Chemistry

Combinatorics

COLLEGE MATHEMATICS

Solving Problems in Finite Mathematics and Calculus

COLLEGE MATHEMATICS

Solving Problems in Finite Mathematics and Calculus

Bill Armstrong
Lakeland Community College

Don Davis
Lakeland Community College

Prentice Hall

Pearson Education, Inc.
Upper Saddle River, New Jersey 07458

Library of Congress Cataloging-in-Publication Data

Armstrong, Bill (William A.)
College mathematics : solving problems in finite mathematics and calculus / Bill Armstrong, Don Davis.
p. cm.
Includes index.
ISBN 0-13-089131-2
1. Mathematics. I. Davis, Don, 1958- II. Title.

QA39.3 .A75 2003
510--dc21

2002030803

Editor-in-Chief/Acquisitions Editor: Sally Yagan
Executive Project Manager: Ann Heath
Assistant Editor: Vince Jansen
Vice President/Director of Production and Manufacturing: David W. Riccardi
Executive Managing Editor: Kathleen Schiaparelli
Senior Managing Editor: Linda Mihatov Behrens
Production Editor: Brittney Corrigan-McElroy
Editor in Chief, Development: Carol Trueheart
Development Editor: Tony Palermino
Manufacturing Buyer: Alan Fischer
Manufacturing Manager: Trudy Pisciotti
Marketing Manager: Patrice Lumumba Jones
Marketing Assistant: Rachel Beckman
Assistant Managing Editor, Math Media Production: John Matthews
Editorial Assistant/Supplements Editor: Joanne Wendelken
Art Director: Maureen Eide
Interior/Cover Designer: Joseph Sengotta
Creative Director: Carole Anson
Art Editor: Thomas Benfatti
Director of Creative Services: Paul Belfanti
Director, Image Resource Center: Melinda Reo
Manager, Rights and Permissions: Zina Arabia
Interior Image Specialist: Beth Boyd-Brenzel
Cover Image Specialist: Karen Sanatar
Image Permission Coordinator: Debbie Latronica
Cover Photo: David Lissy/EStock Photography LLC Leo De Wys
Art Studio: Laserwords

Prentice Hall

Pearson Education, Inc.
Upper Saddle River, New Jersey 07458

Printed in the United States of America
10 9 8 7 6 5 4 3 2 1

ISBN: 0-13-089131-2

Pearson Education LTD., *London*
Pearson Education Australia PTY, Limited, *Sydney*
Pearson Education Singapore, Pte. Ltd
Pearson Education North Asia Ltd., *Hong Kong*
Pearson Education Canada, Ltd., *Toronto*
Pearson Educación de Mexico, S.A. de C.V.
Pearson Education – Japan, *Tokyo*
Pearson Education Malaysia, Pte. Ltd.

Our book is dedicated . . .

In memory of Mr. James Young.
His love of mathematics and the teaching of mathematics continues to inspire 20 years later.

and

To Dr. David Keck, whose mathematical and teaching skills were a true inspiration.

Contents

Chapter 6 Sets and the Fundamentals of Probability 307

Chapter 7 Graphical Data Description 397

Chapter 8 Probability Distributions 461

Chapter 9 Markov Chains 525

Chapter 10 Game Theory 567

Chapter 11 Functions, Models, and Average Rate of Change 601

Chapter 12 Limits, Instantaneous Rate of Change, and the Derivative 729

Chapter 13 Applications of the Derivative 825

Chapter 14 Additional Differentiation Techniques 863

Preface

Audience

In preparing to write the two components that comprise this text, we talked with many colleagues who teach finite mathematics and a brief or applied calculus course to find out if they experienced the same difficulties in teaching these subjects that we have encountered. What we learned is that while there is very little similarity in the topics covered or in how much time is spent on each area, there is remarkable uniformity in the needs of students who enroll in these diverse courses. Professors at Community Colleges, Universities, and Liberal Arts Colleges all told us that their students are generally unmotivated, unsure of their algebra skills, uncomfortable with translating English into mathematics, and unschooled in how to set up problems for solution. Armed with this knowledge, we prepared *College Mathematics: Solving Problems in Finite Mathematics and Calculus* to address these fundamental needs.

As with other college mathematics textbooks, our text is designed for a two-term course for students majoring in economics, business, social or behavioral sciences. The book is organized into two parts. The first ten chapters cover the topics most frequently taught in a finite mathematics course. The second part of the book consists of eight chapters of applied calculus. We have organized the topics for maximum flexibility within each part so that the text may be adapted to any college or university's curriculum. However, that is where the similarity ends. We have crafted this book around six key principles designed to address student's needs:

- Present the mathematics in language that students can read and understand
- Teach good problem solving techniques and provide ample practice
- Use real data applications to keep it interesting
- Provide timely reinforcement of algebra and other essential skills
- Let Instructor's decide whether to incorporate technology
- Present Probability and Statistics with clarity and precision

Present the Mathematics in Language That Students Can Read and Understand

By writing this text in a conversational, easy-to-read style, we strive to evoke the one-on-one communication of a tutorial session. When students find that they can understand the clear presentation and follow the interesting, real world examples, we believe they will get into the habit of reading the text. Although we have written a text that is accessible, we have been careful not to sacrifice the proper depth of coverage and necessary rigor required of finite mathematics. We are confident both objectives have been met.

Teach Good Problem Solving Techniques and Provide Ample Practice

Problem Solving Sections

Before explaining how to solve an entire class of problems, we provide a special section that demonstrates how to apply the appropriate mathematical tools to analyze a given type of problem. For example, the first section of the book, *Problem Solving: Linear Equations*, introduces our general approach to problem solving and then addresses the challenge of analyzing a word problem and translating the words into mathematical expressions; after which we examine the more complex situation of writing a word problem as a system of linear equations. The other dedicated problem solving sections introduce strategies and techniques for solving each type of problem.

- Section 3.1, *Problem Solving: Linear Inequalities* reviews writing linear models and solving linear inequalities, both of which are vital skills for solving linear programming problems.
- Section 6.1, *Problem Solving: Probability and Statistics* explains some basic concepts in probability and statistics such as probability experiment, mutual exclusivity, and ordered samples.
- Section 11.2, *Introduction to Problem Solving* extends our general approach to problem solving to mathematical models and their properties, and explains how numerical solutions to mathematical models are interpreted.
- Section 12.3, *Problem Solving: Rates of Change* connects the average rate of change and secant line slope to instantaneous rate of change and tangent line slope using numerical, algebraic, and graphical techniques.
- Section 16.4, *Problem Solving: Integral Calculus* explains how the definite integral can be used, given a rate of change function, to determine a continuous sum.

Many of the exercises in these problem solving sections prepare students for subsequent sections by introducing exercises that are solved later in the textbook.

Problem Solving Method

To help students solve problems in finite mathematics and brief calculus, we developed a problem solving method that is used throughout the entire text. Many reviewers commented that this is the single most distinctive, and user-friendly feature of our book as it provides students with a tool for learning and understanding the mathematics at hand. Frequently, applied mathematics instructors hear students comment that "I don't even know how to begin this problem." or "If the problem was just set up for me, I could solve it." Because skills such as setting up problems and writing the solution in its proper context can be a major challenge for applied calculus students, we have integrated a **problem solving method** throughout the text. The steps of the problem-solving method are referenced throughout the text. We use the phrases "Understand the Situation" and "Interpret the Solution" to identify two critical steps in problem solving.

- When solving an application example, we first need to **Understand the Situation**. In these clearly labeled paragraphs we state the quantity we are trying to determine and identify the method used to determine the solution. Frequently we reveal the thought processes necessary to attack the problem.
- After determining a numerical solution in a traditional solution step, we **Interpret the Solution**. In the interpretation step, we write a concluding sentence that conveys the meaning of the numerical answer in the context of the application. To reinforce these techniques, we often ask students to solve and interpret the solution to various applied problems in the exercise sets.

Exercises

The comprehensive exercises sets are the heart of our textbook. The typical section exercise set contains numerous skill builder problems, a generous selection of applications from many different disciplines. Another lesson learned from our market research is that many applied calculus textbooks fail to provide enough exercises for the student to grasp the course content. With **more than 6700 exercises**, we are confident our textbook has more than enough exercises to meet student's needs.

Use Real Data Applications to Keep It Interesting

Students in this course tend to be very pragmatic; they want to know why they must learn the mathematical content in this course. Including many **real data modeling** applications (denoted in text by 🌐) in examples and exercises helps to answer their unstated question and provides motivation and interest. Many of the models, parameters, and scenarios in the examples and exercises are based on data gathered from the U.S. Statistical Abstract, the Census Bureau, and other reliable sources for which URLs are always given. For example, we include real data modeling applications such as determining planting schedules in California (*Source:* U.S. Department of Agriculture, www.usda.gov/nass) in the linear programming chapters. Other examples include:

- The out-of-pocket hospital care and physician services expenditures by patients and by Medicare used as an application of matrix operations (*Source:* U.S. Census Bureau, www.census.gov).
- The percentage of American adults who exercised for 30 minutes or more at least 5 times per week, classified by gender, as an application of conditional probability. (*Source:* National Center for Chronic Disease Prevention, www.cdc.gov/nccdphp).
- The typical amount of financial debt held by American families, classified by age group, as an application of frequency distributions (*Source:* U.S. Census Bureau, www.census.gov).
- The number of milligrams of cholesterol consumed each day per person in the U.S. (*Source:* U.S. Department of Agriculture, www.usda.gov).
- The annual total expenditures on pollution abatement in the U.S. (*Source:* U.S. Statistical Abstract, www.census.gov/statab/www).
- The rate of change in the number of arrests for drug abuse violations by American adults (*Source:* U.S. Census Bureau, www.census.gov).

The quantity, quality, and variation of these types of applications are simply not found in other applied calculus textbooks. We believe that real data modeling applications not only keep the course content relevant and fresh, but compel students to interpret the numerical solution in the context of the problem they have solved.

Provide Timely Reinforcement of Algebra and Other Essential Skills

One of the major challenges faced by students, and frustrations encountered by instructors, is weak preparation in algebra and other essential skills. Even students who have proficient algebra skills are often rusty and unsure of which algebraic tool to apply. Further, the content demands of a finite mathematics and an applied calculus course does not allow for extensive time spent on review. In an effort to address this pervasive problem, we have developed the "From Your Toolbox" feature.

From Your Toolbox

When appropriate, the "From Your Toolbox" feature directs students to read background material in the Algebra Review or other appendices. In addition, this feature is used to review previously introduced definitions, theorems, or properties as needed. By providing a brief review when it is needed, students stay on task and do not need to flip back to hunt through previous sections for key information.

Let Instructor's Decide Whether to Incorporate Technology

As graduates and instructors of the Ohio State University, we began using graphing calculator technology in the classroom long before it was fashionable to do so. Based on our years of experience in this area, we have seen the strengths of using a graphing calculator and the drawbacks as well. Our philosophy is to let the instructor, rather than the textbook, determine how much or how little graphing calculators, spreadsheets, or other desktop applications are used in classroom. Consequently, we have developed the **Technology Option** in our text, which allows each instructor to decide whether or not to use technology in the curriculum.

Technology Option

These shaded, optional parts are easy to find, or to skip, and typically follow examples. The content of these parts mirrors the traditional presentation but shows how the answer to a particular example may be found using a graphing calculator. Although keystroke commands are not given, we provide answers to general questions students may have. All screen shots included in the text are from the Texas Instruments TI-83 calculator. Keystrokes and commands for various models and brands of calculator are found at the online graphing calculator manual found at the companion website at www.prenhall.com/armstrong.

Exercises That Assume Technology

Exercises that assume the use of a graphing calculator are clearly marked with a symbol so they can be assigned or skipped as desired by the instructor. We recognize that the graphing calculator is simply a tool to be used in the understanding of mathematics. We have been very careful to introduce the technology only where it is appropriate and not to let its use overshadow the mathematics.

Present Probability and Statistics with Clarity and Precision

Our experiences in the classroom, and supported by the market research, indicated that probability is one of the most difficult topics for the student to comprehend.

Because probability is so challenging, we believe it is important to prepare students for the study of this topic. Students need to know the types of problems they will encounter and whether or not order is important. They also need to know when events are independent and when events are mutually exclusive. Addressing these basic concepts and introducing the language of probability in Section 6.1, *Problem Solving: Probability and Statistics*, provides students with a solid foundation. Additional strengths in the probability and statistics chapters include:

- Clear and well-paced development of permutations and combinations (Section 6.3). The *Flashback* to Section 6.1 helps students understand these often confusing topics.
- Introducing frequency distributions in connection with empirical probability in Section 6.4 helps students distinguish the difference between meanings of probability in various contexts.
- Chapter 7 incorporates more graphics and illustrations than competitors and focuses on interpretation. Students are encouraged to understand and interpret different representations of mathematical ideas by way of algebra, graphs, data, and verbal descriptions and to easily translate from one representation to another.
- Many Finite Math books incorrectly use a single notation, X or x, to represent both a random variable and the values that the random variable can assume. We believe that one need not sacrifice accuracy and precision to be accessible, therefore we use X to represent the random variable and x to represent the values that the random variable can assume. Students quickly become comfortable with this distinction.
- The coverage of random variables, expectation, and standard deviation are divided into two sections (8.1 and 8.2). This division allows for a slower paced development of these complex ideas and better prepares students for probability distributions.
- We have written these chapters in consultation with experts in this field to ensure that our goal of clarity and precision was being met. Saeed Ghahramani, Dean, School of Arts and Sciences Western New England College, served as the expert reviewer for the Probability and Statistics chapters.

Content Features and Highlights

Finite Math Coverage (Chapters 1–10). In addition to exceptionally strong coverage of probability and statistics (Chapters 6, 7 and 8), the following sections have been praised by reviewers:

- Reviewers found the motivation in Section 2.3, Matrix Multiplication, to be particularly effective. Especially interpreting the entries of a matrix product; when entries make sense as well as when the entries are meaningless.
- Several reviewers commented that the explanations in Section 2.5, Leontief Input-Output Models, were exceptionally clear.
- The section on Applications of Linear Programming (Section 3.4) drew rave reviews from reviewers. Most liked the approach of the section in tying together all of Chapter 3 and summarizing the process. Some reviewers found the Flashback in this section to be a wonderful transition to applications of linear programming.
- Reviewers lauded the first two sections of Chapter 4 (Linear Programming and the Simplex Method) as being exceptionally clear and concise, yet thorough.
- Chapter 5 (Mathematics of Finance) was another favorite for our review panel. From the Chapter Opening page through the Chapter Project, these are topics that all reviewers said were necessary and appropriate for students.
- Reviewers who said they covered Markov and Game Theory were extremely impressed with Chapters 9 and 10. All applauded the flow of material in these chapters and how students should find the material quite easy to follow.

Brief Calculus Coverage (Chapters 11–18)

- **Rate of Change Theme** Because we believe it is important for students to understand that calculus is the study of rates of change, we have highlighted this theme in Chapters 11–18. Beginning in Chapter 11, the basic algebraic and transcendental functions are reviewed in a concise and comprehensive manner. As each type of function is introduced, the average rate of change of the function on a closed interval is presented in examples and exercises. From the onset, appropriate units and an emphasis on interpreting, rather than merely finding a numerical answer, are stressed.
    - For example, the solution to an exercise in Section 11.4 reads, "From 1980 to 1990, the number of lawyers in private practice increased at an average rate of 17.07 thousand per year." This requires students to determine the average rate of change of a nonlinear model on a closed interval. We have found that by introducing the difference quotient in this manner early and often in Chapter 11, we create a smooth transition to computing the instantaneous rate of change and the derivative in Chapter 12.
    - The remaining Chapters continue to highlight the rate of change theme, along with the importance of proper units and a sound interpretation of the solution. If a student who uses *College Mathematics: Solving Problems in Finite Mathematics and Calculus* is asked, "What is calculus?" we are confident that the student will confidently proclaim, "It is the study of rates of change!"

- **Use of the Differential** To supplement the rate of change theme, we have paid particular attention to the use of the differential in applied calculus topics. The differential is introduced early in Chapter 13 and is used as a mathematical tool to introduce new topics in later sections.
  - In Section 13.2 for example, marginal analysis is introduced by means of the differential and then we show that the marginal business functions are a type of differential.
  - In Section 13.3, relative errors and the relative rate of change are explained by the use of the differential.
  - In Chapter 14, the differential is used to motivate the formula for elasticity of demand.
  - In Chapter 16, we revisit the differential again when we introduce integration by u-substitution. In contrast to many applied calculus texts which teach the differential as an isolated topic, we present it as a powerful tool that is useful in both differential and integral calculus.
- **The Definite Integral as a Continuous Sum** We have found that many of the applications of the definite integral in applied calculus textbooks tend to be contrived, esoteric, and difficult to interpret. To address this shortcoming, we have made the use of the definite integral to compute a total accumulation over a continuous interval. Introduced in the Chapter 16 problem solving section, students learn that integrating a rate function (that is, a derivative), on an interval gives the total accumulation of the dependent variable values on that interval. For example, the total difference in cost $C(b) - C(a)$ can be determined by integrating the marginal cost function $MC(x)$ over the interval $[a, b]$. This central idea is used throughout the integral calculus unit resulting in applications that are richer and more interesting that is found in most texts.

Chapter Features

Chapter Openers. The first page of each chapter lists the sections included in that chapter, along with a photo and representative graphs or figures that foreshadow the fundamental ideas presented in the chapter in the context of an application. "What We Know" reiterates what information has been learned in previous chapters and "Where Do We Go" explains what topics will be covered. The chapter opener helps create a roadmap to guide the student through the book and underscore the connections between topics.

Flashbacks. Selected examples used earlier in the textbook are revisited in the **Flashback** feature. The Flashback carries over an applied example from a previous section and then extends the content of the example by considering new questions. In this manner, new topics are introduced in a more natural way within a familiar context. Moreover, the Flashback often reviews the necessary skills and concepts from previous chapters. We believe that this pedagogical technique of using applications previously discussed allows students to concentrate on new topics using familiar applications.

Interactive Activities. Extensions to completed examples are included in the **interactive activities** that appear after many examples. The interactive activities may ask students to solve a problem using a different method, to discover a pattern that can lead to a mathematical property, or to explore additional properties of recently introduced topics. Interactive Activities may be used in a number of ways in the classroom. Instructors may assign them as critical thinking exercises, they may be used as a springboard for classroom discussion, or they may provide a vehicle for collaborative activities. Interactive Activities that include the Web icon have solutions provided on the companion website at www.prenhall.com/armstrong.

Checkpoints. At strategic points in each section, an example is followed by a **checkpoint**. Each checkpoint directs a student to complete a selected odd numbered problem in that section's exercise set. Guiding the student to a parallel exercise to the example problem helps to reinforce the topic at hand and ensures that the recently introduced skill or concept is practiced immediately. This pedagogical tool promotes interaction between the text and the student and helps students to develop good study habits. Students who use the checkpoints will quickly learn to take ownership of the course material.

Notes. Another popular feature is the **Notes** that appear after many definitions, theorems, and properties. Notes are used to clarify a mathematical idea verbally and to provide students with additional insight into the material. Many times these notes echo what a professor might state in the classroom to help the student understand the definition, theorem or property.

Section Projects. At the end of each exercise set, a **Section Project** presents a series of questions that ask students to explore the idea that is presented. Some of the projects are based on real data and ask students to use the regression capabilities of their calculator to determine a model. Others give a step-by-step procedure that can be used to solve classic problems in finite mathematics. Instructors can use section projects as standard hand-in assignments, collaborative activities, or for demos in graphing calculator usage.

Section Summary. At the end of each section, a short summary highlights the important concepts of the section. Section summaries are a convenient resource that students may use while completing exercises, or to offer a springboard for test review.

End of Chapter Material. Each chapter concludes with three components; first is a summary called **Why We Learned It**. This narrative outlines the major topics of each chapter and describes how the topics are used in various careers. An extensive set of **Chapter Review Exercises** is designed to augment the exercises in each section of the chapter. Concluding the chapter is the **Chapter Project**. These extensive exercises use real data modeling to explore and extend the ideas discussed in each chapter. They may be assigned as individual homework or as a group project.

Supplements

Instructor's Solutions Manual (0-13-066018-3). Written by Denyse Kerr and accuracy checked by the authors and Priscilla Gathoni (Finite Mathematics) and John Cruthirds of North Georgia College and State University (Brief Calculus) these volumes contain complete worked out solutions to all of the exercises, and the section and chapter projects.

Student's Solutions Manual (0-13-066010-8). Written by Denyse Kerr and accuracy checked by the authors and Priscilla Gathoni (Finite Mathematics) and

John Cruthirds of North Georgia College and State University (Brief Calculus) these volumes contain complete worked out solutions to all of the odd numbered section and review exercises in the textbook.

Companion Website www.prenhall.com/armstrong**.** This site is designed to supplement the textbook by offering a variety of teaching and learning resources for each chapter including a net search of topics relevant to the chapter, links to related chapter topics, solutions to selected interactive activities, quizzes, program downloads for various models of graphing calculators, and an online graphing calculator reference manual.

MathPro5 for Finite Math (0-13-038569-7). MathPro5 is online, customizable tutorial and assessment software that is fully integrated with *Finite Math* at the section level. Loaded with many support tools, students work algorithmically-generated tutorial exercises for review and concept mastery and get step-by-step guidance as needed.

The easy-to-use Course Management System enables instructors to track and assess student performance on tutorial work, quizzes, and tests. Instructors can create any number of algorithmic, multi-form quizzes, pre-tests, post-tests, chapter tests and cumulative reviews with the **integrated TestGen-EQ software**. A **robust reports wizard** provides a grade book, individual student reports and class summaries. **A messaging system** enhances communication between instructors and students from wherever they are working in MathPro5. The **customizable course syllabus** allows instructors to remove and reorganize topics, and add website links and tests to maximize integration of the classroom and tutorial experience.

The combination of MathPro5's richly integrated tutorial and testing and robust course management tools provides an unparalleled tutorial experience for students, and new assessment and time-saving tools for instructors.

TestGen-EQ for Windows and Macintosh (0-13-066022-1). This algorithmic software-testing program allows instructors to create tests quickly and efficiently using the supplied questions or personalize tests using the built-in editing features.

Test Item File (0-13-066017-5). Hard copy of the algorithmic computerized testing materials provides a quick reference for the testing software.

Acknowledgements

We owe a debt of gratitude to many individuals who helped us shape and refine the Finite Mathematics chapters of this text. The reviewers included:

Brenda Ammons, Pellissippi State Tech and CC; Paul Boisvert, Oakton Community College; Dale Buske, St. Cloud State; Elias Deeba, University of Houston—Downtown; Jennifer Fowler, UT Knoxville; Lou Hoelzlel, Bucks County Community College; Frank Hummer, Iowa State University; Stephanie Kurtz, Louisiana State University; Mike McCraith, Joliet Junior College David Miller, William Patterson University; and Bill Siebler.

We owe special thanks to Saeed Ghahramani, Dean, School of Arts and Sciences Western New England College who served as our expert reviewer for the Probability and Statistics chapters.

Ed Bolger, Miami University of Ohio, was our expert reviewer for the Markov Process and Game Theory chapters. Sarah Street served as an additional reviewer for Chapter 8.

Many reviewers gave is invaluable direction on both the first and second edition of the Brief Calculus chapters, including pre-revision reviewers Michele Clement (Louisiana State University), Karabi Datta (Northern Illinois University), and Richard Witt (University of Wisconsin—Eau Claire) and first edition reviewers Martin Bonsangue (California State University at Fullerton), Fred Bakenhus (St. Philips College), Biswa Datta (Northern Illinois University), Matthew Hudock (St. Philips College), Anthony Macula (SUNY Geneseo), Thomas Ordayne (University of South Carolina at Spartanburg), Rene Barrientos (Miami-Dade Community College), Mark Burtch (Arizona State University), Adrienne Goldstein (Miami-Dade Community College), John Grima (Glendale Community College), Michael Kirby (Tidewater Community College), Zhuangyi Lui (University of Minnesota at Duluth), and Martha Pratt (Mississippi State University).

A special thank you is owed to our accuracy checkers Priscilla Gathoni and John Cruthirds (North Georgia College and State University). The Chapter Projects and Solutions to Interactive Activities were provided by Mike McCraith, who deserves special thanks. We offer a very special thank you to an extremely valuable team member, Denyse Kerr. Her professional, high-quality work in creating both the Instructor's and Student's Solutions Manuals was valued for its precision and careful proofreading.

We owe special thanks to Tony Palermino. Besides serving as our Developmental Editor, Tony was also our mentor, coach, advisor, advocate and valued team member. Tony's top quality professional work was critical in preparing the manuscript. Tony deserves much credit for helping create this text.

Many people involved in the management and production of the text deserve recognition. Brittney Corrigan-McElroy has provided high quality, professional production abilities. Maureen Eide worked through several iterations of the design. Vince Jansen contributed his expertise on the website and the TestGen EQ. We appreciate the quality services provided by Laser Words in preparing the graphs, tables and art in the text and of Laurel Technical Services in writing the review and test bank exercises. Patrice Jones continues to provide effective marketing techniques. Thanks to Joanne Wendelken for routing are numerous telephone calls. A very special thanks to the Prentice Hall sales staff for their sales efforts and enthusiasm.

Special thanks to Sally Yagan for continuing to believe in this project. Sally continues to impress us in her level of dedication to our project and her can-do attitude is refreshing and contagious.

Ann Heath deserves special recognition for the many hats she wears. Without Ann on our team, this project would not be at the high level of quality it currently is. Ann is a master of crisis management as well as providing an ear at any time to hear our frustrations.

Finally, we owe personal thanks to our families. Our wives, Lisa and Melissa, were supportive and patient during the entire process. To our children Austin and Dylan; Randy, Rusty and Ronnie who understood why sometimes Daddy could not come out and play in the backyard.

Bill Armstrong

Don Davis

A Guide to Using This Text

Opening Application

These engaging real-world applications appear at the beginning of every chapter. They include compelling photos and graphs that illustrate how the math introduced and covered in the chapter is used outside the classroom.

Chapter 6

Sets and the Fundamentals of Probability

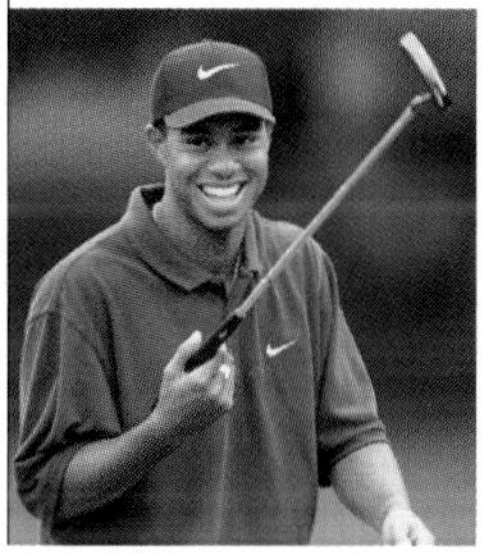

Odds to Probability

$$P(A) = \frac{a}{a+b}$$

Probability of Tiger Woods Winning the 2001 British Open

$$P(A) = \frac{a}{a+b} = \frac{3}{3+2} = \frac{3}{5}$$

We frequently hear about the odds of something occurring. But how do odds relate to probability? In 2001, golf pro Tiger Woods was "hot" and the odds of his winning the British Open were 3 to 2. Using the formula for converting odds to probability, we can translate odds into the probability of a given event occurring. In this case, the probability of Tiger Woods winning the Open was 3/5. Although the probability of a win was high, Tiger actually suffered a stunning defeat. Probability is a valuable tool for predicting likely outcomes, but it does not insure that a given event will occur.

What We Know

So far in this text, we have seen many applications involving matrices and linear programming, as well as the mathematics of finance. We have seen how an understanding of these topics can be important for individuals in many diverse areas.

Where Do We Go

In this chapter we begin our journey into the

Page 307

Problem Solving

Teaching problem-solving skills is central to the approach of the text. The pedagogy and organization encourage students to apply sound problem-solving principles and give them practice applying them.

Section 1.1 **Problem Solving: Linear Equations**

Most of us use or hear the word *problem* in our daily conversation. We may say or hear statements such as

- "I'm having a problem loading this software."
- "The problem is that management doesn't communicate with us."
- "I had a problem getting tickets for the concert."

Many problems that we encounter can be solved easily, but others require a more thoughtful approach. In finite mathematics we are often confronted with the challenge of solving problems that require a thoughtful approach. As a result, we need to review a general problem-solving strategy and the accompanying skills that will allow us to solve a variety of problems. The skills reviewed in this section will be used throughout the text.

English to Mathematics

From prior mathematics classes we have seen how a **variable** can be used to represent an unknown quantity. For example, suppose that we hear the following report on the local news:

"CleanAir produces 1475 of two types of vacuums, Mod I and Mod II, each week."

Page 2

Examples Reinforce Good Problem-Solving Skills

The solutions to many examples are broken into three steps.

- The first step is to **Understand the Situation.** Using straightforward language, the problem situation is analyzed. The quantity to be determined is identified, the method used to find the solution is discussed, and the thought process necessary to attack the problem is presented.
- Then the numerical solution is found in a traditional solution step.
- Finally, students see how to **Interpret the Solution** in the concluding sentence that conveys the meaning of the numerical answer in the context of the application.

Many of the exercises and examples in the text use **Real Data** to make them more interesting and relevant to students.

Example 4 Using the Addition Rule with Tabular Data

The source of payment for patients discharged from U.S. hospitals during 1996 for two age groups is shown in Table 6.5.3. The number of discharges is given in thousands. For the labeled events *A*, *B*, *C*, and *D*, determine the following probabilities and interpret.

(a) $P(B \cap D)$ **(b)** $P(A \cup C)$

Table 6.5.3

Age Group	Private (*A*)	Government (*B*)	Total
Less than 44 years old (C)	6,090	6,442	12,532
45 years or older (D)	4,636	13,337	17,973
Total	10,726	19,779	30,505

SOURCE: U.S. National Center for Health Statistics, www.cdc.gov/nchs

Solution

(a) Understand the Situation: First notice that event *B* is the event that the source of payment for a discharged patient in 1996 was the government. Event *D* is the event that a discharged patient was 45 years old or older. So when we compute $P(B \cap D)$, we are computing the probability that a randomly selected patient discharged from a hospital in 1996 was 45 years old or older and the source of payment was the government.

By inspecting Table 6.5.3, we see that if we follow down the event *B* column and across the event *D* row they intersect at the cell with the value 13,337 thousand. Since the total number of patients discharged in 1996 was 30,505 thousand, we have

$$P(B \cap D) = \frac{\text{value where the event } B \text{ column and event } D \text{ row intersect}}{\text{total patients discharged}}$$

$$= \frac{13,337}{30,505} \approx 0.44$$

Interpret the Solution: This means that if we select a discharged patient at random the probability that he or she had government health insurance **and** was 45 year or older is about 0.44 or about 44%.

(b) Understand the Situation: Since event *A* is the event that the source of payment was private and event *C* is the event that the discharged patient was less than 44 years old, we are being asked to determine the probability that a randomly selected patient has private health insurance, is 44 years old or younger, or both. Table 6.5.3 indicates that

$$P(A) = \frac{\text{total in event } A \text{ column}}{\text{total patients discharged}} = \frac{10,726}{30,505},$$

$$P(C) = \frac{\text{total in event } C \text{ row}}{\text{total patients discharged}} = \frac{12,532}{30,505},$$

and where event *A* column and event *C* row intersect gives us

$$P(A \cap C) = \frac{6090}{30,505}.$$

Page 359

Technology Option

We can evaluate definite integrals on a graphing calculator using the FNINT command. To check our work in Example 1a, we begin by entering $f(x) = x^2 - 3x$ into Y_1. Figure 16.3.2 shows what must be entered into the calculator and Figure 16.3.3 shows the result when we execute the command.

```
fnInt(Y₁,X,-2,0)
```

```
fnInt(Y₁,X,-2,0)
              8.666666667
Ans▶Frac
                     26/3
```

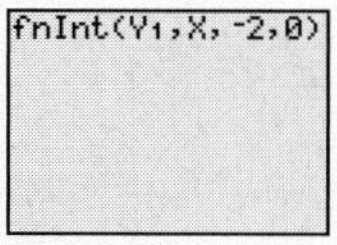

Figure 16.3.2   **Figure 16.3.3**

For more information on the FNINT command or to see how to use this command on your calculator, consult the online graphing calculator manual at www.prenhall.com/armstrong.

To see how the Fundamental Theorem of Calculus can be used to determine an area, let's take another look at an example from Section 16.2.

Page 1027

Technology Option

The text's technology coverage is thorough yet optional. Presented in shaded boxes, technology coverage is easy to find, or to skip, and typically follows examples. The contents of these boxes mirrors the traditional presentation but shows how the answer to a particular example may be found using a TI-83 graphing calculator. For keystroke level instruction on a variety of different calculators, students are directed to the online graphing calculator manuals on the companion website at **www.prenhall.com/armstrong**

   From Your Toolbox

- $\frac{d}{dx}[e^x] = e^x$.
- If f is a differentiable function of x, then, by the Chain Rule,
$$\frac{d}{dx}e^{f(x)} = e^{f(x)} \cdot f'(x)$$

Integration of Exponential Functions

Now we turn our attention from integrals yielding logarithmic functions to those involving exponential functions. Let's consult the Toolbox to the left to recall some facts on exponential functions.

From the first bulleted fact, we know that $\frac{d}{dx}[e^x] = e^x$. This, along with the fact that differentiation and integration are inverse operations, gives the following integration rule.

■ **Integration Formula for e^x**

For the exponential function $f(x) = e^x$,

$$\int e^x \, dx = e^x + C$$

Page 1061

From Your Toolbox

The "From Your Toolbox" makes study time more efficient. The feature pulls the appropriate review material into a given discussion or example. This timely review helps students stay on task, reducing the need to flip back through the text, hunting through previous sections for key information.

■ **Linear Cost Function**

The cost $C(x)$ of producing x units of a product is given by the linear **cost function**

$$C(x) = (\text{variable costs}) \cdot (\text{units produced}) + (\text{fixed costs})$$

▷ **Note:** Some cost functions are not linear because the variable cost is a function itself. However, every cost function is made up of a variable cost expression and a fixed cost.

Page 1022

Notes

synthesize the ideas presented into everyday language—similar to how an instructor would explain in class.

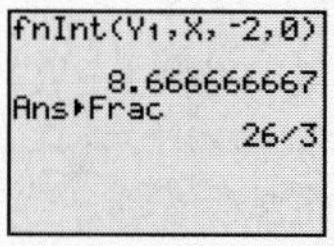

Flashbacks

allow new topics to be introduced in a more natural way within a familiar context. The Flashback carries over an applied example from a previous section and then extends the content of the example by considering new questions.

Interactive Activities

accompany many examples and provide an opportunity for students to explore the example in greater depth. When the icon appears, a worked solution to the Interactive Activity may be found on the companion website. Interactive Activities are ideal for sparking classroom discussion or assigning as an independent or group project.

We conclude this section by taking a Flashback to an example last seen in Section 12.3.

Interactive Activity

Compare the result of the Flashback to the result of Example 4 in Section 12.3.

Flashback

U.S. Imports From China Revisited

The U.S. imports from China can be modeled by

$$f(x) = 0.32x^2 + 1.64x + 3.98 \qquad 1 \le x \le 10$$

where x represents the number of years since 1986 and $f(x)$ represents the dollar value of goods imported from China, measured in billions (*Source:* U.S. Census Bureau, www.census.gov). Determine $f'(x)$. Also, evaluate and interpret $f'(7)$.

Flashback Solution

Understand the Situation: To determine $f'(x)$ we use the definition of the derivative $f'(x) = \lim_{h \to 0} \frac{f(x+h) - f(x)}{h}$. Once we have this general derivative, we will substitute $x = 7$ and evaluate to determine $f'(7)$. Note that $x = 7$ corresponds to 1993.

So we have

$$\begin{aligned} f'(x) &= \lim_{h \to 0} \frac{f(x+h) - f(x)}{h} \\ &= \lim_{h \to 0} \frac{[0.32(x+h)^2 + 1.64(x+h) + 3.98] - [0.32x^2 + 1.64x + 3.98]}{h} \\ &= \lim_{h \to 0} \frac{[0.32(x^2 + 2xh + h^2) + 1.64x + 1.64h + 3.98] - 0.32x^2 - 1.64x - 3.98}{h} \\ &= \lim_{h \to 0} \frac{[0.32x^2 + 0.64xh + 0.32h^2 + 1.64x + 1.64h + 3.98] - 0.32x^2 - 1.64x - 3.98}{h} \\ &= \lim_{h \to 0} \frac{0.64xh + 0.32h^2 + 1.64h}{h} \\ &= \lim_{h \to 0} \frac{h(0.64x + 0.32h + 1.64)}{h} \\ &= \lim_{h \to 0} (0.64x + 0.32h + 1.64) = 0.64x + 1.64 \end{aligned}$$

Interpret the Solution: We have determined that $f'(x) = 0.64x + 1.64$. We evaluate $f'(7)$ by substituting 7 for each occurrence of x. This gives us

$$\begin{aligned} f'(7) &= 0.64(7) + 1.64 \\ &= 6.12 \end{aligned}$$

This means that in 1993, U.S. imports from China were increasing at a rate of $6.12 \frac{\text{billion dollars}}{\text{year}}$.

Page 780

Interactive Activity

Do you ever wonder where the term **zeros** of a function comes from? To find out, graph the function in Example 2 in the viewing window $[-5, 5]$ by $[-10, 5]$; then use the VALUE command on the calculator to evaluate $f(-2)$ and $f\left(\frac{1}{2}\right)$. What are the y-values?

We can check our solutions by evaluating $f(x) = 4x^2 + 6x - 4$ at both solutions to see if the function value is zero. Checking at $x = -2$, we get

$$\begin{aligned} f(-2) &= 4(-2)^2 + 6(-2) - 4 \\ &= 16 - 12 - 4 = 0 \end{aligned}$$

At the second zero $x = \frac{1}{2}$, we get

$$\begin{aligned} f\left(\frac{1}{2}\right) &= 4\left(\frac{1}{2}\right)^2 + 6\left(\frac{1}{2}\right) - 4 \\ &= 1 + 3 - 4 = 0 \end{aligned}$$

So the zeros of the function $f(x) = 4x^2 + 6x - 4$ are $x = -2$ and $x = \frac{1}{2}$. ■

Many times, the quadratic that we wish to solve is not easily factorable. To algebraically solve these types, we use the **quadratic formula**. The derivation of this formula appears in Appendix C.

Page 644

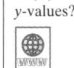

Exercise Sets

The section exercise set contains numerous skill builder problems, applications from many different disciplines, and an abundant supply of real data modeling applications, denoted by this symbol 🌐. The text contains over 6700 exercises.

- Exercises marked with a ✓ are called out in the body of the text as "Checkpoint" problems that relate closely to a worked in-text example.
- Building on the "Connections" approach, we use these symbols ⟸⟹ to indicate a problem that uses a scenario or concept that has been or will be revisited.
- The symbol 🖩 denotes problems that are best solved using a graphing calculator.

30. The import rate of crude oil into the U.S. can be modeled by

$$g(x) = -2.73x^2 + 35.08x + 22.18 \qquad 1 \le x \le 10$$

where x is the number of years since 1989 and $g(x)$ is the rate of imports annually measured in $\frac{\text{millions of barrels}}{\text{year}}$ (*Source:* U.S. Energy Information Administration, www.eia.doe.gov). Determine the total increase (or decrease) in imports of crude oil into the U.S. from 1992 to 1997.

31. The Alaskan oil field production rate can be modeled by

$$g(x) = -0.81x^2 + 5x - 27.56 \qquad 1 \le x \le 10$$

where x is the number of years since 1989 and $g(x)$ is the rate of oil production in the Alaskan oil field measured in $\frac{\text{millions of barrels}}{\text{year}}$ (*Source:* U.S. Energy Information Administration, www.eia.doe.gov). Determine the total increase (or decrease) in Alaskan oil field production from 1992 to 1997.

32. From past records, a botanist knows that a certain species of tree has a rate of growth that can be modeled by

$$f(x) = \frac{3}{2}x^{-\frac{1}{2}} \qquad 1 \le x \le 4$$

where x is the age of the tree in years and $f(x)$ is the growth rate in $\frac{\text{feet}}{\text{year}}$. Determine the total increase in the growth of the tree from the age of 1 year to 4 years.

33. From past records, a botanist knows that a certain species of tree has a rate of growth that can be modeled by

$$f(x) = 2x^{-\frac{1}{2}} \qquad 1 \le x \le 4$$

where x is the age of the tree in years and $f(x)$ is the growth rate in $\frac{\text{feet}}{\text{year}}$. Determine the total increase in the growth of the tree from the age of 1 year to 4 years.

In Exercises 34–39, the velocity function $v(t)$ for an object is given, measured in feet per second.
(a) Evaluate $v(t)$ at the point $t = t_1$.
(b) Evaluate and interpret $\int_{t_1}^{t_2} v(t)\,dt$.

34. $v(t) = 4t;\ t_1 = 2, t_2 = 8$

35. $v(t) = 0.3t;\ t_1 = 10, t_2 = 20$

36. $v(t) = 3t + 6;\ t_1 = 1, t_2 = 5$

37. $v(t) = 2t + 10;\ t_1 = 3, t_2 = 5$

39. $v(t) = -9.8t + 100;\ t_1 = 2, t_2 = 8$

40. The velocity of an object is given by

$$v(t) = -32t + 88$$

where t is the time in seconds and $v(t)$ is the velocity measured in $\frac{\text{feet}}{\text{second}}$.
(a) Evaluate $v(1)$ and interpret.
(b) Determine the total movement, or displacement, of the object between 0 and 2 seconds.

✓ **41.** The velocity of an object is given by

$$v(t) = -32t + 120$$

where t is the time in seconds and $v(t)$ is the velocity measured in $\frac{\text{feet}}{\text{second}}$.
(a) Evaluate $v(1)$ and interpret.
(b) Determine the total movement, or displacement, of the object between 0 and 3 seconds.

In Exercises 42–47, you are given a rate function that, at the present time, you are unable to antidifferentiate. (In Sections 6.5 and 6.6 we will present techniques that will allow us to determine an antiderivative.) As a result, you will need to use the FNINT *command or $\int f(x)\,dx$ on your graphing calculator.*

🖩 ⟹ 6.5 **42.** The CustomKey Company determines that the marginal cost of producing sterling silver key fobs is given by

$$MC(x) = \frac{10x}{\sqrt{x^2 + 10{,}000}} \qquad 0 \le x \le 100$$

where x represents the number of fobs produced daily and $MC(x)$ is the marginal cost in dollars. Use your calculator to approximate the integral $\int_0^{75} \frac{10x}{\sqrt{x^2 + 10{,}000}}\,dx$ and interpret.

🖩 ⟹ 6.5 **43.** The TightNut Company determine that the marginal cost of producing center punches is given by

$$MC(x) = 0.003x\sqrt{x+1} \qquad 0 \le x \le 50$$

where x represents the number of center punches produced in a work shift and $MC(x)$ represents the marginal cost in dollars. Use your calculator to approximate the integral $\int_{35}^{48} 0.003x\sqrt{x+1}\,dx$ and interpret.

🖩 🌐 ⟹ 6.6 **44.** The rate of change function for foreign student enrollment in U.S. colleges can be modeled by

$$g(x) = \frac{92.43}{x} \qquad 1 \le x \le 22$$

where x represents the number of years since 1975 and $g(x)$

Page 1045

🖩 🌐 **SECTION PROJECT**

The rate of change in the average price of natural gas in the United States from 1995 to 1999 is shown in Table 16.4.2.

(a) Use your calculator to determine a cubic regression model in the form

$$g(x) = ax^3 + bx^2 + cx + d \qquad 1 \le x \le 5$$

where x represents the number of years since 1994 and $g(x)$ is the rate of change in the price of natural gas in the U.S. measured in $\frac{\text{dollars per 1000 cubic feet}}{\text{year}}$. Round all coefficients to the nearest thousandth.

(b) Evaluate and interpret $g(3)$ and explain why this answer is the same as the rate of change in price given in Table 16.4.2.

Table 16.4.2

Year	Rate of Change in Price
1995	0.561
1996	0.495
1997	−0.171
1998	−0.333
1999	1.113

SOURCE: U.S. Energy Information Administration, www.eia.doe.gov

(c) Compute $g'(x)$ and use it to determine when the rate of change in price was decreasing most rapidly.

(d) Evaluate and interpret $\int_1^5 g(x)\,dx$.

Page 1046

Section Projects

conclude each exercise set and provide an opportunity for students to explore the main chapter ideas in greater depth. Instructors can use section projects as standard hand-in assignments, collaborative activities, or as demos for graphing calculator usage.

Why We Learned It

Why We Learned It parallels the chapter-opening "road map" and explains how the chapter topic is used in the real world.

■ Why We Learned It

In this chapter we were introduced to a new type of rate of change called an instantaneous rate of change. We learned about limits in the beginning of the chapter so that we would have the necessary mathematical tools to understand how to determine an instantaneous rate of change. We saw that instantaneous rate of change appears in many applications. For example, if we want to know at what rate the total carbon content of carbon dioxide emissions in the United States was increasing or decreasing in a certain year, say 1994, we need to compute an instantaneous rate of change. This instantaneous rate of change is determined by first computing the derivative of the function that models the situation. We also saw how the derivative can tell us the rate at which profits are growing (or falling) at a certain production level. The derivative was used to also determine the rate at which a bacteria colony was growing at a given instance in time. There are many diverse applications from many disciplines that use the derivative as a tool to measure an instantaneous rate of change. This chapter was an introduction to these many, varied applications.

Chapter Review Exercises

Chapter Review Exercises conclude each chapter and offer the same range of skill builder and applied problems as do the section projects. These exercises also serve to tie together ideas from several sections.

CHAPTER REVIEW EXERCISES

1. For $f(x) = \dfrac{1 - \sqrt{1 - 2x - x^2}}{x}$, complete the table to numerically estimate the following.

(a) $\lim_{x \to 0^-} f(x)$ **(b)** $\lim_{x \to 0^+} f(x)$ **(c)** $\lim_{x \to 0} f(x)$

x	-0.1	-0.01	-0.001	0	0.001	0.01	0.1
$f(x)$				?			

2. For $f(x) = (\sqrt{4 - x})^5$, complete the table to numerically estimate the following.

(a) $\lim_{x \to 4^-} f(x)$ **(b)** $\lim_{x \to 4^+} f(x)$ **(c)** $\lim_{x \to 4} f(x)$

x	3	3.9	3.99	4	4.01	4.1	5
$f(x)$				?			

For Exercises 3 and 4, use your calculator to graph the given function. Use the ZOOM and TRACE commands to graphically estimate the indicated limits. Verify your estimate numerically.

3. $f(x) = x^3 - 2x$; $\lim_{x \to 3.1} f(x)$

4. $f(x) = \sqrt{x} + \dfrac{1}{|x|}$; $\lim_{x \to 2.5} f(x)$

For Exercises 5–10, determine the indicated limit algebraically.

5. $\lim_{x \to 2} (7x^3 - 10x)$

6. $\lim_{x \to -1} \dfrac{x^2 - 1}{x + 1}$

7. $\lim_{x \to -3} |x - 5|$

8. $\lim_{x \to 0} \sqrt{36 - 8x}$

9. $\lim_{x \to 10} \dfrac{x + 10}{x^2 - 100}$

10. $\lim_{x \to 2.2} (x + 9)^3$

11. Use the graph of f to find the following:

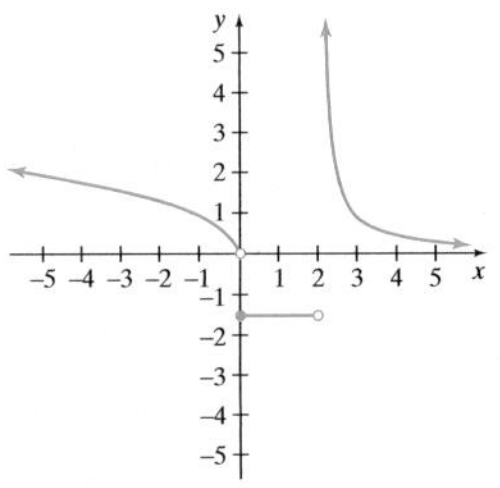

(a) $\lim_{x \to -4} f(x)$ **(b)** $\lim_{x \to 0^-} f(x)$

(c) $f(0)$ **(d)** $\lim_{x \to 2} f(x)$

(e) $\lim_{x \to 3} f(x)$ **(f)** $f(3)$

Chapter Project

A **Chapter Project** rounds out each chapter. These extensive exercises use real data modeling to explore and extend the ideas discussed in each chapter. They may be assigned as individual homework or as a group project.

CHAPTER 2 PROJECT

SOURCE: U.S. Census Bureau, www.census.gov

In Chapter 2 we were introduced to the key elements of differential calculus such as limits, continuity, and the derivative. In this project, we will explore these topics for a piecewise defined regression function model as well as foreshadow an important application of the derivative that appears in Chapter 3. The number of live births in the United States from 1970–1998 can be modeled by

$$L(x) = \begin{cases} 24.29x^2 - 287.00x + 3{,}974.20 & 1 \le x < 11 \\ 0.94x^3 - 37.68x^2 + 518.39x + 1{,}216.66 & 11 \le x < 21 \\ 5.65x^2 - 315.62x + 8{,}316.56 & 21 \le x \le 29 \end{cases}$$

where x is the number of years since 1969 and $L(x)$ represents the number of live births in thousands.

1. Evaluate and interpret $L(10)$, $L(11)$, and $L(21)$.
2. Complete the following table to estimate $\lim_{x \to 11} L(x)$

x	10.9	10.99	10.999	11	11.001	11.01	11.1
$L(x)$				?			

3. Is L continuous at $x = 11$? Explain.
4. Evaluate and interpret $L'(10)$, $L'(11)$, and $L'(21)$.
5. Add $L(11)$ from 1 and $L'(11)$ from 4 and compare your answer to $L(12)$. Notice that it seems that the sum is approximately equal to $L(12)$.
6. Can we use the model, L, to determine the number of live births in 1999? Explain.

The $\lim_{h \to 0^-} \left(\dfrac{f(x+h) - f(x)}{h} \right)$, when it exists, is called the **left-hand derivative** of f at x. This allows us to determine if a function is differentiable at the right endpoint of a closed interval. This means that a function, f, defined on a closed interval $[a, b]$ is differentiable at b if its left-hand derivative exists at b. Let us see if our model L is differentiable at $x = 29$ by determining if the left-hand derivative exists at $x = 29$.

7. Numerically estimate if the left-hand derivative of L exists at $x = 29$ by completing the following table:

h	-0.1	-0.001	-0.00001	-0.0000001
$\dfrac{f(29+h) - L(29)}{h}$				

8. Graphically support your numerical work by graphing the difference quotient.

Companion Website

complements the text by offering a variety of teaching and learning resources including: a chapter quiz, solutions to many of the interactive activities, the online technology manuals referenced in the Technology Options, links to other useful resources, and a wealth of additional support material.

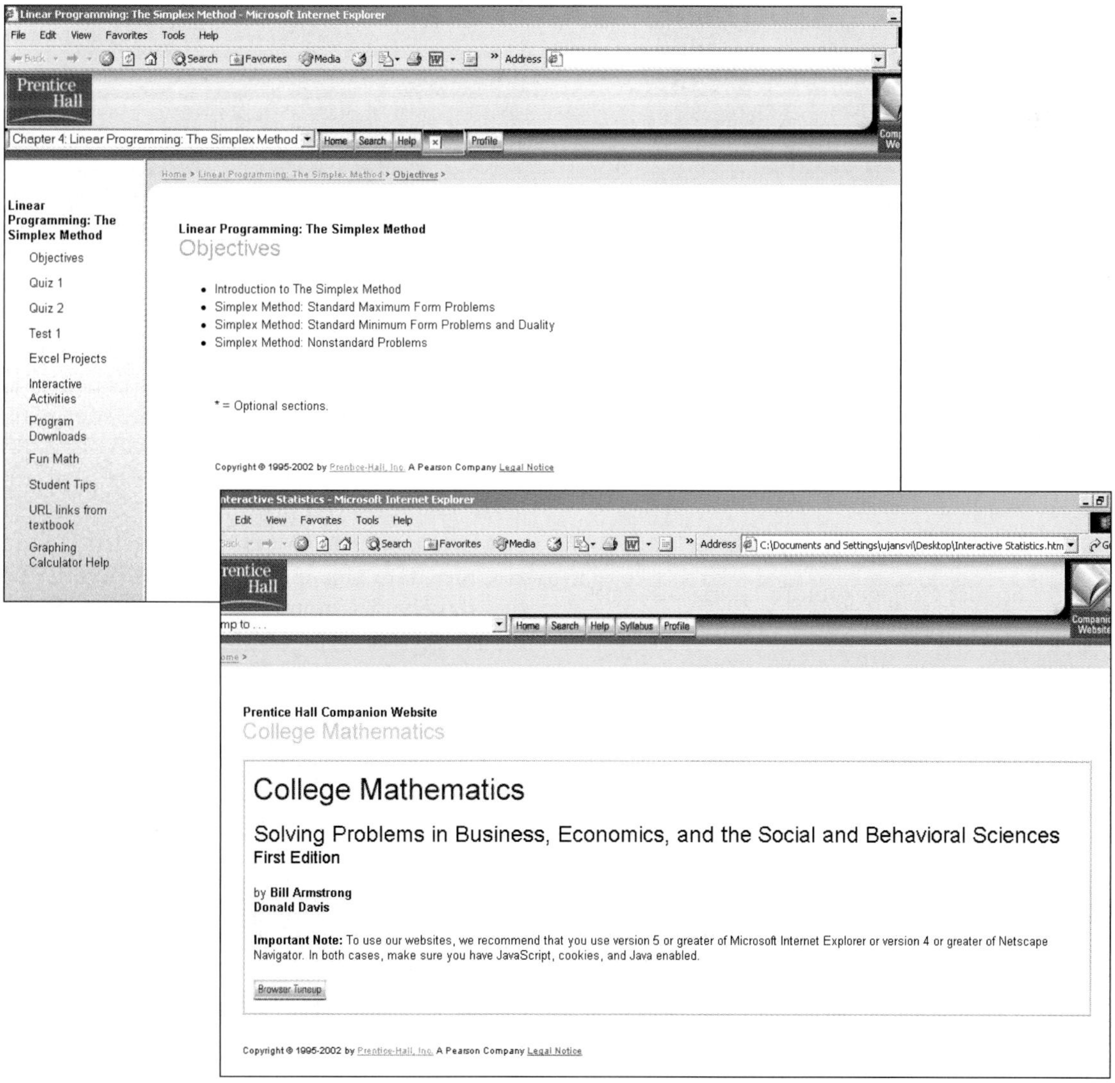

About the Authors

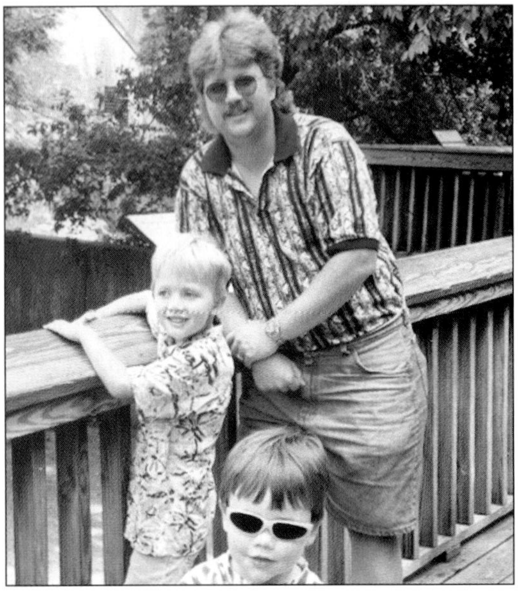

Bill Armstrong

Bill is a native of Ohio and became a hard-core Buckeyes fan after earning both his Bachelors and Masters degrees in Mathematics at The Ohio State University.

Bill taught at numerous colleges including Ohio State and Phoenix College, before taking his current position at Lakeland Community College near Cleveland, Ohio. Bill enjoys working with students at all levels and teaches courses ranging from Algebra to Differential Equations. He employs various teaching strategies to interest and motivate students including using humor to lighten the subject matter and inviting comments from students. His enjoyment of teaching and constant interaction with students has earned him a reputation as an innovative, enthusiastic, and effective teacher.

When Bill is not teaching, tutoring during office hours, or writing, he enjoys playing pool and golf, coaching his sons' little league baseball teams, and hanging out at home with his wife... not necessarily in that order!

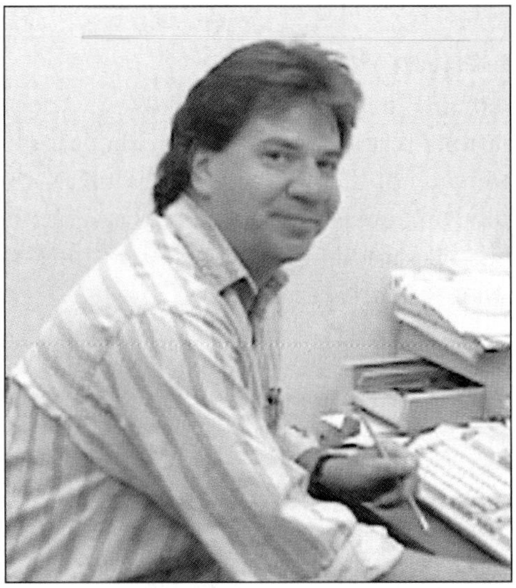

Don Davis

Also a native of Ohio, Don earned a Bachelor of Science degree in Education, specializing in Political Science, Economics, and Mathematics from Bowling Green State University in Bowling Green, Ohio. After teaching at the high school level, Don received his Master of Science degree in Mathematics from Ohio University. He taught at Ohio State University – Newark before joining the Lakeland Community College faculty. With his background in Economics, Don is always searching for ways to apply mathematics to the "real world" and to connect mathematical concepts to the various courses his students are taking. By using technology, along with interesting, practical applications, Don brings his many mathematics classes to life for students. Like Bill, one typically finds Don in his office helping students understand the concepts of his courses. His dedication to his students makes him a sought-after teacher.

In his free time, Don is often found playing with his three children or scurrying after one of the many pets currently in residence at his home including a rat, two parakeets, a turtle, a lizard, a dog, and two cats. Occasionally, he enjoys a quiet night at home.

Chapter 1

Linear Models and Systems of Linear Equations

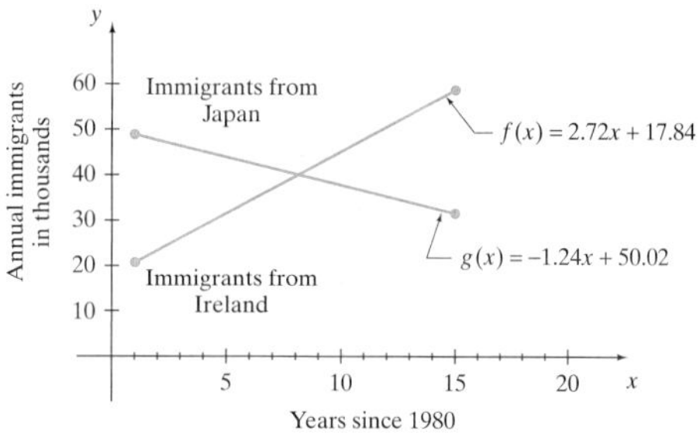

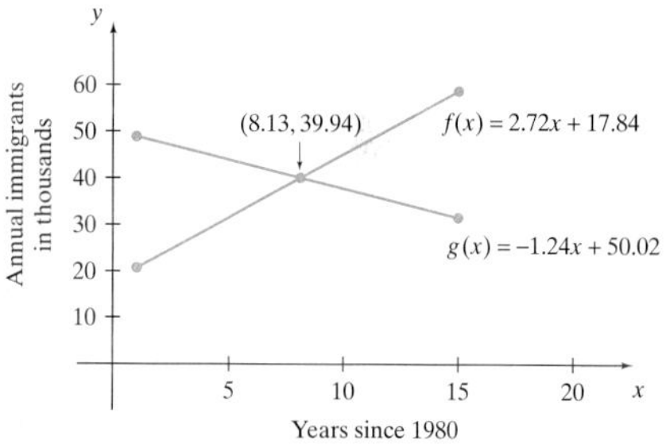

The Statue of Liberty has been a beacon of hope for immigrants to the United States for many years. Although the flow of immigrants slowed by the end of the 20th century, America continued to open her doors to men and women from across the globe. By using linear models and graphing techniques we can compare patterns of immigration and predict future trends, as seen in these graphs of Japanese and Irish immigrants. In the graph to the right, the intersection point shows that the number of immigrants from Japan and Ireland were equal in 1988.

What We Know

We begin our study of finite mathematics having learned the basic skills in algebra. These skills include solving, factoring, and graphing equations.

Where Do We Go

In this chapter we will review the concept of problem solving, explore properties of linear functions, and study the average rate of change. Then we will see how to solve two linear equations simultaneously using graphic and algebraic techniques.

Section 1.1 Problem Solving: Linear Equations

Most of us use or hear the word *problem* in our daily conversation. We may say or hear statements such as

- "I'm having a problem loading this software."
- "The problem is that management doesn't communicate with us."
- "I had a problem getting tickets for the concert."

Many problems that we encounter can be solved easily, but others require a more thoughtful approach. In finite mathematics we are often confronted with the challenge of solving problems that require a thoughtful approach. As a result, we need to review a general problem-solving strategy and the accompanying skills that will allow us to solve a variety of problems. The skills reviewed in this section will be used throughout the text.

English to Mathematics

From prior mathematics classes we have seen how a **variable** can be used to represent an unknown quantity. For example, suppose that we hear the following report on the local news:

"CleanAir produces 1475 of two types of vacuums, Mod I and Mod II, each week."

From this statement we do not know how many of each vacuum is made each week, but from the report we do know that

The number of Mod I produced plus the number of Mod II produced is 1475.

Now, if we let x represent the number of Mod I produced and let y represent the number of Mod II produced, it is more convenient to write the prior English sentence as the mathematical sentence

$$x + y = 1475$$

Converting English to mathematics is an important skill. An essential part of this skill is called **defining the variable**. This is when we state what the variable represents. Defining the variable occurs only after we carefully read the problem and understand what we are asked to do. Example 1 eases us into this process.

Example 1 Translating English to Mathematics

A certain number tripled plus seven is nineteen. Translate this English sentence to a mathematical sentence.

Solution

After reading the first sentence, we realize that we need to define a variable. So we let x represent the certain number. By "a certain number tripled," we mean the product of 3 and a certain number. Thus $3x$ represents this number tripled.

So we have

A certain number tripled plus seven is nineteen.	English sentence
$3x + 7 = 19$	Mathematical sentence

Checkpoint 1 Now work Exercise 5.

The mathematical sentence we came up with in Example 1,

$$3x + 7 = 19$$

is more commonly called an **equation**. Once we have an equation, we can solve it and answer the original problem. Solving the equation in Example 1 gives us

$3x + 7 = 19$	Given equation
$3x + 7 - 7 = 19 - 7$	Subtract 7 from both sides
$3x = 12$	Simplify
$x = 4$	Divide both sides by 3

So the "certain number" in Example 1 was $x = 4$. This can be checked by going back to the original English sentence.

A certain number tripled plus seven is nineteen.

We found that the certain number is $x = 4$. Four tripled is 12; $12 + 7$ is 19.

Example 1 and the discussion following Example 1 illustrate the general strategy for analyzing and solving word problems.

Strategy for Solving Word Problems

1. Carefully **read** the problem and **understand the situation**. As you analyze a problem, you may need to
   - **Define the variables** to represent all the unknowns.
   - **Construct** a mathematical model to represent the problem or write an equation (or equations) to represent the problem.
2. **Solve** the equation. Make sure to check the solution in the original equation.
3. **Interpret the solution.**

Example 2 **Solving a Word Problem**

The Dirt Rocket Company produces spokes for off road motorcycles. Each spoke costs \$3 to produce. The company also spends \$1250 for advertisements in a certain motorcycle enthusiast magazine.

(a) Determine the total cost for advertising and producing 400 spokes.

(b) Write an equation for the costs incurred for advertisement and the production of any number of spokes.

Solution

(a) **Understand the Situation:** If The Dirt Rocket Company produces no spokes, they still have to pay the \$1250 for advertising. Such a cost is called a **fixed cost**. Since each spoke costs \$3 to produce, the cost to produce 400 spokes is given by

$$(\text{cost per spoke}) \cdot (\text{number of spokes})$$
$$3(400) = \$1200$$

Letting C represent the total cost of producing 400 spokes and paying for the advertisements gives us

$$\begin{aligned} C &= (\text{cost per spoke}) \cdot (\text{number of spokes}) + (\text{advertisement cost}) \\ &= 3 \cdot 400 + 1250 \\ &= 1200 + 1250 \\ &= \$2450 \end{aligned}$$

Interpret the Solution: The total cost, C, of producing 400 spokes and the advertisements is \$2450.

(b) **Understand the Situation:** Let C represent the total cost of advertising and producing any number of spokes. Since we do not know how many spokes are produced, we let x represent the number of spokes produced.

Using the variables C and x, we can write an equation to model the problem. This gives us

$$\begin{aligned} C &= (\text{cost per spoke}) \cdot (\text{number of spokes}) + (\text{advertisement cost}) \\ &= 3x + 1250 \end{aligned}$$

■

✓ Checkpoint 2 Now work Exercise 13.

Systems of Linear Equations

In the first part of this text we often need to represent a situation with a **system of linear equations**. (A system of linear equations is basically two or more linear equations with at least one variable.) We will review the techniques for solving systems of linear equations in Sections 1.4 and 1.5. For now, we focus on representing a word problem with a system of linear equations. Example 3 illustrates how to set up linear equations that form a system. You will solve this system in Exercise 67 of Section 2.1.

Example 3 Writing Multiple Linear Equations

Max George is a craftsman who makes end tables and night stands. To make one end table takes 5 hours of his labor, and to make one night stand takes 4 hours of his labor. The materials for one end table cost him \$200 and the materials for one night stand cost him \$80. Suppose that for 1 week Max works 40 hours and spends \$1200 on materials.

(a) Write an equation that represents Max's costs.

(b) Write an equation that represents Max's time (labor).

Solution

(a) **Understand the Situation:** We know that materials for each end table cost Max \$200 and materials for each night stand cost Max \$80. We also know that Max spent \$1200 on materials. What we don't know is how many end tables and how many night stands he made. So we let x represent the number of end tables made in 1 week and let y represent the number of night stands made in 1 week.

This means that he spent

$$\begin{array}{ccc}(\text{cost per end table}) \cdot (\text{number of end tables}) & & \\ 200 \quad\cdot\quad x & = 200x\end{array}$$

dollars on end tables. We also know that he spent

$$\begin{array}{ccc}(\text{cost per night stand}) \cdot (\text{number of night stands}) & & \\ 80 \quad\cdot\quad y & = 80y\end{array}$$

dollars on night stands. Since he spent a total of \$1200 on materials, an equation that represents Max's costs is

$$\begin{pmatrix}\text{dollars spent}\\ \text{on end tables}\end{pmatrix} + \begin{pmatrix}\text{dollars spent}\\ \text{on night stands}\end{pmatrix} = \begin{pmatrix}\text{total dollars}\\ \text{spent on materials}\end{pmatrix}$$

$$200x + 80y = 1200$$

(b) **Understand the Situation:** We know that to make each end table takes Max 5 hours and to make each night stand takes Max 4 hours. We also know that Max spent a total of 40 hours working on these items in 1 week. What we don't know is how many end tables and how many night stands he made. As in part (a), we let x represent the number of end tables made in 1 week and let y represent the number of night stands made in 1 week.

Then we know that he spent

$$\begin{array}{c}(\text{hours per end table}) \cdot (\text{number of end tables}) \\ 5x\end{array}$$

hours on end tables. We also know that he spent

$$\begin{array}{c}(\text{hours per night stand}) \cdot (\text{number of night stands}) \\ 4y\end{array}$$

hours on night stands. Since he spent a total of 40 hours working on these items, an equation that represents Max's time is

$$\begin{pmatrix}\text{hours spent}\\ \text{on end tables}\end{pmatrix} + \begin{pmatrix}\text{hours spent}\\ \text{on night stands}\end{pmatrix} = \begin{pmatrix}\text{total hours}\\ \text{spent on both}\end{pmatrix}$$

$$5x + 4y = 40$$

■

Checkpoint 3 Now work Exercise 15.

In Example 3 notice that in both parts (a) and (b) we let x represent the number of end tables made in 1 week and let y represent the number of night stands made in 1 week. Since each equation had the same variables defined to represent the same unknown quantities, we can write the equations as the **linear system**

$$\begin{aligned}200x + 80y &= 1200\\ 5x + 4y &= 40\end{aligned}$$

Solving this linear system requires finding a value for x and a value for y that makes both equations true. As stated earlier, you will learn how to solve systems later in Chapter 1.

Example 4 Writing Multiple Linear Equations

Gonzalez Office Supplies manufactures three types of office chairs: Secretarial, Executive, and Presidential. Each chair requires time on three different machines. These times and the total available time for each machine are given in Table 1.1.1.

Table 1.1.1

	Secretarial	Executive	Presidential	Total Weekly Hours
Hours on Machine A	1	1	2	30
Hours on Machine B	2	3	2	53
Hours on Machine C	1	2	3	47

Assuming that all available time for each machine must be used each week, do the following:

(a) Write an equation that represents the total number of hours that all three chairs spend on Machine A.

(b) Write an equation that represents the total number of hours that all three chairs spend on Machine B.

(c) Write an equation that represents the total number of hours that all three chairs spend on Machine C.

Solution

(a) **Understand the Situation:** Let x represent the number of Secretarial chairs made in 1 week, y represent the number of Executive chairs made in 1 week, and z represent the number of Presidential chairs made in 1 week. Since x represents the number of Secretarial chairs made in 1 week, and it takes 1 hour of machine time on Machine A, the weekly hours on Machine A for x Secretarial chairs is given by

$$\begin{pmatrix}\text{hours per}\\\text{Secretarial chair}\end{pmatrix}\cdot\begin{pmatrix}\text{number of}\\\text{Secretarial chairs}\end{pmatrix}=\begin{pmatrix}\text{hours on Machine A}\\\text{spent on Secretarial chairs}\end{pmatrix}$$

$$1 \cdot x = x$$

Similarly, the weekly hours on Machine A for y Executive chairs is given by

$$\begin{pmatrix}\text{hours per}\\\text{Executive chair}\end{pmatrix}\cdot\begin{pmatrix}\text{number of}\\\text{Executive chairs}\end{pmatrix}=\begin{pmatrix}\text{hours on Machine A}\\\text{spent on Executive chairs}\end{pmatrix}$$

$$1 \cdot y = y$$

We know that it takes 2 hours on Machine A for each Presidential chair. So for z Presidential chairs, we have

$$\begin{pmatrix}\text{hours per}\\\text{Presidential chair}\end{pmatrix}\cdot\begin{pmatrix}\text{number of}\\\text{Presidential chairs}\end{pmatrix}=\begin{pmatrix}\text{hours on Machine A}\\\text{spent on Presidential chairs}\end{pmatrix}$$

$$2 \cdot z = 2z$$

So if we have a total of 30 hours available on Machine A each week that must be used, we get the equation

$$\begin{pmatrix}\text{hours spent on}\\ \text{Secretarial chairs}\end{pmatrix} + \begin{pmatrix}\text{hours spent on}\\ \text{Executive chairs}\end{pmatrix} + \begin{pmatrix}\text{hours spent on}\\ \text{Presidential chairs}\end{pmatrix} = \begin{pmatrix}\text{total weekly hours}\\ \text{on Machine A}\end{pmatrix}$$

$$x + y + 2z = 30$$

(b) For Machine B, we know that each Secretarial chair takes 2 hours, each Executive chair takes 3 hours, and each Presidential chair takes 2 hours. So for a total of 53 hours available on Machine B that must be used, we get the equation

$$\begin{pmatrix}\text{hours spent on}\\ \text{Secretarial chairs}\end{pmatrix} + \begin{pmatrix}\text{hours spent on}\\ \text{Executive chairs}\end{pmatrix} + \begin{pmatrix}\text{hours spent on}\\ \text{Presidential chairs}\end{pmatrix} = \begin{pmatrix}\text{total weekly hours}\\ \text{on Machine B}\end{pmatrix}$$

$$2x + 3y + 2z = 53$$

(c) For Machine C, we know that each Secretarial chair takes 1 hour, each Executive chair takes 2 hours and each Presidential chair takes 3 hours. So for a total of 47 hours available on Machine C that must be used, we get the equation

$$\begin{pmatrix}\text{hours spent on}\\ \text{Secretarial chairs}\end{pmatrix} + \begin{pmatrix}\text{hours spent on}\\ \text{Executive chairs}\end{pmatrix} + \begin{pmatrix}\text{hours spent on}\\ \text{Presidential chairs}\end{pmatrix} = \begin{pmatrix}\text{total weekly hours}\\ \text{on Machine C}\end{pmatrix}$$

$$x + 2y + 3z = 47$$ ■

Since x, y, and z represent the number of Secretarial, Executive, and Presidential chairs, respectively, for each of the three linear equations, we can express the time used on all three machines with the system of linear equations

$$\begin{aligned} x + y + 2z &= 30 \\ 2x + 3y + 2z &= 53 \\ x + 2y + 3z &= 47 \end{aligned}$$

We will solve and interpret this system in a Flashback in Section 1.5.

SUMMARY

In this section, we learned how to translate English sentences into mathematical sentences. Regardless of the application, we use the **Strategy for Solving Word Problems**:

1. Carefully **read** the problem and **understand the situation**. As you analyze a problem, you may need to
   - **Define the variables** to represent all the unknowns.
   - **Construct** a mathematical model to represent the problem or write an equation (or equations) to represent the problem.
2. **Solve** the equation. Make sure to check the solution in the original equation.
3. **Interpret the solution.**

SECTION 1.1 EXERCISES

In our exercise sets, we often revisit applications. When a number appears under the ⇒ symbol, it means that the application will be seen again or was previously seen in that section.

In Exercises 1–10, translate the English sentence into a mathematical sentence.

1. A certain number increased by ten is 14.

2. A certain number increased by four is six.

3. A certain number decreased by seven is 28.

4. A certain number decreased by one is five.

✓ **5.** The product of five and a certain number is 45.

6. The product of two and a certain number is 112.

7. The product of six and a certain number minus one is 29.

8. The product of four and a certain number minus three is five.

9. The product of three and a certain number plus 11 is 44.

10. The product of eight and a certain number plus nine is 57.

⇒1.4 **11.** Adel starts a business of manufacturing picture frames for small 3 by 5 inch photographs. He determines the fixed cost to be \$252 and the cost to produce each frame is \$1.50.

(a) Determine the total cost for producing 10 picture frames.

(b) Determine an equation for the total cost of producing any number of picture frames.

⇒1.4 **12.** Sally makes dream catchers as a hobby and wants to sell them at craft shows. She knows that the fixed costs are \$90 and the cost to produce each dream catcher is \$2.50.

(a) Determine the total cost for producing 18 dream catchers.

(b) Determine an equation for the total cost of producing any number of dream catchers.

⇒1.4 ✓ **13.** The Lullaby Company knows that the fixed costs to manufacture their portable crib are \$250 and the cost to produce each crib is \$100.

(a) Determine the total cost for producing 110 portable cribs.

(b) Determine an equation for the total cost of producing any number of portable cribs.

⇒1.4 **14.** Clark and Florence make bumper stickers. They know that their fixed costs are \$200 and the cost of each bumper sticker is \$3.

(a) Determine the total cost for producing 200 bumper stickers.

(b) Determine an equation for the total cost of producing any number of bumper stickers.

⇒2.1 ✓ **15.** Martha George is a craftswoman who makes coffee tables and night stands. To make 1 coffee table takes 6 hours of her labor, and to make 1 night stand takes 4 hours of her labor. The materials for 1 coffee table cost her \$220, and the materials for 1 night stand cost her \$80. Suppose that for 1 week Martha works 40 hours and spends \$1400 on materials. Let x represent the number of coffee tables made in 1 week and let y represent the number of night stands made in 1 week.

(a) Write an equation that represents Martha's costs.

(b) Write an equation that represents Martha's time (labor).

⇒2.1 **16.** The WeDo Wood Canoes Company makes two types of canoes: a two-person model and a four-person model. Each two-person model requires 1 hour in the cutting department and 1.5 hours in the assembly department. Each four-person model requires 1 hour in the cutting department and 2.75 hours in the assembly department. The cutting department has a maximum of 640 hours available each week, while the assembly department has a maximum of 1080 hours available each week. Let x represent the number of two-person canoes made each week and let y represent the number of four-person canoes made each week.

(a) Write an equation that represents the number of hours that both models of canoe spend in the cutting department each week, assuming that all available hours are used.

(b) Write an equation that represents the number of hours that both models of canoe spend in the assembly department each week, assuming that all available hours are used.

⇒2.1 **17.** (*continuation of Exercise 16*) Suppose that a breakthrough in the assembly process decreases the time required in the assembly department for each two-person model to $\frac{3}{4}$ hour and for each four-person model to 2 hours. The cutting department still has a maximum of 640 hours available each week, while the assembly department has a maximum of 1080 hours available each week. Let x represent the number of two-person canoes made each week and let y represent the number of four-person canoes made each week.

(a) Write an equation that represents the number of hours that both models of canoe spend in the cutting department each week, assuming that all available hours are used.

(b) Write an equation that represents the number of hours that both models of canoe spend in the assembly department each week, assuming that all available hours are used.

18. The SnackPower company has two ingredients to combine to make their new ZapSnack energy bar. Ingredient I contains 3 grams of carbohydrate and 2 grams of protein per ounce. Ingredient II contains 5 grams of carbohydrate and 3 grams of protein per ounce. They want to combine the two ingredients so that the energy bar has 73 grams of carbohydrate and 46 grams of protein per ounce. Let x represent the number of ounces of ingredient I and let y represent the number of ounces of ingredient II.

(a) Write an equation that represents the number of ounces of carbohydrates in the two ingredients.

(b) Write an equation that represents the number of ounces of protein in the two ingredients.

⇒2.4 **19.** The Drama Department in a local high school is performing a play for three consecutive nights. The auditorium at the high school has 1600 seats, and tickets are broken into two categories: \$3 student tickets and \$5 adult tickets. Table 1.1.2 gives the data for ticket revenue and total tickets sold for each play night.

Table 1.1.2

	Thursday	**Friday**	**Saturday**
Tickets sold	1600	1600	1600
Revenue	\$5760	\$5980	\$5500

(a) Define the variables and write an equation representing the number of tickets sold on Thursday night, and write an equation representing the revenue generated on Thursday night.

(b) Define the variables and write an equation representing the number of tickets sold on Friday night, and write an equation representing the revenue generated on Friday night.

(c) Define the variables and write an equation representing the number of tickets sold on Saturday night, and write an equation representing the revenue generated on Saturday night.

⇒2.4 **20.** (*continuation of Exercise 19*) Because each night was a sellout, the next semester the Drama Department moved the play to a local community college. The auditorium at the community college has 2800 seats. The ticket prices remain the same. Table 1.1.3 gives the data for ticket revenue and total tickets sold for each play night.

Table 1.1.3

	Thursday	Friday	Saturday
Tickets sold	2800	2800	2800
Revenue	\$11,400	\$10,180	\$10,760

(a) Define the variables and write an equation representing the number of tickets sold on Thursday night, and write an equation representing the revenue generated on Thursday night.

(b) Define the variables and write an equation representing the number of tickets sold on Friday night, and write an equation representing the revenue generated on Friday night.

(c) Define the variables and write an equation representing the number of tickets sold on Saturday night, and write an equation representing the revenue generated on Saturday night.

21. The Second City Bank has two types of checking accounts. The City account charges \$5 per month plus \$0.15 for each check that is written. The Urban account charges \$4 per month plus \$0.20 for each check that is written.

(a) Define the variables and write an equation that represents the monthly cost of the City checking account.

(b) Define the variables and write an equation that represents the monthly cost of the Urban checking account.

22. Membership at the Iron Wrist Tennis Club costs \$1000 per year plus \$3 per hour of court usage. The Line-Out! Tennis Club has an \$800 yearly fee with \$5 per hour court fees.

(a) Define the variables and write an equation that represents the total yearly cost of playing at the Iron Wrist Tennis Club.

(b) Define the variables and write an equation that represents the total yearly cost of playing at the Line-Out! Tennis Club.

23. The Frank & Earnst Law Firm charges a \$1500 flat fee plus 25% of the settlement for litigating a civil case. The Lopez & Mackenzie Law Firm charges a \$2000 flat fee plus 20% of the settlement for litigating a civil case.

(a) Define the variables and write an equation that represents the cost that the Frank & Earnst Law Firm charges for litigating a civil case.

(b) Define the variables and write an equation that represents the cost that the Lopez & Mackenzie Law Firm charges for litigating a civil case.

24. The Yorkville School District has two property tax proposals. Under proposal D, a homeowner pays \$500 in addition to 15% of the assessed value of the house. Under proposal E, a homeowner pays \$2000 in addition to 4% of the assessed value of the house.

(a) Define the variables and write an equation for the property tax under proposal D.

(b) Define the variables and write an equation for the property tax under proposal E.

⇒2.1 **25.** The Stitz Tent Company makes three types of tents: two- , four- , and six-person models. Each tent requires the services of three departments as shown in Table 1.1.4. The fabric cutting, assembly, and packaging departments have available a maximum of 450, 390, and 190 hours each week, respectively.

Table 1.1.4

	Two-person (hr)	Four-person (hr)	Six-person (hr)
Fabric cutting	0.5	0.8	1.1
Assembly	0.6	0.7	0.9
Packaging	0.2	0.3	0.6

Let x represent the number of two-person tents made in 1 week, y represent the number of four-person tents made in 1 week, and z represent the number of six-person tents made in 1 week.

(a) Write an equation that represents the number of hours that the three models spend in the fabric cutting department per week, assuming that all available hours are used.

(b) Write an equation that represents the number of hours that the three models spend in the assembly department per week, assuming that all available hours are used.

(c) Write an equation that represents the number of hours that the three models spend in the packaging department per week, assuming that all available hours are used.

⇒2.1 **26.** Rework Exercise 25(a), (b), and (c), except here assume that the fabric cutting, assembly, and packaging departments have available a maximum of 420, 390, and 185 hours each week, respectively.

⇒2.1 **27.** Last year the Hannus Corporation took out three loans totaling \$50,000 to fund the purchase of new equipment. Some of the money was financed at 6%, \$1000 more than $\frac{1}{2}$ the amount of the 6% loan was financed at 8%, and the rest was financed at 7%. The total annual interest on the loans was \$3410. Let x represent the amount borrowed at 6%, y represent the amount borrowed at 7%, and z represent the amount borrowed at 8%.

(a) Write an equation that represents the first sentence.

(b) Write an equation that represents "\$1000 more than $\frac{1}{2}$ the amount of the 6% loan was financed at 8%."

(c) Write an equation that represents the third sentence.

28. Rework Exercise 27(a), (b), and (c), except here assume that the three loans totaled \$60,000 and the total annual interest on the loans was \$4085.

⇒ 2.1 **29.** The data in Table 1.1.5 give the amounts of nitrogen, phosphate, and potash (all in lb/acre) needed to grow asparagus, cauliflower, and celery for an entire growing season in California.

Table 1.1.5

	Asparagus	Cauliflower	Celery
Nitrogen	240	215	336
Phosphate	148	98	171
Potash	42	69	147

SOURCE: U.S. Department of Agriculture, www.usda.gov/nass

Juan has 280 acres, 65,790 pounds of nitrogen, 31,860 pounds of phosphate and 21,630 pounds of potash. Let x represent the number of acres planted in asparagus, y represent the number of acres planted in cauliflower, and z represent the number of acres planted in celery. Assume that Juan wants to use all his resources completely.

(a) Write an equation that represents the total amount of nitrogen needed for the three crops.

(b) Write an equation that represents the total amount of phosphate needed for the three crops.

(c) Write an equation that represents the total amount of potash needed for the three crops.

⇒ 2.1 **30.** The data in Table 1.1.6 give the amounts of nitrogen, phosphate, and potash (all in lb/acre) needed to grow head lettuce, cantaloupe, and bell peppers for an entire growing season in California.

Table 1.1.6

	Head lettuce	Cantaloupe	Bell peppers
Nitrogen	173	175	212
Phosphate	122	161	108
Potash	76	39	72

SOURCE: U.S. Department of Agriculture, www.usda.gov/nass

Zach has 200 acres, 38,430 pounds of nitrogen, 27,845 pounds of phosphate, and 12,985 pounds of potash. Let x represent the number of acres planted in head lettuce, y represent the number of acres planted in cantaloupe, and z represent the number of acres planted in bell peppers. Assume that Zach wants to use all his resources completely.

(a) Write an equation that represents the total amount of nitrogen needed for the three crops.

(b) Write an equation that represent the total amount of phosphate needed for the three crops.

(c) Write an equation that represent the total amount of potash needed for the three crops.

⇒ 2.1 **31.** Chris O'Neill, a dietitian in a local hospital, is to arrange a special diet for a patient that is made up of three basic foods. The diet is to include 340 units of calcium, 205 units of vitamin A, and 225 units of vitamin C. The number of units per ounce for each basic food is given in Table 1.1.7.

Table 1.1.7

	Basic!	Vaca	Secra
Calcium	20	10	30
Vitamin A	10	5	20
Vitamin C	20	15	10

Assume that the dietitian wants to exactly meet the diet requirements.

(a) Define the variables and write an equation that represents the total amount of calcium from the three foods.

(b) Define the variables and write an equation that represents the total amount of vitamin A from the three foods.

(c) Define the variables and write an equation that represents the total amount of vitamin C from the three foods.

⇒ 2.1 **32.** Rework Exercise 31 if the patient's diet is changed so that it includes 480 units of calcium, 285 units of vitamin A, and 325 units of vitamin C.

⇒ 2.4 **33.** Davram manufactures bicycle frames at three different plants. Due to different technologies at each plant, the production rate (in frames/hr) differs for each plant, as shown in Table 1.1.8.

Table 1.1.8

	Plant I	Plant II	Plant III
Racing frame	18	20	16
Hybrid frame	12	9	11
Mountain frame	22	18	17

Shannon, the production manager at Davram, has received an order to produce 5900 racing frames, 3580 hybrid frames, and 6350 mountain frames. Shannon needs to fill this order exactly.

(a) Define the variables and write an equation that represents the total number of racing frames from the three plants.

(b) Define the variables and write an equation that represents the total number of hybrid frames from the three plants.

(c) Define the variables and write an equation that represents the total number of mountain frames from the three plants.

⇒ 2.4 **34.** (*continuation of Exercise 33*) Shannon has received another order requesting 6040 racing frames, 3780 hybrid frames, and 6695 mountain frames. With everything else from Exercise 33 remaining the same, rework parts (a), (b), and (c).

⇒ 2.1

SECTION PROJECT

During rush hour, traffic congestion may be encountered at certain intersections. For a network of four one-way streets in a certain city, the rush hour flow is as shown in Figure 1.1.1. To help to speed the flow of traffic, traffic engineers need to coordinate the signals at these intersections. To do this, data are collected giving the hourly flow of traffic during rush hour at certain locations. Figure 1.1.1 shows the data that were collected. Notice that the figure indicates that a total of 1530 cars per hour enter the intersection of 1st Street and 5th Avenue. Of this 1530 cars, x of these cars leave the intersection on 5th Avenue and w of these cars leave the intersection on 1st Street. Since the total number of cars leaving this (or any) intersection equals the number entering, we have

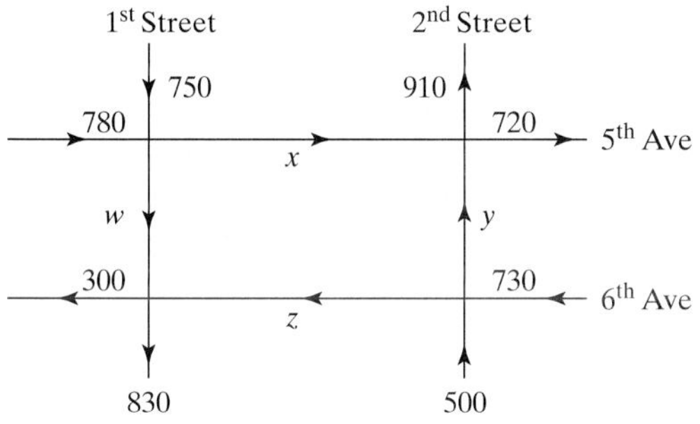

Figure 1.1.1

$$x + w = 1530$$

This equation gives the flow at the intersection of 1st Street and 5th Avenue.

(a) Determine the equations for the traffic flow at the other three intersections. Remember that the total entering the intersection equals the total leaving the intersection, and vice versa.

(b) Write a system of four equations representing the traffic flow for this network of four one-way streets.

Section 1.2 Introduction to Linear Functions

Many times in life, what people desire is some measure of predictability. There is a certain solace in knowing that if today is Tuesday, then it is garbage day. This same kind of idea is also sought in mathematics. We want to know that if we are given a value x we can determine a value for y, without ambiguity. This, in part, is what the **function concept** is about. In this section we will learn how functions are defined, what makes up functions, and the usefulness of functions. We will also explore properties of a special type of function called the **linear function**.

The Function Concept

Suppose that Manuel takes a typing class and, during the course of the semester, takes a series of tests to determine how fast he is typing. His results are shown in Table 1.2.1.

Table 1.2.1

Week of class	0	3	5	7	9	12
Typing speed in words per minute	10	20	28	45	50	54

To make a picture of these data, we can use the **Cartesian plane**. This rectangular coordinate system is made up of two perpendicular number lines called **axes**. Usually, the horizontal number line is called the ***x*-axis**, and the vertical number line is called the ***y*-axis**. See Figure 1.2.1.

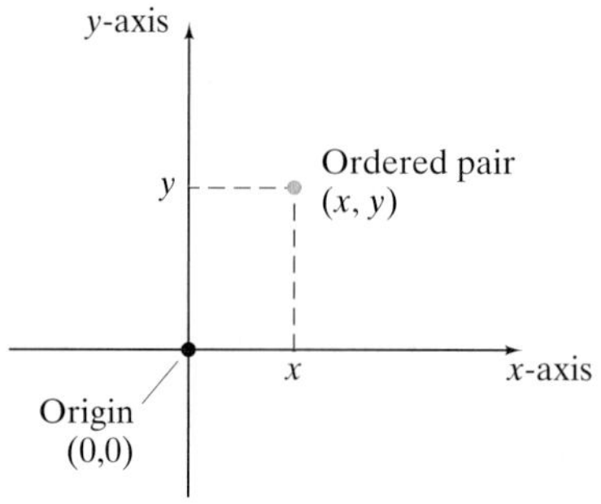

Figure 1.2.1

Since we want to investigate the relationship between the two sets of data, we can plot the data in the table by making **ordered pairs** consisting of x- and y-**coordinates** in the form (x, y). For Manuel, if we let x represent the week of typing class and let y represent typing speed in words per minute, we can construct ordered pairs of the form (x, y) as shown in Table 1.2.2.

Table 1.2.2

Week of class, x	0	3	5	7	9	12
Typing speed in words per minute, y	10	20	28	45	50	54
Ordered pair (x, y)	$(0, 10)$	$(3, 20)$	$(5, 28)$	$(7, 45)$	$(9, 50)$	$(12, 54)$

Now we can plot each ordered pair or **point** by moving to the right x units and up y units. The graph of the typing speed data is given in Figure 1.2.2. Note the labels on the axes.

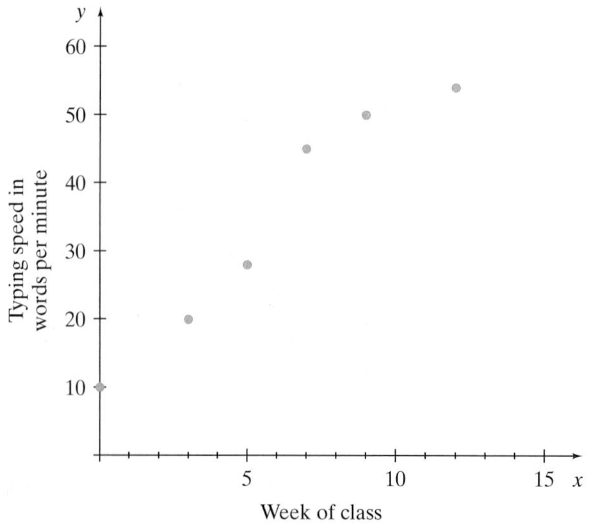

Figure 1.2.2

The choice of letting x represent the number of weeks in class and y the typing speed was not done arbitrarily. We commonly call x the **independent** variable and y the **dependent** variable. Since the typing speed presumably *depends* on the length of time that Manuel is in the typing class, we assigned y to the typing speed.

Notice that if we know what week of class it is we can determine the typing speed. This is much of what the function concept is all about! Now let's apply a mathematical definition to the function concept.

Function

A **function** is a rule, or a set of ordered pairs, that assigns a unique y-value to each x-value. The set of all possible x-values is called the **domain**, and the set of all possible y-values is called the **range**.

Example 1 Plotting Ordered Pairs and Identifying the Domain and Range

Plot the data in Table 1.2.3 on the Cartesian plane and write the sets for the domain and range.

Table 1.2.3

x	−2	−1	0	1	2
y	3	−2	4	5	0

Solution

Notice here that some of the x-values are negative, so we must include negative numbers on the axes on our Cartesian plane. The data are plotted in Figure 1.2.3.

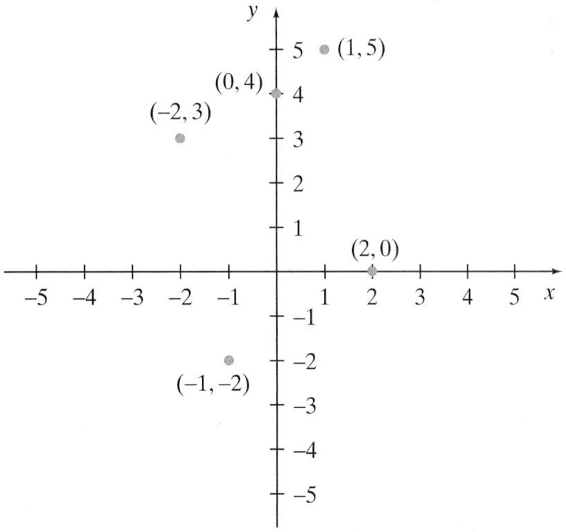

Figure 1.2.3

Since the domain is the set of all possible x-values, the domain is the set $\{-2, -1, 0, 1, 2\}$. Since the range is the set of all possible y-values, the range is the set $\{-2, 0, 3, 4, 5\}$. Note that we write the elements of these sets in ascending order, that is, from smallest to largest. ■

Checkpoint 1

Now work Exercise 11.

We often desire to write a convenient algebraic expression to represent a function. We denote a function by writing $f(x)$, read "f of x," where f is the name of the function and x is the independent variable. This notation allows us to write functions such as

$$f(x) = 3x - 4 \qquad g(x) = \frac{9x + 6}{4} \qquad h(t) = t^2 - 2t$$

Notice that we can designate functions with names like f, g, and h. The letters x and t are used to denote independent variables. **This means that $f(x)$, $g(x)$, and $h(t)$ are all different names for the dependent variable y.** To get ordered pairs from these functions, we must substitute (or "plug in") a value for the independent variable and compute the corresponding value of the dependent variable. This process of substituting values is called **evaluating** a function. When we write $f(4)$, for example, this means to replace every occurrence of the independent variable with 4 and then simplify. The process of evaluating functions is shown in Example 2.

Example 2 **Evaluating a Function**

For the function $f(x) = 2x - 3$, evaluate $f(-1)$, $f(0)$, $f(2)$, and $f(4)$ and then make a table of values including ordered pairs.

Solution

Evaluating this function, we get

$$f(-1) = 2(-1) - 3 = -2 - 3 = -5$$
$$f(0) = 2(0) - 3 = 0 - 3 = -3$$
$$f(2) = 2(2) - 3 = 4 - 3 = 1$$
$$f(4) = 2(4) - 3 = 8 - 3 = 5$$

The values and ordered pairs are shown in Table 1.2.4.

Table 1.2.4

x	-1	0	2	4
y	-5	-3	1	5
Ordered pairs (x, y)	$(-1, -5)$	$(0, -3)$	$(2, 1)$	$(4, 5)$

Linear Functions

Many of the topics that we study in finite mathematics focus on the study of **linear functions**. When graphed, a linear function has the shape of a line. But in what ways are linear functions used? Let's say that Ken has a pickup truck with a 12-gallon gas tank. From experience, he knows that the truck consumes $\frac{1}{20}$ of a gallon of gas for each mile driven. (This is another way to say that the truck gets 20 miles per gallon.) This means, for example, that if he drives the truck 60 miles he uses $60\left(\frac{1}{20}\right) = 3$ gallons of gas. We can say that the amount of gas in the tank has **decreased** by 3 gallons. Now let's suppose that Ken starts a road trip with a full tank of gas and wants to know how many gallons of gas remain in the tank as he takes his trip. To solve this problem, let's begin by **defining the variables**.

Define the Variables: Let x represent the number of miles traveled and $y = f(x)$ represent the number of gallons of gas remaining in the tank.

To see how a function can be written, we make a table to represent the number of gallons left in the tank as various mileages are driven. See Table 1.2.5.

Table 1.2.5

Miles driven, x	Calculation	Gallons remaining, $y = f(x)$
0	$12 - \frac{1}{20}(0)$	12
60	$12 - \frac{1}{20}(60)$	9
120	$12 - \frac{1}{20}(120)$	6
180	$12 - \frac{1}{20}(180)$	3
240	$12 - \frac{1}{20}(240)$	0

The general form for the function is

$$\begin{pmatrix}\text{gallons remaining}\\ \text{in the tank}\end{pmatrix} = \begin{pmatrix}\text{gas tank}\\ \text{capacity}\end{pmatrix} - \begin{pmatrix}\text{rate of gas}\\ \text{consumption}\end{pmatrix}\begin{pmatrix}\text{miles}\\ \text{driven}\end{pmatrix}$$

In terms of our defined variables, we have

$$f(x) = 12 - \frac{1}{20}x, \qquad 0 \le x \le 240$$

The graph of this function is shown in Figure 1.2.4. Notice that the graph only applies for values of x between 0 and 240, inclusive. This is why we wrote $0 \le x \le 240$ to the right of the function. We call this the **reasonable domain** of the function.

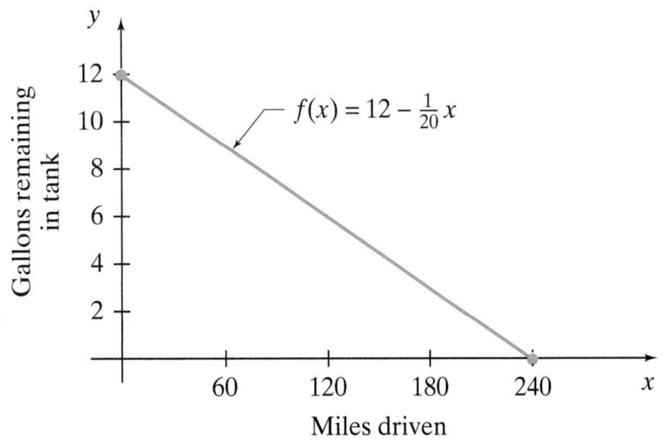

Figure 1.2.4

We know that Ken's truck uses gas at a **rate** of $\frac{1}{20}$ gallon per mile. This tells us how quickly the fuel in the tank decreases. We call this the **slope**, or **average rate of change**, of the linear function. Now let's define the linear function.

Linear Function

A **linear function** has the form

$$y = f(x) = mx + b$$

where m is the slope or the average rate of change of the line and b is a constant.

We could have written the linear function for the gas remaining in the tank in the form $f(x) = -\frac{1}{20}x + 12$. In this form we observe that the slope is $m = -\frac{1}{20}$, which tells us that the amount of gas in the tank is **decreasing** at a rate of $\frac{1}{20}$ gallon per mile, on average. Also, we observe that $b = 12$. This is simply the capacity of the fuel tank or, equivalently, the number of gallons remaining when 0 miles have been driven.

Often we are given ordered pairs and are not explicitly given the slope. To compute the slope, we use the following formula.

Slope

The **slope** of a line, denoted by m, is a measurement of the steepness of the line. Given two ordered pairs (x_1, y_1) and (x_2, y_2), the slope is determined by computing

$$m = \frac{y_2 - y_1}{x_2 - x_1}$$

The slope of a line gives the **average rate of change** of y with respect to x.

Note: The units of measure associated with slope are $\frac{\text{dependent units}}{\text{independent units}}$.

A graphical description of the slope is given in Figure 1.2.5. In addition to an average rate of change, the following are ways to interpret the slope of a line.

$$\text{slope} = m = \frac{\text{difference in } y\text{-values}}{\text{difference in } x\text{-values}} = \frac{\text{change in } y}{\text{change in } x}$$
$$= \frac{\text{rise}}{\text{run}} = \frac{\Delta y}{\Delta x} = \text{steepness of the line}$$

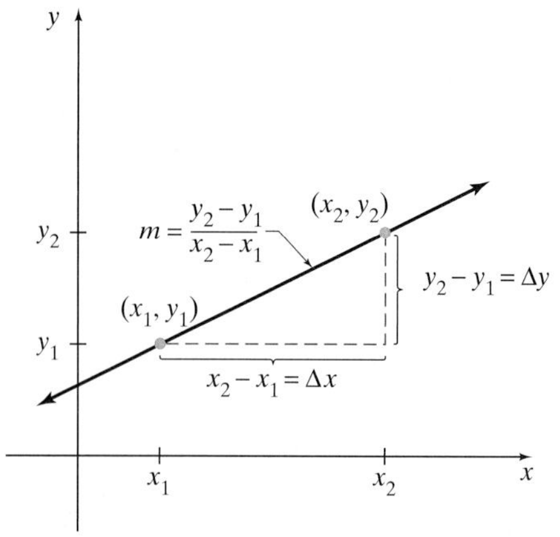

Figure 1.2.5

Now let's see how we can compute the slope if we are given some ordered pairs.

Example 3 Computing the Slope between Two Points

For each of the following, compute the slope m of the line passing through the points that represent the data. Also plot the points and sketch the line through the points.

(a) Table 1.2.6

x	y
−2	−3
3	1

(b) $(0, 4)$ and $(5, 1)$

Solution

(a) For the values in Table 1.2.6, we let $(x_1, y_1) = (-2, -3)$ and $(x_2, y_2) = (3, 1)$. Then the slope is

$$m = \frac{y_2 - y_1}{x_2 - x_1} = \frac{1 - (-3)}{3 - (-2)} = \frac{1 + 3}{3 + 2} = \frac{4}{5}$$

See Figure 1.2.6 for the graph of the line through the two points.

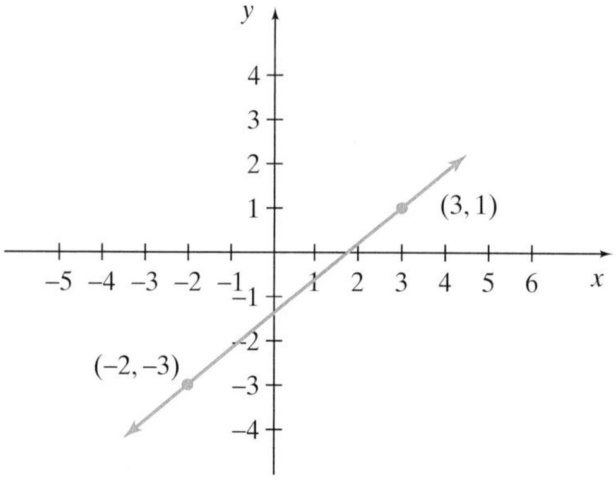

Figure 1.2.6

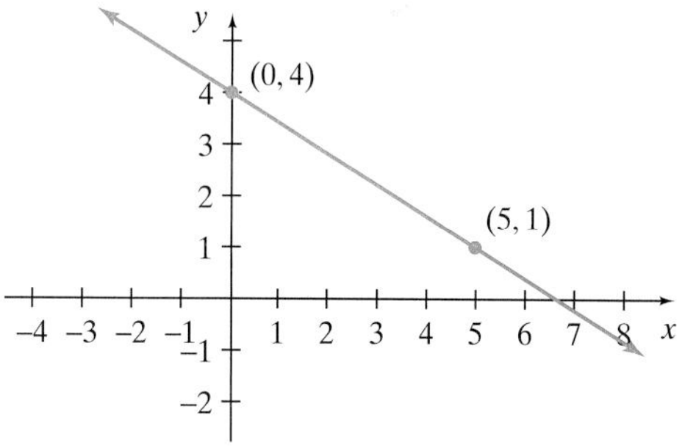

Figure 1.2.7

(b) If we let $(x_1, y_1) = (0, 4)$ and $(x_2, y_2) = (5, 1)$ we get

$$m = \frac{1 - 4}{5 - 0} = \frac{-3}{5}$$

The line through these points is shown in Figure 1.2.7. ■

Interactive Activity

Recompute the slope for part (b) of Example 3, letting $(x_1, y_1) = (5, 1)$ and $(x_2, y_2) = (0, 4)$. This illustrates that **the slope of a line is unique**.

Before we leave the slope concept, let's discuss a few of its properties. Note in part (a) of Example 3 that the slope is a positive number. We can see in Figure 1.2.6 that the line is rising or **increasing** as we follow the graph from the left to the right. In part (b) of Example 3 the slope has a negative value. We can see in Figure 1.2.7 that the line is falling or **decreasing** as we follow it from left to right. It appears that there is an association between the value of the slope and the type of line that we have. This is summarized in Figures 1.2.8 through 1.2.11.

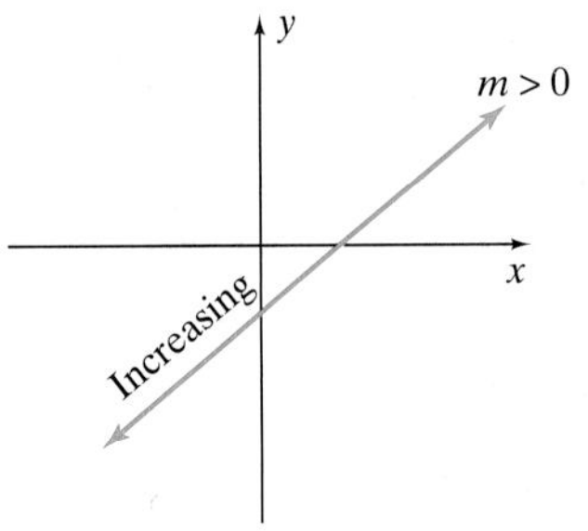

Slope is Positive

Figure 1.2.8

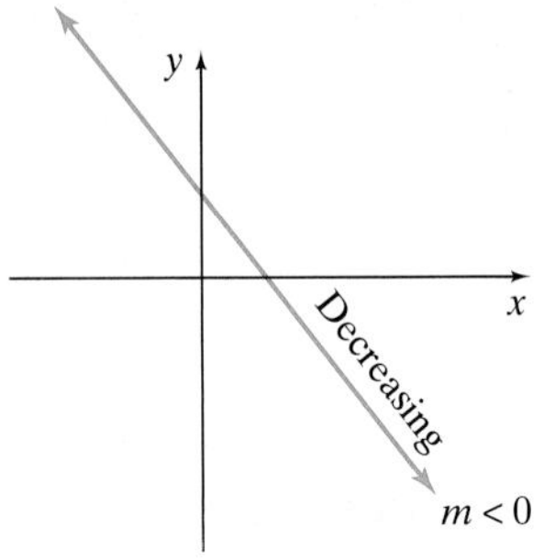

Slope is Negative

Figure 1.2.9

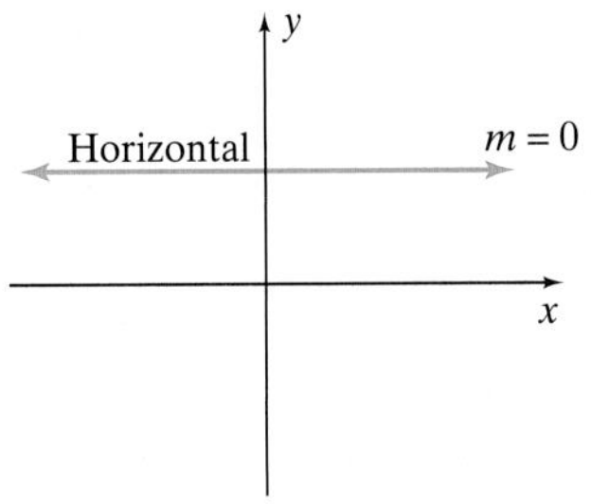

Slope is Zero

Figure 1.2.10

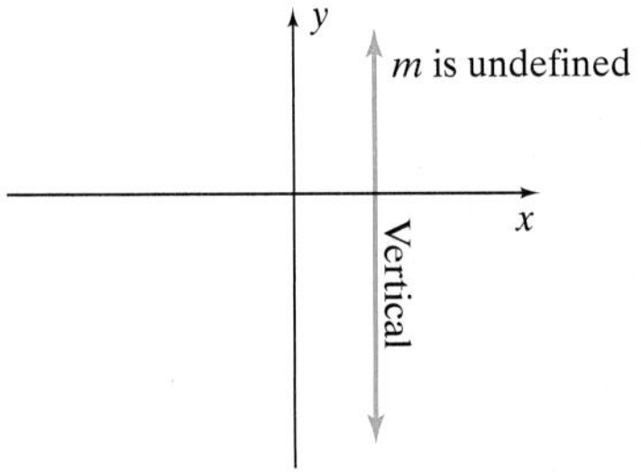

Slope is Undefined

Figure 1.2.11

Now let's see how we can interpret slope in an application.

Example 4 Applying Slope

The per capita bottled water consumption, measured in gallons per day, in the United States from 1990 to 1998 is shown in Table 1.2.7. Compute the slope of the water consumption with respect to time and interpret the result.

Solution

Understand the Situation: We need to use the formula $m = \dfrac{y_2 - y_1}{x_2 - x_1}$ to compute the slope.

If we let $(x_1, y_1) = (1990, 8.0)$ and $(x_2, y_2) = (1998, 13.1)$, we get

$$m = \frac{\text{change in consumption}}{\text{change in years}} = \frac{13.1 - 8.0}{1998 - 1990} = \frac{5.1}{8} \approx 0.64 \frac{\text{gallon}}{\text{year}}$$

Interpret the Solution: This means that from 1990 to 1998 the per capita consumption of bottled water in the United States was increasing at an average rate of 0.64 gallon per day each year. ■

Table 1.2.7

Year	Per Capita Water Consumption
1990	8.0
1998	13.1

SOURCE: U.S. Census Bureau, www.census.gov

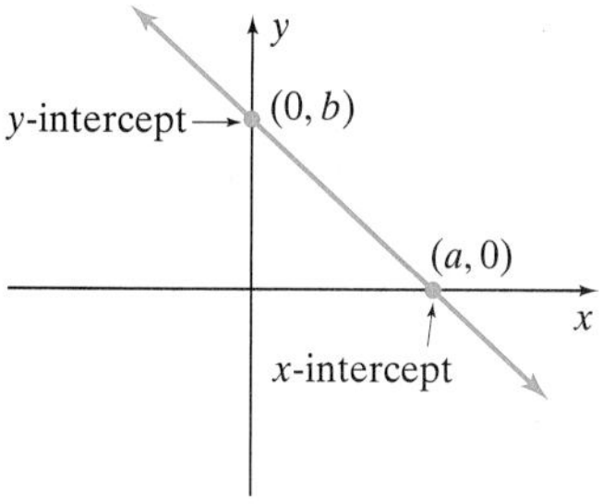

Figure 1.2.12

Intercepts

Two important points on the graph of a linear function are its **intercepts**. These are the points where the graph crosses the x (independent variable) and y (dependent variable) axes, as shown in Figure 1.2.12. The figure also suggests how to find the intercepts.

- To determine the y-intercept, we can replace x with zero in the equation of the line and solve the resulting equation for y.
- To determine the x-intercept, we replace y with zero in the equation of the line and solve the resulting equation for x.

Example 5 Finding the Intercepts of a Linear Function

Determine the x- and y-intercepts of the graph of the linear function $y = \frac{2}{3}x + \frac{11}{3}$.

Solution

To find the y-intercept, we replace x with zero and solve for y. This gives

$$y = \frac{2}{3}(0) + \frac{11}{3} = 0 + \frac{11}{3} = \frac{11}{3}$$

So the y-intercept is $\left(0, \frac{11}{3}\right)$.

To get the x-intercept, we replace y with zero and solve for x to get

From Your Toolbox

For more on how to solve linear equations, consult Appendix B.3.

$$\begin{aligned} 0 &= \frac{2}{3}x + \frac{11}{3} \\ 0 &= 2x + 11 && \text{Multiply both sides by 3.} \\ -11 &= 2x && \text{Subtract 11 from both sides.} \\ \frac{-11}{2} &= x && \text{Divide both sides by 2.} \end{aligned}$$

Thus the x-intercept is $\left(\frac{-11}{2}, 0\right)$. These points are shown on the graph in Figure 1.2.13.

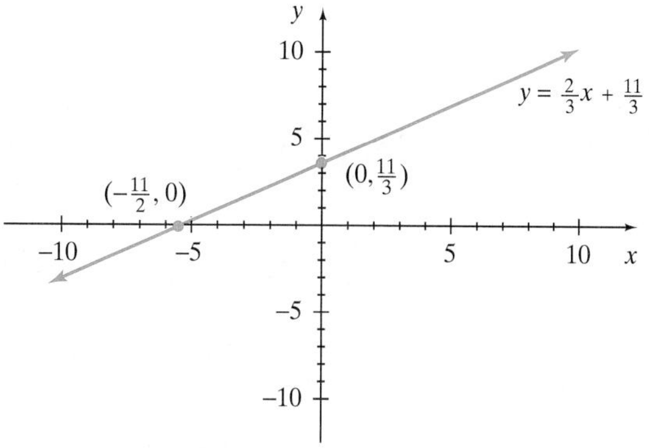

Figure 1.2.13

Checkpoint 2 Now work Exercise 49.

Linear Cost Functions

In business, managers are usually very interested in the cost of producing items. This is often expressed as a **linear cost function**. A cost function consists of two parts. The first part is called the **fixed cost**. A fixed cost is what the business pays no matter how many items are produced. Utilities, rent, and other monthly costs are fixed costs. The second part changes as the number of items produced changes. Economists call this part the **variable costs**. We can put fixed costs together with the variable costs and get the cost function.

Linear Cost Function

The cost $C(x)$ of producing x units of a product is given by the **linear cost function**

$$C(x) = \begin{pmatrix}\text{cost per}\\ \text{unit}\end{pmatrix} \cdot x + \begin{pmatrix}\text{fixed}\\ \text{costs}\end{pmatrix}$$

Note: Some cost functions are not linear because the cost per unit is a variable itself. In the applications that we will study in this text, all cost functions will be linear.

Example 6 Determining a Cost Function

A printing job has fixed costs of \$5 and per unit costs of \$0.10 per copy.

(a) Determine the cost function for the printing job.

(b) Use the result of part (a) to evaluate and interpret $C(100)$.

Solution

(a) **Understand the Situation:** Let x represent the number of copies made and $C(x)$ represent the cost of the printing job.

Then, using the model given in the definition of linear cost function, we have

$$\begin{aligned} C(x) &= \begin{pmatrix}\text{cost per}\\ \text{unit}\end{pmatrix} \cdot x + \begin{pmatrix}\text{fixed}\\ \text{costs}\end{pmatrix} \\ &= 0.10x + 5 \end{aligned}$$

(b) To evaluate $C(100)$, we simply substitute 100 for each occurrence of x. This gives

$$C(x) = 0.10x + 5$$
$$C(100) = 0.10(100) + 5 = 15$$

Interpret the Solution: The total cost of printing 100 copies is \$15. ■

Checkpoint 3 Now work Exercise 61.

SUMMARY

In this section we examined the definition of function and explored the properties of lines, including the slope of a line.

- A **function** is a rule, or a set of ordered pairs, that assigns a unique y-value to each x-value. The set of all possible x-values is called the **domain**, and the set of all possible y-values is called the **range**.
- A **linear function** has the form $y = f(x) = mx + b$.
- Given two ordered pairs (x_1, y_1) and (x_2, y_2), the **slope** of the line through the two points is determined by computing $m = \dfrac{y_2 - y_1}{x_2 - x_1}$.
- We also found that the value of the slope indicates the behavior of the line.

  If $m > 0$, then the line is **increasing**.
  If $m < 0$, then the line is **decreasing**.
  If $m = 0$, then the line is **horizontal**.
  If m is undefined, then the line is **vertical**.
- The **intercepts** are the points where a graph crosses the x- (independent variable) and y- (dependent variable) axes.
- The cost $C(x)$ of producing x units of a product is given by the **linear cost function**

$$C(x) = \begin{pmatrix}\text{cost per}\\ \text{unit}\end{pmatrix} \cdot x + \begin{pmatrix}\text{fixed}\\ \text{costs}\end{pmatrix}$$

SECTION 1.2 EXERCISES

In Exercises 1–8, use the table to construct ordered pairs (x, y). Plot the ordered pairs on the Cartesian plane.

1.

x	y
0	1
3	5
6	7

2.

x	y
1	3
4	7
7	10

3.

x	y
−2	3
−1	5
4	−4

4.

x	y
−2	0
−1	−3
−2	−5

5.

x	−4	−3.5	−2	2
y	−5	−6	−8	−11

6.

x	0	−1	−2	−3
y	0	1	5	6

7.

x	0	1	2	3
y	1.1	1.5	1.8	2

8.

x	1	2	3	4
y	1.5	2	2.5	3.5

For Exercises 9–16, write the sets for the domain and range for the values in the table:

9. In Exercise 1. **10.** In Exercise 2.

11. In Exercise 3. **12.** In Exercise 4.

13. In Exercise 5. **14.** In Exercise 6.

15. In Exercise 7. **16.** In Exercise 8.

For the functions in Exercises 17–26, complete the following:

(a) Identify the independent and dependent variables.

(b) Complete the table. Note that in Exercises 25 and 26 the dependent variable value is given.

17. $f(x) = x - 5$

x	$f(x)$
2	
−3	

18. $f(x) = 3 - x$

x	$f(x)$
1	
−4	

19. $g(x) = 3x + \frac{1}{2}$

x	$g(x)$
−1	
0	
5	

20. $g(x) = 2x + \frac{1}{4}$

x	$g(x)$
−2	
1	
4	

21. $y = 0.3x + 6$

x	y
0	
5	
10	

22. $y = 0.2x - 1.1$

x	y
1	
6	
11	

23. $f(t) = \frac{1}{2}(7 - 0.3t)$

t	$f(t)$
0	
1	
3	
6	

24. $f(t) = \frac{1}{3}(10 - 0.4t)$

t	$f(t)$
0	
4	
8	
12	

25. $f(x) = \dfrac{2x+3}{5}$

x	$f(x)$
1	
4	
	3

26. $f(x) = \dfrac{3x-1}{4}$

x	$f(x)$
3	
4	
	5

In Exercises 27–30, determine which of the tables represents function.

27.

Domain	Range
0	1
2	1
3	2
4	5

28.

Domain	Range
−5	−2
−3	1
−2	5
0	10

29.

Domain	Range
−3	1
−2	0
−2	3
−1	4
0	5
1	6

30.

Domain	Range
2	1
−2	3
0	0
−1	2
−3	7
2	4

In Exercises 31–34, determine if the graph represents a function.

31.

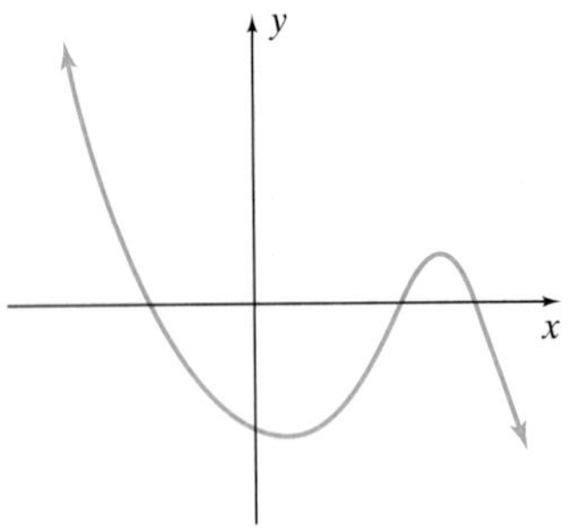

32.

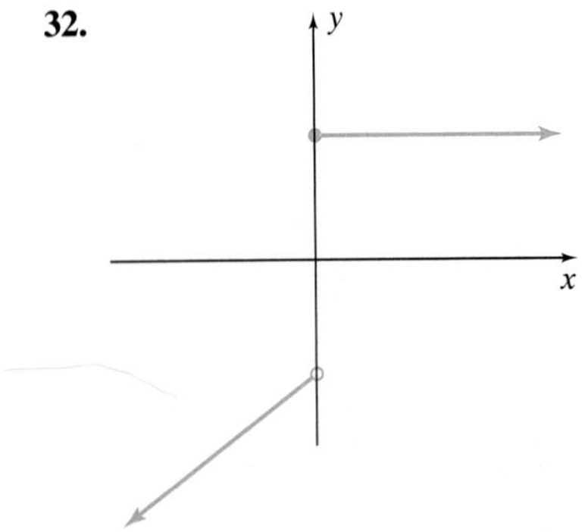

33.

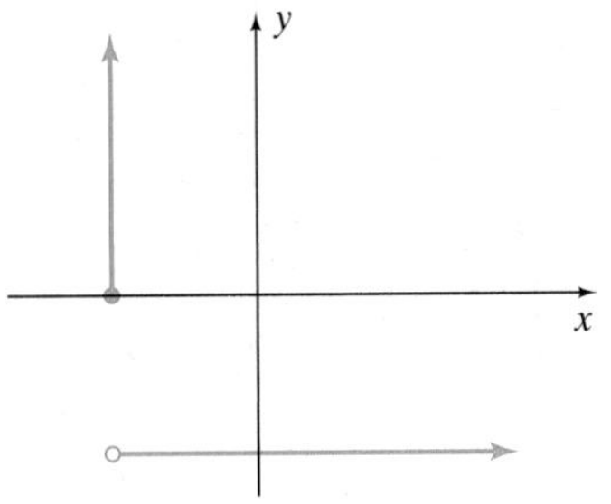

34.

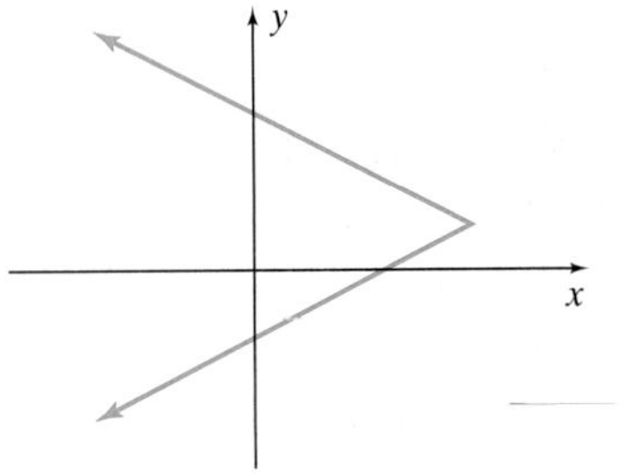

In Exercises 35–38, a table of values is given.

(a) Complete the table.

(b) Write a linear function for the data in the table.

35.

x	expression	y
0		
2	$3(2)+1$	
		4

36.

x	expression	y
	$2(0)+4$	4
		8
3		

37.

x	expression	$f(x)$
0		6
	$6-3(1)$	
		0
3		

38.

x	expression	$f(x)$
	$10-\frac{1}{2}(0)$	10
2		
		8
	$10-\frac{1}{2}(6)$	

In Exercises 39–48, compute the slope of the line passing through the two points. Then classify the line as increasing, decreasing, horizontal, or vertical.

39.

x	y
−6	1
4	4

40.

x	y
−4	−2
−1	7

41.

x	$f(x)$
−1	−4
−3	2

42.

x	$f(x)$
−2	9
1	−3

43. (3, 8) and (−3, 8)

44. (−4, 7) and (3, 7)

45. (2, −3) and (2, 4)

46. (5, −2) and (5, 3)

47.

x	0.5	2.5
y	4.1	1.1

48.

x	1.6	4.8
y	0.3	1.5

In Exercises 49–58, determine the x- and y-intercepts of the graph of the linear function.

49. $y = 2x - 10$

50. $y = 3x - 9$

51. $2x - 3y = 6$

52. $x - 4y = 16$

53. $y = \frac{1}{2}x - 4$

54. $y = \frac{1}{3}x - 12$

55. $g(x) = \frac{10x - 5}{2}$

56. $g(x) = \frac{15x - 3}{5}$

57. $f(x) = 0.2x + 10.3$

58. $f(x) = 1.4x + 6.6$

Applications

In Exercises 59–62, write the linear cost function for the given information in the form $C(x) = mx + b$.

59. The cost per unit is \$70 and the fixed costs are \$10,000.

60. The cost per unit is \$147.75 and the fixed costs are \$2700.

61. The fixed costs are \$14,500 and the cost per unit is \$134.

62. The fixed costs are \$3580 and the cost per unit is \$80.

63. The Dirt Rocket Company produces spokes for offroad motorcycles and determines that the daily cost in dollars of producing x quick-replace spokes is $C(x) = 1250 + 3x$.

(a) What are the fixed and per unit costs for producing the spokes?

(b) Evaluate and interpret $C(400)$.

(c) How many spokes can be produced for \$1520? That is, what is x when $C(x) = 1520$?

64. The Fontana Vitamin Company determines that the cost of producing x bottles of a new supplement is given by the linear cost function $C(x) = 1.25x + 550$.

(a) What are the fixed and per unit costs for producing the supplement?

(b) Evaluate and interpret $C(70)$.

(c) How many bottles of the supplement can be produced for \$693.75? That is, what is x when $C(x) = 693.75$?

65. The Get-It-Now florist determines that the cost, in dollars, of arranging and delivering x get-well bouquets is given by the cost function $C(x) = 10 + 32x$.

(a) What are the fixed and per units costs for producing the bouquets?

(b) Evaluate and interpret $C(15)$.

(c) How many bouquets can be produced and delivered for \$8010?

66. The Never Die Light Bulb Company determines that the cost, in dollars, of producing x units of their new brand of bulb is $C(x) = 2.5x + 1050$.

(a) What are the fixed and per units costs for producing the light bulbs?

(b) Evaluate and interpret $C(260)$.

(c) How many bulbs can be produced for \$1480?

67. The data in Table 1.2.8 represent the public expenditures at private universities and colleges in the United States from 1989 to 1999.

Table 1.2.8

Year	Public Expenditures
1989	\$68.4 billion
1999	\$92.6 billion

SOURCE: U.S. Census Bureau, www.census.gov

Compute the average rate of change in public expenditures at private universities and colleges from 1989 to 1999 and interpret. Be sure to include proper units.

68. The data in Table 1.2.9 represent the number of visits to national parks from 1990 to 1998.

Table 1.2.9

Year	Number of Visits
1990	58.7 million
1998	64.5 million

SOURCE: U.S. Census Bureau, www.census.gov

Compute the average rate of change in the number of visits to national parks from 1990 to 1998 and interpret. Be sure to include proper units.

SECTION PROJECT

Determine if each of the following can be represented by a function. Explain your reasoning.

(a) A person's astrological sign is a function of his or her birth date.

(b) A person's income is a function of how long the person works each day.

(c) The voter turnout in an election is a function of the weather.

(d) The distance a person is from a pizzeria is a function of the deliveries made.

(e) The number of U.S. congress members from a state is a function of its population.

Section 1.3 Linear Models

In Section 1.2 we defined the function and explored the properties of the average rate of change for linear functions. In this section we will see the different ways in which we can write linear functions and examine the concept of **linear models**.

Equations of Lines

When we defined the linear function, we wrote it in the form $f(x) = mx + b$. But there are other ways to mathematically express a line. One way that we will see frequently is the **standard form** of a line.

Standard Form

In the Cartesian plane, the **standard form** of a line is given by

$$ax + by = c$$

where a, b, and c are real numbers, $a \geq 0$, and a and b are not both zero.

Note: If a, b, and c are rational numbers, they are often converted to integers by multiplying both sides of the equation by the least common multiple of the denominators.

Example 1 Writing a Linear Equation in Standard Form

For the linear equation $y = \frac{2}{3}x + \frac{11}{3}$, write the equation in the standard form $ax + by = c$, where a, b, and c are integers.

Solution

First, we collect the terms with x and y on the left side and leave the constant on the right.

$$-\frac{2}{3}x + y = \frac{11}{3}$$

Since we need $a \geq 0$, we multiply both sides of the equation by -1 to get

$$\frac{2}{3}x - y = -\frac{11}{3}$$

Finally, we can multiply both sides by 3 so that no fractions are present.

$$2x - 3y = -11$$

In addition to standard form, there are other ways to write an equation of a line. Let's say that we know the slope of a line and the coordinates of one point on the line. For example, suppose that a line has a slope of $m = -2$ and passes through the point $(-3, 5)$. Now call some other point on the line (x, y). See

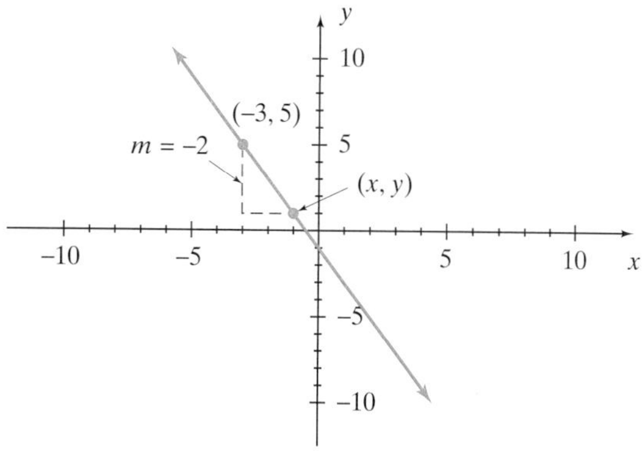

Figure 1.3.1 The slope of the line through $(-3, 5)$ and (x, y) is $m = -2$.

Figure 1.3.1. Using the definition of slope, $m = \dfrac{y_2 - y_1}{x_2 - x_1}$, we can write

$$-2 = \frac{y-5}{x-(-3)} \quad \text{or} \quad -2 = \frac{y-5}{x+3}$$

where $(x_1, y_1) = (-3, 5)$ and $(x_2, y_2) = (x, y)$. Multiplying both sides of the equation by $x + 3$ gives us

$$y - 5 = -2(x + 3)$$

This is an equation of the line with slope $m = -2$ passing through the point $(-3, 5)$. Since this equation uses a known slope and a known point, we call it the **point–slope form**.

Point–Slope Form

An equation of the line with known slope m and known point (x_1, y_1) can be written in the **point–slope form** given by

$$y - y_1 = m(x - x_1)$$

We can use this form to write an equation of any line if we are given either one point on the line and its slope or if we are given two points on the line. This is illustrated in Example 2.

Example 2 Writing an Equation in Point–Slope Form

For the points $(2, 4)$ and $(3, 7)$, complete the following:

(a) Determine the slope m of the line passing through the two points.

(b) Use the point $(2, 4)$ to write an equation for the line passing through the points.

(c) Rewrite the equation in standard form.

Solution

(a) If we let $(x_1, y_1) = (2, 4)$ and let $(x_2, y_2) = (3, 7)$ we get

$$m = \frac{y_2 - y_1}{x_2 - x_1} = \frac{7-4}{3-2} = \frac{3}{1} = 3$$

(b) Using $m = 3$ from part (a) and knowing that $(x_1, y_1) = (2, 4)$, we get

$$y - 4 = 3(x - 2)$$

(c) Here we need to rewrite the equation in the standard form $ax + by = c$, where $a \geq 0$ (that is, a is nonnegative).

$y - 4 = 3(x - 2)$	Point–slope form
$y - 4 = 3x - 6$	Distribute the 3.
$-3x + y - 4 = -6$	Subtract $3x$ from both sides.
$-3x + y = -2$	Add 4 to both sides.
$3x - y = 2$	Multiply both sides by -1. ■

Interactive Activity*

Redo Example 2, but let $(x_1, y_1) = (3, 7)$. Do you get the same result as was found in part (c)? This is why $ax + by = c$ is called the *standard* form.

WWW

From Your Toolbox

In part (c) of Example 2, we used the distributive property. For more practice on using this property, consult the Appendix B.1.

Let's say that we want to write the equation of a line in point–slope form with slope m and y-intercept $(0, b)$. Using $(x_1, y_1) = (0, b)$, we get

$$y - b = m(x - 0)$$

Solving this equation for y gives us

$$y = m(x - 0) + b \quad \text{or} \quad y = mx + b$$

This is another way to write the equation of a line and is called the **slope–intercept form.**

Slope Intercept Form

An equation of a line with slope m and y-intercept $(0, b)$ can be written in the **slope–intercept form** given by

$$y = mx + b$$

Note: The slope–intercept form is how we defined the linear function in Section 1.2. The slope–intercept form is useful when we are given an average rate of change (that is, a slope) along with some y-value when $x = 0$. This is sometimes called an **initial value**. Now let's practice writing a linear equation in slope–intercept form.

Example 3 Writing a Linear Equation in Slope–Intercept Form

A line passes through the points $(-2, 6)$ and $(0, 1)$.

(a) Write an equation of the line in slope–intercept form.

(b) Rewrite the equation in standard form.

*Interactive Activities that have the WWW icon have worked out solutions at the companion website at www.prenhall.com/armstrong.

Solution

(a) We begin by computing the slope of the line using the two given points.

$$m = \frac{y_2 - y_1}{x_2 - x_1} = \frac{1-6}{0-(-2)} = \frac{-5}{2} = -\frac{5}{2}$$

Knowing that the y-intercept is $(0, b) = (0, 1)$, we have $b = 1$. So we get the slope–intercept form

$$y = mx + b$$
$$y = -\frac{5}{2}x + 1$$

(b) Here we need to write the equation found in part (a) in the form $ax + by = c$, where $a \geq 0$.

$$y = -\frac{5}{2}x + 1 \quad \text{Given equation}$$
$$\frac{5}{2}x + y = 1 \quad \text{Add } \frac{5}{2}x \text{ to both sides.}$$
$$5x + 2y = 2 \quad \text{Multiply both sides by 2.}$$

■

Checkpoint 1 Now work Exercise 3.

Linear Models

Often, in our study of finite mathematics, we wish to take real data and use them to write a function that shows the general trends of these data. We call this process **modeling**. Many times, this model starts with a given value and then either steadily increases or decreases. We call these types of models **linear models**.

Linear Model

A **linear model** is a type of linear function that is derived from a real-world situation or from real-world data.

Let's start by examining a linear model in an example. In this section we will assume that all the applications discussed can be modeled by a linear function.

Example 4 Evaluating and Interpreting a Linear Model

The average annual pay of Americans from 1990 to 2000 can be represented by a linear model that contains the points shown in Table 1.3.1. In the table, x represents the number of years since 1990 and y represents the average annual pay in thousands of dollars.

Table 1.3.1

x	y
1	24.28
3	26.40

SOURCE: U.S. Census Bureau, www.census.gov

(a) Compute the average rate of change and interpret.

(b) Write an equation of the line containing the two points in point–slope form.

(c) Write a linear model in slope–intercept form

$$f(x) = mx + b, \qquad 0 \leq x \leq 10$$

for the average annual pay of Americans from 1990 to 2000.

(d) Evaluate $f(5)$ and interpret.

Solution

(a) **Understand the Situation:** Recall that the average rate of change is simply the slope, m, of the line through two points.

Using the ordered pairs $(x_1, y_1) = (1, 24.28)$ and $(x_2, y_2) = (3, 26.40)$, we get

$$m = \frac{y_2 - y_1}{x_2 - x_1} = \frac{26.40 - 24.28}{3 - 1} = \frac{2.12}{2} = 1.06$$

Interpret the Solution: This means that from 1990 to 2000 the average annual pay for Americans increased at an average rate of about 1.06 thousand dollars per year.

(b) Using the point $(x_1, y_1) = (1, 24.28)$, we get

$$y - y_1 = m(x - x_1)$$
$$y - 24.28 = 1.06(x - 1)$$

A graph of this equation is given in Figure 1.3.2.

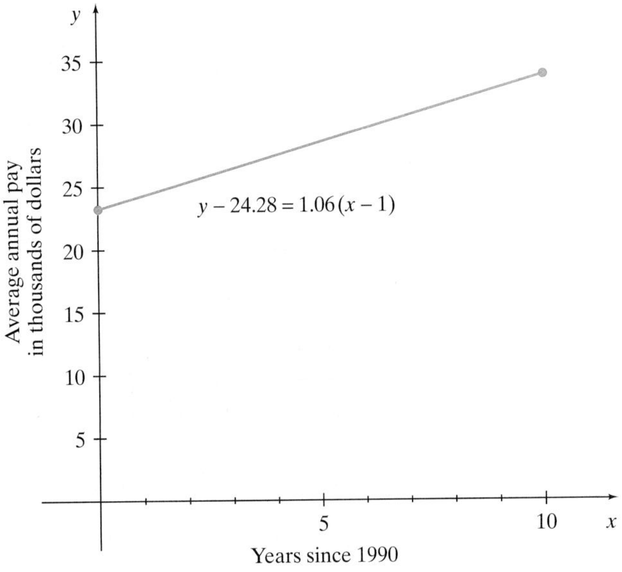

Figure 1.3.2

(c) To get the result of part (b) in the form $f(x) = mx + b$, we solve the equation for y and then replace y with $f(x)$.

$y - 24.28 = 1.06(x - 1)$	Given equation
$y = 1.06(x - 1) + 24.28$	Add 24.28 to both sides.
$y = 1.06x - 1.06 + 24.28$	Distribute 1.06.
$y = 1.06x + 23.22$	Simplify.
$f(x) = 1.06x + 23.22$	Replace y with $f(x)$.

(d) **Understand the Situation:** Since x represents the number of years since 1990, we know that $x = 5$ corresponds to the year $1990 + 5 = 1995$.

Now we need to evaluate the linear model when $x = 5$ and interpret the result.

$$f(5) = 1.06(5) + 23.22 = 5.3 + 23.22 = 28.52$$

Interactive Activity

Write the result of part (b) in Example 4 in standard form. (*Hint:* Multiply both sides of the equation by 100 to eliminate the decimals.)

WWW

Interpret the Solution: This means that in 1995 the average annual pay of Americans was about 28.52 thousand dollars (that is, \$28,520).

Checkpoint 2

Now work Exercise 51.

Another application of linear modeling involves the **linear depreciation model**. When the value of a product decreases as time goes on, we say it depreciates. The amount of value that the product loses each year is called the **depreciation rate**. Many companies consider linear depreciation models in their fiscal plans and for planning upgrades in equipment. Example 5 illustrates how to construct a linear depreciation model.

Example 5 Writing a Linear Depreciation Model

The Clark Clipboard Company recently purchased a metal press for \$15,000. The press is estimated to depreciate at a rate of $2500\frac{\text{dollars}}{\text{year}}$.

(a) Write a linear model in the form $f(x) = mx + b$ for the value of the metal press.

(b) Determine the x-intercept and interpret.

Solution

(a) **Understand the Situation:** Since the value of the press appears to depend on how old it is, we let x represent the age of the press in years and $y = f(x)$ represent the value of the press in dollars.

We know that when the press is new, meaning that the age is $x = 0$, its value is \$15,000. So $f(0) = 15{,}000$, which means that the point $(0, 15{,}000)$ is on the graph. Notice that this point is the y-intercept. (Recall that this is also known as the initial value.) Since the rate at which the value of the press decreases is $2500\frac{\text{dollars}}{\text{year}}$, the value of the press has an average rate of change of $-2500\frac{\text{dollars}}{\text{year}}$. So knowing that $m = -2500$ and the initial value is $b = 15{,}000$, we get the model

$$f(x) = mx + b$$
$$f(x) = -2500x + 15{,}000$$

(b) To determine the x-intercept, we let $f(x) = 0$ and solve for x.

$$0 = -2500x + 15{,}000 \qquad \text{Set } f(x) = 0.$$
$$2500x = 15{,}000 \qquad \text{Add } 2500x \text{ to each side.}$$
$$x = \frac{15{,}000}{2500} \qquad \text{Divide by 2500.}$$
$$= 6 \qquad \text{Simplify.}$$

Interpret the Solution: This means that after 6 years the metal press will have no value.

Interactive Activity

Determine the reasonable domain for the model in Example 5a.

WWW

Checkpoint 3

Now work Exercise 37.

Technology Option

Calculator-Friendly Form

Many times, we want to use our calculator to graph a line with a known point and known slope. We can do this by solving the point–slope form $y - y_1 = m(x - x_1)$ for y. This gives us the calculator-friendly (or C-F) form.

> ### Calculator-Friendly or C-F Form
>
> A linear function with a known slope m and known point (x_1, y_1) can be written in the **calculator-friendly form** given by
>
> $$y = m(x - x_1) + y_1$$

This form is particularly useful in applications, as is illustrated in Example 6.

Example 6   **Using the Calculator-Friendly Form**

In 1994 there were 25.27 million children enrolled in the National School Lunch Program. In 1999, this number had risen to 26.82 million. Assume that the number of children enrolled in the lunch program grew linearly from 1990 to 1999 (*Source:* U.S. Census Bureau, www.census.gov).

(a) Let x represent the number of years since 1990 and y represent the number of children enrolled in the National School Lunch Program in millions. Determine the average rate of change in the number of enrollees from 1990 to 1999.

(b) Write a linear model in the **calculator-friendly form** given by

$$y = m(x - x_1) + y_1, \qquad 0 \le x \le 9$$

to represent the number of children enrolled in the National School Lunch Program in millions from 1990 to 1999.

(c) During what year did the number of enrollees exceed 25 million?

Solution

(a) **Understand the Situation:** If we let x represent the number of years since 1990, we see that 1994 corresponds to $x = 4$ and 1999 corresponds to $x = 9$. So we have to determine the average rate of change, or slope, between the points $(x_1, y_1) = (4, 25.27)$ and $(x_2, y_2) = (9, 26.82)$.
This gives us

$$m = \frac{26.82 - 25.27}{9 - 4} = \frac{1.55}{5} = 0.31$$

Interpret the Solution: This means that from 1990 to 1999 the number of children enrolled in the National School Lunch Program increased at an average rate of 0.31 million enrollees per year.

(b) Knowing that $m = 0.31$ and $(x_1, y_1) = (4, 25.27)$, we get

$$y = 0.31(x - 4) + 25.27$$

Figure 1.3.3 is a graph of the model.

Figure 1.3.3

(c) Here we need the x-values so that $y = 25$. This gives us the equation

$$y = 0.31(x - 4) + 25.27$$
$$25 = 0.31(x - 4) + 25.27$$
$$-0.27 = 0.31(x - 4)$$
$$\frac{-0.27}{0.31} = x - 4$$
$$\frac{-0.27}{0.31} + 4 = x$$
$$3.13 \approx x$$

So we get $x \approx 3.13$. Since x represents the number of years since 1990, we see that during 1993 the number of children enrolled in the National School Lunch Program exceeded 25 million. Figure 1.3.4 shows how to check our algebraic solution by using the TRACE command. For more information on using the TRACE command, consult the online calculator manual at www.prenhall.com/armstrong.

Figure 1.3.4

▌ SUMMARY

In this section we learned four ways to write linear equations.

- **Standard form:** $ax + by = c$
- **Point–slope form:** $y - y_1 = m(x - x_1)$
- **Slope–intercept form:** $y = mx + b$
- **Calculator-friendly form:** $y = m(x - x_1) + y_1$

SECTION 1.3 EXERCISES

In Exercises 1–14, write an equation of the line in standard form (if possible).

1. A line with a slope of $m = \dfrac{2}{3}$ that passes through the point $(2, -5)$.

2. A line with a slope of $m = -\dfrac{1}{2}$ that passes through the point $(2, -2)$.

✓ **3.** The line that passes through the points $(2, 6)$ and $(5, -3)$.

4. The line that passes through the points $(3, 2)$ and $(-1, -2)$.

5. The line is given by the values in Table 1.3.2.

Table 1.3.2

x	y
3	5
5	13

6. The line is given by the values in Table 1.3.3.

Table 1.3.3

x	y
2	4
0	4

7. The line is given by the values in Table 1.3.4.

Table 1.3.4

x	1	5
$f(x)$	5.1	10.6

8. The line is given by the values in Table 1.3.5.

Table 1.3.5

x	0	3
$f(x)$	8.8	4.3

9. The line has a y-intercept at -3 and passes through the origin.

10. The line has an x-intercept at 2 and a y-intercept at 4.

11. The line is vertical and passes through the point $(-2, 9)$.

12. The line is horizontal and passes through the point $(-3, 1)$.

13. The line has an average rate of change of $-\dfrac{2}{3}$ and an x-intercept at 3.

14. The line has an average rate of change of 0.3 and an x-intercept at -4.

In Exercises 15–30, write a function for the line in $f(x) = mx + b$ form.

15. A line with a slope of $m = 3$ that passes through the point $(8, 4)$.

16. A line with a slope of $m = \dfrac{5}{6}$ that passes through the point $(-1, 6)$.

17. A linear function has a slope of $m = -\dfrac{5}{8}$ and $f(1) = 3$.

18. A linear function has a slope of $m = 4$ and $f(-2) = 4$.

19. A linear function has a slope of $m = 5$ and a y-intercept at $(0, -13)$.

20. A linear function has a slope of $m = \dfrac{3}{5}$ and a y-intercept at $(0, -10)$.

21. A linear function with $f(1) = 3$ and $f(-10) = 1$.

22. A linear function with $f(7) = 3$ and $f(-3) = 6$.

23. A linear function $f(x)$ has function values of $f(5) = 1$ and $f(-7) = -1.19$.

24. A linear function $g(x)$ has function values of $g(0) = 6$ and $g(4) = 1.3$.

25. A linear function contains the points in Table 1.3.6.

Table 1.3.6

x	$f(x)$
-1	2
3	1

26. A linear function contains the points in Table 1.3.7.

Table 1.3.7

x	$f(x)$
0	2
-3	4

27. A line has an x-intercept at $(-12, 0)$ and y-intercept at $\left(0, \dfrac{3}{2}\right)$.

28. A line has an x-intercept at $(-5, 0)$ and a y-intercept at $(0, 9)$.

29. A linear function contains the points in Table 1.3.8.

Table 1.3.8

x	1	3
$f(x)$	25.3	10.8

30. A linear function contains the points in Table 1.3.9.

Table 1.3.9

x	1	6
$f(x)$	13.7	16.9

In Exercises 31–34, write an equation in slope–intercept form for the given graph.

31.

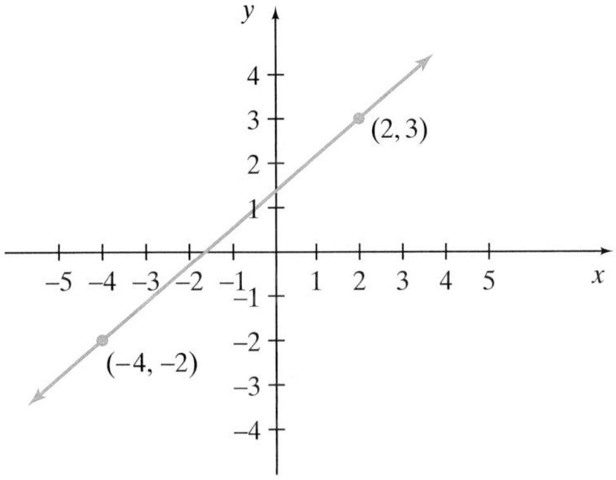

32.

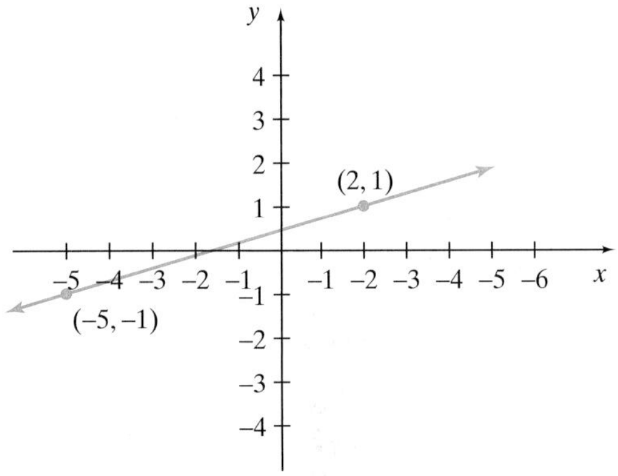

33.

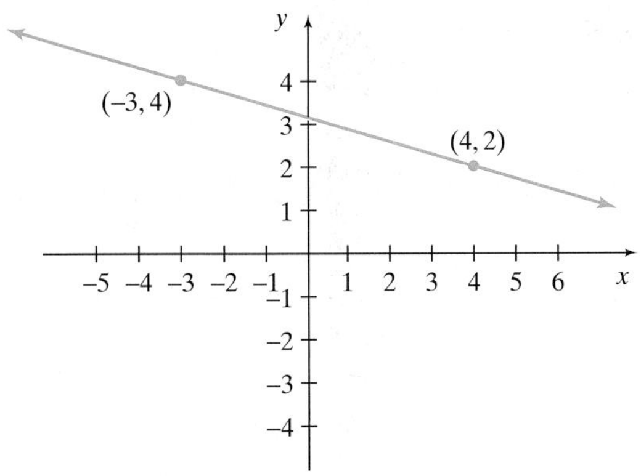

34.

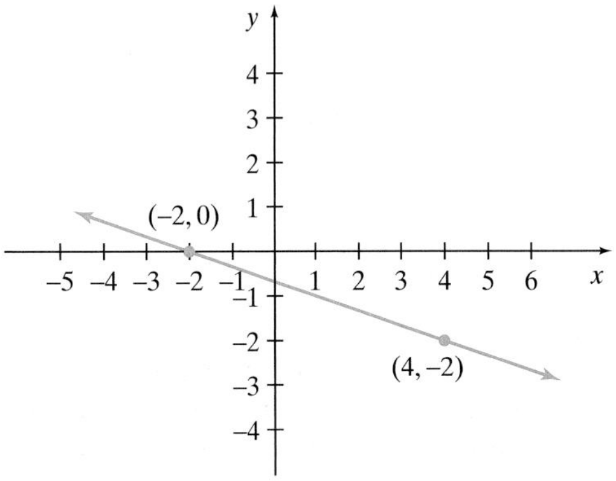

Applications

In Exercises 35–38, answer by using a linear depreciation model.

35. A new photocopier initially costs \$25,000 and depreciates at a rate of \$1250 per year.

(a) Define the variables and determine the depreciation function.

(b) In how many years will the photocopier be worth \$10,000?

36. A new multimedia computer costs \$4000 and depreciates at a rate of $150 \dfrac{\text{dollars}}{\text{year}}$.

(a) Define the variables and determine the depreciation function.

(b) When will the computer depreciate to the point that it has no worth?

✓ **37.** The Westbrook Company invests \$90,000 into a new machine press that depreciates at a rate of \$6000 per year.

(a) Define the variables and determine the depreciation function.

(b) How long will it take for the press to be worth half of its original cost?

38. The XQ-383 sports cars costs \$40,000 and depreciates \$1500 each year.

(a) Define the variables and determine the depreciation function.

(b) How much will the car be worth in 5 years?

39. The annual total expenditures on pollution abatement in the United States can be modeled by

$$f(x) = 2x + 48.17, \qquad 0 \le x \le 20$$

where x represents the number of years since 1972 and $f(x)$ represents the expenditures in billions of constant 1987 dollars (*Source:* U.S. Bureau of Economic Analysis, www.bea.doc.gov).

(a) Is the amount of money spent on pollution abatement generally increasing or decreasing? At what rate?

(b) Evaluate and interpret $f(12)$.

(c) In what year did the annual expenditures approach 90 billion dollars?

40. The percentage of waste generated in the United States that is wood can be modeled by

$$f(x) = 0.23x + 5.95, \qquad 0 \le x \le 9$$

where x represents the number of years since 1990 and $f(x)$ represents the percentage of waste generated that is wood (*Source:* U.S. Bureau of Economic Analysis, www.bea.doc.gov).

(a) Is the percentage of waste generated that is wood generally increasing or decreasing? At what rate?

(b) Evaluate and interpret $f(6)$.

(c) In what year did the percentage approach 7.4%?

41. The population of the United States can be modeled by

$$f(x) = 1.92x + 67.15, \qquad 0 \le x \le 100$$

where x represents the number of years since 1900 and $f(x)$ represents the U.S. population in millions (*Source:* U.S. Census Bureau, www.census.gov).

(a) Evaluate and interpret $f(50)$.

(b) Solve $f(x) = 150$ and interpret.

42. The percentage of pleasure trips taken by U.S. households with incomes above \$40,000 can be modeled by

$$f(x) = 2.66x + 23.62, \qquad 1 \le x \le 14$$

where x represents the number of years since 1984 and $f(x)$ represents the percentage of pleasure trips taken by U.S. households with incomes above \$40,000 (*Source:* U.S. Census Bureau, www.census.gov).

(a) Evaluate and interpret $f(10)$.

(b) Solve $f(x) = 40$ and interpret.

43. The amount spent annually in college bookstores in the United States can be modeled by

$$f(x) = 0.19x + 1.67, \qquad 0 \le x \le 15$$

where x represents the number of years since 1982 and $f(x)$ represents the amount spent in billions of dollars (*Source:* U.S. National Center for Education Statistics, www.nces.ed.gov).

(a) On average, by how much does the amount of spending increase in college bookstores each year?

(b) According to the model, how much was spent in college bookstores in 1985?

(c) According to the model, how much was spent in college bookstores in 1990?

(d) How much more was spent in college bookstores in 1990 compared to 1985?

44. The average annual salary of counselors in the U.S. public school system can be modeled by

$$f(x) = 1557.29x + 19{,}647.80, \qquad 0 \le x \le 20$$

where x represents the number of years since 1980 and $f(x)$ represents the average salary in dollars (*Source:* U.S. National Center for Education Statistics, www.nces.ed.gov).

(a) On average, by how much does a counselor's average salary increase each year?

(b) According to the model, what was a counselor's average salary in 1985?

(c) According to the model, what was a counselor's average salary in 1995?

(d) How much more was a counselor's average salary in 1995 compared to 1985?

45. In 1990 there were 4.24 million children in the United States with physical disabilities and the number was increasing at an average rate of 0.12 million per year (*Source:* U.S. Census Bureau, www.census.gov).

(a) Let x represent the number of years since 1990. Write a linear model of the form

$$g(x) = y = mx + b, \qquad 0 \le x \le 10$$

to represent the number of children in the United States with physical disabilities from 1990 to 2000.

(b) Evaluate and interpret $g(4)$.

46. In 1980 there were 28.3 million Americans enrolled in the Medicare system. The number of enrollees was increasing at an average rate of 0.61 enrollees per year (*Source:* U.S. Census Bureau, www.census.gov).

(a) Let x represent the number of years since 1980. Write a linear model of the form

$$g(x) = y = mx + b, \qquad 0 \le x \le 20$$

to represent the number of Medicare enrollees from 1980 to 2000.

(b) Evaluate and interpret $g(15)$.

47. From 1990 to 1999 the amount spent to purchase a car exported to the United States increased at an average rate of \$1776.71 per year. In 1995, the average amount spent to purchase a car exported to the United States was \$25,483.95 (*Source:* U.S. Census Bureau, www.census.gov).

(a) Let x represent the number of years since 1990. Write a linear model in the calculator-friendly form

$$y = m(x - x_1) + y_1, \qquad 0 \le x \le 9$$

to represent the average amount spent to purchase a car exported to the United States from 1990 to 1999.

(b) Use the model to determine in what year the average price exceeded \$30,000.

48. From 1990 to 2000 the percentage of married women in the U.S. labor force was increasing at an average rate of 0.23% each year. In 1996, 48.1% of the labor force was comprised of married women (*Source:* U.S. Census Bureau, www.census.gov).

(a) Let x represent the number of years since 1990. Write a linear model in the calculator-friendly form

$$y = m(x - x_1) + y_1, \qquad 0 \le x \le 10$$

to represent the percentage of married women in the U.S. labor force from 1990 to 2000.

(b) Use the model to determine in what year the percentage of married women in the U.S. labor force exceeded 48%.

49. In 1989 the median price of a single-family house in the southern United States was \$96.4 thousand. In 1999 the median price had risen to \$140 thousand (*Source:* U.S. Census Bureau, www.census.gov).

(a) Let x represent the number of years since 1989 and y represent the median price of a single-family house in the southern United States in thousands. Determine the average rate of change in the median price from 1989 to 1999.

(b) Write a linear model in the slope–intercept form

$$y = mx + b, \qquad 0 \le x \le 10$$

to represent the median price of a single-family house in the southern United States from 1989 to 1999.

(c) According to the model, during what year was the median price of a single-family house in the southern United States \$118.2 thousand?

50. The annual amount of Social Security payments from 1990 to 2000 can be represented by a linear model that contains the points shown in Table 1.3.10, where x represents the number of years since 1990 and $f(x)$ represents the annual amount of Social Security payments measured in billions of dollars (*Source:* U.S. Census Bureau, www.census.gov).

Table 1.3.10

x	$f(x)$
0	250.63
1	266.18

(a) Compute the slope m and interpret.

(b) Use the information in the table to write a linear model in the slope–intercept form

$$f(x) = mx + b, \qquad 0 \le x \le 10$$

for the annual amount of Social Security payments from 1990 to 2000.

(c) Evaluate and interpret $f(5)$.

✓ **51.** The size of the adult civilian labor force in the United States from 1990 to 2000 can be represented by a linear model that contains the points shown in Table 1.3.11, where x represents the number of years since 1990 and $f(x)$ represents the civilian labor force in millions (*Source:* U.S. Census Bureau, www.census.gov).

Table 1.3.11

x	$f(x)$
0	125.50
1	127.02

(a) Compute the slope m and interpret.

(b) Use the information in the table to write a linear model in slope–intercept form

$$f(x) = mx + b, \qquad 0 \le x \le 10$$

for the size of the adult civilian labor force in the United States from 1990 to 2000.

(c) Evaluate and interpret $f(4)$.

52. The average annual pay of Americans from 1990 to 2000 can be represented by a linear model that contains the points shown in Table 1.3.12, where x represents the number of years since 1990 and $f(x)$ represents the average annual pay in thousands of dollars (*Source:* U.S. Census Bureau, www.census.gov).

Table 1.3.12

x	$f(x)$
1	24.28
3	26.40

(a) Compute the slope m and interpret.

(b) Use the information in the table to write a linear model in slope–intercept form

$$f(x) = mx + b, \qquad 0 \le x \le 10$$

for the average annual pay of Americans from 1990 to 2000.

(c) Determine the y-intercept and interpret.

53. The number of U.S. motor vehicle registrations from 1990 to 2000 can be represented by a linear model that contains the points shown in Table 1.3.13, where x represents the number of years since 1990 and $f(x)$ represents the number of motor vehicle registrations in millions (*Source:* U.S. Census Bureau, www.census.gov).

Table 1.3.13

x	$f(x)$
1	191.28
4	200.1

(a) Compute the slope m and interpret.

(b) Use the information in the table to write a linear model in slope–intercept form

$$f(x) = mx + b, \qquad 0 \le x \le 10$$

for the number of U.S. motor vehicle registrations from 1990 to 2000.

(c) Determine the y-intercept and interpret.

54. The number of students enrolled in elementary (K–8) school annually in the United States from 1990 to 2000 can be represented by a linear model that contains the points shown in Table 1.3.14, where x represents the number of years since 1990 and y represents the number of students enrolled in elementary (K–8) school in millions (*Source:* U.S. Census Bureau, www.census.gov).

Table 1.3.14

x	y
2	35.09
6	37.01

(a) Use the information in the table to write a linear model in standard form

$$ax + by = c, \qquad 0 \leq x \leq 10$$

for the number of students enrolled in elementary (K–8) school annually in the United States from 1990 to 2000. (*Hint:* Multiply both sides of the equation by 100 so that a, b, and c are integers.)

(b) Determine x when $y = 37.97$ and interpret.

55. The number of students enrolled in higher education institutions annually in the United States from 1990 to 2000 can be represented by a linear model that contains the points shown in Table 1.3.15, where x represents the number of years since 1990 and y represents the number of students enrolled in higher education institutions in millions (*Source:* U.S. Census Bureau, www.census.gov).

Table 1.3.15

x	y
2	13.97
6	14.41

(a) Use the information in the table to write a linear model in standard form

$$ax + by = c, \qquad 0 \leq x \leq 10$$

for the number of students enrolled in higher education institutions from 1990 to 2000. (*Hint:* Multiply both sides of the equation by 100 so that a, b, and c are integers.)

(b) Determine x when $y = 14.3$ and interpret.

SECTION PROJECT

Consider the data in Table 1.3.16, where the values in the right column show the number of Americans covered by private or government health insurance, measured in millions of people.

Table 1.3.16

Year	Number Insured
1987	241
1989	246
1991	251
1993	260
1995	264
1997	269

SOURCE: U.S. Healthcare Financing Administration, www.hcfa.gov

Let x represent the number of years since 1985 and let $f(x)$ represent the number of Americans covered by private or government health insurance, in millions. Then the data can be represented by the linear model

$$f(x) = 2.9x + 234.87, \qquad 2 \leq x \leq 12$$

(a) Evaluate and interpret $f(6)$.

(b) Determine the difference between the value in the table in 1991 and the answer in part (a) by computing

$$\text{(value in table)} - [\text{answer to part (a)}]$$

This difference is called the **residual**.

(c) Now graph the scatterplot of the data in the table along with the given linear model. Is the data point for $x = 6$ (1991) above or below the graph of the linear model? How does this relate to the result of part (b)? Explain.

Section 1.4 Solving Systems of Linear Equations Graphically

There are instances when we need more than just one equation with one variable to find a solution. For example, let's say that an amusement park charges \$6 for children under 6 and \$22 for anyone 7 years or older. On a certain day, the cashier reports that the park took in \$140,000 in gate receipts and that 8000 people visited the park. If we want to know how many child and how many adult tickets were sold, the best way to find the solution is by solving a **system of**

equations. In this section we will learn how to solve systems of equations using graphing techniques and how these systems are used in applications.

Fundamentals of Systems of Linear Equations

A **system of linear equations** is a collection of two or more linear equations with each having at least one variable. The following are examples of systems of equations common in this text.

$$\begin{aligned} 2x + y &= 6 \\ x - y &= 6 \end{aligned} \qquad \begin{aligned} 2x + y - z &= 2 \\ x + 3y + 2z &= 1 \\ x + y + z &= 2 \end{aligned} \qquad \begin{aligned} y &= 3x - 5 \\ y &= 1 - 4x \end{aligned}$$

Note that the exponents on all the variables in these systems are 1. This is why they are called **linear systems** of equations. Linear systems are our focus in the remainder of this chapter and in Chapter 2.

A **solution** to a system of linear equations is the values of the variables that make each equation true. Solutions to systems of equations with two or three variables are written as ordered pairs (x, y) or ordered triples (x, y, z). Let's see how we can verify that an ordered pair is a solution to a system of equations.

Example 1 Checking a Solution to a System of Equations

Determine if the ordered pair $(4, -2)$ is a solution to the system

$$\begin{aligned} 2x + y &= 6 \\ x - y &= 6 \end{aligned}$$

Solution

Here we need to show that if we substitute $x = 4$ and $y = -2$ into each of the equations both of the equations are true. Substituting into the first equation gives

$$\begin{aligned} 2x + y &= 6 \\ 2(4) + (-2) &= 6 \\ 8 + (-2) &= 6 \\ 6 &= 6 \end{aligned}$$

which is true. In the second equation, we have

$$\begin{aligned} x - y &= 6 \\ 4 - (-2) &= 6 \\ 6 &= 6 \end{aligned}$$

which is also true. Thus $(4, -2)$ is a solution to the system. ■

Checkpoint 1 Now work Exercise 7.

To see why we want to write solutions as ordered pairs, let's graph each equation in Example 1. We see in the first equation $2x + y = 6$ that the x-intercept is at $(3, 0)$ and the y-intercept is at $(0, 6)$. In the second equation, $x - y = 6$, the x- and y-intercepts are at $(6, 0)$ and $(0, -6)$, respectively. The graphs are shown in

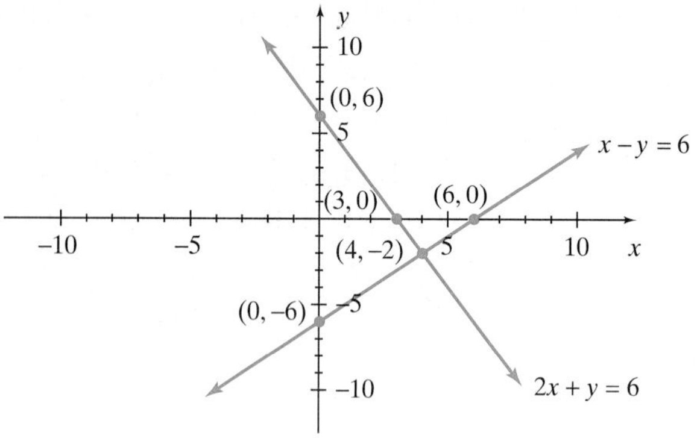

Figure 1.4.1

Figure 1.4.1. Notice that the solution $(4, -2)$ is the point of intersection of the two graphs. This illustrates a fundamental fact behind solving systems of equations.

The solution to a linear system of two equations in two variables corresponds to the point(s) of intersection of their graphs.

Let's use this fact to solve a system of equations graphically.

Example 2 Solving a System of Equations by Graphing

For the system

$$x + 3y = 18$$
$$2x - 6y = 12$$

determine the intercepts of each equation, graph the equations, and write the coordinates of the point of intersection.

Interactive Activity

Numerically verify that (12, 2) is an exact solution in Example 2 by using the technique outlined in Example 1.

Solution

For the first equation, $x + 3y = 18$, the x-intercept is $(18, 0)$ and the y-intercept is $(0, 6)$. For the second equation, $2x - 6y = 12$, the x and y intercepts are at $(6, 0)$ and $(0, -2)$, respectively. The graphs of these equations are shown in Figure 1.4.2. We can see by inspecting the graph that the point of intersection appears to be $(12, 2)$. So the ordered pair $(12, 2)$ is an approximate solution to the system.

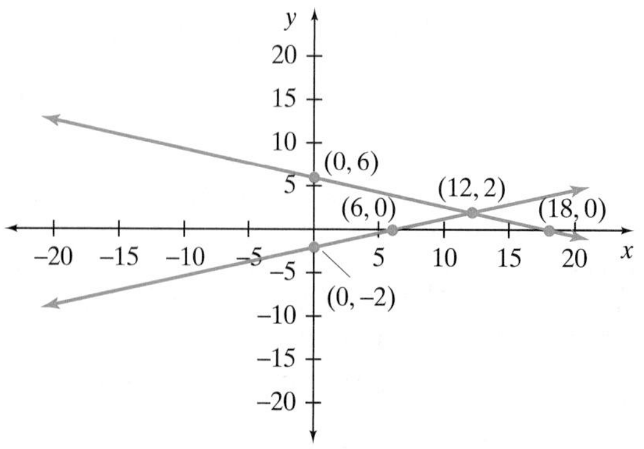

Figure 1.4.2

■

Solutions to Systems

We were able to find a solution to the system in Example 2 because the two equations had exactly **one point of intersection**. To see what else can happen, consider the system

$$x - y = 2$$
$$3x - 3y = -9$$

The graph of this system is shown in Figure 1.4.3. It appears from the graph that the lines are **parallel**, meaning that they never intersect. If the lines are parallel and never intersect, then the system has **no solution**. But how do we know for certain that the lines are parallel? First, we write each of the equations in slope–intercept form (that is, solve each equation for y) and get

$$x - 2 = y \qquad \qquad y = x - 2$$
$$\frac{-9 - 3x}{-3} = y \quad \text{or in slope–intercept form} \quad y = x + 3$$

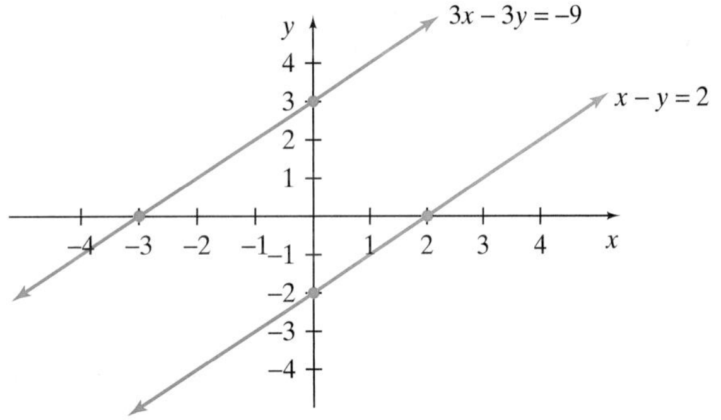

Figure 1.4.3

Notice that each of these equations has a slope of $m = 1$. It stands to reason that if the steepness of the lines are the same then they will not intersect, making the lines parallel. This illustrates that ***parallel lines have equal slopes***. Linear systems that do not have an ordered pair that satisfy both equations are called **inconsistent**.

Inconsistent Linear Systems

Linear systems that have no solution are called **inconsistent systems**.

Let's now take a look at another type of system. The system

$$y = -2x + 1$$
$$6x + 3y = 3$$

is graphed in Figure 1.4.4. Notice that the two equations really represent the same line! We say that the two lines **coincide**. Thus any ordered pair that satisfies the first equation also satisfies the second equation. This means that the system has **infinitely many solutions**. If we write the first equation in standard form, we get

$$2x + y = 1$$
$$6x + 3y = 3$$

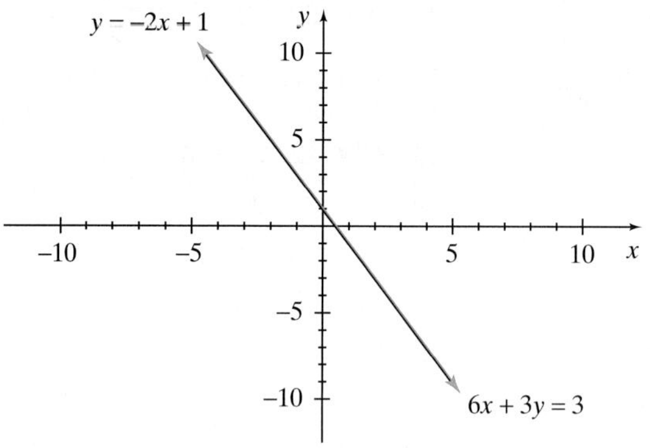

Figure 1.4.4

Notice that we can obtain the second equation by multiplying the first equation by 3. In mathematics, we say that the equations are **linearly dependent**.

Dependent Linear Systems

Linear systems whose graphs coincide are called **dependent systems**. Ordered pairs that satisfy either equation are solutions to the system.

Table 1.4.1 summarizes the types of solutions possible for systems of linear equations.

Table 1.4.1

Position of Graphs	Number of Solutions	Characteristics of Equations
Lines intersect at one point	One (see Figure 1.4.5(a))	Slopes are not equal
Lines are parallel	None (see Figure 1.4.5(b))	Slopes are equal and y-intercepts are not equal
Lines coincide	Infinite (see Figure 1.4.5(c))	Slopes are equal and y-intercepts are equal

From Table 1.4.1 we see that we can determine the system type if it is written in slope–intercept form. In slope–intercept form we can compare the values of the slopes and the y-intercepts.

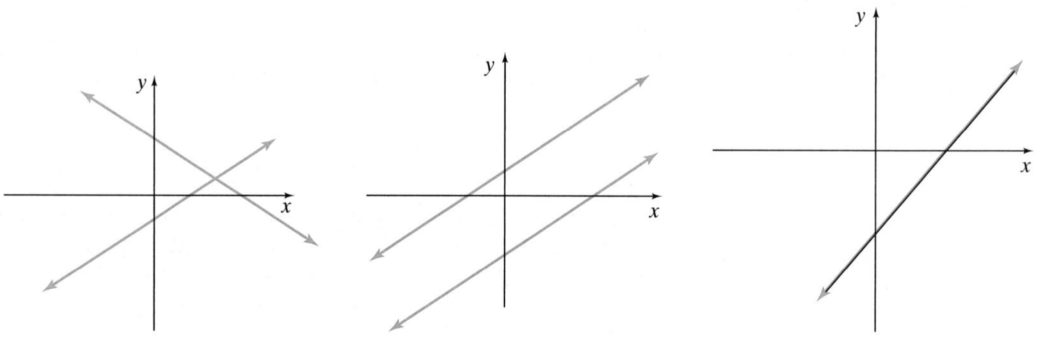

(a) **(b)** **(c)**

Figure 1.4.5 **(a)** Lines intersect at one point. The system has one solution. **(b)** Lines are parallel. The system has no solution. **(c)** Lines coincide. The system has infinitely many solutions.

Example 3 **Determining the Number of Solutions of a Linear System**

Determine the number of solutions for each system and describe the position of the lines without graphing the system.

(a) $2y - x = 4$
$2x + 8 = 4y$

(b) $2x + 2y = 4$
$2x - y = 3$

(c) $2x + 3y = 6$
$4x + 6y = 24$

Solution

Understand the Situation: To determine the system type without graphing, we will write each equation in slope–intercept form and compare the values of m and b.

(a) Solving each equation for y, we get

$$y = \frac{x+4}{2} \qquad\qquad y = \frac{1}{2}x + 2$$
$$\text{or}$$
$$\frac{2x+8}{4} = y \qquad\qquad y = \frac{1}{2}x + 2$$

Since each equation has a slope of $m = \frac{1}{2}$ and y-intercept at $b = 2$, we see that the graphs of the lines coincide, and this system has an infinite number of solutions.

(b) In the slope–intercept form, this system is

$$y = \frac{4-2x}{2} \qquad\qquad y = -x + 2$$
$$\text{or}$$
$$2x - 3 = y \qquad\qquad y = 2x - 3$$

Here the slopes are not equal. This means that the graphs of the lines intersect at a point and the system has one solution.

(c) Isolating y gives us the slope–intercept form.

$$y = \frac{6-2x}{3} \qquad\qquad y = -\frac{2}{3}x + 2$$
$$\text{or}$$
$$y = \frac{24-4x}{6} \qquad\qquad y = -\frac{2}{3}x + 4$$

In this case the slopes are the same at $m = -\frac{2}{3}$, yet the y-intercepts are different. This means that the graphs of the lines are parallel and this linear system has no solution. ■

✓ Checkpoint 2 Now work Exercise 41.

Applications of Linear Systems

We can see that graphing systems of linear equations gives us a visualization of the solution. However, it has drawbacks. For example, we saw that in part (b) of Example 3 the system is dependent and has one solution. However, this solution turns out to be $\left(\frac{5}{3}, \frac{1}{3}\right)$. It is very difficult to determine the exact solution just by graphing. For now, we will approximate the coordinates of the point of intersection. Let's take a look at a few of these solutions in the context of applications.

One application of systems of equations common to economics is the **Law of Supply and Demand.**

Law of Supply and Demand

As the price of a product increases,

- the **demand** for this product decreases and
- the **supply** of this product increases.

The demand for the product decreases when the price increases because some consumers are not willing to pay higher prices. The supply of a product increases because the manufacturers of that product are willing to supply more units of the product, allowing more to be sold. A simplificd version of this relationship between the supply, given by the function s and the demand, denoted by the function d, is shown in Figure 1.4.6. Notice that x represents the number of units of the product sold.

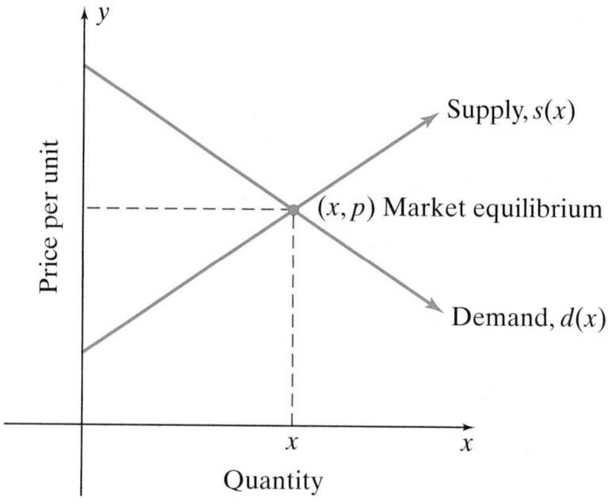

Figure 1.4.6

The point of intersection of the graphs is called the **market equilibrium.** This represents the point at which the quantity produced is equal to the quantity demanded. Let's see how this law works in an application.

Example 4 Applying the Law of Supply and Demand

The Dress-4-Success Company determines that the demand for their new office-appropriate skirt is given by

$$d(x) = 90 - 0.5x$$

while the related supply function is given by

$$s(x) = 0.45x$$

where x is the daily quantity and d and s are measured in dollars for each skirt.

(a) Determine the x- and y-intercepts of the graph of the demand function d.

(b) Determine the supply s when $x = 0$ and $x = 160$ and then graph d and s on the same Cartesian plane.

(c) Approximate the point of intersection rounded to the nearest 5 skirts and 5 dollars and interpret.

Solution

(a) The y-intercept of d can be found by letting $x = 0$ to get

$$d(0) = 90 - 0.5(0) = 90$$

So the y-intercept is at $(0, 90)$. The x-intercept is found by setting $d(x) = 0$ and solving for x.

$$0 = 90 - 0.5x$$
$$0.5x = 90$$
$$x = \frac{90}{0.5} = 180$$

So the x-intercept is at the point $(180, 0)$.

(b) When $x = 0$, the supply is $s(0) = 0.45(0) = 0$, and when $x = 160$, the supply is $s(160) = 0.45(160) = 72$. The graphs of the demand and supply functions are shown in Figure 1.4.7.

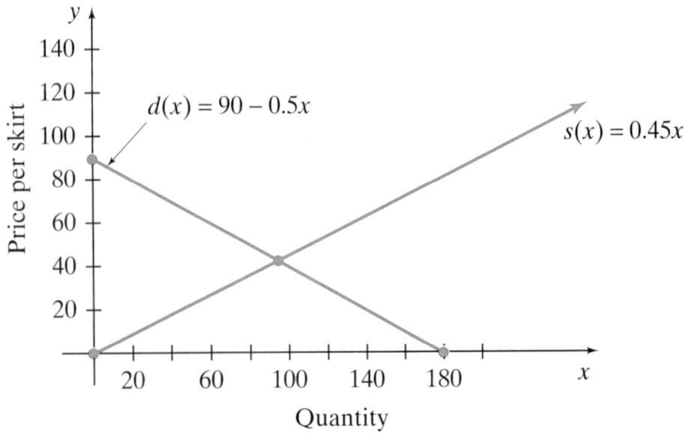

Figure 1.4.7

(c) From inspecting the graph, we can see that the market equilibrium is at about the point $(95, 40)$.

Interpret the Solution: This means that the daily supply will meet the demand when 95 skirts are supplied at a price of \$40 per skirt. ■

Technology Option

By using the INTERSECT command on the calculator, we can get a more accurate approximation of the point of intersection. By graphing the demand and supply functions in Example 4 and using the INTERSECT command, we get an intersection point of about $(94.74, 42.63)$. See Figure 1.4.8. This means that at a daily supply of about 95 skirts the price that consumers are willing

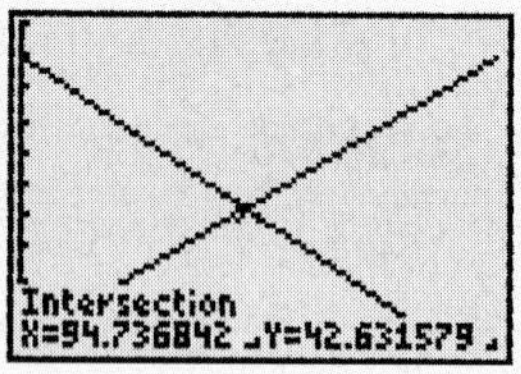

Figure 1.4.8

to pay is about \$42.63. For more on how to use the INTERSECT command on your calculator, consult the online graphing calculator manual at www.prenhall.com/armstrong.

Another type of application of systems in economics is **break-even analysis**. This compares the cost of producing a product to the revenue that the product generates. To see how this process works, let's start by reviewing the **cost function** in the Toolbox below.

From Your Toolbox

The cost of producing x units of a product is given by

$$\begin{pmatrix}\text{cost of}\\ \text{producing } x \text{ units}\end{pmatrix} = \begin{pmatrix}\text{cost per}\\ \text{unit}\end{pmatrix} \cdot x + \begin{pmatrix}\text{fixed}\\ \text{costs}\end{pmatrix}$$

For x units produced at cost per unit m and fixed costs b, the **cost function**, denoted by C, is $C(x) = mx + b$. Recall that fixed costs are costs incurred no matter how many units are produced.

We define the **revenue function** as follows:

Revenue Function

The revenue generated by producing and selling x units of a product at a certain price is

$$\begin{pmatrix}\text{revenue from producing}\\ \text{and selling } x \text{ units}\end{pmatrix} = \begin{pmatrix}\text{quantity}\\ \text{sold}\end{pmatrix} \cdot \begin{pmatrix}\text{unit}\\ \text{price}\end{pmatrix}$$

For x units sold at a price p, the **revenue function** R is

$$R(x) = x \cdot p$$

When the revenue is equal to the costs (that is, when money coming in is equal to money going out), we have reached the break-even point. Mathematically, this is the point where $R(x) = C(x)$. See Figure 1.4.9. Notice in the figure that when $R(x) > C(x)$ a profit occurs. When $R(x) < C(x)$, a loss occurs. Now let's try an application.

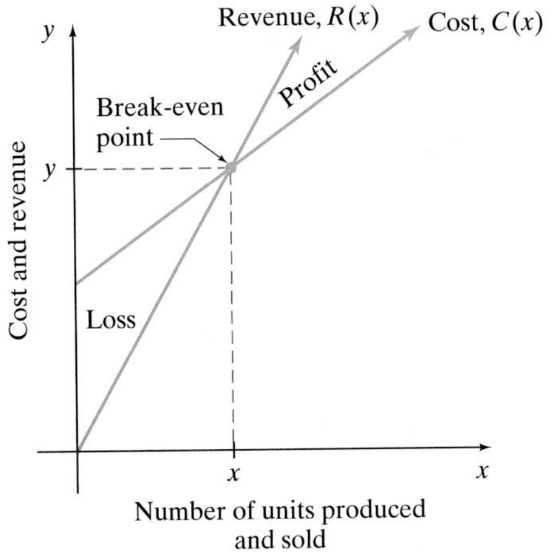

Figure 1.4.9

Example 5 Determining the Break-even Point

The Outdoors Always Company determines that the fixed costs for their new pop-up mini canopy are \$2500 and the cost to produce each canopy is \$25. If they plan to sell the mini canopies for \$50 each, what is the break-even point?

Solution

Understand the Situation: Since the fixed costs are \$2500 and the per unit cost is \$25, we get the cost function $C(x) = 25x + 2500$. The selling price for the mini canopy is \$50, so the revenue is given by the function $R(x) = 50x$.

This means that we must solve the system

$$C(x) = y = 25x + 2500$$
$$R(x) = y = 50x$$

The graphs of these functions are shown in Figure 1.4.10. From inspecting the figure, we believe that the point of intersection is at about (100, 5000).

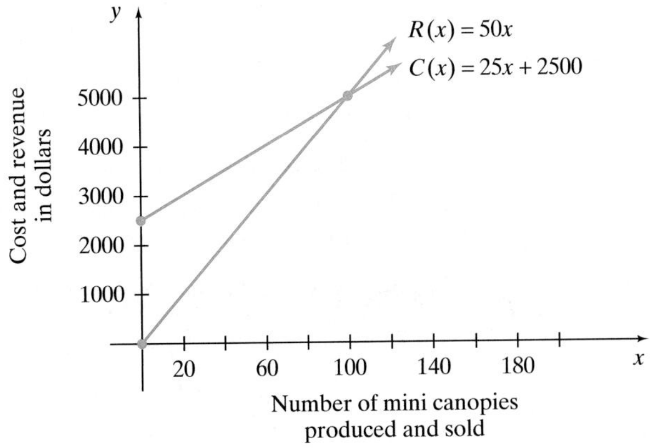

Figure 1.4.10

Interpret the Solution: This means that the company will have to produce and sell 100 mini canopies for \$5000 in revenue in order to break even.

Checkpoint 3 Now work Exercise 51.

Technology Option

If we graph the cost and revenue functions given in Example 5 and use the INTERSECT command, we get the result shown in Figure 1.4.11.

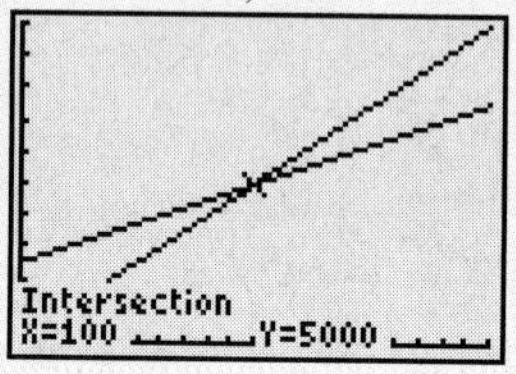

Figure 1.4.11

We see that the point of intersection is at (100, 5000). This confirms our approximation for the break-even point found in Example 5.

We can also use graphical methods when examining **systems of linear models**. These are pairs of models much like the ones we studied in Section 1.3. One of these systems is given in Example 6.

Example 6 Solving a System of Linear Models

From 1990 to 1997, the percent of low-birth-weight babies (less than 5 pounds, 8 ounces) born in Indiana can be modeled by

$$f(x) = -0.038x + 7.621, \qquad 0 \le x \le 7$$

where x represents the number of years since 1990 and $f(x)$ represents the percentage of low-birth-weight babies born. During the same time period, the percent of low-birth-weight babies born in Missouri can be modeled by

$$g(x) = 0.088x + 7.113, \qquad 0 \le x \le 7$$

where x represents the number of years since 1990 and $g(x)$ represents the percentage of low-birth-weight babies born.

(a) Evaluate each model at $x = 0$ and $x = 7$ and graph the system of equations.

(b) Approximate the point of intersection rounded to the nearest half-unit and interpret each coordinate.

Solution

(a) Table 1.4.2 shows the values of each model evaluated at $x = 0$ and $x = 7$.

Table 1.4.2

x	$f(x)$	$g(x)$
0	$-0.038(0) + 7.621 = 7.621$	$0.088(0) + 7.113 = 7.113$
7	$-0.038(7) + 7.621 = 7.355$	$0.088(7) + 7.113 = 7.729$

Using the coordinates in Table 1.4.2, we get the graph of the system shown in Figure 1.4.12.

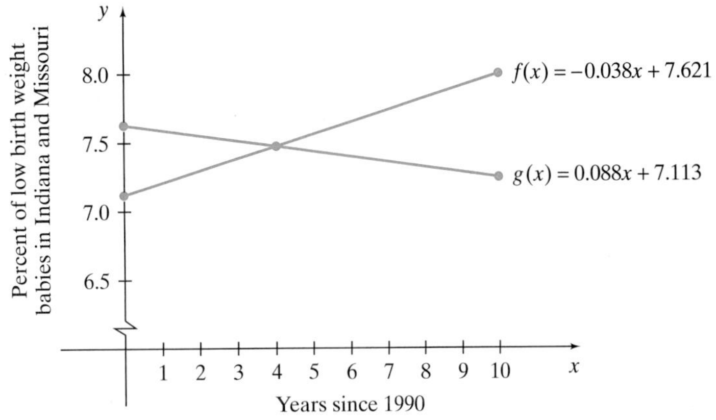

Figure 1.4.12

(b) By inspecting the graph, we approximate the point of intersection to be about $(4, 7.5)$.

Interpret the Solution: This means that in 1994 both Indiana and Missouri had low-birth-weight percentages of about 7.5%. ■

SUMMARY

In this section, we were introduced to systems of linear equations. We learned that one way we could determine the solution to a system of linear equations is by **graphing** each equation in the system. The **point of intersection** of the graphs gives us the solution. We also found that we could have three types of solution for linear systems of equations. Linear systems can produce

- a point of intersection and **one solution**,
- parallel lines and **no solution**,
- the same line and **infinitely many solutions**.

We also learned that some of the applications of linear systems of equations included

- Law of Supply and Demand
- Break-even analysis
- Systems of linear models

SECTION 1.4 EXERCISES

In Exercises 1–10, determine if the given ordered pair is a solution to the system.

1. $(1, -1)$
$x + 4y = -3$
$5x - y = 6$

2. $(2, 2)$
$2x - y = 2$
$x + 3y = 8$

3. $(2, -1)$
$2x + y = 5$
$x + y = 3$

4. $(3, 2)$
$x - y = 5$
$2x + y = 10$

5. $(-1, 1)$
$2x + y = 1$
$x - 2y = 3$

6. $(-2, 2)$
$3x + 2y = -2$
$2x - y = -6$

✓ **7.** $(0, 2)$
$-2x + y = 0$
$7x - 2y = 3$

8. $(3, -3)$
$x - 4y = 15$
$3x - 2y = 15$

9. $(3, 13)$
$x + y = 16$
$x - y = 10$

10. $(-1, 0)$
$x - y = 1$
$3x + y = -1$

In Exercises 11–20, solve the system by determining the x- and y-intercepts of each equation. Then graph the system on the same Cartesian plane. If the system has no solution *or* infinite solutions, *then say so and state the reason why.*

11. $x + y = 2$
$x - y = 4$

12. $x + y = 6$
$x - y = 2$

13. $y = 4 - x$
$y = x + 4$

14. $y = 1 - x$
$y = x + 3$

15. $-x + y = -3$
$3x - y = 3$

16. $4x + 3y = 12$
$2x - 3y = 6$

17. $6x + 2y = 12$
$-3x - y = -3$

18. $x + y = 6$
$-x - y = -5$

19. $-4x - 2y = -8$
$2x + y = 4$

20. $-x + y = 3$
$3x - 3y = -9$

In Exercises 21–24, solve the system by inspecting the graph and approximating the point of intersection.

21. $2x + y = 3$
$-x + y = -3$

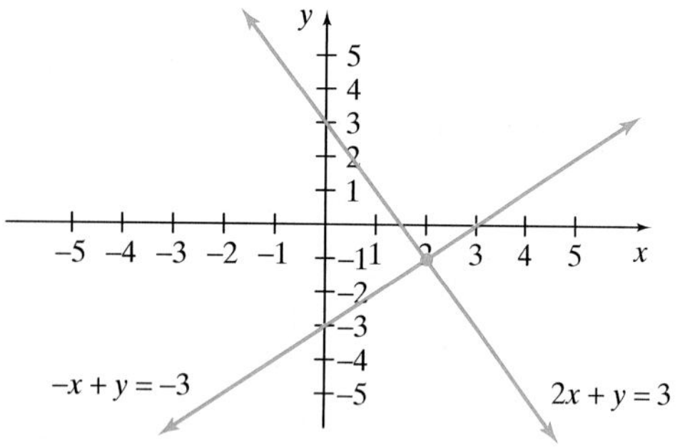

22. $4x - y = -3$
$x - y = 0$

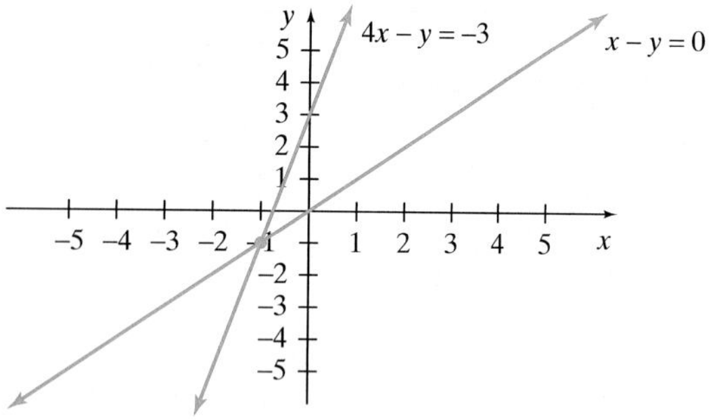

23. $4x + y = 4$
$3x - y = 3$

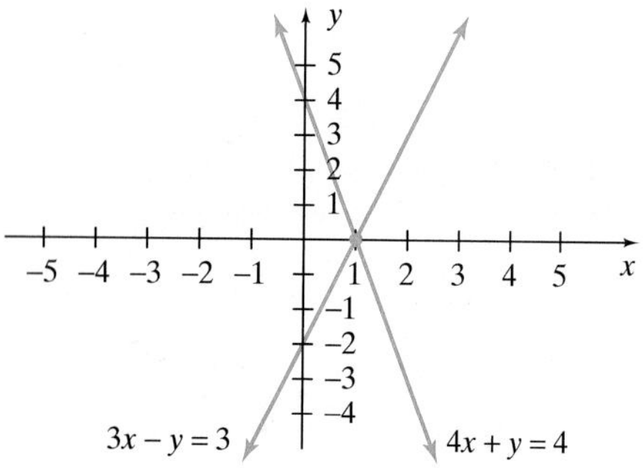

24. $2x + y = 4$
$5x - y = 10$

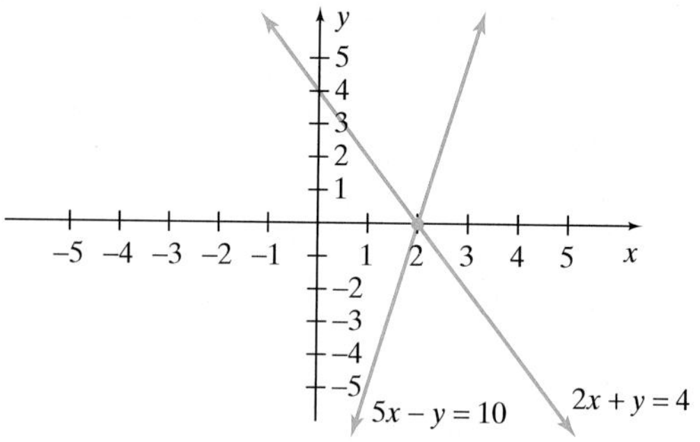

In Exercises 25–32, graph the equations on the same Cartesian plane and then use the graphs to estimate the solutions.

25. $x + y = 4$
$y = -2$

26. $x + y = 10$
$y = 5$

27. $x - y = 8$
$x = 4$

28. $x - y = 6$
$x = 3$

29. $x = 3$
$y = -2$

30. $y = -1$
$x = 4$

31. $y = 3 - 2x$
$y = 1 - x$

32. $y = 3x - 4$
$y = 1 - 2x$

In Exercises 33–38, solve the system by graphing each equation on your calculator and using the INTERSECT command. If the system has no solution *or* infinite solutions, *then say so.*

33. $3y = 2x$
$y - x = 1$

34. $2x + y = 3$
$2y = 3x + 6$

35. $3x + 2y = 8$
$3x = 10 - 2y$

36. $3x + 4y = 5$
$10x - 6y = 3$

37. $2x + y = 3$
$2y - 3x = 6$

38. $2x + 3y = 6$
$4x + 2y = 4$

In Exercises 39–44, determine the value of k that satisfies the given condition for each system.

39. k so that $\begin{array}{l} y = kx + 8 \\ y = 3x - 1 \end{array}$ has no solution.

40. k so that $\begin{array}{l} 2x - y = -2 \\ kx - y = 1 \end{array}$ has no solution.

41. k so that $\begin{array}{l} 4x - 2y = 7 \\ 12x - ky = 21 \end{array}$ has infinitely many solutions.

42. k so that $\begin{array}{l} y = 3x - 4 \\ 4y = kx - 16 \end{array}$ has infinitely many solutions.

43. k so that $\begin{array}{l} y = -6x - 4 \\ y = -3x - k \end{array}$ has no solution.

44. k so that $\begin{array}{l} y = 8x + 5 \\ y = 4x + k \end{array}$ has no solution.

Applications

45. The Say-It-Now Company develops communications software and determines that the demand for their teleconference mini suite is given by $d(x) = 80 - \frac{3}{5}x$ with a supply function given by $s(x) = \frac{2}{5}x$, where x is the daily quantity and $d(x)$ and $s(x)$ are measured in dollars per minisuite.

(a) Determine the x- and y-intercepts for the graph of the demand function d.

(b) Determine the supply s when $x = 0$ and $x = 130$. Graph d and s on the same Cartesian plane.

(c) Approximate the point of intersection and interpret.

46. The SunMaker Company has determined that the demand for their patio sets is given by $d(x) = 630 - 0.75x$, while the supply is given by $s(x) = 0.75x$, where x is the quantity and $d(x)$ and $s(x)$ are in dollars per patio set.

(a) Determine the x- and y-intercepts for the graph of the demand function d.

(b) Determine the supply s when $x = 0$ and $x = 830$. Graph d and s on the same Cartesian plane.

(c) Approximate the point of intersection and interpret.

47. The Omniview Company makes hand-held color televisions and determines that the demand for their entry-level TK-2 model is $d(x) = 150 - 0.2x$, while the supply is given by $s(x) = 0.4x$, where x is the quantity and $d(x)$ and $s(x)$ are in dollars per TK-2 television.

(a) Determine the x- and y-intercepts for the graph of the demand function d.

(b) Determine the supply s when $x = 0$ and $x = 400$. Graph d and s on the same Cartesian plane.

(c) Approximate the point of intersection and interpret.

48. The Big Sound Company has determined that the demand for their Faster Blaster boom box is $d(x) = 450 - 0.3x$, while the supply is given by $s(x) = 0.15x$, where x is the quantity and $d(x)$ and $s(x)$ are measured in dollars per boom box.

(a) Determine the x- and y-intercepts for the graph of the demand function d.

(b) Determine the supply s when $x = 0$ and $x = 2000$. Graph d and s on the same Cartesian plane.

(c) Approximate the point of intersection and interpret.

49. The Ryovic Company determines that the demand for their single-person pup tent is given by $d(x) = 75 - \frac{1}{4}x$, while the supply is given by $s(x) = \frac{1}{2}x$, where x is the quantity and $d(x)$ and $s(x)$ are measured in dollars per tent.

(a) Write a system of equations to determine the equilibrium point and graph the system in the viewing window [0, 200] by [0, 100].

(b) Determine the point of intersection via the INTERSECT command and interpret each coordinate.

50. The TechnoPic Company determines that the demand for their new picture-in-a-picture TV set is given by $d(x) = 1800 - \frac{3}{5}x$, while the supply is given by $s(x) = \frac{3}{10}x$, where x is the quantity and $d(x)$ and $s(x)$ are measured in dollars per TV set.

(a) Write a system of equations to determine the equilibrium point and graph the system in the viewing window [0, 3000] by [0, 1200].

(b) Determine the point of intersection via the INTERSECT command and interpret each coordinate.

⇐ 1.1 **51.** Adel starts a business of manufacturing picture frames for small 3 by 5 inch photographs. He determines that the fixed cost is \$252 and the unit cost is \$1.50. He sells each frame for \$5.50.

(a) Determine a system of linear equations for the cost and the revenue that Adel makes from the picture frames.

(b) Use Figure 1.4.13 to approximate the break-even point and interpret each coordinate. Round the cost and revenue to the nearest \$10 and the number of frames sold to the nearest 10.

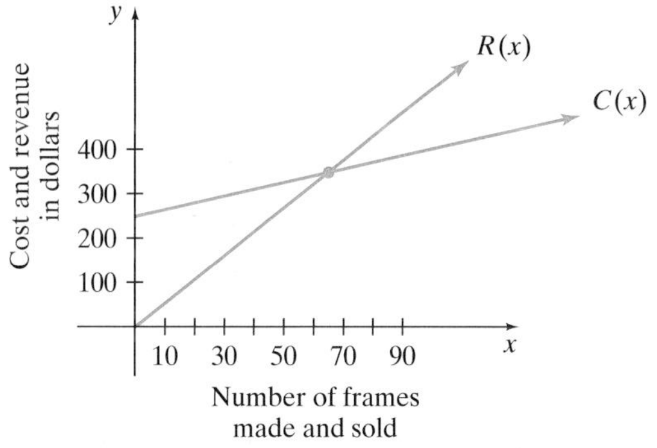

Figure 1.4.13

⇐ 1.1 **52.** Sally makes dream catchers as a hobby and wants to sell them at craft shows. She knows that the fixed costs are \$90 and the cost to produce each dream catcher is \$2.50. She plans to set a price of \$3 for each dream catcher.

(a) Determine a system of equations for the cost and the revenue that Sally makes from the dream catchers.

(b) Use Figure 1.4.14 to approximate the break-even point and interpret each coordinate.

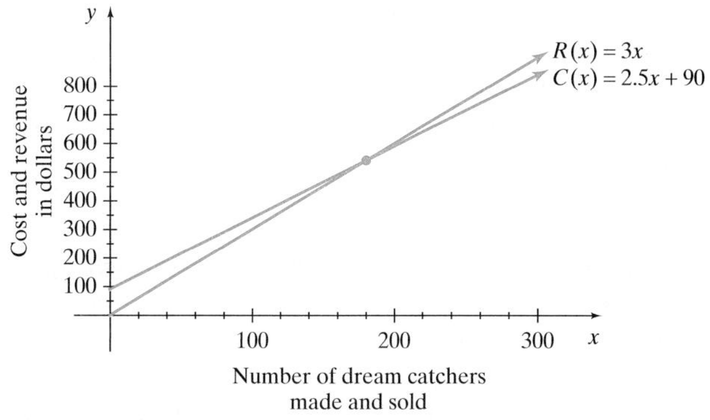

Figure 1.4.14

⇐ 1.1 **53.** The Lullaby Company knows that the fixed costs to manufacture their portable crib are \$250 and the cost to produce each crib is \$100. The sale price for each crib is \$125. Define the variables and write a system of equations for the cost and the revenue that the Lullaby Company makes from the portable cribs.

54. For the system determined in Exercise 53, graph C and R in the same viewing window. Use the INTERSECT command to determine the break-even point and interpret each coordinate.

1.1 **55.** Clark and Florence make bumper stickers. They know that their fixed costs are \$200 and the per unit cost is \$3. They sell the bumper stickers for \$5 each. Define the variables and write a system of equations for the cost and the revenue that Clark and Florence make from the bumper stickers.

56. For the system determined in Exercise 55, graph C and R in the same viewing window. Use the INTERSECT command to determine the break-even point and interpret each coordinate.

57. The Stay-Connected Company determines that the fixed costs for their Family Radio Service radios are \$20,000 and it costs \$30 to manufacture each unit. They sell each radio for \$50. Define the variables and write a system of equations for the cost and the revenue that the Stay-Connected Company makes from the Family Radio Service radios.

58. For the system determined in Exercise 57, graph C and R in the same viewing window. Use the INTERSECT command to determine the break-even point and interpret each coordinate.

59. The Squeaky Clean Company has fixed costs of \$100,000 and unit costs of \$150 for their pressure washers. Each pressure washer sells for \$250. Define the variables and write a system of equations for the cost and the revenue that the Squeaky Clean Company makes from the pressure washers.

60. For the system determined in Exercise 59, graph C and R in the same viewing window. Use the INTERSECT command to determine the break-even point and interpret each coordinate.

61. From 1990 to 1998, the number of U.S. children under 5 years of age who were of Hispanic origin can be modeled by

$$f(x) = 0.12x + 2.51, \qquad 0 \le x \le 8$$

where x represents the number of years since 1990 and $f(x)$ represents the number of children under 5 years of age who were of Hispanic origin, in millions. During the same time period, the number of U.S. children under 5 years of age who were African American can be modeled by

$$g(x) = -0.02x + 2.84, \qquad 0 \le x \le 8$$

where x represents the number of years since 1990 and $g(x)$ represents the number of children under 5 years of age who were African American, in millions (*Source:* U.S. Census Bureau, www.census.gov).

(a) Evaluate each model at $x = 0$ and $x = 8$ and graph the system of equations.

(b) Approximate the point of intersection rounded to the nearest half-unit and interpret each coordinate.

62. For the models given in Exercise 61, graph f and g in the viewing window [0, 8] by [2, 4]. Determine the point of intersection via the INTERSECT command and interpret each coordinate.

63. From 1990 to 1998, the number of people 25 to 29 years of age who were of Hispanic origin in the United States can be modeled by

$$f(x) = 0.024x + 2.335, \qquad 0 \le x \le 8$$

where x represents the number of years since 1990 and $f(x)$ represents the number of people 25 to 29 years of age who were of Hispanic origin, in millions. During the same time period, the number of people 25 to 29 years of age who were African American in the United States can be modeled by

$$g(x) = -0.023x + 2.629, \qquad 0 \le x \le 8$$

where x represents the number of years since 1990 and $g(x)$ represents the number of people 25 to 29 years of age who were African American, in millions (*Source:* U.S. Census Bureau, www.census.gov).

(a) Evaluate each model at $x = 0$ and $x = 8$ and graph the system of equations.

(b) Approximate the point of intersection rounded to the nearest half-unit and interpret each coordinate.

64. For the models given in Exercise 63, graph f and g in the viewing window [0, 8] by [2, 3]. Determine the point of intersection via the INTERSECT command and interpret each coordinate.

SECTION PROJECT

The populations (measured in thousands) of the metropolitan areas of the cities of Austin, Texas, and Dayton, Ohio, are shown in Table 1.4.3.

Table 1.4.3

City	1990 Population	1998 Population
Austin	846	1106
Dayton	951	949

SOURCE: U.S. Census Bureau, www.census.gov

(a) Write linear models in slope–intercept form for the populations of Austin and Dayton.

(b) State the average rate of change for the populations of each city.

(c) Graph each of the models by hand, and estimate the point of intersection and interpret each coordinate.

(d) Graph each model with your calculator, determine the point of intersection, and interpret each coordinate.

Section 1.5 Solving Systems of Linear Equations Algebraically

In the previous section we saw how we could solve systems of linear equations using the **graphical method**. This method provides a good visualization of the solution, but we have seen that the method has some drawbacks. If the solution contains a fraction, for example, the graphing method makes it difficult to pinpoint the solution. In this section we will learn two algebraic methods for solving systems of linear equations. We will also examine more applications of systems of linear equations.

The Substitution Method

The challenge in solving systems of linear equations is that there are two equations with two unknowns (that is, variables). Much of what we studied in algebra focused on how to solve one equation with one variable. With the **substitution method**, we can alter the system in such a way that there is only one linear equation written in terms of one variable. To see how this works, let's Flashback to an example from Section 1.4.

Flashback

Revisiting Systems of Equations

In Example 3, part (b) of the previous section, we were given the system

$$2x + 2y = 4$$
$$2x - y = 3$$

Graph the system and verify that the solution is $\left(\frac{5}{3}, \frac{1}{3}\right)$.

Flashback Solution

The graph of the system is shown in Figure 1.5.1. If the solution to the system is $\left(\frac{5}{3}, \frac{1}{3}\right)$, then the lines intersect at $\left(\frac{5}{3}, \frac{1}{3}\right)$, as shown in Figure 1.5.1.

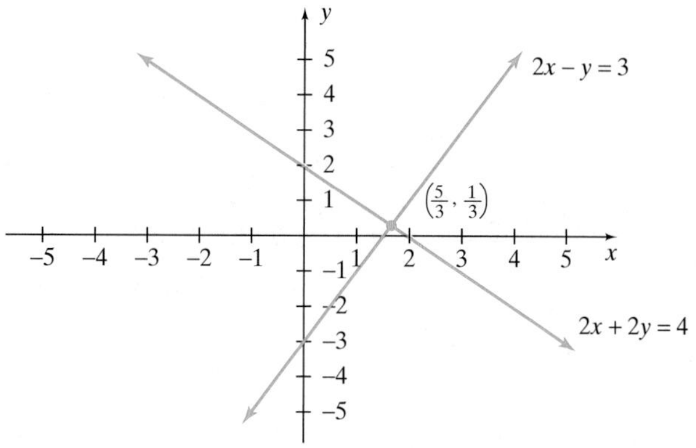

Figure 1.5.1

To verify the solution, we replace x with $\frac{5}{3}$ and y with $\frac{1}{3}$ in each equation and show that the equations are true. In the first equation we get

$$\begin{aligned} 2x + 2y &= 4 \\ 2\left(\frac{5}{3}\right) + 2\left(\frac{1}{3}\right) &= 4 \\ \frac{10}{3} + \frac{2}{3} &= 4 \\ \frac{12}{3} &= 4 \end{aligned}$$

which is true. The second equation yields

$$\begin{aligned} 2x - y &= 3 \\ 2\left(\frac{5}{3}\right) - \left(\frac{1}{3}\right) &= 3 \\ \frac{10}{3} - \frac{1}{3} &= 3 \\ \frac{9}{3} &= 3 \end{aligned}$$

which is true. So the solution is verified.

In the Flashback we verified that the solution to the system is $\left(\frac{5}{3}, \frac{1}{3}\right)$, although it is difficult to see this from looking at the graph. Now let's try an algebraic approach. For the system in the Flashback

$$\begin{aligned} 2x + 2y &= 4 \\ 2x - y &= 3 \end{aligned}$$

let's begin by solving the second equation for y. This gives us

$$\begin{aligned} 2x + 2y &= 4 \\ 2x - 3 &= y \end{aligned}$$

Since the solution to a system of linear equations is made up of the values that make each equation true, we can replace or **substitute** the expression $2x - 3$ for every occurrence of y in the first equation. This gives us

$$2x + 2(2x - 3) = 4 \qquad \text{First equation, with } 2x - 3 \text{ replacing } y$$

Now we have one equation written in terms of one variable, x. This allows us to solve for x to get

$$\begin{aligned} 2x + 4x - 6 &= 4 && \text{Distribute the 2.} \\ 6x &= 10 && \text{Combine like terms.} \\ x = \frac{10}{6} &= \frac{5}{3} && \text{Divide both sides by 6.} \end{aligned}$$

To find y, we can replace $\frac{5}{3}$ for x in either equation of the system. If we let $x = \frac{5}{3}$ in the system's first equation, we get

$$\begin{aligned} 2\left(\frac{5}{3}\right) + 2y &= 4 && \text{Substitute } x \text{ with } \frac{5}{3}. \\ \frac{10}{3} + 2y &= 4 && \text{Simplify.} \\ 2y &= 4 - \frac{10}{3} = \frac{2}{3} && \text{Subtract } \frac{10}{3} \text{ from both sides.} \\ y &= \frac{1}{3} && \text{Divide both sides by 2.} \end{aligned}$$

So the solution to the system is $(x, y) = \left(\frac{5}{3}, \frac{1}{3}\right)$.

We have just illustrated an algebraic technique known as the **substitution method**. We now list the steps used in the substitution method.

Substitution Method

Given a system of linear equations and two variables, we can solve the system using the substitution method by the following steps.

1. Choose one of the equations and solve for one of the variables, x or y. A good choice is a variable with a coefficient of 1 or –1.
2. Substitute the resulting expression from step 1 into the other equation for the appropriate variable.
3. Solve the resulting equation.
4. Find the value of the remaining variable by substituting the value of the variable found in step 3 into either equation containing both variables.

Example 1 Solving a System of Linear Equations by the Substitution Method

Solve the system using the substitution method.

$$\begin{aligned} 2x + 3y &= 1 \\ x - y &= 3 \end{aligned}$$

Solution

For this example, let's follow the four steps in the substitution process.

Step 1: Since the coefficient of x in the second equation is 1, we will solve the second equation for x to get

$$\begin{aligned} x - y &= 3 \\ x &= y + 3 && \text{Add } y \text{ to both sides.} \end{aligned}$$

Step 2: Substituting the expression $y + 3$ for x in the first equation gives us

$$\begin{aligned} 2x + 3y &= 1 \\ 2(y + 3) + 3y &= 1 \end{aligned}$$

Step 3: Now we want to solve the equation found in step 2 for y to get

$$\begin{aligned} 2y+6+3y&=1 && \text{Distribute the 2.}\\ 5y+6&=1 && \text{Combine like terms.}\\ 5y&=-5 && \text{Subtract 6 from both sides.}\\ y&=-1 && \text{Divide both sides by 5.} \end{aligned}$$

Step 4: We can now substitute $y=-1$ into either equation to determine the x-value.

Using the first equation in the system gives us

$$\begin{aligned} 2x+3(-1)&=1\\ 2x-3&=1\\ 2x&=4\\ x&=2 \end{aligned}$$

So the solution to the system is $(x, y) = (2, -1)$. ■

Checkpoint 1

Now work Exercise 5.

Many of our systems of linear models are written in terms of the dependent variable, whether it be f, g, y, and so on. We can use the substitution method to solve these systems of linear models by simply setting the expressions containing the independent variables equal to each other. Some call this method of using substitution the **comparison method**.

Example 2 Solving a System of Linear Models

The number of immigrants from Ireland to the United States from 1981 to 1995 can be modeled by

$$y = f(x) = 2.72x + 17.84, \qquad 1 \le x \le 15$$

where x represents the number of years since 1980 and $f(x)$ represents the annual number of immigrants from Ireland to the United States, in thousands. During the same time period, the number of immigrants from Japan to the United States can be modeled by

$$y = g(x) = -1.24x + 50.02, \qquad 1 \le x \le 15$$

where x represents the number of years since 1980 and $g(x)$ represents the annual number of immigrants from Japan to the United States, in thousands. Solve the corresponding system of equations and interpret the meaning of each coordinate (*Source:* U.S. Immigration and Naturalization Service, www.ins.usdoj.gov).

Solution

Understand the Situation: Here we have to solve the system of equations

$$\begin{aligned} y = f(x) &= 2.72x + 17.84\\ y = g(x) &= -1.24x + 50.02 \end{aligned}$$

Since each equation is solved for y, we can set the expressions involving x equal to each other to get

$$\begin{aligned} 2.72x + 17.84 &= -1.24x + 50.02 \\ 3.96x &= 32.18 \\ x &= \frac{32.18}{3.96} \\ x &\approx 8.13 \end{aligned}$$

Interactive Activity

Using your calculator, enter $y_1 = 2.72x + 17.84$ and $y_2 = -1.24x + 50.02$. Graph y_1 and y_2 in the viewing window [0,15] by [10,50] and use the INTERSECT command to verify the result of Example 2.

Substituting this result into the second model gives us

$$y = -1.24(8.13) + 50.02 \approx 39.94$$

So the solution to the system is about (8.13, 39.94).

Interpret the Solution: This means that during the year 1988 the number of immigrants from Ireland and from Japan to the United States was approximately equal at about 39.94 thousand, or about 39,940. ■

The Elimination Method

The **elimination method** is a powerful method used to solve systems of equations and works well when none of the coefficients of the variables is 1 or −1. The method will also lead us to the method for solving systems of equations using matrices in Chapter 2. The fundamental idea behind the elimination method is to continue replacing the original systems of equations by a system of **equivalent equations** until the solution is reached.

Consider the system

$$\begin{aligned} 3x - 5y &= 2 \\ 2x + 5y &= 13 \end{aligned}$$

Notice that the coefficients of the variable y in the two equations are opposite, so if we simply add the like terms of the two equations, we get

$$\begin{array}{r} 3x - 5y = 2 \\ \underline{2x + 5y = 13} \\ 5x = 15 \end{array}$$

Solving for x gives us

$$\begin{aligned} 5x &= 15 \\ x &= 3 \end{aligned}$$

If we substitute $x = 3$ into the first equation, we get

$$\begin{aligned} 3(3) - 5y &= 2 \\ 9 - 5y &= 2 \\ -5y &= -7 \\ y &= \frac{7}{5} \end{aligned}$$

So the solution is $(x, y) = \left(3, \frac{7}{5}\right)$.

From Your Toolbox

A skill from algebra that is necessary for solving systems by the elimination method is combining monomial terms. For more practice on this skill, see Appendix B.

We have just outlined the elimination method. But what if the coefficients of a variable are not opposites? Then we use one of the following three operations for obtaining equivalent equations.

Equivalent Equation Operations

We can transform a system of linear equations into an equivalent system of linear equations using these three operations.

1. Interchange two equations (this is sometimes called *swapping* equations).
2. Multiply an equation by a nonzero constant.
3. Add a multiple of one equation to another equation.

Usually, the system's equations are written in the standard form $ax + by = c$ before performing any equivalent equation operations. We will see that the third of the operations is used often. Now let's use these operations to solve another system.

Example 3 Solving a System of Equations by the Elimination Method

Solve the system using the elimination method.

$$\begin{aligned} 3x + 2y &= 11 \\ -x + y &= 3 \end{aligned}$$

Solution

We can make the coefficients of x opposites if we multiply the second equation by 3. This gives us the equivalent system

$$\begin{aligned} 3x + 2y &= 11 \\ -3x + 3y &= 9 \end{aligned}$$

Now we can add the two equations to get

$$\begin{array}{r} 3x + 2y = 11 \\ \underline{-3x + 3y = 9} \\ 5y = 20 \\ y = 4 \end{array}$$

If we replace the y-value in our given system's first equation, we have

$$\begin{aligned} 3x + 2(4) &= 11 \\ 3x + 8 &= 11 \\ 3x &= 3 \\ x &= 1 \end{aligned}$$

So the solution is (1, 4). This is illustrated in Figure 1.5.2.

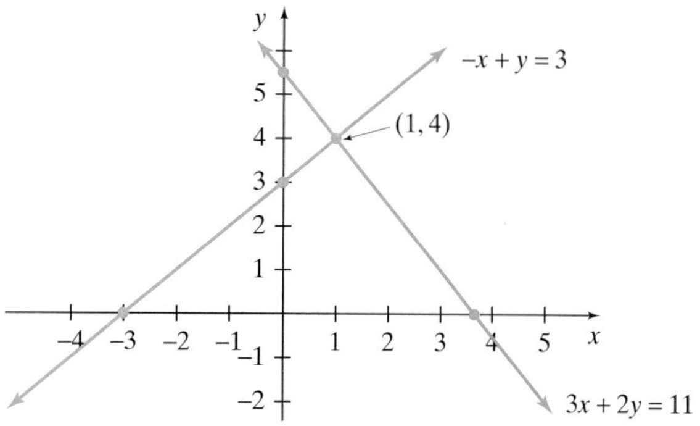

Figure 1.5.2

■

Interactive Activity

In Example 3, replace the y-value in our given system's second equation. Do you get the same result?

WWW

In Section 1.4, we saw that a system of equations can have either one solution, no solution, or infinitely many solutions. If a system has an infinite number of solutions, we can express the solution by using what is called a **parameter**. Let's see how we can express a solution to a system with an infinite number of solutions by using a parameter.

Example 4 Using a Parameter in Writing a Linear System's Solution

Solve the system by elimination.

$$\begin{aligned} x - y &= 3 \\ -2x + 2y &= -6 \end{aligned}$$

Solution

We begin by multiplying the first equation by 2 and adding the equations to get

$$\begin{aligned} 2x - 2y &= 6 \\ -2x + 2y &= -6 \\ \hline 0 &= 0 \end{aligned}$$

Notice that all the variables have disappeared and we are left with the equation $0 = 0$. This is sometimes called an **identity** and indicates that the system has an infinite number of solutions. We will express the solutions in terms of y, where y can represent any real number. In this case, y is the parameter. (We could have also chosen x to be our parameter.) When we solve the first equation for x, we get

$$\begin{aligned} x - y &= 3 \\ x &= y + 3 \end{aligned}$$

So we can say that all ordered pairs of the form

$$(y + 3, y)$$

are solutions to the original system. ■

Interactive Activity

If both linear equations in Example 4 are graphed on one Cartesian plane, would the lines intersect in a point, be parallel, or coincide? Explain.

WWW

Example 4 illustrated that when using an algebraic technique to solve a system and all variables disappear, the system has an infinite number of solutions if we are left with an equation that is an identity. To see what happens when we have no solution, consider the following system:

$$\begin{aligned} x - y &= 4 \\ -2x + 2y &= -6 \end{aligned}$$

We begin by multiplying the first equation by 2 and then add the equations to get

$$\begin{aligned} 2x - 2y &= 8 \\ -2x + 2y &= -6 \\ \hline 0 &= 2 \end{aligned}$$

Notice that all the variables have disappeared and we are left with the equation $0 = 2$. This is sometimes called a **contradiction** and indicates that the system has no solution.

Solving a System with Three Linear Equations and Three Variables

One reason that elimination is so powerful is that we can use it to solve systems of three linear equations and three variables. This is sometimes called a **3 by 3 system**. Similarly, a system of two equations and two unknowns is called a **2 by 2 system**. The general strategy behind solving these systems is to reduce the system to a system of two equations and two variables. Then we can use the techniques we have learned so far to determine the variable's values.

Solving a System of Three Equations and Three Variables

1. Choose any two equations and eliminate a variable.
2. Choose a different pair of equations and eliminate the same variable.
3. Use the results of steps 1 and 2 to write a system of two equations and two variables.
4. Solve the system from step 3.
5. Replace the value found in step 4 into the system of two equations and two variables to determine the second variable's value.
6. Replace the known values into any equation in the original system to determine the third variable's value.

Notice that, since the system has three variables, we will get an **ordered triple** for a solution. Let's look at an example of how to solve one of these systems.

Example 5 Solving a System of Three Linear Equations and Three Variables

Solve the 3 by 3 system using the elimination method.

$$\begin{aligned} x + 2y + 2z &= 9 \\ 2x + y + 4z &= 12 \\ 3x - 3y - 2z &= 1 \end{aligned}$$

Solution

In this example, let's follow the six steps for solving 3 by 3 systems.

Step 1: Notice that if we add the first and third equations, we can eliminate z.

$$\begin{array}{r} x + 2y + 2z = \ \ 9 \\ 3x - 3y - 2z = \ \ 1 \\ \hline 4x - y = 10 \end{array}$$

Step 2: Now we need to eliminate z from another pair of equations in our given system. If we multiply the third equation by 2 and add it to the second equation, we get

$$\begin{array}{r} 2x + y + 4z = 12 \\ 3x - 3y - 2z = \ \ 1 \end{array} \quad \xrightarrow{\text{Multiply bottom equation by 2 and add.}} \quad \begin{array}{r} 2x + \ \ y + 4z = 12 \\ 6x - 6y - 4z = \ \ 2 \\ \hline 8x - 5y = 14 \end{array}$$

Step 3: This gives us the 2 by 2 system

$$\begin{aligned} 4x - y &= 10 \\ 8x - 5y &= 14 \end{aligned}$$

Step 4: We can now eliminate y by multiplying the top equation by -5 and adding the equations. This yields

$$\begin{array}{r} 4x - y = 10 \\ 8x - 5y = 14 \end{array} \quad \xrightarrow{\text{Multiply top equation by } -5 \text{ and add.}} \quad \begin{array}{r} -20x + 5y = -50 \\ 8x - 5y = \ \ \ 14 \\ \hline -12x = -36 \\ x = \ \ \ \ 3 \end{array}$$

Step 5: We can now determine the y-value by replacing x by 3 in either equation of the 2 by 2 system. Substituting x with 3 in the second equation gives us

$$\begin{aligned} 8(3) - 5y &= 14 \\ 24 - 5y &= 14 \\ -5y &= -10 \\ y &= 2 \end{aligned}$$

Step 6: We can now determine the z-value by substituting $x = 3$ and $y = 2$ into any equation in our original 3 by 3 system. If we use the first equation, we find

$$\begin{aligned} x + 2y + 2z &= 9 \\ 3 + 2(2) + 2z &= 9 \\ 3 + 4 + 2z &= 9 \\ 7 + 2z &= 9 \\ 2z &= 2 \\ z &= 1 \end{aligned}$$

So the solution to the system is the ordered triple $(x, y, z) = (3, 2, 1)$. ■

Checkpoint 2 Now work Exercise 41.

We learned in Section 1.4 that the graph of a system of two linear equations and two variables was generally two lines in the Cartesian plane. But what is the graph of three linear equations with three variables? The graph of an equation of the form $ax + by + cz = d$ is a surface in three dimensions called a **plane**. Just like 2 by 2 systems, the 3 by 3 system can have either one solution, no solution, or infinitely many solutions. This depends on how the three planes given by the three equations intersect. Figure 1.5.3 shows this relationship.

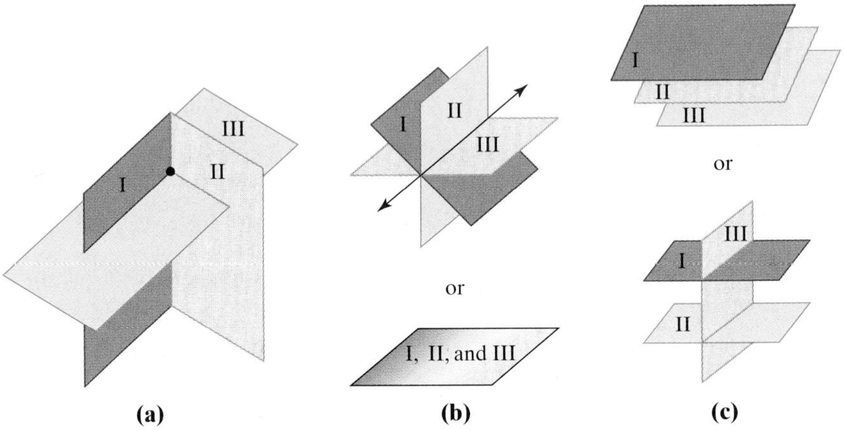

Figure 1.5.3 **(a)** Three planes intersect at a point. System has one solution. **(b)** Three planes intersect in a line or are the same plane. System has infinitely many solutions. **(c)** Three parallel planes or two parallel planes intersected by a third plane. No point or points in common. System has no solution.

Example 6 Solving a 3 by 3 System of Linear Equations

Solve the system by elimination.

$$\begin{aligned} 2x + 3y - 2z &= 4 \\ 3x - 3y + 2z &= 16 \\ 6x - 12y + 8z &= 5 \end{aligned}$$

Solution

If we multiply the first equation by 4 and add it to the third equation, we get

$$\begin{aligned} 2x + 3y - 2z &= 4 \\ 6x - 12y + 8z &= 5 \end{aligned} \quad \xrightarrow{\text{Multiply top equation by 4 and add.}} \quad \begin{aligned} 8x + 12y - 8z &= 16 \\ 6x - 12y + 8z &= 5 \\ \hline 14x &= 21 \\ x &= \frac{3}{2} \end{aligned}$$

Now if we add the first and second equations of our original system, we get

$$\begin{aligned} 2x + 3y - 2z &= 4 \\ 3x - 3y + 2z &= 16 \\ \hline 5x &= 20 \\ x &= 4 \end{aligned}$$

So we see that two different values are obtained for x, which is a contradiction. This indicates that the system has no solution. ■

In Example 4 we saw that the solutions to a 2 by 2 system of linear equations with an infinite number of solutions can be expressed by a parameter. Parameters can also be written to express infinite solutions to 3 by 3 systems of linear equations.

Application

Systems of three equations and three variables can also be used in applications, as is shown in the following Flashback.

Flashback

System of Three Equations and Three Variables Revisited

In Section 1.1 we saw that Gonzalez Office Supplies manufactures three types of office chairs: Secretarial, Executive, and Presidential. Each chair requires time on three different machines. These times and the total available time for each machine are given in Table 1.5.1. If all available time for each machine must be used each week, determine how many of each type of chair can be made each week.

Table 1.5.1

Type:	**Secretarial**	**Executive**	**Presidential**	**Total Weekly Hours**
Hours on Machine A	1	1	2	30
Hours on Machine B	2	3	2	53
Hours on Machine C	1	2	3	47

Flashback Solution

Understand the Situation: Here we let x represent the number of Secretarial chairs made each week, y represent the number of Executive chairs made each week, and z represent the number of Presidential chairs made each week. Then we have to solve the system

$$\begin{aligned} x + y + 2z &= 30 \\ 2x + 3y + 2z &= 53 \\ x + 2y + 3z &= 47 \end{aligned}$$

We can eliminate x by multiplying the first equation by -1 and adding it to the third equation.

$$\begin{aligned} x + y + 2z &= 30 \\ x + 2y + 3z &= 47 \end{aligned} \quad \text{Multiply top equation by } -1 \text{ and add.} \quad \rightarrow \quad \begin{aligned} -x - y - 2z &= -30 \\ x + 2y + 3z &= 47 \\ \hline y + z &= 17 \end{aligned}$$

We can eliminate x from the second and third equations by multiplying the third equation by -2 and adding it to the second.

$$\begin{aligned} 2x + 3y + 2z &= 53 \\ x + 2y + 3z &= 47 \end{aligned} \quad \text{Multiply bottom equation by } -2 \text{ and add.} \quad \rightarrow \quad \begin{aligned} 2x + 3y + 2z &= 53 \\ -2x - 4y - 6z &= -94 \\ \hline -y - 4z &= -41 \end{aligned}$$

This gives us the 2 by 2 system

$$\begin{aligned} y + z &= 17 \\ -y - 4z &= -41 \end{aligned}$$

We can now eliminate y just by adding the two equations. This gives

$$\begin{aligned} y + z &= 17 \\ -y - 4z &= -41 \\ \hline -3z &= -24 \\ z &= 8 \end{aligned}$$

If we replace z with 8 in the first equation of our 2 by 2 system, we find

$$\begin{aligned} y + 8 &= 17 \\ y &= 9 \end{aligned}$$

If we replace z with 8 and y with 9 in the first equation of our given 3 by 3 system, we have

$$\begin{aligned} x + y + 2z &= 30 \\ x + 9 + 2(8) &= 30 \\ x + 9 + 16 &= 30 \\ x &= 5 \end{aligned}$$

So we get the ordered triple (5, 9, 8).

Interpret the Solution: This means that if we use all the available time for each machine each week we can make 5 Secretarial chairs, 9 Executive chairs, and 8 Presidential chairs.

SUMMARY

In this section we learned two new ways to solve systems of equations. The first was the substitution method and the second was the elimination method. We can now examine the advantages and disadvantages of each method.

Method	*Advantages*	*Disadvantages*
Graphical	• We can visualize the solution	• Difficult to determine the solution if the coordinates are not integers
Substitution	• Gives exact solutions • Can be used for systems of linear models	• Solutions cannot be visualized
Elimination	• Can be used to solve 3 by 3 systems • Gives exact solutions	• Difficult to use with systems of linear models with decimal coefficients. • Solutions cannot be visualized

SECTION 1.5 EXERCISES

In Exercises 1–20, solve the system by the substitution method. If the system has no solution, then say so. If the system has an infinite number of solutions, use a parameter to represent the infinite number of solutions as shown in Example 4.

1. $y = x$
$x + y = 4$

2. $x = y$
$x - y = 6$

3. $x + y = 4$
$3x - y = 0$

4. $x + y = 6$
$2x - y = 0$

✓ **5.** $2x - y = 9$
$x + 3y = 8$

6. $2x - y = -7$
$2x - 3y = -13$

7. $2x + 3y = 11$
$x = 2 + 2y$

8. $2x + y = 20$
$x = -2y - 12$

9. $4x - 3y = 2$
$y = -2x - 4$

10. $5x + 6y = 14$
$y = 4x - 17$

11. $x + 2y = 13$
$x = -2y + 9$

12. $2x - y = -1$
$y = 2x + 3$

13. $y = 2x + 1$
$y = x + 4$

14. $y = 4x - 7$
$y = 2x + 1$

15. $-3x + y = -5$
$21x - 7y = 35$

16. $-3x + y = -1$
$12x - 4y = 4$

17. $x - 2y = -2$
$2x - y = 3$

18. $-2x + y = 3$
$x + y = 4$

19. $2x - 3y = 1.3$
$-x + y = -0.5$

20. $1.5x + y = 4.05$
$2x - 3y = -3.7$

In Exercises 21–40, solve the system by the elimination method. If the system has no solution, then say so. If the system has an infinite number of solutions, use a parameter to represent the infinite number of solutions as shown in Example 4.

21. $x - y = -2$
$x + y = 6$

22. $x - y = 3$
$x + y = 1$

23. $x - y = 3$
$x + y = 7$

24. $x - y = 7$
$x + y = 1$

25. $2x + y = -10$
$2x - y = -6$

26. $x + 2y = -9$
$x - 2y = -1$

27. $2x + y = 4$
$3x + 2y = 3$

28. $x - y = -3$
$2x - 5y = -21$

29. $7x - y = 4$
$3x - 2y = -3$

30. $2x + y = 1$
$-3x + y = -4$

31. $5x - 10y = -2$
$-5x + 10y = -2$

32. $3x - y = 1$
$-3x + y = -2$

33. $2x + 2y = 3$
$-5x + 4y = 15$

34. $x + 4y = 2$
$x - 8y = -7$

35. $3x - 2y = 6$
$x + y = 3$

36. $4x - y = 3$
$x - 3y = -4$

37. $4x - 7y = -13$
$-7x + 3y = -5$

38. $9x + 7y = 21$
$x + 12y = 36$

39. $3x - 4y - 1 = 0$
$4x + 3y - 1 = 0$

40. $4x - 2y - 5 = 0$
$2x + 3y - 4 = 0$

In Exercises 41–50, solve the system of three equations and three unknowns by the elimination method. If the system has no solution, then say so. If the system has an infinite number of solutions, use a parameter to represent the infinite number of solutions as shown in Example 4.

41. $-x + 3y + z = 10$
$3x + 2y - 2z = 3$
$2x - y - 4z = -7$

42. $x + y - 3z = 8$
$3x + 4y - 2z = 20$
$2x - 3y + z = -6$

43. $x + y + z = 1$
$x + 3y + 7z = 13$
$x + 2y + 3z = 4$

44. $x + y + z = 2$
$2x + y + 2z = 3$
$3x - y + z = 4$

45. $x + 2y + 3z = 9$
$-3x + 5y - 4z = -7$
$3x - y + 2z = -1$

46. $2x + 2y + 3z = 10$
$3x + y - z = 0$
$x + y + 2z = 6$

47. $2x + y - z = 1$
$x + 2y + 2z = 2$
$4x + 5y + 3z = 3$

48. $2x + 3y + 4z = 6$
$2x - 5y - 4z = -4$
$4x + 6y + 8z = 12$

49. $x - 2y = 14$
$y + 2z = 11$
$2x + z = 16$

50. $y + z = -3$
$x - 2z = -5$
$3x + 2y = -5$

Applications

51. Victor purchased 2 coats and 1 pair of slacks for \$120 at the Men's Suit Palace. The next week, he returned and purchased 1 coat and 2 pairs of slacks also for \$120.

(a) Define the variables and write a system of equations for the price of the coat and slacks.

(b) Solve the system in part (a) and interpret.

52. An office assistant purchased 5 cups of coffee and 6 bagels for \$24 from the DaMatta Deli. The next day the assistant purchased 3 cups of coffee and 5 bagels for \$18.

(a) Define the variables and write a system of equations for the price of the coffee and bagels.

(b) Solve the system in part (a) and interpret.

53. From 1980 to 1995 the number of households in the District of Columbia can be modeled by

$$f(x) = -1.14x + 255.31, \qquad 0 \le x \le 15$$

where x represents the number of years since 1980 and $f(x)$ represents the number of households in thousands. During the same time period, the number of households in South Dakota can be modeled by

$$g(x) = 1.75x + 242.58, \qquad 0 \le x \le 15$$

where x represents the number of years since 1980 and $g(x)$ represents the number of households in thousands. Determine the point of intersection and interpret each coordinate (*Source:* U.S. Census Bureau, www.census.gov).

54. From 1980 to 1997 the number of Americans aged 25 to 29 who owned a home can be modeled by

$$f(x) = -0.06x + 9.81, \qquad 0 \le x \le 17$$

where x represents the number of years since 1980 and $f(x)$ represents the number of Americans, in millions, aged 25 to 29 who owned a home. During the same time period, the number of Americans aged 30 to 34 who owned a home can be modeled by

$$g(x) = 0.09x + 9.60, \qquad 0 \le x \le 17$$

where x represents the number of years since 1980 and $g(x)$ represents the number of Americans, in millions, aged 30 to 34 who owned a home. Determine the point of intersection and interpret each coordinate (*Source:* U.S. Census Bureau, www.census.gov).

55. From 1990 to 2000 the population of the District of Columbia can be modeled by

$$f(x) = -10.30x + 604.64, \qquad 0 \le x \le 10$$

where x represents the number of years since 1990 and $f(x)$ represents the District of Columbia population in thousands. During the same time period, the population of Alaska can be modeled by

$$g(x) = 33.58x + 394.78, \qquad 0 \le x \le 10$$

where x represents the number of years since 1990 and $g(x)$ represents the Alaska population in thousands. Determine the point of intersection and interpret each coordinate (*Source:* U.S. Census Bureau, www.census.gov).

56. From 1990 to 2000 the population of Utah can be modeled by

$$f(x) = 46.49x + 1732.49, \qquad 0 \le x \le 10$$

where x represents the number of years since 1990 and $f(x)$ represents the Utah population in thousands. During the same time period, the population of West Virginia can be modeled by

$$g(x) = 1.76x + 1802.56, \qquad 0 \le x \le 10$$

where x represents the number of years since 1990 and $g(x)$ represents the West Virginia population in thousands. Determine the point of intersection and interpret each coordinate (*Source:* U.S. Census Bureau, www.census.gov).

57. The Clip-it Company manufacturers three types of paper clips: standard, large, and jumbo. The daily cost of manufacturing a box of each type of paper clip, along with the sales price for each box, is shown in Table 1.5.2.

Table 1.5.2

	Type		
	Standard	**Large**	**Jumbo**
Cost per box	\$4	\$6	\$7
Sales price per box	\$6	\$9	\$12

The company knows that the daily cost of manufacturing 100 total boxes is \$520 and the daily revenue from the sales is \$810. Let x, y, and z represent the daily number of standard, large, and jumbo boxes manufactured, respectively.

(a) Write an equation for the total number of boxes manufactured daily.

(b) Write equations for the daily total cost and for the sales revenue.

(c) Solve the system and interpret the result.

58. In a blind study the number of grams of nutrients per serving of three foods is as shown in Table1.5.3.

Table 1.5.3

	Food ID			
	I	**II**	**III**	**Total**
Fat	1	2	2	11
Carbohydrate	1	1	1	6
Protein	2	1	2	10

Let x, y, and z represent the number of servings of foods I, II, and III, respectively. The study is intended to see how many servings of each food must be used to provide 11 grams of fat, 6 grams of carbohydrate, and 10 grams of protein.

(a) Write a system of equations to determine the number of servings of each food that is needed to provide the necessary number of grams of fat, carbohydrate, and protein.

(b) Solve the system and interpret the result.

59. In a blind study the number of grams of nutrients per serving of three foods is as shown in Table 1.5.4.

Table 1.5.4

	Food ID			
	A	**B**	**C**	**Total**
Fat	2	3	1	14
Carbohydrate	1	2	1	9
Protein	2	1	2	9

Let x, y, and z represent the number of servings of foods A, B, and C, respectively. The study is intended to see how many servings of each food must be used to provide 14 grams of fat, 9 grams of carbohydrate, and 9 grams of protein.

(a) Write a system of equations to determine the number of servings of each food that is needed to provide the necessary number of grams of fat, carbohydrate, and protein.

(b) Solve the system and interpret the result.

60. The Crash-em-Up body repair shop fixes three types of cars: compact, midsize, and luxury. If a damaged car fender arrives, the times required, in hours, to fill, sand, and paint the fender for the three types of cars is as shown in Table 1.5.5.

Table 1.5.5

	Type			
	Compact	**Midsize**	**Luxury**	**Time Available**
Filling	2	2	1	14
Sanding	1	2	2	15
Painting	3	2	2	21

Let x, y, and z, represent the number of compact, midsize, and luxury car fenders repaired, respectively. The total time available to complete each job is in the right column.

(a) Write a system of equations to determine the number of fenders of each type that can be repaired using all available labor hours.

(b) Solve the system and interpret the result.

SECTION PROJECT

Systems of equations can also be used to perform curve fitting, meaning that if we are given some coordinates of a function we can determine the function itself. Let's say that a line of the form $y = ax + b$ contains the coordinates $(-1, 4)$ and $(2, 6)$. We can write the following system of equations to determine the constants a and b.

$$\begin{aligned} 4 &= a(-1) + b \\ 6 &= a(2) + b \end{aligned} \quad \text{or in simplified form} \quad \begin{aligned} 4 &= -a + b \\ 6 &= 2a + b \end{aligned}$$

(a) Solve the given system to determine the constants a and b and write the equation in slope–intercept form. Graph and verify that the two given points are on the line.

(b) Determine a and b so that the line $y = ax + b$ contains the points $(1, 1)$ and $(-2, 4)$.

We can also use a system to perform curve fitting on parabolas resulting from quadratic equations. Let's say that we have a quadratic equation that is symmetric about the vertical line $x = 2$. Then the quadratic equation has the form $y = a(x-2)^2 + k$. Suppose that we know that the points $(1, -1)$ and $(4, 8)$ are on the parabola. Then, to determine the values of a and k, we can solve the system

$$\begin{aligned} -1 &= a(1-2)^2 + k \\ 8 &= a(4-2)^2 + k \end{aligned} \quad \text{or in simplified form} \quad \begin{aligned} -1 &= a + k \\ 8 &= 4a + k \end{aligned}$$

(c) Solve the given system to determine the constants a and k and write the quadratic equation in the form $y = a(x-2)^2 + k$. Graph and verify that the two given points are on the parabola.

(d) Determine a and k so that the parabola $y = a(x-2)^2 + k$ contains the points $(1, -3)$ and $(4, 0)$. Note that this parabola is also symmetric about the line $x = 2$.

Why We Learned It

In Chapter 1 we reviewed the properties of linear functions and linear systems of equations. The study of linear functions is very important to the study of finite mathematics. We saw that linear models could be used to express applications in the business, social, and life sciences. For example, a researcher could examine trends in immigration or birth weight over time. The amount that these values increase or decrease over time is given by the average rate of change. We saw that systems of equations are often used in business, particularly in applications such as the Law of Supply and Demand and for determining break-even points of cost and revenue functions. Finally, we saw that we could use graphical or algebraic techniques to solve systems of linear equations.

CHAPTER REVIEW EXERCISES

In Exercises 1–5, translate the English sentence into a mathematical sentence.

1. A certain number increased by 7 is 12.

2. A certain number decreased by 4 is five.

3. The product of three and a certain number is 36.

4. The product of two and a certain number minus three is seven.

5. The product of nine and a certain number plus five is 41.

6. Molly makes wool sweaters and sells them at craft shows. She knows that her fixed costs are \$65 and the cost to produce each sweater is \$14.

(a) Determine the total cost for producing 15 sweaters.

(b) Determine an equation for the total cost of producing any number of sweaters.

7. The Drink Power company has two ingredients, zap and wham, to combine to make their new power drink. Zap contains 7 grams of carbohydrate and 4 grams of protein per ounce. Wham contains 5 grams of carbohydrate and 4 grams of protein per ounce. They want to combine the two ingredients so that the power drink has 70 grams of carbohydrates and 48 grams of protein. Let x represent the number of ounces of zap and y represent the number of ounces of wham.

(a) Write an equation that represents the number of grams of carbohydrate in the power drink.

(b) Write an equation that represents the number of grams of protein in the power drink.

8. A local ski shop sells discounted lift tickets for a nearby ski resort. Tickets cost \$30 per day for children and \$42 per day for adults. Table R.1 gives the data for the number of tickets sold and revenue generated over one weekend.

Table R.1

	Friday	Saturday	Sunday
Tickets sold	20	30	15
Revenue	\$696	\$720	\$570

(a) Define the variables and write an equation representing the number of tickets sold on Friday, and write an equation representing the revenue generated on Friday.

(b) Define the variables and write an equation representing the number of tickets sold on Saturday, and write an equation representing the revenue generated on Saturday.

(c) Define the variables and write an equation representing the number of tickets sold on Sunday, and write an equation representing the revenue generated on Sunday.

9. Harold invested a total of \$20,000 in three different investments. Some of the money was invested in a certificate of deposit earning 6%, \$8000 more than $\frac{1}{4}$ the amount invested in the certificate of deposit earning 6% was invested in a mutual fund earning 7%, and the rest was placed in a savings account earning 5%. The total annual return on Harold's investment was \$1280. Let x represent the amount invested in the certificate of deposit earning 6%, y represent the amount invested in the mutual fund earning 7%, and z represent the amount invested in the savings account earning 5%.

(a) Write an equation that represents the first sentence.

(b) Write an equation that represents "\$8000 more than $\frac{1}{4}$ the amount invested in the certificate of deposit earning 6% was invested in a mutual fund earning 7%."

(c) Write an equation that represents the third sentence.

10. The Tasty Snack company makes three brands of snack mixes made up of three different kinds of dried fruit. The proportions of the ingredients of the three brands are shown in Table R.2.

Table R.2

	Cranberry Lovers (%)	Banana Lovers (%)	Apricot Lovers (%)
Cranberries	70	20	10
Bananas	10	70	20
Apricots	20	10	70

Tasty Snack wants to create a new snack mix that contains 6.6 ounces of cranberries, 3.4 ounces of bananas, and 6 ounces of apricots.

(a) Define the variables and write an equation that represents the total amount of each snack mix that can be combined to produce 6.6 ounces of cranberries in the new mix.

(b) Define the variables and write an equation that represents the total amount of each snack mix that can be combined to produce 3.4 ounces of bananas in the new mix.

(c) Define the variables and write an equation that represents the total amount of each snack mix that can be combined to produce 6 ounces of apricots in the new mix.

For Exercises 11 and 12, use the given table to construct ordered pairs and plot them on the Cartesian plane.

11.

x	y
−2	4
−1	0
2	3

12.

x	y
−1	−2
1	3
−3	4

For Exercises 13 and 14, write the sets for the domain and range.

13. For the values in the table in Exercise 11.

14. For the values in the table in Exercise 12.

For the functions in Exercises 15 and 16, complete the following.

(a) Identify the independent and dependent variables.

(b) Complete the table.

15. $g(x) = 2x + \frac{3}{4}$

x	$g(x)$
−2	
4	
10	

16. $f(x) = \frac{2x - 1}{5}$

x	$f(x)$
5	
0	
	7

In Exercises 17 and 18, determine which of the tables represent functions.

17.

Domain	Range
2	5
−1	6
3	4
2	0

18.

Domain	Range
−3	4
3	2
0	1
5	1

In Exercises 19 and 20, determine if the graph represents a function.

19.

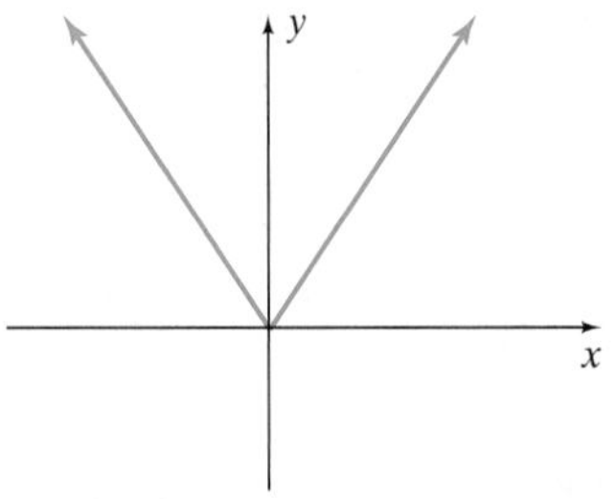

20.

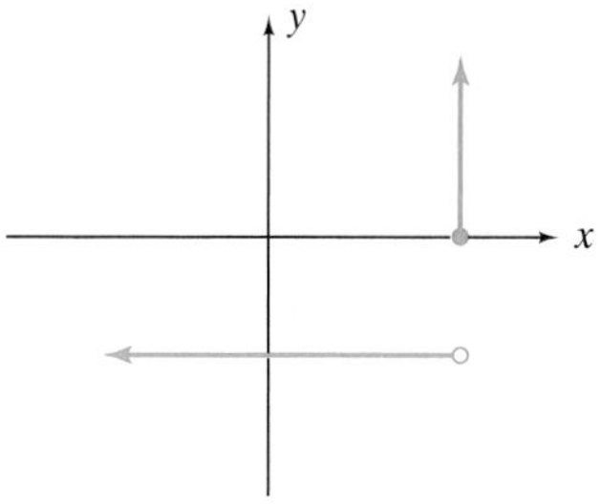

In Exercises 21 and 22, a table of values is given.

(a) Complete the table.

(b) Write a linear function for the data in the table.

21.

x	Expression	y
0		
4	$2(4) - 1$	
		0
2		

22.

x	Expression	$f(x)$
	$3 + 4(-2)$	-5
2		
		15
	$3 + 4(0)$	

In Exercises 23 and 24, compute the slope of the line passing through the two ordered pairs. Then classify the line as increasing, decreasing, horizontal, or vertical.

23.

x	y
-3	4
5	6

24. $(-3, 4)$ and $(-3, 7)$

In Exercises 25 and 26, determine the x- and y-intercepts of the graph of the linear function.

25. $y = 2x - 14$

26. $g(x) = \dfrac{5x + 4}{2}$

In Exercises 27 and 28, write the linear cost function for the given information in the form $C(x) = mx + b$.

27. The cost per unit is \$17.33 and the fixed costs are \$210.

28. The fixed costs are \$35,200 and the cost per unit is \$410.

29. The Spencer Baseball Company determines that the cost of producing x baseballs is given by the linear cost function $C(x) = 1.50x + 700$.

(a) What are the fixed and per unit costs for producing the baseballs?

(b) Evaluate and interpret $C(50)$.

(c) How many baseballs can be produced for \$812.50? That is, what is x when $C(x) = 812.50$?

30. The data in Table R.3 show the enrollment of a small college in 1995 and 2000.

Table R.3

Year	Enrollment
1995	3210
2000	5600

Compute the average rate of change in enrollment at the small college from 1995 to 2000 and interpret. Be sure to include proper units.

In Exercises 31–34, write an equation of the line in standard form (if possible).

31. The line has slope $m = \dfrac{3}{4}$ and passes through the point $(-1, 4)$.

32. The line is given by the values in Table R.4.

Table R.4

x	0	4
$f(x)$	-3	-19

33. The line is vertical and passes through the point $(-5, 12)$.

34. The line has an average rate of change of $\dfrac{1}{5}$ and an x-intercept at 3.

In Exercises 35–38, write a function for the line in slope–intercept form.

35. The line has a slope of $m = 2$ and passes through the point $(1, -4)$.

36. The linear function $f(x)$ has a slope of $m = -\dfrac{3}{7}$ and $f(7) = 1$.

37. The linear function $g(x)$ has function values of $g(0) = -5$ and $g(2) = -11$.

38. The line has an x-intercept at $(-3, 0)$ and a y-intercept at $(0, 4)$.

In Exercises 39 and 40, write an equation in slope–intercept form for the given graph.

39.

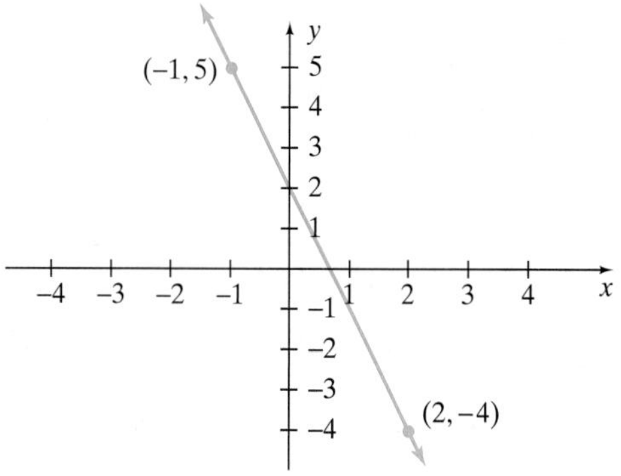

40.

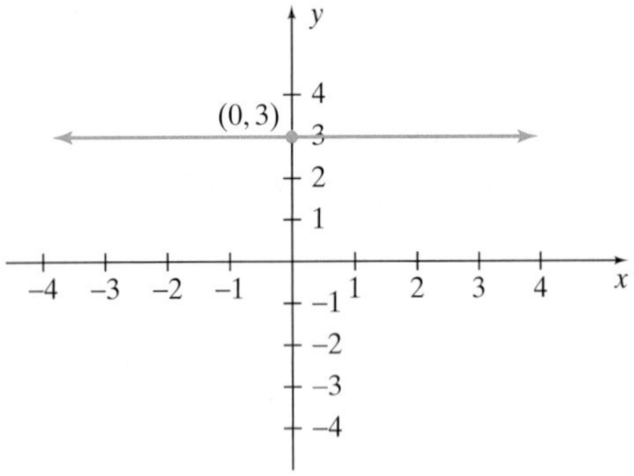

In Exercises 41 and 42, answer using a depreciation model.

41. A new \$15,525 car depreciates \$800 each year.

(a) Define the variables and determine the depreciation function.

(b) How much will the car be worth in 4 years?

42. A new computer costs \$3200 and depreciates at a rate of \$400 per year.

(a) Define the variables and determine the depreciation function.

(b) When will the computer depreciate to the point that it has no worth?

43. The total pounds of garbage produced per year in a growing housing community since 1995 can be modeled by

$$f(x) = 50x + 5200, \qquad 0 \le x \le 5$$

where x represents the number of years since 1995 and $f(x)$ represents the total amount of garbage in pounds.

(a) Is the total pounds of garbage produced per year increasing or decreasing and by how much?

(b) How many pounds of garbage were produced in 1995?

(c) Evaluate and interpret $f(5)$.

44. Suppose that the population of the housing community, from Exercise 43, stabilizes in the year 2000 and the community begins to recycle. Since the recycling concept takes some time to catch on, over the next 5 years the community will reduce the amount of garbage that they produce by 375 pounds per year.

(a) Assuming that the community produced 5450 pounds of garbage in the year 2000, write a linear model of the form

$$g(x) = y = mx + b, \qquad 0 \le x \le 5$$

to represent the predicted amount of garbage that the community will produce per year x years after 2000.

(b) Evaluate and interpret $g(5)$.

45. From 1990 to 2000 the average price of a resident fishing license was increasing at an average rate of \$2.07 per year. In 1992 the average cost for a license was \$20.15.

(a) Let x represent the number of years since 1990. Write a linear model in the calculator-friendly form

$$y = m(x - x_1) + y_1, \qquad 0 \le x \le 10$$

to represent the average cost of a license from 1990 to 2000.

(b) At the beginning of what year did the average cost per license exceed \$24.50?

In Exercises 46 and 47, determine if the given ordered pair is a solution to the system.

46. $(2, -3)$
$$2x - 3y = 13$$
$$x + 5y = -2$$

47. $(1, 4)$
$$3x - 5y = -17$$
$$x + 4y = 17$$

In Exercises 48 and 49, solve the system by determining the x- and y-intercepts of the equation and then graphing the equations to determine their point of intersection. If the system has no solution *or* infinite solutions, *then say so and state a reason why.*

48.
$$y = 3x - 2$$
$$y = 5x - 11$$

49.
$$3x + 4y = 1$$
$$9x + 12y = 3$$

In Exercises 50 and 51, graph the equations on the same Cartesian plane and then use the graphs to estimate the solutions.

50.
$$x - y = 3$$
$$x = 4$$

51.
$$y = -3$$
$$x = 2$$

In Exercises 52 and 53, solve the system by graphing each equation on the calculator and using the INTERSECT command. If the system has no solution *or* infinite solutions, *then say so.*

52.
$$4x - 3y = 2$$
$$2y = 4x$$

53.
$$x + 2y = -3$$
$$3x + 6y = 4$$

In Exercises 54 and 55, determine the value of k that satisfies the given condition for each system.

54. k so that $\begin{aligned} y &= kx + 5 \\ y &= 2x - 1 \end{aligned}$ has no solution.

55. k so that $\begin{aligned} 3x - 4y &= 6 \\ 9x - ky &= 18 \end{aligned}$ has infinitely many solutions.

56. The Quinn Outdoor Company determines that the demand for their new dog backpacks is given by $d(x) = 80 - 0.75x$, with a supply function given by $s(x) = 0.34x$, where x is the daily quantity and $d(x)$ and $s(x)$ are measured in dollars per backpack.

(a) Determine the x- and y-intercepts for the graph of the demand function d.

(b) Determine the supply s when $x = 0$ and $x = 50$ and then graph d and s on the same Cartesian plane.

(c) Approximate the point of intersection and interpret.

57. The All Gear Company determines that the demand for their new adventure running shoe is given by $d(x) = 100 - \frac{1}{5}x$, while the supply is given by $s(x) = \frac{3}{5}x$, where x is the quantity and $d(x)$ and $s(x)$ are measured in dollars per shoe.

(a) Write a system of equations to determine the equilibrium point and graph the system in the viewing window [0, 500] by [0, 100].

(b) Determine the point of intersection via the INTERSECT command and interpret each coordinate.

58. The All Gear Company determines that the demand function for their new kayak passes through the points (0, 800) and (7, 797), and the supply function passes through the points (0, 0) and (14, 12).

(a) Use the given information to find demand and supply functions of the form $d(x) = mx + b$ and $s(x) = mx + b$, where x is the quantity and $d(x)$ and $s(x)$ are measured in dollars per kayak.

(b) Write a system of equations to determine the equilibrium point and graph the system in the viewing window [0, 2000] by [0, 1000].

(c) Determine the point of intersection via the INTERSECT command and interpret each coordinate. Round each coordinate to the nearest hundredth.

59. Molly makes wool sweaters and sells them at craft shows. She knows that her fixed costs are \$65 and the cost to produce each sweater is \$14. She sells each sweater for \$30.

(a) Define the variables and determine a system of linear equations for the cost and the revenue that Molly makes from the sweaters.

(b) Graph the system determined in part (a) in the same viewing window, use the INTERSECT command to determine the break-even point, and interpret each coordinate. Round each coordinate to the nearest whole number.

60. From July 1984 to July 1999 the number of U.S. children under 5 years of age can be modeled by

$$f(x) = -151.6x + 19{,}700, \qquad 0 \le x \le 5$$

where x represents the number of years since July 1994 and $f(x)$ represents the number of U.S. children under 5 in thousands. During the same time period, the number of U.S. children who were 10 to 14 years old can be modeled by

$$g(x) = 166.4x + 18{,}716, \qquad 0 \le x \le 5$$

where x represents the number of years since July 1994 and $g(x)$ represents the number of U.S. children 10 to 14 years old in thousands (*Source:* U.S. Census Bureau, www.census.gov).

(a) Evaluate each model at $x = 0$ and $x = 5$.

(b) Graph f and g in the viewing window [0, 5] by [18,000, 20,000] and determine the point of intersection via the INTERSECT command. Round each coordinate to the nearest thousandth.

(c) Interpret each of the coordinates from your result in part (b).

In Exercises 61–64, solve the system by the substitution method. If the system has no solution, then say so. If the system has an infinite number of solutions, use a parameter to represent the infinite number of solutions.

61. $3x + y = -5$, $x + 4y = 2$

62. $x + 3y = 10$, $x = -3y + 4$

63. $-2x + 4y = -12$, $3x - 2y = 6$

64. $y = 4x + 1$, $y = x + 13$

In Exercises 65–68, solve the system by the elimination method. If the system has no solution, then say so. If the system has an infinite number of solutions, use a parameter to represent the infinite number of solutions.

65. $x + y = 8$, $x - y = 12$

66. $x + 5y = -9$, $4x + y = -17$

67. $x - 3y = 7$, $-4x + 12y = -28$

68. $2x - 5y + 14 = 0$, $4x + 3y - 24 = 0$

In Exercises 69 and 70, solve the system of three equations and three unknowns by the elimination method. If the system has no solution, then say so. If the system has an infinite number of solutions, use a parameter to represent the infinite number of solutions.

69. $-x + 2y + z = -1$, $3x + 2y - 2z = 9$, $x + 4y + 3z = -1$

70. $x + 3y = 1$, $y = 3z - 8$, $2x + z = -1$

71. Alicia purchased 3 T-shirts and 2 pairs of jeans for \$90 at Casual Wear. The next week she purchased 1 T-shirt and 4 pairs of jeans for \$130.

(a) Define the variables and write a system of equations for the price of the T-shirts and jeans.

(b) Solve the system in part (a) and interpret.

72. An office assistant purchased 5 boxes of pens and 2 boxes of staples for \$9. The next day the assistant purchased 2 boxes of pens and 8 boxes of staples for \$9.

(a) Define the variables and write a system of equations for the boxes of pens and boxes of staples.

(b) Solve the system in part (a) and interpret.

73. From July 1990 to July 1999 the population of Colorado can be modeled by

$$f(x) = 0.084x + 3.30, \qquad 0 \le x \le 9$$

where x represents the number of years since July 1990 and $f(x)$ represents the population of Colorado in millions. During the same time period, the population of Kentucky can be modeled by

$$g(x) = 0.030x + 3.69, \qquad 0 \le x \le 9$$

where x represents the number of years since July 1990 and $g(x)$ represents the population of Kentucky in millions. Determine the point of intersection and interpret each coordinate (*Source:* U.S. Census Bureau, www.census.gov).

74. Papa Harold's sandwich shop makes three sizes of sandwiches: large, extra-large, and Papa size. The cost of making each sandwich and the sales price for each sandwich are shown in Table R.5.

Table R.5

	Type		
	Large	**Extra-large**	**Papa size**
Cost per sandwich	\$1	\$2	\$3
Sales price per sandwich	\$4	\$5	\$7

The shop knows that the daily cost of making 200 sandwiches is \$360 and the daily revenue from the sales is \$990. Let x, y, and z represent the daily number of large, extra-large, and Papa size sandwiches made, respectively.

(a) Write an equation for the total number of sandwiches made daily.

(b) Write equations for the daily cost and for the sales revenue.

(c) Solve the system and interpret the results.

75. In a blind study the number of grams of nutrients per serving of three foods is as shown in Table R.6.

Table R.6

	Food ID			
	A	**B**	**C**	**Total**
Fat	4	2	1	14
Carbohydrate	3	1	2	15
Protein	2	2	3	18

Let x, y, and z represent the number of servings of foods A, B, and C, respectively. The study is intended to see how many servings of each food must be used to provide grams of fat, grams of carbohydrates, and grams of protein.

(a) Write a system of equations to determine the number of servings of each food that is needed to provide the necessary number of grams of fat, carbohydrates, and protein.

(b) Solve the system and interpret the results.

CHAPTER 1 PROJECT

SOURCE: Centers for Disease Control and Prevention, www.cdc.gov

As you have seen in this chapter, there are many examples of two quantities having a linear relationship. However, some relationships can only be *approximated* by a linear equation. To compute this correlation, we can use regression.

Table 1 offers data collected by the Centers for Disease Control and Prevention concerning deaths by race and sex from 1985 to 1997 in the United States. The selected races are Native Americans[1] and Asian or Pacific Islanders[2] and the sex is female for both groups.

1. Let x represent the number of years since 1984. Let y_1 be the number of deaths of American Indian females. Enter the data into your calculator and find a linear regression model of the form $y_1 = ax + b$ to model the data. Round coefficients to two decimal places. Interpret the slope of the line.

2. Again, let x be as defined in question 1 and let y_2 be the number of deaths of Asian or Pacific Islander females. Enter the data into your calculator and find a linear regression model of the form $y_2 = ax + b$ to model the data. Round coefficients to two decimal places. Interpret the slope of the line.

3. Plot the data along with the regression lines found in questions 1 and 2 in the same viewing window. According to the regression lines, during what year did the number of deaths for American Indian females equal the number of deaths for Asian or Pacific Islander females?

4. Confirm your answer from question 3 by using the substitution method.

5. According to the models, predict for each race the number of deaths in the years 2000, 2005, and 2010.

6. For the year 1986, compute the residual. That is, compute the difference between the value in the table and the value given by the linear regression line from questions 1 and 2.

Table 1 Crude Rates on an Annual Basis per 100,000 Population

Year	American Indian Females	Asian or Pacific Islander Females
1985	2,973	6,646
1986	2,936	6,719
1987	3,170	7,193
1988	3,300	7,808
1989	3,548	8,354
1990	3,439	8,916
1991	3,673	9,446
1992	3,772	10,092
1993	4,145	10,854
1994	4,140	11,695
1995	4,423	12,364
1996	4,564	12,958
1997	4,591	13,696

[1]Includes Aleuts and Eskimos (Inuits).

[2]Includes Chinese, Filipino, Hawaiian, Japanese, and other Asian or Pacific Islanders.

Chapter 2

Systems of Linear Equations and Matrices

Although the concept of a matrix may seem complex, the fact is that matrices are part of our everyday experience. Matrices are used in such diverse areas as agriculture and urban design. Farmers use systems of equations and matrices to keep track of crops and to determine planting schedules. Traffic engineers help control gridlock in busy cities by using matrices and systems of equations to sequence traffic signals.

What We Know

In Chapter 1 we reviewed linear functions and their graphs. Solving systems of linear equations was discussed using both graphical and analytical methods. The analytical methods included the substitution method and the elimination method.

Where Do We Go

In this chapter we will learn how to use *matrices* to solve systems of linear equations. We begin with *augmented matrices* and the *Gauss–Jordan* method and end with *inverse matrices* and matrix equations. Along the way, we will learn operations with matrices and encounter many applications.

Section 2.1 Matrices and Gauss–Jordan

In this section we will learn what a **matrix**, an **augmented matrix**, **reduced row echelon form**, and the **Gauss–Jordan method** are all about. The Gauss–Jordan method is simply a generalization of the elimination method studied in Section 1.5. Its main advantage is that it works well for systems of any size. Before we learn the power of the Gauss–Jordan method, we must first analyze the matrix.

Basic Matrix Vocabulary

A **matrix** (plural is **matrices**) is a rectangular arrangement of numbers enclosed by brackets. Table 2.1.1 gives the predicted high temperature for selected cities in Texas for four days in June.

Table 2.1.1

	Saturday	Sunday	Monday	Tuesday
Amarillo	86	85	88	91
Austin	87	88	87	91
Brownsville	88	89	89	91
Dallas	84	88	88	90
Houston	84	88	86	89

SOURCE: *The Plain Dealer*

These data can be written as the matrix T as follows:

$$T = \begin{bmatrix} 86 & 85 & 88 & 91 \\ 87 & 88 & 87 & 91 \\ 88 & 89 & 89 & 91 \\ 84 & 88 & 88 & 90 \\ 84 & 88 & 86 & 89 \end{bmatrix}$$

The horizontal listings of any matrix are called **rows**, and the vertical listings are called **columns**. In any matrix each number is called an **element** or an **entry**.

The number of rows and the number of columns give the **dimension** of a matrix. If a matrix has m rows and n columns, we say that its dimension is $m \times n$ (read "m by n"). Hence, since matrix T has 5 rows and 4 columns, we say that it has dimension 5×4. We can also say that matrix T is a 5×4 matrix.

It is a convention to use uppercase letters to name a matrix and lowercase letters with a double subscript to label entries within a matrix. For example, $t_{4,3}$ refers to the entry in matrix T that is in the 4th row and 3rd column. Notice that $t_{4,3}$ refers to the number 88.

Example 1 Determining the Dimension and Entries of a Matrix

For the given matrix A, determine the dimension of A and determine $a_{2,3}$.

$$A = \begin{bmatrix} -3 & 2 & 0 \\ 4 & -1 & 3 \end{bmatrix}$$

Solution

Since A has 2 rows and 3 columns, it is a 2×3 matrix. Also, $a_{2,3}$ refers to the entry in the 2nd row and the 3rd column. This means that $a_{2,3} = 3$. ■

✓ Checkpoint 1

Now work Exercises 5 and 11.

Technology Option

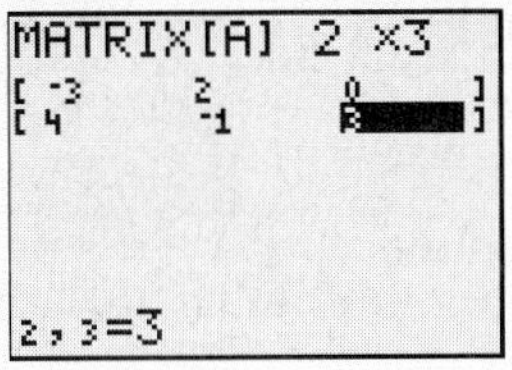

Figure 2.1.1

Matrices can be entered on a graphing calculator. Figure 2.1.1 shows a screen shot of what matrix A from Example 1 looks like on a graphing calculator. Note that the entry $a_{2,3}$ is highlighted and the lower-left corner of the screen shows the value of this entry.

To learn how to enter matrices on your calculator, consult the online graphing calculator manual at www.prenhall.com/armstrong

Augmented Matrices and Solving Systems of Equations

At the beginning of this section we stated that the **Gauss–Jordan** method is a generalization of the elimination method studied in Section 1.5. To aid our transition to this new method for solving systems of linear equations, let's take a Flashback to a problem encountered in Sections 1.1 and 1.5.

Flashback

Gonzalez Office Supplies Revisited

In Chapter 1 we learned that Gonzalez Office Supplies manufactures three types of office chairs: Secretarial, Executive, and Presidential. Each chair requires time on three different machines. These times and the total available time for each machine are given in Table 2.1.2. If all available time for each machine must be used each week, set up a system of equations that can be solved to determine how many of each type can be made each week.

Table 2.1.2

	Secretarial	Executive	Presidential	Total hours each week
Hours on Machine A	1	1	2	30
Hours on Machine B	2	3	2	53
Hours on Machine C	1	2	3	47

Flashback Solution

We let x represent the number of Secretarial chairs made each week, y represent the number of Executive chairs made each week, and z represent the number of Presidential chairs made each week. The total number of hours each week for all three chairs on Machine A is given by $x + y + 2z$ and must equal 30. This gives the equation

$$x + y + 2z = 30$$

In a similar fashion, the total number of hours for Machine B is given by

$$2x + 3y + 2z = 53$$

and the total number of hours for Machine C is given by

$$x + 2y + 3z = 47$$

These three equations form the system of equations

$$\begin{aligned} x + y + 2z &= 30 \\ 2x + 3y + 2z &= 53 \\ x + 2y + 3z &= 47 \end{aligned}$$

In Section 1.5 we solved the system in the Flashback solution using the elimination method. Our task at this time is to develop a new method. For any linear system the matrix having as its entries the coefficients of the variables is called the **coefficient matrix** of the system. For example, the system of equations in the Flashback solution has the coefficient matrix

$$\begin{bmatrix} 1 & 1 & 2 \\ 2 & 3 & 2 \\ 1 & 2 & 3 \end{bmatrix}$$

If we add another column to the coefficient matrix whose entries are the constants from the right side of each equation in the system, we form what is called an **augmented matrix** of the system. The augmented matrix from the Flashback solution is

$$\left[\begin{array}{ccc|c} 1 & 1 & 2 & 30 \\ 2 & 3 & 2 & 53 \\ 1 & 2 & 3 & 47 \end{array}\right]$$

We use a vertical line to separate the constants in the last column from the coefficients of the variables.

Example 2 Writing an Augmented Matrix

Write an augmented matrix for the system

$$\begin{aligned} 2x - 3y &= 7 \\ 5x + y &= 8 \end{aligned}$$

Solution

The augmented matrix is

$$\left[\begin{array}{cc|c} 2 & -3 & 7 \\ 5 & 1 & 8 \end{array}\right]$$

■

✓ Checkpoint 2 Now work Exercise 19.

Since an augmented matrix is just a compact form of a system of equations, we can manipulate the rows of an augmented matrix in the same way as we did the equations in the system in Section 1.5. In fact, the manipulations of systems of equations given in Section 1.5 are now given as **row operations** for an augmented matrix.

Row Operations

For an augmented matrix, the following row operations produce an augmented matrix of an equivalent system:

1. Two rows may be interchanged (denoted $R_i \leftrightarrow R_j$).
2. A row is multiplied by a nonzero constant (denoted $cR_i \rightarrow R_i$).
3. Add a constant multiple of one row to another (denoted $cR_i + R_j \rightarrow R_j$).

Note: We use the arrow $\rightarrow$ to mean *replaces*. For example, $3R_1 + R_2 \rightarrow R_2$ means "3 times row 1 plus row 2 replaces row 2."

Example 3 Solving a System Using Augmented Matrices

Solve the system

$$8x + 9y = 69$$
$$2x + 3y = 21$$

using augmented matrices and the row operations.

Solution

We begin by writing the augmented matrix corresponding to the system

$$\left[\begin{array}{cc|c} 8 & 9 & 69 \\ 2 & 3 & 21 \end{array}\right]$$

Our goal is to use the row operations to transform our augmented matrix into an augmented matrix of the form

$$\left[\begin{array}{cc|c} 1 & 0 & c \\ 0 & 1 & d \end{array}\right]$$

where c and d are real numbers. This form gives the solution $x = c$ and $y = d$ because **a row in an augmented matrix corresponds to an equation in a linear system**. In other words,

The first row of $\left[\begin{array}{cc|c} \mathbf{1} & \mathbf{0} & \boldsymbol{c} \\ 0 & 1 & d \end{array}\right]$ corresponds to $x + 0y = c$ or $x = c$.

The second row of $\left[\begin{array}{cc|c} 1 & 0 & c \\ \mathbf{0} & \mathbf{1} & \boldsymbol{d} \end{array}\right]$ corresponds to $0x + y = d$ or $y = d$.

We proceed by getting a 1 in the upper-left corner. We multiply R_1 by $\frac{1}{8}$ (row operation 2) to get

$$\left[\begin{array}{cc|c} 8 & 9 & 69 \\ 2 & 3 & 21 \end{array}\right] \tfrac{1}{8}R_1 \rightarrow R_1 \left[\begin{array}{cc|c} \frac{1}{8}(8) & \frac{1}{8}(9) & \frac{1}{8}(69) \\ 2 & 3 & 21 \end{array}\right] \rightarrow \left[\begin{array}{cc|c} 1 & \frac{9}{8} & \frac{69}{8} \\ 2 & 3 & 21 \end{array}\right]$$

To get a 0 in the lower-left corner, we multiply $-2R_1$ and add to R_2 (row operation 3). This gives

$$\left[\begin{array}{cc|c} 1 & \frac{9}{8} & \frac{69}{8} \\ 2 & 3 & 21 \end{array}\right] -2R_1 + R_2 \rightarrow R_2 \left[\begin{array}{cc|c} 1 & \frac{9}{8} & \frac{69}{8} \\ -2(1)+2 & -2(\frac{9}{8})+3 & -2(\frac{69}{8})+21 \end{array}\right]$$

$$\rightarrow \left[\begin{array}{cc|c} 1 & \frac{9}{8} & \frac{69}{8} \\ 0 & \frac{3}{4} & \frac{15}{4} \end{array}\right]$$

Notice how the last row operation left row 1 unaffected. We now get a 1 in the 2nd row, 2nd column by multiplying R_2 by $\frac{4}{3}$. This gives

$$\left[\begin{array}{cc|c} 1 & \frac{9}{8} & \frac{69}{8} \\ 0 & \frac{3}{4} & \frac{15}{4} \end{array}\right] \frac{4}{3}R_2 \rightarrow R_2 \left[\begin{array}{cc|c} 1 & \frac{9}{8} & \frac{69}{8} \\ \frac{4}{3}(0) & \frac{4}{3}(\frac{3}{4}) & \frac{4}{3}(\frac{15}{4}) \end{array}\right] \rightarrow \left[\begin{array}{cc|c} 1 & \frac{9}{8} & \frac{69}{8} \\ 0 & 1 & 5 \end{array}\right]$$

Finally, to get a 0 in the 1st row, 2nd column, we multiply $-\frac{9}{8}R_2$ and add to R_1. This yields

$$\left[\begin{array}{cc|c} 1 & \frac{9}{8} & \frac{69}{8} \\ 0 & 1 & 5 \end{array}\right] -\frac{9}{8}R_2 + R_1 \rightarrow R_1 \left[\begin{array}{cc|c} -\frac{9}{8}(0)+1 & -\frac{9}{8}(1)+\frac{9}{8} & -\frac{9}{8}(5)+\frac{69}{8} \\ 0 & 1 & 5 \end{array}\right]$$

$$\rightarrow \left[\begin{array}{cc|c} 1 & 0 & 3 \\ 0 & 1 & 5 \end{array}\right]$$

Interactive Activity

Verify that $x = 3$ and $y = 5$ is the solution to the original system given in Example 3.

WWW

This last augmented matrix corresponds to the linear system

$$x = 3$$
$$y = 5$$

So the solution to our original system is $x = 3$ and $y = 5$. ■

Checkpoint 3

Now work Exercise 43.

Technology Option

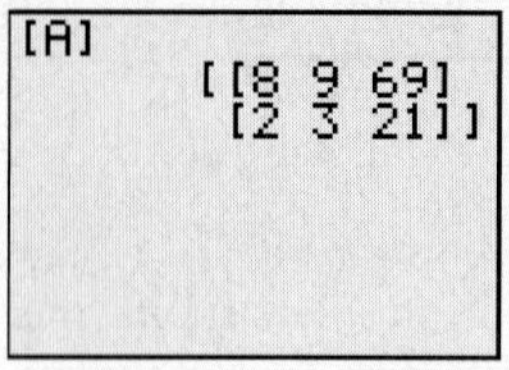

Figure 2.1.2

The row operations in Example 3 can be performed on most graphing calculators.

Figure 2.1.2 shows the augmented matrix for the original system in Example 3.

Figure 2.1.3(a) shows how to get a 1 in the upper-left corner by using the *row(command, and Figure 2.1.3(b) shows the resulting matrix with fractions instead of decimals.

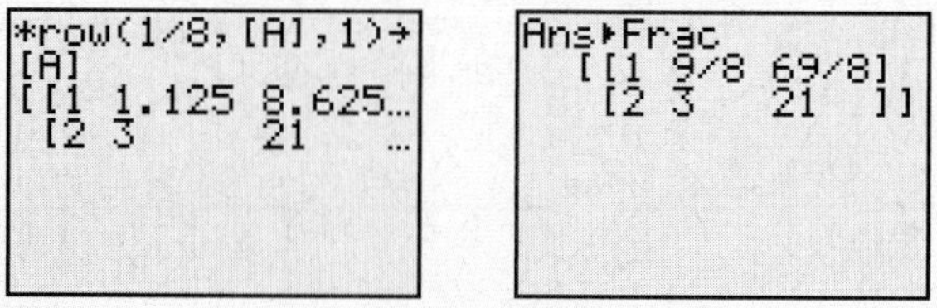

(a) **(b)**

Figure 2.1.3

Figure 2.1.4(a) shows how to get a 0 in the lower-left corner by using the *row+(command, and Figure 2.1.4(b) shows the resulting matrix with fractions instead of decimals.

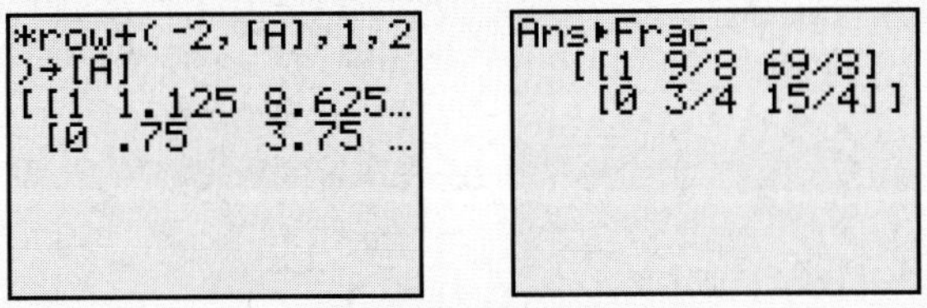

(a) (b)

Figure 2.1.4

To learn how row operations are done on your calculator, consult the online graphing calculator manual at www.prenhall.com/armstrong

Reduced Row Echelon Form and Gauss–Jordan

We found the final matrix in Example 3 to be $\left[\begin{array}{cc|c} 1 & 0 & 3 \\ 0 & 1 & 5 \end{array}\right]$. This is an example of a matrix that is in **reduced row echelon form**. Let's see how to determine when a matrix is in this form.

Reduced Row Echelon Form

For an $m \times n$ matrix to be in **reduced row echelon form**, it must satisfy the following:

1. All rows containing only zeros are at the bottom of the matrix.
2. The first (leftmost) nonzero entry in a nonzero row is a 1. (This is called the **leading 1** of its row).
3. The leading 1 of each nonzero row lies left of the leading 1 of any lower row.
4. For any column that has a leading 1, the other entries in the column are zeros.

From the preceding definition we can see that matrices such as $\left[\begin{array}{cc|c} 1 & 0 & 5 \\ 0 & 1 & 6 \end{array}\right]$ and $\left[\begin{array}{ccc|c} 1 & 0 & 0 & 6 \\ 0 & 1 & 0 & 9 \\ 0 & 0 & 0 & 0 \end{array}\right]$ are in reduced row echelon form. On the other hand, matrices such as $\left[\begin{array}{cc|c} 0 & 1 & -2 \\ 1 & 0 & 6 \end{array}\right]$ and $\left[\begin{array}{ccc|c} 1 & 0 & 0 & 7 \\ 0 & 0 & 0 & 0 \\ 0 & 0 & 1 & 8 \end{array}\right]$ are not in reduced row echelon form. These matrices do not meet requirements 3 and 1, respectively. The process we went through in Example 3 to solve the system of linear equations was really our first example of using the **Gauss–Jordan method**. Let's examine the procedure used to transform a matrix into reduced row echelon form using the Gauss–Jordan method.

Gauss–Jordan Method

1. **Write the system** so that all variables (in the same order) are on the left side of the equations.
2. **Write the augmented matrix** that corresponds to the system.
3. **Use the row operations** to transform the augmented matrix into reduced row echelon form. (This is called **pivoting**.)
4. **Read the solution** once in reduced row echelon form.

There are many ways to do step 3 in the Gauss–Jordan method. In Example 3 we illustrated a routine procedure that always works. It is called **pivoting** and basically it effectively produces the requisite 1s and 0s in reduced row echelon form.

Pivoting Procedure

1. Change the appropriate entry of the 1st column to a 1 by multiplying the row containing that entry by a nonzero constant (row operation 2).
2. Once the 1 is secured, change all entries in the column to a 0 by adding a multiple of the row containing the 1 to the row that we wish to have a 0 (row operation 3).
3. Repeat the above steps with the next column.

Example 4 Using the Gauss–Jordan Method

Solve the following system by using the Gauss–Jordan method.

$$\begin{aligned} x - y &= 2 \\ 5x + 2y &= 17 \end{aligned}$$

Solution

Understand the Situation: We begin by writing the augmented matrix for the system and use the row operations to transform the augmented matrix into reduced row echelon form.

$$\left[\begin{array}{rr|r} 1 & -1 & 2 \\ 5 & 2 & 17 \end{array}\right] \begin{array}{c} -5R_1 + R_2 \to R_2 \end{array} \left[\begin{array}{rr|r} 1 & -1 & 2 \\ 0 & 7 & 7 \end{array}\right]$$

$$\tfrac{1}{7}R_2 \to R_2 \left[\begin{array}{rr|r} 1 & -1 & 2 \\ 0 & 1 & 1 \end{array}\right]$$

$$R_2 + R_1 \to R_1 \left[\begin{array}{rr|r} 1 & 0 & 3 \\ 0 & 1 & 1 \end{array}\right]$$

So the solution to the system is $x = 3$ and $y = 1$. ■

In Section 1.5 we solved some systems of linear equations that had no solutions and some that had an infinite number of solutions. In Examples 5 and 6 we analyze these two scenarios using the Gauss-Jordan method.

Example 5 Using the Gauss–Jordan Method (No Solution)

Use Gauss–Jordan to solve the linear system

$$\begin{aligned} x - y &= 3 \\ 2x - 2y &= 7 \end{aligned}$$

Solution

The augmented matrix for the system is

$$\left[\begin{array}{cc|c} 1 & -1 & 3 \\ 2 & -2 & 7 \end{array}\right]$$

We try to put this matrix into reduced row echelon form as follows:

$$\left[\begin{array}{cc|c} 1 & -1 & 3 \\ 2 & -2 & 7 \end{array}\right] -2R_1 + R_2 \to R_2 \left[\begin{array}{cc|c} 1 & -1 & 3 \\ 0 & 0 & 1 \end{array}\right]$$

Observe that this transformed augmented matrix corresponds to the system

$$\begin{aligned} x - y &= 3 \\ 0x + 0y &= 1 \end{aligned}$$

Since the second equation states $0 = 1$, which is not true, we know that this system is inconsistent and has no solution. Notice how the last row $[0 \quad 0 \mid 1]$ of the matrix is our clue that the system is **inconsistent**. ■

Example 6 Using the Gauss–Jordan Method (Infinite Number of Solutions)

Use the Gauss–Jordan method to solve the linear system

$$\begin{aligned} x + 2y + z &= 3 \\ 2x - 3y + z &= 2 \\ -2x - 4y - 2z &= -6 \end{aligned}$$

Solution

We first write the augmented matrix for the system. Since the first entry in row 1 is a 1, we begin with row operations in order to get 0s for all the other entries in the first column.

$$\left[\begin{array}{ccc|c} 1 & 2 & 1 & 3 \\ 2 & -3 & 1 & 2 \\ -2 & -4 & -2 & -6 \end{array}\right] \begin{array}{c} -2R_1 + R_2 \to R_2 \\ 2R_1 + R_3 \to R_3 \end{array} \left[\begin{array}{ccc|c} 1 & 2 & 1 & 3 \\ 0 & -7 & -1 & -4 \\ 0 & 0 & 0 & 0 \end{array}\right]$$

$$-\tfrac{1}{7}R_2 \to R_2 \left[\begin{array}{ccc|c} 1 & 2 & 1 & 3 \\ 0 & 1 & \frac{1}{7} & \frac{4}{7} \\ 0 & 0 & 0 & 0 \end{array}\right]$$

$$-2R_2 + R_1 \to R_1 \left[\begin{array}{ccc|c} 1 & 0 & \frac{5}{7} & \frac{13}{7} \\ 0 & 1 & \frac{1}{7} & \frac{4}{7} \\ 0 & 0 & 0 & 0 \end{array}\right]$$

Although the first two columns are in the correct form, we cannot get a zero in row 1, column 3 without changing the structure of the 1st two columns. So we write the system that corresponds to this last matrix:

$$\begin{aligned} x + \tfrac{5}{7}z &= \tfrac{13}{7} \\ y + \tfrac{1}{7}z &= \tfrac{4}{7} \\ 0 &= 0 \end{aligned}$$

Interactive Activity

Give three solutions to the system in Example 6 corresponding to $z = 2$, $z = -3$, and $z = 5$.

WWW

As in Section 1.5, since the first two equations contain a z, we choose z to be the parameter. Notice that the system has an infinite number of solutions for the infinite number of values of z that could be selected. The last row of the matrix, $[0\ \ 0\ \ 0 \mid 0]$, is our clue to this. Solving the first two equations for x and y, respectively, yields

$$x = \frac{13 - 5z}{7} \quad \text{and} \quad y = \frac{4 - z}{7}$$

We can write the solution as $\left(\frac{13-5z}{7}, \frac{4-z}{7}, z\right)$. ■

Application

We conclude this section by using the Gauss–Jordan method to solve the application in the Flashback from earlier in the section.

Example 7 Applying the Gauss–Jordan Method

Solve the system

$$\begin{aligned} x + y + 2z &= 30 \\ 2x + 3y + 2z &= 53 \\ x + 2y + 3z &= 47 \end{aligned}$$

that was set up in the Flashback solution earlier in this section.

Solution

Again, we write the augmented matrix for the system and use row operations.

$$\left[\begin{array}{ccc|c} 1 & 1 & 2 & 30 \\ 2 & 3 & 2 & 53 \\ 1 & 2 & 3 & 47 \end{array}\right] \begin{array}{c} -2R_1 + R_2 \to R_2 \\ -1R_1 + R_3 \to R_3 \end{array} \left[\begin{array}{ccc|c} 1 & 1 & 2 & 30 \\ 0 & 1 & -2 & -7 \\ 0 & 1 & 1 & 17 \end{array}\right]$$

$$\begin{array}{c} -1R_2 + R_1 \to R_1 \\ -1R_2 + R_3 \to R_3 \end{array} \left[\begin{array}{ccc|c} 1 & 0 & 4 & 37 \\ 0 & 1 & -2 & -7 \\ 0 & 0 & 3 & 24 \end{array}\right]$$

$$\tfrac{1}{3}R_3 \to R_3 \left[\begin{array}{ccc|c} 1 & 0 & 4 & 37 \\ 0 & 1 & -2 & -7 \\ 0 & 0 & 1 & 8 \end{array}\right]$$

$$\begin{array}{c} 2R_3 + R_2 \to R_2 \\ -4R_3 + R_1 \to R_1 \end{array} \left[\begin{array}{ccc|c} 1 & 0 & 0 & 5 \\ 0 & 1 & 0 & 9 \\ 0 & 0 & 1 & 8 \end{array}\right]$$

So the solution is $x = 5$, $y = 9$, and $z = 8$.

Interpret the Solution: This means that Gonzalez Office Supplies should manufacture 5 Secretarial, 9 Executive, and 8 Presidential chairs each week. ■

Technology Option

In Figure 2.1.5 we have the augmented matrix for the system in Example 7, and Figure 2.1.6 is the result of using the rref(command on a graphing calculator. Notice that Figure 2.1.6 supports our work in Example 7.

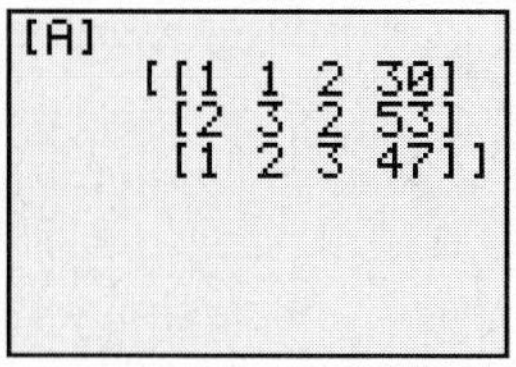

Figure 2.1.5

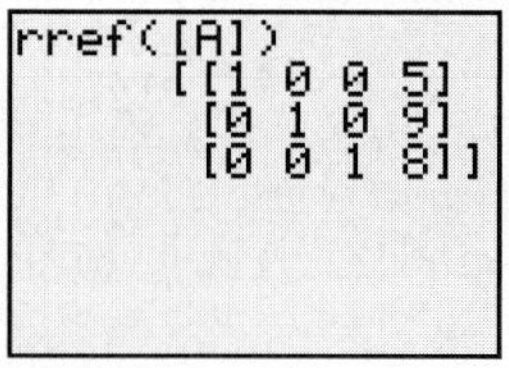

Figure 2.1.6

To see if your calculator transforms matrices into reduced row echelon form, consult the online graphing calculator manual at www.prenhall.com/armstrong

SUMMARY

In this section we learned much about matrices. We first defined a **matrix** and identified the **rows**, **columns**, and **dimension** of a matrix. We then discussed **augmented matrices** and their relationship to systems of linear equations. We saw that the **row operations** used to produce equivalent matrices are analogous to the elimination method presented in Section 1.5. The row operations are used to put an augmented matrix into **reduced row echelon form**. A systematic procedure to achieve this, called **pivoting**, was given. We then saw how all of this built up to the **Gauss–Jordan** method for solving systems of linear equations.

SECTION 2.1 EXERCISES

In Exercises 1–8, find the dimension of the given matrix.

1. $A = \begin{bmatrix} 2 & 1 & 3 \\ 4 & 3 & 2 \end{bmatrix}$

2. $B = \begin{bmatrix} -1 & 3 \\ 2 & 7 \\ 5 & -7 \end{bmatrix}$

3. $C = \begin{bmatrix} -5 & 2 \\ 1 & -3 \\ 4 & 5 \end{bmatrix}$

4. $D = \begin{bmatrix} 1 & -1 & 2 \\ 3 & 4 & -2 \\ 7 & -5 & 5 \end{bmatrix}$

5. $E = \begin{bmatrix} 7 & -1 & -6 & 1 \\ 2 & 4 & -7 & -3 \\ 3 & 5 & 0 & -2 \end{bmatrix}$

6. $F = \begin{bmatrix} -2 & -1 & 3 \\ 1 & 0 & 7 \\ 8 & -3 & 4 \\ -4 & 5 & -6 \end{bmatrix}$

7. $G = \begin{bmatrix} 1 \\ -1 \\ 3 \\ 4 \end{bmatrix}$

8. $H = \begin{bmatrix} 7 & -1 & 2 & 3 \end{bmatrix}$

In Exercises 9–16, find the indicated entry for the matrices given in Exercises 1–8.

9. $b_{3,1}$ **10.** $a_{1,3}$ **11.** $d_{2,3}$

12. $c_{2,1}$ **13.** $f_{2,3}$ **14.** $e_{3,2}$

15. $h_{1,3}$ **16.** $g_{3,1}$

In Exercises 17–24, write an augmented matrix for the given system. **Do not solve.**

17. $3x - 7y = 4$
$2x + 5y = 8$

18. $-2x + 4y = -7$
$3x - 2y = 14$

19. $x + 3y = -11$
$2x + 5y = -17$

20. $-x - y = 13$
$4x + y = 7$

21. $x + y + z = 1$
$3x + y - z = -2$
$2x - y + 3z = 4$

22. $-x - y - z = -3$
$2x - y + 4z = 8$
$-2x + 3y + 5z = 7$

23. $3x + y + 10z = 38$
$5x + 10y + z = 22$
$7x - y - 12z = -15$

24. $11x - y + 12z = 42$
$-x + 13y - 5z = 17$
$51x - 17y + z = -13$

In Exercises 25–32, write the linear system corresponding to each augmented matrix.

25. $\left[\begin{array}{cc|c} 2 & -3 & 7 \\ -2 & 4 & 3 \end{array}\right]$

26. $\left[\begin{array}{cc|c} -1 & 1 & 7 \\ 3 & -1 & -4 \end{array}\right]$

27. $\left[\begin{array}{cc|c} 1 & 0 & 8 \\ 0 & 1 & -3 \end{array}\right]$

28. $\left[\begin{array}{cc|c} 1 & 0 & -2 \\ 0 & 1 & 4 \end{array}\right]$

29. $\left[\begin{array}{ccc|c} 3 & -1 & 4 & 7 \\ 2 & 8 & -3 & 2 \\ -1 & 4 & 5 & -3 \end{array}\right]$

30. $\left[\begin{array}{ccc|c} 4 & -4 & -3 & -2 \\ -5 & 18 & 4 & 22 \\ 7 & 31 & -2 & 37 \end{array}\right]$

31. $\left[\begin{array}{ccc|c} 1 & 0 & 0 & 8 \\ 0 & 1 & 0 & -2 \\ 0 & 0 & 1 & 7 \end{array}\right]$

32. $\left[\begin{array}{ccc|c} 1 & 0 & 0 & -3 \\ 0 & 1 & 0 & 15 \\ 0 & 0 & 1 & 9 \end{array}\right]$

In Exercises 33–40, perform the indicated row operations to change each augmented matrix.

33. $-1R_1 \to R_1$

$\left[\begin{array}{cc|c} -1 & 1 & 7 \\ 3 & -1 & -4 \end{array}\right]$

34. $\frac{1}{2}R_1 \to R_1$

$\left[\begin{array}{cc|c} 2 & -3 & 7 \\ -2 & 4 & 3 \end{array}\right]$

35. $2R_1 + R_2 \to R_2$

$\left[\begin{array}{cc|c} 1 & 3 & -2 \\ -2 & 4 & 3 \end{array}\right]$

36. $-5R_1 + R_2 \to R_2$

$\left[\begin{array}{cc|c} 1 & -8 & 7 \\ 5 & 2 & -3 \end{array}\right]$

37. $\frac{1}{4}R_1 \to R_1$

$\left[\begin{array}{ccc|c} 4 & -4 & -3 & -2 \\ -5 & 18 & 4 & 22 \\ 7 & 31 & -2 & 37 \end{array}\right]$

38. $\frac{1}{3}R_1 \to R_1$

$\left[\begin{array}{ccc|c} 3 & 8 & -2 & 5 \\ -7 & -1 & 1 & 8 \\ 2 & 5 & 8 & 13 \end{array}\right]$

39. $-2R_1 + R_3 \to R_3$

$\left[\begin{array}{ccc|c} 1 & -1 & -1 & -2 \\ 0 & 3 & 2 & 5 \\ 2 & -8 & 1 & 5 \end{array}\right]$

40. $3R_1 + R_3 \to R_3$

$\left[\begin{array}{ccc|c} 1 & 4 & 5 & 3 \\ 0 & -1 & 4 & -7 \\ -3 & 1 & 4 & 8 \end{array}\right]$

In Exercises 41–58, use the Gauss–Jordan method to solve the given system of linear equations.

41. $x - y = 2$
$-2x + 3y = 7$

42. $x + y = 4$
$-3x - 2y = -5$

43. $3x - 5y = 8$
$-2x + 3y = -7$

44. $-4x + 5y = 9$
$3x - 4y = -2$

45. $-5x + 3y = -11$
$2x - y = 5$

46. $-3x + 5y = -13$
$2x - 2y = 5$

47. $-7x - 2y = 9$
$4x + 5y = -2$

48. $7x - 8y = -9$
$-2x - 5y = 6$

49. $x + 3y - 6z = 7$
$2x - y + 2z = 0$
$x + y + 2z = -1$

50. $2x + 3y + 5z = 21$
$x - y - 5z = -2$
$2x + y - z = 11$

51. $x + 3y - z = -3$
$2x + y - 2z = 4$
$3x + 4y - z = 7$

52. $x + 2y + 2z = 11$
$x - y - z = -4$
$2x + 5y + 9z = 39$

53. $x - 3y + z = 5$
$-2x + 7y - 6z = -9$
$x + 2y - 3z = 6$

54. $3x - 6y + 9z = 0$
$4x - 6y + 8z = -4$
$-2x - y + z = 7$

55. $x + 2y - 4z = 5$
$3x - y + z = 3$
$x + 2y + z = -3$

56. $-2x + 3y + 4z = 10$
$3x - 5y + 2z = 7$
$4x + 2y - 3z = -2$

57. $x + 3y + 2z - w = 2$
$-x - y - 4z + 2w = 1$
$2x + y - z + 3w = -4$
$x - 2y - z - 3w = 4$

58. $x - 3y + z + 2w = 3$
$-2x + y + 3z + w = 0$
$-x + 2y - 3z + 2w = -3$
$x + 2y - z + 3w = 4$

In Exercises 59–66, use the Gauss–Jordan method to solve the given system of equations using your calculator's matrix capabilities (either step-by-step row operations or rref(command). Round answers to four decimal places.

59. $2.3x - 5.7y = 12.3$
$-13.1x - 18.2y = 7.1$

60. $-3.1x + 2.9y = -5.8$
$6.1x + 3.5y = -15.2$

61. $1.53x + 3.57y = 2.68$
$7.28x + 9.85y = 13.41$

62. $-3.75x - 4.23y = 7.82$
$7.28x + 8.65y = 4.86$

63. $1.1x - 3.4y + 7.2z = 5.9$
$-8.9x - 4.7y - 3.1z = -7.1$
$0.8x + 0.3y + 1.2z = 0.9$

64. $-2.3x + 3.5y - 2.1z = 5.7$
$3.6x - 0.5y - 0.3z = -5.3$
$-8.2x - 1.1y + 3.7z = -2.2$

65. $-2.31x + 4.21y - 7.28z + 3.42w = 8.17$
$7.91x + 8.99y + 4.01z - 4.21w = -2.71$
$-13.11x - 5.07y - 0.39z + 1.22w = 0.77$
$8.71x + 3.72y - 2.73z - 4.81w = 3.14$

66. $0.72x - 1.13y + 4.27z - 3.12w = 5.77$
$-1.11x + 0.93y - 2.21z + 5.19w = -2.29$
$-0.04x + 1.98y + 7.61z + 4.23w = 5.13$
$1.92x - 5.03y + 3.41z - 2.71w = 8.28$

Applications

67. Max George is a craftsman who makes end tables and night stands. To make one end table takes 5 hours of his labor, and to make one night stand takes 4 hours of his labor. The materials for one end table cost him $200, and the materials for one night stand cost him $80. Suppose that for 1 week Max works 40 hours and spends $1200 on materials.

(a) Let x represent the number of end tables made in 1 week and y represent the number of night stands made in 1 week. Write a system of two linear equations in which one equation relates costs and the other relates time.

(b) Solve the system in part (a) to determine how many end tables and how many night stands Max can make in 1 week, assuming that he works 40 hours and spends $1200 on materials.

1.1 **68.** Martha George is a craftswoman who makes coffee tables and night stands. To make one coffee table takes 6 hours of her labor, and to make one night stand takes 4 hours of her labor. The materials for one coffee table cost her $220 and the materials for one night stand cost her $80. Suppose that for 1 week Martha works 40 hours and spends $1400 on materials.

(a) Let x represent the number of coffee tables made in 1 week and y represent the number of night stands made in 1 week. Write a system of two linear equations in which one equation relates costs and the other relates time.

(b) Solve the system in part (a) to determine how many coffee tables and how many night stands Martha can make in 1 week, assuming that she works 40 hours and spends $1400 on materials.

1.1 **69.** The WeDo Wood Canoes Company makes two types of canoes: a two-person model and a four-person model. Each two-person model requires 1 hour in the cutting department and 1.5 hours in the assembly department. Each four-person model requires 1 hour in the cutting department and 2.75 hours in the assembly department. The cutting department has a maximum of 640 hours available each week, while the assembly department has a maximum of 1080 hours available each week. How many of each type of canoe should be produced each week for the company to operate at full capacity, that is, use all available hours in each department?

1.1 **70.** (*continuation of Exercise 69*) Suppose that a breakthrough in the assembly process decreases the time required in the assembly department for each two-person model to $\frac{3}{4}$ hour and for each four-person model to 2 hours. Now how many of each type of canoe should be produced each week for the company to operate at full capacity, that is, use all available hours in each department?

71. Each order from the TidyTrader wholesaler contains 10 boxes of lighters and 36 boxes of coffee stirrers, while each order from the Trade-it-Now wholesaler contains 30 boxes of lighters and 12 boxes of coffee stirrers. How many orders should the Gas-N-Go convenient store get from each wholesaler to get exactly 120 boxes of lighters and 240 boxes of coffee stirrers?

72. The Java-Go-Go coffee shop carries two blends of coffee. The first blend contains 20% Mexican and 80% Colombian coffee. The second blend contains 60% Mexican and 40% Colombian coffee. How many ounces of each blend should be combined to get a new blend that has 100 ounces of Mexican and 100 ounces of Colombian coffee?

1.1 **73.** The Stitz Tent Company makes three types of tents: two-person, four-person and, six-person models. Each tent requires the services of three departments, as shown in Table 2.1.3. The fabric cutting, assembly, and packaging departments have available a maximum of 450, 390, and 190 hours each week, respectively.

Table 2.1.3

	Two-person	Four-person	Six-person
Fabric cutting (hr)	0.5	0.8	1.1
Assembly (hr)	0.6	0.7	0.9
Packaging (hr)	0.2	0.3	0.6

(a) Let x represent the number of two-person tents made in 1 week, y represent the number of four-person tents made in 1 week, and z represent the number of six-person tents made in 1 week. Write a system of three linear equations using this information and the given department maximum weekly times.

(b) Solve the system in part (a) to determine how many two-, four-, and six-person tents can be made in 1 week, assuming that the company operates at full capacity.

1.1 **74.** Rework Exercise 73(a) and (b), except here assume that the fabric cutting, assembly, and packaging departments have available a maximum of 420, 390, and 185 hours each week, respectively.

1.1 **75.** Last year the Hannus Corporation took out three loans totaling $50,000 to fund the purchase of new equipment. Some of the money was at 6%, $1000 more than $\frac{1}{2}$ the amount of the 6% loan was at 8%, and the rest was at 7%. The total annual interest on the loans was $3410.

(a) Let x represent the amount borrowed at 6%, y represent the amount borrowed at 7%, and z represent the amount borrowed at 8%. Write a system of three linear equations that can be solved using the Gauss–Jordan method to determine the amount borrowed at each rate.

(b) Solve the system in part (a) to determine the amount borrowed at each rate.

76. Debbie Ashton, a certified financial planner, is to invest $80,000 for a client in three different areas: certificates of deposit earning 6%, a mutual fund earning 12%, and speculative junk bonds earning 17%. The client is a conservative investor and stipulates that four times as much be invested in CDs as in junk bonds. The client also wishes that $7700 be produced each year on these investments. Assuming these rates and adhering to her client's wishes, how much should Debbie place in each area to produce exactly $7700 each year on the investments?

77. Tim Kling, chairman of the Chemistry Department at a local community college, wants to make 2 quarts of an 18% acid solution. He has available a solution that is 21% acid and a solution that is 14% acid. How many quarts of each type should he use to produce the 18% solution?

78. Rework Exercise 77 assuming that Tim made an error and really desired 2 quarts of a 16% acid solution.

1.1 **79.** Chris O'Neill, a dietitian in a local hospital, is to arrange a special diet for a patient that is made up of three basic foods, Basic!, Vaca, and Secra. The diet is to include 340 units of calcium, 205 units of vitamin A, and 225 units of vitamin C. The number of units per ounce for each basic food is given in Table 2.1.4. How many ounces of each food must be used to exactly meet the diet requirements?

Table 2.1.4

	Basic!	Vaca	Secra
Calcium	20	10	30
Vitamin A	10	5	20
Vitamin C	20	15	10

1.1 **80.** Rework Exercise 79 if the patient's diet is changed so that it includes 480 units of calcium, 285 units of vitamin A, and 325 units of vitamin C.

1.1 **81.** The data in Table 2.1.5 give the amounts of nitrogen, phosphate, and potash (all in lb/acre) needed to grow asparagus, cauliflower, and celery for an entire growing season in California.

Table 2.1.5

	Asparagus	Cauliflower	Celery
Nitrogen	240	215	336
Phosphate	148	98	171
Potash	42	69	147

SOURCE: U.S. Department of Agriculture, www.usda.gov/nass

(a) If Juan has 280 acres, 20,000 pounds of nitrogen, 31,860 pounds of phosphate, and 21,630 pounds of potash, is it possible for him to use all resources completely. If so, how many acres should he allot for each crop?

(b) Suppose that everything is the same as in part (a), except now Juan has 65,790 pounds of nitrogen. Is it possible for him to use all resources completely? If so, how many acres should he allot for each crop?

1.1 **82.** The data in Table 2.1.6 give the amounts of nitrogen, phosphate, and potash (all in lb/acre) needed to grow head lettuce, cantaloupe, and bell peppers for an entire growing season in California.

Table 2.1.6

	Head Lettuce	Cantaloupe	Bell Peppers
Nitrogen	173	175	212
Phosphate	122	161	108
Potash	76	39	72

SOURCE: U.S. Department of Agriculture, www.usda.gov/nass

(a) If Zach has 200 acres, 38,430 pounds of nitrogen, 27,845 pounds of phosphate, and 6125 pounds of potash, is it possible for him to use all resources completely. If so, how many acres should he allot for each crop?

(b) Suppose that everything is the same as in part (a), except now Zach has 12,985 pounds of potash. Is it possible for him to use all resources completely? If so, how many acres should he allot for each crop?

83. Tammy K's has a 1-month intensive training program for receptionist/secretaries for three types of office positions: medical field, small business, and executive secretarial. Each individual receives training in three departments at Tammy K's: word processing, telephone and LAN (local area network), and ethics. The data in Table 2.1.7 show the number of units of training that each type of receptionist/secretary receives.

Table 2.1.7

	Medical Field	Small Business	Executive Secretary
Word processing	0.8	1.1	1.7
Telephone and LAN	0.6	1.3	1.5
Ethics	1.2	0.8	1.3

Each month at Tammy K's, word processing can give 32.1 units of word processing, telephone and LAN can give 29.7 units, and ethics can give 33 units. To completely utilize their training capacity, how many of each type of receptionist/secretary should Tammy K's admit to their training program each month?

84. Due to a strong demand for their services, Tammy K's from Exercise 83 has added more trainers. This results in Tammy K's being able to deliver each month 41.9 units of word processing, 38.9 units of telephone and LAN training, and 41 units of ethics training. To completely utilize their training capacity, how many of each type of receptionist/secretary should Tammy K's admit to their training program each month?

⇐ 1.1

SECTION PROJECT

During rush hour, traffic congestion may be encountered at certain intersections. For a network of four one-way streets in a certain city, the rush hour flow is as shown in Figure 2.1.7.

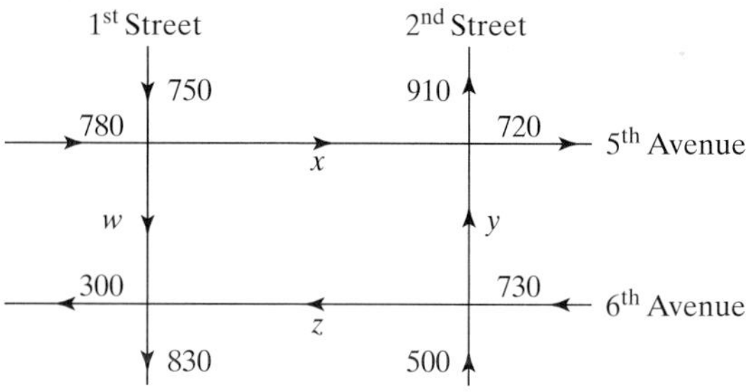

Figure 2.1.7

To help to speed the flow of traffic, traffic engineers need to coordinate the signals at these intersections. To do this, data are collected giving the hourly flow of traffic during rush hour at certain locations. Figure 2.1.7 shows the data that were collected. Notice that the figure indicates that a total of 1530 cars per hour enter the intersection of 1st Street and 5th Avenue. Of these 1530 cars, x cars leave the intersection on 5th Avenue and w cars leave the intersection on 1st Street. Since the total number of cars leaving this (or any) intersection equals the number entering, we have

$$x + w = 1530$$

This equation gives the flow at the intersection of 1st Street and 5th Avenue.

(a) Determine the equations for the traffic flow at the other three intersections. Remember that the total entering the intersection equals the total leaving the intersection, and vice versa.

(b) Write a system of four equations representing the traffic flow for this network of four one-way streets.

(c) Write the augmented matrix for the system in part (b). Note that some entries must be zero.

(d) Use the Gauss–Jordan method to solve. Here we use w as the parameter. If necessary, see Example 6.

(e) Using the result of part (d), what is the smallest and largest value for the parameter w.

(f) Using the results of parts (d) and (e), determine the smallest and largest values for the variables x, y, and z.

Section 2.2 Matrices and Operations

In Section 2.1 we introduced matrices as a means of helping us solve systems of linear equations using the Gauss–Jordan method. In the next three sections we take a more in-depth look at matrices and operations with matrices. We will also see how matrices can be used to solve problems in the business, life, and social sciences.

Matrix Arithmetic

One beauty of matrices is its ability to represent data. For example, Staminly's has three furniture stores, East, West and Metro, in a large metropolitan area. Tables 2.2.1, 2.2.2, and 2.2.3 give the sales of their Country Style Contemporary sofa for each store during July.

Table 2.2.1 Staminly's East

	60″	72″	84″
Two cushion	8	13	5
Three cushion	9	16	7

Table 2.2.2 Staminly's Metro

	60″	72″	84″
Two cushion	13	7	3
Three cushion	6	8	4

Table 2.2.3 Staminly's West

	60″	72″	84″
Two cushion	4	11	13
Three cushion	6	9	8

With the understanding that rows refer to the sofa type (two or three cushion) and columns refer to the lengths (60″, 72″, or 84″), we can use matrices to represent the information in Tables 2.2.1 through 2.2.3 as follows:

$$E = \begin{bmatrix} 8 & 13 & 5 \\ 9 & 16 & 7 \end{bmatrix} \qquad M = \begin{bmatrix} 13 & 7 & 3 \\ 6 & 8 & 4 \end{bmatrix} \qquad W = \begin{bmatrix} 4 & 11 & 13 \\ 6 & 9 & 8 \end{bmatrix}$$

Matrix for Table 2.2.1 — Matrix for Table 2.2.2 — Matrix for Table 2.2.3

Now suppose that we wanted to know the total number of two-cushion 60″ sofas sold at Staminly's East and Metro. Since 8 were sold at East and 13 were sold at Metro, we conclude that a total of $8 + 13 = 21$ two-cushion 60″ sofas were sold at Staminly's East and Metro. Notice that $e_{1,1} + m_{1,1} = 21$. If we continue to sum corresponding entries for E and M, we get a new matrix, call it S, that is given by

$$S = E + M = \begin{bmatrix} 8 & 13 & 5 \\ 9 & 16 & 7 \end{bmatrix} + \begin{bmatrix} 13 & 7 & 3 \\ 6 & 8 & 4 \end{bmatrix}$$

$$= \begin{bmatrix} 8+13 & 13+7 & 5+3 \\ 9+6 & 16+8 & 7+4 \end{bmatrix} = \begin{bmatrix} 21 & 20 & 8 \\ 15 & 24 & 11 \end{bmatrix}$$

Matrix S gives the total sales of Staminly's Country Style Contemporary sofa for both styles and all three lengths for their East and Metro stores during July. The manner in which E and M were added illustrates the following definition for the addition of matrices.

Interactive Activity

For the matrices M and W in Example 1, determine $W + M$. Compare the result with the result of Example 1. This illustrates that matrix addition is **commutative**; that is, $M + W = W + M$.

Addition of Matrices

The **sum** of two (or more) $m \times n$ matrices A and B is the $m \times n$ matrix $A + B$, where each entry in $A + B$ is the sum of the corresponding entries of A and B.

Note: We can only add matrices that have the same dimension.

Example 1 Adding Matrices

Use matrix addition to find the total sales of Staminly's Country Style Contemporary sofa for their Metro and West stores during July.

Solution

The total sales for Metro and West are

$$M + W = \begin{bmatrix} 13 & 7 & 3 \\ 6 & 8 & 4 \end{bmatrix} + \begin{bmatrix} 4 & 11 & 13 \\ 6 & 9 & 8 \end{bmatrix}$$

$$= \begin{bmatrix} 13+4 & 7+11 & 3+13 \\ 6+6 & 8+9 & 4+8 \end{bmatrix} = \begin{bmatrix} 17 & 18 & 16 \\ 12 & 17 & 12 \end{bmatrix}$$ ■

Checkpoint 1

Now work Exercise 1.

Example 2 Adding Matrices and Interpreting an Entry

For Staminly's, define matrix $T = E + M + W$. Find matrix T and interpret $t_{2,3}$.

Solution

Matrix T is given by

$$T = E + M + W = \begin{bmatrix} 8 & 13 & 5 \\ 9 & 16 & 7 \end{bmatrix} + \begin{bmatrix} 13 & 7 & 3 \\ 6 & 8 & 4 \end{bmatrix} + \begin{bmatrix} 4 & 11 & 13 \\ 6 & 9 & 8 \end{bmatrix}$$

$$= \begin{bmatrix} 25 & 31 & 21 \\ 21 & 33 & 19 \end{bmatrix}$$

By looking in the 2nd row, 3rd column we have $t_{2,3} = 19$.

Interpret the Solution: This entry means that the total number of three-cushion 84″ Country Style Contemporary sofas sold in July at all three Staminly stores was 19. ■

Interactive Activity

For matrices E, M, and W from Example 2, do the following: Find $(E + M) + W$ and compare to $E + (M + W)$. This will illustrate that matrix addition is **associative**; that is, $(E + M) + W = E + (M + W)$.

Technology Option

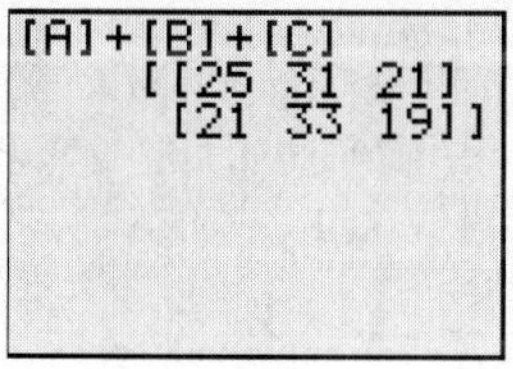

Figure 2.2.1

Matrix operations can be performed on graphing calculators as well. Figure 2.2.1 shows the result of $E + M + W$ in Example 2. Matrices E, M, and W were entered into the graphing calculator as A, B, and C, respectively.

The way that matrices are added and subtracted can vary from calculator to calculator. To learn how matrix operations are performed on your calculator, consult the online graphing calculator manual at www.prenhall.com/armstrong

Scalar Multiplication

The product of a real number k and a matrix A, denoted by kA, is a matrix in which each entry is the product of each entry in A by k. When a matrix is multiplied by a real number, we call this **scalar multiplication**.

Scalar Multiplication

For a matrix $A = \begin{bmatrix} a & b \\ c & d \end{bmatrix}$ and a real number k, the scalar product kA is given by the **scalar multiplication**

$$kA = k\begin{bmatrix} a & b \\ c & d \end{bmatrix} = \begin{bmatrix} ka & kb \\ kc & kd \end{bmatrix}$$

Note: This definition can be extended to a matrix of any dimension.

Example 3 Performing a Scalar Multiplication

Given matrix $A = \begin{bmatrix} -1 & 2 & 2 \\ 4 & 3 & -7 \end{bmatrix}$, find $3A$.

Solution

We multiply each entry in A by 3 to get

$$3A = 3\begin{bmatrix} -1 & 2 & 2 \\ 4 & 3 & -7 \end{bmatrix} = \begin{bmatrix} 3(-1) & 3(2) & 3(2) \\ 3(4) & 3(3) & 3(-7) \end{bmatrix} = \begin{bmatrix} -3 & 6 & 6 \\ 12 & 9 & -21 \end{bmatrix}$$ ■

Technology Option

Figure 2.2.2

Figure 2.2.2 shows the scalar multiplication in Example 3 on a graphing calculator.

Matrix subtraction is defined similarly to real number subtraction. This means that if A and B are two matrices with the same dimension then $A - B = A + (-B)$.

Example 4 Subtracting Scalar Multiplied Matrices

Given $A = \begin{bmatrix} -2 & 3 \\ 4 & -1 \end{bmatrix}$ and $B = \begin{bmatrix} 7 & 2 \\ -1 & 3 \end{bmatrix}$, find $3A - 2B$.

Solution

We have

$$\begin{aligned} 3A - 2B &= 3\begin{bmatrix} -2 & 3 \\ 4 & -1 \end{bmatrix} - 2\begin{bmatrix} 7 & 2 \\ -1 & 3 \end{bmatrix} \\ &= 3\begin{bmatrix} -2 & 3 \\ 4 & -1 \end{bmatrix} + (-2)\begin{bmatrix} 7 & 2 \\ -1 & 3 \end{bmatrix} \\ &= \begin{bmatrix} -6 & 9 \\ 12 & -3 \end{bmatrix} + \begin{bmatrix} -14 & -4 \\ 2 & -6 \end{bmatrix} \\ &= \begin{bmatrix} -20 & 5 \\ 14 & -9 \end{bmatrix} \end{aligned}$$

■

Checkpoint 2 Now work Exercise 13.

Technology Option

Figure 2.2.3

Figure 2.2.3 shows the result of $3A - 2B$ from Example 4 on a graphing calculator.

In Example 4 matrix A and matrix B are examples of **square matrices**. In general, an $n \times n$ matrix, that is, a matrix having the same number of rows as columns, is called a **square matrix**. A matrix with one row is called a **row matrix**, and a matrix with one column is called a **column matrix**.

Equality of Matrices

We have to satisfy two conditions in order for two matrices A and B to be equal.

Equality of Matrices

Two matrices are **equal** if they have the same dimension and if corresponding entries are equal.

Example 5 Determining Matrix Equality

Determine the values for x, y, z, and w that make the following matrices equal.

(a) $\begin{bmatrix} -2 & x & 3 \\ y+2 & 4 & 7 \end{bmatrix} = \begin{bmatrix} z & 1 & 3 \\ -5 & 4 & w+1 \end{bmatrix}$

(b) $\begin{bmatrix} 3 & x & 1 \\ y & -7 & 4 \end{bmatrix} = \begin{bmatrix} 5 & w \\ z & -1 \\ 3 & 4 \end{bmatrix}$

Solution

(a) Both matrices are 2×3, so they are equal if corresponding entries are equal. Hence

$$x = 1, \qquad y + 2 = -5, \qquad z = -2, \qquad w + 1 = 7$$

This means that $x = 1$, $y = -7$, $z = -2$, and $w = 6$ makes the matrices equal.

(b) The matrix on the left is a 2×3, and the one on the right is a 3×2; hence they cannot be equal no matter the values for x, y, z, and w. ■

Checkpoint 3 Now work Exercise 35.

Zero Matrix and Transpose of a Matrix

In the real number system, the number zero is called the **additive identity**, because any real number plus zero produces the same real number. There is a matrix that behaves the same way, the **zero matrix**. The zero matrix is simply a matrix in which all entries are zero. Given a matrix A, the zero matrix O has the property that

$$A + O = O + A = A$$

as long as A and matrix O have the same dimension. For example, if matrix A is any 2×3 matrix, then the zero matrix for any 2×3 matrix is

$$O = \begin{bmatrix} 0 & 0 & 0 \\ 0 & 0 & 0 \end{bmatrix}$$

In Section 4.3, we will use the **transpose** of a matrix to solve some applications. For a given matrix A, if we interchange the rows and columns we produce what is called the **transpose of A** and denote it with A^T.

Transpose of a Matrix

If A is a $m \times n$ matrix with entries $a_{i,j}$, then the **transpose** of A, denoted A^T, is an $n \times m$ matrix with entries $a_{j,i}$.

Example 6 Finding a Transpose

For matrix A, $A = \begin{bmatrix} 3 & -1 & 2 \\ 4 & 7 & -2 \end{bmatrix}$, determine A^T and the dimension of A^T.

Solution

Interchanging the rows and columns gives us

$$A^T = \begin{bmatrix} 3 & 4 \\ -1 & 7 \\ 2 & -2 \end{bmatrix}$$

The dimension of matrix A is 2×3, and the dimension of A^T is 3×2. ■

Checkpoint 4 Now work Exercise 23.

Technology Option

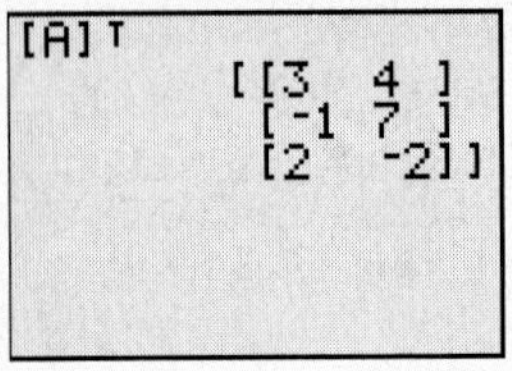

Figure 2.2.4

Figure 2.2.4 shows the result of Example 6 on a graphing calculator.

Our final example returns us to the Staminly furniture stores from the beginning of the section.

Example 7 Predicting Monthly Sales

Joyce, the CEO of Staminly's, wants to see an across the board increase in sales for the month of August. She wants the East and West stores to increase monthly sales by 15% and the Metro store to increase monthly sales by 10%. Determine a matrix giving the total sales that Joyce expects for all three stores in August. Round all entries to the nearest whole number.

Solution

Recall that the July sales for the East, West, and Metro stores, given by matrices E, W, and M, respectively, are

$$E = \begin{bmatrix} 8 & 13 & 5 \\ 9 & 16 & 7 \end{bmatrix} \quad W = \begin{bmatrix} 4 & 11 & 13 \\ 6 & 9 & 8 \end{bmatrix} \quad M = \begin{bmatrix} 13 & 7 & 3 \\ 6 & 8 & 4 \end{bmatrix}$$

The matrix for expected August sales at the East store is computed to be

$$E+0.15E=\begin{bmatrix}8&13&5\\9&16&7\end{bmatrix}+0.15\begin{bmatrix}8&13&5\\9&16&7\end{bmatrix}$$
$$=\begin{bmatrix}8&13&5\\9&16&7\end{bmatrix}+\begin{bmatrix}1.2&1.95&0.75\\1.35&2.4&1.05\end{bmatrix}=\begin{bmatrix}9.2&14.95&5.75\\10.35&18.4&8.05\end{bmatrix}$$

Similarly, the expected August sales at the West store are

$$W+0.15W=\begin{bmatrix}4&11&13\\6&9&8\end{bmatrix}+0.15\begin{bmatrix}4&11&13\\6&9&8\end{bmatrix}$$
$$=\begin{bmatrix}4&11&13\\6&9&8\end{bmatrix}+\begin{bmatrix}0.6&1.65&1.95\\0.9&1.35&1.2\end{bmatrix}=\begin{bmatrix}4.6&12.65&14.95\\6.9&10.35&9.2\end{bmatrix}$$

Finally, the expected August sales at the Metro store are

$$M+0.10M=\begin{bmatrix}13&7&3\\6&8&4\end{bmatrix}+0.10\begin{bmatrix}13&7&3\\6&8&4\end{bmatrix}$$
$$=\begin{bmatrix}13&7&3\\6&8&4\end{bmatrix}+\begin{bmatrix}1.3&0.7&0.3\\0.6&0.8&0.4\end{bmatrix}=\begin{bmatrix}14.3&7.7&3.3\\6.6&8.8&4.4\end{bmatrix}$$

Interactive Activity

For the matrices given in Example 7, determine $1.15E+1.15W+1.1M$ and compare to the final result.

WWW

So the matrix giving the total sales that Joyce expects for all three stores in August is the sum of

$$(E+0.15E)+(W+0.15W)+(M+0.10M)$$
$$=\begin{bmatrix}9.2&14.95&5.75\\10.35&18.4&8.05\end{bmatrix}+\begin{bmatrix}4.6&12.65&14.95\\6.9&10.35&9.2\end{bmatrix}+\begin{bmatrix}14.3&7.7&3.3\\6.6&8.8&4.4\end{bmatrix}$$
$$=\begin{bmatrix}28.1&35.3&24\\23.85&37.55&21.65\end{bmatrix}\approx\begin{bmatrix}28&35&24\\24&38&22\end{bmatrix}$$

■

SUMMARY

In this section we saw how to add, subtract, and multiply matrices by a scalar. We also learned how to determine when two matrices are equal. The zero matrix was presented along with how to determine the transpose of a matrix. We saw how matrices can be used to represent data and analyze problems in an applied setting.

SECTION 2.2 EXERCISES

In Exercises 1–16, refer to the following matrices to perform the indicated operations.

$$A=\begin{bmatrix}-2&0&3\\1&-3&4\end{bmatrix}\qquad B=\begin{bmatrix}1&-3\\4&-2\end{bmatrix}$$

$$C=\begin{bmatrix}-4&1\\3&5\end{bmatrix}\qquad D=\begin{bmatrix}4&-1&-3\\0&7&1\end{bmatrix}$$

$$E=[1\quad -1\quad 4\quad -3]\qquad F=\begin{bmatrix}-2\\1\\0\\3\end{bmatrix}$$

$$G=\begin{bmatrix}-1&4\\3&-7\\-5&-2\end{bmatrix}\qquad H=\begin{bmatrix}4&-2\\3&-1\\-2&5\end{bmatrix}$$

✓ **1.** $A+D$

2. $G+H$

3. $A+B$

4. $D+G$

5. $B-C$

6. $C-B$

7. $-2A$

8. $-3B$

9. $4E$

10. $2F$

11. $3A+2D$

12. $-4A+3D$

✓ **13.** $-3G-4H$

14. $-2G-7H$

15. $4H+2G$

16. $5H-3G$

To answer Exercises 17–30, refer to the matrices A through H given for Exercises 1–16.

17. Which matrices are square matrices?

18. Identify the column matrix.

19. Identify the row matrix.

20. Explain why $A+B$ cannot be done.

21. Explain why $D+G$ cannot be done.

22. Determine A^T, the transpose of A, and give the dimension of A^T.

✓ 23. Determine D^T, the transpose of D, and give the dimension of D^T.

24. Determine B^T, the transpose of B, and give the dimension of B^T.

25. Determine C^T, the transpose of C, and give the dimension of C^T.

26. Determine $(A+D)^T$, the transpose of $A+D$.

27. Write the zero matrix for A.

28. Write the zero matrix for B.

29. Write the zero matrix for G.

30. Write the zero matrix for E.

In Exercises 31–38, find the values of the variables in each equation.

31. $\begin{bmatrix} 3 \\ x \\ y \end{bmatrix} = \begin{bmatrix} z \\ -1 \\ 4 \end{bmatrix}$

32. $\begin{bmatrix} -4 \\ x \\ 3 \end{bmatrix} = \begin{bmatrix} y \\ 2 \\ z \end{bmatrix}$

33. $\begin{bmatrix} -2 & x \\ y & 1 \end{bmatrix} = \begin{bmatrix} -2 & 5 \\ 3 & z \end{bmatrix}$

34. $\begin{bmatrix} x & 1 \\ y & 4 \end{bmatrix} = \begin{bmatrix} -9 & 1 \\ 3 & z \end{bmatrix}$

✓ 35. $\begin{bmatrix} x+3 & y-2 \\ 4 & 3 \end{bmatrix} = \begin{bmatrix} 8 & 11 \\ 4 & z \end{bmatrix}$

36. $\begin{bmatrix} 3 & 8 \\ x & 1 \end{bmatrix} = \begin{bmatrix} y-4 & z+3 \\ 2 & 1 \end{bmatrix}$

37. $\begin{bmatrix} -3+x & 4y & 3z \\ 5w & 3 & 4 \end{bmatrix} + \begin{bmatrix} -6 & 3y & 2 \\ 7 & 2 & 9 \end{bmatrix} = \begin{bmatrix} 18 & 28 & 8 \\ 22 & 5 & 26p \end{bmatrix}$

38. $\begin{bmatrix} x+1 & 3y-1 & 4z \\ -7w & -3 & 5 \end{bmatrix} + \begin{bmatrix} 2x & 7y & 5z-1 \\ -3w & 8 & 13 \end{bmatrix} = \begin{bmatrix} 16 & 91 & 8 \\ 20 & -11 & 54p \end{bmatrix}$

In Exercises 39–48, use a graphing calculator and its matrix capabilities, and refer to the given matrices to perform the indicated operations.

$$A = \begin{bmatrix} 1.32 & 4.67 & 1.19 \\ 3.14 & 2.78 & 5.61 \end{bmatrix} \qquad B = \begin{bmatrix} 8.11 & 9.23 & 1.41 \\ 4.15 & 3.14 & 4.59 \end{bmatrix}$$

39. $A+B$

40. $A-B$

41. $2A+3B$

42. $4A-2B$

43. $1.3A$

44. $2.4B$

45. $2.37A-0.17B$

46. $7.81B+2.19A$

47. Verify that $1.7(A+B) = 1.7A+1.7B$.

48. Verify that $(6.3 \cdot 1.2)A = 6.3(1.2A)$.

Applications

49. Kirzan Auto has the following data for sales of the popular Blink sedan during June, July, and August. The data are from Kirzan's Northside dealership. (Manual and automatic refer to transmission type.)

June Sales

	Manual	Automatic
Two door	13	7
Four door	5	12

July Sales

	Manual	Automatic
Two door	12	9
Four door	7	14

August Sales

	Manual	Automatic
Two door	14	11
Four door	3	19

(a) If the rows refer to the sedan type (two door/four door) and the columns refer to the transmission type (manual/automatic), write three matrices to represent the data. Let J, Y, and A be the matrices for June, July, and August, respectively.

(b) Determine a matrix giving the total sales for Kirzan's Northside dealership during June, July, and August.

50. Kirzan Auto has the following data for sales of the popular Blink sedan during June, July, and August. The data are from Kirzan's Southside dealership. (Manual and automatic refer to transmission type.)

June Sales

	Manual	Automatic
Two door	11	8
Four door	6	9

July Sales

	Manual	Automatic
Two door	15	6
Four door	7	11

August Sales

	Manual	Automatic
Two door	16	13
Four door	2	13

(a) If the rows refer to the sedan type (two door/four door) and the columns refer to the transmission type (manual/automatic), write three matrices to represent the data. Let J, Y, and A be the matrices for June, July, and August, respectively.

(b) Determine a matrix giving the total sales for Kirzan's Southside dealership during June, July, and August.

51. For J, Y, and A in Exercise 49, compute $\frac{1}{3}(J + Y + A)$ and interpret.

52. For J, Y, and A in Exercise 50, compute $\frac{1}{3}(J + Y + A)$ and interpret.

53. The Basich Company has two plants, one on the East Coast and one on the West Coast, that manufacture 2D and 3D graphing calculators. The production costs, in dollars, for each calculator are summarized in the following matrices:

$$E = \begin{array}{c} \\ \end{array} \begin{array}{cc} \text{2D} & \text{3D} \\ \begin{bmatrix} 25 & 35 \\ 10 & 25 \end{bmatrix} & \begin{array}{l} \text{Materials} \\ \text{Labor} \end{array} \end{array}$$
East Coast

$$W = \begin{array}{cc} \text{2D} & \text{3D} \\ \begin{bmatrix} 28 & 39 \\ 15 & 27 \end{bmatrix} & \begin{array}{l} \text{Materials} \\ \text{Labor} \end{array} \end{array}$$
West Coast

(a) Find and interpret $\frac{1}{2}(E + W)$.

(b) If both labor and materials at the East Coast plant increase by 10% and both labor and materials at the West Coast plant are decreased by 5%, determine new matrices to represent the new production costs.

54. Shawna Wells has two types of IRA retirement accounts, certificates of deposit (CD) and Grade A municipal bonds (GMB), held in two banks, First Century (FC) and Money National (MN). The dollar amounts in each at the beginning of 2002 are given in matrix A.

$$A = \begin{array}{cc} \text{CD} & \text{GMB} \\ \begin{bmatrix} 15{,}000 & 25{,}000 \\ 40{,}000 & 35{,}000 \end{bmatrix} & \begin{array}{l} \text{FC} \\ \text{MN} \end{array} \end{array}$$

(a) Assuming a 7% growth rate on each account, determine a matrix I giving the earnings on the various accounts during 2002.

(b) Determine a matrix that represents the value of each account at the end of 2002.

55. Nick Dominic has two types of IRA retirement accounts, treasury notes (TN) and municipal bonds (MB), held in two banks, Valley Bank (VB) and Hilltop Bank (HB). The dollar amounts in each at the beginning of 2002 are given in matrix A.

$$A = \begin{array}{cc} \text{TN} & \text{MB} \\ \begin{bmatrix} 20{,}000 & 25{,}000 \\ 15{,}000 & 35{,}000 \end{bmatrix} & \begin{array}{l} \text{VB} \\ \text{HB} \end{array} \end{array}$$

(a) Assuming an 8% growth rate on each account, determine a matrix I giving the earnings on the various accounts during 2002.

(b) Determine a matrix that represents the value of each account at the end of 2002.

56. The hospital care and physician services expenditures, in billions of dollars, by patients (out of pocket) and Medicare are given in Table 2.2.4.

Table 2.2.4

	Hospital Care		Physician Services	
	Out of pocket	Medicare	Out of pocket	Medicare
1980	5.3	26.4	14.7	8.0
1990	11.1	68.7	32.2	29.2
1995	11.5	108.9	30.0	39.9
1996	11.8	116.3	31.1	42.8
1997	12.4	123.7	34.1	46.4

SOURCE: U.S. Census Bureau, www.census.gov

The data for hospital care are summarized in matrix H.

$$H = \begin{array}{c} \begin{array}{cc} \text{Out of} & \\ \text{pocket} & \text{Medicare} \end{array} \\ \begin{bmatrix} 5.3 & 26.4 \\ 11.1 & 68.7 \\ 11.5 & 108.9 \\ 11.8 & 116.3 \\ 12.4 & 123.7 \end{bmatrix} \begin{array}{l} 1980 \\ 1990 \\ 1995 \\ 1996 \\ 1997 \end{array} \end{array}$$

(a) Write a matrix M to summarize the data for physician services.

(b) Find $H + M$ and interpret the entries.

57. The number of people, in millions, covered by private or government health insurance is given in Table 2.2.5.

Table 2.2.5

Year	Private	Government	
		Medicare	Medicaid
1990	182.1	32.3	24.3
1992	181.5	33.2	29.4
1993	182.4	33.1	31.7
1994	184.3	33.9	31.6
1995	185.9	34.7	31.9
1996	187.4	35.2	31.5
1997	188.5	35.6	29.0

SOURCE: U.S. Census Bureau, www.census.gov

The data for people with private insurance is summarized in matrix P.

$$P = \begin{bmatrix} 182.1 \\ 181.5 \\ 182.4 \\ 184.3 \\ 185.9 \\ 187.4 \\ 188.5 \end{bmatrix} \begin{array}{l} 1990 \\ 1992 \\ 1993 \\ 1994 \\ 1995 \\ 1996 \\ 1997 \end{array}$$

(a) Write a matrix MCR to summarize the data for people with Medicare.

(b) Write a matrix MCD to summarize the data for people with Medicaid.

(c) Let $TG = MCR + MCD$. Determine TG and interpret the entries.

58. (*continuation of Exercise 57*)

(a) Let $TC = TG + P$. Matrix TC gives the total covered by private or government insurance.

(b) Matrix F gives the U.S. population in millions (*Source:* U.S. Census Bureau, www.census.gov).

Population

$$F = \begin{bmatrix} 248.9 \\ 256.8 \\ 259.8 \\ 262.1 \\ 264.3 \\ 266.8 \\ 269.1 \end{bmatrix} \begin{matrix} 1990 \\ 1992 \\ 1993 \\ 1994 \\ 1995 \\ 1996 \\ 1997 \end{matrix}$$

Compute $F - TC$ and interpret the entries.

59. Matrix M gives the percentage of U.S. male smokers, and matrix F gives the percentage of U.S. female smokers (*Source:* U.S. Census Bureau, www.census.gov).

$$\begin{matrix} & \text{18–24} & \text{25–34} & \text{35–44} & \text{45–64} & \text{65 and over} \end{matrix}$$
$$M = \begin{bmatrix} 28.0 & 38.2 & 37.6 & 33.4 & 19.6 \\ 26.6 & 31.6 & 34.5 & 29.3 & 14.6 \\ 29.8 & 31.4 & 33.2 & 28.3 & 13.2 \end{bmatrix} \begin{matrix} 1985 \\ 1990 \\ 1994 \end{matrix}$$

$$\begin{matrix} & \text{18–24} & \text{25–34} & \text{35–44} & \text{45–64} & \text{65 and over} \end{matrix}$$
$$F = \begin{bmatrix} 30.4 & 32.0 & 31.5 & 29.9 & 13.5 \\ 22.5 & 28.2 & 24.8 & 24.8 & 11.5 \\ 25.2 & 28.8 & 26.8 & 22.8 & 11.1 \end{bmatrix} \begin{matrix} 1985 \\ 1990 \\ 1994 \end{matrix}$$

(a) Interpret the meaning of entry $m_{3,2}$ and entry $f_{3,2}$.

(b) Using M and F, write a matrix in which each entry shows how much more (or less) smoking is done by males than females.

60. The data in Table 2.2.6 give the educational attainment in the United States of people of Hispanic origin 25 years old and over (entries are measured in percentages). HS means high school and CG means college.

Table 2.2.6

	Puerto Rican		Cuban	
	4 yr HS	**4 yr CG or more**	**4 yr HS**	**4 yr CG or more**
1990	55.5	9.7	63.5	20.2
1995	61.3	10.7	64.7	19.4
1996	60.4	11.0	63.8	18.8
1997	61.1	10.7	65.2	19.7
1998	63.8	11.9	67.8	22.2

SOURCE: U.S. Census Bureau, www.census.gov

(a) Write a matrix P for the educational attainment of Puerto Ricans and a matrix C for the educational attainment of Cubans.

(b) Use the matrices from part (a) to write a matrix in which each entry shows how much more (or less) education Cubans have attained than Puerto Ricans.

61. The data in Table 2.2.7 give the percentage of U.S. students in grades 9–12 that participated on a sports team in 1997.

Table 2.2.7

	Male		Female	
Run by:	**School**	**Other organization**	**School**	**Other organization**
Grade 9	57.2	51.3	48.5	36.8
Grade 10	58.0	47.3	45.0	34.7
Grade 11	54.0	41.6	40.7	26.4
Grade 12	53.4	42.6	35.7	21.9

SOURCE: U.S. Census Bureau, www.census.gov

(a) Write a 4×2 matrix M for the participation of males on a sports team.

(b) Write a 4×2 matrix F for the participation of females on a sports team.

(c) Use the matrices from parts (a) and (b) to write a matrix in which each entry shows how much more (or less) participation on sports teams males have than females.

SECTION PROJECT

In the text two Interactive Activities illustrated that matrix addition is commutative and associative. In this project you will discover and verify another property of matrices. Given matrices A and B, do the following:

$$A = \begin{bmatrix} -2 & 3 \\ 4 & 7 \end{bmatrix} \quad B = \begin{bmatrix} 8 & -1 \\ 2 & 3 \end{bmatrix}$$

(a) $4A$

(b) $4B$

(c) $4A + 4B$

(d) $A + B$

(e) $4(A + B)$

(f) Based on the results to parts (c) and (e), do you think scalar multiplication is distributive?

(g) To verify that scalar multiplication is distributive in the general case, consider

$$A=\begin{bmatrix} a & b \\ c & d \end{bmatrix} \qquad B=\begin{bmatrix} e & f \\ g & h \end{bmatrix}$$

and real number k. Also, a, b, c, d, e, f, g, and h are real numbers. To show that

$$k(A+B)=kA+kB$$

determine a matrix for $k(A+B)$ and a matrix for $kA+kB$. Then use the equality of matrices definition to verify that the matrices are equal.

Section 2.3 Matrix Multiplication

In Section 2.2 we saw how to do scalar multiplication, that is, multiplying a real number k times a matrix. In this section we learn how to multiply two matrices. **Matrix multiplication** has several applications, as we will see in this section. Matrix multiplication will also reappear later in this chapter, as well as in later chapters.

Matrix Multiplication

Multiplying two matrices is more involved than scalar multiplication. To understand the meaning behind matrix multiplication, it may be helpful to take a Flashback to the Staminly's furniture store from Section 2.2.

Flashback

Staminly's Furniture Revisited

In Section 2.2 we were introduced to Staminly's and its three furniture stores. In Example 2 from Section 2.2, we found a matrix T that gave the total number of two-cushion and three-cushion Country Style Contemporary sofas (of lengths 60″, 72″, and 84″) sold in July for all three stores. This matrix is

$$\begin{matrix} & \begin{matrix} 60'' & 72'' & 84'' \end{matrix} & \\ T=\!\!\! & \begin{bmatrix} 25 & 31 & 21 \\ 21 & 33 & 19 \end{bmatrix} & \begin{matrix} \text{Two cushion} \\ \text{Three cushion} \end{matrix} \end{matrix}$$

From matrix T, write a matrix A for the total number of two-cushion models of each length that were sold and a matrix B for the total number of three-cushion models of each length that were sold.

Flashback Solution

The total number of two-cushion models sold is given by the matrix

$$\begin{matrix} & \begin{matrix} 60'' & 72'' & 84'' \end{matrix} \\ A= & [25 \quad 31 \quad 21] \end{matrix} \qquad \text{(a } 1\times 3 \text{ row matrix)}$$

and the total number of three-cushion models sold is given by the matrix

$$\begin{matrix} & \begin{matrix} 60'' & 72'' & 84'' \end{matrix} \\ B= & [21 \quad 33 \quad 19] \end{matrix} \qquad \text{(a } 1\times 3 \text{ row matrix)}$$

Now suppose that the selling price of the two-cushion models of lengths 60″, 72″, and 84″ is \$600, \$800, and \$1100, respectively. We can represent this information in the following column matrix, P:

$$P = \begin{bmatrix} 600 \\ 800 \\ 1100 \end{bmatrix} \begin{matrix} 60'' \\ 72'' \\ 84'' \end{matrix} \qquad \text{(a } 3 \times 1 \text{ column matrix)}$$

The first entry in the row matrix A in the Flashback gives the number of 60″ two-cushion sofas sold, and the first entry in the column matrix P gives the selling price for each 60″ two-cushion sofa. The product of these entries, $25 \cdot 600$, gives the revenue generated from the sale of the 60″ two-cushion sofas in July. Similarly, the second entry in matrix A multiplied by the second entry in matrix P gives the revenue realized from the sale of 72″ two-cushion sofas. Also, the third entry in matrix A multiplied by the third entry in P yields the revenue made from the sale of 84″ two-cushion sofas. So the revenue generated for each type of two-cushion sofa is

$$\begin{aligned} 60'':&\quad 25 \cdot 600 = \$15{,}000 \\ 72'':&\quad 31 \cdot 800 = \$24{,}800 \\ 84'':&\quad 21 \cdot 1100 = \$23{,}100 \end{aligned}$$

The total revenue for July for all three types of two-cushion sofas is the sum of these products, which gives us

$$(25 \cdot 600) + (31 \cdot 800) + (21 \cdot 1100) = \$62{,}900$$

Notice that the revenue, \$62,900, was simply found by taking the entries in matrix A and multiplying them by the entries in matrix P and then adding the products. We have actually computed the following matrix multiplication:

$$\begin{aligned} A \cdot P &= [25 \quad 31 \quad 21] \cdot \begin{bmatrix} 600 \\ 800 \\ 1100 \end{bmatrix} \\ &= (25 \cdot 600) + (31 \cdot 800) + (21 \cdot 1100) = 62{,}900 \end{aligned}$$

This suggests that if we have a **row matrix** A with dimension $1 \times n$,

$$A = [a_1 \quad a_2 \quad a_3 \quad a_4 \quad \ldots \quad a_n]$$

and a **column matrix** B with dimension $n \times 1$,

$$B = \begin{bmatrix} b_1 \\ b_2 \\ b_3 \\ b_4 \\ \vdots \\ b_n \end{bmatrix}$$

then the **matrix product** of A and B, AB, is given by

$$\begin{aligned} AB &= [a_1 \quad a_2 \quad a_3 \quad a_4 \quad \ldots \quad a_n] \begin{bmatrix} b_1 \\ b_2 \\ b_3 \\ b_4 \\ \vdots \\ b_n \end{bmatrix} \\ &= a_1 b_1 + a_2 b_2 + a_3 b_3 + a_4 b_4 + \cdots + a_n b_n \end{aligned}$$

Example 1 Multiplying a Row Matrix and a Column Matrix

The total number of three-cushion Country Style Contemporary sofas (lengths 60″, 72″, and 84″) sold in July by Staminly's is given by matrix B, and the selling price for each of these sofas is given by matrix C.

$$\begin{array}{c} \quad\;\; 60'' \;\; 72'' \;\; 84'' \\ B = [21 \quad 33 \quad 19] \end{array} \qquad C = \begin{bmatrix} 650 \\ 850 \\ 1200 \end{bmatrix} \begin{array}{l} 60'' \\ 72'' \\ 84'' \end{array}$$

Determine the total revenue for July for all three types of three-cushion sofas.

Solution

The total revenue is given by

$$BC = [21 \quad 33 \quad 19]\begin{bmatrix} 650 \\ 850 \\ 1200 \end{bmatrix} = 21(650) + 33(850) + 19(1200) = 64{,}500$$

So the total revenue for July for all three types of three-cushion sofas is \$64,500. ■

If we look closer at the matrix product BC in Example 1, we notice that the number of columns of matrix B is equal to the number of rows of matrix C. Also, the matrix product BC has dimension 1×1 (we can think of a real number as being a 1×1 matrix). See Figure 2.3.1. In general, if matrix A has dimension $m \times n$ and matrix B has dimension $n \times p$, then the matrix product of A and B, AB, is defined and has dimension $m \times p$. See Figure 2.3.2.

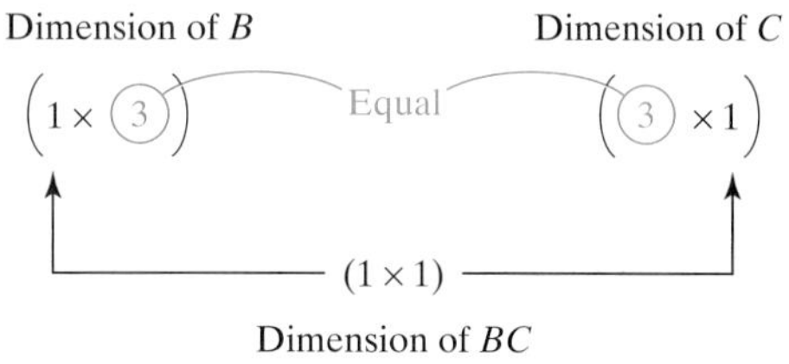

Figure 2.3.1

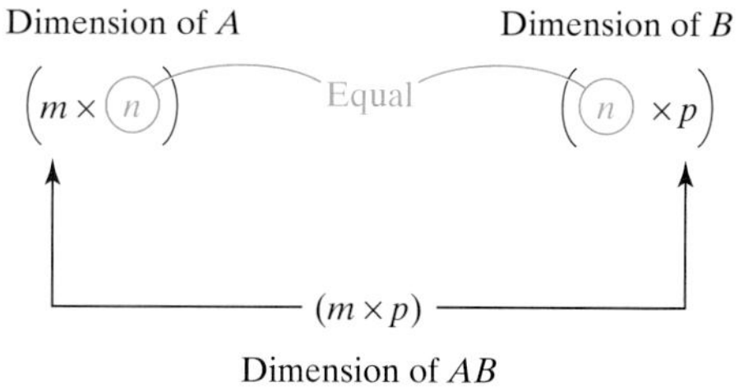

Figure 2.3.2

Example 2 Determining the Dimension of a Matrix Product

Given matrices A, B, and C, determine if the following matrix products are defined. If so, give the dimension of the matrix product.

$$A = \begin{bmatrix} 2 & 3 & -1 & 7 \\ 5 & 6 & 2 & 1 \end{bmatrix} \qquad B = \begin{bmatrix} 1 & 3 \\ -1 & 4 \\ 2 & -2 \end{bmatrix} \qquad C = \begin{bmatrix} 2 & 1 \\ -1 & 3 \\ 4 & -2 \\ 7 & -3 \end{bmatrix}$$

(a) AB **(b)** BA **(c)** AC

Interactive Activity

At this time we do not have a procedure to compute matrix products. However, based on Example 2(a) and (b), do you believe matrix multiplication is commutative? That is, do you think that $AB = BA$ for any matrices A and B?

Solution

(a) A is a 2×4 matrix and B is a 3×2 matrix. Since the number of columns of A *does not equal* the number of rows of B, the matrix product AB does not exist.

(b) B is a 3×2 matrix and A is a 2×4 matrix. Since the number of columns of B *does equal* the number of rows of A, the matrix product BA is defined and would yield a 3×4 matrix.

(c) A is a 2×4 matrix and C is a 4×2 matrix. As in part (b), we see that the matrix product AC is defined and would yield a 2×2 matrix. ■

Checkpoint 1 Now work Exercise 5.

Example 2(b) indicates that BA is defined and yields a 3×4 matrix. If we call the matrix product D, we have

$$BA = D = \begin{bmatrix} d_{1,1} & d_{1,2} & d_{1,3} & d_{1,4} \\ d_{2,1} & d_{2,2} & d_{2,3} & d_{2,4} \\ d_{3,1} & d_{3,2} & d_{3,3} & d_{3,4} \end{bmatrix}$$

$$\begin{bmatrix} 1 & 3 \\ -1 & 4 \\ 2 & -2 \end{bmatrix}\begin{bmatrix} 2 & 3 & -1 & 7 \\ 5 & 6 & 2 & 1 \end{bmatrix} = D = \begin{bmatrix} d_{1,1} & d_{1,2} & d_{1,3} & d_{1,4} \\ d_{2,1} & d_{2,2} & d_{2,3} & d_{2,4} \\ d_{3,1} & d_{3,2} & d_{3,3} & d_{3,4} \end{bmatrix}$$

To compute the entries of D, we do the following:

$$d_{1,1} = [\,b_{1,1} \quad b_{1,2}\,]\begin{bmatrix} a_{1,1} \\ a_{2,1} \end{bmatrix} = [1 \quad 3]\begin{bmatrix} 2 \\ 5 \end{bmatrix} = 1(2) + 3(5) = 17$$

$$d_{1,2} = [\,b_{1,1} \quad b_{1,2}\,]\begin{bmatrix} a_{1,2} \\ a_{2,2} \end{bmatrix} = [1 \quad 3]\begin{bmatrix} 3 \\ 6 \end{bmatrix} = 1(3) + 3(6) = 21$$

The other entries in D are computed in a similar fashion. Matrix products are summarized in the following definition.

Matrix Product

If A is an $m \times n$ matrix and B is an $n \times p$ matrix, then the **matrix product**, AB, is an $m \times p$ matrix. To find the entry of the ith row and jth column of AB, multiply each entry in the ith row of A by the corresponding entry in the jth column of B and add the products.

Note: Observe that determining the entries in a matrix product is analogous to taking a row matrix times a column matrix, as we did in Example 1.

Example 3 Computing a Matrix Product

Determine AB given

$$A = \begin{bmatrix} 2 & -1 & -3 \\ 4 & 1 & 6 \end{bmatrix} \qquad B = \begin{bmatrix} 3 & 1 \\ -1 & 4 \\ 5 & 6 \end{bmatrix}$$

Solution

The first thing that we should always check is whether the matrix product is defined. Since A is a 2×3 and B is a 3×2, the matrix product is defined and yields a 2×2 matrix. If we call the matrix product C, we have

$$AB = C$$

$$\begin{bmatrix} 2 & -1 & -3 \\ 4 & 1 & 6 \end{bmatrix} \begin{bmatrix} 3 & 1 \\ -1 & 4 \\ 5 & 6 \end{bmatrix} = C = \begin{bmatrix} c_{1,1} & c_{1,2} \\ c_{2,1} & c_{2,2} \end{bmatrix}$$

$c_{1,1}$ is found by taking the 1st row of A and multiplying by the 1st column of B.

$$c_{1,1} = [2 \quad -1 \quad -3] \begin{bmatrix} 3 \\ -1 \\ 5 \end{bmatrix} = 2(3) + -1(-1) + -3(5) = -8$$

$c_{1,2}$ is found by taking the 1st row of A and multiplying by the 2nd column of B.

$$c_{1,2} = [2 \quad -1 \quad -3] \begin{bmatrix} 1 \\ 4 \\ 6 \end{bmatrix} = 2(1) + -1(4) + -3(6) = -20$$

$c_{2,1}$ is found by taking the 2nd row of A and multiplying by the 1st column of B.

$$c_{2,1} = [4 \quad 1 \quad 6] \begin{bmatrix} 3 \\ -1 \\ 5 \end{bmatrix} = 4(3) + 1(-1) + 6(5) = 41$$

$c_{2,2}$ is found by taking the 2nd row of A and multiplying by the 2nd column of B.

$$c_{2,2} = [4 \quad 1 \quad 6] \begin{bmatrix} 1 \\ 4 \\ 6 \end{bmatrix} = 4(1) + 1(4) + 6(6) = 44$$

So we have

$$AB = \begin{bmatrix} -8 & -20 \\ 41 & 44 \end{bmatrix}$$

■

Technology Option

Figure 2.3.3

Figure 2.3.3 shows the matrix product from Example 3 on a graphing calculator. Many calculators perform matrix multiplication. To learn how your calculator multiplies matrices, consult the online graphing calculator manual at www.prenhall.com/armstrong

✔ Checkpoint 2

Now work Exercise 15.

Example 4 Computing a Matrix Product

Determine the matrix products AB and BA for the matrices

$$A = \begin{bmatrix} 2 & 3 & 1 \\ -1 & 2 & 3 \\ 3 & 4 & 1 \end{bmatrix} \qquad B = \begin{bmatrix} 4 & -1 & 3 \\ 2 & 1 & 4 \\ 1 & 3 & 2 \end{bmatrix}$$

Solution

We have

$$AB = \begin{bmatrix} 2 & 3 & 1 \\ -1 & 2 & 3 \\ 3 & 4 & 1 \end{bmatrix} \begin{bmatrix} 4 & -1 & 3 \\ 2 & 1 & 4 \\ 1 & 3 & 2 \end{bmatrix}$$

$$= \begin{bmatrix} 2(4)+3(2)+1(1) & 2(-1)+3(1)+1(3) & 2(3)+3(4)+1(2) \\ -1(4)+2(2)+3(1) & -1(-1)+2(1)+3(3) & -1(3)+2(4)+3(2) \\ 3(4)+4(2)+1(1) & 3(-1)+4(1)+1(3) & 3(3)+4(4)+1(2) \end{bmatrix}$$

$$= \begin{bmatrix} 15 & 4 & 20 \\ 3 & 12 & 11 \\ 21 & 4 & 27 \end{bmatrix}$$

For BA we have

$$BA = \begin{bmatrix} 4 & -1 & 3 \\ 2 & 1 & 4 \\ 1 & 3 & 2 \end{bmatrix} \begin{bmatrix} 2 & 3 & 1 \\ -1 & 2 & 3 \\ 3 & 4 & 1 \end{bmatrix}$$

Interactive Activity

By hand, perform the details of computing BA in Example 4. Check AB and BA using a graphing calculator.

WWW

We save the details of this multiplication for you in the Interactive Activity to the left. The matrix product is

$$BA = \begin{bmatrix} 18 & 22 & 4 \\ 15 & 24 & 9 \\ 5 & 17 & 12 \end{bmatrix}$$

■

Example 5 Applying Matrix Multiplication

Leslie, the director of KiddieTime a local day-care center, needs to get the monthly billing under control. Children at the center are there either 2, 4, or 8 hours. Some attend daily, for different hours each day, and some are there only certain days of the week. To further complicate matters, months do not have the same number of Mondays, Tuesdays, and so on, and this is further complicated for months with holidays. Leslie's task of creating a monthly billing system must account for the individual schedule of each child, the number of Mondays, Tuesdays, and so on, in a given month, and the cost per hour for day care. (The latter will probably rise over time.) Leslie believes that matrices can help her, and she selected three children and their schedules for the month of November 2000 to see. Matrix C is the children's daily schedule in which the entries represent the number of hours spent in day care for a particular day. Matrix M gives the number of Mondays, Tuesdays, and so on, that the day care was open in November

2000. Currently, the cost is \$4 per hour. How can Leslie use this information to create a billing system?

$$C = \begin{array}{c} \begin{matrix} \text{M} & \text{T} & \text{W} & \text{Th} & \text{F} \end{matrix} \\ \begin{bmatrix} 4 & 0 & 4 & 0 & 2 \\ 8 & 4 & 8 & 4 & 2 \\ 0 & 8 & 2 & 8 & 0 \end{bmatrix} \end{array} \begin{matrix} \text{Don} \\ \text{Ron} \\ \text{Sarah} \end{matrix} \qquad M = \begin{bmatrix} 4 \\ 4 \\ 5 \\ 4 \\ 3 \end{bmatrix} \begin{matrix} \text{M} \\ \text{T} \\ \text{W} \\ \text{Th} \\ \text{F} \end{matrix}$$

Solution

Understand the Situation: Leslie needs to use the given information and determine the total number of hours that each child was in day care. Then she can multiply by \$4 to find the bill for each child.

The matrix product CM will give the total hours in November 2000 for each child.

$$CM = \begin{bmatrix} 4 & 0 & 4 & 0 & 2 \\ 8 & 4 & 8 & 4 & 2 \\ 0 & 8 & 2 & 8 & 0 \end{bmatrix} \begin{bmatrix} 4 \\ 4 \\ 5 \\ 4 \\ 3 \end{bmatrix}$$

$$= \begin{bmatrix} 4(4) + 0(4) + 4(5) + 0(4) + 2(3) \\ 8(4) + 4(4) + 8(5) + 4(4) + 2(3) \\ 0(4) + 8(4) + 2(5) + 8(4) + 0(3) \end{bmatrix} = \begin{bmatrix} 42 \\ 110 \\ 74 \end{bmatrix} \begin{matrix} \text{Don} \\ \text{Ron} \\ \text{Sarah} \end{matrix}$$

To determine the November 2000 bill for each child, we simply multiply the matrix product CM by \$4. This gives

$$4CM = 4 \begin{bmatrix} 42 \\ 110 \\ 74 \end{bmatrix} = \begin{bmatrix} 168 \\ 440 \\ 296 \end{bmatrix} \begin{matrix} \text{Don} \\ \text{Ron} \\ \text{Sarah} \end{matrix}$$

Interpret the Solution: For November 2000, Don owes \$168, Ron owes \$440, and Sarah owes \$296. It appears that Leslie can enter into matrix C all the children in the day-care center and their daily hours of attendance. In matrix M she can put the number of Mondays, Tuesdays, and so on, that the center is open. The hourly rate multiplied by CM will produce the monthly bill for each child. ■

In our next example we compute a matrix product in which some entries are meaningful while other entries are meaningless. It is important to always remember the definition of matrix multiplication so that we know when a matrix product has meaning in an application.

Example 6 Applying Matrix Multiplication

Recall that for Staminly's three furniture stores, the total sales in July for its two- and three-cushion Country Style Contemporary sofa (lengths of 60″, 72″, and 84″) are represented in matrix T.

$$T = \begin{array}{c} \begin{matrix} 60'' & 72'' & 84'' \end{matrix} \\ \begin{bmatrix} 25 & 31 & 21 \\ 21 & 33 & 19 \end{bmatrix} \end{array} \begin{matrix} \text{Two cushion} \\ \text{Three cushion} \end{matrix}$$

The selling price for each is given in matrix S.

$$S = \begin{array}{c} \\ \\ \end{array}\begin{matrix} \text{Two cushion} & \text{Three cushion} \end{matrix}$$

$$S = \begin{bmatrix} 600 & 650 \\ 800 & 850 \\ 1100 & 1200 \end{bmatrix} \begin{matrix} 60'' \\ 72'' \\ 84'' \end{matrix}$$

Compute TS and interpret the entries.

Solution

The matrix product TS is

$$TS = \begin{bmatrix} 25 & 31 & 21 \\ 21 & 33 & 19 \end{bmatrix} \begin{bmatrix} 600 & 650 \\ 800 & 850 \\ 1100 & 1200 \end{bmatrix}$$

$$= \begin{bmatrix} 62{,}900 & 67{,}800 \\ 59{,}900 & 64{,}500 \end{bmatrix}$$

Interpret the Solution: $ts_{1,1}$ gives the total revenue for the three lengths of the two-cushion model to be \$62,900. Now $ts_{1,2}$ is the product of the number of two-cushion models sold times the prices for the three-cushion models; hence this entry has no meaning. For the same reason, $ts_{2,1}$ is also meaningless. Finally, $ts_{2,2}$ gives the total revenue for the three lengths of the three-cushion model to be \$64,500. ■

Technology Option

Figure 2.3.4

In Example 6, if we use a graphing calculator to do the matrix multiplication, with matrix T called A and matrix S called B, we get the result in Figure 2.3.4.

Whether matrix multiplication is done by hand or on a graphing calculator, it is imperative to know the definition of matrix multiplication so that we can interpret the entries meaningfully!

Representing Systems with a Matrix Equation

Our final topic in this section is to lay the groundwork for Section 2.4. In Example 7 we show how to use matrices to rewrite a system of linear equations in a more compact form called a **matrix equation**.

Example 7 Writing a Matrix Equation for a System of Equations

Write the following system of linear equations as a matrix equation.

$$\begin{aligned} 2x + y + 3z &= 4 \\ x - 2y + z &= -3 \\ 4x - y - z &= -1 \end{aligned}$$

Solution

We start by writing

$$A = \begin{bmatrix} 2 & 1 & 3 \\ 1 & -2 & 1 \\ 4 & -1 & -1 \end{bmatrix} \quad X = \begin{bmatrix} x \\ y \\ z \end{bmatrix} \quad B = \begin{bmatrix} 4 \\ -3 \\ -1 \end{bmatrix}$$

Observe that A is a 3×3 matrix of coefficients, X is a 3×1 column matrix of the variables, and B is a 3×1 column matrix of the constants. The matrix equation for this system has the form

$$AX = B$$

$$\begin{bmatrix} 2 & 1 & 3 \\ 1 & -2 & 1 \\ 4 & -1 & -1 \end{bmatrix} \begin{bmatrix} x \\ y \\ z \end{bmatrix} = \begin{bmatrix} 4 \\ -3 \\ -1 \end{bmatrix}$$

To verify that this is true, we perform the matrix product AX and get

$$AX = \begin{bmatrix} 2 & 1 & 3 \\ 1 & -2 & 1 \\ 4 & -1 & -1 \end{bmatrix} \begin{bmatrix} x \\ y \\ z \end{bmatrix} = \begin{bmatrix} 2x + y + 3z \\ x - 2y + z \\ 4x - y - z \end{bmatrix}$$

Equating this product with matrix B gives

$$\begin{bmatrix} 2x + y + 3z \\ x - 2y + z \\ 4x - y - z \end{bmatrix} = \begin{bmatrix} 4 \\ -3 \\ -1 \end{bmatrix}$$

which via matrix equality is obviously the same as the original system. ■

Checkpoint 3 Now work Exercise 39.

SUMMARY

The section started with multiplying a row matrix by a column matrix, and that lead us to defining **matrix multiplication** in general. It is important to know when matrix multiplication is defined and the dimension of the resulting product. For example, if A is $m \times n$ and B is $n \times p$, then the product AB is defined and has dimension $m \times p$, whereas the product BA is not defined. We saw several applications of matrix multiplication, and the section concluded with a preview of the next section: representing systems of linear equations with a matrix equation.

SECTION 2.3 EXERCISES

In Exercises 1–8, the dimension of matrices A and B are given.

(a) Find the dimension of the matrix product AB, if it exists.

(b) Find the dimension of the matrix product BA, if it exists.

1. A is a 3×2 and B is a 2×3

2. A is a 2×4 and B is a 4×2

3. A is a 2×4 and B is a 4×2

4. A is a 5×2 and B is a 2×3

5. A is a 3×4 and B is a 5×3

6. A is a 3×2 and B is a 4×3

7. A is a 5×2 and B is a 3×3

8. A is a 6×2 and B is a 4×2

In Exercises 9–24, find the matrix product, if it exists.

9. $\begin{bmatrix} 3 & 1 \\ 4 & 2 \end{bmatrix} \begin{bmatrix} -2 \\ 3 \end{bmatrix}$

10. $\begin{bmatrix} -3 & 2 \\ 4 & 1 \end{bmatrix} \begin{bmatrix} 2 \\ 3 \end{bmatrix}$

11. $\begin{bmatrix} 1 & 3 & -2 \\ -1 & 2 & 4 \end{bmatrix} \begin{bmatrix} -2 \\ 2 \\ 1 \end{bmatrix}$

12. $\begin{bmatrix} 6 & -1 & 2 \\ 3 & 4 & 1 \end{bmatrix}\begin{bmatrix} 0 \\ 4 \\ 5 \end{bmatrix}$

13. $\begin{bmatrix} 2 & -2 \\ 1 & -3 \end{bmatrix}\begin{bmatrix} 4 & -3 \\ -1 & 2 \end{bmatrix}$

14. $\begin{bmatrix} -1 & 2 \\ 3 & -4 \end{bmatrix}\begin{bmatrix} -2 & 3 \\ -1 & 5 \end{bmatrix}$

15. $\begin{bmatrix} 3 & 2 & 0 \\ -1 & -2 & 4 \end{bmatrix}\begin{bmatrix} 6 & 1 \\ 5 & -1 \\ 2 & 4 \end{bmatrix}$

16. $\begin{bmatrix} 1 & 6 \\ -1 & -5 \\ 2 & -4 \end{bmatrix}\begin{bmatrix} -3 & 1 & -1 \\ 1 & 4 & -2 \end{bmatrix}$

17. $\begin{bmatrix} 2 & -3 \\ 4 & 1 \\ 0 & -4 \\ -1 & 5 \end{bmatrix}\begin{bmatrix} 2 \\ 1 \\ 4 \\ 3 \end{bmatrix}$

18. $\begin{bmatrix} -1 & 3 & 2 & -4 \end{bmatrix}\begin{bmatrix} -1 & 1 \\ 2 & 3 \end{bmatrix}$

19. $\begin{bmatrix} 5 & -6 \\ -2 & 1 \end{bmatrix}\begin{bmatrix} 1 & 0 \\ 0 & 1 \end{bmatrix}$

20. $\begin{bmatrix} 1 & 0 \\ 0 & 1 \end{bmatrix}\begin{bmatrix} 5 & -6 \\ -2 & 1 \end{bmatrix}$

21. $\begin{bmatrix} 6 & -1 & 4 \\ 0 & 3 & 2 \\ 1 & 4 & 5 \end{bmatrix}\begin{bmatrix} 1 & 0 & 0 \\ 0 & 1 & 0 \\ 0 & 0 & 1 \end{bmatrix}$

22. $\begin{bmatrix} 1 & 0 & 0 \\ 0 & 1 & 0 \\ 0 & 0 & 1 \end{bmatrix}\begin{bmatrix} 6 & -1 & 4 \\ 0 & 3 & 2 \\ 1 & 4 & 5 \end{bmatrix}$

23. $\begin{bmatrix} 10 & 20 & 15 & 5 \\ 6 & 8 & 6 & 4 \\ 5 & 5 & 15 & 30 \\ 20 & 18 & 22 & 16 \end{bmatrix}\begin{bmatrix} 5 \\ 8 \\ 11 \\ 13 \end{bmatrix}$

24. $\begin{bmatrix} 6 & 9 & 18 \\ 15 & 12 & 10 \\ 8 & 6 & 9 \end{bmatrix}\begin{bmatrix} 4 \\ 6 \\ 7 \end{bmatrix}$

In Exercises 25–32, use a graphing calculator to find the matrix products, if possible.

25. $\begin{bmatrix} 1.3 & 1.4 \\ 2.8 & 3.2 \end{bmatrix}\begin{bmatrix} 5.4 & 6.1 \\ 8.9 & 3.5 \end{bmatrix}$

26. $\begin{bmatrix} 0.7 & 1.3 \\ 0.4 & 9.1 \end{bmatrix}\begin{bmatrix} 8.4 & 7.3 \\ 6.1 & 5.9 \end{bmatrix}$

27. $\begin{bmatrix} \frac{4}{11} & -\frac{3}{22} \\ -\frac{3}{11} & \frac{5}{22} \end{bmatrix}\begin{bmatrix} 5 & 3 \\ 6 & 8 \end{bmatrix}$

28. $\begin{bmatrix} -1.5 & 2 \\ -0.5 & 1 \end{bmatrix}\begin{bmatrix} -2 & 4 \\ -1 & 3 \end{bmatrix}$

29. $\begin{bmatrix} 21 & 33 & 58 \\ 17 & 25 & 13 \\ 42 & 20 & 37 \end{bmatrix}\begin{bmatrix} 700 \\ 800 \\ 900 \end{bmatrix}$

30. $\begin{bmatrix} 19 & 17 & 24 \\ 64 & 33 & 36 \\ 56 & 90 & 18 \end{bmatrix}\begin{bmatrix} 20 \\ 50 \\ 82 \end{bmatrix}$

31. $\begin{bmatrix} 2 & 3 & 2 \\ -1 & 5 & 6 \\ 2 & -3 & 4 \end{bmatrix}\begin{bmatrix} \frac{19}{55} & -\frac{9}{55} & \frac{4}{55} \\ \frac{8}{55} & \frac{2}{55} & -\frac{7}{55} \\ -\frac{7}{110} & \frac{6}{55} & \frac{13}{110} \end{bmatrix}$

32. $\begin{bmatrix} 1 & 5 & -2 \\ 2 & 3 & 4 \\ -1 & -3 & 1 \end{bmatrix}\begin{bmatrix} -\frac{5}{3} & -\frac{1}{9} & -\frac{26}{9} \\ \frac{2}{3} & \frac{1}{9} & \frac{8}{9} \\ \frac{1}{3} & \frac{2}{9} & \frac{7}{9} \end{bmatrix}$

33. Consider $A = \begin{bmatrix} 0.65 & 0.35 \\ 0.71 & 0.29 \end{bmatrix}$

(a) Use a graphing calculator to compute A^2, A^5, A^{10}, and A^{20}.

(b) Describe any pattern observed in part (a)

34. Consider $B = \begin{bmatrix} 0.37 & 0.63 \\ 0.25 & 0.75 \end{bmatrix}$

(a) Use a graphing calculator to compute B^2, B^5, B^{10}, and B^{20}.

(b) Describe any pattern observed in part (a)

In Exercises 35–42 write the system of linear equations as a matrix equation. See Example 7.

35. $3x - 2y = 14$
$2x - 5y = 8$

36. $x + 3y = -11$
$2x - 7y = 4$

37. $x + y = 4$
$-3x - 2y = -5$

38. $7x - 8y = -9$
$-2x - 5y = 6$

39. $x + 2y + 2z = 11$
$x - y - z = -4$
$2x + 5y + 9z = 39$

40. $x - 3y + z = 5$
$-2x + 7y - 6z = -9$
$x + 2y - 3z = 6$

41. $-2x + 3y + 4z = 10$
$3x - 5y + 2z = 7$
$4x + 2y - 3z = -2$

42. $x + 2y - 4z = 5$
$3x - y + z = 3$
$x + 2y + z = -3$

Applications

43. Hilltop Motors has two dealerships, East and West. The inventory for its three most popular selling sports utility vehicles (simply called A, B, and C) at each dealership is given in matrix M.

$M =$

	A	B	C
East	15	22	19
West	17	18	23

The wholesale (W) and retail (R) values of each model of vehicle are given in matrix P.

$$P = \begin{array}{c} \\ \\ \\ \end{array} \begin{array}{cc} W & R \\ \left[\begin{array}{cc} 15{,}000 & 22{,}000 \\ 17{,}500 & 26{,}100 \\ 18{,}700 & 28{,}000 \end{array}\right] & \begin{array}{c} A \\ B \\ C \end{array} \end{array}$$

(a) Find MP.

(b) Interpret the entries in MP.

44. Holobinko's Auto has three dealerships, North, South, and Metro. The inventory for its four most popular selling vehicles (we will call them simply A, B, C, and D) at each dealership is given in matrix M.

$$M = \begin{array}{cccc} A & B & C & D \\ \left[\begin{array}{cccc} 10 & 12 & 8 & 15 \\ 13 & 7 & 11 & 12 \\ 17 & 11 & 5 & 13 \end{array}\right] & & & \begin{array}{c} \text{North} \\ \text{South} \\ \text{Metro} \end{array} \end{array}$$

The wholesale (W) and retail (R) values of each model of car are given in matrix P.

$$P = \begin{array}{cc} W & R \\ \left[\begin{array}{cc} 8000 & 11{,}000 \\ 8900 & 12{,}200 \\ 10{,}100 & 15{,}600 \\ 9500 & 13{,}000 \end{array}\right] & \begin{array}{c} A \\ B \\ C \\ D \end{array} \end{array}$$

(a) Find MP.

(b) Interpret the entries in MP.

45. The Stumpf Company has two manufacturing plants, one on the East Coast and one on the West Coast, that produce three types of tents. The labor hours and wage requirements for the manufacture of these three types of tents are given in matrices L and W, respectively.

$$L = \begin{array}{cccc} \text{Fabric} & & \\ \text{cutting} & \text{Assembly} & \text{Packaging} \\ \left[\begin{array}{ccc} 0.5 & 0.6 & 0.2 \\ 0.8 & 0.7 & 0.3 \\ 1.1 & 0.9 & 0.6 \end{array}\right] & & \begin{array}{c} \text{Two person} \\ \text{Four person} \\ \text{Six person} \end{array} \end{array}$$

and

$$W = \begin{array}{cc} \text{East} & \text{West} \\ \text{Coast} & \text{Coast} \\ \left[\begin{array}{cc} 13 & 14 \\ 15 & 15 \\ 10 & 11 \end{array}\right] & \begin{array}{c} \text{Fabric cutting} \\ \text{Assembly} \\ \text{Packaging} \end{array} \end{array}$$

The entries in L are in hours and the entries in W are in dollars.

(a) Find LW.

(b) Determine the labor costs for a two-person tent manufactured on the West Coast.

(c) Determine the labor costs for a six-person tent manufactured on the East Coast.

(d) Interpret the entries in LW.

46. (*continuation of Exercise 45*) Dave, the CEO of the Stumpf Company, decides to give all employees a 10% raise.

(a) Use this information to construct a new wage matrix W_2.

(b) Find LW_2.

(c) Determine the labor costs for a two-person tent manufactured on the West Coast.

(d) Determine the labor costs for a six-person tent manufactured on the East Coast.

47. Aida, a dietitian in a local hospital, plans a special diet for a patient that is made up of three basic foods, Guhde, Vita, and Zaca. The number of units of calcium, vitamin A, and vitamin C, per ounce, is given in matrix F.

$$F = \begin{array}{cccc} \text{Guhde} & \text{Vita} & \text{Zaca} \\ \left[\begin{array}{ccc} 20 & 10 & 30 \\ 10 & 5 & 20 \\ 20 & 15 & 10 \end{array}\right] & & \begin{array}{c} \text{Calcium} \\ \text{Vitamin A} \\ \text{Vitamin C} \end{array} \end{array}$$

At lunch the patient had 8 ounces of Guhde, 5 ounces of Vita, and 9 ounces of Zaca. For dinner the patient had 4 ounces of Guhde, 5 ounces of Vita, and 7 ounces of Zaca.

(a) Write a 3×1 matrix A that represents the amount of each food consumed (in ounces) at lunch.

(b) Write a 3×1 matrix B that represents the amount of each food consumed (in ounces) at dinner.

(c) Find FA and interpret the entries.

(d) Find FB and interpret the entries.

48. Manuel, a dietitian in a local hospital, plans a special diet for a patient that is made up of three basic foods, Veni, Gala, and Mogi. The number of units of calcium, vitamin A, vitamin C, and fiber, per ounce, is given in matrix F.

$$F = \begin{array}{cccc} \text{Veni} & \text{Gala} & \text{Mogi} \\ \left[\begin{array}{ccc} 10 & 15 & 30 \\ 5 & 20 & 10 \\ 10 & 30 & 15 \\ 10 & 5 & 5 \end{array}\right] & & \begin{array}{c} \text{Calcium} \\ \text{Vitamin A} \\ \text{Vitamin C} \\ \text{Fiber} \end{array} \end{array}$$

At lunch the patient had 8 ounces of Veni, 9 ounces of Gala, and 5 ounces of Mogi. For dinner the patient had 3 ounces of Veni, 5 ounces of Gala, and 7 ounces of Mogi.

(a) Write a 3×1 matrix A that represents the amount of each food consumed (in ounces) at lunch.

(b) Write a 3×1 matrix B that represents the amount of each food consumed (in ounces) at dinner.

(c) Find FA and interpret the entries.

(d) Find FB and interpret the entries.

49. The amounts of nitrogen, phosphate, and potash needed to grow artichokes, cabbage, and carrots for an entire growing season in California are given in matrix F, in lb/acre (*Source:* U.S. Department of Agriculture, www.usda.gov/nass).

$$\begin{array}{c} \\ \\ F = \end{array}\begin{array}{c} \text{Arti-} \\ \text{chokes} \quad \text{Cabbage} \quad \text{Carrots} \\ \begin{bmatrix} 197 & 203 & 249 \\ 123 & 86 & 228 \\ 123 & 78 & 67 \end{bmatrix} \end{array}\begin{array}{l} \text{Nitrogen} \\ \text{Phosphates} \\ \text{Potash} \end{array}$$

Matrix P represents the number of acres of each crop that Francis wishes to plant.

$$P = \begin{bmatrix} 30 \\ 40 \\ 130 \end{bmatrix} \begin{array}{l} \text{Artichokes} \\ \text{Cabbage} \\ \text{Carrots} \end{array}$$

(a) Find FP.

(b) How many pounds each of nitrogen, phosphate, and potash will Francis need for the entire growing season?

50. (*continuation of Exercise 49*) Matrix T gives the number of acres of each crop planted in 1990 in California (*Source:* U.S. Department of Agriculture, www.usda.gov/nass).

$$T = \begin{bmatrix} 10{,}000 \\ 11{,}200 \\ 56{,}100 \end{bmatrix} \begin{array}{l} \text{Artichokes} \\ \text{Cabbage} \\ \text{Carrots} \end{array}$$

(a) Find FT.

(b) In 1990, how many pounds each of nitrogen, phosphate, and potash were used in California for an entire growing season to grow these three crops?

51. The amounts of three pollutants, measured in thousands of tons, released into the atmosphere during 1985, 1990, and 1995 by the United States are given in matrix A (*Source:* U.S. Census Bureau, www.census.gov).

$$\begin{array}{c} \\ A = \end{array}\begin{array}{c} \quad 1985 \quad\; 1990 \quad\; 1995 \\ \begin{bmatrix} 45{,}584 & 29{,}844 & 26{,}760 \\ 23{,}230 & 23{,}678 & 19{,}189 \\ 23{,}488 & 23{,}436 & 23{,}768 \end{bmatrix} \end{array}\begin{array}{l} \text{Particulates} \\ \text{Sulfur dioxide} \\ \text{Nitrogen oxides} \end{array}$$

Matrix B gives the percent, as a decimal, of these pollutants due to electric utilities.

$$\begin{array}{ccc} \text{Parti-} & \text{Sulfur} & \text{Nitrogen} \\ \text{culates} & \text{dioxide} & \text{oxides} \end{array}$$
$$B = [\,0.01 \quad 0.64 \quad 0.26\,]$$

Here for example, 0.01 means 1% of particulate air pollution is caused by electric companies. Compute BA and interpret the entries.

52. The amounts of waste generated, in million of tons, for various items in the United States during 1990, 1992, 1994, and 1996 is given in matrix A (*Source:* U.S. Census Bureau, www.census.gov).

$$\begin{array}{c} \\ A = \end{array}\begin{array}{c} 1990 \quad 1992 \quad 1994 \quad 1996 \\ \begin{bmatrix} 72.7 & 74.3 & 80.8 & 79.7 \\ 2.8 & 2.9 & 3.0 & 3.0 \\ 13.1 & 13.1 & 13.4 & 12.3 \\ 17.1 & 18.4 & 19.3 & 19.8 \end{bmatrix} \end{array}\begin{array}{l} \text{Paper and paper board} \\ \text{Aluminum} \\ \text{Glass} \\ \text{Plastics} \end{array}$$

Matrix B gives the percent, as a decimal, of these items that were recovered or recycled over these 4 years.

$$\begin{array}{cccc} \text{Pp} & \text{A} & \text{G} & \text{Pl} \end{array}$$
$$B = [\,0.35 \quad 0.37 \quad 0.23 \quad 0.04\,]$$

Here 0.35 means 35% of paper and paperboard was recovered. Compute BA and interpret the entries.

53. Denyse and Jamie's stock holdings for the companies CGC, BA, and DD are given in matrix A. The entries in A represent the number of shares owned.

$$\begin{array}{c} \\ A = \end{array}\begin{array}{c} \text{CGC} \quad \text{BA} \quad \text{DD} \\ \begin{bmatrix} 250 & 300 & 150 \\ 300 & 350 & 200 \end{bmatrix} \end{array}\begin{array}{l} \text{Denyse} \\ \text{Jamie} \end{array}$$

At the close of trading on December 31, 2000, the prices in dollars per share of the stocks are as given in matrix P.

$$P = \begin{bmatrix} 27 \\ 81 \\ 67 \end{bmatrix} \begin{array}{l} \text{CGC} \\ \text{BA} \\ \text{DD} \end{array}$$

(a) Find AP.

(b) Use the entries of AP to determine the value of Denyse's stock holdings and the value of Jamie's stock holdings on December 31, 2000.

54. (*continuation of Exercise 53*) Suppose that on December 31, 2001, Denyse and Jamie noticed that the prices of their stocks had increased by 15%.

(a) Determine a new matrix P representing the prices in dollars per share of their stocks.

(b) Assuming A from Exercise 53 and P from part (a), determine AP.

(c) Use the result from part (b) to determine the value of Denyse's stock holdings and the value of Jamie's stock holdings on December 31, 2001.

55. Peggy, a mathematics professor, has given three tests to a class of five students. Peggy records their scores in matrix S.

$$\begin{array}{c} \\ \\ S = \end{array}\begin{array}{c} \text{Test} \\ 1 \quad 2 \quad 3 \\ \begin{bmatrix} 88 & 78 & 82 \\ 89 & 81 & 91 \\ 92 & 93 & 89 \\ 96 & 66 & 98 \\ 72 & 92 & 81 \end{bmatrix} \end{array}\begin{array}{l} \text{Josh} \\ \text{Juan} \\ \text{Maria} \\ \text{Diane} \\ \text{Zach} \end{array}$$

Let $C = \begin{bmatrix} 1 \\ 1 \\ 1 \end{bmatrix}$ and $R = [\,1 \quad 1 \quad 1 \quad 1 \quad 1\,]$.

(a) Compute $\frac{1}{3}(SC)$ and interpret the entries.

(b) Compute $\frac{1}{5}(RS)$ and interpret the entries.

56. Susan, a biology professor, has given three tests to a class of six students. Susan records their scores in matrix S.

$$S = \begin{array}{c} \text{Test} \\ \begin{array}{ccc} 1 & 2 & 3 \end{array} \\ \left[\begin{array}{ccc} 82 & 78 & 91 \\ 80 & 76 & 82 \\ 94 & 92 & 89 \\ 82 & 92 & 98 \\ 76 & 86 & 96 \\ 90 & 76 & 72 \end{array}\right] \end{array} \begin{array}{l} \\ \\ \text{Paul} \\ \text{Sue} \\ \text{Brent} \\ \text{Megan} \\ \text{Teri} \\ \text{Steve} \end{array}$$

Let $C = \begin{bmatrix} 1 \\ 1 \\ 1 \end{bmatrix}$ and $R = [1 \quad 1 \quad 1 \quad 1 \quad 1 \quad 1]$.

(a) Compute $\frac{1}{3}(SC)$ and interpret the entries.

(b) Compute $\frac{1}{6}(RS)$ and interpret the entries.

57. Bill is a beagle breeder. He uses two feeds, Octav and Zoso, to make three different mixtures of the perfect beagle food. The three mixtures are shown in matrix F (the entries in F are in ounces).

$$F = \begin{array}{c} \begin{array}{ccc} \text{Mix I} & \text{Mix II} & \text{Mix III} \end{array} \\ \left[\begin{array}{ccc} 30 & 15 & 10 \\ 10 & 15 & 20 \end{array}\right] \end{array} \begin{array}{l} \\ \text{Octav} \\ \text{Zoso} \end{array}$$

The nutritional value of each of the two feeds is given in matrix N (the entries are in grams/ounce).

$$N = \begin{array}{c} \begin{array}{cc} \text{Octav} & \text{Zoso} \end{array} \\ \left[\begin{array}{cc} 6 & 4 \\ 15 & 12 \\ 20 & 15 \end{array}\right] \end{array} \begin{array}{l} \\ \text{Protein} \\ \text{Carbohydrates} \\ \text{Fat} \end{array}$$

(a) Find NF.

(b) Use part (a) to determine the amount of protein in mix II.

(c) Use part (a) to determine the amount of carbohydrates in mix I.

(d) Interpret the entries in NF.

58. (*continuation of Exercise 57*) Bill, the beagle breeder, introduces a new feed, Zipz, and creates a fourth mixture. The four mixtures are shown in matrix F (the entries are in ounces).

$$F = \begin{array}{c} \begin{array}{cccc} \text{Mix I} & \text{Mix II} & \text{Mix III} & \text{Mix IV} \end{array} \\ \left[\begin{array}{cccc} 15 & 10 & 10 & 20 \\ 10 & 10 & 10 & 5 \\ 5 & 10 & 10 & 5 \end{array}\right] \end{array} \begin{array}{l} \\ \text{Octav} \\ \text{Zoso} \\ \text{Zipz} \end{array}$$

The nutritional values of each of the three feeds are given in matrix N (the entries are in grams/ounce).

$$N = \begin{array}{c} \begin{array}{ccc} \text{Octav} & \text{Zoso} & \text{Zipz} \end{array} \\ \left[\begin{array}{ccc} 6 & 4 & 8 \\ 15 & 12 & 15 \\ 20 & 15 & 12 \end{array}\right] \end{array} \begin{array}{l} \\ \text{Protein} \\ \text{Carbohydrates} \\ \text{Fat} \end{array}$$

(a) Find NF.

(b) Use part (a) to determine the amount of carbohydrates in mix III.

(c) Use part (a) to determine the amount of fat in mix II.

(d) Interpret the entries in NF.

59. Matrix A gives the percentage, as a decimal, of eligible voters in the United States in 1994 classified according to party affiliation and age group. (*Note:* Rows may not sum to 100% due to not including parties such as Libertarian, Reform, Green, and the like.) (*Source:* U.S. Census Bureau, www.census.gov.)

$$A = \begin{array}{c} \begin{array}{ccc} \text{Democratic} & \text{Independent} & \text{Republican} \end{array} \\ \left[\begin{array}{ccc} 0.51 & 0.10 & 0.37 \\ 0.44 & 0.12 & 0.43 \\ 0.47 & 0.07 & 0.44 \\ 0.53 & 0.08 & 0.385 \end{array}\right] \end{array} \begin{array}{l} \\ \text{18–24} \\ \text{25–44} \\ \text{45–64} \\ \text{Over 65} \end{array}$$

The population of eligible voters, in millions, in 1994 in the United States according to age group is given in matrix B.

$$B = \begin{array}{c} \begin{array}{cccc} \text{18–24} & \text{25–44} & \text{45–64} & \text{Over 65} \end{array} \\ [25.2 \quad 83 \quad 50.9 \quad 31.1] \end{array}$$

If all eligible voters vote along party lines, find a matrix giving the total number of eligible voters in the United States in 1994 who will vote Democratic, Independent, and Republican.

60. Matrix A gives the percentage, as a decimal, of the total vote for five states for the Democratic, Republican, and Reform party candidates in the 1996 presidential election (*Source:* U.S. Census Bureau, www.census.gov).

$$A = \begin{array}{c} \begin{array}{ccc} \text{Democratic} & \text{Republican} & \text{Reform} \end{array} \\ \left[\begin{array}{ccc} 0.511 & 0.382 & 0.07 \\ 0.48 & 0.423 & 0.091 \\ 0.474 & 0.41 & 0.107 \\ 0.44 & 0.498 & 0.056 \\ 0.498 & 0.373 & 0.089 \end{array}\right] \end{array} \begin{array}{l} \\ \text{California} \\ \text{Florida} \\ \text{Ohio} \\ \text{South Carolina} \\ \text{Washington} \end{array}$$

Matrix B gives the total number of votes, in thousands, cast in each state.

$$B = \begin{array}{c} \begin{array}{ccccc} \text{California} & \text{Florida} & \text{Ohio} & \text{South Carolina} & \text{Washington} \end{array} \\ [10{,}019 \quad 5304 \quad 4534 \quad 1152 \quad 2254] \end{array}$$

Compute BA and interpret each entry.

SECTION PROJECT

For this section project, let

$$A = \begin{bmatrix} 2 & -1 \\ 3 & 4 \end{bmatrix} \quad B = \begin{bmatrix} -1 & 3 \\ 1 & 2 \end{bmatrix} \quad C = \begin{bmatrix} 4 & -2 \\ -1 & 3 \end{bmatrix}$$

(a) Let $D = AB$. Find D.

(b) Find DC.

(c) Let $E = BC$. Find E.

(d) Find AE.

(e) Compare the results of parts (b) and (d). They should be equal. This illustrates the **associative law** of matrix multiplication, $(AB)C = A(BC)$. Of course, the associative law is true only if the matrix products are defined. Using the letters m, n, p, and q and given that matrix A (not the A given above, but any matrix A) has dimensions $m \times n$, determine the dimensions of B and C (not the B and C given above) so that the associative law is true.

For parts (f) through (i), use the matrices given above.

(f) Let $S = B + C$. Find S.

(g) Find AS.

(h) Let $D = AB$ and $P = AC$. Find $D + P$.

(i) Compare the results of parts (g) and (h). They should be equal, illustrating the **distributive law** of matrix multiplication, $A(B + C) = AB + AC$. Of course, the distributive law is true only if the matrix sums and products are defined. For any matrix A having dimensions $m \times n$, what must be true about the dimensions of any matrices B and C?

Section 2.4 Matrix Inverses and Solving Systems of Linear Equations

> **From Your Toolbox**
>
> If $b \neq 0$, then $b^{-1} = \frac{1}{b}$ is the multiplicative inverse (reciprocal) of b, and $b \cdot b^{-1} = b \cdot \frac{1}{b} = 1$.

Sections 2.2 and 2.3 have laid the groundwork for matrix arithmetic. In this section, we learn how to determine the **inverse of a matrix** and how to use the inverse matrix to solve a **matrix equation** that represents a system of linear equations. We will see that a **matrix inverse** is analogous to the reciprocal of a real number. See the Toolbox to the left.

Since $b \cdot 1 = b$ for any real number b, the number 1 is called the **multiplicative identity**. Before studying the inverse of a matrix, we need to discuss another important matrix called the **identity matrix**.

Identity Matrix

> **From Your Toolbox**
>
> A matrix is a **square matrix** if the number of rows equals the number of columns. That is, its dimension is $n \times n$.

An **identity matrix** is a matrix that behaves like the multiplicative identity 1. We know that, for any real number b, $b \cdot 1 = 1 \cdot b = b$. If I is an identity matrix and A is another matrix, then the matrix products (assuming that they exist) AI and IA must equal A. For this reason, an identity matrix exists only for square matrices.[1] (See the Toolbox to the left for a review of square matrices.) In general, the identity matrix of dimension $n \times n$ has 1s along the diagonal from upper left to lower right (called the **main diagonal**) and 0s elsewhere.

Identity Matrix

The **identity matrix** with dimension $n \times n$ is given by

$$I = \begin{bmatrix} 1 & 0 & \cdots & 0 \\ 0 & 1 & \cdots & 0 \\ \vdots & \vdots & \ddots & \vdots \\ 0 & 0 & \cdots & 1 \end{bmatrix}$$

[1]Although some nonsquare matrices may have an identity matrix, only square matrices satisfy $AI = IA = A$.

Example 1 Verifying an Identity Matrix

Given the matrices

$$A = \begin{bmatrix} 2 & 1 & 3 \\ -1 & 3 & -2 \\ 4 & 1 & 7 \end{bmatrix} \qquad I = \begin{bmatrix} 1 & 0 & 0 \\ 0 & 1 & 0 \\ 0 & 0 & 1 \end{bmatrix}$$

show that $AI = A$ and $IA = A$. (We can express these equalities as $AI = IA = A$.)

Solution

$$AI = \begin{bmatrix} 2 & 1 & 3 \\ -1 & 3 & -2 \\ 4 & 1 & 7 \end{bmatrix}\begin{bmatrix} 1 & 0 & 0 \\ 0 & 1 & 0 \\ 0 & 0 & 1 \end{bmatrix} = \begin{bmatrix} 2 & 1 & 3 \\ -1 & 3 & -2 \\ 4 & 1 & 7 \end{bmatrix} = A$$

$$IA = \begin{bmatrix} 1 & 0 & 0 \\ 0 & 1 & 0 \\ 0 & 0 & 1 \end{bmatrix}\begin{bmatrix} 2 & 1 & 3 \\ -1 & 3 & -2 \\ 4 & 1 & 7 \end{bmatrix} = \begin{bmatrix} 2 & 1 & 3 \\ -1 & 3 & -2 \\ 4 & 1 & 7 \end{bmatrix} = A$$

So $AI = IA = A$. This means that I is an identity matrix. ■

The identity matrix in Example 1 is the identity matrix for 3×3 matrices. The identity matrix for 2×2 matrices and 4×4 matrices are, respectively

$$I = \begin{bmatrix} 1 & 0 \\ 0 & 1 \end{bmatrix} \quad \text{and} \quad I = \begin{bmatrix} 1 & 0 & 0 & 0 \\ 0 & 1 & 0 & 0 \\ 0 & 0 & 1 & 0 \\ 0 & 0 & 0 & 1 \end{bmatrix}$$

Inverse of a Square Matrix

From Your Toolbox

To see more review on negative exponents, consult Section A.3 in Appendix A.

As stated in the Toolbox on page 108, the **multiplicative inverse** (reciprocal) of a real number b ($b \neq 0$) is the unique real number b^{-1} such that

$$b \cdot b^{-1} = b^{-1} \cdot b = 1$$

For example, 3^{-1} is the multiplicative inverse of 3 since $3^{-1} \cdot 3 = \frac{1}{3} \cdot 3 = 1$. We use this same idea to define the **inverse of a matrix**.

■ **Inverse of a Matrix**

Let A be a square matrix of dimension $n \times n$. The square matrix A^{-1} of dimension $n \times n$, where

$$A^{-1} \cdot A = A \cdot A^{-1} = I$$

is called the **inverse of A**. Note that I is the identity matrix of dimension $n \times n$.

Note: A^{-1} does *not* mean $\frac{1}{A}$. When working with matrices, A^{-1} is simply notation for the multiplicative inverse, or simply inverse, of matrix A.

Example 2 Verifying an Inverse Matrix

Show that matrix A has as its inverse matrix A^{-1}.

$$A = \begin{bmatrix} 2 & 3 \\ -1 & 4 \end{bmatrix} \qquad A^{-1} = \begin{bmatrix} \frac{4}{11} & -\frac{3}{11} \\ \frac{1}{11} & \frac{2}{11} \end{bmatrix}$$

Solution

We need to verify that $A \cdot A^{-1} = A^{-1} \cdot A = I$. We have

$$A \cdot A^{-1} = \begin{bmatrix} 2 & 3 \\ -1 & 4 \end{bmatrix} \begin{bmatrix} \frac{4}{11} & -\frac{3}{11} \\ \frac{1}{11} & \frac{2}{11} \end{bmatrix}$$

$$= \begin{bmatrix} \frac{8}{11} + \frac{3}{11} & -\frac{6}{11} + \frac{6}{11} \\ -\frac{4}{11} + \frac{4}{11} & \frac{3}{11} + \frac{8}{11} \end{bmatrix} = \begin{bmatrix} 1 & 0 \\ 0 & 1 \end{bmatrix} = I$$

$$A^{-1} \cdot A = \begin{bmatrix} \frac{4}{11} & -\frac{3}{11} \\ \frac{1}{11} & \frac{2}{11} \end{bmatrix} \begin{bmatrix} 2 & 3 \\ -1 & 4 \end{bmatrix}$$

$$= \begin{bmatrix} \frac{8}{11} + \frac{3}{11} & \frac{12}{11} + \left(-\frac{12}{11}\right) \\ \frac{2}{11} + \left(-\frac{2}{11}\right) & \frac{3}{11} + \frac{8}{11} \end{bmatrix} = \begin{bmatrix} 1 & 0 \\ 0 & 1 \end{bmatrix} = I$$

Since $AA^{-1} = A^{-1}A = I$, we see that A^{-1} is the inverse of A. ■

Checkpoint 1

Now work Exercise 1.

Before we outline a process to determine inverse matrices, we must first address whether or not every square matrix has an inverse matrix. We know that every nonzero real number has an inverse (reciprocal); however, unlike real numbers, this is not true for matrices. A matrix that does not have an inverse matrix is said to be **singular**. For example, matrix A does not have an inverse matrix; hence it is singular.

$$A = \begin{bmatrix} 0 & 0 \\ 2 & 0 \end{bmatrix}$$

If A had an inverse, it would be of the form

$$A^{-1} = \begin{bmatrix} a & b \\ c & d \end{bmatrix}$$

where a, b, c, and d are real numbers. By the definition of an inverse, we must have $AA^{-1} = I$, that is,

$$\begin{bmatrix} 0 & 0 \\ 2 & 0 \end{bmatrix} \begin{bmatrix} a & b \\ c & d \end{bmatrix} = \begin{bmatrix} 1 & 0 \\ 0 & 1 \end{bmatrix}$$

Multiplying on the left side of the equation gives

$$\begin{bmatrix} 0a + 0c & 0b + 0d \\ 2a + 0c & 2b + 0d \end{bmatrix} = \begin{bmatrix} 1 & 0 \\ 0 & 1 \end{bmatrix}$$

$$\begin{bmatrix} 0 & 0 \\ 2a & 2b \end{bmatrix} = \begin{bmatrix} 1 & 0 \\ 0 & 1 \end{bmatrix}$$

Recalling that in order for two matrices to be equal they must have the same dimension and corresponding entries must be equal, we see that if these two matrices are equal, then, by examining their entries in the 1st row, 1st column, we must have $0 = 1$, which is not possible. This impossibility tells us that A does not have an inverse; that is, A is singular.

Finding the Inverse of a Square Matrix

To discover a step-by-step procedure to find the inverse of a matrix, let's find the inverse of matrix B

$$B = \begin{bmatrix} 1 & -3 \\ -1 & 4 \end{bmatrix}$$

Let's assume that B^{-1} exists and has the form

$$B^{-1} = \begin{bmatrix} a & b \\ c & d \end{bmatrix}$$

where a, b, c, and d are real numbers. By definition we know that $BB^{-1} = I$, which gives us the equation

$$\begin{bmatrix} 1 & -3 \\ -1 & 4 \end{bmatrix} \begin{bmatrix} a & b \\ c & d \end{bmatrix} = \begin{bmatrix} 1 & 0 \\ 0 & 1 \end{bmatrix}$$

Performing the multiplication on the left side of the equation yields

$$\begin{bmatrix} a - 3c & b - 3d \\ -a + 4c & -b + 4d \end{bmatrix} = \begin{bmatrix} 1 & 0 \\ 0 & 1 \end{bmatrix}$$

From equality of matrices, this equation is the same as the following two systems of linear equations.

$$\begin{aligned} a - 3c &= 1, \qquad & b - 3d &= 0 \\ -a + 4c &= 0, \qquad & -b + 4d &= 1 \end{aligned}$$

From Section 2.1 we can write each system as an augmented matrix.

$$\left[\begin{array}{cc|c} 1 & -3 & 1 \\ -1 & 4 & 0 \end{array}\right] \quad \text{and} \quad \left[\begin{array}{cc|c} 1 & -3 & 0 \\ -1 & 4 & 1 \end{array}\right]$$

Notice that the matrices of coefficients (that is, each matrix left of the vertical line) for both augmented matrices are identical. This means that we would use the same row operations on each augmented matrix to transform it into reduced row echelon form. So we can save some time and effort by combining the two augmented matrices into one augmented matrix.

$$\left[\begin{array}{cc|cc} 1 & -3 & 1 & 0 \\ -1 & 4 & 0 & 1 \end{array}\right]$$

We now use the Gauss–Jordan method to get

$$\left[\begin{array}{cc|cc} 1 & -3 & 1 & 0 \\ -1 & 4 & 0 & 1 \end{array}\right] R_1 + R_2 \to R_2 \left[\begin{array}{cc|cc} 1 & -3 & 1 & 0 \\ 0 & 1 & 1 & 1 \end{array}\right]$$

$$3R_2 + R_1 \to R_1 \left[\begin{array}{cc|cc} 1 & 0 & 4 & 3 \\ 0 & 1 & 1 & 1 \end{array}\right]$$

The numbers in the 1st column to the right of the vertical bar give the values of a and c, and the 2nd column gives the values of b and d. Hence $a = 4$, $b = 3$, $c = 1$, and $d = 1$, which means that B^{-1} is

$$B^{-1} = \begin{bmatrix} 4 & 3 \\ 1 & 1 \end{bmatrix}$$

We use the prior discussion as motivation for the following algorithm, which is used to compute the inverse of a square matrix.

Interactive Activity

For B and B^{-1} in the above discussion, multiply BB^{-1} and $B^{-1}B$ to show that both products equal I. This verifies that B^{-1} is the inverse of B.

Finding an Inverse Matrix

Given an $n \times n$ matrix A, we find A^{-1} (when it exists) by doing the following:

1. Make the augmented matrix $[A \mid I]$, where I is the $n \times n$ identity matrix.
2. Use row operations to reduce $[A \mid I]$ to the form $[I \mid B]$, if possible.
3. Matrix B is the inverse of A; that is, $B = A^{-1}$.

Example 3 Finding an Inverse Matrix

Given $A = \begin{bmatrix} 1 & 1 & 2 \\ 2 & 1 & 3 \\ 1 & -1 & 2 \end{bmatrix}$, find A^{-1}.

Solution

We write the augmented matrix $[A \mid I]$.

$$\left[\begin{array}{rrr|rrr} 1 & 1 & 2 & 1 & 0 & 0 \\ 2 & 1 & 3 & 0 & 1 & 0 \\ 1 & -1 & 2 & 0 & 0 & 1 \end{array}\right]$$

We now use the Gauss–Jordan method to transform it to the form $[I \mid B]$.

$$\left[\begin{array}{rrr|rrr} 1 & 1 & 2 & 1 & 0 & 0 \\ 2 & 1 & 3 & 0 & 1 & 0 \\ 1 & -1 & 2 & 0 & 0 & 1 \end{array}\right] \begin{array}{l} -2R_1 + R_2 \to R_2 \\ -1R_1 + R_3 \to R_3 \end{array} \left[\begin{array}{rrr|rrr} 1 & 1 & 2 & 1 & 0 & 0 \\ 0 & -1 & -1 & -2 & 1 & 0 \\ 0 & -2 & 0 & -1 & 0 & 1 \end{array}\right]$$

$$-2R_2 + R_3 \to R_3 \left[\begin{array}{rrr|rrr} 1 & 1 & 2 & 1 & 0 & 0 \\ 0 & -1 & -1 & -2 & 1 & 0 \\ 0 & 0 & 2 & 3 & -2 & 1 \end{array}\right]$$

$$\tfrac{1}{2}R_3 \to R_3 \left[\begin{array}{rrr|rrr} 1 & 1 & 2 & 1 & 0 & 0 \\ 0 & -1 & -1 & -2 & 1 & 0 \\ 0 & 0 & 1 & \frac{3}{2} & -1 & \frac{1}{2} \end{array}\right]$$

$$R_2 + R_1 \to R_1 \left[\begin{array}{rrr|rrr} 1 & 0 & 1 & -1 & 1 & 0 \\ 0 & -1 & -1 & -2 & 1 & 0 \\ 0 & 0 & 1 & \frac{3}{2} & -1 & \frac{1}{2} \end{array}\right]$$

$$\begin{array}{l} -1R_3 + R_1 \to R_1 \\ R_3 + R_2 \to R_2 \end{array} \left[\begin{array}{rrr|rrr} 1 & 0 & 0 & -\frac{5}{2} & 2 & -\frac{1}{2} \\ 0 & -1 & 0 & -\frac{1}{2} & 0 & \frac{1}{2} \\ 0 & 0 & 1 & \frac{3}{2} & -1 & \frac{1}{2} \end{array}\right]$$

$$-1R_2 \to R_2 \left[\begin{array}{rrr|rrr} 1 & 0 & 0 & -\frac{5}{2} & 2 & -\frac{1}{2} \\ 0 & 1 & 0 & \frac{1}{2} & 0 & -\frac{1}{2} \\ 0 & 0 & 1 & \frac{3}{2} & -1 & \frac{1}{2} \end{array}\right]$$

So the inverse of A is

$$A^{-1} = \begin{bmatrix} -\frac{5}{2} & 2 & -\frac{1}{2} \\ \frac{1}{2} & 0 & -\frac{1}{2} \\ \frac{3}{2} & -1 & \frac{1}{2} \end{bmatrix}$$

We leave the verification of this result to you in the Interactive Activity to the left.

Interactive Activity

For A in Example 3, compute AA^{-1} and $A^{-1}A$.

✓ Checkpoint 2 Now work Exercise 13.

Technology Option

Inverses of matrices can be found using a graphing calculator. In Figure 2.4.1(a), we have entered matrix A from Example 3 into the calculator and used the calculator to find A^{-1}. Figure 2.4.1(b) shows the entries as fractions.

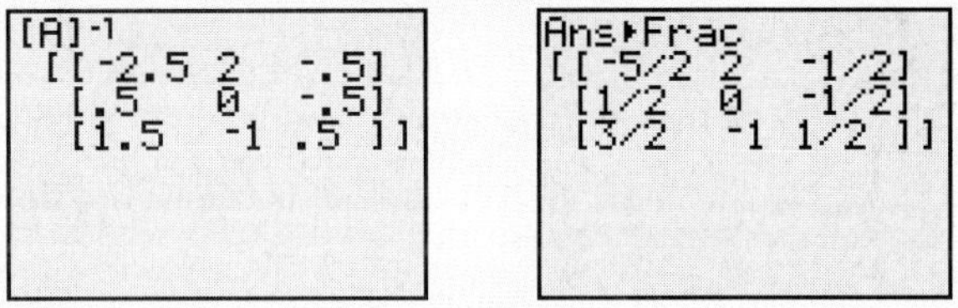

(a) **(b)**

Figure 2.4.1

Example 4 Finding an Inverse Matrix

Given $A = \begin{bmatrix} 1 & 2 & 2 \\ 2 & 1 & 2 \\ 3 & 3 & 4 \end{bmatrix}$, find A^{-1}.

Solution

We form the augmented matrix

$$\left[\begin{array}{ccc|ccc} 1 & 2 & 2 & 1 & 0 & 0 \\ 2 & 1 & 2 & 0 & 1 & 0 \\ 3 & 3 & 4 & 0 & 0 & 1 \end{array}\right]$$

and use the Gauss–Jordan method as in Example 3.

$$\left[\begin{array}{ccc|ccc} 1 & 2 & 2 & 1 & 0 & 0 \\ 2 & 1 & 2 & 0 & 1 & 0 \\ 3 & 3 & 4 & 0 & 0 & 1 \end{array}\right] \begin{array}{l} -2R_1 + R_2 \to R_2 \\ -3R_1 + R_3 \to R_3 \end{array} \left[\begin{array}{ccc|ccc} 1 & 2 & 2 & 1 & 0 & 0 \\ 0 & -3 & -2 & -2 & 1 & 0 \\ 0 & -3 & -2 & -3 & 0 & 1 \end{array}\right]$$

$$-1R_2 + R_3 \to R_3 \left[\begin{array}{ccc|ccc} 1 & 2 & 2 & 1 & 0 & 0 \\ 0 & -3 & -2 & -2 & 1 & 0 \\ 0 & 0 & 0 & -1 & -1 & 1 \end{array}\right]$$

Since the last row has all zeros to the left of the vertical bar, there is no way we can continue and get the reduced form $[I \mid B]$. This means that matrix A is singular and has no inverse. ■

Whenever we encounter a matrix like the final augmented matrix in Example 4 that has all zeros in a row to the left of the vertical bar, this tells us that the matrix has no inverse.

Formula for the Inverse of 2 × 2 Matrices

Before we look at how inverse matrices can be used to solve matrix equations, let's examine an alternative way to find the inverse of a 2 × 2 matrix. The derivation of this formula is outlined in Exercise 27.

Formula for the Inverse of a 2 × 2 Matrix

Let $A = \begin{bmatrix} a & b \\ c & d \end{bmatrix}$. Suppose that $\det(A) = ad - bc$ and $\det(A) \neq 0$. Then A^{-1} is given by

$$A^{-1} = \frac{1}{\det(A)} \begin{bmatrix} d & -b \\ -c & a \end{bmatrix}$$

Note: $\det(A)$ is called the **determinant of A**.

Example 5 Using the Formula to Find A^{-1} for a 2 × 2 Matrix

Given $A = \begin{bmatrix} 3 & -1 \\ 4 & -2 \end{bmatrix}$, find A^{-1} by using the formula.

Solution

First we compute $\det(A)$. Since $a = 3$, $b = -1$, $c = 4$, and $d = -2$, we have

$$\det(A) = 3(-2) - (-1)4 = -6 - (-4) = -2$$

Next we write the matrix

$$\begin{bmatrix} d & -b \\ -c & a \end{bmatrix} = \begin{bmatrix} -2 & 1 \\ -4 & 3 \end{bmatrix}$$

Finally, we multiply this matrix by $\frac{1}{\det(A)} = \frac{1}{-2}$ to get A^{-1}.

$$A^{-1} = \frac{1}{-2}\begin{bmatrix} -2 & 1 \\ -4 & 3 \end{bmatrix} = \begin{bmatrix} 1 & -\frac{1}{2} \\ 2 & -\frac{3}{2} \end{bmatrix}$$ ■

Interactive Activity

Use a graphing calculator to find A^{-1} for matrix A in Example 5.

WWW

Checkpoint 3

Now work Exercise 19.

Using Inverses to Solve Systems of Linear Equations

When we have a system of linear equations in which the number of equations equals the number of variables, we can write a corresponding matrix equation and use the inverse of the coefficient matrix to solve the system. For example, consider the system

$$3x - y = 6$$
$$4x - 2y = 7$$

If we let A be the coefficient matrix, X be the variable matrix, and B be the constants matrix, we have

$$A = \begin{bmatrix} 3 & -1 \\ 4 & -2 \end{bmatrix} \qquad X = \begin{bmatrix} x \\ y \end{bmatrix} \qquad B = \begin{bmatrix} 6 \\ 7 \end{bmatrix}$$

Notice that

$$AX = B$$

$$\begin{bmatrix} 3 & -1 \\ 4 & -2 \end{bmatrix}\begin{bmatrix} x \\ y \end{bmatrix} = \begin{bmatrix} 6 \\ 7 \end{bmatrix}$$

In Example 5 we found that $A^{-1} = \begin{bmatrix} 1 & -\frac{1}{2} \\ 2 & -\frac{3}{2} \end{bmatrix}$. We can use A^{-1} to solve as follows. (We will show both the numerical and general solution simultaneously.)

$$\begin{bmatrix} 3 & -1 \\ 4 & -2 \end{bmatrix}\begin{bmatrix} x \\ y \end{bmatrix} = \begin{bmatrix} 6 \\ 7 \end{bmatrix} \qquad AX = B$$

$$\begin{bmatrix} 1 & -\frac{1}{2} \\ 2 & -\frac{3}{2} \end{bmatrix}\begin{bmatrix} 3 & -1 \\ 4 & -2 \end{bmatrix}\begin{bmatrix} x \\ y \end{bmatrix} = \begin{bmatrix} 1 & -\frac{1}{2} \\ 2 & -\frac{3}{2} \end{bmatrix}\begin{bmatrix} 6 \\ 7 \end{bmatrix} \qquad A^{-1}AX = A^{-1}B$$

$$\begin{bmatrix} 1 & 0 \\ 0 & 1 \end{bmatrix}\begin{bmatrix} x \\ y \end{bmatrix} = \begin{bmatrix} \frac{5}{2} \\ \frac{3}{2} \end{bmatrix} \qquad IX = A^{-1}B$$

$$\begin{bmatrix} x \\ y \end{bmatrix} = \begin{bmatrix} \frac{5}{2} \\ \frac{3}{2} \end{bmatrix} \qquad X = A^{-1}B$$

So the solution to the system is $x = \frac{5}{2}$ and $y = \frac{3}{2}$. The generalization on the right side is summarized as follows:

Using Matrix Inverses to Solve Systems of Equations

Given a system of linear equations $AX = B$, where A is the coefficient matrix, X is the variable matrix, and B is the constants matrix, we solve the system by first finding A^{-1}. Then $X = A^{-1}B$ gives the solution.

Example 6 Using a Matrix Inverse to Solve a System

Write the given system in the form $AX = B$ and use A^{-1} to solve the system.

$$\begin{aligned} x + y + 2z &= -3 \\ 2x + y + 3z &= 4 \\ x - y + 2z &= 5 \end{aligned}$$

Solution

We have

$$A = \begin{bmatrix} 1 & 1 & 2 \\ 2 & 1 & 3 \\ 1 & -1 & 2 \end{bmatrix} \qquad X = \begin{bmatrix} x \\ y \\ z \end{bmatrix} \qquad B = \begin{bmatrix} -3 \\ 4 \\ 5 \end{bmatrix}$$

So

$$AX = B$$

$$\begin{bmatrix} 1 & 1 & 2 \\ 2 & 1 & 3 \\ 1 & -1 & 2 \end{bmatrix}\begin{bmatrix} x \\ y \\ z \end{bmatrix} = \begin{bmatrix} -3 \\ 4 \\ 5 \end{bmatrix}$$

Interactive Activity

To verify the result of Example 6, enter matrices A and B into your calculator. Calculate $A^{-1} * B$ and compare with the result of Example 6.

WWW

In Example 3, we determined that $A^{-1} = \begin{bmatrix} -\frac{5}{2} & 2 & -\frac{1}{2} \\ \frac{1}{2} & 0 & -\frac{1}{2} \\ \frac{3}{2} & -1 & \frac{1}{2} \end{bmatrix}$,

so the solution is given by

$$X = A^{-1}B = \begin{bmatrix} -\frac{5}{2} & 2 & -\frac{1}{2} \\ \frac{1}{2} & 0 & -\frac{1}{2} \\ \frac{3}{2} & -1 & \frac{1}{2} \end{bmatrix} \begin{bmatrix} -3 \\ 4 \\ 5 \end{bmatrix} = \begin{bmatrix} 13 \\ -4 \\ -6 \end{bmatrix}$$

So $x = 13$, $y = -4$, and $z = -6$ is the solution to the system. ■

Checkpoint 4

Now work Exercise 41.

Applications

For our first application, let's revisit the Flashback from Section 2.1.

Flashback

Gonzalez Office Supplies Revisited

In Section 2.1 we revisited Gonzalez Office Supplies and saw that they manufacture three types of office chairs: Secretarial, Executive, and Presidential. Each chair requires time on three different machines. These times and the total available time for each machine are given in Table 2.4.1.

Table 2.4.1

	Secretarial	**Executive**	**Presidential**	**Total hours each week**
Hours on Machine A	1	1	2	30
Hours on Machine B	2	3	2	53
Hours on Machine C	1	2	3	47

We assume that all available time for each machine must be used each week. If x represents the number of Secretarial chairs made each week, y represents the number of Executive chairs made each week, and z represents the number of Presidential chairs made each week, then the system of equations that can be solved to determine how many of each type of chair can be made each week is

$$\begin{aligned} x + y + 2z &= 30 \\ 2x + 3y + 2z &= 53 \\ x + 2y + 3z &= 47 \end{aligned}$$

Solve this system of linear equations by using the inverse of the coefficient matrix.

Flashback Solution

For this system we have

$$A = \begin{bmatrix} 1 & 1 & 2 \\ 2 & 3 & 2 \\ 1 & 2 & 3 \end{bmatrix} \qquad X = \begin{bmatrix} x \\ y \\ z \end{bmatrix} \qquad B = \begin{bmatrix} 30 \\ 53 \\ 47 \end{bmatrix}$$

To find A^{-1}, we start with

$$[A \mid I] = \left[\begin{array}{ccc|ccc} 1 & 1 & 2 & 1 & 0 & 0 \\ 2 & 3 & 2 & 0 & 1 & 0 \\ 1 & 2 & 3 & 0 & 0 & 1 \end{array}\right]$$

and transform it to get $[I \mid A^{-1}]$. Omitting the details, the result is

$$A^{-1} = \begin{bmatrix} \frac{5}{3} & \frac{1}{3} & -\frac{4}{3} \\ -\frac{4}{3} & \frac{1}{3} & \frac{2}{3} \\ \frac{1}{3} & -\frac{1}{3} & \frac{1}{3} \end{bmatrix}$$

The solution is given by

$$X = A^{-1}B = \begin{bmatrix} \frac{5}{3} & \frac{1}{3} & -\frac{4}{3} \\ -\frac{4}{3} & \frac{1}{3} & \frac{2}{3} \\ \frac{1}{3} & -\frac{1}{3} & \frac{1}{3} \end{bmatrix} \begin{bmatrix} 30 \\ 53 \\ 47 \end{bmatrix} = \begin{bmatrix} 5 \\ 9 \\ 8 \end{bmatrix}$$

So the solution to the system is $x = 5$, $y = 9$, and $z = 8$.

Interpret the Solution: Five Secretarial, nine Executive, and eight Presidential chairs should be made each week to use all available time on each machine.

Our final example utilizes the formula for determining an inverse for a 2×2 matrix.

Example 7 Determining a Planting Schedule

The data in Table 2.4.2 give the amounts of nitrogen and phosphate needed to grow cauliflower and head lettuce for an entire growing season in California (all entries are in lb/acre). If Fernando has 150 acres and uses 28,890 pounds of nitrogen and 16,620 pounds of phosphate, how many acres did he allot for each crop to use all the resources completely?

Table 2.4.2

	Cauliflower	Head Lettuce
Nitrogen	215	173
Phosphate	98	122

SOURCE: U.S. Department of Agriculture, www.usda.gov/nass

Solution

Understand the Situation: We need to determine the number of acres of cauliflower and the number of acres of head lettuce that Fernando planted. To do this, we let x represent the number of acres of cauliflower and y represent the number of acres of head lettuce. The number of acres of cauliflower times the number of pounds per acre of nitrogen plus the number of acres of head lettuce times the number of pounds per acre of nitrogen must equal 28,890. This gives

$$215x + 173y = 28{,}890$$

Similarly, for phosphates we get

$$98x + 122y = 16{,}620$$

So the system is

$$\begin{aligned} 215x + 173y &= 28{,}890 \\ 98x + 122y &= 16{,}620 \end{aligned}$$

Letting $A = \begin{bmatrix} 215 & 173 \\ 98 & 122 \end{bmatrix}$, $X = \begin{bmatrix} x \\ y \end{bmatrix}$, and $B = \begin{bmatrix} 28{,}890 \\ 16{,}620 \end{bmatrix}$, we have the system in the form $AX = B$. Since $X = A^{-1}B$, we need to find A^{-1}. Here A is a 2×2 matrix, so we find A^{-1} by using the formula

$$A^{-1} = \frac{1}{\det(A)} \begin{bmatrix} d & -b \\ -c & a \end{bmatrix}$$

We have $\det(A) = 215(122) - 173(98) = 9276$, so

$$A^{-1} = \frac{1}{9276} \begin{bmatrix} 122 & -173 \\ -98 & 215 \end{bmatrix} = \begin{bmatrix} \frac{122}{9276} & -\frac{173}{9276} \\ -\frac{98}{9276} & \frac{215}{9276} \end{bmatrix}$$

Therefore,

$$X = A^{-1}B = \begin{bmatrix} \frac{122}{9276} & -\frac{173}{9276} \\ -\frac{98}{9276} & \frac{215}{9276} \end{bmatrix} \begin{bmatrix} 28{,}890 \\ 16{,}620 \end{bmatrix} = \begin{bmatrix} 70 \\ 80 \end{bmatrix}$$

So the solution to the system is $x = 70$ and $y = 80$.

Interpret the Solution: This means that Fernando planted 70 acres of cauliflower and 80 acres of head lettuce to use all his resources completely. ■

SUMMARY

We began this section with a brief discussion on an **identity matrix**, I, and quickly proceeded to determining the **inverse of a square matrix**. We saw how the Gauss–Jordan method is key in finding an inverse matrix by hand. We also learned a formula for the inverse of a 2×2 matrix that uses the **determinant**. The section concluded with a discussion on using inverse matrices to solve systems of equations, provided the number of equations equals the number of variables, and using this new method of solving systems in an applied setting.

- **Finding an Inverse Matrix** Given an $n \times n$ matrix A, we find A^{-1} (when it exists) by:

  1. Writing the augmented matrix $[A \mid I]$, where I is the $n \times n$ identity matrix.

2. Using row operations to reduce $[A \mid I]$ to the form $[I \mid B]$, if possible.
3. Matrix B is the inverse of A; that is, $B = A^{-1}$.

- **Formula for the Inverse of a 2×2 Matrix** Let $A = \begin{bmatrix} a & b \\ c & d \end{bmatrix}$. Suppose that $\det(A) = ad - bc$ and $\det(A) \neq 0$. Then A^{-1} is given by

$$A^{-1} = \frac{1}{\det(A)} \begin{bmatrix} d & -b \\ -c & a \end{bmatrix}.$$

- **Using Matrix Inverses to Solve Systems of Equations** Given a system of linear equations $AX = B$, where A is the coefficient matrix, X is the variable matrix, and B is the constants matrix, we solve the system by first finding A^{-1}. Then $X = A^{-1}B$ gives the solution.

SECTION 2.4 EXERCISES

In Exercises 1–8, determine if the given matrices are inverses of each other. (Hint: See if their product is the identity matrix I.)

1. $\begin{bmatrix} -2 & -3 \\ 1 & 1 \end{bmatrix}$ and $\begin{bmatrix} 1 & 3 \\ -1 & -2 \end{bmatrix}$

2. $\begin{bmatrix} 2 & -1 \\ 3 & 1 \end{bmatrix}$ and $\begin{bmatrix} \frac{1}{5} & \frac{1}{5} \\ -\frac{3}{5} & \frac{2}{5} \end{bmatrix}$

3. $\begin{bmatrix} 1 & 1 \\ 2 & 3 \end{bmatrix}$ and $\begin{bmatrix} -3 & 1 \\ 2 & -1 \end{bmatrix}$

4. $\begin{bmatrix} -3 & 1 \\ 5 & 1 \end{bmatrix}$ and $\begin{bmatrix} \frac{1}{8} & -\frac{1}{8} \\ -\frac{5}{8} & -\frac{3}{8} \end{bmatrix}$

5. $\begin{bmatrix} 1 & 3 & 2 \\ 4 & -2 & -1 \\ 0 & 4 & 3 \end{bmatrix}$ and $\begin{bmatrix} \frac{1}{3} & \frac{1}{6} & -\frac{1}{6} \\ 2 & -\frac{1}{2} & -\frac{3}{2} \\ -\frac{8}{3} & \frac{2}{3} & \frac{7}{3} \end{bmatrix}$

6. $\begin{bmatrix} -1 & 2 & 3 \\ 0 & 2 & -1 \\ 2 & 4 & 5 \end{bmatrix}$ and $\begin{bmatrix} -\frac{7}{15} & -\frac{1}{15} & \frac{4}{15} \\ \frac{1}{15} & \frac{11}{30} & \frac{1}{30} \\ \frac{2}{15} & -\frac{4}{15} & \frac{1}{15} \end{bmatrix}$

7. $\begin{bmatrix} -1 & 0 & 2 \\ 0 & -1 & 4 \\ 3 & 0 & 5 \end{bmatrix}$ and $\begin{bmatrix} \frac{5}{11} & 0 & \frac{2}{11} \\ \frac{12}{11} & -1 & \frac{4}{11} \\ \frac{3}{11} & 0 & -\frac{1}{11} \end{bmatrix}$

8. $\begin{bmatrix} 1 & 2 & 7 \\ 0 & -1 & 1 \\ 2 & 1 & 2 \end{bmatrix}$ and $\begin{bmatrix} -\frac{1}{5} & \frac{1}{5} & -\frac{3}{5} \\ \frac{2}{15} & -\frac{4}{5} & -\frac{1}{15} \\ -\frac{2}{15} & \frac{1}{5} & -\frac{1}{15} \end{bmatrix}$

In Exercises 9–18, find the inverse for each matrix. See Example 3.

9. $\begin{bmatrix} 1 & 3 \\ 4 & 2 \end{bmatrix}$

10. $\begin{bmatrix} -1 & -2 \\ 3 & 2 \end{bmatrix}$

11. $\begin{bmatrix} -2 & 7 \\ -4 & -1 \end{bmatrix}$

12. $\begin{bmatrix} 3 & 4 \\ -2 & -3 \end{bmatrix}$

13. $\begin{bmatrix} 1 & -1 & 2 \\ 0 & 2 & 3 \\ 3 & -1 & 4 \end{bmatrix}$

14. $\begin{bmatrix} 1 & 2 & 7 \\ 0 & -1 & 1 \\ 2 & 1 & 2 \end{bmatrix}$

15. $\begin{bmatrix} 2 & 1 & 0 \\ -1 & 1 & 2 \\ 4 & 2 & 3 \end{bmatrix}$

16. $\begin{bmatrix} 3 & 2 & 0 \\ -2 & -1 & -2 \\ 1 & 0 & 1 \end{bmatrix}$

17. $\begin{bmatrix} 1 & 2 & -2 & 3 \\ 0 & 2 & -3 & 1 \\ 2 & 0 & 1 & 2 \\ -3 & 1 & 2 & 5 \end{bmatrix}$

18. $\begin{bmatrix} 1 & -2 & 3 & 4 \\ 0 & 1 & -1 & 2 \\ -2 & 1 & 3 & -2 \\ 3 & 2 & 1 & 5 \end{bmatrix}$

In Exercises 19–26, use the formula for the inverse of a 2×2 matrix to find the inverse for the given matrix, if it exists.

19. $\begin{bmatrix} 2 & -3 \\ 4 & 8 \end{bmatrix}$

20. $\begin{bmatrix} -3 & 4 \\ 6 & 3 \end{bmatrix}$

21. $\begin{bmatrix} 2 & -4 \\ -2 & 4 \end{bmatrix}$

22. $\begin{bmatrix} 3 & 7 \\ -6 & -14 \end{bmatrix}$

23. $\begin{bmatrix} 7 & 3 \\ -1 & -2 \end{bmatrix}$

24. $\begin{bmatrix} -1 & -2 \\ 3 & 2 \end{bmatrix}$

25. $\begin{bmatrix} 2 & -1 \\ -4 & 7 \end{bmatrix}$

26. $\begin{bmatrix} -3 & -2 \\ 1 & 4 \end{bmatrix}$

27. Given that $A = \begin{bmatrix} a & b \\ c & d \end{bmatrix}$ is a 2×2 matrix, where a, b, c, and d are real numbers, $a \neq 0$, derive the formula, $A^{-1} = \frac{1}{\det(A)} \begin{bmatrix} d & -b \\ -c & a \end{bmatrix}$, $\det(A) \neq 0$, by doing the following:

(a) Write the matrix $[A \mid I]$ and use row operations until you have $[I \mid A^{-1}]$.

(b) From A^{-1}, factor out $\frac{1}{ad-bc}$. This should leave you with $\frac{1}{ad-bc} \begin{bmatrix} d & -b \\ -c & a \end{bmatrix}$.

28. Given that $A = \begin{bmatrix} 3 & 2 \\ -1 & 1 \end{bmatrix}$ and $B = \begin{bmatrix} 1 & -2 \\ 3 & 4 \end{bmatrix}$:

(a) Find A^{-1} and B^{-1}.
(b) Show that $(A^{-1})^{-1} = A$ and $(B^{-1})^{-1} = B$.
(c) Find $(AB)^{-1}$.
(d) Find $A^{-1}B^{-1}$ and $B^{-1}A^{-1}$.
(e) Which is true? $(AB)^{-1} = A^{-1}B^{-1}$ or $(AB)^{-1} = B^{-1}A^{-1}$.

In Exercises 29–36, use a graphing calculator to find the inverse for each matrix, if it exists. Where possible, express entries as fractions.

29. $\begin{bmatrix} 1.1 & 2.1 & 1.1 \\ 3.1 & 2.1 & 2.1 \\ 4.1 & 3.1 & 4.1 \end{bmatrix}$

30. $\begin{bmatrix} 2.2 & 3.2 & 1.2 \\ 4.2 & 1.2 & 5.2 \\ 1.2 & 0.2 & 3.2 \end{bmatrix}$

31. $\begin{bmatrix} 1 & 3 & -1 & 2 \\ 4 & 7 & -4 & -1 \\ 0 & 3 & 8 & 1 \\ 2 & 1 & 4 & 3 \end{bmatrix}$

32. $\begin{bmatrix} 1 & -1 & -2 & 3 \\ 7 & 1 & 8 & 2 \\ 3 & -2 & -1 & -4 \\ 0 & 3 & 5 & 7 \end{bmatrix}$

33. $\begin{bmatrix} 7 & -1 & 3 & 4 \\ 5 & 3 & -2 & 5 \\ 0 & 2 & -3 & -1 \\ 1 & -1 & 4 & 3 \end{bmatrix}$

34. $\begin{bmatrix} 2 & 1 & 7 & 8 \\ -3 & -1 & -4 & -1 \\ 5 & 9 & -1 & 3 \\ 0 & 1 & 2 & 1 \end{bmatrix}$

35. $\begin{bmatrix} 1 & 5 & 4 & 3 & 2 \\ 0 & 4 & -1 & 4 & 5 \\ -1 & -2 & 6 & -1 & 0 \\ 7 & -3 & 3 & 2 & 1 \\ 8 & 1 & 4 & 7 & 9 \end{bmatrix}$

36. $\begin{bmatrix} 1 & 0 & -1 & 7 & 8 \\ 5 & 4 & -2 & -3 & 1 \\ 4 & -1 & 6 & 3 & 4 \\ 3 & 4 & -1 & 2 & 7 \\ 2 & 5 & 0 & 1 & 9 \end{bmatrix}$

In Exercises 37–46, solve each system of linear equations by using the inverse of the coefficient matrix. See Example 6. (All inverses were found in Exercises 9–18.)

37. $\begin{aligned} x + 3y &= 7 \\ 4x + 2y &= 9 \end{aligned}$
(See Exercise 9)

38. $\begin{aligned} -x - 2y &= 8 \\ 3x + 2y &= -2 \end{aligned}$
(See Exercise 10)

39. $\begin{aligned} -2x + 7y &= 3 \\ -4x - y &= -5 \end{aligned}$
(See Exercise 11)

40. $\begin{aligned} 3x + 4y &= -1 \\ -2x - 3y &= 5 \end{aligned}$
(See Exercise 12)

41. $\begin{aligned} x - y + 2z &= 3 \\ 2y + 3z &= -1 \\ 3x - y + 4z &= 2 \end{aligned}$
(See Exercise 13)

42. $\begin{aligned} x + 2y + 7z &= -2 \\ -y + z &= -1 \\ 2x + y + 2z &= -3 \end{aligned}$
(See Exercise 14)

43. $\begin{aligned} 2x + y &= 3 \\ -x + y + 2z &= -2 \\ 4x + 2y + 3z &= 4 \end{aligned}$
(See Exercise 15)

44. $\begin{aligned} 3x + 2y &= -1 \\ -2x - y - 2z &= 2 \\ x + z &= 3 \end{aligned}$
(See Exercise 16)

45. $\begin{aligned} x + 2y - 2z + 3w &= 3 \\ 2y - 3z + w &= -1 \\ 2x + z + 2w &= 1 \\ -3x + y + 2z + 5w &= 4 \end{aligned}$
(See Exercise 17)

46. $\begin{aligned} x - 2y + 3z + 4w &= -2 \\ y - z + 2w &= 3 \\ -2x + y + 3z - 2w &= 1 \\ 3x + 2y + z + 5w &= -3 \end{aligned}$
(See Exercise 18)

In Exercises 47–52, solve each system of linear equations by using the inverse of the coefficient matrix and a graphing calculator. (All inverse were found in Exercises 29–36.)

47. $\begin{aligned} 1.1x + 2.1y + 1.1z &= 13.1 \\ 3.1x + 2.1y + 2.1z &= 5.1 \\ 4.1x + 3.1y + 4.1z &= 8.1 \end{aligned}$
(See Exercise 29)

48. $\begin{aligned} 2.2x + 3.2y + 1.2z &= 4.2 \\ 4.2x + 1.2y + 5.2z &= 1.2 \\ 1.2x + 0.2y + 3.2z &= 6.2 \end{aligned}$
(See Exercise 30)

49. $\begin{aligned} 7x - y + 3z + 4w &= 8 \\ 5x + 3y - 2z + 5w &= -1 \\ 2y - 3z - w &= 2 \\ x - y + 4z + 3w &= 3 \end{aligned}$
(See Exercise 33)

50. $\begin{aligned} 2x + y + 7z + 8w &= 2 \\ -3x - y - 4z - w &= -4 \\ 5x + 9y - z + 3w &= 5 \\ y + 2z + w &= 3 \end{aligned}$
(See Exercise 34)

51. $\begin{aligned} x + 5y + 4z + 3v + 2w &= 1 \\ 4y - z + 4v + 5w &= -2 \\ -x - 2y + 6z - v &= 2 \\ 7x - 3y + 3z + 2v + w &= -3 \\ 8x + y + 4z + 7v + 9w &= 4 \end{aligned}$
(See Exercise 35)

52.
$$\begin{aligned} x - z + 7v + 8w &= 4 \\ 5x + 4y - 2z - 3v + w &= -2 \\ 4x - y + 6z + 3v + 4w &= 3 \\ 3x + 4y - z + 2v + 7w &= 1 \\ 2x + 5y + v + 9w &= 2 \end{aligned}$$

(See Exercise 36)

Applications

53. Mary Lou Smith makes three types of bread, lemon poppy seed glaze (Type I), pumpkin glaze (Type II), and banana walnut glaze (Type III). Each type of bread requires the same basic ingredients as shown in matrix A.

$$A = \begin{array}{c} \begin{matrix} \text{I} & \text{II} & \text{III} \end{matrix} \\ \begin{bmatrix} 1 & 2 & 2 \\ 3 & 2 & 3 \\ 2 & 2 & 1 \end{bmatrix} \end{array} \begin{matrix} \text{Flour (in cups)} \\ \text{Sugar (in cups)} \\ \text{Eggs} \end{matrix}$$

Mary Lou has regular customers, so she makes these breads daily. To fill her daily orders for these three breads, the amount of basic ingredients used daily is given in matrix B.

$$B = \begin{bmatrix} 24 \\ 45 \\ 27 \end{bmatrix} \begin{matrix} \text{Flour (in cups)} \\ \text{Sugar (in cups)} \\ \text{Eggs} \end{matrix}$$

(a) Let the number of daily orders for the three bread types (Types I, II, and III) be represented by x, y, and z, respectively. Write a 3×1 matrix X with entries x, y, and z that represents the daily orders. Write a matrix equation in the form $AX = B$.

(b) Use A^{-1} to solve the equation in part (a).

54. During the holiday season, Mary Lou Smith from Exercise 53 makes more bread on a daily basis than usual. While matrix A remains the same, matrix B changes to

$$B = \begin{bmatrix} 30 \\ 56 \\ 34 \end{bmatrix} \begin{matrix} \text{Flour (in cups)} \\ \text{Sugar (in cups)} \\ \text{Eggs} \end{matrix}$$

(a) Let the number of daily orders for the three bread types (Types I, II, and III) be represented by x, y, and z, respectively. Write a 3×1 matrix X with entries x, y, and z that represents the daily orders. Write a matrix equation in the form $AX = B$.

(b) Use A^{-1} to solve the equation in part (a).

1.1 **55.** The Drama Department in a local high school is performing a play for three consecutive nights. The auditorium at the high school has 1600 seats, and tickets are broken into two categories: \$3 student tickets and \$5 adult tickets. Table 2.4.3 gives the data for ticket revenue and total tickets sold for each play night.

Table 2.4.3

	Thursday	Friday	Saturday
Tickets sold	1600	1600	1600
Revenue	\$5760	\$5980	\$5500

Let x represent the number of \$3 student tickets sold and let y represent the number of \$5 adult tickets sold.

(a) Write a system of equations for each night, Thursday, Friday, and Saturday.

(b) For each system in part (a), let matrix A be the coefficient matrix, matrix X the variable matrix, and matrix B the constants matrix. Write a matrix equation in the form $AX = B$ for each system in part (a).

(c) Use A^{-1} to solve each system in part (b) and interpret. (Notice that A^{-1} is the same for all three equations.)

1.1 **56.** (*continuation of Exercise 55*) Because each night was a sellout, the next semester the Drama Department moved the play to a local community college. The auditorium at the community college has 2800 seats. The ticket prices remain the same. Table 2.4.4 gives the data for ticket revenue and total tickets sold for each play night.

Table 2.4.4

	Thursday	Friday	Saturday
Tickets sold	2,800	2,800	2,800
Revenue	\$11,400	\$10,180	\$10,760

(a) Write a system of equations for each night, Thursday, Friday, and Saturday.

(b) For each system in part (a), let matrix A be the coefficient matrix, matrix X the variable matrix, and matrix B the constants matrix. Write a matrix equation in the form $AX = B$ for each system in part (a).

(c) Use A^{-1} to solve each system in part (b) and interpret. (Notice that A^{-1} is the same for all three equations.)

1.1 **57.** Davram manufactures bicycle frames at three different plants. Due to different technologies at each plant, the production rate (in frames/hour) differs for each plant, as shown in Table 2.4.5.

Table 2.4.5

	Plant I	Plant II	Plant III
Racing frame	18	20	16
Hybrid frame	12	9	11
Mountain frame	22	18	17

Shannon, the production manager at Davram, has received an order to produce 5900 racing frames, 3580 hybrid frames, and 6350 mountain frames. To fill this order exactly, Shannon lets x, y, and z represent the number of hours that plants I, II, and III, respectively, should be scheduled to operate.

(a) Write a system of equations to model this situation.

(b) Write a matrix equation in the form $AX = B$ for the system in part (a).

(c) Use A^{-1} to solve the equation in part (b) and interpret.

1.1 **58.** (*continuation of Exercise 57*) Shannon has received another order requesting 6040 racing frames, 3780 hybrid frames, and 6695 mountain frames. With everything else from Exercise 57 remaining the same, rework parts (a), (b), and (c). Notice that A^{-1} is the same as it was in Exercise 57.

59. The Basich Company makes three types of tents: two-person, four-person, and six-person models. Each tent requires the services of three departments, as shown in Table 2.4.6. (The entries in the table are in hours.) The fabric cutting, assembly, and packaging departments have available a maximum of 450, 390, and 190 hours each week, respectively.

Table 2.4.6

Department:	Two-person	Four-person	Six-person
Fabric cutting	0.5	0.8	1.1
Assembly	0.6	0.7	0.9
Packaging	0.2	0.3	0.6

Let x, y, and z represent the number of two-, four-, and six-person tents made in 1 week, respectively. Assuming that the company operates at full capacity, write a matrix equation in the form $AX = B$ that models this situation. Use

$$A^{-1} = \begin{bmatrix} -6 & 6 & 2 \\ \frac{36}{5} & -\frac{16}{5} & -\frac{42}{5} \\ -\frac{8}{5} & -\frac{2}{5} & \frac{26}{5} \end{bmatrix}$$ to solve the matrix equation and interpret.

60. Rework Exercise 59, except here assume that the fabric cutting, assembly, and packaging departments have available a maximum of 420, 390, and 185 hours each week, respectively.

61. Last year the Yagan Corporation took out three loans totaling \$120,000 to fund the purchase of new computer equipment. Some of the money was at 6%, \$5000 more than $\frac{1}{3}$ the amount of the 6% loan was at 8.5%, and the rest was at 7.5%. The total annual interest on the loans was \$8350. Let x represent the amount borrowed at 6%, y the amount borrowed at 7.5%, and z the amount borrowed at 8.5%.

(a) Write a matrix equation in the form $AX = B$ that models this scenario.

(b) Use A^{-1} to solve the matrix equation and interpret.

62. A dog breeder has available three commercial dog food mixes, Grava, Zita, and Blando. These mixes containing the percentages, as a decimal, of protein, carbohydrates, and fat are summarized in Table 2.4.7.

Table 2.4.7

	Grava	Zita	Blando
Protein	0.25	0.10	0.20
Carbohydrates	0.10	0.15	0.05
Fat	0.20	0.15	0.15

How many ounces of each mix should be used to prepare a diet for a cocker spaniel that requires 10.25 ounces of protein, 4.75 ounces of carbohydrates, and 8.75 ounces of fat?

63. Mona is a dietitian for a national weight loss program. Using only three basic foods, Lacto, Fracto, and Hima, she needs to arrange a restrictive diet for a new patient. The diet is to include 280 units of calcium, 175 units of vitamin B_{12}, and 225 units of vitamin B_6. The number of units per ounce for each basic food is given in Table 2.4.8. How many ounces of each food must be used to exactly meet the diet requirements?

Table 2.4.8

	Lacto	Fracto	Hima
Calcium	20	10	20
Vitamin B_{12}	10	5	15
Vitamin B_6	15	20	10

64. The data in Table 2.4.9 give the amounts of nitrogen, phosphate, and potash (all in lb/acre) needed to grow three crops for an entire growing season in California.

Table 2.4.9

	Asparagus	Cauliflower	Celery
Nitrogen	240	215	336
Phosphate	148	98	171
Potash	42	69	147

SOURCE: U.S. Department of Agriculture, www.usda.gov/nass

If George has 280 acres and uses 65,790 pounds of nitrogen, 31,860 pounds of phosphate, and 21,630 pounds of potash, how many acres did he allot for each crop in order to use all his resources completely?

65. The data in Table 2.4.10 give the amounts of nitrogen, phosphate, and potash (all in lb/acre) needed to grow three crops for an entire growing season in California.

Table 2.4.10

	Head Lettuce	Cantaloupe	Bell Peppers
Nitrogen	173	175	212
Phosphate	122	161	108
Potash	76	39	72

SOURCE: U.S. Department of Agriculture, www.usda.gov/nass

If Zach has 300 acres and uses 55,220 pounds of nitrogen, 39,380 pounds of phosphate, and 18,780 pounds of potash, how many acres did he allot for each crop in order to use all his resources completely?

SECTION PROJECT

This section project illustrates how matrices and inverse matrices can be used in **cryptography**, that is, encoding and decoding messages. To start, we assign the numbers 1 through 26 to the letters of the alphabet (A = 1, B = 2, C = 3, ..., Y = 25, Z = 26) and assign the number 27 to a blank so that we can put a space between words. For example, the message GET LOST corresponds to

7 5 20 27 12 15 19 20

Now any square matrix whose entries are positive integers (and whose inverse exists) can be used to encode the message. Such a matrix is called an **encoding matrix**. For example, we can use encoding matrix E:

$$E = \begin{bmatrix} 1 & 2 \\ 3 & 4 \end{bmatrix}$$

to encode the message. To do this, we put our sequence of numbers into matrix A.

$$A = \begin{bmatrix} 7 & 20 & 12 & 19 \\ 5 & 27 & 15 & 20 \end{bmatrix}$$

Notice that we entered the sequence going down each column. Matrix product EA encodes the message.

$$EA = \begin{bmatrix} 1 & 2 \\ 3 & 4 \end{bmatrix}\begin{bmatrix} 7 & 20 & 12 & 19 \\ 5 & 27 & 15 & 20 \end{bmatrix} = \begin{bmatrix} 17 & 74 & 42 & 59 \\ 41 & 168 & 96 & 137 \end{bmatrix}$$

Hence the original message has been coded as

17 41 74 168 42 96 59 137

To decode the message, the receiver of the message needs E^{-1}, the inverse matrix of the encoding matrix. E^{-1} is called the **decoding matrix**. Here E^{-1} is

$$E^{-1} = \begin{bmatrix} -2 & 1 \\ \frac{3}{2} & -\frac{1}{2} \end{bmatrix}$$

To decode, put the coded message into matrix form and multiply on the left by E^{-1}.

(a) You have received the following sequence of a coded message

31 70 21 62 36 81 46 119 37 90

It was encoded with matrix $E = \begin{bmatrix} 1 & 1 \\ 2 & 3 \end{bmatrix}$. Decode the message.

(b) Use the encoding matrix in part (a) to encode the message MEET ME AT NOON.

(c) Determine an encoding matrix E and encode any message that you desire. Give the encoded message and matrix E to a friend and see if he/she can decode it.

Section 2.5 Leontief Input–Output Models

In this concluding section of Chapter 2, we introduce one of the most important applications of matrices and their inverses and systems of equations to economics. It is called the **Leontief input–output** model, named after Wassily Leontief. Wassily Leontief was awarded the Nobel prize in economics in 1973 because of this matrix theory applied to economics. What Leontief did was organize the 1958 U.S. economy into an 81 × 81 matrix. The 81 sectors of the U.S. economy, such as steel, agriculture, manufacturing, and transportation, each represented resources that relied on input from the output of other resources. For example, production of agriculture requires an input of manufacturing, transportation, and even some agriculture. Our discussion will not be nearly as in depth as Leontief's; however, the ideas that we present can be generalized to an economy of any size.

Input–Output Models

Input–output analysis deals with the relationship between the inputs and outputs of different sectors of an economy. To help us understand the concepts at hand, let's consider a simple model of an economy in which only three goods are produced: petroleum, transportation, and chemicals. These three industries are called **sectors** in a three-sector economy. Any industry that does not produce petroleum, transportation, or chemicals is said to be outside the economy. However, industries (or governments) outside the economy may have demands on the outputs of these three sectors. These types of demands are called **external demands**. Now the situation becomes a little more complicated because the three sectors tend to use their own outputs. For example, petroleum may demand as an input the output from transportation. These types of demands among the three sectors are referred to as **internal demands**.

Internal Demands

To get a handle on internal demands, and the most important matrix that it leads to, consider the following for our three-sector economy:

- One unit of petroleum (P) requires 0.1 unit of itself, 0.2 unit of transportation, and 0.4 unit of chemicals.
- One unit of transportation (T) requires 0.6 unit of petroleum, 0.1 unit of itself, and 0.25 unit of chemicals.
- One unit of chemicals (C) requires 0.2 unit of petroleum, 0.3 unit of transportation, and 0.2 unit of chemicals.

Units are usually measured in dollars. Hence we can view the above as saying that the production of $1 worth of petroleum requires $0.10 of petroleum, $0.20 of transportation, and $0.40 of chemicals. The information above can be nicely summarized in what is the central idea of input–output analysis: the input–output or **technological matrix**, matrix T.

Output (amount produced)

		Petr	Tran	Chem
$T =$ Input (amount used in production)	Petr	0.1	0.6	0.2
	Tran	0.2	0.1	0.3
	Chem	0.4	0.25	0.2

Notice that in matrix T the 1st column read top to bottom tells us the requirements to produce 1 unit of petroleum.

Example 1 Determining an Internal Demand

If the three-sector economy produces $900 million of petroleum, $850 million of transportation, and $800 million of chemicals, determine how much of this production is consumed internally. This internal consumption is also known as the **internal demand**.

Solution

We begin by organizing the information in a 3×1 matrix P.

$$P = \begin{bmatrix} 900 \\ 850 \\ 800 \end{bmatrix} \begin{matrix} \text{Petr} \\ \text{Tran} \\ \text{Chem} \end{matrix}$$

Matrix P is called the **production matrix** or the **total output matrix**. The internal demand is simply the matrix product TP. This gives

$$TP = \begin{bmatrix} 0.1 & 0.6 & 0.2 \\ 0.2 & 0.1 & 0.3 \\ 0.4 & 0.25 & 0.2 \end{bmatrix} \begin{bmatrix} 900 \\ 850 \\ 800 \end{bmatrix} = \begin{bmatrix} 760 \\ 505 \\ 732.5 \end{bmatrix} \begin{matrix} \text{Petr} \\ \text{Tran} \\ \text{Chem} \end{matrix}$$

Interpret the Solution: The entries of TP tell us that the internal demand for petroleum, transportation, and chemicals is \$760 million, \$505 million, and \$732.5 million, respectively. ■

Checkpoint 1

Now work Exercise 5.

External Demands and Leontief

In Example 1 we saw that the production (or total output) matrix was $P = \begin{bmatrix} 900 \\ 850 \\ 800 \end{bmatrix}$ and the internal demand was $\begin{bmatrix} 760 \\ 505 \\ 732.5 \end{bmatrix}$. To determine the **external demand** matrix D, we simply subtract internal demand from matrix P. That is,

$$\text{external demand} = D = \begin{bmatrix} 900 \\ 850 \\ 800 \end{bmatrix} - \begin{bmatrix} 760 \\ 505 \\ 732.5 \end{bmatrix} = \begin{bmatrix} 140 \\ 345 \\ 67.5 \end{bmatrix}$$

Observe that this leads to the following:

$$\text{production} = \text{internal demand} + \text{external demand}$$

$$\begin{bmatrix} 900 \\ 850 \\ 800 \end{bmatrix} = \begin{bmatrix} 760 \\ 505 \\ 732.5 \end{bmatrix} + \begin{bmatrix} 140 \\ 345 \\ 67.5 \end{bmatrix}$$

This is a fundamental observation of input–output analysis.

$$\text{Production} = \text{Internal demand} + \text{External demand}$$

or

$$\text{Total output} = \text{Internal demand} + \text{External demand}$$

Using the result of Example 1 and the discussion following Example 1, we can use matrices and write the prior boxed equation as

$$\begin{aligned} \text{Total output} &= \text{Internal demand} + \text{External demand} \\ P &= TP + D \end{aligned} \qquad \text{(I)}$$

This simple model will be our guide as we develop a technique to attack the major goal of input–output analysis. We now present this major goal.

Basic Input–Output Problem

Given the technological matrix for an economy and the external demands, determine the total output matrix (production matrix) of each sector that will meet both the internal and external demands.

Our basic input–output problem suggests that usually we are not given matrix P in equation (I). This is what we need to determine! Since these are usually our unknowns, let's work through our three-sector economy and see how matrices (and matrix inverses) will aid us. First let's call the output of the three sectors

$$x_1 = \text{dollar value of output of petroleum}$$
$$x_2 = \text{dollar value of output of transportation}$$
$$x_3 = \text{dollar value of output of chemicals}$$

This means that matrix P now has the form

$$P = \begin{bmatrix} x_1 \\ x_2 \\ x_3 \end{bmatrix}$$

Since this is a variable matrix, let's rename it X. This means equation (I) becomes

$$X = TX + D \tag{II}$$

Example 2 Solving for Total Output

Using the technological matrix T for our three-sector economy and external demand matrix D, use equation (II) and solve for X.

Solution

Substituting the given information into equation (II) and solving for X gives

$$\begin{bmatrix} x_1 \\ x_2 \\ x_3 \end{bmatrix} = \begin{bmatrix} 0.1 & 0.6 & 0.2 \\ 0.2 & 0.1 & 0.3 \\ 0.4 & 0.25 & 0.2 \end{bmatrix} \begin{bmatrix} x_1 \\ x_2 \\ x_3 \end{bmatrix} + \begin{bmatrix} 140 \\ 345 \\ 67.5 \end{bmatrix} \qquad X = TX + D$$

$$\begin{bmatrix} x_1 \\ x_2 \\ x_3 \end{bmatrix} - \begin{bmatrix} 0.1 & 0.6 & 0.2 \\ 0.2 & 0.1 & 0.3 \\ 0.4 & 0.25 & 0.2 \end{bmatrix} \begin{bmatrix} x_1 \\ x_2 \\ x_3 \end{bmatrix} = \begin{bmatrix} 140 \\ 345 \\ 67.5 \end{bmatrix} \qquad X - TX = D$$

$$\left(\begin{bmatrix} 1 & 0 & 0 \\ 0 & 1 & 0 \\ 0 & 0 & 1 \end{bmatrix} - \begin{bmatrix} 0.1 & 0.6 & 0.2 \\ 0.2 & 0.1 & 0.3 \\ 0.4 & 0.25 & 0.2 \end{bmatrix} \right) \begin{bmatrix} x_1 \\ x_2 \\ x_3 \end{bmatrix} = \begin{bmatrix} 140 \\ 345 \\ 67.5 \end{bmatrix} \qquad (I - T)X = D$$

$$\begin{bmatrix} 0.9 & -0.6 & -0.2 \\ -0.2 & 0.9 & -0.3 \\ -0.4 & -0.25 & 0.8 \end{bmatrix} \begin{bmatrix} x_1 \\ x_2 \\ x_3 \end{bmatrix} = \begin{bmatrix} 140 \\ 345 \\ 67.5 \end{bmatrix} \qquad \text{Simplification}$$

We now multiply both sides of the equation, on the left, by the inverse of the coefficient matrix.

$$\begin{bmatrix} \frac{1290}{661} & \frac{1060}{661} & \frac{720}{661} \\ \frac{560}{661} & \frac{1280}{661} & \frac{620}{661} \\ \frac{820}{661} & \frac{930}{661} & \frac{1380}{661} \end{bmatrix} \begin{bmatrix} 0.9 & -0.6 & -0.2 \\ -0.2 & 0.9 & -0.3 \\ -0.4 & -0.25 & 0.8 \end{bmatrix} \begin{bmatrix} x_1 \\ x_2 \\ x_3 \end{bmatrix} = \begin{bmatrix} \frac{1290}{661} & \frac{1060}{661} & \frac{720}{661} \\ \frac{560}{661} & \frac{1280}{661} & \frac{620}{661} \\ \frac{820}{661} & \frac{930}{661} & \frac{1380}{661} \end{bmatrix} \begin{bmatrix} 140 \\ 345 \\ 67.5 \end{bmatrix}$$

$$\begin{bmatrix} 1 & 0 & 0 \\ 0 & 1 & 0 \\ 0 & 0 & 1 \end{bmatrix} \begin{bmatrix} x_1 \\ x_2 \\ x_3 \end{bmatrix} = \begin{bmatrix} 900 \\ 850 \\ 800 \end{bmatrix}$$

$$\begin{bmatrix} x_1 \\ x_2 \\ x_3 \end{bmatrix} = \begin{bmatrix} 900 \\ 850 \\ 800 \end{bmatrix}$$

Interpret the Solution: An output of $900 million in petroleum, $850 million in transportation, and $800 million in chemicals will satisfy internal and external demands.

The result of Example 2 is no surprise since the results were given to us in Example 1. There are two reasons we went through the calculations in Example 2. The first is to illustrate that we *can* find the outputs x_1, x_2, and x_3, and the second reason is to show numerically how to solve for matrix X. We now generalize our work in Example 2 by algebraically solving $X = TX + D$ for X.

$$X = TX + D \qquad \text{Given equation}$$
$$X - TX = D \qquad \text{Subtract } TX \text{ from both sides.}$$
$$IX - TX = D \qquad I \text{ is the identity matrix.}$$
$$(I - T)X = D \qquad \text{Factor out } X.$$
$$(I - T)^{-1}(I - T)X = (I - T)^{-1}D \qquad \text{Provided } I - T \text{ has an inverse}$$
$$IX = (I - T)^{-1}D \qquad (I - T)^{-1}(I - T) = I$$
$$X = (I - T)^{-1}D \qquad \text{Simplification}$$

We now summarize these results.

Leontief Input–Output Model

In a Leontief input–output model, the matrix equation giving the total output of each sector that meets the internal and external demands is

$$X = TX + D$$

where X is the **total output matrix**, T is the **technological matrix**, and D is the **external demand matrix**. The solution of this equation is

$$X = (I - T)^{-1}D \qquad (\text{provided } I - T \text{ has an inverse})$$

Example 3 Determining Total Output

Bucksville has three main industries: a coal-mining operation, an electric power-generating plant, and a local railroad. To mine $1 of coal, the mining operation must purchase $0.35 of electricity and $0.15 of transportation. To produce $1 of electricity, the generating plant requires $0.60 of coal, $0.10 of its own electricity, and $0.05 of transportation. To provide $1 of transportation, the railroad needs $0.50 of coal and $0.15 of electricity. In a certain week, the coal-mining operation receives an order for $50,000 of coal outside Bucksville and the electric generating plant receives an order for $30,000 of electricity outside Bucksville. For this week, there is no external demand for the local railroad. How much must each industry produce in this week to exactly meet the internal and external demands?

Solution

Understand the Situation: We begin by determining the technological matrix T and the external demand matrix D.

$$T = \text{Input} \begin{array}{c} \text{Coal} \\ \text{Elec.} \\ \text{RR} \end{array} \begin{array}{c} \overset{\text{Output}}{\overset{\text{Coal Elec. RR}}{\begin{bmatrix} 0 & 0.60 & 0.50 \\ 0.35 & 0.10 & 0.15 \\ 0.15 & 0.05 & 0 \end{bmatrix}}} \end{array} \qquad D = \begin{bmatrix} 50{,}000 \\ 30{,}000 \\ 0 \end{bmatrix} \begin{array}{c} \text{Coal} \\ \text{Elec.} \\ \text{RR} \end{array}$$

We need to find

$$X = \begin{bmatrix} x_1 \\ x_2 \\ x_3 \end{bmatrix}$$

where x_1, x_2, and x_3 represent the value of coal products, electricity, and the railroad, respectively.

From the Leontief input–output model we know that

$$X = (I - T)^{-1} D$$

To determine $(I - T)^{-1}$, we first compute $I - T$.

$$I - T = \begin{bmatrix} 1 & 0 & 0 \\ 0 & 1 & 0 \\ 0 & 0 & 1 \end{bmatrix} - \begin{bmatrix} 0 & 0.60 & 0.50 \\ 0.35 & 0.10 & 0.15 \\ 0.15 & 0.05 & 0 \end{bmatrix} = \begin{bmatrix} 1 & -0.6 & -0.5 \\ -0.35 & 0.9 & -0.15 \\ -0.15 & -0.05 & 1 \end{bmatrix}$$

Using the methods in Section 2.4, we compute $(I - T)^{-1}$ to be

$$(I - T)^{-1} = \begin{bmatrix} \frac{3570}{2371} & \frac{2500}{2371} & \frac{2160}{2371} \\ \frac{1490}{2371} & \frac{3700}{2371} & \frac{1300}{2371} \\ \frac{610}{2371} & \frac{560}{2371} & \frac{2760}{2371} \end{bmatrix}$$

Finally, we compute X to be

$$X = (I - T)^{-1} D = \begin{bmatrix} \frac{3570}{2371} & \frac{2500}{2371} & \frac{2160}{2371} \\ \frac{1490}{2371} & \frac{3700}{2371} & \frac{1300}{2371} \\ \frac{610}{2371} & \frac{560}{2371} & \frac{2760}{2371} \end{bmatrix} \begin{bmatrix} 50{,}000 \\ 30{,}000 \\ 0 \end{bmatrix}$$

$$\approx \begin{bmatrix} 106{,}916.91 \\ 78{,}237.03 \\ 19{,}949.39 \end{bmatrix} \quad \text{(entries rounded to nearest hundredth)}$$

Interactive Activity

Determine the internal demand of the three industries in Example 3.

WWW

Interpret the Solution: To fulfill internal and external demand, \$106,916.91 worth of coal, \$78,237.03 worth of electricity, and \$19,949.39 worth of railroad transportation should be produced. ■

Checkpoint 2 Now work Exercise 25.

SUMMARY

This section saw the presentation of one of the most important applications of matrices and their inverses to economics: the Leontief input–output model. Our presentation revolved around a simple three-sector economy; however, the concepts in this section easily extend to any n-sector economy. This includes the 81 sectors that Wassily Leontief analyzed in the 1958 U.S. economy. (You would want to use a computer and a spreadsheet for 81 sectors.) The most important part of Leontief input–output is the **technological matrix**. Other important components include **internal demands** and **external demands**.

- **Basic input–output problem** Given the technological matrix for an economy and the external demands, determine the total output matrix (production matrix) of each sector that will meet both the internal and external demands.
- **Leontief input–output model** In a Leontief input–output model, the matrix equation giving the total output of each sector that meets the internal and external demands is

$$X = TX + D$$

where X is the total output matrix, T is the technological matrix, and D is the external demand matrix. The solution of this equation is

$$X = (I - T)^{-1} D \quad \text{(provided } I - T \text{ has an inverse)}$$

SECTION 2.5 EXERCISES

Exercises 1–4 use the following technological matrix T. Here agr, manu, trd, and srv represent agriculture, manufacturing, trade, and service, respectively.

$$T = \begin{array}{c} \\ \text{agr} \\ \text{manu} \\ \text{trd} \\ \text{srv} \end{array} \begin{bmatrix} \overset{\text{agr}}{0.27} & \overset{\text{manu}}{0.39} & \overset{\text{trd}}{0} & \overset{\text{srv}}{0.02} \\ 0.05 & 0.15 & 0 & 0.01 \\ 0.06 & 0.07 & 0.36 & 0.15 \\ 0.12 & 0.13 & 0.2 & 0.17 \end{bmatrix}$$

Determine the number of units of agriculture, manufacturing, trade, and service required to produce:

1. One unit of agriculture

2. One unit of manufacturing

3. One unit of trade

4. One unit of service

In Exercises 5–8, a technological matrix T and a total output matrix X for each sector is given. Determine the internal demand. Assume that all entries represent dollars. See Example 1.

✓ **5.** $T = \begin{array}{c} \text{Util} \\ \text{Trans} \end{array} \begin{bmatrix} \overset{\text{Util}}{0.25} & \overset{\text{Trans}}{0.40} \\ 0.30 & 0.10 \end{bmatrix}$ $X = \begin{bmatrix} 25{,}000 \\ 30{,}000 \end{bmatrix} \begin{array}{c} \text{Util} \\ \text{Trans} \end{array}$

6. $T = \begin{array}{c} \text{Util} \\ \text{Trans} \end{array} \begin{bmatrix} \overset{\text{Util}}{0.20} & \overset{\text{Trans}}{0.45} \\ 0.30 & 0.05 \end{bmatrix}$ $X = \begin{bmatrix} 30{,}000 \\ 40{,}000 \end{bmatrix} \begin{array}{c} \text{Util} \\ \text{Trans} \end{array}$

7. $T = \begin{array}{c} \text{Agr} \\ \text{Cons} \\ \text{Util} \end{array} \begin{bmatrix} \overset{\text{Agr}}{0.05} & \overset{\text{Cons}}{0.03} & \overset{\text{Util}}{0.04} \\ 0.01 & 0.23 & 0.09 \\ 0.15 & 0.17 & 0.17 \end{bmatrix}$

$X = \begin{bmatrix} 30{,}000 \\ 50{,}000 \\ 45{,}000 \end{bmatrix} \begin{array}{c} \text{Agr} \\ \text{Cons} \\ \text{Util} \end{array}$

8. $T = \begin{array}{c} \text{Agr} \\ \text{Cons} \\ \text{Util} \end{array} \begin{bmatrix} \overset{\text{Agr}}{0.04} & \overset{\text{Cons}}{0.02} & \overset{\text{Util}}{0.01} \\ 0.02 & 0.29 & 0.16 \\ 0.23 & 0.21 & 0.19 \end{bmatrix}$

$X = \begin{bmatrix} 25{,}000 \\ 60{,}000 \\ 50{,}000 \end{bmatrix} \begin{array}{c} \text{Agr} \\ \text{Cons} \\ \text{Util} \end{array}$

Exercises 9–14 utilize the following Leontief input–output model: An economy is based on two sectors, construction (C) and services (S). The technological matrix T and the external demand matrices D_1, D_2, and D_3, in millions of dollars, are

$$T = \begin{array}{c} \text{C} \\ \text{S} \end{array} \begin{bmatrix} \overset{\text{C}}{0.2} & \overset{\text{S}}{0.1} \\ 0.4 & 0.6 \end{bmatrix} \qquad D_1 = \begin{bmatrix} 5 \\ 3 \end{bmatrix} \begin{array}{c} \text{C} \\ \text{S} \end{array}$$

$$D_2 = \begin{bmatrix} 6 \\ 5 \end{bmatrix} \begin{array}{c} \text{C} \\ \text{S} \end{array} \qquad D_3 = \begin{bmatrix} 8 \\ 6 \end{bmatrix} \begin{array}{c} \text{C} \\ \text{S} \end{array}$$

9. Determine the number of units (dollars) of C and S required to produce \$1 worth of C.

10. Determine the number of units (dollars) of C and S required to produce \$1 worth of S.

11. Determine $I - T$ and $(I - T)^{-1}$.

12. Determine the total output for each sector that is required to satisfy D_1. (In other words, find the total output matrix X.)

13. Determine the total output for each sector that is required to satisfy D_2. (In other words, find the total output matrix X.)

14. Determine the total output for each sector that is required to satisfy D_3. (In other words, find the total output matrix X.)

Exercises 15–20 utilize the following Leontief input–output model: An economy is based on three sectors, agriculture (A), manufacturing (M), and transportation (T). The technological matrix T and the external demand matrices D_1 and D_2, in millions of dollars, are

$$T = \begin{array}{c} \text{A} \\ \text{M} \\ \text{T} \end{array} \begin{bmatrix} \overset{\text{A}}{0.2} & \overset{\text{M}}{0.2} & \overset{\text{T}}{0.1} \\ 0.4 & 0.5 & 0.2 \\ 0.1 & 0 & 0.2 \end{bmatrix}$$

$$D_1 = \begin{bmatrix} 9 \\ 12 \\ 16 \end{bmatrix} \begin{array}{c} \text{A} \\ \text{M} \\ \text{T} \end{array} \qquad D_2 = \begin{bmatrix} 8 \\ 10 \\ 15 \end{bmatrix} \begin{array}{c} \text{A} \\ \text{M} \\ \text{T} \end{array}$$

15. Determine the number of units (dollars) of A, M, and T required to produce \$1 worth of M.

16. Determine the number of units (dollars) of A, M, and T required to produce \$1 worth of T.

17. Find $(I - T)$.

18. Given $(I - T)^{-1} = \begin{bmatrix} \frac{400}{247} & \frac{160}{247} & \frac{90}{247} \\ \frac{340}{247} & \frac{630}{247} & \frac{200}{247} \\ \frac{50}{247} & \frac{20}{247} & \frac{320}{247} \end{bmatrix}$, determine the

total output for each sector required to satisfy D_1.

19. Repeat Exercise 18 for D_2.

20. Explain why the entries in the columns of any technological matrix should not sum to greater than 1.

Applications

21. An economy has two main industries: agriculture and energy. Production of \$1 worth of agriculture requires \$0.10 from the agriculture sector and \$0.35 from the energy sector. Production of \$1 worth of energy requires \$0.05 from the agriculture sector and \$0.25 from the energy sector. Find the total outputs for each sector that are needed to meet the external demand of \$5 billion for agriculture and \$12 billion for energy.

22. An economy has two main industries: agriculture and chemicals. Production of \$1 worth of agriculture requires \$0.10 from the agriculture sector and \$0.45 from the chemicals sector. Production of \$1 worth of chemicals requires \$0.05 from the agriculture sector and \$0.35 from the chemicals sector. Find the total outputs for each sector that are needed to meet the external demand of \$5 billion for agriculture and \$11 billion for chemicals.

23. A much simplified version of the U.S. economy in 1996, which deals with only two sectors, has the following technological matrix:

$$T = \begin{array}{r} \\ \text{Agri} \\ \text{Manu} \end{array} \begin{array}{c} \begin{array}{cc} \text{Agri} & \text{Manu} \end{array} \\ \begin{bmatrix} 0.24 & 0.04 \\ 0.17 & 0.36 \end{bmatrix} \end{array}$$

The external demand (in millions of dollars) as given by exports is

$$D = \begin{bmatrix} 27{,}066 \\ 465{,}357 \end{bmatrix} \begin{array}{l} \text{Agri} \\ \text{Manu} \end{array}$$

Find the total outputs of each sector that are needed to meet the external demands. (*Source:* Bureau of Economic Analysis, www.bea.gov.)

24. A much simplified version of the U.S. economy in 1996, which deals with only two sectors, has the following technological matrix:

$$T = \begin{array}{r} \\ \text{Minerals} \\ \text{Const} \end{array} \begin{array}{c} \begin{array}{cc} \text{Minerals} & \text{Const} \end{array} \\ \begin{bmatrix} 0.18 & 0.007 \\ 0.02 & 0.0008 \end{bmatrix} \end{array}$$

The external demand (in millions of dollars) as given by exports is

$$D = \begin{bmatrix} 8123 \\ 97 \end{bmatrix} \begin{array}{l} \text{Minerals} \\ \text{Const} \end{array}$$

Find the total outputs of each sector that are needed to meet the external demands. (*Source:* Bureau of Economic Analysis, www.bea.gov.)

25. An economy has three main industries: agriculture (A), manufacturing (M), and trade (T). Production of \$1 worth of agriculture requires \$0.27 from the agriculture sector, \$0.05 from the manufacturing sector, and \$0.06 from the trade sector. Production of \$1 worth of manufacturing requires \$0.39 from the agriculture sector, \$0.15 from the manufacturing sector, and \$0.07 from the trade sector. Production of \$1 worth of trade requires nothing from agriculture and manufacturing and \$0.36 from the trade sector.

(a) Write the technological matrix T for this economy.

(b) The external demand is \$8 billion for agriculture, \$23 billion for manufacturing, and \$10 billion for trade. Write the external demand matrix D.

(c) Given that $(I - T)^{-1} = \begin{bmatrix} 1.41 & 0.65 & 0 \\ 0.08 & 1.21 & 0 \\ 0.14 & 0.19 & 1.56 \end{bmatrix}$, (entries rounded to the nearest hundredth), determine the total outputs that are needed to meet the external demand.

26. An economy has three main industries: manufacturing (M), trade (T), and service (S). Production of \$1 worth of manufacturing requires \$0.15 from the manufacturing sector, \$0.07 from the trade sector, and \$0.13 from the service sector. Production of \$1 worth of trade requires nothing from manufacturing, \$0.36 from the trade sector, and \$0.20 from the service sector. Production of \$1 worth of service requires \$0.01 from the manufacturing sector, \$0.15 from the trade sector, and \$0.17 from the service sector.

(a) Write the technological matrix T for this economy.

(b) The external demand is \$12 billion for manufacturing, \$20 billion for trade, and \$8 billion for service. Write the external matrix D.

(c) Given that $(I - T)^{-1} = \begin{bmatrix} 1.18 & 0.05 & 0.02 \\ 0.18 & 1.66 & 0.30 \\ 0.23 & 0.40 & 1.28 \end{bmatrix}$, (entries rounded to the nearest hundredth), determine the total outputs that are needed to meet the external demand.

27. A much simplified version of the U.S. economy in 1996, which deals with only three sectors, has the following technological matrix:

$$T = \begin{array}{r} \\ \text{Agri} \\ \text{Manu} \\ \text{Trade} \end{array} \begin{array}{c} \begin{array}{ccc} \text{Agri} & \text{Manu} & \text{Trade} \end{array} \\ \begin{bmatrix} 0.24 & 0.04 & 0.0009 \\ 0.17 & 0.36 & 0.05 \\ 0.05 & 0.06 & 0.02 \end{bmatrix} \end{array}$$

The external demand (in millions of dollars) as given by exports is

$$D = \begin{bmatrix} 27{,}066 \\ 465{,}357 \\ 66{,}786 \end{bmatrix} \begin{array}{l} \text{Agri} \\ \text{Manu} \\ \text{Trade} \end{array}$$

Find the total output of each sector that is needed to meet the external demands. (*Source:* Bureau of Economic Analysis, www.bea.gov.)

28. A much simplified version of the U.S. economy in 1996, which deals with only three sectors, has the following technological matrix:

$$T = \begin{array}{r} \\ \text{Minerals} \\ \text{Const} \\ \text{Manu} \end{array} \begin{array}{c} \begin{array}{ccc} \text{Minerals} & \text{Const} & \text{Manu} \end{array} \\ \begin{bmatrix} 0.18 & 0.007 & 0.03 \\ 0.02 & 0.0008 & 0.007 \\ 0.08 & 0.30 & 0.36 \end{bmatrix} \end{array}$$

The external demand (in millions of dollars) as given by exports is

$$D = \begin{bmatrix} 8123 \\ 97 \\ 465{,}357 \end{bmatrix} \begin{array}{l} \text{Minerals} \\ \text{Const} \\ \text{Manu} \end{array}$$

Find the total output of each sector that is needed to meet the external demands. (*Source:* Bureau of Economic Analysis, www.bea.gov.)

SECTION PROJECT

We began this section by stating that Wassily Leontief organized the 1958 U.S. economy into an 81 × 81 matrix. He also aggregated these 81 sectors into six groups. This section project deals with four of these six groups. The technological matrix for the four groups is

$$T = \begin{array}{c} \text{FM} \\ \text{BM} \\ \text{BN} \\ \text{E} \end{array} \overset{\begin{array}{cccc} \text{FM} & \text{BM} & \text{BN} & \text{E} \end{array}}{\begin{bmatrix} 0.295 & 0.018 & 0.002 & 0.004 \\ 0.173 & 0.46 & 0.007 & 0.011 \\ 0.037 & 0.021 & 0.403 & 0.011 \\ 0.001 & 0.039 & 0.025 & 0.358 \end{bmatrix}}$$

FM represents final metals, such as construction machinery and household goods.

BM represents basic metal, such as mining and machine shop products.

BN represents basic nonmetal, such as glass, wood, and textiles.

E represents energy, such as coal, petroleum, electricity, and gas.

Matrix D gives the estimated external demands for the 1958 U.S. economy, in millions of dollars.

$$D = \begin{bmatrix} 75{,}548 \\ 14{,}444 \\ 33{,}501 \\ 23{,}527 \end{bmatrix} \begin{array}{c} \text{FM} \\ \text{BM} \\ \text{BN} \\ \text{E} \end{array}$$

(a) Use a graphing calculator to determine the total output for each sector.

(b) If an energy crisis increased the cost of energy by 20% in each sector, what would the total output for each sector need to be?

Why We Learned It

In this chapter we learned that matrices are common in business, life science, and other occupations. Knowing the characteristics of matrices, as well as operations used with matrices, is vital for applying them to the many diversified applications. For example, we saw that a business may use matrices to determine the maximum production capacity of a product, and we saw that a farmer can use matrix multiplication to determine the amount of nutrients that have to be added to crops for a growing season. Another example of matrix operations is that matrices can be used to keep track of inventories and the sales of certain items and the revenue generated from these sales at different stores. Matrix multiplication quickly led us to learn about inverse matrices. Inverse matrices were used to solve systems of equations. For example, we saw how a commercial baker could use matrix inverses to determine how much of each ingredient is needed to fill an order. Finally, Leontief input–output models are an application of matrices and inverse matrices, frequently used by economists, to analyze the relationship between production and demands for various sectors in an economy.

CHAPTER REVIEW EXERCISES

In Exercises 1 and 2, determine the dimension of the given matrix.

1. $A = \begin{bmatrix} 0 & 5 \\ -1 & 7 \\ 2 & 8 \end{bmatrix}$

2. $B = [-3 \quad 4 \quad 6 \quad -1]$

In Exercises 3 and 4, determine the indicated entry for the matrices given in Exercises 1 and 2.

3. $a_{2,1}$

4. $b_{1,3}$

5. Write an augmented matrix for the given system. **Do not solve.**

$$\begin{aligned} -2x + 3y - z &= 5 \\ x + y + z &= 13 \\ 4x - 5y + 2z &= -1 \end{aligned}$$

6. Write the linear system corresponding to the augmented matrix.

$$\left[\begin{array}{cc|c} -2 & 5 & -7 \\ 1 & 4 & 3 \end{array}\right]$$

In Exercises 7 10, use the Gauss–Jordan method to solve the given system of linear equations.

7. $-3x + 4y = 11$
$x + 5y = 9$

8. $-2x + 5y - z = 6$
$x - 10y + 4z = 0$
$3x + y - 2z = -11$

9. $x + 3y - z = 1$
$2x + 3y + 4z = 5$
$-3x - 9y + 3z = -3$

10. $x - 2y + z - 3w = -10$
$3x - y + 2z + w = -1$
$-x + 2y - z - 3w = -8$
$2x - 3y + 3z - w = -4$

In Exercises 11 and 12, use the Gauss–Jordan method to solve the given system of equations using your calculator's matrix capabilities (either step-by-step row operators or rref(command). Round answers to four decimal places.

11. $-3.7x + 5.1y = 12.1$
$0.5x - 14.2y = 3.8$

12. $-2.17x + 5.13y - 0.42z + 4.12w = 5.67$
$5.89x - 3.24y + 8.53z + 1.02w = -3.29$
$-0.05x + 1.07y - 0.43z + 8.59w = 0.78$
$1.14x + 2.29y + 1.92z - 7.38w = 1.11$

13. Grace Richards makes 4-weight and 5-weight fly-fishing rods. To make a 4-weight rod takes 6 hours of labor and to make a 5-weight rod takes 8 hours of labor. The materials for one 4-weight rod cost her \$60 and the materials for one 5-weight rod cost her \$80. Suppose that for 1 week Grace works 40 hours and spends \$400 on materials.

(a) Let x represent the number of 4-weight rods made in 1 week and y represent the number of 5-weight rods made in 1 week. Write a system of two linear equations in which one equation relates costs and the other relates time.

(b) Solve the system in part (a) to determine how many 4- and 5-weight rods Grace can make in 1 week, assuming that she works 40 hours and spends \$400 on materials.

14. The Smoothie-Go-Go drink stand carries two blends of smoothies. The first blend contains 70% fruit and 30% yogurt. The second blend contains 85% fruit and 15% yogurt. How many ounces of each blend should be combined to get a 16-ounce drink that is 75% fruit and 25% yogurt?

15. The Quinn Outdoor Company makes three types of day backpacks: small, medium, and large. Each size requires the resources of three departments, as shown in Table R.1. The fabric cutting, assembly, and packing departments have available a maximum of 310, 405, and 150 hours each week, respectively.

Table R.1

	Small (hr)	Medium (hr)	Large (hr)
Fabric cutting	0.3	0.4	0.6
Assembly	0.4	0.5	0.8
Packaging	0.1	0.2	0.3

(a) Let x represent the number of small day backpacks made in 1 week, y represent the number of medium day backpacks made in 1 week, and z represent the number of large day backpacks made in 1 week. Write a system of three linear equations using this information and the given department maximum weekly times.

(b) Solve the system in part (a) to determine how many small, medium, and large day backpacks can be made in 1 week, assuming that the company operates at full capacity.

16. Pat wishes to invest \$15,000 in three different areas: a savings account earning 3%, a certificate of deposit earning 5%, and a mutual fund earning 8%. She decides that twice as much should be invested in the mutual fund as in the savings account. Assuming these rates, how much should she place in each area to produce exactly \$3000 each year on the investments?

17. A dietitian arranged a special diet for a patient that is made up of three basic foods, Alive, Gday, and Cardio. The diet is to include 270 units of iron, 340 units of vitamin C, and 126 units of vitamin B. The number of units per ounce for each basic food is given in Table R.2. How many ounces of each food must be used to exactly meet the diet requirements?

Table R.2

	Alive	Gday	Cardio
Iron	15	10	8
Vitamin C	25	5	10
Vitamin B	5	7	4

Refer to the following matrices to perform the indicated operations in Exercises 18–23.

$$A = \begin{bmatrix} -2 & 0 \\ 1 & 7 \end{bmatrix} \quad B = \begin{bmatrix} 4 & 5 \\ -1 & 3 \end{bmatrix}$$

$$C = \begin{bmatrix} 2 & 1 & 5 \\ -1 & 0 & 4 \\ 3 & 1 & 2 \end{bmatrix} \quad D = \begin{bmatrix} -1 & -3 & 5 \\ 0 & 8 & 4 \\ -2 & 1 & 3 \end{bmatrix}$$

18. $A + B$ **19.** $C - D$

20. $2A$ **21.** $-3D$

22. $4A + 2B$ **23.** $-2C - 3D$

Refer to matrices A through D given for Exercises 18–23 to answer Exercises 24 and 25.

24. Explain why $A + D$ cannot be done.

25. Determine $(C + D)^T$, the transpose of $C + D$.

In Exercises 26 and 27, find the values of the variables in each equation.

26. $\begin{bmatrix} 3 & 4 \\ y & x \end{bmatrix} = \begin{bmatrix} z+3 & 4 \\ 7 & -3 \end{bmatrix}$

27. $\begin{bmatrix} -x+1 & 3y & z+2 \\ -2w & 5 & 2 \end{bmatrix} + \begin{bmatrix} 3x & 4y & 1 \\ w-1 & 3 & 4 \end{bmatrix} = \begin{bmatrix} 11 & 21 & -4 \\ -8 & 8 & 24p \end{bmatrix}$

In Exercises 28 and 29, use a graphing calculator and its matrix capabilities and refer to the given matrices to perform the indicated operations.

$$A = \begin{bmatrix} 5.9 & 3.8 \\ 1.2 & -4.1 \\ -0.4 & 2.1 \end{bmatrix} \qquad B = \begin{bmatrix} -2.6 & -1.1 \\ 7.3 & 2.7 \\ -0.5 & 1.6 \end{bmatrix}$$

28. $3.8A + 0.6B$ **29.** $-2.2A - 4.7B$

30. Coulombe's Snowboards has the following data for sales of a popular board during December, January, and February (150″ and 160″ refer to the size of the board and Step-in and Traditional refer to two different types of bindings).

December Sales

	Step-in	Traditional
150″	5	8
160″	4	6

January Sales

	Step-in	Traditional
150″	10	12
160″	7	9

February Sales

	Step-in	Traditional
150″	9	7
160″	7	12

(a) If the rows refer to the length of the board (150″ or 160″) and the columns refer to the binding type (Step-in or Traditional), write three matrices to represent the data. Let D, J, and F be the matrices for December, January, and February, respectively.

(b) Determine a matrix giving the total sales for Coulombe's Snowboards during December, January, and February.

31. Quin Jackson has two types of IRA retirement accounts, certificates of deposit (CD) and treasury notes (TN), held in two banks, First Century (FC) and Hilltop Bank (HB). The dollar amounts in each at the beginning of 2001 are given in matrix A.

$$A = \begin{array}{c} \\ \end{array}\begin{matrix} \text{CD} & \text{TD} \end{matrix}$$
$$A = \begin{bmatrix} 25{,}000 & 20{,}000 \\ 15{,}000 & 30{,}000 \end{bmatrix} \begin{matrix} \text{FC} \\ \text{HB} \end{matrix}$$

(columns: CD, TD)

(a) Assuming a 9% growth rate on each account, determine a matrix I giving the earnings on the various accounts during 2001.

(b) Determine a matrix that represents the value of each account at the end of 2001.

(c) Assuming that the CDs and TNs at both banks have the same growth rate in 2002, what would those growth rates have to be for Quin Jackson to earn \$3379 in interest from both accounts at FC and \$3597 in interest from both accounts at HB at the end of 2002?

32. For D, J, and F in Exercise 30, compute $\frac{1}{3}(D + J + F)$ and interpret.

33. For matrices A and I in Exercise 31, find $a_{2,2}$ and $i_{1,2}$ and interpret.

34. The data in Table R.3 give the percentage of U.S. male smokers and the percentage of U.S. female smokers.

Table R.3

	18–24	25–34	35–44	45–64	65 and over
			Male		
1985	28.0	38.2	37.6	33.4	19.6
1990	26.6	31.6	34.5	29.3	14.6
1994	29.8	31.4	33.2	28.3	13.2
			Female		
1985	30.4	32.0	31.5	29.9	13.5
1990	22.5	28.2	24.8	24.8	11.5
1994	25.2	28.8	26.8	22.8	11.1

SOURCE: U.S. Census Bureau, www.census.gov

(a) Write a matrix M and a matrix F showing the percentage of U.S. male smokers and U.S. female smokers, respectively.

(b) Find and interpret $m_{2,3}$ and $f_{1,4}$.

(c) Using M and F, write a matrix S in which each entry shows how much more (or less) smoking is done by females than males.

(d) During which year and in what age group did a higher percentage of females than males smoke?

In Exercises 35 and 36, the dimension of matrices A and B are given.

(a) Find the dimension of the matrix product AB, if it exists.

(b) Find the dimension of the matrix product BA, if it exists.

35. A is a 5×2 and B is a 2×5.

36. A is a 3×4 and B is a 2×3.

In Exercises 37–41 find the matrix product.

37. $\begin{bmatrix} -3 & 1 \\ 7 & 4 \end{bmatrix}\begin{bmatrix} -2 \\ 3 \end{bmatrix}$

38. $\begin{bmatrix} -1 & 4 \\ 5 & 3 \end{bmatrix}\begin{bmatrix} 1 & 0 \\ 0 & 1 \end{bmatrix}$

39. $\begin{bmatrix} -2 & 1 \\ 4 & 4 \\ 3 & -2 \end{bmatrix}\begin{bmatrix} 7 & 6 & -2 \\ 1 & 4 & 3 \end{bmatrix}$

40. $\begin{bmatrix} -2 & 0 & 5 \end{bmatrix}\begin{bmatrix} 7 & -8 \\ 2 & -6 \end{bmatrix}$

41. $\begin{bmatrix} 1 & -2 & 4 & 3 \\ 7 & 8 & -5 & 2 \\ 16 & -1 & -12 & 4 \\ 8 & 9 & -13 & 7 \end{bmatrix}\begin{bmatrix} 2 \\ -3 \\ 5 \\ 7 \end{bmatrix}$

42. Use a graphing calculator to find the matrix product.

$$\begin{bmatrix} 27 & 35 & 80 \\ 42 & 51 & 17 \\ 14 & 37 & 63 \end{bmatrix}\begin{bmatrix} 200 \\ 500 \\ 700 \end{bmatrix}$$

43. Consider $A = \begin{bmatrix} 0.45 & 0.55 \\ 0.34 & 0.66 \end{bmatrix}$.

(a) Use a graphing calculator to compute A^2, A^5, A^{10}, and A^{20}.

(b) Describe any pattern observed in part (a).

In Exercises 44 and 45, write the system of linear equations as a matrix equation.

44.
$$\begin{aligned} 4x - 5y &= 8 \\ -2x + y &= -3 \end{aligned}$$

45.
$$\begin{aligned} x + 3y - 2z &= 4 \\ -2x + 4y - 5z &= -1 \\ -x - y + z &= 7 \end{aligned}$$

46. The total sales for January for three models of snowboards (lengths of 160″ and 170″), Racing, All-mountain, and Backcountry, at Coulombe's Snowboards is represented in matrix T.

$$T = \begin{array}{c} \begin{array}{cc} 160'' & 160'' \end{array} \\ \begin{bmatrix} 5 & 4 \\ 20 & 10 \\ 7 & 10 \end{bmatrix} \end{array} \begin{array}{l} \text{Racing} \\ \text{All-Racing} \\ \text{Backcountry} \end{array}$$

The selling price in dollars for each is given in the matrix S.

$$S = \begin{array}{c} \begin{array}{ccc} \text{Racing} & \text{All-mountain} & \text{Back-country} \end{array} \\ \begin{bmatrix} 600 & 500 & 550 \\ 700 & 600 & 650 \end{bmatrix} \end{array} \begin{array}{l} 160'' \\ 170'' \end{array}$$

(a) Find TS.

(b) Interpret the entries in TS.

47. Kris, a dietitian in a local hospital, plans a special diet for a patient that is made up of three basic foods, Zor, Daan, and Cretin. The number of units of iron, vitamin C, and vitamin B, per ounce, is given in matrix F.

$$F = \begin{array}{c} \begin{array}{ccc} \text{Zor} & \text{Daan} & \text{Cretin} \end{array} \\ \begin{bmatrix} 15 & 10 & 8 \\ 25 & 5 & 10 \\ 5 & 7 & 4 \end{bmatrix} \end{array} \begin{array}{l} \text{Iron} \\ \text{Vitamin C} \\ \text{Vitamin B} \end{array}$$

At breakfast the patient had 5 ounces of Zor, 8 ounces of Daan, and 12 ounces of Cretin. For lunch the patient had 6 ounces of Zor, 10 ounces of Daan, and 16 ounces of Cretin.

(a) Write a 3×1 matrix A that represents the amount of each food consumed (in ounces) at breakfast.

(b) Write a 3×1 matrix B that represents the amount of each food consumed (in ounces) at lunch.

(c) Find FA and interpret the entries.

(d) Find FB and interpret the entries.

48. The data in matrix F give the amounts of nitrogen, phosphate, and potash (in lb/acre) needed to grow celery, cabbage, and carrots for an entire growing season in California. (*Source:* U.S. Department of Agriculture, www.usda.gov/nass.)

$$F = \begin{array}{c} \begin{array}{ccc} \text{Celery} & \text{Cabbage} & \text{Carrots} \end{array} \\ \begin{bmatrix} 336 & 203 & 249 \\ 171 & 86 & 228 \\ 147 & 78 & 67 \end{bmatrix} \end{array} \begin{array}{l} \text{Nitrogen} \\ \text{Phosphate} \\ \text{Potash} \end{array}$$

Matrix P represents the number of acres of each crop that Rebecca wishes to plant.

$$P = \begin{bmatrix} 20 \\ 40 \\ 15 \end{bmatrix} \begin{array}{l} \text{Celery} \\ \text{Cabbage} \\ \text{Carrots} \end{array}$$

(a) Find FP.

(b) How many pounds of nitrogen, phosphate, and potash will Rebecca need for the entire growing season?

49. Matrix B gives the percentage of people who live in three different buildings of an apartment complex, A, B, and C, who recycle and don't recycle.

$$R = \begin{array}{c} \begin{array}{ccc} \text{A} & \text{B} & \text{C} \end{array} \\ \begin{bmatrix} 0.80 & 0.40 & 0.50 \\ 0.20 & 0.60 & 0.50 \end{bmatrix} \end{array} \begin{array}{l} \text{Recycle} \\ \text{Don't recycle} \end{array}$$

(a) If 500 people live in building A, 400 in building B, and 600 in building C, write a 3×1 matrix P that represents the amount of people in each building.

(b) Find RP.

(c) How many people total, among the three buildings, recycle?

(d) How many people total, among the three buildings, don't recycle?

50. Lindsay, a mathematics teacher, has given three tests to two different geometry sections, Sections 1 and 2. Each section consist of five students. Lindsay records their scores in matrices S and T.

$$S = \begin{bmatrix} 89 & 95 & 97 \\ 92 & 99 & 100 \\ 84 & 76 & 79 \\ 85 & 87 & 88 \\ 74 & 78 & 81 \end{bmatrix} \begin{matrix} \text{Paul} \\ \text{Beth} \\ \text{Dave} \\ \text{Kris} \\ \text{Joe} \end{matrix} \qquad T = \begin{bmatrix} 90 & 92 & 87 \\ 71 & 93 & 74 \\ 81 & 87 & 78 \\ 96 & 99 & 100 \\ 72 & 83 & 85 \end{bmatrix} \begin{matrix} \text{Teri} \\ \text{Cindy} \\ \text{Mika} \\ \text{Lera} \\ \text{Rob} \end{matrix}$$

(Columns: Test 1, 2, 3. Left: Section 1; right: Section 2.)

Let $C = \begin{bmatrix} 1 \\ 1 \\ 1 \end{bmatrix}$ and $R = [1 \quad 1 \quad 1 \quad 1 \quad 1]$. In parts (a) through (e), round all entries to the nearest whole number.

(a) Compute $\frac{1}{3}(SC)$ and interpret the entries.

(b) Compute $\frac{1}{3}(TC)$ and interpret the entries.

(c) Compute $\frac{1}{5}(RS)$ and interpret the entries.

(d) Compute $\frac{1}{5}(RT)$ and interpret the entries.

(e) Compute $\frac{1}{10}(RS + RT)$ and interpret the entries.

In Exercises 51 and 52, determine if the given matrices are inverses of each other.

51. $\begin{bmatrix} -1 & 2 \\ 7 & 6 \end{bmatrix}$ and $\begin{bmatrix} 3 & 5 \\ 2 & -1 \end{bmatrix}$

52. $\begin{bmatrix} -1 & 2 & 4 \\ 2 & 6 & 0 \\ 0 & 3 & 3 \end{bmatrix}$ and $\begin{bmatrix} -3 & -1 & 4 \\ 1 & \frac{1}{2} & -\frac{4}{3} \\ -1 & -\frac{1}{2} & \frac{5}{3} \end{bmatrix}$

In Exercises 53 and 54, find the inverse of each matrix.

53. $\begin{bmatrix} -1 & 0 & 3 \\ 2 & 1 & 2 \\ 0 & -2 & 4 \end{bmatrix}$ **54.** $\begin{bmatrix} 1 & 0 & 3 & -1 \\ 0 & -2 & 1 & 4 \\ 0 & 0 & 1 & -3 \\ -2 & -3 & 1 & 4 \end{bmatrix}$

In Exercises 55 and 56, use the formula for the inverse of a 2×2 matrix to find the inverse for the given matrix, if it exists.

55. $\begin{bmatrix} 3 & 4 \\ -2 & 6 \end{bmatrix}$ **56.** $\begin{bmatrix} 2 & -3 \\ -2 & 3 \end{bmatrix}$

In Exercises 57 and 58, use a graphing calculator to find the inverse for each matrix, if it exists. Where possible, express entries as fractions.

57. $\begin{bmatrix} 3.3 & 2.3 & 1.3 \\ 0.3 & 3.3 & 1.3 \\ 4.3 & 6.3 & 5.3 \end{bmatrix}$ **58.** $\begin{bmatrix} 1 & 0 & 2 & -2 & 3 \\ 1 & 5 & -2 & 6 & -1 \\ 0 & 0 & 5 & -3 & -2 \\ 8 & 4 & -4 & -2 & 5 \\ 1 & 0 & 1 & -1 & 2 \end{bmatrix}$

In Exercises 59 and 60, solve each system of linear equations by using the inverse of the coefficient matrix.

59. $3x + 4y = 13$
$-2x + 6y = 26$
(See Exercise 55)

60. $-x + 3z = -5$
$2x + y + 2z = 10$
$-2y + 4z = 20$
(See Exercise 53)

In Exercises 61 and 62, solve each system of linear equations by using the inverse of the coefficient matrix and a graphing calculator.

61. $3.3x + 2.3y + 1.3z = 4.3$
$0.3x + 3.3y + 1.3z = 5.3$
$4.3x + 6.3y + 5.3z = 8.3$
(See Exercise 57)

62. $x + 2z - 2v + 3w = 1$
$x + 5y - 2z + 6v - w = -2$
$5z - 3v - 2w = 7$
$8x + 4y - 4z - 2v + 5w = 4$
$x + z - v + 2w = 3$
(See Exercise 58)

63. A small pizza shop makes three different sizes of pizza: small, medium, and large. Each pizza requires a certain amount (in cups) of flour, cheese, and sauce, as shown in matrix A.

$$A = \begin{bmatrix} 1 & 3 & 3 \\ 2 & 3 & 4 \\ 1 & 2 & 3 \end{bmatrix} \begin{matrix} \text{Flour} \\ \text{Cheese} \\ \text{Sauce} \end{matrix}$$

(Columns: Small, Medium, Large.)

To fill the daily orders for the three sizes of pizza, the amounts of ingredients used daily are given in matrix B.

$$B = \begin{bmatrix} 135 \\ 180 \\ 125 \end{bmatrix} \begin{matrix} \text{Flour} \\ \text{Cheese} \\ \text{Sauce} \end{matrix}$$

(a) Let the number of daily orders for small, medium, and large pizzas be represented by x, y, and z, respectively. Write a 3×1 matrix X with entries x, y, and z that represents the daily orders. Write a matrix equation $AX = B$.

(b) Use A^{-1} to solve the equation in part (a).

(c) How many large pizzas does the shop sell daily?

(d) Using X from part (b), verify that $AX = B$.

64. Freeman's Mountain, a local ski resort, sells seasons passes for children and adults at \$350 and \$500, respectively. Table R.4 gives the revenue generated for the 1998–1999, 1999–2000, and 2000–2001 seasons and the number of season passes sold.

Table R.4

	1998–1999	1999–2000	2000–2001
Tickets sold	1200	1300	1550
Revenue	\$525,000	\$560,000	\$670,000

Let x represent the number of \$350 children season passes sold, and let y represent the number of \$500 adult season passes sold.

(a) Write a system of equations for each season: 1998–1999, 1999–2000, and 2000–2001.

(b) For each system in part (a), let matrix A be the coefficient matrix, matrix X the variable matrix, and matrix B the constant matrix. Write a matrix equation in the form $AX = B$ for each system in part (a).

(c) Use A^{-1} to solve each system in part (b) and interpret. (Notice that A^{-1} is the same for all three equations.)

65. Snowsports manufactures snowshoes at three different plants. Due to different technologies at each plant, the production rate (in snowshoes/hour) differs for each plant, as shown in Table R.5.

Table R.5

	Plant I	Plant II	Plant III
Running snowshoes	10	12	8
Hiking snowshoe	15	7	12
Backpacking snowshoes	6	12	15

Kerri, the production manager at Snowsports, has received an order to produce 2800 running snowshoes, 2880 hiking snowshoes, and 2970 backpacking snowshoes. To fill this order exactly, Kerri lets x, y, and z represent the number of hours that plants I, II, and III, respectively, should be scheduled to operate.

(a) Write a system of equations to model this situation.

(b) Write a matrix equation in the form $AX = B$ for the system in part (a).

(c) Use A^{-1} to solve the equation in part (b) and interpret.

66. (*continuation of Exercise 65*) Kerri has received another order requesting 2980 running snowshoes, 3370 hiking snowshoes, and 3390 backpacking snowshoes. With everything else from Exercise 65 remaining the same, do the following.

(a) Write a system of equations to model this situation.

(b) Set up an augmented matrix that represents the system in part (a).

(c) Use your graphing calculator to find the row reduced echelon form of the matrix in part (b) and interpret.

(d) Write a matrix equation in the form $AX = B$ for the system in part (a).

(e) Use A^{-1} to solve the equation in part (d) and compare the results to those in part (c).

67. Matrix A gives the percentage, as a decimal, of the total vote for three states, California, Florida, and Ohio, for the Democratic, Republican, and Reform party candidates in the 1996 presidential election (*Source:* U.S. Census Bureau, www.census.gov).

$$A = \begin{array}{c} \\ \\ \\ \end{array}\begin{array}{c} \text{CA} \quad \text{FL} \quad \text{OH} \\ \begin{bmatrix} 0.511 & 0.48 & 0.474 \\ 0.382 & 0.423 & 0.41 \\ 0.07 & 0.091 & 0.107 \end{bmatrix} \end{array} \begin{array}{l} \\ \text{Democratic} \\ \text{Republican} \\ \text{Reform} \end{array}$$

Matrix B represents the total number of Democratic, Republican, and Reform party votes, in thousands, for the three states combined.

$$B = \begin{bmatrix} 9815 \\ 7930 \\ 1669 \end{bmatrix} \begin{array}{l} \text{Democratic} \\ \text{Republican} \\ \text{Reform} \end{array}$$

(a) Let the total number of votes in California, Florida, and Ohio be represented by x, y, and z, respectively. Write a matrix equation $AX = B$.

(b) Use your graphing calculator to solve the equation in part (a) and interpret. Round entries to the nearest whole number.

Exercises 68–70 use the following technology matrix T. Here A, E, and C represent agriculture, energy, and chemicals, respectively.

$$T = \begin{array}{c} \text{A} \\ \text{E} \\ \text{C} \end{array} \begin{array}{c} \text{A} \quad\; \text{E} \quad\; \text{C} \\ \begin{bmatrix} 0.10 & 0.18 & 0.05 \\ 0.25 & 0.15 & 0.20 \\ 0.34 & 0 & 0.20 \end{bmatrix} \end{array}$$

68. Determine the number of units of agriculture, energy, and chemicals required to produce 1 unit of agriculture.

69. Determine the number of units of agriculture, energy, and chemicals required to produce 1 unit of energy.

70. Determine the number of units of agriculture, energy, and chemicals required to produce 1 unit of chemicals.

In Exercises 71 and 72 you are given a technology matrix T and a total output matrix X for each sector. Determine the internal demand. Assume that all entries represent dollars.

71. $T = \begin{array}{c} \text{Util} \\ \text{Trans} \end{array} \begin{array}{c} \text{Util} \quad \text{Trans} \\ \begin{bmatrix} 0.15 & 0.20 \\ 0.35 & 0.40 \end{bmatrix} \end{array} \qquad X = \begin{bmatrix} 30{,}000 \\ 45{,}000 \end{bmatrix} \begin{array}{l} \text{Util} \\ \text{Trans} \end{array}$

72. $T = \begin{array}{c} \text{Agr} \\ \text{Cons} \\ \text{Util} \end{array} \begin{array}{c} \text{Agr} \quad \text{Cons} \quad \text{Util} \\ \begin{bmatrix} 0.10 & 0.01 & 0.08 \\ 0.04 & 0.30 & 0.09 \\ 0.23 & 0.15 & 0.24 \end{bmatrix} \end{array} \qquad X = \begin{bmatrix} 15{,}000 \\ 25{,}000 \\ 20{,}000 \end{bmatrix} \begin{array}{l} \text{Agr} \\ \text{Cons} \\ \text{Util} \end{array}$

Exercises 73–77 utilize the following Leontief input–output model: An economy is based on three sectors, construction (C), manufacturing (M), and transportation (T). The technology matrix T and the external demand matrices D_1 and D_2, in millions of dollars, are

$$T = \begin{array}{c} \\ C \\ M \\ T \end{array}\begin{array}{c} \begin{array}{ccc} C & M & T \end{array} \\ \begin{bmatrix} 0.1 & 0.3 & 0.1 \\ 0.2 & 0.1 & 0.3 \\ 0.4 & 0.1 & 0.5 \end{bmatrix} \end{array} \qquad D_1 = \begin{bmatrix} 6 \\ 8 \\ 13 \end{bmatrix} \begin{array}{c} C \\ M \\ T \end{array}$$

$$D_2 = \begin{bmatrix} 7 \\ 4 \\ 10 \end{bmatrix} \begin{array}{c} A \\ M \\ T \end{array}$$

73. Determine the number of units (dollars) of C, M, and T required to produce \$1 worth of C.

74. Determine the number of units (dollars) of C, M, and T required to produce \$1 worth of M.

75. Find $(I - T)$.

76. Given $(I - T)^{-1} = \begin{bmatrix} \frac{210}{137} & \frac{80}{137} & \frac{90}{137} \\ \frac{110}{137} & \frac{205}{137} & \frac{145}{137} \\ \frac{190}{137} & \frac{105}{137} & \frac{375}{137} \end{bmatrix}$, determine the total output for each sector required to satisfy D_1.

77. Repeat Exercise 76 for D_2.

78. An economy has two main industries: agriculture and manufacturing. Production of \$1 worth of agriculture requires \$0.15 from the agriculture sector and \$0.45 from the manufacturing sector. Production of \$1 worth of manufacturing requires \$0.20 from the agriculture sector and \$0.25 from the manufacturing sector. Find the total output for each sector needed to meet the external demand of \$4 billion for agriculture and \$10 billion for manufacturing.

79. A simplified version of an economy that deals with two sectors, agriculture (A) and energy (E), has the following technology matrix:

$$T = \begin{array}{c} \\ A \\ E \end{array}\begin{array}{c} \begin{array}{cc} A & E \end{array} \\ \begin{bmatrix} 0.10 & 0.14 \\ 0.34 & 0.21 \end{bmatrix} \end{array}$$

The external demand (in millions of dollars) is given by

$$D = \begin{bmatrix} 8 \\ 15 \end{bmatrix} \begin{array}{c} A \\ E \end{array}$$

Find the total output of each sector needed to meet the external demands.

80. (*continuation of Exercise 79*) Repeat Exercise 79 if the external demand for agriculture is \$10 million and \$15 million for energy.

81. An economy has three main industries: agriculture (A), chemicals (C), and energy (E). Production of \$1 worth of agriculture requires \$0.15 from the agriculture sector, \$0.04 from the chemical sector, and \$0.05 from the energy sector. Production of \$1 worth of chemicals requires \$0.02 from the agriculture sector, \$0.12 from the chemical sector, and \$0.25 from the energy sector. Production of \$1 worth of energy requires \$0.04 from the agriculture sector, \$0.18 from the chemical sector, and \$0.10 from the energy sector.

(a) Write the technological matrix T for this economy.

(b) The external demand is \$12 billion for agriculture, \$8 billion for chemicals, and \$15 billion for energy. Write the external demand matrix D.

(c) Given $(I - T)^{-1} = \begin{bmatrix} 1.18 & 0.04 & 0.06 \\ 0.07 & 1.21 & 0.24 \\ 0.09 & 0.34 & 1.18 \end{bmatrix}$ (entries to the nearest hundredth), determine the total output needed to meet the external demand.

82. An economy that deals with three sectors has the following technological matrix:

$$T = \begin{array}{c} \\ \text{Agri} \\ \text{Manu} \\ \text{Trade} \end{array}\begin{array}{c} \begin{array}{ccc} \text{Agri} & \text{Manu} & \text{Trade} \end{array} \\ \begin{bmatrix} 0.18 & 0.24 & 0.03 \\ 0.12 & 0.16 & 0.14 \\ 0.05 & 0.20 & 0.22 \end{bmatrix} \end{array}$$

The total output of each sector, in millions of dollars, needed to meet the external demand is given by

$$X = \begin{bmatrix} 16.2 \\ 4.2 \\ 8.7 \end{bmatrix}$$

(a) Find $I - T$.

(b) Determine the external demand matrix D. Round entries to the nearest tenth.

(c) Verify the given external demand matrix X by using your calculator to compute $(I - T)^{-1}D$, where D is the matrix found in part (b).

CHAPTER 2 PROJECT

SOURCE: Palm, Inc., www.palm.com

Tech Mart, Electrifun, Gadgets-4-U, and Plug-n-Play are selling the same type of Palm Pilots: the m100, m105, m500, and m505. The retail prices of these models are \$129.00, \$199.00, \$399.00, and \$449.00, respectively.

Suppose that Tech Mart sells 2 m100s, 3 m105s, 7 m500s, and 6 m505s; Electrifun sells 4 m100s, 4 m105s, 2 m500s, and 5 m505s; Gadgets-4-U sells 3 m100s, 5 m105s, 8 m500s, and 9 m505s; and Plug-n-Play sells 3 m100s, 3 m105s, 5 m500s, and 5 m505s. All sales are for 1 month's time.

1. Create a 4×4 matrix A for the amount sold by each store in which the rows represent the name of the store and the columns represent the model name.

2. Create a 4×1 matrix B for the price of each model.

3. Find AB and interpret the results. Which store generated the most revenue? By referring to the number of Palms sold and the price per model, can you explain why this store generated the most revenue?

Suppose that a manufacturer makes the Palm m100, m500, and m505. The m100 has a volume of 10.47 cubic inches, weighs 4.4 ounces, and contains 2 MB of memory. The m500 has a volume of 5.58 cubic inches, weighs 4 ounces, and contains 8 MB of memory. The m505 has a volume of 6.98 cubic inches, weighs 4.9 ounces, and contains 8 MB of memory.

Consider a store, McMikes, that will sell a specified number of each type of Palm Pilot in one bundled package. Suppose that the bundled package holds 72.58 cubic inches and can carry 45.3 ounces, and that there's a tag on the box that says "a total of 68 MB inside."

4. Create a 3×3 matrix C where the rows represent the size, weight and amount of memory and the columns represent Palm m100, m500, and m505, respectively.

5. Create a 3×1 matrix D for the limits of the package.

6. Find $C^{-1}D$ and interpret the results.

7. Using the result of question 6 and the price of each Palm Pilot as listed above, how much revenue can McMikes hope to generate from selling all the Palms in this shipping box?

8. Repeat questions 1 through 7 if a different shipping box was used that could hold 70.49 cubic inches and carry 34.9 ounces, and the limit of memory is 34 MB.

Chapter 3 Linear Programming: The Graphical Method

Minimize
$$C = 2.09x + 1.59y$$

Subject to :
$$28x + 8y \geq 116$$
$$2x + 6y \geq 30$$
$$x \geq 0, \ y \geq 0$$

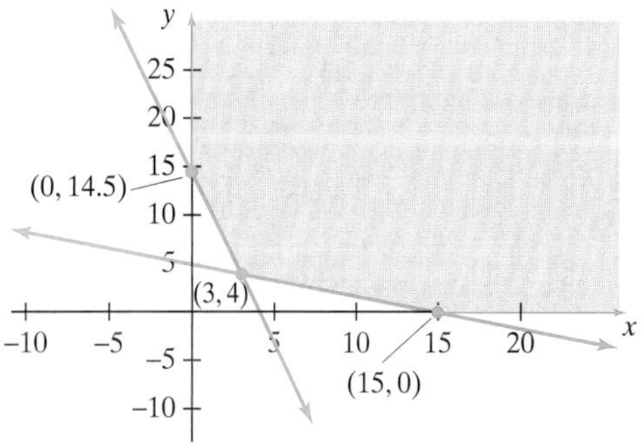

Hospital dietitians are often charged with minimizing food costs while meeting certain nutrient requirements. One technique for meeting both goals is to develop a mathematical model involving constraints and then to use linear programming methods to solve the model. Individuals can employ the same process even when eating at a fast food restaurant such as McDonald's. The figure shows the graph of the linear inequalities that form the problem constraints. One of the corner points of the feasible region minimizes the objective function.

What We Know

In Chapter 1 we learned how to solve systems of linear equations graphically and algebraically. In Chapter 2 we analyzed how matrices could be used to solve systems of linear equations.

Where Do We Go

In this chapter we will discuss how to graph linear inequalities and systems of linear inequalities and how this relates to a powerful tool called *linear programming*. Linear programming is a relatively new field of mathematics, and we will use it to solve a variety of applications.

Section 3.1 Problem Solving: Linear Inequalities

In Section 1.1 we examined techniques for solving linear equations and writing systems of equations. Now we turn our attention to the problem-solving techniques associated with **linear inequalities**. This is a skill necessary for solving linear programming problems.

Expressing Inequalities

We begin our study of **inequalities** with a review of the inequality symbols used most frequently in this text.

Inequality Symbols

Symbol	*Meaning*
$<$	Less than
$>$	Greater than
$\leq$	Less than or equal to
$\geq$	Greater than or equal to

Note: The symbols $<$ and $>$ are sometimes called **strict inequalities**.

Any value that makes an inequality a true statement is a **solution** to the inequality. Example 1 illustrates how to determine if a value is a solution to an inequality.

Example 1 Checking Solutions to Inequalities

For the following inequalities, complete Tables 3.1.1 and 3.1.2 by stating whether the inequality is *true* or *false*.

(a) $x > -7$ **(b)** $x \leq 12$

Table 3.1.1

x	$x > -7$	**Statement**
−10		
−7		
0		
3		

Table 3.1.2

x	$x \leq 12$	**Statement**
0		
9		
12		
15		

Solution

(a) The inequality statements are listed in Table 3.1.3.

(b) The inequality statements are shown in Table 3.1.4.

Notice that $12 \leq 12$ is a true statement since $12 = 12$.

Table 3.1.3

x	$x > -7$	Statement
-10	$-10 > -7$	False
-7	$-7 > -7$	False
0	$0 > -7$	True
3	$5 > -7$	True

Table 3.1.4

x	$x \le 12$	Statement
0	$0 \le 12$	True
9	$9 \le 12$	True
12	$12 \le 12$	True
15	$15 \le 12$	False

Since the solution to an inequality is the set of all values that make the inequality true, one of the most efficient ways to represent solutions to inequalities is by graphing them on a number line. To aid us in graphing, we utilize the following:

- $a < b$ means that a is to the left of b on a number line.
- $a > b$ means that a is to the right of b on a number line.

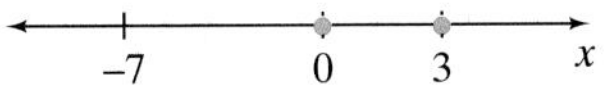

Figure 3.1.1

Figure 3.1.2

In Example 1 (a) we verified that $x = 0$ and $x = 3$ are two solutions to the inequality $x > -7$. These solutions are shown in Figure 3.1.1. These are not the only solutions to $x > -7$. Any value for x *to the right of* -7 on the number line is a solution to $x > -7$. So we graph all solutions to $x > -7$ as shown in Figure 3.1.2. We use parentheses, (or,) to show when an endpoint is not a solution. We use brackets, [or,] to show when an endpoint is a solution. Inequalities and their graphs are summarized below.

Inequalities and Graphs

For a real number a:

Inequality	Graph	Inequality	Graph
$x < a$	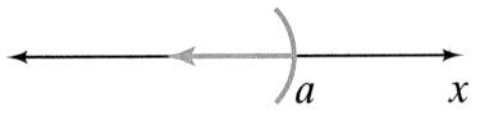	$x \le a$	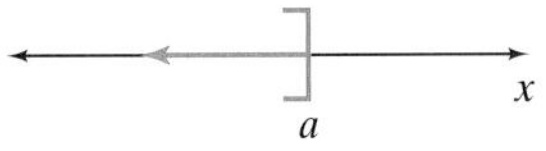
$x > a$	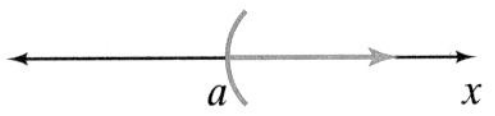	$x \ge a$	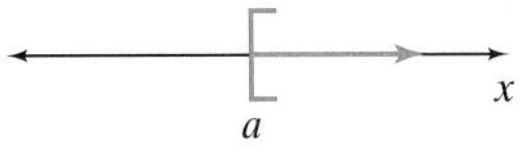

Solving Linear Inequalities

Before we discuss how to solve a linear inequality, let's see how we can translate an English sentence into a mathematical sentence using an inequality symbol.

Example 2 Translating an English Sentence into an Inequality

The product of five and a certain number plus two is greater than 37. Define the variable and translate the English sentence into a mathematical sentence.

Solution

Understand the Situation: Let x represent the certain number. The phrase "the product of five and a certain number" means to multiply 5 by x. This gives the term $5x$. So we get

The product of five and a certain number plus two is greater than 37. English sentence

$5x + 2 > 37$ Mathematical sentence ■

Checkpoint 1 Now work Exercise 9.

The mathematical sentence we came up with in Example 2 is a **linear inequality**. In general, a linear inequality is defined as follows:

Linear Inequality

An inequality of the form

$$ax + b \leq c$$

is called a **linear inequality**, where a, b, and c are constants.

Note: When working with inequalities, we tend generally to use the symbol $\leq$ in definitions and theorems. However, remember that the symbols $\geq$, $<$, or $>$ could be used in place of $\leq$.

Solving linear inequalities is much like solving linear equations. The following inequality properties are used in solving linear inequalities.

Inequality Properties

When given an inequality, an equivalent inequality is produced and the inequality symbol **remains the same** if

1. Each side of the original inequality has the same real number added to or subtracted from it.
2. Each side of the original inequality is multiplied or divided by the same positive real number.

An equivalent inequality is produced and the inequality symbol **reverses direction** if

3. Each side of the original inequality is multiplied or divided by the same negative real number.

Note: 1. As with solving linear equations, multiplying by zero and dividing by zero are not allowed.
2. These properties illustrate that we solve linear inequalities in the same manner as we solve linear equations. The exception is when multiplying or dividing the inequality by a negative coefficient. In this case, reverse the inequality symbol.

Example 3 Solving Linear Inequalities

Solve and graph the solutions to the following.

(a) $2x - 3 \leq 7$ **(b)** $3 - 7x < 17$

Solution

(a) To solve this inequality, we isolate the variable x. This gives us

$2x - 3 \leq 7$	Given inequality
$2x \leq 10$	Add 3 to both sides.
$x \leq 5$	Divide both sides by 2.

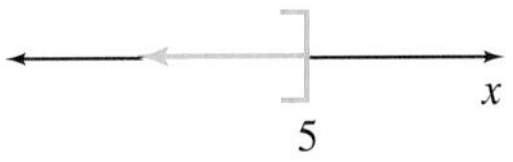

Figure 3.1.3

So any value of x that is less than or equal to 5 is a solution to the inequality. The solution is graphed, using a bracket on the endpoint 5, in Figure 3.1.3.

(b) Proceeding as in part (a) yields

$3 - 7x < 17$	Given inequality
$-7x < 14$	Subtract 3 from both sides.
$x > 2$	Divide by -7 and reverse inequality symbol.

Figure 3.1.4

Notice in the final step that we had to reverse the inequality sign from $<$ to $>$ because we divided by a negative number. The solution is shown graphically in Figure 3.1.4 using a parenthesis on the endpoint 2. ■

Checkpoint 2 Now work Exercise 23.

Writing Linear Inequalities with More Than One Variable

Many applications in this text require us to work with *linear inequalities with more than one variable*. This means that many times we will need to write inequalities to represent a situation. Typically, linear inequalities have many solutions. We will study solutions of linear inequalities, as well as systems of linear inequalities, later in Chapter 3 and in Chapter 4. For now our goal is to represent situations with linear inequalities. Example 4 illustrates how to set up linear inequalities. Before we proceed with Example 4, let's recall the Strategy for Solving Word Problems first presented in Section 1.1.

Strategy for Solving Word Problems

1. Carefully **read** the problem and **understand the situation**. As you analyze a problem, you may need to:
   - **Define the variables** to represent all of the unknowns.
   - **Construct** a mathematical model or write one or more inequalities to represent the problem.
2. **Solve** the inequality. Be sure to check the solution in the original inequality.
3. **Interpret the solution.**

Note: The only difference we see in the above strategy from the one presented in Section 1.1 is that we are now working with inequalities.

Example 4 Writing Inequalities

Table 3.1.5 gives nutritional information in grams for selected items from McDonald's menu.

Table 3.1.5

Item	Protein (g)	Fat (g)
Small French fries	3	10
Garden salad	7	6

SOURCE: McDonald's, www.mcdonalds.com

Julia wishes to eat a lunch consisting of small French fries and garden salad. She wants the lunch to contain at least 12 grams of protein. Let x represent the number of orders of small French fries and let y represent the number of orders of garden salad. Write inequalities that give appropriate restrictions on x and y.

Solution

Here we have already had our variables defined for us. Since x represents the number of orders of small French fries, and Julia cannot eat a negative number of small French fries, we have the restriction of x being greater than or equal to zero. This English sentence can be expressed as an inequality by writing

$$x \geq 0$$

Likewise, since y represents the number of orders of garden salad, we have

$$y \geq 0$$

Since each order of small French fries has 3 grams of protein, if Julia has x orders of small French fries, she would consume

$$3x$$

grams of protein. Since each order of garden salad has 7 grams of protein, if Julia has y orders of garden salad, she would consume

$$7y$$

grams of protein. So the number of grams of protein in a meal consisting of small French fries and garden salad is given by

$$3x + 7y$$

But Julia wants **at least** 12 grams of protein in her meal. Her meal could have 12, 14, or even 20 grams of protein, but not 11, 9, or 7 grams of protein. She wants at least 12 grams! The equation $3x + 7y = 12$ represents a meal with *exactly* 12 grams of protein. However, this equation excludes a meal of 14 or 20 grams. The inequality $3x + 7y > 12$ represents a meal having *more than* 12 grams of protein. However, it excludes a meal of exactly 12 grams. Combining the equation and the inequality gives us

$$3x + 7y \geq 12$$

which satisfies the condition that her meal have 12 or more grams of protein. ■

✔ **Checkpoint 3** Now work Exercise 29.

Example 5 Writing Inequalities with More than One Variable

Using the information from Table 3.1.5 in Example 4, let's say that Julia wishes to eat a lunch consisting of orders of small French fries and orders of garden salad that contains at most 25 grams of fat. Let x represent the number of small French fries and let y represent the number of garden salads. Write inequalities that give appropriate restrictions on x and y.

Solution

In this case we still have the restrictions $x \geq 0$ and $y \geq 0$, since Julia cannot eat a negative number of French fries or garden salads. Since each order of small French fries has 10 grams of fat, if Julia has x orders of small French fries, she would consume

$$10x$$

grams of fat. Since each order of garden salad has 6 grams of fat, if Julia has y orders of garden salad, she would consume

$$6y$$

grams of fat. So the number of grams of fat in a meal consisting of small French fries and garden salad is given by

$$10x + 6y$$

Since Julia wants **at most** 25 grams of fat, this means that she could have any number of fat grams that are *less than or equal to* 25. This gives us the inequality

$$10x + 6y \leq 25$$

■

We can see in Examples 4 and 5 that we get a total of four inequalities. In all these inequalities, x represents the number of orders of small French fries and y represents the number of orders of garden salad. If we want to express the number of orders of small French fries and garden salads that Julia can order so that she consumes both at least 12 grams of protein and at most 25 grams of fat, we can write the **system of inequalities**

$$\begin{aligned} x &\geq 0 \\ y &\geq 0 \\ 3x + 7y &\geq 12 \\ 10x + 6y &\leq 25 \end{aligned}$$

In Section 3.2 we will study how to graph systems of linear inequalities.

Examples 4 and 5 illustrated how inequalities can be used to place restrictions on the variables so that we only consider reasonable values for the variables. This is true of many applications that we will encounter in the next two chapters, particularly in linear programming. In Example 6 we continue to practice how to use inequalities to place restrictions on the variables.

Example 6 Using Inequalities to Express Reasonable Values

Dub, Inc., is a small manufacturing firm that produces two microwave switches, switch A and switch B. Each switch passes through two departments: assembly and testing. Switch A requires 4 hours of assembly and 1 hour of testing, and switch B requires 3 hours of assembly and 2 hours of testing. The maximum number of hours available each week for assembly and testing is 240 hours and 140 hours, respectively.

(a) Define the variables and write inequalities to express reasonable values for the variables.

(b) Write an inequality to express the time restriction of the assembly department.

(c) Write an inequality to express the time restriction of the testing department.

Solution

(a) **Understand the Situation:** Since we do not know how many units of switch A and switch B are made each week, we let x represent the number of units of switch A made each week and let y represent the number of units of switch B made each week.

We know that there cannot be a negative number of switches made. This means that x and y must be either zero or greater than zero. This gives us the two inequalities

$$x \geq 0 \quad \text{and} \quad y \geq 0$$

(b) **Understand the Situation:** To see what kind of time restriction is made, we carefully examine the assembly department. It takes 4 hours to assemble each switch A and 3 hours to assemble each switch B. So the total amount of time that it takes to assemble x switch A's and y switch B's is given by the expression

$$\begin{pmatrix}\text{hours to assemble}\\ \text{each switch A}\end{pmatrix}\begin{pmatrix}\text{number of switch}\\ \text{A's assembled}\end{pmatrix} + \begin{pmatrix}\text{hours to assemble}\\ \text{each switch B}\end{pmatrix}\begin{pmatrix}\text{number of switch}\\ \text{B's assembled}\end{pmatrix}$$

$$4x + 3y$$

We know that a maximum of 240 hours is available each week for assembly. Thus the total number of hours cannot be more than 240. This is another way to say that the number of hours available is less than or equal to 240. So the restriction on the assembly department is given by

$$4x + 3y \leq 240$$

(c) **Understand the Situation:** For the testing department, we know that it takes 1 hour to test each switch A and 2 hours to test each switch B. Using the same kind of reasoning that we used for the assembly department in part (b), the total amount of time that it takes to test x switch A's and y switch B's is given by

$$x + 2y$$

We know that the maximum number of hours available each week for testing is 140 hours. So the restriction on the testing department is given by

$$x + 2y \leq 140$$

■

The restrictions in Example 6 (a), $x \geq 0$ and $y \geq 0$, are commonly called the **nonnegative constraints**. We will study constraints in more detail as we learn the techniques of linear programming in Chapters 3 and 4. We conclude this section with an example that involves four variables.

Example 7 Using Inequalities to Express Restrictions

Skinner Bikes sends bicycles from its two plants, plant A and plant B, to its bicycle retailers I and II located in Columbus, Ohio. Plant A has a total of 210 bicycles to send and plant B has 60. Retailer I needs at least 170 bicycles and retailer II needs at least 100.

(a) Define variables and express the nonnegative constraints.

(b) Write inequalities to express restrictions on the availability and transportation costs for the bicycles.

Solution

(a) **Understand the Situation:** Since we do not know how many bicycles are sent from each plant to each retailer, we let x represent the number of bicycles sent from plant A to retailer I, y represent the number of bicycles sent from plant A to retailer II, z represent the number of bicycles sent from plant B to retailer I, and w represent the number of bicycles sent from plant B to retailer II.

Since all these variables represent quantities of bicycles, all must be greater than or equal to zero. This gives the nonnegative constraints

$$x \geq 0, \qquad y \geq 0, \qquad z \geq 0, \qquad w \geq 0$$

(b) Since plant A has 210 bicycles to ship and we know that the total number of bicycles sent from plant A to retailer I is x and that the total of the number of bicycles sent from plant A to retailer II is y, we get the inequality

$$\left(\begin{array}{c}\text{bicycles sent from}\\ \text{plant A to retailer I}\end{array}\right) + \left(\begin{array}{c}\text{bicycles sent from}\\ \text{plant A to retailer II}\end{array}\right)\left(\begin{array}{c}\text{cannot}\\ \text{exceed}\end{array}\right)\left(\begin{array}{c}\text{total shipped from}\\ \text{plant A}\end{array}\right)$$

$$x + y \leq 240$$

Similarly, since plant B has a total of 60 bicycles to ship, we get the inequality

$$\left(\begin{array}{c}\text{bicycles sent from}\\ \text{plant B to retailer I}\end{array}\right) + \left(\begin{array}{c}\text{bicycles sent from}\\ \text{plant B to retailer II}\end{array}\right)\left(\begin{array}{c}\text{cannot}\\ \text{exceed}\end{array}\right)\left(\begin{array}{c}\text{total shipped from}\\ \text{plant B}\end{array}\right)$$

$$z + w \leq 60$$

Since retailer I needs at least 170 bicycles, this means that no less than 170 bicycles must be delivered. So we get

$$\left(\begin{array}{c}\text{bicycles sent from}\\ \text{plant A to retailer I}\end{array}\right) + \left(\begin{array}{c}\text{bicycles sent from}\\ \text{plant B to retailer I}\end{array}\right)\left(\begin{array}{c}\text{must be at}\\ \text{least}\end{array}\right)\left(\begin{array}{c}\text{number of}\\ \text{bicycles needed}\end{array}\right)$$

$$x + z \geq 170$$

Likewise, we know that retailer II needs at least 100 bicycles, giving us the inequality

$$\left(\begin{array}{c}\text{bicycles sent from}\\ \text{plant A to retailer II}\end{array}\right) + \left(\begin{array}{c}\text{bicycles sent from}\\ \text{plant B to retailer II}\end{array}\right)\left(\begin{array}{c}\text{must be at}\\ \text{least}\end{array}\right)\left(\begin{array}{c}\text{number of}\\ \text{bicycles needed}\end{array}\right)$$

$$y + w \geq 100$$

■

Notice that if we collect the solutions from Example 7 we get the **system of inequalities**

$$x + y \leq 210$$
$$z + w \leq 60$$
$$x + z \geq 170$$
$$y + w \geq 100$$
$$x \geq 0, \quad y \geq 0, \quad z \geq 0, \quad w \geq 0$$

We will study how to solve systems of inequalities like this one in Chapter 4.

SUMMARY

- **Inequality Symbols**

Symbol	*Meaning*
$<$	Less than
$>$	Greater than
$\leq$	Less than or equal to
$\geq$	Greater than or equal to

- A **linear inequality** has the form $ax + b \leq c$, where a, b, and c are constants.
- When **solving linear inequalities**, we use the same process as for solving linear equations. The exception is when multiplying or dividing the inequality by a negative coefficient. In this case, reverse the inequality symbol.
- When translating English sentences into mathematical sentences using inequalities, we use the Strategy for Solving Word Problems.

SECTION 3.1 EXERCISES

For the inequalities in Exercises 1–6, complete the tables by stating whether the inequality is true *or* false.

1. $x > 6$

x	$x > 6$	Statement
−12		
2		
6		
10		

2. $x > 0$

x	$x > 0$	Statement
−1		
0		
1		
6		

3. $x \leq -4$

x	$x \leq -4$	Statement
−8		
−6		
−4		
0		

4. $x \leq 5$

x	$x \leq 5$	Statement
−5		
0		
5		
15		

5. $x \geq -1$

x	$x \geq -1$	Statement
−2		
−1		
0		
1		

6. $x \geq -10$

x	$x \geq -10$	Statement
−15		
−12		
−10		
0		

In Exercises 7–16, translate the English sentence into a mathematical sentence.

7. A certain number increased by five is greater than four.

8. A certain number increased by seven is greater than six.

✓ **9.** A certain number decreased by eight is less than -18.

10. A certain number decreased by two is less than negative six.

11. The product of nine and a certain number is greater than or equal to 54.

12. The product of three and a certain number is greater than or equal to one.

13. The product of three and a certain number minus 15 is less than or equal to -20.

14. The product of four and a certain number minus four is less than or equal to -10.

15. The product of eight and a certain number plus 1 is not equal to 14.

16. The product of six and a certain number plus one is not equal to 97.

In Exercises 17–28, solve and graph the solutions.

17. $x - 3 > 4$

18. $x - 1 > 5$

19. $x + 4 < 9$

20. $x + 6 < 4$

21. $2x + 2 \leq 18$

22. $3x + 5 \leq 20$

✓ **23.** $2x - 3 \geq 7$

24. $3x + 2 \geq 14$

25. $-2x - 3 > 7$

26. $-4x + 3 > 15$

27. $14 - 3x \leq 5$

28. $3 - 7x \leq 17$

Applications

⇒ 3.2 ✓ **29.** McSavaney's trains administrative assistants for two types of office positions: medical field and executive secretarial. Each individual receives training in two departments at McSavaney's: word processing along with telephone and LAN (local area networks). An individual in the medical field program receives 2 units of word processing and 2.5 units of telephone and LAN training. An individual in the executive secretarial program receives 3 units of word processing and 2.5 units of telephone and LAN training. Each month at McSavaney's, corporate training can give no more than 32 units of word processing training and no more than 30 units of telephone and LAN training. Let x represent the number admitted each month to the medical field program, and let y represent the number admitted each month to the executive secretarial program. Write linear inequalities that give appropriate restrictions on x and y.

⇒ 3.2 **30.** (*continuation of Exercise 29*) Feedback from employers has indicated that McSavaney's needs to provide another area of training, ethics. McSavaney's adds an ethics training department. It has been determined that an individual in the medical field program receive 3.5 units of ethics training and an individual in the executive secretarial program receive 2.5 units of ethics training. Each month at McSavaney's, corporate training can give no more than 35 units of ethics training. Assuming that everything else from Exercise 29 remains the same, write linear inequalities that give appropriate restrictions on x and y.

⇒ 3.2 **31.** Sandkuhl, Inc., is a small manufacturing firm that produces two microwave switches, switch A and switch B. Each switch passes through two departments: assembly and testing. Switch A requires 4 hours of assembly and 1 hour of testing; switch B requires 3 hours of assembly and 2 hours of testing. The maximum number of hours available each week for assembly and testing is 240 hours and 140 hours, respectively. Let x represent the number of units of switch A manufactured each week, and let y represent the number of units of switch B manufactured each week. Write linear inequalities that give appropriate restrictions on x and y.

⇒ 3.2 **32.** (*continuation of Exercise 31*) Due to a new contract commitment, Sandkuhl, Inc., must manufacture at least 10 units of switch A each week. Assuming that everything else from Exercise 31 remains the same, write linear inequalities that give appropriate restrictions on x and y.

⇒ 3.2 **33.** Klosterman's owns 200 acres of tillable farmland that is used to grow corn and wheat. On average, each acre of corn and wheat yields 110 and 35 bushels, respectively. At least 11,000 bushels of corn and 2205 bushels of wheat are required due to prior contracts. Let x represent the number of acres of corn planted and let y represent the number of acres of wheat planted. Write linear inequalities that give appropriate restrictions on x and y.

⇒ 3.2 **34.** Mix for Chix, Inc., a local feed store, manufactures chicken feed by mixing two ingredients. Each ingredient contains two key nutrients: protein and fat. The number of units of each nutrient in 1 pound of the two basic ingredients is shown in Table 3.1.6.

Table 3.1.6

Ingredient	Protein	Fat
1	25	11
2	40	8

A local chicken farmer desires at most 500 pounds of a mix in which each pound of feed has a maximum of 35 units of protein and a maximum of 10 units of fat. Let x represent the number of pounds of ingredient 1 used and let y represent the number of pounds of ingredient 2 used. Write linear inequalities that give appropriate restrictions on x and y.

⇒ 3.2 **35.** The Stitz Tent Company makes two types of tents: two-person and six-person models. Each tent requires the services of two departments, in hours, as shown in Table 3.1.7.

Table 3.1.7

Department	Two-person	Six-person
Fabric cutting	6	8
Assembly and packaging	2	4

Fabric cutting has a maximum of 580 hours available each week and assembly and packaging has a maximum of 260 hours available.

(a) Define the variables and write linear inequalities to express reasonable values for the variables.

(b) Write linear inequalities to express the time restrictions of the fabric cutting and of the assembly and packaging departments.

⇒4.1 **36.** Due to new regulations approved by the Environmental Protection Agency (EPA), the Henderson Chemical Company was required to implement a new process to reduce the pollution created when making a particular chemical. The older process releases 4 grams of sulfur and 3 grams of particulates for each gallon of chemical produced. The newer process releases 2 grams of sulfur and 1 gram of particulates for each gallon of chemical produced. The new EPA regulations do not allow for more than 10,000 grams of sulfur and 7000 grams of particulates to be released daily.

(a) Define the variables and write linear inequalities to express reasonable values for the variables.

(b) Write linear inequalities to express the EPA restrictions on sulfur and particulates.

⇒4.1 **37.** Randy and Rusty have decided to enter the field of personal digital assistants (PDAs). Initially, they plan to design two types of PDAs, the WOW 1000 and the WOW 1500. Because of interest in PDAs, they can sell all that they can produce. However, they need to determine a production rate so as to satisfy various limits with a small production crew. These include a maximum of 150 hours per week for assembly and 100 hours per week for testing. The WOW 1000 requires 3 hours of assembly and 3 hours of testing, and the WOW 1500 requires 6 hours of assembly and 3.5 hours of testing.

(a) Define the variables and write linear inequalities to express reasonable values for the variables.

(b) Write linear inequalities to express the time restrictions of the assembly and testing departments.

⇒4.1 **38.** Table 3.1.8 gives data for growing artichokes and asparagus for an entire growing season in California (entries are in lb/acre).

Table 3.1.8

	Artichokes	Asparagus
Nitrogen	197	240
Potash	123	42

SOURCE: U.S. Department of Agriculture, www.usda.gov/nass

Franco has available 150 acres suitable for cultivating crops. He has purchased a contract for a total of 31,880 pounds of nitrogen and 9120 pounds of potash.

(a) Define the variables and write linear inequalities to express reasonable values for the variables.

(b) Write linear inequalities to express the restrictions of land, nitrogen, and potash.

⇒4.1 **39.** Table 3.1.9 gives data for growing cauliflower and head lettuce for an entire growing season in California (entries are in lb/acre).

Table 3.1.9

	Cauliflower	Head Lettuce
Nitrogen	215	173
Phosphate	98	122

SOURCE: U.S. Department of Agriculture, www.usda.gov/nass

Sonja has 210 acres available for cultivating crops. She has purchased a contract for a total of 38,380 pounds of nitrogen and 22,240 pounds of phosphate.

(a) Define the variables and write linear inequalities to express reasonable values for the variables.

(b) Write linear inequalities to express the restrictions of land, nitrogen, and phosphate.

⇒4.1 **40.** The Smolko Tent Company makes three types of tents: two-person, four-person and six-person models. Each tent requires the services, in hours, of three departments, as shown in Table 3.1.10. The fabric cutting, assembly, and packaging departments have available a maximum of 420, 310, and 180 hours each week, respectively.

Table 3.1.10

	Two-person	Four-person	Six-person
Fabric cutting	0.5	0.8	1.1
Assembly	0.6	0.7	0.9
Packaging	0.2	0.3	0.6

(a) Define the variables and write linear inequalities to express reasonable values for the variables.

(b) Write linear inequalities to express the time restrictions of the fabric cutting, assembly, and packing departments.

⇒4.1 **41.** Kathy's Secretarial Services has a 1-month intensive training program for receptionist/secretaries for three types of office positions: medical field, small business, and executive secretarial. Each individual receives training in all three departments at Kathy's: word processing, telephone and LAN (local area network), and ethics. Table 3.1.11 shows the number of units of training that each type of receptionist/secretary receives.

Table 3.1.11

	Medical Field	Small Business	Executive Secretary
Word processing	0.8	1.1	1.7
Telephone and LAN	0.6	1.3	1.5
Ethics	1.2	0.8	1.3

Each month, word processing, telephone and LAN, and ethics have available a maximum of 32.1 units, 29.7 units, and 33 units, respectively.

(a) Define the variables and write linear inequalities to express reasonable values for the variables.

(b) Write linear inequalities to express the time restrictions of the word processing, telephone and LAN, and ethics departments.

⇒ 4.1 **42.** The data in Table 3.1.12 give the amount of nitrogen, phosphate, and potash (all in lb/acre) needed to grow asparagus, cauliflower, and celery for an entire growing season in California.

Table 3.1.12

	Asparagus	Cauliflower	Celery
Nitrogen	240	215	336
Phosphate	148	98	171
Potash	42	69	147

SOURCE: U.S. Department of Agriculture, www.usda.gov/nass

Eldrick has 280 acres and can get no more than 65,970 pounds of nitrogen, 31,860 pounds of phosphate, and 21,630 pounds of potash.

(a) Define the variables and write linear inequalities to express reasonable values for the variables.

(b) Write linear inequalities to express the restrictions on nitrogen, phosphate, and potash.

⇒ 3.2

SECTION PROJECT

The Sierra Refining Company produces a grade of unleaded gasoline, grade A, which it supplies to its chain of service stations. The grade A unleaded gasoline is blended from Sierra's inventory of gasoline components and must meet the specifications given in the table.

	Minimum Octane Rating	Maximum Demand (barrels/week)	Minimum Deliveries (barrels/week)
Grade A	87	80,000	60,000

The characteristics of the components in inventory are given in the next table.

Gasoline Component	Octane Rating	Inventory (barrels)
1	86	70,000
2	96	60,000

Let x represent the number of barrels of component 1 blended into grade A gasoline per week, and let y represent the number of barrels of component 2 blended into grade A gasoline per week.

(a) Verify that the total weekly amount of grade A gasoline is given by $x + y$.

(b) Determine two linear inequality restrictions for $x + y$.

Section 3.2 Graphing Systems of Linear Inequalities

Many realistic applications involve **inequalities** rather than equations. For example, a dietitian may need to plan a diet for a patient that has *at least* 30 grams of calcium. Let's say that one food provides 15 grams per serving while another provides 7 grams per serving. If the dietitian has two (or more) foods to select from and two (or more) dietary requirements, he can use a **system of linear inequalities** to find the right amount of each food. One of the most convenient ways to represent a solution to a system of linear inequalities with two variables is with a graph. Before we can do this, however, we must first learn how to graph a single linear inequality.

Graphing Linear Inequalities

In Chapter 1 we stated that the **standard form** of a line is given by

$$ax + by = c$$

where a, b, and c are constants, $a \geq 0$, and a and b are not both zero. We can use this standard form to define a **linear inequality**.

Linear Inequality

A **linear inequality** in two variables is given by

$$ax + by \leq c$$

where a, b, and c are constants, with a and b not both zero.

Example 1 Graphing a Linear Inequality

Graph the linear inequality $2x + y \leq 6$.

From Your Toolbox

To determine the x-intercept, we let $y = 0$, and to determine the y-intercept, we let $x = 0$.

Solution

Recall that the symbol $\leq$ is read "less than **or** equal to." The coordinates of any point (x, y) in the plane must satisfy $2x + y < 6$ **or** $2x + y = 6$ to be a solution to the inequality, $2x + y \leq 6$. Using the techniques from Chapter 1, we begin by graphing $2x + y = 6$. To graph $2x + y = 6$, we determine its intercepts. (Consult the Toolbox to the left.) The x-intercept is at (3, 0) and the y-intercept is at (0, 6). We use these points to graph $2x + y = 6$, as shown in Figure 3.2.1.

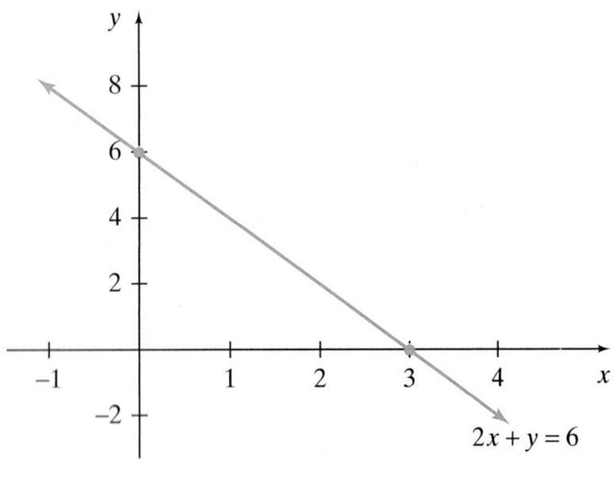

Figure 3.2.1

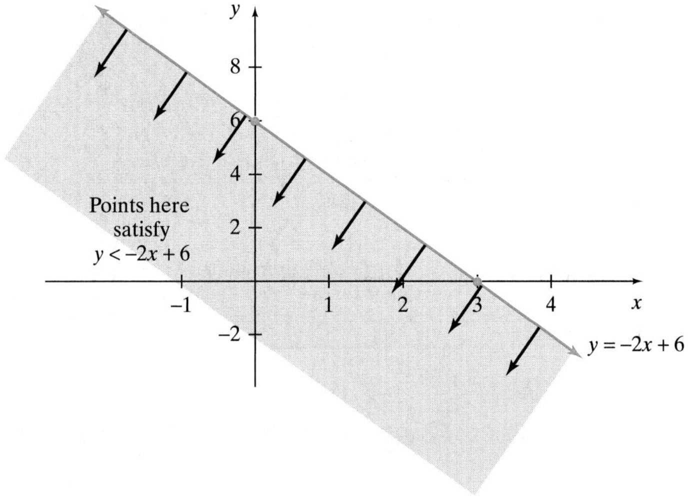

Figure 3.2.2

Before continuing, we must emphasize that each point on the line is a solution to $2x + y = 6$. Now, to determine the points that satisfy (or are solutions to) the inequality $2x + y < 6$, we can solve the inequality for y to get

$$2x + y < 6$$
$$y < -2x + 6$$

Figure 3.2.2 shows that points below the line $2x + y = 6$ (or in slope–intercept form, $y = -2x + 6$) satisfy the inequality $y < -2x + 6$. Since points on the line satisfy $y = -2x + 6$, visually it appears that when below the line the values of y

decrease the farther we move away from the line. What this means is that our inequality $2x + y \le 6$ has as its solution **all points on or below the line** $2x + y = 6$. We indicate the points below the line by shading, as shown in Figure 3.2.3. The line and shaded region in Figure 3.2.3 give us the graph of $2x + y \le 6$.

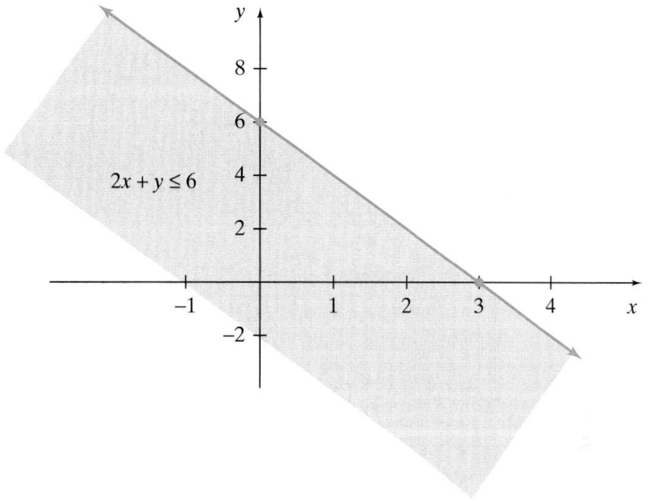

Figure 3.2.3 Graph of $2x + y \le 6$ ■

The line $2x + y = 6$ in Figure 3.2.3 is called the **boundary**. Quite simply, a boundary is a line that separates the points in the solution from those **not** in the solution. Also, the line $2x + y = 6$ has split the Cartesian plane into two planes called the **lower half-plane** and the **upper half-plane**. Figure 3.2.4 summarizes these important features that we find in graphing linear inequalities.

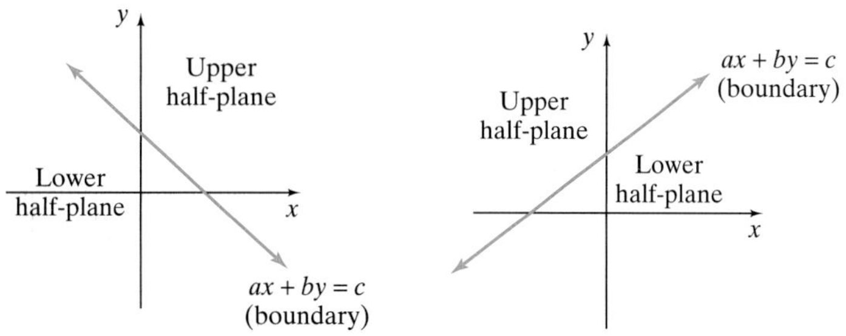

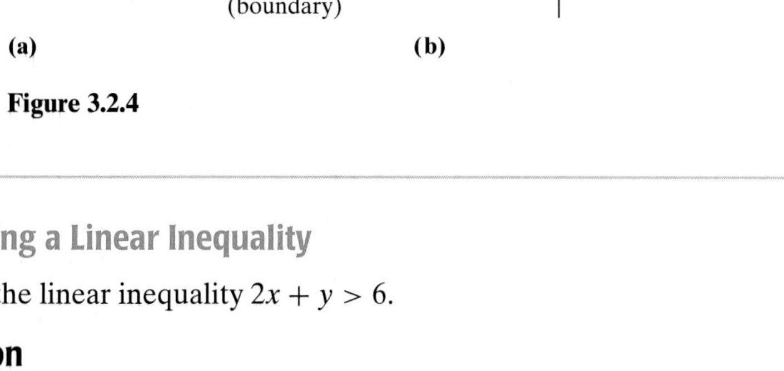

Figure 3.2.4

Example 2 Graphing a Linear Inequality

Graph the linear inequality $2x + y > 6$.

Solution

We have the same boundary as in Example 1, $2x + y = 6$. However, since our inequality symbol (>) indicates that points on the line itself do **not** satisfy the inequality, we use a dashed line. As in Example 1, to determine the points that satisfy $2x + y > 6$, we solve for y and get $y > -2x + 6$. Since points on the line satisfy $y = -2x + 6$, from Figure 3.2.2 it appears that, when above the line, the values of y increase the farther we move away from the line. This means that we must shade the upper half-plane. Figure 3.2.5 shows the graphical solution to $2x + y > 6$.

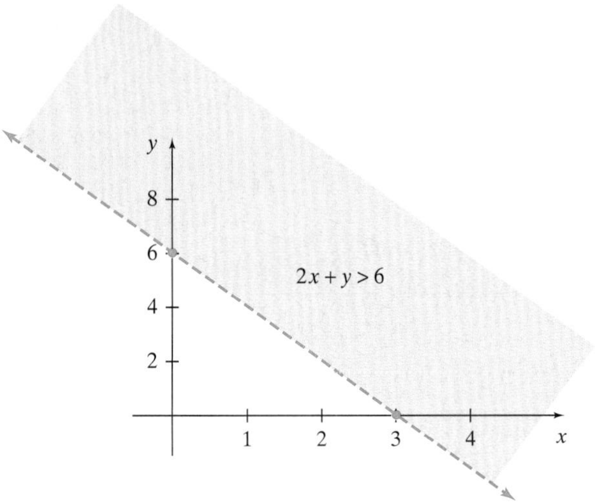

Figure 3.2.5

There is an alternative method for finding the correct region to shade when graphing a linear inequality. After graphing the boundary, select a **test point** that is **not** on the boundary. For example, in Examples 1 and 2 such a point is the origin (0, 0). We can then substitute the coordinates of the test point into the given inequality. If, after substituting, the resulting inequality is true, shade the region containing the test point. If the resulting inequality is false, shade the region not containing the test point. For example, using the test point (0, 0) in Example 1 would give us

$2x + y \leq 6$	Given inequality
$2(0) + 0 \leq 6$	Substitute (0, 0) for (x, y).
$0 \leq 6$	Inequality is true.

Since the result is a true statement, we would shade the region containing the test point (0, 0). This discussion, along with Examples 1 and 2, supplies the following procedure for graphing linear inequalities.

Graphing a Linear Inequality

1. Draw the graph of the boundary line. If the inequality involves $\leq$ or $\geq$, make the line solid. If the inequality involves $<$ or $>$, make the line dashed.
2. Pick a test point (x, y) that is in one of the half-planes. Substitute the values of x and y into the given inequality and simplify. (*Hint:* Whenever possible, pick the origin.)
3. If the test point satisfies the original inequality (gives a true statement), shade the region containing the test point. If the test point does not satisfy the original inequality (gives a false statement), shade the region that does not contain the test point.

Note: When choosing a test point (x, y), be sure it is not on the boundary line.

Example 3 Graphing a Linear Inequality

Graph the following linear inequalities.

(a) $3x + 2y > 6$ **(b)** $x \geq -2$

Solution

(a) We begin by graphing the boundary line $3x + 2y = 6$. The x-intercept is $(2, 0)$ and the y-intercept is $(0, 3)$. Since the inequality symbol is $>$, the boundary line is dashed, as shown in Figure 3.2.6. Our test point is $(0, 0)$. Substituting into the original inequality gives

$$3x + 2y > 6 \quad \text{Given inequality}$$
$$3(0) + 2(0) > 6 \quad \text{Substitute } (0, 0) \text{ for } (x, y).$$
$$0 > 6 \quad \text{Inequality is false.}$$

Since $0 > 6$ is false, we shade the half-plane that does *not* contain the test point, as shown in Figure 3.2.6.

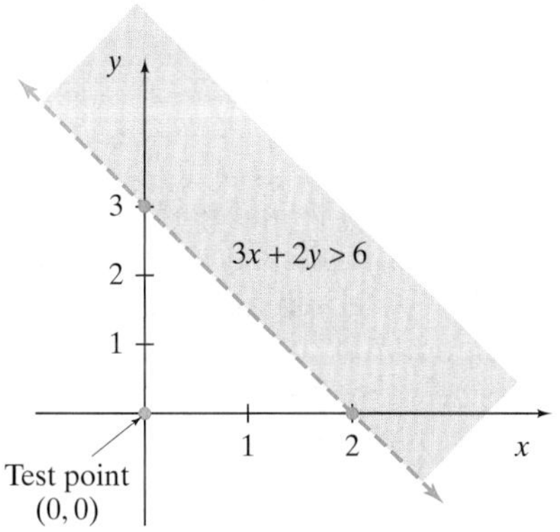

Figure 3.2.6

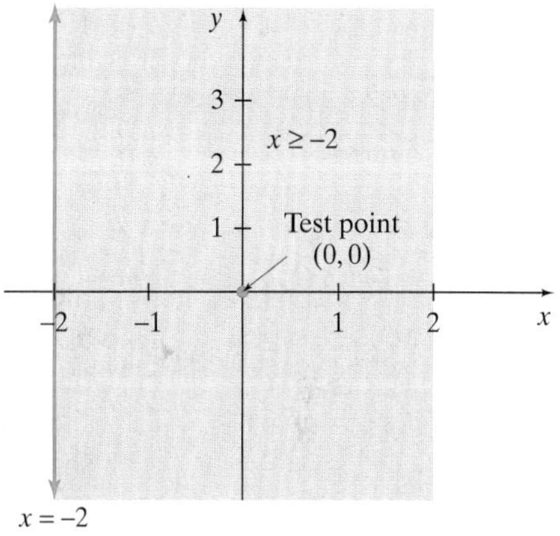

Figure 3.2.7

(b) The graph of $x = -2$ is a vertical line. Since the inequality symbol is $\geq$, we make the boundary line solid, as shown in Figure 3.2.7. Our test point is $(0, 0)$, and substituting into the given inequality yields

$$x \geq -2 \quad \text{Given inequality}$$
$$0 \geq -2 \quad \text{Inequality is true when substituting 0 for } x.$$

Since it is true, we shade the half-plane containing the test point, as shown in Figure 3.2.7. ■

Checkpoint 1 Now work Exercise 15.

Graphing Systems of Linear Inequalities

In Chapter 1 we saw that a solution to a system of linear equations with two variables was a pair of values for the variables that made all the equations in the

Technology Option

Graphing calculators can shade regions showing solutions to linear inequalities. Figure 3.2.8 gives two different shadings options for Example 3a. Figure 3.2.8a uses the DRAW SHADE command, and Figure 3.2.8b uses a feature on the Y = editor screen.

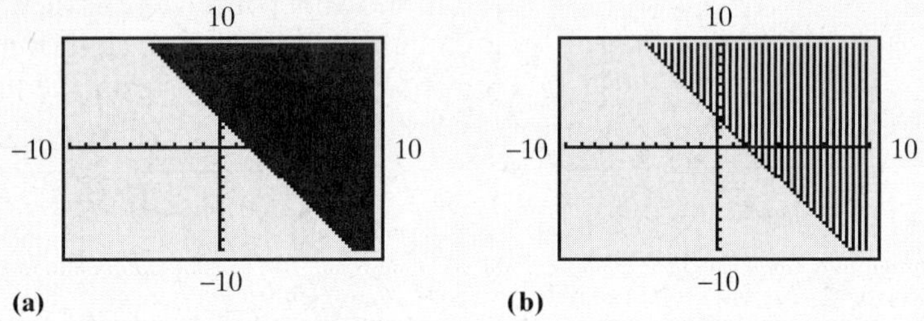

Figure 3.2.8

To learn how to shade inequalities on your calculator, consult the online graphing calculator manual at www.prenhall.com/armstrong

system true. A solution to a **system of linear inequalities** is the set of all points in the plane that satisfies (makes true) all the inequalities in the system. We can determine these points by graphing all inequalities in the system and determining the region that is common to all the graphs of the inequalities. To avoid confusion (and a mess!), we will use arrows on each boundary line to indicate the shading for a single linear inequality and will shade only the solution to the system. Example 4 illustrates this process.

Example 4 Graphing a Solution to a System of Linear Inequalities

Graph the system

$$\begin{aligned} 2x + y &\geq 6 \\ x - y &\leq 0 \end{aligned}$$

Solution

We begin by graphing each boundary line, $2x + y = 6$ and $x - y = 0$, as shown in Figure 3.2.9. Using the test point $(0, 0)$ for $2x + y \geq 6$ gives us $2(0) + 0 \geq 6$ or $0 \geq 6$, which is *false*. Hence we shade the region not containing $(0, 0)$. This is indicated by the arrows on the boundary line. Since $(0, 0)$ is on the boundary line $x - y = 0$ and not in a half-plane, we cannot use $(0, 0)$ as a test point. Choosing $(0, 1)$, which is not on the boundary line, as our test point for $x - y \leq 0$ gives us $0 - 1 \leq 0$ or $-1 \leq 0$, which is *true*. Hence we shade the region containing $(0, 1)$. This is indicated by the arrows on the boundary line. The region common to both inequalities is shaded in Figure 3.2.9.

Interactive Activity

Pick three points in the solution set (shaded region) in Example 4 and verify that each point satisfies both inequalities.

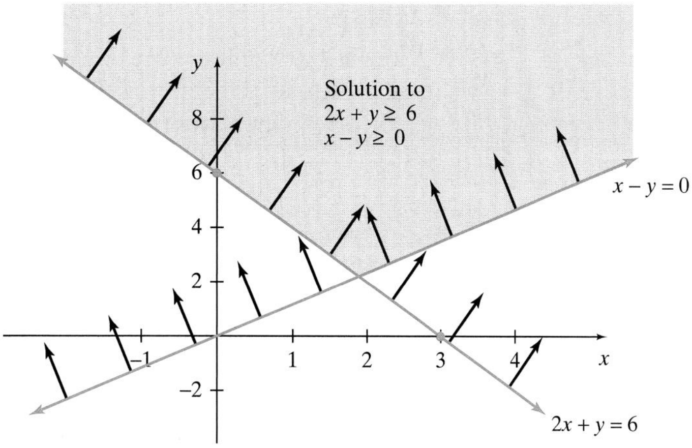

Figure 3.2.9

Technology Option

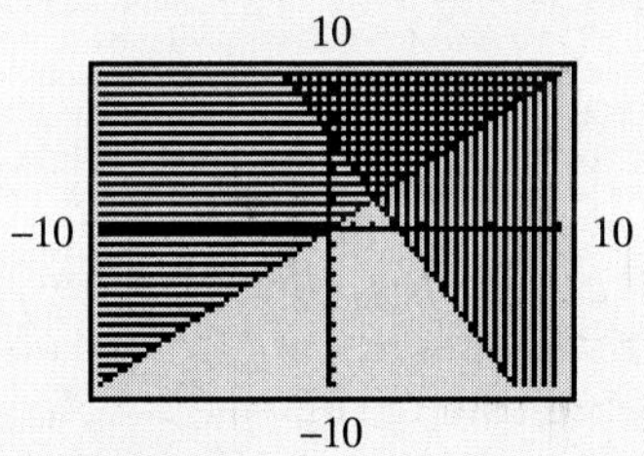

Figure 3.2.10

The graph of the system in Example 4 is shown in Figure 3.2.10. Notice that the solution to the system is given by the region where the two shaded regions overlap.

Checkpoint 2

Now work Exercise 27.

The shaded region in Figure 3.2.9 from Example 4, which gives the solutions to the system of linear inequalities, is also called the **feasible region**. It has this name since it gives all points that satisfy all inequalities in the system. We will use this phrase frequently in the remainder of Chapter 3.

Feasible Region

The set of all points that satisfies a system of linear inequalities is called the **feasible region**.

Example 5 Graphing a Feasible Region

Graph the feasible region for the system

$$x + y \leq 6$$
$$x - y \leq 2$$
$$x \geq 0,\ y \geq 0$$

Solution

We begin by examining the last two inequalities. Observe that $x \geq 0$ corresponds to all points to the right of the y-axis plus all points on the y-axis itself. Also, $y \geq 0$ corresponds to all points above the x-axis plus all points on the x-axis itself. The half-plane solutions for the first two inequalities are shown by arrows in Figure 3.2.11. The graph of the feasible region is shaded in Figure 3.2.11.`

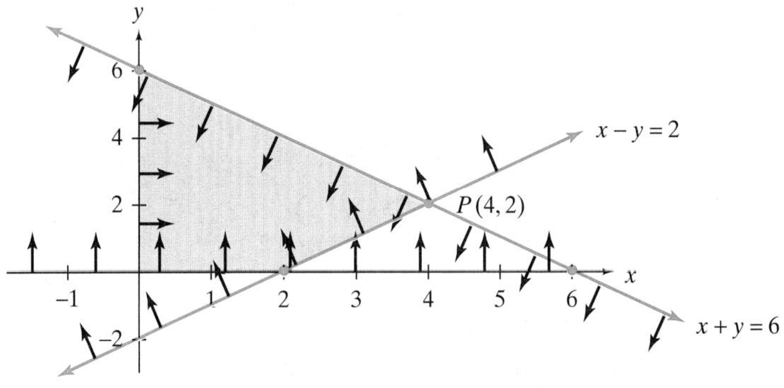

Figure 3.2.11

Interactive Activity

How did we determine the coordinates of point P in Figure 3.2.11 to be (4, 2)?

WWW

The feasible region in Example 5, the shaded region, is said to be **bounded**. This simply means that it is enclosed or it could be enclosed by some circle. If we draw a circle centered at the origin with a radius of, say, 15, then the shaded region would be entirely inside the circle. The feasible region in Example 4 is said to be **unbounded**. This means that it cannot be enclosed by a circle.

Applications of Systems of Linear Inequalities

In this section we focus on writing systems of linear inequalities to represent a problem situation and graphing and interpreting the corresponding feasible region. In the remaining two sections of Chapter 3 we learn how to apply this necessary skill to solving applications.

Example 6 Writing a System of Linear Inequalities

The WeDo Wood Company makes two types of canoes: a two-person model and a four-person model. Each two-person model requires 1 hour in the cutting department and 1.5 hours in the assembly department. Each four-person model requires 1 hour in the cutting department and 2.75 hours in the assembly department. The cutting department has a maximum of 640 hours available each week, and the assembly department has a maximum of 1080 hours available each week. Define the variables and write a system of linear inequalities that expresses these restrictions. Graph the feasible region.

Solution

Understand the Situation: We do not know how many two-person canoes or four-person canoes are made each week. So we let x represent the number of two-person canoes made each week and let y represent the number of four-person canoes made each week.

Since we cannot make a negative number of canoes, we immediately have the linear inequalities

$$x \geq 0 \quad \text{and} \quad y \geq 0$$

Now x units of two-person canoes requires $1 \cdot x = x$ hours in the cutting department, and y units of four-person canoes requires $1 \cdot y = y$ hours in the cutting department. Since the cutting department has available a maximum of 640 hours, we have the linear inequality

$$x + y \leq 640 \qquad \text{Cutting department restrictions}$$

Similarly, x units of two-person canoes requires $1.5x$ hours in the assembly department, and y units of four-person canoes requires $2.75y$ hours in the assembly department. Since the assembly department has available a maximum of 1080 hours, this gives the inequality

$$1.5x + 2.75y \leq 1080 \qquad \text{Assembly department restrictions}$$

So the system of linear inequalities that expresses the restrictions is

$$\begin{aligned} x &\geq 0 \\ y &\geq 0 \\ x + y &\leq 640 \\ 1.5x + 2.75y &\leq 1080 \end{aligned}$$

A graph of the feasible region is given in Figure 3.2.12. Observe that every point in the feasible region satisfies all inequalities in the system. Thus every point in the feasible region is a feasible solution to the problem.

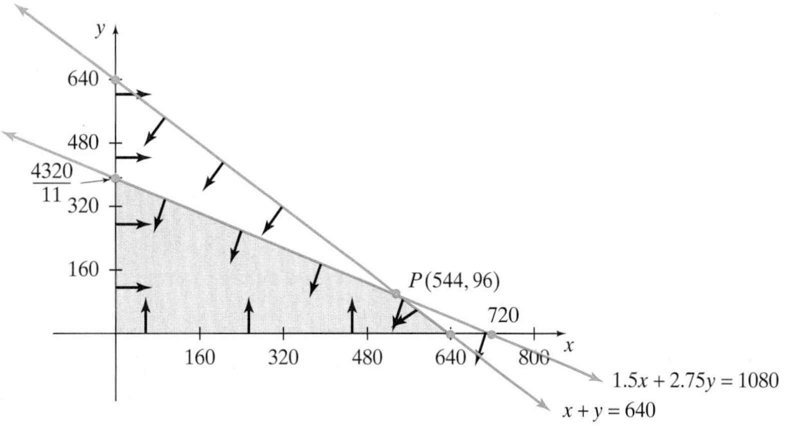

Figure 3.2.12

■

Checkpoint 3 Now work Exercise 51.

The linear inequalities in Example 6 are also known as the problem **constraints**. Each point in the feasible region in Figure 3.2.12 is a feasible solution given the problem constraints. We will see more on constraints in the last two sections of Chapter 3.

SUMMARY

In this section we learned what a linear inequality is and what its graph looks like. The latter gave rise to a **boundary** line as well as **half-planes**. A three-step procedure was given to graph linear inequalities. We then discussed graphing systems of linear inequalities. We saw that the common region of the graph of a system gives the solution to the system. We called this region the **feasible region**. We concluded the section with an application on writing a system of linear inequalities, given some restrictions.

- A **linear inequality** in two variables is given by $ax + by \le c$, where a, b, and c are constants, with a and b not both zero.
- **Graphing a Linear Inequality**

1. Draw the graph of the boundary line. If the inequality involves $\le$ or $\ge$, make the line solid. If the inequality involves $<$ or $>$, make the line dashed.
2. Pick a test point (x, y) that is in one of the half-planes. Substitute the values of x and y into the given inequality and simplify. (*Hint:* Whenever possible, pick the origin.)
3. If the test point satisfies the original inequality (gives a true statement), shade the region containing the test point. If the test point does not satisfy the original inequality (gives a false statement), shade the region that does not contain the test point.

SECTION 3.2 EXERCISES

In Exercises 1–4 match the graph with the correct inequality.

1. $x + y < 4$ **2.** $x + y > 4$ **3.** $x + y \ge 4$ **4.** $x + y \le 4$

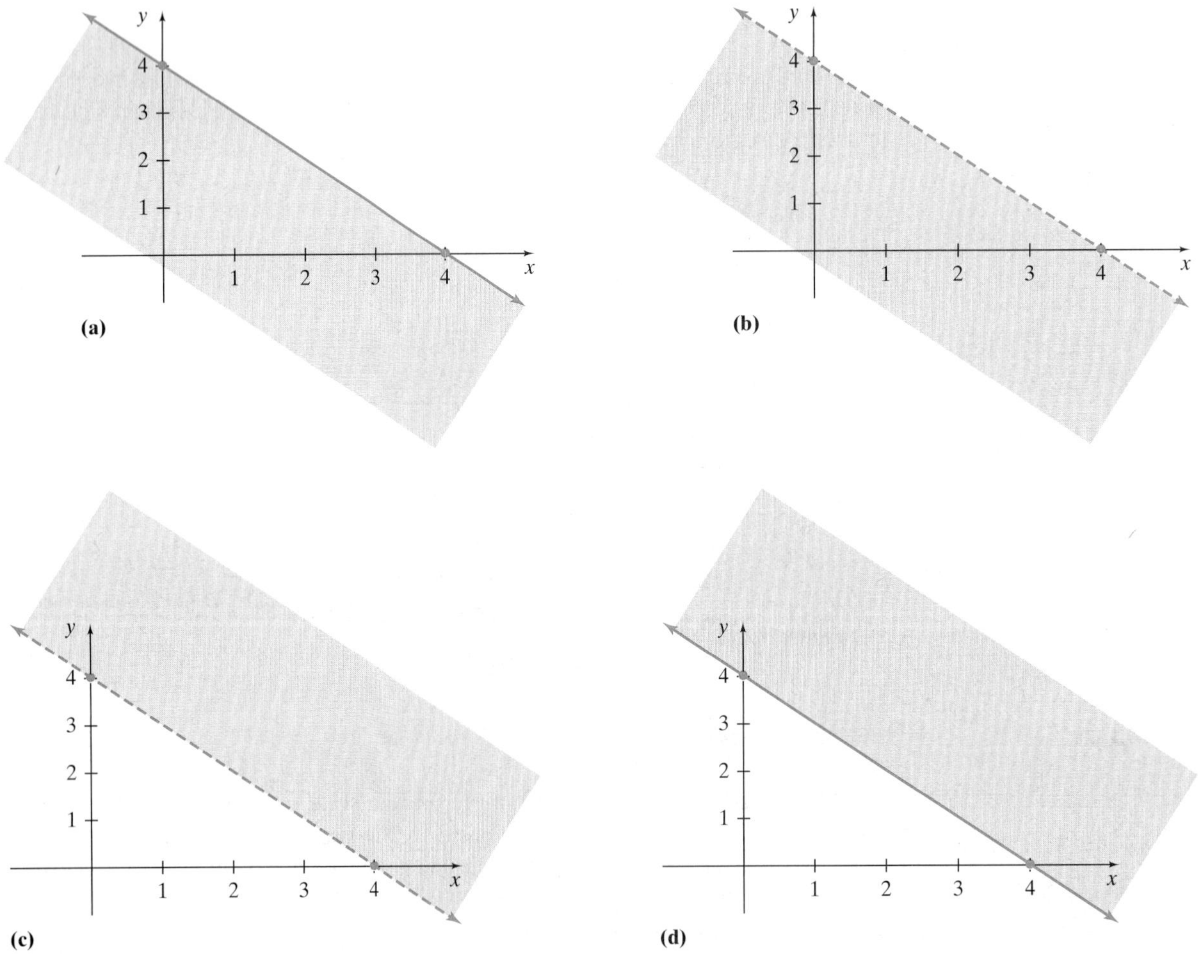

In Exercises 5–24, graph each linear inequality.

5. $y \le x + 2$

6. $y \le x - 3$

7. $y \ge -2x + 1$

8. $y \ge -3x + 2$

9. $y < x - 2$

10. $y < x + 5$

11. $y > 3x - 2$

12. $y > -2x + 5$

13. $3x - y \le 7$

14. $4x - y \le 6$

15. $2x + 4y \ge 8$

16. $3x + 6y \ge 12$

17. $x \ge -1$

18. $y \ge -3$

19. $-2x - y \le 12$

20. $-3x - y \le 15$

21. $2x + 3y \le 0$

22. $4x + 5y \le 0$

23. $3x - 2y \ge 0$

24. $5x - 4y \ge 0$

In Exercises 25–28, use Figure 3.2.13 to match the correct feasible region with the corresponding system of linear inequalities.

25. $x - y \le 0$
$x + 2y \ge 6$

26. $x - y \ge 0$
$x + 2y \ge 6$

27. $x - y \le 0$
$x + 2y \le 6$

28. $x - y \ge 0$
$x + 2y \le 6$

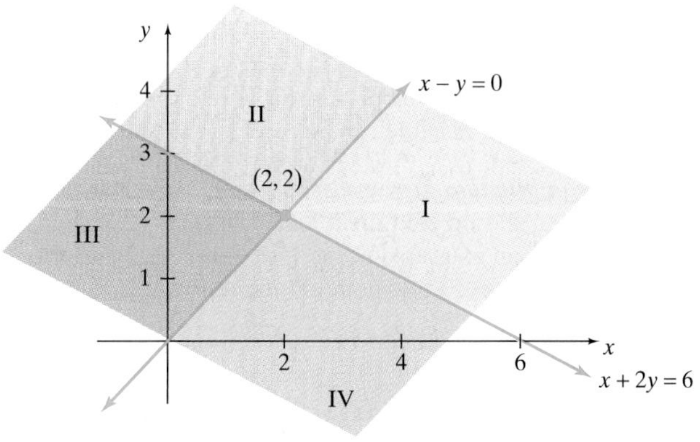

Figure 3.2.13

In Exercises 29–42, graph the feasible region for each system of linear inequalities. Indicate if the region is bounded or unbounded.

29. $x + y \le 2$
$x - y \ge 3$

30. $x - y \le 4$
$x + y \ge -2$

31. $2x + 3y \le 6$
$-x + 2y < 4$

32. $3x + 2y > 6$
$x + 3y \ge 9$

33. $2x - y \ge 4$
$4x - 2y \le -2$

34. $3x - y \le 9$
$2x + y \ge 8$

35. $x + y \le 7$
$2x - y \le 4$
$x \ge 0,\ y \ge 0$

36. $x + y \le 7$
$2x - y \ge 4$
$x \ge 0,\ y \ge 0$

37. $-2x + y \le 3$
$5x + 2y \le 10$
$x \ge 0,\ y \ge 0$

38. $-2x + y \le 3$
$5x + 2y \ge 10$
$x \ge 0,\ y \ge 0$

39. $3x + y \le 7$
$4x + y \ge 6$
$x \ge 0,\ y \ge 0$

40. $3x + 6y \le 12$
$5x + 4y \ge 8$
$x \ge 0,\ y \ge 0$

41. $5x - y \le 10$
$2x + y \ge 3$
$x + 2y \le 4$
$x \ge 0,\ y \ge 0$

42. $6x + 7y \le 42$
$4x - 9y \le 18$
$6x - 7y \le 14$
$x \ge 0,\ y \ge 0$

In Exercises 43–48, use a graphing calculator to graph the following.

43. $y \le 6 - x$

44. $y \ge 3 + x$

45. $2x + 3y \ge 7$

46. $4x - y \le 8$

47. $3x + 2y \ge 6$
$x - 3y \ge 9$

48. $2x - y \ge 4$
$4x - 2y \le -2$

Applications

3.1 **49.** McSavaney's trains administrative assistants for two types of office positions: medical field and executive secretarial. Each individual receives training in two departments at McSavaney's: word processing along with telephone and LAN (local area networks). An individual in the medical field program receives 2 units of word processing and 2.5 units of telephone and LAN training. An individual in the executive secretarial program receives 3 units of word processing and 2.5 units of telephone and LAN training. Each month at McSavaney's, corporate training can give no more than 32 units of word processing training and no more than 30 units of telephone and LAN training. Let x represent the number admitted each month to the medical field program, and let y represent the number admitted each month to the executive secretarial program.

(a) Write a system of linear inequalities that gives appropriate restrictions on x and y.

(b) Graph the feasible region for the system in part (a). Interpret what the feasible region represents.

3.1 **50.** (*continuation of Exercise 49*) Feedback from employers has indicated that McSavaney's needs to provide another area of training, ethics. McSavaney's adds an ethics training department. It has been determined that an individual in the medical field program receives 3.5 units of ethics training and an individual in the executive secretarial program receives 2.5 units of ethics training. Each month at McSavaney's, corporate training can give no more than 35 units of ethics training. Assuming that everything else from Exercise 49 remains the same, complete the following:

(a) Write a system of linear inequalities that gives appropriate restrictions on x and y.

(b) Graph the feasible region for the system in part (a). Interpret what the feasible region represents.

3.1 **51.** Sandkuhl, Inc., is a small manufacturing firm that produces two microwave switches, switch A and switch B. Each switch passes through two departments: assembly and testing. Switch A requires 4 hours of assembly and 1 hour of testing, and switch B requires 3 hours of assembly and 2 hours of testing. The maximum number of hours available each week for assembly and testing is 240 hours and 140 hours, respectively.

(a) Define variables and write a system of linear inequalities that gives appropriate restrictions on x and y.

(b) Graph the feasible region for the system in part (a). Interpret what the feasible region represents.

3.1 **52.** (*continuation of Exercise 51*) Due to a new contract commitment, Sandkuhl, Inc., must manufacture at least 10 units of switch A each week. Assuming that everything else from Exercise 51 remains the same, complete the following:

(a) Write a system of linear inequalities that gives appropriate restrictions on x and y.

(b) Graph the feasible region for the system in part (a). Interpret what the feasible region represents.

3.1 **53.** Klosterman's owns 200 acres of tillable farmland that is used to grow corn and wheat. On average, each acre of corn and wheat yields 110 and 35 bushels, respectively. At least 11,000 bushels of corn and 2205 bushels of wheat are required due to prior contracts.

(a) Define the variables and write a system of linear inequalities that gives appropriate restrictions on x and y.

(b) Graph the feasible region for the system in part (a). Interpret what the feasible region represents.

3.1 **54.** Mix for Chix, Inc., a local feed store, manufactures chicken feed by mixing two ingredients. Each ingredient contains two key nutrients: protein and fat. The number of units of each nutrient in 1 pound of the two basic ingredients is shown in Table 3.2.1.

Table 3.2.1

Ingredient	Protein	Fat
1	25	11
2	40	8

A local chicken farmer desires at most 500 pounds of a mix in which each pound of the feed has a maximum of 35 units of protein and a maximum of 10 units of fat.

(a) Define the variables and write a system of linear inequalities that gives appropriate restrictions on x and y.

(b) Graph the feasible region for the system in part (a). Interpret what the feasible region represents.

55. Table 3.2.2 gives nutritional information, in grams, for selected items from McDonald's menu.

Table 3.2.2

Item	Protein	Fat
Small French fries	3	10
Garden salad	7	6

SOURCE: McDonald's, www.mcdonalds.com

Julia wishes to eat a lunch consisting of small French fries and garden salad. She wants the lunch to contain at least 12 grams of protein.

(a) Define the variables and write a system of linear inequalities that gives appropriate restrictions on x and y.

(b) Graph the feasible region for the system in part (a). Interpret what the feasible region represents.

56. (*continuation of Exercise 55*) Now suppose that Julia wishes to eat a lunch consisting of small French fries and garden salad that contains no more than 25 grams of fat. Use the data in Table 3.2.2 to answer the following:

(a) Define the variables and write a system of linear inequalities that gives appropriate restrictions on x and y.

(b) Graph the feasible region for the system in part (a). Interpret what the feasible region represents.

57. (*continuation of Exercises 55 and 56*) Julia decides that her lunch should contain at least 12 grams of protein, but no more than 25 grams of fat. Use Table 3.2.2 and the results of Exercises 55 and 56 to do the following:

(a) Write a system of linear inequalities that gives appropriate restrictions on x and y.

(b) Graph the feasible region for the system in part (a). Interpret what the feasible region represents.

3.1 **58.** The Stitz Tent Company makes two types of tents: two-person and six-person models. Each tent requires the services, in hours, of two departments as shown in Table 3.2.3.

Table 3.2.3

Department	Two-person	Six-person
Fabric cutting	6	8
Assembly and packaging	2	4

Fabric cutting has a maximum of 580 hours available each week, and assembly and packaging has a maximum of 260 hours available.

(a) Define the variables and write a system of linear inequalities that gives appropriate restrictions on x and y.

(b) Graph the feasible region for the system in part (a). Interpret what the feasible region represents.

59. (*continuation of Exercise 58*) Suppose that the Stitz Tent Company makes a profit of \$40 on each two-person model and a profit of \$60 on each six-person model.

(a) Verify that the point (20, 20) is in the feasible region found in Exercise 58.

(b) If the company makes and sells 20 two-person models and 20 six-person models, the weekly profit is \$2000. The equation $40x + 60y = 2000$ gives *all* production schedules that result in a \$2000 weekly profit. Graph this line in the feasible region from Exercise 58 and verify that the point (20, 20) is on the line.

(c) Determine an equation that gives all production schedules that result in a \$2400 weekly profit.

(d) Verify that the graph of the equation found in part (c) is parallel to the graph of the equation given in part (b).

⇐ 3.1

SECTION PROJECT

The Sierra Refining Company produces a grade of unleaded gasoline, grade A, which it supplies to its chain of service stations. The grade A unleaded gasoline is blended from Sierra's inventory of gasoline components and must meet the specifications in Table 3.2.4.

Table 3.2.4

	Minimum Octane Rating	Maximum Demand (barrels/week)	Minimum Deliveries (barrels/week)
Grade A	87	80,000	60,000

The characteristics of the components in inventory are given in Table 3.2.5

Table 3.2.5

Gasoline Component	Octane Rating	Inventory (barrels)
1	86	70,000
2	96	60,000

Let x represent the number of barrels of component 1 blended into grade A gasoline per week, and let y represent the number of barrels of component 2 blended into grade A gasoline per week.

(a) Verify that the total weekly amount of grade A gasoline is given by $x + y$.

(b) Use Table 3.2.4 to determine two linear inequality restrictions for $x + y$.

(c) The minimum octane rating is obtained by computing the weighted average of the octane in the mixture. This means to divide the total octane in the mixture by the number of barrels in the mixture. Mathematically, this gives

$$\frac{86x + 96y}{x + y} \geq 87 \qquad \text{Restriction for minimum octane rating for grade A}$$

Multiply both sides by $(x + y)$ and simplify so that 0 is on one side of the inequality. (How do we know that $x + y$ is not a negative number?)

(d) Determine any other restrictions on x and y.

(e) Write the system of linear inequalities that gives all restrictions on x and y. (*Hint:* You should have seven total inequalities.)

(f) Graph the feasible region for the system in part (e).

Section 3.3 Solving Linear Programming Problems Graphically

In Section 3.2 we laid the groundwork for our next task, solving the linear programming problem. In the areas of management science, quantitative analysis, and operations research, linear programming is the most used and best known mathematical process. Linear programming allows us to maximize the profit of a product or maximize the protein in a diet, given certain restrictions. It also allows us to minimize the costs in the manufacture of a product or minimize the calories in a diet, given certain restrictions. For example, a company that makes two types of canoes may have restrictions as given by the number of labor hours available in the assembly and cutting departments. If the company knows the profit made on each type of canoe, it would want to determine how to utilize the available labor hours so that the profit is maximized. In Section 3.4 we will address applications such as these. In this section we will focus exclusively on how to solve linear programming problems graphically so that this skill can be used effectively in the next section.

Linear Programming

Many applications in the real world ask us to determine an **optimal value** (maximum or minimum) of a function that is subject to certain **constraints**. For example, the agribusiness in Chang Qing County in China is made up of crop farming, livestock husbandry, forestry, fisheries, and food processing. County officials wanted to increase the net profit while having no adverse effects on the environment. (*Source:* A Case Write Up: Agriculture in China, www.informs.org). In a **linear programming** problem the function that we are trying to find an optimal value for is called an **objective function**. In Chang Qing County, the objective function was the profit function for the agribusinesses, and the constraints were the restrictions on environmental impact. The constraints that the objective function is subject to are usually given by linear inequalities. When only two variables are involved, we can use the techniques of Section 3.2. Before solving applications, let's examine the method behind linear programming.

Example 1 Maximizing an Objective Function

Maximize the objective function $P = 2x + y$ subject to the constraints

$$x + y \leq 6$$
$$x - y \leq 2$$
$$x \geq 0,\ y \geq 0$$

Solution

The first thing we do is **graph the feasible region** as given by the system of linear inequalities using the techniques discussed in Section 3.2. See Figure 3.3.1. The points labeled A, B, C, and D occur where boundary lines intersect. Such points are called **corner points**. Soon we will see that corner points are crucial in maximizing (or minimizing) an objective function. Hence we need to **determine the coordinates** of the corner points. A, B, and C are fairly easy to determine since

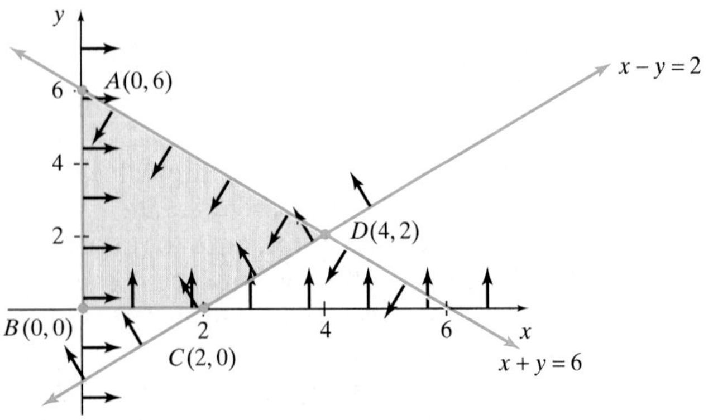

Figure 3.3.1

they are simply intercepts. We determine the coordinates of point A to be (4, 2) by solving the system

$$x - y = 2$$
$$x + y = 6$$

Recall that this system can be solved by using the substitution method (Chapter 1) or the elimination method (Chapter 1), graphically on a calculator, and by using the INTERSECT command (Chapter 1 and 2), Gauss–Jordan (Chapter 2), or matrix inverses (Chapter 2).

Now every point in the feasible region satisfies all the constraints given by the system of linear inequalities. However, we want to **find the point that gives the largest value for the objective function**. To see how we can do this, we will add lines to Figure 3.3.1 that represent our objective function $P = 2x + y$ for various values of P. We arbitrarily pick 0, 4, 8, and 12 for P. This produces the equations

$$0 = 2x + y, \qquad 4 = 2x + y, \qquad 8 = 2x + y, \qquad 12 = 2x + y$$

We now add the graphs of these four lines to Figure 3.3.1 as shown in Figure 3.3.2. The four lines are parallel since they all have a slope of -2. Observe that P cannot have a value of 12 since the graph for $P = 12$ is outside the feasible region. Every point on the line $P = 8$ that is within the feasible region gives a value for x and y such that $2x + y = 8$. It appears that a line parallel to these four lines, somewhere between the lines representing the objective function when $P = 8$ and $P = 12$, will yield the maximum value for P. The values of x and y that maximize P **and** still satisfy all constraints will be coordinates (x, y) of a point on this parallel line **if** the line just touches the feasible region. We observe that this happens at the **corner point** (4, 2). The value of P at this point is

$$P = 2x + y = 2(4) + 2 = 10$$

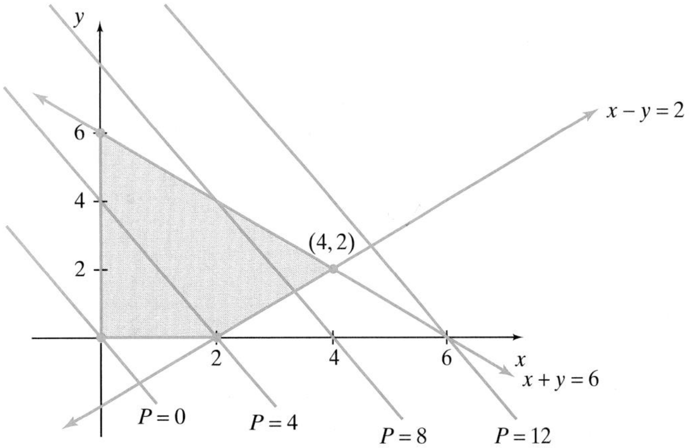

Figure 3.3.2 ■

Interactive Activity

Conjecture what point minimizes the objective function in Example 1. What is the minimum value?

Before doing another example, we want to stress that of all the points in the feasible region in Example 1, the corner point (4, 2) gives the largest possible value of P. Also, recall from Section 3.2 that the feasible region in Example 1 is said to be **bounded**.

Example 2 Minimizing an Objective Function

Minimize the objective function $C = 2x + 2y$ subject to the constraints

$$x + y \geq 8$$
$$5x + 7y \geq 50$$
$$x \geq 0,\ y \geq 0$$

Solution

Figure 3.3.3 shows the feasible region along with lines when C in the objective function is replaced with 0, 8, 16, and 24. The smallest value of z for which a line touches the feasible region is $C = 16$. We have two corner points, $(0, 8)$ and $(3, 5)$, on this line. This means that both corner points, $(0, 8)$ and $(3, 5)$, as well as **all points on the boundary line between them**, yield the same minimum value for C. So there are an infinite number of values for x and y that produce the minimum value of 16 for the objective function.

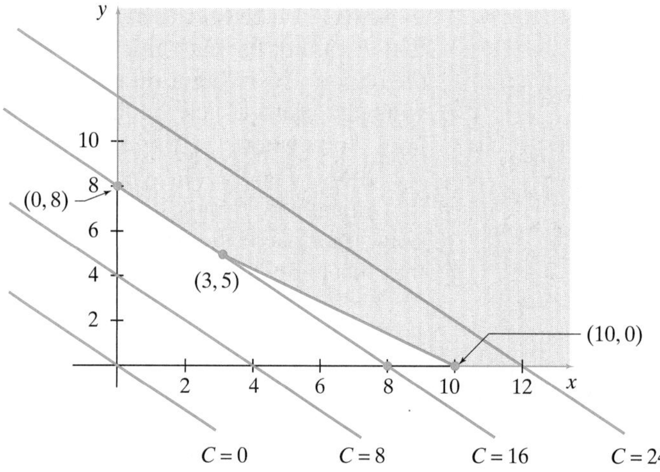

Figure 3.3.3

Interactive Activity

Explain why the lines for $C = 0$ and $C = 8$ in Example 2 are not candidates for minimizing C. Also, do you believe that there is a maximum value of the objective function in Example 2? Explain your reasoning.

WWW

Example 2 is an example of a **multiple optimal solution**. In general,

Multiple Optimal Solution

If two corner points are both optimal solutions to a linear programming problem, then any point that is on the line segment joining them is also an optimal solution. We call the collection of these points the **multiple optimal solution**.

Recall from Section 3.2 that the feasible region in Example 2 is said to be **unbounded**. The feasible region in Example 1 was bounded and, as Example 1 showed, along with the Interactive Activity following Example 1, linear programming problems with bounded regions always have optimal solutions (a maximum and a minimum). This is not true for the unbounded region in Example 2. It had a minimum, but there is no value for x and y that will maximize the objective function.

We now summarize the prior discussion as follows:

Existence of Optimal Solution

- If the feasible region for a linear programming problem is bounded, then both a maximum value and a minimum value of the objective function exist.
- If the feasible region for a linear programming problem is unbounded, then carefully examine the feasible region and objective function to see if the maximum or minimum value exists.

Example 2 illustrates bullet item 2. Examples 1 and 2 illustrate that the optimal value of an objective function occurs at a corner point. These examples suggests the following theorem.

Corner Point Theorem

If the optimal value (either maximum or minimum) of the objective function exists, then it will occur at one (or more) of the corner points of the feasible region.

The Corner Point Theorem greatly simplifies the task of finding an optimal value. We now supply a general strategy for locating optimal values.

Solving Linear Programming Problems with Two Variables Graphically

Step 1: Graph the feasible region. Take note if the region is bounded or unbounded.

Step 2: Determine the coordinates of all corner points.

Step 3: Evaluate each corner point in the objective function.

Step 4: Determine the optimal solution from the results of step 3.

Example 3 Solving a Linear Programming Problem

$$\begin{aligned} \text{Maximize:}\quad & z = x - 3y \\ \text{Subject to:}\quad & 2x - 3y \le 6 \\ & -x + 4y \le 4 \\ & x \ge 0,\ y \ge 0 \end{aligned}$$

Solution

We will utilize the step-by-step procedure given prior to Example 3.

Step 1: The graph of the feasible region is shown in Figure 3.3.4. Notice that the feasible region is bounded, so the existence of a maximum value for the objective function is guaranteed.

Step 2: Corner point A has coordinates $(3, 0)$, corner point B has coordinates $(0, 0)$, and corner point C has coordinates $(0, 1)$. The coordinates for corner point D are found by solving the system

$$\begin{aligned} 2x - 3y &= 6 \\ -x + 4y &= 4 \end{aligned}$$

We leave it to you to decide which method to use to solve the system and verify that the coordinates for corner point D are $\left(\frac{36}{5}, \frac{14}{5}\right)$.

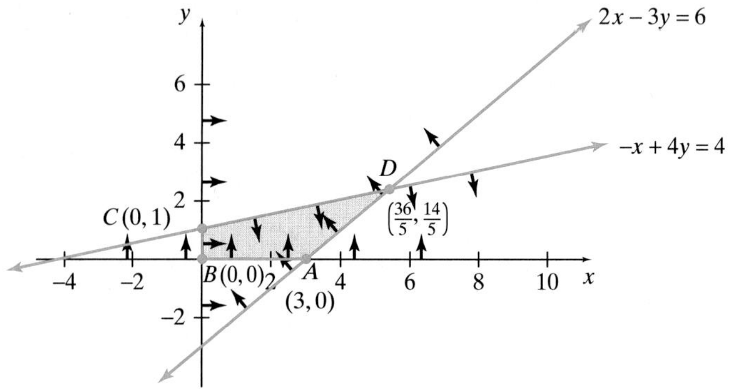

Figure 3.3.4

Step 3: Here we evaluate each corner point in the objective function. We use Table 3.3.1 to organize our work.

Table 3.3.1

Corner Point	Value of $z = x - 3y$	
$(3, 0)$	$3 - 3(0) = 3$	**Maximum**
$(0, 0)$	$0 - 3(0) = 0$	
$(0, 1)$	$0 - 3(1) = -3$	
$\left(\frac{36}{5}, \frac{14}{5}\right)$	$\frac{36}{5} - 3\left(\frac{14}{5}\right) = -\frac{6}{5}$	

Step 4: The maximum value of $z = x - 3y$ is 3 at the corner point $(3, 0)$. ■

Checkpoint 1 Now work Exercise 9.

In the next example we determine both a maximum and a minimum.

Example 4 Solving a Linear Programming Problem

$$\text{Maximize and minimize:} \quad z = 2x + y$$

$$\begin{aligned} \text{Subject to:} \quad x + y &\geq 2 \\ 6x + 4y &\leq 36 \\ 4x + 2y &\leq 20 \\ x \geq 0, y &\geq 0 \end{aligned}$$

Solution

As in Example 3, we utilize the step-by-step procedure.

Step 1: Figure 3.3.5 is a graph of the feasible region.

Step 2: The corner points are shown in Figure 3.3.5. Notice that four corner points are intercepts, while the fifth corner point is the solution to the system

$$\begin{aligned} 6x + 4y &= 36 \\ 4x + 2y &= 20 \end{aligned}$$

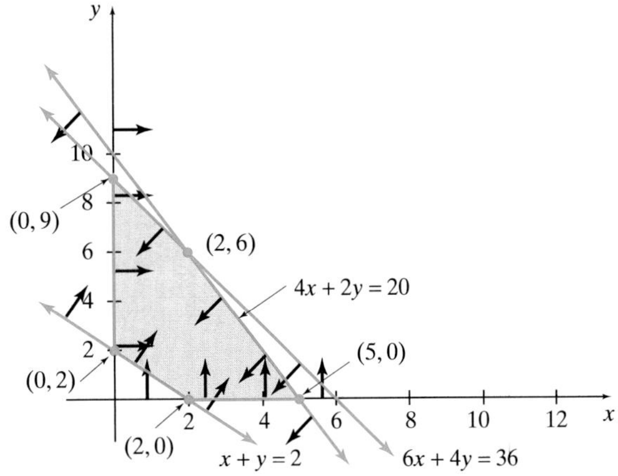

Figure 3.3.5

Step 3: We evaluate each corner point in the objective function. As in Example 3, we use a table to organize our work. See Table 3.3.2.

Table 3.3.2

Corner Point	Value of $z = 2x + y$	
$(5, 0)$	$2(5) + 0 = 10$	**Maximum**
$(2, 0)$	$2(2) + 0 = 4$	
$(0, 2)$	$2(0) + 2 = 2$	**Minimum**
$(0, 9)$	$2(0) + 9 = 9$	
$(2, 6)$	$2(2) + 6 = 10$	**Maximum**

Step 4: The minimum value of $z = 2x + y$ is 2 at the corner point $(0, 2)$. The maximum value of $z = 2x + y$ is 10 at the corner points $(5, 0)$ and $(2, 6)$. As stated earlier, this is a multiple optimal solution, and any point on the line segment connecting $(5, 0)$ and $(2, 6)$ also produces a maximum value of $z = 2x + y$. ■

✓ Checkpoint 2 Now work Exercise 15.

Our final example supplies a general strategy when dealing with an unbounded feasible region.

Example 5 Solving a Linear Programming Problem

Maximize and minimize: $z = x + y$

Subject to:
$$9x - 3y \le 6$$
$$4x + 2y \ge 8$$
$$x \ge 0,\ y \ge 0$$

Solution

Using our four-step process yields:

Step 1: Figure 3.3.6 is a graph of the feasible region. Notice that the feasible region is unbounded.

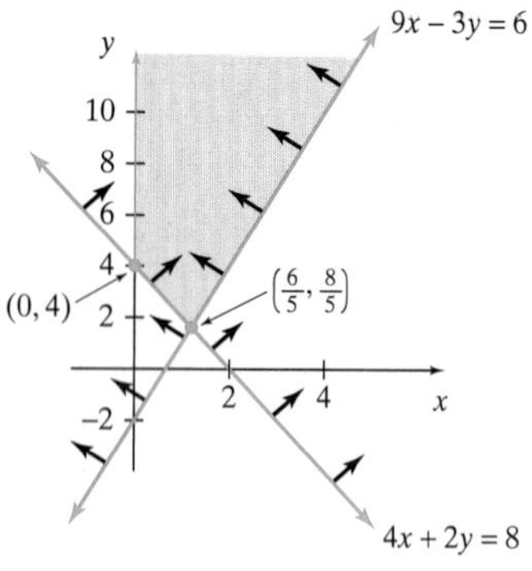

Figure 3.3.6

Step 2: The corner points are shown in Figure 3.3.6.

Step 3: We evaluate each corner point in the objective function and record the results in Table 3.3.3.

Table 3.3.3

Corner Point	Value of $z = x + y$	
$(0, 4)$	$0 + 4 = 4$	
$\left(\frac{6}{5}, \frac{8}{5}\right)$	$\frac{6}{5} + \frac{8}{5} = \frac{14}{5} = 2.8$	**Minimum**

Step 4: The minimum value of $z = x + y$ is 2.8 at the corner point $\left(\frac{6}{5}, \frac{8}{5}\right)$. For a maximum value, we need to take care since the feasible region is unbounded. To help us, we will add two lines to Figure 3.3.6, representing our objective function $z = x + y$ for $z = 2.8$ and $z = 4$, the values from Table 3.3.3. See Figure 3.3.7, which suggests that as z increases the lines move farther away from the origin. Since the feasible region is unbounded, the region continues indefinitely; there is no line that produces a maximum value of z.

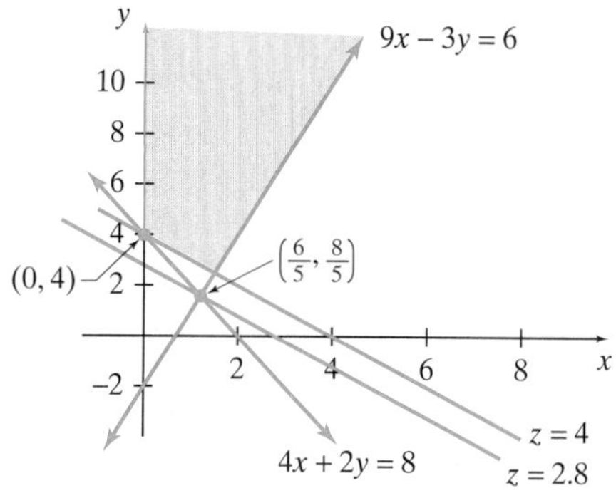

Figure 3.3.7

Example 5 illustrates how we need to take care when dealing with an unbounded feasible region. It also illustrates the following fact when confronted with an unbounded feasible region.

Optimal Values of an Unbounded Feasible Region

If the feasible region is unbounded and all the coefficients of the objective function are positive, then a minimum value of the objective function exists, but a maximum value does not.

Note: This fact is true only when the unbounded feasible region is in the first quadrant. In this text, all applications considered will be restricted to the first quadrant.

SUMMARY

In this section we introduced the methods of solving a **linear programming problem** graphically. We determined the **optimal value** of an **objective function**, given certain **constraints**. We saw how **corner points** were the key to the linear programming problem.

- **Corner Point Theorem** If the optimal value (either maximum or minimum) of the objective function exists, then it will occur at one (or more) of the corner points of the feasible region.

- **Solving Linear Programming Problems with Two Variables Graphically**

  Step 1: Graph the feasible region. Take note if it is bounded or unbounded.

  Step 2: Determine the coordinates of all corner points.

  Step 3: Evaluate each corner point in the objective function.

  Step 4: Determine the optimal solution from the results of step 3.

SECTION 3.3 EXERCISES

In Exercises 1–8, the graph of a feasible region is given. Use these regions to determine the maximum and minimum values of the given objective function.

1. $z = 2x + 3y$

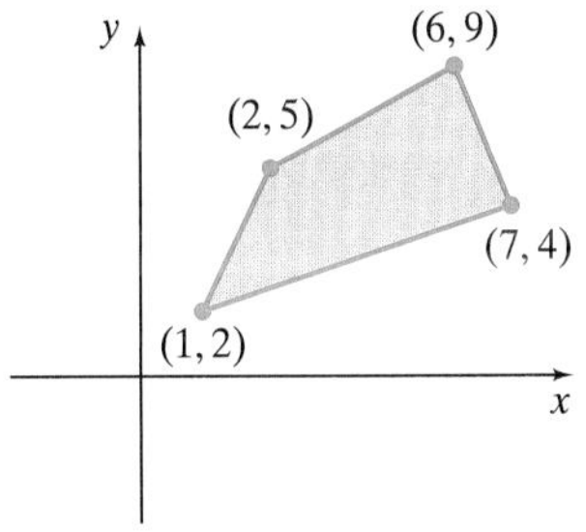

2. $z = 3x + 2y$

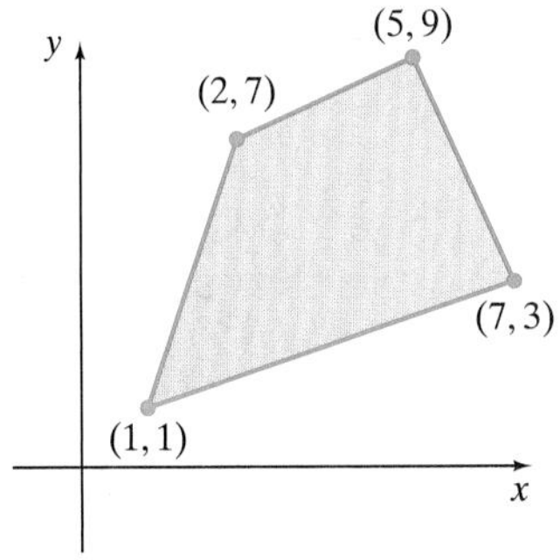

3. $z = 2x + 5y$

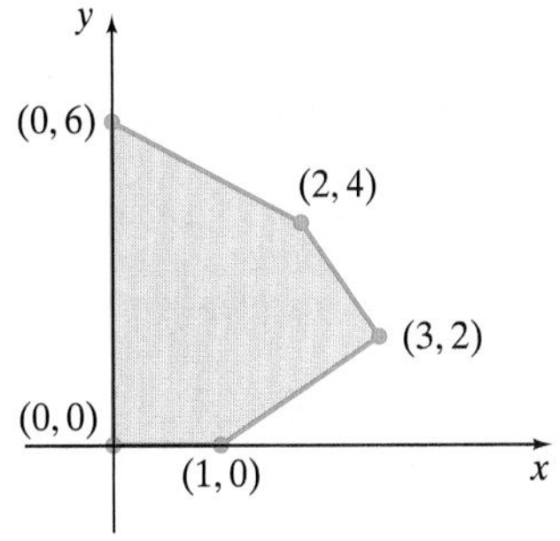

4. $z = 4x + y$

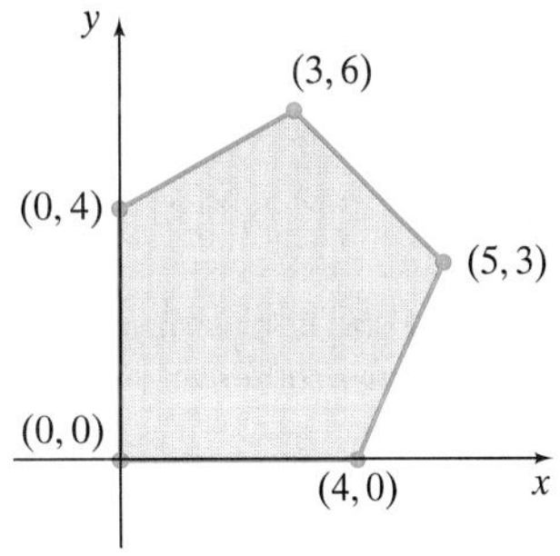

5. $z = x + 4y$

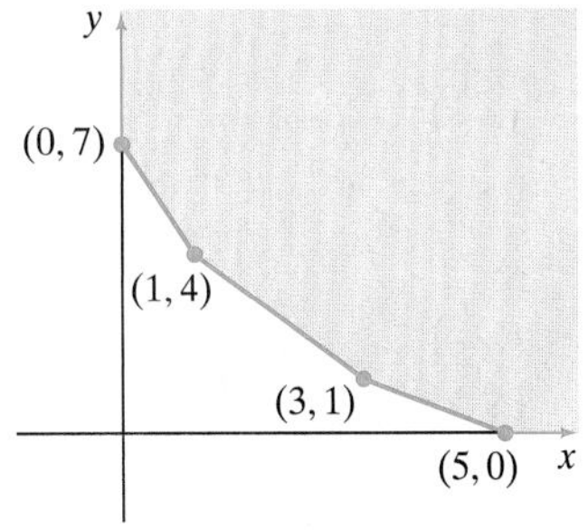

6. $z = 2x + 4y$

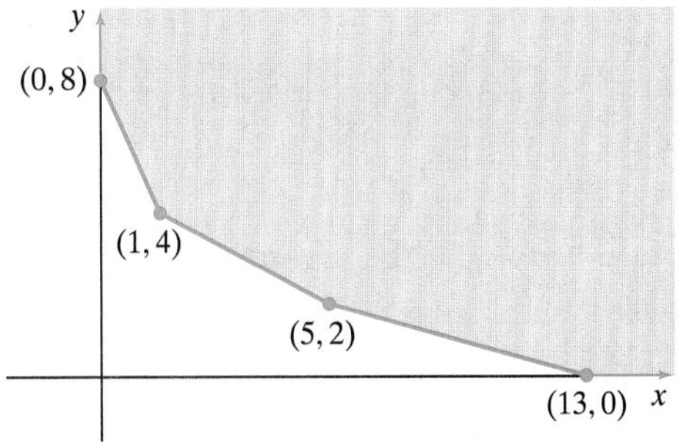

7. $z = 0.4x + 0.3y$

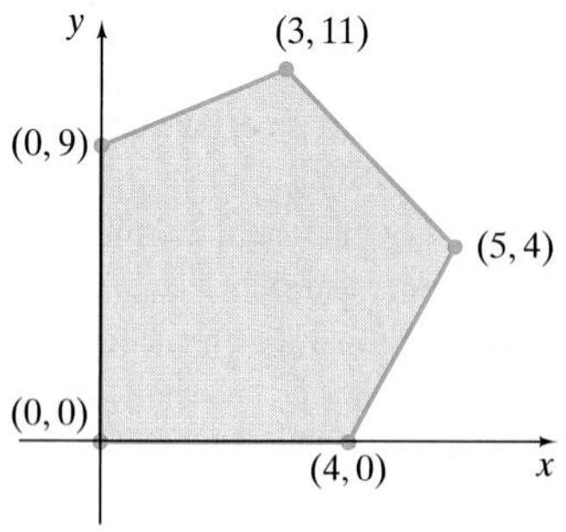

8. $z = 0.2x + 0.7y$

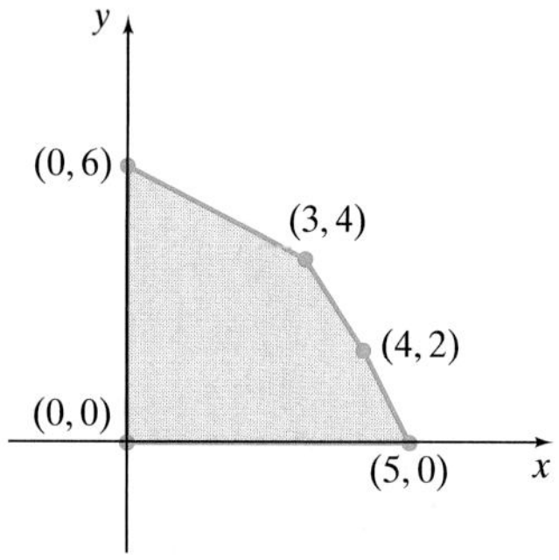

In Exercises 9–22, use the graphical and corner point method to solve the given linear programming problem.

9. Maximize: $z = 4x + 3y$
Subject to: $3x - y \leq 5$
$x + 2y \leq 4$
$x \geq 0,\ y \geq 0$

10. Maximize: $z = 3x + y$
Subject to: $4x + 5y \leq 9$
$2x - y \geq 1$
$x \geq 0,\ y \geq 0$

11. Minimize: $z = x + 2y$
Subject to: $x + y \leq 7$
$2x + 3y \leq 16$
$x \geq 0, y \geq 0$

12. Minimize: $z = 2x + 5y$
Subject to: $x + y \leq 7$
$2x + 3y \geq 16$
$x \geq 0, y \geq 0$

13. Maximize: $z = 2x + 2y$
Subject to: $x + y \leq 7$
$x - 2y \leq 16$
$x \geq 0, y \geq 0$

14. Maximize: $z = x + 3y$
Subject to: $x + y \leq 3$
$x - 2y \geq 0$
$x \geq 0, y \geq 0$

15. Minimize and maximize: $z = 3x + 2y$
Subject to: $x + y \leq 5$
$3x - 2y \leq 0$
$2x + y \leq 6$
$x \geq 0, y \geq 0$

16. Minimize and maximize: $z = 2x + 3y$
Subject to: $x + 2y \leq 2$
$3x - y \leq 3$
$x \geq 0, y \geq 0$

17. Minimize and maximize: $z = x + 4y$
Subject to: $2x + y \leq 32$
$x + y \leq 18$
$x + 3y \leq 36$
$x \geq 0,\ y \geq 0$

18. Minimize and maximize: $z = 5x + y$
Subject to: $-x + y \geq 0$
$x + 3y \leq 9$
$4x + 3y \leq 12$
$x \geq 0,\ y \geq 0$

19. Minimize and maximize: $z = 2x + 5y$
Subject to: $x + y \geq 2$
$2x + 3y \leq 12$
$3x + y \leq 12$
$x \geq 0,\ y \geq 0$

20. Minimize and maximize: $z = 5x + 2y$
Subject to: $x + y \geq 2$
$x + y \leq 8$
$2x + y \leq 10$
$x \geq 0,\ y \geq 0$

21. Minimize and maximize: $z = 3x + 5y$
Subject to: $-x + 2y \leq 10$
$2x + y \leq 15$
$5x + y \geq 8$
$-x + y \geq -1$
$x \geq 0,\ y \geq 0$

22. Minimize and maximize: $z = 5x + y$
Subject to: $-x + 2y \leq 10$
$2x + y \leq 15$
$5x + y \geq 8$
$-x + y \geq -5$
$x \geq 0,\ y \geq 0$

SECTION PROJECT

Minimize and maximize: $z = \frac{1}{2}x + \frac{1}{3}y$

Subject to:

$$\begin{aligned} 2x + 3y &\le 18 \\ 3x + 2y &\le 18 \\ 8x + 4y &\le 32 \\ 9x + 2y &\ge 16 \\ 0 \le x &\le 5 \\ 0 \le y &\le 6 \end{aligned}$$

Section 3.4 Applications of Linear Programming

This section combines the skills learned in the first three sections of this chapter. This includes writing systems of linear inequalities, representing problem constraints as inequalities, graphing the feasible region, and maximizing (or minimizing) an objective function for given constraints. In other words, in this section we will solve applications using linear programming. We'll start with a Flashback.

Flashback

WeDo Wood Canoes Company Revisited

In Section 3.2 we saw that the WeDo Wood Canoes Company makes two types of canoes: a two-person model and a four-person model. Each two-person model requires 1 hour in the cutting department and 1.5 hours in the assembly department. Each four-person model requires 1 hour in the cutting department and 2.75 hours in the assembly department. The cutting department has a maximum of 640 hours available each week, and the assembly department has a maximum of 1080 hours available each week. We saw in Section 3.2 that if we let x represent the number of two-person canoes made each week and let y represent the number of four-person canoes made each week then the following system of linear inequalities expresses the constraints.

$$\begin{aligned} x + y &\le 640 \\ 1.5x + 2.75y &\le 1080 \\ x \ge 0,\ y &\ge 0 \end{aligned}$$

A graph of this system, showing the feasible region and corner points, is given in Figure 3.4.1.

A profit of \$40 on each two-person canoe and \$60 on each four-person canoe is realized by WeDo Wood Canoes. Using the definition of the variables given, write an equation to represent the weekly profit and determine values for x and y that maximize weekly profit.

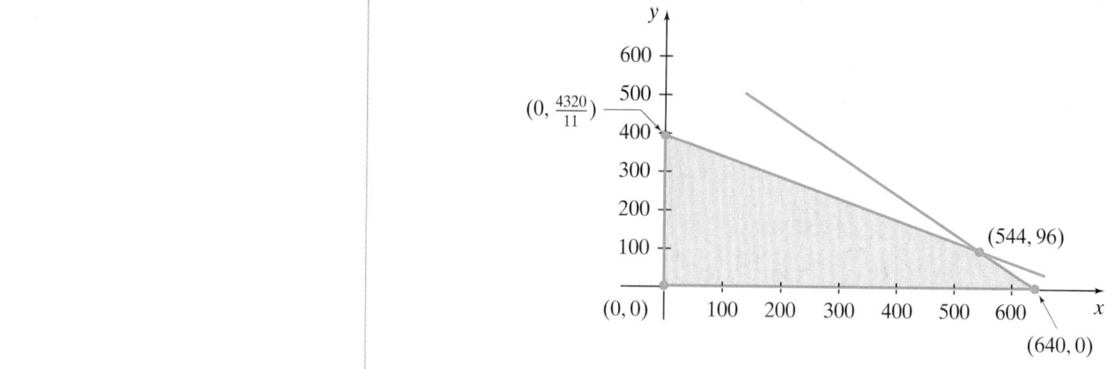

Figure 3.4.1

Flashback Solution

Understand the Situation: We have already defined x to represent the number of two-person canoes produced and sold and to let y represent the number of four-person canoes produced and sold. Since each two-person canoe yields a profit of \$40, WeDo Wood Canoes realizes a profit of $40x$ dollars on the sale of x two-person canoes. Similarly, a profit of $60y$ dollars is realized on the sale of y four-person canoes. So the weekly profit is represented by

$$P = 40x + 60y$$

Determining the values for x and y that maximize weekly profit sounds similar to what we did in Section 3.3. Actually, it is identical to what we did in Section 3.3! We are really solving the following linear programming problem:

$$\begin{aligned} \text{Maximize:} \quad & P = 40x + 60y \\ \text{Subject to:} \quad & x + y \le 640 \\ & 1.5x + 2.75y \le 1080 \\ & x \ge 0,\ y \ge 0 \end{aligned}$$

Figure 3.4.1 shows the feasible region along with the corner points, so all we need to do is evaluate the objective function at each corner point, as shown in Table 3.4.1.

Table 3.4.1

Corner Point	Value of $P = 40x + 60y$	
$(0, 0)$	$40(0) + 60(0) = 0$	
$(0, \frac{4320}{11})$	$40(0) + 60\left(\frac{4320}{11}\right) \approx 23{,}563.64$	
$(544, 96)$	$40(544) + 60(96) = 27{,}520$	**Maximum**
$(640, 0)$	$40(640) + 60(0) = 25{,}600$	

Interpret the Solution: Table 3.4.1 indicates that producing and selling 544 two-person canoes and 96 four-person canoes each week maximizes the weekly profit at \$27,520.

In the Flashback, the variables x and y are known as **decision variables**. Our goal was to determine values of the decision variables that maximized the **objective function**, $P = 40x + 60y$, given the **problem constraints**. Our problem constraints, as given by the system of linear inequalities, were

Cutting department constraint:	$x + y \leq 640$
Assembly department constraint:	$1.5x + 2.75y \leq 1080$
Nonnegative constraints:	$x \geq 0,\ y \geq 0$

Since our decision variables, x and y, represented the number of items produced, the feasible region in Figure 3.4.1 actually gave us the **feasible region for production schedules**. In other words, we could choose any production schedule (x, y) in the feasible region and use the objective function $P = 40x + 60y$ to compute the profit at production level (x, y). In the Flashback we used the techniques from Section 3.3 to find the production level that maximized profit.

Before we consider another example, let's combine our solving strategy from Section 3.3 with some new information to get a strategy for solving applied linear programming problems.

Solving an Applied Linear Programming Problem with Two Decision Variables

Step 1: Construct a mathematical model for the problem. This includes:

- Define the decision variables.
- Write a linear objective function.
- Write the problem constraints.

Step 2: Graph the feasible region. Note if it is bounded or unbounded.

Step 3: Determine the coordinates of all corner points.

Step 4: Evaluate each corner point in the objective function.

Step 5: Determine the optimal solution from the result of step 4.

Step 6: Interpret the optimal solution in terms of the original problem.

Note: Steps 2 through 5 are taken from the steps for solving linear programming problems in Section 3.3.

Example 1 Solving an Applied Linear Programming Problem

Dub, Inc., is a small manufacturing firm that produces two microwave switches, switch A and switch B. Each switch passes through two departments: assembly and testing. Switch A requires 4 hours of assembly and 1 hour of testing, and switch B requires 3 hours of assembly and 2 hours of testing. The maximum number of hours available each week for assembly and testing are 240 and 140 hours, respectively. The per unit profit for switch A is \$20, whereas the per unit profit for switch B is \$30. How many units of each switch should be made and sold each week to maximize profit?

Solution

We use the six-step strategy outlined prior to Example 1.

Step 1: Understand the Situation: Since we do not know how many of each switch is made, we let x represent the number of units of switch A made each week and let y represent the number of units of switch B made each week. Since the profit must be maximized, we determine an objective function for profit. The objective function is

$$P = 20x + 30y$$

The problem constraints are

Assembly department constraint:	$4x + 3y \leq 240$
Testing department constraint:	$x + 2y \leq 140$
Nonnegative assembly and testing constraints:	$x \geq 0,\ y \geq 0$

So the mathematical model is

$$\begin{aligned} \text{Maximize:} \quad & P = 20x + 30y \\ \text{Subject to:} \quad & 4x + 3y \leq 240 \\ & x + 2y \leq 140 \\ & x \geq 0,\ y \geq 0 \end{aligned}$$

Step 2: The feasible region is shown in Figure 3.4.2. Since the region is bounded, a maximum value is guaranteed to exist.

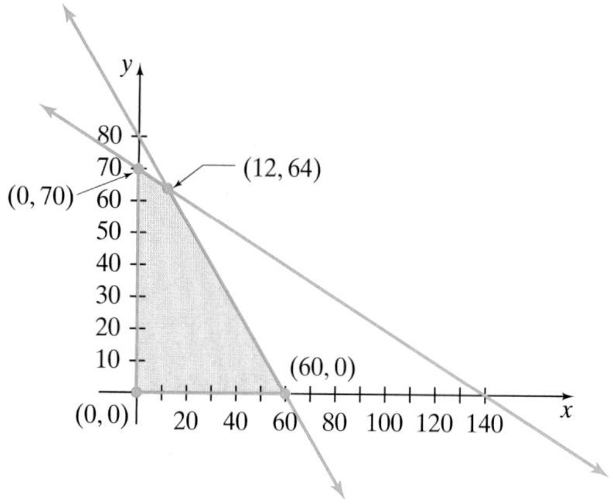

Figure 3.4.2

Step 3: The corner points are shown in Figure 3.4.2.

Step 4: Table 3.4.2 shows the result of each corner point evaluated in the objective function.

Table 3.4.2

Corner Point	Value of $P = 20x + 30y$	
(0, 0)	$20(0) + 30(0) = 0$	
(0, 70)	$20(0) + 30(70) = 2100$	
(12, 64)	$20(12) + 30(64) = 2160$	**Maximum**
(60, 0)	$20(60) + 30(0) = 1200$	

Step 5: The optimal (maximum) value shown in Table 3.4.2 is 2160 and occurs at the corner point (12, 64).

Step 6: Interpret the Solution: To maximize weekly profit, Dub, Inc., should produce and sell 12 units of switch A and 64 units of switch B. This generates a maximum weekly profit of $2160. ■

Checkpoint 1 Now work Exercise 1.

In Example 2 we turn our attention to minimizing costs.

Example 2 Solving an Applied Linear Programming Problem

A beagle breeder can buy a special food, DogsLife, at $0.30 per pound and another special food, King Kibble, at $0.40 per pound. Each pound of DogsLife has 210 units of protein and 240 units of fat. Each pound of King Kibble has 150 units of protein and 380 units of fat. If the minimum daily requirements for the beagles (collectively) are 2115 units of protein and 3460 units of fat, how many pounds of each food mix should be used daily to minimize daily food costs while still meeting (or even exceeding) the minimum daily requirements?

Solution

Step 1: Understand the Situation: Since we do not know how many pounds of each special food are used, we let x represent the number of pounds of DogsLife used daily and let y represent the number of pounds of King Kibble used daily. Since the daily food costs must be minimized, we determine an objective function for daily food costs. The objective function is

$$C = 0.30x + 0.40y$$

The problem constraints are

Protein constraint:	$210x + 150y \geq 2115$
Fat constraint:	$240x + 380y \geq 3460$
Nonnegative food constraints:	$x \geq 0,\ y \geq 0$

So the mathematical model is

$$\begin{aligned} \text{Minimize:} \quad & C = 0.30x + 0.40y \\ \text{Subject to:} \quad & 210x + 150y \geq 2115 \\ & 240x + 380y \geq 3460 \\ & x \geq 0,\ y \geq 0 \end{aligned}$$

Step 2: The feasible region is shown in Figure 3.4.3. The region is unbounded. However, since all coefficients in the objective function are positive, a minimum optimal value exists. (Consult the Toolbox on p. 178.)

Step 3: The corner points are shown in Figure 3.4.3.

Step 4: The results of the corner points being evaluated in the objective function are shown in Table 3.4.3.

Step 5: The optimal (minimum) value shown in Table 3.4.3 is 3.95 and occurs at corner point (6.5, 5).

From Your Toolbox

If the feasible region is unbounded and all the coefficients of the objective function are positive, then a minimum value of the objective function exists, but a maximum value does not.

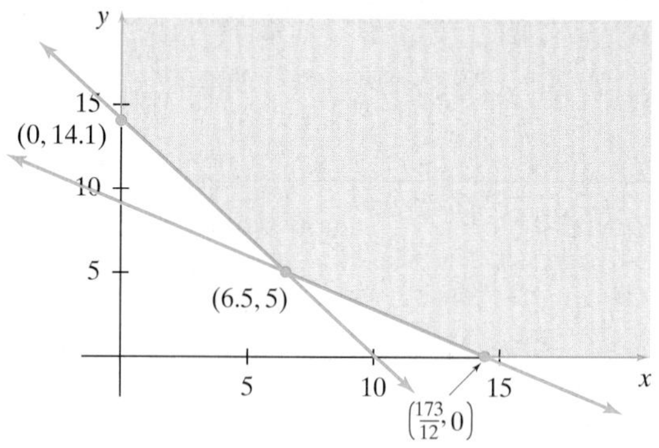

Figure 3.4.3

Table 3.4.3

Corner Point	Value of $C = 0.30x + 0.40y$	
(0, 14.1)	$0.30(0) + 0.40(14.1) = 5.64$	
(6.5, 5)	$0.30(6.5) + 0.40(5) = 3.95$	**Minimum**
$\left(\frac{173}{12}, 0\right)$	$0.30\left(\frac{173}{12}\right) + 0.40(0) = 4.325 \approx 4.33$	

Step 6: Interpret the Solution: To minimize daily food costs, while still meeting the minimum daily nutritional requirements, the beagle breeder should use 6.5 pounds of DogsLife and 5 pounds of King Kibble. The daily minimum cost is \$3.95. ■

Checkpoint 2 Now work Exercise 19.

Example 3 Solving an Applied Linear Programming Problem

Table 3.4.4 gives nutritional information, in grams, and current prices for selected items from Wendy's menu.

Table 3.4.4

Item	Protein	Fiber	Price
Breaded Chicken Sandwich	28	2	\$2.89
Biggie® Fries	7	6	\$1.19

SOURCE: Wendy's, www.wendys.com

Joshua desires to eat nothing but Breaded Chicken Sandwiches and Biggie® Fries. He wants at least 112 grams of protein and at least 30 grams of fiber daily. If Joshua eats nothing but Breaded Chicken Sandwiches and Biggie® Fries and wishes to meet his minimum nutrient requirements, how much of each food should he eat to minimize daily food costs?

Solution

Step 1: **Understand the Situation:** Let x represent the number of orders of Breaded Chicken Sandwiches and let y represent the number of orders of Biggie® Fries. The objective function is

$$C = 2.89x + 1.19y$$

The problem constraints are

Protein constraint:	$28x + 7y \geq 112$
Fiber constraint:	$2x + 6y \geq 30$
Nonnegative constraints:	$x \geq 0,\ y \geq 0$

So the mathematical model is

$$\begin{aligned} \text{Minimize:} \quad & C = 2.89x + 1.19y \\ \text{Subject to:} \quad & 28x + 7y \geq 112 \\ & 2x + 6y \geq 30 \\ & x \geq 0,\ y \geq 0 \end{aligned}$$

Step 2: Figure 3.4.4 shows the feasible region. The region is unbounded but, as in Example 2, since all coefficients of the objective function are positive, a minimum optimal value exists.

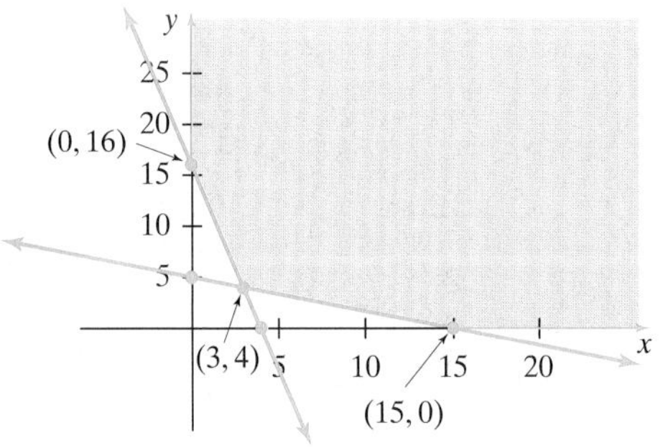

Figure 3.4.4

Step 3: The corner points are shown in Figure 3.4.4.

Step 4: Table 3.4.5 shows the results of each corner point evaluated in the objective function.

Table 3.4.5

Corner Point	Value of $C = 2.89x + 1.19y$	
(0, 16)	$2.89(0) + 1.19(16) = 19.04$	
(3, 4)	$2.89(3) + 1.19(4) = 13.43$	**Minimum**
(15, 0)	$2.89(15) + 1.19(0) = 43.35$	

Step 5: The optimal (minimum) value shown in Table 3.4.5 is 13.43 and occurs at the corner point (3, 4).

Step 6: **Interpret the Solution:** To minimize daily food costs, while still meeting the daily nutrient requirements, Joshua should eat 3 Breaded Chicken Sandwiches and 4 orders of Biggie® Fries. His daily minimum cost is \$13.43. ■

Checkpoint 3

Now work Exercise 21.

Before we consider our final example, we hope that the first three examples have illustrated that the skills learned in Sections 3.1 and 3.2 (writing a system of linear inequalities and graphing the feasible region), along with the skills learned in Section 3.3 (solving linear programming problems using corner points), are simply being combined in this section.

Example 4 Solving an Applied Linear Programming Problem

The Hoofer Mining Company owns two different mines that produce ore, which is crushed and graded into three classes: high, medium, and low grade. The company has a contract to provide a smelting plant with at least 13 tons of high-grade, 9 tons of medium-grade, and 24 tons of low-grade ore per week. The two mines have different operating characteristics as given in Table 3.4.6.

Table 3.4.6

Mine	Cost per Day (in 1000s)	Production (tons per day)		
		High	Medium	Low
I	12	3	3	4
II	10	2	1	6

Each mine operates at most 5 days per week. How many days per week should each mine operate to meet (or exceed) the smelting plant contract and minimize the total cost?

Solution

Step 1: **Understand the Situation:** Let x represent the number of days per week that mine I is operated, and let y represent the number of days per week that mine II is operated. The objective function is

$$C = 12x + 10y \qquad \text{In thousands}$$

The problem constraints are

High-grade ore constraint:	$3x + 2y \geq 13$
Medium-grade ore constraint:	$3x + y \geq 9$
Low-grade ore constraint:	$4x + 6y \geq 24$
Operating days constraint:	$x \leq 5,\ y \leq 5$
Nonnegative constraints:	$x \geq 0,\ y \geq 0$

So the mathematical model is

$$\begin{aligned} \text{Minimize:} \quad & C = 12x + 10y \\ \text{Subject to:} \quad & 3x + 2y \geq 13 \\ & 3x + y \geq 9 \\ & 4x + 6y \geq 24 \\ & 0 \leq x \leq 5, 0 \leq y \leq 5 \end{aligned}$$

Step 2: Figure 3.4.5 shows the feasible region.

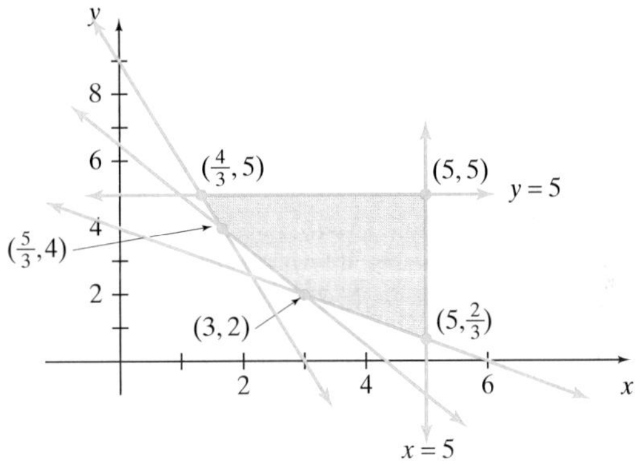

Figure 3.4.5

Step 3: The corner points are shown in Figure 3.4.5.

Step 4: Table 3.4.7 shows the results of each corner point evaluated in the objective function.

Table 3.4.7

Corner Point	Value of $C = 12x + 10y$	
$(5, \frac{2}{3})$	$12(5) + 10\left(\frac{2}{3}\right) = 66\frac{2}{3}$	
$(3, 2)$	$12(3) + 10(2) = 56$	**Minimum**
$(\frac{5}{3}, 4)$	$12\left(\frac{5}{3}\right) + 10(4) = 60$	
$(\frac{4}{3}, 5)$	$12\left(\frac{4}{3}\right) + 10(5) = 66$	
$(5, 5)$	$12(5) + 10(5) = 110$	

Interactive Activity

Determine how many total tons of high-, medium-, and low-grade ore are produced for a week by the Hoofer Mining Company at the optimal value found in Example 4.

Step 5: The optimal (minimum) value in Table 3.4.7 is 56 and occurs at the corner point (3, 2).

Step 6: Interpret the Solution: To minimize the total cost, while still meeting the smelting plant contract, Hoofer Mining Company should operate mine I 3 days per week and mine II 2 days per week. The minimum weekly cost is $56,000. ■

SUMMARY

In this section we learned how to solve applications of linear programming problems. Basically, this combined the skills acquired in the first three sections of the chapter. We also used a six-step process to solve these applications.

- **Solving an Applied Linear Programming Problem with Two Decision Variables**

  Step 1: Construct a mathematical model for the problem. This includes:
  - Define the decision variables.
  - Write a linear objective function.
  - Write the problem constraints.

  Step 2: Graph the feasible region. Note if it is bounded or unbounded.

  Step 3: Determine the coordinates of all corner points.

  Step 4: Evaluate each corner point in the objective function.

  Step 5: Determine the optimal solution from the result of step 4.

  Step 6: Interpret the optimal solution in terms of the original problem.

SECTION 3.4 EXERCISES

✓ **1.** The Yuen Tent Company makes two types of tents: two-person and six-person models. Each tent requires the services of two departments, as shown in Table 3.4.8.

Table 3.4.8

	Labor (hours per unit)	
Department	**Two-person**	**Six-person**
Fabric cutting	6	8
Assembly and packaging	2	4

Fabric cutting has a maximum of 580 hours available each week and assembly and packaging has a maximum of 260 hours available each week. Each two-person tent yields \$40 in profit and each six-person tent yields \$60 in profit.

(a) Define the variables and write the objective function and the problem constraints.

(b) How many of each type of tent should be made and sold to maximize weekly profit?

2. (*continuation of Exercise 1*) Irene, the head of research and development at the Yuen Tent Company, has discovered a way to reduce the amount of time needed for the six-person tent in the fabric cutting department to 7 hours. Assume that everything else from Exercise 1 remains the same.

(a) Define the variables and write the objective function and the problem constraints.

(b) How many of each type of tent should be made and sold to maximize weekly profit?

3. Denyse and Mike, two recent graduates, have decided to enter the field of personal digital assistants. Initially, they plan to make two types of personal digital assistants, Zap 1000 and Zap 1200. Because of the interest in personal digital assistants, they can sell all that they can produce. However, they need to determine a production rate so as to satisfy various limits with a small production crew. These limits include a maximum of 150 hours per week for assembly and 100 hours per week for testing. The Zap 1000 requires 3 hours of assembly and 3 hours of testing, and the Zap 1200 requires 6 hours of assembly and 3.5 hours of testing. Profit for the Zap 1000 is estimated at \$300 per unit; profit for the Zap 1200 is estimated at \$450 per unit.

(a) Define the variables and write the objective function and the problem constraints.

(b) How many of each type of microcomputer should be made and sold to maximize weekly profit?

4. (*continuation of Exercise 3*) Due to a favorable response from the market, Denyse and Mike decide to hire more employees and increase production. By doing so, they now have a maximum of 210 hours per week for assembly and 160 hours per week for testing. Everything else from Exercise 3 remains the same.

(a) Define the variables and write the objective function and the problem constraints.

(b) How many of each type of microcomputer should be made and sold to maximize weekly profit?

5. HotRubber, Inc., produces two types of tires for racing cars: slicks and rain. Each slick tire requires 4 minutes to make and costs \$50; each rain tire requires 12 minutes to make and costs \$60. The company has contracts that mandate that they must make a total of at least 2000 tires each day, of which at least 800 must be rain tires. The labor union contract with HotRubber dictates that they use at least 300 hours each day making these two types of tires.

(a) How many of each type of tire should be made to meet the conditions and minimize costs?

(b) What are the minimum costs?

6. (*continuation of Exercise 5*) HotRubber, Inc., has a machine that is crucial to the production process break down. As a result it now takes 5 minutes to make a slick tire and 15 minutes to make a rain tire. The costs per tire are still the same as in Exercise 5. The company still has to make a total of at least 2000 tires each day, but now at least 600 must be rain tires. Given that the labor union contract is the same as in Exercise 5, answer the following:

(a) How many of each type of tire should be made to meet the conditions and minimize costs?

(b) What are the minimum costs?

For Exercises 7 and 8, use the fact that

return on an investment
= (amount invested) · (percentage rate)

7. Kevan has received a \$50,000 bonus from his employer and wishes to invest it in a growth-oriented mutual fund and an income mutual fund. The growth mutual fund has a historical average yearly return of 12%, and the income fund has a historical average yearly return of 9%. Each fund has a minimum investment of \$5000. Kevan desires that at least twice as much be invested in the income fund as in the growth fund. Assuming the historical average yearly returns, how much should Kevan invest in each fund to maximize his return?

8. Brendan has received a \$30,000 bonus from his employer and wishes to invest it in an international mutual fund and a growth and income mutual fund. The international fund has a historical average yearly return of 8% and the growth and income fund has a historical average yearly return of 11%. The international fund has a minimum investment of \$2500 and the growth and income fund has a minimum investment of \$1000. Brendan wishes that at least three times as much be invested in the growth and income fund as in the international fund. Assuming the historical average yearly returns, how much should Brendan invest in each fund to maximize his return?

9. Hoofer Oil has two refineries, one in Findlay, Ohio, and one in Ft. Wayne, Indiana. Each refinery can refine the amounts of fuel, in barrels, in a 24-hour period as shown in Table 3.4.9.

Table 3.4.9

Maximum Production of:	Findlay	Ft. Wayne
Low-octane gasoline	800	1000
High-octane gasoline	500	600
Diesel	1000	600

Daily operating costs for the Findlay refinery are \$28,000, and daily operating costs for the Ft. Wayne refinery are \$33,000. Zachary, the production manager of Hoofer Oil, receives an order for 60,000 barrels of low-octane gasoline, 120,000 barrels of high-octane gasoline, and 150,000 barrels of diesel.

(a) How many days should Hoofer Oil operate each refinery to fulfill the order while minimizing cost?

(b) What is the minimum cost?

10. Shannon Oil has two refineries, one in Lima, Ohio, and one in Detroit, Michigan. Each refinery can refine the amounts of fuel, in barrels, in a 24-hour period as shown in Table 3.4.10.

Table 3.4.10

Maximum Production of:	Lima	Detroit
Low-octane gasoline	900	1100
High-octane gasoline	600	700
Diesel	1100	700

Daily operating costs for the Lima refinery are \$35,000, and daily operating costs for the Detroit refinery are \$30,000. Joshua, the production manager of Shannon Oil, receives an order for 50,000 barrels of low-octane gasoline, 90,000 barrels of high-octane gasoline, and 130,000 barrels of diesel.

(a) How many days should Shannon Oil operate each refinery to fulfill the order while minimizing cost?

(b) What is the minimum cost?

11. Due to new regulations approved by the Environmental Protection Agency (EPA), the Roskos Chemical Company was required to implement a new process to reduce the pollution created when making a particular chemical. The older process releases 4 grams of sulfur and 3 grams of particulates for each gallon of chemical produced. The newer process releases 2 grams of sulfur and 1 gram of particulates for each gallon of chemical produced. The company realizes a profit of \$0.40 for each gallon produced under the old process and \$0.16 for each gallon produced under the new process. The new EPA regulations do not allow for more than 13,000 grams of sulfur and 7000 grams of particulates to be released daily.

(a) How many gallons of this chemical should Roskos Chemical Company produce using the old process and how many gallons should be produced using the new process to maximize profits and still satisfy the new EPA regulations?

(b) What is the maximum daily profit?

12. (*continuation of Exercise 11*) Phil, the plant manager for the Roskos Chemical Company, received word that the EPA is going to change the new regulations so that no more than 10,000 grams of sulfur can be released daily. Assuming that all other data remain the same as in Exercise 11, answer the following:

(a) How many gallons of this chemical should Roskos Chemical Company produce using the old process and how many gallons should be produced using the new process to maximize profits and still satisfy the new EPA regulations?

(b) What is the maximum daily profit?

13. Michael Ende, a famous German author, wrote *The Neverending Story*. Amazon.com currently provides two versions of the book: hardback and softback. The hardback has a volume of 69.90 in.^{3}, and the softback has a volume of

25.87 in.^{3} Suppose that a shipping box that holds 5184 in.^{3} will be used to ship the books to stores and that Amazon.com wants to ship at most 39 hardback versions. Currently, Amazon.com sells the hardback version for $21.99 and the softback version for $6.99 (*Source:* www.amazon.com).

(a) How many books of each should be shipped to maximize revenue? (Round to the nearest integer so that the shipping box restrictions are not violated.)

(b) What is the maximum revenue?

14. (*continuation of Exercise 13*) Suppose that Amazon.com desires to ship at most 56 hardback versions. Assuming that all other data from Exercise 13 remain the same, answer the following:

(a) How many books of each should be shipped to maximize revenue? (Round to the nearest integer so that the shipping box restrictions are not violated.)

(b) What is the maximum revenue?

15. Compaq makes many types of laptops. Currently, the 1200 Series sells for $1099, while the 1700 Series sells for $1499. The 1200 Series weighs 12.3 pounds and the 1700 Series weighs 11.0 pounds (both in their respective packaging). Suppose that the company was planning to ship the laptops in a box, but the shipper requires the package to weigh no more than 89.3 pounds. Also, suppose that Compaq wants at least one of each laptop in the package (*Source:* www.compaq.com).

(a) How many of each laptop should be packaged to maximize the potential revenue in a package?

(b) What is the maximum potential revenue?

16. (*continuation of Exercise 15*) Suppose that Compaq has a new shipper that requires the package to weigh no more than 78.3 pounds. With this new shipper, Compaq desires to ship at least one 1200 Series and two 1700 Series in a package. Assuming that all the other data in Exercise 15 remain the same, answer the following:

(a) How many of each laptop should be packaged to maximize the potential revenue in a package?

(b) What is the maximum potential revenue?

17. Dominic plans to make pancakes and crepes for breakfast at his restaurant. Pancakes require 1 egg, $\frac{3}{4}$ cup milk, 1 cup flour, and $3\frac{1}{2}$ teaspoons baking powder. Crepes require 2 eggs, 2 cups milk, $1\frac{1}{2}$ cups flour, and $\frac{1}{2}$ teaspoon baking powder. Pancakes sell for $3.00 and crepes sell for $5.50. Dominic only has 3 dozen eggs, 4 gallons of milk (64 cups), 40 cups of flour, and 35 teaspoons of baking powder.

(a) How many pancakes and how many crepes can Dominic make to maximize his revenue?

(b) What is the maximum revenue?

18. (*continuation of Exercise 17*) If Dominic makes the amount of pancakes and crepes determined in Exercise 17(a), how much of each ingredient is left over?

19. Anita, a dietitian at a local hospital, is to prepare a diet to contain at least 400 units of vitamins C and E, 500 units of minerals (such as zinc and iron), and 1400 calories. She has two foods, HiGlo and Sheen, available. An ounce of HiGlo contains 2 units of vitamin C and E, 1 unit of minerals, and 4 calories and costs $0.05. An ounce of Sheen contains 1 unit of vitamins C and E, 2 units of minerals, and 4 calories and costs $0.03.

(a) How many ounces of each food should be used to minimize daily food cost?

(b) What is the minimum daily food cost?

20. Lisa, a dietitian at a local hospital, is to prepare two foods to meet certain nutritional requirements. Each pound of JumpStart contains 200 units of vitamin C, 80 units of vitamin B_{12}, and 20 units of vitamin E and costs $0.40. Each pound of MainTain contains 20 units of vitamin C, 160 units of vitamin B_{12}, and 10 units of vitamin E and costs $0.30. To meet the patients' nutritional needs, the combination of the two foods must have at least 520 units of vitamin C, 640 units of vitamin B_{12}, and 100 units of vitamin E.

(a) How many pounds of each food should be used to minimize daily food cost?

(b) What is the minimum daily food cost?

21. Table 3.4.11 gives nutritional information, in grams, and current prices for selected items from Burger King's menu.

Table 3.4.11

Item	Protein	Fiber	Price
Cheeseburger	22	2	$1.19
Small onion rings	3	2	$0.99

SOURCE: Burger King, www.burgerking.com

Mike desires to eat nothing but cheeseburgers and orders of small onion rings. He wants at least 96 grams of protein and at least 26 grams of fiber daily. Also, he wishes to eat at least 1 cheeseburger and at least 1 order of small onion rings. If Mike eats nothing but cheeseburgers and small onion rings and meet his minimum nutrient requirements, answer the following:

(a) How much of each food should Mike eat to minimize his daily food cost?

(b) What is his minimum daily food cost?

22. (*continuation of Exercise 21*). Mike from Exercise 21 found out that Burger King is having a promotion where cheeseburgers sell for $0.49 on Wednesdays. Assuming all other data in Exercise 21 remains the same, answer the following:

(a) How much of each food should Mike eat to minimize his daily food cost?

(b) What is his minimum daily food cost?

23. Table 3.4.12 gives nutritional information, in grams, for selected items from McDonald's menu.

Table 3.4.12

Item	Protein	Fiber	Total Fat
Filet-o-fish™	15	1	26
Large French fries	8	6	26

SOURCE: McDonald's, www.mcdonalds.com

Austin wants to eat nothing but Filet-o-fish and large French fries. He wants at least 122 grams of protein and at least 30 grams of fiber daily. If Austin eats nothing but Filet-o-fish and large French fries and meets his minimum nutrient requirements, answer the following:

(a) How much of each food should Austin eat to minimize his daily total fat intake?

(b) What is his minimum daily total fat intake?

24. (*continuation of Exercise 23*) A Filet-o-fish contains 470 calories and a large French fries has 540 calories.

(a) If Austin is to consume no more than 2500 calories in a day, would he meet this restriction with the optimal solution found in Exercise 23(a)? If yes, by how many calories is he below 2500? If no, by how many calories would he exceed this restriction?

(b) Is it possible for Austin to meet the calorie restriction and still satisfy his nutrient requirements in Exercise 23? Explain. (*Hint:* Write an inequality for his calorie constraint and graph with the constraints in Exercise 23. Is there a feasible region?)

25. Table 3.4.13 gives data (in lb/acre) for growing artichokes and asparagus for an entire growing season in California.

Table 3.4.13

	Artichokes	Asparagus
Nitrogen	197	240
Potash	123	42

SOURCE: U.S. Department of Agriculture, www.usda.gov/nass

Frances has 150 acres available suitable for cultivating crops. She has purchased a contract for a total of 31,880 pounds of nitrogen and 9120 pounds of potash. She expects to make a profit of \$140/acre on artichokes and \$160/acre on asparagus.

(a) How many acres of each should she plant to maximize her profit?

(b) What is the maximum profit?

26. Table 3.4.14 gives data (in lb/acre) for growing cauliflower and head lettuce for an entire growing season in California.

Table 3.4.14

	Cauliflower	Head lettuce
Nitrogen	215	173
Phosphate	98	122

SOURCE: U.S. Department of Agriculture, www.usda.gov/nass

Sonja has 210 acres available for cultivating crops. She has purchased a contract for a total of 38,380 pounds of nitrogen and 22,240 pounds of phosphate. She expects to make a profit of \$90/acre on cauliflower and \$100/acre on head lettuce.

(a) How many acres of each should she plant to maximize her profit?

(b) What is the maximum profit?

SECTION PROJECT

Table 3.4.15 gives nutritional information and current prices for selected items from McDonald's menu.

Table 3.4.15

Item	Calories	Cholesterol (mg)	Price
Garden salad	100	75	\$2.29
Small French fries	210	0	\$1.29

SOURCE: McDonald's, www.mcdonalds.com

Frank is on a diet of 2500 (or fewer) calories per day, and it is recommended that his daily total cholesterol intake be less than 300 mg. Frank wants to eat nothing but fries and garden salad and not exceed the calorie and cholesterol requirement.

(a) Form the mathematical model for this situation in which the objective function is cost.

(b) Graph the feasible region. Label the corner points.

(c) Evaluate the objective function at each corner point.

(d) From part (c), determine the optimal (minimum) value. Interpret including the number of garden salads, small French fries, calories, and milligrams of cholesterol that Frank would consume.

(e) From part (c), determine the optimal (maximum) value. Interpret including the number of garden salads, small French fries, calories, and milligrams of cholesterol that Frank would consume.

(f) Based on parts (d) and (e), what would you tell Frank? In other words, how many small French fries and how many garden salads should he expect to eat each day and what is the daily cost?

Why We Learned It

In this chapter we learned how to graph systems of linear inequalities and a graphical method for solving linear programming problems. We saw that many managers in business can use linear programming to maximize profit, to minimize costs, or to determine optimum manufacturing times. For example, suppose that a company has three departments that its products must pass through. Due to certain restrictions, each department has a maximum number of hours available each week. Assuming that the company knows the profit it receives on each product that it produces and sells, it would want to determine how many units of each product to make in a week (given the time constraints of each department) in order to maximize profit. We also learned that dietitians can use linear programming to maximize the healthy nutritional elements of a diet, such as vitamins. Dietitians can also use linear programming to minimize elements of a diet, such as fat intake. We also saw that farmers can use linear programming to maximize profit, given restrictions of land and the availability of various fertilizers.

CHAPTER REVIEW EXERCISES

For the inequalities in Exercises 1 and 2, complete the tables by stating whether the inequality is true *or* false.

1. $x > 5$

x	$x > 5$	Statement
−14		
5		
7		
10		

2. $x \leq -3$

x	$x \leq -3$	Statement
−7		
−5		
−3		
0		

In Exercises 3–6, translate the English sentence into a mathematical sentence.

3. A certain number increased by eight is twelve.

4. A certain number decreased by three is greater than negative one.

5. The product of five and a certain number minus seven is less than or equal to nine.

6. The product of three and a certain number plus two is not equal to ten.

In Exercises 7–10, solve and graph the solution to the following.

7. $x - 5 > 3$

8. $x - 2 < 6$

9. $4x - 3 \geq 13$

10. $5 - 2x \leq 3$

11. Music Notes offers introductory music workshops for drums and guitar. Each student attends an individual music theory session and a music lesson. Drum students attend a 2-hour music theory session and a 3-hour music lesson. Guitar students attend a 3-hour music theory session and a 4-hour music lesson. Each month instructors can offer no more than 37 hours of music theory sessions and 51 hours of music lessons. Let x represent the number of students admitted each month for the drum workshop, and let y represent the number of students admitted each month for the guitar workshop. Write linear inequalities that give appropriate restrictions on x and y.

12. (*continuation of Exercise 11*) Feedback from students has indicated that Music Notes needs to provide another workshop: music recording. Guitar and drum students will both receive 1 hour of instruction in music recording. Each month instructors can give no more than 14 hours of instruction in music recording. Assuming that everything else from Exercise 11 remains the same, write linear inequalities that give appropriate restrictions on x and y.

13. Sopris Mountaineering makes two types of backpacks: a 3500-cubic-inch day pack and a 5500-cubic-inch multiday pack. Each backpack requires the services, in hours, of two departments as shown in Table R.1.

Table R.1

Department	Day Pack	Multiday Pack
Fabric cutting	4	8
Assembly and packaging	2	6

Fabric cutting has a maximum of 340 hours available each week and assembly and packaging has a maximum of 230 hours available.

(a) Define the variables and write linear inequalities to express reasonable values for the variables.

(b) Write linear inequalities to express the time restrictions of the fabric cutting and assembly and packaging departments.

14. The data in Table R.2 give the amount of nitrogen, phosphate, and potash (all in lb/acre) needed to grow head lettuce, cantaloupe, and bell peppers for an entire growing season.

Table R.2

	Head Lettuce	Cantaloupe	Bell Peppers
Nitrogen	173	175	212
Phosphate	122	161	108
Potash	76	39	72

SOURCE: U.S. Department of Agriculture, www.usda.gov/nass

Mika has 250 acres and can get no more than 46,140 pounds of nitrogen, 32,640 pounds of phosphate, and 15,760 pounds of potash.

(a) Define the variables and write linear inequalities to express reasonable values for the variables.

(b) Write linear inequalities to express the restrictions on nitrogen, phosphate, and potash.

15. Blue's Music offers 1-month introductory music workshops for drums, guitar, and piano. Each student attends an individual music theory session, music lesson, and recording session. Each student attends each of the three workshops. Table R.3 shows the number of hours of instruction that each student receives.

Table R.3

	Drum Student	Guitar Student	Piano Student
Music theory	2	3	5
Music lesson	4	5	2
Recording session	1	3	1

Each month, instructors can give a maximum of 57 hours, 54 hours, and 24 hours of instruction in music theory, music lessons, and recording, respectively.

(a) Define the variables and write linear inequalities to express reasonable values for the variables.

(b) Write linear inequalities to express the time restrictions of the music theory, music lessons, and recording workshops.

In Exercises 16 and 17, match the graph with the correct inequality.

16. $x + y \leq 5$ **17.** $x + y \geq 5$

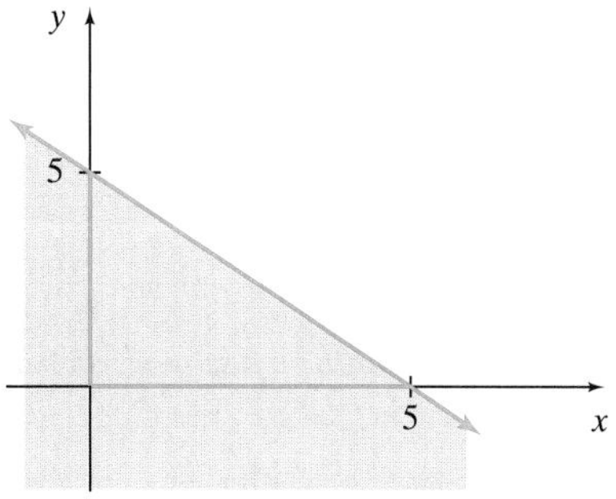

(a)

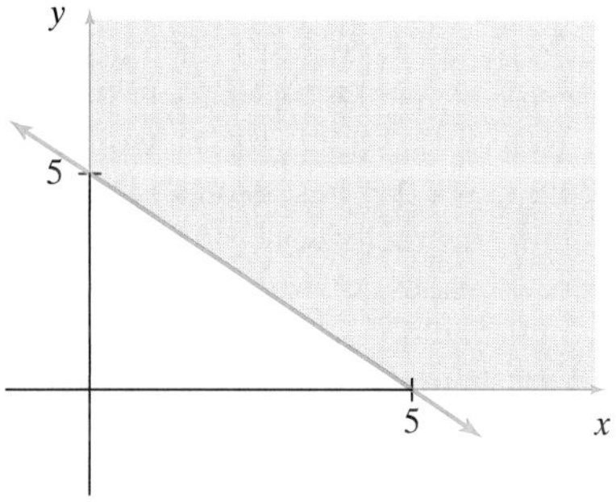

(b)

In Exercises 18–21, graph each linear inequality.

18. $y \leq 2x + 6$ **19.** $y > -4x + 2$

20. $-2x - y \geq 5$ **21.** $x > -2$

In Exercises 22 and 23, use Figure R.1 to match the correct feasible region with the corresponding system of linear inequalities.

22. $x - 2y \leq 0$
$2x + y \geq 5$

23. $x - 2y \geq 0$
$2x + y \geq 5$

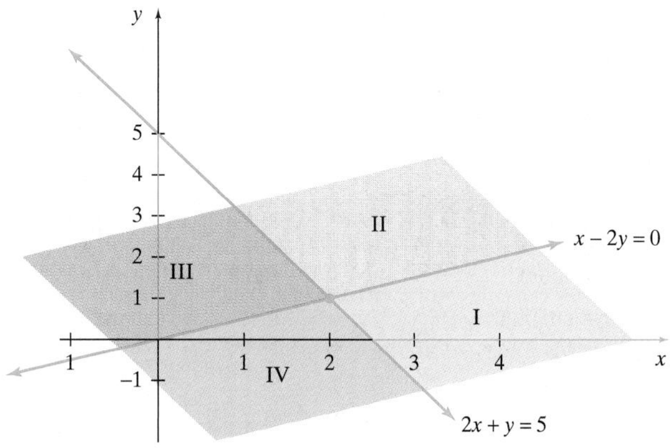

Figure R.1

In Exercises 24–27, graph the feasible region for each system of linear inequalities. Indicate if the region is bounded or unbounded.

24. $x + y \le 4$
$x - y \ge 2$

25. $3x + 2y \ge 5$
$x + 3y \le 4$

26. $x + y \le 5$
$3x - y \ge 2$
$x \ge 0, y \ge 0$

27. $2x - y \le 6$
$x + 2y \ge 4$
$3x + y \ge 1$
$x \ge 0, y \ge 0$

In Exercises 28 and 29, use a graphing calculator to graph the following:

28. $y \le 7 - 2x$

29. $2x - 3y \ge 4$
$4x + y \ge 3$

30. Music Notes offers introductory music workshops for drums and guitar. Each student attends an individual music theory session and a music lesson. Drum students attend a 2-hour music theory session and a 3-hour music lesson. Guitar students attend a 3-hour music theory session and a 4-hour music lesson. Each month, instructors can offer no more than 37 hours of music theory sessions and 51 hours of music lessons. Let x represent the number of students admitted each month for the drum workshop, and let y represent the number of students admitted each month for the guitar workshop.

(a) Write a system of linear inequalities that gives appropriate restrictions on x and y.

(b) Graph the feasible region for the system in part (a). Interpret what the feasible region represents.

31. (*continuation of Exercise 30*) Feedback from students has indicated that Music Notes needs to provide another workshop: music recording. Guitar and drum students will both receive 1 hour of instruction in music recording. Each month, instructors can give no more than 14 hours of instruction in music recording. Assuming that everything else from Exercise 30 remains the same, complete the following:

(a) Write a system of linear inequalities that gives appropriate restrictions on x and y.

(b) Graph the feasible region for the system in part (a). Interpret what the feasible region represents.

32. Sopris Mountaineering makes two types of backpacks: a 3500-cubic-inch day pack and a 5500-cubic-inch multiday pack. Each backpack requires the services, in hours, of two departments as shown in Table R.4.

Table R.4

Department	Day Pack	Multiday Pack
Fabric cutting	4	8
Assembly and packaging	2	6

Fabric cutting has a maximum of 340 hours available each week, and assembly and packaging has a maximum of 230 hours available.

(a) Define the variables and write a system of linear inequalities that give appropriate restrictions on x and y.

(b) Graph the feasible region for the system in part (a). Interpret what the feasible region represents.

33. Table R.5 gives nutritional information, in grams, for selected items from Subway's menu.

Table R.5

Item	Protein	Fat
6″ Turkey	16	3.5
6″ Veggie Delight	7	2.5

SOURCE: Subway, www.subway.com

Lera wishes to eat a lunch consisting of turkey subs and veggie subs. She wants the lunch to contain at least 25 grams of protein.

(a) Define the variables and write a system of linear inequalities that gives appropriate restrictions on x and y.

(b) Graph the feasible region for the system in part (a). Interpret what the feasible region represents.

34. (*continuation of Exercise 33*) Now suppose that Lera wishes to eat a lunch consisting of turkey subs and veggie subs that contains no more than 8 grams of fat. Use the data in Table R.5 to answer the following:

(a) Define the variables and write a system of linear inequalities that gives appropriate restrictions on x and y.

(b) Graph the feasible region for the system in part (a). Interpret what the feasible region represents.

35. (*continuation of Exercises 33 and 34*) Lera decides that her lunch should contain at least 25 grams of protein, but no more than 8 grams of fat. Use Table R.5 and the results of Exercises 33 and 34 to do the following:

(a) Write a system of linear inequalities that gives appropriate restrictions on x and y.

(b) Graph the feasible region for the system in part (a). Interpret what the feasible region represents.

In Exercises 36–39, the graph of a feasible region is given. Use the regions to determine the maximum and minimum values of the given objective function.

36. $z = 3x + 4y$

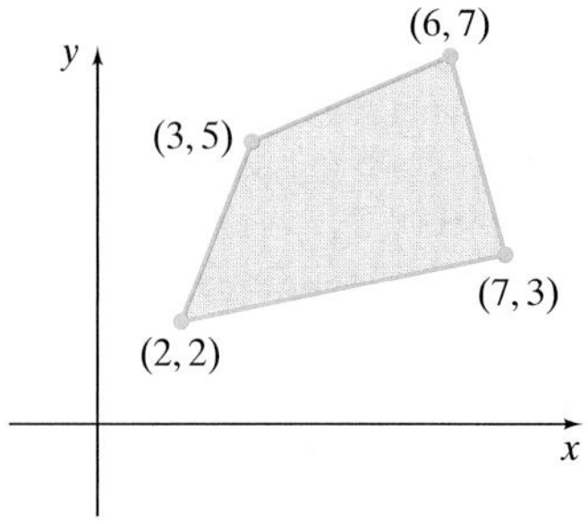

37. $z = 4x + 3y$

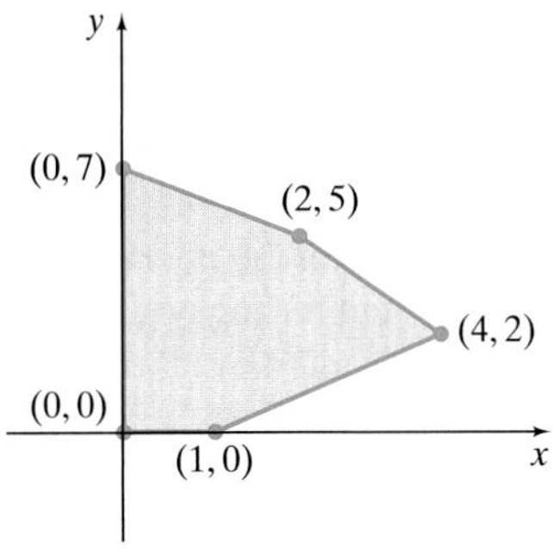

38. $z = x + 2y$

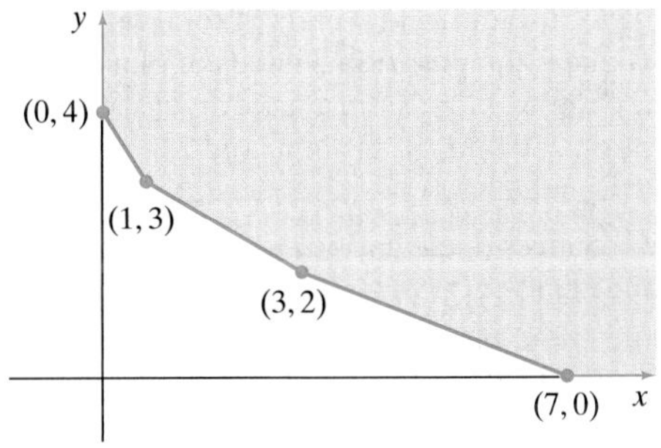

39. $z = 2x + 3y$

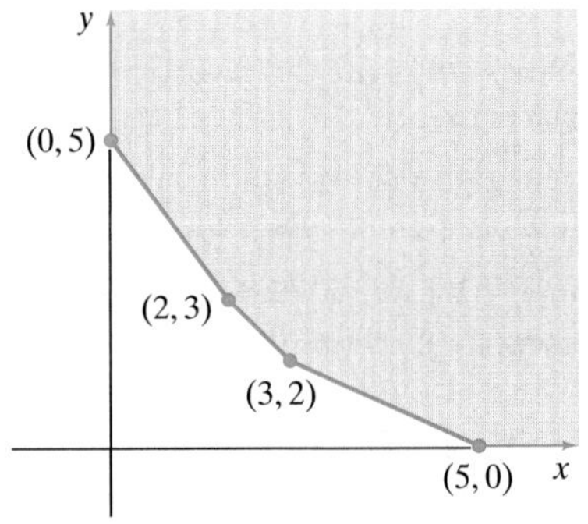

In Exercises 40–45, use the graphical and corner point method to solve the given linear programming problem.

40. Maximize: $z = 2x + y$

Subject to: $2x - 3y \leq 1$

$x + y \leq 3$

$x \geq 0, y \geq 0$

41. Maximize: $z = 3x + y$

Subject to: $6x - 2y \geq 4$

$2x + 2y \leq 4$

$x \geq 0, y \geq 0$

42. Minimize: $z = x + 4y$

Subject to: $x - y \leq 5$

$x + 2y \geq 8$

$x \geq 0, y \geq 0$

43. Minimize: $z = 2x + 5y$

Subject to: $2x + y \geq 6$

$3x - y \leq 2$

$x \geq 0, y \geq 0$

44. Minimize and maximize: $z = x + 6y$

Subject to: $x + y \geq 4$

$2x - 3y \leq 8$

$2x - y \geq 2$

$x \geq 0, y \geq 0$

45. Minimize and maximize: $z = 4x + 5y$

Subject to: $-x + 4y \leq 6$

$x + y \leq 13$

$2x + y \geq 3$

$-x + y \geq -2$

$x \geq 0, y \geq 0$

46. Sopris Mountaineering makes two types of sleeping bags: +15 sleeping bag and −15 sleeping bag (+15 and −15 indicate the temperature ratings in degrees Fahrenheit). Each bag requires the services, in hours, of two departments as shown in Table R.6.

Table R.6

Department	+15 Bag	−15 Bag
Fabric cutting	4	6
Assembly and packaging	2	4

Fabric cutting has a maximum of 320 hours available each week, and assembly and packaging has a maximum of 200 hours available each week. Each +15 bag yields $30 in profit and each −15 bag yields $50 in profit.

(a) Define the variables and write the objective function and problem constraints.

(b) How many of each type of bag should be made and sold to maximize weekly profit?

47. Music Notes offers introductory music workshops for drums and guitar. Each student attends an individual music theory session and a music lesson. Drum students attend a 2-hour music theory session and a 3-hour music lesson. Guitar students attend a 3-hour music theory session and a 4-hour music lesson. Each month, instructors can offer no more than 37 hours of music theory sessions and 51 hours of music lessons. Music notes makes a profit of $25 for each drum student and $30 for each guitar student.

(a) How many of each type of student should be admitted each month to maximize profit?

(b) What is the maximum profit?

48. (*continuation of Exercise 47*) Since Music Notes would really like to offer guitar lessons and students have been demanding a music recording workshop, Music Notes has decided to offer 1 hour of instruction in music recording to both drum and guitar students. Instructors can give no more than 14 hours of instruction in music recording. Assuming that everything else from Exercise 47 remains the same, answer the following questions:

(a) How many of each type of student should be admitted each month to maximize profit?

(b) What is the maximum profit?

(c) Compare your results from part (b) of Exercises 47 and 48. What is your recommendation for Music Notes?

49. Henderson Oil has two refineries, one in Aztec, New Mexico, and one in Buena Vista, Colorado. Each refinery can refine the amounts of fuel, in barrels, in a 24-hour period as shown in Table R.7.

Table R.7

Maximum Production of:	Aztec	Buena Vista
Low-octane gasoline	700	900
High-octane gasoline	400	500
Diesel	900	500

Daily operating costs for the Aztec refinery are $32,000, and daily operating costs for the Buena Vista refinery are $30,000. Lissa, the production manager at Henderson Oil, receives an order for 90,000 barrels of low-octane gasoline, 145,000 barrels of high-octane gasoline, and 150,000 barrels of diesel.

(a) How many days should Henderson Oil operate each refinery to fulfill the order while minimizing cost?

(b) What is the minimum cost?

50. Carolyn, a dietitian for a sports team, is to prepare two foods to meet certain nutritional requirements. Each ounce of PowerSource contains 40 units of carbohydrates, 8 units of protein, and 4 units of total fat and costs $0.12. Each ounce of EnergyMax contains 50 units of carbohydrates, 10 units of protein, and 4 units of total fat and costs $0.14. To meet the nutritional needs of players, the combination of the two foods must have at least 600 units of carbohydrates, 450 units of protein, and 220 units of total fat.

(a) How many ounces of each food should be used to minimize daily food cost?

(b) What is the minimum daily food cost?

CHAPTER 3 PROJECT

SOURCE: EAS, Inc., www.eas.com

EAS produces supplements for anyone who enjoys athletic activities. Two of their products are the SimplyProtein Nutrition Bar and the Myoplex Deluxe Shake. The nutritional information for each of these products is provided in Table 1.

Table 1 Information Is for One Serving

	SimplyProtein Nutrition Bar	Myoplex Deluxe Shake
Protein	33 g	42 g
Calories	300	300
Sugar	13 g	2 g
Sodium	45 mg	450 mg

Suppose that Dave is getting ready to run the Shamrock Shuffle 8k race in Chicago. He is planning on using the SimplyProtein Nutrition Bar and the Myoplex Deluxe Shake in his training. Dave wants to maximize his protein intake while keeping his calories under 1500, sugar under 32 g, and sodium under 1845 mg.

1. Declare the variables and write the objective function and the problem constraints.

2. Rewrite each nonnegative constraint in calculator-friendly form and graph in the same viewing window.

3. Determine the corner points of the feasible region.

4. Use the corner point method to solve the linear programming problem.

5. Using the optimal solution from question 4, determine how many calories, how much sugar, and how much sodium are being consumed.

6. Determine the difference between the amount of calories from question 5 and the limit on calories. Do the same for sugar and sodium. This result is called *slack*.

7. Referring to question 6, explain why some of the differences were approximately 0. (Note the graph and the answer you got for question 4.)

Now suppose that Dave's requirements change as follows: no more than 1800 calories, 56 g of sugar, and 1485 mg of sodium.

8. Repeat questions 1 through 7 for the new constraints.

Chapter

4 Linear Programming: The Simplex Method

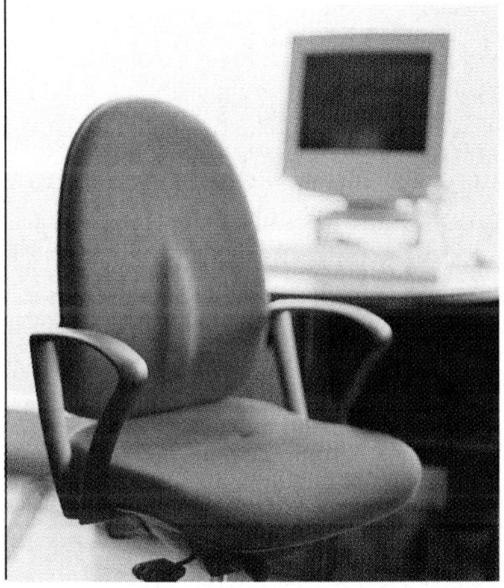

Maximize
$P = 40x_1 + 80x_2 + 70x_3$
Subject to:
$x_1 + x_2 + x_3 \le 30$
$2x_1 + 3x_2 + 2x_3 \le 53$
$x_1 + 2x_2 + 3x_3 \le 47$
$x_1 \ge 0, x_2 \ge 0, x_3 \ge 0$

$$\begin{array}{c} \begin{matrix} x_1 & x_2 & x_3 & s_1 & s_2 & s_3 & P & \end{matrix} \\ \left[\begin{array}{ccccccc|c} \frac{2}{5} & 0 & 0 & 1 & -\frac{1}{5} & -\frac{1}{5} & 0 & 10 \\ \frac{4}{5} & 1 & 0 & 0 & \frac{3}{5} & -\frac{2}{5} & 0 & 13 \\ -\frac{1}{5} & 0 & 1 & 0 & -\frac{2}{5} & \frac{3}{5} & 0 & 7 \\ \hline 10 & 0 & 0 & 0 & 20 & 10 & 1 & 1530 \end{array}\right] \end{array}$$

Modern factories employ specialized machines to manufacture goods. Consider the case of a chair manufacturer who produces several different types of chairs. To maximize profit, the manufacturer must decide how many of each chair to produce each week. Several variables come into play for each chair including the amount of time spent on each machine, the total available time each machine can be run, and the profit per chair. When two or more decision variables must be considered the simplex method is a useful tool. The final tableau of the simplex method gives the solution to the linear programming problem.

What We Know

In Chapter 3 we solved linear programming problems graphically by sketching the feasible region, locating corner points, and then evaluating each corner point in the objective function.

Where Do We Go

In this chapter, we learn a method called the simplex method that allows us to solve linear programming problems without graphing the feasible region. This new method is also used when we have more than two decision variables.

Section 4.1 Introduction to the Simplex Method

In practice, applications of linear programming have more than two decision variables. Some applications may have dozens (or even hundreds) of variables. In 1947, George Dantzig developed an algebraic method to solve problems with many variables called the **simplex method**. In this section we introduce the vocabulary and skills necessary to convert a problem into a form so that the simplex method can be used. We also learn the beginning steps of the method. In Section 4.2 we learn how to complete the method and solve applications. The last two sections of this chapter address special situations involving the simplex method.

Standard Maximum Form

Since the simplex method is used for problems with many variables, it is now more convenient for us to use subscripts to name variables. This means that we will use x_1 (read "x sub one"), x_2, x_3, x_4, and so on, to name the unknowns. Let's quickly practice this new convention through a Flashback.

Flashback

A Maximization Problem Revisited

In Section 3.3, Example 3, we graphically solved the linear programming problem

$$\begin{aligned} \text{Maximize:} \quad & z = x - 3y \\ \text{Subject to:} \quad & 2x - 3y \le 6 \\ & -x + 4y \le 4 \\ & x \ge 0, y \ge 0 \end{aligned}$$

Rewrite this problem using x_1 and x_2 instead of x and y.

Flashback Solution

Using the new notation, this linear programming problem is written as

$$\begin{aligned} \text{Maximize:} \quad & z = x_1 - 3x_2 \\ \text{Subject to:} \quad & 2x_1 - 3x_2 \le 6 \\ & -x_1 + 4x_2 \le 4 \\ & x_1 \ge 0, x_2 \ge 0 \end{aligned}$$

Look carefully at the Flashback solution. In this section and the next, we will use the simplex method for problems written in this form. Problems written in the form shown in the Flashback solution are said to be in **standard maximum form**.

Standard Maximum Form

A linear programming problem is said to be in **standard maximum form** if the following are true:

1. The objective function is linear and is to be **maximized**.
2. All variables are nonnegative.
3. All constraints, except the nonnegative constraints, are written as a linear expression less than or equal to a nonnegative constant. Mathematically, this means

$$a_1x_1 + a_2x_2 + a_3x_3 + \cdots + a_nx_n \le b, \qquad \text{with } b \ge 0$$

Note: In Sections 4.3 and 4.4 we address problems that do not meet these conditions.

Slack Variables and the Initial Simplex Tableau

The first step in using the simplex method to solve a standard maximum form problem, and the key in allowing us to view a linear programming problem algebraically is to turn each constraint (inequality) into an equation. If we can write equations, we can then use matrices to solve the problem. To turn an inequality into an equation requires us to introduce a nonnegative variable called a **slack variable**.

For example, to turn $2x_1 + x_2 \le 12$ into an equation, we add the slack variable s_1 to the left side of the inequality to get

$$2x_1 + x_2 + s_1 = 12, \qquad \text{where } s_1 \ge 0$$

As x_1 and x_2 change, the variable s_1 "picks up the slack" to guarantee that an equality exists. If $2x_1 + x_2$ equals 9, then s_1 is 3. If $2x_1 + x_2$ equals 12, then s_1 is 0. In Example 1 we illustrate how to use slack variables to write constraints as equations.

Example 1 Adapting a Linear Programming Problem

Use slack variables to restate the linear programming problem from the Flashback into equational form.

$$\begin{aligned} \text{Maximize:} \quad & z = x_1 - 3x_2 \\ \text{Subject to:} \quad & 2x_1 - 3x_2 \le 6 \\ & -x_1 + 4x_2 \le 4 \\ & x_1 \ge 0,\ x_2 \ge 0 \end{aligned}$$

Solution

We begin by observing that this problem is in standard maximum form. We can rewrite the two constraints as equations by using the slack variables s_1 and s_2. Using these slack variables, we have the equational form as

$$\begin{aligned} \text{Maximize:} \quad & z = x_1 - 3x_2 \\ \text{Subject to:} \quad & 2x_1 - 3x_2 + s_1 = 6 \\ & -x_1 + 4x_2 + s_2 = 4 \end{aligned}$$

where $x_1 \ge 0$, $x_2 \ge 0$, $s_1 \ge 0$, and $s_2 \ge 0$ ■

Before continuing, let's make some observations about slack variables.

Slack Variables

- Slack variables are not added to the nonnegative constraints $x_1 \geq 0$, $x_2 \geq 0$.
- A different slack variable is needed for each constraint (inequality).
- A slack variable "takes up the slack" to ensure that an equality exists.
- For any point in the feasible region of a linear programming problem, the value of each slack variable is either positive or zero.

To gain some insight into the last bullet item, consult Figure 4.1.1, which shows the feasible region of the linear programming problem considered in the Flashback and again in Example 1.

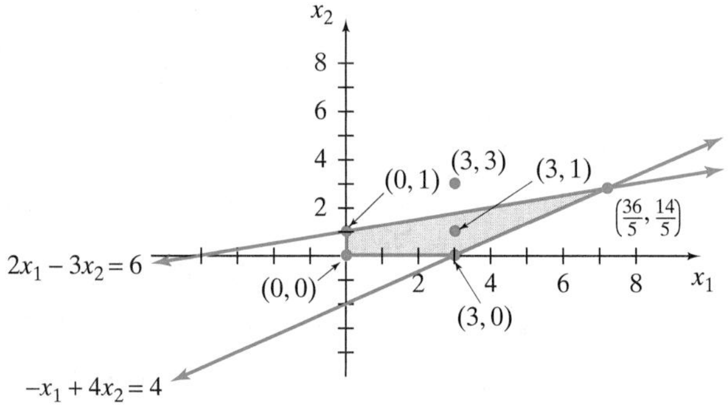

Figure 4.1.1

First observe that when $s_1 = 0$ and $s_2 = 0$, we get the graphs of the lines $2x_1 - 3x_2 = 6$ and $-x_1 + 4x_2 = 4$, respectively. We now pick a point in the feasible region, (3, 1), and a point **not** in the feasible region, (3, 3). These points are of the form (x_1, x_2). Let's use these points to determine s_1 and s_2 as shown in Table 4.1.1.

Table 4.1.1

Point	First Equation	Slack Value, s_1	Second Equation	Slack Value, s_2
	$2x_1 - 3x_2 + s_1 = 6$		$-x_1 + 4x_2 + s_2 = 4$	
(3, 1)	$2(3) - 3(1) + s_1 = 6$	$s_1 = 3$	$-3 + 4(1) + s_2 = 4$	$s_2 = 3$
(3, 3)	$2(3) - 3(3) + s_1 = 6$	$s_1 = 9$	$-3 + 4(3) + s_2 = 4$	$s_2 = -5$

As Table 4.1.1 indicates, for point (3, 1) in the feasible region, the value of each slack variable is positive. For the point (3, 3) which is **not** in the feasible region, there is at least one slack variable (s_2) that is **negative**. This illustrates that the value of each slack variable is either positive or zero for any point in the feasible region.

Back in Example 1, the addition of slack variables converted the constraints into equational form. Hence the slack variables actually converted the linear programming problem into a system of linear equations. From Chapters 1 and 2, we know that for a system of linear equations we should have all variables on one side of the equal sign and the constants on the other side. In Example 1 the only equation that does not satisfy this condition is the objective function, $z = x_1 - 3x_2$. However, we can write this objective function as

$$-x_1 + 3x_2 + z = 0$$

So the standard maximum form from the Flashback

$$\begin{aligned} \text{Maximize:} \quad & z = x_1 - 3x_2 \\ \text{Subject to:} \quad & 2x_1 - 3x_2 \le 6 \\ & -x_1 + 4x_2 \le 4 \\ & x_1 \ge 0, x_2 \ge 0 \end{aligned}$$

has the **equational form**

$$\begin{aligned} \text{Maximize:} \quad & -x_1 + 3x_2 + z = 0 \\ \text{Subject to:} \quad & 2x_1 - 3x_2 + s_1 \qquad = 6 \\ & -x_1 + 4x_2 \qquad + s_2 = 4 \end{aligned}$$

We can use an **augmented matrix** to write the equational form as follows:

$$\begin{array}{c} \overbrace{x_1 \quad x_2}^{\text{Decision variables}} \; \overbrace{s_1 \quad s_2}^{\text{Slack variables}} \; z \\ \left[\begin{array}{rrrrr|r} 2 & -3 & 1 & 0 & 0 & 6 \\ -1 & 4 & 0 & 1 & 0 & 4 \\ \hline -1 & 3 & 0 & 0 & 1 & 0 \end{array}\right] \\ \underbrace{\qquad\qquad\qquad\qquad}_{\text{Indicators}} \end{array}$$

The numbers in the bottom row are from the objective function. These numbers, except for the rightmost 1 and 0, are called **indicators**. This matrix is called the **initial simplex tableau**. Since forming the initial simplex tableau is the first step in the simplex method, let's do an example practicing this necessary skill.

Example 2 Forming an Initial Simplex Tableau

Form the initial simplex tableau for the following standard maximum form linear programming problem.

$$\begin{aligned} \text{Maximize:} \quad & z = 2x_1 - x_2 + x_3 \\ \text{Subject to:} \quad & 2x_1 + x_2 + 3x_3 \le 5 \\ & -x_1 + 3x_2 \qquad \le 7 \\ & x_1 \ge 0, x_2 \ge 0, x_3 \ge 0 \end{aligned}$$

Solution

Understand the Situation: Since we have two constraints (ignoring the non-negative constraints), we need to introduce two slack variables, s_1 and s_2, one for each constraint.

Rewriting the constraints in equational form gives

$$\begin{aligned} 2x_1 + x_2 + 3x_3 + s_1 \qquad &= 5 \\ -x_1 + 3x_2 \qquad\qquad + s_2 &= 7 \end{aligned}$$

Next we write the objective function as

$$-2x_1 + x_2 - x_3 + z = 0$$

We now have the initial simplex tableau as

$$\begin{array}{c} \begin{matrix} x_1 & x_2 & x_3 & s_1 & s_2 & z & \end{matrix} \\ \left[\begin{array}{rrrrrr|r} 2 & 1 & 3 & 1 & 0 & 0 & 5 \\ -1 & 3 & 0 & 0 & 1 & 0 & 7 \\ \hline -2 & 1 & -1 & 0 & 0 & 1 & 0 \end{array}\right] \end{array}$$

Indicators

■

Checkpoint 1 Now work Exercise 15.

Basic Variables, Nonbasic Variables, and Basic Feasible Solutions

In Example 2 we have seen the system of linear inequalities rewritten as the system of linear equations

$$\begin{aligned} 2x_1 + x_2 + 3x_3 + s_1 &= 5 \\ -x_1 + 3x_2 + s_2 &= 7 \end{aligned}$$

This system of two equations in five variables has an infinite number of solutions. For example, if we solve each equation for s_1 and s_2, respectively, we get

$$\begin{aligned} s_1 &= 5 - 2x_1 - x_2 - 3x_3 \\ s_2 &= 7 + x_1 - 3x_2 \end{aligned}$$

Notice that each choice of values for x_1, x_2, and x_3 yields values for s_1 and s_2 that give a solution to the system. However, a solution to the system is said to be **feasible** only if all variables are nonnegative.

When a linear system is written in the form

$$\begin{aligned} s_1 &= 5 - 2x_1 - x_2 - 3x_3 \\ s_2 &= 7 + x_1 - 3x_2 \end{aligned}$$

s_1 and s_2 are called **basic variables** and x_1, x_2, and x_3 are called **nonbasic variables**. In general, we consider the nonbasic variables to be free and set them equal to zero. Doing this for the system under discussion gives us

$$\begin{aligned} s_1 &= 5 - 2(0) - (0) - 3(0) = 5 \\ s_2 &= 7 + (0) - 3(0) = 7 \end{aligned}$$

The solution obtained when setting all nonbasic variables equal to zero is called a **basic solution**. In fact, the basic solution for this system is $x_1 = 0, x_2 = 0, x_3 = 0$, $s_1 = 5$, and $s_2 = 7$. Since all values are nonnegative, we can call it a **basic feasible solution**.

Since the identification of nonbasic variables and basic variables will be important, we offer the following procedure on how to determine basic and nonbasic variables from a tableau.

When a variable heads a column containing exactly one 1 and all other entries are 0, that variable is a **basic variable**. A variable that does not head such a column is a **nonbasic variable** and may be set equal to zero.

Example 3 Identifying Basic and Nonbasic Variables and Reading a Solution

Given the following tableau:

(a) Determine the basic and nonbasic variables.

(b) Determine the solution given by the tableau.

$$\begin{array}{cccccc|c} x_1 & x_2 & x_3 & s_1 & s_2 & z & \\ 3 & 1 & -2 & 3 & 0 & 0 & 13 \\ 1 & 0 & -1 & 1 & 1 & 0 & 22 \\ \hline 4 & 0 & 7 & 0 & 0 & 1 & 87 \end{array}$$

Solution

(a) Since x_2, s_2, and z head columns containing exactly one 1 and all other entries are 0, we have the variables x_2, s_2, and z as our basic variables. This means that x_1, x_3, and s_1 are the nonbasic variables.

(b) Since x_1, x_3, and s_1 are the nonbasic variables, they may be set equal to 0. Recall that the first row of the tableau represents the equation

$$3x_1 + x_2 - 2x_3 + 3s_1 = 13$$

Since x_1, x_3, and s_1 are the nonbasic variables (and may be set equal to zero), this equation becomes

$$3(0) + x_2 - 2(0) + 3(0) = 13$$
$$x_2 = 13$$

Similarly, the second row of the tableau gives us $1 \cdot s_2 = 22$ or simply $s_2 = 22$. The bottom row gives us $1 \cdot z = 87$ or simply $z = 87$. So the solution represented by the tableau is

$$x_1 = 0, \quad x_2 = 13, \quad x_3 = 0, \quad s_1 = 0, \quad s_2 = 22, \quad z = 87$$

Observe how we can determine the solution directly from the tableau by reading down the column that is a basic variable until we hit the 1 and then turning right and reading the number in the last column. ■

✓ Checkpoint 2 Now work Exercise 33.

Pivot

Rarely is the solution read from the initial simplex tableau the optimal solution. Almost always, we need to locate other solutions. To determine these other solutions, we use the row operations for matrices (see the Toolbox below) and **pivot**. Example 4 reviews how to pivot. Keep in mind that pivoting produces another matrix (tableau) that gives another solution to the system of equations. Also, pivoting is the key to using the simplex method, as we will see in Section 4.2.

From Your Toolbox

1. Two rows may be interchanged (denoted $R_i \leftrightarrow R_j$).
2. A row is multiplied by a nonzero constant (denoted $cR_i \leftrightarrow R_i$).
3. Add a constant multiple of one row to another (denoted $cR_i + R_j \rightarrow R_j$).

Note: It will not be necessary to use row operation 1 in this chapter.

Example 4 Pivoting

Pivot about the indicated 2.

$$\begin{array}{c} \begin{matrix} x_1 & x_2 & s_1 & s_2 & z \end{matrix} \\ \left[\begin{array}{rrrrr|r} \textcircled{2} & -3 & 1 & 0 & 0 & 6 \\ -1 & 4 & 0 & 1 & 0 & 4 \\ \hline -1 & 3 & 0 & 0 & 1 & 0 \end{array}\right] \end{array}$$

Solution

We get a 1 where the 2 is by using row operation 2.

$$\left[\begin{array}{rrrrr|r} 2 & -3 & 1 & 0 & 0 & 6 \\ -1 & 4 & 0 & 1 & 0 & 4 \\ \hline -1 & 3 & 0 & 0 & 1 & 0 \end{array}\right] \begin{array}{c} \tfrac{1}{2}R_1 \to R_1 \end{array} \left[\begin{array}{rrrrr|r} 1 & -\frac{3}{2} & \frac{1}{2} & 0 & 0 & 3 \\ -1 & 4 & 0 & 1 & 0 & 4 \\ \hline -1 & 3 & 0 & 0 & 1 & 0 \end{array}\right]$$

(columns: x_1, x_2, s_1, s_2, z)

We now get zeros in the first column with the pivot by using row operation 3.

$$\left[\begin{array}{rrrrr|r} 1 & -\frac{3}{2} & \frac{1}{2} & 0 & 0 & 3 \\ -1 & 4 & 0 & 1 & 0 & 4 \\ \hline -1 & 3 & 0 & 0 & 1 & 0 \end{array}\right] \begin{array}{c} R_1 + R_2 \to R_2 \\ R_1 + R_3 \to R_3 \end{array} \left[\begin{array}{rrrrr|r} 1 & -\frac{3}{2} & \frac{1}{2} & 0 & 0 & 3 \\ 0 & \frac{5}{2} & \frac{1}{2} & 1 & 0 & 7 \\ \hline 0 & \frac{3}{2} & \frac{1}{2} & 0 & 1 & 3 \end{array}\right]$$

(columns: x_1, x_2, s_1, s_2, z)

From this last matrix we can read the solution as $x_1 = 3, x_2 = 0, s_1 = 0, s_2 = 7$, and $z = 3$. ■

Interactive Activity

Is the solution in Example 4 feasible? Explain.

WWW

Checkpoint 3

Now work Exercise 37.

The matrix given in Example 4,

$$\begin{array}{c} \begin{matrix} x_1 & x_2 & s_1 & s_2 & z \end{matrix} \\ \left[\begin{array}{rrrrr|r} 2 & -3 & 1 & 0 & 0 & 6 \\ -1 & 4 & 0 & 1 & 0 & 4 \\ \hline -1 & 3 & 0 & 0 & 1 & 0 \end{array}\right] \end{array}$$

was the initial simplex tableau for the equational form of the linear programming problem presented in the Flashback. We notice from this tableau that x_1 and x_2 are the nonbasic variables (and may be set equal to zero), and s_1, s_2, and z are the basic variables. Observe that this means that the tableau corresponds to the solution $x_1 = 0, x_2 = 0, s_1 = 6, s_2 = 4$, and $z = 0$. In particular, the solution corresponds to the corner point (0, 0). See Figure 4.1.2.

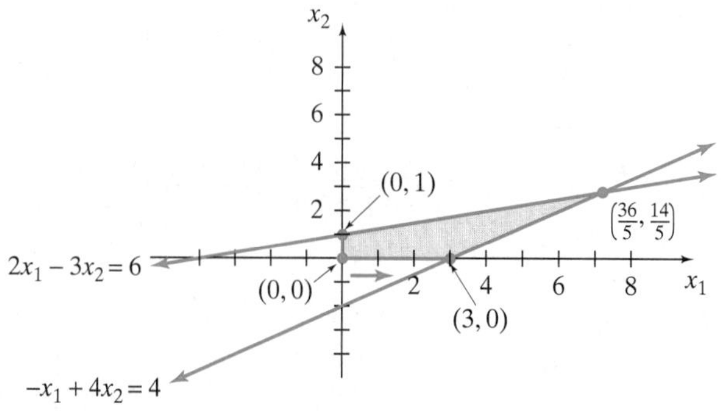

Figure 4.1.2

In Example 4 we actually performed a step in the simplex method (discussed in detail in Section 4.2). Observe that the solution read from the final tableau corresponds to the corner point (3, 0) because $x_1 = 3$ and $x_2 = 0$. This illustrates that the simplex method seeks out an optimal solution by moving from corner point to corner point in the feasible region in a strategic fashion. In Section 4.2 we learn the strategy used to find the optimal solution.

Technology Option

In Chapter 2 we saw that row operations can be performed on a graphing calculator. Figure 4.1.3 shows the initial tableau from Example 4. Notice that the right side of the matrix is not visible, however, it can be seen by pressing the right arrow button. Figure 4.1.4 shows $\frac{1}{2}R_1 \to R_1$. Figures 4.1.5 and 4.1.6 show $R_1 + R_2 \to R_2$ and $R_1 + R_3 \to R_3$, respectively.

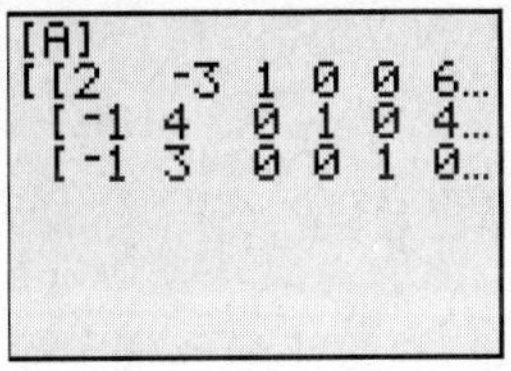

Figure 4.1.3

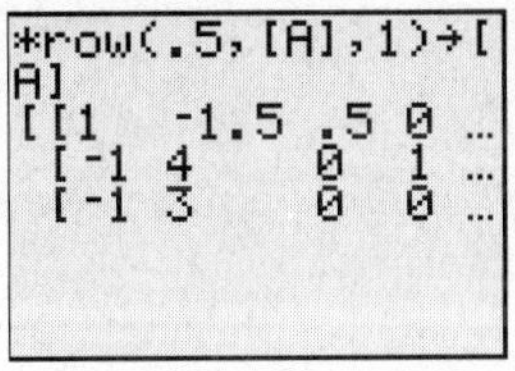

Figure 4.1.4

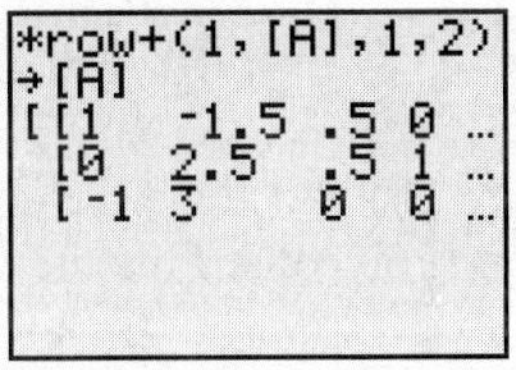

Figure 4.1.5

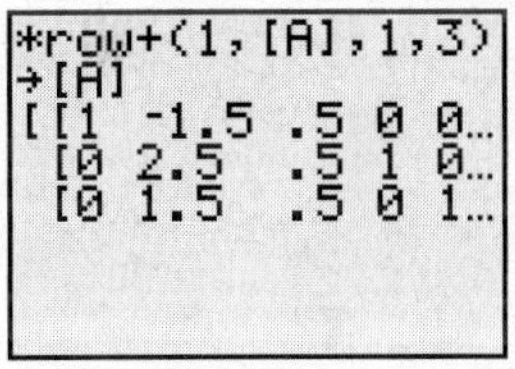

Figure 4.1.6

For more information on how to do matrix operations on your calculator, consult the online graphing calculator manual at www.prenhall.com/armstrong.

Applications

The main thrust of the applications in this section will be to use slack variables to adapt standard maximum form linear programming problems into equational form and then write the initial simplex tableau.

Example 5 Writing an Initial Simplex Tableau

Dub, Inc., is a small manufacturing firm that produces two microwave switches, switch A and switch B. Each switch passes through two departments: assembly and testing. Switch A requires 4 hours of assembly and 1 hour of testing and switch B requires 3 hours of assembly and 2 hours of testing. The maximum numbers of hours available each week for assembly and testing are 240 hours and 140 hours, respectively. The per unit profit for switch A is \$20, whereas the per unit profit of switch B is \$30. To determine how many units of each switch should be made and sold each week to maximize profit, do the following:

(a) Define the variables and write the mathematical model for this problem.

(b) Use slack variables to adapt the mathematical model to equational form.

(c) Write the initial simplex tableau.

Solution

(a) Understand the Situation: Since we do not know how many of each switch are made each week, we let x_1 represent the number of units of switch A made each week and x_2 represent the number of units of switch B made each week.

The objective function is

$$P = 20x_1 + 30x_2$$

where P is the profit in dollars. The problem constraints are

Assembly department constraint:	$4x_1 + 3x_2 \leq 240$
Testing department constraint:	$x_1 + 2x_2 \leq 140$
Nonnegative constraints:	$x_1 \geq 0, x_2 \geq 0$

So the mathematical model is

$$\begin{aligned} \text{Maximize:} \quad & P = 20x_1 + 30x_2 \\ \text{Subject to:} \quad & 4x_1 + 3x_2 \leq 240 \\ & x_1 + 2x_2 \leq 140 \\ & x_1 \geq 0, x_2 \geq 0 \end{aligned}$$

(b) Since we have two constraints, we need two slack variables, s_1 and s_2. Our constraints in equational form are

$$\begin{aligned} 4x_1 + 3x_2 + s_1 &= 240 \\ x_1 + 2x_2 + s_2 &= 140 \end{aligned}$$

Our objective function is

$$-20x_1 - 30x_2 + P = 0$$

(c) The initial simplex tableau is

$$\begin{array}{c} \begin{matrix} x_1 & x_2 & s_1 & s_2 & P & \end{matrix} \\ \left[\begin{array}{rrrrr|r} 4 & 3 & 1 & 0 & 0 & 240 \\ 1 & 2 & 0 & 1 & 0 & 140 \\ \hline -20 & -30 & 0 & 0 & 1 & 0 \end{array}\right] \end{array}$$

In the next section, we will learn a method that will transform this tableau into one that provides the optimal solution to the original problem. ■

Checkpoint 4 Now work Exercise 51.

SUMMARY

In this section we presented many necessary preliminaries pertaining to the simplex method. These included the **standard maximum form** of a linear programming problem, **slack variables**, how to adapt a standard maximum form linear programming problem into **equational form**, the **initial simplex tableau**, **basic variables**, **nonbasic variables**, **basic feasible solution**, and a review of matrix row operations and the **pivot**. The ability to read a simplex tableau was discussed and is very important. It gives a solution (optimal or otherwise) and indicates which variables are basic and which nonbasic.

SECTION 4.1 EXERCISES

In Exercises 1–8, convert the system of linear inequalities into a system of equations by using slack variables.

1. $x_1 + x_2 \le 8$
   $2x_1 - x_2 \le 5$

2. $3x_1 + 2x_2 \le 12$
   $5x_1 - 3x_2 \le 35$

3. $2x_1 + x_2 - x_3 \le 5$
   $-3x_1 + 2x_2 + x_3 \le 7$
   $4x_1 - x_2 + 2x_3 \le 9$

4. $-x_1 + 2x_2 + x_3 \le 1$
   $7x_1 + 4x_2 - 2x_3 \le 8$
   $2x_1 - 3x_2 - 4x_3 \le 7$

5. $2x_1 - x_2 \le 4$
   $-3x_1 + 2x_2 \le 7$
   $x_1 - 4x_2 \le 8$

6. $-3x_1 + 2x_2 \le 16$
   $4x_1 - 7x_2 \le 14$
   $x_1 - x_2 \le 7$

7. $2x_1 + 3x_2 - x_3 \le 8$
   $-4x_1 + x_2 + x_3 \le 5$

8. $-2x_1 + x_2 - 3x_3 \le 12$
   $3x_1 - 2x_2 - 4x_3 \le 31$

In Exercises 9–26, for the given standard maximum form linear programming problem, introduce slack variables and set up the initial simplex tableau.

9. Maximize: $z = 2x_1 + x_2$
   Subject to: $x_1 + x_2 \le 6$
   $x_1 - x_2 \le 2$
   $x_1 \ge 0, x_2 \ge 0$

10. Maximize: $z = 3x_1 + x_2$
    Subject to: $4x_1 + 5x_2 \le 9$
    $2x_1 - x_2 \le 1$
    $x_1 \ge 0, x_2 \ge 0$

11. Maximize: $z = x_1 + 2x_2$
    Subject to: $x_1 + x_2 \le 7$
    $2x_1 + 3x_2 \le 16$
    $x_1 \ge 0, x_2 \ge 0$

12. Maximize: $z = 2x_1 + 3x_2$
    Subject to: $3x_1 + 2x_2 \le 8$
    $x_1 - x_2 \le 7$
    $x_1 \ge 0, x_2 \ge 0$

13. Maximize: $z = 2x_1 + 2x_2$
    Subject to: $x_1 + x_2 \le 3$
    $x_1 - 2x_2 \le 0$
    $x_1 \ge 0, x_2 \ge 0$

14. Maximize: $z = 5x_1 + 7x_2$
    Subject to: $3x_1 - x_2 \le 8$
    $2x_1 + 4x_2 \le 5$
    $x_1 \ge 0, x_2 \ge 0$

15. Maximize: $z = 2x_1 + 2x_2$
    Subject to: $3x_1 - 2x_2 \le 0$
    $x_1 + x_2 \le 5$
    $2x_1 + x_2 \le 6$
    $x_1 \ge 0, x_2 \ge 0$

16. Maximize: $z = x_1 + 4x_2$
    Subject to: $2x_1 + x_2 \le 32$
    $x_1 + x_2 \le 18$
    $x_1 + 3x_2 \le 36$
    $x_1 \ge 0, x_2 \ge 0$

17. Maximize: $z = 5x_1 + x_2$
    Subject to: $-x_1 + 2x_2 \le 10$
    $2x_1 + x_2 \le 15$
    $5x_1 + x_2 \le 8$
    $-x_1 + x_2 \le 1$
    $x_1 \ge 0, x_2 \ge 0$

18. Maximize: $z = 3x_1 + 5x_2$
    Subject to: $-x_1 + 2x_2 \le 10$
    $4x_1 + 2x_2 \le 15$
    $x_1 + 3x_2 \le 8$
    $6x_1 - x_2 \le 4$
    $x_1 \ge 0, x_2 \ge 0$

19. Maximize: $z = 2x_1 + x_2 + 3x_3$
    Subject to: $x_1 - x_2 + 2x_3 \le 4$
    $2x_1 + x_2 - 3x_3 \le 7$
    $x_1 \ge 0, x_2 \ge 0, x_3 \ge 0$

20. Maximize: $z = 3x_1 + 2x_2 + 4x_3$
    Subject to: $2x_1 + x_2 + x_3 \le 10$
    $5x_1 - x_2 + 2x_3 \le 8$
    $x_1 \ge 0, x_2 \ge 0, x_3 \ge 0$

21. Maximize: $z = 3x_1 + x_2 + 5x_3$
    Subject to: $2x_1 + x_2 + 4x_3 \le 8$
    $4x_1 - x_2 + 3x_3 \le 15$
    $12x_1 + 2x_2 + 7x_3 \le 12$
    $x_1 \ge 0, x_2 \ge 0, x_3 \ge 0$

22. Maximize: $z = 2x_1 + 5x_2 + x_3$
    Subject to: $5x_1 + 2x_2 + x_3 \le 20$
    $x_1 + 3x_2 + 5x_3 \le 13$
    $4x_1 + x_2 + 3x_3 \le 15$
    $x_1 \ge 0, x_2 \ge 0, x_3 \ge 0$

23. Maximize: $z = 4x_1 + 2x_2 + 3x_3$

Subject to:
$$\begin{aligned} 2x_1 \quad\quad + 3x_3 &\le 8 \\ 3x_1 + 2x_2 \quad\quad &\le 5 \\ x_1 \ge 0, x_2 \ge 0, x_3 &\ge 0 \end{aligned}$$

24. Maximize: $z = 9x_1 + x_2 + x_3$

Subject to:
$$\begin{aligned} 4x_1 \quad\quad + x_3 &\le 7 \\ 2x_1 + 3x_2 \quad\quad &\le 8 \\ x_1 \ge 0, x_2 \ge 0, x_3 &\ge 0 \end{aligned}$$

25. Maximize: $z = 5x_1 + x_2 + 3x_3$

Subject to:
$$\begin{aligned} 15x_1 + 3x_2 \quad\quad &\le 9 \\ 6x_2 + x_3 &\le 7 \\ x_1 \ge 0, x_2 \ge 0, x_3 &\ge 0 \end{aligned}$$

26. Maximize: $z = 4x_1 + 3x_2 + x_3$

Subject to:
$$\begin{aligned} 6x_1 + x_2 \quad\quad &\le 9 \\ 3x_2 + 4x_3 &\le 24 \\ x_1 \ge 0, x_2 \ge 0, x_3 &\ge 0 \end{aligned}$$

In Exercises 27–36, for each given matrix:

(a) Identify the basic and nonbasic variables.

(b) Determine the solution given by the matrix.

27.
$$\begin{array}{ccccc|c} x_1 & x_2 & s_1 & s_2 & z & \\ 1 & 1 & 0 & 3 & 0 & 5 \\ 0 & 2 & 1 & 1 & 0 & 2 \\ \hline 0 & -5 & 0 & 4 & 1 & 10 \end{array}$$

28.
$$\begin{array}{ccccc|c} x_1 & x_2 & s_1 & s_2 & z & \\ 0 & 2 & 1 & 6 & 0 & 5 \\ 1 & 5 & 0 & 11 & 0 & 11 \\ \hline 0 & -3 & 0 & 4 & 1 & 15 \end{array}$$

29.
$$\begin{array}{ccccc|c} x_1 & x_2 & s_1 & s_2 & z & \\ 2 & -3 & 1 & 0 & 0 & 3 \\ 4 & 6 & 0 & 1 & 0 & 7 \\ \hline -1 & 5 & 0 & 0 & 1 & 9 \end{array}$$

30.
$$\begin{array}{ccccc|c} x_1 & x_2 & s_1 & s_2 & z & \\ 1 & 3 & 0 & 6 & 0 & 4 \\ 0 & -1 & 1 & 2 & 0 & 7 \\ \hline 0 & -2 & 0 & 7 & 1 & 13 \end{array}$$

31.
$$\begin{array}{cccccc|c} x_1 & x_2 & s_1 & s_2 & s_3 & z & \\ 2 & 5 & 1 & 0 & 0 & 0 & 3 \\ 3 & 6 & 0 & 1 & 0 & 0 & 8 \\ 4 & 2 & 0 & 0 & 1 & 0 & 15 \\ \hline -1 & -2 & 0 & 0 & 0 & 1 & 0 \end{array}$$

32.
$$\begin{array}{cccccc|c} x_1 & x_2 & s_1 & s_2 & s_3 & z & \\ -1 & 2 & 0 & 1 & 0 & 0 & 3 \\ 3 & 8 & 1 & 0 & 0 & 0 & 6 \\ 7 & 5 & 0 & 0 & 1 & 0 & 8 \\ \hline -2 & -7 & 0 & 0 & 0 & 1 & 0 \end{array}$$

33.
$$\begin{array}{cccccc|c} x_1 & x_2 & x_3 & s_1 & s_2 & z & \\ 1 & 0 & -3 & 5 & -1 & 0 & 7 \\ 0 & 1 & 2 & 8 & 2 & 0 & 12 \\ \hline 0 & 0 & 4 & 3 & 9 & 1 & 27 \end{array}$$

34.
$$\begin{array}{cccccc|c} x_1 & x_2 & x_3 & s_1 & s_2 & z & \\ 0 & 1 & -2 & 1 & 8 & 0 & 9 \\ 1 & 0 & 3 & 8 & 2 & 0 & 17 \\ \hline 0 & 0 & 1 & 9 & 3 & 1 & 38 \end{array}$$

35.
$$\begin{array}{cccccccc|c} x_1 & x_2 & x_3 & x_4 & s_1 & s_2 & s_3 & z & \\ 2 & 1 & 0 & 2 & 2 & 0 & 5 & 0 & 15 \\ 4 & 0 & 1 & 10 & 1 & 0 & 1 & 0 & 20 \\ 2 & 0 & 0 & 0 & 2 & 1 & 0 & 0 & 10 \\ \hline -4 & 0 & 0 & -2 & 3 & 0 & 3 & 1 & 100 \end{array}$$

36.
$$\begin{array}{cccccccc|c} x_1 & x_2 & x_3 & x_4 & s_1 & s_2 & s_3 & z & \\ 1 & 5 & 0 & 0 & 2 & 5 & 3 & 0 & 9 \\ 0 & 3 & 0 & 1 & -1 & 4 & 1 & 0 & 13 \\ 0 & 2 & 1 & 0 & 3 & 7 & 9 & 0 & 27 \\ \hline 0 & -2 & 0 & 0 & 1 & 0 & 4 & 1 & 80 \end{array}$$

In Exercises 37–46:

(a) Pivot about the circled entry.

(b) Read the solution after pivoting.

37.
$$\begin{array}{ccccc|c} x_1 & x_2 & s_1 & s_2 & z & \\ 1 & 1 & 0 & 3 & 0 & 5 \\ 0 & \textcircled{2} & 1 & 1 & 0 & 2 \\ \hline 0 & -5 & 0 & 4 & 1 & 10 \end{array}$$

38.
$$\begin{array}{ccccc|c} x_1 & x_2 & s_1 & s_2 & z & \\ 0 & 2 & 1 & 6 & 0 & 5 \\ 1 & \textcircled{5} & 0 & 11 & 0 & 11 \\ \hline 0 & -3 & 0 & 4 & 1 & 15 \end{array}$$

39.
$$\begin{array}{ccccc|c} x_1 & x_2 & s_1 & s_2 & z & \\ \textcircled{2} & -3 & 1 & 0 & 0 & 3 \\ 4 & 6 & 0 & 1 & 0 & 7 \\ \hline -1 & 5 & 0 & 0 & 1 & 9 \end{array}$$

40.
$$\begin{array}{ccccc|c} x_1 & x_2 & s_1 & s_2 & z & \\ 1 & \textcircled{3} & 0 & 6 & 0 & 4 \\ 0 & -1 & 1 & 2 & 0 & 7 \\ \hline 0 & -2 & 0 & 7 & 1 & 13 \end{array}$$

41.
$$\begin{array}{cccccc|c} x_1 & x_2 & s_1 & s_2 & s_3 & z & \\ -1 & 2 & 0 & 1 & 0 & 0 & 3 \\ 3 & \textcircled{8} & 1 & 0 & 0 & 0 & 6 \\ 7 & 5 & 0 & 0 & 1 & 0 & 8 \\ \hline -2 & -7 & 0 & 0 & 0 & 1 & 0 \end{array}$$

42.
$$\begin{array}{cccccc|c} x_1 & x_2 & s_1 & s_2 & s_3 & z & \\ 2 & \textcircled{5} & 1 & 0 & 0 & 0 & 3 \\ 3 & 6 & 0 & 1 & 0 & 0 & 8 \\ 4 & 2 & 0 & 0 & 1 & 0 & 15 \\ \hline -1 & -2 & 0 & 0 & 0 & 1 & 0 \end{array}$$

43. $$\begin{array}{cccccc|c} x_1 & x_2 & x_3 & s_1 & s_2 & z & \\ \hline 10 & 4 & 2 & 1 & 0 & 0 & 90 \\ 6 & 3 & \textcircled{4} & 0 & 1 & 0 & 40 \\ \hline -2 & -3 & -8 & 0 & 0 & 1 & 0 \end{array}$$

44. $$\begin{array}{cccccc|c} x_1 & x_2 & x_3 & s_1 & s_2 & z & \\ \hline 3 & 6 & \textcircled{1} & 1 & 0 & 0 & 13 \\ 5 & 2 & 2 & 0 & 1 & 0 & 15 \\ \hline -8 & -4 & -10 & 0 & 0 & 1 & 0 \end{array}$$

45. $$\begin{array}{cccccccc|c} x_1 & x_2 & x_3 & x_4 & s_1 & s_2 & s_3 & z & \\ \hline 1 & 5 & 0 & 0 & 2 & 5 & 3 & 0 & 9 \\ 0 & 3 & 0 & 1 & -1 & 4 & 1 & 0 & 13 \\ 0 & \textcircled{2} & 1 & 0 & 3 & 7 & 9 & 0 & 27 \\ \hline 0 & -2 & 0 & 0 & 1 & 0 & 4 & 1 & 80 \end{array}$$

46. $$\begin{array}{cccccccc|c} x_1 & x_2 & x_3 & x_4 & s_1 & s_2 & s_3 & z & \\ \hline 6 & 0 & 1 & 2 & 2 & 0 & 5 & 0 & 5 \\ \textcircled{2} & 1 & 0 & 10 & 1 & 0 & 2 & 0 & 10 \\ 3 & 0 & 0 & 0 & 2 & 1 & 0 & 0 & 8 \\ \hline -10 & 0 & 0 & -2 & 3 & 0 & 3 & 1 & 20 \end{array}$$

In Exercises 47–50, use a graphing calculator to:

(a) Pivot about the circled entry.

(b) Read the solution after pivoting.

47. Matrix in Exercise 38.

48. Matrix in Exercise 42.

49. Matrix in Exercise 44.

50. Matrix in Exercise 46.

Applications

In Exercises 51–60, perform the setup so that the solution can be determined by the simplex method as illustrated in Example 5. In the Section 4.2 exercises, you will determine optimal solutions for these exercises.

3.1 ✓ **51.** Due to new regulations approved by the Environmental Protection Agency (EPA), the Henderson Chemical Company was required to install a new process to reduce the pollution created when making a particular chemical. The older process releases 4 grams of sulfur and 3 grams of particulates for each gallon of chemical produced. The newer process releases 2 grams of sulfur and 1 gram of particulates for each gallon of chemical produced. The company realizes a profit of \$0.40 for each gallon produced under the old process and \$0.16 for each gallon produced under the new process. The new EPA regulations do not allow for more than 10,000 grams of sulfur and 7000 grams of particulates to be released daily. To determine how many gallons of this chemical should be produced using the old process to maximize profit, do the following:

(a) Define the variables and write the mathematical model for this problem.

(b) Use slack variables to adapt the mathematical model to equational form.

(c) Write the initial simplex tableau.

52. (*continuation of Exercise 51*) Barb, the plant manager for Henderson Chemical Company, has learned that the EPA is going to relax the new regulations so that no more than 13,000 grams of sulfur can be released daily. Assuming that all other data remain the same as in Exercise 51, to determine how many gallons of this chemical should be produced using the old process to maximize profit, do the following:

(a) Define the variables and write the mathematical model for this problem.

(b) Use slack variables to adapt the mathematical model to equational form.

(c) Write the initial simplex tableau.

3.1 **53.** Randy and Rusty have decided to enter the field of personal digital assistants (PDAs). Initially, they plan to design two types of PDAs, the WOW 1000 and the WOW 1500. Because of interest in PDAs, they can sell all that they can produce. However, they need to determine a production rate to satisfy various limits with a small production crew. These include a maximum of 150 hours per week for assembly and 100 hours per week for testing. The WOW 1000 requires 3 hours of assembly and 3 hours of testing, and the WOW 1500 requires 6 hours of assembly and 3.5 hours of testing. Profit for the WOW 1000 is estimated at \$100 per unit, and profit for the WOW 1500 is estimated at \$150 per unit. To determine how many of each type of PDA should be made and sold to maximize weekly profit, do the following:

(a) Define the variables and write the mathematical model for this problem.

(b) Use slack variables to adapt the mathematical model to equational form.

(c) Write the initial simplex tableau.

54. (*continuation of Exercise 53*) Randy and Rusty hire some new employees and can, as a result, increase production. By doing so, they now have a maximum of 210 hours per week for assembly and 160 hours per week for testing. Assuming that all other data remain the same as in Exercise 53, to determine how many of each type of PDA should be made and sold to maximize weekly profit, do the following:

(a) Define the variables and write the mathematical model for this problem.

(b) Use slack variables to adapt the mathematical model to equational form.

(c) Write the initial simplex tableau.

3.1 **55.** Table 4.1.2 gives data for growing artichokes and asparagus for an entire growing season in California (entries are in lb/acre).

Table 4.1.2

	Artichokes	Asparagus
Nitrogen	197	240
Potash	123	42

SOURCE: U.S. Department of Agriculture, www.usda.gov/nass

Franco has 150 acres available suitable for cultivating crops. He has purchased a contract for a total of 31,880 pounds of nitrogen and 9120 pounds of potash. He expects to make a profit of \$140 per acre on artichokes and \$160 per acre on asparagus. To determine how many acres of each he should plant to maximize his profit, do the following:

(a) Define the variables and write the mathematical model for this problem.

(b) Use slack variables to adapt the mathematical model to equational form.

(c) Write the initial simplex tableau.

⇐3.1 **56.** Table 4.1.3 gives data for growing cauliflower and head lettuce for an entire growing season in California (entries are in lb/acre).

Table 4.1.3

	Cauliflower	Head Lettuce
Nitrogen	215	173
Phosphate	98	122

SOURCE: U.S. Department of Agriculture, www.usda.gov/nass

Sonja has 210 acres available for cultivating crops. She has purchased a contract for a total of 38,380 pounds of nitrogen and 22,240 pounds of phosphate. She expects to make a profit of \$90 per acre on cauliflower and \$100 per acre on head lettuce. To determine how many acres of each she should plant to maximize her profit, do the following:

(a) Define the variables and write the mathematical model for this problem.

(b) Use slack variables to adapt the mathematical model to equational form.

(c) Write the initial simplex tableau.

⇐3.1 **57.** The Smolko Tent Company makes three types of tents: two-person, four-person, and six-person models. Each tent requires the services, in hours, of three departments, as shown in Table 4.1.4. The fabric cutting, assembly, and packaging departments have available a maximum of 420, 310, and 180 hours each week, respectively.

Table 4.1.4

	Two-person	Four-person	Six-person
Fabric cutting	0.5	0.8	1.1
Assembly	0.6	0.7	0.9
Packaging	0.2	0.3	0.6

The per unit profit is \$30, \$50, and \$80 for the two-, four-, and six-person tents, respectively. To determine how many of each type of tent should be made and sold to maximize weekly profit, do the following:

(a) Define the variables and write the mathematical model for this problem.

(b) Use slack variables to adapt the mathematical model to equational form.

(c) Write the initial simplex tableau.

⇐3.1 **58.** Kathy's Secretarial Services has a 1-month intensive training program for receptionist/secretaries for three types of office positions: medical field, small business, and executive secretarial. Each individual receives training in three departments at Kathy's: word processing, telephone and LAN (local area network), and ethics. Table 4.1.5 shows the number of units of training that each type of receptionist/secretary receives.

Table 4.1.5

	Medical Field	Small Business	Executive Secretary
Word processing	0.8	1.1	1.7
Telephone and LAN	0.6	1.3	1.5
Ethics	1.2	0.8	1.3

Each month, word processing, telephone and LAN, and ethics have available a maximum of 32.1 units, 29.7 units and 33 units, respectively. The profit that Kathy's realizes from each individual admitted to medical field, small business, and executive secretary is \$1000, \$1000, and \$1400, respectively. To determine how many individuals should be admitted each month to each program to maximize monthly profit, do the following:

(a) Define the variables and write the mathematical model for this problem.

(b) Use slack variables to adapt the mathematical model to equational form.

(c) Write the initial simplex tableau.

⇐3.1 **59.** The data in Table 4.1.6 give the amount of nitrogen, phosphate, and potash (all in lb/acre) needed to grow asparagus, cauliflower, and celery for an entire growing season in California.

Table 4.1.6

	Asparagus	Cauliflower	Celery
Nitrogen	240	215	336
Phosphate	148	98	171
Potash	42	69	147

SOURCE: U.S. Department of Agriculture, www.usda.gov/nass

Eldrick has 280 acres and can get no more than 65,970 pounds of nitrogen, 31,860 pounds of phosphate, and 21,630 pounds of potash. He expects a profit of \$160 per acre for asparagus, \$90 per acre for cauliflower, and \$80 per acre for celery. To determine how many acres he should allot for each crop to maximize profit, do the following:

(a) Define the variables and write the mathematical model for this problem.

(b) Use slack variables to adapt the mathematical model to equational form.

(c) Write the initial simplex tableau.

60. The data in Table 4.1.7 give the amount of nitrogen, phosphate, and potash (all in lb/acre) needed to grow head lettuce, cantaloupe, and bell peppers for an entire growing season in California. Maria has 200 acres and can get no more than 38,430 pounds of nitrogen, 27,845 pounds of phosphate, and 12,985 pounds of potash. She expects a profit of \$100 per acre for head lettuce, \$90 per acre for cantaloupe, and \$85 per acre for bell peppers. To determine how many acres she should allot for each crop to maximize profit, do the following:

Table 4.1.7

	Head Lettuce	Cantaloupe	Bell Peppers
Nitrogen	173	175	212
Phosphate	122	161	108
Potash	76	39	72

SOURCE: U.S. Department of Agriculture, www.usda.gov/nass

(a) Define the variables and write the mathematical model for this problem.

(b) Use slack variables to adapt the mathematical model to equational form.

(c) Write the initial simplex tableau.

SECTION PROJECT

The biggest change to take place in the field of linear programming solution techniques in the time since George Dantzig developed the simplex method in 1947 was the arrival in 1984 of Karmarkar's algorithm, named after its inventor Narendra Karmarkar. Karmarkar's method often takes significantly less computer time to solve very large-scale linear programming problems.

As we saw in Figure 4.1.2, the simplex method moves from corner point to corner point around the feasible region seeking out an optimal solution. In contrast, Karmarkar's method follows a path of points on the inside of the feasible region.

Although it is likely that the simplex method will continue to be used for many linear programming problems, a new generation of linear programming software built around Karmarkar's algorithm is already becoming popular. For example, Delta Air Lines became the first commercial airline to use the Karmarkar program (called KORBX), which was developed and sold by AT&T.

For this section project, research and write a short paper on the Karmarkar method. You should include the following in your paper.

(a) The history of the simplex method and George Dantzig and the history of the Karmarkar method and Narendra Karmarkar.

(b) The advantages and/or disadvantages of each method.

(c) Current available computer programs for the simplex method and the Karmarkar method.

(d) An outline of the basic features of each method.

Section 4.2 Simplex Method: Standard Maximum Form Problems

In Section 4.1 we learned how to prepare a standard maximum form linear programming problem for the simplex method. This preparation included the following:

- Introduction of slack variables, one for each constraint
- Using slack variables to convert linear inequality constraints into equational form
- Using the coefficients of the system of linear equations corresponding to the constraints and objective function to write an augmented matrix called the initial simplex tableau

We also reviewed how to pivot about an element in the initial simplex tableau. We saw that pivoting took us from one corner point to another corner point in the feasible region.

At this time we are ready to extend the pivoting process to obtain an optimal solution for the objective function. To see how this is done, let's Flashback to Example 5 from Section 4.1.

Flashback

Dub, Inc., Revisited

Dub, Inc., is a small manufacturing firm that produces two microwave switches, switch A and switch B. Each switch passes through two departments: assembly and testing. Switch A requires 4 hours of assembly and 1 hour of testing, and switch B requires 3 hours of assembly and 2 hours of testing. The maximum numbers of hours available each week for assembly and testing are 240 hours and 140 hours, respectively. The per unit profit for switch A is \$20, whereas the per unit profit of switch B is \$30. To determine how many units of each switch should be made and sold each week to maximize profit P, we let x_1 represent the number of units of switch A made each week and x_2 represent the number of units of switch B made each week. The mathematical model is

$$\begin{aligned} \text{Maximize:} \quad & P = 20x_1 + 30x_2 \\ \text{Subject to:} \quad & 4x_1 + 3x_2 \le 240 \\ & x_1 + 2x_2 \le 140 \\ & x_1 \ge 0, x_2 \ge 0 \end{aligned}$$

Since we have two constraints, we need two slack variables, s_1 and s_2. Our constraints in equational form are

$$\begin{aligned} 4x_1 + 3x_2 + s_1 \quad\quad &= 240 \\ x_1 + 2x_2 \quad\quad + s_2 &= 140 \end{aligned}$$

Our objective function is

$$-20x_1 - 30x_2 + P = 0$$

The initial simplex tableau is

$$\begin{array}{c} \quad x_1 \quad\; x_2 \;\; s_1 \;\; s_2 \;\; P \\ \left[\begin{array}{rrrrr|r} 4 & 3 & 1 & 0 & 0 & 240 \\ 1 & 2 & 0 & 1 & 0 & 140 \\ \hline -20 & -30 & 0 & 0 & 1 & 0 \end{array}\right] \end{array}$$

Write the solution corresponding to the initial simplex tableau and interpret the value of each variable.

Flashback Solution

Since s_1, s_2, and P head columns containing exactly one 1 and all other entries are 0, these variables are the **basic variables**. They have the values $s_1 = 240$, $s_2 = 140$, and $P = 0$. Since x_1 and x_2 are the **nonbasic variables**, we set them equal to 0. So the corresponding solution to the initial simplex tableau is

$$x_1 = 0, \quad x_2 = 0, \quad s_1 = 240, \quad s_2 = 140, \quad P = 0$$

The interpretation of each variable is

$x_1 = 0$ (make 0 units of switch A)

$x_2 = 0$ (make 0 units of switch B)

$s_1 = 240$ (there are 240 unused hours in the assembly department)

$s_2 = 140$ (there are 140 unused hours in the testing department)

$P = 0$ (there is a profit of 0 dollars)

Intuitively, this makes sense since, if we make and sell 0 units of switch A and switch B, our profit is \$0, and we should still have available the maximum number of hours in each department.

Obviously, a value of zero for Dub, Inc.'s, profit in the Flashback solution is not optimal. Our goal is to find the optimum solution, and we do this by pivoting.

Pivoting and the Simplex Method

After writing the initial simplex tableau, the simplex method requires us to determine on which element in the matrix we perform the pivot operation. To do this we need to determine the **pivot column** and the **pivot row**. To gain some understanding of how we determine the pivot column, consider the objective function in the Flashback, $P = 20x_1 + 30x_2$. Since both coefficients are positive, the value of P can be increased from 0 by increasing x_1 and/or x_2. Since the coefficient of x_2 is larger than the coefficient of x_1, increasing x_2 while holding x_1 constant results in a greater increase in the value of P than increasing x_1 and keeping x_2 constant. Hence we should increase the value of x_2 while keeping x_1 constant. This analysis means that our **pivot column** in the initial simplex tableau is the column headed by x_2. Notice that this column has the most negative of the indicators.

$$\begin{array}{c} \begin{array}{cccccc} x_1 & x_2 & s_1 & s_2 & P & \end{array} \\ \left[\begin{array}{ccccc|c} 4 & 3 & 1 & 0 & 0 & 240 \\ 1 & 2 & 0 & 1 & 0 & 140 \\ \hline -20 & -30 & 0 & 0 & 1 & 0 \end{array}\right] \\ \underbrace{\hspace{4cm}}_{\text{Indicators}} \end{array}$$

Recall from Section 4.1 that the indicators are the numbers in the bottom row except for the rightmost 0 and 1. Also recall that the indicators come from rewriting the objective function $P = 20x_1 + 30x_2$ as $-20x_1 - 30x_2 + P = 0$. This suggests the following for determining the pivot column.

Pivot Column

The column with the most negative indicator is the **pivot column**. If there are two such indicators, use the leftmost indicator.

Now that we know how to determine the pivot column, we must determine which row is the pivot row. This will determine precisely on which entry to pivot. Continuing with the problem from the Flashback, we have found the initial simplex tableau to be

$$\begin{array}{c} \\ \\ \end{array}\begin{matrix} x_1 & x_2 & s_1 & s_2 & P & \end{matrix}$$
$$\left[\begin{array}{ccccc|c} 4 & 3 & 1 & 0 & 0 & 240 \\ 1 & 2 & 0 & 1 & 0 & 140 \\ \hline -20 & -30 & 0 & 0 & 1 & 0 \end{array}\right]$$

↑ Pivot column (under x_2)

The initial simplex tableau corresponds to the system

$$\begin{aligned} 4x_1 + 3x_2 + s_1 \qquad &= 240 \\ x_1 + 2x_2 \qquad + s_2 &= 140 \end{aligned}$$

or

$$\begin{aligned} s_1 &= 240 - 4x_1 - 3x_2 \\ s_2 &= 140 - x_1 - 2x_2 \end{aligned}$$

From Your Toolbox

A solution is a basic solution if all values for the variables are nonnegative.

Since pivoting in the column headed x_2 will result in one 1 and all other entries are 0 in this column, the pivot will turn x_2 into a basic variable while x_1 remains nonbasic. Now if x_1 remains nonbasic, we can set $x_1 = 0$. Substituting this into the last two equations gives us

$$\begin{aligned} s_1 &= 240 - 3x_2 \\ s_2 &= 140 - 2x_2 \end{aligned}$$

We know that s_1 and s_2 must be nonnegative in order to be consistent with a basic feasible solution (consult the Toolbox to the left). Since $s_1 \geq 0$ and $s_1 = 240 - 3x_2$, then $240 - 3x_2 \geq 0$. Solving for x_2 gives us

$$\begin{aligned} 240 - 3x_2 &\geq 0 \\ -3x_2 &\geq -240 \\ x_2 &\leq 80 \end{aligned}$$

This means that x_2 cannot exceed $\frac{240}{3}$ or 80. Similarly, the equation $s_2 = 140 - 2x_2$ implies that x_2 cannot exceed $\frac{140}{2}$ or 70; in other words, $x_2 \leq 70$. Together, this means that x_2 can be increased by at most 70; which would correspond to the corner point (0, 70). More importantly for our purposes at this time, the quotient $\frac{140}{2} = 70$ tells us the **pivot row**. Observe that the quotients $\frac{240}{3}$ and $\frac{140}{2}$ are simply the elements in the last column (column of constants) divided by the corresponding entry in the pivot column. These observations suggest the following on determining the pivot row.

Pivot Row

To determine the **pivot row**, divide each number from the rightmost column (column of constants) by the corresponding number from the pivot column. The smallest nonnegative quotient gives the pivot row. If two quotients are equal (and smallest), arbitrarily choose the row nearest the top of the matrix.

Note: We do not form a quotient for the bottom row, our indicators. The entry where the pivot column and pivot row intersect is the **pivot element**.

Interactive Activity

Consider the initial simplex tableau from the Flashback and to the right. If we pivot on any row that is not the smallest quotient, the value of x_2 will increase too much, causing one of the slack variables to become negative. To illustrate, pivot on the 3 in column 2 of the initial simplex tableau. Why is the resulting tableau undesirable?

Putting together what has been discussed so far and continuing with the problem from the Flashback yields

$$\begin{array}{c} \\ \text{Pivot element} \\ \\ \end{array}\begin{array}{c} x_1 \quad x_2 \quad s_1 \quad s_2 \quad P \\ \left[\begin{array}{rrrrr|r} 4 & 3 & 1 & 0 & 0 & 240 \\ 1 & \textcircled{2} & 0 & 1 & 0 & 140 \\ \hline -20 & -30 & 0 & 0 & 1 & 0 \end{array}\right] \\ \uparrow \\ \text{Pivot column} \\ \text{(most negative indicator)} \end{array} \begin{array}{l} \text{Quotients:} \\ 240/3 = 80 \\ 140/2 = 70 \quad \text{Smaller; so} \\ \qquad\qquad\quad\ \text{pivot row} \end{array}$$

Now that we have our pivot element, let's pivot on the circled 2 using our row operations.

$$\begin{array}{c} x_1 \quad x_2 \quad s_1 \quad s_2 \quad P \\ \left[\begin{array}{rrrrr|r} 4 & 3 & 1 & 0 & 0 & 240 \\ 1 & \textcircled{2} & 0 & 1 & 0 & 140 \\ \hline -20 & -30 & 0 & 0 & 1 & 0 \end{array}\right] \end{array} \begin{array}{c} \\ \frac{1}{2}R_2 \to R_2 \\ \\ \end{array} \begin{array}{c} x_1 \quad x_2 \quad s_1 \quad s_2 \quad P \\ \left[\begin{array}{rrrrr|r} 4 & 3 & 1 & 0 & 0 & 240 \\ \frac{1}{2} & 1 & 0 & \frac{1}{2} & 0 & 70 \\ \hline -20 & -30 & 0 & 0 & 1 & 0 \end{array}\right] \end{array}$$

$$\begin{array}{c} -3R_2 + R_1 \to R_1 \\ 30R_2 + R_3 \to R_3 \end{array} \begin{array}{c} x_1 \quad x_2 \quad s_1 \quad s_2 \quad P \\ \left[\begin{array}{rrrrr|r} \frac{5}{2} & 0 & 1 & -\frac{3}{2} & 0 & 30 \\ \frac{1}{2} & 1 & 0 & \frac{1}{2} & 0 & 70 \\ \hline -5 & 0 & 0 & 15 & 1 & 2100 \end{array}\right] \end{array}$$

This last matrix, the result of our first pivot, is called the **second simplex matrix** or **second simplex tableau.**

Example 1 Interpreting the Second Simplex Tableau

Interpret the entries in the second simplex tableau from the Flashback.

$$\begin{array}{c} x_1 \quad x_2 \quad s_1 \quad s_2 \quad P \\ \left[\begin{array}{rrrrr|r} \frac{5}{2} & 0 & 1 & -\frac{3}{2} & 0 & 30 \\ \frac{1}{2} & 1 & 0 & \frac{1}{2} & 0 & 70 \\ \hline -5 & 0 & 0 & 15 & 1 & 2100 \end{array}\right] \end{array}$$

Solution

The basic variables are x_2, s_1, and P and have values $x_2 = 70$, $s_1 = 30$, and $P = 2100$. The nonbasic variables are x_1 and s_2 and have the value of 0. This corresponds to the solution

$$x_1 = 0, \quad x_2 = 70, \quad s_1 = 30, \quad s_2 = 0, \quad P = 2100$$

Interpret the Solution: This means that producing and selling 0 units of switch A and 70 units of switch B leaves 30 unused hours in the assembly department and 0 unused hours in the testing department and yields a profit of $2100. ■

Does the second simplex tableau, which we just interpreted in Example 1, give the optimal solution? The answer is no and is exhibited by the presence of a negative number (-5) in the bottom row of the second simplex tableau. The last row represents the equation

$$-5x_1 + 0x_2 + 0s_1 + 15s_2 + P = 2100$$

which is equivalent to

$$P = 2100 + 5x_1 - 15s_2$$

This last equation tells us that profit is \$2100 if $x_1 = s_2 = 0$. However, if x_1 is a positive number, then the profit will be larger than \$2100. Hence it may be possible to have a larger profit by changing x_1 from 0 to a positive number. We make this change by repeating the pivot process! Continuing with the pivot process as outlined in this section, we have

$$\text{Pivot element} \quad \begin{array}{c} \begin{matrix} x_1 & x_2 & s_1 & s_2 & P & \end{matrix} \\ \left[\begin{array}{ccccc|c} \textcircled{\tfrac{5}{2}} & 0 & 1 & -\frac{3}{2} & 0 & 30 \\ \frac{1}{2} & 1 & 0 & \frac{1}{2} & 0 & 70 \\ \hline -5 & 0 & 0 & 15 & 1 & 2100 \end{array}\right] \end{array} \quad \begin{array}{l} \text{Quotients:} \\ 30/\frac{5}{2} = 12 \quad \text{Smaller; pivot row} \\ 70/\frac{1}{2} = 140 \end{array}$$

$\uparrow$ Pivot column

$$\begin{array}{c} \begin{matrix} x_1 & x_2 & s_1 & s_2 & P & \end{matrix} \\ \left[\begin{array}{ccccc|c} \frac{5}{2} & 0 & 1 & -\frac{3}{2} & 0 & 30 \\ \frac{1}{2} & 1 & 0 & \frac{1}{2} & 0 & 70 \\ \hline -5 & 0 & 0 & 15 & 1 & 2100 \end{array}\right] \end{array} \quad \frac{2}{5}R_1 \to R_1 \quad \begin{array}{c} \begin{matrix} x_1 & x_2 & s_1 & s_2 & P & \end{matrix} \\ \left[\begin{array}{ccccc|c} 1 & 0 & \frac{2}{5} & -\frac{3}{5} & 0 & 12 \\ \frac{1}{2} & 1 & 0 & \frac{1}{2} & 0 & 70 \\ \hline -5 & 0 & 0 & 15 & 1 & 2100 \end{array}\right] \end{array}$$

$$\begin{array}{l} -\frac{1}{2}R_1 + R_2 \to R_2 \\ 5R_1 + R_3 \to R_3 \end{array} \quad \begin{array}{c} \begin{matrix} x_1 & x_2 & s_1 & s_2 & P & \end{matrix} \\ \left[\begin{array}{ccccc|c} 1 & 0 & \frac{2}{5} & -\frac{3}{5} & 0 & 12 \\ 0 & 1 & -\frac{1}{5} & \frac{4}{5} & 0 & 64 \\ \hline 0 & 0 & 2 & 12 & 1 & 2160 \end{array}\right] \end{array}$$

The result of the second pivot gives the **third simplex tableau**. Interpreting the entries gives

$$x_1 = 12, \quad x_2 = 64, \quad s_1 = 0, \quad s_2 = 0, \quad P = 2160$$

Notice that the last row of the third simplex tableau represents the following:

$$\begin{aligned} 0x_1 + 0x_2 + 2s_1 + 12s_2 + P &= 2160 \\ 0(12) + 0(64) + 2(0) + 12(0) + P &= 2160 \\ P &= 2160 \end{aligned}$$

We have found our optimal (maximum) solution! So each week Dub, Inc., should produce and sell 12 units of switch A and 64 units of switch B to maximize weekly profit at \$2160. Notice that $s_1 = s_2 = 0$ means that all available hours should be used. This result suggests when we stop the pivot process.

The optimal solution has been found when no indicators are negative.

We now summarize what we have done so far in this section. The following steps are used when solving a standard maximum form linear programming problem using the **simplex method**.

Simplex Method for Solving Standard Maximum Form Problems

1. Write the initial simplex tableau. This involves the following:
   - Determining the objective function
   - Determining all constraints
   - Using slack variables to write each constraint in equational form
2. Determine the pivot element by finding the pivot column and constructing the necessary quotients to determine the pivot row.
3. Use row operations to pivot about the element found in step 2.
4. If the indicators are all nonnegative, this is the final tableau. Proceed to step 5. If the indicators are not all nonnegative, go back to step 2.
5. Read the optimal solution(s) from the final tableau.

Before continuing with the simplex method, we want to compare it to the graphical method in Chapter 3. Figure 4.2.1 shows the feasible region corresponding to the mathematical model in the Flashback. In Chapter 3 we had to evaluate the objective function at **each** corner point to determine the optimal value. See Table 4.2.1.

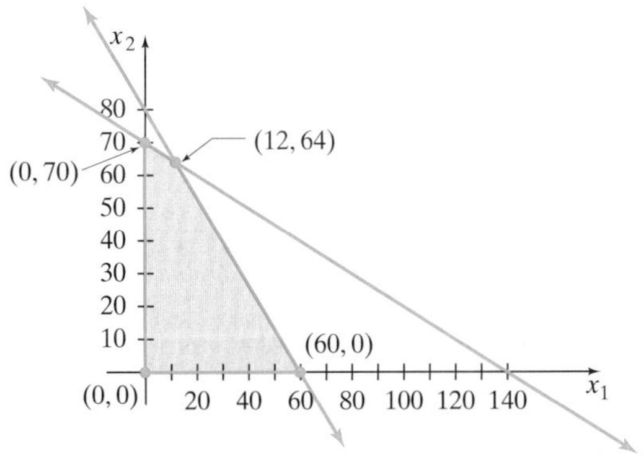

Figure 4.2.1

Table 4.2.1

Corner Point	Value of $P = 20x + 30y$	
(0, 0)	$20(0) + 30(0) = 0$	
(0, 70)	$20(0) + 30(70) = 2100$	
(12, 64)	$20(12) + 30(64) = 2160$	**Maximum**
(60, 0)	$20(60) + 30(0) = 1200$	

Notice how the simplex method evaluated the objective function at the corner points (0, 0), (0, 70), and finally (12, 64). There was no need to test the corner point (60, 0) since the optimum value of P was determined before the point was reached. Figure 4.2.2 shows the order in which corner points were evaluated using the simplex method.

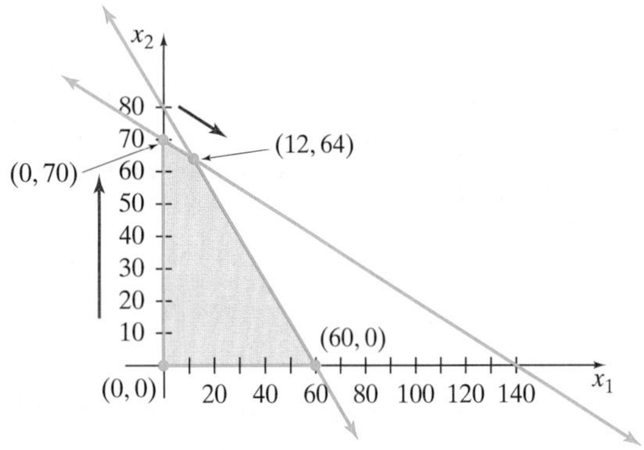

Figure 4.2.2

Example 2 Using the Simplex Method

Use the simplex method to

$$\begin{aligned} \text{Maximize:} \quad & z = 3x_1 + 2x_2 \\ \text{Subject to:} \quad & x_1 + x_2 \le 5 \\ & 2x_1 + x_2 \le 6 \\ & x_1 \ge 0, x_2 \ge 0 \end{aligned}$$

Solution

We use the Five-step process:

Step 1: Since the objective function and constraints are given, we need to introduce two slack variables, s_1 and s_2 (one for each constraint), to write the constraints in equational form. This gives

$$\begin{aligned} x_1 + x_2 + s_1 \qquad &= 5 \\ 2x_1 + x_2 \qquad + s_2 &= 6 \end{aligned}$$

The objective function is written as

$$-3x_1 - 2x_2 + z = 0$$

So our initial simplex tableau is

$$\begin{array}{c} \begin{matrix} x_1 & x_2 & s_1 & s_2 & z \end{matrix} \\ \left[\begin{array}{rrrrr|r} 1 & 1 & 1 & 0 & 0 & 5 \\ 2 & 1 & 0 & 1 & 0 & 6 \\ \hline -3 & -2 & 0 & 0 & 1 & 0 \end{array}\right] \end{array}$$

Step 2: The pivot column, quotients, pivot row, and pivot element are shown next.

$$\begin{array}{c} \begin{matrix} x_1 & x_2 & s_1 & s_2 & z \end{matrix} \\ \left[\begin{array}{rrrrr|r} 1 & 1 & 1 & 0 & 0 & 5 \\ \textcircled{2} & 1 & 0 & 1 & 0 & 6 \\ \hline -3 & -2 & 0 & 0 & 1 & 0 \end{array}\right] \end{array} \begin{array}{l} \text{Quotients:} \\ 5/1 = 5 \\ 6/2 = 3 \quad \text{Pivot row} \\ \end{array}$$

Pivot element: the circled 2. Pivot column: ↑ first column.

Step 3: We pivot about the selected 2 as follows:

$$\begin{array}{c} \begin{matrix} x_1 & x_2 & s_1 & s_2 & z \end{matrix} \\ \left[\begin{array}{rrrrr|r} 1 & 1 & 1 & 0 & 0 & 5 \\ \textcircled{2} & 1 & 0 & 1 & 0 & 6 \\ \hline -3 & -2 & 0 & 0 & 1 & 0 \end{array}\right] \end{array} \xrightarrow{\frac{1}{2}R_2 \to R_2} \begin{array}{c} \begin{matrix} x_1 & x_2 & s_1 & s_2 & z \end{matrix} \\ \left[\begin{array}{rrrrr|r} 1 & 1 & 1 & 0 & 0 & 5 \\ 1 & \frac{1}{2} & 0 & \frac{1}{2} & 0 & 3 \\ \hline -3 & -2 & 0 & 0 & 1 & 0 \end{array}\right] \end{array}$$

$$\begin{array}{r} -R_2 + R_1 \to R_1 \\ 3R_2 + R_3 \to R_3 \end{array} \begin{array}{c} \begin{matrix} x_1 & x_2 & s_1 & s_2 & z \end{matrix} \\ \left[\begin{array}{rrrrr|r} 0 & \frac{1}{2} & 1 & -\frac{1}{2} & 0 & 2 \\ 1 & \frac{1}{2} & 0 & \frac{1}{2} & 0 & 3 \\ \hline 0 & -\frac{1}{2} & 0 & \frac{3}{2} & 1 & 9 \end{array}\right] \end{array}$$

Step 4: Since we still have negative indicators, we go back to step 2 and repeat. Column 2 is the pivot column, and constructing quotients as before indicates that we choose the $\frac{1}{2}$ in row 1, column 2 as the pivot element. This gives us

$$\begin{array}{c} \begin{matrix} x_1 & x_2 & s_1 & s_2 & z \end{matrix} \\ \left[\begin{array}{ccccc|c} 0 & \textcircled{\frac{1}{2}} & 1 & -\frac{1}{2} & 0 & 2 \\ 1 & \frac{1}{2} & 0 & \frac{1}{2} & 0 & 3 \\ \hline 0 & -\frac{1}{2} & 0 & \frac{3}{2} & 1 & 9 \end{array}\right] \end{array} \quad 2R_1 \rightarrow R_1 \quad \begin{array}{c} \begin{matrix} x_1 & x_2 & s_1 & s_2 & z \end{matrix} \\ \left[\begin{array}{ccccc|c} 0 & 1 & 2 & -1 & 0 & 4 \\ 1 & \frac{1}{2} & 0 & \frac{1}{2} & 0 & 3 \\ \hline 0 & -\frac{1}{2} & 0 & \frac{3}{2} & 1 & 9 \end{array}\right] \end{array}$$

$$\begin{array}{r} -\frac{1}{2}R_1 + R_2 \rightarrow R_2 \\ \frac{1}{2}R_1 + R_3 \rightarrow R_3 \end{array} \quad \begin{array}{c} \begin{matrix} x_1 & x_2 & s_1 & s_2 & z \end{matrix} \\ \left[\begin{array}{ccccc|c} 0 & 1 & 2 & -1 & 0 & 4 \\ 1 & 0 & -1 & 1 & 0 & 1 \\ \hline 0 & 0 & 1 & 1 & 1 & 11 \end{array}\right] \end{array}$$

Since the indicators are all nonnegative, this is the final tableau.

Step 5: The solutions are $x_1 = 1$, $x_2 = 4$, $s_1 = 0$, $s_2 = 0$, and $z = 11$. This means that z has a maximum value of 11 when $x_1 = 1$ and $x_2 = 4$ and both slack variables are 0. ■

Checkpoint 1 Now work Exercise 11.

Example 3 Using the Simplex Method in an Application

The Gonzalez Office Supplies Company has determined that the profits are $40, $80, and $70 for each Secretarial, Executive, and Presidential chair produced and sold. Each chair requires time on three different machines. These times and the total available time are given in Table 4.2.2, in hours.

Table 4.2.2

	Secretarial	Executive	Presidential	Total Hours Each Week
Hours on machine A	1	1	1	30
Hours on machine B	2	3	2	53
Hours on machine C	1	2	3	47

How many of each type of chair should Gonzalez Office Supplies make each week in order to maximize its profit?

Solution

Step 1: Understand the Situation: Since we are asked to determine how many of each type of chair should be made, we let x_1, x_2, and x_3 represent the number of Secretarial, Executive, and Presidential chairs to be made, respectively. The mathematical model is

$$\begin{array}{ll} \text{Maximize:} & P = 40x_1 + 80x_2 + 70x_3 \\ \text{Subject to:} & x_1 + x_2 + x_3 \leq 30 \\ & 2x_1 + 3x_2 + 2x_3 \leq 53 \\ & x_1 + 2x_2 + 3x_3 \leq 47 \\ & x_1 \geq 0,\ x_2 \geq 0,\ x_3 \geq 0 \end{array}$$

Since we have three constraints, we need three slack variables, s_1, s_2, and s_3, in order to write the constraints in equational form. This gives

$$\begin{aligned} x_1 + x_2 + x_3 + s_1 \qquad\qquad &= 30 \\ 2x_1 + 3x_2 + 2x_3 \quad + s_2 \qquad &= 53 \\ x_1 + 2x_2 + 3x_3 \qquad\quad + s_3 &= 47 \end{aligned}$$

The objective function is written as

$$-40x_1 - 80x_2 - 70x_3 + P = 0$$

So the initial simplex tableau is

$$\begin{array}{c} \begin{matrix} x_1 & x_2 & x_3 & s_1 & s_2 & s_3 & P & \end{matrix} \\ \left[\begin{array}{ccccccc:c} 1 & 1 & 1 & 1 & 0 & 0 & 0 & 30 \\ 2 & 3 & 2 & 0 & 1 & 0 & 0 & 53 \\ 1 & 2 & 3 & 0 & 0 & 1 & 0 & 47 \\ \hdashline -40 & -80 & -70 & 0 & 0 & 0 & 1 & 0 \end{array}\right] \end{array}$$

Step 2: The pivot column, quotients, pivot row, and pivot element are shown next.

$$\begin{array}{c} \begin{matrix} x_1 & x_2 & x_3 & s_1 & s_2 & s_3 & P & \end{matrix} \\ \text{Pivot element} \left[\begin{array}{ccccccc:c} 1 & 1 & 1 & 1 & 0 & 0 & 0 & 30 \\ 2 & \textcircled{3} & 2 & 0 & 1 & 0 & 0 & 53 \\ 1 & 2 & 3 & 0 & 0 & 1 & 0 & 47 \\ \hdashline -40 & -80 & -70 & 0 & 0 & 0 & 1 & 0 \end{array}\right] \begin{array}{l} \text{Quotients:} \\ 30/1 = 30 \\ 53/3 = 17\tfrac{2}{3} \quad \text{Pivot row} \\ 47/2 = 23.5 \\ \end{array} \end{array}$$

$\uparrow$ Pivot column (under x_2)

Step 3: We pivot about the selected 3 as follows:

$$\begin{array}{c} \begin{matrix} x_1 & x_2 & x_3 & s_1 & s_2 & s_3 & P & \end{matrix} \\ \left[\begin{array}{ccccccc:c} 1 & 1 & 1 & 1 & 0 & 0 & 0 & 30 \\ 2 & \textcircled{3} & 2 & 0 & 1 & 0 & 0 & 53 \\ 1 & 2 & 3 & 0 & 0 & 1 & 0 & 47 \\ \hdashline -40 & -80 & -70 & 0 & 0 & 0 & 1 & 0 \end{array}\right] \end{array}$$

$$\begin{array}{c} \begin{matrix} x_1 & x_2 & x_3 & s_1 & s_2 & s_3 & P & \end{matrix} \\ \tfrac{1}{3}R_2 \to R_2 \quad \left[\begin{array}{ccccccc:c} 1 & 1 & 1 & 1 & 0 & 0 & 0 & 30 \\ \frac{2}{3} & 1 & \frac{2}{3} & 0 & \frac{1}{3} & 0 & 0 & \frac{53}{3} \\ 1 & 2 & 2 & 0 & 0 & 1 & 0 & 47 \\ \hdashline -40 & -80 & -70 & 0 & 0 & 0 & 1 & 0 \end{array}\right] \end{array}$$

$$\begin{array}{c} \begin{matrix} x_1 & x_2 & x_3 & s_1 & s_2 & s_3 & P & \end{matrix} \\ \begin{array}{r} -R_2 + R_1 \to R_1 \\ -2R_2 + R_3 \to R_3 \\ 80R_2 + R_4 \to R_4 \end{array} \left[\begin{array}{ccccccc:c} \frac{1}{3} & 0 & \frac{1}{3} & 1 & -\frac{1}{3} & 0 & 0 & \frac{37}{3} \\ \frac{2}{3} & 1 & \frac{2}{3} & 0 & \frac{1}{3} & 0 & 0 & \frac{53}{3} \\ -\frac{1}{3} & 0 & \frac{5}{3} & 0 & -\frac{2}{3} & 1 & 0 & \frac{35}{3} \\ \hdashline \frac{40}{3} & 0 & -\frac{50}{3} & 0 & \frac{80}{3} & 0 & 1 & \frac{4240}{3} \end{array}\right] \end{array}$$

Step 4: Since we do not have all nonnegative indicators, we need to repeat step 2. Column 3 is the pivot column, and constructing quotients as before indicates that we choose the $\frac{5}{3}$ in row 3, column 3 as the pivot element. Continuing the pivot, we begin by performing the row operation $\frac{3}{5}R_3 \rightarrow R_3$.

$$
\begin{array}{c}
\begin{array}{ccccccc|c} x_1 & x_2 & x_3 & s_1 & s_2 & s_3 & P & \end{array} \\
\left[\begin{array}{ccccccc|c}
\frac{1}{3} & 0 & \frac{1}{3} & 1 & -\frac{1}{3} & 0 & 0 & \frac{37}{3} \\
\frac{2}{3} & 1 & \frac{2}{3} & 0 & \frac{1}{3} & 0 & 0 & \frac{53}{3} \\
-\frac{1}{3} & 0 & \textcircled{\frac{5}{3}} & 0 & -\frac{2}{3} & 1 & 0 & \frac{35}{3} \\
\hline
\frac{40}{3} & 0 & -\frac{50}{3} & 0 & \frac{80}{3} & 0 & 1 & \frac{4240}{3}
\end{array}\right]
\end{array}
$$

$$
\frac{3}{5}R_3 \rightarrow R_3 \quad
\begin{array}{c}
\begin{array}{ccccccc|c} x_1 & x_2 & x_3 & s_1 & s_2 & s_3 & P & \end{array} \\
\left[\begin{array}{ccccccc|c}
\frac{1}{3} & 0 & \frac{1}{3} & 1 & -\frac{1}{3} & 0 & 0 & \frac{37}{3} \\
\frac{2}{3} & 1 & \frac{2}{3} & 0 & \frac{1}{3} & 0 & 0 & \frac{53}{3} \\
-\frac{1}{5} & 0 & 1 & 0 & -\frac{2}{5} & \frac{3}{5} & 0 & 7 \\
\hline
\frac{40}{3} & 0 & -\frac{50}{3} & 0 & \frac{80}{3} & 0 & 1 & \frac{4240}{3}
\end{array}\right]
\end{array}
$$

$$
\begin{array}{r}
-\frac{1}{3}R_3 + R_1 \rightarrow R_1 \\
-\frac{2}{3}R_3 + R_2 \rightarrow R_2 \\
\frac{50}{3}R_3 + R_4 \rightarrow R_4 \\
\
\end{array}
\begin{array}{c}
\begin{array}{ccccccc|c} x_1 & x_2 & x_3 & s_1 & s_2 & s_3 & P & \end{array} \\
\left[\begin{array}{ccccccc|c}
\frac{2}{5} & 0 & 0 & 1 & -\frac{1}{5} & -\frac{1}{5} & 0 & 10 \\
\frac{4}{5} & 1 & 0 & 0 & \frac{3}{5} & -\frac{2}{5} & 0 & 13 \\
-\frac{1}{5} & 0 & 1 & 0 & -\frac{2}{5} & \frac{3}{5} & 0 & 7 \\
\hline
10 & 0 & 0 & 0 & 20 & 10 & 1 & 1530
\end{array}\right]
\end{array}
$$

Since all indicators are nonnegative, this is the final tableau.

Step 5: The optimal solution is $x_1 = 0$, $x_2 = 13$, $x_3 = 7$, $s_1 = 10$, $s_2 = 0$, $s_3 = 0$, and $P = 1530$.

Interpret the Solution: This means that to maximize profit Gonzalez Office Supplies should produce 0 Secretarial, 13 Executive, and 7 Presidential chairs each week. This yields a maximum weekly profit of \$1530. The value of the slack variable $s_1 = 10$ means that 10 hours of available time on machine A is unused. ■

✓ Checkpoint 2

Now work Exercise 35.

We began Section 2.1 with a Flashback to the Gonzalez Office Supplies Company and in Example 7 in Section 2.1 we found that in order to use **all** available time on each machine required Gonzalez Office Supplies to produce 5 Secretarial, 9 Executive, and 8 Presidential chairs each week. Given the profit per chair in Example 3, this production schedule would result in a profit of $40(5) + 80(9) + 70(8) = \$1480$, which is less than the \$1530 we found in Example 3. Example 3 illustrates that through optimal use of the equipment a company can maximize its profit while reducing wear and tear on machines.

Technology Option

Simplex Method on a Calculator

Several computer programs can perform the simplex method. A popular one is LINDO. In Appendix C we supply a calculator program for the TI-83 that will perform the simplex method. This program may be useful, especially when many variables are involved, as in Example 3. If you have a calculator different than a TI-83, consult www.prenhall.com/armstrong for the program for your calculator.

Multiple Solutions and Unbounded Feasible Region

We saw in Chapter 3 that some linear programming problems have an infinite number of solutions. We also saw that if the feasible region is unbounded there may be no optimal (maximum) value. When using the simplex method, here is how to recognize these situations:

- A linear programming problem may have an infinite number of solutions if the bottom row to the left of the vertical dashed line of the final tableau contains a zero in a column that is not a basic variable column.
- A linear programming problem has no solution if the simplex method fails at some point in the process.

An example of the last bullet item is if, when constructing the quotients, we have no nonnegative quotients. In this case, the problem has no solution.

SUMMARY

In this section we continued the process established in Section 4.1. This process of solving linear programming problems is called the **simplex method**. We saw how selecting the **pivot column** led to choosing the correct **pivot row** by constructing quotients. The intersection of the pivot column and the pivot row is the **pivot element**. We then supplied a Five-step process for using the simplex method.

Simplex Method for Solving Standard Maximum Form Problems

1. Write the initial simplex tableau. This involves the following:
   - Determining the objective function
   - Determining all constraints
   - Using slack variables to write each constraint in equational form
2. Determine the pivot element by finding the pivot column and constructing the necessary quotients to determine the pivot row.
3. Use row operations to pivot about the element found in step 2.
4. If the indicators are all nonnegative, this is the final tableau. Proceed to step 5. If the indicators are not all nonnegative, go back to step 2.
5. Read the optimal solution(s) from the final tableau.

SECTION 4.2 EXERCISES

In Exercises 1–10, the initial simplex tableau of a linear programming problem is given. Use the simplex method to solve.

1.
$$\begin{array}{ccccc}x_1 & x_2 & s_1 & s_2 & z\end{array}$$
$$\left[\begin{array}{ccccc|c} 1 & 1 & 1 & 0 & 0 & 5 \\ 1 & 2 & 0 & 1 & 0 & 2 \\ \hline -4 & -5 & 0 & 0 & 1 & 10 \end{array}\right]$$

2.
$$\begin{array}{ccccc}x_1 & x_2 & s_1 & s_2 & z\end{array}$$
$$\left[\begin{array}{ccccc|c} 6 & 2 & 1 & 0 & 0 & 5 \\ 11 & 5 & 0 & 1 & 0 & 11 \\ \hline 4 & -3 & 0 & 0 & 1 & 15 \end{array}\right]$$

3.
$$\begin{array}{ccccc}x_1 & x_2 & s_1 & s_2 & z\end{array}$$
$$\left[\begin{array}{ccccc|c} 1 & 4 & 1 & 0 & 0 & 6 \\ 2 & 7 & 0 & 1 & 0 & 10 \\ \hline -3 & -7 & 0 & 0 & 1 & 14 \end{array}\right]$$

4.
$$\begin{array}{ccccc}x_1 & x_2 & s_1 & s_2 & z\end{array}$$
$$\left[\begin{array}{ccccc|c} 1 & 3 & 1 & 0 & 0 & 4 \\ 2 & 4 & 0 & 1 & 0 & 7 \\ \hline -6 & -2 & 0 & 0 & 1 & 12 \end{array}\right]$$

5.
$$\begin{array}{cccccc}x_1 & x_2 & s_1 & s_2 & s_3 & z\end{array}$$
$$\left[\begin{array}{cccccc|c} -1 & 2 & 1 & 0 & 0 & 0 & 3 \\ 3 & 8 & 0 & 1 & 0 & 0 & 6 \\ 7 & 5 & 0 & 0 & 1 & 0 & 8 \\ \hline -2 & -7 & 0 & 0 & 0 & 1 & 0 \end{array}\right]$$

6.
$$\begin{array}{cccccc}x_1 & x_2 & s_1 & s_2 & s_3 & z\end{array}$$
$$\left[\begin{array}{cccccc|c} 2 & 5 & 1 & 0 & 0 & 0 & 3 \\ 3 & 6 & 0 & 1 & 0 & 0 & 8 \\ 4 & 2 & 0 & 0 & 1 & 0 & 15 \\ \hline -1 & -2 & 0 & 0 & 0 & 1 & 0 \end{array}\right]$$

7.
$$\begin{array}{cccccc}x_1 & x_2 & x_3 & s_1 & s_2 & z\end{array}$$
$$\left[\begin{array}{cccccc|c} 10 & 4 & 2 & 1 & 0 & 0 & 90 \\ 6 & 3 & 4 & 0 & 1 & 0 & 40 \\ \hline -2 & -3 & -8 & 0 & 0 & 1 & 0 \end{array}\right]$$

8.
$$\begin{array}{cccccc}x_1 & x_2 & x_3 & s_1 & s_2 & z\end{array}$$
$$\left[\begin{array}{cccccc|c} 3 & 6 & 1 & 1 & 0 & 0 & 13 \\ 5 & 2 & 2 & 0 & 1 & 0 & 15 \\ \hline -8 & -4 & -10 & 0 & 0 & 1 & 0 \end{array}\right]$$

9.

$$\begin{array}{cccccccc|c} x_1 & x_2 & x_3 & x_4 & s_1 & s_2 & s_3 & z & \\ \hline 1 & 5 & 0 & 0 & 1 & 0 & 0 & 0 & 9 \\ 0 & 3 & 0 & 1 & 0 & 1 & 0 & 0 & 13 \\ 0 & 2 & 1 & 0 & 0 & 0 & 1 & 0 & 27 \\ \hline 2 & -2 & -1 & 0 & 0 & 0 & 0 & 1 & 80 \end{array}$$

10.

$$\begin{array}{cccccccc|c} x_1 & x_2 & x_3 & x_4 & s_1 & s_2 & s_3 & z & \\ \hline 6 & 0 & 1 & 2 & 1 & 0 & 0 & 0 & 5 \\ 2 & 1 & 0 & 10 & 0 & 1 & 0 & 0 & 10 \\ 3 & 0 & 2 & 1 & 0 & 0 & 1 & 0 & 8 \\ \hline -10 & -5 & 0 & -2 & 0 & 0 & 0 & 1 & 20 \end{array}$$

For the standard maximum for linear programming problems in Exercises 11–26:

(a) Introduce slack variables and set up the initial simplex tableau.

(b) Use the simplex method to solve.

✓ **11.** Maximize: $z = 2x_1 + x_2$
Subject to: $x_1 + x_2 \le 6$
$x_1 - x_2 \le 2$
$x_1 \ge 0, x_2 \ge 0$

12. Maximize: $z = 3x_1 + x_2$
Subject to: $4x_1 + 5x_2 \le 9$
$2x_1 - x_2 \le 2$
$x_1 \ge 0, x_2 \ge 0$

13. Maximize: $z = 2x_1 + 3x_2$
Subject to: $3x_1 + 2x_2 \le 8$
$x_1 - x_2 \le 7$
$x_1 \ge 0, x_2 \ge 0$

14. Maximize: $z = x_1 + 2x_2$
Subject to: $x_1 + x_2 \le 7$
$2x_1 + 3x_2 \le 16$
$x_1 \ge 0, x_2 \ge 0$

15. Maximize: $z = x_1 + 4x_2$
Subject to: $2x_1 + x_2 \le 32$
$x_1 + x_2 \le 18$
$x_1 + 3x_2 \le 36$
$x_1 \ge 0, x_2 \ge 0$

16. Maximize: $z = 5x_1 + 7x_2$
Subject to: $3x_1 - x_2 \le 8$
$2x_1 + 4x_2 \le 5$
$x_1 + x_2 \le 7$
$x_1 \ge 0, x_2 \ge 0$

17. Maximize: $z = 2x_1 + 2x_2$
Subject to: $3x_1 - 2x_2 \le 0$
$x_1 + x_2 \le 5$
$2x_1 + x_2 \le 6$
$x_1 \ge 0, x_2 \ge 0$

18. Maximize: $z = 2x_1 + 2x_2$
Subject to: $x_1 + x_2 \le 3$
$x_1 - 2x_2 \le 0$
$2x_1 + x_2 \le 2$
$x_1 \ge 0, x_2 \ge 0$

19. Maximize: $z = 5x_1 + x_2$
Subject to: $-x_1 + 2x_2 \le 10$
$2x_1 + x_2 \le 15$
$5x_1 + x_2 \le 8$
$-x_1 + x_2 \le 1$
$x_1 \ge 0, x_2 \ge 0$

20. Maximize: $z = 3x_1 + 5x_2$
Subject to: $-x_1 + 2x_2 \le 10$
$4x_1 + 2x_2 \le 15$
$x_1 + 3x_2 \le 8$
$6x_1 - x_2 \le 4$
$x_1 \ge 0, x_2 \ge 0$

21. Maximize: $z = 3x_1 + 2x_2 + 4x_3$
Subject to: $2x_1 + x_2 + x_3 \le 10$
$5x_1 - x_2 + 2x_3 \le 8$
$x_1 \ge 0, x_2 \ge 0, x_3 \ge 0$

22. Maximize: $z = 2x_1 + x_2 + 3x_3$
Subject to: $3x_1 - x_2 + 2x_3 \le 4$
$2x_1 + x_2 - x_3 \le 7$
$x_1 \ge 0, x_2 \ge 0, x_3 \ge 0$

23. Maximize: $z = 2x_1 + 5x_2 + x_3$
Subject to: $5x_1 + 2x_2 + x_3 \le 20$
$x_1 + 3x_2 + 5x_3 \le 13$
$4x_1 + x_2 + 3x_3 \le 15$
$x_1 \ge 0, x_2 \ge 0, x_3 \ge 0$

24. Maximize: $z = 3x_1 + x_2 + 5x_3$
Subject to: $2x_1 + x_2 + 4x_3 \le 8$
$4x_1 - x_2 + 3x_3 \le 15$
$12x_1 + 2x_2 + 7x_3 \le 12$
$x_1 \ge 0, x_2 \ge 0, x_3 \ge 0$

25. Maximize: $z = 4x_1 + 2x_2 + 3x_3$
Subject to: $2x_1 + 3x_3 \le 8$
$3x_1 + 2x_2 \le 5$
$x_1 \ge 0, x_2 \ge 0, x_3 \ge 0$

26. Maximize: $z = 9x_1 + x_2 + x_3$
Subject to: $4x_1 + x_3 \le 7$
$2x_1 + 3x_2 \le 9$
$x_1 \ge 0, x_2 \ge 0, x_3 \ge 0$

Applications

In the remainder of the exercises, set up and solve each using the simplex method. Interpret solutions as shown in Example 3. Include an interpretation of any nonzero slack variables.

27. The Young Furniture Company produces tables and chairs. Each table requires 2 hours of painting and 4 hours of carpentry work and yields a profit of \$70. Each chair requires 1 hour painting and 3 hours of carpentry work and yields a profit of \$50. The maximum of painting hours and carpentry hours available each week is 100 and 240, respectively. How many of each should be produced each week to maximize profit?

28. (*continuation of Exercise 27*) Due to weak demand, the Young Furniture Company must lay off some employees. As a result, the maximum number of painting hours and carpentry hours available each week is 50 and 110, respectively. Assuming that all other data remain the same as in Exercise 27, how many tables and chairs should be produced each week to maximize profit?

29. Due to new regulations approved by the Environmental Protection Agency (EPA), the Henderson Chemical Company was required to install a new process to reduce the pollution created when making a particular chemical. The older process releases 4 grams of sulfur and 3 grams of particulates for each gallon of chemical produced. The newer process releases 2 grams of sulfur and 1 gram of particulates for each gallon of chemical produced. The company realizes a profit of \$0.40 for each gallon produced under the old process and \$0.16 for each gallon produced under the new process. The new EPA regulations do not allow for more than 10,000 grams of sulfur and 7000 grams of particulates to be released daily. How many gallons of this chemical should be produced using the old process to maximize profits and still satisfy the new EPA regulations?

30. (*continuation of Exercise 29*) Barb, the plant manager for Henderson Chemical Company, has learned that the EPA is going to relax the new regulations so that no more than 13,000 grams of sulfur can be released daily. Assuming that all other data remain the same as in Exercise 29, how many gallons of this chemical should be produced using the old process to maximize profits and still satisfy the new EPA regulations?

31. Randy and Rusty have decided to enter the field of personal digital assistants (PDAs). Initially, they plan to design two types of PDAs, the WOW 1000 and the WOW 1500. Because of interest in PDAs, they can sell all that they can produce. However, they need to determine a production rate so as to satisfy various limits with a small production crew. These include a maximum of 150 hours per week for assembly and 100 hours per week for testing. The WOW 1000 requires 3 hours of assembly and 3 hours of testing and the WOW 1500 requires 6 hours of assembly and 3.5 hours of testing. Profit for the WOW 1000 is estimated at \$100 per unit and the profit for the WOW 1500 is estimated at \$150 per unit. How many of each type of PDA should be made and sold to maximize weekly profits?

32. (*continuation of Exercise 31*) Randy and Rusty hire some new employees and can, as a result, increase production. By doing so, they now have a maximum of 210 hours per week for assembly and 160 hours per week for testing. Assuming that all other data remain the same as in Exercise 31, how many of each type of PDA should be made and sold to maximize weekly profits?

33. Table 4.2.3 gives data for growing artichokes and asparagus for an entire growing season in California (entries are in lb/acre).

Table 4.2.3

	Artichokes	Asparagus
Nitrogen	197	240
Potash	123	42

SOURCE: U.S. Department of Agriculture, www.usda.gov/nass

Franco has 150 acres available suitable for cultivating crops. He has purchased a contract for a total of 31,880 pounds of nitrogen and 9120 pounds of potash. He expects to make a profit of \$140 per acre on artichokes and \$160 per acre on asparagus.

(a) How many acres of each should he plant to maximize his profit?

(b) What is the maximum profit?

34. Table 4.2.4 gives data for growing cauliflower and head lettuce for an entire growing season in California (entries are in lb/acre).

Table 4.2.4

	Cauliflower	Head Lettuce
Nitrogen	215	173
Phosphate	98	122

SOURCE: U.S. Department of Agriculture, www.usda.gov/nass

Sonja has 210 acres available for cultivating crops. She has purchased a contract for a total of 38,380 pounds of nitrogen and 22,240 pounds of phosphate. She expects to make a profit of \$90 per acre on cauliflower and \$100 per acre on head lettuce.

(a) How many acres of each should she plant to maximize her profit?

(b) What is the maximum profit?

35. The Smolko Tent Company makes three types of tents: two-person, four-person and six-person models.

Each tent requires the services, in hours, of three departments as shown in Table 4.2.5. The fabric cutting, assembly, and packaging departments have available a maximum of 420, 310, and 180 hours each week, respectively.

Table 4.2.5

	Two-person	Four-person	Six-person
Fabric cutting	0.5	0.8	1.1
Assembly	0.6	0.7	0.9
Packaging	0.2	0.3	0.6

If the per unit profit is \$30, \$50, and \$80 for the two-person, four-person and six-person tent, respectively, how many of each type tent should be made and sold to maximize weekly profit?

36. Kathy's Secretarial Services has a 1-month intensive training program for receptionist/secretaries for three types of office positions: medical field, small business, and executive secretarial. Each individual receives training in three departments at Kathy's: word processing, telephone and LAN (local area network), and ethics. Table 4.2.6 shows the number of units of training that each type of receptionist/secretary receives.

Table 4.2.6

	Medical Field	Small Business	Executive Secretary
Word processing	0.8	1.1	1.7
Telephone and LAN	0.6	1.3	1.5
Ethics	1.2	0.8	1.3

Each month, word processing, telephone and LAN, and ethics have available a maximum of 32.1 units, 29.7 units, and 33 units, respectively. The profit that Kathy's realizes from each individual admitted to medical field, small business, and executive secretary is \$1000, \$1000, and \$1400, respectively. How many individuals should be admitted each month to each program to maximize monthly profit?

37. The data in Table 4.2.7 give the amount of nitrogen, phosphate, and potash (all in lb/acre) needed to grow asparagus, cauliflower, and celery for an entire growing season in California.

Table 4.2.7

	Asparagus	Cauliflower	Celery
Nitrogen	240	215	336
Phosphate	148	98	171
Potash	42	69	147

SOURCE: U.S. Department of Agriculture, www.usda.gov/nass

If Eldrick has 280 acres and can get no more than 65,970 pounds of nitrogen, 31,860 pounds of phosphate, and 21,630 pounds of potash, and if he expects a profit of \$160 per acre for asparagus, \$90 per acre for cauliflower, and \$80 per acre for celery, how many acres should he allot for each crop to maximize profit?

38. The data in Table 4.2.8 give the amount of nitrogen, phosphate, and potash (all in lb/acre) needed to grow head lettuce, cantaloupe, and bell peppers for an entire growing season in California.

Table 4.2.8

	Head Lettuce	Cantaloupe	Bell Peppers
Nitrogen	173	175	212
Phosphate	122	161	108
Potash	76	39	72

SOURCE: U.S. Department of Agriculture, www.usda.gov/nass

If Maria has 200 acres and can get no more than 38,430 pounds of nitrogen, 27,845 pounds of phosphate, and 12,985 pounds of potash, and if she expects a profit of \$100 per acre for head lettuce, \$90 per acre for cantaloupe, and \$85 per acre for bell peppers, how many acres should she allot for each crop to maximize profit?

39. Table 4.2.9 gives nutritional information for selected Hershey's bites products.

Table 4.2.9

Product	Serving Size	Calories	Protein	Carbohydrates (grams)	Total fat
Reeses cups	7 pieces	90	1	9	6
Cookies 'n' creme	8 pieces	90	2	10	5
York peppermint	9 pieces	90	0	19	1.5

SOURCE: www.hersheys.com

Meagan desires a snack consisting of these three candies. She wants total calories to not exceed 700, total fat to not exceed 36 grams, and total carbohydrates to not exceed 80 grams. How many servings of each should she have if she wants to maximize the total grams of protein?

40. Justin desires a snack consisting of the three candies given in Table 4.2.9. He wants total calories to not exceed 700, total fat to not exceed 30 grams, and total carbohydrates to not exceed 80 grams. How many servings of each should he have if he wants to maximize the total grams of protein?

41. Julia makes and sells three types of bread, Lemon Poppy Seed Glaze, Pumpkin Glaze, and Banana Walnut

Glaze. A loaf of Lemon Poppy Seed requires 1 cup of flour, 3 cups of sugar, and 2 eggs. A loaf of Pumpkin Glaze requires 2 cups of flour, 2 cups of sugar, and 2 eggs. A loaf of Banana Walnut Glaze requires 2 cups of flour, 3 cups of sugar, and 1 egg. Julia makes a profit of \$1 per loaf of Lemon Poppy Seed Glaze, \$1.50 per loaf of Pumpkin Glaze, and \$1.50 per loaf of Banana Walnut Glaze. If Julia has available 24 cups of flour, 45 cups of sugar, and 27 eggs, how many of each type of bread should she make to maximize her profit?

42. Rework Exercise 40 except now assume that Julia has available 30 cups of flour, 56 cups of sugar, and 34 eggs.

SECTION PROJECT

Although it is not necessary, the graphing calculator program in Appendix C called SIMPLEX *may be beneficial for this section project.*

Shaver Frames manufactures three types of bicycle frames. The types of frames and the number of minutes that each frame requires on three different machines are given in Table 4.2.10.

Table 4.2.10

	Racing Frame	Hybrid Frame	Mountain Frame
Machine A	2	3	4
Machine B	5	6	5
Machine C	10	5	4

In any given week, machine A has available a maximum of 130 hours; machine B, 90 hours; and machine C, 110 hours. Shaver Frames realizes a profit of \$14 for each racing frame, \$13.50 for each hybrid frame, and \$11.25 for each mountain frame. Shaver Frames desires to make no more than 560 mountain frames and no more than 1200 racing and mountain frames combined. Let x_1 represent the number of racing frames produced, x_2 the number of hybrid frames produced, and x_3 the number of mountain frames produced.

(a) Determine the number of each type of frame that Shaver Frames should produce to maximize profit?

(b) What is the maximum profit?

(c) Interpret the values of all variables, including the slack variables. (*Hint:* Five slack variables are needed and you should convert the hours to minutes.)

Section 4.3 Simplex Method: Standard Minimum Form Problems and Duality

In the first two sections of this chapter we have seen how to use the simplex method to solve linear programming problems in standard maximum form. In this section we learn how to solve linear programming problems in **standard minimum form**. This will quickly lead us to study what is called the **dual**, and we will see how duality plays a role in economics using what is known as **sensitivity analysis**.

Standard Minimum Form

Linear programming problems of the form

$$\begin{aligned} \text{Minimize:} \quad & w = 6y_1 + 7y_2 \\ \text{Subject to:} \quad & 3y_1 + y_2 \geq 2 \\ & y_1 + 2y_2 \geq 0 \\ & y_1 \geq 0,\ y_2 \geq 0 \end{aligned}$$

are said to be in **standard minimum form**. Notice that we use $y_1, y_2, \ldots$ as decision variables rather than $x_1, x_2, \ldots$. The reason for this will be apparent later.

Standard Minimum Form

A linear programming problem is said to be in **standard minimum form** if the following are true:

1. The objective function is linear and must be **minimized.**
2. All variables are nonnegative.
3. All constraints, except the nonnegative constraints, are written as a linear expression greater than or equal to a nonnegative constant. Mathematically, this means

$$a_1 y_1 + a_2 y_2 + a_3 y_3 + \cdots + a_n y_n \geq b, \qquad b \geq 0$$

Note: The differences between standard maximum form and standard minimum form are that in a standard minimum form problem the objective function is to be **minimized** and the constraints have the $\geq$ symbol (with the constants to the right).

Constructing the Dual

The note following the box defining standard minimum form linear programming problems shows that standard maximum form and standard minimum form problems are somewhat similar. The similarities do not stop there. In fact, we will soon see that the solution of a standard maximum form problem yields a solution to an associated standard minimum form problem, and the solution to a standard minimum form problem gives a solution to an associated standard maximum form problem. We call these associated problems the **dual** of the other. The power of duals is that we can solve standard minimum form problems using the simplex method on its associated standard maximum form problem. But, in order to do this, we must first see how to construct the dual. To illustrate how we form the dual problem, we consider the following minimization problem:

$$\begin{aligned} \text{Minimize:} \quad & w = 6y_1 + 7y_2 \\ \text{Subject to:} \quad & 3y_1 + y_2 \geq 2 \\ & 4y_1 + 2y_2 \geq 3 \\ & y_1 \geq 0,\ y_2 \geq 0 \end{aligned}$$

In general, the given problem is called the **primal problem**. (The objective function and the inequality constraints associated with the primal problem are called the **primal objective function** and **primal inequalities**, respectively). The first step in forming the dual problem is to rewrite the problem constraints and objective function. We will use this rewritten form to construct an augmented matrix.

Rewritten Form	*Augmented Matrix*
$3y_1 + y_2 \geq 2$ $4y_1 + 2y_2 \geq 3$ $6y_1 + 7y_2 = w$	$A = \left[\begin{array}{cc|c} 3 & 1 & 2 \\ 4 & 2 & 3 \\ \hline 6 & 7 & 0 \end{array}\right]$

Since we have not introduced slack variables, matrix A is **not** a simplex tableau. The solid vertical and horizontal bars are used to make this **important** distinction.

From Your Toolbox

For a given matrix A, if we interchange the rows and columns, we produce what is called the **transpose of A** and denote it with A^T.

Next we form the transpose of A, denoted A^T. Consult the Toolbox to the left.

$$A = \left[\begin{array}{cc|c} 3 & 1 & 2 \\ 4 & 2 & 3 \\ \hline 6 & 7 & 0 \end{array}\right] \qquad A^T = \left[\begin{array}{cc|c} 3 & 4 & 6 \\ 1 & 2 & 7 \\ \hline 2 & 3 & 0 \end{array}\right]$$

Now A^T corresponds to a new linear programming problem that is in standard maximum form. Using different variables, x_1 and x_2, so as to avoid confusion, the linear programming problem A^T corresponds to is

$$\begin{aligned} 3x_1 + 4x_2 &\le 6 \\ x_1 + 2x_2 &\le 7 \\ 2x_1 + 3x_2 &= z \end{aligned} \qquad A^T = \left[\begin{array}{cc|c} 3 & 4 & 6 \\ 1 & 2 & 7 \\ \hline 2 & 3 & 0 \end{array}\right]$$

To recap, we have:

Primal Problem		*Dual Problem*	
Minimize:	$w = 6y_1 + 7y_2$	Maximize:	$z = 2x_1 + 3x_2$
Subject to:	$3y_1 + y_2 \ge 2$	Subject to:	$3x_1 + 4x_2 \le 6$
	$4y_1 + 2y_2 \ge 3$		$x_1 + 2x_2 \le 7$
	$y_1 \ge 0,\ y_2 \ge 0$		$x_1 \ge 0,\ x_2 \ge 0$

We now summarize the procedure for constructing the dual.

Forming the Dual Problem

Given a standard minimum form linear programming problem:

Step 1: Use the coefficients and constants in the primal problem constraints and the objective function to construct an augmented matrix A. Place the coefficients of the objective function in the bottom row.

Step 2: Determine A^T, the transpose of matrix A.

Step 3: Use the rows of A^T to construct a standard maximum form linear programming problem.

Example 1 Constructing the Dual

Form the dual problem for

$$\begin{aligned} \text{Minimize:}\quad & w = 3y_1 + 7y_2 \\ \text{Subject to:}\quad & 4y_1 + y_2 \ge 8 \\ & 5y_1 + 2y_2 \ge 1 \\ & y_1 \ge 0,\ y_2 \ge 0 \end{aligned}$$

Solution

Step 1: We begin by forming matrix A.

$$A = \left[\begin{array}{cc|c} 4 & 1 & 8 \\ 5 & 2 & 1 \\ \hline 3 & 7 & 0 \end{array}\right]$$

Step 2: A^T, the transpose of A, is

$$A^T = \left[\begin{array}{cc|c} 4 & 5 & 3 \\ 1 & 2 & 7 \\ \hline 8 & 1 & 0 \end{array}\right]$$

Step 3: The dual of the given linear programming problem is the standard maximum form problem

Maximize: $z = 8x_1 + x_2$

Subject to:
$$4x_1 + 5x_2 \le 3$$
$$x_1 + 2x_2 \le 7$$
$$x_1 \ge 0, x_2 \ge 0$$

■

✓Checkpoint 1 Now work Exercise 7.

Technology Option

Recall that A^T can be done on a graphing calculator. Figure 4.3.1 shows A on a graphing calculator and Figure 4.3.2 shows A^T.

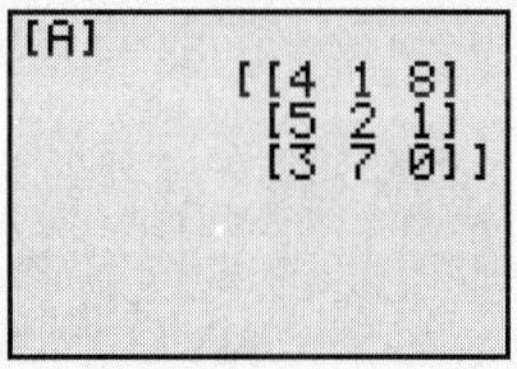

Figure 4.3.1

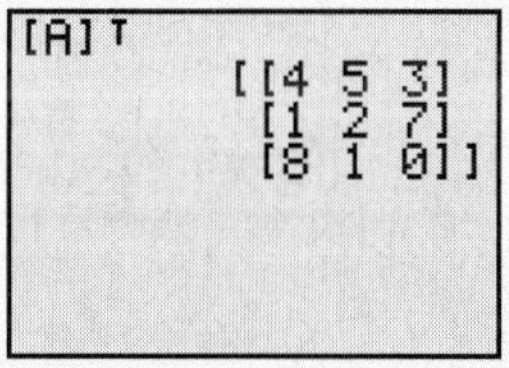

Figure 4.3.2

To review how to perform a matrix transpose on your calculator, consult the online calculator manual at www.prenhall.com/armstrong.

Solutions to Minimization Problems

To gain understanding on why duals are so important, let's consider the primal problem and its dual from the beginning of the section. We solve both in Example 2 using the graphical techniques from Chapter 3

Example 2 Solving Linear Programming Problems Graphically

Solve the following primal and dual problems by graphing the feasible region and evaluating the corner points in the objective function.

Primal Problem		*Dual Problem*	
Minimize:	$w = 6y_1 + 7y_2$	Maximize:	$z = 2x_1 + 3x_2$
Subject to:	$3y_1 + y_2 \ge 2$	Subject to:	$3x_1 + x_2 \le 6$
	$y_1 + 2y_2 \ge 3$		$x_1 + 2x_2 \le 7$
	$y_1 \ge 0, y_2 \ge 0$		$x_1 \ge 0, x_2 \ge 0$

Solution

We begin by graphing the feasible regions and labeling the corner points. See Figure 4.3.3 for the primal problem and Figure 4.3.4 for the dual problem. Notice in Figure 4.3.3 that the axes are labeled y_1 and y_2, whereas in Figure 4.3.4 the axes are labeled x_1 and x_2.

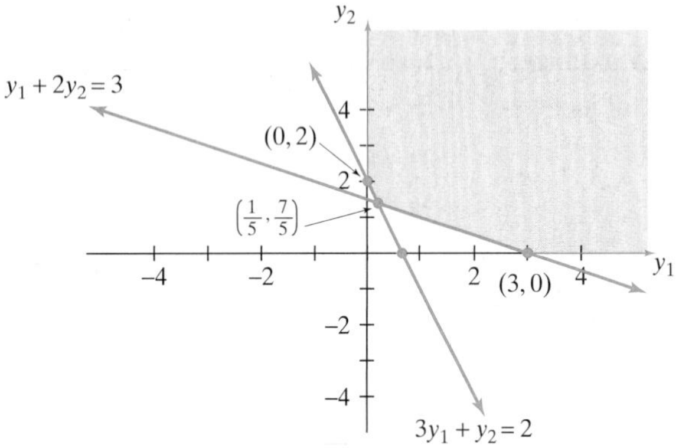

Figure 4.3.3

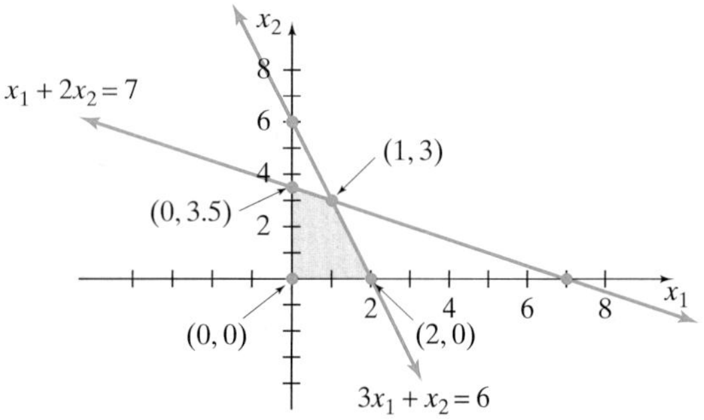

Figure 4.3.4

We now evaluate the objective function at each corner point. See Table 4.3.1 for the primal problem and Table 4.3.2 for the dual problem.

Table 4.3.1 Primal Problem

Corner Point	$w = 6y_1 + 7y_2$	
(0, 2)	$6(0) + 7(2) = 14$	
$(\frac{1}{5}, \frac{7}{5})$	$6(\frac{1}{5}) + 7(\frac{7}{5}) = 11$	**Minimum**
(3, 0)	$6(3) + 7(0) = 18$	

Table 4.3.2 Dual Problem

Corner Point	$z = 2x_1 + 3x_2$	
(0, 0)	$2(0) + 3(0) = 0$	
(0, 3.5)	$2(0) + 3(3.5) = 10.5$	
(1, 3)	$2(1) + 3(3) = 11$	**Maximum**
(2, 0)	$2(2) + 3(0) = 4$	

For the primal problem we have a minimum value of 11 at the corner point $(\frac{1}{5}, \frac{7}{5})$. For the dual problem we have a maximum value of 11 at the corner point (1, 3). ■

Notice in Example 2 that the two feasible regions are different and the corner points are different, but the optimal values for each objective function are equal. Both have optimal values of 11. This is not coincidental! Example 2 suggests the following theorem, which establishes the connection between the solution of a minimum problem and its dual. The proof of the theorem requires advanced techniques and is beyond the scope of this text.

Duality Theorem

The objective function of a minimum linear programming problem has an optimal (minimum) value if and only if the objective function of the associated dual maximum problem has an optimal (maximum) value. Furthermore, the optimal (minimum) value of the minimization problem equals the optimal (maximum) value of the associated dual maximization problem.

The duality theorem tells us that the optimal value of the minimization problem and its dual are equal. As Example 2 illustrated, these optimal values occurred at different corner points. Generally, we cannot determine the variable values that yield the optimal value for the minimum problem by examining the feasible region of the dual problem. But, since the dual is a maximum problem, we can apply the simplex method and find the optimal value and variable values that produce the optimal value to the original primal minimum problem, as we now illustrate in Example 3.

Example 3 Using the Simplex Method on the Dual

Use the simplex method to solve the dual problem from Example 2.

Solution

> **From Your Toolbox**
>
> The column with the most negative indicator is the **pivot column**. If there are two such indicators, use the leftmost indicator. To determine the pivot row, divide each number from the rightmost column (column of constants) by the corresponding number from the pivot column. The smallest nonnegative quotient gives the pivot row. If two quotients are equal (and smallest), arbitrarily choose the row nearest the top of the matrix.

Recall that the dual problem in Example 2 is in standard maximum form. After introducing slack variables, the equational form is

$$\begin{aligned} 3x_1 + x_2 + s_1 \quad &= 6 \\ x_1 + 2x_2 \quad + s_2 \quad &= 7 \\ -2x_1 - 3x_2 \quad\quad + z &= 0 \end{aligned}$$

The initial simplex tableau is

$$\begin{array}{ccccc|c} x_1 & x_2 & s_1 & s_2 & z & \end{array}$$
$$\left[\begin{array}{ccccc|c} 3 & 1 & 1 & 0 & 0 & 6 \\ 1 & 2 & 0 & 1 & 0 & 7 \\ \hline -2 & -3 & 0 & 0 & 1 & 0 \end{array}\right]$$

Using the techniques from Section 4.2 (consult the Toolbox to the left), we pivot on the circled 2.

$$\begin{array}{ccccc|c} x_1 & x_2 & s_1 & s_2 & z & \end{array} \qquad\qquad \begin{array}{ccccc|c} x_1 & x_2 & s_1 & s_2 & z & \end{array}$$
$$\left[\begin{array}{ccccc|c} 3 & 1 & 1 & 0 & 0 & 6 \\ 1 & \textcircled{2} & 0 & 1 & 0 & 7 \\ \hline -2 & -3 & 0 & 0 & 1 & 0 \end{array}\right] \begin{array}{c} \\ \tfrac{1}{2}R_2 \to R_2 \\ \\ \end{array} \left[\begin{array}{ccccc|c} 3 & 1 & 1 & 0 & 0 & 6 \\ \frac{1}{2} & 1 & 0 & \frac{1}{2} & 0 & \frac{7}{2} \\ \hline -2 & -3 & 0 & 0 & 1 & 0 \end{array}\right]$$

$$\begin{array}{ccccc|c} x_1 & x_2 & s_1 & s_2 & z & \end{array}$$
$$\begin{array}{r} -R_2 + R_1 \to R_1 \\ 3R_2 + R_3 \to R_3 \end{array} \left[\begin{array}{ccccc|c} \frac{5}{2} & 0 & 1 & -\frac{1}{2} & 0 & \frac{5}{2} \\ \frac{1}{2} & 1 & 0 & \frac{1}{2} & 0 & \frac{7}{2} \\ \hline -\frac{1}{2} & 0 & 0 & \frac{3}{2} & 1 & \frac{21}{2} \end{array}\right]$$

We now pivot on the circled $\frac{5}{2}$.

$$\begin{array}{ccccc|c} x_1 & x_2 & s_1 & s_2 & z & \end{array} \qquad\qquad \begin{array}{ccccc|c} x_1 & x_2 & s_1 & s_2 & z & \end{array}$$
$$\left[\begin{array}{ccccc|c} \textcircled{$\frac{5}{2}$} & 0 & 1 & -\frac{1}{2} & 0 & \frac{5}{2} \\ \frac{1}{2} & 1 & 0 & \frac{1}{2} & 0 & \frac{7}{2} \\ \hline -\frac{1}{2} & 0 & 0 & \frac{3}{2} & 1 & \frac{21}{2} \end{array}\right] \begin{array}{c} \\ \tfrac{2}{5}R_1 \to R_1 \\ \\ \end{array} \left[\begin{array}{ccccc|c} 1 & 0 & \frac{2}{5} & -\frac{1}{5} & 0 & 1 \\ \frac{1}{2} & 1 & 0 & \frac{1}{2} & 0 & \frac{7}{2} \\ \hline -\frac{1}{2} & 0 & 0 & \frac{3}{2} & 1 & \frac{21}{2} \end{array}\right]$$

$$\begin{array}{ccccc|c} x_1 & x_2 & s_1 & s_2 & z & \end{array}$$
$$\begin{array}{r} -\frac{1}{2}R_1 + R_2 \to R_2 \\ \frac{1}{2}R_1 + R_3 \to R_3 \end{array} \left[\begin{array}{ccccc|c} 1 & 0 & \frac{2}{5} & -\frac{1}{5} & 0 & 1 \\ 0 & 1 & -\frac{1}{5} & \frac{3}{5} & 0 & 3 \\ \hline 0 & 0 & \frac{1}{5} & \frac{7}{5} & 1 & 11 \end{array}\right]$$

As we saw in Example 2, the optimal (maximum) value is $z = 11$ and occurs at $x_1 = 1,\ x_2 = 3,$ and $s_1 = s_2 = 0$. ■

Look carefully at the bottom row in the final tableau in Example 3. In particular, the solution to the **primal minimization problem** in Example 2 is found in the bottom row and in the slack variable columns. As shown in Example 2 and Table 4.3.1, the primal minimization problem has a minimum value of 11 and it occurs at the corner point $(\frac{1}{5}, \frac{7}{5})$, precisely the values in the slack variable columns. This astonishing result is no accident!

An optimal solution to a primal minimum problem can be obtained from the bottom row of the final simplex tableau for the dual maximum problem.

We now have a procedure for solving a minimum problem. It requires applying the simplex method to its dual maximum problem. We summarize this procedure in the following box.

Solving Minimization Problems Using Duals

Given a standard minimum form linear programming problem:

1. Form the dual, which is a standard maximum form problem.
2. Solve the standard maximum form linear programming problem using the simplex method.
3. Read the solution to the minimum problem from the bottom row of the final simplex tableau in step 2. Specifically, the solution occurs in the slack variable columns.

Note: If the dual maximum problem has no optimal solution, then the primal minimum problem has no solution.

Example 4 Solving a Minimum Problem

Solve the following minimum problem by using the dual.

$$\begin{aligned}
\text{Minimize:} \quad & w = 3y_1 + 7y_2 \\
\text{Subject to:} \quad & 4y_1 + y_2 \geq 8 \\
& 5y_1 + 2y_2 \geq 1 \\
& y_1 \geq 0,\ y_2 \geq 0
\end{aligned}$$

Solution

This is the problem from Example 1, where we found the dual to be

$$\begin{aligned}
\text{Maximize:} \quad & z = 8x_1 + x_2 \\
\text{Subject to:} \quad & 4x_1 + 5x_2 \leq 3 \\
& x_1 + 2x_2 \leq 7 \\
& x_1 \geq 0,\ x_2 \geq 0
\end{aligned}$$

Introducing slack variables and converting to equational form yields

$$\begin{aligned}
4x_1 + 5x_2 + s_1 \qquad\qquad &= 3 \\
x_1 + 2x_2 \qquad + s_2 \qquad &= 7 \\
-8x_1 - x_2 \qquad\qquad + z &= 0
\end{aligned}$$

The initial simplex tableau is

$$\begin{array}{c} \begin{matrix} x_1 & x_2 & s_1 & s_2 & z \end{matrix} \\ \left[\begin{array}{rrrrr|r} 4 & 5 & 1 & 0 & 0 & 3 \\ 1 & 2 & 0 & 1 & 0 & 7 \\ \hline -8 & -1 & 0 & 0 & 1 & 0 \end{array}\right] \end{array}$$

We pivot on the 4 as follows:

$$\left[\begin{array}{rrrrr|r} \textcircled{4} & 5 & 1 & 0 & 0 & 3 \\ 1 & 2 & 0 & 1 & 0 & 7 \\ \hline -8 & -1 & 0 & 0 & 1 & 0 \end{array}\right] \xrightarrow{\frac{1}{4}R_1 \to R_1} \left[\begin{array}{rrrrr|r} 1 & \frac{5}{4} & \frac{1}{4} & 0 & 0 & \frac{3}{4} \\ 1 & 2 & 0 & 1 & 0 & 7 \\ \hline -8 & -1 & 0 & 0 & 1 & 0 \end{array}\right]$$

$$\begin{array}{l} -R_1 + R_2 \to R_2 \\ 8R_1 + R_3 \to R_3 \end{array} \left[\begin{array}{rrrrr|r} 1 & \frac{5}{4} & \frac{1}{4} & 0 & 0 & \frac{3}{4} \\ 0 & \frac{3}{4} & -\frac{1}{4} & 1 & 0 & \frac{25}{4} \\ \hline 0 & 9 & 2 & 0 & 1 & 6 \end{array}\right]$$

(columns: x_1, x_2, s_1, s_2, z)

This is our final tableau. We can read from the bottom row that the solution to the minimum problem is

$$\text{Minimum is } w = 6 \text{ at } y_1 = 2,\ y_2 = 0$$ ■

Interactive Activity

Use the final tableau in Example 4 to read the solution to the dual maximum problem.

WWW

Checkpoint 2

Now work Exercise 17.

Applications

Our first application is a Flashback to an example considered in Section 3.4.

Flashback

Wendy's Revisited

Table 4.3.3 gives nutritional information and current prices for selected items from Wendy's menu.

Table 4.3.3

Item	Protein (g)	Fiber (g)	Price
Breaded Chicken Sandwich	28	2	\$2.89
Biggie® fries	7	6	\$1.19

SOURCE: Wendy's, www.wendys.com

Joshua desires to eat nothing but Breaded Chicken Sandwiches and Biggie® fries. He wants at least 112 grams of protein and at least 30 grams of fiber daily. If Joshua eats nothing but Breaded Chicken Sandwiches and

Biggie® fries and wishes to meet his minimum nutrient requirements, how much of each food should he eat to minimize daily food costs?

Flashback Solution

Understand the Situation: We let y_1 represent the number of Breaded Chicken Sandwiches and y_2 represent the number of orders of Biggie® fries. The primal problem is

$$\begin{aligned} \text{Minimize:} \quad & C = 2.89y_1 + 1.19y_2 \\ \text{Subject to:} \quad & 28y_1 + 7y_2 \geq 112 \\ & 2y_1 + 6y_2 \geq 30 \\ & y_1 \geq 0,\ y_2 \geq 0 \end{aligned}$$

To get the dual maximum problem, we need the augmented matrix A for the given primal problem. Remember that the objective function goes on the bottom row.

$$A = \left[\begin{array}{cc|c} 28 & 7 & 112 \\ 2 & 6 & 30 \\ \hline 2.89 & 1.19 & 0 \end{array}\right]$$

We now determine A^T, the transpose of A, to be

$$A^T = \left[\begin{array}{cc|c} 28 & 2 & 2.89 \\ 7 & 6 & 1.19 \\ \hline 112 & 30 & 0 \end{array}\right]$$

This gives the dual maximum problem

$$\begin{aligned} \text{Maximize:} \quad & z = 112x_1 + 30x_2 \\ \text{Subject to:} \quad & 28x_1 + 2x_2 \leq 2.89 \\ & 7x_1 + 6x_2 \leq 1.19 \\ & x_1 \geq 0,\ x_2 \geq 0 \end{aligned}$$

Introducing slack variables and converting to equational form gives us

$$\begin{aligned} 28x_1 + 2x_2 + s_1 \qquad\qquad &= 2.89 \\ 7x_1 + 6x_2 \qquad + s_2 \qquad &= 1.19 \\ -112x_1 - 30x_2 \qquad\qquad + z &= 0 \end{aligned}$$

The initial simplex tableau is

$$\begin{array}{c} \begin{matrix} x_1 & x_2 & s_1 & s_2 & z \end{matrix} \\ \left[\begin{array}{ccccc|c} 28 & 2 & 1 & 0 & 0 & 2.89 \\ 7 & 6 & 0 & 1 & 0 & 1.19 \\ \hline -112 & -30 & 0 & 0 & 1 & 0 \end{array}\right] \end{array}$$

We leave the pivoting details for you in the Interactive Activity on p. 229. The final simplex tableau is

$$\begin{array}{c} \begin{matrix} x_1 & x_2 & s_1 & s_2 & z \end{matrix} \\ \left[\begin{array}{ccccc|c} 1 & 0 & 0.039 & -0.013 & 0 & 0.0971 \\ 0 & 1 & -0.0455 & 0.1818 & 0 & 0.085 \\ \hline 0 & 0 & 3 & 4 & 1 & 13.43 \end{array}\right] \end{array}$$

Entries rounded

Interactive Activity

Verify the final simplex tableau in the Flashback by performing the appropriate pivoting.

We can read from the bottom row that the solution to the primal minimum problem is

$$C = 13.43, \qquad y_1 = 3, \qquad y_2 = 4$$

Interpret the Solution: This means that to minimize daily food costs, while still meeting the daily nutrient requirements, Joshua should eat 3 Breaded Chicken Sandwiches and 4 orders of Biggie® fries. His daily minimum cost is $13.43.

Sensitivity Analysis and Shadow Variables

In some economic applications, the variables in the dual of a linear programming problem measure something useful and are called **shadow variables**. These variables can be used to measure the **sensitivity** of the optimal solution to small changes in the problem requirements.

For example, consider the scenario in the Flashback and Table 4.3.3. If we consider price to be price per unit (that is, price per Breaded Chicken Sandwich or price per Biggie® fries), then x_1 is the value of each gram of protein and x_2 is the value of each gram of fiber. The total value of a Breaded Chicken Sandwich is $28x_1 + 2x_2$, the value of protein plus the value of fiber. This value can be no greater than $2.89, the price of a Breaded Chicken Sandwich. So $28x_1 + 2x_2 \leq 2.89$. Similarly, for Biggie® fries we have $7x_1 + 6x_2 \leq 1.19$.

In the Flashback, Joshua wanted at least 112 grams of protein and at least 30 grams of fiber. It stands to reason that we would want to maximize $112x_1 + 30x_2$, which is the total value of protein and fiber. This means that we want to

$$\begin{aligned} \text{Maximize:} \quad & z = 112x_1 + 30x_2 \\ \text{Subject to:} \quad & 28x_1 + 2x_2 \leq 2.89 \\ & 7x_1 + 6x_2 \leq 1.19 \\ & x_1 \geq 0, x_2 \geq 0 \end{aligned}$$

But this is simply the dual maximum problem that we solved in the Flashback! The final tableau tells us that $x_1 = 0.0971$ and $x_2 = 0.085$. This means that a gram of protein is worth $0.0971 and a gram of fiber is worth $0.085. The maximum nutrient value is $13.43, the same as the minimum daily cost. Observe that the minimum daily cost, $13.43, can be found as follows:

$$\begin{aligned} (\$0.0971 \text{ per gram of protein}) \cdot (112 \text{ grams of protein}) &= \$10.8752 \\ + (\$0.085 \text{ per gram of fiber}) \cdot (30 \text{ grams of fiber}) &= \quad 2.55 \\ &= \$13.4252 \approx \$13.43 \end{aligned}$$

The numbers $x_1 = 0.0971$ and $x_2 = 0.085$ are called **shadow costs** of the nutrients. These two numbers from the dual (see the final tableau in the Flashback) measure the sensitivity of the optimal solution to changes in the requirements of protein and fiber. Each extra gram of protein costs about $0.0971 and each extra gram of fiber costs about $0.085. For example, if Joshua desires 113 grams of protein and 31 grams of fiber, his new total cost is

$$\begin{array}{rl} \$13.43 & \text{(112 grams of protein and 30 grams of fiber)} \\ 0.0971 & \text{(extra gram of protein)} \\ +\ 0.085 & \text{(extra gram of fiber)} \\ \hline \$13.6121 \approx \$13.61 & \end{array}$$

SUMMARY

In this section we saw the **standard minimum form** of a linear programming problem. We saw how the **transpose of a matrix** was key in constructing the **dual**. If the **primal** linear programming problem is a standard minimum form, then the dual is a standard maximum form. The primal and the dual have the same optimal value as stated in the Duality Theorem. We supplied a strategy for solving minimization problems using duals. The section concluded with an application and a look at **shadow variables**.

- **Duality Theorem** The objective function of a minimum linear programming problem has an optimal (minimum) value if and only if the objective function of the associated dual maximum problem has an optimal (maximum) value. Furthermore, the optimal (minimum) value of the minimization problem equals the optimal (maximum) value of the associated dual maximization problem.
- **Solving Minimum Problems** Given a standard minimum form linear programming problem:
  1. Form the dual, which is a standard maximum form problem.
  2. Solve the standard maximum form linear programming problem using the simplex method.
  3. Read the solution to the minimum problem from the bottom row of the final simplex tableau in step 2. Specifically, the solution occurs in the slack variable columns.

SECTION 4.3 EXERCISES

In Exercises 1–6 determine the transpose of the given matrix.

1. $\begin{bmatrix} 2 & 1 & 3 \\ 4 & 1 & 2 \end{bmatrix}$

2. $\begin{bmatrix} 3 & 2 & 1 & 0 & 2 \\ 1 & 4 & 0 & 1 & 0 \end{bmatrix}$

3. $\begin{bmatrix} 5 & 7 & 6 & 1 \\ 3 & 2 & 1 & 0 \\ 0 & 1 & 1 & 2 \end{bmatrix}$

4. $\begin{bmatrix} 1 & 3 \\ 2 & 4 \\ 5 & 7 \end{bmatrix}$

5. $\begin{bmatrix} 4 & 1 & 3 \\ 1 & 2 & 7 \\ 8 & 1 & 9 \end{bmatrix}$

6. $\begin{bmatrix} 2 & 1 \\ 3 & 9 \\ 8 & 0 \\ 5 & 4 \end{bmatrix}$

In Exercises 7–12, form the dual problem for the given linear programming problem. See Example 1.

✓ **7.** Minimize: $w = 2y_1 + 3y_2$
Subject to:
$5y_1 + 2y_2 \geq 4$
$3y_1 + y_2 \geq 7$
$y_1 \geq 0,\ y_2 \geq 0$

8. Minimize: $w = 7y_1 + 3y_2$
Subject to:
$8y_1 + 10y_2 \geq 15$
$y_1 + y_2 \geq 4$
$y_1 \geq 0,\ y_2 \geq 0$

9. Minimize: $w = y_1 + 3y_2 + 4y_3$
Subject to:
$2y_1 + y_2 + 4y_3 \geq 0$
$8y_1 + y_2 + y_3 \geq 7$
$5y_1 + 2y_2 + 7y_3 \geq 11$
$y_1 \geq 0,\ y_2 \geq 0,\ y_3 \geq 0$

10. Minimize: $w = 2y_1 + y_2 + 5y_3$
Subject to:
$y_1 + 3y_2 + y_3 \geq 9$
$2y_1 + 5y_2 + y_3 \geq 8$
$y_1 + 2y_2 + 4y_3 \geq 10$
$y_1 \geq 0,\ y_2 \geq 0,\ y_3 \geq 0$

11. Minimize: $w = 2y_1 + y_2 + 3y_3$
Subject to:
$5y_1 + y_2 \geq 8$
$3y_1 + 8y_3 \geq 7$
$y_2 + y_3 \geq 11$
$y_1 \geq 0,\ y_2 \geq 0,\ y_3 \geq 0$

12. Minimize: $w = 8y_1 + y_2 + 5y_3$
Subject to:
$y_1 + 3y_2 \geq 9$
$2y_1 + 4y_3 \geq 8$
$3y_2 + y_3 \geq 10$
$y_1 \geq 0,\ y_2 \geq 0,\ y_3 \geq 0$

In Exercises 13–16, a linear programming problem is given along with the final simplex tableau for the dual problem. Use the final simplex tableau for the dual to:

(a) Find the solution to the primal problem.

(b) Find the solution to the dual problem.

13. Minimize: $w = y_1 + 5y_2$
Subject to:
$y_1 + 2y_2 \geq 6$
$3y_1 + y_2 \geq 9$
$y_1 \geq 0,\ y_2 \geq 0$

$$\begin{array}{ccccc|c} x_1 & x_2 & s_1 & s_2 & z & \\ \hline 1 & 3 & 1 & 0 & 0 & 1 \\ 0 & -5 & -2 & 1 & 0 & 3 \\ \hline 0 & 9 & 6 & 0 & 1 & 6 \end{array}$$

14. Minimize: $w = 3y_1 + y_2$
Subject to:
$2y_1 + 3y_2 \geq 6$
$y_1 + 2y_2 \geq 4$
$y_1 \geq 0,\ y_2 \geq 0$

$$\begin{array}{ccccc|c} x_1 & x_2 & s_1 & s_2 & z & \\ \hline 0 & -\frac{1}{3} & 1 & -\frac{2}{3} & 0 & \frac{7}{3} \\ 1 & \frac{2}{3} & 0 & \frac{1}{3} & 0 & \frac{1}{3} \\ \hline 0 & 0 & 0 & 2 & 1 & 2 \end{array}$$

15. Minimize: $w = 2y_1 + 3y_2$

Subject to:
$$y_1 + y_2 \geq 12$$
$$3y_1 + y_2 \geq 9$$
$$y_1 + 2y_2 \geq 6$$
$$y_1 \geq 0,\ y_2 \geq 0$$

$$\begin{array}{cccccc|c} x_1 & x_2 & x_3 & s_1 & s_2 & z & \\ \hline 1 & 3 & 1 & 1 & 0 & 0 & 2 \\ 0 & -2 & 1 & -1 & 1 & 0 & 1 \\ \hline 0 & 27 & 6 & 12 & 0 & 1 & 24 \end{array}$$

16. Minimize: $w = 4y_1 + y_2$

Subject to:
$$2y_1 + y_2 \geq 8$$
$$y_1 + 3y_2 \geq 12$$
$$3y_1 + 2y_2 \geq 6$$
$$y_1 \geq 0,\ y_2 \geq 0$$

$$\begin{array}{cccccc|c} x_1 & x_2 & x_3 & s_1 & s_2 & z & \\ \hline 0 & -5 & -1 & 1 & -2 & 0 & 2 \\ 1 & 3 & 2 & 0 & 1 & 0 & 1 \\ \hline 0 & 12 & 10 & 0 & 8 & 1 & 8 \end{array}$$

In Exercises 17–30:

(a) Construct the dual problem.

(b) Use the simplex method on the dual problem to find the solution to the original primal problem.

17. Minimize: $w = 7y_1 + 3y_2$

Subject to:
$$8y_1 + 10y_2 \geq 15$$
$$y_1 + y_2 \geq 4$$
$$y_1 \geq 0,\ y_2 \geq 0$$

18. Minimize: $w = 2y_1 + 3y_2$

Subject to:
$$5y_1 + 2y_2 \geq 4$$
$$3y_1 + y_2 \geq 7$$
$$y_1 \geq 0,\ y_2 \geq 0$$

19. Minimize: $w = y_1 + 4y_2$

Subject to:
$$y_1 + 2y_2 \geq 5$$
$$y_1 + 3y_2 \geq 6$$
$$y_1 \geq 0,\ y_2 \geq 0$$

20. Minimize: $w = 9y_1 + 2y_2$

Subject to:
$$4y_1 + y_2 \geq 13$$
$$3y_1 + y_2 \geq 12$$
$$y_1 \geq 0,\ y_2 \geq 0$$

21. Minimize: $w = 2y_1 + y_2$

Subject to:
$$y_1 + y_2 \geq 8$$
$$y_1 + 2y_2 \geq 4$$
$$y_1 \geq 0,\ y_2 \geq 0$$

22. Minimize: $w = 3y_1 + 9y_2$

Subject to:
$$2y_1 + y_2 \geq 8$$
$$y_1 + 2y_2 \geq 10$$
$$y_1 \geq 0,\ y_2 \geq 0$$

23. Minimize: $w = 4y_1 + 5y_2$

Subject to:
$$2y_1 + 3y_2 \geq 29$$
$$5y_1 + 4y_2 \geq 55$$
$$2y_1 + y_2 \geq 18$$
$$y_1 \geq 0,\ y_2 \geq 0$$

24. Minimize: $w = 5y_1 + 2y_2$

Subject to:
$$y_1 + 5y_2 \geq 15$$
$$2y_1 + 4y_2 \geq 24$$
$$y_1 + y_2 \geq 7$$
$$y_1 \geq 0,\ y_2 \geq 0$$

25. Minimize: $w = 2y_1 + 3y_2$

Subject to:
$$15y_1 + y_2 \geq 29$$
$$y_1 + 6y_2 \geq 20$$
$$8y_1 + 7y_2 \geq 78$$
$$y_1 \geq 0,\ y_2 \geq 0$$

26. Minimize: $w = 2y_1 + 7y_2$

Subject to:
$$y_1 + 4y_2 \geq 40$$
$$3y_1 + y_2 \geq 36$$
$$y_1 + y_2 \geq 5$$
$$y_1 \geq 0,\ y_2 \geq 0$$

27. Minimize: $w = 14y_1 + 8y_2 + 20y_3$

Subject to:
$$y_1 + y_2 + 3y_3 \geq 6$$
$$2y_1 + y_2 + y_3 \geq 9$$
$$y_1 \geq 0,\ y_2 \geq 0,\ y_3 \geq 0$$

28. Minimize: $w = 5y_1 + 7y_2 + 12y_3$

Subject to:
$$y_1 + y_2 + 2y_3 \geq 7$$
$$2y_1 + y_2 + y_3 \geq 4$$
$$y_1 \geq 0,\ y_2 \geq 0,\ y_3 \geq 0$$

29. Minimize: $w = 4y_1 + 10y_2 + 9y_3$

Subject to:
$$y_1 + 3y_2 + y_3 \geq 6$$
$$y_1 + y_2 + 3y_3 \geq 9$$
$$3y_1 + y_2 + y_3 \geq 3$$
$$y_1 \geq 0,\ y_2 \geq 0,\ y_3 \geq 0$$

30. Minimize: $w = 2y_1 + 3y_2 + 4y_3$

Subject to:
$$y_1 + 4y_2 + 2y_3 \geq 16$$
$$y_1 + 3y_2 + 4y_3 \geq 7$$
$$2y_1 + y_2 + 3y_3 \geq 6$$
$$y_1 \geq 0,\ y_2 \geq 0,\ y_3 \geq 0$$

Applications

31. Richie needs to supplement his diet with at least 50 milligrams (mg) of vitamin C and 5 mg of niacin daily. The nutrients are available in two supplement pills, AllDay and Live!. AllDay contains 5 mg of vitamin C and 1 mg of niacin, while Live! contains 10 mg of vitamin C and 2 mg of niacin. Each AllDay pill costs \$0.08 and each Live! pill costs \$0.09. Let y_1 represent the number of AllDay pills taken daily, and let y_2 represent the number of Live! pills taken daily.

(a) Write a linear programming problem to minimize the cost of adding the nutrients to Richie's diet.

(b) Write the dual of the linear programming problem in part (a).

(c) Use the simplex method on the dual problem in part (b) to solve the original problem in part (a). Your answer should include the number of each pill taken and the daily minimum cost.

32. Roberto needs to supplement his diet with at least 30 mg of vitamin C and 10 mg of niacin daily. The nutrients are available in two supplement pills, GoodDay and Now!. GoodDay contains 3 mg of vitamin C and 2 mg of niacin, while Now! contains 6 mg of vitamin C and 1 mg of niacin. Each GoodDay pill costs \$0.06 and each Now! pill costs \$0.08. Let y_1 represent the number of GoodDay pills taken daily, and let y_2 represent the number of Now! pills taken daily.

(a) Write a linear programming problem to minimize the cost of adding the nutrients to Roberto's diet.

(b) Write the dual of the linear programming problem in part (a).

(c) Use the simplex method on the dual problem in part (b) to solve the original problem in part (a). Your answer should include the number of each pill taken and the daily minimum cost.

33. Elle, a beagle breeder, can buy a special food, Shine, at \$0.30 per pound and a special food, Lustre, at \$0.40 per pound. Each pound of Shine has 210 units of protein and 150 units of fat. Each pound of Lustre has 240 units of protein and 380 units of fat. The minimum daily requirements for the beagles (collectively) are 2115 units of protein and 3460 units of fat. Let y_1 represent the number of pounds of Shine used daily and y_2 represent the number of pounds of Lustre used daily.

(a) Write a linear programming problem to minimize Elle's daily food costs while still meeting (or even exceeding) the minimum daily requirements.

(b) Write the dual of the linear programming problem in part (a).

(c) Use the simplex method on the dual problem in part (b) to solve the original problem in part (a). Your answer should include the number of pounds of each mix and the daily minimum cost.

34. (*continuation of Exercise 33*) Ginger, Elle's cousin from Exercise 33, also breeds beagles but in a different state. Ginger can buy Shine at \$0.40 per pound and Lustre at \$0.50 per pound. Everything else from Exercise 33 remains the same.

(a) Write a linear programming problem to minimize Ginger's daily food costs while still meeting (or even exceeding) the minimum daily requirements.

(b) Write the dual of the linear programming problem in part (a).

(c) Use the simplex method on the dual problem in part (b) to solve the original problem in part (a). Your answer should include the number of pounds of each mix and the daily minimum cost.

35. The AJA Mining Company owns two different mines that produce ore, which is crushed and graded into three classes: high, medium, and low grade. The company has a contract to provide a smelting plant with at least 13 tons of high-grade, 9 tons of medium-grade, and 24 tons of low-grade ore per week. The two mines have different operating characteristics as given in Table 4.3.4.

Table 4.3.4

		Production (tons per day)		
Mine	**Cost per Day (in \$1000's)**	**High**	**Medium**	**Low**
I	12	3	3	4
II	10	2	1	6

(a) Define the variables and write a linear programming problem to minimize AJA's total operating cost while still meeting (or even exceeding) the smelting plant contract.

(b) Write the dual of the linear programming problem in part (a).

(c) Use the simplex method on the dual problem in part (b) to solve the original problem in part (a). Your answer should include the number of days that each mine should operate and the minimum weekly cost.

36. (*continuation of Exercise 35*) Suppose that the contract with the smelting plant is changed so that the AJA Mining Company needs to provide at least 15 tons of high-grade, 11 tons of medium-grade, and 30 tons of low-grade ore per week. Everything else from Exercise 35 remains the same.

(a) Write a linear programming problem to minimize AJA's total operating cost while still meeting (or even exceeding) the smelting plant contract.

(b) Write the dual of the linear programming problem in part (a).

(c) Use the simplex method on the dual problem in part (b) to solve the original problem in part (a). Your answer should include the number of days that each mine should operate and the minimum weekly cost.

37. Hoofer Oil has two refineries, one in Findlay, Ohio, and one in Ft. Wayne, Indiana. Each refinery can refine the amounts of fuel, in barrels, in a 24-hour period as shown in Table 4.3.5.

Table 4.3.5

Maximum Production of:	Findlay	Ft. Wayne
Low-octane gasoline	800	1000
High-octane gasoline	500	600
Diesel	1000	600

Daily operating costs for the Findlay refinery are $28,000 and daily operating costs for the Ft. Wayne refinery are $33,000. Zachary, the production manager of Hoofer Oil, receives an order for 60,000 barrels of low-octane gasoline, 120,000 barrels of high-octane gasoline, and 150,000 barrels of diesel.

(a) How many days should Hoofer Oil operate each refinery to fulfill the order while minimizing cost?

(b) What is the minimum cost?

38. Shannon Oil has two refineries, one in Lima, Ohio, and one in Detroit, Michigan. Each refinery can refine the amounts of fuel, in barrels, in a 24-hour period as shown in Table 4.3.6.

Table 4.3.6

Maximum Production of:	Lima	Detroit
Low-octane gasoline	900	1100
High-octane gasoline	600	700
Diesel	1100	700

Daily operating costs for the Lima refinery are $35,000 and daily operating costs for the Detroit refinery are $30,000. Joshua, the production manager of Shannon Oil, receives an order for 50,000 barrels of low-octane gasoline, 90,000 barrels of high-octane gasoline, and 130,000 barrels of diesel.

(a) How many days should Shannon Oil operate each refinery to fulfill the order while minimizing cost?

(b) What is the minimum cost?

39. Anita, a dietitian at a local hospital, is to prepare a diet to contain at least 400 units of vitamins C and E, 500 units of minerals (such as zinc and iron), and 1400 calories. She has two foods available, Flakes and Woday. An ounce of Flakes contains 2 units of vitamin C and E, 1 unit of minerals, and 4 calories and costs $0.05. An ounce of Woday contains 1 unit of vitamins C and E, 2 units of minerals, and 4 calories and costs $0.03.

(a) How many ounces of each food should be used to minimize daily food cost?

(b) What is the minimum daily food cost?

40. Lisa, a dietitian at a local hospital, is to prepare two foods to meet certain nutritional requirements. Each pound of Vaca contains 200 units of vitamin C, 80 units of vitamin B_{12}, and 20 units of vitamin E and costs $0.40. Each pound of Carne contains 20 units of vitamin C, 160 units of vitamin B_{12}, and 10 units of vitamin E and costs $0.30. To meet the nutritional needs of the patients, the combination of the two foods must have at least 520 units of vitamin C, 640 units of vitamin B_{12}, and 100 units of vitamin E.

(a) How many pounds of each food should be used to minimize daily food cost?

(b) What is the minimum daily food cost?

41. Table 4.3.7 gives nutritional information in grams for selected items from Taco Bell's menu.

Table 4.3.7

Item	Protein	Fiber	Total Fat
Burrito Supreme®-Beef	17	9	18
Double Burrito Supreme®-Steak	28	3	18

SOURCE: Taco Bell, www.tacobell.com

Warren wants to eat nothing but Burrito Supreme®-Beef and Double Burrito Supreme®-Steak. He wants at least 118 grams of protein and at least 27 grams of fiber daily. If Warren eats nothing but Burrito Supreme®-Beef and Double Burrito Supreme®-Steak and meets his minimum nutrient requirements, answer the following.

(a) How much of each food should Warren eat to minimize his daily total fat intake?

(b) What is this minimum daily total fat intake?

42. Table 4.3.8 gives the nutritional information in grams for selected items from McDonald's menu.

Table 4.3.8

Item	Carbohydrates	Protein	Fat
Cheeseburger	35	16	13
Large French fries	68	8	26
Chocolate shake	60	11	9

SOURCE: McDonald's, www.mcdonalds.com

Diane wishes to eat nothing but cheeseburgers, large French fries, and chocolate shakes. If she wishes to have at least 450 grams of carbohydrates and 80 grams of protein a day, how many of each should she have to minimize her fat intake and still meet her requirements?

Exercises 43–46 pertain to shadow costs. These exercises refer back to previously worked exercises.

43. (a) Determine the shadow costs of the nutrients in Exercise 31.

(b) Use the shadow costs along with the minimum cost answer in Exercise 31(c) to determine the new total cost if Richie changes his nutrient requirements to 51 mg of vitamin C and 6 mg of niacin daily.

44. (a) Determine the shadow costs of the nutrients in Exercise 32.

(b) Use the shadow costs along with the minimum cost answer in Exercise 32(c) to determine the new total cost if Roberto changes his nutrient requirements to 31 mg of vitamin C and 11 mg of niacin daily.

45. (a) Determine the shadow costs of the nutrients in Exercise 33.

(b) Use the shadow costs along with the minimum cost answer in Exercise 33(c) to determine the new total cost if Elle changes the beagles' nutrient requirements to 2116 units of protein and 3461 units of fat daily.

46. (a) Determine the shadow costs of the nutrients in Exercise 34.

(b) Use the shadow costs along with the minimum cost answer in Exercise 33(c) to determine the new total cost if Ginger changes the beagles' nutrient requirements to 2116 units of protein and 3461 units of fat daily.

SECTION PROJECT

Ty and Mike are management analysts at a Houston, Texas, laboratory. Table 4.3.9 gives information that Ty and Mike need.

Table 4.3.9

Experiment	Biophysicist	Biochemist	Minimum Needed
Test 1	8	4	120
Test 2	4	6	115
Test 3	9	4	116

These numbers mean that a biophysicist can complete 8, 4, and 9 of Tests 1, 2, and 3, respectively, per hour. Similarly, a biochemist can complete 4, 6, and 4 of Tests 1, 2, and 3 per hour. The lab needs a minimum of 120 of Test 1, 115 of Test 2, and 116 of Test 3 done each day. A biophysicist costs \$23 per hour, and a biochemist costs \$18 per hour.

(a) Define the variables and write a linear programming problem to minimize costs.

(b) Write the dual of the problem in part (a).

(c) Use the simplex method on the dual in part (b) to solve the original problem in part (a). Your answer should include the number of hours for each scientist type and the minimum cost each day.

(d) Determine the solutions to the dual problem in part (b).

(e) Interpret the meaning of the dual and its solution.

(f) Are there any shadow variables in the final simplex tableau in part (a)? Explain.

Section 4.4 Simplex Method: Nonstandard Problems

In Section 4.2, all the linear programming problems that we encountered were of the standard maximum form variety, and in Section 4.3 we saw how the dual was instrumental in solving standard minimum form linear programming problems. In this section we discuss **nonstandard problems**, that is, problems with **mixed constraints** (constraints involving $\leq$ and $\geq$). We also discuss another technique for solving minimization problems. This discussion requires the introduction of a new type of variable called a **surplus variable**. We introduce mixed constraints through a Flashback.

Flashback

Dub Inc. Revisited

We were first introduced to Dub, Inc., in Section 3.4 (and revisited in Section 4.2). Recall that Dub, Inc., is a small manufacturing firm that produces two microwave switches, switch A and switch B. Each switch passes through two departments: assembly and testing. Switch A requires 4 hours of assembly and 1 hour of testing, and switch B requires 3 hours of assembly and 2 hours of testing. The maximum numbers of hours available each week for assembly and testing are 240 hours and 140 hours, respectively. The per unit profit for switch A is \$20, whereas the per unit profit for switch B is \$30. Now suppose that the total number of switches made each week must be at least 10. Graphically determine how many of each switch should be made and sold each week to maximize profit.

Flashback Solution

Understand the Situation: Since we do not know how many of each switch should be produced and sold each week, we let x_1 represent the number of units of switch A made each week and x_2 represent the number of units of switch B made each week. Our mathematical model for this linear programming problem is

$$\begin{aligned} \text{Maximize:} \quad & P = 20x_1 + 30x_2 \\ \text{Subject to:} \quad & 4x_1 + 3x_2 \le 240 \\ & x_1 + 2x_2 \le 140 \\ & x_1 + x_2 \ge 10 \\ & x_1 \ge 0, x_2 \ge 0 \end{aligned}$$

The feasible region is shown in Figure 4.4.1.

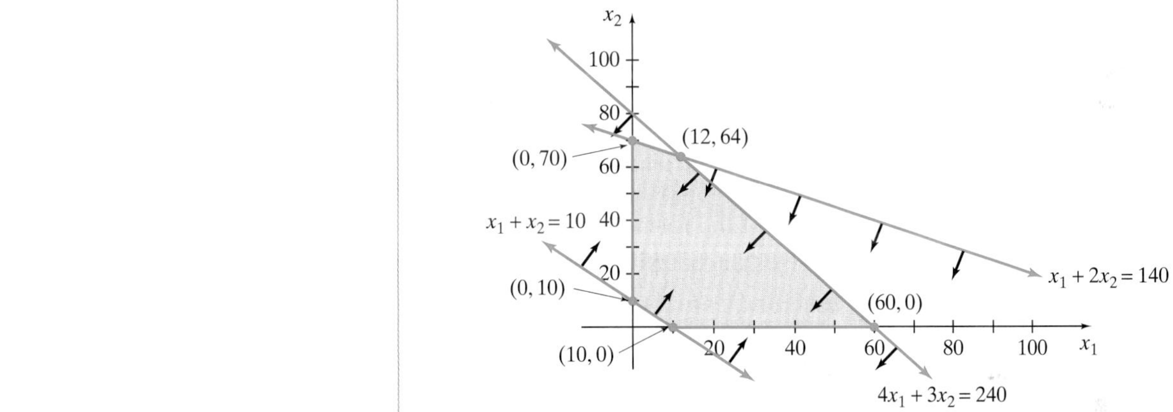

Figure 4.4.1

Table 4.4.1 shows the result of each corner point evaluated in the objective function.

Table 4.4.1

Corner Point	Value of $P = 20x_1 + 30x_2$	
(10, 0)	$20(10) + 30(0) = 200$	
(0, 10)	$20(0) + 30(10) = 300$	
(0, 70)	$20(0) + 30(70) = 2100$	
(12, 64)	$20(12) + 30(64) = 2160$	**Maximum**
(60, 0)	$20(60) + 30(0) = 1200$	

The optimal (maximum) value in Table 4.4.1 is 2160 and occurs at the corner point (12, 64).

Interpret the Solution: This means that to maximize weekly profit Dub, Inc., should produce and sell 12 units of switch A and 64 units of switch B. This generates a maximum weekly profit of \$2160.

Notice that the problem in the Flashback is not a standard maximum form problem since the third inequality, $x_1 + x_2 \geq 10$, does not have the $\leq$ symbol. However, if the total number of switches made is greater than 10 (and it was in the Flashback), then we have some *surplus* switches made. These surplus switches are switches beyond the necessary 10, and we can write

$$(\text{switch A}) + (\text{switch B}) - (\text{surplus switches}) = 10$$

If we let s_3 represent the surplus switches, we can write the above as

$$x_1 + x_2 - s_3 = 10$$

Notice that this equation is simply the third constraint in the Flashback written in equational form. Here s_3 is an example of a **surplus variable**. As with slack variables, surplus variables are nonnegative.

Slack and Surplus Variables

- We **add** a **slack** variable to the left side of $a_1x_1 + a_2x_2 + \cdots + a_nx_n \leq b$ to pick up the slack and create an equation of the form $a_1x_1 + a_2x_2 + \cdots + s_1 = b$.
- We **subtract** a **surplus** variable from the left side of $a_1x_1 + a_2x_2 + \cdots + a_nx_n \geq b$ to take away the excess and create an equation of the form $a_1x_1 + a_2x_2 + \cdots - s_2 = b$.
- Both slack and surplus variables are nonnegative.

Example 1 Using Slack and Surplus Variables

Use slack and surplus variables to write the mathematical model in the Flashback in equational form.

Solution

Using s_1 and s_2 for slack variables and s_3 as the surplus variable, we have

$$\begin{aligned} &\text{Maximize:} && P = 20x_1 + 30x_2 \\ &\text{Subject to:} && 4x_1 + 3x_2 + s_1 = 240 \\ &&& x_1 + 2x_2 + s_2 = 140 \\ &&& x_1 + x_2 - s_3 = 10 \end{aligned}$$

■

Checkpoint 1 Now work Exercise 3.

Let's set up the initial simplex tableau for the system in Example 1. It is

$$\begin{array}{cccccc|c} x_1 & x_2 & s_1 & s_2 & s_3 & P & \\ \hline 4 & 3 & 1 & 0 & 0 & 0 & 240 \\ 1 & 2 & 0 & 1 & 0 & 0 & 140 \\ 1 & 1 & 0 & 0 & -1 & 0 & 10 \\ \hline -20 & -30 & 0 & 0 & 0 & 1 & 0 \end{array}$$

The row operation $-1R_3 \to R_3$ gives us

$$\begin{array}{cccccc|c} x_1 & x_2 & s_1 & s_2 & s_3 & P & \\ \hline 4 & 3 & 1 & 0 & 0 & 0 & 240 \\ 1 & 2 & 0 & 1 & 0 & 0 & 140 \\ -1 & -1 & 0 & 0 & 1 & 0 & -10 \\ \hline -20 & -30 & 0 & 0 & 0 & 1 & 0 \end{array}$$

From Your Toolbox

A solution is feasible if all variables are nonnegative.

This tableau gives the solution

$$x_1 = 0, \quad x_2 = 0, \quad s_1 = 240, \quad s_2 = 140, \quad s_3 = -10, \quad P = 0$$

This solution is not **feasible** (consult the Toolbox to the left) since $s_3 = -10$.

Also, observe that the solution $x_1 = 0$ and $x_2 = 0$ corresponds to the point (0, 0). Consulting Figure 4.4.1, we see that the point (0, 0) is not in the feasible region. Why is this observation important? The answer is that the simplex method seeks out an optimal solution by moving from corner point to corner point in the feasible region! Hence, if our initial simplex tableau does not correspond to a feasible solution, we cannot use the simplex method. In order to use the simplex method, we must first get a feasible solution. This requires us to use what is called the **two-phase method**.

The Two-Phase Method

Most mixed constraint problems, such as Example 1, have two phases when solving the simplex method. These two phases are as follows:

- **Phase I: Locating a Corner Point** Our goal is to locate a feasible solution. To do this we offer the following strategy:
  1. Select a row with a negative entry in the rightmost column.
  2. Pivot on a negative entry in that row.
  3. Repeat this process until our tableau has nonnegative entries in the rightmost column.
- **Phase II: Solving after Phase I** After using Phase I to obtain a tableau that gives a feasible solution, we then apply the simplex method as usual.

Phase I will take us from the origin to a corner point. Phase II then takes us from corner point to corner point until the optimal value is obtained. In Phase I, if more than one row has a negative entry in the rightmost column, pick a row arbitrarily. To find the negative entry in the row to pivot on, we arbitrarily choose the leftmost one.

Example 2 Using the Two-Phase Process

Use the two-phase process on the following tableau to solve the linear programming problem in the Flashback.

$$\begin{array}{cccccc|c} x_1 & x_2 & s_1 & s_2 & s_3 & P & \\ \hline 4 & 3 & 1 & 0 & 0 & 0 & 240 \\ 1 & 2 & 0 & 1 & 0 & 0 & 140 \\ -1 & -1 & 0 & 0 & 1 & 0 & -10 \\ \hline -20 & -30 & 0 & 0 & 0 & 1 & 0 \end{array}$$

Solution

We begin Phase I by pivoting on the -1 in row 3, column 1. Our first step is to use the row operation $-1R_3 \to R_3$.

$$\begin{array}{c} \begin{array}{cccccc} x_1 & x_2 & s_1 & s_2 & s_3 & P \end{array} \\ \left[\begin{array}{cccccc|c} 4 & 3 & 1 & 0 & 0 & 0 & 240 \\ 1 & 2 & 0 & 1 & 0 & 0 & 140 \\ \boxed{-1} & -1 & 0 & 0 & 1 & 0 & -10 \\ \hline -20 & -30 & 0 & 0 & 0 & 1 & 0 \end{array}\right] \end{array} \xrightarrow{-1R_3 \to R_3} \begin{array}{c} \begin{array}{cccccc} x_1 & x_2 & s_1 & s_2 & s_3 & P \end{array} \\ \left[\begin{array}{cccccc|c} 4 & 3 & 1 & 0 & 0 & 0 & 240 \\ 1 & 2 & 0 & 1 & 0 & 0 & 140 \\ 1 & 1 & 0 & 0 & -1 & 0 & 10 \\ \hline -20 & -30 & 0 & 0 & 0 & 1 & 0 \end{array}\right] \end{array}$$

$$\begin{array}{r} -4R_3 + R_1 \to R_1 \\ -R_3 + R_2 \to R_2 \\ 20R_3 + R_4 \to R_4 \end{array} \quad \begin{array}{c} \begin{array}{cccccc} x_1 & x_2 & s_1 & s_2 & s_3 & P \end{array} \\ \left[\begin{array}{cccccc|c} 0 & -1 & 1 & 0 & 4 & 0 & 200 \\ 0 & 1 & 0 & 1 & 1 & 0 & 130 \\ 1 & 1 & 0 & 0 & -1 & 0 & 10 \\ \hline 0 & -10 & 0 & 0 & -20 & 1 & 200 \end{array}\right] \end{array}$$

We now have the solution $x_1 = 10$, $x_2 = 0$, $s_1 = 200$, $s_2 = 130$, $s_3 = 0$, and $P = 200$. Since all variables are nonnegative, this is a basic feasible solution and we move to Phase II. [Observe that we are at the corner point $(10, 0)$ in Figure 4.4.1.] We begin Phase II by pivoting on the 4 in row 1, column 5.

$$\begin{array}{c} \begin{array}{cccccc} x_1 & x_2 & s_1 & s_2 & s_3 & P \end{array} \\ \left[\begin{array}{cccccc|c} 0 & -1 & 1 & 0 & \boxed{4} & 0 & 200 \\ 0 & 1 & 0 & 1 & 1 & 0 & 130 \\ 1 & 1 & 0 & 0 & -1 & 0 & 10 \\ \hline 0 & -10 & 0 & 0 & -20 & 1 & 200 \end{array}\right] \end{array} \xrightarrow{\frac{1}{4}R_1 \to R_1} \begin{array}{c} \begin{array}{cccccc} x_1 & x_2 & s_1 & s_2 & s_3 & P \end{array} \\ \left[\begin{array}{cccccc|c} 0 & -\frac{1}{4} & \frac{1}{4} & 0 & 1 & 0 & 50 \\ 0 & 1 & 0 & 1 & 1 & 0 & 130 \\ 1 & 1 & 0 & 0 & -1 & 0 & 10 \\ \hline 0 & -10 & 0 & 0 & -20 & 1 & 200 \end{array}\right] \end{array}$$

$$\begin{array}{r} -R_1 + R_2 \to R_2 \\ R_1 + R_3 \to R_3 \\ 20R_1 + R_4 \to R_4 \end{array} \quad \begin{array}{c} \begin{array}{cccccc} x_1 & x_2 & s_1 & s_2 & s_3 & P \end{array} \\ \left[\begin{array}{cccccc|c} 0 & -\frac{1}{4} & \frac{1}{4} & 0 & 1 & 0 & 50 \\ 0 & \frac{5}{4} & -\frac{1}{4} & 1 & 0 & 0 & 80 \\ 1 & \frac{3}{4} & \frac{1}{4} & 0 & 0 & 0 & 60 \\ \hline 0 & -15 & 5 & 0 & 0 & 1 & 1200 \end{array}\right] \end{array}$$

Next we pivot on the $\frac{5}{4}$ in row 2, column 2.

$$\begin{array}{c} \begin{array}{cccccc} x_1 & x_2 & s_1 & s_2 & s_3 & P \end{array} \\ \left[\begin{array}{cccccc|c} 0 & -\frac{1}{4} & \frac{1}{4} & 0 & 1 & 0 & 50 \\ 0 & \boxed{\frac{5}{4}} & -\frac{1}{4} & 1 & 0 & 0 & 80 \\ 1 & \frac{3}{4} & \frac{1}{4} & 0 & 0 & 0 & 60 \\ \hline 0 & -15 & 5 & 0 & 0 & 1 & 1200 \end{array}\right] \end{array} \xrightarrow{\frac{4}{5}R_2 \to R_2} \begin{array}{c} \begin{array}{cccccc} x_1 & x_2 & s_1 & s_2 & s_3 & P \end{array} \\ \left[\begin{array}{cccccc|c} 0 & -\frac{1}{4} & \frac{1}{4} & 0 & 1 & 0 & 50 \\ 0 & 1 & -\frac{1}{5} & \frac{4}{5} & 0 & 0 & 64 \\ 1 & \frac{3}{4} & \frac{1}{4} & 0 & 0 & 0 & 60 \\ \hline 0 & -15 & 5 & 0 & 0 & 1 & 1200 \end{array}\right] \end{array}$$

$$\begin{array}{r} \frac{1}{4}R_2 + R_1 \to R_1 \\ -\frac{3}{4}R_2 + R_3 \to R_3 \\ 15R_2 + R_4 \to R_4 \end{array} \quad \begin{array}{c} \begin{array}{cccccc} x_1 & x_2 & s_1 & s_2 & s_3 & P \end{array} \\ \left[\begin{array}{cccccc|c} 0 & 0 & \frac{1}{5} & \frac{1}{5} & 1 & 0 & 66 \\ 0 & 1 & -\frac{1}{5} & \frac{4}{5} & 0 & 0 & 64 \\ 1 & 0 & \frac{2}{5} & -\frac{3}{5} & 0 & 0 & 12 \\ \hline 0 & 0 & 2 & 12 & 0 & 1 & 2160 \end{array}\right] \end{array}$$

We have reached our final tableau. The solution is,

$$x_1 = 12, \quad x_2 = 64, \quad s_1 = 0, \quad s_2 = 0, \quad s_3 = 66, \quad P = 2160$$

■

Interactive Activity

Interpret each variable value in the Example 2 solution in the context of the Flashback and Example 1. In particular, what does $s_3 = 66$ mean?

WWW

Checkpoint 2

Now work Exercise 11.

Before we supply a procedure for solving nonstandard problems, let's examine an alternative for minimization. To gain some insight, consider the following objective functions along with the feasible region in Figure 4.4.2.

$$\text{Minimize} \quad w = 3x + y$$
$$\text{Maximize} \quad z = -w = -3x - y$$

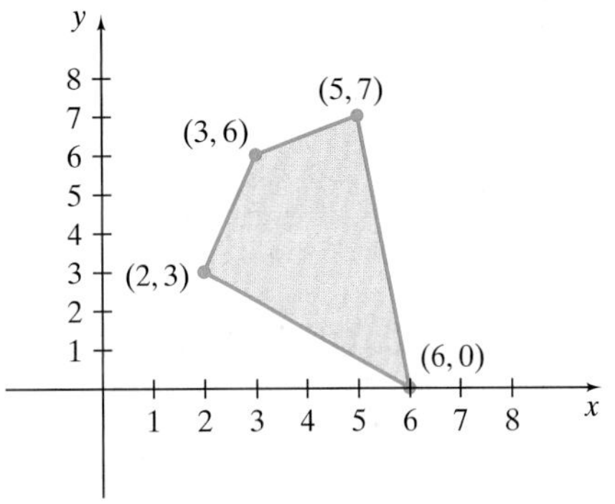

Figure 4.4.2

We evaluate each objective function at the corner points. See Table 4.4.2.

Table 4.2.2

Corner Point	$w = 3x + y$	$z = -3x - y$
(2, 3)	$3(2) + 3 = 9$	$-3(2) - 3 = -9$
(3, 6)	$3(3) + 6 = 15$	$-3(3) - 6 = -15$
(5, 7)	$3(5) + 7 = 22$	$-3(5) - 7 = -22$
(6, 0)	$3(6) + 0 = 18$	$-3(6) - 0 = -18$

Table 4.4.2 indicates that the minimum of w is 9 at the corner point (2, 3). The maximum of z is -9 at the corner point (2, 3). This illustrates the following fact:

The minimum value of an objective function w occurs at the same point as the maximum value of the objective function z, where $z = -w$, provided both have the same constraints.

We illustrate how to use this fact in Example 3.

Example 3 Minimizing With Mixed Constraints

$$\begin{aligned} \text{Minimize:} \quad & w = 2x_1 + 5x_2 \\ \text{Subject to:} \quad & 2x_1 + 3x_2 \leq 5 \\ & 4x_1 - x_2 \geq 3 \\ & x_1 \geq 0, x_2 \geq 0 \end{aligned}$$

Solution

We change this to a maximization problem by letting $z = -w$. We now have

$$\begin{aligned}
\text{Maximize:} \quad & z = -w = -2x_1 - 5x_2 \\
\text{Subject to:} \quad & 2x_1 + 3x_2 \le 5 \\
& 4x_1 - x_2 \ge 3 \\
& x_1 \ge 0, x_2 \ge 0
\end{aligned}$$

Next we add slack variable s_1 and subtract the surplus variable s_2. This gives the equational form

$$\begin{aligned}
2x_1 + 3x_2 + s_1 \quad\quad\quad &= 5 \\
4x_1 - \quad x_2 \quad\quad - s_2 \quad &= 3 \\
2x_1 + 5x_2 \quad\quad\quad\quad + z &= 0
\end{aligned}$$

The initial simplex tableau is

$$\begin{array}{c} \begin{matrix} x_1 & x_2 & s_1 & s_2 & z & \end{matrix} \\ \left[\begin{array}{ccccc|c} 2 & 3 & 1 & 0 & 0 & 5 \\ 4 & -1 & 0 & -1 & 0 & 3 \\ \hline 2 & 5 & 0 & 0 & 1 & 0 \end{array}\right] \end{array} \quad -1R_2 \to R_2 \quad \begin{array}{c} \begin{matrix} x_1 & x_2 & s_1 & s_2 & z & \end{matrix} \\ \left[\begin{array}{ccccc|c} 2 & 3 & 1 & 0 & 0 & 5 \\ -4 & 1 & 0 & 1 & 0 & -3 \\ \hline 2 & 5 & 0 & 0 & 1 & 0 \end{array}\right] \end{array}$$

The solution that this represents, $x_1 = x_2 = 0, s_1 = 5, s_2 = -3, z = 0$, is not feasible, so we move into Phase I. We pivot on the -4 in row 2, column 1.

$$\begin{array}{c} \begin{matrix} x_1 & x_2 & s_1 & s_2 & z & \end{matrix} \\ \left[\begin{array}{ccccc|c} 2 & 3 & 1 & 0 & 0 & 5 \\ \textcircled{-4} & 1 & 0 & 1 & 0 & -3 \\ \hline 2 & 5 & 0 & 0 & 1 & 0 \end{array}\right] \end{array} \quad -\tfrac{1}{4}R_2 \to R_2 \quad \begin{array}{c} \begin{matrix} x_1 & x_2 & s_1 & s_2 & z & \end{matrix} \\ \left[\begin{array}{ccccc|c} 2 & 3 & 1 & 0 & 0 & 5 \\ 1 & -\frac{1}{4} & 0 & -\frac{1}{4} & 0 & \frac{3}{4} \\ \hline 2 & 5 & 0 & 0 & 1 & 0 \end{array}\right] \end{array}$$

$$\begin{array}{l} -2R_2 + R_1 \to R_1 \\ -2R_2 + R_3 \to R_3 \end{array} \quad \begin{array}{c} \begin{matrix} x_1 & x_2 & s_1 & s_2 & z & \end{matrix} \\ \left[\begin{array}{ccccc|c} 0 & \frac{7}{2} & 1 & \frac{1}{2} & 0 & \frac{7}{2} \\ 1 & -\frac{1}{4} & 0 & -\frac{1}{4} & 0 & \frac{3}{2} \\ \hline 0 & \frac{11}{2} & 0 & \frac{1}{2} & 1 & -\frac{3}{2} \end{array}\right] \end{array}$$

The rightmost column has no negatives and there are no negative indicators, so this is our final tableau. (Note that $z = -\frac{3}{2}$. It must be negative since it is the opposite of w, that is, $z = -w$.) To answer the original problem, the solution is $x_1 = \frac{3}{4}$ and $x_2 = 0$. Since $z = -w = -\frac{3}{2}$, we have the minimum value of $w = \frac{3}{2}$. ■

✓ Checkpoint 3 Now work Exercise 17.

To summarize, the following steps are used in solving the nonstandard problems in this section.

Solving Nonstandard Problems

1. If necessary, convert to a maximization problem.
2. Add slack variables and subtract surplus variables.
3. Write the initial simplex tableau.
4. Employ the Two-Phase Method until an optimal solution is found.

Applications

An important application of linear programming is the minimization of the cost of transporting goods. These types of problems are commonly called **transportation problems**. Example 4 is one of these types of problems.

Example 4 Minimizing Transportation Costs

Skinner Bikes sends bicycles from its two plants, Plant A and Plant B, to its bicycle retailers I and II located in Columbus, Ohio. Plant A has a total of 210 bicycles to send and Plant B has 60. Retailer I needs 170 bicycles and retailer II needs 100. Transportation costs per bicycle are \$18 from A to I, \$20 from A to II, \$22 from B to I, and \$16 from B to II. How many bicycles should Skinner Bikes send from each plant to each retailer to minimize transportation costs?

Solution

Understand the Situation: We begin by letting

x_1 represent the number of bicycles sent from A to I

x_2 represent the number of bicycles sent from A to II

x_3 represent the number of bicycles sent from B to I

x_4 represent the number of bicycles sent from B to II

Since Plant A has 210 bicycles to send, this gives us

$$x_1 + x_2 \leq 210$$

Similarly, Plant B has 60 bicycles to send, which means that

$$x_3 + x_4 \leq 60$$

Since retailer I needs 170 bicycles, we have

$$x_1 + x_3 \geq 170$$

Likewise, since retailer II needs 100 bicycles,

$$x_2 + x_4 \geq 100$$

Skinner Bikes wants to minimize transportation costs; hence the objective function is

$$w = 18x_1 + 20x_2 + 22x_3 + 16x_4$$

By letting $z = -w$, we convert the minimization problem to a maximization problem.

$$\begin{aligned}
\text{Maximize:} \quad & z = -w = -18x_1 - 20x_2 - 22x_3 - 16x_4 \\
\text{Subject to:} \quad & x_1 + x_2 \leq 210 \\
& x_3 + x_4 \leq 60 \\
& x_1 + x_3 \geq 170 \\
& x_2 + x_4 \geq 100 \\
& x_1 \geq 0, x_2 \geq 0, x_3 \geq 0, x_4 \geq 0
\end{aligned}$$

Adding slack variables s_1 and s_2 and subtracting surplus variables s_3 and s_4 gives the equational form

$$
\begin{array}{lcl}
x_1 + x_2 + s_1 & = & 210 \\
x_3 + x_4 + s_2 & = & 60 \\
x_1 + x_3 - s_3 & = & 170 \\
x_2 + x_4 - s_4 & = & 100 \\
18x_1 + 20x_2 + 22x_3 + 16x_4 + z & = & 0
\end{array}
$$

The initial simplex tableau is

$$
\begin{array}{ccccccccc|c}
x_1 & x_2 & x_3 & x_4 & s_1 & s_2 & s_3 & s_4 & z & \\
\hline
1 & 1 & 0 & 0 & 1 & 0 & 0 & 0 & 0 & 210 \\
0 & 0 & 1 & 1 & 0 & 1 & 0 & 0 & 0 & 60 \\
1 & 0 & 1 & 0 & 0 & 0 & -1 & 0 & 0 & 170 \\
0 & 1 & 0 & 1 & 0 & 0 & 0 & -1 & 0 & 100 \\
\hline
18 & 20 & 22 & 16 & 0 & 0 & 0 & 0 & 1 & 0
\end{array}
$$

$$
\begin{array}{l}
\\ \\ \\
-1R_3 \to R_3 \\
-1R_4 \to R_4 \\
\\
\end{array}
\begin{array}{ccccccccc|c}
x_1 & x_2 & x_3 & x_4 & s_1 & s_2 & s_3 & s_4 & z & \\
\hline
1 & 1 & 0 & 0 & 1 & 0 & 0 & 0 & 0 & 210 \\
0 & 0 & 1 & 1 & 0 & 1 & 0 & 0 & 0 & 60 \\
-1 & 0 & -1 & 0 & 0 & 0 & 1 & 0 & 0 & -170 \\
0 & -1 & 0 & -1 & 0 & 0 & 0 & 1 & 0 & -100 \\
\hline
18 & 20 & 22 & 16 & 0 & 0 & 0 & 0 & 1 & 0
\end{array}
$$

We are now in Phase I. We pivot on the -1 in row 3, column 1. After the pivot we have

$$
\begin{array}{ccccccccc|c}
x_1 & x_2 & x_3 & x_4 & s_1 & s_2 & s_3 & s_4 & z & \\
\hline
0 & 1 & -1 & 0 & 1 & 0 & 1 & 0 & 0 & 40 \\
0 & 0 & 1 & 1 & 0 & 1 & 0 & 0 & 0 & 60 \\
1 & 0 & 1 & 0 & 0 & 0 & -1 & 0 & 0 & 170 \\
0 & -1 & 0 & -1 & 0 & 0 & 0 & 1 & 0 & -100 \\
\hline
0 & 20 & 4 & 16 & 0 & 0 & 18 & 0 & 1 & -3060
\end{array}
$$

Since $s_4 = -100$, we are still in Phase I. Pivoting on the -1 in row 4, column 2 gives the following:

$$
\begin{array}{ccccccccc|c}
x_1 & x_2 & x_3 & x_4 & s_1 & s_2 & s_3 & s_4 & z & \\
\hline
0 & 0 & -1 & -1 & 1 & 0 & 1 & 1 & 0 & -60 \\
0 & 0 & 1 & 1 & 0 & 1 & 0 & 0 & 0 & 60 \\
1 & 0 & 1 & 0 & 0 & 0 & -1 & 0 & 0 & 170 \\
0 & 1 & 0 & 1 & 0 & 0 & 0 & -1 & 0 & 100 \\
\hline
0 & 0 & 4 & -4 & 0 & 0 & 18 & 20 & 1 & -5060
\end{array}
$$

We are still in Phase I since $s_1 = -60$. We pivot on the -1 in row 1, column 3 and get

$$
\begin{array}{ccccccccc|c}
x_1 & x_2 & x_3 & x_4 & s_1 & s_2 & s_3 & s_4 & z & \\
\hline
0 & 0 & 1 & 1 & -1 & 0 & -1 & -1 & 0 & 60 \\
0 & 0 & 0 & 0 & 1 & 1 & 1 & 1 & 0 & 0 \\
1 & 0 & 0 & -1 & 1 & 0 & 0 & 1 & 0 & 110 \\
0 & 1 & 0 & 1 & 0 & 0 & 0 & -1 & 0 & 100 \\
\hline
0 & 0 & 0 & -8 & 4 & 0 & 22 & 24 & 1 & -5300
\end{array}
$$

We have all variables with nonnegative values, so we move to Phase II. (Recall that z must be negative.) Now we pivot on the 1 in row 1, column 4 and get

$$\begin{array}{ccccccccc}
x_1 & x_2 & x_3 & x_4 & s_1 & s_2 & s_3 & s_4 & z \\
\end{array}$$
$$\left[\begin{array}{ccccccccc|c}
0 & 0 & 1 & 1 & -1 & 0 & -1 & -1 & 0 & 60 \\
0 & 0 & 0 & 0 & 1 & 1 & 1 & 1 & 0 & 0 \\
1 & 0 & 1 & 0 & 0 & 0 & -1 & 0 & 0 & 170 \\
0 & 1 & -1 & 0 & 1 & 0 & 1 & 0 & 0 & 40 \\
\hline
0 & 0 & 8 & 0 & -4 & 0 & 14 & 16 & 1 & -4820
\end{array}\right]$$

Our next pivot is on the 1 in row 2, column 5. This yields

$$\begin{array}{ccccccccc}
x_1 & x_2 & x_3 & x_4 & s_1 & s_2 & s_3 & s_4 & z \\
\end{array}$$
$$\left[\begin{array}{ccccccccc|c}
0 & 0 & 1 & 1 & 0 & 1 & 0 & 0 & 0 & 60 \\
0 & 0 & 0 & 0 & 1 & 1 & 1 & 1 & 0 & 0 \\
1 & 0 & 1 & 0 & 0 & 0 & -1 & 0 & 0 & 170 \\
0 & 1 & -1 & 0 & 0 & -1 & 0 & -1 & 0 & 40 \\
\hline
0 & 0 & 8 & 0 & 0 & 4 & 18 & 20 & 1 & -4820
\end{array}\right]$$

This is the final tableau. The maximum value of z is -4820, so the minimum value of w is 4820. This occurs when $x_1 = 170, x_2 = 40, x_3 = 0, x_4 = 60$, and $s_1 = s_2 = s_3 = s_4 = 0$.

Interpret the Solution: This means that Plant A should send 170 bicycles to retailer I and 40 bicycles to retailer II, while Plant B should send 0 bicycles to retailer I and 60 bicycles to retailer II. This minimizes the cost at \$4820. ■

SUMMARY

In this section we saw how to solve nonstandard problems, that is, problems with **mixed constraints**. We introduced a new type of variable called a **surplus variable**. The key to solving mixed constraint problems was employing the Two-Phase Method. The section concluded with a classic economics problem called a transportation problem.

- **Two-Phase Method**

  Phase I: Locating a Corner Point Our goal is to locate a feasible solution. To do this we offer the following strategy: Select a row with a negative entry in the rightmost column and then pivot on a negative entry in that row. We repeat this process until our tableau has nonnegative entries in the rightmost column.

  Phase II: Solving after Phase I After using Phase I to obtain a tableau that gives a feasible solution, we apply the simplex method as usual.

- The minimum value of an objective function w occurs at the same point as the maximum value of the objective function z, where $z = -w$, provided that both have the same constraints.

SECTION 4.4 EXERCISES

In Exercises 1–6, add slack variables and subtract surplus variables to write each system of inequalities in equational form.

1. $3x_1 + 2x_2 \le 12$
$x_1 + 4x_2 \ge 7$

2. $4x_1 + 2x_2 \le 8$
$3x_1 + 5x_2 \ge 15$

3. $x_1 + 4x_2 + 3x_3 \le 9$
$2x_1 + x_2 + 5x_3 \ge 10$
$3x_1 + 5x_2 + 2x_3 \le 8$

4. $5x_1 + 6x_2 + x_3 \le 20$
$2x_1 + x_2 + 3x_3 \ge 6$
$3x_1 + 4x_2 + x_3 \le 13$

5. $x_1 + 2x_2 + x_3 \le 60$
$2x_1 + 3x_3 \ge 18$
$3x_2 + 2x_3 \ge 30$

6. $2x_1 + 3x_2 + 7x_3 \le 80$
$9x_1 + x_3 \ge 12$
$7x_2 + 4x_3 \ge 20$

In Exercises 7–10, complete the following:

(a) Convert to a maximization problem.

(b) Add slack variables and subtract surplus variables to write in equational form.

(c) Write the initial simplex tableau.

7. Minimize: $w = 8x_1 + 3x_2$
Subject to: $4x_1 + 2x_2 \le 8$
$x_1 + 4x_2 \ge 7$
$x_1 \ge 0, x_2 \ge 0$

8. Minimize: $w = x_1 + 7x_2$
Subject to: $3x_1 + 2x_2 \le 12$
$5x_1 + 3x_2 \ge 15$
$x_1 \ge 0, x_2 \ge 0$

9. Minimize: $w = 2x_1 + x_2 + 3x_3$
Subject to: $5x_1 + 6x_2 + x_3 \le 20$
$2x_1 + x_2 + 5x_3 \ge 10$
$7x_2 + 4x_3 \ge 20$
$x_1 \ge 0, x_2 \ge 0, x_3 \ge 0$

10. Minimize: $w = x_1 + 2x_2 + x_3$
Subject to: $x_1 + 4x_2 + 3x_3 \le 9$
$2x_1 + x_2 + x_3 \ge 7$
$3x_2 + 2x_3 \ge 30$
$x_1 \ge 0, x_2 \ge 0, x_3 \ge 0$

In Exercises 11—24, use the method in this section to solve each linear programming problem.

✓ **11.** Maximize: $z = 5x_1 + 2x_2$
Subject to: $x_1 + x_2 \ge 11$
$2x_1 + 3x_2 \ge 24$
$x_1 + 3x_2 \le 18$
$x_1 \ge 0, x_2 \ge 0$

12. Maximize: $z = 2x_1 + 5x_2$
Subject to: $x_1 + 2x_2 \le 18$
$2x_1 + x_2 \le 21$
$x_1 + x_2 \ge 10$
$x_1 \ge 0, x_2 \ge 0$

13. Maximize: $z = 2x_1 + 3x_2$
Subject to: $3x_1 + x_2 \le 15$
$x_1 + x_2 \le 10$
$4x_1 + 3x_2 \ge 12$
$x_1 \ge 0, x_2 \ge 0$

14. Maximize: $z = 4x_1 + x_2$
Subject to: $x_1 + 3x_2 \le 18$
$2x_1 + x_2 \le 12$
$3x_1 + 4x_2 \ge 12$
$x_1 \ge 0, x_2 \ge 0$

15. Maximize: $z = 2x_1 + x_2 + 3x_3$
Subject to: $2x_1 + 4x_2 + x_3 \le 150$
$3x_1 + 3x_2 + x_3 \le 90$
$-x_1 + 5x_2 + x_3 \ge 120$
$x_1 \ge 0, x_2 \ge 0, x_3 \ge 0$

16. Maximize: $z = x_1 + 2x_2 + 3x_3$
Subject to: $2x_1 + x_2 + x_3 \le 30$
$x_1 + 2x_2 + 2x_3 \le 80$
$x_1 + 2x_2 - x_3 \ge 10$
$x_1 \ge 0, x_2 \ge 0, x_3 \ge 0$

✓ **17.** Minimize: $w = x_1 + 2x_2$
Subject to: $2x_1 + 3x_2 \le 12$
$3x_1 + x_2 \ge 6$
$x_1 \ge 0, x_2 \ge 0$

18. Minimize: $w = 2x_1 + x_2$
Subject to: $x_1 + 2x_2 \le 20$
$2x_1 + 3x_2 \ge 6$
$x_1 \ge 0, x_2 \ge 0$

19. Minimize: $w = 3x_1 + 4x_2$
Subject to: $x_1 + 2x_2 \le 18$
$x_1 + x_2 \le 10$
$x_1 + 3x_2 \ge 3$
$x_1 \ge 0, x_2 \ge 0$

20. Minimize: $w = 4x_1 + x_2$
Subject to: $3x_1 + x_2 \le 15$
$x_1 + x_2 \le 12$
$2x_1 + 3x_2 \ge 6$
$x_1 \ge 0, x_2 \ge 0$

21. Minimize: $w = x_1 + x_2 + 2x_3$
Subject to: $x_1 - x_2 + x_3 \le 8$
$x_1 + 3x_2 + 2x_3 \ge 10$
$-3x_1 + x_2 + 4x_3 \ge 12$
$x_1 \ge 0, x_2 \ge 0, x_3 \ge 0$

22. Minimize: $w = x_1 - 2x_2 + x_3$
Subject to: $x_1 - 2x_2 + 3x_3 \le 10$
$2x_1 + x_2 - 2x_3 \ge 2$
$2x_1 + x_2 + 3x_3 \ge 1$
$x_1 \ge 0, x_2 \ge 0, x_3 \ge 0$

23. Minimize: $w = 2x_1 - x_2 + 3x_3$
Subject to: $2x_1 + x_2 + x_3 \ge 2$
$x_1 + 3x_2 + x_3 \ge 6$
$2x_1 + x_2 + 2x_3 \le 12$
$x_1 \ge 0, x_2 \ge 0, x_3 \ge 0$

24. Minimize: $w = x_1 + 2x_2 + 3x_3$
Subject to: $x_1 + x_2 + x_3 \le 20$
$2x_1 + x_2 + x_3 \ge 10$
$2x_1 + 2x_2 + x_3 \ge 2$
$x_1 \ge 0, x_2 \ge 0, x_3 \ge 0$

Applications

25. The Tentsmith Company makes two types of tents: two-person and six-person models. Each tent requires the services, in hours, of two departments as shown in Table 4.4.3

Table 4.4.3

	Labor	
Department	**Two-person**	**Six-person**
Fabric cutting	6	8
Assembly and packaging	2	4

Fabric cutting has a maximum of 580 hours available each week, and assembly and packaging has a maximum of 260 hours available each week. Due to prior contracts, the total number of tents made each week must be at least 20. Each two-person tent yields $40 in profit, and each six-person tent yields $60 in profit.

(a) How many of each type of tent should be made and sold to maximize weekly profit? What is the maximum weekly profit?

(b) Interpret the value of all slack and surplus variables in the final tableau.

26. (*continuation of Exercise 25*) Tanya, the head of research and development at The Tentsmith Company has discovered a way to reduce the amount of time needed for the six-person tent in the fabric cutting department to 7 hours. Everything else from Exercise 25 remains the same.

(a) How many of each type of tent should be made and sold to maximize weekly profit? What is the maximum weekly profit?

(b) Interpret the value of all slack and surplus variables in the final tableau.

27. Denyse and Mike, two recent graduates, have decided to enter the field of microcomputers. Initially, they plan to make two types of microcomputers, Deluxe 1000 and Deluxe 1200. Because of the interest in microcomputers, they can sell all that they can produce. However, they need to determine a production rate so as to satisfy various limits with a small production crew. These limits include a maximum of 150 hours per week for assembly and 100 hours per week for testing. The Deluxe 1000 requires 3 hours of assembly and 3 hours of testing, and the Deluxe 1200 requires 6 hours of assembly and 3.5 hours of testing. For market demonstration purposes, the total number of microcomputers made each week must be at least 8. Profit for the Deluxe 100 is estimated at $300 per unit, and profit for the Deluxe 1200 is estimated at $450 per unit.

(a) How many of each type of microcomputer should be made and sold to maximize weekly profit? What is the maximum weekly profit?

(b) Interpret the value of all slack and surplus variables in the final tableau.

28. (*continuation of Exercise 27*) Due to a favorable response from the market, Denyse and Mike decide to hire more employees and increase production. By doing so, they now have a maximum of 210 hours per week for assembly and 160 hours per week for testing. Everything else from Exercise 27 remains the same.

(a) How many of each type of microcomputer should be made and sold to maximize weekly profit? What is the maximum weekly profit?

(b) Interpret the value of all slack and surplus variables in the final tableau.

29. Table 4.4.4 gives nutritional information in grams and current prices for selected items from Burger King's menu.

Table 4.4.4

Item	**Protein**	**Fiber**	**Price**
Cheeseburger	22	2	$1.19
Small onion rings	3	2	$0.99

SOURCE: Burger King, www.burgerking.com

Mike desires to eat nothing but cheeseburgers and orders of small onion rings. He wants at least 96 grams of protein and at least 26 grams of fiber daily. Also, he wishes to eat at least 1 cheeseburger and at least 1 order of small onion rings. If Mike eats nothing but cheeseburgers and small onion rings and meets his minimum nutrient requirements, answer the following:

(a) How much of each food should Mike eat to minimize his daily food cost?
(b) What is his minimum daily food cost?

30. (*continuation of Exercise 29*) Mike has found out that Burger King is having a promotion during which cheeseburgers sell for $0.49 on Wednesdays. Assuming that all other data in Exercise 29 remain the same, answer the following:

(a) How much of each food should Mike eat to minimize his daily food cost?
(b) What is his minimum daily food cost?

31. Hoofer Oil has two refineries, one in Findlay, Ohio, and one in Ft. Wayne, Indiana. Each refinery can refine the amounts of fuel, in barrels, in a 24-hour period as shown in Table 4.4.5.

Table 4.4.5

Maximum Production of:	**Findlay**	**Ft. Wayne**
Low-octane gasoline	800	1000
High-octane gasoline	500	600
Diesel	1000	600

Daily operating costs for the Findlay refinery are $28,000, and daily operating costs for the Ft. Wayne refinery are $33,000. Zachary, the production manager of Hoofer Oil, receives an order for 60,000 barrels of low-octane gasoline, 120,000 barrels of high-octane gasoline, and 150,000 barrels of diesel. Due to budgetary issues, the combined total number of days that the refineries operate cannot exceed 300.

(a) How many days should each refinery operate to minimize cost? What is the minimum cost?

(b) Interpret the value of all slack and surplus variables in the final tableau.

32. Shannon Oil has two refineries, one in Lima, Ohio, and one in Detroit, Michigan. Each refinery can refine the amounts of fuel, in barrels, in a 24-hour period as shown in Table 4.4.6.

Table 4.4.6

Maximum Production of:	Lima	Detroit
Low-octane gasoline	900	1100
High-octane gasoline	600	700
Diesel	1100	700

Daily operating costs for the Lima refinery are $35,000, and daily operating costs for the Detroit refinery are $30,000. Joshua, the production manager of Shannon Oil, receives an order for 50,000 barrels of low-octane gasoline, 90,000 barrels of high-octane gasoline, and 130,000 barrels of diesel. Due to budgetary issues, the combined total number of days that the refineries can operate cannot exceed 200.

(a) How many days should each refinery operate to minimize cost? What is the minimum cost?

(b) Interpret the value of all slack and surplus variables in the final tableau.

33. Anita, a dietitian at a local hospital, is to prepare a diet to contain at least 400 units of vitamins C and E, 500 units of minerals (such as zinc and iron), and 1400 calories. She has two foods available, Goli and Zarobila. An ounce of Goli contains 2 units of vitamin C and E, 1 unit of minerals, and 4 calories and costs $0.05. An ounce of Zarobila contains 1 unit of vitamins C and E, 2 units of minerals, and 4 calories and costs $0.03. Due to restricted availability, the total number of ounces of the two foods is no more than 450 ounces.

(a) How many ounces of each food should be used to minimize daily food cost? What is the minimum daily food cost?

(b) Interpret the value of all slack and surplus variables in the final tableau.

34. Lisa, a dietitian at a local hospital, is to prepare two foods to meet certain nutritional requirements. Each pound of Viega contains 200 units of vitamin C, 80 units of vitamin B_{12}, and 20 units of vitamin E and costs $0.40. Each pound of Zella contains 20 units of vitamin C, 160 units of vitamin B_{12}, and 10 units of vitamin E and costs $0.30. To meet the nutritional needs of the patients, the combination of the two foods must have at least 520 units of vitamin C, 640 units of vitamin B_{12}, and 100 units of vitamin E. Due to restricted availability, the total number of pounds of the two foods is no more than 6 pounds.

(a) How many ounces of each food should be used to minimize daily food cost? What is the minimum daily food cost?

(b) Interpret the value of all slack and surplus variables in the final tableau.

35. A DVD player manufacturer needs to fulfill orders from two retailers in Miami, Florida. The manufacturer has the DVD players stored in two warehouses, warehouse A and warehouse B. Warehouse A has a total of 200 DVD players and warehouse B has 240. Retailer I needs 110 DVD players and retailer II needs 150. Transportation costs per DVD player are $16 from warehouse A to retailer I, $24 from warehouse A to retailer II, $26 from warehouse B to retailer I, and $14 from warehouse B to retailer II. How many DVD players should be sent from each warehouse to each retailer to minimize transportation costs? What is the minimum transportation cost?

36. A manufacturer of computer scanners has two plants, Plant A and Plant B, that can produce up to 400 units and 280 units per month, respectively. The units are shipped to two warehouses, warehouse I and warehouse II. Warehouse I needs at least 360 units each month, and warehouse II needs at least 300 units each month. Transportation costs per scanner are $150 from Plant A to warehouse I, $80 from Plant A to warehouse II, $160 from Plant B to warehouse I, and $60 from Plant B to warehouse II. How many scanners should be sent from each warehouse to each retailer to minimize transportation costs? What is the minimum transportation cost?

SECTION PROJECT

Linger Furniture Company manufactures office desks at three locations: Chicago, Detroit, and Fort Lauderdale. The company distributes the desks through regional warehouses located in Phoenix, Boston, and Cleveland. Each month the Chicago plant, Detroit plant, and Fort Lauderdale plant can produce up to 100 units, 300 units, and 300 units, respectively. Each month the Phoenix warehouse, Boston warehouse, and Cleveland warehouse need at least 300 units, 200 units, and 200 units, respectively. The transportation costs per desk are given in Table 4.4.7.

Table 4.4.7

From \ To	Phoenix	Boston	Cleveland
Chicago	$5	$4	$3
Detroit	$8	$4	$3
Fort Lauderdale	$9	$7	$5

(a) How many desks should be sent from each plant to each warehouse to minimize transportation costs?

(b) What is the minimum transportation cost?

Why We Learned It

In this chapter we learned how to solve linear programming problems that have more than two decision variables and more than two constraints. The method discussed to solve these types of linear programming problems is called the simplex method. As in Chapter 3, we saw that many managers in business can use the simplex method to maximize profit and minimize costs or to determine optimum manufacturing times. For example, suppose that a company makes three types of chairs and each chair must spend some time on three different machines. Due to certain restrictions, each machine has a maximum number of hours available each week. Assuming that the company knows the profit that it receives on each chair produced and sold, it would want to determine how many units of each chair to make in a week (given the time constraints of each machine) in order to maximize profit. The simplex method will also show the company how to maximize profit through optimal use of the machines. This means that the company can maximize profit while reducing wear and tear on the machines. We also learned that dietitians can use the simplex method to maximize healthy nutritional elements in a diet, such as vitamins, or minimize elements of a diet, such as total fat intake. Farmers can also use the simplex method to maximize profit, given restrictions of land and the availability of various fertilizers.

CHAPTER REVIEW EXERCISES

In Exercises 1 and 2, convert the system of linear inequalities into a system of equations by using slack variables.

1. $x_1 + x_2 \le 14$
$-2x_1 + 5x_2 \le 6$

2. $2x_1 + x_2 - 5x_3 \le 11$
$-3x_1 + 2x_2 - x_3 \le 10$

In Exercises 3–6, for the given standard maximum form linear programming problem, introduce slack variables and setup the initial simplex tableau.

3. Maximize: $z = x_1 + 3x_2$
Subject to: $x_1 + 2x_2 \le 8$
$x_1 - x_2 \le 4$
$x_1 \ge 0, x_2 \ge 0$

4. Maximize: $z = 3x_1 + 3x_2$
Subject to: $2x_1 - x_2 \le 8$
$x_1 + 3x_2 \le 7$
$5x_1 - 2x_2 \le 4$
$x_1 \ge 0, x_2 \ge 0$

5. Maximize: $z = 2x_1 + 3x_2 + x_3$
Subject to: $x_1 + 2x_2 + 3x_3 \le 14$
$-x_1 + 4x_2 - 2x_3 \le 3$
$x_1 \ge 0, x_2 \ge 0, x_3 \ge 0$

6. Maximize: $z = 3x_1 + 2x_2 + x_3$
Subject to: $x_1 + 2x_2 + 3x_3 \le 9$
$-x_1 + 4x_2 - x_3 \le 4$
$5x_1 - x_2 + 4x_3 \le 12$
$x_1 \ge 0, x_2 \ge 0, x_3 \ge 0$

In Exercises 7 and 8, for each given matrix:

(a) Identify the basic and nonbasic variables.
(b) Determine the solution given by the matrix.

7.
$$\begin{array}{c} \begin{array}{ccccc|c} x_1 & x_2 & s_1 & s_2 & z & \end{array} \\ \left[\begin{array}{ccccc|c} 0 & -3 & 1 & 5 & 0 & 6 \\ 1 & 2 & 0 & -2 & 0 & 14 \\ \hline 0 & 4 & 0 & 7 & 1 & 5 \end{array}\right] \end{array}$$

8.
$$\begin{array}{c} \begin{array}{cccccc|c} x_1 & x_2 & x_3 & s_1 & s_2 & z & \end{array} \\ \left[\begin{array}{cccccc|c} 1 & 0 & -2 & 4 & 5 & 0 & 6 \\ 0 & 1 & 3 & 2 & 1 & 0 & 14 \\ \hline 0 & 0 & -1 & 4 & 8 & 1 & 13 \end{array}\right] \end{array}$$

In Exercises 9 and 10:

(a) Pivot about the circled entry.
(b) Read the solution after pivoting.

9.
$$\begin{array}{c} \begin{array}{cccccc|c} x_1 & x_2 & s_1 & s_2 & s_3 & z & \end{array} \\ \left[\begin{array}{cccccc|c} 2 & 3 & 1 & 0 & 0 & 0 & 3 \\ -1 & \textcircled{4} & 0 & 1 & 0 & 0 & 4 \\ 1 & -2 & 0 & 0 & 1 & 0 & 12 \\ \hline 3 & 5 & 0 & 0 & 0 & 1 & 2 \end{array}\right] \end{array}$$

10.
$$\begin{array}{c} \\ \\ \\ \\ \\ \end{array}\begin{array}{cccccccc|c} x_1 & x_2 & x_3 & x_4 & s_1 & s_2 & s_3 & z & \\ -1 & 1 & 3 & 0 & 0 & 5 & 6 & 0 & 8 \\ 2 & 0 & 7 & 1 & 0 & 1 & 2 & 0 & 14 \\ \textcircled{-3} & 0 & 0 & 0 & 1 & 5 & 4 & 0 & 2 \\ \hline 6 & 0 & 3 & 0 & 0 & 4 & -1 & 1 & 4 \end{array}$$

In Exercises 11–15, perform the setup so that the solution can be determined by the simplex method as illustrated in Section 4.1, Example 5. In Review Exercises 26–30, you will determine optimal solutions for these exercises.

11. Music Notes offers introductory music workshops for drums and guitar. Each student attends an individual music theory session and a music lesson. Drum students attend a 2-hour music theory session and a 4-hour music lesson. Guitar students attend a 4-hour music theory session and a 2-hour music lesson. Each month Music Notes instructors can offer no more than 84 hours of music theory sessions and 78 hours of music lessons. Music Notes makes a profit of \$25 for each drum student and \$30 for each guitar student. To determine how many of each type of student should be admitted each month to maximize profit, do the following:

(a) Define the variables and write the mathematical model for this problem.

(b) Use slack variables to adapt the mathematical model to equational form.

(c) Write the initial simplex tableau.

12. (*continuation of Exercise 11*) Blues Notes, a local competitor of Music Notes, also offers introductory music workshops for drums and guitar. In addition to an individual music theory session and a music lesson, students also attend an individual music recording session. As with Music Notes, drum students attend a 2-hour music session and a 4-hour music lesson, and guitar students attend a 4-hour music theory session and a 2-hour music lesson. In addition, drum students and guitar students both attend a 2-hour music recording session at Blues Notes. Instructors at Blues Notes each month can offer no more than 64 hours of music theory sessions, 56 hours of music lessons, and 42 hours of music recording. Blues Notes makes a profit of \$30 for each drum student and \$35 for each guitar student. To determine how many of each type of student should be admitted each month to maximize profit, do the following:

(a) Define the variables and write the mathematical model for this problem.

(b) Use slack variables to adapt the mathematical model to equational form.

(c) Write the initial simplex tableau.

13. Sopris Mountaineering makes two types of sleeping bags: +15 sleeping bag and −15 sleeping bag (+15 and −15 indicate the temperature rating in degrees Fahrenheit). Each bag requires the services of two departments, in hours, as shown in Table R.1.

Table R.1

Department	**+15 bag**	**−15 bag**
Fabric cutting	3	6
Assembly and packaging	3	3

Fabric cutting has a maximum of 240 hours available each week, and assembly and packaging has a maximum of 180 hours available each week. Each +15 bag yields \$60 in profit and each −15 bag yields \$70 in profit. To determine the number of each type of bag that should be made each week to maximize profit:

(a) Define the variables and write the mathematical model for this problem.

(b) Use slack variables to adapt the mathematical model to equational form.

(c) Write the initial simplex tableau.

14. Table R.2 gives data for growing head lettuce and cantaloupe in California for an entire growing season (entries are in lb/acre).

Table R.2

	Head Lettuce	**Cantaloupe**
Nitrogen	173	175
Phosphate	122	161
Potash	76	39

SOURCE: U.S. Department of Agriculture, www.usda.gov/nass

Amara has 220 acres, and can get no more than 34,700 pounds of nitrogen, 27,845 pounds of phospate, and 13,350 pounds of potash. She expects a profit of \$120 per acre for head lettuce and \$100 per acre for cantaloupe. To determine how many acres she should allot for each crop to maximize profit, do the following:

(a) Define the variables and write the mathematical model for this problem.

(b) Use slack variables to adapt the mathematical model to equational form.

(c) Write the initial simplex tableau.

15. (*continuation of Exercise 14*) Amara wants to consider the possibility of allocating some acres to growing bell peppers. The amounts of nitrogen, phosphate, and potash needed for the entire growing season are 212 pounds per acre, 108 pounds per acre, and 72 pounds per acre, respectively. Amara expects a profit of \$140 per acre for bell peppers. With everything else remaining the same from Exercise 14, to determine how many acres she should allot for each crop to maximize profit, do the following:

(a) Define the variables and write the mathematical model for this problem.

(b) Use slack variables to adapt the mathematical model to equational form.

(c) Write the initial simplex tableau.

In Exercises 16–19, the initial simplex tableau of a linear programming problem is given. Use the simplex method to solve.

16.
$$\begin{array}{ccccc|c} x_1 & x_2 & s_1 & s_2 & z & \\ \hline 4 & 1 & 1 & 0 & 0 & 14 \\ -2 & 1 & 0 & 1 & 0 & 6 \\ \hline -2 & -3 & 0 & 0 & 1 & 12 \end{array}$$

17.
$$\begin{array}{cccccc|c} x_1 & x_2 & s_1 & s_2 & s_3 & z & \\ \hline -3 & 2 & 1 & 0 & 0 & 0 & 8 \\ 1 & 4 & 0 & 1 & 0 & 0 & 5 \\ -2 & -1 & 0 & 0 & 1 & 0 & 7 \\ \hline -4 & -5 & 0 & 0 & 0 & 1 & 0 \end{array}$$

18.
$$\begin{array}{cccccc|c} x_1 & x_2 & x_3 & s_1 & s_2 & z & \\ \hline 4 & 5 & 7 & 1 & 0 & 0 & 20 \\ 3 & 2 & 1 & 0 & 1 & 0 & 40 \\ \hline -3 & -4 & -2 & 0 & 0 & 1 & 0 \end{array}$$

19.
$$\begin{array}{cccccccc|c} x_1 & x_2 & x_3 & x_4 & s_1 & s_2 & s_3 & z & \\ \hline 2 & 5 & 0 & 0 & 1 & 0 & 0 & 0 & 5 \\ 0 & 1 & 2 & 1 & 0 & 1 & 0 & 0 & 14 \\ 2 & 0 & 5 & 0 & 0 & 0 & 1 & 0 & 12 \\ \hline 3 & -2 & -4 & -1 & 0 & 0 & 0 & 1 & 10 \end{array}$$

In Exercises 20–25, for the standard maximum form linear programming problems:

(a) Introduce slack variables and set up the initial simplex tableau.

(b) Use the simplex method to solve.

20. Maximize: $z = 2x_1 + 3x_2$

Subject to: $2x_1 + x_2 \leq 7$

$x_1 - x_2 \leq 4$

$x_1 \geq 0, x_2 \geq 0$

21. Maximize: $z = 2x_1 + 2x_2$

Subject to: $4x_1 - 2x_2 \leq 0$

$x_1 + x_2 \leq 7$

$2x_1 + x_2 \leq 8$

$x_1 \geq 0, x_2 \geq 0$

22. Maximize: $z = 4x_1 + 3x_2 + 5x_3$

Subject to: $3x_1 + x_2 + x_3 \leq 14$

$7x_1 - 2x_2 + 3x_3 \leq 4$

$x_1 \geq 0, x_2 \geq 0, x_3 \geq 0$

23. Maximize: $z = 3x_1 + x_2 + x_3$

Subject to: $x_1 - x_2 + x_3 \leq 5$

$2x_1 + x_2 + 4x_3 \leq 4$

$x_1 \leq 0, x_2 \geq 0, x_3 \geq 0$

24. Maximize: $z = 3x_1 + x_2 + 2x_3$

Subject to: $4x_1 + 2x_2 + x_3 \leq 10$

$x_1 + 3x_2 + 4x_3 \leq 14$

$3x_1 + 2x_2 + 4x_3 \leq 12$

$x_1 \geq 0, x_2 \geq 0, x_3 \geq 0$

25. Maximize: $z = 8x_1 + 2x_2$

Subject to: $x_1 + x_2 \leq 9$

$2x_1 + 2x_2 \leq 4$

$x_1 \geq 0, x_2 \geq 0$

In Exercises 26–30, set up and solve each using the simplex method. Interpret solutions as shown in Section 4.2, Example 3. Include an interpretation of any nonzero slack variables (see Exercises 11–15).

26. Music Notes offers introductory music workshops for drums and guitar. Each student attends an individual music theory session and a music lesson. Drum students attend a 2-hour music theory session and a 4-hour music lesson. Guitar students attend a 4-hour music theory session and a 2-hour music lesson. Each month Music Notes instructors can offer no more than 84 hours of music theory sessions and 78 hours of music lessons. Music Notes makes a profit of \$25 for each drum student and \$30 for each guitar student. Determine how many of each type of student should be admitted each month to maximize profit.

27. (*continuation of Exercise 26*) Blues Notes, a local competitor of Music Notes, also offers introductory music workshops for drums and guitar. In addition to an individual music theory session and a music lesson, students also attend an individual music recording session. As with Music Notes, drum students attend a 2-hour music theory session and a 4-hour music lesson, and guitar students attend a 4-hour music theory session and a 2-hour music lesson. In addition, drum students and guitar students both attend a 2-hour music recording session at Blues Notes. Instructors at Blues Notes each month can offer no more than 64 hours of music theory sessions, 56 hours of music lessons, and 42 hours of music recording. Blues Notes makes a profit of \$30 for each drum student and \$35 for each guitar student.

(a) Determine how many of each type of student should be admitted each month to maximize profit.

(b) Compare the results of Exercise 26 and Exercise 27a.

28. Sopris Mountaineering makes two types of sleeping bags: a +15 sleeping bag and a −15 sleeping bag (+15 and −15 indicate the temperature rating in degrees Fahrenheit). Each bag requires the services, in hours, of two departments as shown in Table R.3.

Table R.3

Department	+15 bag	−15 bag
Fabric cutting	3	6
Assembly and packaging	3	3

Fabric cutting has a maximum of 240 hours available each week, and assembly and packaging has a maximum of 180 hours available each week. Each +15 bag yields \$60 in profit and each −15 bag yields \$70 in profit. Determine the number of each type of bag that should be made each week to maximize profit.

29. Table R.4 gives data for growing head lettuce and cantaloupe in California for an entire growing season (entries are in lb/acre).

Table R.4

	Head Lettuce	**Cantaloupe**
Nitrogen	173	175
Phosphate	122	161
Potash	76	39

SOURCE: U.S. Department of Agriculture, www.usda.gov/nass

Amara has 220 acres and can get no more than 34,700 pounds of nitrogen, 27,845 pounds of phosphate, and 13,350 pounds of potash. She expects a profit for \$120 per acre for head lettuce and \$100 per acre for cantaloupe. Determine how many acres she should allot for each crop to maximize profit.

30. (*continuation of Exercise 29*) Amara wants to consider the possibility of allocating some acres to growing bell peppers. The amounts of nitrogen, phosphate, and potash needed for the entire growing season are 212 pounds per acre, 108 pounds per acre, and 72 pounds per acre, respectively. Amara expects a profit of \$140 per acre for bell peppers.

(a) With everything else remaining the same from Exercise 29, determine how many acres she should allot for each crop to maximize profit.

(b) Compare the results of Exercises 29 and 30a.

In Exercise 31 and 32, determine the transpose of the given matrix.

31. $\begin{bmatrix} 3 & 1 & 0 & 2 \\ 2 & 0 & 3 & 7 \\ 4 & 1 & 8 & 9 \end{bmatrix}$

32. $\begin{bmatrix} 0 & 2 \\ 3 & 4 \\ 1 & 7 \\ 8 & 9 \end{bmatrix}$

In Exercises 33 and 34, form the dual problem for the given linear programming problem.

33. Minimize: $w = 6y_1 + 4y_2$

Subject to: $4y_1 + 2y_2 \geq 10$

$y_1 + 3y_2 \geq 5$

$y_1 \geq 0,\ y_2 \geq 0$

34. Minimize: $w = 3y_1 + 2y_2 + y_3$

Subject to: $y_1 + y_2 + y_3 \geq 9$

$2y_1 + 3y_2 + 8y_3 \geq 10$

$y_1 + 5y_2 + y_3 \geq 2$

$y_1 \geq 0,\ y_2 \geq 0,\ y_3 \geq 0$

In Exercises 35 and 36, a linear programming problem is given along with the final simplex tableau for the dual problem. Use the final simplex tableau for the dual to:

(a) Find the solution to the primal problem.

(b) Find the solution to the dual problem.

35. Minimize: $w = y_1 + 4y_2$

Subject to: $y_1 + 3y_2 \geq 4$

$3y_1 + y_2 \geq 6$

$y_1 \geq 0,\ y_2 \geq 0$

$$\begin{array}{ccccc|c} x_1 & x_2 & s_1 & s_2 & z & \\ \hline 1 & 3 & 1 & 0 & 0 & 1 \\ 0 & -8 & -3 & 1 & 0 & 1 \\ \hline 0 & 6 & 4 & 0 & 1 & 4 \end{array}$$

36. Minimize: $w = 3y_1 + y_2$

Subject to: $4y_1 + y_2 \geq 9$

$y_1 + 2y_2 \geq 4$

$2y_1 + 3y_2 \geq 6$

$y_1 \geq 0,\ y_2 \geq 0$

$$\begin{array}{cccccc|c} x_1 & x_2 & x_3 & s_1 & s_2 & z & \\ \hline 1 & 0 & \frac{1}{7} & \frac{2}{7} & -\frac{1}{7} & 0 & \frac{5}{7} \\ 0 & 1 & \frac{10}{7} & -\frac{1}{7} & \frac{4}{7} & 0 & \frac{1}{7} \\ \hline 0 & 0 & 1 & 2 & 1 & 1 & 7 \end{array}$$

In Exercises 37–40:

(a) Construct the dual problem.

(b) Use the simplex method in the dual problem to find the solution to the original primal problem.

37. Minimize: $w = 3y_1 + 4y_2$

Subject to: $4y_1 + 2y_2 \geq 6$

$3y_1 + 4y_2 \geq 9$

$y_1 \geq 0,\ y_2 \geq 0$

38. Minimize: $w = 3y_1 + 9y_2$

Subject to: $y_1 + 5y_2 \geq 24$

$5y_1 + y_2 \geq 35$

$y_1 + y_2 \geq 8$

$y_1 \geq 0,\ y_2 \geq 0,\ y_3 \geq 0$

39. Minimize: $w = 16y_1 + 7y_2 + 22y_3$

Subject to: $y_1 + 2y_2 + 3y_3 \geq 5$

$2y_1 + y_2 + y_3 \geq 12$

$y_1 \geq 0,\ y_2 \geq 0,\ y_3 \geq 0$

40. Minimize: $w = 4y_1 + 5y_2 + 6y_3$

Subject to:
$$y_1 + 3y_2 + 6y_3 + 18$$
$$y_1 + y_2 + 2y_3 \geq 8$$
$$3y_1 + y_2 + 4y_3 \geq 5$$
$$y_1 \geq 0,\ y_2 \geq 0,\ y_3 \geq 0$$

41. Amara needs to supplement her diet with at least 60 milligrams (mg) of vitamin C and 12 mg of niacin. The nutrients are available in two supplement pills, Everyday and Allday. Everyday contains 6 mg of vitamin C and 2 mg of niacin, and Allday contains 6 mg of vitamin C and 1 mg of niacin. Each Everyday pill cost $0.08 and each Allday pill costs $0.07. Let y_1 represent the number of Everyday pills taken daily, and let y_2 represent the number of Allday pills taken daily.

(a) Write a linear programming problem to minimize the cost of adding the nutrients to Amara's diet.

(b) Write the dual of the linear programming problem in part (a).

(c) Use the simplex method on the dual problem in part (b) to solve the original problem in part (a). Your answer should include the numbers of each pill taken and the daily minimum costs.

42. Niel, a dietitian for a sports team, is to prepare an energy snack from two foods, PowerSource and EnergyMax, to meet certain nutritional requirements. The combination of the two foods must have at least 550 grams of carbohydrates, 400 grams of protein, and 200 grams of total fat. The foods have different nutritional characteristics as given in Table R.5.

Table R.5

		Nutrition (grams per ounce)		
Food	**Cost per Ounce**	**Carbohydrates**	**Protein**	**Fat**
PowerSource	$0.10	30	8	2
EnergyMax	$0.12	40	10	4

(a) Define the variables and write a linear programming problem to minimize Niel's cost while still meeting the nutritional requirements.

(b) Write the dual of the linear programming problem in part (a).

(c) Use the simplex method on the dual problem in part (b) to solve the original problem in part (a). Your answer should include the number of ounces of each type of food to use and the minimum cost.

43. Henderson Oil has two refineries, one in Aztec, New Mexico, and one in Buena Vista, Colorado. Each refinery can refine the amounts of fuel, in barrels, in a 24-hour period as shown in Table R.6.

Table R.6

Maximum Production of:	Aztec	Buena Vista
Low-octane gasoline	800	1200
High-octane gasoline	500	600
Diesel	900	600

Daily operating costs for the Aztec refinery are $36,000, and daily operating costs for the Buena Vista refinery are $32,000. Lissa, the production manager at Henderson Oil, receives an order for 120,000 barrels of low-octane gasoline, 100,000 barrels of high-octane gasoline, and 120,000 barrels of diesel.

(a) How many days should Henderson Oil operate each refinery to fulfill (or even exceed) the order while minimizing costs?

(b) What is the minimum cost?

44. Table R.7 gives nutritional information, in grams, for selected items from Wendy's menu.

Table R.7

Item	Carbohydrates	Protein	Fat
Grilled chicken sandwich	44	30	21
Big Bacon Classic®	45	33	31
Large Frosty™	91	14	14

SOURCE: Wendy's, www.wendys.com

(a) Dave wishes to eat nothing but grilled chicken sandwiches, Big Bacon Classics®, and large Frostys™. He wants at least 650 grams of carbohydrates and 130 grams of protein in a day. How many of each item should he have to eat to minimize his fat intake and still meet the nutritional requirements?

(b) What is Dave's minimum daily fat intake?

45. (a) Determine the shadow costs of the nutrients in Exercise 41.

(b) Use shadow costs along with the minimum cost answer in Exercise 41(c) to determine the new cost if Amara changes her nutrient requirements to 61 mg of vitamin C and 13 mg of niacin daily.

In Exercises 46 and 47, add slack variables and subtract surplus variables to write each system of inequalities in equational form.

46.
$$4x_1 + 3x_2 \leq 9$$
$$x_1 + 5x_2 \geq 4$$

47.
$$4x_1 + x_2 + 3x_3 \leq 12$$
$$x_1 + 2x_2 + 5x_3 \geq 9$$
$$3x_1 + x_2 + 7x_3 \leq 4$$

In Exercises 48 and 49, complete the following:

(a) Convert to a maximization problem.

(b) Add slack variables and subtract surplus variables to write in equational form.

(c) Write the initial simplex tableau.

48. Minimize: $w = x_1 + 4x_2$

Subject to:
$$2x_1 + 5x_2 \le 10$$
$$3x_1 + 4x_2 \ge 9$$
$$x_1 \ge 0, x_2 \ge 0$$

49. Minimize: $w = x_1 + 3x_2 + 8x$

Subject to:
$$x_1 + 2x_2 + 5x_3 \le 7$$
$$x_1 + x_2 + 8x_3 \ge 10$$
$$4x_2 + 2x_3 \ge 24$$
$$x_1 \ge 0, x_2 \ge 0, x_3 \ge 0$$

In Exercises 50–55, use the method in Section 4.4 to solve each linear programming problem.

50. Maximize: $z = 3x_1 + x_2$

Subject to:
$$2x_1 + 5x_2 \le 20$$
$$4x_1 + 3x_2 \le 14$$
$$5x_1 + 6x_2 \ge 10$$
$$x_1 \ge 0, x_2 \ge 0$$

51. Maximize: $z = 4x_1 + x_2$

Subject to:
$$x_1 + x_2 \ge 10$$
$$3x_1 + 4x_2 \ge 26$$
$$x_1 + 4x_2 \le 16$$
$$x_1 \ge 0, x_2 \ge 0$$

52. Maximize: $z = 2x_1 + 3x_2 + 4x_3$

Subject to:
$$3x_1 + x_2 + x_3 \le 25$$
$$x_1 + 4x_2 + 4x_3 \le 55$$
$$x_1 + 3x_2 - x_3 \ge 6$$
$$x_1 \ge 0, x_2 \ge 0, x_3 \ge 0$$

53. Minimize: $w = 3x_1 + 2x_2$

Subject to:
$$x_1 + 4x_2 \le 28$$
$$2x_1 + 5x_2 \ge 10$$
$$x_1 \ge 0, x_2 \ge 0$$

54. Minimize: $w = 6x_1 + 2x_2$

Subject to:
$$2x_1 + x_2 \le 18$$
$$x_1 + x_2 \le 10$$
$$3x_1 + 4x_2 \ge 4$$
$$x_1 \ge 0, x_2 \ge 0$$

55. Minimize: $w = 2x_1 + 3x_2 + 4x_3$

Subject to:
$$x_1 + x_2 + x_3 \le 30$$
$$3x_1 + x_2 + x_3 \ge 15$$
$$4x_1 + 4x_2 + x_3 \ge 5$$
$$x_1 \ge 0, x_2 \ge 0, x_3 \ge 0$$

56. Sopris Mountaineering makes two types of backpacks: 3500-cubic-inch daypacks and 5500-cubic-inch multiday packs. Each backpack requires the services of two departments, in hours, as shown in Table R.8.

Table R.8

	Labor	
Department	**Day Pack**	**Multiday Pack**
Fabric cutting	7	10
Assembly and packaging	3	5

Fabric cutting has a maximum of 540 hours available each week, and assembly and packaging has a maximum of 260 hours available each week. Due to prior contracts, the total number of backpacks made each week must be at least 15. Each daypack yields $20 in profit, and each multiday pack yields $50 in profit.

(a) How many of each type of backpack should be made and sold to maximize weekly profit? What is the maximum weekly profit?

(b) Interpret the value of all slack and surplus variables in the final tableau.

57. (*continuation of Exercise 56*) Lindsay, the head of research and development at Sopris Mountaineering, has discovered a way to reduce the amount of time needed for the day pack to 6 hours in the fabric-cutting department and 2 hours in the assembly and packaging department, while increasing the profit per day pack to $35. Everything else from Exercise 56 remains the same.

(a) How many of each type of backpack should be made and sold to maximize weekly profit? What is the maximum weekly profit?

(b) Interpret the value of all slack and surplus variables in the final tableau.

58. Table R.9 gives nutritional information, in grams, and price for selected items from Subway's menu.

Table R.9

Item	**Protein**	**Fat**	**Price**
6-Inch roasted chicken	25	6	$3.88
6-Inch Subway Club®	22	5	$3.55

SOURCE: Subway, www.subway.com

Kris desires to eat nothing but 6-inch roasted chicken sandwiches and 6-inch Subway Clubs. He wants at least 119 grams of protein, but no more than 28 gram of fat. Also, he wishes to eat at least one 6-inch roasted chicken sandwich and at least one 6-inch Subway Club sandwich. If Kris eats nothing but 6-inch roasted chicken sandwiches and 6-inch Subway Club sandwiches and meets his nutritional requirements, answer the following:

(a) How much of each food should Kris eat to minimize his daily food cost?

(b) What is his minimum daily food cost?

59. (*continuation of Exercise 58*) Kris found out that Subway has increased its price of the 6-inch roasted chicken sandwich to $4.00 and increased the price of the 6-inch Subway Club sandwich to $3.70. Assuming that all other data in Exercise 58 remain the same, answer the following:

(a) How much of each food should Kris eat to minimize his daily food cost?

(b) What is his minimum daily food cost?

(c) Compare the results from parts (a) and (b) of Exercises 58 and 59.

60. Henderson Oil has two refineries, one in Aztec, New Mexico, and one in Buena Vista, Colorado. Each refinery can refine the amounts of fuel, in barrels, in a 24-hour period as shown in Table R.10.

Table R.10

Maximum Production of:	Aztec	Buena Vista
Low-octane gasoline	800	1200
High-octane gasoline	500	600
Diesel	900	600

Daily operating costs for the Aztec refinery are $36,000, and daily operating costs for the Buena Vista refinery are $32,000. Lissa, the production manager at Henderson Oil, receives an order for 40,000 barrels of low-octane gasoline, 100,000 barrels of high-octane gasoline, and 120,000 barrels of diesel. Due to budgetary issues, the total number of days that the refineries can operate cannot exceed 250 days.

(a) How many days should each refinery operate to minimize costs while meeting (or even exceeding) the order? What is the minimum cost?

(b) Interpret the value of all slack and surplus variables in the final tableau.

CHAPTER 4 PROJECT

SOURCE: www.thesims.com and *The Sims: Prima's Official Strategy Guide* by Rick Barba

Maxis is a computer game company known for its Sim games. One of the more recent and very popular computer games is called "The Sims." In the game, the player controls people (called Sims) and helps them to get a job and establish relationships and shows them the way to the bathroom. One of the more strategic points of the game is to make the Sims happy. To do this, the player must purchase accessories for the home, such as a couch or a bubble blower. Most objects will affect the Sims' motive(s): those being Hunger, Comfort, Hygiene, Bladder, Energy, Fun, Social, and Room. Some objects do a better job at satisfying needs than others. For instance, would you rather watch T.V. on a 13-inch black and white screen or on a 53-inch HDTV? Unfortunately, the objects that do a better job cost a lot more. The point of this project is to minimize the amount of money spent while satisfying the given needs.

We will consider six objects and only three motives: Fun, Room, and Comfort. Table 1 gives information about these six objects and the needs that they satisfy.

Suppose that a Sim needed at least 16 Fun points, 6 Room points, and 15 Comfort points.

1. Define the six items using x_1 through x_6 as your variables. Write the objective function and the problem constraints.

2. Write the mathematical model in 1 in equational form. Write the initial simplex tableau.

3. Solve the linear program by using the Two-Phase Method in any way that you wish and interpret your results.

4. Referring to the objective function in 1 and the results of 3, why does it make sense that these items were purchased?

5. Referring to 3, for some items more than one is needed. Suppose that we don't want to purchase more than two items. Repeat 1 and 2 and construct new constraints for those items so that no more than two are purchased. Do not solve the corresponding linear programming problem.

Table 1 Items and Corresponding Price, Fun, Room, and Comfort Values

Item	Price § (in Simoleans)	Fun	Room	Comfort
Super Schlooper Bubble Blower	§710	3	0	2
Liebefunkenmann Piano	§5399	4	6	0
Spazmatronic Plasmatronic Go-Go Cage	§1749	7	3	0
WhirlWizard Hot Tub	§6500	2	0	6
Sky Surfer Inflatable Sofa	§190	0	1	3
Dolce Tutti Frutti Sofa	§1450	0	3	9

Chapter 5

Mathematics of Finance

$$FV = PMT\,\frac{(1+i)^n - 1}{i}$$

$$= 82.50\,\frac{\left(1+\frac{0.0525}{12}\right)^{360} - 1}{\frac{0.0525}{12}}$$

$$\approx 71{,}924.44$$

The old saying "a penny saved is a penny earned" is more than just a cliché. Consider the fact that in 2000, the average price of a pack of cigarettes in the U.S. was $2.75. Assuming a 30-day month, a pack-a-day smoker spent $82.50 per month on cigarettes. If the smoker quit and invested the monthly cigarette budget of $82.50 in a savings account earning 5.25% interest compounded monthly, the value of the account after 30 years would be greater than $70,000.

What We Know

In Chapters 3 and 4 we studied an important family of problems called linear programming problems. Analysts and managers use this to effectively run and operate their businesses.

Where Do We Go

In this chapter we study interest, another important topic for business people as well as those not necessarily in the management sciences. A good understanding of interest is essential for anyone who wants to manage their own finances or those of other people.

Section 5.1 Simple Interest

When we deposit money into a savings account or purchase a certificate of deposit, we have essentially loaned our money to the bank. When we borrow money from the bank to purchase a vehicle or a house, the bank has loaned us money. In either case, the amount of money loaned is called the **principal**. The individual or bank that loans the money receives a fee for lending the money. This fee is called **interest**. See Figure 5.1.1. This interest is classified as either **simple interest** or **compound interest**. We begin our journey into the mathematics of finance with the study of simple interest since the concept of simple interest lays the foundation for much of the material in this chapter.

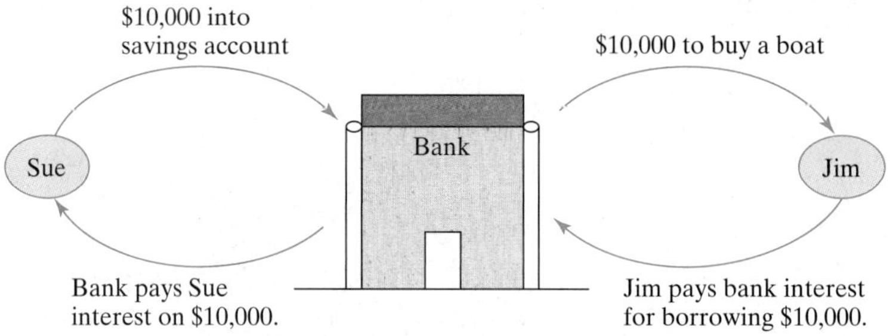

Figure 5.1.1

Simple Interest

From Your Toolbox

Interest rates should be converted to decimals before being used in any formulas. For more review on working with percents, consult the review in Appendix A.1.

When financial institutions advertise **interest rates** on certificates of deposit (CDs) or on loans, it usually means **rate per year**. In general, rates are expressed as percents. Since the word percent means **per hundred**, we can quickly convert from percents to decimals, and vice versa. For our purposes in this chapter, remember the information in the Toolbox to the left.

Simple interest is earned on the amount borrowed and not on any past interest. The **time** that the money is earning interest is given in years. **Simple interest** is simply the product of the principal, rate, and time.

Simple Interest

To compute **simple interest**, use the formula $I = Prt$, where

$P =$ principal (amount invested or borrowed)
$r =$ interest rate per year (as a decimal)
$t =$ time in years
$I =$ interest (earned or paid)

Example 1 Calculating Simple Interest

Lisa borrows $10,000 from her parents for a down payment on a house. She agrees to pay back the loan in 5 years at the rate of 6.5% simple interest. How much interest will Lisa pay?

Solution

Understand the Situation: For this situation, we have that the principal P is $10,000, the interest rate r is $6.5\% = 0.065$, and the time t is 5 years.

So the interest that Lisa will pay is

$$\begin{aligned} I &= Prt \\ &= 10{,}000(0.065)(5) \\ &= \$3250 \end{aligned}$$

Interpret the Solution: Lisa will pay \$3250 in interest in order to use the \$10,000 for 5 years. ■

Checkpoint 1 Now work Exercise 1.

When Lisa from Example 1 pays back the loan, she has to pay back the principal, P, plus the interest, I. So the total amount, A, that Lisa pays back to her parents is

$$A = P + I = \$10{,}000 + \$3250 = \$13{,}250$$

For Lisa, the \$10,000 was the amount that she borrowed now, so this amount is also referred to as a **present value**. The \$13,250 she pays back in 5 years is called the **future value**. In other words, \$13,250 is the future value of \$10,000 at a simple interest rate of 6.5% for 5 years.

In general, if a principal P is borrowed at a simple interest rate r, then after t years the borrower owes the lender an amount A, where

$$\begin{aligned} A &= P + I \\ &= P + Prt \\ &= P(1 + rt) \end{aligned}$$

P is also called the **present value**, and A is also known as the **future value**.

Future Value

The **future value** A of P dollars at interest rate r (as a decimal) for t years is

$$A = P(1 + rt)$$

Example 2 Calculating Total Amount Due

Juanita borrows \$6500 at 9.9% simple interest for 54 months. How much does Juanita owe at the end of the 54 months?

Solution

Interactive Activity

Using the result of Example 2, what is the future value of \$6500 at an interest rate of 9.9% for 4.5 years?

Understand the Situation: Here P is \$6500 and the simple interest rate, as a decimal, is 0.099. Recall that time t must be in years. So 54 months is $\frac{54}{12} = 4.5$ years.

The amount Juanita that owes at the end of 54 months is

$$\begin{aligned} A &= P(1 + rt) \\ &= 6500[1 + 0.099(4.5)] \\ &= \$9395.75 \end{aligned}$$

Interpret the Solution: The total amount due on Juanita's \$6500 loan at 9.9% simple interest for 54 months is \$9395.75. ■

The formula for future value, $A = P(1 + rt)$, has four variables. If we have values for any three of these variables, we can solve for the fourth. Examples 3 and 4 illustrate how we can do this.

Example 3 Determining a Simple Interest Rate

Joshua borrows \$2500 from Zachary and promises to pay back \$2630 in 6 months. What simple interest rate will Joshua pay?

Solution

Understand the Situation: We can use the formula for future value with $A = 2630$, $P = 2500$, and $t = \frac{6}{12} = 0.5$ and solve for r.

This gives us

$$
\begin{aligned}
A &= P(1 + rt) \\
2630 &= 2500[1 + r(0.5)] \\
2630 &= 2500 + 1250r \qquad \text{Distributive property} \\
130 &= 1250r \\
r &= \frac{130}{1250} = 0.104 = 10.4\%
\end{aligned}
$$

Interpret the Solution: This means that Joshua will pay a simple interest rate of 10.4%. ■

✓ Checkpoint 2 Now work Exercise 19.

Example 4 Determining a Present Value

Shannon desires to earn a simple interest rate of 9% on her investment. How much should she pay, to the nearest cent, for a note that will be worth \$7000 in 6 months?

Solution

Understand the Situation: Again we use the formula for future value, $A = P(1 + rt)$. However, here we are interested in finding the principal, P, or the present value. We have $A = 7000$, $r = 0.09$, and $t = \frac{6}{12} = 0.5$.

Using these values gives us

$$
\begin{aligned}
A &= P(1 + rt) \\
7000 &= P[1 + 0.09(0.5)] \\
7000 &= 1.045P \\
P &\approx \$6698.56 \qquad \text{Rounded to the nearest cent}
\end{aligned}
$$

Interpret the Solution: This means that Shannon should pay about \$6698.56 for the note. ■

✓ Checkpoint 3 Now work Exercise 29.

Future Value and Time

Table 5.1.1

Time, *t*, in Years	Amount Owed (future value)
1	\$2140
2	2280
4	2560
5	2700
8	3120
10	3400

Suppose that Gene borrows \$2000 from his Aunt Frances. Frances tells Gene to repay the principal plus 7% simple interest whenever he is able. Using the formula for future value, $A = P(1 + rt)$, Gene computes how much he would owe his aunt if he repaid the loan after various periods of time. Since $P = 2000$ and $r = 0.07$, the formula for future value becomes

$$A = P(1 + rt)$$
$$A = 2000(1 + 0.07t) = 2000 + 140t$$

Table 5.1.1 shows the future values for various values of t.

If we plot the points in Table 5.1.1 and connect them, we get the graph in Figure 5.1.2. Notice in the graph that the future value is a **linear function of time, *t*.**

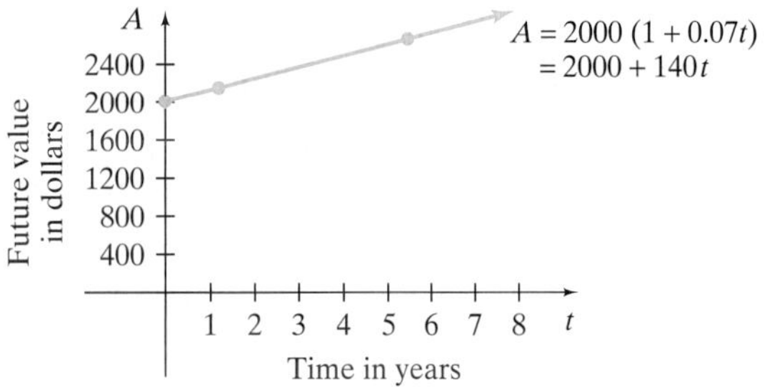

Figure 5.1.2

Technology Option

If we replace A with y and t with x, the equation graphed in Figure 5.1.2 becomes

$$y = 2000 + 140x$$

where y represents the future value and x represents the time in years. We can enter this equation into our graphing calculator and quickly reproduce Table 5.1.1 and Figure 5.1.2. Table 5.1.2 and Figure 5.1.3 show the output.

Table 5.1.2

X	Y1
1	2140
2	2280
4	2560
5	2700
8	3120
10	3400

X=

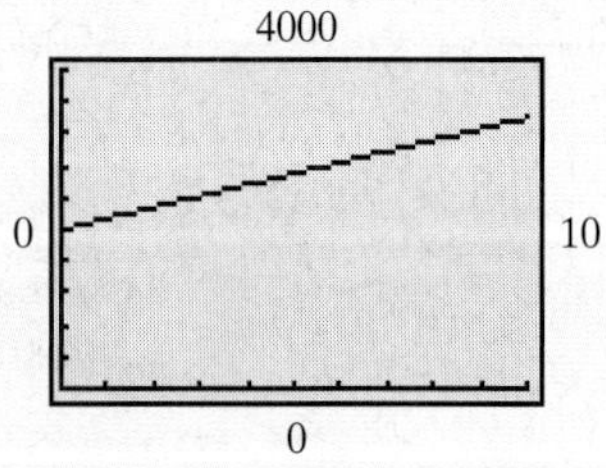

Figure 5.1.3

To learn how to make a table or to use the TABLE command, consult the online graphing calculator manual at www.prenhall.com/armstrong.

Example 5 Analyzing Future Values as a Function of Time

Stefani deposits \$5000 into a savings account that earns 5.5% simple interest.

(a) Write the future value, A, as a function of time, t.

(b) Make a table to determine A for $t = 1, 3, 5, 7$ and 9.

(c) Graph the equation from part (a). Determine the slope and interpret.

Table 5.1.3

t	$A = 5000 + 275t$
1	\$5275
3	5825
5	6375
7	6925
9	7475

Solution

(a) We use the formula for future value, $A = P(1 + rt)$, with $P = 5000$ and $r = 0.055$ to get

$$A = 5000(1 + 0.055t)$$
$$= 5000 + 275t$$

So the future value of Stefani's \$5000 at 5.5% simple interest as a function of time is given by $A = 5000 + 275t$.

(b) Table 5.1.3 shows the results for $t = 1, 3, 5, 7$, and 9.

(c) Figure 5.1.4 shows a graph of $A = 5000 + 275t$. The slope of the line is $m = 275$ (consult the Toolbox to the left).

From Your Toolbox

A linear equation in slope–intercept form $y = mx + b$ has slope given by m.

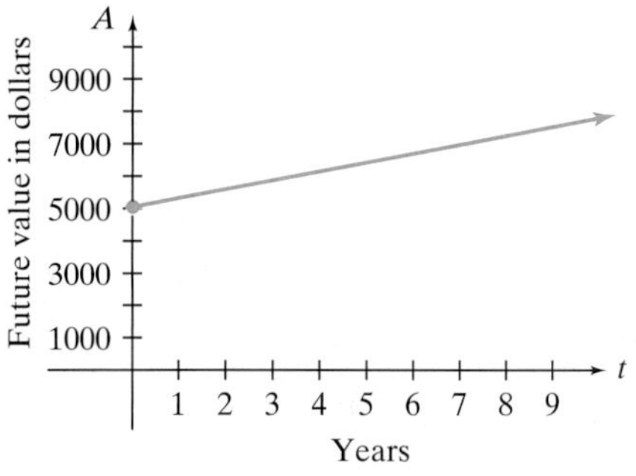

Figure 5.1.4 Graph of $A = 5000 + 275t$

Interpret the Solution: This means that each year Stefani earns \$275 in interest on her \$5000 deposit. ■

Doubling Time

Many times we wish to know how long it will take for our money to double. For example, suppose that Stefani in Example 5 wants to know when her deposit of \$5000 at 5.5% simple interest has a value of \$10,000. Since the future value as a function of time, as found in Example 5(a), is given by

$$A = 5000 + 275t$$

we can algebraically determine the time necessary for her money to double by substituting \$10,000 for A and solving for t.

$$10{,}000 = 5000 + 275t$$
$$5000 = 275t$$
$$t = \frac{5000}{275} \approx 18.18 \text{ years}$$

So it would take about 18.18 years for Stefani's \$5000 to double to \$10,000 at 5.5% simple interest.

Technology Option

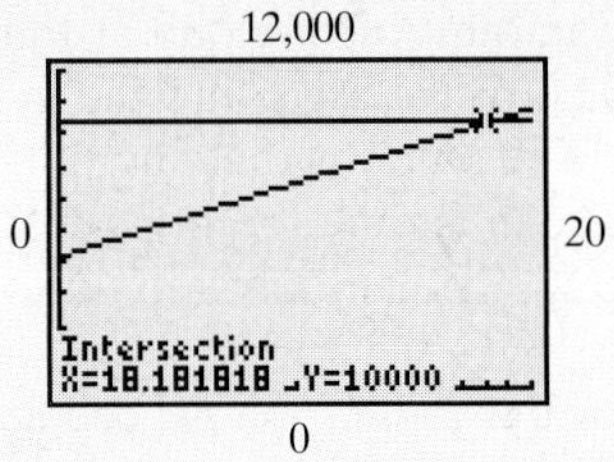

Figure 5.1.5

We can support the algebraic solution for the time required for Stefani's \$5000 to double to \$10,000 by graphing in the same viewing window $y_1 = 5000 + 275t$ and $y_2 = 10{,}000$ and using the INTERSECT command. See Figure 5.1.5.

To learn more about using the INTERSECT command on your calculator, consult the online graphing calculator manual at www.prenhall.com/armstrong.

In general, the time required for an amount of money P to double in value to $2P$ at a simple interest rate r can be determined by solving the following equation for t:

$$2P = P(1 + rt)$$

Dividing both sides by P gives us

$$2 = 1 + rt$$

Subtracting 1 from both sides and then dividing by r yields

$$1 = rt$$
$$t = \frac{1}{r}$$

Doubling Time for Simple Interest

The time required for money to double is given by

$$t = \frac{1}{r}$$

where t is time in years and r is the simple interest rate (as a decimal).

Example 6 Determining a Doubling Time

How long will it take \$8000 to double at 8% simple interest?

Solution

Understand the Situation: The principal (present value) is not important in answering this question. According to the formula, $t = \frac{1}{r}$, we only need to know r to determine the doubling time. Hence we have

$$t = \frac{1}{r}$$
$$= \frac{1}{0.08} = 12.5$$

Interpret the Solution: So at 8% simple interest the doubling time is 12.5 years.

Discounted Loans and Proceeds

Usually, when money is borrowed, it is expected that the amount borrowed (principal) plus interest will have been repaid at the end of the loan term. However, sometimes lenders collect the interest from the amount of the loan at the time that the loan is made. This type of loan is said to be **discounted**, and the interest that is deducted from the loan is the **discount**. The amount that the borrower receives is then called the **proceeds**. When the loan is due, the borrower pays back the amount before the discount was deducted. The amount repaid to the lender is called the **face value** of the loan.

If F represents the amount of the loan (its face value), then the proceeds, p, are simply

$$p = F - D$$

where D is the interest deducted, or the discount. Since the interest is given by

$$D = Frt$$

we have that the proceeds of a discounted loan are

$$p = F - Frt = F(1 - rt)$$

Discounted Loans and Proceeds

The **discount**, D, on a discounted loan of F dollars at annual simple interest rate r over t years is given by

$$D = Frt$$

The **proceeds**, p, of the loan is what the borrower receives and is given by

$$p = F - Frt = F(1 - rt)$$

Example 7 Computing Discount and Proceeds

Franklin signs a note for \$20,000 at 10% annual simple interest rate for 8 months. Compute the discount and the proceeds.

Interactive Activity

Suppose that Franklin in Example 7 actually needs \$20,000, that is, he needs the proceeds to be $p = 20{,}000$. Use the formula $p = F(1 - rt)$ with $p = 20{,}000$ and r and t having the same values as in Example 7 to determine F, the amount Franklin must borrow so that his proceeds are \$20,000.

WWW

Solution

Understand the Situation: We have $F = 20{,}000$, $r = 0.10$, and $t = \frac{8}{12} = \frac{2}{3}$. The discount is

$$\begin{aligned} D &= Frt \\ &= 20{,}000(0.10)\left(\frac{2}{3}\right) \\ &\approx 1333.33 \quad \text{Rounded to the nearest cent} \end{aligned}$$

So the discount is \$1333.33. This amount is deducted from the loan amount, F, to give the proceeds.

$$\begin{aligned} p &= F - D \\ &= 20{,}000 - 1333.33 \\ &= 18{,}666.67 \end{aligned}$$

Interpret the Solution: This means that Franklin receives \$18,666.67. However, after 8 months he must repay the face value of \$20,000. ■

Checkpoint 4 Now work Exercise 65.

SUMMARY

This section focused on discussing **simple interest**. **Future value** and **present value** of money were discussed, as was how to represent future value as a function of time. We analyzed the doubling time for simple interest and concluded the section with a look at **discounted loans** and **proceeds**.

- **Simple Interest:** To compute simple interest, use the formula $I = Prt$, where P = principal (amount invested or borrowed), r = interest rate per year (as a decimal), t = time in years, and I = interest (earned or paid).
- **Future Value:** The **future value** A of P dollars at interest rate r (as a decimal) for t years is $A = P(1 + rt)$.
- **Doubling Time for Simple Interest:** The time required for money to double is given by $t = \frac{1}{r}$, where t is time in years and r is the simple interest rate (as a decimal).
- **Discounted Loans and Proceeds:** The **discount**, D, on a discounted loan of F dollars at annual interest rate r over t years is given by $D = Frt$. The **proceeds**, p, of the loan are what the borrower receives and are given by $p = F - Frt = F(1 - rt)$.

SECTION 5.1 EXERCISES

In Exercises 1–8, the amount of money P is borrowed for period of time t at simple interest rate r. Compute the simple interest owed for the use of the money. Round answers to the nearest cent.

✓ **1.** $P = \$1000, r = 6.25\%, t = 3$ years

2. $P = \$2000, r = 7.25\%, t = 2$ years

3. $P = \$10{,}000, r = 5.5\%, t = 5$ years

4. $P = \$20{,}000, r = 6.5\%, t = 7$ years

5. $P = \$5000, r = 8.5\%, t = 9$ months

6. $P = \$8000, r = 7.5\%, t = 8$ months

7. $P = \$25{,}000, r = 8\%, t = 66$ months

8. $P = \$15{,}000, r = 9\%, t = 78$ months

In Exercises 9–16, the amount of money P is borrowed for period of time t at simple interest rate r. Compute the total amount, A, owed at the end of time t. (Recall that this is the same as future value.) Round answers to the nearest cent.

9. $P = \$2000, r = 7\%, t = 2$ years

10. $P = \$1000, r = 6\%, t = 3$ years

11. $P = \$10{,}000, r = 6.5\%, t = 5$ years

12. $P = \$25{,}000, r = 9.5\%, t = 7$ years

13. $P = \$7000, r = 7.5\%, t = 8$ months

14. $P = \$5000, r = 8.5\%, t = 9$ months

15. $P = \$20{,}000, r = 9.75\%, t = 78$ months

16. $P = \$15{,}000, r = 8.75\%, t = 66$ months

In Exercises 17–24, the amount of money P loaned for a period of time t along with the simple interest I charged are given. Determine the simple interest rate r of the loan. Round percentages to the nearest hundredth. See Example 3.

17. $P = \$2000, I = \$150, t = 1$ year

18. $P = \$3000, I = \$160, t = 1$ year

✓ **19.** $P = \$5000, I = \$900, t = 2$ years

20. $P = \$5000, I = \$880, t = 2$ years

21. $P = \$10{,}000, I = \$4280, t = 4.5$ years

22. $P = \$15{,}000, I = \$6300, t = 4.5$ years

23. $P = \$20{,}000, I = \$14{,}200, t = 6$ years, 9 months

24. $P = \$22{,}000, I = \$13{,}400, t = 5$ years, 9 months

In Exercises 25–32, determine the present value P given the future value A, time t, and annual simple interest rate r. Round to the nearest cent. See Example 4.

25. $A = \$5000, r = 8\%, t = 2$ years

26. $A = \$7500, r = 7\%, t = 3$ years

27. $A = \$10{,}000, r = 5.25\%, t = 5$ years

28. $A = \$12{,}000, r = 6.25\%, t = 6$ years

✓ **29.** $A = \$4000, r = 6.5\%, t = 9$ months

30. $A = \$3000, r = 5.5\%, t = 8$ months

31. $A = \$20{,}000, r = 7.75\%, t = 78$ months

32. $A = \$25{,}000, r = 8.25\%, t = 66$ months

In Exercises 33–36, complete the following:

(a) *Write the future value, A, as a function of time t. See Example 5(a).*
(b) *Graph the equation from part (a). Determine the slope and interpret. See Example 5(c).*

33. Sue deposits \$8000 into a savings account that earns 6.5% simple interest.

34. Cheryl deposits \$5000 into a savings account that earns 5.5% simple interest.

35. Rich deposits \$4000 into a savings account that earns 5.25% simple interest.

36. Renee deposits \$6000 into a savings account that earns 5.75% simple interest.

In Exercises 37–42, determine how long it will take a deposit to double at the given simple interest rate. Round answers to the nearest hundredth.

37. $r = 10\%$ **38.** $r = 6\%$

39. $r = 8.5\%$ **40.** $r = 6.75\%$

41. $r = 5.75\%$ **42.** $r = 7.25\%$

In Exercises 43–48, determine the discount and the proceeds given the loan amount F, annual interest rate r and time t. See Example 7.

43. $F = \$1000, r = 7\%, t = 8$ months

44. $F = \$1500, r = 8\%, t = 9$ months

45. $F = \$5000, r = 6.5\%, t = 2$ years

46. $F = \$6000, r = 8.5\%, t = 3$ years

47. $F = \$10{,}000, r = 8.25\%, t = 42$ months

48. $F = \$15{,}000, r = 7.75\%, t = 66$ months

In Exercises 49–54, determine the loan amount F to produce the given proceeds p at annual interest rate r over time t.

49. $p = \$1000, r = 7\%, t = 8$ months

50. $p = \$1500, r = 8\%, t = 9$ months

51. $p = \$5000, r = 6.5\%, t = 2$ years

52. $p = \$6000, r = 8.5\%, t = 3$ years

53. $p = \$10{,}000, r = 8.25\%, t = 42$ months

54. $p = \$15{,}000, r = 7.75\%, t = 66$ months

Applications

55. Shelby borrows \$25,000 for 8 years at a simple interest rate of 9%. Determine the interest due on the loan.

56. Lena borrows \$22,000 for 7 years at a simple interest rate of 8.5%. Determine the interest due on the loan.

57. Diane borrows \$3000 for 8 months at a simple interest rate of 8%. Determine the interest due on the loan.

58. Jeff borrows \$5000 for 9 months at a simple interest rate of 9%. Determine the interest due on the loan.

59. In June 2000, the World Bank granted China a loan for \$200 million (*Source:* http://suratkabar.com/arsip/6/2/170615.shtml). Assuming a simple interest rate of 9%, how much interest is due on the loan after 1 year?

60. In November 1996, the World Bank granted the Philippines \$281 million in loans (*Source:* www.worldbank.org). Assuming a simple interest rate of 8%, how much interest is due on the loan after 1 year?

61. Ardra borrowed \$22,000 from her aunt to start a beauty salon. She repaid her aunt after 2 years. The annual simple interest rate was 8.25%. Determine the total amount she repaid.

62. Maria borrowed \$15,000 from her uncle to start an auto parts store. She repaid him after 3 years. The annual simple interest rate was 7.25%. Determine the total amount she repaid.

63. Denyse borrowed \$20,000 from her grandfather to start a bakery. She repaid him after 30 months. The annual simple interest rate was 7.25%. Determine the total amount she repaid.

64. Mike borrowed \$18,000 from his mother to start a catering business. He repaid her after 42 months. The annual simple interest rate was 8.75%. Determine the total amount he repaid.

✓ **65.** Brian signs a note for \$7000 at 8.5% annual simple interest rate for 2 years. Assuming that this is a discounted loan, determine the discount and the proceeds.

66. Sandy signs a note for \$6000 at 8.25% annual simple interest rate for 3 years. Assuming that this is a discounted loan, determine the discount and the proceeds.

67. Suppose that Brian in Exercise 65 actually needs \$7000, that is, he needs the proceeds of the discounted loan to be \$7000. Assuming a 8.5% annual simple interest rate for 2 years, determine the amount that Brian must borrow so that his proceeds are \$7000.

68. Suppose that Sandy in Exercise 66 actually needs \$6000, that is, she needs the proceeds of the discounted loan to be \$6000. Assuming a 8.25% annual simple interest rate for 3 years, determine the amount that Sandy must borrow so that her proceeds are \$6000.

69. Melissa wants to buy a \$3000 entertainment center in 2 years. How much should she invest now at 6.75% simple interest to have enough for the entertainment center in 2 years?

70. Don wants to buy a \$40,000 recreational vehicle in 6 years. How much should he invest now at 7.25% simple interest to have enough for the recreational vehicle in 6 years?

71. In 52 months, Bill wants to build a \$20,000 three-season room addition on his house. How much should he invest now at 5.25% simple interest to have enough for the addition in 52 months?

72. In 66 months, Lisa wants to have $8000 worth of remodeling done to her basement. How much should she invest now at 4.75% simple interest to have enough for the remodeling in 66 months?

73. Francine borrows $2500 from Arthur and promises to pay back $2912.50 in 2 years. What interest rate will Francine pay?

74. Brock borrows $1400 from Misty and promises to pay back $1725.50 in 3 years. What interest rate will Brock pay?

75. Lenny borrows $5000 from Robert and promises to pay back $5912.50 in 2.5 years. What interest rate will Lenny pay?

76. Aida borrows $10,000 from Diane and promises to pay back $13,881.25 in 5 years, 9 months. What interest rate will Aida pay?

77. Dylan desires to earn a simple interest rate of 8.5% on his investment. How much should he pay, to the nearest cent, for a note that will be worth $5000 in 3 years?

78. Elle desires to earn a simple interest rate of 7.5% on her investment. How much should she pay, to the nearest cent, for a note that will be worth $8000 in 4 years?

79. As of 8/26/00, BankAtlantic.com offers a 12-month CD (certificate of deposit) at 7.25% given a minimum deposit of $1000 (*Source:* www.bankatlantic.com). Assuming the 7.25% is simple interest, how much, to the nearest cent, should Franco invest to have $2000 after 12 months?

80. As of 8/26/00, BankAtlantic.com offers a 12-month CD (certificate of deposit) at 7.25% given a minimum deposit of $1000 (*Source:* www.bankatlantic.com). Assuming the 7.25% is simple interest, how much, to the nearest cent, should Serena invest to have $3000 after 12 months?

81. Treasury bills (T-bills) can be purchased from the U.S. Treasury Department. At maturity, you are paid the face value of the bill. As of May 17, 2001, you could buy a 13-week T-bill with a face value of $1000 for $990.82 (*Source:* www.publicdebt.treas.gov). What annual simple interest rate does this T-bill earn?

82. Treasury bills (T-bills) can be purchased from the U.S. Treasury Department. At maturity, you are paid the face value of the bill. As of May 17, 2001, you could buy a 26-week T-bill with a face value of $1000 for $981.60 (*Source:* www.publicdebt.treas.gov). What annual simple interest rate does this T-bill earn?

83. If Alicia desires to earn an annual simple interest rate of 9.5% on a 26-week T-bill with a face value of $5000, how much should she pay for the T-bill?

84. If Horace desires to earn an annual simple interest rate of 8.75% on a 26-week T-bill with a face value of $10,000, how much should he pay for the T-bill?

85. Bill is considering a **home equity loan** that has an 8.5% annual simple interest rate. His home has appraised for $191,000 and he still has a balance of $120,000 on his first mortgage.

(a) Compute 80% of the appraisal minus the balance of his first mortgage. This is called the **loan to value** ratio and is the amount that the bank will loan Bill as a home equity loan.

(b) Interest is paid yearly on a home equity loan and is tax deductible. How much will Bill be able to take as a tax deduction in a given year due to interest on the home equity loan?

86. (*continuation of Exercise 85*) Suppose that Bill in Exercise 85 takes the home equity loan as computed in Exercise 85(a).

(a) Assuming the interest rate from Exercise 85, what is the total amount that Bill repays if he has the loan for 6 years?

(b) For the amount in part (a), what would his monthly payment be if he pays the loan off in 6 years with equal monthly payments? (*Hint:* There are 72 months in 6 years.)

SECTION PROJECT

In discounted loans, the actual interest rate can be different from the amount quoted by the bank. The stated rate is called the **nominal rate**, whereas the rate of interest actually paid is called the **effective rate**. For simple interest rate loans, the effective rate and the nominal rates are the same. However, for simple discount notes, the effective rate can be larger than the nominal rate, as this section project illustrates.

Mark Freeman agrees to pay his banker $20,000 in 6 months. The banker subtracts a discount of 8% and gives the rest to Mark.

(a) Compute the discount.
(b) Compute the proceeds.
(c) Using the formula $A = P(1 + rt)$, we can compute the effective rate. Use $A = P(1 + rt)$ with $A = 20{,}000$ (the loan amount), $P =$ the result from part (b), and $t = 0.5$ (6 months) and solve for r. Convert to a percent and this is the effective rate of interest.
(d) Rework parts (a)–(c) if Mark agrees to pay his banker the $20,000 in 1 year. Assume that the banker subtracts the same discount of 8%.

Section 5.2 Compound Interest

The concept of simple interest and how to compute it presented in Section 5.1 laid the foundation for our next topic, **compound interest**. If the interest earned is added to the original principal so that the next time interest is computed it is based on the original principal plus interest, then this type of interest is said to be **compounded**. In other words,

Compound interest is interest earned on the initial principal plus previously earned interest.

Compounding Periods

An important component in working with situations involving compound interest, is the **compounding period** or simply the **period**. The compounding period is the number of times a year that interest is compounded and is denoted by k. This means that at the end of each compounding period interest is calculated on the initial principal plus all interest earned to that point. Table 5.2.1 gives the most popular compounding periods used by most financial institutions.

Table 5.2.1

Compound Period Name	Number of Compounding Periods per Year, k
Annually	1
Semiannually	2
Quarterly	4
Monthly	12
Weekly	52
Daily	365

When interest is compounded k times per year, we need to determine the **interest per compounding period** or the **periodic rate**. The following shows how to compute the interest per compounding period.

Interest per Compounding Period

If r is the annual interest rate and k is the number of compounding periods per year, then the interest rate per compounding period, i, is given by

$$i = \frac{r}{k}$$

Note: We can think of i as giving the rate of interest for each compounding period.

Example 1 **Determining an Interest Rate per Compounding Period**

Suppose that a bank pays an annual interest rate of $r = 8\%$. Determine the interest rate per compounding period, i, when the number of compounding periods per year, k, is

(a) $k = 1$ (compounded annually)
(b) $k = 4$ (compounded quarterly)
(c) $k = 12$ (compounded monthly)

Solution

(a) The interest rate per compounding period is

$$i = \frac{r}{k} = \frac{0.08}{1} = 0.08 \quad (\text{or } 8\%)$$

(b) The interest rate per compounding period is

$$i = \frac{r}{k} = \frac{0.08}{4} = 0.02 \quad (\text{or } 2\%)$$

(c) The interest rate per compounding period is

$$i = \frac{r}{k} = \frac{0.08}{12} \approx 0.006667 \quad (\text{or about } 0.6667\%)$$

■

Interactive Activity

For the annual interest rate given in Example 1, compute i when the number of compounding periods per year, k, is 2, 52, and 365.

Formula for Compound Interest

To aid us in developing a formula for compound interest, let's review the simple interest formula from Section 5.1 and use it as shown in Example 2.

Example 2 Computing Compound Interest

Suppose that a bank pays 5% annual interest compounded quarterly. If we deposit $1000 into a savings account and the quarterly interest earned is left in the account, we can compute how much money is in the account after 1 year. Determine the amount in the account after 1 year.

Solution

Recall from Section 5.1 that the future value A of P dollars at interest rate r (as a decimal) for t years is given by $A = P(1 + rt)$. After 1 quarter (3 months) the amount in the account is

$$\begin{aligned} A &= P(1 + rt) \\ &= 1000\left[1 + 0.05\left(\frac{3}{12}\right)\right] \qquad \text{Note that } t = \frac{3}{12}. \\ &= 1000\left[1 + 0.05\left(\frac{1}{4}\right)\right] = \$1012.50 \end{aligned}$$

To compute the amount after the second quarter, the amount after the first quarter, $1012.50, is now the principal. Hence the amount after the second quarter is

$$\begin{aligned} A &= P(1 + rt) \\ &= 1012.50\left[1 + 0.05\left(\frac{1}{4}\right)\right] \\ &\approx \$1025.16 \qquad \text{Rounded to the nearest cent} \end{aligned}$$

To compute the amount after the third quarter, the amount after the second quarter, \$1025.16, is now the principal. So the amount after the third quarter is

$$\begin{aligned} A &= P(1+rt) \\ &= 1025.16\left[1+0.05\left(\frac{1}{4}\right)\right] \\ &\approx \$1037.97 \quad \text{Rounded to the nearest cent} \end{aligned}$$

Finally, to compute the amount after the fourth quarter, the amount after the third quarter, \$1037.97, is now the principal. Hence the amount after the fourth quarter is

$$\begin{aligned} A &= P(1+rt) \\ &= 1037.97\left[1+0.05\left(\frac{1}{4}\right)\right] \\ &\approx \$1050.94 \quad \text{Rounded to the nearest cent} \end{aligned}$$

Interpret the Solution: This means that after 1 year the total in the savings account is \$1050.94. ■

Interactive Activity

Rework Example 2, except here assume that the bank paid 5% simple interest. In other words, what is the future value of \$1000 at simple interest rate 5% in 1 year?

We would like to generalize the process in Example 2 and determine a formula for compound interest. Our first observation is that each time we computed the amount at the end of the given quarter in Example 2 we used the formula

$$A = P(1+rt)$$

Observe that in each calculation the expression in parenthesis $(1+rt)$ was of the form

$$\left[1+0.05\left(\frac{1}{4}\right)\right] = \left(1+\frac{0.05}{4}\right)$$

Notice that $\frac{0.05}{4} = \frac{r}{k} = i$, the **interest per compounding period**. So, using Example 2 as a model, in general the amount A_1 after the first compounding period is

$$A_1 = P(1+i) = P + Pi$$

We notice that the interest earned is given by Pi. Since A_1 takes the place of the principal P for the second compounding period, the amount A_2 after the second compounding period is

$$\begin{aligned} A_2 = A_1 + A_1 i &= A_1(1+i) \\ &= P(1+i)(1+i) \quad \text{Since } A_1 = P(1+i) \\ &= P(1+i)^2 \end{aligned}$$

Since A_2 takes the place of the principal P for the third compounding period, the amount A_3 after the third compounding period is

$$\begin{aligned} A_3 = A_2 + A_2 i &= A_2(1+i) \\ &= P(1+i)^2(1+i) \quad \text{Since } A_2 = P(1+i)^2 \\ &= P(1+i)^3 \end{aligned}$$

Since A_3 takes the place of the principal P for the fourth compounding period, the amount A_4 after the fourth compounding period is

$$\begin{aligned} A_4 = A_3 + A_3 i &= A_3(1+i) \\ &= P(1+i)^3(1+i) \quad \text{Since } A_3 = P(1+i)^3 \\ &= P(1+i)^4 \end{aligned}$$

Generalizing this pattern, after n periods the amount is given by the following:

Amount of Compound Interest

A principal P earning compound interest at a rate of i percent per compounding period has a **compound amount** (or **future value**) after n periods of

$$A = P(1+i)^n$$

where $i = \frac{r}{k}$ and $n = kt$. Also,

P = principal (present value)
i = interest rate per compounding period
r = annual interest rate
k = number of compounding periods in 1 year
t = time, in years
n = total number of compounding periods
A = compound amount (future value) at end of n periods

Note: Since $i = \frac{r}{k}$ and $n = kt$, an alternative compound interest formula is $A = P\left(1 + \frac{r}{k}\right)^{kt}$.

Example 3 Using the Compound Interest Formula

Suppose that \$2000 is invested at an 8% annual interest rate. What is the amount of the investment, rounded to the nearest cent, after 6 years if the 8% annual interest rate is compounded

(a) Annually
(b) Quarterly
(c) Monthly

Solution

For all three parts of the example, $P = 2000$, $r = 0.08$, and $t = 6$.

(a) For annual compounding, we have $i = \frac{r}{k} = \frac{0.08}{1} = 0.08$ and $n = kt = 1 \cdot 6 = 6$. The amount A is given by

$$\begin{aligned} A &= P(1+i)^n \\ &= 2000(1+0.08)^6 \approx \$3173.75 \end{aligned}$$

(b) For quarterly compounding, we have $i = \frac{r}{k} = \frac{0.08}{4} = 0.02$ and $n = kt = 4 \cdot 6 = 24$. The amount A is given by

$$\begin{aligned} A &= P(1+i)^n \\ &= 2000(1+0.02)^{24} \approx \$3216.87 \end{aligned}$$

(c) For monthly compounding, we have $i = \frac{r}{k} = \frac{0.08}{12}$ and $n = kt = 12 \cdot 6 = 72$. The amount A is given by

$$\begin{aligned} A &= P(1+i)^n \\ &= 2000\left(1 + \frac{0.08}{12}\right)^{72} \approx \$3227.00 \end{aligned}$$

Notice that we did not use a decimal approximation for $i = \frac{0.08}{12}$ in the formula. In general, we save any rounding until the final answer. ■

Checkpoint 1 Now work Exercise 7.

Technology Option

Many calculators have built-in financial commands. Figure 5.2.1(a) shows the input for Example 3(c) on a TI-83 using TVM Solver. Figure 5.2.1(b) shows the result (FV) in the TVM Solver window, and Figure 5.2.1(c) shows the result of tvm_FV. (Note that tvm_FV finds the future value.) To learn more about these commands or to see if your calculator has a financial menu, consult the online graphing calculator manual at www.prenhall.com/armstrong.

(a) (b) (c)

Figure 5.2.1 **(a)** $PV = -2000$ because it is a deposit. **(b)** Place the cursor after the equals sign for FV and press ALPHA ENTER. An indicator square in the left column designates the solution variable.

As Example 3 indicates, the more compounding periods there are, the greater the future value. Table 5.2.2 shows the results of Example 3 along with other compounding periods.

Table 5.2.2

Annual Rate (%)	Periodic Rate	Compounding Method	Total Number of Periods	Future Value after 6 Years	Interest Earned
8	$\frac{0.08}{1}$	Annually	6	\$3173.75	\$1173.75
8	$\frac{0.08}{2}$	Semiannually	12	3202.06	1202.06
8	$\frac{0.08}{4}$	Quarterly	24	3216.87	1216.87
8	$\frac{0.08}{12}$	Monthly	72	3227.00	1227.00
8	$\frac{0.08}{52}$	Weekly	312	3230.96	1230.96
8	$\frac{0.08}{365}$	Daily	2190	3231.98	1231.98

It may be surprising that the difference shown in Table 5.2.2 between compounding weekly and compounding daily, \$1.02, is relatively small. This small difference is because the interest earned approaches a limit. This limit is when the compounding is done **continuously**. (Continuous compound interest is found by using the formula $A = Pe^{rt}$, where $e \approx 2.718281828\ldots$.) Here, if the interest is compounded continuously, the future value after 6 years of the \$2000 principal would be \$3232.15, or \$0.17 more than daily compounding!

Present Value

Just as we did in Section 5.1, we can determine the present value for some future amount by using the formula $A = P(1+i)^n$. Example 4 illustrates the algebraic process.

Example 4 Determining a Present Value

To the nearest cent, determine how much Brad Bundy should invest now at 5.5% interest compounded monthly to have \$10,000 toward the purchase of a minivan in 4 years?

Interactive Activity

Show that a formula for present value can be rewritten as $P = \dfrac{A}{(1+i)^n}$ or as $P = A(1+i)^{-n}$.

WWW

Solution

Understand the Situation: Notice that we are given the future value $A = \$10{,}000$ for a compound investment and we need to find the present value (or principal) P. We know that $i = \dfrac{r}{k} = \dfrac{0.055}{12}$ and $n = kt = 12(4) = 48$.

This gives us

$$A = P(1+i)^n$$

$$10{,}000 = P\left(1 + \frac{0.055}{12}\right)^{48}$$

$$P = \frac{10{,}000}{\left(1 + \dfrac{0.055}{12}\right)^{48}} \approx \$8029.22$$

Interpret the Solution: This means that Brad should invest \$8029.22 now to have \$10,000 in 4 years. ■

Checkpoint 2

Now work Exercise 21.

Technology Option

Figure 5.2.2(a) shows the input for Example 4 on a TI-83 using TVM Solver. Figure 5.2.2(b) shows the result (PV) in the TVM Solver window, and Figure 5.2.2(c) shows the result of tvm_PV. (Note that tvm_PV finds the present value.) The negative sign in Figure 5.2.2(b) and (c) means a cash outflow or an investment.

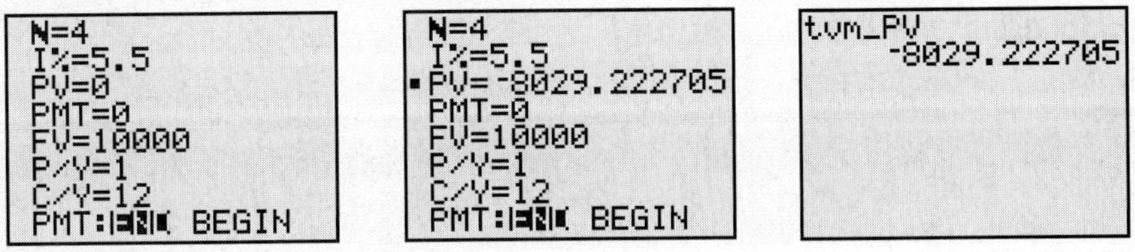

(a) (b) (c)

Figure 5.2.2

To learn more about these commands or to see if your calculator has a financial menu, consult the online graphing calculator manual at www.prenhall.com/armstrong.

Future Value and Time

In Section 5.1 we analyzed the future value of simple interest as a function of time. We can do the same thing with compound interest. Suppose that Sara invests \$2000 at 8% interest compounded monthly. Using the formula $A = P(1+i)^n$, where $i = \frac{r}{k}$ and $n = kt$, we can determine the amount that Sara has for various periods of time. Since $P = 2000$ and $i = \frac{0.08}{12}$, the formula for future value becomes

$$A = P(1+i)^n = P\left(1 + \frac{r}{k}\right)^{kt}$$
$$= 2000\left(1 + \frac{0.08}{12}\right)^{12t}$$

Table 5.2.3 shows the future value for different values of t.

Table 5.2.3

Time, t, in Years	Future Value, $A = 2000\left(1 + \frac{0.08}{12}\right)^{12t}$
1	\$2166.00
2	2345.76
4	2751.33
5	2979.69
8	3784.91
10	4439.28

If we plot the points in Table 5.2.3 and connect them with a smooth curve, we get the graph in Figure 5.2.3.

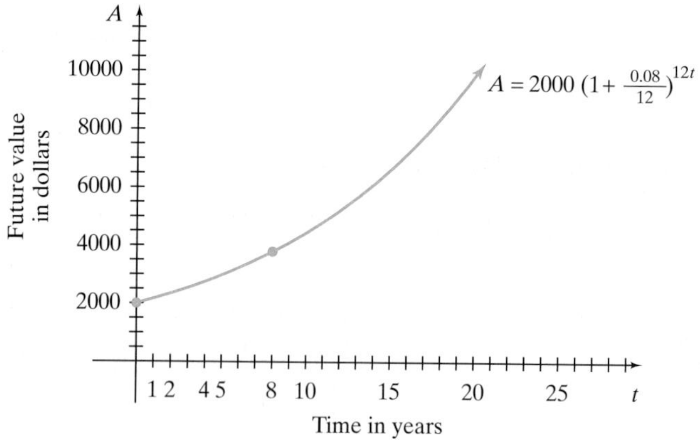

Figure 5.2.3

As we saw in Section 5.1, if the 8% interest was simple interest, we would have Sara's future value given by

$$A = 2000(1 + 0.08t)$$

Figure 5.2.4 shows the graph of the future value of a compound amount along with the future value for simple interest. With simple interest, the future value $A = 2000(1 + 0.08t)$ is a linear function of time, t, whereas with compound interest the future value $A = 2000\left(1 + \frac{0.08}{12}\right)^{12t}$ is an **exponential function of**

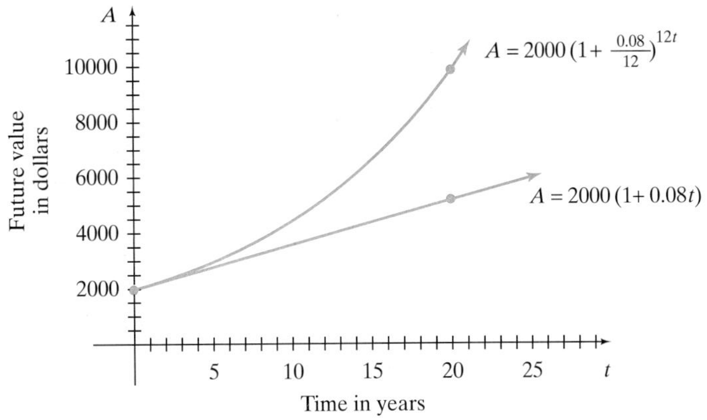

Figure 5.2.4

time, *t*. Observe that as time increases, the graph for the future value of compound interest starts to increase more rapidly than the graph for the future value of simple interest.

Technology Option

If we replace t with x, and let y_1 represent the future value for compound interest and y_2 represent the future value for simple interest, the equations in Figure 5.2.4 become

Compound Interest *Simple Interest*

$$y_1 = 2000\left(1 + \frac{0.08}{12}\right)^{12x} \qquad y_2 = 2000(1 + 0.08x)$$

We can enter these equations in our graphing calculator and analyze these two functions using the TABLE command as well as graphically. See Table 5.2.4 and Figure 5.2.5(a) and (b).

Table 5.2.4

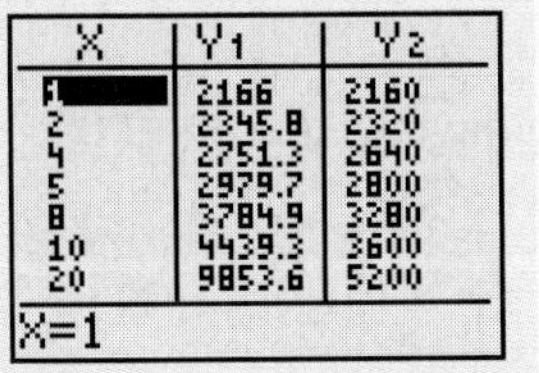

X	Y1	Y2
1	2166	2160
2	2345.8	2320
4	2751.3	2640
5	2979.7	2800
8	3784.9	3280
10	4439.3	3600
20	9853.6	5200

X=1

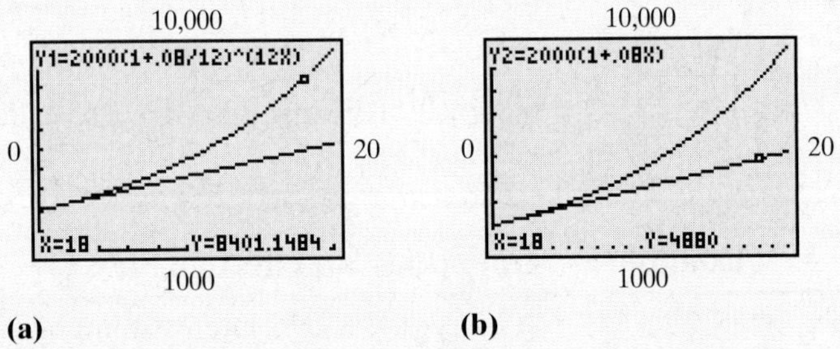

Figure 5.2.5 **(a)** Future value after 18 years with compound interest. **(b)** Future value after 18 years with simple interest.

Effective Rate

A common problem in financial planning is choosing the best investment from two or more investments. For example, is an investment that pays 10% interest compounded monthly better than one that pays 10.2% interest compounded semiannually? A good way to answer this question is to compare their **effective rates**. An effective rate is the simple interest rate that would give the same amount in 1 year as the compounded rate gives. For example, if \$2000 is deposited in a bank that pays 8% interest compounded monthly, then the future value after 1 year is

$$A = P(1+i)^n$$

$$= 2000\left(1+\frac{0.08}{12}\right)^{12} \approx \$2166.00 \qquad \text{Rounded to the nearest cent}$$

Using $A = P(1+rt)$ from Section 5.1 (future value for simple interest formula), we can determine the simple interest rate by substituting in $A = 2166$, $P = 2000$, and $t = 1$ and solving for r. This gives us

$$A = P(1+rt)$$

$$2166 = 2000(1+r)$$

$$\frac{2166}{2000} = 1+r$$

$$r = \frac{2166}{2000} - 1 = 0.083 = 8.3\%$$

The 8% rate is called the **nominal rate**. The 8.3% rate is the **effective rate** of interest. Notice that this effective rate is a **simple interest rate**. To find a formula relating effective rate, denoted r_{eff}, and nominal rate, r, we simply set the equations for future value for simple interest and future value for compound interest equal to each other. (Note that in both $t = 1$ since it is 1 year.)

$$\text{Future value (simple interest) in 1 year} = \text{Future value (compound interest) in 1 year}$$

$$P(1+r_{\text{eff}}) = P(1+i)^k \qquad \text{Since } n = kt \text{ and } t = 1, \text{ we have } n = k.$$

$$1+r_{\text{eff}} = (1+i)^k \qquad \text{Divide both sides by } P.$$

$$r_{\text{eff}} = (1+i)^k - 1$$

Effective Rate

The **effective rate**, r_{eff}, corresponding to the nominal rate i, where $i = \frac{r}{k}$, compounded k times per year, is given by

$$r_{\text{eff}} = (1+i)^k - 1$$

Note: Effective rates are also called **annual percentage yields**, or *APY*.

Example 5 Computing an Effective Rate

A bank pays 6.25% interest compounded quarterly. Determine the effective rate rounded to the nearest thousandth.

Solution

We have

$$r_{\text{eff}} = (1+i)^k - 1$$

$$= \left(1+\frac{0.0625}{4}\right)^4 - 1$$

$$= (1.015625)^4 - 1$$

$$\approx 0.0639801621$$

$$= 6.398\%$$

So the effective rate is 6.398%.

Interpret the Solution: This means that money invested at 6.398% simple interest earns the same amount of interest in 1 year as money invested at 6.25% interest compounded quarterly. Hence the effective rate of 6.25% interest compounded quarterly is 6.398%. ■

Checkpoint 3 Now work Exercise 41.

Example 6 Computing a Nominal Rate

Sam says that he has a CD with an effective rate of 6.5%. What annual nominal rate compounded monthly does this correspond to? Round to the nearest thousandth.

Solution

Understand the Situation: We use the same formula as we did in Example 5. Here we are trying to determine r, the annual nominal rate. Recall that $i = \frac{r}{k}$.

Proceeding, we get

$$r_{\text{eff}} = (1+i)^k - 1$$

$$0.065 = \left(1 + \frac{r}{12}\right)^{12} - 1 \qquad \text{Recall that } i = \frac{r}{k}.$$

$$1.065 = \left(1 + \frac{r}{12}\right)^{12}$$

$$\sqrt[12]{1.065} = 1 + \frac{r}{12}$$

$$\sqrt[12]{1.065} - 1 = \frac{r}{12}$$

$$r = 12(\sqrt[12]{1.065} - 1) \approx 0.0631403313 \quad \text{or} \quad 6.314\%$$

Interpret the Solution: So an annual nominal rate of 6.314% compounded monthly is the same as an effective rate of 6.5%. ■

Technology Option

Figure 5.2.6 shows how to calculate the effective rate for the situation in Example 5 by using the ►Eff(command on the TI-83. This command computes the effective rate given the nominal rate and the number of compounding periods in 1 year. Figure 5.2.7 shows the result of using the ►Nom(command on the TI-83 to answer the situation in Example 6. This command computes the nominal rate given the effective rate and the number of compounding periods in 1 year.

Figure 5.2.6

Figure 5.2.7

To see if your graphing calculator computes effective rates and nominal rates, consult the online graphing calculator manual at www.prenhall.com/armstrong.

Growth of Money

Many times we wish to know how many compounding periods it will take for a principal to reach a certain dollar amount. We can analyze these questions either graphically or algebraically. Algebraically, we need to use logarithms to solve $A = P(1+i)^n$ for n. (For a review on logarithms, see Appendix B.5.) Proceeding, we get

$$A = P(1+i)^n$$

$$\frac{A}{P} = (1+i)^n$$

$$\ln\left(\frac{A}{P}\right) = \ln(1+i)^n \quad \text{Take the natural log of both sides.}$$

$$\ln\left(\frac{A}{P}\right) = n\ln(1+i) \quad \text{Property of logarithms}$$

$$n = \frac{\ln\left(\frac{A}{P}\right)}{\ln(1+i)}$$

Number of Periods for Money to Grow from P to A

The number of compounding periods n for a present value P to reach a future value A with interest rate per compounding period i is given by

$$n = \frac{\ln\left(\frac{A}{P}\right)}{\ln(1+i)}$$

Note: The value for n will be rounded up to the nearest integer to guarantee that the value A is reached.

Example 7 Computing a Growth Time Algebraically

Determine how long it will take \$5000 to grow to at least \$8000 if it is invested at 6% interest compounded quarterly.

Solution

Understand the Situation: We begin by using the formula $n = \dfrac{\ln\left(\frac{A}{P}\right)}{\ln(1+i)}$ and note that $A = 8000$, $P = 5000$, and $i = \dfrac{0.06}{4}$.

This gives us

$$n = \frac{\ln\left(\frac{8000}{5000}\right)}{\ln\left(1+\frac{0.06}{4}\right)} = \frac{\ln 1.6}{\ln 1.015} \approx 31.56799396$$

Interpret the Solution: This means that it will take 32 quarters, or 8 years, for \$5000 to grow to at least \$8000 at 6% interest compounded quarterly. ■

Technology Option

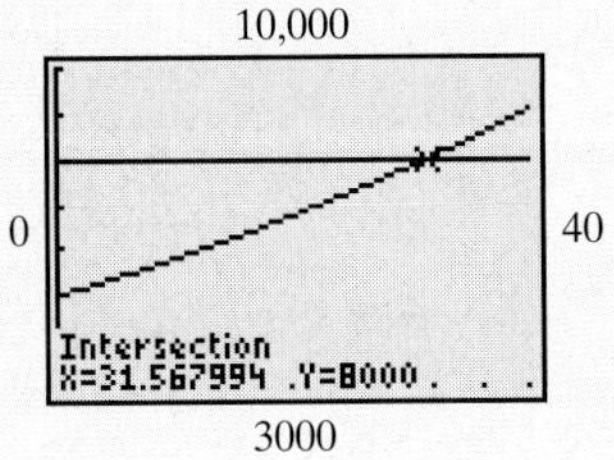

Figure 5.2.8

To analyze Example 7 graphically, we need to graph both sides of the equation

$$A = P(1+i)^n$$

$$8000 = 5000\left(1+\frac{0.06}{4}\right)^n$$

We enter $y_1 = 8000$ and $y_2 = 5000\left(1+\frac{0.06}{4}\right)^x$ into our graphing calculator, graph both in the same viewing window and then use the INTERSECT command. See Figure 5.2.8. Graphically, we reach the same result of 32 quarters for \$5000 to grow to \$8000.

Checkpoint 4

Now work Exercise 37.

SUMMARY

In this section we learned how to compute the future amount when the interest is **compounded**. We derived a formula for compound interest and used it to compute future and present values. We saw that the future value (when interest is compounded) as a function of time is an exponential function. We learned a formula for effective rates, and we finished the section with a formula for computing the number of compounding periods for money to grow from *P* to *A*.

SECTION 5.2 EXERCISES

In Exercises 1–10, compute the future value for the given investments at the end of 5 and 8 years. Round answers to the nearest cent.

	Principal or Present Value, *P*	Annual Interest Rate, *r*	Compounding Type
1.	\$3000	5%	Annually
2.	\$5000	6%	Annually
3.	\$2000	5.5%	Semiannually
4.	\$1500	6.5%	Semiannually
5.	\$8000	7.25%	Quarterly
6.	\$6000	7.75%	Quarterly
✓ **7.**	\$15,000	6.75%	Monthly
8.	\$10,000	6.25%	Monthly
9.	\$9700	8.125%	Daily
10.	\$8200	7.125%	Daily

In Exercises 11–14, use a graphing calculator's built-in financial package to compute the future value for the given investments at the end of 3 and 6 years. Round answers to the nearest cent.

	Principal or Present Value, *P*	Annual Interest Rate, *r*	Compounding Type
11.	\$8000	5.5%	Quarterly
12.	\$5000	6.5%	Monthly
13.	\$10,000	5.75%	Monthly
14.	\$9000	6.25%	Quarterly

In Exercises 15–24, compute the present value for the given future values. Round answers to the nearest cent.

	Future Value, *A*	Annual Interest Rate, *r*	Time, *t*	Compounding Type
15.	\$5000	5%	4	Annually
16.	\$4000	6%	6	Annually
17.	\$3000	5.5%	5	Semiannually

	Future Value, A	Annual Interest Rate, r	Time, t	Compounding Type
18.	$8000	6.5%	7	Semiannually
19.	$8500	7.25%	6	Quarterly
20.	$6500	7.75%	8	Quarterly
✓ 21.	$13,000	6.75%	7	Monthly
22.	$9000	6.25%	9	Monthly
23.	$8200	8.125%	10	Daily
24.	$7800	7.125%	15	Daily

In Exercises 25–28, use a graphing calculator's built-in financial package to compute the present value for the given future values. Round answers to the nearest cent.

	Future Value, A	Annual Interest Rate, r	Time, t	Compounding Type
🖩 25.	$10,000	6%	6	Quarterly
🖩 26.	$8000	5%	5	Monthly
🖩 27.	$9000	6.25%	7	Monthly
🖩 28.	$13,000	5.75%	8	Quarterly

In Exercises 29–32, match the given description with the correct graph, (a), (b), (c), or (d).

29. Which graph models the future value of $2000 invested in a simple interest rate savings account?

30. Which graph models the future value of $1000 invested in a simple interest rate savings account?

31. Which graph models the future value of $1000 invested in a compound interest rate savings account?

32. Which graph models the future value of $2000 invested in a compound interest rate savings account?

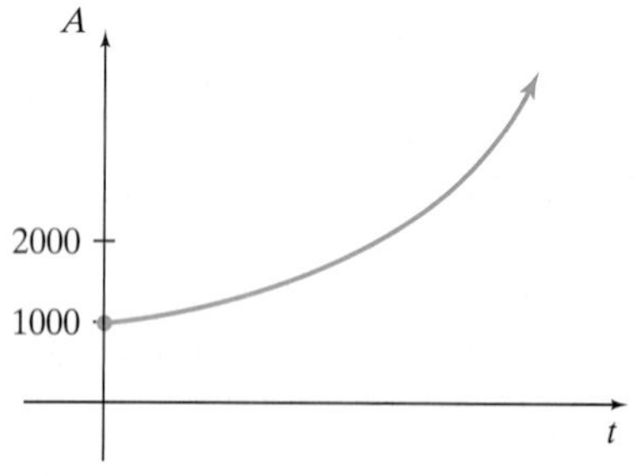

(a)

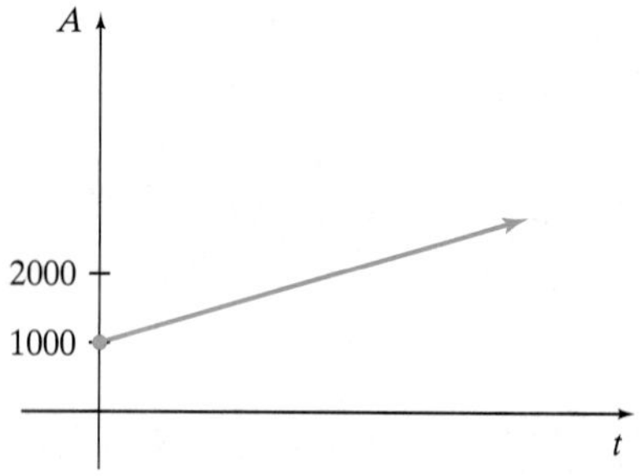

(b)

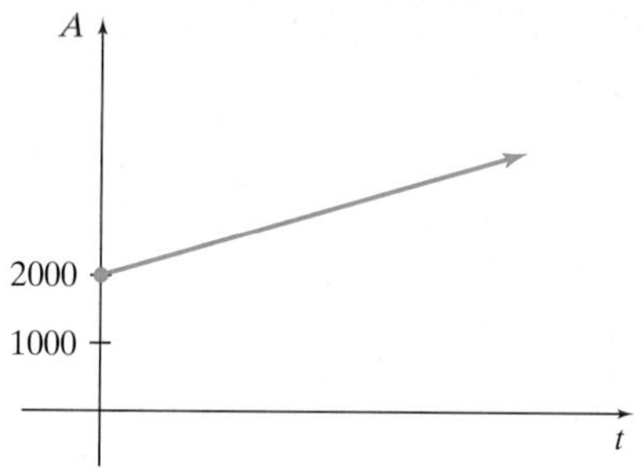

(c)

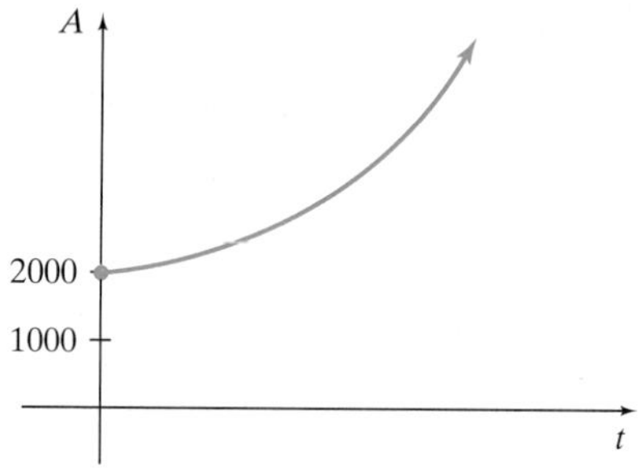

(d)

In Exercises 33–40, determine how long (that is, the number of compounding periods) it will take the given present value P to grow to the given future value A for the stated interest rate and compounding type. Round the answer up to the nearest integer.

	Present Value, P	Future Value, A	Annual Interest rate, r	Compounding Type
33.	$4000	$5000	6%	Annually
34.	$1000	$1500	5%	Annually
35.	$8000	$13,000	5.5%	Semiannually
36.	$5000	$9000	6.5%	Semiannually
✓ 37.	$10,000	$16,000	5.25%	Quarterly
38.	$13,000	$23,000	5.75%	Quarterly
39.	$12,000	$20,000	6.75%	Monthly
40.	$11,000	$19,000	6.25%	Monthly

In Exercises 41–48, determine the effective rate corresponding to each of the following nominal rates. Round the percentages to the nearest thousandth.

✓ **41.** 8.5% compounded quarterly

42. 7.5% compounded quarterly

43. 6% compounded monthly

44. 7% compounded monthly

45. 5.25% compounded semiannually

46. 6.25% compounded semiannually

47. 6.75% compounded monthly

48. 5.75% compounded monthly

In Exercises 49–56, use a graphing calculator and its built-in financial package to compute the effective rate.

49. 7.5% compounded quarterly

50. 8.5% compounded quarterly

51. 7% compounded monthly

52. 6% compounded monthly

53. 6.25% compounded semiannually

54. 5.25% compounded semiannually

55. 5.75% compounded monthly

56. 6.75% compounded monthly

In Exercises 57–60, for the given effective rate determine the nominal rate if the compounding is (a) compounded quarterly; or (b) compounded monthly.

57. 6.25% **58.** 6.75%

59. 5.5% **60.** 6.5%

In Exercises 61–64, use a graphing calculator and its financial package to determine, for the given effective rate, the nominal rate if the compounding is (a) compounded quarterly; or (b) compounded monthly.

61. 5.75% **62.** 6.25%

63. 7.5% **64.** 5.5%

Applications

In Exercises 65–68, use effective rates to determine which is the better investment.

65. 8% compounded monthly or 8.25% compounded annually?

66. 7% compounded monthly or 7.25% compounded quarterly?

67. 8.25% compounded quarterly or 8.125% compounded monthly?

68. 7.875% compounded quarterly or 7.75% compounded monthly?

69. The Consumer Price Index (CPI) is a measure of inflation. The CPI for all urban consumers in the United States during 1999 was 2.7%. If this rate holds true for the next 5 years, what will a $2.29 gallon of milk cost in 5 years if the rate is compounded annually? (*Source:* Bureau of Labor Statistics, http://stats.bls.gov)

70. Assuming the rate and compounding in Exercise 69, what will a $5.99 pound of coffee cost in 5 years?

71. Gene received a $2500 holiday bonus from his employer. He placed the bonus in an account earning 4.25% interest compounded monthly. How much is in his account after 4 years?

72. Kathy won $1000 in a raffle. She placed her winnings in an account earning 4.5% interest compounded quarterly. How much is in her account after 5 years?

73. Susan just purchased a new house costing $135,000. Houses in the area are appreciating at a rate of 3% per year. If this rate holds steady for 7 years, what will the house be worth assuming that this 3% rate is compounded annually?

74. Omar just purchased a new house costing $129,000. Houses in the area are appreciating at a rate of 3.25% per year. If this rate holds steady for 8 years, what will the house be worth assuming that this 3.25% rate is compounded annually?

75. Alec and Lexi compute that they will need $20,000 in 4 years to remodel their kitchen. How much should they invest now at 5.5% interest compounded quarterly to have the $20,000 in 4 years?

76. Lee and Jordan compute that they will need $13,000 in 6 years to build an addition on their house. How much should they invest now at 4.75% interest compounded daily to have the $13,000 in 6 years?

77. Sam and Diane have recently married. They compute that they will need $25,000 in 5 years as a down payment for their first home. How much should they invest now at 8.5% interest compounded monthly to have the $25,000 down payment in 5 years?

78. Brad and Tammy want to buy a new house in 7 years. They estimate that they will need $20,000 as a down payment. How much should they invest now at 7.75% interest compounded quarterly to have the $20,000 down payment in 7 years?

79. Jorge and Jenna have $10,000 to invest toward the purchase of a $16,000 ski boat. How many compounding periods will it take for the $10,000 to grow to at least $16,000 if it is invested at 5.25% compounded quarterly?

80. Doug and Patty have $8000 to invest toward the purchase of a $14,000 fishing boat. How many compounding periods will it take for the $8000 to grow to at least $14,000 if it is invested at 4.8% compounded monthly?

81. As of May 17, 2001, First Internet Bank of Indiana offered a money market account paying 4.50% interest compounded monthly. NetBank offered a money market account paying 4.40% interest compounded daily (*Source:* www.bankrate.com).

(a) Compute the effective rate for the First Internet Bank of Indiana account.

(b) Compute the effective rate for the NetBank account.

(c) Is one account a better investment than the other? Explain.

82. As of August 29, 2000, Key Bank USA offered an interest-bearing checking account paying 5.85% interest compounded daily. Compute the effective rate for this account.

Exercises 83–86 refer to **zero coupon bonds.** *A zero coupon bond is a bond that is sold now at a discount. When it matures at some time in the future, it pays its face value. No interest payments are made.*

83. Paige's grandparents are considering purchasing a \$45,000 face value zero coupon bond at her birth to help to finance her college education in 18 years. If money is worth 7.85% compounded annually, how much should they pay for this bond?

84. Repeat Exercise 83, except here assume a \$40,000 face value and 8.125% compounded annually.

85. If Beth and Frank pay \$7500 for a \$40,000 zero coupon bond that matures in 18 years, what is their annual compound rate of return?

86. If Todd and Laurie pay \$6235 for a \$38,000 zero coupon bond that matures in 17 years, what is their annual compound rate of return?

87. According to an advertisement in the September 2000 issue of *Money* magazine, \$10,000 invested in the T. RowePrice Science and Technology Fund on 9/30/87 would have increased in value to \$174,440 by 6/30/00. What interest rate compounded annually would account for this increase? (*Hint:* Use $t = 12.75$ years.)

88. According to an advertisement in the August 2000 issue of *Money* magazine, \$10,000 invested in the T. RowePrice Mid-Cap Growth Fund on 6/30/92 would have increased in value to \$54,518 by 3/31/00. What interest rate compounded annually would account for this increase? (*Hint:* Use $t = 7.75$ years.)

SECTION PROJECT

In July 2000, the Ohio Lottery started a game called Super Lotto PLUS™. At the time of ticket purchase, you must choose how you would like to receive your winnings (assuming that you win the jackpot!). Your options are yearly equal installments for 30 years or a one-time lump-sum payment. According to the Ohio Lottery Web site (www.ohiolottery.com), a jackpot of \$10,000,000 results in either \$333,333.33 per year for 30 years or a lump sum of \$4,232,055.00. (Both amounts are before taxes.)

(a) Assume that you won this jackpot and took the equal yearly installment plan. If you place the \$333,333.33 into a savings account earning 5.5% interest compounded monthly, how much interest would the account earn in 1 month?

(b) How much would the first installment of \$333,333.33 grow to in this account in 30 years?

(c) Assume that you won the jackpot and took the lump-sum option. If you place the \$4,232,055 into a savings account earning 5.5% interest compounded monthly, how much interest would the account earn in 1 month?

(d) How much would the \$4,232,055 grow to in 30 years?

(e) Which option (yearly installment or lump sum) would you take and why?

Section 5.3 Future Value of an Annuity

In Section 5.2 we saw how to find the future value of money when the money was invested at a fixed interest rate that was compounded. Many times, however, people and/or companies do not invest money and then sit back and watch it grow. Instead, they invest smaller amounts at regular intervals. There are many examples of this strategy. Some examples are annual life insurance premiums, loans paid off in monthly installments, and monthly contributions to a 403(b), 401(k), or an IRA account. This section explains the financial tool used by individuals as well as companies to fund large expenses, such as a child's college education or purchasing a large piece of equipment. This tool is called an **annuity**.

Future Value of an Annuity

An **annuity** is simply a sequence of equal periodic payments. The interval between payments, called the **payment period**, can be a year, every 6 months, every quarter, every month, and so on. For example, if you deposit \$500 into a savings account at the end of each year and the account earns 8% interest compounded annually, you have created an annuity for which the \$500 is the payment and the payment period is every year. Notice that each \$500 payment is made at the end of each year; hence the payment period is every year. If payments are made at the end of each period, then the annuity is called an **ordinary annuity**. (In this text we only study ordinary annuities. However, the Section Project in this section does analyze an **annuity due**.) In this section we concern ourselves with computing the amount, or **future value**, of an annuity.

Amount or Future Value of an Annuity

The amount, or **future value**, of an annuity is the sum of all payments made plus all interest earned.

Let's compute the future value of an ordinary annuity and see if it sheds some light on a possible formula. In Example 1, observe how we make use of the compound interest formula from Section 5.2.

Example 1 Computing the Future Value of an Annuity

Angela plans on depositing \$500 into a savings account at the end of each year for the next 5 years. The bank pays an interest rate of 8% compounded annually. What is the future value of Angela's account at the end of 5 years?

Solution

We first observe that, since Angela is depositing an equal amount (\$500) periodically (end of each year) into this account, she is constructing an annuity. Figure 5.3.1 illustrates the payment schedule for Angela.

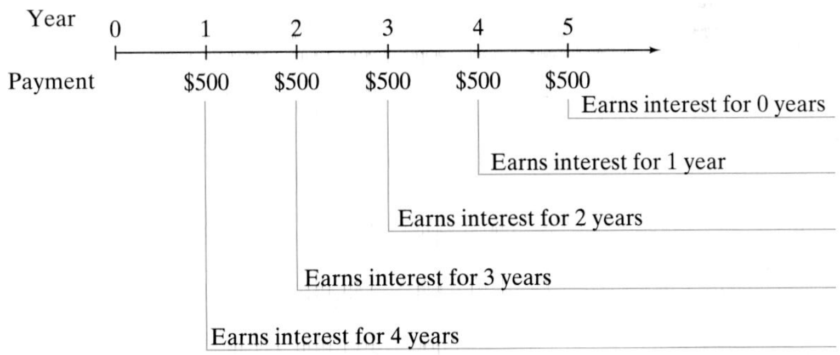

Figure 5.3.1

To determine the future value of Angela's account, we need to use the compound interest formula, $A = P(1+i)^n$, and determine the future value of each deposit. For example, at the end of the first year Angela deposits \$500 and it earns 8% interest compounded annually for 4 years. The future value of this amount, at the end of the fifth year, is given by

$$A = 500\left(1 + \frac{0.08}{1}\right)^4 = 500(1.08)^4$$

At the end of the second year, Angela deposits \$500 and it earns 8% interest compounded annually for 3 years. The future value of this amount, at the end of the fifth year, is given by

$$A = 500\left(1 + \frac{0.08}{1}\right)^3 = 500(1.08)^3$$

Similarly, the third, fourth, and fifth deposits at the end of years 3, 4, and 5 will earn 8% interest compounded annually for 2, 1, and 0 years, respectively. Their future values are given by

3rd deposit at the end of year 3: $A = 500(1.08)^2$
4th deposit at the end of year 4: $A = 500(1.08)^1$
5th deposit at the end of year 5: $A = 500(1.08)^0 = 500$

So the future value, FV, of Angela's account after 5 deposits, or at the end of 5 years, is

$$FV = 500(1.08)^4 + 500(1.08)^3 + 500(1.08)^2 + 500(1.08)^1 + 500$$
$$= \$2933.30 \quad \text{Rounded to the nearest cent}$$

The second to last line in Example 1 is disguising a useful pattern. If we write this line and factor out the 500, we get

$$FV = 500(1.08)^4 + 500(1.08)^3 + 500(1.08)^2 + 500(1.08)^1 + 500$$
$$= 500[(1.08)^4 + (1.08)^3 + (1.08)^2 + (1.08)^1 + 1]$$

Recognizing this pattern allows us to write an expression for the future value of Angela's account at the end of 8 years, if she continues to deposit \$500 at the end of each year. It is

$$FV = 500[(1.08)^7 + (1.08)^6 + (1.08)^5 + (1.08)^4 + (1.08)^3 + (1.08)^2 + (1.08)^1 + 1]$$

Following this pattern, if an amount PMT is deposited into an account at the end of each interest earning period, with i representing the interest rate per period, then the FV (future value) of the annuity at the end of the nth payment period is given by

$$FV = PMT[(1+i)^{n-1} + (1+i)^{n-2} + (1+i)^{n-3} + \cdots + (1+i) + 1]$$

From Your Toolbox

The sum of the first n terms of the geometric sequence with common ratio r is $1 + r + r^2 + r^3 + \cdots + r^{n-1} = \frac{1 - r^n}{1 - r}$. For a more detailed discussion on geometric sequences and the derivation of the sum formula, see Appendix B.6.

The expression inside the brackets is the sum of a **geometric sequence** with n terms and common ratio $(1 + i)$. A well-known algebraic formula (consult the Toolbox to the left) gives its sum as

$$(1+i)^{n-1} + (1+i)^{n-2} + \cdots + (1+i) + 1 = \frac{1 - (1+i)^n}{1 - (1+i)}$$
$$= \frac{1 - (1+i)^n}{-i} = \frac{(1+i)^n - 1}{i}$$

Hence the future value, FV, is given by the following:

Future Value of an Ordinary Annuity

If PMT represents the amount of each payment (or deposit) made at the end of each payment period and i is the interest rate per period, then the future value FV after n payment periods is

$$FV = PMT\frac{(1+i)^n - 1}{i}$$

Note: Some books represent $\frac{(1+i)^n - 1}{i}$ with $s_{\overline{n}|i}$, read "*S* angle *n* at *i*." Before the readily available technology we have today, tables were constructed listing values of $s_{\overline{n}|i}$ for various values of *n* and *i*. Thanks to calculators, we can handle many more situations for *n* and *i* than any table.

Example 2 Computing a Future Value of an Annuity

Kyle decides to deposit \$100 at the end of each month into a savings account earning a guaranteed interest rate of 4.5% compounded monthly. What is the value of this annuity at the end of 10 years?

Solution

Interactive Activity

How much of the future value computed in Example 2 is interest?

Understand the Situation: The periodic payment, *PMT*, is \$100. So $PMT = 100$, $r = 0.045$, and $k = 12$. The interest rate per period, *i*, is $i = \frac{r}{k} = \frac{0.045}{12} = 0.00375$. The number of payment periods, *n*, is $n = kt = 12(10) = 120$.

So the future value of this annuity is given by

$$FV = PMT\frac{(1+i)^n - 1}{i}$$

$$= 100\frac{(1+0.00375)^{120} - 1}{0.00375}$$

$$\approx \$15{,}119.81 \quad \text{Rounded to the nearest cent}$$

Interpret the Solution: This means that after 10 years of depositing \$100 at the end of each month into an account that earns 4.5% interest compounded monthly Kyle will have about \$15,119.81 in the account. ■

Checkpoint 1

Now work Exercise 5.

Technology Option

As stated in Section 5.2, some graphing calculators have a special FINANCE menu. To use this feature to solve Example 2, Figure 5.3.2(a) shows what is entered on the TI-83 using the TVM Solver, and Figure 5.3.2(b) shows the result when we solve for FV.

To learn more about the FINANCE menu and TVM Solver, or to learn if your calculator has this feature, consult the online graphing calculator manual at www.prenhall.com/armstrong. In case your graphing calculator does not have a FINANCE menu, we have supplied a program for the graphing calculator in Appendix C that will compute the future value. Consult the online graphing calculator manual at www.prenhall.com/armstrong for the program for your graphing calculator.

(a)

(b)

Figure 5.3.2 (a) *PMT* = −100 because it is a cash outflow. **(b)** Place the cursor after the equals sign for *FV* and press ALPHA ENTER. An indicator square in the left column designates the solution variable.

Example 3 Computing a Future Value of an Annuity

The winnings of Super Lotto PLUS™, an Ohio Lottery game, can be taken as a lump-sum distribution or as a 30-year annuity. A jackpot of \$22,000,000 has an after-taxes yearly payment (for 30 years) of \$502,333.34. (*Source:* www.ohiolottery.com) Assume that Seth wins this jackpot and takes the 30-year annuity option. Assume that he places \$250,000 of his yearly payment into an account that earns 6% interest compounded annually. What is the value of this account at the end of 30 years?

Solution

Here we have $PMT = \$250{,}000$, $i = \frac{0.06}{1} = 0.06$, and $n = 30$. So the future value is given by

$$\begin{aligned} FV &= PMT\frac{(1+i)^n - 1}{i} \\ &= 250{,}000\frac{(1+0.06)^{30} - 1}{0.06} \\ &= \$19{,}764{,}546.55 \end{aligned}$$

Observe that Seth would have deposited a total of $250{,}000(30) = \$7{,}500{,}000$, which means that he would earn $\$19{,}764{,}546.55 - \$7{,}500{,}000 = \$12{,}264{,}546.55$ in interest! Also notice that Seth would be able to spend \$252,333.34 of his winnings each year. ■

Sinking Funds

In many instances, individuals or companies know of an expense that will arise in the future. Examples include college expenses that parents will realize for a newborn baby or the replacement of delivery vehicles for a delivery company. When individuals or companies know of a future expense, many times they make regular payments into a fund to meet the future financial obligation. These funds are special types of annuities called **sinking funds**.

Sinking Fund

In general, any account that is established for the purpose of generating funds to cover a future financial obligation is called a **sinking fund**.

Example 4 Computing the Payment for a Sinking Fund

Craig and Laurie decide to start a college fund for their newborn son Jared. They estimate that they will need \$100,000. They decide to deposit money in a bank account that pays 6% interest compounded monthly. How much should they deposit every month to have generated \$100,00 after 17 years?

Solution

Understand the Situation: Here we want the future value, FV, to be \$100,000. We are trying to find PMT given $FV = 100{,}000$, $i = \frac{0.06}{12} = 0.005$, and $n = 12(17) = 204$.

Interactive Activity

Solve $FV = PMT\dfrac{(1+i)^n - 1}{i}$ for PMT to get a formula for the sinking fund payment.

WWW

Substituting into the formula for future value of an ordinary annuity and solving for PMT gives

$$FV = PMT\frac{(1+i)^n - 1}{i}$$

$$100{,}000 = PMT\frac{(1+0.005)^{204} - 1}{0.005}$$

Using a calculator, we get

$$PMT \approx \$283.10 \quad \text{Rounded to the nearest cent}$$

Interpret the Solution: So Craig and Laurie should deposit $283.10 each month into the account. Observe that the total amount that they invest is $283.10(204) = $57,752.40, which means that the interest earned is $100,000 − $57,752.40 = $42,247.60. ■

Checkpoint 2

Now work Exercise 21.

Technology Option

Figure 5.3.3(a) shows the setup to solve Example 4 using TVM Solver on a TI-83, and Figure 5.3.3(b) shows the result when we solve for PMT.

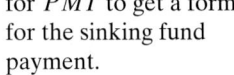

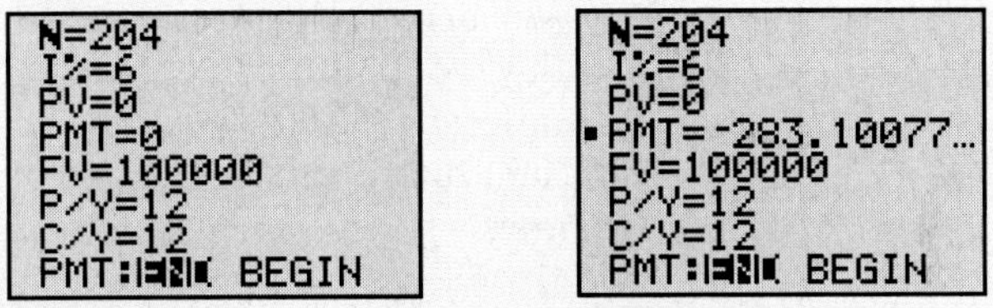

(a) (b)

Figure 5.3.3

In case your graphing calculator does not have a FINANCE menu, we have supplied a program for the graphing calculator in Appendix C that will compute the payment to fund a sinking fund. Consult the online graphing calculator manual at www.prenhall.com/armstrong for the program for your graphing calculator.

Determining an Interest Rate Using a Graphing Calculator

The last example of this section uses a graphing calculator to determine an interest rate that will achieve an individual's financial goal. This material may be skipped without loss of continuity.

Example 5 Determining an Interest Rate

Carl has started a new job and he decides to deposit $200 each month in an account that pays interest compounded monthly. Carl wishes to have at least

\$250,000 after 30 years. What annual interest rate compounded monthly would provide Carl with this amount?

Solution

Here we have $PMT = \$200$, $FV = \$250{,}000$, and $n = 12(30) = 360$. For Carl, the periodic interest rate i is $i = \frac{r}{12}$. We are trying to solve for r, the annual interest rate, in the equation

$$FV = PMT\frac{(1+i)^n - 1}{i}$$

$$250{,}000 = 200\frac{\left(1+\frac{r}{12}\right)^{360} - 1}{\frac{r}{12}}$$

This equation is very difficult to solve algebraically! We opt to solve it graphically by letting x represent the annual interest rate r. We then graph

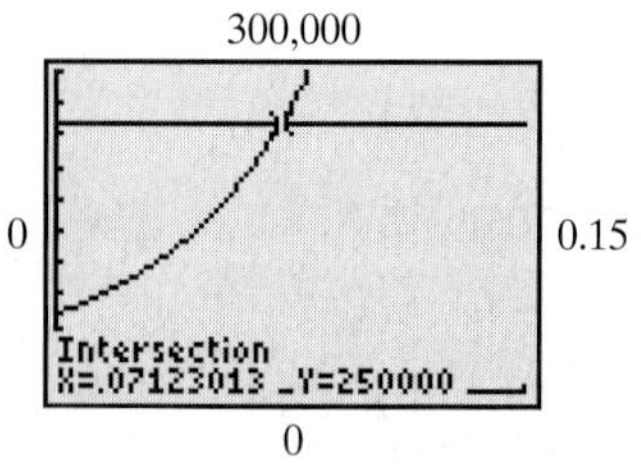

Figure 5.3.4

$$y_1 = 200\frac{\left(1+\frac{x}{12}\right)^{360} - 1}{\frac{x}{12}}$$

$$y_2 = 250{,}000$$

in the same viewing window and use the INTERSECT command as shown in Figure 5.3.4.

So an annual interest rate, to the nearest hundredth, of 7.13% compounded monthly would provide Carl with \$250,000. We rounded to 7.13% to guarantee that at least \$250,000 would be accumulated. A rate of 7.12% would not give a sufficient future value. We can check this by comparing the future values at the two rates.

At 7.12%: $$FV = PMT\frac{(1+i)^n - 1}{i} = 200\frac{\left(1+\frac{0.0712}{12}\right)^{360} - 1}{\frac{0.0712}{12}} = \$249{,}850.89$$

At 7.13%: $$FV = PMT\frac{(1+i)^n - 1}{i} = 200\frac{\left(1+\frac{0.0713}{12}\right)^{360} - 1}{\frac{0.0713}{12}} = \$250{,}346.20$$

■

We can also use the TVM Solver on the TI-83 to solve Example 5. Figure 5.3.5(a) shows the setup, and Figure 5.3.5(b) shows the result when we solve for I%.

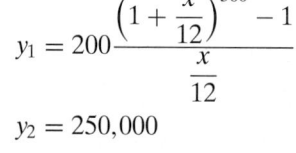

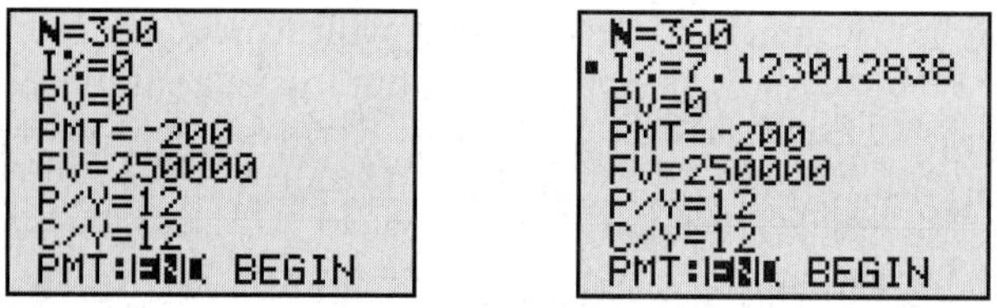

(a) (b)

Figure 5.3.5

SUMMARY

This section focused on computing the **future value of an ordinary annuity**. The formula that we developed used the compound interest formula from Section 5.2. We used the future value of an ordinary annuity formula to compute future values and to examine **sinking funds**.

- **Future Value of an Ordinary Annuity** If PMT represents the amount of each payment (or deposit) made at the end of each payment period, and i is the interest rate per period, then the future value FV after n payment periods is $FV = PMT\frac{(1+i)^n - 1}{i}$.
- **Sinking Fund** In general, any account that is established for the purpose of generating funds to cover a future financial obligation is called a **sinking fund**.

SECTION 5.3 EXERCISES

In Exercises 1–6, determine the amount of each annuity.

1. A deposit of $150 monthly for 10 years at an interest rate of 6% compounded monthly

2. A deposit of $250 monthly for 15 years at an interest rate of 7% compounded monthly

3. A deposit of $2000 annually for 30 years at an interest rate of 7.5% compounded annually

4. A deposit of $1800 annually for 25 years at an interest rate of 6.5% compounded annually

✓ **5.** A deposit of $500 quarterly for 30 years at an interest rate of 6.75% compounded quarterly

6. A deposit of $400 monthly for 25 years at an interest rate of 7.25% compounded quarterly

In Exercises 7–12, determine the amount of each annuity by using the FINANCE menu on your graphing calculator or the ANNFV calculator program found in Appendix C. Consult www.prenhall.com/armstrong *for the program for your calculator.*

7. A deposit of $250 monthly for 15 years at an interest rate of 7% compounded monthly

8. A deposit of $150 monthly for 10 years at an interest rate of 6% compounded monthly

9. A deposit of $1800 annually for 25 years at an interest rate of 6.5% compounded annually

10. A deposit of $2000 annually for 30 years at an interest rate of 7.5% compounded annually

11. A deposit of $400 quarterly for 25 years at an interest rate of 7.25% compounded quarterly

12. A deposit of $500 quarterly for 30 years at an interest rate of 6.75% compounded quarterly

In Exercise 13–18, determine how much of the amount (or future value) of each annuity is interest.

13. The annuity in Exercise 1

14. The annuity in Exercise 2

15. The annuity in Exercise 3

16. The annuity in Exercise 4

17. The annuity in Exercise 5

18. The annuity in Exercise 6

In Exercises 19–24, determine the periodic payment amount required for each sinking fund.

19. The amount needed is $20,000 in 10 years at 6% interest compounded annually. What is the annual payment?

20. The amount needed is $30,000 in 15 years at 7% interest compounded annually. What is the annual payment?

✓ **21.** The amount needed is $80,000 in 15 years at 6.5% interest compounded monthly. What is the monthly payment?

22. The amount needed is $60,000 in 12 years at 7.5% interest compounded monthly. What is the monthly payment?

23. The amount needed is $30,000 in 8 years at 7.25% interest compounded quarterly. What is the quarterly payment?

24. The amount needed is $40,000 in 7 years at 6.75% interest compounded quarterly. What is the quarterly payment?

In Exercises 25–30, determine the payment required for each sinking fund by using the FINANCE menu on your graphing calculator or the ANNPMT calculator program found in Appendix C. Consult www.prenhall.com/armstrong *for the program for your calculator.*

25. The amount needed is $30,000 in 15 years at 7% interest compounded annually. What is the annual payment?

26. The amount needed is $20,000 in 10 years at 6% interest compounded annually. What is the annual payment?

27. The amount needed is $60,000 in 12 years at 7.5% interest compounded monthly. What is the monthly payment?

28. The amount needed is $80,000 in 15 years at 6.5% interest compounded monthly. What is the monthly payment?

29. The amount needed is $40,000 in 7 years at 6.75% interest compounded quarterly. What is the quarterly payment?

30. The amount needed is \$30,000 in 8 years at 7.25% interest compounded quarterly. What is the quarterly payment?

In Exercises 31–36, refer back to Exercises 19–24 to answer the following. See Example 4.

(a) How much was invested?

(b) How much interest was earned?

31. The sinking fund in Exercise 19

32. The sinking fund in Exercise 20

33. The sinking fund in Exercise 21

34. The sinking fund in Exercise 22

35. The sinking fund in Exercise 23

36. The sinking fund in Exercise 24

In Exercises 37–42, determine the annual interest rate (to the hundredth) that guarantees the given future value of the annuity. See Example 5.

37. The future value is \$100,000, where \$200 is deposited each month for 20 years into an account paying interest compounded monthly.

38. The future value is \$150,000, where \$250 is deposited each month for 20 years into an account paying interest compounded monthly.

39. The future value is \$125,000, where \$2000 is deposited each year for 25 years into an account paying interest compounded annually.

40. The future value is \$175,000, where \$2000 is deposited each year for 30 years into an account paying interest compounded annually.

41. The future value is \$500,000, where \$1000 is deposited each quarter for 30 years into an account paying interest compounded quarterly.

42. The future value is \$600,000, where \$1250 is deposited each quarter for 25 years into an account paying interest compounded quarterly.

Applications

43. Teri deposits \$100 per month into a savings account that pays 4.75% interest compounded monthly. Determine the value of this account after 10 years.

44. Steve deposits \$80 per month into a savings account that pays 4.25% interest compounded monthly. Determine the value of this account after 15 years.

45. For Teri in Exercise 43, how much of the future value was money that she deposited and how much of the future value was interest earned?

46. For Steve in Exercise 44, how much of the future value was money that he deposited and how much of the future value was interest earned?

47. In 2000, the average price for a pack of cigarettes in the United States was \$2.75 (*Source:* www.legislature.state.tn.us/bills/currentga/BILL/SB0192.pdf). Assuming a 30-day month, a typical pack-a-day smoker spends about \$82.50 per month on cigarettes. Suppose that a smoker quits smoking and invests the monthly amount spent on cigarettes in a savings account earning 5.25% interest compounded monthly. Determine the value of the account after 30 years.

48. A Senate Office of Research Analysis estimates that California households with gross incomes at or below \$15,000 per year with one smoker spend an average of \$870 per year on tobacco products (*Source:* www.cbp.org/brief/bb980903.html). If this annual amount were invested in a savings account earning 5.75% interest compounded annually, determine the value of the account after 25 years.

49. From September 1987 through June 2000, the S&P 500 Index had an average annual return of 15.54% (*Source: Money* magazine, September 2000 issue). If Amanda invests \$2000 a year in a S&P 500 Index mutual fund (a fund that imitates the S&P 500) for 10 years, how much is her investment worth after 10 years assuming the 15.54% average annual rate?

50. For the 10-year period from 6/30/90 to 6/30/00, the T. RowePrice Science and Technology Fund had an average annual return of 27.43% (*Source: Money* magazine, September 2000 issue). Suppose that Gene invested \$2000 a year in this fund during this 10-year period of time. Assuming the 27.43% average annual rate, what is his investment worth after the 10-year period?

51. Jack and Diane are saving money for a down payment on a new house. They decide to deposit \$350 per month into an account paying 5.75% interest compounded monthly. Determine the value of their account after 4 years.

52. Hank and Peggy are saving money to purchase a new pickup truck. They decide to deposit \$100 per month into an account paying 5.25% interest compounded monthly. Determine the value of their account after 5 years.

53. Maury and Hannah would like to have \$30,000 in 6 years in order to purchase a new minivan. What amount should they deposit each month into a savings account paying 5.125% interest compounded monthly?

54. Ted and Kathy would like to have \$20,000 in 5 years as a down payment on a new house. What amount should they deposit each month into a savings account paying 5.375% interest compounded monthly?

55. Tamika and Tim have established a sinking fund in order to have \$120,000 in 17 years for their twin daughters' college education. How much should they deposit each quarter into a savings account paying 6.125% interest compounded quarterly?

56. Sally and Juan have established a sinking fund in order to have \$180,000 in 15 years for their children's college education. How much should they deposit each quarter into a savings account paying 6.5% interest compounded quarterly?

57. For her retirement, Irene wishes to have \$200,000 in a 403(b) account 30 years from now. How much should she deposit each month in an account paying 5.75% interest compounded monthly?

58. For his retirement, Dave wishes to have \$250,000 in a 403(b) account 25 years from now. How much should he deposit each month in an account paying 5.25% interest compounded monthly?

59. As of October 1999, tuition at 4-year public institutions was increasing at a rate of 3.4% per year (*Source:* www.dailybruin.ucla.edu). For 1998–1999, the average tuition at 4-year public institutions was \$3243 (*Source:* www.collegeboard.org).

(a) Assuming the 3.4% rate and the average tuition cost of \$3243, how much will tuition at a 4-year public institution be in 15 years? Round to the nearest dollar.

(b) Multiply the answer in part (a) by 4 to get an approximation for 4 years of tuition for a child starting college in 15 years.

(c) If parents wish to establish a sinking fund now to cover 4 years of tuition, how much should they deposit each month in an account paying 5% interest compounded monthly to have the amount in part (b) in 15 years?

60. As of October 1999, tuition at 4-year private schools was increasing at a rate of 4.7% per year (*Source:* www.dailybruin.ucla.edu). From 1998–1999, the average tuition at 4-year private schools was \$14,508 (*Source:* www.collegeboard.org).

(a) Assuming the 4.7% rate and the average tuition cost of \$14,508, how much will tuition at a 4-year private institution be in 15 years?

(b) Multiply the answer in part (a) by 4 to get an approximation for 4 years of tuition for a child starting college in 15 years.

(c) If parents wish to establish a sinking fund now to cover 4 years of tuition, how much should they deposit each month in an account paying 5% interest compounded monthly to have the amount in part (b) in 15 years?

61. On her 25th birthday, Greg and Chris opened an IRA for their daughter Erika and deposited \$2000. On every birthday up to and including her 55th, Erika deposits \$2000 into her IRA. How much will be in the account on her 55th birthday if the IRA earns

(a) 7% interest compounded annually?

(b) 9% interest compounded annually?

(c) 11% interest compounded annually?

62. (*continuation of Exercise 61*) Suppose that the \$2000 deposit on her 55th birthday is the last that Erika makes. Assume that she leaves the money in the IRA for 10 more years. On her 65th birthday, what is the value of her account if the IRA had earned and continues to earn

(a) 7% interest compounded annually?

(b) 9% interest compounded annually?

(c) 11% interest compounded annually?

63. (a) Debbie deposits \$2000 a year into an IRA earning 8% interest compounded annually. She makes her first deposit on her 23rd birthday and her last deposit on her 32nd birthday, a total of 10 deposits. Debbie makes no additional deposits. She leaves the amount accumulated from the 10 deposits in the account, earning 8% interest compounded annually until her 65th birthday. How much is in Debbie's account on her 65th birthday?

(b) Bubba deposits \$2000 a year into an IRA earning 8% interest compounded annually. He makes his first deposit on his 33rd birthday and continues to deposit \$2000 on every birthday until he is 65, a total of 33 deposits. How much is in his account when he makes his last deposit on his 65th birthday?

(c) Compare the results of parts (a) and (b). In your comparison, determine how much Debbie deposited versus how much Bubba deposited and the amount in each account on their 65th birthday. Are you surprised? Is there an investing moral here?

64. The after-taxes yearly installment for a \$10,000,000 Super Lotto PLUS™ jackpot in the Ohio Lottery is \$228,333.34 (*Source:* www.ohiolottery.com). This yearly installment is for 30 years and assumes a single winner. Suppose that Kahrin wins such a jackpot. Each year when she receives her annual installment she invests \$28,333.34 in an account earning 5.5% interest compounded annually. How much will be in this account in 30 years when she receives her final winning installment?

65. The after-taxes lump-sum payment for a \$10,000,000 Super Lotto PLUS™ jackpot in the Ohio Lottery is \$2,898,957.68. Suppose that Kyle wins such a jackpot. Kyle deposits his winnings in a savings account paying 5.5% interest compounded monthly.

(a) Kyle quits his job and decides to live off the interest. Each month he has the interest earned in his savings account transferred to his checking account. How much is this monthly amount?

(b) Kyle realizes he can live quite comfortably on \$10,000 each month. Of the amount found in part (a), he has \$10,000 put into his checking account and the rest invested each month in an account earning 6% interest compounded monthly. How much is this monthly investment?

(c) For the amount invested at 6% interest compounded monthly found in part (b), compute the value of this account after 30 years.

SECTION PROJECT

The formula $FV = PMT\frac{(1+i)^n - 1}{i}$ is for ordinary annuities for which payments are made at the *end* of each time period. Annuities in which the payments are made at the *beginning* of each time period are called **annuities due**. The **future value of an annuity due** of n payments of PMT dollars each at the beginning of each period, with i being the interest rate per period, is given by

$$FV = PMT\frac{(1+i)^{n+1} - 1}{i} - PMT$$

(a) Compare and contrast the formulas for ordinary annuities and annuities due.

(b) Rich and Cheryl deposit \$200 at the end of each month in an account paying 6.5% interest compounded monthly. This is an ordinary annuity. Compute the value of the account after 30 years.

(c) Suppose that Rich and Cheryl deposit \$200 at the beginning of each month in an account paying 6.5% interest compounded monthly. This is an annuity due. Compute the value of the account after 30 years.

(d) How much more money would Rich and Cheryl have with an annuity due versus an ordinary annuity? Consult the results from parts (b) and (c).

Section 5.4 Present Value of an Annuity

In Section 5.2 we studied compound interest and learned the compound interest formula, $A = P(1+i)^n$. Recall that A was called the **future value**. We also learned how to find the **present value**, which is the amount of money needed now to obtain the amount A in the future. A similar idea holds for annuities. For example, suppose that you want to withdraw \$500 per month each month for the next 4 years from an account that earns 6.5% compounded monthly. The amount of money initially required in the account to allow this to happen is the sum of the present values of each of these \$500 withdrawals. We will return to this scenario in Example 1, but let's first take a Flashback to Example 1 from Section 5.3.

Flashback

Angela's Annuity Revisited

In Example 1 in Section 5.3, we saw that Angela deposits \$500 at the end of each year into a savings account that pays an interest rate of 8% compounded annually. The future value of Angela's account at the end of 5 years can be found by using the **future value of an ordinary annuity** formula

$$FV = PMT\frac{(1+i)^n - 1}{i}$$

Using this formula, the future value of Angela's account is

$$FV = 500\frac{(1+0.08)^5 - 1}{0.08} \approx \$2933.30 \qquad \text{Rounded to the nearest cent}$$

How much would Angela have to invest today, in one lump sum, at 8% interest compounded annually to yield the same amount of \$2933.30 at the end of 5 years?

Flashback Solution

Understand the Situation: This is nothing more than computing a present value using the compound interest formula discussed in Section 5.2.

Using the compound interest formula,

$$A = P(1+i)^n$$

with $A = 2933.30$, $i = \dfrac{0.08}{1} = 0.08$, and $n = 5$, and solving for P gives us

$$2933.30 = P(1+0.08)^5$$

$$2933.30 = P(1.08)^5$$

$$P = \frac{2933.30}{(1.08)^5} \approx \$1996.35 \qquad \text{Rounded to the nearest cent}$$

Interpret the Solution: So Angela would have to deposit \$1996.35 now at 8% interest compounded annually to have \$2933.30 at the end of 5 years.

Present Value of an Annuity

What we did in the Flashback was determine the **present value of an annuity**. If we look back at what we did in the Flashback, we should be able to determine a formula for the present value of an annuity. The future value of Angela's annuity was determined to be \$2933.30 and was found by using

$$FV = PMT\frac{(1+i)^n - 1}{i}$$

The present value of \$1996.35 was found by using the formula

$$A = P(1+i)^n$$

where A was \$2933.30. Since we are trying to find the present value so that $FV = A$, we can simply equate the two equations as follows:

$$P(1+i)^n = PMT\frac{(1+i)^n - 1}{i}$$

To solve for P, we multiply both sides by $(1+i)^{-n}$ and get

$$P = PMT(1+i)^{-n}\frac{(1+i)^n - 1}{i}$$

Distributing $(1+i)^{-n}$ on the right side gives us

$$P = PMT\frac{(1+i)^{-n}(1+i)^n - (1+i)^{-n}}{i} = PMT\frac{1-(1+i)^{-n}}{i}$$

We have now developed a formula to determine the **present value of an annuity**.

From Your Toolbox

Algebra facts used in developing the formula for present value:

$b^0 = 1, \quad b \neq 0$

$b^n \cdot b^m = b^{n+m}$

$b^n \cdot b^{-n} = b^0 = 1, \quad b \neq 0$

Present Value of an Annuity

The present value PV of an annuity with n payments of PMT dollars made at the end of each payment period with i as the interest rate per period is given by

$$PV = PMT\frac{1-(1+i)^{-n}}{i}$$

Note: Some books represent $\dfrac{1-(1+i)^{-n}}{i}$ with $a_{\overline{n}|i}$, read "*a* angle *n* at *i*." Before the readily available technology we have today, tables were constructed listing values of $a_{\overline{n}|i}$ for various values of n and i. Thanks to calculators, we can handle many more situations for n and i than can any table.

Example 1 Computing a Present Value of an Annuity

Determine the present value of an annuity that pays \$500 per month for 4 years at 6.5% interest compounded monthly. (Note that this is the scenario described prior to the Flashback.)

Solution

Interactive Activity

For the situation in Example 1, how much interest has been earned over the 4-year period?

Understand the Situation: We use the present value for an annuity formula with $PMT = 500$, $i = \dfrac{0.065}{12}$, and $n = 4(12) = 48$.

This gives us

$$PV = PMT\frac{1-(1+i)^{-n}}{i}$$

$$= 500\frac{1-\left(1+\dfrac{0.065}{12}\right)^{-48}}{\dfrac{0.065}{12}}$$

$$\approx \$21{,}083.74 \qquad \text{Rounded to the nearest cent}$$

Interpret the Solution: This means that to receive \$500 per month each month for 4 years requires that an initial amount of \$21,083.74 be placed in the account that earns 6.5% interest compounded monthly. ■

Checkpoint 1 Now work Exercise 3.

Technology Option

As we have seen in this chapter, if your calculator has a FINANCE menu, it can be used to solve present value problems such as Example 1. To use these commands to solve Example 1, consult Figure 5.4.1(a), which shows what is entered on the TI-83 using the TVM Solver, and Figure 5.4.1(b), which shows the result when we solve for PV (present value).

(a) (b)

Figure 5.4.1

If your graphing calculator does not have a FINANCE menu, we have supplied a program for the graphing calculator that will compute the present value. See Appendix C. Consult the online graphing calculator manual at www.prenhall.com/armstrong for the program for your calculator.

Example 2 Computing Retirement Income

Sasha Jones starts her career at the age of 25 and makes equal annual deposits for 35 years into an annuity that earns 6.5% interest compounded annually. She plans on retiring at the age of 60 and would like to withdraw \$40,000 annually for 25 years, bringing the account balance to \$0. How much should Sasha deposit annually to accumulate enough to provide her with these \$40,000 payments?

Solution

Understand the Situation: To answer this question, we need to address both future and present values. As Figure 5.4.2 illustrates, the future value of Sasha's account after 35 years of investing must equal the present value to supply 25 years of yearly \$40,000 payments.

We begin by finding the present value required to supply these withdrawals. Using the formula for the present value of an annuity with $PMT = 40{,}000$, $i = 0.065$, and $n = 25$ gives us

$$\begin{aligned} PV &= PMT\frac{1-(1+i)^{-n}}{i} \\ &= 40{,}000\frac{1-(1+0.065)^{-25}}{0.065} \\ &\approx \$487{,}915.07 \quad \text{Rounded to the nearest cent} \end{aligned}$$

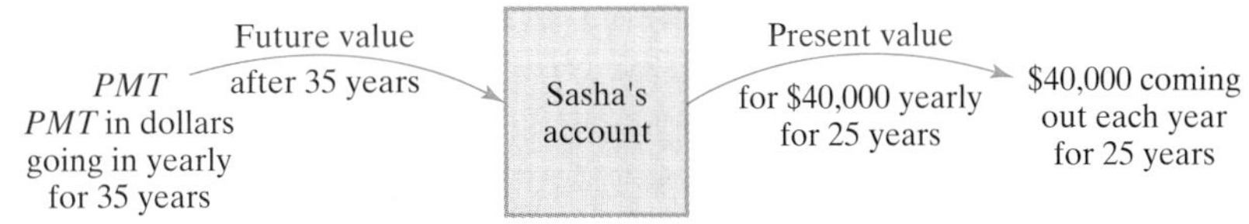

Figure 5.4.2

Now we need to find the annual deposit that Sasha must make to yield a future value of \$487,915.07 in 35 years. Using the formula for future value of an annuity from Section 5.3, with $FV = \$487{,}915.07$, $i = 0.065$, and $n = 35$, we get

$$\begin{aligned} FV &= PMT\frac{(1+i)^n-1}{i} \\ \$487{,}915.07 &= PMT\frac{(1+0.065)^{35}-1}{0.065} \\ PMT &= \$3933.70 \quad \text{Rounded to the nearest cent} \end{aligned}$$

Interpret the Solution: So Sasha needs to deposit \$3933.70 annually for 35 years in order to get 25 annual withdrawals of \$40,000. ■

Amortization

One of the most important applications of the present value of an annuity formula occurs in the area of purchasing a home. We introduce this topic in Example 3.

Example 3 Determining the Cost of a Home

Frank and Beth have put \$30,000 down toward the purchase of a house. To finance the balance of the purchase price of the house, they agree to pay a bank \$950 per month for 30 years at 8% annual interest charged monthly to pay off the mortgage (loan) on the house. (Mortgage rates are given as **annual percentage**

rates, ***APR***, but it is understood that the monthly rate is $\frac{1}{12}$ of the quoted annual rate and that the interest is compounded monthly.) Determine how much the house originally cost and how much interest was paid.

Solution

Understand the Situation: This scenario is the same as Example 1. In other words, we are finding the present value of an annuity of \$950 per month at 8% for 30 years.

Using the present value of an annuity formula with $PMT = 950$, $i = \frac{0.08}{12}$, and $n = 30(12) = 360$ gives us

$$\begin{aligned} PV &= PMT\frac{1-(1+i)^{-n}}{i} \\ &= 950\frac{1-\left(1+\frac{0.08}{12}\right)^{-360}}{\frac{0.08}{12}} \\ &\approx \$129{,}469.32 \quad \text{Rounded to the nearest cent} \end{aligned}$$

So the original cost of the house is

$$\$129{,}469.32 + \$30{,}000 = \$159{,}469.32$$

Since Beth and Frank are paying \$950 each month for 30 years (360 months), they pay a total of

$$950(360) = \$342{,}000$$

for the house. So the total interest paid is

$$\text{Total paid} - \text{Present value} = \text{Total interest paid}$$
$$\$342{,}000 - \$129{,}469.32 = \$212{,}530.68$$

This shockingly large amount of interest is typical of what happens with a long mortgage. ■

There are two different ways we can look at the loan in Example 3.

1. If Frank and Beth pays \$950 per month for 30 years (360 months) with an interest rate of 8% compounded monthly, then the house is theirs.
2. The loan is the same as an annuity that the bank bought from Frank and Beth. The bank paid Frank and Beth \$129,469.32 (a present value) for an annuity that then paid the bank \$950 per month for 360 months at 8% interest compounded monthly.

No matter how we look at it, the loan that Frank and Beth had in Example 3 had a fixed rate of interest and the monthly payments of \$950 included both principal and interest. In general, a loan is said to be **amortized** if both principal and interest are paid by a sequence of equal payments over equal time periods. Many times, we are interested in computing the equal payments that amortize the loan. If we solve the present value of an annuity formula for *PMT*, we obtain the following formula:

Payments for an Amortized Loan

A loan of *PV* dollars at interest rate i per period is **amortized** in n equal payments of *PMT* dollars, where *PMT* is given by

$$PMT = PV\frac{i}{1-(1+i)^{-n}}$$

Example 4 Determining a Monthly House Payment

As of Thursday 5/17/01, Ohio Savings Bank was offering a 30-year fixed-rate mortgage of 7.25% (*Source: The Plain Dealer*). Juan and Andrea Martinez purchase a house for \$140,000. After putting 20% down as a down payment, they finance the balance with Ohio Savings Bank.

(a) Determine the size of the down payment and the amount financed.

(b) Determine the monthly payment needed to amortize the loan.

Solution

(a) The down payment is 20% of \$140,000. This dollar amount is

$$0.20(140{,}000) = \$28{,}000$$

So with a down payment of \$28,000, Juan and Andrea need to finance

$$\$140{,}000 - \$28{,}000 = \$112{,}000$$

(b) The monthly payment needed to amortize the \$112,000 loan is determined by using the payments for an amortized loan formula. Here $PV = 112{,}000$, $i = \dfrac{0.0725}{12}$, and the number of monthly payments is $n = 12(30) = 360$. This gives us

$$\begin{aligned} PMT &= PV\frac{i}{1-(1+i)^{-n}} \\ &= 112{,}000\frac{\frac{0.0725}{12}}{1-\left(1+\frac{0.0725}{12}\right)^{-360}} \\ &\approx \$764.04 \quad \text{Rounded to the nearest cent} \end{aligned}$$

Interpret the Solution: This means that monthly payments of \$764.04 are needed to amortize the loan. ■

Interactive Activity

Find the total interest paid when the loan in Example 4 is amortized over 30 years. Now determine the monthly payment if the loan in Example 4 is to be amortized in 15 years and compare to the monthly payment in Example 4. Find the total interest paid when the loan in Example 4 is amortized over 15 years. Comment on your results.

WWW

Technology Option

Figure 5.4.3(a) and (b) shows the result of Example 4b using the FINANCE menu on a graphing calculator.

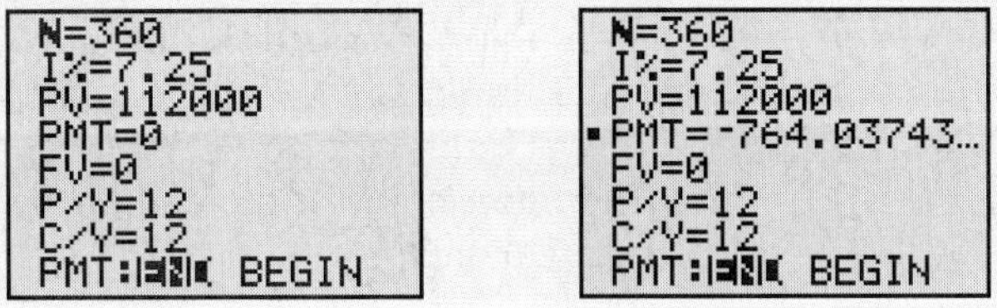

(a) (b)

Figure 5.4.3

✔ Checkpoint 2 Now work Exercise 45.

Amortization Schedules

When amortizing a loan such as the 30-year mortgage in Example 4, the situation may arise where at some point in time we wish to pay off the remainder of the debt in one lump-sum payment. Every time a home with an outstanding mortgage is sold, such a situation arises. To understand what happens in a situation like this, we need a closer analysis of the amortization process. In Example 5 we offer a simple example that allows us to analyze the effect that each payment has on an amortized loan.

Example 5 Constructing an Amortization Schedule

Manny borrows \$1000 that he agrees to repay in 12 equal monthly payments at an interest rate of 9%. How much of each monthly payment goes toward reducing the unpaid balance (or principal) of the loan and how much goes toward interest?

Solution

First we observe that an interest rate of 9% has a periodic rate of $i = \frac{0.09}{12} = 0.0075$ per month. Next we use the formula for payments of an amortized loan to determine the monthly payment. With $PV = 1000$, $i = 0.0075$, and $n = 12$, this gives

$$\begin{aligned} PMT &= PV\frac{i}{1-(1+i)^{-n}} \\ &= 1000\frac{0.0075}{1-(1+0.0075)^{-12}} \\ &\approx \$87.45 \quad \text{Rounded to the nearest cent} \end{aligned}$$

Now, at the end of the first month, interest is due on the entire \$1000 since this is the unpaid balance. This interest due is

$$\$1000(0.0075) = \$7.50$$

The monthly payment of \$87.45 is split into two parts: payment of the interest and reduction of the unpaid balance (or principal). This gives

Monthly payment		Interest due		Reduction of unpaid balance
\$87.45	=	\$7.50	+	\$79.95

So the unpaid balance for the next month (month 2) is

Prior unpaid balance		Reduction of unpaid balance		New unpaid balance
\$1000	−	\$79.95	=	\$920.05

The interest due at the end of the second month is based on the \$920.05 and is computed to be

$$\$920.05(0.0075) \approx \$6.90 \quad \text{Rounded to the nearest cent}$$

So the monthly payment of \$87.45 is split as follows:

Monthly payment		Interest due		Reduction of unpaid balance
\$87.45	=	\$6.90	+	\$80.55

Hence the unpaid balance for the next month (month 3) is

Prior unpaid balance		Reduction of unpaid balance		New unpaid balance
$920.05	−	$80.05	=	$839.50

We continue this process until all payments have been made and the unpaid balance is $0. Table 5.4.1 shows the entire process month by month and is known as an **amortization schedule**.

Table 5.4.1

Payment Number	Amount of Payment	Interest for Period	Reduction of Unpaid Balance	Unpaid Balance (principal)
0				$1000
1	$87.45	$7.50	$79.95	$920.05
2	87.45	6.90	80.55	839.50
3	87.45	6.30	81.15	758.35
4	87.45	5.69	81.76	676.59
5	87.45	5.07	82.38	594.21
6	87.45	4.46	82.99	511.22
7	87.45	3.83	83.62	427.60
8	87.45	3.21	84.24	343.36
9	87.45	2.58	84.87	258.49
10	87.45	1.94	85.51	172.98
11	87.45	1.30	86.15	86.83
12	87.48	0.65	86.83	0

Observe that the last payment needs to be increased by $0.03 in order to reduce the unpaid balance to $0. This minor change is due to rounding off in the computations and is typical of an amortized loan. ■

Example 6 Finding an Unpaid Loan Balance

In Example 4, we saw that Juan and Andrea Martinez financed a balance of $112,000 at 7.25% over 30 years to purchase a house. The monthly payments needed to amortize this loan were computed to be $764.04. After 10 years they wish to sell their house and purchase a new one. What is the unpaid balance on the mortgage after 10 years?

Solution

Understand the Situation: One way to answer this question is by constructing an amortization schedule like the one in Table 5.4.1. However, this table would be 120 lines long! Luckily, there is an easier method to answer this question. After 10 years (or 120 payments) the unpaid balance is equivalent to the amount of the loan that can be paid off with the remaining 240 payments (or 20 years). As we stated earlier in this section, the bank views mortgages as an annuity bought from individuals. Hence, for Juan and Andrea, the unpaid balance of their loan with 240 remaining payments is the same as the present value of an annuity with monthly payments of $764.04 for 20 years.

Using the present value of an annuity formula with $PMT = 764.04$, $n = 240$, and $i = \frac{0.0725}{12}$, we get

$$PV = PMT\frac{1-(1+i)^{-n}}{i}$$

$$= 764.04\frac{1-\left(1+\frac{0.0725}{12}\right)^{-240}}{\frac{0.0725}{12}}$$

$$\approx \$96{,}667.92 \quad \text{Rounded to the nearest cent}$$

Interpret the Solution: This means that after 10 years of making monthly payments of \$764.04 Juan and Andrea still owe \$96,667.92 on their \$112,000 mortgage. ■

The scenario in Example 6 illustrates the following for computing an unpaid balance of an amortized loan.

Unpaid Balance of an Amortized Loan

The unpaid balance of an amortized loan is equal to the present value of the remaining (unpaid) payments.

Checkpoint 3 Now work Exercise 49.

The unpaid loan balance in Example 6 may seem too large an amount to owe after making monthly payments for 10 years. But if we compute the amount of interest due on the first payment, we would get

$$\$112{,}000\frac{0.0725}{12} \approx \$676.67 \quad \text{Rounded to the nearest cent}$$

So the first monthly payment of \$764.04 was divided as

Monthly payment		Interest due		Reduction of unpaid balance
\$764.04	=	\$676.67	+	\$87.37

Only \$87.37 was used to reduce the unpaid balance with the first payment. Again, it is typical in long-term amortization loans to start out with small reductions in the unpaid balance.

SUMMARY

This section concluded our discussion of the mathematics of finance. Here we learned how to compute the present value of an annuity and applied this concept to retirement planning and loans. We also supplied a formula to compute payments for an amortized loan. We discussed how to construct an amortization schedule, a task ideally suited for a spreadsheet. The section concluded with an example illustrating how to compute an unpaid balance on a loan.

- **Present Value of an Annuity:** The present value PV of an annuity with n payments of PMT dollars made at the end of each payment period, with i as the interest rate per period, is $PV = PMT\frac{1-(1+i)^{-n}}{i}$.
- **Payments for an Amortized Loan:** A loan of PV dollars at interest rate i per period is **amortized** in n equal payments of PMT dollars, where PMT is given by $PMT = PV\frac{i}{1-(1+i)^{-n}}$.

SECTION 5.4 EXERCISES

In Exercises 1–6, determine the present value of each annuity.

1. Payments of $800 monthly for 8 years at an interest rate of 6% compounded monthly

2. Payments of $600 monthly for 15 years at an interest rate of 7% compounded monthly

✓ **3.** Payments of $900 monthly for 20 years at an interest rate of 7.5% compounded monthly

4. Payments of $700 monthly for 25 years at an interest rate of 6.5% compounded monthly

5. Payments of $2000 yearly for 30 years at an interest rate of 8.25% compounded annually

6. Payments of $1500 yearly for 20 years at an interest rate of 7.25% compounded annually

In Exercises 7–12, determine the present value of each annuity by using the FINANCE menu on your graphing calculator or the ANNPV calculator program found in Appendix C. Consult www.prenhall.com/armstrong *for the program for your graphing calculator.*

7. Payments of $600 monthly for 15 years at an interest rate of 7% compounded monthly

8. Payments of $800 monthly for 8 years at an interest rate of 6% compounded monthly

9. Payments of $700 monthly for 25 years at an interest rate of 6.5% compounded monthly

10. Payments of $900 monthly for 20 years at an interest rate of 7.5% compounded monthly

11. Payments of $1500 yearly for 20 years at an interest rate of 7.25% compounded annually

12. Payments of $2000 yearly for 30 years at an interest rate of 8.25% compounded annually

In Exercises 13–18, find the payment required to amortize each loan.

13. $25,000 at 7.5% compounded monthly for 10 years

14. $20,000 at 8.5% compounded monthly for 5 years

15. $100,000 at 7.75% compounded monthly for 30 years

16. $100,000 at 7.75% compounded monthly for 20 years

17. $13,000 at 6.25% compounded quarterly for 5 years

18. $20,000 at 7.25% compounded quarterly for 10 years

In Exercises 19–24, find the payment required to amortize each loan by using the FINANCE menu on your graphing calculator or the AMORTPAY calculator program found in Appendix C. Consult www.prenhall.com/armstrong *for the program for your graphing calculator.*

19. $20,000 at 8.5% compounded monthly for 5 years

20. $25,000 at 7.5% compounded monthly for 10 years

21. $100,000 at 7.75% compounded monthly for 20 years

22. $100,000 at 7.75% compounded monthly for 30 years

23. $20,000 at 7.25% compounded quarterly for 10 years

24. $13,000 at 6.25% compounded quarterly for 5 years

In Exercises 25–30, refer back to Exercises 13–18 and answer the following:

(a) Determine the total amount paid over the life of the loan.

(b) Determine how much interest was paid over the life of the loan.

25. The loan in Exercise 13

26. The loan in Exercise 14

27. The loan in Exercise 15

28. The loan in Exercise 16

29. The loan in Exercise 17

30. The loan in Exercise 18

Applications

31. Teri deposits $100 per month in a savings account that pays 4.75% interest compounded monthly for 10 years. Find the lump sum that Teri needs to deposit now to yield the same amount as the annuity after 10 years. (*Hint:* See Exercise 43, Section 5.3, for the future value of Teri's annuity.)

32. Steve deposits $80 per month in a savings account that pays 4.25% compounded monthly for 15 years. Find the lump sum that Steve needs to deposit now to yield the same amount as the annuity after 15 years. (*Hint:* See Exercise 44, Section 5.3, for the future value of Steve's annuity.)

33. Al and Tootie purchased a powerboat costing $34,000. They paid 20% down and amortized the rest with equal monthly payments over a 10-year period. They agree to pay 10% annual interest compounded monthly.

(a) Determine the monthly payment.

(b) Determine the total amount of interest paid over the 10 years.

34. Rework parts (a) and (b) from Exercise 33 keeping everything the same, except here compute for a 5-year period.

35. Irene deposits $200 each month into a retirement account earning 6.5% interest compounded monthly for 30 years. After the 30 years, she retires and withdraws equal monthly payments from the account for 25 years, at which time the account balance is $0. What monthly payment will Irene receive for the last 25 years?

36. Nick deposits $2400 each year into a retirement account earning 6.5% interest compounded annually for

30 years. After the 30 years, he retires and withdraws equal yearly payments from the account for 25 years at which time the account balance is $0. What yearly payments will Nick receive for the last 25 years?

37. (*continuation of Exercises 35 and 36*)

(a) How much did Irene in Exercise 35 deposit over the 30 years?

(b) How much did Nick in Exercise 36 deposit over the 30 years?

(c) How much do Irene and Nick receive on an annual basis once they retire and begin withdrawing money? Why aren't their annual amounts equal?

38. Rework Exercise 36, except here assume that Nick can earn 7% interest compounded annually.

39. Malika Agwani starts her career at the age of 25 and makes equal monthly deposits for 35 years into an annuity that earns 6.25% compounded monthly. She plans on retiring at the age of 60 and would like to withdraw $3000 monthly for 25 years, bringing the account balance to $0. How much should Malika deposit monthly to accumulate enough to provide her with these $3000 monthly payments?

40. Kahuna Smolko starts his career at the age of 30 and makes equal monthly deposits for 30 years into an annuity that earns 6.25% interest compounded monthly. He plans on retiring at the age of 60 and would like to withdraw $3000 monthly for 25 years, bringing the account balance to $0. How much should Kahuna deposit monthly to accumulate enough to provide him with these $3000 monthly payments?

41. In the Ohio Super Lotto PLUS™ lottery game, the smallest jackpot is $4 million. If there is one winner of this jackpot, he or she can take 30 annual payments (before taxes) of $133,333.33 or a lump sum that is equivalent to the present-day cash value (*Source:* www.ohiolottery.com). This present-day cash value is determined by the interest rate at which the money could be invested.

(a) Determine the present value of a $4 million lottery jackpot distributed in 30 annual payments (before taxes) of $133,333.33 using an interest rate of 6% compounded annually.

(b) Determine the present value of a $4 million lottery jackpot distributed in 30 annual payments (before taxes) of $133,333.33 using an interest rate of 7% compounded annually.

42. In the Lotto Virginia lottery game, a jackpot of $10 million has 26 annual payments (before taxes) of about $384,620 (*Source:* www.valottery.com). A lump sum that is equivalent to the present-day cash value can be determined by using the interest rate at which the money could be invested.

(a) Determine the present value of a $10 million lottery jackpot distributed in 26 annual payments (before taxes) of $384,620 using an interest rate of 6% compounded annually.

(b) Determine the present value of a $10 million lottery jackpot distributed in 26 annual payments (before taxes) of $384,620 using an interest rate of 7% compounded annually.

43. Sue and Tim state at a dinner party that they paid $20,000 down toward a new house. They also state that they have a 30-year mortgage for which they pay $870 per month. If interest is 8% compounded monthly, what did the house that Sue and Tim purchased originally cost?

44. Laurie and Craig have purchased a new home. They paid $25,000 down and have a 15-year mortgage for which they pay $980 per month. If interest is 8.25% compounded monthly, what did the house that Laurie and Craig purchased originally cost?

✓ **45.** As of Thursday 5/17/01, Firstar Bank was offering a 30-year fixed-rate mortgage of 7.38% (*Source: The Plain Dealer*). George and Debbie Ashton purchase a house for $180,000. After putting 20% down as a down payment, they finance the balance with Firstar Bank.

(a) Determine the size of the down payment and the amount financed.

(b) Determine the monthly payment needed to amortize the loan.

46. (*continuation of Exercise 45*) As of Thursday 5/17/01, Firstar Bank was offering a 15-year fixed-rate mortgage of 7.00% (*Source: The Plain Dealer*). Repeat Exercise 45, except here assume that George and Debbie Ashton take out a 15-year mortgage.

47. (*continuation of Exercises 45 and 46*)

(a) Find the total interest that George and Debbie Ashton will pay if they take out a 30-year mortgage.

(b) Find the total interest that George and Debbie Ashton will pay if they take out a 15-year mortgage.

48. As of Thursday 5/17/01, Dollar Bank was offering a 30-year fixed-rate mortgage of 7.25% (*Source: The Plain Dealer*). Mike and Leslie Williams have taken out a $150,000 30-year mortgage with Dollar Bank.

(a) Determine the monthly payment needed to amortize the loan.

(b) Determine the total interest paid on this 30-year mortgage.

✓ **49.** For the Ashton family in Exercise 45, determine the unpaid balance on their mortgage after

(a) 10 years **(b)** 15 years **(c)** 20 years

50. For the Williams family in Exercise 48, determine the unpaid balance on their mortgage after

(a) 10 years **(b)** 15 years **(c)** 20 years

Exercises 51 and 52 use the following:

It can be shown that the unpaid balance on a 30-year mortgage after x payments can be approximated by the function

$$y = PMT\left[\frac{1-(1+i)^{-(360-x)}}{i}\right]$$

where PMT is the monthly payment and i is the periodic rate.

51. (a) For the Ashton family in Exercise 45, write the function that approximates the unpaid balance after x payments.

(b) In the viewing window [0, 360] by [0, 150,000], graph the function from part (a).

(c) Use the graph to approximate the unpaid balance after 10 years ($x = 120$), 15 years ($x = 180$), and 20 years ($x = 240$). Compare the results with Exercise 49.

52. (a) For the Williams family in Exercise 48, write the function that approximates the unpaid balance after x payments.

(b) In the viewing window [0, 360] by [0, 160,000], graph the function from part (a).

(c) Use the graph to approximate the unpaid balance after 10 years ($x = 120$), 15 years ($x = 180$), and 20 years ($x = 240$). Compare the results with Exercise 50.

53. A loan of \$10,000 with interest of 9% compounded monthly is repaid with equal monthly payments over 1 year. Construct the amortization schedule for this loan.

54. A loan of \$4800 with interest of 12% compounded monthly is repaid with equal monthly payments over 2 years. Construct the amortization schedule for this loan.

SECTION PROJECT

Bill and Lisa wish to take out a **home equity loan** to finance their sons' college education. A local bank offers them a home equity loan called an **80% LTV ratio**. (LTV means loan-to-value and is typical of many home equity loans.) To compute the amount available as a loan, Bill and Lisa need to know the current value of their house and the unpaid balance on their mortgage. The amount available in an 80% LTV home equity loan is then computed to be

(80% of home's value) – (balance on mortgage)

Bill and Lisa purchased their home for \$160,000 and put 20% down. They took out a 30-year fixed-rate mortgage at 6.75%.

(a) Determine the size of the down payment.

(b) Determine the amount financed.

(c) Determine the monthly payment needed to amortize the loan.

(d) Many things determine the current value of a home. With all items factored in, assume that the home appreciates in value at a rate of 2.75% per year. Determine the value of the home in 18 years.

(e) Determine the unpaid balance on the mortgage after 18 years.

(f) Determine the maximum amount that Bill and Lisa can receive in a home equity loan assuming an 80% LTV ratio loan.

Why We Learned It

In this chapter we learned the basics of the mathematics of finance. To make good financial decisions, companies and individuals need to have an understanding of finance. For example, suppose that a company knows that in 5 years it will need to replace a piece of machinery that costs \$1 million. Instead of coming up with the \$1 million all at once in 5 years, the company could make fixed monthly deposits into an account with the goal that the future value of the account will be \$1 million in 5 years. Or the company could simply make a lump-sum deposit now that would grow in value to \$1 million in 5 years. The first situation requires an understanding of future value of an annuity, while the second situation uses knowledge of zero coupon bonds. Individuals can also use annuities to plan for retirement. For example, suppose that an individual wants to withdraw \$1000 per month each month for 25 years to supplement his or her retirement income. If this individual is just beginning his or her career and has 30 years of investing, he or she would need to determine how much money is needed in the account in 30 years to finance \$1000 monthly withdraws for 25 years. This situation would require a knowledge of future value as well as present value of an annuity.

CHAPTER REVIEW EXERCISES

In Exercises 1 and 2, the amount of money P borrowed for period of time t at simple interest rate r is given.

(a) Compute the simple interest owed for the use of the money.

(b) Compute the total amount, A, owed at the end of time t.

1. $P = \$12{,}000, r = 8.25\%, t = 3$ years

2. $P = \$8000, r = 7.75\%, t = 9$ months

In Exercises 3 and 4, the amount of money P loaned for period of time t along with the simple interest I charged is given. Determine the simple interest rate r of the loan. Round percentages to the nearest hundredth.

3. $P = \$5000, I = \$750, t = 2$ years

4. $P = \$18{,}000, I - \$6600, t = 5$ years, 9 months

In Exercises 5 and 6, determine the present value P given the future value A, time t, and annual simple interest rate r. Round to the nearest cent.

5. $A = \$8500, r = 8\%, t = 4$ years

6. $A = \$17{,}000, r = 7.25\%, t = 75$ months

7. Yoshii deposits \$6000 into a savings account that earns 6.25% simple interest.

(a) Write the future value A as a function of time t.

(b) Graph the equation from part (a).

(c) Determine the slope and interpret.

In Exercises 8 and 9, determine how long it will take a deposit to double at the given simple interest rate. Round answers to the nearest hundredth.

8. $r = 7.75\%$

9. $r = 6.25\%$

In Exercises 10 and 11, determine the discount and the proceeds given the loan amount F, annual interest rate r, and time t.

10. $F = \$1800, r = 7\%, t = 9$ months

11. $F = \$12{,}000, r = 8.75\%, t = 4$ years

In Exercises 12 and 13, determine the loan amount F to produce the given proceeds p at annual interest rate r over time t. Round to the nearest cent.

12. $P = \$4000, r = 6\%, t = 2$ years

13. $P = \$18{,}000, r = 7.25\%, t = 66$ months

14. Troy borrows \$4000 for 9 months at a simple interest rate of 7.5%. Determine the interest due on the loan.

15. Nicole borrowed \$16,000 from her father to start an accounting firm. She repaid him after 54 months. The annual simple interest rate was 6.75%. Determine the total amount that she repaid.

16. Rebecca wants to buy a \$10,000 sailboat in 3 years. How much should she invest now, to the nearest cent, at 7.25% simple interest to have enough for the sailboat in 3 years?

17. Quinn borrows \$1500 from Melissa and promises to pay back \$1800 in 2 years. What simple interest rate will Quinn pay?

18. If Mark desires to earn an annual simple interest rate of 8.5% on a 26-week T-bill with a face value of \$15,000, how much should he pay for the T-bill?

In Exercises 19 and 20, compute the future value for the given principal or present value P, annual interest rate r, and given compounding type at the end of 4 and 9 years. Round answers to the nearest cent.

19. $P = \$7000, r = 6.75\%$, compounded semiannually

20. $P = \$11{,}000, r = 7.25\%$, compounded monthly

In Exercises 21 and 22, use a graphing calculator's built-in financial package to compute the future value for the given principal or present value P, annual interest rate r, and given compounding type at the end of 2 and 5 years. Round answers to the nearest cent.

21. $P = \$9000, r = 5.75\%$, compounded annually

22. $P = \$12{,}000, r = 6.5\%$, compounded quarterly

In Exercises 23 and 24, compute the present value for the given future value A, annual interest rate r, time t, and given compounding type. Round answers to the nearest cent.

23. $A = \$7500, r = 6.75\%, t = 6$ years, compounded monthly

24. $A = \$4500, r = 6.5\%, t = 4$ years, compounded daily

In Exercises 25 and 26, use a graphing calculator's built-in financial package to compute the present value for the given future value A, annual interest rate r, time t, and given compounding type. Round answers to the nearest cent.

25. $A = \$12{,}000, r = 6\%, t = 5$ years, compounded semiannually

26. $A = \$8500, r = 7.25\%, t = 3$ years, compounded quarterly

In Exercises 27 and 28, determine how long (that is, is the number of compounding periods) it will take the given present value P to grow to the given future value A for the stated annual interest rate r and compounding type. Round the answer to the nearest integer.

27. $P = \$4000, A = \$7500, r = 6.5\%$, compounded annually

28. $P = \$10{,}000, A = \$12{,}000, r = 6.75\%$, compounded quarterly

In Exercises 29 and 30, determine the effective rate corresponding to each of the following nominal rates. Round the percentages to the nearest thousandth.

29. 7.75% compounded quarterly

30. 6.5% compounded semiannually

In Exercises 31 and 32, for the given effective rate determine the nominal rate if the compounding is (a) quarterly; or (b) monthly.

31. 5.75% **32.** 7.25%

33. The Consumer Price Index (CPI) is a measure of inflation. The CPI for all urban consumers in 2000 was 3.4%. If this rate holds true for the next 3 years, what will a $1.69 half-gallon of orange juice cost in 3 years if the rate is compounded annually (*Source:* Bureau of Labor Statistics, http://stats.bls.gov).

34. Kris and Rachel purchased a new condominium costing $159,000. Condominiums in the area are appreciating at a rate of 2.25% per year. If this rate holds steady for 6 years, what will the condominium be worth, assuming that this 2.25% rate is compounded annually? Round to the nearest thousand.

35. Mike wants to buy a new boat in 2 years. He estimates that he will need $8000 for the boat. How much should he invest now, to the nearest cent, at 6.75% interest compounded quarterly to have the $8000 in 2 years?

36. Paul wishes to invest some money in a money market account. He finds a bank that has an account paying 4.60% interest compounded monthly and another bank with an account paying 4.38% compounding daily.

(a) Compute the effective rate for the account paying 4.60% interest compounded monthly to the nearest hundredth.

(b) Compute the effective rate for the account paying 4.38% interest compounded daily to the nearest hundredth.

(c) Is one account a better investment than the other? Explain.

37. Beth's parents are considering purchasing a $50,000 face value zero coupon bond at her birth to help to finance her college education in 17 years. If money is worth 7.65% compounded annually, how much, to the nearest cent, should they pay for this bond?

In Exercises 38 and 39, determine the amount of each annuity.

38. A deposit of $200 monthly for 10 years at an interest rate of 6.25% compounded monthly

39. A deposit of $3000 annually for 30 years at an interest rate of 6.75% compounded annually

40. Use the FINANCE menu on your graphing calculator or the ANNFV calculator program found in Appendix C to find the amount of a deposit of $300 monthly for 20 years at an interest rate of 7.25% compounded monthly. (Consult www.prenhall.com/armstrong for the program for your calculator.)

41. Determine how much of the amount (or future value) of the annuity in Exercise 39 is interest.

In Exercises 42 and 43, determine the periodic payment amount required for each sinking fund.

42. The amount needed is $20,000 in 15 years at 6.75% interest compounded monthly. What is the monthly payment?

43. The amount needed is $50,000 in 10 years at 7.25% interest compounded quarterly. What is the quarterly payment?

44. Use the FINANCE menu on your graphing calculator or the ANNPMT calculator program found in Appendix C to determine the annual payment required for a sinking fund for which the amount needed is $20,000 in 8 years at 6% interest compounded annually. (Consult www.prenhall.com/armstrong for the program for your calculator.)

45. Refer back to Exercise 42 to answer the following:

(a) How much was invested?

(b) How much interest was earned?

In Exercises 46 and 47, determine the annual interest rate (to the nearest hundredth) that guarantees the future value of the annuity.

46. The future value is $130,000; $300 is deposited each month for 20 years into an account paying interest compounded monthly.

47. The future value is $500,000; $2000 is deposited each quarter for 30 years into an account paying interest compounded quarterly.

48. Kerri deposits $150 per month into a savings account that pays 4.5% interest compounded monthly.

(a) Determine the value of this account after 5 years.

(b) How much of the future value was money that Kerri deposited and how much of the future value was interest earned?

49. Meagan estimates that she spends $150 per month on cigarettes. She decides that she will quit and invest the monthly amount spent on cigarettes in an account earning 6.25% interest compounded monthly. At the end of 10 years, she will donate the interest earned on her account to her favorite charity. How much money can the charity expect to get in 10 years?

50. Jen and Mide would like to have $25,000 in 5 years as a down payment on a new house. What amount should they deposit each month into a savings account paying 6.375% interest compounded quarterly?

51. For his retirement, Kris wishes to have $225,000 in a 403(b) account 25 years from now. How much should he deposit each month in an account paying 6.25% interest compounded monthly?

52. (a) Grace deposits $3000 a year into an IRA earning 7% interest compounded annually. She makes her first deposit on her 25th birthday and her last deposit on her 35th birthday, a total of 11 deposits. Grace makes no additional deposits. She leaves the amount accumulated from the 11 deposits in the account, earning 7% interest compounded annually until her 65th birthday. How much is in Grace's account on her 65th birthday?

(b) Ted, Grace's friend, encouraged Grace to deposit $1500 into the same IRA account from part (a) each year beginning on her 25th birthday and ending with a deposit on her 65th birthday, a total of 41 deposits. How much would be in Grace's account when she makes her last deposit on her 65th birthday if she had taken Ted's advice?

(c) Compare the results of parts (a) and (b). In your comparison, determine how much Grace actually deposited and how much she would have deposited had she taken Ted's advice.

In Exercises 53 and 54, determine the present value of each annuity.

53. Payments of \$750 monthly for 20 years at an interest rate of 7.25% compounded monthly

54. Payments of \$1000 yearly for 15 years at an interest rate of 8.25% compounded annually

In Exercises 55 and 56, determine the present value of each annuity by using the FINANCE menu on your graphing calculator or the ANNPV calculator program found in Appendix C. (Consult www.prenhall.com/armstrong *for the program for your graphing calculator.)*

55. Payments of \$800 monthly for 6 years at an interest rate of 7.5% compounded monthly

56. Payments of \$2500 yearly for 30 years at an interest rate of 8.75% compounded annually

In Exercises 57 and 58, find the payment required to amortize each loan.

57. \$30,000 at 6.75% compounded monthly for 15 years

58. \$25,000 at 8.25% compounded quarterly for 20 years

In Exercises 59 and 60, find the payment required to amortize each loan by using the FINANCE menu on your graphing calculator or the AMORTPAY calculator program found in Appendix C. (Consult: www.prenhall.com/armstrong *for the program for your graphing calculator.)*

59. \$75,000 at 7.25% compounded monthly for 30 years

60. \$18,000 at 8.5% compounded quarterly for 15 years

61. Refer back to Exercise 57 to answer the following:

(a) Determine the total amount paid over the life of the loan.

(b) Determine how much interest was paid over the life of the loan.

62. Lena deposits \$200 per month in a savings account that pays 4.5% interest compounded monthly for 8 years. Find the lump sum that Lena needs to deposit now to yield the same amount as the annuity after 8 years.

63. Al and Peggy purchased 15 acres of land costing \$40,000. They paid 25% down and amortized the rest with equal monthly payments over a 15-year period. They agree to pay 8.25% annual interest compounded monthly.

(a) Determine the monthly payment.

(b) Determine the total amount of interest paid over the 15 years.

64. Lindsay deposits \$240 each month into a retirement account earning 6.75% interest compounded monthly for 25 years. After 25 years, she retires and withdraws equal monthly payments from the account for 20 years, at which time the account balance is \$0.

(a) What monthly payment will Lindsay receive for the last 20 years?

(b) How much did Lindsay deposit over the 25 years?

65. (*continuation of Exercise 64*) Suppose that Lindsay deposits \$400 each month into a retirement account earning 6.75% interest compounded monthly for 15 years. She then leaves the amount accumulated after 15 years in the account, still earning 6.75% interest compounded monthly, for another 10 years without making any further deposits. She then withdraws equal monthly payments from the account for 20 years, at which time the account balance is \$0.

(a) What monthly payment will Lindsay receive for the last 20 years?

(b) Compare your results from Exercise 64(a) and 65(a). In your comparison include the total amount that Lindsay deposits in each case.

66. Amy and Mike have purchased a new home. They paid \$20,000 down and have a 30-year mortgage for which they pay \$725 per month; the interest is 7.25% compounded monthly.

(a) What did the house that Amy and Mike purchased originally cost?

(b) Determine the total interest paid on this 30-year mortgage.

(c) Suppose that Amy and Mike decide to sell their home after 15 years. Determine the unpaid balance on their mortgage.

67. A loan of \$6000 with interest of 8% compounded monthly is repaid with equal monthly payments over 1 year. Construct the amortization schedule for this loan.

In Exercises 68–71, do the following:

(a) Determine if the problem is a simple interest, compound interest, future value of an annuity, or present value of an annuity problem.

(b) Solve the problem.

68. Spencer's parents have established a sinking fund in order to have \$150,000 in 17 years for his college education. How much should they deposit each quarter into a savings account paying 5.75% interest compounded quarterly?

69. Ryan borrows \$3000 from Kris and promises to pay back \$3400 in 3 years. What interest rate, to the nearest hundredth, will Ryan pay?

70. Lindsay and Mark purchase a house for \$130,000. After putting 20% down, they finance the balance with a 30-year fixed-rate mortgage of 7.45% and make equal monthly payments to pay off the balance. If Lindsay and Mark decide to sell their home in 10 years, what is their unpaid balance?

71. How much should Ron invest now at 8.25% interest compounded quarterly to have \$15,000 in 5 years to purchase a new boat?

CHAPTER 5 PROJECT

In this project we analyze credit cards and debt. We will learn how the finance charge is calculated, how to find the minimum payment due, and the importance of paying off the debt as soon as possible.

Kahuna Smolko just graduated from Old State University and wishes to reward himself with a new television. Recently, he received a Big Bucks credit card in the mail that offered a 12.99% fixed APR (annual percentage rate). He used the card to purchase a 65-inch wide screen HDTV from Sony for $10,000.00 (including sales tax) on the first of the month (*Source:* Sony Style Electronics: www.sonystyle.com). After this purchase, Kahuna cut up his credit card. He decides to pay only the minimum payment due for each bill. The credit card company receives his payments on the 16th day of the month.

To compute the finance charge, many credit card companies use a process called a *two-cycle average daily balance* since it results in a higher interest payment by the consumer. For simplicity, we will assume 30-day months and 360 days in a year.

On the first bill, the balance will be $10,000.00. Big Bucks computes the minimum payment due as 2.5% of the balance. Thus, the minimum payment due will be $10{,}000.00 \times 0.025 = 250.00$. Normally, the minimum payment is truncated (meaning if the decimal were actually $250.43, $250 will still be the minimum payment). Thus the remaining balance will be $10{,}000.00 - 250.00 = 9750.00$ (remember, this payment is received by Big Bucks on the 16th). We will assume that Kahuna must pay at least $25.00 per month.

Since this is a two-cycle average daily balance, Kahuna will be charged interest for the previous month *plus* the current month (on a one-cycle method, only the current month is charged interest). To calculate the finance charge for the current bill, we use the formula

$$\begin{pmatrix}\text{Finance charge of}\\ \text{previous month}\end{pmatrix} + \underbrace{\begin{pmatrix}\text{Average daily balance}\\ \text{of current month}\end{pmatrix} \times \frac{\text{APR}}{360} \times 30}_{\text{Finance charge for the current month}}$$

As mentioned, the previous month's average daily balance (ADB) was $10,000 (since no payment was made in the first month). To compute the ADB for the second month, we use

$$(\underbrace{10{,}000}_{\substack{\text{Balance before}\\ \text{payment made}}} \times \underbrace{15}_{\substack{\text{Number of days}\\ \text{balance was held}}} + \underbrace{9750}_{\substack{\text{Balance after}\\ \text{payment made}}} \times \underbrace{15}_{\substack{\text{Number of}\\ \text{days remaining}\\ \text{in month}}}) \div \underbrace{30}_{\substack{\text{Number of}\\ \text{days in billing cycle}}} = 9875$$

The finance charge for the first month is $10{,}000 \times \frac{0.1299}{360} \times 30 = 108.25$. The finance charge for the second month is $9875 \times \frac{0.1299}{360} \times 30 = 106.896875$. The sum of the finance charges will be $108.25 + 106.896875 = 215.146875 \approx 215.14$. Thus, the new balance will be

$$\underbrace{9750}_{\substack{\text{Balance after}\\ \text{payment made}}} + \underbrace{215.14}_{\substack{\text{Finance charge}\\ \text{for current bill}}} = 9965.14$$

The new monthly payment will be 2.5% of this new balance, or $0.025 \times 9965.14 = 249.00$ after truncating.

1. Calculate the next two cycles. Be sure to find the new balance after the last minimum monthly payment was made, the new ADB, the new finance charge, and the new minimum monthly payment.

2. Visit www.prenhall.com/armstrong and input the program called Credit into your calculator. Check your answers to question 1 with this program.

3. Using the calculator program, determine the number of cycles necessary to pay off the credit card completely. (As you may have guessed, it will take many cycles.) Also, determine the total amount paid.

4. Suppose that Kahuna pays $300 a month instead of the given minimum monthly balance. Compute the first three cycles by hand and compare your results to question 1.

5. Visit www.prenhall.com/armstrong and input the program called Credit2 into your calculator. Check your answers to question 4 with this program. Use the program to determine the number of cycles necessary to pay off the credit card completely and the total amount paid. Compare total amount paid here with the total amount paid in question 3.

Chapter

6 Sets and the Fundamentals of Probability

Odds to Probability

$$P(A) = \frac{a}{a+b}$$

Probability of Tiger Woods Winning the 2001 British Open

$$P(A) = \frac{a}{a+b} = \frac{3}{3+2} = \frac{3}{5}$$

We frequently hear about the odds of something occurring. But how do odds relate to probability? In 2001, golf pro Tiger Woods was "hot" and the odds of his winning the British Open were 3 to 2. Using the formula for converting odds to probability, we can translate odds into the probability of a given event occurring. In this case, the probability of Tiger Woods winning the Open was 3/5. Although the probability of a win was high, Tiger actually suffered a stunning defeat. Probability is a valuable tool for predicting likely outcomes, but it does not ensure that a given event will occur.

What We Know

So far in this text, we have seen many applications involving matrices and linear programming, as well as the mathematics of finance. We have seen how an understanding of these topics can be important for individuals in many diverse areas.

Where Do We Go

In this chapter we begin our journey into the study of probability. Areas as diverse as insurance policy analysis, weather forecasting, medical research, and financial planning utilize concepts of probability. Since probability is different from the mathematics of the first part of this text, we ease into its study with a section dedicated to problem-solving techniques in probability.

Section 6.1 Problem Solving: Probability and Statistics

The idea of chance is a part of our lives everyday. Many of the decisions that we make are influenced, in part, by chance. When we use the word *chance* we mean *the likelihood that an event will occur*. For example, an individual may decide whether or not to take an umbrella with him or her to work based on the chance that it will rain that day. In this scenario, the **event** in question is "it will rain that day." Another example is taking a chance that guessing C as an answer for a question on a multiple-choice test is the correct answer. Or we may take a chance on whether the balding tires on our car will blow out this summer. All these events and many others are based on **chance or probability**. You probably have heard about games of chance, that is, games that involve probability, such as a state lottery, a card game, or slot machines. However, probability is a necessary tool in areas such as insurance policy analysis, weather forecasting, medical research, and financial planning. In general terms, we can say that probability is the branch of mathematics that produces a number that expresses the likelihood of occurrence of a specific event.

The challenge that many students face when beginning the study of probability is the use of language in probability applications. This means that you must **read the problems closely** and **think carefully** about what you are asked to find. An author takes a great deal of care to select just the right words for a probability problem. You need to read the problem slowly and carefully, and you may need to read it more than once so that you understand what the author is saying. Then you must think about the problem in a careful, step-by-step fashion to figure out what is needed and how you can achieve the correct results. In this section, we will introduce some of the key concepts that are necessary to solve the application problems commonly encountered in the study of probability. We will also demonstrate how to read closely and think carefully when confronted with an application. The ideas in this section can be applied to the remainder of Chapter 6 and also in Chapter 8.

Thinking about Populations and Samples

We begin with two terms fundamental to the study of probability: population and sample. We will revisit these definitions later in the text.

Population and Sample

- A **population** is the collection of **all** outcomes, responses, measurements, or counts that are of interest.
- A **sample** is a portion (or subset) of the population.

When we study statistics, the companion topic to probability, we take a sample at random and use the results to make estimates of the population as a whole. By at random, we mean that every item in the population has an equal chance of being included in the sample. We call this branch of statistics **inferential statistics**. In the study of **probability**, we do the reverse. That is, we use the information about a population to make inference about the possible nature of the sample. This contrast is illustrated in Figure 6.1.1(a) and (b).

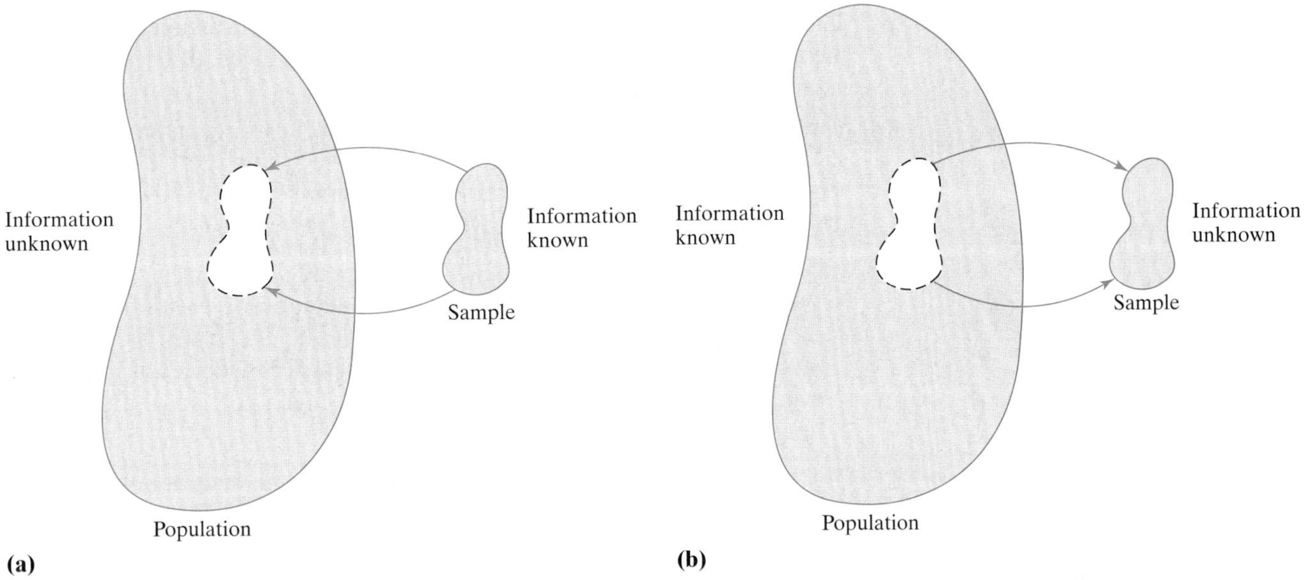

Figure 6.1.1 (a) Inferential statistics involves making conclusions about a population based on information gathered in a sample. **(b)** Probability involves making statements about a sample based on information that is known about a population.

Example 1 Distinguishing between Statistical Inference and Probability

Classify each of the following as an example of probability or of statistical inference.

(a) Past records show that about 15% of the population of Austinburg use the emergency 911 number each year. In a random sample of 50 Austinburg citizens, the chances are really high that at least one used the 911 emergency number last year.

(b) A survey shows that in the town of Austinburg, about 1 out of every 14 residents say that they use public transportation at least once per week.

Solution

(a) **Understand the Situation:** The first task when solving a probability problem is to read the problem slowly and repeatedly until you recognize exactly what the writer is talking about. Make sure that you understand exactly what the writer says. In this case, there are two statements. In the first statement, we know that 15% of Austinburg citizens use 911. In the second statement, we are told that if we take a random sample of 50 citizens the chances are really high that one has called 911 during the past year.

Now let's think about these two statements.

- The statement "15% of Austinburg citizens use 911" is a statement about a population (the people of Austinburg).
- The statement "In a random sample of 50 Austinburg citizens, there is a good chance that at least one used the 911 emergency number last year" is a statement about a sample (portion) of the population. We can make this statement because we know that 15% of the people (or 15 out of 100) used 911 last year.

The second statement shows that we are drawing a conclusion about a sample from information known about a population. Hence, this scenario is an example of probability.

(b) **Understand the Situation:** In this case there is only one statement to consider. In the statement, we are told that a survey shows that in the town of Austinburg about 1 out of every 14 residents says that they use public transportation at least once per week.

Now let's think about this statement.

- We are not given any facts about the population of Austinburg. Here, there are no past records, as was the case for the 911 calls in part (a), simply a survey. Under normal circumstances, a survey does not imply that every resident was contacted. It does imply that a portion of the population, that is, a sample, was contacted.

This statement indicates that we are estimating some facts about a population based on some sample results. Hence, this scenario is an example of statistical inference. ■

✓ Checkpoint 1 Now work Exercise 3.

Thinking about Probability Experiments

In order to collect information that can be used in situations involving probability, we conduct **probability experiments**. A probability experiment can be as simple as flipping a coin or as complex as a medical research study.

Probability Experiment

A **probability experiment** is a process that leads to well-defined results, called **outcomes**. All possible outcomes are predictable, but any given outcome cannot be predicted with certainty.

Notice from the definition that a probability experiment must have two conditions satisfied. The first condition is that the outcomes must be unambiguous. In other words, the result of each individual outcome must be able to be easily placed into a specific category. The second condition is that all possible outcomes are predictable, but a specific outcome cannot be predicted with certainty. For example, when flipping a coin we know that all possible outcomes are heads or tails, yet we cannot predict with certainty if the flip will yield heads or tails.

Example 2 Determining If a Process Is a Probability Experiment

Determine whether each of the following processes represents a probability experiment.

(a) 100 people were called on the telephone and asked, "How was your day?"

(b) 100 people were called on the telephone and asked, "How was your day?
1. Good 2. Fair 3. Bad 4. None of your business 5. No comment"

(c) A bag contains three marbles painted with the numbers 1, 2, and 3. A marble is drawn from the bag, its number is recorded, and it is then returned to the bag. Another marble is drawn and its number is recorded.

Solution

(a) **Understand the Situation:** To ask someone a general question such as "How was your day?" could elicit many different kinds of responses. This is typical of an open-ended question. Since open-ended questions can lead to many different kinds of responses, this indicates that the outcomes (responses) are not well defined. So this process is not a probability experiment.

(b) **Understand the Situation:** At first glance, this seems the same as part (a). But when we read the scenario carefully, we see that it is different. Specifically, people surveyed are asked to pick one of five possible responses. This means that we do not have an open-ended question! Each outcome is well defined (all possible outcomes are Good, Fair, Bad, None of your business, or No comment). We know that the outcome of a single telephone call is going to be recorded as one of these five responses, but the outcome of each telephone call is not predictable with certainty. We have no way of knowing exactly how an individual will respond to the question. Since the outcomes are well defined, but any given outcome is not predictable with certainty, this process is an example of a probability experiment.

(c) **Understand the Situation:** In this case, each individual outcome is well defined (all possible outcomes are the whole numbers 1, 2, or 3). We know that the outcome of the first draw is going to be a 1, 2, or 3. We know that the outcome of the second draw is also going to be a 1, 2, or 3. However, the outcome of each draw is not predictable with certainty. This means that the process is a probability experiment. ■

Thinking about Ordered and Unordered Samples

When studying the concepts of probability, we commonly have a need to count the number of different outcomes that we can have for a probability experiment. In Example 2(c), we can see that there are nine possible outcomes. These nine possible outcomes are (1, 1), (1, 2), (1, 3), (2, 1), (2, 2), (2, 3), (3, 1), (3, 2), and (3, 3). But there are times when counting the number of possible outcomes can be more complicated. If we want to know the number of ways that a 12-member board of trustees can choose a president, vice-president, and CEO, for example, we need additional counting techniques. One of the keys that enables us to count these possible outcomes depends on whether order is important. By order being important, we mean that:

- The order of arrangement makes a difference, or
- There is a ranking of the members of the arrangement.

For example, let's say that 12 horses are entered in this year's Jasper Derby. If we want to count the number of ways that the horses can finish first, second, and third, then the order of finish is important. In other words, there is a ranking of the first three finishers. We will see how to compute the actual number of outcomes for a probability experiment like this one in Section 6.3. For now, we are going to focus on determining whether order is important.

Example 3 Determining If Order Is Important

For each of the following, determine whether order is important.

(a) There are 24 members in the New Strawsburg Junior High School Computer Club. When computing how many ways the club can elect a president, vice-president, secretary, and treasurer, is order important?

(b) The trauma unit at Gothicheath Hospital has 44 staff members. When computing how many ways a committee of three can be formed to investigate salary discrepancies, is order important?

Solution

(a) Understand the Situation: The names that we associate with the president, vice-president, secretary, and treasurer make a difference. This indicates that order is important.

(b) Understand the Situation: The hospital staff wants to form a committee of **any** three from among the possible 44 staff members. Notice that no ranking of committee members is involved. This means that, in this case, order is not important. ■

Checkpoint 2 Now work Exercise 21.

Thinking about Simultaneous Events

Many times in the study of probability we want to know if it is possible for two things to happen at the same time. These things can be regarded as a collection of outcomes of a probability experiment. We call this collection of outcomes **events**. (We will discuss events in more depth in Section 6.4.)

For example, if a six-sided die is tossed in a probability experiment, then an event could be that the outcome of the roll is an odd number. The outcomes associated with this event are the numbers 1, 3, and 5. Another event could be that the outcome of the roll is an even number. The outcomes associated with this event are the numbers 2, 4, and 6. Since a single roll of a die cannot produce **both** an odd number **and** an even number, these events cannot happen at the same time.

By contrast, consider the event that a randomly chosen student from Berliner College is registered for a chemistry class this semester, and the other event is that the randomly chosen student from Berliner College is also registered for a Spanish class this semester. It is possible for a student to sign up for a chemistry class and for a Spanish class during the same semester. Hence, this is an example of two events that can occur simultaneously.

Before proceeding, let's pause for a word of caution. When we ask questions such as if two events can happen at the same time, we are asking the question to be considered under normal circumstances.

A common error made by many individuals who are just beginning to study probability is trying to read too much information into the questions in an attempt to find some exception.

So when posed questions in the study of probability, such as if two events can happen at the same time, we mean for the question to be considered under everyday, usual, commonsense circumstances. For example, when flipping a coin, the only reasonable outcomes are heads or tails. Some may say that the coin could land on its edge, but this is not a usual or common sense result.

Example 4 Determining If Events Can Happen at the Same Time

Determine if the following events can happen at the same time.

(a) The event that Sandy gets a speeding ticket and the event that Sandy gets a ticket for not wearing a seat belt.

(b) The event that, during the same at bat, a softball player, Ariel, hits a single and the event that Ariel hits a triple.

(c) The event that during a single roll of a six-sided die Don rolls a 2 and the event that Don rolls a 4.

Solution

(a) Understand the Situation: To help us determine what is happening in this case, we will list the two events.

- Event 1: Sandy gets a speeding ticket.
- Event 2: Sandy gets a ticket for not wearing her seat belt.

It appears that these events, receiving a speeding ticket and receiving a ticket for not wearing a seat belt, can happen at the same time.

(b) Understand the Situation: As an aid to help us determine what is happening here, we again will list the two events.

- Event 1: During an at bat, Ariel hits a single.
- Event 2: During the same at bat, Ariel hits a triple.

During the same at bat, a softball player cannot hit a single **and** a triple. This means that these events cannot happen at the same time.

(c) Understand the Situation: As in parts (a) and (b), we list the two events.

- Event 1: During the roll of a die, Don rolls a 2.
- Event 2: During the same roll, Don rolls a 4.

During the same roll of a die, an individual cannot roll a 2 **and** a 4. This means that these events cannot happen at the same time. ■

✓ Checkpoint 3 Now work Exercise 29.

Thinking about Whether One Event Influences Another Event

Many times, a crucial piece of information that we want to know about two successive events is whether the result of one event causes the next event to be more or less likely to occur. For example, let's say that one event is that Alan takes an aspirin and the next event is that Alan's headache is relieved. For most people, taking an aspirin increases the likelihood that the headache will be relieved. This indicates that the first event (taking an aspirin) does influence the second event (headache is relieved).

Now suppose that Michelle oversleeps because she did not set her alarm clock and then tries to start her car only to discover that its battery is dead. We know that oversleeping does not make it more or less likely for her car to have a dead battery. So the outcome of the first event (oversleeping) does not influence the second event (the car having a dead battery).

Once again, we want to point out that we should not read too much into the event(s) that we are given. Even though some may say that Michelle may be suffering from bad karma or perhaps the alarm clock could have somehow drained the car battery or that not setting the alarm clock indicates irresponsibility, hence she may also be irresponsible of car maintenance, we can conclude that under normal, usual circumstances, one event (oversleeping) does not influence the other event (dead car battery).

Example 5 Determining If One Event Influences the Other

Determine if the outcome of one event influences the outcome of the other event.

(a) The event of receiving 9 inches of snow on Sunday night and the event that the local school district cancels classes on Monday morning.

(b) The event that the first toss of a coin is heads and the event that the second toss of the coin is tails.

Solution

(a) **Understand the Situation:** As we did in Example 4, we will list the two events separately as an aid to determine what is happening here.

- Event 1: Receiving 9 inches of snow Sunday night.
- Event 2: Local school district cancels classes Monday morning.

Since getting a large snowfall makes it more likely for school to be canceled, we conclude that the outcome of the first event (receiving 9 inches of snow Sunday night) does influence the second event (local school district cancels classes Monday morning).

(b) **Understand the Situation:** Again, we begin by listing the two events.

- Event 1: First toss of a coin is heads.
- Event 2: Second toss of a coin is tails.

Tossing heads with the first toss does not alter the likelihood of tossing tails on the second toss. So, in this case, the outcome of the first event (first toss of a coin is heads) does not influence the second event (second toss of a coin is tails). ■

✓ Checkpoint 4 Now work Exercise 53.

Solving Applications in Probability

We conclude this section by offering some general tips on solving applications in the area of probability.

Problem Solving Tips for Probability

1. Read the application problem closely and think carefully.
2. For the question that is being asked, ask yourself whether or not the following apply:
   (a) Is order important?
   (b) Can the events occur simultaneously?
   (c) Can the outcome of one event influence the likelihood of another event?
3. Do not conjecture about a problem too much. Answer the question as it would occur under ordinary circumstances.
4. Answer the question being posed using a complete sentence.

SUMMARY

In this section, we explored the unique use of language when studying probability. We found that many of the applications addressed one of three key questions. The first question asks, "Is order important?", meaning does the order of arrangement make a difference, or is there a ranking of the members of the arrangement. The second central question that we examined is "Can events occur simultaneously?", meaning can the outcome of a probability experiment satisfy two events at once. The third of the equations that we addressed is "Can the outcome of one event influence the likelihood of another event?" The ability to answer these questions will make the voyage through the probability and statistics part of this text less arduous.

SECTION 6.1 EXERCISES

In Exercises 1–8, classify each of the following as an example of probability or statistical inference.

⇒ 8.3 **1.** A 1992 report on prisoners and education showed that 49% of the prisoners did not have a high school diploma. In a sample of 10 prisoners selected at random, 6 do not have a high school diploma. (*Source:* National Center for Education Statistics, www.nces.ed.gov.)

⇒ 8.3 **2.** A 1994 report revealed that 32.6% of all births in the United States were to unmarried women. In a random sample of 30 women who gave birth in 1994, nine of them were unmarried. (*Source:* Center for Disease Control, www.cdc.gov.)

✓ **3.** To determine the attitudes of Americans concerning the political leadership in Russia, a June 2001 Gallup poll asked a random sample of 1011 adults 18 years and older if they are more likely to consider Russian President Vladimir Putin friendly or unfriendly to the United States; 435 of them considered him friendly to the United States. (*Source:* The Gallup Organization, www.gallup.com.)

4. A randomly selected sample of 1061 American adults 18 years and older were asked, "Could the entire moon program have been an elaborate deception staged to fool the public?" 944 of them did not believe that the U.S. government staged or faked the *Apollo* moon landing. (*Source:* The Gallup Organization, www.gallup.com.)

5. A May 2001 survey determined that 76% of all Internet users check their e-mail daily. In a random sample of 35 Internet users, 25 said that they check their e-mail daily. (*Source:* UCLA Center for Communication Policy, www.ccp.ucla.edu.)

6. In a report on the adult public, 53.1% viewed American children negatively. In a sample of 20 adults selected at random, 14 viewed children negatively. (*Source:* www.publicagenda.org.)

⇒ 8.3 **7.** A report on minorities and transportation found that 59% of African Americans in New York City had no automobiles. In a random sample of 15 African Americans from New York City, 9 of them did not have an automobile. (*Source:* American Highway Users Alliance, www.highways.org.)

⇒ 8.3 **8.** In 1997, 10.8% of female smokers smoked cigars. In a sample of 20 female smokers selected at random, two of them smoked cigars. (*Source:* American Cancer Society, www.cancer.org.)

In Exercises 9–12, determine whether each of the following processes represents a probability experiment.

⇒ 6.4 **9.** A box contains 5 identical tickets labeled 1 through 5. In a process, a ticket is drawn, its number is recorded, and it is returned to the box. Then a second draw is taken. The sum of the two numbers is recorded.

⇒ 6.4 **10.** The suggestions card at Pappy's restaurant asks customers to rate the service as fair, good, or excellent and then to rate the quality of the food on a scale from 1 to 5, with 1 being superior and 5 being lousy.

11. In a mail survey, 400 people were asked, "What is your least favorite song?"

12. A random sample of 1050 people was asked, "If you could change your name, what would you change it to?"

In Exercises 13 and 14, consult the customer satisfaction card supplied.

	Excellent	Above Average	Average	Below Average	Poor
The Store					
Overall shopping experience	○	○	○	○	○
The Merchandise					
Selection	○	○	○	○	○
In stock	○	○	○	○	○
Value	○	○	○	○	○
The Sales Personnel					
Sales people available	○	○	○	○	○
Sales people knowledgeable	○	○	○	○	○
The Checkout Area					
Speed	○	○	○	○	○
Friendliness	○	○	○	○	○
The Service Desk					
Courtesy	○	○	○	○	○
Helpfulness	○	○	○	○	○

Customer Type Do-It-Yourselfer ○ Contractor / Commercial Customer ○

Additional Comments:____

13. Which item(s) represents a probability experiment. Explain.

14. Which item(s) does not represent a probability experiment. Explain.

In Exercises 15–24, determine whether order is important.

⇒ 6.3 **15.** The board of directors at the SafeSide Company consists of 12 members. When computing how many different ways a chief executive officer, a vice-president and a chief financial officer can be selected, is order important?

⇒ 6.3 **16.** The Chess Club at Dyeston High School has 15 members. When computing how many different ways a president, a vice-president, and a treasurer can be selected, is order important?

⇒ 6.3 **17.** The Miss Farrago County Fair queen contest has 19 entrants. When computing how many different ways a winner, first runner-up, and second runner-up can be selected, is order important?

⇒ 6.3 **18.** Traditionally, 33 cars start the Indianapolis 500. When computing how many different ways the drivers can place first, second, and third, is order important?

⇒ 6.3 **19.** In Crawfords County, 6 judges are running for 3 county judge positions. When computing how many different ways the 3 county judges can be elected, is order important?

⇒ 6.3 **20.** The central office of the Higgens Corporation has to choose 3 of its 8 regional offices for an audit. When computing how many different ways the regional offices can be chosen, is order important?

✓ ⇒ 6.3 **21.** A group of 20 professors in an Archeology department want to choose a research team of 5 to study a new archeological site. When computing how many different ways the research team can be formed, is order important?

6.3 **22.** For a literature class, a student has to read 4 books from a list of 21. When computing how many different ways the 4 books can be chosen, is order important?

6.3 **23.** In the Maine State lottery called *Megabucks,* 6 balls numbered 1 to 42 are drawn from a hopper. When computing how many different ways the 6 numbers can be chosen, is order important? (*Source:* Maine State Lottery Commission, www.mainelottery.com.)

6.3 **24.** Jelly Belly manufactures 40 different flavors of jelly beans. If a person goes to a Jelly Belly store, he can get a sampler cup of 8 jelly beans, in which the 8 jelly beans are all different flavors. When computing how many different sampler cups are possible, is order important? (*Source:* Jelly Belly, www.jellybelly.com.)

In Exercises 25–34, determine if the following events can happen at the same time.

6.5 **25.** A person in a mall is selected at random. The event that this the person has brown hair and the event that this person has blue eyes.

6.5 **26.** A person in a mall is selected at random. The event that the person has a college degree and the event that the person is employed.

6.5 **27.** In a state that requires voters to register for a single party, a voter is selected at random for an exit poll. The event that the person is a registered Republican and the event that the person is a registered Democrat.

6.5 **28.** In a state that uses electronic ballots, a voter is selected at random for an exit poll. The event that the person voted for issue 1 on the ballot and the event that the person voted against issue 1 on the ballot.

✓ 6.5 **29.** A student at Brockston College is selected at random. The event that the student is a biology major and the event that the student is a senior.

6.5 **30.** A student at Brockston College is selected at random. The event that the student receives financial aid and the event that the student is on the Dean's list.

6.5 **31.** A box contains 30 identical tickets numbered 1 to 30. One ticket is drawn at random. The event that the ticket is labeled with an even number and the event that the ticket is labeled with an odd number.

6.5 **32.** A box contains 30 identical tickets numbered 1 to 30. One ticket is drawn at random. The event that the ticket is labeled with a prime number and the event that the ticket is labeled with a composite number.

6.5 **33.** Two respondents from a nationwide survey are selected at random for a follow-up interview. The event that the two respondents are brothers and the event that the two respondents are sisters.

6.5 **34.** Two respondents from a nationwide survey are selected at random for a follow-up interview. The event that the two respondents are brothers and the event that the two respondents are not related.

In Exercises 35–40, an event is given. Write another event so that together the events cannot happen at the same time.

35. If a coin is flipped, the event that the outcome is heads.

36. If a card is drawn from a shuffled, standard 52-card deck of cards, the event that the card is red.

37. If a person wakes up from a nap, the event that it will be dark outside.

38. The event that a randomly selected student will live in a coed dormitory.

39. The event that a randomly selected person has a hearing aid.

40. The event that a randomly selected student taking history passes a history test.

In Exercises 41–46, an event is given. Write another event so that together the events can happen at the same time.

41. In an automobile, the event that a person is arrested for fleeing from the police.

42. At an airport, the event that the flight from Chicago to Washington is delayed due to bad weather in Chicago.

43. During the fall semester, the event that a randomly selected student receives a failing grade in biology.

44. During a meeting of the school's chess club, the event that Thom is elected club vice-president.

45. When drawing a card from a deck, the event that a black card is drawn.

46. When drawing a card from a deck, the event that a numbered (nonface) card is drawn.

In Exercises 47–56, determine if the outcome of one event influences the outcome of the other event.

6.6 **47.** The event that a flash food occurs at Windsor Park on Tuesday afternoon and the event that the Tuesday evening company picnic at Windsor Park is postponed.

6.6 **48.** The event that Rebecca completes her political science course reading assignment every day and the event that Rebecca passes her political science course.

6.6 **49.** The event that Kahrin's car battery is dead and the event that Kahrin's oven light is burned out.

6.6 **50.** The event that the telephone in Rich's office does not work and the event that the vending machine in Rich's daughter's high school is empty.

6.6 **51.** The event that Clifford gets a raise and a promotion and the event that Clifford purchases a new car.

6.6 **52.** The event that Lana is exposed to ultraviolet radiation for three decades and the event that Lana develops skin cancer.

6.6 **53.** A box contains 5 tickets numbered 1 to 5. In an experiment, a ticket is drawn, its number is recorded, and it is replaced. Then a second ticket is drawn. The event that the first ticket is a 3 and the event that the second ticket is a 3.

6.6 **54.** A bag contains 5 green and 4 purple marbles. In an experiment, a marble is drawn, its color is recorded, and it is replaced. Then a second marble is drawn. The event that the first marble is purple and the event that the second marble is purple.

6.6 **55.** The event that a person has brown eyes and the event that the person's son has brown eyes.

6.6 **56.** The event that Jim has received seven speeding tickets and the event that Jim has high car insurance premiums.

In Exercises 57–62, an event is given. Write another event so that together the outcome of one event does not influence the outcome of the other event.

57. The event that it will be sunny on Wednesday.

58. The event that the next president will be a Republican.

59. A child pulls out two cookies from a jar containing five chocolate chip and five oatmeal cookies. The event that the first cookie taken is an oatmeal cookie.

60. If 2 cards are drawn from a shuffled, standard 52-card deck of cards, the event that the first card drawn is an ace.

61. If a die is tossed, the event that the outcome is an even number.

62. Ten identical balls numbered 0 through 9 are placed in a hopper for the Mini Lotto game. If three numbered balls are drawn for the Mini Lotto game, the event that the winning numbers the first week of August are 3–8–9.

In Exercises 63–68, an event is given. Write another event so that together the outcome of one event does influence the outcome of the other event.

63. If a person goes outside to sunbathe, the event that the person does not put on sunscreen.

64. In Olympic skating, the event that Tara Lapinski wins the gold medal in freestyle skating.

65. In a graduation ceremony, the event that Tammy gets the award for valedictorian.

66. At an ice cream parlor, the event that Ben eats four banana splits.

67. For a political science exam, the event that Ron did not go to class or complete the assigned reading to prepare for the exam.

68. At a stop sign down the street from her house, the event that Chandra does not come to a complete stop at the intersection.

SECTION PROJECT

In this section, we have seen that events can have certain properties, such as if the outcome of one event influences another or if two events can occur at the same time. Another important property to know is if the list of events includes every possible outcome of an experiment. If we toss a coin, for example, we can say that the possible events associated with the experiment are to flip a head and to flip a tail. On any given flip, we know that both events cannot occur at the same time. We also know that on any given flip we must either flip a head or flip a tail. So the list of events, flip a head and flip a tail, includes every possible outcome of the experiment.

A standard deck of 52 cards consists of four suits: hearts, diamonds, clubs, and spades. Each suit, in turn, consists of 13 cards: Ace, two, three, . . . , ten, jack, queen, and king. We draw a card at random from the deck. Let's say that one event is that it is a diamond and another event is that it is a heart. Even though these events cannot occur at the same time, the list of these two events does not include every possible outcome. Knowing that two events cannot occur at the same time does not necessarily mean that every possible outcome is accounted for.

Let's test our understanding of these ideas by completing the table by writing *yes* or *no* for each situation.

Card Draws	Can the two events occur at the same time?	Will the two events list all possible outcomes?
(a) Draw a spade and draw a club		
(b) Draw a face card and a numbered card		
(c) Draw a king and a 2		
(d) Draw a spade and a nonspade		
(e) Draw a 9 and a heart		
(f) Draw a black card and a club		

Section 6.2 Sets and Set Operations

We often categorize things in life so that we can make sense of them. We categorize cars by their make, model, color, list of options, and so forth. Creating collections of objects helps us to find order and meaning. When we categorize objects, we are creating **sets**. In this section, we will learn how to write sets of objects, perform operations on sets, and explore some properties of sets. Later, we will apply many of these ideas when we study probability.

Set Notation

A **set** is a well-defined collection of objects. For example, the months January, February, May, and July represent the set of all months of the year that end in the letter y. On the other hand, a list of the five best female tennis players of all time is not a set since it is *not* well defined. This is because naming the five best female tennis players is a subjective matter; different people could have different lists.

Capital letters, such as A, B, C, and so on, are usually used to name sets. The objects in a set are its **elements**. The elements of a set are separated by commas and enclosed in **braces**, { }. For example, the set of all even numbers from 1 to 10, inclusive, can be represented by

$$A = \{2, 4, 6, 8, 10\}$$

Notice that if $B = \{4, 8, 10, 6, 2\}$ then the sets A and B have exactly the same elements, just written in a different order. We call the sets A and B **equal**, and write $A = B$. To indicate that 4 is an element of A, we write

$$4 \in \{2, 4, 6, 8, 10\} \quad \text{or equivalently} \quad 4 \in A$$

The symbol $\in$ means "belongs to" or "is an element of." To indicate that 5 is not an element of A, we write

$$5 \notin \{2, 4, 6, 8, 10\} \quad \text{or equivalently} \quad 5 \notin A$$

Some sets have no elements. For example, the set of U.S. coins in which each coin is worth exactly 7 cents has no elements. We call a set with no elements the **empty set** and designate it with the symbol $\emptyset$ or { }.

There are generally two ways to express the elements in a set. The first is the **listing method** where we simply list all the elements in a set. However, at times this method is not practical to use. Let's say that set C consists of the even numbers between 1 and 1000, inclusive. Instead of listing all 500 numbers in the set, we could use **set-builder notation** and write

$$C = \{x \mid x \text{ is an even number between 1 and 1000, inclusive}\}$$

This is read, "C is the set of all elements x such that x is an even number between 1 and 1000, inclusive." Note that x is the name of the typical element in C and that the vertical bar is read "such that."

Example 1 Writing Sets

Use the listing method to rewrite each of the following sets:

(a) $A = \{x \mid x \text{ is a U.S. coin less than a dollar in value}\}$

(b) $B = \{x \mid x \text{ is an even number and } 10 \le x \le 20\}$

Solution

(a) Here we write the names of the desired coins in set notation.

$$A = \{\text{penny, nickel, dime, quarter, half-dollar}\}$$

(b) For B, the designated numbers are listed and separated by commas.

$$B = \{10, 12, 14, 16, 18, 20\}$$

■

Checkpoint 1

Now work Exercise 5.

At times the elements of one set are contained within another set. For example, if we let $A = \{x | x \text{ is a whole number and } 1 \leq x \leq 30\}$ and $B = \{5, 10, 15, 20\}$, then we observe that every element of B is also an element of A. We call B a **subset** of A, denoted by $B \subseteq A$. In fact, there are elements of A that are not elements of B, so we call B a **proper subset** of A and write $B \subset A$.

Subset and Proper Subset

- If every element of set B is also an element of set A, then we say that B is a **subset** of A, denoted by $B \subseteq A$.
- If set B is a subset of set A and there is at least one element in A that is not in B, then we say that B is a **proper subset** of A, denoted by $B \subset A$.

Note: From the definition of subset, we observe that for any set A and any set B the following are true:

1. The empty set is a subset of every set. That is $\emptyset \subseteq A$ and $\emptyset \subseteq B$.
2. Every set is a subset of itself. So, for example, $A \subseteq A$ and $B \subseteq B$.
3. If $A \subseteq B$ and $A \neq B$, then $A \subset B$.
4. If $A \subseteq B$ and $B \subseteq A$, then $A = B$.
5. If $A \subset B$, then $A \subseteq B$. However, the converse (if $A \subseteq B$, then $A \subset B$) is not true in general.

Example 2 Listing Subsets

List all possible subsets for the set $A = \{3, 4, 5\}$.

Solution

We know from the note just prior to this example that the empty set and A itself are subsets. So the list of all possible subsets is

$$\emptyset, \{3, 4, 5\}, \{3\}, \{4\}, \{5\}, \{3, 4\}, \{3, 5\}, \{4, 5\}$$

■

Interactive Activity

Notice that in Example 2 a set with three elements has eight subsets. Now list all the subsets for the sets $\{3, 4\}$ and $\{3, 4, 5, 6\}$. Do you see a pattern? Complete the following statement. *A set of n distinct elements has ____ subsets.*

Set Operations

In arithmetic, we have four basic **operations** of numbers: addition, subtraction, multiplication, and division with the operation symbols $+$, $-$, $\times$, and $\div$. In sets, we also have operations. One such operation is the **intersection** of sets. Consider the sets $A = \{1, 2, 3, 4, 5, 6\}$ and $B = \{2, 4, 6, 8, 10\}$. Notice that A and B have the elements 2, 4, and 6 in common. We say that the intersection of A and B, denoted by $A \cap B$, is the set $\{2, 4, 6\}$. Visually, we can use **Venn diagrams** to construct graphical representations of set operations. A Venn diagram starts with a

universal set U that contains all the elements under consideration. Let's say that the universal set for A and B is $U = \{x|x \text{ is a whole number and } 1 \le x \le 15\}$. Then the Venn diagram for $A \cap B$ is as shown in Figure 6.2.1.

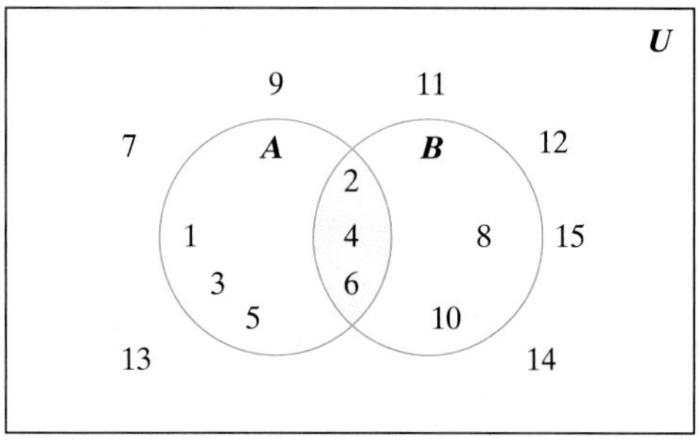

Figure 6.2.1 The shaded region shows the intersection of sets A and B, that is, the elements in both A and B.

Intersection of Sets

For sets A and B, the set of all elements that belong to both A and B is the **intersection** of A and B, denoted $A \cap B$. Mathematically, this is written as

$$A \cap B = \{x|x \in A \textbf{ and } x \in B\}$$

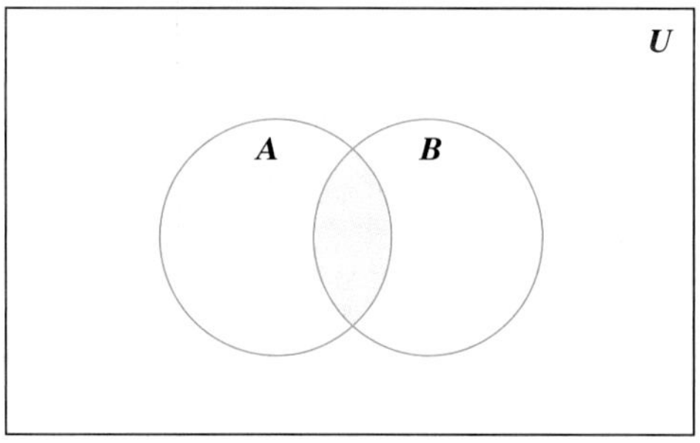

Shaded region shows $A \cap B$.

Example 3 Determining the Intersection of Sets

Let $A = \{1, 5, 7, 9\}$, $B = \{3, 4, 5, 6, 7\}$, and $C = \{2, 3, 4\}$ with $U = \{x|x \text{ is a whole number and } 1 \le x \le 10\}$. Write the desired set and draw a shaded Venn diagram for each of the following:

(a) $A \cap B$ **(b)** $A \cap C$

Solution

(a) For $A \cap B$, we notice that A and B only have the elements 5 and 7 in common. This means that $A \cap B = \{5, 7\}$. This is shown in the Venn diagram in Figure 6.2.2.

(b) Here we see that sets A and C have no elements in common. This means that the intersection of A and C is the empty set; that is, $A \cap C = \emptyset$. Notice in the Venn diagram in Figure 6.2.3 that the two circles representing the sets do not overlap.

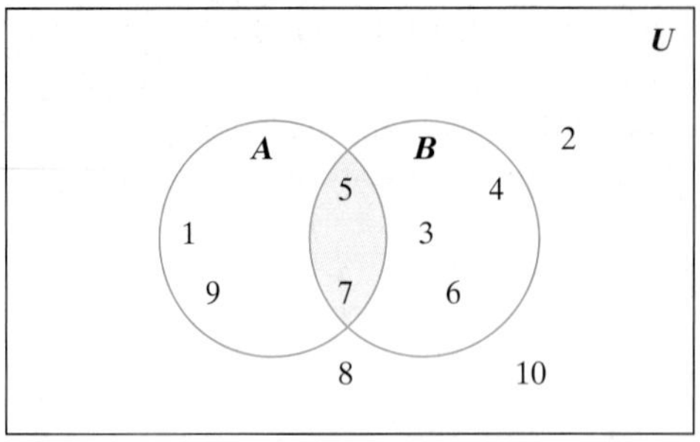

Figure 6.2.2 Shaded region shows $A \cap B$.

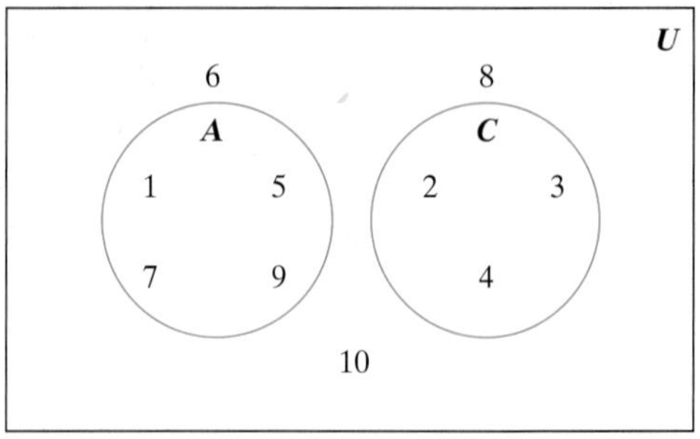

Figure 6.2.3

In part (b) of Example 3, we saw that $A \cap C = \emptyset$. We call sets that have no elements in common **disjoint sets**.

Disjoint Sets

If $A \cap B = \emptyset$, then the sets A and B are **disjoint**.

Sometimes we want to know what makes up the merging of two or more sets. Let's say that we have the sets $A = \{1, 2, 3, 4, 5, 6\}$ and $B = \{2, 4, 6, 8, 10\}$. The merging of A and B is the set $\{1, 2, 3, 4, 5, 6, 8, 10\}$. Mathematically, we call the merging of A and B the **union** of A and B.

Union of Sets

For sets A and B, the set of all elements that belong to at least one of A and B is the **union** of A and B, denoted by $A \cup B$. Mathematically, this is written as

$$A \cup B = \{x | x \in A \textbf{ or } x \in B \textbf{ or } x \in A \cap B\}$$

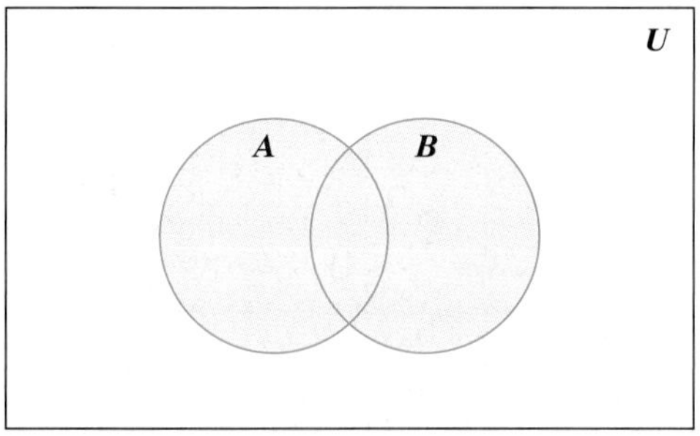

Shaded region shows $A \cup B$.

Note: If an element is a member of set A **or** is a member of set B **or both**, it is in the union of A and B.

Example 4 Determining the Union of Sets

Let $A = \{1, 3, 5, 7, 9\}$, $B = \{3, 4, 5, 6, 7\}$, and $C = \{1, 2, 3, 4\}$ with $U = \{x | x$ is a whole number and $1 \le x \le 10\}$. Write the desired set and draw a shaded Venn diagram for each of the following:

(a) $A \cup B$ **(b)** $(A \cup B) \cap C$

Solution

(a) Here we take the elements that are in set A or in set B or in both to get $A \cup B = \{1, 3, 4, 5, 6, 7, 9\}$. See Figure 6.2.4.

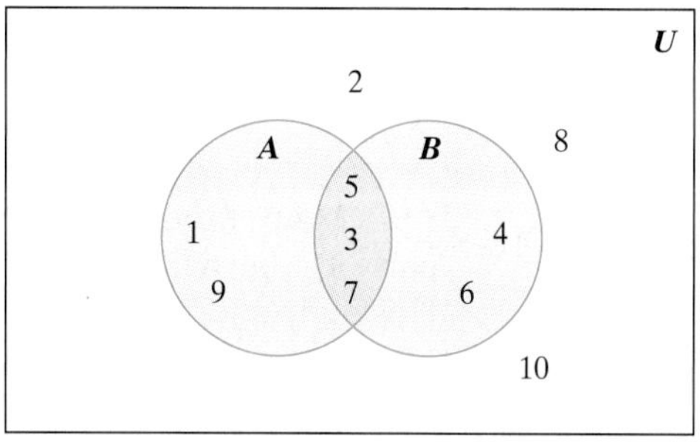

Figure 6.2.4 Shaded region shows $A \cup B$.

(b) When simplifying arithmetic expressions, we know from the order of operations that we must perform the operations in the parentheses first. The same applies to set operations. So the expression $(A \cup B) \cap C$ means that we must perform $A \cup B$ first. We then take this result and intersect with set C. In part (a) we found $A \cup B$. So here we simply take the result of part (a) and intersect with set C. This gives us

$$(A \cup B) \cap C = \{1, 3, 4, 5, 6, 7, 9\} \cap \{1, 2, 3, 4\} = \{1, 3, 4\}$$

This result of $(A \cup B) \cap C$ is shown in the Venn diagram in Figure 6.2.5.

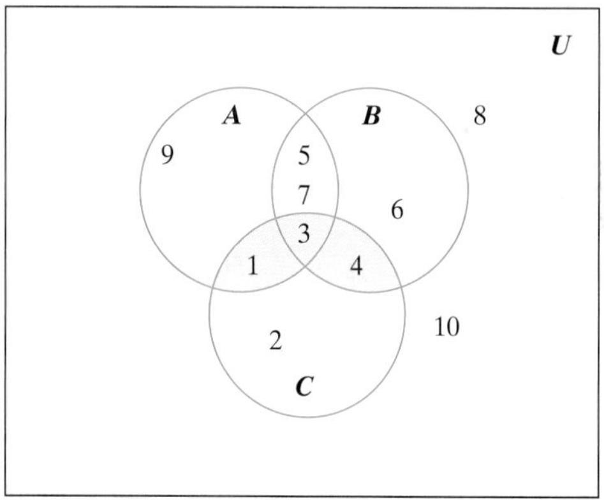

Figure 6.2.5 ■

Checkpoint 2 Now work Exercise 57.

Example 4(b) illustrates that we complete set operations that are in parentheses first, just as we do with arithmetic and algebraic expressions. In fact, the properties associated with set operations are very similar to those in arithmetic and algebra.

Properties of Set Operations

Let U be a universal set. If A, B, and C are subsets of U, then we have the following:

1. Commutative Property for Union of Sets $\quad A \cup B = B \cup A$
2. Commutative Property for Intersection of Sets $\quad A \cap B = B \cap A$

3. Associative Property for Union of Sets $(A \cup B) \cup C = A \cup (B \cup C)$
4. Associative Property for Intersection of Sets $(A \cap B) \cap C = A \cap (B \cap C)$
5. Distributive Properties $A \cup (B \cap C) = (A \cup B) \cap (A \cup C)$
   $A \cap (B \cup C) = (A \cap B) \cup (A \cap C)$

Many times, we are interested in simply knowing what elements are **not** part of a particular set. Let's say we have $U = \{x | x$ is an employee at Farberg Hospital$\}$ and $A = \{x | x$ is a female employee at Farberg Hospital$\}$; then the set of elements *not* in A is $\{x | x$ is a male employee at Farberg Hospital$\}$. We call this the **complement** of A.

Complement of a Set

If U is a universal set and A is a subset of U, then the **complement** of set A, denoted by A^C, is the set of all elements not in set A. Mathematically, this is written as

$$A^C = \{x | x \notin A\}$$

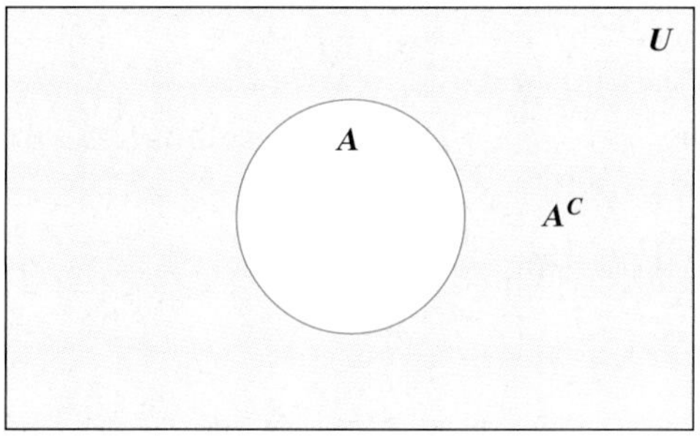

Shaded area represents A^C.

Example 5 **Determining the Complement of a Set**

Let $A = \{1, 5, 7, 9\}$, $B = \{3, 4, 5, 6, 7\}$, and $C = \{2, 3, 4\}$ with $U = \{x | x$ is a whole number and $1 \le x \le 10\}$. Write the desired set and draw a shaded Venn diagram for each of the following:

(a) A^C **(b)** $(A \cap B)^C$ **(c)** $A^C \cup B^C$

Solution

Interactive Activity

Using the sets given in Example 5, determine $(A \cup B)^C$ and then determine $A^C \cap B^C$. Compare the results. What do you notice?

(a) For A^C, we want the elements that are not in A, so we have $A^C = \{2, 3, 4, 6, 8, 10\}$. See Figure 6.2.6.

(b) From Example 3(a), we know that $A \cap B = \{5, 7\}$. The complement of this set is $(A \cap B)^C = \{1, 2, 3, 4, 6, 8, 9, 10\}$. The Venn diagram for $(A \cap B)^C$ is in Figure 6.2.7. In words, this is the set of all elements of U not in the intersection of A and B.

(c) From part (a), we know that $A^C = \{2, 3, 4, 6, 8, 10\}$, and we can see that the set of elements not in B is $B^C = \{1, 2, 8, 9, 10\}$. If we take the union of these

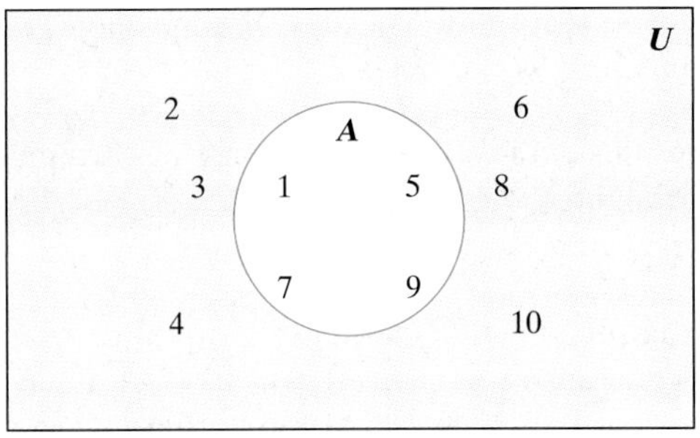

Figure 6.2.6 Shaded region represents A^C.

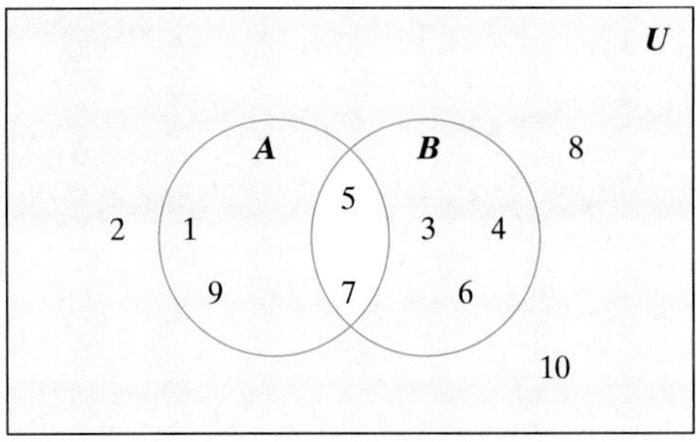

Figure 6.2.7 Shaded region represents $(A \cap B)^C$.

sets we get

$$A^C \cup B^C = \{2, 3, 4, 6, 8, 10\} \cup \{1, 2, 8, 9, 10\} = \{1, 2, 3, 4, 6, 8, 9, 10\}$$

$A^C \cup B^C$ is shown graphically in Figure 6.2.8.

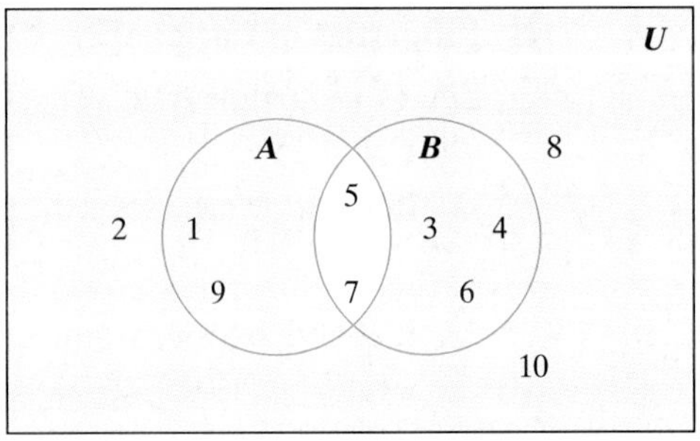

Figure 6.2.8 Shaded region represents $A^C \cup B^C$. ■

Cardinality of Sets

Often our primary interest in a set is to know the number of elements in the set. We call the number of elements in a set A its **cardinality** and denote it with $n(A)$. For example, if $B = \{25, 26, 27, 30\}$, then $n(B) = 4$, since B has four elements. Similarly, if $A = \{15, 20, 25, 30, 35\}$, then $n(A) = 5$.

Example 6 Determining Cardinality

Let $A = \{$Monday, Tuesday, Wednesday, Thursday, Friday$\}$ and $B = \{$Friday, Saturday, Sunday, Monday$\}$. Determine the following:

(a) $A \cap B$ **(b)** $n(A \cap B)$

(c) $A \cup B$ **(d)** $n(A \cup B)$

Solution

(a) For $A \cap B$, we see that A and B only have the elements Monday and Friday in common. This means that $A \cap B = \{$Monday, Friday$\}$.

(b) Since $A \cap B = \{$Monday, Friday$\}$, we have $n(A \cap B) = 2$.

(c) Here we take the elements that are in set A or in set B or in both to get $A \cup B = \{$Monday, Tuesday, Wednesday, Thursday, Friday, Saturday, Sunday$\}$.

(d) Since $A \cup B = \{$Monday, Tuesday, Wednesday, Thursday, Friday, Saturday, Sunday$\}$, we have $n(A \cup B) = 7$. ■

Example 6 illustrates a useful property of cardinality of sets. Observe that $n(A) = 5$, $n(B) = 4$, $n(A \cap B) = 2$, and $n(A \cup B) = 7$. From arithmetic, we have

$$n(A) + n(B) = 5 + 4 = 9$$

It might seem that $n(A) + n(B)$ should be equal to $n(A \cup B)$, but we know that $n(A \cup B) = 7$. The reason that $n(A) + n(B) \neq n(A \cup B)$ relies on the fact that when we summed $n(A) + n(B) = 5 + 4 = 9$, there were two elements common to both A and B. These two elements in $A \cap B$ are being counted twice when we compute $n(A) + n(B)$. So to compute $n(A \cup B)$, it appears that $n(A \cup B)$ is simply the sum of $n(A)$ and $n(B)$ minus $n(A \cap B)$. In other words,

$$n(A \cup B) = n(A) + n(B) - n(A \cap B)$$

This serves as motivation for the first of the following properties of cardinality of sets.

Properties of Cardinality of Sets

Let U be a universal set. If A and B are subsets of U, then we have the following:

1. If A and B are sets with elements in common, then
$$n(A \cup B) = n(A) + n(B) - n(A \cap B)$$
2. If A and B are disjoint sets, then
$$n(A \cup B) = n(A) + n(B)$$
3. $n(A^C) = n(U) - n(A)$
4. $n(\emptyset) = 0$

Note: Property 2 is a special case of property 1. If A and B are disjoint sets, then $n(A \cap B) = 0$ since $n(\emptyset) = 0$. Thus

$$\begin{aligned} n(A \cup B) &= n(A) + n(B) - n(A \cap B) \\ &= n(A) + n(B) - 0 \\ &= n(A) + n(B) \end{aligned}$$

These properties of the cardinality of sets can be used in applications as shown in Example 7.

Example 7 Determining the Cardinality of a Set

The numbers of public elementary and secondary school teachers in the states of Texas and New York during 1998 are shown in Table 6.2.1.

Table 6.2.1

State \ Grade Level:	Elementary	Secondary	Total
Texas	132,500	122,200	254,700
New York	98,800	98,400	197,200
Total	231,300	220,600	451,900

SOURCE: U.S. Department of Education, www.ed.gov

We let A represent the set of elementary school teachers from the states of Texas and New York, B represent the set of secondary school teachers from the states of Texas and New York, C represent the set of elementary and secondary school teachers from Texas, and D represents the set of elementary and secondary school teachers from New York. With these definitions of sets A, B, C, and D, Table 6.2.1 can be written as shown in Table 6.2.2.

Table 6.2.2

Grade Level: / State	Elementary (A)	Secondary (B)	Total
Texas (C)	132,500	122,200	254,700
New York (D)	98,800	98,400	197,200
Total	231,300	220,600	451,900

SOURCE: U.S. Department of Education, www.ed.gov

Determine the following and interpret each.

(a) $n(A)$ **(b)** $n(A \cap C)$ **(c)** $n(A \cup C)$

Solution

(a) **Understand the Situation:** Here we are being asked to determine the number of elements in the set of elementary school teachers from the states of Texas and New York.

If we look down the column represented by Elementary (A), we see a total of 231,300. So $n(A) = 231{,}300$.

Interpret the Solution: This means that during 1998 there were about 231,300 elementary school teachers in the states of Texas and New York.

(b) **Understand the Situation:** We are being asked to find the number of elements common to the set of elementary school teachers from the states of Texas and New York *and* the set of elementary and secondary school teachers from Texas.

If we look down the column represented by Elementary (A) and across the row represented by Texas (C), we see a value of 132,500, meaning that $n(A \cap C) = 132{,}500$.

Interpret the Solution: This means that during 1998 there were about 132,500 elementary school teachers in the state of Texas.

(c) First note that if we look across the row represented by Texas (C) we see that $n(C) = 245{,}700$. So, using the solutions from part (a) and part (b) along with the first property of cardinality of sets, we get

$$\begin{aligned} n(A \cup C) &= n(A) + n(C) - n(A \cap C) \\ &= 231{,}300 + 245{,}700 - 132{,}500 = 344{,}500 \end{aligned}$$

Interpret the Solution: This means that in 1998 there were about 344,500 who were either an elementary school teacher from Texas or New York or a public school teacher in Texas. ■

✓ Checkpoint 3

Now work Exercise 71.

SUMMARY

In this section, we introduced **sets**, which are well-defined collections of objects. We discussed the two methods, **listing** and **set-builder**, for writing sets. We then discussed some of the operations associated with sets.

- The set of all elements that belong to both A and B is the **intersection** of A and B, denoted $A \cap B$. That is, $A \cap B = \{x | x \in A \textbf{ and } x \in B\}$.
- The set of all elements that belong to at least one of A and B is the **union** of A and B, denoted by $A \cup B$. That is, $A \cup B = \{x | x \in A \textbf{ or } x \in B \textbf{ or } x \in A \cap B\}$.
- The **complement** of set A, denoted by A^C, is the set of all elements not in A.
- If A and B are any sets, then $n(A \cup B) = n(A) + n(B) - n(A \cap B)$.
- If A and B are disjoint sets, then $n(A \cup B) = n(A) + n(B)$.

SECTION 6.2 EXERCISES

In Exercises 1–6, rewrite the set using the listing method.

1. $A = \{x | x$ is a whole number and $1 \leq x \leq 7\}$

2. $B = \{x | x$ is a whole number and $2 \leq x \leq 10\}$

3. $C = \{x | x$ is an even number and $2 \leq x \leq 25\}$

4. $D = \{x | x$ is an odd number and $2 \leq x \leq 25\}$

5. $E = \{x | x$ is a month that ends with the letters b-e-r$\}$

6. $F = \{x | x$ is a day of the week that starts with the letter S$\}$

In Exercises 7–12, rewrite the set using set-builder notation.

7. $A = \{2, 3, 4, 5, 6\}$

8. $B = \{12, 13, 14, 15, 16\}$

9. $C = \{5, 10, 15, 20, 25, 30\}$

10. $D = \{20, 30, 40, 50, 60, 70, 80\}$

11. $E = \{$Tuesday, Thursday$\}$

12. $F = \{$January, June, July$\}$

In Exercises 13–18, write all possible subsets for the given set.

13. $\{2\}$

14. $\{5\}$

15. {Honda, Ford}

16. $\{5, 7\}$

17. $\{8, 9, 10\}$

18. {knitting, needlepoint, quilting}

In Exercises 19–24, fill in the blanks with $\subset$, $\subseteq$, $\supset$, or $\supseteq$ so that each statement is true. More than one answer may be possible.

19. $\{1, 2, 3\}$_____$\{2, 3\}$

20. $\{10, 20, 30\}$_____$\{10, 20\}$

21. $\{a, v, r\}$_____$\{r, a, v\}$

22. {Fred, Ruth, Ralph}_____{Ralph, Fred, Ruth}

23. $\emptyset$_____$\emptyset$

24. For any set A, A_____A.

In Exercises 25–46, let $U = \{x | x$ is a whole number and $1 \leq x \leq 15\}$, $A = \{1, 2, 3, 4, 5\}$, $B = \{2, 4, 6, 8, 10\}$, and $C = \{11, 12, 13, 14, 15\}$. Write each of the following sets using the listing method.

25. $A \cap B$

26. $A \cap C$

27. $B \cap C$

28. $B \cap U$

29. $A \cup B$

30. $A \cup C$

31. $B \cup C$

32. $B \cup U$

33. A^C

34. C^C

35. $B \cap C^C$

36. $B^C \cap C$

37. U^C

38. $\emptyset^C$

39. $(A \cap C)^C$

40. $(B \cap C)^C$

41. $A^C \cup C$

42. $B^C \cup C$

43. $A \cup \emptyset$

44. $C \cup \emptyset$

45. $\emptyset \cap B^C$

46. $\emptyset \cap A^C$

In Exercises 47–56, use the Venn diagram to write the desired set.

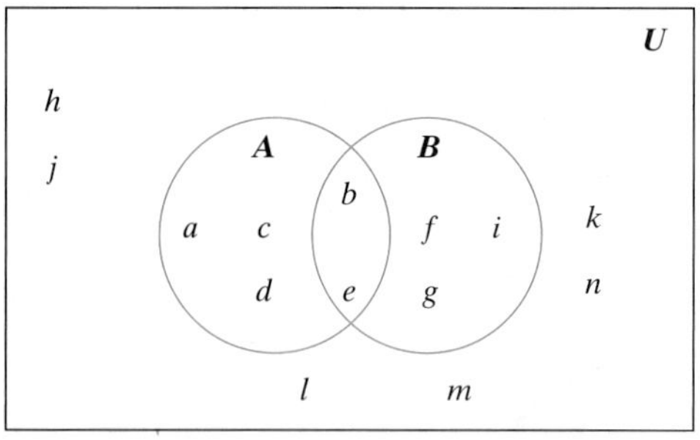

47. A

48. B

49. U

50. $A \cup B$

51. $A \cap B$

52. B^C

53. A^C

54. $(A \cap B)^C$

55. $(A \cup B)^C$

56. $A \cap B^C$

In Exercises 57–66, let $U = \{1, 2, 3, 4, 5, 6, 7, 8, 9\}$, $A = \{2, 4, 6, 8\}$, $B = \{4, 5, 6\}$, *and* $C = \{2, 3, 4, 5, 6\}$. *Write a set and draw a shaded Venn diagram for each of the following:*

57. $A \cup (B \cap C)$

58. $A \cap (B \cup C)$

59. $A \cap (B \cup C)^C$

60. $A^C \cap (B \cup C)$

61. $(A \cup B) \cap (A \cup C)$

62. $(A \cap B) \cup (A \cap C)$

63. $A^C \cup B^C$

64. $A^C \cup C^C$

65. $(A \cap B)^C$

66. $(A \cap C)^C$

For Exercises 67–72, the values of arms transfers, in millions of dollars, from the United States and the United Kingdom to India and Pakistan during 1996 are as shown in the table.

Recipient/ Supplier	United States (A)	United Kingdom (B)	Total
India (C)	110	0	$n(C) = 110$
Pakistan (D)	140	70	$n(D) = 210$
Total	$n(A) = 250$	$n(B) = 70$	$n(U) = 320$

SOURCE: U.S. Statistical Abstract, www.census.gov/statab/www

Here A represents the set of arms supplied by the United States, B represents the set of the arms supplied by the United Kingdom, C represents the set of the arms received by India, and D represents the set of the arms received by Pakistan. Determine the following and interpret each.

67. $n(A)$

68. $n(C)$

69. $n(B \cap D)$

70. $n(A \cap C)$

71. $n(B \cup D)$

72. $n(A \cup C)$

For Exercises 73–78, the number of American households in 1997 in the midwestern and the southern United States, measured in millions, whose household incomes are below and above \$35,000 are as shown in the table.

Region/ Income	Under \$35,000 (A)	\$35,000 and over (B)	Total
Midwest (C)	35.3	13.1	48.4
South (D)	18.5	18.0	36.5
Total	53.8	31.1	84.9

SOURCE: U.S. Statistical Abstract, www.census.gov/statab/www

Here A represents the set of households in the regions with incomes under \$35,000, B represents the set of households in the regions with incomes of \$35,000 and over, C represents the set of households in the Midwest, and D represents the set of households in the South. Determine the following and interpret each.

73. $n(A^C)$

74. $n(C^C)$

75. $n(A \cup D)$

76. $n(A \cup C)$

77. $n(A^C \cup D)$

78. $n(A \cup C^C)$

SECTION PROJECT

In Example 6, we illustrated, for two sets A and B that are not disjoint, that $n(A \cup B) = n(A) + n(B) - n(A \cap B)$. For this section project, consider the following sets:

$$A = \{1, 2, 3, 4, 5\}, \quad B = \{2, 4, 6, 8\}, \quad C = \{1, 3, 5, 7, 9, 11\}$$

(a) Determine $n(A)$.
(b) Determine $n(B)$.
(c) Determine $n(C)$.
(d) Determine $n(A \cap B)$.
(e) Determine $n(A \cap C)$.
(f) Determine $n(B \cap C)$.
(g) Determine $n(A \cap B \cap C)$.
(h) Verify that

$$n(A \cup B \cup C) = n(A) + n(B) + n(C) - n(A \cap B) - n(A \cap C) - n(B \cap C) + n(A \cap B \cap C)$$

Explain why this formula is correct in general.

Section 6.3 Principles of Counting

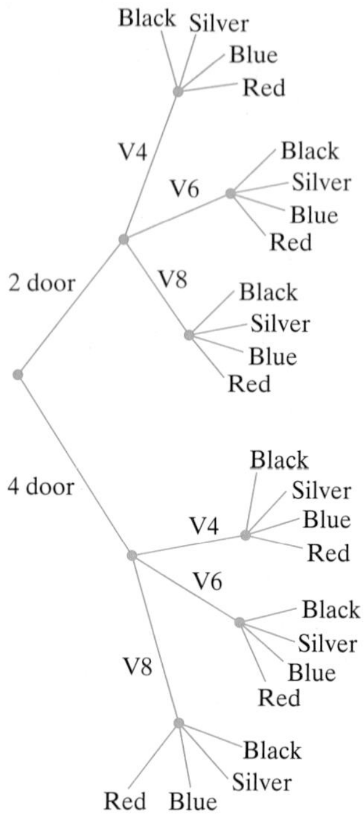

Figure 6.3.1

In the previous section we learned that cardinality was an important idea in the discussion of sets. Now we want to generalize our discussion of counting objects. In this section we will discuss the techniques of counting objects under different conditions and assumptions. The study of counting techniques is called **combinatorics**.

Fundamental Principles

Let's say that a person goes to a new car lot to order a new sport utility vehicle called the Trailmaster. The SUV comes in a two-door and a four-door model. There are three types of engines: V4, V6, and V8. Additionally, the SUV comes in the colors black, silver, red, and blue. To see in how many distinct ways a person can order a Trailmaster, we can draw a **tree diagram** to visually show how many ways the vehicle can be ordered. The person has to make choices in three steps. Choosing the number of doors is the first step, choosing the engine type is the second step, and selection of a color is the third step. We will draw **branches** on the tree diagram for the choices in each step. The tree diagram for ordering the Trailmaster is in Figure 6.3.1. We can see from counting the branches on the right side of the tree diagram that there are 24 ways to order a Trailmaster.

Example 1 Constructing a Tree Diagram

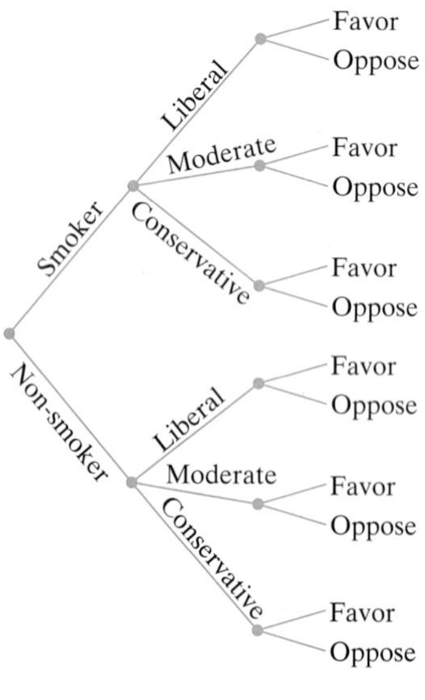

Figure 6.3.2

A survey given by a local pollster has three questions.

Question 1: Are you a smoker?

Question 2: Do you consider yourself to be a liberal, moderate, or conservative?

Question 3: Do you support or oppose capital punishment?

Draw a tree diagram to represent all possible answers in the survey. How many ways could a respondent answer all three questions?

Solution

Understand the Situation: An individual who completes this survey has to respond to three questions. Each question requires the individual to make a choice. The tree diagram for all the choices in the survey is in Figure 6.3.2.

Interpret the Solution: If we count the number of branches on the right side of the tree diagram, we see that there are a total of 12 ways to respond to the three questions. ■

✓ Checkpoint 1

Now work Exercise 3.

Let's inspect Example 1 a little more closely. The first question, "Are you a smoker?", can be answered in 2 ways, the second question, "Do you consider yourself to be liberal, moderate or conservative?", can be answered in 3 ways, and the third question, "Do you support or oppose capital punishment?", can be answered in 2 ways. Notice that if we multiply these values together, we get

Possible answers for Question 1		Possible answers for Question 2		Possible answers for Question 3	
2	·	3	·	2	= 12 ways

This is the same result that we determined by using the tree diagram in Figure 6.3.2! This illustrates one of the fundamental counting techniques, the **Multiplication Principle**.

Multiplication Principle

If a choice consists of k steps, of which the first can be made in n_1 ways, the second in n_2 ways, . . . , and the last step can be made in n_k ways, then the number of different ways that the total choice can be made is given by

$$n_1 \cdot n_2 \cdot n_3 \cdot \cdots \cdot n_k$$

Even though tree diagrams give us a good visual perspective of counting objects, it is not practical for many applications. An application of the Multiplication Principle is given in Example 2. Notice in this example that drawing a tree diagram would be very difficult.

Example 2 Applying the Multiplication Principle

A certain state's automobile license plates are embossed with three letters followed by four digits. Assuming that there are no restrictions on numbers or letters, how many different license plates can be made?

Solution

Understand the Situation: In this case, we have to make choices in seven steps. There is a choice for each of the three letters and there is a choice for each of the four digits.

Since there are 26 letters in the alphabet and there are 10 digits, the total number of ways to make license plates in this state is given by

Choices for:	Letter 1	Letter 2	Letter 3	Digit 1	Digit 2	Digit 3	Digit 4
	26 ·	26 ·	26 ·	10 ·	10 ·	10 ·	10

$= 26^3 \cdot 10^4 = 175{,}760{,}000$

Interpret the Solution: This means that in this particular state a total of 175,760,000 different license plates can be made if there are no restrictions on numbers or letters. ■

Interactive Activity

Let's say in Example 2 that a law was passed so that no letter could be repeated on a license plate. Under the new law, how many different license plates could be made?

WWW

Permutations

Let's explore another counting technique. Suppose that there are 4 people who need to sit for a photograph and the photographer must arrange them in a row. How many ways can the photographer do this? To find out, we can again use the multiplication principle. The first choice can be made in 4 ways since all the people are waiting to be seated. The second choice can be made in 3 ways since the first person chosen is already sitting for the photo. The third choice can be made in 2 ways and the fourth choice can be made in only 1 way, since there is only one person left to be seated. So the total number of ways is

Choice 1		Choice 2		Choice 3		Choice 4	
4	$\cdot$	3	$\cdot$	2	$\cdot$	1	= 12 ways

So there are 12 ways to seat the 4 people for the photograph. For simplicity, special products such as $4 \cdot 3 \cdot 2 \cdot 1$ are given the special name **factorial**. Factorials appear frequently in combinatorics.

Factorial Notation

If n is a natural number, then ***n* factorial,** denoted by $n!$, is given by

$$n! = n(n-1)(n-2) \cdot \cdots \cdot 3 \cdot 2 \cdot 1$$

As a special case, we define $0! = 1$.

Note: We call n the first factor of $n!$, $n-1$ is called the second factor of $n!$, and so forth.

Example 3 Using Factorial Notation

In the Jasper Derby, 12 horses start the race.

(a) In how many ways can the horses finish first, second, and third?

(b) Rewrite the answer in part (a) using factorial notation.

Solution

(a) Of the 12 horses, there are 3 places to de determined. For the first-place horse, there are 12 possibilities. For the second-place horse there are 11 possibilities (the first-place horse is already determined). Finally, for the third-place horse there are 10 possibilities. So the total number of possibilities that these 12 horses can finish first, second, or third is

Choice 1		Choice 2		Choice 3	
12	$\cdot$	11	$\cdot$	10	= 1320

So the horses can finish first, second, and third in 1320 different ways.

(b) We know by the definition of n factorial that

$$12! = 12 \cdot 11 \cdot 10 \cdot 9 \cdot 8 \cdot 7 \cdot 6 \cdot 5 \cdot 4 \cdot 3 \cdot 2 \cdot 1$$

Since we need only the first three factors of 12!, we can rewrite 12! as

$$\begin{aligned} 12! &= (12 \cdot 11 \cdot 10) \cdot (9 \cdot 8 \cdot 7 \cdot 6 \cdot 5 \cdot 4 \cdot 3 \cdot 2 \cdot 1) \\ &= (12 \cdot 11 \cdot 10) \cdot 9! \end{aligned}$$

To get the solution in part (a), we can divide 12! by 9! to get

$$\frac{12!}{9!} = \frac{(12 \cdot 11 \cdot 10) \cdot 9!}{9!} = 12 \cdot 11 \cdot 10 = 1320$$

So we see that by using factorial notation the solution to part (a) can be written as $\frac{12!}{9!}$.

Two aspects to the scenario in Example 3 are critical. First, we see that the **order is important**. Recall from Section 6.1, by order being important we mean that

- The order of arrangement makes a difference, or
- There is a ranking of the members of the arrangement.

Next we see that there is **no repetition**, meaning that, for example, a horse cannot finish both first and third. These conditions are critical components of the counting technique called **permutation**.

Formula for Permutation

The number of ways in which r objects can be selected in a specific order from among n objects is given by the **permutation** ${}_nP_r$ (read "n select r"), where

$${}_nP_r = \frac{n!}{(n-r)!}$$

Interactive Activity

For any value of n, use the definition of factorial to compute ${}_nP_n$ and ${}_nP_0$. What special properties do these permutations have and how can they be interpreted? Also, compare the two quantities and see if there is a relation between them.

Note: Permutations have two important characteristics:

1. Order is important
2. There is no repetition.

Let's take a Flashback to an example first presented in Section 6.1 to practice our newest counting technique.

Flashback

Computer Club Revisited

There are 24 members in the New Strawsburg Junior High School Computer club. How many ways can the club pick a president, a vice-president, a secretary, and a treasurer?

Flashback Solution

In Section 6.1 we saw that order is important and there is no repetition. The names that we associate with the president, vice-president, secretary, and treasurer make a difference. So the total number of ways that the 4 club officers can be chosen is a 24 select 4 permutation, which is calculated by

$${}_{24}P_4 = \frac{24!}{(24-4)!} = \frac{24!}{20!} = 24 \cdot 23 \cdot 22 \cdot 21 = 255{,}024$$

Interpret the Solution: So the Computer club officers can be selected in 255,024 different ways.

Combinations

Often when we count objects, the order is *not* important. For these types of applications, we use a counting technique called the **combination**.

Combination

Unordered selections (order not important) of distinct objects are called **combinations**. An unordered selection using r of the n distinct objects is given by the combination ${}_nC_r$ (read "n choose r"). ${}_nC_r$ is defined to be the number of ways of choosing r distinct objects from a set of n distinct objects without regard to the order of the selection.

Note: 1. For the combination, order is **not** important and there is no repetition.

2. Some textbooks express the combination ${}_nC_r$ by writing $\binom{n}{r}$ or $C_{n,r}$.

Before we supply a formula to determine ${}_nC_r$, let's do an example to fully understand the difference between when order is important and when order is not important, as well as to gain some insight into the formula.

Example 4 Arranging Letters

Consider the letters w, x, y, and z. Determine in how many ways we can choose three without repeating any letter where

(a) Order is important

(b) Order is not important

Solution

(a) If order is important, the total number of ways that 3 letters letters can be chosen is a 4 select 3 permutation calculated by

$$ {}_4P_3 = \frac{4!}{(4-3)!} = 24 $$

All 24 possible selections are shown in the right-hand column of the tree diagram in Figure 6.3.3.

(b) When order is not important, only 4 of the 24 selections determined in part (a) are listed. They are

$$ wxy \quad wxz \quad wyz \quad xyz $$

Notice that when selecting the letters w, x, and y different results are obtained depending on whether order is important. When order is not important, it does not matter how we arrange the letters. However, when order is important, it does matter how the letters are arranged. This can be summarized as follows:

Order Is Not Important	*Order Is Important*
wxy	wxy, wyx, xwy, xyw, ywx, yxw

Notice that the three letters wxy in the "order is not important" list allow for six rearrangements in the "order is important" list. ■

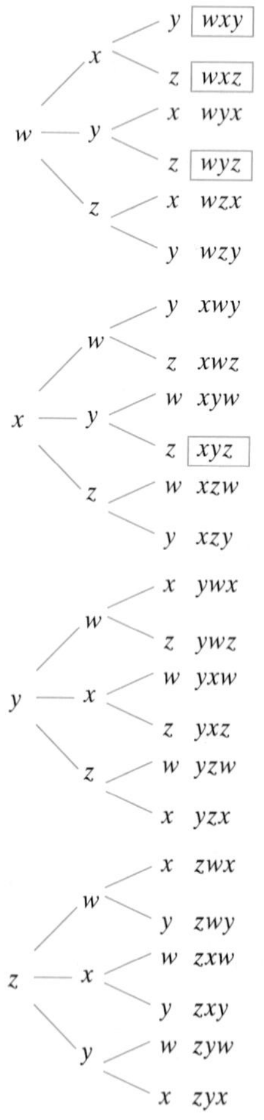

Figure 6.3.3

Observe in Example 4 that the number of possible selections when order is important, 24, is six times the number of possibilities when order is not important, 4. The reason for this is that each unordered selection consisted of three letters. As part (b) of Example 4 illustrated, these three letters allow for 6 (or 3!) rearrangements. We can use this observation to obtain a formula for ${}_nC_r$. In general, **each** unordered selection of r objects gives rise to $r!$ ordered selections. But the number of ordered selections is given by the permutation ${}_nP_r$. So the number of ordered selections, ${}_nP_r$, is $r!$ times the number of unordered selections. Since the number of unordered selections is given by ${}_nC_r$, we have

$${}_nP_r = r! \cdot {}_nC_r$$

Since ${}_nP_r = \dfrac{n!}{(n-r)!}$, we can substitute this and solve for ${}_nC_r$ and get

$${}_nP_r = r! \cdot {}_nC_r$$

$$\frac{n!}{(n-r)!} = r! \cdot {}_nC_r \qquad \text{Substitute.}$$

$${}_nC_r = \frac{n!}{r!(n-r)!} \qquad \text{Divide both sides by } r!.$$

Formula for Combination

The number of different selections of r objects chosen from n distinct objects when no object is repeated and order is not important is given by the **combination** ${}_nC_r$, where ${}_nC_r$ is computed by

$${}_nC_r = \frac{n!}{r!(n-r)!}$$

Example 5 Computing Combinations

A hospital has 44 staff members. In how many ways can a committee of 3 be formed to investigate salary discrepancies?

Solution

Understand the Situation: Here we have $n = 44$ distinct objects. The order is not important since there is no ranking of committee members. Also notice that there is no repetition.

So the number of ways that the committee can be formed is the number of 3 combinations of 44 objects, which is calculated by

$${}_{44}C_3 = \frac{44!}{3!(44-3)!} = \frac{44!}{3! \cdot 41!} = 13{,}224$$

Interpret the Solution: This means that there are 13,224 possible committees of size 3 chosen from the 44 available staff members. ■

✓ Checkpoint 2 Now work Exercise 65.

Technology Option

Calculators can perform permutations and combinations. Figure 6.3.4 shows the result of the Flashback and Figure 6.3.5 shows the result of Example 5.

Figure 6.3.4

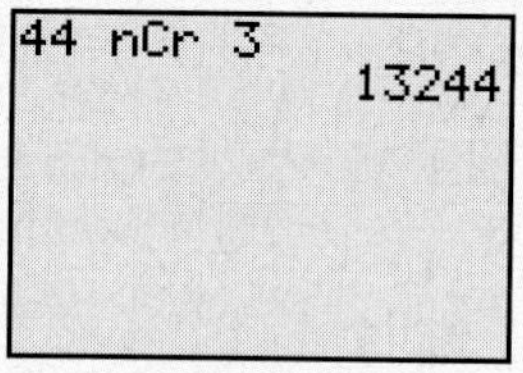

Figure 6.3.5

To learn how to compute permutations and combinations on your calculator, consult the online graphing calculator manual at www.prenhall.com/armstrong.

There are times when the combination and the multiplication principle are used together. Let's say that a city council consists of 6 Democrats and 5 Republicans. The council wants to form a committee of 4 to investigate possible campaign finance irregularities. If the committee is to consist of equal representation from each political party, then we have two tasks ahead of us. Specifically, we need to choose 2 Democrats and 2 Republicans. The number of ways that we can choose the 2 Democrats and the number of ways that we can choose the 2 Republicans are ${}_6C_2$ and ${}_5C_2$, respectively. These choices are illustrated in Figure 6.3.6.

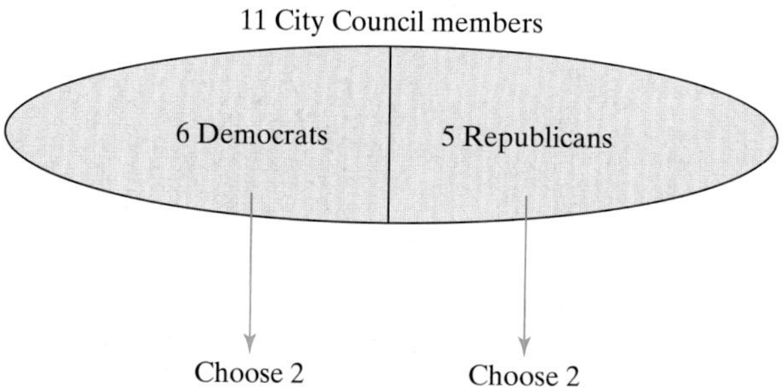

Figure 6.3.6

So the number of ways to form the committee is given by

$$ {}_6C_2 \cdot {}_5C_2 = 15 \cdot 10 = 150 $$

Hence, by the Multiplication Principle, there are 150 different ways to form the committee.

Example 6 Combining the Multiplication Principle and the Combination

A shipment of 20 suits arrives at the Businessman's Discount Warehouse. The shipment contains 7 suits with flaws in the fabric.

(a) Determine ${}_{20}C_5$ and interpret.

(b) In how many ways can 5 suits be chosen from the shipment so that exactly 2 of them have flaws?

Solution

(a) **Understand the Situation:** We are being asked to determine the number of ways of choosing 5 suits, in any order, from the shipment of 20 suits.

The number of ways to choose 5 suits in any order from a shipment of 20 is given by

$$_{20}C_5 = \frac{20!}{5!(20-5)!} = \frac{20!}{5! \cdot 15!} = 15{,}504$$

Interpret the Solution: So there are 15,504 ways that 5 suits can be chosen from the shipment of 20 suits.

(b) **Understand the Situation:** Notice that if we want to choose 5 suits so that exactly 2 of them have flaws this means that the other 3 suits have no flaws. Since 7 of the 20 suits have flaws, we want to choose 2 of the 7 with flaws and choose 3 of the 13 without flaws. Figure 6.3.7 shows the graphical representation of this part of the example.

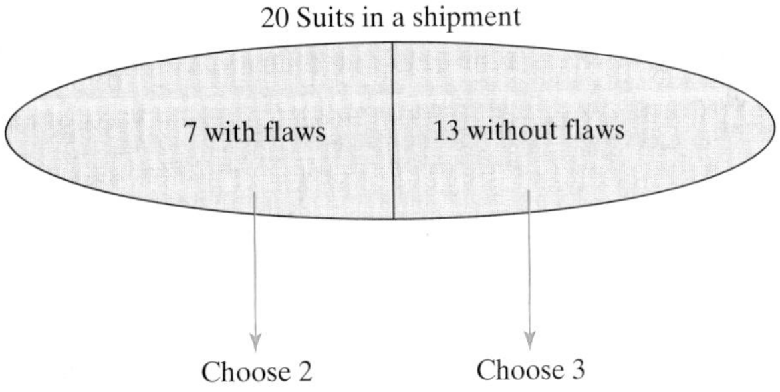

Figure 6.3.7

So the desired number of possibilities is given by

$$_7C_2 \cdot {}_{13}C_3 = \frac{7!}{2!(7-2)!} \cdot \frac{13!}{3!(13-3)!} = \frac{7!}{2!5!} \cdot \frac{13!}{3!10!} = 21 \cdot 286 = 6006$$

Interpret the Solution: This means that there are 6006 ways to choose 5 suits from the shipment so that 2 suits have flaws and 3 suits do not have flaws. ■

SUMMARY

In this section, we learned several counting techniques, such as the **Multiplication Principle**, the **permutation**, and the **combination**.

- **Multiplication Principle** If a choice consists of k steps, of which the first can be made in n_1 ways, the second in n_2 ways, . . . , and the last step can be made in n_k ways, then the number of different ways that the total choice can be made is given by $n_1 \cdot n_2 \cdot n_3 \cdot \cdots \cdot n_k$.
- ***n* factorial** $n! = n(n-1)(n-2) \cdot \cdots \cdot 3 \cdot 2 \cdot 1$
- **Number of Permutations** $_nP_r = \frac{n!}{(n-r)!}$. Order is important and there is no repetition.
- **Number of Combinations** $_nC_r = \frac{n!}{r!(n-r)!}$. Order is not important and there is no repetition.

SECTION 6.3 EXERCISES

1. A two-question phone survey asks a person the following questions:

Question 1: Have you donated to the local Fraternal Order of Police?

Question 2: Do you feel safe or unsafe in your home at night?

(a) Draw a tree diagram to represent all possible answers in the survey.

(b) In how many ways could a respondent answer both questions?

2. A two-question survey by the Find-It Company asks a person the following questions:

Question 1: Have you used an Internet search engine in the last week?

Question 2: Have you made an online purchase in the last month?

(a) Draw a tree diagram to represent all possible answers in the survey.

(b) In how many ways could a respondent answer both questions?

3. A person wants a new bookshelf and finds that there are two color choices: white and oak. There is also a size choice of three, four, or five shelves.

(a) Draw a tree diagram to represent all possible bookshelves.

(b) In how many ways could a person create a bookshelf?

4. A customer who orders a computer from the Z-Dron Company has the choice of three microprocessors: budget, faster, and fastest. The customer also has a choice of a 17-inch or a 19-inch monitor.

(a) Draw a tree diagram to represent all possible computer systems.

(b) In how many ways could a customer create a computer system?

5. The new XZR-200 motorcycle comes in a choice of four colors: black, silver, green, and magenta. There is also an option for a luggage rack.

(a) Draw a tree diagram to represent all possible motorcycle packages.

(b) In how many ways could a customer create a motorcycle package?

6. A ceiling fan can come from one of four factories: East, South, North, and Midwest. Each factory includes the option of a brass or silver fixture.

(a) Draw a tree diagram to represent all possible ceiling fans.

(b) In how many ways could a customer create a ceiling fan?

In Exercises 7–20, use the Multiplication Principle to determine the following:

7. A new concert hall features 12 classical and 16 rock concerts in its premiere season. In how many different ways can a person construct a package deal as a present that consists of a ticket for a classical concert and a ticket for a rock concert?

8. The Blossom Blouse Company has 8 warehouses and 24 retail outlets. In how many different ways can a person order a blouse from the warehouse and another blouse from a retail outlet?

9. A football squad has 21 linemen, 8 receivers, and 9 running backs. A most-improved player award will be given to each position. In how many ways can a most-improved player award be given to a lineman, a receiver, and a running back?

10. The SpectraSpectacle Company makes its Laservision sunglasses with 9 different colored lenses, 6 different colored ear pieces, and 4 different colored nose pieces. In how many different ways can a person order a pair of Laservision sunglasses?

11. The menu at the Top Class restaurant has a choice of 4 appetizers, 6 entrees, 6 beverages, and 5 desserts. In how many different ways can a customer order an appetizer, entree, beverage, and dessert at the restaurant?

12. A child has 6 T-shirts, 7 pairs of pants, 4 pairs of shoes, and 10 pairs of socks. Assuming that the clothes are color coordinated, in how many different ways can the child pick a T-shirt, pants, shoes, and socks to wear?

13. A textbook committee has to choose from among 10 prealgebra, 13 college algebra, 5 precalculus, and 7 calculus books. In how many ways can the committee chose a textbook from each subject area?

14. The call letters of a radio station consists of 4 letters. The first letter must be a K or a W. If the last 3 letters can be repeated, how many different call letters can be made?

15. The Deanol sports car comes in a choice of 5 colors and 3 engine sizes. There is also an air conditioner and a turbocharger option. In how many different ways can the car be ordered?

16. An identification card used at the Linger Corporation consists of two digits from 0 to 4 and then 3 letters. If the digits and letters can be repeated, in how many different ways can an identification card be made?

17. The lock on the Takeaway brief case has 5 digits. If the numbers cannot be repeated, how many different possibilities are there for the lock?

18. The password to gain entrance to the Techway Company is a sequence of 5 letters from A to K. If a letter can be repeated, how many different passwords can be made?

19. A 25-question multiple-choice test in engineering has choice of A, B, C, or D for each question. In how many different ways can a student answer all the questions?

20. A history quiz consists of 8 true–false questions. In how many different ways can a student answer all the questions?

In Exercises 21–28, evaluate the expression without using a calculator.

21. $7!$

22. $10!$

23. ${}_5P_3$

24. ${}_7P_2$

25. ${}_{10}P_4$

26. ${}_9P_5$

27. ${}_8P_0$

28. ${}_8P_8$

In Exercises 29–34, use a calculator to determine the following:

29. ${}_{50}P_{20}$

30. ${}_{30}P_{15}$

31. ${}_{40}P_{15}$

32. ${}_{45}P_{28}$

33. ${}_{90}P_{30}$

34. ${}_{50}P_{41}$

6.1 **35.** The board of directors at the SafeSide Company consists of 12 members. In how many different ways can a chief executive officer, a vice-president and a chief financial officer be selected?

6.1 **36.** The Chess Club at Dyeston High School has 15 members. In how many different ways can a president, vice-president, and treasurer be selected?

6.1 **37.** The Miss Farrago County Fair queen contest has 19 entrants. In how many different ways can a winner, first runner-up, and second runner-up be selected?

6.1 **38.** Traditionally, 33 cars start the Indianapolis 500. In how many different ways can the drivers place first, second, and third?

39. Food critic Hero Maramoto has eaten at 25 restaurants this year. To introduce the top 4 restaurants in his newspaper column, how many choices does he have?

40. In a customer preference survey, a person is asked to try 16 different brands of applesauce and rank the first 5 in order of preference. How many different rankings are possible?

41. Cabin cruisers can communicate with each other by lofting 5 from among 12 flags up a mast. If the message is determined by the order in which the flags are raised, how many different messages are possible?

42. A researcher has found that 13 drugs are effective in treating a certain ailment. She conjectures that the sequencing of the administration of the drugs can affect their effectiveness. In how many different ways can 6 of the drugs be administered?

43. The starting lineup for the Jasper Comets softball team consists of 9 members. In how many ways can the manager arrange the batting order?

44. The 1921 class of Stringsbrug High School consists of 7 surviving members. In how many ways can they be arranged in 7 chairs, in a row, for a class photo?

In Exercises 45–56, evaluate the expression without using a calculator.

45. ${}_{5}C_{3}$

46. ${}_{6}C_{2}$

47. ${}_{10}C_{6}$

48. ${}_{9}C_{5}$

49. ${}_{12}C_{5}$

50. ${}_{11}C_{4}$

51. ${}_{7}C_{1}$

52. ${}_{9}C_{1}$

53. ${}_{13}C_{0}$

54. ${}_{17}C_{0}$

55. ${}_{12}C_{12}$

56. ${}_{14}C_{14}$

In Exercises 57–62, use a calculator to determine the following:

57. ${}_{50}C_{20}$

58. ${}_{30}C_{15}$

59. ${}_{40}C_{15}$

60. ${}_{45}C_{28}$

61. ${}_{90}C_{30}$

62. ${}_{50}C_{41}$

6.1 **63.** In Crawfords County, 6 judges are running for 3 county judge positions. In how many different ways can the 3 county judges be elected?

6.1 **64.** The central office of the Higgens Corporation has to choose 3 of its 8 regional offices for an audit. In how many different ways can the regional offices be chosen?

6.1 ✓ **65.** A group of 20 professors in an Archeology department wants to choose a research team of 5 to study a new archeological site. In how many different ways can the research team be formed?

6.1 **66.** For a literature class, a student has to read 4 books from a list of 21. In how many different ways can the 4 books be chosen?

67. There are 29 companies bidding for the construction of a new building. In how many different ways can 5 of the companies be chosen for hire?

68. How many different 5-card hands can be dealt from a standard deck of 52 cards?

6.1 **69.** In the Maine State lottery called *Megabucks*, 6 balls numbered 1 to 42 are drawn from a hopper. In how many different ways can the 6 numbers be chosen? (*Source:* www.mainelottery.com.)

70. In the Washington State lottery called Lotto, 6 balls numbered 1 to 49 are drawn from a hopper. In how many different ways can the 6 numbers be chosen? (*Source:* www.wa.gov/lot/home.htm.)

6.1 **71.** Jelly Belly manufactures 40 different flavors of jelly beans. If a person goes to a Jelly Belly store, they can get a sampler cup of 8 jelly beans in which the 8 jelly beans are all different flavors. How many different sampler cups are possible? (*Source:* www.jellybelly.com.)

72. Sweet Dreams has 34 different flavors of ice cream. One of its ice cream cones is a triple dip of 3 different flavors for which the order of the 3 different flavors does not matter. How many different kinds of triple-dip ice cream cone can be made?

73. A city council has 5 Democrats and 7 Republicans. In how many different ways can a subcommittee be formed with 2 Democrats and 4 Republicans?

74. The Euratal Company has 4 women and 8 men on its board of directors. In how many different ways can a gender-equality subcommittee of 4 be formed so that the numbers of men and women are equal?

75. In a shipment of 30 car stereos, 8 are known to be defective. In how many ways can a sample of 10 be chosen so that 2 are defective and 8 are not defective?

76. A box of 50 pieces of taffy has 20 that are pink and 30 that are yellow. In how many ways can a handful of 8 pieces of taffy be pulled out so that 2 are pink and 6 are yellow?

77. A group of 20 environmentalists is made up of 6 Republicans, 8 Democrats, and 6 members of the Green Party. In how many different ways can a public relations committee of 7 be formed with 2 Republicans, 2 Democrats, and 3 Green Party members?

SECTION PROJECT

(a) Compute ${}_{10}C_3$ and ${}_{10}C_7$. Compare the solutions.

(b) Prove that for any whole numbers n and r, where $0 \le r \le n$, ${}_nC_r = {}_nC_{n-r}$. (*Hint:* Use the factorial definition to write the left and right sides. Simplify the right side and compare it to the left side.)

(c) Compute the sum ${}_{15}C_3 + {}_{15}C_4$. Compare the solution to ${}_{16}C_4$.

(d) Prove that for any whole numbers n and r, where $0 \le r \le n$, ${}_nC_r + {}_nC_{r+1} = {}_{n+1}C_{r+1}$. Use the hint given in part (b).

Section 6.4 Introduction to Probability

Probability is a topic that many people experience in their everyday lives, yet know little about. When a weather forecaster states that there is an 80% chance of rain tomorrow or when a physician asserts that a certain surgical procedure has a 97% chance of success, they are often stating a probability. **Probability** can be thought of as the branch of mathematics that produces a number that expresses the likelihood of occurrence of a specific event. The study of probability is called **probability theory**. In this section, we will introduce the terminology and notation of probability theory and examine three fundamental ways in which probabilities can be computed.

Probability Experiments

To illustrate some of the terminology of probability theory, let's start by examining an activity that involves probability. Recall from Section 6.1 that a **probability experiment** is a process that leads to well-defined results called **outcomes**. All possible outcomes are predictable, but any given outcome cannot be predicted with certainty. Let's take a Flashback to an example from Section 6.1 to refamiliarize ourselves with probability experiments.

Flashback

Marbles in a Bag Revisited

A bag contains three marbles that are painted with the numbers 1, 2, and 3. A marble is drawn randomly from the bag, its number is recorded, and it is then returned to the bag. Another marble is drawn randomly and its number is recorded. Construct a **tree diagram** showing all possible outcomes of this experiment.

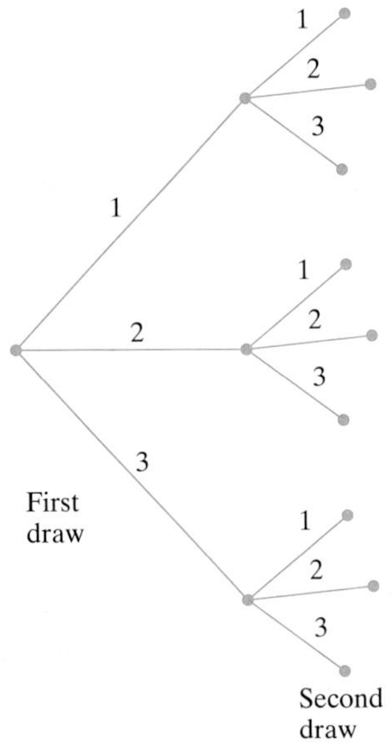

Figure 6.4.1

Flashback Solution

First, since we can observe each pair of numbers that are drawn, this is an example of a probability experiment. Each result of an experiment or a **trial**, such as drawing the marbles and recording the numbers, is called an outcome. We can display all the outcomes of this experiment several ways. One way is to construct a tree diagram as shown in Figure 6.4.1. We can see from the right side of the tree diagram that there are a total of nine possible outcomes for the experiment.

As stated at the end of the Flashback solution, there are nine possible outcomes for the experiment. These nine outcomes are **equally likely**, since the marbles are identical and the ones drawn are selected randomly. By **equally likely**, we mean that each outcome has an equal chance of occurring. We can write these nine possible equally likely outcomes for the experiment in the Flashback using **set notation** as shown in set S.

$$S = \{(1,1), (1,2), (1,3), (2,1), (2,2), (2,3), (3,1), (3,2), (3,3)\}$$

Set S is a special set in probability called a **sample space**.

Sample Space

For a probability experiment, the **sample space**, denoted by S, is the set of all possible outcomes.

From Your Toolbox

The cardinality of a set is its number of elements.

Many times we are interested in the number of elements in a sample space. Recall that in Section 6.2 we called this the **cardinality** of a set. Consult the Toolbox to the left. For our experiment of drawing the marbles from the bag, the size of the sample space, denoted by $n(S)$, is $n(S) = 9$.

Example 1 Identifying the Sample Space of a Probability Experiment

Consider a box that contains 4 identical tickets. Suppose that a ticket is labeled with a star, a ticket is labeled with a triangle, a ticket is labeled with a square, and a ticket is labeled with a circle. Also consider a bag that contains 2 marbles identical

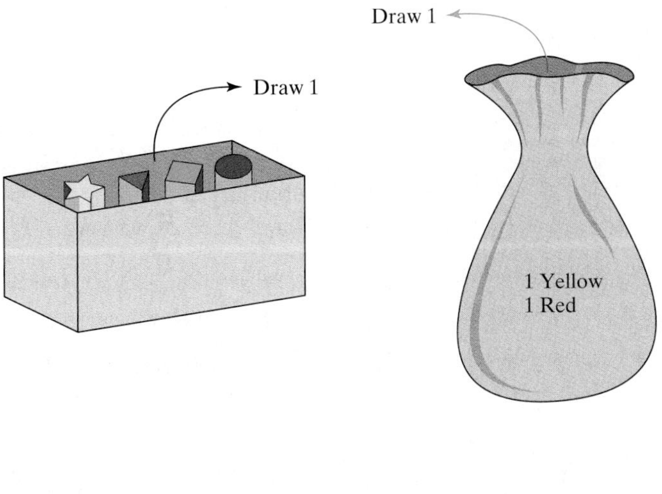

in every way except color. One is a yellow and the other is red. In an experiment, one random draw is made from the box and one random draw is made from the bag and the outcome is recorded. Write the sample space S and determine $n(S)$.

Solution

The draws from the box can be represented by the symbols ★, △, ■, and ○. For the bag, let Y indicate that the yellow marble is drawn and R indicate that the red marble is drawn. This gives us the sample space

$$S = \{(\bigstar, Y), (\bigstar, R), (\triangle, Y), (\triangle, R), (\blacksquare, Y), (\blacksquare, R), (\bigcirc, Y), (\bigcirc, R)\}$$

By counting the outcomes, we see that $n(S) = 8$. ■

Checkpoint 1

Now work Exercise 3.

Note in Example 1 that we really did not have to list the whole sample space to determine $n(S)$. We could have also used the **Multiplication Principle**, which was discussed in Section 6.3. Since the draw from the box can be made in 4 ways and the draw from the bag can be made in 2 ways, we get

Choice 1		*Choice 2*	
Draw from box		Draw from bag	
4	·	2	$= 8 = n(S)$

Now let's return to the experiment of drawing a marble numbered 1, 2, or 3 from the bag, replacing it, and drawing again. Recall that in this experiment the set of all possible outcomes (that is, the sample space) is

$$S = \{(1, 1), (1, 2), (1, 3), (2, 1), (2, 2), (2, 3), (3, 1), (3, 2), (3, 3)\}$$

Suppose that we want to list the outcomes that are associated when a 2 is drawn. If we label this set of outcomes with A, we get

$$A = \{(1, 2), (2, 1), (2, 2), (2, 3), (3, 2)\}$$

We call A an **event**. In words, we can say that A represents the event that at least one 2 is drawn. Notice that $A \subset S$.

Event

An **event** is simply a subset of the sample space.

Note: We usually label events with capital letters such as A, B, C, and D or B_1, B_2, and so on.

Types of Probability

There are several fundamental ways to compute probability. The first way is one that almost everyone has done at some point in his or her life and is called **subjective probability**. For example, when a student frowns and says, "I don't have much of a chance of passing this test," the student is stating a subjective probability.

Subjective Probability

A probability that is based on estimates, educated guesses, intuition, emotion, or other indirect information is called a **subjective probability**.

Although subjective probability can be useful in sorting out certain decisions that a person can make, it has little or no mathematical basis. So we now turn to another type of probability that uses many of the mathematical tools that we have already studied. This type is called **classical probability**.

Classical Probability

Assume that S is a sample space that has $n(S)$ outcomes, where each outcome is equally likely. If A is an event associated with S, then the probability of event A occurring, denoted by $P(A)$, is given by

$$\begin{array}{c}\text{Probability of}\\ \text{event } A \text{ occurring}\end{array} = P(A) = \frac{n(A)}{n(S)} = \frac{\text{number of outcomes of event } A}{\text{total number of possible outcomes}}$$

Note: 1. Recall that by equally likely we mean that every element in the sample space has an equal chance of being selected.
2. Probabilities can be expressed as either fractions, decimals, or percents.

Example 2 Computing Classical Probability

In Example 1, we were given the sample space

$$S = \{(\bigstar, Y), (\bigstar, R), (\triangle, Y), (\triangle, R), (\blacksquare, Y), (\blacksquare, R), (\bigcirc, Y), (\bigcirc, R)\}$$

Let A represent the event that a ticket with a ★ is drawn and let B represent the event that a red marble is drawn.

(a) Determine $P(A)$ and interpret.

(b) Determine $P(B)$ and interpret.

Solution

(a) **Understand the Situation:** When we are asked to determine $P(A)$, we are being asked to determine the probability of A, the event that a ticket with a ★ is drawn. To do this, we need to know the number of outcomes associated with event A as well as the total number of equally likely outcomes, that is, the cardinality of the sample space.

First, recall from Example 1 that the total number of possible outcomes is given by $n(S) = 8$. The outcomes associated with event A are $(\bigstar, Y)$, $(\bigstar, R)$. Therefore,

$$A = \{(\bigstar, Y), (\bigstar, R)\}$$

So, using classical probability, we get

$$P(A) = \frac{\text{number of ways to draw a } \bigstar}{\text{total number of possible outcomes}} = \frac{n(A)}{n(S)} = \frac{2}{8} = \frac{1}{4}$$

Interpret the Solution: This means that the probability of having an outcome that includes a ticket with a ★ is $\frac{1}{4}$.

(b) **Understand the Situation:** When we are asked to determine $P(B)$, we are being asked to determine the probability of B, the event that a red marble is drawn. To do this, we need to know the number of outcomes associated with event B as well as the total number of equally likely outcomes, that is, the cardinality of the sample space.

The outcomes associated with event B are elements of the following set:

$$B = \{(\bigstar, R), (\triangle, R), (\blacksquare, R), (\bigcirc, R)\}$$

So the probability of event B is

$$P(B) = \frac{\text{number of ways to draw a red marble}}{\text{total number of possible outcomes}} = \frac{n(B)}{n(S)} = \frac{4}{8} = \frac{1}{2}$$

Interpret the Solution: This means that the probability of having an outcome that includes a red marble is $\frac{1}{2}$. ■

Checkpoint 2 Now work Exercise 13.

When computing classical probability, we can also use some of the counting techniques discussed in Section 6.3. Example 3 illustrates how to use the multiplication principle.

Example 3 Computing Classical Probability

In an experiment, Conor tosses a six-sided die and records the number. He tosses it again and records the number from the second toss. Let A represent the event that the sum of the two tosses is an 8. Determine $P(A)$ and interpret.

Solution

Understand the Situation: When we are asked to determine $P(A)$, we are being asked to determine the probability of A, the event that the sum of the two tosses is 8. To do this, we need to know the number of outcomes associated with event A as well as the total number of equally likely outcomes, that is, the cardinality of the sample space.

We begin by determining the total number of possible outcomes. Since any of the numbers 1, 2, 3, 4, 5, or 6 is possible on the first toss as well as the second toss, we can employ the multiplication principle to determine the total number of possible outcomes. This gives

Possibilities for first toss		Possibilities for second toss	
6	·	6	= 36

Table 6.4.1

First Toss	Second Toss	Sum
2	6	8
3	5	8
4	4	8
5	3	8
6	2	8

So the sample space has 36 equally likely outcomes. This means that $n(S) = 36$. Now the number of ways to have the sum of the two tosses equal 8 is shown in Table 6.4.1, which indicates that there are 5 ways to sum the results of the two tosses of the die and get a sum of 8. So the probability of event A is

$$P(A) = \frac{\text{number of ways to sum the tosses to get 8}}{\text{total number of possible outcomes}} = \frac{n(A)}{n(S)} = \frac{5}{36} \approx 0.139$$

Interpret the Solution: This means that the probability of having the sum of the two tosses of a six-sided die equal 8 is $\frac{5}{36} \approx 0.139$. That is, there is a 13.9% chance that the sum of the two tosses is 8. ■

Permutations and combinations, two counting techniques discussed in Section 6.3, can also be used in computing classical probability. The use of combinations to compute classical probability is shown in Example 4.

Example 4 Computing Classical Probability with Combinations

The Garforth County Board of Trustees is made up of 4 Republicans and 5 Democrats. The Board has agreed to form a subcommittee of 4 to investigate the impact of commercial development in the county. The 4 will be chosen at random.

(a) In how many ways can a subcommittee of 4 be formed?

(b) In how many ways can a subcommittee be formed that has 2 Republicans and 2 Democrats?

(c) Let A represent the event that the subcommittee is formed randomly with 2 Republicans and 2 Democrats. Determine $P(A)$ and interpret.

Solution

(a) **Understand the Situation:** Here we have to choose 4 trustees from among a total of 9 trustees. Since order is not important and there is no repetition, the total number of ways that the subcommittee can be formed is the number of 4 combinations of 9 objects.

The total number of ways that the subcommittee can be formed is given by

$$_9C_4 = \frac{9!}{4!(9-4)!} = \frac{9 \cdot 8 \cdot 7 \cdot 6}{4 \cdot 3 \cdot 2 \cdot 1} = 126$$

So the subcommittee of 4 can be formed in 126 ways.

Interactive Activity

For Example 4, determine the probability that the subcommittee is randomly formed with all members being Republicans.

WWW

(b) Using the techniques introduced in Section 6.3 and consulting Figure 6.4.2, we see that the number of ways to form the subcommittee with 2 Republicans and 2 Democrats is given by

Choice 1		*Choice 2*	
Republicans		Democrats	
$_4C_2$	$\cdot$	$_5C_2$	
6	$\cdot$	10	$= 60$

where $_4C_2 = \frac{4!}{2!2!} = 6$ and $_5C_2 = \frac{5!}{3!2!} = 10$. So there are 60 ways to form a subcommittee of 4 with 2 Republicans and 2 Democrats.

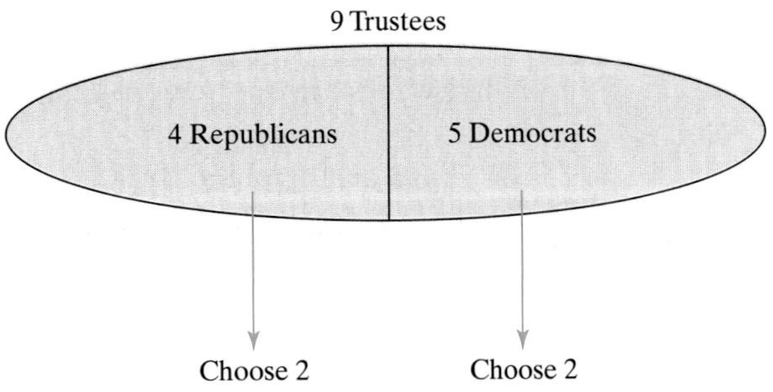

Figure 6.4.2

(c) Using classical probability along with the results of parts (a) and (b), we get

$$P(A) = \frac{\begin{array}{c}\text{number of ways to form a subcommittee}\\ \text{with 2 Republicans and 2 Democrats}\end{array}}{\begin{array}{c}\text{total number of ways to}\\ \text{form a subcommittee of 4}\end{array}} = \frac{{}_4C_2 \cdot {}_5C_2}{{}_9C_4} = \frac{60}{126} \approx 0.476$$

Interpret the Solution: This means that there is about a 47.6% probability that the subcommittee is formed with 2 Republicans and 2 Democrats. ■

Checkpoint 3 Now work Exercise 25.

The third type of probability, called **empirical probability**, makes use of frequency distributions of qualitative data. Frequency distributions and qualitative data will be studied in more depth in Chapter 7. For our purposes at this time, think of a **frequency**, denoted by f, as telling us how often something occurs or how many of a certain item there are. A **frequency distribution** is a table that summarizes the frequencies. Think of **qualitative data** as being data that are organized by categories. (Some even call qualitative data categorical data.) See Table 6.4.2, which gives the distribution of the U.S. population by race for 1998.

Table 6.4.2

Race	Population in millions, f
White	195.44
Black	32.72
American Indian, Eskimo	2.00
Asian	9.90
Hispanic	30.25

SOURCE: U.S. Census Bureau, www.census.gov

In Table 6.4.2, race is the qualitative data and is shown in the left column. The frequency for each racial category is given in the right column. If we sum the numbers (frequencies) in the right column, we get

$$195.44 + 32.72 + 2.00 + 9.90 + 30.25 = 270.31$$

Since these numbers are in millions, the sum represents 270.31 million. This number represents the population of the United States in 1998. Mathematically, we use the symbol $\sum$ (Greek capital letter sigma) to communicate sum. Hence, a more compact way to represent summing the numbers in the right column of Table 6.4.2 is by writing

$$\sum f = 270.31$$

Now that we have addressed some terminology common to empirical probability, let's look at how to compute it.

Empirical Probability

Given a frequency distribution of qualitative data, if A is an event associated with a qualitative data item, then the probability of event A is given by

$$P(A) = \frac{\text{frequency of event } A}{\text{sum of all frequencies}} = \frac{f}{\sum f}$$

Example 5 Computing Empirical Probability

The distribution of the U.S. population by race in 1998 is shown in Table 6.4.2. If a person in the United States is selected at random for a survey, determine the probability that the person is Hispanic.

Solution

Understand the Situation: We need to determine the U.S. population in 1998 and the total number of the U.S. Hispanic population in 1998.

To determine the U.S. population, we must add the frequencies in the table to get

$$\sum f = 195.44 + 32.72 + 2.00 + 9.90 + 30.25 = 270.31$$

So in 1998 the U.S. population was about 270.31 million. Table 6.4.2 also indicates that there were 30.25 million U.S. Hispanics in 1998. Now we let A represent the event that the randomly selected person is Hispanic. To determine $P(A)$ using empirical probability, we compute

$$P(A) = \frac{\text{frequency of Hispanics}}{\text{sum of frequencies}} = \frac{30.25}{270.31} \approx 0.112$$

Interpret the Solution: This means that in 1998 the probability that a randomly selected person in the United States is Hispanic is about 0.112 or about 11.2%.

Checkpoint 4 Now work Exercise 33.

So far in this section we have seen many examples that include the phrase **selected at random**. We will use this phrase often in this chapter. When we use this phrase, we mean that all elements in the sample space are equally likely to be selected. Now let's turn our attention to the basic rules of probability.

Basic Rules of Probability

For sample space S, assume that $A \subseteq S$ is an event and let $P(A)$ represent the probability of A.

1. The probability of an event has a value between 0 and 1, inclusive. In other words, for any event A, $0 \le P(A) \le 1$. Equivalently, using percentages, $0\% \le P(A) \le 100\%$.
2. If event A is certain to occur, then $P(A) = 1$. If event A is certain not to occur, then $P(A) = 0$.

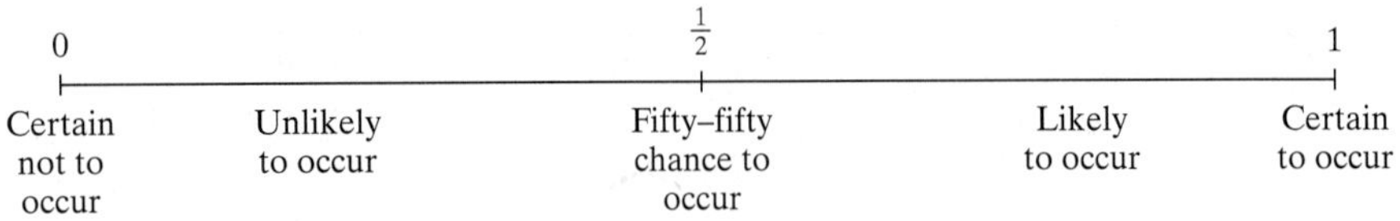

3. The event that A does *not* occur is the **complement of A**, denoted by A^C. If $P(A) = a$, then $P(A^C) = 1 - a$. That is, $P(A^C) = 1 - P(A)$.

In Example 6 we begin to apply some of these rules of probabilities.

Example 6 Using Probability Rules

Table 6.4.3

Marital Status	Percentage (%)
Never married	25
Married	55
Widowed	10
Divorced	10

SOURCE: U.S. Census Bureau, www.census.gov

The marital status of U.S. females 15 years or older in 2000 is given by the distribution in Table 6.4.3. For a study, a U.S. female 15 years or older was selected at random. Let W represent the event that she is widowed. Determine $P(W^C)$ and interpret.

Solution

Understand the Situation: Since W represents the event that a female 15 years or older is widowed, W^C (the complement of W) is the event that the female is *not* widowed.

We can see from reading Table 6.4.3 that $P(W) = 10\%$ or 0.10. From the basic rules of probability, we know that

$$P(W^C) = 1 - P(W) = 1 - 0.10 = 0.90$$

Interpret the Solution: This means that in 2000 the probability that a 15 year old or older female selected at random in the United States was *not* widowed is about 0.90 or 90%. ■

SUMMARY

In this section, we learned that there are three types of probabilities: **subjective**, **classical**, and **empirical**. We also learned some terminology concerning probability theory and the basic rules of probability.

- For a probability experiment, the **sample space**, denoted by S, is the set of all possible outcomes.
- For a probability experiment, an **event** is simply a subset of the sample space.
- **Subjective Probability** A probability that is based on estimates, educated guesses, intuition, emotion, or other information.
- **Classical Probability**

$$\text{Probability of event } A \text{ occurring} = P(A) = \frac{n(A)}{n(S)} = \frac{\text{number of outcomes of event } A}{\text{total number of possible outcomes}}$$

- **Empirical Probability**

$$\text{Probability of event } A \text{ occurring} = P(A) = \frac{\text{frequency of event } A}{\text{sum of all frequencies}} = \frac{f}{\sum f}$$

- **Basic Rules of Probability**
  1. For any event A, $0 \le P(A) \le 1$.
  2. If event A is certain to occur, then $P(A) = 1$. If event A is certain not to occur, then $P(A) = 0$.
  3. The event that A does *not* occur is the **complement of A**, denoted by A^C. If $P(A) = a$, then $P(A^C) = 1 - a$. That is, $P(A^C) = 1 - P(A)$.

SECTION 6.4 EXERCISES

In Exercises 1–10, write the sample space S using set notation and then determine n(S).

1. A box contains 9 identical tickets labeled 1 through 9. In an experiment, a ticket is drawn at random and its number is recorded.

2. A jar contains 11 identical marbles labeled 1 through 11. In an experiment, a marble is drawn at random and its number is recorded.

6.1 ✓ **3.** A box contains 5 identical tickets labeled 1 through 5. In an experiment, a ticket is drawn at random, its number is recorded, and it is returned to the box. Then a second ticket is drawn at random. The sum of the two numbers is recorded.

4. A box contains 5 identical tickets labeled 1 through 5. In an experiment, a ticket is drawn at random, its number is recorded, and it is returned to the box. Then a second ticket is drawn at random. The difference of the first minus the second number is recorded.

5. A pollster asks a respondent to classify her political ideology as conservative, moderate, or liberal.

6. The Holster Company asks 500 users of their eyeliner if their eye color is blue, brown, green, or hazel.

7. A warranty card for a new camping tent asks two questions. The first is to rate the quality of the tent as good, excellent, or superior, and the second asks to rate how waterproof the tent is on a scale from 1 to 4, with 1 being extremely waterproof and 4 being not very waterproof at all.

⇐ 6.1 **8.** The suggestions card at Pappy's restaurant asks customers to rate the service as fair, good, or excellent and then to rate the quality of the food on a scale from 1 to 5, with 1 being superior and 5 being lousy.

9. A pollster asks the question "Do you think that the quality of television programming is better or worse than it was 15 years ago?" The pollster then asks, "What do you listen to the most, AM or FM radio?"

10. A pollster asks the question "Is your personal financial situation better, worse, or about the same as compared to 5 years ago?" The pollster then asks, "Have you ever declared bankruptcy?"

In Exercises 11–20, use classical probability to compute $P(A)$ and $P(B)$ for the indicated events A and B.

11. For the experiment in Exercise 1, a ticket is drawn at random and its number is recorded. Let A represent the event that the outcome is an odd number and B represent the event that the outcome is an even number.

12. For the experiment in Exercise 2, a marble is drawn at random and its number is recorded. Let A represent the event that the outcome is an odd number and B represent the event that the outcome is an even number.

✓ **13.** For the experiment in Exercise 3, a ticket is drawn at random, its number is recorded, and it is returned to the box. Then a second ticket is drawn randomly. The sum of the two numbers is recorded. Let A represent the event that the sum is even and B represent the event that the sum is odd.

14. For the experiment in Exercise 4, a ticket is drawn randomly, its number is recorded, and it is returned to the box. Then a second ticket is drawn at random. The difference of the first minus the second number is recorded. Let A represent the event that the difference is positive and B represent the event that the difference is negative. Explain why $P(A) + P(B) \neq 1$.

15. A bag contains 10 identical marbles painted with the numbers 1 through 10. In an experiment, 2 marbles are drawn in succession and randomly, and the first marble is not replaced before the second draw. The numbers are recorded. Let A represent the event that the numbers are the same and B represent the event that a 2 is drawn.

16. A box contains 10 tickets numbered 1 through 10. In an experiment, 2 tickets are drawn in succession and randomly and the first ticket is not replaced before the second draw. The numbers are recorded. Let A represent the event that a 3 is drawn and B represent the event that the numbers add up to 5.

17. For the experiment in Exercise 7, a warranty card is drawn at random. Let A represent the event that the tent quality is rated as excellent and B represent the event that the tent's waterproof rating is a 2.

18. For the experiment in Exercise 8, a suggestions card is drawn at random. Let A represent the event that the service is rated as good and B represent the event that the food quality is rated as a 1.

19. A box contains 4 identical tickets numbered 1 through 4. In an experiment, a ticket is drawn at random, its number recorded, the ticket is returned to the box, and a second ticket is drawn randomly. Let A represent the event that the numbers are the same and B represent the event that a 4 is drawn.

20. A bag contains 4 identical marbles with 1 painted red, 1 painted yellow, 1 painted green, and 1 painted blue. In an experiment, a marble is drawn randomly, its color is recorded, the marble is returned to the bag, and a second marble is drawn at random. Let A represent the event that the marbles have the same color and B represent the event that a blue marble is drawn.

21. A cookie jar contains 15 identical chocolate chip cookies and 15 identical oatmeal raisin cookies. Kim takes two cookies out of the jar without looking at what kind they are. Let S be the sample space.

(a) Determine $n(S)$.

(b) If A represents the event that both cookies that Kim takes are oatmeal raisin, determine $n(A)$.

(c) Use classical probability to determine $P(A)$ and interpret.

22. Mrs. Perez mixes 10 plain and 10 peanut M&M's in a sack for Chris's school lunch. After finishing the rest of his lunch, Chris turns over the sack and two M&M's fall out. Let S be the sample space.

(a) Determine $n(S)$.

(b) If A represents the event that both M&M's are plain, determine $n(A)$.

(c) Use classical probability to determine $P(A)$ and interpret.

23. A box contains 30 identical marbles of which 15 are painted red and 15 are painted blue. Cole reaches into the box and pulls out 3 marbles in a single draw at random.

(a) Determine $n(S)$.

(b) If A represents the event that all three marbles are painted blue, determine $n(A)$.

(c) Use classical probability to determine $P(A)$ and interpret.

24. A box contains 20 identical marbles of which 10 are painted with an **X** and 10 are painted with a **Y**. Lexi reaches into the box and pulls out 4 marbles in a single draw at random.

(a) Determine $n(S)$.

(b) If A represents the event that all four marbles are labeled with an $\boldsymbol{X}$, determine $n(A)$.

(c) Use classical probability to determine $P(A)$ and interpret.

25. The city council in Staten City is comprised of 5 Democrats and 7 Republicans.

(a) In how many different ways can a subcommittee of 6 council persons be formed?

(b) In how many different ways can a subcommittee be formed with 2 Democrats and 4 Republicans?

(c) If A represents the event that a subcommittee is formed randomly with 2 Democrats and 4 Republicans, determine $P(A)$ and interpret.

26. The Euratal Company has 4 women and 8 men on its board of directors.

(a) In how many different ways can a gender-equality subcommittee of 4 be formed?

(b) In how many different ways can a gender-equality subcommittee of 4 be formed so that the number of men and women are equal?

(c) If A represents the event that a gender-equality subcommittee of 4 is formed at random so that the number of men and women are equal, determine $P(A)$ and interpret.

27. In a shipment of 30 car stereos, 8 are known to be defective.

(a) In how many different ways can a sample of 10 car stereos be chosen?

(b) In how many different ways can a sample of 10 be chosen so that 2 are defective and 8 are not defective?

(c) If A represents the event that in a sample of 10 car stereos 2 are defective and 8 are not defective, determine $P(A)$ and interpret.

28. In a box of 50 pieces of taffy, 20 pieces are pink and 30 pieces are yellow. If A represents the event that a handful of 8 pieces of taffy is pulled out at random so that 2 are pink and 6 are yellow, determine $P(A)$ and interpret.

29. A group of 20 environmentalists is made up of 6 Republicans, 8 Democrats, and 6 members of the Green Party. If A represents the event that a public relations committee of 7 is formed randomly with 2 Republicans, 2 Democrats, and 3 Green Party members, determine $P(A)$ and interpret.

30. In the Washington State lottery game called Lotto, 6 balls numbered 1 to 49 are drawn from a hopper. (*Source:* www.wa.gov/lot/home.htm.) If A represents the event that a person buys one ticket and wins, determine $P(A)$ and interpret. Write the solution as a proper fraction.

In Exercises 31–40, use empirical probability to answer the following questions.

31. The number and type of U.S. periodicals published in 1980 is shown in the table. A journalism professor selects a periodical from 1980 at random for a paper on reporting quality.

Type	Number
Weekly	1716
Semimonthly	645
Monthly	3985
Bimonthly	1114
Quarterly	1444

SOURCE: U.S. Census Bureau, www.census.gov

(a) Determine the probability that the periodical was a quarterly.

(b) Determine the probability that the periodical was a weekly.

32. The number and type of U.S. periodicals published in 1998 is shown in the table. A journalism professor selects a periodical from 1998 at random for a paper on reporting quality.

Type	Number
Weekly	364
Semimonthly	156
Monthly	3363
Bimonthly	2168
Quarterly	3309

SOURCE: U.S. Census Bureau, www.census.gov

(a) Determine the probability that the periodical was a quarterly.

(b) Determine the probability that the periodical was a weekly.

33. The type of robbery crimes, in thousands, reported in the United States in 1996 is shown in the table. A criminal justice professor selects a reported robbery crime from 1996 at random for a case study.

Type of Crime	Crimes (in thousands)
Street	274
Commercial	72
Gas station	13
Convenience store	32
Residence	57
Bank	11

SOURCE: U.S. Bureau of Crime Statistics, www.ojp.osdoj.gov

(a) Determine the probability that the crime involved a gas station.

(b) Determine the probability that the crime involved a bank.

34. The type of weapon, in thousands, used to commit crimes in 1995 is shown in the table. A criminal justice professor selects a reported robbery crime from 1995 at random for a case study.

Weapon	Number (in thousands)
Firearm	218
Knife	48
Other dangerous weapon	62
Strongarm	207

SOURCE: U.S. Bureau of Crime Statistics, www.ojp.osdoj.gov

(a) Determine the probability that the crime involved a firearm.

(b) Determine the probability that the crime involved a knife.

35. The number of U. S. enlisted military personnel, in thousands, in 1997 is shown in the table. In 1997 an enlisted military personnel was selected at random for a survey.

Branch of Service	Thousands Enlisted
Army	162.3
Navy	93.3
Marines	50.0
Air Force	74.5

SOURCE: U.S. Census Bureau, www.census.gov

(a) Determine the probability that the person was in the Army.

(b) Determine the probability that the person was in the Navy.

36. The number of recipients, in thousands, of master's degrees in science-related areas during 1995 and 1996 is shown in the table.

Degree Field	Master's Recipients (in thousands)
Computer science	18.2
Mathematics	7.9
Life science	15.3
Physical science	9.7
Social science	25.1

SOURCE: U.S. Department of Education, www.ed.gov

(a) If a master's degree recipient from 1995 or 1996 is selected at random, determine the probability that the degree was in the social science field.

(b) If a master's degree recipient from 1995 or 1996 is selected at random, determine the probability that the degree was in the life science fields.

37. The number of scientists, in thousands, employed in the United States in 1996 is shown in the table. In 1996 a scientist was selected at random for a survey.

Type of Scientist	Number (in thousands)
Social scientist	263.5
Physical scientist	206.7
Life scientist	180.0
Mathematical scientist	15.6

SOURCE: U.S. Census Bureau, www.census.gov

(a) Determine the probability that the scientist is a life scientist.

(b) Determine the probability that the scientist is not a life scientist.

38. The number of scientists, in thousands, employed by the U.S. government in 1996 is shown in the table. In 1996 a U.S. government scientist was selected at random for a survey.

Type of Scientist	Number (in thousands)
Social scientist	74.7
Physical scientist	71.2
Life scientist	47.2
Mathematical scientist	4.8

SOURCE: U.S. Census Bureau, www.census.gov

(a) Determine the probability that the scientist is a life scientist.

(b) If a U.S. government scientist from 1996 is selected at random, determine the probability that the scientist is not a life scientist.

39. The number of U.S. houses sold at various prices, in dollars, in 1998 is shown in the table. For an upcoming story, a national magazine selects at random a house that sold in 1998.

Sales Price	Number of Houses
Under $100,000	140,000
$100,000 to $119,999	112,000
$120,000 to $149,999	183,000
$150,000 to $199,999	208,000
$200,000 and over	248,000

SOURCE: U.S. Department of Housing and Urban Development, www.hud.gov

(a) Determine the probability that it sold for anywhere between $120,000 to $149,999.

(b) Determine the probability that it sold for $120,000 or more.

40. The percentage of U.S. houses built in 1999 of various sizes is shown in the table. For an upcoming story, a national magazine selects at random a house built in 1999.

Area (ft^2)	Houses (%)
Under 1200	7
1200 to 1599	19
1600 to 1999	22
2000 to 2399	18
2400 and over	34

SOURCE: U.S. Department of Housing and Urban Development, www.hud.gov

(a) Determine the probability that the house had an area of 1600 to 1999 square feet.

(b) Determine the probability that the house had an area of 1600 square feet or more.

In Exercises 41–46, classify the following as an example of classical, empirical, *or* subjective *probability.*

41. Samantha claims that the probability that the telephone rings just when she is getting out of the shower seems close to 80%.

42. A girl tells a boy at the local high school, "There is zero chance of me going on a date with you!"

43. A survey shows that 45% of all Americans watched one NASCAR race in 1999. (*Source:* www.harrisinteractive.com.)

44. A recent poll shows that 81% of Americans feel that the choice of a vice-president has no influence on who they vote for president of the United States. (*Source:* www.harrisinteractive.com.)

45. The probability of winning a drawing when 500 tickets have been sold is $\frac{1}{500}$.

46. The probability of tossing a quarter and having the flip be a tail is 50%.

SECTION PROJECT

A traditional example of classical probability is computing what is commonly known as the *Birthday Problem*. Suppose that a group of r randomly selected people are together in a room. What is the probability that at least two of them have the same birthday? Assume that by the same birthday we mean month and day (and not necessarily the year) and we ignore the effect of leap years. To find out, let A represent the event that at least two of the r people in the room have the same birthday. Determining $P(A)$ directly can be a difficult task, but what we can do is compute $P(A^C)$, where A^C represents the event that no two people in the room have the same birthday (that is, all people in the room have different birthdays).

(a) Write an equation to show how $P(A)$ and $P(A^C)$ are related.

(b) To determine $P(A^C)$ using classical probability, we need to compute

$$P(A^C) = \frac{\text{number of ways that } r \text{ randomly selected people can have different birthdays}}{\text{the number of all possible sets of birthdays for } r \text{ randomly selected people}}$$

For the denominator, we know that the first person can have a birthday 365 ways, the second person can have a birthday 365 ways, the third person can have a birthday 365 ways, and so on. Use the multiplication principle to determine the denominator of $P(A^C)$.

(c) For the numerator of $P(A^C)$, we need the number of ways that r randomly selected people can have different birthdays. This means that the first person can have a birthday 365 ways, the second person 364 ways, the third 363 ways, and so on. Note that the order is important and there is no repetition. Write an expression for the numerator of $P(A^C)$. Then write a fraction that gives a formula to compute $P(A^C)$.

(d) Complete the table below to determine the probability that at least two people in a room of r randomly selected people have the same birthday.

Number of People, r	Probability $P(A)$
5	
20	
30	
40	
50	
60	

(e) Using the table in part (d) as a guide, determine how many randomly selected people should be in a room so that there is at least a 50% chance that at least two people in the room have the same birthday.

Section 6.5 Computing Probabilities Using the Addition Rule

In the study of probability theory, we frequently need to determine a probability involving two or more events. For example, suppose that you roll a six-sided die and want to know the probability that on a single roll the result is a 1 **or** a 6. Or suppose that a medical researcher wants to know the probability that a newborn has low birth weight **or** whether the mother takes prenatal vitamins. These two situations illustrate what are called **compound events**. In this section we will learn a rule called the **Addition Rule**, which we can use to compute the probability of certain kinds of compound events. We will also examine the notion of **mutually exclusive events** and learn how the probabilities of such events are determined using the Addition Rule.

Mutually Exclusive Events

In Section 6.1 we looked at whether or not two events could happen at the same time. For example, in Section 6.1, Example 4(a), we saw that the event that Sandy gets a speeding ticket and the event that Sandy gets a ticket for not wearing a seat belt could happen at the same time. Whereas in Section 6.1, Example 4(b), we saw that the event that, during the same at bat, a softball player, Ariel, hits a single and the event that Ariel hits a triple were two events that could not happen at the same time. Events that have no outcomes in common, such as Ariel hitting a single and a triple in the same at bat, are what we call **mutually exclusive events**.

Mutually Exclusive Events

Two events A and B are **mutually exclusive** if A and B cannot occur at the same time. In other words, $A \cap B = \emptyset$.

Note: Recall in Section 6.1 that we offered the following warning. A common error that many who are just beginning to study probability make is trying to read too much information into the questions in an attempt to find some exception.

The Venn diagrams in Figure 6.5.1 contrast mutually exclusive and non-mutually exclusive events.

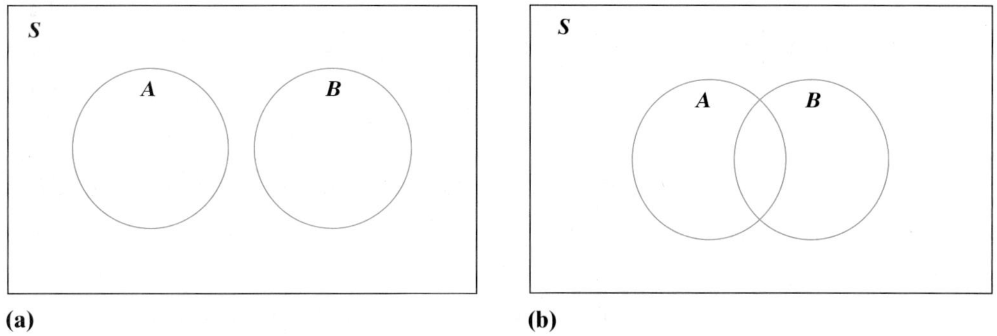

Figure 6.5.1 **(a)** A and B are mutually exclusive events. **(b)** A and B are not mutually exclusive events.

Note: In Section 6.2, we defined **disjoint sets** as sets with no elements in common. Mutually exclusive events have no outcomes in common, so they are disjoint subsets of the sample space.

Example 1 Determining If Events Are Mutually Exclusive

For the following, determine if the events A and B are mutually exclusive.

(a) A blood donor is selected at random. Let A represent the event that the donor has type A blood and B represent the event that the donor has type B blood.

(b) A student from Harperston Business College is selected at random. Let A represent the event that the student is an accounting major and B represent the event that the student is a female.

Solution

(a) Since the person cannot have two different blood types, we see that events A and B are mutually exclusive.

(b) In this case, we know that there must be students who are accounting majors, since it is a business college. It also seems reasonable that some accounting majors are female. Since these events can happen at the same time, events A and B are not mutually exclusive. ■

Checkpoint 1

Now work Exercise 7.

Table 6.5.1

Branch of Service	Enlistment
Army (A)	162,300
Navy (B)	93,300
Marines (C)	50,000
Air Force (D)	74,500

SOURCE: U.S. Department of Defense, www.defenselink.mil

Now suppose that we want to determine the probability that, given two events, **either** of the events occur. To see how we compute this type of probability, consider Table 6.5.1, which shows the distribution of enlisted U.S. military personnel in 1997. Notice that the events A, B, C, and D are assigned to the events that an enlisted person is in the Army, Navy, Marines, or Air Force, respectively.

First notice that if we add the enlistments in each of the branches we get $\sum f = 380{,}100$. Suppose that a researcher selects an enlisted person at random. To determine the probability that the enlisted person is in the Navy *or* the Air Force, we begin by determining the total number that is enlisted in the Navy or Air Force. This sum is $93{,}300 + 74{,}500 = 167{,}800$. We have already determined that the total number of enlisted personnel is 380,100. So the probability that the randomly selected enlisted person is in the Navy **or** Air Force is

$$\begin{aligned} P(\text{Navy } \textbf{or} \text{ Air Force}) &= P(B \cup D) \\ &= \frac{\text{number enlisted in the Navy or Air Force}}{\text{total enlisted military personnel}} \\ &= \frac{167{,}800}{380{,}100} \approx 0.441 \end{aligned}$$

This means that in 1997 the probability that a randomly selected enlisted person is a member of the Navy or the Air Force is about 44.1%.

Notice that the probability that the enlisted person is in the Navy is given by $P(B) = \dfrac{93{,}300}{380{,}100}$, while the probability that the enlisted person is in the Air

Force is given by $P(D) = \frac{74,500}{380,100}$. If we sum these probabilities, we get

$$P(B) + P(D) = \frac{93,300}{380,100} + \frac{74,500}{380,100} = \frac{167,800}{380,100} = P(B \cup D)$$

So it appears for this situation that $P(B \cup D) = P(B) + P(D)$. This is not a coincidence! Since events B and D are mutually exclusive, these calculations illustrate the Addition Rule for Mutually Exclusive Events.

Addition Rule for Mutually Exclusive Events

If A and B are mutually exclusive events (this means that only one of them can occur or, equivalently, $A \cap B = \emptyset$), then the probability that either A occurs **or** B occurs is given by

$$P(A \textbf{ or } B) = P(A \cup B) = P(A) + P(B)$$

Note: 1. In Section 6.2 we discussed the cardinality of sets and showed that, if sets A and B are disjoint, then $n(A \cup B) = n(A) + n(B)$.

2. If A and B are mutually exclusive events, then $P(A \cap B) = 0$ since $A \cap B = \emptyset$.

Example 2 Applying the Addition Rule for Mutually Exclusive Events

Table 6.5.2 shows the number of successful space launches for all countries that launched during 1999. (Success is defined as attainment of Earth orbit or Earth escape.) Notice that the events A, B, C, D, E, and F are assigned to the events that a launch was from Russia, United States, ESA, China, India, or Ukraine, respectively. A space exploration researcher selects a 1999 successful space launch at random. Determine $P(B \cup C)$ and interpret.

Table 6.5.2

Country	Number of Successful Launches
Russia (A)	26
United States (B)	30
ESA (European Space Agency) (C)	10
China (D)	4
India (E)	1
Ukraine (F)	2

SOURCE: Statistical Abstract of the United States, www.census.gov/statab/www/

Interactive Activity

For the experiment in Example 2 determine $P(D \cup E)$.

WWW

Solution

Understand the Situation: We begin by noting that events B and C are mutually exclusive. The same space launch cannot take place from the United States (event B) and the ESA (event C) at the same time. This means that we can use the Addition Rule for Mutually Exclusive Events. Summing the numbers in the right-hand column of Table 6.5.2 determines that in 1999 a total of 73 successful space launches took place. This tells us that $P(B) = \frac{30}{73}$ and $P(C) = \frac{10}{73}$.

Since we know that events B and C are mutually exclusive, we have

$$P(B \cup C) = P(B) + P(C)$$
$$= \frac{30}{73} + \frac{10}{73} = \frac{40}{73}$$

Interpret the Solution: In 1999 the probability that a randomly selected successful space launch came either from the United States or the ESA is $\frac{40}{73}$. ■

✓ Checkpoint 2

Now work Exercise 23.

The General Addition Rule

We have seen how to compute $P(A \cup B)$ if A and B are mutually exclusive events. The next step is to see how we compute $P(A \cup B)$ if A and B are **not** mutually exclusive. For example, suppose that you roll a six-sided die and are interested in the probability of rolling a number that is less than 3 or rolling an even number. In Figure 6.5.2 we can see that these events are not mutually exclusive, since 2 is an outcome of both events.

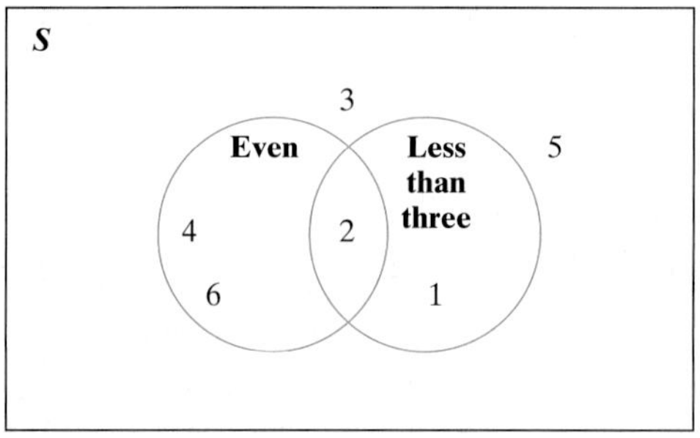

Figure 6.5.2

We know that there are 6 possible outcomes on a single roll of a die. Figure 6.5.2 illustrates that 4 of these 6 are less than 3 or even. This means that

$$P(\text{less than 3 or even}) = \frac{4}{6} = \frac{2}{3}$$

Notice that it appears that the probability of rolling a number that is less than three or an even number can be expressed as

$$P(\text{less than 3 or even}) = P(\text{less than 3}) + P(\text{even}) - P(\text{less than 3 and even})$$
$$= \frac{2}{6} + \frac{3}{6} - \frac{1}{6} = \frac{4}{6} = \frac{2}{3} \approx 0.667$$

To generalize what we just did, we need to consult the Toolbox below to recall a useful rule from Section 6.2, as well as a type of probability that we explored in Section 6.4.

From Your Toolbox

- If A and B are any sets, then $n(A \cup B) = n(A) + n(B) - n(A \cap B)$.
- **Classical Probability**

$$\text{Probability of event } A \text{ occurring} = P(A) = \frac{n(A)}{n(S)} = \frac{\text{number of outcomes of event } A}{\text{total number of possible outcomes}}$$

We can let A and B represent two events where the numbers of outcomes for these events are given by $n(A)$ and $n(B)$, respectively. For a sample space S, we can use classical probability to get

$$P(A \cup B) = \frac{\text{number of ways } A \text{ or } B \text{ can occur}}{\text{total number of possible outcomes}} = \frac{n(A) + n(B) - n(A \cap B)}{n(S)}$$

$$= \frac{n(A)}{n(S)} + \frac{n(B)}{n(S)} - \frac{n(A \cap B)}{n(S)} = P(A) + P(B) - P(A \cap B)$$

We call this rule the General Addition Rule.

General Addition Rule

For any events A and B, the probability that at least one of A and B occurs (that is, A occurs or B occurs or both A and B occur) is given by

$$P(A \cup B) = P(A) + P(B) - P(A \cap B)$$

Note: Notice that the Addition Rule for Mutually Exclusive Events is just a special case of the General Addition Rule. If A and B are mutually exclusive, then $P(A \cap B) = 0$.

Venn diagrams are commonly used to visualize applications of the General Addition Rule. These diagrams can either show the number of outcomes associated with given events or can display the probabilities assigned to these events. This is illustrated in Example 3.

Example 3 Applying the General Addition Rule

The curator of a science museum surveys 1100 patrons at random and finds that 800 saw the space probe exhibit, 500 attended the 3D movie, and 300 saw both the exhibit and the movie. Let A represent the event that a patron saw the space probe exhibit and B represent the event that he or she saw the 3D movie.

(a) Draw a Venn diagram to represent the sample space and the events A and B.

(b) Determine $P(A \cup B)$ and interpret.

(c) Determine $P(A \cup B)^C$ and interpret.

Solution

(a) The Venn diagram in Figure 6.5.3 shows the number of outcomes associated with events A, B, $A \cap B$ and sample space S. Let's make a few observations

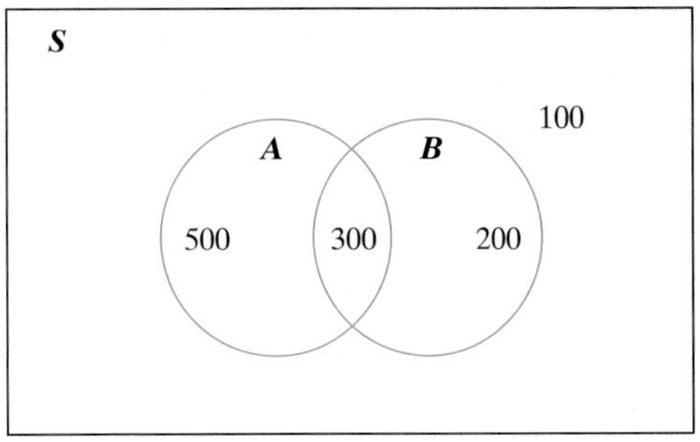

Figure 6.5.3

about the Venn diagram. Notice that if we sum all of the outcomes in the circle associated with event A we get $500 + 300 = 800$, and if we sum the outcomes in the circle associated with event B, we get $300 + 200 = 500$. Also notice that if we sum all of the values in the diagram we get $500 + 300 + 200 + 100 = 1100$, which is the size of the sample space.

(b) **Understand the Situation:** First note that

$$P(A) = \frac{\text{number surveyed who saw the space probe exhibit}}{\text{total number surveyed}} = \frac{800}{1100} = \frac{8}{11} \quad \text{and}$$

$$P(B) = \frac{\text{number surveyed who saw the 3D movie}}{\text{total number surveyed}} = \frac{500}{1100} = \frac{5}{11}.$$

Since 300 patrons saw both the exhibit and the movie, we know that $P(A \cap B) = \frac{300}{1100} = \frac{3}{11}$.

So the General Addition Rule yields

$$P(A \cup B) = P(A) + P(B) - P(A \cap B) = \frac{8}{11} + \frac{5}{11} - \frac{3}{11} = \frac{10}{11}$$

Interpret the Solution: This means that if a surveyed patron is selected at random the probability that she or he saw either the space probe exhibit, the 3D movie, or both is $\frac{10}{11}$.

(c) From the definition of complementary events, we know that

$$P(A \cup B)^C = 1 - P(A \cup B)$$

So, using the solution from part (b), we get

$$P(A \cup B)^C = 1 - P(A \cup B) = 1 - \frac{10}{11} = \frac{1}{11}$$

Interpret the Solution: This means that if a surveyed patron is selected at random the probability that she or he attended neither the space probe exhibit nor the 3D movie is $\frac{1}{11}$. ■

There is a useful observation embedded within Example 3. If we take each number shown in Figure 6.5.3 and divide it by 1100 (the size of the sample space), we get what is called a **probability Venn diagram**. This diagram is shown in Figure 6.5.4. Note that the sum of all the values in the probability Venn diagram add up to 1.

The General Addition Rule can also be used to compute probabilities based on data that are in tabular form. In practice, many probabilities involving two or more events are computed using this method.

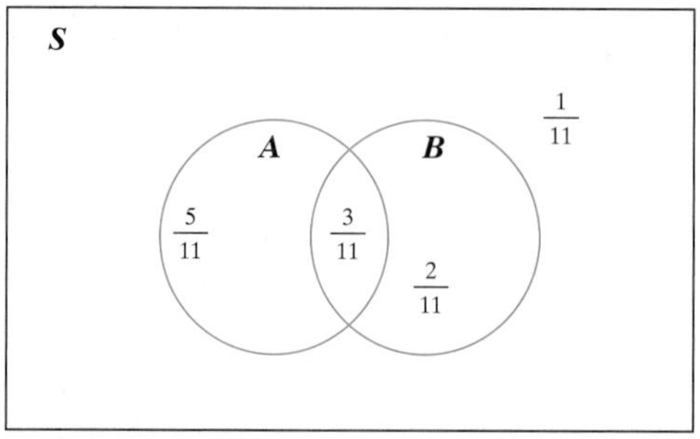

Figure 6.5.4 Probability Venn Diagram.

Example 4 Using the Addition Rule with Tabular Data

The source of payment for patients discharged from U.S. hospitals during 1996 for two age groups is shown in Table 6.5.3. The number of discharges is given in thousands. For the labeled events A, B, C, and D, determine the following probabilities and interpret.

(a) $P(B \cap D)$ **(b)** $P(A \cup C)$

Table 6.5.3

Age Group	Private (*A*)	Government (*B*)	Total
Less than 44 years old (C)	6,090	6,442	12,532
45 years or older (D)	4,636	13,337	17,973
Total	10,726	19,779	30,505

SOURCE: U.S. National Center for Health Statistics, www.cdc.gov/nchs

Solution

(a) Understand the Situation: First notice that event B is the event that the source of payment for a discharged patient in 1996 was the government. Event D is the event that a discharged patient was 45 years old or older. So when we compute $P(B \cap D)$, we are computing the probability that a randomly selected patient discharged from a hospital in 1996 was 45 years old or older and the source of payment was the government.

By inspecting Table 6.5.3, we see that if we follow down the event B column and across the event D row they intersect at the cell with the value 13,337 thousand. Since the total number of patients discharged in 1996 was 30,505 thousand, we have

$$P(B \cap D) = \frac{\text{value where the event } B \text{ column and event } D \text{ row intersect}}{\text{total patients discharged}}$$

$$= \frac{13{,}337}{30{,}505} \approx 0.44$$

Interpret the Solution: This means that if we select a discharged patient at random the probability that he or she had government health insurance **and** was 45 year or older is about 0.44 or about 44%.

(b) Understand the Situation: Since event A is the event that the source of payment was private and event C is the event that the discharged patient was less than 44 years old, we are being asked to determine the probability that a randomly selected patient has private health insurance, is 44 years old or younger, or both. Table 6.5.3 indicates that

$$P(A) = \frac{\text{total in event } A \text{ column}}{\text{total patients discharged}} = \frac{10{,}726}{30{,}505},$$

$$P(C) = \frac{\text{total in event } C \text{ row}}{\text{total patients discharged}} = \frac{12{,}532}{30{,}505},$$

and where event A column and event C row intersect gives us

$$P(A \cap C) = \frac{6090}{30{,}505}.$$

The General Addition Rule gives us

$$P(A \cup C) = P(A) + P(C) - P(A \cap C)$$

$$= \frac{\text{total in event } A \text{ column}}{\text{total patients discharged}} + \frac{\text{total in event } C \text{ row}}{\text{total patients discharged}}$$

$$- \frac{\text{value where the event } A \text{ column and event } C \text{ row intersect}}{\text{total patients discharged}}$$

$$= \frac{10{,}726}{30{,}505} + \frac{12{,}532}{30{,}505} - \frac{6090}{30{,}505} = \frac{17{,}168}{30{,}505} \approx 0.56$$

Interpret the Solution: This means that if we select a discharged patient at random the probability that he or she either has private health insurance, is 44 years or younger, or both is about 0.56 or about 56%.

Checkpoint 3

Now work Exercise 35.

Odds

We conclude this section with a brief discussion of **odds**. Sometimes, when we know the probability of an event A, we speak of the odds for (or against) the event A, rather than the probability of event A. This is particularly true in cases involving betting on sporting events. For example, we may hear that "The odds of Tiger Woods winning the British Open are 3 to 2." We will return to this example later, but first we introduce a definition that ties together the concept of odds and probabilities.

Probability to Odds

If $P(A)$ is the probability of an event A occurring, then

1. The **odds in favor of A** are "$P(A)$ to $1 - P(A)$" or, equivalently, "$P(A)$ to $P(A^C)$." This is represented by

$$\frac{P(A)}{1 - P(A)} = \frac{P(A)}{P(A^C)}, \qquad P(A) \neq 1$$

2. The **odds against A** are "$1 - P(A)$ to $P(A)$" or, equivalently, "$P(A^C)$ to $P(A)$." This is represented by

$$\frac{1 - P(A)}{P(A)} = \frac{P(A^C)}{P(A)}, \qquad P(A) \neq 0$$

Note: When possible, odds are expressed as ratios of whole numbers.

Example 5 Converting from Probability to Odds

Dillon rolls a six-sided die once. The probability of rolling a 2 is $\frac{1}{6}$. Determine the odds in favor of rolling a 2.

Solution

Understand the Situation: We let A represent the event that Dillon rolls a 2. This means that we are given $P(\text{rolling a 2}) = P(A) = \frac{1}{6}$. This also gives us

Interactive Activity

For the situation in Example 5, determine the odds against rolling a 2.

$P(A^C) = \frac{5}{6}$. The odds in favor of A (rolling a 2) are $\frac{1}{6}$ to $\frac{5}{6}$ or, equivalently,

$$\frac{P(A)}{1 - P(A)} = \frac{\frac{1}{6}}{1 - \frac{1}{6}} = \frac{\frac{1}{6}}{\frac{5}{6}} = \frac{1}{5}$$

Interpret the Solution: So the odds in favor of rolling a 2 are $\frac{1}{5}$. We could read this as "Odds in favor of rolling a 2 are 1 to 5." ■

If the situation in Example 5 were in a gaming situation where money was wagered, we would call it a **fair game** if the following occurred:

- If Dillon bet \$1 on a 2 turning up, he would lose his \$1 if any number other than a 2 turns up.
- If Dillon bet \$1 on a 2 turning up, he would win \$5 (and have his bet of \$1 returned) if a 2 turns up.

Generally, if the odds in favor of an event A are "a to b", the **game is fair** if a bet of \$$a$ is lost if event A does not occur, but a win of \$$b$ (as well as returning the bet of \$$a$) is realized if event A does occur.

We have seen how to convert probability to odds. Now suppose that we are given the odds for an event, but wish to determine the probability of the event. If the odds in favor of event A occurring are "a to b," using part 1 of the Probability to Odds formula box allows us to write

$$\frac{a}{b} = \frac{P(A)}{1 - P(A)}$$

Solving for $P(A)$ (we leave the details for Exercise 75), yields the following:

Odds to Probability

If the odds for event A are a to b, then the probability of event A occurring is

$$P(A) = \frac{a}{a+b}$$

Example 6 Converting Odds to Probability

As of July 9, 2001, the odds of Tiger Woods winning the 2001 British Open are 3 to 2. Express these odds as a probability. (*Source:* www.vegasinsider.com/u/futures.)

Solution

Understand the Situation: Let A be the event that Tiger Woods wins the British Open. The odds in favor of A are given as 3 to 2. So we have $a = 3$ and $b = 2$.

Using the formula for converting odds to probability gives us

$$P(A) = \frac{a}{a+b} = \frac{3}{3+2} = \frac{3}{5}$$

Interpret the Solution: If the odds of Tiger Woods winning the British Open are 3 to 2, then the probability that he will win the British Open is $\frac{3}{5}$. ■

✓ Checkpoint 4

Now work Exercise 69.

SUMMARY

In this section we learned how to compute compound events of the form $P(A \cup B)$. We saw that the rule that we used to compute this probability depended on the events being mutually exclusive. We also learned the relationship between probability and odds.

- **Mutually Exclusive Events** Two events A and B are **mutually exclusive** if A and B cannot occur at the same time; that is, $A \cap B = \emptyset$.
- **Addition Rule for Mutually Exclusive Events** $P(A \cup B) = P(A) + P(B)$
- **General Addition Rule** $P(A \cup B) = P(A) + P(B) - P(A \cap B)$
- If $P(A)$ is the probability of an event A occurring, then

1. The **odds in favor of A** are "$P(A)$ to $1 - P(A)$" or, equivalently, "$P(A)$ to $P(A^C)$." This is represented by

$$\frac{P(A)}{1 - P(A)} = \frac{P(A)}{P(A^C)}, \qquad P(A) \neq 1$$

2. The **odds against A** are "$1 - P(A)$ to $P(A)$" or, equivalently, "$P(A^C)$ to $P(A)$." This is represented by

$$\frac{1 - P(A)}{P(A)} = \frac{P(A^C)}{P(A)}, \qquad P(A) \neq 0$$

- **Odds to Probability** If the odds for event A are a to b, then the probability of event A occurring is $P(A) = \dfrac{a}{a+b}$.

SECTION 6.5 EXERCISES

In Exercises 1–10, determine if the events A and B are mutually exclusive.

6.1 **1.** A person in a mall is selected at random. A represents the event that the person has brown hair and B represents the event that the person has blue eyes.

6.1 **2.** A person in a mall is selected at random. A represents the event that the person has a college degree and B represents the event that the person is employed.

6.1 **3.** In a state that requires voters to register for a single party, a voter is selected at random for an exit poll. The event that the person is a registered Republican and the event that the person is a registered Democrat.

6.1 **4.** In a state that uses electronic ballots, a voter is selected at random for an exit poll. The event that the person voted for issue 1 on the ballot and the event that the person voted against issue 1 on the ballot.

6.1 **5.** A student at Brockston College is selected at random. A represents the event that the student is a biology major and B represents the event that the student is a senior.

6.1 **6.** A student at Brockston College is selected at random. A represents the event that the student receives financial aid and B represents the event that the student is on the dean's list.

6.1 ✓ **7.** A box contains 30 identical tickets numbered 1 to 30. One ticket is drawn at random. A represents the event that the ticket is labeled with an even number and B represents the event that the ticket is labeled with an odd number.

6.1 **8.** A box contains 30 identical tickets numbered 1 to 30. One ticket is drawn at random. A represents the event that the ticket is labeled with a prime number and B represents the event that the ticket is labeled with a composite number.

6.1 **9.** Two respondents from a nationwide survey are selected at random for a follow-up interview. A represents the event that the two respondents are brothers and B represents the event that the two respondents are sisters.

6.1 **10.** Two respondents from a nationwide survey are selected at random for a follow-up interview. A represents the event that the two respondents are brothers and B represents the event that the two respondents are not related.

In Exercises 11–14, consider a box that contains 10 identical tickets numbered 1 to 10. If a draw is made at random, determine the following probabilities.

11. A ticket labeled with a 1 or a 6 is drawn.

12. A ticket labeled with a 2 or a 3 is drawn.

13. A ticket labeled with a 5 or a 12 is drawn.

14. A ticket labeled with a 4 or a 14 is drawn.

In Exercises 15–18, consider a bag that contains 6 red, 5 blue, and 4 yellow marbles of identical shapes. If a draw is made at random, determine the following probabilities.

15. A red or a blue marble is drawn.

16. A red or a yellow marble is drawn.

17. A black or a blue marble is drawn.

18. A red or a white marble is drawn.

For Exercises 19–24, consider a survey of all 500 workers at the Canal Junction water treatment plant. Let A represent the event that the worker took a 3-week vacation in the last year and B represent the event that the worker did more than 5 weeks of overtime in the last year. The results are shown in the Venn diagram. Assuming that a worker is selected at random, use the diagram to answer the following:

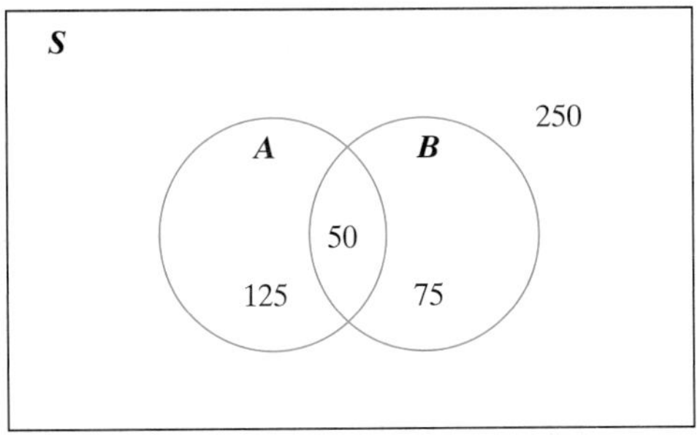

19. Determine $P(A)$ and interpret.

20. Determine $P(B)$ and interpret.

21. Determine the probability that the worker took a 3-week vacation in the last year and did more than 5 weeks of overtime in the last year.

22. Determine the probability that the worker did not take a 3-week vacation in the last year and did not do more than 5 weeks of overtime in the last year.

✓ **23.** Determine $P(A \cup B)$ and interpret.

24. Determine $P(A \cup B)^C$ and interpret.

In Exercises 25–30, consider a survey of a group of seniors, selected at random, at Clayborne College who were asked about their attitudes concerning academic probation. Let A represent the event that the student had been placed on academic probation for at least one semester and B represent the event that the person had been placed on the dean's list for at least one semester. The results are shown in the probability Venn diagram. Assuming that a senior in the survey group is selected at random, use the diagram to answer the following:

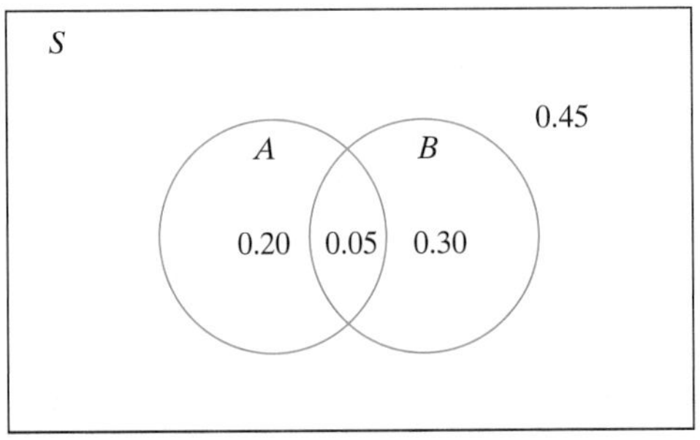

25. Determine $P(B)$ and interpret.

26. Determine $P(A)$ and interpret.

27. Determine $P(A \cap B)^C$ and interpret.

28. Determine $P(A \cap B)$ and interpret.

29. Determine the probability that the student had been placed on academic probation for at least one semester or had been placed on the dean's list for at least one semester.

30. Determine the probability that the student had not been placed on academic probation for at least one semester or had not been placed on the dean's list for at least one semester.

For Exercises 31–40 consider the following. The number of doctorates conferred in the social sciences (anthropology, sociology, economics, and political science) and the physical sciences (astronomy, physics, and chemistry) during 1998 is shown in the following table. The data are classified as being a U.S. citizen or a foreign-born citizen.

	Social Sciences (A)	Physical Sciences (B)	Total
U.S. citizen (C)	2353	2415	4768
Foreign-born citizen (D)	760	1111	1871
Total	3113	3526	6639

SOURCE: U.S. Census Bureau, www.census.gov

For the events labeled in the table as A, B, C, and D, answer the following:

31. Determine $P(A)$ and interpret.

32. Determine $P(B)$ and interpret.

33. Determine $P(A^C)$ and interpret.

34. Determine $P(B^C)$ and interpret.

✓ **35.** Determine the probability that the doctorate is in social sciences and is conferred on a U.S. citizen.

36. Determine the probability that the doctorate is in physical sciences and is conferred on a foreign-born citizen.

37. Determine $P(A \cap B)$ and interpret.

38. Determine $P(C \cap D)$ and interpret.

39. Determine the probability that the doctorate is in social sciences or is conferred on a foreign-born citizen.

40. Determine the probability that the doctorate is in physical sciences or is conferred on a U.S. citizen.

For Exercises 41–50, consider the following. In 1996, the percentage of American adults who exercised for 30 minutes or more at least 5 times per week is shown in the table. The percentages are classified by gender.

	Did Not Exercise (A)	Exercised Regularly (B)	Total (%)
Men (C)	22.0	25.6	47.6
Women (D)	25.6	26.8	52.4
Total	47.6	52.4	100

SOURCE: National Center for Chronic Disease Prevention, www.cdc.gov/nccdphp

For the events labeled in the table as A, B, C, and D, answer the following:

41. Determine $P(B)$ and interpret.

42. Determine $P(A)$ and interpret.

43. Determine $P(B^C)$ and interpret.

44. Determine $P(A^C)$ and interpret.

45. Determine the probability that the person exercised regularly and is female.

46. Determine the probability that the person did not exercise regularly and is male.

47. Determine $P(C \cap D)$ and interpret.

48. Determine $P(A \cap B)$ and interpret.

49. Determine the probability that the person exercised regularly or is male.

50. Determine the probability that the person did not exercise regularly or is female.

51. The probability of event A occurring is 0.8.

(a) Determine the odds in favor of A occurring.

(b) Determine the odds against A occurring.

52. The probability of event A occurring is 0.2.

(a) Determine the odds in favor of A occurring.

(b) Determine the odds against A occurring.

53. The probability of event A not occurring is 0.7.

(a) Determine the odds in favor of A occurring.

(b) Determine the odds against A occurring.

54. The probability of event A not occurring is 0.6.

(a) Determine the odds in favor of A occurring.

(b) Determine the odds against A occurring.

55. The odds in favor of event A occurring are 5 to 3. Determine the probability of A occurring.

56. The odds in favor of event A occurring are 7 to 5. Determine the probability of A occurring.

57. The odds against event A occurring are 2 to 5. Determine the probability of A not occurring.

58. The odds against event A occurring are 4 to 7. Determine the probability of A not occurring.

In Exercises 59–62, determine the odds in favor of obtaining the specified event.

59. A tail in a single toss of a coin

60. An even number in a single roll of a die

61. At least 1 tail when a coin is tossed 2 times

62. At least 1 tail when a coin is tossed 3 times

In Exercises 63–66, determine the odds against the specified event.

63. A head in a single toss of a coin

64. A number divisible by 3 in a single roll of a die

65. Two tails when a coin is tossed 2 times

66. An even number in a single roll of a die

67. Latasha hears that the probability of rain tomorrow is 30%.

(a) Determine the odds that it will rain tomorrow.

(b) Determine the odds that it will not rain tomorrow.

68. Frank, a carpet sales representative, believes that the odds are 9 to 4 that he will close a deal of carpeting the offices of a certain company. Determine the probability that Frank will make the sale.

69. As of April 23, 2001, the odds in favor of the Dallas Cowboys winning the 2002 Super Bowl were listed as 1 to 75. As of April 23, 2001, determine the probability that the Dallas Cowboys will win the 2002 Super Bowl. (*Source:* www.vegasinsider.com/u/futures.)

70. As of April 23, 2001, the odds in favor of the St. Louis Rams winning the 2002 Super Bowl were listed as 1 to 4. As of April 23, 2001, determine the probability that the St. Louis Rams will win the 2002 Super Bowl. (*Source:* www.vegasinsider.com/u/futures.)

71. As of July 2, 2001, the odds in favor of the Los Angeles Dodgers winning the 2001 World Series were listed as 1 to 25. As of July 2, 2001, determine the probability that the Los Angeles Dodgers will win the 2001 World Series. (*Source:* www.vegasinsider.com/u/futures.)

72. As of July 2, 2001, the odds in favor of the Cleveland Indians winning the 2001 World Series were listed as 1 to 6. As of July 2, 2001, determine the probability that the Cleveland Indians will win the 2001 World Series. (*Source:* www.vegasinsider.com/u/futures.)

73. As of July 2, 2001, the odds in favor of the Los Angeles Lakers winning the 2002 NBA Championship were listed as 5 to 7. As of July 2, 2001, determine the probability that the Los Angeles Lakers will win the 2002 NBA Championship. (*Source:* www.vegasinsider.com/u/futures.)

74. As of July 2, 2001, the odds in favor of the Charlotte Hornets winning the 2002 NBA Championship were listed as 5 to 7. As of July 2, 2001, determine the probability that the Charlotte Hornets will win the 2002 NBA Championship. (*Source:* www.vegasinsider.com/u/futures.)

75. If the odds in favor of event A occurring are a to b, show that

$$P(A) = \frac{a}{a+b}$$

Hint: Start with the equation $\frac{a}{b} = \frac{P(A)}{1 - P(A)}$ and solve for $P(A)$.

SECTION PROJECT

For the following probability Venn diagram, $P(A) = 0.45$, $P(B) = 0.45$, and $P(C) = 0.40$. Determine the values of w, x, y, and z.

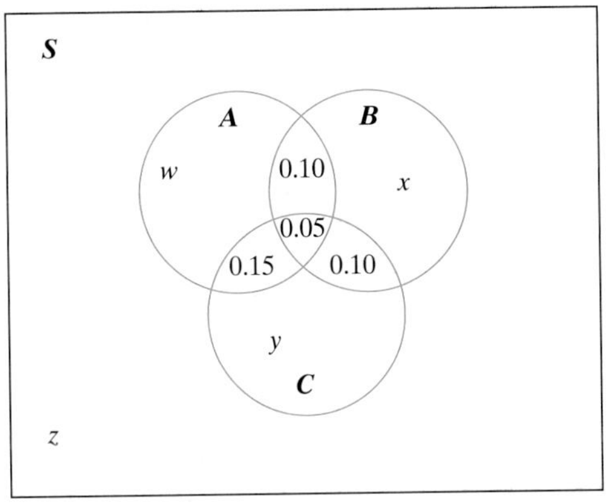

Section 6.6 Computing Probabilities Using the Multiplication Rule

In the previous section we used the Addition Rule to determine the probability that **either** of two events would occur. To compute this probability, we first had to know if the events were mutually exclusive. In this section we will use a new rule called the **Multiplication Rule**, which determines the probability, given two events, that **both** events occur. To compute this probability, we often have to determine if the events have a condition called **independence**. We will also examine a new type of probability called **conditional probability**.

Independent Events

Recall that in Section 6.1 we saw that often the occurrence of an event has no particular effect on the occurrence of some other event. For example, in Section 6.1 Example 5(b), we analyzed the event that the first toss of a coin is heads and the event that the second toss of the coin is tails. The likelihood of tossing heads or tails is the same, so tossing heads with the first toss does not alter the likelihood of tossing tails on the second toss. So, in this case, the outcome of the first event (first toss of a coin is heads) does not influence the second event (second toss of the coin is tails). We call events of this type **independent events**.

Independent Events

Two events A and B are **independent** if the occurrence (or nonoccurrence) of one event does not affect the other's chance of occurring. If A and B are not independent, we say that they are **dependent events**.

Note: We again offer a warning first presented in Section 6.1. A common error that many who are just beginning to study probability make is trying to read too much information into the questions in an attempt to find some exception.

Example 1 Determining If Events Are Independent

For each of the following, determine if the events A and B are independent.

(a) A represents the event that Austin practices golf 6 hours each day for 10 years and B represents the event that Austin becomes a professional golfer.

(b) Each week, 5 numbers from 1 to 40 are randomly selected from a bin in the Ultra-Lotto game. A represents the event that the five winning numbers in the first Ultra-Lotto held this year are 7, 10, 19, 29, and 33, and B represents the event that the five winning numbers in the second Ultra-Lotto held this year are 7, 10, 19, 29, and 33.

Solution

(a) **Understand the Situation:** If Austin practices golf 6 hours each day for 10 years, his chances of becoming a professional golfer are increased. So in this case, events A and B are dependent.

(b) **Understand the Situation:** Since all the numbers drawn in a lottery are placed back into the bin for the next drawing, the occurrence or nonoccurrence of A does not change the probability of B. Thus, A and B are independent. ■

Checkpoint 1 Now work Exercise 3.

Now let's see how independence can be used to find probabilities of two or more events occurring simultaneously. Suppose that a box contains 3 identical tickets colored red, blue, and green. In an experiment, a ticket is drawn at random, its color is recorded, and it is returned to the box. Then a second ticket is drawn at random from the box. If we let R, B, and G represent that a red, blue, or green ticket is drawn respectively, we can make a tree diagram for the experiment as shown in Figure 6.6.1.

If we write the sample space for this experiment, we get

$$S = \{RR, RB, RG, BR, BB, BG, GR, GB, GG\}$$

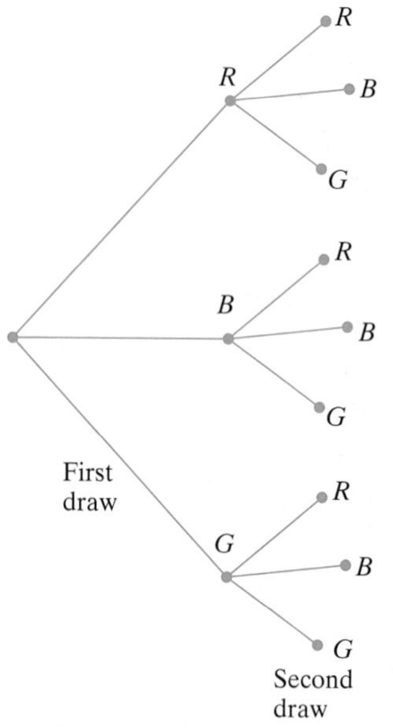

Figure 6.6.1

Suppose that we want to find the probability that we get a blue ticket on the first draw *and* a green ticket on the second draw. To do this we let

A represent the event that the blue ticket is drawn on the first draw
B represent the event that the green ticket is drawn on the second draw

The only associated outcome of this experiment is BG. So, using classical probability, we get

$$P(\text{blue first } \textbf{and} \text{ green second}) = P(A \cap B)$$

$$= \frac{\text{number of ways to draw a blue ticket and then a green ticket}}{\text{total number of possible outcomes}}$$

$$= \frac{n(A \cap B)}{n(S)} = \frac{1}{9}$$

Also notice that, on any given draw, the probability that a blue ticket is drawn is $P(A) = \frac{1}{3}$ and the probability that a green ticket is drawn is $P(B) = \frac{1}{3}$. Notice that if we multiply these two probabilities together we get

$$P(A) \cdot P(B) = \frac{1}{3} \cdot \frac{1}{3} = \frac{1}{9} = P(A \cap B)$$

It is no coincidence that these probabilities are the same. This illustrates the **Multiplication Rule for Independent Events**.

Multiplication Rule for Independent Events

If two events A and B are independent, then the probability of both A and B occurring is given by

$$P(A \textbf{ and } B) = P(A \cap B) = P(A) \cdot P(B)$$

Note: Make sure that you understand the difference between *independent events* and *mutually exclusive events*. Independent events pertain to how the occurrence of one event affects the occurrence of another event. Mutually exclusive events pertain to whether events can occur at the same time.

Example 2 Applying the Multiplication Rule for Independent Events

In a July 2000 Harris poll, it was found that 47% of the American public had independently read or heard something about the missile defense system that the Pentagon was developing. If two Americans are selected at random, determine the probability that both had heard about the missile defense system. (*Source:* www.harrisinteractive.com.)

Solution

Understand the Situation: Let A represent the event that the first randomly selected American had heard about the missile defense system and B represent the event that the second randomly selected American had heard about the missile defense system. The events A and B are independent since whether the first person has heard about the system does not affect whether the second person has heard about the system.

Since events A and B are independent, we can use the Multiplication Rule for Independent Events. This gives us

$$P(A \cap B) = P(A) \cdot P(B) = (0.47)(0.47) \approx 0.22$$

Interpretation: This means that if two Americans are selected at random there is about a 22% probability that both of them have read or heard something about the missile defense system. ■

Interactive Activity

Recompute Example 2, except here determine the probability that neither of the two randomly selected Americans had read or heard anything about the missile defense system.

WWW

Conditional Probability

We have seen that events are independent if the occurrence of one event does not affect the occurrence of the other. Our next step is to determine the probability $P(A \cap B)$ if the events are dependent. If the events are dependent, we must compute a new type of probability called **conditional probability** denoted by $\boldsymbol{P(B \mid A)}$. We read $P(B \mid A)$ as "the probability of B given A." This means that we compute the probability of B given that A has occurred.

To see how conditional probability works, let's consider a bag that contains 3 red, 4 white and 3 green marbles. In an experiment, a marble is drawn at random, its color recorded, and then a second marble is drawn **without replacing** the first. Let A represent the event that a white marble is drawn on the first draw and B represent the event that a green marble is drawn on the second draw. Suppose that we want to determine $P(B \mid A)$. This is the probability that we draw a green marble on the second draw, given that the first draw produced a white marble. If the first marble drawn is white, then there are 9 marbles left in the bag and 3 of those are green. So for the second draw this gives us

$$P(B \mid A) = \frac{\begin{array}{c}\text{number of green marbles left in the}\\ \text{bag after a white marble is drawn}\end{array}}{\begin{array}{c}\text{number of marbles in}\\ \text{the bag after the first draw}\end{array}} = \frac{3}{9} = \frac{1}{3}$$

Frequently, sample spaces are often much larger than the number of marbles that were in the bag for the experiment that we just considered. Hence, computing conditional probabilities using the method just outlined is not always practical. Fortunately, there is a formula that we can use to compute conditional probabilities.

Conditional Probability

If A and B are events with $P(A) \neq 0$, then the **conditional probability** that B occurs, given that A has occurred, is given by

$$P(B \mid A) = \frac{P(A \cap B)}{P(A)}$$

We can think of the conditional probability $P(B \mid A)$ as the probability that event B occurs if the sample space is restricted to the outcomes associated with event A. This idea is illustrated visually in Figure 6.6.2.

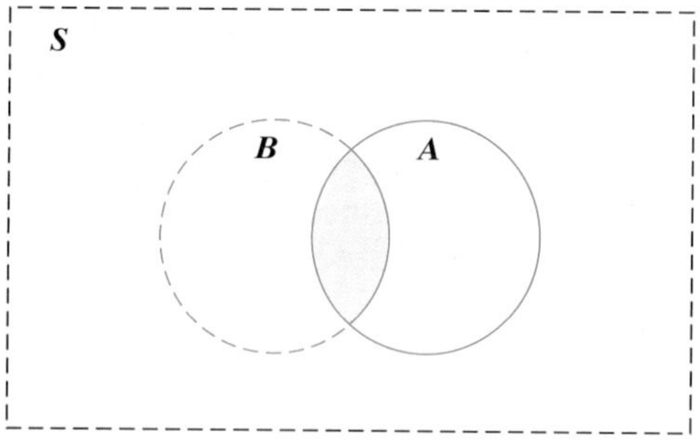

Figure 6.6.2

Example 3 Determining a Conditional Probability Using a Venn Diagram

In a random sample of 700 business people, 350 said that they had an airline flight delayed in the last month, 200 said that they had an airline flight canceled in the last month, and 75 said that they had both a flight delay and a cancellation in the last month. Let D represent the event that a businessperson had a flight delayed and C represent the event that the businessperson had a flight canceled.

(a) Draw a Venn diagram to represent the business people sampled with events D and C. Compute $P(D \cap C)$ and interpret.

(b) Determine $P(C \mid D)$ and interpret.

Solution

(a) The Venn diagram for the sample is shown in Figure 6.6.3. To compute $P(D \cap C)$, we see that there are 75 business people who had both a cancellation and a flight delay and there were 700 business people surveyed. This gives us

$$P(D \cap C) = \frac{\text{business people who had both a cancellation and a delay}}{\text{total number surveyed}} = \frac{75}{700} = \frac{3}{28} \approx 0.107$$

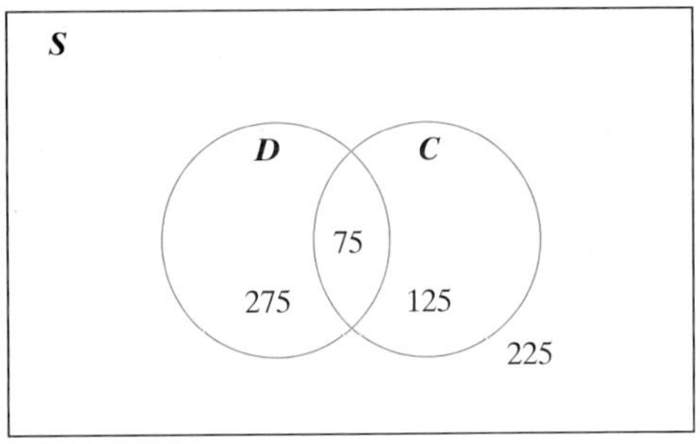

Figure 6.6.3

Interpret the Solution: This means that if one of the business people is selected at random, the probability that the person had **both** a cancellation **and** a flight delay in the last month is about 10.7%.

(b) First note that $P(D) = \frac{350}{700} = \frac{1}{2}$. Using the formula for computing conditional probability, we get

$$P(C \mid D) = \frac{P(D \cap C)}{P(D)} = \frac{\frac{75}{700}}{\frac{350}{700}} = \frac{75}{350} = \frac{3}{14} \approx 0.214$$

Interpret the Solution: This means that the probability that a randomly chosen businessperson had a flight canceled, given that he or she had a delayed flight, is about 21.4%. ■

Checkpoint 2 Now work Exercise 37.

We need to make some observations based on Example 3. First note that

$$P(D \mid C) = \frac{P(C \cap D)}{P(C)} = \frac{\frac{75}{700}}{\frac{200}{700}} = \frac{75}{200} = \frac{3}{8}$$

which is different from the result in part (b). This is evidence that conditional probability is not commutative, meaning that generally, for events A and B,

$$\boldsymbol{P(B \mid A) \neq P(A \mid B)}$$

Also note that, when we computed $P(C \mid D)$ in part (b) of Example 3, we got the fraction $\frac{75}{350}$. This implies that we can think of conditional probability as a way of restricting the sample space. For $P(C \mid D)$ in Example 3, we can say that we determined the probability that a businessperson had a canceled flight as we restrict the sample space just to those who had delayed flights. This is displayed graphically in Figure 6.6.4.

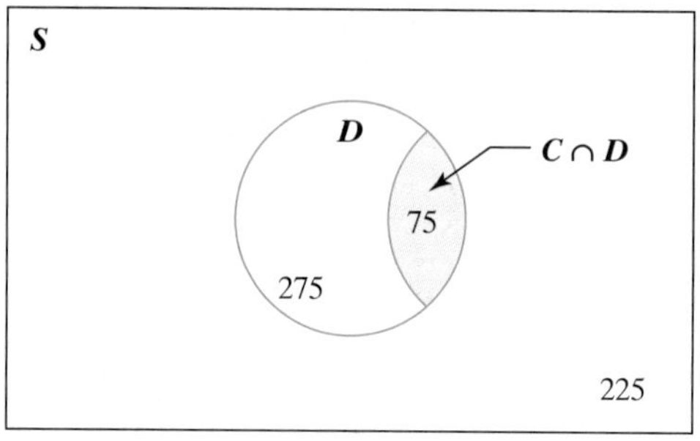

Figure 6.6.4 $P(C \mid D) = \frac{75}{350} = \frac{3}{14}$

General Multiplication Rule

Using the formula for conditional probability, we can derive a general formula for computing $P(A \cap B)$.

$$P(B \mid A) = \frac{P(A \cap B)}{P(A)} \quad \text{Formula for conditional probability}$$

$$P(A) \cdot P(B \mid A) = P(A \cap B) \quad \text{Multiply both sides by } P(A),\ P(A) \neq 0.$$

This gives us the **General Multiplication Rule.**

General Multiplication Rule

For any events A and B, the probability that A and B **both** occur is given by

$$P(A \cap B) = P(A) \cdot P(B \mid A)$$

Recall that the Multiplication Rule for Independent Events A and B is given by

$$P(A \cap B) = P(A) \cdot P(B)$$

By comparing the General Multiplication Rule to the Multiplication Rule for Independent Events A and B, we can see that if A and B are independent events then $P(B \mid A) = P(B)$. This means that the probability of B given A equals the probability of B. Another way of looking at it is that the fact that the event A has occurred does not affect the probability of event B. This gives us the following alternative for independent events.

B* Is Independent of *A

Let A and B be two events of a sample space S, where $P(B) > 0$. The event B is independent of A if and only if

$$P(B \mid A) = P(B)$$

Example 4 Applying the General Multiplication Rule

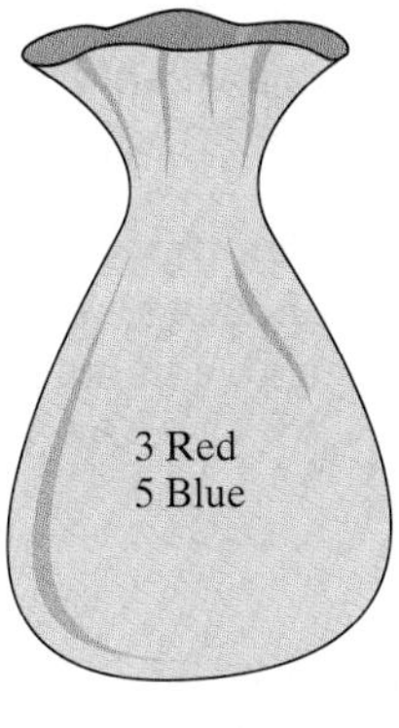

A bag contains 3 red and 5 blue marbles. In an experiment, a marble is drawn at random, its color is recorded, and then a second marble is drawn at random and *without* replacing the first. Determine the probability that first a red and then a blue marble are drawn.

Solution

Understand the Situation: Let R_1 represent the event that first a red marble is drawn. Let B_2 represent the event that the second draw is a blue marble.

By the General Multiplication Rule, we have

$$P(R_1 \cap B_2) = \begin{matrix}\text{probability that the}\\ \text{first marble is red}\end{matrix} \cdot \begin{matrix}\text{probability that the second marble is}\\ \text{blue, given that the first was red}\end{matrix}$$
$$= P(R_1) \cdot P(B_2 \mid R_1) = \frac{3}{8} \cdot \frac{5}{7} = \frac{15}{56}$$

Interpret the Solution: So the probability that first a red and then a blue marble is drawn is $\frac{15}{56}$. ■

Interactive Activity

In Example 4, we computed $P(R_1 \cap B_2) = \frac{15}{56}$. Compute $P(B_1 \cap R_2)$ and compare to the solution of Example 4.

WWW

We can also use tabular data to determine conditional probabilities and probabilities for other compound events. When using data from tables, it is important to remember that we get a compound fraction when computing a conditional probability. This is illustrated in Example 5.

Example 5 Computing Conditional Probability Using a Table

In 1987, the number of men and women enrolled in 2- and 4-year colleges and universities, measured in thousands, is shown in Table 6.6.1.

Table 6.6.1

	2-Year (A)	4-Year (B)	Total
Male (M)	1522	3356	4,878
Female (F)	2127	3299	5,426
Total	3649	6655	10,304

SOURCE: U.S. Census Bureau, www.census.gov

A 2- or 4-year college student is selected at random. For the labeled events A, B, M, and F, determine the following probabilities and interpret.

(a) Determine the probability that a college student selected at random is female.

(b) Given that the randomly selected student is female, determine the probability that she attends a 4-year college.

Solution

(a) Understand the Situation: Here we are trying to determine $P(F)$.

If we follow across the row associated with event F, we get a total of 5426. So dividing by the total number of college enrollees (in thousands), we get

$$P(F) = \frac{5426}{10{,}304} \approx 0.527$$

Interpret the Solution: This means that if a 2- or 4-year college student is selected at random there is about a 52.7% chance that the enrollee is female.

(b) **Understand the Situation:** We are really being asked to determine the probability that the randomly chosen student attends a 4-year college, *given* that she is a female. Mathematically, this means we are trying to compute

$$P(B \mid F)$$

The formula for conditional probability gives us

$$P(B \mid F) = \frac{P(F \cap B)}{P(F)} = \frac{\frac{3299}{10{,}304}}{\frac{5426}{10{,}304}} = \frac{3299}{5426} \approx 0.608$$

Interpret the Solution: This means that if a college student is selected at random the probability that the student attends a 4-year college, given that he or she is a female, is about 60.8%. ■

Checkpoint 3 Now work Exercise 53.

SUMMARY

In this section we studied how to use the multiplication rule to determine $P(A \cap B)$. We saw that the formula we used depended on whether the events were independent.

- Two events A and B are **independent** if the occurrence (or nonoccurrence) of one event does not affect the other's chance of occurring.
- **Multiplication Rule for Independent Events** $P(A \cap B) = P(A) \cdot P(B)$
- **Conditional Probability** $P(B \mid A) = \frac{P(A \cap B)}{P(A)}$, $P(A) \neq 0$
- **General Multiplication Rule** $P(A \cap B) = P(A) \cdot P(B \mid A)$

SECTION 6.6 EXERCISES

In Exercises 1–10, determine if the events A and B are independent or dependent. Explain your reasoning.

6.1 **1.** A represents the event that a flash food occurs at Windsor Park on Tuesday afternoon and B represents the event that the Tuesday evening company picnic at Windsor Park is postponed.

6.1 **2.** A represents the event that Rebecca completes her political science course reading assignment every day and B represents the event that Rebecca passes her political science course.

6.1 ✓ **3.** A represents the event that Kahrin's car battery is dead and B represents the event that Kahrin's oven light is burned out.

6.1 **4.** A represents the event that the telephone in Rich's office does not work and B represents the event that the vending machine in Rich's daughter's high school is empty.

6.1 **5.** A represents the event that Clifford gets a raise and a promotion and B represents the event that Clifford purchases a new car.

6.1 **6.** A represents the event that Lana is exposed to ultraviolet radiation for three decades and B represents the event that Lana develops skin cancer.

6.1 **7.** A box contains 5 tickets numbered 1 to 5. In an experiment, a ticket is drawn at random, its number is recorded, and it is replaced. Then a second ticket is drawn randomly. A represents the event that the first ticket is a 3 and B represents the event that the second ticket is a 3.

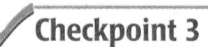

6.1 **8.** A bag contains 5 green and 4 purple marbles. In an experiment, a marble is drawn at random, its color is recorded, and it is replaced. Then a second marble is drawn randomly. A represents the event that the first marble is purple and B represents the event that the second marble is purple.

6.1 **9.** A represents the event that Jim has received seven speeding tickets and B represents the event that Jim has high car insurance premiums.

6.1 **10.** A represents the event that a person has brown eyes and B represents the event that the person's son has brown eyes.

11. A box contains 3 tickets numbered 1 to 3. In an experiment, a ticket is drawn at random, its number is recorded, and it is returned to the box. Then a second ticket is drawn randomly. Determine the probability that both tickets are labeled with the number 2.

12. A box contains 2 tickets numbered 1 and 2. In an experiment, a ticket is drawn at random, its number is recorded, and it is returned to the box. Then a second ticket is drawn randomly. Determine the probability that both tickets are labeled with the number 1.

13. A bag contains 5 green and 4 blue marbles. In an experiment, a marble is drawn at random, its color is recorded, and then it is returned to the bag. Then a second marble is drawn randomly. Determine the probability that two green marbles are drawn.

14. A bag contains 3 red and 8 yellow marbles. In an experiment, a marble is drawn at random, its color is recorded, and then it is returned to the bag. Then a second marble is drawn randomly. Determine the probability that two yellow marbles are drawn.

15. A June 2000 Gallup poll showed that 14% of American adults said independently that the decline of ethics, morals, and family values was the most important problem facing the country. A sociologist selects two Americans at random for an interview. Let A represent the event that the first person believes that the decline of ethics, morals, and family values is the most important problem facing the country and B represent the event that the second person believes that the decline of ethics, morals, and family values is the most important problem facing the country. (*Source:* www.gallup.com.)

(a) Are events A and B independent? Explain.
(b) Determine $P(A \cap B)$ and interpret.
(c) Determine $P(A^C)$ and $P(B^C)$ and interpret each.
(d) Determine $P(A^C \cap B^C)$ and interpret.

16. A February 1999 Gallup poll showed that 34% of adult Americans said independently of each other that AIDS was the most urgent health problem facing the United States. A medical researcher selects two Americans at random for a report. Let A represent the event that the first person believes that AIDS is the most urgent health problem facing the United States and B represent the event that the second person believes that AIDS is the most urgent health problem facing the United States. (*Source:* www.gallup.com.)

(a) Are events A and B independent? Explain.
(b) Determine $P(A \cap B)$ and interpret.
(c) Determine $P(A^C)$ and $P(B^C)$ and interpret each.
(d) Determine $P(A^C \cap B^C)$ and interpret.

17. A March 2000 Gallup poll showed that 25% of adult Americans reported that their doctors diagnosed a condition or suggested a treatment that turned out to be wrong. If an HMO selects four adult Americans at random, determine the probability that all four had their doctors make such an error. (*Source:* www.gallup.com.)

18. An April 2000 Gallup poll showed that 46% of American adults said independently that the federal government should be more involved in public education than it currently is. If an educational political action committee selects four adult Americans at random, determine the probability that all four believe the federal government should be more involved in public education. (*Source:* www.gallup.com.)

19. In 1996 the U.S. Census Bureau reported that 35 million Americans owned personal computers. Among the 9.9 million Americans who used their computers more than 16 hours per week, 4.7 million said that they used their computer for personal use only, 2.1 million said that they used their computer for business use only, and 3.1 million said that they used their computer for both personal and business use. (*Source:* U.S. Census Bureau, www.census.gov.)

(a) A representative from a computer manufacturer selects at random a computer owner who uses his or her computer for more than 16 hours per week. Let A represent the event that the computer was for personal use and B represent the event that the computer was for business use. Draw a Venn diagram to represent the computer owners who use their computer for more than 16 hours per week with events A and B.
(b) Compute $P(A \cap B)$ and interpret.
(c) Compute $P(B \mid A)$ and interpret.

In Exercises 20–23, a Venn diagram is given. Determine $P(A \cap B)$ and $P(B \mid A)$.

20.

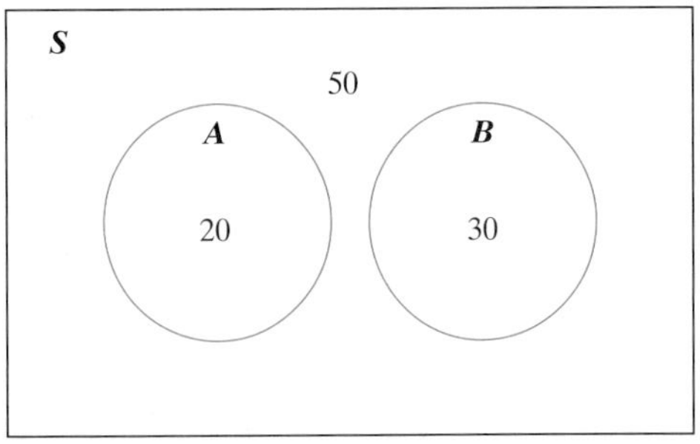

21.

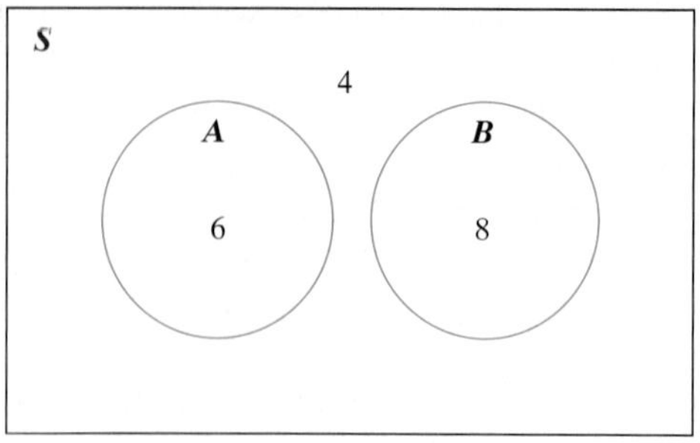

22.

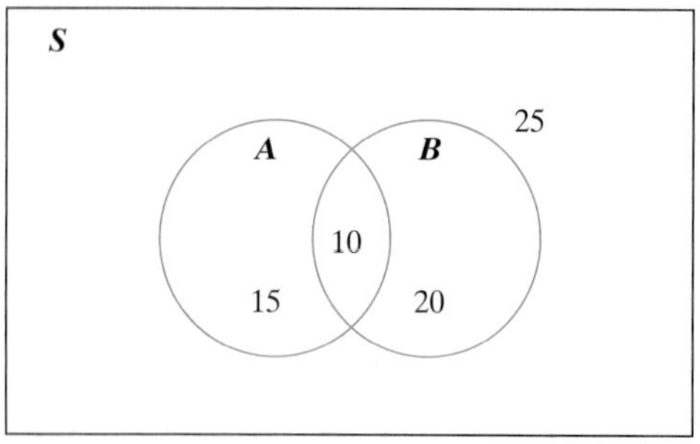

23.

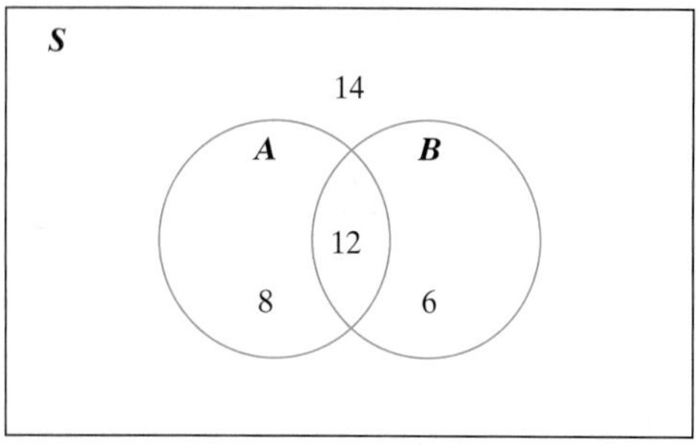

In Exercise 24–27, a probability Venn diagram is given. Determine $P(A \cap B)$ and $P(B \mid A)$.

24.

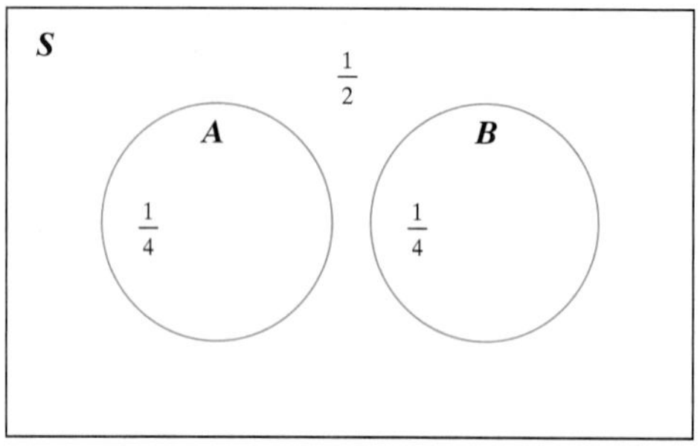

25.

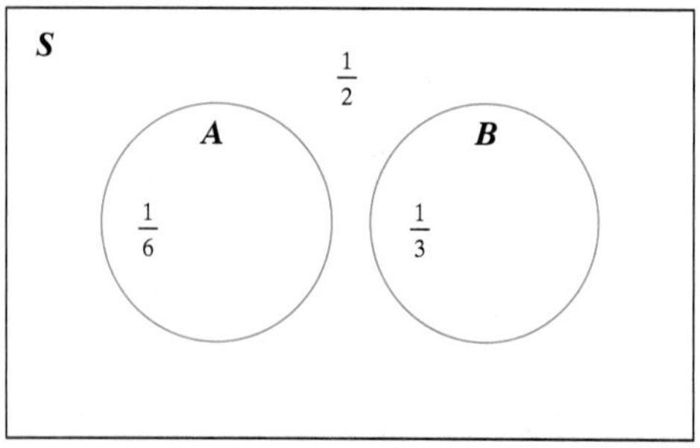

26.

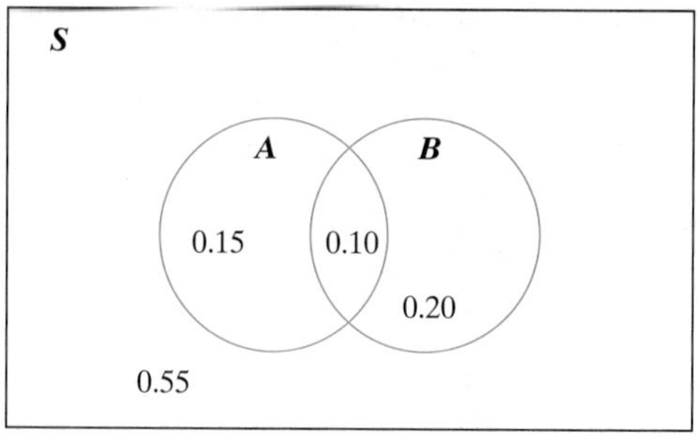

27.

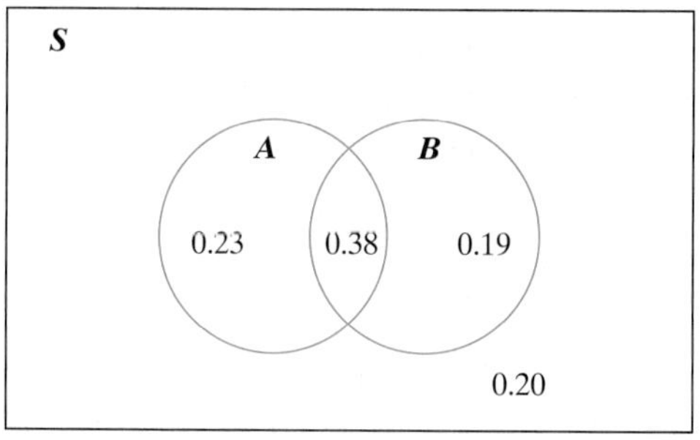

In Exercises 28–31, a table of data is given. Determine $P(A \cap C)$ and $P(C \mid A)$.

28.

	A	*B*	Total
C	10	20	30
D	15	5	20
Total	25	25	50

29.

	A	*B*	Total
C	40	60	100
D	55	45	100
Total	95	105	200

30.

	A	*B*	Total
C	26	33	59
D	18	48	66
Total	44	81	125

31.

	A	*B*	Total
C	87	52	139
D	61	50	111
Total	148	102	250

In Exercises 32–35, a table of probabilities is given. Determine $P(B \cap D)$ and $P(D \mid B)$.

32.

	A	*B*	Total
C	$\frac{1}{4}$	$\frac{1}{4}$	$\frac{1}{2}$
D	$\frac{1}{8}$	$\frac{3}{8}$	$\frac{1}{2}$
Total	$\frac{3}{8}$	$\frac{5}{8}$	1

33.

	A	*B*	Total
C	$\frac{1}{10}$	$\frac{3}{10}$	$\frac{2}{5}$
D	$\frac{1}{5}$	$\frac{2}{5}$	$\frac{3}{5}$
Total	$\frac{3}{10}$	$\frac{7}{10}$	1

34.

	A	*B*	Total
C	0.22	0.15	0.37
D	0.42	0.21	0.63
Total	0.64	0.36	1.00

35.

	A	*B*	Total
C	0.17	0.24	0.41
D	0.20	0.39	0.59
Total	0.37	0.63	1.00

36. A bag contains 5 yellow and 6 green marbles. In an experiment, a marble is drawn at random, its color is recorded, and a second marble is drawn randomly without replacing the first.

(a) Determine the probability that the first marble is green and the second marble is yellow.

(b) Determine the probability that both marbles drawn are green.

✓ **37.** A bag contains 6 red and 4 blue marbles. In an experiment, a marble is drawn at random, its color is recorded, and a second marble is drawn randomly without replacing the first.

(a) Determine the probability that the first marble is red and the second marble is blue.

(b) Determine the probability that both marbles drawn are blue.

38. A box contains 10 tickets with 4 of them labeled with a △ and 6 tickets labeled with a ○. In an experiment, 3 tickets are drawn consecutively without replacement and at random.

(a) Determine the probability that all three tickets are labeled with a △.

(b) Determine the probability that the first ticket is labeled with a △ and the second and third tickets are labeled with a ○.

39. A box contains 10 tickets with 7 tickets labeled with a ★ and 3 tickets labeled with a ◆. In an experiment, 3 tickets are drawn consecutively without replacement and at random.

(a) Determine the probability that all three tickets are labeled with a ◆.

(b) Determine the probability that the first ticket is labeled with a ◆ and the second and third tickets are labeled with a ★.

For Exercises 40–47, consider the following: The number of doctorates conferred in the social sciences (anthropology, sociology, economics, and political science) and the physical sciences (astronomy, physics, and chemistry) during 1998 is shown in the following table. The data are classified as being a U.S. citizen or a foreign-born citizen.

	Social Sciences (*A*)	Physical Sciences (*B*)	Total
U.S. citizen (*C*)	2353	2415	4768
Foreign-born citizen (*D*)	760	1111	1871
Total	3113	3526	6639

SOURCE: U.S. Census Bureau, www.census.gov

For the events labeled in the table as *A, B, C,* and *D* answer the following:

40. Determine $P(A \cap B)$ and interpret.

41. Determine $P(C \cap D)$ and interpret.

42. Determine the probability that the doctorate is in social sciences and is conferred on a U.S. citizen.

43. Determine the probability that the doctorate is in physical sciences and is conferred on a foreign-born citizen.

44. Determine $P(C \mid A)$ and interpret.

45. Determine $P(D \mid B)$ and interpret.

46. Determine the probability that the doctorate is in physical sciences, given that the recipient is a U.S. citizen.

47. Determine the probability that the doctorate is in social sciences, given that the recipient is a foreign-born citizen.

For Exercises 48–55, consider the following: In 1996, the percentage of American adults who exercised for 30 minutes or more at least 5 times per week is shown in the table. The percentages are classified by gender.

	Did Not Exercise (*A*)	Exercised Regularly (*B*)	Total
Men (*C*)	22.0%	25.6%	47.6%
Women (*D*)	25.6%	26.8%	52.4%
Total	47.6%	52.4%	100%

SOURCE: National Center for Chronic Disease Prevention, www.cdc.gov/nccdphp

For the events labeled in the table as A, B, C, and D answer the following.

48. Determine $P(C \cap D)$ and interpret.

49. Determine $P(A \cap B)$ and interpret.

50. Determine $P(D \cap B)$ and interpret.

51. Determine $P(A \cap C)$ and interpret.

52. Determine the probability that the person is a female given that the person exercises regularly.

✓ **53.** Determine the probability that the person is a male given that the person does not exercise regularly.

54. Determine $P(A \mid D)$ and interpret.

55. Determine $P(B \mid C)$ and interpret.

SECTION PROJECT

We can use the Multiplication Rule for Independent Events, along with the complement of an event, to determine the probability that **at least one** outcome occurs. For example, let's say that a student takes a six-question multiple-choice test, with each question having choices A, B, C, and D. The student guesses on each question. If we let A represent the event that the student gets *at least one* question correct, then the complement of the event, A^C, represents the event that the student gets *no* questions correct. Since the trials are independent and the probability of getting a question wrong is $\frac{3}{4}$, we get

$$P(A^C) = \frac{3}{4} \cdot \frac{3}{4} \cdot \frac{3}{4} \cdot \frac{3}{4} \cdot \frac{3}{4} \cdot \frac{3}{4} = \frac{3^6}{4^6} \approx 0.178$$

This means the probability of getting at least one question correct is

$$P(A) = 1 - P(A^C) = 1 - 0.178 = 0.822$$

Thus, the probability that the student gets at least one question correct is about 82.2%. Use this technique to answer the following:

In an August 1999 Harris poll, 24% of all Americans knew that Trent Lott was the U.S. Senate Majority Leader. If eight Americans are chosen at random, determine the probability that at least one knows that Lott was the U.S. Senate Majority Leader. (*Source:* www.harrisinteractive.com.)

Section 6.7 Bayes' Theorem and Its Applications

In the previous section we saw that conditional probability is an important tool for computing probabilities by the Multiplication Rule. Now we want to expand our discussion of conditional probability. First, we want to see if there is a relationship between the probabilities $P(B \mid A)$ and $P(A \mid B)$. This will lead us to discover a powerful formula called **Bayes' Theorem**, which can be used to compute certain conditional probabilities.

The Relationship between $P(B \mid A)$ and $P(A \mid B)$

In Section 6.6 we showed that, generally, $P(B \mid A) \neq P(A \mid B)$. To see how the probabilities $P(B \mid A)$ and $P(A \mid B)$ are related, however, we must first understand the difference between these two expressions. This difference is illustrated in Example 1.

Example 1 Computing Conditional Probabilities

In a January 1999 Harris poll, a random sample of American adults were asked the question "By the year 2020, do you think the quality of life in your community is more likely to get better or worse?" The percentages of the results, separated by gender, are shown in Table 6.7.1.

Table 6.7.1

	Male (A)	Female (B)	Total
Better (C)	31%	30%	61%
Worse (D)	19%	20%	39%
Total	50%	50%	100%

SOURCE: www.harrisinteractive.com

Let's select an American adult at random. For the labeled events A, B, C, and D, determine the following probabilities and interpret.

(a) $P(B \mid C)$ **(b)** $P(C \mid B)$

Solution

(a) **Understand the Situation:** We are being asked to determine the probability that we select a female (event B), given that the response to the question is "better" (event C).

Using the definition of conditional probability, we need to compute $P(B \mid C) = \dfrac{P(C \cap B)}{P(C)}$. Since the values in the table are in percentages, we can write them in decimal form to get

$$P(B \mid C) = \frac{P(C \cap B)}{P(C)} = \frac{0.30}{0.61} \approx 0.492$$

Interpret the Solution: This means that if we select an American adult at random the probability is about 49.2% that the person is female, given that the adult's opinion is that the community's quality of life will get better.

(b) **Understand the Situation:** In this case, we are being asked to compute the probability that the response to the question is "better" (event C) given that the selected person is a female (event B).

Using the definition of conditional probability, we have

$$P(C \mid B) = \frac{P(B \cap C)}{P(B)} = \frac{0.30}{0.50} = 0.60$$

Interpret the Solution: This means that if we select an American adult at random the probability is 60% that the person thinks his or her community's quality of life will get better, given that the selected individual is female. ■

Checkpoint 1

Now work Exercise 3.

Example 1 illustrates that, if we are given a complete table of probabilities, computing a particular conditional probability is not difficult. But there are times when we do not have a complete table of values. To see how $P(B \mid A)$ and $P(A \mid B)$ are related, let's turn to our definition of conditional probability. First, we know that $P(B \mid A) = \dfrac{P(A \cap B)}{P(A)}$, so after multiplying both sides by $P(A)$ [note that $P(A) \neq 0$], we get

$$P(A) \cdot P(B \mid A) = P(A \cap B)$$

We also know that $P(A \mid B) = \dfrac{P(B \cap A)}{P(B)}$, which, after multiplying both sides by $P(B)$ [note that $P(B) \neq 0$], yields

$$P(B) \cdot P(A \mid B) = P(B \cap A)$$

Since we know that, for any events A and B, $P(B \cap A) = P(A \cap B)$, we can set the two equations equal to each other to get

$$P(A) \cdot P(B \mid A) = P(B) \cdot P(A \mid B)$$

If we divide both sides of the equation by $P(A)$, we get

$$P(B \mid A) = \frac{P(B) \cdot P(A \mid B)}{P(A)}$$

We now have the following relationship between $P(A \mid B)$ and $P(B \mid A)$.

Relationship between $P(A \mid B)$ and $P(B \mid A)$

The conditional probabilities $P(A \mid B)$ and $P(B \mid A)$ are related by

$$P(B \mid A) = \frac{P(B) \cdot P(A \mid B)}{P(A)}$$

where $P(A) \neq 0$.

The relationship between events involving conditional probabilities are often expressed in the form of tree diagrams like the one shown in Figure 6.7.1. Notice that the tree diagram has two **branches**. There is one branch for B and another for its complement B^C. We will soon see that we use B^C to guarantee that all elements of the sample space are accounted for.

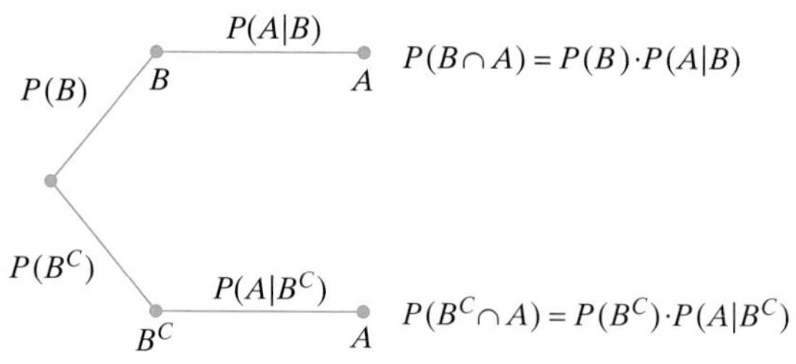

Figure 6.7.1

Let's say that we want to compute $P(A)$. To do this, we must follow the first branch to compute $P(B \cap A)$ **and** we must follow the second branch to compute $P(B^C \cap A)$. From the General Multiplication Rule, the probability associated with the top branch, $P(B \cap A)$, is $P(B) \cdot P(A \mid B)$, and the probability associated with the second branch, $P(B^C \cap A)$, is $P(B^C) \cdot P(A \mid B^C)$. Since B and B^C, along with their associated branches, are mutually exclusive, we can use the Addition Rule for Mutually Exclusive Events to get

$$\begin{aligned} P(A) &= P((B \cap A) \cup (B^C \cap A)) \\ &= P(B \cap A) + P(B^C \cap A) \\ &= P(B) \cdot P(A \mid B) + P(B^C) \cdot P(A \mid B^C) \end{aligned}$$

Informally, we can say that **$P(A)$ is equal to the sum of the branches**. Let's see how these formulas can be used in an application.

Example 2 Determining $P(B \mid A)$ If $P(A \mid B)$ Is Given

In a March 2000 survey, a Harris poll queried 1149 adults; 43% of the adults had disabilities and the remainder did not have disabilities. The survey found that 30% of the respondents who had disabilities accessed the Internet from their home and 48% of the respondents without disabilities accessed the Internet from their home. Let B represent the event that a respondent had disabilities and A represent the event that the respondent accessed the Internet from home. (*Source:* www.harrisinteractive.com.)

(a) Construct a tree diagram for the events A, B, and B^C.

(b) Determine $P(A)$ and interpret.

(c) Determine $P(B \mid A)$ and interpret.

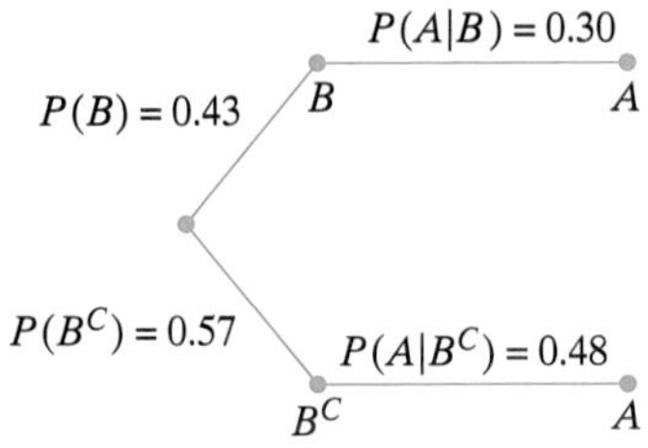

Figure 6.7.2

Solution

(a) First notice that we are given that $P(B) = 0.43$ and $P(B^C) = 1 - 0.43 = 0.57$, where B^C represents the event that a respondent does not have disabilities. We are also given that $P(A \mid B) = 0.30$ and $P(A \mid B^C) = 0.48$. These events and probabilities are shown in the tree diagram in Figure 6.7.2.

(b) To calculate $P(A)$, we must compute the sum of the branches to get

$$P(A) = P(B) \cdot P(A \mid B) + P(B^C) \cdot P(A \mid B^C)$$
$$= (0.43)(0.30) + (0.57)(0.48) \approx 0.403$$

Interpret the Solution: This means that if a respondent is selected at random there is about a 40.3% probability that she or he accesses the Internet from home.

(c) To determine $P(B \mid A)$, we need to use the relationship between $P(A \mid B)$ and $P(B \mid A)$ formula.

$$P(B \mid A) = \frac{\text{first branch}}{\text{sum of branches}} = \frac{P(B) \cdot P(A \mid B)}{P(A)}$$
$$= \frac{P(B) \cdot P(A \mid B)}{P(B) \cdot P(A \mid B) + P(B^C) \cdot P(A \mid B^C)}$$
$$= \frac{(0.43)(0.30)}{(0.43)(0.30) + (0.57)(0.48)} \approx 0.320$$

Interpret the Solution: This means that if a respondent is selected at random the probability that she or he has a disability, given that they access the Internet from home, is about 32%. ■

Interactive Activity

Use the method shown in part (c) of Example 2 to compute and interpret $P(B^C \mid A)$.

www

Checkpoint 2

Now work Exercise 9.

The formula that we have derived and used in Example 2 is a version of a theorem called **Bayes' Theorem**, which is named after Reverend Thomas Bayes who, during the 18th century, studied calculations much like those computed in Example 2. Let's now summarize the formula that we have seen thus far.

Bayes' Theorem for Two Branches

For events A and B with complement B^C,

- The tree diagram for the events A, B and B^C is as follows:

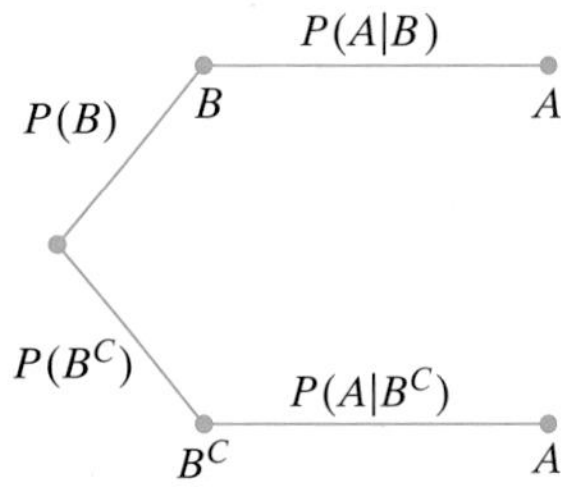

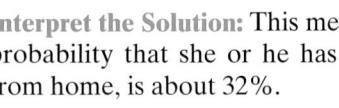

- $P(A) = \text{sum of branches} = P(B) \cdot P(A \mid B) + P(B^C) \cdot P(A \mid B^C)$
- $P(B \mid A) = \dfrac{\text{first branch}}{\text{sum of branches}} = \dfrac{P(B) \cdot P(A \mid B)}{P(B) \cdot P(A \mid B) + P(B^C) \cdot P(A \mid B^C)}$
- $P(B^C \mid A) = \dfrac{\text{second branch}}{\text{sum of branches}} = \dfrac{P(B^C) \cdot P(A \mid B^C)}{P(B) \cdot P(A \mid B) + P(B^C) \cdot P(A \mid B^C)}$

In Example 3, we apply Bayes' Theorem in a factory setting.

Example 3 Computing Probabilities Using Bayes' Theorem

Suppose that $\frac{7}{10}$ of the computer keyboards made by the TouchRight Company come from factory A, while factory B produces the rest. Quality control records show that $\frac{1}{5}$ of the keyboards from factory A do not pass inspection and $\frac{1}{3}$ of the keyboards from factory B do not pass inspection. Determine the probability that if a randomly selected keyboard does not pass inspection, the keyboard came from factory B.

Solution

Understand the Situation: To apply Bayes' Theorem, we need to first assign events. We let

A represent the event that the keyboard came from factory A
A^C represent the event that the keyboard came from factory B
D represent the event that the keyboard did not pass inspection

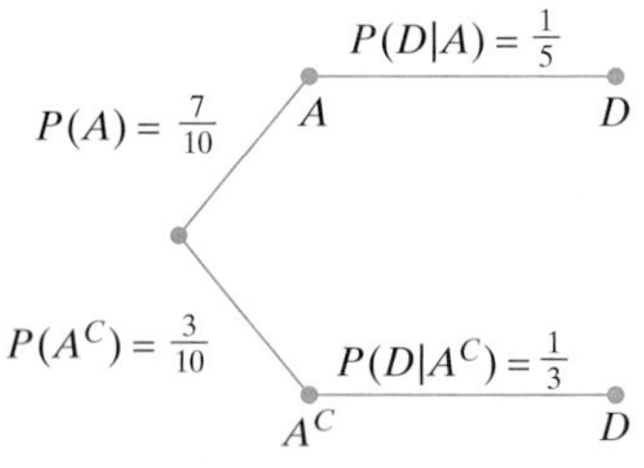

Figure 6.7.3

We have been given the probabilities $P(A) = \frac{7}{10}$, $P(A^C) = 1 - \frac{7}{10} = \frac{3}{10}$, $P(D \mid A) = \frac{1}{5}$, and $P(D \mid A^C) = \frac{1}{3}$. These probabilities are shown in the tree diagram in Figure 6.7.3.

To determine the probability that if a randomly selected keyboard does not pass inspection the keyboard came from factory B, we need to compute $P(A^C \mid D)$. This gives us

$$P(A^C \mid D) = \frac{\text{second branch}}{\text{sum of branches}} = \frac{P(A^C) \cdot P(D \mid A^C)}{P(A) \cdot P(D \mid A) + P(A^C) \cdot P(D \mid A^C)}$$

$$= \frac{\frac{3}{10} \cdot \frac{1}{3}}{\frac{7}{10} \cdot \frac{1}{5} + \frac{3}{10} \cdot \frac{1}{3}} = \frac{\frac{1}{10}}{\frac{7}{50} + \frac{1}{10}} = \frac{\frac{1}{10}}{\frac{6}{25}} = \frac{5}{12}$$

Interpret the Solution: This means that there is a $\frac{5}{12}$ probability that a randomly selected keyboard that did not pass inspection was manufactured at factory B. ■

The Generalized Bayes' Theorem

So far we have seen how to apply the Bayes' Theorem when we have mutually exclusive events B and B^C, which results in two branches in a tree diagram. In some applications, however, there can be many branches in the tree diagram. In order to use Bayes' Theorem in these cases, we need to have all the events take up the whole sample space. We call events of this type **collectively exhaustive**.

Collectively Exhaustive Events

Events $B_1, B_2, B_3, \ldots, B_k$ are **collectively exhaustive** if $B_1 \cup B_2 \cup B_3 \cup \cdots \cup B_k = S$, where S is the sample space.

Note: 1. In terms of their probabilities, we can see that the events $B_1, B_2, B_3, \ldots, B_k$ are collectively exhaustive if $P(B_1 \cup B_2 \cup B_3 \cup \cdots \cup B_k) = 1$ and they are mutually exclusive.
2. The events $B_1, B_2, B_3, \ldots, B_k$ are mutually exclusive if any two of them are mutually exclusive.

We can now state the Generalized Bayes' Theorem. This generalization of the theorem was developed by the mathematician Pierre LaPlace in 1814, several years after Bayes' death.

Generalized Bayes' Theorem

For mutually exclusive and collectively exhaustive events $B_1, B_2, B_3, \ldots, B_k$ and for any event A with $i = 1, 2, 3, \ldots, k$, we have the following:

- The tree diagram for the events is as follows:

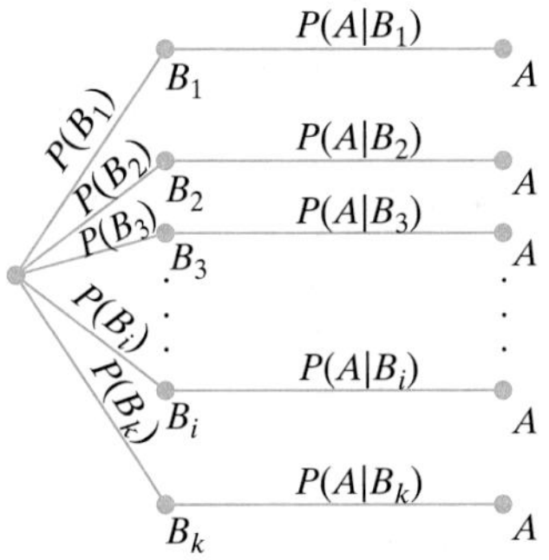

- $P(A) = P(B_1) \cdot P(A \mid B_1) + P(B_2) \cdot P(A \mid B_2) + \cdots + P(B_k) \cdot P(A \mid B_k)$
  $= \text{sum of branches}$
- For any value i where $1 \leq i \leq k$, we have

$$P(B_i \mid A) = \frac{P(B_i) \cdot P(A \mid B_i)}{P(B_1) \cdot P(A \mid B_1) + P(B_2) \cdot P(A \mid B_2) + \cdots + P(B_k) \cdot P(A \mid B_k)}$$

$$= \frac{i\text{th branch}}{\text{sum of branches}}$$

Now let's see how we can apply the Generalized Bayes' Theorem.

Example 4 Applying the Generalized Bayes' Theorem

In a September 1999 Harris poll, a random sample of 1009 Americans were asked about the aftermath of the fire at the Branch Dividian compound in

Waco, Texas. In the sample, 43% of the respondents identified themselves as Republicans, 36% as Democrats, and 21% as Independents. They were asked, "Do you think Attorney General Janet Reno should resign because of her handling of what happened in Waco?" The results showed that 33% of the Republicans, 10% of the Democrats, and 25% of the Independents said that she should resign. Let B_1, B_2, and B_3 represent the event that a randomly selected respondent was a Republican, Democrat, or Independent, respectively, and let A represent the event that the respondent said that Reno should resign. (*Source:* www.harrisinteractive.com.)

(a) Construct a tree diagram of the events.

(b) Compute and interpret $P(A)$.

(c) Compute and interpret $P(B_3 \mid A)$.

Solution

(a) **Understand the Situation:** Let's begin by organizing our information in tabular form as shown in Table 6.7.2. We can use the table to construct the tree diagram in Figure 6.7.4, where the percentages have been converted to decimal form.

Table 6.7.2

Party Affiliation	Percentage of Sample	Percentage Favoring Resignation
Republican	43%	33%
Democrat	36%	10%
Independent	21%	25%

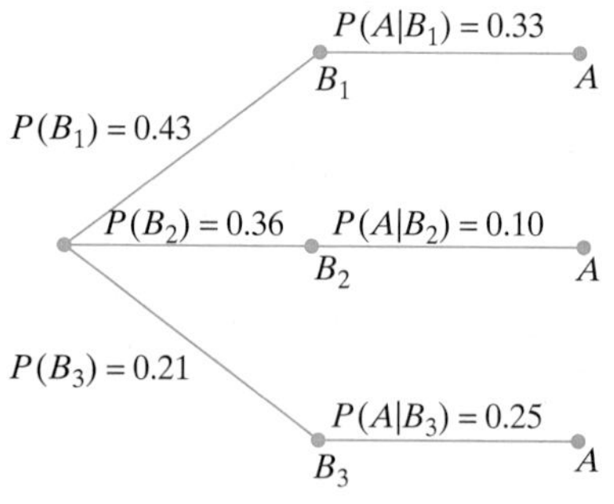

Figure 6.7.4

(b) To determine $P(A)$, we use the Generalized Bayes' Theorem and compute the sum of the three branches to get

$$\begin{aligned} P(A) &= (B_1) \cdot P(A \mid B_1) + P(B_2) \cdot P(A \mid B_2) + P(B_3) \cdot P(A \mid B_3) \\ &= (0.43)(0.33) + (0.36)(0.10) + (0.21)(0.25) \approx 0.230 \end{aligned}$$

Interpret the Solution: This means that if we select a respondent at random there is about a 23% probability that she or he feels that Attorney General Reno should resign.

(c) To determine $P(B_3 \mid A)$ we need to compute

$$P(B_3 \mid A) = \frac{\text{third branch}}{\text{sum of branches}}$$
$$= \frac{P(B_3)\cdot P(A \mid B_3)}{P(B_1)\cdot P(A \mid B_1) + P(B_2)\cdot P(A \mid B_2) + P(B_3)\cdot P(A \mid B_3)}$$
$$= \frac{(0.21)(0.25)}{(0.43)(0.33) + (0.36)(0.10) + (0.21)(0.25)} \approx 0.228$$

Interpret the Solution: This means that if we select a respondent at random the probability that the respondent is an Independent, given that she or he favored Reno's resignation, is about 23%. ■

SUMMARY

In this section we learned about the association between the probabilities $P(B \mid A)$ and $P(A \mid B)$. We then generalized this relationship with a theorem called Bayes' Theorem.

- **Bayes' Theorem for Two Branches** For events A, B, and B^C:

1. $P(A) = \text{sum of branches}$
$= P(B)\cdot P(A \mid B) + P(B^C)\cdot P(A \mid B^C)$

2. $$P(B \mid A) = \frac{\text{first branch}}{\text{sum of branches}} = \frac{P(B)\cdot P(A \mid B)}{P(B)\cdot P(A \mid B) + P(B^C)\cdot P(A \mid B^C)}$$

3. $$P(B^C \mid A) = \frac{\text{second branch}}{\text{sum of branches}} = \frac{P(B^C)\cdot P(A \mid B^C)}{P(B)\cdot P(A \mid B) + P(B^C)\cdot P(A \mid B^C)}$$

- The events $B_1, B_2, B_3, \ldots, B_k$ are mutually exclusive if any two of them are mutually exclusive.
- Events $B_1, B_2, B_3, \ldots, B_k$ are **collectively exhaustive** if $B_1 \cup B_2 \cup B_3 \cup \cdots \cup B_k = S$
- **Generalized Bayes' Theorem** For mutually exclusive and collectively exhaustive events $B_1, B_2, B_3, \ldots, B_k$ and for any event A:

1. $P(A) = P(B_1)\cdot P(A \mid B_1) + P(B_2)\cdot P(A \mid B_2) + \cdots + P(B_k)\cdot P(A \mid B_k)$
2. For any value i, where $1 \le i \le k$, we have

$$P(B_i \mid A) = \frac{P(B_i)\cdot P(A \mid B_i)}{P(B_1)\cdot P(A \mid B_1) + P(B_2)\cdot P(A \mid B_2) + \cdots + P(B_k)\cdot P(A \mid B_k)} = \frac{i\text{th branch}}{\text{sum of branches}}$$

SECTION 6.7 EXERCISES

In Exercises 1–4, a table of data is given. Compute the values of $P(D \mid B)$ and $P(B \mid D)$.

1.

	A	B	Total
C	10	20	30
D	15	5	20
Total	25	25	50

2.

	A	B	Total
C	40	60	100
D	55	45	100
Total	95	105	200

✓ **3.**

	A	B	Total
C	26	33	59
D	18	48	66
Total	44	81	125

4.

	A	B	Total
C	87	52	139
D	61	50	111
Total	148	102	250

In Exercises 5–8, a table of probabilities is given. Compute the values of $P(C \mid A)$ and $P(A \mid C)$.

5.

	A	B	Total
C	$\frac{1}{4}$	$\frac{1}{4}$	$\frac{1}{2}$
D	$\frac{1}{8}$	$\frac{3}{8}$	$\frac{1}{2}$
Total	$\frac{3}{8}$	$\frac{5}{8}$	1

6.

	A	B	Total
C	$\frac{1}{10}$	$\frac{3}{10}$	$\frac{2}{5}$
D	$\frac{1}{5}$	$\frac{2}{5}$	$\frac{3}{5}$
Total	$\frac{3}{10}$	$\frac{7}{10}$	1

7.

	A	B	Total
C	0.22	0.15	0.37
D	0.42	0.21	0.63
Total	0.64	0.36	1.00

8.

	A	B	Total
C	0.17	0.24	0.41
D	0.20	0.39	0.59
Total	0.37	0.63	1.00

9. The number of doctorates conferred in the social sciences (anthropology, sociology, economics, and political science) and the physical sciences (astronomy, physics, and chemistry) during 1998 is shown in the following table. The data are classified as being a U.S. citizen or a foreign-born citizen.

	Social Sciences (A)	Physical Sciences (B)	Total
U.S. citizen (C)	2353	2415	4768
Foreign-born citizen (D)	760	1111	1871
Total	3113	3526	6639

SOURCE: U.S. Census Bureau, www.census.gov

For the events labeled in the table as A, B, C, and D, compute and interpret the following probabilities:

(a) $P(D \mid B)$

(b) $P(B \mid D)$

10. In 1996, the percentage of American adults who exercised for 30 minutes or more at least 5 times per week is shown in the table. The percentages are classified by gender.

	Did Not Exercise (A)	Exercised Regularly (B)	Total
Men (C)	22.0%	25.6%	47.6%
Women (D)	25.6%	26.8%	52.4%
Total	47.6%	52.4%	100%

SOURCE: National Center for Chronic Disease Prevention, www.cdc.gov/nccdphp

For the events labeled in the table as A, B, C, and D, compute and interpret the following probabilities:

(a) $P(C \mid A)$

(b) $P(A \mid C)$

In Exercises 11–16, a tree diagram is given. Determine $P(A)$, $P(B \mid A)$, and $P(B^C \mid A)$.

11.

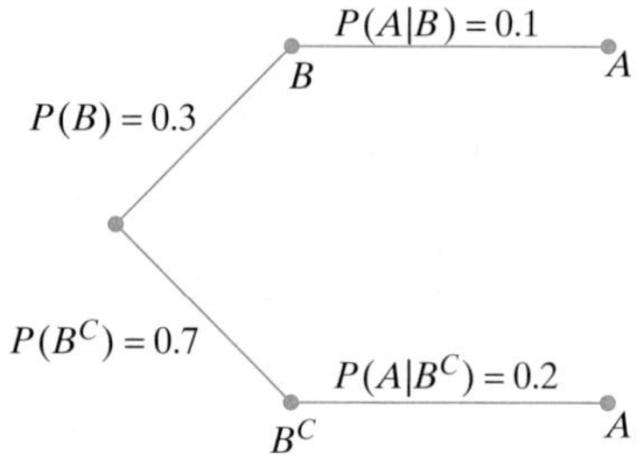

12.

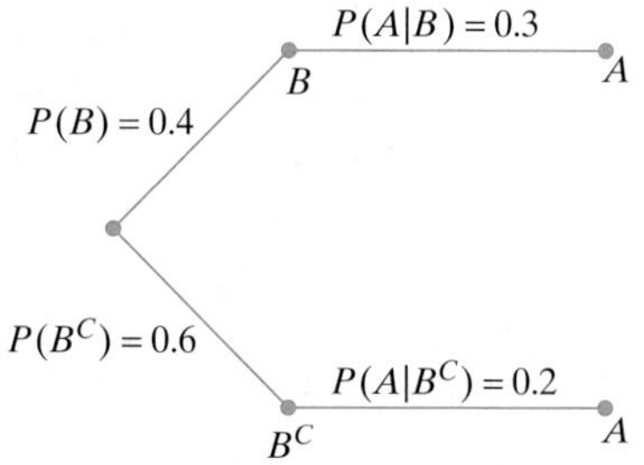

13.

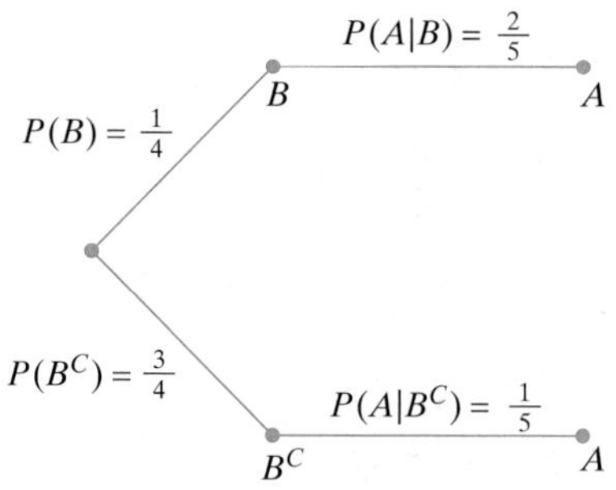

14.

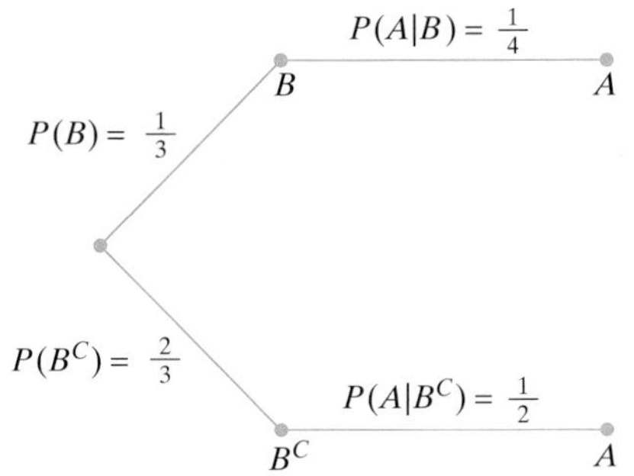

15.

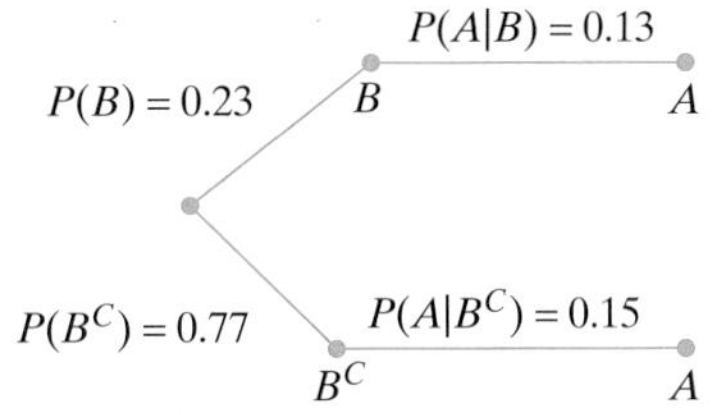

16.

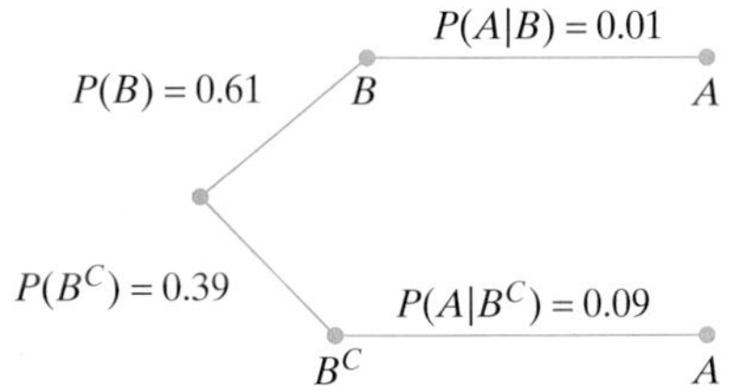

17. The Garforth Company runs a day and a night shift at its factory. Of the 500 workers, 300 work the day shift and the remainder work the night shift. When asked whether they would like to have a paid holiday for St. Patrick's Day or for Columbus Day next year, 150 of the workers on the day shift favored St. Patrick's Day and 150 of the workers on the night shift favored St. Patrick's Day. A union leader selects a worker at random. Let B represent the event that the worker is on the day shift, B^C represent the event that the worker is on the night shift, and A represent the event that the worker favored a paid holiday for St. Patrick's Day.

(a) Determine $P(A)$ and interpret.

(b) Determine $P(B \mid A)$ and interpret.

18. The Clanford Mk V device is made at two factories. The northern plant manufactures 40% of the devices and the southern plant manufacturers the remainder of the devices. Records show that 2% of the devices from the northern plant are defective and 6% of the devices from the southern plant are defective. A quality control engineer selects a Clanford Mk V device at random. Let B present the event that the device came from the northern plant, B^C represent the event that the device came from the southern plant, and A represent the event that the device is defective.

(a) Determine $P(A)$ and interpret.

(b) Determine $P(B \mid A)$ and interpret.

19. At Ketterson College, 66% of the student body is underclass persons and the remainder is upperclass persons. In a survey, 33% of the underclass persons favored switching from quarters to semesters and 12% of the upperclass persons favored switching from quarters to semesters. A prospective student visiting the campus selects a Ketterson College student at random. Let B represent the event that the selected student is an underclass person, B^C represent the event that the selected student is an upperclass person, and A represent the event that the selected student favored switching from quarters to semesters.

(a) Determine $P(A)$ and interpret.

(b) Determine $P(B \mid A)$ and interpret.

20. In a May 2000 Gallup poll, 54% of the respondents said that they had money invested in the stock market and the remaining respondents did not. They were asked the question, "If you had \$1000 to spend, do you think investing it in the stock market would be a good idea." Of the stockholders, 81% said that it was a good idea and 45% of the nonstockholders said that it was a good idea. A national money magazine selects a respondent at random. Let B represent the event that the respondent was a stockholder and let A represent the event that he or she thought that investing \$1000 in the stock market would be a good idea. (*Source:* www.gallup.com.)

(a) Determine $P(A)$ and interpret.

(b) Determine $P(B \mid A)$ and interpret.

(c) Determine $P(B^C \mid A)$ and interpret.

21. In a March 2000 Gallup poll, 26% of the respondents called themselves fans of professional golf and the remaining respondents did not. When asked who they felt was the greatest active golfer, 81% of the golf fans replied that Tiger Woods was the greatest, while 64% of those who said they were not a fan said that Tiger Woods was the greatest. A national golf magazine selects a respondent at random. Let B represent the event that the respondent was a golf fan and let A represent the event that he or she felt that Tiger Woods was the greatest active golfer. (*Source:* www.gallup.com.)

(a) Determine $P(A)$ and interpret.

(b) Determine the probability that the respondent is a golf fan given that he or she felt that Tiger Woods is the greatest active golfer.

(c) Determine $P(B^C \mid A)$ and interpret.

22. In a November 1999 Gallup poll, 45% of the respondents said that they were baseball fans and the remaining respondents said that they were not. Former Major League player Pete Rose is ineligible for baseball's Hall of Fame due to charges that he bet on Major League baseball games. When asked if Rose should be eligible for the Hall of Fame, 62% of the baseball fans said that he should be eligible, while 57% of those who were not baseball fans said that he should be eligible. Select a respondent at random and let B represent the event that the respondent was a baseball fan and A represent the event that the respondent said that Rose should be eligible for the Hall of Fame. (*Source:* www.gallup.com.)

(a) Determine $P(A)$ and interpret.

(b) Determine the probability that the respondent is a baseball fan given that the respondent felt that Rose should be eligible for the Hall of Fame.

(c) Determine $P(B^C \mid A)$ and interpret.

23. In a November 1999 Gallup poll, 45% of the respondents said that they were baseball fans and the remaining respondents said that they were not. Former Major League player Shoeless Joe Jackson is ineligible for baseball's Hall of Fame due to charges that he took money from gamblers in exchange for fixing the 1919 World Series. When asked if Jackson should be eligible for the Hall of Fame, 44% of the baseball fans said that he should be eligible, while 33% of those who were not baseball fans said that he should be eligible. A national sports magazine selects a respondent at random. Let B represent the event that the respondent was a baseball fan and let A represent the event that the respondent said that Jackson should be eligible for the Hall of Fame. (*Source:* www.gallup.com.)

(a) Determine $P(A)$ and interpret.

(b) Determine the probability that the respondent is a baseball fan, given that the respondent felt that Jackson should be eligible for the Hall of Fame.

(c) Determine $P(B^C \mid A)$ and interpret.

24. In a July 1999 Gallup poll, half of the respondents called themselves a regular viewer of the Cable News Network (CNN) and the other half said that they were not regular viewers. When asked if they were interested in the results of polls about political campaigns and elections, 50% of the CNN regular viewers said that they were very interested, while 39% of those who were not regular CNN viewers said that they were very interested. A reporter selects a respondent at random. Let B represent the event that the respondent was a regular CNN viewer and let A represent the event that the respondent said that she or he was very interested in the results of polls about political campaigns and elections. (*Source:* www.gallup.com.)

(a) Determine $P(A)$ and interpret.

(b) Determine the probability that the respondent is a regular CNN viewer, given that the respondent said that she or he is very interested in the results of polls about political campaigns and elections.

(c) Determine $P(B^C \mid A)$ and interpret.

25. In a February 2000 poll, a group of adults were asked if they ate organic foods on a regular basis and if they were willing to spend higher prices for organic foods. The results are summarized in the following table:

Eat Organic Foods Regularly	Percent	Willing to Pay Higher Prices for Organic Food
Yes	46	65%
No	54	34%

SOURCE: www.abcnews.go.com

Select a respondent at random and let B represent the event that the respondent is a regular eater of organic foods and A represent the event that the respondent is willing to pay higher process for organic foods.

(a) Determine $P(A)$ and interpret.

(b) Determine the probability that the respondent is a regular eater of organic foods, given that the respondent is willing to pay higher prices for organic foods.

(c) Determine $P(B^C \mid A)$ and interpret.

26. In a September 2000 poll, a group of adults were asked if they were gun owners and if the National Rifle Association (NRA) had too much influence on the issue of gun control. The results are summarized in the following table:

Gun Owner	Percent	Too Much NRA Influence
Yes	29	31%
No	71	55%

SOURCE: www.abcnews.go.com

Select a respondent at random and let B represent the event that the respondent was a gun owner and A represent the event that the respondent believed that the NRA had too much influence on the issue of gun control.

(a) Determine $P(A)$ and interpret.

(b) Determine the probability that the respondent is a gun owner, given that the respondent believes that the NRA has too much influence on the issue of gun control.

(c) Determine $P(B^C \mid A)$ and interpret.

27. In a fall 1998 survey, a sample of adults was asked if they were either parents or teachers and if they felt that kids whose parents are not involved in school sometimes get shortchanged. The results are summarized in the following table:

Group	Percent	Kids Get Shortchanged
Parents	55	72%
Teachers	45	48%

SOURCE: www.publicagenda.org

Select a respondent at random and let B_1 represent the event that the respondent is a parent, B_2 represent the event that the respondent is a teacher, and A represent the event that the respondent felt that kids whose parents are not involved in school sometimes get shortchanged.

(a) Determine $P(A)$ and interpret.

(b) Determine the probability that the respondent is a parent, given that the respondent felt that kids whose parents are not involved in school sometimes get shortchanged.

(c) Determine $P(B_2 \mid A)$ and interpret.

In Exercises 28–33, a tree diagram is given. Determine if the events are collectively exhaustive.

28.

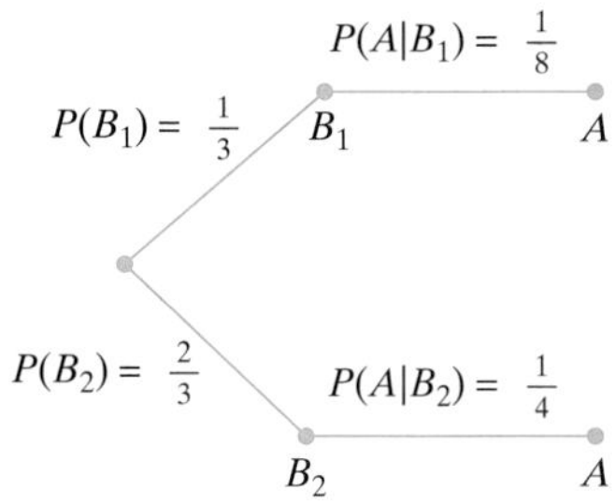

29.

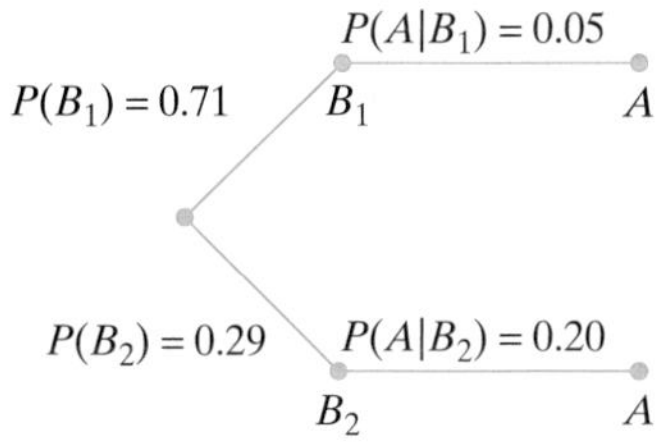

30.

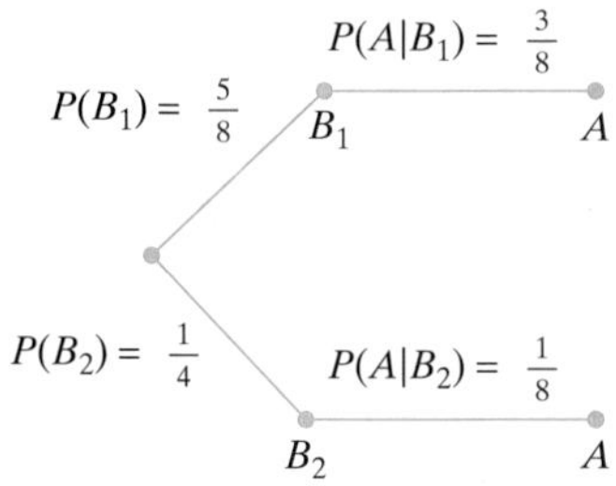

31.

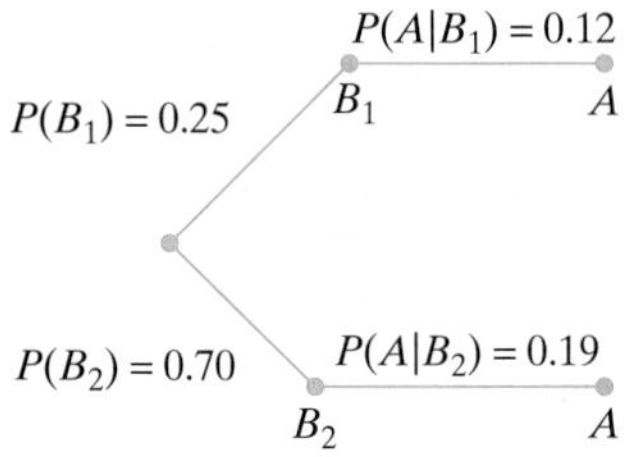

32.

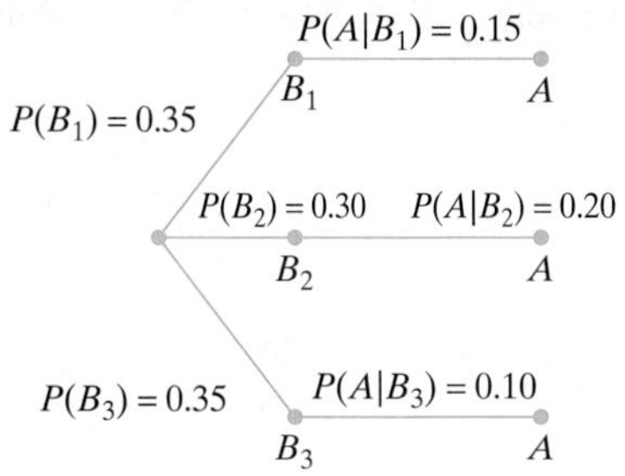

33.

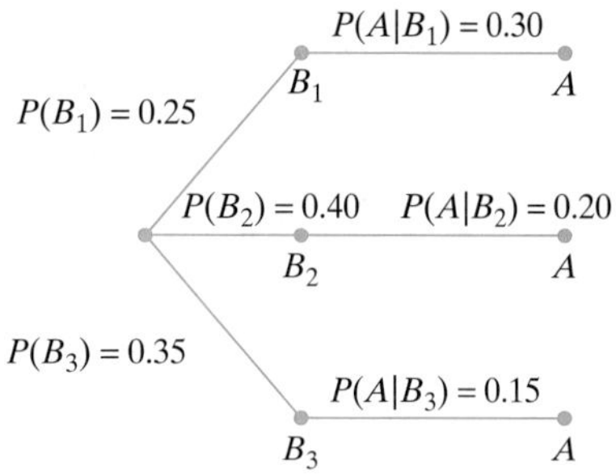

In Exercises 34–39, a tree diagram is given. Determine $P(A)$, $P(B_1 \mid A)$, and $P(B_3 \mid A)$.

34.

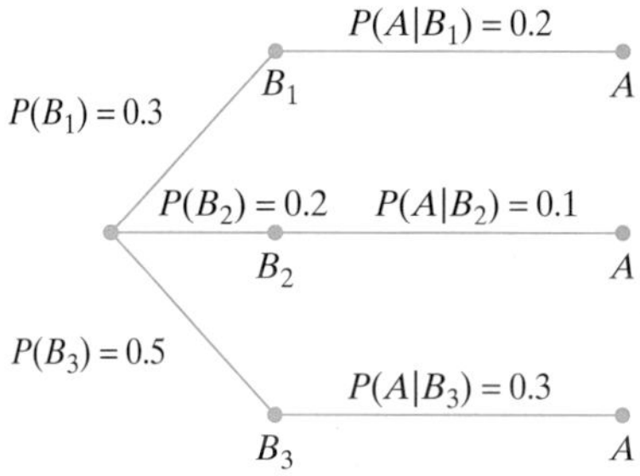

35.

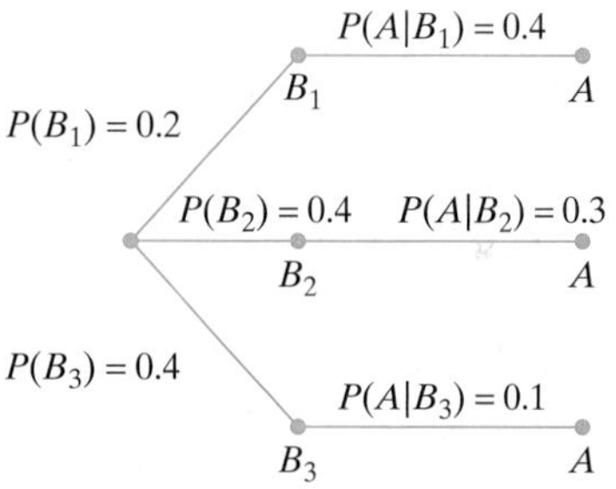

36.

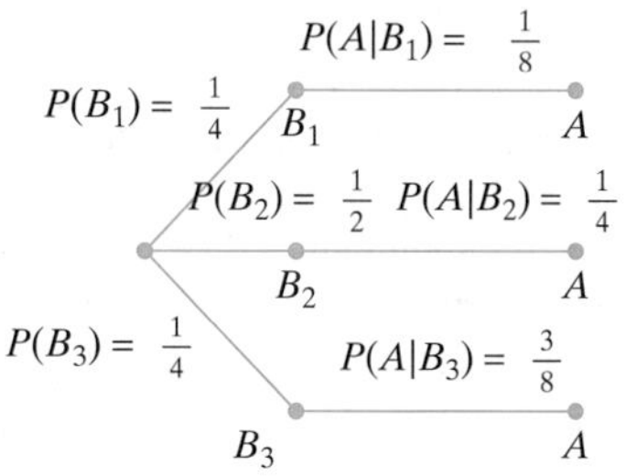

37.

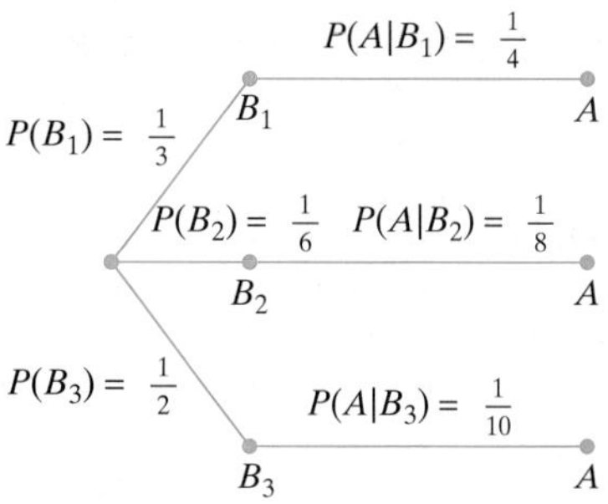

38.

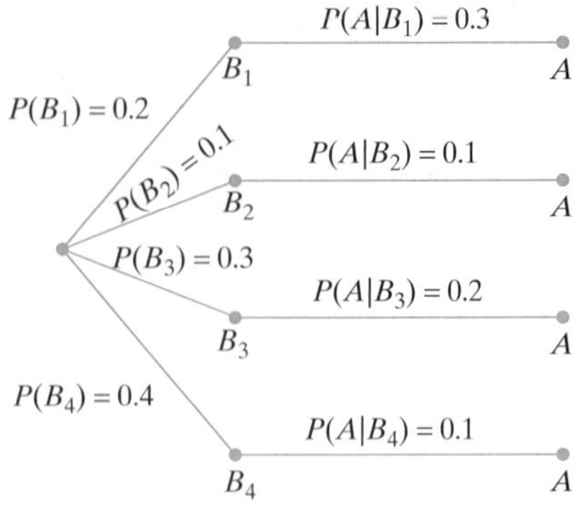

39.

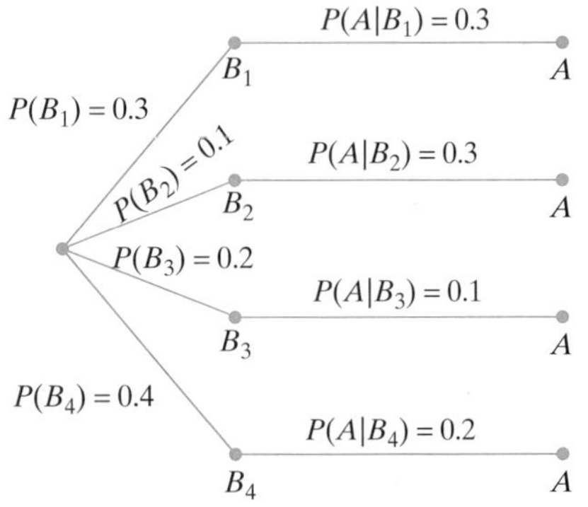

40. In Binfort County, the local election board keeps records of the number of registered voters from each party for school board elections. The voter turnout for the last school board election is shown in the table.

Party Affiliation	Percent of Electorate	Percent Turnout
Republican	35	20
Democrat	40	25
Independent	25	10

A newspaper reporter selects a registered voter at random. Let B_1, B_2, and B_3 represent the event that the voter is a Republican, Democrat, or Independent, respectively, and let A represent the event that the voter turned out to vote for the last school board election.

(a) Construct a tree diagram.

(b) Determine $P(A)$ and interpret.

(c) Determine the probability that the voter is a Republican, given that the voter turned out to vote for the last school board election.

(d) Determine $P(B_2 \mid A)$ and interpret.

41. The Juniper Company makes light fixtures at three factories. From past records they know the proportion of output from each factory and the percentage of defects that come from each factory. This information is shown in the table.

Factory	Percent Output	Percent Defects
I	40	1
II	25	2
III	35	4

A quality control engineer selects a Juniper Company light fixture at random. Let B_1, B_2, and B_3 represent the events that the light fixture comes from factory I, factory II, or factory III, respectively, and let A represent the event that the light fixture is defective.

(a) Construct a tree diagram.

(b) Determine $P(A)$ and interpret.

(c) Determine the probability that the light fixture came from factory II, given that the light fixture is defective.

(d) Determine $P(B_3 \mid A)$ and interpret.

42. In the United States in 1998, the income of parents and the percentage of children under 18 living with their father only is shown in the table.

Income	Percent of Population	Percent Living with Father Only
Under \$25,000 (B_1)	17	6
\$25,000 to \$40,000 (B_2)	30	7
Over \$40,000 (B_3)	53	3

SOURCE: U.S. Census Bureau, www.census.gov

A researcher selects an American child under 18 at random. Let B_1, B_2, and B_3 represent the events labeled in the table and A represent the event that a child is living alone with his or her father.

(a) Construct a tree diagram.

(b) Determine $P(A)$ and interpret.

(c) Determine the probability that parental income is under \$25,000, given that the child is living alone with his or her father.

(d) Determine $P(B_3 \mid A)$ and interpret.

43. In the United States in 1995, the percentage of women from 15 to 44 years of age and the percentage of women who had never been married or cohabited is shown in the table.

Age	Percent of Study	Percent Never Married or Cohabited
15 to 24 (B_1)	30	67
25 to 34 (B_2)	34	16
35 to 44 (B_3)	36	6

SOURCE: U.S. Census Bureau, www.census.gov

A medical researcher selects an American woman aged 15 to 44 at random. Let B_1, B_2, and B_3 represent the events labeled in the table and A represent the event that a woman had never been married or cohabited.

(a) Construct a tree diagram.

(b) Determine $P(A)$ and interpret.

(c) Determine the probability the woman is from 25 to 34 years of age, given that the woman has never been married or cohabited.

(d) Determine $P(B_3 \mid A)$ and interpret.

SECTION PROJECT

Some applications of the Bayes Theorem can be somewhat daunting. For example, the following table shows the percent of remarriages of women in the United States for several age groups.

Age	Percent of Population	Percent remarried
Under 20 (B_1)	21.1	1.7
20 to 24 (B_2)	37.1	15.3
25 to 29 (B_3)	18.7	26.7
30 to 34 (B_4)	9.3	20.6
35 to 44 (B_5)	7.8	20.8
45 to 64 (B_6)	5	14.3
65 and over (B_7)	1	2.9

SOURCE: U.S. Census Bureau, www.census.gov

A medical researcher selects an American woman at random. Let B_1, B_2, B_3, B_4, B_5, B_6, and B_7 represent the events labeled in the table, and let A represent the event that the woman has been remarried. Use the BAYES calculator program in Appendix C in answering parts (b) through (e).

(a) Construct a tree diagram.

(b) Determine $P(A)$ and interpret.

(c) Determine $P(B_4 \mid A)$ and interpret.

(d) Determine $P(B_5 \mid A)$ and interpret.

(e) Determine $P(B_7 \mid A)$ and interpret.

Why We Learned It

In this chapter, we learned the fundamentals of counting and the fundamentals of probability. State lottery officials use the counting technique combination to help determine rules for lottery games. We saw that many researchers use probability to help understand the nature of a sample if some information about its population is known. For example, researchers use probability theory to study diverse areas as stock market fluctuations, information theory, coding theory, and weather forecasting just to name a few. Quality control managers use probability to help determine if a given item has a defect. These are just a few of the ways in which probability can be used.

CHAPTER REVIEW EXERCISES

In Exercises 1 and 2, classify each of the following as an example of probability or statistical inference.

1. An April 2000 Gallup poll asked a random sample of 1004 adults if they had voluntarily recycled newspapers, glass, aluminum, motor oil, or other items in the past year; 90% said that they had voluntarily recycled these items in the past year. (*Source:* www.publicagenda.org.)

2. In a 1998 report on the American teen-age public, 72% of American teen-agers said that they can always trust their parents to be there for them when needed. In a sample of 30 American teen-agers selected at random, 23 said that they could trust their parents to be there when needed. (*Source:* www.publicagenda.org.)

In Exercises 3–5, determine whether each of the following processes represents a probability experiment.

3. A teacher evaluation asks students to rate their teacher's energy level on a scale from 1 to 5, with 1 being lethargic and 5 being tireless.

4. A random sample of 2010 people were asked, "If you could live anywhere in the world, where would it be?"

5. A survey asked respondents to fill in their political party affiliation as Republican, Democratic, or Independent.

In Exercises 6–8, determine whether order is important.

6. There are 24 students who attend a field trip to a science museum. When computing in how many ways the teacher can select a pilot, copilot, and navigator for a flight simulator, is order important?

7. A music festival has 8 different musicians playing. When computing in how many different ways the musicians can be first, second, and third to take the stage, is order important?

8. From a group of 8 backpackers, 2 are to be chosen to guide the group on their backpacking trip in Escalante, Utah. When computing in how many different ways the team of backpackers can choose their leaders, is order important?

In Exercises 9–11, determine if the following events can happen at the same time.

9. A respondent from a nationwide survey is selected at random for an interview. The event that the respondent was born in Michigan and New Hampshire.

10. A resident of Colorado Springs, Colorado, is selected at random. The event that the resident is a student and has hiked Pikes Peak.

11. A spectator at a soccer match is selected at random. The event that the person is seated in section A and the event that the person is seated in section H.

In Exercise 12, an event is given.

(a) Write another event so that together the events cannot happen at the same time.

(b) Write another event so that together the events can happen at the same time.

12. The event that a student is a senior.

In Exercises 13 and 14, determine if the outcome of one event influences the outcome of the other event.

13. A box contains 5 blue chips and 4 red chips. In an experiment, a chip is drawn at random, its color is recorded, and it is replaced. Then a second chip is drawn randomly. The event that the first chip is blue and that the second chip is blue.

14. The event that Rebecca's car breaks down and the event that Rebecca rides her bike to work the next day.

In Exercise 15, an event is given.

(a) Write another event so that together the outcome of one event does not influence the outcome of another event.

(b) Write another event so that together the outcome of one event influences the outcome of the other event.

15. On a science exam, the event that Ron has not prepared.

In Exercises 16 and 17, rewrite the set using the listing method.

16. $A = \{x|x \text{ is a prime number and } 1 \leq x \leq 20\}$

17. $B = \{x|x \text{ is a state that starts with the word "New"}\}$

In Exercises 18 and 19, rewrite the set using set-builder notation.

18. $A = \{2, 4, 6, 8, 10\}$

19. $B = \{\text{Alabama, Alaska, Arizona, Arkansas}\}$

In Exercises 20 and 21, write all possible subsets for the given set.

20. $\{10, 12\}$

21. {McKinley, Pikes Peak, Washington}

22. Let $A = \{x|x$ is a Review Exercise that is an example of probability and $1 \leq x \leq 2\}$ and $B = \{x|x$ is a Review Exercise that is an example of statistical inference and $1 \leq x \leq 2\}$.

(a) Rewrite sets A and B using the listing method.

(b) Find $A \cap B$.

(c) Are A and B disjoint? Explain.

In Exercises 23–25, let $U = \{x|x$ is a whole number and $16 \leq x \leq 30\}$, $A = \{16, 17, 18, 19, 20\}$, $B = \{17, 19, 21, 23, 25, 27, 29\}$, and $C = \{26, 27, 28, 29, 30\}$. Write each of the following sets using the listing method.

23. $A \cap B$

24. $A \cap C^C$

25. $(A \cup C)^C$

In Exercises 26–28, use the Venn diagram to write the desired set.

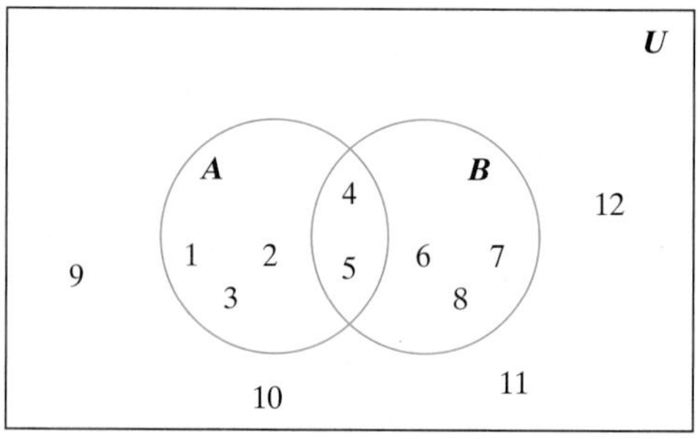

26. B

27. $A^C \cap B$

28. $(A \cap B)^C$

29. Let $U = \{1, 2, 3, 4, 5, 6\}$, $A = \{1, 3, 5\}$, $B = \{1, 2, 3\}$, and $C = \{2, 3, 4, 5\}$. Write a set and draw a shaded Venn diagram for $A^C \cap (B \cup C)$.

In Exercises 30 and 31, the number of students enrolled in public and private schools, in thousands, in the United States in grades 9 to 12 and college in 1997 is shown in the table.

Type of School	Grades 9–12 (*A*)	College (*B*)	Total
Public (*C*)	13,054	11,146	24,200
Private (*D*)	1,308	3,199	4,507
Total	14,362	14,345	28,707

SOURCE: U.S. National Center for Education Statistics, www.nces.ed.gov

Here A represents the set of students in grades 9 to 12 at public and private schools, B represents the set of students in college at public and private schools, C represents the set of students in grades 9 to 12 and college at public schools, and D represents the number of students in grades 9 to 12 and college at private schools. Determine the following and interpret each.

30. $n(D^C)$ **31.** $n(A \cup C)$

32. A three-question phone survey of registered voters asks the following questions:

Question 1: Do you worry about the loss of open spaces and natural habitats for wildlife?

Question 2: Do you favor or oppose President Bush's decision to drill in the Arctic National Wildlife Refuge?

Question 3: Are you registered Republican, Democrat, or Independent?

Draw a tree diagram to represent all possible answers in the survey. How many ways could a respondent answer all three questions?

In Exercises 33–35, determine the following:

33. A new sports pack feature offered through a ticket agency allows you to construct a package deal that includes one ticket for a professional soccer game and one ticket for a professional football game. If there are 33 games in the soccer season and 16 games in a football season, how many ways can a person construct a package deal?

34. A password consists of 4 letters and 3 digits. If letters and digits can be repeated, in how many different ways can a password be created?

35. A science quiz consists of ten true–false questions. In how many different ways can a student answer all the questions?

In Exercises 36 and 37, evaluate the expression without using a calculator.

36. 6! **37.** ${}_6P_2$

In Exercises 38 and 39, use a calculator to determine the following:

38. ${}_{35}P_{15}$ **39.** ${}_{60}P_{10}$

In Exercises 40 and 41, evaluate the expression without using a calculator.

40. ${}_6C_3$ **41.** ${}_8C_4$

In Exercises 42 and 43, use a calculator to determine the following:

42. ${}_{40}C_{20}$ **43.** ${}_{60}C_{40}$

44. There are 24 students who attend a field trip to a science museum. In how many different ways can the teacher select a pilot, copilot, and navigator for a flight simulator?

45. From a group of 8 backpackers, 2 are to be chosen to guide the group on their backpacking trip in Escalante, Utah. In how many different ways can the team of backpackers choose their leaders?

46. A music festival has 8 different musicians playing. In how many different ways can the musicians be first, second, and third to take the stage?

47. In a shipment of 40 CD players, 6 are known to be defective. In how many ways can a sample of 8 be chosen so that 3 are defective and 5 are not defective?

48. A city council has 6 Democrats, 5 Republicans, and 3 Independents. In how many different ways can a subcommittee be formed with 3 Democrats, 2 Republicans, and 1 Independent?

In Exercises 49–51, write the sample space S using set notation and then determine n(S).

49. A box contains 12 cards labeled *A* through *L*. In an experiment, a card is drawn out and its letter is recorded.

50. A box contains 1 yellow, 1 red, and 1 blue marble. A bag contains a ticket labeled with a 1 and a ticket labeled with a 3. In an experiment, one draw is made from the box and one draw is made from the bag and the outcome is recorded.

51. A pollster asks the question, "Have you ever received a student loan?" The pollster then asks, "Did you attend a public or private school?"

In Exercises 52–55, use classical probability to compute P(A) and P(B) for the indicated events A and B.

52. For the experiment in Exercise 50, one draw is made from the box and one draw is made from the bag and the outcome is recorded. Let *A* represent the event that a yellow marble is drawn and *B* represent the event that a ticket labeled with a 3 is drawn.

53. In an experiment, Rebecca performs a pH test on a sample of drinking water and records the result as acidic, neutral, or basic. She then performs a nitrate test on the same sample of water and records the result as normal or excessive. Let *A* represent the event that the water is acidic and *B* represent the event that the water is neutral and the nitrate level is normal.

54. For the experiment in Exercise 51, a pollster asks two questions and records the results. Let *A* represent the event that the respondent answered "yes" to question 1 and "private" to question 2 and *B* represent the event that the respondent answered "no" to question 1.

55. In an experiment, Chad tosses a six-sided die and records the number. He tosses it again and records the number from the second toss. Let *A* represent the event that the numbers on each toss are the same and *B* represent the event that the sum of the two tosses is a 6.

56. A fruit basket contains 10 oranges and 10 apples. Kris takes two pieces of fruit from the basket without looking at what kind they are.

(a) Determine $n(S)$.

(b) If A represents the event that both pieces of fruit are apples, determine $n(A)$.

(c) Use classical probability to determine $P(A)$ and interpret.

57. The city council in Platt City is comprised of 6 Democrats and 8 Republicans.

(a) In how many different ways can a subcommittee of 5 council persons be formed?

(b) In how many different ways can a subcommittee be formed with 3 Democrats and 2 Republicans?

(c) If A represents the event that a subcommittee is formed with 3 Democrats and 2 Republicans, determine $P(A)$ and interpret.

58. In a shipment of 40 pairs of waders, 6 are known to leak.

(a) In how many different ways can a sample of 8 pairs of waders be chosen?

(b) In how many different ways can a sample of 8 be chosen so that 3 pairs of waders leak and 5 do not leak?

(c) If A represents the event that in a sample of 8 pairs of waders 3 leak and 5 do not leak, determine $P(A)$ and interpret.

59. The number of workers 16 years old or older in the United States and their means of transportation to work in 2000 is shown in the table.

Transportation	Number (in millions)
Car, truck, or van (drove alone)	97.2
Car, truck, or van (carpooled)	14.3
Public transportation	6.6
Walked	3.4
Other means	1.8
Worked at home	4.1

SOURCE: U.S. Census Bureau, www.census.gov

In 2000, a working individual 16 years old or older was selected at random.

(a) Determine the probability that a working individual 16 years old or older drives alone to work in a car, truck, or van.

(b) Determine the probability that a working individual 16 years old or older does not drive alone to work in a car, truck, or van.

In Exercises 60 and 61, classify the following as an example of classical, empirical, *or* subjective *probability.*

60. A local weatherman uses the fact that it has rained 35 days out of 50 days that have shown the same weather patterns as today to state that there is a 70% chance of rain today.

61. The probability of rolling a 4 on one toss of a die is $\frac{1}{6}$.

In Exercises 62 and 63, determine if the events A and B are mutually exclusive.

62. A respondent from a nationwide survey is selected at random for an interview. A represents the event that the respondent was born in Michigan and B represents the event that the respondent was born in New Hampshire.

63. A resident of Colorado Springs, Colorado, is selected at random. A represents the event that the resident is a student and B represents the event that the resident has hiked Pikes Peak.

64. A box contains 12 tickets numbered 1 to 12. If a draw is made at random, determine the probability that the ticket is labeled with a 1 or an even number.

65. A bowl of cereal contains 15 red, 8 green, 14 yellow, and 10 orange pieces. If a piece of cereal is selected at random from the bowl, determine the probability that a red or orange piece is drawn.

For Exercises 66–71, consider a survey of a group of 400 students at Osceola College. Let A represent the event that the student has season football tickets and B represent the event that the student is an upperclassperson. The results are shown in the Venn diagram. Assuming that a student in the survey group is selected at random, use the diagram to answer the following:

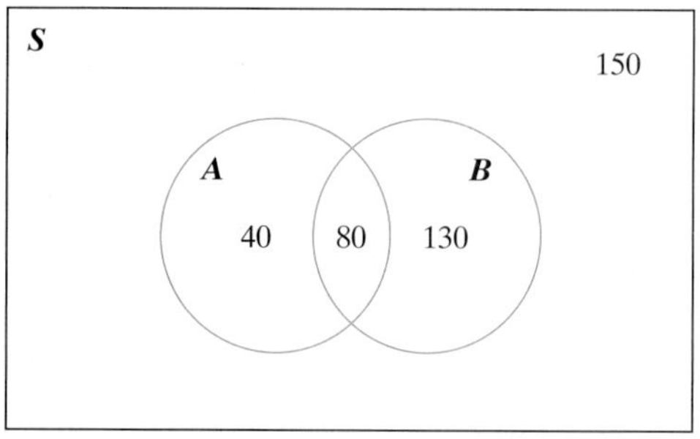

66. Determine $P(A)$ and interpret.

67. Determine $P(B)$ and interpret.

68. Determine $P(A \cup B)$ and interpret.

69. Determine $P(A \cap B)^C$ and interpret.

70. Determine the probability that the student has season football tickets and is an upperclassperson.

71. Determine the probability that the student does not have season football tickets and is not an upperclassperson.

For Exercises 72–74, consider the following. In 1998, the percentage of persons living in counties in California or Texas that do not meet U.S. Environmental Protection

Agency air quality standards is shown in the following table.

	Meeting Standards, % (*A*)	Not Meeting Standards, % (*B*)	Total (%)
California (C)	6.7	55.7	62.4
Texas (D)	18.8	18.8	37.6
Total	25.5	74.5	100

SOURCE: National Center of Chronic Disease Prevention, www.cdc.gov, U.S. Census Bureau, www.census.gov

72. Determine $P(A)$ and interpret.

73. Determine the probability that a randomly selected person from California or Texas is living in California and in a county that is not meeting the U.S. Environmental Protection Agency air quality standards.

74. Determine the probability that a randomly selected person from California or Texas is living in Texas or in a county that is not meeting U.S. Environmental Protection Agency air quality standards.

75. The probability of event A occurring is 0.4.

(a) Determine the odds in favor of A occurring.

(b) Determine the odds against A occurring.

76. The odds in favor of event A occurring are 7 to 2. Determine the probability of A occurring.

77. Determine the odds in favor of a prime number on a single roll of a ten-sided die containing the whole numbers 1 to 10.

78. Determine the odds against a number divisible by 3 on a single roll of a ten-sided die containing the whole numbers 1 to 10.

79. As of July 11, 2001, the odds in favor of the Detroit Red Wings winning the 2002 Stanley Cup are listed as 1 to 5. Determine the probability that the Detroit Red Wings will win the 2002 Stanley Cup. (*Source:* www.vegasinsider.com/u/futures.)

In Exercises 80 and 81, determine if events A and B are independent or dependent. Explain your reasoning.

80. A box contains 5 blue chips and 4 red chips. In an experiment, a chip is drawn at random, its color is recorded, and it is replaced. Then a second chip is drawn randomly. Let A represent the event that the first chip is blue and B represent the event that the second chip is blue.

81. A is the event that Rebecca's car breaks down and B is the event that Rebecca rides her bike to work the next day.

82. A March 2001 Gallup poll showed that 18% of adult Americans fear flying on an airplane. If an airline chooses 4 adult Americans at random to receive a free promotional flight, what is the probability that all 4 fear flying? (*Source:* www.publicagenda.org.)

83. In a sample of 25 students, 14 said that they studied at least 3 hours for their Biology exam, 18 said that they received a grade of 80% or higher, and 12 said that they studied at least 3 hours and received a grade of 80% or higher. Let A represent the event that a student studied at least 3 hours and B represent the event that a student received a grade of 80% or higher.

(a) Draw a Venn diagram to represent the students sampled with events A and B.

(b) Compute $P(A \cap B)$ and interpret.

(c) Determine $P(B \mid A)$ and interpret.

84. Use the probability Venn diagram to determine $P(A \cap B)$ and $P(B \mid A)$.

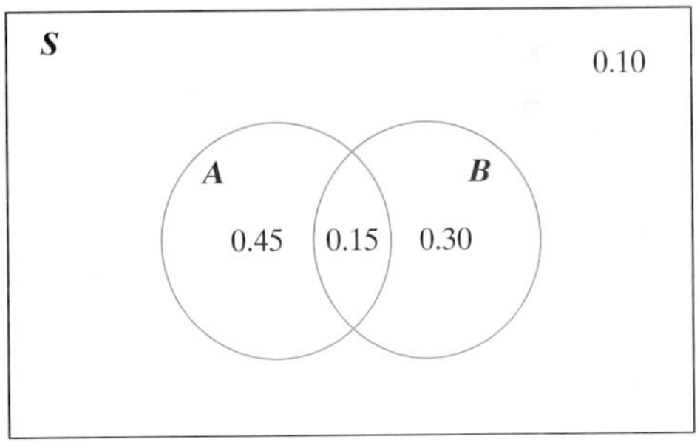

85. Use the table of data to determine $P(A \cap C)$ and $P(C \mid A)$.

	A	*B*	Total
C	15	20	35
D	25	45	70
Total	40	65	105

86. Use the table of probabilities to determine $P(B \cap D)$ and $P(D \mid B)$.

	A	*B*	Total
C	0.34	0.18	0.52
D	0.28	0.20	0.48
Total	0.62	0.38	1.00

87. A bag contains 6 yellow and 8 blue marbles. In an experiment, a marble is drawn at random, its color is recorded, and a second marble is drawn randomly without replacing the first.

(a) Determine the probability that the first marble is blue and the second marble is yellow.

(b) Determine the probability that both marbles are yellow.

For Exercises 88–92, consider the following: In 1998, the percentage of persons living in counties in California or Texas that do not meet U.S. Environmental Protection Agency air quality standards is shown in the following table:

	Meeting Standards, % (A)	Not Meeting Standards, % (B)	Total (%)
California (C)	6.7	55.7	62.4
Texas (D)	18.8	18.8	37.6
Total	25.5	74.5	100

SOURCE: National Center of Chronic Disease Prevention, www.cdc.gov, U.S. Census Bureau, www.census.gov

88. Determine $P(A \cap B)$ and interpret.

89. Determine $P(C \cap D)$ and interpret.

90. Determine $P(B \mid C)$ and interpret.

91. Determine the probability that a randomly selected person lives in a county not meeting U.S. Environmental Protection Agency air quality standards, given that they are from Texas.

92. Determine $P(C \mid B)$.

93. Use the table of data to compute the values of $P(D \mid B)$ and $P(B \mid D)$.

	A	*B*	Total
C	20	30	50
D	15	10	25
Total	35	40	75

94. Use the table of probabilities to compute the values of $P(C \mid A)$ and $P(A \mid C)$.

	A	*B*	Total
C	$\frac{1}{3}$	$\frac{1}{3}$	$\frac{2}{3}$
D	$\frac{1}{4}$	$\frac{1}{12}$	$\frac{1}{3}$
Total	$\frac{7}{12}$	$\frac{5}{12}$	1

95. A pollster conducts a phone survey of 500 adult men and 516 adult women. The pollster asks each person if he or she is afraid of heights. The results are shown in the table.

	Afraid (*A*)	Not Afraid (*B*)	Total
Men (C)	312	188	500
Women (D)	289	227	516
Total	601	415	1016

For the events labeled in the table as *A*, *B*, *C*, and *D*, compute and interpret the following probabilities.

(a) $P(C \mid A)$ **(b)** $P(A \mid C)$

In Exercises 96 and 97, a tree diagram is given. Determine $P(A)$, $P(B \mid A)$ and $P(B^C \mid A)$.

96.

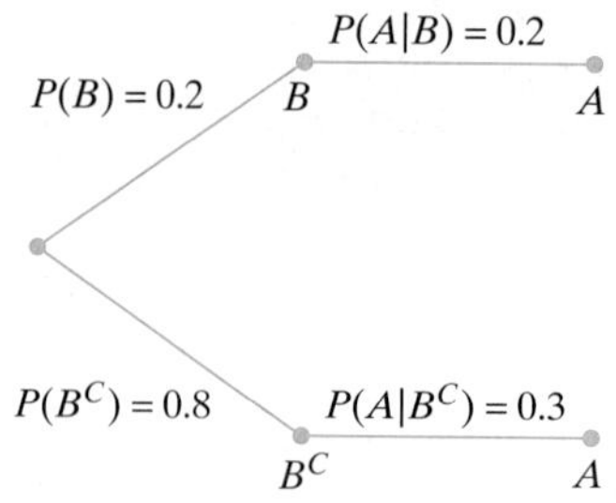

97.

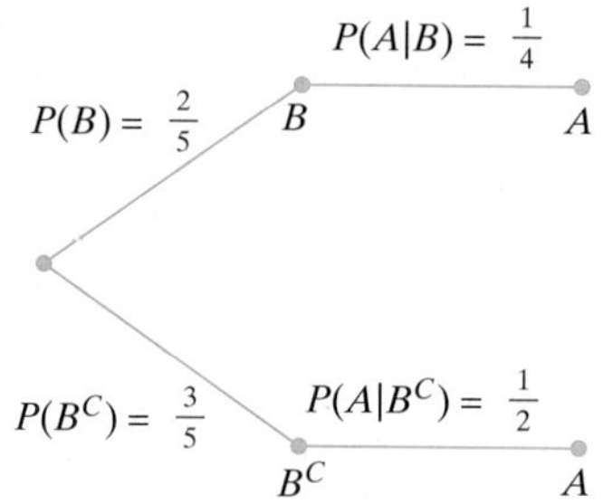

98. At Osceola College, 58% of the student body is underclasspeople and the remainder is upperclasspeople. In a survey, 60% of the underclasspeople favored the construction of a new parking structure and 30% of the upperclasspeople favored the construction of a new parking structure. A prospective student visiting the campus selects an Osceola College student at random. Let *B* represent the event that the selected student is an underclassperson, B^C represent the event that the selected student is an upperclassperson, and *A* represent the event that the selected student favored the construction of a new parking garage.

(a) Determine $P(A)$ and interpret.

(b) $P(B \mid A)$ and interpret.

(c) Determine $P(B^C \mid A)$ and interpret.

99. In a March 1999 poll, a random sample of American adults were asked if they were aware that American and British air forces were continuing to attack Iraqi air defenses and radar stations in no-fly zones. This sample group was then asked if they support the attacks. (For those who answered "no" to the first question, they were asked if they support the attacks now that they are aware of them.) The results are summarized in the following table:

Aware of Attacks	Percent	Support Attacks
Yes	69%	61%
No	31%	48%

SOURCE: www.harrisinteractive.com

Suppose that a respondent was selected at random. Let *B* represent the event that the respondent is aware of the attacks and *A* represent the event that the respondent supports the attacks.

(a) Determine $P(A)$ and interpret.

(b) Determine the probability that the respondent is aware of the attacks, given that the respondent supports the attacks.

(c) Determine $P(B^C \mid A)$ and interpret.

In Exercises 100 and 101, a tree diagram is given. Determine if the events are collectively exhaustive.

100.

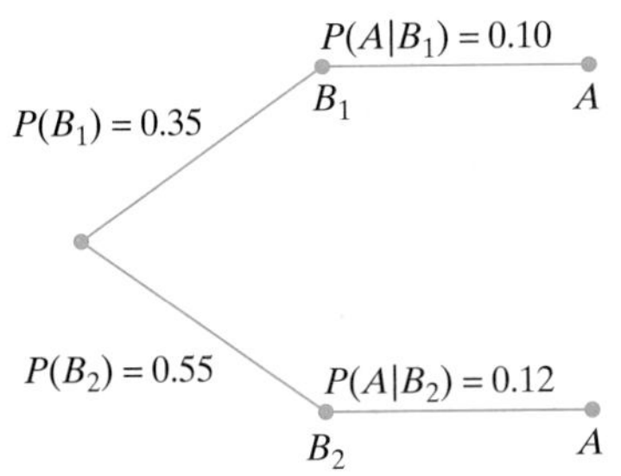

101.

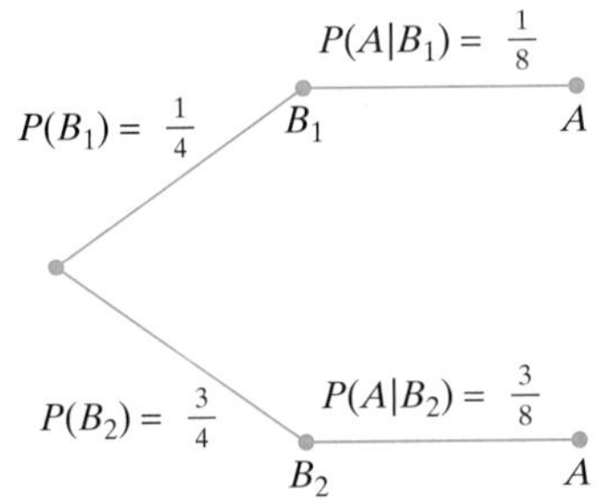

In Exercises 102 and 103 a tree diagram is given. Determine $P(A)$, $P(B_1 | A)$, and $P(B_3 | A)$.

102.

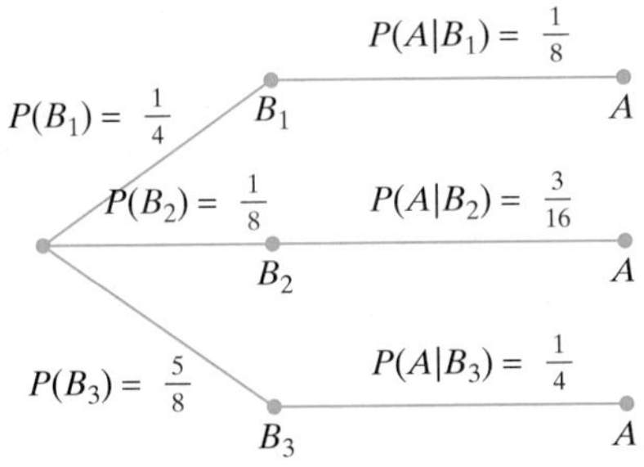

103.

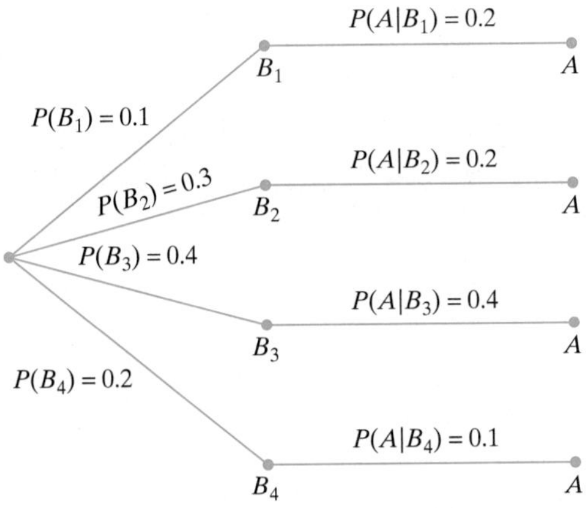

104. At Monadrock High School, the student council handed out surveys to determine if students favored adding lacrosse to the athletic program. The percent of surveys returned from each class is shown in the table.

Class	Student Body (%)	Surveys Returned (%)
Freshman	30	75
Sophomore	25	50
Junior	20	40
Senior	25	10

A student working for the school newspaper selects a student at random. Let B_1, B_2, B_3, and B_4 represent the event that the student is a freshman, sophomore, junior, or senior, respectively, and let A represent the event that the student returned the survey.

(a) Construct a tree diagram.

(b) Determine $P(A)$ and interpret.

(c) Determine the probability that the voter is a freshman, given that the voter returned the survey.

(d) Determine $P(B_3 \mid A)$ and interpret.

105. Sopris Mountaineering Company makes ski bindings at three factories. From past records, they know the proportion of output from each factory and the percentage of defects that come from each factory. The information is shown in the table.

Factory	Output (%)	Defects (%)
I	30	2
II	25	1
III	45	3

A quality control engineer selects a ski binding at random. Let B_1, B_2, and B_3 represent the event that the binding comes from factory I, factory II, or factory III, respectively, and let A represent the event that the binding has a defect.

(a) Construct a tree diagram.

(b) Determine $P(A)$ and interpret.

(c) Determine the probability that the ski binding came from factory II, given that the ski binding is defective.

(d) Determine $P(B_3 \mid A)$ and interpret.

CHAPTER 6 PROJECT

SOURCE: University of Minnesota: www.umn.edu

In this project, we will consider the enrollment of the Spring 2001 Semester at the University of Minnesota. We will use these data to compute different types of probabilities, such as conditional probabilities. Consider the Tables 1 and 2 below.

For a case study on university life, suppose that a national education magazine selects a student at random.

1. Using Table 1, let M stand for medical student, G stand for graduate student, U stand for undergraduate.

(a) Find the probability that the student is a medical student. Find the probability that the student is an undergraduate Medical student.

(b) Determine the probability that the student is between the ages of 21 and 24, given that the student is a medical student.

(c) Determine the probability that the student is a medical student, given that the student is between the ages of 21 and 24.

(d) Find $P(M \mid G)$ and interpret.

(e) Find $P(G \mid M)$ and interpret.

(f) Compare the results from parts (a) and (b) and also parts (d) and (e). Why do the answers differ?

2. Using Table 2, let W stand for white, AA stand for African American, and TC stand for twin cities campus.

(a) Find the probability that the student attends the Duluth campus.

(b) Find the probability that the student goes to the Morris campus, given that the student is American Indian.

(c) Find $P(W \cup AA)$ and interpret.

(d) Find $P[(W \cup AA)^C]$ and interpret.

(e) Find the probability that the student attends the Crookston campus or the Twin Cities campus, given that she or he is an International student. Assume that a student cannot attend both colleges.

3. There are a total of 1448 white medical students at the Twin Cities campus. Using Tables 1 and 2, find the probability that the student is a white medical student, given that he or she attends the Twin Cities campus. Let WMS stand for white medical student.

Table 1 Number of Students Attending the University of Minnesota's Twin Cities Campus

	Undergraduate				Graduate				Professional				Nondegree				
Age	Under 21	21–24	25–34	Other	Under 21	21–24	25–34	Other	Under 21	21–24	25–34	Other	Under 21	21–24	25–34	Other	Total
Medical	6	64	30	14	0	95	174	41	0	195	537	45	0	3	519	182	1,897
Total	11,873	9918	2476	1023	5	1809	5840	2333	8	930	1417	185	1035	959	1967	1616	43,394

Table 2 Number of Students Attending the University of Minnesota Classified by Race

	African American	American Indian	Asian/ Pacific	Chicano/ Latino	White	International	N/A	Total
Crookston	40	20	19	30	1,736	32	440	2,317
Duluth	72	110	149	69	7,659	153	291	8,503
Morris	96	102	49	21	1,444	8	54	1,774
Twin Cities	1432	273	2917	755	31,890	3245	2882	43,394
Total	1640	505	3134	875	42,729	3438	3667	55,988

Chapter 7

Graphical Data Description

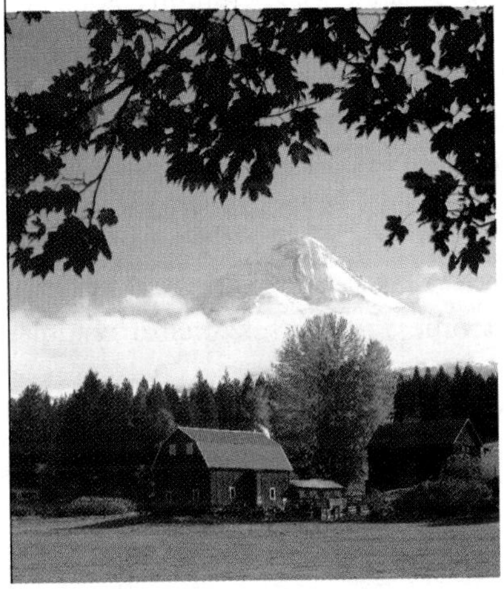

Type of asset	Value of asset in billions of dollars
Real estate	823
Machinery	89
Livestock	62
Financial assets	55
Crops	30

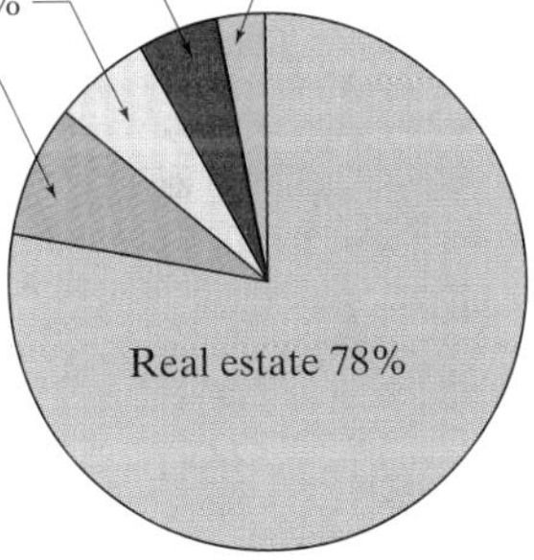

It is often said that a picture is worth a thousand words. This adage is never more apparent than when trying to make sense of a set of data. Consider the five categories of total assets of farms in the U.S. in 1995, as listed in the table. Although the data shows that a farms' greatest asset is the land itself, the statistical graph, or pie chart, makes this information perfectly clear in a quick visual summary.

What We Know

In the previous chapter we learned how to construct sets and how to find the number of elements in a set. This led us to some counting principles, which then led us to an introductory study of probability.

Where Do We Go

In this chapter we will study ways to communicate important information on sets of data, including the representation of data in graphical form and interpreting data that are presented in graphical form. We will also study the different ways of measuring the centrality of a data set and the overall dispersion of the data.

Section 7.1 Graphing Qualitative Data

Often in newspapers and magazines and on television, we see graphs and charts that summarize data. These graphs are used because they provide a quick, easy, and effective way to summarize raw data. In this section we will explore the properties of these graphs and study how they can be constructed.

Types of Data

We often hear statements such as "In 1998, 12.7% of the U.S. population was at least 65 years old," or "From 1980 to 1988, there were 2.62 residents per U.S. household." These statements are made after analyzing **data**. Data are made up of information coming from responses to surveys, from making observations, and from taking measurements. A collection of data is called a **data set**. Data sets come from two sources, called populations and samples. Recall from Section 6.1 that we defined **population** and **sample** as follows:

Population and Sample

- A **population** is the collection of *all* outcomes, responses, measurements, or counts that are of interest.
- A **sample** is a portion (or subset) of the population.

Example 1 Identifying a Population and a Sample

In an April 29, 2000, poll conducted by CNN/USA Today and the Gallup Organization, 1003 adult Americans were asked if they would favor or oppose the registration of all handguns; 76% of the adults favored registration. Identify the population and sample. Estimate the number in the data set who favor registration and the number who oppose registration (*Source:* www.gallup.com).

Solution

The **population** consists of the responses, or possible responses, of all American adults. The **sample** is made up of the 1003 American adults who participated in the survey. Since 76% of 1003 is about 762, the data set consists of the 762 responses favoring registration and the 241 opposing responses.

Checkpoint 1 Now work Exercise 11.

Data can be classified in various ways. For now, we will classify data as either **qualitative** or **quantitative**.

Qualitative and Quantitative Data

- Data that can be placed into distinct categories according to some characteristic, attribute, or nonnumerical label are called **qualitative data**. Since this type of data can be placed into distinct categories, they are sometimes referred to as **categorical data**.
- Data that are numerical in nature and can be ordered or ranked are called **quantitative data**.

Example 2 Classifying Data as Qualitative or Quantitative

The average gasoline price, in dollars per gallon, in Phoenix, Arizona, on June 23, 2000, for several types of gasoline is shown in Table 7.1.1. Which data are qualitative data and which are quantitative data? Explain.

Table 7.1.1

Type	Price in Dollars per Gallon
Regular	1.42
Premium	1.56
Diesel	1.55

SOURCE: www.gallup.com

Solution

The data in this table can be separated into two sets. One set consists of the type of gasoline and the other set consists of the average per gallon price. The gasoline types are nonnumerical and are separated into **categories**, so these are **qualitative** data. The average prices are numerical and can be ordered, so they make up the **quantitative** data. ■

We will call a table like Table 7.1.1 a **data table**. The remainder of this section will analyze the graphical techniques describing qualitative data.

Bar Graphs

One of the most widely used graphs for qualitative data is the **bar graph**. This type of graph is popular because bar graphs are easy to make and easy to interpret. Consider the bar graph in Figure 7.1.1, which represents the top five U.S. cities visited, in millions, by overseas travelers in 1997. From the graph, we see that New York City had the most overseas travelers at about 5 million. Los Angeles had just under 4 million overseas visitors. Note that the cities are listed in descending order by the number of overseas travelers. The horizontal axis lists the qualitative data values, that is, the cities. The vertical bars show the number of overseas travelers. The number of overseas travelers visiting these U.S. cities

Top 5 United States Cities Visited by Overseas Travelers, 1997

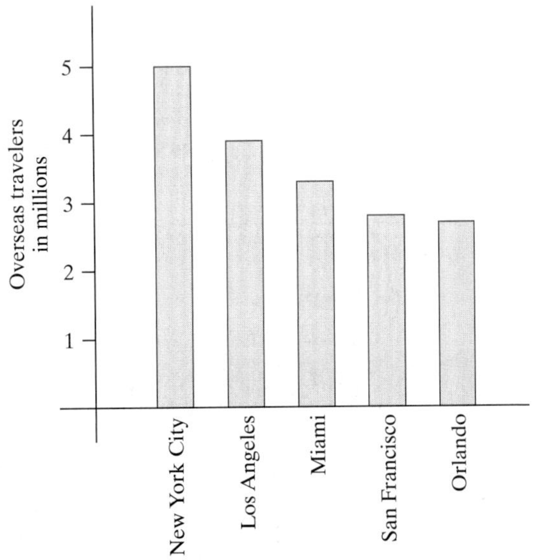

Figure 7.1.1

SOURCE: U.S. Census Bureau, www.census.gov

is called the **frequency**. The bars on the graph are all of equal width, and there is an equal space between each bar. We now list some of the features of the bar graph.

Properties of Bar Graphs

1. The horizontal axis contains the qualitative data values.
2. The vertical axis is labeled with the frequencies.
3. The data are arranged from largest to smallest in value as the graph is read from left to right.
4. The height of the bars is determined by the frequency.
5. The bars are of equal width and have an equal space between them.
6. The bar graph includes labels and units in the axes and a descriptive title.

Note: Sometimes the axes in items 1 and 2 are interchanged. However, in this text we will use the outline above unless otherwise indicated.

Example 3 Constructing a Bar Graph

The data table in Table 7.1.2 shows the amount spent, in billions of dollars, in the United States in 1997 on various recreation services.

Table 7.1.2

Type of Recreation Service	Amount Spent (in billions of dollars)
Physical fitness facilities	5.7
Public golf courses	4.3
Video arcades	3.7
Amusement parks	7.3
Sports clubs	7.7

SOURCE: U.S. Census Bureau, www.census.gov

Construct a bar graph for the data.

Solution

We begin by ranking the recreation services in descending order by the amount spent. The types of recreation services are labeled on the horizontal axis since they are the qualitative values. The vertical axis is labeled with the amount spent in billions of dollars. See Figure 7.1.2.

Now we construct the vertical bars with equal width and with equal spacing between them. The height of the bars is determined by the amounts spent, which are the frequencies. The completed bar graph is in Figure 7.1.3. Observe that, when we start with the leftmost bar and count to the right, physical fitness facilities rank third in total amount spent in 1997.

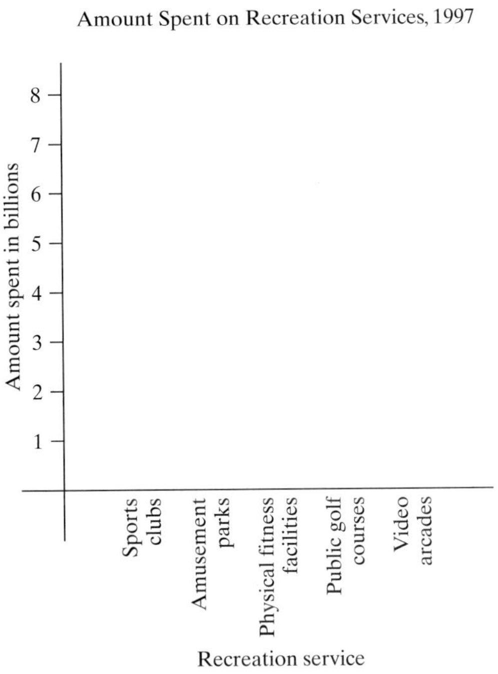

Figure 7.1.2

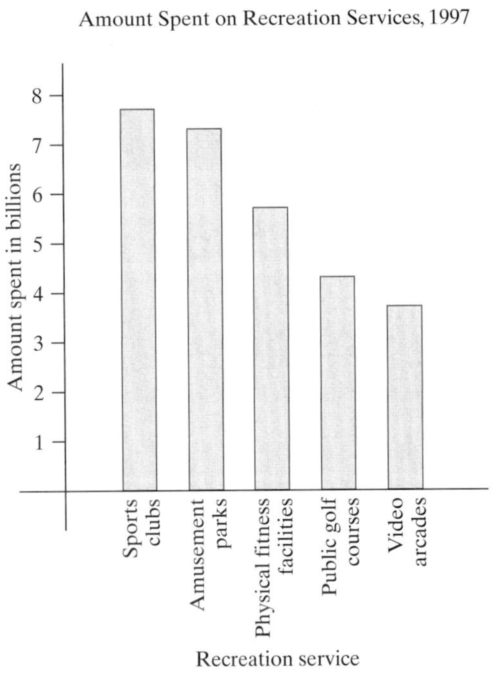

Figure 7.1.3 ■

Technology Option

Bar graphs can be made using computer spreadsheet programs such as Excel or statistical packages such as Minitab. Figure 7.1.4 shows the bar graph obtained for the data given in Example 3 using Excel.

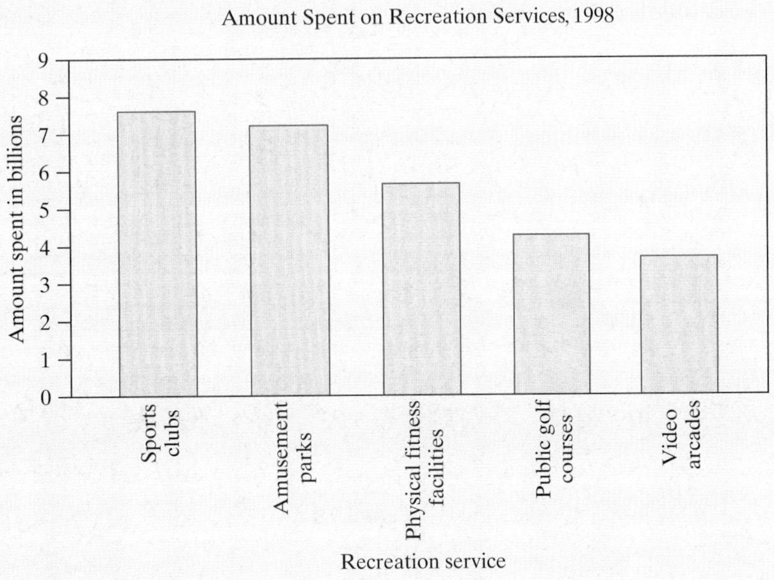

Figure 7.1.4

Interactive Activity

Sometimes bar graphs are constructed with horizontal bars and the axes are interchanged. Redraw the bar graph in Example 3 so that the bars are horizontal.

Bar graphs can also be used to display two sets of data. Table 7.1.3 shows the highest educational attainment of U.S. men and women in 1998, measured by percent.

Table 7.1.3

Educational Attainment	Percent Male	Percent Female
No high school diploma	17.2	17.1
High school graduate	32.3	35.2
Some college	17.1	17.3
Associate degree	6.9	8.0
Bachelor's degree	17.1	15.8
Advanced college degree	9.4	6.6

SOURCE: The U.S. Statistical Abstract, www.census.gov/statab/www/

We can compare the highest educational attainment by gender if we draw dual adjacent bars for each educational attainment level. When we have two separate sets of data, ordering them as in Example 3 is difficult, if not impossible. In this case we will organize the data by the listed educational attainment. The bar graph is shown in Figure 7.1.5. Note that the **legend** in the upper-right corner of the graph differentiates the educational attainment of men and of women.

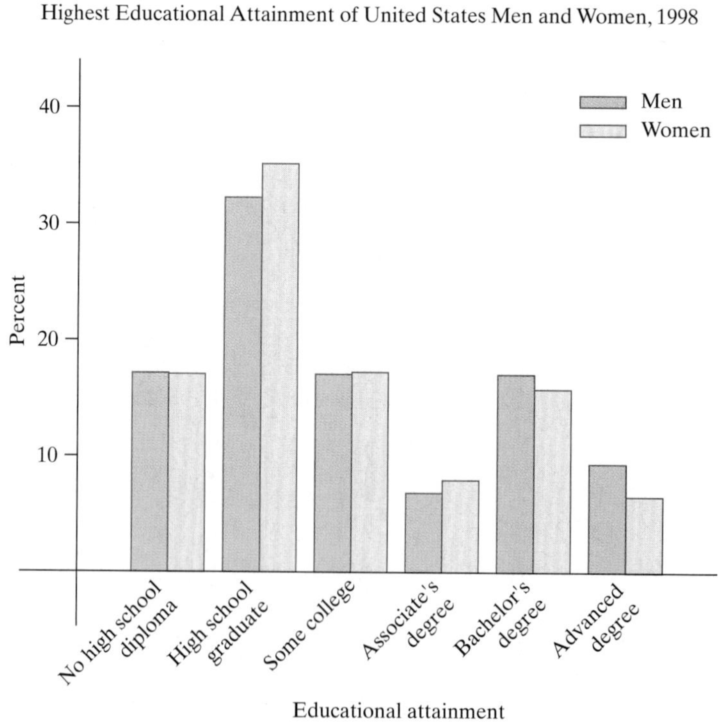

Figure 7.1.5

Pie Charts

One type of graph that is used extensively in descriptive statistics is the **pie chart**. This chart is ideally suited to show how a whole quantity is divided into several parts. The pie chart is made up of a circle divided into various slices called **sectors**. The size of each sector is determined by the percentage that each qualitative data value is of the whole. Let's take a look at an example of a pie chart. Table 7.1.4 shows the marital status of U.S. females 15 years or older in 2000. The corresponding pie chart for this data table is shown in Figure 7.1.6.

Table 7.1.4

Marital Status	Percentage of Females
Never married	25
Married	55
Widowed	10
Divorced	10

SOURCE: U.S. Census Bureau, www.census.gov

Marital Status of United States Females 15 Years or Older, 2000

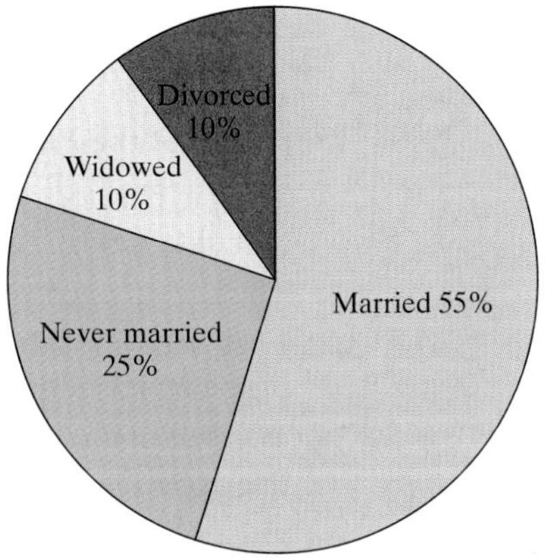

Figure 7.1.6

The graph clearly shows that the majority of females age 15 or older were married in 2000, while a relatively small percentage were divorced. This illustrates how effectively and quickly we can access data from a pie chart as compared to a numerical table.

When constructing a pie chart by hand, one key is to determine how large each sector will be. To determine its size, we need to find the measure, in degrees, of its **central angle**. The central angle is shown in Figure 7.1.7.

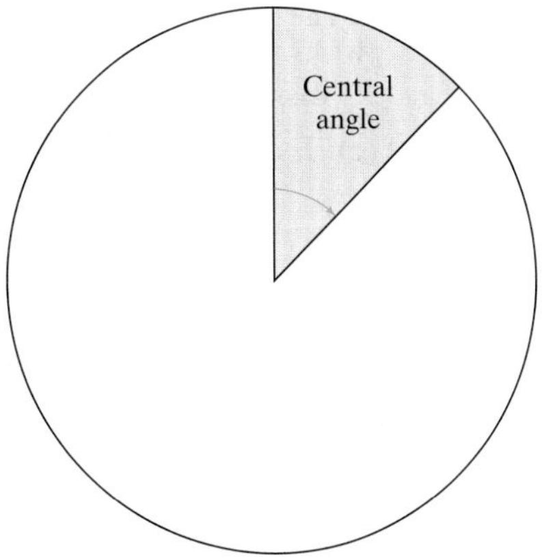

Figure 7.1.7

If we are given percentages as was done in Table 7.1.4, we can multiply the percentage, converted to decimal form, by 360°, the number of degrees in a circle. The measure of each central angle of the pie chart in Figure 7.1.6 is shown in Table 7.1.5. The data have been ordered from the highest to the lowest percentage.

Table 7.1.5

Marital Status	Percentage of Females	Degrees
Married	55	0.55(360°) = 198°
Never married	25	0.25(360°) = 90°
Widowed	10	0.10(360°) = 36°
Divorced	10	0.10(360°) = 36°

Now let's take a look at some of the properties of the pie chart.

Properties of the Pie Chart

1. The pie chart shows the part of the whole for each qualitative (categorical) data item.
2. The size of each sector in the pie chart corresponds to its frequency.
3. Each sector is labeled and includes its frequency or a percentage. Every pie chart has a title.
4. The sectors are arranged with the largest in the upper-right portion of the circle and, then clockwise in descending order.

Our main focus with pie charts will be reading and interpreting what they tell us. This is illustrated in Example 4.

Example 4 Interpreting a Pie Chart

The total assets of U.S. farms in 1998, divided into five categories, are shown in the pie chart in Figure 7.1.8. Note that, since the labels are too large to fit inside the smaller sectors, we use **leader lines** to connect the label with its sector (*Source:* U.S. Statistical Abstract, www.census.gov/statab/www).

(a) Which type of asset is the largest and which is the smallest?

(b) If total assets for these five categories were $1059 billion, estimate the assets for livestock.

(c) Estimate the assets that were not real estate.

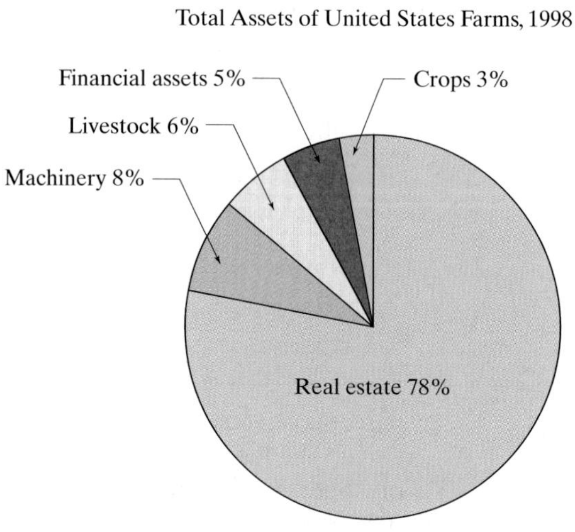

Figure 7.1.8

Solution

(a) Since real estate is the largest sector and has the highest percentage, it is the largest asset. On the other hand, crops is the smallest sector and has the lowest percentage, which means that it is the smallest asset.

(b) From Figure 7.1.8 we know that livestock is about 6% of the total assets. Since the total assets for these five categories in 1998 totaled \$1059 billion, the assets for livestock are estimated by

$$0.06(1059) = 63.54$$

So we can estimate that the assets for livestock in 1998 were about \$63.54 billion.

(c) Figure 7.1.8 indicates that about 78% of the total assets were in real estate. This means that 100% − 78% = 22% was in the other four categories. Our estimate for the assets not in real estate is found by

$$0.22(1059) = 232.98$$

So we can estimate that the assets not in real estate in 1998 were about \$232.98 billion. ■

Technology Option

Pie charts can be made using computer spreadsheets such as Excel or statistical packages such as Minitab. Figure 7.1.9 shows the pie chart for Example 4 using Excel.

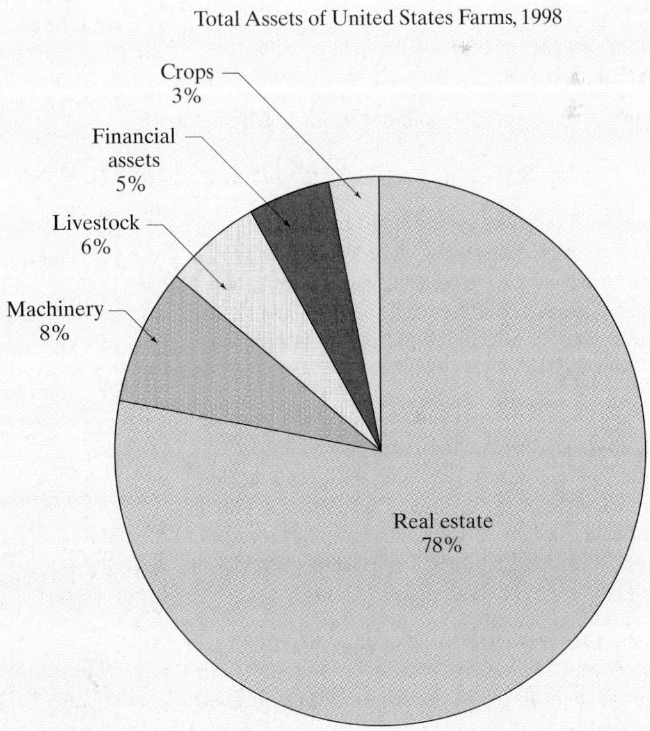

Figure 7.1.9

✓ Checkpoint 2 Now work Exercise 37.

SUMMARY

In this section, we studied two types of commonly used graphs, the **bar graph** and the **pie chart**. We explored the properties, construction techniques, and how to interpret these graphs.

- **Data** are made up of information coming from responses to surveys, from making observations, and from taking measurements.
- A **population** is the collection of *all* outcomes, responses, measurements, or counts that are of interest.
- A **sample** is a portion (or subset) of the population.
- **Qualitative data** are data that can be placed into distinct categories according to some characteristic, attribute, or nonnumerical label. These are sometimes called **categorical data**.
- **Quantitative data** are data that are numerical in nature and can be ordered or ranked.

SECTION 7.1 EXERCISES

In Exercises 1–8, determine whether the data set is a population or a sample. Explain your reasoning.

1. The age of every member of the U.S. Senate.

2. The annual salary of every employee at a certain corporation.

3. The party affiliation of every third voter exiting a polling booth.

4. A survey of 50 workers at a factory that employs 3000.

5. The annual income of all homeowners in California.

6. The number of children of all families who own home computers.

7. The number of years of education of those who responded to a mail survey.

8. The fuel prices at selected gasoline stations in Michigan.

In Exercises 9–14, identify the population and the sample.

9. In a Spring 2000 survey conducted by the Gallup Organization, 568 adult Americans were asked if they would favor or oppose a proposal that would allow people to put a portion of their Social Security payroll taxes into personal retirement accounts; 65% of the adults favored the proposal (*Source:* www.gallup.com).

10. In a June 1999 survey conducted by NBC and the *Wall Street Journal*, 491 adult Americans were asked, "In terms of setting Medicare policy dealing with prescription drug benefits, who do you trust more: President Clinton or companies that make prescription drugs?" Fifty-seven percent of the respondents said that they trusted President Clinton more (*Source:* NBC/*Wall Street Journal*).

✓ **11.** In an October 1998 survey conducted by the Gallup Organization, 467 baseball fans were asked, "Who is the greatest baseball player in the game today?" Eleven percent responded Sammy Sosa (*Source:* www.gallup.com).

12. In a March 2000 survey, 205 golf fans were asked by the Gallup Organization, if they could play any golf course in the world, which one would it be? Thirty-seven percent of the respondents said Pebble Beach (*Source:* www.gallup.com).

13. CNN conducted an on-line, interactive quick vote on May 20, 1999. They asked 19,026 respondents, "A study found that bioengineered corn can harm butterflies; should such crops be put on hold pending more study?" Seventy-five percent said there should be a hold on genetically engineered crops until further study had been conducted (*Source:* www.cnn.com).

14. In a summer 2000 *USA Today* online survey, 1909 readers were asked, "Which activist, leader, or politician has done the most to raise the visibility of women?" to which 34.6% responded by saying Mother Teresa (*Source:* www.usatoday.com).

In Exercises 15–20, determine whether the data are qualitative or quantitative.

15. The gender of the members of a book reading club.

16. The religious preference of the citizens of a small town.

17. The number of student desks in a classroom.

18. The length of bass caught in Lake Superior.

19. The colors of blazers in a ladies clothing store.

20. The ranking of service (excellent, good, poor) at a suburban restaurant.

In Exercises 21–30, a table of data is given. Construct a bar graph for the data.

21. The total cash receipts, in billions of dollars, from U.S. farm marketings for four of the top farm livestock commodities in 1998 are shown in the table.

Commodity	Receipts (in billions of dollars)
Cattle	33.7
Hogs	9.4
Chickens	15.1
Turkeys	2.7

SOURCE: U.S. Statistical Abstract, www.census.gov/statab/www

22. The total cash receipts, in billions of dollars, from U.S. farm marketings for four of the top farm crop commodities in 1998 are shown in the table.

Commodity	Receipts (in billions of dollars)
Corn	17.1
Soybeans	15.4
Wheat	7.0
Cotton	6.0

SOURCE: U.S. Statistical Abstract, www.census.gov/statab/www

23. The number of establishments, in thousands, for selected U.S. service industries in 1992 is shown in the table.

Establishment	Number (in thousands)
Social services	140.8
Personal services	197.1
Business services	306.6
Automotive repair	172.0
Health services	465.3

SOURCE: U.S. Statistical Abstract, www.census.gov/statab/www/

24. The number of establishments, in thousands, for selected U.S. service industries in 1997 is shown in the table.

Establishment	Number (in thousands)
Social services	162.3
Personal services	204.6
Business services	398.2
Automotive repair	192.3
Health services	498.9

SOURCE: U.S. Statistical Abstract, www.census.gov/statab/www/

25. The number of units, in millions, of various recorded audio media shipped in the United States in 1990 is shown in the table.

Media Type	Number Shipped (in millions)
Compact disks	286.5
Cassettes	442.2
Cassette singles	87.4
Vinyl albums	39.9

SOURCE: U.S. Census Bureau, www.census.gov

26. The number of units, in millions, of various recorded audio media shipped in the United States in 1997 is shown in the table.

Media Type	Number Shipped (in millions)
Compact disks	847.0
Cassettes	158.5
Cassette singles	26.4
Vinyl Albums	8.8

SOURCE: U.S. Census Bureau, www.census.gov

27. The number of titles of new books imported to the United States in various subjects during 1990 is shown in the table.

Subject	New Titles Imported
Education	234
History	329
Fiction	241
Medicine	588
Technology	546

SOURCE: U.S. Census Bureau, www.census.gov

28. The number of titles of new books imported to the United States in various subjects during 1997 is shown in the table.

Subject	New Titles Imported
Education	190
Fiction	273
History	512
Medicine	706
Technology	501

SOURCE: U.S. Census Bureau, www.census.gov

29. The number of foreign students, in thousands, who came to study in the United States in 1995 from various countries is shown in the table.

Country	Foreign Students (in thousands)
Canada	22.7
Mexico	9.0
Japan	45.2
South Korea	33.6
Spain	5.1
United Kingdom	7.8
Turkey	6.7

SOURCE: U.S. Department of Education, www.ed.gov

30. The number of foreign students, in thousands, who went to study in the United Kingdom in 1995 from various countries is shown in the table.

Country	Foreign Students (in thousands)
Canada	1.7
France	8.9
Japan	2.6
Germany	9.5
Greece	10.3
United States	6.2
Ireland	9.8

SOURCE: U.S. Department of Education, www.ed.gov

In Exercises 31–36, answer the questions associated with the given bar graph.

31. The number of cigarettes, in billions, exported by various countries in 1998 is shown in the bar graph on the next page (*Source:* U.S. Census Bureau, www.census.gov).

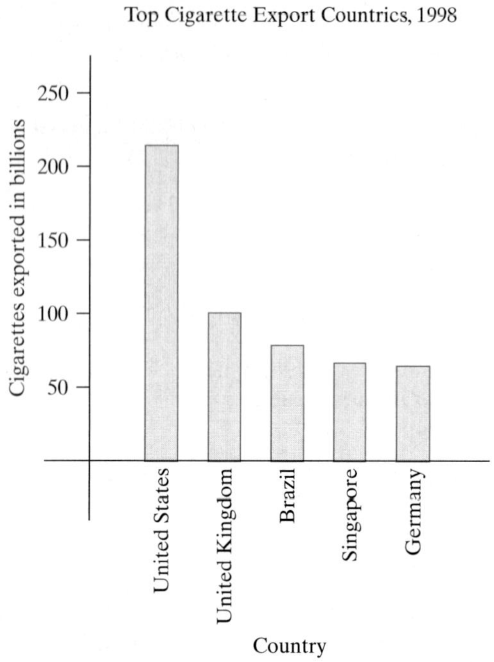

(a) Which countries exceeded 100 billion cigarettes exported in 1998?

(b) What countries had exports that were about the same amount?

32. The number of cigarettes, in billions, imported by various countries in 1998 is shown in the bar graph (*Source:* U.S. Census Bureau, www.census.gov).

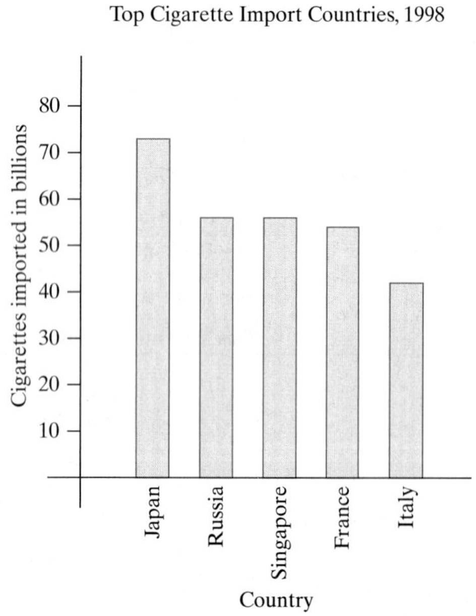

(a) Which countries exceeded 50 billion cigarettes imported in 1998?

(b) What countries had imports that were about the same amount?

33. The number of monitoring stations in the U.S. for various forms of air pollution in 1998 is shown in the bar graph (*Source:* U.S. Environmental Protection Agency, www.epa.gov).

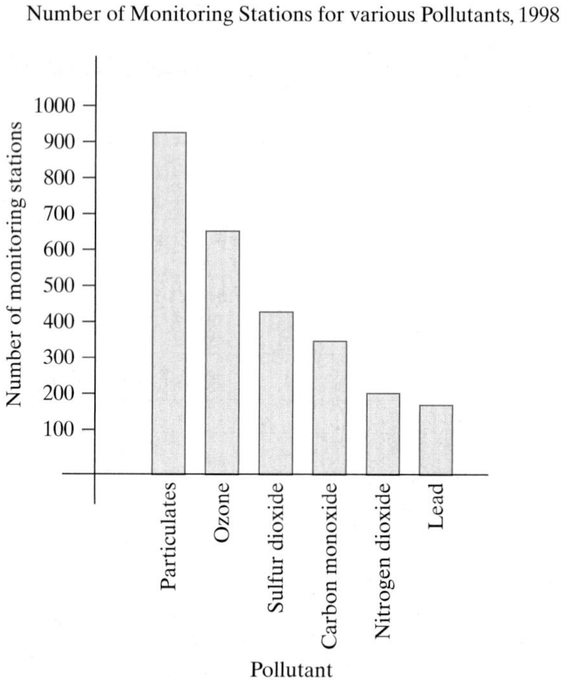

(a) Which pollutants have less than 500 monitoring stations?

(b) About how many more monitoring stations are there for particulates than there are for lead?

34. The amount of U.S. municipal waste generated, in millions of tons, by various items in 1997 is shown in the bar graph (*Source:* U.S. Census Bureau, www.census.gov).

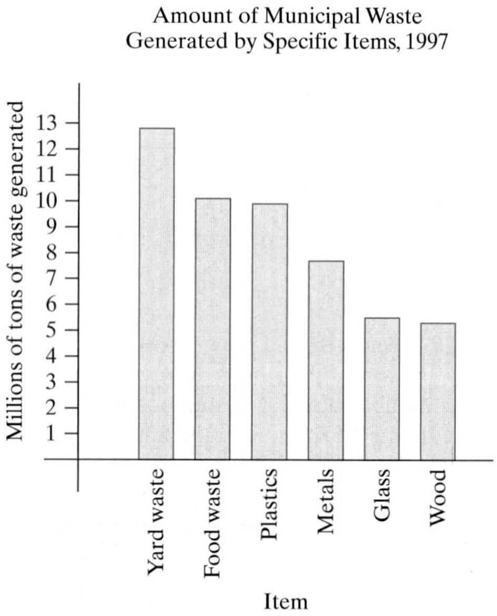

(a) About how many millions of tons of yard waste was generated in 1997?

(b) About how many millions of tons of waste was generated by food and yard wastes combined?

35. The number of U.S. households, in thousands, who went on a leisure vacation of various types in 1995 is shown in the bar graph (*Source:* U.S. Census Bureau, www.census.gov).

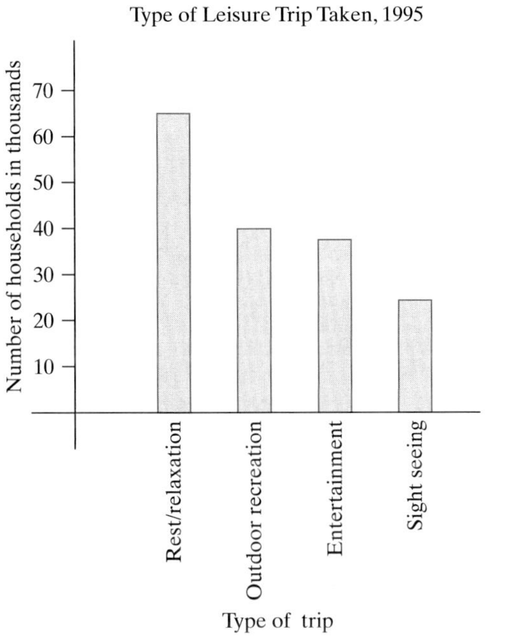

(a) About how many households took trips for rest and relaxation in 1995?

(b) About how many more rest and relaxation trips were taken in 1995 than sightseeing trips?

36. The number of traffic fatalities involving a large truck in 1997 from various states in the United States is shown in the bar graph (*Source:* U.S. Department of Transportation, www.dot.gov).

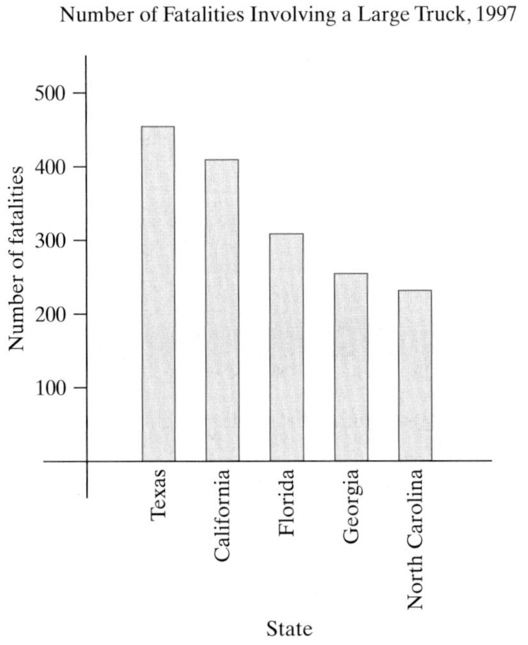

(a) Which states had more than 300 traffic fatalities involving a truck?

(b) Would the data from the bar graph be suitable for a pie chart? Explain.

In Exercises 37–46, answer the questions associated with the given pie chart.

37. The percent of the amount of debt from various sources held by American families in 1995 is shown in the pie chart (*Source:* U.S. Census Bureau, www.census.gov).

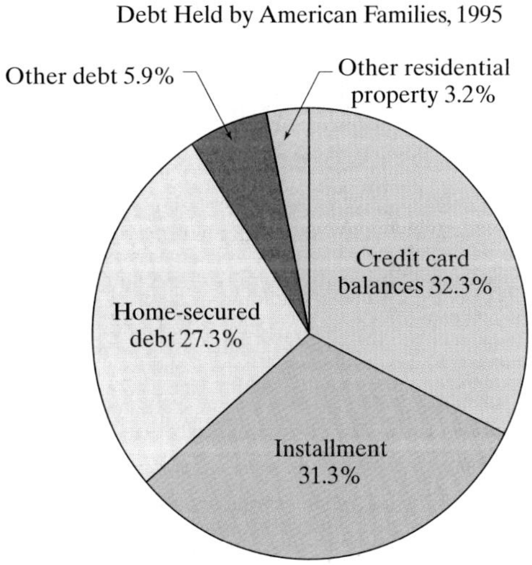

(a) What percentage of family debt is from home-secured debt and installment?

(b) What percentage of family debt is from credit-card balances and installment?

38. The percent of the amount of debt from various sources held by American families in 1998 is shown in the pie chart (*Source:* U.S. Census Bureau, www.census.gov).

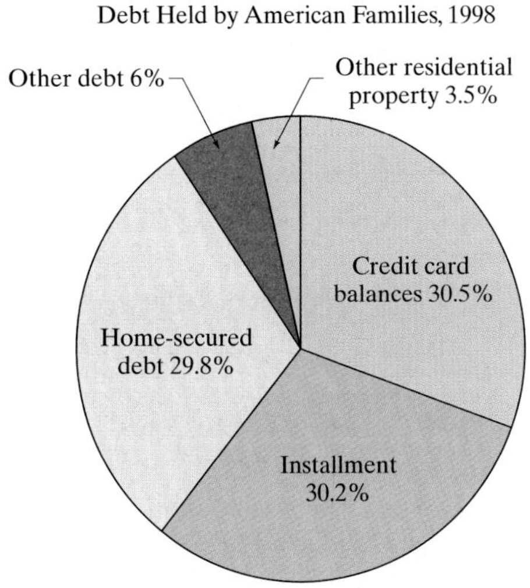

(a) What percentage of family debt is from home-secured debt and installment?

(b) What was the third largest source of family debt in 1995?

39. The types of motorized transportation used by the public in France during 1996 are shown in the pie chart. Numbers are in thousands (*Source:* U.S. Census Bureau, www.census.gov).

Types of Motorized Transportation in France, 1996

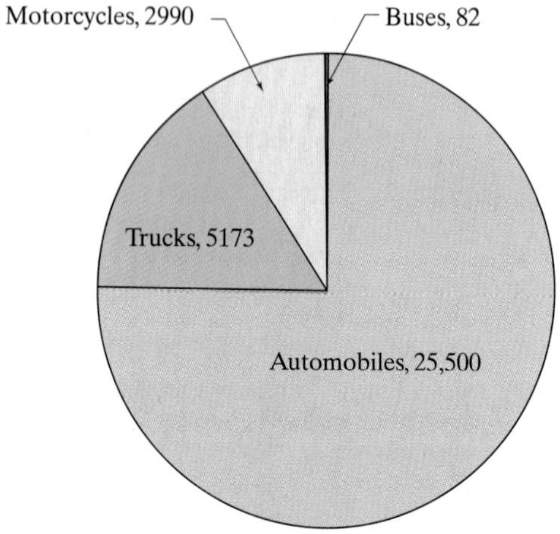

(a) What type of motorized transportation was used the least in France in 1996?

(b) What percentage of all transportation vehicles were motorcycles? Round to the nearest tenth.

40. The types of motorized transportation used by the public in the United Kingdom during 1996 are shown in the pie chart. Numbers are in thousands (*Source:* U.S. Census Bureau, www.census.gov).

Types of Motorized Transportation in the United Kingdom, 1996

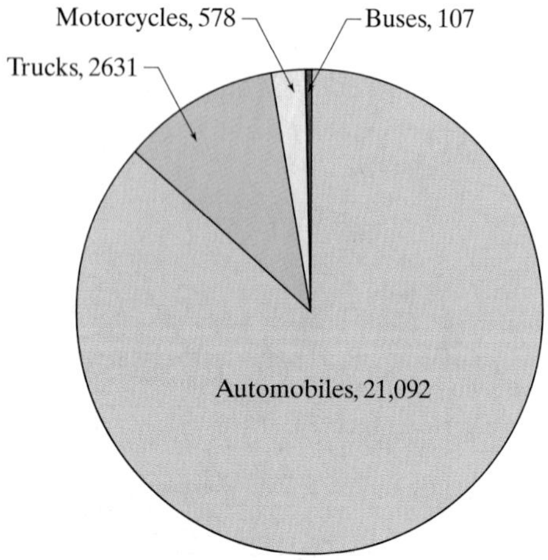

(a) What type of motorized transportation was used the least in the United Kingdom in 1996?

(b) What percentage of all motorized transportation were trucks? Round to nearest tenth.

41. The number of scientists, in thousands, employed in the United States in 1996 is shown in the pie chart (*Source:* U.S. Census Bureau, www.census.gov).

Scientists Employed in the United States, 1996

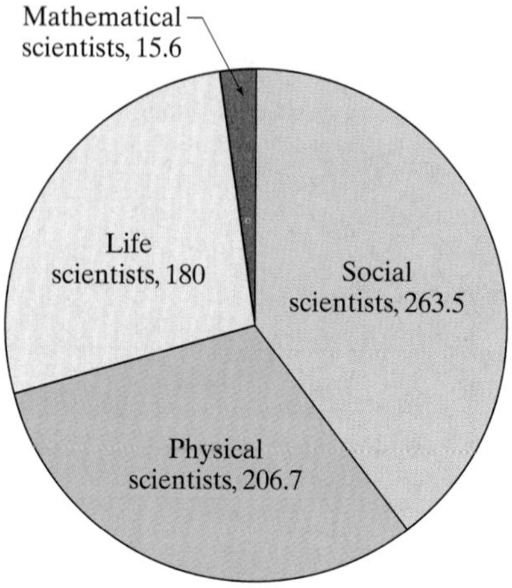

(a) What type of scientist ranked third in terms of employment?

(b) What percentage of all scientists employed in the United States were social scientists? Round to nearest tenth.

42. The number of scientists, in thousands, employed by the U.S. government in 1996 is shown in the pie chart (*Source:* U.S. Census Bureau, www.census.gov).

Scientists Employed by the United States Government, 1996

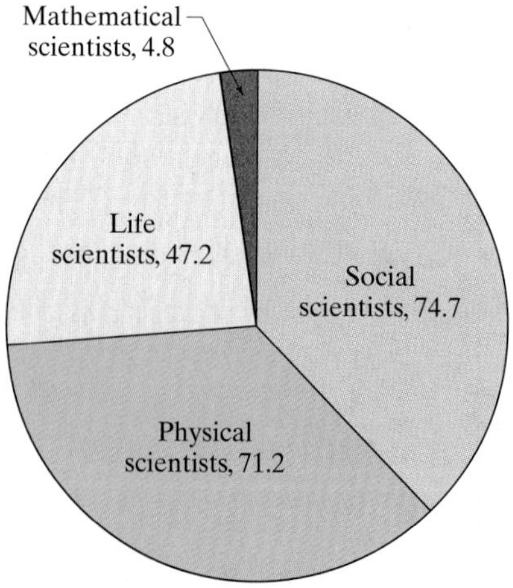

(a) What type of scientist ranked third in terms of employment?

(b) What percentage of U.S. government scientists were social scientists? Round to nearest tenth.

43. The number of hours per person per year that Americans watched various forms of television in 1992 is shown in the pie chart (*Source:* U.S. Census Bureau, www.census.gov).

Television Viewing (hours per person per year) in the United States, 1992

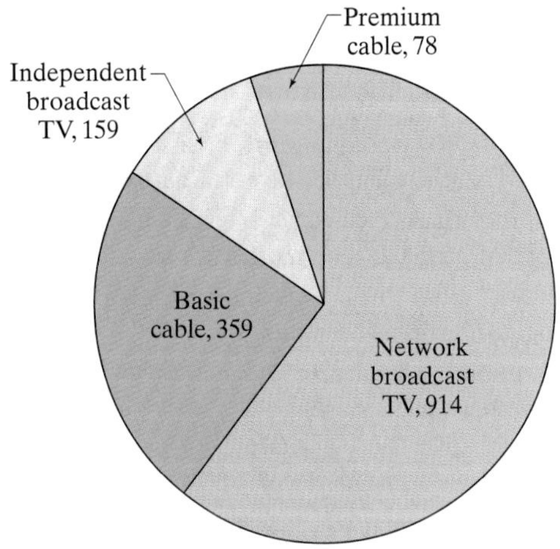

(a) What form of television was least watched in terms of hours per person per year?

(b) What percentage of television viewing was basic cable? Round to nearest tenth.

44. The number of hours per person per year that Americans watched various forms of television in 2000 is shown in the pie chart (*Source:* U.S. Census Bureau, www.census.gov).

Television Viewing (hours per person per year) in the United States, 2000

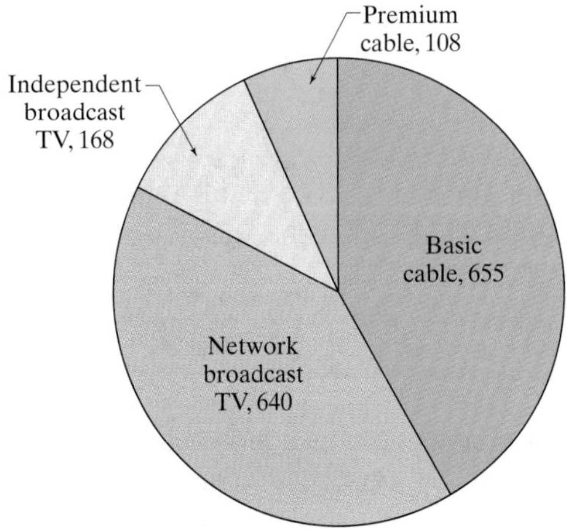

(a) What form of television was least watched in terms of hours per person per year?

(b) What percentage of television viewing was basic cable? Round to nearest tenth.

45. The types of robbery crimes, in thousands, reported in the United States in 1996 are shown in the pie chart (*Source:* U.S. Bureau of Crime Statistics, www.ojp.usdoj.gov/bjs).

Reported Robbery Crimes in the United States, 1996

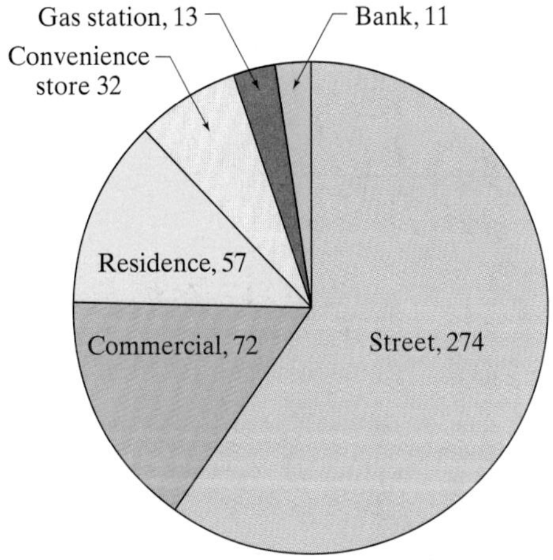

(a) About how many gas station and convenience store crimes were reported in 1996?

(b) What percentage of reported robbery crimes were residence robbery crimes? Round to nearest tenth.

46. The types of weapon used, in thousands, to commit crimes in 1995 in the United States are shown in the pie chart (*Source:* U.S. Bureau of Crime Statistics, www.ojp.usdoj.gov/bjs).

Weapons Used in Crimes in the United States, 1995

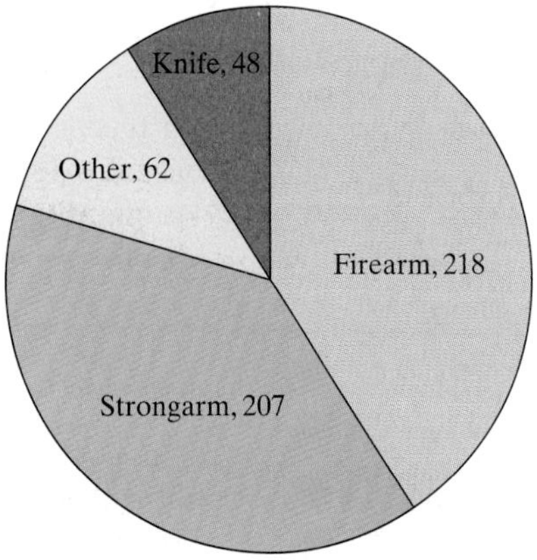

(a) About how many crimes in 1995 involved firearms and knives?

(b) What percentage of crimes involved firearms? Round to nearest tenth.

SECTION PROJECT

There are times when a **stacked bar graph** shows not only the total frequency, but also the frequency for a certain portion of the whole group. The stacked bar graph that follows shows the total births in 1993 for five of the 20 largest U.S. metropolitan areas. Births to teen mothers are also displayed. The frequencies are shown at the top of each bar for accuracy (*Source:* U.S. National Center for Health Statistics, www.cdc.gov/nchs).

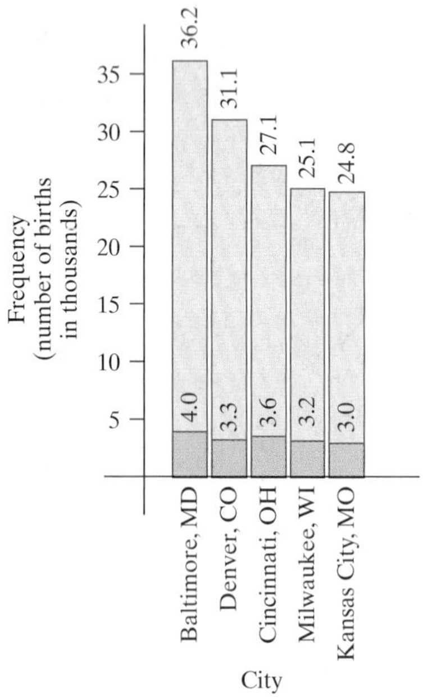

Use the graph to answer the following.

(a) Determine the percentage of births of teen mothers in Cincinnati. Use the formula

$$\frac{\text{number of births to teen mothers}}{\text{total births}} \cdot 100\%.$$

(b) Determine the percentage of births to teen mothers in Baltimore. Use the formula shown in part (a).

(c) Of all the cities displayed, which city had the highest percentage of births to teen mothers?

(d) Of all the cities displayed, which city had the lowest percentage of births to teen mothers?

(e) Determine the total percentage of births to teen mothers in these five cities.

Section 7.2 Graphing Quantitative Data

In the previous section we saw how we could use bar graphs and pie charts to visually display qualitative data. But much of the data that we study are numeric or **quantitative data**. In this section we will examine the techniques of organizing, displaying, and interpreting this type of data.

The Frequency Distribution

When a set of quantitative data in raw and ungrouped form has many values, it is difficult to see trends and patterns. We can organize the data by the use of the **frequency distribution**.

Frequency Distribution

A table that categorizes raw data into **classes** is called a **frequency distribution**. The number of data values that falls into each class is called the class **frequency**, denoted by f.

By raw data, we mean data that are given in their original form. Let's examine a frequency distribution to see how it summarizes data. Table 7.2.1 displays a frequency distribution for the number of community hospitals in the 50 states of the United States in 1990.

Table 7.2.1

Class (number of hospitals)	Frequency, f (number of states)
8 to 62	19
63 to 117	13
118 to 172	10
173 to 227	4
228 to 282	2
283 to 337	0
338 to 392	0
393 to 447	2

SOURCE: U.S. Department of Health and Human Services, www.hhs.gov

We can see that there are eight classes in this frequency distribution. The first class has 8 to 62 hospitals, the second class from 63 to 117 hospitals, and so on. The frequency of the first class is 19. This means that 19 states in the United States had from 8 to 62 hospitals in 1990. We can also see from inspecting Table 7.2.1 that in 1990 no states had from 338 to 392 community hospitals.

The leftmost number in each class is called the **lower class limit**. For example, in Table 7.2.1 the lower limit of the first class is 8 and the lower limit of the third class is 118. The rightmost number in each class is called the **upper class limit**. In Table 7.2.1, 62 and 172 are the upper limits for the first and third classes, respectively. Notice that if we take the difference between any two lower class limits we get the same number.

$$63 - 8 = 55, \quad 118 - 63 = 55, \quad 173 - 118 = 55, \quad 228 - 173 = 55, \quad \text{and so on}$$

This value is called the **class width**. The class width of the frequency distribution given in Table 7.2.1 is 55. If we add up all the frequencies, we get 50 as expected. In Chapter 6 we used the capital Greek letter $\sum$ (sigma) to signify "to sum." Hence, we can denote the sum of the frequencies by writing $\sum f = 50$. We call this sum the **sample size** and denote it with the letter n. In other words, $\sum f = n$. Note that the **population** is the set of *all* hospitals in the 50 states, not only the community hospitals.

One useful type of frequency is the **relative frequency**. This tells us what portion or percent falls into each class. To compute the relative frequency of a class, we take the frequency f and divide it by the sample size n.

Relative Frequency

The **relative frequency** of a class is the portion of observations that falls into that class and is given by

$$\text{relative frequency} = \frac{\text{frequency of a class}}{\text{sum of frequencies}} = \frac{f}{n} = \frac{f}{\sum f}$$

Note: Relative frequencies can be expressed as either whole number percents or decimals rounded to the hundredths place.

Example 1 Computing Relative Frequency

For the frequency distribution in Table 7.2.1, include a third column that gives the relative frequency of each class. Write the frequencies in decimal form.

Solution

We have already seen that, for the frequency distribution in Table 7.2.1, $n = 50$. For the first class, the relative frequency is

$$\frac{\text{frequency of first class}}{\text{sum of frequencies}} = \frac{f}{n} = \frac{19}{50} = 0.38$$

Continuing in this manner, we get the relative frequencies shown in Table 7.2.2.

Interactive Activity

Sum the relative frequencies in Table 7.2.2. How do you explain the value of the sum? Will the sum of relative frequencies always be this number? Explain.

WWW

Table 7.2.2

Class (number of hospitals)	Frequency, f (number of states)	Relative Frequency
8 to 62	19	0.38
63 to 117	13	0.26
118 to 172	10	0.20
173 to 227	4	0.08
228 to 282	2	0.04
283 to 337	0	0.00
338 to 392	0	0.00
393 to 447	2	0.04

Constructing Frequency Distributions

Now that we have seen some of the features of the frequency distribution, we can use the guidelines that follow to construct a frequency distribution from raw data.

Constructing a Frequency Distribution

Step 1: Decide on the number of classes that you would like the frequency distribution to have. In general, there should be between 5 and 15 classes. (Many times, the number of classes to be used is given.)

Step 2: Determine the class width, denoted by CW. This is given by

$$CW = \frac{\text{largest data value} - \text{smallest data value}}{\text{number of classes}}$$

As a convention, we round CW up to the next whole number.

Step 3: Set the class limits. The smallest data value is the lower limit of the first class. The other lower class limits are determined by adding the class width to the lower limit of the preceding class. The upper limits are one unit less than the lower limit of the next class.

Step 4: Tally the data to determine the frequency of each class. Sorting the data in ascending order can simplify this task.

Note: When setting the class limits, we should keep in mind the following:

1. Each data item can fall into only one class. That is, the classes cannot overlap.

2. Each class should have the same width. (An exception could occur if there are outliers.)
3. Every data item must fall into a class. If this does not occur, classes must be added or the class width must be increased.

Example 2 Constructing a Frequency Distribution

The average annual expenditures in dollars on vehicle purchases in 28 U.S. metropolitan areas during 1995 are shown in Table 7.2.3.

Table 7.2.3

1225	1755	1815	1926	1967	2240	2270
2339	2401	2454	2640	2656	2684	2828
2833	2931	3108	3184	3235	3241	3286
3339	3521	3521	3598	4225	5136	5814

SOURCE: U.S. Statistical Abstract, www.census.gov/statab/www/

Construct a frequency distribution using six classes.

Solution

For this example, we will follow the four steps in the frequency distribution guidelines.

Step 1: Here we are given that we should have six classes.

Step 2: Since the largest data value is 5814 and the smallest data value is 1225, we determine that the approximate class width is

$$\frac{5814 - 1225}{6} \approx 764.83$$

Rounding up to the next whole number, we get $CW = 765$.

Step 3: The smallest data value is 1225, so this gives us the lower limit of the first class. The lower limits of the successive classes are found by adding the class width to the lower limit of the previous class. This means that the lower limit of the second class is $1225 + 765 = 1990$, the lower limit of the third class is $1990 + 765 = 2755$, and so on. For the upper class limits, we know that the upper limit of the first class is 1989, which is one less than the lower limit of the second class. The upper limit of the second class is 2754, which is one less than the upper limit of the third class, and so on. The classes and the class limits are given in Table 7.2.4.

Table 7.2.4

Class Number	Lower Limit	Upper Limit
1	1225	$1990 - 1 = 1989$
2	$1225 + 765 = 1990$	2754
3	2755	3519
4	3520	4284
5	4285	5049
6	5050	5814

Step 4: Now we tally the number of data items that fall into each class. This is the class frequency. For example, since there are five data values between 1225 and 1989 inclusive, the frequency of the first class is 5. The completed frequency distribution is shown in Table 7.2.5.

Table 7.2.5

Class (expenditure on vehicle purchases)	Frequency, f (number of metro areas)
1225 to 1989	5
1990 to 2754	8
2755 to 3519	9
3520 to 4284	4
4285 to 5049	0
5050 to 5814	2

■

Checkpoint 1 Now work Exercise 5.

The Histogram

One of the most popular graphical data representations of a frequency distribution is the **histogram**. To plot histograms, we first need to compute the **class boundaries**.

Class Boundary

For the given class limits of a frequency distribution, we have the following:

1. The **lower class boundary** is the lower class limit minus one-half unit. That is,

$$\text{lower boundary} = \text{lower limit} - 0.5$$

2. The **upper class boundary** is the upper class limit plus one-half unit. That is,

$$\text{upper boundary} = \text{upper limit} + 0.5$$

In Table 7.2.6, we add a column of class boundaries to the frequency distribution of the state community hospitals given in Table 7.2.1. Notice in the table

Table 7.2.6

Class (number of hospitals)	Class Boundaries	Frequency, f (number of states)
8 to 62	7.5 to 62.5	19
63 to 117	62.5 to 117.5	13
118 to 172	117.5 to 172.5	10
173 to 227	172.5 to 227.5	4
228 to 282	227.5 to 282.5	2
283 to 337	282.5 to 337.5	0
338 to 392	337.5 to 392.5	0
393 to 447	392.5 to 447.5	2

that the upper boundary of a class is equal to the lower boundary of the next class. We can see from the properties of the histogram that follow that this is no coincidence. We want the rectangles in a histogram to have no gaps between them, as was seen with the bar graph.

Properties of the Histogram

1. The horizontal axis is labeled with the class boundaries.
2. The vertical axis is labeled with the frequency.
3. Adjacent bars are constructed with the width determined by the class boundaries and the height determined by the class frequency.
4. The histogram can also be constructed using the relative frequencies.
5. Every histogram is given a title.

Example 3 Constructing a Histogram

Construct a histogram for the frequency distribution in Example 2.

Interactive Activity

Add a column to the frequency distribution in Table 7.2.7 for the relative frequency of each class. Then construct a histogram using the relative frequencies as the label on the vertical axis. How does the shape of this histogram compare to the histogram in Figure 7.2.1?

Solution

First we need to add a column containing the class boundaries to the frequency distribution. This is shown in Table 7.2.7.

Figure 7.2.1 shows the histogram for the data.

Table 7.2.7

Expenditure on Vehicle Purchases	Class Boundaries	Frequency, f
1225 to 1989	1224.5 to 1989.5	5
1990 to 2754	1989.5 to 2754.5	8
2755 to 3519	2754.5 to 3519.5	9
3520 to 4284	3519.5 to 4284.5	4
4285 to 5049	4284.5 to 5049.5	0
5050 to 5814	5049.5 to 5814.5	2

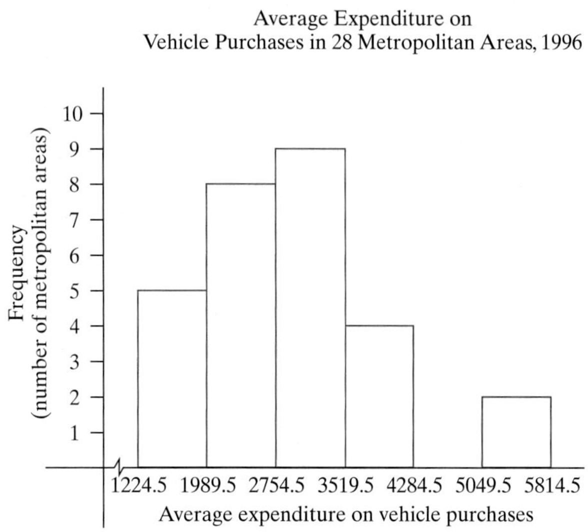

Figure 7.2.1

Checkpoint 2 Now work Exercise 15.

Technology Option

Histograms can also be made using a graphing calculator. Figure 7.2.2(a) shows the histogram for Example 3 made on a graphing calculator, and Figure 7.2.2(b) shows the histogram for Example 3 made in Excel.

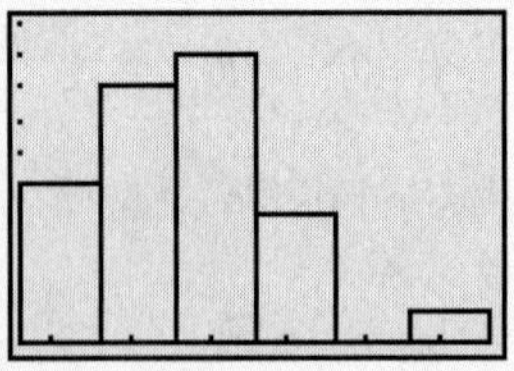

(a)

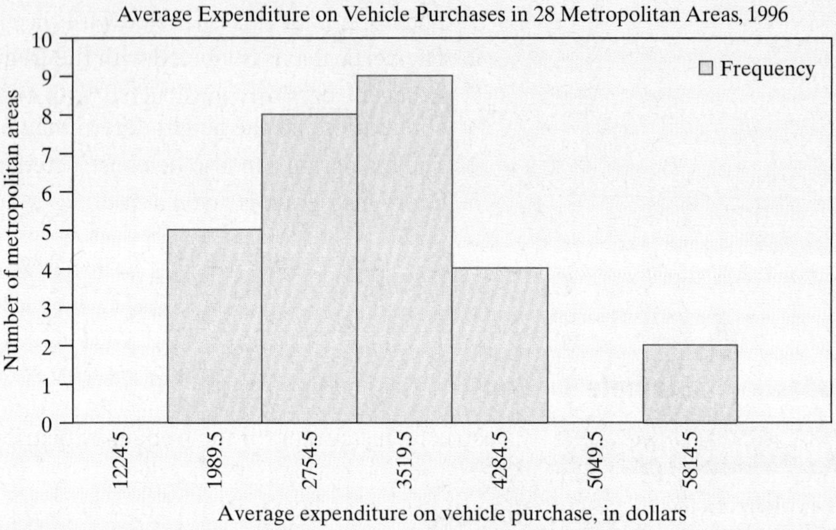

(b)

Figure 7.2.2

For more on making histograms on your calculator, consult the online graphing calculator manual at www.prenhall.com/armstrong.

The Frequency Polygon

Another way to graph a frequency distribution is to construct a **frequency polygon**. This type of graph uses the **midpoints** of the classes on the horizontal axis.

Class Midpoint

Given a frequency distribution, the **class midpoint**, denoted by m, is given by

$$m = \frac{\text{lower class limit} + \text{upper class limit}}{2}$$

The frequency polygon is usually, but not always, associated with the classes being some measure of time, with the units being minutes, hours, or years.

Properties of the Frequency Polygon

1. The horizontal axis is labeled with the class midpoints.
2. The vertical axis is labeled with the frequency or the relative frequency.
3. Points are plotted using the midpoints and the frequencies.
4. The points are connected with line segments.

5. Extra *tie-down* points are added to the beginning and end of the graph. The tie-down points have a frequency of zero. The left tie-down point is the first class midpoint minus the class width. The right tie-down point is the last class midpoint plus the class width.

Example 4 Constructing a Frequency Polygon

The number of Americans, measured in millions, aged 5 to 74 who were covered by private or government health insurance in 1997 is shown in the frequency distribution in Table 7.2.8.

Table 7.2.8

Class (age group)	Frequency, f (number covered in millions)
5 to 14	38
15 to 24	35
25 to 34	40
35 to 44	45
45 to 54	34
55 to 64	22
65 to 74	32

SOURCE: U.S. Department of Health and Human Services, www.hhs.gov

(a) Compute the class width.

(b) Add a column to the frequency distribution that contains the class midpoints.

(c) Construct a frequency polygon for the distribution.

Solution

(a) The difference between the lower class limits of the first two classes is $15 - 5 = 10$, which is also true for each successive class. So the class width is $CW = 10$.

(b) The midpoint of the first class is $\frac{5+14}{2} = 9.5$. Continuing in this fashion, we get the distribution shown in Table 7.2.9. Notice that the difference between any two consecutive class midpoints is also the class width.

Table 7.2.9

Class (age group)	Class Midpoint, m	Frequency, f (number covered in millions)
5 to 14	9.5	38
15 to 24	$\frac{15+24}{2} = 19.5$	35
25 to 34	29.5	40
35 to 44	39.5	45
45 to 54	49.5	34
55 to 64	59.5	22
65 to 74	69.5	32

(c) Before we construct the frequency polygon, we determine the tie-down points. To get the left tie-down point, we subtract the class width from the first class midpoint to get $9.5 - 10 = -0.5$. The right tie-down point is found by adding the class width to the last class limit, which gives us $69.5 + 10 = 79.5$. The completed frequency polygon is shown in Figure 7.2.3. Note that, even though it may not make sense to have a negative value on the horizontal axis, the idea is to include tie-down points so that a closed polygon is formed.

Americans Covered by Private or Government Health Insurance, 1997

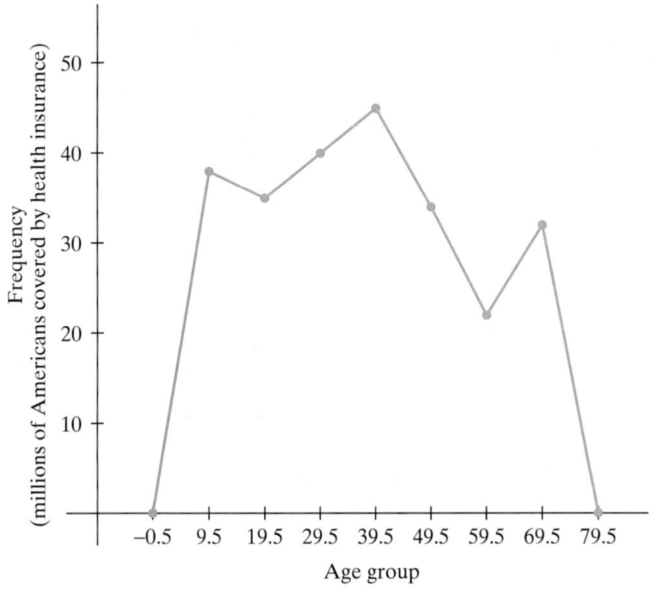

Figure 7.2.3

Checkpoint 3

Now work Exercise 27.

The Ogive

Many times we want to know how many data values are either below or above a certain quantitative value or even between two given values. This can be done by constructing an **ogive** (pronounced ójive). An ogive is a graph of the **cumulative frequencies** of each class.

Cumulative Frequency

The **cumulative frequency** of a class is the frequency of that class added to the frequencies of the previous classes.

We can see from the definition that the cumulative frequency of the first class is just that class's frequency. The cumulative frequency of the last class in a frequency distribution is the same as the sample size n.

Properties of the Ogive

1. The horizontal axis contains the upper class boundaries.
2. The vertical axis contains the cumulative frequencies.
3. Points are plotted with the upper class boundaries and the cumulative frequencies. The points are connected with line segments.
4. A leftmost point is plotted with a cumulative frequency of zero. The point is plotted on the horizontal axis and is given by the first upper class boundary minus the class width.

Example 5 Constructing an Ogive

For the frequency distribution in Example 3, complete the following:

(a) Compute the cumulative frequency of each class.

(b) Construct an ogive for the distribution.

Solution

(a) The cumulative frequency of the first class is 5 since there are no classes that come before it. The cumulative frequency of the second class is $5 + 8 = 13$. Continuing in this fashion, we get the distribution shown in Table 7.2.10.

Table 7.2.10

Class	Class Boundaries	Frequency, f	Cumulative Frequency
1225 to 1989	1224.5 to 1989.5	5	5
1990 to 2754	1989.5 to 2754.5	8	$5 + 8 = 13$
2755 to 3519	2754.5 to 3519.5	9	$13 + 9 = 22$
3520 to 4284	3519.5 to 4284.5	4	$22 + 4 = 26$
4285 to 5049	4284.5 to 5049.5	0	$26 + 0 = 26$
5050 to 5814	5049.5 to 5814.5	2	$26 + 2 = 28$

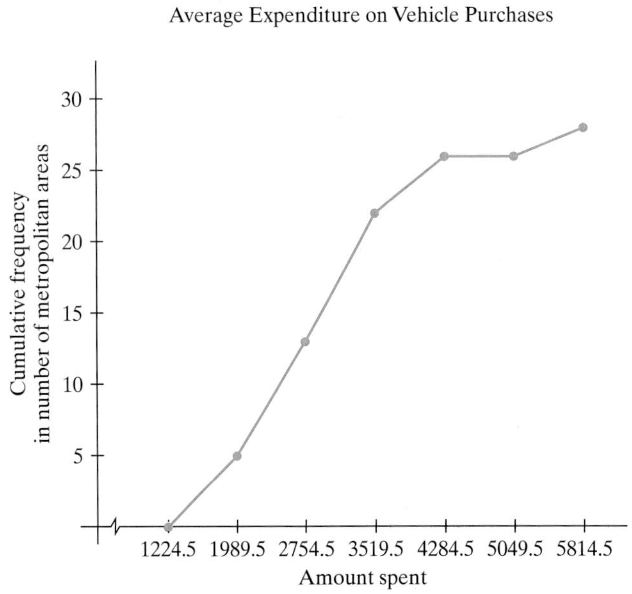

Figure 7.2.4

(b) Since the class width is 765, we see that the left tie-down point is $1989.5 - 765 = 1224.5$. Note that this is also the lower boundary for the first class. The ogive is shown in Figure 7.2.4. Among other observations, we can see from the graph that about 20 of the U.S. metropolitan areas had average vehicle purchases of about \$3500 or less. ■

Checkpoint 4 Now work Exercise 35.

SUMMARY

In this section, we saw that we could organize a set of quantitative data by constructing a **frequency distribution**. From this distribution, we could make the **histogram**, **frequency polygon**, and the **ogive**.

- The **relative frequency** of a class is given by

$$\text{(relative frequency)} = \frac{\text{frequency of a class}}{\text{sum of frequencies}} = \frac{f}{n} = \frac{f}{\sum f}.$$

- The **class width**, denoted by CW, is given by

$$CW = \frac{\text{largest data value} - \text{smallest data values}}{\text{number of classes}},$$

where CW is rounded up to the next whole number.

- The **lower class boundary** is the lower class limit minus one-half unit.
- The **upper class boundary** is the upper class limit plus one-half unit.
- The **class midpoint**, denoted by m, is given by

$$m = \frac{\text{lower class limit} + \text{upper class limit}}{2}.$$

- The **cumulative frequency** of a class is the frequency of that class added to the frequencies of the previous classes.

SECTION 7.2 EXERCISES

In Exercises 1–10, a frequency distribution is given. Answer the following.

1. The amount of square footage for single-family homes, in thousands, in the United States during 1997 is shown in the frequency distribution.

Class (square feet)	Frequency (in thousands)
501 to 999	9,764
1000 to 1499	17,030
1500 to 1999	15,007
2000 to 2499	10,680
2500 to 2999	5,944
3000 to 3499	3,282
3500 to 3999	2,188

SOURCE: U.S. Census Bureau, www.census.gov

(a) What is the class width?

(b) How many houses had square footage from 2000 to 2499 square feet?

2. The amount of square footage for *vacant* single family homes, in thousands, in the United States during 1997 is shown in the frequency distribution.

Class (square feet)	Frequency (in thousands)
501 to 999	799
1000 to 1499	1130
1500 to 1999	741
2000 to 2499	373
2500 to 2999	238
3000 to 3499	127
3500 to 3999	85

SOURCE: U.S. Census Bureau, www.census.gov

(a) What is the class width?

(b) How many vacant houses had square footage from 2000 to 2499 square feet?

3. The average amount of financial debt, in thousands of dollars, held by U.S. families during 1995 is classified by age group in the frequency distribution.

Class (age group)	Frequency (in thousands of dollars)
25 to 34	15.0
35 to 44	37.6
45 to 54	40.7
55 to 64	21.5
65 to 74	7.6
75 to 84	1.9

SOURCE: U.S. Census Bureau, www.census.gov

(a) What is the class width?

(b) Which age group had the smallest amount of financial debt? Which age group had the largest amount of financial debt?

4. The average amount of *credit-card* debt, rounded to the nearest hundred dollars, held by U.S. families during 1995 is classified by age group in the frequency distribution.

Class (age group)	Frequency (in dollars)
25 to 34	1300
35 to 44	1900
45 to 54	2000
55 to 64	1300
65 to 74	800
75 to 84	400

SOURCE: U.S. Census Bureau, www.census.gov

(a) What is the class width?

(b) Which age group had the smallest amount of credit-card debt? Which age group had the largest amount of credit-card debt?

5. The number of farms, in thousands, owned by Americans during 1992 is classified by age group in the following frequency distribution.

Class (age group)	Frequency (in thousands of farms)
15 to 24	28
25 to 34	179
35 to 44	382
45 to 54	429
55 to 64	430
65 to 74	478

SOURCE: U.S. Department of Agriculture, www.usda.gov

(a) How many farms were owned by people who were 55 years or older?

(b) How many farms were owned by people who were 25 to 54 years old?

6. The number of farms, in thousands, owned by Americans during 1997 is classified by age group in the following frequency distribution.

Class (age group)	Frequency (in thousands of farms)
15 to 24	21
25 to 34	128
35 to 44	371
45 to 54	467
55 to 64	427
65 to 74	497

SOURCE: U.S. Department of Agriculture, www.usda.gov

(a) How many farms were owned by people who were 55 years or older?

(b) How many farms were owned by people who were 25 to 54 years old?

7. The number of Americans, in millions, aged 15 to 74 covered by private or government health insurance by age group in 1997 is shown in the frequency distribution.

Class (age group)	Frequency (in millions)
15 to 24	32.3
25 to 34	39.4
35 to 44	44.5
45 to 54	34.1
55 to 64	22.3
65 to 74	32.1

SOURCE: U.S. Department of Health and Human Services, www.hhs.gov

(a) According to the frequency distribution, how many Americans aged 15 to 74 were covered by private or government health insurance in 1997?

(b) Which age group had the most Americans covered by private or government health insurance in 1997?

8. The number of Americans, in millions, covered by private or government health insurance classified by household income in 1997 is shown in the frequency distribution.

Class (income in dollars)	Frequency (in millions)
1 to 25,000	72.2
25,001 to 50,000	80.4
50,001 to 75,000	56.2
75,001 to 100,000	60.3

SOURCE: U.S. Department of Health and Human Services, www.hhs.gov

(a) According to the frequency distribution, how many Americans with household incomes of \$100,000 or less were covered by private or government health insurance in 1997?

(b) How many Americans with household incomes of \$50,000 or less were covered by private or government health insurance in 1997?

(c) How many Americans with household incomes of \$50,001 or more were covered by private or government health insurance in 1997?

9. The costs for homeowners, in thousands, in the northeastern United States whose monthly housing costs were under \$1000 are shown in the frequency distribution.

Class (cost in dollars)	Frequency (in thousands of homeowners)
200 to 299	1923
300 to 399	1493
400 to 499	1339
500 to 599	967
600 to 699	849
700 to 799	765
800 to 899	482
900 to 999	722

SOURCE: U.S. Census Bureau, www.census.gov

(a) How many northeastern homeowners had monthly costs of less than \$500?

(b) How many northeastern homeowners had monthly costs of \$300 or more, but less than \$700?

10. The costs for homeowners, in thousands, in the midwestern United States whose monthly housing costs were under \$1000 are shown in the frequency distribution.

Class (cost in dollars)	Frequency (in thousands of homeowners)
200 to 299	4912
300 to 399	2186
400 to 499	1430
500 to 599	1379
600 to 699	1241
700 to 799	1078
800 to 899	700
900 to 999	1052

SOURCE: U.S. Census Bureau, www.census.gov

(a) How many midwestern homeowners had monthly costs of less than $500?

(b) How many midwestern homeowners had monthly costs of $300 or more, but less than $700?

In Exercises 11–16, complete the following:

(a) Determine the class boundaries.

(b) Construct a histogram.

11. Use the frequency distribution in Exercise 1.

12. Use the frequency distribution in Exercise 2.

13. Use the frequency distribution in Exercise 3.

14. Use the frequency distribution in Exercise 4.

✓ **15.** Use the frequency distribution in Exercise 5.

16. Use the frequency distribution in Exercise 6.

In Exercises 17–20, complete the following:

(a) Determine the class boundaries.

(b) Determine the relative frequency for each class. Round to the nearest hundredth.

(c) Construct a relative frequency histogram.

17. Use the frequency distribution in Exercise 7.

18. Use the frequency distribution in Exercise 8.

19. Use the frequency distribution in Exercise 9.

20. Use the frequency distribution in Exercise 10.

For the data in Exercises 21–26, complete the following:

(a) Write a frequency distribution. See Example 2.

(b) Use your calculator to construct a histogram.

21. The amount in dollars, spent on new clothes at the Happening Zone for a sample of 30 preteens is shown in the table. Use five classes.

48	345	127	33	48	260
189	388	190	286	370	100
247	255	400	130	189	263
0	144	188	266	241	188
378	330	249	287	274	91

22. The time, in minutes, spent watching the local news during one day for a sample of 28 citizens of Lewisburg is shown in the table. Use five classes.

45	120	25	10	35	29	39
15	51	14	38	210	180	185
125	60	35	40	81	36	110
85	66	90	41	50	35	15

23. The area, measured in square miles, of selected major bodies of water in the Gulf Coast area of the United States is shown in the table. Use six classes.

813	733	631	616	511
310	271	253	245	236
212	189	151	146	129
122	118	112	112	104
101	99	94	93	89
85	83	79	75	

SOURCE: U.S. Department of the Interior, www.doi.gov

24. The area, measured in square miles, of selected major bodies of water in Alaska is shown in the table. Use six classes.

1559	1382	1199	1022	792	791
791	702	640	463	447	436
433	403	393	348	345	324
324	322	310	310	301	295
293	241	229	225	210	206

SOURCE: U.S. Department of the Interior, www.doi.gov

25. The length, in miles, of the 25 longest rivers in the United States is shown in the table. Use five classes.

2540	2340	1980	1900	1900
1460	1450	1420	1310	1290
1280	1240	1040	990	926
906	886	862	800	774
743	724	692	659	649

SOURCE: U.S. Department of the Interior, www.doi.gov

26. The number of visitors, in thousands, by overseas travelers to the top 35 traveled states in the United States in 1996 is shown in the table. Use six classes.

22,658	6004	5710	4804	3059	2062	1292	1156	1178	974
897	634	657	702	612	566	363	385	498	408
340	363	385	498	408	340	363	249	227	272
272	227	181	159	159					

SOURCE: U.S. Census Bureau, www.census.gov

In Exercises 27–34, a frequency distribution is given. Complete the following:

(a) Determine the midpoint of each class.

(b) Construct a frequency polygon.

✓ **27.** The ages of the employees at the NewFinish Corporation are shown in the frequency distribution.

Class (age)	Frequency (in employees)
20 to 29	15
30 to 39	50
40 to 49	56
50 to 59	30
60 to 69	18

28. The pretest scores of 24 entering freshmen at Festivelle High School are shown in the frequency distribution.

Class (test score)	Frequency (in students)
96 to 102	2
103 to 109	3
110 to 116	3
117 to 123	4
124 to 130	5
131 to 137	1
138 to 144	4
145 to 151	1
152 to 158	1

29. The weights, in pounds, of 100 sixth-grade students at P.H. Johnson elementary school are shown in the frequency distribution.

Class (pounds)	Frequency (in students)
59 to 61	4
62 to 64	8
65 to 67	12
68 to 70	13
71 to 73	21
74 to 76	15
77 to 79	12
80 to 82	9
83 to 85	4
86 to 88	2

30. The total fines, in dollars, for outstanding parking tickets in the village of Kettersville are shown in the frequency distribution.

Class (dollars)	Frequency
10 to 24	5
25 to 39	8
40 to 54	9
55 to 69	15
70 to 84	5
85 to 99	2

31. Use the frequency distribution in Exercise 1.

32. Use the frequency distribution in Exercise 2.

33. Use the frequency distribution in Exercise 3.

34. Use the frequency distribution in Exercise 4.

In Exercises 35–40, complete the following:

(a) Determine the cumulative frequency for each class.

(b) Construct an ogive.

✓ **35.** Use the frequency distribution in Exercise 27.

36. Use the frequency distribution in Exercise 28.

37. Use the frequency distribution in Exercise 29.

38. Use the frequency distribution in Exercise 30.

39. Use the frequency distribution in Exercise 6.

40. Use the frequency distribution in Exercise 7.

*For Exercises 41–44, use the histogram for U.S. mothers aged 15 to 39 of newborns in 1990 to answer the following (*Source: *U.S. Census Bureau,* www.census.gov*).*

Age of Mother to Newborns, 1990

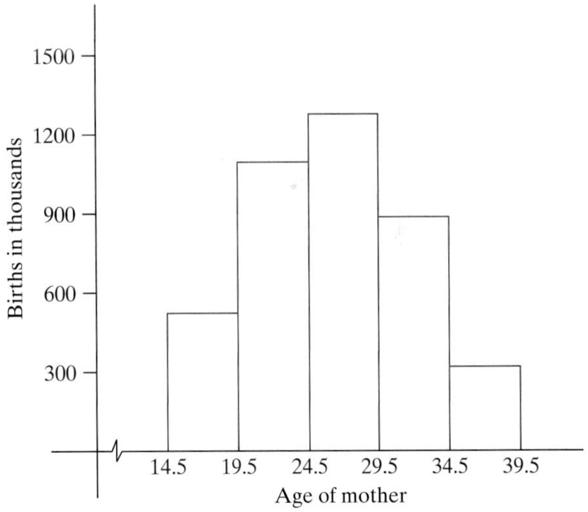

41. What age group had the most newborns?

42. What age group had the least newborns?

43. What is the class width?

44. What are the class *limits* of the third class?

*For Exercises 45–48, use the histogram (see on the next page) for U.S. mothers aged 15 to 44 of newborns in 1998 to answer the following (*Source: *U.S. Census Bureau,* www.census.gov*).*

45. What age group had the most newborns?

46. What age group had the least newborns?

47. What is the class width?

48. What are the class *limits* of the third class?

Age of Mother to Newborns, 1998

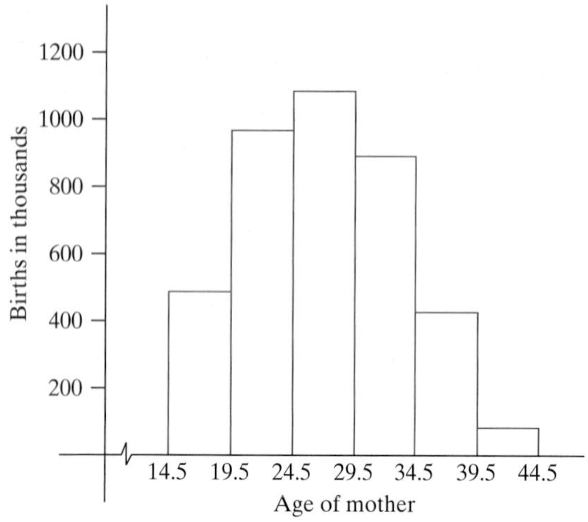

For Exercises 49–52, use the frequency polygon for the amount spent on child care and education to raise a child for husband–wife families whose incomes are between $36,000 and $60,600 in 1998 to answer the following (Source: U.S. Department of Health and Human Services, www.hhs.gov).

Annual Cost of Childcare and Education, 1998

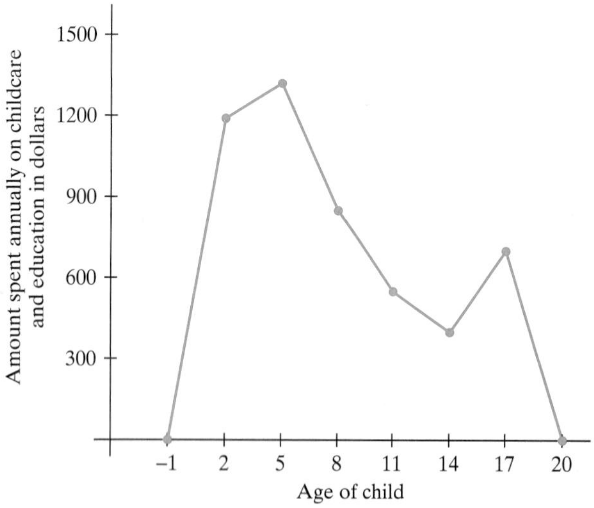

49. What is the class *width*?

50. What are the *limits* of the fourth class?

51. In which class do parents spend over $1300 annually on child care and education?

52. In which class do parents spend about $400 annually on child care and education?

For Exercises 53–56, use the frequency polygon for the number of deaths, in thousands, caused by cancer in the United States in 1997 to answer the following (Source: U.S. National Center for Health Statistics, www.cdc.gov/nchs).

53. What is the class *width*?

54. What are the class *limits* of the third class?

Deaths Caused by Cancer, 1997

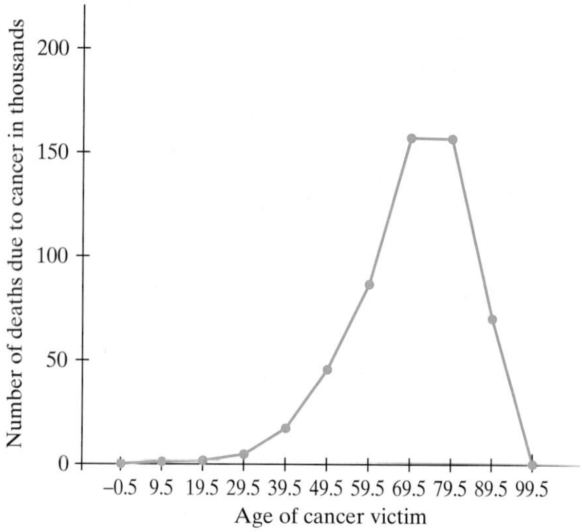

55. In which class(es) is the number of deaths caused by cancer close to 160,000?

56. In which class is the number of deaths caused by cancer close to 70,000?

For Exercises 57–60, use the ogive for the number of U.S. drivers, in thousands, involved in alcohol-related fatal crashes in 1998 to answer the following (Source: U.S. Department of Transportation, www.dot.gov).

Alcohol Involvement in Fatal Auto Crashes, 1998

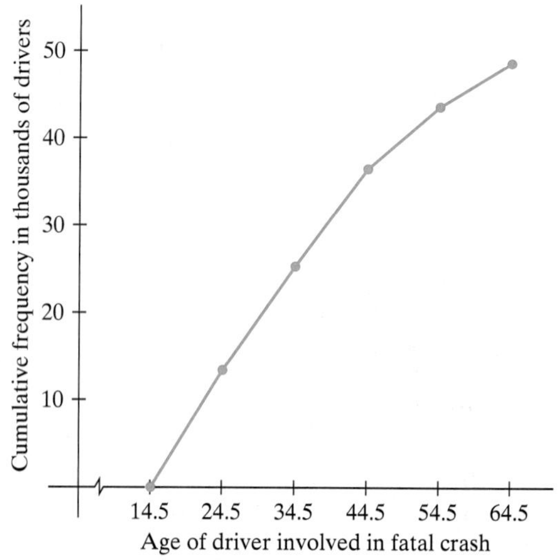

57. What are the class *limits* of the first class?

58. What is the class *width*?

59. About how many drivers aged 34.5 years and below were involved with alcohol-related fatal crashes in 1998?

60. From inspecting the ogive, which class has the greatest frequency? Explain.

For Exercises 61–64, use the ogive for the number of U.S. deaths, in thousands, caused by accidental and adverse effects in 1997 (Source: *U.S. National Center for Health Statistics,* www.cdc.gov/nchs).

Deaths Caused by Accidental and Adverse Effects, 1997

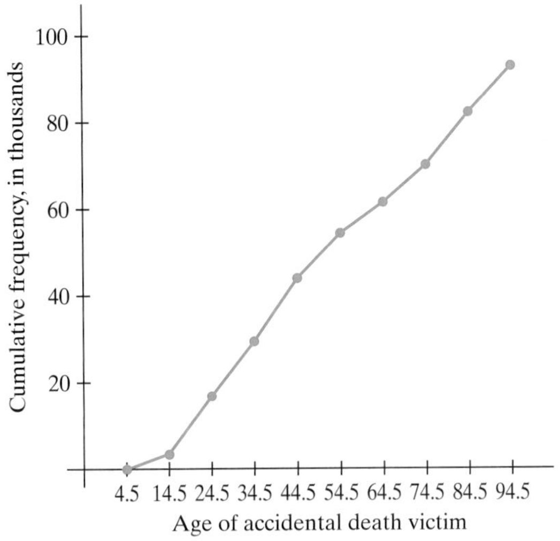

61. What are the class *limits* of the first class?

62. What is the class *width*?

63. About how many individuals aged 54.5 years and below died from accidental and adverse effects in 1997?

64. From inspecting the ogive, which class has the greatest frequency? Explain.

SECTION PROJECT

Another quick way to obtain a graph similar to a histogram is to use a **stemplot** (sometimes called a stem and leaf plot) to plot the data. On the stemplot, the data are separated by a **stem** (the leftmost digits) and a **leaf** (the rightmost digit) by a vertical bar. For example, consider the following data for the test results for 20 students taking a philosophy final exam.

22	40	51	52	62	62	68	69	69	74
78	79	80	82	85	86	88	90	94	99

The stemplot for the data is given in Figure 7.2.5.

Stem	Leaves
2	2
3	
4	0
5	1 2
6	2 2 8 9 9
7	4 8 9
8	0 2 5 6 8
9	0 4 9

Figure 7.2.5

The average fuel consumption, in miles per gallon, for U.S. automobiles for selected years from 1970 to 1996 is shown in the table (*Source:* U.S. Department of Transportation, www.dot.gov).

13.5	14.0	16.0	16.4	16.8	17.0	17.3	17.5	17.4	18.0
18.7	18.9	20.2	21.1	20.9	20.5	20.7	21.1	21.3	21.4

(a) Construct a stemplot for the average fuel consumption for automobiles. Use the digits to the left of the decimal point for the stems and the digit to the right of the decimal point for the leaves.

(b) Which row has the greatest number of leaves? How would you interpret this?

(c) Are there any rows that have no leaves? What does this mean?

(d) If a histogram of the fuel consumption were constructed, how would its shape compare to that of the stemplot?

Section 7.3 Measures of Centrality

Many times we are given a set of data to inspect, but there are so many numbers, or the numbers are so diverse, that it is difficult to analyze the data. To make sense of sets of data, we often want to know what mathematical characteristics data items possess. These characteristics are often referred to as **measures of centrality**, because they give us the value that represents the center or core of the data. When we hear statements such as "The average salary of American workers in 1997 was $29,261" or "The median sales price of a home in the United States in 1996 was $115,800," we are hearing measures of centrality. In this section we will see how these measures are computed and interpreted, the properties of these measures, and how to approximate a measure of centrality even if we do not have all the data at our disposal.

The Mean

One of the most frequently used measures in statistics is the **mean**, which is commonly called the **average**. We can determine the mean of a sample of n data items by determining its sum and dividing the sum by the sample size. To express the sum of the data items, we will use sigma notation and write $\sum x$, which we read as "the sum of all x-values."

From Your Toolbox

A **population** is the set of all elements under consideration. A **sample** is a subset of the population.

Mean

- For a **sample** of n data items, the **sample mean**, denoted by $\bar{x}$ (read "x-bar") is given by

$$\bar{x} = \frac{\sum x}{n} = \frac{\text{sum of data}}{\text{sample size}}$$

- If we have a **population** of N data items, then the **population mean**, denoted by μ (the Greek lowercase letter mu), is given by

$$\mu = \frac{\sum x}{N} = \frac{\text{sum of data}}{\text{population size}}$$

Note: Observe that we use n for the sample size and N for population size.

We can see from the definition of the mean and the note following the definition that we write $\bar{x}$ if we have a sample and μ if the data are regarded as a population. Since one is divided by n and the other by N, we must use two different notations.

Example 1 Determining a Mean

The seven most popular recreational activities of Americans in 1997 are shown in Table 7.3.1. Determine the mean and interpret.

Table 7.3.1

Activity	Participation (in millions)
Exercise walking	76.3
Swimming	59.5
Exercising with equipment	47.9
Camping	46.6
Bicycle riding	45.1
Bowling	44.8
Fishing	39.0

SOURCE: U.S. Census Bureau, www.census.gov

Solution

Understand the Situation: Notice that the data in Table 7.3.1 are for the seven most popular recreational activities. The table does not list all the recreational activities. This means that the data in Table 7.3.1 are a sample of all recreational activities. As a result, we are to compute a sample mean, $\bar{x}$.

Here we have a sample with size $n = 7$. The formula for the sample mean yields

$$\bar{x} = \frac{\sum x}{n} = \frac{76.3 + 59.5 + 47.9 + 46.6 + 45.1 + 44.8 + 39.0}{7} = \frac{359.2}{7} \approx 51.3$$

Interpret the Solution: Since inevitably some people participate in more than one activity, this does not mean that in 1997 an average of 51.3 million Americans participated in the top seven recreational activities. What it does mean is that in 1997 an average of 51.3 million participations were made in the top seven recreational activities by Americans. Some of the participations may have been made by the same people. ■

✓ Checkpoint 1 Now work Exercise 7.

Technology Option

All graphing calculators compute the mean for a given set of data, but the way that the data are entered and used for statistical computations varies from model to model. In Figure 7.3.1(a), we have entered the data from Example 1 into a list. In Figure 7.3.1(b), we have used the 1-Variable Stats menu. Notice in Figure 7.3.1(b)

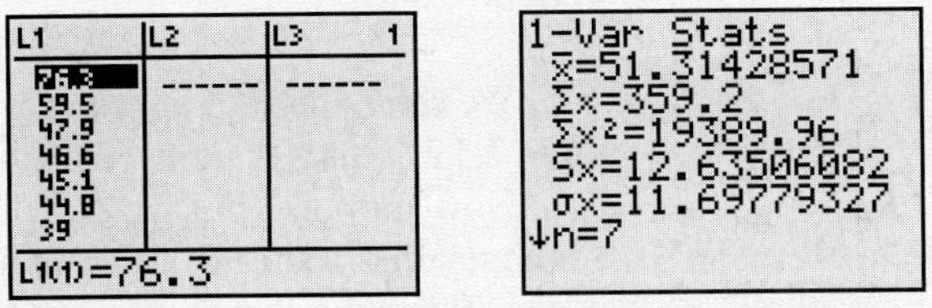

(a) (b)

Figure 7.3.1

that $\bar{x} \approx 51.3$, the same result achieved in Example 1. Other items of interest in Figure 7.3.1(b) are $\sum x = 359.2$ and $n = 7$. Both of these were also determined in Example 1.

To see how your calculator computes the mean, consult the online graphing calculator manual at www.prenhall.com/armstrong.

We learned in Section 7.2 that when samples of data are relatively large we can categorize the data into classes and summarize them in a **frequency distribution**. However, if the data have already been summarized in a frequency distribution, it is still possible to estimate the mean. Before we illustrate how this is done, let's consider another example of computing a mean.

Example 2 Determining a Mean

The length in miles, of the 25 longest U.S. rivers is shown in Table 7.3.2. Determine the mean and interpret.

Table 7.3.2

2540	2340	1980	1900	1900
1460	1450	1420	1310	1290
1280	1240	1040	990	926
906	886	862	800	774
743	724	692	659	649

SOURCE: U.S. Department of the Interior, www.doi.gov

Solution

Understand the Situation: Table 7.3.2 shows the lengths of the 25 longest U.S. rivers. The table does not contain all U.S. rivers, which indicates that we have a sample. As a result, we need to compute a sample mean, $\bar{x}$.

Here we have a sample with size $n = 25$. The formula for the sample mean yields

$$\bar{x} = \frac{\sum x}{n} = \frac{2540 + 2340 + 1980 + \cdots + 692 + 659 + 649}{25} = \frac{30{,}761}{25} = 1230.44$$

Interpret the Solution: This means that the average length of the 25 longest U.S. rivers is about 1230 miles. ■

Table 7.3.3

Class	Frequency, f
649 to 1027	12
1028 to 1406	5
1407 to 1785	3
1786 to 2164	3
2165 to 2543	2

We can construct a frequency distribution, using five classes, for the data in Table 7.3.2. This frequency distribution is shown in Table 7.3.3.

Now suppose that we did not have the actual data and were only given the frequency distribution shown in Table 7.3.3. To see how we can approximate $\bar{x}$, let's examine the distribution further. We know, for example, that the first class contains 12 data items with values ranging anywhere from 649 to 1027. If we do not know what these 12 values are, the best guess would be the value in the middle of the class. In Section 7.2, we called this value the **class midpoint**, x_m. The midpoint of each class is shown in Table 7.3.4.

From Your Toolbox

The **midpoint** of a class is determined by computing

$$\frac{\text{lower limit} + \text{upper limit}}{2} = x_m.$$

Table 7.3.4

Class	Midpoint, x_m	Frequency, f
649 to 1027	$\frac{649+1027}{2} = 838$	12
1028 to 1406	$\frac{1028+1406}{2} = 1217$	5
1407 to 1785	1596	3
1786 to 2164	1975	3
2165 to 2543	2354	2

To approximate $\bar{x}$, we first multiply the class midpoint by the class frequency. This operation gives us an approximation for the sum of the values in a class. Then we add the products and divide by the sum of the frequencies to determine our estimate. For the distribution in Table 7.3.4, we get

$$\bar{x} \approx \frac{838(12) + 1217(5) + 1596(3) + 1975(3) + 2354(2)}{25} = 1262.48$$

This means, using the frequency distribution, that the mean of the 25 longest U.S. rivers is about 1262 miles. Notice that this value differs from the actual mean found in Example 2 by only about 32 miles.

Mean from a Frequency Distribution

Given a frequency distribution, an approximation for the **mean** of the summarized data is given by

$$\bar{x} \approx \frac{\sum x_m \cdot f}{n}$$

where

x_m is the midpoint of each class
f is the class frequency
n is the sample size, where $n = \sum f$

The Median

Table 7.3.5

2	1	12	3	2	3	2

Even though the mean is the most commonly used measure of centrality, it often does not give us enough information about the data. To see how this can happen, let's suppose that a group of seven students from Vanderburg Junior High School are participants in the summer *Got the Need to Read* program. The number of books read by each participant is shown in Table 7.3.5.

The mean of the data is $\bar{x} = \frac{\sum x}{n} = \frac{2+1+12+3+2+3+2}{7} = \frac{25}{7} \approx 3.6.$

This number is somewhat misleading. Notice that the mean is about 3.6, yet there are six values less than 3.6 and only one greater than 3.6. The reason that the mean is so large is because one motivated student read 12 books; nine more than any other student in the group. We call this value of 12 an **outlier** because it "lies outside" all the other data items. Since the mean uses every data element when it is calculated, it is susceptible to being distorted by outliers. When outliers are present, and in other cases as well, we may find another measure of centrality called the **median** useful.

Interactive Activity

We saw that if $n = 7$ then the median was the fourth item of the ordered data. Which item in the ordered data would be the median if $n = 11$ or $n = 23$? Write a general expression telling which item in the ordered data would be the median for a sample of size n, where n is an odd number.

Median

For a set of data, the **median**, denoted by M, is the middle value of the set of data after the data have been ordered from smallest to largest.

Note: For a set of data, we can think of the median as being similar to the "median" in a divided highway.

If we order the data from the summer reading program from smallest to largest, we get

$$1, \quad 2, \quad 2, \quad 2, \quad 3, \quad 3, \quad 12$$

Since there are $n = 7$ data items, the median is the fourth value in the order, meaning that $M = 2$.

Computing the median when we have an odd number of data items is fairly casy, but what happens when we have an even number of data? Example 3 illustrates how to handle this situation.

Example 3 Determining the Median

The total number of votes cast, in millions, in U.S. presidential elections from 1960 to 1996 is shown in Table 7.3.6. Determine the median M.

Table 7.3.6

Year	Total Votes Cast (in millions)
1960	68.8
1964	70.6
1968	73.2
1972	77.7
1976	81.6
1980	86.5
1984	92.7
1988	91.6
1992	104.4
1996	96.3

SOURCE: U.S. Census Bureau, www.census.gov

Solution

First we must rank the data from smallest to largest. The ordered data are

$$68.8, 70.6, 73.2, 77.7, 81.6, 86.5, 91.6, 92.7, 96.3, 104.4$$

Since there are $n = 10$ data items, an even number of data items, no single data item is the middle. In this situation we must find the mean (average) of the fifth and sixth values.

This yields

$$M = \frac{81.6 + 86.5}{2} = 84.05$$

Interpret the Solution: This means that between 1960 and 1996 in half of the presidential elections less than 84.05 million Americans voted and in the other half more than 84.05 million Americans voted. ■

Technology Option

Graphing calculators with statistical capabilities can compute the median of a set of entered data. Scientific calculators generally do not compute the median. For Figures 7.3.2(a) and (b), we have entered the data in Example 3 into a list. We then used the 1-Variable Stats menu. In Figure 7.3.2(a), can you determine what the mean for the set of data is? Figure 7.3.2(b) shows us that Med = 84.05, the same result attained in Example 3. It also tells us that $n = 10$.

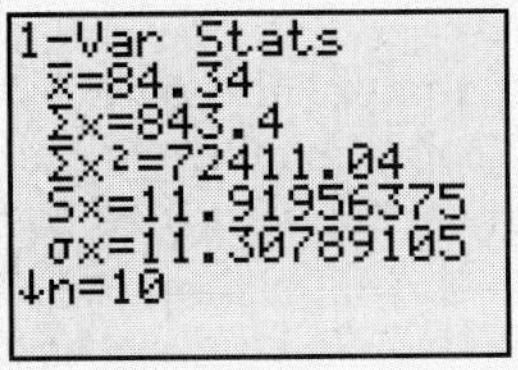

(a)

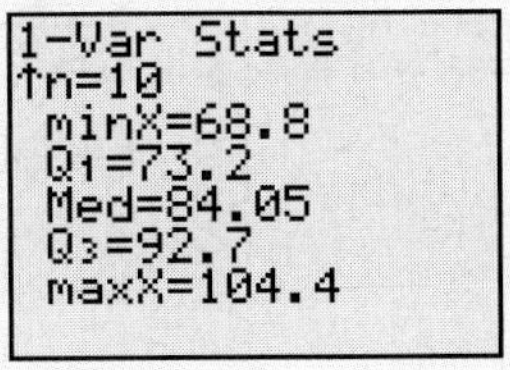

(b)

Figure 7.3.2

To see how your calculator computes the median, consult the online graphing calculator manual at www.prenhall.com/armstrong.

Since the median is computed by determining the position of an item in a set of data, rather than by the data values themselves, the median is not affected by outliers. Because it is more resistant to extreme values, the median is sometimes called a **resistant** measure of centrality.

The Mode

So far, we have seen how to determine measures of centrality for quantitative data. If the data are qualitative (categorical), however, the mean and median cannot be determined. The **mode** is a measure of centrality that sometimes is useful in analyzing qualitative data.

Mode

For a set of data, either quantitative or qualitative, the **mode**, denoted by M_0, is the data item or data items that occur most frequently.

Note: 1. If no data items are repeated, then the set of data has no mode.
2. If more than one data item has the highest frequency, then each of these data items is a mode.

Example 4 Determining the Mode

ActionNews 16 conducts a focus group to determine which newscast, morning, noon, or evening, these viewers watch most frequently. The results are shown in Table 7.3.7.

Table 7.3.7

evening	morning	evening	noon
morning	evening	morning	morning
evening	noon	morning	evening

Determine the mode, M_0, and interpret.

Table 7.3.8

Morning	5
Noon	2
Evening	5

Solution

We begin to analyze the data by constructing a frequency table as shown in Table 7.3.8. From the table we see that both the morning and evening newscasts have the highest frequency of 5. So both morning and evening newscasts are modes.

Interpret the Solution: This means that the morning and the evening newscasts are most frequently viewed by members of this focus group. Since we have *two* modes in this data set, we would say that it is **bimodal**. ■

✓ Checkpoint 2

Now work Exercise 29.

Technology Option

Most graphing calculators do not explicitly compute the mode of a set of entered quantitative data. The mode can be determined by first sorting the quantitative data in ascending order and then inspecting the data for the item with the highest frequency. For example, the quantitative data 3, 1, 4, 1, 5, 9, 2 are stored in a list as shown in Figure 7.3.3(a). In Figure 7.3.3(b), they are sorted in ascending order and we can see that the mode is 1.

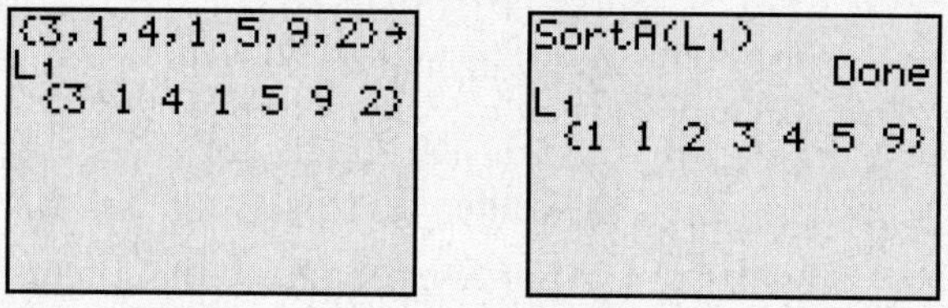

(a) (b)

Figure 7.3.3

To see how data are sorted on your calculator, consult the online graphing calculator manual at www.prenhall.com/armstrong.

The Midrange

There are times when we are not given the complete set of data, but just the extreme values of a set of data. For example, the local weather report usually gives the low and high temperatures for the day, but no other temperatures. In situations like these, we can still determine a measure of centrality by computing the **midrange** of the data. The type of measure of centrality that the midrange gives us is similar to the mean (or average).

Midrange

For a set of data, the **midrange**, denoted by MR, is given by

$$MR = \frac{(\text{minimum data value}) + (\text{maximum data value})}{2}$$

Example 5 Determining the Midrange

When researching the total market value of major league baseball franchises in the United States and Canada in 1996, it was found that the New York Yankees had the highest value at \$241 million and the Minnesota Twins had the lowest value at \$77 million (*Source:* www.infoplease.com). Determine MR.

Solution

For these data we have a minimum value of 77 and a maximum value of 241, so the midrange is

$$MR = \frac{77 + 241}{2} = \frac{318}{2} = 159$$

This means that the midrange of the major league baseball franchises in the United States and Canada in 1996 was about \$159 million. ■

Technology Option

Generally, calculators do not have built-in capabilities to compute the midrange. However, many compute the minimum and maximum values from which the midrange can be found. To see how to get the minimum and maximum values for a set of data on your calculator, consult the online graphing calculator manual at www.prenhall.com/armstrong.

When looking at the market values of **every** major league baseball franchise in 1996, it can be shown that the median franchise market value is $M = \$128$ million and the mean is $\mu = \$134$ million. The reason that the midrange is so much different from the median or mean is because the New York Yankees' market value was significantly more than the rest of the baseball teams in 1996. This illustrates that, even though the midrange can be used to determine a measure of centrality when little information is given, it is very susceptible to outliers.

Skewness

Frequency polygons and histograms can have many shapes. Some graphs have an important feature called **skewness**. Let's say that in a frequency distribution the higher frequencies occur in the beginning, or lower classes. Then we say the graph is **skewed right**. On the other hand, if the higher frequencies occur in the ending, or higher classes, then the graph is **skewed left**. If the higher frequencies occur in the middle classes, we call the graph **symmetric**. Examples of these cases are shown in Figure 7.3.4(a)–(c).

The graph in Figure 7.3.4(a) is skewed right because the data *tails off* or *falls off* to the right. We see that the skewness is determined by the direction of the tail, and not where the data are clustered. The severity of the skewness is often determined by how much the mean and median differ in value.

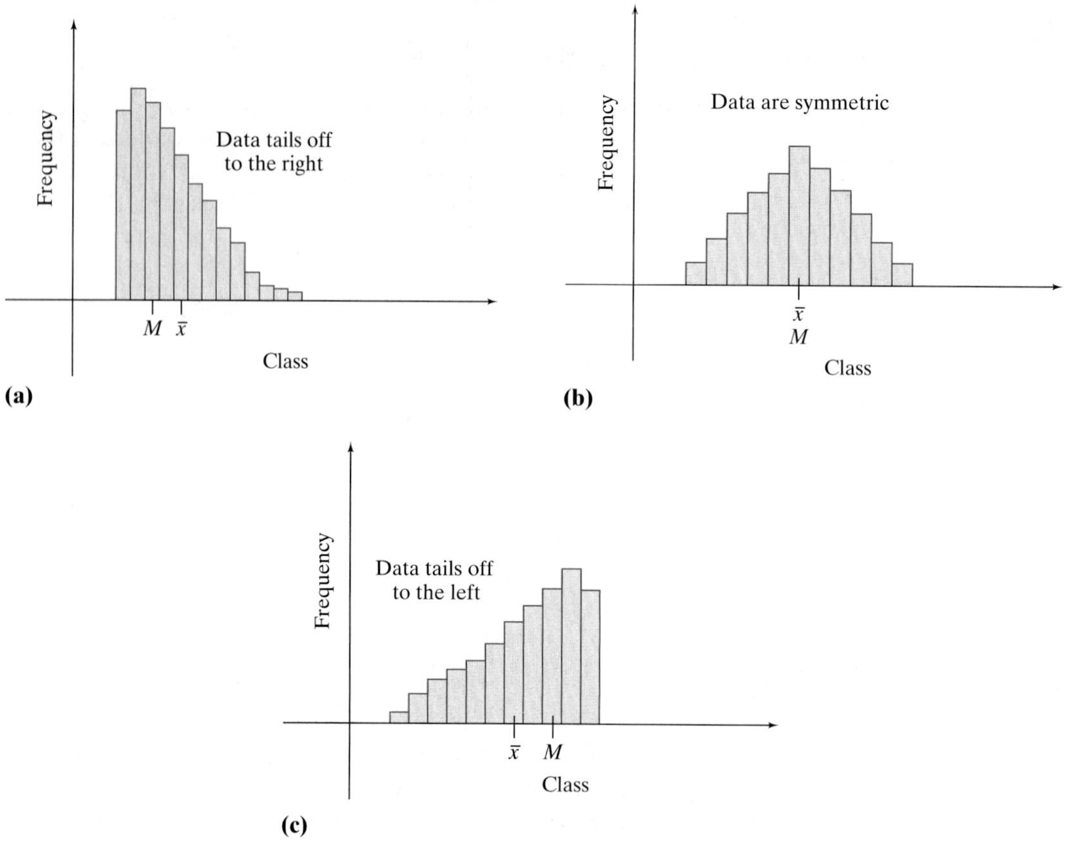

Figure 7.3.4 **(a)** Histogram is skewed right. **(b)** Histogram is symmetric. **(c)** Histogram is skewed left.

Example 6 Determining the Skewness of a Distribution

To decide if a new parking garage is needed, managers at the William Welsh Office Tower survey the employees to see how many had at least one parking ticket in the last 3 months. The distribution of those who had at least one ticket in the last 3 months is shown in Table 7.3.9.

(a) Determine the mean number of tickets given to those who received tickets.

(b) Construct a histogram of the number of tickets given in the 3-month period to those who received tickets.

(c) Classify the histogram as skewed left, skewed right, or symmetric.

Table 7.3.9

Number of Tickets	Frequency, f
1	23
2	35
3	20
4	8
5	4
6	1

Solution

(a) First note that the size of the sample is $n = \sum f = 23 + 35 + \cdots + 1 = 91$. To find the mean, we can use the formula $\bar{x} \approx \dfrac{\sum x_m \cdot f}{n}$, where x_m represents the number of tickets.

$$\bar{x} \approx \frac{(1)23 + (2)35 + (3)20 + (4)8 + (5)4 + 6(1)}{91} = \frac{211}{91} \approx 2.32$$

So, among those receiving parking tickets in the 3-month period, the mean number of tickets received was about 2.32.

(b) Here we have six classes, yet there are no class boundaries. We can construct the histogram by drawing the rectangles directly over the values on the horizontal axis, as shown in Figure 7.3.5.

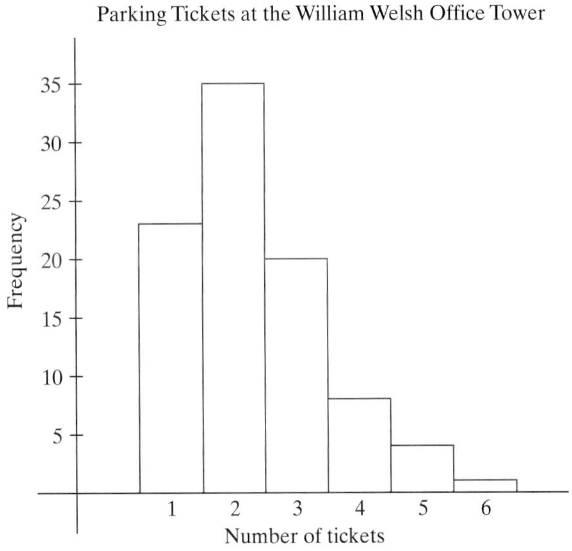

Figure 7.3.5

(c) We can see in Figure 7.3.5 that the histogram tails off to the right. This tells us that the histogram is skewed right. ■

SUMMARY

- For a sample of n data items, the **sample mean** is given by $\bar{x} = \dfrac{\sum x}{n} = \dfrac{\text{sum of data}}{\text{sample size}}$.
- Given a frequency distribution, an approximation for the **mean** of the summarized data is given by $\bar{x} \approx \dfrac{\sum x_m \cdot f}{n}$, where x_m is the midpoint of each class and f is the class frequency.
- For a set of data, the **median**, denoted by M, is the middle value of the set of data after the data have been ordered from smallest to largest.
- For a set of data, either quantitative or qualitative, the **mode**, denoted by M_0, is the data item (or items) that occurs most frequently.
- For a set of data, the **midrange**, denoted by MR, is given by

$$MR = \frac{\left(\begin{array}{c}\text{minimum data}\\ \text{value}\end{array}\right) + \left(\begin{array}{c}\text{maximum data}\\ \text{value}\end{array}\right)}{2}.$$

SECTION 7.3 EXERCISES

In Exercises 1–10, a set of data is given. Determine the sample mean $\bar{x}$ and the mode M_0.

1.

6	3	6	5	9	8

2.

9	8	4	5	8	1

3.

4	8	6	1	1
2	3	4	7	0

4.

5	1	4	0	3
4	0	7	8	2

5.

3	17	18	14	9	7
16	2	15	4	19	12

6.

21	15	25	2	13	8
22	17	18	9	24	7

✓ **7.**

14.3	15.8	17.4	18.3	19.1
15.8	11.2	9.0	11.2	19.5

8.

7.1	6.8	4.9	5.2	6.3
4.9	7.0	6.3	7.1	5.5

9.

0.25	0.59	0.63	0.58	0.21
0.11	0.78	0.65	0.93	0.27

10.

0.55	0.73	0.18	0.69	0.71
0.83	0.63	0.78	0.68	0.03

In Exercises 11–16, determine the population mean μ for the given population data.

11.

6	2	4	8	9
2	5	2	11	8

12.

4	7	8	1	25
5	6	2	3	14

13.

135	120	148	165	123
140	156	187	203	162
154	186	155	139	147

14.

248	291	254	211	214
258	265	247	251	283
281	265	263	267	288

15.

50.5	20.1	38.7	71.2
67.7	59.1	39.8	47.5
82.6	41.4	61.1	58.6
60.5	48.8	61.7	55.3

16.

7.7	8.3	6.9	1.1
6.8	9.3	7.1	8.5
6.3	5.7	4.9	3.8
1.3	7.6	8.5	4.1

In Exercises 17–26, a set of data is given. Determine the median M and the midrange MR.

17.

6	3	5	11	4

18.

5	11	2	9	0

19.

3	9	8	4	5	5

20.

9	8	4	7	6	9

21.

18	16	17	14	21
11	29	25	27	29

22.

55	68	51	63	54
59	64	63	63	57

23.

0.36	0.25	0.81	0.93	0.66
0.31	0.58	0.33	0.47	0.44
0.58	0.36	0.09	0.87	0.25

24.

1.7	5.0	3.6	8.4	2.2
7.3	3.9	4.5	8.6	3.1
4.7	7.5	6.8	7.7	8.6

25.

168	173	189	201
159	256	547	209
256	163	211	177
198	253	261	256

26.

530	691	598	603
552	641	538	614
540	390	548	626
584	572	629	540

In Exercises 27–30, determine the mode for the qualitative data.

27.

Medium	Small	Medium
Small	Medium	Small
Medium	Large	Medium

28.

Good	Fair	Poor
Fair	Poor	Good
Poor	Fair	Poor

29.

Blue	Brown	Brown	Black
Blue	Blue	Brown	Blue
Black	Black	Brown	Black
Green	Brown	Green	Blue

30.

Maple	Oak	Birch	Oak
Oak	Oak	Pine	Maple
Oak	Maple	Maple	Oak
Birch	Maple	Birch	Maple

In Exercises 31–38, a frequency distribution is given. Estimate the mean $\bar{x}$.

31.

Class	Frequency, f
0 to 4	4
5 to 9	7
10 to 14	3

32.

Class	Frequency, f
0 to 9	4
10 to 19	10
20 to 19	5

33.

Midpoint, x_m	Frequency, f
15	5
25	8
35	8
45	5

34.

Midpoint, x_m	Frequency, f
30	2
40	6
50	6
60	2

35.

Class	Frequency, f
2.1 to 2.5	1
2.6 to 3.0	1
3.1 to 3.5	6
3.6 to 4.0	7
4.1 to 4.5	10

36.

Class	Frequency, f
42.5 to 45.4	15
45.5 to 48.4	12
48.5 to 51.4	5
51.5 to 54.4	1
54.5 to 57.4	1

37.

Midpoint, x_m	Frequency, f
25	6
50	8
75	12
100	22
125	10
150	9

38.

Midpoint, x_m	Frequency, f
13	3
18	5
23	9
28	10
33	8
38	6

In Exercises 39–46, classify the distribution as skewed right, skewed left, or symmetric.

39.

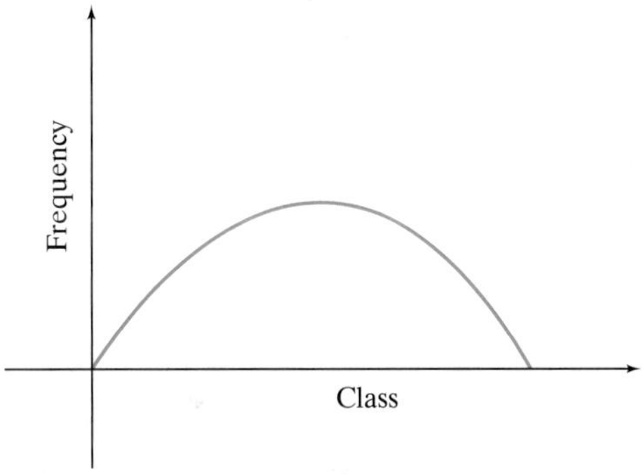

40.

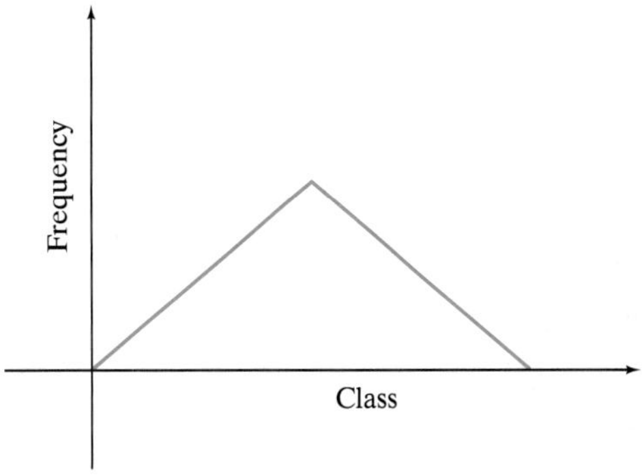

41.

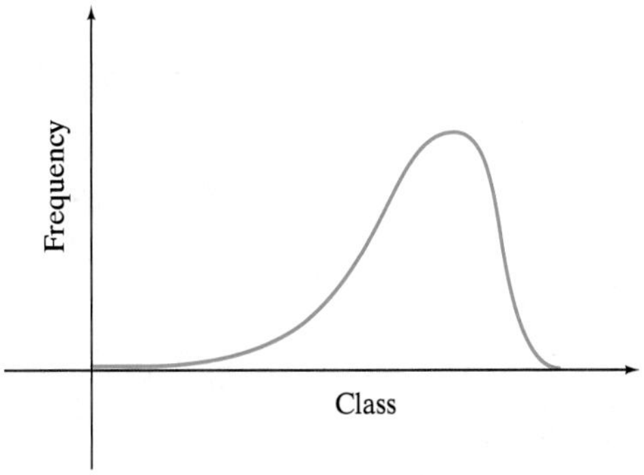

42.

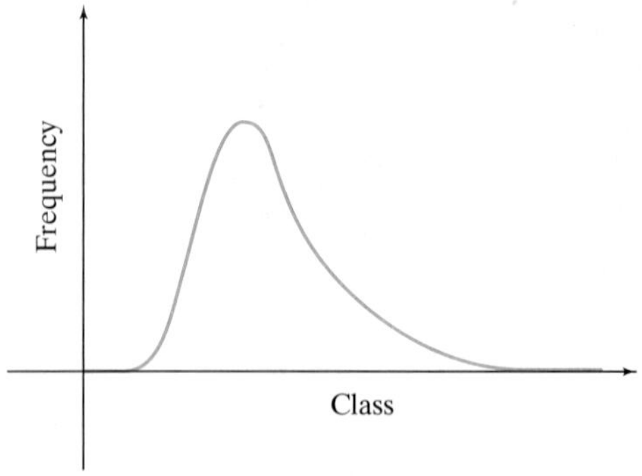

43.

Class	Frequency
100 to 109	100
110 to 119	150
120 to 129	300
130 to 139	850
140 to 149	1025
150 to 159	925

44.

Class	Frequency
0.20 to 0.29	7
0.30 to 0.39	18
0.40 to 0.49	26
0.50 to 0.59	26
0.60 to 0.69	19
0.70 to 0.79	6

45.

Class	Frequency
0.10 to 0.19	3
0.20 to 0.29	14
0.30 to 0.39	19
0.40 to 0.49	19
0.50 to 0.59	13
0.60 to 0.69	2

46.

Class	Frequency
10 to 19	50
20 to 29	60
30 to 39	90
40 to 49	300
50 to 59	635
60 to 69	440

47. The top (most visited) six Web sites in May 2000 are shown in the table.

Web Site	Visitors (in millions)
AOL Network	59.2
Microsoft	49.3
Yahoo	48.9
Lycos	32.5
Excite@Home	28.8
Go Network	23.1

SOURCE: www.infoplease.com

(a) Determine $\bar{x}$ and interpret.

(b) Determine M and interpret.

48. The top (most visited) news, information, and entertainment Web sites in May 2000 are shown in the table.

Web Site	Visitors (in millions)
ZDnet	9.8
CNET	9.6
About.com	9.4
MSNBC.com	8.9
Weather.com	7.6
Disney Online	6.5

SOURCE: www.infoplease.com

(a) Determine $\bar{x}$ and interpret.

(b) Determine M and interpret.

49. The nine largest oil reserves by country in 1999 are shown in the table.

Country	Reserves (in billions of barrels)
Saudi Arabia	259
Iraq	113
Kuwait	93
Adu Dhabi	93
Iran	90
Venezuela	73
Russia	49
Mexico	39
Libya	30

SOURCE: ogj.pennnet.com

(a) Determine $\bar{x}$ and interpret.

(b) Determine M_0 and interpret.

50. The nine largest oil reserves by country in 2000 are shown in the table.

Country	Reserves (in billions of barrels)
Saudi Arabia	261
Iraq	113
Kuwait	94
Adu Dhabi	92
Iran	90
Venezuela	73
Russia	49
Libya	29
Mexico	29

SOURCE: ogj.pennnet.com

(a) Determine $\bar{x}$ and interpret.

(b) Determine M_0 and interpret.

51. The top 15 countries in personal computer usage in 1998 are shown in the table.

Country	PC's in Use (in millions)
United States	129.0
Japan	32.8
Germany	21.1
United Kingdom	18.3
France	15.4
Canada	11.8
Italy	10.6
China	8.3
Australia	7.7
South Korea	6.7
Spain	5.7
Russia	5.6
Brazil	5.2
Netherlands	5.1
Mexico	4.6

SOURCE: www.c-i-a.com

(a) Determine M and interpret.

(b) Determine MR and interpret.

52. The top 15 countries in Internet usage in 1998 are shown in the table.

Country	Weekly Internet Users (in millions)
United States	76.0
Japan	9.8
United Kingdom	8.1
Germany	7.1
Canada	6.5
Australia	4.4
France	2.8
Sweden	2.6
Italy	2.1
Taiwan	2.1
South Korea	2.0
Spain	2.0
Netherlands	2.0
China	1.6
Finland	1.6

SOURCE: www.c-i-a.com

(a) Determine M_0 and interpret.

(b) Determine MR and interpret.

53. The number of emergency room visits in the United States during 1997, classified by age from 5 to 84, is shown in the frequency distribution.

Age	Number of visits (in millions)
5 to 14	15.7
15 to 24	14.4
25 to 34	14.8
35 to 44	14.6
45 to 54	12.1
55 to 64	9.1
65 to 74	6.2
75 to 84	8.6

SOURCE: U.S. National Center for Health Statistics, www.cdc.gov/nchs

(a) Determine $n = \sum f$ and interpret.

(b) Estimate the mean of the frequency distribution and interpret.

54. The ages of U.S. prisoners sentenced to death in 1980 are shown in the table.

Age	Number of Prisoners
15 to 24	184
25 to 34	334
35 to 44	144
45 to 54	42
55 to 64	10

SOURCE: U.S. Census Bureau, www.census.gov

(a) Estimate the mean of the frequency distribution and interpret.

(b) Determine if the distribution of the data is generally skewed right, skewed left, or symmetric.

55. The ages of U.S. prisoners sentenced to death in 1997 are shown in the table.

Age	Number of Prisoners
15 to 24	289
25 to 34	1075
35 to 44	1050
45 to 54	768
55 to 64	153

SOURCE: U.S. Census Bureau, www.census.gov

(a) Estimate the mean of the frequency distribution and interpret.

(b) Determine if the distribution of the data is generally skewed right, skewed left, or symmetric.

SECTION PROJECT

Let's examine some of the properties of the measures of centrality. Historically, the traditional holiday bonuses given to six employees at the Happy Healthcare Company are

250	450	600	400	650	600

(a) Determine the mean, median, and mode of the data.

(b) Now suppose that the company CEO decided to give an additional $300 in bonuses to each employee last year. Determine the mean, median, and mode of this new set of data.

(c) Compare the measures from parts (a) and (b). What effect does adding the same number to a set of data have on the mean, median, and mode?

(d) Now suppose that unexpected company profits result in the CEO doubling the traditional bonuses this year. Determine the mean, median, and mode of this new set of data.

(e) Compare the measures from parts (a) and (d). What effect does multiplying every number in a set of data by a constant have on the mean, median, and mode?

Section 7.4 Measures of Dispersion

In the previous section we saw that the measures of centrality can help us to analyze a set of data. But there is another piece of information that is useful when summarizing data numerically. We also want to know how the data are spread out. For example, if we survey the ages of patrons at a fast-food restaurant, the ages can vary greatly, because anyone from infants to senior citizens could frequent the restaurant. By contrast, if we survey the ages of guests at a senior citizen party, the ages would be relatively the same and not very spread out. In this section we will study **measures of dispersion**, which are numerical values associated with how much the data in a set are spread out.

The Range

The simplest measure of dispersion to compute is the **range**. The range is the difference between the minimum and maximum values in a set of data. The advantage of using the range is that it is easy to compute. The disadvantage is that it, much like the midrange, is very susceptible to outliers, because only the minimum and maximum data values are used. The range can be useful in comparing the dispersions of two sets of data.

Range

For a set of data, the **range**, denoted by R, is given by

$$R = (\text{maximum data value}) - (\text{minimum data value})$$

Example 1 Comparing the Ranges of Data

In Example 5 in Section 7.3, we saw that in 1996 the New York Yankees had the highest franchise market value at \$241 million and the Minnesota Twins had the lowest value at \$77 million. During the same year in the National Football League, the Dallas Cowboys had the highest franchise market value at \$320 million and the Indianapolis Colts had the lowest value at \$170 million. Compare the ranges of the two sets of data (*Source:* www.infoplease.com).

Solution

For the major league baseball franchises, the range is

$$R = 241 - 77 = 164$$

and for the National Football League franchises, the range is

$$R = 320 - 170 = 150$$

So it appears from inspecting the ranges that the spread of the market value of the major league baseball franchises is greater than that of the National Football League franchises. ■

Checkpoint 1 Now work Exercise 5.

The Standard Deviation

A drawback of using the range to determine how much the data are spread out is that it is always affected by outliers. The **standard deviation** is a more accurate measure of dispersion. In general terms, the standard deviation gives an "average" distance that each data item lies away from the mean. To see how the standard deviation is computed, consider the data in Table 7.4.1, which gives the number, in thousands, of Persian Gulf War veterans from each state in the western region of the United States. By defining the western region to be the six states shown in Table 7.4.1, these data can be considered population data.

Table 7.4.1

State	Gulf War Veterans (in thousands)
California	172
Oregon	28
Washington	46
Idaho	11
Arizona	30
Nevada	9

SOURCE: U.S. Census Bureau, www.census.gov

The population mean of the data is

$$\mu = \frac{\sum x}{N} = \frac{176 + 28 + 46 + 11 + 30 + 9}{6} = \frac{296}{6} \approx 49.33$$

Now let's plot μ along with each data item and its distance from the mean. See Figure 7.4.1. These distances are shown numerically in Table 7.4.2. Since these distances are obtained by subtracting the mean from the data value, we denote this difference by writing $x - \mu$.

We can see that the value of $x - \mu$ is positive when the data item is greater than the mean and $x - \mu$ is negative when the data item is less than the mean. Since

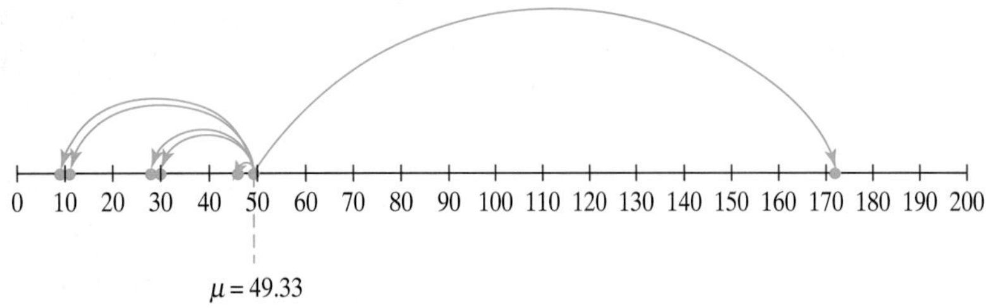

Figure 7.4.1

Table 7.4.2

State	x	$x - \mu$
California	172	$172 - 49.33 = 122.67$
Oregon	28	$28 - 49.33 = -21.33$
Washington	46	-3.33
Idaho	11	-38.33
Arizona	30	-19.33
Nevada	9	-40.33

we are trying to determine the "average" distance that each data item lies away from the mean, and distances are always nonnegative, let's square each $x - \mu$ value so that none of them will be negative. This squaring is shown in Table 7.4.3.

Table 7.4.3

State	x	$x - \mu$	$(x - \mu)^2$
California	172	$172 - 49.33 = 122.67$	$(122.67)^2 = 15{,}047.9289$
Oregon	28	$28 - 49.33 = -21.33$	$(-21.33)^2 = 454.9689$
Washington	46	-3.33	11.0889
Idaho	11	-38.33	1469.1889
Arizona	30	-19.33	373.6489
Nevada	9	-40.33	1626.5089

If we compute the sum of these squares, we get $\sum(x - \mu)^2 = 18{,}983.3334$. Dividing this sum by the population size gives us $\dfrac{\sum(x-\mu)^2}{N} = \dfrac{18{,}983.3334}{6} \approx 3163.89$. Now to "undo" the squaring of the distance from the mean, we take the the square root of this value to get $\sqrt{\dfrac{\sum(x-\mu)^2}{N}} = \sqrt{3163.89} \approx 56.25$. This value tells us that the number of Persian Gulf veterans in the six western states varied on average by about 56.25 thousand from the mean. We have just seen how the **population standard deviation** is computed. Since we primarily use sample data in our computations, let's define the formula for the **sample standard deviation**.

Standard Deviation

- Given a sample of n data items, the **sample standard deviation**, denoted by s, is given by

$$s = \sqrt{\frac{\sum(x - \bar{x})^2}{n - 1}}$$

- For a population of size N, the **population standard deviation**, denoted by σ, is given by

$$\sigma = \sqrt{\frac{\sum(x - \mu)^2}{N}}$$

Note: The main difference between the sample standard deviation and the population standard deviation is the denominator. In the sample standard deviation, we divide by $n - 1$; in the population standard deviation, we divide by N. Since we often use the sample standard deviation to estimate the population standard deviation, for some technical mathematical reasons that are beyond the scope of this text, dividing by $n - 1$ rather than N produces a more accurate estimate.

Example 2 Determining the Sample Standard Deviation

The 1990 population size of seven randomly selected cities in the Boston metropolitan area is shown in Table 7.4.4. Use the formula for sample standard deviation to determine s.

Table 7.4.4

City	Population (in thousands)	City	Population (in thousands)
Brockton	236	Manchester	174
Fitchburg	138	New Bedford	176
Lawrence	353	Worcester	478
Lowell	281		

SOURCE: U.S. Census Bureau, www.census.gov

Solution

Let's start by computing the mean.

$$\bar{x} = \frac{236 + 138 + 353 + 281 + 174 + 176 + 478}{7} = \frac{1836}{7} \approx 262.29$$

Using the formula for sample standard deviation, we get

$$s = \sqrt{\frac{(236 - 262.29)^2 + (138 - 262.29)^2 + \cdots + (176 - 262.29)^2 + (478 - 262.29)^2}{7 - 1}}$$

$$= \sqrt{\frac{86{,}489.4287}{6}} \approx 120.06$$

Interpret the Solution: This means that the populations of the seven cities varied on average by about 120,060 from the mean population of 262,290 people. ■

Interactive Activity

Recompute the standard deviation in Example 2 as if the data were a population. Which value is larger, s or σ? For any given set of data, will this always be true? Explain.

WWW

Technology Option

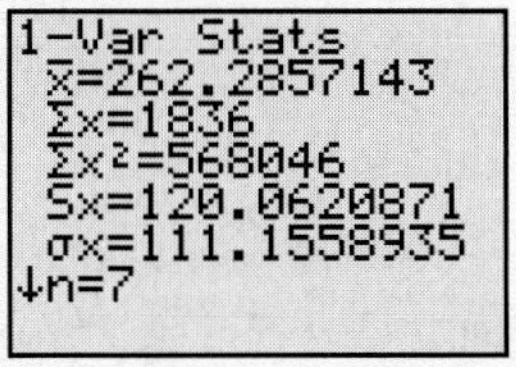

Figure 7.4.2

Almost all scientific and graphing calculators compute the standard deviation of a set of data. On many graphing calculators, the sample standard deviation is denoted by Sx and the population standard deviation is denoted by σx. For Figure 7.4.2, the data in Example 2 have been entered in a list and the 1-Variable Stats menu has been utilized. Notice that $Sx \approx 120.06$, the same value that we obtained in Example 2. Also, notice that Figure 7.4.2 contains other items that we found in Example 2.

To see how your calculator computes the standard deviation, consult the online graphing calculator manual at www.prenhall.com/armstrong.

Standard Deviation from a Frequency Distribution

In Section 7.3 we learned that we can approximate the mean of a set of values summarized in a frequency distribution. We can also approximate the standard deviation of such summarized data. To see how this is done, let's return to the frequency distribution of the 25 longest U.S. rivers, in miles, that we used in Section 7.3. This distribution is given in Table 7.4.5.

Table 7.4.5

Class	Midpoint, x_m	Frequency, f
649 to 1027	838	12
1028 to 1406	1217	5
1407 to 1785	1596	3
1786 to 2164	1975	3
2165 to 2543	2354	2

To approximate the standard deviation, we must first compute how much each class midpoint varies from the mean. Recall that in Section 7.3 we *approximated* the mean $\bar{x}$ to be about 1262. But we must also account for the class frequencies. Knowing that the sum of the frequencies is $n = \sum f = 25$, we get

$$s \approx \sqrt{\frac{(838 - 1262)^2 \cdot 12 + (1217 - 1262)^2 \cdot 5 + (1596 - 1262)^2 \cdot 3 + \cdots + (2354 - 1262)^2 \cdot 2}{25 - 1}} = \sqrt{\frac{6{,}412{,}140}{24}}$$

$$\approx 516.89$$

So, using the frequency distribution for the 25 longest U.S. rivers, an approximation for the mean is 1262 miles and an approximation for the standard deviation is about 517 miles. We now generalize how to approximate the standard deviation from a frequency distribution.

Standard Deviation from a Frequency Distribution

Given a frequency distribution, an approximation for the **sample standard deviation** of the summarized data is given by

$$s = \sqrt{\frac{\sum (x_m - \bar{x})^2 \cdot f}{n - 1}}$$

where

x_m is the midpoint of each class
f is the frequency of each class
n is the sample size
$\bar{x}$ is the sample mean

Example 3 Estimating the Mean and Standard Deviation from a Frequency Distribution

The number of U.S. males living alone aged 15 to 84 in 1998 is given in Table 7.4.6. Estimate $\bar{x}$ and s and interpret each.

Table 7.4.6

Age Group	Number Living Alone (in millions)
15 to 24	0.72
25 to 34	2.22
35 to 44	2.56
45 to 54	1.90
55 to 64	1.26
65 to 74	1.11
75 to 84	1.23

SOURCE: U.S. Census Bureau, www.census.gov

Table 7.4.7

Midpoint, x_m	Frequency, f
$\frac{15+24}{2} = 19.5$	0.72
29.5	2.22
39.5	2.56
49.5	1.90
59.5	1.26
69.5	1.11
79.5	1.23

Solution

First, we must determine the midpoint of each class as shown in Table 7.4.7. Using the formula $\bar{x} = \frac{\sum x_m \cdot f}{n}$, where $n = \sum f = 0.72 + 2.22 + 2.56 + \cdots + 1.23 = 11$, we get

$$\begin{aligned}\bar{x} &= \frac{19.5(0.72) + 29.5(2.22) + 39.5(2.56) + \cdots + 79.5(1.23)}{11} \\ &= \frac{14.04 + 65.49 + 101.12 + \cdots + 97.785}{11} = \frac{524.6}{11} \approx 47.69\end{aligned}$$

Now, to get the standard deviation, we must compute

$$\begin{aligned}s &= \sqrt{\frac{\sum (x_m - \bar{x})^2 \cdot f}{n-1}} \\ &= \sqrt{\frac{(19.5 - 47.69)^2 \cdot 0.72 + (29.5 - 47.69)^2 \cdot 2.22 + (39.5 - 47.69)^2 \cdot 2.56 + \cdots + (79.5 - 47.69)^2 \cdot 1.23}{11 - 1}} \\ &\approx 18.53\end{aligned}$$

Interpret the Solution: According to the frequency distribution, the mean age of U.S. males living alone in 1998 was about 47.69 years with a standard deviation of about 18.53 years. This means that the age of U.S. males living alone in 1998 varied on average by about 18.53 years from the mean of 47.69 years. ■

Technology Option

The program MEANSD in Appendix C approximates the sample mean and sample standard deviation of a frequency distribution, where the class midpoints are entered in L_1 and the class frequencies are entered in L_2. The program is written for the Texas Instruments TI-83. The input and output from executing the MEANSD calculator program for the data in Example 3 are shown in Figure 7.4.3.

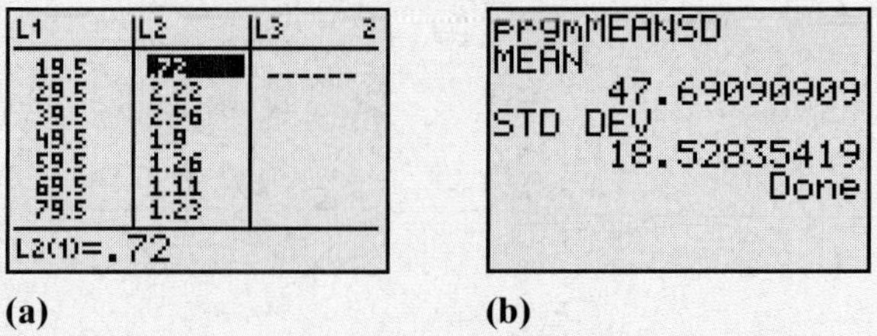

Figure 7.4.3

To see the program listing for your calculator, consult the textbook's companion Web site at www.prenhall.com/armstrong.

Checkpoint 2

Now work Exercise 31.

Pearson Skewness Coefficient

Our final topic in this section revisits the concept of skewness that was introduced in Section 7.3. We revisit this concept through a Flashback.

Flashback

Parking Tickets Revisited

In Example 6 in Section 7.3, we saw that in an effort to decide if a new parking garage is needed the managers at the William Welsh Office Tower surveyed employees to see how many had at least one parking ticket in the last 3 months. The distribution of those who had at least one ticket in the last 3 months is shown in Table 7.4.8 (to the left), and a histogram for the data is given in Figure 7.4.4.

Table 7.4.8

Number of Tickets	Frequency, f
1	23
2	35
3	20
4	8
5	4
6	1

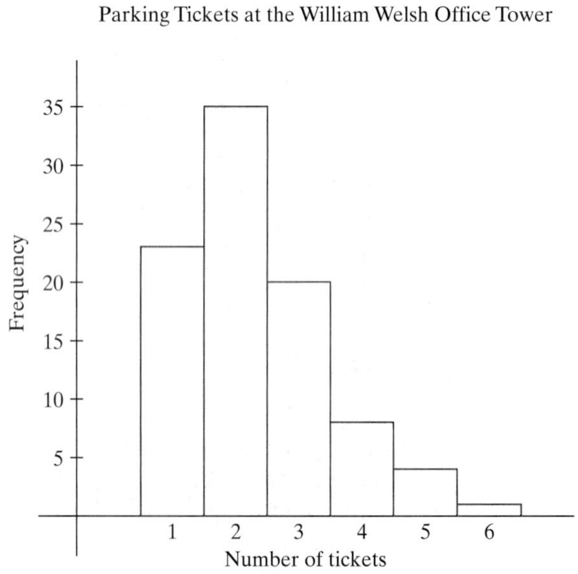

Figure 7.4.4

As Figure 7.4.4 indicates, the histogram is skewed right since the histogram tails off or falls off to the right. In Example 6 in Section 7.3, we determined that the mean number of tickets received in the 3-month period is $\bar{x} \approx 2.32$. Determine the standard deviation s.

Flashback Solution

We know that the formula for standard deviation is

$$s = \sqrt{\frac{\sum(x-\bar{x})^2}{n-1}}$$

Table 7.4.8 indicates that we have $n = \sum f = 91$ data items. In using the formula for standard deviation, the quantity $(x-\bar{x})^2 = (1-2.32)^2$ would occur 23 times, since Table 7.4.8 shows the frequency of one ticket is 23. Similarly, $(x-\bar{x})^2 = (2-2.32)^2$ would occur 35 times, $(x-\bar{x})^2 = (3-2.32)^2$ would occur 20 times, and so on. Rather than list all these occurrences, we can use multiplication and simplify our work as follows:

$$\begin{aligned} s &= \sqrt{\frac{(1-2.32)^2\cdot 23 + (2-2.32)^2\cdot 35 + (3-2.32)^2\cdot 20 + (4-2.32)^2\cdot 8 + (5-2.32)^2\cdot 4 + (6-2.32)^2\cdot 1}{91-1}} \\ &= \sqrt{\frac{117.7584}{90}} \\ &\approx 1.14 \end{aligned}$$

So the standard deviation for the number of parking tickets is 1.14.

It can be shown that the median, M, for the number of parking tickets in the Flashback is $M = 2$. There exists a numerical value using M, $\bar{x}$, and s that tells us if the distribution of the data is skewed left, skewed right, or symmetric. This numerical value is called the **Pearson Skewness Coefficient**.

Pearson Skewness Coefficient

A measure to determine the skewness of a distribution is called the **Pearson Skewness Coefficient**, denoted SK. The formula for SK is

$$SK = \frac{3(\bar{x}-M)}{s}$$

where $\bar{x}$ is the mean, M is the median, and s is the standard deviation. The values for SK usually range from -3 to 3.

- If $SK < 0$, then the distribution is skewed left.
- If $SK > 0$, then the distribution is skewed right.
- If $SK \approx 0$, then the distribution is symmetrical.

Example 4 Determining the Pearson Skewness Coefficient

For the data in the Flashback, determine the Pearson Skewness Coefficient given that $\bar{x} \approx 2.32$, $M = 2$, and $s \approx 1.14$.

Solution

We compute the Pearson Skewness Coefficient to be

$$SK = \frac{3(\bar{x} - M)}{s} = \frac{3(2.32 - 2)}{1.14} \approx 0.84$$

Since $SK \approx 0.84$, that is, $SK > 0$, the distribution is skewed right. This can be visually verified by consulting Figure 7.4.4. ■

Checkpoint 3 Now work Exercise 43.

Final Word on Standard Deviation

The standard deviation can give us more information about a set of data than already discussed. For example, let's draw a curve through the midpoints of the tops of the rectangles for the histogram in Figure 7.4.5. Assume that the histogram is for a relatively large frequency distribution. If the resulting curve is roughly bell shaped, then it can be shown that about 68% of the data lies within 1 standard deviation of the mean. In other words, 68% of the data lies in the interval $(\bar{x} - s, \bar{x} + s)$. About 95% of the data lies within 2 standard deviations of the mean, that is, about 95% of the data lies in the interval $(\bar{x} - 2s, \bar{x} + 2s)$. We will have more to say about bell-shaped curves and this way of looking at standard deviation in Chapter 8.

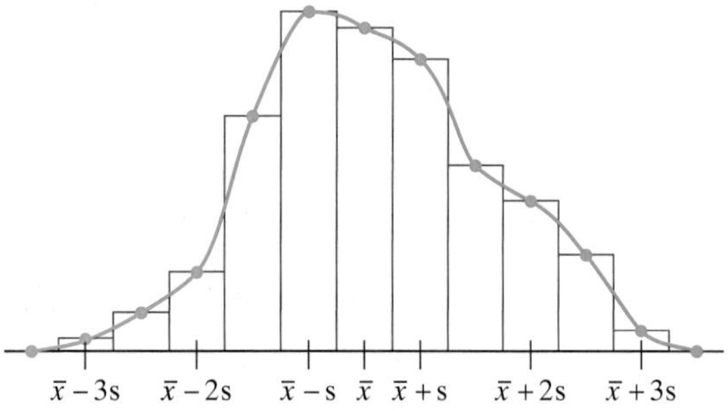

Figure 7.4.5

SUMMARY

- The **range**, denoted by R, is given by $R = \begin{pmatrix}\text{maximum data}\\ \text{value}\end{pmatrix} - \begin{pmatrix}\text{minimum data}\\ \text{value}\end{pmatrix}$.
- The **sample standard deviation**, denoted by s, is given by $s = \sqrt{\frac{\sum(x - \bar{x})^2}{n-1}}$.
- The **population standard deviation**, denoted by σ, is given by $\sigma = \sqrt{\frac{\sum(x - \mu)^2}{N}}$.
- The approximation for the **sample standard deviation** for a given frequency distribution is given by $s \approx \sqrt{\frac{\sum(x_m - \bar{x})^2 \cdot f}{n-1}}$.
- The **Pearson Skewness Coefficient**, denoted SK, is given by $SK = \frac{3(\bar{x} - M)}{s}$, where $\bar{x}$ is the mean, M is the median, and s is the standard deviation. The values for SK usually range from −3 to 3. If $SK < 0$, then the distribution is skewed left. If $SK > 0$, then the distribution is skewed right. If $SK \approx 0$, then the distribution is symmetrical.

SECTION 7.4 EXERCISES

In Exercises 1–10, a set of data is given. Determine the range R.

1.

6	3	6	5	9	8

2.

9	8	4	5	8	1

3.

4	8	6	1	1
2	3	4	7	0

4.

5	1	4	0	3
4	0	7	8	2

✓ **5.**

3	17	18	14	9	7
16	2	15	4	19	12

6.

21	15	25	2	13	8
22	17	18	9	24	7

7.

14.3	15.8	17.4	18.3	19.1
15.8	11.2	9.0	11.2	19.5

8.

7.1	6.8	4.9	5.2	6.3
4.9	7.0	6.3	7.1	5.5

9.

0.25	0.59	0.63	0.58	0.21
0.11	0.78	0.65	0.93	0.27

10.

0.55	0.73	0.18	0.69	0.71
0.83	0.63	0.78	0.68	0.03

In Exercises 11–16, determine the population standard deviation for the given population data. Use the formula

$$\sigma = \sqrt{\frac{\sum(x-\mu)^2}{N}}.$$

11.

6	2	4	8	9
2	5	2	11	8

12.

4	7	8	1	25
5	6	2	3	14

13.

135	120	148	165	123
140	156	187	203	162
154	186	155	139	147

14.

248	291	254	211	214
258	265	247	251	283
281	265	263	267	288

15.

50.5	20.1	38.7	71.2
67.7	59.1	39.8	47.5
82.6	41.4	61.1	58.6
60.5	48.8	61.7	55.3

16.

7.7	8.3	6.9	1.1
6.8	9.3	7.1	8.5
6.3	5.7	4.9	3.8
1.3	7.6	8.5	4.1

In Exercises 17–26, a set of data is given. Determine the sample standard deviation s.

17.

6	3	5	11	4

18.

5	11	2	9	0

19.

3	9	8	4	5	5

20.

9	8	4	7	6	9

21.

18	16	17	14	21
11	29	25	27	29

22.

55	68	51	63	54
59	64	63	63	57

23.

0.36	0.25	0.81	0.93	0.66
0.31	0.58	0.33	0.47	0.44
0.58	0.36	0.09	0.87	0.25

24.

1.7	5.0	3.6	8.4	2.2
7.3	3.9	4.5	8.6	3.1
4.7	7.5	6.8	7.7	8.6

25.

168	173	189	201
159	256	547	209
256	163	211	177
198	253	261	256

26.

530	691	598	603
552	641	538	614
540	390	548	626
584	572	629	540

In Exercises 27–34, a frequency distribution is given. Estimate the standard deviation using the formula

$$s = \sqrt{\frac{\sum(x_m - \bar{x})^2 \cdot f}{n-1}}.$$

27.

Class	Frequency, *f*
0 to 4	4
5 to 9	7
10 to 14	3

28.

Class	Frequency, f
0 to 9	4
10 to 19	10
20 to 29	5

29.

Midpoint, x_m	Frequency, f
15	5
25	8
35	8
45	5

30.

Midpoint, x_m	Frequency, f
30	2
40	6
50	6
60	2

✓ 31.

Class	Frequency, f
2.1 to 2.5	1
2.6 to 3.0	1
3.1 to 3.5	6
3.6 to 4.0	7
4.1 to 4.5	10

32.

Class	Frequency, f
42.5 to 45.4	15
45.5 to 48.4	12
48.5 to 51.4	5
51.5 to 54.4	1
54.5 to 57.4	1

33.

Midpoint, x_m	Frequency, f
25	6
50	8
75	12
100	22
125	10
150	9

34.

Midpoint, x_m	Frequency, f
13	3
18	5
23	9
28	10
33	8
38	6

35. The area of selected major bodies of water in the Pacific Coast region of the United States is shown in the table.

Body of Water	Area (in square miles)
Puget Sound, WA	808
San Francisco Bay, CA	264
Willapa Bay, WA	125
Hood Canal, WA	117

SOURCE: U.S. Census Bureau, www.census.gov

(a) Determine the range and interpret.

(b) Determine the sample standard deviation and interpret.

36. The area of selected major bodies of water in the Atlantic Coast region of the United States is shown in the table.

Body of Water	Area (in square miles)
Chesapeake Bay, MD	2747
Pamlico Sound, NC	1622
Long Island Sound, NY	914
Delaware Bay, DE	614
Cape Cod, MA	598
Albemarle Sound, NC	492

SOURCE: U.S. Census Bureau, www.census.gov

(a) Determine the range and interpret.

(b) Determine the sample standard deviation and interpret.

37. The ages of U.S. prisoners sentenced to death in 1980 are shown in the table. Estimate the standard deviation of the frequency distribution and interpret.

Age	Number of Prisoners
15 to 24	184
25 to 34	334
35 to 44	144
45 to 54	42
55 to 64	10

SOURCE: U.S. Census Bureau, www.census.gov

38. The ages of U.S. prisoners sentenced to death in 1997 are shown in the table. Estimate the standard deviation of the frequency distribution and interpret.

Age	Number of Prisoners
15 to 24	289
25 to 34	1075
35 to 44	1050
45 to 54	768
55 to 64	153

SOURCE: U.S. Census Bureau, www.census.gov

39. The number of emergency room visits in the United States during 1997, classified by age from 5 to 84, is shown in the frequency distribution. Estimate the standard deviation of the frequency distribution and interpret.

Age	Number of visits (in millions)
5 to 14	15.7
15 to 24	14.4
25 to 34	14.8
35 to 44	14.6
45 to 54	12.1
55 to 64	9.1
65 to 74	6.2
75 to 84	8.6

SOURCE: U.S. National Center for Health Statistics, www.cdc.gov/nchs

In Exercises 40–45, determine the Pearson Skewness Coefficient and classify the distribution as skewed right, skewed left, or symmetric.

40.

Summary	Statistics
$\bar{x}$	35.7
M	36
s	0.93

41.

Summary	Statistics
$\bar{x}$	103.5
M	108
s	2.1

42. A sample of data has a sample mean $\bar{x} = 48$ with a median $M = 48$ and standard deviation of $s = 4.01$.

43. A sample of data has a sample mean $\bar{x} = 7.5$ with a median $M = 7.5$ and standard deviation of $s = 1.03$.

44. The ages of U.S. females in 1990 had a mean of 36.1 years, with a median age of 34.0 and standard deviation of 22.9 years (*Source:* U.S. Census Bureau, www.census.gov).

45. The ages of U.S. males in 1990 had a mean of 33.4 years, with a median age of 31.6 and standard deviation of 21.6 years (*Source:* U.S. Census Bureau, www.census.gov).

SECTION PROJECT

Let's examine some of the properties of the measures of variation. Suppose that the traditional holiday bonuses given to six employees at the Happy Healthcare Company are

250	450	600	400	650	600

(a) Determine the range and sample standard deviation of the data.

(b) Now suppose that the company CEO decided to give an additional $300 in bonuses this year to the six employees. Determine the range and sample standard deviation of this new set of data.

(c) Compare the measures from parts (a) and (b). What effect does adding the same number to a set of data have on the range and sample standard deviation?

(d) Now suppose that in an effort to have a consistent bonus policy the company CEO gives each of the six employees a $500 bonus. Determine the range and sample standard deviation of this new set of data.

(e) What can we say about the value of the range and sample standard deviation of a set of data that is comprised of a single numerical value?

Why We Learned It

In this chapter, we learned how to analyze data sets both numerically and graphically. We looked at pie charts and saw that the ability to read a pie chart can give much useful information. For example, a social scientist given a pie chart listing the number of hours per person per year watching various forms of television can sum the numbers given in the pie chart to determine the total hours per person per year watching television. Once this is done, the scientist can also determine the percentages watching the various forms of television. We also saw how the histogram, bar graph, frequency polygon, and ogive can visually represent a data set. The latter part of this chapter was dedicated to measures of central tendency, along with ways to measure how spread out the data are. For example, if a criminal justice professor has the ages of all U.S. prisoners sentenced to death, he or she may be interested in the mean (average) age of this group of prisoners.

To measure how spread out the ages are, the professor could then compute the standard deviation. The professor could also compute the median of the ages and the mode. If the professor constructed a histogram of the data, he or she may be interested in whether the histogram is skewed. This can be analyzed visually or numerically by using the Pearson Skewness Coefficient.

CHAPTER REVIEW EXERCISES

In Exercises 1 and 2, identify the population and the sample.

1. In a March 2001 survey conducted by the Gallup Organization, 1060 adult Americans were asked if the quality of the environment as a whole is getting better or getting worse; 36% of adults responded that the overall quality of the environment is getting better (*Source:* www.gallup.com).

2. In a November 2000 survey, 501 college football fans were asked by the Gallup Organization what team they considered to be the number 1 team in the country; 28% responded that the number 1 team was Oklahoma (*Source:* www.gallup.com).

In Exercises 3–5, determine whether the data are qualitative or quantitative.

3. The number of students per class in each course offered at a university

4. The type of fish caught during a fishing tournament

5. The movies playing at the local cinema

In Exercises 6–8, a table of data is given. Construct a bar graph for the data.

6. The total numbers of females, in thousands, participating in various NCAA sports during the 1997–1998 academic year are shown in the table.

Sport	Number of Females Participating (in thousands)
Basketball	13.8
Soccer	16.0
Cross country	10.5
Tennis	8.1

SOURCE: U.S. Statistical Abstract, www.census.gov/statab/www

7. The total numbers of pounds, in millions, of various domestic fish and shellfish caught in 1998 are shown in the table.

Species	Number of Pounds (in millions)
Lobsters	79.6
Halibut	73.2
Atlantic cod	24.5
Sablefish	43.5
Sea scallops	13.1

SOURCE: U.S. Statistical Abstract, www.census.gov/statab/www

8. The numbers of adults, in millions, attending various sporting events one or more times a month in 1998 are shown in the table.

Sporting Event	Number of Adults Attending (in millions)
Auto racing	3.6
Baseball	8.6
College football	5.0
Ice hockey	2.9
Soccer	3.4
Golf	2.2

SOURCE: U.S. Statistical Abstract, www.census.gov/statab/www

In Exercises 9–11, answer the questions associated with the given bar graph.

9. The average yearly incomes, in thousands, of 25- to 34-year-olds by highest degree earned for 1998 is shown in the bar graph (*Source:* U.S. Statistical Abstract, www.census.gov/statab/www).

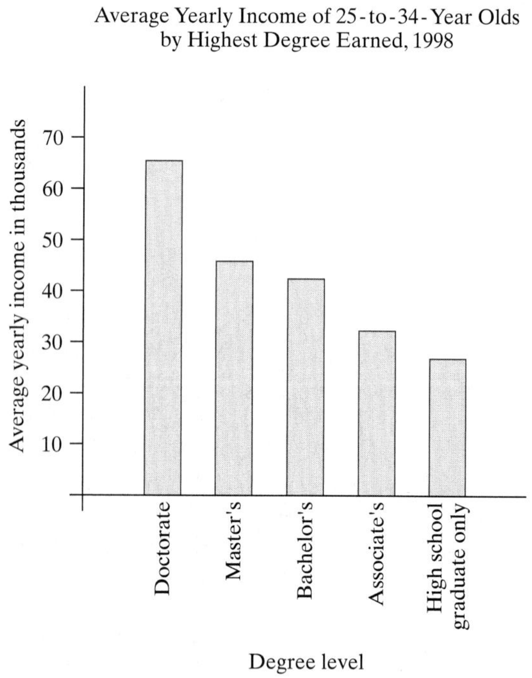

(a) Which degrees were held by 25- to 34-year-olds who earned at least $40,000 in 1998?

(b) About how much more was the average yearly income of a 25- to 34-year-old in 1998 holding a doctorate degree than someone in this age group who was only a high school graduate?

10. The numbers of people, in millions, using home computers in 1997 by various age groups is shown in the bar graph (*Source:* U.S. Statistical Abstract, www.census.gov/statab/www).

Use of Home Computers, 1997

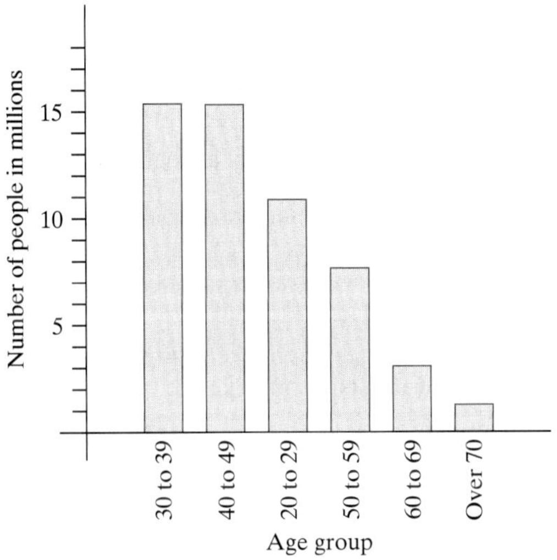

(a) Which age groups had about the same number of users?

(b) Which age groups had under 6 million users in 1997?

11. The numbers of people 5 years old and over, in thousands, speaking various languages other than English at home in 1990 are shown in the bar graph (*Source:* U.S. Statistical Abstract, www.census.gov/statab/www).

Persons 5 Years Old and Over Speaking a Language Other Than English at Home, 1990

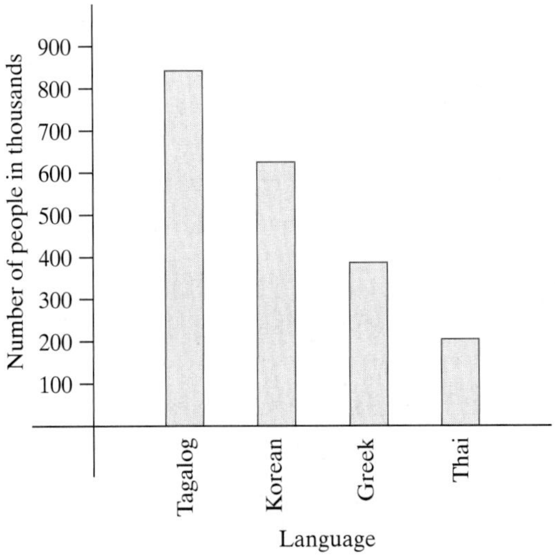

(a) About how many people 5 years old and over spoke Thai at home in 1990?

(b) About how many more people 5 years old and over spoke Tagalog than Thai at home in 1990?

In Exercises 12–14, answer the questions associated with the given pie chart.

12. The percent of toxic chemical releases in 1998 by industry is shown in the pie chart (*Source:* U.S. Statistical Abstract, www.census.gov/statab/www).

Toxic Chemical Releases by Industry, 1998

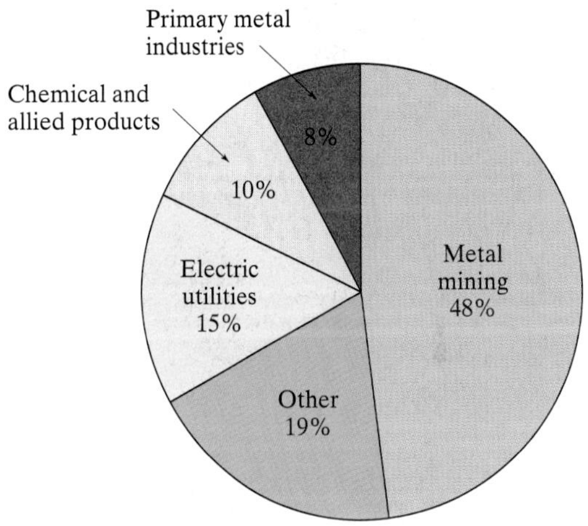

(a) What industry was the largest source of toxic chemical releases in 1998?

(b) If 7,307,000,000 pounds of toxic chemicals was released by industries in 1998, how many pounds were released by electric utilities?

13. Juvenile arrests, in thousands, for drug possession in 1998 are shown in the pie chart (*Source:* U.S. Statistical Abstract, www.census.gov/statab/www).

Juvenile Arrest for Drug Possession, 1998

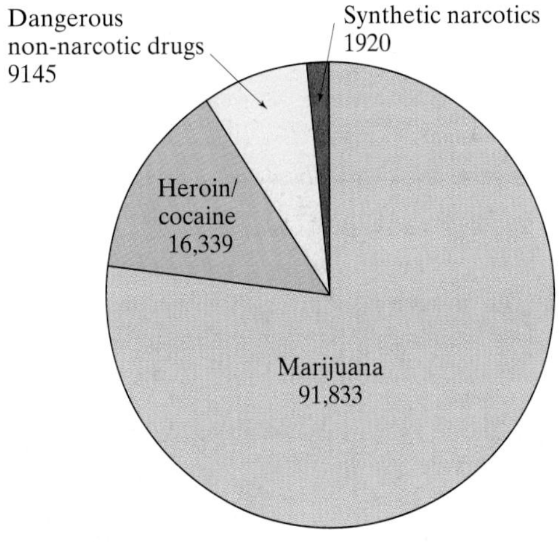

(a) What type of possession ranked third in terms of arrests?

(b) If there were 119,237 thousand arrests in 1998 for drug possession, what percentage of those were for possession of marijuana? Round to the nearest tenth.

14. The number of lottery ticket sales, in billions of dollars, by various types of games in 1999 is shown in the pie chart (*Source:* U.S. Statistical Abstract, www.census.gov/statab/www).

Lottery Ticket Sales, 1999

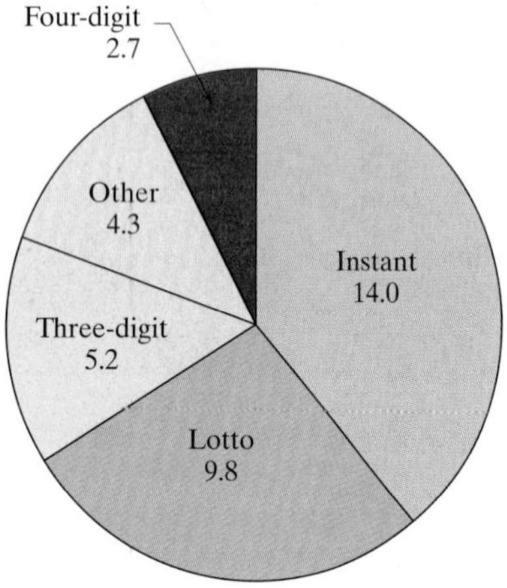

(a) Which type of game had the lowest number of ticket sales?

(b) If total ticket sales in 1999 were $36 billion, what percent of sales was from instant tickets and 3-digit tickets? Round to the nearest tenth.

In Exercises 15 and 16, answer the questions associated with the given frequency distribution.

15. The total number of adults, in millions, who attended at least one movie in 1996 is classified by age group in the frequency distribution.

Class (age group)	Frequency (in millions)
25 to 34	31.7
35 to 44	33.1
45 to 54	21.9
55 to 64	9.6
65 to 74	7.4

SOURCE: U.S. Statistical Abstract, www.census.gov/statab/www

(a) What is the class width?

(b) How many adults age 45 to 54 attended at least one movie in 1996?

16. The number of adults, in millions, who had read the newspaper in the week prior to a spring 2000 survey is classified by household income in the frequency distribution.

Class (income in dollars)	Frequency (in millions)
10,000 to 19,999	16.7
20,000 to 29,999	26.1
30,000 to 39,999	18.6
40,000 to 49,999	17.3

SOURCE: U.S. Statistical Abstract, www.census.gov/statab/www

(a) How many adults whose income was between $10,000 and $19,999 in spring 2000 read the newspaper at least once in the week prior to the survey?

(b) If there were 24,055,000 adults whose income was between $30,000 and $39,999 in spring 2000, what percentage of them read the newspaper at least once in the week prior to the survey? Round to the nearest tenth.

17. Use Exercise 15 to complete the following.

(a) Determine the class boundaries.

(b) Construct a frequency histogram.

18. Use Exercise 16 to complete the following.

(a) Determine the class boundaries.

(b) Determine the relative frequency for each class.

(c) Construct a relative frequency histogram.

For the data in Exercises 19 and 20, complete the following:

(a) Write a frequency distribution.

(b) Use your calculator to construct a histogram.

19. The time, in minutes, spent exercising during one day for a sample of 30 adults is shown in the table. Use seven classes.

0	15	25	0	45	30
60	120	15	35	40	45
45	30	30	0	0	60
90	0	120	0	45	60
15	30	20	40	80	0

20. The amount, in dollars, spent on recreation per month for a sample of 18 teen-agers is shown in the table. Use five classes.

40	60	15	10	5	30
97	25	20	18	30	12
80	20	40	42	37	26

In Exercises 21 and 22, complete the following:

(a) Determine the midpoint of each class.

(b) Construct a frequency polygon.

21. Use the frequency distribution in Exercise 15.

22. Use the frequency distribution in Exercise 16.

In Exercises 23 and 24, complete the following:

(a) Determine the cumulative frequency for each class.

(b) Construct an ogive.

23. Use the frequency distribution in Exercise 15.

24. Use the frequency distribution in Exercise 16.

For Exercises 25–27, use the frequency polygon for death rates from cancer per 100,000 females in the given age groups for ages 25 to 74 in 1998 (Source: *U.S. Statistical Abstract,* www.census.gov/statab/www).

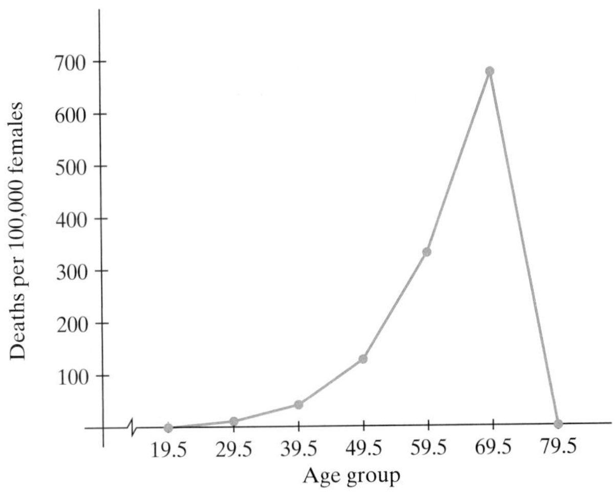

25. What is the class *width*?

26. What are the *limits* of the third class?

27. In which class are there about 330 deaths per 100,000 females?

For Exercises 28–30, use the ogive for the number of deaths reported due to injuries at work for persons age 15 to 64 in 1999 (Source: *U.S. National Center for Health Statistics,* www.cdc.gov/nchs).

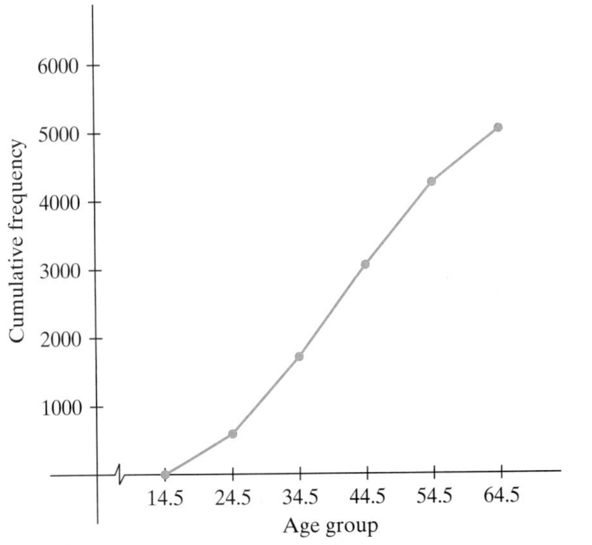

28. What are the class *limits* of the second class?

29. What is the class *width*?

30. About how many individuals age 15 to 44.5 years old died from work-related injuries in 1999?

In Exercises 31 and 32, a sample of data is given. Determine the sample mean $\bar{x}$ and the mode M_0.

31.

5	8	9	2	2
3	7	6	1	8

32.

0.22	0.53	0.77	0.64	0.21
0.14	0.08	0.27	0.89	0.47

In Exercises 33 and 34, determine the population mean μ for the given population data.

33.

122	128	134	188	133
118	121	126	150	145
115	119	141	143	112

34.

2.6	3.4	1.1	3.3
4.1	5.6	5.3	3.4
1.8	9.8	1.9	2.7

In Exercises 35 and 36, a sample of data is given. Determine the median M and the midrange MR.

35.

10	13	27	29	19
11	31	18	21	15

36.

320	325	415	584
608	409	353	305
223	287	619	387
298	521	409	508

In Exercises 37 and 38, determine the mode for the qualitative data.

37.

First	First	Third
Second	Second	First
Third	Second	First

38.

Biology	Math	Music	Physics
Physics	Biology	Music	Music
Physics	Math	Math	Physics
Math	Math	Biology	Biology

In Exercises 39 and 40, a frequency distribution is given. Estimate the mean $\bar{x}$.

39.

Class	Frequency, f
0 to 4	5
5 to 9	6
10 to 14	4
15 to 19	3

40.

Midpoint, x_m	Frequency, f
20	2
30	5
40	4
50	3
60	8

In Exercises 41 and 42, classify the distribution as skewed right, skewed left, or symmetric.

41.

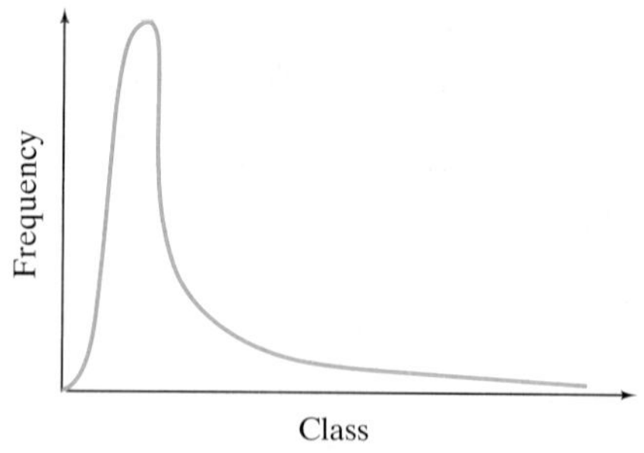

42.

Class	Frequency
0.10 to 0.14	2
0.15 to 0.19	3
0.20 to 0.24	4
0.25 to 0.29	10
0.30 to 0.34	12

43. The top (highest grossing) six summer films of all time are shown in the table.

Film	Domestic gross (in millions)
Star Wars	461.0
Star Wars: Episode One—The Phantom Menace	431.1
E.T.	399.8
Jurassic Park	357.1
Forrest Gump	329.7
The Lion King	312.9

SOURCE: www.infoplease.com

(a) Determine $\bar{x}$ and interpret.

(b) Determine M and interpret.

44. The top nine hockey teams to make the most appearances in the Stanley Cup finals are given in the table along with their total number of appearances.

Team	Appearances
Montreal Canadiens	32
Toronto Maple Leafs	21
Detroit Red Wings	21
Boston Bruins	17
New York Rangers	10
Chicago Blackhawks	10
Philadelphia Flyers	7
Edmonton Oilers	6
New York Islanders	5

SOURCE: www.infoplease.com

(a) Determine $\bar{x}$ and interpret.

(b) Determine M_0 and interpret.

(c) Determine MR and interpret.

45. The total number of arrests in 1999 of 10- to 18-year-olds is given in the table.

Age	Number of Arrests
10 to 12	117,679
13 to 15	662,847
16 to 18	1,244,361

SOURCE: www.infoplease.com

(a) Estimate the mean of the frequency distribution and interpret.

(b) Determine if the distribution of the data is generally skewed right, skewed left, or symmetric.

In Exercises 46–48, a sample of data is given. Determine the range R.

46.

3	2	7	12	5

47.

4	18	25	27	38	20
10	12	42	16	5	31

48.

0.10	0.57	0.32	0.19	0.89	0.63
0.22	0.64	0.05	0.67	0.74	0.44

In Exercises 49–51, determine the population standard deviation for the given population data. Use the formula

$$\sigma = \sqrt{\frac{\sum(x-\mu)^2}{N}}.$$

49.

3	12	10	17	6
5	8	12	13	14

50.

138	94	87	126	312
220	119	184	155	175
198	123	146	158	202

51.

3.4	2.8	5.1	3.2
2.7	8.9	9.9	1.6
3.4	5.7	1.8	9.5
3.8	7.2	7.8	9.4

In Exercises 52–54, a sample of data is given. Determine the sample standard deviation s.

52.

2	8	3	1	9

53.

74	71	70	68	77
62	65	70	75	76

54.

3.8	4.7	2.6	8.9	5.6
2.1	1.5	3.7	7.4	2.3
4.1	1.8	1.7	2.3	6.6

In Exercises 55 and 56, a frequency distribution is given. Estimate the standard deviation using the formula

$$s = \sqrt{\frac{\sum(x_m - \bar{x})^2 \cdot f}{n-1}}.$$

55.

Class	Frequency, f
13 to 17	2
18 to 22	6
23 to 27	8
28 to 32	3

56.

Midpoint, x_m	Frequency, f
10	12
20	15
30	8
40	7
50	20
60	16

57. The 1999 population size of seven randomly selected counties in New Hampshire is shown in the table.

County	Population (in thousands)
Belknap	54
Coos	33
Grafton	79
Hillsbourough	367
Merrimack	130
Stafford	111
Sullivan	40

SOURCE: U.S. Census Bureau, www.census.gov

(a) Determine the range and interpret.

(b) Determine the sample standard deviation and interpret.

58. The total number of arrests in 1999 of 10- to 18-year-olds is given in the table. Estimate the standard deviation of the frequency distribution and interpret.

Age	Number of Arrests
10 to 12	117,679
13 to 15	662,847
16 to 18	1,244,361

SOURCE: www.infoplease.com

In Exercises 59 and 60, determine the Pearson Skewness Coefficient and classify the distribution as skewed right, skewed left, or symmetric.

59.

Summary	Statistics
$\bar{x}$	42.8
M	38
s	0.74

60. A sample of data has a sample mean $\bar{x} = 32$ with a median $M = 32$ and standard deviation of $s = 3.98$.

CHAPTER 7 PROJECT

SOURCE: www.nielsenmedia.com and 7metasearch.com

One application of the standard deviation is Chebyshev's Theorem. The theorem states that for any set of data, population or sample, *at least* $p = 1 - \frac{1}{k^2}$ of the data lie within k standard deviations of the mean, where $k > 1$. For a sample of data, this means that at least p% of the data lies between $\bar{x} - ks$ and $\bar{x} + ks$.

Consider the following table of a sample of the number of unique visitors to random Internet sites during September 2001.

8169	7903	7255	6517	7556	7350	8295
6100	6618	6968	6216	5054	4907	6275
6188	6660	6510	6009	6111	6076	6639
6387	6401	5866	6405	5229	5729	6069
5264	5502	5316	5176	5092	5796	6048
5530	6104	5439	5379	6261	5943	5414
5722	5428	5551	5971	6069	5432	5764
5939						

1. Determine the sample mean and the sample standard deviation. Round to hundredths place.

2. According to Chebyshev's Theorem, at least 75% of the data lies within k standard deviations of the mean where $1 - \frac{1}{k^2} = \frac{3}{4}$ (that is 75%). Solve this equation to determine the proper k value.

3. Determine the values $n_1 = \bar{x} - ks$ and $n_2 = \bar{x} + ks$ using the $\bar{x}$- and s-values found in step 1 along with the k-value found in step 2.

4. What number represents 75% of the data in the table? That is, what is 75% of the sample size?

5. Tally the data to determine the number of unique visitors to random Internet sites during September 2001 that are between n_1 and n_2.

6. How do you account for the difference in the solutions between steps 4 and 5?

7. Repeat steps 2 through 6 for at least 88.9% $\left(\text{that is, } \frac{8}{9}\right)$ of the data.

Chapter

8 Probability Distributions

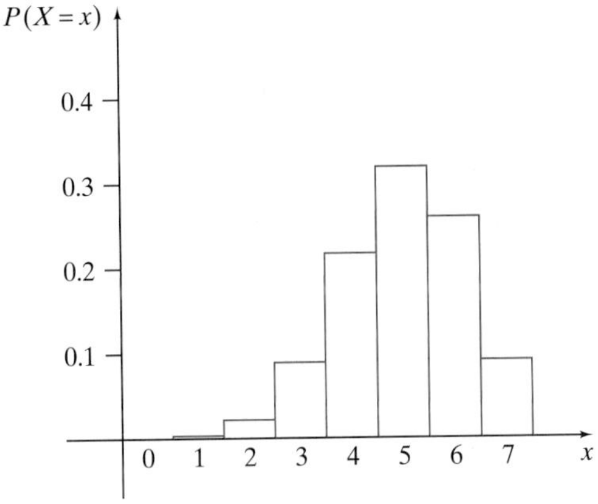

$$P(X = x) = {}_7C_x(0.71)^x\,(0.29)^{7-x}$$

$$P(X = 5) = {}_7C_5(0.71)^5\,(0.29)^2 \approx 0.319$$

The Internet has opened a world of information to anyone who can go online. This is especially true for people seeking medical information. An August 2000 Harris poll found that 71% of Americans who are online use the Internet to access health and medical information. The probability that any number of 7 randomly selected online Americans use the Internet to access this information can be shown using a probability histogram. The binomial distribution gives the probability that 5 of 7 randomly selected Americans go online to access health and medical information. This probability is about 0.319 or 31.9%.

What We Know

In Chapter 6 we learned the fundamentals of probability, and in Chapter 7 we learned about describing data graphically. We also learned about measures of central tendency and measures of dispersion.

Where Do We Go

In this chapter we will study random variables as a precursor to the study of probability distributions. This allows us to group outcomes that share similar attributes and to compute probabilities. We will also revisit the measure of central tendency and the measure of dispersion.

Section 8.1 Discrete Random Variables

Thus far in our study of probability we have learned how to determine the probability of individual events. Now we want to find how to determine the probabilities of many events with specific common characteristics. To do this we use **random variables**. We will also examine numerical tables of values, sometimes called **probability distributions**, that can be used to determine probabilities.

Discrete Random Variables

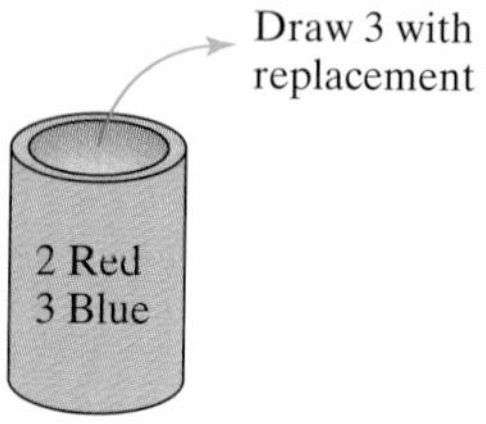

Figure 8.1.1

Let's begin with a typical example from Chapter 6 and see how we can use the results already known to assign a random variable to the experiment and to construct a probability distribution. Suppose that a jar contains 2 red and 3 blue marbles. In an experiment, we pull out a marble and record its color and then return it to the jar. We repeat this two more times until we have made three draws (Figure 8.1.1). The sample space for the experiment is

$$S = \{RRR, RRB, RBR, BRR, RBB, BRB, BBR, BBB\}$$

To compute the probabilities for these outcomes, we can use the classical probability formula from Chapter 6. (Consult the Toolbox below.)

From Your Toolbox

Classical Probability: Assume that S is a sample space that has $n(S)$ outcomes for which each outcome is equally likely. If A is an event associated with S, then the probability of event A occurring, denoted by $P(A)$, is given by

$$\text{probability of event } A \text{ occurring} = P(A) = \frac{n(A)}{n(S)} = \frac{\text{number of outcomes of event } A}{\text{total number of possible outcomes}}$$

We notice that on any given draw the probability of drawing a red marble, denoted by $P(R)$, can be found by dividing the number of red marbles in the jar by the total number of marbles in the jar. Hence, for any given single draw,

$$P(R) = \frac{\text{number of red marbles in the jar}}{\text{total number of marbles in the jar}} = \frac{2}{5}$$

The probability of drawing a blue marble on any single draw is

$$P(B) = \frac{\text{number of blue marbles in the jar}}{\text{total number of marbles in the jar}} = \frac{3}{5}$$

From Your Toolbox

Two events A and B are **independent** if the outcome of one event does not affect the other's chance of occurring. If two events A and B are independent, then the probability of both A and B occurring is given by $P(A \textbf{ and } B) = P(A \cap B) = P(A) \cdot P(B)$.

Since the marble is replaced before the next draw, we know this means that the trials are independent. (Consult the Toolbox to the left.) So, to determine the probability of drawing three marbles, we can apply the multiplication rule for independent events. (Consult the Toolbox to the left.) If we compute this for all eight possible outcomes, we get the results shown in Table 8.1.1.

We can see that some of the probability values in Table 8.1.1 are the same. Let's find a way to group them. If we let the variable X represent the number of

Table 8.1.1

Outcome	*P* (outcome)
RRR	$\frac{2}{5} \cdot \frac{2}{5} \cdot \frac{2}{5} = \frac{8}{125} = 0.064 = 6.4\%$
RRB	$\frac{2}{5} \cdot \frac{2}{5} \cdot \frac{3}{5} = \frac{12}{125} = 0.096 = 9.6\%$
RBR	$\frac{2}{5} \cdot \frac{3}{5} \cdot \frac{2}{5} = \frac{12}{125} = 0.096 = 9.6\%$
BRR	$\frac{3}{5} \cdot \frac{2}{5} \cdot \frac{2}{5} = \frac{12}{125} = 0.096 = 9.6\%$
BBR	$\frac{3}{5} \cdot \frac{3}{5} \cdot \frac{2}{5} = \frac{18}{125} = 0.144 = 14.4\%$
BRB	$\frac{3}{5} \cdot \frac{2}{5} \cdot \frac{3}{5} = \frac{18}{125} = 0.144 = 14.4\%$
RBB	$\frac{2}{5} \cdot \frac{3}{5} \cdot \frac{3}{5} = \frac{18}{125} = 0.144 = 14.4\%$
BBB	$\frac{3}{5} \cdot \frac{3}{5} \cdot \frac{3}{5} = \frac{27}{125} = 0.216 = 21.6\%$

blue marbles drawn, we see that the possible values of X can be $x = 0$, $x = 1$, $x = 2$, and $x = 3$, because anywhere from zero to three blue marbles can be drawn. If we associate the eight outcomes with the values of X, we get Table 8.1.2.

Table 8.1.2

x-value	$x = 0$	$x = 1$	$x = 2$	$x = 3$
Associated outcomes	*RRR*	*RRB*, *RBR*, *BRR*	*BBR*, *BRB*, *RBB*	*BBB*

If we assign probabilities to these x-values, we get the table of values shown in Table 8.1.3. Notice that for the values of $x = 1$ and $x = 2$ we must multiply 0.096 and 0.144, respectively, by 3, because they are associated with three different outcomes.

Table 8.1.3

$X = x$	$P(X = x)$
$X = 0$	$P(X = 0) = 0.064$
$X = 1$	$P(X = 1) = 3(0.096) = 0.288$
$X = 2$	$P(X = 2) = 3(0.144) = 0.432$
$X = 3$	$P(X = 3) = 0.216$

We want Table 8.1.3 to illustrate that X assigns a probability to each value of x. We call X a **random variable**. Since there is a finite number of possible values of X for this experiment, we specifically call X a **discrete random variable**.

Discrete Random Variable

- A **random variable**, X, is a rule that assigns a real number to each outcome of a probability experiment.
- A random variable is **discrete** if its set of values is finite or countably infinite.

Note: 1. X represents the random variable for the experiment, whereas x represents the possible value for the random variable.

2. By *countably infinite*, we mean that all the possible values of the random variable can be arranged so that one is called the first, one the second, one the third, and so on. That is, they can be listed in some order.

Since Table 8.1.3 is a table of a random variable and its possible values and their corresponding probabilities, some call this type of table a **probability distribution**. (We will use *table of values* and *probability distribution* interchangeably in this text.) Notice that in our probability distribution all the probability values are between 0 and 1 and the sum of the probabilities is 1. That is,

$$\sum_{x=0}^{3} P(X=x) = P(X=0) + P(X=1) + P(X=2) + P(X=3)$$
$$= 0.064 + 0.288 + 0.432 + 0.216 = 1$$

Note that in the expression $\sum_{x=0}^{3} P(X=x)$ we use an **index of summation** below and above the $\sum$ symbol to show where we start and stop adding probabilities. Now let's list the two conditions that define a discrete probability distribution.

Discrete Probability Distribution

A distribution of a discrete random variable X is a **discrete probability distribution** if:

1. $0 \le P(X=x) \le 1$ That is, all the probability values are between 0 and 1, inclusive.
2. $\sum_{\text{all } x \text{ values}} P(X=x) = 1$ The sum of the probabilities equals 1.

Sometimes we are given a function that is used to determine the probabilities for given values of the random variable. We determine these probabilities by simply evaluating the function. Such functions are called **probability mass functions**. A probability mass function will help us to compute the probabilities for given values of X.

Probability Mass Function

A **probability mass function** is a discrete probability distribution for which the probabilities are computed using a function.

Note: A probability mass function refers to a discrete random variable.

For a function to be a probability mass function, it must meet the two criteria for a discrete probability distribution. Example 1 illustrates how to determine if a function is a probability mass function.

Example 1 Determining If a Function Is a Probability Mass Function

For the function

$$P(X = x) = \begin{cases} \frac{1}{10}x, & x = 1, 2, 3, 4 \\ 0, & \text{otherwise} \end{cases}$$

complete the following:

(a) Write a table of values for $P(X = x)$.

(b) Determine if $P(X = x)$ is a probability mass function.

Solution

Table 8.1.4

$X = x$	$P(X = x)$
$X = 1$	$P(X = 1) = \frac{1}{10} = 0.1$
$X = 2$	$P(X = 2) = \frac{1}{5} = 0.2$
$X = 3$	$P(X = 3) = \frac{3}{10} = 0.3$
$X = 4$	$P(X = 4) = \frac{2}{5} = 0.4$

(a) To make a table of values for $P(X = x)$, we substitute the values of x and evaluate as shown in Table 8.1.4. The bottom part of $P(X = x)$ that reads "0 otherwise" means that, if any values other than $x = 1, 2, 3, 4$ are used, then $P(X = x) = 0$.

(b) To determine if $P(X = x)$ is a probability mass function, we must see if it meets the two criteria outlined in the definition of discrete probability distribution.

1. From Table 8.1.4, we see that the probability values for $P(X = x)$ are 0.1, 0.2, 0.3, and 0.4. So all the probability values are between 0 and 1.
2. The sum of the probabilities is

$$\sum_{x=1}^{4} P(X = x) = P(X = 1) + P(X = 2) + P(X = 3) + P(X = 4)$$
$$= 0.1 + 0.2 + 0.3 + 0.4 = 1$$

Since $P(X = x)$ meets the two criteria, it is a probability mass function. ■

✓ Checkpoint 1

Now work Exercise 15.

In addition to making a numerical table of probabilities, we can also make a graph of the probability values. In Chapter 7 we learned how to construct histograms. Here we will construct what we call **probability histograms**. The probability histogram for $P(X = x)$ in Example 1 is shown in Figure 8.1.2. In a

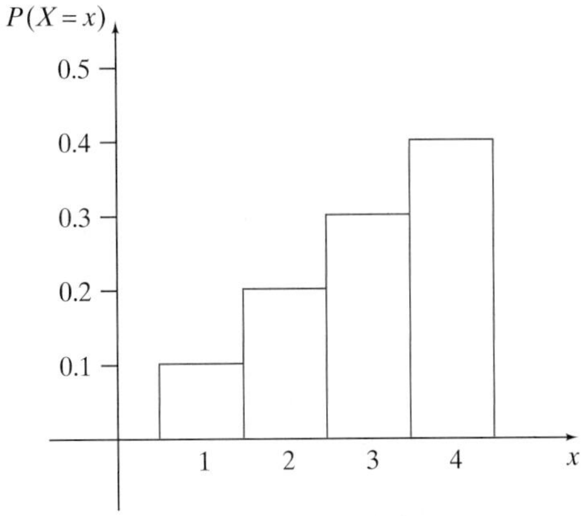

Figure 8.1.2 Probability histogram for Table 8.1.4.

probability histogram, the length of the base of each rectangle is 1, and x is the midpoint of the base. The height of each rectangle is equal to the probability value for the specified value of x.

Example 2 Determining the Value That Makes $P(X = x)$ a Probability Mass Function

For the function

$$P(X = x) = \begin{cases} \frac{1}{k}x, & x = 2, 3, 4, 5 \\ 0, & \text{otherwise} \end{cases}$$

determine the value of k so that $P(X = x)$ is a probability mass function.

Solution

Using the definition of probability mass function, we need a value of k such that $\sum_{x=2}^{5} P(X = x) = 1$. The sum gives

$$\sum_{x=2}^{5} \frac{1}{k}x = \frac{1}{k} \cdot 2 + \frac{1}{k} \cdot 3 + \frac{1}{k} \cdot 4 + \frac{1}{k} \cdot 5 = \frac{1}{k}(2 + 3 + 4 + 5) = \frac{1}{k} \cdot 14$$

Setting this sum equal to 1 yields the equation $\frac{1}{k} \cdot 14 = 1$. Solving for k, we see that $14 = k$. This gives the probability mass function

$$P(X = x) = \begin{cases} \frac{1}{14}x, & x = 2, 3, 4, 5 \\ 0, & \text{otherwise} \end{cases}$$

A quick check show that the values of $P(X = x)$ are between 0 and 1 and that $\sum_{x=2}^{5} \frac{1}{14}x = 1$. This verifies that $P(X = x)$ is a probability mass function. ■

To simplify the writing of a probability mass function, we will use the following agreement throughout the remainder of this chapter.

We will only write a probability mass function for values of x for which it is positive and assume that $P(X = x) = 0$ where it is not positive. For example, this means that we can write the solution to Example 2 as

$$P(X = x) = \frac{1}{14}x, \qquad x = 2, 3, 4, 5$$

Probability between Two Values

Let's return to the probability experiment that started this section. Recall that a jar contains 2 red and 3 blue marbles. We pull out a marble and record its color and then return it to the jar. We repeat this two more times until we have made three draws. Suppose that we want to compute the probability that we draw either 2 or 3 blue marbles from the jar. To do this, we can simply add the probability values for $P(X=2)$ and $P(X=3)$. Recall that the values for $P(X=2)$ and $P(X=3)$ are in Table 8.1.3. Symbolically, this is

$$P(2 \leq X \leq 3) = \sum_{x=2}^{3} P(X=x) = P(X=2) + P(X=3)$$
$$= 0.432 + 0.216 = 0.648$$

This means that, as the experiment is repeated, the probability that we draw either 2 or 3 blue marbles from the jar is 0.648 or 64.8%. We can use the probability histogram in Figure 8.1.3 to get this same result. To do this, we simply add the areas of the shaded rectangles that correspond to $P(X=2)$ and $P(X=3)$. (Recall that the base of each rectangle is 1 and the height is simply the probability value.)

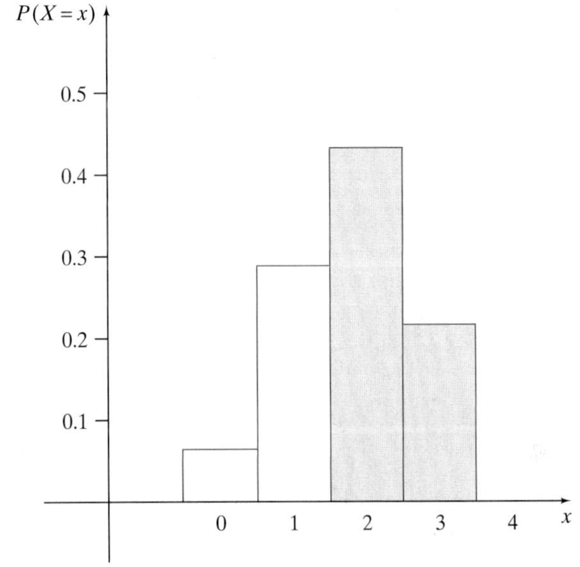

Figure 8.1.3 The sum of the areas of the two shaded rectangles gives $P(2 \leq X \leq 3)$.

Computing the Probability between Two Values

For a probability mass function $P(X=x)$, $P(a \leq X \leq b) = \sum_{x=a}^{b} P(X=x)$.

Example 3 Computing the Probability between Two Values

In a probability experiment, a coin is tossed 4 times and the result is recorded. Let X represent the total number of heads recorded. Figure 8.1.4 shows all 16 possible outcomes along with the value of X for each outcome.

(a) Construct a probability histogram of the distribution.

(b) Determine $P(1 \leq X \leq 3)$ and interpret.

		HTTH		
		HTHT		
	HTTT	THTH	HHHT	
	THTT	HHTT	HHTH	
	TTHT	THHT	HTHH	
TTTT	TTTH	TTHH	THHH	HHHH
$X = 0$	$X = 1$	$X = 2$	$X = 3$	$X = 4$

Figure 8.1.4

Solution

(a) **Understand the Situation:** To aid us in drawing the probability histogram, we begin by making a table of values for $P(X = x)$. See Table 8.1.5.

Using these values, we construct the probability histogram shown in Figure 8.1.5.

Table 8.1.5

$X = x$	$P(X = x)$
$X = 0$	$P(X = 0) = \frac{1}{16} = 0.0625$
$X = 1$	$P(X = 1) = \frac{4}{16} = 0.25$
$X = 2$	$P(X = 2) = \frac{6}{16} = 0.375$
$X = 3$	$P(X = 3) = \frac{4}{16} = 0.25$
$X = 4$	$P(X = 4) = \frac{1}{16} = 0.0625$

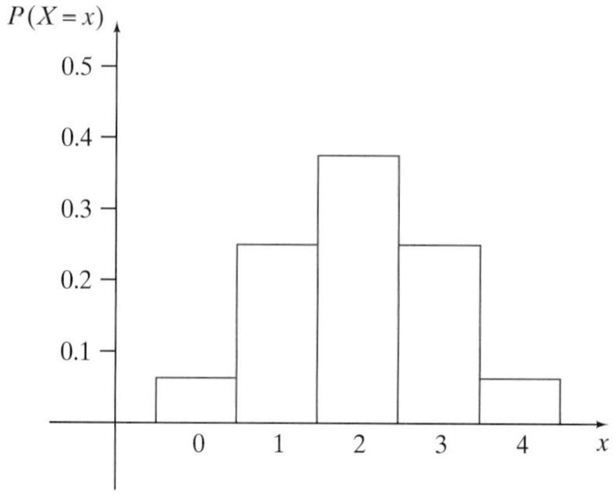

Figure 8.1.5 Probability histogram for the number of heads in four tosses of a coin.

Interactive Activity

Compute $P(X = 0)$ for the situation in Example 3. Also, compute $1 - P(X = 0)$ and interpret what this means.

WWW

(b) **Understand the Situation:** When asked to determine $P(1 \le X \le 3)$, we are being asked to determine the probability that 4 tosses of a coin will result in 1, 2, or 3 heads.

We compute this probability to be

$$\begin{aligned} P(1 \le X \le 3) &= P(X = 1) + P(X = 2) + P(X = 3) \\ &= 0.25 + 0.375 + 0.25 \\ &= 0.875 \end{aligned}$$

Interpret the Solution: When a coin is tossed 4 times, the probability that 1, 2, or 3 heads occurs is 0.875, or 87.5%. ■

Checkpoint 2

Now work Exercise 33.

Notice that the result of Example 3b corresponds to the sum of the areas of the shaded rectangles for the probability histogram as shown in Figure 8.1.6.

Probability Distributions from Empirical Probabilities

Our final topic in this introductory section to Chapter 8 revisits empirical probabilities. To recall what empirical probabilities are, consult the Toolbox on p. 469.

From Your Toolbox

Empirical Probability: Given a frequency distribution of qualitative data, if A is an event associated with a qualitative data item, then the probability of event A is given by $P(A)$

$$= \frac{\text{frequency of event } A}{\text{sum of all frequencies}} = \frac{f}{\sum f}.$$

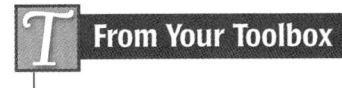

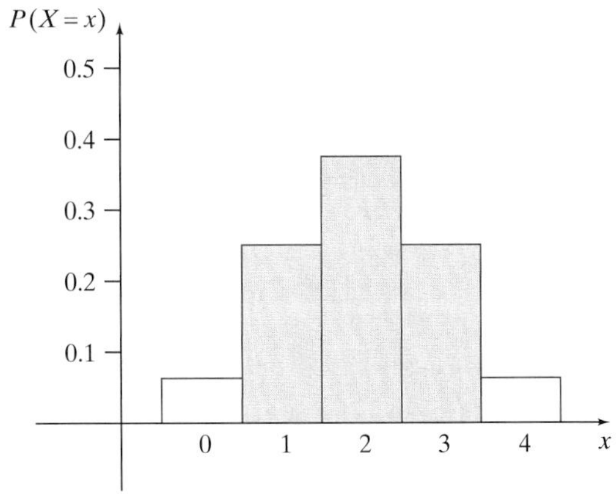

Figure 8.1.6 The sum of the areas of the three shaded rectangles equals $P(1 \leq X \leq 3)$.

Now let's examine an application in which we use empirical probabilities to construct a table of values (probability distribution).

Example 4 Constructing a Probability Distribution Using Relative Frequencies

Between 1993 and 1997, the number of two-bedroom rental units, in thousands, under construction at various monthly rent levels, in dollars, in the United States is shown in Table 8.1.6.

Table 8.1.6

Monthly Rent (rounded to the nearest hundred)	Number of Units (in thousands)
100	62
200	377
300	205
400	233
500	187
600	198
700	236

SOURCE: U.S. Census Bureau, www.census.gov

(a) Construct a table of values for the data.

(b) If a housing unit under construction is chosen at random for a housing study, determine the probability that its rent will be \$500 or more per month.

Solution

(a) **Understand the Situation:** To construct a table of values, we must first find the sum of the frequencies, f, in the second column of Table 8.1.6. The total number of units is

$$\sum f = 62 + 377 + 205 + 233 + 187 + 198 + 236 = 1498$$

Now, if we let X be a discrete random variable representing the monthly rent, in dollars, we can construct the table of values for $P(X = x)$.

The table of values for $P(X = x)$ is shown in Table 8.1.7. Note that the decimals for probability values are rounded to the nearest thousandth. As a result of this rounding, the decimal values may not sum to exactly 1.

Table 8.1.7

$X = x$	$P(X = x)$
$X = 100$	$P(X = 100) = \frac{62}{1498} \approx 0.041$
$X = 200$	$P(X = 200) = \frac{377}{1498} \approx 0.252$
$X = 300$	$P(X = 300) = \frac{205}{1498} \approx 0.137$
$X = 400$	$P(X = 400) = \frac{233}{1498} \approx 0.156$
$X = 500$	$P(X = 500) = \frac{187}{1498} \approx 0.125$
$X = 600$	$P(X = 600) = \frac{198}{1498} \approx 0.132$
$X = 700$	$P(X = 700) = \frac{236}{1498} \approx 0.158$

Interactive Activity

Use the result of $P(X \geq 500)$ from Example 4b to determine $1 - P(X \geq 500)$. Explain what this result represents.

(b) Understand the Situation: Since X is a random variable representing monthly rent, we want to determine $P(500 \text{ or more})$, which symbolically is $P(X \geq 500)$.

Continuing, we get

$$P(X \geq 500) = P(500) + P(600) + P(700)$$
$$\approx 0.125 + 0.132 + 0.158 = 0.415$$

Interpret the Solution: The probability that a randomly selected housing unit under construction will have a monthly rent of \$500 or more is 0.415, or 41.5%.

■

Checkpoint 3 Now work Exercise 47.

SUMMARY

- A **random variable** is a rule that assigns a real number to each outcome of a probability experiment.
- A random variable is **discrete** if its set of values is finite or countably infinite.
- A function P for a discrete random variable X is a **probability mass function** if $0 \leq P(X = x) \leq 1$ and $\sum_{\substack{\text{all } x \\ \text{values}}} P(X = x) = 1$.
- For a probability mass function $P(X = x)$, $P(a \leq X \leq b) = \sum_{x=a}^{b} P(X = x)$.

SECTION 8.1 EXERCISES

In Exercises 1–10, determine if the table meets the two criteria of a probability distribution.

1.

$X = x$	$P(X = x)$
$X = 0$	$P(X = 0) = \frac{1}{2}$
$X = 1$	$P(X = 1) = \frac{1}{4}$
$X = 2$	$P(X = 2) = \frac{1}{4}$

2.

$X = x$	$P(X = x)$
$X = 1$	$P(X = 1) = \frac{1}{3}$
$X = 2$	$P(X = 2) = \frac{2}{3}$
$X = 3$	$P(X = 3) = \frac{1}{3}$

3.

$X = x$	$P(X = x)$
$X = 0$	$P(X = 0) = \frac{1}{10}$
$X = 1$	$P(X = 1) = \frac{1}{20}$
$X = 2$	$P(X = 2) = \frac{1}{5}$
$X = 3$	$P(X = 3) = \frac{2}{5}$
$X = 4$	$P(X = 4) = \frac{1}{4}$

4.

$X = x$	$P(X = x)$
$X = 1$	$P(X = 1) = \frac{3}{32}$
$X = 2$	$P(X = 2) = \frac{3}{16}$
$X = 3$	$P(X = 3) = \frac{5}{32}$
$X = 4$	$P(X = 4) = \frac{3}{8}$
$X = 5$	$P(X = 5) = \frac{3}{16}$

5.

$X = x$	$P(X = x)$
$X = -2$	$P(X = -2) = 0.03$
$X = -1$	$P(X = -1) = 0.13$
$X = 0$	$P(X = 0) = 0.40$
$X = 1$	$P(X = 1) = 0.35$
$X = 2$	$P(X = 2) = 0.09$

6.

$X = x$	$P(X = x)$
$X = -1$	$P(X = -1) = 0.174$
$X = 0$	$P(X = 0) = 0.111$
$X = 2$	$P(X = 2) = 0.260$
$X = 4$	$P(X = 4) = 0.300$
$X = 6$	$P(X = 6) = 0.155$

7.

$X = x$	$P(X = x)$
$X = 10$	$P(X = 10) = -0.1$
$X = 20$	$P(X = 20) = 0.3$
$X = 30$	$P(X = 30) = 0.6$
$X = 40$	$P(X = 40) = 0.2$

8.

$X = x$	$P(X = x)$
$X = 5$	$P(X = 5) = 0$
$X = 10$	$P(X = 10) = 1.2$
$X = 25$	$P(X = 25) = 0.1$
$X = 35$	$P(X = 35) = -0.3$

9.

$X = x$	$P(X = x)$
$X = 8$	$P(X = 8) = \frac{1}{5}$
$X = 9$	$P(X = 9) = \frac{1}{5}$
$X = 10$	$P(X = 10) = \frac{1}{10}$
$X = 11$	$P(X = 11) = \frac{3}{20}$
$X = 12$	$P(X = 12) = \frac{7}{20}$

10.

$X = x$	$P(X = x)$
$X = 0$	$P(X = 0) = \frac{11}{20}$
$X = 1$	$P(X = 1) = \frac{1}{4}$
$X = 2$	$P(X = 2) = \frac{1}{20}$
$X = 3$	$P(X = 3) = \frac{1}{10}$
$X = 4$	$P(X = 4) = \frac{1}{20}$

In Exercises 11–20, a formula for $P(X = x)$ is given. Complete the following:

(a) Write a table of values for $P(X = x)$.

(b) Determine if the table meets the two criteria of a probability distribution. Remember that $P(X = x) = 0$ when the x-values are not defined.

11. $P(X = x) = \dfrac{1}{6}x, \quad x = 0, 1, 2, 3$

12. $P(X = x) = 0.01x, \quad x = 10, 20, 30, 40$

13. $P(X = x) = \dfrac{1}{4}, \quad x = 1, 2, 3, 4$

14. $P(X = x) = \dfrac{1}{5}, \quad x = 0, 1, 2, 3, 4$

✓ **15.** $P(X = x) = \dfrac{1}{15}x^2, \quad x = 0, 1, 2, 3, 4$

16. $P(X = x) = \dfrac{1}{18}x^3, \quad x = 0, 1, 2, 3, 4$

17. $P(X = x) = 0.05x, \quad x = 2, 3, 4, 6$

18. $P(X = x) = \dfrac{1}{55}x, \quad x = 5, 10, 15, 20$

19. $P(X = x) = \dfrac{x + 2}{10}, \quad x = 0, 1, 2, 3$

20. $P(X = x) = \dfrac{x - 4}{14}, \quad x = 6, 7, 8, 9$

In Exercises 21–26, determine the values of k so that $P(X = x)$ is a probability mass function.

21. $P(X = x) = \dfrac{1}{k}x, \quad x = 0, 1, 2, 3$

22. $P(X = x) = \dfrac{2}{k}x, \quad x = 1, 2, 3, 4$

23. $P(X = x) = kx, \quad x = 5, 10, 15, 20$

24. $P(X = x) = kx, \quad x = 3, 6, 9, 12, 15$

25. $P(X = x) = \dfrac{1}{k}x^2, \quad x = 0, 1, 2, 4$

26. $P(X = x) = \dfrac{1}{k}x^3, \quad x = 0, 1, 2, 3, 5$

In Exercises 27–32, a table of frequencies is given. Construct the corresponding probability distribution with $X = x$ in the left column and $P(X = x)$ in the right column. See Example 4.

27.

Value	Frequency
1	5
2	5
3	10

28.

Value	Frequency
0	6
1	10
2	14

29.

Value	Frequency
1	10
2	20
3	15
4	5
5	25

30.

Value	Frequency
0	55
10	45
20	15
30	10

31.

Value	Frequency
0	150
1	600
2	400
3	200

32.

Value	Frequency
0	1500
10	0
20	1750
30	2000

Applications

✓ **33.** The Videoshack has determined that the number of videos that a customer rents on any given visit can be represented by a discrete random variable X with the following probability distribution.

$X = x$	$P(X = x)$
$X = 0$	$P(X = 0) = 0.05$
$X = 1$	$P(X = 1) = 0.50$
$X = 2$	$P(X = 2) = 0.30$
$X = 3$	$P(X = 3) = 0.15$

(a) Construct a probability histogram of the distribution.
(b) Determine $P(1 \leq X \leq 2)$ and interpret.

34. The MoccaMania coffee shop has found that its number of daily customers can be represented by a discrete random variable X with the following probability distribution.

$X = x$	$P(X = x)$
$X = 45$	$P(X = 45) = 0.10$
$X = 46$	$P(X = 46) = 0.30$
$X = 47$	$P(X = 47) = 0.25$
$X = 48$	$P(X = 48) = 0.20$
$X = 49$	$P(X = 49) = 0.05$
$X = 50$	$P(X = 50) = 0.10$

(a) Construct a probability histogram of the distribution.
(b) Determine $P(46 \leq X \leq 49)$ and interpret.

35. Milford College hires the next four staff employees without regard to gender. The personnel office knows that the pool of applicants is large and there is an equal chance of each applicant being hired. Records show that the number of women hired can be represented by a discrete random variable X with the following probability distribution.

$X = x$	$P(X = x)$
$X = 0$	$P(X = 0) = 0.1$
$X = 1$	$P(X = 1) = 0.2$
$X = 2$	$P(X = 2) = 0.4$
$X = 3$	$P(X = 3) = 0.2$
$X = 4$	$P(X = 4) = 0.1$

(a) Construct a probability histogram of the distribution.
(b) Determine $P(2 \leq X \leq 4)$ and interpret.

36. When four different households are surveyed on Tuesday night, the National Rating Service Company determines that the discrete random variable X represents the number of households who are watching the new drama "Alien from Within."

$X = x$	$P(X = x)$
$X = 0$	$P(X = 0) = 0.2$
$X = 1$	$P(X = 1) = 0.3$
$X = 2$	$P(X = 2) = 0.3$
$X = 3$	$P(X = 3) = 0.1$
$X = 4$	$P(X = 4) = 0.1$

(a) Construct a probability histogram of the distribution.
(b) Determine $P(0 \leq X \leq 2)$ and interpret.

37. The Longberg Elementary School PTA has determined that the number of commercials per show aired during afterschool television shows can be represented by a discrete random variable X with the following probability distribution.

$X = x$	$P(X = x)$
$X = 0$	$P(X = 0) = 0.20$
$X = 1$	$P(X = 1) = 0.15$
$X = 2$	$P(X = 2) = 0.25$
$X = 3$	$P(X = 3) = 0.20$
$X = 4$	$P(X = 4) = 0.05$
$X = 5$	$P(X = 5) = 0.10$
$X = 6$	$P(X = 6) = 0.05$

(a) Construct a probability histogram of the distribution.
(b) Determine $P(X \leq 3)$ and interpret.

38. The North-Am racing series has found that the number of yellow flag caution laps, rounded to the nearest 10, during any given race can be represented by a discrete random variable X with the following probability distribution.

$X = x$	$P(X = x)$
$X = 0$	$P(X = 0) = 0.20$
$X = 10$	$P(X = 10) = 0.25$
$X = 20$	$P(X = 20) = 0.38$
$X = 30$	$P(X = 30) = 0.10$
$X = 40$	$P(X = 40) = 0.07$

(a) Construct a probability histogram of the distribution.

(b) Determine $P(X \geq 10)$ and interpret.

39. Parmiter College has found that the number of students who drop out of introductory psychology can be represented by a discrete random variable X with the following probability distribution.

$X = x$	$P(X = x)$
$X = 12$	$P(X = 12) = 0.09$
$X = 13$	$P(X = 13) = 0.18$
$X = 14$	$P(X = 14) = 0.38$
$X = 15$	$P(X = 15) = 0.20$
$X = 16$	$P(X = 16) = 0.15$

(a) Determine the probability that at most 14 students will drop out of introductory psychology during any given quarter.

(b) According to the distribution, what is the least likely number of students to drop out of introductory psychology during any given quarter?

40. The Supreme Pie pizzeria determines that its number of late-night pizza deliveries on any given Wednesday can be represented by a discrete random variable X with the following probability distribution.

$X = x$	$P(X = x)$
$X = 4$	$P(X = 4) = 0.32$
$X = 5$	$P(X = 5) = 0.11$
$X = 6$	$P(X = 6) = 0.40$
$X = 7$	$P(X = 7) = 0.12$
$X = 8$	$P(X = 8) = 0.05$

(a) Determine the probability that at least 5 late-night deliveries will be made on any given Wednesday night.

(b) According to the distribution, what is the least likely number of late-night pizza deliveries on any given Wednesday?

41. Officials of Stone Temple City have determined that the size of family, including homeowner and all relatives, can be represented by a discrete random variable X. The probability distribution for family size is shown in the following table. A city official chooses a family at random to complete a questionnaire about civil planning.

$X = x$	$P(X = x)$
$X = 2$	$P(X = 2) = 0.15$
$X = 3$	$P(X = 3) = 0.23$
$X = 4$	$P(X = 4) = 0.21$
$X = 5$	$P(X = 5) = 0.15$
$X = 6$	$P(X = 6) = 0.14$
$X = 7$	$P(X = 7) = 0.12$

(a) Determine the probability that the size of the family will be at least 4.

(b) According to the distribution, what is the least likely size that the chosen family will be?

42. Researchers at the Hobart Sleep Institute have determined that the number of hours of sleep (rounded to the nearest hour) that the subjects of a large study get during a sleep study can be represented by a random variable X that has the probability distribution shown in the following table. A sleep study subject is chosen at random to answer follow-up questions.

$X = x$	$P(X = x)$
$X = 4$	$P(X = 4) = 0.11$
$X = 5$	$P(X = 5) = 0.15$
$X = 6$	$P(X = 6) = 0.20$
$X = 7$	$P(X = 7) = 0.33$
$X = 8$	$P(X = 8) = 0.12$
$X = 9$	$P(X = 9) = 0.09$

(a) Determine the probability that the subject slept at least 6 hours.

(b) According to the distribution, what is the least likely number of hours that a subject slept during the study?

43. The enrollment at Hagerty College is given in the following frequency table.

Academic Rank	Enrollment
Freshmen	7800
Sophomores	5500
Juniors	3200
Seniors	2500

(a) Let X be a discrete random variable representing the number of years of college completed by a randomly selected student from Hagerty College, meaning that $x = 0, 1, 2, 3$. Construct a probability distribution for the data.

(b) Compute $P(X = 3)$ and interpret.

44. The number of defective blue jeans discovered after inspection by the Saxsworth Company is shown in the following table:

Number of Inspections	Number of Defects
1	1500
2	800
3	200
4	50

(a) Let X be a discrete random variable where X represents the number of inspections completed. Construct a probability distribution for the data.

(b) Compute $P(X = 2)$ and interpret.

45. The number of passengers who miss the overnight flight from Tokyo to Honolulu, as indicated by the annual records of Pacific Rim Airlines, are shown in the table.

Passengers Who Miss the Flight	Number of Flights
0	53
1	127
2	280
3	128
4	77

(a) Let X be a discrete random variable representing the number of passengers who miss the overnight flight. Construct a probability distribution for the data.

(b) Compute $P(X = 3)$ and interpret.

46. A marketing research firm determines that if groups of 5 youths are chosen the number (of the 5) who own the new Dream Boy hand-held video game are as shown in the table.

Number of Youths	Number Who Own Dream Boy
0	233
1	207
2	170
3	118
4	82
5	30

(a) Let X be a discrete random variable representing the number of youths who own the Dream Boy. Construct a probability distribution for the data.

(b) Compute $P(X = 2)$ and interpret.

47. The number of drivers aged 16 to 24 who were in automobile accidents in 1996 is shown in the table.

Age	Number of Accidents (in millions)
16	1.6
17	2.2
18	2.4
19	2.7
20	2.8
21	2.9
22	2.9
23	3.1
24	3.6

SOURCE: National Safety Council, www.nsc.org

(a) Let X be a discrete random variable representing the age of a driver in an accident. Construct a probability distribution for the data.

(b) Compute $P(X = 20)$ and interpret.

(c) Compute $P(X \leq 20)$ and interpret.

48. The distribution of weights of American males aged 70 to 79 from 1988 to 1994 is shown in the table.

Weight (in pounds)	Percentage	Weight (in pounds)	Percentage
100–109	0.2	200–209	9.5
110–119	0.2	210–219	5.2
120–129	1.3	220–229	5.4
130–139	4.1	230–239	2.5
140–149	7.4	240–249	1.0
150–159	10.1	250–259	0.7
160–169	11.9	260–269	1.1
170–179	13.2	270–279	0.9
180–189	11.1	280–289	0.4
190–199	13.6	290 and above	0.2

SOURCE: U.S. Center for Health Statistics, www.cdc.gov/nchs

(a) Let X be a discrete random variable representing the weight of an American male aged 70 to 79 from 1988 to 1994. Construct a probability distribution for the data.

(b) Compute $P(X < 160)$ and interpret.

(c) Compute $P(X \geq 200)$ and interpret.

49. In 1997, the number of school activities in which at least one parent of a child in grades K–5 participated is shown in the following table:

Activities Attended	Percentage
0	3.5
1	7.4
2	19.7
3	32.0
4 or more	37.4

SOURCE: U.S. Census Bureau, www.census.gov

Let X be a discrete random variable representing the number of school activities that at least one parent attended in 1997.

(a) Compute $P(X = 2)$ and interpret.

(b) Compute $P(X \geq 2)$ and interpret.

50. In 1997, the number of school activities in which at least one parent of a child in grades 6–8 participated is shown in the following table:

Activities Attended	Percentage
0	7.3
1	11.7
2	25.8
3	32.2
4 or more	23.0

SOURCE: U.S. Census Bureau, www.census.gov

Let X be a discrete random variable representing the number of school activities that at least one parent attended in 1997.

(a) Compute $P(X = 3)$ and interpret.

(b) Compute $P(X \geq 3)$ and interpret.

SECTION PROJECT

Suppose that a mathematics department has nine women and five men as full-time faculty members.

(a) In how many ways can a committee of three women and one man be selected from the department?

(b) In how many ways can a committee of four be selected from the department.

(c) Using the results from (a) and (b), determine the probability that a committee of four from the department will consist of three women and one man.

The result of part (c) illustrates a special probability distribution called the **hypergeometric distribution**. The hypergeometric distribution is a distribution of a random variable that has two outcomes when selection is done without replacement. The distribution is described as follows: Given a population with only two types (males and females, for example) such that there are a items of one kind and b items of another kind and $a+b$ equals the total population, the probability $P(X=x)$ of selecting without replacement a sample of size n with X items of type a and $n-X$ items of type b is $P(X)=\dfrac{({}_aC_X)\cdot({}_bC_{n-X})}{{}_{(a+b)}C_n}$.

(d) For the mathematics department with nine women and five men in (a)–(c), if we were to use the hypergeometric distribution to determine the probability that a committee of four consists of three women and one man, determine the values of a, b, X, and n that would be used in the hypergeometric formula. Use the hypergeometric formula to verify your answer to part (c).

(e) Use the hypergeometric distribution to determine the following: A recent housing study by a major homeowners insurance company found that exactly four out of nine houses on a certain street were underinsured. If five houses are selected from the nine houses, determine the probability that two are underinsured.

Section 8.2 Expected Value and Standard Deviation of a Discrete Random Variable

Given a particular probability distribution or a probability mass function, we often want to know what the *average* outcome would be if we repeat the experiment over and over many times. Since we are working with a probability distribution, this average is computed in a different way from computing the mean of a set of data.

Expected Value

To discover how an average of a probability distribution can be computed, let's start with a familiar experiment of tossing a six-sided die. Suppose that we roll a die 4 times and the results of the rolls are 2, 3, 6, and 1. Then the average of the rolls is $\dfrac{4+3+6+2}{4}=\dfrac{15}{4}=3.75$. It is important to note that the average of the 4 rolls is not a value that occurs in a single roll. It is not possible to roll a 3.75!

Table 8.2.1

$X=x$	$P(X=x)$
$X=1$	$P(X=1)=\frac{1}{6}$
$X=2$	$P(X=2)=\frac{1}{6}$
$X=3$	$P(X=3)=\frac{1}{6}$
$X=4$	$P(X=4)=\frac{1}{6}$
$X=5$	$P(X=5)=\frac{1}{6}$
$X=6$	$P(X=6)=\frac{1}{6}$

If we wanted to roll a die, say, 300 times, the average of the rolls is computed in a similar fashion. To determine the average, we can sum the 300 outcomes and divide the result by 300. To see how we can approximate this sum, we let X represent the outcome on any roll of the die, where the possible values of X are $x=1,2,3,4,5,6$. Since the outcomes of the rolls are equally likely, we get the values of the probability distribution $P(X=x)$ shown in Table 8.2.1.

Let's examine the next to last row of the table. Since $P(X=5)=\frac{1}{6}$, this tells us that if we roll a die 300 times we expect that about $\frac{1}{6}$ of the 300 rolls will produce a 5. So, in 300 rolls the number of times that we expect to see a 5 is found by $\frac{1}{6}\cdot 300=50$. Hence, we expect to roll a 5 about 50 times if we toss the die a total of 300 times. Notice that Table 8.2.1 indicates that we would expect to roll **any** number $\frac{1}{6}$ of the time. This means that in 300 rolls we expect to see each number occur 50 times. We summarize this information in Table 8.2.2.

Table 8.2.2

$X = x$	$P(X = x)$	Number of Times $X = x$ Is Expected
$X = 1$	$P(X = 1) = \frac{1}{6}$	$\frac{1}{6} \cdot 300 = 50$
$X = 2$	$P(X = 2) = \frac{1}{6}$	$\frac{1}{6} \cdot 300 = 50$
$X = 3$	$P(X = 3) = \frac{1}{6}$	$\frac{1}{6} \cdot 300 = 50$
$X = 4$	$P(X = 4) = \frac{1}{6}$	$\frac{1}{6} \cdot 300 = 50$
$X = 5$	$P(X = 5) = \frac{1}{6}$	$\frac{1}{6} \cdot 300 = 50$
$X = 6$	$P(X = 6) = \frac{1}{6}$	$\frac{1}{6} \cdot 300 = 50$

To determine the expected sum of the 300 rolls, we do the following:

$$\begin{aligned}\text{sum of rolls} &= 1 \cdot \begin{pmatrix}\text{number of}\\ \text{1's expected}\end{pmatrix} + 2 \cdot \begin{pmatrix}\text{number of}\\ \text{2's expected}\end{pmatrix}\\ &\quad + 3 \cdot \begin{pmatrix}\text{number of}\\ \text{3's expected}\end{pmatrix} + \cdots + 6 \cdot \begin{pmatrix}\text{number of}\\ \text{6's expected}\end{pmatrix}\\ &= 1 \cdot 50 + 2 \cdot 50 + 3 \cdot 50 + 4 \cdot 50 + 5 \cdot 50 + 6 \cdot 50\\ &= 1050\end{aligned}$$

Dividing this sum by the number of rolls, 300, gives us our expected average. This is

$$\text{average of rolls} = \frac{\text{sum of rolls}}{\text{number of rolls}} = \frac{1050}{300} = 3.5$$

Notice that if we write this average in the form of a fraction we get

$$\begin{aligned}\text{average of rolls} &= \frac{\text{sum of rolls}}{\text{number of rolls}}\\ &= \frac{1 \cdot \begin{pmatrix}\text{number of}\\ \text{1's expected}\end{pmatrix} + 2 \cdot \begin{pmatrix}\text{number of}\\ \text{2's expected}\end{pmatrix} + 3 \cdot \begin{pmatrix}\text{number of}\\ \text{3's expected}\end{pmatrix} + \cdots + 6 \cdot \begin{pmatrix}\text{number of}\\ \text{6's expected}\end{pmatrix}}{300}\\ &= \frac{1 \cdot [(\frac{1}{6})(300)] + 2 \cdot [(\frac{1}{6})(300)] + 3 \cdot [(\frac{1}{6})(300)] + \cdots + 6 \cdot [(\frac{1}{6})(300)]}{300}\\ &= \frac{300\,[1 \cdot (\frac{1}{6}) + 2 \cdot (\frac{1}{6}) + 3 \cdot (\frac{1}{6}) + \cdots + 6 \cdot (\frac{1}{6})]}{300}\\ &= 1 \cdot \left(\frac{1}{6}\right) + 2 \cdot \left(\frac{1}{6}\right) + 3 \cdot \left(\frac{1}{6}\right) + \cdots + 6 \cdot \left(\frac{1}{6}\right)\\ &= \frac{21}{6} = 3.5\end{aligned}$$

The value 3.5 tells us that, if we repeat the experiment over and over many times, the average of the outcomes is expected to be about 3.5. Again, we note that the average is not a value that occurs in a single roll. What we have just determined is the **mean** or average value of a random variable X. Often we call the mean of a random variable X the **expected value** of X. (The expected value can also be thought of as a probability weighted average, since each outcome is *weighted* by its probability.)

Expected Value of a Discrete Random Variable

For a discrete random variable X having values $x_1, x_2, \ldots, x_n$, the **expected value, $E(X)$**, (or **mean, μ_X**) of the distribution is

$$E(X) = \mu_X = x_1 P(X = x_1) + x_2 P(X = x_2) + \cdots + x_n P(X = x_n)$$

Note: The formula in the preceding box is for the case of a finite number of random variable values. In the case that X is countably infinite, the general definition is $\mu_X = \sum_{\substack{\text{all } x \\ \text{values}}} x \cdot P(X = x)$, assuming that the sum exists.

Before we leave the experiment of tossing a die, let's make a final observation. When we compute the expected value as a fraction, notice that the 300 for the number of rolls was factored from the numerator and then canceled with the 300 in the denominator. This tells us that the actual number of trials does not matter, because the expected value represents the **long-run average** of the outcomes. In Example 1 we determine the expected value for another probability distribution.

Example 1 Determining an Expected Value

In Example 3 of Section 8.1, we conducted a probability experiment of tossing a coin 4 times where X represented the total number of heads recorded. The table of values for $P(X = x)$ is shown in Table 8.2.3. Determine the expected value and interpret.

Table 8.2.3

$X = x$	$P(X = x)$
$X = 0$	$P(X = 0) = \frac{1}{16}$
$X = 1$	$P(X = 1) = \frac{4}{16}$
$X = 2$	$P(X = 2) = \frac{6}{16}$
$X = 3$	$P(X = 3) = \frac{4}{16}$
$X = 4$	$P(X = 5) = \frac{1}{16}$

Solution

Using the formula for expected value, we get

$$E(X) = \mu_X = 0 \cdot P(X = 0) + 1 \cdot P(X = 1) + 2 \cdot P(X = 2) + 3 \cdot P(X = 3) + 4 \cdot P(X = 4)$$

$$= 0 \cdot \frac{1}{16} + 1 \cdot \frac{4}{16} + 2 \cdot \frac{6}{16} + 3 \cdot \frac{4}{16} + 4 \cdot \frac{1}{16}$$

$$= 0 + \frac{4}{16} + \frac{12}{16} + \frac{12}{16} + \frac{4}{16} = \frac{32}{16} = 2$$

Interpret the Solution: This means that if the experiment of tossing a coin 4 times is repeated over time we can expect on average that 2 of the 4 tosses will be heads.

If we make a probability histogram for the distribution in Table 8.2.3 and look at where the expected value, $E(X) = 2$, found in Example 1 occurs, we can obtain a geometric meaning of expected value. The expected value is similar to a fulcrum (a balancing point) that would perfectly "balance" the probability histogram. See Figure 8.2.1.

Example 1 had proper fractions for all the probability values. In applications, many of the probabilities are given as decimals. Example 2 shows this type of application.

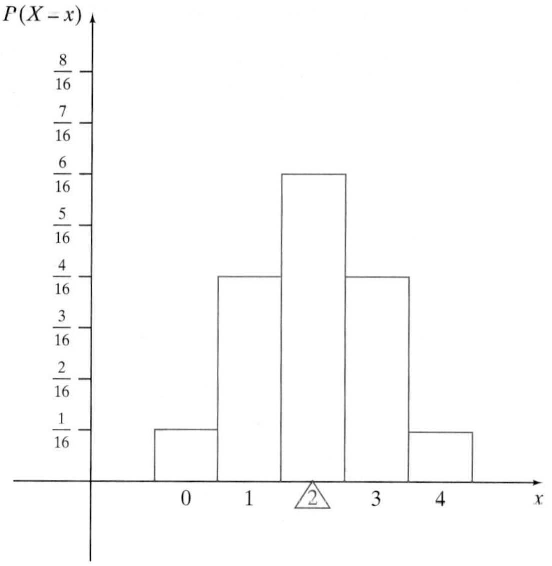

Figure 8.2.1

Example 2 Applying an Expected Value

Between 1993 and 1997, the number of two-bedroom rental units, in thousands, under construction at various monthly rent levels X, in dollars, in the United States can be represented by the probability distribution in Table 8.2.4.

Table 8.2.4

$X = x$	$P(X = x)$
$X = 100$	$P(X = 100) \approx 0.041$
$X = 200$	$P(X = 200) \approx 0.252$
$X = 300$	$P(X = 300) \approx 0.137$
$X = 400$	$P(X = 400) \approx 0.156$
$X = 500$	$P(X = 500) \approx 0.125$
$X = 600$	$P(X = 600) \approx 0.132$
$X = 700$	$P(X = 700) \approx 0.158$

SOURCE: U.S. Census Bureau, www.census.gov

Compute the expected value and interpret.

Solution

Using the formula for expected value, we get

$$E(X) = \mu_X = 100(0.041) + 200(0.252) + 300(0.137) + 400(0.156)$$
$$+ 500(0.125) + 600(0.132) + 700(0.158)$$
$$= 410.30$$

Interpret the Solution: This means that, if many two-bedroom housing units were selected randomly between 1993 and 1997 in the United States, the average of the monthly rent of all selected rentals would have been about \$410.30. ■

Checkpoint 1

Now work Exercise 41.

Example 3 Applying an Expected Value

The Ashton Wrestling Club is having a fund-raising raffle. One thousand tickets have been sold for \$5 each. There is 1 first-place prize of \$1000, 3 second-place prizes of \$500 each, 5 third-place prizes of \$200 each, and 20 consolation prizes of \$50 each. Let X represent the net winnings, that is, the winnings minus the cost of purchasing a ticket. Determine $E(X)$ and interpret.

Table 8.2.5

Type of Ticket	Net Winnings
Losing	$0 - 5 = -5$
Consolation	$50 - 5 = 45$
Third place	$200 - 5 = 195$
Second place	$500 - 5 = 495$
First place	$1000 - 5 = 995$

Solution

Understand the Situation: We need to determine the possible values of X, that is, the value of a losing ticket, a first-place ticket, a second-place ticket, a third-place ticket, and a consolation prize ticket. After we determine the probability of having each type of ticket, we can then construct a table of values for $P(X = x)$. We can then use the table to determine $E(X)$.

The net winnings of any ticket, which give us the values of X, is simply

$$\text{winnings} - \text{cost of a ticket} = \text{net winnings}$$

We summarize the net winnings of each type of ticket in Table 8.2.5.

Since there is 1 first-place prize, the probability of having a first-place ticket is

$$P(\text{first place}) = \frac{1}{1000}$$

We know that there are 3 second-place tickets, which means the probability of having a second-place ticket is

$$P(\text{second place}) = \frac{3}{1000}$$

Continuing in this manner allows us to make the table of values for $P(X = x)$ shown in Table 8.2.6, from which we can compute $E(X)$ to be

$$\begin{aligned} E(X) &= -5(0.971) + 45(0.02) + 195(0.005) + 495(0.003) + 995(0.001) \\ &= -0.5 \end{aligned}$$

Table 8.2.6

$X = x$	$P(X = x)$
$X = -5$	$P(X = -5) = \frac{971}{1000} = 0.971$
$X = 45$	$P(X = 45) = \frac{20}{1000} = 0.02$
$X = 195$	$P(X = 195) = \frac{5}{1000} = 0.005$
$X = 495$	$P(X = 495) = \frac{3}{1000} = 0.003$
$X = 995$	$P(X = 995) = \frac{1}{1000} = 0.001$

Interpret the Solution: The expected value gives the long-run average loss for an owner of one ticket. In other words, if someone participated in such a raffle by purchasing just one ticket each time, in the long run this individual can expect to lose \$0.50 per raffle on average. ■

Standard Deviation

In addition to knowing the expected value, or mean, of a random variable X, we also want to know how spread out or *dispersed* the probability distribution is.

Consider the two distributions whose probability histograms are given in Figure 8.2.2. Figure 8.2.2(a) shows a probability distribution that is fairly concentrated about its expected value (mean), whereas Figure 8.2.2(b) is more dispersed or spread about the expected value (mean).

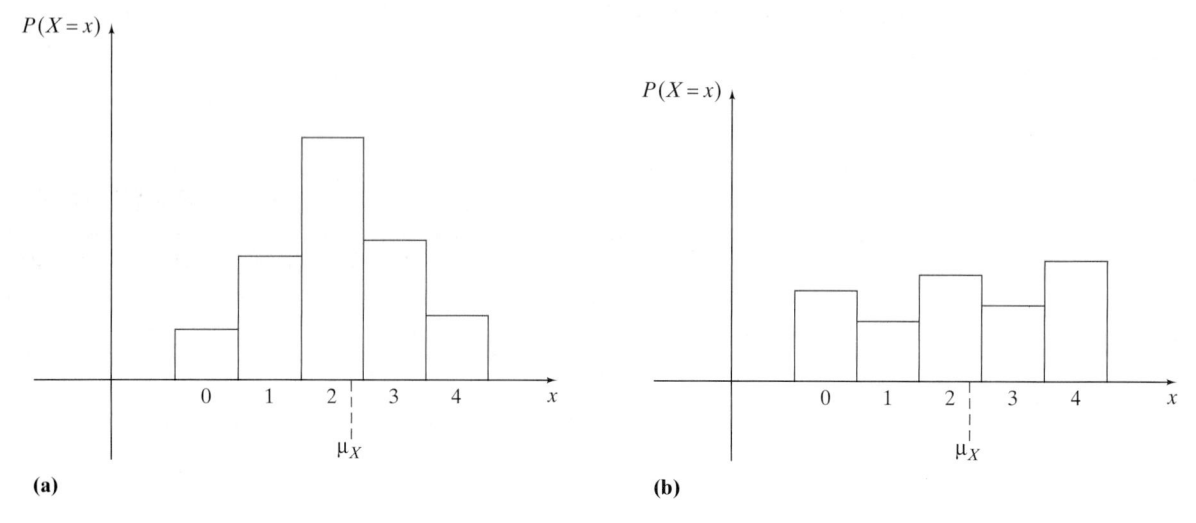

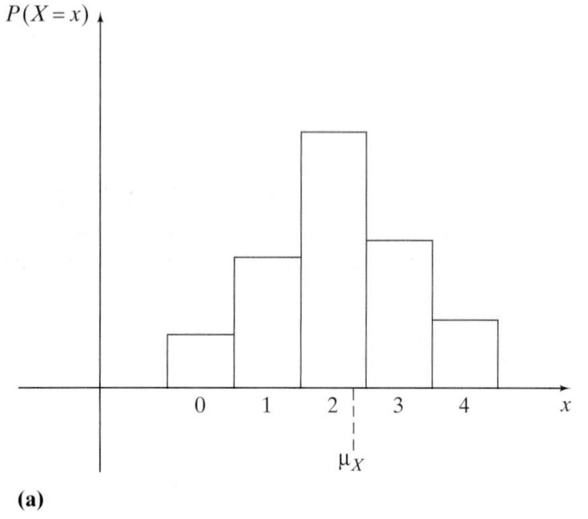

Figure 8.2.2

We can give a numerical value to this dispersion by computing the **standard deviation** of the probability distribution. The standard deviation for a probability distribution is denoted by the Greek lowercase letter σ (pronounced *sigma*). We can think of the standard deviation as the average distance that each random variable value lies away from the mean, when accounting for probability.

Standard Deviation of a Discrete Random Variable

For a discrete random variable X having values $x_1, x_2, \ldots, x_n$, the **standard deviation** is

$$\sigma_X = \sqrt{\sum_{i=1}^{n} x_i^2 P(x = x_i) - \mu_X^2}$$

Note: The formula in the preceding box is for the case of a finite number of random variable values. In the case when X is countably infinite, the general definition is

$$\sigma_X = \sqrt{\sum_{\substack{\text{all } x \\ \text{values}}} x^2 \cdot P(X = x) - \mu_X^2}$$

Using the formula to compute the standard deviation can appear to be somewhat perplexing. To simplify computations, we can compute the standard deviation in three steps:

Step 1: Compute

$$\sum_{i=1}^{n} x_i^2 P(X = x_i) = x_1^2 P(X = x_1) + x_2^2 P(X = x_2) + \cdots + x_n^2 P(X = x_n)$$

Step 2: Subtract the square of the mean, expected value, from the result in step 1 to get

$$\sum_{i=1}^{n} x_i^2 P(X = x_i) - \mu_X^2$$

This quantity is called the **variance** and is denoted by σ_X^2.

Step 3: Take the square root of the result in step 2 to get

$$\sigma_X = \sqrt{\sum_{i=1}^{n} x_i^2 P(X = x_i) - \mu_X^2}$$

Example 4 illustrates how these three steps are computed.

Example 4 Computing the Standard Deviation of a Discrete Random Variable

For the probability distribution of monthly rents in Example 2, compute the variance and standard deviation.

Solution

We will follow the three steps in computing the standard deviation.

Step 1: $\sum_{i=1}^{7} x_i^2 P(X = x_i) \approx 100^2(0.041) + 200^2(0.252) + 300^2(0.137)$

$$+ 400^2(0.156) + 500^2(0.125) + 600^2(0.132) + 700^2(0.158) = 203{,}970$$

Step 2: Now we square the expected value $E(X)$, or mean μ_X, that we found in Example 2 and subtract it from the value found in step 1.

$$\sum_{i=1}^{7} x_i^2 P(X = x_i) - \mu_X^2 = 203{,}970 - (410.30)^2 = 35{,}623.91$$

So the variance is $\sigma_X^2 = 35{,}623.91$.

Step 3: To determine the standard deviation, we simply take the square root of the variance. This gives us

$$\sigma_X = \sqrt{\sigma_X^2} = \sqrt{35{,}623.91} \approx 188.74$$

Interpret the Solution: So the standard deviation of the monthly rents is about \$188.74. This means that the monthly rents varied on average by about \$188.74 from the mean of \$410.30. ■

Checkpoint 2 Now work Exercise 31.

Discrete Probability Distribution Models

At times we may know some specific properties of a probability experiment. In some cases we can use a predetermined function called a **probability distribution**

model to determine $P(X = x)$. For example, suppose that eight tickets labeled with the numbers 1 through 8 are placed into a box. In an experiment, one ticket is drawn out randomly. If we let X represent the outcome of any draw, then the possible values of X are $x = 1, 2, 3, 4, 5, 6, 7, 8$. Since each outcome is equally likely, the probability of getting any of these outcomes is $\frac{1}{8}$. This is shown in the probability histogram in Figure 8.2.3.

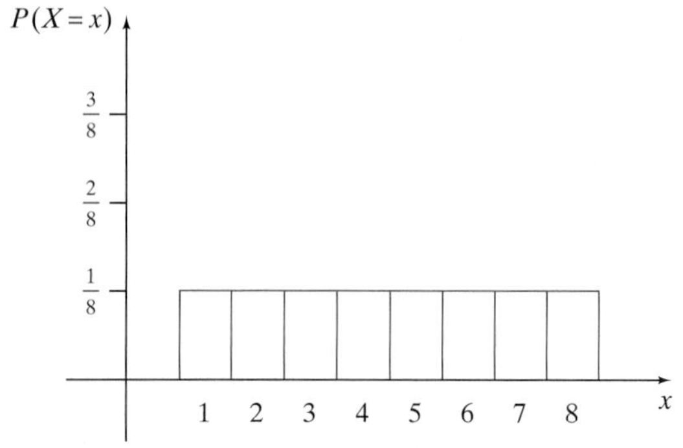

Figure 8.2.3

The probability mass function for X is

$$P(X = x) = \frac{1}{8}, \qquad x = 1, 2, 3, 4, 5, 6, 7, 8$$

Since the rectangles in the probability histogram are evenly or *uniformly* dispersed and each rectangle has the same height, this type of distribution is called the **discrete uniform distribution**.

Discrete Uniform Distribution

Let X be a discrete random variable having values $x = 1, 2, \ldots, N$. If all the probabilities associated with X are equal, then X is a **discrete uniform random variable** with probability mass function

$$P(X = x) = \begin{cases} \frac{1}{N}, & x = 1, 2, \ldots, N \\ 0, & \text{otherwise} \end{cases}$$

Example 5 Determining Probabilities with the Discrete Uniform Distribution

A new radio station in Evansburg called Red Hot Radio airs a promotion for their Red Hot Chili Cookoff contest. The station manager knows that the number of times that the promotion airs for the contest is a random variable uniformly distributed between 1 and 5 times per hour, for any given hour.

(a) Determine the probability mass function $P(X = x)$ for the distribution. Sketch a probability histogram for the distribution.

(b) Determine $P(X > 3)$ and interpret.

Solution

(a) Here X is a discrete uniform random variable for which the possible values of X are $x = 1, 2, 3, 4, 5$. So, knowing that $N = 5$, the probability mass function for the distribution is

$$P(X = x) = \frac{1}{5}, \qquad x = 1, 2, 3, 4, 5$$

Figure 8.2.4 shows the probability histogram for the distribution.

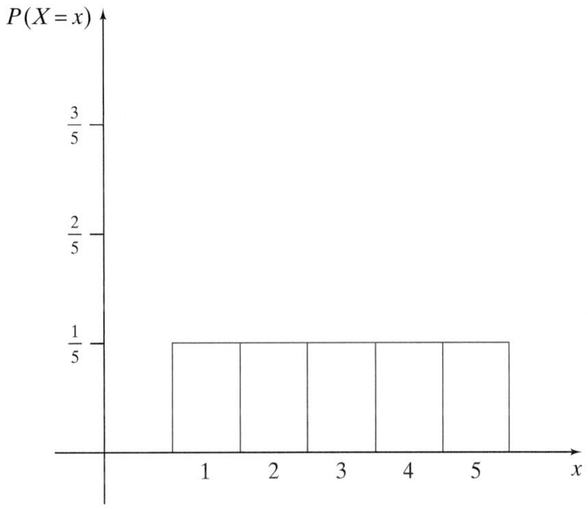

Figure 8.2.4

(b) **Understand the Situation:** To compute $P(X > 3)$, we need to add the probability values that are associated with X being greater than 3. This means that we can have $X = 4$ or $X = 5$, or $P(X \geq 4)$.

So we get

$$\begin{aligned} P(X > 3) &= P(X \geq 4) = P(X = 4) + P(X = 5) \\ &= \frac{1}{5} + \frac{1}{5} = \frac{2}{5} \end{aligned}$$

Interpret the Solution: This means that, during any given hour, there is a $\frac{2}{5}$ or 40% chance that the promotion will be aired more than 3 times. ■

Checkpoint 3

Now work Exercise 57 (a) and (d).

Observe in Example 5(b) that $P(X > 3) = P(X \geq 4)$ can be represented graphically by summing the areas of the shaded rectangles for the probability histogram shown in Figure 8.2.5 on p. 484.

The mean, or expected value, and standard deviation can be computed for the discrete uniform distribution using the techniques outlined in this section. But discrete probability distribution models, including the discrete uniform distribution, also have shortcut formulas that can be used to determine the mean (expected value) and standard deviation. The derivation of these formulas requires more advanced mathematics.

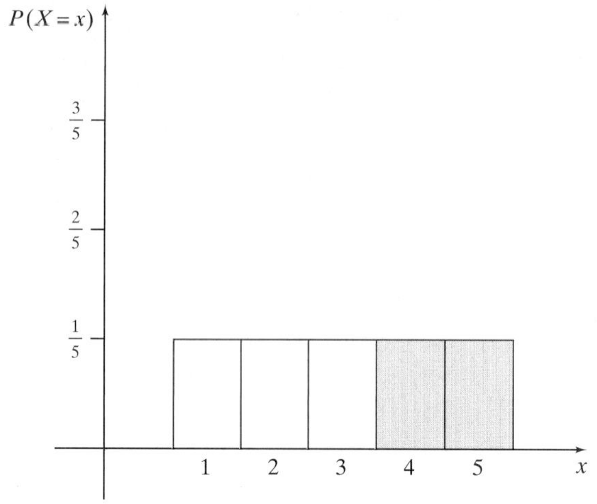

Figure 8.2.5 The sum of the areas of the shaded rectangles equals $P(X \geq 4)$.

Mean and Standard Deviation of the Discrete Uniform Distribution

For discrete uniform distributions, the **mean**, or **expected value**, and **standard deviation** are given by the formulas

$$E(X) = \mu_X = \frac{N+1}{2} \quad \text{and} \quad \sigma_X = \sqrt{\frac{(N-1)(N+1)}{12}}$$

Example 6 Determining the Mean and Standard Deviation for the Discrete Uniform Distribution

In Example 5 we determined that the probability mass function for the number of times that the Red Hot Chili Cookoff airs per hour is given by

$$P(X = x) = \frac{1}{5}, \qquad x = 1, 2, 3, 4, 5$$

(a) Determine the mean using the formula

$$E(X) = \mu_X = x_1 P(X = x_1) + x_2 P(X = x_2) + \cdots + x_n P(X = x_n)$$

and verify by using the shortcut formula $\mu_X = \dfrac{N+1}{2}$.

(b) Determine the standard deviation, σ_X, of the distribution using the shortcut formula.

Solution

(a) Using the definition of expected value, we get

$$\mu_X = 1 \cdot \frac{1}{5} + 2 \cdot \frac{1}{5} + 3 \cdot \frac{1}{5} + 4 \cdot \frac{1}{5} + 5 \cdot \frac{1}{5}$$

$$= \frac{1}{5} + \frac{2}{5} + \frac{3}{5} + \frac{4}{5} + \frac{5}{5} = \frac{15}{5} = 3$$

With the shortcut formula using $N = 5$, we get

$$\mu_X = \frac{N+1}{2} = \frac{5+1}{2} = \frac{6}{2} = 3$$

Interpret the Solution: Using either formula, we see that the promotion airs, on average, 3 times an hour.

(b) Using the shortcut formula, we get

$$\sigma_X = \sqrt{\frac{(N-1)(N+1)}{12}} = \sqrt{\frac{(5-1)(5+1)}{12}} = \sqrt{\frac{(4)(6)}{12}} = \sqrt{\frac{24}{12}}$$
$$= \sqrt{2} \approx 1.41$$

Interpret the Solution: So the standard deviation of the distribution is about 1.41. This means that the number of times that the promotion airs varied on average by about 1.41 times per hour from the mean of 3 times per hour. ■

SUMMARY

In this section we learned how to determine the long-run average of a discrete probability distribution. We also learned how to compute the standard deviation of these distributions.

- For a discrete random variable, the **expected value** or **mean** is

$$E(X) = \mu_X = x_1 P(X = x_1) + x_2 P(X = x_2) + \cdots + x_n P(X = x_n)$$

- For a discrete random variable, the **standard deviation** is

$$\sigma_X = \sqrt{\sum_{i=1}^{n} x_i^2 P(x) - \mu^2}$$

- If all the probabilities associated with X are equal, then X is a **discrete uniform random variable** with probability mass function

$$P(X = x) = \begin{cases} \frac{1}{N}, & x = 1, 2, \ldots, N \\ 0, & \text{otherwise} \end{cases}$$

- For the discrete uniform distribution, the **mean (expected value)** and **standard deviation** are given by the formulas

$$\mu_X = \frac{N+1}{2} \quad \text{and} \quad \sigma_X = \sqrt{\frac{(N-1)(N+1)}{12}}$$

SECTION 8.2 EXERCISES

In Exercises 1–14, a table of values for the probability distribution $P(X = x)$ is given. Determine the expected value $E(X)$ of the random variable X having the given discrete probability distribution.

1.

$X = x$	$P(X = x)$
$X = 0$	$P(X = 0) = \frac{1}{2}$
$X = 1$	$P(X = 1) = \frac{1}{4}$
$X = 2$	$P(X = 2) = \frac{1}{4}$

2.

$X = x$	$P(X = x)$
$X = 1$	$P(X = 1) = \frac{1}{3}$
$X = 2$	$P(X = 2) = \frac{1}{3}$
$X = 3$	$P(X = 3) = \frac{1}{3}$

3.

$X = x$	$P(X = x)$
$X = 0$	$P(X = 0) = \frac{1}{10}$
$X = 1$	$P(X = 1) = \frac{1}{20}$
$X = 2$	$P(X = 2) = \frac{1}{5}$
$X = 3$	$P(X = 3) = \frac{2}{5}$
$X = 4$	$P(X = 4) = \frac{1}{4}$

4.

$X = x$	$P(X = x)$
$X = 1$	$P(X = 1) = \frac{3}{32}$
$X = 2$	$P(X = 2) = \frac{3}{16}$
$X = 3$	$P(X = 3) = \frac{5}{32}$
$X = 4$	$P(X = 4) = \frac{3}{8}$
$X = 5$	$P(X = 5) = \frac{3}{16}$

5.

$X = x$	$P(X = x)$
$X = -2$	$P(X = -2) = 0.03$
$X = -1$	$P(X = -1) = 0.13$
$X = 0$	$P(X = 0) = 0.40$
$X = 1$	$P(X = 1) = 0.35$
$X = 2$	$P(X = 2) = 0.09$

6.

$X = x$	$P(X = x)$
$X = -1$	0.174
$X = 0$	0.111
$X = 2$	0.260
$X = 4$	0.300
$X = 6$	0.155

7.

$X = x$	$P(X = x)$
$X = 10$	$P(X = 10) = 0.1$
$X = 20$	$P(X = 20) = 0.1$
$X = 30$	$P(X = 30) = 0.6$
$X = 40$	$P(X = 40) = 0.2$

8.

$X = x$	$P(X = x)$
$X = 5$	$P(X = 5) = 0.3$
$X = 10$	$P(X = 10) = 0.2$
$X = 25$	$P(X = 25) = 0.1$
$X = 35$	$P(X = 35) = 0.4$

9.

$X = x$	$P(X = x)$
$X = 8$	$P(X = 8) = \frac{1}{5}$
$X = 9$	$P(X = 9) = \frac{1}{5}$
$X = 10$	$P(X = 10) = \frac{1}{10}$
$X = 11$	$P(X = 11) = \frac{3}{20}$
$X = 12$	$P(X = 12) = \frac{7}{20}$

10.

$X = x$	$P(X = x)$
$X = 0$	$P(X = 0) = \frac{11}{20}$
$X = 1$	$P(X = 1) = \frac{1}{4}$
$X = 2$	$P(X = 2) = \frac{1}{20}$
$X = 3$	$P(X = 3) = \frac{1}{10}$
$X = 4$	$P(X = 4) = \frac{1}{20}$

11.

$X = x$	$P(X = x)$
$X = 0$	$P(X = 0) = 0.5$
$X = 1$	$P(X = 1) = 0.2$
$X = 2$	$P(X = 2) = 0.1$
$X = 3$	$P(X = 3) = 0.2$

12.

$X = x$	$P(X = x)$
$X = 0$	$P(X = 0) = 0.2$
$X = 2$	$P(X = 2) = 0.4$
$X = 3$	$P(X = 3) = 0.1$
$X = 4$	$P(X = 4) = 0.3$

13.

$X = x$	$P(X = x)$
$X = 0$	$P(X = 0) = 0.05$
$X = 2$	$P(X = 2) = 0.05$
$X = 3$	$P(X = 3) = 0.10$
$X = 4$	$P(X = 4) = 0.20$
$X = 5$	$P(X = 5) = 0.60$

14.

$X = x$	$P(X = x)$
$X = 0$	$P(X = 0) = 0.15$
$X = 1$	$P(X = 1) = 0.15$
$X = 2$	$P(X = 2) = 0.25$
$X = 3$	$P(X = 3) = 0.10$
$X = 4$	$P(X = 4) = 0.35$

In Exercises 15–26, a probability mass function $P(X = x)$ is given. Determine the expected value.

15. $P(X = x) = \frac{1}{6}x, \quad x = 0, 1, 2, 3$

16. $P(X = x) = 0.01x, \quad x = 10, 20, 30, 40$

17. $P(X = x) = \frac{1}{8}x, \quad x = 0, 1, 2, 5$

18. $P(X = x) = \frac{1}{100}x, \quad x = 10, 30, 60$

19. $P(X = x) = \frac{1}{4}, \quad x = 1, 2, 3, 4$

20. $P(X = x) = \frac{1}{5}, \quad x = 0, 1, 2, 3, 4$

21. $P(X = x) = \frac{1}{14}x^2, \quad x = 1, 2, 3$

22. $P(X = x) = \frac{1}{36}x^3, \quad x = 1, 2, 3$

23. $P(X = x) = 0.05x, \quad x = 2, 3, 4, 5, 6$

24. $P(X = x) = \frac{1}{50}x, \quad x = 5, 10, 15, 20$

25. $P(X = x) = \frac{x + 2}{10}, \quad x = -1, 0, 1, 2$

26. $P(X = x) = \frac{x - 4}{14}, \quad x = 6, 7, 8, 9$

In Exercises 27–40, determine the standard deviation, σ_X, for the following:

27. For the discrete probability distribution in Exercise 1

28. For the discrete probability distribution in Exercise 2

29. For the discrete probability distribution in Exercise 3

30. For the discrete probability distribution in Exercise 4

✓ **31.** For the discrete probability distribution in Exercise 7

32. For the discrete probability distribution in Exercise 8

33. For the probability mass function in Exercise 15

34. For the probability mass function in Exercise 16

35. For the probability mass function in Exercise 19

36. For the probability mass function in Exercise 20

37. For the probability mass function in Exercise 21

38. For the probability mass function in Exercise 22

39. For the probability mass function in Exercise 25

40. For the probability mass function in Exercise 26

Applications

✓ **41.** From prior records the owner of the Video4U has determined that the number of videos that a customer rents on any given visit can be represented by a discrete random variable X with the following probability distribution.

$X = x$	$P(X = x)$
$X = 0$	$P(X = 0) = 0.05$
$X = 1$	$P(X = 1) = 0.50$
$X = 2$	$P(X = 2) = 0.30$
$X = 3$	$P(X = 3) = 0.15$

Determine the expected value, $E(X)$, and interpret.

42. From prior records the manager of Javah! coffee shop has determined that its total number of daily customers can be represented by a discrete random variable X with the following probability distribution.

$X = x$	$P(X = x)$
$X = 45$	$P(X = 45) = 0.10$
$X = 46$	$P(X = 46) = 0.30$
$X = 47$	$P(X = 47) = 0.25$
$X = 48$	$P(X = 48) = 0.20$
$X = 49$	$P(X = 49) = 0.05$
$X = 50$	$P(X = 50) = 0.10$

Determine the expected value, $E(X)$, and interpret.

43. The Franklinton Group, a consortium of educators, has determined that the number of commercials per show aired during afterschool television shows on a certain network can be represented by a discrete random variable X with the following probability distribution.

$X = x$	$P(X = x)$
$X = 0$	$P(X = 0) = 0.20$
$X = 1$	$P(X = 1) = 0.15$
$X = 2$	$P(X = 2) = 0.25$
$X = 3$	$P(X = 3) = 0.20$
$X = 4$	$P(X = 4) = 0.05$
$X = 5$	$P(X = 5) = 0.10$
$X = 6$	$P(X = 6) = 0.05$

(a) Determine the expected value and interpret.

(b) What is the most frequently occurring number of commercials on any given show? (That is, what x-value yields the greatest $P(X = x)$-value?) This value is called the **mode** of the distribution.

44. The North-Am racing series has determined that the number of yellow flag caution laps, rounded to the nearest 10, during any given race can be represented by a discrete random variable X having the following probability distribution.

$X = x$	$P(X = x)$
$X = 0$	$P(X = 0) = 0.20$
$X = 10$	$P(X = 10) = 0.25$
$X = 20$	$P(X = 20) = 0.38$
$X = 30$	$P(X = 30) = 0.10$
$X = 40$	$P(X = 40) = 0.07$

(a) Determine the expected value and interpret.

(b) What is the most frequently occurring number of yellow flag caution laps run during any given race? (That is, what x-value yields the greatest $P(X = x)$-value?) This value is called the **mode** of the distribution.

45. Parmiter College has determined that the number of students who drop out of introductory psychology can be represented by a discrete random variable X having the following probability distribution.

$X = x$	$P(X = x)$
$X = 12$	$P(X = 12) = 0.09$
$X = 13$	$P(X = 13) = 0.18$
$X = 14$	$P(X = 14) = 0.38$
$X = 15$	$P(X = 15) = 0.20$
$X = 16$	$P(X = 16) = 0.15$

According to the distribution, how many students can we expect to drop out of introductory psychology?

46. Linger's Pizzeria management team has determined that the number of late-night pizza deliveries on any given Wednesday can be represented by a discrete random variable X having the following probability distribution.

$X = x$	$P(X = x)$
$X = 4$	$P(X = 4) = 0.32$
$X = 5$	$P(X = 5) = 0.11$
$X = 6$	$P(X = 6) = 0.40$
$X = 7$	$P(X = 7) = 0.12$
$X = 8$	$P(X = 8) = 0.05$

According to the distribution, how many late-night pizzas can the pizzeria expect to deliver on any given Wednesday night?

47. The Sedgewick Baseball Club is having a fund-raising raffle. Twenty thousand raffle tickets have been sold for $10 each. Tickets are drawn at random and cash prizes are awarded as follows: 1 prize of $10,000; 2 prizes of $5000; 10 prizes of $1000, and 20 prizes of $500. Determine the expected value of this raffle if 1 ticket is purchased.

48. The Library Society is having a fund-raising raffle. Fifteen thousand raffle tickets have been sold for \$5 each. Tickets are drawn at random and cash prizes are awarded as follows: 1 prize of \$5000, 3 prizes of \$1000, 5 prizes of \$750, and 10 prizes of \$300. Determine the expected value of this raffle if 1 ticket is purchased.

49. The number of defective springs discovered after inspection by the Springsworth Company is shown in the following table.

Number of Inspections	Number of Defects
1	1500
2	800
3	200
4	50

(a) Determine the expected value and interpret.

(b) Determine the standard deviation of the distribution and interpret.

50. The number of drivers aged 16 to 24 who were in automobile accidents in 1996 is shown in the table.

Age	Number of Accidents (in millions)
16	1.6
17	2.2
18	2.4
19	2.7
20	2.8
21	2.9
22	2.9
23	3.1
24	3.6

SOURCE: National Safety Council, www.nsc.org

(a) Determine the expected value and interpret.

(b) Determine the standard deviation of the distribution and interpret.

51. The distribution of weights of American males aged 70 to 79 from 1988 to 1994 is shown in the table.

Weight (in pounds)	Percentage	Weight (in pounds)	Percentage
100–109	0.2	200–209	9.5
110–119	0.2	210–219	5.2
120–129	1.3	220–229	5.4
130–139	4.1	230–239	2.5
140–149	7.4	240–249	1.0
150–159	10.1	250–259	0.7
160–169	11.9	260–269	1.1
170–179	13.2	270–279	0.9
180–189	11.1	280–289	0.4
190–199	13.6	290 and over	0.2

SOURCE: U.S. Center for Health Statistics, www.cdc.gov/nchs

(a) According to the distribution, what can we expect the average weight of an American male aged 70 to 79 from 1988 to 1994 to be?

(b) Determine the standard deviation of the distribution and interpret.

52. John Spears wishes to purchase a 7-year term-life insurance policy that pays the beneficiary \$100,000 in the event that his death occurs during the next 7 years. From inspecting life insurance tables, John determines that the probability that he will live another 7 years is about 0.97. To calculate the minimum premium (payment) for this insurance, determine when the insurance company's profit is zero.

53. Chris Thomas purchased a \$100,000 1-year term-life insurance policy for \$276. Assuming that the probability that she will live another year is 0.981, determine the company's expected gain.

54. Chenush Gas is a natural gas drilling company. After testing and analysis by geologists, Chenush Gas is considering drilling on two different sites. The geologists have estimated that site I will net \$50 million if successful (probability 0.3) and lose \$5 million if not successful (probability 0.7). Site II is estimated to net \$80 million if successful (probability 0.2) and lose \$6 million if not successful (probability 0.8).

(a) Determine the expected return for each site.

(b) Based on part (a), which site would you advise the company to select.

(c) Which site has less risk? (This corresponds to which has a smaller variance.)

55. Chenush Oil is an oil drilling company. After testing and analysis by geologists, Chenush Oil is considering drilling on two different sites. The geologists have estimated that site I will net \$40 million if successful (probability 0.25) and lose \$4 million if not successful (probability 0.75). Site II is estimated to net \$70 million if successful (probability 0.12) and lose \$5 million if not successful (probability 0.88).

(a) Determine the expected return for each site.

(b) Based on part (a), which site would you advise the company to select.

(c) Which site has less risk? (This corresponds to which has a smaller variance.)

56. The number of times that KTTV radio gives the time of day each hour is a random variable uniformly distributed between 1 and 5, inclusive. Let X be a discrete uniform random variable representing the number of times that the time of day is given each hour.

(a) Write a probability mass function for X.

(b) Determine μ_X and interpret.

(c) Compute σ_X for the distribution and interpret.

(d) Determine the probability that the time of day will be given two or more times in a given hour.

57. The Kronix Company knows that the number of times that its old printing machine breaks down during a shift is a random variable uniformly distributed between 1 and 10 times inclusive. Let X be a discrete uniform random variable representing the number of times that the printing machine breaks down during a shift.

(a) Write a probability mass function for X.

(b) Determine μ_X and interpret.

(c) Compute σ_X for the distribution and interpret.

(d) Determine the probability that during any given shift the printing machine will break down three or fewer times.

58. Manufacturers of Mushy Slushy drinks know that the machine that dispenses the Slushys keeps the drink mixture at a temperature (rounded to the nearest degree) that is a random variable uniformly distributed between 1 and 6 degrees Celsius, inclusive. Let X be a discrete uniform random variable representing the temperature in degrees Celsius.

(a) Write a probability mass function for X.

(b) Determine μ_X and interpret.

(c) Compute σ_X for the distribution and interpret.

(d) Determine the probability that if the Mushy Slushy drink dispenser is checked the temperature will be at most 3 degrees Celsius.

59. The Rockford Company manufactures plastic automobile parts and uses the Rockford hardness scale to determine the hardness of the plastic pieces manufactured. The scale and the manufacturing process can be represented by a discrete uniform random variable from 1 to 25. Let X be a discrete uniform random variable representing the hardness on the Rockford hardness scale.

(a) Write a probability mass function for X.

(b) Determine μ_X and interpret.

(c) Compute σ_X for the distribution and interpret.

(d) Determine the probability that if a plastic piece is selected at random the piece will have a Rockford hardness scale value of more than 20.

SECTION PROJECT

Let's explore some of the properties of the expected value and standard deviation of discrete random variables. Consider the following three probability distributions:

Distribution A

$X = x$	$P(X = x)$
$X = 10$	$P(X = 10) = 0.1$
$X = 20$	$P(X = 20) = 0.4$
$X = 30$	$P(X = 30) = 0.3$
$X = 40$	$P(X = 40) = 0.2$

Distribution B

$X = x$	$P(X = x)$
$X = 15$	$P(X = 15) = 0.1$
$X = 25$	$P(X = 25) = 0.4$
$X = 35$	$P(X = 35) = 0.3$
$X = 45$	$P(X = 45) = 0.2$

Distribution C

$X = x$	$P(X = x)$
$X = 90$	$P(X = 90) = 0.1$
$X = 180$	$P(X = 180) = 0.4$
$X = 270$	$P(X = 270) = 0.3$
$X = 360$	$P(X = 360) = 0.2$

(a) How are the x-values between Distribution A and Distribution B related? That is, how can we get the x-values for Distribution B using the x-values from Distribution A?

(b) How are the x-values between Distribution A and Distribution C related? That is, how can we get the x-values for Distribution C using the x-values from Distribution A?

(c) Compute the mean μ_X of each distribution. How is the mean of Distribution A related to the mean of Distribution B? How is the mean of Distribution A related to the mean of Distribution C?

(d) Compute the standard deviation σ_X of each distribution. How is the standard deviation of Distribution A related to the standard deviation of Distribution B? How is the standard deviation of Distribution A related to the standard deviation of Distribution C?

Section 8.3 The Binomial Distribution

Many of the types of probability that we study have exactly two possible outcomes. For example, a person whose television viewing habits are being monitored can either watch or not watch a specific program. A student taking a multiple-choice test can get a question either correct or incorrect. A college football team can either win or lose a particular game. In this section we will study the **binomial distribution**, which is used to determine probabilities of events concerning repeated experiments that have only two possible outcomes in each experiment. We will also explore the properties of the binomial distribution, including the mean and standard deviation.

The Binomial Distribution

Generally, the trials associated with probability experiments that have exactly two outcomes, like the ones we mentioned earlier, are called **Bernoulli trials**. Named after the Swiss mathematician Jakob Bernoulli, these experiments are regarded as having an outcome of **success** (watching a particular television show, getting a test question correct, or winning a football game) or **failure**. If we combine a series of independent Bernoulli trials, we get a **binomial experiment**.

Binomial Experiment

A **binomial experiment** is a probability experiment that satisfies the following four assumptions:

1. Each trial has exactly two possible outcomes: success or failure.
2. The probability of success for each trial is the same. We call this fixed probability of success p.
3. The experiment has a finite number of trials, denoted by n.
4. Each trial is independent of the others.

Example 1 Identifying a Binomial Experiment

A 1995 report revealed that 31.4% of African Americans were smokers (*Source:* www.ymn.org/newstat). Suppose that a sample of 20 African Americans is selected at random. Determine if this meets the assumptions of a binomial experiment and, if so, identify the values of p and n.

Solution

Understand the Situation: Let's check the four assumptions for a binomial experiment.

1. There are two possible outcomes. These are defined as

   Success: member of the sample is a smoker

   Failure: member of the sample is not a smoker
2. The probability of success for each trial is the same, with the probability of success written as $p = 0.314$ in decimal form.
3. The experiment has a fixed number of trials with $n = 20$.
4. Since the fact that one member of the sample being a smoker does not affect whether the other members of the sample smoke, the trials are independent. Since all four assumptions are satisfied, we have a binomial experiment. ■

Checkpoint 1

Now work Exercise 11.

Example 1 shows a peculiar aspect of binomial experiments. We defined a success as identifying a person from the sample as a smoker. Even though smoking is not regarded as a "success," since we are trying to determine probabilities concerning smokers in the sample, this is how we would define a success. We must be aware that success and failure are merely mathematical conventions to characterize binomial random variables. Sometimes what in real life is clearly a failure is, for mathematical convention, defined to be a success.

The possible results of binomial experiments and probabilities associated with them give us a type of discrete probability distribution called the **binomial distribution**. This important distribution has its own probability mass function, based on the values of n and p.

Binomial Distribution

In a binomial experiment, the probability of having x successes in n trials is given by the probability mass function.

$$P(X = x) = {}_nC_x \cdot p^x(1-p)^{n-x}, \qquad x = 0, 1, 2, \ldots, n$$

where

n is the number of trials
p is the fixed probability of success
x is the number of successes in n trials

Note: 1. The binomial distribution is a special type of discrete distribution. This means that the information presented in Sections 8.1 and 8.2 can be used here.

2. Some textbooks write the binomial probability mass function as $P(X = x) = {}_nC_x \cdot p^x \cdot q^{n-x}$, where $q = 1 - p$ represents the probability of failure.

Now let's see how the binomial distribution can be used to calculate probabilities in some classic probability experiments.

Example 2 Determining a Binomial Probability

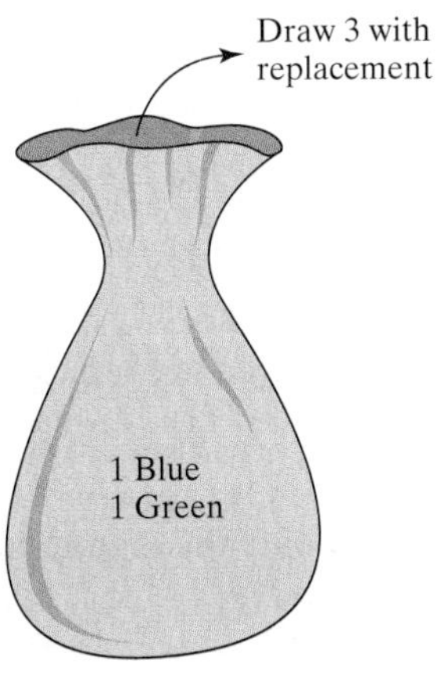

Figure 8.3.1

A bag contains 1 blue and 1 green marble. In an experiment, a marble is drawn at random, its color is recorded, and it is returned to the bag. Two more draws are done in the same way, for a total of three draws (see Figure 8.3.1).

(a) Use classical probability to compute the probability that in 3 draws exactly 2 will be green marbles.

(b) Use the binomial distribution to determine the probability that in 3 draws exactly 2 will be green marbles.

Solution

(a) The sample space for the experiment has 8 elements:

$$S = \{BBB, BBG, BGB, BGG, GBG, GGB, GBB, GGG\}$$

The outcomes associated with drawing exactly 2 green marbles are BGG, GBG, GGB. So, using classical probability, we get

$$P(\text{2 green marbles}) = \frac{\text{number of ways to get 2 green marbles}}{\text{total number of ways to draw 3 marbles}} = \frac{3}{8}$$

(b) **Understand the Situation:** We first note that we have two possible outcomes: drawing a blue marble or drawing a green marble. Since we want to determine the probability that in 3 draws exactly 2 will be green marbles, let's define success as drawing a green marble. The probability of success for each trial is

$$p = \frac{\text{number of green marbles in the bag}}{\text{total number of marbles in the bag}} = \frac{1}{2}$$

Since the number of draws is 3, this means that there are a fixed number of trials, $n = 3$. We replace the marble before the next draw, which means that the trials are independent. So we have a binomial experiment, with $p = \frac{1}{2}$ and $n = 3$, and the probability mass function is

$$P(X = x) = {}_nC_x \cdot p^x(1-p)^{n-x}, \qquad x = 0, 1, 2, \ldots, n$$

$$= {}_3C_x \left(\frac{1}{2}\right)^x \left(1 - \frac{1}{2}\right)^{3-x}, \qquad x = 0, 1, 2, 3$$

where x represents the number of green marbles drawn in 3 trials.

From Your Toolbox

$${}_nC_r = \frac{n!}{r!(n-r)!}$$

To determine the probability of drawing 2 green marbles, we need to compute $P(X = x)$ for $x = 2$. Consulting the Toolbox to the left for evaluating ${}_3C_2$, we get

$$P(X = 2) = {}_3C_2 \left(\frac{1}{2}\right)^2 \left(1 - \frac{1}{2}\right)^{3-2}$$

$$= \frac{3!}{2!(3-2)!} \left(\frac{1}{2}\right)^2 \left(\frac{1}{2}\right)^1 = 3 \cdot \frac{1}{4} \cdot \frac{1}{2} = \frac{3}{8}$$

Interactive Activity

Repeat Example 2 if there are 5 draws from the bag.

Interpret the Solution: The probability of drawing 2 green marbles is $\frac{3}{8}$. Notice that this is the same result obtained in part (a).

■

One way that we use the binomial distribution is to examine the number of successes, x, in a **simple random sample** of size n. By a simple random sample of size n, we mean that the sample selected from the population has n elements and every set of n elements has an equal chance to be in the sample actually selected. Let's see how we can use the binomial distribution to approximate probabilities of a simple random sample.

Example 3 Applying the Binomial Distribution

An August 2000 Harris poll found that 71% of Americans who are on-line use the Internet to access health and medical information (*Source:* www.harrisinteractive.com). A simple random sample of 7 Americans who are on-line is selected.

(a) Make a table of values for $P(X = x)$. Construct a probability histogram for the distribution.

(b) Determine $P(X = 5)$ and interpret.

(c) Determine $P(X \geq 5)$ and interpret.

Solution

(a) **Understand the Situation:** To make a table of values, we define success as the selected on-line American uses the Internet to access health and medical information. Since 71% of all on-line Americans polled use the Internet to access health and medical information, we have a probability of success

$p = 0.71$. There are a fixed number of trials $n = 7$, so the probability mass function for the distribution is

$$P(X = x) = {}_nC_x \cdot p^x(1-p)^{n-x}, \qquad x = 0, 1, 2, \ldots, n$$
$$= {}_7C_x(0.71)^x(1-0.71)^{7-x}$$
$$= {}_7C_x(0.71)^x(0.29)^{7-x}, \qquad x = 0, 1, 2, \ldots, 7$$

Using this formula, we get the table of values for $P(X = x)$ shown in Table 8.3.1. The probabilities are rounded to three decimal places.

Table 8.3.1

$X = x$	$P(X = x)$
$X = 0$	$P(X = 0) = {}_7C_0(0.71)^0(0.29)^7 \approx 0.000$
$X = 1$	$P(X = 1) = {}_7C_1(0.71)^1(0.29)^6 \approx 0.003$
$X = 2$	$P(X = 2) = {}_7C_2(0.71)^2(0.29)^5 \approx 0.022$
$X = 3$	$P(X = 3) = {}_7C_3(0.71)^3(0.29)^4 \approx 0.089$
$X = 4$	$P(X = 4) = {}_7C_4(0.71)^4(0.29)^3 \approx 0.217$
$X = 5$	$P(X = 5) = {}_7C_5(0.71)^5(0.29)^2 \approx 0.319$
$X = 6$	$P(X = 6) = {}_7C_6(0.71)^6(0.29)^1 \approx 0.260$
$X = 7$	$P(X = 7) = {}_7C_7(0.71)^7(0.29)^0 \approx 0.091$

Using the values in Table 8.3.1, we get the probability histogram shown in Figure 8.3.2.

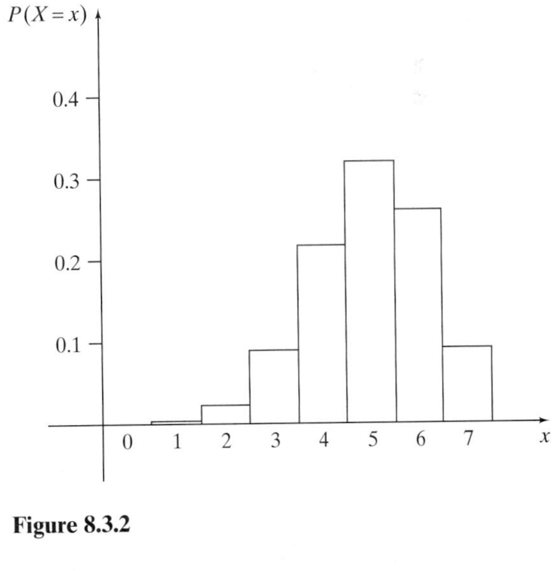

Figure 8.3.2

(b) Using Table 8.3.1, we determine that $P(X = 5) \approx 0.319$.

Interpret the Solution: This means that if 7 Americans who are on-line are selected at random, the probability that 5 of them use the Internet to access health and medical information is about 31.9%.

(c) To compute $P(X \geq 5)$, we have

$$P(X \geq 5) = P(X = 5) + P(X = 6) + P(X = 7)$$
$$\approx 0.319 + 0.260 + 0.091 = 0.670$$

Interpret the Solution: This means that if we select 7 Americans who are on-line at random, there is about a 67% chance that 5 or more of them use the Internet to access health and medical information. Notice that this corresponds to summing the areas of the shaded rectangles of the probability histogram, as shown in Figure 8.3.3.

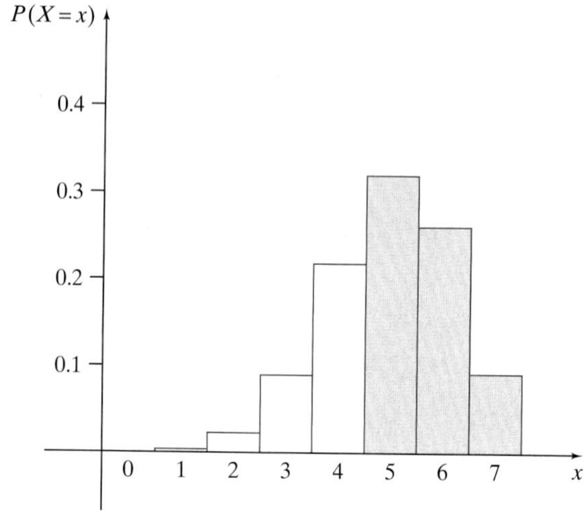

Figure 8.3.3

■

Checkpoint 2 Now work Exercise 45.

Technology Option

The output from using the BINOMPDF command on a graphing calculator to compute, say, $P(X = 5)$ in part (a) of Example 3 is shown in Figure 8.3.4(a).

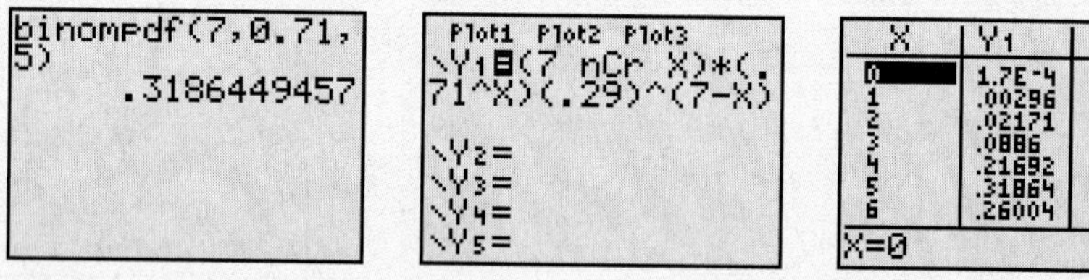

(a) **(b)** **(c)**

Figure 8.3.4

To quickly reproduce the table of values in Table 8.3.1, we can enter the probability mass function $P(X = x)$ in our $y =$ editor [see Figure 8.3.4(b)] and use the TABLE command [see Figure 8.3.4(c)].

To learn about these commands for your calculator, consult the on-line graphing calculator manual at www.prenhall.com/armstrong.

Two phrases frequently used in probability applications are "at most" and "at least." Example 4 demonstrates how these phrases are used in applications.

Example 4 Using the Phrases "at most" and "at least"

A May 2000 CBS News/NY Times poll asked "Will Social Security have money for you?" Thirty percent of American adults aged 45 years and younger responded yes (*Source:* www.CBSnews.com). Suppose that a random sample of 9 American adults 45 years and younger is selected at random.

(a) Determine the probability that at least 5 respond yes to the Social Security question.

(b) Determine the probability that at most 5 respond yes to the Social Security question.

Solution

Understand the Situation: To aid us in both parts of this example, we construct a table of values for $P(X = x)$. Here we have a binomial random variable X with $n = 9$ and $p = 0.30$. This means that

$$P(X = x) = {}_9C_x \cdot (0.30)^x(1 - 0.30)^{9-x}, \qquad x = 0, 1, 2, \ldots, 9$$
$$= {}_9C_x \cdot (0.30)^x(0.70)^{9-x}, \qquad x = 0, 1, 2, \ldots, 9$$

See Table 8.3.2.

Table 8.3.2

$X = x$	$P(X = x)$
$X = 0$	$P(X = 0) = {}_9C_0 \cdot (0.30)^0(1 - 0.30)^{9-0} \approx 0.040$
$X = 1$	$P(X = 1) = {}_9C_1 \cdot (0.30)^1(1 - 0.30)^{9-1} \approx 0.156$
$X = 2$	$P(X = 2) \approx 0.267$
$X = 3$	$P(X = 3) \approx 0.267$
$X = 4$	$P(X = 4) \approx 0.172$
$X = 5$	$P(X = 5) \approx 0.074$
$X = 6$	$P(X = 6) \approx 0.021$
$X = 7$	$P(X = 7) \approx 0.004$
$X = 8$	$P(X = 8) \approx 0.000$
$X = 9$	$P(X = 9) \approx 0.000$

(a) To say "at least 5" means that we could have 5 or more respond yes to the question. So we compute

$$\begin{aligned} P(\text{at least } 5) &= P(X \geq 5) \\ &= P(X = 5) + P(X = 6) + P(X = 7) + P(X = 8) + P(X = 9) \\ &\approx 0.074 + 0.021 + 0.004 + 0.000 + 0.000 = 0.099 \end{aligned}$$

Interpret the Solution: This means that there is about a 9.9% chance that at least 5 will respond yes to the question.

(b) To say "at most 5" means that we could have 5 or less respond yes to the question. So we compute

$$\begin{aligned} P(\text{at most } 5) &= P(X \leq 5) \\ &= P(X = 0) + P(X = 1) + P(X = 2) + P(X = 3) \\ &\quad + P(X = 4) + P(X = 5) \\ &\approx 0.040 + 0.156 + 0.267 + 0.267 + 0.172 + 0.074 = 0.976 \end{aligned}$$

Interactive Activity

Why is it that the probabilities found in parts (a) and (b) of Example 4 are not complements?

WWW

Interpret the Solution: This means that there is about a 97.6% chance that at most 5 will respond yes to the question. ■

Checkpoint 3

Now work Exercise 49.

Mean and Standard Deviation of the Binomial Distribution

Probability distribution models such as the binomial distribution have specific formulas for the mean and standard deviation. To discover what the formulas are, let's first revisit an example presented in Section 8.1.

Flashback

Coin Toss Revisited

In Section 8.1, Example 3, we examined a probability experiment in which a coin is tossed 4 times and the result is recorded. If we let X represent the total number of heads recorded, show that the experiment can be considered a binomial experiment and determine its probability mass function $P(X = x)$.

Flashback Solution

To show that the experiment can be considered a binomial experiment, we check the four assumptions for a binomial experiment given at the beginning of this section.

1. There are two possible outcomes that we define as

   Success: toss is a head

   Failure: toss is a tail
2. The probability of success for each trial is the same, with the probability of success written as $p = 0.5$.
3. The experiment has a fixed number of trials, with $n = 4$.
4. Since the result of one toss does not affect any other toss, the trials are independent.

Since all four assumptions are met, we have a binomial experiment. We have determined that $p = 0.5$ and $n = 4$, so the probability mass function for the binomial distribution is given by

$$\begin{aligned} P(X = x) &= {}_4C_x \cdot (0.5)^x(1 - 0.5)^{4-x}, \qquad x = 0, 1, 2, 3, 4 \\ &= {}_4C_x \cdot (0.5)^x(0.5)^{4-x}, \qquad x = 0, 1, 2, 3, 4 \end{aligned}$$

In the Flashback, we determined that X is a binomial random variable with $n = 4$ and $p = 0.5$. Using the probability mass function,

$$P(X = x) = {}_4C_x \cdot (0.5)^x(0.5)^{4-x}, \qquad x = 0, 1, 2, 3, 4$$

we can make a table of values for $P(X = x)$ as shown in Table 8.3.3.

Table 8.3.3

$X = x$	$P(X = x)$
$X = 0$	$P(X = 0) = 0.0625$
$X = 1$	$P(X = 1) = 0.25$
$X = 2$	$P(X = 2) = 0.375$
$X = 3$	$P(X = 3) = 0.25$
$X = 4$	$P(X = 4) = 0.0625$

Using the formula from Section 8.2 to compute the mean (consult the Toolbox on p. 497), we get

$$\begin{aligned} \mu_X = E(X) &= 0(0.0625) + 1(0.25) + 2(0.375) + 3(0.25) + 4(0.0625) \\ &= 0 + 0.25 + 0.75 + 0.75 + 0.25 = 2 \end{aligned}$$

Observe that this is the same as computing $np = 4(0.5) = 2$. This is no coincidence! This suggests that for the binomial distribution the mean of the distribution is $\mu_X = np$.

From Your Toolbox

For any discrete probability distribution, the mean (expected value) is given by

$$E(X) = \mu_X = x_1 P(X = x_1) + x_2 P(X = x_2) + \cdots + x_n P(X = x_n)$$

with standard deviation

$$\sigma_X = \sqrt{\sum_{i=1}^{n} x_i^2 P(X = x_i) - [E(X)]^2}$$

Now let's compute the standard deviation for the same distribution. Using the formula from Section 8.2 to compute the standard deviation (consult the Toolbox above) and following our three steps outlined in Section 8.2, we start by computing the value of $\sum x^2 P(X = x)$.

$$\sum x^2 P(X = x) = 0^2(0.0625) + 1^2(0.25) + 2^2(0.375) + 3^2(0.25) + 4^2(0.0625) = 5$$

Subtracting the square of the mean gives us

$$\sum x^2 P(X = x) - \mu_X^2 = 5 - (2)^2 = 1$$

Taking the square root of this result gives us the standard deviation.

$$\sigma_X = \sqrt{\sum x^2 P(X) - \mu^2} = \sqrt{1} = 1$$

Observe that this is the same as computing

$$\sigma_X = \sqrt{np(1-p)} = \sqrt{4 \cdot (0.5)(1 - 0.5)} = \sqrt{4 \cdot (0.5) \cdot (0.5)} = \sqrt{1} = 1$$

There is no coincidence here either! We have just illustrated that the mean and standard deviation of the binomial distribution are based on the values of n and p.

Mean and Standard Deviation of the Binomial Random Variable

Let X be a binomial random variable with number of trials n and probability of success p. Then the **mean**, or expected value, of the distribution is

$$E(X) = \mu_X = np$$

with **standard deviation** given by

$$\sigma_X = \sqrt{np(1-p)}$$

Example 5 Determining the Mean and Standard Deviation

In Example 1 we saw that 31.4% of African Americans were smokers and a sample of 20 African Americans is selected at random. Compute μ_X and σ_X and interpret each.

Solution

Understand the Situation: Here we have a binomial random variable with $n = 20$ and $p = 0.314$.

So the mean is

$$E(X) = \mu_X = np = 20(0.314) = 6.28$$

with a standard deviation

$$\sigma_X = \sqrt{np(1-p)} = \sqrt{20(0.314)(1-0.314)} = \sqrt{20(0.314)(0.686)} \approx 2.08$$

Interpret the Solution: This means that, as we repeat the experiment over time, taking many samples of 20, in the long run we would expect about 6.28 of those in the sample to be smokers. The number of smokers varied, on average, by about 2.08 from the mean of 6.28. ■

Checkpoint 4 Now work Exercise 53.

SUMMARY

- A **binomial experiment** is a probability experiment such that each trial has exactly two possible outcomes: success or failure. The probability of success for each trial is the same and is denoted by p. The experiment has a finite number of trials, denoted by n, and each trial is independent.
- In a binomial experiment the probability of having x successes in n trials is given by the binomial **probability mass function:** $P(X=x) = {}_nC_x \cdot p^x(1-p)^{n-x}$, $x = 0, 1, 2, \ldots, n$.
- The **mean**, or expected value, of the binomial distribution is $\mu_X = np$.
- The **standard deviation** of the binomial distribution is $\sigma_X = \sqrt{np(1-p)}$.

SECTION 8.3 EXERCISES

In Exercises 1–10, evaluate the binomial probability mass function $P(X=x) = {}_nC_x \cdot p^x(1-p)^{n-x}$ for the given values of n, p, and x.

1. $n = 4$, $p = \frac{1}{2}$, $x = 2$

2. $n = 3$, $p = \frac{1}{3}$, $x = 2$

3. $n = 5$, $p = 0.2$, $x = 0$

4. $n = 4$, $p = 0.3$, $x = 0$

5. $n = 8$, $p = \frac{2}{3}$, $x = 6$

6. $n = 7$, $p = \frac{3}{5}$, $x = 5$

7. $n = 10$, $p = 0.08$, $x = 5$

8. $n = 10$, $p = 0.11$, $x = 6$

9. $n = 20$, $p = 0.50$, $x = 10$

10. $n = 12$, $p = 0.6$, $x = 6$

In Exercises 11–20, determine if the assumptions of a binomial experiment are reasonably satisfied. If so, identify the values for p and n.

✓ **11.** A bag contains 4 blue and 4 white marbles. A marble is drawn at random, its color is recorded and it is returned to the bag. This is repeated for a total of 30 draws.

12. Fifty people are independently surveyed at a grocery store to see if they like the store brand of baked beans.

13. In a survey, 200 people selected at random and independently are asked which brand of motor oil they use.

14. In a public opinion poll, respondents are asked in which state they would most like to retire.

15. Ninety percent of the students at Plymouth High School pass the mandatory competency test. A random sample of 12 students is asked if they passed the test.

16. Twenty-three percent of the citizens in Klossington feel that the local police department is too tough on crime. A random sample of 50 Klossington citizens is asked if the local police department is too tough on crime.

17. A new brand of breakfast cereal called Hearty Flakes is liked by 11% of the market. A telemarketer conducts a phone survey and calls consumers until a consumer says that he or she likes Hearty Flakes.

18. Fifteen percent of the shoppers at the Manson Mall buy an item at the food court. A mall employee stands at the door and asks people if they intend to go to the food court until she gets an affirmative response.

19. One out of every five of the households in Classton call 911 during any given year. 150 Classton households randomly selected are asked in a direct mailing if they called 911 within the last year.

20. Two out of every seven children in Sherman Elementary School wear glasses. In a mail questionnaire, a local health organization asks a random sample of 15 students if they wear glasses.

In Exercises 21–26, a binomial probability mass function is given. Complete the following.

(a) Make a table of values for $P(X = x)$.

(b) Construct a probability histogram for the distribution found in part (a).

21. $P(X = x) = {}_3C_x \cdot \left(\frac{1}{2}\right)^x \left(1 - \frac{1}{2}\right)^{3-x}, \quad x = 0, 1, 2, 3$

22. $P(X = x) = {}_2C_x \cdot \left(\frac{1}{4}\right)^x \left(1 - \frac{1}{4}\right)^{2-x}, \quad x = 0, 1, 2$

23. $P(X = x) = {}_5C_x \cdot (0.3)^x (1 - 0.3)^{5-x}, \quad x = 0, 1, 2, 3, 4, 5$

24. $P(X = x) = {}_4C_x \cdot (0.6)^x (1 - 0.6)^{4-x}, \quad x = 0, 1, 2, 3, 4$

25. $P(X = x) = {}_5C_x \cdot (0.7)^x (1 - 0.7)^{5-x}, \quad x = 0, 1, 2, 3, 4, 5$

26. $P(X = x) = {}_4C_x \cdot (0.4)^x (1 - 0.4)^{4-x}, \quad x = 0, 1, 2, 3, 4$

In Exercises 27–32, match the probability statement with the appropriately shaded probability histogram.

(a)

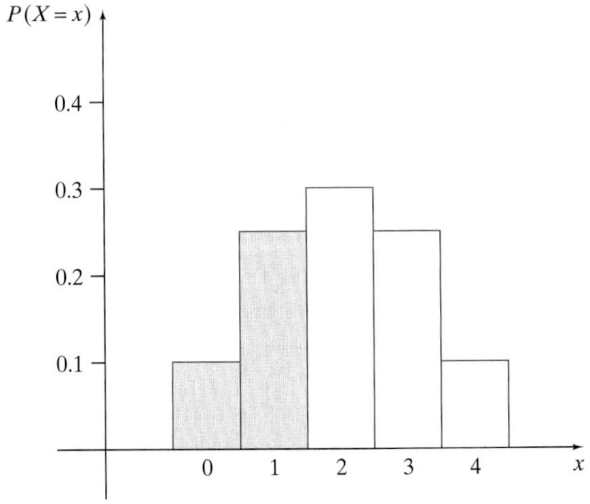

(b)

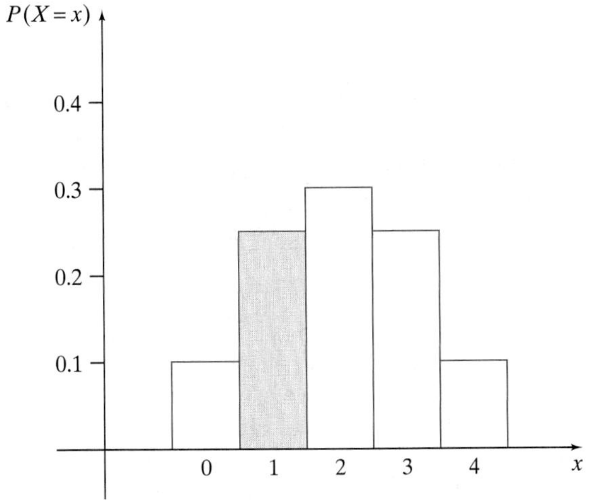

(c)

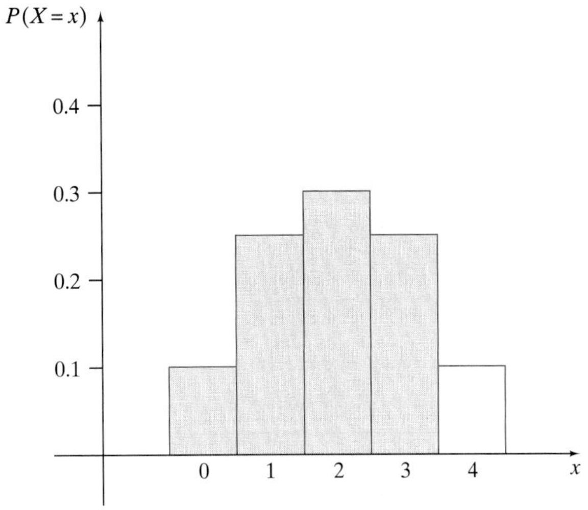

(d)

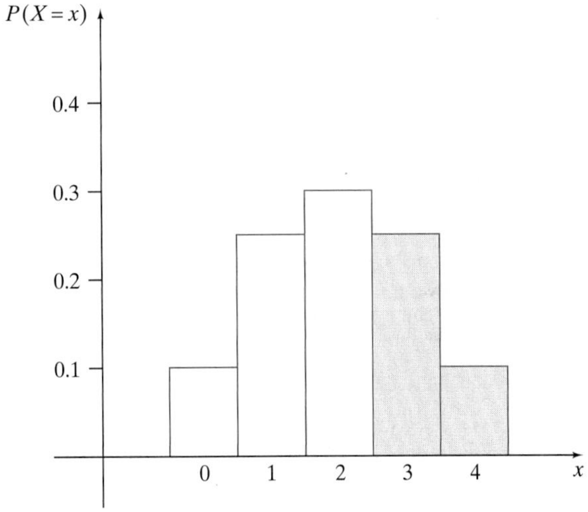

(e)

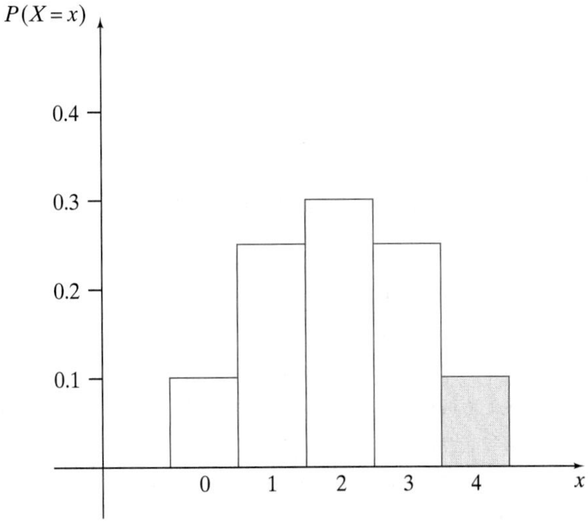

(f)

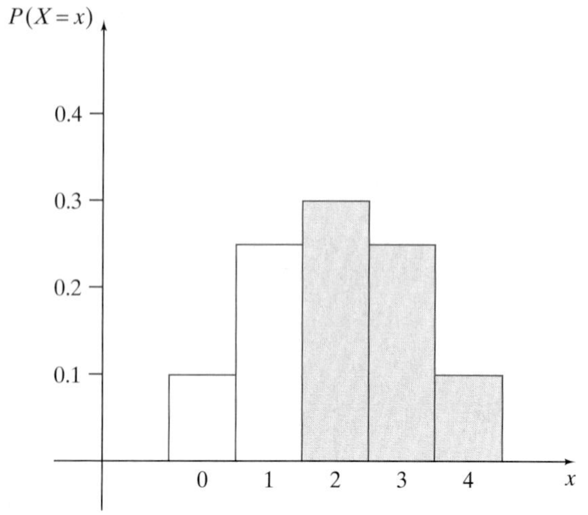

27. $P(X = 1)$

28. $P(X = 4)$

29. $P(X \leq 3)$

30. $P(X \leq 1)$

31. $P(X \geq 2)$

32. $P(X \geq 3)$

In Exercises 33–40, assume that X represents a binomial random variable. Determine the mean $E(X) = \mu_X$ and standard deviation σ_X of the distribution.

33. $n = 12,\ p = \frac{1}{2}$

34. $n = 18,\ p = \frac{1}{3}$

35. $n = 20,\ p = 0.25$

36. $n = 16,\ p = 0.3$

37. $P(X = x) = {}_6C_x \cdot \left(\frac{1}{4}\right)^x \left(1 - \frac{1}{4}\right)^{6-x}$,

$x = 0, 1, 2, 3, 4, 5, 6$

38. $P(X = x) = {}_8C_x \cdot \left(\frac{1}{10}\right)^x \left(1 - \frac{1}{10}\right)^{8-x}$,

$x = 0, 1, 2, 3, \ldots, 7, 8$

39. $P(X = x) = {}_{18}C_x \cdot \left(\frac{1}{8}\right)^x \left(1 - \frac{1}{8}\right)^{18-x}$,

$x = 0, 1, 2, 3, \ldots, 17, 18$

40. $P(X = x) = {}_{22}C_x \cdot \left(\frac{1}{10}\right)^x \left(1 - \frac{1}{10}\right)^{22-x}$,

$x = 0, 1, 2, 3, \ldots, 21, 22$

Applications

In Exercises 41–58, assume that all applications can be expressed by a binomial random variable X.

41. The Vector Tech Company gives a preemployment screening exam and finds that 40% of those who take the exam pass. A group of 7 prospective employees takes the exam.

(a) Verify that the assumptions of a binomial experiment are reasonably satisfied.

(b) Evaluate $P(X = 5)$ and interpret.

42. A survey at Harperston College reveals that 70% of the students sold their philosophy book back to the bookstore at the end of the course. The bookstore selects a group of 6 philosophy students at random.

(a) Verify that the assumptions of a binomial experiment are reasonably satisfied.

(b) Evaluate $P(X = 3)$ and interpret.

43. The Direct Satisfaction Company knows from experience that 40% of the people who enter their sweepstakes also order a magazine. A new national magazine selects a group of 12 sweepstakes entrants at random.

(a) Determine $P(X = 4)$ and interpret.

(b) Determine $P(4 \leq X \leq 5)$ and interpret.

44. In the town of Litherville, 30% of the households say that they are satisfied with the garbage service. The mayor's office selects 10 houses at random.

(a) Determine $P(X = 3)$ and interpret.

(b) Determine $P(3 \leq X \leq 4)$ and interpret.

45. When asked, "Who do you think was the greatest American president?" 23% of the Republicans responded Abraham Lincoln (*Source:* www.CBSNews.com). A sample of 15 Republicans is selected at random.

(a) Determine $P(X = 3)$ and interpret.

(b) Determine $P(X \leq 3)$ and interpret.

46. When asked, "Who do you think was the greatest American president?" 26% of the Democrats responded John F. Kennedy (*Source:* www.CBSNews.com). A sample of 12 Democrats is selected at random.

(a) Determine $P(X = 4)$ and interpret.

(b) Determine $P(X \leq 4)$ and interpret.

47. In January 1999, a study showed that 75% of all American households had cable TV (*Source:* www.infoplease.com). A satellite TV company selects 8 American households at random.

(a) Determine the probability that 7 have cable TV.

(b) Determine the probability that at most 5 houses have cable TV.

48. A 1997 mayoral report on curfews stated that 88% of the cities with curfews felt that they made the city safer (*Source:* U.S. Conference of Mayors, www.usmayors.org). Suppose that a Justice Department official selects 15 cities with curfews at random.

(a) Determine the probability that 12 mayors felt that the curfew made the city safer.

(b) Determine the probability that at least 12 mayors felt that the curfew made the city safer.

49. In February 2000, 2.8% of Colorado's labor force was unemployed (*Source:* U.S. Bureau of Labor Statistics, www.bls.gov). A national money magazine selects 50 people from Colorado's labor force at random.

(a) Determine the probability that 1 was unemployed.

(b) Determine the probability that at most 1 was unemployed.

50. In February 2000, 2.2% of Iowa's labor force was unemployed (*Source:* U.S. Bureau of Labor Statistics, www.bls.gov). A national money magazine selects 60 people from Iowa's labor force at random.

(a) Determine the probability that 1 was unemployed.

(b) Determine the probability that at least 1 was unemployed.

51. In 1997, 31.2% of male smokers smoked cigars (*Source:* American Cancer Society, www.cancer.org). A cigar importer selects 13 male smokers at random.

(a) Determine the probability that 3 were cigar smokers.

(b) Determine the probability that at most 3 were cigar smokers.

6.1 **52.** In 1997, 10.8% of female smokers smoked cigars (*Source:* American Cancer Society, www.cancer.org). A cigar importer selects 20 female smokers at random.

(a) Determine the probability that 2 were cigar smokers.

(b) Determine the probability that at most 2 were cigar smokers.

6.1 **53.** A report on minorities and transportation found that 59% of African Americans in New York City had no automobiles (*Source:* American Highway Users Alliance, www.highways.org). A poll selects 10 African Americans from New York City at random.

(a) Construct a probability histogram for the distribution of New York City African Americans with no automobiles.

(b) Compute μ_X and σ_X for the distribution and interpret each.

54. A report on minorities and transportation found that 40% of African Americans in Baltimore had no automobiles (*Source:* American Highway Users Alliance, www.highways.org). A poll selects 10 African Americans from Baltimore at random.

(a) Construct a probability histogram for the distribution of Baltimore African Americans with no automobiles.

(b) Compute μ_X and σ_X for the distribution and interpret each.

6.1 **55.** A 1992 survey on prisoners and education showed that 49% of the prisoners did not have a high school diploma (*Source:* National Center for Education Statistics, www.nces.ed.gov). A social worker selects 20 prisoners at random.

(a) How many are expected not to have a high school diploma?

(b) Construct a table of values for $P(X = x)$ and determine the most likely number of prisoners that did not have a high school diploma. That is, find the value x that corresponds to the greatest probability.

56. A 1992 survey on prisoners and education showed that 36% of the parents of prisoners did not have a high school diploma (*Source:* National Center for Education Statistics, www.nces.ed.gov). A social worker selects parents of 10 prisoners at random.

(a) How many are expected not to have a high school diploma?

(b) Construct a table of values for $P(X = x)$ and determine the most likely number of parents of prisoners that did not have a high school diploma. That is, find the value x that corresponds to the greatest probability.

6.1 **57.** A 1994 report revealed that 32.6% of all U.S. births were to unmarried women (*Source:* Centers for Disease Control, www.cdc.gov). A parenting magazine selects 30 women who gave birth in 1994 at random.

(a) How many are expected to be unmarried?

(b) What is the probability that at most 10 of the women were unmarried?

58. In a survey of adults, 53.1% viewed American children negatively (*Source:* www.publicagenda.org). A parenting magazine selects 20 adults at random.

(a) How many are expected to view American children negatively?

(b) What is the probability that at most 10 view American children negatively?

SECTION PROJECT

Consider a binomial random variable X with $n = 8$ and $p = \frac{1}{2}$.

(a) Construct a table of values for $P(X = x)$. Do you see a pattern in the probabilities? Explain.

(b) Draw a probability histogram for the distribution found in part (a). Classify the probability histogram as skewed left, skewed right, or symmetric.

(c) For the equation $P(x) = P(k - x)$, find the value of k that satisfies this equation for this distribution.

Section 8.4 The Normal Distribution

Thus far we have examined probability distributions that correspond to discrete random variables. The possible x-values of the random variable X up until now have been a finite subset of the whole numbers $0, 1, 2, \ldots, n$ or have been countably infinite. But let's say that we have a probability experiment in which we test the life length, in hours, of a new rechargeable battery. If we were interested in the probability that a battery lasted more than 5.5 hours, for example, we would have fractions or irrational numbers for the random variable values. In other words, the possible values of x are in the interval $[5.5, \infty)$, which in no way can be ordered. This means that the possible values of x are uncountable. To accommodate this situation, we need a different kind of random variable, called a **continuous random variable**. In this section we introduce continuous random variables. This will allow us to examine a continuous probability distribution called the **normal distribution**. This is the best known and most used of all probability distribution models. We will also see how we can use the normal distribution in applications.

Normal Random Variables

In Section 8.3 we saw that the binomial random variable had a corresponding probability mass function that was used to compute probabilities for known values of x. If the random variable is continuous, the probabilities are found using a function called a **probability density function**.

Continuous Random Variable

- If a random variable X can assume values corresponding to any of the points contained in one or more intervals of real numbers, then X is a **continuous random variable**.
- The probability function associated with a continuous random variable is called a **probability density function**.
- The graph of a probability density function is called a **density curve**.

Note: In Section 8.1 we noted that a probability mass function was associated with a discrete random variable. Here we note that a **probability density function** is associated with a **continuous random variable**.

The **normal distribution** is a particular type of continuous probability distribution with a particular type of probability density function. It is sometimes called the Gaussian distribution after one of its developers, German mathematician Karl Frederick Gauss.

Normal Random Variable

If X is a continuous random variable with probability density function

$$f(x) = \frac{1}{\sigma\sqrt{2\pi}} e^{-\frac{1}{2}\left(\frac{x-\mu}{\sigma}\right)^2}$$

then X is a **normal random variable**. The possible values of X are the set of real numbers.

Note: In this section we will use μ for μ_X and σ for σ_X for simplicity.

We will not be directly computing with this rather complicated looking formula, but instead will often refer to its graph, or **density curve**. This normal density curve, shown in Figure 8.4.1, is sometimes called the **bell curve** (or **normal curve**), because of its distinctive shape.

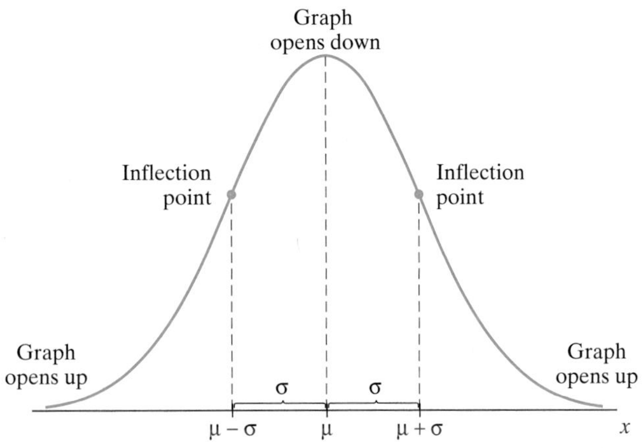

Figure 8.4.1

Figure 8.4.1 shows that the two important values that define the shape of the normal density curve are μ and σ. We can also see from the graph in Figure 8.4.1 that if we move the distance of 1 standard deviation to the left of the mean we get to the point on the graph where it switches from opening up (or bowing up) to opening down. This point is called an **inflection point**. If we move the distance of 1 standard deviation to the right of the mean, we get to the point on the curve where the graph switches from opening down to opening up again (another inflection point).

To compute probabilities with the normal distribution, we will transform the normal distribution into a simpler distribution called the **standard normal distribution**. First, let's define the **standard normal random variable**.

Standard Normal Random Variable

If Z is a continuous random variable with probability density function

$$f(Z) = \frac{1}{\sqrt{2\pi}} e^{-\frac{1}{2}z^2}$$

then Z is called a **standard normal random variable**. The possible values of Z are the set of real numbers.

Note: A standard normal random variable is a normal random variable with mean $\mu = 0$ and standard deviation $\sigma = 1$.

To compute probabilities with normal distributions, we first use the following rule to transform the normal distribution to a standard normal distribution.

Let X be a normal random variable with mean μ and standard deviation σ; then

$$Z = \frac{X - \mu}{\sigma}$$

is a standard normal random variable.

▶ **Note:** We will refer to the value $z = \dfrac{x - \mu}{\sigma}$ as a ***z*-score**.

The standard normal distribution allows us to determine probabilities associated with the normal distribution all in the same way. Since we will be using the z-scores and the standard normal curve for all our normal distribution applications, let's look at some of the properties of the standard normal distribution.

Properties of the Standard Normal Distribution

1. The graph of the standard normal curve is shown in Figure 8.4.2.

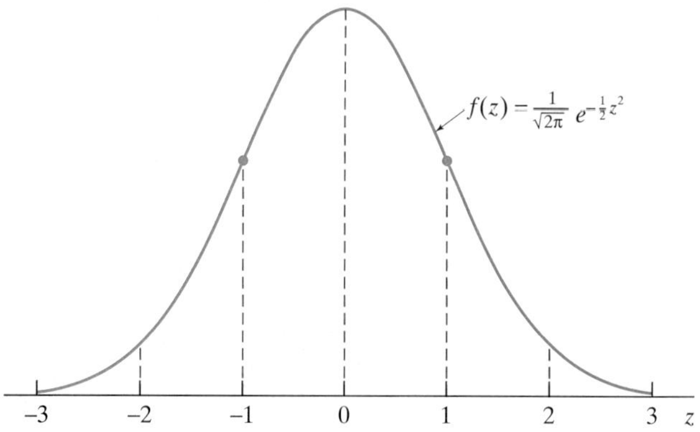

Figure 8.4.2

2. The total area under the standard normal curve is 1. That is, $P(-\infty < Z < \infty) = 1$.
3. The curve is bell shaped.
4. The mean, median, and mode are equal and centered in the middle of the graph.
5. The graph is symmetric about the mean. This means that the shape of the curve is the same on both sides of a vertical line through the center of the distribution, that is, at $z = 0$. This line divides the area under the curve into two equal parts.
6. The maximum occurs at $\mu = 0$.
7. About 68% of the area under the curve is within 1 standard deviation of the mean $\mu = 0$. That is, about 68% of the area under the curve is in the interval $(-1, 1)$. About 95% of the area under the curve is within 2 standard deviations of the mean $\mu = 0$, or in the interval $(-2, 2)$. About 99.7% of the area under the curve is within 3 standard deviations of the mean $\mu = 0$, or in the interval $(-3, 3)$.

The key to using the standard normal distribution is the computation of the z-scores. The z-score tells us how many standard deviations a value lies away from the mean. For example, a z-score of $z = 1$ means that the data value is 1 standard deviation greater than the mean μ. A z-score of $z = -1$ means that the data value is 1 standard deviation less than the mean μ. Let's see how z-scores can be interpreted in a familiar setting.

Example 1 Interpreting z-scores

The results of Rachel's midterms for Chemistry and English classes are shown in Table 8.4.1, along with the class mean and standard deviation. Assume that the distribution of scores in each class is normal. Compute Rachel's z-score for each class and interpret.

Table 8.4.1

Midterm Results	Chemistry	English
Midterm score, x	72	70
Mean, μ	68	76
Standard deviation, σ	3.2	8.0

Solution

Interactive Activity

In Example 1 suppose that, Rachel's English midterm was graded incorrectly and she really earned a score of 81. Recompute her z-score for the English midterm. With this new score, relative to the rest of the students, on which midterm did she do better.

WWW

Let's call the z-score for the Chemistry midterm z_1. Using the z-score formula, we get

$$z_1 = \frac{x - \mu}{\sigma} = \frac{72 - 68}{3.2} = 1.25$$

Interpret the Solution: In Chemistry, her score was 1.25 standard deviations greater than the mean.

The z-score for the English midterm, which we will call z_2, is

$$z_2 = \frac{70 - 76}{8.0} = -0.75$$

Interpret the Solution: In English, her score was 0.75 standard deviation less than the mean. This shows that Rachel's grade in Chemistry was much better than her grade in English *relative* to other students.

The z-scores are plotted on the standard normal curve in Figure 8.4.3.

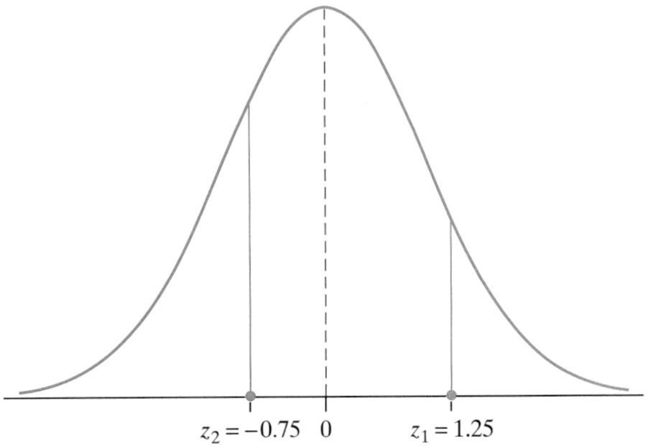

Figure 8.4.3

Checkpoint 1 Now work Exercise 3.

Finding Areas under the Standard Normal Curve

When using the binomial distribution in Section 8.3, we could add up the area of the rectangles in the probability histogram to determine probabilities. For the normal distribution we must compute the amount of area under the standard normal curve. Since we have a continuous random variable, determining the area requires techniques of calculus. However, tables have been created using calculus techniques, and we can use these tables to determine these areas. A table used to calculate these areas, called the Standard Normal Table, is in Appendix D. This table gives the area from $z = 0$ to some given z-score. See Figure 8.4.4.

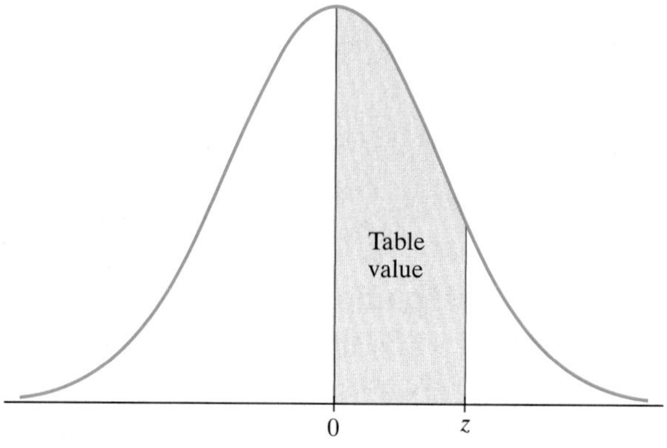

Figure 8.4.4

A table value is all that is needed to compute a probability. To understand how to use this table, let's say we want to compute $P(0 < Z < 1.24)$. In this case we need the table value at $z = 1.24$. To get this value, we look down the left column until we find 1.2 (the z-score accurate to the tenths place). Then we look across the row until we find 0.04 at the top of the page (the hundredths place digit of the z-score). From the table, we get $P(0 < Z < 1.24) = (\text{value at } z = 1.24) = 0.3925$. See Figure 8.4.5. The area that $z = 1.24$ corresponds to is shown on the standard normal curve in Figure 8.4.6. This means that the probability that the z-score value is between 0 and 1.24 is about 39.25%. Note that we said that the "z-score value is between 0 and 1.24," but did not specify if the values of 0 and 1.24 were included or not. This is because lines have length and not width, so the area between these two z-scores does not change whether we include these

z	.00	.01	.02	.03	.04	.05
0.0	0.0000	0.0040	0.0080	0.0120	0.0160	0.0199
0.1	0.0398	0.0438	0.0478	0.0517	0.0557	0.0596
0.2	0.0793	0.0832	0.0871	0.0910	0.0948	0.0987
0.3	0.1179	0.1217	0.1255	0.1293	0.1331	0.1368
0.4	0.1554	0.1591	0.1628	0.1664	0.1700	0.1736
0.5	0.1915	0.1950	0.1985	0.2019	0.2054	0.2088
0.6	0.2257	0.2291	0.2324	0.2357	0.2389	0.2422
0.7	0.2580	0.2611	0.2642	0.2673	0.2704	0.2734
0.8	0.2881	0.2910	0.2939	0.2967	0.2995	0.3023
0.9	0.3159	0.3186	0.3212	0.3238	0.3264	0.3289
1.0	0.3413	0.3438	0.3461	0.3485	0.3508	0.3531
1.1	0.3643	0.3665	0.3686	0.3708	0.3729	0.3749
1.2	0.3849	0.3869	0.3888	0.3907	0.3925	0.3944
1.3	0.4032	0.4049	0.4066	0.4082	0.4099	0.4115

Figure 8.4.5

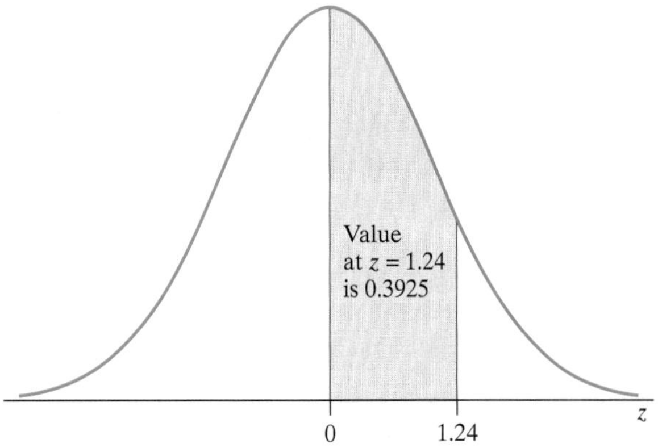

Figure 8.4.6 The value at $z = 1.24$ gives the area of the shaded region.

endpoints or not. So, for continuous random variables, including the normal random variable, we can use the following notation rule:

Notation for Endpoints of Continuous Random Variables

For real numbers a and b and a continuous random variable X, such as a normally distributed random variable, we have the following:

1. $P(a < X < b) = P(a \leq X \leq b)$
2. $P(X < a) = P(X \leq a)$
3. $P(X > b) = P(X \geq b)$

In Example 2 we illustrate how to use the Standard Normal Table to compute other probabilities, as well as use some properties of the standard normal distribution.

Example 2 Computing Areas under the Standard Normal Curve

Determine the following areas under the standard normal curve.

(a) $P(Z \geq 1.68)$ **(b)** $P(-1.31 < Z < 0.50)$

Solution

(a) From Property 2 of Properties of the Standard Normal Distribution, we know that the total area under the standard normal curve is 1. Using this fact along with Property 5 (graph is symmetric about the mean, $z = 0$), we know that $P(Z \geq 0) = 0.5$. This is shown in Figure 8.4.7 on p. 508.

To determine $P(Z \geq 1.68)$, we can simply take away the area for $P(0 < Z < 1.68)$ from $P(Z \geq 0)$. See Figure 8.4.8 on p. 508.

The table value at $z = 1.68$ is 0.4535 (see Standard Normal Table in Appendix D). So we have

$$\begin{aligned} P(Z \geq 1.68) &= P(Z \geq 0) - P(0 < Z < 1.68) \\ &= 0.50 - (\text{value at } z = 1.68) = 0.50 - 0.4535 \\ &= 0.0465 \end{aligned}$$

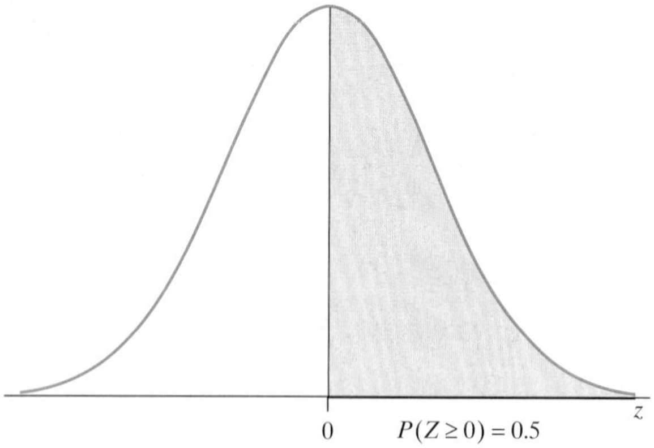

Figure 8.4.7

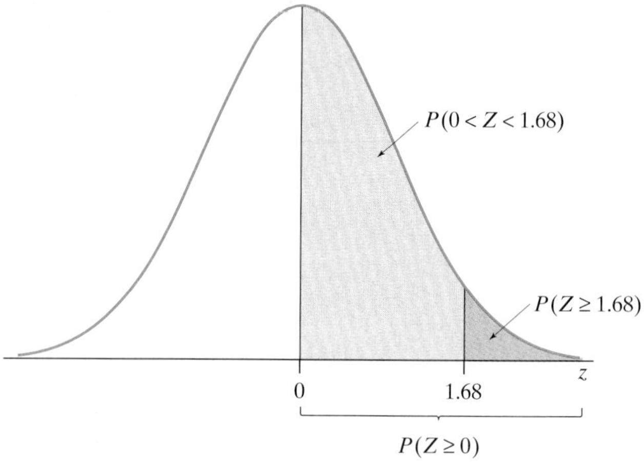

Figure 8.4.8 $P(Z \geq 1.68) = P(Z \geq 0) - P(0 < Z < 1.68)$.

(b) To compute $P(-1.31 < Z < 0.50)$ corresponds to finding the area shown in Figure 8.4.9.

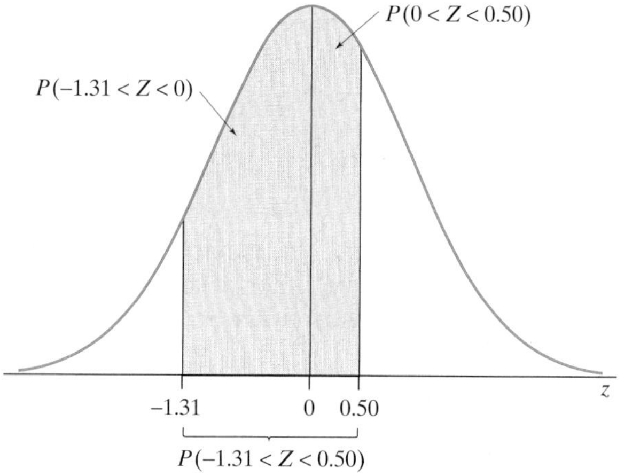

Figure 8.4.9

Notice in Figure 8.4.9 that

$$P(-1.31 < Z < 0.50) = P(-1.31 < Z < 0) + P(0 < Z < 0.50)$$

$P(0 < Z < 0.50)$ is immediately found by consulting the Standard Normal Table in Appendix D. Since the value at $z = 0.50$ is 0.1915, we have

$$P(0 < Z < 0.50) = 0.1915$$

To determine $P(-1.31 < Z < 0)$, we again utilize Property 5 (graph is symmetric about the mean, $z = 0$). The area shaded in Figure 8.4.10 corresponds to $P(0 < Z < 1.31)$ and is identical to $P(-1.31 < Z < 0)$, thanks to symmetry. This means that the table values for $z = -1.31$ and $z = 1.31$ are the same.

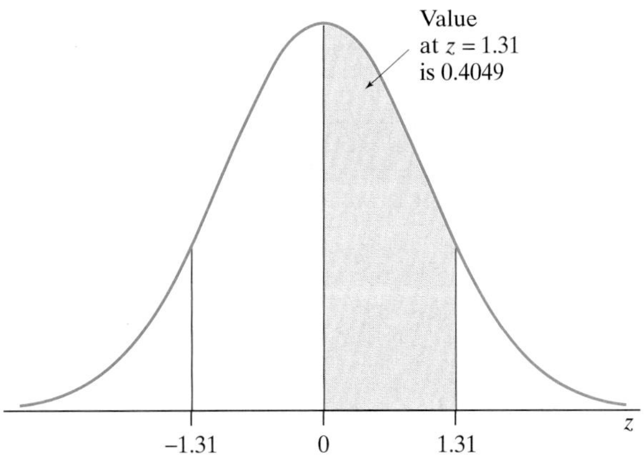

Figure 8.4.10

We can now compute

$$\begin{aligned} P(-1.31 < Z < 0.50) &= P(-1.31 < Z < 0) + P(0 < Z < 0.50) \\ &= 0.4049 + 0.1915 \\ &= 0.5964 \end{aligned}$$

Observe that in computing $P(-1.31 < Z < 0.50)$ one z-score was negative and the other was positive. What we ended up doing was simply adding the table value for $z = 1.31$ and $z = 0.50$. ■

In Table 8.4.2 on p. 510 we summarize how to determine areas under the standard normal curve for different z-values. We will use Table 8.4.2 frequently.

Applications of the Normal Distribution

The normal distribution can be used on a wide variety of applications. The only requirement is that the distribution have the distinctive bell-shaped curve. We can assume that in this section all the applications come from data that are normally distributed. To solve applications of the normal distribution, we will use the following six-step process.

Solving Normal Distribution Applications

1. After reading the application problem, identify the mean μ and the standard deviation σ.
2. Write the desired probability in terms of the normal random variable X.
3. Convert the given x-values to standard z-scores using the formula $z = \dfrac{x - \mu}{\sigma}$.
4. Sketch the standard normal curve and shade the desired area.
5. Use Table 8.4.2 and the Standard Normal Table in Appendix D to determine the probability.
6. Interpret the solution.

Table 8.4.2

Description of Area	Graph	Computation of Area
1. Between 0 and any value	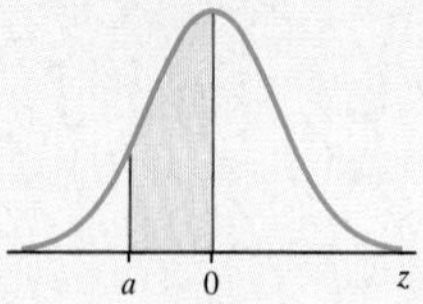	$P(a < Z < 0) = (\text{value at } z = a)$
	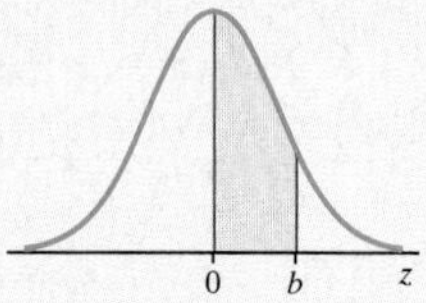	$P(0 < Z < b) = (\text{value at } z = b)$
2. A left or right tail	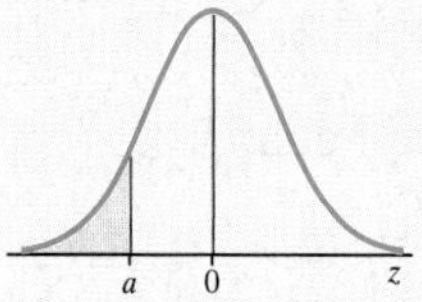	$P(Z < a) = 0.50 - (\text{value at } z = a)$
	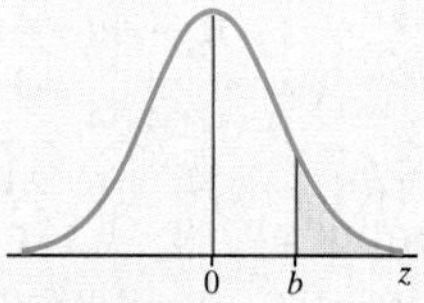	$P(Z > b) = 0.50 - (\text{value at } z = b)$
3. Between any two values where the z-scores have the same sign	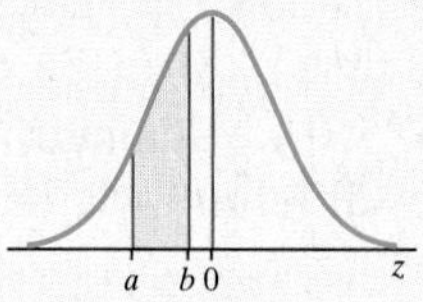	$P(a < Z < b) = (\text{value at } z = a) - (\text{value at } z = b)$
	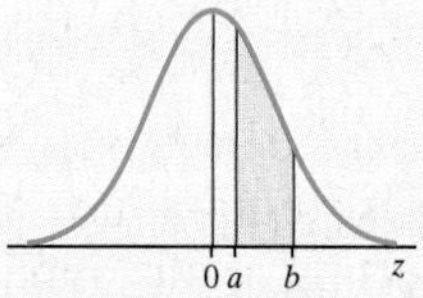	$P(a < Z < b) = (\text{value at } z = b) - (\text{value at } z = a)$
4. Between any two values where the z-scores have opposite sign	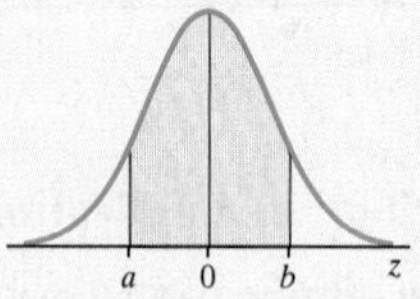	$P(a < Z < b) = (\text{value at } z = a) + (\text{value at } z = b)$
5. Less than any z-score to the right of the mean	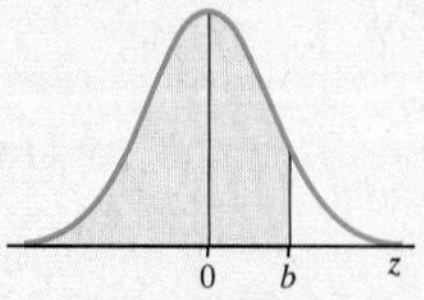	$P(Z < b) = 0.50 + (\text{value at } z = b)$
6. Greater than any z-score to the left of the mean	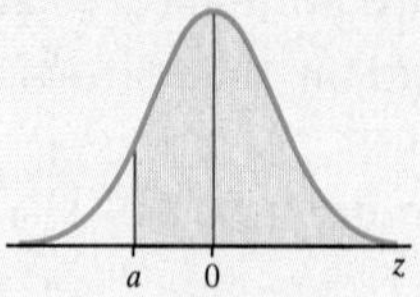	$P(Z > a) = 0.50 + (\text{value at } z = a)$

Now let's try this procedure in an application. **For all applications in this section, assume that the quantity discussed can be represented by a normal random variable**. When computing z-scores, always round to the nearest hundredths place so that you can use the Standard Normal Table in Appendix D.

Example 3 Applying the Normal Distribution

The amount of floor area in a new, privately owned, single-family house built in 1970 had an average of 1500 square feet with a standard deviation of 683 square feet (*Source:* U.S. Census Bureau, www.census.gov). For a special featured article, a national home improvement magazine selects a house built in 1970 at random.

(a) Determine the probability that the amount of floor space in the house is less than 1200 square feet.

(b) Determine the probability that the amount of floor space in the house is between 1600 and 2000 square feet.

Solution

For part (a) of this example, we will list the six steps.

(a)

Step 1: We let X be a normal random variable. We are given the mean $\mu = 1500$ and $\sigma = 683$.

Step 2: Here we need the probability

$$P(\text{house has less than 1200 square feet}) = P(X < 1200)$$

Step 3: Converting this value to a z-score gives us

$$P(X < 1200) = P\left(Z < \frac{1200 - 1500}{683}\right) \approx P(Z < -0.44)$$

Step 4: This region is shown in Figure 8.4.11.

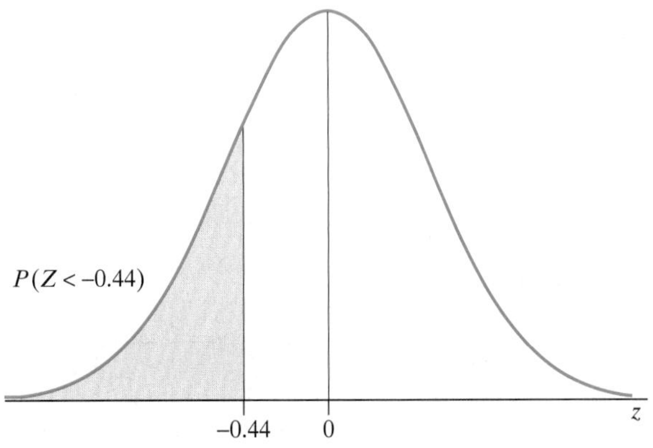

Figure 8.4.11

Step 5: Table 8.4.2 indicates that to determine $P(Z < -0.44)$ we do the following:

$$P(Z < -0.44) = 0.50 - (\text{value at } z = 0.44)$$

The Standard Normal Table in Appendix D gives us the value at $z = 0.44$. Continuing, we get

$$P(Z < -0.44) = 0.50 - (\text{value at } z = 0.44) = 0.50 - 0.1700 = 0.330$$

Step 6: This means that there is about a 33% probability that a randomly chosen house built in 1970 will have a floor area of less than 1200 square foot.

(b) Now we need to find the probability that the square footage is between 1600 and 2000. In terms of our random variable, this is $P(1600 < X < 2000)$. Converting to a z-score gives us

$$P(1600 < X < 2000) = P\left(\frac{1600 - 1500}{683} < Z < \frac{2000 - 1500}{683}\right)$$
$$\approx P(0.15 < Z < 0.73)$$

The area corresponding to this probability is shown in Figure 8.4.12. To find this area, we use Table 8.4.2 and the Standard Normal Table for the values of $z = 0.73$ and $z = 0.15$ to get

$$P(0.15 < Z < 0.73) = (\text{value at } z = 0.73) - (\text{value at } z = 0.15)$$
$$= 0.2673 - 0.0596 = 0.2077$$

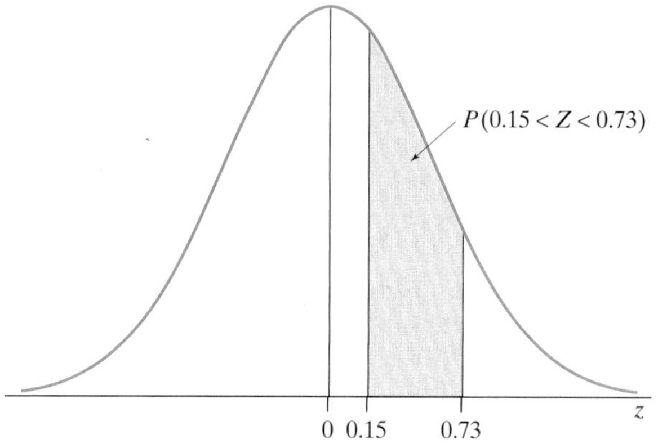

Figure 8.4.12

This means that there is about a 21% chance that a randomly chosen house built in 1970 will have floor area between 1600 and 2000 square feet. ■

Checkpoint 2 Now work Exercise 45.

Technology Option

Figure 8.4.13

The solution to part (b) of Example 3, using the NORMALCDF command on a graphing calculator, is shown in Figure 8.4.13. The difference in the answers is due to the rounding of the z-score to the hundredths place.

To see how your calculator computes normal probabilities, consult the on-line calculator manual at www.prenhall.com/armstrong.

Let's try another application in order to practice computing probabilities using the standard normal distribution. In this application we want an area to the right of a given z-score and also two z-scores that *straddle* the mean.

Example 4 Applying the Normal Distribution

The average age of a farmer in 1992 was 51.9 years with a standard deviation of 8.7 years (*Source:* U.S. Census Bureau, www.census.gov). An agriculture economics professor selects a farmer from 1992 at random for a case study paper.

(a) Determine the probability that the farmer's age exceeded 40 years.

(b) Determine the probability that the farmer's age was between 45 and 55 years.

Solution

(a) We are given the mean $\mu = 51.9$ and the standard deviation $\sigma = 8.7$. We seek the probability that a farmer is older than 40. So we want $P(X > 40)$. Converting this value to a z-score gives

$$P(X > 40) = P\left(Z > \frac{40 - 51.9}{8.7}\right) \approx P(Z > -1.37)$$

This area is shown in Figure 8.4.14.

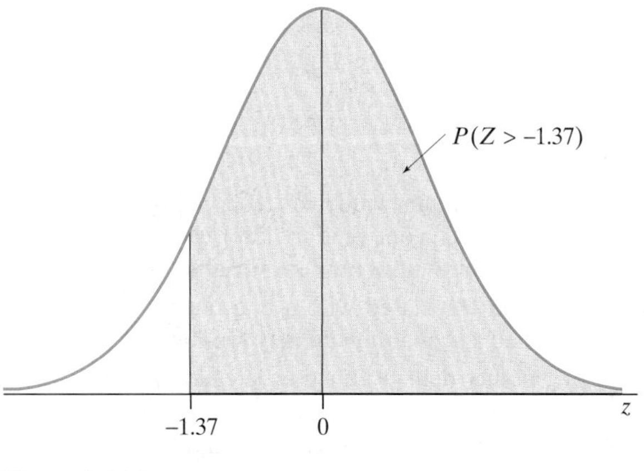

Figure 8.4.14

Using Table 8.4.2 and the Standard Normal Table in Appendix D for $z = 1.37$ gives us

$$P(Z > -1.37) = (\text{value at } z = 1.37) + 0.50 = 0.4147 + 0.50 = 0.9147$$

Interpret the Solution: This means that if a farmer is randomly chosen there is about a 91% chance that the farmer is older than 40.

(b) Here we want the probability that the farmer is between 45 and 55 years of age. So we need to compute

$$P(45 < X < 55) = P\left(\frac{45 - 51.9}{8.7} < Z < \frac{55 - 51.9}{8.7}\right) \approx P(-0.79 < Z < 0.36)$$

This area is shown in Figure 8.4.15.

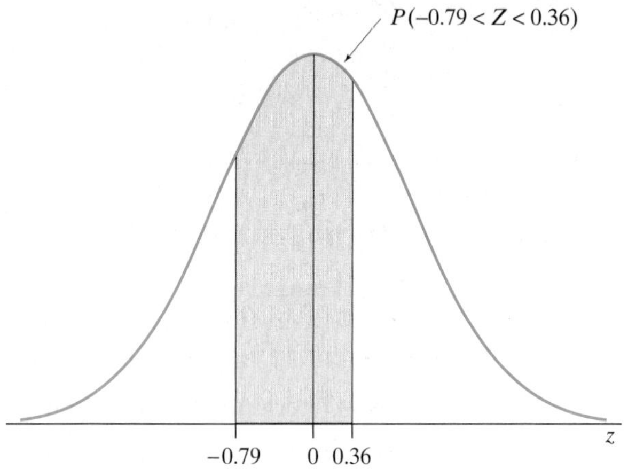

Figure 8.4.15

Using Table 8.4.2 and the Standard Normal Table in Appendix D for $z = 0.79$ and $z = 0.36$ gives us

$$P(-0.79 < Z < 0.36) = (\text{value at } z = 0.79) + (\text{value at } z = 0.36)$$
$$= 0.2852 + 0.1406 = 0.4258$$

Interpret the Solution: This means that if a farmer is chosen a random there is about a 43% chance that the farmer is between 45 and 55 years of age. ■

Checkpoint 3 Now work Exercise 49.

SUMMARY

- If a random variable X can assume values corresponding to any of the points contained in one or more intervals of real numbers, then X is a **continuous random variable**.
- The probability function associated with a continuous random variable is called a **probability density function**.
- The graph of a probability density function is called a **density curve**.
- If X is a continuous random variable with probability density function

$$f(x) = \frac{1}{\sigma\sqrt{2\pi}} e^{-\frac{1}{2}\left(\frac{x-\mu}{\sigma}\right)^2}$$

then X is a **normal random variable**. The possible values of x are the set of real numbers.
- If X is a normal random variable with mean μ and standard deviation σ, then we define Z as a **standard normal random variable**, where $Z = \dfrac{X-\mu}{\sigma}$.
- The value $z = \dfrac{x-\mu}{\sigma}$ is called a ***z*-score**.
- The mean of the standard normal distribution is $\boldsymbol{\mu = 0}$ with a standard deviation of $\boldsymbol{\sigma = 1}$.

SECTION 8.4 EXERCISES

Assume that each of the exercises in this section can be represented by a normal random variable X.

1. The mean and standard deviation of the gas mileages for the Jupiter and the Atlantis mid-sized cars are shown in the table.

	Jupiter	Atlantis
μ	23.5	26.0
σ	3.5	5.1

(a) In a test, a Jupiter gets 27 mpg, while an Atlantis gets 29.1 mpg. Compute the z-score for each car's gas mileage.

(b) Graph the z-scores from part (a) on the standard normal curve. Relative to the rest of the cars of the same make, which car got better mileage?

2. The mean and standard deviation of the wages of those who work on first and second shift at the Homburg Manufacturing Plant are shown in the table.

	First Shift	Second Shift
μ	22.0	19.5
σ	1.1	1.9

(a) A randomly chosen worker from the first shift makes \$23.87 per hour, while a randomly chosen worker from the second shift makes \$23.11. Compute the z-score for each of the worker's wages.

(b) Graph the z-scores from part (a) on the standard normal curve. Relative to the rest of the workers in their respective shifts, which shift worker made higher wages?

✓ **3.** The mean and standard deviation of the entrance exam scores for the Sunny Day Prep School for the 1999–2000 and 2000–2001 academic years are shown in the table.

	1999–2000	2000–2001
μ	153	142
σ	10.6	8.7

(a) A randomly chosen student who took the test in 1999–2000 scored 137, while a randomly chosen student who took the test in 2000–2001 scored 131. Compute the z-score for each student's exam.

(b) Graph the z-scores from part (a) on the standard normal curve. Relative to the rest of the students taking the test that year, which student did better?

4. The mean and standard deviation of the annual snowfall in the neighboring communities of Nashbrook and Hampton are shown in the table.

	Nashbrook	Hampton
μ	36	41
σ	1.5	0.8

(a) Last year Nashbrook received 35.1 inches of snow, while Hampton received 40.6 inches of snow. Compute the z-score for each community's snowfall.

(b) Graph the z-scores from part (a) on the standard normal curve. Relative to the rest of the years, which community received less snow?

5. The mean and standard deviation of the number of miles traveled each month by the sales divisions in the MegaStar and HemelGraph companies are shown in the table.

	MegaStar	HemelGraph
μ	1570	1645
σ	53	42

(a) Last month the sales division of MegaStar traveled 1702.5 miles, while the sales division of HemelGraph traveled 1741.6 miles. Compute the z-score for each company's mileages.

(b) Graph the z-scores from part (a) on the standard normal curve. Relative to the rest of the months, which company's sales division traveled less?

6. The mean and standard deviation of the number of bags lost each week by AirCom and BlueSky airlines are shown in the table.

	AirCom	BlueSky
μ	120	78
σ	2.8	2.7

(a) Last week Aircom lost 114 bags and BlueSky lost 74 bags. Compute the z-score for each airline's lost bags.

(b) Graph the z-scores from part (a) on the standard normal curve. Relative to the rest of the weeks, which company lost fewer bags?

In Exercises 7–16, find the shaded area using Table 8.4.2 and the Standard Normal Table in Appendix D.

7.

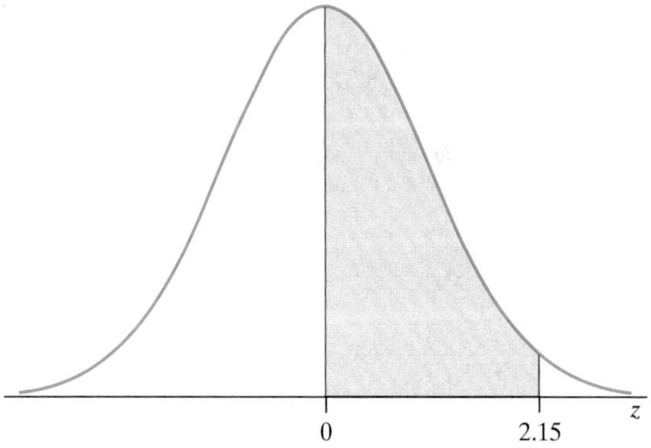

8.

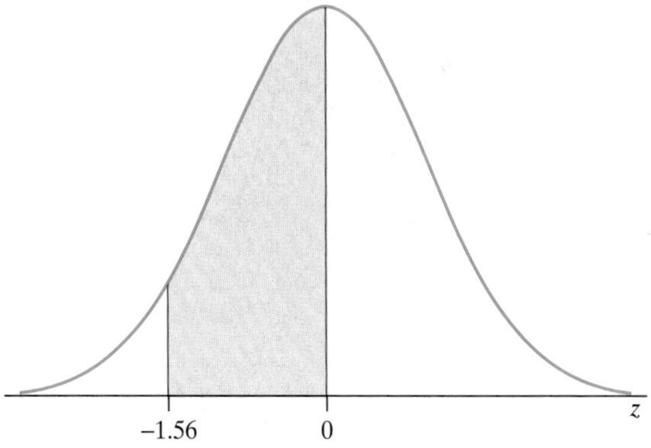

9.

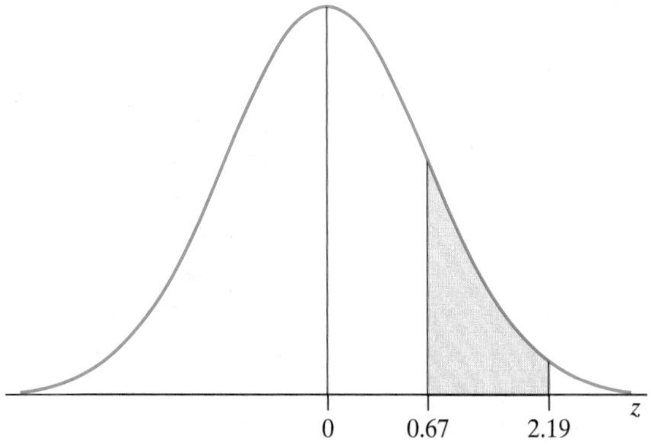

10.

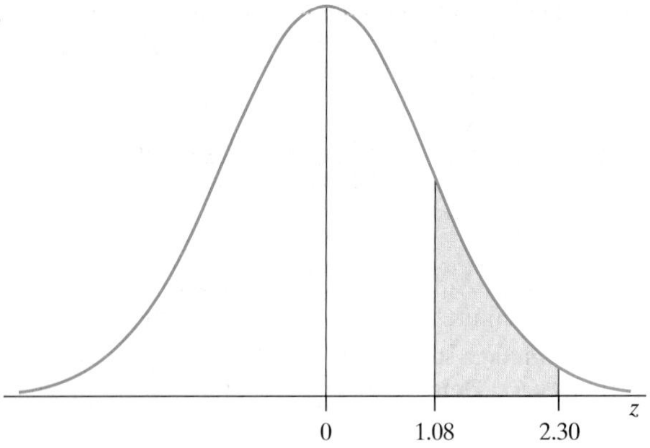

11.

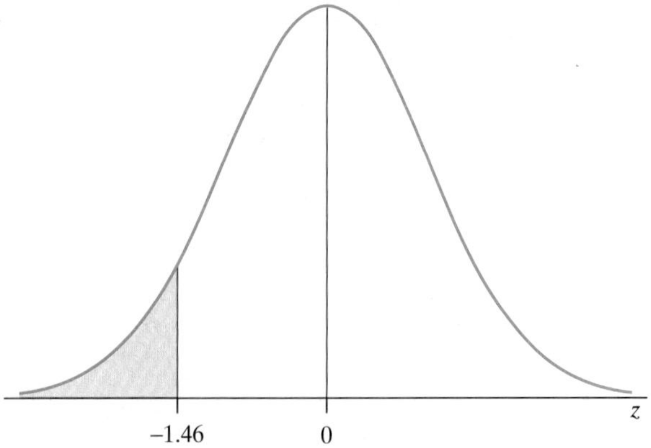

12.

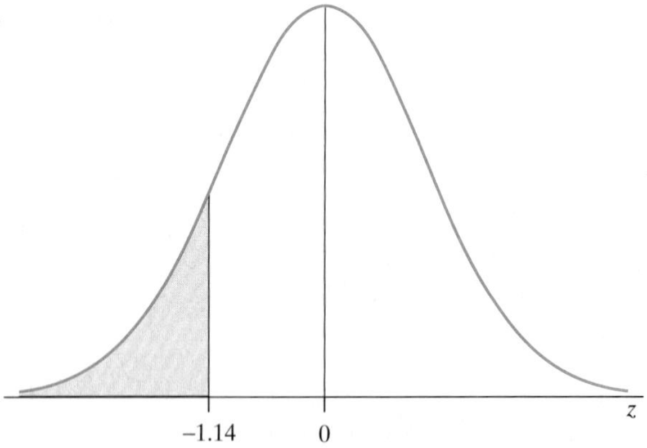

13.

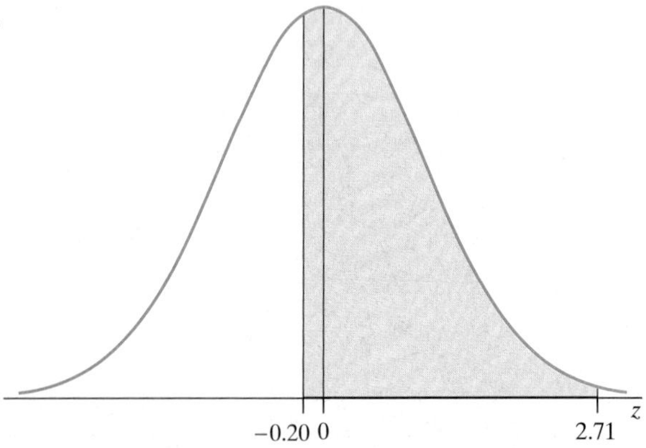

14.

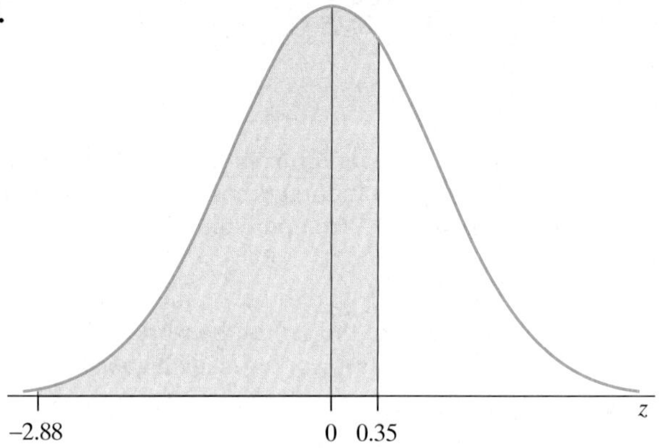

15.

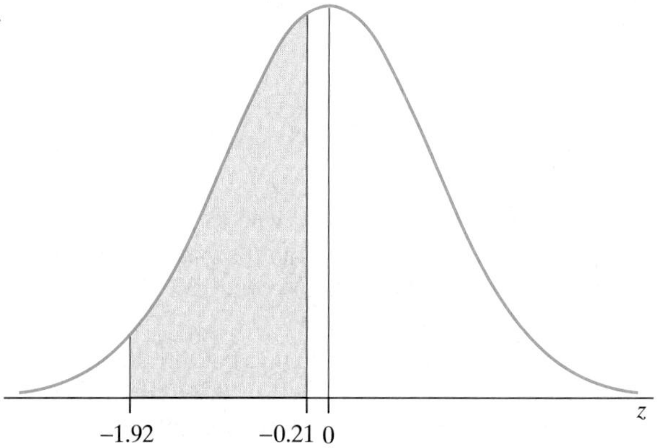

16.

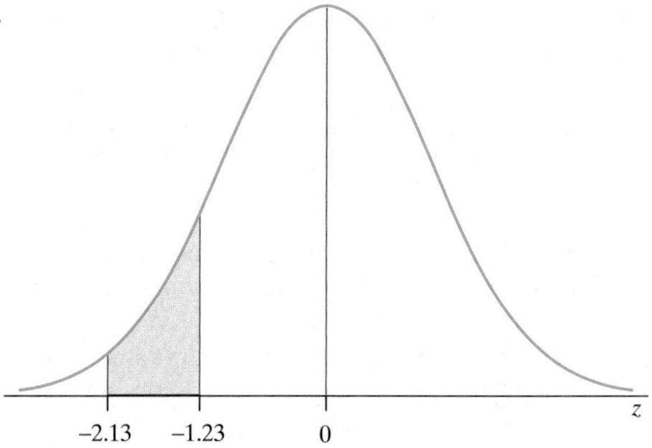

In Exercises 17–26, find the area between the given z-scores using Table 8.4.2 and the Standard Normal Table in Appendix D.

17. Between $z = 0$ and $z = 1.50$

18. Between $z = 0$ and $z = 1.70$

19. Between $z = 0$ and $z = -0.65$

20. Between $z = 0$ and $z = -0.85$

21. Between $z = -1.23$ and $z = 0.61$

22. Between $z = -0.75$ and $z = 1.45$

23. Between $z = 0.53$ and $z = 2.11$

24. Between $z = 0.67$ and $z = 2.04$

25. Between $z = -1.25$ and $z = -0.19$

26. Between $z = -1.98$ and $z = -0.51$

In Exercises 27–40, determine the given probabilities using Table 8.4.2 and the Standard Normal Table in Appendix D.

27. $P(Z > 0)$

28. $P(Z < 0)$

29. $P(Z > 1.30)$

30. $P(Z > 1.60)$

31. $P(Z \leq -1.76)$

32. $P(Z \leq -1.88)$

33. $P(Z \geq -0.73)$

34. $P(Z \geq -0.91)$

35. $P(Z < 1.96)$

36. $P(Z < 2.18)$

37. $P(-0.50 < Z < 0.50)$

38. $P(-0.38 < Z < 0.38)$

39. $P(-0.91 \leq Z \leq -0.12)$

40. $P(-0.87 \leq Z \leq -0.32)$

Applications

In the remainder of the exercises, assume that X represents a normal random variable.

41. The LandTrans Company runs shuttle buses from the airport to the local hotels. Records show that the mean number of runs for each bus each day is $\mu = 15$ with a standard deviation of $\sigma = 0.5$. For an audit, the chief financial officer at LandTrans Company selects a shuttle bus at random.

(a) Determine $P(X > 15)$ and interpret.

(b) Determine $P(X < 15)$ and interpret.

42. The Lock-It-Now Company finds that, on any given day, a mean of $\mu = 24{,}000$ locks is produced with a standard deviation of $\sigma = 2000$. For an audit, the chief financial officer at Lock-It-Now Company selects a manufacturing day at random.

(a) Determine $P(X \geq 24{,}000)$ and interpret.

(b) Determine $P(X \leq 24{,}000)$ and interpret.

43. The We-Make-Tools machine shop manufactures wire and finds that its standard-gauge wire has a mean of $\mu = 0.12$ inch with a standard deviation of $\sigma = 0.02$ inch. For an inspection, a quality control engineer selects a wire at random.

(a) Determine $P(0.12 \leq X \leq 0.16)$ and interpret.

(b) Determine $P(0.08 \leq X \leq 0.12)$ and interpret.

44. The electronic life of the MindReader personal digital assistant is found to have a mean of $\mu = 4.01$ years with a standard deviation of $\sigma = 0.25$ year. For an upcoming article, a computer magazine selects a MindReader owner at random.

(a) Determine $P(4.01 \leq X \leq 4.26)$ and interpret.

(b) Determine $P(3.76 \leq X \leq 4.01)$ and interpret.

45. The weight of players on the 2000 Ohio State University spring training football roster had a mean of $\mu = 234.7$ pounds with a standard deviation of $\sigma = 48.8$ pounds (*Source:* www.ohiostatebuckeyes.fansonly.com). A nutritionist selects a player from the spring training football roster at random.

(a) Determine $P(X \leq 260)$ and interpret.

(b) Determine $P(200 \leq X \leq 260)$ and interpret.

46. From January 1, 1996, to August 19, 1996, the weekly mean retail price of a gallon of gasoline was \$1.37 with a standard deviation of \$0.07 (*Source:* U.S. Department of Energy, www.eia.doe.gov). For an upcoming story in a motor magazine, a week during this time period is chosen at random.

(a) Determine $P(X \leq 1.40)$ and interpret.

(b) Determine $P(1.30 \leq X \leq 1.40)$ and interpret.

47. The average monthly temperature (in degrees Fahrenheit) in northwestern Pennsylvania during the month of March from 1895 to 1921 had a mean of 33.7 degrees with a standard deviation of 4.8 degrees (*Source:* National Weather Service, www.nws.noaa.gov). In a study on global warming, a climatologist selects a March monthly temperature from this time period at random.

(a) Determine the probability that the average monthly temperature was below 30 degrees.

(b) Determine the probability that the average monthly temperature was between 30 and 40 degrees.

48. In 1998 the mean school district enrollment in the Cuyahoga County, Ohio, school system was 6395.3 students in each school district with a standard deviation of 560.1 students (*Source:* www.nsrs.com). For a new testing procedure, the state superintendent selects a school district at random.

(a) Determine the probability that the school district has fewer than 6000 students.

(b) Determine the probability that the school district has between 6000 and 7500 students.

49. From 1943 to 1974, the U.S. minimum wage had a mean of \$1.00 per hour with a standard deviation of \$0.46 (*Source:* U.S. Department of Labor, www.dol.gov). An economics professor selects a minimum-wage earner from this time period at random.

(a) Determine the probability that the selected minimum-wage earner made between \$1.10 per hour and \$1.50 per hour.

(b) Determine the probability that the selected minimum-wage earner made between \$0.90 per hour and \$1.50 per hour.

50. In 1998 the number of acres per state of nonfederal developed land in the continental United States had a mean of 1.84 million acres with a standard deviation of 1.44 millions acres (*Source:* U.S. Census Bureau, www.census.gov).

Suppose that in 1998 a state in the continental United States is chosen at random by a political science student for a report.

(a) Determine $P(1.5 \leq X \leq 1.7)$ and interpret.

(b) Determine $P(1.3 \leq X \leq 1.7)$ and interpret.

51. In 1995 the mean age of a person receiving outpatient care in the United States was 45.73 years with a standard deviation of 18.68 years (*Source:* National Center for Health Statistics, www.cdc.gov/nchs). An HMO selects an outpatient from 1995 at random.

(a) Determine the probability that the patient is at least 40 years old.

(b) Determine the probability that the patient is at most 50 years old.

52. In 1995 the mean age of a person receiving emergency room care in the United States was 42.92 years with a standard deviation of 19.57 years (*Source:* National Center for Health Statistics, www.cdc.gov/nchs). An HMO selects an emergency room patient from 1995 at random.

(a) Determine the probability that the patient is at least 38 years old.

(b) Determine the probability that the patient is at most 45 years old.

53. In 1996 the mean age of a person getting the flu in the United States was 27.43 years with a standard deviation of 20.98 years (*Source:* National Center for Health Statistics, www.cdc.gov/nchs). An HMO selects a person who had the flu in 1996 at random.

(a) Determine the probability that the patient was between 16 and 19 years old.

(b) Determine the probability that the patient was between 20 and 40 years old.

54. Between 1981 and 1998 the age of a person with AIDS had a mean of 37.26 years with a standard deviation of 10.24 years (*Source:* National Center for Health Statistics, www.cdc.gov/nchs). For a national news report, a person with AIDS during this time period is chosen at random.

(a) Determine the probability that the age of the person was between 30 and 35 years.

(b) Determine the probability that the age of the person was between 38 and 48 years.

55. In 1995 the age of a person in a nursing home in the United States had a mean of 81.77 years with a standard deviation of 7.25 years (*Source:* National Center for Health Statistics, www.cdc.gov/nchs). A psychology professor selects a nursing home resident from 1995 at random.

(a) Determine the probability that the resident was older than 85.

(b) Determine the probability that the resident was older than 72.

SECTION PROJECT

The distribution of weights of American females aged 40 to 49 from 1988 to 1994 is shown in the table.

Weight (in pounds)	Percent	Weight (in pounds)	Percent
100–109	1.6	230–239	1.7
110–119	3.9	240–249	2.4
120–129	7.6	250–259	0.3
130–139	12.5	260–269	0.5
140–149	12.3	270–279	0.3
150–159	12.0	280–289	0.3
160–169	9.9	290–299	0.3
170–179	10.2	300–309	0.2
180–189	8.6	310–319	0.1
190–199	4.0	320–329	0.1
200–209	4.8	330–339	0.1
210–219	3.1	340–349	0.1
220–229	2.7	350 and over	0.4

SOURCE: U.S. Census Bureau, www.census.gov

(a) According to the table, what percent of American females aged 40 to 49 weighed anywhere between 140 and 170 pounds?

(b) Use the MEANSD program in Appendix C to estimate the mean and the standard deviation of the distribution.

(c) Using μ and σ found in part (b), determine $P(140 \leq X \leq 170)$ and interpret.

(d) How do you account for the difference in the answers in part (a) and in part (c)?

Why We Learned It

In Chapter 8 we saw that random variables are powerful tools that can be used in various disciplines. We saw many situations in which random variables, probability distributions, expected value, and standard deviation were applied. Discrete random variables and some probability distributions associated with them included the uniform distribution and the binomial distribution. We saw that the binomial distribution had many varied applications. For example, a social worker who desires to do a case study on the educational attainment of parents of prisoners can use the binomial distribution to predict the probability that, say, 5 parents in a group of 20 have a high school diploma. The social worker could also compute the expected value (mean) to determine how many are expected to have a high school diploma. Finally, the social worker could compute the standard deviation of the distribution to measure the dispersion. Continuous random variables gave rise to the normal distribution and the standard normal distribution. A quality control engineer in a machine shop could use the standard normal distribution to determine the probability that the diameter of wire produced is between 0.12 and 0.16 inch given that the mean is 0.12 inch and the standard deviation is 0.02 inch.

CHAPTER REVIEW EXERCISES

In Exercises 1–3, determine if the table meets the two criteria of a probability distribution.

1.

$X = x$	$P(X = x)$
$X = 0$	$P(X = 0) = -\frac{1}{2}$
$X = 1$	$P(X = 1) = \frac{3}{4}$
$X = 2$	$P(X = 2) = \frac{3}{4}$

2.

$X = x$	$P(X = x)$
$X = -3$	$P(X = -3) = 0.125$
$X = -2$	$P(X = -2) = 0.138$
$X = -1$	$P(X = -1) = 0.225$
$X = 0$	$P(X = 0) = 0.118$
$X = 1$	$P(X = 1) = 0.206$
$X = 2$	$P(X = 2) = 0.134$
$X = 3$	$P(X = 3) = 0.054$

3.

$X = x$	$P(X = x)$
$X = 5$	$P(X = 5) = \frac{1}{10}$
$X = 10$	$P(X = 10) = \frac{3}{5}$
$X = 15$	$P(X = 15) = \frac{1}{8}$
$X = 20$	$P(X = 20) = \frac{1}{9}$

In Exercises 4–6, a formula for $P(X = x)$ is given. Complete the following.

(a) Write a table of values for $P(X = x)$.

(b) Determine if the table meets the two criteria of a probability distribution. Remember that $P(X = x) = 0$, where the x-values are not defined.

4. $P(X = x) = \frac{1}{7}x, \quad x = 1, 2, 3, 4$

5. $P(X = x) = \frac{1}{30}x^2, \quad x = 1, 2, 3, 4$

6. $P(X = x) = \frac{x+4}{34}, \quad x = 3, 4, 5, 6$

In Exercises 7 and 8, determine the value of k so that $P(X = x)$ is a probability mass function.

7. $P(X = x) = \frac{4}{k}x, \quad x = 2, 3, 4, 5$

8. $P(X = x) = \frac{2}{k}x^2, \quad x = 1, 2, 3, 4$

In Exercises 9 and 10, a table of frequencies is given. Construct the corresponding probability distribution with $X = x$ in the left column and $P(X = x)$ in the right column.

9.

Value	Frequency
1	10
2	12
3	18
4	14

10.

Value	Frequency
0	220
5	300
10	150
15	170

11. Salem High School has determined that the number of students per class can be represented by a discrete random variable X with the following probability distribution:

$X = x$	$P(X = x)$
$X = 25$	$P(X = 25) = 0.10$
$X = 26$	$P(X = 26) = 0.20$
$X = 27$	$P(X = 27) = 0.35$
$X = 28$	$P(X = 28) = 0.10$
$X = 29$	$P(X = 29) = 0.20$
$X = 30$	$P(X = 30) = 0.05$

(a) Construct a probability histogram of the distribution.

(b) Determine $P(26 \leq X \leq 28)$ and interpret.

12. Sopris Ski Club has determined that the number of snowboarders that they have on any given night during the ski season can be presented by a discrete random variable X having the following probability distribution:

$X = x$	$P(X = x)$
8	$P(X = 8) = 0.06$
9	$P(X = 9) = 0.28$
10	$P(X = 10) = 0.36$
11	$P(X = 11) = 0.20$
12	$P(X = 12) = 0.10$

(a) Construct a probability histogram of the distribution.

(b) Determine $P(X \geq 10)$ and interpret.

13. Hillsbourough Health Club determines that the number of new memberships during any given month can be represented by a discrete random variable X having the following probability distribution:

$X = x$	$P(X = x)$
$X = 4$	$P(X = 4) = 0.22$
$X = 5$	$P(X = 5) = 0.15$
$X = 6$	$P(X = 6) = 0.26$
$X = 7$	$P(X = 7) = 0.28$
$X = 8$	$P(X = 8) = 0.09$

(a) Determine the probability that there will be at least 6 new memberships during any given month.

(b) According to the distribution, what is the least likely number of new memberships during any given month?

14. The enrollment at Saco River Kayaking School for last June is given in the following frequency table.

Course	Enrollment
Beginning	25
Intermediate	12
Advanced	8
Expert	5

(a) Let X be a discrete random variable, where X represents the level of the course in which a person is enrolled, meaning $x = 1, 2, 3, 4$ for beginning, intermediate, advanced, and expert, respectively. Construct a probability distribution for the data.

(b) Compute $P(X = 4)$ and interpret.

15. The number of unprovoked worldwide shark attacks for the years 1997 to 2000 is shown in the table.

Year	Number of Unprovoked Attacks
1997	56
1998	51
1999	58
2000	79

SOURCE: Florida Museum of Natural History, www.flmnh.ufl.edu

(a) Let X be a discrete random variable, where X represents a year between 1997 and 2000. Construct a probability distribution for the data.

(b) Compute $P(X = 2000)$ and interpret.

(c) Compute $P(X \geq 1999)$ and interpret.

In Exercises 16–18, a table of values for the probability distribution $P(X = x)$ is given. Determine the expected value $E(X)$ of the random variable X having the given discrete probability distribution.

16.

$X = x$	$P(X = x)$
$X = 0$	$P(X = 0) = \frac{1}{8}$
$X = 1$	$P(X = 1) = \frac{1}{4}$
$X = 2$	$P(X = 2) = \frac{1}{4}$
$X = 3$	$P(X = 3) = \frac{3}{8}$

17.

$X = x$	$P(X = x)$
$X = -1$	$P(X = -1) = 0.227$
$X = 0$	$P(X = 0) = 0.314$
$X = 1$	$P(X = 1) = 0.186$
$X = 2$	$P(X = 2) = 0.273$

18.

$X = x$	$P(X = x)$
$X = 5$	$P(X = 5) = 0.2$
$X = 10$	$P(X = 10) = 0.1$
$X = 15$	$P(X = 15) = 0.3$
$X = 20$	$P(X = 20) = 0.1$
$X = 25$	$P(X = 25) = 0.3$

In Exercises 19–21, a probability mass function $P(X = x)$ is given. Determine the expected value.

19. $P(X = x) = \dfrac{1}{6}, \quad x = 1, 2, 3, 4, 5, 6$

20. $P(X = x) = \frac{1}{55}x^2, \quad x = 1, 2, 3, 4, 5$

21. $P(X = x) = \frac{x - 2}{6}, \quad x = 3, 4, 5$

In Exercises 22–24, determine the standard deviation, σ_X, for the following:

22. For the discrete probability distribution in Exercise 16

23. For the discrete probability distribution in Exercise 17

24. For the probability mass function in Exercise 20

25. Salem High School has determined that the number of students per class can be represented by a discrete random variable X with the following probability distribution:

$X = x$	$P(X = x)$
$X = 25$	$P(X = 25) = 0.10$
$X = 26$	$P(X = 26) = 0.20$
$X = 27$	$P(X = 27) = 0.35$
$X = 28$	$P(X = 28) = 0.10$
$X = 29$	$P(X = 29) = 0.20$
$X = 30$	$P(X = 30) = 0.05$

Determine the expected value $E(X)$ and interpret.

26. Sopris Ski Club has determined that the number of snowboarders that they have on any given night during the ski season can be presented by a discrete random variable X having the following probability distribution:

$X = x$	$P(X = x)$
8	$P(X = 8) = 0.06$
9	$P(X = 9) = 0.28$
10	$P(X = 10) = 0.36$
11	$P(X = 11) = 0.20$
12	$P(X = 12) = 0.10$

(a) Determine the expected value and interpret.

(b) What is the most frequently occurring number of snowboarders on any given night? That is, what x-value yields the greatest $P(X)$-value?

27. Hillsbourough Health Club determines that the number of new memberships during any given month can be represented by a discrete random variable X having the following probability distribution:

$X = x$	$P(X = x)$
$X = 4$	$P(X = 4) = 0.22$
$X = 5$	$P(X = 5) = 0.15$
$X = 6$	$P(X = 6) = 0.26$
$X = 7$	$P(X = 7) = 0.28$
$X = 8$	$P(X = 8) = 0.09$

(a) According to the distribution, how many new memberships can the health club expect during a month?

(b) Determine the standard deviation of the distribution.

28. Chenush Gas is a natural gas drilling company. After testing and analysis by geologists, Chenush Gas is considering drilling on two different sites. The geologists have estimated that Site I will net \$75 million if successful (probability 0.2) and lose \$7 million if not successful (probability 0.8). Site II is estimated to net \$60 million if successful (probability 0.3) and lose \$5 million if not successful (probability 0.7).

(a) Determine the expected return for each site.

(b) Based on part (a), which site would you advise the company to select?

(c) Which site has less risk? (This corresponds to which has a smaller variance.)

29. The number of times that KTTV radio gives the weather each hour is uniformly distributed between 1 and 4, inclusive. Let X be a discrete uniform random variable representing the number of times that the weather is given each hour.

(a) Write a probability mass function for X.

(b) Determine μ_X and interpret.

(c) Compute σ_X for the distribution.

(d) Determine the probability that the weather will be given 3 or more times in a given hour.

In Exercises 30–32, evaluate the binomial probability mass function $P(X = x) = {}_nC_x \cdot p^x(1 - p)^{n-x}$ for the given values of n, p, and x.

30. $n = 4, p = \frac{1}{4}, x = 2$

31. $n = 3, p = 0.2, x = 0$

32. $n = 10, p = 0.40, x = 4$

In Exercises 33 and 34, determine if the assumptions of a binomial experiment are reasonably satisfied. If so, identify the values for p and n.

33. In a survey, 100 people are asked which brand of toothpaste they use.

34. Eighty-five percent of the students at Safeway Driving School pass their written driver's examination. A sample of 15 students is asked if they passed the test.

In Exercises 35 and 36, a binomial probability mass function is given. Complete the following:

(a) Make a table of values for $P(X = x)$.

(b) Construct a probability histogram for the distribution in part (a).

35. $P(X = x) = {}_3C_x \left(\frac{1}{8}\right)^x \left(1 - \frac{1}{8}\right)^{3-x}, \quad x = 0, 1, 2, 3$

36. $P(X = x) = {}_5C_x(0.4)^x(1 - 0.4)^{5-x}, \quad x = 0, 1, 2, 3, 4, 5$

In Exercises 37 and 38, match the probability with the appropriately shaded probability histogram.

(a)

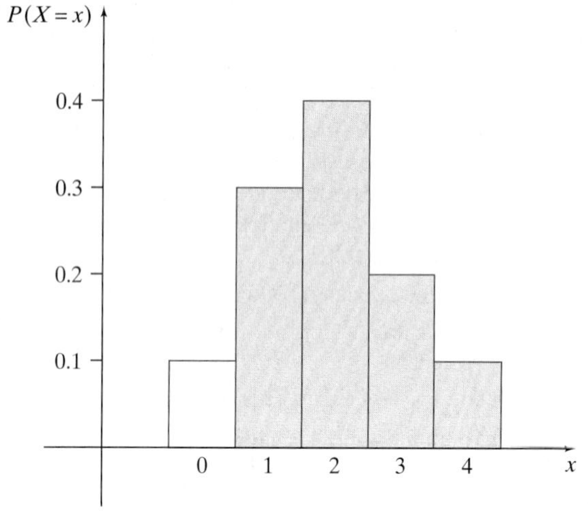

(b)

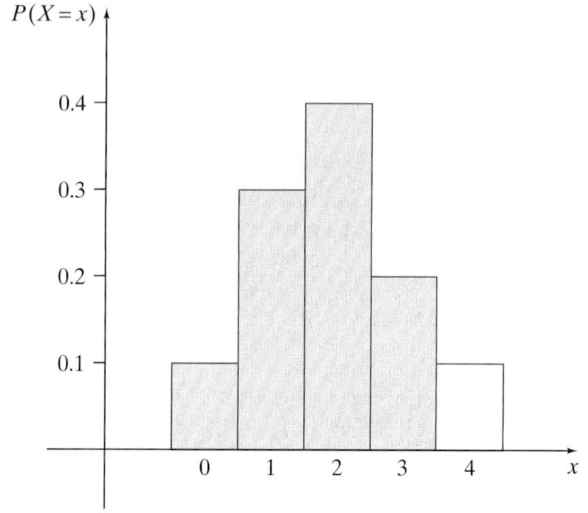

37. $P(X > 0)$ **38.** $P(X \leq 3)$

In Exercises 39 and 40, assume that X represents a binomial random variable. Determine the mean $E(X) = \mu_X$ and the standard deviation σ_X of the distribution.

39. $n = 14,\ p = \frac{1}{4}$

40. $P(X = x) = {}_7C_x \cdot \left(\frac{1}{5}\right)^x \left(1 - \frac{1}{5}\right)^{7-x}$,

$x = 0, 1, 2, 3, 4, 5, 6, 7$

In Exercises 41–45, assume that all applications can be represented by a binomial random variable.

41. A survey at Hawthorne Middle School reveals that 75% of the students pass their physical fitness test. A group of 8 students is selected at random.

(a) Verify that the assumptions of a binomial experiment are reasonably satisfied.

(b) Evaluate $P(X = 5)$ and interpret.

42. The Direct Satisfaction Company knows from experience that 20% of the people who receive a free trial issue of *Backcountry* magazine will sign up for a 1-year subscription. The marketing department selects a group of 14 people at random who received a free trial issue of *Backcounty*.

(a) Determine $P(X = 6)$ and interpret.

(b) Determine $P(6 \leq X \leq 7)$ and interpret.

43. A 1998 study showed that 49% of Americans dined out at least once within 12 months of the survey (*Source:* U.S. Statistical Abstract, www.census.gov/statab). A sample of 15 Americans is selected at random.

(a) Determine the probability that 8 people dined out at least once within 12 months of the survey.

(b) Determine the probability that at most 7 of them dined out at least once within 12 months of the survey.

44. In a 1998 report on the American teen-age public, 72% of American teen-agers said that they can always trust their parents to be there for them when needed (*Source:* www.publicagenda.org). A poll selects 10 American teen-agers at random.

(a) Construct a probability histogram for the distribution.

(b) Compute μ_X and σ_X for the distribution and interpret each.

45. In 1998, 29% of the residents in Wisconsin lived in counties not meeting U.S. Environmental Protection Agency air quality standards (*Source:* National Center for Chronic Disease Prevention, www.cdc.gov). Eight residents of Wisconsin are selected at random.

(a) How many are expected to live in counties not meeting U.S. Environmental Protection Agency air quality standards?

(b) What is the probability that at most 3 of the residents live in counties not meeting U.S. Environmental Protection Agency air quality standards?

In Exercises 46–60, assume that each exercise can be represented by a normal random variable X.

46. The mean and standard deviation of the final exam scores for Kris's Biology and Physics classes are shown in the table.

	Biology	Physics
μ	78	73
σ	2.1	6.2

(a) Kris scored an 85 on the Biology exam and an 88 on the Physics exam. Compute Kris's z-scores for each exam.

(b) Graph the z-scores from part (a) on the standard normal curve. Relative to the rest of the students taking the Biology and Physics final exams, on which exam did Kris do better?

In Exercises 47–49, find the shaded area using Table 8.4.2 and the Standard Normal Table in Appendix D.

47.

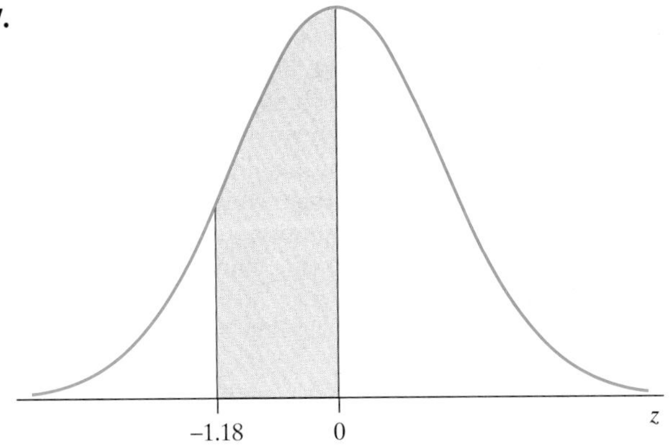

48.

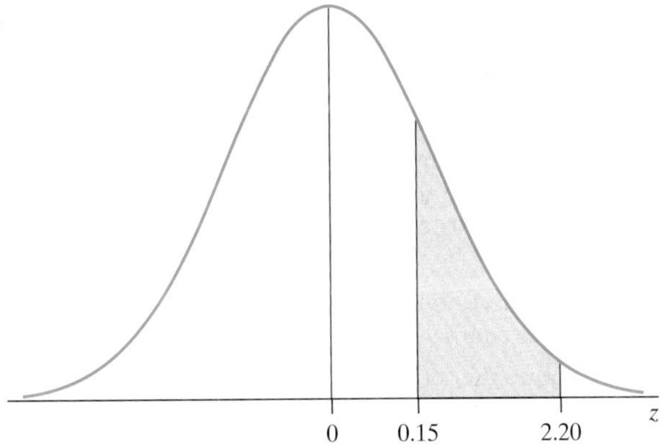

49.

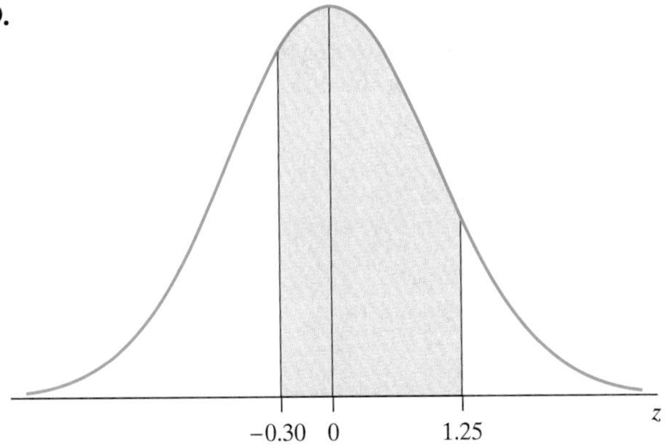

In Exercises 50–52, find the area between the given z-scores using Table 8.4.2 and the Standard Normal Table in Appendix D.

50. Between $z = -1.30$ and $z = 1.46$

51. Between $z = 0.27$ and $z = 2.18$

52. Between $z = -2.24$ and $z = -1.19$

In Exercises 53–55, determine the given probabilities using Table 8.4.2 and the Standard Normal Table in Appendix D.

53. $P(z > 1.28)$

54. $P(-0.45 \leq z \leq 0.48)$

55. $P(-1.85 < z < -0.35)$

56. The LandTrans Company runs shuttle buses from the ski resort to local hotels. Records show that the mean number of runs for each bus each day is $\mu = 20$ with a standard deviation of $\sigma = 1.4$. For an audit, the chief financial officer at LandTrans Company selects a shuttle bus at random.

(a) Determine $P(X > 17)$ and interpret.

(b) Determine $P(X < 17)$ and interpret.

57. Speedy Printing Company finds that, on any given day, a mean of $\mu = 10{,}000$ copies are made with a standard deviation of $\sigma = 800$ copies. The chief financial officer at Speedy Printing Company selects a business day at random.

(a) Determine $P(8000 \leq X \leq 12{,}000)$ and interpret.

(b) Determine $P(10{,}000 \leq X \leq 12{,}000)$ and interpret.

58. The weights of players at a soccer tournament had a mean of $\mu = 165.3$ pounds with a standard deviation of $\sigma = 17.2$ pounds. The athletic trainer selects a player from the roster at random.

(a) Determine the probability that the player weighed at least 150 pounds.

(b) Determine the probability that the player weighed between 150 and 170 pounds.

59. The mean school district enrollment for the 1999–2000 academic year in the Claremont school system was 3485.4 students in each school district with a standard deviation of 285.9 students. The superintendent selects a school district at random.

(a) Determine the probability that the school district has fewer than 3000 students.

(b) Determine the probability that the school district has between 3000 and 4000 students.

60. During the 1999–2000 ski season, the average age of a skier at a local ski resort was 29.7 years with a standard deviation of 14.3 years. The local newspaper selects a skier at random for an interview.

(a) Determine the probability that the skier is at least 18 years old.

(b) Determine the probability that the skier is between 18 and 35 years old.

CHAPTER 8 PROJECT

SOURCE: U.S. Census Bureau, www.census.gov

In this project we will see that we can use the normal distribution to approximate values for the binomial distribution.

According to the U.S. Census Bureau (www.census.gov), about 19% of Americans who are 65 years old and over are taking adult education courses. Suppose that a sample of 10 Americans 65 years old and over is selected at random.

1. Identify n and p.

2. Determine $P(X = 2)$ and interpret.

3. Determine the probability that either 1 or 2 of the 10 Americans in the sample are taking adult education courses.

Now suppose that group of 100 Americans that are 65 years old and over are selected at random. In this case, with $n = 100$, the calculations of the binomial distribution formula can be cumbersome. We can simplify the calculations however if we use the **normal approximation to the binomial distribution**. Statisticians use the following rule of thumb: The normal distribution can be used to approximate the binomial distribution if

$$\text{(i) } np \geq 5$$

$$\text{(ii) } n(1 - p) \geq 5$$

The values for the mean and the standard deviation are $\mu = np$ and $\sigma = \sqrt{npq} = \sqrt{np(1-p)}$, respectively.

4. Verify that the two conditions are satisfied for the values of n and p given in step 1.

5. Use the values of n and p to determine μ and σ for the normal distribution.

Since the binomial distribution is a discrete probability distribution and the normal distribution is a continuous probability distribution, we need to use a **continuity correction**. Let's say that we want to compute $P(X = 5)$ using the normal approximation for the binomial distribution. On a probability histogram, the rectangle over $X = 5$ would begin at 4.5 and end at 5.5. So when we apply the continuity correction to the binomial probability $P(X = 5)$, we get $P(4.5 \leq X \leq 5.5)$. We can then convert these values to z-scores and use the computation techniques from Section 8.4. We can use the following guideline to compute the continuity correction.

Binomial Probability	Normal Probability with Continuity Correction
$P(X = a)$	$P(a - 0.5 \leq X \leq a + 0.5)$
$P(a \leq X \leq b)$	$P(a - 0.5 \leq X \leq b + 0.5)$

6. Compute $P(19.5 \leq X \leq 20.5)$ using the mean and the standard deviation values found in step 5.

7. Now let's determine the probability that anywhere from 10 to 20 of those in the sample of 100 are taking adult education courses. Write a binomial probability to determine this value.

8. Apply the continuity correction to the probability that you wrote in step 7.

9. Compute the probability value that you found in step 8 using the mean and the standard deviation values found in step 5. Interpret.

10. Using statistical software, we can determine that the value of $P(10 \leq X \leq 20)$ using $n = 100$ and $p = 0.19$ is 0.6524. Compare this answer to the one you got in step 9. How do you account for the difference?

Chapter 9 Markov Chains

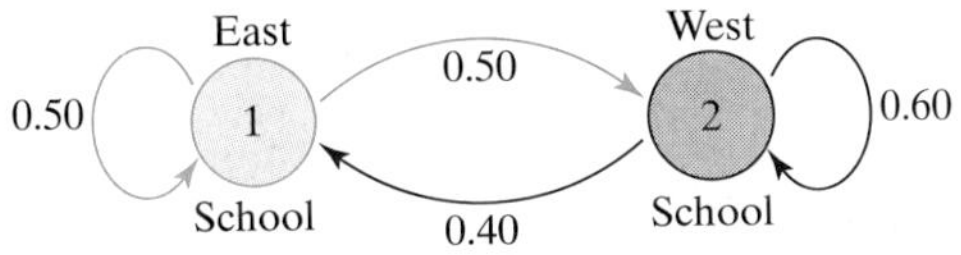

$$S_n = S_0 P^n$$

$$= [9000 \quad 13{,}000] \cdot \begin{bmatrix} \frac{6}{11} & \frac{5}{11} \\ \frac{6}{11} & \frac{5}{11} \end{bmatrix}$$

$$= [12{,}000 \quad 10{,}000]$$

One of the more controversial ideas in the debate over how to improve and fund public education is the use of school vouchers. To analyze the impact of vouchers, economists and public education officials use various models. One such model shows that enrollment changes resulting from school vouchers can be thought of as a system associated with a Markov chain. A state transition diagram is a visual representation of a Markov chain. A matrix product can reveal the long term enrollment pattern in a given school as a result of school vouchers.

What We Know

In Chapters 6, 7, and 8 we studied probability and its applications. This included conditional probability and Bayes' Theorem.

Where Do We Go

In this chapter we will extend the idea of conditional probability and use matrix operations to compute a sequence of experimental trials. This sequence of trials is called a Markov Chain.

Section 9.1 Introduction to Markov Chains

Many real-life applications involve modeling a process that changes from one category to the next, depending on given probabilities. For example, let's say that in a voting precinct in a certain city a voter can cast a vote for a Democrat, Republican, or Independent congressional candidate. These three outcomes associated with the party affiliation can be thought of as **states**, since they are the categories into which a voter can be identified. The voting precinct, relative to all congressional elections now, in the past, and in the future, can be thought of as a **system**. We call an individual voter in the precinct an **element** of the system. If the system changes from one state to another based on chance, then the sequence can be thought of as a **Markov chain**. This is part of a branch of the study of probability called **stochastic theory**, in which we study mathematical models whose experimental outcomes are based on outcomes of previous experiments. Markov chains are named after the Russian mathematician A. A. Markov who began the study of stochastic theory in the early twentieth century. In this section we will learn how to express Markov chains with diagrams and matrices.

From Your Toolbox

In Section 6.4 we saw that the event A does *not* occur is given by the **complement of A**, denoted by A^C. If $P(A) = a$, then $P(A^C) = 1 - a$.

Diagramming Markov Chains

We begin our study of Markov chains by developing ways to express states of systems visually. States of Markov chains are sometimes expressed as given events and complements of given events. To review how complements are expressed in a system, let's flashback to an example from Chapter 6. For a review of the complement of an event, consult the Toolbox to the left.

Flashback

Harris Poll Revisited

In Example 1 of Section 6.7, we saw that a sample of American adults was asked, "By the year 2020, do you think the quality of life in your community will be better or worse?" The percentages of results, separated by gender, are shown in Table 9.1.1.

Table 9.1.1

	Male	Female	Total
Worse	31%	30%	61%
Better	19%	20%	39%
Total	50%	50%	100%

SOURCE: www.harrisinteractive.com

Assume that we just use the results from the female respondents. Answer the following:

(a) Describe the system and the two states associated with the poll.

(b) If A represents the event that a female responded that the quality of life would be better, write the values of $P(A)$ and $P(A^C)$, and draw a tree diagram for the possible outcomes.

Flashback Solution

(a) The system is the opinions of all American females on the quality of life in their community and the two states are

State 1: Responded that the quality of life would be better

State 2: Responded that the quality of life would be worse

(b) To determine the probabilities, we first note that $P(A) = \frac{0.20}{0.50} = 0.40$. This means that $P(A^C) = 1 - 0.40 = 0.60$. The tree diagram for the female's opinion is shown in Figure 9.1.1.

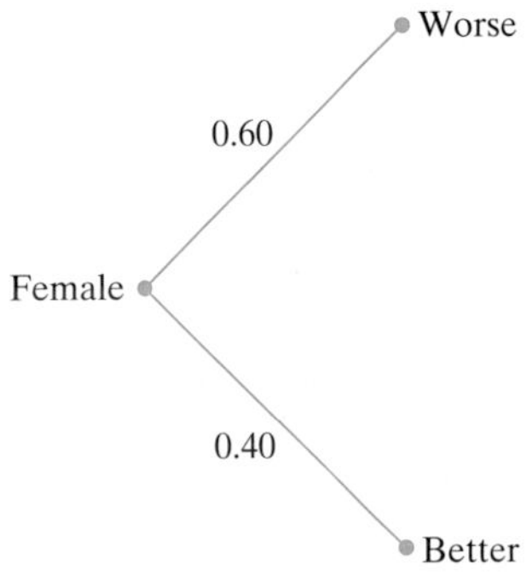

Figure 9.1.1

Note that we had a finite number of outcomes, that is, two, and that the poll was taken so that the respondents in the poll could not change their opinions once they had given their opinions. The Flashback exhibits many of the characteristics of a Markov chain.

Markov Chain

A sequence of trials associated with an experiment is a **Markov chain** if the outcome of an experiment meets the following two requirements:

1. The outcome is one of a finite set of defined states.
2. The outcome depends only on the present states and not any previous states.

A distinguishing component of a Markov chain is the **time step**. To see the impact of time steps on a Markov chain, let's examine another scenario. Suppose that a market analyst for Galaxy Communications is interested in whether those customers who sign a 1-year agreement for wireless telephone service will renew their contracts the next year. (Here we consider only those who renew with Galaxy or another company. Those who discontinue service altogether are not included.) A market analysis shows that if a wireless phone customer currently has a 1-year agreement with Galaxy Communications there is a 70% chance that he or she will renew this agreement for the next year. If a wireless phone customer does not have an agreement with Galaxy, then there is a 20% chance that he or she will switch to Galaxy when it comes time to renew the agreement. Here the two states are

State 1: User has a 1-year agreement with Galaxy Communications.

State 2: User has a 1-year agreement with another wireless phone provider.

The percentages given are called **transient probabilities** and correspond to the following conditional probabilities as we transcend from one state to another.

$$P(\text{state } 1 \mid \text{state } 1) = 0.70$$
$$P(\text{state } 1 \mid \text{state } 2) = 0.20$$

To denote this probability, we will use the letter p to represent the probability and use subscripts to identify the states. So we can write the conditional probabilities as

$$p_{1,1} = 0.70$$
$$p_{2,1} = 0.20$$

A graphical representation of these states with their probabilities is called a **state transition diagram**. We use labeled circles for the numbered states and arrows to represent the probabilities. The part of the diagram that represents $p_{1,1}$ is shown in Figure 9.1.2, which shows that there is a 70% chance that an element in state 1 will remain in state 1 when it comes time to renew the agreement next year. Observe that the time step for this system is 1 year. The number inside each circle corresponds to the number of each state. The notation $p_{1,2}$ represents the probability that a wireless user who currently has an agreement with Galaxy will switch to a different provider next year. Since we know that 70% of the current Galaxy customers will renew, this means that 30% of the customers will not. So we can write this complement as

$$p_{1,2} = 1 - p_{1,1} = 1 - 0.70 = 0.30$$

So the probability of going from state 1 to state 2 is 0.30. This is illustrated in Figure 9.1.3.

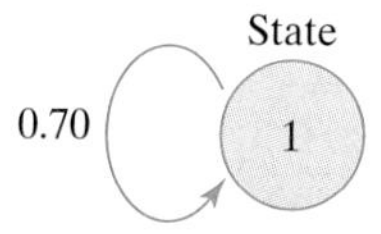

Figure 9.1.2

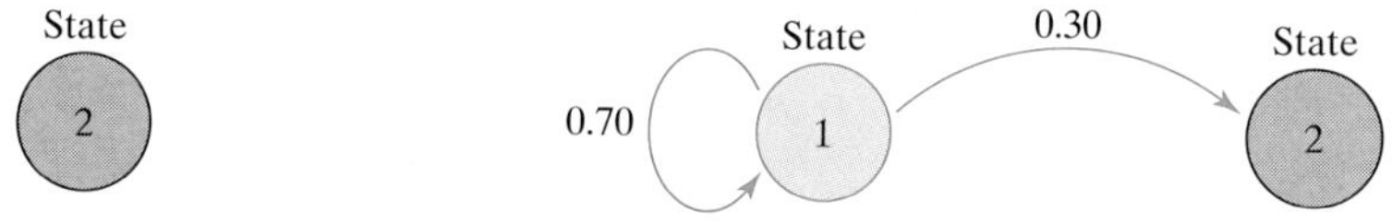

Figure 9.1.3

Market research indicated that $p_{2,1} = 0.20$, which means that the probability of going from state 2 to state 1 is 0.20. Figure 9.1.4 adds this probability to the state transition diagram. (Recall that this assumes that all the customers continue with wireless service.)

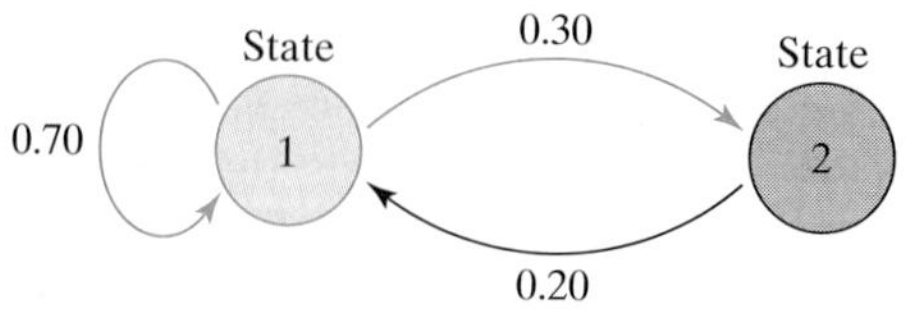

Figure 9.1.4

To complete the diagram, we see that the probability that a wireless customer who has an agreement with a provider other than Galaxy will also have an agreement with a provider other than Galaxy next year is $p_{2,2} = 1 - 0.20 = 0.80$. The complete state transition diagram for the system is shown in Figure 9.1.5.

Before we continue, let's list some properties of state transition diagrams.

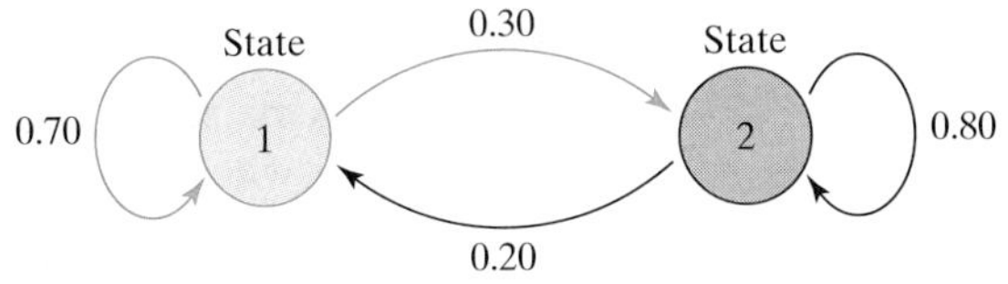

Figure 9.1.5

Properties of State Transition Diagrams

- There must be one circle to represent each state.
- Every state must have at least one outgoing arrow.
- The sum of the probabilities labeled on the arrows coming from a state must be 1.

Example 1 Drawing a State Transition Diagram

At Urbanix College, records show that 15% of all education majors changed their major between their first and second years. Of all the noneducation majors, 10% changed their major to education, while the rest did not.

(a) List the system and the states.

(b) Draw the state transition diagram for the system.

Solution

(a) The system is represented by all students who have attended, are attending, or will attend Urbanix College. The two states are

State 1: The student is an education major.

State 2: The student is not an education major.

(b) The state transition diagram for the system is shown in Figure 9.1.6. Notice that the corresponding probabilities for the system are $p_{1,1} = 0.85$, $p_{1,2} = 0.15$, $p_{2,1} = 0.10$, and $p_{2,2} = 0.90$.

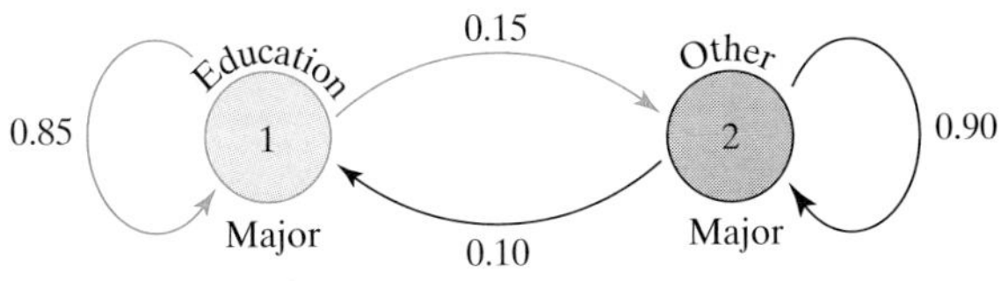

Figure 9.1.6

The subscripts that we are using are not accidental. Notice that they look just like those used when naming elements of a matrix. This means that we can write a matrix corresponding to the Urbanix College system as

$$P = \begin{bmatrix} 0.85 & 0.15 \\ 0.10 & 0.90 \end{bmatrix}$$

We call a matrix such as P a **transition matrix**.

Transition Matrix

The **transition matrix** P for a given system associated with a Markov chain with two states whose corresponding state transition diagram is

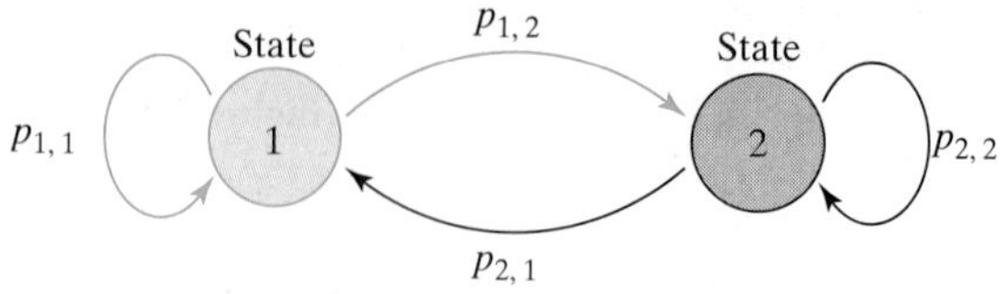

is given by

$$P = \begin{bmatrix} p_{1,1} & p_{1,2} \\ p_{2,1} & p_{2,2} \end{bmatrix} \quad \begin{matrix} \text{Arrows that come from state 1} \\ \text{Arrows that come from state 2} \end{matrix}$$

Example 2 Writing Transition Matrices

Write a transition matrix for each of the state transition diagrams in Figures 9.1.7 and 9.1.8.

(a)

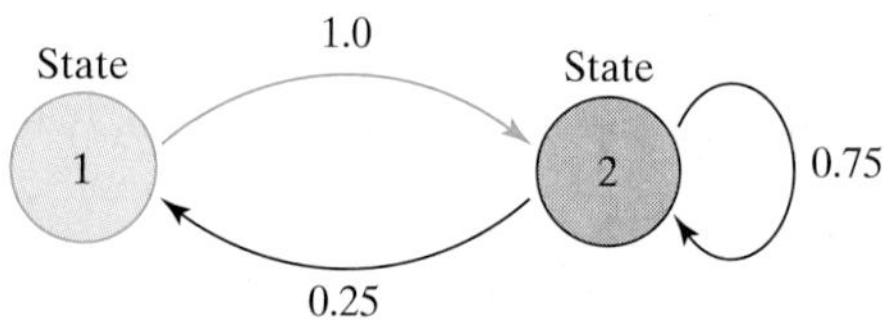

Figure 9.1.7

(b)

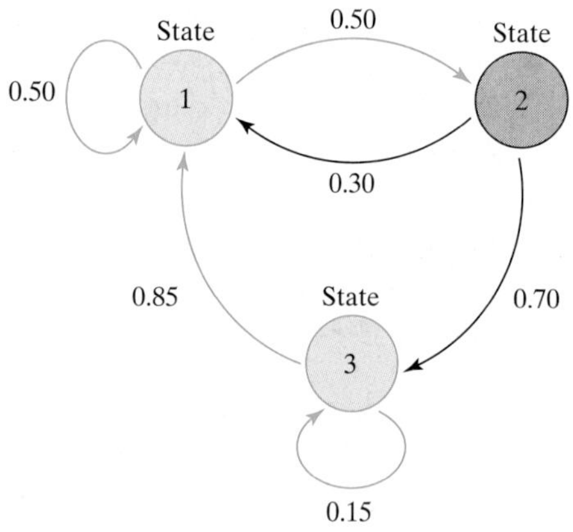

Figure 9.1.8

> **Interactive Activity**
>
> Draw the state transition diagram for the transition matrix
>
> $$P = \begin{bmatrix} 1 & 0 & 0 \\ 0 & 1 & 0 \\ 0 & 0 & 1 \end{bmatrix}$$
>
> What special type of matrix is P? In general, what will the state transition diagram for any square matrix of this form look like?
>
> WWW

Solution

(a) Since there is only one arrow originating from state 1, we have $p_{1,2} = 1$. This means that $p_{1,1} = 0$. Knowing that $p_{2,1} = 0.25$ and $p_{2,2} = 0.75$, we get the transition matrix

$$P = \begin{bmatrix} 0 & 1 \\ 0.25 & 0.75 \end{bmatrix}$$

(b) Even though this diagram has three states, the transition matrix is formed using the same method as in part (a). Here the transition matrix will have three rows and three columns. We have two arrows originating from state 1. These arrows indicate that $p_{1,1} = 0.50$, $p_{1,2} = 0.50$, and $p_{1,3} = 0$. From state 2 we get $p_{2,1} = 0.30$, $p_{2,2} = 0$, and $p_{2,3} = 0.70$. For state 3 we get the values $p_{3,1} = 0.85$, $p_{3,2} = 0$, and $p_{3,3} = 0.15$. This gives us the transition matrix

$$P = \begin{bmatrix} 0.50 & 0.50 & 0 \\ 0.30 & 0 & 0.70 \\ 0.85 & 0 & 0.15 \end{bmatrix}$$

■

Checkpoint 1 Now work Exercise 23.

Example 2 illustrates some properties of transition matrices. We can now generalize the results and list some of the properties of transition matrices for any number of states.

Properties of Transition Matrices

For a system of n states associated with a Markov chain, we have the following:

1. The size of the transition matrix is $n \times n$. That is, the transition matrix has the form

$$P = \begin{bmatrix} p_{1,1} & p_{1,2} & p_{1,3} & \cdots & p_{1,n} \\ p_{2,1} & p_{2,2} & p_{2,3} & \cdots & p_{2,n} \\ \vdots & \vdots & \vdots & \vdots & \vdots \\ p_{n,1} & p_{n,2} & p_{n,3} & \cdots & p_{n,n} \end{bmatrix}$$

2. The elements of a transition matrix are values between 0 and 1, inclusive. That is, $0 \le p_{i,j} \le 1$ for $i = 1, 2, \ldots, n$ and $j = 1, 2, \ldots, n$.
3. The sum of the elements in any row is 1.

Initial State Vectors

Once we get the transition matrix for a system, we can use known values associated with the system to make time step estimations. To see how, let's suppose that the Milson City School System has had a tax levy on the ballot for the last two elections, with the levy failing each time. Exit polls reveal the following facts:

- Sixteen percent of those who opposed the levy in the first election favored the levy in the second election, while the rest voted against the levy in both elections.
- On the other hand, 7% of those who favored the levy in the first election voted against it in the second election, while the rest favored the levy in both elections.

The system associated with this Markov chain has two states:

State 1: Opposed the levy

State 2: Favored the levy

The state transition diagram for the system is shown in Figure 9.1.9. This gives us the transition matrix $P = \begin{bmatrix} 0.84 & 0.16 \\ 0.07 & 0.93 \end{bmatrix}$. Now suppose that in the first election,

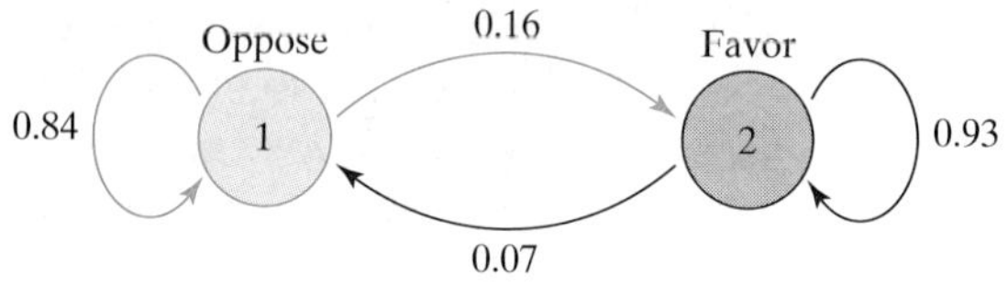

Figure 9.1.9

of the approximately 15,000 votes cast, 9000 were against the levy, while 6000 voted for the levy. To approximate how many voted against the levy in the second election, we can compute

$$\begin{array}{c}\text{number opposed in}\\ \text{second election}\end{array} = \begin{pmatrix}\text{number}\\ \text{opposed in}\\ \text{the first}\\ \text{election}\end{pmatrix}\begin{pmatrix}\text{percent who}\\ \text{continued to}\\ \text{oppose in the}\\ \text{second election}\end{pmatrix} + \begin{pmatrix}\text{number who}\\ \text{favored in}\\ \text{first}\\ \text{election}\end{pmatrix}\begin{pmatrix}\text{percent who}\\ \text{switched to being}\\ \text{opposed in}\\ \text{second election}\end{pmatrix}$$

$$= (9000)(0.84) + (6000)(0.07) = 7980$$

The number who favored the levy in the second election is given by

$$\begin{array}{c}\text{number favored in}\\ \text{second election}\end{array} = \begin{pmatrix}\text{number}\\ \text{opposed in}\\ \text{the first}\\ \text{election}\end{pmatrix}\begin{pmatrix}\text{percent who}\\ \text{switched to}\\ \text{favoring in}\\ \text{second election}\end{pmatrix} + \begin{pmatrix}\text{number who}\\ \text{favored in}\\ \text{first}\\ \text{election}\end{pmatrix}\begin{pmatrix}\text{percent who}\\ \text{continued to}\\ \text{favor in the}\\ \text{second election}\end{pmatrix}$$

$$= (9000)(0.16) + (6000)(0.93) = 7020$$

From Your Toolbox

If A is an $m \times n$ matrix and B is an $n \times p$ matrix, then the matrix product is an $m \times p$ matrix. To find the entry of the ith row and jth column of AB, multiply each entry in the ith row of A by the corresponding entry in the jth column of B and add the products.

The pattern used in the calculations is one that we first saw in Chapter 2. (Consult the Toolbox to the left.)

Notice that if we let $[9000 \quad 6000]$ be a 1×2 matrix and multiply by the 2×2 transition matrix, we get

$$[9000 \quad 6000]\begin{bmatrix}0.84 & 0.16\\ 0.07 & 0.93\end{bmatrix} = [7980 \quad 7020]$$

Here we call the matrix $[9000 \quad 6000]$ an **initial state vector**. A **vector** is a name given to any matrix that is made up of a single row or a single column. If a system associated with a Markov chain has n states, then the size of the initial state vector is $1 \times n$.

Initial State Vector

For a system associated with a Markov chain with n states, the **initial state vector** S_0 is a $1 \times n$ row vector whose elements are the system's values in its beginning state.

Notes: 1. The matrix product $S_0 P = S_1$ gives us the **first state vector**.
2. We can also use probability values for initial state vectors.

Using the definition for the initial state vector, the initial vector S_0 for the Milson City School System tax levy is $S_0 = [9000 \quad 6000]$. The first state vector is given by

$$\begin{aligned} S_1 &= S_0 P\\ &= [9000 \quad 6000] \cdot \begin{bmatrix}0.84 & 0.16\\ 0.07 & 0.93\end{bmatrix}\\ &= [7980 \quad 7020]\end{aligned}$$

Let's assume that the percentages of voters who will either change their minds or vote the same way remain the same in the upcoming election. To approximate how many will oppose or favor the tax levy the next time it is on the ballot, we can repeat the matrix multiplication using the first state vector to get

$$S_2 = S_1 P = [\,7980 \quad 7020\,]\begin{bmatrix} 0.84 & 0.16 \\ 0.07 & 0.93 \end{bmatrix} = [\,7194.6 \quad 7805.4\,]$$

This means that by the third time the levy is placed on the ballot we can expect the levy to pass with approximately 7195 opposed and approximately 7805 favoring.

Now let's see how probability values can be used in initial state vectors. For example, since 15,000 people voted in the first election and 9000 opposed the levy, we see that $\frac{9000}{15,000} = 60\%$ of the voters opposed the levy and $\frac{6000}{15,000} = 40\%$ favored the levy. This means that we can write $S_0 = [\,0.60 \quad 0.40\,]$ as our initial state vector and compute the first state vector as follows:

$$S_1 = S_0 P = [\,0.60 \quad 0.40\,]\begin{bmatrix} 0.84 & 0.16 \\ 0.07 & 0.93 \end{bmatrix} = [\,0.532 \quad 0.468\,]$$

We can now compute the second state vector, S_2. This is

$$S_2 = S_1 P = [\,0.532 \quad 0.468\,]\begin{bmatrix} 0.84 & 0.16 \\ 0.07 & 0.93 \end{bmatrix} \approx [\,0.480 \quad 0.520\,]$$

This tells use that by the third time the levy goes on the ballot we can expect about 48% to oppose the levy and about 52% to favor the levy. When we use probabilities as elements in our vectors, we call these **probability distribution vectors**.

Computing State Vectors

In general, we call S_k the ***k*th state vector**, where $S_k = S_{k-1}P$. But we do not have to compute all the lower state vectors first in order to get S_k. Using the properties of matrix multiplication, we have

$$\text{first state vector} = S_1 = S_0 P$$
$$\text{second state vector} = S_2 = S_1 P = (S_0 P)P = S_0 P^2$$
$$\text{third state vector} = S_3 = S_2 P = (S_0 P^2)P = S_0 P^3$$

This pattern illustrates that we can use powers of the transition matrix to determine the kth state vector.

*k*th State Vector

For a system associated with a Markov chain with a transition matrix P and initial state vector S_0, the ***k*th state vector** is given by

$$S_k = S_0 P^k$$

where $k = 0, 1, 2, \ldots$.

Example 3 Determining the *k*th State Vector

The zoning commission in Chesterfield identifies properties as being zoned in one of three categories: public, residential, or commercial. During the last 5 years, 85% of the land zoned as public has remained public, while 15% of the land has been rezoned as residential. Seventy percent of the land zoned as residential has remained residential, with 10% rezoned as public and 20% rezoned as commercial. The commission also found that 60% of the land zoned as commercial has remained commercial, while 25% has been rezoned as public and 15% rezoned as residential. Chesterfield initially had 2000 acres zoned public, 24,000 acres zoned residential, and 6000 acres zoned commercial.

(a) Identify the time step, draw a state transition diagram, and write a transition matrix for the zoned properties.

(b) Compute S_4 and interpret.

Solution

(a) The time step for the system is 5 years. The system has three states:

State 1: Land is zoned public.

State 2: Land is zoned residential.

State 3: Land is zoned commercial.

The state transition diagram for the zoned property is shown in Figure 9.1.10.

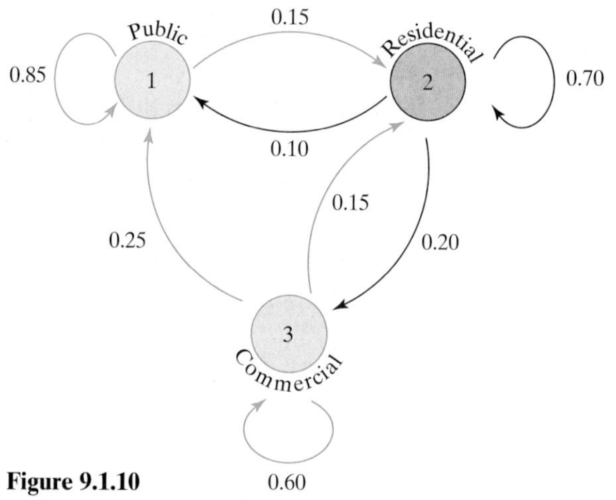

Figure 9.1.10

The diagram indicates that the 3×3 transition matrix for the system is

$$P = \begin{bmatrix} 0.85 & 0.15 & 0 \\ 0.10 & 0.70 & 0.20 \\ 0.25 & 0.15 & 0.60 \end{bmatrix}$$

(b) **Understand the Situation:** The initial state vector is

$$S_0 = [\,2000 \quad 24000 \quad 6000\,]$$

Recall that the time step for the system is 5 years, so S_4 represents the states in 20 years after the land was initially zoned. To determine S_4, we need to compute $S_4 = S_0 P^4$.

So we get

$$S_4 = S_0 P^4 = [\,2000 \quad 24{,}000 \quad 6000\,] \begin{bmatrix} 0.85 & 0.15 & 0 \\ 0.10 & 0.70 & 0.20 \\ 0.25 & 0.15 & 0.60 \end{bmatrix}^4$$

$$= [\,12{,}661.85 \quad 11{,}886.75 \quad 7451.40\,]$$

Interpret the Solution: This means that 20 years after initially zoning the land Chesterfield can expect to have about 12,662 acres zoned public, 11,887 acres zoned residential, and 7451 acres zoned commercial. ■

Technology Option

Figure 9.1.11

We can compute S_4 on our calculator by assigning S_0 to matrix A and P to matrix B and then computing $A * B\hat{}4$. See Figure 9.1.11.

To learn more on using your calculator to perform matrix operations, consult the on-line graphing calculator manual at www.prenhall.com/armstrong

Checkpoint 2 Now work Exercises 31 and 41.

SUMMARY

- A sequence of trials associated with an experiment is a **Markov chain** if the outcome of an experiment is one of a finite set of defined states and the outcomes depend only on the present states and not any previous states.
- A graphic depiction of a Markov system is a **state transition diagram**.
- A **transition matrix** shows the transitional probabilities from one state to another.
- The **initial state vector** S_0 is a $1 \times n$ row vector whose elements are the system's values in its beginning state.
- For a system associated with a Markov chain with a transition matrix P and initial state vector S_0, the ***k*th state vector** is given by $S_k = S_0 P^k$.

SECTION 9.1 EXERCISES

In Exercises 1–10, determine if the matrices are transition matrices. Supply an explanation for your answer.

1. $\begin{bmatrix} 0.1 & 0.9 \\ 0.3 & 0.7 \end{bmatrix}$

2. $\begin{bmatrix} 0.5 & 0.5 \\ 0.6 & 0.4 \end{bmatrix}$

3. $\begin{bmatrix} 0.4 & 0.6 \\ 0.8 & 0.3 \end{bmatrix}$

4. $\begin{bmatrix} 0.1 & 0.8 \\ 0.9 & 0.1 \end{bmatrix}$

5. $\begin{bmatrix} 0 & -0.9 & 0.1 \\ 0 & 1 & 0 \\ 0 & 0 & 1 \end{bmatrix}$

6. $\begin{bmatrix} 0.5 & -0.1 & 0.6 \\ 0.4 & 0.3 & 0.3 \\ 0.9 & 0 & 0.1 \end{bmatrix}$

7. $\begin{bmatrix} 0.3 & 0 & 0.7 \\ 0.1 & 0.5 & 0.4 \\ 0.7 & 0.3 & 0 \end{bmatrix}$

8. $\begin{bmatrix} 0.6 & 0.4 & 0 \\ 0.9 & 0.1 & 0 \\ 0 & 0 & 1 \end{bmatrix}$

9. $\begin{bmatrix} 0.35 & 0.60 & 0.05 \\ 0.20 & 0.35 & 0.45 \\ 0.50 & 0.15 & 0.35 \end{bmatrix}$

10. $\begin{bmatrix} 0.33 & 0.36 & 0.31 \\ 0.64 & 0.20 & 0.16 \\ 0.91 & 0.02 & 0.07 \end{bmatrix}$

In Exercises 11–20, write the transition matrix for the state transition diagrams.

11.

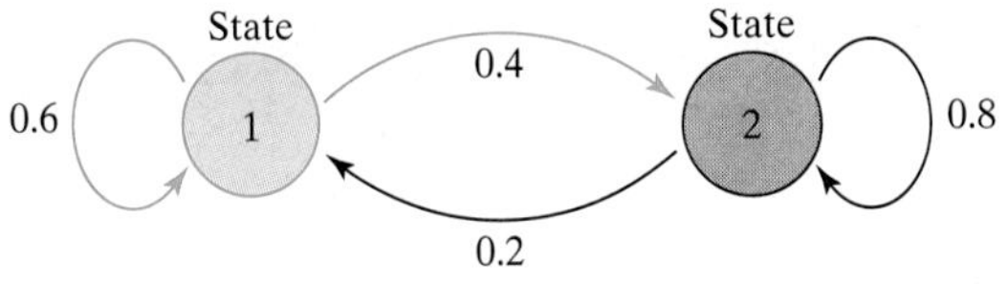

12.

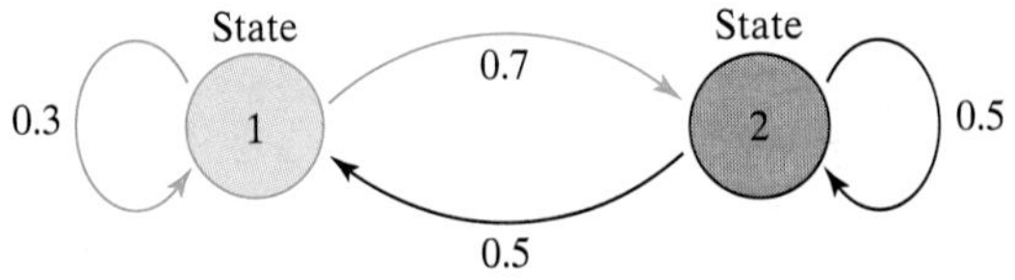

13.

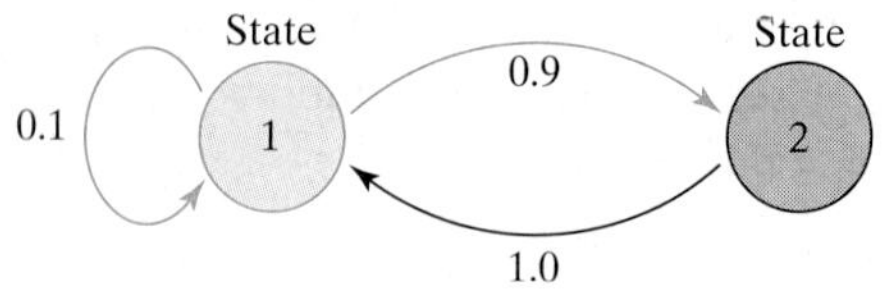

14.

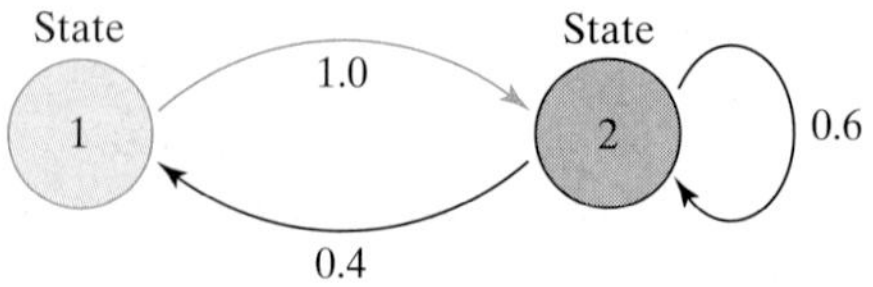

15.

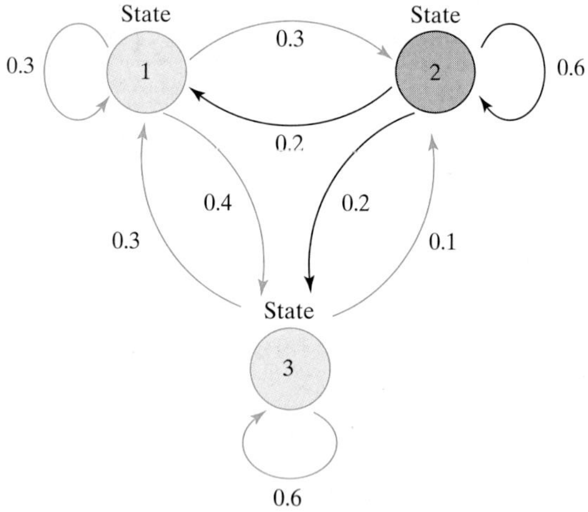

16.

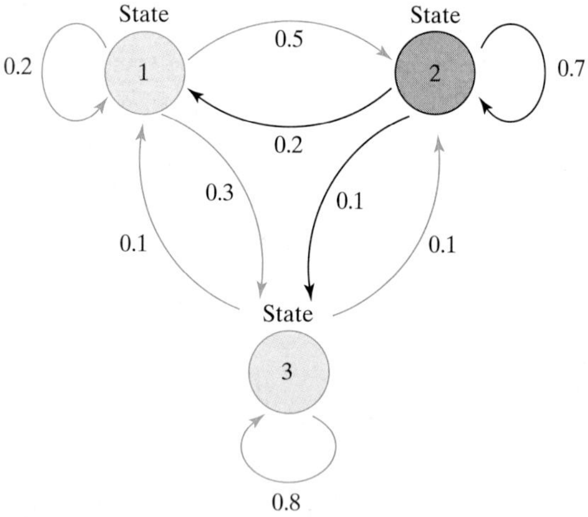

17.

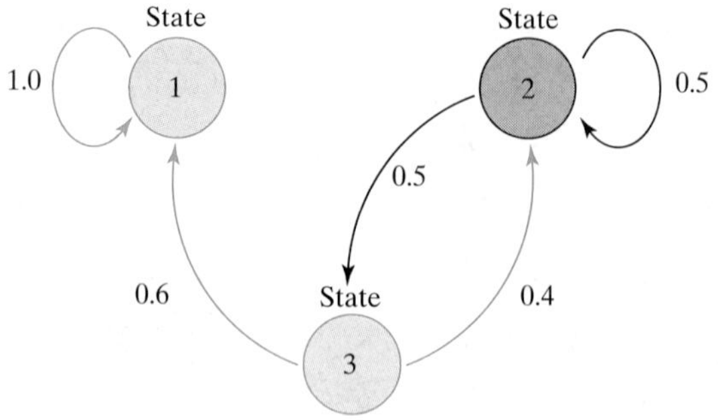

18.

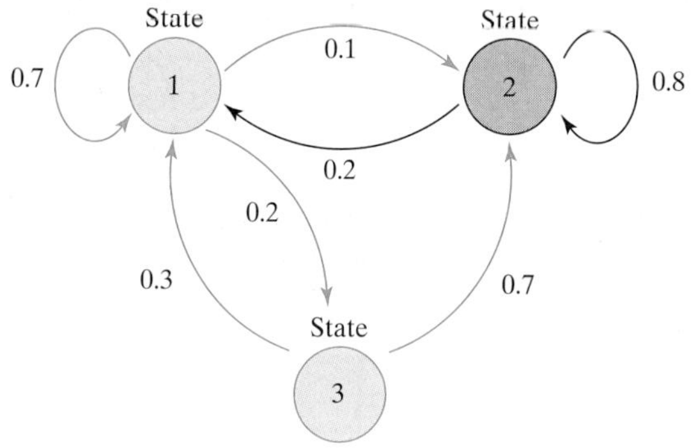

19.

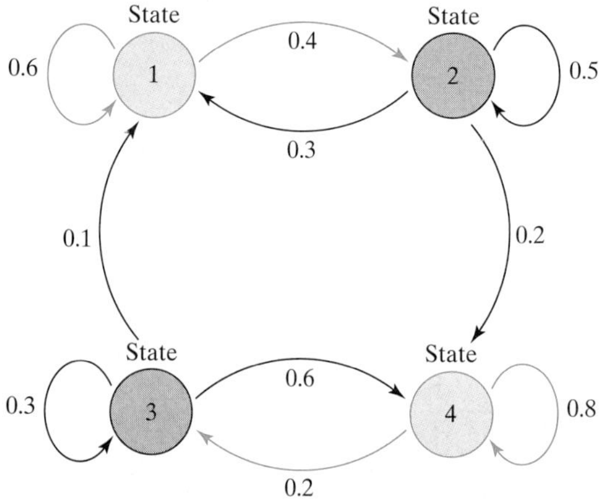

20.

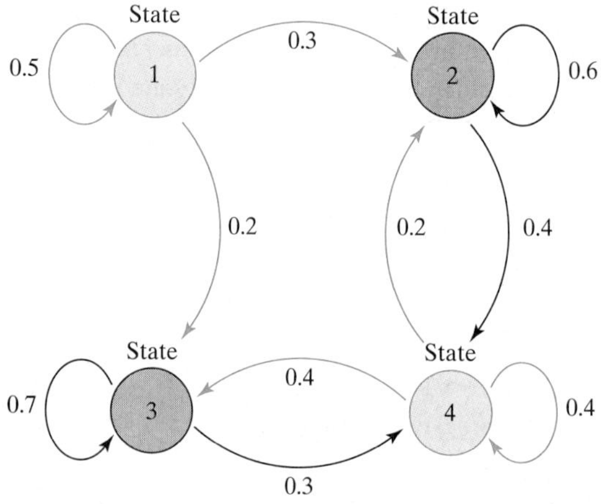

In Exercises 21–30, draw a state transition diagram for the given transition matrix.

21. $P = \begin{bmatrix} 0.3 & 0.7 \\ 0.1 & 0.9 \end{bmatrix}$

22. $P = \begin{bmatrix} 0.5 & 0.5 \\ 0.8 & 0.2 \end{bmatrix}$

23. $P = \begin{bmatrix} 0 & 1 \\ 0.4 & 0.6 \end{bmatrix}$

24. $P = \begin{bmatrix} 0.7 & 0.3 \\ 0.1 & 0.9 \end{bmatrix}$

25. $P = \begin{bmatrix} 0.2 & 0.3 & 0.5 \\ 0.8 & 0.1 & 0.1 \\ 0.5 & 0.1 & 0.4 \end{bmatrix}$

26. $P = \begin{bmatrix} 0.7 & 0.2 & 0.1 \\ 0.3 & 0.5 & 0.2 \\ 0.6 & 0.3 & 0.1 \end{bmatrix}$

27. $P = \begin{bmatrix} 0 & 0.5 & 0.5 \\ 0.63 & 0.30 & 0.07 \\ 0 & 0.6 & 0.4 \end{bmatrix}$

28. $P = \begin{bmatrix} 0.69 & 0.23 & 0.08 \\ 0.53 & 0 & 0.47 \\ 0.66 & 0 & 0.34 \end{bmatrix}$

29. $P = \begin{bmatrix} 0.22 & 0.15 & 0.60 & 0.03 \\ 0.48 & 0.11 & 0.03 & 0.38 \\ 0.06 & 0.37 & 0.31 & 0.26 \\ 0.52 & 0 & 0.31 & 0.17 \end{bmatrix}$

30. $P = \begin{bmatrix} 0.36 & 0.05 & 0.08 & 0.51 \\ 0 & 0.62 & 0.33 & 0.05 \\ 0.09 & 0.02 & 0.59 & 0.30 \\ 0.81 & 0 & 0.18 & 0.01 \end{bmatrix}$

For Exercises 31–40, complete the following:

(a) Identify the states.

(b) Draw the state transition diagram.

(c) Write the transition matrix P.

✓ **31.** A market survey on activity-related sunglasses shows that 30% of the owners of Blade Shades did not like the product and switched to Optifilters the next summer, while the rest continued to purchase Blade Shades. Conversely, 40% of the owners of Optifilters did not like their sunglasses and switched to Blade Shades the next summer, while the rest continued to purchase Optifilters.

32. Following a new advertising campaign, a market analysis shows that 20% of the consumers who do not currently drink SportSquirt will purchase it the next time that they buy a sport drink. Ninety percent of the consumers who currently drink SportSquirt will purchase it the next time that they buy a sport drink.

33. The *Suburban Almanac* presumes that the severity of hot temperatures of one summer can be determined by the severity of the previous summer. Records show that if the last year's summer was severely hot, then there is a 45% chance that the next summer will be severely hot. On the other hand, if the last year's summer was moderate, then there is a 40% chance that the next summer will be moderate.

34. The *Suburban Almanac* presumes that the chance of rain the previous day can indicate if it will rain the next day. Records show that if it rained the previous day there is a 40% chance that it will rain again the next day. If it was sunny the previous day, there is an 80% chance that it will be sunny the next day.

35. The town of Minsterton has an election for mayor every 2 years. Voting records show that if a Democrat is in office then there is a 60% chance that a Democrat will be elected in the next election and a 40% chance that a Republican will be elected. If a Republican is in office, there is a 65% chance that a Republican will be elected in the next election and a 35% chance that a Democrat will be elected.

36. A regional health study shows that, from one year to the next, 75% of the people in the region who smoked will continue to smoke, while 25% of the people who smoked will stop smoking the next year. Eight percent of the people who did not smoke will start smoking the next year, while 92% of those who did not smoke will continue to be nonsmokers.

37. The Move-It-Safe Company rents moving trucks in the cities of Adamstown, Breston, and Carpersville. They have both round-trip and point-to-point rentals. Records show that if a truck is rented in Adamstown, there is a 70% chance that it will be returned to Adamstown, a 15% chance that it will be returned to Breston, and a 15% chance that it will be returned to Carpersville. If a truck is rented in Breston, there is a 65% chance that it will be returned to Breston, a 10% chance that it will be returned to Adamstown, and a 25% chance that it will be returned to Carpersville. If a truck is rented in Carpersville, there is an 85% chance that it will be returned to Carpersville, a 5% chance that it will be returned to Adamstown, and a 10% chance that it will be returned to Breston.

38. In the LaDona metropolitan area, commuters either drive alone, carpool, or take public transportation. A study completed by the transportation department shows that 80% of those who drive alone intend to drive alone a year from now, while 15% intend to switch to carpooling and 5% intend to use public transportation. Ninety percent of those who carpool intend to continue carpooling the next year, while 5% intend to drive alone and 5% will begin to use public transportation. Sixty percent of those who use public transportation will continue to use it next year, while 30% intend to drive alone and 10% will begin to carpool.

39. In a particular county in the northern part of the country, voters can declare themselves as members of the Republican Party, Democrat Party or Green Party. A public opinion poll shows that 85% of current Republicans intend to continue with their party affiliation, while 13% intend to switch to the Democrat Party in the next election and 2% to the Green Party. Seventy percent of current Democrats intend to continue their party affiliation, while 10% intend to switch to the Republican Party in the next election and 20% to the Green Party. Ninety-five percent of current Green Party members plan to continue with their party affiliation, while 1% intend to switch to the Republican party in the next election and 4% to the Democrat Party.

40. The Armont Conservatory has been keeping records on their own Armont carnation and finds that the flowering color of the next generation of plant is related to the current generation. If a carnation is red, there is a 60%

chance that the next generation will be red, a 40% chance that the plant will be pink, and no chance of the plant being white. If a carnation is pink, there is a 60% chance that the next generation will be pink, a 30% chance that it will be red, and a 10% chance that it will be white. If a carnation is white, there is a 50% chance that the next generation will be white, a 50% chance that it will be pink, and no chance that it will be red.

✓ **41.** Refer to Exercise 31.

(a) Currently, Blade Shade holds 60% of the market share and Optifilters holds 40%. Write the probability distribution vector S_0 for the market share.

(b) Determine S_3 and interpret.

42. Refer to Exercise 32.

(a) Currently, about 35% of those who drink sports drinks purchase SportSquirt. Write the probability distribution vector S_0 for the market share.

(b) Determine S_3 and interpret.

43. Refer to Exercise 33.

(a) If the probability distribution vector is given to be $S_0 = [0.4 \quad 0.6]$, interpret the values of these two matrix elements.

(b) Determine S_2 and interpret.

44. Refer to Exercise 34.

(a) If the probability distribution vector is given to be $S_0 = [0.7 \quad 0.3]$, interpret the values of these two matrix elements.

(b) Determine S_2 and interpret.

45. Refer to Exercise 35.

(a) Interpret each coordinate of the probability distribution vector $S_0 = [0.4 \quad 0.6]$.

(b) Determine S_5 and interpret.

46. Refer to Exercise 36.

(a) Currently, there are about 60,000 smokers and 250,000 nonsmokers in the region. Write the corresponding initial state vector S_0.

(b) Determine S_4 and interpret.

47. Refer to Exercise 37.

(a) Currently, 800 Move-It-Safe Company moving trucks have been rented from Adamstown, 1200 from Breston, and 400 from Carpersville. Write the corresponding initial state vector S_0.

(b) Determine S_2 and interpret.

48. Refer to Exercise 38.

(a) Interpret each value of the probability distribution vector $S_0 = [0.8 \quad 0.1 \quad 0.1]$.

(b) Determine S_2 and interpret.

49. Refer to Exercise 39.

(a) Interpret each value of the probability distribution vector $S_0 = [0.62 \quad 0.37 \quad 0.01]$.

(b) Determine S_4 and interpret.

50. Refer to Exercise 40.

(a) Interpret each value of the probability distribution vector $S_0 = [0.3 \quad 0.6 \quad 0.1]$.

(b) Determine S_8 and interpret.

SECTION PROJECT

To explore the computing of powers of matrices, consider the transition matrix

$$P = \begin{bmatrix} \frac{1}{5} & \frac{4}{5} \\ \frac{3}{10} & \frac{7}{10} \end{bmatrix}$$

(a) Compute $10P$ using scalar multiplication. What type of elements does this matrix have?

(b) Compute $(10P)^5$, the fifth power of the scalar product $10P$.

(c) Multiply the result of part (b) by the scalar $\frac{1}{10^5}$, to get the expression $\frac{1}{10^5}(10P)^5$.

(d) Use your calculator to compute P^5. Compare the answer to the answer in part (c). What do you notice?

Section 9.2 Regular Markov Chains

In the previous section we learned that Markov chains can be used to analyze trends of data for a given increment of time called a time step. Now we will turn our attention to the long-range behavior of Markov chains and kth state matrices. We will also see what conditions are necessary to study these long-range behaviors.

Steady-State Transition Matrices

So far we have studied how we can use powers of transition matrices to determine the kth state vector S_k. But suppose that we want to determine the outcome of a Markov chain over a very long period of time. This means we would need to find a kth state vector for a large value of k. For example, consider the transition matrix $P = \begin{bmatrix} 0.2 & 0.8 \\ 0.7 & 0.3 \end{bmatrix}$. If we raise P to higher and higher powers, we get (after rounding to four decimal places)

$$P^2 = \begin{bmatrix} 0.6 & 0.4 \\ 0.35 & 0.65 \end{bmatrix}, \qquad P^4 = \begin{bmatrix} 0.5 & 0.5 \\ 0.4375 & 0.5625 \end{bmatrix}$$

$$P^{16} \approx \begin{bmatrix} 0.4667 & 0.5333 \\ 0.4667 & 0.5333 \end{bmatrix}, \qquad P^{32} \approx \begin{bmatrix} 0.4667 & 0.5333 \\ 0.4667 & 0.5333 \end{bmatrix}$$

Let's make two observations about these matrices:

- The two rows of the higher-power matrices have the same entries.
- As the power of P gets relatively large, we see that no change occurs in the elements of P^n.

For the transition matrix $P = \begin{bmatrix} 0.2 & 0.8 \\ 0.7 & 0.3 \end{bmatrix}$, we can show for large values of n that P^n is approximately $\begin{bmatrix} \frac{7}{15} & \frac{8}{15} \\ \frac{7}{15} & \frac{8}{15} \end{bmatrix}$. Then, for any initial probability vector S_0 and $S = \begin{bmatrix} \frac{7}{15} & \frac{8}{15} \end{bmatrix}$, it can be shown that

$$S_0 P^n \approx S \qquad \text{for large values of } n$$

It is also true that $SP = S$. We call S a **steady-state vector**. The steady-state vector is sometimes called a long-run distribution. The matrix P^n, for large values of n, can also be called a **limiting matrix**. (Observe that the steady-state vector, S, gives the rows of the limiting matrix P^n.) Now let's see how steady-state vectors can be used in applications.

Example 1 Using Steady-State Vectors in Applications

Snellburg has two schools on opposite ends of the city, East School and West School. The city school system adopts a voucher system that allows a student to choose the school that he or she attends every 4 years. After the first 4 years, it was found that half of the students who attend East School opted to stay at that school while the other half chose to go to West School. Forty percent of the students who attended West School opted to stay at West, while 60% used their vouchers to enroll at East School. When the voucher system began, 9000 students attended East School and 13,000 students attended West School.

(a) Draw a state transition diagram and write the transition matrix for the East and West School enrollments.

(b) Write the initial state matrix S_0. Given that, for large values of n, $P^n \approx \begin{bmatrix} \frac{6}{11} & \frac{5}{11} \\ \frac{6}{11} & \frac{5}{11} \end{bmatrix}$, compute $S_n = S_0 P^n$ and interpret.

Solution

(a) The system has two states:

State 1: Student is enrolled at East School.

State 2: Student is enrolled at West School.

The state transition diagram for the system is shown in Figure 9.2.1. This gives us the transition matrix $P = \begin{bmatrix} 0.5 & 0.5 \\ 0.6 & 0.4 \end{bmatrix}$.

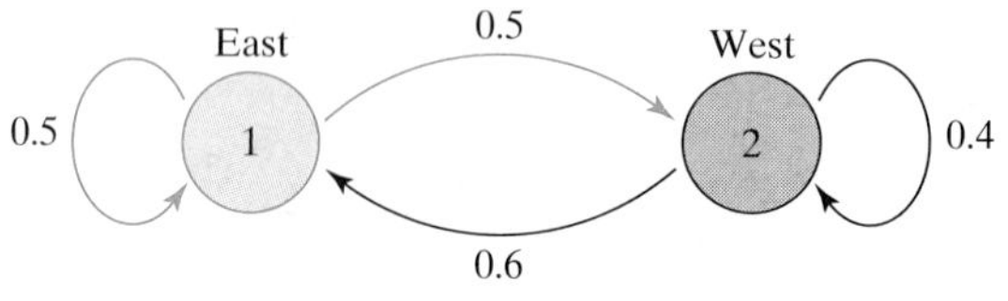

Figure 9.2.1

(b) The initial state matrix is $S_0 = [\,9000 \quad 13{,}000\,]$. Multiplying by $P^n \approx \begin{bmatrix} \frac{6}{11} & \frac{5}{11} \\ \frac{6}{11} & \frac{5}{11} \end{bmatrix}$ gives us

$$S_n = S_0 P^n \approx [\,9000 \quad 13{,}000\,] \begin{bmatrix} \frac{6}{11} & \frac{5}{11} \\ \frac{6}{11} & \frac{5}{11} \end{bmatrix} = [\,12{,}000 \quad 10{,}000\,]$$

Interpret the Solution: Assuming that the enrollment trends continue, this means that, enrollment at East School will increase to 12,000, while enrollment at West School will decrease to 10,000, in the long run.

Technology Option

We can use the calculator to help us to determine steady-state vectors. To determine the steady-state vector in Example 1, we store the transition matrix $P = \begin{bmatrix} 0.5 & 0.5 \\ 0.6 & 0.4 \end{bmatrix}$ in our calculator as matrix A. Figure 9.2.2 shows the output of computing P^{16}.

Using the FRAC command gives the result shown in Figure 9.2.3.

Figure 9.2.2

Figure 9.2.3

For more information on performing matrix operations on your calculator, consult the on-line graphing calculator manual at www.prenhall.com/armstrong

Regular Transition Matrices

For a transition matrix to have a steady-state vector, it is sufficient that it be what is called a **regular transition matrix**.

Regular Transition Matrix

Let P be a transition matrix corresponding to a Markov chain. P is a **regular transition matrix** if some power of P has only positive elements.

Note: Every regular transition matrix has a steady-state vector.

Let's examine some techniques that can be used to determine if a transition matrix is a regular transition matrix.

Example 2 Determining Regular Matrices

Determine whether the following transition matrices are regular transition matrices.

(a) $P = \begin{bmatrix} 0.6 & 0.4 \\ 0.1 & 0.9 \end{bmatrix}$ **(b)** $P = \begin{bmatrix} 0.5 & 0 & 0.5 \\ 0 & 1 & 0 \\ 0 & 0 & 1 \end{bmatrix}$

Solution

(a) Notice that all the elements of P are positive. When we multiply to get higher powers of P, we will always be multiplying and adding positive numbers, which will give us a positive result. This means that P is a regular transition matrix.

(b) Let's compute some powers of P to see if a pattern emerges.

$$P^2 = P \cdot P = \begin{bmatrix} 0.5 & 0 & 0.5 \\ 0 & 1 & 0 \\ 0 & 0 & 1 \end{bmatrix}\begin{bmatrix} 0.5 & 0 & 0.5 \\ 0 & 1 & 0 \\ 0 & 0 & 1 \end{bmatrix} = \begin{bmatrix} 0.25 & 0 & 0.75 \\ 0 & 1 & 0 \\ 0 & 0 & 1 \end{bmatrix}$$

$$P^3 = P^2 \cdot P = \begin{bmatrix} 0.25 & 0 & 0.75 \\ 0 & 1 & 0 \\ 0 & 0 & 1 \end{bmatrix}\begin{bmatrix} 0.5 & 0 & 0.5 \\ 0 & 1 & 0 \\ 0 & 0 & 1 \end{bmatrix} = \begin{bmatrix} 0.125 & 0 & 0.875 \\ 0 & 1 & 0 \\ 0 & 0 & 1 \end{bmatrix}$$

$$P^4 = P^3 \cdot P = \begin{bmatrix} 0.125 & 0 & 0.875 \\ 0 & 1 & 0 \\ 0 & 0 & 1 \end{bmatrix}\begin{bmatrix} 0.5 & 0 & 0.5 \\ 0 & 1 & 0 \\ 0 & 0 & 1 \end{bmatrix} = \begin{bmatrix} 0.0625 & 0 & 0.9375 \\ 0 & 1 & 0 \\ 0 & 0 & 1 \end{bmatrix}$$

Notice that for each power of P the zero elements seem to always be in the same location. It appears that all higher powers of P will also have zero elements in the same location. This means that P is not a regular transition matrix. ■

✓ Checkpoint 1 Now work Exercise 7.

Example 2 has illustrated some ways to inspect transition matrices in order to determine whether they are regular. Before analyzing a more direct way to determine steady-state matrices, let's first examine some properties of regular transition matrices.

Properties of Regular Transition Matrices

Let P be a regular transition matrix. Then the following properties apply.

1. There is a unique vector S such that $S_0 P^n \approx S$, for large values of n.
2. The rows of the limiting matrix P^n are the same, for large values of n. The elements of S are the row elements of P^n, for large values of n.
3. If a transition matrix has all positive elements, then the matrix is a regular transition matrix.
4. The steady-state vector, S, corresponding to a regular transition matrix can be determined by solving a system of equations.

To find the steady-state vector for a given regular transition matrix, we can use the equation $SP = S$ and write the corresponding system of equations. We will also use the condition that the sum of the elements in a row of a transition matrix is 1. This allows us to solve the resulting system of equations using the substitution method shown in Chapter 1 or Gauss–Jordan Elimination shown in Chapter 2. The solution to this system gives us the **steady-state vector**. The elements of this vector are the row elements of our limiting matrix P^n, for large values of n.

Example 3 Computing the Steady-State Vector and Limiting Matrix

Solve a system of equations to determine the steady-state vector for the transition matrix

$$P = \begin{bmatrix} 0.1 & 0.9 \\ 0.7 & 0.3 \end{bmatrix}$$

Solution

Note that, since P has only positive elements, P is a regular transition matrix. To determine the steady-state vector, we let $S = [x \quad y]$ and solve the matrix equation

$$[x \quad y]\begin{bmatrix} 0.1 & 0.9 \\ 0.7 & 0.3 \end{bmatrix} = [x \quad y]$$

If we perform the matrix multiplication, we get a system of two equations and two unknowns.

$$0.1x + 0.7y = x$$
$$0.9x + 0.3y = y$$

If we write the equations in standard form, we get

$$-0.9x + 0.7y = 0$$
$$0.9x - 0.7y = 0$$

Since we know that all the row entries must equal 1, we get a third equation:

$$-0.9x + 0.7y = 0$$
$$0.9x - 0.7y = 0$$
$$x + \quad y = 1$$

From the third equation, we know that $x = 1 - y$. Substituting this expression into the second equation gives us

$$\begin{aligned} 0.9(1-y) - 0.7y &= 0 \\ 0.9 - 0.9y - 0.7y &= 0 \\ 0.9 - 1.6y &= 0 \\ 0.9 &= 1.6y \\ \frac{9}{16} &= y \end{aligned}$$

Substituting this value into the third equation gives

$$\begin{aligned} x + \frac{9}{16} &= 1 \\ x = 1 - \frac{9}{16} &= \frac{7}{16} \end{aligned}$$

Interactive Activity

Verify the solution to Example 3 by computing P^{16}, P^{32}, and P^{64} on your calculator and using the FRAC key.

So we get the steady-state vector $[x \quad y] = [\frac{7}{16} \quad \frac{9}{16}]$. Since the rows of a limiting matrix are made up of the elements of the steady-state vector $[x \quad y]$, we have the limiting matrix, P^n, that corresponds to the transition matrix $P = \begin{bmatrix} 0.1 & 0.9 \\ 0.7 & 0.3 \end{bmatrix}$ is

$$P^n \approx \begin{bmatrix} \frac{7}{16} & \frac{9}{16} \\ \frac{7}{16} & \frac{9}{16} \end{bmatrix} \quad \text{for large } n$$ ■

Applications of Steady-State Vectors

Two key pieces of information result from computing the limiting matrix P^n. First, the steady-state vector gives us the probability that an element will stay in a particular state in the long run. In Example 1 the limiting matrix was $P^n \approx \begin{bmatrix} \frac{6}{11} & \frac{5}{11} \\ \frac{6}{11} & \frac{5}{11} \end{bmatrix}$. This tells us that the steady-state vector for the system is $[x \quad y] = [\frac{6}{11} \quad \frac{5}{11}]$. This means that, in the long run, about $\frac{6}{11} \approx 55\%$ of the students will enroll in East School, while about $\frac{5}{11} \approx 45\%$ will attend West School.

The other piece of information is gathered from multiplying the initial state vector S_0 by the limiting matrix P^n to get $S_0 P^n \approx S_n$. This gives us the number of elements that will stay in a particular state in the long run. This is what we computed in part (b) of Example 1. Now let's try some of the techniques that we have learned in an application.

Example 4 Applying the Steady-State Vector

In Suffox County deregulation of gas and electricity utilities has had many customers reevaluating whether they will use electricity, natural gas, or heating oil as their primary heating energy source. Many are switching from one source to another based on their particular needs. A consumer group has surveyed the households of the county and has found that, during the past year:

- 70% of those who use electricity remained using electricity, while 20% have switched to natural gas and 10% to heating oil.
- 80% of those who use natural gas continued using it, while 10% have switched to electricity and 10% to heating oil.
- 60% of those who use heating oil continued to use heating oil, while 30% have switched to natural gas and 10% to electricity.

(a) Draw a state transition diagram and write the transition matrix P for the system.

(b) Let $S = [x \quad y \quad z]$ be a steady-state vector. Solve the matrix equation $SP = S$ to determine the values of the steady-state vector and interpret the results.

(c) Prior to deregulation, about 35,000 households used electricity, 45,000 households used natural gas, and about 15,000 used heating oil. About how many households will be using each of these energy sources in the long run?

Solution

(a) Here we have a system with three states:

State 1: Household uses electricity.
State 2: Household uses natural gas.
State 3: Household uses heating oil.

The state transition diagram for the system is shown in Figure 9.2.4. This gives us the transition matrix $P = \begin{bmatrix} 0.7 & 0.2 & 0.1 \\ 0.1 & 0.8 & 0.1 \\ 0.1 & 0.3 & 0.6 \end{bmatrix}$.

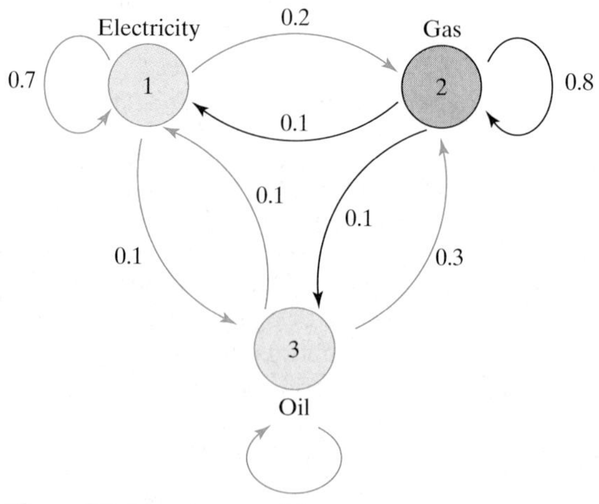

Figure 9.2.4

(b) To determine the steady-state vector, we need to get the values of $[x \quad y \quad z]$ by solving the matrix equation

$$[x \quad y \quad z] \begin{bmatrix} 0.7 & 0.2 & 0.1 \\ 0.1 & 0.8 & 0.1 \\ 0.1 & 0.3 & 0.6 \end{bmatrix} = [x \quad y \quad z]$$

Using matrix multiplication, we get the system of equations

$$\begin{aligned} 0.7x + 0.1y + 0.1z &= x \\ 0.2x + 0.8y + 0.3z &= y \\ 0.1x + .01y + 0.6z &= z \end{aligned}$$

Knowing that the sum of the rows of the steady-state vector must equal 1 and writing the equations in standard form give us

$$\begin{aligned} -0.3x + 0.1y \ + 0.1z &= 0 \\ 0.2x - 0.2y \ + 0.3z &= 0 \\ 0.1x + 0.01y - 0.4z &= 0 \\ x + \quad y + \quad z &= 1 \end{aligned}$$

Using the techniques of Chapter 2, we need to row reduce the augmented matrix

$$\left[\begin{array}{ccc|c} -0.3 & 0.1 & 0.1 & 0 \\ 0.2 & -0.2 & 0.3 & 0 \\ 0.1 & 0.1 & -0.4 & 0 \\ 1 & 1 & 1 & 1 \end{array}\right]$$

Interactive Activity

Use the Gauss–Jordan method on

$$\left[\begin{array}{ccc|c} -0.3 & 0.1 & 0.1 & 0 \\ 0.2 & -0.2 & 0.3 & 0 \\ 0.1 & 0.1 & -0.4 & 0 \\ 1 & 1 & 1 & 1 \end{array}\right]$$

to verify that the row reduced augmented matrix is

$$\left[\begin{array}{ccc|c} 1 & 0 & 0 & \frac{1}{4} \\ 0 & 1 & 0 & \frac{11}{20} \\ 0 & 0 & 1 & \frac{1}{5} \\ 0 & 0 & 0 & 0 \end{array}\right]$$

WWW

We leave the details on using Gauss–Jordan to you in the Interactive Activity to the left. We urge you to refer to Chapter 2 to review the Gauss–Jordan method. The row reduced augmented matrix is

$$\left[\begin{array}{ccc|c} 1 & 0 & 0 & \frac{1}{4} \\ 0 & 1 & 0 & \frac{11}{20} \\ 0 & 0 & 1 & \frac{1}{5} \\ 0 & 0 & 0 & 0 \end{array}\right]$$

This gives us the steady-state vector values $[x \quad y \quad z] = [\frac{1}{4} \quad \frac{11}{20} \quad \frac{1}{5}]$.

Interpret the Solution: This means that in the long run about $\frac{1}{4} = 25\%$ of the households will use electricity as their heating energy source, about $\frac{11}{20} = 55\%$ will use natural gas, and about $\frac{1}{5} = 20\%$ will use heating oil.

(c) From part (b), we know that the limiting matrix is

$$P^n \approx \begin{bmatrix} 0.25 & 0.55 & 0.20 \\ 0.25 & 0.55 & 0.20 \\ 0.25 & 0.55 & 0.20 \end{bmatrix} \qquad \text{for large values of } n$$

From the given information, the initial state vector is $S_0 = [35{,}000 \quad 45{,}000 \quad 15{,}000]$. Multiplying the initial state vector S_0 by the limiting matrix P^n, we get

$$\begin{aligned} S_n = S_0 P^n &= [35{,}000 \quad 45{,}000 \quad 15{,}000] \begin{bmatrix} 0.25 & 0.55 & 0.20 \\ 0.25 & 0.55 & 0.20 \\ 0.25 & 0.55 & 0.20 \end{bmatrix} \\ &= [23{,}750 \quad 52{,}250 \quad 19{,}000] \end{aligned}$$

Interpret the Solution: This means that, in the long run, we would expect about 23,750 to use electricity as their heating energy source, while about 52,250 will use natural gas and about 19,000 will use heating oil. ■

Technology Option

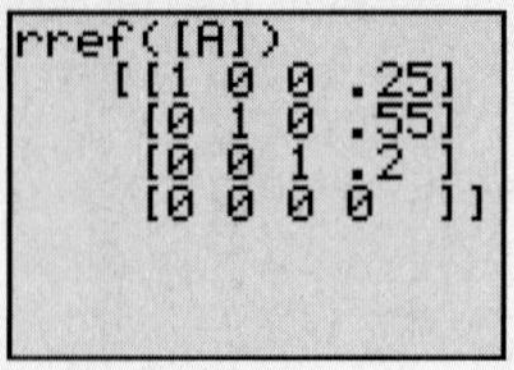

Figure 9.2.5

The rref(command on the calculator takes a matrix that has been entered in the matrix editor and transforms it into reduced row echelon form. For the matrix

$$\left[\begin{array}{ccc|c} -0.3 & 0.1 & 0.1 & 0 \\ 0.2 & -0.2 & 0.3 & 0 \\ 0.1 & 0.1 & -0.4 & 0 \\ 1 & 1 & 1 & 1 \end{array}\right]$$

shown in part (b) of Example 4, we get the row reduced echelon form shown in Figure 9.2.5.

For more information about the rref(command, consult the on-line graphing calculator manual at www.prenhall.com/armstrong

✓ Checkpoint 2 Now work Exercise 39.

SUMMARY

In this section, we saw that we could determine high-state Markov chains using the steady-state vector. We saw that we can use these special vectors to determine long-run trends.

- P is a **regular transition matrix** if some power of P has only positive elements.
- Solving $SP = S$ for S gives us the **steady-state vector**.
- If P is a regular transition matrix, then $S_0 P^n \approx S$ for large values of n.

SECTION 9.2 EXERCISES

In Exercises 1–10, compute P^4, P^{16}, P^{32}, and P^{64} on a calculator to determine if the given transition matrix P is a regular transition matrix. If the matrix is regular, approximate the limiting matrix P^n accurate to four decimal places.

1. $P = \begin{bmatrix} 0.2 & 0.8 \\ 0.9 & 0.1 \end{bmatrix}$ **2.** $P = \begin{bmatrix} 0.7 & 0.3 \\ 0.5 & 0.5 \end{bmatrix}$

3. $P = \begin{bmatrix} \frac{1}{3} & \frac{2}{3} \\ \frac{4}{5} & \frac{1}{5} \end{bmatrix}$ **4.** $P = \begin{bmatrix} \frac{1}{2} & \frac{1}{2} \\ \frac{1}{8} & \frac{7}{8} \end{bmatrix}$

5. $P = \begin{bmatrix} 1 & 0 \\ 0.65 & 0.35 \end{bmatrix}$ **6.** $P = \begin{bmatrix} 0.55 & 0.45 \\ 0 & 1 \end{bmatrix}$

✓ **7.** $P = \begin{bmatrix} 0.1 & 0.1 & 0.8 \\ 0.2 & 0.2 & 0.6 \\ 0.1 & 0.2 & 0.7 \end{bmatrix}$

8. $P = \begin{bmatrix} 0.5 & 0.2 & 0.3 \\ 0.1 & 0.4 & 0.5 \\ 0.2 & 0.7 & 0.1 \end{bmatrix}$

9. $P = \begin{bmatrix} 0 & 1 & 0 \\ 0.4 & 0.2 & 0.4 \\ 1 & 0 & 0 \end{bmatrix}$

10. $P = \begin{bmatrix} 0.3 & 0.5 & 0.2 \\ 1 & 0 & 0 \\ 0.5 & 0.4 & 0.1 \end{bmatrix}$

In Exercises 11–18, solve the matrix equation $SP = S$ for S to get the steady-state vector. Solve a system of equations using either the substitution or elimination method to determine the values for the steady-state vector $[x \quad y]$. Also, use the steady-state vector to determine the limiting matrix P^n.

11. $P = \begin{bmatrix} 0.2 & 0.8 \\ 0.1 & 0.9 \end{bmatrix}$ **12.** $P = \begin{bmatrix} 0.5 & 0.5 \\ 0.6 & 0.4 \end{bmatrix}$

13. $P = \begin{bmatrix} 0.1 & 0.9 \\ 0.7 & 0.3 \end{bmatrix}$ **14.** $P = \begin{bmatrix} 0.7 & 0.3 \\ 0.6 & 0.4 \end{bmatrix}$

15. $P = \begin{bmatrix} 0.8 & 0.2 \\ 1 & 0 \end{bmatrix}$ **16.** $P = \begin{bmatrix} 0 & 1 \\ 0.5 & 0.5 \end{bmatrix}$

17. $P = \begin{bmatrix} 0.25 & 0.75 \\ 0.85 & 0.15 \end{bmatrix}$ **18.** $P = \begin{bmatrix} 0.55 & 0.45 \\ 0.65 & 0.35 \end{bmatrix}$

In Exercises 19–26, solve the matrix equation $SP = S$ for S to get the steady-state vector. Use the Gauss–Jordan method *to determine the values for the steady-state vector* $[x \;\; y \;\; z]$. *Also, use the steady-state vector to determine the limiting matrix* P^n.

19. $P = \begin{bmatrix} 0 & 1 & 0 \\ 0.3 & 0.1 & 0.6 \\ 0.5 & 0.5 & 0 \end{bmatrix}$ **20.** $P = \begin{bmatrix} 0 & 1 & 0 \\ 0 & 0 & 1 \\ 0.5 & 0 & 0.5 \end{bmatrix}$

21. $P = \begin{bmatrix} 0 & \frac{1}{2} & \frac{1}{2} \\ \frac{1}{2} & \frac{1}{2} & 0 \\ 0 & \frac{1}{4} & \frac{3}{4} \end{bmatrix}$ **22.** $P = \begin{bmatrix} \frac{1}{4} & 0 & \frac{3}{4} \\ \frac{1}{2} & \frac{1}{4} & \frac{1}{4} \\ 0 & \frac{1}{4} & \frac{3}{4} \end{bmatrix}$

23. $P = \begin{bmatrix} \frac{1}{2} & \frac{1}{2} & 0 \\ \frac{1}{2} & \frac{1}{4} & \frac{1}{4} \\ \frac{1}{8} & \frac{3}{8} & \frac{1}{2} \end{bmatrix}$ **24.** $P = \begin{bmatrix} \frac{1}{10} & \frac{1}{10} & \frac{4}{5} \\ \frac{1}{2} & 0 & \frac{1}{2} \\ \frac{1}{2} & 0 & \frac{1}{2} \end{bmatrix}$

25. $P = \begin{bmatrix} 0.1 & 0.2 & 0.7 \\ 0.1 & 0.3 & 0.6 \\ 0.1 & 0.4 & 0.5 \end{bmatrix}$ **26.** $P = \begin{bmatrix} 0.09 & 0.05 & 0.86 \\ 0.28 & 0.16 & 0.56 \\ 0.43 & 0.45 & 0.12 \end{bmatrix}$

Applications

In Exercises 27–40, you can use your calculator to approximate the limiting matrix P^n *by computing* P^{64} *or solve the matrix equation* $SP = S$ *for* S *using the methods in Exercises 11–26.*

27. The Osborn–Wilcox Company manufactures paint pigments and tests the pigments that are made in a quality control lab. Records show that if a sample passed quality test then there is an 80% chance that the next sample will pass and a 20% chance that the next sample will fail. If the sample failed the test the last time it was taken, there is a 30% chance that the next sample will fail and a 70% chance that it will pass.

(a) Label the two states and write the transition matrix.

(b) Determine the limiting matrix P^n, for large values of n.

(c) Use the solution from part (b) to determine the probability that, in the long run, a test will pass.

28. Histronics Incorporated manufactures gauges for uses in aeronautics and calibrates each altimeter before it is shipped to see if the altimeter is within specifications. The calibration department's data show that if an altimeter is not in proper specification then there is a 60% chance that the next altimeter will not be within specification and a 40% chance that the next altimeter will be within specification. If an altimeter is within specification, there is a 75% chance that the next altimeter will be within specification and a 25% chance that the next altimeter will not be within specification.

(a) Label the two states and write the transition matrix.

(b) Determine the limiting matrix P^n, for large values of n.

(c) Use the solution from part (b) to determine the probability that, in the long run, an altimeter will be within specification.

29. The Rock Steady Insurance Company classifies their insured motorists as low risk or high risk based on whether they have had an accident in the last 3 years. Their research shows that 40% of the motorists who are classified as low risk remain classified as low risk 3 years later and 60% are reclassified as high risk. Fifty percent of the motorists who are classified as high risk remain classified as high risk 3 years later and the other 50% are reclassified as low risk.

(a) Draw a state transition diagram and write the transition matrix for the motorist's insurance classifications.

(b) Determine the limiting matrix P^n, for large values of n.

(c) The company currently insures 12,000 low-risk motorists and 8000 high-risk motorists. In the long run, about how many of the motorists will be classified as low risk and how many will be classified as high risk?

30. Metro Area Temps is an employment agency for temporary employees and classifies their temporary employees as either dependable or questionable based on whether they were cited as being late to work or absent without notice from work in their last job. They find that 90% of the employees classified as dependable will be classified the same way in their next job and 10% will be reclassified as questionable. Eighty-five percent of those who are classified as questionable will be classified the same way in their next job, while 15% will be reclassified as dependable.

(a) Draw a state transition diagram and write the transition matrix for the temporary employee classifications.

(b) Determine the limiting matrix P^n, for large values of n.

(c) The company currently has 900 employees classified as dependable and 400 classified as questionable. In the long run, about how many will be classified as dependable and how many will be classified as questionable?

31. (*Refer to Exercise 31 in Section 9.1.*) A market survey on sunglasses shows that 30% of the owners of Blade Shades did not like the product and switched to Optifilters the next summer, while the rest continued to purchase Blade Shades. Conversely, 40% of the owners of Optifilters did not like their sunglasses and switched to Blade Shades the next summer, while the rest continued to purchase Optifilters.

(a) Determine the limiting matrix P^n, for large values of n.

(b) Assuming that Blade Shades and Optifilters are the two primary manufacturers of activity-related sunglasses, how much of the market share will each company have in the long run?

32. (*Refer to Exercise 32 in Section 9.1.*) Following a new advertising campaign, a market analysis shows that 20% of the consumers who did not currently drink SportSquirt will purchase it the next time that they buy a sport drink, and 90% of the consumers who currently drink SportSquirt will purchase it the next time that they buy a sport drink.

(a) Determine the limiting matrix P^n, for large values of n.

(b) How much of the market share would SportSquirt expect to have in the long run?

33. At V. F. Rashford Business College, students are declared as either business majors or general studies majors. The Research and Development Office finds that 63% of the students who are declared business majors switch to general studies majors the next year, while 37% remain business majors. Eighty-two percent of those declared as general studies majors remain general studies majors the next year, while 18% switch to being business majors.

(a) Draw the state transition diagram and write the transition matrix for the student's declared majors.

(b) Determine the limiting matrix P^n, for large values of n, and interpret the result.

34. The city of Galston in Montburg County undergoes a renovation program to deter what they interpret as being urban flight. A year after the program is completed, a consulting firm for the city determines that 89% of the residents in Galston remain in the city, while 11% relocate to the suburbs in Montburg County. They also find that 6% of those who live in a Montburg County suburb relocate in Galston, while 94% remain in the suburbs.

(a) Draw the state transition diagram and write the transition matrix for the public's migration.

(b) Determine the limiting matrix P^n, for large values of n, and interpret the result.

35. (*Refer to Exercise 35 in Section 9.1.*) The town of Minsterton has an election for mayor every 2 years. Voting records show that if a Democrat is in office, there is a 60% chance that a Democrat will be elected in the next election and a 40% chance that a Republican will be elected. If a Republican is in office, there is a 65% chance that a Republican will be elected in the next election and a 35% chance that a Democrat will be elected.

(a) Determine the limiting matrix P^n, for large values of n.

(b) What percentage of Democrats would be expected to win the mayoral election in the long run?

36. (*Refer to Exercise 36 in Section 9.1.*) A regional health study shows that, from one year to the next, 75% of the people who smoked will continue to smoke, while 25% of the people who smoke will stop smoking the next year. Eight percent of the people who did not smoke will start smoking the next year, while 92% of those who did not smoke will continue to be nonsmokers.

(a) Determine the limiting matrix P^n, for large values of n.

(b) What percentage of smokers would the region expect to have in the long run?

37. (*Refer to Exercise 37 in Section 9.1.*) The Move-It-Safe Company rents moving trucks in the cities of Adamstown, Breston, and Carpersville. They have both round-trip and point-to-point rentals. Records show that if a truck is rented in Adamstown, there is a 70% chance that it will be returned to Adamstown, a 15% chance that it will be returned to Breston, and a 15% chance that it will be returned to Carpersville. If a truck is rented in Breston, there is a 65% chance that it will be returned to Breston, a 10% chance that it will be returned to Adamstown, and a 25% chance that it will be returned to Carpersville. If a truck is rented in Carpersville, there is an 85% chance that it will be returned to Carpersville, a 5% chance that it will be returned to Adamstown, and a 10% chance that it will be returned to Breston.

(a) Determine the limiting matrix P^n, for large values of n.

(b) In the long run, what percentage of trucks would be returned to Adamstown, Breston, and Carpersville, respectively?

38. (*Refer to Exercise 38 in Section 9.1.*) In the LaDona metropolitan area, commuters either drive alone, carpool, or take public transportation. A study completed by the transportation department show that 80% of those who drive alone intend to drive alone a year from now, while 15% intend to switch to carpooling and 5% intend to use public transportation. Ninety percent of those who carpool intend to continue carpooling the next year, while 5% intend to drive alone and 5% will begin to use public transportation. Sixty percent of those who use public transportation will continue to use it next year, while 30% intend to drive alone and 10% will begin to carpool.

(a) Determine the limiting matrix P^n, for large values of n.

(b) In the long run, what percentage of LaDona commuters would be expected to either drive alone, carpool, or take public transportation?

✓ **39.** (*Refer to Exercise 39 in Section 9.1.*) In a particular county in the northern part of the country, voters can declare themselves as members of the Republican Party, Democrat Party, or Green Party. A public opinion poll shows that 85% of the Republicans intend to continue with their party affiliation, while 13% intend to switch to the Democrat Party in the next election and 2% to the Green Party. Seventy percent of the Democrats intend to continue their party affiliation, while 10% intend to switch to the Republican Party in the next election and 20% to the Green Party. Ninety-five percent of the Green Party members plan to continue with their party affiliation, while 1% intend to switch to the Republican Party in the next election and 4% to the Democratic Party.

(a) Determine the limiting matrix P^n, for large values of n.

(b) In the long run, what percentage of voters in the county is expected to be members of the Green Party?

40. (*Refer to Exercise 40 in Section 9.1.*) The Armont Conservatory has been keeping records on their own Armont carnations and find that the flowering color of the next generation of plant is related to the current generation. If a carnation is red, there is a 60% chance that the next generation will be red, a 40% chance that the plant will be pink, and no chance of the plant being white. If a carnation is pink, there is a 60% chance that the next generation will be pink, 30% that it will be red, and a 10% chance that it will be white. If a carnation is white, there is a 50% chance that the next generation will be white, a 50% chance that it will be pink, and no chance that it will be red.

(a) Determine the limiting matrix P^n, for large values of n.

(b) In the long run, what percentage of Armont carnations are expected to be pink?

SECTION PROJECT

To generalize some of the properties of regular transition matrices, consider the following:

(a) Let the transition matrix for a Markov chain be $P = \begin{bmatrix} p & 1-p \\ q & 1-q \end{bmatrix}$ for $0 < p < 1$ and $0 < q < 1$. Show that the limiting matrix, P^n, is

$$P^n = \begin{bmatrix} \frac{q}{1+q-p} & \frac{1-p}{1+q-p} \\ \frac{q}{1+q-p} & \frac{1-p}{1+q-p} \end{bmatrix} \quad \text{for large values of } n$$

(b) Determine the limiting matrix P^n for large values of n, if possible, for the transition matrix

$$P = \begin{bmatrix} p & 1-p \\ 1 & 0 \end{bmatrix}$$

for $0 < p < 1$. If the limiting matrix P^n cannot be determined, explain why not.

(c) Determine the limiting matrix P^n for large values of n, if possible, for the transition matrix

$$P = \begin{bmatrix} 0 & 1 & 0 \\ 0 & 0 & 1 \\ p & q & 1-p-q \end{bmatrix}$$

for $0 < p + q < 1$. If the limiting matrix P^n cannot be determined, explain why not.

Section 9.3 Absorbing Markov Chains

In Section 9.2 we saw that if a transition matrix associated with a Markov chain was a regular matrix then the matrix reached a steady state from which we could make long-term estimations. However, some Markov chains involve entering certain states, but never leaving these states. We call these **absorbing states**. In this section we will examine the transition matrix that is associated with these absorbing states and see how we can make long-term estimations of systems that have these states.

Absorbing Systems

To begin to see how these absorbing states can occur, let's examine the state transition diagram in Figure 9.3.1. The transition matrix associated with the system shown is

$$P = \begin{bmatrix} 0.2 & 0.8 & 0 \\ 0.1 & 0.6 & 0.3 \\ 0 & 0 & 1 \end{bmatrix}$$

State 3 is an example of an **absorbing state**, since the arrows in the diagram only enter state 3. There are no arrows leaving this state. In states 1 and 2 we see that there is at least one arrow leaving each of these states. Since states 1 and 2 are not absorbing, we call these states **nonabsorbing states**. We can see in the transition

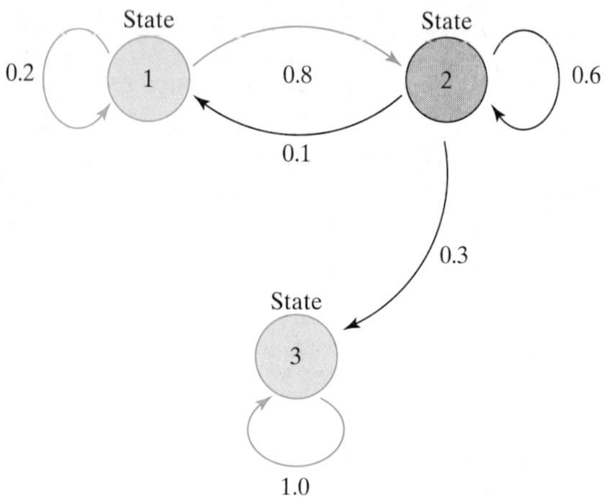

Figure 9.3.1

matrix for the system that the third row contains a 1 on the main diagonal and has 0s elsewhere. This illustrates how absorbing states can be identified in transition matrices.

Absorbing State

- For a system associated with a Markov chain, an **absorbing state** is a state with the characteristic that, once the state is reached, it is not possible for an element to leave the state. If a state is not an absorbing state, we call it a **nonabsorbing state**.
- A Markov chain is an **absorbing Markov chain** if
  1. The system associated with the Markov chain has at least one absorbing state.
  2. It is possible for an element to go from any nonabsorbing state to an absorbing state in a finite number of time steps.

In an absorbing Markov chain, nonabsorbing states can also be called **transient**.

Note: Ways to identify absorbing states in a system include:

- In the state transition diagram, no arrows leave the absorbing state.
- In the transition matrix, a row has a 1 on the main diagonal and 0s elsewhere. In this case the matrix row number and the absorbing state number are the same.

Notice in the second part of the preceding definition that simply having absorbing states in a system does not guarantee that we have an absorbing Markov chain. To see how this can happen, let's examine the state transition diagram in Figure 9.3.2. We see that state 3 is an absorbing state. Yet it is not possible to follow the arrows from state 1 or state 2 to absorbing state 3. To have an absorbing Markov chain, an element in the system must be able to go from each nonabsorbing state to the absorbing state in a finite number of time steps. This means that the system depicted in the diagram in Figure 9.3.2 is not associated with an absorbing Markov chain.

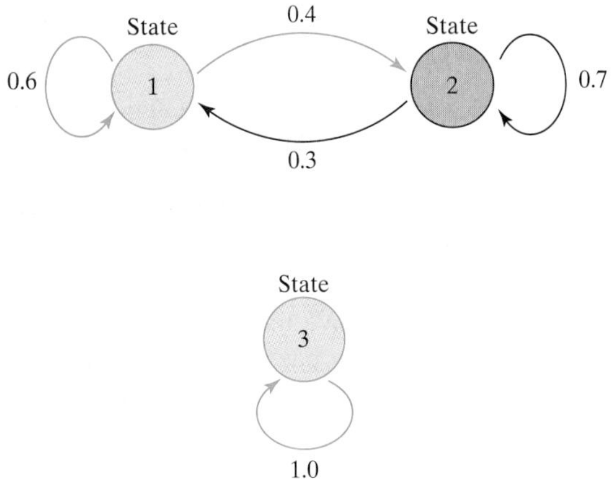

Figure 9.3.2

Example 1 Identifying Absorbing Markov Chains

For each of the following, identify the absorbing states and determine if the associated system represents an absorbing Markov chain.

(a)

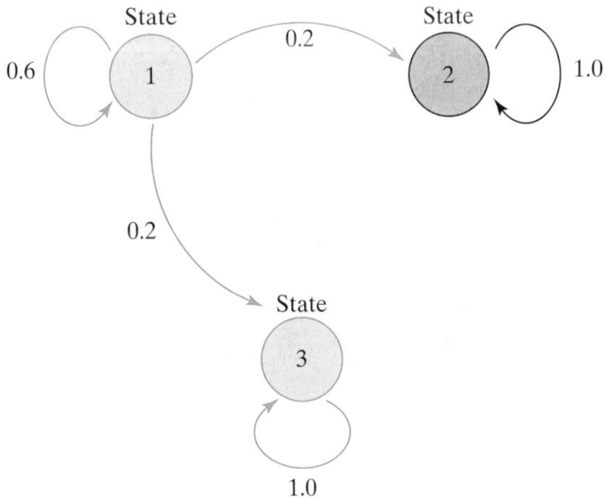

(b) $P = \begin{bmatrix} 0.2 & 0.8 & 0 \\ 0.3 & 0.7 & 0 \\ 0 & 0 & 1 \end{bmatrix}$

Solution

(a) Since states 2 and 3 have arrows entering, but not leaving, the states, we conclude that states 2 and 3 are absorbing states. Since it is possible to go from state 1 (a nonabsorbing state) to state 2 or 3 (the absorbing states), the associated system represents an absorbing Markov chain.

(b) Since the final entry in the third row of the transition matrix is 1 and all the other entries in the row are 0s, we see that state 3 is an absorbing state. Notice that all the other entries in the third column in the transition matrix are zero. That is, $p_{1,3} = 0$ and $p_{2,3} = 0$. This means that it is not possible to go from either state 1 or from state 2 to state 3. Thus, the associated system does not represent an absorbing Markov chain. ■

Checkpoint 1 Now work Exercise 13.

To be able to make long-term estimates using absorbing Markov chains, we use the following convention:

In an absorbing Markov system, we will always number the absorbing states last.

We follow this convention because the transition matrix for an absorbing system has a particular structure. To see what this structure looks like, consider the state transition diagram in Figure 9.3.3.

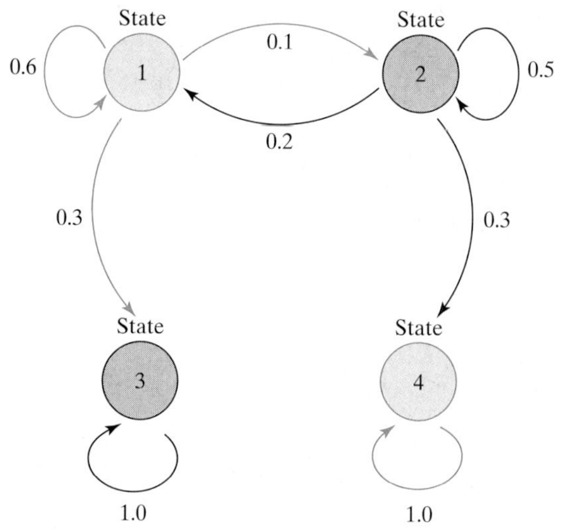

Figure 9.3.3

First, notice that two states, states 3 and 4, are absorbing states. The transition matrix associated with this system is

$$P = \begin{bmatrix} 0.6 & 0.1 & 0.3 & 0 \\ 0.2 & 0.5 & 0 & 0.3 \\ 0 & 0 & 1 & 0 \\ 0 & 0 & 0 & 1 \end{bmatrix}$$

Notice that there is a square 2×2 identity matrix in the lower-right corner resulting from the two absorbing states. Let's now **partition** the transition matrix so that we can see the identity matrix more clearly.

$$P = \left[\begin{array}{cc|cc} 0.6 & 0.1 & 0.3 & 0 \\ 0.2 & 0.5 & 0 & 0.3 \\ \hline 0 & 0 & 1 & 0 \\ 0 & 0 & 0 & 1 \end{array}\right]$$

In addition to the 2×2 identity matrix in the lower-right corner, there is a zero matrix in the lower-left corner. Generally, if a system has n states of which k of them are absorbing states, then the transition matrix has the partitioned form

$$P = \left[\begin{array}{c|c} Q & R \\ \hline 0 & I \end{array}\right]$$

where I is a k by k identity matrix, 0 is a k by $(n - k)$ zero matrix, and Q is an $(n - k)$ by $(n - k)$ matrix. For the transition matrix associated with Figure 9.3.3, the four partitioned matrices are

$$Q = \begin{bmatrix} 0.6 & 0.1 \\ 0.2 & 0.5 \end{bmatrix} \qquad R = \begin{bmatrix} 0.3 & 0 \\ 0 & 0.3 \end{bmatrix}$$

$$0 = \begin{bmatrix} 0 & 0 \\ 0 & 0 \end{bmatrix} \qquad I = \begin{bmatrix} 1 & 0 \\ 0 & 1 \end{bmatrix}$$

Since matrix Q is associated with states 1 and 2, this matrix gives the probabilities of going from one nonabsorbing state to another nonabsorbing state. But we know that the probability exists for us to go from a nonabsorbing to an absorbing state. The number of time steps it is expected to take to go from a nonabsorbing state to an absorbing state is given by the **fundamental matrix**.

Fundamental Matrix

Let P be a transition matrix associated with an absorbing Markov chain, where P has the partitioned form

$$P = \left[\begin{array}{c|c} Q & R \\ \hline 0 & I \end{array}\right]$$

Then the **fundamental matrix F** gives the expected number of time steps that the Markov chain will spend in each nonabsorbing state and is given by

$$F = (I - Q)^{-1}$$

Note that the identity matrix I that is in matrix P and matrix F are not necessarily the same. That is, they may not have the same dimension. In matrix F, I must have the same dimension as Q.

Note: We can make the following two interpretations with regard to the fundamental matrix.

1. The i, jth element of F, that is $f_{i,j}$, gives the expected number of time steps that an element that begins in the ith nonabsorbing state will be in the jth nonabsorbing state before it reaches an absorbing state.
2. The sum of the elements in the ith row of the fundamental matrix gives the expected number of time steps that an element that begins in the ith state will take to reach an absorbing state.

Example 2 Computing the Fundamental Matrix

The proline Communications Cable Company has an accounting system in which each customer is given 3 months to pay his or her current bill before cable service is discontinued. The cable company classifies its accounts as current (less than 1 month due), 1 month overdue, 2 months overdue, 3 months overdue, and service discontinued. The company's past records show the following:

- 75% of the accounts are paid within 1 month.
- 60% of the accounts that are 1 month overdue are paid at the end of the month.

- 40% of the accounts that are 2 months overdue are paid at the end of the month.
- 80% of the accounts that are 3 months overdue have service discontinued, while the remaining 20% pay their bill at the end of the month.

(a) Draw a state transition diagram and write a transition matrix P for the system.

(b) Compute the fundamental matrix F and interpret the element in the first row, fourth column (that is, $f_{1,4}$).

Solution

(a) Here we have six states, with two being absorbing states:

State 1: Account is less than 1 month due.
State 2: Account is 1 month overdue.
State 3: Account is 2 months overdue.
State 4: Account is 3 months overdue.
State 5: Account is paid.
State 6: Service is discontinued.

The state transition diagram for the accounts is shown in Figure 9.3.4.

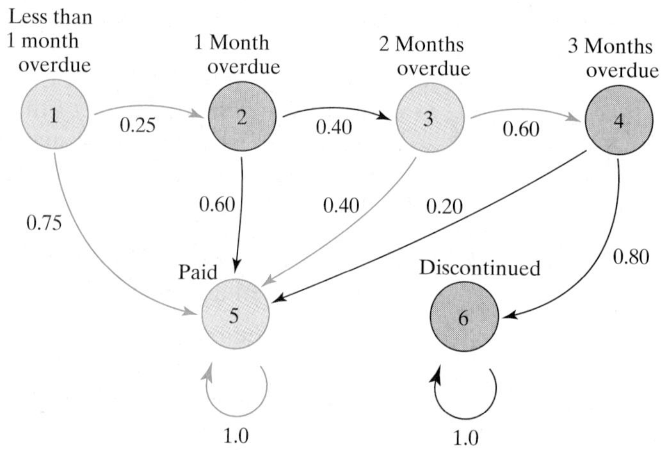

Figure 9.3.4

The corresponding partitioned transition matrix for the system is

$$P = \left[\begin{array}{cccc|cc} 0 & 0.25 & 0 & 0 & 0.75 & 0 \\ 0 & 0 & 0.40 & 0 & 0.60 & 0 \\ 0 & 0 & 0 & 0.60 & 0.40 & 0 \\ 0 & 0 & 0 & 0 & 0.20 & 0.80 \\ \hline 0 & 0 & 0 & 0 & 1 & 0 \\ 0 & 0 & 0 & 0 & 0 & 1 \end{array}\right]$$

So we see that

$$Q = \begin{bmatrix} 0 & 0.25 & 0 & 0 \\ 0 & 0 & 0.40 & 0 \\ 0 & 0 & 0 & 0.60 \\ 0 & 0 & 0 & 0 \end{bmatrix}$$

(b) Using the definition of the fundamental matrix, we get

$$F=(I-Q)^{-1}=\left(\begin{bmatrix}1&0&0&0\\0&1&0&0\\0&0&1&0\\0&0&0&1\end{bmatrix}-\begin{bmatrix}0&0.25&0&0\\0&0&0.40&0\\0&0&0&0.60\\0&0&0&0\end{bmatrix}\right)^{-1}$$

$$=\begin{bmatrix}1&-0.25&0&0\\0&1&-0.40&0\\0&0&1&-0.6\\0&0&0&1\end{bmatrix}^{-1}=\begin{bmatrix}1&0.25&0.1&0.06\\0&1&0.40&0.24\\0&0&1&0.60\\0&0&0&1\end{bmatrix}$$

Interpret the Solution: We see that the element in the first row, fourth column is $f_{1,4}=0.06$. This means that an account that is less than 1 month due (state 1) will spend about 0.06 time step as an account that is 3 months overdue (state 4) before it reaches an absorbing state (state 5 or state 6). ■

Technology Option

Graphing calculators with matrix menus can readily compute the inverse of a matrix that has been entered in the matrix editor. If we enter matrix Q,

$$\begin{bmatrix}0&0.25&0&0\\0&0&0.40&0\\0&0&0&0.60\\0&0&0&0\end{bmatrix}$$

in our calculator as matrix A, we can compute $(I-Q)^{-1}$ by entering $(I-A)^{-1}$ on the calculator. See Figure 9.3.5.

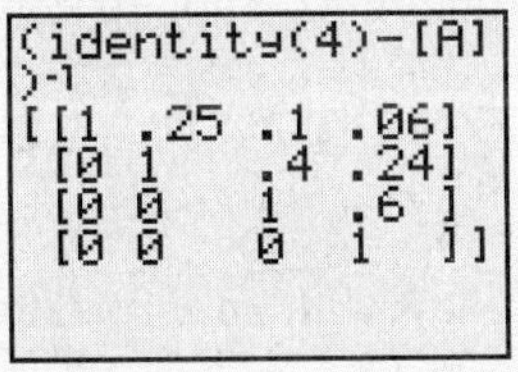

Figure 9.3.5

For more information about using your calculator to perform these types of operations, consult the on-line graphing calculator manual at www.prenhall.com/armstrong

Interactive Activity

Verify the matrix inverse to get the fundamental matrix in Example 2 by using the Gauss–Jordan method to write the augmented matrix

$$\left[\begin{array}{cccc|cccc}1&-0.25&0&0&1&0&0&0\\0&1&-0.40&0&0&1&0&0\\0&0&1&-0.6&0&0&1&0\\0&0&0&1&0&0&0&1\end{array}\right]$$

in reduced row echelon form.

WWW

In our partitioning of the transition matrix P, it appears that we have ignored the matrix R. However, this matrix can be useful, too, since the matrix product FR gives the probability that, if an element in the system was originally in a nonabsorbing state, it will go to a specific absorbing state. In Example 2 we found matrices F and R to be

$$F = \begin{bmatrix} 1 & 0.25 & 0.1 & 0.06 \\ 0 & 1 & 0.40 & 0.24 \\ 0 & 0 & 1 & 0.60 \\ 0 & 0 & 0 & 1 \end{bmatrix} \quad \text{and} \quad R = \begin{bmatrix} 0.75 & 0 \\ 0.60 & 0 \\ 0.40 & 0 \\ 0.20 & 0.80 \end{bmatrix}$$

The matrix product FR is

$$FR = \begin{bmatrix} 1 & 0.25 & 0.1 & 0.06 \\ 0 & 1 & 0.40 & 0.24 \\ 0 & 0 & 1 & 0.60 \\ 0 & 0 & 0 & 1 \end{bmatrix} \begin{bmatrix} 0.75 & 0 \\ 0.60 & 0 \\ 0.40 & 0 \\ 0.20 & 0.80 \end{bmatrix} = \begin{bmatrix} 0.952 & 0.048 \\ 0.808 & 0.192 \\ 0.520 & 0.480 \\ 0.200 & 0.800 \end{bmatrix}$$

We can see by inspecting row 2 of the matrix product FR, for example, that if an account is 1 month overdue there is an 80.8% chance of the bill being paid and a 19.2% chance that the service will be discontinued. Now let's see what kind of information the sum of the entries of the fundamental matrix can provide.

Example 3 Interpreting the Row Sums of the Fundamental Matrix

The fundamental matrix is often used to analyze enrollment trends in colleges. To illustrate how this is done, consider the following scenario. Bearstate Junior College has found by studying historical data that any one of its students can be classified in one of the following four states:

State 1: Student is a freshman.

State 2: Student is a sophomore.

State 3: Student transfers to a 4-year school or graduates with an associate's degree.

State 4: Student drops out.

The state transition diagram for the college is shown in Figure 9.3.6.

(a) Determine the fundamental matrix F for the system.

(b) Compute the sum of the entries of the rows of the fundamental matrix and interpret.

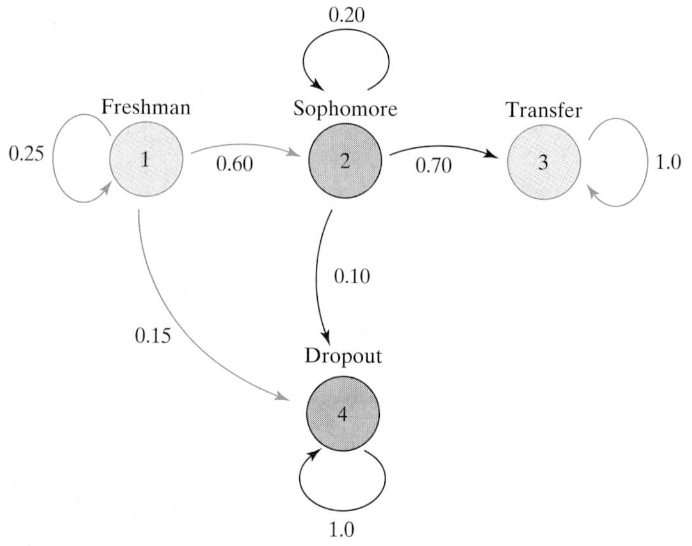

Figure 9.3.6

Solution

(a) From the state transition diagram in Figure 9.3.6, we see that the partitioned transition matrix for the system is

$$P = \left[\begin{array}{cc|cc} 0.25 & 0.60 & 0 & 0.15 \\ 0 & 0.20 & 0.70 & 0.10 \\ \hline 0 & 0 & 1 & 0 \\ 0 & 0 & 0 & 1 \end{array}\right] = \left[\begin{array}{c|c} Q & R \\ \hline 0 & I \end{array}\right]$$

Knowing that $Q = \begin{bmatrix} 0.25 & 0.60 \\ 0 & 0.20 \end{bmatrix}$, we get the fundamental matrix:

$$F = (I - Q)^{-1} = \left(\begin{bmatrix} 1 & 0 \\ 0 & 1 \end{bmatrix} - \begin{bmatrix} 0.25 & 0.60 \\ 0 & 0.20 \end{bmatrix} \right)^{-1}$$

$$= \begin{bmatrix} 0.75 & -0.60 \\ 0 & 0.80 \end{bmatrix}^{-1} = \begin{bmatrix} \frac{4}{3} & 1 \\ 0 & \frac{5}{4} \end{bmatrix}$$

(b) The sum of the entries in row 1 is $\frac{4}{3} + 1 = 2\frac{1}{3}$, and the sum of the entries in row 2 is $0 + \frac{5}{4} = 1\frac{1}{4}$. This means that we can expect a freshman to be at Bearstate Junior College for about $2\frac{1}{3}$ years before dropping out, transferring, or getting an associate's degree, and we can expect a sophomore to be at the college about $1\frac{1}{4}$ years before dropping out, transferring, or getting an associate's degree. ■

SUMMARY

- An **absorbing state** is a state with the characteristic that, once the state is reached, it is not possible for an element to leave the state.
- If a state is not an absorbing state, we call it a **nonabsorbing state**.
- A Markov chain is an **absorbing Markov chain** if the system associated with the Markov chain has at least one absorbing state and it is possible for an element to go from any nonabsorbing state to an absorbing state in a finite number of time steps.
- For the transition matrix P in the partitioned form $P = \left[\begin{array}{c|c} Q & R \\ \hline 0 & I \end{array}\right]$, the **fundamental matrix** $F = (I - Q)^{-1}$ gives the expected number of time steps that the Markov chain will spend in each nonabsorbing state.
- The matrix product FR gives the probability that, if an element in the system was originally in a nonabsorbing state, it will go to a specific absorbing state.

SECTION 9.3 EXERCISES

In Exercises 1–10, a transition matrix is given. Identify the absorbing states, if any, and draw a state transition diagram corresponding to the matrix.

1. $P = \begin{bmatrix} \frac{1}{2} & \frac{1}{2} \\ 1 & 0 \end{bmatrix}$

2. $P = \begin{bmatrix} 0 & 1 \\ \frac{1}{2} & \frac{1}{2} \end{bmatrix}$

3. $P = \begin{bmatrix} 0.3 & 0.7 & 0 \\ 0 & 1 & 0 \\ 0 & 0 & 1 \end{bmatrix}$

4. $P = \begin{bmatrix} 0 & 1 & 0 \\ 0.3 & 0.5 & 0.2 \\ 0 & 0 & 1 \end{bmatrix}$

5. $P = \begin{bmatrix} 0.5 & 0.5 & 0 \\ 0 & 0.5 & 0.5 \\ 0 & 0 & 1 \end{bmatrix}$

6. $P = \begin{bmatrix} 0.5 & 0 & 0.5 \\ 0.5 & 0.5 & 0 \\ 0 & 0.6 & 0.4 \end{bmatrix}$

7. $P = \begin{bmatrix} 1 & 0 & 0 \\ 0.4 & 0.4 & 0.2 \\ 0 & 0 & 1 \end{bmatrix}$

8. $P = \begin{bmatrix} 0 & 0 & 1 \\ 1 & 0 & 0 \\ 0 & 1 & 0 \end{bmatrix}$

9. $P = \begin{bmatrix} 0.9 & 0 & 0.1 & 0 \\ 0.3 & 0.3 & 0.4 & 0 \\ 0.5 & 0 & 0 & 0.5 \\ 0 & 0 & 0 & 1 \end{bmatrix}$

10. $P = \begin{bmatrix} 0.25 & 0.25 & 0.3 & 0.2 \\ 0.25 & 0.25 & 0.25 & 0.25 \\ 0 & 0 & 1 & 0 \\ 0 & 0 & 0 & 1 \end{bmatrix}$

In Exercises 11–20, a state transition diagram is given. Identify the absorbing states and determine if the diagram represents an absorbing Markov chain.

11.

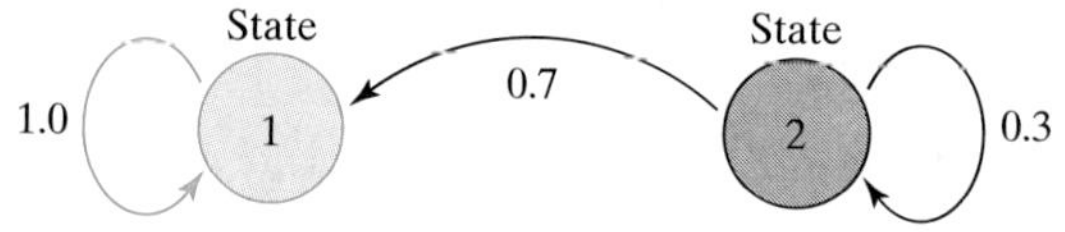

12.

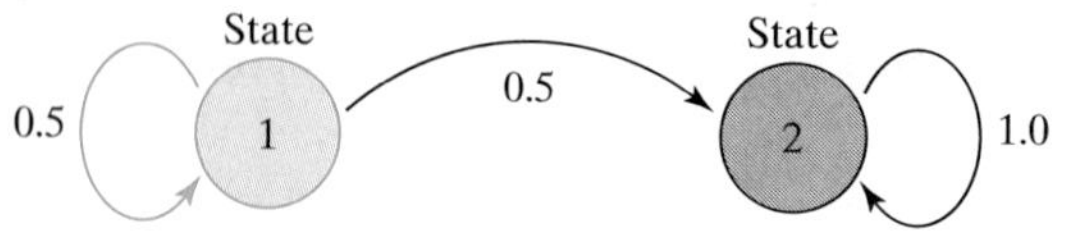

✓ **13.**

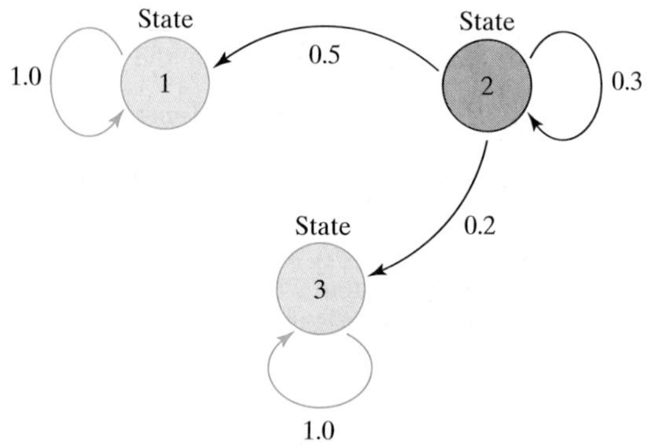

14.

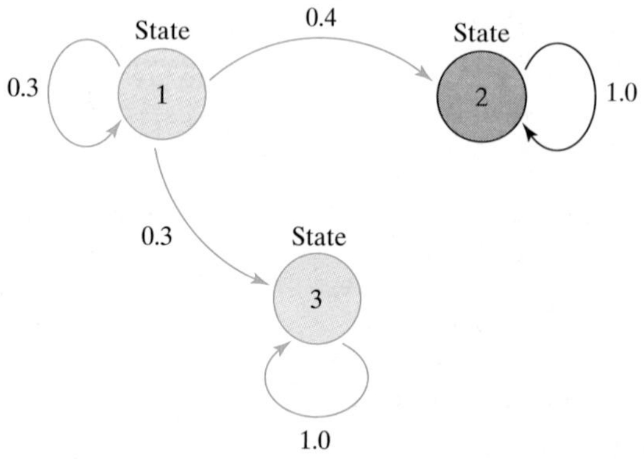

15.

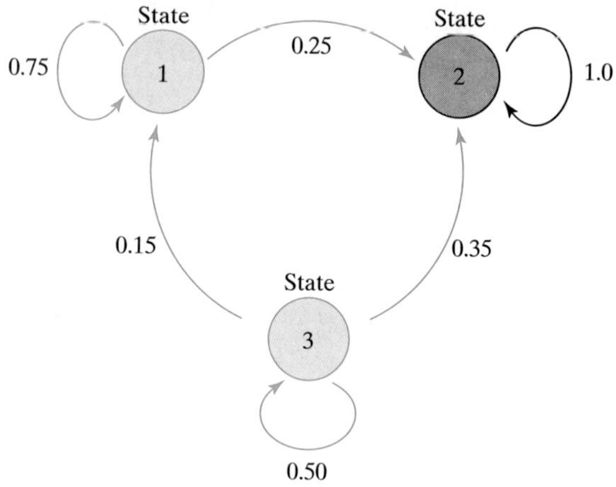

16.

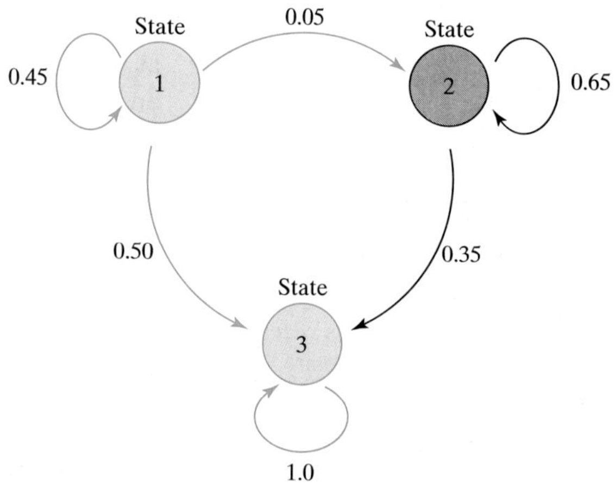

17.

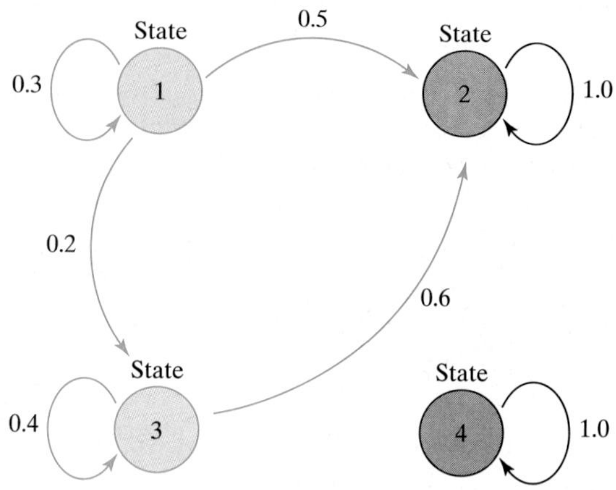

18.

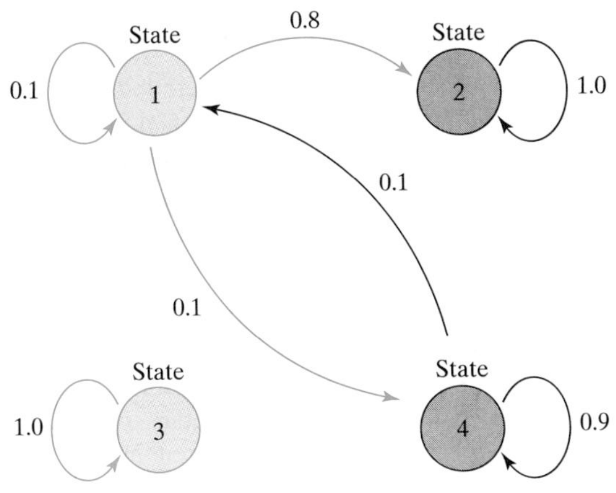

19.

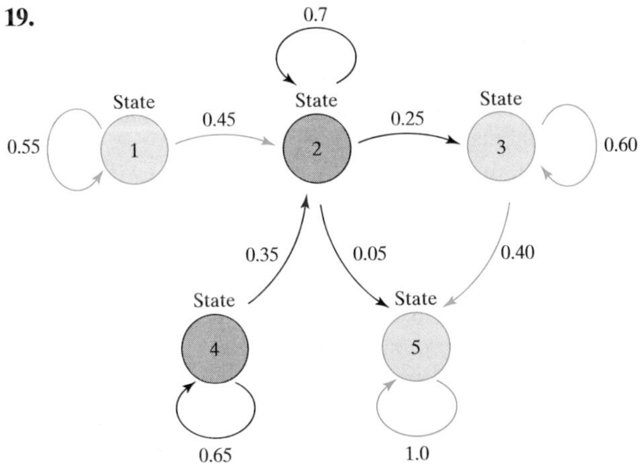

20.

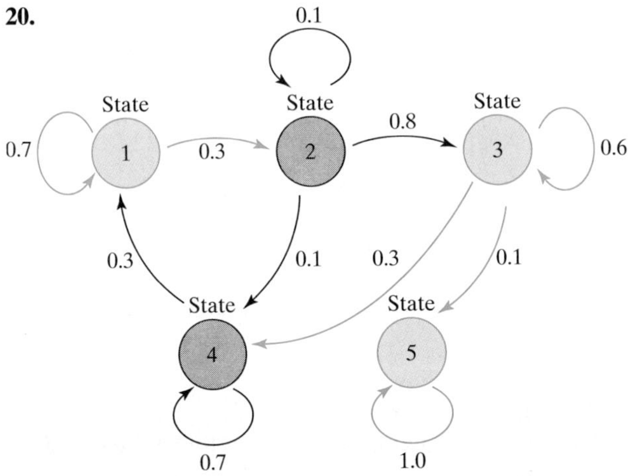

In Exercises 21–36:

(a) Determine the fundamental matrix F, if possible.

(b) Determine FR, if possible.

21. $P = \begin{bmatrix} 0.5 & 0 & 0.5 \\ 0.75 & 0.25 & 0 \\ 0 & 0 & 1 \end{bmatrix}$ **22.** $P = \begin{bmatrix} 0.3 & 0 & 0.7 \\ 0.2 & 0.6 & 0.2 \\ 0 & 0 & 1 \end{bmatrix}$

23. $P = \begin{bmatrix} \frac{1}{2} & 0 & \frac{1}{2} \\ \frac{3}{4} & \frac{1}{4} & 0 \\ 0 & 0 & 1 \end{bmatrix}$ **24.** $P = \begin{bmatrix} \frac{1}{4} & 0 & \frac{3}{4} \\ 0 & \frac{3}{4} & \frac{1}{4} \\ 0 & 0 & 1 \end{bmatrix}$

25. $P = \begin{bmatrix} 0.2 & 0.7 & 0.1 \\ 0.5 & 0 & 0.5 \\ 0 & 0 & 1 \end{bmatrix}$ **26.** $P = \begin{bmatrix} 0.1 & 0.6 & 0.3 \\ 0.9 & 0 & 0.1 \\ 0 & 0 & 1 \end{bmatrix}$

27. $P = \begin{bmatrix} 0.9 & 0.1 & 0 \\ 0.1 & 0.9 & 0 \\ 0 & 0 & 1 \end{bmatrix}$ **28.** $P = \begin{bmatrix} 0.4 & 0.6 & 0 \\ 0.6 & 0.4 & 0 \\ 0 & 0 & 1 \end{bmatrix}$

29. $P = \begin{bmatrix} 0.3 & 0 & 0.7 \\ 1 & 0 & 0 \\ 0 & 0 & 1 \end{bmatrix}$ **30.** $P = \begin{bmatrix} 0.4 & 0 & 0.6 \\ 1 & 0 & 0 \\ 0 & 0 & 1 \end{bmatrix}$

31. $P = \begin{bmatrix} \frac{1}{2} & \frac{1}{2} & 0 & 0 \\ \frac{3}{4} & \frac{1}{4} & 0 & 0 \\ 0 & 0 & 1 & 0 \\ 0 & 0 & 0 & 1 \end{bmatrix}$ **32.** $P = \begin{bmatrix} \frac{2}{3} & \frac{1}{3} & 0 & 0 \\ \frac{1}{5} & \frac{4}{5} & 0 & 0 \\ 0 & 0 & 1 & 0 \\ 0 & 0 & 0 & 1 \end{bmatrix}$

33. $P = \begin{bmatrix} 0.2 & 0.2 & 0.2 & 0.4 \\ 0.5 & 0.2 & 0 & 0.3 \\ 0 & 0 & 1 & 0 \\ 0 & 0 & 0 & 1 \end{bmatrix}$

34. $P = \begin{bmatrix} 0 & 0.5 & 0.1 & 0.4 \\ 0.3 & 0.3 & 0.3 & 0.1 \\ 0 & 0 & 1 & 0 \\ 0 & 0 & 0 & 1 \end{bmatrix}$

35. $P = \begin{bmatrix} 0.15 & 0.25 & 0.10 & 0.3 & 0.2 \\ 0.10 & 0.10 & 0.25 & 0.25 & 0.3 \\ 0.20 & 0.25 & 0.15 & 0.2 & 0.2 \\ 0 & 0 & 0 & 1 & 0 \\ 0 & 0 & 0 & 0 & 1 \end{bmatrix}$

36. $P = \begin{bmatrix} 0.2 & 0.3 & 0.5 & 0 & 0 \\ 0.1 & 0.4 & 0.2 & 0.3 & 0 \\ 0.3 & 0.3 & 0.3 & 0 & 0.1 \\ 0 & 0 & 0 & 1 & 0 \\ 0 & 0 & 0 & 0 & 1 \end{bmatrix}$

Applications

37. The zoning commission in the town of Gresden zones parcels of land as public, commercial, or residential. Over the last 5 years, records show that 75% of the land zoned as residential remained that way, while the remaining 25% was rezoned as commercial. Seventy-five percent of the land zoned as commercial was rezoned that way, with 15% being rezoned as residential and 10% as public. All the public land zoned as public remained zoned as public.

(a) Identify the three states and draw a state transition diagram for the zoning records.

(b) Write the transition matrix, and determine the fundamental matrix F.

(c) Interpret $f_{1,2}$.

38. The CorpNet software company gives each of its employees the title of programmer or project manager. An annual evaluation shows that, during any given year, 70% of the programmers are retitled as programmers, 20% are retitled as project managers, and 10% are asked to leave the company. Ninety percent of the project managers are retitled as project managers, and 10% are asked to leave the company.

(a) Identify the three states and draw a state transition diagram for the zoning records.

(b) Write the transition matrix, and determine the fundamental matrix F.

(c) Interpret $f_{1,2}$.

39. At DeRosa State Community College, records show that 15% of the freshmen at the school will remain freshmen after 1 year, 65% will become sophomores, and 20% will leave the school. During the same year, 25% of the sophomores will remain sophomores and 75% will leave the school (either by dropping out or graduating). Assume that none of the students who leave the school re-enroll.

(a) Identify the three states and draw a state transition diagram for the student enrollment.

(b) Write the transition matrix, and determine the fundamental matrix F.

(c) Interpret $f_{1,2}$.

40. The OilFirst Company employs geologists to conduct their fieldwork. Before being given their field assignment, however, the geologists first must pass a rigorous classroom program. Company records show that, during any given year, 45% of those enrolled in the classroom training will be promoted to field assignments, while the remaining 55% leave the company. Eighty percent of those involved in field assignments continue to do that work, while 20% of the field employees leave the company.

(a) Identify the three states and draw a state transition diagram for the company's training.

(b) Write the transition matrix, and determine the fundamental matrix F.

41. Drive-On Auto Sales gives short-term loans and classifies its clients as being in good standing if they make their payments by the due date or as probationary if they periodically miss payments. The accounting office records show that, during any given month, 75% of the clients in good standing remain in good standing, 20% pay off their loans, and 5% have their cars repossessed. Forty percent of the probationary clients remain probationary, 30% are reclassified as being in good standing, 10% pay off their loans, and 20% have their cars repossessed.

(a) Identify the four states and draw a state transition diagram for the loan classifications.

(b) Write the transition matrix, and determine the fundamental matrix F.

(c) Compute the sum of the entries in each row of the fundamental matrix F and interpret.

(d) Compute FR and interpret the entries in the first row of this matrix product.

42. During any given week, Piloton Metropolitan Library has books that are soon to be due or overdue. Thirty percent of the books that are due remain due, 50% are returned, and 20% are relabeled as overdue. Sixty-five percent of the overdue books remain overdue, while 20% of the books are returned and 15% of the patrons who borrow books have their library cards voided.

(a) Identify the four states and draw a state transition diagram for the library book labelings.

(b) Write the transition matrix, and determine the fundamental matrix F.

(c) Compute the sum of the entries in each row of the fundamental matrix F and interpret.

43. At the Public Water Works Laboratories, employees can have the titles of trainee, technician, or supervisor. During any given year, 20% of the trainees remain trainees, 60% are promoted to technician, and 20% are dismissed or quit. Fifty-five percent of the technicians remain technicians, 15% are promoted to supervisor, and 30% are dismissed or quit. Supervisor is the highest title that an employee can have at the lab.

(a) Identify the four states and draw a state transition diagram for the employee classifications.

(b) Write the transition matrix, and determine the fundamental matrix F.

(c) How many years can an employee expect to hold one of the two lowest positions?

44. Many homeowners who have their mortgages at the Clipston Financial Trust refinance their loans to take advantage of falling interest rates. The banks offers variable-rate loans, 30-year fixed-rate mortgages, and 15-year fixed-rate mortgages. Their records reveal the following facts:

- 55% of the borrowers with variable-rate mortgages remain with variable-rate mortgages, 35% refinance to get a 30-year fixed-rate mortgage, 5% pay off their loan, and 5% foreclose.
- 54% of the borrowers with 30-year fixed-rate loans remain with those loans, 15% refinance to get a variable-rate mortgage, 25% refinance to get a 15-year fixed-rate loan, 5% pay off their loans, and 1% foreclose.
- 75% of those with 15-year fixed-rate mortgages remain with those loans, while 20% refinance to have a

variable-rate mortgage, 4% pay off their loans, and 1% foreclose.

(a) Identify the five states and draw a state transition diagram for the bank loans.

(b) Write the transition matrix, and determine the fundamental matrix F.

(c) In the long run, what percentage of borrowers who now have variable-rate loans are expected to have variable-rate loans, 30-year fixed-rate mortgages, or 15-year fixed-rate mortgages?

45. At Edwards and Smith University, past experience shows that students move through the undergraduate ranks as follows:

- 65% of the freshmen will become sophomores after 1 year, 10% will remain freshmen, and 25% will withdraw from the university.
- 65% of the sophomores will become juniors after 1 year, 15% will remain sophomores, and 20% will withdraw from the university.
- 75% of the juniors will become seniors after 1 year, 5% will remain juniors, and 20% will withdraw from the university.
- 80% of the seniors will graduate from the university, 15% will remain as seniors, and 5% will withdraw from the university.

(a) Identify the six states and draw a state transition diagram for the student ranking.

(b) Write the transition matrix, and determine the fundamental matrix F.

(c) Determine the number of years that a freshman, sophomore, junior, and senior can expect to be at the university.

SECTION PROJECT

Let's explore some of the more general properties of absorbing Markov chains.

(a) Consider an absorbing Markov system whose transition matrix is given by

$$P = \begin{bmatrix} p & 1-p \\ 0 & 1 \end{bmatrix}$$

where $0 < p < 1$. Compute the fundamental matrix F and show that, if the Markov chain starts in state 1, then the expected number of times that it is in state 1 before being absorbed is given by $\frac{1}{p}$.

(b) Consider an absorbing Markov system whose transition matrix is given by

$$P = \begin{bmatrix} p & q & 1-p-q \\ 0 & 1 & 0 \\ 0 & 0 & 1 \end{bmatrix}$$

where $0 < p+q < 1$. Compute the fundamental matrix F. If the Markov chain starts in state 1, determine the expected number of time steps needed before an element is absorbed. Determine the probability that it is absorbed in state 3.

Why We Learned It

In this chapter we offered a brief introduction to Markov chains. We analyzed many diverse applications of Markov chains. For example, in the area of education some applications included school officials in a local school district analyzing vouchers, a community college analyzing the length of time that a student spends as a freshman or a sophomore, and a university analyzing the switching of majors by students. Long-term projections can be made by schools officials once a steady-state vector and/or a limiting matrix is determined. We also saw that zoning commissions can analyze how land is being zoned and project the long-term partition of land through a Markov process. Political parties can even use Markov chains to analyze voter registration and project the long-term consequences of voters changing party affiliations. Finally, we saw how botanists could use the Markov process to analyze a long-term projection of floral types. These are just a few of the applications of Markov chains. A more thorough study of Markov chains would yield even more areas of application for this branch of mathematics.

CHAPTER REVIEW EXERCISES

In Exercises 1–3, determine if the matrices are transition matrices.

1. $\begin{bmatrix} 0.4 & 0.6 \\ 0.1 & 0.9 \end{bmatrix}$

2. $\begin{bmatrix} 0.3 & -0.2 & 0.9 \\ 0 & 0.8 & 0.2 \\ 0.1 & 0.3 & 0.6 \end{bmatrix}$

3. $\begin{bmatrix} 0.32 & 0.41 & 0.27 \\ 0.15 & 0.28 & 0.57 \\ 0.05 & 0.78 & 0.17 \end{bmatrix}$

In Exercises 4–6, write the transition matrix for the state transition diagrams.

4.

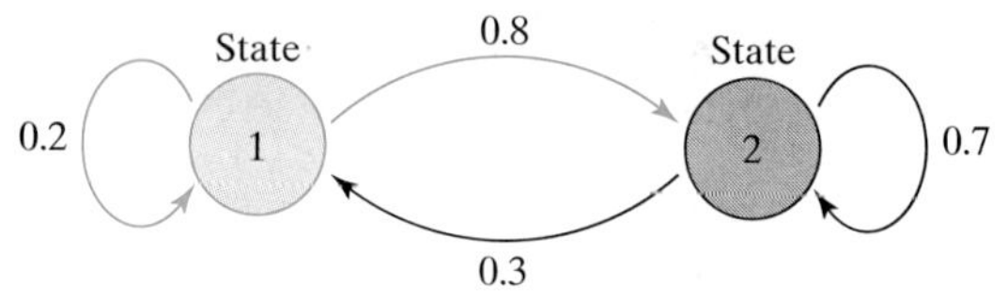

5.

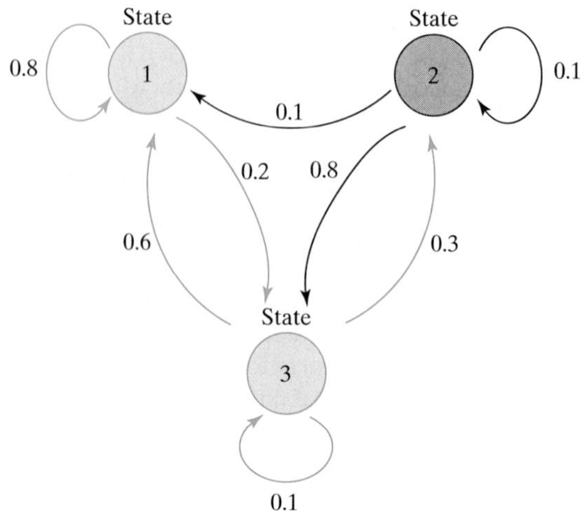

6.

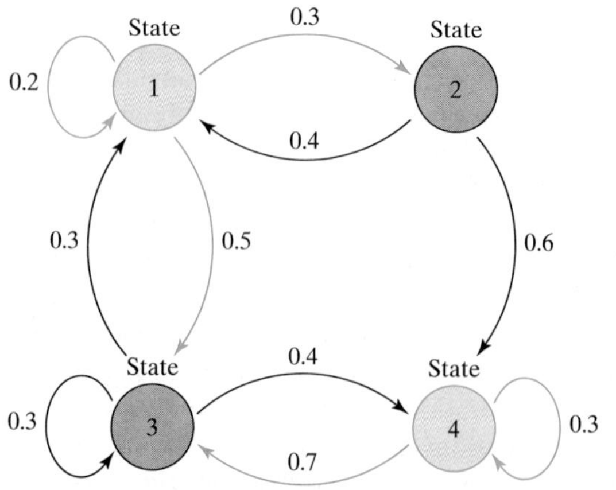

In Exercises 7–9, draw a state transition diagram for the given transition matrix.

7. $\begin{bmatrix} 0.1 & 0.9 \\ 0.2 & 0.8 \end{bmatrix}$

8. $\begin{bmatrix} 0.1 & 0.2 & 0.7 \\ 0.3 & 0.2 & 0.5 \\ 0.2 & 0.2 & 0.6 \end{bmatrix}$

9. $\begin{bmatrix} 0.34 & 0 & 0.28 & 0.38 \\ 0.10 & 0.34 & 0 & 0.56 \\ 0.66 & 0 & 0.19 & 0.15 \\ 0 & 0 & 0.25 & 0.75 \end{bmatrix}$

In Exercises 10–12, complete the following:

(a) Identify the states.

(b) Draw the state transition diagram.

(c) Write the transition matrix P.

10. The town of Plymouth has two elite soccer teams, Metro and Juventus. The director of parks and recreation has determined that there is a 30% chance that a player on Metro will play for Juventus next summer, while the rest continue to play for Metro. Conversely, 15% of the Juventus players will play for Metro next summer, while the rest continue to play for Juventus.

11. Goffstown has a community bike program in which a resident may pick up a bike at one of three locations and ride the bike anywhere provided that the bike is returned to one of the pickup locations the next morning. The pickup locations are the library, the YMCA, and the grocery store. The director of the program has determined that if a bike is picked up at the library there is a 10% chance that it will be returned to the library, a 25% chance that it will be returned to the YMCA, and a 65% chance that it will be returned to the grocery store. If a bike is picked up at the YMCA, there is an 80% chance that it will be returned to the YMCA, a 10% chance that it will be returned to the library, and a 10% chance that it will be returned to the grocery store. If a bike is picked up at the grocery store, there is a 5% chance that it will be returned to the grocery store, a 75% chance that it will be returned to the library, and a 20% chance that it will be returned to the YMCA.

12. In a particular county in New England, voters can declare themselves as Republican, Democrat, or Independent. A public opinion poll shows that 80% of the Republicans intend to continue with their party affiliation, while 12% intend to switch to Democrat in the next election and 8% intend to switch to Independent. Ninety percent of Democrats intend to continue with their party affiliation, while 6% intend to switch to Republican in the next election and 4% intend to switch to Independent. Ninety-six percent of Independents intend to continue with their party affiliation, while 1% intends to switch to Republican in the next election and 3% intend to switch to Democrat.

13. Refer to Exercise 11.

(a) On Monday, 20 bikes were picked up from the library, 24 bikes were picked up from the YMCA, and 18 bikes were picked up from the grocery store. Write the corresponding initial state vector S_0.

(b) Determine S_2 and interpret.

14. Refer to Exercise 12.

(a) Interpret each value of the probability distribution vector $S_0 = [0.58 \quad 0.40 \quad 0.02]$.

(b) Determine S_4 and interpret.

In Exercises 15–17, compute P^4, P^{16}, P^{32}, and P^{64} on a calculator to determine if the given transition matrix P is a regular transition matrix. If the matrix is regular, approximate the limiting matrix P^n accurate to four decimal places.

15. $P = \begin{bmatrix} 0.3 & 0.7 \\ 0.1 & 0.9 \end{bmatrix}$ **16.** $P = \begin{bmatrix} 0.65 & 0.35 \\ 0 & 1 \end{bmatrix}$

17. $P = \begin{bmatrix} 0.2 & 0.2 & 0.6 \\ 0.1 & 0.3 & 0.6 \\ 0.5 & 0.2 & 0.3 \end{bmatrix}$

In Exercises 18–20, solve the matrix equation $SP = S$ for S to determine the steady-state vector. Solve a system of equations using either the substitution or elimination method to determine the values for the steady state vector $[x \quad y]$. Also, use the steady-state vector to determine the limiting matrix P^n.

18. $P = \begin{bmatrix} 0.4 & 0.6 \\ 0.7 & 0.3 \end{bmatrix}$ **19.** $P = \begin{bmatrix} 0 & 1 \\ 0.4 & 0.6 \end{bmatrix}$

20. $P = \begin{bmatrix} 0.22 & 0.78 \\ 0.34 & 0.66 \end{bmatrix}$

In Exercises 21–23, solve the matrix equation $SP = S$ for S to determine the steady-state vector. Use the Gauss–Jordan method to determine the values for the steady-state vector $[x \quad y \quad z]$. Also, use the steady-state vector to determine the limiting matrix P^n.

21. $\begin{bmatrix} 0 & 1 & 0 \\ 0 & 0 & 1 \\ 0.8 & 0 & 0.2 \end{bmatrix}$ **22.** $\begin{bmatrix} \frac{1}{2} & 0 & \frac{1}{2} \\ \frac{1}{2} & \frac{1}{2} & 0 \\ 0 & \frac{1}{5} & \frac{4}{5} \end{bmatrix}$

23. $\begin{bmatrix} 0.10 & 0 & 0.90 \\ 0.55 & 0.15 & 0.30 \\ 0.45 & 0.25 & 0.30 \end{bmatrix}$

In Exercises 24–28, you can use your calculator to approximate the limiting matrix P^n or to solve the matrix equation $SP = S$ for S using the methods of Exercises 18–23.

24. A professional sports league gives its players a monthly drug test. The league has determined that if a player passes a monthly test there is an 80% chance that the player will pass the test next month and a 20% chance that the player will fail the test next month. If a player fails a monthly test, there is a 40% chance that the player will fail the test next month and a 60% chance that the player will pass the test.

(a) Label the two states and write the transition matrix.

(b) Determine the limiting matrix P^n, for large values of n.

(c) Use the solution from part (b) to determine the probability that, in the long run, a player will pass the test.

25. Students at Hillsbourough High School are put on probation if they are truant in any of their classes a total of five or more times a week. The attendance office has found that 90% of the students who are not on probation will remain off probation the next week, while 10% will go on probation. Eighty-five percent of the students who are on probation will remain on probation the next week, while 15% will go off probation.

(a) Draw a state transition diagram and write the transition matrix.

(b) Determine the limiting matrix P^n, for large values of n.

(c) Hillsbourough High School currently has 900 students not on probation and 300 students on probation. In the long run, about how many students will be off probation and how many will be on probation?

26. The town of Plymouth has two elite soccer teams, Metro and Juventus. The director of parks and recreation has determined that there is a 30% chance that a player on Metro will play for Juventus next summer, while the rest continue to play for Metro. Conversely, 15% of Juventus players will play for Metro next summer, while the rest continue to play for Juventus.

(a) Determine the limiting matrix P^n, for large values of n.

(b) If Metro originally has a roster of 24 players and Juventus has a roster of 18 players, in the long run how many players will be on each team?

27. Goffstown has a community bike program in which a resident may pick up a bike at one of three locations and ride the bike anywhere provided that the bike is returned to one of the pickup locations the next morning. The pickup locations are the library, the YMCA, and the grocery store. The director of the program has determined that if a bike is picked up at the library there is a 10% chance that it will be returned to the library, a 25% chance that it will be returned to the YMCA, and a 65% chance that it will be returned to the grocery store. If a bike is picked up at the YMCA, there is an 80% chance that it will be returned to the YMCA, a 10% chance that it will be returned to the library, and a 10% chance that it will be returned to the grocery store. If a bike is picked up at the grocery store, there is a 5% chance that it will be returned to the grocery store, a 75% chance that it will be returned to the library, and a 20% chance that it will be returned to the YMCA.

(a) Determine the limiting matrix P^n, for large values of n.

(b) In the long run, what percentage of the bikes will be at the YMCA?

28. In a particular county in New England, voters can declare themselves as Republican, Democrat, or Independent. A public opinion poll shows that 80% of the Republicans intend to continue with their party affiliation, while 12% intend to switch to Democrat in the next election and 8% intend to switch to Independent. Ninety percent of Democrats intend to continue with their party affiliation, while 6% intend to switch to Republican in the next election and 4% intend to switch to Independent. Ninety-six percent of Independents intend to continue with their party affiliation, while 1% intends to switch to

Republican in the next election and 3% intend to switch to Democrat.

(a) Determine the limiting matrix P^n, for large values of n.

(b) In the long run, what percentage of voters will be declared as Independents?

In Exercises 29–31, a transition matrix is given. Identify the absorbing states, if any, and draw a state transition diagram corresponding to the matrix.

29. $P = \begin{bmatrix} \frac{1}{4} & \frac{3}{4} \\ 0 & 1 \end{bmatrix}$

30. $P = \begin{bmatrix} 0.4 & 0 & 0.6 \\ 0.3 & 0.7 & 0 \\ 0 & 0.2 & 0.8 \end{bmatrix}$

31. $P = \begin{bmatrix} 0.6 & 0.3 & 0.1 & 0 \\ 0.2 & 0.4 & 0.2 & 0.2 \\ 0 & 0 & 1 & 0 \\ 0 & 0 & 0 & 1 \end{bmatrix}$

In Exercises 32–35, a state transition diagram is given. Identify the absorbing states and determine if the diagram represents an absorbing Markov chain.

32.

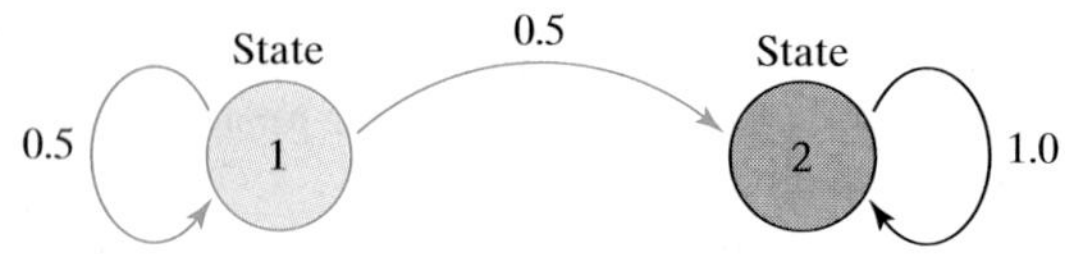

33.

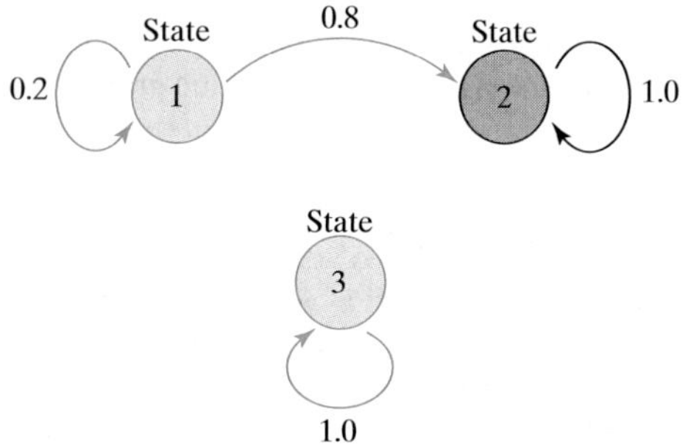

34.

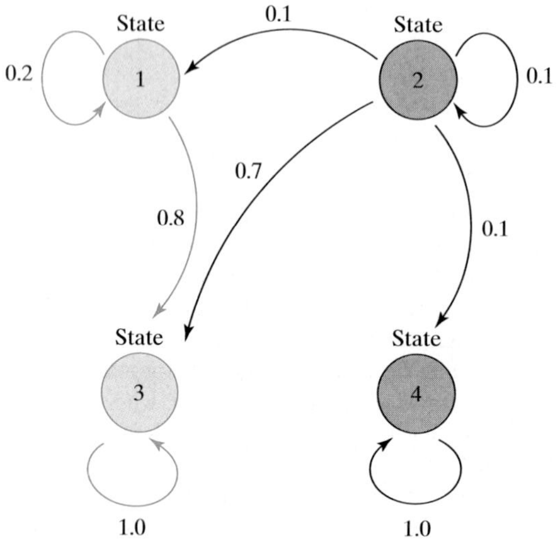

35.

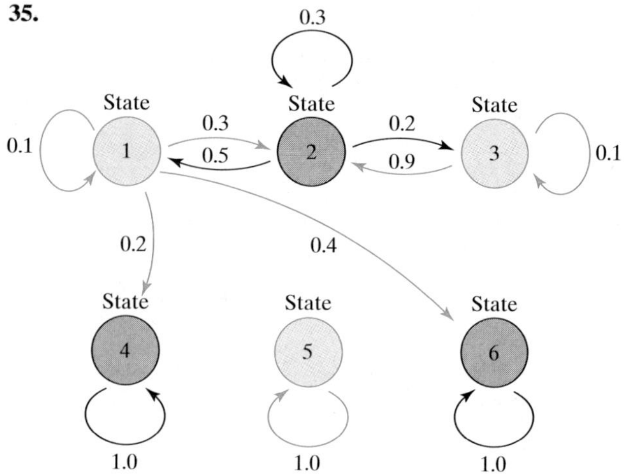

In Exercises 36–40, determine the fundamental matrix F.

36. $P = \begin{bmatrix} \frac{2}{5} & 0 & \frac{3}{5} \\ \frac{3}{10} & \frac{7}{10} & 0 \\ 0 & 0 & 1 \end{bmatrix}$

37. $P = \begin{bmatrix} 0.2 & 0 & 0.8 \\ 0.1 & 0.6 & 0.3 \\ 0 & 0 & 1 \end{bmatrix}$

38. $P = \begin{bmatrix} 0.3 & 0.7 & 0 \\ 0.7 & 0.3 & 0 \\ 0 & 0 & 1 \end{bmatrix}$

39. $P = \begin{bmatrix} \frac{1}{10} & \frac{3}{5} & \frac{1}{5} & \frac{1}{10} \\ \frac{3}{10} & \frac{2}{5} & \frac{3}{10} & 0 \\ 0 & 0 & 1 & 0 \\ 0 & 0 & 0 & 1 \end{bmatrix}$

40. $P = \begin{bmatrix} 0.10 & 0.20 & 0.35 & 0.10 & 0.25 \\ 0.25 & 0.15 & 0.10 & 0.25 & 0.25 \\ 0.30 & 0.10 & 0.20 & 0.15 & 0.25 \\ 0 & 0 & 0 & 1 & 0 \\ 0 & 0 & 0 & 0 & 1 \end{bmatrix}$

41. Players on Metro's soccer team are classified as first string or second string. Records from past seasons show that there is a 75% chance that a first-string player will remain first string the next season, a 20% chance that the player will be on the second string next season, and a 5% chance that the player will leave the team. There is a 40% chance that a second-string player will remain second string the next season, a 20% chance that the player will become first string the next season, and a 40% chance that the player will leave the team. Assume that none of the players who leave the team return.

(a) Identify the three states and draw a state transition diagram for a player's classification.

(b) Write the transition matrix, and determine the fundamental matrix F.

(c) Interpret $f_{1,2}$.

42. At Hillsbourough Community College, records show that 10% of the freshmen at the school will remain freshmen after 1 year, 70% will become sophomores, and 20% will leave the school. During the same year, 20% of sophomores will remain sophomores and 80% will leave the school (either by dropping out or graduating). Assume that none of the students who leave the school re-enroll.

(a) Identify the three states and draw a state transition diagram for the student enrollment.

(b) Write the transition matrix, and determine the fundamental matrix F.

(c) Compute FR and interpret the entries in the first row of this matrix product.

43. At Bedford Hospital's Physical Therapy Center patients are placed in a two-part program. Results of a weekly strength test determine in which stage a patient is placed. Stage I is for patients who test at less than half of their normal strength levels, and Stage II is for patients who test at greater than half of their normal strength levels. When a patient tests at 100%, the patient completes the program. Past records show that there is a 40% chance that a patient in Stage I remains in Stage I after the next testing, a 40% chance that the patient will move to Stage II, and a 20% chance that the patient will transfer to another program. There is a 45% chance that a patient in Stage II will remain in Stage II after the next testing, a 15% chance that the patient returns to Stage I, indicating reinjury, a 5% chance that the patient will transfer to another program, and a 35% chance that the patient will test 100% and complete the program. Assume that a patient that transfers or completes the program does not return.

(a) Identify the four states and draw a state transition diagram for the patient's status.

(b) Write the transition matrix, and determine the fundamental matrix F.

(c) Compute the sum of the entries in each row of the fundamental matrix F and interpret.

44. Vertical Climbing Gym offers beginner, intermediate, and advanced courses. There is a 20% chance that a student in the beginner course will remain in that course next session, a 60% chance that the student moves into the intermediate course, and a 20% chance that the student does not take another course. There is a 30% chance that a student in the intermediate course remains there next session, a 50% chance that the student moves into the advanced course, and a 20% chance that the student does not take another course. The advanced course is the highest course offered, and there is a 25% chance that a student remains in this course next session.

(a) Identify the four states and draw a state transition diagram for the student's course.

(b) Write the transition matrix, and determine the fundamental matrix F.

(c) If a student begins in the intermediate course, about how many sessions will the student take?

45. At Keller and Harris College, past records show that ranks of undergraduates progress as follows:

- 70% of the freshmen will become sophomores after 1 year, 15% will remain freshmen, and 15% will withdraw from the college.
- 70% of the sophomores will become juniors after 1 year, 10% will remain sophomores, and 20% will withdraw from the college.
- 65% of the juniors will become seniors after 1 year, 25% will remain juniors, and 10% will withdraw from the college.
- 85% of the seniors will graduate from the university, 10% will remain seniors, and 5% will withdraw from the university.

(a) Identify the six states and draw a state transition diagram for the ranks of undergraduates.

(b) Write the transition matrix, and determine the fundamental matrix F.

(c) Determine the number of years that a freshman, sophomore, junior, or senior can expect to be at the college.

(d) Determine the probability that a freshman will eventually graduate.

(e) Determine the probability that a sophomore will eventually withdraw.

CHAPTER 9 PROJECT

SOURCE: U.S. Census Bureau, www.census.gov

A Markov chain is a way to describe and predict travel through *states*. In this project, we'll consider the mobility of Americans within the United States. Consider the following information (in thousands):

178 moved from the West to the Midwest.
101 moved from the West to the Northeast.
400 moved from the West to the South.
98 moved from the Midwest to the Northeast.
556 moved from the Midwest to the South.
253 moved from the Midwest to the West.
383 moved from the Northeast to the South.
146 moved from the Northeast to the West.
94 moved from the Northeast to the Midwest.
334 moved from the South to the West.
463 moved from the South to the Midwest.
262 moved from the South to the Northeast.

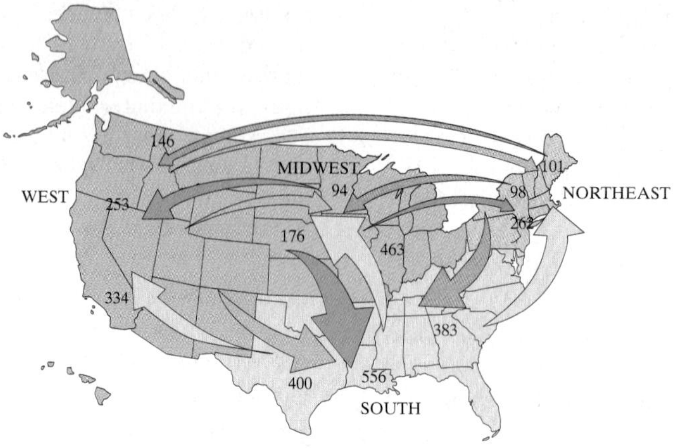

Figure R.1 Census Regions and Divisions of the United States

Let x_1 = population in the West, x_2 = population in the Midwest, x_3 = population in the Northeast, and x_4 = population in the South. Assume that no one enters the United States and no one leaves.

1. In 1998 there were 60,263,000 people in the West, 62,951,000 people in the Midwest, 51,686,000 people in the Northeast, and 95,349,000 people in the South. Find the percentage of each transition. Round to five decimal places.

2. Determine the percentage of people who stayed within their own region. Use the decimals from question 1 to find the total percentage of people that left each region. Subtract this answer from 1 to get the percentage of people who stayed in the region.

3. Draw a transition diagram.

4. Use the diagram in question 3 to form a transition matrix, P. Make sure that your rows add up to 1.

5. Given that the initial state vector is $S_0 = [\,60{,}263 \quad 62{,}951 \quad 51{,}686 \quad 95{,}349\,]$, find and interpret S_2. Round answers to the nearest whole number.

6. Solve $SP = S$ for S using a system of equations in which P is the matrix found in question 4 and S is the steady-state matrix. Round to five decimal places.

7. Interpret the entries in S found in question 6.

8. Assuming that the U. S. population maintains its 1998 level of 270,249 (in thousands), multiply each entry of S by 270,249 and interpret the results.

9. Use the result from question 6 to determine the limiting matrix, P^n.

10. Find $S_0 P^n$ and compare your answer to that of question 8.

Chapter

10 Game Theory

	Economy strong	Recession
Remodel	300,000	−100,000
Don't Remodel	−150,000	−125,000

The decision to remodel a store can be thought of as a game between the store and the economy. Although a dated looking store can discourage customers, spending too much money on the remodel can bankrupt the store. Managers can use matrices and game theory as an aid in determining which course of action to follow. The above payoff matrix gives the expected payoff or loss in dollars. The −100,000 is the saddle point which gives the store its best strategy in this strictly determined game.

What We Know

We have learned many things in this text. Linear systems, matrices, linear programming, the simplex method, probability, and expected value are a few topics that we have covered so far.

Where Do We Go

In this chapter, we study game theory. Game theory provides an exceptional review, similar to a capstone, for all the topics listed above. At times, you may need to refer to other chapters in the text to review specific items.

Section 10.1 Strictly Determined Games

When we hear the word **game**, we tend to think of competitors working to determine a strategy that allows one of the competitors to win the game. Many games, such as a chess match, a football game, or a wrestling match, are played by only two competitors. We call these types of games **two-person games**. For our purposes, the number of actual players does not matter since we may be able to think of two-person games as involving two competing forces. However, some games may have several competitors. For example, a jewelry store in a mall might have several jewelry stores in the mall competing for business.

In this chapter we offer a brief exposure to competitive **game theory**, a branch of mathematics invented by John von Neumann. This exposure is restricted to games involving only two competitors, that is, two-person games. Also, we restrict our work in this chapter to games where the amount "gained" or "won" by one competitor is equal to the amount "lost" by the opposing competitor. This means that at the end of one play of the game the sum of the winnings of the two players is zero. Hence, we denote a loss by one competitor with a negative number. The types of games that we have just described are known as **zero-sum games**. In this chapter we analyze **two-person**, **zero-sum games**. Example 1 introduces us to one such game.

Example 1 Introducing a Two-Person, Zero-Sum Game

The city of Bundyville has two digital camera stores, Digicity and ClearVu. Both stores offer discounts of 25% or 35%, depending on how aggressively they want business. When both stores offer the same discount, Digicity gets 65% of the business. When Digicity discounts more than ClearVu, Digicity gets 75% of the business. When ClearVu discounts more than Digicity, Digicity gets 55% of the business. Determine the percentage of business that ClearVu "loses":

(a) When both stores have the same discount.

(b) When Digicity discounts more than ClearVu.

Solution

(a) When both stores offer the same discount, Digicity gets 65% of the business. This means that ClearVu loses 65% of the business.

(b) When Digicity discounts more than ClearVu, it gets 75% of the business. Hence, ClearVu loses 75% of the business. ■

Example 1 illustrates a **two-person**, **zero-sum game**, since the sales gain of one store comes at the expense of the other store. (The game in Example 1 can actually be viewed as a *constant-sum* game, since the sum of Digicity's percentage and ClearVu's percentage is 100%.) In the game in Example 1, each store has a choice of how much to discount. Each choice leads to different percentages of the market share for the stores. For example, if Digicity's strategy is to discount 25% and ClearVu's strategy is to discount 35%, then Digicity would have a market share of 55%, while ClearVu would have a market share of 45%. This scenario outlines succinctly what this chapter covers:

1. Each player has a choice on a course of action, and these choices affect the outcome of the game.
2. Any choice made by a player is called a **strategy** for that player.

3. How does each player decide on a best strategy after considering all the choices that the other player has?

Determining the best or **optimal strategy** for each player is called **solving the game** and this is, of course, our ultimate goal. To help us in solving a game, it is advantageous to organize information in what is known as a **payoff matrix**.

Payoff Matrix

Continuing with the scenario in Example 1, the choice made by Digicity or ClearVu is called a **strategy** and each strategy has a consequence, which is called a **payoff**. We can summarize the payoff for each strategy in what is known as a **payoff matrix**. Each entry in the payoff matrix gives the payoff for a particular pair of choices by Digicity and ClearVu. To facilitate our organization, we arbitrarily select one store, Digicity, and use the rows of the matrix to represent its gains, as a percentage of the market share written as a decimal. In this setup, Digicity is called the **row player** and ClearVu is called the **column player**.

$$\begin{array}{ccc} & & \text{ClearVu} \\ & & \begin{array}{cc} 25\% & 35\% \end{array} \\ \text{Digicity} & \begin{array}{c} 25\% \\ 35\% \end{array} & \begin{bmatrix} 0.65 & 0.55 \\ 0.75 & 0.65 \end{bmatrix} \end{array}$$

Payoff matrix where Digicity is the row player.

Example 2 Constructing a Two-Person, Zero-Sum Payoff Matrix

Construct a payoff matrix for Digicity and ClearVu in which ClearVu is the row player.

Solution

Using the information given in Example 1, the payoff matrix is given by

$$\begin{array}{ccc} & & \text{Digicity} \\ & & \begin{array}{cc} 25\% & 35\% \end{array} \\ \text{ClearVu} & \begin{array}{c} 25\% \\ 35\% \end{array} & \begin{bmatrix} 0.35 & 0.25 \\ 0.45 & 0.35 \end{bmatrix} \end{array}$$

■

✓ **Checkpoint 1** Now work Exercise 9.

Example 3 Understanding Entries in a Payoff Matrix

Zach and Josh are playing a penny-matching game. They each choose a side of a coin without knowing the other's choice. Then, when their friend Diane blows a whistle, both players show their choices simultaneously. The payoff matrix associated with this two-person, zero-sum game is given by

$$\begin{array}{ccc} & & \text{Josh's choice} \\ & & \begin{array}{cc} H & T \end{array} \\ \text{Zach's choice} & \begin{array}{c} H \\ T \end{array} & \begin{bmatrix} 4 & -5 \\ 3 & 2 \end{bmatrix} \end{array}$$

Given that the entries are in dollars, supply an interpretation of this payoff matrix.

Solution

Understand the Situation: Since this is a two-person, zero-sum game, a gain for one player equals a loss of the same size for the other player.

If Zach chooses heads (row 1 of the payoff matrix), he wins \$4 if Josh chooses heads (column 1), whereas Zach loses \$5 if Josh chooses tails (column 2). If Zach chooses tails (row 2), he wins \$3 if Josh chooses heads (column 1), and Zach wins \$2 if Josh chooses tails (column 2). ■

Strictly Determined Games and Optimal Strategies

Looking more closely at the game and payoff matrix in Example 3, we may wonder how Zach and Josh should play the game. If Zach is greedy, he may choose row 1 because of the \$4 payoff. However, we assume that Josh is not stupid and as a result would try to avoid column 1 to avoid losing \$4. If Josh plays column 2, then he might win \$5. But Zach, who we also assume is not stupid, might anticipate Josh's rationale and in turn he might play row 2, preventing Josh from winning at all!

To help us to determine a best play for Josh and Zach, as well as illustrate a general approach for a two-person, zero-sum game with an $m \times n$ payoff matrix, we utilize a worst-case scenario mentality. For Zach, the row player, the worst that could happen in row 1 is a \$5 loss and in row 2 a \$2 gain. We circle these entries as shown below.

$$\text{Zach's choice} \quad \begin{matrix} & \text{Josh's choice} \\ & \begin{matrix} \text{H} & \text{T} \end{matrix} \\ \begin{matrix} \text{H} \\ \text{T} \end{matrix} & \begin{bmatrix} 4 & \textcircled{-5} \\ 3 & \textcircled{2} \end{bmatrix} \end{matrix}$$

Zach's **best strategy** is to select row 2, since it has the *largest* of these minimum values. (Some call this the **maximin strategy**.) By choosing row 2, Zach is guaranteed a \$2 win no matter what Josh does.

For Josh, the column player, the worst that could happen in column 1 is a \$4 loss and in column 2 a \$2 loss. (Remember, for Josh, the column player, a positive number indicates a loss to Zach, the row player.) We place squares around these entries as shown below.

$$\text{Zach's choice} \quad \begin{matrix} & \text{Josh's choice} \\ & \begin{matrix} \text{H} & \text{T} \end{matrix} \\ \begin{matrix} \text{H} \\ \text{T} \end{matrix} & \begin{bmatrix} \boxed{4} & \textcircled{-5} \\ 3 & \boxed{\textcircled{2}} \end{bmatrix} \end{matrix}$$

Josh's **best strategy** is to select column 2, since it has the *smallest* of the maximum values. (Some call this the **minimax strategy**.) By choosing column 2, Josh is guaranteed no worse than a \$2 loss no matter what Zach does. Unfortunately for Josh, this is the best he can do as long as Zach continues to make his best move.

Notice that the 2 in row 2, column 2 is enclosed by a circle and a square. This means that it is both the smallest entry in its row and the largest entry in its column. This entry is called a **saddle value** or a **saddle point**. Just like a horse saddle, the saddle point is a minimum in one direction and a maximum from the other direction. Since saddle points are crucial in game theory, we now formalize the procedure on how to find them.

Locating Saddle Points

1. Circle the minimum value in each row. This may occur in more than one place.
2. Place squares around the maximum value in each column. This may occur in more than one place.
3. Any entry enclosed by a circle and a square is a **saddle point**. If more than one occurs, they will be equal.

Example 4 Finding a Saddle Point

Locate the saddle point for the following payoff matrix from a two-person, zero-sum game.

$$\begin{bmatrix} 2 & -2 & -6 \\ -1 & 1 & -2 \\ 4 & 10 & -8 \end{bmatrix}$$

Solution

To locate the saddle point, we use the three steps in the Locating Saddle Points box.

Step 1: Here we circle the minimum value in each row to get

$$\begin{bmatrix} 2 & -2 & \textcircled{-6} \\ -1 & 1 & \textcircled{-2} \\ 4 & 10 & \textcircled{-8} \end{bmatrix}$$

Step 2: Using the matrix that resulted from step 1, we now place squares around the maximum value in each column to get

$$\begin{bmatrix} 2 & -2 & \textcircled{-6} \\ -1 & 1 & \boxed{\textcircled{-2}} \\ \boxed{4} & \boxed{10} & \textcircled{-8} \end{bmatrix}$$

Step 3: The resulting matrix in step 2 shows that there is one entry enclosed by a circle and a square. This means that there is one saddle point and its value is −2. ■

Checkpoint 2 Now work Exercise 13a.

Let's return to Zach and Josh and the payoff matrix for their game.

		Josh's choice H	Josh's choice T
Zach's choice	H	$\boxed{4}$	$\textcircled{-5}$
	T	3	$\boxed{\textcircled{2}}$

We know that 2 is the saddle point. We also know that this means that it is the smallest entry in row 2 and the largest entry in column 2. This means that Zach should always play row 2 and Josh should always play column 2. This results in a win of \$2 for Zach every time. This is the **optimal strategy** for both Zach and Josh. Notice that if Zach keeps playing row 2 and Josh changes to column 1 Zach would win more money (bad for Josh). If Josh keeps playing column 2 and Zach changes to row 1, then Josh wins $5 (bad for Zach). This type of game is said to be **strictly determined**, since Zach must always play row 2 to maximize his winnings and Josh must always play column 2 to minimize his losses. Optimal strategy and strictly determined games can be determined by the payoff matrix and saddle points as follows:

Strictly Determined Games

A game is called **strictly determined** if its payoff matrix has a saddle point. In a strictly determined game, the **optimal strategy** for row player R and column player C is as follows:

1. R should choose any row that contains a saddle point.
2. C should choose any column that contains a saddle point.

A saddle point is also called the **value of the game**.

Notes:
- If the value of the game is positive, then the game favors the row player.
- If the value of the game is negative, then the game favors the column player.
- If the value of the game is zero, then the game is called a **fair game**.

Example 5 Determining Optimal Strategies

Sam's Shoes and Louie's Loafers are two competing shoe stores. Each must decide how to price a popular brand of hiking boots. Each store will charge either $45 or $49 for the boots. If Sam's charges $45 and Louie's charges $49, Sam's will get 70% of the business for these boots. If Sam's charges $49 and Louie's charges $45, Sam's will get 40% of the business. If both stores charge the same amount, each will get 50% of the business.

(a) Construct the payoff matrix for this game writing percentages as decimals.

(b) Verify that the game is strictly determined.

(c) Determine the optimal strategy for each company.

Solution

T From Your Toolbox

Recall that an entry in a matrix is given by $a_{i,j}$, where i is the row and j is the column.

(a) Arbitrarily assigning Sam's Shoes as the row player, the payoff matrix is

$$\begin{array}{cc} & \text{Louie's Loafers} \\ & \begin{array}{cc} \$45 & \$49 \end{array} \\ \text{Sam's Shoes} \begin{array}{c} \$45 \\ \$49 \end{array} & \begin{bmatrix} 0.50 & 0.70 \\ 0.40 & 0.50 \end{bmatrix} \end{array}$$

(b) A game is strictly determined if it has a saddle point. Using the three steps in locating a saddle point, we have

$$\begin{array}{cc} & \text{Louie's Loafers} \\ & \begin{array}{cc} \$45 & \$49 \end{array} \\ \text{Sam's Shoes} \begin{array}{c} \$45 \\ \$49 \end{array} & \begin{bmatrix} \boxed{0.50} & \boxed{0.70} \\ 0.40 & 0.50 \end{bmatrix} \end{array}$$

The payoff matrix has a saddle point (enclosed by both a circle and a square) and hence is a strictly determined game.

(c) The entry $a_{1,1} = 0.50$ is the saddle point, so Sam's Shoes' optimal strategy is to choose row 1 and Louie's Loafers' optimal strategy is to choose column 1. In other words, Sam's Shoes and Louie's Loafers should both charge $45 for the hiking boots. ■

✓ Checkpoint 3 Now work Exercise 29.

SUMMARY

In this section we introduced **two-person, zero-sum games**. Central to our introduction to game theory was the **payoff matrix**. The payoff matrix allows us to determine if the game is **strictly determined**, to locate **saddle points**, and to find an **optimal strategy** for each player. The saddle point also gives the value of the game, which in turn tells us if the game favors one player over the other or if it is a fair game.

- **Locating Saddle Points**
  1. Circle the minimum value in each row. This may occur in more than one place.
  2. Place squares around the maximum value in each column. This may occur in more than one place.
  3. Any entry enclosed by a circle and a square is a **saddle point**. If more than one occurs, they will be equal.
- **Strictly Determined Games** A game is called **strictly determined** if its payoff matrix has a saddle point. In a strictly determined game, the **optimal strategy** for row player R and column player C is as follows:
  1. R should choose any row that contains a saddle point.
  2. C should choose any column that contains a saddle point.

A saddle point is also called the **value of the game**.

SECTION 10.1 EXERCISES

Exercises 1–8, use the following payoff matrix for a two-person, zero-sum game. The entries represent dollars.

$$\text{Jenna}\ \overset{\text{Elle}}{\begin{bmatrix} 2 & -2 & 3 \\ -1 & 4 & 6 \\ 1 & -3 & 2 \end{bmatrix}}$$

Determine the payoff when the following strategies are used.

1. Jenna plays row 1; Elle plays column 1.

2. Jenna plays row 1; Elle plays column 3.

3. Jenna plays row 1; Elle plays column 2.

4. Jenna plays row 2; Elle plays column 1.

5. Jenna plays row 2; Elle plays column 3.

6. Jenna plays row 3; Elle plays column 1.

7. Jenna plays row 3; Elle plays column 2.

8. Jenna plays row 3; Elle plays column 3.

In Exercises 9–12, you are given a payoff matrix for a game. Construct a new payoff matrix for the game where the row player becomes the column player and the column player becomes the row player.

✓ **9.** $\text{Bill}\ \overset{\text{Don}}{\begin{bmatrix} 3 & -3 \\ -2 & -4 \end{bmatrix}}$

10. $\text{Lisa}\ \overset{\text{Melissa}}{\begin{bmatrix} -3 & 4 \\ 7 & 1 \end{bmatrix}}$

11. $\text{Eric}\ \overset{\text{Patrice}}{\begin{bmatrix} -3 & 1 & 2 \\ 4 & -6 & 3 \\ 4 & 6 & -8 \end{bmatrix}}$

12. $\text{Ann}\ \overset{\text{Sally}}{\begin{bmatrix} 3 & -2 & -1 \\ -1 & 2 & -7 \\ -2 & -3 & 5 \end{bmatrix}}$

In Exercises 13–24, you are given a payoff matrix for a two-person, zero-sum game. If the game is strictly determined (if it is not, say so) determine:

(a) The saddle point(s).

(b) The optimal strategy for each player.

(c) The value of the game, and state whether the game favors one player over the other.

✓ **13.** $\text{Bill}\ \overset{\text{Al}}{\begin{bmatrix} 2 & 3 \\ -2 & 5 \end{bmatrix}}$

14. $\text{George}\ \overset{\text{Dick}}{\begin{bmatrix} 3 & -4 \\ -7 & -5 \end{bmatrix}}$

15. Monica

Hillary $$\begin{bmatrix} 3 & -7 \\ -2 & 1 \end{bmatrix}$$

16. Linda

Ken $$\begin{bmatrix} 7 & -6 \\ -3 & -1 \end{bmatrix}$$

17. Scoonie

Khalid $$\begin{bmatrix} 2 & 4 & 2 \\ 0 & 3 & 0 \\ -1 & -2 & 1 \end{bmatrix}$$

18. Maria

Nina $$\begin{bmatrix} 1 & 2 & 3 \\ -1 & 2 & -1 \\ 1 & 6 & 3 \end{bmatrix}$$

19. Dylan

Austin $$\begin{bmatrix} 4 & 3 & 0 \\ 5 & 7 & 8 \\ 3 & -1 & 9 \end{bmatrix}$$

20. Rusty

Randy $$\begin{bmatrix} 1 & 0 & 7 \\ -3 & 2 & 0 \\ 2 & -3 & 8 \end{bmatrix}$$

21. Ash

Brock $$\begin{bmatrix} -5 & 1 \\ -2 & 0 \\ 4 & 3 \\ -3 & 1 \end{bmatrix}$$

22. Misty

Erika $$\begin{bmatrix} 3 & 1 & -4 & 5 \\ -2 & 5 & 0 & 3 \end{bmatrix}$$

23. Mike

Brian $$\begin{bmatrix} 5 & -2 & 3 & -8 \\ 3 & 0 & 2 & 1 \\ 1 & -1 & -7 & 2 \\ 4 & 0 & 1 & 3 \end{bmatrix}$$

24. Luther

Denyse $$\begin{bmatrix} 1 & 4 & 0 & 1 \\ -8 & 3 & -3 & 2 \\ 4 & 1 & 0 & 4 \\ -2 & 3 & -7 & 0 \end{bmatrix}$$

Applications

25. The Chasm, a clothing retail store, is contemplating remodeling to modernize its physical appearance. A market research firm determines that the main determining factor in The Chasm's decision is the economy and whether it will remain strong or slip into recession. The firm determines that if The Chasm remodels and the economy is strong The Chasm will earn \$300,000. However, if The Chasm remodels and the economy slips into recession, the business will lose \$100,000. If The Chasm does not remodel and the economy remains strong, the business will lose \$150,000 due to customers going to other stores. Finally, if The Chasm does not remodel and there is a recession, the business will lose \$125,000. The market research firm summarizes this information in the following payoff matrix, where the two-person game is The Chasm against the economy. Find the saddle point and determine the best strategy for The Chasm.

The Chasm	Economy: Strong	Economy: Recession
Remodel	300,000	−100,000
Don't remodel	−150,000	−125,000

26. You Do It and I Did It, two home improvement stores, are each exploring opening new stores at one of two possible locations: Eastown Mall and Westown Mall. A market research firm has developed the following payoff matrix, where each entry represents the annual amounts (in millions of dollars) either gained or lost by one store from or to the other as a result of the location selected. Find the saddle point and determine the best strategy for the management of each store.

You Do It	I Did It: Eastown	I Did It: Westown
Eastown	1	−1
Westown	1.5	−2

27. Big Lakes Bank and Bank Two are two banks that serve a certain region. Each bank is exploring opening a new branch that will be located in one of the three following towns: Jasonville (J), Kerrton (K), or Heathton (H). Depending on where each bank locates its new branch, Bank Two will either gain or lose deposits at the expense of Big Lakes Bank as indicated in the following payoff matrix (entries are in thousands of dollars):

Bank Two	Big Lakes Bank: J	K	H
J	80	0	−20
K	50	25	40
H	−30	−20	100

(a) Is this a strictly determined game?

(b) Find the saddle point and determine the best strategy for each bank.

(c) Determine the value of the game. Does this game favor one bank over the other? If so, which bank has the advantage?

28. C Me Play and R You Game are two competing sporting goods stores. On a weekly basis, each store selects one (and only one) of the following types of advertisement: television, radio, or mail circulars. A survey by a market research company yields the following payoff matrix. Entries represent the percentages of the market gain or loss for each choice of action by the stores. (Assume that a gain by C Me Play is a loss for R You Game, and vice versa.)

		C Me Play		
		TV	Radio	Mail
R You Game	TV	2	2	3
	Radio	−3	−2	0
	Mail	2	2	−2

(a) Is this a strictly determined game?

(b) Find the saddle point and determine the best strategy for each store.

(c) Determine the value of the game. Does this game favor one store over the other? If so, which store has the advantage?

✓ **29.** Two competing auto parts stores, Redan and Erog, are each planning to add another store in a certain city. It is possible for each store to be located centrally in the city or peripherally in a large suburb of the city. If both stores decide to build in the city, Redan will have a monthly profit of \$200 more than the profit of Erog. If both build in the suburb, then Redan's monthly profit will be \$400 less than Erog's. If Redan builds in the suburb while Erog builds in the city, Redan's monthly profit will be \$800 more than Erog's. If Redan builds in the city and Erog builds in the suburb, then Redan's monthly profit will be \$600 less than Erog's. Let Redan be the row player and Erog be the column player.

(a) Construct the payoff matrix for this game, where the entries are in hundreds of dollars.

(b) Verify that the game is strictly determined.

(c) Determine the optimal strategy for each store.

30. (*continuation of Exercise 29*)

(a) Determine the value of the game in Exercise 29.

(b) Does the game in Exercise 29 favor one store over the other? If so, which store has the advantage?

31. Two health clubs, Gutbuster and Noflab, are considering expanding the special fitness programs that they offer. Gutbuster is considering offering a middle-age spread buster program, a senior citizen toning program, and a twenty-something fit-for-life program. Noflab is considering a middle-age spread buster program and a senior citizen toning program. A survey by a market research firm determines the following. If Gutbuster starts a middle-age spread buster program, it will gain 5% of the market if Noflab starts a middle-age spread buster program, but will lose 5% of the market if Noflab starts a senior citizen toning program. If Gutbuster starts a senior citizen toning program, it will gain 5% of the market if Noflab starts a middle-age spread buster program, but see no gain if Noflab starts a senior citizen toning program. If Gutbuster starts a twenty-something program, it will gain 10% of the market if Noflab starts a middle-age spread buster program and gain 15% of the market if Noflab starts a senior citizen toning program. (Assume that a gain by Gutbuster is a loss for Noflab, and vice versa.) Let Gutbuster be the row player and Noflab be the column player.

(a) Construct the payoff matrix for this game, where the entries are percentages of the market.

(b) Verify that the game is strictly determined.

(c) Determine the optimal strategy for each health club.

32. (*continuation of Exercise 31*)

(a) Determine the value of the game in Exercise 31.

(b) Does the game in Exercise 31 favor one health club over the other? If so, which health club has the advantage?

33. A candidate for governor of a certain state has informed his campaign manager about a prior, potentially damaging indiscretion. The campaign manager has to decide whether to reveal this "skeleton in the closet." The campaign manager has determined that if the candidate reveals the skeleton and the news media are critical, the candidate will lose 35,000 votes. However, if the news media are not critical, the candidate will gain 7000 votes. If the candidate does not reveal the skeleton and the media learns of it first and are critical, the candidate will lose 150,000 votes. Yet if the candidate does not reveal it and the media does not learn about it (it is not revealed during the campaign), the candidate will neither gain nor lose votes. This is a two-person game between the candidate and the media. Let the candidate be the row player and the media be the column player.

(a) Construct the payoff matrix for this game, where the entries are votes gained or lost in thousands.

(b) Verify that the game is strictly determined.

(c) Determine the optimal strategy for the candidate.

SECTION PROJECT

Brian and Jo play a game of matching fingers. When Jamie blows a whistle, both players extend, at the same time, 1, 2 or 3 fingers. If the sum of the extended fingers is 2 or 3, then Brian receives from Jo an amount in dollars equal to that sum. If the sum of the extended fingers is 5 or 6, then Jo receives from Brian an amount in dollars equal to that sum. If the sum of the extended fingers is 4, then neither player wins any money. Let Brian be the row player and Jo the column player.

(a) Construct the payoff matrix for this game.

(b) Is the game strictly determined?

(c) What is the optimal strategy for Brian and Jo?

(d) What is the value of the game?

(e) Does the game favor Brian or Jo or is it a fair game?

Section 10.2 Mixed Strategies

In Section 10.1 we studied strictly determined games and saw that, in a strictly determined game, both the row player and the column player choose a row and a column, respectively, that contains a saddle point. This is the optimal strategy for each player. To review how to find saddle points and determine the corresponding optimal strategy, let's take a brief Flashback.

Flashback

Penny-Matching Game Revisited

In Section 10.1, Example 3, we saw that Zach and Josh are playing a penny-matching game. They each choose a side of a coin without knowing the other's choice. Then, when their friend Diane blows a whistle, both players show their choices simultaneously. The payoff matrix associated with this two-person, zero-sum game is given by

$$\text{Zach's choice}\quad \begin{array}{c} \\ H \\ T \end{array}\overset{\text{Josh's choice}}{\begin{array}{c} \begin{matrix} H & T \end{matrix} \\ \begin{bmatrix} 4 & -5 \\ 3 & 2 \end{bmatrix} \end{array}}$$

Recall that entries are in dollars. Find the saddle point and optimal strategy for each player.

Flashback Solution

Using the circle and square method given in Section 10.1, we have

$$\text{Zach's choice}\quad \begin{array}{c} \\ H \\ T \end{array}\overset{\text{Josh's choice}}{\begin{array}{c} \begin{matrix} H & T \end{matrix} \\ \begin{bmatrix} \boxed{4} & \text{(}-5\text{)} \\ 3 & \boxed{\text{(}2\text{)}} \end{bmatrix} \end{array}}$$

The saddle point is 2; hence the optimal strategy for Zach is row 2 and the optimal strategy for Josh is column 2. Also recall that the saddle point gives the value of the game. Here the value of the game is 2. Since the value of the game is a positive number, the game favors the row player, Zach.

In repeated plays of the game in the Flashback, Zach or Josh having knowledge of the other's strategy is of no advantage provided that the opponent is using an optimal strategy. If Josh figures out that Zach is always playing row 2, it does him no good to change to column 1, since he would then lose more money to Zach. Likewise, if Zach discovers that Josh is always playing column 2, it would do him no good to change to row 1, since he would then lose money to Josh. So the optimal strategy for both in repeated plays of the game is to make the same move over and over again. Such strategies are called **pure strategies**.

Nonstrictly Determined Games

Let's consider the following variation of the payoff matrix for Josh and Zach in the Flashback, while keeping everything else the same.

$$\begin{array}{cc} & \text{Josh's choice} \\ & \begin{array}{cc} \text{H} & \text{T} \end{array} \\ \text{Zach's choice} \begin{array}{c} \text{H} \\ \text{T} \end{array} & \begin{bmatrix} 4 & -5 \\ -3 & 2 \end{bmatrix} \end{array}$$

To see if Zach or Josh has an advantage, as well as to see if there exists a pure strategy, we begin by checking to see if the game is strictly determined. The circle and square method allows us to determine this.

$$\begin{array}{cc} & \text{Josh's choice} \\ & \begin{array}{cc} \text{H} & \text{T} \end{array} \\ \text{Zach's choice} \begin{array}{c} \text{H} \\ \text{T} \end{array} & \begin{bmatrix} \boxed{4} & \textcircled{-5} \\ \textcircled{-3} & \boxed{2} \end{bmatrix} \end{array}$$

Since no entry is enclosed by both a circle and a square, there are no saddle points. This means that the game is *not* strictly determined. So how should Zach and Josh play this game?

If Zach consistently plays row 1 because of its large payoff of \$4, it would not take Josh too long to figure out Zach's strategy, and as a result Josh would then play column 2. But then Zach would play row 2. This would cause Josh to then play column 1. This brief analysis indicates that neither player should consistently play the same row or column. Instead, each player should play each row and each column in some mixed pattern unknown to the other player. Strategies such as this are called **mixed strategies**.

Mixed Strategies

How should Zach or Josh choose a mixed pattern? The best way to select a mixed pattern is to use a probability distribution and some sort of chance device to determine the distribution. For example, Zach might choose row 1 with probability $\frac{1}{2}$ and row 2 with probability $\frac{1}{2}$. This could be determined by flipping an unbiased coin before each move, with heads corresponding to playing row 1 and tails corresponding to playing row 2. Similarly, Josh might choose column 1 with probability $\frac{3}{13}$ and column 2 with probability $\frac{10}{13}$. These could be determined by drawing a card at random from a standard 52-card deck, where selecting a Jack, Queen, or King corresponds to playing column 1 and selecting any other card corresponds to playing column 2. (The selected card would need to be replaced after each play.)

Since payoffs for a game are expressed in a payoff matrix, it makes mathematical sense to use matrices for mixed strategies. For Zach, the row player, and the aforementioned strategy of flipping a coin, his strategy is represented by the **row matrix**

$$P = \begin{bmatrix} \frac{1}{2} & \frac{1}{2} \end{bmatrix}$$

For Josh, the column player, and the aforementioned strategy of drawing a card, his strategy is given by the **column matrix**

$$Q = \begin{bmatrix} \frac{3}{13} \\ \frac{10}{13} \end{bmatrix}$$

Before we illustrate exactly why row and column matrices are used for strategies, let's summarize what we have done so far.

Interactive Activity

If an entry in P (or Q) is 1 and the other is 0, then the strategy is a pure strategy. Why?

WWW

Strategies for the Row Player and Column Player

Given the payoff matrix

$$A = \begin{bmatrix} a & b \\ c & d \end{bmatrix}$$

the row player strategy is

$$P = [\,p_1 \quad p_2\,]$$

where $p_1 \geq 0$, $p_2 \geq 0$.

$$p_1 + p_2 = 1$$

the column player strategy is

$$Q = \begin{bmatrix} q_1 \\ q_2 \end{bmatrix}$$

where $q_1 \geq 0$, $q_2 \geq 0$.

$$q_1 + q_2 = 1$$

In Zach and Josh's game we saw that pure strategies were not the best choices since it is a **nonstrictly determined game**. This led us to using mixed strategies. However, is there an optimal mixed strategy for each player? The answer is yes and it uses the idea of **expected value** discussed in Section 8.2.

Expected Value of a Game

Let's assume for Zach (row player) and Josh (column player) the payoff matrix A, row matrix P, and column matrix Q introduced earlier.

$$A = \begin{bmatrix} a & b \\ c & d \end{bmatrix} = \text{Zach's choice} \begin{matrix} \\ \text{H} \\ \text{T} \end{matrix} \overset{\text{Josh's choice}}{\begin{matrix} \text{H} & \text{T} \\ \end{matrix}} \begin{bmatrix} 4 & -5 \\ -3 & 2 \end{bmatrix}$$

Suppose that in repeated plays of the game Zach uses the mixed strategy

$$P = [\,p_1 \quad p_2\,] = [\,\tfrac{1}{2} \quad \tfrac{1}{2}\,]$$

and Josh uses mixed strategy

$$Q = \begin{bmatrix} q_1 \\ q_2 \end{bmatrix} = \begin{bmatrix} \frac{3}{13} \\ \frac{10}{13} \end{bmatrix}$$

This means that Zach plays row 1 with probability $\frac{1}{2}$ and row 2 with probability $\frac{1}{2}$, while Josh plays column 1 with probability $\frac{3}{13}$ and column 2 with probability $\frac{10}{13}$. Each play of the game yields four possible outcomes or payoffs, as given by the payoff matrix A. Each of these payoffs has a probability of occurring that we can determine. For example, the probability of \$4 occurring is simply the probability associated with Zach playing row 1 and Josh playing column 1. Since the choice of moves is made by a player without knowing the other's choice, each outcome (or payoff) makes up a pair of independent events. Reviewing the information in the Toolbox to the left, the probability of \$4 occurring is simply the product $\frac{1}{2}\left(\frac{3}{13}\right) = \frac{3}{26}$. That is,

$$P(\text{row 1 and column 1}) = P(\text{row 1}) \cdot P(\text{column 1})$$

$$P(a) = p_1 \cdot q_1$$

$$P(4) = \frac{1}{2} \cdot \frac{3}{13} = \frac{3}{26}$$

From Your Toolbox

Two events are independent if the outcome of one does not affect the other from occurring. If two events A and B are independent, the probability that both occur is $P(A) \cdot P(B)$.

In a similar fashion, we can compute the probability of the other three payoffs. This is summarized in Table 10.2.1.

Table 10.2.1

Payoff	Probability
$a = 4$	$p_1q_1 = \frac{1}{2} \cdot \frac{3}{13} = \frac{3}{26}$
$b = -5$	$p_1q_2 = \frac{1}{2} \cdot \frac{10}{13} = \frac{10}{26}$
$c = -3$	$p_2q_1 = \frac{1}{2} \cdot \frac{3}{13} = \frac{3}{26}$
$d = 2$	$p_2q_2 = \frac{1}{2} \cdot \frac{10}{13} = \frac{10}{26}$

From Your Toolbox

Expected value is computed as follows:

$$E(X) = x_1 \cdot P(X = x_1) + x_2 \cdot P(X = x_2) + \cdots + x_n \cdot P(X = x_n)$$

Now using the concept of expected value from Section 8.2 (consult the Toolbox to the left), the **expected value of the game *E*** is

$$\begin{aligned} E &= ap_1q_1 + bp_1q_2 + cp_2q_1 + dp_2q_2 \\ &= 4\left(\frac{1}{2}\right)\left(\frac{3}{13}\right) + (-5)\left(\frac{1}{2}\right)\left(\frac{10}{13}\right) + (-3)\left(\frac{1}{2}\right)\left(\frac{3}{13}\right) + 2\left(\frac{1}{2}\right)\left(\frac{10}{13}\right) \\ &= -\frac{27}{26} \approx -1.04 \end{aligned}$$

This means that in the long run, when Zach and Josh use the indicated mixed strategies, Zach will lose about \$1.04 per game on average. Hence, Josh will win about \$1.04 per game on average.

Before we address whether this is the best that Zach and Josh can do, let's summarize what we have done.

Expected Value of a Game

Given the payoff matrix

$$A = \begin{bmatrix} a & b \\ c & d \end{bmatrix}$$

and strategies

$$P = [\,p_1 \quad p_2\,] \quad \text{and} \quad Q = \begin{bmatrix} q_1 \\ q_2 \end{bmatrix}$$

for the row player and the column player, respectively, the **expected value of the game**, E, is given by

$$E = PAQ$$

Note: To show that $E = PAQ$, use matrix multiplication to multiply PAQ and show that $PAQ = ap_1q_1 + bp_1q_2 + cp_2q_1 + dp_2q_2$. This is left for you to do in Exercise 29.

Example 1 Computing an Expected Value of a Game

Given the payoff matrix A for Zach and Josh

$$A = \text{Zach's choice} \begin{array}{c} \\ \text{H} \\ \text{T} \end{array} \overset{\text{Josh's choice}}{\begin{array}{c} \begin{array}{cc} \text{H} & \text{T} \end{array} \\ \begin{bmatrix} 4 & -5 \\ -3 & 2 \end{bmatrix} \end{array}}$$

and mixed strategies

$$P = \begin{bmatrix} \frac{1}{2} & \frac{1}{2} \end{bmatrix} \quad \text{and} \quad Q = \begin{bmatrix} \frac{1}{2} \\ \frac{1}{2} \end{bmatrix}$$

determine the expected value, E, of the game and interpret.

Solution

Since $E = PAQ$, we have

$$\begin{aligned} E &= \begin{bmatrix} \frac{1}{2} & \frac{1}{2} \end{bmatrix} \cdot \begin{bmatrix} 4 & -5 \\ -3 & 2 \end{bmatrix} \cdot \begin{bmatrix} \frac{1}{2} \\ \frac{1}{2} \end{bmatrix} \\ &= \begin{bmatrix} \frac{1}{2} & -\frac{3}{2} \end{bmatrix} \cdot \begin{bmatrix} \frac{1}{2} \\ \frac{1}{2} \end{bmatrix} \\ &= \begin{bmatrix} -\frac{1}{2} \end{bmatrix} \quad \text{or simply} \quad -\frac{1}{2} \end{aligned}$$

Interpret the Solution: This means that in the long run, when Zach and Josh use the indicated mixed strategies, Zach will lose about \$0.50 per game on average. ■

Checkpoint 1 Now work Exercise 3.

Technology Option

Figure 10.2.1

Recall from Chapter 2 that matrix multiplication can be performed on a graphing calculator. In Figure 10.2.1, we have entered matrices P, A, and Q into a calculator and, for the calculator's sake, we have renamed them matrices A, B, and C. Figure 10.2.1 shows the result of the matrix multiplication. Note that it is the same result that we obtained in Example 1.

For more information on how to use your calculator with matrices, consult the on-line graphing calculator manual at www.prenhall.com/armstrong

Solutions to a Game

In Section 10.1 we saw that the optimal strategy (pure strategy) for a strictly determined game was the row and column associated with a saddle point. This optimal strategy gave us the **solution** to a strictly determined game. For **nonstrictly determined games** a player needs a mixed strategy, since a pure strategy would be detected by the opponent. Unfortunately, there are an infinite number of mixed strategies for each player in a nonstrictly determined game. So how can an optimal mixed strategy be determined for each player? (An optimal mixed strategy would maximize the expected payoff for the row player, while the column player simultaneously wants to minimize it).

Before we supply a formula to determine the optimal mixed strategy, let's consider the thought processes of each player. The row player assumes that any mixed strategy he uses will be countered by the column player using a strategy that minimizes the row player's payoff. Hence, the row player will use the mixed strategy for which his expected payoff is maximized when the column player offers his best counterstrategy.

Likewise, the column player assumes that the row player will use a counterstrategy that maximizes the row player's payoff. Hence, the column player will use the mixed strategy that minimizes the payoff to the row player, when the row player uses his best counterstrategy.

The following formulas, which give the optimal strategies and therefore the solution to a nonstrictly determined game, are supplied without proof. In Section 10.3 we show how to solve a game by converting it to a linear programming problem and using the simplex method, along with duality.

Optimal Mixed Strategy for Nonstrictly Determined Games

For payoff matrix

$$A = \begin{bmatrix} a & b \\ c & d \end{bmatrix}$$

let $S = (a + d) - (b + c)$.

The **optimal mixed strategy for the row player** is

$$P_{\text{opt}} = [\, p_1 \quad p_2 \,]$$

where $p_1 = \dfrac{d - c}{S}$ and $p_2 = 1 - p_1$.

The **optimal mixed strategy for the column player** is

$$Q_{\text{opt}} = \begin{bmatrix} q_1 \\ q_2 \end{bmatrix}$$

where $q_1 = \dfrac{d - b}{S}$ and $q_2 = 1 - q_1$.

The **value of the game**, v, is equivalent to the expected value $E = P_{\text{opt}} A Q_{\text{opt}}$. This can be simplified to $v = \dfrac{ad - bc}{S}$.

Note: In Section 2.4 we defined the determinant of a 2×2 matrix, $\det(A)$, as $\det(A) = ad - bc$. Notice that the value of the game, v, can be restated as $v = \dfrac{\det(A)}{S}$.

Example 2 Finding Optimal Mixed Strategies

Solve the Zach and Josh penny-matching game where the payoff matrix is

$$A = \text{Zach's choice} \begin{matrix} \\ \text{H} \\ \text{T} \end{matrix} \overset{\text{Josh's choice}}{\begin{matrix} \text{H} & \text{T} \\ \end{matrix}} \begin{bmatrix} 4 & -5 \\ -3 & 2 \end{bmatrix}$$

Solution

Understand the Situation: To solve this game means to determine the optimal mixed strategy for both players.

From the payoff matrix, we have $a = 4$, $b = -5$, $c = -3$, and $d = 2$. So S is

$$S = (a + d) - (b + c) = (4 + 2) - (-5 + -3) = 14$$

So p_1 is

$$p_1 = \frac{d-c}{S} = \frac{2-(-3)}{14} = \frac{5}{14}$$

and p_2 is

$$p_2 = 1 - p_1 = 1 - \frac{5}{14} = \frac{9}{14}$$

This gives us Zach's optimal mixed strategy:

$$P_{\text{opt}} = [\,p_1 \quad p_2\,] = [\,\tfrac{5}{14} \quad \tfrac{9}{14}\,]$$

For Josh we have

$$q_1 = \frac{d-b}{S} = \frac{2-(-5)}{14} = \frac{7}{14} = \frac{1}{2}$$

and q_2 is

$$q_2 = 1 - q_1 = 1 - \frac{1}{2} = \frac{1}{2}$$

Josh's optimal mixed strategy is

$$Q_{\text{opt}} = \begin{bmatrix} q_1 \\ q_2 \end{bmatrix} = \begin{bmatrix} \frac{1}{2} \\ \frac{1}{2} \end{bmatrix}$$

■

Checkpoint 2

Now work Exercise 13.

Example 3 Finding the Value of a Game

Find the value of the penny-matching game in Example 2 and determine if it favors one player over the other.

Solution

Recall from Example 2 that $a = 4$, $b = -5$, $c = -3$, and $d = 2$. We also determined that $S = 14$. Since the value of a game, v, is given by $v = \dfrac{ad - bc}{S}$, we determine that $ad - bc = 4(2) - (-5)(-3) = -7$. So the value of the game is

$$v = \frac{ad - bc}{S} = \frac{-7}{14} = -\frac{1}{2}$$

Interpret the Solution: Since the value of the game is negative, this game favors Josh, the column player, over Zach, the row player. Over the long run, in repeated plays of this game, when each player uses his optimal mixed strategy, Josh is expected to win $0.50 per game on average. ■

Checkpoint 3

Now work Exercise 23.

When doing the applications in the exercises, be sure to write down the payoff matrix (if it is not supplied) and employ the techniques of this section if the payoff matrix represents a nonstrictly determined game. If the payoff matrix represents a strictly determined game, use the techniques from Section 10.1.

SUMMARY

In this section we saw that **pure strategies** are used in strictly determined games and **mixed strategies** are used in nonstrictly determined games. Mixed strategies saw us revisit some old topics: probability, expected value, and independent events. We also revisited row matrices and column matrices, as well as matrix multiplication. We provided formulas for determining the optimal strategy (solution) for nonstrictly determined games.

- **Optimal Mixed Strategy for Nonstrictly Determined Games** For payoff matrix

$$A = \begin{bmatrix} a & b \\ c & d \end{bmatrix}$$

let $S = (a + d) - (b + c)$.

The **optimal mixed strategy for the row player** is

$$P_{\text{opt}} = [\, p_1 \quad p_2 \,]$$

where $p_1 = \dfrac{d - c}{S}$ and $p_2 = 1 - p_1$.

The **optimal mixed strategy for the column player** is

$$Q_{\text{opt}} = \begin{bmatrix} q_1 \\ q_2 \end{bmatrix}$$

where $q_1 = \dfrac{d - b}{S}$ and $q_2 = 1 - q_1$.

The **value of the game**, v, is equivalent to the expected value $E = P_{\text{opt}} A Q_{\text{opt}}$. This can be simplified to $v = \dfrac{ad - bc}{S}$.

SECTION 10.2 EXERCISES

In Exercises 1–10, determine the expected value E for each game whose payoff matrix and strategies P and Q (row and column player, respectively) are given.

1. $\begin{bmatrix} 1 & -2 \\ -2 & 3 \end{bmatrix}$, $P = [\, \frac{1}{2} \quad \frac{1}{2} \,]$, $Q = \begin{bmatrix} \frac{1}{3} \\ \frac{2}{3} \end{bmatrix}$

2. $\begin{bmatrix} -4 & 2 \\ 3 & 1 \end{bmatrix}$, $P = [\, \frac{1}{3} \quad \frac{2}{3} \,]$, $Q = \begin{bmatrix} \frac{3}{4} \\ \frac{1}{4} \end{bmatrix}$

3. $\begin{bmatrix} 5 & -4 \\ -3 & 6 \end{bmatrix}$, $P = [\, \frac{3}{8} \quad \frac{5}{8} \,]$, $Q = \begin{bmatrix} \frac{1}{2} \\ \frac{1}{2} \end{bmatrix}$

4. $\begin{bmatrix} -3 & 2 \\ 5 & -4 \end{bmatrix}$, $P = [\, \frac{2}{3} \quad \frac{1}{3} \,]$, $Q = \begin{bmatrix} \frac{1}{4} \\ \frac{3}{4} \end{bmatrix}$

5. $\begin{bmatrix} 4 & 1 & -3 \\ 3 & -2 & 0 \\ 4 & 2 & -1 \end{bmatrix}$, $P = [\, 0.3 \quad 0.4 \quad 0.3 \,]$, $Q = \begin{bmatrix} 0.2 \\ 0.5 \\ 0.3 \end{bmatrix}$

6. $\begin{bmatrix} 5 & -3 & 1 \\ 1 & 6 & -2 \\ 2 & -4 & 3 \end{bmatrix}$, $P = [\, 0.1 \quad 0.4 \quad 0.5 \,]$, $Q = \begin{bmatrix} 0.6 \\ 0.1 \\ 0.3 \end{bmatrix}$

7. $\begin{bmatrix} 3 & -1 & 4 \\ -2 & 2 & 5 \end{bmatrix}$, $P = [\, 0.75 \quad 0.25 \,]$, $Q = \begin{bmatrix} 0.2 \\ 0.2 \\ 0.6 \end{bmatrix}$

8. $\begin{bmatrix} -1 & 3 & -3 \\ 1 & -2 & 4 \end{bmatrix}$, $P = [\, 0.15 \quad 0.85 \,]$, $Q = \begin{bmatrix} 0.25 \\ 0.35 \\ 0.4 \end{bmatrix}$

9. $\begin{bmatrix} -1 & 3 \\ 2 & -2 \\ 3 & -3 \end{bmatrix}$, $P = [\, 0.2 \quad 0.4 \quad 0.4 \,]$, $Q = \begin{bmatrix} 0.7 \\ 0.3 \end{bmatrix}$

10. $\begin{bmatrix} 3 & -2 \\ -1 & 1 \\ -3 & 2 \end{bmatrix}$, $P = [\, 0.3 \quad 0.4 \quad 0.3 \,]$, $Q = \begin{bmatrix} 0.2 \\ 0.8 \end{bmatrix}$

11. A certain game has a payoff matrix

$$\begin{bmatrix} 2 & -3 \\ -2 & 4 \end{bmatrix}$$

Compute the expected value E for the pairs of strategies given in (a) to (d). Which of these strategies is most favorable to the row player?

(a) $P = [\, \frac{1}{2} \quad \frac{1}{2} \,]$, $Q = \begin{bmatrix} \frac{1}{2} \\ \frac{1}{2} \end{bmatrix}$

(b) $P = [\, \frac{1}{3} \quad \frac{2}{3} \,]$, $Q = \begin{bmatrix} \frac{1}{4} \\ \frac{3}{4} \end{bmatrix}$

(c) $P = [\, \frac{2}{5} \quad \frac{3}{5} \,]$, $Q = \begin{bmatrix} \frac{2}{7} \\ \frac{5}{7} \end{bmatrix}$

(d) $P = [\, \frac{3}{8} \quad \frac{5}{8} \,]$, $Q = \begin{bmatrix} \frac{2}{9} \\ \frac{7}{9} \end{bmatrix}$

12. A certain game has a payoff matrix

$$\begin{bmatrix} 4 & 1 & -3 \\ 3 & -2 & 0 \\ 4 & 2 & -1 \end{bmatrix}$$

Compute the expected value E for the pairs of strategies given in (a) to (d). Which of these strategies is most favorable to the row player?

(a) $P = [\, 0.3 \quad 0.4 \quad 0.3 \,]$, $Q = \begin{bmatrix} 0.2 \\ 0.5 \\ 0.3 \end{bmatrix}$

(b) $P = [0.4 \quad 0.2 \quad 0.4]$, $Q = \begin{bmatrix} 0.6 \\ 0.3 \\ 0.1 \end{bmatrix}$

(c) $P = [1 \quad 0 \quad 0]$, $Q = \begin{bmatrix} 1 \\ 0 \\ 0 \end{bmatrix}$

(d) $P = [0.25 \quad 0.45 \quad 0.3]$, $Q = \begin{bmatrix} 0.55 \\ 0.25 \\ 0.2 \end{bmatrix}$

In Exercises 13–22, determine the optimal strategies P_{opt} and Q_{opt} for the row and column players, respectively. Be sure to check for saddle points since some may be strictly determined games.

✓ **13.** $\begin{bmatrix} -4 & 5 \\ 3 & -4 \end{bmatrix}$ **14.** $\begin{bmatrix} -1 & 3 \\ 2 & 1 \end{bmatrix}$

15. $\begin{bmatrix} 3 & -4 \\ -7 & -5 \end{bmatrix}$ **16.** $\begin{bmatrix} 3 & -7 \\ -2 & 1 \end{bmatrix}$

17. $\begin{bmatrix} 0 & 3 \\ 4 & 0 \end{bmatrix}$ **18.** $\begin{bmatrix} 4 & 0 \\ 0 & 5 \end{bmatrix}$

19. $\begin{bmatrix} 5 & -4 \\ -3 & 6 \end{bmatrix}$ **20.** $\begin{bmatrix} -3 & 2 \\ 5 & -4 \end{bmatrix}$

21. $\begin{bmatrix} -\frac{1}{2} & \frac{1}{2} \\ \frac{2}{3} & -\frac{1}{3} \end{bmatrix}$ **22.** $\begin{bmatrix} \frac{5}{4} & -\frac{3}{4} \\ -\frac{7}{8} & \frac{7}{4} \end{bmatrix}$

In Exercises 23–28:

(a) Determine the value of the game.

(b) Determine which player (if any) the game favors, assuming that the row player and the column player use their optimal strategies.

✓ **23.** The game in Exercise 13

24. The game in Exercise 14

25. The game in Exercise 17

26. The game in Exercise 18

27. The game in Exercise 21

28. The game in Exercise 22

29. Let $A = \begin{bmatrix} a & b \\ c & d \end{bmatrix}$ be a payoff matrix and $P = [\, p_1 \quad p_2 \,]$ and $Q = \begin{bmatrix} q_1 \\ q_2 \end{bmatrix}$ be the strategies for the row player and the column player, respectively.

(a) Use matrix multiplication to multiply PA.

(b) Use matrix multiplication to multiply the result from part (a) by Q.

(c) Part (b) is really the product PAQ. Verify that this is equal to E, the expected value, given in the text.

30. Let $A = \begin{bmatrix} a & b \\ c & d \end{bmatrix}$ be a payoff matrix.

(a) Compute $P_{opt} = [\, p_1 \quad p_2 \,]$, the optimal mixed strategy for the row player.

(b) Compute $Q_{opt} = \begin{bmatrix} q_1 \\ q_2 \end{bmatrix}$, the optimal mixed strategy for the column player.

(c) Compute $E = P_{opt}\, AQ_{opt}$, the expected value of the game. Simplify the result to verify that $E = P_{opt}\, AQ_{opt} = \dfrac{ad - bc}{(a + d) - (b + c)}$.

Applications

31. The Stumpf Company introduces its new product on the market with a major television and newspaper advertising campaign. Dave, the CEO of the Stumpf Company, finds out that its major competitor, the Stitz Company, is planning a major advertising campaign for a comparable product. A market research firm has determined the following payoff matrix, where each entry shows the sales (in millions of dollars) either gained or lost by one company to the other.

		Stumpf	
		TV	Paper
Stitz	TV	1.5	−0.8
	Paper	−0.7	1.2

(a) Find the optimum strategies for the Stumpf Company and the Stitz Company.

(b) Find the value of the game assuming that both use the optimal mixed strategy. Does the game favor anyone? If so, who?

32. The Stumpf Company introduces its new product on the market with a radio and mail circular advertising campaign. Dave, the CEO of the Stumpf Company, finds out that its major competitor, the Stitz Company, is planning a major advertising campaign for a comparable product. A market research firm has determined the following payoff matrix, where each entry shows the sales (in millions of dollars) either gained or lost by one company to the other.

		Stumpf	
		Radio	Mail
Stitz	Radio	1.5	−0.8
	Mail	−0.7	1.2

(a) Find the optimum strategies for the Stumpf Company and the Stitz Company.

(b) Find the value of the game, assuming that both use the optimal mixed strategy. Does the game favor anyone? If so, who?

33. The Bridge, a clothing retail store, is contemplating remodeling to modernize its physical appearance. A market research firm determines that the main determining factor in The Bridge's decision is the economy and whether it will remain strong or slip into recession. The firm determines that if The Bridge remodels and the economy is strong The Bridge will earn $400,000. However, if The Bridge remodels and the economy slips into recession, the business will lose $150,000. If The Bridge does not remodel and the economy remains strong, the business will lose $125,000 due to customers going to other stores. Finally, if The Bridge does not remodel and there is a recession, the business will lose $100,000. Let The Bridge be the row player and the economy the column player.

(a) Construct the payoff matrix for this game, where the entries are in thousands of dollars.

(b) Determine the optimal strategy for The Bridge.

(c) If you were the CEO of The Bridge, what would you do?

34. (*continuation of Exercise 33*)

(a) Assuming the optimal strategies in Exercise 33, determine the value of the game.

(b) Does the game in Exercise 33 favor anyone? If so, who?

35. The Riblets have decided to invest $50,000 in the stocks of two companies listed on the New York Stock Exchange (NYSE). One of the companies gets its revenue mainly from the technology sector, whereas the other is a pharmaceutical company. Market analysts believe that if the economy is growing the technology stock will have an annual gain of 35%, whereas the pharmaceutical stock will have an annual gain of 15%. However, if the economy experiences a recession, the technology stock will have an annual loss of 15% and the pharmaceutical stock will have an annual gain of 16%. Let each stock choice be the row player and the economy be the column player.

(a) Construct the payoff matrix for this game, where entries are percentages of increase or decrease in the value of each investment.

(b) Determine the optimal strategy for the Riblet's investment of $50,000.

(c) If the Riblet's use their optimal investment strategy, what profit can they expect to make on their investments after 1 year?

36. The Hannuses have decided to invest $80,000 into two types of mutual funds: a stock fund and a bond fund. Their financial planner believes that if the economy is growing the stock fund will have an annual gain of 30%, whereas the bond fund will have an annual gain of 10%. However, if the economy experiences a recession, the stock fund will have an annual loss of 15% and the bond fund will have an annual gain of 12%. Let the fund choice be the row player and the economy be the column player.

(a) Construct the payoff matrix for this game, where entries are percentages of increase or decrease in the value of each investment.

(b) Determine the optimal strategy for the Hannuses investment of $80,000.

(c) If the Hannuses use their optimal investment strategy, what profit can they expect to make on their investments after 1 year?

37. Beth Toth and Mike Williams are running for governor of a certain state. The state has several urban areas and rural areas. If both candidates campaign only in urban areas, Toth is expected to get 65% of the vote, and if both campaign only in rural areas, Toth is expected to get 60% of the vote. If Toth campaigns solely in the urban areas and Williams campaigns solely in the rural areas, Toth is expected to get 35% of the vote. If Toth campaigns solely in the rural areas and Williams campaigns solely in the urban areas, Toth is expected to get 40% of the vote. Let Toth be the row player and Williams the column player.

(a) Construct the payoff matrix for this game, where the entries are percentages written as decimals.

(b) Determine the optimal strategy for Toth and Williams.

38. (*continuation of Exercise 37*) A scandal in the Toth versus Williams governor race has been reported and it is affecting the Toth campaign. If both candidates campaign only in urban areas, Toth is expected to get 55% of the vote, and if both campaign only in rural areas, Toth is expected to get 65% of the vote. If Toth campaigns solely in urban areas and Williams campaigns solely in rural areas, Toth is expected to get 45% of the vote. If Toth campaigns solely in rural areas and Williams campaigns solely in urban areas, Toth is expected to get 35% of the vote. Let Toth be the row player and Williams the column player.

(a) Construct the payoff matrix for this game, where the entries are percentages written as decimals.

(b) Determine the optimal strategy for Toth and Williams.

SECTION PROJECT

Maria and Gloria are playing a game called **matching fingers** in which they show either 1 or 2 fingers at the same time. If there is a match (both show 1 or 2 fingers), Maria wins an amount of dollars equal to the total number of fingers shown. If there is no match, Gloria wins an amount of dollars equal to the number of fingers shown. Let Maria be the row player and Gloria the column player.

(a) Write the payoff matrix for this game.

(b) Find the optimal strategies for Maria and Gloria.

(c) Determine the value of this game, assuming that both players use their optimal strategies.

(d) Does the game favor one player over the other? If so, who?

(e) Repeat parts (a) to (d), except now the game is one in which Maria and Gloria show either 1 or 3 fingers at the same time. Assume the same type of money payoff as described earlier.

Section 10.3 Game Theory and Linear Programming

From Your Toolbox

A matrix is square if the number of rows equals the number of columns.

All games that we discussed in Section 10.2 used mixed strategies and had a square matrix for the payoff matrix. (Consult the Toolbox to the left.) Specifically, all games had 2×2 payoff matrices. In this section we will see how the techniques of linear programming and the simplex method can be utilized to solve games in which the row player and column player have an arbitrary (but finite) number of choices.

The Fundamental Theorem of Game Theory

Before we get to the heart of this section, we state without proof the **fundamental theorem of game theory**.

Fundamental Theorem of Game Theory

For a two-person, zero-sum game with a finite payoff matrix A, there exist strategies P_{opt} and Q_{opt} (not necessarily unique) for the row player and column player, respectively, and a unique number v such that

$$P_{\text{opt}} AQ \geq v$$

for every strategy Q of the column player and

$$PAQ_{\text{opt}} \leq v$$

for every strategy P of the row player.

Some observations on the fundamental theorem of game theory follow:

- v is the **value of the game**.
- P_{opt} and Q_{opt} are the optimal strategies for the row player and the column player, respectively.
- $P_{\text{opt}} AQ$ is the expected value of the game when the row player uses the optimal strategy and the column player uses any strategy. PAQ_{opt} is the expected value of the game when the column player uses the optimal strategy and the row player uses any strategy.
- The theorem says that every game has optimal strategies for the row player and the column player. Finding these strategies, as we did in Section 10.2, is called **solving the game**.
- As we saw in Section 10.2, when both players use optimal strategies, the expected value of the game is v.

We delayed introducing the fundamental theorem of game theory until now since it has some structures in it (the $\geq$ and $\leq$) that lead us to linear programming.

Linear Programming and Game Theory

Now that we have introduced the fundamental theorem of game theory, let's look at an example in which the optimal strategy for the row player can be determined by rewriting it as a linear programming problem.

Example 1 Using Linear Programming to Find an Optimal Strategy

Determine a linear programming problem that will yield the row player's optimal strategy for the payoff matrix

$$A = \begin{bmatrix} 1 & 4 \\ 3 & 2 \end{bmatrix}$$

Solution

In general, any strategy for the row and column player, respectively, is given by

$$P = [\, p_1 \quad p_2 \,] \quad \text{and} \quad Q = \begin{bmatrix} q_1 \\ q_2 \end{bmatrix}$$

This means that the expected value of the game, in general, is

$$\begin{aligned} E = PAQ &= [\, p_1 \quad p_2 \,] \cdot \begin{bmatrix} 1 & 4 \\ 3 & 2 \end{bmatrix} \cdot \begin{bmatrix} q_1 \\ q_2 \end{bmatrix} \\ &= [\, p_1 + 3p_2 \quad 4p_1 + 2p_2 \,] \cdot \begin{bmatrix} q_1 \\ q_2 \end{bmatrix} \\ &= (p_1 + 3p_2)q_1 + (4p_1 + 2p_2)q_2 \end{aligned}$$

If the column player always chooses column 1, then $q_1 = 1$ and $q_2 = 0$, and the expected value is

$$E = p_1 + 3p_2 \qquad \text{(I)}$$

However, if the column player always chooses column 2, then $q_1 = 0$ and $q_2 = 1$ and the expected value is

$$E = 4p_1 + 2p_2 \qquad \text{(II)}$$

The goal of the row player is to find values for p_1 and p_2 (where $p_1 + p_2 = 1$) that maximize the expected value of the game. Let v be the smaller of the two expected values in Equations (I) and (II). In either case, we have $E \geq v$. This means that if we can maximize v then we maximize E. Also, since all entries in the payoff matrix A are positive, we can claim that v is positive, that is, $v > 0$. We can now write Equations (I) and (II) as

$$\begin{aligned} p_1 + 3p_2 &\geq v \\ 4p_1 + 2p_2 &\geq v \end{aligned}$$

Since $v > 0$, we can divide both sides of each inequality by v and not have to worry about the inequality symbol reversing direction. Doing this gives

$$\begin{aligned} \frac{p_1}{v} + 3\left(\frac{p_2}{v}\right) &\geq 1 \\ 4\left(\frac{p_1}{v}\right) + 2\left(\frac{p_2}{v}\right) &\geq 1 \end{aligned}$$

To simplify notation, we substitute $x_1 = \dfrac{p_1}{v}$ and $x_2 = \dfrac{p_2}{v}$ ($x_1 \geq 0$ and $x_2 \geq 0$). The inequalities can be written more simply as

$$\begin{aligned} x_1 + 3x_2 &\geq 1 && \text{(III)} \\ 4x_1 + 2x_2 &\geq 1 && \text{(IV)} \\ x_1 \geq 0,\ x_2 &\geq 0 && \text{(V)} \end{aligned}$$

Inequalities (III), (IV), and (V) look like the constraints for a linear programming problem. All we need is to determine the objective function. The objective function comes from the fact that $p_1 + p_2 = 1$. Dividing both sides of $p_1 + p_2 = 1$ by v yields

$$\frac{p_1}{v} + \frac{p_2}{v} = \frac{1}{v} \quad \text{or} \quad x_1 + x_2 = \frac{1}{v}$$

Remember that we are trying to **maximize** v. This happens if $\frac{1}{v}$ is **minimized.** (As v gets larger, $\frac{1}{v}$ gets smaller.) Hence, we now have the linear programming problem as

$$\begin{aligned} \text{Minimize:} \quad & y = \frac{1}{v} = x_1 + x_2 \\ \text{Subject to:} \quad & x_1 + 3x_2 \geq 1 \\ & 4x_1 + 2x_2 \geq 1 \\ & x_1 \geq 0, x_2 \geq 0 \end{aligned}$$

Notice that once the values for x_1 and x_2 have been found we can use $\frac{1}{v} = x_1 + x_2$ to find v. Quite simply, v is

$$v = \frac{1}{x_1 + x_2}$$

Also, the substitutions $x_1 = \frac{p_1}{v}$ and $x_2 = \frac{p_2}{v}$ allow us to write the optimal row strategy as

$$p_1 = vx_1 \quad \text{and} \quad p_2 = vx_2$$ ■

Using Example 1 as a guide, we now summarize the procedure for finding the optimal row strategy when given a 2×2 payoff matrix. This can be further generalized to larger payoff matrices, as we show later in Example 3.

Using Linear Programming to Determine Optimal Row Strategy

Given a game with payoff matrix

$$A = \begin{bmatrix} a & b \\ c & d \end{bmatrix}$$

where $a, b, c,$ and d are positive numbers, we can find $P_{\text{opt}} = [\,p_1 \quad p_2\,]$, the optimal row strategy, by solving the linear programming problem

$$\begin{aligned} \text{Minimize:} \quad & y = \frac{1}{v} = x_1 + x_2 \\ \text{Subject to:} \quad & ax_1 + cx_2 \geq 1 \\ & bx_1 + dx_2 \geq 1 \\ & x_1 \geq 0, x_2 \geq 0 \end{aligned}$$

The expected value of the game is $v = \frac{1}{x_1 + x_2}$. The optimal row strategy is $P_{\text{opt}} = [\,p_1 \quad p_2\,]$, where $p_1 = vx_1$ and $p_2 = vx_2$.

From Your Toolbox

For a given matrix A, if we interchange the rows and columns, we produce the transpose of A, denoted A^T.

Before continuing, notice how the coefficients of x_1 and x_2 in the main constraints are the rows of the **transpose** of the original payoff matrix A. (Consult the Toolbox to the left.)

Going through a similar analysis as we did in Example 1, except for the column player, we would reach the following for finding the optimal column strategy.

Using Linear Programming to Determine Optimal Column Strategy

Given a game with payoff matrix

$$A = \begin{bmatrix} a & b \\ c & d \end{bmatrix}$$

where a, b, c, and d are positive numbers, we can find $Q_{\text{opt}} = \begin{bmatrix} q_1 \\ q_2 \end{bmatrix}$, the optimal column strategy, by solving the linear programming problem

$$\begin{aligned} &\text{Maximize:} && z = \frac{1}{v} = y_1 + y_2 \\ &\text{Subject to:} && ay_1 + by_2 \leq 1 \\ & && cy_1 + dy_2 \leq 1 \\ & && y_1 \geq 0,\ y_2 \geq 0 \end{aligned}$$

The expected value of the game is $v = \dfrac{1}{y_1 + y_2}$. The optimal column strategy is $Q_{\text{opt}} = \begin{bmatrix} q_1 \\ q_2 \end{bmatrix}$, where $q_1 = vy_1$ and $q_2 = vy_2$.

Here observe that the coefficients of y_1 and y_2 in the main constraints are simply the rows of the payoff matrix A. Since the coefficients of x_1 and x_2 in the main constraints in the Optimal Row Strategy box are the transpose of payoff matrix A and the coefficients of y_1 and y_2 in the main constraints in the Optimal Column Strategy box are simply the payoff matrix A, it appears that the linear programming problems for row and column strategies are actually **dual problems** (see Section 4.3). This means that we can determine both optimal strategies by using the simplex method to solve the linear programming problem for the column player's optimal strategy. The concept of duality then allows us to read the row player's optimal strategy from the bottom row of the final tableau. Example 2 illustrates.

Example 2 Solving a Game Using the Simplex Method

Solve the game that has payoff matrix

$$A = \begin{bmatrix} 1 & 4 \\ 3 & 2 \end{bmatrix}$$

Solution

Understand the Situation: This is the same payoff matrix given in Example 1. The row player's optimal strategy is found by solving

$$\begin{aligned} &\text{Minimize:} && y = \frac{1}{v} = x_1 + x_2 \\ &\text{Subject to:} && x_1 + 3x_2 \geq 1 \\ & && 4x_1 + 2x_2 \geq 1 \\ & && x_1 \geq 0,\ x_2 \geq 0 \end{aligned}$$

The column player's optimal strategy is found by solving

$$\begin{aligned} \text{Maximize:} \quad & z = \frac{1}{v} = y_1 + y_2 \\ \text{Subject to:} \quad & y_1 + 4y_2 \le 1 \\ & 3y_1 + 2y_2 \le 1 \\ & y_1 \ge 0,\ y_2 \ge 0 \end{aligned}$$

From Your Toolbox

The column with the most negative indicator is the pivot column. If there are two, use the leftmost indicator. To find the pivot row, divide each number from the rightmost column by the corresponding number from the pivot column. The smallest nonnegative quotient gives the pivot row. For more details, see Section 4.2.

We use the simplex method on the column player's linear programming problem. The initial tableau, with first pivot element (consult the Toolbox to the left) circled, is

$$\begin{array}{c} \begin{matrix} y_1 & y_2 & s_1 & s_2 & z \end{matrix} \\ \left[\begin{array}{rrrrr|r} 1 & 4 & 1 & 0 & 0 & 1 \\ \textcircled{3} & 2 & 0 & 1 & 0 & 1 \\ \hline -1 & -1 & 0 & 0 & 1 & 0 \end{array}\right] \begin{matrix} \text{Quotients:} \\ 1/1 \\ 1/3 \\ \end{matrix} \\ \underbrace{\qquad\qquad\qquad}_{\text{Indicators}} \end{array}$$

Pivoting gives us

$$\begin{matrix} y_1 & y_2 & s_1 & s_2 & z \end{matrix} \qquad\qquad \begin{matrix} y_1 & y_2 & s_1 & s_2 & z \end{matrix}$$

$$\left[\begin{array}{rrrrr|r} 1 & 4 & 1 & 0 & 0 & 1 \\ \textcircled{3} & 2 & 0 & 1 & 0 & 1 \\ \hline -1 & -1 & 0 & 0 & 1 & 0 \end{array}\right] \xrightarrow{\frac{1}{3}R_2 \to R_2} \left[\begin{array}{rrrrr|r} 1 & 4 & 1 & 0 & 0 & 1 \\ 1 & \frac{2}{3} & 0 & \frac{1}{3} & 0 & \frac{1}{3} \\ \hline -1 & -1 & 0 & 0 & 1 & 0 \end{array}\right]$$

$$\begin{array}{r} -1R_2 + R_1 \to R_1 \\ R_2 + R_3 \to R_3 \end{array} \quad \begin{matrix} y_1 & y_2 & s_1 & s_2 & z \end{matrix} \left[\begin{array}{rrrrr|r} 0 & \frac{10}{3} & 1 & -\frac{1}{3} & 0 & \frac{2}{3} \\ 1 & \frac{2}{3} & 0 & \frac{1}{3} & 0 & \frac{1}{3} \\ \hline 0 & -\frac{1}{3} & 0 & \frac{1}{3} & 1 & \frac{1}{3} \end{array}\right]$$

The next pivot element is the $\frac{10}{3}$ in row 1, column 2.

$$\begin{matrix} y_1 & y_2 & s_1 & s_2 & z \end{matrix} \qquad\qquad \begin{matrix} y_1 & y_2 & s_1 & s_2 & z \end{matrix}$$

$$\left[\begin{array}{rrrrr|r} 0 & \textcircled{\tfrac{10}{3}} & 1 & -\frac{1}{3} & 0 & \frac{2}{3} \\ 1 & \frac{2}{3} & 0 & \frac{1}{3} & 0 & \frac{1}{3} \\ \hline 0 & -\frac{1}{3} & 0 & \frac{1}{3} & 1 & \frac{1}{3} \end{array}\right] \xrightarrow{\frac{3}{10}R_1 \to R_1} \left[\begin{array}{rrrrr|r} 0 & 1 & \frac{3}{10} & -\frac{1}{10} & 0 & \frac{1}{5} \\ 1 & \frac{2}{3} & 0 & \frac{1}{3} & 0 & \frac{1}{3} \\ \hline 0 & -\frac{1}{3} & 0 & \frac{1}{3} & 1 & \frac{1}{3} \end{array}\right]$$

$$\begin{array}{r} -\frac{2}{3}R_1 + R_2 \to R_2 \\ \frac{1}{3}R_1 + R_3 \to R_3 \end{array} \quad \begin{matrix} y_1 & y_2 & s_1 & s_2 & z \end{matrix} \left[\begin{array}{rrrrr|r} 0 & 1 & \frac{3}{10} & -\frac{1}{10} & 0 & \frac{1}{5} \\ 1 & 0 & -\frac{1}{5} & \frac{2}{5} & 0 & \frac{1}{5} \\ \hline 0 & 0 & \frac{1}{10} & \frac{3}{10} & 1 & \frac{2}{5} \end{array}\right]$$

We now have our final tableau. From this we find that $y_1 = \frac{1}{5}$ and $y_2 = \frac{1}{5}$. This means that $y_1 + y_2 = \frac{2}{5}$, giving the expected value of the game to be $v = \frac{5}{2}$. Also, we have

$$\begin{aligned} q_1 &= vy_1 & \text{and} \quad q_2 &= vy_2 \\ &= \frac{5}{2}\left(\frac{1}{5}\right) & &= \frac{5}{2}\left(\frac{1}{5}\right) \\ &= \frac{1}{2} & &= \frac{1}{2} \end{aligned}$$

Hence the column player's optimal strategy is $Q_{\text{opt}} = \begin{bmatrix} q_1 \\ q_2 \end{bmatrix} = \begin{bmatrix} \frac{1}{2} \\ \frac{1}{2} \end{bmatrix}$.

Now the solution to the minimum linear programming problem (the row player's strategy) is obtained from the bottom row of the final tableau, precisely in the slack variables columns. In other words, the final tableau shows us that $x_1 = \frac{1}{10}$ and $x_2 = \frac{3}{10}$. This means that $x_1 + x_2 = \frac{4}{10} = \frac{2}{5}$, from which we again find that the expected value of the game is $v = \frac{5}{2}$. Also, we have

$$\begin{aligned} p_1 &= vx_1 & \text{and} \quad p_2 &= vx_2 \\ &= \frac{5}{2}\left(\frac{1}{10}\right) & &= \frac{5}{2}\left(\frac{3}{10}\right) \\ &= \frac{1}{4} & &= \frac{3}{4} \end{aligned}$$

So the row player's optimal strategy is $P_{\text{opt}} = [\,p_1 \quad p_2\,] = [\,\frac{1}{4} \quad \frac{3}{4}\,]$. As a final check, the value of the game $v = \frac{5}{2}$, should be the result of $P_{\text{opt}} \cdot A \cdot Q_{\text{opt}}$. Evaluating, we get

$$\begin{aligned} P_{\text{opt}} \cdot A \cdot Q_{\text{opt}} &= [\,\tfrac{1}{4} \quad \tfrac{3}{4}\,] \cdot \begin{bmatrix} 1 & 4 \\ 3 & 2 \end{bmatrix} \cdot \begin{bmatrix} \frac{1}{2} \\ \frac{1}{2} \end{bmatrix} \\ &= [\,\tfrac{5}{2} \quad \tfrac{5}{2}\,] \cdot \begin{bmatrix} \frac{1}{2} \\ \frac{1}{2} \end{bmatrix} = 2.5 = \frac{5}{2} \end{aligned}$$

■

✓ Checkpoint 1 Now work Exercise 1.

Before considering another example, we need to comment on the requirement for all entries in payoff matrix A to be positive. Recall that in Example 1 this was necessary to preserve the direction of the inequalities. However, we can relax this restriction from this point forward. If some of the entries in a payoff matrix A are negative, we can add a positive number n to each entry so that we create a new payoff matrix B in which all entries are positive. Matrix B and matrix A will have exactly the same optimal strategies. To find the value of the game with matrix A, we find the value of the game with matrix B and then subtract n. All this can be summarized as follows:

Optimal strategies of a game do not change if a positive constant n is added to each entry in the payoff matrix. If w is the value of the new game, then the value of the original game is $v = w - n$.

In the Flashback that follows, we illustrate how to use this technique.

Flashback

Zach and Josh Revisited

Recall the penny-matching game that Zach and Josh are playing with the payoff matrix

$$\text{Zach's choice}\;\begin{array}{c} \\ \text{H} \\ \text{T}\end{array}\!\!\begin{array}{c}\text{Josh's choice}\\ \begin{array}{cc}\text{H} & \text{T}\end{array}\\ \begin{bmatrix} 4 & -5 \\ -3 & 2 \end{bmatrix}\end{array}$$

Use the simplex method and duals to find the optimal strategy for each player and the value of the game.

Flashback Solution

We begin by adding 6 to each entry in the payoff matrix to yield an alternative payoff matrix B with all positive entries.

$$B = \begin{bmatrix} 10 & 1 \\ 3 & 8 \end{bmatrix}$$

From Josh's viewpoint, the column player, the linear programming problem has the following standard maximum form:

$$\begin{aligned} \text{Maximize:}\quad & z = \frac{1}{v} = y_1 + y_2 \\ \text{Subject to:}\quad & 10y_1 + y_2 \le 1 \\ & 3y_1 + 8y_2 \le 1 \\ & y_1 \ge 0,\ y_2 \ge 0 \end{aligned}$$

The initial simplex tableau with pivot element circled is

$$\begin{array}{c} \begin{array}{cccccc} y_1 & y_2 & s_1 & s_2 & z & \end{array} \\ \left[\begin{array}{ccccc:c} \textcircled{10} & 1 & 1 & 0 & 0 & 1 \\ 3 & 8 & 0 & 1 & 0 & 1 \\ \hdashline -1 & -1 & 0 & 0 & 1 & 0 \end{array}\right] \end{array}$$

Pivoting gives us

$$\left[\begin{array}{ccccc:c} \textcircled{10} & 1 & 1 & 0 & 0 & 1 \\ 3 & 8 & 0 & 1 & 0 & 1 \\ \hdashline -1 & -1 & 0 & 0 & 1 & 0 \end{array}\right] \xrightarrow{\frac{1}{10}R_1 \to R_1} \left[\begin{array}{ccccc:c} 1 & \frac{1}{10} & \frac{1}{10} & 0 & 0 & \frac{1}{10} \\ 3 & 8 & 0 & 1 & 0 & 1 \\ \hdashline -1 & -1 & 0 & 0 & 1 & 0 \end{array}\right]$$

(columns: y_1, y_2, s_1, s_2, z)

$$\begin{array}{c} -3R_1 + R_2 \to R_2 \\ R_1 + R_3 \to R_3 \end{array} \left[\begin{array}{ccccc:c} 1 & \frac{1}{10} & \frac{1}{10} & 0 & 0 & \frac{1}{10} \\ 0 & \frac{77}{10} & -\frac{3}{10} & 1 & 0 & \frac{7}{10} \\ \hdashline 0 & -\frac{9}{10} & \frac{1}{10} & 0 & 1 & \frac{1}{10} \end{array}\right]$$

(columns: y_1, y_2, s_1, s_2, z)

Next we pivot on the $\frac{77}{10}$ in row 2, column 2.

$$\begin{array}{c} \quad y_1 \quad y_2 \quad s_1 \quad s_2 \quad z \\ \left[\begin{array}{ccccc:c} 1 & \frac{1}{10} & \frac{1}{10} & 0 & 0 & \frac{1}{10} \\ 0 & \boxed{\frac{77}{10}} & -\frac{3}{10} & 1 & 0 & \frac{7}{10} \\ \hdashline 0 & -\frac{9}{10} & \frac{1}{10} & 0 & 1 & \frac{1}{10} \end{array}\right] \end{array} \xrightarrow{\frac{10}{77}R_2 \to R_2} \begin{array}{c} \quad y_1 \quad y_2 \quad s_1 \quad s_2 \quad z \\ \left[\begin{array}{ccccc:c} 1 & \frac{1}{10} & \frac{1}{10} & 0 & 0 & \frac{1}{10} \\ 0 & 1 & -\frac{3}{77} & \frac{10}{77} & 0 & \frac{1}{11} \\ \hdashline 0 & -\frac{9}{10} & \frac{1}{10} & 0 & 1 & \frac{1}{10} \end{array}\right] \end{array}$$

$$\begin{array}{l} -\frac{1}{10}R_2 + R_1 \to R_1 \\ \frac{9}{10}R_2 + R_3 \to R_3 \end{array} \quad \begin{array}{c} \quad y_1 \quad y_2 \quad s_1 \quad s_2 \quad z \\ \left[\begin{array}{ccccc:c} 1 & 0 & \frac{8}{77} & -\frac{1}{77} & 0 & \frac{1}{11} \\ 0 & 1 & -\frac{3}{77} & \frac{10}{77} & 0 & \frac{1}{11} \\ \hdashline 0 & 0 & \frac{5}{77} & \frac{9}{77} & 1 & \frac{2}{11} \end{array}\right] \end{array}$$

From the final tableau, we find that $y_1 = \frac{1}{11}$ and $y_2 = \frac{1}{11}$. This means that $y_1 + y_2 = \frac{2}{11}$ and the value of the game is $v = \frac{11}{2}$. Also

$$q_1 = vy_1 = \frac{11}{2}\left(\frac{1}{11}\right) = \frac{1}{2} \qquad \text{and} \qquad q_2 = vy_2 = \frac{11}{2}\left(\frac{1}{11}\right) = \frac{1}{2}$$

So Josh's optimal strategy is $Q_{\text{opt}} = \begin{bmatrix} \frac{1}{2} \\ \frac{1}{2} \end{bmatrix}$. The row player Zach's optimal strategy is found by noting that $x_1 = \frac{5}{77}$ and $x_2 = \frac{9}{77}$. So $x_1 + x_2 = \frac{14}{77} = \frac{2}{11}$, which means again that the value of the game is $v = \frac{11}{2}$. Also,

$$p_1 = vx_1 = \frac{11}{2}\left(\frac{5}{77}\right) = \frac{5}{14} \qquad \text{and} \qquad p_2 = vx_2 = \frac{11}{2}\left(\frac{9}{77}\right) = \frac{9}{14}$$

Interpret the Solution: So Zach's optimal strategy is $P_{\text{opt}} = \begin{bmatrix} \frac{5}{14} & \frac{9}{14} \end{bmatrix}$. The value of the original game is $v = \frac{11}{2}$ minus the 6, which was added to the original payoff matrix. This gives $v = -\frac{1}{2}$.

Interactive Activity

Compare the result of the Flashback with Example 2 and Example 3 in Section 10.2.

✓ Checkpoint 2

Now work Exercise 11.

Our final example illustrates how the linear programming techniques and the simplex method can be applied to payoff matrices of any size.

Example 3 Analyzing Larger Payoff Matrices

Use linear programming and the simplex method to find the optimal strategy for the column player for a game that has payoff matrix

$$A = \begin{bmatrix} 1 & 4 & 7 \\ 5 & 3 & 8 \end{bmatrix}$$

Solution

The linear programming problem for the column player is

$$\begin{aligned} &\text{Maximize:} && z = \frac{1}{v} = y_1 + y_2 + y_3 \\ &\text{Subject to:} && y_1 + 4y_2 + 7y_3 \le 1 \\ &&& 5y_1 + 3y_2 + 8y_3 < 1 \\ &&& y_1 \ge 0,\ y_2 \ge 0,\ y_3 \ge 0 \end{aligned}$$

The initial tableau with pivot element circled is

$$\begin{array}{c} \begin{matrix} y_1 & y_2 & y_3 & s_1 & s_2 & z \end{matrix} \\ \left[\begin{array}{cccccc:c} 1 & 4 & 7 & 1 & 0 & 0 & 1 \\ \textcircled{5} & 3 & 8 & 0 & 1 & 0 & 1 \\ \hdashline -1 & -1 & -1 & 0 & 0 & 1 & 0 \end{array}\right] \end{array}$$

Pivoting gives us

$$\begin{array}{r} \frac{1}{5}R_2 \to R_2 \\ \text{Then } -1R_2 + R_1 \to R_1 \\ R_2 + R_3 \to R_3 \end{array} \begin{array}{c} \begin{matrix} y_1 & y_2 & y_3 & s_1 & s_2 & z \end{matrix} \\ \left[\begin{array}{cccccc:c} 0 & \frac{17}{5} & \frac{27}{5} & 1 & -\frac{1}{5} & 0 & \frac{4}{5} \\ 1 & \frac{3}{5} & \frac{8}{5} & 0 & \frac{1}{5} & 0 & \frac{1}{5} \\ \hdashline 0 & -\frac{2}{5} & \frac{3}{5} & 0 & \frac{1}{5} & 1 & \frac{1}{5} \end{array}\right] \end{array}$$

Pivoting on the $\frac{17}{5}$ in row 1, column 2 yields

$$\begin{array}{r} \frac{5}{17}R_1 \to R_1 \\ \text{Then } -\frac{3}{5}R_1 + R_2 \to R_2 \\ \frac{2}{5}R_1 + R_3 \to R_3 \end{array} \begin{array}{c} \begin{matrix} y_1 & y_2 & y_3 & s_1 & s_2 & z \end{matrix} \\ \left[\begin{array}{cccccc:c} 0 & 1 & \frac{27}{17} & \frac{5}{17} & -\frac{1}{17} & 0 & \frac{4}{17} \\ 1 & 0 & \frac{11}{17} & -\frac{3}{17} & \frac{4}{17} & 0 & \frac{1}{17} \\ \hdashline 0 & 0 & \frac{21}{17} & \frac{2}{17} & \frac{3}{17} & 1 & \frac{5}{17} \end{array}\right] \end{array}$$

From our final tableau, we find that $y_1 = \frac{1}{17}$, $y_2 = \frac{4}{17}$, and $y_3 = 0$. So $y_1 + y_2 + y_3 = \frac{5}{17}$, which means that $v = \frac{17}{5}$. This gives us

$$\begin{aligned} q_1 &= vy_1 & q_2 &= vy_2 & q_3 &= vy_3 \\ &= \frac{17}{5}\left(\frac{1}{17}\right) & &= \frac{17}{5}\left(\frac{4}{17}\right) & &= \frac{17}{5}(0) \\ &= \frac{1}{5} & &= \frac{4}{5} & &= 0 \end{aligned}$$

So the column player's optimal strategy is

$$Q_{\text{opt}} = \begin{bmatrix} q_1 \\ q_2 \\ q_3 \end{bmatrix} = \begin{bmatrix} \frac{1}{5} \\ \frac{4}{5} \\ 0 \end{bmatrix}$$

■

Interactive Activity

Find the row player's optimal strategy in Example 3.

WWW

SUMMARY

We introduced this section with the fundamental theorem of game theory and used it to help to set up a linear programming problem to determine optimal strategies. We saw how the linear programming problems to determine P_{opt} and Q_{opt} are actually **dual problems**. The techniques outlined in Section 4.3 allowed us to solve and use duals to determine each player's optimal strategy. We also saw how to modify a payoff matrix so that it has no negative entries and how this has no effect on the optimal strategies. The section concluded with an example of how to use linear programming to solve a game with any size of payoff matrix.

- **Fundamental Theorem of Game Theory** For a game with payoff matrix A, there exist strategies P_{opt} and Q_{opt} (not necessarily unique) for the row player and column player, respectively, and a unique number v such that $P_{\text{opt}}AQ \geq v$ for every strategy Q of the column player and $PAQ_{\text{opt}} \leq v$ for every strategy P of the row player.

SECTION 10.3 EXERCISES

In Exercises 1–10, the payoff matrix for a two-person, zero-sum game is given (some are strictly determined).

(a) Solve each game. For those that are nonstrictly determined, solve using the techniques of this section.

(b) Determine the value of each game.

✓ **1.** $\begin{bmatrix} 1 & 3 \\ 7 & 2 \end{bmatrix}$ **2.** $\begin{bmatrix} 2 & 6 \\ 4 & 1 \end{bmatrix}$

3. $\begin{bmatrix} 8 & 11 \\ 9 & 7 \end{bmatrix}$ **4.** $\begin{bmatrix} 3 & 5 \\ 9 & 4 \end{bmatrix}$

5. $\begin{bmatrix} 2 & 6 \\ 1 & 3 \end{bmatrix}$ **6.** $\begin{bmatrix} 3 & 2 \\ 1 & 1 \end{bmatrix}$

7. $\begin{bmatrix} 4 & 2 \\ 5 & 1 \\ 3 & 4 \end{bmatrix}$ **8.** $\begin{bmatrix} 5 & 1 \\ 7 & 3 \\ 2 & 6 \end{bmatrix}$

9. $\begin{bmatrix} 3 & 6 & 2 \\ 2 & 1 & 4 \end{bmatrix}$ **10.** $\begin{bmatrix} 2 & 4 & 6 \\ 4 & 5 & 3 \end{bmatrix}$

In Exercises 11–20, the payoff matrix for a two-person, zero-sum game is given. Use the duality of a linear programming problem to determine the row and column player's optimal strategies.

✓ **11.** $\begin{bmatrix} -3 & 1 \\ 2 & -2 \end{bmatrix}$ **12.** $\begin{bmatrix} 3 & -2 \\ -3 & 4 \end{bmatrix}$

13. $\begin{bmatrix} 3 & -5 \\ -9 & 4 \end{bmatrix}$ **14.** $\begin{bmatrix} -4 & 3 \\ 5 & -2 \end{bmatrix}$

15. $\begin{bmatrix} 5 & 3 & -2 \\ 2 & -2 & 3 \end{bmatrix}$ **16.** $\begin{bmatrix} -3 & 1 & 2 \\ -1 & -2 & 4 \end{bmatrix}$

17. $\begin{bmatrix} 2 & -2 & 4 \\ 3 & 4 & -3 \\ -3 & 5 & 3 \end{bmatrix}$ **18.** $\begin{bmatrix} 3 & -1 & 8 \\ 1 & 2 & -2 \\ -3 & 4 & 5 \end{bmatrix}$

19. $\begin{bmatrix} 5 & 2 \\ -3 & 4 \\ 2 & 6 \end{bmatrix}$ **20.** $\begin{bmatrix} -4 & 3 \\ 5 & -1 \\ -3 & -2 \end{bmatrix}$

Applications

21. The Hoffman Company introduces its new product on the market with a major television and radio advertising campaign. Ty, the CEO of the Hoffman Company, finds out that its major competitor, the Baker Company, is planning a major advertising campaign for a comparable product. A market research firm has determined the following payoff matrix, where each entry shows the sales (in millions of dollars) either gained or lost by one company to the other.

		Hoffman	
		TV	Radio
Baker	TV	2	−1
	Radio	−1	1

(a) Use the techniques of this section to determine the optimal strategy for each company.

(b) Does the game favor anyone? If so, who?

22. The Hoffman Company introduces its new product on the market with a newspaper and mail circular advertising campaign. Ty, the CEO of the Hoffman Company, finds out that its major competitor, the Baker Company, is planning a major advertising campaign for a comparable product. A market research firm has determined the following payoff matrix, where each entry shows the sales (in millions of dollars) either gained or lost by one company to the other.

		Hoffman	
		Paper	Mail
Baker	Paper	1	−1
	Mail	−2	3

(a) Use the techniques of this section to determine the optimal strategy for each company.

(b) Does the game favor anyone? If so, who?

23. Jessie and Jamie play the following game: Jessie hides a coin in one of her hands. If Jamie guesses which hand the coin is in, Jessie gives Jamie $4. If Jamie does not guess correctly, Jamie gives Jessie $3 if the coin is in the right hand and $6 if the coin is in the left hand. Let Jessie be the row player and Jamie the column player.

(a) Write the payoff matrix associated with this game.

(b) Use the techniques of this section to determine the optimal strategy for each player.

(c) Does this game favor anyone? If so, who?

24. **Two-finger morra** is a game in which two players, at a predetermined signal, simultaneously show 1 or 2 fingers. Jose and Maria are playing a version of two-finger morra in which if the total number of fingers shown is even Maria pays Jose \$2. However, if the total number of fingers shown is odd, Jose pays Maria \$3. Let Maria be the row player and Jose the column player.

(a) Write the payoff matrix associated with this game.

(b) Use the techniques of this section to determine the optimal strategy for each player.

(c) Does this game favor anyone? If so, who?

25. Based on preliminary tests, a physician determines that Jacques has one of two possible diseases, labeled 1 and 2. The doctor can prescribe one of three treatments, labeled A, B, and C. Each treatment is effective on certain diseases but less so for others. The payoff matrix for this scenario is

$$\begin{array}{cc} & \text{Diseases} \\ & \begin{array}{cc} 1 & 2 \end{array} \\ \text{Treatment} \begin{array}{c} A \\ B \\ C \end{array} & \begin{bmatrix} 4 & 2 \\ -1 & 1 \\ -2 & 3 \end{bmatrix} \end{array}$$

Determine the optimal strategy for the physician.

SECTION PROJECT

The game rock, paper, scissors is well known in many parts of the world. In this game, two players simultaneously show a hand in the form of a fist (rock), open hand (paper), or two fingers extended (scissors). The payoff for this game is one unit according to the following rules: paper beats rock, scissors beats paper, and rock beats scissors. If the players present the same choice, the game is a tie and the payoff is 0.

(a) Set up the payoff matrix for this game.

(b) Using the methods of this section, determine the optimum strategies for the row player and the column player.

(c) Determine the value of this game. Are you surprised?

Why We Learned It

In Chapter 10 we saw that game theory was an interesting topic that reviewed many topics in this text. We saw that probability, expected value, matrices, linear programming, and the simplex method are used in the study of game theory. Situations in which game theory arises are many and varied. A political campaign manager may employ game theory and consider the game to be the candidate for office and the media. This is especially true if there is some potentially damaging information in the candidate's past. The decision on whether to remodel a store can become a game between the store manager and the economy. If market research can determine expected profits, or losses, for the store based on the decision of remodeling or not and whether the economy remains strong or slips into recession, techniques of game theory can help the store owner to make a more informed decision. Game theory can also be applied to games such as matching fingers or two-finger morra.

CHAPTER REVIEW EXERCISES

In Exercises 1–4, use the following payoff matrix for a two-person, zero-sum game. The entries represent dollars.

$$\begin{array}{cc} & \text{Grace} \\ \text{Amy} & \begin{bmatrix} -3 & -2 & 4 \\ 1 & 6 & 2 \\ -5 & 1 & 4 \end{bmatrix} \end{array}$$

Determine the payoff when the following strategies are used.

1. Amy plays row 1, Grace plays column 1.

2. Amy plays row 1, Grace plays column 3.

3. Amy plays row 2, Grace plays column 2.

4. Amy plays row 3, Grace plays column 1.

In Exercises 5 and 6, you are given a payoff matrix for a game. Construct a new payoff matrix for the game where the row player becomes the column player and the column player becomes the row player.

5. $$\begin{array}{cc} & \text{Sharon} \\ \text{Ron} & \begin{bmatrix} 2 & -5 \\ -1 & -3 \end{bmatrix} \end{array}$$

6. Quinn (rows), Nelle (columns)

$$\text{Quinn}\ \begin{bmatrix} -2 & 1 & -3 \\ 3 & -4 & 2 \\ 2 & 3 & -5 \end{bmatrix}$$

In Exercises 7–10, you are given a payoff matrix for a two-person, zero-sum game that is strictly determined.

(a) Determine the saddle point(s).

(b) Determine the optimal strategy for each player.

(c) Determine the value of the game and state which person the game favors.

7. Mike (rows), Jen (columns)

$$\text{Mike}\ \begin{bmatrix} 3 & -5 \\ 4 & 2 \end{bmatrix}$$

8. Fran (rows), Larry (columns)

$$\text{Fran}\ \begin{bmatrix} 5 & 4 & 3 \\ 1 & 4 & 1 \\ -1 & -3 & 2 \end{bmatrix}$$

9. Lindsay (rows), Mark (columns)

$$\text{Lindsay}\ \begin{bmatrix} -2 & -1 \\ -1 & 0 \\ 0 & -3 \\ 3 & 5 \end{bmatrix}$$

10. Troy (rows), Nicole (columns)

$$\text{Troy}\ \begin{bmatrix} -1 & -3 & 6 & 8 \\ 0 & -2 & 1 & 4 \\ 4 & -5 & -5 & 7 \\ 3 & -2 & -1 & 0 \end{bmatrix}$$

11. A professional indoor soccer team is contemplating moving to another city. The team owner determines that the main determining factor in the team's decision is whether they will have a winning or losing season. If the team moves and has a winning season, they will generate \$120,000 more than last season due to increased ticket sales. If the team moves and has a losing season, they will generate \$40,000 less than last season due to decreased ticket sales. If the team does not move and they have a winning season, they will generate \$50,000 more than last season due to increased ticket sales. If the team does not move and they have a losing season, they will generate \$8000 less than last season due to decreased ticket sales. The team's owner summarizes this information in the following payoff matrix, where the two-person game is the team against their season. Find the saddle point and determine the best strategy for the team.

Team	Season: Winning	Season: Losing
Move	120,000	−40,000
Don't move	50,000	−8000

12. Mark's Auto and Jim's Auto are two competing used car dealerships. On a weekly basis each dealership selects one (and only one) of the following types of advertisement: television, radio, or newspaper. A survey by a market research company yields the following payoff matrix. Entries represent the percentage of the market gain or loss for each choice of action by the dealerships. (Assume that a gain by Mark's Auto is a loss for Jim's Auto, and vice versa.)

Mark's Auto	Jim's Auto: TV	Radio	Newspaper
TV	3	2	1
Radio	−2	−1	0
Newspaper	−1	2	−3

(a) Is this a strictly determined game?

(b) Find the saddle point and determine the best strategy for each dealership.

(c) Determine the value of the game. Does this game favor one dealership over the other? If so, which dealership has the advantage?

13. Two competing ski resorts, Willard Mountain and Gorham Mountain, are considering expanding the type of daily ski passes that they offer. Both mountains currently only offer full-day passes. Willard Mountain is considering offering a 2-hour, 3-hour, or 4-hour pass and Gorham Mountain is considering offering a 2-hour or 3-hour pass. Each day both mountains decide on one of the passes to offer. If Willard Mountain offers a 2-hour pass, it will lose 2% of the market if Gorham Mountain offers a 2-hour pass and gain 4% of the market if Gorham Mountain offers a 3-hour pass. If Willard Mountain offers a 3-hour pass, it will see no gain if Gorham Mountain offers a 2-hour pass and a 1% gain of the market if Gorham Mountain offers a 3-hour pass. If Willard Mountain offers a 4-hour pass, it will lose 1% of the market if Gorham Mountain offers a 2-hour pass and lose 2% of the market if Gorham Mountain offers a 3-hour pass. (Assume that a gain by Willard Mountain is a loss for Gorham Mountain, and vice versa.) Let Willard Mountain be the row player and Gorham Mountain be the column player.

(a) Construct the payoff matrix for this game, where the entries are percentages of the market.

(b) Verify that the game is strictly determined.

(c) Determine the optimal strategy for each ski resort.

14. (*continuation of Exercise 13*)

(a) Determine the value of the game in Exercise 13.

(b) Does the game in Exercise 13 favor one ski resort over the other? If so, which ski resort has the advantage?

15. A bookstore is considering expanding to include a music section and a coffee bar. A market research firm determines that the main factor in the bookstore's decision is the economy and whether it remains strong or slips into recession. The firm determines that if the bookstore expands and the economy is strong, the store will earn

$50,000. However, if the store expands and the economy slips into recession, the store will lose $100,000. If the store does not expand and the economy is strong, the store will lose $75,000 due to customers going to other stores. If the store does not expand and there is a recession, the store will lose $80,000. This is a two-person game between the store and the economy. Let the store be the row player and the economy be the column player.

(a) Construct the payoff matrix for the game.

(b) Verify that the game is strictly determined.

(c) Determine the optimal strategy for the bookstore.

In Exercises 16–18, determine the expected value E for each game whose payoff matrix and strategies P and Q (row and column player, respectively) are given.

16. $\begin{bmatrix} -2 & 3 \\ 5 & 1 \end{bmatrix}, \quad P = [\,\frac{1}{4} \quad \frac{3}{4}\,], \quad Q = \begin{bmatrix} \frac{5}{6} \\ \frac{1}{6} \end{bmatrix}$

17. $\begin{bmatrix} -2 & 0 & 1 \\ 3 & -1 & 2 \\ 4 & -5 & 6 \end{bmatrix}, \quad P = [\,0.1 \quad 0.3 \quad 0.6\,],$

$Q = \begin{bmatrix} 0.2 \\ 0.3 \\ 0.5 \end{bmatrix}$

18. $\begin{bmatrix} 1 & -1 & -2 \\ -2 & 3 & 2 \end{bmatrix}, \quad P = [\,0.25 \quad 0.75\,],$

$Q = \begin{bmatrix} 0.15 \\ 0.35 \\ 0.50 \end{bmatrix}$

19. A certain game has a payoff matrix $\begin{bmatrix} 2 & -1 \\ -4 & 3 \end{bmatrix}$.

Compute the expected value E for the pairs of strategies given in parts (a) to (d). Which of these strategies is most favorable to the column player?

(a) $P = [\,\frac{1}{2} \quad \frac{1}{2}\,], \quad Q = \begin{bmatrix} \frac{1}{2} \\ \frac{1}{2} \end{bmatrix}$

(b) $P = [\,\frac{1}{4} \quad \frac{3}{4}\,], \quad Q = \begin{bmatrix} \frac{1}{6} \\ \frac{5}{6} \end{bmatrix}$

(c) $P = [\,\frac{1}{3} \quad \frac{2}{3}\,], \quad Q = \begin{bmatrix} \frac{2}{3} \\ \frac{1}{3} \end{bmatrix}$

(d) $P = [\,\frac{1}{8} \quad \frac{7}{8}\,], \quad Q = \begin{bmatrix} \frac{1}{2} \\ \frac{1}{2} \end{bmatrix}$

In Exercises 20–22, determine the optimal strategies P_{opt} and Q_{opt} for the row and column players, respectively. Be sure to check for saddle points since some may be strictly determined games.

20. $\begin{bmatrix} -2 & 0 \\ 3 & -1 \end{bmatrix}$

21. $\begin{bmatrix} 2 & -3 \\ -8 & -6 \end{bmatrix}$

22. $\begin{bmatrix} -\frac{1}{3} & \frac{2}{3} \\ \frac{1}{2} & -\frac{1}{4} \end{bmatrix}$

In Exercises 23–25, use the specified payoff matrix for a two-person, zero-sum game given in Exercises 20–22.

(a) Determine the value of the game.

(b) Determine which player (if any) the game favors assuming that the row player and the column player use their optimal strategies.

23. Use the game in Exercise 20.

24. Use the game in Exercise 21.

25. Use the game in Exercise 22.

26. Sopris Mountaineering is deciding to introduce new guided trips to two new destinations, either Mount Cook or Denali. To advertise, Sopris Mountaineering will launch a major magazine and newspaper campaign. Greg, the CEO of Sopris Mountaineering, finds out that its major competitor, Mountain Adventures, is also planning an advertising campaign for comparable trips. A market research firm has determined the following payoff matrix, where each entry shows the sales (in dollars) either gained or lost by one company to the other.

		Mountain Adventures	
		Magazine	Newspaper
Sopris Mountaineering	Magazine	20,000	−12,000
	Newspaper	−8000	15,000

(a) Find the optimal strategies for Sopris Mountaineering and Mountain Adventures.

(b) Find the value of the game assuming that both use the optimal mixed strategy. Does the game favor anyone? If so, who?

27. A bookstore is considering expanding to include a music section and a coffee bar. A market research firm determines that the main factor in the bookstore's decision is the economy and whether it remains strong or slips into recession. The firm determines that if the bookstore expands and the economy is strong the store will earn $50,000. However, if the store expands and the economy slips into recession, the store will lose $100,000. If the store does not expand and the economy is strong, the store will lose $125,000 due to customers going to other stores. If the store does not expand and there is a recession, the store will lose $80,000. This is a two-person game between the store and the economy. Let the store be the row player and the economy be the column player.

(a) Construct the payoff matrix for this game where the entries are in thousands of dollars.

(b) Determine the optimal strategy for the bookstore.

(c) If you were the owner of the bookstore, what would you do?

28. (*continuation of Exercise 27*)

(a) Assuming the optimal strategies in Exercise 27, determine the value of the game.

(b) Does the game in Exercise 27 favor anyone? If so, who?

29. Kris and Rachel are running for senior class president. They may deliver their campaign speech at all school meetings and football games. If both candidates campaign only at all school meetings, Kris is expected to get 55% of the vote, and if both campaign only at football games, Kris is expected to get 60% of the vote. If Kris campaigns solely at all school meetings and Rachel campaigns solely at football games, Kris is expected to get 40% of the vote. If Kris solely campaigns at football games and Rachel solely campaigns at all school meetings, Kris is expected to get 45% of the vote. Let Kris be the row player and Rachel be the column player.

(a) Construct the payoff matrix for this game where the entries are percentages written as decimals.

(b) Determine the optimal strategy for Kris and Rachel.

30. (*continuation of Exercise 29*) The school newspaper has printed some information that has affected Kris's campaign. If both candidates campaign only at all school meetings, Kris is expected to get 40% of the vote, and if both campaign only at football games, Kris is expected to get 55% of the vote. If Kris campaigns solely at all school meetings and Rachel campaigns solely at football games, Kris is expected to get 35% of the vote. If Kris campaigns solely at football games and Rachel solely campaigns at all school meetings, Kris is expected to get 45% of the vote. Let Kris be the row player and Rachel be the column player.

(a) Construct the payoff matrix for this game where the entries are percentages written as decimals.

(b) Determine the optimal strategy for Kris and Rachel.

In Exercises 31–35, the payoff matrix for a two-person, zero-sum game is given (some are strictly determined).

(a) Solve the game. For those that are not strictly determined, solve using the techniques of Section 10.3.

(b) Determine the value of each game.

31. $\begin{bmatrix} 3 & 7 \\ 8 & 4 \end{bmatrix}$ **32.** $\begin{bmatrix} 3 & 5 \\ 4 & 6 \end{bmatrix}$

33. $\begin{bmatrix} 3 & 1 \\ 4 & 3 \\ 2 & 4 \end{bmatrix}$ **34.** $\begin{bmatrix} 3 & 1 & 4 \\ 2 & 3 & 5 \end{bmatrix}$

35. $\begin{bmatrix} 2 & 3 & 4 \\ 1 & 4 & 6 \end{bmatrix}$

In Exercises 36–40, the payoff matrix for a two-person, zero-sum game is given. Use the duality of a linear programming problem to determine the row and column players' optimal strategies.

36. $\begin{bmatrix} -2 & 1 \\ 5 & -4 \end{bmatrix}$ **37.** $\begin{bmatrix} 1 & -6 \\ -4 & 8 \end{bmatrix}$

38. $\begin{bmatrix} 2 & -3 & 5 \\ -3 & 1 & 4 \end{bmatrix}$

39. $\begin{bmatrix} -2 & 2 & 4 \\ 0 & 1 & -3 \\ 1 & -1 & 3 \end{bmatrix}$

40. $\begin{bmatrix} -3 & -4 \\ -2 & 0 \\ 1 & -3 \end{bmatrix}$

41. The Babcock Company introduces its new product on the market with a major newspaper and magazine campaign. Grace, the CEO of the Babcock Company, finds that its major competitor, the Mayhew Company, is planning a major newspaper and magazine campaign for a comparable product. A market research firm has determined the following payoff matrix, where each entry shows the sales (in millions of dollars) either gained or lost by one company to the other.

		Mayhew	
		Newspaper	Magazine
Babcock	Newspaper	3	−2
	Magazine	−1	1

(a) Use the techniques of Section 10.3 to determine the optimal strategy for each company.

(b) Does the game favor anyone? If so, who?

42. (*continuation of Exercise 41*) The Mayhew Company has decided to launch a television campaign in addition to the newspaper and magazine campaign. Based on this new campaign, a market research firm has determined the following payoff matrix, where each entry shows the sales (in millions of dollars) either gained or lost by one company to the other.

		Mayhew		
		Newspaper	Magazine	TV
Babcock	Newspaper	3	−2	−3
	Magazine	−1	1	−5

(a) Determine the optimal strategy for each company.

(b) Does the game favor anyone? If so, who?

43. **(a)** Use the techniques of Section 10.2 to verify the solutions to Exercise 41(a).

(b) What is the value of the game?

44. Two-finger morra is a game in which two players, at a predetermined signal, simultaneously show 1 or 2 fingers. Lissa and Lera are playing a version of the two-finger morra in which, if the total number of fingers shown is odd, Lissa pays Lera $2. However, if the total number of fingers shown is even, Lera pays Lissa $1. Let Lissa be the row player and Lera be the column player.

(a) Write the payoff matrix associated with this game.

(b) Use the techniques of Section 10.3 to determine the optimal strategy for each player.

(c) Verify the solution to part (b) by using the techniques of Section 10.2.

(d) What is the value of the game?

(e) Does the game favor anyone? If so, who?

45. Based on preliminary tests, a physician determines that Mika has one of two possible diseases, labeled 1 and 2. The doctor can prescribe one of three treatments, labeled A, B, and C. Each treatment is effective on certain diseases, but less so for others. The payoff matrix for this scenario is

$$\begin{array}{cc} & \text{Disease} \\ & \begin{array}{cc} 1 & 2 \end{array} \\ \text{Treatment} \begin{array}{c} A \\ B \\ C \end{array} & \begin{bmatrix} -1 & 2 \\ -2 & 4 \\ 3 & -2 \end{bmatrix} \end{array}$$

Determine the optimal strategy for the physician.

CHAPTER 10 PROJECT

Kids love to receive money, even if they have to do chores to earn it. Consider a two-person game between brothers in which the goal is to make the most money from the completed chores. Suppose that Donald can either do the laundry or wash cars, while Matthew can vacuum, do the dishes, or mow the lawn. Consider the following values for the chores:

If Donald does the laundry and if Matthew vacuums, Donald receives \$3.

If Donald does the laundry and if Matthew does the dishes, Matthew receives \$5.

If Donald does the laundry and if Matthew mows the lawn, Donald receives \$4.

If Donald washes the cars and if Matthew vacuums, Donald receives \$6.

If Donald washes the cars and if Matthew does the dishes, Donald receives \$7.

If Donald washes the cars and if Matthew mows the lawn, Matthew receives \$6.

Suppose that Nicole, the boys' older sister, realizes that this is actually a two-person game between Donald and Matthew. Nicole is in charge of distributing the money between the boys. She knows that if Donald makes any money it is considered a loss to Matthew. Also, if Matthew makes any money, it is considered a loss to Donald. Only one brother can be paid based on whether his chore was more valuable.

1. Determine the payoff matrix, A, for this game for which each entry is the dollar value of the chore. Have Donald be the row player and Matthew the column player. Let L stand for does laundry, W for washes cars, V for vacuums, D for does dishes, and M for mows lawn.

2. Determine the location of the saddle point for the payoff matrix (if any). Is the game strictly determined? Why or why not?

3. Does the payoff matrix have all nonnegative entries? If not, add a positive number, n, to each entry.

4. Find a linear programming problem to determine the optimal (row) strategy for Donald.

5. Find a linear programming problem to determine the optimal (column) strategy for Matthew.

6. If both linear programming problems in questions 4 and 5 were solved, what would be true about their optimal values? Why?

7. Use the Simplex Method to solve the linear programming problem in question 5.

8. Determine the optimal (row) strategy for Donald, the optimal (column) strategy for Matthew, and the expected value of the game. Be sure to subtract n from the expected value of the game (where n is the number added in 3).

9. In the long run, to whom would Nicole pay the most?

Chapter 11 Functions, Models, and Average Rate of Change

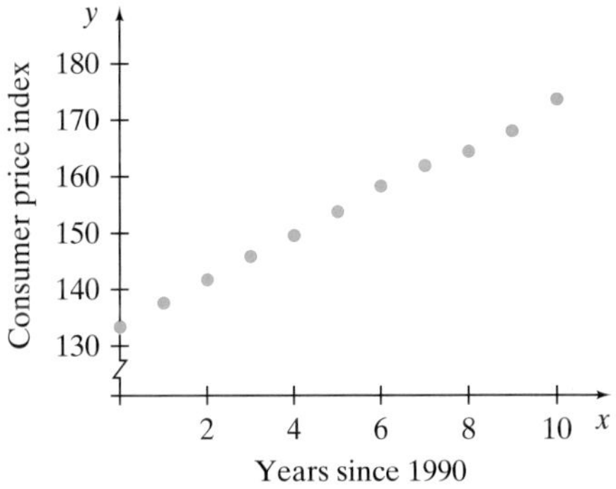

(b)

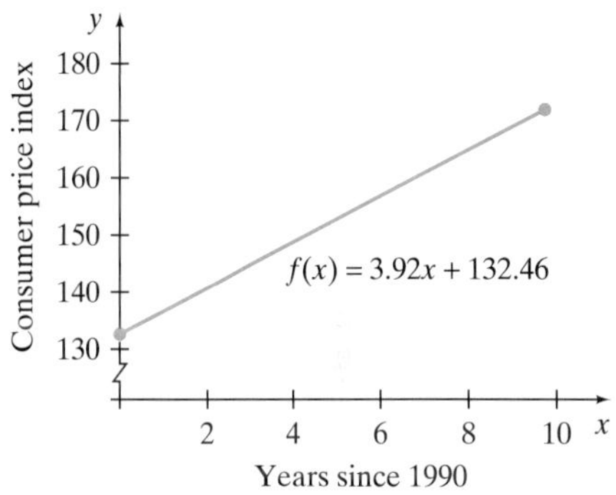

(c)

During the prosperous 1990's shopping at "the mall" proved to be a popular activity for many suburban teens. A question one might ask is whether \$100 bought as many goods at the end of the decade as it did at the beginning? One source of this information is the Consumer Price Index (CPI) which is a measure of prices facing consumers. Figure b graphs the data for the CPI as a scatterplot and Figure c gives the function modeling that data. These figures show that the CPI rose steadily throughout the decade.

What We Know

We begin our study of calculus having learned the basics of algebra. These include equation solving, factoring, and the graphing of functions.

Where Do We Go

In this chapter, we will review the function concept, the properties of various types of functions, and the average rate of change of functions over an interval. We will see how data can be used to form or model functions.

Section 11.1 The Coordinate System and Functions

Plotting Points

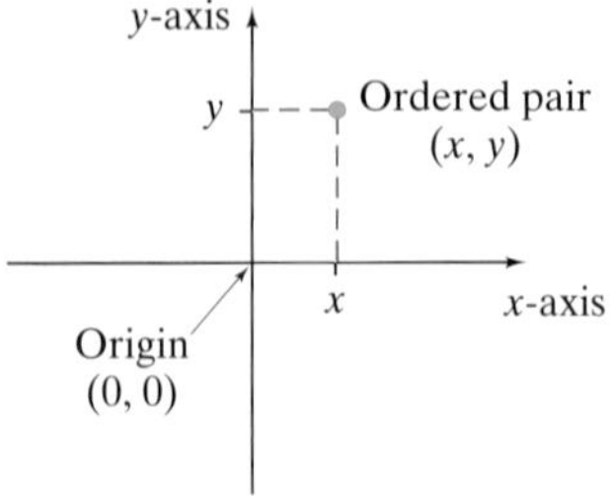

Figure 11.1.1

We start our study of functions by examining how they appear visually. The graph on which functions are plotted is called the **Cartesian plane.**

The Cartesian plane can be thought of as two number lines that are perpendicular to one another, as shown in Figure 11.1.1. The point at which the lines cross is called the **origin**. The horizontal number line, or horizontal axis, locates the values of the **independent variable**, which is usually denoted by x. The vertical axis locates the values of the **dependent variable**, which is usually denoted by y. These are commonly called the x and y axes, respectively. Points are plotted as **ordered pairs** in the form (independent variable, dependent variable), which, most of the time, are of the form (x, y). Before we continue, let's plot some points on a Cartesian plane. Figure 11.1.2 shows the graph of some ordered pairs. To plot the ordered pair $(-3, 1)$, we start at the origin and move to the left 3 units and then up 1 unit. In a similar fashion, we plotted the remainder of the points shown in Figure 11.1.2.

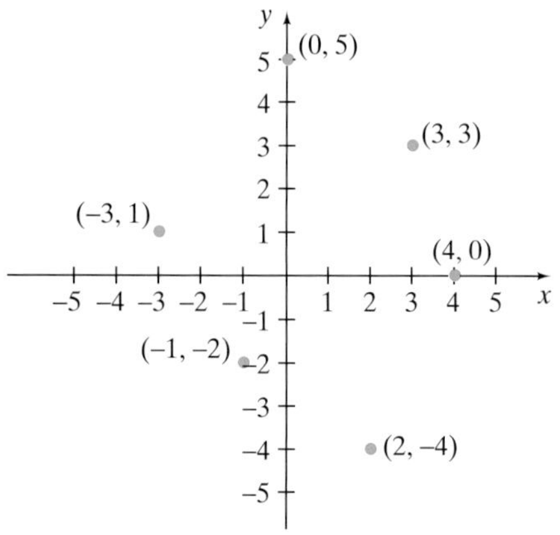

Figure 11.1.2

In the study of applied calculus, we often use tabular data as a basis for forming a mathematical model. When we plot data, we are producing what is called a **scatterplot**. Frequently, the values of the independent variable are a measure of time, primarily in years. The dependent variable then represents some phenomenon in business, life, or social science that is related to time.

Example 1 Creating a Scatterplot

The data in Table 11.1.1 represent the U.S. imports of products from France for the years 1990 to 1994. The value of the imports is in billions of U.S. dollars. Let x represent the number of years since 1990, and let y represent the value in billions of dollars of U.S. imports from France. Construct ordered pairs to represent the data, and plot the data.

Solution

To avoid working with large values of the independent variable x, it is convenient to let x represent the number of years since 1990 (see Table 11.1.2). The

Table 11.1.1

Year	Imports
1990	10.7
1991	13.3
1992	14.8
1993	15.3
1994	16.8

SOURCE: U.S. Census Bureau
www.census.gov

Table 11.1.2

Years since 1990	Imports
0	10.7
1	13.3
2	14.8
3	15.3
4	16.8

year 1990 corresponds with $x = 0$, 1991 corresponds to $x = 1$, and so on. Notice that we get these values by taking the year and subtracting 1990. This process is called **standardizing the values**. So constructing these ordered pairs gives (0, 10.7), (1, 13.3), (2, 14.8), (3, 15.3), and (4, 16.8). Using these ordered pairs, we get the scatterplot shown in Figure 11.1.3.

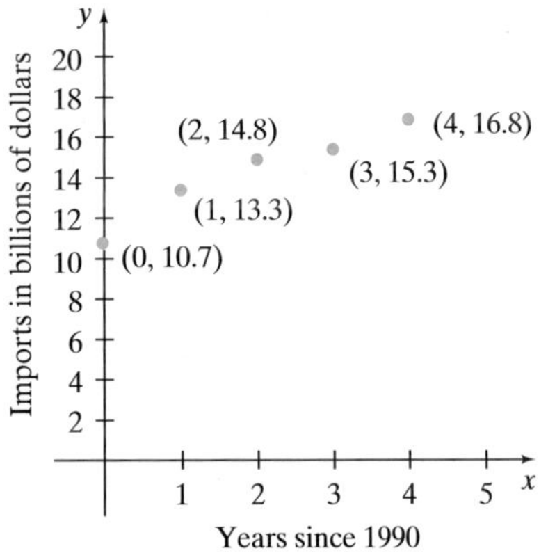

Figure 11.1.3

■

Checkpoint 1

Now work Exercise 11.

Notice in Example 1 that the possible values of x came from the set $\{0, 1, 2, 3, 4\}$ and the possible values of y came from the set $\{10.7, 13.3, 14.8, 15.3, 16.8\}$. These sets of numbers have special names in mathematics. The set of all possible values of the independent variable, in this case x, is called the **domain**; the set of all possible values of the dependent variable, in this case y, is called the **range**. These two sets of numbers are critical in the study of functions.

Function Notation and Evaluating Functions

Many times, a dependent relationship exists between two phenomena. The price of a concert ticket may **depend** on the popularity of the band, the cost of manufacturing may depend on the quantity manufactured, and the life expectancy of a person may depend on the year in which she or he was born. These examples exhibit a direct relationship, or correspondence, between **independent variable** values (price, quantity, year) and the **dependent variable** values (popularity, cost, life length). We write these relationships between the independent and dependent variables using **functions**.

A function f can be thought of as a process in which an independent value x in the domain is mapped to a dependent value y in the range. This is illustrated in Figure 11.1.4.

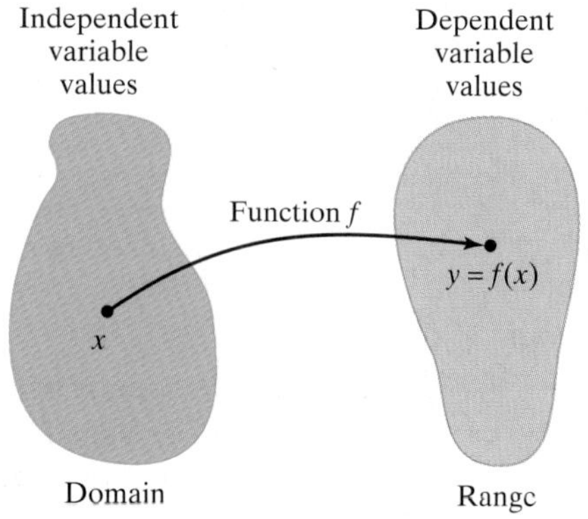

Figure 11.1.4

Function, Domain, and Range

A **function** is a rule, or a set of ordered pairs, that assigns to each element in the domain one and only one element in the range. The **domain** is the set of all possible independent variable values. The **range** is the set of all possible dependent variable values.

Functions are usually expressed by first naming the function with a letter like f or g, and then writing the independent variable letter in parentheses. For example, in the expression $f(x)$ (read "f of x"), f represents the function and $f(x)$ represents the value of f at x, where x is the independent variable. It is often convenient to represent $f(x)$ with the dependent variable y. This means that on our Cartesian plane we can now label the vertical axis with $f(x)$, $g(x)$, or any other function name that we wish to use.

Functions are not only expressed as tables and graphs, but also as mathematical expressions. For example, the function $f(x) = \frac{x+2}{5}$ means: "Take the independent value and add 2; then divide that sum by 5."

The process of "plugging" in various x-values into a function is called **evaluating** the function. To evaluate the function $f(x) = \frac{x+2}{5}$ at $x = 13$, denoted by writing $f(13)$, we substitute 13 for every x. This gives

$$f(x) = \frac{x+2}{5}$$

$$f(13) = \frac{13+2}{5} = \frac{15}{5} = 3$$

Example 2 Evaluating Functions

For the function $g(x) = -2x + 6$, evaluate $g(-2)$, $g(0)$, and $g(3)$, and write the results as ordered pairs.

Solution

Evaluating the function, we get $g(-2) = -2(-2) + 6 = 4 + 6 = 10$, which produces the ordered pair $(-2, 10)$.

Now $g(0) = -2(0) + 6 = 0 + 6 = 6$. This gives us the ordered pair $(0, 6)$.

And $g(3) = -2(3) + 6 = -6 + 6 = 0$, which produces the third ordered pair $(3, 0)$. ■

Now let's take a look at a relationship that is not a function. Figure 11.1.5 shows the graph of relationship R. Since $(11, 3)$ and $(11, -3)$ are on the graph, R assigns both 3 and -3 to the value 11. Applying our definition of function, we see that R is not a function. A visual way to tell whether a graph represents a function is to use the **vertical line test**.

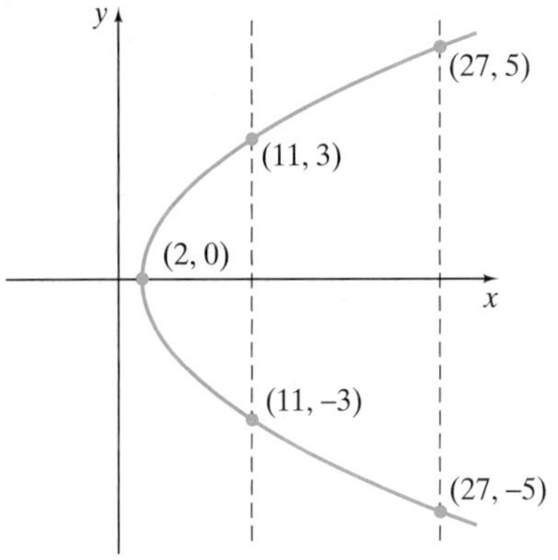

Figure 11.1.5

Vertical Line Test

If every vertical line drawn through a graph intersects the graph at only one point, then the graph represents a function.

Example 3 Determining if a Graph Represents a Function

Use the vertical line test to determine which of the graphs in Figures 11.1.6a, b, and c do not represent functions.

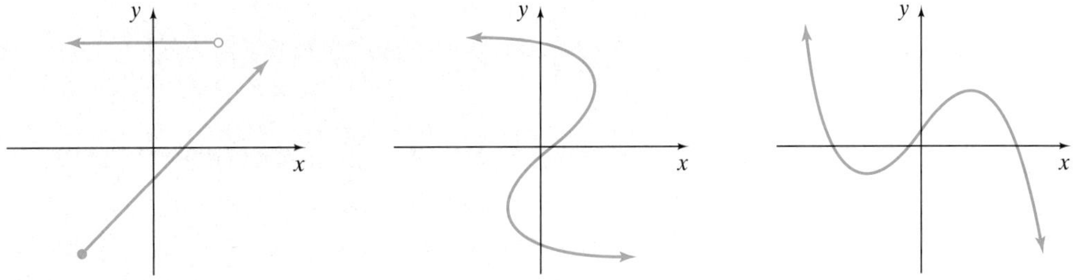

(a) **(b)** **(c)**

Figure 11.1.6

Solution

We see in Figure 11.1.6c that vertical lines will intersect the graph at only one place, but in Figures 11.1.6a and b they will intersect at more than one point. So the first two graphs do not represent functions, while the third graph is a function. ■

Checkpoint 2

Now work Exercise 21.

Many times, we need to express the domain and range of a function as some portion of the real number line. To write this in a convenient form, we use **interval notation**. Let's say that we have the inequality notation $2 \le x < 5$; that is, all the x-values from 2 to 5, including 2. We can conveniently write this section of the real number line using interval notation as $[2, 5)$. Intervals with two endpoints are called **finite intervals**. If the endpoints are included, the interval is **closed**. If the endpoints are not included, the interval is **open**.

Finite Intervals

Number Line	Interval Notation	Inequality Notation	Interval Type
a b	$[a, b]$	$a \le x \le b$	Closed
a b	(a, b)	$a < x < b$	Open
a b	$(a, b]$	$a < x \le b$	Half-open
a b	$[a, b)$	$a \le x < b$	Half-open

Intervals that extend indefinitely in at least one direction are called **infinite intervals**. We use the **infinity** symbol ∞ to indicate that the numbers extend positively (or to the right on the number line) without bound. We use $-\infty$ to represent **negative infinity** to show that the numbers extend negatively (or to the left on the number line) without bound. Since ∞ and $-\infty$ are only **concepts** and not real numbers themselves, we always write these endpoints as open, using parentheses.

Infinite Intervals

Number Line	Interval Notation	Inequality Notation	Interval Type
a	$[a, \infty)$	$x \ge a$	Unbounded closed
a	(a, ∞)	$x > a$	Unbounded open
b	$(-\infty, b]$	$x \le b$	Unbounded closed
b	$(-\infty, b)$	$x < b$	Unbounded open
a	$(-\infty, \infty)$	All x in $\mathbb{R}$	Number line

Now that we have reviewed interval notation, let's return to the analysis of functions, domains, and ranges. We may think of the domain as the set of all

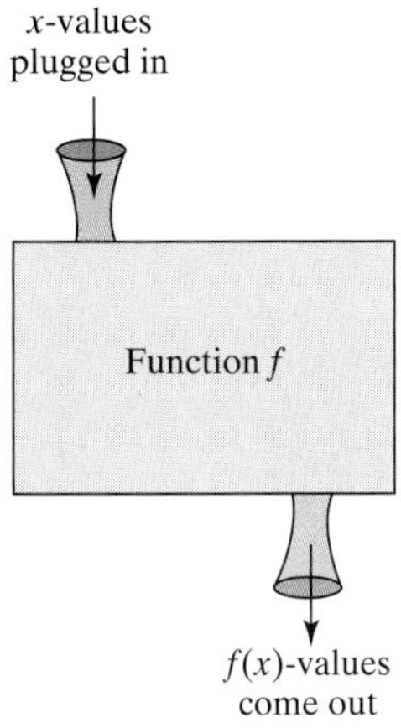

Figure 11.1.7

possible values that can be *plugged in* for the independent variable of the function. The range is the set of numbers that *come out* of the function. An illustration of this *plugged-in* and *come out* process is given in Figure 11.1.7.

To find numbers that cannot be in the domain, think of the number properties learned in algebra. We cannot divide a number by zero; for example, $\frac{3}{0}$ is not defined. And we cannot take the square root of a negative number (that is, $\sqrt{-9}$ does not have a *real* number solution). (You might have heard of complex numbers, but in applied calculus we consider only real numbers.) So, generally, when we look for domain values of functions, we exclude values that can make the denominator of a fraction zero or make the expression under a square root (or any even index root) function negative. (Another restriction will arise in Section 11.7, when we study logarithmic functions.)

Example 4 Finding the Domain of a Function Algebraically

Determine the domain of the following functions. Show the domain on a real number line and by writing in interval notation.

(a) $f(x) = \sqrt{5x - 2}$ **(b)** $g(x) = \dfrac{\sqrt{x-2}}{x^2 - 4x}$

Solution

(a) To find the domain for f, we need the x-values so that the radicand (the expression under the square root symbol) is greater than or equal to zero. As an inequality, this means that we need

$$5x - 2 \geq 0$$
$$5x \geq 2$$
$$x \geq \frac{2}{5}$$

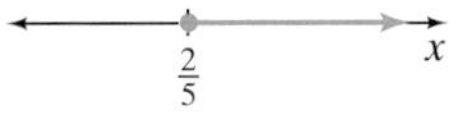

Figure 11.1.8

This domain is shown on the number line in Figure 11.1.8. In interval notation, the domain is $\left[\frac{2}{5}, \infty\right)$.

(b) This function requires us to examine the restrictions on both the numerator and denominator. In the numerator, we see that we need values of x such that

$$x - 2 \geq 0$$
$$x \geq 2$$

For the denominator, we can factor x from each term to get

$$x^2 - 4x = x(x - 4)$$

If we set the factored form equal to zero and solve, we get

$$x(x - 4) = 0$$
$$x = 0 \quad \text{or} \quad x = 4$$

So we must also exclude the values of 0 and 4 from the domain, since they make the denominator of g zero. The resulting domain is shown on the

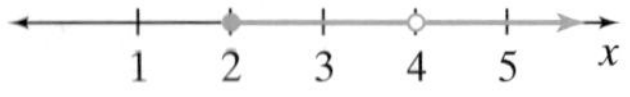

Figure 11.1.9

number line in Figure 11.1.9. Since the domain of g is the x-values for which both the numerator and denominator are defined, the domain of g is $[2, 4) \cup (4, \infty)$. ■

Checkpoint 3 Now work Exercise 43.

Before continuing, let's summarize how to determine the domain of a function algebraically.

Determining the Domain of a Function

To determine the domain of a function, we exclude independent variable values that

1. Make the denominator of a fraction zero.
2. Produce a negative result under an even-indexed radical.

Using purely algebraic techniques, finding the range of a function can be difficult. Many functions require techniques learned in calculus to determine the range, but the range is relatively easy to determine when we see the graph of the function. We illustrate how in Example 5.

Example 5 Determining Domains and Ranges From a Graph

Determine visually the domain and range for the function in Figure 11.1.10, and write your answer using interval notation.

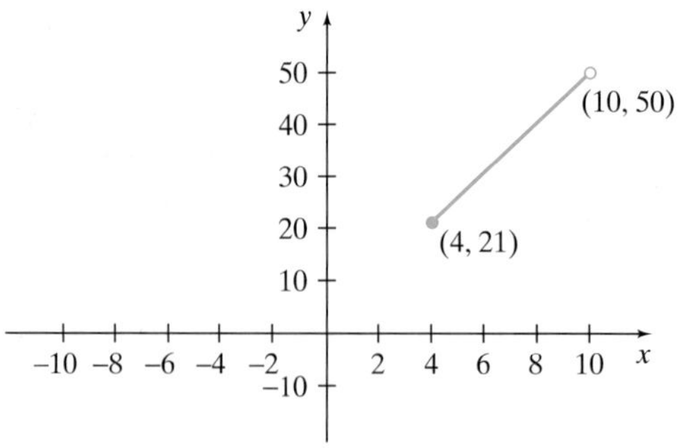

Figure 11.1.10

Solution

First, notice that the left endpoint of the function is a solid dot and the right endpoint is a hollow dot. This means that the domain is the half-open interval $[4, 10)$. Notice that the lowest point on the graph is at the ordered pair $(4, 21)$ and the highest point is near $(10, 50)$. So we see that the range of the function is $[21, 50)$. ■

Now let's introduce a function that we will see frequently in this text, the **price–demand function**.

Price–Demand Function

The **price–demand function** p gives us the price at which people buy exactly x units of product.

Generally, as the price of a product decreases, more and more people will buy the product. So for the price–demand function as the x-values increase, the p-values decrease. This is shown in Figure 11.1.11.

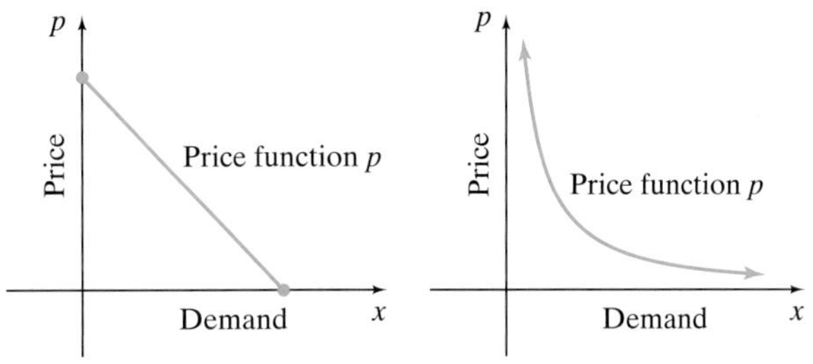

Figure 11.1.11

Example 6 Determining the Domain and Range of a Price–Demand Function

The price–demand function for electronic organs sold at Red River Mall is given by

$$p(x) = 16{,}000 - 320x$$

The graph of p is given in Figure 11.1.12. Here, x represents the number of electronic organs that people buy at Red River Mall weekly and $p(x)$ represents the price per organ in dollars.

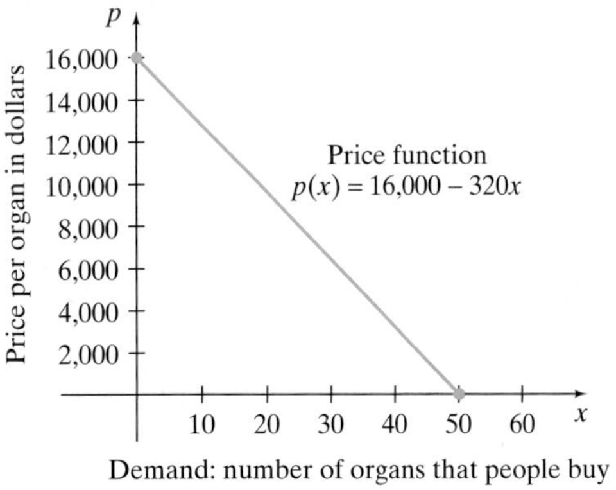

Figure 11.1.12

(a) Determine the domain and the range of the function.

(b) Evaluate $p(15)$ and interpret.

Solution

(a) Notice that the function applies only for x-values up to 50 units. For other values, $p(x)$ is not defined. We may express this restriction on x using interval notation by writing

$$p(x) = 16{,}000 - 320x \qquad [0, 50]$$

We can see that the possible values for x are in the interval [0, 50], which gives us our domain. Since the range is the set of all possible function values, we can see from the graph that the range is [0, 16,000].

(b) Evaluating the price–demand function at $x = 15$, we get

$$p(15) = 16{,}000 - 320(15) = 16{,}000 - 4800 = 11{,}200$$

This means that when the weekly demand for organs is 15, the price of an organ is \$11,200. ■

SUMMARY

In this section we have reviewed the fundamental properties of functions.

- A **function** is a rule that assigns to each element in the domain one and only one element in the range.
- The **domain** of a function is the set of all possible independent variable values.
- The **range** of a function is the set of all possible dependent variable values.
- The **price–demand function** p gives us the price $p(x)$ at which people buy exactly x units of product.

SECTION 11.1 EXERCISES

For Exercises 1–4, use the table to construct ordered pairs and plot them in the Cartesian plane.

1.

x	$f(x)$
−2	3
1	5
4	4

2.

x	$g(x)$
1	2.1
2	2.3
3	2.9

3.

x	−4	−3.5	−2	2
$f(x)$	−5	−6	−8	−11

4.

x	0	−1	−2
$f(x)$	0	1.1	1.5

For Exercises 5–10, make a table and write the ordered pairs based on the scatterplots.

5.

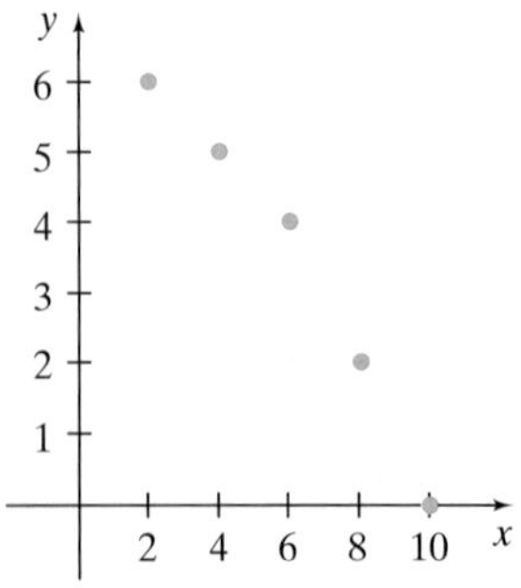

6.

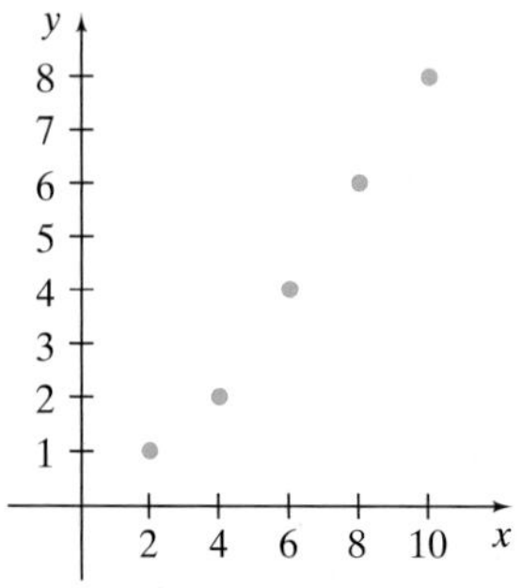

7.

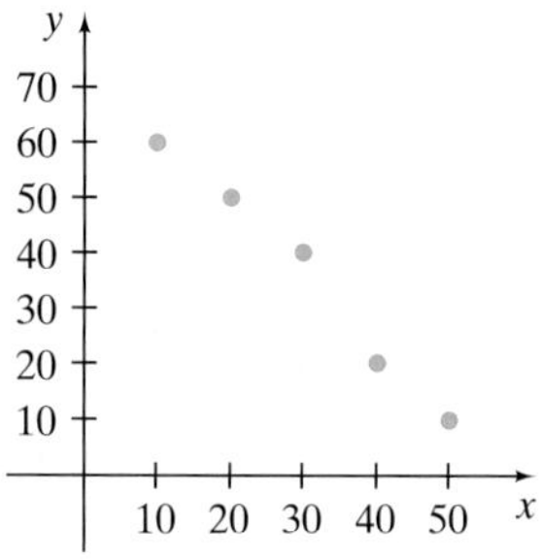

8.

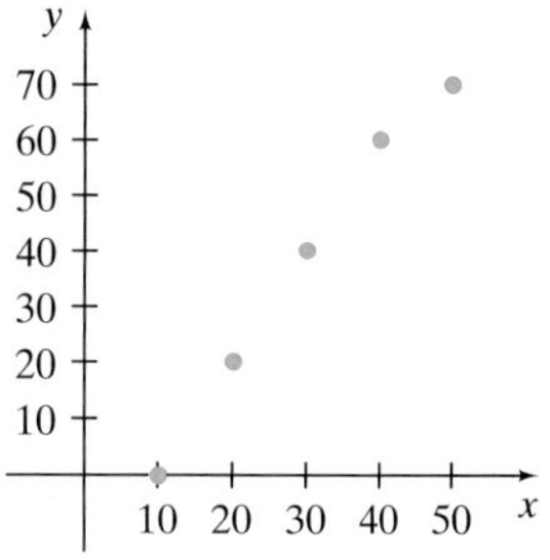

9.

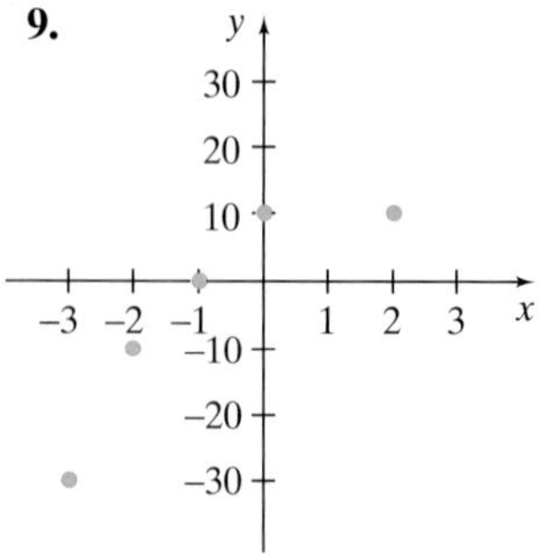

10.

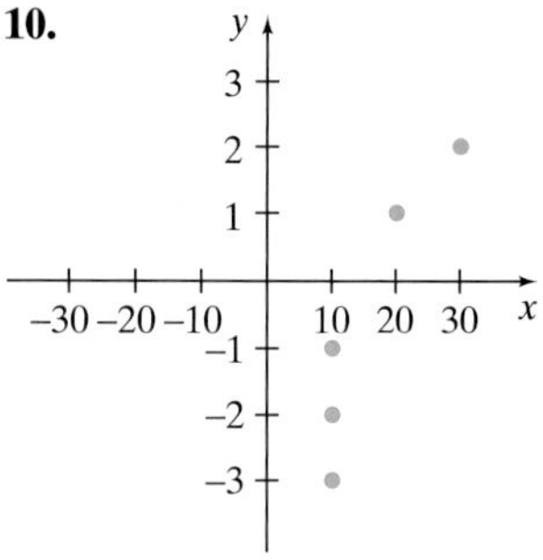

For Exercises 11–14, standardize the values of the independent variable based on its definition and then make a scatterplot of the data.

11. Let x represent the number of years since 1980.

Year	Expenditures at Public Colleges and Universities (in billions of dollars)
1980	88
1985	99
1990	125
1993	133
1996	141
1999	154

SOURCE: U.S. National Center for Education Statistics, www.nces.ed.gov

12. Let x represent the number of years since 1990.

Year	Number of Recreational Visits to National Parks (in millions)
1994	63
1995	64.8
1996	63.1
1997	65.3
1998	64.5

SOURCE: U.S. Statistical Abstract, www.census.gov/statab/www

13. Let x represent the elapsed time in seconds.

Time	Speed (in miles per hour)
0	0
1	40
2	120
3	160
4	180

14. Let x represent the number of years since 1980.

Year	Per Capita Bottled Water Consumption (in gallons per year)
1980	2.4
1985	4.5
1990	8.0
1993	9.4
1995	11.6
1997	13.1

SOURCE: U.S. Statistical Abstract, www.census.gov/statab/www

For Exercises 15–18, determine which of the tables represents a function.

15.

Domain	Range
0	1
2	1
3	2
4	5

16.

Domain	Range
−5	−2
−3	1
−2	5
−1	7
0	10
1	6

17.

Domain	Range
−3	1
−2	0
−2	3
−1	4
0	5
1	6

18.

Domain	Range
2	1
−2	3
0	0
−1	2
−3	7
2	4

For Exercises 19–24, determine if the graph represents a function.

19.

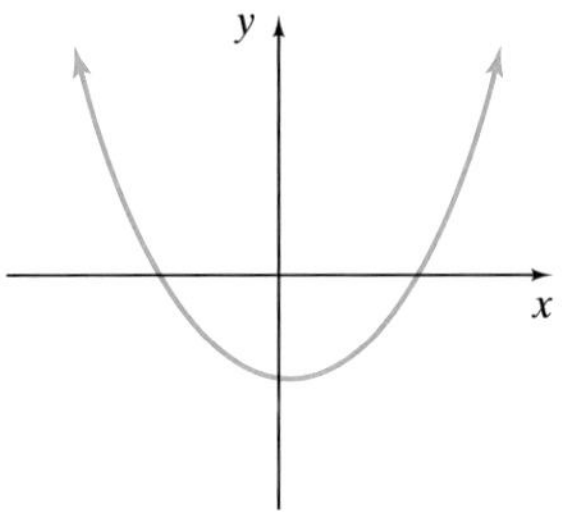

20.

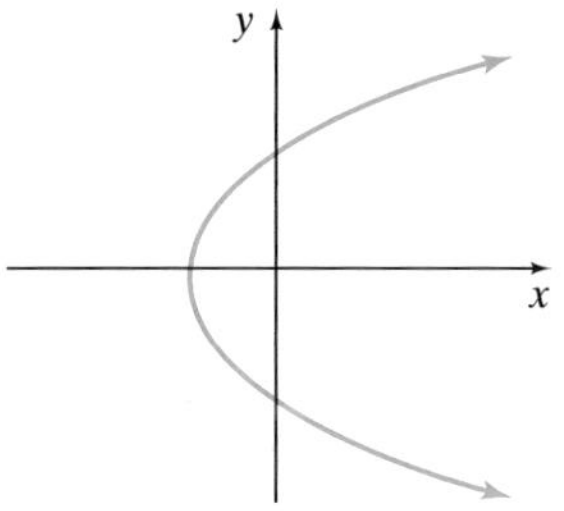

21.

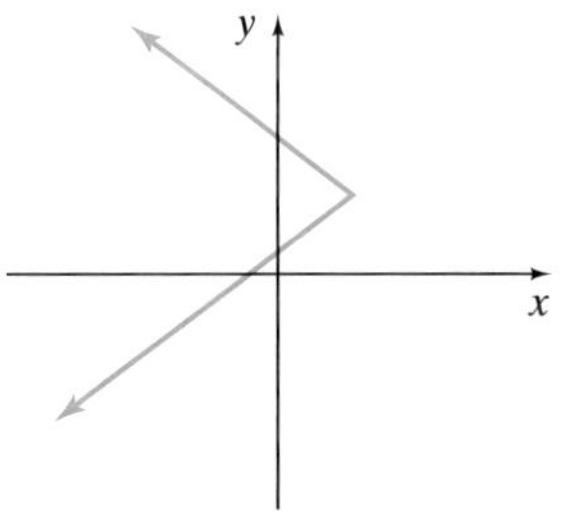

22.

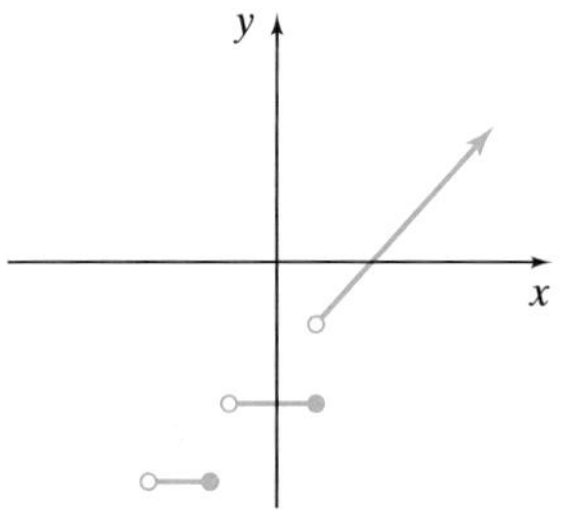

23.

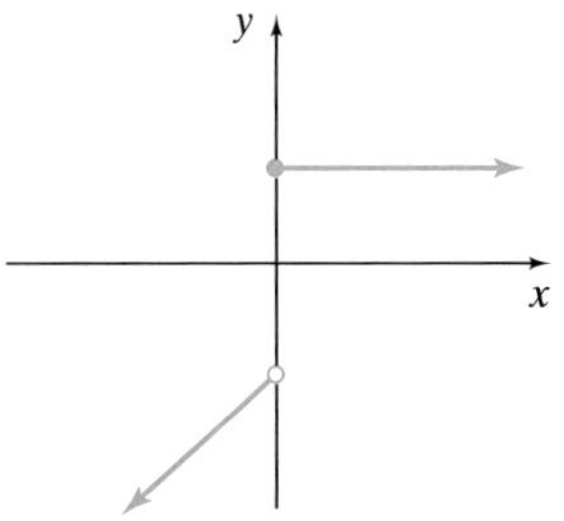

24.

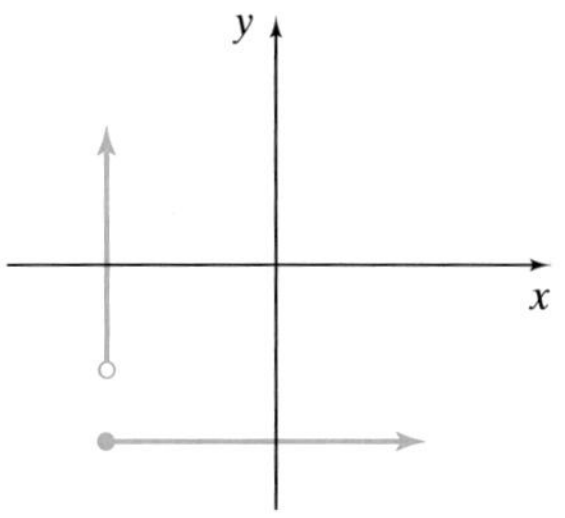

For Exercises 25–35, let $f(x) = 2x - 1$ and $g(x) = x^2$. Evaluate each expression and write the solution as an ordered pair.

25. $f(2)$

26. $f(-1)$

27. $f(1)$

28. $f(-2)$

29. $g(3)$

30. $g(1)$

31. $g(0)$

32. $f\left(\frac{1}{2}\right)$

33. $g\left(\frac{1}{2}\right)$

34. $f(-0.5)$

35. $g(-0.25)$

36. Use the graph of the function f to answer parts (a) through (f).

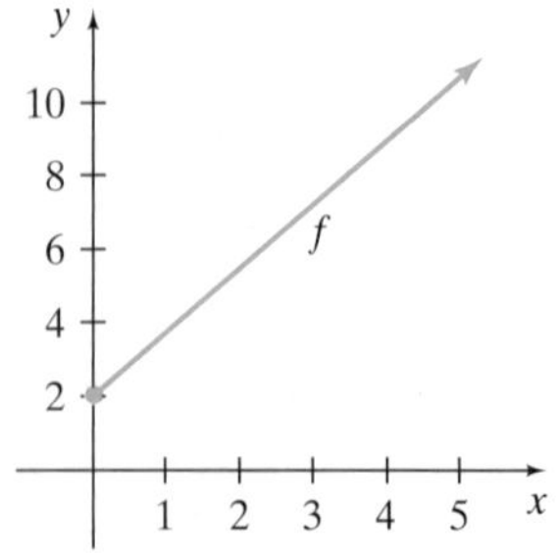

(a) What is the independent variable?
(b) What is the dependent variable?
(c) What is the value of $f(x)$ when $x = 0$?
(d) What is $f(3)$?
(e) What is the domain of f?
(f) What is the range of f?

37. Use the graph of the function f to answer parts (a) through (f).

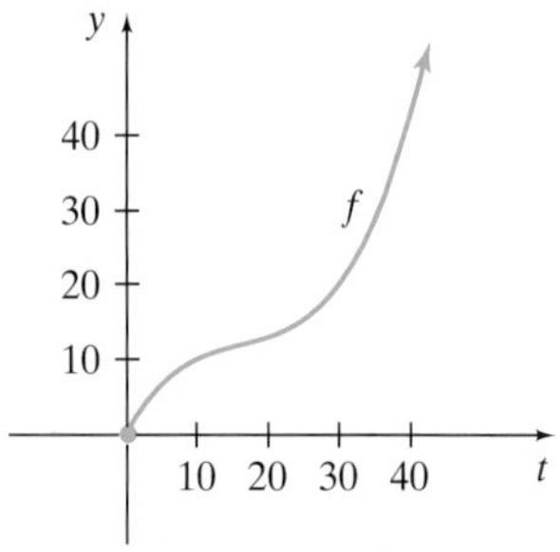

(a) What is the independent variable?
(b) What is the dependent variable?
(c) What is the value of $f(t)$ when $t = 10$?
(d) What is $f(30)$?
(e) What is the domain of f?
(f) What is the range of f?

38. Use the graph of the function g to answer parts (a) through (g).

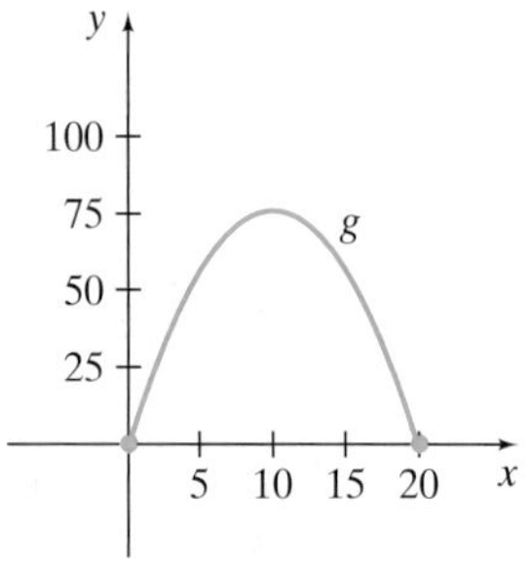

(a) What is the independent variable?
(b) What is the dependent variable?
(c) What is the value of $g(x)$ when $x = 20$?
(d) What is $g(10)$?
(e) What are the values of x when $g(x) = 50$?
(f) What is the domain of g?
(g) What is the range of g?

39. Complete the following table.

Interval Notation	Inequality Notation	Number Line
$[-\frac{1}{2}, 5)$		x
	$x \geq -2$	x
		-3 x
	$-1 < x < 10$	x
$(-\infty, 3) \cup (3, \infty)$		x

For Exercises 40–51, determine the domain of the function algebraically, and write the domain using interval notation.

40. $f(x) = 2x$

41. $g(x) = -10x$

42. $f(x) = \dfrac{x^2 - 36}{x - 6}$

43. $g(x) = 3x^2 - 10$

44. $f(x) = x^3 - 6x^2$

45. $y = \dfrac{x - 10}{5 - x}$

46. $g(x) = \sqrt{x - 6}$

47. $f(x) = \sqrt{6 - x}$

48. $f(x) = \dfrac{\sqrt{x - 1}}{x - 4}$

49. $y = \sqrt{x^2 + 3}$

50. $y = \dfrac{10x}{x^2 - 25}$

51. $g(x) = \dfrac{x - 1}{2x^2 - 4x}$

In Exercises 52–55, graph the function on your calculator and use the graph to determine the domain and range of the function. Write your solutions using interval notation.

52. $g(x) = x^2 - 4$

53. $f(x) = \dfrac{x^2 + 5}{10}$

54. $f(x) = \sqrt{9 - x}$

55. $y = \dfrac{x - 2}{x^2 + 6}$

56. The price–demand function for the new Hanford hand-held game unit is given by

$$p(x) = 120 - 0.1x \qquad 0 \leq x \leq 80$$

Here x represents the number of units sold per day and $p(x)$ represents the price of the unit.

(a) Complete the following table of values. (In other words, determine the values for the price when $x = 0, 10, 20, 30, \ldots, 80$.)

x	0	10	20	30	40	50	60	70	80
$p(x)$									

(b) Make a graph of p on the interval [0, 80].

(c) If [0, 80] is the domain, what is the range of the function?

(d) Evaluate $p(45)$ and interpret the answer.

57. The price–demand function for the new Teddy Bear line of designer lingerie is given by

$$p(x) = \frac{165 - x}{4} \qquad 0 \le x \le 35$$

Here x represents the number of units sold per day, and $p(x)$ represents the price of the unit.

(a) Complete the following table of values.

x	0	5	10	15	20	25	30	35
$p(x)$								

(b) Make a graph of p on the interval [0, 35].

(c) If [0, 35] is the domain, what is the range of the function?

(d) Evaluate $p(22)$ and interpret the answer.

58. For the price–demand function in Exercise 56, determine the desired price for each unit if the number sold is 65 units per day.

59. For the price–demand function in Exercise 57, determine the desired price for each unit if the number sold is 18 units per day.

For Exercises 60–65, write the domain and range of the function using interval notation.

60.

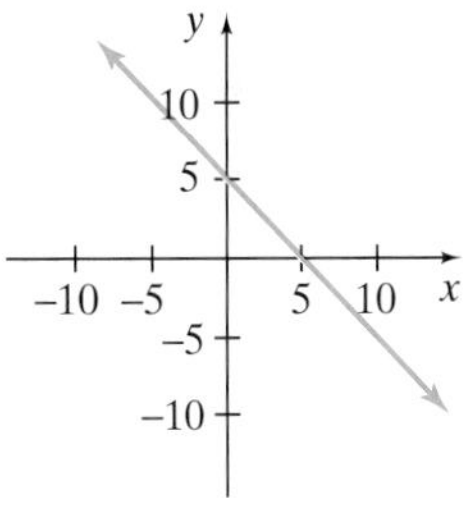

61.

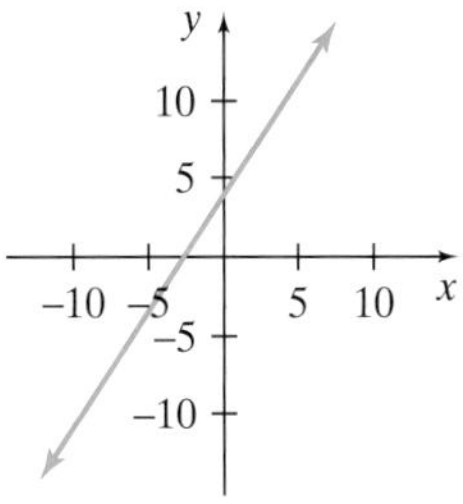

62.

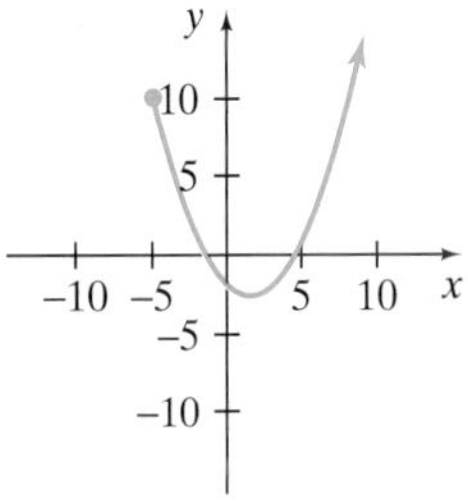

63.

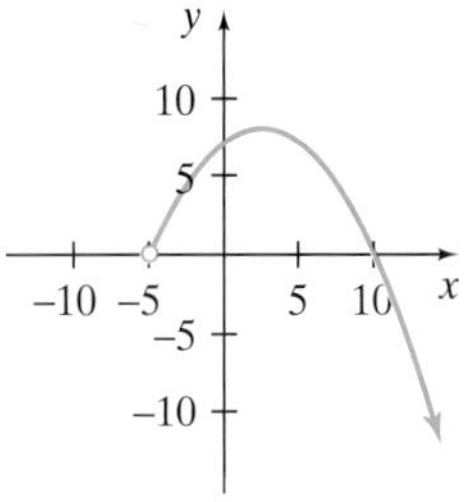

64.

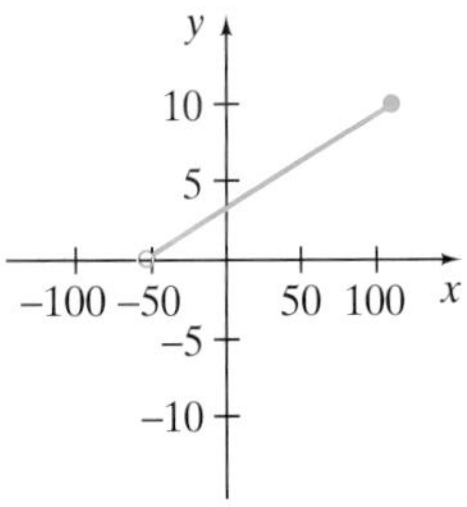

65.

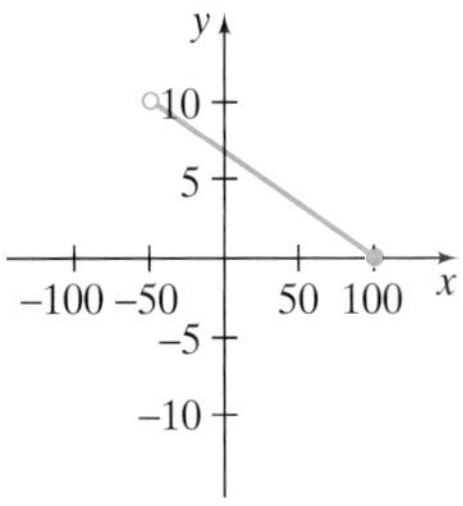

SECTION PROJECT

Determine if the following can be represented by functions.

(a) A person's astrological sign is a function of his or her birth date.

(b) A person's income is a function of how long he or she works each day.

(c) The voter turnout for an election is a function of the weather.

(d) The distance a delivery person is from a pizzeria is a function of the number of deliveries made.

(e) The number of U.S. Congressmen a state has is a function of its population.

Section 11.2 Introduction to Problem Solving

Students who study mathematics often find that solving applications can be one of the most challenging tasks that they encounter. These applications (sometimes called "word problems") can be difficult, and at times frustrating, because there does not appear to be a consistent way to solve these particular problems. This section reintroduces the problem solving method that is used throughout this textbook to organize and solve various applications.

Thinking about Problem Solving

When solving applications, students most often ask themselves questions such as, "What do I do first?" or "What am I trying to find?" The key in beginning to solve any type of application is to **Understand the Situation**. This can include a statement on what formula or method can be used to solve the problem or defining a quantity such as x that represents a solution to the problem. The other often-asked question goes something like, "I've found this answer; now what does it mean?" We address this question in our problem solving method when we **Interpret the Solution**. This involves writing a concluding sentence that states what the numerical answer means in the context of the problem being asked. This concluding sentence also includes proper labeling of the units used in the problem.

Problem Solving Method

When encountering an application, we can use the following method to organize our work.

1. **Understand the Situation**: This involves stating any particular methods or formulas that can be used to solve the application, defining the proper unknown quantities, or stating how the problem will be solved.
2. After the problem is organized and set up, the application is then solved using the methods discussed in the applicable section.
3. **Interpret the Solution**: This involves writing a concluding sentence that states what the numerical answer means in the context of the problem being asked. Proper units are included in the interpretation.

We will see that this method can be used in a variety of topics and that all three steps are not always used. In Example 1, we illustrate how to use the problem solving method.

Example 1 Using the Problem Solving Method

The Fontana Vitamin Company determines that the cost, in dollars, of producing x bottles of a new supplement is given by

$$C(x) = 1.25x + 550.$$

(a) How much does it cost to produce 70 bottles of the new supplement?

(b) How many bottles are produced if the cost is \$703.75?

Solution

(a) Understand the Situation: We are given the equation that relates cost, C, as a function of the number of bottles produced, x. When we are asked to determine the cost to produce 70 bottles of the new supplement, we are being told that $x = 70$. To answer this question, we simply need to evaluate $C(70)$.

We evaluate $C(70)$ by substituting 70 for x as follows.

$$C(x) = 1.25x + 550$$
$$C(70) = 1.25(70) + 550 = 637.5$$

Interpret the Solution: The Fontana Vitamin Company realizes a cost of \$637.50 to produce 70 bottles of a new supplement.

(b) Understand the Situation: Here we are asked to determine the number of bottles (an x-value) produced when the cost is \$703.75. Since $C(x)$

represents the cost in dollars, we are being asked to find the x-value when $C(x) = \$703.75$.

The equation we want to solve is

$$703.75 = 1.25x + 550$$

Solving this equation for x gives us

$703.75 = 1.25x + 550$	Given equation
$703.75 - 550 = 1.25x$	Subtract 550 from both sides
$153.75 = 1.25x$	Simplify
$\dfrac{153.75}{1.25} = x$	Divide both sides by 1.25
$123 = x$	Simplify

Interpret the Solution: This means that the Fontana Vitamin Company can produce 123 bottles of the new supplement for a cost of \$703.75. ■

Thinking about Reasonable Domain

Many of the applications that we study are based on mathematical models. These models are based on functions that are derived from data which is commonly given in the form of a table.

Mathematical Model

A function that is derived from a real-world situation is called a **mathematical model**.

Table 11.2.1

Year	Average Family Income (in thousands of dollars)
1990	22.23
1995	25.88
1997	30.33
1998	30.66
1999	32.90

SOURCE: U.S. Census Bureau, www.census.gov

Let's examine a typical example of a mathematical model. The average family income for Puerto Ricans from 1990 to 1999 is shown in Table 11.2.1. Note that the data was not available for every year during this decade.

The data in Table 11.2.1 can be modeled by

$$f(x) = 1.17x + 21.61 \qquad 0 \le x \le 9$$

where x represents the number of years since 1990 and $f(x)$ represents the average family income, in thousands of dollars, of Puerto Ricans. An important part of this model is the interval $0 \le x \le 9$ seen to the right of the model. *Every mathematical model in this text that is based on data will have an interval similar to this, written to the right of the model.* Understanding what this interval means is an important component when attempting to **Understand the Situation**.

Example 2 Understanding Mathematical Models

The average family income for Puerto Ricans can be modeled by

$$f(x) = 1.17x + 21.61 \qquad 0 \le x \le 9$$

where x represents the number of years since 1990 and $f(x)$ represents the average family income, in thousands of dollars, of Puerto Ricans. Interpret the meaning of the interval $0 \le x \le 9$.

Solution

Understand the Situation: We know that the independent variable x represents the number of years since 1990. So $x = 0$ corresponds to the year $1990 + 0 = 1990$ whereas $x = 9$ corresponds to the year $1990 + 9 = 1999$.

Interpret the Solution: Our analysis indicates that the interval $0 \le x \le 9$ corresponds to the years 1990 to 1999. This means that our mathematical model is defined only for the time frame 1990 to 1999. ■

For the model in Example 2, the interval $0 \le x \le 9$ gives the x-values for which the model is defined. We call intervals such as this the **reasonable domain** of the model.

Reasonable Domain

For a mathematical model, the **reasonable domain** gives the independent variable values for which the model is defined.

Note: In Section 11.1, we learned that the domain of a linear function is the set of all real numbers. The reasonable domain is only the portion of the real numbers for which our model is defined.

Example 3 Understanding the Reasonable Domain

The number of cable TV subscribers in the U.S. can be modeled by

$$f(x) = 1.77x + 49.71 \qquad 0 \le x \le 10$$

where x represents the number of years since 1990 and $f(x)$ represents the number of cable TV subscribers in the U.S. in millions (*Source:* U.S. Statistical Abstract, www.census.gov/statab/www).

(a) Interpret the reasonable domain of the model.

(b) Explain why the model cannot be used to determine the number of cable TV subscribers in 2002.

Solution

(a) **Understand the Situation:** The reasonable domain for the model is $0 \le x \le 10$. Since x represents the number of years since 1990, we see that $x = 0$ corresponds to the year 1990, $x = 1$ corresponds to 1991 and so on until we get to $x = 10$ which corresponds to 2000.

Interpret the Solution: So the reasonable domain, $0 \le x \le 10$, corresponds to the years 1990 to 2000.

(b) From the analysis in part (a), we know that $x = 12$ corresponds to the year 2002 since $1990 + 12 = 2002$. Because our model is defined on the reasonable domain $0 \le x \le 10$, we cannot use the model to estimate the number of cable TV subscribers in 2002, since $x = 12$ is not in the reasonable domain. ■

Checkpoint 1 Now work Exercise 15.

Examples 2 and 3 illustrate that the reasonable domain is an important part of the problem solving method. Once we understand the reasonable domain, we quite often evaluate the function for a value that is in the reasonable domain. Once this evaluation is done and we have a numerical result, one of the most common questions focuses on what does the numerical answer mean. We address this often asked question when we **Interpret the Solution**. By interpreting, we mean to write a sentence that explains the numerical answer written in the context of the given mathematical model, as we did in Example 1. Example 4 illustrates how to interpret a solution and incorporate the reasonable domain as well.

Example 4 Using the Problem Solving Method

The number of cellular telephone subscribers in the U.S. can be modeled by

$$f(x) = 0.84x^2 + 1.33x + 5.17 \qquad 0 \le x \le 9$$

where x represents the number of years since 1990 and $f(x)$ represents the number of cellular telephone subscribers in millions. Evaluate and interpret $f(8)$ (*Source:* Cellular Telecommunications Industry Association, www.tiaonline.org).

Interactive Activity

The tabular data for the number of cellular telephone subscribers in the U.S. from 1990 to 1999 shows that there were 65.89 million cellular telephone subscribers in the U.S. in 1998. How much does this answer differ from the solution to Example 4. Why do you think there is a difference?

Solution

Understand the Situation: Here we need to substitute 8 for x in the given model $f(x)$ and simplify. Also note that $x = 8$ corresponds to 1998.

Evaluating, we get

$$f(8) = 0.84(8)^2 + 1.33(8) + 5.17 = 53.76 + 10.64 + 5.17 = 69.57$$

Interpret the Solution: This means that in 1998 there were about 69.57 million cellular telephone subscribers in the U.S. ■

Our final example illustrates how the problem solving method can be used to attack another application.

Example 5 Using the Problem Solving Method

In Example 3 we saw that the number of cable TV subscribers in the U.S. can be modeled by

$$f(x) = 1.77x + 49.71 \qquad 0 \le x \le 10$$

where x represents the number of years since 1990 and $f(x)$ represents the number of cable TV subscribers in the U.S. in millions. Use the model to determine the year that the number of cable TV subscribers in the U.S. was about 76.26 million (*Source:* U.S. Statistical Abstract, www.census.gov/statab/www).

Solution

Understand the Situation: Here we are asked to find the year (an x-value) when the number of cable TV subscribers was about 76.26 million. Since $f(x)$ represents the number of cable TV subscribers in millions, we are being asked to find the x-value when $f(x) = 76.26$.

The equation we want to solve is

$$1.77x + 49.71 = 76.26$$

Solving this equation for x gives us

$1.77x + 49.71 = 76.26$	Given equation
$1.77x = 76.26 - 49.71$	Subtract 49.71 from both sides
$1.77x = 26.55$	Simplify
$x = \frac{26.55}{1.77}$	Divide both sides by 1.77
$x = 15$	Simplify

Notice that $x = 15$ corresponds to the year 2005.

Interpret the Solution: Since our solution, $x = 15$, does not fall in the reasonable domain, $0 \le x \le 10$, there is no year in this interval in which the number of cable TV subscribers is 76.26 million.

In Example 5, it may be tempting to say that the solution $x = 15$ means that in 2005 the number of cable TV subscribers will be 76.26 million. This is not correct since the value $x = 15$ is not in the reasonable domain. Also, there is a big danger in using a model to make predictions for values outside the reasonable domain. Many things could change after 2000 (the last year the model in Example 5 is defined for). For example, more people may start to subscribe to a satellite service, which could affect the number of cable TV subscribers.

Checkpoint 2 Now work Exercise 29.

SUMMARY

- **The Problem Solving Method**:
  1. **Understand the Situation**: This involves stating any particular methods or formulas that can be used to solve the application, defining the proper unknown quantities, or stating how the problem will be solved.
  2. After the problem is organized and set up, the application is then solved using the methods discussed in the applicable section.
  3. **Interpret the Solution**: This involves writing a concluding sentence that states what the numerical answer means in context of the problem being asked. Proper units are included in the interpretation.
- A function that is derived from a real-world situation is called a **mathematical model**.
- For a mathematical model, the **reasonable domain** gives the independent variable values for which the model is defined.

SECTION 11.2 EXERCISES

In this Exercise set, the number under the ⇒ symbol indicates the section where the application is used.

⇒ 11.3 **1.** The Dirt Rocket Company produces spokes for off-road motorcycles. It determines that the daily cost in dollars of producing x quick-replace spokes is $C(x) = 1250 + 3x$. Determine the cost of producing 300 quick-replace spokes.

⇒ 11.3 **2.** The Get-It-Now florist determines that the cost in dollars of arranging and delivering x get-well bouquets is given by the cost function $C(x) = 10 + 32x$. Determine the cost of arranging and delivering 20 get-well bouquets.

⇒ 12.1 **3.** The cost, in dollars, of producing x units of a product is given by $C(x) = 22{,}500 + 7.35x$.

(a) Determine the cost of producing 30 units.

(b) How many units can be produced if the cost is $24,197.85?

⇒ 12.1 **4.** The cost, in dollars, of producing x units of a product is given by $C(x) = 33{,}125 + 6.38x$.

(a) Determine the cost of producing 85 units.

(b) How many units can be produced if the cost is $33,546.08?

⇒ 12.2 **5.** The total cost of producing q units of a product is given by $C(q) = 13{,}700 + 6.85q$, where $C(q)$ is in dollars.

(a) Evaluate $C(110)$ and interpret.

(b) How many units can be produced if the cost is $16,714?

12.2 **6.** The total cost of producing q units of a product is given by $C(q) = 15{,}200 + 4.85q$, where $C(q)$ is in dollars.

(a) Evaluate $C(120)$ and interpret.

(b) How many units can be produced if the cost is \$15,539.50?

In Exercises 7–12, a table of data is given and the independent variable x is defined. Write the reasonable domain for the model that represents the data.

11.9 **7.** The variable x represents the number of years since 1980.

Year	Multiple Births (twins, triplets, etc.) per 1000 Live Births
1980	19.3
1985	21
1990	23.3
1991	23.9
1992	24.4
1993	25.2
1994	25.7
1995	26.1
1996	27.4
1997	28.6
1998	30

SOURCE: U.S. National Center for Health Statistics, www.cdc.gov/nchs

11.9 **8.** The variable x represents the number of years since 1975.

Year	Number of U.S. Golf Facilities
1975	11,370
1980	12,005
1985	12,346
1990	12,846
1992	13,210
1993	13,439
1994	13,683
1995	14,074
1999	14,900

SOURCE: National Golf Foundation, www.ngf.org

11.9 **9.** The variable x represents the number of years since 1989.

Year	Average Expenditure for a New Car (in \$)
1990	15,926
1991	16,650
1992	17,825
1993	18,585
1994	19,463
1995	19,757

SOURCE: U.S. Bureau of Economic Analysis, www.bea.doc.gov

11.9 **10.** The variable x represents the number of years since 1969.

Year	Percentage of 3- to 5-Year-Olds in Preschool
1970	37.5
1975	48.6
1980	52.5
1985	54.6
1990	59.4
1992	55.5
1993	55.1
1994	61
1995	61.8
1998	64.5

SOURCE: U.S. Census Bureau, www.census.gov

12.2 **11.** The variable x represents the number of years since 1899.

Year	*y*, Population in Thousands
1900	1,485.053
1910	2,377.549
1920	3,426.861
1930	5,677.251
1940	6,907.387
1950	10,586.223
1960	15,717.204
1970	19,971.069
1980	23,667.764
1990	29,760.021
1999	33,145.000

SOURCE: U.S. Census Bureau, www.census.gov

12.6 **12.** The variable x represents the number of years since 1980.

Year	National Debt (in billions of dollars)	U.S. Civilian Population (in millions)
1980	909.1	225.621
1982	1137.3	229.995
1984	1564.7	234.110
1986	2120.6	238.412
1988	2601.3	242.817
1990	3206.6	247.758
1992	4002.1	253.426
1994	4643.7	258.960
1996	5207.3	263.998

SOURCE: U.S. Census Bureau, www.census.gov

13.1 **13.** The percentage of males and females who lived alone in the U.S. and were between the ages of 45 and 64 years old can be modeled by

$$f(x) = 0.03x^2 - 0.5x + 25.56 \qquad 1 \le x \le 20$$

where x represents the number of years since 1979 and $f(x)$ represents the percentage of males and females who lived alone and were between the ages of 45 and 64 years old (*Source:* U.S. Census Bureau, www.census.gov). Interpret the reasonable domain of the model.

13.1 **14.** The percentage of males who lived alone in the U.S. and were between the ages of 45 and 64 years old can be modeled by

$$f(x) = 0.014x^2 - 0.09x + 9.08 \qquad 1 \le x \le 20$$

where x represents the number of years since 1979 and $f(x)$ represents the percentage of males who lived alone and were between the ages of 45 and 64 years old (*Source:* U.S. Census Bureau, www.census.gov). Interpret the reasonable domain of the model.

13.3 ✓ **15.** The poverty threshold in the United States can be modeled by

$$f(x) = 0.004x^2 + 0.057x + 1.280 \qquad 0 \le x \le 35$$

where x represents the number of years since 1960 and $f(x)$ represents the U.S. poverty threshold in thousands of dollars (*Source:* U.S. Census Bureau, www.census.gov). Interpret the reasonable domain of the model.

13.3 **16.** The population of Florida can be modeled by

$$f(x) = 0.003x^2 + 0.168x + 4.908 \qquad 0 \le x \le 35$$

where x represents the number of years since 1960 and $f(x)$ represents the population in millions (*Source:* U.S. Census Bureau, www.census.gov). Interpret the reasonable domain of the model.

14.2 **17.** The annual per capita consumption of light and skim milk in the United States can be modeled by

$$f(x) = 10.12 + 2\ln x \qquad 1 \le x \le 16$$

where x represents the number of years since 1979 and $f(x)$ represents the annual per capita consumption of light and skim milk in gallons (*Source:* U.S. Department of Agriculture, www.usda.gov). Interpret the reasonable domain of the model.

14.2 **18.** The average expenditure for a new domestic car in the United States can be modeled by

$$f(x) = 15{,}241.24 + 1517.65\ln x \qquad 1 \le x \le 8$$

where x represents the number of years since 1991 and $f(x)$ represents the average expenditure for a new domestic car in dollars (*Source:* U.S. Bureau of Economic Analysis, www.bea.doc.gov). Interpret the reasonable domain of the model.

11.3 **19.** The annual total expenditures on pollution abatement in the United States can be modeled by

$$f(x) = 2x + 48.17 \qquad 0 \le x \le 20$$

where x represents the number of years since 1972 and $f(x)$ represents the expenditures in billions of constant 1987 dollars (*Source:* U.S. Statistical Abstract, www.census.gov/statab/www).

(a) Evaluate $f(12)$ and interpret.

(b) Explain why the model cannot be used to determine the annual total expenditures on pollution abatement in 1997.

11.3 **20.** The amount spent annually in college bookstores in the United States can be modeled by

$$f(x) = 0.19x + 1.67 \qquad 0 \le x \le 15$$

where x represents the number of years since 1982 and $f(x)$ represents the amount spent in billions of dollars (*Source:* U.S. Census Bureau, www.census.gov).

(a) Evaluate $f(13)$ and interpret.

(b) Explain why the model cannot be used to determine the amount spent annually in college bookstores in 2001.

11.4 **21.** The number of milligrams of cholesterol consumed each day per person in the United States can be modeled by

$$f(x) = 0.11x^2 - 4.04x + 445.02 \qquad 1 \le x \le 23$$

where x represents the number of years since 1974 and $f(x)$ represents the number of milligrams of cholesterol consumed each day per person (*Source:* U.S. Department of Agriculture, www.usda.gov).

(a) Evaluate $f(20)$ and interpret.

(b) Explain why the model cannot be used to determine the number of milligrams of cholesterol consumed each day per person in 1974.

11.4 **22.** The number of lawyers in private practice in the U.S. can be modeled by

$$f(x) = 0.37x^2 - 1.43x + 190.34 \qquad 0 \le x \le 30$$

where x represents the number of years since 1960 and $f(x)$ represents the number of lawyers in private practice measured in thousands (*Source:* U.S. Statistical Abstract, www.census.gov/statab/www).

(a) Evaluate $f(22)$ and interpret.

(b) Explain why the model cannot be used to determine the number of lawyers in private practice in 1993.

11.5 **23.** The percent of people taking the SAT whose intended area of study was engineering can be modeled by

$$f(x) = 0.002x^3 - 0.11x^2 + 1.38x + 6.7 \qquad 0 \le x \le 22$$

where x represents the number of years since 1975 and $f(x)$ represents the percent of people taking the SAT whose intended area of study was engineering (*Source:* U.S. Statistical Abstract, www.census.gov/statab/www).

(a) Evaluate $f(20)$ and interpret.

(b) Explain why the model cannot be used to determine the percent of people taking the SAT whose intended area of study was engineering in 1999.

11.5 **24.** The percent of people taking the SAT whose intended area of study was business can be modeled by

$$f(x) = -0.08x^2 + 1.77x + 11.59 \qquad 0 \le x \le 22$$

where x represents the number of years since 1975 and $f(x)$ represents the percent of people taking the SAT whose intended area of study was business (*Source:* U.S. Statistical Abstract, www.census.gov/statab/www).

(a) Evaluate $f(20)$ and interpret.

(b) Explain why the model cannot be used to determine the percent of people taking the SAT whose intended area of study was business in 1974.

25. For the mathematical model given in Exercise 19, use the model to determine the year that the annual total expenditures on pollution abatement was about 58.17 billion dollars.

26. For the mathematical model given in Exercise 20, use the model to determine the year that the amount spent annually in college bookstores was about 3 billion dollars.

27. The total payments (passenger fares and expenditures abroad) by U.S. travelers to foreign countries can be modeled by

$$f(x) = 2.78x + 33.81 \qquad 1 \le x \le 9$$

where x represents the number of years since 1986 and $f(x)$ represents the total payments, in billions of dollars, by U.S. travelers to foreign countries. Use the model to determine the year that the total payments by U.S. travelers was about 53.27 billion dollars (*Source:* U.S. Department of Commerce, www.doc.gov).

12.6 **28.** The average hourly earnings, in dollars, of employees in finance, insurance, and real estate in the United States are given by

$$g(t) = 0.48t + 9.44 \qquad 1 \le t \le 7$$

where t is the number of years since 1989 and $g(t)$ is the average hourly earnings in dollars per hour. Use the model to determine the year that the average hourly earnings was about \$11.84 per hour (*Source:* U.S. Bureau of Labor Statistics, www.bls.gov).

12.6 **29.** The average weekly earnings, in dollars, of employees in finance, insurance, and real estate in the United States are given by

$$f(t) = 17.5t + 336.86 \qquad 1 \le t \le 7$$

where t is the number of years since 1989 and $f(t)$ is the average weekly earnings in dollars. Use the model to determine the year that the average weekly earnings was about \$599.36 (*Source:* U.S. Bureau of Labor Statistics, www.bls.gov).

12.6 **30.** The U.S. civilian population can be modeled by

$$CP(t) = 2.57t + 232.91 \qquad 1 \le t \le 11$$

where t is the number of years since 1984 and $CP(t)$ is the U.S. civilian population in millions. Use the model to determine the year that the U.S. civilian population was about 284.31 million (*Source:* U.S. Census Bureau, www.census.gov).

SECTION PROJECT

The U.S. Census Bureau has projected the number of people in the U.S. who are 100 years old and over (called centenarians) from 2000 to 2040 in five year increments. The results are shown in Table 11.2.2.

(a) Let x represent the number of years since 2000. Rewrite the table with the standardized variable values.

(b) Construct a scatterplot of the table in part (a). Use values from 0 to 600 on the vertical axis.

(c) Estimate which of the four models best fits the data in the scatterplot. Explain.

(i) $f(x) = 12.06x + 24.06$

(ii) $f(x) = 120.6x + 24.06$

(iii) $f(x) = 12.06x + 240.6$

(iv) $f(x) = 12.06x + 2406$

Table 11.2.2

Year	Projected Centenarian Population (in thousands)
2000	65
2005	96
2010	129
2015	177
2020	235
2025	313
2030	381
2035	441
2040	551

SOURCE: U.S. Census Bureau, www.census.gov

Section 11.3 Linear Functions and Average Rate of Change

The first type of function that we will study is called a **linear function**. When we say things like "The cost of a phone call is a function of the length of the call" or "The price of a cab ride is a function of the distance to the destination," we are referring to linear functions.

Another critical concept that we address in this section is **average rate of change**. This kind of rate is a common component of linear functions. If we know the average rate of change of a linear function (also known as the **slope** of the line), it is usually a short process to write the function itself. Statements such as "On average, the price of used cars is going up \$100 per year" or "My computer equipment is depreciating at about \$80 each month" are using the average rate of change concept.

Linear Functions

Table 11.3.1

Number of Copies	Cost
100	\$15
500	\$55
1000	\$105

We begin our investigation of linear functions with an example of how it is used in an everyday application. Suppose that The Fashion Mystique wants to print promotional flyers for its annual sidewalk sale. The store manager goes to the local copy center and finds that there is a \$5 setup charge and a 10 cent copying fee for each flyer produced. The manager is handed the pricing schedule shown in Table 11.3.1.

There must be a functional relationship between the values in the two columns since the cost of the copies is a **function** of the number of copies made. Since the cost of the job **depends** on the number of copies produced, we believe that the number of copies is the independent variable, and the cost is the dependent variable.

This function consists of two parts. The first is the setup charge, which is a flat fee that is assessed no matter how many copies are made. It is what economists refer to as a **fixed cost**. The other part is the copying fee, which changes as the number of copies changes. Economists call this type of cost **variable costs**. We can use this combination of fixed and variable costs to verify the cost of making 500 flyers. For example, the \$5 setup fee plus 10 cents times 500 for the copying fee gives

$$\begin{aligned}\text{Cost for 500} &= (\text{price per copy}) \cdot (\text{number of copies}) + (\text{setup fee}) \\ &= (0.10) \cdot 500 + 5 \\ &= 50 + 5 = 55 \text{ or } \$55\end{aligned}$$

This is an exact numerical representation of what a linear cost model looks like. We now define the general form of a linear cost function.

Linear Cost Function

The cost $C(x)$ of producing x units of a product is given by the linear **cost function**

$$C(x) = (\text{variable costs}) \cdot (\text{units produced}) + (\text{fixed costs})$$

Note: Some cost functions are not linear because the variable cost is a function itself. However, every cost function is made up of a variable cost expression and a fixed cost.

Example 1 Determining a Cost Function

Determine a cost function, C, for the printing job just described.

Solution

The variable costs are 10 cents per copy and the fixed costs are 5 dollars, which means that the cost function for the printing job can be written as

$$C(x) = 0.10x + 5$$

where x represents the number of copies made and $C(x)$ represents the cost, in dollars, of making the copies. Figure 11.3.1 shows a graph of the cost function and the points from the pricing schedule given in Table 11.3.1.

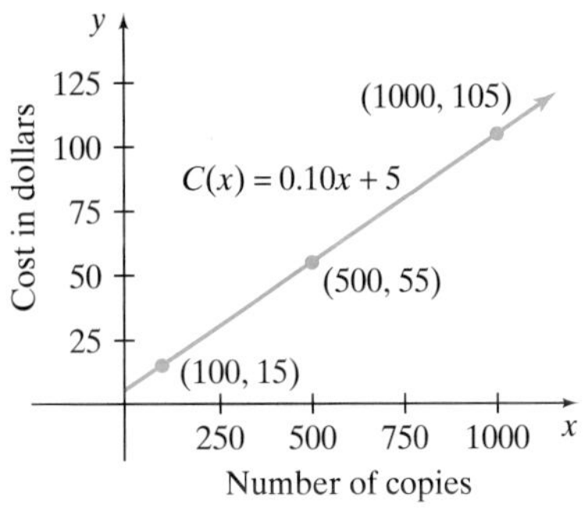

Figure 11.3.1

The cost function in Example 1 is one example of a linear function. Notice that the three points of the graph all *line up*. This is precisely why we call functions of this type **linear**. A linear function is made up of two key parts: a fixed number called a constant and a variable part that is the product of a number and a variable.

Linear Function

A **linear function** has the form

$$f(x) = mx + b$$

where m is the **average rate of change** or **slope** of the line and b is a **constant**, where m and b are real numbers.

The number given by m has a very important meaning. Let's return to the printing job described in Example 1 and use the graph to compute the cost per copy, denoted $\frac{\text{cost}}{\text{copy}}$. To do this, we calculate the difference between two of the points (500, 55) and (1000, 105) as follows:

$$\begin{aligned}\frac{\text{cost}}{\text{copy}} &= \frac{\text{difference in cost}}{\text{difference in copies}}\\ &= \frac{105 - 55}{1000 - 500} = \frac{50}{500} = \frac{1}{10} = 0.1\end{aligned}$$

We say that the cost of this printing job is increasing at an average rate of \$0.10 per copy. If we think of the cost as the variable y and the copies as x, we have the following definition.

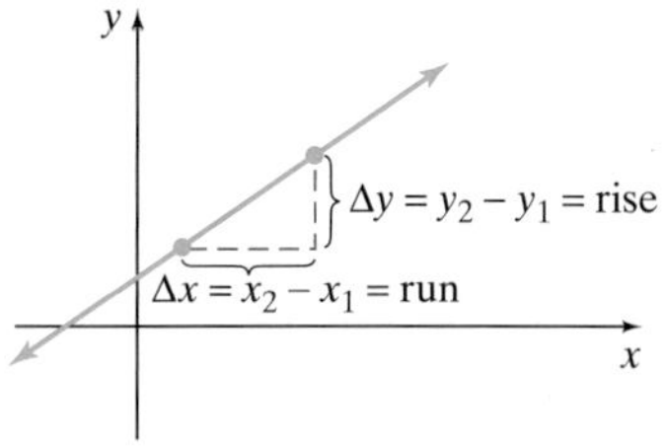

Figure 11.3.2
$$\text{slope} = \frac{\text{rise}}{\text{run}} = \frac{\Delta y}{\Delta x} = \frac{y_2 - y_1}{x_2 - x_1}$$

Average Rate of Change

The **average rate of change** of y with respect to x is

$$\frac{\text{difference in } y}{\text{difference in } x} = \frac{y_2 - y_1}{x_2 - x_1}$$

(Notice that the average rate of change in the cost of the copies is 0.10, which is exactly the same as what we called the variable costs.) When referring to a linear function, the average rate of change is also called the **slope** of the line. See Figure 11.3.2.

Slope

The **slope** of a line, denoted by m, is a measurement of the steepness of the line. Given two points on a line, (x_1, y_1) and (x_2, y_2), the slope of the line is computed by

$$m = \frac{y_2 - y_1}{x_2 - x_1}$$

The slope of the line also gives the average rate of change of y with respect to x.

Note: All of the following are ways to interpret slope:

$$\text{Average rate of change} = m = \frac{y_2 - y_1}{x_2 - x_1}$$
$$= \frac{\text{rise}}{\text{run}} = \frac{\text{difference in } y}{\text{difference in } x}$$
$$= \frac{\text{change in } y}{\text{change in } x} = \frac{\Delta y}{\Delta x}$$

Example 2 Computing Average Rate of Change

For the points $P = (2, -3)$ and $Q = (4, 2)$:

(a) Compute the average rate of change between the points.

(b) If the independent variable value increases by 1 unit, how will this affect the dependent value?

Solution

(a) Understand the Situation: Here we will let P be the first point and Q be the second point. This means that (x_1, y_1) is the ordered pair $(2, -3)$ and (x_2, y_2) is the ordered pair $(4, 2)$.

Using the definition of average rate of change we get

$$m = \frac{y_2 - y_1}{x_2 - x_1} = \frac{2 - (-3)}{4 - 2} = \frac{5}{2} = 2.5 \frac{\text{dependent units}}{\text{independent unit}}$$

(b) Since the average rate of change is $m = 2.5$, we see that if x, the independent value, increases by 1 unit then the dependent value will increase by 2.5 or $\frac{5}{2}$ units. This change is illustrated in Figure 11.3.3.

Interactive Activity*

Rework Example 2a, except let P be the second point and Q be the first point.

WWW

* Interactive Activities marked with a (WWW) have solutions provided at the companion website at www.prenhall.com/armstrong.

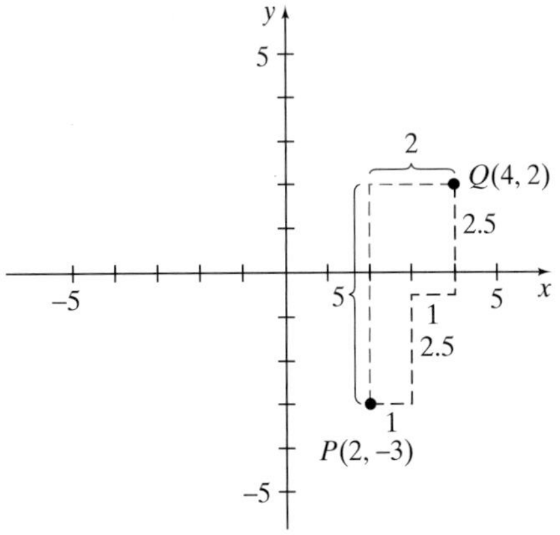

Figure 11.3.3

Checkpoint 1 Now work Exercise 3.

The Interactive Activity on page 624 illustrates that the average rate of change between two points has only one numerical value. In other words, the **slope of a line is unique**.

Equations of Lines

There are several ways to write equations of lines. The form in the definition of a linear function, $f(x) = mx + b$, is called the **slope–intercept form**. Since linear equations are critical in our study of graphing and calculus, let's explore other ways to write a linear function.

We said at the beginning of this section that if the average rate of change of a line is known, it is a short process to write its linear function. To show how easy it is, let's say that we know that the slope of a line is $m = 3$ and a point on the line is (4, 2). Now we call any other point on the line (x, y). See Figure 11.3.4.

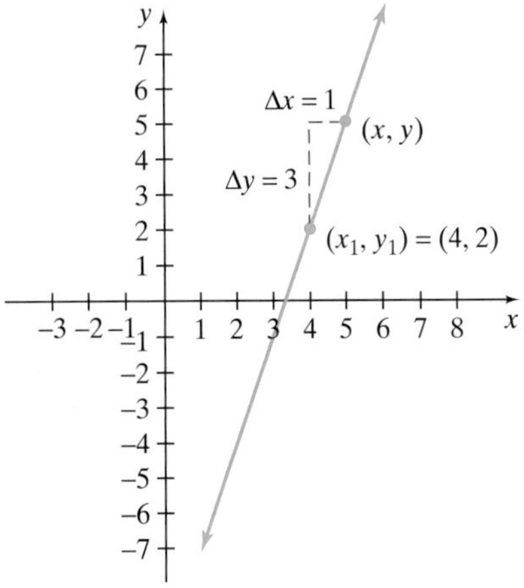

Figure 11.3.4

Using the definition of average rate of change, $m = \dfrac{y_2 - y_1}{x_2 - x_1}$, gives us

$$3 = \frac{y - 2}{x - 4}$$

where $(x_1, y_1) = (4, 2)$ and $(x_2, y_2) = (x, y)$. Multiplying each side of the equation by $x - 4$ yields

$$y - 2 = 3(x - 4)$$

We now have an equation for the line with slope $m = 3$ through the point (4, 2). Since this equation uses a known slope and a known point, it is called the **point–slope form**.

Point–Slope Form

The equation of a line with known slope m and known point (x_1, y_1) is given by the **point–slope form**

$$y - y_1 = m(x - x_1)$$

Technology Option

Many times, we want to use our calculator to graph a line with a known point and a known slope. We can do this by solving the point-slope form equation for y.

$$y - y_1 = m(x - x_1)$$
$$y = m(x - x_1) + y_1$$

We call this form the **calculator friendly** or simply **CF form** of a line.

Calculator Friendly or CF Form

A linear function with a known slope m and a known point (x_1, y_1) is given by

$$y = m(x - x_1) + y_1$$

Example 3 Writing a Linear Function in Point–Slope Form

For the points (2, –3) and (4, 2):

(a) Write an equation in point–slope form using the slope found in Example 2.

(b) Write the function in the slope–intercept form $f(x) = mx + b$.

Solution

(a) In Example 2, we found the slope between these two points to be $m = 2.5$. Replacing (x_1, y_1) with (2, –3), we get

$$y - (-3) = 2.5(x - 2)$$

or simply

$$y + 3 = 2.5(x - 2)$$

(b) To get the slope–intercept form $f(x) = mx + b$, we just need to simplify the function that is given in the point–slope form.

$$y + 3 = 2.5(x - 2)$$
$$y + 3 = 2.5x - 5$$
$$y = 2.5x - 8$$

Interactive Activity

Rework Example 3, except use the point (4, 2) when using the point–slope form.

WWW

Since y is just another name for $f(x)$, we see that the slope–intercept form is $f(x) = 2.5x - 8$. ■

Technology Option

If we solve the equation in Example 3 part (a) for y, we get the CF form of the line to be $y = 2.5(x - 2) - 3$. Figure 11.3.5 has the graph of the line.

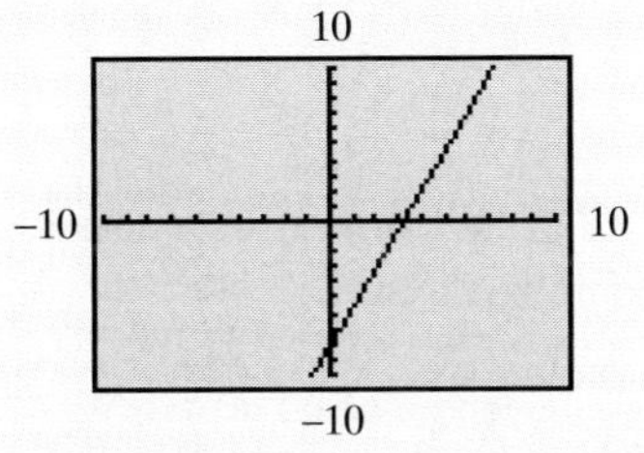

Figure 11.3.5

Throughout the textbook, the viewing window for calculator-generated graphs will appear on the perimeter of the graph. In text, we denote the viewing window by writing $[X\text{min}, X\text{max}]$ by $[Y\text{min}, Y\text{max}]$.

Intercepts

Two important points on the graph of a linear function are its **intercepts**. These are the points where the graph crosses the x (independent) and y (dependent) axes, as shown in Figure 11.3.6.

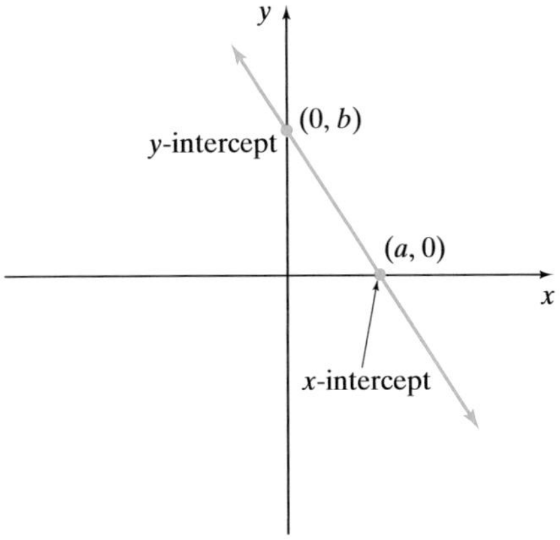

Figure 11.3.6

The figure also suggests how to find the intercepts. To get the y-intercept, we can replace x with zero and solve the resulting equation for y. To get the x-intercept, we replace y with zero and solve the resulting equation for x.

Example 4 Finding the Intercepts of a Function

Determine the x- and y-intercepts of the graph of the linear function

$$y = \frac{2}{3}(x+1)+3.$$

Solution

To find the y-intercept, we replace x with zero and solve for y. This gives

$$y = \frac{2}{3}(0+1)+3 = \frac{2}{3}+3 = \frac{11}{3}$$

So the y-intercept is $\left(0, \frac{11}{3}\right)$.

To get the x-intercept, we replace y with zero and solve for x to get

$$0 = \frac{2}{3}(x+1)+3$$

$$-3 = \frac{2}{3}(x+1)$$

$$\frac{-9}{2} = x+1$$

$$\frac{-11}{2} = x$$

Thus, the x-intercept is $\left(\frac{-11}{2}, 0\right)$. These points are highlighted on the graph in Figure 11.3.7.

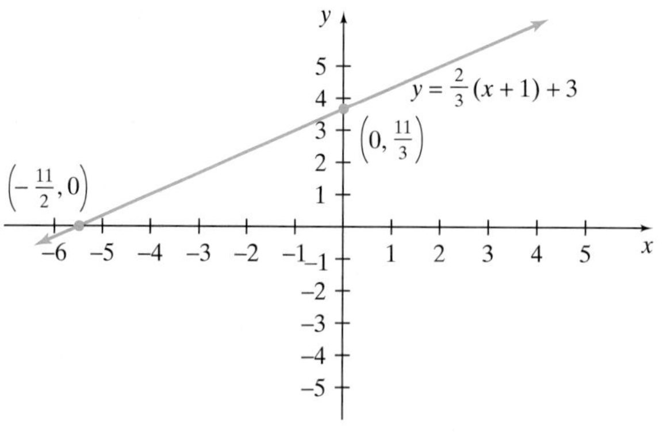

Figure 11.3.7

Notice that the y-intercept can be written in the form $(0, b)$. If we evaluate the linear function $f(x) = mx + b$ when $x = 0$, we get $f(0) = m(0) + b = b$. This means that the y-intercept of the graph of a linear function is the same as the constant b. This is why the form $f(x) = mx + b$ is called the **slope–intercept form** of a line. Now let's look at how intercepts can be used in an application.

Example 5 Writing a Depreciation Model and Interpreting Intercepts

The value of many products is said to **depreciate**, meaning that they decrease in value, as time goes on. The amount that the value of the product loses each year is called the **depreciation rate**. Many businesses incorporate depreciation models into their fiscal plans and for scheduling upgrades of equipment. For example, the Clark Clipboard Company recently purchased a metal press for \$15,000. The press is estimated to depreciate at a rate of $2500 \frac{\text{dollars}}{\text{year}}$.

(a) Write a linear function in the form $f(x) = mx + b$ for the value of the press after x years.

(b) Determine the intercepts and interpret each.

Solution

(a) Understand the Situation: We know that when the press is new, meaning that its age is $x = 0$, its value is \$15,000. Thus, $f(0) = 15{,}000$, which means that the point $(0, 15{,}000)$ is on the graph. Notice that this point is the y-intercept. Now, since the rate at which the value of the press **decreases** is $2500 \frac{\text{dollars}}{\text{year}}$, the average rate of change in the press's value is $-2500 \frac{\text{dollars}}{\text{year}}$. Since the average rate of change is the same as the slope of a line, we have $m = -2500$.

So the value of the press after x years is $f(x) = -2500x + 15{,}000$.

(b) From part (a) we know that the y-intercept is $(0, 15{,}000)$. To find the x-intercept, we set $f(x)$ equal to zero and solve the equation for x.

$$0 = -2500x + 15{,}000$$
$$-15{,}000 = -2500x$$
$$\frac{-15{,}000}{-2500} = x$$
$$6 = x$$

Interpret the Solution: The y-intercept, at the point $(0, 15{,}000)$, means that when the press has just been purchased its value is \$15,000. The x-intercept of $(6, 0)$ means that after 6 years of use the press has no value. Notice that the reasonable domain of $f(x)$ is $[0, 6]$. A graph of this depreciation function is shown in Figure 11.3.8.

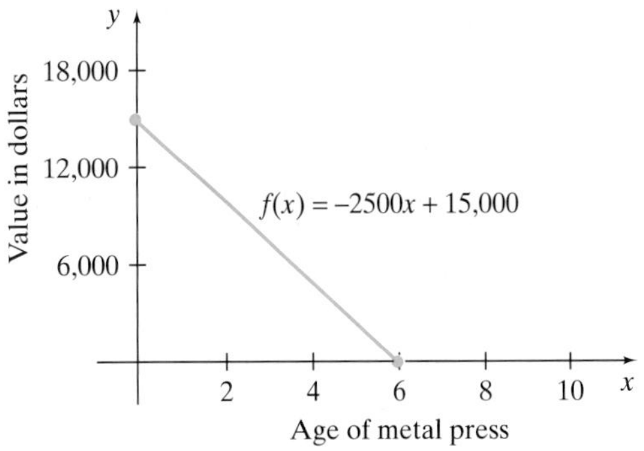

Figure 11.3.8

■

Checkpoint 2 Now work Exercise 41.

We found in Example 5 that, as time went on, the value of the metal press decreased. On the other hand, as the x-values increased in the function in Example 4, so did the function values. These examples serve as motivation for the following important behaviors of functions.

Increasing and Decreasing Functions

1. A function is said to be **increasing** if the function values $y = f(x)$ increase as the x-values increase.
2. A function is said to be **decreasing** if the function values $y = f(x)$ decrease as the x-values increase.

Note: In simple terms, this means that a line is increasing if it runs uphill when viewed from left to right, and is decreasing if it runs downhill when viewed from left to right.

Example 6 Identifying Increasing and Decreasing Functions

Graph each line and classify the function as increasing, decreasing, or neither. State the slope of the line.

(a) $h(x) = 2x - 1$

(b) $g(x) = -3x + 5$

(c) $f(x) = 4$

Solution

(a)

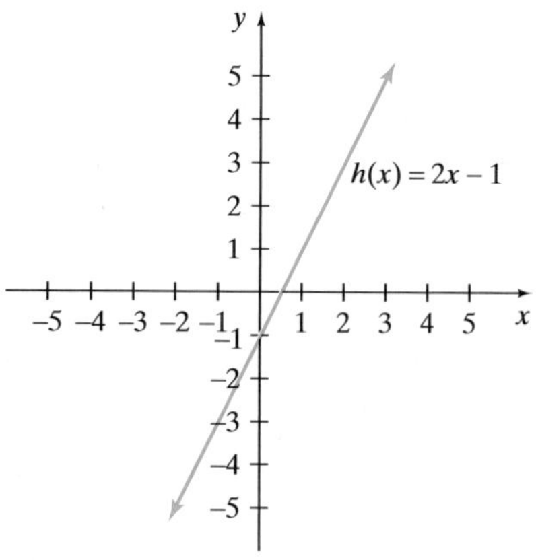

Figure 11.3.9

The graph "runs uphill" when viewed from left to right (Figure 11.3.9), so the function values increase as the x-values increase. This means that the function is increasing. The function has the slope–intercept form $h(x) = mx + b$, so we can directly read that the slope of the line is $m = 2$.

(b)

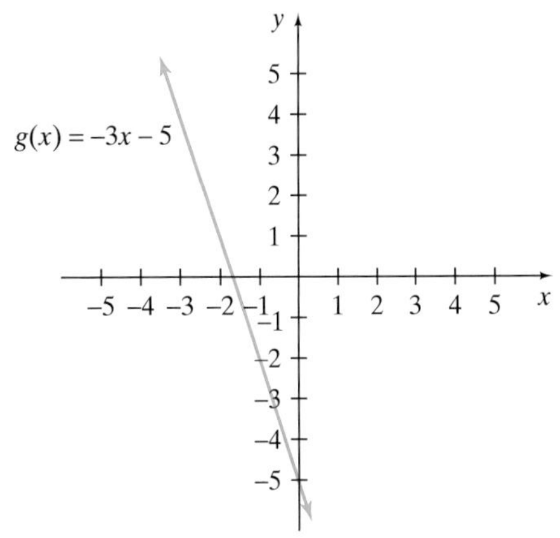

Figure 11.3.10

The graph "runs downhill" when viewed from left to right (Figure 11.3.10), so the function values decrease as the x-values increase. This means that the function is decreasing. The function has the slope–intercept form $g(x) = mx + b$, so the slope of the line is $m = -3$.

(c)

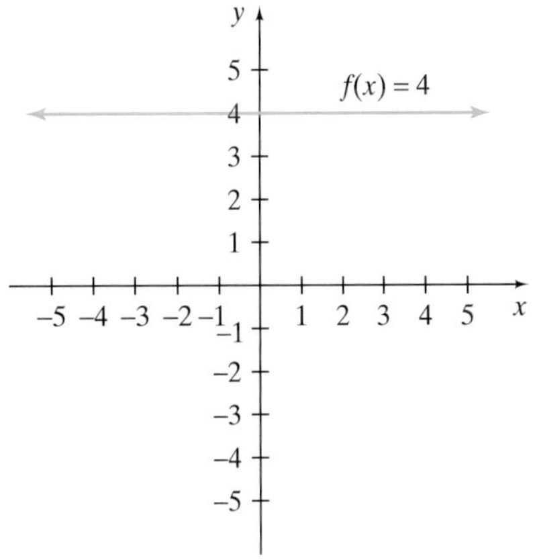

Figure 11.3.11

For this function, the function values stay the same as the x-values increase (Figure 11.3.11). We call this function a **constant function**, since the function is made up of only a constant. Notice that if we write the function in the form $f(x) = mx + b$ it would look like $f(x) = 0 \cdot x + b$. This means that the slope of the line is $m = 0$. ■

A natural question to ask is "What is the slope of a vertical line?" A line of this type has the form $x = a$, where a is a constant. Since the change in x is zero, the formula for slope, $m = \dfrac{y_2 - y_1}{x_2 - x_1}$, would have zero in the denominator. Since division by zero is not defined, we can say that **vertical lines have undefined slope**. We summarize lines and their graphs in Table 11.3.2 on page 632.

Marginal Cost

Often in economics it is desirable to know the cost of producing the $(x + 1)^{\text{st}}$ item of a product. This cost is commonly called the **marginal cost**.

TABLE 11.3.2

Type of Line	Value of Slope m	Diagram
Increasing	Positive	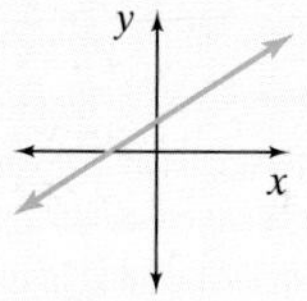
Decreasing	Negative	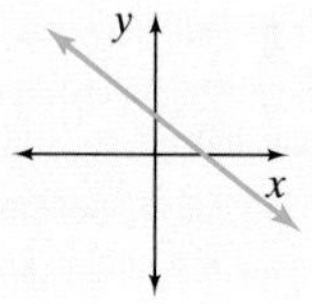
Horizontal	Zero	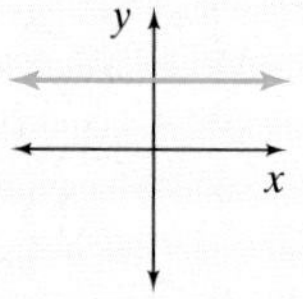
Vertical	Undefined	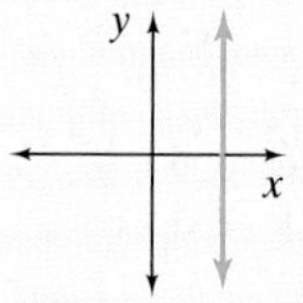

At the beginning of the section, we found that the cost function of copying promotional flyers for The Fashion Mystique was given by $C(x) = 0.10x + 5$, where x represents the number of copies produced and $C(x)$ represents the cost of producing x flyers. The cost of producing 110 flyers is given by $C(110)$. This cost is

$$C(110) = 0.10(110) + 5 = 11 + 5 = 16$$

So the cost of producing 110 flyers is \$16. The cost of producing 111 flyers is found by evaluating $C(111)$. This is

$$C(111) = 0.10(111) + 5 = 11.1 + 5 = 16.1$$

The value of $C(111) - C(110)$ is then $16.1 - 16 = 0.1$, which is the difference between the cost of producing 110 and 111 flyers. In other words, we have found that the cost of producing the 111th flyer is \$0.10, or 10 cents.

Recall, on page 624, that we discovered the cost is increasing at an average rate of \$0.10 per copy. Thus, the marginal cost for the 111th copy is the same as the average rate of change for the cost of the copies. The fact that the marginal cost and the slope (given by the variable costs) are the same is no coincidence.

For linear cost functions, variable costs and marginal costs are equal.

We will study marginal functions in more depth in Chapter 13.

Example 7 Determining Marginal Costs

The ProAudio Company manufactures DVD disks. It determines that the weekly fixed costs are \$14,000 and the variable costs are \$2.60 per disk.

(a) Determine the linear cost function C and interpret $C(1500)$.

(b) Identify the marginal cost. At a 1500 per week production level, what is the cost of manufacturing the 1501st disk?

Solution

(a) Understand the Situation: The manufacturing process has fixed costs of \$14,000 and variable costs of \$2.60, so the linear cost function is $C(x) = 2.60x + 14{,}000$.

Evaluating at $x = 1500$ gives

$$C(1500) = 2.60(1500) + 14{,}000 = 3900 + 14{,}000 = 17{,}900$$

Interpret the Solution: Thus, the weekly cost of producing 1500 DVD disks is \$17,900.

(b) Since the marginal cost and the variable cost are the same for a linear function, the marginal cost for the manufacturing process is \$2.60. Thus, at a 1500 disk per week production level, the cost of manufacturing the 1501st disk is \$2.60. ■

Piecewise-defined Functions

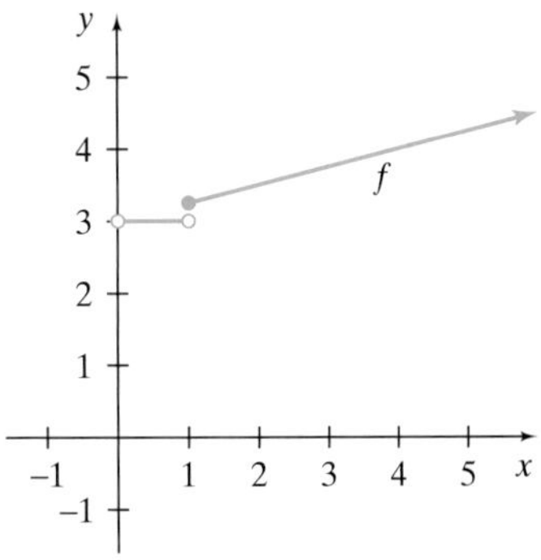

Figure 11.3.12 Graph of $f(x) = \begin{cases} 3, & 0 < x < 1 \\ 0.25x + 3, & x \geq 1 \end{cases}$

Often one simple function alone cannot represent everyday applications. For example, a taxi may charge a flat fee of \$3 up to the first mile driven and an additional \$0.25 for 1 mile and for each mile afterward. To express a function for the cost of the taxi, we must write the function in two parts. Up to the first mile, the x-values go from 0 to 1, and the cost is given by the constant function, which we will call $y_1 = 3$. For x-values greater than or equal to 1, we have to add 25 cents for each mile ridden to the \$3. So for the x-values greater than or equal to 1, the cost is given by a second piece that we will call $y_2 = 0.25x + 3$. So we can join the two pieces y_1 and y_2 to get the cost of riding x miles in the taxi as

$$f(x) = \begin{cases} 3, & 0 < x < 1 \\ 0.25x + 3, & x \geq 1 \end{cases}$$

A graph of f is given in Figure 11.3.12. This kind of function, which is defined for specific intervals of the domain, is called a **piecewise-defined function**. Let's try some examples of graphing this new type of function.

Example 8 Graphing a Piecewise-defined Function

For the piecewise-defined function $f(x) = \begin{cases} x + 2, & x < 1 \\ 3 - x, & x \geq 1 \end{cases}$

(a) Evaluate $f(0)$, $f(1)$, and $f(3)$.

(b) Make an accurate graph of the function.

Solution

(a) Since the x-value 0 is in the interval $(-\infty, 1)$, we evaluate using the "top" piece to get $f(0) = 0 + 2 = 2$. To get $f(1)$, we see that $x = 1$ is the left end point of our interval $[1, \infty)$, so we use the "bottom" piece to evaluate and get $f(1) = 3 - 1 = 2$. Finally, $f(3)$ is in the interval $[1, \infty)$, so we use the bottom piece again to get $f(3) = 3 - 3 = 0$.

(b) On the interval of x-values $(-\infty, 1)$, we plot the linear equation $y = x + 2$ shown in Figure 11.3.13a. Then, on the interval $[1, \infty)$, we plot the equation $y = 3 - x$ as in Figure 11.3.13b. Finally, the graph of f is found by plotting the two graphs shown in Figures 11.3.13a and b on the same Cartesian plane, as shown in Figure 11.3.13c.

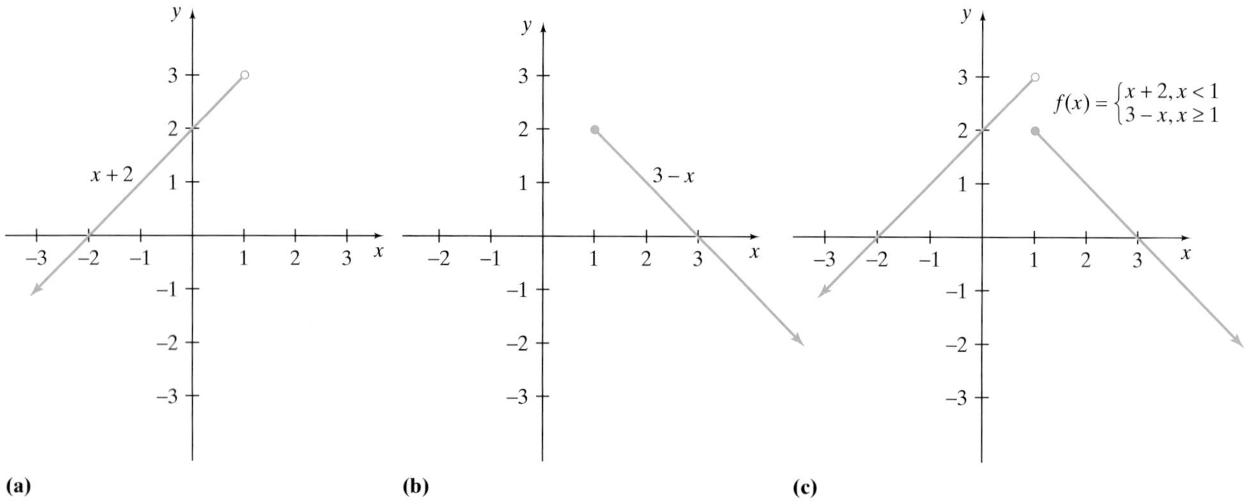

Figure 11.3.13

We use a solid dot to show which end point is included when $x = 1$ and a hollow dot to indicate that the endpoint is not included when $x = 1$. ■

Absolute Value Function

A piecewise-defined function that is often used is the **absolute value function**. Recall from algebra that the absolute value of any real number x, denoted by $|x|$, is defined by

$$|x| = \begin{cases} -x, & x < 0 \\ x, & x \geq 0 \end{cases}$$

In words, when taking the absolute value of any number, just write that number if the value is positive or zero, but if the number is negative, take its opposite. For example, $|8| = 8$, since $8 > 0$; and $|-5| = -(-5) = 5$, since $-5 < 0$. Notice how the definition of absolute value looks like a piecewise-defined function. It seems natural to define the absolute value function as a piecewise-defined function.

Absolute Value Function

The **absolute value function** $f(x) = |x|$ is defined by

$$f(x) = |x| = \begin{cases} -x, & x < 0 \\ x, & x \geq 0 \end{cases}$$

A graph of $f(x) = |x|$ is shown in Figure 11.3.14. Notice that the domain of the function is $(-\infty, \infty)$, the set of real numbers. Since an absolute value cannot be negative, the range of the absolute value function is $[0, \infty)$.

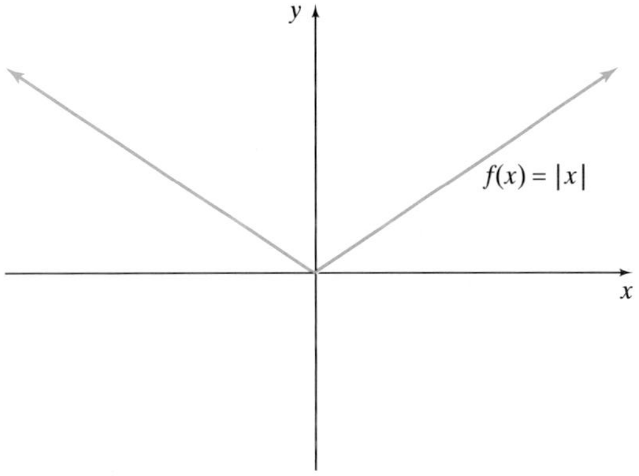

Figure 11.3.14

There are two key points to recognize about this type of function. First, the graph of the function $f(x) = |x|$ has a sharp "corner" at the origin. Second, the variable x inside the absolute value bars can be replaced by **any** algebraic expression.

Example 9 Rewriting and Graphing Piecewise-defined Functions

For the function $f(x) = |x - 5|$, graph the function and rewrite $f(x) = |x - 5|$ as a piecewise-defined function.

Solution

A graph of the function $f(x) = |x - 5|$ is given in Figure 11.3.15 on page 636. We see from the graph that the corner has shifted 5 units to the right to the point (5, 0). Also from the graph, the piecewise-defined function will have one piece for x-values less than 5 and another for x-values greater than or equal to 5. For $x \geq 5$, we see that $|x - 5| = (x - 5) = x - 5$. For the x-values less than 5, we have to take the opposite of $x - 5$ to get $|x - 5| = -(x - 5) = -x + 5$. This application of the definition of absolute value allows us to write the function in piecewise-defined form as

$$f(x) = |x - 5| = \begin{cases} -x + 5, & x < 5 \\ x - 5, & x \geq 5 \end{cases}$$

Interactive Activity

Use your calculator to verify the solution of Example 9 by comparing the graphs of the absolute value function to the piecewise-defined function.

WWW

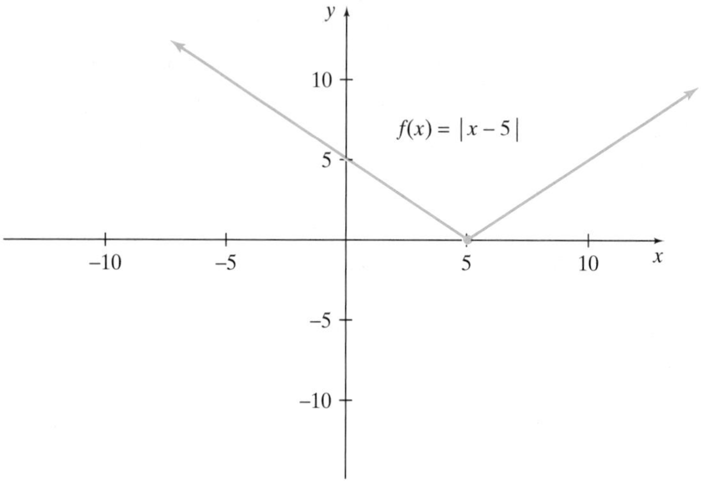

Figure 11.3.15

Technology Option

Most graphing calculators use abs(to denote the absolute value operator. For the function given in Example 9, Figure 11.3.16a shows what is entered in the $y =$ editor to produce the graph shown in Figure 11.3.16b. To learn how to enter absolute value on your calculator, consult the online graphing calculator manual at www.prenhall.com/armstrong.

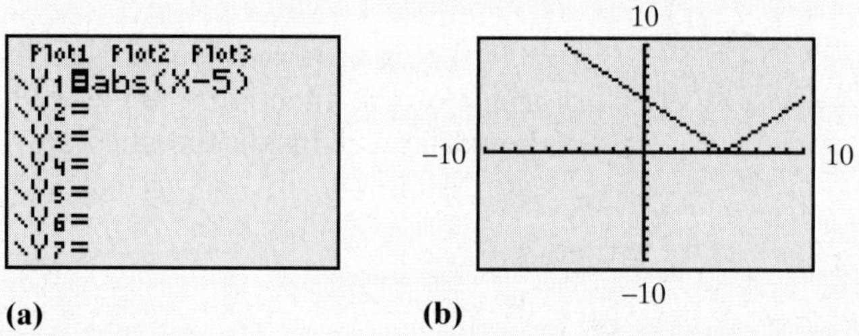

(a) (b)

Figure 11.3.16

SUMMARY

In this section, we learned that the slope of a line is also its **average rate of change** and is computed by

$$m = \frac{y_2 - y_1}{x_2 - x_1}$$

We also found that the value of the slope indicated whether the linear function was increasing or decreasing:

- If $m > 0$, the function is increasing.
- If $m < 0$, the function is decreasing.
- If $m = 0$, the function is the constant function.
- If m is undefined, the line is vertical.

The slope, along with a point on the line, can be used to write **linear equations** in two main forms.

- **Slope–intercept** form: $f(x) = mx + b$
- **Point–slope** form: $y - y_1 = m(x - x_1)$

The linear cost function has the form $f(x) = mx + b$, where m is the variable costs and b represents the fixed costs. This same model, $f(x) = mx + b$, can be used as a depreciation model, where b represents the cost of a new item and m is the depreciation rate.

Functions that have different rules based on the x-values are called **piecewise-defined functions**, and a special type of piecewise-defined function is the **absolute value function**.

SECTION 11.3 EXERCISES

For Exercises 1–8, find the average rate of change between the two points.

1. (4, 8) and (5, 3) **2.** (4, 8) and (3, 5)

3. (2, 3) and (−4, 8) **4.** (2, 2) and (−4, 4)

5. (5, 6.1) and (7, 8.3) **6.** (0, 1.25) and (4, 5)

7. $(a-1, b-1)$ and (a, b) where a and b are constants

8. $(1-a, 1-b)$ and $(a-1, b)$ where a and b are constants

For Exercises 9–14, determine the average rate of change for the given graphs.

9.

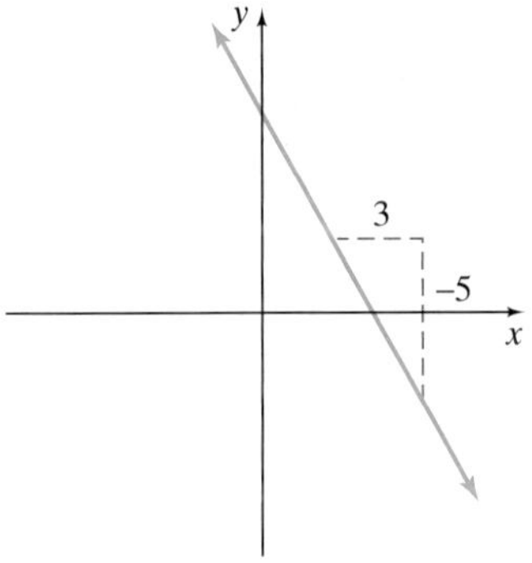

10.

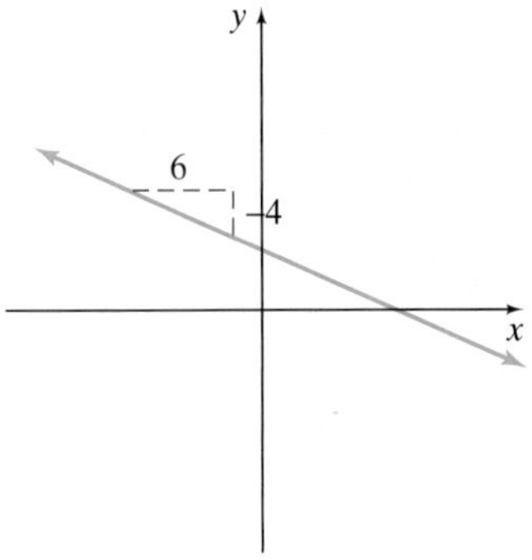

11.

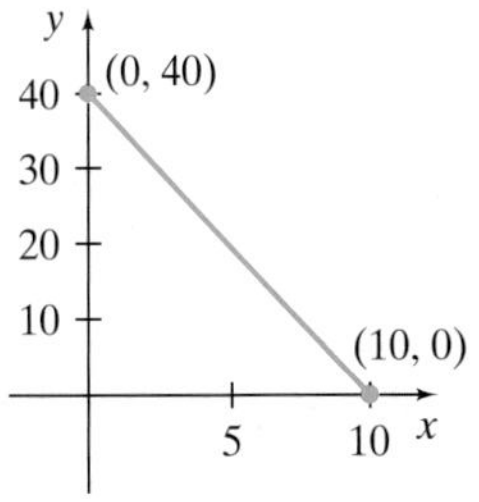

12.

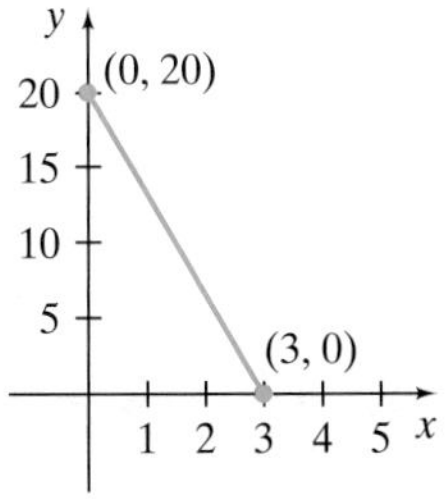

13.

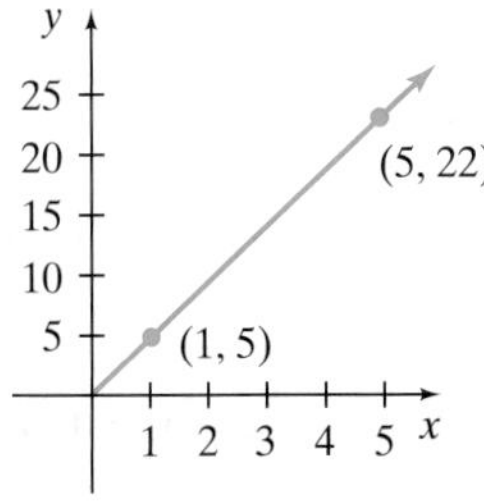

14.

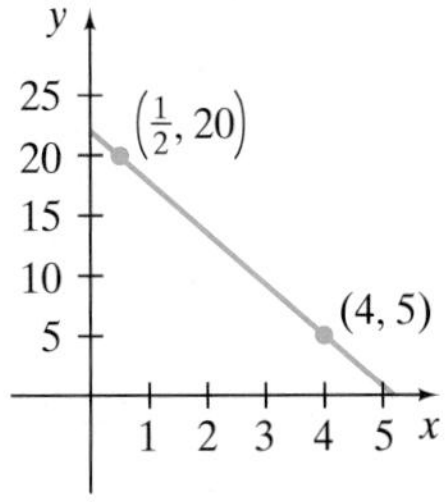

For Exercises 15–22, write an equation of the line in point–slope form if possible and graph.

15. The line contains the points (2, 6) and (5, −3).

16. The line contains the points (3, 2) and (−1, −2).

17. The line is given by the values in Table 11.3.3.

Table 11.3.3

x	y
3	5
5	13

18. The line is given by the values in Table 11.3.4.

Table 11.3.4

x	y
2	4
0	4

19. The line has a y-intercept at −3 and passes through the origin.

20. The line has an x-intercept at 2 and a y-intercept at 4.

21. The line is vertical and passes through the point (−2, 9).

22. The line is horizontal and passes through the point (−3, 1).

For Exercises 23–33, write an equation of the line in CF form.

23. A line has a slope of 3 and passes through the point (8, 4).

24. A line has a slope $-\frac{5}{8}$ and $f(0) = 3$.

25. A line has a slope of 5 and a y-intercept at −13.

26. A line has an average rate of change of $-\frac{2}{3}$ and an x-intercept at 3.

27. A line has an average rate of change of 0.3 and an x-intercept at −4.

28. The graph that contains the data points (1, 3) and (−10, 1).

29. The graph that contains the data points in Table 11.3.5.

Table 11.3.5

x	$f(x)$
−1	2
3	1

30. The graph that contains the data points in Table 11.3.6.

Table 11.3.6

x	$f(x)$
0	2
−3	4

31. A line has an x-intercept at $(-12, 0)$ and a y-intercept of $(0, \frac{3}{2})$.

32. A linear function f has function values of $f(5) = 1$ and $f(-7) = -1$.

33. A linear function g has function values of $g(0) = 6$ and $g(4) = 1.3$.

For Exercises 34–40, determine the x- and y-intercepts for each graph.

34.

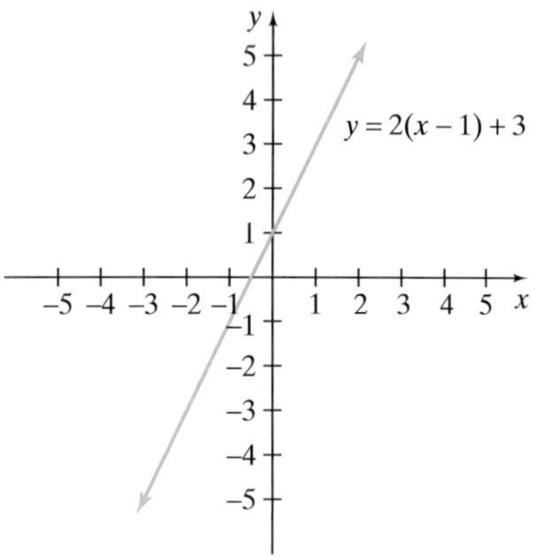

35.

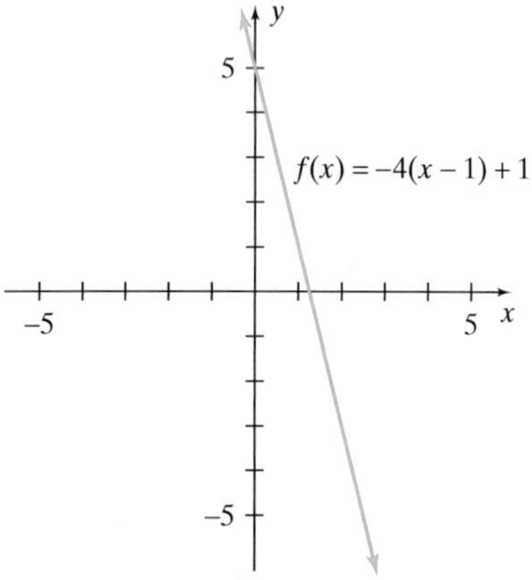

36. The graph of the function $f(x) = \dfrac{x}{2} - 4$.

37. The graph of the function $g(x) = 120 - 0.3x$.

38. The graph of the function $f(x) = \dfrac{750 - 35x}{10}$.

39. The graph of the function $g(x) = 0.2(x - 5.1) + 10$.

40. The graph of the linear equation $y - 3 = -2(x + 4)$.

Answer Exercises 41–44 by using the depreciation model.

41. A new photocopy machine initially costs \$25,000 and depreciates at a rate of \$1250 per year.

(a) Determine an equation for the depreciation function.

(b) In how many years will the machine be worth \$10,000?

42. A new multimedia computer costs \$800 and depreciates at a rate of $150 \dfrac{\text{dollars}}{\text{year}}$.

(a) Determine an equation for the depreciation function.

(b) When will the computer depreciate to the point that it has no worth?

43. A factory invests \$90,000 in a new machine press that depreciates at a rate of \$6000 per year.

(a) Determine an equation for the depreciation function.

(b) How long will it take for the press to have half its original value?

44. A XQ-383 sports car costs \$40,000 and depreciates \$3000 per year.

(a) Determine an equation for the depreciation function.

(b) How much will the car be worth in five years?

For Exercises 45–53, classify each of the linear functions as increasing or decreasing, and identify the slope of the line.

45. $f(x) = -4.5(x - 2)$

46. $g(x) = \dfrac{100 - 5x}{2}$

47. $f(x) = \dfrac{500 + 60x}{15}$

48. $g(x) = 4(x - 6) + 8$

49. $f(x) = 5 + 4x$

50. The graph of the function containing the points in Table 11.3.7.

Table 11.3.7

x	$f(x)$
1	3
7	5

51. The linear function containing the points in Table 11.3.8.

Table 11.3.8

x	y
10	5
1	2

52.

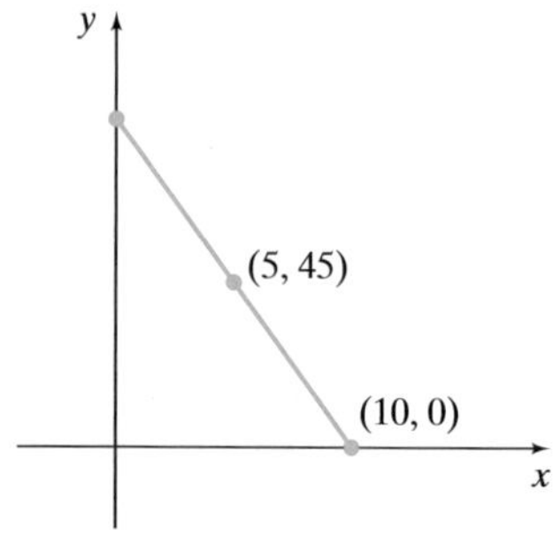

53.

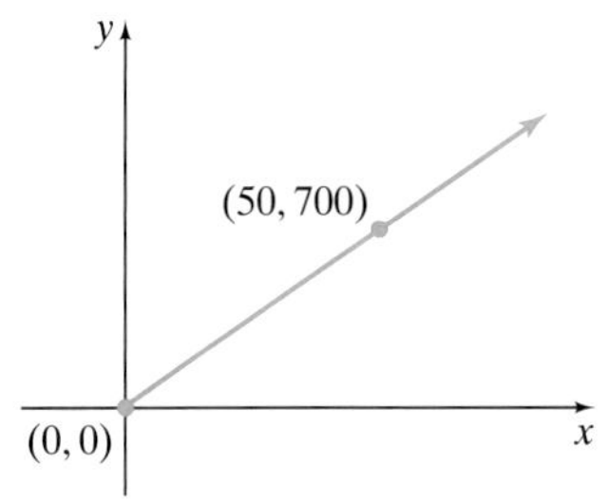

In Exercises 54–59, write the linear cost function for the given information in the slope–intercept form $C(x) = mx + b$.

54. The variable costs are $70 \dfrac{\text{dollars}}{\text{unit}}$, and the fixed costs are \$10,000.

55. The variable costs are \$147.75 per unit, and the fixed costs are \$2700.

56.

Number of Units	Cost
5	\$1250
23	\$5580

57.

Quantity	Cost
10	\$7000
50	\$30,000

58. The fixed costs are \$14,500, and it costs \$100,100 to produce 75 units of product.

59. The variable costs are $80 \dfrac{\text{dollars}}{\text{unit}}$, and the cost of producing 40 units is \$3850.

11.2 **60.** The Dirt Rocket Company produces spokes for off road motorcycles. It determines that the daily cost in dollars of producing x quick-replace spokes is $C(x) = 1250 + 3x$.

(a) What are the fixed and variable costs for producing the spokes?

(b) Evaluate $C(400)$ and interpret.

(c) How many spokes can be produced for \$1820?

(d) What is the marginal cost for the spokes?

(e) If the current daily production level is 100 spokes, what is the cost in dollars of producing the 101st spoke?

61. The Fontana Vitamin Company determines that the cost in dollars of producing x bottles of a new supplement is given by the linear cost function $C(x) = 1.25x + 550$.

(a) What are the fixed and variable costs for producing the supplement?

(b) Evaluate $C(70)$ and interpret.

(c) How many bottles of the supplement can be produced for \$693.75?

(d) What is the marginal cost for the supplement?

(e) If the current production level is 70 bottles, what is the cost of producing the 71st bottle of the supplement?

11.2 **62.** The Get-It-Now florist determines that the cost in dollars of arranging and delivering x get-well bouquets is given by the cost function $C(x) = 10 + 32x$.

(a) What are the fixed and variable costs for producing the bouquets?

(b) Evaluate $C(15)$ and interpret.

(c) How many bouquets can be produced and delivered for \$8010?

(d) What is the marginal cost for the bouquets?

(e) If the current production level is 40 bouquets, what is the cost of producing the 41st bouquet?

For Exercises 63–68, make an accurate graph of the piecewise-defined functions.

63. $f(x) = \begin{cases} 4 - 0.2x, & x < 5 \\ 5, & x \geq 5 \end{cases}$

64. $g(x) = \begin{cases} 2x, & x \leq 0 \\ -3x, & x > 0 \end{cases}$

65. $y = \begin{cases} -3, & x < 4 \\ 3, & x > 4 \end{cases}$

66. $f(x) = \begin{cases} \dfrac{x+6}{2}, & x \leq 1 \\ \dfrac{7}{2}, & x > 1 \end{cases}$

67. $g(x) = \begin{cases} -4, & x < 5 \\ x - 2, & x \geq 5 \end{cases}$

68. $f(x) = \begin{cases} -2, & x < 0 \\ 2x + 4, & 0 \leq x \leq 5 \\ -x, & x > 5 \end{cases}$

For Exercises 69–74, graph the absolute value function then rewrite the function in piecewise form.

69. $f(x) = |x + 1|$

70. $y = |x - 3|$

71. $g(x) = |6 - 2x|$

72. $f(x) = |10 - x|$

73. $g(x) = |3x - 15|$

74. $f(x) = |4x + 12|$

75. The Lesky Truck Rental Company charges a flat fee of \$20 plus 15 cents per mile driven for renting a moving truck.

(a) What are the fixed costs? What are the variable costs?

(b) Write a linear function C, where x represents the number of miles driven and $C(x)$ represents the rental costs.

(c) What would be a realistic domain for this function?

(d) Evaluate $C(120)$ and interpret.

(e) What is the cost of the 121st mile driven? What is this cost called?

76. Consider the daily cost schedule (Table 11.3.9) for the cost of producing x Never Die light bulbs.

Table 11.3.9

Bulbs Produced	Cost (in dollars)
0	1050
100	1250
215	1480

(a) What are the fixed costs? Explain how you determined this from Table 11.3.9.

(b) How much does the cost increase for each additional bulb produced? How do you classify this kind of cost?

(c) Write a linear cost function C, where x represents the number of bulbs produced and $C(x)$ represents the cost, in dollars.

(d) What would the cost be if 260 bulbs were produced?

77. Consider the price schedule for the repair cost of a Ripen microwave oven (Table 11.3.10).

Table 11.3.10

Hours Labor	Repair Cost (in dollars)
0	50
1	65
2	80

(a) What are the fixed costs? Explain how you determined this from Table 11.3.10.

(b) How much does the repair cost increase for each additional hour of labor? How do you classify this kind of cost?

(c) Write a linear repair cost function C, where x represents hours of labor and $C(x)$ represents the repair cost, in dollars.

(d) What would the repair cost be if the microwave oven took 4.5 hours to fix?

11.2 **78.** The annual total expenditures on pollution abatement in the United States can be modeled by

$$f(x) = 2x + 48.17 \qquad 0 \le x \le 20$$

where x represents the number of years since 1972 and $f(x)$ represents the expenditures in billions of constant 1987 dollars (*Source:* U.S. Statistical Abstract, www.census.gov/statab/www).

(a) Is the amount of money spent on pollution abatement generally increasing or decreasing? At what rate?

(b) Interpret the reasonable domain.

(c) Evaluate $f(12)$ and interpret.

(d) In what year did the annual expenditures approach 80 billion dollars?

11.2 **79.** The amount spent annually in college bookstores in the United States can be modeled by

$$f(x) = 0.19x + 1.67 \qquad 0 \le x \le 15$$

where x represents the number of years since 1982 and $f(x)$ represents the amount spent in billions of dollars (*Source:* U.S. Census Bureau, www.census.gov).

(a) How much does the amount of spending increase in college bookstores each year?

(b) According to the model, how much was spent in college bookstores in 1985?

(c) According to the model, how much was spent in college bookstores in 1995?

(d) How much more was spent in college bookstores in 1995 compared to 1985?

SECTION PROJECT

The number of disabled Korean War veterans receiving compensation is shown in Table 11.3.11.

Table 11.3.11

Year	Number of Disabled Korean War Veterans (in thousands)
1990	209
1999	175

SOURCE: U.S. Census Bureau, www.census.gov

(a) If x represents the number of years since 1990 and $f(x)$ represents the number of disabled Korean War veterans (in thousands), compute the average rate of change between the points (0, 209) and (9, 175). Write the average rate of change using proper units.

(b) Is the number of disabled Korean War veterans receiving compensation increasing or decreasing?

(c) Write a linear mathematical model for the number of disabled Korean War veterans receiving compensation since 1990.

(d) Using the model from part (c), evaluate $f(7)$ and interpret.

(e) Compute the x-intercept of the graph of the function. Is the x-intercept in the reasonable domain?

Section 11.4 Quadratic Functions and Average Rate of Change on an Interval

In this section, we turn our attention to quadratic functions. This family of functions has many applications in the physical laws of motion, economics, and the social sciences. We then extend our knowledge of the average rate of change by discussing the secant line slope of a function on an interval.

Properties of Quadratic Functions

After the linear function, the next most common type of function we study is the **quadratic function**. The quadratic function may be called a **second-degree function** since the largest exponent for the independent variable is 2.

Quadratic Function

A function of the form $f(x) = ax^2 + bx + c$ is called a **quadratic function**, where a, b, and c are real numbers and $a \neq 0$.

Since there are no x-values that can make a denominator zero or give us the square root of a negative number, the domain of a quadratic function is the set of real numbers $(-\infty, \infty)$. The shape of a quadratic function is called a **parabola**.

Example 1 Graphing and Evaluating a Quadratic Function

The percentage of people taking the Scholastic Aptitude Test (SAT) who intend to study business and commerce can be modeled by

$$f(x) = -0.09x^2 + 2.01x + 9.63 \qquad 1 \leq x \leq 22$$

where $f(x)$ is the percentage and x is the number of years since 1974. Evaluate $f(10)$ and interpret the result (*Source:* U.S. Census Bureau, www.census.gov).

Solution

Understand the Situation: Since x represents the number of years since 1974, $x = 10$ corresponds to the year 1984. To evaluate $f(10)$ requires us to substitute 10 for every occurrence of x in the model.

Evaluating the model at $x = 10$ gives

$$\begin{aligned} f(10) &= -0.09(10)^2 + 2.01(10) + 9.63 \\ &= -9 + 20.1 + 9.63 \\ &= 20.73 \end{aligned}$$

Interpret the Solution: Thus, in 1984 about 20.73% of all the people taking the SAT intended to study business and commerce. See Figure 11.4.1.

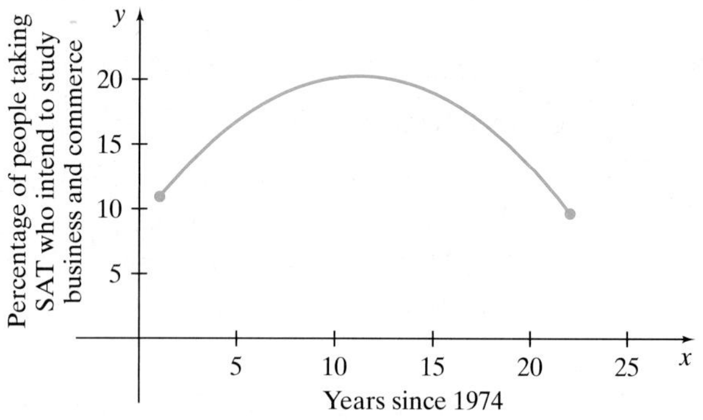

Figure 11.4.1 Graph of $f(x) = -0.09x^2 + 2.01x + 9.63$ on [1, 22].

Figure 11.4.1 shows that the graph appears to have a peak or maximum value when x has the value of about 11. This peak is called the **vertex** of the graph. To the left of the vertex, the graph appears to be increasing. To the right, the graph appears to be decreasing.

The vertex of a parabola is one of its most important features, since it gives us the maximum or minimum value of the function. Usually, finding a maximum or minimum value requires tools from calculus, but in the case of quadratic functions, we can use a simple formula to find the x-coordinate of the vertex. (You will derive this property in Chapter 15.)

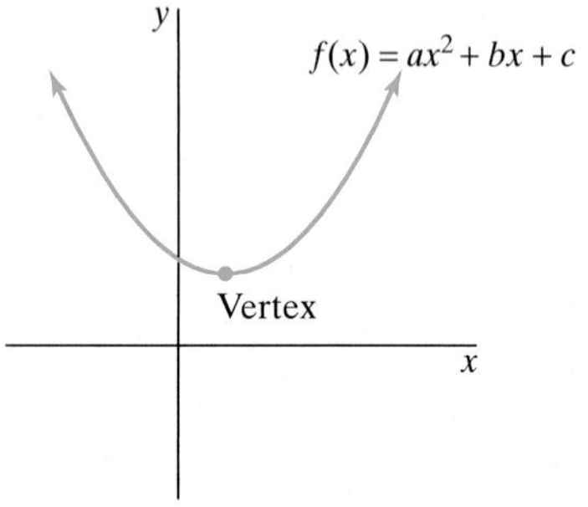

(a)

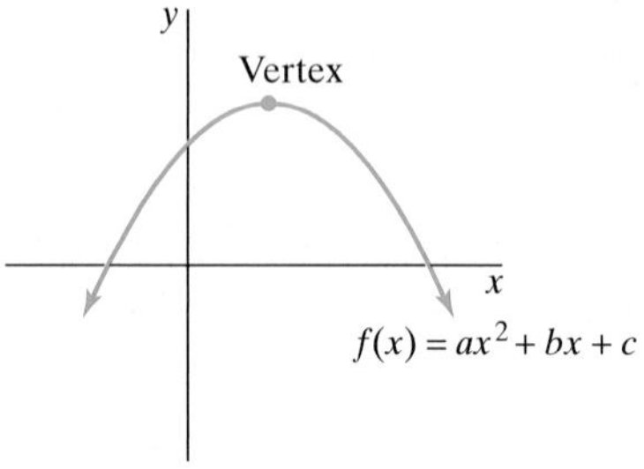

(b)

Figure 11.4.2 **(a)** If $a > 0$ then the parabola opens up and the vertex is a minimum. **(b)** If $a < 0$ then the parabola opens down and the vertex is a maximum.

Vertex of a Parabola

The x-coordinate of the **vertex** of a parabola given by $f(x) = ax^2 + bx + c$ is

$$x = \frac{-b}{2a}$$

The y-coordinate can be found by evaluating $f\left(\frac{-b}{2a}\right)$.

The way that the parabola $f(x) = ax^2 + bx + c$ opens is determined by the value of a, which is called the **leading coefficient**. If a is a positive number, the graph opens up, and if a is negative, the graph opens downward. If the graph of a parabola opens up, the vertex is a **minimum**, and if the graph opens down, the vertex is a **maximum**. See Figures 11.4.2a and b. Consequently, if we know the y-coordinate of the vertex and the direction in which the graph opens, we can readily determine the range of any quadratic function. For example, consider the function $f(x) = -x^2 + x + 6$. Since the leading coefficient of the quadratic function is $-1 < 0$, we know that the graph opens down. This means that the vertex is a maximum point. The x-coordinate of the vertex is given by $x = \frac{-b}{2a} = \frac{-1}{2(-1)} = \frac{1}{2}$. The y-coordinate of the vertex is

$$f\left(\frac{1}{2}\right) = -\left(\frac{1}{2}\right)^2 + \left(\frac{1}{2}\right) + 6$$
$$= -\frac{1}{4} + \frac{1}{2} + 6 = \frac{25}{4}$$

Since the maximum function value is $y = \frac{25}{4}$, we see that the range of the function is $\left(-\infty, \frac{25}{4}\right]$. This is shown in Figure 11.4.3.

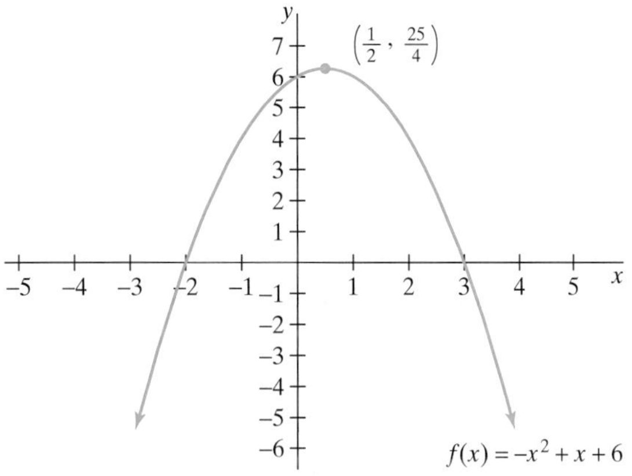

Figure 11.4.3

Determining Zeros of Quadratic Functions

The real **zeros** of a function f are given by the x-intercepts of its graph. They can also be called the **roots** of the function, since they are the solutions to the equation $f(x) = 0$.

Many times, the zeros, or roots, of a function can be found algebraically by the process of **factoring**. We just set $f(x)$ equal to zero and factor, as shown in Example 2.

Example 2 Determining the Zeros of a Quadratic Function by Factoring

Determine the zeros of the function $f(x) = 4x^2 + 6x - 4$. Check the result numerically.

Solution

To find the zeros for this function, we need to solve the quadratic equation $4x^2 + 6x - 4 = 0$. So, by factoring, we get

$4x^2 + 6x - 4 = 0$	Given equation
$2(2x^2 + 3x - 2) = 0$	Factor out the constant 2
$2x^2 + 3x - 2 = 0$	Divide both sides by the constant 2
$(x + 2)(2x - 1) = 0$	Factor the expression $2x^2 + 3x - 2$
$(x + 2) = 0$ or $(2x - 1) = 0$	Set each factor equal to zero
$x = -2$ or $x = \frac{1}{2}$	Solve for x

Interactive Activity

Do you ever wonder where the term **zeros** of a function comes from? To find out, graph the function in Example 2 in the viewing window [−5, 5] by [−10, 5]; then use the VALUE command on the calculator to evaluate $f(-2)$ and $f\left(\frac{1}{2}\right)$. What are the y-values?

We can check our solutions by evaluating $f(x) = 4x^2 + 6x - 4$ at both solutions to see if the function value is zero. Checking at $x = -2$, we get

$$f(-2) = 4(-2)^2 + 6(-2) - 4$$
$$= 16 - 12 - 4 = 0$$

At the second zero $x = \frac{1}{2}$, we get

$$f\left(\frac{1}{2}\right) = 4\left(\frac{1}{2}\right)^2 + 6\left(\frac{1}{2}\right) - 4$$
$$= 1 + 3 - 4 = 0$$

So the zeros of the function $f(x) = 4x^2 + 6x - 4$ are $x = -2$ and $x = \frac{1}{2}$. ■

Many times, the quadratic that we wish to solve is not easily factorable. To algebraically solve these types, we use the **quadratic formula**. The derivation of this formula appears in Appendix E.

Quadratic Formula

The zeros of the quadratic function $f(x) = ax^2 + bx + c$ are given by the **quadratic formula**

$$x = \frac{-b \pm \sqrt{b^2 - 4ac}}{2a}$$

$b^2 - 4ac$ is called the **discriminant**.

If $b^2 - 4ac > 0$, the graph looks like

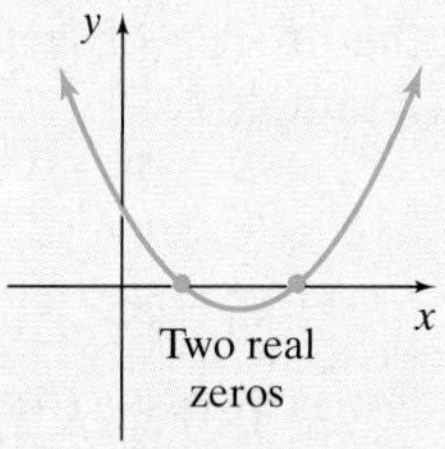

or

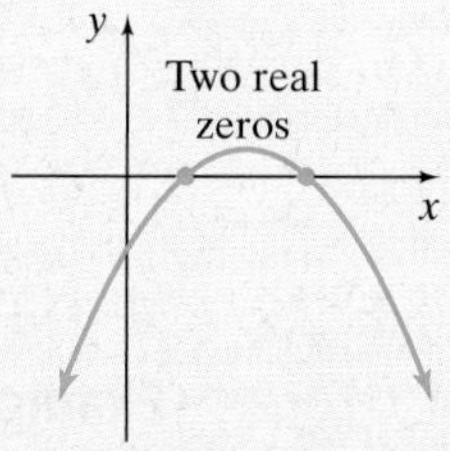

If $b^2 - 4ac = 0$, the graph looks like

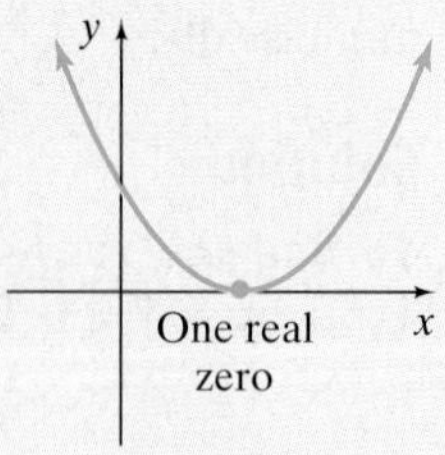

or

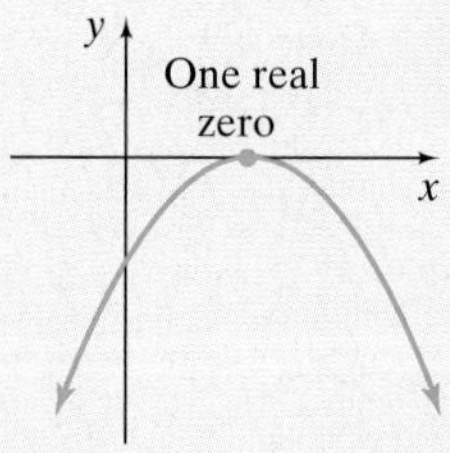

If $b^2 - 4ac < 0$, the graph looks like

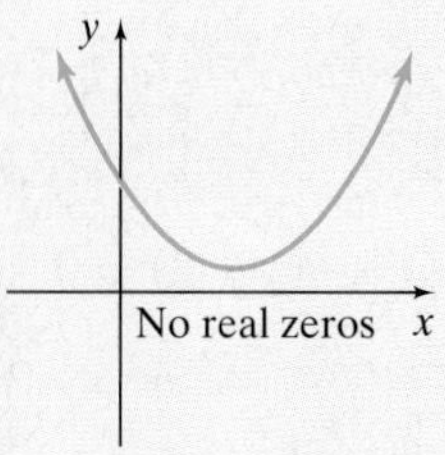

or

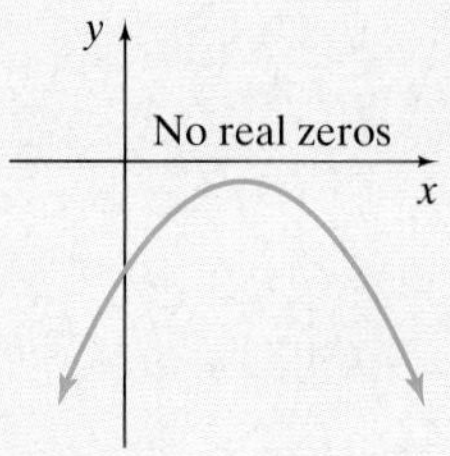

Example 3 Determining Zeros with the Quadratic Formula

Determine the zeros of the quadratic function $f(x) = 2x^2 - 6x + 3$.

Solution

To find the zeros of the function, we need to solve the quadratic equation $2x^2 - 6x + 3 = 0$. Since this equation is not easily factorable, we determine the roots by evaluating the quadratic formula using $a = 2$, $b = -6$, and $c = 3$. This gives us

$$\begin{aligned} x &= \frac{-(-6) \pm \sqrt{(-6)^2 - 4(2)(3)}}{2(2)} \\ &= \frac{6 \pm \sqrt{36 - 24}}{4} = \frac{6 \pm \sqrt{12}}{4} \\ &= \frac{6 \pm 2\sqrt{3}}{4} = \frac{2(3 \pm \sqrt{3})}{4} = \frac{3 \pm \sqrt{3}}{2} \end{aligned}$$

So the two real zeros of the function $f(x) = 2x^2 - 6x + 3$ are $x = \frac{3 - \sqrt{3}}{2}$ and $x = \frac{3 + \sqrt{3}}{2}$. ■

Checkpoint 1 Now work Exercise 19.

Technology Option

Sometimes even the quadratic formula can be unwieldy when locating zeros of quadratic functions. In some situations, we use a calculator to approximate the solutions to quadratic equations. To approximate the zeros of $f(x) = 2x^2 - 6x + 3$, we can use the ZERO or ROOT command on our calculator. Figure 11.4.4a shows the approximation of the zero of the leftmost x-intercept and the rightmost is shown in Figure 11.4.4b. So the approximations for the zeros of the function are $x \approx 0.63$ and $x \approx 2.37$, rounded to the nearest hundredth.

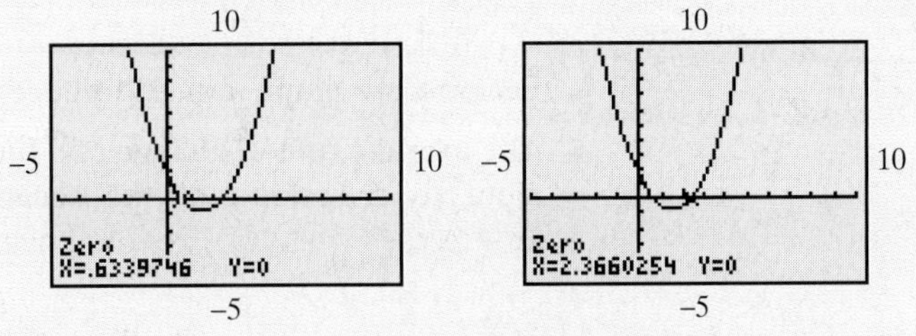

(a) (b)

Figure 11.4.4

To see how to use the ZERO or ROOT command on your calculator, consult the online graphing calculator manual at www.prenhall.com/armstrong.

Secant Line Slope and the Difference Quotient

From Your Toolbox

The **slope** of a line, denoted by m, is a measurement of the steepness of the line. Given two points on a line, (x_1, y_1) and (x_2, y_2), the slope is computed by

$$m = \frac{y_2 - y_1}{x_2 - x_1} = \frac{\text{change in } y}{\text{change in } x} = \frac{\Delta y}{\Delta x}$$

The slope of the line also gives the **average rate of change** of y with respect to x.

In Section 11.3, we introduced the average rate of change of a line. This definition is reviewed in the Toolbox to the left.

At times, we want to know what the average rate of change would be for a **nonlinear function** over a specified interval. We can find this rate by computing the slope of a **secant line**, which is a line that passes through two points on a curve. To determine the slope of the secant line, let's suppose that a curve defined by a nonlinear function f has two distinct points, $P(x_1, y_1)$ and $Q(x_2, y_2)$. The average rate of change, or slope, of a line passing through P and Q is

$$m = \frac{y_2 - y_1}{x_2 - x_1} = \frac{f(x_2) - f(x_1)}{x_2 - x_1}$$

We call the change from P to Q in the x-coordinate Δx (read "delta x"). This means that $\Delta x = x_2 - x_1$. If we let $x_1 = x$, then $x_2 = x + \Delta x$. See Figure 11.4.5.

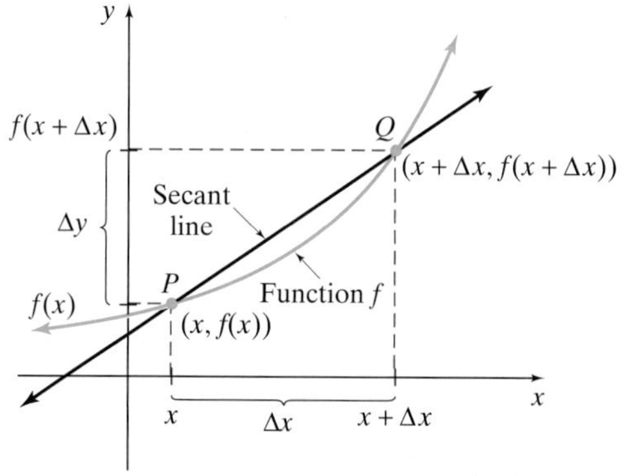

Figure 11.4.5

We can see from the graph that the coordinates of these two points are $P(x, f(x))$ and $Q(x + \Delta x, f(x + \Delta x))$. Now, determining the $\frac{\text{change in } y}{\text{change in } x}$, we get

$$\text{Secant line slope} = \frac{\text{change in } y}{\text{change in } x} = \frac{f(x + \Delta x) - f(x)}{(x + \Delta x) - x} = \frac{f(x + \Delta x) - f(x)}{\Delta x}$$

This fraction represents the **average rate of change** between the points P and Q. The fraction $\frac{f(x + \Delta x) - f(x)}{\Delta x}$ is also called the **difference quotient**.

Secant Line Slope and Difference Quotient

The **average rate of change** of a function f on an interval $[x, x + \Delta x]$ is equivalent to the **slope of the secant line** through the points $(x, f(x))$ and $(x + \Delta x, f(x + \Delta x))$, denoted by m_{sec}, and is given by the **difference quotient**

$$m_{\text{sec}} = \frac{f(x + \Delta x) - f(x)}{\Delta x}$$

where $\Delta x \neq 0$. The units of this average rate of change are $\frac{\text{units of } f}{\text{unit of } x}$.

Note: The denominator of the difference quotient, Δx, is also called the **increment in x**. The numerator $f(x + \Delta x) - f(x)$ is the same as Δy and is called the **increment in y**.

Example 4 Determining the Secant Line Slope

Find the average rate of change of $f(x) = 2x^2 + 1$ on the interval [2, 5].

Solution

The first x-value is $x = 2$, and the increment in x is $\Delta x = 5 - 2 = 3$. Using the difference quotient gives us

$$m_{\text{sec}} = \frac{f(x + \Delta x) - f(x)}{\Delta x} = \frac{f(2+3) - f(2)}{3} = \frac{f(5) - f(2)}{3}$$
$$= \frac{(2(5)^2 + 1) - (2(2)^2 + 1)}{3} = \frac{(51) - (9)}{3} = \frac{42}{3} = 14$$

Since the slope of the secant line on [2, 5] is 14, we conclude that the average rate of change of f on [2, 5] is 14 or $14 \dfrac{\text{units of } f}{\text{unit of } x}$. The graph of f and this secant line is shown in Figure 11.4.6.

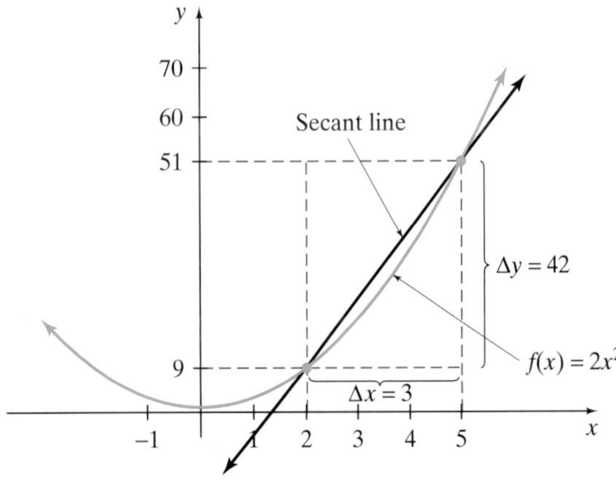

Figure 11.4.6 ■

In applications, the secant line slope can be an indicator of the general trend in the rate of change of some phenomena on some interval. This is illustrated in Example 5.

Example 5 Interpreting the Secant Line Slope

The U.S. imports from China for the years 1987 to 1996 can be modeled by

$$f(x) = 0.32x^2 + 1.64x + 3.98 \qquad 1 \le x \le 10$$

where x represents the number of years since 1986 and $f(x)$ represents the dollar value, in billions, of goods imported (*Source:* U.S. Census Bureau, www.census.gov).

(a) Make a table of function values for $x = 1, 2, 3, \ldots, 10$. Use these values when calculating parts (b) and (c).

(b) Determine the average rate of change in U.S. imports from China from 1987 to 1991 and interpret. Include appropriate units.

(c) Determine the average rate of change in U.S. imports from China from 1991 to 1996 and interpret. Include appropriate units.

(d) Compare the results from parts (b) and (c).

Solution

(a) A numerical table of values for the model is shown in Table 11.4.1.

Table 11.4.1

x	1	2	3	4	5	6	7	8	9	10
$f(x)$	5.94	8.54	11.78	15.66	20.18	25.34	31.14	37.58	44.66	52.38

(b) **Understand the Situation:** First we see that the year 1987 corresponds to $x = 1$ and the year 1991 corresponds to $x = 1991 - 1986 = 5$. This means that the increment in x is $\Delta x = 5 - 1 = 4$. So we need to compute the difference quotient

$$m_{\text{sec}} = \frac{f(x + \Delta x) - f(x)}{\Delta x} = \frac{f(1 + 4) - f(1)}{4} = \frac{f(5) - f(1)}{4}$$

The values from Table 11.4.1 give us

$$m_{\text{sec}} = \frac{20.18 - 5.94}{4} = \frac{14.24}{4} = 3.56 \quad \text{or} \quad 3.56 \frac{\text{billion dollars}}{\text{year}}$$

Interpret the Solution: This means that during the period from 1987 to 1991, U.S. imports from China increased at an average rate of $3.56 \frac{\text{billion dollars}}{\text{year}}$. Notice that the amount of imports increased, since the secant line slope is positive.

(c) **Understand the Situation:** For the years from 1991 to 1996, the corresponding x-values are $x = 5$ and $x = 10$, respectively. Thus, $\Delta x = 10 - 5 = 5$.

Now we compute the difference quotient as

$$m_{\text{sec}} = \frac{f(5 + 5) - f(5)}{5} = \frac{f(10) - f(5)}{5}$$

$$= \frac{52.38 - 20.18}{5} = \frac{32.2}{5}$$

$$= 6.44 \quad \text{or} \quad 6.44 \frac{\text{billion dollars}}{\text{year}}$$

Interpret the Solution: This means that during the period from 1991 to 1996, U.S. imports from China increased at an average rate of $6.44 \frac{\text{billion dollars}}{\text{year}}$.

(d) By comparing the rates in part (b) to those in part (c), we can see that total U.S. imports from China, on average, were increasing much faster between 1991 to 1996 than between 1987 to 1991. This comparison is shown graphically in Figure 11.4.7.

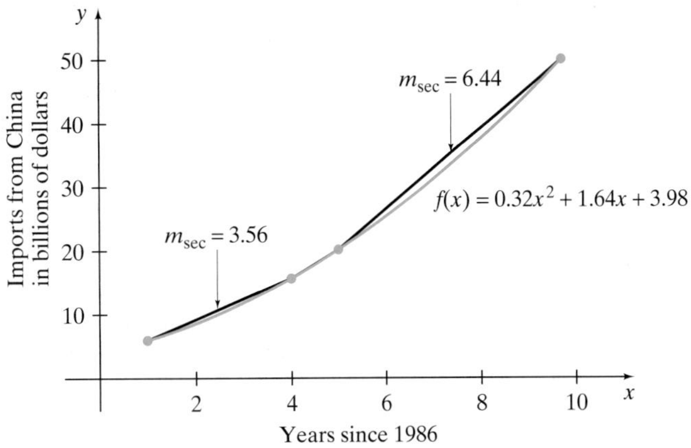

Figure 11.4.7

The difference quotient is a fundamental component of our understanding of differential calculus, so we will continue to compute the secant line slope throughout the remainder of Chapter 11 and into Chapter 12.

SUMMARY

This section featured the introduction of the **quadratic function**, $f(x) = ax^2 + bx + c$, which is a second-degree function with $a \neq 0$. We found that every graph of a quadratic function is a parabola, and the **vertex of the parabola** was found at the coordinates $\left(\frac{-b}{2a}, f\left(\frac{-b}{2a}\right)\right)$. Roots, also known as zeros of the quadratic function, or x-intercepts of the graph, can be found in three ways:

- Algebraic methods like factoring
- The **quadratic formula**, $x = \dfrac{-b \pm \sqrt{b^2 - 4ac}}{2a}$
- The capabilities of the calculator

Finally, we introduced the **slope of the secant line** and found that this type of average rate of change could be computed by the difference quotient

$$m_{sec} = \frac{f(x + \Delta x) - f(x)}{\Delta x}$$

We found that the **difference quotient** could determine the average rate of change over an interval for a nonlinear function. As you will see in upcoming sections, the difference quotient will be used for more than just quadratic functions.

SECTION 11.4 EXERCISES

For Exercises 1–10:

(a) Determine the coordinates of the vertex algebraically.

(b) Use the result from part (a) to determine the intervals where the function is increasing and where it is decreasing.

1. $f(x) = x^2 + 6x + 5$

2. $f(x) = x^2 - 4x + 2$

3. $g(x) = -x^2 + 6x + 6$

4. $g(x) = -2x^2 - 4x + 5$

5. $g(x) = 5x^2 + 6x - 3$

6. $g(x) = 5x^2 - 7x + 2$

7. $f(x) = 0.14x^2 + 0.5x - 0.3$

8. $f(x) = 0.81x^2 + 3.24x - 0.4$

9. $f(x) = -0.9x^2 - 1.8x + 0.5$

10. $f(x) = -0.35x^2 + 2.8x - 0.3$

For Exercise 11–18, find the zeros of the quadratic function by factoring.

11. $f(x) = x^2 - 16$

12. $f(x) = x^2 - 49$

13. $f(x) = -2x^2 - 4x$

14. $f(x) = -13x^2 - 39x$

15. $f(x) = x^2 - 5x + 6$

16. $f(x) = x^2 + 2x - 8$

17. $g(x) = 6x^2 - 5x - 50$

18. $g(x) = 8x^2 + 14x + 3$

In Exercises 19–24, find the real number roots of the quadratic function by the quadratic formula or write no real roots.

✓ 19. $f(x) = x^2 - x - 1$

20. $f(x) = x^2 - 3x - 2$

21. $f(x) = 11x^2 - 7x + 1$

22. $f(x) = 4x^2 - 12x + 11$

23. $g(x) = 2x^2 + 2x + 1$

24. $f(x) = 9x^2 - 12x + 8$

In Exercises 25–30, use the ZERO or ROOT capabilities of your calculator to approximate the zeros of the quadratic function. Round to the nearest hundredth. If the graph of the function has no x-intercepts, write no real zeros.

25. $f(x) = x^2 - 2x - 24$

26. $f(x) = x^2 - 2x - 15$

27. $f(x) = 4x^2 - 12x + 9$

28. $g(x) = 9x^2 + 24x + 16$

29. $g(x) = x^2 - 5x + 8$

30. $f(x) = 1.53x^2 - 3x + 2.67$

In Exercises 31–36, use the difference quotient to determine the slope of the secant line for each function over the indicated intervals.

31. $f(x) = -5x + 1$

(a) x changes from $x = 0$ to $x = 6$.
(b) x changes from $x = 0$ to $x = 3$.
(c) x changes from $x = 3$ to $x = 4$.

32. $f(x) = 3x - 5$

(a) x changes from $x = 0$ to $x = 8$.
(b) x changes from $x = 0$ to $x = 4$.
(c) x changes from $x = 4$ to $x = 8$.

33. $f(x) = x^2 + x$

(a) x changes from $x = 0$ to $x = 4$.
(b) x changes from $x = 0$ to $x = 2$.
(c) x changes from $x = 2$ to $x = 4$.

34. $f(x) = x^2 - 2x$

(a) x changes from $x = 0$ to $x = 6$.
(b) x changes from $x = 0$ to $x = 3$.
(c) x changes from $x = 3$ to $x = 6$.

35. $f(x) = x^2 - x + 3$

(a) If $x = 1$ and $\Delta x = 2$.
(b) If $x = 1$ and $\Delta x = 1$.
(c) If $x = 1$ and $\Delta x = 0.5$.

36. $f(x) = 2x^2 - 3$

(a) If $x = 2$ and $\Delta x = 3$.
(b) If $x = 2$ and $\Delta x = 1$.
(c) If $x = 2$ and $\Delta x = 0.5$.

Applications

37. The number of parking citations that the Sampsonburg Police Department gave over a five-year period can be modeled by

$$f(x) = \frac{1}{5}x^2 + 100x + 30 \qquad 0 \le x \le 5$$

where x represents the number of years since record taking of the citations began and $f(x)$ represents the number of citations given. Determine $\frac{f(5) - f(0)}{5}$ and interpret.

38. The number of unbreakable sunglasses sold at the U-C-Me specialty store during a long-term sales promotion can be modeled by

$$g(x) = \frac{1}{10}x^2 + 50x + 10 \qquad 0 \le x \le 10$$

where x represents the number of months since the promotion began and $g(x)$ represents the number of sunglasses sold. Determine $\frac{g(10) - g(0)}{10}$ and interpret.

39. If a rock is thrown from the ground with an initial velocity of 80 feet per second, then its height can be modeled by

$$s(t) = -16t^2 + 80t \qquad 0 \le t \le 5$$

where t represents the number of seconds since the rock was thrown and $s(t)$ represents the rock's height in feet.

(a) Evaluate $s(3)$ and interpret.
(b) Determine the average rate of change in the rock's height for $t = 1$ to $t = 3$ and interpret.

40. If a model rocket is launched from a 3-foot platform with an initial velocity of 150 feet per second, then its height can be modeled by

$$s(t) = -16t^2 + 150t + 3$$

where t represents the number of seconds since launch and $s(t)$ represents the height in feet.

(a) Evaluate $s(8)$ and interpret.

(b) Determine the average rate of change in the rocket's height for $t = 5$ to $t = 8$ and interpret.

41. The number of bacteria in a colony after t hours is given by

$$g(t) = t^2 + 8t + 2000 \qquad 0 \le t \le 24$$

where t is the number of hours since the colony was established and $g(t)$ represents the number of bacteria.

(a) Evaluate $g(3)$ and interpret.

(b) Determine the average rate of change in the increase of the colony's population for $t = 3$ to $t = 6$ and interpret.

42. The number of seeds dispersed by a plot of dandelions each day in April can be modeled by

$$f(x) = 40x - x^2 \qquad 1 \le x \le 30$$

where x represents the number of days in April and $f(x)$ represents the number of seeds dispersed.

(a) Evaluate $f(7)$ and interpret. Does the graph of the model open up or down? How do you know this?

(b) On which day(s) of the month is the dispersion of seeds the greatest?

(c) Determine the average rate of change in the dispersion of the seeds between $x = 24$ and $x = 29$ and interpret.

43. The annual number of aggravated assaults in the United States can be modeled by

$$f(x) = -3.08x^2 + 40.35x + 305.89 \qquad 0 \le x \le 10$$

where x represents the number of years since 1986 and $f(x)$ represents the number of aggravated assaults per 100,000 people (*Source:* U.S. Federal Bureau of Investigation, www.fbi.gov).

(a) Evaluate $f(2)$ and interpret. Does the graph of the model open up or down? How do you know this?

(b) Use your calculator to make a graph of the annual aggravated assault rate for $0 \le x \le 10$. During what year was the number of aggravated assaults per 100,000 people the greatest from 1986 to 1996?

(c) Determine the average rate of change in the number of aggravated assaults per 100,000 people between $x = 4$ and $x = 9$.

44. Courthouse records show that the average property tax on a three-bedroom home in Independence, during the 1990s can be modeled by

$$f(x) = 22x^2 + 35x + 500 \qquad 0 \le x \le 9$$

where x represents the number of years since 1990 and $f(x)$ is the average property tax, measured in dollars.

(a) Does the graph of the model open up or down? How do you know this?

(b) The domain of the model consists of x-values in the interval $[0, 9]$. What is the corresponding range?

(c) Determine the average rate of change in the property tax from 1991 to 1995.

11.2 **45.** The number of milligrams of cholesterol consumed each day per person in the United States can be modeled by

$$f(x) = 0.11x^2 - 4.04x + 445.02 \qquad 1 \le x \le 23$$

where x represents the number of years since 1974 and $f(x)$ represents the number of milligrams of cholesterol consumed each day per person (*Source:* U.S. Department of Agriculture, www.usda.gov).

(a) Evaluate $f(8)$ and interpret.

(b) Determine the vertex and interpret each coordinate. Round the coordinates to the nearest whole number.

(c) Determine the average rate of change in the number of milligrams of cholesterol consumed each day per person on the interval $[1, 15]$ and interpret.

(d) Determine the average rate of change in the number of milligrams of cholesterol consumed each day per person on the interval $[19, 23]$ and interpret.

46. *(continuation of Exercise 45)*

(a) Make a graph of the model in Exercise 45 in the window $[0, 25]$ by $[390, 450]$.

(b) Use the MINIMUM command to determine the coordinates of the vertex.

(c) During what year did the cholesterol consumption per day first dip below 430 milligrams?

11.2 **47.** The number of lawyers in private practice in the United States can be modeled by

$$f(x) = 0.37x^2 - 1.43x + 190.34 \qquad 0 \le x \le 30$$

where x represents the number of years since 1960 and $f(x)$ represents the total number of lawyers in private practice, in thousands (*Source:* U.S. Statistical Abstract, www.census.gov/statab/www).

(a) Evaluate $f(22)$ and interpret.

(b) Make a table of the function $f(x)$ for the values $x = 0$, 5, 10, 15, 25, and 30.

(c) Determine the average rate of change in the number of lawyers in private practice from 1970 to 1980 and interpret using appropriate units.

(d) Determine the average rate of change in the number of lawyers in private practice from 1980 to 1990 and compare the result to part (c).

48. *(continuation of Exercise 47)*

(a) Make a graph of the model in Exercise 47 in the window $[0, 30]$ by $[180, 500]$.

(b) During what year did the number of lawyers in private practice exceed 300,000?

SECTION PROJECT

The number of bowling alleys in the United States can be modeled by

$$f(x) = -4.2x^2 + 2.2x + 8622.97 \qquad 0 \le x \le 20$$

Here x represents the number of years since 1975, and $f(x)$ represents the number of bowling alleys open for business in the United States that year (*Source:* U.S. Census Bureau, www.census.gov).

(a) According to the model, how many bowling alleys were open for business in the United States in 1978?

(b) Does the graph of the function open up or down? How do you know this from inspecting the model?

(c) Make a graph of the model in the viewing window [0, 20] by [7000, 8700]. Is the graph consistent with your answer to part (b)?

(d) Determine the average rate of change in the number of bowling alleys in the United States from 1975 to 1985 and interpret.

(e) Determine the average rate of change in the number of bowling alleys in the United States from 1985 to 1995 and interpret.

(f) Does the average rate of change during the time periods in parts (d) and (e) indicate a steady decline or a sharp decline in bowling establishments?

Section 11.5 Operations on Functions

In this section we extend our discussion to functions of higher degree. In general, we can call this family of functions **polynomial functions**. While continuing to build up our tools to prepare for the study of calculus, we will also discuss **operations on functions**, which are commonly used to determine the **business functions**. Finally, we will revisit the difference quotient and examine more applications of the secant line.

Operations on Functions

The basics of arithmetic with real numbers—addition, subtraction, multiplication, and division—are called mathematical **operations**. These operations can also be performed on functions.

Operations on Functions

Let f and g be functions. Then, for all x-values for which both f and g exist, we have

1. The **sum** of f and g is $(f+g)(x) = f(x) + g(x)$.
2. The **difference** of f and g is $(f-g)(x) = f(x) - g(x)$.
3. The **product** of f and g is $(f \cdot g)(x) = f(x) \cdot g(x)$.
4. The **quotient** of f and g is $\left(\frac{f}{g}\right)(x) = \frac{f(x)}{g(x)}$, provided that $g(x) \neq 0$.

Example 1 Evaluating Operations of Functions

Let $f(x) = x^2 + x - 3$ and $g(x) = 4x + 2$. Evaluate the following:

(a) $(f+g)(1)$

(b) $(f-g)(1)$

(c) $(f \cdot g)(1)$

(d) $\left(\frac{f}{g}\right)(1)$

Solution

We first evaluate the two functions at $x = 1$ and get

$$f(1) = (1)^2 + 1 - 3 = -1 \quad \text{and} \quad g(1) = 4(1) + 2 = 6$$

(a) From the definition of sum of functions, we get

$$(f + g)(1) = f(1) + g(1) = -1 + 6 = 5$$

(b) From the definition of difference of functions, we get

$$(f - g)(1) = f(1) - g(1) = -1 - 6 = -7$$

(c) From the definition of the product of functions, we get

$$(f \cdot g)(1) = f(1) \cdot g(1) = (-1)(6) = -6$$

(d) From the definition of the quotient of functions, we get

$$\left(\frac{f}{g}\right)(1) = \frac{f(1)}{g(1)} = \frac{-1}{6}$$

■

Checkpoint 1 Now work Exercise 11.

The critical restriction when using operations with functions occurs when using the quotient $\left(\frac{f}{g}\right)(x)$. We must make sure to exclude all values from the domain that can make the denominator zero.

Functions of Business

The operations on functions are commonly used in the functions of business, such as price–demand and cost functions. Before we extend this important family of functions, let's review a few business functions that we have already studied.

From Your Toolbox

1. The **price–demand function** p gives us the price $p(x)$ at which people buy exactly x units of product.
2. The cost $C(x)$ of producing x units of a product is given by the **cost function**

$$C(x) = (\text{variable costs}) \cdot (\text{units produced}) + (\text{fixed costs})$$

Other types of business functions are illustrated in everyday life. Let's say that a youngster has a lemonade stand in her front yard and that she charges 50 cents for each cup. If 35 cups are sold in an afternoon, the amount that the youngster makes is $\$0.50(35) = \$17.50$. This amount earned represents the **revenue**. We computed this revenue by multiplying the price per cup, called the **unit price**, times the quantity sold. We can generalize this notion to form the **revenue function**.

Revenue Function

The revenue generated by selling a certain quantity of a product at a certain price is

$$(\text{Total revenue}) = (\text{quantity sold}) \cdot (\text{unit price})$$

For x units sold at a price given by the price function $p(x)$, the **revenue function** R is

$$R(x) = x \cdot [p(x)]$$

Example 2 Determining a Revenue Function

The Professional Bookstore knows from past sales records that the weekly number of units demanded, x, of a certain pack of computer tutorial software at price p is as shown in Table 11.5.1.

Table 11.5.1

Demand x	Price $p(x)$
10	\$125
30	\$75

(a) Use Table 11.5.1 to find a linear price function, p, written in slope–intercept form for the tutorial software.

(b) Use the result from part (a) to determine the revenue function R and graph R.

(c) Determine the vertex of the graph of R and interpret each coordinate.

Solution

(a) To find the slope–intercept form, we must compute the slope m of the graph of the price function given by the table,

$$m = \frac{75 - 125}{30 - 10} = \frac{-50}{20} = -\frac{5}{2}$$

Using the point–slope form with $(x_1, y_1) = (10, 125)$, we find that the price function is

$$y - y_1 = m(x - x_1)$$

$$y - 125 = -\frac{5}{2}(x - 10)$$

$$y - 125 = -\frac{5}{2}x + 25$$

$$y = -\frac{5}{2}x + 150$$

Substituting $p(x)$ for y gives us

$$p(x) = -\frac{5}{2}x + 150$$

(b) To get the revenue function R, we need to multiply the price function by the quantity x. This gives us

$$R(x) = x \cdot [p(x)] = x \cdot \left[-\frac{5}{2}x + 150\right] = -\frac{5}{2}x^2 + 150x$$

Notice that this revenue function is a parabola that opens down (Figure 11.5.1). This is a common shape for revenue functions.

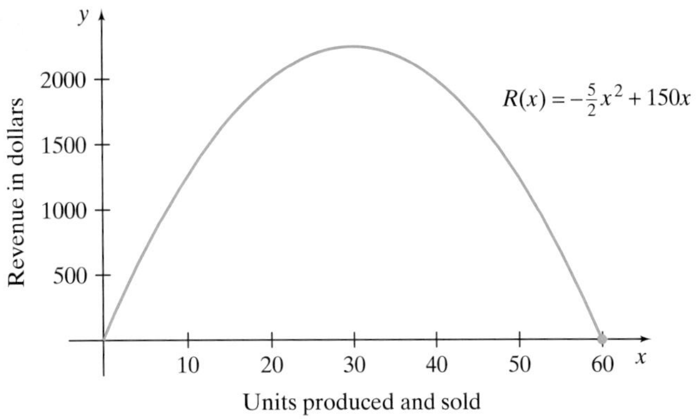

Figure 11.5.1

(c) **Understand the Situation:** The point where the revenue is a maximum is at the vertex. Recall that the x-coordinate of the vertex is given by $x = \frac{-b}{2a}$.

For this function, we know that $a = -\frac{5}{2}$ and $b = 150$, so we get

$$x = \frac{-150}{2\left(-\frac{5}{2}\right)} = \frac{-150}{-5} = 30$$

Evaluating this number in the revenue function, we get

$$R(30) = -\frac{5}{2}(30)^2 + 150(30) = -2250 + 4500 = 2250$$

Interpret the Solution: So the weekly revenue is at its maximum of \$2250 dollars when 30 packs of tutorial software are sold. ■

Let's return to our young entrepreneur. If the cost of making the lemonade was \$4.25, then the **profit** that is made during that afternoon is \$17.50 – \$4.25 = \$13.25. This same model applies no matter how small or large the business venture may be. The profit made is just revenue minus the cost. This relationship between these two quantities allows us to form the **profit function**.

Profit Function

The profit generated after producing and selling a certain quantity of a product is given by

$$\text{Profit} = \text{revenue} - \text{cost}$$

For x units of a product, the **profit function** P is

$$P(x) = R(x) - C(x)$$

where R and C are the revenue and cost functions, respectively.

Note: Notice that the last two definitions use operations on functions. The revenue function is a product of functions, and the profit function is the difference of functions.

Example 3 Determining a Profit Function

The FlashBroc Company makes designer ties. It determines that the weekly cost and revenue functions, in dollars, for producing and selling x designer ties are

$$C(x) = 30x + 50 \quad \text{and} \quad R(x) = 90x - x^2$$

respectively.

(a) Determine the profit function P and graph P.

(b) Find the vertex of the graph of P and interpret each coordinate.

Solution

(a) Since the definition of the profit function is $P(x) = R(x) - C(x)$, we get

$$\begin{aligned} P(x) &= R(x) - C(x) \\ &= (90x - x^2) - (30x + 50) \\ &= -x^2 + 60x - 50 \end{aligned}$$

So the profit function for producing and selling x units is $P(x) = -x^2 + 60x - 50$. A graph of P is given in Figure 11.5.2.

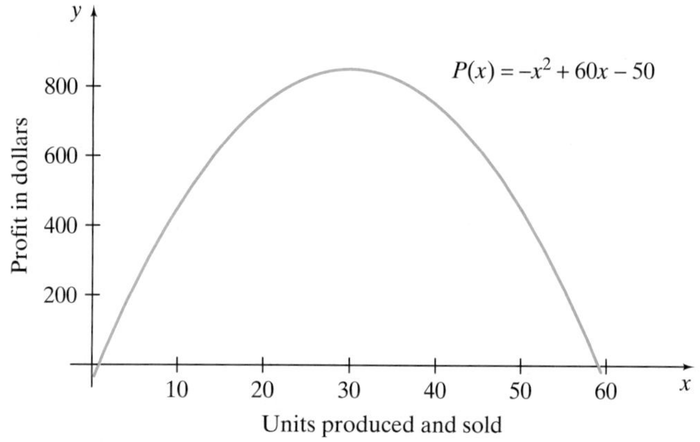

Figure 11.5.2

(b) The x-coordinate of the vertex is $x = \frac{-b}{2a} = \frac{-60}{2(-1)} = 30$. The y-coordinate is given by

$$P(30) = -30^2 + 60(30) - 50 = -900 + 1800 - 50 = 850$$

Interpret the Solution: This means that a maximum profit of \$850 will be realized when 30 designer ties are produced and sold each week. ■

Interactive Activity

Using the cost and revenue functions in Example 3, graph the following in the viewing window [0, 100] by [0, 2500].

$$y_1 = 90x - x^2$$
$$y_2 = 30x + 50$$
$$y_3 = y_1 - y_2$$

Without using the TRACE command, determine which function matches which graph. How do you know this?

WWW

Intuitively, when the revenue is equal to the cost (that is, when money coming in is equal to money going out), we have reached the **break-even point**. Mathematically, this occurs at the point where the graphs of the cost and revenue functions intersect. See Figure 11.5.3. Algebraically, this means that the break-even point occurs where $R(x) = C(x)$. If the cost is more than the revenue, the result is a **loss**, and when the revenue is greater than the cost, a **profit** is the result. Let's try an example of how we can use these functions in practice.

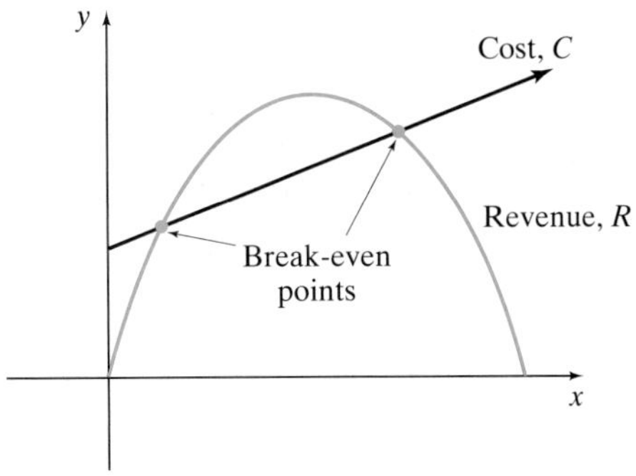

Figure 11.5.3

Example 4 Determining the Break-even Points

In Example 2, we showed that the revenue function for the tutorial software packs was $R(x) = -\frac{5}{2}x^2 + 150x$. Now assume that the fixed costs for the software are \$1505, with variable costs of $7.5\ \frac{\text{dollars}}{\text{pack}}$.

(a) Write a linear cost function C for the tutorial software.

(b) Find the break-even points(s) and interpret.

(c) Determine the profit function P.

Solution

(a) Using the definition of the linear cost function with variable costs of $m = 7.5\ \frac{\text{dollars}}{\text{pack}}$ and fixed costs of \$1505, we get $C(x) = 7.5x + 1505$.

(b) **Understand the Situation:** To determine the break-even point(s), we need the x-values such that $R(x) = C(x)$.

Setting $R(x) = C(x)$ and solving for x yields

$$
\begin{aligned}
R(x) &= C(x) \\
-\frac{5}{2}x^2 + 150x &= 7.5x + 1505 \\
-\frac{5}{2}x^2 + 142.5x - 1505 &= 0 \\
-5x^2 + 285x - 3010 &= 0 \\
x^2 - 57x + 602 &= 0
\end{aligned}
$$

Since this quadratic is not easily factorable, we use the quadratic formula to get

$$
\begin{aligned}
x &= \frac{57 \pm \sqrt{(-57)^2 - 4(1)(602)}}{2(1)} \\
&= \frac{57 \pm \sqrt{841}}{2} = \frac{57 \pm 29}{2}
\end{aligned}
$$

So $x = \frac{57+29}{2}$ or $x = \frac{57-29}{2}$. These solutions simplify to $x = 43$ and $x = 14$. So the x-coordinate of the break-even point(s) are $x = 43$ and $x = 14$. The y-coordinates are determined by evaluating either the cost function C or the revenue function R at these two x-values. (Recall from algebra that a point of intersection satisfies both equations.) Using the cost function C we have

$$C(x) = 7.5x + 1505$$
$$C(14) = 7.5(14) + 1505 = 1610$$

and

$$C(43) = 7.5(43) + 1505 = 1827.5$$

So the two break-even points are (14, 1610) and (43, 1827.5).

Interpret the Solution: When 14 packs of tutorial software are sold the cost and revenue are equal at \$1610. When 43 packs of tutorial software are sold the cost and revenue are equal at \$1827.50.

(c) Using our definition of the profit function, we have

$$\begin{aligned} P(x) &= R(x) - C(x) \\ &= \left(-\frac{5}{2}x^2 + 150x\right) - (7.5x + 1505) \\ &= -\frac{5}{2}x^2 + 142.5x - 1505 \end{aligned}$$

So the profit function for the tutorial software packs is

$$P(x) = -\frac{5}{2}x^2 + 142.5x - 1505.$$

Using the cost, revenue, and profit functions in Example 4, we see that the break-even points on the cost and revenue functions lie directly above the zeros of the profit function P. See Figure 11.5.4

Interactive Activity

Graphically verify the solution to part (b) of Example 4 in two ways. First graph R and C in the same viewing window and use the INTERSECT command on your calculator to find the break-even points. Then graph the profit function P and use the ZERO or ROOT command on your calculator to find the x-intercepts.

WWW

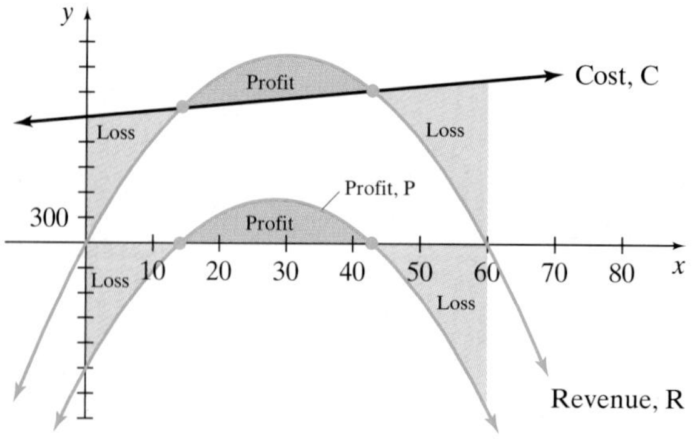

Figure 11.5.4

Thus, another way to find the break-even points is to find the zeros of the profit function. A profit is realized in the interval where $P(x)$ lies above the x-axis (that is, $P(x) > 0$). Let's summarize the relationship between P, R, and C before continuing.

- Break-even occurs when $R(x) = C(x)$. This is equivalent to $P(x) = 0$. Graphically, $P(x) = 0$ at the x-intercepts.
- Profit occurs when $R(x) > C(x)$. This is equivalent to $P(x) > 0$, when the graph of P is above the x-axis.
- A loss occurs when $R(x) < C(x)$. This is equivalent to $P(x) < 0$, when the graph of P is below the x-axis.

We can also apply the difference quotient to business functions to approximate the average rate of change on specific intervals.

Example 5 Determining the Secant Line Slope of a Business Function

The Midwest Manufacturing Company manufactures titanium lock washers. Its revenue function is given by

$$R(x) = -0.02x^2 + 8x$$

where x represents the number of titanium lock washers sold and $R(x)$ represents the amount of revenue generated in dollars. Find the average rate of change in revenue as the number of washers sold increases from 50 to 150 washers and interpret.

Solution

Understand the Situation: To find the desired average rate of change, we need to compute the difference quotient

$$m_{\text{sec}} = \frac{R(x + \Delta x) - R(x)}{\Delta x}$$

With $x = 50$ and $\Delta x = 150 - 50 = 100$, we get

$$\begin{aligned} m_{\text{sec}} &= \frac{R(50 + 100) - R(50)}{100} = \frac{R(150) - R(50)}{100} \\ &= \frac{-0.02(150)^2 + 8(150) - (-0.02(50)^2 + 8(50))}{100} \\ &= \frac{750 - 350}{100} \\ &= \frac{400}{100} = 4 \end{aligned}$$

Interpret the Solution: This means that as the number of titanium lock washers sold increases from 50 to 150, the revenue increases at an average rate of $4 \frac{\text{dollars}}{\text{washer}}$. See Figure 11.5.5.

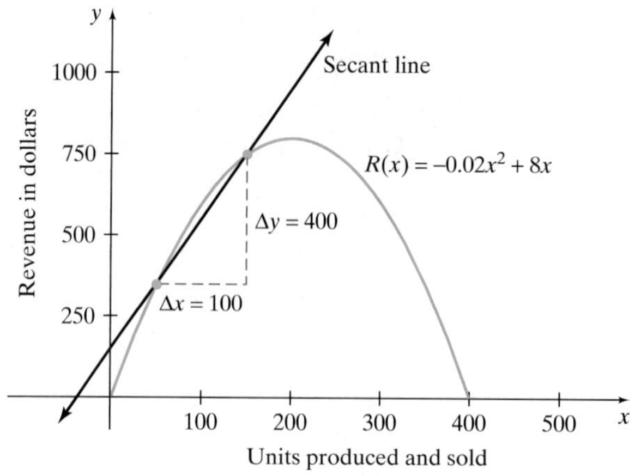

Figure 11.5.5

Checkpoint 2 Now work Exercise 69.

Polynomial Functions

So far we have studied linear, or first-degree, functions and quadratic, or second-degree, functions. Now let's take a look at functions of higher degree. We call this general family of functions **polynomial functions**.

Polynomial Function

A **polynomial function** of degree n has the form

$$f(x) = a_n x^n + a_{n-1}x^{n-1} + \cdots + a_1 x + a_0$$

where $a_0, a_1, a_2, \ldots, a_n$ are real number constants where $a_n \neq 0$ and n is a whole number. The **degree** of the polynomial function is given by n.

Using this definition, we see that $f(x) = 2x^3 - 4x^2 - x + 5$ is a third-degree polynomial function, also called a **cubic** function. The functions $f(x) = \sqrt{x} + 3$ and $g(x) = \frac{4}{x} + 8$ are not polynomial functions because neither can be written with whole number exponents.

A few properties of polynomial functions make them very convenient to use mathematically.

- The domain of every polynomial function is the set of real numbers, $(-\infty, \infty)$. This means that there are no restrictions on the numbers evaluated in polynomial functions.
- There are no holes or breaks in the graphs of polynomial functions.
- We say that polynomial functions are **continuous**. This means their graphs are smooth and unbroken.

Example 6 Using Polynomial Models

The personal income per capita in the United States, in constant 1992 dollars, can be modeled by

$$f(x) = 3.75x^3 - 115.23x^2 + 1229.81x + 16{,}025.65 \qquad 1 \le x \le 15$$

where x represents the number of years since 1979 and $f(x)$ represents the personal income per capita in the United States in constant 1992 dollars (*Source:* U.S. Bureau of Economic Analysis, www.bea.doc.gov).

(a) Use the difference quotient to determine the average rate of change in the personal income per capita where $x = 1$ and $\Delta x = 10$. Interpret the result.

(b) Use the difference quotient to determine the average rate of change in the personal income per capita where $x = 1$ and $\Delta x = 5$. Interpret the result. Compare it with the result from part (a).

Solution

(a) Evaluating the difference quotient with $x = 1$ and $\Delta x = 10$, we get

$$m_{\text{sec}} = \frac{f(x+\Delta x) - f(x)}{\Delta x} = \frac{f(1+10) - f(1)}{10} = \frac{f(11) - f(1)}{10}$$

$$= \frac{[3.75(11)^3 - 115.23(11)^2 + 1229.81(11) + 16{,}025.65] - [3.75(1)^3 - 115.23(1)^2 + 1229.81(1) + 16{,}025.65]}{10}$$

$$= \frac{20{,}601.98 - 17{,}143.98}{10} = \frac{3458}{10} = 345.8$$

Interpret the Solution: Here $x = 1$ corresponds to the year 1980, and $x + \Delta x = 11$ corresponds to 1990. Thus, during the period from 1980 to 1990, the personal per capita income was increasing at an average rate of $345.8\,\frac{\text{dollars}}{\text{year}}$.

(b) Evaluating the difference quotient with $x = 1$ and $\Delta x = 5$, we get

$$m_{\text{sec}} = \frac{f(1+5) - f(1)}{5} = \frac{f(6) - f(1)}{5}$$

$$= \frac{20{,}066.23 - 17{,}143.98}{5}$$

$$= \frac{2922.25}{5} = 584.45$$

Interpret the Solution: This tells us that in the period between 1980 and 1985, the personal per capita income was increasing at an average rate of $584.45\,\frac{\text{dollars}}{\text{year}}$. Knowing that the income data are adjusted to constant 1992 dollars, it appears that per capita income grew faster in the first half of the

1980s as compared to the decade as a whole. As can be seen in Figure 11.5.6, the personal income leveled off during the late 1980s. This is why the average rate of change is larger in part (b).

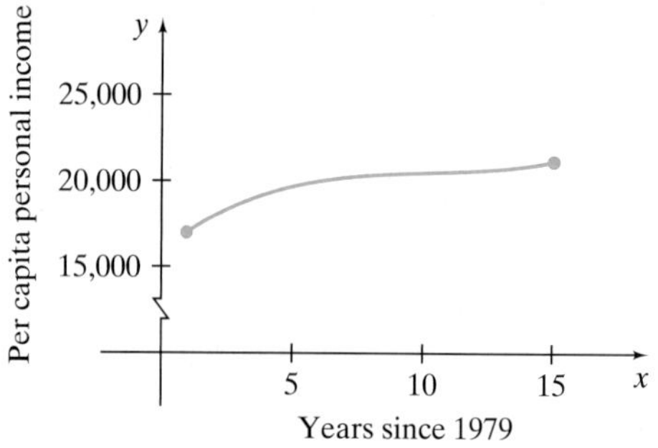

Figure 11.5.6 Graph of
$f(x) = 3.75x^3 - 115.23x^2 + 1229.81x + 16{,}025.65$. ■

End Behavior of Functions

At times we want to know the values of a function for extremely large positive and extremely large negative values of the independent variable x. This determines what is called the **end behavior** of a function. To begin to see what happens, let's take a look at the quadratic function $f(x) = 2x^2 - 6x + 3$. Table 11.5.2 shows the function values for extremely large positive and negative values of x. From this table, it appears that as the x-values grow both negatively and positively large the function values grow positively large. Table 11.5.2 illustrates just what end behavior is all about. We commonly describe the end behavior symbolically. We write as $\boldsymbol{x \to -\infty}$, $\boldsymbol{f(x) \to \infty}$ to denote that, as we move from right to left on the x-axis, the $f(x)$-values grow large positively. (This is read "as x approaches negative infinity, $f(x)$ approaches infinity"). We write as $\boldsymbol{x \to \infty}$, $\boldsymbol{f(x) \to \infty}$ to denote that, as we move from left to right on the x-axis, the $f(x)$-values grow large positively. (This is read "as x approaches infinity, $f(x)$ approaches infinity".)

Table 11.5.2

x	$f(x) = 2x^2 - 6x + 3$
−100,000	20,000,600,003
−10,000	200,060,003
−1000	2,006,003
−100	20,603
0	3
100	19,403
1000	1,994,003
10,000	199,940,003
100,000	19,999,400,003

To determine the end behavior of a polynomial function, all we need do is examine the leading term, $a_n x^n$, of the polynomial. We can summarize the results with the **leading coefficient test**.

Leading Coefficient Test

1. Assume that $f(x)$ is an **even-degree** polynominal function with leading term $a_n x^n$ and leading coefficient a_n.

If the leading coefficient is	as $x \to -\infty$	as $x \to \infty$	The typical graph is
Positive ($a_n > 0$)	$f(x) \to \infty$	$f(x) \to \infty$	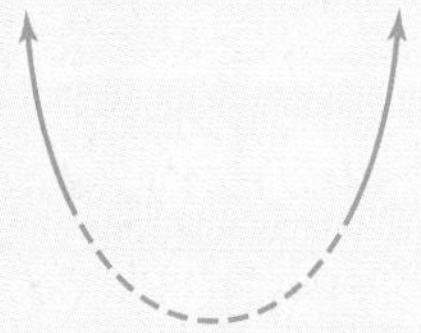
Negative ($a_n < 0$)	$f(x) \to -\infty$	$f(x) \to -\infty$	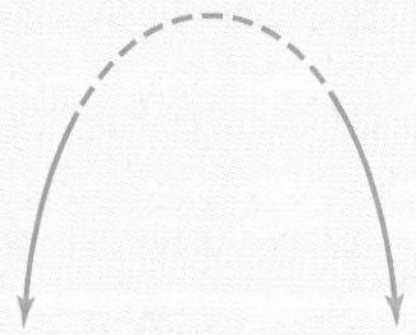

2. Assume that $f(x)$ is an **odd-degree** polynominal function with leading term $a_n x^n$ and leading coefficient a_n.

If the leading coefficient is	as $x \to -\infty$	as $x \to \infty$	The typical graph is
Positive ($a_n > 0$)	$f(x) \to -\infty$	$f(x) \to \infty$	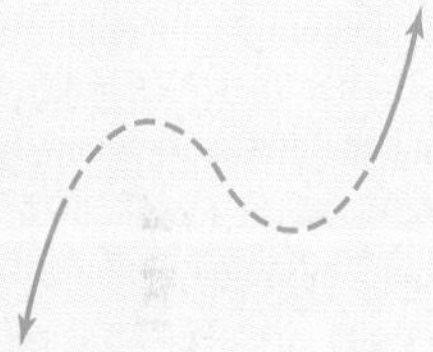
Negative ($a_n < 0$)	$f(x) \to \infty$	$f(x) \to -\infty$	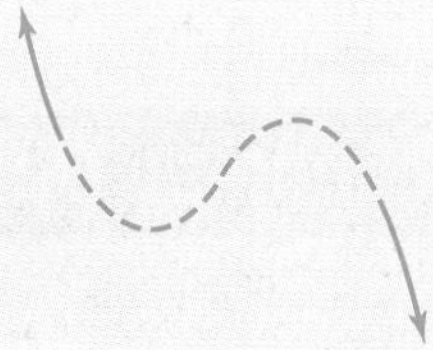

Example 7 Determining the End Behavior of a Polynomial Function

For the cubic function $f(x) = x^3 - 3x^2 - 6x + 8$:

(a) Determine the end behavior of the function.

(b) Use the graph of $f(x) = x^3 - 3x^2 - 6x + 8$ in Figure 11.5.7 to approximate the *peaks* and *valleys* of the graph of the function. Use these approximations to determine the intervals where the function increases and where it decreases.

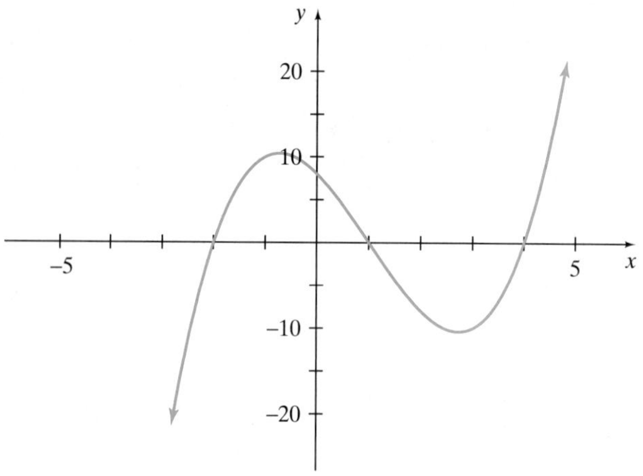

Figure 11.5.7 Graph of $f(x) = x^3 - 3x^2 - 6x + 8$.

Solution

(a) From case 2 of the Leading Coefficient Test, since f is a third, or odd, degree polynomial function and the leading coefficient, $a_n = 1$, is positive, we know that as $x \to -\infty$, $f(x) \to -\infty$ and as $x \to \infty$, $f(x) \to \infty$. This is shown graphically in Figure 11.5.7 with the arrowheads at the beginning and end of the graph indicating that the graph continues in the same direction.

(b) Using Figure 11.5.7, we estimate a peak or maximum point at $(-1, 10)$ and a valley or minimum point at $(3, -10)$. We can see from the graph that $f(x) = x^3 - 3x^2 - 6x + 8$ is increasing, or going uphill, for all the x-values to the left of $x \approx -1$ and to the right of $x \approx 3$. So, using interval notation, we write that f is increasing on the intervals $(-\infty, -1) \cup (3, \infty)$ and is decreasing on the interval $(-1, 3)$. Keep in mind that these are **estimates** of the maximum and minimum values. To find the **exact** values, we need the power of calculus. ■

Technology Option

We can use the MAXIMUM and MINIMUM commands on the graphing calculator to determine approximations for the **peak** and **valley** of the graph of the function given in Example 7. Figure 11.5.8a shows the result of using the MAXIMUM command while Figure 11.5.8b shows the result of using the MINIMUM command.

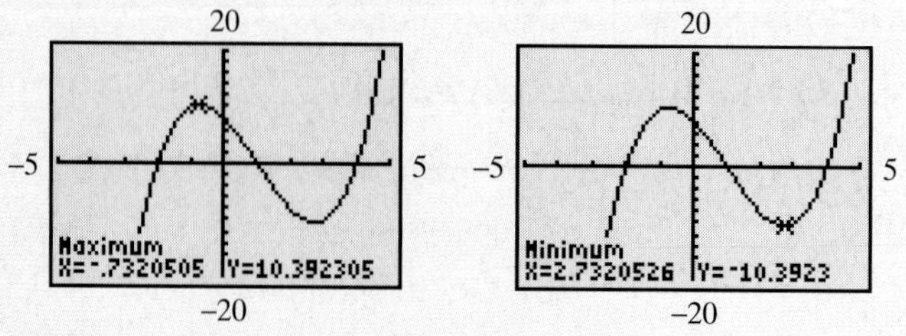

(a) **(b)**

Figure 11.5.8

Please note that these are still approximations to the maximum and minimum values. We need calculus to determine the maximum and minimum exactly. To see how to use the MAXIMUM and MINIMUM commands on your calculator, consult the online graphing calculator manual at www.prenhall.com/armstrong.

Checkpoint 3

Now work Exercise 73.

Interactive Activity

Here are graphs of two cubic polynomial functions. Are they complete graphs of the functions, or is there a piece missing? How do you know this?

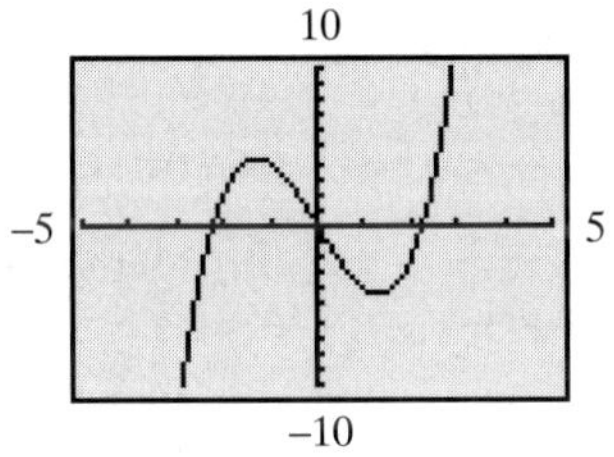

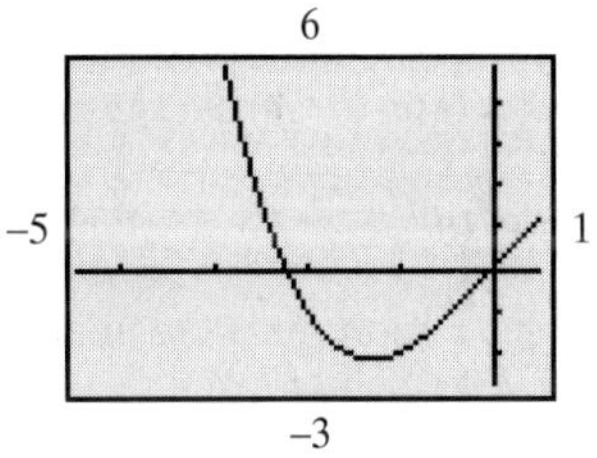

SUMMARY

This section began with operations on functions. We learned that we can add, subtract, multiply, and divide functions, much as we do with real numbers. Then we discussed the functions of business:

- The **revenue function** $R(x) = x \cdot [p(x)]$
- The **profit function** $P(x) = R(x) - C(x)$
- The **break-even** point(s) occurs when $R(x) = C(x)$, or equivalently, where $P(x) = 0$.

Then we turned our attention to the properties of polynomial functions. These functions all have the form

$$f(x) = a_n x^n + a_{n-1}x^{n-1} + \cdots + a_1 x + a_0$$

where $a_0, a_1, a_2, \ldots, a_n$ are real number constants with $a_n \neq 0$ and n is a whole number. The end behavior can be found by using the leading coefficient test.

Leading Coefficient Test

	***n* Is Odd**	***n* Is Even**
$a_n > 0$	As $x \to -\infty$, $f(x) \to -\infty$ and as $x \to \infty$, $f(x) \to \infty$	As $x \to \pm\infty$, $f(x) \to \infty$
$a_n < 0$	As $x \to -\infty$, $f(x) \to \infty$ and as $x \to \infty$, $f(x) \to -\infty$	As $x \to \pm\infty$, $f(x) \to -\infty$

SECTION 11.5 EXERCISES

For Exercises 1–4, simplify and write the domain for the following operations.

(a) $(f + g)(x)$ **(b)** $(f - g)(x)$

(c) $(f \cdot g)(x)$ **(d)** $\left(\frac{f}{g}\right)(x)$

1. $f(x) = 6x + 3, g(x) = 4x - 1$

2. $f(x) = -2x + 9, g(x) = -5x - 2$

3. $f(x) = \sqrt{x + 5}, g(x) = x^2 + 5$

4. $f(x) = \sqrt{x - 3}, g(x) = x - 5$

For Exercises 5–12, let $f(x) = 2x^2 - 4x$ and $g(x) = 5x + 1$. Evaluate each of the following:

5. $(f + g)(0)$

6. $(f + g)(-3)$

7. $(f \cdot g)(4)$

8. $(f \cdot g)(-2)$

9. $(f - g)(2)$

10. $(f - g)(-1)$

✓ **11.** $\left(\frac{f}{g}\right)(5)$

12. $\left(\frac{f}{g}\right)(-1)$

Applications

13. The Handi-Neighbor hardware store sells PowerDriver hammers and finds that the amount made in dollars for the weekly sales is

$$f(x) = 9.75x \qquad 0 \le x \le 50$$

The competing U-Do-It hardware store finds that the weekly sales for selling PowerDriver hammers is

$$g(x) = \frac{4}{5}x^2 + 6x \qquad 0 \le x \le 50$$

In both functions, x represents the number of PowerDriver hammers sold.

(a) Evaluate $(f - g)(20)$ and interpret.

(b) Evaluate $(f + g)(20)$ and interpret.

14. The Nuthin'-But-Socks specialty store finds that the monthly expenses for operation during its first year of business is given by

$$f(x) = \frac{1}{5}x^2 + 300 \qquad 1 \le x \le 12$$

where $x = 1$ corresponds to January, $x = 2$ corresponds to February, and so on, and $f(x)$ is the monthly expenses in hundreds of dollars. The monthly revenue during the first year of business is given by

$$g(x) = 27.5x \qquad 1 \le x \le 12$$

where $x = 1$ corresponds to January, $x = 2$ corresponds to February, and so on, and $g(x)$ is the monthly revenue in hundreds of dollars.

(a) Evaluate $(g - f)(6)$ and interpret.

(b) Did the store break even before the end of the first year?

15. The number of arrests for arson in the town of Waterbridge from 1960 to 1970 can be modeled by

$$f(x) = 4x + 45 \qquad 0 \le x \le 10$$

where x represents the number of years since 1960, and $f(x)$ represents the number of arrests. The number of arson convictions in Waterbridge during the same time period is given by

$$g(x) = x + 35 \qquad 0 \le x \le 10$$

(a) Determine the arson conviction rate during 1967.

(b) Determine a model for the arson *conviction rate* in Waterbridge during the decade. The conviction rate is the number of convictions divided by the number of arrests.

Exercises 16 and 17 use the following. The number of drug **convictions** in the United States can be modeled by the quartic function

$$y_1 = f(x) = 6.13x^4 - 207.23x^3 + 2127.9x^2 - 6622.41x + 15{,}219.92 \qquad 1 \le x \le 11$$

where x represents the number of years since 1984, and $f(x)$ represents the number of drug convictions in the United States. The number of drug **arrests** in the United States can be modeled by

$$y_2 = g(x) = 27.82x^4 - 710.56x^3 + 5963.14x^2 - 1695.09x + 27{,}423.98 \qquad 1 \le x \le 11$$

where x represents the number of years since 1984 and $g(x)$ represents the number of drug arrests in the United States (*Source for both models:* U.S. Census Bureau, www.census.gov).

16. (a) Evaluate $f(10)$ and $g(10)$ and interpret each.

(b) Determine $\left(\frac{f}{g}\right)(10)$ using the solutions to part (a). Interpret the result.

17. (a) Use your calculator to graph $y_3 = \frac{y_1}{y_2} = \left(\frac{f}{g}\right)(x)$ in the viewing window [1, 11] by [0, 1]. This quotient of f and g represents the **conviction rate** for drug arrests in the United States from 1985 to 1995.

(b) Over which two-year period of time did the conviction rate decrease the most—[1, 3], [5, 7], or [9, 11]?

Exercises 18 and 19 use the following. The annual Medicaid **payments** in the United States can be modeled by

$$y_1 = f(x) = 413.48x^2 + 185.72x + 24{,}031.95 \quad 0 \le x \le 15$$

where x represents the number of years since 1980 and $f(x)$ represents the amount of Medicaid payments in millions of dollars. The annual number of Medicaid **recipients** in the United States can be modeled by

$$y_2 = g(x) = 0.004x^4 + 0.02x^3 + 0.01x^2 - 0.24x + 21.66 \qquad 0 \le x \le 15$$

where x represents the number of years since 1980 and $g(x)$ represents the number of Medicaid recipients in millions of people (*Source for both models:* Center for Medicare and Medicaid Services, www.hcfa.gov).

18. (a) Evaluate $f(5)$ and $g(5)$ and interpret each.

(b) Determine $\left(\frac{f}{g}\right)(5)$ using the solutions to part (a). Interpret the result.

19. (a) Use your calculator to graph $y_3 = \frac{y_1}{y_2} = \left(\frac{f}{g}\right)(x)$ in the viewing window [0, 15] by [400, 1400]. This quotient of f and g is the **amount paid per recipient** of Medicaid each year from 1980 to 1995.

(b) By using the TRACE command on your calculator, find the year in which the amount paid per Medicaid recipient dropped below \$900.

For Exercises 20–27, the price–demand function p is given. Determine the revenue function R.

20. $p(x) = 2.55$

21. $p(x) = 87.1$

22. $p(x) = -3.1x$

23. $p(x) = -0.19x$

24. $p(x) = -0.3x + 20$

25. $p(x) = \dfrac{-x}{2000} + 3$

26. $p(x) = -0.1x + 50$

27. $p(x) = -0.12x + 30$

For Exercises 28–31, the revenue and cost functions are given.

(a) Determine the break-even point(s).

(b) Determine how much revenue must be generated to reach the break-even point.

28. $R(x) = 50x$; $C(x) = 20x + 170$

29. $R(x) = 8x$; $C(x) = 5000$

30. $R(x) = 25x - 0.25x^2$; $C(x) = 2x + 5$

31. $R(x) = 100x - x^2$; $C(x) = 20x + 4$

For Exercises 32–35, the revenue and cost functions are given.

(a) Graph R and C in the same viewing window.

(b) Use the INTERSECT command on your calculator to determine the break-even point(s).

(c) Determine how much revenue must be generated to reach the break-even point.

32. $R(x) = 25x - 0.25x^2$; $C(x) = 2x + 5$

33. $R(x) = 100x - x^2$; $C(x) = 20x + 4$

34. $R(x) = -2.1x^2 + 500x$; $C(x) = 80x + 9500$

35. $R(x) = 95.2x - 5x^2$; $C(x) = 155 + 20x$

For Exercises 36–39, the revenue and cost functions are given.

(a) Determine the profit function $P(x) = R(x) - C(x)$.

(b) Determine the zeros of the profit function and interpret.

(c) Determine the vertex of the graph of the profit function P, and interpret each coordinate of the vertex.

36. $R(x) = 120x - 6x^2$; $C(x) = 240 + 2x$

37. $R(x) = 1200x - 37x^2$; $C(x) = 4300 + 148x$

38. $R(x) = 95.2x - 5x^2$; $C(x) = 155 + 20x$

39. $R(x) = -2.1x^2 + 500x$; $C(x) = 80x + 9500$

For Exercises 40–45, the price–demand function p, along with the fixed and variable costs, is given.

(a) Determine the profit function $P(x) = R(x) - C(x)$.

(b) Find the vertex of the graph of the profit function P, and interpret each coordinate of the vertex.

40. $p(x) = 105.7 - 0.89x$; variable costs are \$80 per unit and fixed costs are \$61.80.

41. $p(x) = 37.8 - x$; variable costs are \$29 per unit and fixed costs are \$10.15.

42. $p(x) = 50 - x$; variable costs are \$20 per unit and fixed costs are \$200.

43. $p(x) = -10x + 1040$; variable costs are \$500 per unit and fixed costs are \$6650.

44. $p(x) = 40 - x$; variable costs are \$20 per unit and fixed costs are \$80.

45. $p(x) = -x + 30$; variable costs are \$5 per unit and fixed costs are \$60.

For Exercises 46–49, use the ZERO or ROOT command on your calculator to determine the zeros of the profit function P. Also, supply an interpretation for the zeros.

46. The profit function determined in Exercise 40.

47. The profit function determined in Exercise 41.

48. The profit function determined in Exercise 44.

49. The profit function determined in Exercise 45.

For Exercises 50–55, determine if the given function is a polynomial function. Identify each polynomial as linear, quadratic or cubic.

50. $f(x) = 7$

51. $f(x) = -30$

52. $y = 3x - \dfrac{5}{x}$

53. $g(x) = \dfrac{4}{x^2} + 3x + \dfrac{1}{2}$

54. $f(x) = x^2 + 5x + 2$

55. $f(x) = x^3 + x + 1$

In Exercises 56–65, determine the end behavior of the function.

56. $f(x) = 3x$

57. $f(x) = -17x$

58. $g(x) = -2.1x$

59. $g(x) = 3.8x$

60. $f(x) = -3x^2 + 10x$

61. $g(x) = 5.5x^2 + 4$

62. $f(x) = -x^3 + 2x - 1$

63. $f(x) = 11x + 0.1x^3$

64. $f(x) = -x^6 - x$

65. $g(x) = x^6 - 2x^3 - x^2$

Applications

66. If a stone is thrown straight up from ground level with an initial velocity of 48 feet per second, then the height s of the stone above the ground after t seconds is given by

$$s(t) = -16t^2 + 48t$$

where t is the number of seconds after the stone has been thrown and $s(t)$ is the height of the stone in feet.

(a) Evaluate $s(3)$ and interpret.

(b) Find the average rate of change for $t = 1$ and $\Delta t = 2$. Interpret the result.

67. A girl throws a ball straight up from a 48-foot-tall building with an initial velocity of 30 feet per second. The height s of the ball above the ground after t seconds is given by

$$s(t) = -16t^2 + 30t + 48$$

where t is the number of seconds after the ball has been thrown and $s(t)$ is the height of the ball in feet.

(a) Evaluate $s(2)$ and interpret.

(b) Find the average rate of change for $t = 0$ and $\Delta t = 2.5$. Interpret the result.

68. The ActivLife Company finds that the revenue, in dollars, generated by selling x exercise machines is

$$R(x) = -\frac{x^2}{900} + 10x \qquad x \geq 0$$

(a) Evaluate $R(30)$ and interpret.

(b) Find the average rate of change for $x = 10$ and $\Delta x = 90$ and interpret.

✓ **69.** The BackStreet Company determines that the revenue, in dollars, generated by selling x of its Boomer headphones is given by

$$R(x) = 11x - \frac{x^2}{730} \qquad x \geq 0$$

(a) Evaluate $R(250)$ and interpret.

(b) Find the average rate of change for $x = 100$ and $\Delta x = 200$ and interpret.

11.2 **70.** The percent of people taking the SAT whose intended area of study was engineering can be modeled by

$$f(x) = 0.002x^3 - 0.11x^2 + 1.38x + 6.7 \qquad 0 \leq x \leq 22$$

where x represents the number of years since 1975 and $f(x)$ represents the percent of people taking the SAT whose intended area of study was engineering (*Source:* U.S. Statistical Abstract, www.census.gov/statab/www).

(a) Evaluate $f(6)$ and interpret.

(b) Find the average rate of change for $x = 2$ and $\Delta x = 10$ and interpret.

(c) Find the average rate of change from 1987 to 1997 and interpret.

(d) Compare the results from parts (b) and (c).

11.2 **71.** The percent of people taking the SAT whose intended area of study was business can be modeled by

$$f(x) = -0.08x^2 + 1.77x + 11.59 \qquad 0 \leq x \leq 22$$

where x represents the number of years since 1975 and $f(x)$ represents the percent of people taking the SAT whose intended area of study was business (*Source:* U.S. Statistical Abstract, www.census.gov/statab/www).

(a) Evaluate $f(7)$ and interpret.

(b) Find the average rate of change for $x = 2$ and $\Delta x = 10$ and interpret.

(c) Find the average rate of change from 1987 to 1997 and interpret.

(d) Compare the results from parts (b) and (c).

In Exercises 72–75, use the given graph to determine the following.

(a) Determine the end behavior of the function.

(b) Approximate the peaks and valleys of the graph.

(c) Use the solution from part (b) to determine intervals where f increases and where it decreases.

72. $f(x) = 3x + 4$

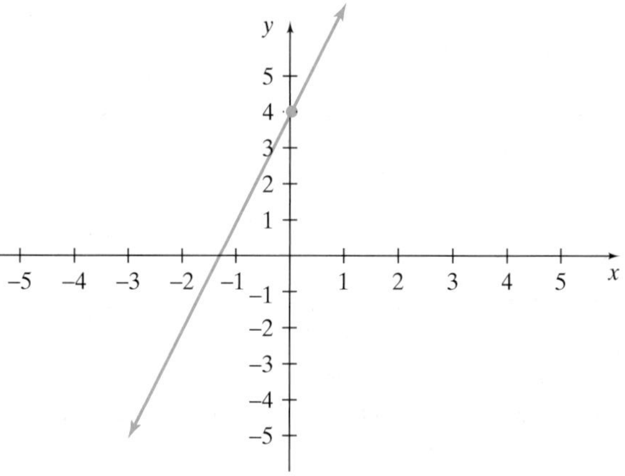

✓ **73.** $f(x) = -x^2 - 10x + 21$

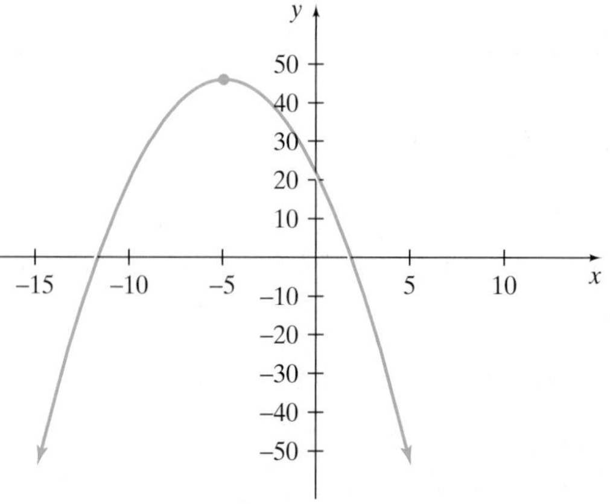

74. $f(x) = x^3 - 3x^2 - 9x + 15$

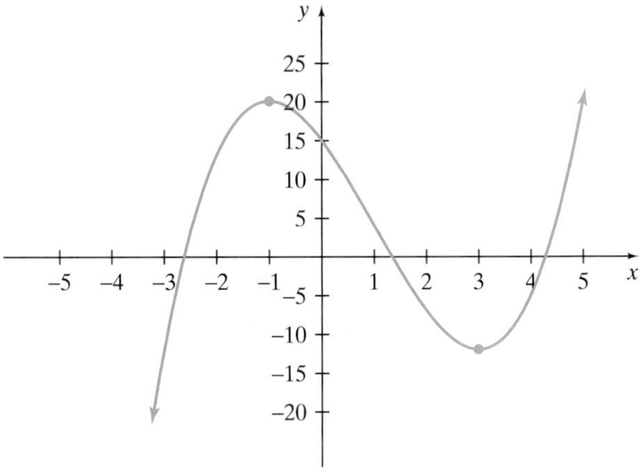

75. $f(x) = -x^4 + 8x + 12$

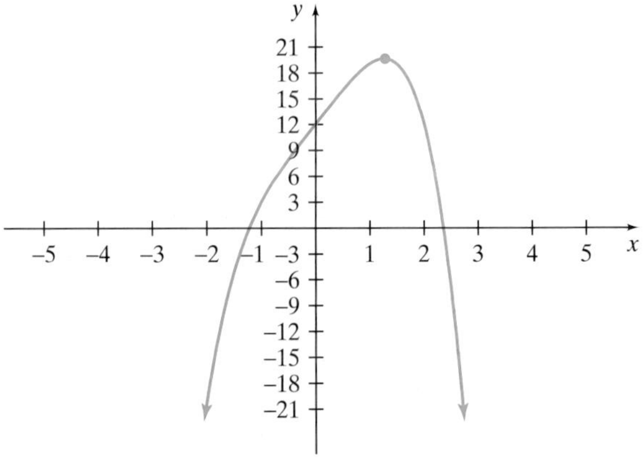

For Exercises 76–79, consider the given polynomial functions.

(a) Determine the end behavior of the function.

(b) Use the MAXIMUM and MINIMUM commands on your calculator to approximate the peaks and valleys of the graph. Estimate the points to the nearest hundredth.

(c) Use the solution from part (b) to determine intervals where f increases and where it decreases.

76. $f(x) = -3x + 6$

77. $f(x) = x^2 - 6x + 7$

78. $f(x) = -2x^3 - 7x^2 + 4x - 2$

79. $f(x) = \frac{1}{2}x^4 - x^2$

SECTION PROJECT

The annual total **debt** in the U.S. farming sector can be modeled by

$$y_1 = f(x) = 0.36x^3 - 4.71x^2 + 42.53x + 717.04$$
$$1 \le x \le 9$$

where x represents the number of years since 1986, and $f(x)$ represents the total farm debt in billions of dollars. The annual total **assets** in the U.S. farming sector can be modeled by

$$y_2 = g(x) = -0.02x^3 + 0.88x^2 - 6.14x + 149.25$$
$$1 \le x \le 9$$

where x represents the number of years since 1986, and $f(x)$ represents the total farm assets in billions of dollars (*Source for both models:* U.S. Department of Agriculture, www.usda.gov).

(a) Evaluate $f(4)$ and $g(4)$ and interpret each.

(b) Determine $\left(\frac{f}{g}\right)(4)$ using the solutions to part (a). Interpret the result.

(c) Use your calculator to graph $y_3 = \frac{y_1}{y_2} = \left(\frac{f}{g}\right)(x)$ in the viewing window [1, 9] by [0, 10]. This quotient of f and g gives us what the U.S. Department of Agriculture calls the **farm debt/asset ratio**.

(d) From inspecting the graph, would you conclude that the debt/asset ratio generally increased, decreased, or remained constant from 1987 to 1996? Explain.

Section 11.6 Rational, Radical, and Power Functions

Thus far, the type of functions that we have studied are polynomial functions. However, many of the functions that we study in calculus are not polynomials. In this section, we study **rational functions**. These functions are made up of polynomials, but their graphs do not resemble those that we have seen. Then we will examine **radical functions**, which are functions involving square roots, cube roots, and the like, and having restricted domains and ranges. Finally, we will investigate **power functions**, which have fractions, not only whole numbers, in their exponents. Once again we will see how these functions are used in applications, and we will calculate the slope of a secant line over an interval and interpret it as an average rate of change.

Rational Functions

Expressions for **rational functions** have fractions whose numerator and denominator are polynomials. Here are some examples of rational functions:

$$f(x) = \frac{x}{x+1}, \qquad g(x) = \frac{2x^2 - 1}{x - 2}, \qquad y = \frac{x - 3}{x^2 + x + 1}$$

A function like $f(x) = \dfrac{\sqrt{x+2}}{x+5}$ is not a rational function, since the numerator is not a polynomial.

Rational Function

A **rational function** has the form

$$y = \frac{f(x)}{g(x)}$$

where f and g are polynomials and $g(x) \neq 0$.

Notice that the domain of a rational function is determined by the set of all real numbers that make the denominator not equal to zero. Since there are usually x-values that make the denominator zero, rational functions frequently have breaks in their graphs.

Example 1 Finding Vertical Asymptotes Using Tables

Consider the rational function $f(x) = \dfrac{2x}{x-3}$.

(a) Determine the domain.

(b) Complete Table 11.6.1 and describe what happens to the function values.

Table 11.6.1

x	2	2.9	2.99	2.999	2.9999	3.0001	3.001	3.01	3.1	4
$f(x)$										

Solution

(a) We see that the denominator, $x - 3$, is zero when $x = 3$. So the domain of f is the set of all real numbers except 3 or, using interval notation, $(-\infty, 3) \cup (3, \infty)$.

(b) **Table 11.6.2**

x	2	2.9	2.99	2.999	2.9999	3.0001	3.001	3.01	3.1	4
$f(x)$	−4	−58	−598	−5998	−59,998	60,002	6002	602	62	8

Table 11.6.2 shows that for x-values very close but less than 3, the function values get negatively large or, symbolically, $f(x) \to -\infty$. Also, for x-values very close but greater than 3, the function values grow positively large, or $f(x) \to \infty$. Since x cannot equal 3, the graph of $f(x) = \dfrac{2x}{x-3}$ will never cross the vertical line $x = 3$. This line is called a **vertical asymptote**. A graph of f is given in Figure 11.6.1.

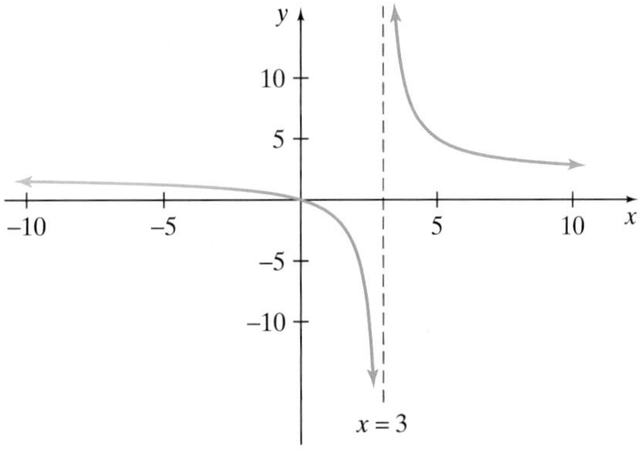

Figure 11.6.1 Graph of $f(x) = \dfrac{2x}{x-3}$. ■

Technology Option

Most graphing calculators graph functions in either DOT mode or CONNECTED mode. Figure 11.6.2a shows the graph of the function in Example 1 when the calculator is in DOT mode. Figure 11.6.2b shows the graph of the function in Example 1 when the calculator is in CONNECTED mode.

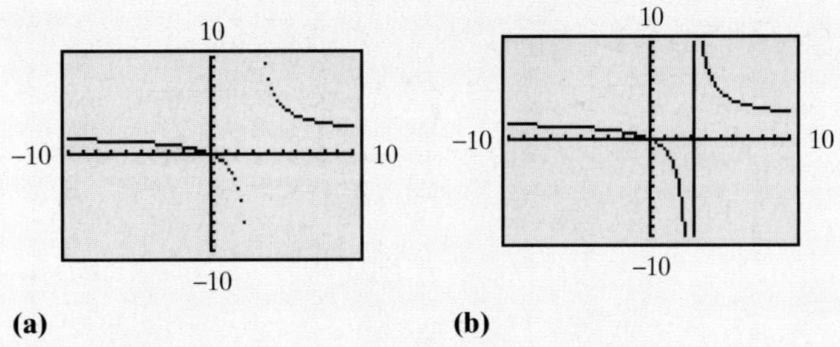

Figure 11.6.2

The graph in Figure 11.6.2b looks as if the vertical asymptote has been drawn, however this is not the case. It is a limitation of the calculator when it is in CONNECTED mode. To learn how to have your calculator graph in DOT or CONNECTED mode, consult the online graphing calculator manual at www.prenhall.com/armstrong.

Making numerical tables is a sound, yet time-consuming, way to determine vertical asymptotes. Algebraically, we see that for the function $f(x) = \frac{2x}{x-3}$, the value $x = 3$ makes the denominator equal to zero and yet makes the numerator a nonzero real number (in this case $2(3) = 6$). This is the common procedure for identifying vertical asymptotes.

Finding Vertical Asymptotes

Consider the rational function $h(x) = \frac{f(x)}{g(x)}$, where f and g are polynomials. If there is a value c that makes the denominator zero, yet the numerator is not zero, then the vertical line $x = c$ is a vertical asymptote.

The graphs of some rational functions have more than one vertical asymptote, as illustrated in Example 2.

Example 2 Determining Vertical Asymptotes

Find the vertical asymptotes of the graph of the function $f(x) = \frac{x-4}{x^2+x-2}$ and graph.

Solution

We start by factoring the denominator. This gives us

$$f(x) = \frac{x-4}{x^2+x-2} = \frac{x-4}{(x+2)(x-1)}$$

We see that the denominator of the rational function is zero when $x = -2$ and $x = 1$. Since neither of these values makes the numerator equal to zero, the graph of $f(x) = \frac{x-4}{x^2+x-2}$ has vertical asymptotes at the vertical lines $x = -2$ and $x = 1$. See Figure 11.6.3.

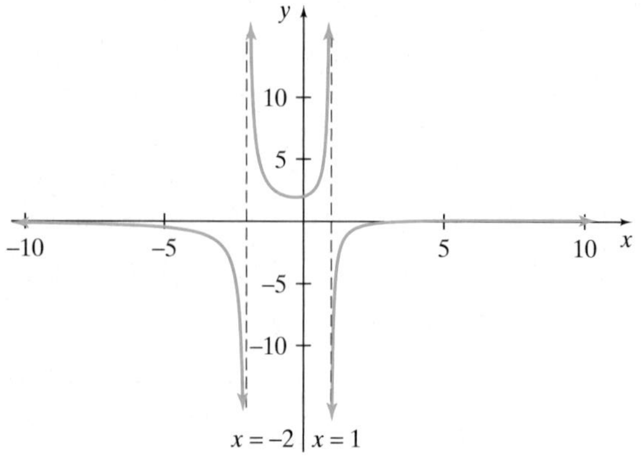

Figure 11.6.3 Graph of $f(x) = \frac{x-4}{x^2+x-2}$. ■

Interactive Activity

Sometimes the calculator does not display the hole clearly. To see this shortcoming, graph the function $f(x) = \frac{x^2 - 4}{x - 2}$ in the standard viewing window. Do you see a hole? Now use the VALUE command of your calculator to evaluate $f(2)$. What is the output? Regraph the function in the viewing window $[-9.4, 9.4]$ by $[-6.2, 6.2]$. Now do you see a hole?

WWW

We stated that vertical asymptotes are identified by values that make the denominator zero, yet the numerator is not zero. But what if a value makes both the numerator and denominator zero? Let's say we have the function $f(x) = \frac{x^2 - 4}{x - 2}$. We see that the value $x = 2$ is excluded from the domain since this number makes the denominator zero. Since the factored form of the function is $f(x) = \frac{(x+2)(x-2)}{x-2}$, we can algebraically reduce the numerator and denominator by $(x - 2)$ and rewrite the function as

$$f(x) = x + 2 \qquad x \neq 2$$

The result is a line, $f(x) = x + 2$, with a "hole" in the graph when $x = 2$. The graph of f is given in Figure 11.6.4.

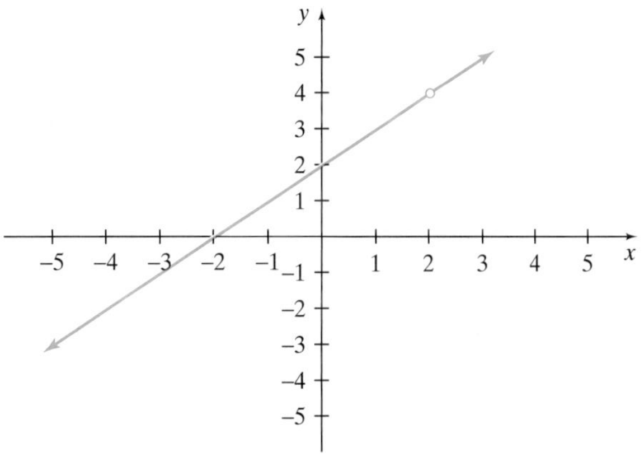

Figure 11.6.4 Graph of $f(x) = \frac{x^2 - 4}{x - 2}$ with a hole when $x = 2$.

Checkpoint 1

Now work Exercise 3.

End Behavior of Rational Functions

Table 11.6.3

x	$f(x)$
$-100,000$	1.99994 ...
$-10,000$	1.9994 ...
$-1,000$	1.994 ...
-100	1.94 ...
0	0
100	2.061855 ...
1,000	2.006018 ...
10,000	2.000600 ...
100,000	2.000060 ...

In Section 11.5, we said that the end behavior of a polynomial function describes the type of numerical values that a function takes on for extreme values of the independent variable x. To see how end behavior works for rational functions, we return to the rational function $f(x) = \frac{2x}{x - 3}$. The numerical output for extreme x-values is shown in Table 11.6.3. The ... at the end of the number in the second column means that the decimal values continue on, but are not written.

From Table 11.6.3, it appears that as x gets very large positively and very large negatively, the function values are getting close to 2. Figure 11.6.5 shows a graph of $f(x) = \frac{2x}{x - 3}$, and we notice that $y = 2$ is a **horizontal asymptote** for the graph.

Horizontal asymptotes can be determined quickly by inspecting the degree of the numerator and the degree of the denominator. Notice that both the numerator and denominator of f have a degree of 1, and the ratio of the leading coefficients is $\frac{2}{1} = 2$. It is not a coincidence that this ratio is the horizontal

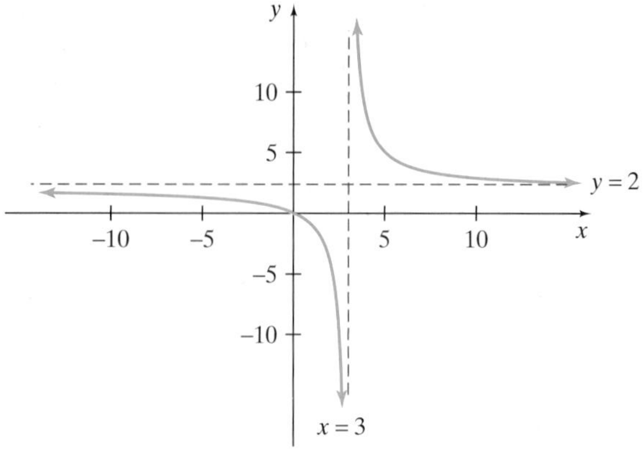

Figure 11.6.5 Graph of $f(x) = \dfrac{2x}{x-3}$ has a vertical asymptote at $x = 3$ and a horizontal asymptote at $y = 2$.

asymptote, as we state in the following. (A more rigorous analysis of horizontal asymptotes is presented in Chapter 12.)

Finding Horizontal Asymptotes

Consider the rational function $h(x) = \dfrac{f(x)}{g(x)}$, where f and g are polynomials.

1. If the degree of the numerator is largest, then the graph of the rational function has no horizontal asymptote.
2. If the degree of the denominator is largest, then the graph of the rational function has a horizontal asymptote at the line $y = 0$ (the x-axis).
3. If the degrees of the numerator and denominator are the same, then the horizontal asymptote is determined by the fraction made up of the coefficients of the highest-degree terms.

Note: The graph of a rational function can have at most one horizontal asymptote.

Example 3 Determining Horizontal Asymptotes

Find the horizontal asymptote of the graph of $f(x) = \dfrac{2x^2 - 5x - 3}{x^2 - 16}$.

Solution

For the function $f(x) = \dfrac{2x^2 - 5x - 3}{x^2 - 16}$, the numerator and denominator are both quadratic, so the horizontal asymptote is given by the coefficients of the highest-degree terms, $y = \dfrac{2}{1} = 2$.

Checkpoint 2 Now work Exercise 13.

Another key feature of the graphs of rational functions are the intercepts. The process for determining intercepts is the same as we have shown for other functions, as illustrated in Example 4.

Example 4 Determining the Intercepts of the Graph of a Rational Function

Determine the x- and y-intercepts of the graph of $f(x) = \dfrac{2x^2 - 5x - 3}{x^2 - 16}$.

Solution

To find the y-intercept, we substitute 0 for x. This yields

$$f(0) = \frac{2(0)^2 - 5(0) - 3}{(0)^2 - 16} = \frac{-3}{-16} = \frac{3}{16}$$

So the y-intercept is at the point $\left(0, \dfrac{3}{16}\right)$.

To find the x-intercept, we substitute 0 for y [or here $f(x)$] and solve for x. This gives

$$0 = \frac{2x^2 - 5x - 3}{x^2 - 16}$$

Factoring gives us

$$0 = \frac{(2x+1)(x-3)}{(x+4)(x-4)}$$

Now the only time a fraction equals 0 is when the numerator equals 0. Thus,

$$0 = (2x+1)(x-3)$$

for $x = -\dfrac{1}{2}$ and $x = 3$. (It is important to check that neither of these x-values makes the denominator equal 0.) We conclude that the x-intercepts occur at $\left(-\dfrac{1}{2}, 0\right)$ and $(3, 0)$. ■

Now that we know how to find vertical asymptotes, horizontal asymptotes, and intercepts, let's put it all together and sketch a graph of a rational function by hand.

Example 5 Sketching a Rational Function

Sketch a graph of $f(x) = \dfrac{2x^2 - 5x - 3}{x^2 - 16}$. Label all asymptotes and intercepts.

Solution

The factored form of $f(x)$ is

$$f(x) = \frac{2x^2 - 5x - 3}{x^2 - 16} = \frac{(2x+1)(x-3)}{(x+4)(x-4)}$$

Vertical asymptotes are at $x = -4$ and $x = 4$, while the horizontal asymptote is at $y = 2$. In Example 4, we determined that the y-intercept is at $\left(0, \dfrac{3}{16}\right)$ and the

x-intercepts are at $\left(-\frac{1}{2}, 0\right)$ and $(3, 0)$. By plotting a couple of other points, $f(-5) = 8$ and $f(5) = \frac{22}{9}$, we get the graph in Figure 11.6.6.

Interactive Activity

Use your calculator to graphically verify the sketch of the rational function $f(x) = \dfrac{2x^2 - 5x - 3}{x^2 - 16}$ in Example 5.

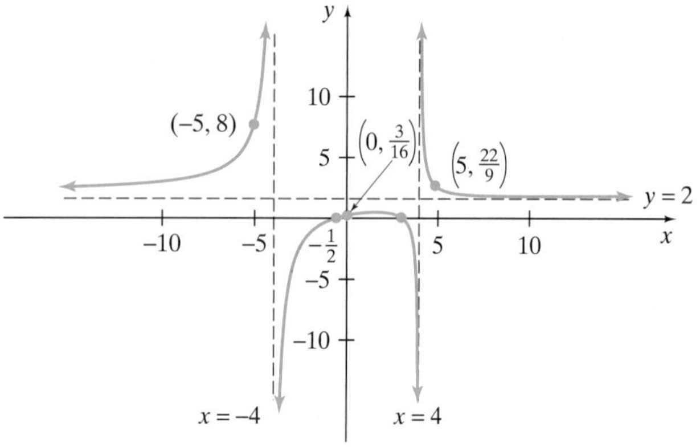

Figure 11.6.6 Graph of $f(x) = \dfrac{2x^2 - 5x - 3}{x^2 - 16}$. Vertical asymptotes at $x = -4$ and $x = 4$. Horizontal asymptote at $y = 2$. ■

The intercepts of a rational function are often used in applications.

Example 6 Applying Rational Functions in Economics

During the 1980s, the controversial economist Arthur Laffer promoted the idea of supply-side economics, which later gained the moniker "trickle-down theory." This theory centered on the **Laffer curve**. According to this curve, an increase in the tax rate can actually produce a reduction in government revenue. Suppose that a supply-side economist develops a model of the Laffer curve based on the rational function

$$f(x) = \frac{80x - 8000}{x - 110} \qquad 30 \le x \le 100$$

where x represents the tax rate percentage and $f(x)$ represents the government tax revenue in tens of billions of dollars. See Figure 11.6.7.

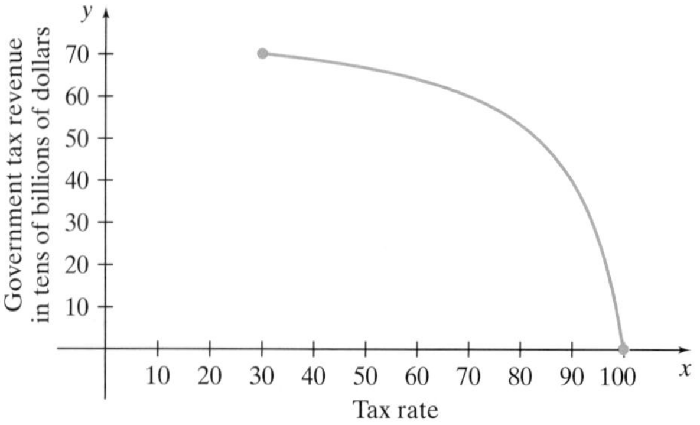

Figure 11.6.7 Graph of Laffer Curve given by the function $f(x) = \dfrac{80x - 8000}{x - 110}$.

(a) Evaluate $f(45)$ and interpret.

(b) Find the x-intercept and interpret.

(c) Find the average rate of change for $x = 45$ and $\Delta x = 30$ and interpret.

Solution

(a) **Understand the Situation:** To evaluate $f(45)$ requires that we substitute 45 for every occurrence of x. We also note that $x = 45$ corresponds to a tax rate percentage of 45%.

Evaluating the model at $x = 45$ yields

$$f(45) = \frac{80(45) - 8000}{45 - 110} = \frac{-4400}{-65} \approx 67.69$$

Interpret the Solution: This means that at a tax rate of 45%, the government tax revenue generated will be 67.69 times 10 billion dollars, or \$676.9 billion.

(b) Setting $f(x) = 0$ and solving gives us

$$\begin{aligned} \frac{80x - 8000}{x - 110} &= 0 \\ 80x - 8000 &= 0 \\ 80x &= 8000 \\ x &= 100 \end{aligned}$$

Interpret the Solution: This means that at a tax rate of 100%, zero dollars in tax revenue is generated. This is a common agreement among economic theorists, since no one will earn income if all of it would be taken by the government.

(c) **Understand the Situation:** Recall that an average rate of change over an interval is given by the slope of the secant line over the interval. So we utilize the difference quotient to determine the slope of the secant line.

$$\begin{aligned} m_{\text{sec}} &= \frac{f(45+30) - f(45)}{30} = \frac{f(75) - f(45)}{30} \\ &= \frac{\left[\frac{80(75) - 8000}{75 - 110}\right] - \left[\frac{80(45) - 8000}{45 - 110}\right]}{30} \\ &\approx \frac{57.14 - 67.69}{30} = \frac{-10.55}{30} \approx -0.35 \end{aligned}$$

Interpret the Solution: This means that as the tax rate increases from 45% to 75%, the tax revenue will **decrease** at an average rate of $3.5 \dfrac{\text{billion dollars}}{\text{percent}}$. ■

> **From Your Toolbox**
>
> A **rational exponent** has the form $\frac{a}{b}$, where a and b are integers, and has the property
>
> $$x^{a/b} = \sqrt[b]{x^a} = (\sqrt[b]{x})^a$$
>
> where $\sqrt{\ }$ is the **radical sign**, x is the **radicand**, and b is the **root index**.

Radical Functions

To begin to understand radical functions, let's recall a definition from algebra as shown in the Toolbox to the left.

Notice that the Toolbox implies that any function written in radical form can be written as a function with a rational exponent, and vice versa. Knowing how to rewrite a function between radical and rational exponent form is essential for success in calculus. The following definition relates these two families of functions.

Radical/Rational Exponent Functions

A function of the form $f(x) = \sqrt[b]{[g(x)]^a}$ is called a **radical function**, where g is called the **radicand.** Rewriting $f(x) = \sqrt[b]{[g(x)]^a}$ as $f(x) = [g(x)]^{a/b}$ produces what we call the **rational exponent function.**

Many times, $g(x)$ is a polynomial expression like $5x - 1$ or $8 - x$. For example, for the function $g(x) = x^{2/3}$, the root index is 3, so we can rewrite the function in radical form as $g(x) = \sqrt[3]{x^2}$. Similarly, for the function $f(x) = (2x + 3)^{1/2}$, the function has a rational exponent of $\frac{1}{2}$, so we can rewrite f as $f(x) = \sqrt{2x + 3}$. This kind of function is called a **square root function.**

Checkpoint 3

Now work Exercise 51.

The domains for radical/rational exponent functions are determined by the value of the root index. Let's say that we have the function $f(x) = \sqrt[3]{x - 1}$. Since we can take the cube root of any real number, the domain of this function is the set of real numbers $(-\infty, \infty)$. This can be verified visually in Figure 11.6.8.

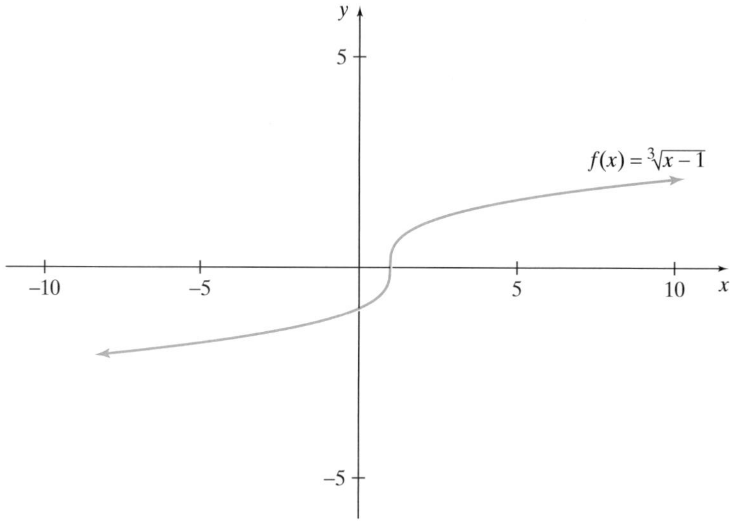

Figure 11.6.8

If the function is a square root function, we have to be more cautious. Recall from Section 11.1 that we found that the domain of $g(x) = \sqrt{2x - 5}$ is $\left[\frac{5}{2}, \infty\right)$ by determining the x-values that make $2x - 5$ greater than or equal to zero. Thus, the domain of a radical function depends on whether the root index is even or odd.

Domain of a Radical/Rational Exponent Function

1. The domain of a function of the form $f(x) = \sqrt[b]{[g(x)]^a}$ or $f(x) = [g(x)]^{a/b}$ is all real numbers $(-\infty, \infty)$ if b is an odd number.
2. The domain is determined by solving $g(x) \geq 0$ if b is even.

In practice, many of the radical functions that we study are square root functions. Let's take a look at an application of this class of radical functions.

Example 7 Using the Square Root Function

A sapling of a certain kind of tree grows according to the mathematical model

$$g(x) = 10\sqrt{x} + 0.75$$

where x represents the number of years since the sapling has been planted, and $g(x)$ represents the height of the tree in feet after x years.

(a) How tall will the sapling be 20 years after being planted? Round the solution to the nearest hundredth of a foot.

(b) How long will it take for the tree to be 30 feet tall?

Solution

Interactive Activity

Graphically verify the solution to part (b) of Example 7 by letting $y_1 = 10\sqrt{x} + 0.75$ and $y_2 = 30$, then determining the point of intersection.

WWW

(a) **Understand the Situation:** To determine the height of the tree after 20 years, we need to evaluate $g(20)$.

This gives us

$$g(20) = 10\sqrt{20} + 0.75 \approx 45.47$$

Interpret the Solution: So, according to the model, the tree will be about 45.47 feet tall in 20 years.

(b) **Understand the Situation:** In this case, we have to find the x-value(s) so that the $g(x)$-value is 30. That is, we need to find x such that $g(x) = 30$.

Substituting 30 for $g(x)$ and solving for x gives us

$$10\sqrt{x} + 0.75 = 30$$
$$10\sqrt{x} = 29.25$$
$$\sqrt{x} = 2.925$$
$$x = (2.925)^2 \approx 8.56$$

Interpret the Solution: So, according to the model, the tree will be 30 feet tall in about 8.56 years. ■

Power Functions

Some rational exponent functions are easier to study mathematically when the exponents are written as decimals, such as $f(x) = x^{0.145}$. We will call this class of functions the **power functions**.

Power Function

A function of the form

$$f(x) = a \cdot x^b$$

is called a **power function**, where a and b are real numbers.

The power function has various properties based on the values of a and b. We will study these properties in two cases. If a is a positive number, then the

Table 11.6.4

x	$f(x)$	$g(x)$
0	0	0
1	1	1
2	2.83	1.62
3	5.20	2.16
4	8	2.64
5	11.18	3.09
6	14.70	3.51
7	18.52	3.90
8	22.63	4.29
9	27	4.66
10	31.62	5.01

Some values are rounded to the nearest hundredth.

power function appears much like the polynomial functions that we studied in Section 11.5. Much of the shape of the graph of these functions is determined by whether b is between zero and 1 or b is greater than 1.

For example, consider the table of values and graphs of the power functions $f(x) = x^{1.5}$ and $g(x) = x^{0.7}$, where f and g are defined for $x \geq 0$ (Table 11.6.4). The graph (Figure 11.6.9) supports the numerical notion that f increases more rapidly than g. Finally, we observe that both functions are increasing.

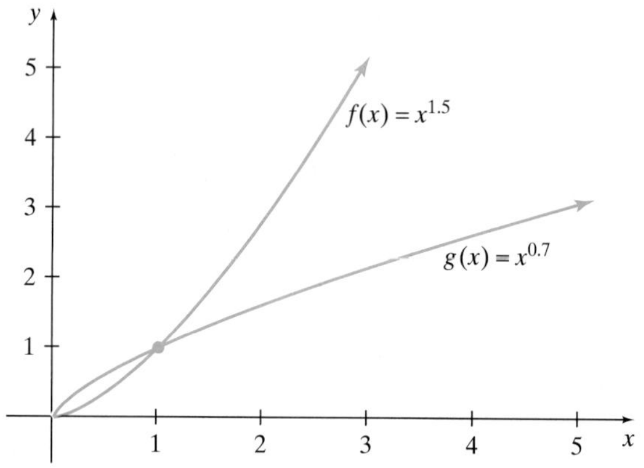

Figure 11.6.9

Positive Exponent Power Functions

For the power function of the form $f(x) = a \cdot x^b$ with $a > 0$, we have

If the value of b is	The graph	Diagram
$0 < b < 1$	Opens down	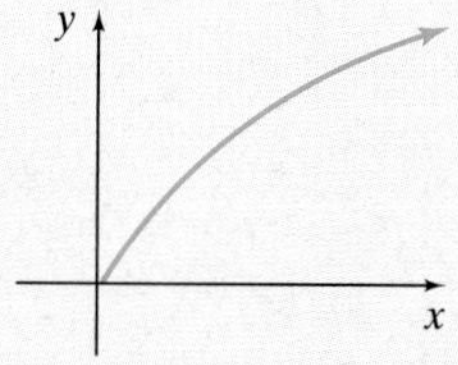
$b > 1$	Opens up	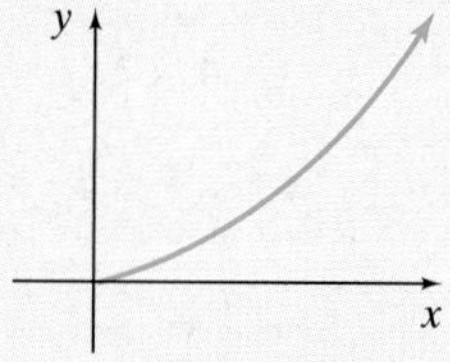

Example 8 Using Positive Exponent Power Functions in Applications

The median sale price of new single-family homes in the U.S. Midwest can be modeled by

$$f(x) = 63.25\sqrt[4]{x} \quad \text{or} \quad f(x) = 63.25x^{0.25} \qquad 1 \leq x \leq 14$$

where x represents the number of years since 1981, and $f(x)$ represents the sale price in thousands of dollars (*Source:* U.S. Census Bureau, www.census.gov).

(a) Evaluate $f(11)$ and interpret.

(b) Determine the slope of the secant line of the graph of f for $x = 1$ and $\Delta x = 10$ and interpret.

Solution

(a) **Understand the Situation:** First, note that $x = 11$ corresponds to the year $1981 + 11 = 1992$.

Evaluating the model at $x = 11$, we get

$$f(11) = 63.25(11)^{0.25} \approx 63.25(1.82116) \approx 115.188$$

Interpret the Solution: This means that in 1992 the median sale price of a new single-family house in the U.S. Midwest was about \$115.188 thousand or \$115,188.

(b) Evaluating the secant line slope, we get

$$m_{\text{sec}} = \frac{f(x + \Delta x) - f(x)}{\Delta x} = \frac{f(11) - f(1)}{10} \approx \frac{115.188 - 63.25}{10} = 5.1938$$

Interpret the Solution: This means that between the years 1982 and 1992 the median sale price of a new single-family house in the Midwest increased at an average rate of about \$5193.80 per year. ■

> **Interactive Activity**
>
> Contrast the graphs of $f(x) = 3 \cdot x^{2.5}$ and $g(x) = -3 \cdot x^{2.5}$. What do you think you can generally say about the relationship of the graphs of the functions $f(x) = a \cdot x^b$ and $g(x) = -a \cdot x^b$?
>
> WWW

In Example 8, note that when we evaluated $63.25(11)^{0.25}$ we raised 11 to the 0.25 power first and *then* multiplied by 63.25. Remember from the order of operations in algebra: exponentiation first, then multiplication.

Now let's examine power functions for which the exponent is a negative number. The domain used for these functions is $(0, \infty)$ and a is a positive number. Consider the power functions $f(x) = x^{-1.8}$ and $g(x) = x^{-0.2}$, where f and g are defined for $x > 0$. Rewriting with positive exponents, we get $f(x) = x^{-1.8} = \dfrac{1}{x^{1.8}}$ and $g(x) = x^{-0.2} = \dfrac{1}{x^{0.2}}$. Since $0.2 = \dfrac{1}{5}$, we could write g as $g(x) = \dfrac{1}{\sqrt[5]{x}}$. Notice that when $x = 0$ the denominators of f and g are zero. This is why zero is not part of the domain of these functions. The graphs of $f(x) = x^{-1.8}$ and $g(x) = x^{-0.2}$ are shown in Figure 11.6.10. Note that since $x = 0$ is not in the domain, the graphs of f and g have the y-axis as a vertical asymptote. We can see from the graph that both functions are decreasing.

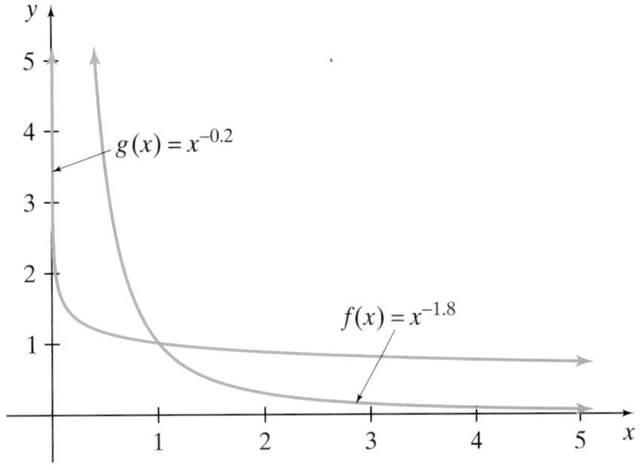

Figure 11.6.10

So when a power function has a negative exponent, the domain is the set of all positive real numbers. The functions are decreasing and their graphs have vertical asymptotes. This kind of power function is frequently used as a model for a decreasing phenomenon.

SUMMARY

In this section, we discussed the rational functions, which included finding the intercepts and asymptotes of their graphs. Then we examined radical functions and saw the relationship between radical and rational exponent functions. Finally, we examined the power functions. The investigation of all these functions is key in our understanding of calculus.

The types of functions we discussed in this section include:

- **Rational function**: $h(x) = \dfrac{f(x)}{g(x)}$, f and g are polynomials, $g(x) \neq 0$
- **Radical/rational exponent function**: $f(x) = \sqrt[b]{[g(x)]^a}$ or $f(x) = [g(x)]^{a/b}$
- **Square root function**: $f(x) = a\sqrt{x} + b$
- **Power function**: $f(x) = a \cdot x^b$

Vertical Asymptote Rule for the Rational Function $\dfrac{f(x)}{g(x)}$

For $h(x) = \dfrac{f(x)}{g(x)}$, if there is a value c so that $g(c) = 0$, yet $f(c) \neq 0$, then $x = c$ is a vertical asymptote.

Horizontal Asymptote Rule for the Rational Function $\dfrac{f(x)}{g(x)}$

- If the degree of f is largest, then there is no horizontal asymptote.
- If the degree of g is largest, then the x-axis is the horizontal asymptote.
- If the degrees of f and g are the same, then the horizontal asymptote is the reduced form of the leading terms of f and g.

SECTION 11.6 EXERCISES

For Exercises 1–10,

(a) Write the domain using interval notation.

(b) Identify any holes or vertical asymptotes in the graph.

1. $f(x) = \dfrac{12x^3}{3x}$ **2.** $f(x) = \dfrac{15x^2}{25x^3}$

✓ **3.** $f(x) = \dfrac{3x - 12}{x - 2}$ **4.** $f(x) = \dfrac{2x - 16}{x - 5}$

5. $f(x) = \dfrac{3x + 6}{x + 2}$ **6.** $g(x) = \dfrac{x^2 - 1}{x + 1}$

7. $g(x) = \dfrac{x^2 + 2x + 1}{x^2 + 4x + 3}$ **8.** $g(x) = \dfrac{6x^2 + x - 2}{8x^2 + 2x - 3}$

9. $f(x) = \dfrac{4x^2 + 8x + 3}{2x^2 - x - 6}$ **10.** $f(x) = \dfrac{x^2 + x - 30}{x^2 - x - 20}$

For Exercises 11–20, identify any horizontal asymptotes for the following.

11. The function in Exercise 1

12. The function in Exercise 2

✓ **13.** The function in Exercise 3

14. The function in Exercise 4

15. The function in Exercise 5

16. The function in Exercise 6

17. The function in Exercise 7

18. The function in Exercise 8

19. The function in Exercise 9

20. The function in Exercise 10

For Exercises 21–34, determine the x- and y-intercepts.

21. $g(x) = \dfrac{x}{2x + 6}$ **22.** $f(x) = \dfrac{x - 1}{3x + 9}$

23. $f(x) = \dfrac{6x - 3}{2x + 4}$ **24.** $g(x) = \dfrac{7x + 4}{3x - 1}$

25. $f(x) = \dfrac{x^2 - x - 6}{x^2 - 7x + 12}$ **26.** $f(x) = \dfrac{x^2 - 5x + 4}{x^2 + 2x - 3}$

27. $g(x) = \dfrac{x - 2}{x^2 + 6x + 9}$ **28.** $g(x) = \dfrac{x^2 + 2x}{3x^2 + 18x + 24}$

29. $f(x) = \dfrac{x^2 - 4}{x^2 + 2x - 3}$ **30.** $g(x) = \dfrac{4x^2 - 9}{x^2 + 3x + 4}$

For Exercises 31–36, determine the x- and y-intercepts graphically.

31. $f(x) = \dfrac{x^2 - 2x + 5}{x + 1}$ **32.** $f(x) = \dfrac{x^2 - 2x + 1}{x - 2}$

33. $f(x) = \dfrac{2x^3 - x^2 + 3x - 2}{x^3 + 3}$ **34.** $f(x) = \dfrac{x^3 - 2x + 1}{x - 1}$

35. $f(x) = \dfrac{2x^3 - x^2 + 1}{2 - x}$ **36.** $f(x) = \dfrac{3x^2 - x + 5}{x^2 - 4}$

In Exercises 37–42, sketch a graph by hand for each function on the interval $1 \le x \le 10$. Label any asymptotes and intercepts.

37. $f(x) = \dfrac{x}{x+2}$ **38.** $f(x) = \dfrac{2x}{x-4}$

39. $f(x) = \dfrac{x+4}{x-4}$ **40.** $f(x) = \dfrac{x-2}{x+2}$

41. $f(x) = \dfrac{6-3x}{x-6}$ **42.** $f(x) = \dfrac{4-4x}{x-2}$

Applications

43. Environmental scientists and municipal planners often are guided by **cost–benefit models**. These mathematical models estimate the cost of removing a pollutant from the atmosphere as a function of the percentage of pollutant removed. Let's suppose that a cost-benefit function is given by

$$f(x) = \frac{30x}{100 - x} \qquad 0 \le x < 100$$

where x represents the percentage of the pollutant removed and $f(x)$ represents the associated cost in millions of dollars.

(a) Evaluate $f(85)$ and interpret.

(b) Complete the following table.

x	5	50	70	90	95
$f(x)$					

(c) Many cost–benefit functions exhibit the very high cost of removing the final percentage of a pollutant. To calculate this behavior, evaluate $f(99.9) - f(95)$. This represents the approximate cost of removing the final 5% of the pollutant.

(d) Why can we not compute $f(100) - f(95)$ to get the *actual* cost of removing the final 5% of the pollutant?

44. The cost–benefit function for removing a certain pollutant from the atmosphere is given by

$$f(x) = \frac{20x}{100 - x} \qquad 0 \le x < 100$$

where x represents the percentage of the pollutant removed and $f(x)$ represents the associated cost in millions of dollars.

(a) Evaluate $f(70)$ and interpret.

(b) Complete the following table.

x	5	50	70	90	95
$f(x)$					

(c) Evaluate $f(99.9) - f(95)$. This represents the approximate cost of removing the final 5% of the pollutant.

(d) Why can we not compute $f(100) - f(95)$ to get the *actual* cost of removing the final 5% of the pollutant?

45. The students in an anatomy class were asked to memorize a list of 20 parts of the human body. After each class, a student was chosen and asked to write as many of these anatomical parts as she could. The average number of parts remembered is given by the mathematical model

$$f(x) = \frac{20x - 18}{x} \qquad x \ge 4$$

where x represents the number of days since the list was distributed and $f(x)$ represents the average number of items that were remembered.

(a) Evaluate $f(10)$ and interpret.

(b) Evaluate $f(100)$ and interpret.

46. The technicians at the Arp Brothers auto parts factory are given a checklist of 40 items to inspect for quality control. Over the next three months, a technician is selected and asked to write the checklist from memory. The mathematical model for the technician's performance is given by

$$f(x) = \frac{80x - 36}{2x} \qquad 1 \le x \le 120$$

where x represents the number of days since the checklist was distributed and $f(x)$ represents the average number of inspection items that was remembered.

(a) Evaluate $f(2)$ and $f(30)$ and interpret each.

(b) Find the horizontal asymptote and interpret its meaning.

(c) The State Regulatory Agency requires technicians to remember a minimum of 25 items after three months. According to the model, will the technicians meet this requirement?

47. The Slaybaugh Satellite Company is manufacturing a new low-cost satellite dish and promotes its sales through an aggressive sales campaign. The income from sales is given by the sales function

$$S(x) = \frac{120x^2 - 600x + 3}{2x^2 - 10x + 1} \qquad x \ge 5$$

where x represents the number of dollars spent on advertising in thousands of dollars and $S(x)$ represents the income from sales measured in hundreds of thousands of dollars.

(a) Evaluate $S(10)$ and interpret.

(b) Find the horizontal asymptote and interpret its meaning.

For Exercises 48–55, rewrite the radical functions as rational exponent functions. Also, write the domain using interval notation.

48. $f(x) = \sqrt[4]{x-1}$ **49.** $f(x) = \sqrt{2x+3}$

50. $f(x) = \sqrt[3]{4-x}$ **51.** $f(x) = \sqrt[3]{5x-8}$

52. $g(x) = \sqrt{(x-4)^3}$

53. $g(x) = \sqrt[4]{(6x+1)^5}$

54. $f(x) = \dfrac{1}{\sqrt{3x+4}}$

55. $f(x) = \dfrac{3}{\sqrt{7x-2}}$

For Exercises 56–63, rewrite the rational exponent functions as radical functions. Also, write the domain using interval notation.

56. $f(x) = (4x+3)^{1/2}$

57. $f(x) = (8x-9)^{1/2}$

58. $f(x) = (3x+1)^{1/3}$

59. $f(x) = (7x+9)^{1/3}$

60. $y = (6x-1)^{3/2}$

61. $f(x) = (4x-5)^{3/2}$

62. $g(x) = (6x+3)^{-1/2}$

63. $g(x) = (7x-2)^{-1/3}$

64. A slightly banked highway corner will safely accommodate traffic at a speed given by the model

$$f(x) = \frac{29}{20}\sqrt{x}$$

where x represents the radius of the corner in feet and $f(x)$ represents the speed at which the traffic can travel safely in miles per hour.

(a) Evaluate $f(20)$ and interpret.

(b) If the highway planners expect the traffic to travel at a speed of 64 miles per hour, what radius should the corner be?

65. In a forest fire tower, the distance that an observer is able to see into the forest is related to the height of the fire tower via the function

$$g(x) = \frac{7}{5}\sqrt{x}$$

where x represents the height of the tower in feet and $g(x)$ represents the distance that the observer can see in miles.

(a) Evaluate $g(70)$ and interpret.

(b) If the observer is required to see 29 miles into the forest, how high must the tower be?

66. The distance that a person can see from an airplane to the horizon on a cloudless day is given by

$$g(x) = \frac{61}{50}\sqrt{x}$$

where x represents the altitude of the plane in feet and $g(x)$ is the distance that a person can see in miles.

(a) Evaluate $g(32{,}000)$ and interpret.

(b) Find the slope of the secant line for $x = 36{,}000$ and $\Delta x = 4000$ and interpret.

67. A biologist has shown that the number of plant species in the South American rainforest is related to the area of the land studied via the model

$$g(x) = 28.1\sqrt[3]{x}$$

where x represents the area studied in square miles and $g(x)$ represents the number of plant species.

(a) Evaluate $g(300)$ and interpret.

(b) Find the slope of the secant line for $x = 1728$ and $\Delta x = 469$ and interpret.

68. A city planner has projected that the population of a newly developed suburb will grow for the next four years according to the mathematical model

$$f(x) = 10{,}000 + 20\sqrt{x^3} + 30x \qquad 0 \le x \le 48$$

where x is the number of months from the present and $f(x)$ is the suburb's population.

(a) Evaluate $f(12)$ and interpret.

(b) Find the slope of the secant line for $x = 12$ and $\Delta x = 24$ and interpret.

69. The number of grocery stores in the United States can be modeled by the power function

$$f(x) = 17{,}311x^{-0.028} \qquad 1 \le x \le 6$$

where x represents the number of years since 1989 and $f(x)$ represents the number of grocery stores in the United States (*Source:* U.S. Statistical Abstract, www.census.gov/statab/www).

(a) Evaluate $f(3)$ and interpret.

(b) Evaluate $\dfrac{f(6)-f(3)}{6-3}$ and interpret this result.

70. The amount of lead emitted by the United States into the atmosphere as air pollution can be modeled by

$$f(x) = \frac{13.97}{\sqrt{x}} \qquad 1 \le x \le 10$$

where x represents the number of years since 1984 and $f(x)$ is the number of tons of lead emitted (*Source:* U.S. Environmental Protection Agency, www.epa.gov).

(a) Evaluate $f(9)$ and interpret.

(b) Compute the average rate of change of the model for $x = 1$ and $\Delta x = 8$ and interpret.

71. The number of retired workers in the United States receiving Social Security benefits can be modeled by

$$f(x) = 19.14x^{0.112} \qquad 1 \le x \le 20$$

where x represents the number of years since 1979 and $f(x)$ represents the number of retired workers, in millions, receiving Social Security benefits (*Source:* U.S. Census Bureau, www.census.gov).

(a) Evaluate $f(19)$ and interpret.

(b) Compute the average rate of change of the model for $x = 12$ and $\Delta x = 7$ and interpret.

72. *(continuation of Exercise 70)*

(a) Graph the model given in Exercise 70 in the viewing window [1, 10] by [0, 15]. From the graph, do you think that the amount of lead air pollution is generally increasing or decreasing? Explain.

(b) Graphically approximate the year in which the amount of lead emitted first fell below 8 tons.

73. *(continuation of Exercise 71)*

(a) Graph the model given in Exercise 71 in the viewing window [1, 20] by [19, 30]. From the graph, do you think that the number of retired workers receiving Social Security benefits is generally increasing or decreasing? Explain.

(b) Graphically approximate the year in which the number of retired workers receiving Social Security benefits first exceeded 25 million.

SECTION PROJECT

The expected lifespan of African-American males in the U.S. can be modeled by

$$f(x) = 60.95x^{0.033} \qquad 1 \le x \le 27$$

where x represents the number of years since 1969 and $f(x)$ represents the expected lifespan of African-American males born during the xth year (*Source:* U.S. National Center for Health Statistics, www.cdc.gov/nchs).

(a) Evaluate $f(20)$ and interpret.

(b) During what year did the expected lifespan of African-American males exceed 65 years?

(c) Find the average rate of change of the expected lifespan of African-American males for $x = 2$ and $\Delta x = 5$ and interpret.

The expected lifespan of African-American females in the U.S. can be modeled by

$$g(x) = 69.27x^{0.028} \qquad 1 \le x \le 27$$

where x represents the number of years since 1969 and $f(x)$ represents the expected lifespan of African-American females born during the xth year (*Source:* U.S. National Center for Health Statistics, www.cdc.gov/nchs).

(d) Evaluate $g(20)$ and interpret.

(e) During what year did the expected lifespan of African-American females exceed 75 years?

(f) Find the average rate of change of the expected lifespan of African-American females for $x = 2$ and $\Delta x = 5$ and interpret.

(g) Compare the answers of parts (c) and (f).

Section 11.7 Exponential Functions

Often business people make a proclamation like "Our sales are growing exponentially!" This may not be exactly true, but it is an effective way to communicate that the sales are growing very rapidly. In this section, we will study **general exponential functions** that typically have function values that either increase or decrease rapidly. Then we will discuss compound interest, which is a direct application of exponential functions. After introducing the irrational number ***e***, we will explore additional applications in which exponential functions arise, such as population growth, radioactive decay, and learning curves. Finally, we will introduce a special type of function called the **logistic function**.

General Exponential Functions

Exponential functions are different from the functions studied so far in that the independent variable is in the exponent. The following are examples of exponential functions:

$$f(x) = 3^x, \qquad y = \left(\frac{1}{5}\right)^x, \qquad g(x) = 2 \cdot 4^x, \qquad f(x) = 1.6 \cdot (1.09)^x$$

General Exponential Function

A **general exponential function** has the form

$$f(x) = a \cdot b^x$$

where a is a real number **constant** with $a \neq 0$, and b is a real number with $b > 0, b \neq 1$. Here b is called the **base**.

Note: Do not confuse exponential functions with power functions. For an exponential function, the base is a constant and the exponent is the independent variable. The reverse is true for power functions. For example, $f(x) = x^3$ is a power function, whereas $f(x) = 3^x$ is an exponential function.

Example 1 Graphing Exponential Functions

(a) Make a table of values for the general exponential functions $f(x) = 2^x$ and $g(x) = \left(\frac{1}{2}\right)^x$ for $x = -5, -4, -3, -2, \ldots, 4, 5$.

(b) Graph the functions f and g and write their domains using interval notation.

Interactive Activity

For Table 11.7.1, make another row of the table that calculates the **successive ratios** of the function values by dividing each value by the preceding value. For example, for the function $g(x) = \left(\frac{1}{2}\right)^x$, the ratios would begin $\frac{16}{32}, \frac{8}{16}, \ldots$. What pattern do you notice?

WWW

Solution

(a) The values for the functions are shown in Table 11.7.1.

Table 11.7.1

x	−5	−4	−3	−2	−1	0	1	2	3	4	5
$f(x) = 2^x$	$\frac{1}{32}$	$\frac{1}{16}$	$\frac{1}{8}$	$\frac{1}{4}$	$\frac{1}{2}$	1	2	4	8	16	32
$g(x) = \left(\frac{1}{2}\right)^x$	32	16	8	4	2	1	$\frac{1}{2}$	$\frac{1}{4}$	$\frac{1}{8}$	$\frac{1}{16}$	$\frac{1}{32}$

(b) From the table and the graphs in Figures 11.7.1a and b, we see that any real number can be substituted for x, so the domain of the functions $f(x) = 2^x$ and $g(x) = \left(\frac{1}{2}\right)^x$ is $(-\infty, \infty)$.

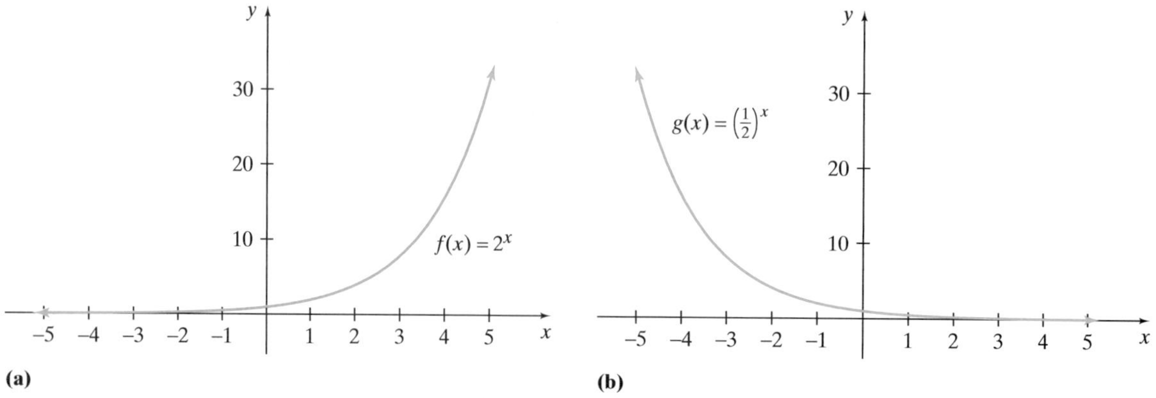

Figure 11.7.1

Notice that $f(x) = 2^x$ is increasing on its domain and as $x \to \infty$, $f(x) \to \infty$. Consequently, we call this kind of function an **exponential growth function**. Whereas, notice that $g(x) = \left(\frac{1}{2}\right)^x$ is decreasing on its domain and, as $x \to \infty$, $g(x) \to 0$. We call this kind of function an **exponential decay function**. (See Properties of General Exponential Functions at the bottom of this page.)

For the function $f(x) = a \cdot b^x$, the value of a can affect the end behavior. First, we see that with any value of a, $f(0) = a \cdot b^0 = a \cdot 1 = a$. This means that the y-intercept of the graph of $f(x) = a \cdot b^x$ is at $(0, a)$. To see the effect of a, compare the graphs of $f(x) = 2 \cdot 3^x$ and $g(x) = -2 \cdot 3^x$ in Figure 11.7.2.

From Figure 11.7.2, we see that switching the sign of a from positive to negative results in a reflection about the x-axis. Now let's see how these exponential models are used in applications.

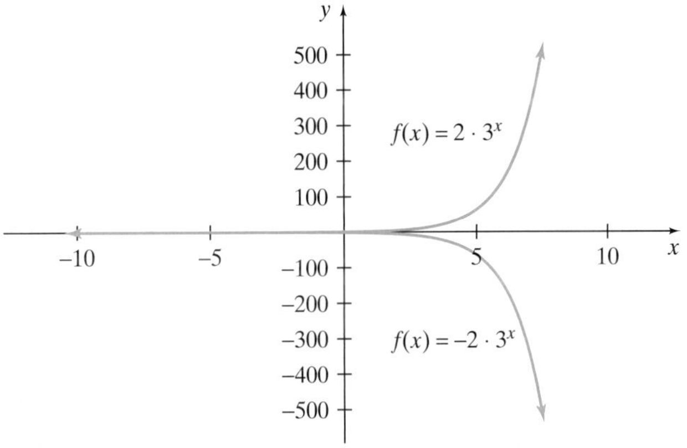

Figure 11.7.2

Properties of General Exponential Functions

- The domain is the set of all real numbers $(-\infty, \infty)$.
- The y-intercept is at $(0, 1)$.
- The behavior of the function is shown in the table.

Function Type	**Definition**	**Graph**
Exponential growth function	$f(x) = b^x, b > 1$	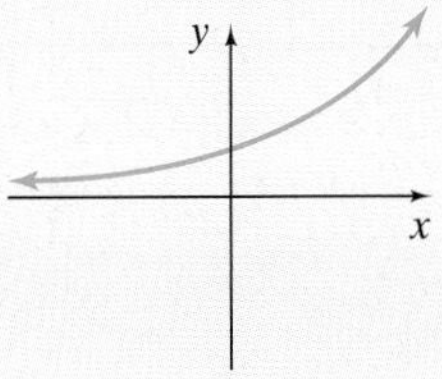
Exponential decay function	$f(x) = b^x, 0 < b < 1$	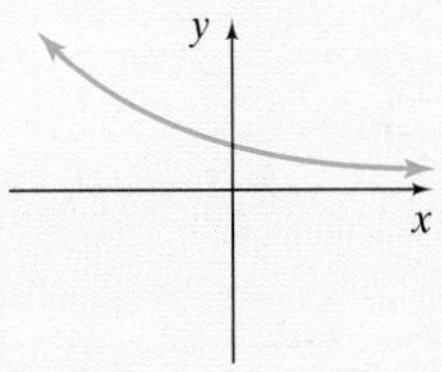

Example 2 Using General Exponential Decay Models

The number of disabled World War I veterans receiving compensation for service-connected disabilities can be modeled by

$$f(x) = 47.55(0.78)^x \qquad 1 \le x \le 16$$

where x represents the number of years since 1979 and $f(x)$ represents the number of World War I veterans receiving compensation, measured in thousands (*Source:* U.S. Census Bureau, www.census.gov).

(a) Evaluate $f(6)$ and interpret.

(b) Graph f and classify the function as an exponential growth or exponential decay function.

(c) What is the behavior of f as the x-values increase? Interpret.

Solution

(a) **Understand the Situation:** We begin by noting that $x = 6$ corresponds to the year $1979 + 6 = 1985$.

Evaluating, we get

$$f(6) = 47.55(0.78)^6 \approx 10.71$$

Interpret the Solution: This means that in 1985 there were approximately 10.71 thousand, or about 10,710 disabled World War I veterans receiving compensation.

(b) The graph of f is shown in Figure 11.7.3. For the model $f(x) = 47.55(0.78)^x$, the value of $b = 0.78$ is between zero and one. This means f is an exponential decay function. This is supported by Figure 11.7.3.

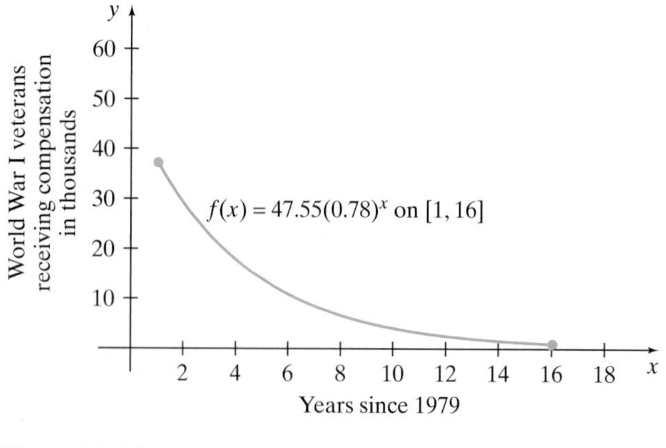

Figure 11.7.3

(c) We can see from the graph that as the x-values, representing the years, increase the function values decrease. Since x represents the number of years since 1979, this means that as we move farther away from 1979 the number of World War I veterans receiving compensation is decreasing. We would expect this trend because the number of World War I veterans has been declining since 1979. ■

Compound Interest

A common application of the general exponential function in the banking industry is *compound interest*. To understand where the compound interest formula comes from, we begin by looking at simple interest.

Simple Interest Formula

If P dollars is deposited in an account that earns interest at a rate of r (written as a decimal), then the **interest** I accumulated after time t is

$$I = Prt$$

$$\text{Interest} = \text{Principal} \cdot \text{rate} \cdot \text{time}$$

Note: See Section 5.1 for a more in depth discussion on simple interest.

Compound interest means that interest is paid not only on the money deposited, called the **principal**, but also on interest that has already accumulated. Usually, a bank computes the interest earned and deposits the interest in your account. The next time that the interest is computed the bank computes interest on the original principal **plus** previously earned interest! The amount in an account that earns compound interest is given by the **compound interest formula**.

Compound Interest Formula

If a principal of P dollars is invested in an account earning an annual interest rate of r (in decimal form) compounded k times per year, then the amount A in the account at the end of t years is given by the **compound interest formula**

$$A = P\left(1 + \frac{r}{k}\right)^{kt}$$

Example 3 Calculating Compound Interest

Suppose that \$20,000 is deposited into an account that yields an annual interest rate of 6.5% compounded quarterly.

(a) How much will be in the account after three years? Round the solution to the nearest cent.

(b) How much interest was earned at the end of the three-year period?

Solution

(a) Using the compound interest formula with $P = 20{,}000$, $r = 0.065$, $k = 4$, and $t = 3$ we get

$$A = P\left(1 + \frac{r}{k}\right)^{kt} = 20{,}000\left(1 + \frac{0.065}{4}\right)^{4(3)} = 20{,}000(1.01625)^{12} \approx 24{,}268.15$$

So there will be about \$24,268.15 in the account in three years.

(b) To determine the interest earned, we take the amount A and subtract the principal P from it. This gives us \$24,268.15 – \$20,000 = \$4268.15. So about \$4268.15 is earned in interest in three years. ■

Checkpoint 1 Now work Exercise 27.

If we hold the values of P, r, and k constant, we can think of the compound interest formula as a function of time t. Writing this gives us the **compound interest function**: $A(t) = P\left(1 + \frac{r}{k}\right)^{kt}$. Notice here that the independent variable is time, denoted with a t. For the account in Example 3, we can write the compound interest function

$$A(t) = 20{,}000\left(1 + \frac{0.065}{4}\right)^{4t}$$

To find the value of the account after five years, we evaluate $A(5)$.

$$A(5) = 20{,}000\left(1 + \frac{0.065}{4}\right)^{4(5)} = 20{,}000(1.01625)^{20} \approx 27{,}608.40$$

This means that after five years, there will be about $27,608.40 in the account.

The Number *e*

At the beginning of this section, we saw that we could write an exponential function $f(x) = a \cdot b^x$ with any base b as long as $b > 0, b \neq 1$. However, the most frequently used base is the famous number e. This number is so widely used that it is found on nearly every scientific and graphing calculator. To see what this number is, let's numerically examine the expression $\left(1 + \frac{1}{k}\right)^k$, where k is an independent variable. See Table 11.7.2.

Table 11.7.2

k	$\left(1+\frac{1}{k}\right)^k$
1	2
10	2.59374...
100	2.70481...
1,000	2.71692...
10,000	2.71815...
100,000	2.71827...
1,000,000	2.71828...

We see that as the k-values become very large the values of the expression $\left(1 + \frac{1}{k}\right)^k$ seem to get close to a single value called ***e***. Many applications in the physical, life, and social sciences use e as the base of their exponential functions. As a matter of fact, this is referred to as *the* exponential function.

The Exponential Function

The **exponential function** has the form

$$f(x) = ae^{bx}$$

where a and b are real number constants.

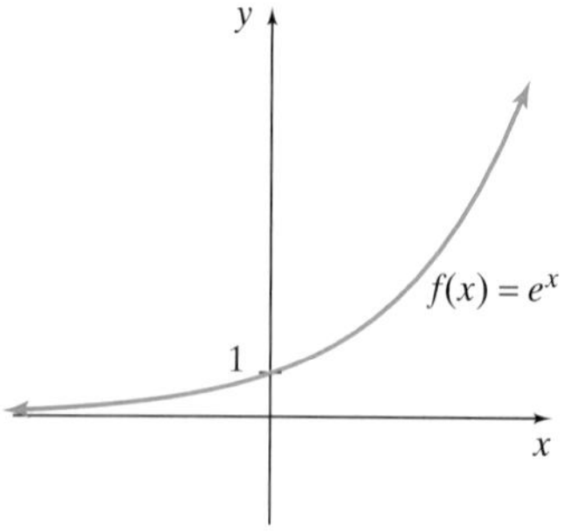

Figure 11.7.4

Note: The simplest form of the exponential function is $y = e^x$. See Figure 11.7.4.

Example 4 **Applying the Exponential Function**

The annual per capita consumption of whole milk in the United States can be modeled by

$$f(x) = 18.04e^{-0.046x} \qquad 1 \leq x \leq 16$$

where x represents the number of years since 1979 and $f(x)$ represents the number of gallons consumed per capita (*Source:* U.S. Department of Agriculture, www.usda.gov).

(a) Evaluate $f(3)$ and interpret.

(b) Compute the difference quotient for $x = 3$ and $\Delta x = 10$ and interpret.

Solution

(a) Understand the Situation: First, we notice that $x = 3$ corresponds to the year $1979 + 3 = 1982$.

Evaluating the function at $x = 3$ yields

$$f(3) = 18.04e^{-0.046(3)} = 18.04e^{-0.138}$$
$$\approx 18.04(0.87109869) \approx 15.71$$

Interpret the Solution: This means that in 1982 the annual per capita consumption of whole milk was about 15.71 gallons.

(b) Computing the difference quotient for $x = 3$ and $\Delta x = 10$, we get

$$m_{\text{sec}} = \frac{f(x+\Delta x) - f(x)}{\Delta x} = \frac{f(3+10) - f(3)}{10}$$
$$= \frac{f(13) - f(3)}{10} \approx \frac{9.92 - 15.71}{10} = -0.58$$

Interpret the Solution: This means that from 1982 to 1992 the annual per capita whole milk consumption decreased at an average rate of $0.58 \frac{\text{gallons per capita}}{\text{year}}$. See Figure 11.7.5.

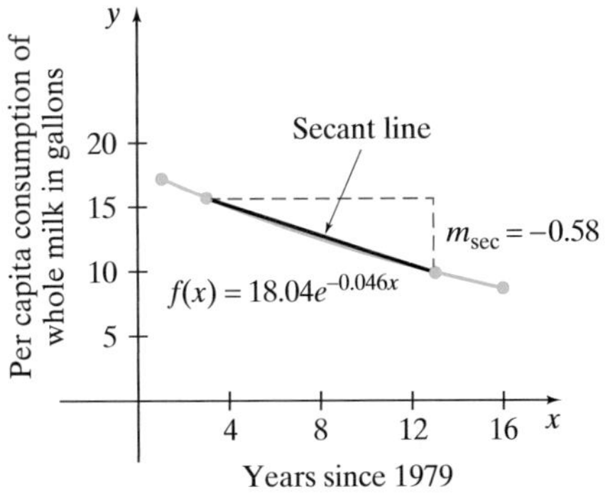

Figure 11.7.5 Slope of secant line on interval [3, 13] gives the average rate of change from 1982 to 1992.

■

Checkpoint 2 Now work Exercise 45.

Since the function in Example 4 is a decreasing function, we can classify it as an **exponential decay function**. The growth or decay property of an exponential function $f(x) = a \cdot e^{bx}$ is determined by the sign of the constant b.

Properties of Exponential Functions

Given the exponential function $f(x) = a \cdot e^{bx}$, with $a > 0$ and $b \neq 0$, we have

- The domain is the set of all real numbers $(-\infty, \infty)$.
- The y-intercept is at $(0, a)$.
- If $b > 0$, then f is an exponential growth function.
- If $b < 0$, then f is an exponential decay function.

The exponential function $f(x) = a \cdot e^{bx}$ is used often in the sciences, as well as the business sciences. Let's suppose that a banker with a strong entrepreneurial spirit wants to offer customers an account that accumulates interest **continuously** over time. To find a formula for this kind of interest, recall that the compound interest formula is

$$A = P\left(1 + \frac{r}{k}\right)^{kt}$$

First, it can be shown that, as $k \to \infty$, $\left(1 + \frac{r}{k}\right)^{k} = e^{r}$. So, rewriting the compound interest formula, we get

$$A = P\left[\left(1 + \frac{r}{k}\right)^{k}\right]^{t}$$

For $k \to \infty$, we can substitute e^r for the expression in brackets and get $A = Pe^{rt}$. This is the **continuous compound interest** formula.

Continuous Compound Interest

If a principal of P dollars is invested into an account earning an annual interest rate r (in decimal form) compounded continuously, then the amount A in the account at the end of t years is given by the **continuous compound interest formula**

$$A = Pe^{rt}$$

As a function of time, we can define the **continuous compound interest function** as $A(t) = Pe^{rt}$.

Example 5 Computing Compound Interest

Suppose that we have \$5000 deposited in an account that earns 5.3% interest compounded continuously. Find the continuous compound interest function for this account, and evaluate $A(3)$. Round the solution to the nearest cent.

Solution

Knowing that $r = 5.3\% = 0.053$ and $P = 5000$, we get the function

$$A(t) = 5000e^{0.053t}$$

Evaluating $A(t)$ for $t = 3$, we get

$$A(3) = 5000e^{0.053(3)} = 5000e^{0.159} \approx \$5861.69$$

So after 3 years, there would be \$5861.69 in the account. ■

Logistic Curves

One type of function that is used in the study of areas such as population growth and the spread of disease is the **logistic function**. The derivation of this function will be discussed in Chapter 16; for now, we simply define this new type of function.

Logistic Function

A **logistic function** has the form

$$f(t) = \frac{L}{1 + a \cdot e^{-kLt}}$$

where L is the limit of growth (that is, a horizontal asymptote) and a and k are real number constants.

Notice that the independent variable of this function is t, since the logistic function is often a function of time. The typical S-shape of the graph of a logistic function is shown in Figure 11.7.6.

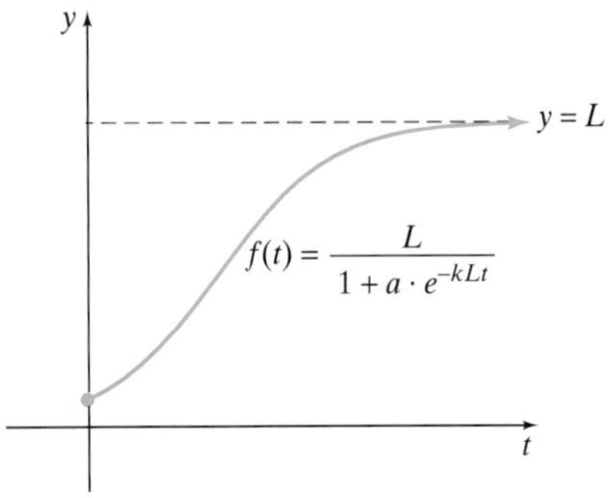

Figure 11.7.6

Example 6 Using Logistic Functions

The number of people affected by a flu epidemic in the isolated town of Spring Point (population 1800) can be modeled by

$$f(t) = \frac{1800}{1 + 359e^{-0.9t}}$$

where t represents the number of weeks since the flu was first detected and $f(t)$ represents the number of people who are affected (see Figure 11.7.7).

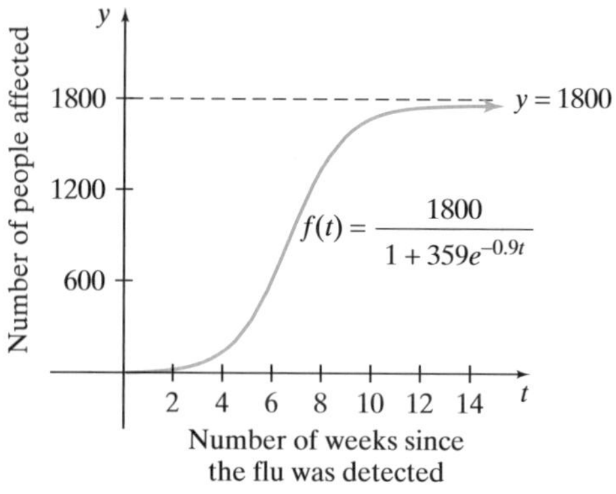

Figure 11.7.7

(a) Evaluate $f(1)$ and interpret.

(b) Compute the difference quotient for $t = 1$ and $\Delta t = 20$ and interpret.

Solution

(a) Evaluating the function at $t = 1$, we get

$$f(1) = \frac{1800}{1+359e^{-0.9(1)}} = \frac{1800}{1+359e^{-0.9}} \approx 12.25$$

According to the model, this means that after one week about 12 people were affected.

(b) Computing the difference quotient for $t = 1$ and $\Delta t = 20$, we find that

$$m_{\text{sec}} = \frac{f(1+20) - f(1)}{20} = \frac{f(21) - f(1)}{20}$$

$$= \frac{\dfrac{1800}{1+359e^{-0.9(21)}} - \dfrac{1800}{1+359e^{-0.9(1)}}}{20} \approx \frac{1799.99 - 12.25}{20} \approx 89.39$$

Interpret the Solution: So during the interval from 1 week to 21 weeks, the number of people affected increased at an average rate of about $89.39 \dfrac{\text{affected people}}{\text{week}}$. ■

Interactive Activity

Use the INTERSECT command on your calculator to determine the week that the number of people who were affected by the flu epidemic in Spring Point exceeded 500.

WWW

SUMMARY

In this section, we examined two types of exponential functions: the general exponential function $f(x) = a \cdot b^x$ and the exponential function $f(x) = a \cdot e^{bx}$. We found that the value of b determines whether f is an exponential growth or exponential decay function. Then we reviewed the simple and compound interest formulas. The definitions that this section highlighted were the following:

- **General exponential function**: $f(x) = a \cdot b^x$
- **Simple interest formula**: $I = Prt$
- **Compound interest formula**: $A = P\left(1 + \dfrac{r}{k}\right)^{kt}$
- **The number *e***: $\left(1 + \dfrac{1}{k}\right)^k = e$ as $k \to \infty$
- **Exponential function**: $f(x) = a \cdot e^{bx}$
- **Compound interest formula**: $A = Pe^{rt}$
- **Logistic function**: $f(t) = \dfrac{L}{1 + a \cdot e^{-kLt}}$

SECTION 11.7 EXERCISES

For Exercises 1–14, answer the following:

(a) Make a table of function values for $x = -5$ to $x = 5$, similar to Table 11.7.1.

(b) Graph f and classify the functions as exponential growth or decay.

1. $f(x) = \left(\frac{1}{3}\right)^x$ **2.** $f(x) = \left(\frac{1}{4}\right)^x$

3. $f(x) = 4^x$ **4.** $f(x) = 3^x$

5. $f(x) = (2.3)^x$ **6.** $f(x) = (4.5)^x$

7. $f(x) = (0.7)^x$ **8.** $f(x) = (0.22)^x$

9. $f(x) = e^{2x}$ **10.** $f(x) = e^{7x}$

11. $f(x) = e^{0.3x}$ **12.** $f(x) = e^{0.92x}$

13. $f(x) = e^{-1.6x}$ **14.** $f(x) = e^{-3.1x}$

15. The healing of a 200-square-centimeter wound depends on the amount of time since the wound occurred and can be modeled by

$$f(x) = 200\left(\frac{4}{5}\right)^x \qquad x \geq 0$$

where x represents the number of days since the wound occurred and $f(x)$ represents the size of the wound in square centimeters.

(a) Evaluate $f(7)$ and interpret.

(b) How many days will it take for the wound to be half its original size?

(c) Find the slope of the secant line for $x = 1$ and $\Delta x = 7$ and interpret.

16. The Desert Fox Dune Buggy Company launches a sales blitz for its new Family Buggy vehicle, offering a 30% discount during the first week. The percentage of customers who respond to the sale can be modeled by

$$f(x) = 70 - 100\left(\frac{3}{5}\right)^x \qquad 1 \leq x \leq 7$$

where x is the day of the sales blitz and $f(x)$ represents the percentage of customers who respond to the sale.

(a) Evaluate $f(1)$ and interpret.

(b) Find the slope of the secant line for $x = 2$ and $\Delta x = 2$ and interpret.

17. The number of students at a college who hear the rumor that the college president will resign can be modeled by

$$g(x) = 8000 - 8000\left(\frac{43}{50}\right)^x \qquad 1 \leq x \leq 14$$

where x represents the number of days since the rumor started and $g(x)$ represents the number of students who have heard the rumor.

(a) Evaluate $g(5)$ and interpret.

(b) Find the slope of the secant line for $x = 1$ and $\Delta x = 4$ and interpret.

18. The pupil-teacher ratio in U.S. elementary and secondary schools can be modeled by

$$f(x) = 25.34(0.987)^x \qquad 1 \leq x \leq 36$$

where x represents the number of years since 1959 and $f(x)$ represents the pupil-teacher ratio (*Source:* U.S. National Center for Education Statistics, www.nces.ed.gov).

(a) Evaluate $f(17)$ and interpret.

(b) Classify the model as exponential growth or exponential decay.

(c) Compute the difference quotient for $x = 31$ and $\Delta x = 5$ and interpret.

19. *(continuation of Exercise 18)*

(a) Graph the model in Exercise 18 in the viewing window [1, 36] by [0, 26].

(b) Use the INTERSECT command on your calculator to determine the year that the pupil-teacher ratio became less than $21\,\frac{\text{pupils}}{\text{teacher}}$.

20. The percentage of waste generated in the United States that is yard waste can be modeled by

$$f(x) = 19.12(0.944)^x \qquad 1 \leq x \leq 6$$

where x represents the number of years since 1990 and $f(x)$ represents the percentage of waste generated that is yard waste (*Source:* U.S. Statistical Abstract, www.census.gov/statab/www).

(a) Graph f in the viewing window [1, 6] by [10, 20]. Based upon the graph, classify the model as exponential growth or exponential decay.

(b) Use the INTERSECT command on your calculator to determine the year that the percentage of waste generated is yard waste became less than 15%.

21. *(continuation of Exercise 20)*

(a) For the model given in Exercise 20, evaluate $f(2)$ and interpret.

(b) Compute the difference quotient for $x = 1$ and $\Delta x = 5$ and interpret.

22. The annual total assets in mutual funds in the United States can be modeled by

$$f(x) = 126.67(1.217)^x \qquad 1 \leq x \leq 16$$

where x represents the number of years since 1979 and $f(x)$ represents the annual total assets in mutual funds in billions of dollars (*Source:* U.S. Statistical Abstract, www.census.gov/statab/www).

(a) Evaluate $f(9)$ and interpret.

(b) Classify the model as exponential growth or exponential decay.

(c) Compute the difference quotient for $x = 5$ and $\Delta x = 10$ and interpret.

23. *(continuation of Exercise 22)*

(a) Graph the model in Exercise 22 in the viewing window [1, 16] by [100, 3000].

(b) Use the INTERSECT command on your calculator to determine the year that the annual total assets in mutual funds exceeded $1000 billion (1 trillion dollars).

24. Use the simple interest formula to find the total interest earned for an account in which $500 is deposited at a 4% interest rate for 6 years.

25. Use the simple interest formula to find the total interest earned for an account in which $2000 is deposited at a 6.5% interest rate for 4 years.

26. If Kevin deposits $3000 into an account that yields 6% interest compounded annually, how much will be in the account after 3 years?

27. If Shuna deposits $10,000 into an account that yields 6.5% interest compounded semiannually, how much will be in the account after 4 years?

28. If Lucy deposits $1800 into an account that yields 7.1% interest compounded monthly, how much will be in the account after 5 years?

29. If Lana deposits $800 into an account that yields 7% interest compounded annually, how much will be in the account after 8 years?

30. If Joe deposits $2000 into an account that yields 8% annual interest, how much will be in the account after 6 years if the interest is compounded

(a) annually?

(b) semiannually (twice a year)?

(c) quarterly?

31. If Mrs. Johanson deposits $5000 into an account that yields 6.5% annual interest, how much will be in the account after 10 years if the interest is compounded

(a) annually?

(b) monthly (12 times a year)?

(c) weekly (52 times a year)?

32. If the Klien family deposits $15,000 into an account that yields 5.9% annual interest, how much will be in the account after 16 years if the interest is compounded

(a) annually?

(b) monthly (12 times a year)?

(c) weekly (52 times a year)?

33. If the LaDukes deposit $8000 into an account that yields 6.1% annual interest, how much will be in the account after 10 years if the interest is compounded

(a) annually?

(b) semiannually (twice a year)?

(c) quarterly?

For Exercises 34–38, use the compound interest function $A(t) = P\left(1 + \frac{r}{k}\right)^{kt}$ *to evaluate* $A(5)$ *and interpret. How much interest was made at the end of the 5-year period?*

34. $5500 is invested at a 5.5% interest rate compounded quarterly (four times a year).

35. $6400 is invested at a 6.5% interest rate compounded quarterly (four times a year).

36. $10,000 is invested at a 7.25% interest rate compounded monthly (12 times a year).

37. $7500 is invested at a 6.25% interest rate compounded monthly (12 times a year).

38. $2500 is invested at a 4.75% interest rate compounded semiannually (2 times a year).

39. Suppose that Elizabeth deposits $4000 at a 5.75% annual interest rate compounded monthly.

(a) Write the compound interest function A for the given information.

(b) Graph A in the viewing window [0, 16] by [4000, 10,000].

(c) Use your calculator to graphically find the amount of time that it takes the account to accumulate a total balance of $8000. This is called the **doubling time** of the account.

40. Now assume that Elizabeth deposits $6000 at a 5.75% annual interest rate compounded monthly.

(a) Write and graph the compound interest function A for the given information.

(b) Find the doubling time of this new account. Compare your answer to part (c) of Exercise 39.

41. Under certain conditions, the spread of the *E. coli* strain of bacteria can be modeled by

$$f(t) = N_0 e^{0.23t}$$

where t represents time in minutes, N_0 is the size of the colony when it starts, and $f(t)$ represents the size of the colony after time t. If $N_0 = 1{,}200{,}000$, answer the following:

(a) Evaluate $f(5)$ and interpret.

(b) Find the slope of the secant line for $t = 0$ and $\Delta t = 10$ and interpret.

42. The amount of a new experimental drug present in a patient's bloodstream can be modeled by

$$g(t) = 5e^{-0.3t} \qquad t \geq 0$$

where t represents the time since the drug was administered in hours and $g(t)$ represents the amount of drug in the bloodstream, measured in milligrams.

(a) Evaluate $g(2)$ and interpret.

(b) Find the slope of the secant line for $t = 2$ and $\Delta t = 3$ and interpret.

43. The sales of the new Flashfast Cigarette Lighters can be modeled by

$$S(t) = 10{,}000(2 - e^{-0.2t})$$

where t represents the number of years that the lighter has been on the market and $S(t)$ represents the number of sales.

(a) Evaluate $S(5)$ and interpret.

(b) Find the slope of the secant line for $t = 0$ and $\Delta t = 5$ and interpret.

44. The annual new import car sales in the United States can be modeled by

$$f(x) = 2.72e^{-0.07x} \qquad 1 \le x \le 6$$

where x represents the number of years since 1989 and $f(x)$ represents the new import car sales in millions of cars (*Source:* U.S. Census Bureau, www.census.gov).

(a) Evaluate $f(2)$ and interpret.

(b) Classify the model as exponential growth or exponential decay.

(c) Determine the slope of the secant line for $x = 1$ and $\Delta x = 4$ and interpret.

45. The number of U.S. World War II veterans receiving compensation for service-connected disabilities can be modeled by

$$f(x) = 1285.34e^{-0.037x} \qquad 1 \le x \le 16$$

where x represents the number of years since 1979 and $f(x)$ represents the number of World War II veterans receiving compensation, measured in thousands (*Source:* U.S. Census Bureau, www.census.gov).

(a) Evaluate $f(12)$ and interpret.

(b) Classify the model as exponential growth or exponential decay.

(c) Compute the difference quotient for $x = 1$ and $\Delta x = 10$ and interpret.

46. The U.S. federal government *receipts* can be modeled by

$$f(x) = 37.14e^{0.081x} \qquad 1 \le x \le 46$$

where x represents the number of years since 1949 and $f(x)$ represents the federal government receipts in billions of dollars (*Source:* U.S. Office of Management and Budget, www.whitehouse.gov/omb).

(a) Evaluate $f(21)$ and interpret.

(b) Classify the model as exponential growth or exponential decay.

(c) Compute the difference quotient for $x = 31$ and $\Delta x = 10$ and interpret.

47. The U.S. federal government *outlays* can be modeled by

$$g(x) = 37.52e^{0.084x} \qquad 1 \le x \le 46$$

where x represents the number of years since 1949 and $g(x)$ represents the U.S. federal government outlays in billions of dollars (*Source:* U.S. Office of Management and Budget, www.whitehouse.gov/omb).

(a) Classify the model as exponential growth or exponential decay.

(b) Evaluate $g(21)$ and interpret.

(c) Compute the difference quotient for $x = 31$ and $\Delta x = 10$ and interpret.

48. Using function f from Exercise 46 and function g from Exercise 47, complete the following.

(a) Evaluate the difference $f(41) - g(41)$ and interpret.

(b) Graph $f - g$ in the viewing window $[1, 46]$ by $[-250, 50]$. What does this difference of functions represent?

(c) Use the INTERSECT command on your calculator to determine the year in which the difference $f - g$ dipped below -100.

For Exercises 49–54, assume that the interest is compounded continuously. Use the continuous compound interest function to evaluate $A(7)$ and interpret.

49. \$1000 is invested at a $5\frac{1}{2}\%$ interest rate.

50. \$800 is invested at a 4.75% interest rate.

51. \$20,000 is invested at a 5.9% interest rate.

52. \$10,000 is invested at a 7.1% interest rate.

53. \$8000 is invested at a 8.1% interest rate.

54. \$75,000 is invested at a $5\frac{3}{4}\%$ interest rate.

55. The proportion of U.S. households that owns a DVD player can be modeled by

$$f(x) = \frac{0.8}{1 + 6e^{-0.3x}} \qquad x \ge 0$$

where x represents the number of years since 1997 and $f(x)$ represents the proportion of U.S. households that own a DVD player (*Source:* U.S. Census Bureau, www.census.gov).

(a) Evaluate $f(3)$ and interpret.

(b) Compute the difference quotient for $x = 3$ and $\Delta x = 2$ and interpret.

56. The proportion of home computers sold that have the latest 3-D WOW sound technology can be modeled by

$$g(t) = \frac{0.9}{1 + 4e^{-.3t}} \qquad t \ge 0$$

where t represents the number of months that the sound

technology has been on the market and $g(t)$ represents the number of home computers sold with the sound technology.

(a) Evaluate $g(7)$ and interpret.

(b) Compute the difference quotient for $t = 5$ and $\Delta t = 10$ and interpret.

57. Suppose that the spread of a certain kind of influenza at a rural school is given by the logistic model

$$f(t) = \frac{105}{1 + 20.08e^{-t}} \qquad 0 \le t \le 10$$

where t represents the number of days that healthy students are exposed to infected ones and $f(t)$ represents the number of students afflicted.

(a) Evaluate $f(0)$ and interpret.

(b) Compute the difference quotient for $t = 3$ and $\Delta t = 4$ and interpret.

58. Suppose that the spread of measles at Grover Elementary School is given by the mathematical model

$$f(t) = \frac{200}{1 + 200.34e^{-t}} \qquad 0 \le t \le 10$$

where t represents the number of days since the first student was diagnosed and $f(t)$ represents the number of affected students.

(a) Evaluate $f(4)$ and interpret.

(b) Compute the difference quotient for $t = 4$ and $\Delta t = 5$ and interpret.

59. Conservationists wish to repopulate a rare breed of lizard by transporting a seed colony to a wildlife preserve. They estimate that the lizard population will grow according to the logistic model

$$f(t) = \frac{1000}{1 + 121.51e^{-0.72t}} \qquad 0 \le t \le 15$$

where t represents the number of years since the seeding and $f(t)$ represents the lizard population.

(a) According to the model, what was the initial size of the seed colony?

(b) How many lizards will be in the population in 8 years?

(c) Compute the average rate of change for $t = 0$ and $\Delta t = 8$ and interpret.

60. In a biological experiment, a seed colony of paramecia were placed in a petri dish along with a nutritional medium. The number of paramecia in the dish is given by the logistic model

$$f(t) = \frac{380}{1 + 181.22e^{-2.1t}} \qquad 0 \le t \le 10$$

where t represents the number of days since the start of the experiment and $f(t)$ represents the population of paramecia.

(a) According to the model, how many paramecia were initially placed in the petri dish?

(b) How many paramecia will be in the petri dish in 2 days?

(c) Compute the average rate of change for $t = 2$ and $\Delta t = 8$ and interpret.

SECTION PROJECT

Investors compare accounts that compound interest differently using an effective rate. For example, an interest rate of 6.25% compounded *monthly* produces the same amount of interest as an interest rate of 6.43% compounded *annually*. Here 6.43% is called the **effective rate** and 6.25% is called the **nominal rate**. For the interest rate r compounded k times per year, the effective rate is

$$\text{Effective rate} = \left(1 + \frac{r}{k}\right)^k - 1$$

Compute the effective rates for the following accounts and interpret the results.

(a) A 4% nominal rate compounded weekly (52 times a year)

(b) A 7.5% nominal rate compounded monthly

(c) A 6.1% nominal rate compounded daily (365 times a year)

(d) A 5.75% nominal rate compounded semiannually

For interest compounded continuously, the effective rate is given by the formula

$$\text{Effective rate} = e^r - 1$$

Compute the effective rates for the following accounts and interpret the results.

(e) A 4% nominal rate compounded continuously

(f) A 7.5% nominal rate compounded continuously

(g) A 6.1% nominal rate compounded continuously

(h) A 5.75% nominal rate compounded continuously

Section 11.8 Logarithmic Functions

In the previous section, we studied exponential functions and their properties. Now we want to develop some mathematical tools that undo the process of exponentiation. This undoing is accomplished with **logarithms**. To understand logarithms, we need to introduce some other useful ideas. The first of these is the composition of functions. We will see that **composite functions** occur frequently. Next we discuss **inverse functions**, which allow us to define functions such as logarithms.

Composite Functions

In Section 11.1, we learned that when we apply a function to an independent variable value a dependent variable value is returned. Now let's suppose that the dependent value of the one function becomes the independent value of another. To illustrate, suppose that a refinery's underwater supply line ruptures, resulting in an oil spill that is fairly circular. The petroleum company, the local fisheries, and those who are concerned about the environment would like to know how much area the spill will cover as time passes.

Assume that past experiments show that the radius of the greasy, circular slick increases at a rate of about 0.7 feet per second. As a function of time t, the radius of the slick is given by

$$r(t) = 0.7t$$

Since the area of the oil spill depends on its radius, the area is given by the function

$$A(r) = \pi r^2$$

Since the radius is given by $r(t) = 0.7t$, to find the area of the slick as a function of time, we can substitute $r(t) = 0.7t$ into $A(r) = \pi r^2$. This yields

$$A(t) = \pi(0.7t)^2 = \pi \cdot 0.49 \cdot t^2 = 0.49\pi t^2$$

Figure 11.8.1 suggests that we are nesting the functions, meaning that we could write $A(t) = A(r(t))$. This is exactly what a composition of functions is—a *nesting* of one function inside another.

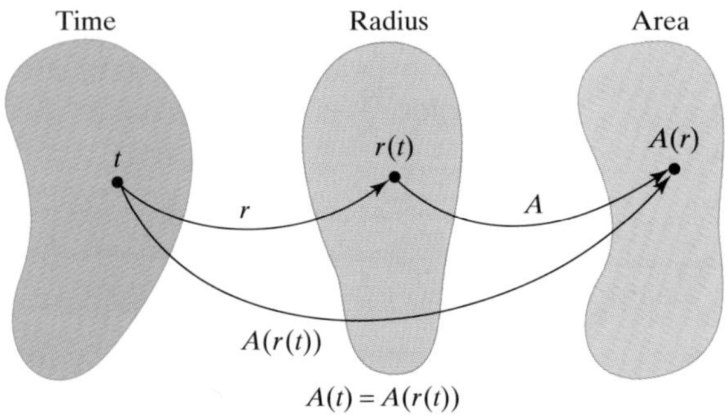

Figure 11.8.1

Composite Function

A function h is a **composition** of the functions f and g if

$$h(x) = f(g(x))$$

Note: The composition of functions $f(g(x))$ can also be denoted as $(f \circ g)(x)$.

Example 1 Evaluating Composite Functions

Consider the functions $f(x) = \frac{3}{x}$ and $g(x) = 2x - 1$. Evaluate $f(g(6))$ and $g(f(6))$.

Solution

Starting from the inside of the expression $f(g(6))$ and working our way out, we first evaluate

$$g(6) = 2(6) - 1 = 11$$

To evaluate $f(g(6))$, we substitute 11 for $g(6)$ to get

$$f(g(6)) = f(11) = \frac{3}{11}$$

In the case of $g(f(6))$, we first start by evaluating $f(6)$ to get

$$f(6) = \frac{3}{6} = \frac{1}{2}$$

So, to evaluate $g(f(6))$, we get

$$g(f(6)) = g\left(\frac{1}{2}\right) = 2 \cdot \frac{1}{2} - 1 = 0$$

■

Checkpoint 1 Now work Exercise 3.

Notice in Example 1 that $f(g(6))$ and $g(f(6))$ were different. This shows that *generally* $\boldsymbol{f(g(x)) \neq g(f(x))}$. We will soon see a special case when these compositions are the same.

When determining the domain of a composition of functions, we have to take care that the dependent values of the *inside* function are all members of the domain of the second. Knowing the range values of the inside function takes on added importance.

Building up functions through composition is a very exacting process. However, **decomposing** a function into simpler parts offers several options. This is illustrated in Example 2.

Example 2 Decomposing Functions

Determine functions f and g for each composition so that $f(g(x)) = h(x)$; then check your answer.

(a) $h(x) = \sqrt{(x+5)^3}$

(b) $h(x) = (5x^3 + 6)^2$

Solution

(a) Here we could let the *inside function* $g(x) = (x+5)^3$ and the *outside function* $f(x) = \sqrt{x}$. Checking the solution yields

$$f(g(x)) = f[(x+5)^3] = \sqrt{(x+5)^3} = h(x)$$

Interactive Activity

Check the solution to part (a) of Example 2 on the calculator by entering the following functions in the $y =$ menu and comparing the graphs of the functions y_3 and y_4.

$$y_1 = (x+5)^3$$
$$y_2 = \sqrt{(x)}$$
$$y_3 = y_2(y_1(x))$$
$$y_4 = \sqrt{((x+5)^3)}$$

WWW

(b) For this function we could let $g(x) = 5x^3 + 6$ and $f(x) = x^2$. Checking the solution, we get

$$f(g(x)) = f[(5x^3 + 6)]$$
$$= (5x^3 + 6)^2 = h(x)$$

■

Inverse Functions

Addition and subtraction are examples of **inverse operations**. We know that if we start with some number like 3 and add 7, then subtract 7, we get back the original number 3. Similarly, functions such as $f(x) = \frac{x}{3}$, and $g(x) = 3x$ are inverses of one another. To show that these functions undo each other, choose a number like $x = 12$. Then in the function f we get

$$f(12) = \frac{12}{3} = 4$$

Taking this result and calculating $g(4)$, we get

$$g(4) = 3(4) = 12$$

which is the value that we started with. In terms of composition of functions, we have shown that

$$g(f(12)) = 12$$

Furthermore, we can show that for the functions $f(x) = \frac{x}{3}$ and $g(x) = 3x$ that

$$g(f(x)) = g\left(\frac{x}{3}\right)$$
$$= 3\left(\frac{x}{3}\right) = x$$

These are examples of inverse functions. All inverse functions have a special property: they are all one-to-one functions.

One-to-One Function

A function f is a **one-to-one function** if, for elements a and b in the domain of f,

$$a \neq b \quad \text{implies that} \quad f(a) \neq f(b)$$

Finding these values for some functions can be rather difficult, so there is another way to determine if a function is 1-1. It is called the **horizontal line test**.

Horizontal Line Test

If every horizontal line intersects the graph of a function f in no more than one place, then f is a one-to-one function.

Now that we know what kind of functions have inverses, we present the definition of an inverse function.

Inverse Function

Two one-to-one functions f and g are **inverses** of each other if

$$f(g(x)) = x \quad \text{for every } x\text{-value in the domain of } g$$

and

$$g(f(x)) = x \quad \text{for every } x\text{-value in the domain of } f$$

A special notation is often used for the inverse function. If g is the inverse of f, then g may be written as $\boldsymbol{f^{-1}}$ (read "f inverse"). Do *not* confuse this with a negative exponent. $f^{-1}(x)$ does not represent $\dfrac{1}{f(x)}$; it represents the inverse function of f. Figure 11.8.2 shows the general relationship between a function and its inverse function.

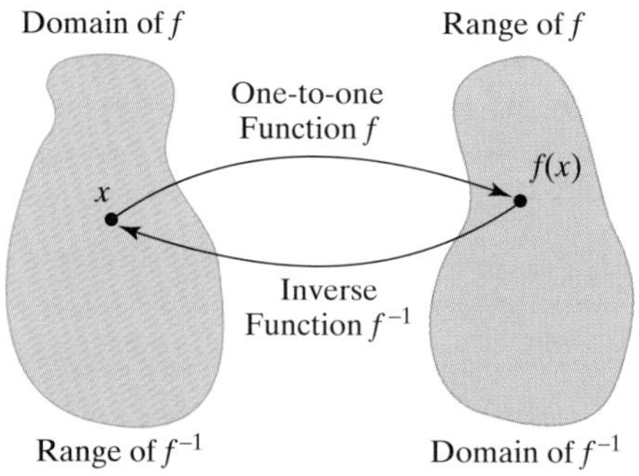

Figure 11.8.2

Example 3 Showing That Two Functions Are Inverses

Use the definition of inverse functions to show that $f(x) = 3x - 2$ and $g(x) = \dfrac{x+2}{3}$ are inverses of each other.

Solution

We need to show that $f(g(x)) = x$ and $g(f(x)) = x$. Completing the first composition, we get

$$f(g(x)) = f\left(\frac{x+2}{3}\right)$$

Replacing the expression $\dfrac{x+2}{3}$ at every occurrence of x in $f(x) = 3x - 2$, we get

$$f(g(x)) = 3\left(\frac{x+2}{3}\right) - 2 = (x+2) - 2 = x$$

In the Interactive Activity to the left, showing that $g(f(x)) = x$ will verify that $f(x) = 3x - 2$ and $g(x) = \dfrac{x+2}{3}$ are inverses of each other. ■

The Interactive Activity to the left shows that there is not only an algebraic but also a graphical relationship between two inverse functions.

Interactive Activity

Simplify the composition $g(f(x)) = x$ to verify that $f(x) = 3x - 2$ and $g(x) = \dfrac{x+2}{3}$ are inverses. Then graph both functions in the viewing window $[-9.4, 9.4]$ by $[-6.2, 6.2]$. Do you see a relationship in the appearance of the graphs? Now let $y_3 = y_1(y_2(x))$ and regraph the functions. What is this new function y_3?

Graphical Relationship between Inverse Functions f and g

If f is a one-to-one function, then the graph of its inverse function g is a reflection about the line $y = x$.

Logarithmic Functions

Now we want to look at a special family of inverse functions. Their importance can be seen in the Flashback that follows.

Flashback

Disabled World War I Veterans Revisited

In Section 11.7, we saw that the number of disabled World War I veterans receiving compensation for service-connected disabilities can be modeled by

$$f(x) = 47.55(0.78)^x \qquad 1 \le x \le 16$$

where x represents the number of years since 1979 and $f(x)$ represents the number of disabled World War I veterans, measured in thousands. Use tables to numerically estimate the time x when the number of disabled World War I veterans in the United States dipped below 12,000. Write the answer accurate to the half-year (*Source:* U.S. Census Bureau, www.census.gov).

Flashback Solution

We start by making a table of values for $x = 1, 2, 3, \ldots$. The result is shown in Table 11.8.1. Numerically, we see that at some time between the values of $x = 5$ and $x = 6$ the $f(x)$-values dipped below 12. To get a better estimate, let's make another table using increments of 0.5 for x. This output is shown in Table 11.8.2. This table tells us that the $f(x)$-value dipped below 12 in the interval $5.5 < x < 6.0$. Since $f(5.5) = 12.125$, our estimate accurate to 0.5 is $x = 5.5$.

Table 11.8.1

x	$f(x)$
1	$47.55(0.78)^1 = 37.089$
2	$47.55(0.78)^2 = 28.929$
3	$47.55(0.78)^3 = 22.565$
4	$47.55(0.78)^4 = 17.601$
5	$47.55(0.78)^5 = 13.729$
6	$47.55(0.78)^6 = 10.708$
7	$47.55(0.78)^7 = 8.352$

Table 11.8.2

x	$f(x)$
5.0	13.729
5.5	12.125
6.0	10.708

We could continue to get a more accurate solution to the Flashback by continuing with this numerical method or even by using the calculator's INTERSECT command. To get an accurate solution algebraically, we need a way to undo the exponential function. This is why we study the **logarithm**. The inverse association between logarithm and exponent is illustrated in the definition of a logarithm.

Logarithm

We say that

$$y = \log_b x \quad \text{if and only if} \quad b^y = x$$

where $b > 0, b \neq 1, x > 0$ ($\log_b x$ is read "the log of x, base b" or "the log base b of x").

This definition tells us, for example, that

$$6^2 = 36 \quad \text{means} \quad \log_6 36 = 2$$

or

$$3^{-2} = \frac{1}{9} \quad \text{means} \quad \log_3\left(\frac{1}{9}\right) = -2$$

To illustrate that the logarithm function is an inverse function, let's recall the table of values of $y = 2^x$ that was given in Section 11.7. These results are repeated on the left-hand side of Table 11.8.3. Now, if we use the definition of the logarithm, we obtain the values on the right-hand side. For example, for $x = \frac{1}{32}$, we see that since $\frac{1}{32} = 2^{-5}$, we can say that $\log_2\left(\frac{1}{32}\right) = -5$.

Table 11.8.3

x	$f(x) = 2^x$	x	$g(x) = \log_2 x$
-5	$\frac{1}{32}$	$\frac{1}{32}$	-5
-4	$\frac{1}{16}$	$\frac{1}{16}$	-4
-3	$\frac{1}{8}$	$\frac{1}{8}$	-3
-2	$\frac{1}{4}$	$\frac{1}{4}$	-2
-1	$\frac{1}{2}$	$\frac{1}{2}$	-1
0	1	1	0
1	2	2	1
2	4	4	2
3	8	8	3
4	16	16	4
5	32	32	5

The table shows that the x- and y-coordinates of $f(x) = 2^x$ and $g(x) = \log_2 x$ are interchanged. Thus, the graph of $g(x) = \log_2 x$ is a reflection of $f(x) = 2^x$ about the line $y = x$. See Figure 11.8.3. So it seems that the logarithm undoes exponentiation. That is, **$f(x) = b^x$ and $g(x) = \log_b x$ are inverse functions**. The table and graph also suggest that, since the range of the exponential function is the set of positive numbers, the domain of the logarithmic function is also the set of positive numbers. In other words, we cannot take the logarithm of a negative number. This is all summarized in the definition of a logarithm function.

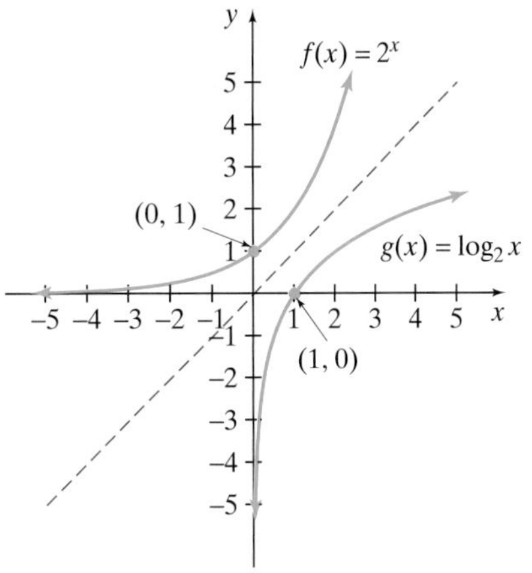

Figure 11.8.3

Logarithm Function

$$f(x) = \log_b x$$

is the **logarithm function**, where $b > 0$, $b \neq 1$, and the domain is $(0, \infty)$.

The definition states that the base b of a logarithm can be any positive number not equal to 1, but two bases are the most widely used.

Logarithm Notation

1. The **common logarithm** is $\log_{10} x$ and is denoted by $\log x$.
2. The **natural logarithm** is $\log_e x$ and is denoted by $\ln x$.

The properties of logarithms are much like those for exponents. Some of these properties are proved in Appendix E.

Properties of Logarithms

Let m and n be positive numbers. Then the following are true:

1. $\log_b(mn) = \log_b m + \log_b n$
2. $\log_b\left(\frac{m}{n}\right) = \log_b m - \log_b n$
3. $\log_b m^n = n \cdot \log_b m$
4. $\log_b 1 = 0$
5. $\log_b b = 1$
6. $x = b^{\log_b x}$
7. $x = \log_b b^x$
8. $\log_b m = \log_b n$ if and only if $m = n$

Example 4 Using the Properties of Logarithms

Use the properties of logarithms to rewrite the following.

(a) $\log(6 \cdot 17)$ **(b)** $\log_5\left(\frac{10}{3}\right)$ **(c)** $\log_6 \sqrt{11}$

Solution

(a) Using property 1, we get $\log(6 \cdot 17) = \log 6 + \log 17$.

(b) Property 2 tells us that we can rewrite $\log_5\left(\frac{10}{3}\right)$ as $\log_5 10 - \log_5 3$.

(c) Before we apply property 3, we rewrite $\sqrt{11}$ as $11^{1/2}$,

$$\log_6 \sqrt{11} = \log_6 11^{1/2} = \frac{1}{2}\log_6 11$$

■

Checkpoint 2 Now work Exercise 43.

In our study of calculus, we primarily use the **natural logarithm**. Since the natural logarithm is key to solving exponential equations, we note that all properties of logarithms are true for $\ln x$. These properties allow us to *take the log* of both sides of an equation so that the independent variable is no longer in the exponent. We outline this procedure to solve the general exponential equation $y = a \cdot b^x$ in Example 5.

Example 5 Solving Exponential Equations

In the Flashback, we numerically approximated the solution to the exponential equation $12 = 47.55(0.78)^x$. Use the properties of logarithms to solve the equation and round the solution to the nearest hundredth.

Solution

We start solving the equation $12 = 47.55(0.78)^x$ by dividing both sides by 47.55.

$$12 = 47.55(0.78)^x$$

$$\frac{12}{47.55} = (0.78)^x \qquad \text{Divide by 47.55.}$$

$$\ln\left(\frac{12}{47.55}\right) = \ln(0.78)^x \qquad \text{Take the ln of both sides, Property 8.}$$

$$\ln\left(\frac{12}{47.55}\right) = x \cdot \ln(0.78) \qquad \text{Property 3.}$$

$$\frac{1}{\ln(0.78)} \cdot \ln\left(\frac{12}{47.55}\right) = x \qquad \text{Divide by } \ln(0.78).$$

$$5.54 \approx x$$

■

Note in the example that no rounding takes place until the final step. Rounding earlier in the process can throw off the final result.

Applications with Logarithmic Functions

Many phenomena that we study start with rapid growth and then the growth begins to level off. This type of behavior is well suited to the **logarithmic model** of the form $f(x) = a + b \ln x$, where a and b are constants. Since we cannot take the logarithm of a negative number, the domain of this function is the set of positive numbers $(0, \infty)$. In practice, we usually start with an x-value of 1.

Example 6 Applying Logarithmic Models

The percentage of U.S. households with cable TV can be modeled by

$$f(x) = 18.32 + 15.94 \ln x \qquad 1 \le x \le 17$$

where x represents the number of years since 1979 and $f(x)$ represents the percentage of households in the United States with cable TV (*Source:* U.S. Statistical Abstract, www.census.gov/statab/www).

(a) Evaluate $f(6)$ and interpret.

(b) Use the result of part (a) to compute the difference quotient for $x = 6$ and $\Delta x = 5$ and interpret.

Interactive Activity

Compute the difference quotient for $x = 11$ and $\Delta x = 5$ and compare the result with part (b) of Example 6.

WWW

Solution

(a) Understand the Situation: Recall that to evaluate $f(6)$, we substitute 6 in the function for every occurrence of x. Also note that $x = 6$ corresponds to the year 1985.

Evaluating, we get

$$f(6) = 18.32 + 15.94 \ln(6) \approx 46.88$$

Interpret the Solution: This means that about 46.88% of the households in the United States had cable TV in 1985.

(b) Computing the difference quotient with $x = 6$ and $\Delta x = 5$, we get

$$m_{\text{sec}} = \frac{f(6+5) - f(6)}{5} = \frac{f(11) - f(6)}{5}$$

$$\approx \frac{56.54 - 46.88}{5} \approx 1.93$$

Interpret the Solution: This means that from 1985 to 1990 the number of U.S. households with cable TV was growing at an average rate of about $1.93 \frac{\text{percent}}{\text{year}}$. ■

SUMMARY

This section featured several new topics, including the definition and properties of the composite and inverse functions. Then we used this information to discover that the exponential and logarithmic functions were inverses.

- A function h is a **composition** of the functions f and g if $h(x) = f(g(x))$.
- Two **one-to-one** functions f and g are **inverses** of each other if $f(g(x)) = x$ for every x-value in the domain of g and $g(f(x)) = x$ for every x-value in the domain of f.
- We say that $y = \log_b x$ if and only if $b^y = x$, where $b > 0$, $b \neq 1$, $x > 0$.
- The logarithmic model is $f(x) = a + b \ln x$.

Properties of Logarithms

- $\log_b(mn) = \log_b m + \log_b n$
- $\log_b\left(\frac{m}{n}\right) = \log_b m - \log_b n$
- $\log_b m^n = n \cdot \log_b m$
- $\log_b 1 = 0$
- $\log_b b = 1$
- $x = b^{\log_b x}$
- $x = \log_b b^x$
- $\log_b m = \log_b n$ if and only if $m = n$

SECTION 11.8 EXERCISES

For Exercises 1–4, answer the following:

1. Let $f(x) = 3x$ and $g(x) = 6x$. Evaluate $f(g(3))$ and $f(g(-2))$.

2. Let $f(x) = 3x - 1$ and $g(x) = x^2 + 1$. Evaluate $g(f(2))$ and $g(f(3))$.

✓ **3.** Let $f(x) = 3x^2 - x$ and $g(x) = 6x + 1$. Evaluate $f(g(0))$ and $g(f(0))$.

4. Let $f(x) = \sqrt{2x}$ and $g(x) = 3x^2$. Evaluate $f(g(2))$ and $g(f(10))$.

For Exercises 5–12, find $g(f(x))$ and $f(g(x))$.

5. $f(x) = 3x - 1$; $g(x) = 8x + 2$

6. $f(x) = 6x - 9$; $g(x) = 7x + 5$

7. $f(x) = 5x - 3$; $g(x) = x^2 + 3x + 4$

8. $f(x) = 3x^2 + x + 5$; $g(x) = x - 1$

9. $f(x) = x^3$; $g(x) = \dfrac{1}{x}$

10. $f(x) = \dfrac{2}{x}$; $g(x) = x + 3$

11. $f(x) = \sqrt{x+1}$; $g(x) = x^2 + 2$

12. $f(x) = \sqrt{x+2}$; $g(x) = 4x - 3$

For Exercises 13–20, determine f and g so that $f(g(x)) = h(x)$ for each composite function h; then check your answer. There is more than one correct answer.

13. $h(x) = (x+3)^3$

14. $h(x) = (5x+6)^4$

15. $h(x) = \left(\dfrac{1}{x+3}\right)^2$

16. $h(x) = \left(\dfrac{5}{x-1}\right)^4$

17. $h(x) = \sqrt[4]{x-2}$

18. $h(x) = \sqrt[3]{2x+3}$

19. $h(x) = 2 - 3\sqrt{x}$

20. $h(x) = 9 + 2\sqrt{x}$

21. Suppose that a spherical balloon is being inflated with the radius increasing at an average rate of 1.3 inches per second.

(a) Write the radius $r(t)$ as a function of time t. (Assume that when $t = 0$ then $r = 0$.)

(b) The volume of a sphere is a function of its radius and is given by $V(r) = \frac{4}{3}\pi r^3$. Simplify $V(t) = V(r(t))$ and interpret the meaning of this composition.

(c) Evaluate $V(t)$ when $t = 6$ and interpret.

22. The Top Drug Company has developed a new type of pill in the shape of a ball. The spherical pill is dropped into a glass of water and dissolves. Tests show that when the 2-centimeter pill is dropped into the water its radius decreases at an average rate of $0.003\frac{\text{cm}}{\text{sec}}$.

(a) Write the radius $r(t)$ as a function of time t. In this case, $r(t) = 2$ when $t = 0$.

(b) Knowing that the surface area of a sphere as a function of its radius is $S(r) = 4\pi r^2$, simplify $S(t) = S(r(t))$ and interpret the meaning of this composition.

(c) Evaluate $S(t)$ when $t = 20$ and interpret.

For the one-to-one functions f and g in Exercises 23–30, show that $f(g(x)) = x$.

23. $f(x) = 8x$; $g(x) = \dfrac{1}{8}x$

24. $f(x) = \dfrac{3}{4}x$; $g(x) = \dfrac{4}{3}x$

25. $f(x) = 7x - 10$; $g(x) = \dfrac{x+10}{7}$

26. $f(x) = \dfrac{x-3}{4}$; $g(x) = 4x + 3$

27. $f(x) = x^3 + 6$; $g(x) = \sqrt[3]{x-6}$

28. $f(x) = x^5 - 9$; $g(x) = \sqrt[5]{x+9}$

29. $f(x) = \sqrt{x-1}$; $g(x) = x^2 + 1$, for $x \ge 0$

30. $f(x) = \sqrt[4]{x-1}$; $g(x) = x^4 + 1$, for $x \ge 0$

In Exercises 31–42, rewrite the equation in the logarithmic form $y = \log_b x$.

31. $2^5 = 32$

32. $3^4 = 81$

33. $2^{-3} = \dfrac{1}{8}$

34. $5^{-2} = \dfrac{1}{25}$

35. $\left(\dfrac{1}{2}\right)^2 = \dfrac{1}{4}$

36. $\left(\dfrac{1}{3}\right)^{-1} = 3$

37. $81^{3/4} = 27$

38. $64^{4/3} = 256$

39. $10^5 = 100{,}000$

40. $10^3 = 1000$

41. $e^1 = e$

42. $e^0 = 1$

In Exercises 43–50, use the properties of logarithms to rewrite the following as the sum and/or difference of logarithms.

✓ **43.** $\log_2\left(\dfrac{3}{5}\right)$

44. $\log_6\left(\dfrac{4}{7}\right)$

45. $\log(8 \cdot 20)$

46. $\log(17 \cdot 11)$

47. $\ln\sqrt{26}$

48. $\ln\sqrt[3]{11}$

49. $\log_3 \dfrac{4\sqrt{3}}{9}$

50. $\log_{16} \dfrac{8\sqrt[3]{6}}{11}$

For Exercises 51–58, solve the exponential equations algebraically and round the solution to the hundredths place.

51. $6 = 3^x$

52. $12 = 4^x$

53. $8 \cdot 2^x = 51$

54. $6 \cdot 9^x = 126$

55. $2e^x = 62$

56. $4e^x = 81$

57. $1.21(0.3)^x = 42$

58. $0.33(1.27)^x = 58$

59. The percentage of unemployed Hispanics in the U.S. labor force can be modeled by

$$f(x) = 11.39(0.94)^x \qquad 1 \le x \le 7$$

where x represents the number of years since 1992 and $f(x)$ represents the percentage of unemployed Hispanics in the U.S. labor force (*Source:* U.S. Bureau of Labor Statistics, www.bls.gov).

(a) Evaluate $f(4)$ and interpret.

(b) Solve the exponential equation $11.39(0.94)^x = 10$ and interpret.

60. The percentage of unemployed Caucasians in the U.S. labor force can be modeled by

$$f(x) = 7.17(0.91)^x \qquad 1 \le x \le 7$$

where x represents the number of years since 1989 and $f(x)$ represents the percentage of unemployed Caucasians in the U.S. labor force (*Source:* U.S. Bureau of Labor Statistics, www.bls.gov).

(a) Evaluate $f(4)$ and interpret.

(b) Solve the exponential equation $7.17(0.91)^x = 5.4$ and interpret.

61. The total annual personal expenditures for admission to spectator sports can be modeled by

$$f(x) = 4.21 + 1.29 \ln x \qquad 1 \le x \le 9$$

where x represents the number of years since 1989 and $f(x)$ represents the total annual personal expenditures for admission to spectator sports in billions of dollars (*Source:* U.S. Bureau of Economic Analysis, www.bea.doc.gov).

(a) Make a table of values for $x = 1, 3, 5, 7, 9$.

(b) Evaluate $f(6)$ and interpret.

(c) Determine the average rate of change in personal expenditures for admission to spectator sports from 1990 to 1998 and interpret.

62. The percentage of U.S. households that have video cassette recorders (VCRs) can be modeled by

$$f(x) = 21.88 + 25.34 \ln x \qquad 1 \le x \le 17$$

where x represents the number of years since 1979 and $f(x)$ represents the percentage of households in the United States with VCRs (*Source:* U.S. Statistical Abstract, www.census.gov/statab/www).

(a) Make a table of values for $x = 1, 6, 11, 16$.

(b) Use the values from part (a) to compute the difference quotient for $x = 1$ and $\Delta x = 5$ and interpret.

63. The total school expenditures in higher education by the federal government can be modeled by

$$f(x) = 18.48 + 3.31 \ln x \qquad 1 \le x \le 13$$

where x represents the number of years since 1984 and $f(x)$ represents the total school expenditures in higher education by the federal government in billions of constant 1997–1998 dollars (*Source:* U.S. National Center for Education Statistics, www.nces.ed.gov).

(a) Evaluate $f(10)$ and interpret.

(b) Compute the difference quotient for $x = 10$ and $\Delta x = 3$ and interpret.

64. The total school expenditures in higher education by *state* governments can be modeled by

$$g(x) = 49.25 + \ln x \qquad 1 \le x \le 4$$

where x represents the number of years since 1993 and $g(x)$ represents the total school expenditures in higher education by *state* governments in billions of constant 1997–1998 dollars (*Source:* U.S. National Center for Education Statistics, www.nces.ed.gov).

(a) Evaluate $g(1)$ and interpret.

(b) Compute the difference quotient for $x = 1$ and $\Delta x = 3$ for the function g and compare with the result of Exercise 63 part (b).

SECTION PROJECT

The annual per capita consumption of *light* and *skim* milk in the United States can be modeled by

$$f(x) = 10.12 + 2 \ln x \qquad 1 \le x \le 16$$

where x represents the number of years since 1979 and $f(x)$ represents the annual per capita consumption of light and skim milk in gallons (*Source:* U.S. Department of Agriculture, www.usda.gov).

(a) Evaluate $f(12)$ and interpret.

(b) Compute the difference quotient for $x = 12$ and $\Delta x = 2$ and interpret.

The annual per capita consumption of *whole* milk in the United States can be modeled by

$$g(x) = 18.04(0.96)^x \qquad 1 \le x \le 16$$

where x represents the number of years since 1979 and $g(x)$ represents the annual per capita consumption of whole milk in gallons (*Source:* U.S. Department of Agriculture, www.usda.gov).

(c) Evaluate $g(12)$ and interpret.

(d) Evaluate $f(12) - g(12)$ and interpret.

(e) Compute the difference quotient for $x = 12$ and $\Delta x = 2$ for the function g and compare with part (b).

*Section 11.9 Modeling Data with Functions (Optional Section–Requires Graphing Calculator)

Many of the applications in this book have the 🌐 icon indicating that the function is derived from data gathered on the Internet. But how is this done? The answer lies within a statistical process called **regression analysis**. In this section, we will show how computers and, particularly, graphing calculators can be used to develop a **mathematical model** to study real data. We will also discuss the shortcomings of overreliance on these regression techniques.

Mathematical Models

We often want to find a nice, compact function to explain complex events in the real world. This function, called a **mathematical model**, is a function that is derived from a real-world situation. Let's say that a local meteorologist makes an observation such as "the height of the Bend River started at 4 feet and has been steadily climbing half a foot each day for the last 10 days." We can make a simple mathematical model for the water level such as

$$w(t) = \frac{1}{2}t + 4 \qquad 0 \le t \le 10$$

where t represents the number of days and $w(t)$ represents the water level.

This simple, compact function numerically explains what the water level has been for each day of the 10-day period. Unfortunately, few phenomena in the business, life, and social sciences are so straightforward. This is why we need to gather data and use them to create a mathematical model via **regression** techniques. (See Figure 11.9.1.) We can then use this model to study the phenomenon.

Table 11.9.1

Year	CPI
1987	113.6
1988	118.3
1989	124.0
1990	130.7
1991	136.2
1992	140.3
1993	144.5
1994	148.2
1995	152.4
1996	156.9

SOURCE: U.S. Bureau of Labor Statistics, www.bls.gov

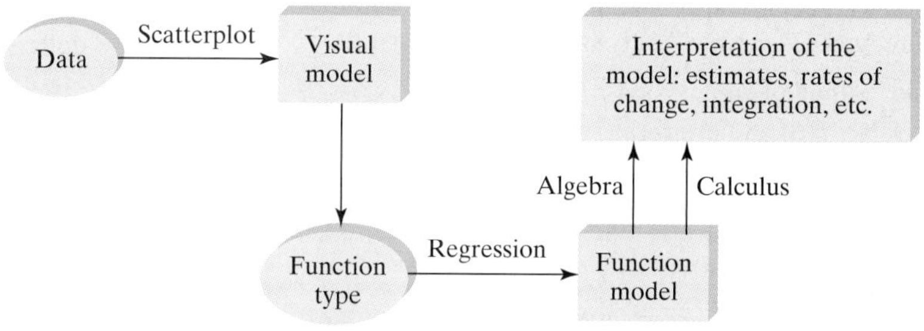

Figure 11.9.1

Regression Model Process

Finding a proper fit for data can be almost as much art and experience as it is mathematics. There are some common steps to follow. To see how such a procedure works, let's consider the Bureau of Labor Statistics data on the consumer price index (CPI) for the years 1987 to 1996 in Table 11.9.1.

As we mentioned in Section 11.1, we begin the process by **standardizing the independent variable** (Table 11.9.2). The next step is to **make a scatterplot** of the CPI data, as shown in Figure 11.9.2.

Note: To see how to make scatterplots for your data, check the online calculator manual at www.prenhall.com/armstrong.

Table 11.9.2

Year	x	y, the CPI
1987	0	113.6
1988	1	118.3
1989	2	124.0
1990	3	130.7
1991	4	136.2
1992	5	140.3
1993	6	144.5
1994	7	148.2
1995	8	152.4
1996	9	156.9

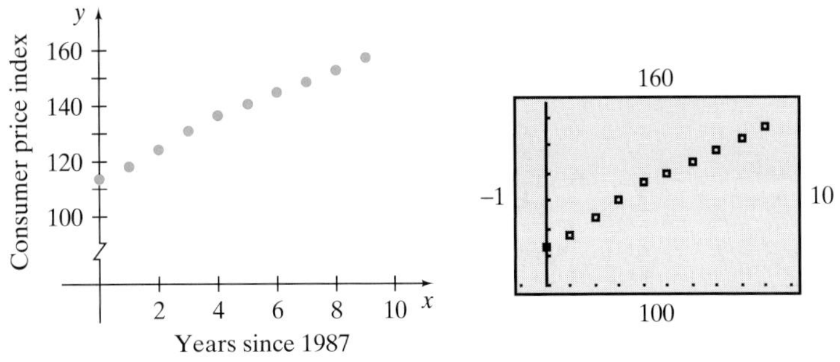

(a) **(b)**

Figure 11.9.2

We can now **inspect the dependent values** to see if there is a pattern in the data. We can see from the scatterplot of the CPI data that the points appear to be fairly linear. This is a good clue that the regression model that we want to use is the linear model. But there is another way to inspect the data for a pattern. Let's compute the annual differences of the dependent values by taking a dependent value and subtracting the one before it. The results of this analysis are shown in Table 11.9.3.

Table 11.9.3

Year	x-value	CPI	Successive Difference
1987	0	113.6	
1988	1	118.3	118.3 − 113.6 = 4.7
1989	2	124.0	5.7
1990	3	130.7	6.7
1991	4	136.2	5.5
1992	5	140.3	4.1
1993	6	144.5	4.2
1994	7	148.2	3.7
1995	8	152.4	4.2
1996	9	156.9	156.9 − 152.4 = 4.5

Even though the differences vary a little, their average value is about 4.8. We can conclude that the average rate of change is nearly constant. For each year, there is about a 4.8 increase in the CPI. This constant difference is a clue that a linear relationship exists between the independent and dependent variable values. This analysis is all part of **deciding which model is the best**.

After selecting the model, the next step is to use a computer or calculator to **compute the regression**. The common method used to compute linear regression is called the **least-squares method**, developed by mathematician G. Yule, who used a technique developed nearly a century before by Adrien Lagendre. The linear regression model takes the form $f(x) = ax + b$. For the CPI data, the calculator output of the regression is similar to that shown in Figure 11.9.3.

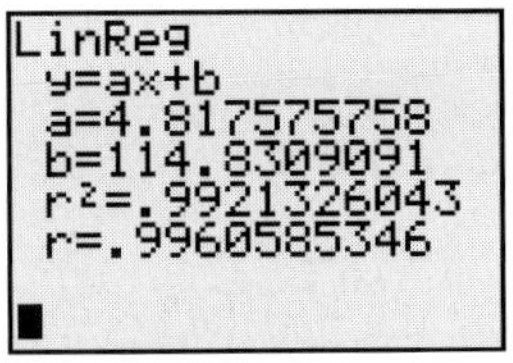

Figure 11.9.3

Note: The DIAGNOSTICS ON indicator may need to be activated to get r^2 and r. Consult the online calculator manual at www.prenhall.com/armstrong.

Rounding to the nearest hundredth, we obtain the linear regression model $f(x) = 4.82x + 114.83$ for the data. Note that the slope of the line is $m = 4.82$, which is close to the average annual difference of 4.8. The graph of this model, along with the scatterplot, is shown in Figure 11.9.4. Notice that the line does not fit exactly through the points. Some of the points are above and some are below the regression line.

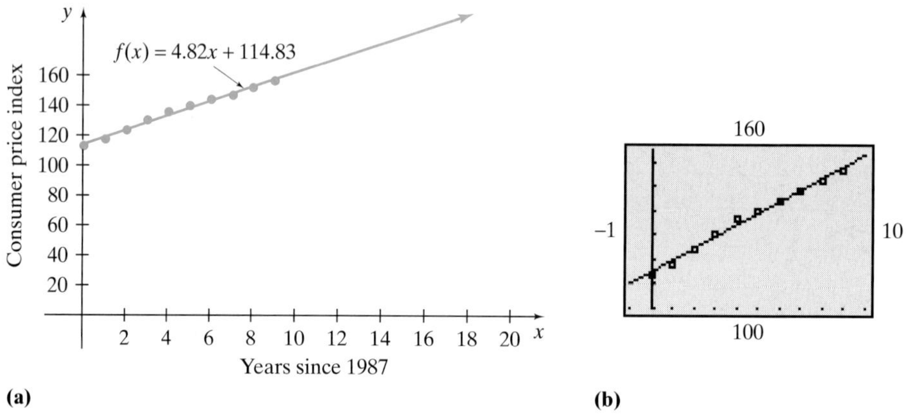

Figure 11.9.4

The final step in the curve-fitting process is to check for **goodness of fit** of the model to the tabular data. Notice in Figure 11.9.3 that values for r and r^2 are included in the regression output. These values measure the quality of fit. The measure used for the linear regression model is the **correlation coefficient *r*.**

Correlation Coefficient

The **correlation coefficient** r, which ranges from -1 to 1, measures the strength and direction of the linear relationship between x and y.

The interpretation of the correlation coefficient for linear regression is summarized in Table 11.9.4. (See page 713.)

Since the CPI linear model has a correlation coefficient of $r = 0.996$, there is a strong positive relationship between x and y. Calling a relationship strong means that the data points on the scatterplot generally line up. A positive r means that as the x-values increase so do the y-values.

The r^2-value is a measure of goodness of fit for nonlinear functions, which we will discuss next. Before continuing, let's summarize the process used in curve fitting tabular data.

The Regression Model Process

1. **Standardize the independent variable** x by setting the first independent value equal to 0 or 1 and labeling all the other independent values accordingly.
2. **Make a scatterplot** of the data.

3. **Inspect the dependent values** to see if there is a pattern in the data that appears to be consistent with a familiar type of function. Use this information to **decide which curve-fitting model** to use.
4. Use a computer or graphing calculator to **compute the regression**. Record the constants along with the r and r^2 values.
5. Verify for **goodness of fit** by checking the r- and r^2-values. The closer that these numbers are to -1 and 1 the better the model fits the data.

Table 11.9.4 Interpretation of r for Linear Models

If the scatterplot looks like	Then x and y Have	AND The r Value Is
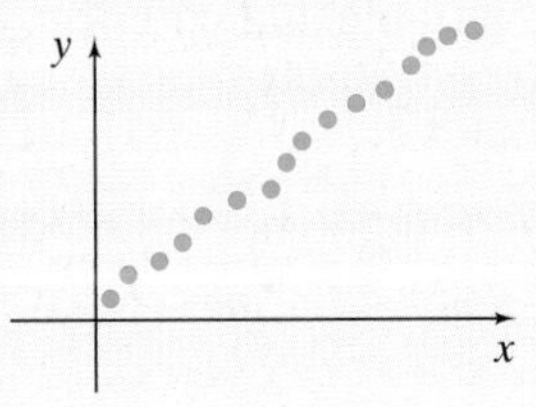	A strong positive association	close to 1
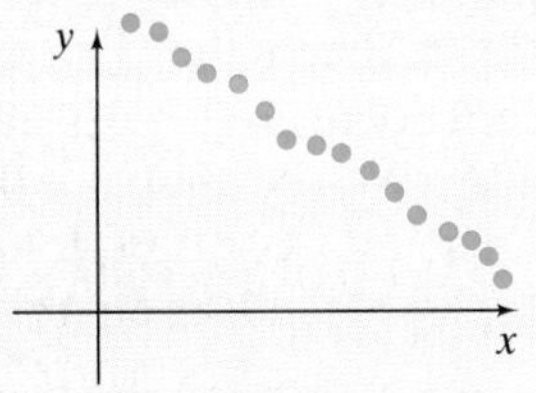	A strong negative association	close to –1
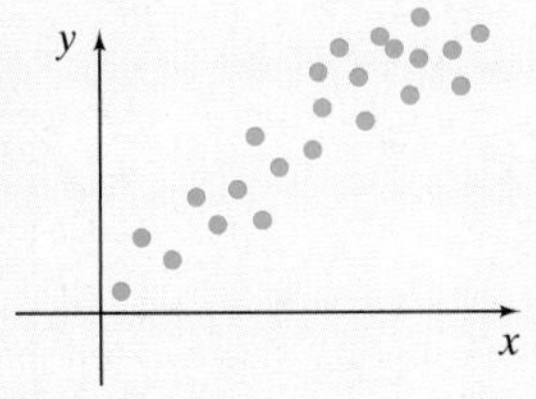	A weak positive association	between 0 and 1
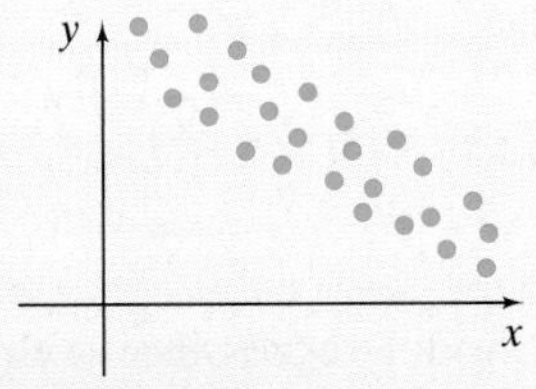	A weak negative association	between –1 and 0
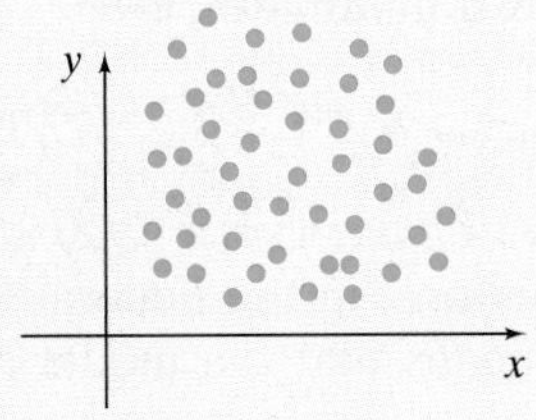	Little or no association	close to zero

Analyzing Nonlinear Regression Models

Many of the mathematical models that we have seen in this chapter are nonlinear models, such as quadratic, cubic, exponential, or logarithmic models. The curve-fitting process is done the same way for these models. For quadratic, cubic, and quartic (fourth-degree) functions, the r^2 value is used in lieu of the correlation coefficient. The closer that the r^2-value is to 1 the better the model fits the data.

Example 1 Fitting a Polynomial Model

Table 11.9.5

Year	Participants
1990	20.1
1992	25.4
1993	27
1994	27.5
1995	26.6
1996	25.5

SOURCE: U.S. Census Bureau, www.census.gov

Table 11.9.5 shows the number of participants (in millions) in the Federal Food Stamp Program from 1990 to 1996.

(a) Standardize the independent variable and enter the data in a table, make a scatterplot of the data, decide which polynomial model to use, and explain why.

(b) Compute the regression to get the polynomial model. Record the r^2-value and interpret.

Solution

Table 11.9.6

Number of Years Since 1990	Participants
0	20.1
2	25.4
3	27
4	27.5
5	26.6
6	25.5

(a) We can standardize the independent variable values by letting x represent the number of years since 1990. Then the data table looks like Table 11.9.6. Standardizing the independent variable values this way simplifies the interpretation of the results. We see that $x = 0$ corresponds to 1990, $x = 2$ corresponds to 1992, and so on. A scatterplot of the data is shown in Figure 11.9.5. Since the scatterplot resembles a parabola, we will try a quadratic model.

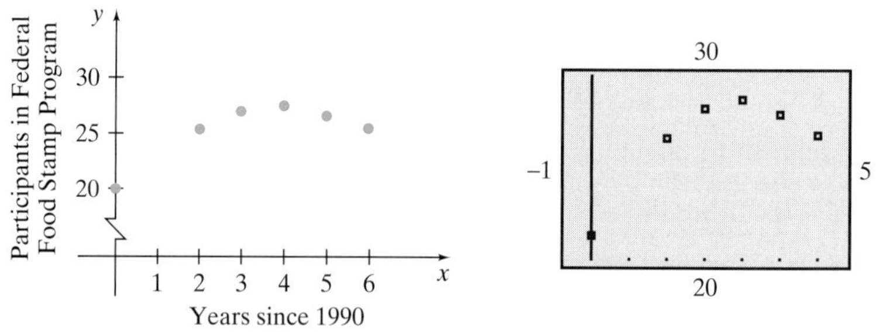

(a) **(b)**

Figure 11.9.5

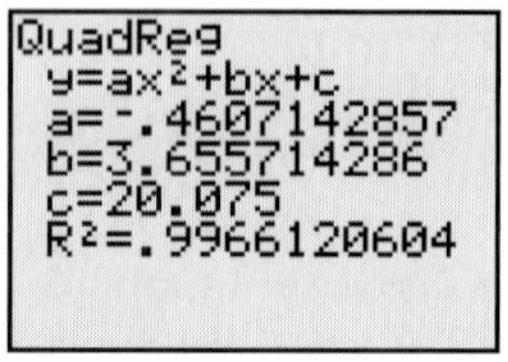

Figure 11.9.6

(b) The result of the quadratic model regression is shown in Figure 11.9.6. From the output, we see that the quadratic model of the data, with all the constants rounded to the nearest hundredth, is

$$f(x) = -0.46x^2 + 3.66x + 20.08 \qquad 0 \le x \le 6$$

where x represents the number of years since 1990 and $f(x)$ represents the number of participants, in millions, in the Federal Food Stamp Program. We have $r^2 = 0.996$, which means that the model fits the data very well since 0.996 is close to 1. ■

Interactive Activity

Repeat Example 1, but choose a **quartic** regression model.

Curve-fitting Polynomial Models

Most graphing calculators compute regression models for first-, second-, third-, and fourth-degree polynomials. Many times, the choice of polynomial used is determined by the number of peaks and valleys in the data. Following are some tips for each model. Remember that these are loose guidelines; when in doubt, try several models and compare the r- and r^2-values.

- **Linear Model** The form of the linear model is $f(x) = ax + b$, where a and b are real number constants. A linear model is a good choice when the dependent values have approximately equal successive differences and the values are steadily increasing or decreasing Figure 11.9.7a has a strong positive correlation, and Figure 11.9.7b has a strong negative association.

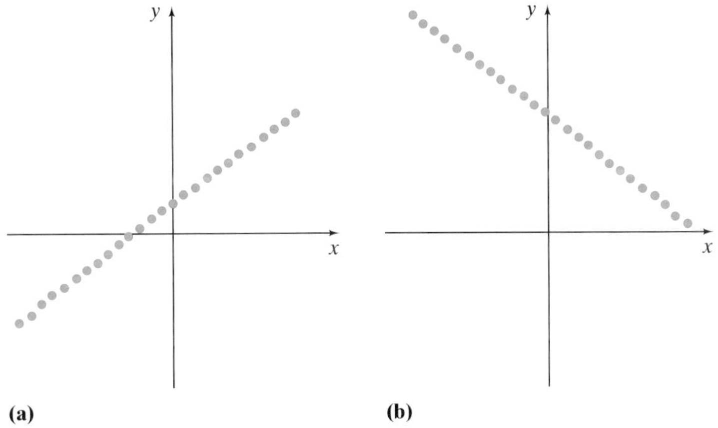

Figure 11.9.7

- **Quadratic Model** The form of the quadratic model is $f(x) = ax^2 + bx + c$, where a, b, and c are real number constants. The quadratic model lends itself to data whose scatterplot has a distinct peak or a distinct valley. See Figure 11.9.8.

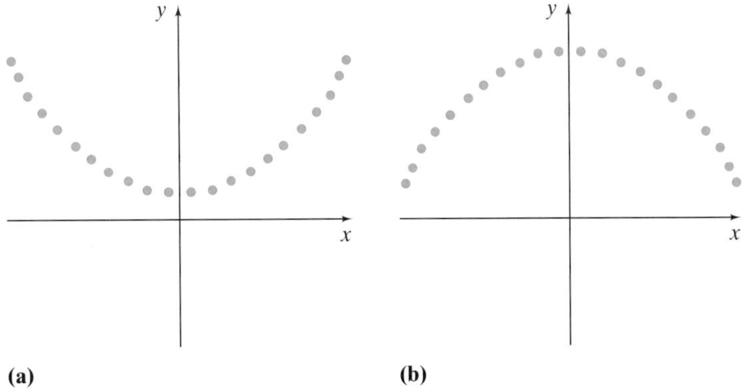

Figure 11.9.8

- **Cubic Model** The form of the cubic model is $f(x) = ax^3 + bx^2 + cx + d$, where a, b, c, and d are real number constants. The cubic model lends itself to data whose scatterplot has values with one distinct peak and one distinct valley. See Figure 11.9.9.

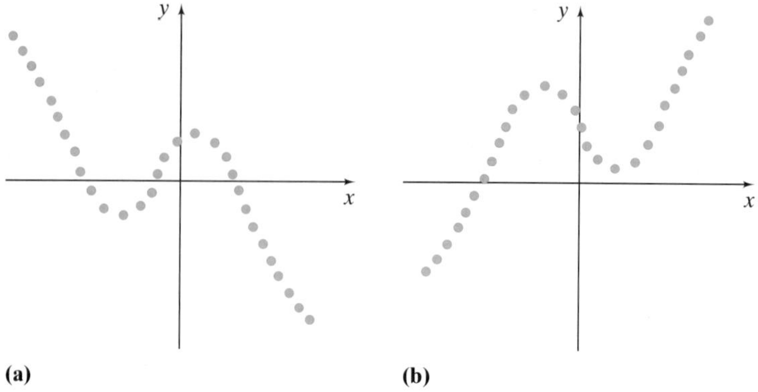

Figure 11.9.9

- **Quartic Model** The form of the quartic model is

$$f(x) = ax^4 + bx^3 + cx^2 + dx + e,$$

where a, b, c, d, and e are real number constants. The cubic model lends itself to data whose scatterplot has a distinct peak between two valleys, or vice versa. See Figure 11.9.10.

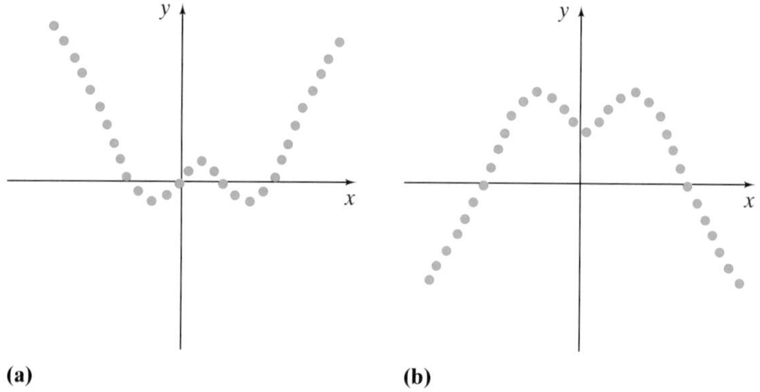

Figure 11.9.10

- **Power Model** The form of the power model is $f(x) = a \cdot x^b$ where a and b are real number constants. The power model lends itself to data whose scatterplot resembles Figure 11.9.11.

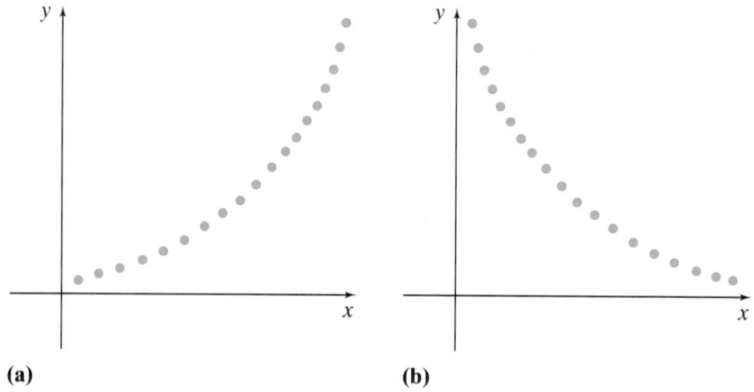

Figure 11.9.11

Modeling with Exponential and Logarithmic Functions

Most graphing calculators compute regressions for general exponential, logarithmic, and logistic functions. A key step when curve fitting with this family of models is standardizing the independent variable. When computing a regression model of the form $y = a \cdot b^x$, the calculator rewrites the equation using logarithms. Since the domain of the logarithmic function is the set of positive numbers, **zero must not be a value of the independent variable.**

Following is a summary of common logarithmic and exponential models.

- **General Exponential Model** The form of the general exponential model is $f(x) = a \cdot b^x$, where a and b are real number constants, with $b > 0$, $b \neq 1$. The exponential model lends itself to data whose scatterplot is always increasing or decreasing on the domain. See Figure 11.9.12.

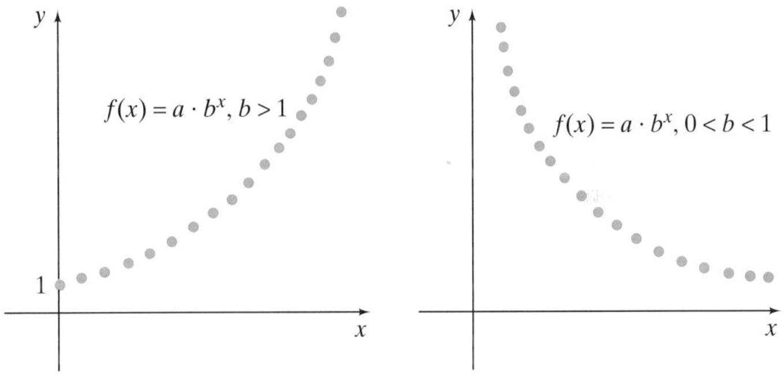

Figure 11.9.12 **(a)** **(b)**

From Your Toolbox

Recall property 6 of the properties of the natural logarithm:

$$e^{\ln x} = x$$

One other property of the general exponential model requires an explanation. First, we need to revisit a property of logarithms as shown in the Toolbox to the left.

Notice that we can replace x with a base like b to get $b = e^{\ln b}$. This allows us to rewrite any general exponential model $f(x) = a \cdot b^x$ in exponential form by writing $\boldsymbol{f(x) = a \cdot b^x}$ **as** $\boldsymbol{f(x) = a \cdot e^{\ln b \cdot x}}$.

- **Logarithmic Model** The form of the logarithmic model is $f(x) = a + b \ln x$, where a and b are real number constants. The logarithmic model lends itself to data whose scatterplot initially increases rapidly and then begins to level off. See Figure 11.9.13.
- **Logistic Model** The form of the logistic model is $f(x) = \dfrac{c}{1 + a \cdot e^{-bx}}$ where a, b, and c are real number constants. The logistic model lends itself to data whose scatterplot increases (decreases) slowly, then increases (decreases) rapidly, and then levels off. See Figure 11.9.14.

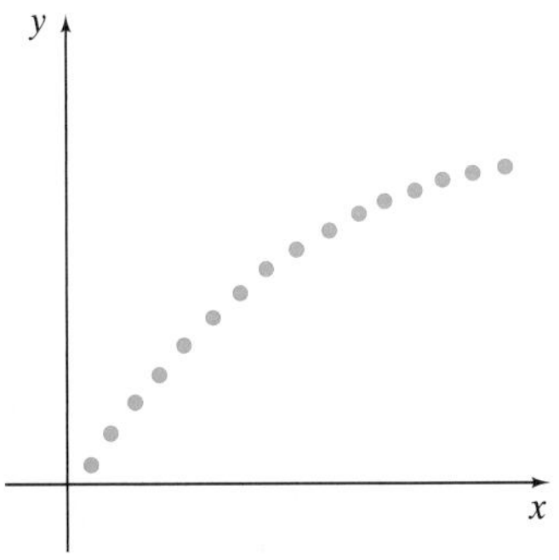

Figure 11.9.13

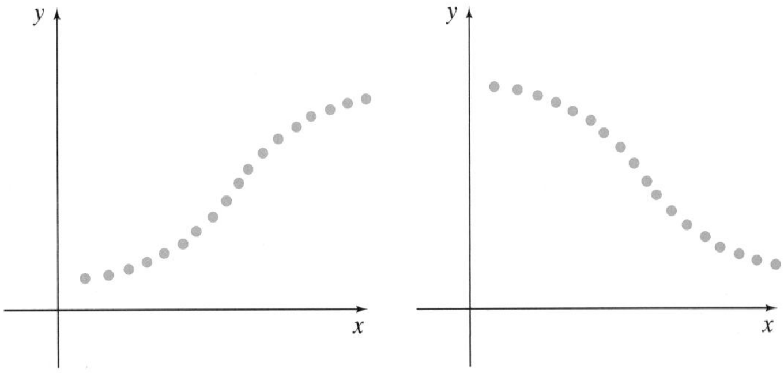

Figure 11.9.14 **(a)** **(b)**

Example 2 Fitting an Exponential/Logarithmic Model

Table 11.9.7

Year	Number Employed
1980	462.5
1990	1174.3
1994	1274.0
1995	1299.6
1996	1306.1

SOURCE: U.S. Census Bureau, www.census.gov

Table 11.9.7 shows the number of employees in the environmental industry (measured in thousands of employees) in the United States.

(a) Standardize the independent variable, make a scatterplot of the data, and explain which, exponential or logarithmic, model to use.

(b) Compute the regression to determine the model. Record the r^2-value and interpret.

Solution

(a) Since the data start in the year 1980, and we want to start our independent variable values at $x = 1$, let's assign x to represent the number of years since 1979. This gives us the data shown in Table 11.9.8.

A scatterplot of the data is shown in Figure 11.9.15. Since the data in this scatterplot increase rapidly at first and then begin to level off a bit, we suspect that a logarithmic model may be a good choice.

Table 11.9.8

Years Since 1979	Number Employed
1	462.5
11	1174.3
15	1274
16	1299.6
17	1306.1

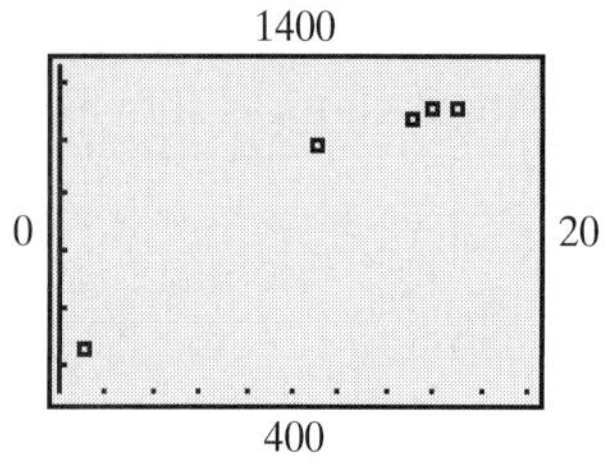

Figure 11.9.15

(b) The result of the logarithmic model regression is shown in Figure 11.9.16. From the output, we see that the logarithmic model of the data, with the coefficients rounded to the nearest hundredth, is

$$f(x) = 461.87 + 299.40 \ln x \qquad 1 \le x \le 17$$

where x represents the number of years since 1979 and $f(x)$ represents the number of employees in the environmental industry (measured in thousands of employees). Since $r^2 \approx 0.999$, this means that the model fits the data very well, since this number is close to 1. The model and the scatterplot are shown in Figure 11.9.17.

Figure 11.9.16

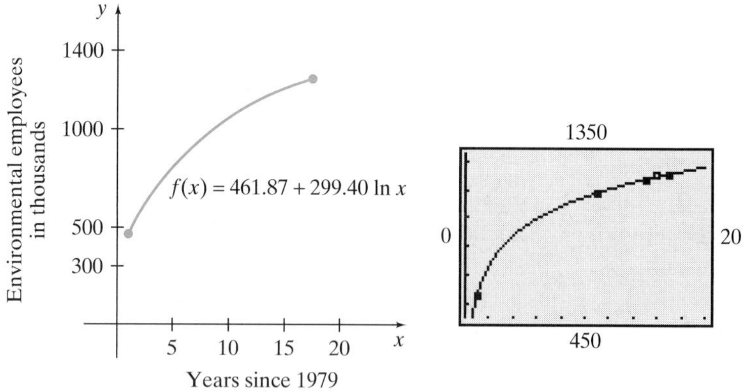

(a) **(b)**

Figure 11.9.17

■

Shortcomings and Cautions

Curve fitting is a powerful tool that can be used to enhance the study of calculus. However, it does have some limitations. One of these is that regression models cannot be used to make long-term predictions. Remember that with many models, particularly polynomial models, the end behavior is given by either $-\infty$ or ∞. However, almost all phenomena in the business, life, and social sciences have some kind of physical, legal, or practical limitation. This is why almost all the regression models that we use are defined on closed intervals using the reasonable domain.

We also must be cautious of particular data points that appear to be much different from the rest of the data. Called **outliers**, they can throw off the accuracy of the resulting regression model, particularly if they occur for large independent values. If the outlier occurs as either the first or last data point, it can be excluded in order to make the model more representative of the entire set of data.

SUMMARY

This section built the framework for data modeling that is seen throughout this textbook. We discussed the curve-fitting procedure that derives a function model from real-world data. Then we outlined the following types of regression models:

- **Linear model**: $f(x) = ax + b$
- **Quadratic model**: $f(x) = ax^2 + bx + c$
- **Cubic model**: $f(x) = ax^3 + bx^2 + cx + d$
- **Quartic model**: $f(x) = ax^4 + bx^3 + cx^2 + dx + e$
- **General Exponential model**: $f(x) = a \cdot b^x$
- **Exponential model**: $f(x) = a \cdot e^{\ln b \cdot x}, b > 0, b \neq 1$
- **Logarithmic model**: $f(x) = a + b \ln x$
- **Logistic model**: $f(x) = \dfrac{c}{1 + a \cdot e^{-bx}}$

SECTION 11.9 EXERCISES

For Exercises 1–10, determine which type of regression model would be a good choice based on the scatterplot. Answers may vary.

1.

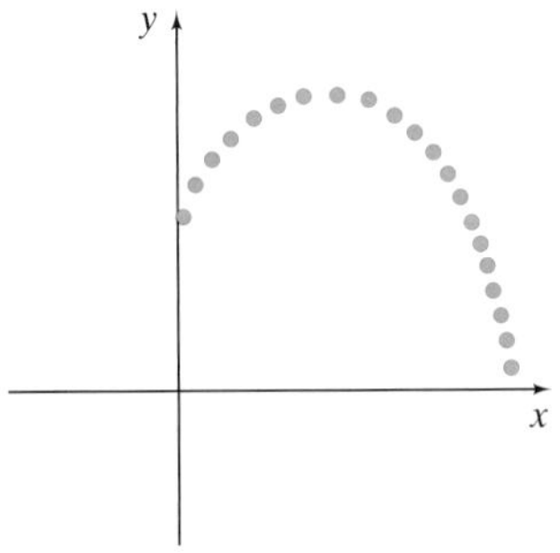

2.

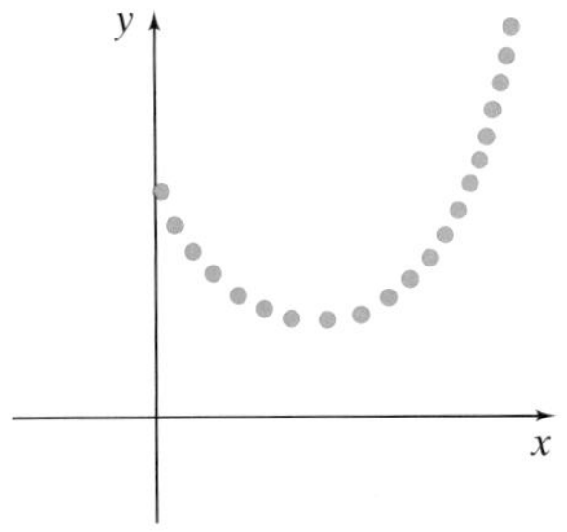

3.

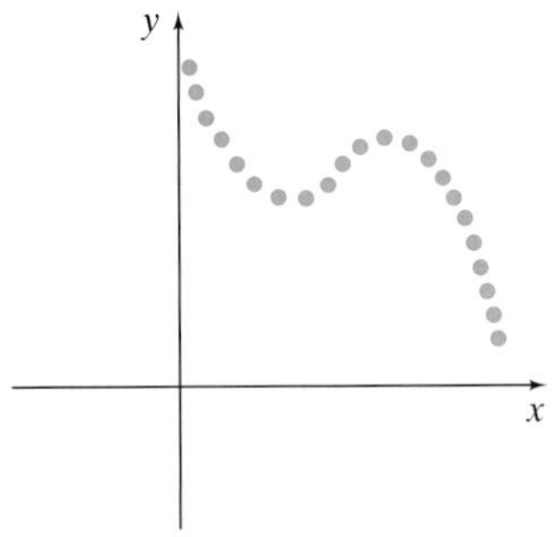

4.

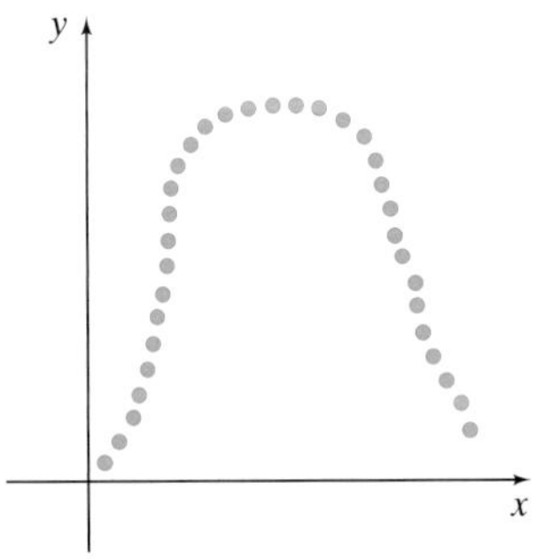

5.

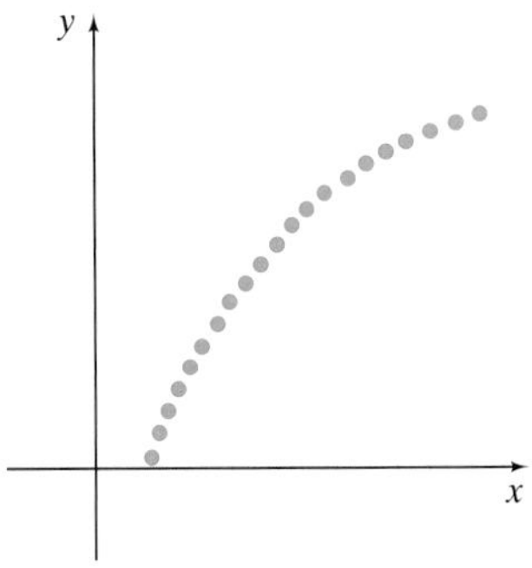

6.

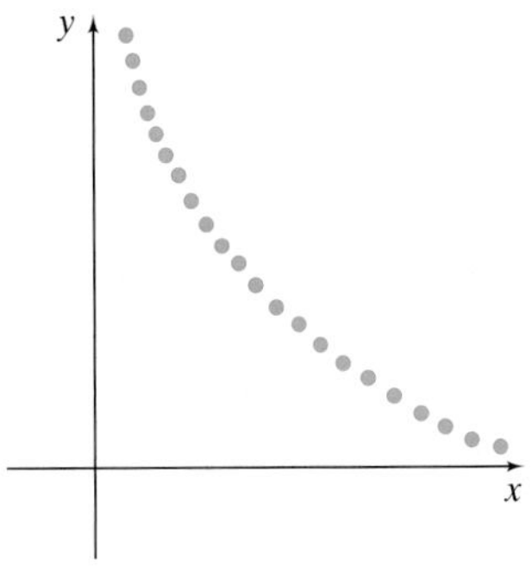

7.

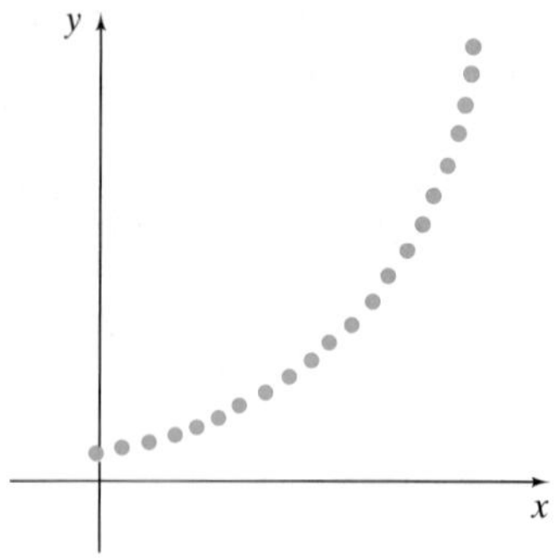

8.

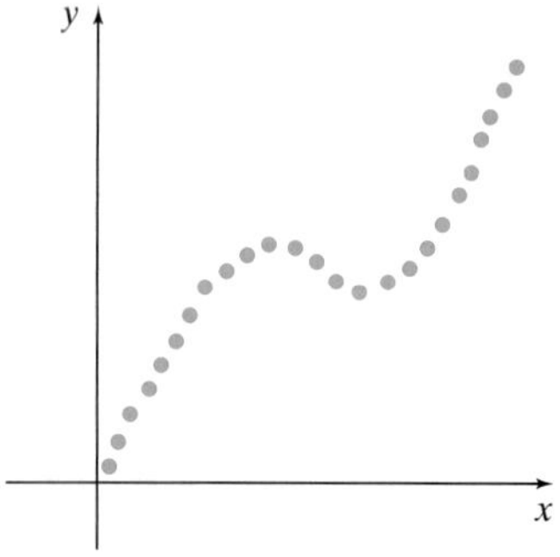

9.

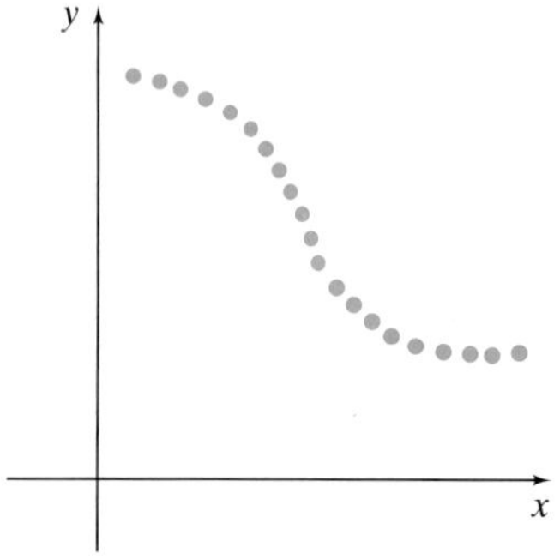

10.

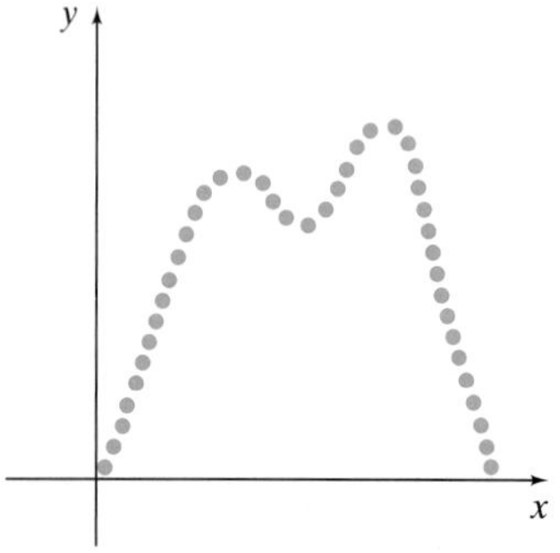

For Exercises 11–14, follow the curve-fitting process to find an appropriate model for the data.

11.2 **11.** Use a linear model, where x represents the number of years since 1980.

Year	Multiple Births (twins, triplets, etc.) Per 1000 Live Births
1980	19.3
1985	21
1990	23.3
1991	23.9
1992	24.4
1993	25.2
1994	25.7
1995	26.1
1996	27.4
1997	28.6
1998	30

SOURCE: U.S. National Center for Health Statistics, www.cdc.gov/nchs

11.2 **12.** Use a cubic model, where x represents the number of years since 1975.

Year	Number of U.S. Golf Facilities
1975	11,370
1980	12,005
1985	12,346
1990	12,846
1992	13,210
1993	13,439
1994	13,683
1995	14,074
1999	14,900

SOURCE: National Golf Foundation, www.ngf.org

11.2 **13.** Use a logarithmic model, where x represents the number of years since 1989.

Year	Average Expenditure for a New Car (in $)
1990	15,926
1991	16,650
1992	17,825
1993	18,585
1994	19,463
1995	19,757

SOURCE: U.S. Bureau of Economic Analysis, www.bea.doc.gov

11.2 **14.** Use a power function model, where x represents the number of years since 1969.

Year	Percentage of 3- to 5-Year-Olds in Preschool
1970	37.5
1975	48.6
1980	52.5
1985	54.6
1990	59.4
1992	55.5
1993	55.1
1994	61
1995	61.8
1998	64.5

SOURCE: U.S. Census Bureau, www.census.gov

15. To see the shortcomings of regression modeling, consider the mathematical model of the percentage of births to unmarried mothers.

$$f(x) = -0.01x^3 + 0.14x^2 + 0.46x + 21.46 \qquad 1 \le x \le 14$$

Here, x represents the number of years since 1984 and $f(x)$ is the percentage of births to unmarried mothers (*Source:* U.S. National Center for Health Statistics, www.cdc.gov/nchs).

(a) Evaluate $f(10)$ and interpret.

(b) Evaluate $f(22)$ and interpret.

(c) Why does the solution to part (b) not make sense?

Why We Learned It

In this chapter, we supplied a review of functions and the different kinds of functions. Included in this review was the concept that the slope of a secant line over an interval gives the average rate of change of the function over the interval. The average rate of change concept was applied to many diversified applications. For example, suppose that we have a function that models the annual amount spent by the United States on pollution abatement from 1985–2000. If we want to know the average rate of change in the amount spent on pollution abatement by the United States from 1990–2000, we would find the slope of the secant line over the interval that corresponds to these years. In nearly every section of this chapter we encountered the average rate of change concept. As we will see in the next chapter, a firm understanding of the average rate of change concept eases the understanding of the instantaneous rate of change concept.

CHAPTER REVIEW EXERCISES

1. Use the table to construct ordered pairs and plot them in the Cartesian plane.

x	-1	2	2.8	4
$f(x)$	3.5	1	-1.7	2

2. Make a table and write the ordered pairs based on the scatterplot.

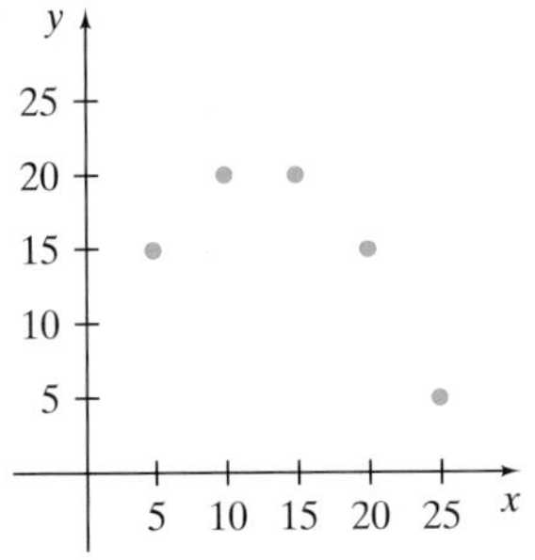

3. For the function $A = \pi r^2$, identify the independent and the dependent variable.

4. Let x represent the number of years since 1990. Standardize the values of the independent variable based on this definition, and make a scatterplot of the data.

Year	U.S. Population (in millions)
1910	92.0
1920	105.7
1930	122.8
1940	131.7
1950	150.7
1960	179.3
1970	203.3
1980	226.5
1990	248.7
1999	272.9

SOURCE: U.S. Census Bureau, www.census.gov

5. Determine if the table represents a function.

Domain	1	2	3	4	5
Range	5	3	1	3	5

6. Determine if the graph represents a function.

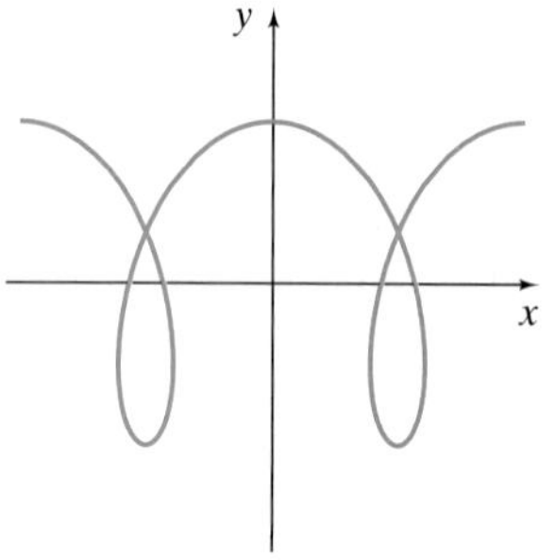

For Exercises 7 and 8, let $f(x) = x^2 - 4$ and $g(x) = 5x + 6$. Evaluate each expression and write the solution as an ordered pair.

7. $f(3)$

8. $g(-3)$

9. Use the function $y = f(x)$ to answer parts (a) through (f).

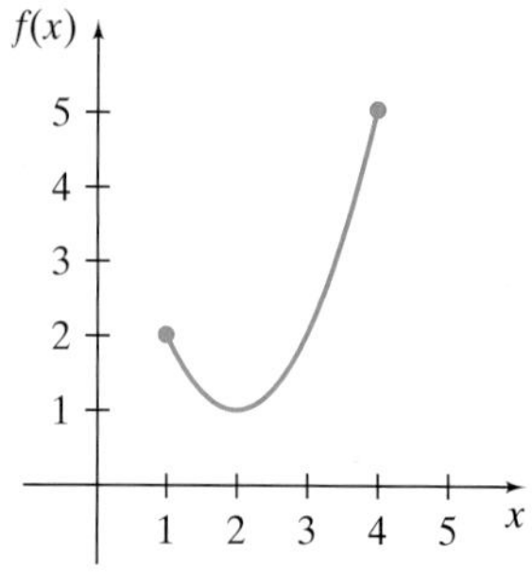

(a) What is the independent variable?
(b) What is the dependent variable?
(c) What is the value of $f(x)$ when $x = 3$?
(d) What is $f(1)$?
(e) What is the domain of f?
(f) What is the range of f?

10. Solve the linear inequality $3x - 7 \geq 20$, and write the solution using interval notation.

For Exercises 11 and 12, determine the domain of the function algebraically, and write the domain using interval notation.

11. $f(x) = \sqrt{-4 + 2x}$

12. $g(x) = \dfrac{x + 3}{x^2 + 3x - 10}$

13. Graph the function $f(x) = \dfrac{x + 3}{x^2 - 4}$ on your calculator, and determine the domain and range of the function. Write your solutions using interval notation.

14. The price–demand function for the new EasyType computer keyboard is given as

$$p(x) = 35 - 0.2x \qquad 0 \leq x \leq 50$$

Here x represents the number of units sold per day and $p(x)$ represents the price of one unit in dollars.

(a) Complete the following table of values.

x	0	10	20	30	40	50
$p(x)$						

(b) Make a graph of p on the interval $[0, 50]$.
(c) If $[0, 50]$ is the domain, what is the range of the function?
(d) Evaluate $p(35)$ and interpret the answer.

15. Using the graph, write the domain and range of the function using interval notation.

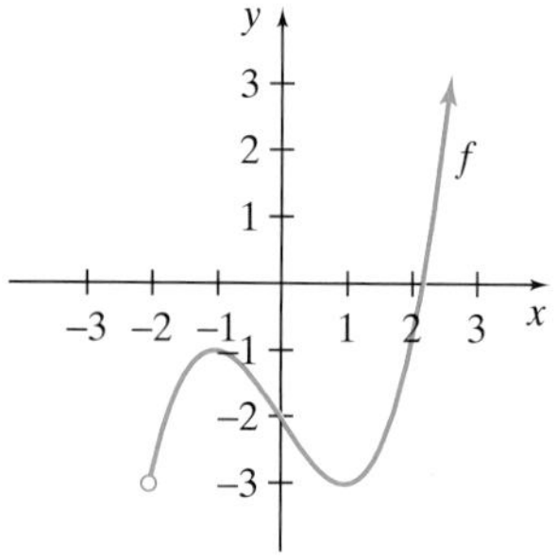

16. Find the average rate of change between the points (3, 7) and (−2, 14).

17. Given the graph, find the average rate of change.

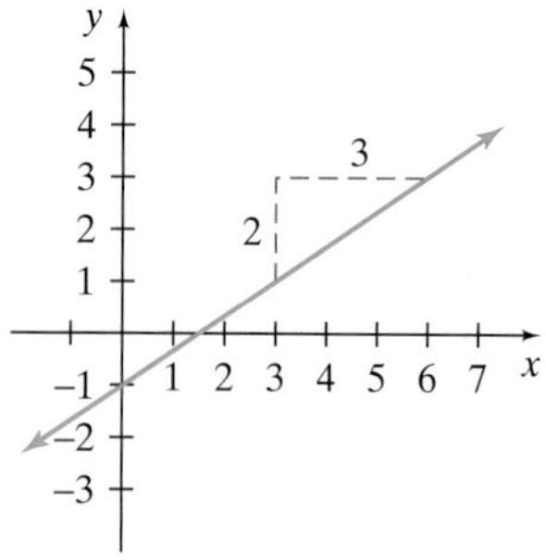

18. A line contains the points (2, 1) and (8, 19). Write the equation of the line in point–slope form and graph.

19. A line has an x-intercept at −4 and a y-intercept at 5. Write the equation of the line in point–slope form and graph.

20. A line has a slope of $\frac{3}{7}$ and passes through the point (−3, 2). Write the equation of the line in CF form.

21. A linear function contains the data points in the table. Write the equation of the function in slope–intercept form.

x	y
−4	−10
12	2

22. Determine the x- and y-intercepts for the graph of the function $g(x) = \frac{60 - 9x}{4}$.

23. A new bicycle costs \$800 and depreciates at the rate of \$40 per year. Answer the following using the depreciation model.

(a) Determine the equation of the depreciation function.

(b) In how many years will the bicycle be worth \$600?

24. Classify the function given by $f(x) = -3(2 - 5x)$ as increasing or decreasing, and identify the slope of the line.

25. For a certain product, the fixed costs are \$2600 and it costs \$3950 to produce 18 units. Write the equation of the linear cost function in the slope–intercept form $C(x) = mx + b$.

26. The manager of Pizza Pizzazz determines that the cost of preparing and delivering x pizzas is given by the cost function $C(x) = 300 + 6.5x$.

(a) What are the fixed and variable costs for producing the pizzas?

(b) Evaluate $C(36)$ and interpret the answer.

(c) How many pizzas can be prepared and delivered for \$703?

(d) What is the marginal cost for the pizzas?

(e) If the current production level is 75 pizzas, what is the cost of preparing and delivering the 76th pizza?

27. Make an accurate graph of the piecewise-defined function.

$$f(x) = \begin{cases} 1 - 0.5x, & x \le 1 \\ 3, & x > 1 \end{cases}$$

28. Graph the function $g(x) = |12 - 4x|$ and then rewrite the function in piecewise form.

29. Consider the daily cost schedule for producing x CasaVerde ceiling fans.

Fans Produced	Cost (in dollars)
0	2,500
50	5,500
220	15,700

(a) What are the fixed costs? Explain how you determined this from the table.

(b) How much does the cost increase for each additional ceiling fan produced? How do you classify this kind of cost?

(c) Write a cost function C, where x represents the number of ceiling fans produced and $C(x)$ represents the cost.

(d) What would the cost be if 84 ceiling fans were produced?

For Exercises 30 and 31:

(a) Determine the coordinates of the vertex algebraically.

(b) Write the range of the function using interval notation.

30. $f(x) = x^2 - 4x + 3$

31. $f(x) = -2x^2 - 6x + 12$

For Exercises 32 and 33:

(a) Determine the vertex.

(b) Use the result from part (a) to determine the intervals where the function is increasing and where it is decreasing.

32. $g(x) = 5x^2 + 20x$ **33.** $g(x) = -x^2 + 3.2x - 7.8$

For Exercises 34 and 35, find the zeros of the quadratic function by factoring.

34. $f(x) = x^2 - 144$ **35.** $f(x) = 2x^2 + 3x - 20$

36. Find the real number roots of the quadratic function $f(x) = 3x^2 - 6x + 2$ by using the quadratic formula. If the discriminant is a negative number, write *no real roots.*

37. Use the ZERO or ROOT capabilities of your calculator to approximate the zeros of the quadratic function $g(x) = 2.4x^2 + 1.7x - 5.3$. Round to the nearest hundredth. If the graph has no x-intercepts, write *no real roots.*

38. Consider the function $f(x) = -1.8x + 5.3$. Use the difference quotient to determine the slope of the line over the following intervals:

(a) As x changes from $x = 0$ to $x = 2$

(b) As x changes from $x = 0$ to $x = 5$

(c) As x changes from $x = 4$ to $x = 7$

39. Consider the function $g(x) = 2x^2 - 3x$. Use the difference quotient to determine the slope of the secant line over the following intervals:

(a) If $x = 0$ and $\Delta x = 1$

(b) If $x = 0$ and $\Delta x = 2$

(c) If $x = 0$ and $\Delta x = 4$

40. The average homeowner's property tax in a certain community can be modeled by

$$f(x) = -15x^2 + 300x + 1200 \qquad 0 \le x \le 9$$

where x represents the number of years since 1990 and $f(x)$ is the average property tax.

(a) Does the graph of the model open up or down? How do you know this?

(b) The domain of the model consists of x-values in the interval $[0, 9]$. What is the corresponding range?

(c) Determine the average rate of change in the property tax from 1992 to 1997.

41. The amount of time that an average American spends per year watching premium cable television programming can be modeled by

$$f(x) = 0.49x^2 - 3.41x + 88.50 \qquad 0 \le x \le 10$$

where x represents the number of years since 1990 and $f(x)$ is the number of hours spent watching premium cable television programming (*Source:* U.S. Statistical Abstract, www.census.gov/statab/www).

(a) Evaluate $f(4)$ and interpret.

(b) Make a graph of the model in the window [0, 10] by [75, 110].

(c) The domain of the model consists of x-values in the interval [0, 10]. What is the corresponding range?

(d) Make a table of values for $x = 0, 1, 2, 3, \ldots, 10$.

(e) According to the model, during what years did the average American spend less than 85 hours watching premium cable television programming?

For Exercises 42 and 43, simplify and write the domain for the following functions.

(a) $(f + g)(x)$

(b) $(f - g)(x)$

(c) $(f \cdot g)(x)$

(d) $\left(\frac{f}{g}\right)(x)$

42. $f(x) = x^2 + 4,\ g(x) = 2x + 3$

43. $f(x) = \sqrt{x + 2},\ g(x) = x - 2$

For Exercises 44–47, let $f(x) = 3x - 5$ and $g(x) = (x + 2)^2$. Evaluate each of the following:

44. $(f + g)(2)$

45. $(f - g)(4)$

46. $(f \cdot g)(-2)$

47. $\left(\frac{f}{g}\right)(3)$

48. The population of Eastbrook from 1980 to 1990 can be modeled by

$$f(x) = 80{,}300 + 800x \qquad 0 \le x \le 10$$

where x represents the number of years since 1980 and $f(x)$ represents the population. The population of Westbrook during the same time period is given by

$$g(x) = 62{,}600 - 200x \qquad 0 \le x \le 10$$

(a) Evaluate $(f + g)(4)$ and interpret.

(b) Evaluate $(f - g)(8)$ and interpret.

49. For the price function $p(x) = 145 - 0.15x$, determine the revenue function R.

50. Suppose that the revenue and cost functions for a certain product are given by $R(x) = 18x - 0.08x^2$ and $C(x) = 5x + 275$.

(a) Graph R and C in the same viewing window.

(b) Use algebra or the INTERSECT command on your calculator to determine the break-even point.

(c) Determine how much revenue must be generated to reach the break-even point.

51. For the revenue and cost functions $R(x) = 32x - 0.08x^2$ and $C(x) = 16x + 400$, determine the profit function P, find the vertex, and interpret.

52. For a certain product, the price–demand function is $p(x) = 400 - 0.1x$. The variable costs are \$225 per unit, and the fixed costs are \$3000.

(a) Determine the linear cost function C.

(b) Determine the revenue function R.

(c) Determine the profit function $P(x) = R(x) - C(x)$.

(d) Use the ZERO or ROOT command on your calculator to find the zeros of the profit function P. These are the break-even points.

(e) Find the vertex of the graph of the profit function P.

(f) Determine the demand level that yields the maximum profit, and find the maximum profit.

53. Determine if each function is a polynomial function. Identify each polynomial as linear, quadratic, cubic, or quartic.

(a) $f(x) = x^2 + 5x - 3$

(b) $f(x) = 17$

(c) $g(x) = \frac{1}{3}x - 4$

(d) $g(x) = \frac{3}{4}x^4 - \frac{2}{3}x^3 + \frac{1}{2}x^2 + x$

For Exercises 54 and 55, determine the end behavior of each function.

54. $f(x) = 5x - 2x^3$

55. $g(x) = 3x^2 - 5x$

56. The Disma Department Store finds that the revenue generated by selling x dresses is given by

$$R(x) = 68x - 0.3x^2 \qquad x \ge 0$$

(a) Evaluate $R(42)$ and interpret.

(b) Find the average rate of change for $x = 40$ and $\Delta x = 10$ and interpret.

57. Consider the polynomial function

$$f(x) = x^3 - 6x^2 + 6x + 4.$$

(a) Determine the end behavior of the function.

(b) Use the MAXIMUM and MINIMUM commands on the calculator to approximate the peaks and valleys of the function. Estimate the points to the nearest hundredth.

(c) Use the solution from part (b) to determine the intervals where f increases and where it decreases.

For Exercises 58–61:

(a) Write the domain using interval notation.

(b) Identify any holes or vertical asymptotes in the graph.

(c) Identify any horizontal asymptotes.

58. $f(x) = \dfrac{x+3}{x-1}$

59. $g(x) = \dfrac{x^2}{2x+3}$

60. $g(x) = \dfrac{x^2 - x - 6}{x^2 + 3x + 2}$

61. $f(x) = \dfrac{2x^2 - x - 15}{x^2 - 9}$

For Exercises 62–65, determine the x- and y-intercepts for the graph of the given function.

62. $g(x) = \dfrac{2x}{x-5}$

63. $f(x) = \dfrac{x^2 - 8x + 16}{x^2 - 9}$

64. $f(x) = \dfrac{x^2 + 5x + 6}{x^2 + x - 2}$

65. $g(x) = \dfrac{12}{x^2 + 6}$

66. For the function $f(x) = \dfrac{3x^2 - 5x + 2}{x+1}$, determine the x- and y-intercepts graphically.

67. Sketch a graph by hand for $f(x) = \dfrac{x+3}{x-2}$ on the interval $1 \le x \le 10$. Label the asymptotes and the intercepts.

68. Consider the cost–benefit function for removing a certain pollutant from the atmosphere.

$$f(x) = \frac{26x}{100 - x} \qquad 0 \le x < 100$$

where x represents the percentage of the pollutant removed and $f(x)$ represents the associated cost in millions of dollars.

(a) Evaluate $f(60)$ and interpret.

(b) Evaluate $f(95)$ and interpret.

69. Rewrite the radical function $f(x) = \sqrt[4]{(x-2)^3}$ as a rational exponent function. Also, write the domain using interval notation.

70. Rewrite the rational exponent function $f(x) = (2x - 7)^{3/5}$ as a radical function. Also, write the domain using interval notation.

71. The time required for an object to fall a distance $f(x)$ is given by the model

$$f(x) = \frac{1}{4}\sqrt{x}$$

where x represents the distance in feet and $f(x)$ represents the time in seconds.

(a) Evaluate $f(64)$ and interpret.

(b) How long does it take an object to fall a distance of 144 feet?

72. A biologist has shown that the number of known insect species in a certain desert is related to the area of the land studied by the model

$$g(x) = 16.3\sqrt[3]{x}$$

where x represents the area studied in square miles and $g(x)$ represents the number of insect species.

(a) Evaluate $g(343)$ and interpret.

(b) Find the slope of the secant line for $x = 1000$ and $\Delta x = 331$ and interpret.

73. The total amount spent annually on research and development in the United States is given by the mathematical model

$$f(x) = 19.66x^{0.74} \qquad 5 \le x \le 21$$

where x represents the number of years after 1975 and $f(x)$ represents the amount spent in billions of dollars (*Source:* U.S. Census Bureau, www.census.gov).

(a) Evaluate $f(12)$ and interpret.

(b) Find the average rate of change in the amount spent on R&D for $x = 8$ and $\Delta x = 4$ and interpret.

For Exercises 74–77:

(a) Complete the following table.

x	−5	−4	−3	−2	−1	0	1	2	3	4	5
$f(x)$											

(b) Graph f.

(c) Classify the function as exponential growth or decay.

74. $f(x) = \left(\dfrac{2}{5}\right)^x$

75. $f(x) = 2.3^x$

76. $f(x) = e^{0.8x}$

77. $f(x) = e^{-2.7x}$

78. The SuperValue Grocery chain announced a two-week sale in conjunction with the grand opening of its new location. The percentage of customers responding to the sale can be modeled by

$$f(x) = 65 - 80(0.71)^x$$

where x is the day of the sale and $f(x)$ represents the percentage of customers who respond to the sale.

(a) Evaluate $f(1)$ and interpret.

(b) Find the slope of the secant line for $x = 4$ and $\Delta x = 3$ and interpret.

(c) When does the percentage of customers responding to the sale first exceed 50%?

79. Suppose that Paul deposits \$1700 into an account that pays interest at a rate of 5.3% compounded monthly. How much will be in the account after 7 years?

80. Suppose that Yushu deposits \$20,000 into an account that pays interest at a rate of 7.5%. How much will be in the account after 12 years if the interest is compounded

(a) Annually?

(b) Monthly (12 times a year)?

(c) Weekly (52 times a year)?

81. Suppose that \$8000 is invested at a 6.75% interest rate compounded quarterly (four times a year).

(a) Write the compound interest function

$$A(t) = P\left(1 + \frac{r}{n}\right)^{nt}$$

for the given information and interpret.

(b) Evaluate $A(6)$ and interpret.

(c) Calculate how much interest was made at the end of the 6-year period.

82. Suppose that Frederick deposits \$5000 at a 6.2% interest rate compounded monthly.

(a) Write the compound interest function A for the given information.

(b) Graph A in the viewing window [0, 15] by [5000, 15,000].

(c) Use your calculator to graphically find the doubling time of the account (that is, the amount of time that it takes the account to accumulate a total balance of \$10,000).

83. The population of Granaco City can be modeled by

$$f(x) = 64{,}000e^{0.04x} \qquad 0 \le x \le 20$$

where x represents the number of years since 1980 and $f(x)$ represents the number of people living in Granaco City.

(a) Evaluate $f(4)$ and interpret.

(b) Find the slope of the secant line for $x = 5$ and $\Delta x = 3$ and interpret.

(c) In what year did the population first exceed 110,000?

84. Suppose that \$2500 is invested at a 5.7% interest rate compounded continuously.

(a) Write the continuous compound interest function $A(t) = Pe^{rt}$ for the given information.

(b) Evaluate $A(4)$ and interpret.

(c) Use your calculator to graphically find the doubling time of the account.

85. For each account, compute the effective rate and interpret the result.

(a) An 8.2% nominal rate compounded monthly

(b) A 6.7% nominal rate compounded continuously

86. The population of leopards in a certain region is given by the logistic model

$$f(x) = \frac{160}{1 + 42.29e^{-0.1x}} \qquad 0 \le x \le 60$$

where x represents the number of years since 1940 and $f(x)$ represents the number of leopards.

(a) Evaluate $f(35)$ and interpret.

(b) Compute the difference quotient for $x = 47$ and $\Delta x = 3$ and interpret.

For Exercises 87 and 88, let $f(x) = x^2 - 3x$ and $g(x) = 2x - 5$. Evaluate each expression.

87. $f(g(4))$

88. $g(f(-2))$

For Exercises 89 and 90:

(a) State the domain of each of the functions f and g.

(b) Algebraically simplify $f(g(x))$.

(c) Algebraically simplify $g(f(x))$.

89. $f(x) = 5x + 3$; $g(x) = x^2$

90. $f(x) = \sqrt{3x - 2}$, $g(x) = 3x^2 + 6$

For Exercises 91–94, determine the functions f and g so that $f(g(x)) = h(x)$ for each composite function h; then check your answer. There is more than one correct answer.

91. $h(x) = (x^2 + 2)^5$

92. $h(x) = x^2 + 5$

93. $h(x) = \sqrt{6 - x}$

94. $h(x) = \dfrac{3x - 4}{3x + 10}$

95. Graph the function $g(x) = \dfrac{1}{x}$, and use the horizontal line test to determine if g is a one-to-one function.

For the functions in Exercises 96 and 97:

(a) Show that $f(g(x)) = x$.

(b) Graph f, g, and the line $y = x$ in the same Cartesian plane.

96. $f(x) = 4 - 3x$; $g(x) = \dfrac{-x + 4}{3}$

97. $f(x) = \sqrt[3]{x + 5}$; $g(x) = x^3 - 5$

For Exercises 98 and 99, rewrite the equation in logarithmic form $y = \log_b x$.

98. $3^5 = 243$

99. $10^{-4} = 0.0001$

For Exercises 100 and 101, use the properties of logarithms to rewrite the expression.

100. $\ln 18^3$

101. $\log_2 \frac{5^2}{7}$

For Exercises 102 and 103, solve the exponential equations algebraically and round the solution to the hundredths place.

102. $5 \cdot 4^x = 65$

103. $6.3(1.7)^x = 24$

104. The annual revenues for a certain business can be modeled by

$$f(x) = 1.38 + 2.4 \ln x \qquad 1 \le x \le 10$$

where x represents the number of years since 1990 and $f(x)$ represents the annual revenues, in millions of dollars.

(a) Complete the following table:

x	1	3	5	10
$f(x)$				

(b) Evaluate $f(8)$ and interpret.

(c) Use the table in part (a) to determine the average rate of change in the revenues between 1991 and 1995.

(d) Use the table in part (a) to compute the difference quotient for $x = 5$ and $\Delta x = 5$ and interpret.

For Exercises 105 and 106, determine which type of regression model would be a good choice based on the scatterplot. Answers may vary.

105.

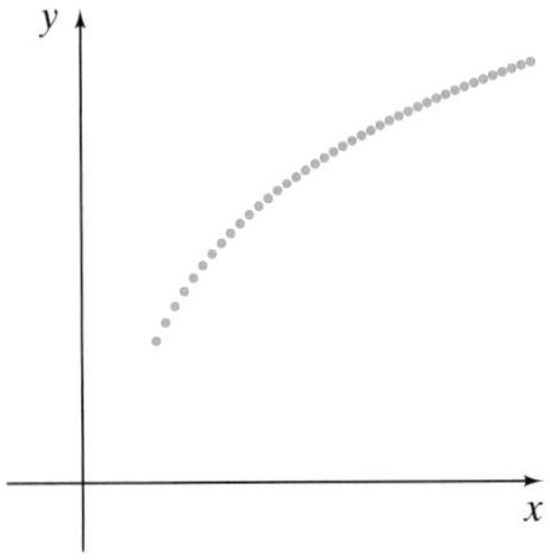

106.

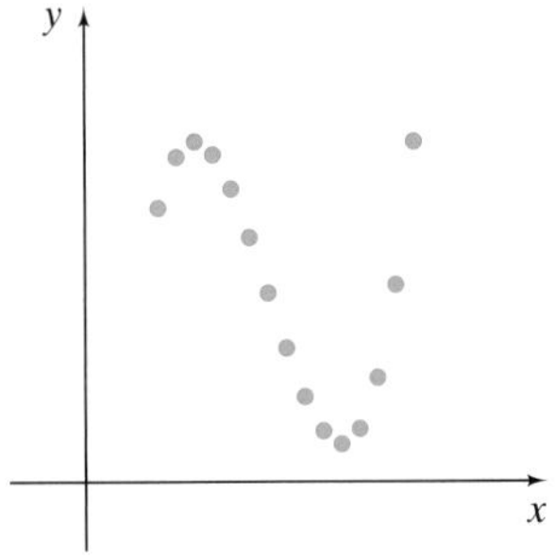

For Exercises 107–109, follow the curve-fitting process to find an appropriate model for the data.

107. Use a linear function model, where x represents the number of years since 1985.

Year	Number of Americans Covered by Private or Government Health Insurance (millions)
1987	241
1989	246
1991	251
1993	260
1995	264
1997	269

SOURCE: U.S. National Center for Health Statistics, www.cdc.gov/nchs

108. Use a quadratic function model, where x represents the number of years since 1960.

Year	Number of U.S. Workers Who Worked at Home (millions)
1960	4.66
1970	2.69
1980	2.18
1990	3.41

SOURCE: U.S. Census Bureau, www.census.gov

109. Use a power function model, where x represents the number of years since 1900.

Year	Average Cost of a New House (in $ thousands)
1963	19.3
1968	26.6
1973	35.5
1978	62.5
1983	89.8
1988	138.3
1993	147.7
1998	181.9

SOURCE: U.S. Census Bureau, www.census.gov

CHAPTER 11 PROJECT

SOURCE: U.S. Census Bureau, www.census.gov

In this chapter, we discussed the properties of various types of functions. One aspect that we have discovered is that data may be modeled by more than one type of function. In this project, we'll explore some of these functions and their accuracy when predicting population growths. We'll use different functions for different periods of time to model the population of the United States between the years of 1960 to 1999. Consider the following.

Years	Type of Function	Population Model ($f(x)$ in thousands)
1960–1979	Quadratic	$f_1(x) = -14.913x^2 + 2{,}577.691x + 178{,}836.684$
	Cubic	$f_2(x) = 2.000x^3 - 77.909x^2 + 3{,}119.858x + 177{,}774.149$
	Quartic	$f_3(x) = -0.095x^4 + 5.998x^3 - 132.812x^2 + 3{,}391.177x + 177{,}427.318$
	Exponential	$f_4(x) = 180{,}848.012(1.011^x)$
1980–1999	Power	$f_5(x) = 94{,}672.892(x^{0.285})$
1960–1999	Logistic	$f_6(x) = \dfrac{560{,}561.048}{1 + 2.101e^{-0.017x}}$

Note: Let x be the number of years since 1959. All decimals are rounded to 3 decimal places. The functions' units are in thousands.

1. Compute and interpret.

(a) The average rate of change from 1965–1975 using the Quadratic function

(b) The average rate of change from 1985–1995 using the Power function

(c) The average rate of change from 1985–1995 using the Logistic function

2. For each of the models, predict the population of the United States in the year 2050. The United States Census Bureau predicts that the population for the year 2050 will be between 313,546 and 552,757 thousand people.

3. Which answers in Exercise 2 were within the predicted interval? Which answers were not? Explain your reasoning using your knowledge of the graph of each type of function.

4. Evaluate each function for large values of x, that is, as x approaches infinity. Which result makes sense?

5. Overall, which function provides the best approximation for the United States population?

Chapter

12 Limits, Instantaneous Rate of Change, and the Derivative

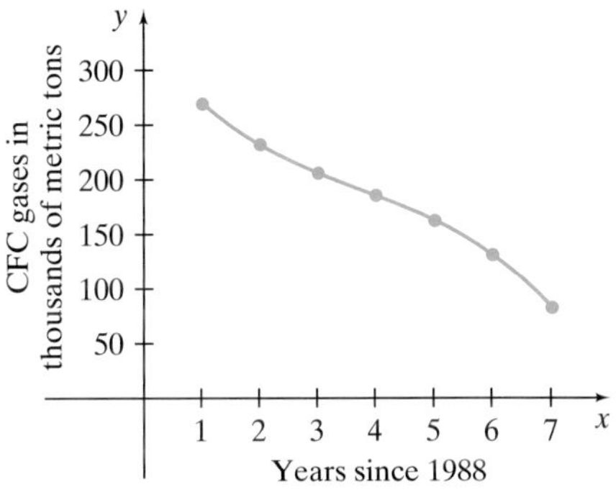

(b)

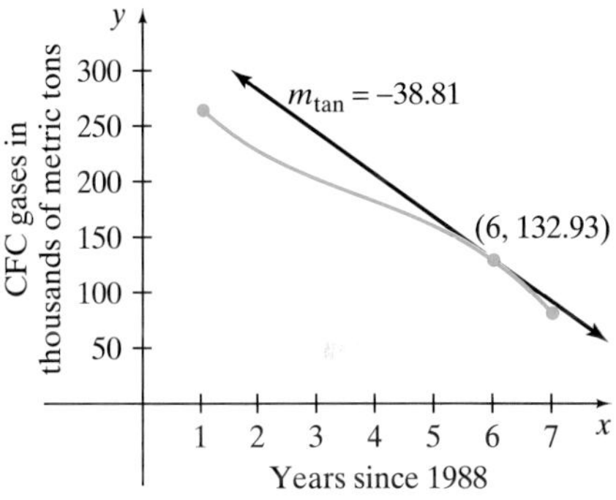

(c)

One of the serious environmental challenges facing the world is that of global warming caused by the burning of fossil fuels. Many nations in the industrialized world, including the United States, have worked to decrease the amount of chlorofluorocarbons (CFCs) released into the air. Figure b, shows a model of the release of CFC's in the atmosphere by the U.S. from 1989–1995. From Figure c, we see that in 1994, the amount of CFC's released by the U.S. in the atmosphere was decreasing at a rate of 38.81 thousand metric tons/year.

What We Know

In Chapter 11, we reviewed algebra and also learned about an important rate of change called an average rate of change. We saw how an average rate of change is equivalent to the slope of a secant line over an interval.

Where Do We Go

In this chapter, we will see how the limit concept is used to introduce a new type of rate of change called an instantaneous rate of change. We will see how an instantaneous rate of change is equivalent to the slope of a tangent line at a specific point.

Section 12.1 Limits

The two main branches of calculus, differential calculus and integral calculus, depend on the limit concept. To help grasp the ideas that calculus is based on, we first provide a practical introduction to limits. We use a combined numerical, graphical, and algebraic approach to examine the concept of a limit.

Determining Limits Numerically and Graphically

We begin our journey by considering the function $f(x) = \dfrac{x^2 - 4}{x - 2}$ and its graph shown in Figure 12.1.1. Since $x = 2$ is not in the domain of the function, in other words $f(2)$ is undefined, there appears to be a "hole" in the graph. However, what is the behavior of $f(x) = \dfrac{x^2 - 4}{x - 2}$ as x gets very, very close to the value of 2? By

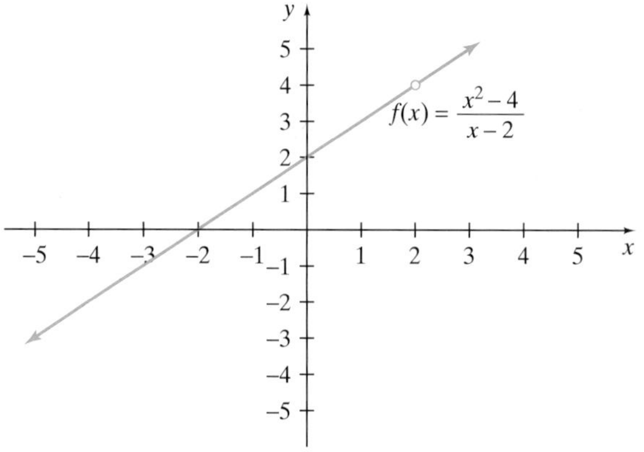

Figure 12.1.1

behavior, we mean what is happening to the function values (or the y-values if you prefer) as x approaches 12. One way to answer this question is by constructing a table to numerically analyze the behavior of f as x gets closer and closer to 2. Since we could approach 2 from the *left side* of 2 or from the *right side* of 2, we must include values of x less than 2 and values greater than 2. See Table 12.1.1.

Table 12.1.1

	$\longrightarrow$ **x Approaches 2 from the Left**						$\longleftarrow$ **x Approaches 2 from the Right**				
x	1	1.9	1.99	1.999	1.9999	2	2.0001	2.001	2.01	2.1	3
$f(x)$	3	3.9	3.99	3.999	3.9999		4.0001	4.001	4.01	4.1	5

Note: We intentionally left a blank below $x = 2$ for two reasons. First, the function is not defined at $x = 2$. Second, we wish to emphasize that in the limit process, we do not care what is happening at $x = 2$, but only in the behavior of the function as x gets close to 2.

It appears from Table 12.1.1 that if we start to the left of $x = 2$ or to the right of $x = 2$, as we allow x to approach 2, our functional values are approaching 4.

We can say that,

$$\text{"The limit of } \frac{x^2 - 4}{x - 2}\text{, as } x \text{ approaches 2, is 4."}$$

Using an arrow for the word *approaches* and *lim* as shorthand for the word limit, the mathematical notation for this English sentence is

$$\lim_{x \to 2} \frac{x^2 - 4}{x - 2} = 4$$

You should interpret this limit notation to mean that as x gets closer and closer to 2, from both sides of 2, $\frac{x^2 - 4}{x - 2}$ gets closer and closer to 4. Notice that Figure 12.1.1 graphically supports our numerical work in Table 12.1.1. At this time, numerically and graphically we believe that $\lim_{x \to 2} \frac{x^2 - 4}{x - 2} = 4$.

Technology Option

Table 12.1.2

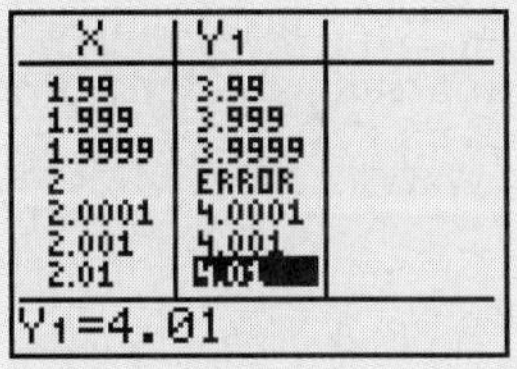

X	Y1
1.99	3.99
1.999	3.999
1.9999	3.9999
2	ERROR
2.0001	4.0001
2.001	4.001
2.01	4.01

Y1=4.01

Once we enter the function $f(x) = \frac{x^2 - 4}{x - 2}$ into the $y =$ editor of our graphing calculator, we can numerically and graphically analyze a limit. Table 12.1.2 shows the result of using the TABLE command on the graphing calculator to numerically analyze $\lim_{x \to 2} \frac{x^2 - 4}{x - 2}$.

Figure 12.1.2 shows a graph of $f(x) = \frac{x^2 - 4}{x - 2}$ and Figure 12.1.3 is the result of utilizing the ZOOM IN command. Notice in Figures 12.1.3b and c that we have used the TRACE command to get as close to 2 as possible from the left side of 2 and from the right side of 2, respectively.

Due to the limitations of the graphing calculator, you may not see a hole in the graph at $x = 2$. Using ZDECIMAL or selecting the x-axis window so that $x = 2$ is the midpoint of the graphing interval should show the hole. See Figure 12.1.2. For information on these features on your graphing calculator consult the online graphing calculator manual at www.prenhall.com/armstrong.

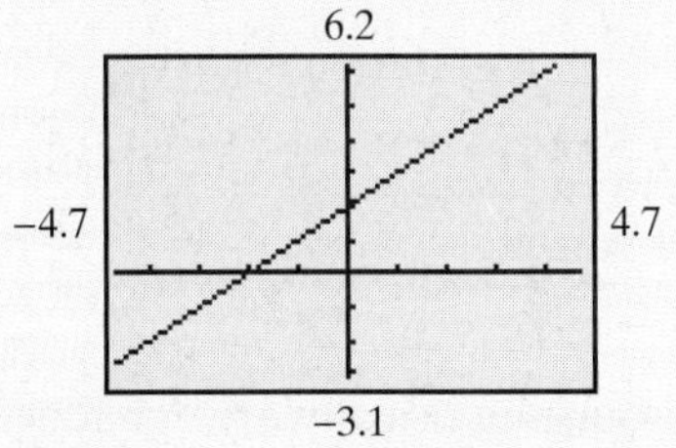

Figure 12.1.2

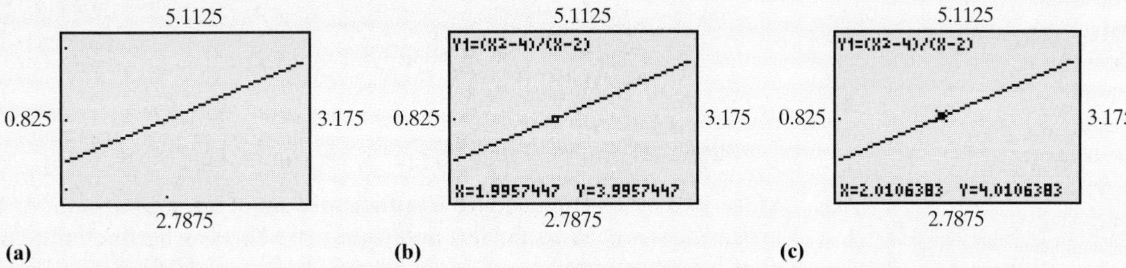

Figure 12.1.3

Example 1 Analyzing a Limit

For $f(x) = \dfrac{x^2 - 1}{x - 1}$, construct a table of values around $x = 1$ and guess the value of $f(x)$ as $x \to 1$. Use the graph of f in Figure 12.1.4 to graphically support your numerical work.

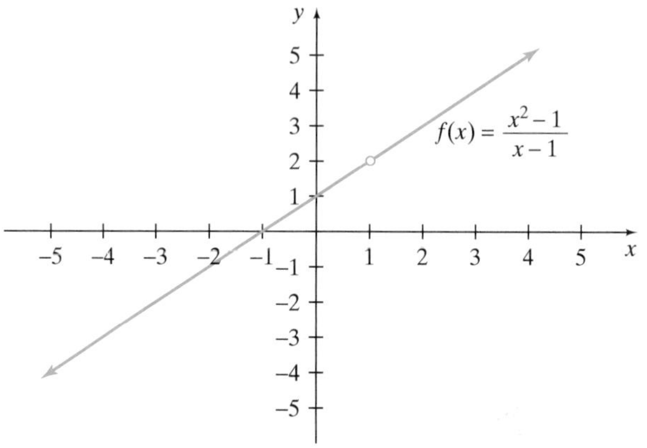

Figure 12.1.4

Solution

Using a calculator, we can quickly construct Table 12.1.3. We notice that as $x \to 1$ from both sides (left and right) of 1, $f(x)$ is approaching 2. Hence, we conjecture that the limit is 2, and we write

$$\lim_{x \to 1} \frac{x^2 - 1}{x - 1} = 2$$

Interactive Activity

Use a graphing calculator to reproduce the values given in Table 12.1.3. Also, graph the function given in Example 1 in the window [−4.7, 4.7] by [−3.1, 6.2]. Use the ZBOX command to capture that part of the graph where $x = 1$. After using the ZBOX command, use the TRACE command to get as close as possible to $x = 1$ from both sides of $x = 1$, to graphically support the numerical work. For more information on the ZBOX command, consult the online calculator manual at www.prenhall.com/armstrong.

Table 12.1.3

	$x \to 1$ from Left					$x \to 1$ from Right			
x	0	0.9	0.99	0.999	1	1.001	1.01	1.1	2
$f(x)$	1	1.9	1.99	1.999		2.001	2.01	2.1	3

The graph of f in Figure 12.1.4 appears to support our numerical work. That is, from the graph it looks like the closer we get to $x = 1$, from both sides of $x = 1$, the closer the function values get to 2. So numerically and graphically we believe that

$$\lim_{x \to 1} \frac{x^2 - 1}{x - 1} = 2.$$

By table building and graphing, we can make $f(x) = \dfrac{x^2 - 1}{x - 1}$ as close to 2 as we like by restricting x to a sufficiently small interval *around* 1. By interval around 1, we mean to the left and right of 1. This is what the limit concept is all about! Before continuing, do Checkpoint 1 to ensure that you understand the process.

Checkpoint 1 Now work Exercise 7.

Left-hand and Right-hand Limits

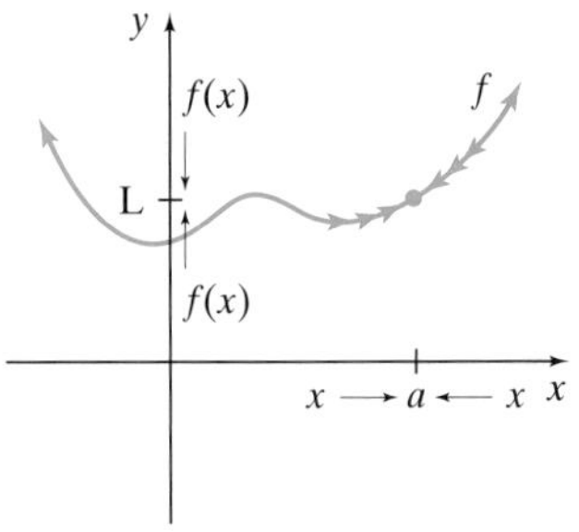

Figure 12.1.5 Illustrating the definition of a limit.

Notice that in Example 1, our table was constructed so that we approached $x = 1$ both from the left side and from the right side. We also made sure that the graph of f in Figure 12.1.4 showed x-values to the left and right of $x = 1$. Because this **left-side** and **right-side** analysis is so important in the limit process, we introduce notation for it. Specifically,

Left-hand and Right-hand Limits

$\lim_{x \to a^-} f(x)$ means "the limit as x approaches a from the left side of a" and is called the **left-hand limit**.

$\lim_{x \to a^+} f(x)$ means "the limit as x approaches a from the right side of a" and is called the **right-hand limit**.

Using this notation, we are ready for the following important definition.

Limit

For any function f, $\lim_{x \to a} f(x) = L$ means that, as x approaches a, $f(x)$ approaches L. Alternatively, if $\lim_{x \to a^-} f(x) = L$ and $\lim_{x \to a^+} f(x) = L$, then $\lim_{x \to a} f(x) = L$.

Notes:
1. If the left-hand limit does not equal the right-hand limit, then there is no limit. In this case, we say that the limit *does not exist*. See Figure 12.1.6a.
2. The existence of $\lim_{x \to a} f(x)$ does *not* depend on whether $f(a)$ is defined. See Figure 12.1.6b.
3. The existence of $\lim_{x \to a} f(x)$ does not depend on the value of $f(a)$ if $f(a)$ *is* defined. See Figure 12.1.6c.

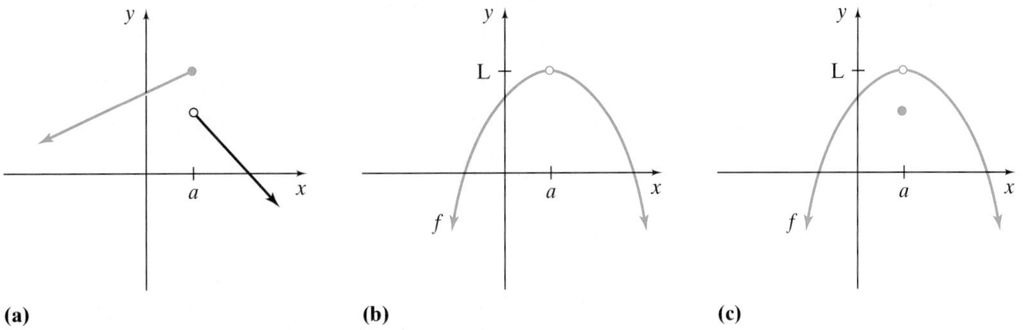

Figure 12.1.6 **(a)** $\lim_{x \to a} f(x)$ does not exist since $\lim_{x \to a^-} f(x) \neq \lim_{x \to a^+} f(x)$. **(b)** $\lim_{x \to a} f(x)$ exists even though $f(a)$ is undefined. **(c)** $f(a)$ is defined, but does *not* equal $\lim_{x \to a} f(x)$.

Example 2 illustrates that the existence of $\lim_{x \to a} f(x)$ does not depend on the value of $f(a)$ if $f(a)$ is defined.

Example 2 Analyzing a Limit

Estimate $\lim_{x \to 2} f(x)$ for $f(x) = \begin{cases} 1 - x, & x \le 2 \\ 4, & x > 2 \end{cases}$.

Solution

From Table 12.1.4 and from the graph of f in Figure 12.1.7, we conclude that

$$\lim_{x \to 2^-} f(x) = -1, \quad \text{whereas} \quad \lim_{x \to 2^+} f(x) = 4$$

Since the left-hand limit **does not equal** the right-hand limit, we conclude that $\lim_{x \to 2} f(x)$ does not exist. Notice that graphically this corresponds to a jump in the graph. See Figure 12.1.7.

Table 12.1.4

	$x \to 2^-$						$x \to 2^+$		
x	1	1.9	1.99	1.999	2	2.001	2.01	2.1	3
$f(x)$	0	−0.9	−0.99	−0.999		4	4	4	4

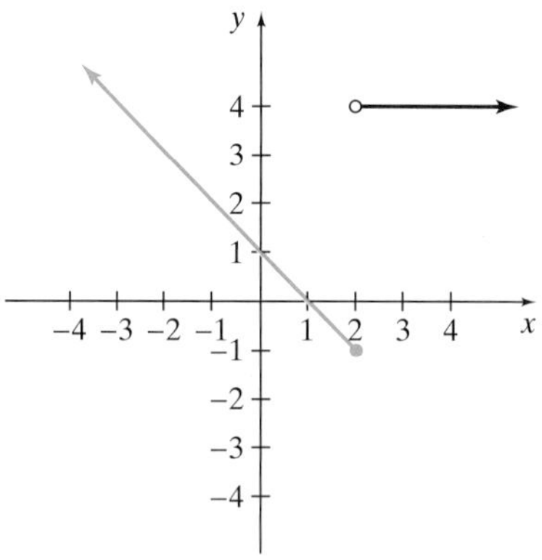

Figure 12.1.7

Again, we want to point out that in Example 2, $f(2) = -1$, yet $\lim_{x \to 2} f(x)$ does not exist. Remember that the existence of the limit $\lim_{x \to a} f(x)$ does *not* depend on the value of $f(a)$.

Algebraically Determining Limits

So far, our approach to limits has been very informal. At this time, we will briefly state limit theorems to aid us in the algebraic evaluation of limits, as well as provide some mathematical validity. Later in the text, we will occasionally use some of these theorems.

Limit Theorems

If a, c, and n are real numbers, then

1. $\lim_{x\to a} c = c$
2. $\lim_{x\to a} x = a$
3. $\lim_{x\to a}[c \cdot f(x)] = c \cdot \lim_{x\to a} f(x)$
4. $\lim_{x\to a}[f(x) \pm g(x)] = \lim_{x\to a} f(x) \pm \lim_{x\to a} g(x)$
5. $\lim_{x\to a}[f(x) \cdot g(x)] = \lim_{x\to a} f(x) \cdot \lim_{x\to a} g(x)$
6. $\lim_{x\to a} \dfrac{f(x)}{g(x)} = \dfrac{\lim_{x\to a} f(x)}{\lim_{x\to a} g(x)}$, provided that $\lim_{x\to a} g(x) \neq 0$
7. $\lim_{x\to a}[f(x)]^n = [\lim_{x\to a} f(x)]^n$

Notes: 1. The first limit theorem simply states that the limit of a constant is that constant.

2. Instead of memorizing these theorems, we suggest the following alternative, which we call the **Substitution Principle**.

Substitution Principle

When attempting to algebraically determine a limit, first try direct substitution. In other words, when attempting to find $\lim_{x\to a} f(x)$, first try to compute $f(a)$ (substitute a for x).

Example 3 Analyzing Limits Algebraically

Determine the following limits.

(a) $\lim_{x\to 2} 7$ **(b)** $\lim_{x\to 1}(2x^2 - 3x + 5)$ **(c)** $\lim_{x\to 3} \sqrt{(2x-1)}$

Solution

(a) Since 7 is a constant, we utilize Limit Theorem 1 and have

$$\lim_{x\to 2} 7 = 7$$

(b) Here, we simply substitute and get

$$\lim_{x\to 1}(2x^2 - 3x + 5) = 2(1)^2 - 3(1) + 5 = 4$$

(c) Again, we simply substitute and have

$$\lim_{x\to 3} \sqrt{(2x-1)} = \sqrt{(2(3)-1)} = \sqrt{5}$$

■

Checkpoint 2 Now work Exercise 31.

At this time you may be asking yourself, "When does the Substitution Principle fail?" The answer to this question can be found in Examples 1 and 2. In Example 2, the Substitution Principle fails, since we have a piecewise-defined function. In Example 1, the Substitution Principle fails, since substituting produces a fraction of the form $\frac{0}{0}$. We call the fraction $\frac{0}{0}$ an **indeterminate form.** When we try substitution and get the indeterminate form $\frac{0}{0}$, we have to use other techniques to determine the limit.

Recall that in Example 1, we determined numerically and graphically the limit

$$\lim_{x \to 1} \frac{x^2 - 1}{x - 1} = 2$$

To determine this limit algebraically, we can do the following.

$$\begin{aligned} \lim_{x \to 1} \frac{x^2 - 1}{x - 1} &= \lim_{x \to 1} \frac{(x-1)(x+1)}{x - 1} && \text{Factor} \\ &= \lim_{x \to 1} (x+1) && \text{Cancel, provided that } x \neq 1 \\ &= 1 + 1 = 2 && \text{Substitution Principle} \end{aligned}$$

Interactive Activity

Complete Table 12.1.3 in Example 1 for $g(x) = x + 1$ and graph $g(x) = x + 1$ to verify that numerically and graphically $g(x) = x + 1$ and $f(x) = \frac{x^2 - 1}{x - 1}$ are equivalent for all values of x, except $x = 1$.

What we have really done using algebra is to determine another function, $g(x) = x + 1$, that is equal to $f(x) = \frac{x^2 - 1}{x - 1}$, for all values of x, except $x = 1$. In other words, we have

$$f(x) = \frac{x^2 - 1}{x - 1} = \frac{(x-1)(x+1)}{x - 1} = x + 1 = g(x), \qquad \text{provided that } x \neq 1$$

The Interactive Activity to the left reinforces that the graphs of $f(x) = \frac{x^2 - 1}{x - 1}$ and $g(x) = x + 1$ are identical for all x close to 1, but not equal to 1. We can write

$$\lim_{x \to 1} \frac{x^2 - 1}{x - 1} = \lim_{x \to 1} (x + 1) = 2$$

Example 4 Analyzing a Limit Involving $\frac{0}{0}$

Determine $\lim_{x \to 3} g(x)$, where $g(x) = \frac{x^2 - 9}{x - 3}$.

Solution

We try substituting, which gives

$$\lim_{x \to 3} \frac{x^2 - 9}{x - 3} = \frac{(3)^2 - 9}{3 - 3} = \frac{0}{0}$$

Since we have the indeterminate form $\frac{0}{0}$, we decide to try a little algebra. We will factor the numerator and cancel as follows:

$$\begin{aligned} \lim_{x \to 3} \frac{x^2 - 9}{x - 3} &= \lim_{x \to 3} \frac{(x-3)(x+3)}{x - 3} && \text{Factor} \\ &= \lim_{x \to 3} (x+3) && \text{Cancel, provided that } x \neq 3 \\ &= (3 + 3) = 6 && \text{Substitute} \end{aligned}$$

Once again, notice that even though $g(3)$ is undefined, we determined that $\lim_{x\to 3} g(x)$ exists. This reinforces the fact that the limit as x approaches 3 is *not* dependent on $g(3)$. See Figure 12.1.8.

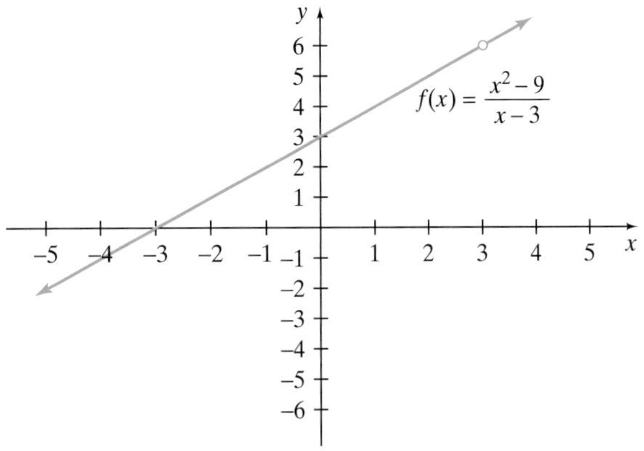

Figure 12.1.8

✓ **Checkpoint 3**

Now work Exercise 39.

Example 5 Analyzing a Limit Involving $\frac{0}{0}$

Determine $\lim_{x\to 0} \frac{|x|}{x}$.

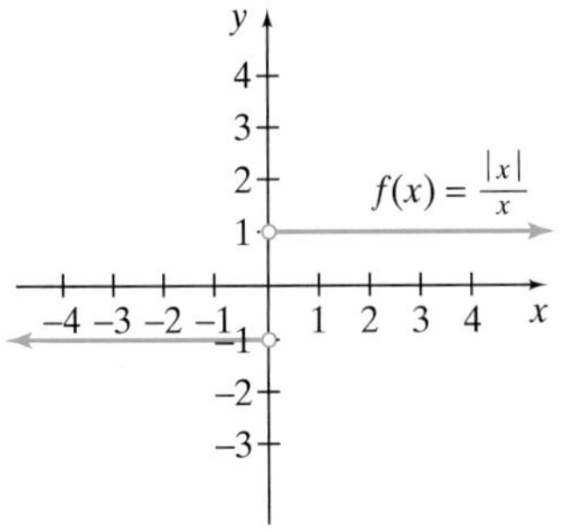

Figure 12.1.9

Solution

If we try substituting, we get

$$\lim_{x\to 0} \frac{|x|}{x} = \frac{0}{0}$$

By letting $f(x) = \frac{|x|}{x}$, we can use a table and a graph to determine this limit. See Table 12.1.5 and Figure 12.1.9.

Table 12.1.5

	$x \to 0^-$					$x \to 0^+$			
x	−1	−0.1	−0.01	−0.001	0	0.001	0.01	0.1	1
$f(x)$	−1	−1	−1	−1		1	1	1	1

As we can see numerically and graphically, since the left-hand limit does not equal the right-hand limit, we conclude that $\lim_{x\to 0} \frac{|x|}{x}$ does not exist. ■

Interactive Activity

Recall that $|x|$ is defined to be

$$|x| = \begin{cases} x, & \text{if } x \geq 0 \\ -x, & \text{if } x < 0 \end{cases}.$$

Use this definition to algebraically verify that $\lim_{x\to 0} \frac{|x|}{x}$ does not exist.

For Example 6, we need to recall Limit Theorem 1, which states that the limit of a constant is that constant.

Example 6 Analyzing Limits Involving Two Variables

Determine the following limits.

(a) $\lim_{h \to 0}(3x + 2h)$

(b) $\lim_{h \to 0} \dfrac{5xh + 2h^2}{h}$

Solution

(a) **Understand the Situation:** First, we notice that two variables, x and h, are involved. Remember that as $h \to 0$, the variable x acts as a **constant**; only the value of h changes.

With this in mind, we try the Substitution Principle, which yields

$$\begin{aligned}\lim_{h \to 0}(3x + 2h) &= 3x + 2(0) \\ &= 3x\end{aligned}$$

(b) Again, two variables are involved, x and h. Proceeding as in part (a) gives

$$\begin{aligned}\lim_{h \to 0} \frac{5xh + 2h^2}{h} &= \frac{5x(0) + 2(0)^2}{0} \\ &= \frac{0}{0}\end{aligned}$$

Since this is an indeterminate form, we try some algebra.

$$\begin{aligned}\lim_{h \to 0} \frac{5xh + 2h^2}{h} &= \lim_{h \to 0} \frac{h(5x + 2h)}{h} && \text{Factor} \\ &= \lim_{h \to 0}(5x + 2h) && \text{Cancel, provided that } h \neq 0 \\ &= 5x + 2(0) = 5x && \text{Substitute}\end{aligned}$$

■

Checkpoint 4 Now work Exercise 49.

Before we conclude this section, we would like to look at another limit that will be quite important in Section 12.3.

Example 7 Analyzing a Limit of a Difference Quotient

For $f(x) = x^2$, compute $\lim_{h \to 0} \dfrac{f(2 + h) - f(2)}{h}$.

Solution

Here we have

$$\lim_{h \to 0} \frac{f(2 + h) - f(2)}{h} = \lim_{h \to 0} \frac{(2 + h)^2 - (2)^2}{h}$$

Substituting 0 for h produces the indeterminate form $\frac{0}{0}$. But, if we try some algebra, we get

$$\begin{aligned}\lim_{h\to 0}\frac{f(2+h)-f(2)}{h} &= \lim_{h\to 0}\frac{(2+h)^2-(2)^2}{h}\\ &= \lim_{h\to 0}\frac{(4+4h+h^2)-4}{h} && \text{Recall that } (2+h)^2 = 4+4h+h^2\\ &= \lim_{h\to 0}\frac{4h+h^2}{h} && \text{Still indeterminate}\\ &= \lim_{h\to 0}\frac{h(4+h)}{h} && \text{Factor}\\ &= \lim_{h\to 0}(4+h) && \text{Cancel, provided that } h\neq 0\\ &= 4+0=4 && \text{Substitute}\end{aligned}$$

■

Applications

The main thrust of this section has been to lay the groundwork for the calculus. Although the number of applications in this section is limited, there will be more in the next section when limits at infinity and infinite limits are discussed. At this time, we offer the following application.

Example 8 Applying Limits

The graph in Figure 12.1.10 shows the cost $C(x)$, in dollars, to make x copies at Cody's Copy Center. Use the information in the graph to estimate the following:

(a) $C(10)$ **(b)** $\lim_{x\to 10} C(x)$ **(c)** $C(15)$ **(d)** $\lim_{x\to 15} C(x)$

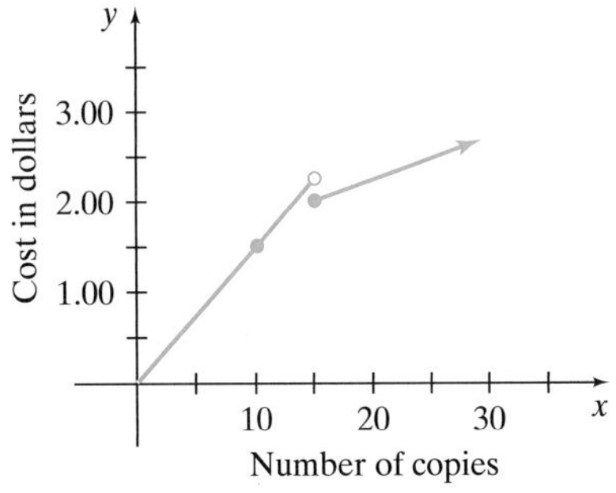

Figure 12.1.10

Solution

(a) From the graph, we estimate that $C(10) = \$1.50$.

(b) Since $\lim_{x\to 10^-} C(x) = 1.50$ and $\lim_{x\to 10^+} C(x) = 1.50$, we estimate that $\lim_{x\to 10} C(x) = 1.50$.

(c) From the graph we estimate that $C(15) = \$2.00$.

(d) Since $\lim_{x\to 15^-} C(x) = 2.25$ and $\lim_{x\to 15^+} C(x) = 2.00$, we conclude that $\lim_{x\to 15} C(x)$ does not exist. ■

SUMMARY

In this introductory section on limits, we have presented the limit concept numerically, graphically, and algebraically. When analyzing a limit of a function, it is important to remember that in order for $\lim_{x \to a} f(x) = L$, we must have the **left-hand limit**, $\lim_{x \to a^-} f(x)$, and the **right-hand limit**, $\lim_{x \to a^+} f(x)$, both equal to L. We used **limit theorems** and the Substitution Principle to determine limits. In addition, we used algebra to find the limits of an indeterminate form $\frac{0}{0}$.

- The limit $\lim_{x \to a} f(x) = L$ means that as x approaches a, $f(x)$ approaches L. Alternatively, if $\lim_{x \to a^-} f(x) = L$ and $\lim_{x \to a^+} f(x) = L$, then $\lim_{x \to a} f(x) = L$.
- **Substitution Principle**: When trying to find $\lim_{x \to a} f(x)$, first try to compute $f(a)$.

SECTION 12.1 EXERCISES

For Exercises 1–10, complete the given tables to numerically estimate the following:

(a) $\lim_{x \to a^-} f(x)$ **(b)** $\lim_{x \to a^+} f(x)$ **(c)** $\lim_{x \to a} f(x)$

1. $f(x) = 2 - 3x;\quad a = 1$

x	0	0.9	0.99	0.999	1	1.001	1.01	1.1	2
f(x)					?				

2. $f(x) = 2x - 3;\quad a = -2$

x	−3	−2.1	−2.01	−2.001	−2	−1.999	−1.99	−1.9	−1
f(x)					?				

3. $f(x) = 2x^4 - 3x^3 + 2x - 4;\quad a = -2$

x	−3	−2.1	−2.01	−2.001	−2	−1.999	−1.99	−1.9	−1
f(x)					?				

4. $f(x) = -2x^3 + 3x - 2;\quad a = 1$

x	0	0.9	0.99	0.999	1	1.001	1.01	1.1	2
f(x)					?				

5. $f(x) = \dfrac{x^2 - 1}{x - 1};\quad a = 1$

x	0	0.9	0.99	0.999	1	1.001	1.01	1.1	2
f(x)					?				

6. $f(x) = \dfrac{x^2 - 3x}{x^2 - 9};\quad a = 3$

x	2	2.9	2.99	2.999	3	3.001	3.01	3.1	4
f(x)					?				

7. $f(x) = \dfrac{x^3 + 8}{x + 2}; \quad a = -2$

x	-3	-2.1	-2.01	-2.001	-2	-1.999	-1.99	-1.9	-1
$f(x)$					?				

8. $f(x) = \dfrac{x^3 - 1}{x - 1}; \quad a = 1$

x	0	0.9	0.99	0.999	1	1.001	1.01	1.1	2
$f(x)$					?				

9. $f(x) = \begin{cases} 3x + 1, & x > 0 \\ x^2 + 1, & x \le 0 \end{cases}; \quad a = 0$

x	-1	-0.1	-0.01	-0.001	0	0.001	0.01	0.1	1
$f(x)$					?				

10. $f(x) = \begin{cases} x^2 + 1, & x \ge 0 \\ 2x^2 - 1, & x < 0 \end{cases}; \quad a = 0$

x	-1	-0.1	-0.01	-0.001	0	0.001	0.01	0.1	1
$f(x)$					?				

For Exercises 11–18, use your calculator to graph the given function. Use the ZOOM IN and TRACE commands to graphically estimate the indicated limits. Verify your estimate numerically.

11. $f(x) = x^2 - 5x - 2; \quad \lim_{x\to 1} f(x)$

12. $f(x) = x^2 - 8x + 15; \quad \lim_{x\to 3} f(x)$

13. $g(x) = \dfrac{x^3 + 1}{x + 1}; \quad \lim_{x\to -1} g(x)$

14. $g(x) = \dfrac{x + 1}{x^3 + 1}; \quad \lim_{x\to -1} g(x)$

15. $f(x) = (x - 2)(x + 1); \quad \lim_{x\to 3} f(x)$

16. $f(x) = \dfrac{x - 2}{x + 1}; \quad \lim_{x\to 3} f(x)$

17. $f(x) = \dfrac{\sqrt{x} - 2}{x - 4}; \quad \lim_{x\to 4} f(x)$

18. $f(x) = x^3 - 2x^2 - 5x + 6; \quad \lim_{x\to 1.5} f(x)$

For Exercises 19–44, determine the indicated limit algebraically by using techniques in this section. Verify your answers numerically and graphically.

19. $\lim_{x\to -2} (3x + 1)$

20. $\lim_{x\to 2} (-2x^2 + 50x)$

21. $\lim_{x\to 2} (2x^2 + 3x - 1)$

22. $\lim_{x\to -4} \sqrt{x^2 + 9}$

23. $\lim_{x\to 3} \dfrac{x^2 - 9}{x - 3}$

24. $\lim_{x\to 1} (2 - 3x)$

25. $\lim_{x\to 0} |x|$

26. $\lim_{x\to 1} |x + 1|$

27. $\lim_{x\to -1} |x + 1|$

28. $\lim_{x\to -1} \dfrac{x + 1}{|x + 1|}$

29. $\lim_{x\to -1} \dfrac{x^2 - 1}{x + 1}$

30. $\lim_{x\to -1} \dfrac{x + 1}{x^2 - 1}$

31. $\lim_{x\to 0} \sqrt{2x + 3}$

32. $\lim_{x\to 5} \dfrac{x^2 - 25}{x - 5}$

33. $\lim_{x\to -5} \dfrac{x^2 - 25}{x - 5}$

34. $\lim_{x\to 5} \dfrac{x - 5}{x^2 - 25}$

35. $\lim_{x\to 1} \dfrac{x^2 - 1}{x + 1}$

36. $\lim_{x\to 1} \dfrac{x - 1}{x^2 - 1}$

37. $\lim_{x\to 2} (x + 1)^2 \cdot (3x - 1)^3$

38. $\lim_{x\to -1} (x + 2)^3(3x + 2)$

39. $\lim_{x\to 3} \dfrac{x^2 - x - 6}{x - 3}$

40. $\lim_{x\to 3} \dfrac{x - 3}{x^2 - x - 6}$

41. $\lim_{x\to -2} \dfrac{x + 2}{x^2 + 5x + 6}$

42. $\lim_{x\to -2} \dfrac{x^2 + 5x + 6}{x + 2}$

43. $\lim_{x\to 4} \dfrac{\sqrt{x} - 2}{x - 4}$ (*Hint:* Algebraically, try multiplying the numerator and denominator by $\sqrt{x} + 2$.)

44. $\lim_{x\to 9} \dfrac{\sqrt{x} - 3}{x - 9}$ (*Hint:* Algebraically, try multiplying the numerator and denominator by $\sqrt{x} + 3$.)

45. Use the graph of f in Figure 12.1.11 to answer the following:

(a) $\lim_{x\to -1} f(x)$ **(b)** $f(-1)$ **(c)** $\lim_{x\to 3^-} f(x)$

(d) $\lim_{x\to 3^+} f(x)$ **(e)** $\lim_{x\to 3} f(x)$ **(f)** $f(1)$

(g) $\lim_{x\to 1} f(x)$

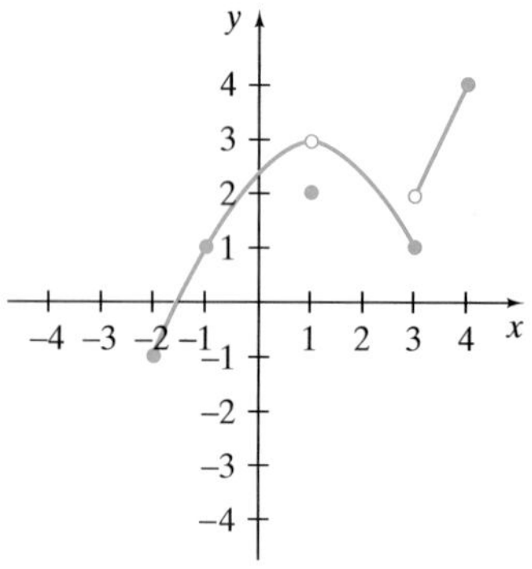

Figure 12.1.11

46. Use the graph of g in Figure 12.1.12 to answer the following:

(a) $\lim_{x\to 2^-} g(x)$ **(b)** $\lim_{x\to 2^+} g(x)$ **(c)** $\lim_{x\to 2} g(x)$

(d) $g(2)$ **(e)** $\lim_{x\to -2} g(x)$ **(f)** $g(-2)$

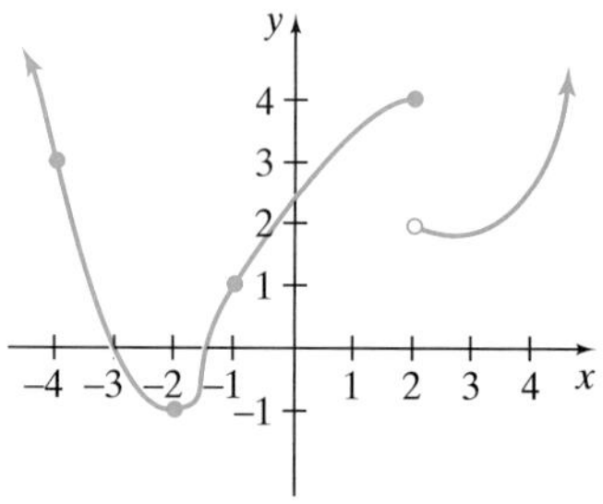

Figure 12.1.12

Determine the indicated limit in Exercises 47–53. See Example 6.

47. $\lim_{h\to 0}(2x+h)$

48. $\lim_{h\to 0}(3x^2+3h)$

49. $\lim_{h\to 0}\dfrac{3xh+h^2}{h}$

50. $\lim_{h\to 0}\dfrac{2xh+h^2}{h}$

51. $\lim_{h\to 0}\dfrac{4x^2h+2h^2}{h}$

52. $\lim_{h\to 0}\dfrac{4x^3h^2+3xh^3+5h^4}{h}$

53. $\lim_{h\to 0}\dfrac{-3x^2h+6h^2}{h}$

For each function in Exercises 54–62 (see Example 7):

(a) Determine $f(1+h)$.

(b) Compute $\lim_{h\to 0}\dfrac{f(1+h)-f(1)}{h}$.

54. $f(x)=2x+3$

55. $f(x)=3x-1$

56. $f(x)=x^2-1$

57. $f(x)=x^2+2$

58. $f(x)=x^2-2x+3$

59. $f(x)=x^2+3x-1$

60. $f(x)=|x|$

61. $f(x)=\dfrac{1}{x}$

62. $f(x)=\sqrt{x}$

Applications

63. The number of dollars spent on advertising for a product influences the number of items of the product that will be purchased by consumers. If x is the number of dollars spent, in thousands, then $N(x)$, in hundreds, is the number of items sold, as given by the function

$$N(x)=2000-\frac{520}{x}$$

(a) Determine $N(10)$ and interpret.

(b) Determine $\lim_{x\to 10} N(x)$ and interpret.

64. Repeat all parts of Exercise 63, except use

$$N(x)=2200-\frac{750}{x}.$$

11.2 **65.** The cost, in dollars, of producing x units of a product is given by $C(x)=22{,}500+7.35x$. The **average cost**, AC, is given by

$$AC(x)=\frac{C(x)}{x}$$
$$=\frac{22{,}500+7.35x}{x}$$

(a) Determine $AC(10)$ and interpret.

(b) Determine $\lim_{x\to 10} AC(x)$ and interpret.

11.2 **66.** Repeat all parts of Exercise 65, except use

$$C(x)=33{,}125+6.38x.$$

67. Lisa's Lease-a-Car charges \$25 per day plus \$0.05 per mile to rent one of its medium-sized cars. Figure 12.1.13 shows the graph of the cost in dollars, $c(m)$, realized by

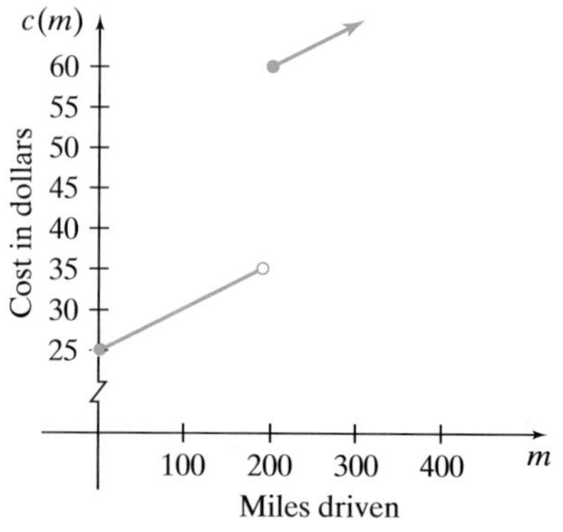

Figure 12.1.13

Sofia as a function of the miles, m, that she has driven the car.

(a) Determine $c(100)$ and interpret.

(b) Determine $\lim_{m \to 100} c(m)$.

(c) Determine $c(200)$ and interpret.

(d) Determine $\lim_{m \to 200} c(m)$.

(e) Give an explanation for the jump in the graph at $m = 200$.

68. Dylan's Truck Rental Company charges \$20 per day plus \$0.10 per mile to rent a moving truck. Figure 12.1.14 shows the graph of the cost in dollars, $c(m)$, realized by Austin as a function of the miles, m, that he has driven the truck.

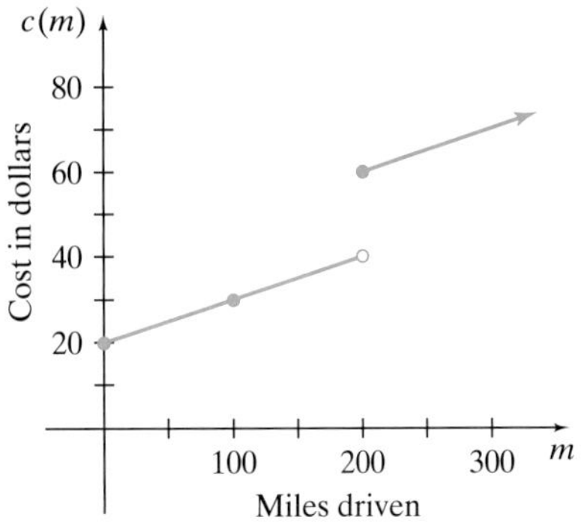

Figure 12.1.14

(a) Determine $c(50)$ and interpret.

(b) Determine $\lim_{m \to 50} c(m)$.

(c) Determine $c(100)$ and interpret.

(d) Determine $\lim_{m \to 100} c(m)$.

(e) Give an explanation for the jump in the graph at $m = 200$.

SECTION PROJECT

Finer Foods is running the following advertisement:

> Hamburger
>
> \$1.50/pound for packages less than 2 pounds
> \$1.00/pound for packages of 2 pounds or more

(a) Let x represent the number of pounds of hamburger in a package. Determine a piecewise-defined function, $C(x)$, that models the cost of a package of hamburger.

(b) Graph C where $0 < x \leq 6$.

(c) Determine $C(1.5)$ and interpret.

(d) Determine $C(2)$ and interpret.

(e) Does $\lim_{x \to 2} C(x)$ exist? Explain.

(f) If Luther needs 2 pounds of hamburger for a cookout, how much money would Luther save by buying one 2-pound package versus two 1-pound packages?

Section 12.2 Limits and Asymptotes

In Section 12.1, we introduced the limit concept and analyzed limits numerically, graphically, and algebraically. In this section, we look at the behavior of functions via limits when the independent variable increases without bound or decreases without bound. Graphically, this is important to the understanding of **horizontal asymptotes**. We also analyze situations when the dependent variable increases and decreases without bound. This situation will be closely related to **vertical asymptotes**.

Infinite Limits and Vertical Asymptotes

Example 1 begins our discussion of the relationship between infinite limits and vertical asymptotes.

Example 1 Analyzing a Limit

Consider $f(x) = \dfrac{1}{x}$. Numerically and graphically estimate $\lim_{x \to 0} \dfrac{1}{x}$.

Solution

We begin our analysis by constructing a table of values as shown in Table 12.2.1. Numerically, it appears that the left-hand limit and the right-hand limit are not equal, which means that the limit does not exist. As $x \to 0^+$, $\frac{1}{x}$ is positive and appears to be increasing forever or increasing without bound. We will indicate this "increasing without bound" behavior by writing

$$\lim_{x \to 0^+} \frac{1}{x} = \infty$$

Table 12.2.1

	$x \to 0^-$					$x \to 0^+$			
x	−1	−0.1	−0.01	−0.001	0	0.001	0.01	0.1	1
$f(x)$	−1	−10	−100	−1000		1000	100	10	1

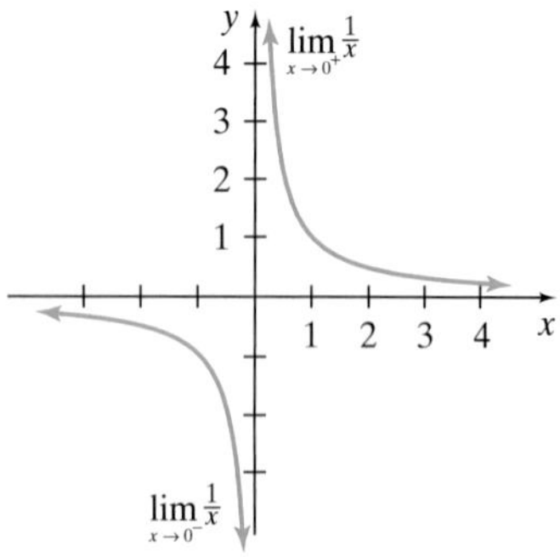

Figure 12.2.1

Likewise, as $x \to 0^-$, $\frac{1}{x}$ is negative and appears to be decreasing forever or decreasing without bound. We will indicate this "decreasing without bound" behavior by writing

$$\lim_{x \to 0^-} \frac{1}{x} = -\infty$$

These behaviors are shown graphically in Figure 12.2.1. The graph seems to support our numerical work. Since both the table and the graph suggest that $\lim_{x \to 0^-} \frac{1}{x} \neq \lim_{x \to 0^+} \frac{1}{x}$, we conclude that $\lim_{x \to 0} \frac{1}{x}$ does not exist. ■

Note: Technically speaking, in neither case does the limit, $\lim_{x \to 0^+} \frac{1}{x}$ nor $\lim_{x \to 0^-} \frac{1}{x}$, exist, since ∞ and $-\infty$ are not real numbers; they are merely concepts. To ensure that these concepts are understood, let's consider another example.

Example 2 Analyzing an Infinite Limit

Consider $f(x) = \frac{1}{x^2}$. Use Table 12.2.2 and Figure 12.2.2 to determine the following:

(a) $\lim_{x \to 0^-} \frac{1}{x^2}$ **(b)** $\lim_{x \to 0^+} \frac{1}{x^2}$ **(c)** $\lim_{x \to 0} \frac{1}{x^2}$

Table 12.2.2

	$x \to 0^-$					$x \to 0^+$			
x	−1	−0.1	−0.01	−0.001	0	0.001	0.01	0.1	1
$f(x)$	1	100	10,000	1,000,000		1,000,000	10,000	100	1

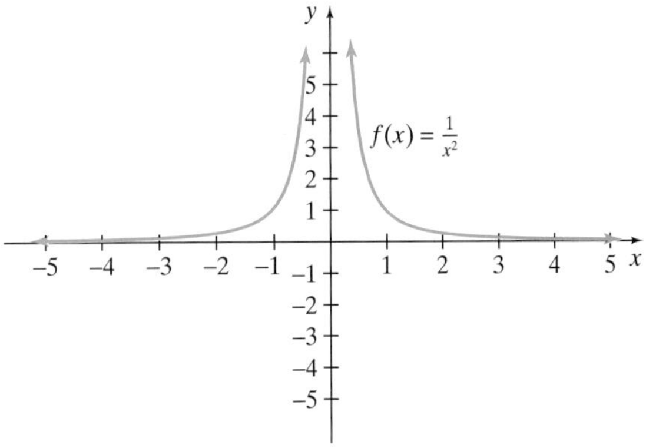

Figure 12.2.2

Solution

(a) Table 12.2.2 and Figure 12.2.2 both suggest that, as $x \to 0^-$, $\frac{1}{x^2}$ increases without bound. So we write

$$\lim_{x \to 0^-} \frac{1}{x^2} = \infty$$

(b) Table 12.2.2 and Figure 12.2.2 both suggest that, as $x \to 0^+$, $\frac{1}{x^2}$ increases without bound. So we write

$$\lim_{x \to 0^+} \frac{1}{x^2} = \infty$$

(c) From our knowledge of left-hand and right-hand limits, the results from parts (a) and (b) suggest that

$$\lim_{x \to 0} \frac{1}{x^2} = \infty$$ ■

Interactive Activity

In your own words, explain why we say that $\lim_{x \to 0} \frac{1}{x}$ does not exist (Example 1), yet we say that $\lim_{x \to 0} \frac{1}{x^2} = \infty$ (Example 2).

WWW

It is worth mentioning one more time that in Example 2, none of the limits really exists, since $-\infty$ and ∞ are not real numbers. We are using the symbols ∞ and $-\infty$ to describe the behavior of the function near $x = 0$.

Notice in Examples 1 and 2 that the line $x = 0$, also known as the y-axis, serves as a **vertical asymptote** for the graphs. We now summarize the connection between infinite limits and vertical asymptotes.

Infinite Limits and Vertical Asymptotes

The line $x = a$ is a **vertical asymptote** of the graph of f if *any* of the following are true.

- If as $x \to a^-$ the function values $f(x)$ increase or decrease without bound, that is, $\lim_{x \to a^-} f(x) = \infty$ or $\lim_{x \to a^-} f(x) = -\infty$, then $x = a$ is a vertical asymptote.
- If as $x \to a^+$ the function values $f(x)$ increase or decrease without bound, that is, $\lim_{x \to a^+} f(x) = \infty$ or $\lim_{x \to a^+} f(x) = -\infty$, then $x = a$ is a vertical asymptote.

Note: If both the left- and the right-hand limits exhibit the same behavior, we say that $\lim_{x \to a} f(x) = \infty$ (or $-\infty$).

Example 3 Analyzing Limits at Excluded Domain Values

Consider $f(x) = \dfrac{x-1}{x^2-1}$. The domain for f is all real numbers except -1 and 1.

(a) Determine $\lim\limits_{x\to 1} \dfrac{x-1}{x^2-1}$ algebraically.

(b) Determine $\lim\limits_{x\to -1} \dfrac{x-1}{x^2-1}$.

Solution

(a) Utilizing the Substitution Principle produces the indeterminate form $\dfrac{0}{0}$, so we need to employ some algebraic manipulation.

$$\lim_{x\to 1} \frac{x-1}{x^2-1} = \lim_{x\to 1} \frac{x-1}{(x-1)(x+1)} \qquad \text{Factor}$$

$$= \lim_{x\to 1} \frac{1}{x+1} \qquad \text{Cancel, provided that } x \neq 1$$

$$= \frac{1}{1+1} = \frac{1}{2} \qquad \text{Substitute}$$

(b) Again substitution yields $\dfrac{0}{0}$, so proceeding with the same algebra as in part (a) gives

$$\lim_{x\to -1} \frac{x-1}{x^2-1} = \lim_{x\to -1} \frac{x-1}{(x-1)(x+1)} \qquad \text{Factor}$$

$$= \lim_{x\to -1} \frac{1}{x+1} \qquad \text{Cancel, provided that } x \neq -1$$

We attempt to substitute at this time, but this gives us $\dfrac{1}{0}$, which we know is undefined. We decide to make a table and see if numerically some light can be shed on what is happening here. See Table 12.2.3.

Table 12.2.3

	$x \to -1^-$					$x \to -1^+$			
x	-2	-1.1	-1.01	-1.001	-1	-0.999	-0.99	-0.9	0
$f(x)$	-1	-10	-100	-1000		1000	100	10	1

Since the table suggests that

$$\lim_{x\to -1^-} \frac{x-1}{x^2-1} \neq \lim_{x\to -1^+} \frac{x-1}{x^2-1},$$

we conclude that $\lim\limits_{x\to -1} \dfrac{x-1}{x^2-1}$ does not exist. Figure 12.2.3 supports our conjecture.

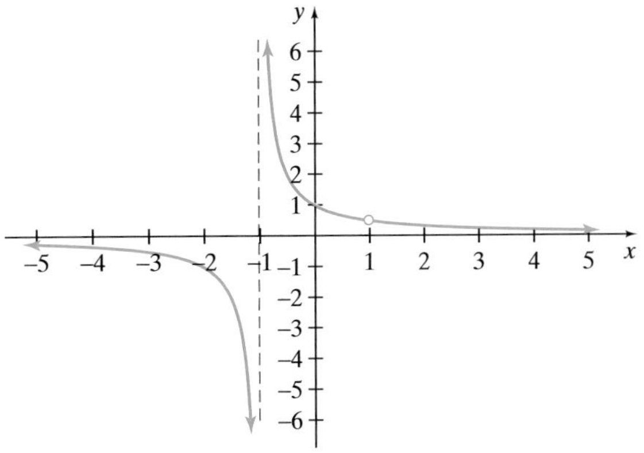

Figure 12.2.3 Graph of $f(x) = \dfrac{x-1}{x^2-1}$ has a vertical asymptote at $x = -1$. ■

Checkpoint 1 Now work Exercise 13.

Example 4 Analyzing a Limit in an Applied Setting

The cost, $C(x)$, in thousands of dollars of removing $x\%$ of a city's pollutants discharged into a lake is given by

$$C(x) = \frac{113x}{100 - x}$$

(a) Determine the reasonable domain for C and graph.

(b) Evaluate $C(50)$ and interpret.

(c) Determine $\lim_{x \to 100^-} C(x)$ and interpret.

Solution

(a) Understand the Situation: Since our independent variable, x, represents a percentage, the reasonable domain for C is $[0, 100)$.

A graph of C is given in Figure 12.2.4.

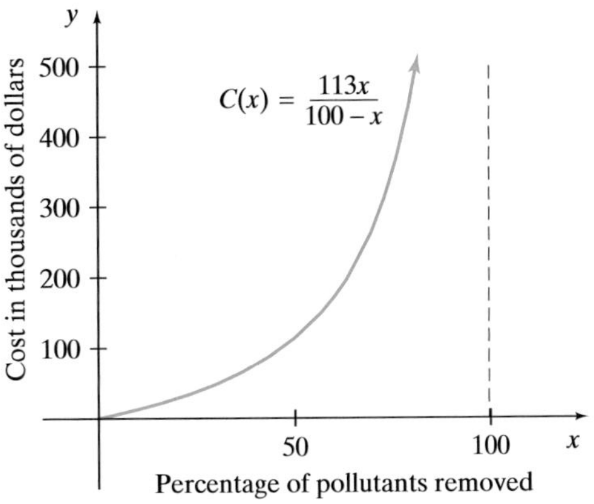

Figure 12.2.4

(b) $C(50) = \dfrac{113(50)}{100-50} = 113$. This means that it will cost the city \$113,000 to remove 50% of the pollutants being discharged into the lake.

(c) Figure 12.2.4 suggests that $\lim_{x\to 100^-} C(x) = \infty$. We construct Table 12.2.4 to support our graphical belief that $\lim_{x\to 100^-} C(x) = \infty$.

Table 12.2.4

	$x \to 100^-$				
x	90	99	99.9	99.99	99.999
$C(x)$	1017	11,187	112,887	1,129,887	11,229,887

Interpret the Solution: This means that as the city tries to remove *all* the pollutants (100%) the cost to do so is "increasing without bound." ■

Checkpoint 2

Now work Exercise 45.

Limits at Infinity and Horizontal Asymptotes

We now turn our attention to analyzing the behavior of a function as the independent variable values increase without bound (heads to ∞) and as the independent variable values decrease without bound (heads to $-\infty$). To introduce this concept, we return to a problem first encountered in Section 11.7.

Flashback

Spring Point Epidemic Revisited

In Section 11.7 we modeled a flu epidemic in Spring Point, an isolated town with a population of 1800, with the function

$$f(t) = \frac{1800}{1 + 359e^{-0.9t}}$$

where $f(t)$ represented the number of people affected by the flu after t weeks. See Figure 12.2.5. How many people were affected after 10 weeks?

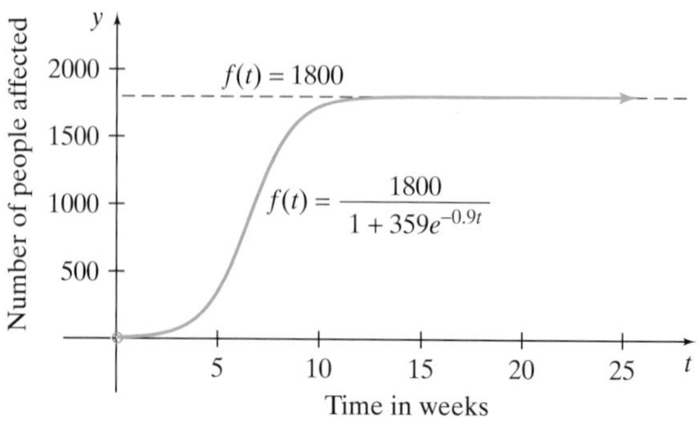

Figure 12.2.5

Flashback Solution

To determine the number of people affected after 10 weeks, we simply substitute 10 for t and get

$$f(10) = \frac{1800}{1 + 359e^{-0.9(10)}} \approx 1723.6358$$

So after 10 weeks approximately 1724 people were affected by the flu.

Interactive Activity

Construct a table of values of $f(t)$ for $t = 1, 5, 10, 20, 50,$ and 100 to numerically confirm that

$$\lim_{t\to\infty} \frac{1800}{1 + 359e^{-0.9t}} = 1800$$

Notice in the Flashback that after 10 weeks 1724 out of 1800 people had been affected by the flu. Intuitively, for any value of t the largest value that $f(t)$ can have is 1800, the entire population of the town! Also, it seems reasonable that as t increases (that is, as the number of weeks since the flu outbreak increases) the number of people affected by the flu, $f(t)$, should increase. Putting this information together, namely that as t increases $f(t)$ increases, yet $f(t)$ cannot exceed Spring Point's population of 1800, leads us to claim that

$$\lim_{t\to\infty} \frac{1800}{1 + 359e^{-0.9t}} = 1800$$

Notice how this is supported by the **horizontal asymptote** in Figure 12.2.5.

Example 5 Analyzing Limits at Infinity

Consider $f(x) = 2^x$.

(a) Use Table 12.2.5 and Figure 12.2.6 to estimate $\lim_{x\to\infty} 2^x$.

(b) Use Table 12.2.6 and Figure 12.2.6 to estimate $\lim_{x\to-\infty} 2^x$.

Table 12.2.5

$x \to \infty$	
x	$f(x)$
1	2
10	1024
50	1.1×10^{15}
100	1.3×10^{30}

Table 12.2.6

$x \to \infty$	
x	$f(x)$
−1	0.5
−10	9.8×10^{-4}
−50	8.9×10^{-16}
−100	7.9×10^{-31}

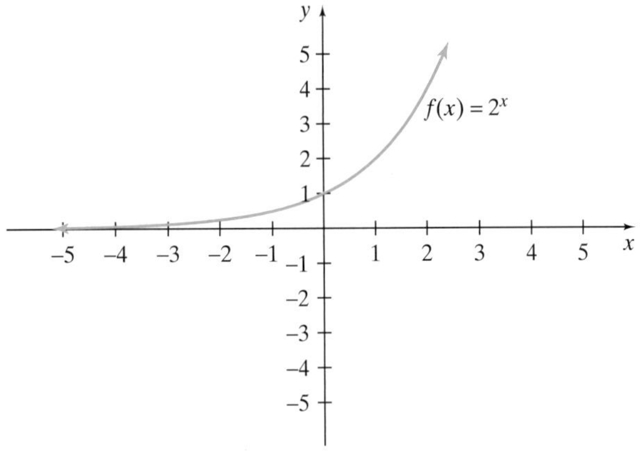

Figure 12.2.6

Solution

(a) Table 12.2.5 and Figure 12.2.6 suggest that $\lim_{x\to\infty} 2^x = \infty$. In other words, as x increases without bound, 2^x also increases without bound.

(b) Table 12.2.6 and Figure 12.2.6 suggest that $\lim_{x\to-\infty} 2^x = 0$. In other words, as x decreases without bound, 2^x gets close to 0. ■

Technology Option

As we saw in Section 12.1, we can use the TABLE command on our graphing calculator to construct tables as an aid to numerically determine a limit. See Tables 12.2.7 and 12.2.8.

Table 12.2.7

X	Y1
1	2
10	1024
50	1.1E15
100	1.3E30
1000	ERROR
X=	

Table 12.2.8

X	Y1
-1	.5
-10	9.8E-4
-50	9E-16
-100	8E-31
-1000	0
X=	

In Table 12.2.7, $x = 1000$ produces an error, and in Table 12.2.8, $x = -1000$ produces a value of 0 due to the limitations of our calculator. Basically, 2^{1000} is too large and 2^{-1000} is too close to 0 for our calculator.

Recall from Chapter 11 that Example 5 illustrates what is known as **exponential growth**. Example 6 illustrates **exponential decay**.

Example 6 Applying Limits at Infinity

Pharmacological studies have determined that the amount of medication present in the body is a function of the amount given and how much time has elapsed since the medication was administered. For a certain medication, the amount present in milliliters, $A(t)$, can be approximated by the function

$$A(t) = 3e^{-0.123t}$$

where t is the number of hours since the medication was administered.

(a) Determine $A(0)$ and interpret.

(b) Determine $\lim_{t\to\infty} A(t)$ and interpret.

Solution

(a) $A(0) = 3e^{-0.123(0)} = 3.$

Interpret the Solution: This means that initially, 3 milliliters of the medication was administered.

(b) Using Table 12.2.9 and Figure 12.2.7, we conclude that $\lim_{t\to\infty} A(t) = 0.$

Table 12.2.9

$t \to \infty$

t	0	1	10	100	1000
$A(t)$	3	2.65	0.88	1.4×10^{-5}	1×10^{-53}

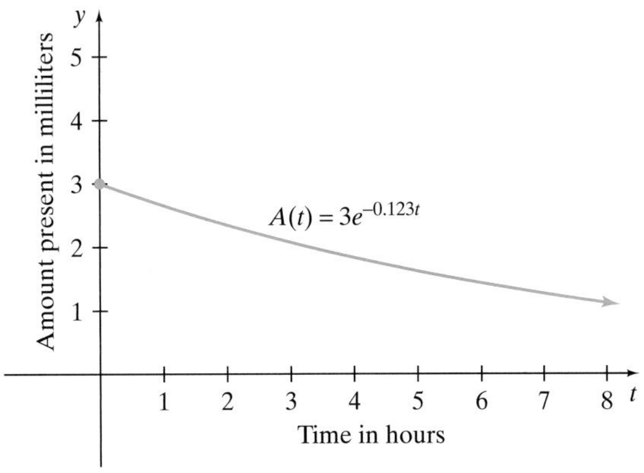

Figure 12.2.7

Interpret the Solution: This means that, as the number of hours since administering the medication increases without bound, the amount present in the body approaches 0 milliliters. It should seem reasonable that the result is 0. Since $t = 1000$ hours is about 42 days, we would hope that, by then, our body would be rid of this medication. ■

Checkpoint 3 Now work Exercise 53.

As seen in the Flashback and in Examples 5 and 6, for some functions there is a relationship between limits at infinity and horizontal asymptotes.

Limits at Infinity and Horizontal Asymptotes

For any function f, if $\lim_{x \to \pm\infty} f(x) = L$, then the line $y = L$ is a **horizontal asymptote** for the graph of f.

Consider $f(x) = \frac{1}{x}$, $g(x) = \frac{1}{x^{5/3}}$ and $h(x) = \frac{1}{x^{1/2}}$ and their graphs in Figures 12.2.8, 12.2.9, and 12.2.10, respectively. Figure 12.2.8 suggests that $\lim_{x \to \infty} \frac{1}{x} = 0$ and $\lim_{x \to -\infty} \frac{1}{x} = 0$, and the line $y = 0$ is a horizontal asymptote for the

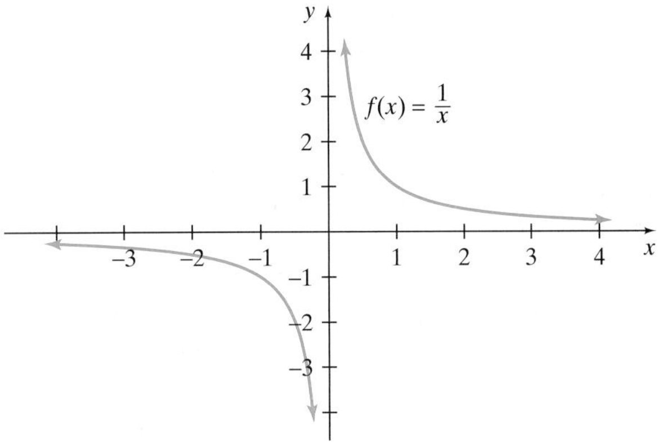

Figure 12.2.8

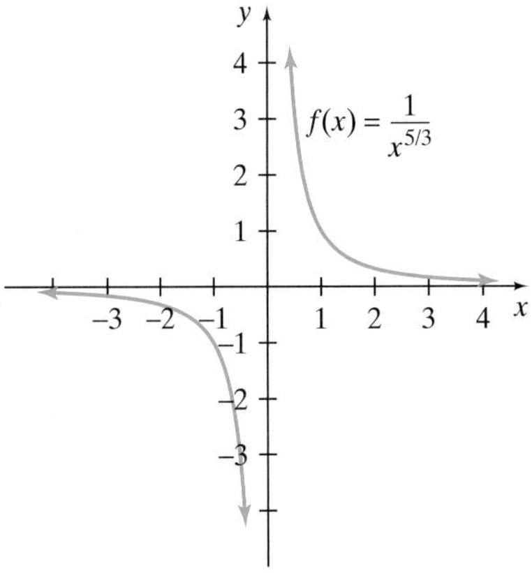

Figure 12.2.9

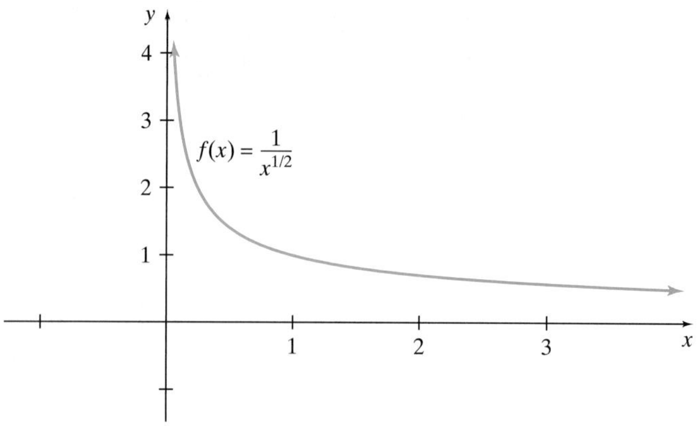

Figure 12.2.10

graph. Figure 12.2.9 suggests that $\lim_{x\to\infty} \frac{1}{x^{5/3}} = 0$ and $\lim_{x\to-\infty} \frac{1}{x^{5/3}} = 0$, and the line $y = 0$ is a horizontal asymptote. Figure 12.2.10 suggests that $\lim_{x\to\infty} \frac{1}{x^{1/2}} = 0$, whereas $\lim_{x\to-\infty} \frac{1}{x^{1/2}}$ does not exist, since negative numbers are not in the domain. Also, the line $y = 0$ is a horizontal asymptote here as well. These three functions and their respective graphs demonstrate the following.

Special Limits at Infinity

1. For n, a positive real number, $\lim_{x\to\infty} \frac{1}{x^n} = 0$.
2. For n, a positive real number, $\lim_{x\to-\infty} \frac{1}{x^n} = 0$, provided that x^n is a real number for negative values of x.

Note: All Limit Theorems from Section 12.1 are true for limits at infinity.

Example 7 illustrates the usefulness of these special limits of infinity while confirming a graphing technique first presented in Section 11.5.

Example 7 Determining Horizontal Asymptotes

For $f(x) = \frac{x^2 + 1}{2x^2 - 1}$, determine $\lim_{x\to\infty} f(x)$ and $\lim_{x\to-\infty} f(x)$ algebraically, and determine the horizontal asymptote for the graph of f.

Solution

Since we have a rational function (the numerator and the denominator are polynomial functions), the algebra manipulation for problems of this type is rather straightforward.

Divide every term in the numerator and denominator by x^b, where b is the larger degree of the numerator and the denominator.

For this problem, since both degrees are 2, we have $b = 2$. Next we divide every term in the numerator and denominator by x^2. This gives

$$f(x) = \frac{x^2+1}{2x^2-1} = \frac{(x^2+1) \div x^2}{(2x^2-1) \div x^2}$$
$$= \frac{\frac{x^2}{x^2} + \frac{1}{x^2}}{\frac{2x^2}{x^2} - \frac{1}{x^2}} = \frac{1 + \frac{1}{x^2}}{2 - \frac{1}{x^2}}$$

Interactive Activity

Make a table and graph $y = \frac{x^2+1}{2x^2-1}$ to numerically and graphically support the result of Example 7.

WWW

So, to determine the limit algebraically, we have

$$\lim_{x\to\infty} \frac{x^2+1}{2x^2-1} = \lim_{x\to\infty} \frac{1 + \frac{1}{x^2}}{2 - \frac{1}{x^2}} = \frac{\lim_{x\to\infty}\left(1 + \frac{1}{x^2}\right)}{\lim_{x\to\infty}\left(2 - \frac{1}{x^2}\right)}$$
$$= \frac{\lim_{x\to\infty} 1 + \lim_{x\to\infty} \frac{1}{x^2}}{\lim_{x\to\infty} 2 - \lim_{x\to\infty} \frac{1}{x^2}} = \frac{1+0}{2-0} = \frac{1}{2}$$

The algebraic verification for $\lim_{x\to-\infty} \frac{x^2+1}{2x^2-1}$ is identical. We immediately conclude that the horizontal asymptote is the line $y = \frac{1}{2}$. ■

Checkpoint 4 Now work Exercise 33.

Example 8 Analyzing Average Cost

The total cost, in dollars, to produce x units of a certain product is given by $C(x) = 22{,}500 + 7.35x$. The **average cost**, AC, is given by

$$AC(x) = \frac{C(x)}{x} = \frac{22{,}500 + 7.35x}{x}$$

Determine $\lim_{x\to\infty} AC(x)$ and interpret.

Solution

Interactive Activity

Numerically and graphically verify the result in Example 8.

Since $AC(x) = \frac{22{,}500 + 7.35x}{x}$ is already written as a fraction with a single term in the denominator, we may break it apart as follows:

$$\lim_{x\to\infty} \frac{22{,}500 + 7.35x}{x} = \lim_{x\to\infty}\left(\frac{22{,}500}{x} + \frac{7.35x}{x}\right) = \lim_{x\to\infty}\left(\frac{22{,}500}{x} + 7.35\right)$$
$$= 0 + 7.35 = 7.35$$

Interpret the Solution: This means that as the number of units produced increases without bound, the average total cost is approaching \$7.35. In other words, the cost per unit is approaching \$7.35 as the number of units produced increases without bound. ■

Checkpoint 5 Now work Exercise 51.

SUMMARY

In this section we continued our study of limits. We introduced the symbol ∞ (and $-\infty$) to represent an increasing without bound (decreasing without bound) behavior. We examined the relationship between **infinite limits** and **vertical asymptotes**. We also examined the relationship between **limits at infinity** and **horizontal asymptotes**.

- If as $x \to a^-$ the function values $f(x)$ increase or decrease without bound, that is, $\lim_{x\to a^-} f(x) = \infty$ or $\lim_{x\to a^-} f(x) = -\infty$, then the line $x = a$ is a vertical asymptote.
- If as $x \to a^+$ the function values $f(x)$ increase or decrease without bound, that is, $\lim_{x\to a^+} f(x) = \infty$ or $\lim_{x\to a^+} f(x) = -\infty$, then the line $x = a$ is a vertical asymptote.
- For any function f, if $\lim_{x\to\pm\infty} f(x) = L$, then the line $y = L$ is a horizontal asymptote for the graph of f.
- For n, a positive real number, $\lim_{x\to\infty} \frac{1}{x^n} = 0$.
- For n, a positive real number, $\lim_{x\to-\infty} \frac{1}{x^n} = 0$, provided that x^n is a real number for negative values of x.

SECTION 12.2 EXERCISES

For Exercises 1–8, complete the given table to numerically estimate the following limits:

(a) $\lim_{x\to a^-} f(x)$ **(b)** $\lim_{x\to a^+} f(x)$ **(c)** $\lim_{x\to a} f(x)$

Use the symbols ∞ and $-\infty$ where applicable.

1. $f(x) = \frac{1}{x^3}$; $a = 0$

x	−0.1	−0.01	−0.001	−0.00001	0	0.00001	0.001	0.01	0.1
$f(x)$									

2. $f(x) = \frac{2}{x^3}$; $a = 0$

x	−0.1	−0.01	−0.001	−0.00001	0	0.00001	0.001	0.01	0.1
$f(x)$									

3. $f(x) = \frac{1}{(x-1)^2}$; $a = 1$

x	0.9	0.99	0.999	0.99999	1	1.00001	1.001	1.01	1.1
$f(x)$									

4. $f(x) = \frac{3}{(x-1)^2}$; $a = 1$

x	0.9	0.99	0.999	0.99999	1	1.00001	1.001	1.01	1.1
$f(x)$									

5. $f(x) = \frac{x+2}{x^2 - x - 6}$; $a = -2$

x	−2.1	−2.001	−2.0001	−2	−1.9999	−1.999	−1.9
$f(x)$							

6. $f(x) = \dfrac{x+2}{x^2 - x - 6}$; $a = 3$

x	2.9	2.99	2.999	2.99999	3	3.00001	3.001	3.01	3.1
$f(x)$									

7. $f(x) = \dfrac{13{,}250 + 2.35x}{x}$; $a = 0$

x	−0.1	−0.01	−0.001	−0.00001	0	0.00001	0.001	0.01	0.1
$f(x)$									

8. $f(x) = \dfrac{4.25x - 25{,}350}{x}$; $a = 0$

x	−0.1	−0.01	−0.001	−0.00001	0	0.00001	0.001	0.01	0.1
$f(x)$									

In Exercises 9–16, determine the limit algebraically.

9. $\lim\limits_{x \to 0} \dfrac{2}{x^3}$

10. $\lim\limits_{x \to 0} \dfrac{13{,}250 + 2.35x}{x}$

11. $\lim\limits_{x \to -2} \dfrac{x+2}{x^2 - x - 6}$

12. $\lim\limits_{x \to 3} \dfrac{x+2}{x^2 - x - 6}$

13. $\lim\limits_{x \to 3} \dfrac{x-3}{x^2 - 9}$

14. $\lim\limits_{x \to -3} \dfrac{x-3}{x^2 - 9}$

15. $\lim\limits_{x \to -1} \dfrac{x^2 - x + 1}{x^3 + 1}$

16. $\lim\limits_{x \to 3} \dfrac{x^2 - 9}{x - 3}$

For each function in Exercises 17–24:

(a) Classify f as an exponential growth or exponential decay function.

(b) Determine $\lim\limits_{x \to \infty} f(x)$ and determine $\lim\limits_{x \to -\infty} f(x)$.

17. $f(x) = 3^x$

18. $f(x) = 3^{-x}$

19. $f(x) = e^x$

20. $f(x) = e^{-x}$

21. $f(x) = e^{-0.215x}$

22. $f(x) = e^{0.215x}$

23. $f(x) = (0.987)^x$

24. $f(x) = (0.987)^{-x}$

In Exercises 25–32, determine the indicated limit algebraically using the method in Example 7. Verify your results numerically.

25. $\lim\limits_{x \to -\infty} \dfrac{2x+5}{x-1}$

26. $\lim\limits_{x \to \infty} \dfrac{2x+5}{x-1}$

27. $\lim\limits_{x \to \infty} \dfrac{3x^2 - x + 2}{2x^2 + x - 5}$

28. $\lim\limits_{x \to -\infty} \dfrac{3x^2 - x + 2}{2x^2 + x - 5}$

29. $\lim\limits_{x \to \infty} \dfrac{2x^2 + 2x + 1}{5x^3 + 3x - 5}$

30. $\lim\limits_{x \to \infty} \dfrac{3x^2 - 2x + 5}{2x^3 + x^2 - 2x + 3}$

31. $\lim\limits_{x \to -\infty} \dfrac{2x^2 + 2x + 1}{5x^3 + 3x - 5}$

32. $\lim\limits_{x \to -\infty} \dfrac{3x^2 - 2x + 5}{2x^3 + x^2 - 2x + 3}$

In Exercises 33–36, determine an equation for the horizontal asymptote for the graph of f. Consult your work performed in Exercises 25–32.

33. $f(x) = \dfrac{2x+5}{x-1}$

34. $f(x) = \dfrac{3x^2 - x + 2}{2x^2 + x - 5}$

35. $f(x) = \dfrac{2x^2 + 2x + 1}{5x^3 + 3x - 5}$

36. $f(x) = \dfrac{3x^2 - 2x + 5}{2x^3 + x^2 - 2x + 3}$

Use the graph of f in Figure 12.2.11 to determine the indicated limit in Exercises 37–43.

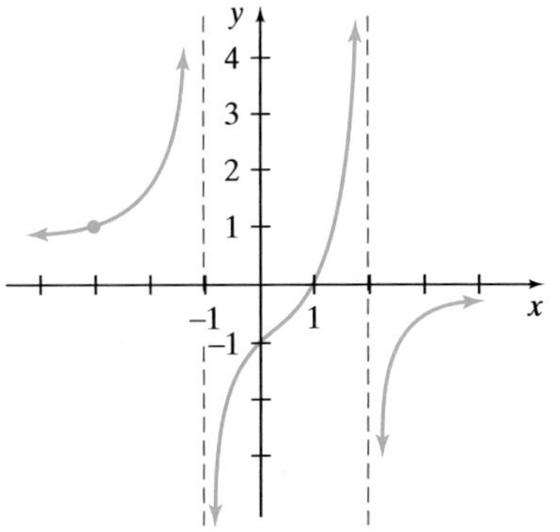

Figure 12.2.11

37. $\lim\limits_{x \to 2^+} f(x)$

38. $\lim\limits_{x \to 2^-} f(x)$

39. $\lim\limits_{x \to 2} f(x)$

40. $\lim\limits_{x \to -1} f(x)$

41. $\lim\limits_{x \to -3} f(x)$

42. $\lim\limits_{x \to \infty} f(x)$

43. $\lim\limits_{x \to -\infty} f(x)$

Applications

44. The cost $C(x)$, in thousands of dollars, of removing $x\%$ of a city's pollutants discharged into a lake is given by

$$C(x) = \frac{223x}{100 - x}$$

(a) Determine the reasonable domain for C.

(b) Evaluate $C(40)$ and interpret.

(c) Determine $\lim_{x \to 100^-} C(x)$ and interpret.

45. The cost $C(x)$, in thousands of dollars, of removing $x\%$ of a city's pollutants discharged into a river is given by

$$C(x) = \frac{93x}{100 - x}$$

(a) Determine the reasonable domain for C.

(b) Evaluate $C(40)$ and interpret.

(c) Determine $\lim_{x \to 100^-} C(x)$ and interpret.

46. The number of dollars spent on advertising for a product influences the number of items of the product that will be purchased by consumers. If x is the number of dollars spent, in thousands, then $N(x)$, in hundreds, is the number of items sold, as given by the function

$$N(x) = 2000 - \frac{520}{x}$$

Determine $\lim_{x \to \infty} N(x)$ and interpret.

47. The number of dollars spent on advertising for a product influences the number of items of the product that will be purchased by consumers. If x is the number of dollars spent, in thousands, then $N(x)$, in hundreds, is the number of items sold, as given by the function

$$N(x) = 3200 - \frac{750}{x}$$

Determine $\lim_{x \to \infty} N(x)$ and interpret.

48. *(continuation of Exercise 46)* Graph the model given in Exercise 46, $N(x) = 2000 - \frac{520}{x}$ in the viewing window [0, 40] by [1000, 2200]. Use the graph and the capabilities of your graphing calculator to estimate an interval for x that would result in $N(x)$ being between 1500 and 1800.

49. *(continuation of Exercise 47)* Graph the model given in Exercise 47, $N(x) = 3200 - \frac{750}{x}$ in the viewing window [0, 40] by [2000, 3400]. Use the graph and the capabilities of your graphing calculator to estimate an interval for x that would result in $N(x)$ being between 2700 and 3000.

11.2 **50.** The total cost of producing q units of a product is given by $C(q) = 13{,}700 + 6.85q$, where $C(q)$ is in dollars. Recall that the average cost, AC, is given by

$$AC(q) = \frac{C(q)}{q} = \frac{13{,}700 + 6.85q}{q}$$

Determine $\lim_{q \to \infty} AC(q)$ and interpret.

11.2 **51.** The total cost of producing q units of a product is given by $C(q) = 15{,}200 + 4.85q$, where $C(q)$ is in dollars. Recall that the average cost, AC, is given by

$$AC(q) = \frac{C(q)}{q} = \frac{15{,}200 + 4.85q}{q}$$

Determine $\lim_{q \to \infty} AC(q)$ and interpret.

52. This exercise uses the following data for the population of Florida.

x, Years Since 1899	y, Population in Thousands
1	528.542
11	752.619
21	968.470
31	1,468.211
41	1,897.414
51	2,771.305
61	4,951.560
71	6,791.418
81	9,746.961
91	12,937.926
100	15,111.000

SOURCE: U.S. Census Bureau, www.census.gov

Bob has modeled the data for the population of Florida, where x represents the number of years since 1899 and y represents the population in thousands, with the following models:

$$y = 2.01x^2 - 56.6x + 939.93$$

$$y = 490.23e^{0.04x}$$

$$y = \frac{23{,}976.11}{1 + 84.95e^{-0.05x}}$$

Each does a good job of modeling the given data. Aware of the dangers of using any model to predict too far into the future, Bob still wants to try to determine an upper limit for the population of Florida, but he does not know how. Help Bob by doing the following:

(a) Determine $\lim_{x \to \infty} (2.01x^2 - 56.6x + 939.93)$ and interpret.

(b) Determine $\lim_{x \to \infty} (490.23e^{0.04x})$ and interpret.

(c) Determine $\lim_{x \to \infty} \left(\frac{23{,}976.11}{1 + 84.95e^{-0.05x}} \right)$ and interpret.

(d) Which of the three models gives the most reasonable estimate for an upper limit of Florida's population and why?

53. Pharmacological studies have determined that the amount of medication present in the body is a function of the amount given and how much time has elapsed since the medication was administered. For a certain medication, the amount present in milliliters, $A(t)$, can be approximated by the function

$$A(t) = 2.5e^{-0.2t} \qquad t \geq 0$$

where t is the number of hours since the medication was administered.

(a) Determine $A(0)$ and interpret.

(b) Determine $\lim_{t \to \infty} A(t)$ and interpret.

54. Pharmacological studies have determined that the amount of medication present in the body is a function of the amount given and how much time has elapsed since the medication was administered. For a certain medication, the amount present in milliliters, $A(t)$, can be approximated by the function

$$A(t) = 3.5e^{-0.3t} \qquad t \geq 0$$

where t is the number of hours since the medication was administered.

(a) Determine $A(0)$ and interpret.

(b) Determine $\lim_{t\to\infty} A(t)$ and interpret.

55. *(continuation of Exercise 53)* Graph the model given in Exercise 53, $A(t) = 2.5e^{-0.2t}$ in the viewing window [0, 5] by [0, 5]. If a patient is to receive more medication when the amount present drops below 1 milliliter, graphically approximate how many hours after the initial administration of the medication more medication should be given.

56. *(continuation of Exercise 54)* Graph the model given in Exercise 54, $A(t) = 3.5e^{-0.3t}$ in the viewing window [0, 5] by [0, 5]. If a patient is to receive more medication when the amount present drops below 1 milliliter, graphically approximate how many hours after the initial administration of the medication more medication should be given.

57. The local game commission decided to stock a lake with bass. To do this, 100 bass were introduced into the lake. The population of the bass can be approximated by

$$P(t) = \frac{10(10 + 7t)}{1 + 0.02t} \qquad t \geq 0$$

where t is time in months since the lake was stocked and $P(t)$ is the population after t months.

(a) Determine the population in 1 year.

(b) Determine $\lim_{t\to\infty} P(t)$ and interpret. The result is called the *limiting size* of the population.

58. The local game commission decided to stock a lake with trout. To do this, 200 trout were introduced into the lake. The population of the trout can be approximated by

$$P(t) = \frac{20(10 + 7t)}{1 + 0.02t} \qquad t \geq 0$$

where t is time in months since the lake was stocked and $P(t)$ is the population after t months.

(a) Determine the population in 1 year.

(b) Determine $\lim_{t\to\infty} P(t)$ and interpret. The result is called the *limiting size* of the population.

59. *(continuation of Exercise 57)* Graph the model given in Exercise 57, $P(t) = \dfrac{10(10 + 7t)}{1 + 0.02t}$ in the viewing window [0, 60] by [0, 3000]. Use the graph to support your answers for Exercise 57 parts (a) and (b).

60. *(continuation of Exercise 58)* Graph the model given in Exercise 58, $P(t) = \dfrac{20(10 + 7t)}{1 + 0.02t}$ in the viewing window [0, 60] by [0, 5000]. Use the graph to support your answers for Exercise 58 parts (a) and (b).

61. The spread of influenza in Surfside, an isolated town with a population of 1000, can be modeled by

$$D(t) = \frac{1000}{1 + 999e^{-0.3t}} \qquad t \geq 0$$

where $D(t)$ is the number of people affected by influenza after t days.

(a) How many people have been affected by influenza after 3 days?

(b) Determine $\lim_{t\to\infty} D(t)$ and interpret.

62. When a new product is introduced on the market, sales often follow a pattern of rapid initial growth followed by some leveling off. For a certain new candy bar, this pattern is modeled by

$$S(t) = 4000 - 4000e^{-0.25t}$$

where $S(t)$ is the number of units sold after t months.

(a) How many units were sold after 2 months?

(b) Determine $\lim_{t\to\infty} S(t)$ and interpret.

11.2

SECTION PROJECT

Use the following data for the population of California.

x, Years Since 1899	*y*, Population in Thousands	*x*, Years Since 1899	*y*, Population in Thousands
1	1,485.053	61	15,717.204
11	2,377.549	71	19,971.069
21	3,426.861	81	23,667.764
31	5,677.251	91	29,760.021
41	6,907.387	100	33,145.000
51	10,586.223		

SOURCE: U.S. Census Bureau, www.census.gov

Let x represent the number of years since 1899 and let y represent the population, in thousands, of California. Enter the data into your calculator and determine the following regression models:

(a) Quadratic model: round coefficients to the nearest hundredth.

(b) Exponential model: round coefficients to the nearest hundredth.

(c) Logistic model: round coefficients to the nearest hundredth.

(d) Plot the data and graph each model to determine if each model gives a good fit to the data.

(e) Determine $\lim_{x\to\infty}$ for the each model and interpret.

(f) Aware of the dangers of using regression models to predict too far into the future, determine which of these three models gives the most reasonable estimate for an upper limit to the population of California.

Section 12.3 Problem Solving: Rates of Change

The study of calculus is the study of rates of change. So far, we have focused on the **average rate of change**. Now we wish to study the **instantaneous rate of change**. This type of rate of change is one of the central topics of differential calculus. In this section, we will see how these rates of change are different from one another. We will also see how we can approximate the instantaneous rate of change for a given function.

Thinking about Rates of Change

In Chapter 11, we saw that the average rate of change of a function f on a closed interval was given by the **slope of the secant line** for a given point $(x, f(x))$ and a given increment in x denoted by Δx. **To simplify computations in the remainder of this chapter, we will use h in the place of Δx.** This gives us the secant line slope formula shown in the Toolbox below. We first saw this formula in Section 11.3.

From Your Toolbox

The **slope of the secant line** between the points $(x, f(x))$ and $(x+h, f(x+h))$, denoted by m_{sec}, is given by the difference quotient

$$m_{\text{sec}} = \frac{f(x+h) - f(x)}{h}$$

where $h = \Delta x$ is the increment in x.

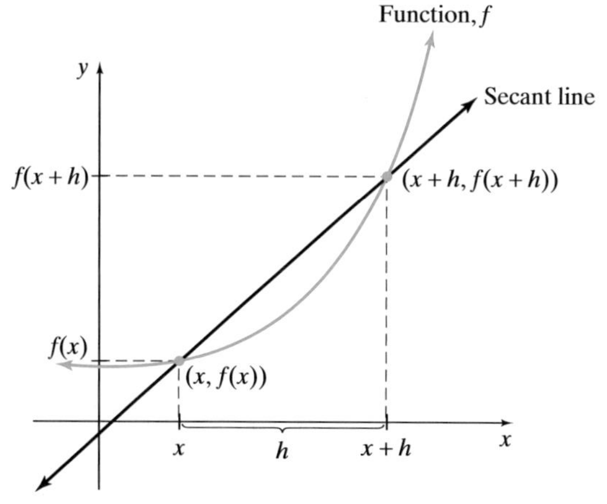

Figure 12.3.1

One of our main uses of the secant line slope was to use it to give us the average rate of change of a function over a closed interval. In Example 1 we review how to do this.

Example 1 Determining an Average Rate of Change

Compute the average rate of change of the function $f(x) = x^2 + 3$ over the interval [2, 5].

Solution

Understand the Situation: Since the average rate of change is given by the slope of the secant line over an interval, we use the formula $m_{sec} = \dfrac{f(x+h) - f(x)}{h}$ to compute the slope of the secant line. We are given the interval [2, 5], which means that $x = 2$ and $h = 5 - 2 = 3$.

So we get

$$m_{sec} = \frac{f(x+h) - f(x)}{h} = \frac{f(2+3) - f(2)}{3}$$
$$= \frac{f(5) - f(2)}{3} = \frac{28 - 7}{3} = \frac{21}{3} = 7$$

Interpret the Solution: The slope of the secant line on the interval [2, 5] is $m_{sec} = 7$. This means that the average rate of change of $f(x) = x^2 + 3$ over the interval [2, 5] is $7 \dfrac{\text{dependent units}}{\text{independent unit}}$. See Figure 12.3.2.

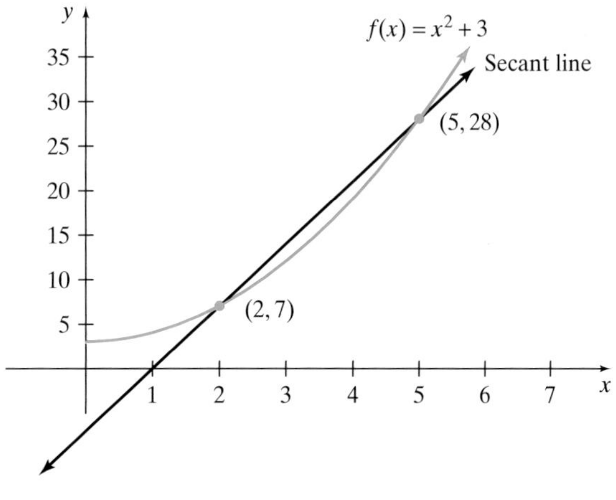

Figure 12.3.2

■

Thinking about the Instantaneous Rate of Change

We now pose the following question: At what rate is the function in Example 1, $f(x) = x^2 + 3$, changing when $x = 2$? We are not asking for the average rate of change over some interval as was done in Example 1. Instead the question asks for the rate of change at a particular *point*. This type of rate of change is called an **instantaneous rate of change**, and is equivalent to the slope of the line **tangent** to the curve at a specific point. What do we mean by **tangent line**?

Consider the curve in Figure 12.3.3.

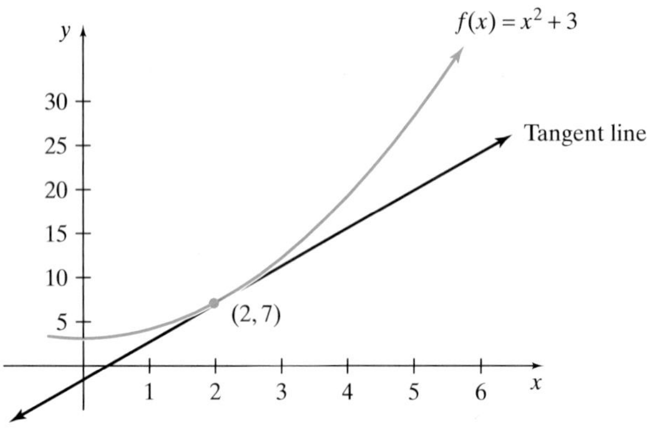

Figure 12.3.3

The steepness of the curve varies from point to point. The measure of the steepness of the curve at the point (2, 7) is equivalent to finding the slope of the tangent line through the point. (We call the point through which the tangent line passes the **fixed point** or the **point of tangency**.) With this in mind, the following serves as a working definition of tangent line.

Tangent Line

The **tangent line** to a curve at a point A is the line through A whose slope matches the steepness of the curve at point A. We can say that the tangent line slope is equal to the slope of the curve at A.

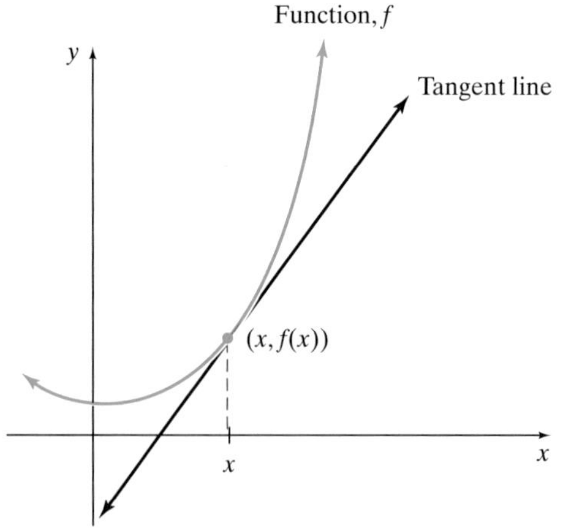

Figure 12.3.4

Figures 12.3.1 and 12.3.4 illustrate the main differences between slope of a secant line and slope of a tangent line. The slope of the **secant line** is determined by two points on the curve and it gives an **average rate of change**. The slope of the **tangent line** is associated with a single point on the curve and it gives an

instantaneous rate of change. Before we begin the task of approximating tangent line slope, Example 2 contrasts these two rates of change.

Example 2 Classifying Rates of Change

Classify each of the following as exhibiting average rate of change or instantaneous rate of change.

(a) The line l_1 that is shown in Figure 12.3.5.

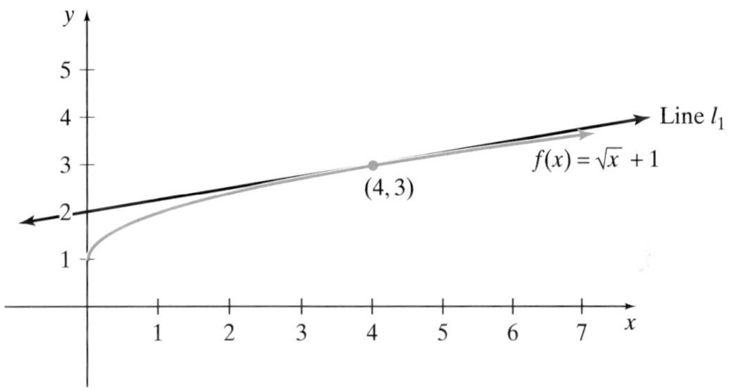

Figure 12.3.5

(b) The line l_2 that is shown in Figure 12.3.6.

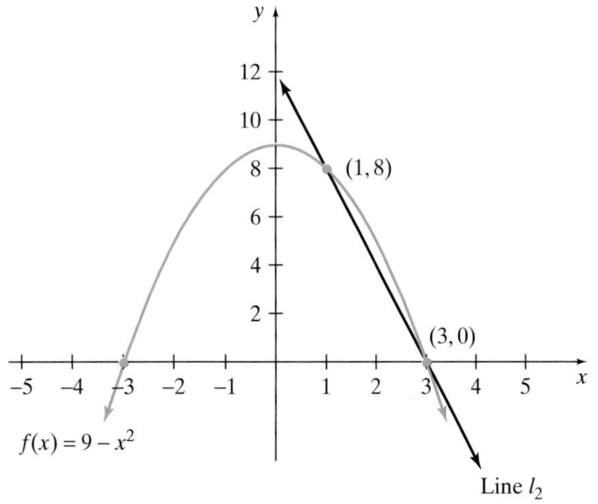

Figure 12.3.6

(c) The Country Day Company determines that the daily cost of producing a lawn tractor tire is given by

$$C(x) = 100 + 40x - 0.001x^2 \qquad 0 \leq x \leq 300$$

where x represents the number of tires produced each day and $C(x)$ represents the daily cost in dollars. Suppose we want the rate of change in cost when the production changes from $x = 100$ to $x = 150$.

(d) The Kelomata Company determines the daily cost of producing patio swings can be modeled by

$$C(x) = 15{,}000 + 100x - 0.001x^2 \qquad 0 \le x \le 200$$

where x represents the number of patio swings produced each day and $C(x)$ represents the daily cost in dollars. Suppose we want the rate of change in cost at a production level of $x = 75$.

Solution

Understand the Situation: In order to distinguish between the two types of rates of change, we can examine the number of points on the curve that a line would pass through. If two points are given, we have a secant line and its slope is associated with an average rate of change. If one point is given, we have a tangent line and its slope is associated with an instantaneous rate of change.

(a) Here we have a line touching the graph of the function $f(x) = \sqrt{x} + 1$ at the point (4, 3). So the line l_1 is a tangent line and its slope is associated with an instantaneous rate of change.

(b) In this case the line intersects the function $f(x) = 9 - x^2$ at the points $(1, 8)$ and (3, 0). So the line l_2 is a secant line and its slope is associated with an average rate of change.

(c) Since we are considering two production levels $x = 100$ and $x = 150$ for this cost function, we are seeking an average rate of change.

(d) For the given cost function, we want the rate of change in cost at a single production level. So in this case we seek the instantaneous rate of change.

■

Checkpoint 1 Now work Exercises 9 and 11.

Approximating Instantaneous Rate of Change

Now that we understand the difference between average rate of change and instantaneous rate of change and the types of lines that each is associated with, we turn our attention to approximating the instantaneous rate of change. Since instantaneous rate of change arises in many situations, the following is the process we use in approximating this rate:

- To find the slope of a tangent line, determine the point of tangency.
- Select points on the curve that are progressively closer to the point of tangency. This is achieved by letting h, the increment in x reviewed at the beginning of the section, approach 0. That is, we are going to let $h \to 0$.
- Calculate the slope of the secant line through each of these points and the point of tangency. The slope of the tangent line is the limit of the slopes of these secant lines.

The third bullet item revisits a concept introduced in the first two sections of this chapter, the limit concept. To ensure that the limit exists, the points we select that are progressively closer to the point of tangency will be points to the left and to the right of the point of tangency. This left and right analysis is done by letting $h \to 0^-$ and by letting $h \to 0^+$. As an aid to streamline the process listed in the three bullet items, we offer the following as a way to organize our work.

Approximating the Instantaneous Rate of Change

To approximate the instantaneous rate of change of a function f at a given point $(x, f(x))$, we use the following steps.

1. Complete a table with the following columns for $h \to 0^+$.

h	Interval, $[x, x+h]$	$m_{sec} = \dfrac{f(x+h) - f(x)}{h}$
1		
0.1		
0.01		
0.001		

2. Complete a table with the following columns for $h \to 0^-$.

h	Interval, $[x+h, x]$	$m_{sec} = \dfrac{f(x+h) - f(x)}{h}$
-1		
-0.1		
-0.01		
-0.001		

3. Examine the values of m_{sec} as $h \to 0^+$ and as $h \to 0^-$ to approximate the instantaneous rate of change.

Example 3 Approximating an Instantaneous Rate of Change

Approximate the instantaneous rate of change of $f(x) = x^2 + 3$ at $x = 2$.

Solution

Understand the Situation: We need to compute two tables of values. One for $h = 1, 0.1, 0.01, 0.001$ (that is $h \to 0^+$) and one for $h = -1, -0.1, -0.01, -0.001$ (that is $h \to 0^-$).

The result of letting $h \to 0^+$ is shown in Table 12.3.1. The secant lines over the intervals in Table 12.3.1 are shown in Figure 12.3.7.

Table 12.3.1

h	Interval, $[2, 2+h]$	$m_{sec} = \dfrac{f(2+h) - f(2)}{h}$
1	$[2, 3]$	$\dfrac{f(2+h) - f(2)}{h} = \dfrac{f(3) - f(2)}{1} = \dfrac{12 - 7}{1} = 5$
0.1	$[2, 2.1]$	$\dfrac{f(2+h) - f(2)}{h} = \dfrac{f(2.1) - f(2)}{0.1} = \dfrac{7.41 - 7}{0.1} = 4.1$
0.01	$[2, 2.01]$	$\dfrac{f(2+h) - f(2)}{h} = \dfrac{f(2.01) - f(2)}{0.01} = \dfrac{7.0401 - 7}{0.01} = 4.01$
0.001	$[2, 2.001]$	$\dfrac{f(2+h) - f(2)}{h} = \dfrac{f(2.001) - f(2)}{0.001} = \dfrac{7.004001 - 7}{0.001} = 4.001$

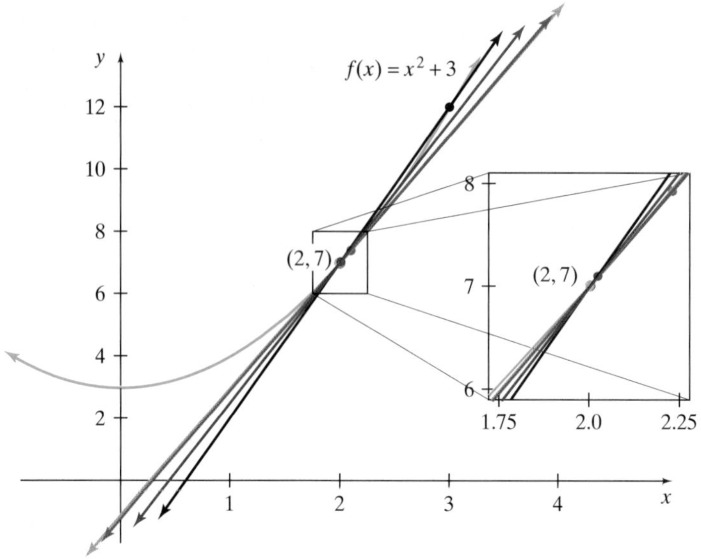

Figure 12.3.7

The result of letting $h \to 0^-$ is shown in Table 12.3.2. The secant lines over the intervals in Table 12.3.2 are shown in Figure 12.3.8.

Table 12.3.2

h	Interval, $[2 + h, 2]$	$m_{sec} = \dfrac{f(2+h) - f(2)}{h}$
-1	$[1, 2]$	$\dfrac{f(2+h) - f(2)}{h} = \dfrac{f(1) - f(2)}{-1} = \dfrac{4 - 7}{-1} = 3$
-0.1	$[1.9, 2]$	$\dfrac{f(2+h) - f(2)}{h} = \dfrac{f(1.9) - f(2)}{-0.1} = \dfrac{6.61 - 7}{-0.1} = 3.9$
-0.01	$[1.99, 2]$	$\dfrac{f(2+h) - f(2)}{h} = \dfrac{f(1.99) - f(2)}{-0.01} = \dfrac{6.9601 - 7}{-0.01} = 3.99$
-0.001	$[1.999, 2]$	$\dfrac{f(2+h) - f(2)}{h} = \dfrac{f(1.999) - f(2)}{-0.001} = \dfrac{6.996001 - 7}{-0.001} = 3.999$

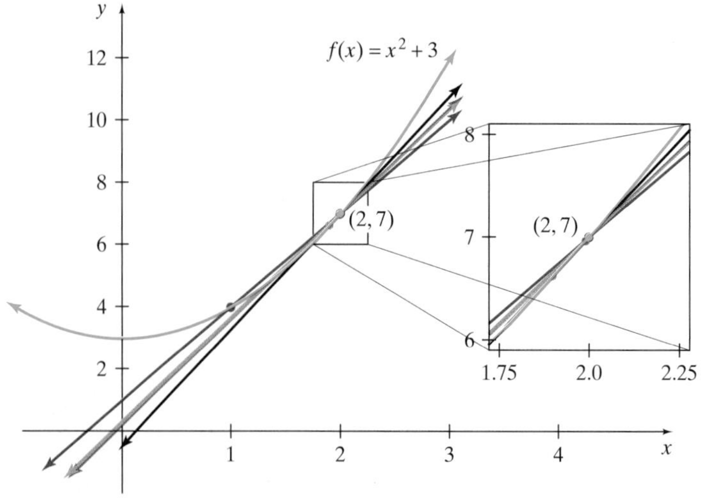

Figure 12.3.8

Interpret the Solution: We can see from inspecting the values of the two tables that, as h becomes smaller and smaller, the slope of the secant lines get closer and closer to 4. So we approximate the instantaneous rate of change of $f(x) = x^2 + 3$ at $x = 2$ is 4. This rate of change, which is the slope of the tangent line of $f(x) = x^2 + 3$ at the point $(2, f(2))$ is shown graphically in Figure 12.3.9.

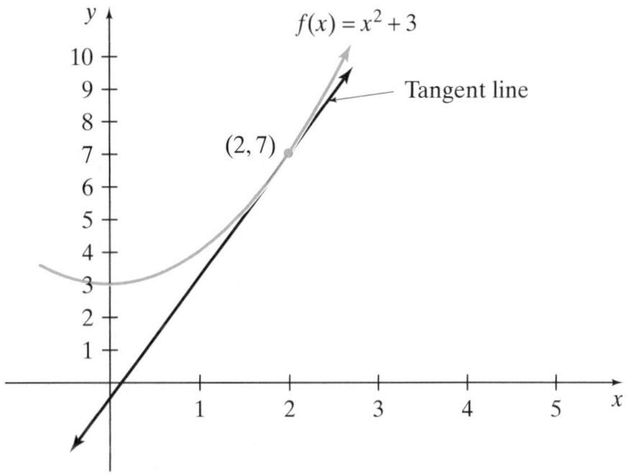

Figure 12.3.9 Slope of tangent line at (2, 7) is approximated as 4. This also gives the instantaneous rate of change of $f(x) = x^2 + 3$ at $x = 2$.

■

Checkpoint 2 Now work Exercise 33.

Technology Option

The computations for approximating the instantaneous rate of change of a function can be tedious, so we can use the program MSEC in Appendix C to help complete the two tables. For the program, the function f is entered in Y_1 and the x-value is entered when the program is executed. The results of Example 3 using the MSEC program are shown in Figure 12.3.10a and in Figure 12.3.10b.

(a) **(b)**

Figure 12.3.10

Thinking about Applying Instantaneous Rates of Change

To see how we can approximate the instantaneous rate of change in an application, let's return to a problem first encountered in Section 11.3. At that time we saw the U.S. imports from China for the years 1987 to 1996 can be modeled by

$$f(x) = 0.32x^2 + 1.64x + 3.98 \qquad 1 \leq x \leq 10$$

where x represents the number of years since 1986 and $f(x)$ represents the dollar value of goods imported from China, measured in billions (*Source:* U.S. Census Bureau, www.census.gov). To determine the average rate of change for the U.S. imports from China from 1990 to 1993 we need to compute the slope of the secant line. For this scenario we have $x = 1990 - 1986 = 4$ and $h = 1993 - 1990 = 3$. So the slope of the secant line is

$$m_{\text{sec}} = \frac{f(x+h) - f(x)}{\Delta x} = \frac{f(4+3) - f(4)}{3}$$
$$= \frac{f(7) - f(4)}{3} = \frac{31.14 - 15.66}{7 - 4} = 5.16$$

This means that from 1990 to 1993, the dollar value of goods imported from China increased at an average rate of 5.16 billion dollars per year. See Figure 12.3.11.

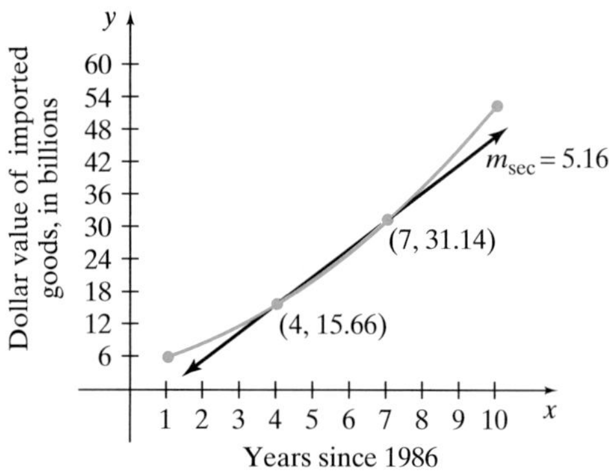

Figure 12.3.11 Slope of the secant line, m_{sec}, gives the average rate of change.

Now suppose we want to determine at what rate the U.S. imports from China was changing in 1993. We are not asking for the average rate of change over some time interval. Instead, the question asks for the rate of change at a particular **point**, namely $(7, f(7))$. Here we need to determine the **instantaneous rate of change**. We now know that this is given by the slope of the line **tangent** to the curve at the given point. We approximate the tangent line slope the same way we did in Example 3.

Example 4 Applying an Instantaneous Rate of Change

The U.S. imports from China can be modeled by

$$f(x) = 0.32x^2 + 1.64x + 3.98 \qquad 1 \le x \le 10$$

where x represents the number of years since 1986 and $f(x)$ represents the dollar value of goods imported from China, measured in billions (*Source:* U.S. Census Bureau, www.census.gov). Approximate the instantaneous rate of change of f at the value $x = 7$ and interpret the approximation.

Solution

Understand the Situation: First note that the value $x = 7$ corresponds to the year 1993. Here we will complete tables just as we did in Example 3.

Completing the values of m_{sec} as $h \to 0^+$ gives us the results shown in Table 12.3.3.

Table 12.3.3

h	Interval, $[7, 7+h]$	$m_{\text{sec}} = \dfrac{f(7+h)-f(7)}{h}$
1	$[7, 8]$	$\dfrac{f(7+h)-f(7)}{h} = \dfrac{f(8)-f(7)}{1} = \dfrac{37.58-31.14}{1} = 6.44$
0.1	$[7, 7.1]$	$\dfrac{f(7+h)-f(7)}{h} = \dfrac{f(7.1)-f(7)}{0.1} = \dfrac{31.7552-31.14}{0.1} = 6.152$
0.01	$[7, 7.01]$	$\dfrac{f(7+h)-f(7)}{h} = \dfrac{f(7.01)-f(7)}{0.01} = \dfrac{31.201232-31.14}{0.01} = 6.1232$
0.001	$[7, 7.001]$	$\dfrac{f(7+h)-f(7)}{h} = \dfrac{f(7.001)-f(7)}{0.001} = \dfrac{31.14612032-31.14}{0.001} = 6.12032$

The secant lines associated with these m_{sec} values are shown graphically in Figure 12.3.12.

The values of m_{sec} as $h \to 0^-$ give us the results shown in Table 12.3.4.

The secant lines associated with these m_{sec} values are shown graphically in Figure 12.3.13.

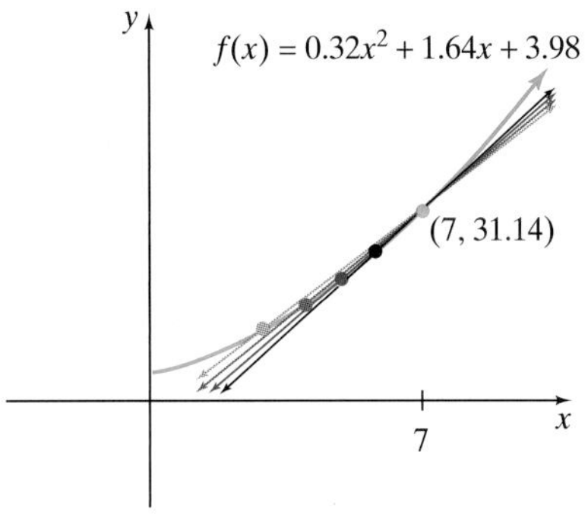

Figure 12.3.12 Secant lines through fixed point (7, 31.14) and as $x \to 7^-$.

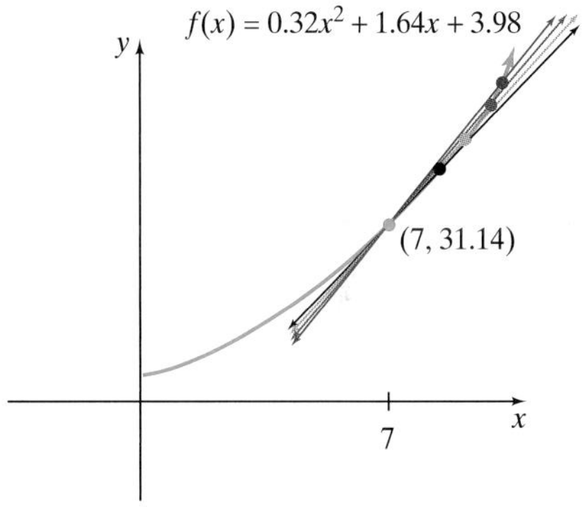

Figure 12.3.13 Secant lines through fixed point (7, 31.14) and as $x \to 7^+$.

Table 12.3.4

h	Interval, $[7+h, 7]$	$m_{\text{sec}} = \dfrac{f(7+h)-f(7)}{h}$
-1	$[6, 7]$	$\dfrac{f(7+h)-f(7)}{h} = \dfrac{f(6)-f(7)}{-1} = \dfrac{25.34-31.14}{-1} = 5.8$
-0.1	$[6.9, 7]$	$\dfrac{f(7+h)-f(7)}{h} = \dfrac{f(6.9)-f(7)}{-0.1} = \dfrac{30.5312-31.14}{-0.1} = 6.088$
-0.01	$[6.99, 7]$	$\dfrac{f(7+h)-f(7)}{h} = \dfrac{f(6.99)-f(7)}{-0.01} = \dfrac{31.078832-31.14}{-0.01} = 6.1168$
-0.001	$[6.999, 7]$	$\dfrac{f(7+h)-f(7)}{h} = \dfrac{f(6.999)-f(7)}{-0.001} = \dfrac{31.13388032-31.14}{-0.001} = 6.11968$

Tables 12.3.3 and 12.3.4 indicate that the approximate instantaneous rate of change is about 6.12.

Interpret the Solution: This means that in 1993, the U.S. imports from China were increasing at a rate of about 6.12 billion dollars per year. ■

SUMMARY

- The **slope of the secant line** between the points $(x, f(x))$ and $(x+h, f(x+h))$, denoted by m_{sec}, is given by the difference quotient

$$m_{\text{sec}} = \frac{f(x+h) - f(x)}{h}$$

where $h = \Delta x$ is the increment in x.

- For a curve given by the function f, the **tangent line** to the curve at point $(x, f(x))$ is the line whose slope is the same as the slope of the curve at $(x, f(x))$.

- To **approximate the instantaneous rate of change** of a function f at a given point $(x, f(x))$, we use the following steps:
  1. Complete a table for $h \to 0^+$.
  2. Complete a table for $h \to 0^-$.
  3. Examine the values of m_{sec} as $h \to 0^+$ and as $h \to 0^-$ to approximate the instantaneous rate of change.

SECTION 12.3 EXERCISES

In Exercises 1–12, classify each of the following as exhibiting average rate of change or instantaneous rate of change.

1.

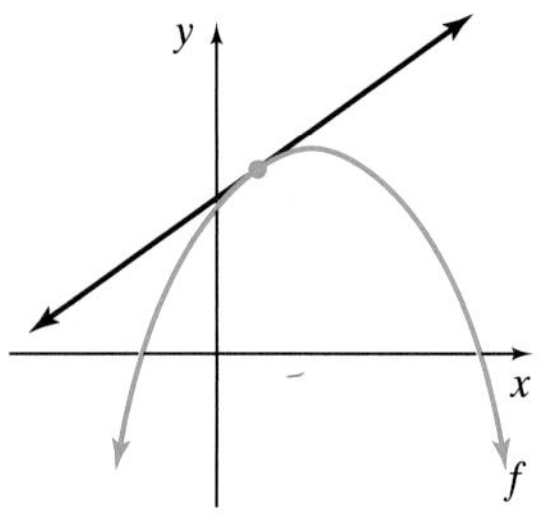

2.

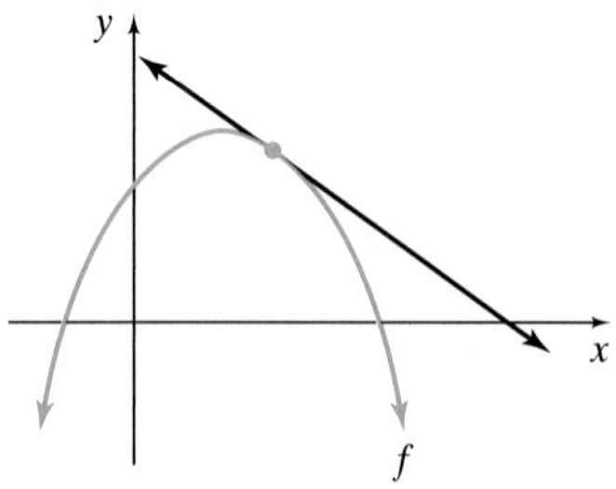

3.

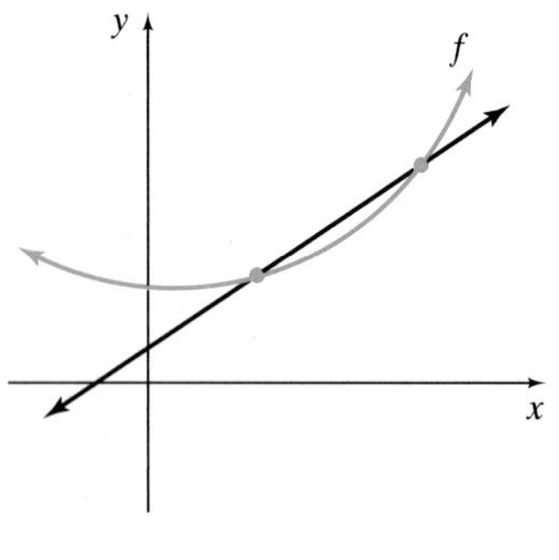

4.

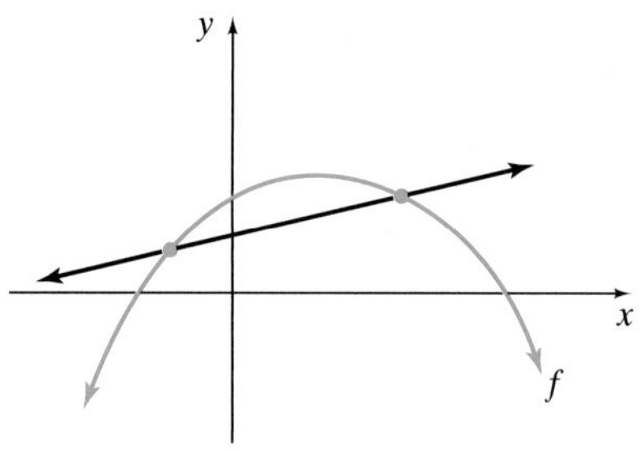

5.

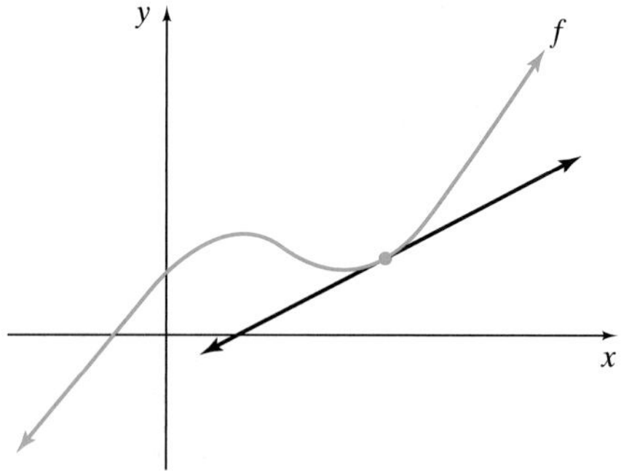

6.

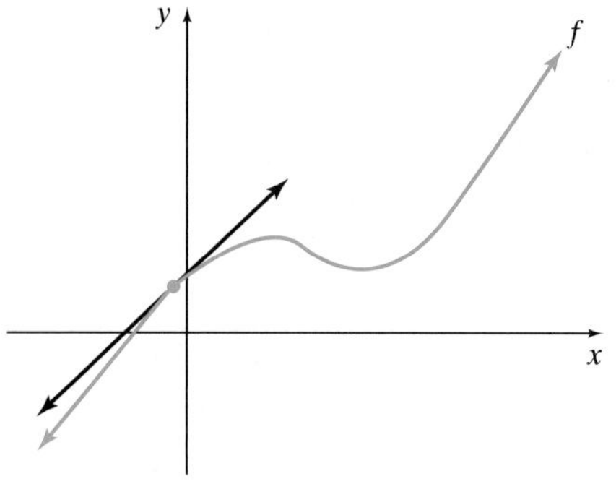

14.1 **7.** The StopCop Company determines that the cost to produce auto antitheft devices can be modeled by

$$C(x) = (3x + 6)^{1.5} + 30 \qquad 0 \le x \le 50$$

where x represents the number of auto antitheft devices produced in hundreds and $C(x)$ represents the production costs in thousands of dollars. The rate of change in cost when the production changes from $x = 10$ to $x = 15$.

14.1 **8.** The SnapPic Company determines that its cost for producing disposable cameras can be modeled by

$$C(x) = 60 + \sqrt{3x + 5} \qquad 0 \le x \le 40$$

where x represents the number of disposable cameras produced each shift and $C(x)$ represents the cost of production in hundreds of dollars. The rate of change in cost when the production changes from $x = 20$ to $x = 35$.

12.4 ✓ **9.** The U.S. imports of all merchandise from all counties can be modeled by

$$f(x) = 2.39x^2 + 47.05x + 432.2 \qquad 1 \le x \le 6$$

where x represents the number of years since 1990 and $f(x)$ is the dollar value in billions of U.S. imports from all counties (*Source:* U.S. Bureau of Economic Analysis, www.bea.doc.gov). The rate of change in the dollar value of imports when x changes from 2 to 5.

12.4 **10.** The U.S. exports of all merchandise to all counties can be modeled by

$$f(x) = 5.18x^2 + 5.86x + 410.6 \qquad 1 \le x \le 6$$

where x represents the number of years since 1990 and $f(x)$ is the dollar value in billions of U.S. exports to all counties (*Source:* U.S. Bureau of Economic Analysis, www.bea.doc.gov). The rate of change in the dollar value of exports when x changes from 1 to 3.

12.5 ✓ **11.** The total carbon content of carbon dioxide emitted into the atmosphere by the U.S. can be modeled by

$$f(x) = -0.55x^3 + 9.03x^2 - 21.8x + 1361.01 \qquad 1 \le x \le 9$$

where x is the number of years since 1989 and $f(x)$ is the total carbon content of carbon dioxide, in millions of metric tons, emitted into the atmosphere by the U.S. (*Source:* U.S. Statistical Abstract, www.census.gov/statab/www). The rate of change of total carbon content of carbon dioxide emitted into the atmosphere when $x = 7$.

12.5 **12.** The number of cases of tuberculosis reported in the U.S. can be modeled by

$$f(x) = -0.014x^3 + 0.39x^2 - 3.11x + 30.29 \qquad 1 \le x \le 19$$

where x represents the number of years since 1979 and $f(x)$ represents the number of cases in thousands (*Source:* U.S. Centers for Disease Control and Prevention, www.cdc.gov). The rate of change in the number of cases of tuberculosis reported when $x = 10$.

In Exercises 13–26, compute the average rate of change of the function over the indicated interval.

13. $f(x) = x^2 + x \quad [1, 4]$

14. $f(x) = x^2 - x \quad [1, 4]$

15. $g(x) = 2x^3 + 3x + 1 \quad [2, 4]$

16. $g(x) = x^3 - 2x - 3 \quad [2, 5]$

17. $f(x) = 2x^2 + x - 3 \quad [-2, 1]$

18. $f(x) = 3x^2 - 2x + 2 \quad [-3, 0]$

19. $g(x) = 3\sqrt{x} \quad [1, 9]$

20. $g(x) = 2\sqrt{x} \quad [1, 16]$

21. $f(x) = 2.1e^{0.2x} \quad [0, 20]$

22. $f(x) = 1.3e^{0.3x} \quad [0, 25]$

23. $f(x) = 2.1e^{-0.2x} \quad [0, 15]$

24. $f(x) = 3.2e^{-0.4x} \quad [0, 5]$

25. $f(x) = 1 + \ln x \quad [1, 10]$

26. $g(x) = 2.1 + \ln x \quad [1, 8]$

27. Between which pairs of consecutive points on the given curve is the average rate of change positive? Negative? Zero?

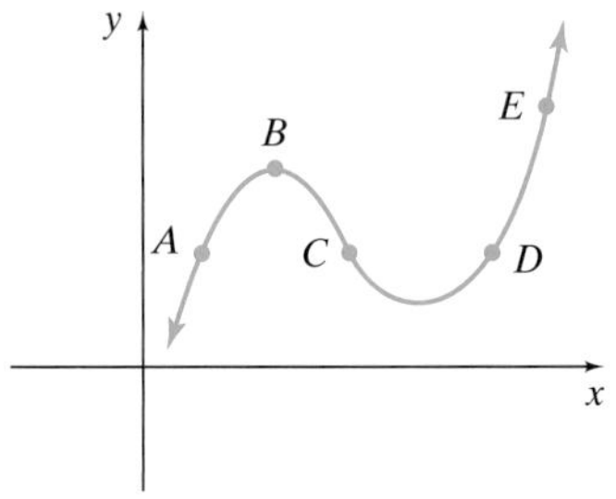

(a)

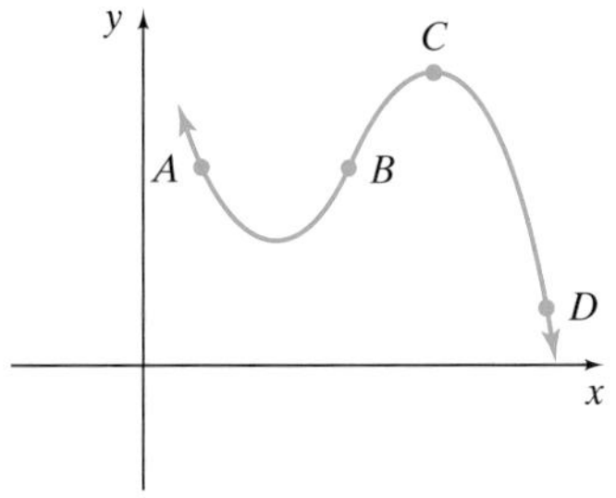

(b)

28. At which points on the given curve is the tangent line slope positive? Negative? Zero?

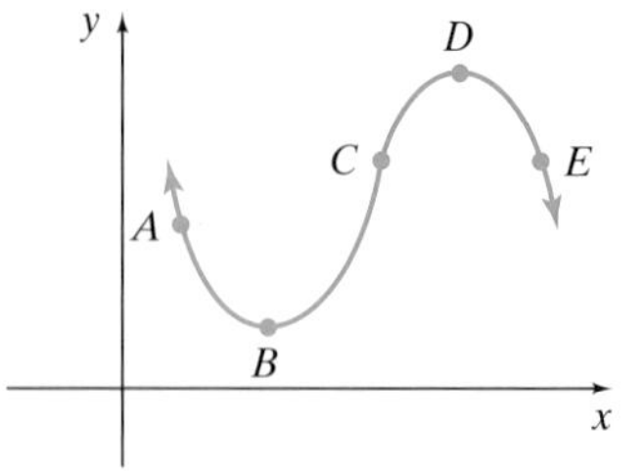

(a)

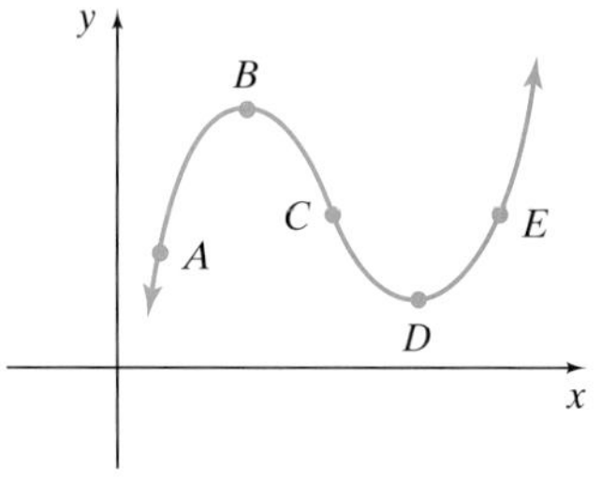

(b)

In Exercises 29–40, estimate the instantaneous rate of change of the given function at the indicated x-value by completing a table for $h = 1, 0.1, 0.01, 0.001$, and another for $h = -1, -0.1, -0.01, -0.001$. See Example 3.

29. $f(x) = 3x$; $x = 2$

30. $f(x) = 4x$; $x = 2$

31. $f(x) = x^2 + x$; $x = 3$

32. $f(x) = x^2 - x$; $x = -3$

✓ **33.** $f(x) = 2x^3 + 3x + 1$; $x = 4$

34. $f(x) = x^3 - 2x - 3$; $x = -1$

35. $f(x) = 3\sqrt{x}$; $x = 4$

36. $f(x) = 2\sqrt{x}$; $x = 9$

37. $f(x) = 2.1e^{0.2x}$; $x = 2$

38. $f(x) = 1.3e^{0.3x}$; $x = 4$

39. $f(x) = 2.1e^{-0.2x}$; $x = 3$

40. $f(x) = 2.1e^{-0.2x}$; $x = 5$

12.5 **41.** Membership in the Girl Scouts of America can be modeled by

$$g(t) = 0.086t^4 - 3.83t^3 + 55.97t^2 - 243.23t + 2972.93 \qquad 1 \le t \le 20$$

where t is the number of years since 1979 and $g(t)$ is the membership in thousands (*Source:* U.S. Statistical Abstract, www.census.gov/statab/www).

(a) Determine the average rate of change in the membership in the Girl Scouts of America from 1980 to 1990 and interpret.

(b) Approximate the instantaneous rate of change of f at $t = 11$ and interpret.

12.5 **42.** Membership in the Boy Scouts of America can be modeled by

$$f(t) = 0.72t^3 - 22.15t^2 + 263.28t + 4043.3 \qquad 1 \le t \le 20$$

where t is the number of years since 1979 and $f(t)$ is the membership in thousands (*Source:* U.S. Statistical Abstract, www.census.gov/statab/www).

(a) Determine the average rate of change in the membership in the Boy Scouts of America from 1980 to 1990 and interpret.

(b) Approximate the instantaneous rate of change of f at $t = 11$ and interpret.

13.2 **43.** The Seas Beginning Company determines that the daily cost function of producing its Rain Forest ultra rain coat can be modeled by

$$C(x) = 1000 + 35x - 0.01x^2 \qquad 0 \le x \le 300$$

where x represents the number of rain coats produced each day and $C(x)$ is the total cost, in dollars, of producing the rain coats.

(a) Determine the average rate of change in cost when production changes from $x = 100$ to $x = 150$ and interpret.

(b) Approximate the instantaneous rate of change in cost at a production level of $x = 200$ and interpret.

13.2 **44.** The Kranky Company determines that its weekly cost function of producing automatic transmissions for automobiles can be modeled by

$$C(x) = 55{,}000 + 600x - 1.25x^2 \qquad 0 \le x \le 230$$

where x represents the number of automatic transmissions produced each week and $C(x)$ is the total cost, in dollars, of producing the automatic transmissions.

(a) Determine the average rate of change in cost when production changes from $x = 180$ to $x = 210$ and interpret.

(b) Approximate the instantaneous rate of change in cost at a production level of $x = 210$ and interpret.

13.3 **45.** The poverty threshold in the United States can be modeled by

$$f(x) = 0.004x^2 + 0.057x + 1.280 \qquad 0 \le x \le 35$$

where x represents the number of years since 1960 and $f(x)$ represents the U.S. poverty threshold in thousands of dollars (*Source:* U.S. Census Bureau, www.census.gov). Approximate the instantaneous rate of change of f at $x = 30$ and interpret.

13.3 **46.** The population of Florida can be modeled by

$$f(x) = 0.003x^2 + 0.168x + 4.908 \qquad 0 \le x \le 35$$

where x represents the number of years since 1960 and $f(x)$ represents the population in millions (*Source:* U.S. Census Bureau, www.census.gov). Approximate the instantaneous rate of change of f at $x = 20$ and interpret.

$\overset{\Rightarrow}{13.2}$ **47.** A telemarketer determines that the monthly profit made from selling magazine subscriptions can be modeled by

$$P(x) = 5x + x^{\frac{1}{2}} \qquad 0 \le x \le 100$$

where x represents the number of magazine subscriptions sold each month and $P(x)$ represents the profit in dollars. Approximate the instantaneous rate of change of profit when there are 50 subscriptions sold and interpret.

$\overset{\Rightarrow}{13.2}$ **48.** A newspaper courier determines the monthly profit from a typical newspaper route can be modeled by

$$P(x) = 2x - x^{\frac{1}{2}} \qquad 0 \le x \le 200$$

where x represents the number of subscribers and $P(x)$ represents the monthly profit in dollars. Approximate the instantaneous rate of change of profit when there are 75 subscribers and interpret.

SECTION PROJECT

There is another method for approximating the instantaneous rate of change of a function. We start with an alternative formula for the average rate of change of f between $(x, f(x))$ and $(c, f(c))$ given by

$$m_{\text{sec}} = \frac{f(x) - f(c)}{x - c} \qquad x \neq c$$

Then we can compute m_{sec} for values of c that get close to x in value. For example, some m_{sec} values of the function $f(x) = \sqrt{x} + 5$ at $x = 4$ and starting with $c = 5$ are shown in Table 12.3.5.

(a) Complete Table 12.3.5.

(b) Explain in a sentence what the values in the right hand column of Table 12.3.5 represent.

(c) Complete Table 12.3.6.

(d) Explain in a sentence what the values in the right hand column of Table 12.3.6 represent.

(e) From examining the values in Table 12.3.5 and in Table 12.3.6, approximate the instantaneous rate of change of $f(x) = \sqrt{x} + 5$ at $x = 4$.

Table 12.3.5

c	$x - c$	$m_{\text{sec}} = \dfrac{f(x) - f(c)}{x - c}$
5	$4 - 5 = -1$	$\dfrac{f(4) - f(5)}{4 - 5} = \dfrac{(\sqrt{4} + 5) - (\sqrt{5} + 5)}{4 - 5} = \dfrac{2 - \sqrt{5}}{-1} \approx 0.236$
4.1	$4 - 4.1 = -0.1$	$\dfrac{f(4) - f(4.1)}{4 - 4.1} = \dfrac{(\sqrt{4} + 5) - (\sqrt{4.1} + 5)}{-0.1} \approx 0.248$
4.01		
4.001		

Table 12.3.6

c	$x - c$	$m_{\text{sec}} = \dfrac{f(x) - f(c)}{x - c}$
3	$4 - 3 = 1$	$\dfrac{f(4) - f(3)}{4 - 3} = \dfrac{(\sqrt{4} + 5) - (\sqrt{3} + 5)}{4 - 3} = \dfrac{2 - \sqrt{3}}{1} \approx 0.268$
3.9	$4 - 3.9 = 0.1$	$\dfrac{f(4) - f(3.9)}{4 - 3.9} = \dfrac{(\sqrt{4} + 5) - (\sqrt{3.9} + 5)}{0.1} \approx 0.252$
3.99		
3.999		

Section 12.4 The Derivative

In Section 12.3, we analyzed the difference between an average rate of change and an instantaneous rate of change. We saw that an instantaneous rate of change is given by the slope of a tangent line. The process we outlined to approximate this type of rate of change employed table building and computing secant line slope over smaller and smaller intervals. This process actually utilized the limit concept presented in Sections 12.1 and 12.2. In this section, we formalize and generalize the process of determining an instantaneous rate of change. This generalization will allow us to compute an instantaneous rate of change for **any** value of x and it will also give us the **exact** value of the instantaneous rate of change.

Difference Quotient and the Derivative

Suppose that we wish to determine the instantaneous rate of change at x for the function graphed in Figure 12.4.1. In Section 12.3 we had a specific value for x. Here, we consider *any* value for x.

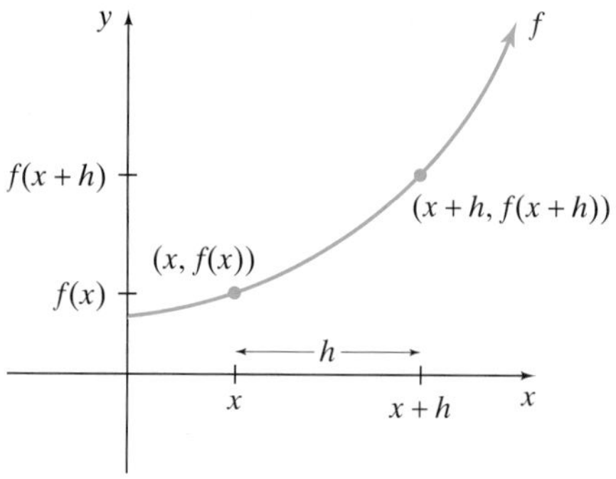

Figure 12.4.1

Using the process outlined in Section 12.3, we apply the limit concept by letting h get smaller and smaller. In other words, we take the limit of the difference quotient as $h \to 0$ and this limit gives the instantaneous rate of change. For any value of x, we can formalize the process as follows.

Instantaneous Rate of Change/Slope of a Tangent Line

The **instantaneous rate of change** of a function, f, at x is equal to the **slope of the line tangent** to the graph of f at x and is given by

$$\text{Instantaneous rate of change} = m_{\tan} = \lim_{h \to 0} \frac{f(x+h) - f(x)}{h}$$

provided that the limit exists. Its units of measurement are units of f per unit of x.

Note: $m_{\tan}$ means slope of the tangent line.

Example 1 Determining an Instantaneous Rate of Change

For $f(x) = x^2 + 3$, determine the instantaneous rate of change at $x = 2$.

Solution

From the preceding definition, we know that

$$\text{Instantaneous rate of change} = m_{\tan} = \lim_{h \to 0} \frac{f(x+h) - f(x)}{h}$$

Substituting $x = 2$ gives us

$$\begin{aligned}
\text{Instantaneous rate of change} = m_{\tan} &= \lim_{h \to 0} \frac{f(2+h) - f(2)}{h} \\
&= \lim_{h \to 0} \frac{[(2+h)^2 + 3] - [2^2 + 3]}{h} \\
&= \lim_{h \to 0} \frac{[4 + 4h + h^2 + 3] - [7]}{h} \\
&= \lim_{h \to 0} \frac{7 + 4h + h^2 - 7}{h} \\
&= \lim_{h \to 0} \frac{4h + h^2}{h} \\
&= \lim_{h \to 0} \frac{h(4+h)}{h} \\
&= \lim_{h \to 0} (4 + h) \\
&= 4
\end{aligned}$$

So the instantaneous rate of change of $f(x) = x^2 + 3$ at $x = 2$ is 4. This is the exact slope of the line tangent to the graph of $f(x) = x^2 + 3$ at $x = 2$. See Figure 12.4.2.

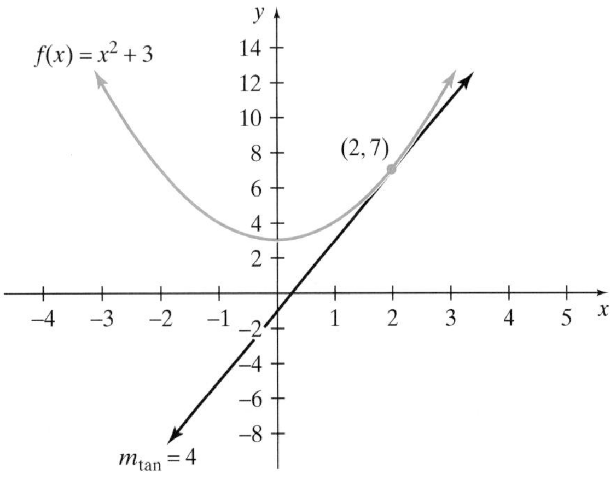

Figure 12.4.2 Slope of the line tangent to the graph of $f(x) = x^2 + 3$ at $x = 2$ is 4. ■

The definition given before Example 1 is so important that it has been given a special name, the *derivative*. We now present one of the most important definitions in all of mathematics, the definition of the derivative.

Derivative

For a function f, the **derivative of f at x**, denoted $f'(x)$ (this is read "f prime of x"), is defined to be

$$f'(x) = \lim_{h \to 0} \frac{f(x+h) - f(x)}{h}$$

provided that the limit exists. The units of $f'(x)$ are units of f per unit of x.

Note: Notice that the derivative of f at x gives an instantaneous rate of change of f at x.

Example 2 Computing a Derivative

Compute $f'(2)$ for $f(x) = x^2 + x$. Interpret the result.

Solution

Understand the Situation: To compute $f'(2)$, we use the definition of the derivative

$$f'(x) = \lim_{h \to 0} \frac{f(x+h) - f(x)}{h}$$

where x has a value of 2.

Proceeding, this gives

$$\begin{aligned} f'(2) &= \lim_{h \to 0} \frac{f(2+h) - f(2)}{h} \\ &= \lim_{h \to 0} \frac{[(2+h)^2 + (2+h)] - [2^2 + 2]}{h} \\ &= \lim_{h \to 0} \frac{[4 + 4h + h^2 + 2 + h] - 6}{h} \\ &= \lim_{h \to 0} \frac{h^2 + 5h}{h} \end{aligned}$$

Now we rely on the algebraic techniques learned in Sections 12.1 and 12.2 to evaluate this limit. Employing the Substitution Principle yields the indeterminate form $\frac{0}{0}$, so we factor the numerator and cancel:

$$\begin{aligned} f'(2) &= \lim_{h \to 0} \frac{h(h+5)}{h} \\ &= \lim_{h \to 0} (h+5) = 5 \end{aligned}$$

Interpret the Solution: We have now determined that $f'(2) = 5$, which means that the instantaneous rate of change of $f(x) = x^2 + x$ at $x = 2$ is 5. Alternatively, we could say that the slope of the curve given by $f(x) = x^2 + x$ at $x = 2$ is 5, or we could even say that the slope of the line tangent to the graph of $f(x) = x^2 + x$ at $x = 2$ is 5. See Figure 12.4.3.

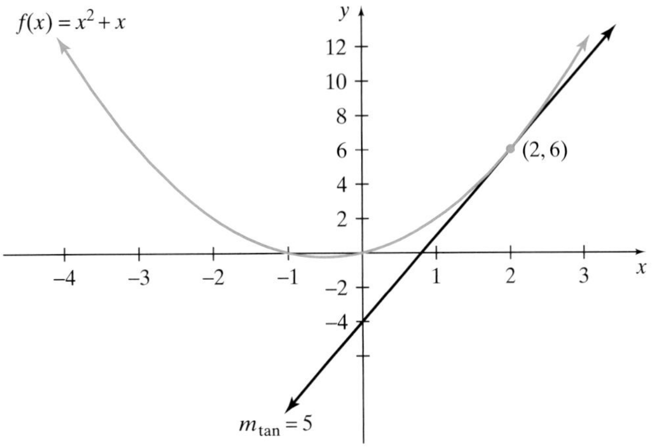

Figure 12.4.3 The notation m_{tan} represents "slope of tangent line."

■

Checkpoint 1

Now work Exercise 3.

WWW

Interactive Activity

In Example 2, we algebraically determined $f'(2) = 5$. We used the definition of the derivative at $x = 2$ to get $f'(2) = \lim_{h \to 0} \frac{f(2+h) - f(2)}{h}$, which we eventually simplified to $f'(2) = \lim_{h \to 0} \frac{h^2 + 5h}{h}$. To examine this numerically and graphically, we change the h's to x's and enter into the calculator $y = \frac{x^2 + 5x}{x}$. Do this, then numerically and graphically determine $\lim_{x \to 0} \frac{x^2 + 5x}{x}$ and see if it supports the result of Example 2. Use Table 12.4.1 for the numerical part.

Table 12.4.1

	$x \to 0^-$					$x \to 0^+$			
x	−1	−0.1	−0.001	−0.0001	0	0.0001	0.001	0.01	1
$\frac{x^2 + 5x}{x}$									

Technology Option

Many graphing calculators have a DRAW TANGENT command. Figure 12.4.4 was generated using such a command. Notice that in the lower left-hand corner of the screen is displayed the equation of the tangent line in slope-intercept form. From this, we can see that the slope of the tangent line is 5 which supports our work in Example 2.

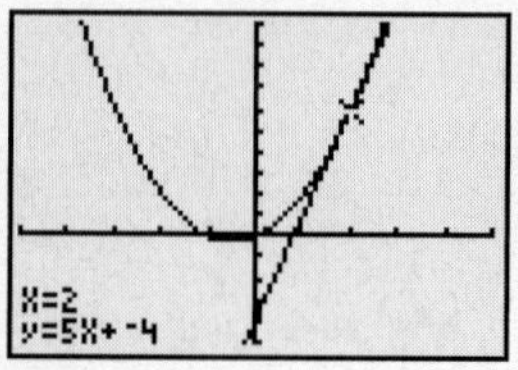

Figure 12.4.4

For more information on this command consult the online graphing calculator manual at www.prenhall.com/armstrong.

Example 3 Computing a Derivative in an Applied Setting

The total cost of producing q recreational vehicles can be modeled by

$$C(q) = 100 + 60q + 3q^2$$

where q is the number of vehicles produced and $C(q)$ is in hundreds of dollars.

(a) Compute $C(5)$ and interpret.

(b) Compute $C'(5)$ and interpret.

Solution

(a) **Understand the Situation:** To compute $C(5)$, we simply substitute 5 for q. This gives

$$C(5) = 100 + 60(5) + 3(5)^2 = 475$$

Interpret the Solution: This means the total cost to produce 5 vehicles is \$47,500.

(b) **Understand the Situation:** To compute $C'(5)$, we will use the definition of the derivative

$$C'(q) = \lim_{h \to 0} \frac{C(q+h) - C(q)}{h}$$

where q has the value of 5.

Proceeding, this gives

$$\begin{aligned} C'(5) &= \lim_{h \to 0} \frac{C(5+h) - C(5)}{h} \\ &= \lim_{h \to 0} \frac{[100 + 60(5+h) + 3(5+h)^2] - [100 + 60(5) + 3(5)^2]}{h} \\ &= \lim_{h \to 0} \frac{100 + 300 + 60h + 3(25 + 10h + h^2) - 475}{h} \\ &= \lim_{h \to 0} \frac{100 + 300 + 60h + 75 + 30h + 3h^2 - 475}{h} \\ &= \lim_{h \to 0} \frac{90h + 3h^2}{h} \end{aligned}$$

As in Example 2, employing the Substitution Principle yields the indeterminate form $\frac{0}{0}$, so we factor the numerator and cancel.

$$C'(5) = \lim_{h \to 0} \frac{h(90 + 3h)}{h}$$
$$= \lim_{h \to 0}(90 + 3h) = 90$$

Interpret the Solution: So we have determined that $C'(5) = 90$, which means that the instantaneous rate of change is \$9000 per vehicle when the production level is 5 vehicles. Alternatively, we could say that the total costs increase approximately \$9000 when production increases from 5 vehicles to 6 vehicles. Figure 12.4.5 shows the graph of C along with the line tangent at the point $(5, 475)$.

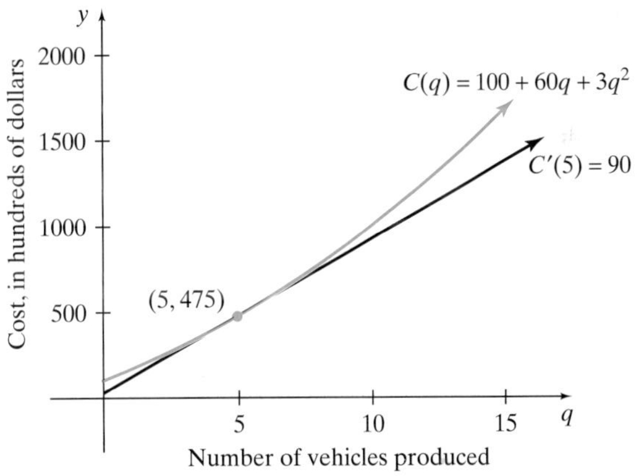

Figure 12.4.5

The derivative of a cost function such as the one in Example 3 is called **marginal cost**. The marginal concept will be discussed in full detail in Chapter 13.

Derivatives in General

Before we continue with how to compute a derivative in general, we need to present some facts about the derivative.

Facts about the Derivative

1. Recall that the derivative of a function f can be found for any value x where the derivative exists. Thus, the derivative of a function, f', is itself a function. This means that it has a domain for which it is defined.
2. For any c in the domain of f', $f'(c)$ is a number that gives the slope of a line tangent to the graph of f at $x = c$. Also, $f'(c)$ gives the instantaneous rate of change of f at $x = c$.
3. The process of determining $f'(x)$ is called **differentiation**.

By taking full advantage of the definition of the derivative coupled with the fact that the derivative is a function, we can compute a formula for the derivative.

In other words, recall the total cost function in Example 3, which was given by $C(q) = 100 + 60q + 3q^2$. Suppose that we were asked to compute $C'(3)$, $C'(7)$, $C'(13)$, and $C'(15)$. At first glance, it appears that we would have the long and tedious task of duplicating our work in Example 3 for these four different values of q. However, the definition of the derivative tells us that the derivative is a *function*, which means that we should be able to determine a rule for this function. In Example 4, we illustrate how to use the definition of the derivative to do this.

Example 4 Computing a Derivative Formula

Compute $C'(q)$ for $C(q) = 100 + 60q + 3q^2$.

Solution

The process is the same as in Example 3, except here we have our independent variable q in place of 5.

$$\begin{aligned} C'(q) &= \lim_{h\to 0} \frac{C(q+h) - C(q)}{h} \\ &= \lim_{h\to 0} \frac{[100 + 60(q+h) + 3(q+h)^2] - [100 + 60(q) + 3(q)^2]}{h} \\ &= \lim_{h\to 0} \frac{[100 + 60q + 60h + 3(q^2 + 2qh + h^2)] - (100 + 60q + 3q^2)}{h} \\ &= \lim_{h\to 0} \frac{100 + 60q + 60h + 3q^2 + 6qh + 3h^2 - 100 - 60q - 3q^2}{h} \\ &= \lim_{h\to 0} \frac{60h + 6qh + 3h^2}{h} \\ &= \lim_{h\to 0} \frac{h(60 + 6q + 3h)}{h} \\ &= \lim_{h\to 0}(60 + 6q + 3h) = 60 + 6q. \end{aligned}$$

So $C'(q) = 60 + 6q$. A huge advantage in having this formula for the derivative is that we can now evaluate it for different values of our independent variable very quickly. For example,

$$\begin{aligned} C'(3) &= 60 + 6(3) = 78 \\ C'(5) &= 60 + 6(5) = 90 \\ C'(7) &= 60 + 6(7) = 102 \end{aligned}$$

■

✓ Checkpoint 2 Now work Exercise 23.

Example 5 Computing and Using a Derivative Formula

Determine $f'(x)$ for $f(x) = x^2 + 3x$. Also, determine the slope of the line tangent to the graph of f at $x = -2$.

Solution

Interactive Activity

Algebraically find an equation for the line through $(-2, -2)$ with slope of -1 to verify that the equation of the tangent line in Figure 12.4.6 is $y = -x - 4$.

WWW

Utilizing the definition of the derivative gives

$$\begin{aligned} f'(x) &= \lim_{h\to 0} \frac{f(x+h) - f(x)}{h} \\ &= \lim_{h\to 0} \frac{[(x+h)^2 + 3(x+h)] - (x^2 + 3x)}{h} \\ &= \lim_{h\to 0} \frac{x^2 + 2xh + h^2 + 3x + 3h - x^2 - 3x}{h} \\ &= \lim_{h\to 0} \frac{2xh + h^2 + 3h}{h} \\ &= \lim_{h\to 0} \frac{h(2x + h + 3)}{h} \\ &= \lim_{h\to 0} (2x + h + 3) = 2x + 3 \end{aligned}$$

So $f'(x) = 2x + 3$. Knowing that the derivative gives the slope of the tangent line at any point, we can quickly compute that at $x = -2$

$$\begin{aligned} m_{\tan} &= f'(-2) \\ &= 2(-2) + 3 \\ &= -1 \end{aligned}$$

Figure 12.4.6 has a graph of f and the line tangent to f at $x = -2$.

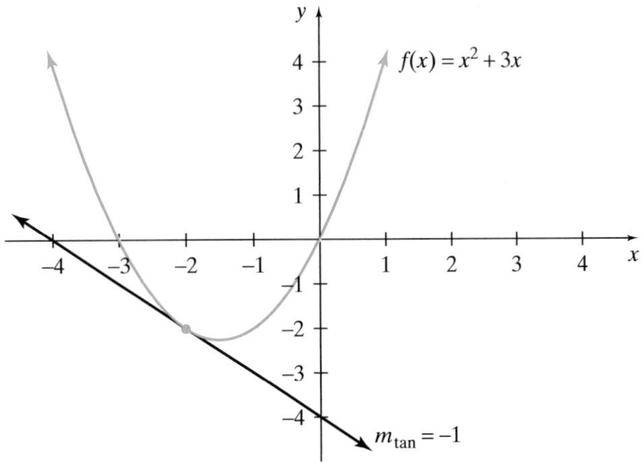

Figure 12.4.6

■

Checkpoint 3

Now work Exercise 35.

Example 5 communicates that the derivative of a function, in its most general form, is very useful when determining tangent line slopes or instantaneous rates of change at *any* point on the graph of the given function.

We conclude this section by taking a Flashback to an example last seen in Section 12.3.

Flashback

U.S. Imports From China Revisited

The U.S. imports from China can be modeled by

$$f(x) = 0.32x^2 + 1.64x + 3.98 \qquad 1 \le x \le 10$$

where x represents the number of years since 1986 and $f(x)$ represents the dollar value of goods imported from China, measured in billions (*Source:* U.S. Census Bureau, www.census.gov). Determine $f'(x)$. Also, evaluate and interpret $f'(7)$.

Flashback Solution

Interactive Activity

Compare the result of the Flashback to the result of Example 4 in Section 12.3.

Understand the Situation: To determine $f'(x)$ we use the definition of the derivative $f'(x) = \lim_{h \to 0} \frac{f(x+h) - f(x)}{h}$. Once we have this general derivative, we will substitute $x = 7$ and evaluate to determine $f'(7)$. Note that $x = 7$ corresponds to 1993.

So we have

$$\begin{aligned}
f'(x) &= \lim_{h \to 0} \frac{f(x+h) - f(x)}{h} \\
&= \lim_{h \to 0} \frac{[0.32(x+h)^2 + 1.64(x+h) + 3.98] - [0.32x^2 + 1.64x + 3.98]}{h} \\
&= \lim_{h \to 0} \frac{[0.32(x^2 + 2xh + h^2) + 1.64x + 1.64h + 3.98] - 0.32x^2 - 1.64x - 3.98}{h} \\
&= \lim_{h \to 0} \frac{[0.32x^2 + 0.64xh + 0.32h^2 + 1.64x + 1.64h + 3.98] - 0.32x^2 - 1.64x - 3.98}{h} \\
&= \lim_{h \to 0} \frac{0.64xh + 0.32h^2 + 1.64h}{h} \\
&= \lim_{h \to 0} \frac{h(0.64x + 0.32h + 1.64)}{h} \\
&= \lim_{h \to 0} (0.64x + 0.32h + 1.64) = 0.64x + 1.64
\end{aligned}$$

Interpret the Solution: We have determined that $f'(x) = 0.64x + 1.64$. We evaluate $f'(7)$ by substituting 7 for each occurrence of x. This gives us

$$\begin{aligned}
f'(7) &= 0.64(7) + 1.64 \\
&= 6.12
\end{aligned}$$

This means that in 1993, U.S. imports from China were increasing at a rate of $6.12 \frac{\text{billion dollars}}{\text{year}}$.

SUMMARY

We have presented some very powerful mathematical concepts in this section and the last section. The main ideas have been rates of change, specifically introducing the **instantaneous rate of change**. In Table 12.4.2, we compare and contrast average rate of change with instantaneous rate of change.

We calculated **derivatives** for specific values of x and for any arbitrary x. The derivative is a function, which means that it has a domain and can be evaluated at any value c in the domain. Remember when computing a derivative, whether at a specific value of c or for an arbitrary x, that we are finding a limit. Therefore, the limit techniques learned in Sections 12.1 and 12.2 can be employed.

Table 12.4.2

Average Rate of Change	**Instantaneous Rate of Change**
Change over an interval	Change at a point
Slope of a secant line	Slope of a tangent line; slope of a curve
$\frac{f(x+h)-f(x)}{h}$	$\lim_{h\to 0}\frac{f(x+h)-f(x)}{h}$
Measured in units of f per unit of x	Measured in units of f per unit of x

SECTION 12.4 EXERCISES

In Exercises 1-10, compute $f'(c)$ by using the definition of the derivative:

$$f'(c) = \lim_{h\to 0}\frac{f(c+h)-f(c)}{h}$$

See Example 2.

1. $f(x) = 2x + 1;\quad c = 2$

2. $f(x) = 3x - 4;\quad c = 3$

✓ **3.** $f(x) = x^2 - 2x;\quad c = -2$

4. $f(x) = x^2 - 3x;\quad c = 2$

5. $f(x) = -x^2 + 2x;\quad c = 4$

6. $f(x) = -2x^2 + x;\quad c = 4$

7. $f(x) = 2\sqrt{x};\quad c = 4$

8. $f(x) = 2.3x^2 - 0.2x + 3.2;\quad c = 8$

9. $f(x) = -1.2x^2 + 3.2x - 2.1;\quad c = 7$

10. $f(x) = \frac{1}{x};\quad c = 3$

Exercises 11–16 use the results of Exercises 1–10. For each function and given value for x determine an equation for the line tangent to the graph of the function at the indicated value for x.

11. $f(x) = x^2 - 2x$ at $x = -2$. See Exercise 3.

12. $f(x) = x^2 - 3x$ at $x = 2$. See Exercise 4.

13. $f(x) = -x^2 + 2x$ at $x = 4$. See Exercise 5.

14. $f(x) = 2\sqrt{x}$ at $x = 4$. See Exercise 7.

15. $f(x) = -1.2x^2 + 3.2x - 2.1$ at $x = 7$. See Exercise 9.

16. $f(x) = \frac{1}{x}$ at $x = 3$. See Exercise 10.

In Exercises 17–32, compute a formula for the derivative by using the definition of the derivative

$$f'(x) = \lim_{h\to 0}\frac{f(x+h)-f(x)}{h}$$

as was illustrated in Examples 4, 5, and 6.

17. $f(x) = 2x - 5$

18. $f(x) = 3x + 1$

19. $g(x) = -2x + 3$

20. $g(x) = -3x - 1$

21. $f(x) = x^2$

22. $f(x) = x^2 - 3x$

✓ **23.** $h(x) = x^2 - 2x + 3$

24. $h(x) = x^2 - 3x + 1$

25. $g(x) = 2.1x^2 + 3.2x$

26. $g(x) = 1.3x^2 - 2.1x$

27. $f(x) = -2x^2 + 3x$

28. $f(x) = -3x^2 + 2x$

29. $f(x) = -2.3x^2 + 3.1x$

30. $f(x) = x^3$

31. $g(x) = x^3 + x^2$

32. $g(x) = \sqrt{x}$

Exercises 33–42 use the results of Exercises 17–32. Determine the slope of the line tangent to the graph of the given function at the indicated values of x. See Example 5.

33. $f(x) = 2x - 5$ at $x = -1$ and $x = 2$. See Exercise 17.

34. $g(x) = -3x - 1$ at $x = -1$ and $x = 2$. See Exercise 20.

✓ **35.** $f(x) = x^2 - 2x + 3$ at $x = -3$ and $x = 2$. See Exercise 23.

36. $h(x) = x^2 - 3x + 1$ at $x = -2$ and $x = 3$. See Exercise 24.

37. $g(x) = 2.1x^2 + 3.2x$ at $x = 0$ and $x = 2$. See Exercise 25.

38. $g(x) = 1.3x^2 - 2.1x$ at $x = 1$ and $x = 4$. See Exercise 26.

39. $f(x) = -2x^2 + 3x$ at $x = 3$ and $x = 6$. See Exercise 27.

40. $f(x) = -3x^2 + 2x$ at $x = 2$ and $x = 7$. See Exercise 28.

41. $f(x) = x^3 + x^2$ at $x = -1$ and $x = 1$. See Exercise 31.

42. $g(x) = \sqrt{x}$ at $x = 4$ and $x = 9$. See Exercise 32.

Applications

43. The Network Standards Company has determined from data that it has collected that the profit function is given by

$$P(q) = 1000q - 2q^2 \qquad 0 \le q \le 300$$

where q is the number of modems produced and sold and $P(q)$ is the profit in dollars.

(a) Use the definition of the derivative to determine $P'(q)$. Here you will use

$$P'(q) = \lim_{h \to 0} \frac{P(q+h) - P(q)}{h}$$

(b) Use the result from part (a) to compute $P'(200)$ and $P'(300)$, and interpret each using appropriate units.

44. The Fore Link Company has determined from data that it has collected that the profit function is given by

$$P(q) = 1000q - q^2 \qquad 0 \le q \le 800$$

where q is the number of golf clubs produced and sold and $P(q)$ is the profit in dollars.

(a) Use the definition of the derivative to determine $P'(q)$.

(b) Use the result from part (a) to compute $P'(400)$ and $P'(600)$, and interpret each using appropriate units.

45. The Fore Link Company has determined that the revenue, in dollars, from the sale of q sets of Junior Golf Clubs can be estimated by

$$R(q) = 200q - q^2 \qquad 0 \le q \le 200$$

where $R(q)$ is in dollars.

(a) Compute $R(50)$ and $R(150)$ and interpret each.

(b) Compute $R'(q)$ by using the definition of the derivative. Here you will use

$$R'(q) = \lim_{h \to 0} \frac{R(q+h) - R(q)}{h}$$

(c) Use the result from part (b) to determine $R'(50)$ and $R'(150)$, and interpret each using appropriate units.

46. The Network Standards Company has determined that the revenue, in dollars, from the sale of q modems can be estimated by

$$R(q) = 300q - q^2 \qquad 0 \le q \le 300$$

where $R(q)$ is in dollars.

(a) Compute $R(100)$ and $R(200)$ and interpret each.

(b) Compute $R'(q)$ by using the definition of the derivative.

(c) Use the result from part (b) to determine $R'(100)$ and $R'(200)$, and interpret each using appropriate units.

47. The FrezMore Company has determined that the cost of producing x refrigerators is modeled by

$$C(x) = x^2 + 15x + 1500 \qquad 0 \le x \le 200$$

where x is the number of refrigerators produced and $C(x)$ is the weekly cost in dollars.

(a) Use the definition of the derivative to determine $C'(x)$. Here you will use

$$C'(x) = \lim_{h \to 0} \frac{C(x+h) - C(x)}{h}$$

(b) Use the result from part (a) to compute $C'(40)$ and $C'(100)$, and interpret each using appropriate units.

48. The FrezMore Company has determined that the cost of producing x dorm room-sized refrigerators is modeled by

$$C(x) = \frac{1}{2}x^2 + 3x + 15 \qquad 0 \le x \le 300$$

where x is the number of refrigerators produced and $C(x)$ is the weekly cost in dollars.

(a) Use the definition of the derivative to determine $C'(x)$.

(b) Use the result from part (a) to compute $C'(40)$ and $C'(100)$, and interpret each using appropriate units.

12.3 **49.** The U.S. imports of all merchandise from all countries can be modeled by

$$f(x) = 2.39x^2 + 47.05x + 432.2 \qquad 1 \le x \le 6$$

where x is the number of years since 1990 and $f(x)$ is the dollar value in billions of all U.S. imports from all countries (*Source:* U.S. Bureau of Economic Analysis, www.bea.doc.gov).

(a) Determine $f'(x)$.

(b) Use the result from part (a) to compute $f'(3)$, $f'(4)$, and $f'(5)$, and interpret each.

(c) Using your knowledge that the derivative gives the slope of a tangent line, does the result from part (b) indicate that the function is increasing or decreasing on its reasonable domain? Explain.

12.3 **50.** The U.S. exports of all merchandise to all countries can be modeled by

$$f(x) = 5.18x^2 + 5.86x + 410.6 \qquad 1 \le x \le 6$$

where x is the number of years since 1990 and $f(x)$ is the dollar value in billions of all U.S. exports to all countries (*Source:* U.S. Bureau of Economic Analysis, www.bea.doc.gov).

(a) Determine $f'(x)$.

(b) Use the result from part (a) to compute $f'(3)$, $f'(4)$, and $f'(5)$, and interpret each.

(c) Using your knowledge that the derivative gives the slope of a tangent line, does the result from part (b) indicate that the function is increasing or decreasing on its reasonable domain? Explain.

51. The per capita consumption of diet carbonated soft drinks in the U.S. can be modeled by

$$f(x) = -0.02x^2 + 0.8x + 3.94 \qquad 1 \le x \le 18$$

where x is the number of years since 1979 and $f(x)$ is the number of gallons of diet carbonated soft drinks consumed per year per capita (*Source:* U.S. Department of Agriculture, www.usda.gov).

(a) Compute $f'(x)$.

(b) Use the result from part (a) to compute $f'(5)$, $f'(10)$, and $f'(15)$, and interpret each.

(c) Using your knowledge that the derivative gives the slope of a tangent line, does the result from part (b) indicate that the function is increasing or decreasing on its reasonable domain? Explain.

52. If \$1000 is deposited in a bank account earning 5% interest compounded annually, then the amount in the account after t years is given by

$$A(t) = 1000(1.05)^t$$

(a) Determine $A(10)$ and interpret.

(b) To approximate $A'(10)$ to the nearest cent, use the definition of the derivative,

$$A'(10) = \lim_{h \to 0} \frac{A(10+h) - A(10)}{h}$$

and employ tables as outlined in the Interactive Activity on p. 775. Interpret $A'(10)$.

53. If \$2000 is deposited in a bank account earning 6.5% interest compounded annually, then the amount in the account after t years is given by

$$A(t) = 2000(1.065)^t$$

(a) Determine $A(7)$ and interpret.

(b) To approximate $A'(7)$ to the nearest cent, use the definition of the derivative,

$$A'(7) = \lim_{h \to 0} \frac{A(7+h) - A(7)}{h}$$

and employ tables as outlined in the Interactive Activity on p. 775. Interpret $A'(7)$.

54. In marketing, the number of items sold can depend on the amount of money spent on advertising. Suppose that

$$N(x) = 2000 - \frac{520}{x} \qquad [0.26, 100]$$

is a function where x is the amount spent in thousands of dollars to sell $N(x)$ items, in hundreds.

(a) Given that $N'(x) = \dfrac{520}{x^2}$ compute $N(20)$ and $N'(20)$ and interpret each.

(b) Compute $N(36)$ and $N'(36)$ and interpret each.

(c) As the amount of dollars, x, spent on advertising increases, what happens to $N'(x)$, the rate of change of items sold per thousands of dollars spent? Is it increasing or decreasing?

(d) Graph N and N', and determine if your result in part (c) seems reasonable.

55. In marketing, the number of items sold can depend on the amount of money spent on advertising. Suppose that

$$N(x) = 1500 - \frac{630}{x} \qquad [0.42, 100]$$

is a function where x is the amount spent in thousands of dollars to sell $N(x)$ items, in hundreds.

(a) Given that $N'(x) = \dfrac{630}{x^2}$, compute $N(15)$ and $N'(15)$ and interpret each.

(b) Compute $N(25)$ and $N'(25)$ and interpret each.

(c) As the amount of dollars, x, spent on advertising increases, what happens to $N'(x)$, the rate of change of items sold per thousands of dollars spent? Is it increasing or decreasing?

(d) Graph N and N', and determine if your result in part (c) seems reasonable.

56. The total inmate population in the Federal Bureau of Prisons can be modeled by

$$f(x) = \frac{1}{4}x^2 + \frac{9}{25}x + 24.5 \qquad 1 \le x \le 17$$

where x is the number of years since 1979 and $f(x)$ represents the total inmate population in thousands (*Source:* Federal Bureau of Prisons, www.bop.gov).

(a) Determine $f(11)$ and $f(16)$ and interpret each.

(b) Determine $f'(x)$.

(c) Compute $f'(11)$ and $f'(16)$ and interpret each.

57. Assume that the mathematical model for the growth of a locust tree in its first century of life is given by

$$h(t) = \sqrt{3t} \qquad [0, 100]$$

where t is the age of the tree in years and $h(t)$ is the height of the tree in feet. Figure 12.4.7 is a graph of h along with two tangent lines labeled t_1 and t_2. Line t_1 is tangent to the graph of h at $t = 25$, and line t_2 is tangent to the graph of h at $t = 64$. Which tangent line is associated with the faster growth rate? Explain in a complete sentence.

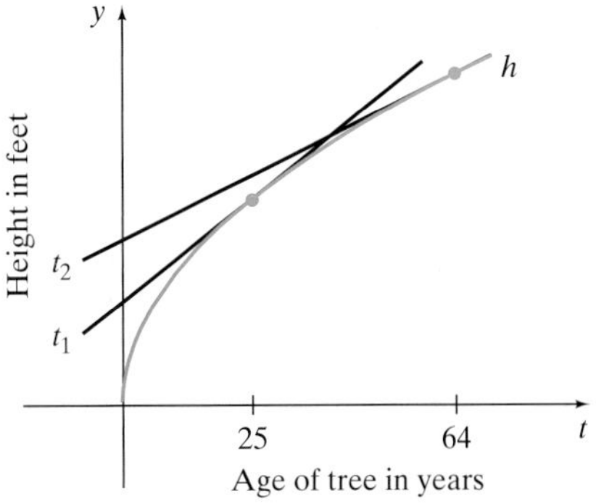

Figure 12.4.7

SECTION PROJECT

Use Figure 12.4.8 to answer the following:

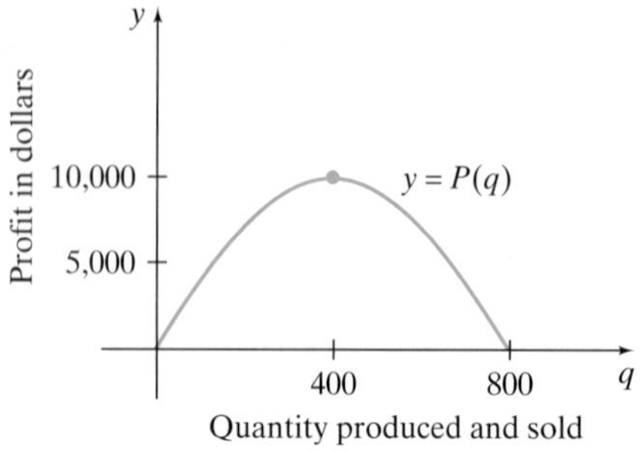

Figure 12.4.8

(a) For what value of q is $P(q)$ a maximum?

(b) For what value of q does $P'(q) = 0$?

(c) What is the maximum profit?

(d) Using interval notation, determine for what values of q is $P(q)$ increasing.

(e) Using interval notation, determine for what values of q is $P'(q) > 0$.

(f) Using interval notation, determine for what values of q is $P(q)$ decreasing.

(g) Using interval notation, determine for what values of q is $P'(q) < 0$.

(h) In your own words, describe the connection between the results of parts (a) and (b).

(i) In your own words, describe the connection between the results of parts (d) and (e).

(j) In your own words, describe the connection between the results of parts (f) and (g).

Section 12.5 Derivatives of Constants, Powers, and Sums

In Section 12.4 we defined the *derivative* of a function and used it to find slopes of tangent lines as well as instantaneous rates of change. However, computing the derivative of a function from the definition can be quite involved. Calculus would not be very useful if all derivatives had to be calculated from the definition.

In this section and the next, we present several **rules of differentiation**, or shortcuts if you like, that will greatly simplify differentiation. So why did we painstakingly compute derivatives via the definition if these shortcut rules exist? Quite simply, these rules are *derived* from the definition! Also, we want to make sure that the *concept* of instantaneous rate of change/tangent line slope was understood for what it is . . . a limit.

Before we delve into differentiation rules, we need to present alternative ways that can be used to represent the derivative.

Alternative Notation for the Derivative

For $y = f(x)$, the following may be used to represent the derivative:

$$f'(x), \quad y', \quad \frac{dy}{dx}, \quad \frac{d}{dx}[f(x)]$$

Each notation has its own advantage, and we will use each of these where appropriate.

Derivative of a Constant Function

The first rule that we present shows how to differentiate a constant function. Geometrically, this rule is fairly obvious. The graph of a constant function,

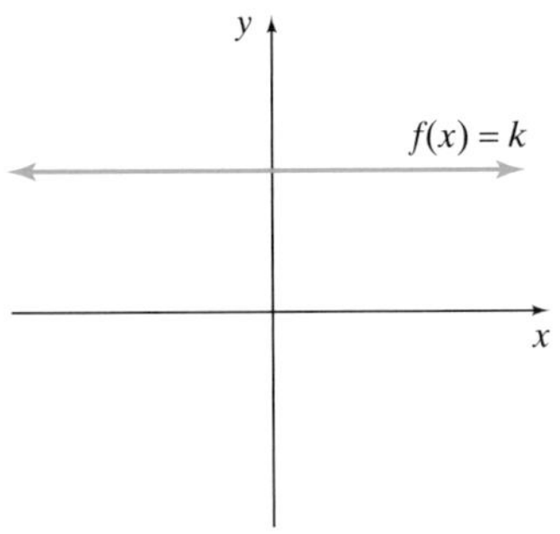

Figure 12.5.1

$f(x) = k$, is a horizontal line like the one shown in Figure 12.5.1. We know that the slope of a horizontal line is 0. This leads us to believe that the instantaneous rate of change at any point on the graph of a horizontal line is 0 or, in other words, for $f(x) = k$ in Figure 12.5.1 we believe that $f'(x) = 0$.

Rule 1: Constant Function Rule

For any constant k, if $f(x) = k$, then $f'(x) = 0$. Equivalently, we have $\frac{d}{dx}[k] = 0$, and if $y = k$, then $\frac{dy}{dx} = 0$.

In English, this rule states:

The derivative of a constant is 0.

To show that the Constant Function Rule is true, we simply apply the definition of the derivative as follows. If $f(x) = k$, $f(x + h) = k$ and $f(x) = k$. From the definition of the derivative, we have

$$f'(x) = \lim_{h \to 0} \frac{f(x+h) - f(x)}{h} = \lim_{h \to 0} \frac{k - k}{h} = \lim_{h \to 0} \frac{0}{h} = \lim_{h \to 0} 0 = 0$$

Example 1 Differentiating Constant Functions

Differentiate the following.

(a) $f(x) = 7$ **(b)** $g(x) = \sqrt[3]{2}$

Solution

(a) Since 7 is a constant, we have $f'(x) = 0$.

(b) Since $\sqrt[3]{2}$ is a constant, we have $g'(x) = 0$. ■

Power Rule

We now present one of the most useful differentiation rules in calculus. It is used to differentiate functions of the form $f(x) = x^n$, where n is any real number. Functions of this form are collectively called **power functions**. The definition of the derivative can be used to determine derivatives for power functions, although the algebra may become cumbersome quickly. For example, the definition of the derivative can be used to derive the following:

For $f(x) = x^3$, we have $f'(x) = 3x^2$.
For $f(x) = x^6$, we have $f'(x) = 6x^5$.
For $f(x) = x^7$, we have $f'(x) = 7x^6$.

There appears to be a pattern with the derivatives of power functions. It appears that the power of $f(x)$ becomes the coefficient of $f'(x)$, and the power of $f'(x)$ is 1 less than the power of $f(x)$. This observation leads us to the Power Rule. The proof of the Power Rule is in Appendix E.

Rule 2: Power Rule

If $f(x) = x^n$, where n is any real number, then $f'(x) = nx^{n-1}$. Equivalently, $\frac{d}{dx}[x^n] = nx^{n-1}$, and if $y = x^n$, then $\frac{dy}{dx} = nx^{n-1}$.

In English, this rule states:

To differentiate x^n, simply bring the exponent out front as a coefficient and then decrease the exponent by 1.

To use the Power Rule, we must have the function in the form $f(x) = x^n$. To do this, sometimes we need some rules from algebra, as Example 2 illustrates.

Example 2 Differentiating Power Functions

Differentiate the following functions.

(a) $f(x) = x^4$ **(b)** $g(x) = x^{1.32}$ **(c)** $y = \sqrt{x}$ **(d)** $g(x) = \frac{1}{x^3}$

Solution

(a) Thanks to the Power Rule, we quickly compute the derivative to be

$$f'(x) = 4x^{4-1} = 4x^3$$

(b) Again, using the Power Rule, we have

$$g'(x) = 1.32x^{1.32-1} = 1.32x^{0.32}$$

(c) Here we need to use a little algebra before differentiating. We need to recall that $\sqrt[n]{x} = x^{1/n}$. This allows us to rewrite $y = \sqrt{x}$ as $y = x^{1/2}$. Now we can apply the Power Rule to get

$$y' = \frac{1}{2}x^{1/2-1} = \frac{1}{2}x^{-1/2}$$

Utilizing the algebraic fact that $x^{-n} = \frac{1}{x^n}$ allows us to write the simplified version of the derivative as

$$y' = \frac{1}{2x^{1/2}} = \frac{1}{2\sqrt{x}}$$

(d) We begin by using a little algebra to rewrite the function. That is, by utilizing the algebraic fact that $x^{-n} = \frac{1}{x^n}$, the function can be rewritten as

$$g(x) = \frac{1}{x^3} = x^{-3}$$

Now we can apply the Power Rule and get

$$g'(x) = -3x^{-3-1} = -3x^{-4} = -\frac{3}{x^4}$$ ■

✓ Checkpoint 1 Now work Exercise 5.

Constant Multiple Rule

The Constant Multiple Rule extends the Power Rule to differentiating functions that are of the form $k \cdot g(x)$, a constant times a function. The definition of the derivative can be used to derive the following:

$$\text{For } f(x) = 2x^3, \text{ we have } f'(x) = 6x^2.$$
$$\text{For } f(x) = 3x^5, \text{ we have } f'(x) = 15x^4.$$
$$\text{For } f(x) = -2x^4, \text{ we have } f'(x) = -8x^3.$$

Again, there appears to be a pattern in these derivatives. It appears that the power of $f(x)$ gets multiplied by the coefficient of $f(x)$, and this product becomes the coefficient of $f'(x)$, while the power of $f'(x)$ is 1 less than the power of $f(x)$. This observation leads to the Constant Multiple Rule.

Rule 3: Constant Multiple Rule

If $f(x) = k \cdot g(x)$, where k is any real number, then $f'(x) = k \cdot g'(x)$, assuming that g is differentiable. Equivalently, $\frac{d}{dx}[k \cdot g(x)] = k \cdot \frac{d}{dx}[g(x)] = k \cdot g'(x)$.

In English, this rule states:

The derivative of a constant times a function is simply the constant times the derivative of the function.

The proof of this rule is given in Appendix E.

Example 3 Differentiating a Constant Times a Function

Differentiate the following.

(a) $g(x) = 1.2x^5$ **(b)** $y = \frac{1}{7x^3}$ **(c)** $f(x) = \frac{2}{3}\sqrt[5]{x}$

Solution

(a) Applying the Constant Multiple Rule yields

$$g'(x) = 1.2 \cdot (5x^{5-1}) = 6x^4$$

(b) Rewriting the function via algebra gives

$$y = \frac{1}{7}x^{-3}$$

Now we apply the Constant Multiple Rule to get

$$y' = \frac{1}{7} \cdot (-3x^{-3-1}) = \frac{1}{7}(-3x^{-4})$$
$$= -\frac{3}{7}x^{-4} = \frac{-3}{7x^4}$$

(c) We begin by rewriting the function, using some algebra, as

$$f(x) = \frac{2}{3}x^{1/5}$$

Now we apply the Constant Multiple Rule, which gives

$$f'(x) = \frac{2}{3}\left(\frac{1}{5}x^{1/5-1}\right) = \frac{2}{15}x^{-4/5}$$

Simplifying this derivatives yields

$$f'(x) = \frac{2}{15x^{4/5}} \quad \text{or} \quad \frac{2}{15\sqrt[5]{x^4}}$$

■

Checkpoint 2 Now work Exercise 13.

Sum and Difference Rule

Consider two functions $f(x) = 2x^2$ and $g(x) = 3x + 1$. We can define a new function, S, as follows.

$$S(x) = f(x) + g(x) = 2x^2 + 3x + 1$$

From the differentiation rules developed so far in this section, we know that

$$f'(x) = 4x \quad \text{and} \quad g'(x) = 3$$

However, we have no rule to help us determine $S'(x)$. We can appeal to the definition of the derivative to help us compute $S'(x)$.

$$\begin{aligned}
S'(x) &= \lim_{h\to 0} \frac{S(x+h) - S(x)}{h} \\
&= \lim_{h\to 0} \frac{[2(x+h)^2 + 3(x+h) + 1] - [2x^2 + 3x + 1]}{h} \\
&= \lim_{h\to 0} \frac{[2(x^2 + 2xh + h^2) + 3x + 3h + 1] - 2x^2 - 3x - 1}{h} \\
&= \lim_{h\to 0} \frac{2x^2 + 4xh + 2h^2 + 3x + 3h + 1 - 2x^2 - 3x - 1}{h} \\
&= \lim_{h\to 0} \frac{4xh + 2h^2 + 3h}{h} \\
&= \lim_{h\to 0} \frac{h(4x + 2h + 3)}{h} \\
&= \lim_{h\to 0} (4x + 2h + 3) \\
&= 4x + 3
\end{aligned}$$

So we have determined that $S'(x) = 4x + 3$. Recall that we defined $S(x)$ as

$$S(x) = f(x) + g(x) = 2x^2 + 3x + 1$$

Also recall that we knew that $f'(x) = 4x$ and $g'(x) = 3$. It appears that we have discovered that

$$S'(x) = 4x + 3 = f'(x) + g'(x).$$

This is not a coincidence! A rule that handles functions that are sums or differences of other functions is the Sum and Difference Rule.

Rule 4: Sum and Difference Rule

If $h(x) = f(x) \pm g(x)$, where f and g are both differentiable functions, then $h'(x) = f'(x) \pm g'(x)$. Equivalently,

$$\frac{d}{dx}[f(x) \pm g(x)] = \frac{d}{dx}[f(x)] \pm \frac{d}{dx}[g(x)] = f'(x) \pm g'(x).$$

In English this rule states:

To differentiate a sum/difference of two (or more) functions, just differentiate the functions separately and add/subtract the results.

A proof of the Sum and Difference Rule is given in Appendix E.

Example 4 Differentiating Sums and Differences

Differentiate the following functions.

(a) $f(x) = 3x^2 + 2x - 1$

(b) $g(x) = \frac{1}{2}x^3 - \frac{3}{2}x^{-1}$

(c) $y = 3x^4 + 2\sqrt{x} - \frac{2}{x^2}$

Solution

(a) According to the Sum and Difference Rule,

$$f'(x) = \frac{d}{dx}[3x^2 + 2x - 1] = \frac{d}{dx}[3x^2] + \frac{d}{dx}[2x] - \frac{d}{dx}[1]$$
$$= 6x + 2 - 0 = 6x + 2$$

Interactive Activity

Algebraically simplify the derivatives in Example 4b and 4c so that there are no negative exponents. Then determine the domain of each derivative. What is the behavior of the original function at these excluded domain values?

WWW

(b) Applying the Sum and Difference Rule yields

$$g'(x) = \frac{d}{dx}\left[\frac{1}{2}x^3 - \frac{3}{2}x^{-1}\right] = \frac{d}{dx}\left[\frac{1}{2}x^3\right] - \frac{d}{dx}\left[\frac{3}{2}x^{-1}\right]$$
$$= \frac{3}{2}x^2 - \left(-\frac{3}{2}x^{-2}\right) = \frac{3}{2}x^2 + \frac{3}{2}x^{-2}$$

(c) First, we rewrite the function as

$$y = 3x^4 + 2x^{1/2} - 2x^{-2}$$

Now we differentiate to get

$$\frac{dy}{dx} = \frac{d}{dx}[3x^4 + 2x^{1/2} - 2x^{-2}]$$
$$= \frac{d}{dx}[3x^4] + \frac{d}{dx}[2x^{1/2}] - \frac{d}{dx}[2x^{-2}]$$
$$= 12x^3 + x^{-1/2} + 4x^{-3}$$

■

Notice in Example 4 that we not only used the Sum and Difference Rule, but we also used the Constant Function Rule, the Power Rule, and the Constant Multiple Rule. In short, we used all the rules presented in this section.

Checkpoint 3 Now work Exercise 23.

Applications

Example 5 Analyzing Greenhouse Gas Emissions

Table 12.5.1 displays data for the amount of chlorofluorocarbons (CFCs), in thousands of metric tons, released into the atmosphere by the United States from 1989 to 1995.

Table 12.5.1

Year	1989	1990	1991	1992	1993	1994	1995
CFC Gases	272	231	210	187	166	133	87

SOURCE: U.S. Census Bureau, www.census.gov

If we let x represent the number of years since 1988 and $f(x)$ represent the CFCs in thousands of metric tons released into the atmosphere, the data can be modeled by

$$f(x) = -1.2x^3 + 13.37x^2 - 69.65x + 328.71 \qquad 1 \le x \le 7$$

Figure 12.5.2 gives a graph of the data along with the cubic model. Determine $f'(x)$. Evaluate $f'(6)$ and interpret. Use appropriate units in the answer.

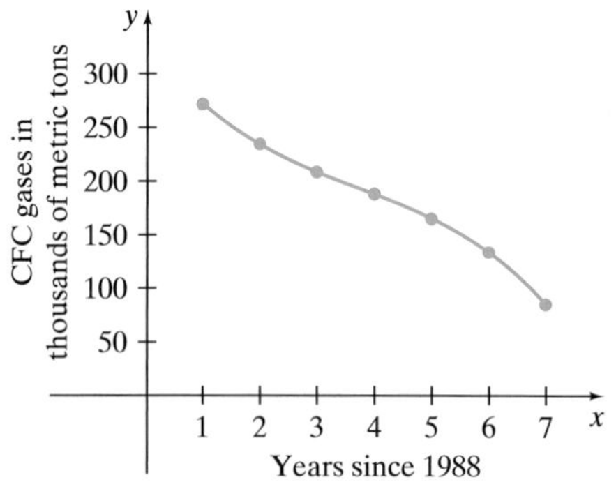

Figure 12.5.2 Graph of $f(x) = -1.2x^3 + 13.37x^2 - 69.65x + 328.71$ along with data points.

Solution

Understand the Situation: We can use the differentiation techniques learned in this section to quickly determine $f'(x)$. To evaluate $f'(6)$ we will simply substitute $x = 6$ into the derivative $f'(x)$. Note that $x = 6$ corresponds to the year 1994.

Using the techniques learned in this section, we get

$$f'(x) = -1.2(3x^2) + 13.37(2x) - 69.65(1) + 0$$
$$f'(x) = -3.6x^2 + 26.74x - 69.65$$

To determine $f'(6)$, we substitute $x = 6$ into the derivative.

$$f'(6) = -3.6(6)^2 + 26.74(6) - 69.65$$
$$= -38.81 \frac{\text{thousand metric tons}}{\text{year}}$$

Interpret the Solution: This means that in 1994 U.S. emission of CFCs into the atmosphere was decreasing at a rate of $38.81 \frac{\text{thousand metric tons}}{\text{year}}$. See Figure 12.5.3.

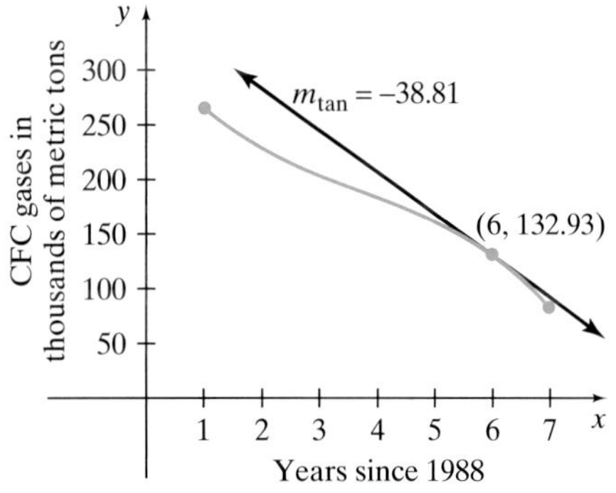

Figure 12.5.3

Checkpoint 4

Now work Exercise 69.

The next example deals with what is known as a **position function** and a **velocity function**. These functions are commonly used in physics.

Example 6 Analyzing Velocity

If a coconut falls from a tree that is 75 feet tall, its height above the ground after t seconds is given by

$$s(t) = 75 - 16t^2$$

where $s(t)$ is measured in feet. This function is called a **position function** since it gives the position of the coconut above the ground as a function of time.

(a) The derivative of a position function, $s'(t)$, is called the **velocity function** and is denoted by $v(t)$. Determine $v(t)$.

(b) Compute $s(2)$ and $v(2)$ and interpret each.

(c) When does the coconut hit the ground?

Solution

(a) Since we have been told that $s'(t) = v(t)$, we determine $v(t)$ as follows.

$$v(t) = s'(t) = -32t$$

(b) We have

$$s(2) = 75 - 16(2)^2$$
$$= 11 \text{ feet}$$

Interpret the Solution: This means that after 2 seconds the coconut is 11 feet above the ground.

We also have

$$v(2) = -32(2)$$
$$= -64\frac{\text{feet}}{\text{sec}}$$

Interpret the Solution: This means that after 2 seconds the velocity of the falling coconut is $-64\frac{\text{feet}}{\text{sec}}$. The negative sign indicates that the coconut is falling toward the ground.

(c) **Understand the Situation:** The coconut hits the ground when it is 0 feet above the ground. Hence, we need to solve $s(t) = 0$.

Setting $s(t) = 0$ and solving yields

$$75 - 16t^2 = 0$$
$$75 = 16t^2$$
$$\frac{75}{16} = t^2$$
$$\pm\sqrt{\frac{75}{16}} = t$$
$$t \approx \pm 2.17 \text{ seconds}$$

Since -2.17 seconds does not make sense (we cannot yet travel back in time), we conclude that the coconut hits the ground in about 2.17 seconds. ■

Technology Option

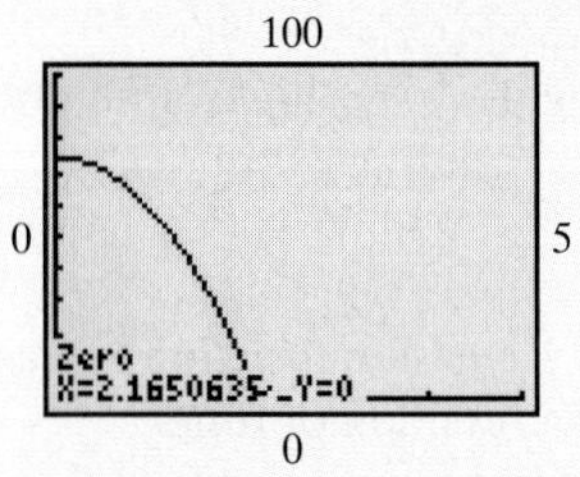

Figure 12.5.4 Result of ZERO command.

We can graphically support the result of Example 6c by graphing the position function and utilizing the ZERO command on the graphing calculator. See Figure 12.5.4.

Example 7 Analyzing a Profit Function

The Cool Air company has determined that the total cost function for the production of its refrigerators can be described by

$$C(x) = 2x^2 + 15x + 1500$$

where x is the weekly production of refrigerators and $C(x)$ is the total cost in dollars. The revenue function for these refrigerators is given by

$$R(x) = -0.3x^2 + 460x$$

where x is the number of refrigerators sold and $R(x)$ is in dollars.

(a) Determine P, the profit function, and determine $P'(x)$.

(b) Evaluate $P(20)$ and $P'(20)$ and interpret each.

Solution

(a) In Chapter 11 we learned that $P(x) = R(x) - C(x)$. Thus,

$$\begin{aligned} P(x) &= (-0.3x^2 + 460x) - (2x^2 + 15x + 1500) \\ &= -2.3x^2 + 445x - 1500 \end{aligned}$$

Since $P(x) = -2.3x^2 + 445x - 1500$, $P'(x)$ is

$$P'(x) = -2.3(2x) + 445(1) - 0 = -4.6x + 445$$

(b) We have $P(20) = -2.3(20)^2 + 445(20) - 1500 = 6480$, or \$6480. This means that when the Cool Air company makes and sells 20 refrigerators, a profit of \$6480 is realized. On the other hand, $P'(20) = -4.6(20) + 445 = 353$, or \$353. This means that the instantaneous rate of change is \$353 per refrigerator when the production level is 20 refrigerators (Figure 12.5.5). In other words, at a weekly production of 20 refrigerators, the profit is increasing at a rate of \$353 per refrigerator. Alternatively, we could say that it means the profit increases approximately \$353 to make and sell the 21st refrigerator.

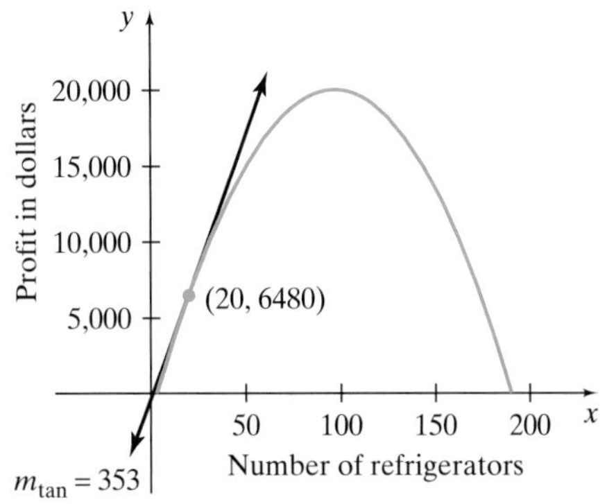

Figure 12.5.5

■

The derivative of a profit function is called **marginal profit**, which will be discussed more thoroughly in Chapter 13.

SUMMARY

In this section we introduced four differentiation rules to allow us to compute derivatives more quickly:

Name	Rule	Example
Constant Function Rule	If $f(x) = k$, then $f'(x) = 0$.	If $f(x) = 7$, then $f'(x) = 0$.
Power Rule	If $f(x) = x^n$, then $f'(x) = nx^{n-1}$.	If $f(x) = x^8$, then $f'(x) = 8x^7$.
Constant Multiple Rule	If $f(x) = k \cdot g(x)$, then $f'(x) = k \cdot g'(x)$.	If $f(x) = 3x^4$, then $f'(x) = 12x^3$.
Sum and Difference Rule	If $h(x) = f(x) \pm g(x)$, then $h'(x) = f'(x) \pm g'(x)$.	If $h(x) = x^3 + x^2$, then $h'(x) = 3x^2 + 2x$.

Even though we now have some tools to allow us to calculate a derivative more quickly than by using the definition of the derivative, do not lose sight of what the derivative tells us: It is a limit that gives an instantaneous rate of change as well as the slope of a tangent line.

SECTION 12.5 EXERCISES

In Exercises 1–16, determine the derivative for the given function. Where appropriate, simplify the derivative so that there are no negative or fractional exponents. Some algebra rules to keep in mind are $x^{-n} = \frac{1}{x^n}$ *and* $\sqrt[n]{x^m} = x^{m/n}$.

1. $f(x) = 3$

2. $g(x) = 5$

3. $f(t) = -2$

4. $g(t) = -7$

✓ **5.** $f(x) = x^6$

6. $f(x) = x^{10}$

7. $y = -3x^4$

8. $y = -5x^7$

9. $g(x) = 2x^{2/3}$

10. $g(x) = 3x^{4/5}$

11. $f(x) = -3x^{-1/3}$

12. $f(x) = -2x^{-2/5}$

✓ **13.** $h(x) = \frac{2}{3}x^4$

14. $h(x) = \frac{3}{4}x^8$

15. $y = -\frac{2}{5}x^{5/3}$

16. $y = -\frac{5}{6}x^{6/5}$

In Exercises 17–48, determine the derivative for the given function. Where appropriate, simplify the derivative so that there are no negative or fractional exponents.

17. $f(x) = 2x^3 + 4x^2 - 7x + 1$

18. $f(x) = 4x^3 - 3x^2 + 5x - 3$

19. $g(x) = 3x^2 - 2x + 6$

20. $g(x) = 3x^2 + 5x - 2$

21. $y = -5x^2 - 6x + 2$

22. $y = -3x^2 + 9x - 1$

✓ **23.** $y = -5x^3 + 7x - 5$

24. $y = -8x^3 - 8x + 3$

25. $f(x) = \frac{1}{2}x^3 + \frac{3}{5}x^2 - \frac{2}{3}x + \frac{2}{5}$

26. $f(x) = \frac{2}{5}x^3 - \frac{2}{3}x^2 + \frac{1}{2}x - 5$

27. $g(x) = 1.31x^2 + 2.05x - 3.9$

28. $g(x) = 3.15x^2 - 1.13x + 5.2$

29. $d(x) = -0.2x^2 + 3.5x^3 - 0.4x^4$

30. $d(x) = 0.3x^2 - 0.67x^3 + 0.8x^4$

31. $y = 1.15x^3 - 2.3x^2 + 2.53x - 7.1$

32. $y = 2.35x^3 + 3.56x^2 - 63.25x + 365.3$

33. $f(x) = 3\sqrt{x} + \frac{1}{2}x - 5x^2$

34. $f(x) = 5\sqrt{x} - \frac{1}{2}x + 7x^2$

35. $y = \sqrt[3]{x} + x^2 - 3x^3$

36. $y = \sqrt[3]{x} - 2x^3 + 5x^4$

37. $g(x) = \sqrt[3]{x^2} - \frac{4}{\sqrt{x}}$

38. $g(x) = \sqrt[3]{x^2} + \frac{3}{\sqrt{x}}$

39. $f(x) = 2.35x^{1.35}$

40. $f(x) = 3.2x^{1.14}$

41. $AC(x) = 2000 + \frac{5}{x^2}$

42. $AC(x) = 3300 + \frac{13}{x^2}$

43. $y = 2x^2 + \frac{1}{x}$

44. $y = -3x^2 - \frac{1}{x}$

45. $f(x) = 3x^{-3/2} - 4x^{1/2} + 5$

46. $f(x) = 4x^{-3/2} - 2x^{1/2} + x^{-1/2}$

47. $y = 2.35x^{-1/2} - 2.3x^{-2/3}$

48. $y = 3.52x^{-2/5} + 3.2x^{-1/2}$

For Exercises 49–56:

(a) Determine the domain of the given function. Write in interval notation.

(b) Use the addition rule for fractions, $\frac{a+b}{c} = \frac{a}{c} + \frac{b}{c}$, to rewrite the function and simplify. For example,

$$f(x) = \frac{x^3 + 3x^2 - x + 2}{x} = \frac{x^3}{x} + \frac{3x^2}{x} - \frac{x}{x} + \frac{2}{x}$$
$$= x^2 + 3x - 1 + 2x^{-1}$$

(c) Compute the derivative using differentiation rules.
(d) Determine the domain of the derivative. Write in interval notation.

49. $f(x) = \dfrac{3x^3 - 9x^2 + 4}{12}$

50. $f(x) = \dfrac{x^2 + 4x + 3}{2}$

51. $f(x) = \dfrac{2x^3 + 3x^2 - x + 3}{x}$

52. $f(x) = \dfrac{-3x^3 - 4x^2 + 2x - 7}{x}$

53. $y = \dfrac{7x^4 - 50x^2 + x}{x^2}$

54. $y = \dfrac{-5x^4 + 25x^2 - x}{x^2}$

55. $h(x) = \dfrac{4x^3 - 14x^2 + 3}{2\sqrt{x}}$

56. $g(x) = \dfrac{x^3 + 2x^2}{\sqrt[3]{x^2}}$

For Exercises 57–62:

(a) Determine the derivative for the given function using differentiation rules.

(b) Determine the slope of the line tangent to the graph of the function at the indicated x-value.

(c) Determine an equation for the line tangent to the graph of the function at the indicated x-value.

(d) Graph the function and the tangent line in the same viewing window.

57. $f(x) = x^3$ at $x = -1$ **58.** $f(x) = \sqrt{x}$ at $x = 4$

59. $y = \dfrac{1}{x}$ at $x = 3$ **60.** $y = \dfrac{1}{x^2}$ at $x = 3$

61. $f(x) = x^{2/3}$ at $x = 8$ **62.** $g(x) = x^{-1/3}$ at $x = 1$

Applications

63. The number of dollars spent on advertising for a new CD influences the number of CDs consumers purchase. If x is the number of dollars spent, in thousands, then $N(x)$, measured in hundreds, is the number of CDs sold, as given by

$$N(x) = 2500 - \frac{520}{x} \qquad x \geq 0.208$$

(a) Compute $N'(x)$.
(b) Compute $N(10)$ and $N'(10)$ and interpret each.

64. The number of dollars spent on advertising for a new surround-sound stereo receiver influences the number of these stereo receivers that consumers purchase. If x is the number of dollars spent, in thousands, then $N(x)$, measured in hundreds, is the number of stereo receivers sold, as given by

$$N(x) = 3600 - \frac{700}{x} \qquad x \geq 0.195$$

(a) Compute $N'(x)$.
(b) Compute $N(5)$ and $N'(5)$ and interpret each.

Exercises 65 and 66 use the following: Recall that if the total cost of making x units of a product is given by $C(x)$ then the average total cost, $AC(x)$, is simply given by

$$AC(x) = \frac{C(x)}{x}$$

Suppose that from past experience a manufacturer of coats knows that the total cost in dollars of making x coats per week is given by

$$C(x) = 3000 + 11x - 7\sqrt{x} + 0.03x^{3/2}$$

65. Determine $C(300)$ and $C'(300)$ and interpret each.

66. **(a)** Determine $AC(x)$.
(b) Determine $AC(300)$ and $AC'(300)$ and interpret each.

67. An electric generating plant burns coal, and data collected from air samples downwind from the plant indicate that the amount of sulfur dioxide, $SD(x)$, x miles downwind is given by

$$SD(x) = \frac{93.21}{x^2} \qquad x > 0$$

where $SD(x)$ is in parts per million (ppm). Compute $SD(1)$ and $SD'(1)$ and interpret each.

68. An electric generating plant burns coal, and data collected from air samples downwind from the plant indicate that the amount of sulfur dioxide, $SD(x)$, x miles downwind is given by

$$SD(x) = \frac{78.35}{x^2} \qquad x > 0$$

where $SD(x)$ is in parts per million (ppm). Compute $SD(2)$ and $SD'(2)$ and interpret each.

12.3 **69.** The total carbon content of carbon dioxide emitted into the atmosphere by the U.S. can be modeled by

$$f(x) = -0.55x^3 + 9.03x^2 - 21.8x + 1361.01 \qquad 1 \leq x \leq 9$$

where x is the number of years since 1989 and $f(x)$ is the total carbon content of carbon dioxide, in millions of metric tons, emitted into the atmosphere by the U.S. (*Source:* U.S. Statistical Abstract, www.census.gov/statab/www).

(a) Compute $f(3)$ and $f(7)$ and interpret each.
(b) Compute $f'(3)$ and $f'(7)$ and interpret each. Use appropriate units.

70. The size of a certain bacteria culture at time t, in minutes, is approximated by $N(t) = 4t^{7/2}$, where $N(t)$ is in milligrams. Compute $N'(9)$ and interpret using appropriate units.

71. The size of a certain bacteria culture at time t, in minutes, is approximated by $N(t) = 6t^{5/2}$, where $N(t)$ is in milligrams.

(a) Compute the average rate of change from $t = 1$ to $t = 4$. Interpret using appropriate units.
(b) Compute $N'(4)$ and interpret using appropriate units.

72. If a watermelon is dropped from a building 200 feet tall, its height above the ground after t seconds is given by $s(t) = 200 - 16t^2$, where $s(t)$ is measured in feet.

(a) Determine an expression for the velocity function, $v(t)$. (Recall that $v(t) = s'(t)$.)

(b) Determine $s(1)$ and $s(3)$ and interpret each using appropriate units.

(c) Determine the average velocity from $t = 1$ to $t = 3$.

(d) Determine $v(1)$ and $v(3)$ and interpret each.

(e) When does the watermelon hit the ground? Round your answer to the nearest tenth of a second.

73. If an egg is dropped from a building 150 feet tall, its height above the ground after t seconds is given by $s(t) = 150 - 16t^2$, where $s(t)$ is measured in feet.

(a) Determine an expression for the velocity function, $v(t)$. (Recall that $v(t) = s'(t)$.)

(b) Determine $s(1)$ and $s(3)$ and interpret each using appropriate units.

(c) Determine the average velocity from $t = 1$ to $t = 3$.

(d) Determine $v(1)$ and $v(3)$ and interpret each.

(e) When does the egg hit the ground? Round your answer to the nearest tenth of a second.

12.3 Exercises 74 and 75 use the data in Table 12.5.2 which gives the number of cases of tuberculosis (in thousands) reported in the U.S. from 1980 to 1998. If we let x represent the number of years since 1979, then the number of cases of tuberculosis reported can be modeled by

$$f(x) = -0.014x^3 + 0.39x^2 - 3.11x + 30.29 \qquad 1 \le x \le 19$$

where $f(x)$ is the number of cases in thousands.

Table 12.5.2

Year	Tuberculosis	Year	Tuberculosis
1980	27.7	1992	26.7
1985	22.2	1993	25.3
1989	23.5	1994	24.4
1990	25.7	1995	22.9
1991	26.3	1999	18.4

SOURCE: U.S. Centers for Disease Control and Prevention, www.cdc.gov

74. (a) Using the model, compute the average rate of change of cases of tuberculosis reported from 1985 to 1992. Interpret the result.

(b) Using the model, compute the average rate of change of cases of tuberculosis reported from 1992 to 1999. Interpret the result.

75. (a) Determine $f'(x)$.

(b) Compute $f'(12)$ and $f'(17)$ and interpret each. Make sure to use appropriate units.

76. Consider any linear function of the form $f(x) = mx + b$, where m and b are real numbers. Compute $f'(x)$ and interpret.

77. The number of milligrams of cholesterol consumed each day per person in the United States can be modeled by

$$C(x) = 0.11x^2 - 4.04x + 445.02 \qquad 1 \le x \le 23$$

where x represents the number of years since 1974 and $C(x)$ represents the number of milligrams of cholesterol consumed each day per person (*Source:* U.S. Department of Agriculture, www.usda.gov).

(a) Compute $C(11)$ and $C(21)$ and interpret each.

(b) Compute $C'(11)$ and $C'(21)$ and interpret each using appropriate units.

Exercises 78 and 79 utilize the following: Double D Cola ran a 10-week ad campaign for its new caffeine-free cola. The total weekly sales, $S(t)$, where $S(t)$ is in thousands of dollars, t weeks after the ad campaign began can be modeled by

$$S(t) = -0.51t^2 + 20.20t + 49.49 \qquad 1 \le t \le 23$$

78. (a) Compute $S(10)$ and interpret.

(b) Compute $S'(10)$ and interpret using appropriate units.

(c) Using the results from parts (a) and (b), complete the following:

Ten weeks after the ad campaign began, weekly sales of the new cola were ____________. Weekly sales were ____________ (increasing/decreasing) at a rate of ____________.

79. (a) Compute $S(23)$ and interpret.

(b) Compute $S'(23)$ and interpret using appropriate units.

(c) Using the results from parts (a) and (b), complete the following:

Twenty-three weeks after the ad campaign began, weekly sales of the new cola were ____________. Weekly sales were ____________ (increasing/decreasing) at a rate of ____________.

80. Psychologists have found that, when learning a new task, people learn quickly in the beginning and then the rate of learning slows. Eventually, an individual reaches a level that cannot be exceeded. For example, to test learning, a psychologist asks people to memorize a long sequence of digits and checks with each person every few minutes to see how many digits have been memorized. Assume that the learning can be described by the equation $y = 22(1 - e^{-0.2x})$, where y is the number of digits memorized and x is the time in minutes. A graph of this equation is called a **learning curve** and will be studied in Chapter 14.

(a) Determine $\lim_{x\to\infty} [22(1 - e^{-0.2x})]$ and interpret.

(b) Graph $y = 22(1 - e^{-0.2x})$, use your calculator's $\frac{dy}{dx}$ or DRAW TANGENT command to determine $\frac{dy}{dx}$ at $x = 3$ and $x = 10$, and interpret each.

(c) Does the result from part (b) indicate that the rate of learning slows down? Explain.

12.3 **81.** Membership in the Girl Scouts of America can be modeled by

$$g(t) = 0.086t^4 - 3.83t^3 + 55.97t^2 - 243.23t + 2972.93 \qquad 1 \le t \le 20$$

where t is the number of years since 1979 and $g(t)$ is the membership in thousands (*Source:* U.S. Statistical Abstract, www.census.gov/statab/www).

(a) Determine $g(14)$ and interpret.

(b) Determine $g'(t)$.

(c) Determine $g'(14)$ and interpret.

12.3 **82.** Membership in the Boy Scouts of America can be modeled by

$$f(t) = 0.72t^3 - 22.15t^2 + 263.28t + 4043.3 \qquad 1 \le t \le 20$$

where t is the number of years since 1979 and $f(t)$ is the membership in thousands (*Source:* U.S. Statistical Abstract, www.census.gov/statab/www).

(a) Determine $f(14)$ and interpret.

(b) Determine $f'(t)$.

(c) Determine $f'(14)$ and interpret.

83. Comparing the results from Exercises 81c and 82c, whose membership was increasing at a faster rate at the beginning of 1993? Explain.

SECTION PROJECT

In Table 12.5.3, expenditures refers to federal government expenditures, in billions of dollars, for health services and supplies.

Table 12.5.3

Year	Expenditures	Year	Expenditures
1980	42.4	1990	115.1
1985	68.4	1991	136.2
1987	76.3	1992	159.7
1988	84.1	1993	181.1
1989	96.3	1994	190.6

SOURCE: U.S. Health Care Financing Administration, www.hcfa.gov

(a) Let x represent the number of years since 1979, and let y represent the amount, in billions, spent on health services and supplies by the federal government. Enter the data into your calculator, and perform a cubic regression on the data to model the data on the interval [1, 15]. Plot the data and graph the model in the same viewing window.

(b) Use the model to compute the average rate of change of federal government expenditures for health services and supplies from 1983 to 1990.

(c) Compute the derivative of the model.

(d) Evaluate the derivative at $x = 4$, and interpret using appropriate units.

(e) Evaluate the derivative at $x = 11$, and interpret using appropriate units.

(f) To the nearest tenth, graphically determine a value for x in the interval [4, 11], where the derivative at this value is equivalent to the average rate of change found in part (b). In other words, find a value for x in [4, 11], where the slope of the tangent line is equal to the slope of the secant line that passes through the points determined by $x = 4$ and $x = 11$. (*Hint:* Two lines have the same slope if they are parallel.)

Section 12.6 Derivatives of Products and Quotients

In Section 12.5 we learned four useful rules for computing derivatives. In this section, we illustrate how to differentiate the **product** and **quotient** of two functions. These two rules are not as simple as the ones in Section 12.5.

Product Rule

To differentiate the product of two functions, we use the Product Rule.

Rule 5: Product Rule

If $h(x) = f(x) \cdot g(x)$, where f and g are differentiable functions, then the derivative is $h'(x) = f'(x) \cdot g(x) + f(x) \cdot g'(x)$.

In English, the Product Rule states:

Take the derivative of the first function times the second function plus the first function times the derivative of the second function.

A proof of the Product Rule is given in Appendix E.

Example 1 Differentiating Products

Differentiate $h(x) = 3x^3(x^4 + 2)$ by using the Product Rule.

Solution

Let $f(x) = 3x^3$ and $g(x) = x^4 + 2$. By the Product Rule, we compute the derivative to be

$$\begin{aligned} h'(x) &= f'(x) \cdot g(x) + f(x) \cdot g'(x) \\ &= 9x^2(x^4 + 2) + 3x^3(4x^3) \\ &= 9x^6 + 18x^2 + 12x^6 \\ &= 21x^6 + 18x^2 \\ &= 3x^2(7x^4 + 6) \end{aligned}$$

Interactive Activity

The derivative of h in Example 1 could have been computed if we had first multiplied out the function and then applied the rules learned in Section 12.5. Do this to verify our result in Example 1.

Example 2 Differentiating Products

Differentiate the following by using the Product Rule.

(a) $f(x) = (2x^2 + 4x + 5)(5x - 4)$

(b) $y = \sqrt{x}(3x^3 - 4x^2 + 8x)$

Solution

(a) Applying the Product Rule yields

$$\begin{aligned} f'(x) &= (4x + 4)(5x - 4) + (2x^2 + 4x + 5)(5) \\ &= (20x^2 - 16x + 20x - 16) + (10x^2 + 20x + 25) \\ &= 30x^2 + 24x + 9 \end{aligned}$$

(b) Before differentiating, we rewrite the function as

$$y = x^{1/2}(3x^3 - 4x^2 + 8x)$$

Now, using the Product Rule, we get

$$y' = \left(\frac{1}{2}x^{-1/2}\right)(3x^3 - 4x^2 + 8x) + x^{1/2}(9x^2 - 8x + 8)$$

Writing the derivative without negative or fractional exponents yields

$$y' = \frac{3x^3 - 4x^2 + 8x}{2\sqrt{x}} + \sqrt{x} \cdot (9x^2 - 8x + 8)$$ ■

Checkpoint 1 Now work Exercise 13.

Example 3 Differentiating Products

Differentiate $f(x) = (6x^{4/3} + 2x)(3x^{5/3} + 4x - 1)$.

Solution

By the Product Rule, we immediately compute the derivative to be

$$f'(x) = (8x^{1/3} + 2)(3x^{5/3} + 4x - 1) + (6x^{4/3} + 2x)(5x^{2/3} + 4)$$ ■

Example 4 Determining an Equation for a Tangent Line

Determine an equation for the line tangent to $y = (x^4 - x^2)(x^3 - x + 2)$ at $x = 1$.

Solution

Understand the Situation: Since the derivative gives the slope of a tangent line at any point, our first task is to determine the derivative of the function by using the Product Rule.

This gives

$$\frac{dy}{dx} = (4x^3 - 2x)(x^3 - x + 2) + (x^4 - x^2)(3x^2 - 1)$$

We then evaluate $\frac{dy}{dx}$ at $x = 1$ to get the slope of the tangent line. This yields, at $x = 1$,

$$\frac{dy}{dx} = (4 \cdot 1^3 - 2 \cdot 1)(1^3 - 1 + 2) + (1^4 - 1^2)(3 \cdot 1^2 - 1) = 4$$

So, at $x = 1$, $m_{\tan} = 4$. Also, when $x = 1$, we determine the point of tangency by substituting 1 for x in the original function. This gives

$$y = (1^4 - 1^2)(1^3 - 1 + 2) = 0$$

So our point of tangency is (1, 0). Since we know a point on the tangent line and we know that the slope of the tangent line is 4, we use the point–slope form of a line to determine an equation for the tangent line. This results in

$$y - y_1 = m(x - x_1)$$
$$y - 0 = 4(x - 1)$$
$$y = 4x - 4$$

Figure 12.6.1 shows a graph of the function along with the tangent line.

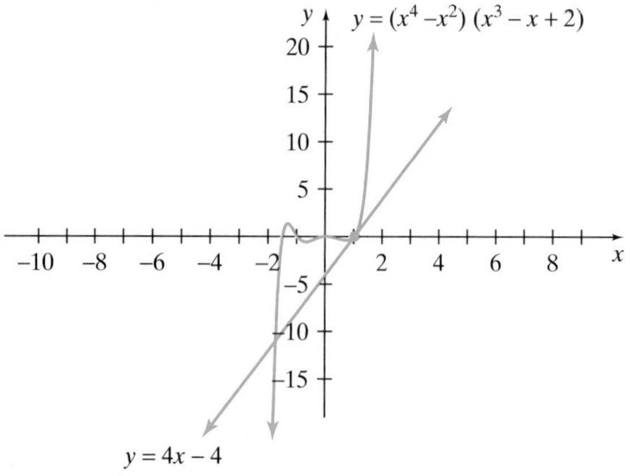

Figure 12.6.1

■

Checkpoint 2 Now work Exercise 39.

Technology Option

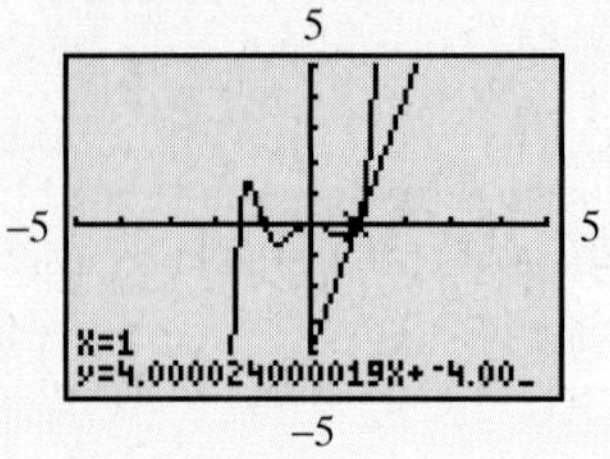

Figure 12.6.2

In Figure 12.6.2 we have used a graphing calculator to graph the function in Example 4. We have also used the DRAW TANGENT command. Notice that the DRAW TANGENT command does not give the same equation for the tangent line as we found in Example 4. This is a limitation of the calculator. Even though it is very close, we needed calculus for the **exact** answer.

Example 5 Applying the Product Rule

Extensive market research has determined that for the next five years the price of a certain mountain bike is predicted to vary according to $p(t) = 300 - 30t + 7.5t^2$, where t is time in years and $p(t)$ is the price in dollars. The number of mountain bikes sold annually by Skinner's Bikes is expected to follow $q(t) = 3000 + 90t - 15t^2$, where $q(t)$ is the number sold and t is time in years.

(a) Determine $R(t)$ and $R'(t)$.

(b) Compute $R'(1)$ and interpret.

(c) Compute $R'(4)$ and interpret.

Solution

(a) Since revenue equals price times quantity, we have

$$R(t) = p(t) \cdot q(t) = (300 - 30t + 7.5t^2) \cdot (3000 + 90t - 15t^2)$$

By the Product Rule, we compute the derivative to be

$$R'(t) = \underbrace{(-30 + 15t)}_{p'(t)}\underbrace{(3000 + 90t - 15t^2)}_{q(t)} + \underbrace{(300 - 30t + 7.5t^2)}_{p(t)}\underbrace{(90 - 30t)}_{q'(t)}$$

We will not simplify $R'(t)$ so that we may see separately the effects on $R(t)$ caused by changing prices and sales. Notice in the derivative that the first term is $p'(t) \cdot q(t)$. This term describes the rate at which $R(t)$ is changing as the price changes. The second term in the derivative is $p(t) \cdot q'(t)$. This term describes the rate at which $R(t)$ is changing as the number of bikes sold changes.

(b) We compute $R'(1)$ to be

$$\begin{aligned} R'(1) &= (-30 + 15)(3000 + 90 - 15) + (300 - 30 + 7.5)(90 - 30) \\ &= (-15)(3075) + (277.5)(60) \\ &= -46{,}125 + 16{,}650 = -29{,}475 \end{aligned}$$

Interpret the Solution: The negative sign indicates that after one year the revenue is decreasing at a rate of \$29,475 per year. Notice that the effect of falling prices represented by the first term, $p'(1) \cdot q(1) = -46{,}125$, is greater than the effect of rising sales given by the second term, $p(1) \cdot q'(1) = 16{,}650$.

(c) We have

$$\begin{aligned} R'(4) &= (-30 + 60)(3000 + 360 - 240) + (300 - 120 + 120)(90 - 120) \\ &= (30)(3120) + (300)(-30) \\ &= 93{,}600 - 9000 = 84{,}600 \end{aligned}$$

Interpret the Solution: Since this is positive, we claim that after four years the revenue is increasing at a rate of \$84,600 per year. Notice that the effect of rising prices represented by the first term, $p'(4) \cdot q(4) = 93{,}600$, is greater than the effect of falling sales given by the second term, $p(4) \cdot q'(4) = -9000$. Figure 12.6.3 shows a graph of R.

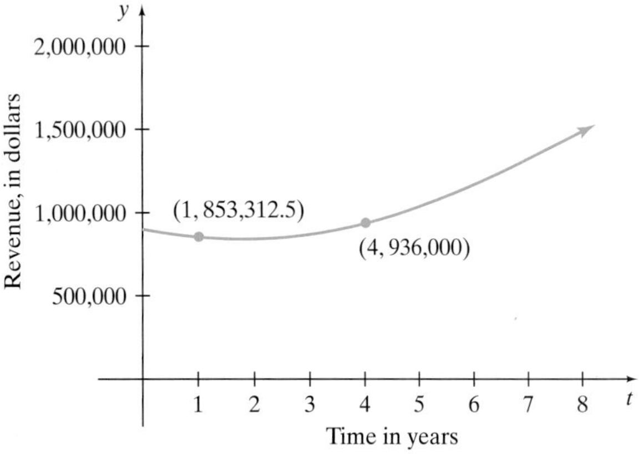

Figure 12.6.3 Graph of $R(t) = (300 - 30t + 7.5t^2)(3000 + 90t - 15t^2)$. ■

Quotient Rule

The final differentiation rule of this section, the Quotient Rule, shows how to differentiate the quotient of two functions.

Rule 6: Quotient Rule

If $h(x) = \dfrac{f(x)}{g(x)}$, where f and g are differentiable functions, then

$$h'(x) = \frac{f'(x) \cdot g(x) - f(x) \cdot g'(x)}{[g(x)]^2}, \quad \text{where } g(x) \neq 0.$$

In English, the Quotient Rule states:

Take the derivative of the numerator times the denominator minus the numerator times the derivative of the denominator, all of this over the denominator squared.

A proof of the Quotient Rule is given in Appendix E.

Example 6 Differentiating Quotients

Differentiate the following using the Quotient Rule.

(a) $h(x) = \dfrac{x+3}{x-2}$

(b) $y = \dfrac{x^4 - 3x}{x^2 + 1}$

Solution

(a) Using the Quotient Rule, we have

$$\begin{aligned} h'(x) &= \frac{f'(x)g(x) - f(x)g'(x)}{[g(x)]^2} \\ &= \frac{1 \cdot (x-2) - (x+3) \cdot 1}{(x-2)^2} \\ &= \frac{x - 2 - x - 3}{(x-2)^2} = \frac{-5}{(x-2)^2} \end{aligned}$$

(b) Again, by the Quotient Rule, the derivative is

$$\begin{aligned} y' &= \frac{f'(x)g(x) - f(x)g'(x)}{[g(x)]^2} \\ &= \frac{(4x^3 - 3)(x^2 + 1) - (x^4 - 3x)(2x)}{(x^2+1)^2} \\ &= \frac{(4x^5 + 4x^3 - 3x^2 - 3) - 2x^5 + 6x^2}{(x^2+1)^2} \\ &= \frac{2x^5 + 4x^3 + 3x^2 - 3}{(x^2+1)^2} \end{aligned}$$

Checkpoint 3 Now work Exercise 27.

The next example requires us to use both the Product Rule and the Quotient Rule.

Example 7 Differentiating Using the Product Rule and Quotient Rule

Differentiate $y = \dfrac{(2x+1)(3x-2)}{x+1}$.

Solution

Notice that when applying the Quotient Rule we must use the Product Rule when differentiating the numerator. To aid us in this computation, let $f(x) = 2x + 1$ and $g(x) = 3x - 2$. We then have

$$f'(x) = 2 \quad \text{and} \quad g'(x) = 3$$

Now, when applying the Quotient Rule, the derivative of the numerator is

$$2 \cdot (3x - 2) + (2x + 1) \cdot 3 = 6x - 4 + 6x + 3 = 12x - 1$$

Putting this all together using the Quotient Rule gives

$$\begin{aligned}\frac{dy}{dx} &= \frac{(12x-1)(x+1) - (2x+1)(3x-2) \cdot 1}{(x+1)^2} \\ &= \frac{12x^2 + 11x - 1 - (6x^2 - x - 2)}{(x+1)^2} \\ &= \frac{12x^2 + 11x - 1 - 6x^2 + x + 2}{(x+1)^2} \\ &= \frac{6x^2 + 12x + 1}{(x+1)^2}\end{aligned}$$

■

Checkpoint 4 Now work Exercise 31.

Example 8 Applying the Quotient Rule

Researchers have determined through experimentation that the percent concentration of a certain medication can be approximated by

$$p(t) = \frac{200t}{2t^2 + 5} - 4 \qquad [0.25, 20]$$

where t is the time in hours after administering the medication and $p(t)$ is the percent concentration.

(a) Evaluate $p'(1)$ and interpret.

(b) Evaluate $p'(6)$ and interpret.

Solution

Understand the Situation: In order to evaluate $p'(1)$ in part (a) and $p'(6)$ in part (b), we first need to determine $p'(t)$. Once we have $p'(t)$, we will substitute $t = 1$ and $t = 6$ to determine the value of $p'(1)$ and $p'(6)$, respectively.

(a) Utilizing the Quotient Rule, along with the Sum and Difference Rule, we compute the derivative to be

$$p'(t) = \frac{200(2t^2 + 5) - (200t)(4t)}{(2t^2 + 5)^2} = \frac{-200(2t^2 - 5)}{(2t^2 + 5)^2}$$

Evaluating $p'(1)$ gives

$$p'(1) = \frac{-200(2(1)^2 - 5)}{(2(1)^2 + 5)^2} \approx 12.24$$

Interpret the Solution: This means that at the end of one hour the concentration of the medication is increasing at a rate of about 12.24% per hour.

(b) Evaluating $p'(t)$ for $t = 6$ gives us

$$p'(6) = \frac{-200(2(6)^2 - 5)}{(2(6)^2 + 5)^2} \approx -2.26$$

Interpret the Solution: Thus, at the end of six hours the concentration of the medication is decreasing at the rate of about 2.26% per hour. Do these results seem reasonable based on Figure 12.6.4?

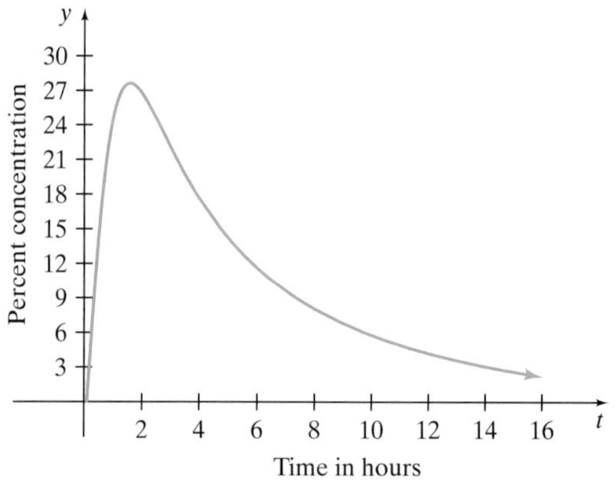

Figure 12.6.4

SUMMARY

In this section we presented the **Product Rule** and the **Quotient Rule** for computing derivatives. The differentiation rules learned so far are listed in Table 12.6.1. The letters k and n represent constants, while f and g represent differentiable functions of x.

Table 12.6.1

Name	Rule
Constant Function	$\frac{d}{dx}[k] = 0$
Power	$\frac{d}{dx}[x^n] = nx^{n-1}$
Constant Multiple	$\frac{d}{dx}[k \cdot f(x)] = k \cdot f'(x)$
Sum and Difference	$\frac{d}{dx}[f(x) \pm g(x)] = f'(x) \pm g'(x)$
Product	$\frac{d}{dx}[f(x) \cdot g(x)] = f'(x) \cdot g(x) + f(x) \cdot g'(x)$
Quotient	$\frac{d}{dx}\left[\frac{f(x)}{g(x)}\right] = \frac{f'(x) \cdot g(x) - f(x) \cdot g'(x)}{[g(x)]^2}, \quad g(x) \neq 0$

SECTION 12.6 EXERCISES

In Exercises 1–22, use the Product Rule to determine the derivative for the given function.

1. $f(x) = x^2(2x + 1)$

2. $f(x) = x^2(3x - 5)$

3. $f(x) = x^3(3x^2 + 2x - 5)$

4. $f(x) = x^3(5x^2 - 6x + 3)$

5. $y = 3x^4(2x^2 - 9x + 1)$

6. $y = 5x^3(3x^2 - 6x + 2)$

7. $y = -5x^2(3x^3 + 5x - 7)$

8. $y = -7x^3(2x^3 - 3x^2 + 8)$

9. $f(x) = (3x + 4)(2x - 1)$

10. $f(x) = (4x - 1)(x + 6)$

11. $y = (5x + 3)(3x^3 + 2x^2 + 1)$

12. $y = (2x - 1)(x^2 - 2x + 3)$

✓ **13.** $g(x) = (3x^2 - 2x + 1)(2x^2 + 5x - 7)$

14. $g(x) = (2x^2 + 5x - 1)(3x^2 - 7x + 3)$

15. $y = (2\sqrt{x} + 4x - 3)(3x - 4)$

16. $y = (3\sqrt{x} - 2x + 1)(5x + 2)$

17. $f(x) = (3x^{6/5} - 5x)(4x^{5/3} + 2x - 5)$

18. $f(x) = (2x^{4/3} + 3x)(-2x^{7/3} + 2x - 5)$

19. $f(x) = (3\sqrt{x} - 5)\left(2\sqrt{x} - \frac{1}{x^3}\right)$

20. $f(x) = (4\sqrt{x} + 2x - 6)(3\sqrt{x} + 6x)$

21. $h(x) = (x^{2/3} + x + 1)(x^{-1} + x^{-2})$

22. $h(x) = (6x^{4/3} - 2x + 3)(3x^{-1} - 4x^{-2})$

In Exercises 23–36, determine the derivative for the given function.

23. $f(x) = \dfrac{x + 2}{x + 1}$

24. $f(x) = \dfrac{3x - 4}{x - 1}$

25. $y = \dfrac{4x - 3}{2x + 1}$

26. $y = \dfrac{5x - 11}{3x - 4}$

✓ **27.** $f(x) = \dfrac{3x^2 - 5x + 1}{5x^2 + 3x + 2}$

28. $f(x) = \dfrac{-2x^2 + 6x - 5}{3x^2 + 5x + 2}$

29. $f(x) = \dfrac{3\sqrt{x} - 5}{6x - 1}$

30. $f(x) = \dfrac{4\sqrt{x} + 3}{2x + 7}$

✓ **31.** $y = \dfrac{(x^2 + 2)(x - 3)}{x - 1}$

32. $y = \dfrac{(x + 2)(x^3 - 3x^2 + 1)}{x - 2}$

33. $f(x) = \dfrac{(5x^4 + 2)(x^2 + 3)}{x - 4}$

34. $f(x) = \dfrac{(6x^3 - 2x^2 + 1)(3x - 5)}{2x + 1}$

35. $g(x) = \dfrac{1}{2}(4x^3 + 2x^2 - 3x - 5)$

36. $g(x) = \dfrac{2x^3 + 3x^2 - 7x + 1}{5}$

For Exercises 37–44:

(a) Determine the derivative.
(b) Determine an equation of the line tangent to the graph of the function at the indicated x-value.

37. $f(x) = x^2(x^2 - 5)$ at $x = 1$

38. $y = -3x^2(2x + 3)$ at $x = 3$

✓ **39.** $f(x) = (x^2 + 1)(x^3 + 1)$ at $x = 1$

40. $f(x) = (2x - 3)(x^2 + 3)$ at $x = 5$

41. $y = \dfrac{x + 2}{x - 1}$ at $x = 2$

42. $y = \dfrac{x^2 + 1}{x}$ at $x = -1$

43. $g(x) = \dfrac{3x^2 - 2x}{-2x + 3}$ at $x = -1$

44. $g(x) = \dfrac{-2x^2 + 3x}{3x - 5}$ at $x = 3$

For Exercises 45–50:

(a) Determine an equation of the line tangent to the graph of the function at the indicated x-value.
(b) Graph the function and the tangent line in the same viewing window.
(c) Use the $\dfrac{dy}{dx}$ command or DRAW TANGENT command of your calculator to check your work.

45. $f(x) = x^2(x^2 - 3)$ at $x = 2$

46. $y = -2x^2(3x + 1)$ at $x = 1$

47. $f(x) = (x^3 + 2)(x^2 + 2)$ at $x = -1$

48. $f(x) = (3x - 1)(x^2 + 1)$ at $x = -2$

49. $y = \dfrac{4x^3 - 3x}{2x + 1}$ at $x = 2$

50. $y = \dfrac{-2x^2 + 5x}{4x - 1}$ at $x = -1$

The Product Rule presented in this section can be extended to functions made up of any finite number of differentiable functions. For example, if $k(x) = f(x) \cdot g(x) \cdot h(x)$, then $k'(x) = f'(x) \cdot g(x) \cdot h(x) + f(x) \cdot g'(x) \cdot h(x) + f(x) \cdot g(x) \cdot h'(x)$. Use this extension of the Product Rule to compute derivatives for each function in Exercises 51–56.

51. $f(x) = (x+1)(x-2)(x+5)$

52. $f(x) = x^2(x^3 - 3x^2 + 1)(x - 4)$

53. $f(x) = (x+1)(2x^2 - 3)(3x + 4)$

54. $f(x) = (x-4)(3x^2 - 5)(2x - 9)$

55. $y = \sqrt{x} \cdot (2x - 1)(3x^2 + 2)$

56. $y = \sqrt[3]{x} \cdot (3x + 1)(2x^3 - 3)$

Applications

57. The monthly sales of a new computer are given by $q(t) = 30t - 0.5t^2$ hundred units per month t months after it hits the market, where $0 \le t \le 7$. Compute $q(3)$ and $q'(3)$ and interpret each.

58. Suppose that the computer described in Exercise 51 has a retail price, in dollars, given by $p(t) = 2200 - 34t^2$ in t months after it hits the market, where $0 \le t \le 7$. Compute $p(3)$ and $p'(3)$ and interpret each.

59. For the computer described in Exercises 57 and 58:

(a) Determine the revenue function $R(t)$. Do not simplify.

(b) Determine the rate of change of revenue $R'(t)$.

(c) Compute $R(3)$ and $R'(3)$ and interpret each.

60. The monthly sales of a new computer are given by $q(t) = 30t - 0.5t^2$ hundred units per month t months after it hits the market, where $0 \le t \le 7$. Compute $q(6)$ and $q'(6)$ and interpret each.

61. Suppose that the computer described in Exercise 60 has a retail price, in dollars, given by $p(t) = 2200 - 34t^2$ in t months after it hits the market, where $0 \le t \le 7$. Compute $p(6)$ and $p'(6)$ and interpret each.

62. For the computer described in Exercises 60 and 61:

(a) Determine the revenue function $R(t)$. Do not simplify.

(b) Determine the rate of change of revenue $R'(t)$.

(c) Compute $R(6)$ and $R'(6)$ and interpret each.

63. **(a)** Verify that upon multiplying and simplifying $R'(t)$ in Exercise 59 we get

$$R'(t) = 68t^3 - 3060t^2 - 2200t + 66{,}000$$

(b) Graph $R'(t)$ and use the ZERO or ROOT or SOLVE command to find where $R'(t) = 0$ in the interval $[0, 7]$.

(c) Graph R and use the $\frac{dy}{dx}$ or DRAW TANGENT command to calculate the slope of the line tangent to the graph of R at the value found in part (b). (*Hint:* It should be close to zero.)

(d) Is *revenue* maximized or minimized at the t-value determined in part (b)? Explain.

64. The monthly sales of a new DVD drive are given by $q(t) = 30t - 0.5t^2$ hundred units per month t months after being introduced on the market, where $0 \le t \le 5$. Compute $q(3)$ and $q'(3)$ and interpret each.

65. The retail price, in dollars, of the DVD drive in Exercise 64 is given by $p(t) = 220 - t^2$, where t is the number of months after being introduced on the market, where $0 \le t \le 5$. Compute $p(3)$ and $p'(3)$ and interpret each.

66. For the DVD drive described in Exercises 64 and 65:

(a) Determine the revenue function $R(t)$. Do not simplify.

(b) Determine the rate of change of revenue $R'(t)$.

(c) Compute $R(3)$ and $R'(3)$ and interpret each.

11.2 **67.** The average hourly earnings, in dollars, of employees in finance, insurance, and real estate in the United States are given by

$$g(t) = 0.48t + 9.44 \qquad 1 \le t \le 7$$

where t is the number of years since 1989. The average weekly earnings, in dollars, of employees in finance, insurance, and real estate in the United States are given by

$$f(t) = 17.5t + 336.86 \qquad 1 \le t \le 7$$

where t is the number of years since 1989 (*Source:* U.S. Bureau of Labor Statistics, www.bls.gov).

(a) Compute $\frac{f(3)}{g(3)}$ and interpret. (*Hint*: Watch the units.)

(b) Let $h(t) = \frac{f(t)}{g(t)}$. What does this function represent?

(c) Compute $h(3)$ and $h'(3)$ and interpret each.

68. The cost $C(x)$ in thousands of dollars of removing x% of a city's pollutants discharged into a lake is given by

$$C(x) = \frac{113x}{100 - x} \qquad 0 \le x < 100$$

Compute $C(50)$ and $C'(50)$ and interpret each.

69. The cost $C(x)$ in thousands of dollars of removing x% of a city's pollutants discharged into a lake is given by

$$C(x) = \frac{50x}{100 - x} \qquad 0 \le x < 100$$

Compute $C(50)$ and $C'(50)$ and interpret each.

70. The local game commission decides to stock a lake with bass. To do this, 100 bass are introduced into the lake. The population of the bass is approximated by

$$P(t) = \frac{10(10 + 7t)}{1 + 0.02t} \qquad t \ge 0$$

where t is time in months. Compute $P(5)$ and $P'(5)$ and interpret each.

71. The local game commission decides to stock a lake with bass. To do this, 200 bass are introduced into the lake. The population of the bass is approximated by

$$P(t) = \frac{20(10 + 7t)}{1 + 0.02t} \qquad t \geq 0$$

where t is time in months. Compute $P(5)$ and $P'(5)$ and interpret each.

72. A cost analyst for Jetway Airlines is trying to estimate the rate at which the total fuel cost will change during the next three years. The number of gallons of jet fuel that Jetway Airlines expects to use annually is given by $N(t) = 300 + 60t$, where $N(t)$ is in millions of gallons per year for t, the time in years, in the interval $[0, 3]$. The price per gallon of the jet fuel is expected to decrease according to $p(t) = \dfrac{100}{100 + 2t^2}$, where $p(t)$ is dollars per gallon for t, the time in years, in the interval $[0, 3]$. Determine the rate at which the airline's annual fuel costs are changing when

(a) $t = 0$ **(b)** $t = 1$ **(c)** $t = 2$

11.2 **73.** The amount of sulfur dioxide emitted into the atmosphere by the United States can be modeled by

$$SD(t) = -0.02t^3 + 0.27t^2 - 1.22t + 24.12 \qquad 1 \leq t \leq 11$$

where t is the number of years since 1984 and $SD(t)$ is the amount of sulfur dioxide, in millions of tons, released into the atmosphere by the United States. The U.S. civilian population can be modeled by

$$CP(t) = 2.57t + 232.91 \qquad 1 \leq t \leq 11$$

where t is the number of years since 1984 and $CP(t)$ is the U.S. civilian population in millions (*Sources:* U.S. Statistical Abstract, www.census.gov/statab/www and U.S. Census Bureau, www.census.gov).

(a) Compute $SD(5)$ and $CP(5)$ and interpret each.

(b) Compute $\dfrac{SD(5)}{CP(5)}$ and interpret.

(c) Define $H(t) = \dfrac{SD(t)}{CP(t)}$. This gives the per capita emission of sulfur dioxide. Compute $H'(t)$.

(d) Compute $H'(5)$ and interpret.

74. *(continuation of Exercise 73)* Graph the function H in Exercise 73 in the viewing window $[0, 12]$ by $[0.05, 0.10]$. Does H appear to be increasing or decreasing?

SECTION PROJECT

11.2 The table gives the U.S. national debt in billions of dollars and the U.S. civilian population in millions from 1980 to 1996.

Year	National Debt	U.S. Civilian Population
1980	909.1	225.621
1982	1137.3	229.995
1984	1564.7	234.110
1986	2120.6	238.412
1988	2601.3	242.817
1990	3206.6	247.758
1992	4002.1	253.426
1994	4643.7	258.960
1996	5207.3	263.998

SOURCE: U.S. Census Bureau, www.census.gov

(a) Let x represent the number of years since 1980 and y represent the national debt in billions. Enter the data into your calculator, and determine a quadratic regression model for the data.

(b) Round the coefficients to the nearest hundredth and store the regression model in y_1. Plot the data points with the regression model to see how well the curve fits the data.

(c) Compute the derivative of the quadratic equation. Evaluate the derivative at $x = 6$, $x = 12$, and $x = 16$, and interpret each.

(d) Let x represent the number of years since 1980 and y the U.S. civilian population in millions. Enter the data into your calculator, and determine a linear regression model for the data.

(e) Round the coefficients to the nearest hundredth and store the linear model in y_2. Plot the data points with the linear model to see how well the line fits the data.

(f) Compute the derivative for the linear equation. Evaluate the derivative at $x = 6$, $x = 12$, and $x = 16$ and interpret each.

(g) If we take the national debt and divide it by the U.S. civilian population, we get the per capita national debt. Define $y_3 = \dfrac{y_1}{y_2}$ and graph y_3. Notice its increasing behavior.

(h) Determine the derivative for y_3. Evaluate the derivative at $x = 6$, $x = 12$, and $x = 16$ and interpret each.

Section 12.7 Continuity and Nondifferentiability

In this final section of Chapter 12, we address what it means for a function to be *continuous* at a point (and over an interval) and when a function does not have a derivative at a point. These two topics nicely tie together the main concepts in this chapter—the limit concept and the derivative.

Continuity

A function is said to be **continuous** if its graph has no breaks in it such as holes, gaps, or jumps. This means that the graph of the function can be drawn without lifting a pencil off the paper. If the graph of a function has a hole, gap, or jump at $x = a$, we say that the function is **discontinuous** at $x = a$. There are many ways for a function to be discontinuous at $x = a$. Rather than list every possibility, we now supply a definition to determine whether a function is continuous at $x = a$.

Continuity

A function, f, is said to be continuous at the point $x = a$ if all the following are true:

(1) $f(a)$ is defined (2) $\lim_{x \to a} f(x)$ exists (3) $\lim_{x \to a} f(x) = f(a)$

Note: A function is continuous on open interval (b, c) if it is continuous for all x in the interval.

Figure 12.7.1 shows a graph of three different functions, each discontinuous at $x = a$. Notice that the function in Figure 12.7.1a violates condition 1 of our continuity definition, the function in Figure 12.7.1b violates condition 2, and the function in Figure 12.7.1c violates condition 3.

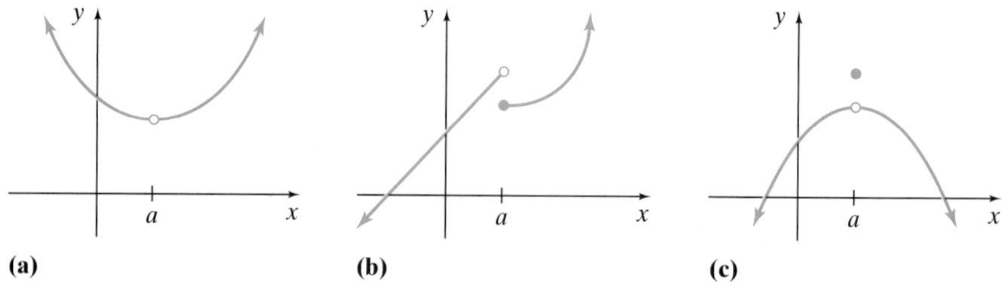

Figure 12.7.1 **(a)** $f(a)$ is not defined. **(b)** $\lim_{x \to a} f(x)$ does not exist. **(c)** $\lim_{x \to a} f(x) \neq f(a)$.

Example 1 Determining Continuity

Determine the continuity of $g(x) = \dfrac{x^2 - 4}{x - 2}$ at $x = 1$ and at $x = 2$.

Solution

According to the definition, three items must be true for continuity to exist. For $x = 1$,

1. Is $g(1)$ defined? Yes; $g(1) = \dfrac{1^2 - 4}{1 - 2} = 3$.

2. Does $\lim_{x \to 1} \frac{x^2 - 4}{x - 2}$ exist? Yes; the Substitution Principle shows that

$$\lim_{x \to 1} \frac{x^2 - 4}{x - 2} = \frac{(1)^2 - 4}{1 - 2} = 3$$

3. Does $\lim_{x \to 1} \frac{x^2 - 4}{x - 2} = g(1)$? Yes; $3 = 3$.

So we conclude that $g(x) = \frac{x^2 - 4}{x - 2}$ is continuous at $x = 1$.
For $x = 2$,

1. Is $g(2)$ defined? No!

Since $g(2)$ is not defined, hence violating part 1 of the definition of continuity, we conclude that $g(x) = \frac{x^2 - 4}{x - 2}$ is not continuous at $x = 2$. ■

Checkpoint 1

Now work Exercise 13.

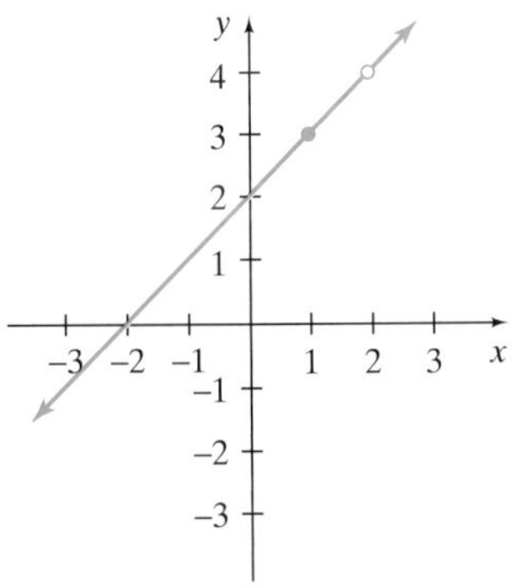

Figure 12.7.2 Graph of $g(x) = \frac{x^2 - 4}{x - 2}$.

Notice in Example 1 that $\lim_{x \to 2} \frac{x^2 - 4}{x - 2}$ exists, as seen in Figure 12.7.2. In fact, in Section 12.1 we determined that $\lim_{x \to 2} \frac{x^2 - 4}{x - 2} = 4$. But since $g(2)$ is not defined, there is a hole in the graph at the point $(2, 4)$. This type of discontinuity is said to be *removable*.

Continuity for Specific Functions

From Chapter 11 and our work thus far in Chapter 12, you may have a hunch that continuity can be determined for certain families of functions just by recognition. In fact, we can determine continuity for certain functions as indicated in the following theorem.

Theorem 12.1

1. A polynomial function is continuous for all real x.
2. A rational function is continuous for all real x except those values of x for which the denominator is 0.
3. An exponential function is continuous for all real x.
4. The natural logarithmic function is continuous for all real x in its domain.
5. A radical function, or rational exponent function, is continuous for all real x in its domain.

Example 2 Determining Intervals of Continuity

Use Theorem 12.1 to determine intervals where the following functions are continuous.

(a) $f(x) = \frac{2x + 1}{(x - 1)(2x + 1)}$ **(b)** $g(x) = (1.21)^x$ **(c)** $h(x) = \sqrt{x - 7}$

Solution

(a) Since this is a rational function, we conclude that it is continuous everywhere except at $x = 1$ and at $x = -\frac{1}{2}$, the values of x that make the denominator 0. See Figure 12.7.3. The function is continuous on

$$\left(-\infty, -\frac{1}{2}\right) \cup \left(-\frac{1}{2}, 1\right) \cup (1, \infty).$$

(b) Since this is an exponential function, it is continuous for all reals. See Figure 12.7.4.

(c) The domain of this radical function is $[7, \infty)$, so we conclude that the function is continuous on $[7, \infty)$. See Figure 12.7.5.

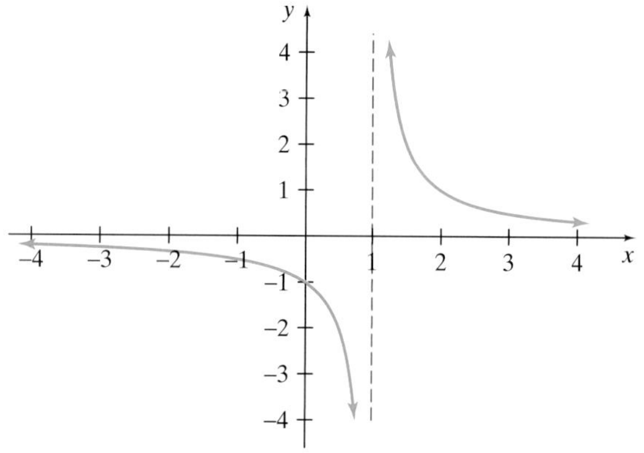

Figure 12.7.3 Graph of $f(x) = \dfrac{2x+1}{(x-1)(2x+1)}$.

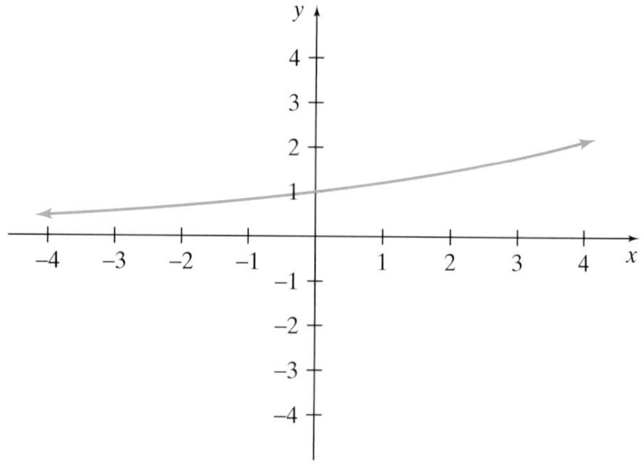

Figure 12.7.4 Graph of $g(x) = (1.21)^x$.

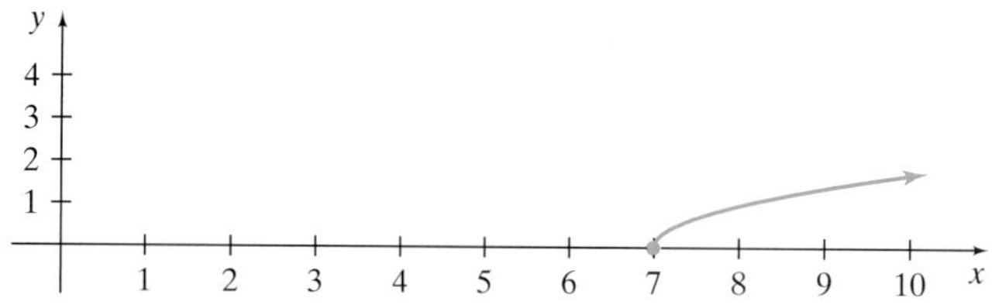

Figure 12.7.5 Graph of $h(x) = \sqrt{x - 7}$. ■

Checkpoint 2 Now work Exercise 21.

Nondifferentiable Functions

So far we have presented six basic rules for differentiation, and we will see a few more in Chapter 14. In spite of all the rules for differentiation, some functions are **nondifferentiable**. That is, some functions cannot be differentiated at certain values. (You may have noticed in some of the rules presented so far the phrase, "where f is differentiable.") We conclude this section by determining where these functions are not differentiable. We begin by analyzing the **absolute value function**, $f(x) = |x|$, whose graph is shown in Figure 12.7.6.

Recall that the definition for $f(x) = |x|$ is

$$f(x) = |x| = \begin{cases} x, & x \geq 0 \\ -x, & x < 0 \end{cases}$$

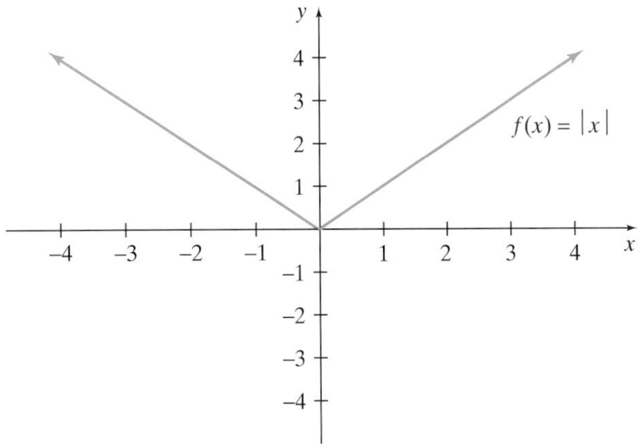

Figure 12.7.6

Example 3 Analyzing Nondifferentiability at a Point

Show that $f(x) = |x|$ is not differentiable at $x = 0$.

Solution

Since we have not given a rule for differentiating the absolute value function, we appeal to the definition of the derivative that was presented in Section 12.4. (See the Toolbox to the left.) Let's try to compute $f'(0)$ from the definition. We have

$$f'(0) = \lim_{h\to 0} \frac{f(0+h) - f(0)}{h} = \lim_{h\to 0} \frac{f(h) - f(0)}{h}$$
$$= \lim_{h\to 0} \frac{|h| - 0}{h} = \lim_{h\to 0} \frac{|h|}{h}$$

From Your Toolbox

For a function f, the derivative of f at x, denoted $f'(x)$, is defined to be

$$f'(x) = \lim_{h\to 0} \frac{f(x+h) - f(x)}{h}$$

provided that the limit exists.

In Section 12.1, we learned that if we try substitution here it produces the indeterminate form $\frac{0}{0}$. In Table 12.7.1, we numerically compute the limit.

Table 12.7.1

	$h \to 0^-$					$h \to 0^+$			
h	−1	−0.1	−0.01	−0.001	0	0.001	0.01	0.1	1
$\frac{\lvert h\rvert}{h}$	−1	−1	−1	−1		1	1	1	1

Table 12.7.1 suggests that the limit does not exist. We confirm this belief algebraically by appealing to the definition of absolute value. Specifically,

$$\lim_{h\to 0^-} \frac{|h|}{h} = \lim_{h\to 0^-} \frac{-h}{h} \qquad |h| = -h, \quad \text{for } h < 0$$
$$= \lim_{h\to 0^-} (-1) = -1 \qquad \text{Limit of a constant}$$

Interactive Activity

Graphically support that $\lim_{h\to 0}\frac{|h|}{h}$ does not exist by graphing $y=\frac{|x|}{x}$ and analyzing the graph as $x\to 0$.

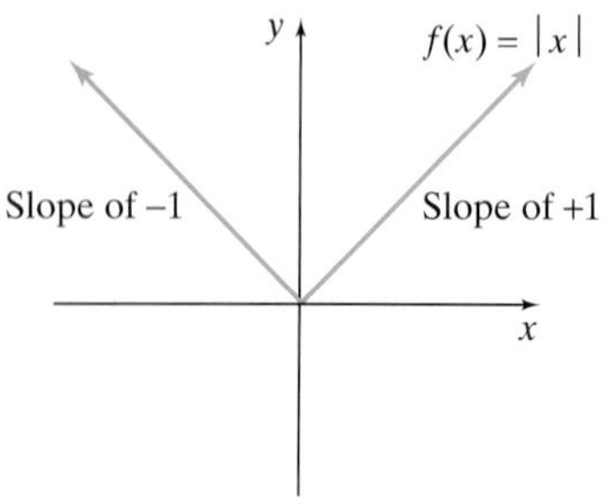

Figure 12.7.7

Similarly,

$$\lim_{h\to 0^+}\frac{|h|}{h}=\lim_{h\to 0^+}\frac{h}{h} \qquad |h|=h,\quad \text{for } h\ge 0$$

$$=\lim_{h\to 0^+}(1)=1 \qquad \text{Limit of a constant}$$

Since the *left-hand limit* does not equal the *right-hand limit*, $\lim_{h\to 0^-}\frac{|h|}{h}\neq\lim_{h\to 0^+}\frac{|h|}{h}$, we conclude that $\lim_{h\to 0}\frac{|h|}{h}$ *does not exist*. Since $\lim_{h\to 0}\frac{|h|}{h}$ does not exist and since $f'(0)=\lim_{h\to 0}\frac{|h|}{h}$, we conclude that the derivative does not exist. This is exactly what we wanted to show—that the absolute value function is not differentiable at $x=0$. Notice that $f(x)=|x|$ is continuous at $x=0$. ■

Geometrically, the reason $f(x)=|x|$ is not differentiable at $x=0$ is due to its graph consisting of two lines with slopes of -1 and $+1$ that meet at the origin. See Figure 12.7.7. The two conflicting slopes make it impossible to define a single slope at the origin.

Technology Option

To gain a different perspective on why $f(x)=|x|$ is not differentiable at $x=0$, consider the sequence of graphs in Figures 12.7.8a through e. Notice that after four applications of the ZOOM IN command the graph of $f(x)=|x|$ still has a corner at the origin. The graph has not straightened out. If a function is differentiable at a point, several applications of the ZOOM IN command will show that the graph "straightens" out and practically becomes its own tangent line.

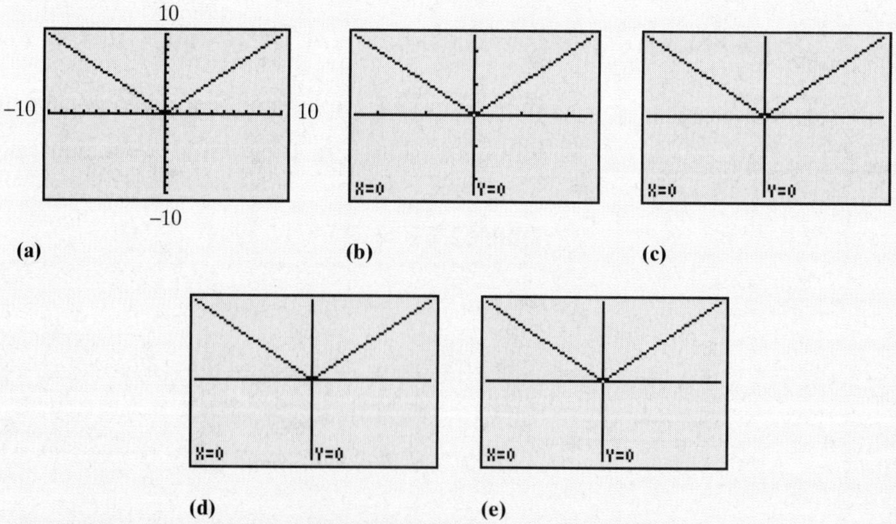

Figure 12.7.8 **(b)** After one ZOOM IN. **(c)** After second ZOOM IN. **(d)** After third ZOOM IN. **(e)** After fourth ZOOM IN.

Example 3 and Figure 12.7.7 show that if a graph has a sharp turn or a corner at a point, then the function is not differentiable at that point. In Example 4, we analyze another scenario in which a function is non-differentiable.

Example 4 Analyzing Nondifferentiability at a Point

Show that $f(x) = \sqrt[3]{x}$ is not differentiable at $x = 0$.

Solution

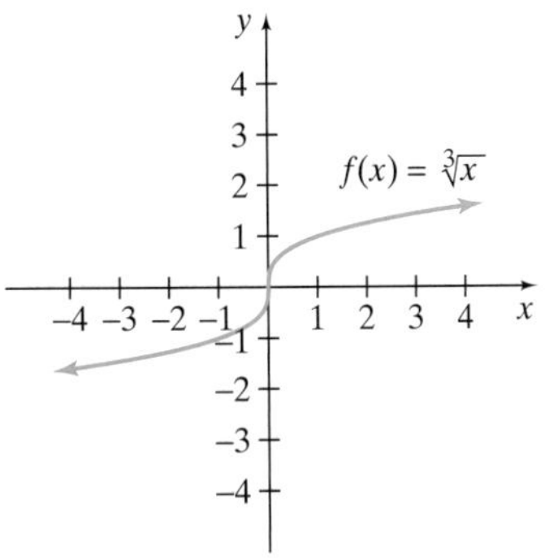

Figure 12.7.9

By looking at the graph of $f(x) = \sqrt[3]{x}$ in Figure 12.7.9, it appears that the tangent line at $x = 0$ would be a vertical line. We know the slope of a vertical line is undefined, which leads us to believe that here, $f'(0)$ is undefined. To analytically verify this, we compute the derivative by using the Power Rule.

$$f(x) = \sqrt[3]{x} = x^{1/3}$$

$$f'(x) = \frac{1}{3}x^{-2/3} = \frac{1}{3x^{2/3}} = \frac{1}{3\sqrt[3]{x^2}}$$

Remembering that f' is a function, we see that its domain is $(-\infty, 0) \cup (0, \infty)$. In other words, the domain of the derivative *excludes* zero. Since $f'(0)$ is not defined, the derivative is not defined at zero, and so our function is not differentiable at zero. Notice that $f(x) = \sqrt[3]{x}$ is continuous at $x = 0$. ■

Example 4 reminds us that the derivative is also a function. Using that fact (and being able to take a function's derivative) can aid us in determining where functions are nondifferentiable.

So far we have seen that a function is not differentiable at $x = c$ if

- The graph of the function has a sharp turn or corner at $x = c$.
- The graph of the function has a vertical tangent at $x = c$.

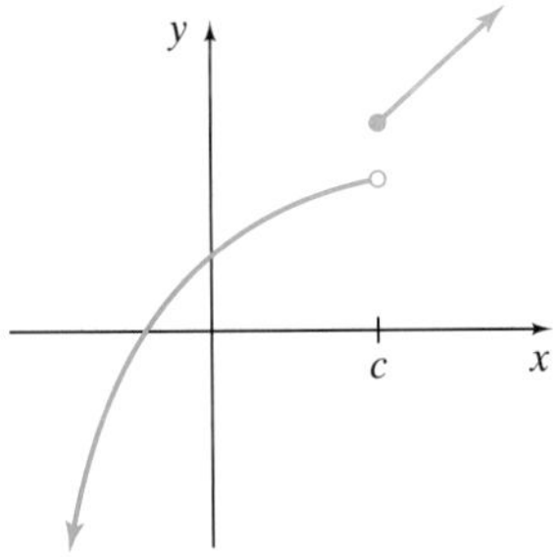

Figure 12.7.10

The third situation for which a function is not differentiable is wherever the function is *discontinuous*. The function graphed in Figure 12.7.10 is not differentiable at $x = c$ since it is discontinuous at $x = c$.

Now both functions analyzed in Examples 3 and 4 were continuous at $x = 0$, but in both cases $f'(0)$ was undefined. At this time we present the following astounding theorem.

Theorem: Differentiability Implies Continuity
If f is differentiable at $x = c$, then f is continuous at $x = c$.

A proof of this theorem is given in Appendix E. We could condense the contents of this theorem to the following:

> Differentiability implies continuity, but continuity does *not* imply differentiability.

Note that Examples 3 and 4 illustrate continuous functions that are not differentiable.

So we have determined that a function is not differentiable at $x = c$ if

- The graph of the function has a **sharp turn** or **corner** at $x = c$.
- The graph of the function has a **vertical tangent** at $x = c$.
- The graph of the function is **not continuous** at $x = c$.

Example 5 Determining Points of Nondifferentiability

For what values of x is the function graphed in Figure 12.7.11 not differentiable?

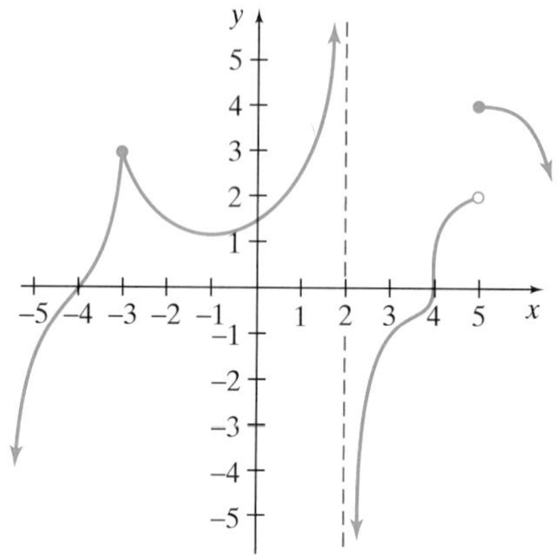

Figure 12.7.11

Solution

$f(x)$ is not differentiable at

$x = -3$	Sharp turn or corner
$x = 2$	Not continuous
$x = 4$	Vertical tangent
$x = 5$	Not continuous

■

Checkpoint 3 Now work Exercise 35.

SUMMARY

In this section we formalized the concept of **continuity**. We concluded this section by looking at **nondifferentiable functions** or, more specifically, by locating values where a function is not differentiable.

- A function, f, is said to be continuous at the point $x = a$ if all the following are true: (1) $f(a)$ is defined; (2) $\lim_{x \to a} f(x)$ exists; (3) $\lim_{x \to a} f(x) = f(a)$.

We then presented Theorem 12.1, which tells us the continuity for certain functions quickly.

- Theorem 12.1
  1. A polynomial function is continuous for all real x.
  2. A rational function is continuous for all real x except those values of x where the denominator is 0.
  3. An exponential function is continuous for all real x.
  4. The natural logarithmic function is continuous for all real x in its domain.
  5. A radical function, or rational exponent function, is continuous for all real x in its domain.
- A function is not differentiable at $x = c$ if
  1. The function has a sharp turn or corner at $x = c$.
  2. The function has a vertical tangent at $x = c$.
  3. The function is not continuous at $x = c$.

SECTION 12.7 EXERCISES

For Exercises 1–4, use the graph of f in Figure 12.7.12 to determine if f is continuous at the indicated point. If f is not continuous at the indicated point, state which of the three conditions given in the definition of continuity is violated.

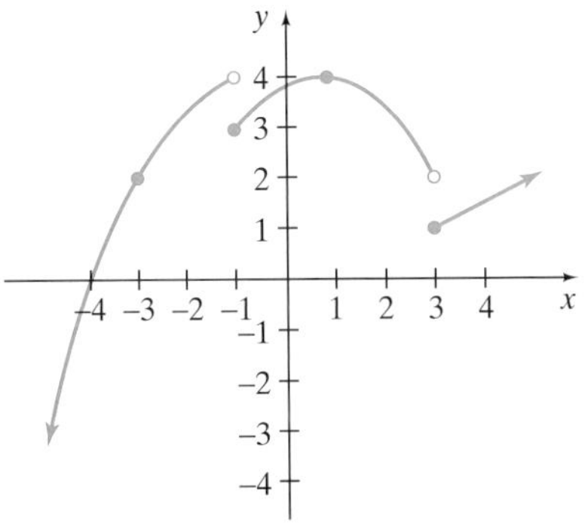

Figure 12.7.12

1. Is f continuous at $x = 1$?

2. Is f continuous at $x = -1$?

3. Is f continuous at $x = 3$?

4. Is f continuous at $x = -3$?

For Exercises 5–8, use the graph of f in Figure 12.7.13.

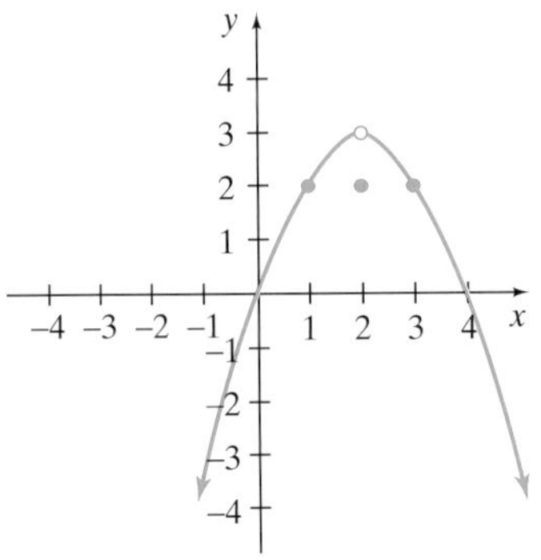

Figure 12.7.13

5. Is f continuous at $x = 1$? Explain.

6. Is f continuous at $x = 2$? Explain.

7. Is f continuous at $x = 3$? Explain.

8. Determine $f(2)$.

In Exercises 9–18, determine the continuity of the given function at the indicated point.

9. $g(x) = 3x - 2, \quad x = -2$

10. $f(x) = 2x + 6, \quad x = -3$

11. $h(x) = x^2 - x - 6, \quad x = 3$

12. $h(x) = 2x^3 + 3x^2 - 5, \quad x = 2$

13. $g(x) = \dfrac{x+1}{x-3}, \quad x = 3$

14. $f(x) = \dfrac{x-5}{x+2}, \quad x = -2$

15. $f(x) = \dfrac{x^2-25}{x-5}, \quad x = 5$

16. $g(x) = \dfrac{x^2-25}{x-5}, \quad x = 0$

17. $f(x) = \begin{cases} x+2, & x \le 1 \\ x^2+3, & x > 1 \end{cases}, \quad x = 1$

18. $f(x) = \begin{cases} x-3, & x \le 0 \\ x^2+x-3, & x > 0 \end{cases}, \quad x = 0$

In Exercises 19–32, determine intervals where the following functions are continuous. Refer to Theorem 12.1 and Example 2.

19. $f(x) = 7x^2 - 3.2x + 10.5$

20. $f(x) = 4x^3 - 2x^2 + 1.3x - 5$

21. $g(x) = \dfrac{2x-3}{x+5}$

22. $g(x) = \dfrac{3x-7}{2x+1}$

23. $f(x) = \dfrac{x+1}{(x+1)(x-3)}$

24. $f(x) = \dfrac{2x-3}{(2x-3)(x+4)}$

25. $f(x) = (1.221)^x$

26. $g(x) = (0.958)^x$

27. $f(x) = -2.1 - \ln(x+3)$

28. $f(x) = 3.2 + \ln(x+1)$

29. $h(x) = e^{-0.254x}$

30. $g(x) = e^{0.21x}$

31. $f(x) = \sqrt{2x+3}$

32. $g(x) = \sqrt{3x-2}$

For each function graphed in Exercises 33–40, state the x-values for which the derivative does not exist and explain why.

33.

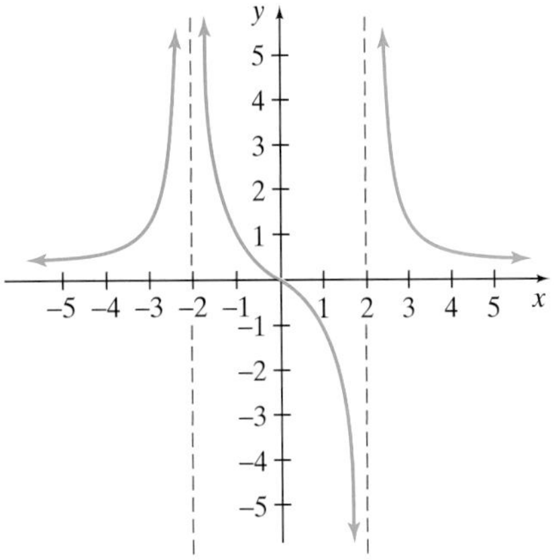

34.

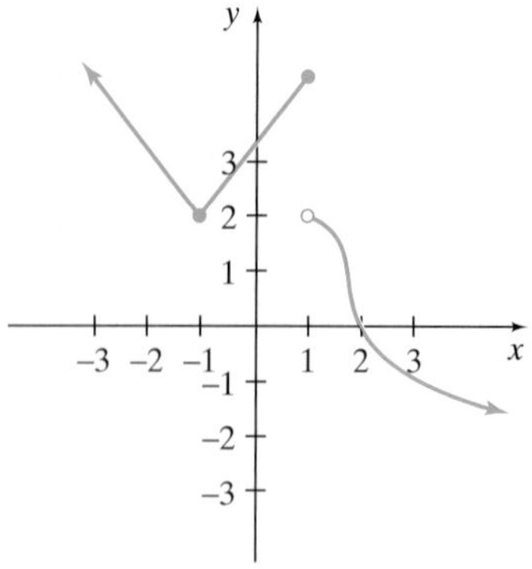

35.

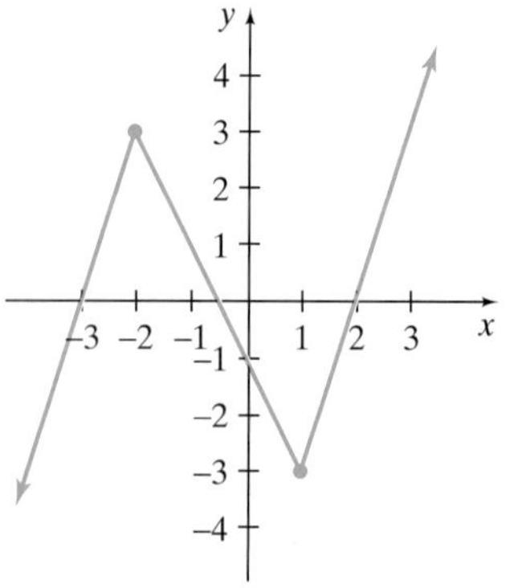

36.

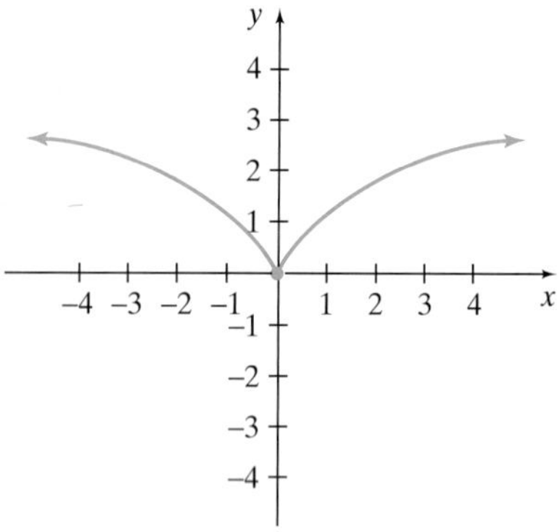

37.

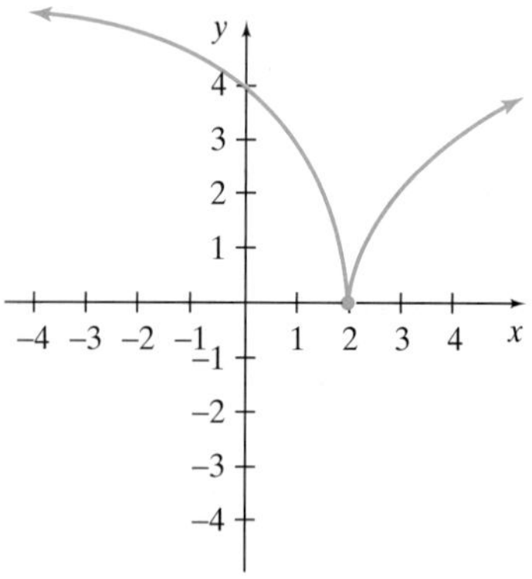

38.

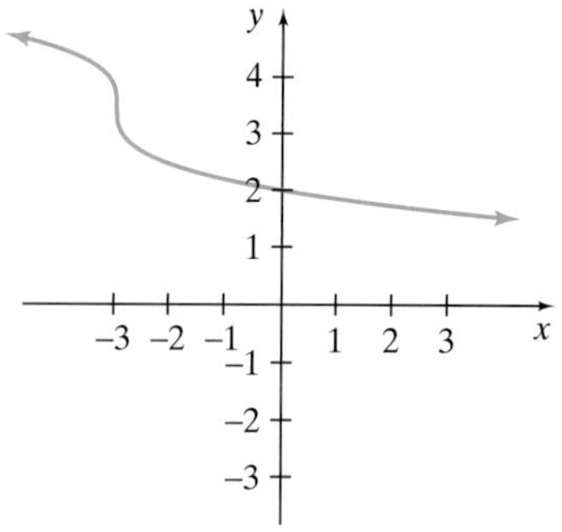

39.

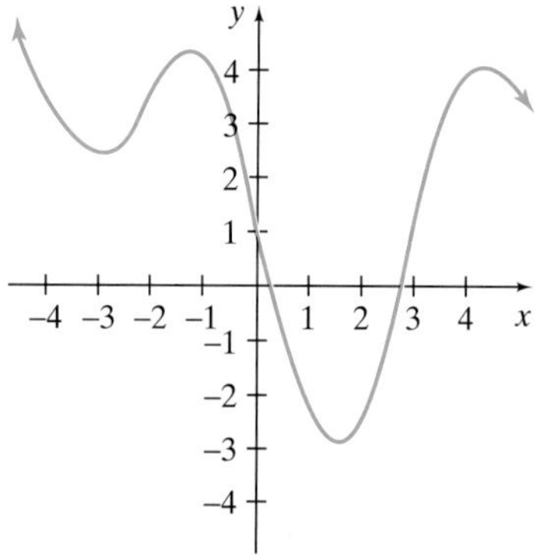

40.

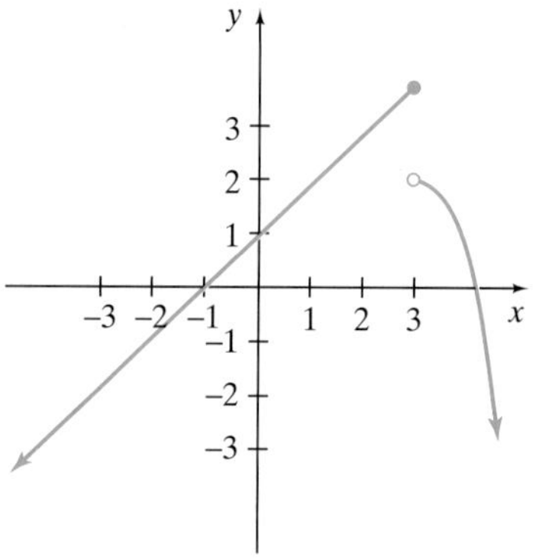

For Exercises 41–48:

(a) Determine the domain of f.

(b) Determine the derivative.

(c) Are there any x-values in the domain of f that makes the derivative undefined?

(d) For the x-values where the derivative is undefined, does the graph of the function have a sharp turn or a vertical tangent or is it discontinuous?

41. $f(x) = \sqrt{x}$

42. $f(x) = \sqrt[4]{x}$

43. $f(x) = x^{2/3}$

44. $f(x) = x^{3/5}$

45. $f(x) = \dfrac{x^2 - 1}{x + 1}$

46. $f(x) = \dfrac{x^2 - 1}{x - 1}$

47. $f(x) = \dfrac{x^2 - 4}{x - 2}$

48. $f(x) = \dfrac{x^2 - 4}{x + 2}$

Applications

49. Lisa's Lease-a-Car charges \$25 per day plus \$0.05 per mile to rent one of its medium-sized cars. Figure 12.7.14 shows the graph of the cost in dollars, $c(m)$, realized by Sofia as a function of the miles, m, that she has driven the car. Give a reason for the discontinuity at $m = 200$.

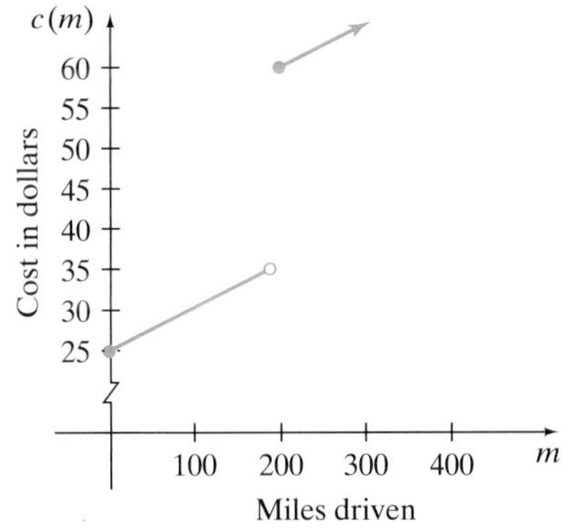

Figure 12.7.14

50. Dylan's Truck Rental Company charges \$20 per day plus \$0.10 per mile to rent a moving truck. Figure 12.7.15 shows the graph of the cost in dollars, $c(m)$, realized by Austin as a function of the miles, m, that he has driven the truck.

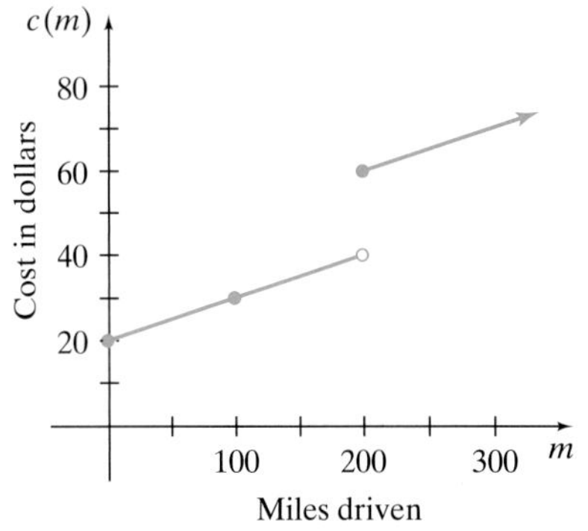

Figure 12.7.15

(a) Determine $c(100)$ and interpret.

(b) Is c continuous at $m = 100$?

(c) Determine $c(200)$ and interpret.

(d) Is c continuous at $m = 200$?

(e) Give a reason for the discontinuity at $m = 200$.

51. Finer Foods is running the following advertisement:

Hamburger

\$1.50/pound for packages less than 2 pounds
\$1.00/pound for packages of 2 pounds or more

(a) Let x represent the number of pounds of hamburger in a package. Determine a piecewise-defined function, C, that models the cost of a package of hamburger.

(b) Graph C, where $0 < x \le 6$.

(c) Determine $C(1.5)$ and interpret.

(d) Determine $C(2)$ and interpret.

(e) Does $\lim_{x \to 2} C(x)$ exist? Explain.

(f) Is C continuous at $x = 2$?

SECTION PROJECT

Consider $f(x) = \sqrt{x}$.

(a) Use the definition of the derivative to show that

$$f'(0) = \lim_{h \to 0} \frac{\sqrt{h}}{h}.$$

(b) Numerically, attempt to find the limit in part (a) by making a table for $h = 0.1, 0.001, 0.00001, 0.0000001$, and 0.00000001.

(c) Recall from Section 12.3 that what you did in part (b) was compute secant line slopes. The limit of the secant line slope is the slope of the tangent line. From part (b), do you believe that the limit exists?

(d) Graphically support your conjecture in part (b) by graphing $y = \dfrac{\sqrt{x}}{x}$ and use the ZOOM IN and TRACE commands. Does this graphical analysis support your numerical conjecture?

(e) Using the results of parts (b), (c), and (d), does the derivative of $f(x) = \sqrt{x}$ at $x = 0$ exist? Explain.

(f) Graph $f(x) = \sqrt{x}$. Explain why the slope at $x = 0$ does not exist.

Why We Learned It

In this chapter we were introduced to a new type of rate of change called an instantaneous rate of change. We learned about limits in the beginning of the chapter so that we would have the necessary mathematical tools to understand how to determine an instantaneous rate of change. We saw that instantaneous rate of change appears in many applications. For example, if we want to know at what rate the total carbon content of carbon dioxide emissions in the United States was increasing or decreasing in a certain year, say 1994, we need to compute an instantaneous rate of change. This instantaneous rate of change is determined by first computing the derivative of the function that models the situation. We also saw how the derivative can tell us the rate at which profits are growing (or falling) at a certain production level. The derivative was used to also determine the rate at which a bacteria colony was growing at a given instance in time. There are many diverse applications from many disciplines that use the derivative as a tool to measure an instantaneous rate of change. This chapter was an introduction to these many, varied applications.

CHAPTER REVIEW EXERCISES

1. For $f(x) = \dfrac{1 - \sqrt{1 - 2x - x^2}}{x}$, complete the table to numerically estimate the following.

(a) $\lim_{x \to 0^-} f(x)$ **(b)** $\lim_{x \to 0^+} f(x)$ **(c)** $\lim_{x \to 0} f(x)$

x	−0.1	−0.01	−0.001	0	0.001	0.01	0.1
$f(x)$				?			

2. For $f(x) = (\sqrt{4 - x})^5$, complete the table to numerically estimate the following.

(a) $\lim_{x \to 4^-} f(x)$ **(b)** $\lim_{x \to 4^+} f(x)$ **(c)** $\lim_{x \to 4} f(x)$

x	3	3.9	3.99	4	4.01	4.1	5
$f(x)$				?			

For Exercises 3 and 4, use your calculator to graph the given function. Use the ZOOM and TRACE commands to graphically estimate the indicated limits. Verify your estimate numerically.

3. $f(x) = x^3 - 2x;\quad \lim_{x \to 3.1} f(x)$

4. $f(x) = \sqrt{x} + \dfrac{1}{|x|};\quad \lim_{x \to 2.5} f(x)$

For Exercises 5–10, determine the indicated limit algebraically.

5. $\lim_{x \to 2} (7x^3 - 10x)$

6. $\lim_{x \to -1} \dfrac{x^2 - 1}{x + 1}$

7. $\lim_{x \to -3} |x - 5|$

8. $\lim_{x \to 0} \sqrt{36 - 8x}$

9. $\lim_{x \to 10} \dfrac{x + 10}{x^2 - 100}$

10. $\lim_{x \to 2.2} (x + 9)^3$

11. Use the graph of f to find the following:

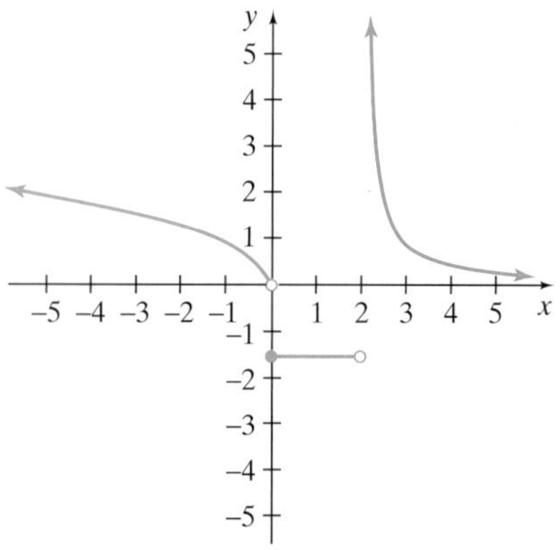

(a) $\lim_{x \to -4} f(x)$ **(b)** $\lim_{x \to 0^-} f(x)$

(c) $f(0)$ **(d)** $\lim_{x \to 2} f(x)$

(e) $\lim_{x \to 3} f(x)$ **(f)** $f(3)$

Determine the indicated limit in Exercises 12–14.

12. $\lim_{h\to 0}(x+2h)^2$

13. $\lim_{h\to 0}\dfrac{2x^2h-9h}{h}$

14. $\lim_{h\to 0}\dfrac{6x^3h^2+h}{h}$

15. For $f(x)=3x^2$, find

(a) $f(2+h)$ **(b)** $\lim_{h\to 0}\dfrac{f(2+h)-f(2)}{h}$

16. For $f(x)=\dfrac{9}{x}$, find

(a) $f(4+h)$ **(b)** $\lim_{h\to 0}\dfrac{f(4+h)-f(4)}{h}$

17. Acme Stuffed Animals, Inc., is introducing a new line of teddy bears. The total cost of producing Scare Bear (with glowing eyes) is projected to be $C(x)=36{,}000+\sqrt{10{,}000x}$, where x is the number of units made and $C(x)$ is the cost in dollars.

(a) Find and interpret $C(400)$.

(b) Find and interpret $\lim_{x\to 100} C(x)$.

(c) Find and interpret $AC(x)=\dfrac{C(x)}{x}$ for $x=25$.

18. A toy speedboat moves away from a dock, and its distance from the dock is given by

$$d(t)=\begin{cases}2t^2, & 0\le t\le 3\\ 12t-18, & 3<t\end{cases}$$

where $d(t)$ is in feet and t is in seconds.

(a) Graph d for $0\le t\le 6$.

(b) Find $\lim_{t\to 3^-} d(t)$.

(c) Find $\lim_{t\to 3^+} d(t)$.

(d) Find $\lim_{t\to 3} d(t)$.

(e) Describe the behavior of the toy speedboat.

19. A pastry chef in a commercial test kitchen is fine-tuning a pie-filling recipe. Colleagues acting as tasters have rated various recipes on a scale of 1 to 10. The average rating $R(s)$ appears to be a function of the sugar content s

$$R(s)=8-\frac{s^2-48s+512}{50}$$

where s is the number of tablespoons of sugar.

(a) Graph R in the viewing window [0, 40] by [0, 10].

(b) Find $R(20)$, $R(25)$, and $R(30)$.

(c) Find $\lim_{h\to 0}\dfrac{R(24+h)-R(24)}{h}$.

(d) How much sugar seems to be optimal?

20. A furnace switches on in such a way that the temperature of the heating element is a function of time, as shown by the graph.

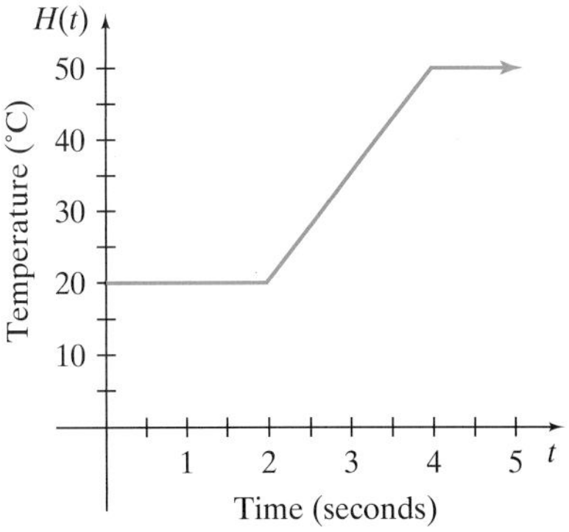

(a) Write an algebraic definition of $H(t)$.

(b) Find $\lim_{h\to 0}\dfrac{H(5+h)-H(5)}{h}$, if possible.

(c) Find $\lim_{h\to 0}\dfrac{H(2+h)-H(2)}{h}$, if possible.

(d) Explain why the middle segment of the curve is not vertical.

21. For $f(x)=\dfrac{2}{(x+3)^2}$, complete the table to numerically estimate

(a) $\lim_{x\to -3^-} f(x)$ **(b)** $\lim_{x\to -3^+} f(x)$ **(c)** $\lim_{x\to -3} f(x)$

Use the symbols ∞ and $-\infty$ as appropriate.

x	−3.1	−3.01	−3.001	−3	−2.999	−2.99	−2.9
$f(x)$							

22. For $f(x)=\dfrac{x+1}{x-2}$, complete the table to numerically estimate

(a) $\lim_{x\to 2^-} f(x)$ **(b)** $\lim_{x\to 2^+} f(x)$ **(c)** $\lim_{x\to 2} f(x)$

Use the symbols ∞ and $-\infty$ as appropriate.

x	1.9	1.99	1.999	2	2.001	2.01	2.1
$f(x)$							

For the functions given in Exercises 23 and 24:

(a) Determine the domain.

(b) Determine the indicated limit.

23. For $f(x)=\dfrac{1250+3.2x}{x}$; $\lim_{x\to 0} f(x)$

24. For $f(x)=\dfrac{x+5}{x^2-25}$; $\lim_{x\to -5} f(x)$

For the functions in Exercises 25–27:

(a) Classify f as modeling exponential growth or decay.

(b) Determine $\lim_{x\to\infty} f(x)$.

(c) Determine $\lim_{x\to -\infty} f(x)$.

25. $f(x) = 4^x$

26. $f(x) = (0.72)^x$

27. $f(x) = (2.72)^x$

In Exercises 28–31, determine the indicated limit algebraically. Verify your results numerically.

28. $\lim_{x\to-\infty} \frac{x+2}{2x-3}$

29. $\lim_{x\to\infty} \frac{-2x^2+5x-1}{x^2-13}$

30. $\lim_{x\to\infty} \frac{-4x^2-3x+11}{8x^3-5}$

31. $\lim_{x\to-\infty} \frac{3x^4-27}{x+3}$

32. Find an equation for the horizontal asymptote of the graph of $f(x) = \frac{2x^2-9x+9}{6x^2+x+11}$.

33. Find an equation for the horizontal asymptote of the graph of $f(x) = \frac{7x+9}{3x^2+2}$.

For Exercises 34–36, use the following graph of f:

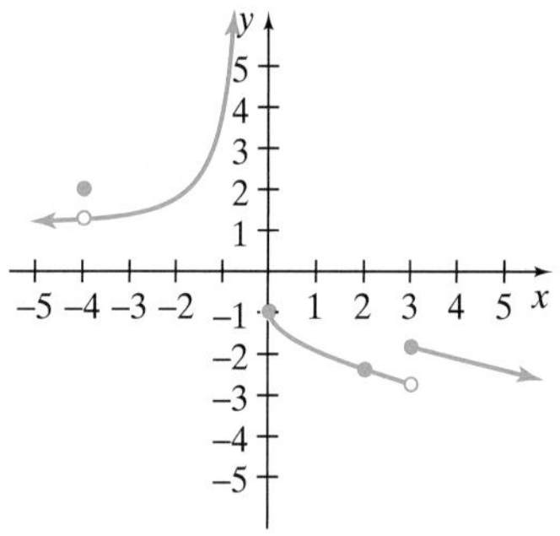

34. Is f continuous at $x = -4$? If not, why not?

35. Is f continuous at $x = 0$? If not, why not?

36. Is f continuous at $x = 2$? If not, why not?

37. Is $g(x) = \frac{x^2-9}{x+3}$ continuous at $x = 9$? At $x = -3$? Explain.

38. Is $h(x) = \begin{cases} x-8, & x \le 2 \\ \frac{6}{x-3}, & x > 2 \end{cases}$ continuous at $x = 2$? At $x = 3$? Explain.

39. Find the intervals where the function $f(x) = \frac{x-2}{(x-5)(x+3)}$ is continuous.

40. Find the intervals where the function $g(x) = 6.4 - \ln(x-9) + \sqrt{x}$ is continuous.

41. The annual cost $C(x)$, in thousands of dollars, of getting the reliability rate of an automated production quality control system up to $x\%$ is given by

$$C(x) = \begin{cases} 0, & 0 \le x \le 80 \\ \frac{400-5x}{x-100}, & 80 < x < 100 \end{cases}$$

(a) Evaluate $\lim_{x\to 80} C(x)$.

(b) Evaluate $\lim_{x\to 100^-} C(x)$.

42. A woman creates lithographs depicting the healing auras surrounding crystals. Producing q copies of a lithograph costs $C(q) = 238 + 5q$ dollars. The average cost, $AC(q)$, is given by $AC(q) = \frac{C(q)}{q}$. Find and interpret $\lim_{q\to 40} AC(q)$ and $\lim_{q\to\infty} AC(q)$.

43. In a swampy area, researchers model the springtime mosquito population using the function $N(t) = 2.4e^{0.008t}$, where t is the number of days after April 1 and $N(t)$ is population in millions.

(a) Find $N(0)$ and interpret.

(b) Find $\lim_{t\to\infty} N(t)$ and interpret. Comment on this result.

(c) How long does it take for the population to reach 5 million?

44. A bowl of soup taken out of a microwave has a temperature given by the function $C(t) = 70e^{-0.065t} + 20$, where t is time in minutes and $C(t)$ is in degrees Celsius.

(a) What is the initial temperature?

(b) What is the temperature at $t = 15$ minutes?

(c) What is the temperature of the surrounding air? (*Hint:* What temperature does the soup tend toward in the long run?)

45. A peach grower takes steps to improve the quality of his product over several years. The annual profits follow the function

$$P(t) = \frac{25(t+4)}{t+5}$$

where t is time in years and $P(t)$ is profit in thousands of dollars.

(a) Find the annual profit 3 years after the start of the improvement program.

(b) Find and interpret $\lim_{t\to\infty} P(t)$.

In Exercises 46–49, compute the average rate of change of the given function over the indicated interval.

46. $f(x) = x^3 - 6x + 1$; $[3, 7]$

47. $g(x) = 2\sqrt{4x}$; $[4, 9]$

48. $h(x) = 1.7e^{0.2x}$; $[-2, 7]$

49. $f(x) = \frac{3}{x-2}$; $[6.2, 7.9]$

50. Consider the curve shown.

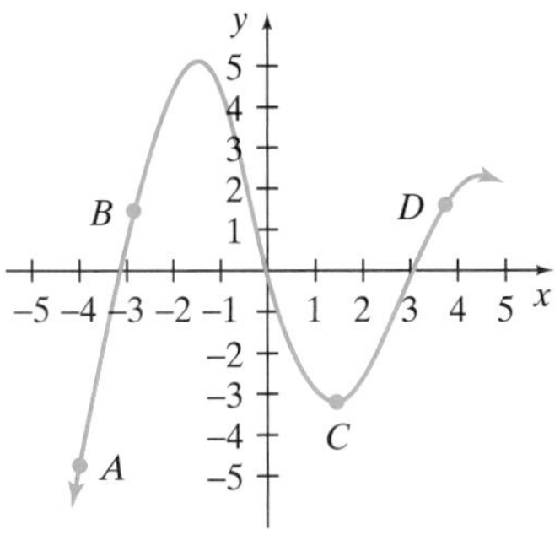

(a) Between which pairs of points is the average rate of change positive? Negative? Zero?

(b) At which points shown is the tangent line slope positive? Negative? Zero?

In Exercises 51–53, estimate the instantaneous rate of change of the function at the indicated value by using $\lim_{h\to 0}\frac{f(x+h)-f(x)}{h}$ *and completing the following tables:*

	$h \to 0^-$				
h	−1	−0.1	−0.01	−0.001	−0.0001
$\frac{f(x+h)-f(x)}{h}$					

	$h \to 0^+$				
h	1	0.1	0.01	0.001	0.0001
$\frac{f(x+h)-f(x)}{h}$					

51. $f(x) = 2.5x;\quad x = 4$

52. $f(x) = 3x^3 + 3x - 25;\quad x = -2$

53. $f(x) = 1.4e^{0.6x};\quad x = -1$

54. Find the slope of the tangent line to the graph of $f(x) = 3\sqrt{x+2}$ at $x = 4$ by estimating $\lim_{h\to 0} f(x)$ using successively smaller values for h.

In Exercises 55–57, compute $f'(c)$ *by using the definition of the derivative*

$$f'(c) = \lim_{h\to 0}\frac{f(c+h)-f(c)}{h}$$

55. $f(x) = x^2 - 2x;\quad c = 7$

56. $f(x) = x^3 + 1;\quad c = 4$

57. $f(x) = \frac{1}{x^2};\quad c = 3$

58. For $f(x) = -3x^2 - 2x$:

(a) Find an equation for the line tangent to the graph of the function at $x = -1$.

(b) Graph the function and tangent line in the same coordinate system.

In Exercises 59–62, find a formula for the general derivative by using the definition of the derivative

$$f'(x) = \lim_{h\to 0}\frac{f(x+h)-f(x)}{h}$$

59. $f(x) = 3x - 7$

60. $g(x) = x^4$

61. $h(x) = 3.4x^2 + 1.9x$

62. $f(x) = \frac{1}{x^2}$

63. Find the slope of the tangent line to the graph of $f(x) = 6x$ at $x = 4$ by first finding a general formula for the derivative.

64. Find the slope of the tangent line to the graph of $g(x) = 3x^2$ at $x = 9$ by first finding a general formula for the derivative.

65. Find the slope of the tangent line to the graph of $h(x) = \frac{1}{x}$ at $x = 7$ by first finding a general formula for the derivative.

66. The graph of a profit function is shown.

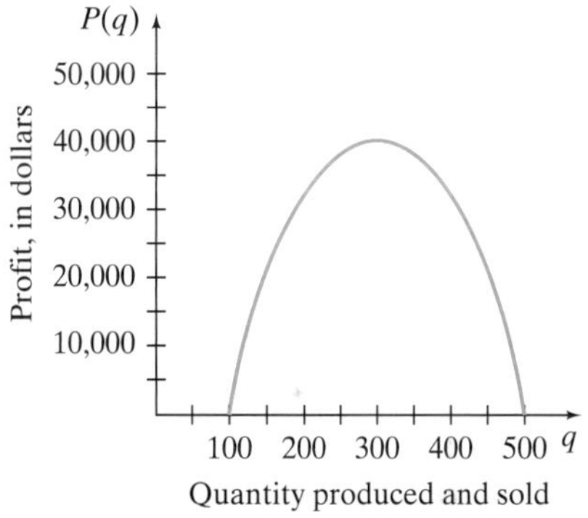

(a) For what value of q is $p(q)$ a maximum?

(b) For what value of q does $P'(q) = 0$?

(c) What is the maximum profit?

67. If \$3000 is deposited in a bank account earning 8.5% interest compounded annually, then the amount in the account after t years is given by $A(t) = 3000(1.085)^t$.

(a) Find and interpret $A(6)$.

(b) Use the definition of the derivative to approximate and interpret $A'(6)$.

68. The number of people attending a high school play is a function of the effort put into publicity, according to the equation $N(w) = 800 - \frac{8000}{w+10}$, where w is the number of person-hours (1 person working 1 hour = 1 person-hour).

(a) Given $N'(w) = \frac{8000}{(w+10)^2}$, compute $N(20)$ and $N'(20)$ and interpret each.

(b) How many people will attend if no effort at all is put into publicity?

(c) What is the maximum number of people that can be expected to attend with heavy publicity?

69. Assume that a mathematical model for the growth over time of a fish population is given by $P(t) = 20 + \sqrt{400t}$, where t is in years.

(a) Use an approximation method to find $P'(t)$ at $t = 4$.

(b) Write an equation for the line tangent to the graph of the population model at $t = 4$.

(c) Interpret the slope of the tangent line.

In Exercises 70–81, determine the derivative for the given function. Where appropriate, simplify the derivative so that there are no negative or fractional exponents.

70. $f(x) = x^8$

71. $g(x) = 3x^{1/5}$

72. $h(x) = -4x^{-3/7}$

73. $f(x) = -\frac{4}{5}x^{7/6}$

74. $g(x) = 6x^2 + 5x - 11$

75. $h(x) = -9x^3 + x + 12$

76. $f(x) = \frac{1}{2}x^3 - 3x^2 + \frac{2}{3}x + 5$

77. $g(x) = 6.23x^2 + 1.98x - 3.34$

78. $h(x) = -7.10x^3 - 5.02x^2 + 11.19x - 16.07$

79. $f(x) = 2\sqrt{x} + 7x^2 - \frac{1}{2}x^3$

80. $g(x) = \sqrt[5]{x} - \sqrt{x} + \frac{1}{x}$

81. $h(x) = 2.08x^{3.79}$

For Exercises 82 and 83:

(a) Write the domain of f in interval notation.

(b) Use the addition rule for fractions to rewrite and simplify f.

(c) Find $f'(x)$ using the differentiation rules.

(d) Write the domain of f' in interval notation.

82. $f(x) = \dfrac{3x^3 - 2x^2 + 6x + 2}{x}$

83. $f(x) = \dfrac{6x^4 + 25x^3 - 9x + 11}{x^2}$

For Exercises 84 and 85:

(a) Determine the derivative for the given function using the differentiation rules.

(b) Determine the slope of the line tangent to the graph of the function at the indicated x-value.

(c) Determine an equation for the line tangent to the graph of the function at the indicated x-value.

(d) Graph the function and the tangent line in the same viewing window.

84. $f(x) = \frac{1}{x^2}$ at $x = 4$

85. $f(x) = x^{3/4}$ at $x = 16$

86. The total amount of energy that an appliance has used since time $t = 0$ is given by

$$E(t) = 6t + 9\sqrt{t} + 0.02t^{3/2}$$

where t is in hours and $E(t)$ is in watt-hours. Find and interpret $E'\left(\frac{1}{3600}\right)$.

87. A rocket taking off has an altitude from the ground of $h(t) = 200t^2$, where $h(t)$ is in feet and t is in seconds. Find and interpret $h'(4)$.

88. Over the course of one year, the average price of homes for sale in a certain area follows the function $P(t) = -30t^2 + 2100t + 163{,}250$, where $P(t)$ is in dollars and t is in weeks after January 1.

(a) Find the average rate of change of the price from $t = 10$ weeks to $t = 30$ weeks.

(b) Graph the function on a calculator and find the maximum price and the time at which that price holds.

89. It has been said that from 1980 through the late 1990s the clock speed of commercially available home computer microprocessors doubled every 18 months. (This is called Moore's law.)

(a) Assuming that the clock speed was 1.172 MHz in 1985 ($t = 0$), write an equation for the clock speed $S(t)$ at t years after 1985.

(b) Find and interpret $S'(13)$ by using successively smaller values of h.

In Exercises 90–98, find the derivative of the given function.

90. $f(x) = x^3(6x^2 - 3x + 8)$

91. $g(x) = -3x^2(6x^3 - 2x^2 + 9x + 10)$

92. $h(x) = (2x^2 - 5x + 1)(3x^2 + 4x - 1)$

93. $f(x) = (3x^{1/2} + 5x)(-2x^{2/5} + x - 9)$

94. $g(x) = (5\sqrt{x} - 3x - 1)(4\sqrt{x} - 7x)$

95. $h(x) = \dfrac{3x + 2}{x - 1}$

96. $f(x) = \dfrac{2x^2 + 2x - 5}{3x^2 - x + 9}$

97. $g(x) = \dfrac{3\sqrt{x} - 2}{3x + 1}$

98. $h(x) = \frac{4}{3}(6x^3 - x^2 + 2x + 11)$

For Exercises 99 and 100:

(a) Determine the derivative.

(b) Determine an equation of the line tangent to the graph of the function at the indicated x-value.

(c) Graph the function and the tangent line in the same viewing window.

(d) Use the $\frac{dy}{dx}$ or DRAW TANGENT command on your calculator to check your work.

99. $f(x) = -3x^2(3x + 5)$ at $x = 2$

100. $g(x) = \dfrac{2x^2 - 3}{x}$ at $x = 2$

For each function graphed in Exercises 101 and 102, state the x-values for which the derivative does not exist.

101.

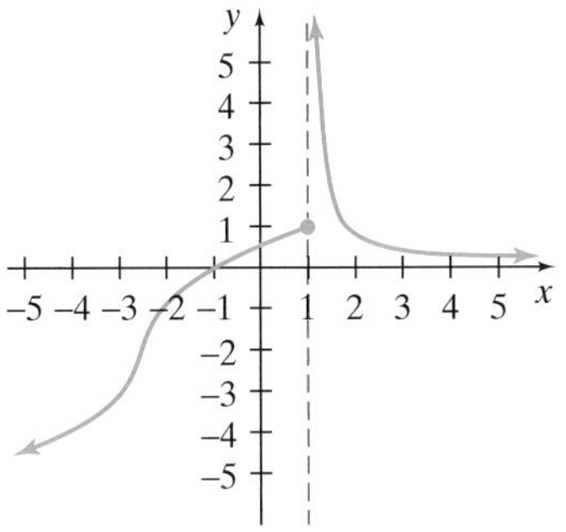

102.

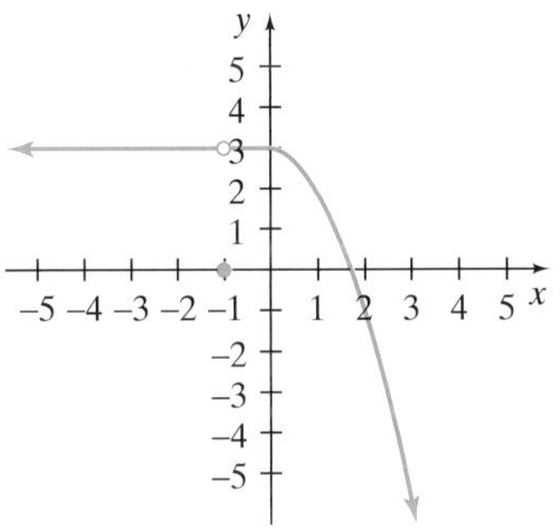

For Exercises 103 and 104:

(a) Determine the domain of f.

(b) Determine the derivative.

(c) Are there any x-values in the domain of f that makes the derivative undefined?

(d) For the x-values for which the derivative is undefined, does the graph of the function have a sharp turn or a vertical tangent or is it discontinuous?

103. $f(x) = x^{5/8}$

104. $f(x) = \dfrac{x^2 - 9}{3 - x}$

105. For $f(x) = \sqrt[4]{x^3} \cdot (3x - 1)(2x + 9)$, find $f'(x)$.

106. A musical instrument manufacturer's monthly sales of clarinets are given by $q(t) = 200e^{-0.3t}$, where t is in months after January 1.

(a) Compute $q(4)$ and interpret.

(b) Compute $q'(4)$ and interpret.

CHAPTER 12 PROJECT

SOURCE: U.S. Census Bureau, www.census.gov

In Chapter 12 we were introduced to the key elements of differential calculus such as limits, continuity, and the derivative. In this project, we will explore these topics for a piecewise defined regression function model as well as foreshadow an important application of the derivative that appears in Chapter 13. The number of live births in the United States from 1970–1998 can be modeled by

$$L(x) = \begin{cases} 24.29x^2 - 287.00x + 3{,}974.20 & 1 \le x < 11 \\ 0.94x^3 - 37.68x^2 + 518.39x + 1{,}216.66 & 11 \le x < 21 \\ 5.65x^2 - 315.62x + 8{,}316.56 & 21 \le x \le 29 \end{cases}$$

where x is the number of years since 1969 and $L(x)$ represents the number of live births in thousands.

1. Evaluate and interpret $L(10)$, $L(11)$, and $L(21)$.
2. Complete the following table to estimate $\lim_{x \to 11} L(x)$

x	10.9	10.99	10.999	11	11.001	11.01	11.1
$L(x)$				?			

3. Is L continuous at $x = 11$? Explain.
4. Evaluate and interpret $L'(10)$, $L'(11)$, and $L'(21)$.
5. Add $L(11)$ from 1 and $L'(11)$ from 4 and compare your answer to $L(12)$. Notice that it seems that the sum is approximately equal to $L(12)$.
6. Can we use the model, L, to determine the number of live births in 1999? Explain.

The $\lim_{h \to 0^-} \left(\dfrac{f(x+h) - f(x)}{h} \right)$, when it exists, is called the **left-hand derivative** of f at x. This allows us to determine if a function is differentiable at the right endpoint of a closed interval. This means that a function, f, defined on a closed interval $[a, b]$ is differentiable at b if its left-hand derivative exists at b. Let us see if our model L is differentiable at $x = 29$ by determining if the left-hand derivative exists at $x = 29$.

7. Numerically estimate if the left-hand derivative of L exists at $x = 29$ by completing the following table:

h	-0.1	-0.001	-0.00001	-0.0000001
$\dfrac{f(29+h) - L(29)}{h}$				

8. Graphically support your numerical work by graphing the difference quotient.

Chapter

13 Applications of the Derivative

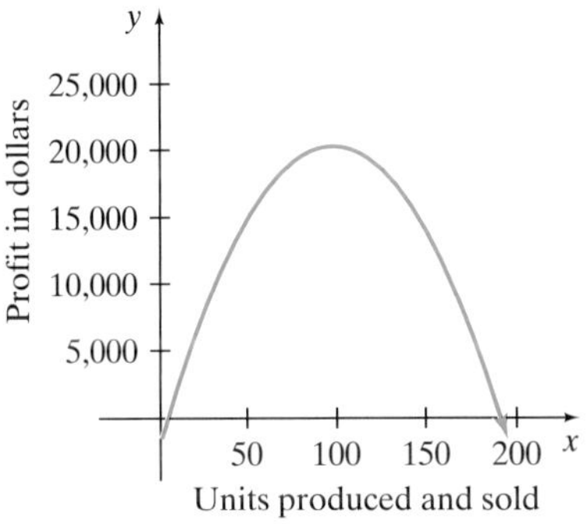

(b)

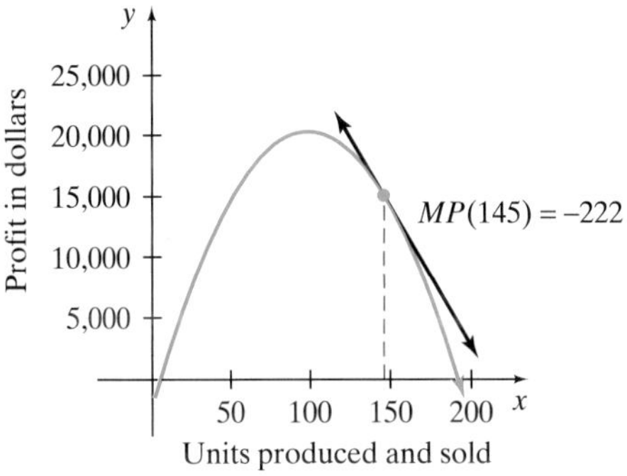

(c)

One of the factors that must be taken into consideration by business managers when determining whether or not to increase production of an item is that of marginal profit. While the goal for any successful business manager is for profit to increase, there is a point at which profit is maximized and the production of a single additional unit will result in a loss of gain in profits. Consider the case of a refrigerator manufacturer. The graph of the profit function in Figure c illustrates the profit lost by producing and selling the 146th refrigerator.

What We Know

In Chapter 12, we found that the instantaneous rate of change gave us the slope of the line tangent to the graph of a function. The way we found this new rate of change was through the derivative. We also saw applications of the derivative.

Where Do We Go

In this chapter, we will use the derivative to study a new quantity called the differential. We will use the differential to approximate solutions, to study the functions of business, and to measure errors.

Section 13.1 The Differential and Linear Approximations

How often have you heard on TV or read in the newspapers a statement such as "If crime keeps growing at this rate. . . ." The problem with this statement is that rates do indeed change. This is the reason that we study calculus—to see the impact of these fluctuating rates of change. In this section, we will examine the mathematics of the statement made above, and how to use the **differential** to make short-term conclusions about applications. Then we will study the idea of **linear approximations**, which can be used to approximate the change in the dependent variable as changes in the independent variable values are made.

The Differential

The differential is a tool used to study the relationship between changes in the independent and dependent variable values. So let's begin by revisiting some tools that we introduced with the difference quotient.

From Your Toolbox

Recall that for a function f on a closed interval $[x_1, x_2]$:

- $\Delta x = x_2 - x_1$ is called the *increment in x*. (Remember, in Chapter 12, we called this quantity h.)
- $\Delta y = f(x_2) - f(x_1)$ is called the *increment in y*.
- The slope of a secant line, m_{sec}, is given by $m_{\text{sec}} = \dfrac{\Delta y}{\Delta x} = \dfrac{f(x + \Delta x) - f(x)}{\Delta x}$.

  Recall that $\dfrac{f(x + \Delta x) - f(x)}{\Delta x}$ is called the *difference quotient*.

Now we can establish a context for our discussion through a Flashback.

Flashback

Prison Inmate Population Revisited

In Section 12.4, we modeled the total U.S. Federal Bureau of Prisons inmate population with the function

$$f(x) = \frac{1}{4}x^2 + \frac{9}{25}x + 24.5 \qquad 1 \le x \le 17$$

where x represents the number of years since 1979 and $f(x)$ represents the total inmate population in thousands (*Source:* Federal Bureau of Prisons, www.bop.gov).

(a) Evaluate $f(7) - f(6)$ and interpret this result.

(b) Determine the difference quotient with $x = 6$ and $\Delta x = 1$ and interpret. Compare this result with part (a).

Flashback Solution

(a) First, note that the values $x = 6$ and $x = 7$ correspond to the years 1985 and 1986, respectively.

$$\Delta y = f(7) - f(6)$$
$$= \left(\frac{1}{4}(7)^2 + \frac{9}{25}(7) + 24.5\right) - \left(\frac{1}{4}(6)^2 + \frac{9}{25}(6) + 24.5\right)$$
$$= 39.27 - 35.66 = 3.61$$

So from 1985 to 1986 the U.S. federal prison inmate population increased by 3.61 thousand (or 3610 inmates).

(b) Computing the difference quotient with $x = 6$ and $\Delta x = 1$,

$$m_{\text{sec}} = \frac{f(x + \Delta x) - f(x)}{\Delta x} = \frac{f(6+1) - f(6)}{1}$$
$$= \frac{f(7) - f(6)}{1} = \frac{39.27 - 35.66}{1} = 3.61$$

This means that between the years 1985 and 1986, the U.S. federal prison inmate population increased by an average rate of $3610 \frac{\text{inmates}}{\text{year}}$. Notice that the result is the same as in part (a). In other words, **when $\Delta x = 1$, then $\Delta y = 3.61 = m_{\text{sec}}$**, where m_{sec} is the slope of the secant line.

Now let's mathematically try the scenario that we suggested at the beginning of this section. What if the prison population continued to grow at a rate constant with that of 1985? Will this assumption give an accurate prediction of the 1986 U.S. federal prison population? Using the derivative and the tangent line, we can find out. First, we find the growth rate in 1985 ($x = 6$) using the derivative f', since we know that $f'(6)$ is the instantaneous rate of change at $x = 6$.

$$f'(x) = \frac{d}{dx}\left(\frac{1}{4}x^2 + \frac{9}{25}x + 24.5\right) = \frac{1}{2}x + \frac{9}{25}$$

$$f'(6) = \frac{1}{2}(6) + \frac{9}{25} = \frac{84}{25} = 3.36$$

Thus, the U.S. federal prison population was growing at a rate of $3.36 \frac{\text{thousand inmates}}{\text{year}}$ in 1985. Assuming that the rate of change is constant during 1985, we use the equation of the line tangent to the curve at $x = 6$ to predict the 1986 federal inmate population. Example 1 illustrates the process.

Example 1 Using the Tangent Line Equation

Determine an equation of the line tangent to the graph of the model $f(x) = \frac{1}{4}x^2 + \frac{9}{25}x + 24.5$ when $x = 6$. On the tangent line, determine y when $x = 7$ and interpret what this means.

Solution

Understand the Situation: We will use the derivative to determine the slope of the tangent line. Specifically, $f'(6)$ will give us the slope of the line tangent to the graph of f at $x = 6$. The point of tangency is $(6, f(6))$. Once we have the slope of the tangent line and the point of tangency, we will use the point–slope form to determine an equation of the tangent line.

Interactive Activity

What would the prison population have been in 1994 if it kept growing at the 1985 rate? To find out, graph the model $f(x) = \frac{1}{4}x^2 + \frac{9}{25}x + 24.5$ and the tangent line equation $y = 3.36(x - 6) + 35.66$ in the viewing window [0, 17] by [0, 105]. Evaluate the model and the tangent line value y when $x = 15$. Compare the results. Explain the shortcoming of using the tangent line to make long-term predictions.

WWW

Knowing that $f(6) = \frac{1}{4}(6)^2 + \frac{9}{25}(6) + 24.5 = \frac{1783}{50} = 35.66$ and $f'(6) = 3.36$, we use the point–slope form of a line to find an equation of the tangent line at $x = 6$.

$$y - y_1 = m(x - x_1)$$
$$y - 35.66 = 3.36(x - 6)$$
$$y = 3.36(x - 6) + 35.66$$

Evaluating the tangent line equation when $x = 7$ (that is, the year 1986), we get

$$y = 3.36(7 - 6) + 35.66 = 3.36(1) + 35.66 = 39.02$$

Interpret the Solution: This means that if the U.S. federal prison inmate population continued growing at the 1985 rate, the 1986 inmate population would be about 39.02 thousand. The number of inmates in 1986 using the model was $f(7) = 39.27$, meaning that the difference is only about 0.25 or 250 inmates. See Figure 13.1.1.

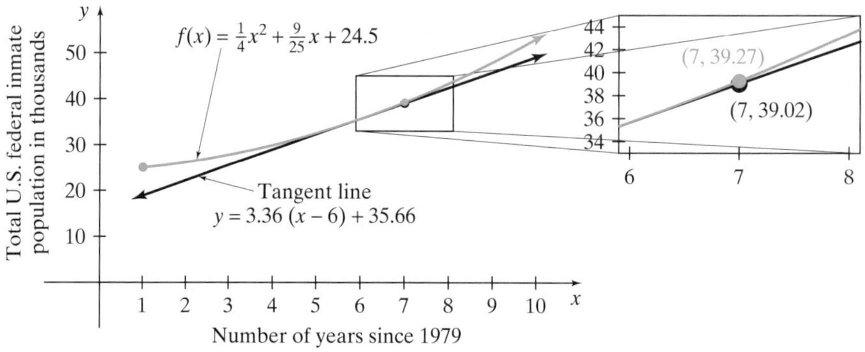

Figure 13.1.1 Graph of the line tangent to the graph of $f(x) = \frac{1}{4}x^2 + \frac{9}{25}x + 24.5$ at $x = 6$. ■

Checkpoint 1

Now work Exercise 47.

We can make two important observations from the results of the Flashback and from Example 1.

1. From part (a) of the Flashback, we found that the actual change as x changed from 6 to 7 ($\Delta x = 1$) was $\Delta y = 3.61$ thousand prisoners. We also found that $f'(6)$, the instantaneous rate of change at $x = 6$, was $3.36 \frac{\text{thousand inmates}}{\text{year}}$. Thus, for $\Delta x = 1$, Δy is approximately equal to $f'(6)$.

2. Example 1 showed that $f(7) = 39.27$ and the y-value on the tangent line at $x = 7 (y = 39.02)$ are very close. We display these observations graphically in Figure 13.1.2.

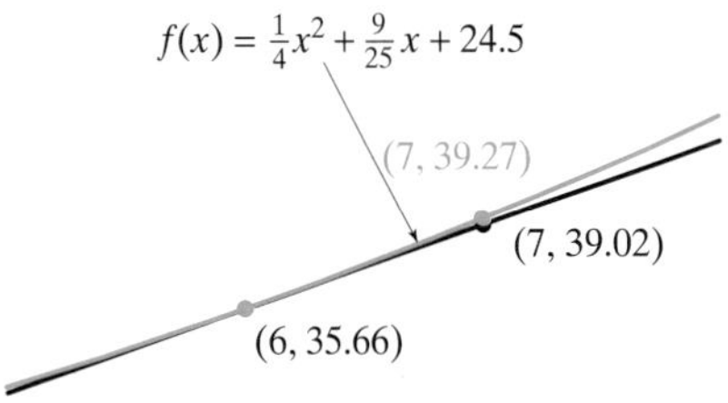

Figure 13.1.2

The concept of the *differential* is imbedded in observation 1, while the concept of a *linear approximation* is imbedded in part 2. We will discuss the linear approximation later in this section, but before that, let's examine the differential and its applications.

The Differential

If $y = f(x)$ where f is a differentiable function then we define the following:

- The **differential in x**, denoted by dx, of the independent variable is given by

$$dx = \Delta x$$

- The **differential in y**, denoted by dy, of the dependent variable is given by

$$dy = f'(x)\,dx$$

Note: Δy and dy are not the same. Δy is the *actual* change in the dependent variable values, whereas dy is an *approximation* of this change. If dx is small, then $dy \approx \Delta y$.

Many disciplines think of dy as the change in y on the tangent line. We show the difference between Δy and dy in Figure 13.1.3.

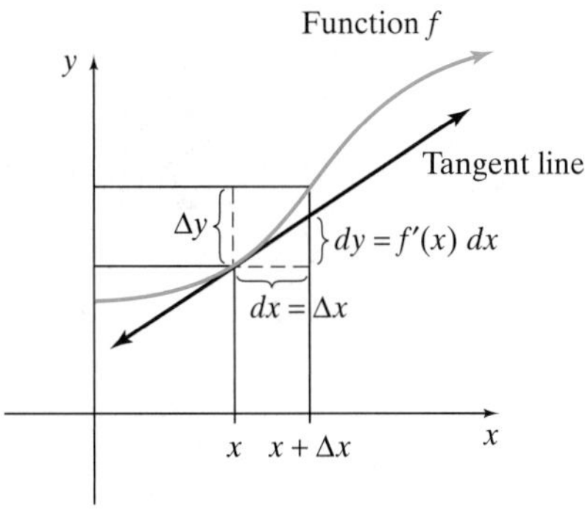

Figure 13.1.3

Example 2 Computing the Differential

For the function $f(x) = x^2 + 3x - 8$, evaluate Δy and dy for $x = 2$ and $\Delta x = dx = 0.1$.

Solution

We can compute the change in y, Δy, by evaluating $f(x + \Delta x) - f(x)$.

$$\begin{aligned}\Delta y &= f(2 + 0.1) - f(2)\\ &= f(2.1) - f(2)\\ &= ((2.1)^2 + 3(2.1) - 8) - ((2)^2 + 3(2) - 8)\\ &= 2.71 - 2 = 0.71\end{aligned}$$

Since $f'(x) = 2x + 3$, the differential in y for the function is

$$dy = f'(x)\,dx$$
$$dy = (2x + 3)\,dx$$

So at $x = 2$ and $\Delta x = dx = 0.1$ we get

$$dy = (2(2) + 3)(0.1) = 7(0.1) = 0.7$$

Notice that $dy \approx \Delta y$ in this example. Again, this is true when dx is small. ■

Checkpoint 2 Now work Exercise 17.

Our next example illustrates how differentials can be used in business applications.

Example 3 Using Differentials in Applications

The Garland Toddler Company determined that the price–demand function for their new pacifier/thermometer is given by

$$p(x) = 15 - 0.2\sqrt{x}$$

where x represents the quantity demanded and $p(x)$ represents the unit price in dollars.

(a) Compute Δp, the actual change in price, for $x = 100$ and $\Delta x = dx = 1$.

(b) Determine the differential dp for the price–demand function. Use the differential dp to approximate the change in price that would cause the quantity demanded to increase from 100 to 101 units.

Solution

(a) Using $x = 100$ and $\Delta x = dx = 1$, we get the increment in p as

$$\begin{aligned}\Delta p &= p(x + \Delta x) - p(x)\\ &= p(100 + 1) - p(100)\\ &= p(101) - p(100)\\ &\approx 12.99002 - 13\\ &= -0.00998\end{aligned}$$

(b) For the function $p(x) = 15 - 0.2\sqrt{x}$, the differential in the dependent variable p is

$$dp = p'(x)\,dx = \frac{d}{dx}(15 - 0.2\sqrt{x})\,dx$$
$$= -0.2\left(\frac{1}{2}x^{-1/2}\right)dx$$
$$= (-0.1x^{-1/2})\,dx$$

Evaluating dp when $x = 100$ and $\Delta x = dx = 101 - 100 = 1$, we get

$$dp = (-0.1(100)^{-1/2})(1)$$
$$= (-0.1)(0.1)(1) = -0.01$$

Interpret the Solution: This means that as the quantity demanded changes from 100 to 101 units, the unit price of the pacifier/thermometer would decrease by about 1 cent. ■

Notice how close the numerical solutions to parts (a) and (b) are in Example 3. This is evidence that **if dx is small, then $dy \approx \Delta y$.**

Linear Approximations

To get a better picture of why linear approximations are used, let's return to the model for the U.S. Federal Bureau of Prisons total inmate population

$$f(x) = \frac{1}{4}x^2 + \frac{9}{25}x + 24.5 \qquad 1 \le x \le 17$$

Let's investigate the graph of f around $x = 16$. We graph f in smaller and smaller intervals as shown in Figure 13.1.4. Notice that as we zoom in on f for values close to $x = 16$ the graph appears to straighten out and look like a line. Let's see if we can take advantage of this characteristic to approximate function values.

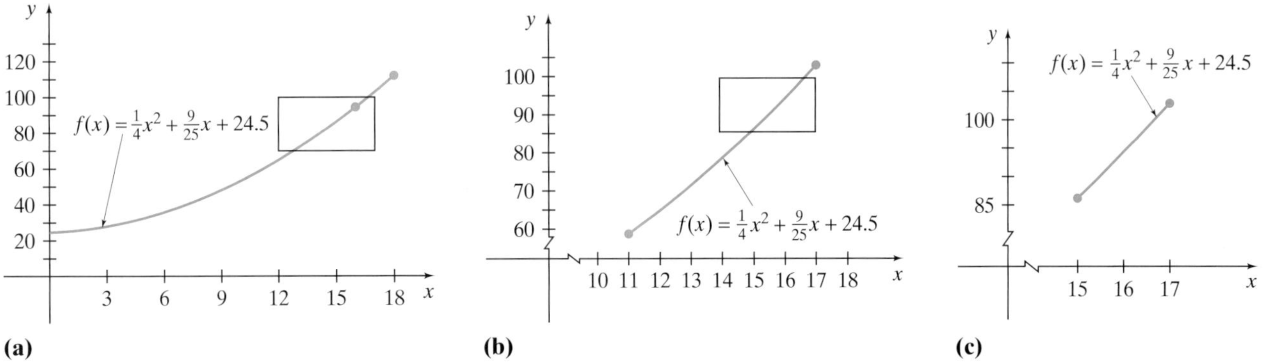

Figure 13.1.4 (a) We zoom in on that portion of the graph shown in the rectangle and redraw the graph as shown in **Figure 13.1.4 (b)** We again zoom in on that portion of the graph shown in the rectangle and redraw the graph as shown in **Figure 13.1.4 (c)** On this smaller interval the graph of f appears to be nearly linear.

Example 4 Computing a Linear Approximation

For the model of the total U.S. Federal Bureau of Prisons inmate population,

(a) Determine dy and evaluate when $x = 15$ and $\Delta x = dx = 1$.

(b) Evaluate $f(16)$ and interpret.

(c) Add $f(15)$ to the result of part (a) and compare to part (b).

Solution

(a) The differential in y for the model is

$$dy = f'(x)\,dx = \frac{d}{dx}\left(\frac{1}{4}x^2 + \frac{9}{25}x + 24.5\right)dx = \left(\frac{1}{2}x + \frac{9}{25}\right)dx$$

So for $x = 15$ and $dx = 1$, the differential in y is

$$\begin{aligned} dy &= f'(15)(1) \\ &= \left(\frac{1}{2}(15) + \frac{9}{25}\right)(1) = \frac{393}{50} = 7.86 \end{aligned}$$

(b) **Understand the Situation:** We evaluate $f(16)$ by substituting 16 for every occurrence of x in the function. Note that $x = 16$ corresponds to the year 1995.

Evaluating, we get

$$f(16) = \frac{1}{4}(16)^2 + \frac{9}{25}(16) + 24.5 = 94.26$$

Interpret the Solution: This means that in 1995 there were about 94,260 inmates in U.S. federal prisons.

(c) Evaluating the model at $x = 15$ gives

$$f(15) = \frac{1}{4}(15)^2 + \frac{9}{25}(15) + 24.5 = 86.15$$

Adding this result to part (a) gives us

$$f(15) + dy = f(15) + f'(15)(1) = 86.15 + 7.86 = 94.01$$

Notice how close this result is to part (b). In other words, $f(16) \approx f(15) + f'(15)(1)$. Numerically, the difference between this result and the result of part (b) is

$$f(16) - [f(15) + f'(15)(1)] = 94.26 - 94.01 = 0.25$$

■

Notice in Example 4 that we can rewrite $f(16)$ as $f(15+1)$. So it appears, for a small value of dx, that $f(15+1) \approx f(15) + f'(15)(1)$. This is exactly what a linear approximation is. We generalize this new tool in the following definition.

Linear Approximation

For a differentiable function f, where $y = f(x)$, the **linear approximation** of f is given by

$$f(x + dx) \approx f(x) + dy = f(x) + f'(x)\,dx$$

when the value of dx is small.

Both the linear approximation and the differential will be the cornerstone of the remainder of this chapter. Let's now try a classic application of linear approximations.

Example 5 Using Linear Approximations to Make Estimates

Use a linear approximation to estimate $\sqrt[3]{63}$, and compare with the calculator value rounded to four decimal places.

Solution

Understand the Situation: At first, it appears that we have nothing to work with to get the differential. But, on closer inspection, we see that a number very close to $\sqrt[3]{63}$ has an integer cube root, that is, $\sqrt[3]{64} = 4$. Thus, we can use the linear approximation to make an estimate of $\sqrt[3]{63}$ by letting $f(x) = \sqrt[3]{x}$, $x = 64$ (a perfect cube), and $dx = 63 - 64 = -1$.

So we want to find

$$f(x + dx) \approx f(x) + f'(x)\,dx$$
$$f(64 + (-1)) \approx f(64) + f'(64)(-1)$$

Knowing that $f(x) = \sqrt[3]{x} = x^{1/3}$, we get $f'(x) = \frac{1}{3}x^{-(2/3)} = \frac{1}{3\sqrt[3]{x^2}}$. So the linear approximation is

$$\begin{aligned}\sqrt[3]{63} &= f(64 + (-1)) \approx f(64) + f'(64)(-1)\\ &= \sqrt[3]{64} + \frac{1}{3\sqrt[3]{(64)^2}}(-1)\\ &= 4 + \frac{1}{3(16)}(-1)\\ &= 4 - \frac{1}{48} = 3\frac{47}{48}\end{aligned}$$

As a decimal number, $3\frac{47}{48} \approx 3.9792$, while $\sqrt[3]{63} \approx 3.9791$. Thus, we find that the difference using the linear approximation is only about one ten-thousandth. ■

Checkpoint 3 Now work Exercise 31.

SUMMARY

This section began with a discussion of the **differential in x** denoted by dx and the **differential in y**, where

- $dy = f'(x)\,dx$

We said that if dx is small in value then $dy \approx \Delta y$. Then we introduced the **linear approximation** and stated that, when dx is small,

- $f(x + dx) \approx f(x) + dy = f(x) + f'(x)\,dx$

SECTION 13.1 EXERCISES

In Exercises 1–16, find dy for the given function by determining $f'(x)\,dx$.

1. $y = 6x$

2. $y = -3x$

3. $f(x) = -3x^2 + 2x$

4. $f(x) = 7x^3 + 3x^2 - 13$

5. $y = \frac{5}{x - 1}$

6. $y = \frac{-2}{x + 3}$

7. $f(x) = \frac{x}{x + 1}$

8. $f(x) = \frac{2x}{x - 3}$

9. $y = \sqrt{x} + \frac{2}{x}$

10. $y = \sqrt[3]{x} - \frac{3}{x^2}$

11. $y = \frac{1}{\sqrt{x}} + \sqrt[3]{x^2}$

12. $y = \frac{1}{2\sqrt{x}} - \sqrt{x}$

13. $f(x) = \frac{x^2+1}{x^2-1}$

14. $f(x) = \frac{x^2+3}{x^2-3}$

15. $y = 3x^{1.7} + 7x^{0.8} + 3$

16. $y = 4x^{2.2} - 6x^{0.7} + 7$

In Exercises 17–28 evaluate Δy and dy for each function at the indicated values.

✓ **17.** $y = x^2 - 2x - 1, \quad x = 2, \quad \Delta x = dx = 0.1$

18. $y = x^2 + 5x, \quad x = 1, \quad \Delta x = dx = 0.2$

19. $y = 750 + 5x - 2x^2, \quad x = 50, \quad \Delta x = dx = 2$

20. $y = 1000 + 2x - 3x^2, \quad x = 100, \quad \Delta x = dx = 1$

21. $y = 100 - \frac{270}{x}, \quad x = 9, \quad \Delta x = dx = 0.5$

22. $y = 75 - \frac{150}{x}, \quad x = 5, \quad \Delta x = dx = 0.5$

23. $y = \sqrt{x}, \quad x = 2, \quad \Delta x = dx = 0.1$

24. $y = 3\sqrt{x}, \quad x = 1.5, \quad \Delta x = dx = 0.1$

25. $f(x) = \frac{x^2+1}{x^2-1}, \quad x = 2$ and $\Delta x = dx = 0.1$

26. $f(x) = \frac{x^2-5}{2x^2+1}, \quad x = 3$ and $\Delta x = dx = 0.1$

27. $y = 2x^2(3x^2 - 2x), \quad x = 1$ and $\Delta x = dx = 0.1$

28. $y = x^3(x^2 - 1), \quad x = 2$ and $\Delta x = dx = 0.1$

For Exercises 29–36, use the linear approximation to estimate the values of the given numbers. Compare to the calculator value when rounded to four decimal places.

29. $\sqrt{26}$

30. $\sqrt{8.9}$

✓ **31.** $\sqrt[3]{26}$

32. $\sqrt[3]{124}$

33. $\sqrt[4]{15.8}$

34. $\sqrt[4]{15}$

35. $\frac{2}{\sqrt{50}}$

36. $\frac{3}{\sqrt[3]{7}}$

Applications

37. The radius of a circle increases from an initial value of $r = 5$ inches by an amount $\Delta r = dr = 0.2$ inch. Estimate the corresponding increase in the circle's area by evaluating dA. The area of a circle is given by the function $A(r) = \pi r^2$.

38. The radius of a circle increases from an initial value of $r = 8$ inches by an amount $\Delta r = dr = 0.1$ inch. Estimate the corresponding increase in the circle's area by evaluating dA. The area of a circle is given by the function $A(r) = \pi r^2$.

39. The Dakorn Company determines that the annual cost of covering its employees' eye and dental insurance can be modeled by the function

$$y = 1000 + 110x^{1/2}$$

where x represents the number of employees covered and y represents the annual insurance cost in dollars. Determine dy and use the differential to approximate the increase in cost if the number of employees increases from $x = 250$ to 254.

40. In a study conducted by the Northern Aid Organization, the number of people identified as having incomes below the poverty level in a certain region of the country can be modeled by the function

$$y = 10 + 707x^{1/2}$$

where x is the total population of the region in thousands and y represents the number of people identified as having incomes below the poverty level in thousands. Determine dy and use the differential to approximate the increase in the number of people below the poverty level if the population in the region increases from $x = 20$ to 22.

41. The Paulson Motor Company estimates that the weekly cost of producing its most popular custom sport utility vehicle can be modeled by

$$y = 0.22x^3 - 2.35x^2 + 14.32x + 10.22$$

where x represents the number of custom sport utility vehicles produced each week and y represents the total cost in thousands of dollars. Determine dy and use the differential to approximate the increase in cost if the weekly production is increased from $x = 30$ to 33 vehicles.

42. The A & D Publishing Company finds that its cost of printing textbooks can be modeled by

$$y = 0.02x^3 - 0.6x^2 + 9.15x + 98.43$$

where x is the number of textbooks printed in thousands and y is the cost of printing the textbooks in thousands. Determine dy and use the differential to approximate the increase in printing cost if the number of textbooks printed each day is increased from $x = 19$ to 20.

43. Using past records, the PowerSet Company has estimated that the association between its monthly sales of volleyballs and its advertising can be modeled by

$$y = 120x - 2.4x^2$$

where x represents the amount spent on advertising in thousands of dollars and y is the number of volleyballs sold in hundreds. Determine dy and use the differential to approximate the increase in sales caused by increasing advertising from $x = 10$ to 11.

44. Repeat Exercise 43, except use the model

$$y = 189.24x - 3.5x^2.$$

45. Compute the actual change Δy in sales in Exercise 43 and compare Δy to the approximation dy.

For each of the models in Exercises 46–48:

(a) Write an equation of the tangent line at $x = 16$.
(b) Find the y-value on the tangent line when $x = 17$. Interpret what this estimate means. See Example 1.

11.2 **46.** The percentage of males and females who lived alone in the U.S. and were between the ages of 45 to 64 years old can be modeled by

$$f(x) = 0.03x^2 - 0.5x + 25.56 \qquad 1 \le x \le 20$$

where x represents the number of years since 1979 and $f(x)$ represents the percentage of males and females who lived alone and were between the ages of 45 to 64 years old (*Source:* U.S. Census Bureau, www.census.gov).

11.2 **47.** The percentage of males who lived alone in the U.S. and were between the ages of 45 to 64 years old can be modeled by

$$f(x) = 0.014x^2 - 0.09x + 9.08 \qquad 1 \le x \le 20$$

where x represents the number of years since 1979 and $f(x)$ represents the percentage of males who lived alone and were between the ages of 45 to 64 years old (*Source:* U.S. Census Bureau, www.census.gov).

48. The percentage of females who lived alone in the U.S. and were between the ages of 45 to 64 years old can be modeled by

$$f(x) = 0.019x^2 - 0.41x + 16.48 \qquad 1 \le x \le 20$$

where x represents the number of years since 1979 and $f(x)$ represents the percentage of females who lived alone and were between the ages of 45 to 64 years old (*Source:* U.S. Census Bureau, www.census.gov).

SECTION PROJECT

The average salary of major league baseball players can be modeled by

$$f(x) = 0.28x^4 - 9.27x^3 + 100.9x^2 - 314.87x + 612.87 \qquad 1 \le x \le 14$$

where x represents the number of years since 1984 and $f(x)$ represents the average salary of major league baseball players in thousands of dollars (*Source:* U.S. Statistical Abstract, www.census.gov/statab/www).

(a) Compute $f(13) - f(12)$ and interpret.

(b) Determine the differential dy for the salary function.

(c) Use the differential to approximate the change in salary as x changes from 12 to 13 and interpret.

(d) Determine an equation of the tangent line at $x = 12$.

(e) Find the y-value of the tangent line when $x = 13$ and interpret.

Section 13.2 Marginal Analysis

Many decisions made by managers in business involve analyzing the effect on the dependent variable when a small change is made to a specific independent value. For example, a company may wish to consider changing the price of an item and examining how this change affects the revenue or profit of the product. **Marginal analysis** can be defined as the study of the amount of change in the dependent variable that results from a single unit change in an independent variable. A **unit change** means a change of one single unit. This change in the dependent variable is a direct application of our now familiar tool—the derivative.

Marginal Analysis Concept

Let's start this discussion of business functions by reviewing their definitions.

From Your Toolbox

1. The *price–demand function* p gives us the price $p(x)$ at which people buy exactly x units of product.
2. The cost $C(x)$ of producing x units of a product is given by the *cost function:*

$$C(x) = (\text{variable costs}) \cdot (\text{units produced}) + (\text{fixed cost})$$

   Note that since variable costs are often expressed as a function, $C(x)$ may be a higher-order polynomial function.
3. The total revenue R generated by producing and selling x units of product at price $p(x)$ is given by the *revenue function:*

$$R(x) = (\text{quantity sold}) \cdot (\text{unit price}) = x \cdot [p(x)]$$

4. The profit P generated after producing and selling x units of a product is given by the *profit function:*

$$P(x) = \text{revenue} - \text{cost} = R(x) - C(x)$$

We referred to marginal analysis as the study of the dependent variable if the independent variable had a single unit change. Let's say that we want to study the marginal cost at a production level x, given a cost function C. We can start by determining the *actual change* in cost, denoted by ΔC, when the number of units produced is increased by 1.

$$\begin{aligned}(\text{actual change in cost}) = &(\text{cost to produce } x + 1 \text{ units})\\ &- (\text{cost to produce } x \text{ units})\end{aligned}$$

In terms of the cost function C, this is

$$\Delta C = C(x + 1) - C(x)$$

This relationship is illustrated in Figure 13.2.1.

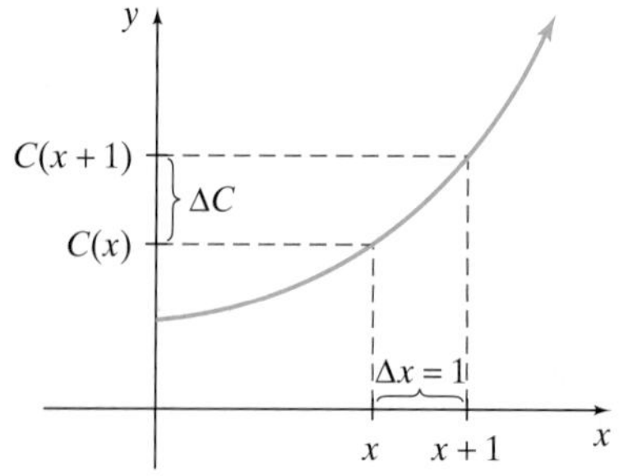

Figure 13.2.1 ΔC is the actual change in cost.

Let's try an example of computing this actual change. In Section 12.4 we found that the cost of producing x units of a recreation vehicle can be modeled by

$$C(x) = 100 + 60x + 3x^2$$

where x represents the number of vehicles produced and $C(x)$ is the cost in hundreds of dollars. To find ΔC for $x = 5$ and $\Delta x = 1$, we seek the difference in cost where

$$\begin{aligned}\Delta C &= C(x + \Delta x) - C(x)\\ &= C(5 + 1) - C(5)\\ &= C(6) - C(5)\\ &= 568 - 475 = 93 \text{ hundred dollars}\end{aligned}$$

Since $C(6)$ is the cost of producing 6 vehicles and $C(5)$ is the cost of producing the first 5 vehicles, then $\Delta C = C(6) - C(5)$ must represent the cost of producing the sixth vehicle. Thus, the cost of producing the sixth vehicle is \$9300.

An exact value can always be computed by evaluating $\Delta C = C(x + 1) - C(x)$, but in Section 13.1 we found that, when dx is small, $\Delta C \approx dy$, where dy is the differential in y given by $dy = C'(x)\,dx$. Thus, for dx being small,

$$(\text{actual change in cost, } \Delta C) \approx (\text{differential in } C\text{, called } dC)$$

But for marginal analysis, $dx = 1$. This gives us

$$\begin{pmatrix}\text{marginal cost at}\\ \text{production level } x\end{pmatrix} \approx \begin{pmatrix}\text{differential in } C,\\ \text{where } dx = 1\end{pmatrix} = C'(x) \cdot (1) = C'(x)$$

Consequently, the marginal cost function, denoted by MC is simply the derivative of the cost function. Also note that MC is the differential in the cost function where $dx = 1$. This is sometimes called a *unit differential.* The relationship of $C(x)$, ΔC, and $MC(x)$ is shown in Figure 13.2.2.

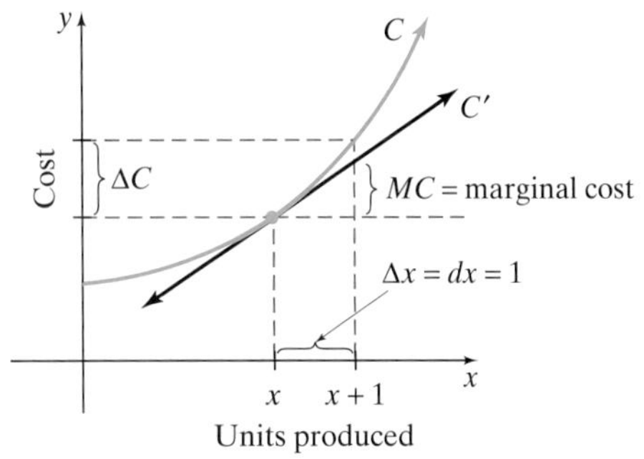

Figure 13.2.2

Marginal Cost Function

The **marginal cost function**, MC, given by $MC(x) = C'(x)$, is the approximate cost of producing one additional unit at a production level x.

Normally, when we write the units for a rate of change, we write them as $\frac{\text{dependent units}}{\text{independent units}}$. But, since $dx = 1$ in these cases, we just use the dependent units, which are usually measured in dollars.

Example 1 Computing a Marginal Cost Function

The cost of producing x units of a certain recreational vehicle can be modeled by

$$C(x) = 100 + 60x + 3x^2$$

where x represents the number of vehicles produced and $C(x)$ is the cost in hundreds of dollars. Compute the marginal cost $MC(x) = C'(x)$. Evaluate $MC(5)$ and interpret.

Solution

Computing the derivative of $C(x)$ with respect to x, we get

$$MC(x) = C'(x) = \frac{d}{dx}(100 + 60x + 3x^2) = 60 + 6x$$

Evaluating the marginal cost function at $x = 5$ yields

$$MC(5) = 60 + 6(5) = 90$$

Interpret the Solution: So the approximate cost of producing the next, or sixth vehicle is 90 hundred dollars, or \$9000. Earlier in this section, we showed that the *exact cost* of producing the sixth vehicle is \$9300, so the error of the marginal cost approximation is \$300. ■

Checkpoint 1

Now work Exercise 3.

Other Marginal Business Functions

At the beginning of this section, we stated that marginal analysis focused on the change in cost, revenue, and profit with a single unit change of the independent variable. This small change can have a significant impact on profit. This is because a change in price can cause changes in the quantity produced and sold, the cost and revenue, and the profit. See Figure 13.2.3.

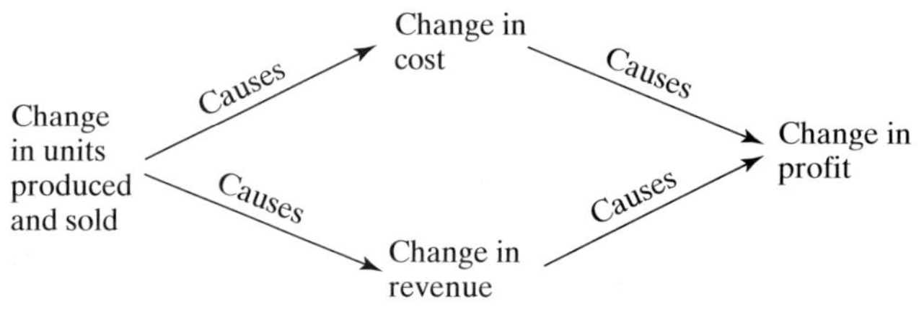

Figure 13.2.3

So it seems logical that we need to examine the marginal functions that are associated with the remainder of the business functions.

Marginal Business Functions

- The **marginal revenue function** MR, given by $MR(x) = R'(x)$, is the approximate loss or gain in revenue by producing one additional unit at a production level x.
- The **marginal profit function** MP, given by $MP(x) = P'(x)$, is the approximate loss or gain in profit by producing one additional unit at a production level x.

These definitions show that differentiating the business functions gives us the marginal business functions. Let's see how these functions can be used to affect profit.

Example 2 Using the Marginal Profit Function

The FrezMore Company has determined that its cost of producing x refrigerators can be modeled by

$$C(x) = 2x^2 + 15x + 1500 \qquad 0 \le x \le 200$$

where x is the number of refrigerators produced each week and $C(x)$ represents the weekly cost in dollars. The company also determines that the price–demand function for the refrigerators is

$$p(x) = -0.3x + 460$$

(a) Determine the profit function P for the refrigerators.

(b) Determine the marginal profit function.

(c) Compute $MP(60)$ and $MP(145)$ and interpret the results.

Solution

(a) Understand the Situation: In order to determine the profit function, we must first find the revenue function. Since the price–demand function for producing and selling x refrigerators is $p(x) = -0.3x + 460$, the revenue from selling x refrigerators is

$$R(x) = x[p(x)] = x(-0.3x + 460) = -0.3x^2 + 460x$$

The profit function can now be written as

$$\begin{aligned}
P(x) &= R(x) - C(x) \\
&= (-0.3x^2 + 460x) - (2x^2 + 15x + 1500) \\
&= -0.3x^2 + 460x - 2x^2 - 15x - 1500 \\
&= -2.3x^2 + 445x - 1500
\end{aligned}$$

(b) Differentiating the profit function to determine the marginal profit function yields

$$MP(x) = \frac{d}{dx}(-2.3x^2 + 445x - 1500) = -4.6x + 445$$

(c) Evaluating the marginal profit at $x = 60$ gives

$$MP(60) = -4.6(60) + 445 = 169$$

Interpret the Solution: This means that at a production level of $x = 60$ refrigerators each week there is about $169 profit for manufacturing and selling the 61st refrigerator.

Finally, evaluating $MP(x)$ at $x = 145$, we get

$$MP(145) = -4.6(145) + 445 = -222$$

Interpret the Solution: This means that at a production level of $x = 145$ refrigerators each week there is about $222 loss in profit for manufacturing and selling the 146th refrigerator. Figure 13.2.4 shows that the profit at $x = 145$ is decreasing at a rate of $222 \dfrac{\text{dollars}}{\text{refrigerator}}$ because the slope of the tangent line at $x = 145$ is -222.

Interactive Activity

For Example 2, use your calculator to:

- Approximate the maximum number of refrigerators that a company can make in a week and still realize a profit.

- Determine the production level x at which the profit is at a maximum. Now find the x-intercept of the graph of MP. Make a conjecture about the association between the production level that yields maximum profit and the x-intercept of the graph of MP.

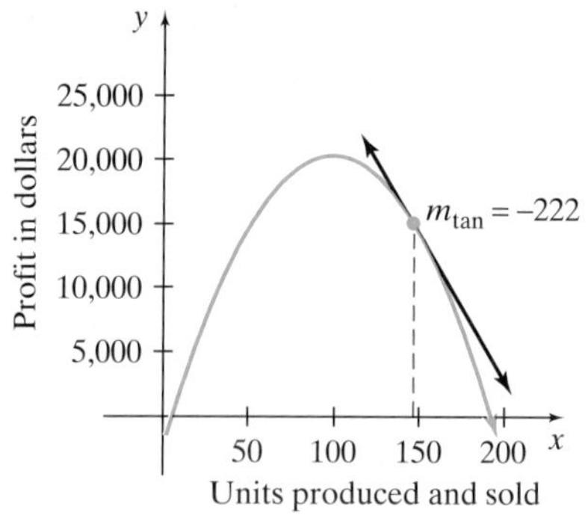

Figure 13.2.4

An extension of the marginal concept is the use of linear approximation for business functions. Recall from Section 13.1 that for a differentiable function f where $y = f(x)$, the *linear approximation* of f is given by

$$f(x + dx) \approx f(x) + dy = f(x) + f'(x)\,dx$$

when the value of dx is small.

In the case of marginal analysis, $dx = 1$. This means that we can write a linear approximation for a cost function with $f(x) = C(x)$ as

$$C(x + 1) \approx C(x) + C'(x)$$

or, in terms of the marginal cost function, as

$$C(x + 1) \approx C(x) + MC(x)$$

Consequently, we can apply the linear approximation concept to marginal analysis to easily make approximations for cost, revenue, and profit functions with information that is already known. This is particularly useful if the data given are tabular or some partial results are already available. Let's apply this new piece of information on differentials to find a linear approximation for a revenue function.

Example 3 Computing Linear Approximations for a Revenue Function

The revenue function for the production of x refrigerators per week in Example 2 was given as $R(x) = -0.3x^2 + 460x$.

(a) Compute $R(110)$ and interpret.

(b) Compute $MR(110)$ and interpret.

(c) Use the solution from parts (a) and (b) to get a linear approximation for $R(111)$. Compare your answer to the exact value of $R(111)$.

Solution

(a) For the revenue function $R(x) = -0.3x^2 + 460x$, we evaluate to get

$$\begin{aligned} R(110) &= -0.3(110)^2 + 460(110) = -0.3(12{,}100) + 50{,}600 \\ &= -3630 + 50{,}600 = 46{,}970 \end{aligned}$$

Interpret the Solution: So, when producing and selling 110 refrigerators per week, the revenue realized is \$46,970.

(b) **Understand the Situation:** We differentiate the revenue function, R, to get the marginal revenue function, MR. We then substitute $x = 110$ into the marginal revenue function.

We differentiate the revenue function to determine the marginal revenue function as follows.

$$MR(x) = R'(x) = \frac{d}{dx}(-0.3x^2 + 460x)$$
$$= -0.6x + 460$$

So the marginal revenue at a production level of $x = 110$ is

$$MR(110) = -0.6(110) + 460$$
$$= -66 + 460 = 394$$

Interpret the Solution: This means that the revenue gained from producing and selling one additional refrigerator at a production level of $x = 110$ is about \$394.

(c) **Understand the Situation:** To get an approximation for $R(111)$, use the linear approximation for the revenue function

$$R(x+1) \approx R(x) + MR(x)$$

Using this, along with the result from parts (a) and (b), we can get an approximation for $R(111)$.

This approximation is

$$R(111) = R(110+1) \approx R(110) + MR(110)$$
$$= 46{,}970 + 394 = 47{,}364$$

So our approximation for $R(111)$ is $R(111) \approx \$47{,}364$. The exact revenue using the given revenue function is $R(111) = \$47{,}363.70$, so the error from using the approximation is only \$0.30! The key to the closeness of this linear approximation is that the change in the independent variable is relatively small. ■

Average Business Functions

Many times in business situations, financial reports are simplified so that the numerical results can be easily understood. Often managers are interested in the *per unit* cost of a product, which is usually easier to work with than the total cost for the production of x units of a product. For example, it is much easier for a manager to think of the production costs at a music production company to be \$5.25 per CD or for a cycling company to think of it costing \$215 to produce each bike.

To find this per unit or *average* for the cost function, we take the total cost and divide it by the number of items produced. In words,

$$\text{per unit cost} = \frac{\text{total cost to produce } x \text{ items}}{\text{number of items produced, } x}$$

Let's see how this works through a Flashback.

Flashback

Fashion Mystique Revisited

In Section 11.3, we found that the cost function of copying promotional flyers for The Fashion Mystique was given by

$$C(x) = 0.10x + 5$$

where x represents the number of copies produced and $C(x)$ represents the cost of producing x flyers.

(a) Compute $C(100)$ and interpret.

(b) Evaluate $\frac{C(100)}{100}$ and interpret.

Flashback Solution

(a) Evaluating the cost function at $x = 100$, we get

$$C(100) = 0.10(100) + 5 = 10 + 5 = 15 \text{ dollars}$$

This means that the cost of producing 100 promotional flyers is \$15.

(b) We use the solution from part (a) to evaluate the expression $\frac{C(100)}{100}$.

$$\frac{C(100)}{100} = \frac{15}{100} = 0.15$$

Thus, at a production level of $x = 100$ flyers, the average cost is 15 cents per flyer.

The Flashback leads us to the following definition.

Average Business Functions

The **average cost function**, AC, which gives the per unit cost of producing x items, is given by

$$AC(x) = \frac{C(x)}{x}$$

The **average profit function**, AP, which gives the per unit profit of producing and selling x, is given by

$$AP(x) = \frac{P(x)}{x}$$

Note: Statistics students may remember that the average of a set of data is $\bar{x} = \frac{\text{sum of data}}{\text{number of data items}}$. This is why some textbooks denote the average cost function as $\overline{C}$.

Notice that the average revenue function is omitted from the average business functions definiton. This is because the average revenue function is $AR(x) = \frac{R(x)}{x} = \frac{x \cdot p(x)}{x} = p(x)$. So we see that the average revenue function is just another name for the price function. The average profit function is discussed in more detail in Exercises 23 and 24. Now let's take a look at an application of an average cost function.

Example 4 Analyzing an Average Cost Function

The Ventoux Athletic Shoe Manufacturing Company knows that, for its Stampeder model basketball shoes, the daily cost function can be modeled by

$$C(x) = 700\sqrt{x} + 5000 \qquad 0 \le x \le 500$$

where x is the number of pairs of shoes produced daily and $C(x)$ is the daily cost in dollars.

(a) Determine AC, the average cost function.

(b) Evaluate and interpret $C(400)$.

(c) Evaluate and interpret $AC(400)$.

Solution

(a) Using the definition of average cost function, we get

$$AC(x) = \frac{C(x)}{x} = \frac{700\sqrt{x} + 5000}{x}$$

(b) Evaluating the cost function when $x = 400$ yields

$$\begin{aligned} C(400) &= 700\sqrt{400} + 5000 \\ &= 700(20) + 5000 \\ &= 19{,}000 \end{aligned}$$

Interpret the Solution: This means that the total cost of producing 400 pairs of shoes a day is \$19,000.

(c) Evaluating $AC(x)$ when $x = 400$, we get

$$AC(400) = \frac{700\sqrt{400} + 5000}{400} = 47.5$$

Interpret the Solution: This means that the average cost of producing 400 pairs of shoes a day is \$47.50 for each pair. ■

✓ Checkpoint 2

Now work Exercise 17a.

A manager may also wish to apply marginal analysis techniques to average business functions. The result is the **marginal average business functions**. These are found by computing the derivative of each of the average business functions, respectively. The result is a function that approximates the per unit cost (or profit) of producing one more item. Now let's define these marginal average business functions.

Marginal Average Business Functions

The **marginal average cost function** approximates the per unit cost for producing an additional item of a product and is given by

$$MAC(x) = \frac{d}{dx}\left(\frac{C(x)}{x}\right)$$

The **marginal average profit function** approximates the per unit profit for producing and selling an additional item of a product and is given by

$$MAP(x) = \frac{d}{dx}\left(\frac{P(x)}{x}\right)$$

Note: To determine these marginal average profit functions, we first determine the average business function and then compute the derivative *in that order.*

The marginal average profit function is discussed in more detail in Exercises 23 and 24. Now let's take a look at an application of marginal average business functions.

Example 5 Computing a Marginal Average Cost Function

In Example 4, we found that the cost function for producing x pairs of Stampeder model basketball shoes was

$$C(x) = 700\sqrt{x} + 5000 \qquad 0 \le x \le 500$$

(a) Compute the marginal average cost function MAC.

(b) Evaluate $MAC(400)$. Round your answer to the nearest hundredth and interpret.

Solution

(a) In Example 4 we computed the average cost function as

$AC(x) = \dfrac{700\sqrt{x} + 5000}{x}$. Before differentiating, we simplify to

$$AC(x) = \frac{700\sqrt{x}}{x} + \frac{5000}{x} = \frac{700}{\sqrt{x}} + \frac{5000}{x} = 700x^{-1/2} + 5000x^{-1}.$$

Now, differentiating this average cost function with respect to x, we get

$$\begin{aligned} MAC(x) &= \frac{d}{dx}(700x^{-1/2} + 5000x^{-1}) \\ &= -350x^{-3/2} - 5000x^{-2} \end{aligned}$$

Simplifying the rational and negative exponents, we get the marginal average cost function

$$MAC(x) = -\frac{350}{\sqrt{x^3}} - \frac{5000}{x^2}$$

(b) Evaluating the result of part (a) at $x = 400$, we get

$$MAC(400) = -\frac{350}{\sqrt{(400)^3}} - \frac{5000}{(400)^2} = -0.075 \approx -0.08$$

Interpret the Solution: This means that when 400 pairs of basketball shoes have been produced, the average cost per pair decreases by about \$0.08 for an additional pair produced. Note that the negative answer tells us that the per unit cost is decreasing. ■

SUMMARY

In this section, we revisited the business functions and from them derived the marginal business functions. These functions are found by differentiation, and they approximate the cost, revenue, or profit for producing one more item of a product. Then we discussed the average business functions, which were found by taking the business function and dividing by the independent variable. Finally, we discussed the marginal average business functions, which were the derivatives of the average business functions.

Important Functions

- **Marginal cost function**: $MC(x) = C'(x)$
- **Marginal revenue function**: $MR(x) = R'(x)$
- **Marginal profit function**: $MP(x) = P'(x)$
- **Average cost function**: $AC(x) = \dfrac{C(x)}{x}$
- **Average profit function**: $AP(x) = \dfrac{P(x)}{x}$
- **Marginal average cost function**:

$$MAC(x) = \frac{d}{dx}\left(\frac{C(x)}{x}\right)$$

- **Marginal average profit function**:

$$MAP(x) = \frac{d}{dx}\left(\frac{P(x)}{x}\right)$$

SECTION 13.2 EXERCISES

For Exercises 1–6, assume $C(x)$ is in dollars and complete the following:

(a) Determine the marginal cost function MC.
(b) For the given production level x, evaluate $MC(x)$ and interpret.
(c) Evaluate the actual change in cost by evaluating $C(x+1) - C(x)$ and compare with the answer to part (b).

1. $C(x) = 23x + 5200; \quad x = 10$

2. $C(x) = 14x + 870; \quad x = 12$

✓ **3.** $C(x) = \frac{1}{2}x^2 + 12.7x + 2100; \quad x = 11$

4. $C(x) = \frac{1}{2}x^2 + 27x + 1200; \quad x = 20$

5. $C(x) = 0.2x^3 - 3x^2 + 50x + 20; \quad x = 30$

6. $C(x) = 0.08x^3 - 2x^2 + 10x + 70; \quad x = 90$

In Exercises 7–12, the cost function C and the price–demand function p are given. Assume $C(x)$ and $p(x)$ are in dollars.

(a) Determine the revenue function R and the profit function P.
(b) Determine the marginal cost function MC and the marginal profit function MP.

7. $C(x) = 5x + 500; \quad p(x) = 6$

8. $C(x) = 12x + 4500; \quad p(x) = 15$

9. $C(x) = \frac{x^2}{100} + 7x + 1000; \quad p(x) = \frac{-x}{20} + 15$

10. $C(x) = \frac{1}{100}x^2 + \frac{1}{2}x + 8; \quad p(x) = \frac{-x}{200} + 1$

11. $C(x) = -0.001x^3 + 4x + 100, 0 \le x \le 70;$
$p(x) = -0.005x + 7$

12. $C(x) = -0.002x^3 + 0.01x^2 + 2x + 50, 0 \le x \le 40;$
$p(x) = -x^{-1/2} + 5$

Applications

13. The Country Day Company determines that the daily cost of producing lawn tractor tires can be modeled by

$$C(x) = 100 + 40x - 0.001x^2 \qquad 0 \le x \le 300$$

where x represents the number of tires produced each day and $C(x)$ is the total cost, in dollars, of producing the tires. Determine MC, the marginal cost function. Evaluate and interpret $MC(200)$.

14. The Kelomata Company determines that the daily cost of producing patio swings can be modeled by

$$C(x) = 15{,}000 + 100x - 0.001x^2 \qquad 0 \le x \le 200$$

where x represents the number of patio swings produced each day and $C(x)$ represents the daily production cost in dollars. Determine MC, the marginal cost function. Evaluate and interpret $MC(100)$.

12.3 **15.** The Seas Beginning Company determines that the daily cost function of producing its Rain Forest ultra rain coat can be modeled by

$$C(x) = 1000 + 35x - 0.01x^2 \qquad 0 \le x \le 300$$

where x represents the number of rain coats produced each day and $C(x)$ is the total cost, in dollars, of producing the rain coats. Determine MC, the marginal cost function. Evaluate and interpret $MC(200)$.

12.3 **16.** The Kranky Company determines that its weekly cost function of producing automatic transmissions for automobiles can be modeled by

$$C(x) = 55{,}000 + 600x - 1.25x^2 \qquad 0 \le x \le 230$$

where x represents the number of automatic transmissions produced each week and $C(x)$ is the total cost, in dollars, of producing the automatic transmissions. Determine MC, the marginal cost function. Evaluate and interpret $MC(210)$.

17. For the cost function in Exercise 13:

(a) Determine AC, the average cost function. Evaluate and interpret $AC(200)$.

(b) Determine MAC, the marginal average cost function. Evaluate and interpret $MAC(200)$.

18. For the cost function in Exercise 14:

(a) Determine AC, the average cost function. Evaluate and interpret $AC(100)$.

(b) Determine MAC, the marginal average cost function. Evaluate and interpret $MAC(100)$.

19. The Hanash Corporation determines that the weekly profit from producing and selling its Jog-R-Radio can be modeled by

$$P(x) = -0.01x^2 + 12x - 2000 \qquad 0 \le x \le 1000$$

where x represents the number of radios produced and sold each week and $P(x)$ is the weekly profit in dollars. Determine MP, the marginal profit function. Evaluate and interpret $MP(500)$.

20. The Tesch Company determines that the weekly profit from producing and selling steering wheels for automobiles can be modeled by

$$P(x) = -0.001x^2 + 8x - 4000 \qquad 0 \le x \le 7000$$

where x represents the number of steering wheels produced and sold each week and $P(x)$ is the weekly profit in dollars. Determine MP, the marginal profit function. Evaluate and interpret $MP(3000)$.

12.3 **21.** A telemarketer determines that the monthly profit from selling magazine subscriptions can be modeled by

$$P(x) = 5x + x^{1/2} \qquad 0 \le x \le 100$$

where x is the number of magazine subscriptions sold per month and $P(x)$ is the profit in dollars. Determine MP, the marginal profit function. Evaluate and interpret $MP(55)$.

12.3 **22.** A newspaper courier determines that the monthly profit from a typical newspaper route can be modeled by

$$P(x) = 2x - x^{1/2} \qquad 0 \le x \le 200$$

where x represents the number of subscribers and $P(x)$ represents the monthly profit in dollars. Determine MP, the marginal profit function. Evaluate and interpret $MP(110)$.

23. For the profit function in Exercise 21:

(a) Determine AP, the average profit function. Evaluate and interpret $AP(55)$.

(b) Determine MAP, the marginal average profit function. Evaluate and interpret $MAP(55)$.

24. For the profit function in Exercise 22:

(a) Determine AP, the average profit function. Evaluate and interpret $AP(110)$.

(b) Determine MAP, the marginal average profit function. Evaluate and interpret $MAP(110)$.

25. The GlobalTalk Company manufactures pocket pagers and finds that the fixed and variable costs to produce q pagers are \$1200 and \$12, per pager respectively.

(a) Find the cost function C in the form $C(q) = mq + b$, and use calculus to compute the marginal cost function $MC(q) = \frac{d}{dq}C(q)$.

(b) Evaluate $MC(100)$ and $MC(150)$ and interpret these results.

(c) Why are the answers in part (b) equal?

26. The NewJoy Company toy manufacturer has just produced a new doll action set that it sells to wholesalers for \$20 each. The cost $C(x)$ in dollars to produce x action sets is given by the function $C(x) = 0.001x^2 + 4x + 5000$.

(a) Algebraically derive the profit function P and simplify it.

(b) Evaluate $P(1000)$ and interpret.

(c) Evaluate $MP(1000)$ and interpret.

27. The NewJoy Company hires a consulting firm to audit their books and consequently revise their price and cost functions to $p(x) = 23$ and $C(x) = \frac{x^2}{95} + \frac{7}{2}x + 5500$.

(a) Algebraically derive the profit function P and simplify it.

(b) Evaluate $P(500)$ and interpret.

(c) Evaluate $MP(500)$ and interpret.

28. The CraftEz Company determines that the price–demand function (in dollars) for their new make-at-home picture frame is

$$p(x) = \frac{-x}{30} + 200$$

with a cost function of

$$C(x) = 60x + 72{,}000$$

where x represents the number of frames produced and $C(x)$ is the cost in dollars.

(a) Determine the revenue function R.

(b) Determine the profit function P. Find the smallest and largest production levels x so that the company realizes a profit. (That is, find the smallest and largest independent variable values so that the revenue is greater than the cost.)

(c) Evaluate $P'(3000)$ and interpret.

29. The Vroncom Company determines that the price–demand function for their handheld computer device is

$$p(x) = \frac{-x}{30} + 300$$

They know that their fixed costs are \$150,000 and variable cost is $30\ \frac{\text{dollars}}{\text{device}}$.

(a) Determine the revenue function R and the cost function C in the form $C(x) = mx + b$.

(b) Determine the profit function P. Find the smallest and largest production levels x so that the company realizes a profit. (That is, find the smallest and largest independent values so that the revenue is greater than the cost.)

(c) Evaluate $P'(1000)$ and interpret.

For Exercises 30 and 31, consider a stockholder's report that lists the following information:

q	$C(q)$
1000	42,500
2000	57,070

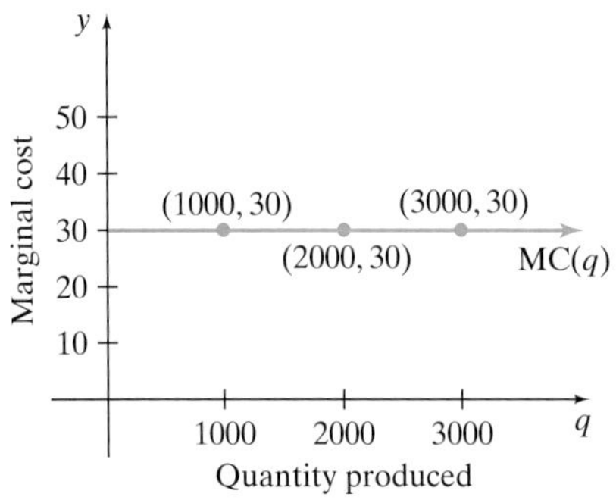

Use the table and the graph of $MC(q)$ to get a linear approximation for the given cost function. Use the approximation $C(q+1) \approx C(q) + MC(q)$.

30. $C(1001)$

31. $C(2001)$

SECTION PROJECT

Consider the following data table for the cost and revenues at various production levels for a new brand of fuel injector.

Number of Injectors Produced, x	Cost, $C(x)$	Revenue, $R(x)$
100	\$5,500	\$3,050
200	\$9,600	\$24,100
300	\$13,500	\$81,200
400	\$17,600	\$192,200

(a) Use your calculator to determine a linear regression model for the cost of producing the fuel injector in the form

$$C(x) = ax + b \qquad 100 \le x \le 400$$

where x represents the number of fuel injectors produced and $C(x)$ represents the cost in dollars.

(b) Use your calculator to determine a cubic regression model for the revenue of producing and selling the fuel injectors in the form

$$R(x) = ax^3 + bx^2 + cx + d \qquad 100 \le x \le 400$$

where x represents the number of fuel injectors produced and sold and $R(x)$ represents the revenue in dollars.

(c) Compute $MAC(x)$ and simplify the result.

(d) Evaluate $MAC(250)$ and interpret.

(e) Compute $MAP(x)$ and simplify the result.

(f) Evaluate $MAP(250)$ and interpret.

Section 13.3 Measuring Rates and Errors

Many times values are computed with differentials, particularly when those values are computed with regression models, there is a difference between the actual measurement and the computed one. In this section, we will quantify these differences using the tools of differential calculus by examining the association between the instantaneous rate of change and the error measurement by using the **relative error** and **percentage error**. We will also measure the amount of error resulting from using a linear approximation. Then we will introduce the **relative rate of change**, a measure that can be used to compare rates of different mathematical models.

Relative Error

How often do you see public opinion polls in the newspapers, on television, or on the Internet and wonder just what is meant by the *margin of error* measurement displayed below the polling results? To see how the margin of error works, let's say that the Concord Computer Company announces that its sales forecast for the next quarter is about 600,000 computers. In reality, the company does not really expect to generate *exactly* this number of sales. To account for uncertainty and volatility in the market, they assume that the actual number of sales will be in the range from 555,000 to 645,000 computers. Computing the difference between the range values and the given forecast, we get

$$555{,}000 - 600{,}000 = -45{,}000$$

and

$$645{,}000 - 600{,}000 = 45{,}000$$

This means that the error in the forecast can be written as $\pm 45{,}000$. So, if the sales forecast is 600,000 units, the error in measurement relative to the sales forecast can be computed as

$$\frac{\text{(maximal error in the forecast)}}{\text{(given value of the forecast)}} = \pm\frac{45{,}000}{600{,}000} = \pm 0.075$$

We call this number the **relative error**. To make this error measurement more informative to managers, analysts, and stockholders, we can write it as a percentage by multiplying by 100%.

$$\pm 0.075 \cdot 100\% = \pm 7.5\%$$

So the error in the forecast is read as "plus or minus 7.5%." Thus, the sales could be 7.5% more than or 7.5% less than the 600,000 units forecast. This relationship is illustrated in Figure 13.3.1.

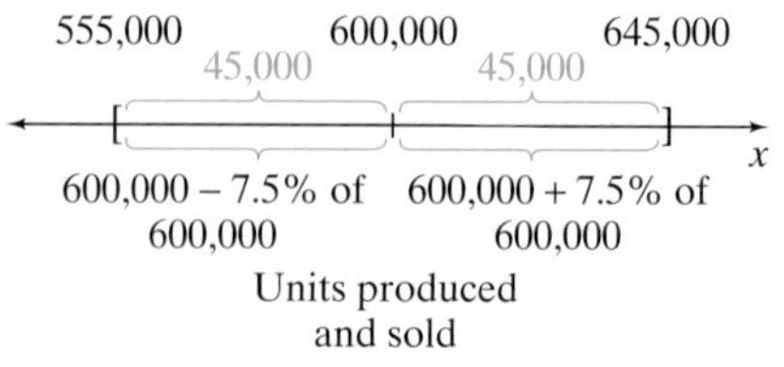

Figure 13.3.1 Sales forecast of the Concord Computer Company.

Notice in this example that we were measuring the error in the quantity produced and sold (that is, the independent variable x). By generalizing the idea illustrated in the computer company sales forecast, we can define the **relative error** and **percentage error** of the independent variable.

Relative and Percentage Error

- The **relative error** in the independent variable x, denoted by ε_x, is defined as

$$\varepsilon_x = \pm\frac{\text{(maximal error in the forecast)}}{\text{(given value of the forecast)}} = \pm\frac{dx}{x}$$

Note that, just as in Section 13.1, dx represents a change in the independent variable x.

- The **percentage error** or margin of error in the independent variable is given by

$$\varepsilon_x \cdot 100\%$$

Note: The ε is the Greek letter epsilon.

Let's take a look at an example of everyday use of this relative error in x.

Example 1 Determining a Range of Values from the Relative Error

Table 13.3.1

Should	31%
Should not	62%

SOURCE: CNN, www.cnn.com

A Gallup poll conducted in June 1997 asked the question, "Do you think the U.S. government should pass legislation that officially apologizes to American blacks for the fact that slavery was practiced before the Civil War in this country?" The results of the poll are shown in Table 13.3.1.

The methodology of the poll states that the margin of error for the poll was ±3.5%. Determine the highest and lowest values of the measure of those who think the government should not pass this legislation. Interpret the result.

Solution

Understand the Situation: Since the figures are already written as percentages, we see that in this case the relative and percentage error are the same. We are given $x = 62\%$ and a maximal error of $dx = 3.5\%$.

This gives us the lowest value, x_a, determined by the margin of error as

$$x_a = x - dx = 62\% - 3.5\% = 58.5\%$$

and the highest value, x_b, is

$$x_b = x + dx = 62\% + 3.5\% = 65.5\%$$

Interpret the Solution: This means that, based on the results of the poll, somewhere between 58.5% and 65.5%, or [58.5%, 65.5%], of the American public think that the government should not pass the legislation. ■

Interactive Activity

Using Example 1, determine the highest and lowest values of the measure of those who think that the government *should* pass legislation that officially apologizes to American blacks for the fact that slavery was practiced before the Civil War in this country.

Since we have defined the relative error for the independent variable, it makes sense to examine how the relative error affects the corresponding dependent variable. Let's say that the relative change in the dependent variable is the differential in y divided by the y-value. This quantity is

$$\varepsilon_y = \frac{dy}{y} = \frac{f'(x) \cdot dx}{f(x)}$$

Many times, in practice, the value of the relative error in x is already given, so to write this expression in terms of ε_x, we can multiply the numerator and denominator of the fraction by x to get

$$\varepsilon_y = \frac{f'(x)\cdot dx\cdot x}{f(x)\cdot x} = \frac{f'(x)\cdot x}{f(x)}\cdot\frac{dx}{x} = \frac{f'(x)\cdot x}{f(x)}\cdot\varepsilon_x$$

Now we can define the relative error in the dependent variable y in terms of ε_x.

Relative Error of the Dependent Variable

With a given value x and the relative error in the independent variable ε_x, for a differentiable function f, the **relative error in the dependent variable** $y = f(x)$, denoted by ε_y, is given by

$$\varepsilon_y = \frac{f'(x)\cdot x}{f(x)}\cdot\varepsilon_x$$

or, alternatively, without the given relative error ε_x, we can write

$$\varepsilon_y = \frac{f'(x)\cdot dx}{f(x)}$$

The **percentage error** in the dependent variable is given by

$$\varepsilon_y\cdot 100\%$$

Let's revisit the Concord Computer Company and see if we can compute a relative error for a dependent variable.

Example 2 Computing the Relative Error of the Dependent Variable

Assume that the revenue function for the sale of x computers is

$$R(x) = 0.0001x^2 + 2x$$

where $R(x)$ is in dollars.

(a) Compute the predicted revenue from the predicted sales forecast of $x = 600{,}000$ and interpret.

(b) Knowing that $\varepsilon_x = \pm 0.075$, compute ε_y, the percentage error, and interpret.

Solution

(a) The predicted revenue based on the forecast of $x = 600{,}000$ units is

$$R(600{,}000) = 0.0001(600{,}000)^2 + 2(600{,}000) = \$37{,}200{,}000$$

Interpret the Solution: This means that if the sales forecast is correct, the company will generate \$37.2 million in revenue.

(b) Knowing that the derivative of the function is $R'(x) = 0.0002x + 2$, we get the relative error based on the predicted sales.

$$\varepsilon_y = \frac{R'(x)\cdot x}{R(x)}\cdot\varepsilon_x = \frac{R'(600{,}000)\cdot 600{,}000}{R(600{,}000)}(\pm 0.075)$$

$$= \pm\frac{(0.0002(600{,}000)+2)(600{,}000)}{0.0001(600{,}000)^2 + 2(600{,}000)}(0.075)$$

$$= \pm\frac{(122)(600{,}000)}{37{,}200{,}000}(0.075) = \pm\frac{73{,}200{,}000}{37{,}200{,}000}(0.075) \approx \pm 0.148$$

As a percentage error, this value is $\pm 0.148\cdot 100\% = \pm 14.8\%$.

Interpret the Solution: This indicates that the revenues will range from 14.8% below to 14.8% above the predicted \$37.2 million, that is, in the interval [31,694,400, 42,705,600], if the sales are in the predicted range from 7.5% above to 7.5% below the 600,000 forecast, that is, in the interval [555,000, 645,000]. ■

Note: We use the "plus or minus" symbol, $\pm$, to denote the numerical *range* of the error. This sign is not used in exactly the same way that it is used in the quadratic formula. In that case, one term is added and then subtracted from another.

Many relative errors in the dependent variable result from basing estimates of cost, revenue, and profit on models, many of which are computed through means of regression. Such models lend themselves to the study of the relative error of these dependent variable values.

Interactive Activity

Graph these functions on your calculator in the viewing window [0, 200] by [−0.1, 0.1].

$$y_1 = \frac{(4x+15)(x)}{2x^2+15x+1500}(0.04)$$

$$y_2 = -y_1$$

with reference to the Flashback, what do these two functions represent? Evaluate $y_1(100)$ and $y_2(100)$ and interpret.

WWW

Flashback

Refrigerator Cost Function Revisited

In Section 13.2, we saw that the FrezMore Company finds that its cost of producing x refrigerators can be modeled by

$$C(x) = 2x^2 + 15x + 1500 \qquad 0 \le x \le 200$$

where x is the number of refrigerators produced and $C(x)$ represents the weekly cost in dollars. Assume that this cost function was modeled based on regression methods and the company's analysts determine that the percentage error in the quantity produced is 4% (that is, $\varepsilon_x = \pm 0.04$). Find the relative and percentage errors in the cost when $x = 150$ and interpret.

Flashback Solution

The relative error for the cost function is $\varepsilon_y = \dfrac{C'(x)\cdot x}{C(x)} \cdot \varepsilon_x$. Knowing that $\varepsilon_x = \pm 0.04$ and $C'(x) = 4x + 15$ with $x = 150$, we find that

$$\varepsilon_y = \frac{C'(x)\cdot x}{C(x)} \cdot \varepsilon_x = \frac{C'(150)\cdot(150)}{C(150)}(\pm 0.04)$$

To avoid confusion, we can move the $\pm$ sign to the front of the fraction to get

$$= \pm\frac{C'(150)\cdot(150)}{C(150)}(0.04) = \pm\frac{(615)(150)}{48,750}(0.04)$$

$$= \pm\frac{92,250}{48,750}(0.04) \approx \pm 0.076$$

As a percentage error, this value is $\pm 7.6\%$, which means that if the production is 4% more than predicted the production costs will rise 7.6%. On the other hand, if the production is 4% less than predicted, the production costs will decrease 7.6%.

Relative Errors and Linear Approximations

In the previous section on marginal analysis, we found that we could use a linear approximation to estimate the cost, revenue, or profit of producing the $(x+1)^{\text{st}}$ unit of a product. We can use the relative error here to see just how "approximate" this linear approximation really is. Remember, in the beginning of this section we stated that

$$\text{Relative error} = \frac{\text{(maximal error in measurement)}}{\text{(given value of measurement)}}$$

We apply this idea in Example 3.

Example 3 Computing the Relative Error of a Linear Approximation

The StopFirst Brake Company determines that their annual profit function for producing and selling brakes is given by

$$P(x) = 4\sqrt[5]{x} \qquad 0 \le x \le 100$$

where x represents the number of brakes sold in thousands and $P(x)$ represents the profit in hundred thousand dollars.

(a) Evaluate $P(55)$ and $MP(55)$ and interpret. Round to the nearest ten-thousandth.

(b) Use the solutions from part (a) to get a linear approximation for $P(56)$.

(c) Evaluate $P(56)$ and compute the relative error in measure of the approximation found in part (b).

Solution

(a) Evaluating $P(x) = 4\sqrt[5]{x}$ when $x = 55$, we get

$$P(55) = 4\sqrt[5]{55} \approx 8.9152$$

Interpret the Solution: This means that when 55 thousand brakes are produced and sold the profit realized is about 8.9152 hundred thousand dollars, or $891,520.

The marginal profit function for the brake company is

$$MP(x) = \frac{d}{dx}(4\sqrt[5]{x}) = \frac{d}{dx}(4x^{1/5}) = \frac{4}{5}x^{-4/5} = \frac{4}{5x^{4/5}} = \frac{4}{5\sqrt[5]{x^4}}$$

Evaluating this function at $x = 55$ yields

$$MP(55) = \frac{4}{5\sqrt[5]{(55)^4}} \approx 0.0324$$

Interpret the Solution: This means that at a production level of 55 thousand units the profit from producing and selling one additional thousand units is about 0.0324 hundred thousand dollars, or $3240.

(b) A linear approximation for the profit function (with $dx = 1$) has the form

$$P(x+1) \approx P(x) + MP(x)$$

So for $x = 55$ we get the approximation of $P(56)$ as

$$P(56) = P(55+1) \approx P(55) + MP(55)$$
$$\approx 8.9152 + 0.0324 = 8.9476$$

(c) Since the actual value at $x = 56$ is $P(56) = 4\sqrt[5]{56} \approx 8.9474$, the error in measurement is

$$\Delta P = 8.9476 - 8.9474 = 0.0002$$

So, knowing that relative error $= \dfrac{\text{(maximal error in measurement)}}{\text{(given value of measurement)}}$, we find the relative error is $\dfrac{0.0002}{8.9474} \approx 0.00002$. This tells us that the relative error made by the linear approximation is 0.00002 or 0.002%. ■

Checkpoint 1 Now work Exercise 23.

Relative Rates of Change

Ideas about rates of change make more sense at times if they are given as a percentage. If someone says that the amount of ozone in the atmosphere decreased at an annual rate of $7\frac{1}{2}\%$ in 1992, this seems to make more sense than saying, "the amount of ozone in the atmosphere decreased at an annual rate of 0.008 parts per million" (*Source:* www.census.gov). This annual rate of $7\frac{1}{2}\%$ in 1992 can be found by dividing the instantaneous rate of change by the current function value. We call this result the **relative rate of change**.

Relative Rate of Change

If f is a differentiable function, then the **relative rate of change**, denoted by $Rel(x)$, is given by

$$Rel(x) = \frac{f'(x)}{f(x)}, \qquad \text{provided that } f(x) \neq 0$$

Note: We can determine the **percent rate of change** by multiplying the relative rate of change by 100. Relative rates of change can be compared between different functions and models, as shown in Example 4.

Example 4 Comparing Relative Rates of Change

Membership in the Boy Scouts of America can be modeled by

$$f(t) = 0.72t^3 - 22.15t^2 + 263.28t + 4043.3 \qquad 1 \le t \le 20$$

where t is the number of years since 1979 and $f(t)$ is the membership in thousands. During the same time period, the membership in the Girl Scouts of America can be modeled by

$$g(t) = 0.086t^4 - 3.83t^3 + 55.97t^2 - 243.23t + 2972.93 \qquad 1 \le t \le 20$$

where t is the number of years since 1979 and $g(t)$ is the membership in thousands (*Source:* U.S. Statistical Abstract, www.census.gov/statab/www).

(a) Compute $Rel(15)$ for the model f and interpret.

(b) Compute $Rel(15)$ for the model g and compare to the result of part (a).

Solution

Interactive Activity

Graph the models f and g in the viewing window [0, 20] by [2000, 8000]. At $t = 14$, how can you tell which function will have the greater relative rate?

WWW

(a) Understand the Situation: We need to compute $Rel(t) = \frac{f'(t)}{f(t)}$ for $t = 15$. Note that $t = 15$ corresponds to the year 1994.

The derivative of the model f is

$$f'(t) = 2.16t^2 - 44.30t + 263.28$$

So for $t = 15$ we have

$$Rel(15) = \frac{f'(15)}{f(15)} = \frac{84.78}{5438.75} \approx 0.016$$

Interpret the Solution: This means that during 1994, membership in the Boy Scouts of America was growing at a rate of about 1.6%.

(b) Understand the Situation: Here, we need to compute $Rel(t) = \frac{g'(t)}{g(t)}$ for $t = 15$. Again, note that $t = 15$ corresponds to the year 1994.

The derivative of the model g is

$$g'(t) = 0.344t^3 - 11.49t^2 + 111.94t - 243.23$$

So for $t = 15$ we have

$$Rel(15) = \frac{g'(15)}{g(15)} = \frac{11.62}{3345.23} \approx 0.003$$

Interpret the Solution: This means that in 1994, membership in the Girl Scouts of America was growing at a rate of about 0.3%. So Boy Scout membership was growing about five times faster than that of the Girl Scouts in 1994. ■

SUMMARY

In this section, we studied errors in estimates and the impact that calculus has on them. We defined the relative and percentage errors and how to determine the range of values based on the maximal error. We applied the relative error to linear approximations of business functions. Then we discussed the relative rates of change and how they can be used to compare rates of growth.

Important Formulas

- **Relative error (independent variable)**:

$$\varepsilon_x = \pm\frac{\text{(maximal error in the forecast)}}{\text{(given value of the forecast)}} = \pm\frac{dx}{x}$$

- **Percentage error (independent variable)**: $\varepsilon_x \cdot 100\%$
- **Relative error (dependent variable)**: $\varepsilon_y = \frac{f'(x) \cdot x}{f(x)} \cdot \varepsilon_x$
- **Percentage error (dependent variable)**: $\varepsilon_y \cdot 100\%$
- **Relative rate of change**: $Rel(x) = \frac{f'(x)}{f(x)}$

SECTION 13.3 EXERCISES

In Exercises 1–6, the range of values $[x_a, x_b]$ is given.

(a) Determine the maximal error of measurement.

(b) Compute the relative error in the independent variable ε_x.

(c) Determine the percentage error.

1. [10, 14]

2. [200, 210]

3. [4000, 4300]

4. [6000, 6100]

5. [42,000, 45,000]

6. [91,000, 92,500]

For Exercises 7–16, find the relative error in the dependent variable ε_y, given $f(x)$, x, and the relative error ε_x.

7. $f(x) = 6x^3$; $x = 1$; $\varepsilon_x = \pm 0.10$

8. $f(x) = 4x^2$; $x = 2$; $\varepsilon_x = \pm 0.10$

9. $f(x) = \dfrac{5}{x-1}$; $x = -3$; $\varepsilon_x = \pm 0.15$

10. $f(x) = \dfrac{3}{x-2}$; $x = 4$; $\varepsilon_x = \pm 0.15$

11. $f(x) = x^2 - 5x - 2$; $x = 2$; $\varepsilon_x = \pm 0.05$

12. $f(x) = x^2 - 6x - 1$; $x = 3$; $\varepsilon_x = \pm 0.08$

13. $f(x) = 3\sqrt{x}$; $x = 2$; $\varepsilon_x = \pm 0.07$

14. $f(x) = 2\sqrt{x^2}$; $x = -3$; $\varepsilon_x = \pm 0.15$

15. $f(x) = \dfrac{3-2x}{x}$; $x = 1$; $\varepsilon_x = \pm 0.19$

16. $f(x) = \dfrac{x^2+1}{2}$; $x = 5$; $\varepsilon_x = \pm 0.15$

Applications

17. A television newscast highlights a pre-election poll showing that two candidates have garnered 52.5% and 47.5% of the vote, respectively. The poll has a margin of error of ±3%. The anchorperson states that the race is a statistical deadheat. Explain her statement.

18. A newspaper poll shows that a school levy is passing 56% to 44%, with a margin of error measured at ±3%. What is the closest that the vote could actually be in the poll and stay within the margin of error?

19. Sales projections for the Griffort Company's new model of ballet shoe have targeted a projected sales of 24,000 pairs of shoes sold in the first year. They know that the sales could actually range from 18,000 to 30,000. Compute the percentage error in the independent variable for the number of pairs of ballet shoes sold in the next year.

20. Sales projections for an author's new romance novel is 70,000. The marketing department knows that the sales could actually range from 65,000 to 75,000. Compute the percentage error in the independent variable for the number of books sold.

For Exercises 21–23, consider the following scenario: The New Alliance Tool and Die factory sells bolts through large government contracts. In the next quarter, the target sales for the bolts are 60,000, but the accounting office knows that the sales could actually range from 55,000 to 65,000.

21. Compute the percentage error in the independent variable for the number of bolts sold in the next quarter.

22. If the cost function for the bolts is $C(x) = 11x^2 + 20x$, where x represents the number of bolts sold in thousands and $C(x)$ is in dollars:

(a) Find the percentage error in the dependent variable ε_y for the cost in next quarter's forecast.

(b) Determine $C(5)$ and $MC(5)$ and interpret each.

(c) Use the solutions from part (b) to get a linear approximation for $C(6)$ by evaluating $C(5) + MC(5)$.

(d) Evaluate $C(6)$ and compute the relative error in measure of the approximation found in part (c).

✓ **23.** If the revenue function for the bolts is $R(x) = 200\sqrt{x^3}$, where x represents the number of bolts sold in thousands and $R(x)$ is in dollars:

(a) Find the percentage error in the dependent variable ε_y for the revenue in next quarter's forecast.

(b) Determine $R(5)$ and $MR(5)$ and interpret each.

(c) Use the solutions from part (b) to get a linear approximation for $R(6)$.

(d) Evaluate $R(6)$ and compute the relative error in measure of the approximation found in part (c).

24. The results of an August 23, 1998, Gallup/CNN poll follow:

"The United States recently launched military attacks against suspected terrorist sites in Afghanistan and Sudan. Do you approve or disapprove of these attacks?"

Approve:	75 percent
Disapprove:	18 percent

SOURCE: CNN, www.cnn.com

Knowing that the margin of error of the poll was ±3.5%, determine the lowest and highest number of those who approved of the military attacks.

For Exercises 25–32, find the relative rate of change and then evaluate $Rel(x)$ at the given value of x.

25. $f(x) = 2x^2$; $x = 5$

26. $f(x) = 5x^3$; $x = 2$

27. $f(x) = \dfrac{x-100}{10}$; $x = 125$

28. $f(x) = \dfrac{x-1}{x^2}$; $x = 2$

29. $f(x) = 20\sqrt{x}$; $x = 8$

30. $f(x) = 5\sqrt[3]{x}$; $x = 1$

31. $f(x) = 3x^2 + 5x$; $x = 2$

32. $f(x) = 4 + \dfrac{3}{x}$; $x = -3$

⇐ 11.2 12.3 **33.** The poverty threshold in the United States can be modeled by

$$f(x) = 0.004x^2 + 0.057x + 1.280 \qquad 0 \le x \le 35$$

where x represents the number of years since 1960 and $f(x)$ represents the U.S. poverty threshold in thousands of dollars (*Source:* U.S. Census Bureau, www.census.gov).

(a) Evaluate $f(30)$ and interpret.
(b) Compute $f'(x)$. Evaluate $f'(30)$ and interpret.
(c) Evaluate $Rel(30)$ and interpret.

11.2 12.3 **34.** The population of Florida can be modeled by

$$f(x) = 0.003x^2 + 0.168x + 4.908 \qquad 0 \le x \le 35$$

where x represents the number of years since 1960 and $f(x)$ represents the population in millions (*Source:* U.S. Census Bureau, www.census.gov).

(a) Evaluate $f(30)$ and interpret.
(b) Compute $f'(x)$. Evaluate $f'(30)$ and interpret.
(c) Evaluate $Rel(30)$ and interpret.

SECTION PROJECT

The number of employees (in thousands) at business establishments in the U.S. that have less than 20 employees and between 20 and 99 employees are shown in Table 13.3.1.

Table 13.3.1

Year	x	Less Than 20 Employees	20–99 Employees
1980	1	19,423	21,268
1985	6	21,810	23,539
1988	9	23,583	25,930
1989	10	23,992	26,829
1990	11	24,373	27,414
1991	12	24,482	26,906
1992	13	25,000	27,030
1993	14	25,233	27,443
1994	15	25,373	27,443
1997	18	26,883	30,631

SOURCE: U.S. Census Bureau, www.census.gov

(a) Use the regression capabilities of your calculator to find a cubic model for the number of business establishments that have less than 20 employees in the form

$$f(x) = ax^3 + bx^2 + cx + d \qquad 1 \le x \le 18$$

where x represents the number of years since 1979 and $f(x)$ represents the number of employees at business establishments that have less than 20 employees, in thousands. Round all coefficients to the nearest hundredth.

(b) Use the regression capabilities of your calculator to find a quartic model for the number of business establishments that have between 20 and 99 employees in the form

$$g(x) = ax^4 + bx^3 + cx^2 + dx + e \qquad 1 \le x \le 18$$

where x represents the number of years since 1979 and $g(x)$ represents the number of employees at business establishments that have between 20 and 99 employees, in thousands. Round all coefficients to the nearest hundredth.

(c) Compute $Rel(17)$ for the model f and interpret.

(d) Compute $Rel(17)$ for the model g and interpret.

(e) Compare the rates found in parts (c) and (d).

Why We Learned It

In this chapter we focused exclusively on applications of the derivative. The differential and linear approximations were the first applications that we saw. This laid the groundwork for the marginal business functions. The marginal business functions are used to give an approximation for the next unit. For example, we can use a given cost function to determine the cost of producing, say, 100 units of a product. The marginal cost function evaluated at 100 will give us an approximation of the cost of producing the 101st unit. Similarly, we can use a profit function to determine the profit realized when producing and selling, say, 100 units of a product. The marginal profit function evaluated at 100 will give us an approximation of the profit of producing and selling the 101st unit. We also applied the derivative to measuring relative rates and errors. In the business world, this has many applications. For example, suppose a certain company has a sales projection for its new product of 25,000 units in the first year. If the company knows that the sales could actually range from 19,000 to 31,000 units, then they could compute the percentage error in the independent variable for the number of units sold in that year. Relative rates of change allow us to compare two items to see which one is increasing (or decreasing) more rapidly. For example, if you wanted to compare the price of a pound of lobster to the price of a round of golf, you could compare their respective relative rates of change to determine which is increasing more quickly.

CHAPTER REVIEW EXERCISES

In Exercises 1–12, find dy for the given function by determining $f'(x)$.

1. $y = 4x + 2$

2. $f(x) = 5x^2 - 3x + 2$

3. $y = \dfrac{x+3}{x-5}$

4. $y = \dfrac{7}{x+5}$

5. $y = \sqrt{x} - \dfrac{3}{x^4}$

6. $f(x) = \dfrac{8}{x^2} + \sqrt[3]{x}$

7. $f(x) = x^4 - 2x + \sqrt[3]{x^2}$

8. $y = x^3 - 5x^2 + 2x + 3$

9. $y = 4x^5 - 21x + 4$

10. $y = \dfrac{x^2 - 5}{x^2 + 5}$

11. $y = 2x^{1.7} - 5x^{0.8} + 4$

12. $f(x) = 3x^{4.1} + 7x^{0.6} - 12$

In Exercises 13–18, evaluate Δy and dy for each function for the indicated values.

13. $y = 4x^2 - x + 6$; $x = 3, \Delta x = dx = 0.1$

14. $y = \dfrac{18}{x} + 5$; $x = 2, \Delta x = dx = 0.5$

15. $f(x) = \sqrt[3]{x}$; $x = 8, \Delta x = dx = 1.261$

16. $f(x) = 2x^3 - 7x^2 + 2x$; $x = 4, \Delta x = dx = 0.2$

17. $y = \dfrac{x^2 + 2}{x^2 - 2}$; $x = 2, \Delta x = dx = 0.1$

18. $y = 4x(2x + 5)$; $x = 1, \Delta x = dx = 0.1$

For Exercises 19–22, use the linear approximation to estimate the values of the given numbers. Compare to the calculator value when rounded to four decimal places.

19. $\sqrt{65}$ **20.** $\sqrt{24.6}$

21. $\sqrt[4]{16.3}$ **22.** $\sqrt[3]{62}$

23. The Wild & Wacky T-shirt company has estimated that the association between its monthly T-shirt sales and its advertising can be modeled by

$$y = 90x - 2.7x^2 \qquad 0 \le x \le 8$$

where x represents the amount spent on advertising in hundreds of dollars and y is the number of T-shirts sold in hundreds.

(a) Determine dy.

(b) Approximate the increase in sales if the advertising is increased from \$400 to \$500.

24. Repeat Exercise 23 using the model $y = 82.76x - 1.87x^2$.

25. For Exercises 23 and 24, compute the actual change Δy in sales and compare to the approximation.

26. The number of 21-year-olds living in New York can be modeled by

$$f(x) = 0.83x^3 - 8.11x^2 + 7.66x + 283.96 \qquad 0 \le x \le 7$$

where x represents the number of years since 1990 and $f(x)$ represents the number of 21-year-olds, in thousands (*Source:* U.S. Census Bureau, www.census.gov).

(a) Evaluate $f(3)$ and $f(5)$ and interpret.

(b) Determine $f'(x)$.

(c) Write the equation of the tangent line at $x = 2$.

(d) Find the y-value on the tangent line when $x = 3$. Interpret what this estimate means and compare to the value of $f(3)$ in part (a).

For Exercises 27–32:

(a) Determine the marginal cost function $MC(x)$.

(b) For the given production level x, evaluate $MC(x)$ and interpret.

(c) Determine the actual change in cost by evaluating $C(x+1) - C(x)$ and compare with the answer to part (b).

27. $C(x) = 18x + 642$; $x = 12$

28. $C(x) = 9x + 1460$; $x = 27$

29. $C(x) = 26.7x + 87.4$; $x = 8$

30. $C(x) = \dfrac{1}{2}x^2 + 3x + 16$; $x = 15$

31. $C(x) = \dfrac{1}{4}x^2 + 12x + 47$; $x = 31$

32. $C(x) = \dfrac{1}{3}x^2 + 318x + 1783$; $x = 23$

In Exercises 33–38, the cost function C and the price–demand function p are given.

(a) Determine the revenue function R.

(b) Determine the profit function P.

(c) Differentiate P in part (b) to get the marginal profit function MP.

(d) Determine the marginal cost function MC and the marginal revenue function MR.

(e) Subtract the solutions found in part (d) to get $MR - MC$ and simplify. Compare with the result of part (c).

33. $C(x) = 7x + 250; \quad p(x) = 11$

34. $C(x) = 14x + 1380; \quad p(x) = 21$

35. $C(x) = \frac{1}{10}x^2 + 3x + 850; \quad p(x) = -\frac{x}{15} + 50$

36. $C(x) = \frac{1}{50}x^2 + \frac{1}{4}x + 70; \quad p(x) = -\frac{x}{50} + 5$

37. $C(x) = -0.001x^3 + 8x + 100; \quad p(x) = -0.005x + 10$

38. $C(x) = -0.01x^3 + 0.1x^2 + 4x + 18 \quad p(x) = -0.6x + 15$

For Exercises 39–43, the Wheelex Company manufactures bicycles and finds the price function for the bicycles to be

$$p(q) = -0.02q + 150$$

where q represents the number of bicycles produced and sold and p(q) is the price of the bicycle. Furthermore, the fixed and variable costs to produce q bicycles are \$5600 and \$85 per bicycle, respectively.

39. The cost function follows the linear form $C(q) = mq + b$. Answer the following:

(a) Write the cost C in the linear form $C(q) = mq + b$.

(b) Use calculus to compute the marginal cost function $MC(q) = \frac{d}{dq}C(q)$.

40. Using the information found in Exercise 39, complete part (a) through part (d).

(a) Evaluate $MC(250)$ and $MC(500)$ and interpret these answers.

(b) Why are the answers in part (a) equal?

(c) Algebraically find $AC(q)$ and simplify.

(d) Evaluate $AC(250)$ and interpret.

41. Use the solution from part (c) of Exercise 40 to answer parts (a) and (b).

(a) Use calculus to compute the marginal average cost function $MAC(q)$.

(b) Evaluate $MAC(250)$ and interpret.

42. Use $p(q)$ and the cost function information given in Exercise 39 to complete parts (a) through (c).

(a) Derive the revenue function $R(q)$.

(b) Use calculus to compute $MR(q) = \frac{d}{dq}R(q)$.

(c) Evaluate $MR(500)$ and interpret.

43. Use the solution from part (a) of Exercise 39 with the solution to part (a) of Exercise 42 to complete parts (a) and (b).

(a) Derive the profit function $P(q)$.

(b) If the bicycles must be manufactured in lots of 250, how many bicycles should be manufactured so that the profit is as large as possible? Verify your answer by completing the table.

Produced, q	Total profit, $P(q)$
0	
250	
500	
750	
1000	
1250	
1500	
1750	
2000	

For Exercises 44–48, repeat Exercises 39–43 all parts if the price–demand function is $p(q) = -0.022q + 180$ and the variable costs are \$125 per bicycle and the fixed costs are \$7300.

49. Consider the following scenario. The Colorama Company, a paint manufacturer, has just produced a watercolor set that it sells to wholesalers for \$6 each. The cost $C(x)$ to produce x watercolor sets is given by the function $C(x) = 0.0002x^2 + 2x + 1250$.

(a) Algebraically derive the profit function $P(x)$ and simplify it.

(b) Evaluate $P(3000)$ and interpret the answer.

(c) Use calculus to compute the marginal profit function.

(d) Evaluate $MP(3000)$ and interpret.

(e) Use the solutions from parts (b) and (d) to get a linear approximation for the value of $P(3001)$.

(f) Compute the error of the approximation for $P(3001)$ in part (e).

50. The Colorama Company hires a consulting firm to assess its work and consequently revises its price and cost functions to $p(x) = 6.5$ and $C(x) = \frac{x^2}{5500} + \frac{7}{3}x + 1500$. Redo parts (a) to (f) in Exercise 49 using these revisions.

51. Knowing that $AC(x) = \frac{C(x)}{x}$, use the quotient rule to show that the marginal average cost function can be written as

$$MAC(x) = \frac{MC(x) - AC(x)}{x}$$

52. The Between the Lines Publishing Company determines that the price–demand function for a new book is

$$p(x) = \frac{-x}{500} + 20$$

with fixed costs of \$12,000 and variable costs of $4.5\frac{\text{dollars}}{\text{book}}$.

(a) Find the cost function C and the revenue function R.

(b) Find the profit function P.

(c) Find the smallest and largest production levels x so that the company realizes a profit. (That is, find the smallest and largest independent values so that the revenue is greater than the cost.)

(d) Compute the marginal profit function $P'(x)$.

(e) Evaluate $P'(2500)$ and interpret the result.

53. The Deluxe Furniture Company determines that the price–demand function for its new bookshelf is

$$p(x) = \frac{-x}{75} + 250$$

The fixed costs are \$10,000 and variable costs are $150 \dfrac{\text{dollars}}{\text{unit}}$. Redo parts (a) through (e) in Exercise 52.

For Exercises 54–59, consider a stockholder's report that lists the following information:

q	$C(q)$	$R(q)$
100	22,830	38,000
200	30,830	72,000
300	38,830	102,000

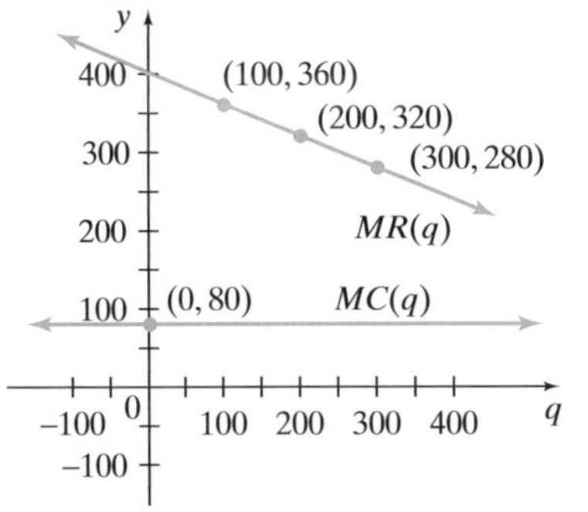

Use the table and the graphs of $MC(q)$ and $MR(q)$ to get a linear approximation for the given cost and revenue functions.

54. $C(101)$ **55.** $C(201)$ **56.** $C(301)$

57. $R(101)$ **58.** $R(201)$ **59.** $R(301)$

60. Consider the following data table for the cost and revenues at various production levels for a new brand of computer printer.

Number of Printers Produced, x	Cost, $C(x)$	Revenue, $R(x)$
1000	243,600	296,950
2000	363,000	575,800
3000	482,800	818,550
4000	603,300	1,007,200

(a) Use your calculator to determine a linear regression model for the cost of producing the printer in the form

$$C(x) = ax + b \qquad 1 \le x \le 4$$

where x represents the number of printers produced in thousands and $C(x)$ represents the cost of production.

(b) Use your calculator to determine a cubic regression model for the revenue of producing and selling the printers in the form

$$R(x) = ax^3 + bx^2 + cx + d \qquad 1 \le x \le 4$$

where x represents the number of printers produced and sold in thousands and $R(x)$ represents the resulting revenue.

(c) Compute $MAC(x)$ and simplify the result.

(d) Evaluate $MAC(1.5)$ and interpret.

61. Use the models found in Exercise 60 parts (a) and (b) to answer parts (a) and (b).

(a) Compute $MAP(x)$ and simplify the result.

(b) Evaluate $MAP(1.5)$ and interpret the answer.

For Exercises 62–67, find ε_y, the relative error in the dependent variable, given $f(x)$, x, and ε_x, the relative error in the independent variable.

62. $f(x) = 4x^2; \quad x = 2; \quad \varepsilon_x = \pm 0.15$

63. $f(x) = x^3 - 2x + 3; \quad x = 1; \quad \varepsilon_x = \pm 0.10$

64. $f(x) = \sqrt{5x}; \quad x = 5; \quad \varepsilon_x = \pm 0.05$

65. $f(x) = \dfrac{2x+3}{3x-5}; \quad x = -4; \quad \varepsilon_x = \pm 0.17$

66. $f(x) = \dfrac{2x^2-7}{15}; \quad x = 4; \quad \varepsilon_x = \pm 0.02$

67. $f(x) = 8\sqrt[4]{x^3}; \quad x = 2; \quad \varepsilon_x = \pm 0.15$

In Exercises 68–73, the range of values $[x_a, x_b]$ is given.

(a) Determine the forecast value in the independent variable x.

(b) Compute ε_x, the relative error in the independent variable.

(c) Determine the percentage error.

68. [13, 17] **69.** [180, 250]

70. [99, 100] **71.** [246, 254]

72. [56,000, 72,000] **73.** [94,150, 94,350]

74. A local newspaper prints pre-election poll results showing that a ballot initiative is losing 48% to 52%. If the margin of error is ±2.5%, could the race be a tie? Explain.

75. A magazine publishes a poll regarding an upcoming mayoral election. The poll shows that 58% of the voters support Candidate Evans and 42% support Candidate Hawthorne, with a margin of error of 4%. If the actual vote falls within the margin of error, what is the closest that the vote could be?

For Exercises 76–78, consider the following scenario:

A high-tech company sells modems to computer manufacturers. In the next quarter, the sales target for the modems is 88,000, but the accounting office knows that the sales could actually range from 85,000 to 91,000.

76. Compute the percentage error in the independent variable for the number of modems sold in the next quarter.

77. The cost function for the modems is $C(x) = 350x^2 + 30{,}000x + 24{,}000$, where x represents the number of modems sold in thousands.

(a) Find ε_y, the percentage error in the dependent variable, for the cost in next quarter's forecast.

(b) Determine $C(88)$ and $MC(88)$ and interpret.

(c) Use the solutions from part (b) to get a linear approximation for $C(89)$.

(d) Evaluate $C(89)$ and compute the relative error in measure of the approximation found in part (c).

78. The revenue function for the modems is $R(x) = 340{,}000\sqrt[3]{x^2}$, where x represents the number of modems sold in thousands. Find the percentage error in the dependent variable ε_y for the revenue in next quarter's forecast.

(a) Find ε_y, the percentage error in the dependent variable, for the revenue in next quarter's forecast.

(b) Determine $R(88)$ and $MR(88)$ and interpret.

(c) Use the solutions from part (b) to get a linear approximation for $R(89)$.

(d) Evaluate $R(89)$ and compute the relative error in measure of the approximation found in part (c).

79. According to a Gallup poll conducted on October 23, 1998, 78% of Americans were generally satisfied with their standard of living, while 22% were generally dissatisfied. The margin of error for this poll was 2%. Determine the lowest and highest values of the measure of those who were generally *satisfied* with their standard of living (*Source:* Gallup Organization, www.gallup.com).

For Exercises 80–87, find the relative rate of change and then evaluate the Rel(x) at the given value of x.

80. $f(x) = 4x^2; \quad x = 7$

81. $f(x) = 7x^3 - 2x; \quad x = 5$

82. $f(x) = \dfrac{x + 18}{5}; \quad x = 12$

83. $f(x) = \dfrac{x - 4}{x^2}; \quad x = 8$

84. $f(x) = 5\sqrt[3]{x}; \quad x = 27$

85. $f(x) = 3\sqrt{x}; \quad x = 2$

86. $f(x) = x^3 + 7x^2 - 3x; \quad x = 1$

87. $f(x) = \dfrac{7}{x} - 1; \quad x = 2$

88. The median income of U.S. one-person households consisting of a female under 65 years old can be modeled by

$$f(x) = -1.33x^3 + 42.80x^2 - 32.31x + 16{,}318.98 \qquad 0 \le x \le 26$$

where x represents the number of years since 1960 and $f(x)$ represents the median income in dollars. Use this model to compute $Rel(22)$ and interpret the answer (*Source:* U.S. Census Bureau, www.census.gov).

89. The number of U.S. households with a male householder (no spouse) and related children under 18 years of age can be modeled by the quadratic function

$$f(x) = 1.52x^2 + 19.90x + 437.52 \qquad 0 \le x \le 26$$

where x represents the number of years since 1970 and $f(x)$ represents the number of households in thousands (*Source:* U.S. Census Bureau, www.census.gov).

(a) Evaluate $f(18)$ and interpret the answer.

(b) Compute $f'(x)$, evaluate $f'(18)$, and interpret the answer.

(c) Evaluate $Rel(18)$ and interpret the answer.

90. The following table shows the estimated total annual retail sales (in million of dollars) for apparel and accessory stores and for automotive dealers.

Year	x	Apparel and Accessory Stores	Automotive Dealers
1986	1	75,626	326,138
1987	2	79,322	342,896
1988	3	85,307	372,570
1989	4	92,341	386,011
1990	5	95,819	387,605
1991	6	97,441	372,647
1992	7	104,212	406,935
1993	8	107,199	457,797
1994	9	109,976	521,768
1995	10	110,936	556,708
1996	11	114,635	599,667
1997	12	117,826	625,682

SOURCE: U.S. Census Bureau, www.census.gov

(a) Use the regression capabilities of your calculator to find a power model for the annual retail sales of apparel and accessory stores in the form

$$f(x) = a \cdot x^b \qquad 1 \le x \le 12$$

where x represents the number of years since 1985 and $f(x)$ represents the annual retail sales of apparel and accessory stores in millions of dollars.

(b) Use the regression capabilities of your calculator to find a quartic model for the annual retail sales of automotive dealers in the form

$$g(x) = ax^4 + bx^3 + cx^2 + dx + e \qquad 1 \le x \le 12$$

where x represents the number of years since 1985 and $g(x)$ represents the annual retail sales of automotive dealers in millions of dollars.

(c) Compute $Rel(5)$ for the model f and interpret.

(d) Compute $Rel(5)$ for the model g and interpret.

(e) Compare the rates found in parts (c) and (d).

CHAPTER 13 PROJECT

This chapter taught us a wonderful application of the derivative: linear approximations. In this project, we will review this and other topics discussed in this chapter while also discovering a different formula to approximate the $(x+1)^{\text{st}}$ item. Consider the following example:

WaterHand Productions manufactures several convenient faucets for modern kitchens. It was recently determined that for one of their latest models, named "Into the Wave," the variable and fixed costs were \$68.50 and \$8,020 respectively. The price–demand function p was found to be

$$p(x) = -\frac{1}{8{,}100}x^3 - \frac{1}{18}x^2 - 3x + 810 \qquad 0 \le x \le 90$$

where x is the number of "Into the Wave" faucets and $p(x)$ is the price of the faucet in dollars.

1. Determine the cost function C.
2. Determine the revenue function R.
3. Using 1 and 2, determine the profit function P.
4. Evaluate and interpret $P(12)$.
5. Evaluate and interpret $MP(12)$.
6. Using 4 and 5, approximate $P(13)$.
7. Evaluate $P(13)$ and compare it to the result in 6. That is, determine the difference in the approximation found in 6 and $P(13)$.
8. Determine the relative error and the percentage error made by the linear approximation.
9. Compute $Rel(12)$ for the model P and interpret. Round to four decimal places.
10. Note that $P(13) \approx [1 + Rel(12)] \times P(12)$. In general, $f(x+1) \approx [1 + Rel(x)] \times f(x)$ for the model f. Explain why this is true.

Chapter

14 Additional Differentiation Techniques

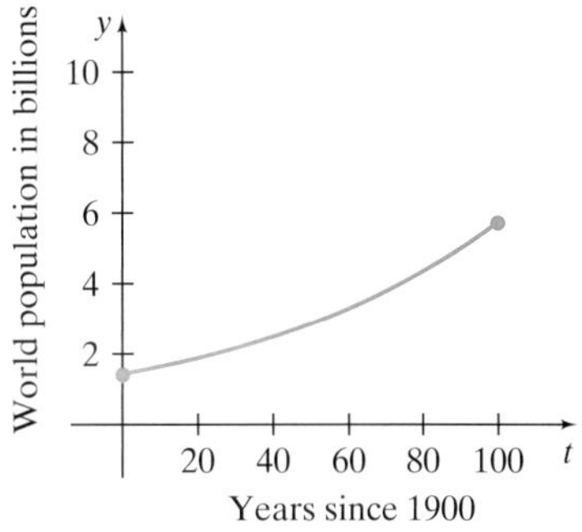

(b)

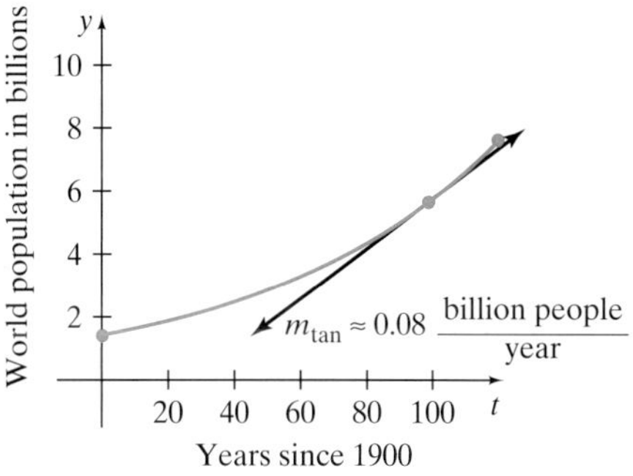

(c)

While the rate of population growth has slowed in most of the industrialized world, it continues to grow at an alarming rate in many places around the world. In Figure b is a graph of a world population model, $p(t) = 1.419e^{0.014t}$, that shows the population in billions during the 20th century. As shown in Figure c, finding the slope of the line tangent to the curve at $t = 99$ (that is, evaluating $p'(99)$) indicates that in 1999 the world's population was growing at a rate of about 80 million people per year.

What We Know

In the previous two chapters, we learned the basic rules of differentiation and their applications. We also learned that the derivative can be used to study marginal analysis of business functions. We found that the role of the differential was central to marginal analysis.

Where Do We Go

In this chapter, we will learn how to differentiate other families of functions for which the rules we have learned so far may not apply. These include composite functions, exponential and logarithmic functions, and implicit functions.

Section 14.1 The Chain Rule

So far, the type of functions that we have differentiated are polynomial functions, rational functions, and power functions. But one family of functions that we have not differentiated is the **composite** function family. If the composite function $f(x) = 1000\sqrt{180 - 2x}$ models the number of college graduates surviving to x years of age, we must find a way to compute the derivative of f in order to determine the rate of change of this function. In this section, we introduce the **Chain Rule**, a powerful technique used to differentiate composite functions.

Chain Rule

We will discuss the differentiation of composite functions that have the form $h(x) = f(g(x))$. A reasonable question to ask at this point is "Can we even determine the derivative of a composite function in the first place?" To answer that question, let's take a Flashback to an example first presented in Chapter 11.

Flashback

Oil Spill Function Revisited

In Section 11.8 we saw that a refinery's underwater supply line ruptures, resulting in a fairly circular oil spill. The radius was modeled by

$$r(t) = 0.7t$$

where t represents the number of seconds since the spill occurred and $r(t)$ represents the radius of the oil slick in feet. The area of the spill was given by

$$A(r) = \pi r^2$$

where r is the radius of the slick in feet and $A(r)$ is the area of the slick in square feet. Simplify the composition $A(r(t))$, which yields $A(t)$, and find the derivative $\frac{d}{dt}(A(t))$. Interpret the resulting function.

Flashback Solution

Simplifying, we get

$$A(r(t)) = A(0.7t)$$

Substituting $0.7t$ for r in the function $A(r) = \pi r^2$, we get the area of the oil slick as a function of time.

$$A(r(t)) = \pi(0.7t)^2 = \pi(0.49 \cdot t^2) = 0.49\pi t^2$$

Differentiating this result yields

$$\frac{d}{dt}(A(t)) = \frac{d}{dt}(0.49\pi t^2) = 0.98\pi t$$

The derivative, $\frac{d}{dt}(A(t)) = 0.98\pi t$, represents the instantaneous rate of change of the area of the oil slick with respect to time.

The Flashback shows that we can differentiate a composite function, yet it does not explicitly show how this is done. Let's examine the functions in the Flashback again and see whether there is another way to find the derivative of $A(t) = A(r(t))$. If we compute the derivatives of the original functions $r(t) = 0.7t$ and $A(r) = \pi r^2$, with respect to t and r respectively, we have

$$r'(t) = 0.7 \quad \text{and} \quad A'(r) = 2\pi r$$

Since we know that $A'(r(t)) = 2\pi(0.7t) = 1.4\pi t$ and $r'(t) = 0.7$, it appears that the derivative $A'(t)$ is equivalent to

$$A'(t) = A'(r(t)) \cdot r'(t) = (1.4\pi t) \cdot (0.7) = 0.98\pi t$$

We have just differentiated a composite function using the Chain Rule! We now state this rule for functions made up of a composition of functions f and g.

Chain Rule

If $y = f(u)$ and $u = g(x)$ are used to define $h(x)$, where $h(x) = f(g(x))$, then

$$h'(x) = f'(g(x)) \cdot g'(x)$$

provided that $f'(g(x))$ and $g'(x)$ exist. Equivalently, using Leibniz notation, we have

$$h'(x) = \frac{dy}{du} \cdot \frac{du}{dx}$$

provided that $\frac{dy}{du}$ and $\frac{du}{dx}$ exist.

To differentiate $h(x) = f(g(x))$, it appears that we can differentiate the "outside" function f, and then chain it to the derivative of the "inside" function g. Many functions that we will study have the form $h(x) = (\text{function})^{\text{power}}$, where the power is some real number. For these types of functions, we can rely on an extension, or corollary, of this rule, called the **Generalized Power Rule**. Notice that this rule looks much like our Power Rule from Chapter 12.

Generalized Power Rule

If u is a differentiable function of x and n is any real number with $f(x) = [u(x)]^n$, then

$$f'(x) = n[u(x)]^{n-1} \cdot u'(x)$$

Example 1 Applying the Generalized Power Rule

Use the Generalized Power Rule to determine the derivatives.

(a) $f(x) = (5x^3 + 3x)^4$ **(b)** $g(x) = (x^2 + 1)^{15}$

Solution

(a) For the function $f(x) = (5x^3 + 3x)^4$, consider $u(x) = 5x^3 + 3x$ and $n = 4$. Applying the Generalized Power Rule gives

$$f'(x) = \frac{d}{dx}[(5x^3 + 3x)^4] = 4(5x^3 + 3x)^{4-1} \cdot \frac{d}{dx}(5x^3 + 3x)$$
$$= 4(5x^3 + 3x)^3(15x^2 + 3)$$

(b) Here, we can consider $u(x) = x^2 + 1$ and $n = 15$ and apply the Generalized Power Rule to get

$$g'(x) = \frac{d}{dx}[(x^2+1)^{15}] = 15(x^2+1)^{14} \cdot \frac{d}{dx}(x^2+1)$$
$$= 15(x^2+1)^{14}(2x) = 30x(x^2+1)^{14}$$

■

Checkpoint 1

Now work Exercise 11.

From Your Toolbox

A function of the form $f(x) = \sqrt[b]{[g(x)]^a}$ is called a *radical function*. Rewriting $f(x) = \sqrt[b]{[g(x)]^a}$ as $f(x) = [g(x)]^{a/b}$ produces what we call the *rational exponent function*.

Notice the power of the Generalized Power Rule in part (b) of Example 1. It would have been possible, yet not at all practical, to algebraically expand the binomial $(x^2+1)^{15}$ in order to apply the differentiation techniques from Chapter 12.

We can use this new technique of differentiation to readily determine derivatives of new families of functions, including the radical and rational exponent functions. To use the Generalized Power Rule with these functions, let's review how these functions can be rewritten as shown in the Toolbox to the left.

The key in differentiating radical functions is to rewrite them in the rational exponent form so that we can apply the Generalized Power Rule. We illustrate this in Example 2.

Example 2 Determining Derivatives of Radical Functions

(a) Use the Generalized Power Rule to determine $f'(x)$ for $f(x) = \sqrt[3]{2x-4}$.

(b) Find an equation of the line tangent to the graph of f at the point $(6, 2)$.

Solution

(a) **Understand the Situation:** Before we can differentiate, we rewrite $f(x) = \sqrt[3]{2x-4}$ using rational exponents as

$$f(x) = \sqrt[3]{2x-4} = (2x-4)^{1/3}$$

Using the Generalized Power Rule with $u(x) = (2x-4)$ and $n = \frac{1}{3}$ gives us

$$f'(x) = \frac{d}{dx}[(2x-4)^{1/3}]$$
$$= \frac{1}{3}(2x-4)^{\frac{1}{3}-1} \cdot \frac{d}{dx}(2x-4) = \frac{1}{3}(2x-4)^{-2/3}(2) = \frac{2}{3}(2x-4)^{-2/3}$$

Writing the derivative without negative or rational exponents, this simplifies to

$$f'(x) = \frac{2}{3} \cdot (2x-4)^{-2/3} = \frac{2}{3(2x-4)^{2/3}} = \frac{2}{3\sqrt[3]{(2x-4)^2}}$$

(b) Since $f'(6)$ gives the slope of the tangent line at $x = 6$, we have

$$f'(6) = \frac{2}{3\sqrt[3]{(2(6)-4)^2}} = \frac{2}{3\sqrt[3]{(8)^2}}$$
$$= \frac{2}{3\sqrt[3]{64}} = \frac{2}{3 \cdot 4} = \frac{1}{6}$$

With a slope of $\frac{1}{6}$ and point (6, 2), the tangent line equation is

$$y - y_1 = m(x - x_1)$$
$$y - 2 = \frac{1}{6}(x - 6)$$
$$y = \frac{1}{6}(x - 6) + 2$$

The graphs of $f(x) = \sqrt[3]{2x - 4}$ and $y = \frac{1}{6}(x - 6) + 2$ are shown in Figure 14.1.1.

Interactive Activity

For what x-value is the function $f(x) = \sqrt[3]{2x - 4}$ not differentiable? Which of the three nondifferentiability characteristics (sharp turn or corner, vertical tangent, or discontinuity) does the graph of f exhibit? To check visually, consult the graph of f in Figure 14.1.1.

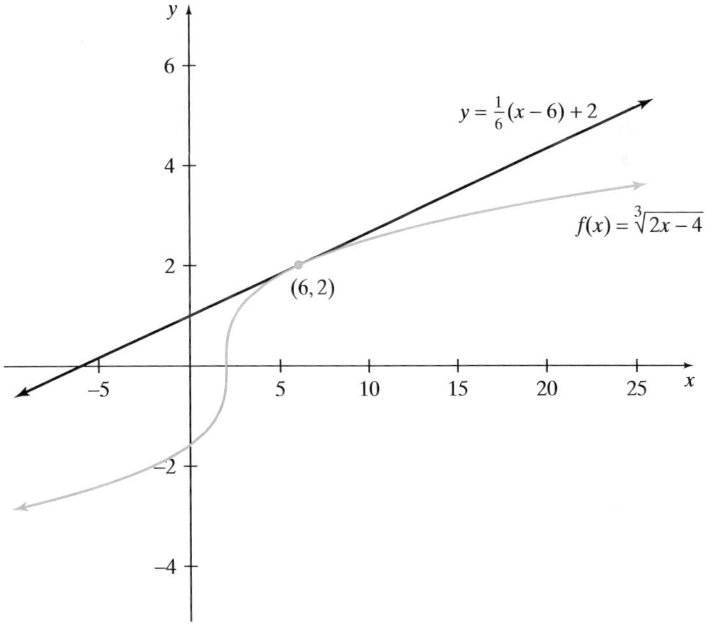

Figure 14.1.1

■

Checkpoint 2 Now work Exercise 37.

Another family of functions that can be differentiated using the Generalized Power Rule are rational functions. Even though the Generalized Power Rule can be used on any rational function, it is particularly useful on rational functions that have constants in their numerators.

Example 3 Determining Derivatives of Rational Functions Two Ways

For the function $h(x) = \frac{5}{(2x - 3)^2}$, determine $h'(x)$ using the Quotient Rule and then determine $h'(x)$ using the Generalized Power Rule.

Solution

Using the Quotient Rule, we get

$$h'(x) = \frac{\frac{d}{dx}[5] \cdot (2x-3)^2 - (5) \cdot \frac{d}{dx}[(2x-3)^2]}{[(2x-3)^2]^2}$$
$$= \frac{0 \cdot (2x-3)^2 - 5 \cdot [2(2x-3)(2)]}{(2x-3)^4}$$
$$= \frac{-20(2x-3)}{(2x-3)^4} = \frac{-20}{(2x-3)^3}$$

When using the Generalized Power Rule, we rewrite the rational function as $h(x) = 5(2x-3)^{-2}$. With help from the Constant Multiple Rule, we determine

$$h'(x) = 5 \cdot \frac{d}{dx}[(2x-3)^{-2}]$$
$$= 5 \cdot (-2)(2x-3)^{-3} \cdot (2)$$
$$= -20(2x-3)^{-3} = \frac{-20}{(2x-3)^3}$$

■

Checkpoint 3 Now work Exercise 33.

Applications

In our first application, we apply the Generalized Power Rule to a rational exponent function.

Example 4 **Applying the Generalized Power Rule**

The death rate caused by heart disease in the United States can be modeled by

$$f(x) = 337.36(x+1)^{-0.07} \qquad 0 \le x \le 17$$

where x represents the number of years since 1980 and $f(x)$ represents the death rate (measured in deaths per 100,000 people) caused by heart disease. Determine $f'(x)$. Evaluate $f'(12)$ and interpret (*Source:* U.S. National Center for Health Statistics, www.cdc.gov/nchs).

Solution

Understand the Situation: In order to evaluate $f'(12)$, we need to determine the derivative, $f'(x)$, and substitute 12 for each occurrence of x. Note that $x = 12$ corresponds to 1992.

Using the Generalized Power Rule, we get

$$f'(x) = \frac{d}{dx}[337.36(x+1)^{-0.07}]$$
$$= (337.36) \cdot (-0.07)(x+1)^{-0.07-1} \cdot (1)$$
$$= -23.6152(x+1)^{-1.07}$$

We now evaluate $f'(12)$ as follows.

$$f'(12) = -23.6152(13)^{-1.07} \approx -1.52$$

Interpret the Solution: So in 1992 the death rate caused by heart disease was decreasing at a rate of about $1.52 \dfrac{\text{deaths per 100,000 people}}{\text{year}}$. See Figure 14.1.2.

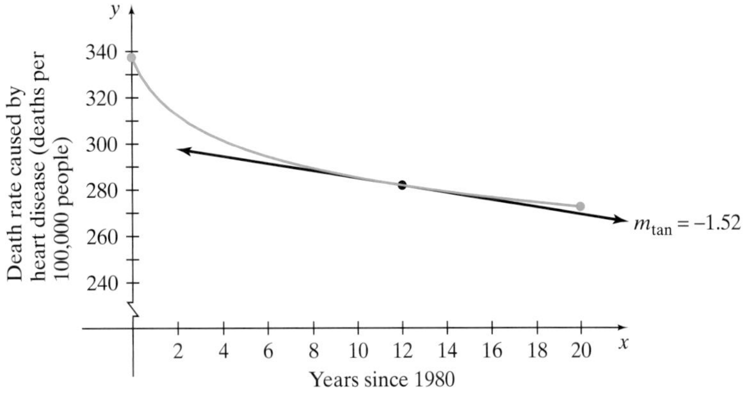

Figure 14.1.2

■

Checkpoint 4 Now work Exercise 83.

Our final example in this section demonstrates how the Generalized Power Rule can be used with business functions.

Example 5 Applying the Generalized Power Rule to Business Functions

The RadRadio Company produces and sells personal stereo devices. Market research has found that the price–demand function is

$$p(x) = 100 - \sqrt{x^2 + 20} \qquad 0 \le x \le 35$$

where x is the number of devices demanded in thousands and $p(x)$ represents the unit price in dollars.

(a) Evaluate $p'(30)$ and interpret.

(b) Determine the marginal revenue function. Evaluate $MR(30)$ and interpret.

Solution

(a) **Understand the Situation:** To aid us in determining $p'(x)$, we rewrite this radical function as a rational exponent function. The rational exponent form of the price–demand function is

$$p(x) = 100 - (x^2 + 20)^{\frac{1}{2}} \qquad 0 \le x \le 35$$

Using the Generalized Power Rule, we get the derivative

$$\begin{aligned} p'(x) &= -\frac{1}{2}(x^2 + 20)^{-1/2} \cdot \frac{d}{dx}(x^2 + 20) \\ &= -\frac{1}{2}(x^2 + 20)^{-1/2} \cdot (2x) \\ &= -x(x^2 + 20)^{-1/2} \\ &= \frac{-x}{(x^2 + 20)^{1/2}} = \frac{-x}{\sqrt{x^2 + 20}} \end{aligned}$$

Evaluating $p'(30)$ yields

$$p'(30) = \frac{-(30)}{\sqrt{(30)^2 + 20}} = \frac{-30}{\sqrt{920}} \approx -0.99 \frac{\text{dollars}}{\text{thousand devices}}$$

Interpret the Solution: This means that at a demand level of 30 thousand personal stereo devices the price is decreasing at a rate of about 99 cents per 1,000 devices sold.

(b) **Understand the Situation:** We need to first determine the revenue function. Knowing that the revenue function is $R(x) = x \cdot [p(x)]$, we use the price–demand function to get

$$R(x) = x \cdot [p(x)] = x \cdot (100 - \sqrt{x^2 + 20})$$
$$= 100x - x\sqrt{x^2 + 20}$$

To determine the marginal revenue function, we determine the derivative $R'(x)$.

In computing $R'(x)$, we must apply the Product Rule to the second term of $R(x)$. This gives us

$$MR(x) = R'(x) = \frac{d}{dx}(100x - x\sqrt{x^2 + 20})$$
$$= 100 - \frac{d}{dx}(x) \cdot \sqrt{x^2 + 20} - x \cdot \frac{d}{dx}[(x^2 + 20)^{1/2}]$$
$$= 100 - (1)\sqrt{x^2 + 20} - x\left(\frac{1}{2}\right)(x^2 + 20)^{-1/2}(2x)$$
$$= 100 - \sqrt{x^2 + 20} - x^2(x^2 + 20)^{-1/2}$$
$$= 100 - \sqrt{x^2 + 20} - \frac{x^2}{\sqrt{x^2 + 20}}$$

Evaluating $MR(x)$ at $x = 30$ produces

$$MR(30) = 100 - \sqrt{(30)^2 + 20} - \frac{(30)^2}{\sqrt{(30)^2 + 20}}$$
$$= 100 - \sqrt{920} - \frac{900}{\sqrt{920}} \approx 40.0 \frac{\text{dollars}}{\text{thousand devices}}$$

Interpret the Solution: Thus, when the production level is 30 thousand, an increase of 1000 personal stereo devices produced and sold increases revenue by about \$40. ■

SUMMARY

In this section, we reviewed composite functions and then introduced the Chain Rule, which is a technique used to differentiate composite functions. We then focused on the Generalized Power Rule, which is an extension of the Chain Rule. It can be used when a composite function has the form $f(x) = (\text{function})^{\text{power}}$, where the power is a real number.

- **Chain Rule**: If $y = f(u)$ and $u = g(x)$ are used to define $h(x)$, where $h(x) = f(g(x))$, then $h'(x) = f'(g(x)) \cdot g'(x)$, provided that $f'(g(x))$ and $g'(x)$ exist.
- **Generalized Power Rule**: If u is a differentiable function of x and n is any real number, with $f(x) = [u(x)]^n$, then $f'(x) = n[u(x)]^{n-1} \cdot u'(x)$.

SECTION 14.1 EXERCISES

In Exercises 1–6, differentiate using the Generalized Power Rule.

1. $f(x) = (x+1)^2$
2. $f(x) = (x+3)^2$
3. $f(x) = (x-5)^3$
4. $f(x) = (x-2)^3$
5. $f(x) = (2-x)^2$
6. $f(x) = (5-x)^2$

In Exercises 7–30, differentiate using the Generalized Power Rule.

7. $g(x) = (2x+4)^3$
8. $g(x) = (3x+3)^3$
9. $f(x) = (5-2x)^5$
10. $f(x) = (10-5x)^4$
✓ **11.** $f(x) = (3x^2+7)^5$
12. $f(x) = (4x^2-3)^3$
13. $f(x) = (x^3-2x^2+x)^2$
14. $f(x) = (2x^3+4x+3)^3$
15. $g(x) = 3(x^3-4)^3$
16. $g(x) = 5(4x^2-10)^6$
17. $f(x) = 5(5x^2-3x-1)^{10}$
18. $f(x) = 10(10x^3+x-9)^8$
19. $g(x) = (4x^2-x-4)^{55}$
20. $g(x) = (8x^2-2x+5)^{94}$
21. $g(x) = (2x-4)^{1/2}$
22. $f(x) = (7x+6)^{1/2}$
23. $g(x) = (x^2+2x)^{1/3}$
24. $g(x) = (x^3+5x)^{1/3}$
25. $f(x) = (5x-2)^{-2}$
26. $f(x) = (4x+3)^{-3}$
27. $g(x) = (x^2+2x+4)^{-1/2}$
28. $g(x) = (3x^2+5x+6)^{-1/2}$
29. $g(x) = (3x^3-x)^{-1/4}$
30. $g(x) = (4x^5+5x^3)^{-1/3}$

For the rational functions in Exercises 31–36:

(a) Determine the derivative using the Quotient Rule.
(b) Determine the derivative using the Generalized Power Rule.

31. $f(x) = \dfrac{1}{3x+4}$
32. $f(x) = \dfrac{1}{7x-5}$
✓ **33.** $f(x) = \dfrac{5}{(x-2)^2}$
34. $f(x) = \dfrac{10}{(2x-1)^3}$
35. $f(x) = \dfrac{2}{x^2+2x+3}$
36. $f(x) = \dfrac{9}{x^3+2x+10}$

In Exercises 37–44, determine an equation of the line tangent to the graph of f at the indicated point.

✓ **37.** $f(x) = (2x-1)^3$; $(1, 1)$
38. $f(x) = (3x-4)^3$; $(1, -1)$
39. $f(x) = (2-x)^4$; $(1, 1)$
40. $f(x) = (x^2-1)^4$; $(1, 0)$
41. $f(x) = (x^3-4x+2)^4$; $(2, 16)$
42. $f(x) = (4x-3)^{1/2}$; $(3, 3)$
43. $f(x) = (2x-4)^{1/2}$; $(2, 0)$
44. $f(x) = (2x+8)^{1/2}$; $(4, 4)$

Use the Generalized Power Rule to differentiate the functions in Exercises 45–52.

45. $f(x) = \sqrt{x^2+5}$
46. $f(x) = \sqrt{3x+6}$
47. $g(x) = \sqrt[3]{2x-1}$
48. $g(x) = \sqrt[3]{4x-3}$
49. $f(x) = \dfrac{5}{\sqrt{2x-8}}$
50. $f(x) = \dfrac{10}{\sqrt{5x+8}}$
51. $f(x) = \dfrac{64}{\sqrt[3]{5x^2-6x+3}}$
52. $f(x) = \dfrac{27}{\sqrt[3]{3x^3+x}}$

In Exercises 53–62, use the Generalized Power Rule, along with the Product and Quotient Rules to find the derivatives of the given functions.

53. $g(x) = x(x-4)^3$
54. $g(x) = x(10-x)^3$
55. $g(x) = x\sqrt{x^2+3x}$
56. $g(x) = x^2\sqrt{2x^2-11}$
57. $f(x) = \dfrac{x^3}{(3x-8)^2}$
58. $f(x) = \dfrac{x^2}{(4x^2-x+5)^3}$
59. $f(x) = (x+3)^3(2x-1)^2$
60. $f(x) = (3x-3)^4(2x-2)^3$
61. $g(x) = \sqrt{\dfrac{x+3}{x-3}}$
62. $g(x) = \sqrt{\dfrac{2x+1}{2x-1}}$

In Exercises 63–68, find the derivatives using the Generalized Power Rule.

63. $y = (4x^2+5x+6)^{0.23}$
64. $y = 3(0.7x^3-0.02x^2)^{0.09}$
65. $g(x) = \left(\dfrac{1}{x+3}\right)^{-1.03}$
66. $g(x) = \left(\dfrac{1}{0.2x+1.7}\right)^{-1.1}$
67. $f(x) = 1.44(x+1)^{1.22}$
68. $f(x) = 67.41(x+1)^{0.97}$

Applications

69. An actuary has determined that for a certain demographic group the number of people surviving over the duration of a century can be modeled by

$$f(x) = 400\sqrt{100-x} \qquad 0 \le x \le 100$$

where x represents the age of the person in years in the group and $f(x)$ represents the number of people surviving. Evaluate and interpret $f'(70)$.

70. During its first season, the number of viewers who watched the new television series "It Ain't Me!" can be modeled by

$$g(x) = \sqrt[3]{(50+2x)^2} \qquad 1 \le x \le 26$$

where x represents the number of weeks that the series has been airing and $g(x)$ is the number of viewers in millions. Evaluate and interpret $g'(13)$.

71. A study by the bursar at Clarksman College determines that the number of students enrolled in the Arts and Science programs during the past decade can be modeled by

$$f(t) = -\frac{10{,}000}{\sqrt{1+0.18t}} + 11{,}000 \qquad 1 \le t \le 11$$

where t represents the number of years since the beginning of the study and $f(t)$ represents the number of students enrolled in the Arts and Science programs. Evaluate and interpret $f'(10)$.

72. Medical researchers studying arteriosclerosis have found that, if the radius of a person's artery is currently 1 centimeter, the amount of fatty tissue called plaque that will build up in the artery can be modeled by

$$g(t) = 0.5t^2(t^2 + 10)^{-1} \qquad 0 \le t \le 10$$

where t represents the number of years since the present time and $g(t)$ represents the thickness of the plaque in the wall of the artery in centimeters. Evaluate and interpret $g'(7)$.

12.3 **73.** The StopCop Company determines that the cost to produce auto antitheft devices is modeled by

$$C(x) = (3x + 6)^{1.5} + 30 \qquad 0 \le x \le 50$$

where x represents the number of auto antitheft devices produced in hundreds and $C(x)$ represents the production costs in thousands of dollars.

(a) Determine the marginal cost function.
(b) Evaluate and interpret $MC(5)$.

74. *(continuation of Exercise 73)*

(a) Determine the average cost function AC.
(b) Determine the marginal average cost function $MAC(x) = \frac{d}{dx}(AC(x))$.
(c) Evaluate and interpret $MAC(5)$.

12.3 **75.** The SnapPic Company determines that its cost for producing disposable cameras is modeled by

$$C(x) = 60 + \sqrt{3x + 5} \qquad 0 \le x \le 40$$

where x represents the number of disposable cameras produced during each shift and $C(x)$ represents the cost of production in hundreds of dollars.

(a) Determine the marginal cost function.
(b) Evaluate and interpret $MC(15)$.

76. *(continuation of Exercise 75)*

(a) Determine the average cost function AC.
(b) Evaluate and interpret $AC(15)$.
(c) Determine the marginal average cost function $MAC(x) = \frac{d}{dx}(AC(x))$.

77. The price–demand function for a collectable doll is found to be

$$p(x) = \sqrt{22{,}500 - 50x} \qquad 0 \le x \le 30$$

where x represents the number of collectable dolls produced in hundreds and $p(x)$ is the price of the dolls in dollars.

(a) Determine $p'(x)$ using the Generalized Power Rule.
(b) Evaluate $p'(25)$ and interpret.

78. *(continuation of Exercise 77)*

(a) Determine the revenue function R.
(b) Determine the marginal revenue function using the Generalized Power Rule.
(c) Evaluate and interpret $MR(15)$.

79. The HotSpark Company has assumed that the price–demand function for their spark plug is

$$p(x) = \frac{125}{\sqrt{2x + 5}} \qquad 0 \le x \le 20$$

where x represents the number of spark plugs manufactured in hundreds and $p(x)$ is the price of the spark plug.

(a) Determine $p'(x)$ using the Generalized Power Rule.
(b) Evaluate $p'(20)$ and interpret.
(c) Determine the revenue function R.
(d) Determine the marginal revenue function using the Generalized Power Rule.
(e) Evaluate and interpret $MR(20)$.

For Exercises 80 and 81, consider the following: In the early 1930s, psychologist L.L. Thurstone determined that the time needed to learn a list of a certain length is given by the model

$$f(x) = ax\sqrt{x - b} \qquad x \ge b$$

where x represents the number of items on the list and $f(x)$ represents the time needed to learn the list, measured in minutes. The constants a and b are different for each subject and are determined by pretesting.

80. Suppose that it has been determined that a freshman subject in a psychology class learns the items on a list according to the model

$$f(x) = \frac{5x}{2}\sqrt{x - 6} \qquad x \ge 6$$

(a) Evaluate $f(20)$ and interpret.
(b) Determine $f'(x)$ using the Generalized Power Rule.
(c) Evaluate $f'(20)$ and interpret.

81. Suppose that it has been determined that a sophomore subject in a psychology class learns the items on a list according to the model

$$f(x) = 2x\sqrt{x - 3} \qquad x \ge 3$$

(a) Evaluate $f(12)$ and interpret.
(b) Determine $f'(x)$ using the Generalized Power Rule.
(c) Evaluate $f'(12)$ and interpret.

82. The amount of toxic material entering Lake Formica is related to the number of years that the Bristine Chemical Company has been operating by the model

$$g(t) = \left(\frac{4}{5}t^{1/5} + 2\right)^4 \qquad 0 \le t \le 30$$

where t represents the number of years that the company has been operating and $g(t)$ represents the amount of toxic material entering the lake in thousands of gallons.

(a) Evaluate $g(15)$ and interpret.
(b) Determine $g'(t)$ using the Generalized Power Rule.
(c) Evaluate $g'(15)$ and interpret.

83. Suppose that an actuary has determined that in Tribble Township the number of people surviving a certain number of years is given by the model

$$P(t) = 400\sqrt{101 - t} \qquad 0 \le t \le 101$$

where t represents the age of the person and $P(t)$ represents the number of people in the township who are still surviving.

(a) Determine $P(75)$ and interpret.
(b) Determine $P'(t)$ using the Generalized Power Rule.
(c) Determine $P'(75)$ and interpret.

84. Suppose that the rural town of Rufusville decides to relax its zoning laws so that more land can be made eligible for commercial use. They find that the town's annual tax base after the zoning changes can be modeled by

$$f(x) = 5x\sqrt{2x + 2} \qquad x \ge 0$$

where x represents the number of years since the zoning laws have changed and $f(x)$ represents the annual tax base in thousands of dollars.

(a) Determine $f(15)$ and interpret.
(b) Determine $f'(x)$ using the Generalized Power Rule.
(c) Determine $f'(15)$ and interpret.

85. Suppose that a lab technician finds that the number of bacteria present in an unidentified culture can be modeled by

$$g(t) = 70(10 + 0.5t)^{2.1} \qquad t \ge 0$$

where t represents the time since the first observation in hours and $g(t)$ represents the number of bacteria present.

(a) Determine $g(8)$ and interpret.
(b) Determine $g'(t)$ using the Generalized Power Rule.
(c) Determine $g'(8)$ and interpret.

SECTION PROJECT

The data in Table 14.1.1 give the percentage of 3- to 5-year-olds enrolled in preschool from 1970 to 1998.

Table 14.1.1

Year	Percent Enrolled
1970	37.5
1975	48.6
1980	52.5
1985	54.6
1990	59.4
1992	55.5
1993	55.1
1994	61.0
1995	61.8
1998	64.5

SOURCE: U.S. Census Bureau, www.census.gov

(a) Use your calculator to determine a power regression model of the form

$$f(x) = a \cdot x^b \qquad 1 \le x \le 29$$

where x represents the number of years since 1969 and $f(x)$ represents the percentage of 3- to 5-year-olds enrolled in preschool. Round the constants a and b to the nearest hundredth.

(b) Use the constants a and b found in part (a) to rewrite the model in the form

$$g(x) = a \cdot (x + 1)^b \qquad x_1 \le x \le x_2$$

where x represents the number of years since 1970. What are the values of x_1 and x_2 for which the model is valid?

(c) Use your calculator to make a table of values for f and g for $x = 0, 1, \ldots, 25$. What pattern do you notice?

(d) Determine $g'(x)$ using the Generalized Power Rule.

(e) Evaluate and interpret $g'(7)$.

(f) Evaluate and interpret $f'(8)$ and compare to part (e).

Section 14.2 Derivatives of Logarithmic Functions

Now we wish to focus our attention on a function first introduced in Section 11.8, the **logarithmic function**. This function has applications in business, social, and life sciences. To discuss its application in calculus, in particular for determining rates of change, we first must learn its derivative. Since the logarithmic function is not a polynomial function, we must develop a new rule for its derivative.

Derivative of the Natural Logarithm Function

From Your Toolbox

- The natural logarithm is $\log_e x$ and is denoted by $\ln x$ for $x > 0$.
- Let m and n be positive numbers. Then the following are true:
  1. $\ln(mn) = \ln m + \ln n$
  2. $\ln\left(\frac{m}{n}\right) = \ln m - \ln n$
  3. $\ln m^n = n \cdot \ln m$
  4. $\ln 1 = 0$
  5. $\ln e = 1$
  6. $e^{\ln x} = x$
  7. $\ln e^x = x$

To set the stage for studying the natural logarithm function, let's review some of its properties that we first encountered in Section 11.8. (See the Toolbox to the left.)

To determine the derivative of $f(x) = \ln x$, we cannot rely on the differentiation rules that we have learned so far. To determine the derivative, we return to the definition of the derivative, which gives

$$f'(x) = \lim_{h \to 0} \frac{\ln(x+h) - \ln x}{h}$$

Continuing we have

$$f'(x) = \lim_{h \to 0} \frac{\ln\left(\frac{x+h}{x}\right)}{h} \qquad \text{Property 2 of logarithms}$$

$$= \lim_{h \to 0} \frac{\ln\left(1 + \frac{h}{x}\right)}{h} \qquad \text{Simplifying } \frac{x+h}{x} \text{ to } 1 + \frac{h}{x}$$

We now let $k = \frac{x}{h}$. Notice that $h \to 0$ implies that $k \to \infty$ (for $h > 0$). We also note that if $k = \frac{x}{h}$ then $\frac{1}{k} = \frac{h}{x}$. So, rewriting the limit in terms of k gives us

$$= \lim_{k \to \infty} \frac{\ln\left(1 + \frac{1}{k}\right)}{\frac{x}{k}} \qquad k = \frac{x}{h} \text{ implies } h = \frac{x}{k}$$

$$= \lim_{k \to \infty} \frac{k}{x}\left[\ln\left(1 + \frac{1}{k}\right)\right] \qquad \text{Dividing by } \frac{x}{k} \text{ is same as multiplying by } \frac{k}{x}$$

$$= \lim_{k \to \infty} \frac{1}{x} \cdot \left[k \ln\left(1 + \frac{1}{k}\right)\right] \qquad \text{Rewriting } \frac{k}{x} \text{ as } \frac{1}{x} \cdot k$$

$$= \lim_{k \to \infty} \frac{1}{x}\left[\ln\left(1 + \frac{1}{k}\right)^k\right] \qquad \text{Property 3 of logarithms}$$

In Section 11.7 we numerically saw that $\lim_{k \to \infty}\left(1 + \frac{1}{k}\right)^k = e$. Using this information we continue and get

$$= \frac{1}{x} \cdot \ln e$$

$$= \frac{1}{x} \qquad \text{Property 5 of logarithms}$$

Derivative of the Natural Logarithm Function

For $f(x) = \ln x$, with $x > 0$, the **derivative of the natural logarithm function** is given by

$$f'(x) = \frac{d}{dx}(\ln x) = \frac{1}{x}$$

Technology Option

Table 14.2.1

X	Y1	Y2
1	0	1
2	.69315	.5
3	1.0986	.33333
4	1.3863	.25
5	1.6094	.2
6	1.7918	.16667
7	1.9459	.14286

X=7

We can use the NDERIV command on the graphing calculator to numerically see that $\frac{d}{dx}(\ln x) = \frac{1}{x}$, for $x > 0$. Table 14.2.1 shows values for $y_1 = \ln x$ for $x = 1, 2, \ldots, 7$ as well as $y_2 = \text{NDERIV}(\ln x)$.

It does appear that each value in column y_2 is the reciprocal of the corresponding x. For more on NDERIV and other ways to compute derivative values with your graphing calculator, consult the online manual at www.prenhall.com/armstrong.

Example 1 Computing Derivatives of Natural Logarithmic Functions

Compute the derivatives of the following functions.

(a) $f(x) = 2\ln x$ **(b)** $g(x) = \ln x^3$ **(c)** $y = 7 - 4\ln x$

Solution

(a) We can apply the Constant Multiple Rule here to get

$$f'(x) = \frac{d}{dx}(2\ln x) = 2 \cdot \frac{d}{dx}(\ln x)$$
$$= 2 \cdot \frac{1}{x} = \frac{2}{x}$$

(b) Before we differentiate, we first take advantage of logarithm property 3, $\ln m^n = n \cdot \ln m$, and rewrite $g(x) = \ln x^3$ as $g(x) = 3\ln x$. Differentiation gives

$$g'(x) = \frac{d}{dx}(\ln x^3) = \frac{d}{dx}(3\ln x)$$
$$= 3 \cdot \frac{d}{dx}(\ln x) = 3 \cdot \frac{1}{x} = \frac{3}{x}$$

(c) For this derivative, we use the Sum and Difference Rule. This gives

$$y' = \frac{d}{dx}(7 - 4\ln x) = \frac{d}{dx}(7) - \frac{d}{dx}(4\ln x)$$
$$= 0 - 4 \cdot \frac{d}{dx}(\ln x)$$
$$= -4 \cdot \frac{1}{x} = \frac{-4}{x}$$

■

✓ Checkpoint 1

Now work Exercise 3.

Example 1c illustrates a special form that we will frequently use for modeling data. The function $y = 7 - 4\ln x$ is an example of the **natural logarithm model** because it has the form $f(x) = a + b \cdot \ln x$, where a and b represent constants. Example 2 illustrates how to calculate the rate of change of this type of function.

Example 2 Differentiating and Interpreting a Natural Logarithm Model

The amount of fish products imported into the U.S. each year for human consumption can be modeled by

$$f(x) = 4.21 + 0.67 \ln x \qquad 1 \le x \le 20$$

where x represents the number of years since 1979 and $f(x)$ represents the annual import of fish products for human consumption in billions of pounds (*Source:* U.S. Census Bureau, www.census.gov).

(a) Evaluate $f'(2)$ and interpret. **(b)** Evaluate $f'(19)$ and interpret.

Solution

Interactive Activity

For the model in Example 2, determine an equation for the line tangent to the graph of f at $x = 2$. Graph the tangent line and the model f in the same viewing window. Determine the value of the tangent line at $x = 19$ and interpret in terms of the amount of fish products imported into the U.S.

WWW

(a) **Understand the Situation:** We need to first determine the derivative and then evaluate it at $x = 2$. Note that $x = 2$ corresponds to the year 1981.

The derivative of the model is

$$f'(x) = \frac{d}{dx}(4.21 + 0.67 \ln x) = \frac{d}{dx}(4.21) + \frac{d}{dx}(0.67 \ln x)$$

$$= 0 + 0.67 \cdot \frac{1}{x} = \frac{0.67}{x}$$

Evaluating $f'(2)$, we get

$$f'(2) = \frac{0.67}{2} = 0.335$$

Interpret the Solution: This means that in 1981, the annual import into the U.S. of fish products for human consumption was increasing at a rate of $0.335 \dfrac{\text{billions of pounds}}{\text{year}}$.

(b) Here, we note that $x = 19$ corresponds to 1998. Evaluating $f'(19)$ gives us

$$f'(19) = \frac{0.67}{19} \approx 0.035$$

Interpret the Solution: This means that in 1998, the annual import into the U.S. of fish products for human consumption was increasing at a rate of $0.035 \dfrac{\text{billions of pounds}}{\text{year}}$. Notice that this is about one-tenth the rate in 1981. See Figure 14.2.1.

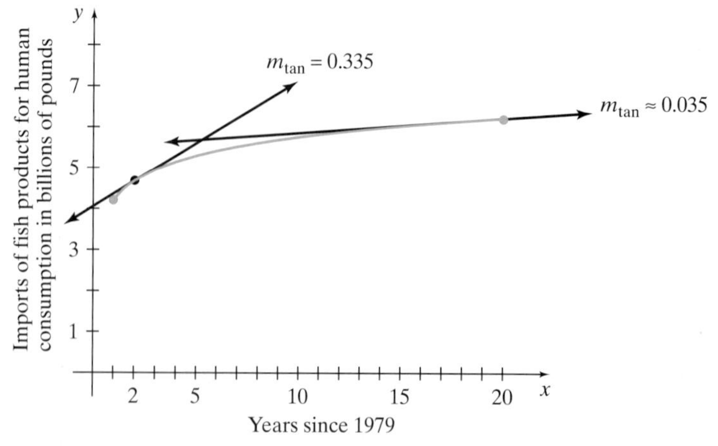

Figure 14.2.1 Comparison of $f'(2)$ and $f'(19)$.

■

Checkpoint 2

Now work Exercise 43.

Sometimes we need to differentiate functions that involve more than just $\ln x$. Let's say that the argument of the logarithmic function was another function, which we will call $u = g(x)$. If we let $f(x) = \ln x$, then we have a composite function $h(x) = f(g(x)) = \ln(g(x)) = \ln u$. We must use the Chain Rule to determine the derivative of $h(x) = f(g(x))$. Since $f'(x) = \frac{1}{x}$ and $g'(x) = \frac{du}{dx}$, we have

$$h'(x) = f'(g(x)) \cdot g'(x)$$
$$= \frac{1}{g(x)} \cdot g'(x) = \frac{g'(x)}{g(x)}$$

Chain Rule for the Natural Logarithm Function

If g is a differentiable function of x and the range of g is $(0, \infty)$, the derivative of the composition of functions $h(x) = \ln[g(x)]$ is given by

$$h'(x) = \frac{1}{g(x)} \cdot g'(x) = \frac{g'(x)}{g(x)}$$

Note: In Section 13.3 we defined the *relative rate of change* of a function f as $Rel(f(x)) = \frac{(\text{derivative of } f(x))}{(\text{given function } f(x))} = \frac{f'(x)}{f(x)}$. This is exactly what the preceding rule states, with $g(x)$ in place of $f(x)$. So we can redefine the relative rate of change of a function as $Rel(f(x)) = \frac{d}{dx}[\ln(f(x))]$ if we restrict the range of f in our definition of $Rel(f(x))$ to $(0, \infty)$.

Example 3 Differentiating the Natural Logarithm Function with the Chain Rule

Determine the derivative of each function.

(a) $y = \ln(x^4 - 2x)$

(b) $h(x) = \ln\sqrt{6x - 1}$

(c) $f(x) = (\ln x)^3$

Solution

(a) Here the inside, or g, function is $g(x) = x^4 - 2x$. So applying the Chain Rule gives us

$$y' = \frac{1}{x^4 - 2x} \cdot \frac{d}{dx}(x^4 - 2x)$$
$$= \frac{1}{x^4 - 2x} \cdot (4x^3 - 2)$$
$$= \frac{4x^3 - 2}{x^4 - 2x}$$

(b) Here, we begin by using some algebra to rewrite the function,

$$h(x) = \ln\sqrt{6x - 1} = \ln(6x - 1)^{1/2} = \frac{1}{2}\ln(6x - 1)$$

Now, we differentiate using the Chain Rule.

$$h'(x) = \frac{d}{dx}\left[\frac{1}{2}\ln(6x-1)\right]$$
$$= \frac{1}{2}\cdot\frac{d}{dx}\ln(6x-1)$$
$$= \frac{1}{2}\cdot\frac{1}{6x-1}\cdot\frac{d}{dx}(6x-1)$$
$$= \frac{1}{2}\cdot\frac{1}{6x-1}\cdot 6 = \frac{3}{6x-1}$$

(c) This is a little different from the previous two parts in that we must apply the Generalized Power Rule to get the derivative.

$$f(x) = (\ln x)^3$$
$$f'(x) = \frac{d}{dx}[(\ln x)^3]$$
$$= 3(\ln x)^2\cdot\frac{d}{dx}(\ln x)$$
$$= 3(\ln x)^2\cdot\frac{1}{x}$$
$$= \frac{3(\ln x)^2}{x}$$

■

The difference between the functions in Example 1b and Example 3c is an important one. In Example 1b, $g(x) = \ln x^3$ means that x is taken to the third power and *then* the logarithm is taken. In Example 3c, $f(x) = (\ln x)^3$ means that the logarithm of x is taken first and *then* the result is cubed. The parentheses are used to make the difference clear. (Some textbooks denote a function like $f(x) = (\ln x)^3$ by writing $f(x) = \ln^3 x$.)

Derivatives of General Logarithmic Functions

The most frequently used logarithm function for applications is the natural logarithm function $f(x) = \ln x$. But what about the derivatives of the logarithm function with other bases? In other words, what is $\frac{d}{dx}(\log_b x)$? The following definition shows how to differentiate a general logarithmic function. The derivation of this result is left to you. (See Interactive Activity to the left.)

Interactive Activity

The general logarithmic function $f(x) = \log_b x$ with $b > 0, b \neq 1$ may be rewritten as

$$f(x) = \log_b x = \frac{\ln x}{\ln b}$$

via what is known as the change of base formula. Differentiate $f(x) = \frac{\ln x}{\ln b}$ to show that for $f(x) = \log_b x$, $f'(x) = \frac{1}{\ln b}\cdot\frac{1}{x}$.

■ **Derivative of the General Logarithmic Function**

- For the general logarithmic function $f(x) = \log_b x$ with $b > 0, b \neq 1$,

$$f'(x) = \frac{1}{\ln b}\cdot\frac{1}{x}$$

- If g is a differentiable function, where the range of g is $(0, \infty)$, then the **Chain Rule** for the general logarithmic function $f(x) = \log_b[g(x)]$ is

$$f'(x) = \frac{1}{\ln b}\cdot\frac{1}{g(x)}\cdot g'(x)$$

Example 4 Determining Derivatives of General Logarithmic Functions

Differentiate the following.

(a) $y = \log_3 x$ **(b)** $f(x) = \log(x^3 + 9)$

Solution

(a) Here we have a general logarithm function with base 3. So the derivative is

$$y' = \frac{d}{dx}(\log_3 x) = \frac{1}{\ln 3} \cdot \frac{1}{x} = \frac{1}{x \ln 3}$$

(b) Since the base is not written, we know that this is a common logarithm with base 10. Applying the Chain Rule, with $g(x) = x^3 + 9$, gives

$$f'(x) = \frac{d}{dx}[\log(x^3 + 9)]$$

$$= \frac{1}{\ln 10} \cdot \frac{1}{(x^3 + 9)} \cdot \frac{d}{dx}(x^3 + 9)$$

$$= \frac{1}{\ln 10} \cdot \frac{1}{(x^3 + 9)} \cdot (3x^2) = \frac{1}{\ln 10} \cdot \frac{3x^2}{(x^3 + 9)}$$

■

Checkpoint 3 Now work Exercise 31.

SUMMARY

In this section we learned how to differentiate the natural logarithmic function, as well as the logarithmic function with any base.

- **Derivative of the natural logarithm function**: For $f(x) = \ln x$, $f'(x) = \frac{d}{dx}(\ln x) = \frac{1}{x}$.
- **Chain rule for the natural logarithm function**: If $h(x) = \ln(g(x))$ then $h'(x) = \frac{1}{g(x)} \cdot g'(x) = \frac{g'(x)}{g(x)}$.
- **Derivative of the general logarithmic function**: For the general logarithmic function $f(x) = \log_b x$, $f'(x) = \frac{1}{\ln b} \cdot \frac{1}{x}$. The **Chain Rule** for the general logarithmic function $f(x) = \log_b[g(x)]$ is $f'(x) = \frac{1}{\ln b} \cdot \frac{1}{g(x)} \cdot g'(x)$.

SECTION 14.2 EXERCISES

In Exercises 1–10, determine the derivative for the following functions.

1. $f(x) = 5 \ln x$
2. $f(x) = -8 \ln x$
✓ **3.** $f(x) = \ln x^6$
4. $f(x) = \ln x^4$
5. $f(x) = 4x^3 \cdot \ln x$
6. $f(x) = 12x^3 \cdot \ln x$
7. $f(x) = \frac{3x^5}{\ln x}$
8. $f(x) = \frac{12}{\ln x}$
9. $f(x) = 10 - 12 \ln x$
10. $f(x) = -2 + 8 \ln x$

In Exercises 11–24, determine the derivative for the following functions.

11. $g(x) = \ln(x + 7)$
12. $g(x) = \ln(2 - x)$
13. $g(x) = \ln(2x - 5)$
14. $g(x) = \ln(3x + 4)$
15. $g(x) = \ln(x^2 + 3)$
16. $g(x) = \ln(3x^3 - 11)$
17. $g(x) = \ln(\sqrt{2x + 5})$
18. $g(x) = \ln(\sqrt[3]{4x + 2})$
19. $g(x) = (\ln x)^6$
20. $g(x) = (\ln x)^4$
21. $g(x) = \sqrt{x} \cdot \ln(\sqrt{x})$
22. $g(x) = 4x^5 \cdot \ln(3x^3)$

23. $g(x) = \dfrac{x^2 + 2x + 3}{\ln(x+5)}$ **24.** $g(x) = \dfrac{4x^3 - x + 2}{\ln(x+7)}$

For Exercises 25–34, determine the derivative for the following functions.

25. $f(x) = \log_{10} x$ **26.** $f(x) = \log_5 x$

27. $f(x) = 6\log_3 x$ **28.** $f(x) = 11\log_4 x$

29. $f(x) = x^2 \log_9 x$ **30.** $f(x) = 2x^5 \log_8 x$

✓ **31.** $f(x) = \log_2(5x+3)$ **32.** $f(x) = \log_5(3x+9)$

33. $f(x) = \log_{10}\left(\dfrac{x+3}{x^2+1}\right)$ **34.** $f(x) = \log_2\left(\dfrac{x^3}{x^2-1}\right)$

For Exercises 35–42, determine an equation for the line tangent to the graph of the function at the given point.

35. $f(x) = \ln x$; $(2, \ln 2)$

36. $f(x) = \ln x$; $(1, 0)$

37. $f(x) = \ln\sqrt{2x-1}$; $(1, 0)$

38. $f(x) = \ln(3x)$; $(2, \ln 6)$

39. $f(x) = 4x^3 \cdot \ln x$; $(1, 0)$

40. $f(x) = 12x^3 \cdot \ln x$; $(2, 96\ln 2)$

41. $y = (\ln x)^6$; $(e, 1)$

42. $y = \ln x^6$; $(e, 6)$

Applications

✓ **43.** A research assistant in biology finds in an experiment that at low temperatures the growth of a certain bacteria culture can be modeled by

$$f(t) = 750 + 12\ln t \qquad t \geq 1$$

where t represents the number of hours since the start of the experiment and $f(t)$ represents the number of bacteria present.

(a) Determine $f'(t)$.

(b) Evaluate and interpret $f(12)$ and $f'(12)$.

44. The city of Plantersville has enacted new zoning laws in order to curb the growth of the city's population. They find that the population can be modeled by

$$P(x) = 10{,}000 + 100\ln x \qquad x \geq 1$$

where x represents the number of years since the laws were adopted and $P(x)$ represents the city's population.

(a) Determine $P'(x)$.

(b) Evaluate and interpret $P(20)$ and $P'(20)$.

45. Prescription drug companies have found that the popularity of the new drug Vectrum has dwindled and can be modeled by

$$f(x) = 150 + 5\log_2 x \qquad x \geq 1$$

where x represents the number of years that the drug has been on the market and $f(x)$ represents the number of prescriptions written for the drug annually in thousands.

(a) Determine $f'(x)$.

(b) Evaluate $f'(2)$ and $f'(10)$ and interpret each.

46. The urban school district of Molisburg has started a new educational campaign in an attempt to reduce the increase in lice found in the elementary school student population. The number of children who contracted lice can be modeled by

$$g(t) = 200 + 8\log_3 t \qquad t \geq 1$$

where t represents the number of years since the new educational campaign has been enacted and $g(t)$ represents the number of students who are diagnosed with lice annually.

(a) Determine $g'(t)$.

(b) Evaluate $g'(2)$ and $g'(7)$ and interpret each.

47. The life expectancy for African-American females in the United States can be modeled by

$$f(x) = 68.41 + 1.75\ln x \qquad 1 \leq x \leq 26$$

where x represents the birth year since 1969 and $f(x)$ represents the life expectancy in years (*Source:* U.S. National Center for Health Statistics, www.cdc.gov/nchs).

(a) Determine $f'(x)$.

(b) Evaluate and interpret $f(3)$ and $f'(3)$.

(c) Write an equation of the tangent line at $x = 3$, and determine y on the tangent line when $x = 15$. Interpret and compare to $f(15)$.

48. The life expectancy for white females in the United States can be modeled by

$$f(x) = 75.32 + 1.29\ln x \qquad 1 \leq x \leq 26$$

where x represents the birth year since 1969 and $f(x)$ represents the life expectancy in years (*Source:* U.S. National Center for Health Statistics, www.cdc.gov/nchs).

(a) Determine $f'(x)$.

(b) Evaluate and interpret $f(3)$ and $f'(3)$. Compare to part (b) in Exercise 47.

(c) Write an equation of the tangent line at $x = 3$, and determine y on the tangent line when $x = 15$. Interpret and compare to $f(15)$. Compare to part (c) in Exercise 47.

11.2 **49.** The annual per capita consumption of light and skim milk in the United States can be modeled by

$$f(x) = 10.12 + 2\ln x \qquad 1 \leq x \leq 16$$

where x represents the number of years since 1979 and $f(x)$ represents the annual per capita consumption of light and skim milk in gallons (*Source:* U.S. Department of Agriculture, www.usda.gov).

(a) Determine $f'(x)$.

(b) Evaluate and interpret $f'(5)$ and compare to $f'(10)$.

11.2 **50.** The average expenditure for a new domestic car in the United States can be modeled by

$$f(x) = 15{,}241.24 + 1517.65 \ln x \qquad 1 \le x \le 8$$

where x represents the number of years since 1991 and $f(x)$ represents the average expenditure, in dollars, for a new domestic car (*Source:* U.S. Bureau of Economic Analysis, www.bea.doc.gov).

(a) Determine $f'(x)$.

(b) Evaluate and interpret $f'(2)$ and compare to $f'(6)$.

51. *(continuation of Exercise 49)*

(a) Graph the model in Exercise 49 in the viewing window [1, 16] by [10, 17].

(b) Use the INTERSECT command to determine in which year the annual per capita consumption of light and skim milk exceeded 15 gallons.

(c) Use the $\frac{dy}{dx}$ command to approximate $f'(5)$ and $f'(10)$. Compare to Exercise 49 (b).

52. *(continuation of Exercise 50)*

(a) Graph the model in Exercise 50 in the viewing window [1, 8] by [15,000, 19,000].

(b) Use the INTERSECT command to determine in which year the average expenditure for a new domestic car exceeded $17,000.

(c) Use the $\frac{dy}{dx}$ command to approximately $f'(2)$ and $f'(6)$. Compare to Exercise 50 (b).

SECTION PROJECT

The revenue, in billions of dollars, generated in the solid waste management industry in the United States from 1980 to 1996 is shown in Table 14.2.2.

Table 14.2.2

Year	x	Revenue (in $billions)
1980	1	52.0
1990	11	146.4
1994	15	172.5
1995	16	180.0
1996	17	184.3

SOURCE: U.S. Census Bureau, www.census.gov

(a) Use your calculator to determine a logarithmic regression model of the form

$$f(x) = a + b \ln x \qquad 1 \le x \le 17$$

where x represents the number of years since 1979 and $f(x)$ represents the revenue generated in the solid waste management industry, in billions of dollars.

(b) Determine $f'(x)$.

(c) Evaluate and interpret $f'(5)$.

(d) Write an equation of the line tangent to the graph of f at $x = 3$. Determine y on the tangent line when $x = 10$ and interpret.

(e) Compare the value found in part (d) to $f(10)$ and interpret.

Section 14.3 Derivatives of Exponential Functions

In this section, we discuss how to differentiate the **exponential function**, $f(x) = e^x$, and the **general exponential function**, $f(x) = b^x$. The exponential function has applications in business as well as in the social and life sciences. One application first introduced in Section 11.7 is exponential growth and decay. Here we analyze the rate of change of the growth or the decay. To discuss these applications, in particular when determining rates of change, we need to learn the derivative. As in Section 14.2, we must determine the derivative utilizing the definition of the derivative.

Derivatives of Exponential Functions with Base e

An important function that has not yet been differentiated is the exponential function with base e, $f(x) = e^x$. To determine this derivative, we must return to the definition of the derivative. By definition, if $f(x) = e^x$, then

$$f'(x) = \lim_{h \to 0} \frac{f(x+h) - f(x)}{h}$$
$$= \lim_{h \to 0} \frac{e^{x+h} - e^x}{h}$$

Using the properties of exponents gives us

$$f'(x) = \lim_{h \to 0} \frac{e^x \cdot e^h - e^x}{h}$$
$$= \lim_{h \to 0} \frac{e^x(e^h - 1)}{h}$$

Since we are computing the limit with respect to h, e^x can be treated as constant.

$$= \lim_{h \to 0} e^x \frac{e^h - 1}{h}$$
$$= e^x \cdot \lim_{h \to 0} \frac{e^h - 1}{h}$$

But what is $\lim_{h \to 0} \frac{e^h - 1}{h}$? If we simply substitute zero for h, we get the indeterminate form $\frac{0}{0}$. To find this limit, we rely on our numerical and graphical methods.

From the graph of $\frac{e^h - 1}{h}$ in Figure 14.3.1, we believe that $\lim_{h \to 0} \frac{e^h - 1}{h} = 1$. Checking numerically, Table 14.3.1 shows the values of $\frac{e^h - 1}{h}$ for h-values close to zero. From the graph and the table, we believe that

$$\lim_{h \to 0} \frac{e^h - 1}{h} = 1$$

We can now take the limit of our difference quotient.

$$f'(x) = e^x \cdot \lim_{h \to 0} \frac{e^h - 1}{h} = e^x \cdot 1 = e^x$$

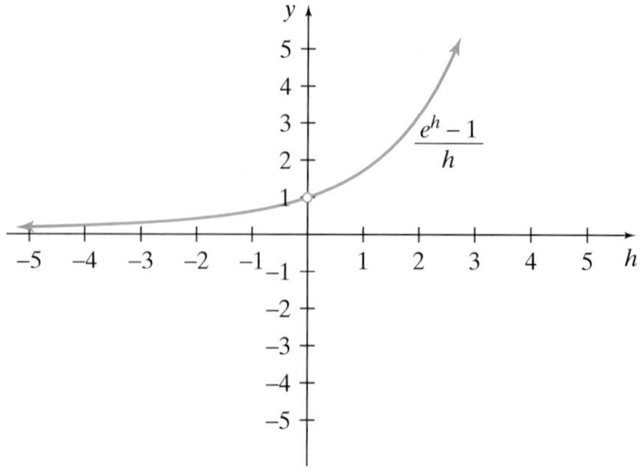

Figure 14.3.1 Graphical determination that $\lim_{h \to 0} \frac{e^h - 1}{h} = 1$.

Interactive Activity

Use the rule to the right along with the constant multiple rule to determine the derivatives for $h(x) = 5e^x$, $g(x) = \frac{1}{2}e^x$, and $y = \sqrt{3} \cdot e^x$. What rule could we conclude for the derivative of $f(x) = ce^x$, where c is any real number?

WWW

Table 14.3.1

h	-0.1	-0.001	-0.0001	0	0.0001	0.001	0.1
$\frac{e^h - 1}{h}$	0.9516	0.9995	0.9999		1.0001	1.0005	1.0517

This surprising result means that the exponential function $f(x) = e^x$ is its own derivative! (The only other function that had a property like this up to now is the trivial function $f(x) = 0$.)

Derivative of the Exponential Function

The **derivative** of the **exponential function** $f(x) = e^x$ is

$$f'(x) = e^x$$

Example 1 Differentiating Exponential Functions

Determine derivatives for the following functions.

(a) $y = \frac{x^2}{e^x}$ **(b)** $g(x) = x^4 \cdot e^x$ **(c)** $f(x) = \log e^x$

Solution

(a) Applying the Quotient Rule, we get

$$y' = \frac{d}{dx}\left(\frac{x^2}{e^x}\right) = \frac{\frac{d}{dx}[x^2] \cdot (e^x) - (x^2) \cdot \frac{d}{dx}[e^x]}{(e^x)^2}$$

$$= \frac{2xe^x - x^2e^x}{e^{2x}} = \frac{e^x(2x - x^2)}{e^{2x}} = \frac{2x - x^2}{e^x}$$

(b) Applying the Product Rule for this function gives

$$g'(x) = \frac{d}{dx}(x^4 \cdot e^x) = \frac{d}{dx}[x^4] \cdot (e^x) + (x^4) \cdot \frac{d}{dx}(e^x)$$

$$= 4x^3e^x + x^4e^x$$

(c) For this common logarithm function, we need to recall the differentiation rule for the general logarithm function from Section 14.2.

$$f'(x) = \frac{d}{dx}(\log e^x) = \frac{1}{\ln 10} \cdot \frac{1}{e^x} \cdot \frac{d}{dx}(e^x)$$

$$= \frac{1}{\ln 10} \cdot \frac{1}{e^x} \cdot (e^x) = \frac{1}{\ln 10}$$

Notice that this derivative is actually a constant. ■

The derivative of e^x by itself has limited value. But the Chain Rule applied to the exponential function is very useful, as it lends itself to many applications. Let's add this extension to our list of differentiation rules.

Chain Rule for the Exponential Function

If f is a differentiable function of x, the derivative of the composition of functions $h(x) = e^{f(x)}$ is given by

$$h'(x) = e^{f(x)} \cdot f'(x)$$

Example 2 Differentiating Composite Exponential Functions

Determine derivatives for the following functions.

(a) $h(x) = e^{3x-3}$ **(b)** $g(x) = e^{6x-(1/2)x^6}$

(c) $y = e^{1-\ln x}$ **(d)** $f(x) = \ln(e^{3x} - 8)$

Solution

(a) In the composite form $h(x) = e^{f(x)}$, $f(x) = 3x - 3$.

$$\begin{aligned} h'(x) &= \frac{d}{dx}(e^{3x-3}) = e^{3x-3} \cdot \frac{d}{dx}(3x-3) \\ &= e^{3x-3} \cdot (3) = 3e^{3x-3} \end{aligned}$$

(b) Applying the Chain Rule for the exponential function gives

$$\begin{aligned} g'(x) &= \frac{d}{dx}(e^{6x-(1/2)x^6}) = (e^{6x-(1/2)x^6}) \cdot \frac{d}{dx}\left(6x - \frac{1}{2}x^6\right) \\ &= (e^{6x-(1/2)x^6}) \cdot (6 - 3x^5) \end{aligned}$$

(c) This function has a natural logarithm term in the exponent.

$$\begin{aligned} y' &= \frac{d}{dx}(e^{1-\ln x}) = (e^{1-\ln x}) \cdot \frac{d}{dx}(1 - \ln x) \\ &= (e^{1-\ln x}) \cdot \left(-\frac{1}{x}\right) = -\frac{e^{1-\ln x}}{x} = -\frac{e}{x^2} \end{aligned}$$

Interactive Activity

Explain how to algebraically simplify

$$-\frac{e^{1-\ln x}}{x} \text{ to } -\frac{e}{x^2}$$

as shown in Example 2c.

(d) This function requires us to start with the Chain Rule for the natural logarithm function.

$$\begin{aligned} f'(x) &= \frac{d}{dx}[\ln(e^{3x} - 8)] = \frac{1}{e^{3x} - 8} \cdot \frac{d}{dx}(e^{3x} - 8) \\ &= \frac{1}{e^{3x} - 8} \cdot (e^{3x} \cdot 3) = \frac{3e^{3x}}{e^{3x} - 8} \end{aligned}$$

■

✓ Checkpoint 1

Now work Exercise 19.

The properties that we studied in earlier chapters can be applied to the exponential functions as well. This is illustrated in Example 3.

Example 3 Applying an Exponential Growth Model

The world's population during the 20th century closely follows the mathematical model

$$p(t) = 1.419e^{0.014t} \qquad 0 \le t \le 100$$

where t represents the number of years since 1900 and $p(t)$ represents the world's population in billions of people. Evaluate and interpret $p'(10)$ and compare to $p'(99)$ (*Source:* U.S. Census Bureau, www.census.gov).

Solution

Differentiating the population model, we get

$$p'(t) = \frac{d}{dt}(1.419e^{0.014t}) = 1.419\frac{d}{dt}(e^{0.014t})$$

$$= 1.419e^{0.014t} \cdot \frac{d}{dt}(0.014t)$$

$$= 1.419e^{0.014t} \cdot (0.014) \approx 0.02e^{0.014t}$$

The year 1910 corresponds to $t = 10$. Evaluating $p'(10)$ gives

$$p'(10) = 0.02e^{(0.014)(10)} \approx 0.023\frac{\text{billion people}}{\text{year}}$$

Interpret the Solution: This means that in 1910 the world's population was increasing at a rate of about 0.023 billion (that is, 23 million) people per year.

The year 1999 corresponds to $t = 99$. Evaluating $p'(99)$ yields

$$p'(99) = 0.02e^{(0.014)(99)} \approx 0.08\frac{\text{billion people}}{\text{year}}$$

Interpret the Solution: Thus, in 1999 the world's population was increasing at a rate of about 0.08 billion (that is, 80 million), people per year. The growth rate in 1999 was nearly 3.5 times the growth rate in 1910. See Figure 14.3.2.

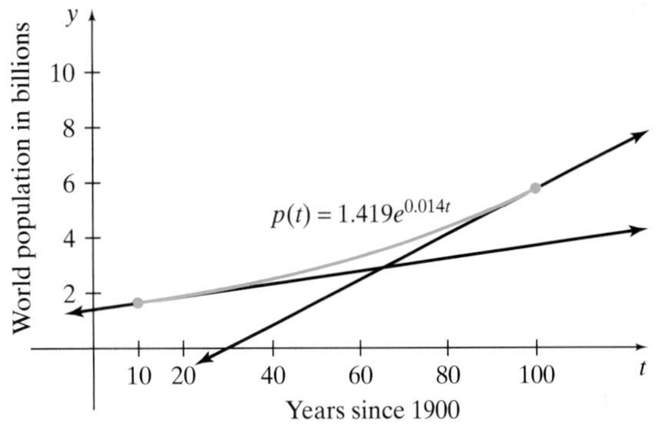

Figure 14.3.2 Growth rate in 1999 was about 3.5 times the growth rate in 1910 as shown by tangent lines. ■

Derivatives of General Exponential Functions

From Your Toolbox

Since the exponential and logarithmic functions are inverses, for $x > 0$,

$$e^{\ln x} = x$$

One type of mathematical model we have not yet differentiated is the **general exponential function**, $f(x) = b^x$. Remember that for these functions b is any positive number other that 1. To find this derivative, let's return to a property that we covered in Chapter 11 as shown in the Toolbox to the left.

This fact in the Toolbox means that we can substitute the value of b for x and write $b = e^{\ln b}$. Thus,

$$b^x = (e^{\ln b})^x = e^{\ln b \cdot x}$$

Using this property, we can now differentiate the function $f(x) = b^x$ using the Chain Rule for the exponential function.

$$\begin{aligned} f'(x) &= \frac{d}{dx}(b^x) \\ &= \frac{d}{dx}(e^{\ln b \cdot x}) \\ &= e^{\ln b \cdot x} \cdot \frac{d}{dx}(\ln b \cdot x) \end{aligned}$$

Since b is a constant, so is $\ln b$, which gives us

$$f'(x) = e^{\ln b \cdot x} \cdot \ln b$$

$$f'(x) = b^x \cdot \ln b \qquad \text{Rewrite } b \text{ for } e^{\ln b}$$

Derivative of the General Exponential Function

If $f(x) = b^x$, with $b > 0$ and $b \neq 1$, then the **derivative** of the **general exponential function** is

$$f'(x) = b^x \cdot \ln b$$

Example 4 Determining Derivatives of General Exponential Functions

Determine the derivatives for the given functions.

(a) $f(x) = 10^x$

(b) $g(x) = \dfrac{3^x}{4^x}$

Solution

(a) Here we use the general exponential function differentiation rule with $b = 10$. So we get

$$\begin{aligned} f'(x) &= \frac{d}{dx}(10^x) \\ &= 10^x \cdot \ln 10 \end{aligned}$$

(b) Instead of using the Quotient Rule, we can use the properties of exponents to rewrite $g(x) = \dfrac{3^x}{4^x}$ as $g(x) = \left(\dfrac{3}{4}\right)^x$. Now we have a general exponential function with $b = \dfrac{3}{4}$. So the derivative is

$$\begin{aligned} g'(x) &= \frac{d}{dx}\left[\left(\frac{3}{4}\right)^x\right] \\ &= \left(\frac{3}{4}\right)^x \cdot \ln\left(\frac{3}{4}\right) = \left(\frac{3}{4}\right)^x (\ln 3 - \ln 4) \end{aligned}$$

■

It seems natural at this point to extend our differentiation capabilities of the general exponential function by adding in the Chain Rule so that we can differentiate composite general exponential functions.

Chain Rule for the General Exponential Function

If f is a differentiable function of x, and $b > 0, b \neq 1$, the derivative of the composition of functions $h(x) = b^{f(x)}$ is given by

$$h'(x) = b^{f(x)} \cdot \ln b \cdot f'(x)$$

Example 5 Differentiating Composite General Exponential Functions

Determine derivatives for the given functions.

(a) $y = 5^{9x-5}$ **(b)** $g(x) = 3^{\ln x+5}$

Solution

(a) In the form $h(x) = b^{f(x)}$, we have $b = 5$ and $f(x) = 9x - 5$. So the derivative is

$$\begin{aligned} y' &= \frac{d}{dx}(5^{9x-5}) \\ &= 5^{9x-5} \cdot \ln 5 \cdot \frac{d}{dx}(9x-5) \\ &= 5^{9x-5} \cdot \ln 5 \cdot 9 \end{aligned}$$

(b) Applying the Chain Rule for the general exponential function yields

$$\begin{aligned} y' &= \frac{d}{dx}(3^{\ln x+5}) \\ &= 3^{\ln x+5} \cdot \ln 3 \cdot \frac{d}{dx}(\ln x + 5) \\ &= 3^{\ln x+5} \cdot \ln 3 \cdot \left(\frac{1}{x}\right) = \frac{3^{\ln x+5} \cdot \ln 3}{x} \end{aligned}$$

■

Interactive Activity

For $y = \ln(3^{2x})$, use the Chain Rule for natural logarithmic functions along with a property of logarithms to show that $y' = \ln 9$.

WWW

Checkpoint 2

Now work Exercise 35.

Applications

Recall that models such as $p(t) = 1.419e^{0.014t}$ from Example 3 are called **exponential growth models**. Let's take another look at the definition of exponential growth and exponential decay by consulting the Toolbox to the left.

Before we look at one final application, we would like to point out that, since $f(x) = b^x$ can be written as $f(x) = e^{\ln b \cdot x}$, any general exponential model may be written as follows:

From Your Toolbox

- The exponential function $f(x) = b^x$, with $b > 1$ (or $f(x) = b^{-x}$, where $0 < b < 1$) models **exponential growth**.
- The exponential function $f(x) = b^x$, where $0 < b < 1$ (or $f(x) = b^{-x}$, where $b > 1$) models **exponential decay**.

Rewriting the General Exponential Model

The general exponential model $f(x) = a \cdot b^x$ can be written in the exponential form

$$g(x) = a \cdot e^{kx}$$

where $k = \ln b$.

For example, the exponential model $f(x) = 3.1(1.92)^x$ can be written as $g(x) = 3.1e^{(\ln 1.92)\cdot x} \approx 3.1e^{0.65x}$. This rewriting technique is useful for modeling, since many calculators will model data only in the general exponential form $f(x) = a \cdot b^x$.

Flashback

Yard Waste Model Revisited

In Section 11.7, we modeled the percentage of waste generated that is yard waste with the general exponential function

$$f(x) = 19.12(0.944)^x \qquad 1 \le x \le 6$$

where x represents the number of years since 1990 and $f(x)$ represents the percentage of waste generated that is yard waste. Determine $f'(2)$ and interpret (*Source:* U.S. Statistical Abstract, www.census.gov/statab/www).

Flashback Solution

The derivative is computed to be

$$\begin{aligned} f'(x) &= \frac{d}{dx}[19.12(0.944)^x] \\ &= 19.12 \cdot (0.944)^x \cdot \ln(0.944) \end{aligned}$$

Evaluating the derivative when $x = 2$ gives us

$$f'(2) = 19.12 \cdot (0.944)^2 \cdot \ln(0.944) \approx -0.98$$

Since $x = 2$ corresponds to 1992, we conclude that in 1992 the percentage of waste generated that is yard waste was decreasing at a rate of about $0.98\ \frac{\text{percent}}{\text{year}}$. See Figure 14.3.3.

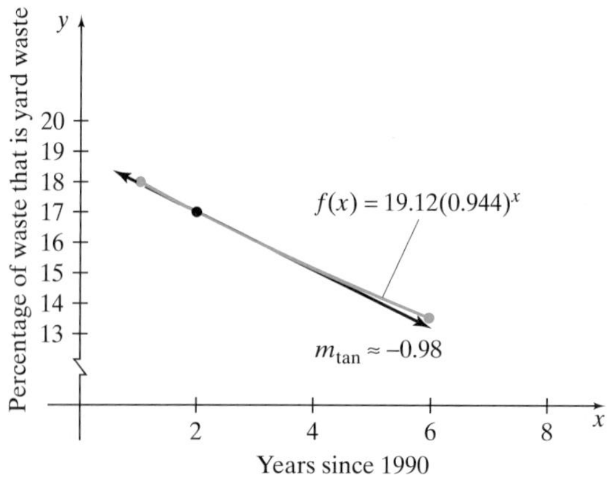

Figure 14.3.3

Checkpoint 3 Now work Exercise 45.

SUMMARY

In this section, we examined the **derivatives of exponential functions**. We also saw how the property $b = e^{\ln b}$ can be used to rewrite $f(x) = b^x$ as $f(x) = e^{\ln b \cdot x}$. This rewriting technique is useful for modeling since many calculators model data only in the general exponential form $f(x) = a \cdot b^x$. We determined the following differentiation rules:

- $\frac{d}{dx}[e^x] = e^x$
- $\frac{d}{dx}[e^{f(x)}] = e^{f(x)} \cdot f'(x)$, where f is a differentiable function of x
- $\frac{d}{dx}[b^x] = b^x \cdot \ln b$
- $\frac{d}{dx}[b^{f(x)}] = b^{f(x)} \cdot \ln b \cdot f'(x)$, where f is a differentiable function of x

SECTION 14.3 EXERCISES

In Exercises 1–16, determine the derivative for the following functions.

1. $g(x) = 7e^x$

2. $g(x) = 10e^x$

3. $g(x) = 2x(4 + e^x)$

4. $g(x) = 10x(e^x + 40)$

5. $g(x) = \dfrac{10}{5 - e^x}$

6. $g(x) = \dfrac{15}{2 + e^x}$

7. $g(x) = 4x^2 e^x$

8. $g(x) = 10x^2 e^x$

9. $g(x) = \sqrt{12 - e^x}$

10. $g(x) = \sqrt[3]{e^x + 5}$

11. $g(x) = \dfrac{e^x - 10}{x^3 - 1}$

12. $g(x) = \dfrac{x^2 + 5}{2 - e^x}$

13. $g(x) = \dfrac{e^x + 1}{e^x - 1}$

14. $g(x) = \dfrac{4 - e^x}{4 + e^x}$

15. $g(x) = 2xe^x - x$

16. $g(x) = 4x^2 e^x - e^x$

In Exercises 17–28, determine the derivative for the following functions.

17. $f(x) = e^{2x-1}$

18. $f(x) = e^{6x+9}$

✓ **19.** $f(x) = e^{\sqrt{x}}$

20. $f(x) = e^{\sqrt[3]{x}}$

21. $f(x) = e^{\ln x}$

22. $f(x) = e^{2 \ln x}$

23. $f(x) = 5x \cdot e^{2x}$

24. $f(x) = 2x^3 \cdot e^{6x^2}$

25. $f(x) = \dfrac{e^{x-1}}{e^{x+1}}$

26. $f(x) = \dfrac{e^{2+x}}{e^{2-x}}$

27. $f(x) = \ln(x^2 + e^{-x})$

28. $f(x) = \ln(x^3 - e^{5x})$

In Exercises 29–34, determine the derivative for the following functions.

29. $g(x) = 10^x$

30. $g(x) = 4^x$

31. $g(x) = \dfrac{5^x}{15^x}$

32. $g(x) = \dfrac{8^x}{2^x}$

33. $g(x) = x^3 \cdot 0.3^x$

34. $g(x) = x^2 \cdot 0.2^x$

In Exercises 35–42, determine the derivative for the following functions.

✓ **35.** $f(x) = 10^{x+3}$

36. $f(x) = 2^{9-x}$

37. $f(x) = 9^{1/x}$

38. $f(x) = 4^{\sqrt{x}}$

39. $f(x) = x \cdot e^x - 5^{2x}$

40. $f(x) = 2x \cdot \ln x + 4^{2x-1}$

41. $f(x) = \ln 5x \cdot 5^{x^2}$

42. $f(x) = \log_3 4x \cdot 4^{x-e}$

Applications

43. A veterinarian finds that when a lab animal specimen is exposed to a new pesticide, the growth of a tumor in the specimen can be modeled by

$$f(t) = 2.1e^{0.2t} \qquad t > 0$$

where t represents the number of days since exposure to the pesticide and $f(t)$ represents the diameter of the tumor in millimeters.

(a) Determine $f'(t)$.

(b) Evaluate and interpret $f'(3)$.

44. Since adding a new herbicide to wheat seed, a farmer finds that the total yield of the crop can be modeled by

$$f(x) = 1500e^{0.15x} \qquad x \geq 0$$

where x represents the number of growing seasons since the farmer started using the herbicide and $f(x)$ represents the crop yield in bushels.

(a) Determine $f'(x)$.

(b) Evaluate and interpret $f'(5)$.

✓ **45.** A zoologist has studied the population growth of a certain species of fish near a recently built lake shore factory. The population can be modeled by

$$p(t) = 12 \cdot (0.8)^t \qquad 0 \leq t \leq 10$$

where t represents the number of years since the factory opened and $p(t)$ represents the population of the fish in hundreds.

(a) Determine $p'(t)$.

(b) Evaluate and interpret $p'(1)$ and compare to $p'(8)$. See Example 3.

46. The director of health services at a large university enacts an immunization drive in order to decrease the number of students contracting the flu. The number of students getting the flu can be modeled by

$$g(x) = 7000 \cdot (0.92)^x \qquad 0 \leq x \leq 12$$

where x represents the number of years since the immunization drive began and $g(x)$ represents the number of students who contract the flu each year.

(a) Determine $g'(x)$.

(b) Evaluate and interpret $g'(2)$ and compare to $g'(11)$. See Example 3.

47. The average salary of players in the National Basketball Association can be modeled by

$$f(x) = 280.8e^{0.16x} \qquad 1 \leq x \leq 14$$

where x represents the number of years since 1984 and $f(x)$ represents the average salary in thousands of dollars (*Source:* U.S. Statistical Abstract, www.census.gov/statab/www).

(a) Classify the function as an exponential growth or decay model. Explain.

(b) Determine $f'(x)$.

(c) Evaluate and interpret $f'(3)$ and compare to $f'(13)$.

48. The tax liability of a single U.S. resident with no dependents who earned at most \$35,000 can be modeled by

$$f(x) = 7042.12e^{-0.03x} \qquad 1 \leq x \leq 11$$

where x represents the number of years since 1984 and $f(x)$ represents the tax liability in dollars (*Source:* U.S. Census Bureau, www.census.gov).

(a) Classify the function as an exponential growth or decay model. Explain.

(b) Determine $f'(x)$.

(c) Evaluate and interpret $f'(1)$ and compare to $f'(10)$.

49. Suppose that \$2000 is invested in an account that earns 6.5% interest compounded continuously. The amount accumulated after t years is given by

$$A(t) = 2000e^{0.065t} \qquad t \geq 0$$

(a) Determine the derivative $A'(t)$.

(b) Evaluate and interpret $A(5)$ and $A'(5)$.

50. Suppose that \$5000 is invested in an account that earns 5% interest compounded continuously. The amount accumulated after t years is given by

$$A(t) = 5000e^{0.05t} \qquad t \geq 0$$

(a) Determine the derivative $A'(t)$.

(b) Evaluate and interpret $A(9)$ and $A'(9)$.

51. *(continuation of Exercise 49)*

(a) Graph A from Exercise 49 in the viewing window [0, 10] by [2000, 4000].

(b) Use the INTERSECT command to determine when the amount accumulated first reaches \$3000.

(c) Use the $\frac{dy}{dx}$ command to approximate $A'(2)$.

52. *(continuation of Exercise 50)*

(a) Graph A from Exercise 50 in the viewing window [0, 10] by [5000, 8500].

(b) Use the INTERSECT command to determine when the amount accumulated first reaches \$7000.

(c) Use the $\frac{dy}{dx}$ command to approximate $A'(2)$.

53. The trading volume on the New York Stock Exchange can be modeled by

$$f(x) = 9.74e^{0.145x} \qquad 1 \leq x \leq 20$$

where x is the number of years since 1979 and $f(x)$ is the number of shares traded in billions (*Source:* U.S. Statistical Abstract, www.census.gov/statab/www).

(a) Determine the derivative $f'(x)$.

(b) Evaluate and interpret $f'(3)$.

(c) Evaluate and interpret $f'(17)$.

54. The number of bond issuers on the New York Stock Exchange can be modeled by

$$f(x) = 1208(0.95)^x \qquad 1 \leq x \leq 20$$

where x is the number of years since 1979 and $f(x)$ is the number of bond issuers listed on the New York Stock Exchange (*Source:* U.S. Statistical Abstract, www.census.gov/statab/www).

(a) Determine the derivative $f'(x)$.

(b) Evaluate and interpret $f'(3)$.

(c) Evaluate and interpret $f'(17)$.

55. The total amount of individual retirement accounts (IRAs) that are held by mutual funds can be modeled by

$$f(x) = 115.9(1.27)^x \qquad 1 \leq x \leq 10$$

where x represents the number of years since 1989 and $f(x)$ represents the total amount of individual retirement accounts invested in mutual funds, in billions of dollars (*Source:* U.S. Statistical Abstract, www.census.gov/statab/www).

(a) Classify the function as an exponential growth or decay model. Explain.

(b) Determine $f'(x)$ for the model. Evaluate and interpret $f'(1)$ and compare to $f'(9)$.

(c) Rewrite the model in the form $g(x) = a \cdot e^{kx}$, where a and k are constants rounded to the nearest hundredth.

56. The number of civil cases going to trial in U.S. district courts can be modeled by

$$f(x) = 12.4 \cdot (0.95)^x \qquad 1 \leq x \leq 12$$

where x represents the number of years since 1984 and $f(x)$ represents the number of civil cases going to trial in U.S. district courts, in thousands (*Source:* U.S. Courts, www.uscourts.gov).

(a) Classify the function as an exponential growth or decay model. Explain.

(b) Determine $f'(x)$ for the model. Evaluate and interpret $f'(3)$.

(c) Rewrite the model in the form $g(x) = a \cdot e^{kx}$, where a and k are constants rounded to the nearest hundredth.

SECTION PROJECT

The percentage of the U.S. labor force who was unemployed between 1992 and 1996 is shown in Table 14.3.2.

Table 14.3.2

Year	x	Percent Unemployed
1992	1	7.5
1993	2	6.9
1994	3	6.1
1995	4	5.6
1996	5	5.4

SOURCE: U.S. Census Bureau, www.census.gov

(a) Use your calculator to determine a general exponential regression model of the form

$$f(x) = a \cdot b^x \qquad 1 \leq x \leq 5$$

where x represents the number of years since 1991 and $f(x)$ represents the percentage of the U.S. labor force who was unemployed.

(b) Classify the function as an exponential growth or decay model.

(c) Graph f in the viewing window [1, 5] by [4, 8].

(d) Determine $f'(x)$ for the model.

(e) Evaluate and interpret $f'(1)$ and compare to $f'(5)$.

(f) Rewrite the model in the form $g(x) = a \cdot e^{kx}$, where a and k are constants rounded to the nearest hundredth.

(g) Evaluate and interpret $g'(1)$ and $g'(5)$ and compare to part (e).

Section 14.4 Implicit Differentiation and Related Rates

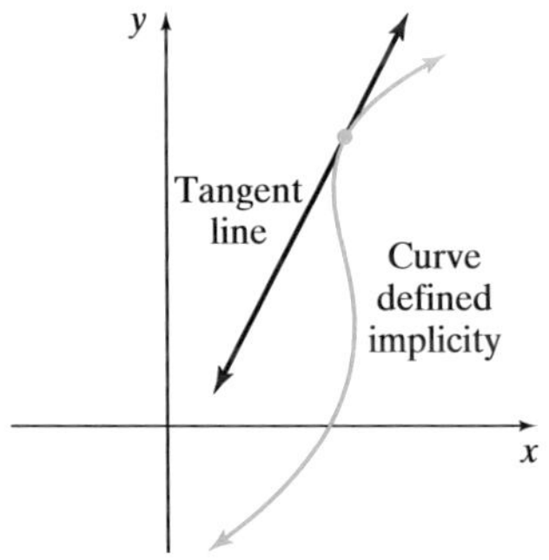

Figure 14.4.1

Until now, all the graphs of functions that we have differentiated are ones that have passed the vertical line test. But what about other graphs like the one in Figure 14.4.1. We see that these graphs have tangent lines, too.

In this section we will learn a new method of differentiation that can be used to find derivatives and, subsequently, rates of change of graphs that may not pass the vertical line test. This method is called **implicit differentiation**. Then we examine applications called **related rates** that use implicit differentiation.

Implicit Differentiation

Thus far, all the functions that we have studied have the form $y = f(x)$. Algebraically, this means that we can solve for y *explicitly* in terms of x, such as $y = x^3 + 3x^2 - 1$. However, not all equations are expressed in this form. For example, for an equation such as

$$xy^4 + y^2 - xy + 3x^2 - 5 = 0$$

solving for y for this implicit function would be both difficult and impractical. The natural question is, "How do we find the derivative $\frac{dy}{dx}$ in this case?" The key is to use implicit differentiation. Since this differentiation method is really an extension of the Chain Rule, let's recall what this rule states. (Consult the Toolbox to the left.)

From Your Toolbox

The Chain Rule

If $y = f(u)$ and $u = g(x)$ are used to define $h(x)$, where $h(x) = f(g(x))$, then

$$h'(x) = f'(g(x)) \cdot g'(x)$$

provided that $f'(g(x))$ and $g'(x)$ exist.

For a function like $y = (x^3 + 3)^4$, we can use the Chain Rule to determine the derivative because our inside function, $g(x) = x^3 + 3$, is given. But what if we were asked to determine

$$\frac{d}{dx}[f(x)^4]?$$

Notice here that our inside function is not explicitly given; it is just represented as $f(x)$. We could still perform the differentiation using the Chain Rule, as follows:

$$\frac{d}{dx}[f(x)^4] = 4 \cdot [f(x)^3] \cdot f'(x)$$

Now, if we replace $f(x)$ with y and $f'(x)$ with $\frac{dy}{dx}$, we have

$$\frac{d}{dx}[f(x)^4] = \frac{d}{dx}[y^4] = 4y^3\frac{dy}{dx}$$

This suggests the following rule:

Implicit Differentiation Rule for y^n

If n is any real number and y is differentiable, then

$$\frac{d}{dx}[y^n] = n(y^{n-1}) \cdot \frac{dy}{dx}$$

This rule allows us to determine derivatives of expressions that involve powers of x, as well as powers of y. The process we use is the following:

- Use the usual differentiation rules for terms in x.
- Use implicit differentiation for terms in y.
- Solve for $\frac{dy}{dx}$.

Example 1 Determining Derivatives of Expressions Containing y^n

Use implicit differentiation to determine $\frac{dy}{dx}$ for the following.

(a) $y^3 - 4x^2 = 7$ **(b)** $x^{1/3} + y^{1/3} - 1 = 0$

Solution

(a) We start by differentiating each term with respect to x.

$$\frac{d}{dx}(y^3) - \frac{d}{dx}(4x^2) = \frac{d}{dx}(7)$$

To differentiate $\frac{d}{dx}(y^3)$, we use the Implicit Differentiation Rule for y^n to get

$$3y^2 \cdot \frac{dy}{dx} - 8x = 0$$

Solving for $\frac{dy}{dx}$ yields

$$3y^2 \cdot \frac{dy}{dx} = 8x$$

$$\frac{dy}{dx} = \frac{8x}{3y^2}$$

(b) Again, we begin by differentiating each term with respect to x.

$$\frac{d}{dx}(x^{1/3}) + \frac{d}{dx}(y^{1/3}) - \frac{d}{dx}(1) = 0$$

$$\frac{1}{3}x^{-2/3} + \frac{1}{3}y^{-2/3} \cdot \frac{dy}{dx} - 0 = 0$$

When writing the equation with positive exponents and in radical form, we get

$$\frac{1}{3\sqrt[3]{x^2}} + \frac{1}{3\sqrt[3]{y^2}} \cdot \frac{dy}{dx} = 0$$

Solving for $\frac{dy}{dx}$ gives

$$\frac{1}{3\sqrt[3]{y^2}} \cdot \frac{dy}{dx} = -\frac{1}{3\sqrt[3]{x^2}}$$

$$\frac{dy}{dx} = -\frac{3\sqrt[3]{y^2}}{3\sqrt[3]{x^2}}$$

$$\frac{dy}{dx} = -\sqrt[3]{\frac{y^2}{x^2}}$$

■

Checkpoint 1 Now work Exercise 5.

Notice that the derivatives of these implicit functions are written in terms of both x and y.

Example 2 Determining $\frac{dy}{dx}$ Implicitly

Determine $\frac{dy}{dx}$ for the following.

(a) $4e^y = x^2$ **(b)** $2x + 1 = \sqrt{2 - y^2}$

Solution

(a) We begin by differentiating each term with respect to x.

$$\frac{d}{dx}[4e^y] = \frac{d}{dx}[x^2]$$

Using techniques from Section 14.3 on the left side of the equation gives us

$$4e^y\frac{dy}{dx} = 2x$$

Solving for $\frac{dy}{dx}$ yields

$$\frac{dy}{dx} = \frac{2x}{4e^y} = \frac{x}{2e^y}$$

(b) Again, we start by differentiating each term with respect to x.

$$\frac{d}{dx}[2x] + \frac{d}{dx}[1] = \frac{d}{dx}[\sqrt{2 - y^2}]$$

$$\frac{d}{dx}[2x] + \frac{d}{dx}[1] = \frac{d}{dx}[(2 - y^2)^{1/2}]$$

$$2 + 0 = \frac{1}{2}(2 - y^2)^{-1/2} \cdot (-2y)\frac{dy}{dx}$$

$$2 = -y(2 - y^2)^{-1/2}\frac{dy}{dx}$$

$$2 = \frac{-y}{\sqrt{2 - y^2}}\frac{dy}{dx}$$

Solving for $\frac{dy}{dx}$ gives

$$\frac{dy}{dx} = \frac{-2\sqrt{2 - y^2}}{y}$$

■

✓ Checkpoint 2

Now work Exercise 25.

Sometimes we encounter situations that contain mixed terms, such as x^2y^3. To apply our implicit differentiation technique to these mixed terms, we must apply the Product Rule and then the Implicit Differentiation Rule for y^n when we differentiate the factor containing y^n. For example, suppose that we want to determine $\frac{d}{dx}(x^2y^3)$. Using the Product Rule, we get

$$\begin{aligned}\frac{d}{dx}(x^2y^3) &= \frac{d}{dx}(x^2) \cdot y^3 + x^2 \cdot \frac{d}{dx}(y^3)\\ &= 2x \cdot y^3 + x^2 \cdot 3y^2\frac{dy}{dx}\end{aligned}$$

Example 3 Determining $\frac{dy}{dx}$ of an Expression Containing x and y Terms

Use implicit differentiation to determine $\frac{dy}{dx}$ for $x^2y^2 - 7x^3 - 5 = 0$.

Solution

Differentiating each term with respect to x gives

$$\frac{d}{dx}(x^2y^2) - \frac{d}{dx}(7y^3) - \frac{d}{dx}(5) = \frac{d}{dx}(0)$$

Using the Product Rule for $\frac{d}{dx}(x^2y^2)$ yields

$$\begin{aligned}\frac{d}{dx}(x^2) \cdot y^2 + x^2 \cdot \frac{d}{dx}(y^2) - \frac{d}{dx}(7y^3) - \frac{d}{dx}(5) &= \frac{d}{dx}(0)\\ 2x \cdot y^2 + x^2 \cdot 2y\frac{dy}{dx} - 21y^2\frac{dy}{dx} - 0 &= 0\\ 2xy^2 + 2x^2y\frac{dy}{dx} - 21y^2\frac{dy}{dx} &= 0\end{aligned}$$

Now we need to solve for $\frac{dy}{dx}$ algebraically.

$$\begin{aligned}2x^2y\frac{dy}{dx} - 21y^2\frac{dy}{dx} &= -2xy^2\\ \frac{dy}{dx}(2x^2y - 21y^2) &= -2xy^2\\ \frac{dy}{dx} &= \frac{-2xy^2}{2x^2y - 21y^2}\end{aligned}$$

■

✓ Checkpoint 3

Now work Exercise 13.

When finding an equation for the line tangent to the graph of an implicit function, we use a procedure that is much the same as what we have always done.

- Compute the derivative.
- Evaluate the derivative at a specified point.
- Write an equation of the tangent line.

The only difference is that for implicit functions we need both the x- and y-coordinates for the tangent point, since many derivatives are expressed in terms of both x and y.

Note: When we are evaluating a derivative of an implicit expression by replacing both x- and y-values, we write $\left.\frac{dy}{dx}\right|_{(x,y)}$

Example 4 Determining an Equation of a Line Tangent to the Graph of an Implicit Function

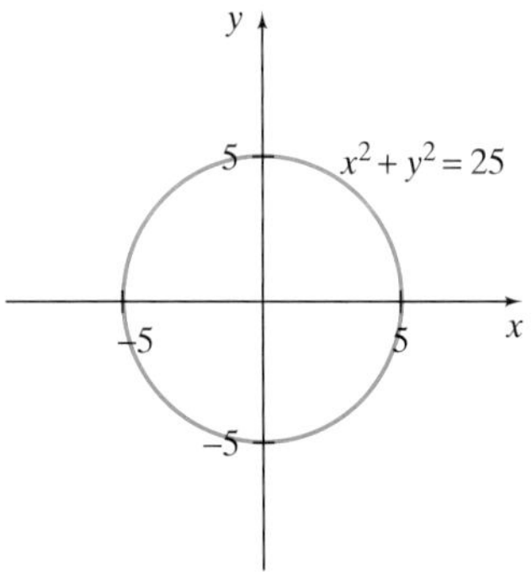

Figure 14.4.2

Consider the equation $x^2 + y^2 = 25$. This equation represents the graph of a circle centered at the origin and has a radius of 5, as shown in Figure 14.4.2.

(a) Use implicit differentiation to determine $\frac{dy}{dx}$ for $x^2 + y^2 = 25$. Evaluate $\left.\frac{dy}{dx}\right|_{(4,-3)}$ and interpret.

(b) Write an equation of the line tangent to the graph of $x^2 + y^2 = 25$ at the point (4, –3).

Solution

(a) Using the Implicit Differentiation Rule for y^n, we get

$$\frac{d}{dx}(x^2) + \frac{d}{dx}(y^2) = \frac{d}{dx}(25)$$

$$2x + 2y\frac{dy}{dx} = 0$$

$$2y\frac{dy}{dx} = -2x$$

$$\frac{dy}{dx} = \frac{-2x}{2y} = \frac{-x}{y}$$

To evaluate $\left.\frac{dy}{dx}\right|_{(4,-3)}$, we replace 4 for x and -3 for y in the derivative $\frac{dy}{dx} = \frac{-x}{y}$ to get

$$\left.\frac{dy}{dx}\right|_{(4,-3)} = \frac{-(4)}{-3} = \frac{4}{3}$$

Interpret the Solution: This means that the slope of the line tangent to the graph of $x^2 + y^2 = 25$ at the point $(4, -3)$ is $m_{\tan} = \frac{4}{3}$.

Interactive Activity

Verify the result to part (b) of Example 4 on your calculator by defining

$y_1 = \sqrt{(25 - x^2)}$

$y_2 = -y_1$

$y_3 = (4/3)(x - 4) - 3$

Do you think all implicit functions can be defined in this way? Explain.

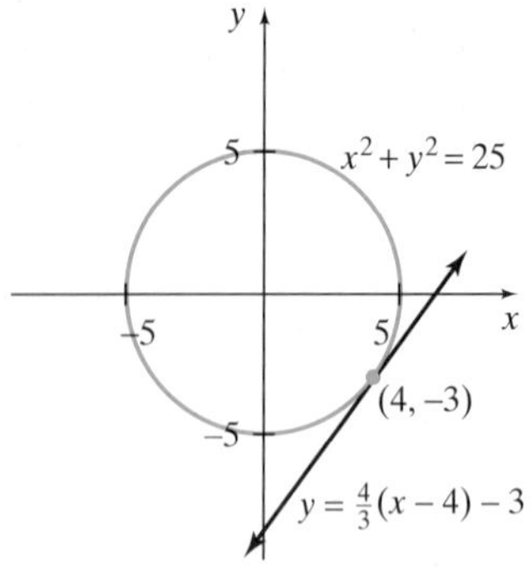

Figure 14.4.3

(b) Using the slope of $\frac{4}{3}$ and the point–slope form of a line, we find that the tangent line equation at the point $(4, -3)$ is

$$y - (-3) = \frac{4}{3}(x - 4)$$

$$y + 3 = \frac{4}{3}(x - 4) \quad \text{or} \quad y = \frac{4}{3}(x - 4) - 3$$

The graph of $x^2 + y^2 = 25$ with tangent line $y = \frac{4}{3}(x - 4) - 3$ is shown in Figure 14.4.3. ■

We can also use implicit differentiation in applications. In the next example we are given a price–demand equation in terms of both demand x and price $p(x)$. To avoid confusion, we will denote the price in such applications by just writing p.

Example 5 Applying Implicit Differentiation

Suppose that the VeeCam Company determines that the price–demand equation for their economy tripod is given by

$$p + 2xp + x^2 = 125 \qquad 0 < x \leq 100$$

where x represents the demand for the tripod in thousands and p represents the price in dollars. Use implicit differentiation to determine $\frac{dp}{dx}$. Evaluate and interpret $\left.\frac{dp}{dx}\right|_{(2.5, 19.5)}$.

Solution

To differentiate $p + 2xp + x^2 = 125$ with respect to x implicitly, we need to use the Product Rule for the term $2xp$.

$$\frac{d}{dx}(p) + \frac{d}{dx}(2xp) + \frac{d}{dx}(x^2) = \frac{d}{dx}(125)$$

$$\frac{d}{dx}(p) + \frac{d}{dx}(2x) \cdot p + 2x \cdot \frac{d}{dx}(p) + \frac{d}{dx}(x^2) = \frac{d}{dx}(125)$$

$$\frac{dp}{dx} + 2p + 2x\frac{dp}{dx} + 2x = 0$$

Solving for $\frac{dp}{dx}$ gives

$$\frac{dp}{dx} + 2x\frac{dp}{dx} = -2x - 2p$$

$$\frac{dp}{dx}(1 + 2x) = -2x - 2p$$

$$\frac{dp}{dx} = \frac{-2x - 2p}{1 + 2x} = \frac{-2(x + p)}{1 + 2x}$$

Evaluating the derivative, when $x = 2.5$ and $p = 19.5$, yields

$$\left.\frac{dp}{dx}\right|_{(2.5,19.5)} = \frac{-2(2.5 + 19.5)}{1 + 2(2.5)} = \frac{-2(22)}{6} \approx -7.33$$

Interpret the Solution: This means that at a demand level of 2.5 thousand tripods and a price of \$19.50, the price is decreasing at a rate of $7.33 \dfrac{\text{dollars}}{\text{thousand tripods}}$. ■

Related Rates

To begin to understand the concept behind related rates, let's suppose that the TastyMix Company determines that the daily revenue function, R for their cupcake mix is given by

$$R(x) = 40x - x^2$$

where x represents the number of boxes of mix produced and sold in hundreds. But instead of having a price–demand equation giving price as a function of x, they find that the amount produced and sold is a function of *time* t or, mathematically, $x = x(t)$. Can we write an expression for $\frac{dR}{dt}$ even if we do not know how the quantity, x, or the revenue, R, are related to t? The answer lies again with the Chain Rule. Since the quantity x is a function of time t and the revenue $R(x)$ is a function of the quantity x, we can write

$$\begin{pmatrix}\text{Change in revenue}\\ \text{with respect to time}\end{pmatrix} = \begin{pmatrix}\text{Change in revenue}\\ \text{with respect to amount}\\ \text{produced and sold}\end{pmatrix} \cdot \begin{pmatrix}\text{Change in amount}\\ \text{produced and sold}\\ \text{with respect to time}\end{pmatrix}$$

$$\frac{dR}{dt} = \frac{dR}{dx} \cdot \frac{dx}{dt}$$

Since $\frac{dR}{dx} = \frac{d}{dx}(40x - x^2) = 40 - 2x$, we get

$$\frac{dR}{dt} = (40 - 2x) \cdot \frac{dx}{dt}$$

$$\frac{dR}{dt} = 40\frac{dx}{dt} - 2x\frac{dx}{dt}$$

This equation is called a **rate of change equation**. It expresses the rate of change of revenue with respect to time, $\frac{dR}{dt}$, in terms of $\frac{dx}{dt}$, the rate of change in the amount produced and sold with respect to time. Hence, we say that $\frac{dR}{dt}$ and $\frac{dx}{dt}$ are **related rates**. Thus, if we know the rate $\frac{dx}{dt}$, then we can find the rate $\frac{dR}{dt}$ using the rate of change equation.

Example 6 Determining a Rate of Change Equation

Determine the rate of change equation by implicitly differentiating $p^2 = x^3 - 300x$ with respect to time, t.

Solution

Using implicit differentiation, we $\frac{d}{dx}$ each term to get

$$2p \cdot \frac{dp}{dt} = 3x^2 \cdot \frac{dx}{dt} - 300 \cdot \frac{dx}{dt}$$

■

Checkpoint 4 Now work Exercise 47.

In the related rates applications that follow, it is necessary to determine a rate of change equation, using implicit differentiation, as we did in Example 6. We then solve for one of the rates, given some information, as illustrated in Example 7.

Example 7 Solving General Related Rates Applications

Assuming that x and p are both functions of t, determine $\frac{dp}{dt}$ for $p^2 = x^3 - 300x$, given $\frac{dx}{dt} = 5$, $x = 2$, and $p = 3$.

Solution

In Example 6 we determined that the rates $\frac{dp}{dt}$ and $\frac{dx}{dt}$ were related by the equation

$$2p \cdot \frac{dp}{dt} = 3x^2 \cdot \frac{dx}{dt} - 300 \cdot \frac{dx}{dt}$$

Substituting the given values and solving for $\frac{dp}{dt}$ gives

$$2(3)\frac{dp}{dt} = 3(2)^2(5) - 300(5)$$

$$6\frac{dp}{dt} = 60 - 1500$$

$$6\frac{dp}{dt} = -1440$$

$$\frac{dp}{dt} = -\frac{1440}{6} = -240$$

■

Applications

The applications of related rates are varied. To aid us in solving related rates applications, we offer the following five-step process.

Solving Related Rates Applications

1. Sketch a diagram, if it can be helpful.
2. Write down all the variables, along with the rates that are given. Be sure to include proper units.
3. Write an equation that relates the variables given.
4. Assuming that the variables are functions of time, t, differentiate the equation implicitly with respect to time, t.
5. Solve for the unknown rate.

Example 8 Applying Related Rates in Geometry Applications

The FreshDay Company delivers its bread in electric-powered trucks. The delivery area is restricted by the range of the trucks to a radius r miles from the company bakery. Through improvements in battery efficiency, the company finds that the radius for the delivery area is increasing at a rate of $3\dfrac{\text{miles}}{\text{year}}$. At the time that the radius is 50 miles, how fast is the delivery area increasing?

Solution

Understand the Situation: For clarity, we will follow the five steps listed in the box on page 898.

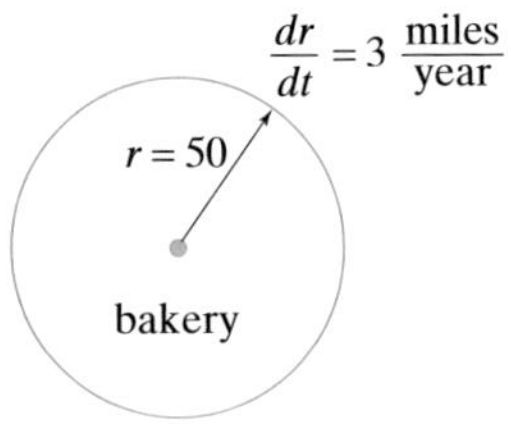

Figure 14.4.4

1. The first step is to sketch a diagram for the application, as shown in Figure 14.4.4.
2. Next, we can label t as the time in years, r as the radius in miles, and A as the delivery area in square miles. When the radius $r = 50$, we know that the radius is increasing at a rate of $\dfrac{dr}{dt} = 3\dfrac{\text{miles}}{\text{year}}$.
3. The radius and the area are related by the equation for a circle, $A = \pi r^2$, where A is the area in square miles and r is the radius in miles.
4. Since r is changing with respect to time t, which means that A is a function of time t, we differentiate the equation $A = \pi r^2$ implicitly with respect to t.

$$\frac{d}{dt}[A] = \frac{d}{dt}[\pi r^2]$$

$$\frac{dA}{dt} = 2\pi r \frac{dr}{dt} \qquad (\pi \text{ is a constant})$$

5. Since we want to know how fast the delivery area is increasing, we seek $\dfrac{dA}{dt}$ when $r = 50$ and $\dfrac{dr}{dt} = 3$. Solving for the unknown rate $\dfrac{dA}{dt}$ yields

$$\begin{aligned}\frac{dA}{dt} &= 2\pi r \frac{dr}{dt} \\ &= 2\pi(50)(3) = 300\pi \approx 942.48\end{aligned}$$

Interpret the Solution: This means that when the radius is 50 miles, the delivery area is increasing at a rate of about $942.48\dfrac{\text{square miles}}{\text{year}}$. ■

Many times, the equation for a related rates application is not given directly. In the previous example, we saw that the geometry formula for the area of a circle was necessary to describe the delivery area. Now let's see how related rates are used in business applications.

Example 9 Applying Related Rates in Business Applications

Past records of the TechTop Company determine that the revenue for the number of software suites produced and sold is given by

$$R = 90x - x^2$$

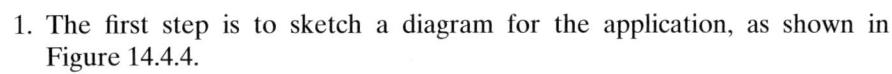

where x is the number of units produced and sold daily and R is the generated revenue in dollars. The company also finds that the software is selling at a rate of 5 suites per day. How fast is the revenue changing when 40 suites are being produced and sold?

Solution

Understand the Situation: Again, we will follow the steps for solving related rates applications.

1. In this application, a diagram is not needed.
2. We have the number of units sold, x, the revenue generated, R, and the time, t. We know that the suites are selling at a rate of $5\frac{\text{suites}}{\text{day}}$, so $\frac{dx}{dt} = 5$ when $x = 40$.
3. In this case, we are given the revenue equation $R = 90x - x^2$.
4. Since we are being asked how revenue, R, is changing with respect to time, t, we differentiate the equation for R with respect to time t.

$$\frac{d}{dt}(R) = \frac{d}{dt}(90x - x^2)$$

$$\frac{dR}{dt} = 90\frac{dx}{dt} - 2x\frac{dx}{dt}$$

5. Since we want to know how fast the revenue is changing, we seek $\frac{dR}{dt}$ when $\frac{dx}{dt} = 5$ and $x = 40$. This gives us

$$\frac{dR}{dt} = 90\frac{dx}{dt} - 2x\frac{dx}{dt}$$

$$= 90(5) - 2(40)(5) = 50$$

Interpret the Solution: Thus, when 40 suites are produced and sold daily, the revenue is increasing at a rate of $50\frac{\text{dollars}}{\text{day}}$. ■

SUMMARY

In this section, we focused on two primary topics. We saw that **implicit differentiation** was a method that could be used to determine derivatives of expressions that were written in terms of x, as well as in terms of both x and y. We introduced a new rule for implicit differentiation. Next we turned our attention to **related rates**. By using implicit differentiation, we found that we could differentiate two quantities with respect to time, t, a third variable. We saw applications of related rates that used geometric formulas and also applications in the business sciences that used our five-step process for solving related rates applications.

- **Implicit differentiation rule for y^n:**

$$\frac{d}{dx}[y^n] = n(y^{n-1}) \cdot \frac{dy}{dx}.$$

- **Solving related rates applications**
  1. Sketch a diagram, if it can be helpful.
  2. Write down all the variables, along with the rates that are given. Be sure to include proper units.
  3. Write an equation that relates the variables given.
  4. Assuming that the variables are functions of time, t, differentiate the equation implicitly with respect to time t.
  5. Solve for the unknown rate.

SECTION 14.4 EXERCISES

Implicit Differentiation

In Exercises 1–10, use implicit differentiation to determine $\frac{dy}{dx}$.

1. $2x + y = 5$ **2.** $x + 3y = 0$

3. $x + 3y^2 = 4$ **4.** $7x^2 - 5y^2 - 100 = 0$

5. $2x^2 + 2y^2 = 32$ **6.** $3x^2 + y^2 = 81$

7. $5x^3 + y^3 - x^4 = 0$ **8.** $4x^2 + 2y^3 = x^3$

9. $x^{1/4} - y^{1/4} = 1$ **10.** $x^{2/3} - y^{2/3} = 4$

In Exercises 11–20, use implicit differentiation to determine $\frac{dy}{dx}$.

11. $x^2 + xy = 6$ **12.** $xy + y^2 = 4$

13. $x^3 - y^3 + 12xy = 0$

14. $2x^3 + 3xy + y^3 = 0$ **15.** $(x + y)^2 - 1 = 7x^2$

16. $(y + 3)^2 = x^2 - 5$ **17.** $y \cdot \ln x = 10 - y$

18. $x + \ln y + 11 = 0$ **19.** $xe^y + x^2 = y^2$

20. $e^x y + y = 4x$

In Exercises 21–26, determine $\frac{dy}{dx}$ *for the following.*

21. $5^y = x^3$ **22.** $6^y = 2x^4$

23. $10^{y-2} = x - 3$ **24.** $7^{3y-1} = x + 1$

25. $\sqrt[3]{y} = x^2 + 6$ **26.** $\sqrt{y} = 3x^4 - 9$

In Exercises 27–32, complete the following.

(a) Use implicit differentiation to determine $\frac{dy}{dx}$.

(b) Solve the equation explicitly for y and differentiate to determine $\frac{dy}{dx}$.

27. $x + 2y = 6$ **28.** $3x - 4y = 12$

29. $xy + 4 = x^4$ **30.** $xy = 2$

31. $\frac{x}{y} - x^2 = 1$ **32.** $\frac{1}{x} + \frac{1}{y} = 3$

For Exercises 33–40, determine an equation of the line tangent to the graph of the given equation at the indicated point.

33. $x^2 + y^2 = 13;\quad (3, 2)$

34. $x^3 + y^3 = 9;\quad (1, 2)$

35. $4x^2 + 9y^2 = 36;\quad (0, 2)$

36. $x^2 + y^2 = 25;\quad (3, 4)$

37. $x \ln y = 2x^3 - 2y;\quad (1, 1)$

38. $x + \ln y - 2y^2 = 0;\quad (2, 1)$

39. $x^2 + y^2 = e^y;\quad (1, 0)$ **40.** $4e^y + y = x^2;\quad (2, 0)$

Applications

41. The Crabtree Company determines that the price–demand equation for their mini picture frames is given by

$$p + x^2 = 150$$

where x represents the demand for the frames in hundreds and p represents the price in dollars.

(a) Use implicit differentiation to determine $\frac{dp}{dx}$.

(b) Evaluate and interpret $\left.\frac{dp}{dx}\right|_{(11,29)}$

42. Suppose that Pottery Plus determines that the price–demand equation for their glazed presidential inauguration mugs is given by

$$\frac{p}{2} + \sqrt{x} = 30$$

where x represents the demand for the mugs and p represents the price in dollars.

(a) Use implicit differentiation to determine $\frac{dp}{dx}$.

(b) Evaluate and interpret $\left.\frac{dp}{dx}\right|_{(400,20)}$

43. The WoodedLife Company has determined that the price–demand equation for their new single-occupant tent is given by

$$px + 2x = 1000 \qquad 10 \le x \le 30$$

where x represents the demand for the tents in hundreds and p represents the price in dollars.

(a) Use implicit differentiation to determine $\frac{dp}{dx}$.

(b) Evaluate and interpret $\left.\frac{dp}{dx}\right|_{(20,48)}$

44. The CorpStyle Company determines that the price–demand equation for their all-silk power ties is given by

$$x^2 = 1000 - p^2 \qquad 10 \le x \le 30$$

where x represents the demand for the ties in thousands and p represents the price in dollars.

(a) Use implicit differentiation to determine $\frac{dp}{dx}$.

(b) Evaluate and interpret $\left.\frac{dp}{dx}\right|_{(18,26)}$

Related Rates

In Exercises 45–52, determine the rate of change equation by implicitly differentiating each of the following with respect to time t.

45. $2x + 3y = 20$ **46.** $x^2 + 2y = 11$

47. $x^2 - 3y = 1$ **48.** $y = x^2 - 3x + 5$

49. $x^2 + y^2 = 5x$ **50.** $x^3 - y^3 = 10y$

51. $5xy + y^4 = x$ **52.** $xy = 4x + 5y + 9$

For Exercises 53–58, assume that x and y are both functions of t and determine the indicated rate.

53. Determine $\frac{dy}{dt}$ for $xy = 7$ given $x = 7$, $y = 1$, and $\frac{dx}{dt} = 2$.

54. Determine $\frac{dx}{dt}$ for $x^2 + 9y^2 = 18$ given $x = 3$, $y = 1$, and $\frac{dy}{dx} = 10$.

55. Determine $\frac{dx}{dt}$ for $y^2 + x = 3$ given $x = 2$, $y = 1$, and $\frac{dy}{dt} = 2$.

56. Determine $\frac{dy}{dt}$ for $y = x^2 - 2x + 3$ given $x = 5$, $y = 18$, and $\frac{dx}{dt} = 3$.

57. Determine $\frac{dy}{dt}$ for $x^2 + y^2 = 25$ given $x = 3$, $y = -4$, and $\frac{dx}{dt} = 2$.

58. Determine $\frac{dx}{dt}$ for $y^2 + xy - 3x = -1$ given $x = 1$, $y = -2$, and $\frac{dy}{dt} = -2$.

Applications

59. The area of a circle with radius r is increasing at a rate of 12 square inches per minute. Find the rate at which the radius is increasing when the radius is 2 inches.

60. The area of a circle with radius r is increasing at a rate of 5 square inches per minute. Find the rate at which the radius is increasing when the radius is 0.8 inch.

61. The area of a square with sides x inches is increasing at a rate of 10 square inches per minute. Find the rate at which a side is increasing when the sides are 3 inches.

62. The area of a square with sides x inches is increasing at a rate of 25 square inches per minute. Find the rate at which a side is increasing when the sides are 4 inches.

63. A circular pan is being heated and the radius of the heated area increases at a rate of 0.02 inch per minute. Find the rate at which the area is increasing when the radius is 8 inches.

64. A 5-foot-tall man is walking away from a 20-foot street lamp at a speed of 6 feet per second. How fast is the tip of his shadow moving along the ground?

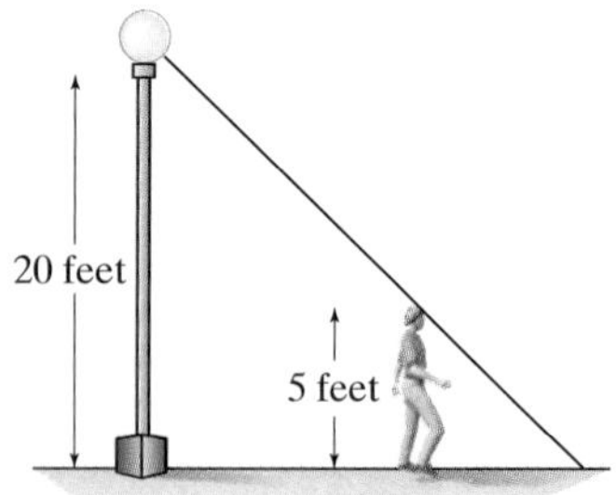

65. A 6-foot-tall woman is walking away from a 24-foot street lamp at a speed of 8 feet per second. How fast is the tip of her shadow moving along the ground?

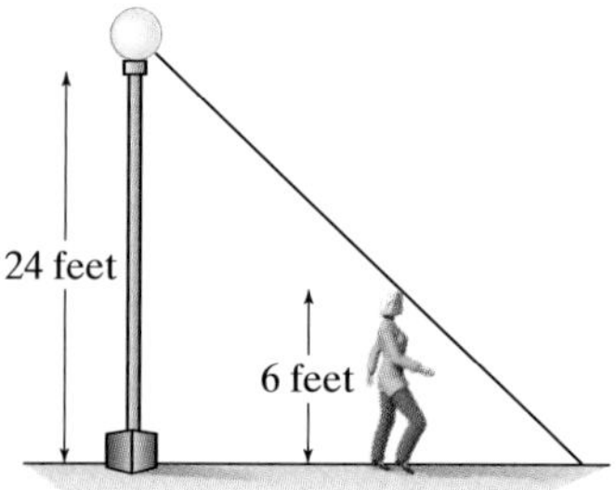

66. The SoftSkirt Company determines that the monthly revenue for a new style of skirt is given by

$$R = 60 - \frac{1}{2}x^2$$

where x is the number of skirts produced and sold in hundreds and R is the revenue, in thousands of dollars, generated from the skirts.

(a) Differentiate the revenue function with respect to time t.

(b) Determine the rate of change in the revenue with respect to time at a production level of $x = 3$ and $\frac{dx}{dt} = 20 \frac{\text{hundred skirts}}{\text{month}}$.

67. The SharpSuit Company determines that monthly revenue for a new type of casual suit is given by

$$R = 250x - \frac{2}{5}x^2$$

where x is the number of suits produced and sold and R is the revenue in dollars.

(a) Differentiate the revenue function with respect to time t.

(b) Determine the rate of change in the revenue with respect to time at a production level of $x = 100$ and $\frac{dx}{dt} = 200 \frac{\text{suits}}{\text{month}}$.

68. A 24-foot ladder is leaning against the wall as shown in the figure below. Assume that the values of x and y are related by the Pythagorean Theorem equation

$$x^2 + y^2 = 24^2$$

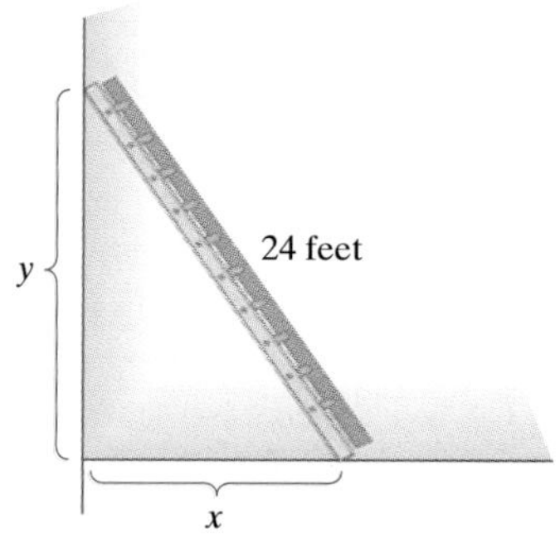

(a) Differentiate each side of the equation with respect to time t.

(b) If the base of the ladder is sliding away from the wall at a rate of $2\frac{\text{feet}}{\text{second}}$, find the rate at which the top of the ladder is sliding down the wall when the top of the ladder is 12 feet from the ground.

69. A 20-foot ladder is leaning against the wall as shown in the figure below. Assume that the values of x and y are related by the Pythagorean Theorem equation

$$x^2 + y^2 = 20^2$$

(a) Differentiate each side of the equation with respect to time t.

(b) If the base of the ladder is sliding away from the wall at a rate of $3\frac{\text{feet}}{\text{second}}$, find the rate at which the top of the ladder is sliding down the wall when the top of the ladder is 8 feet from the ground.

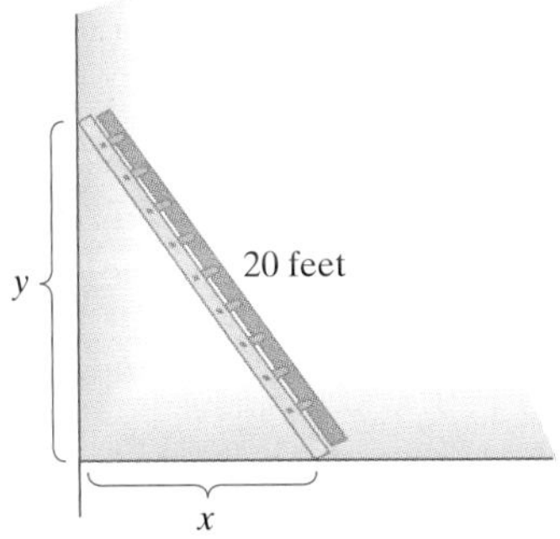

70. A surgeon has a patient with a tumor in the shape of a sphere. Assume that the volume of the tumor is given by the equation

$$V = \frac{4}{3}\pi r^3$$

where r is the radius of the tumor measured in millimeters and V is its volume, measured in cubic millimeters.

(a) Differentiate each side of the equation with respect to time t.

(b) If tests show that the radius of the tumor is growing at a rate of $0.3\frac{\text{millimeter}}{\text{week}}$ determine the rate at which the volume is changing when the radius is 15 millimeters.

71. Assume that the number of bass in a pond is related to the level of polychlorinated biphenyls (or PCBs) in the pond. (PCBs are a group of industrial chemicals used in plasticizers, fire retardants, and other materials.) The bass population is modeled by

$$y = \frac{2500}{1 + x}$$

where x represents the PCB level in parts per million (ppm) and y represents the number of bass in the pond.

(a) Differentiate each side of the equation with respect to time t.

(b) If the level of PCBs is increasing at rate of $40\frac{\text{ppm}}{\text{year}}$, find the rate at which y is changing when there are 100 bass in the pond.

72. The price–demand equation for oranges in the town of Southburg is given by

$$p = \frac{40{,}000}{x^{1.5}}$$

where x is the demand for oranges in bushels per month and p is the wholesale price per bushel. The local farmer's market finds that the demand for oranges is currently 900 bushels each month and is increasing at a rate of $100\frac{\text{bushels}}{\text{month}}$. How fast is the price changing?

73. The price–demand equation of the number of DVD players sold by a local electronics store is given by

$$p = 0.2x^2 + 2x$$

where x is the demand for the DVD players per week and p is the price per player. The store finds that the demand for the players is currently 150 players each week and is increasing at a rate of $10\frac{\text{players}}{\text{week}}$. How fast is the price changing?

SECTION PROJECT

The strategic planning department of the Higbrize Company is given the data in Table 14.4.1 on the cost of producing a new lapel pager. The department is asked to answer the following scenario: The current production level is 7500 pagers and is increasing at a rate of 50 units per day. How is the average cost changing?

(a) Use the regression features of your calculator to determine a linear cost model for the lapel pagers.

(b) Determine the average cost function AC.

(c) Differentiate AC with respect to time t.

(d) What do the numbers 7500 and 50 represent in terms of production x and time t?

(e) Use related rates to find the rate of change for the average cost with respect to time and interpret.

Table 14.4.1

Units Produced, x	Cost, C
1500	14,725
2500	18,125
3500	21,725
4000	23,600
5500	29,525
7000	35,900

Section 14.5 Elasticity of Demand

In Chapter 13, we discussed the concept of marginal analysis, which focused on the effect that a small change in the quantity produced and sold had on cost, revenue, and profit. Now we will take a different approach. We wish to know what effect a change in price has on the quantity demanded. Let's take a look at an illustration. In the 1980s and early 1990s, Japanese microprocessor manufacturers flooded the technology market with low-price microchips in an attempt to force other overseas manufacturers out of business. This huge increase in microchips caused lower prices. The low prices, in turn, stimulated consumer demand, but reduced manufacturers' revenues dramatically. **Elasticity of demand** is a mathematical tool that can be used to measure the impact that a change in price has on the demand for a product. The term **elasticity** generally refers to how sensitive the demand is to a change in price.

Arc Elasticity

Let's begin by looking at some examples of elasticity. We call a product **elastic** if a small change in price produces a significant change in demand. The furniture market is an example of demand that is elastic. This market is very sensitive to changes in price, as furniture shoppers are always seeking value and quality. On the other hand, we call a product **inelastic** if a change in price generally does not affect demand. The heating oil market is an example of a commodity that is inelastic. Since homeowners who use heating oil must have the product in the winter months, changes in price have little effect on the demand.

Elastic and Inelastic Products

- If small changes in the unit price of a product result in significant changes in demand, the product is **elastic**.
- If small changes in the unit price of a product do not result in significant changes in demand, the product is **inelastic**.

Generally, there are two types of elasticity. The first type that economists commonly use is **arc elasticity**. This type of elasticity is computed when the actual changes in price and quantity are known and a price–demand function may not be given. Arc elasticity measures the ratio of relative change in quantity to the relative change in price.

$$(\text{Arc elasticity}) = -\frac{(\text{relative change in quantity demanded})}{(\text{relative change in price})}$$

$$= -\frac{\frac{\Delta q}{q}}{\frac{\Delta p}{p}}$$

The relative changes are simply the change in the quantity or price divided by their original value. The reason for the negative sign is purely a convention used by economists.

Arc Elasticity

For finite changes in quantity and price, the **arc elasticity**, denoted by E_a, is

$$E_a = -\frac{\frac{\Delta q}{q}}{\frac{\Delta p}{p}} = -\frac{\frac{q_2 - q_1}{q_1}}{\frac{p_2 - p_1}{p_1}}$$

where q_1 and q_2 are the original and new quantities, respectively, and p_1 and p_2 are the original and new prices, respectively.

Example 1 Determining Arc Elasticity

The GreenLawn Company can sell 300 mulching lawn mowers each week when the unit price of the mowers is \$200. They find through a promotion that when the price is reduced by 10% to \$180 the number sold rises by 20% to 360 mowers each week. Determine the arc elasticity.

Solution

Understand the Situation: Here we have the original quantity $q_1 = 300$ and original price $p_1 = 200$. We also have the new quantity $q_2 = 360$ and the new price $p_2 = 180$. We will use these values to compute the arc elasticity.

So the arc elasticity is

$$E_a = -\frac{\frac{q_2 - q_1}{q_1}}{\frac{p_2 - p_1}{p_1}} = -\frac{\frac{360 - 300}{300}}{\frac{180 - 200}{200}} = -\frac{\frac{60}{300}}{\frac{-20}{200}} = -\frac{0.2}{-0.1} = 2$$

Interpret the Solution: This means that for the lawn mowers a $0.1 = 10\%$ change in price causes a $0.2 = 20\%$ change in sales. Since the percentage change in sales exceeds the percentage change in price, we say that the demand is *elastic*. That is, if $E_a > 1$ then the demand is elastic. ■

✓ Checkpoint 1

Now work Exercise 3.

Arc elasticity is applicable when we have finite changes in quantity and price. However, we cannot determine the elasticity at a single quantity demanded. For this, we would need the *instantaneous* change in price and quantity. To compute this type of elasticity, we need the derivative. This results in a new type of elasticity, called **point elasticity**.

Point Elasticity

From Your Toolbox

The **price–demand function** p gives us the price $p(x)$ at which people buy exactly x units of product.

To determine this new type of elasticity, we need to rewrite the price–demand function. Let's recall what this function tells us as shown in the Toolbox to the left.

Since the price–demand function is always decreasing on its domain, we know that the function is one-to-one. Thus, we can solve the price–demand function for x to get what economists call the **demand function**. This function plays a key role in determining point elasticity.

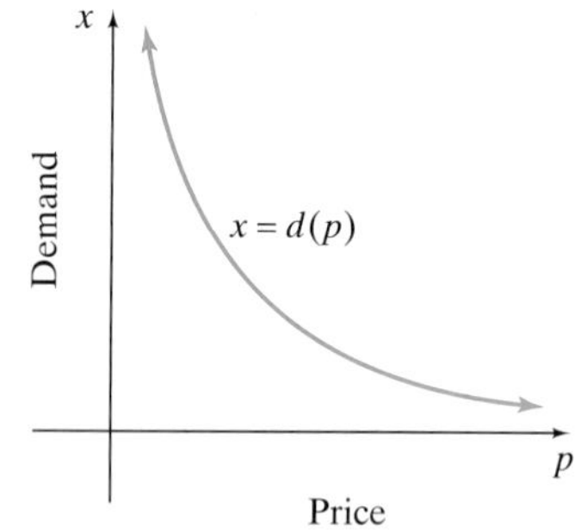

Figure 14.5.1 Graph of demand function $x = d(p)$.

Demand Function

The **demand function**, denoted by $d(p)$, is given by

$$x = d(p)$$

The demand function gives the quantity x of a product demanded by consumers at price p.

A graph of a typical demand function is shown in Figure 14.5.1. Notice that the price p is now on the horizontal axis and the demand x is on the vertical axis. To determine the demand function, we simply need to solve the price–demand function for x. Since p will become the independent variable, we will express $p(x)$ simply as p. Example 2 illustrates this.

Example 2 Determining the Demand Function

Determine the demand function $x = d(p)$ for the price–demand function

$$p = \frac{1000}{x^2} \qquad 10 \le x \le 50$$

Solution

We can get the demand function by solving $p = \frac{1000}{x^2}$ for x.

$$p = \frac{1000}{x^2}$$
$$x^2 \cdot p = 1000$$
$$x^2 = \frac{1000}{p}$$
$$x = \sqrt{\frac{1000}{p}}$$

So the demand function is $x = d(p) = \sqrt{\frac{1000}{p}}$. Notice that we use only the positive square root since both price and demand are represented by positive values. ■

Checkpoint 2 Now work Exercise 13.

To determine point elasticity, suppose that for a demand function the unit price of a product is increased by h dollars to $p + h$ dollars, as shown in Figure 14.5.2. This causes the quantity demanded to decrease from $d(p)$ to $d(p + h)$ units. So we see that the relative change in demand is

$$(\text{relative change in quantity demanded}) = \frac{d(p+h) - d(p)}{d(p)}$$

The corresponding change in unit price is

$$(\text{relative change in price}) = \frac{(p+h) - p}{p} = \frac{h}{p}$$

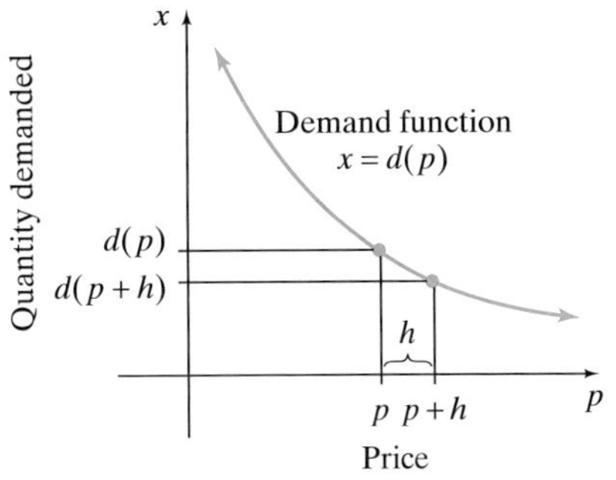

Figure 14.5.2

Knowing from the beginning of this section that elasticity is the ratio of relative change in quantity to the relative change in price, we get

$$-\frac{\text{(relative change in quantity demanded)}}{\text{(relative change in price)}} = -\frac{\dfrac{d(p+h)-d(p)}{d(p)}}{\dfrac{h}{p}}$$

As the change in price becomes small, that is, $h \to 0$, we get the point elasticity formula.

$$\text{(Point elasticity)} = \lim_{h\to 0} -\frac{\dfrac{d(p+h)-d(p)}{d(p)}}{\dfrac{h}{p}}$$

$$= \lim_{h\to 0} -\frac{\dfrac{d(p+h)-d(p)}{h}}{\dfrac{d(p)}{p}} = -\frac{\lim\limits_{h\to 0}\dfrac{d(p+h)-d(p)}{h}}{\dfrac{d(p)}{p}}$$

$$= -\frac{d'(p)}{\dfrac{d(p)}{p}} = -\frac{p \cdot d'(p)}{d(p)}$$

We will simply call this type of elasticity the **elasticity of demand**.

Elasticity of Demand

Given the demand function $x = d(p)$, the **elasticity of demand**, denoted by $E(p)$, of a product at a price p is given by

$$E(p) = \frac{-p \cdot d'(p)}{d(p)}$$

Example 3 Determining the Elasticity of Demand

Determine the elasticity of demand $E(p)$ for each price–demand function.

(a) $p = 12 - 0.4x$ **(b)** $p = \dfrac{50}{2-x}$

Solution

(a) First, we must solve this price–demand function for x to get the demand function, $x = d(p)$.

$$p = 12 - 0.4x$$
$$p - 12 = -0.4x$$
$$\frac{p-12}{-0.4} = x$$
$$-2.5p + 30 = x = d(p)$$

Now, to find $d'(p)$, we differentiate $d(p)$ with respect to p to get

$$d'(p) = \frac{d}{dp}(-2.5p + 30) = -2.5$$

Substituting into the elasticity of demand formula gives

$$E(p) = \frac{-p \cdot d'(p)}{d(p)} = \frac{-p(-2.5)}{-2.5p + 30} = \frac{2.5p}{-2.5p + 30}$$

(b) Solving $p = \dfrac{50}{2-x}$ for x, we get

$$p = \frac{50}{2-x}$$
$$p(2-x) = 50$$
$$2p - xp = 50$$
$$-xp = 50 - 2p$$
$$x = \frac{50-2p}{-p} = -\frac{50}{p} + 2 = 2 - \frac{50}{p} = 2 - 50p^{-1} = d(p)$$

Differentiating with respect to p yields

$$d'(p) = \frac{d}{dp}(2 - 50p^{-1}) = (-1)(-50p^{-2})$$
$$= 50p^{-2} = \frac{50}{p^2}$$

This gives the elasticity of demand $E(p)$ as

$$E(p) = \frac{-p \cdot d'(p)}{d(p)} = \frac{-p\left(\frac{50}{p^2}\right)}{2 - \frac{50}{p}}$$
$$= \frac{\frac{-50}{p}}{\frac{2p-50}{p}} = \frac{-50}{2p-50}$$

■

Notice that the value of $E(p)$ has no units, since the units of demand are canceled out when computing the ratio of the relative changes. Also notice that the point elasticity function requires only one price to compute the elasticity coefficient. The interpretation of the values of $E(p)$ is summarized in Table 14.5.1.

Table 14.5.1 Interpreting the $E(p)$ Value

If the $E(p)$ value is:	then the demand is:	and this means a change in price will cause:
Between 0 and 1	Inelastic	Relatively small changes in demand
Greater than 1	Elastic	Relatively large changes in demand
Equal to 1	Unitary	A relatively equal change in demand

Example 4 Interpreting the Elasticity of Demand

A school's junior business club is holding its annual raffle. Data collected from raffles in the past indicate that the demand function for the tickets follows the model

$$x = d(p) = 36 - p^2$$

where p represents the price of a ticket and $x = d(p)$ represents the number of tickets each member sells each day.

(a) Find the elasticity of demand $E(p)$.

(b) Evaluate $E(3)$ and $E(4)$ and interpret the results.

Solution

(a) To determine $d'(p)$, we differentiate the demand function with respect to p to get

$$d'(p) = \frac{d}{dp}(36 - p^2) = -2p$$

Using the definition of elasticity of demand yields

$$E(p) = \frac{-p \cdot d'(p)}{d(p)} = \frac{-p(-2p)}{36 - p^2} = \frac{2p^2}{36 - p^2}$$

(b) First, evaluating the elasticity of demand at \$3, we get

$$E(3) = \frac{2(3)^2}{36 - (3)^2} = \frac{18}{27} = \frac{2}{3}$$

Interpret the Solution: Since $E(3) = \frac{2}{3} < 1$, we see that the demand for the raffle ticket is **inelastic** at a price of \$3 a ticket. A small increase in price will cause a negligible change in demand.

Evaluating the point elasticity function at \$4, we get

$$E(4) = \frac{2(4)^2}{36 - (4)^2} = \frac{32}{20} = \frac{8}{5}$$

Interpret the Solution: Here $E(4) = \frac{8}{5} > 1$, so the demand for the raffle tickets is **elastic** at a price of \$4 a ticket. So a small increase in price will cause a significant change in demand. ■

Checkpoint 3 Now work Exercise 39.

In Example 4, we found that at a price of \$3 the demand for a ticket was inelastic, so demand is only negligibly sensitive to price changes. Increasing prices slightly will increase the revenue $R(x)$, because the higher price will turn away relatively few buyers. Conversely, when the price was \$4, the demand was elastic, so the demand is sensitive to price changes. Thus, to increase revenue in this case, the club should slightly lower its price. The lower price will increase demand for tickets and will generate more revenue, which will offset the lower price. Furthermore, if the point elasticity was $E(p) = 1$, then prices should not be changed, because an increase in price will result in exactly the same percentage decrease in demand. Thus, when $E(p) = 1$, revenue is at a maximum.

Since $R'(p)$ is the rate of change in the revenue as the price p changes, the equation $R'(p) = d(p)(1 - E(p))$ expresses this relationship between $R'(p)$ and $d(p)$. This is summarized in Table 14.5.2.

Table 14.5.2 Elasticity–Revenue Relationship

If the demand is:	then $E(p)$ is:	and $R'(p)$ is:	so the revenue is:
Elastic	Greater than 1	Negative	Decreasing
Inelastic	Less than 1	Positive	Increasing
Unitary	1	Zero	At a maximum

From Table 14.5.2 and Figure 14.5.3, we can see the following:

- If demand is elastic, $E(p) > 1$, then prices should be lowered to increase revenue.
- If demand is inelastic, $E(p) < 1$, then prices should be raised to increase revenue.
- If demand is unitary, $E(p) = 1$, then the revenue is at its maximum and the price should not be changed.

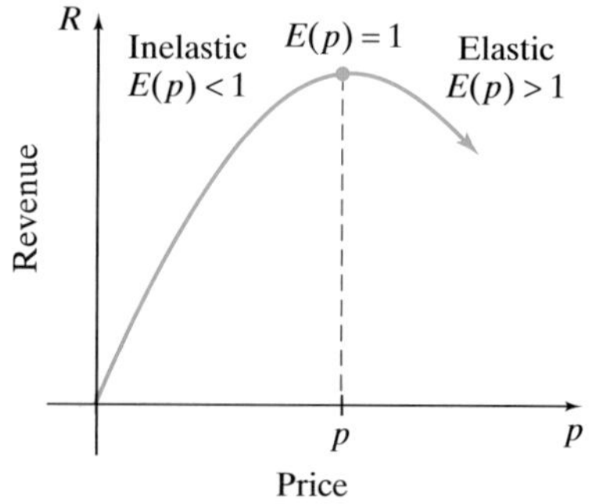

Figure 14.5.3

In Chapter 15, we will focus more on the sign of the derivative and finding maximum values.

Example 5 Determining Price Adjustment to Increase Revenue

The PowerTi Company produces neckties and sells them at a price of \$15. The demand function for the ties is given by

$$x = d(p) = 170{,}000\sqrt{20 - p}$$

where p is the price of a necktie and $d(p)$ is the demand for the ties.

(a) Determine $E(15)$ and state whether the manufacturer should lower or raise the price to increase revenue.

(b) Set $E(p) = 1$ and solve for p to determine at what price the revenue is greatest.

Solution

(a) First, we need the elasticity of demand $E(p)$. Computing the derivative of the demand function $d(p) = 170{,}000(20 - p)^{1/2}$, we get

$$\begin{aligned} d'(p) &= \frac{d}{dp}[170{,}000(20 - p)^{1/2}] \\ &= 170{,}000\left(\frac{1}{2}\right)(20 - p)^{-1/2}(-1) = \frac{-85{,}000}{\sqrt{20 - p}} \end{aligned}$$

Now, using this result in the point elasticity formula,

$$\begin{aligned} E(p) &= \frac{-p \cdot d'(p)}{d(p)} = \frac{-p\left(\dfrac{-85{,}000}{\sqrt{20 - p}}\right)}{170{,}000\sqrt{20 - p}} \\ &= \frac{85{,}000p}{\sqrt{20 - p}} \cdot \frac{1}{170{,}000\sqrt{20 - p}} = \frac{0.5p}{20 - p} \end{aligned}$$

So at a price of $p = 15$, we get

$$E(15) = \frac{0.5(15)}{20 - 15} = \frac{7.5}{5} = 1.5$$

Since $E(15) = 1.5 > 1$, the demand is relatively elastic and the prices should be *lowered* to increase revenue.

(b) To find the price that maximizes revenue, we need to find the value of p that makes $E(p) = 1$. So, solving this equation, we get

$$\begin{aligned} E(p) &= 1 \\ \frac{0.5p}{20 - p} &= 1 \\ 0.5p &= 20 - p \\ 1.5p &= 20 \\ p &= \frac{20}{1.5} \approx 13.33 \end{aligned}$$

Thus, to maximize revenue, the price of the neckties should be \$13.33. ■

✓ Checkpoint 4 Now work Exercise 43.

SUMMARY

In this section we discussed a topic that is specific to economics, **elasticity of demand**. We first computed elasticity for finite changes in price and demand using the **arc elasticity** formula. We saw that products could be classified as **elastic**, **inelastic**, or **unitary**. Then we examined elasticity for instantaneous changes in price and quantity, called **point elasticity**. Finally, we examined the relationship between the **elasticity of demand** and maximizing the revenue.

- **Arc elasticity**: $E_a = -\dfrac{\dfrac{q_2 - q_1}{q_1}}{\dfrac{p_2 - p_1}{p_1}}$
- **Point elasticity**: $E(p) = \dfrac{-p \cdot d'(p)}{d(p)}$

SECTION 14.5 EXERCISES

In Exercises 1–6, compute the arc elasticity E_a.

1. $q_1 = 50, q_2 = 55, p_1 = 110, p_2 = 100$

2. $q_1 = 10, q_2 = 13, p_1 = 8, p_2 = 5$

3. $q_1 = 300, q_2 = 260, p_1 = 40, p_2 = 45$

4. $q_1 = 5000, q_2 = 5800, p_1 = 60, p_2 = 54$

5. $q_1 = 5700, q_2 = 5600, p_1 = 390, p_2 = 400$

6. $q_1 = 420, q_2 = 419, p_1 = 389, p_2 = 395$

7. The 4Com Company determines that the demand function for their mouse pad is

$$x = d(p) = 20 - 2p$$

where p is the unit price of the mouse pad and $d(p)$ is the quantity demanded in hundreds.

(a) Determine $d(5)$ and $d(6)$ and interpret.

(b) Using $q_1 = d(5), q_2 = d(6), p_1 = 5$, and $p_2 = 6$, determine the arc elasticity E_a and interpret.

8. The WetLands Company determines that the demand function for their rain poncho is

$$x = d(p) = 50 - 5p$$

where p is the unit price of the rain poncho and $d(p)$ is the quantity demanded in thousands.

(a) Determine $d(5)$ and $d(7)$ and interpret.

(b) Using $q_1 = d(5), q_2 = d(7), p_1 = 5$, and $p_2 = 7$, determine the arc elasticity E_a and interpret.

9. The 725th Street Movie Theater finds that it can sell 320 tickets per day when the price of each ticket is \$8.50. During a promotion week, the theater finds that by reducing the ticket price to \$7.25, the number of tickets sold increases to 410 tickets per day. Determine the arc elasticity and interpret.

10. The HappyFace Portrait Studio sells 325 family packs of photos monthly at a unit price of \$49. When they offer a sale price of \$39, the number of family packs increases to 330 sold per month. Determine the arc elasticity and interpret.

11. The JemJam Jewelry Store finds that they sell 52 pairs of emerald earrings per week at a price of \$149 each. Under new management, the price is increased to \$159 and the number sold decreases to 45 pairs per week. Determine the arc elasticity and interpret.

12. The Fleetwood Company sells 600 stained glass window ornaments monthly at a unit price of \$19. When they increased the price to \$24 in order to absorb higher fixed costs, they found that the monthly sales decreased to 450 ornaments. Determine the arc elasticity and interpret.

In Exercises 13–22, the price–demand function p is given. Determine the demand function $x = d(p)$.

13. $p = 600 - 100x$

14. $p = 100 - x$

15. $p = \dfrac{300}{x^2} + 10$

16. $p = \dfrac{700}{2x + 1}$

17. $p = 12 - 0.04x$

18. $p = \sqrt{250 - x}$

19. $p = \sqrt{300 - x^2}$

20. $p = 1200 - x^2$

21. $p = 100e^{-0.1x}$

22. $p = 250e^{-0.3x}$

In Exercises 23–38, the demand function $x = d(p)$ and the price p are given. Determine the point elasticity of demand $E(p)$, and classify the demand as elastic, inelastic, or unitary.

23. $d(p) = 220 - 5p; \quad p = 10$

24. $d(p) = 50 - 4p; \quad p = 5$

25. $d(p) = 200 - p^2; \quad p = 8$

26. $d(p) = 250 - p^2; \quad p = 10$

27. $d(p) = \dfrac{200}{p}; \quad p = 15$

28. $d(p) = \dfrac{1000}{7p}; \quad p = 20$

29. $d(p) = \sqrt{150 - 3p}; \quad p = 30$

30. $d(p) = \sqrt{100 - 2p}; \quad p = 25$

31. $d(p) = \dfrac{100}{p^2}; \quad p = 30$

32. $d(p) = \frac{500}{p^3}$; $p = 20$

33. $d(p) = 4500e^{-0.02p}$; $p = 200$

34. $d(p) = 6500e^{-0.04p}$; $p = 100$

35. $d(p) = 100 \ln(1000 - 10p)$; $p = 19$

36. $d(p) = 150 \ln(100 - 2p)$; $p = 5$

37. $d(p) = 100e^{-0.05p}$; $p = 40$

38. $d(p) = 3000e^{-0.03p}$; $p = 100$

Applications

✓ **39.** Currently, about 1800 people ride the MetroTram local public transportation system per day and pay \$4 for each ticket. The number of people x willing to take the public transportation at price p is given by the demand function

$$x = 600(5 - p^{1/2})$$

(a) Is the demand elastic or inelastic at the current \$4 ticket price?

(b) Should the ticket price be raised or lowered to increase revenue?

40. The EconoCommute bus system charges 65 cents for each ride and serves approximately 57,800 riders each day. The demand function for x riders at p cents for each ride is given by

$$x = 2000\sqrt{900 - p}$$

(a) Is the demand elastic or inelastic at the current 65-cent fare?

(b) Should the rider's price be raised or lowered to increase revenue?

41. The SilkTop clothing manufacturer sells blouses at \$15 each and calculates the demand function for the blouses to be

$$d(p) = 60 - 3p$$

where p is the price of a blouse and $d(p)$ represents the demand in hundreds.

(a) Is the demand elastic or inelastic at the current \$15 price?

(b) Should the blouse price be raised or lowered to increase revenue?

42. Suppose that the demand function for a certain brand of cigarettes is

$$d(p) = 4.5p^{-0.73}$$

where p is the price of a pack of cigarettes and $d(p)$ is the demand in thousands.

(a) Determine the point elasticity of demand $E(p)$.

(b) Show that the demand is inelastic for any price $p > 0$. Interpret what this means in terms of raising cigarette prices through manufacturer and tax price increases.

✓ **43.** The IceDream Company determines that the demand function for their frozen yogurt is

$$d(p) = 50 - 2p$$

where p is the price, in dollars, of a quart of frozen yogurt and $d(p)$ represents the demand in hundreds.

(a) Determine $E(4)$ and state whether the manufacturer should lower or raise the price to increase revenue.

(b) Set $E(p) = 1$ and solve for p to determine at what price the revenue is greatest.

44. The PackIt Company determines that the demand function for their lightweight daypack is given by

$$d(p) = -p^2 + 400$$

where p is the unit price, in dollars, of a daypack and $d(p)$ is the demand.

(a) Determine $E(10)$ and state whether the manufacturer should lower or raise the price to increase revenue.

(b) Set $E(p) = 1$ and solve for p to determine at what price the revenue is greatest.

45. The EverGlo Company determines that the demand for their new night light is given by

$$d(p) = 2 - \frac{p^2}{5}$$

where p is the unit price of a night light, in dollars, and $d(p)$ is the demand in thousands.

(a) Determine $E(1.50)$ and state whether the manufacturer should lower or raise the price to increase revenue.

(b) Set $E(p) = 1$ and solve for p to determine at what price the revenue is greatest.

46. The SuperChip Company determines that the demand function for their gourmet potato chips is given by

$$d(p) = 3 - 0.1p - 0.1p^2$$

where p is the unit price, in dollars, of a bag of potato chips and $d(p)$ is the demand in thousands.

(a) Determine $E(4)$ and state whether the manufacturer should lower or raise the price to increase revenue.

(b) Set $E(p) = 1$ and solve for p to determine at what price the revenue is greatest.

47. The MicroTV Company determines that the demand for their 3-inch color TV is

$$d(p) = 100 \ln(150 - p)$$

where p is the unit price, in dollars, of a 3-inch color TV and $d(p)$ is the demand in hundreds.

(a) Determine $E(100)$ and state whether the manufacturer should lower or raise the price to increase revenue.

(b) Set $E(p) = 1$ and solve for p to determine at what price the revenue is greatest.

SECTION PROJECT

Some who study the theory of economics want to generalize the trends that they see relative to demand functions. To begin to understand how this is done, answer the following:

(a) If the price–demand for a product is $d(p) = 300p^{-0.3}$ show that the elasticity of demand $E(p)$ is a constant.

(b) Consider the price–demand function $d(p) = c \cdot p^{-n}$, where c and n are fixed constants. Show that the elasticity $E(p)$ is also a fixed constant.

(c) If the price–demand for a product is $d(p) = \frac{100}{e^{0.2p}}$ show that the elasticity of demand $E(p)$ is a linear function.

(d) Consider the price–demand function $d(p) = \frac{a}{e^{cp}}$, where a and c are fixed constants. Show that the elasticity $E(p)$ is a linear function of the form $E(p) = cp$.

Why We Learned It

In this chapter we learned differentiation formulas that allow us to differentiate composite functions, logarithmic functions, and exponential functions. Since many real world situations involve these types of functions, it was important for us to learn how to differentiate these functions so that we could analyze rates of change. For example, the population of the earth over the last 100 years has followed an exponential growth pattern. In order to compute the rate at which the earth's population was growing in, say, 1995, we need to be able to compute the derivative of the function that models the earth's population. Exponential functions also model money in an account that is earning compound interest. So if we want to compute the rate at which the account balance is growing, in dollars per year, we need the derivative. In the latter part of the chapter, we focused on related rates and elasticity of demand. Related rates allowed us to analyze situations where items were dependent on time. For example, a pizza delivery area may be circular in its coverage and the radius for the delivery area may be growing at a rate of one mile per year. Related rates allow us to determine how fast the delivery area is increasing when the radius is, say, 5 miles. Elasticity of demand is a purely economics topic. Understanding elasticity of demand, and whether or not the demand for a product is sensitive to price changes, is imperative for many businesses.

CHAPTER REVIEW EXERCISES

In Exercises 1–10, differentiate using the Generalized Power Rule.

1. $f(x) = (x + 2)^3$

2. $f(x) = (x - 5)^2$

3. $f(x) = (8 - x)^3$

4. $f(x) = (4x - 3)^3$

5. $f(x) = (2x + 5)^4$

6. $g(x) = (4x^2 + 7)^5$

7. $g(x) = 3(x^2 - 5x + 3)^2$

8. $f(x) = (3x^2 + 7x - 2)^{63}$

9. $f(x) = (2x^2 - 5x + 7)^{1/3}$

10. $g(x) = (3x^2 - 9x - 4)^{-6}$

For the rational functions in Exercises 11–14:

(a) Determine the derivative using the Quotient Rule.

(b) Determine the derivative using the Generalized Power Rule.

11. $f(x) = \frac{3}{2x + 9}$

12. $f(x) = \frac{7}{(x - 7)^2}$

13. $f(x) = \frac{2}{(3x + 5)^3}$

14. $f(x) = \frac{9}{x^2 - 6x + 18}$

In Exercises 15–18, determine an equation of the line tangent to the graph of f at the indicated point.

15. $f(x) = (5x + 3)^5$; $(-1, -32)$

16. $f(x) = (7x - 6)^3$; $(1, 1)$

17. $f(x) = (x^2 - 5x + 8)^4$; $(3, 16)$

18. $f(x) = (6x - 11)^{1/2}$; $(6, 5)$

Use the Generalized Power Rule to differentiate the functions in Exercises 19–22.

19. $g(x) = \sqrt{7x - 12}$

20. $f(x) = \sqrt[3]{8x + 1}$

21. $f(x) = \dfrac{3}{\sqrt{4x + 5}}$

22. $g(x) = \dfrac{8}{\sqrt[3]{2x^3 - 5x + 4}}$

In Exercises 23–26, use the Generalized Power Rule, along with the Product and Quotient Rules, to find the derivatives of the given functions.

23. $f(x) = x(x^2 + 5)^3$

24. $g(x) = 4x\sqrt{x^2 - 2x}$

25. $f(x) = \dfrac{3x - 7}{\sqrt{5x - 6}}$

26. $g(x) = (6x - 5)^7(3x + 2)^4$

In Exercises 27–30, find the derivatives using the Generalized Power Rule.

27. $y = (3x^2 - x + 1)^{0.67}$

28. $y = \left(\dfrac{1}{8.3x - 5.7}\right)^{-2.4}$

29. $f(x) = (x^3 - x^2 + 5x + 1)^{-0.7}$

30. $g(x) = 4.96(x + 1)^{2.78}$

31. The ClearView Window Company determines that the cost to produce a certain kind of window is modeled by

$$C(x) = (10x - 8)^{1.5} + 480 \qquad 0 \le x \le 100$$

where x represents the number of windows produced in hundreds and $C(x)$ represents the production costs in hundreds of dollars.

(a) Determine the marginal cost function, MC.

(b) Evaluate and interpret $MC(7)$.

(c) Determine the average cost function, AC.

(d) Determine the marginal average cost function, MAC.

(e) Evaluate and interpret $MAC(7)$.

In Exercises 32–37, determine the derivatives of the natural logarithm functions.

32. $f(x) = -6 \ln x$

33. $f(x) = \ln x^5$

34. $f(x) = 4x^5 \cdot \ln x$

35. $f(x) = \ln(3x^2 - 5)$

36. $f(x) = (\ln x)^{-2}$

37. $f(x) = \dfrac{x^2 + 5x - 2}{\ln(x + 4)}$

In Exercises 38–41, determine the derivatives of the general logarithm functions.

38. $g(x) = 3 \log_5 x$

39. $g(x) = 4x^3 \log_2 x$

40. $g(x) = \log_{10}(6x - 5)$

41. $g(x) = \log_4 \left(\dfrac{x^2 + 5}{2x - 3}\right)$

In Exercises 42–47, determine the derivatives of the exponential functions.

42. $f(x) = 3e^x + 5$

43. $f(x) = \sqrt{e^x - 5}$

44. $f(x) = 7x^3 e^x - 3e^x$

45. $f(x) = e^{2x-13}$

46. $f(x) = e^{4 \ln x + 5}$

47. $f(x) = \dfrac{e^{2x}}{\ln(x - 4)}$

For Exercises 48–51:

(a) Determine the tangent line equation for the function at the given point.

(b) Check by graphing the function and the tangent line in the same viewing window.

48. $f(x) = 3e^x$; $(2, 3e^2)$

49. $f(x) = \ln x^4$; $(e^4, 16)$

50. $f(x) = \ln \sqrt{2x + 3}$; $(3, \ln 3)$

51. $f(x) = e^x - e^{-x}$; $(0, 0)$

In Exercises 52–55, determine the derivatives of the general exponential functions.

52. $g(x) = 4^x$

53. $g(x) = \dfrac{3^x}{15^x}$

54. $g(x) = x^4 \cdot 0.7^x$

55. $g(x) = \sqrt[3]{10^x}$

In Exercises 56–59, determine the derivatives of the given functions.

56. $f(x) = 5^{x^2 - 1}$

57. $f(x) = 0.4^{\sqrt{x}}$

58. $f(x) = 3xe^{5x} - \ln(x + 4)$

59. $f(x) = \log_7 3x - \log_3 7x$

60. The amount of annual air pollution emissions in the United States can be modeled by

$$f(x) = 13.44 - 3.20 \ln x \qquad 1 \le x \le 26$$

where x represents the number of years since 1969 and $f(x)$ represents the number of millions of tons of particulate matter of less than 10 micrometers (*Source:* U.S. Environmental Protection Agency, www.epa.gov).

(a) Determine f' for the model.

(b) Evaluate and interpret $f'(10)$.

61. The number of commercial FM radio stations in the United States can be modeled by

$$f(x) = 2149.6 \cdot (1.036)^x \qquad 1 \le x \le 26$$

where x represents the number of years since 1969 and $f(x)$ represents the number of commercial FM radio stations (*Source:* U.S. Census Bureau, www.census.gov).

(a) Classify the function as an exponential growth or decay model.

(b) Determine f' for the model.

(c) Evaluate and interpret $f'(7)$ and compare to $f'(12)$.

62. The number of subscribers (in millions) to cellular phone service in the United States between 1989 and 1996 is shown in the following table.

Year	x	Subscribers (millions)
1989	1	3.509
1990	2	5.283
1991	3	7.557
1992	4	11.033
1993	5	16.009
1994	6	24.134
1995	7	33.786
1996	8	44.043

SOURCE: U.S. Census Bureau, www.census.gov

(a) Use your calculator to determine a general exponential regression model of the form

$$f(x) = a \cdot b^x \qquad 1 \le x \le 8$$

where x represents the number of years since 1988 and $f(x)$ represents the number of subscribers (in millions) to cellular phone service in the United States.

(b) Classify the function as an exponential growth or decay model.

(c) Graph f in the viewing window [1, 8] by [0, 50].

(d) Determine f' for the model.

(e) Evaluate and interpret $f'(3)$ and compare to $f'(7)$.

(f) Rewrite the model in the form $g(x) = a \cdot e^{kx}$, where a and k are constants rounded to the nearest hundredth.

In Exercises 63–70, use implicit differentiation to find $\frac{dy}{dx}$.

63. $3x - 5y = 7$

64. $9x^2 + 4y^2 = 36$

65. $5x^6 + y^3 = x^2$

66. $x^{1/3} - y^{1/3} = 8$

67. $y^3 - x^2 y = 5$

68. $x^2 + 6xy + 9y^2 = 4$

69. $x \ln y = y + 4$

70. $\ln(x + y) = e^x + 3y$

For Exercises 71–74:

(a) Use implicit differentiation to determine $\frac{dy}{dx}$.

(b) Solve the equation explicitly for y and determine $\frac{dy}{dx}$.

71. $2x + 3y = 6$

72. $7x - xy = 5x^2$

73. $x^3 y - y = 2x + 2$

74. $\frac{3}{x} - \frac{4}{y} = 5$

For Exercises 75–78, determine the equation of the tangent line for the given equation at the indicated point.

75. $x^2 - y^2 = 9; \quad (5, 4)$

76. $x^2 y + y^2 x = 12; \quad (3, -4)$

77. $2x \ln y = 4x^2 - 4y; \quad (1, 1)$

78. $3x + xe^y = 4 + y; \quad (1, 0)$

79. The BriteWite Light Company has determined that the price–demand equation for a new lamp is

$$px + 10x = 600 \qquad 10 \le x \le 25$$

where x represents the demand for lamps in thousands and p represents the price of one lamp in dollars.

(a) Use implicit differentiation to determine $\frac{dp}{dx}$.

(b) Evaluate and interpret $\left.\frac{dp}{dx}\right|_{(15,30)}$.

In Exercises 80–83, determine the rate of change equation by implicitly differentiating each side of the equation with respect to time t.

80. $x^2 - 3y = 4$

81. $x^3 + y^3 = 6x$

82. $2xy + 5x = 18y$

83. $xy - y^3 = 2x$

For Exercises 84–87, assume that x and y are both functions of t and determine the indicated rate.

84. Determine $\frac{dy}{dt}$ for $2x + y^2 = 7$, given $y = 3$ and $\frac{dx}{dt} = 12$.

85. Determine $\frac{dx}{dt}$ for $xy + 4y - 3x = 7$, given $x = 1$ and $\frac{dy}{dt} = 4$.

86. Determine $\frac{dx}{dt}$ for $y = 2x + 3^{x-4}$, given $x = 4$ and $\frac{dy}{dt} = -6$.

87. Determine $\frac{dy}{dt}$ for $x^2 + 2xy + y^2 = 25$, given $x = 2$, $y = 3$, and $\frac{dx}{dt} = 11$.

88. The Zoot Soot Chimney Service has determined that its monthly revenue is given by

$$R = 150x - \frac{x^2}{300}$$

where x is the number of chimneys cleaned and R is the revenue in dollars.

(a) Differentiate the revenue function with respect to time t.

(b) Determine the rate of change in the revenue if $x = 45$ and $\frac{dx}{dt} = 3\frac{\text{chimneys}}{\text{month}}$.

89. The surface area of a spherical balloon is given by the equation

$$S = 4\pi r^2$$

where r is the radius of the balloon and S is the volume.

(a) Differentiate the equation with respect to time t.

(b) If the balloon is being blown up so that the radius is increasing at a rate of $0.7\frac{\text{centimeter}}{\text{second}}$, determine the rate at which the surface area is changing when the radius is 8 centimeters.

In Exercises 90–93, compute the arc elasticity E_a.

90. $p_1 = 18, p_2 = 20, q_1 = 200, q_2 = 195$

91. $p_1 = 50, p_2 = 55, q_1 = 80, q_2 = 70$

92. $p_1 = 630, p_2 = 650, q_1 = 1300, q_2 = 1250$

93. $p_1 = 15{,}000, p_2 = 15{,}250, q_1 = 75, q_2 = 73$

94. The Macro Software Company determines that the price–demand function for a new FileFinder program is

$$x = d(p) = 350 - 2p$$

where p is the unit price in dollars of the program and x is the quantity demanded in thousands.

(a) Determine $d(8)$ and $d(9)$ and interpret.

(b) Using $q_1 = d(8), q_2 = d(9), p_1 = 8$, and $p_2 = 9$, determine the arc elasticity E_a and interpret.

95. The Thai Way Buffet Restaurant typically has 45 customers for lunch each day when the price for lunch is \$6.95. During a special promotion week, the price is reduced to \$5.50 and the number of lunch customers grows to 58 per day. Determine the arc elasticity and interpret.

In Exercises 96–99, the price–demand function $p(x)$ is given. Determine the demand function $x = d(p)$.

96. $p = 18 - 0.3x$

97. $p = \dfrac{12}{x} + 5$

98. $p = \sqrt{500 - x^2}$

99. $p = 800e^{-0.02x}$

In Exercises 100–103, the demand function $x = d(p)$ and the price p are given.

(a) Determine the point elasticity function $E(p)$.

(b) Classify the demand as elastic, inelastic, or unitary.

100. $d(p) = 180 - 5p;\ p = 20$

101. $d(p) = 400 - p^2;\ p = 12$

102. $d(p) = \sqrt{200 - p};\ p = 100$

103. $d(p) = 18\ln(120 - 3p);\ p = 5$

104. An amusement park charges \$15 per admission and serves approximately 6400 patrons per day. The demand function for the x patrons at p dollars per admission is given by

$$x = 130(35 - p)^{1.3}$$

(a) Determine the point elasticity function $E(p)$.

(b) Is the demand elastic or inelastic at the current \$15 admission price?

(c) Should the ticket price be raised or lowered in order to increase revenue?

105. The FlyingThyme Watch Company determines that the demand for a new watch is given by

$$d(p) = 35\ln(150 - 5p)$$

where p is the unit price, in dollars, of a watch and $d(p)$ is the demand in thousands.

(a) Determine $E(25)$ and state whether the manufacturer should raise or lower the price in order to increase revenue.

(b) Set $E(p) = 1$ and solve for p to determine the watch price that maximizes revenue.

CHAPTER 14 PROJECT

Chapter 14 provided us with new tools for differentiation. Some of these tools, like differentiation of an exponential and a logarithmic function, will be used within this project. We will also revisit some topics discussed in previous chapters, such as linear approximation.

The following table gives the population for the state of Ohio, which will be celebrating its bicentennial in 2003:

Year	Population	Year	Population
1800	42,159	1910	4,767,121
1810	230,760	1920	5,759,394
1820	581,434	1930	6,646,697
1830	937,903	1940	6,907,612
1840	1,519,467	1950	7,946,627
1850	1,980,329	1960	9,706,397
1860	2,339,511	1970	10,652,017
1870	2,665,260	1980	10,797,630
1880	3,198,062	1990	10,847,115
1890	3,672,329	2000	11,353,140
1900	4,157,545		

SOURCE: U.S. Census Bureau, www.census.gov

Let x be the number of years since 1799.

1. Determine a function $y_1(x) = a \cdot b^x$ where y_1 is an exponential regression function for $1 \le x < 141$.

2. Determine a function $y_2(x) = a + b\ln x$ where y_2 is a logarithmic regression function for $141 \le x \le 201$.

3. Form a piecewise-defined function $f(x)$ where

$$f(x) = \begin{cases} y_1(x) & 1 \le x < 141 \\ y_2(x) & 141 \le x \le 201 \end{cases}$$

Evaluate and interpret.

(a) $f(121)$ and $f'(121)$

(b) $f(171)$ and $f'(171)$

4. Use the results from 3b to approximate $f(181)$. Compare this to the actual $f(181)$ and the given Ohio population in 1980. Explain the discrepancy.

5. Compute $f'(201)$. If Ohio's population continued to grow at the 2000 rate, what would the population be during Ohio's bicentennial year?

Chapter

15 Further Applications of the Derivative

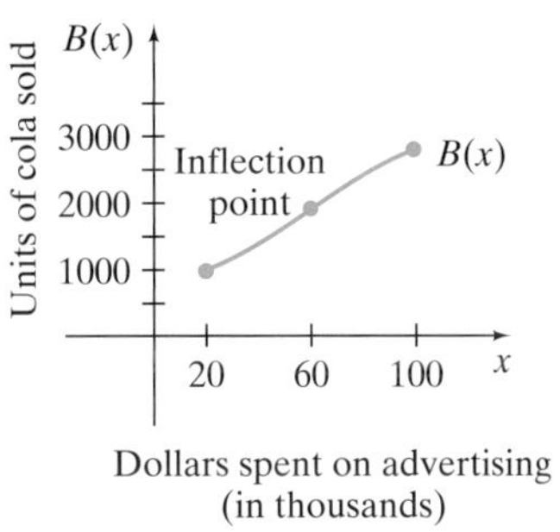

(b)

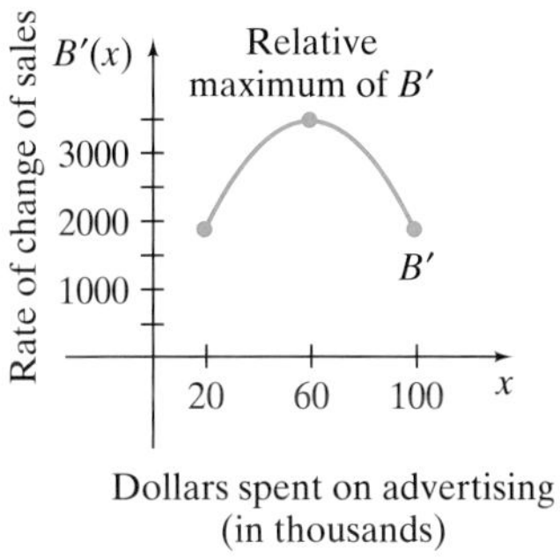

(c)

To gain domination in the lucrative soft drink market, cola makers can spend a fortune on advertising. But how are decisions made on how much money to spend on advertising? In Figure b, $B(x)$ represents the number of units of cola sold after spending x thousand dollars on advertising. Figure c shows that the rate of change of sales $B'(x)$, is maximized at $x = 60$. At $x = 60$ on the graph of $B(x)$ we find what is called an inflection point. In economics, this point is also called the *point of diminishing returns* and indicates where the rate of change of sales started to decrease.

What We Know

In Chapter 14, we continued our study of differentiation and how the derivative gives an instantaneous rate of change. We also examined more applications utilizing the power of the derivative.

Where Do We Go

In this chapter, we will see how the rate of change concept and the derivative tell us the behavior of a function. Specifically, we will see how the derivative tells us where a function is increasing (or decreasing) and where its graph reaches a high point (or low point). We then examine how the increasing (or decreasing) behavior of the derivative gives us the concavity of the graph of the function.

Section 15.1 First Derivatives and Graphs

In this section we examine more deeply the relationship that exists between a function and its derivative. Our study will reveal how the sign of a derivative, whether it is positive or negative, on an interval indicates whether the function is increasing or decreasing. Also, we will see how the derivative can help locate the **relative extrema** for a function.

Intervals of Increase and Decrease for Functions

So far, we have looked at the graphs of many different functions. The graphs of functions generally have portions that are **increasing** and portions that are **decreasing**, such as the graph in Figure 15.1.1.

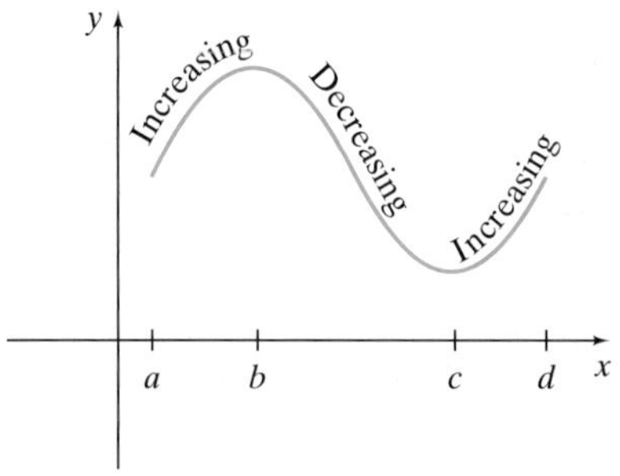

Figure 15.1.1

Flashback

SAT Data Revisited

In Section 11.4, we found that the percentage of people taking the SAT who intended to study business and commerce can be modeled by

$$f(x) = -0.09x^2 + 2.01x + 9.63 \qquad 1 \le x \le 22$$

where $f(x)$ is the percentage of people taking the SAT and x is the number of years since 1974 (*Source:* U.S. Statistical Abstract, www.census.gov/statab/www). See Figure 15.1.2.

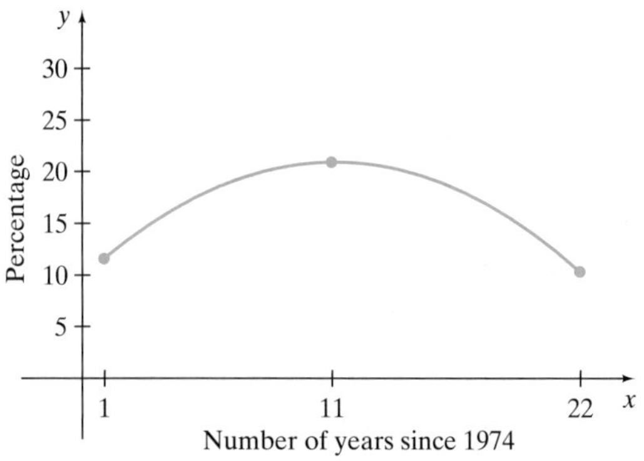

Figure 15.1.2 Graph of $f(x) = -0.09x^2 + 2.01x + 9.63$.

Determine intervals where f is increasing and decreasing. Round to the nearest hundredth and interpret the results.

Flashback Solution

Since the graph of the model is a parabola, to determine intervals where the model is increasing or decreasing, we must find the coordinates of the vertex, the maximum point of the graph. In Chapter 11 it was shown that the x-coordinate of the vertex can be determined algebraically by using $x = -\frac{b}{2a}$. Using this formula, the x-coordinate is

$$x = -\frac{b}{2a} = -\frac{2.01}{2(-0.09)} = -\frac{2.01}{-0.18} \approx 11.17$$

The y-coordinate is found by substituting $x = 11.17$ back into $f(x)$. This gives

$$f(11.17) = -0.09(11.17)^2 + 2.01(11.17) + 9.63 \approx 20.85, \quad \text{or} \quad 20.85\%$$

So we conclude that f is increasing on the interval (1, 11.17) and decreasing on the interval (11.17, 22). Rounding 11.17 to the nearest year gives a value of 11. Thus, the percentage of people taking the SAT who intended to study business and commerce increased from 1975 to 1985 and then it decreased from 1985 to 1996. There was a peak of about 20.85% who intended to study business and commerce, and it occurred in 1985.

Technology Option

You can verify the results in the Flashback by graphing the model on your calculator and using the MAXIMUM command. For more information on the MAXIMUM command, consult the online graphing calculator manual at www.prenhall.com/armstrong.

In the Flashback, we found the intervals where f is increasing and where it is decreasing. Another way to find these intervals is to analyze the sign of f'. (Since $f'(x)$ gives the rate of change of f at x, f' is called a **rate function**.) In Example 1, we observe the relationship between the graph of f and the sign of f'.

Example 1 Analyzing the Behavior of a Function and Its Derivative

The graphs of $f(x) = x^3 - 12x + 4$ and its derivative $f'(x) = 3x^2 - 12$ are shown in Figure 15.1.3 and 15.1.4. Use the graphs to determine where:

(a) f is increasing **(b)** $f' > 0$
(c) f is decreasing **(d)** $f' < 0$
(e) f is constant **(f)** $f' = 0$

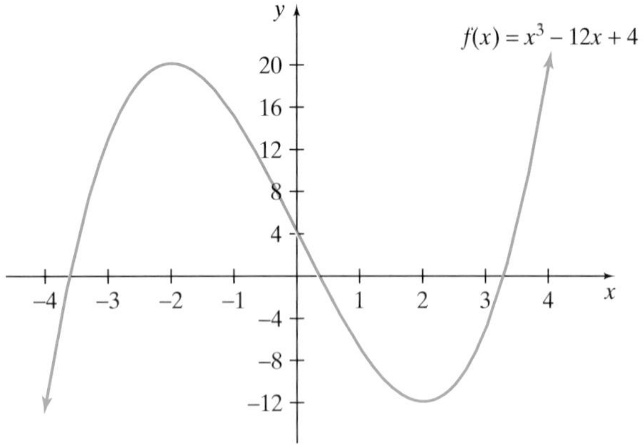

Figure 15.1.3

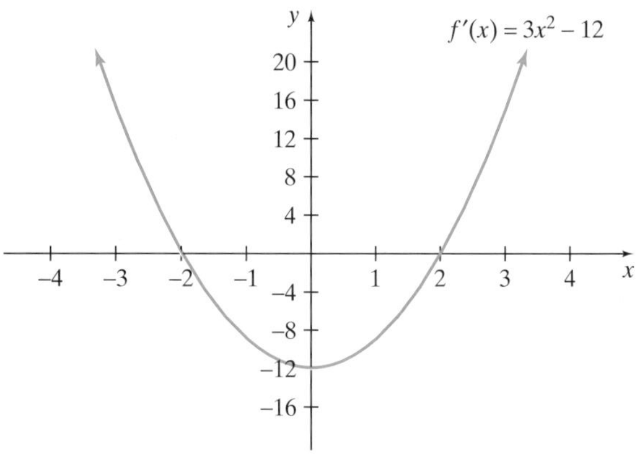

Figure 15.1.4

Solution

(a) f is increasing on $(-\infty, -2) \cup (2, \infty)$.

(b) Recall that a function is greater than zero whenever its graph is *above* the x-axis. Thus, $f' > 0$ on $(-\infty, -2) \cup (2, \infty)$ since this is where the graph of f' is above the x-axis.

(c) f is decreasing on $(-2, 2)$.

(d) Recall that a function is less than zero whenever its graph is *below* the x-axis. Thus, $f' < 0$ on $(-2, 2)$ since this is where the graph of f' is below the x-axis.

(e) f is constant at $x = -2$ and $x = 2$.

(f) Recall that a function is equal to zero whenever its graph *crosses* the x-axis. Thus, $f'(x) = 0$ at $x = -2$ and $x = 2$ since this is where the graph of f' crosses the x-axis. ■

Notice in Example 1 that there appears to be a connection between our answers to parts (a) and (b) as well as parts (c) and (d) and parts (e) and (f). Pairing the results in this manner leads us to make the following observation, which is true for any function and its derivative.

Increasing and Decreasing Functions

On an open interval (a, b) on which f is differentiable and continuous:

1. If $f'(x) > 0$ for all x in (a, b), then f is increasing on (a, b).
2. If $f'(x) < 0$ for all x in (a, b), then f is decreasing on (a, b).
3. If $f'(x) = 0$ for all x in (a, b), then f is constant on (a, b).

▶ **Note:** In English this means that *wherever the derivative of a function is positive, the function is increasing; wherever the derivative of a function is negative, the function is decreasing; and wherever the derivative of a function is zero, the function is constant*. See Figure 15.1.5.

Knowing that the derivative tells us the slope of a tangent line at a point, the preceding properties should come as no surprise. Basically, wherever tangent lines have a positive slope, $f'(x) > 0$, we see that a function is increasing (Figure 15.1.6), and wherever tangent lines have a negative slope, $f'(x) < 0$, we see that a function is decreasing (Figure 15.1.7). Also, where we have *horizontal tangent lines*, $f'(x) = 0$, we say that the function is constant (Figure 15.1.8).

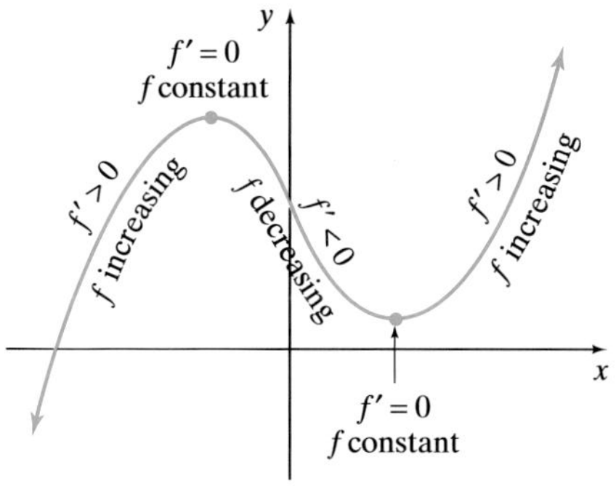

Figure 15.1.5

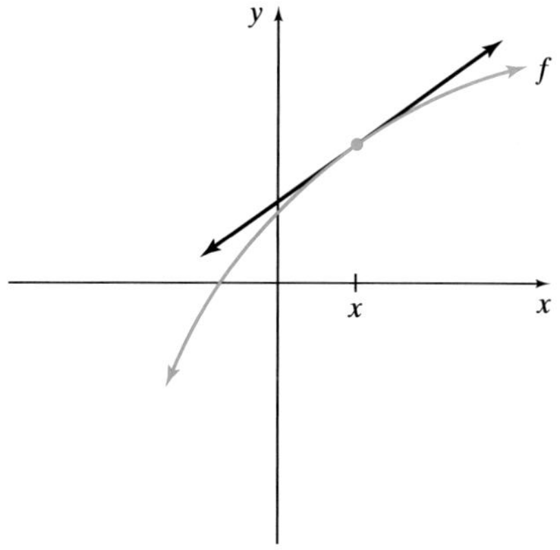

Figure 15.1.6 Tangent line has positive slope; that is, $f'(x) > 0$.

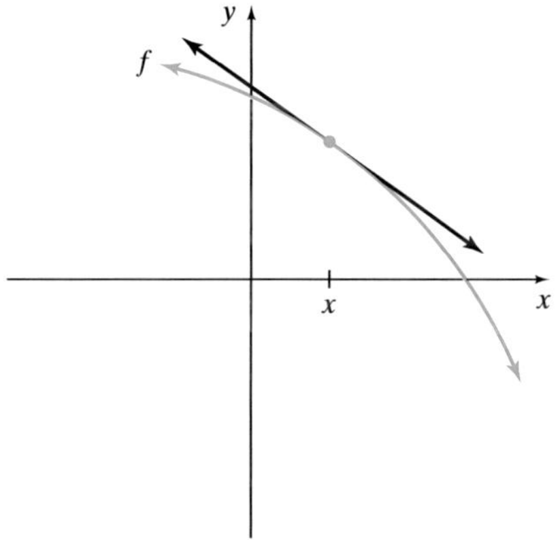

Figure 15.1.7 Tangent line has negative slope; that is, $f'(x) < 0$.

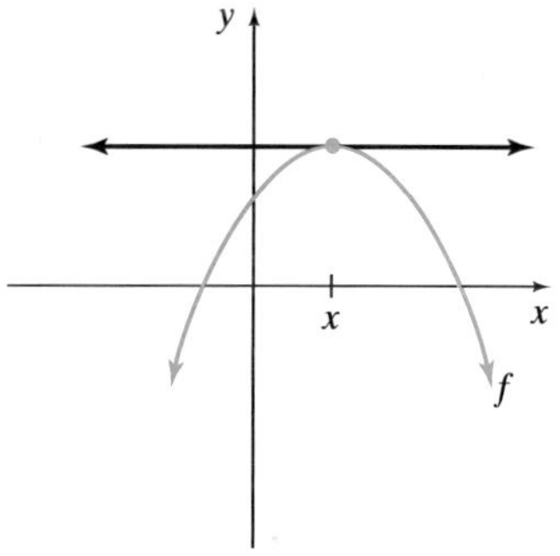

Figure 15.1.8 Tangent line has zero slope; that is, $f'(x) = 0$.

Relative Extrema and Critical Values

In Figure 15.1.5, notice that the graph has a *valley* and a *hill*. In mathematics, the hill is called a **relative maximum**, while the valley is called a **relative minimum**. By using the word *relative* we mean *relative to points nearby*. (In Section 15.4 we will discuss *absolute maximum* and *absolute minimum*). These terms are expressed more formally in the following definition.

Relative Extrema

If f is a continuous function on an open interval containing c, then the point $(c, f(c))$:

- Is a **relative maximum** if $f(c) \geq f(x)$ for all x in the interval.
- Is a **relative minimum** if $f(c) \leq f(x)$ for all x in the interval.

Note: The plural of relative maximum is relative maxima and the plural of relative minimum is relative minima. Collectively, the maxima and minima are called **extrema**.

Example 2 Locating Relative Extrema

Determine where the graphs in Figure 15.1.9 and 15.1.10 have a relative maximum and a relative minimum.

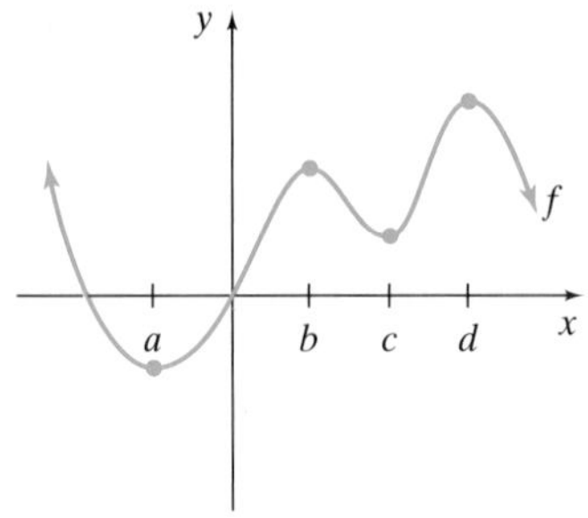

(a)

Figure 15.1.9

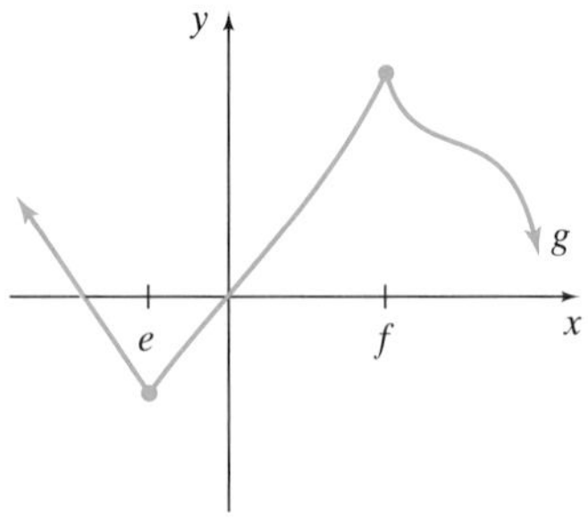

(b)

Figure 15.1.10

Solution

(a) f has a relative maximum at $x = b$ and $x = d$, and f has a relative minimum at $x = a$ and $x = c$.

(b) g has a relative maximum at $x = f$ and a relative minimum at $x = e$. ■

In Example 2a, the relative extrema occurred where the graph has horizontal tangents, that is, where $f'(x) = 0$. In Example 2b the relative extrema occurred where the graph had a corner or made a sharp turn, that is, where $f'(x)$ is undefined. We call points on a curve where $f'(x) = 0$ or where $f'(x)$ is undefined the **critical points** of the function. The x-coordinate of the critical points are called **critical values** or **critical numbers**.

Critical Values

A **critical value** for f is an x-value in the domain of f for which $f'(x) = 0$ *or* $f'(x)$ is undefined.

Example 3 Determining Critical Values

Determine the critical values for $f(x) = 5x^3 + 4x^2 - 12x - 25$.

Solution

Understand the Situation: The key to locating critical values is to first determine the derivative. The derivative is

$$f'(x) = 15x^2 + 8x - 12$$

Since $f'(x)$ is defined for all x, the only critical values occur where $f'(x) = 0$.

So we need to solve

$$15x^2 + 8x - 12 = 0$$

$$(5x + 6)(3x - 2) = 0$$

$$x = -\frac{6}{5} \quad \text{or} \quad x = \frac{2}{3}$$

Thus, the critical values occur at $x = -\frac{6}{5}$ and $x = \frac{2}{3}$. See Figure 15.1.11.

Interactive Activity

Solve $f'(x) = 0$ graphically to support the result in Example 3.

WWW

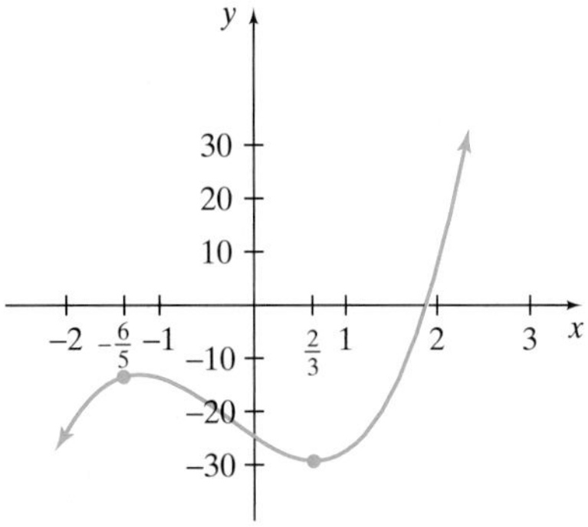

Figure 15.1.11

■

Example 4 Determining Critical Values

Determine the critical values for the following functions.

(a) $f(x) = \sqrt[3]{x}$ **(b)** $y = x^3$

Solution

(a) Again, our first step is to compute the derivative. This yields

$$f(x) = \sqrt[3]{x} = x^{1/3}$$

$$f'(x) = \frac{1}{3}x^{-2/3} = \frac{1}{3x^{2/3}} \quad \text{or} \quad \frac{1}{3\sqrt[3]{x^2}}$$

Here we notice that $f'(x)$ is undefined at $x = 0$ and that $f'(x)$ cannot equal zero; that is, $f'(x) = 0$ has no solution. So we conclude that the only critical value is $x = 0$. See Figure 15.1.12.

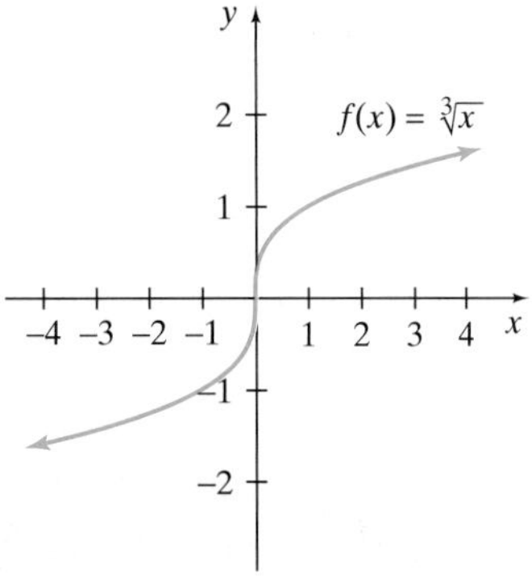

Figure 15.1.12

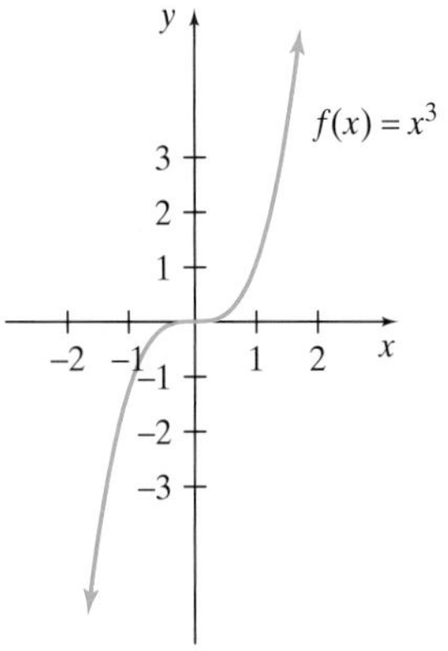

Figure 15.1.13

(b) The derivative is $\frac{dy}{dx} = 3x^2$ and it is defined for all x. We see that $\frac{dy}{dx} = 0$ at $x = 0$, which means that the only critical value is $x = 0$. See Figure 15.1.13.

■

Checkpoint 1 Now work Exercise 17.

Notice that in Example 3 our critical values produced a relative maximum and a relative minimum, whereas in Example 4 our critical values produced neither a relative maximum nor a relative minimum. What this means is that *every relative extreme occurs at a critical value, but not every critical value produces a relative extreme.* So how do we know if a critical value is going to produce a relative maximum or a relative minimum? The answer to this question is found in the First Derivative Test.

First Derivative Test

The First Derivative Test pulls together all the ideas in this section.

First Derivative Test

Let $x = c$ be a critical value of a function f that is continuous on an open interval containing $x = c$. The point $(c,\ f(c))$ can be called:

1. A **relative minimum** if $f' < 0$ to the left of c and $f' > 0$ to the right of c.
2. A **relative maximum** if $f' > 0$ to the left of c and $f' < 0$ to the right of c.

Recalling that the sign of the derivative tells us the increasing-decreasing behavior of the function, we can construct a **sign diagram** that communicates the First Derivative Test in a more visual manner. In the following sign diagrams, the bottom row gives the sign of f' and the top row gives the behavior of f.

First Derivative Test Using a Sign Diagram

Let $x = c$ be a critical value of a function f that is continuous on an open interval containing $x = c$. Place c on a number line.

1. **Relative minimum**

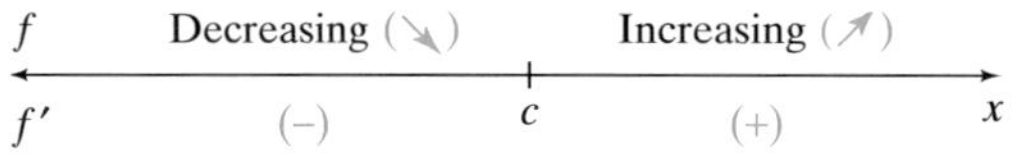

Relative minimum at $(c,\ f(c))$.

2. **Relative maximum**

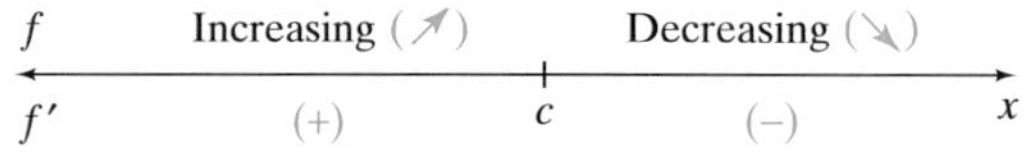

Relative maximum at $(c,\ f(c))$.

Note: If f' has the same sign on both sides of $x = c$, then f has neither a relative maximum nor a relative minimum at $x = c$.

Figure 15.1.14 (a–c) illustrates the First Derivative Test when $x = c$ is a critical value, where $f'(c) = 0$, that is, a horizontal tangent.

Figure 15.1.15 (a–c) illustrates the First Derivative Test when $x = c$ is a critical value, where $f'(c)$ is undefined.

To understand why the sign diagram and the method that we will use in constructing sign diagrams works, we need to quickly review **continuous functions**.

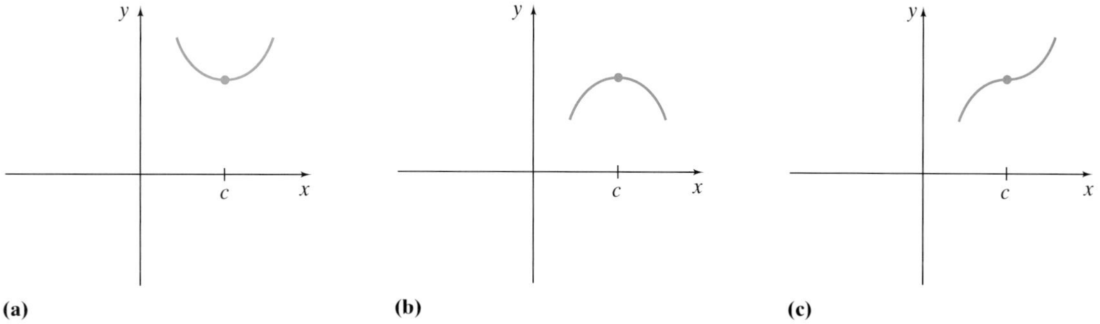

Figure 15.1.14 **(a)** Relative minimum at $x = c$. **(b)** Relative maximum at $x = c$. **(c)** Neither a relative maximum nor a relative minimum at $x = c$.

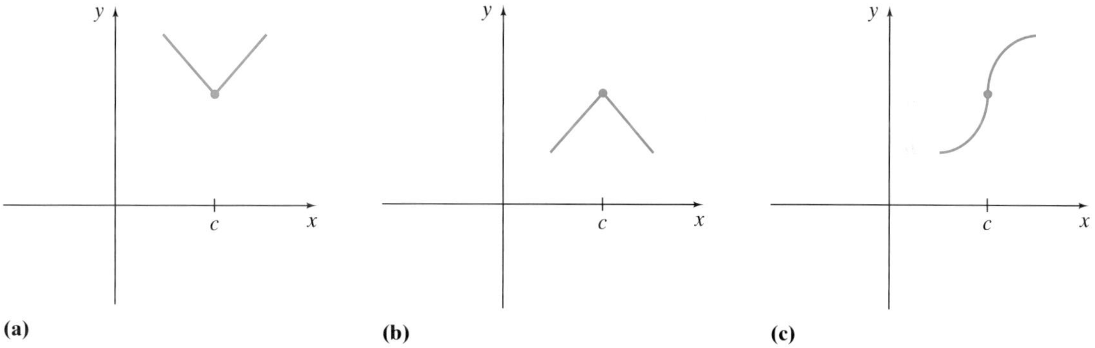

Figure 15.1.15 **(a)** Relative minimum at $x = c$. **(b)** Relative maximum at $x = c$. **(c)** Neither relative maximum nor relative minimum at $x = c$.

Suppose that a function f is continuous on the open interval $(2, 7)$, and $f(x) \neq 0$ for any x in $(2, 7)$. If $f(x)$ is positive for some x, say, $f(3) = 4$, then $f(x)$ is positive for every x in $(2, 7)$. Why? Well, suppose we assume that $f(6)$ is negative, say $f(6) = -3$. Since f is *continuous*, the graph of f could not connect the points $(3, 4)$ and $(6, -3)$ without crossing the x-axis. See Figure 15.1.16. But recall that $f(x) = 0$ at the point where the graph of f crosses the x-axis. Since we agreed that $f(x) \neq 0$ for any x in $(2, 7)$, this assumption has been contradicted, and thus f cannot be negative in $(2, 7)$. This property of continuous functions is stated in the following theorem, which we will use frequently in this chapter.

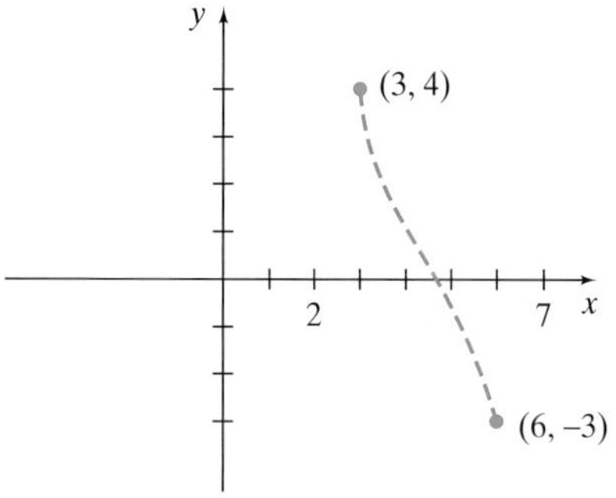

Figure 15.1.16

Theorem 15.1

If f is continuous and $f(x) \neq 0$ for any x in the interval (a, b), then f cannot change sign on (a, b).

Example 5 Applying the First Derivative Test

For $f(x) = 5x^3 + 4x^2 - 12x - 25$, determine intervals where f is increasing and where f is decreasing by making a sign diagram. Locate all relative extrema.

Solution

Understand the Situation: To find intervals where f is increasing and where it is decreasing, we need to determine the critical values for this function. In order to find the critical values, we must first determine the derivative.

In Example 3, we determined that the derivative is $f'(x) = 15x^2 + 8x - 12$ and the critical values are

$$x = -\frac{6}{5} \quad \text{and} \quad x = \frac{2}{3}$$

To determine intervals where the function is increasing and decreasing, we place the critical values on a number line representing the x-axis in Figure 15.1.17.

Figure 15.1.17

Since f' is a polynomial function, it is continuous on the open intervals determined by the critical values. Thanks to Theorem 15.1, f' can change sign only at the critical values. We can determine the sign of the derivative within each interval by evaluating $f'(x)$ for some x within the interval. We call such x-values **test numbers**. In the interval $\left(-\infty, -\frac{6}{5}\right)$, we choose $x = -2$ as our test number.

$$f'(-2) = 15(-2)^2 + 12(-2) - 12 = 24$$

Since $f'(-2) = 24$, which is positive, we conclude that on the interval $\left(-\infty, -\frac{6}{5}\right)$, $f(x) = 5x^3 + 4x^2 - 12x - 25$ is *increasing*. Choosing $x = 0$ as our test number in the interval $\left(-\frac{6}{5}, \frac{2}{3}\right)$ gives

$$f'(0) = 15(0)^2 + 12(0) - 12 = -12$$

So $f'(0) = -12$, a negative result. In a similar fashion, we select $x = 1$ as our test number in the interval $\left(\frac{2}{3}, \infty\right)$ and determine that

$$f'(1) = 15(1)^2 + 12(1) - 12 = 15$$

So $f'(1) = 15$, a positive number. We now list the signs of the derivative on the sign diagram in Figure 15.1.18.

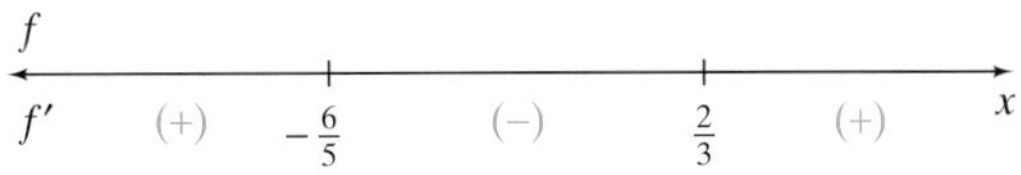

Figure 15.1.18

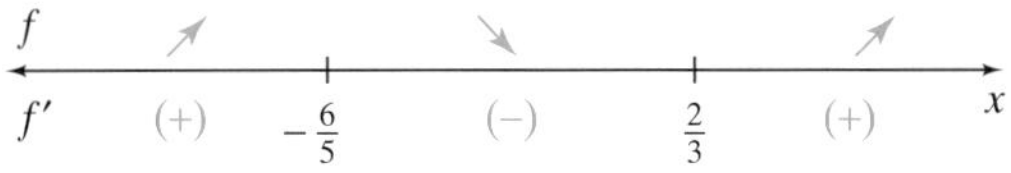

Figure 15.1.19

Now using our knowledge of how the sign of the derivative tells us the behavior of the function, we complete the sign diagram in Figure 15.1.19. This sign diagram indicates that f is increasing on $\left(-\infty, -\frac{6}{5}\right) \cup \left(\frac{2}{3}, \infty\right)$ and decreasing on $\left(-\frac{6}{5}, \frac{2}{3}\right)$. To find the relative extrema, we use the First Derivative Test with a sign diagram. Since f is increasing on $\left(-\infty, -\frac{6}{5}\right)$ and then decreasing on $\left(-\frac{6}{5}, \frac{2}{3}\right)$, we conclude that there is a relative maximum at $x = -\frac{6}{5}$. Since f is decreasing on $\left(-\frac{6}{5}, \frac{2}{3}\right)$ and then increasing on $\left(\frac{2}{3}, \infty\right)$, we conclude that there is a relative minimum at $x = \frac{2}{3}$. Specifically, we have a relative maximum at the point $\left(-\frac{6}{5}, -\frac{337}{25}\right)$ and a relative minimum at the point $\left(\frac{2}{3}, -\frac{803}{27}\right)$. ■

Interactive Activity

Graph $f(x) = 5x^3 + 4x^2 - 12x - 25$ and use the MAXIMUM and MINIMUM commands to verify the result of Example 5.

Example 5 demonstrates the following process, which we can use to locate relative extrema.

To locate relative extrema of a function, we find the critical values of the function and apply the First Derivative Test.

Example 6 Applying the First Derivative Test

For $f(x) = x + \frac{1}{x}$, determine intervals where f is increasing and decreasing. Locate all relative extrema.

Solution

To determine critical values, we need to first determine the derivative.

$$f(x) = x + \frac{1}{x} = x + x^{-1}$$

$$f'(x) = 1 - x^{-2} = 1 - \frac{1}{x^2}$$

Next we set the derivative equal to zero and solve.

$$1 - \frac{1}{x^2} = 0$$

$$1 = \frac{1}{x^2}$$

$$x^2 = 1$$

$$x = \pm 1$$

Notice that, although the derivative is undefined at $x = 0$, we do not list $x = 0$ as a critical value, because $x = 0$ is not in the domain of f. So the only critical values are $x = 1$ and $x = -1$. We proceed as in Example 5 and construct the sign diagram in Figure 15.1.20. Notice that we placed $x = 0$, where the derivative was undefined, in our sign diagram. We did this to take advantage of Theorem 15.1. In the interval $(-\infty, -1)$, we select the test number -2; in $(-1, 0)$, we select $-\frac{1}{2}$; in $(0, 1)$, we pick $\frac{1}{2}$; and in $(1, \infty)$, we choose 2. (Notice that in each of these open intervals our function is continuous, thus allowing us to use Theorem 15.1.) Substituting these test numbers into the derivative yields

$x = -2$:
$$f'(-2) = 1 - \frac{1}{(-2)^2} = 1 - \frac{1}{4} = \frac{3}{4}$$

$x = -\frac{1}{2}$:
$$f'\left(-\frac{1}{2}\right) = 1 - \frac{1}{\left(-\frac{1}{2}\right)^2} = 1 - \frac{1}{\frac{1}{4}} = -3$$

$x = \frac{1}{2}$:
$$f'\left(\frac{1}{2}\right) = 1 - \frac{1}{\left(\frac{1}{2}\right)^2} = 1 - \frac{1}{\frac{1}{4}} = -3$$

$x = 2$:
$$f'(2) = 1 - \frac{1}{(2)^2} = 1 - \frac{1}{4} = \frac{3}{4}$$

Finally, using these results and our knowledge of how the sign of the derivative determines the behavior of the function, we complete the sign diagram as shown in Figure 15.1.21.

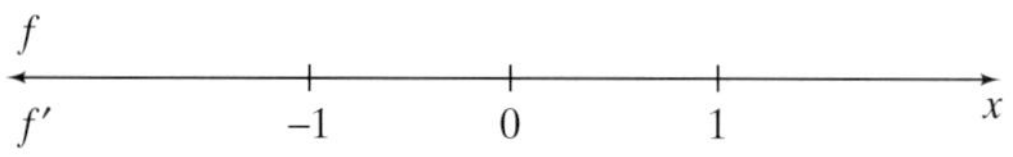

Figure 15.1.20

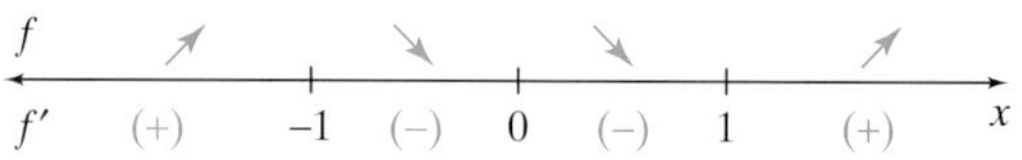

Figure 15.1.21

So $f(x) = x + \frac{1}{x}$ is increasing on the intervals $(-\infty, -1) \cup (1, \infty)$. It is decreasing on the intervals $(-1, 0) \cup (0, 1)$. Using the sign diagram and the First Derivative Test, we have a relative maximum at $x = -1$ and a relative minimum at $x = 1$. Specifically, we have a relative maximum at $(-1, -2)$ and a relative minimum at $(1, 2)$. See Figure 15.1.22.

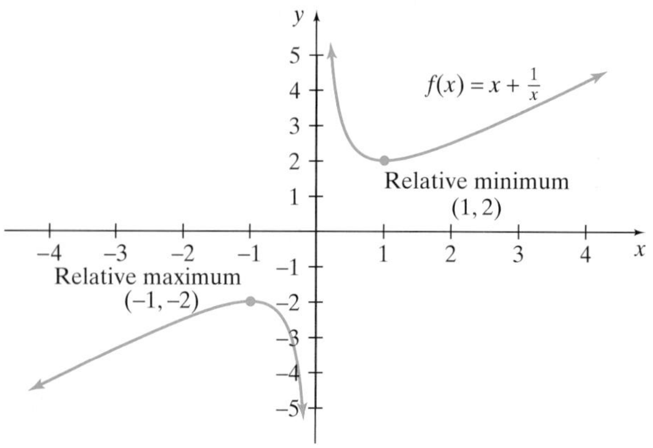

Figure 15.1.22

■

Checkpoint 2

Now work Exercises 31a and b.

Now let's see how the First Derivative Test can be used in an application.

Example 7 Generating Revenue from Marginal Revenue

The owner of Everything Fit to Print has received a shipment of new computer tutorial software. The manufacturer of the software claims that their market research has determined that a retail price of \$74 maximizes revenue. The manufacturer supplied the graph of the marginal revenue function shown in Figure 15.1.23.

(a) Determine what value of q, quantity, will maximize the revenue. Also, determine the maximum revenue.

(b) Sketch a possible graph of R.

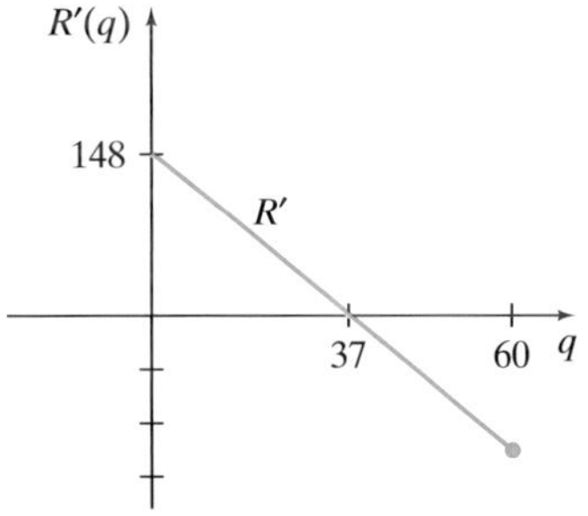

Figure 15.1.23

Solution

(a) Understand the Situation: To find the relative maximum, we determine the critical values of the function and apply the First Derivative Test. Since the critical values of R occur where $R'(q) = 0$, we know from the graph of R' that $q = 37$ is a critical value.

Now we determine the sign of R' to the left and to the right of this critical value.

- The graph R' is above the q-axis on the interval $(0, 37)$. This means that R' is positive and thus R is increasing on this interval.
- The graph R' is below the q-axis on the interval $(37, 60)$. This means that R' is negative and thus R is decreasing on this interval.

We use this information to construct the sign diagram in Figure 15.1.24.

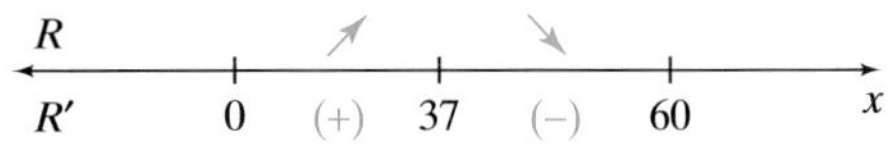

Figure 15.1.24

Since revenue is increasing on the interval $(0, 37)$ and decreasing on the interval $(37, 60)$, we conclude, by the First Derivative Test, that revenue is maximized at $q = 37$. The maximum revenue is then simply

$$p \cdot q = 74 \cdot 37 = \$2738$$

(b) We know two points on the graph, (0, 0) (if nothing is sold, then revenue is zero), and the relative maximum (37, 2738). Using this information, a possible graph is given in Figure 15.1.25.

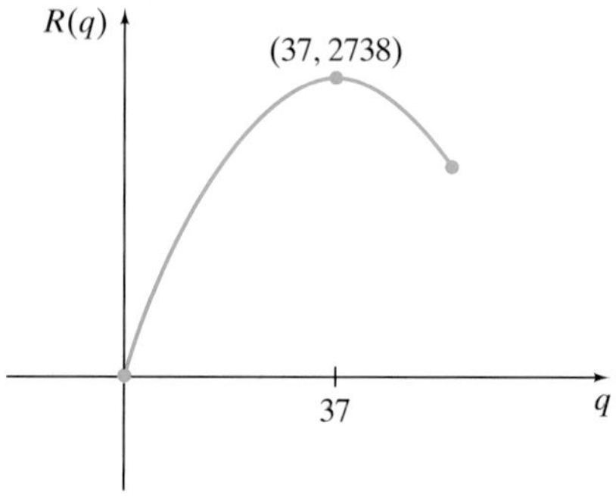

Figure 15.1.25

■

SUMMARY

In this section we saw how the sign of the derivative tells us the behavior of a function. Specifically, on an open interval:

- If $f'(x) > 0$ for all x in (a, b), then f is increasing on (a, b).
- If $f'(x) < 0$ for all x in (a, b), then f is decreasing on (a, b).
- If $f'(x) = 0$ for all x in (a, b), then f is constant on (a, b).

We defined **critical values** as x-values in the domain of the function such that $f'(x) = 0$ or $f'(x)$ is undefined. To find relative extrema, determine critical values and apply the **First Derivative Test**. Use the following process:

1. Determine the derivative and use it to find critical values.
2. Make a sign diagram to determine intervals of increase or decrease.
3. Use the sign diagram and the First Derivative Test to locate relative extrema.
4. Use the calculator as an effective tool to verify your work.

SECTION 15.1 EXERCISES

For each function graphed in Exercises 1–6, determine where the function is increasing, where it is decreasing, and where it is constant.

1.

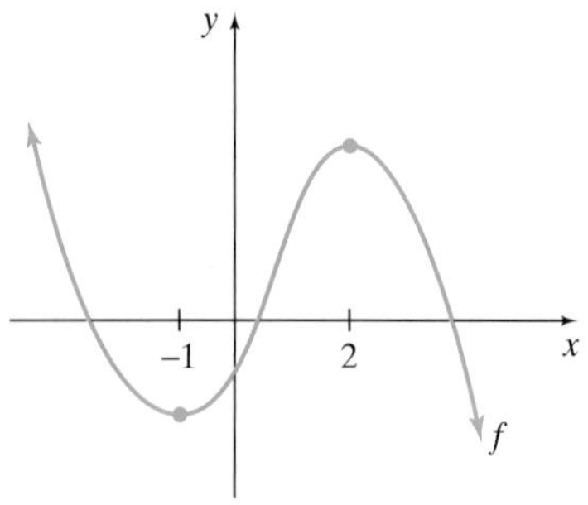

2.

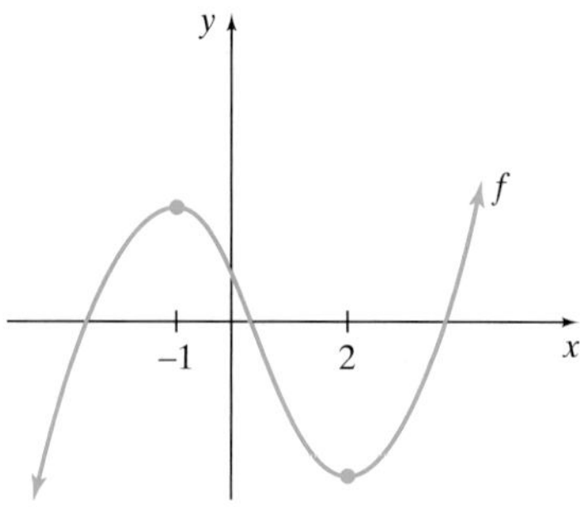

3.

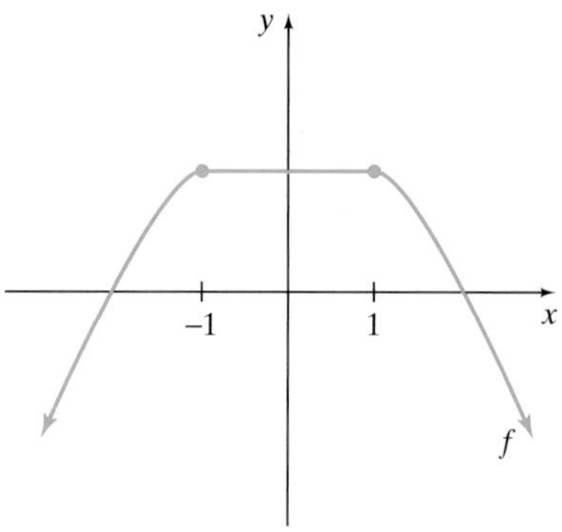

4.

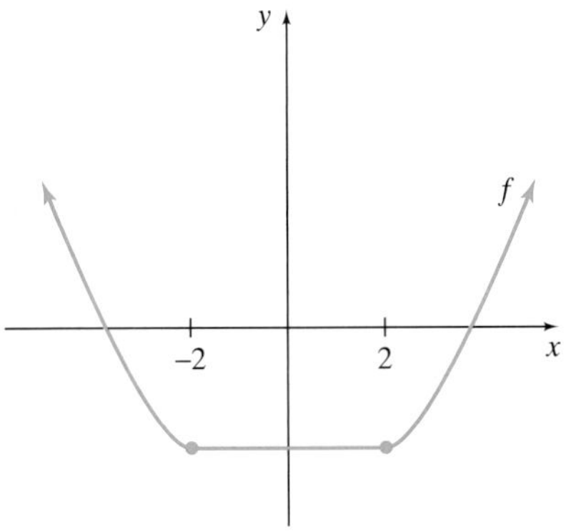

5.

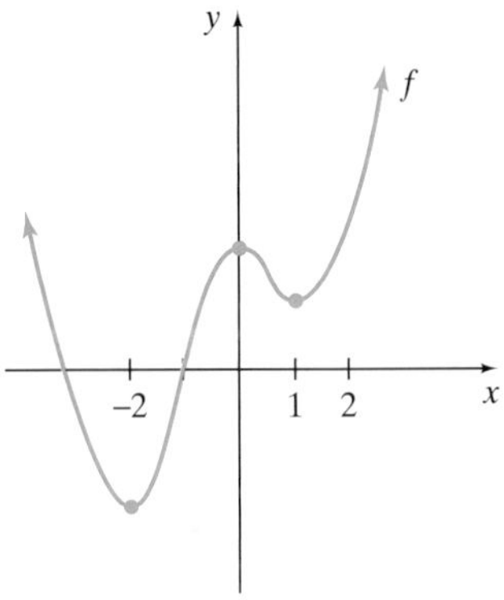

6.

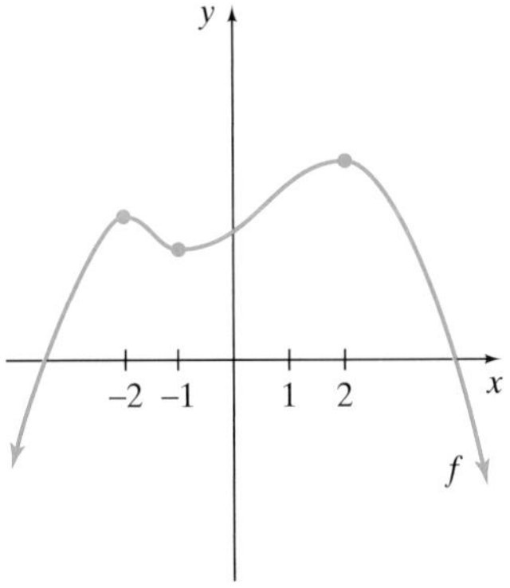

*For each **function** graphed in Exercises 7–12:*

(a) Determine intervals where the derivative is positive.

(b) Determine intervals where the derivative is negative.

(c) List where the derivative is undefined and where the derivative equals zero.

7.

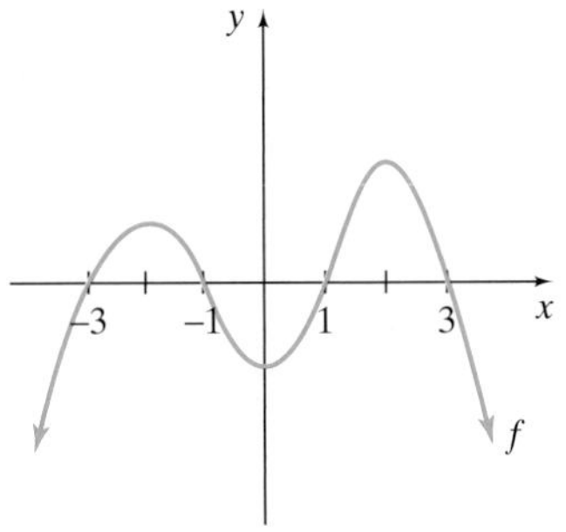

8.

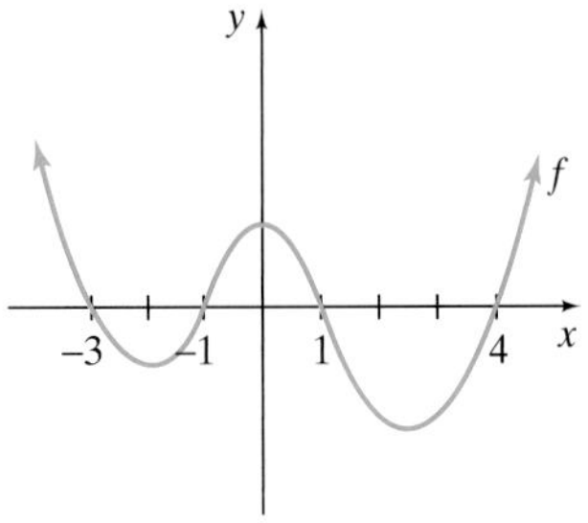

9.

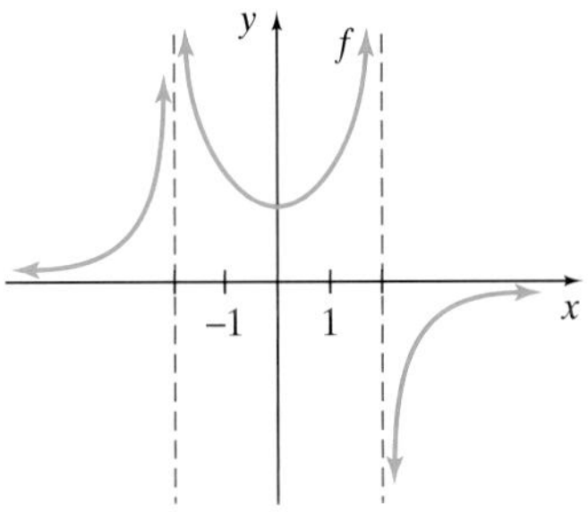

10.

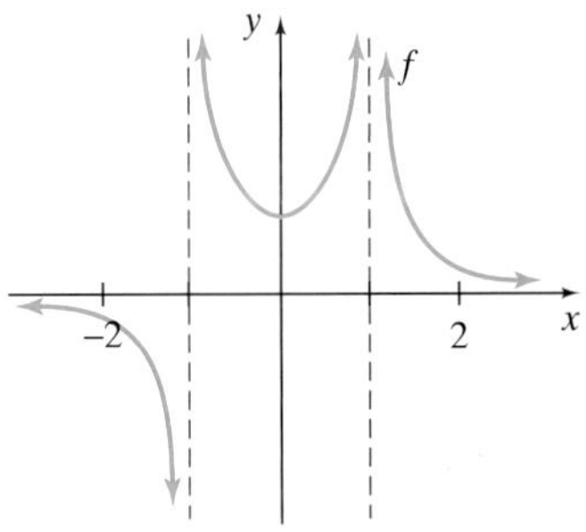

11.

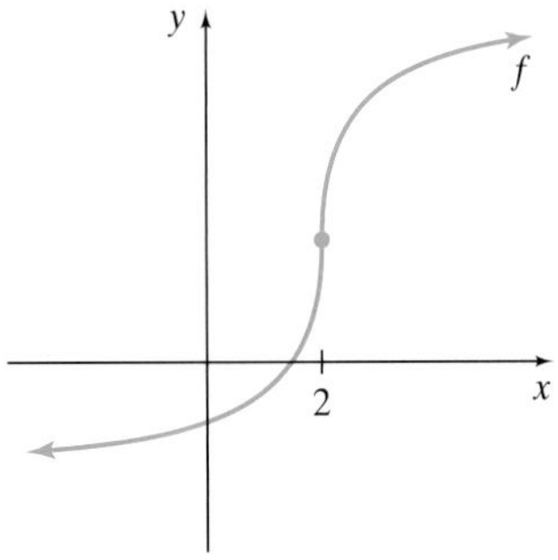

12.

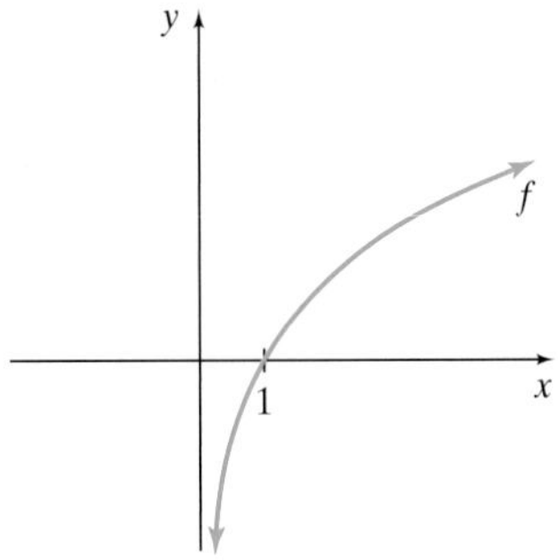

Determine the critical values of each function in Exercises 13–28.

13. $f(x) = 2x - 3$

14. $f(x) = -3x + 5$

15. $g(x) = x^2 - 5x + 6$

16. $f(x) = 2x^2 + 3x - 5$

✓ **17.** $f(x) = \frac{1}{3}x^3 + x^2 - 15x + 3$

18. $g(x) = -x^3 - 3x^2 + 45x - 5$

19. $y = \sqrt{2x+1}$

20. $y = \sqrt{3x-7}$

21. $f(x) = e^x$

22. $g(x) = e^{3x}$

23. $y = x + \ln x$

24. $y = x - \ln x$

25. $f(x) = \sqrt[3]{x+1}$

26. $g(x) = \sqrt[3]{x-1}$

27. $f(x) = \frac{x-2}{2x-3}$

28. $f(x) = \frac{2x+1}{5-3x}$

Exercises 29–38 use the results of Exercises 13–28. For each function:

(a) Determine intervals where the function is increasing and decreasing.

(b) Locate the relative extrema.

(c) Use a calculator to graph the function, and verify your results for parts (a) and (b).

29. The function in Exercise 15.

30. The function in Exercise 16.

✓ **31.** The function in Exercise 17.

32. The function in Exercise 18.

33. The function in Exercise 19.

34. The function in Exercise 20.

35. The function in Exercise 23.

36. The function in Exercise 24.

37. The function in Exercise 25.

38. The function in Exercise 26.

Use the derivatives given in Exercises 39–44 to determine the x-values where f has a relative maximum and/or a relative minimum.

39. $f'(x) = -2x + 5$

40. $f'(x) = -2x - 3$

41. $f'(x) = 3x(x+1)$

42. $f'(x) = -2x(x-4)$

43. $f'(x) = 2(x+1)^2(x-1)(x+3)^3$

44. $f'(x) = 4(x-1)(x+1)^2(x+2)^3$

For Exercises 45–48, sketch a graph of a continuous function that satisfies the given data.

45.

x	$f(x)$	$f'(x)$
0	3	−1
1	2	−1
2	1	Undefined
3	2	1
4	3	1
5	4	1

46.

x	$f(x)$	$f'(x)$
0	2	−2
1	1	−1
2	0.25	0
3	1.25	1
4	2.5	1.5
5	2	−0.5

47.

x	$f(x)$	$f'(x)$
0	0	Undefined
1	1	$\frac{1}{2}$
2	1.4	0.35
3	1.7	0.29
4	2	0.25
5	2.2	0.22

48.

x	$f(x)$	$f'(x)$
0	−7	12
1	0	3
2	1	0
3	2	3
4	9	12
5	28	27

Applications

49. In Table 15.1.1, x represents the number of years since 1989 and $f(x)$ represents the number of liver transplants performed in the U.S.

(a) Sketch a graph of a continuous function for f, on the interval [1, 11], that satisfies the data.

(b) What are the units of $f'(x)$?

Table 15.1.1

x	1	3	5	7	9	11
$f(x)$	2631.7	3358.3	3697.8	3918.5	4288.7	5076.8
$f'(x)$	504.81	244.17	117.69	125.37	267.21	543.21

SOURCE: U.S. Statistical Abstract, www.census.gov/statab/www

Table 15.1.2

x	0	2	4	6	8	10	12	14	16
$f(x)$	31	25	22.2	21.8	23	24.7	25.9	25.6	22.9
$f'(x)$	−4.3	−2.2	−0.7	0.3	0.8	0.81	0.3	−0.7	−2.2

SOURCE: U.S. Centers for Disease Control and Prevention, www.cdc.gov

50. In Table 15.1.2, x represents the number of years since 1980 and $f(x)$ represents the number of cases of tuberculosis, in thousands, reported in the United States.

(a) Sketch a graph of a continuous function, on the interval [0, 16], that satisfies the data.

(b) What are the units of $f'(x)$?

51. The Fore Link Company has determined from data that it has collected that the profit function P is given by

$$P(q) = 1000q - q^2 \qquad 0 \le q \le 1000$$

where q is the number of golf clubs produced and sold and $P(q)$ is the profit in dollars.

(a) Determine intervals where P is increasing and decreasing.

(b) Determine the relative maximum and interpret each coordinate.

52. The Network Standards Company has determined that the revenue, in dollars, from the sale of 56K modems can be estimated by

$$R(q) = 300q - q^2 \qquad 0 \le q \le 300$$

where $R(q)$ is in dollars and q is the number of modems.

(a) Determine intervals where R is increasing and decreasing.

(b) Determine the relative maximum and interpret each coordinate.

53. The U.S. egg consumption can be modeled by

$$f(x) = 0.23x^2 - 6.33x + 279.28 \qquad 1 \le x \le 19$$

where x represents the number of years since 1979 and $f(x)$ is the number of eggs consumed per capita (*Source:* U.S. Department of Agriculture, www.usda.gov).

(a) Determine intervals where f is increasing and decreasing.

(b) Determine the relative minimum and interpret each coordinate.

54. The U.S. imports of crude natural rubber can be modeled by

$$f(x) = -44.33x^2 + 480.83x + 278.66 \qquad 1 \le x \le 10$$

where x is the number of years since 1989 and $f(x)$ is the dollar value, in millions, of U.S. imports of crude natural rubber (*Source:* U.S. Department of Agriculture, www.usda.gov).

(a) Determine intervals where f is increasing and decreasing.

(b) Determine the relative maximum and interpret each coordinate.

55. The U.S. per capita consumption of fish and shellfish can be modeled by

$$f(x) = 0.002x^3 - 0.083x^2 + 0.988x + 11.513 \qquad 1 \le x \le 20$$

where x represents the number of years since 1979 and $f(x)$ is the number of pounds of fish and shellfish consumed per capita (*Source:* U.S. Department of Agriculture, www.usda.gov).

(a) Determine intervals where f is increasing and decreasing.

(b) Determine the relative extrema and interpret each coordinate.

56. The U.S. imports of crude oil can be modeled by

$$f(x) = -0.91x^3 + 17.54x^2 + 22.18x + 2121.58 \qquad 1 \le x \le 10$$

where x represents the number of years since 1989 and $f(x)$ is the number of barrels, in millions, of crude oil imported into the U.S. (*Source:* U.S. Energy Information Administration, www.eia.doe.gov) Show that from 1990 to 1999 U.S. import of crude oil has been increasing.

57. The oil field production of Alaska can be modeled by

$$f(x) = -0.27x^3 + 2.5x^2 - 27.56x + 671.48 \qquad 1 \le x \le 10$$

where x represents the number of years since 1989 and $f(x)$ is the number of barrels, in millions, of Alaskan oil field production (*Source:* U.S. Energy Information Administration, www.eia.doe.gov). Show that from 1990 to 1999 Alaskan oil field production has been decreasing.

58. Linguini's Pizza Palace is starting an all-you-can-eat pizza buffet from 5:00 to 9:00 P.M. on Friday evenings. A survey of local residents produced the price–demand function

$$p(x) = -0.02x + 8.3$$

where x represents the quantity demanded and $p(x)$ represents the price in dollars.

(a) Use the model to determine the price if the demand is 250. Round to the nearest cent.

(b) Determine $R(x)$, revenue as a function of the quantity x demanded.

(c) Use the techniques of this section to determine intervals where R is increasing and where R is decreasing.

(d) Determine the relative maximum and interpret each coordinate.

59. Linguini's Pizza Palace is introducing a line of specialty pizzas, such as a chicken and garlic pizza. Taste tests of this type of pizza produced the price–demand function

$$p(x) = 15.22e^{-0.015x}$$

where x represents the quantity demanded and $p(x)$ represents the price in dollars.

(a) Use the model to determine the price if the demand is 55. Round to the nearest cent.

(b) Determine $R(x)$, revenue as a function of the quantity x demanded.

(c) Use the techniques of this section to determine intervals where R is increasing and where R is decreasing.

(d) Determine the relative maximum. Round the x-coordinate to the nearest whole number and the y-coordinate to the nearest hundredth, and interpret each coordinate.

60. Researchers have determined through experimentation that the percent concentration of a certain medication during the first 20 hours after it has been administered is approximated by

$$p(t) = \frac{200t}{2t^2 + 5} - 4 \qquad [0.25, 20]$$

where t is the time in hours after administration of the medication and $p(t)$ is the percent concentration.

(a) Compute $p'(t)$ and determine the critical value(s).

(b) Determine intervals where p is increasing and decreasing.

(c) Determine the relative maximum and interpret each coordinate.

61. Researchers have determined through experimentation that the percent concentration of a certain medication during the first 20 hours after it has been administered is approximated by

$$p(t) = \frac{230t}{t^2 + 6t + 9} \qquad [0, 20]$$

where t is the time in hours after administration of the medication and $p(t)$ is the percent concentration.

(a) Compute $p'(t)$ and determine the critical values.

(b) Determine intervals where p is increasing and decreasing.

(c) Determine the relative maximum and interpret each coordinate.

62. A manufacturer of Digital Pet, a virtual pet, has the following costs when producing x pets in one day, where $0 \le x \le 200$: Fixed costs are \$150, unit production costs are \$3 per pet, and equipment maintenance is $\frac{2x^2}{30}$. This yields a cost function for manufacturing x pets in one day of

$$C(x) = 150 + 3x + \frac{2x^2}{30} \qquad 0 \le x \le 200$$

(a) Determine the average cost function, AC, and find the critical value(s) for AC.

(b) Determine intervals where AC is increasing and decreasing.

(c) Determine the relative minimum for AC and interpret each coordinate.

63. A manufacturer of coffee mugs has the following costs when producing x mugs in one day, where $0 \le x \le 100$: Fixed costs are \$50, unit production costs are \$1 per mug, and equipment maintenance is $\frac{x^2}{40}$. This yields a cost function for manufacturing x coffee mugs in one day of

$$C(x) = 50 + x + \frac{x^2}{40} \qquad 0 \le x \le 100$$

(a) Determine the average cost function, AC, and find the critical value(s) for AC.

(b) Determine intervals where AC is increasing and decreasing.

(c) Determine the relative minimum for AC and interpret each coordinate.

64. Consider a quadratic function of the form $f(x) = ax^2 + bx + c$, where a, b, and c are real numbers with $a \neq 0$. Use $f'(x)$ to determine the x-coordinate where a horizontal tangent line occurs. (The result should look familiar.)

65. Recall that profit equals revenue minus cost; that is, $P(x) = R(x) - C(x)$.

(a) At a production level x, show that $P(x)$ is maximized when $R'(x) = C'(x)$, that is, when marginal revenue equals marginal cost.

(b) For all x on production interval (a, b), show that profit is increasing when $R'(x) > C'(x)$, that is, when marginal revenue is greater than marginal cost.

(c) For all x on production interval (a, b), show that profit is decreasing when $R'(x) < C'(x)$, that is, when marginal revenue is less than marginal cost.

SECTION PROJECT

Figure 15.1.26 shows a graph of the rate of change for catfish sold in the United States that were raised by aquaculture (*Source:* U.S. Department of Agriculture, www.usda.gov).

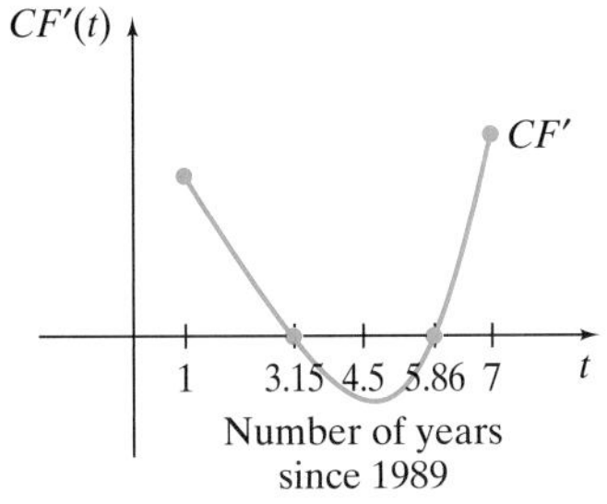

Figure 15.1.26

Given that CF is the function that represents the number of catfish sold, in millions, in the United States that were raised by aquaculture and t represents the number of years since 1989, $1 \leq t \leq 7$, answer the following:

(a) What are the units for $CF'(t)$?

(b) Determine intervals where CF is increasing and decreasing.

(c) Determine the values of t where the rate of change of $CF(t) = 0$.

(d) Determine the t-values that produce relative extrema. Classify each as producing a relative maximum or a relative minimum.

(e) Sketch a possible graph for CF.

Section 15.2 Second Derivatives and Graphs

In Section 15.1, we saw how the derivative of a function, f', allowed us to determine where the function f is increasing and where it is decreasing. Recall that $f'(x)$ tells us an *instantaneous rate of change* at x and that f' is itself a *function*. Since the derivative is a function, we can compute the derivative of $f'(x)$ and use it to determine where f' is increasing and decreasing. The derivative of a derivative is called the **second derivative**.

We just stated that the second derivative tells us where the derivative is increasing and decreasing. This analysis will also give us more information about the behavior of the original function. Specifically, it will give us the intervals of **concavity** and **inflection points**, and other characteristics of the graphs of functions. We will interpret these concepts in the context of rates of change. Before we go much further, let's consider **higher-order derivatives**.

Higher-order Derivatives

If f is some function, $f'(x)$ is the derivative at x which gives the instantaneous rate of change for f at x or, equivalently, the slope of a tangent line at x. We know that f' is used to tell where f is increasing and where it is decreasing. Since f' is itself a function, computing the derivative of f' tells where the derivative is increasing and where it is decreasing. The derivative of f', denoted f'' and read "f double prime", is called the **second derivative**. Table 15.2.1 gives some different notations for the second derivative.

Table 15.2.1

Function	First Derivative	Second Derivative
$f(x)$	$f'(x)$	$f''(x)$
y	y'	y''
y	$\dfrac{dy}{dx}$	$\dfrac{d^2y}{dx^2}$

For most applications, we have no need for a derivative beyond the second derivative. However, any derivative beyond the first derivative is called a **higher-order derivative**.

Example 1 Computing Higher-order Derivatives

Determine the first three derivatives for the following functions.

(a) $f(x) = 2x^5 + 6x^3 - 7x + 1$ **(b)** $y = e^{2x}$

Solution

(a) Applying the rules for differentiation yields

$$f'(x) = 10x^4 + 18x^2 - 7$$

$$f''(x) = \frac{d}{dx}(10x^4 + 18x^2 - 7) = 40x^3 + 36x$$

$$f'''(x) = \frac{d}{dx}(40x^3 + 36x) = 120x^2 + 36$$

(b) Again, applying the rules for differentiation yields

$$y' = 2e^{2x}$$

$$y'' = \frac{d}{dx}(2e^{2x}) = 4e^{2x}$$

$$y''' = \frac{d}{dx}(4e^{2x}) = 8e^{2x}$$ ■

As we can see in Example 1, computing higher-order derivatives is no different than just computing a derivative.

Checkpoint 1 Now work Exercise 3.

Flashback

First Derivative Test Revisited

Determine intervals where $f(x) = 5x^3 + 4x^2 - 12x - 25$ is increasing and where it is decreasing. Also, locate any relative extrema.

Flashback Solution

To find intervals where f is increasing and where it is decreasing, we need to determine the critical values for this function. As we saw in Section 15.1, Example 5, the derivative is given by $f'(x) = 15x^2 + 8x - 12$, and the critical values are $x = \frac{-6}{5}$ and $x = \frac{2}{3}$. Placing the critical values on a number line and choosing test numbers produces the sign diagram in Figure 15.2.1.

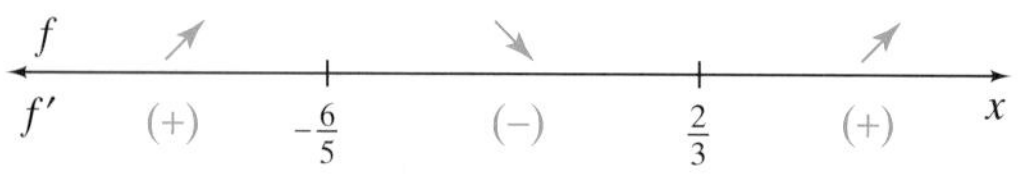

Figure 15.2.1

The sign diagram indicates that $f(x) = 5x^3 + 4x^2 - 12x - 25$ is increasing on $\left(-\infty, \frac{-6}{5}\right) \cup \left(\frac{2}{3}, \infty\right)$ and decreasing on $\left(\frac{-6}{5}, \frac{2}{3}\right)$. To locate the relative extrema, we use the First Derivative Test using a Sign Diagram. This gives a relative maximum at $\left(\frac{-6}{5}, \frac{-337}{25}\right)$ and a relative minimum at $\left(\frac{2}{3}, \frac{-803}{27}\right)$, as shown in Figure 15.2.2.

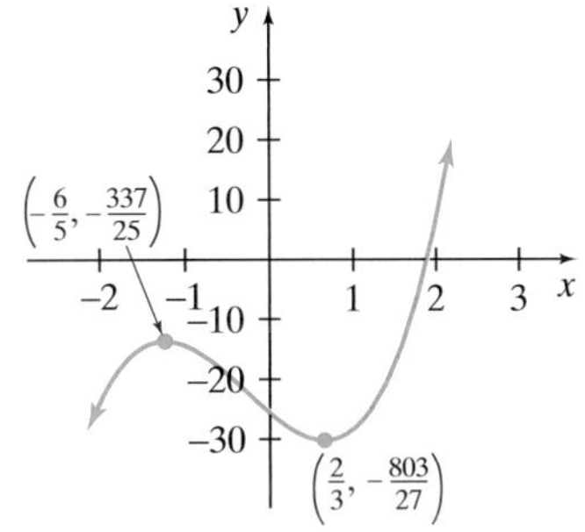

Figure 15.2.2

Recall that, since f' is also a function, we should be able to compute the **second derivative**, f'', and use it to determine the behavior of the derivative, f'. Later we will see how this information is used to determine the shape of the graph of f. First, let's see how we can use f'' to find information about f'.

Example 2 Using f'' to Graph f'

For the function in the Flashback, compute $f''(x)$ and use it to determine where f' is increasing and decreasing. Graph f'.

Solution

In the Flashback we computed the derivative of $f(x) = 5x^3 + 4x^2 - 12x - 25$ to be $f'(x) = 15x^2 + 8x - 12$. So the second derivative is

$$f''(x) = \frac{d}{dx}(15x^2 + 8x - 12) = 30x + 8$$

Since $f''(x)$ is defined for all x, the only critical value for f' is when $f''(x) = 0$. So we solve

$$30x + 8 = 0$$
$$x = \frac{-8}{30} = \frac{-4}{15}$$

So the only critical value for f' is $x = \frac{-4}{15}$.

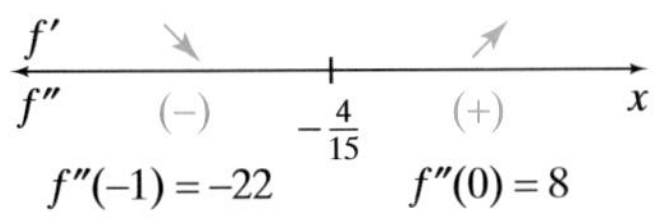

Figure 15.2.3

We employ the same process from Section 15.1 and the Flashback; we place $x = \frac{-4}{15}$ on a number line, select some test numbers, and make a sign diagram. See Figure 15.2.3.

Interactive Activity

From the work in Example 2, notice that a relative minimum for f' occurs at $x = \frac{-4}{15}$. Verify that the y-coordinate is $\frac{-196}{15}$. We know that $f'(x)$ tells us the slope of a line tangent to the graph of f for any x, so at $x = \frac{-4}{15}$ on the graph of f we know that the slope of the tangent line is $\frac{-196}{15}$. What does the fact that this is also the relative minimum for f' mean about the graph of f?

WWW

Notice that f'' and f' are used in the sign diagram. This is exactly what we did in Section 15.1, since we simply have a function and its derivative listed here! From the sign diagram we determine that f' is decreasing on $\left(-\infty, \frac{-4}{15}\right)$ and increasing on $\left(\frac{-4}{15}, \infty\right)$.

Figure 15.2.4 shows a graph of $f'(x) = 15x^2 + 8x - 12$.

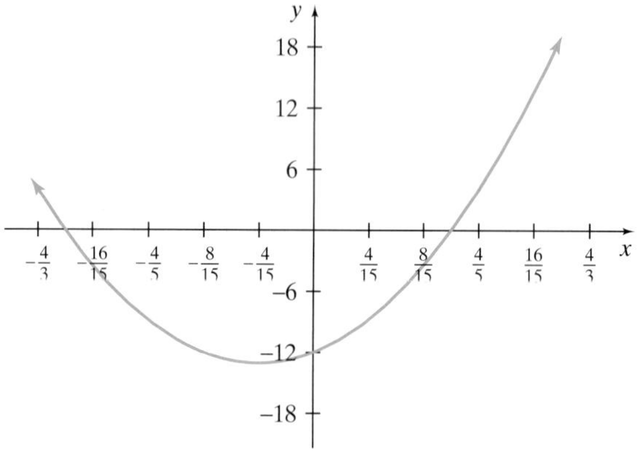

Figure 15.2.4 Graph of $f'(x) = 15x^2 + 8x - 12$. ■

Concavity

In Chapter 12 we saw how the derivative gives the instantaneous rate of change of a function at any point. Example 2 showed how the second derivative tells us where the first derivative is increasing and where it is decreasing. Since the first derivative gives the instantaneous rate of change, the second derivative tells us the increasing or decreasing behavior of the instantaneous rate of change. Knowing intervals where the instantaneous rate of change, f', is increasing and intervals where it is decreasing then tells us the type of **concavity** that the graph of a function, f, has on the interval. Since concavity is important in our continuing analysis of functions and their graphs, we offer the following definition.

Concavity

On an open interval (a, b) where f is differentiable:

1. If f' is increasing, then the graph of f is **concave up.**
2. If f' is decreasing, then the graph of f is **concave down.**

Notice in Figure 15.2.5, on an interval where the graph of a function is concave up, tangent lines to the curve for any point in the interval are below the

Concave up: Tangent lines lie below the curve.

Concave down: Tangent lines lie above the curve.

Figure 15.2.5

curve. Also, on an interval where the graph of a function is concave down, tangent lines to the curve for any point in the interval are above the curve.

Now let's return to our work in the Flashback and Example 2. Recall that

$$f(x) = 5x^3 + 4x^2 - 12x - 25 \qquad \text{(see Figure 15.2.6)}$$
$$f'(x) = 15x^2 + 8x - 12 \qquad \text{(see Figure 15.2.7)}$$
$$f''(x) = 30x + 8 \qquad \text{(see Figure 15.2.8)}$$

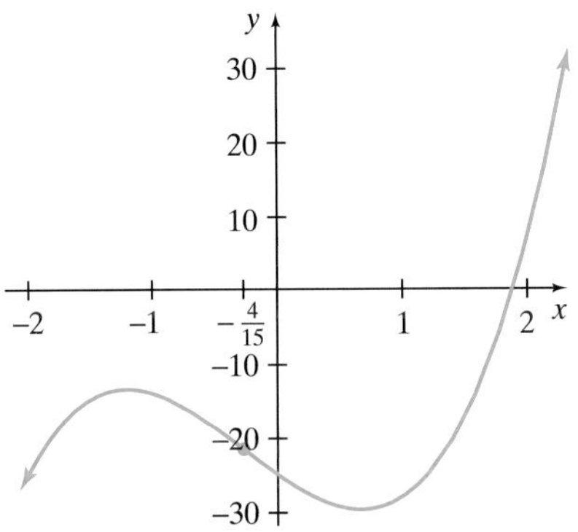

Figure 15.2.6

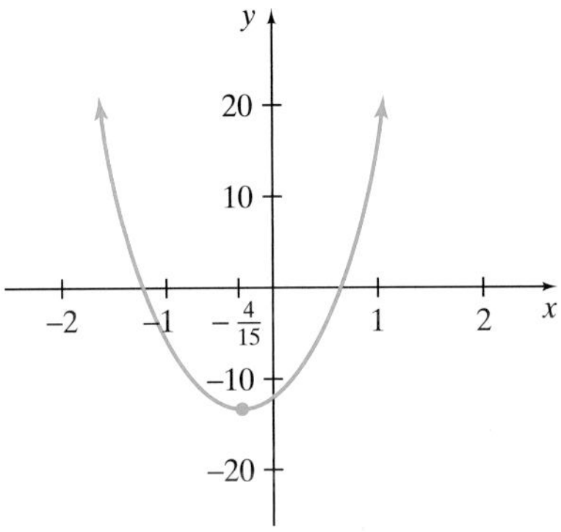

Figure 15.2.7

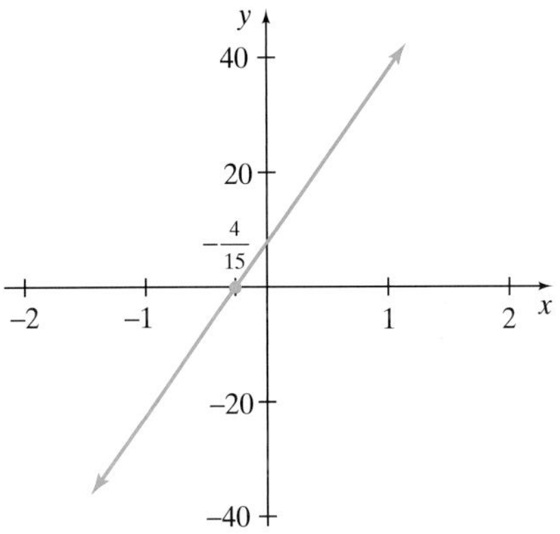

Figure 15.2.8

The Flashback Example 2, and Figures 15.2.6 through 15.2.8 demonstrate the following:

- On the interval $\left(-\infty, \frac{-4}{15}\right)$, $f'' < 0$ (Figure 15.2.8) and f' is decreasing (Figure 15.2.7 and Example 2).
- On the same interval $\left(-\infty, \frac{-4}{15}\right)$, the graph of f is concave down. See Figure 15.2.6.
- On the interval $\left(\frac{-4}{15}, \infty\right)$, $f'' > 0$ (Figure 15.2.8) and f' is increasing (Figure 15.2.7 and Example 2).
- On the same interval $\left(\frac{-4}{15}, \infty\right)$, the graph of f is concave up. See Figure 15.2.6.

These observations can be condensed into the following Tests for Concavity.

Tests for Concavity

For a function f whose second derivative exists on open interval (a, b):

1. If $f''(x) > 0$ for all x on (a, b), then the graph of f is concave up on (a, b).
2. If $f''(x) < 0$ for all x on (a, b), then the graph of f is concave down on (a, b).

Note: In Section 15.1 we saw how the sign of f' determines where f is increasing or decreasing. Here we need to determine the sign of f'' in order to determine the concavity of the graph of f. Hence, the process outlined in Section 15.1 will be used here, as shown in Example 3.

Example 3 Determining Intervals of Concavity

Determine intervals where the graph of $f(x) = x^3 + 3x^2 - 4$ is concave up and where the graph is concave down.

Solution

Understand the Situation: To determine the concavity, we need to determine the sign of the second derivative.

We compute the second derivative to be

$$f'(x) = 3x^2 + 6x$$

$$f''(x) = 6x + 6$$

We note that $f''(x)$ is defined for all values of x, so, using the same process as in Section 15.1, we need to determine where $f''(x) = 0$. Solving this equation yields

$$6x + 6 = 0$$

$$x = -1$$

We place $x = -1$ on a number line and make a sign diagram in Figure 15.2.9 using the same steps as in Section 15.1.

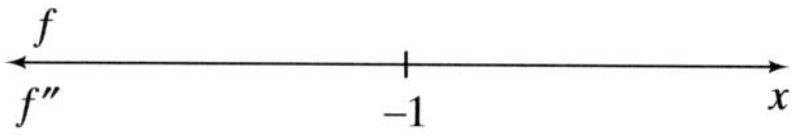

Figure 15.2.9

Selecting $x = -2$ as a test number from the interval $(-\infty, -1)$ and substituting this into $f''(x)$ gives $f''(-2) = 6(-2) + 6 = -6$. Selecting $x = 0$ from the interval $(-1, \infty)$ and substituting into $f''(x)$ gives $f''(0) = 6(0) + 6 = 6$. Putting this information on the sign diagram yields Figure 15.2.10.

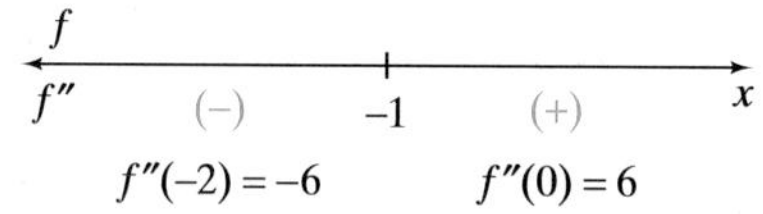

Figure 15.2.10

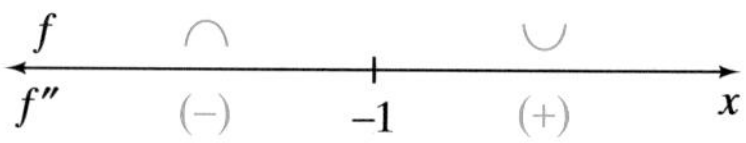

Figure 15.2.11

Knowing that the sign of the second derivative tells the concavity of the graph, we complete the sign diagram as shown in Figure 15.2.11. We use ∩ to mean concave down and ∪ to mean concave up.

The graph of $f(x) = x^3 + 3x^2 - 4$ is concave down on $(-\infty, -1)$ and concave up on $(-1, \infty)$. ■

✓ Checkpoint 2

Now work Exercise 11.

The graph of $f(x) = x^3 + 3x^2 - 4$ is shown in Figure 15.2.12. At the point $(-1, -2)$, the graph switches concavity from concave down to concave up.

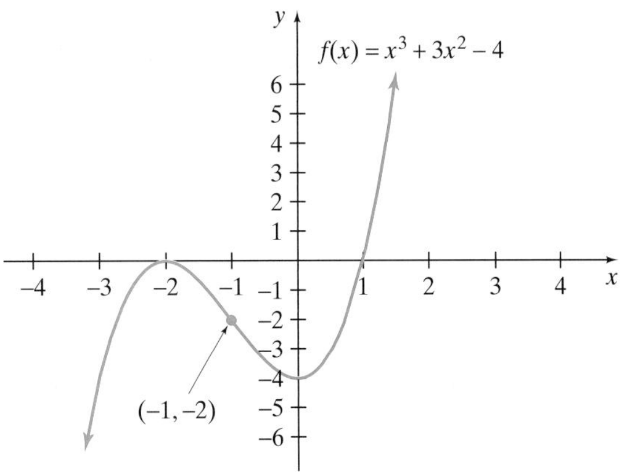

Figure 15.2.12

A point on the graph where concavity changes from concave up to concave down (or from concave down to concave up) is called an **inflection point**. It is also worth noting that the x-coordinate of an inflection point is a critical value for f'.

Example 4 Determining Intervals of Concavity

Determine intervals where the graph of $f(x) = x^4$ is concave up and where it is concave down. Also, locate any inflection points.

Solution

Understand the Situation: We can solve problems that ask for the concavity of the graph of a function by using the following procedure.

To determine intervals of concavity of the graph of a function f, we determine the second derivative f'' and apply the Tests for Concavity.

We begin by computing the second derivative to be

$$f'(x) = 4x^3$$
$$f''(x) = 12x^2$$

Since $f''(x)$ is defined for all x, just like in Example 3, we need to solve $f''(x) = 0$. This results in

$$12x^2 = 0$$
$$x = 0$$

We place $x = 0$ on a number line and make a sign diagram as in Figure 15.2.13. Selecting $x = -1$ as a test number from the interval $(-\infty, 0)$ and substituting this into $f''(x)$ gives $f''(-1) = 12(-1)^2 = 12$. Selecting $x = 1$ from the interval $(0, \infty)$ and substituting into $f''(x)$ gives $f''(1) = 12(1)^2 = 12$. Knowing that the sign of

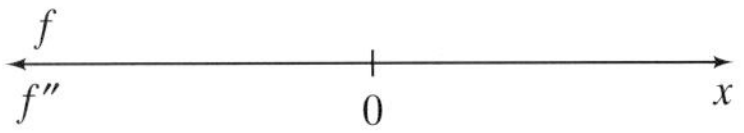

Figure 15.2.13

the second derivative tells the concavity of the graph, we complete the sign diagram as shown in Figure 15.2.14.

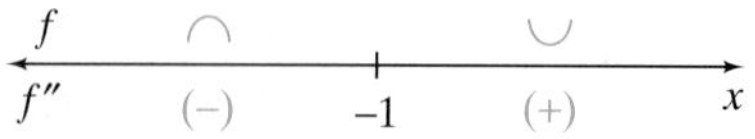

Figure 15.2.14

We conclude that the graph of $f(x) = x^4$ is concave up on $(-\infty, 0)$ and concave up on $(0, \infty)$. Since there is no change in concavity, the graph of the function $f(x) = x^4$ has no inflection points. ■

Example 3 showed that $f''(x)$ may equal zero at an inflection point. There is only one other possibility for an inflection point. If a continuous function has an inflection point at $x = d$, then either $f''(d) = 0$ or $f''(d)$ is undefined. We use the following process to locate inflection points.

Locating Inflection Points

1. Determine the values $x = d$ where $f''(d) = 0$ or where $f''(d)$ is undefined.
2. Place these values on a number line and make a sign diagram.
3. The point $(d, f(d))$ is an inflection point if f'' changes sign at $x = d$ **and** if $x = d$ is in the domain of f.

Note: Example 4 indicates that not every value of x that satisfies $f''(x) = 0$ produces an inflection point. It is worth noting one more time that the process outlined in Examples 3 and 4 is similar to the process that we used in Section 15.1 to locate relative extrema.

Example 5 Locating Inflection Points

Determine any inflection points for the graph of $f(x) = 5x^3 + 4x^2 - 12x - 25$.

Solution

This is the same function from the Flashback and Example 2. In Example 2 we determined that $f''(x) = 30x + 8$ and $f''(x) = 0$ at $x = \frac{-4}{15}$. Placing $x = \frac{-4}{15}$ on a number line and constructing a sign diagram gives Figure 15.2.15. The sign diagram indicates that the graph of f changes concavity at $x = \frac{-4}{15}$. So the point $\left(\frac{-4}{15}, f\left(\frac{-4}{15}\right)\right)$, which is the same as $\left(\frac{-4}{15}, \frac{-14{,}587}{675}\right)$, is an inflection point. See Figure 15.2.16.

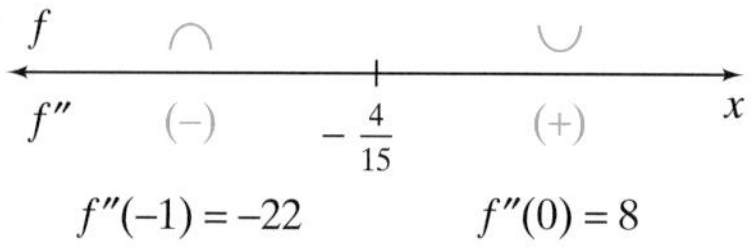

Figure 15.2.15

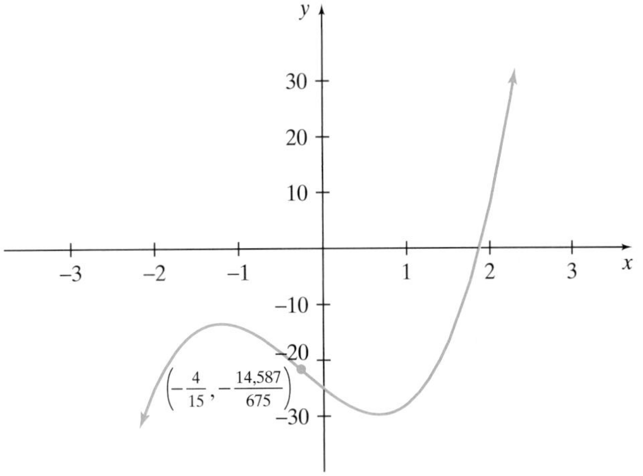

Figure 15.2.16 Graph of $f(x) = 5x^3 + 4x^2 - 12x - 25$ has an inflection point at $\left(\frac{-4}{15}, \frac{-14{,}587}{675}\right)$. ■

Checkpoint 3 Now work Exercise 19.

Inflection Points, Rates of Change, and Applications

In this section, we have learned that the second derivative tells us the intervals where the rate of change is increasing and where it is decreasing. In addition, we know that:

- f' gives the intervals where f is increasing and where it is decreasing.
- f'' gives the intervals where the graph of f is concave up and where it is concave down.

We summarize the results of all this information in Table 15.2.2.

Table 15.2.2

First and Second Derivative	Shape of Graph of f	Behavior of f
$f' > 0$ implies f increasing $f'' > 0$ implies f concave up	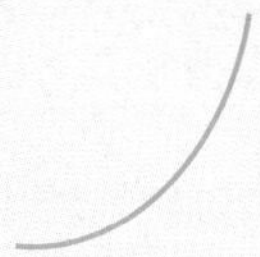	f increasing at a faster rate
$f' > 0$ implies f increasing $f'' < 0$ implies f concave down		f increasing at a slower rate
$f' < 0$ implies f decreasing $f'' < 0$ implies f concave down		f decreasing at a faster rate
$f' < 0$ implies f decreasing $f'' > 0$ implies f concave up	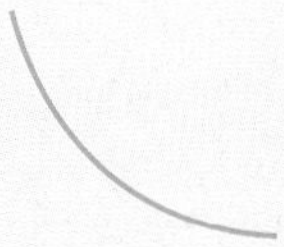	f decreasing at a slower rate

Example 6 Maximizing a Rate of Change

The marketing research department for the Spritz Cola company analyzed data on the number of units of a certain cola that sold after spending x dollars on advertising. They estimate that the company will sell $B(x)$ units of a diet cola after spending x thousand dollars on advertising, according to

$$B(x) = -\frac{1}{3}x^3 + 60x^2 - 110x + 5200 \qquad 20 \le x \le 100$$

(a) Determine where the rate of change of sales is increasing and where it is decreasing.

(b) What level of spending maximizes the rate of change of sales?

Solution

(a) **Understand the Situation:** In order to determine where the rate of change of sales is increasing and where it is decreasing, we must first determine the rate of change of sales, $B'(x)$.

Differentiating $B(x)$ gives us

$$B'(x) = -x^2 + 120x - 110$$

To determine where B' is increasing and decreasing, we compute $B''(x)$, the derivative of $B'(x)$.

$$B''(x) = -2x + 120$$

Since $B''(x)$ is defined for all x, we solve $B''(x) = 0$ and get

$$-2x + 120 = 0$$
$$x = 60$$

Making a sign diagram and analyzing B' and B'' gives Figure 15.2.17. From this diagram we conclude that the rate of change of sales, B', is increasing on the interval (20, 60) and decreasing on the interval (60, 100).

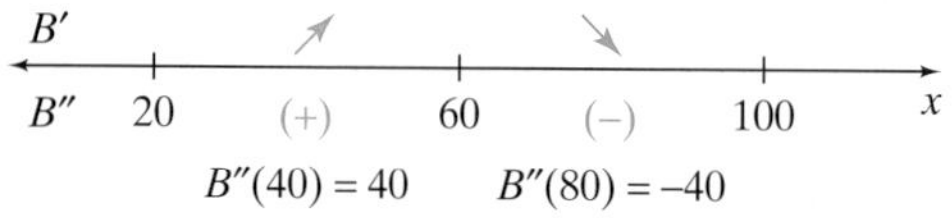

Figure 15.2.17

(b) The sign diagram indicates that a spending level of 60, which is $60,000, for advertising maximizes the rate of change of sales. Figure 15.2.18 shows graphs of B and B' that demonstrate this result.

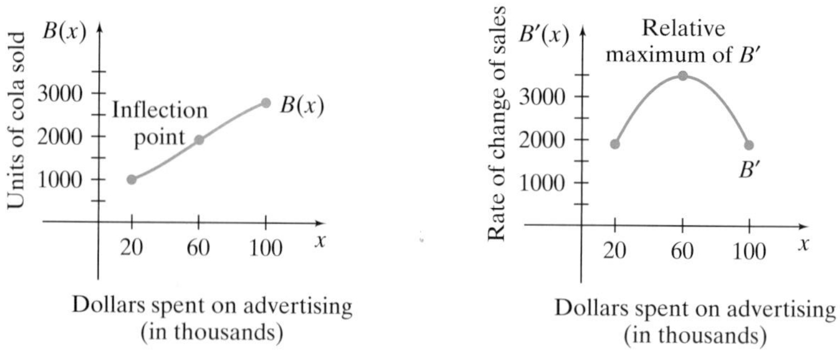

Figure 15.2.18

■

Example 6 and Figure 15.2.18 nicely summarize the key concepts of this section. That is, where B' is increasing, the graph of B is concave up, and where B' is decreasing, the graph of B is concave down. Also, where the rate of change of sales is maximized, at $x = 60$, the graph of B has an inflection point. In Example 6, the inflection point is also known as the **point of diminishing returns**. At the inflection point in Example 6, concavity changes from concave up to concave down. As we analyzed in Example 6, the point of diminishing returns is exactly where the rate of change of sales started to decrease.

Example 7 Interpreting Concavity and Inflection Point

The number of births to women in the United States who were 20 to 24 years old can be modeled by

$$b(x) = 0.38x^3 - 9.19x^2 + 42.85x + 1105 \qquad 1 \le x \le 14$$

where x is the number of years since 1984 and $b(x)$ is the number of births, in thousands (*Source:* U.S. National Center for Health Statistics, www.cdc. gov/nchs).

(a) On the interval (1, 14), determine where the graph of b is concave up and where it is concave down.

(b) Locate and interpret the inflection point.

Solution

(a) **Understand the Situation:** To determine intervals of concavity, we find the second derivative and apply the tests for concavity.

Differentiation gives us

$$b'(x) = 1.14x^2 - 18.38x + 42.85$$
$$b''(x) = 2.28x - 18.38$$

Since $b''(x)$ is defined for all x, we must solve $b''(x) = 0$. This gives

$$2.28x - 18.38 = 0$$
$$x = 8.06 \qquad \text{rounded to the nearest hundredth}$$

We place $x = 8.06$ on a number line and construct a sign diagram as in Figure 15.2.19.

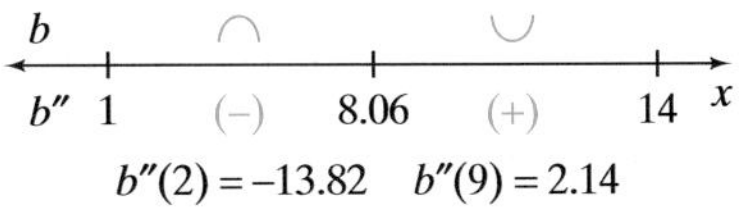

Figure 15.2.19

So the graph of b is concave down on $(1, 8.06)$ and concave up on $(8.06, 14)$.

(b) Since the graph of b changes from concave down to concave up at $x \approx 8.06$, we conclude that there is an inflection point at $(8.06, b(8.06))$ or, specifically, at $(8.06, 1052.33)$. Note that both coordinates have been rounded to the nearest hundredth.

Interpret the Solution: From part (a), we know that the graph of b is concave down on $(1, 8.06)$. This means that the rate of change of births is *decreasing* on this interval. Also from part (a), we know that the graph of b is concave up on $(8.06, 14)$ which means that the rate of change of births is *increasing* on this interval. Hence, we interpret the inflection point as indicating that the rate of change of births was minimized at (8.06, 1052.33). See Figure 15.2.20.

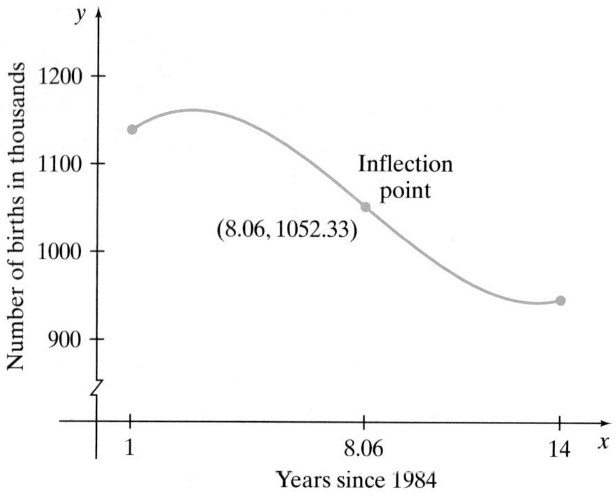

Figure 15.2.20 Rate of change of births is minimized at the inflection point. ■

Checkpoint 4 Now work Exercise 63.

SUMMARY

In this section we saw how the second derivative can be used to determine the increasing and decreasing behavior of the rate of change, or the derivative, of a function. We examined the intervals where the graph of a function is concave up and where it is concave down, and we saw that a point where a graph switches concavity is called an **inflection point**.

To determine the **concavity** of the graph of a function on an interval, apply the Tests for Concavity.

- If $f'' > 0$ on an interval, then the graph of f is concave up on the interval.
- If $f'' < 0$ on an interval, then the graph of f is concave down on the interval.

To locate inflection points:

1. Compute $f''(x)$.
2. Determine where $f''(x)$ is undefined and where $f''(x) = 0$.
3. Place these values on a number line and construct a sign diagram.
4. If concavity changes at any point, then there is an inflection point.

SECTION 15.2 EXERCISES

In Exercises 1–10, compute the first three derivatives for the given function.

1. $f(x) = -4x^5 - 6x^3 + 7x$

2. $f(x) = 5x^4 + 3x^2 - 7x + 1$

✓ **3.** $y = 7x^3 - 3x^2 + 4x + 5$

4. $y = -8x^3 - 7x^2 + 5x + 6$

5. $f(x) = e^x$ **6.** $f(x) = e^{x^2}$

7. $f(x) = \sqrt{x}$ **8.** $f(x) = \sqrt[3]{x}$

9. $y = \ln x$ **10.** $y = \ln 2x$

For each function graphed in Exercises 11–16:

(a) Determine intervals where the graph of function is concave up and where it is concave down.

(b) Determine intervals where the derivative of the function is increasing and where the derivative is decreasing.

✓ **11.**

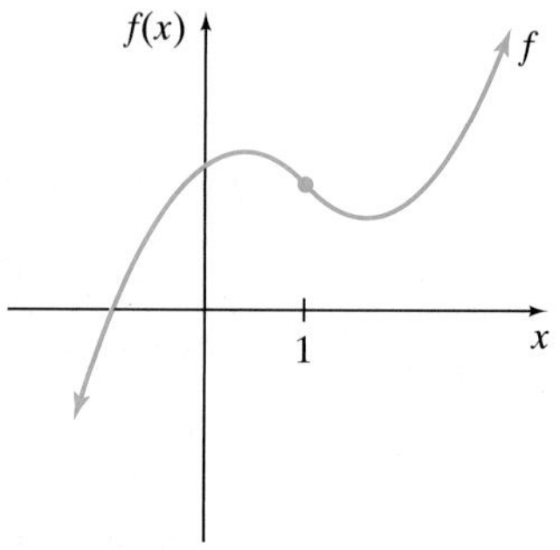

12.

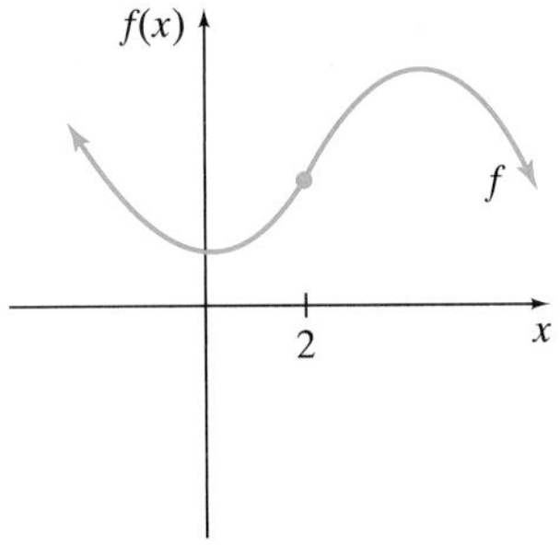

13.

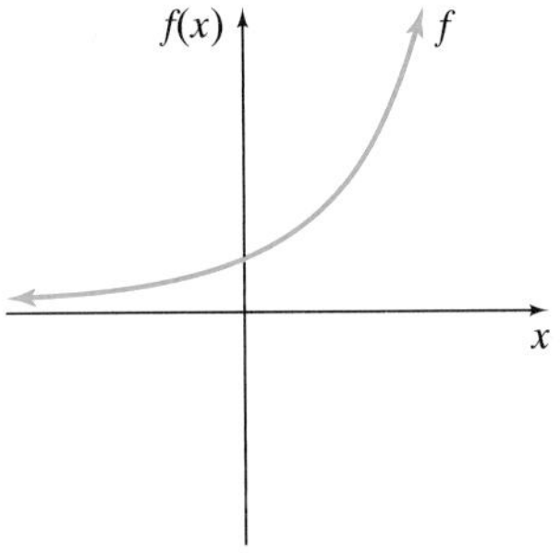

14.

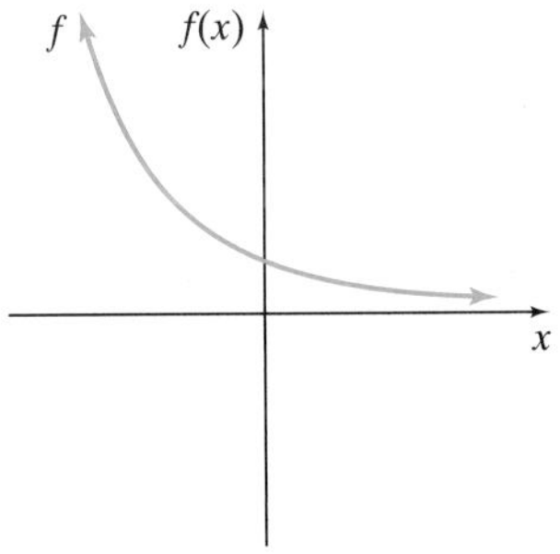

15.

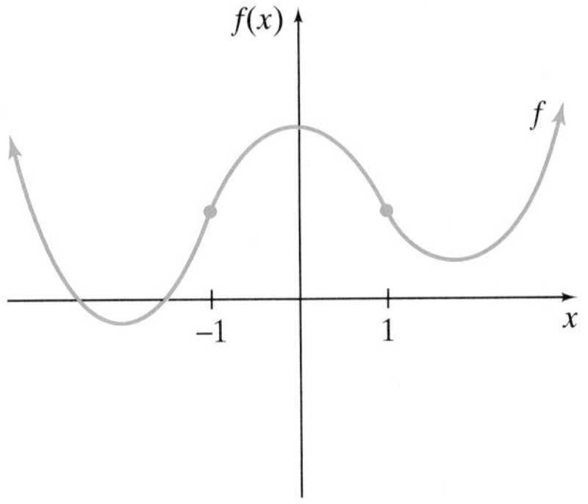

16.

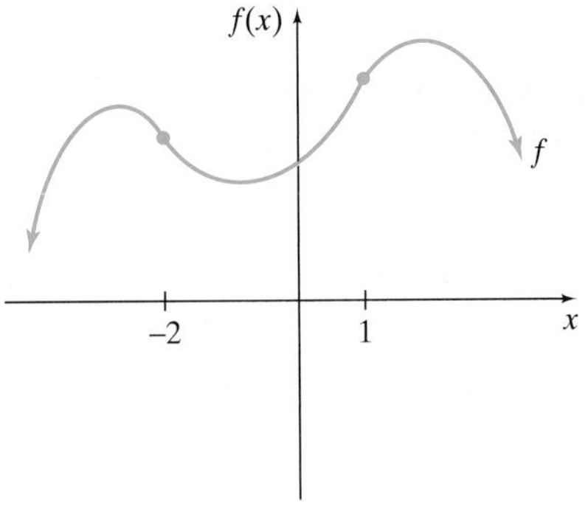

For each function in Exercises 17–34, determine intervals where the graph of the function is concave up and where it is concave down, and locate any inflection points.

17. $y = x^3 + 6x^2 + 18x - 5$

18. $f(x) = x^3 - 6x^2 + 6x - 3$

19. $g(x) = -12x^3 + 6x^2 - 24x - 11$

20. $y = -3x^3 + 5x^2 - 2x + 4$

21. $f(x) = 2x^2 + 3x + 1$

22. $f(x) = 3x^2 - 2x - 5$

23. $f(x) = -3x^2 + 3x - 2$

24. $f(x) = -2x^2 - 5x + 3$

25. $f(x) = x + \dfrac{2}{x}$

26. $f(x) = x + \dfrac{5}{x}$

27. $f(x) = x^{2/3}$

28. $f(x) = (x - 1)^{2/3}$

29. $y = e^{2x}$

30. $y = e^{1.2x}$

31. $y = e^{-2x}$

32. $y = e^{-1.5x}$

33. $g(x) = \ln(x + 1)$

34. $g(x) = \ln(x - 3)$

For each function in Exercises 35–48, determine the following:

(a) Intervals where the function is increasing and where it is decreasing.

(b) The relative extrema.

(c) Intervals where the graph of the function is concave up and where it is concave down.

(d) Any inflection points.

35. $f(x) = 2x^2 + 3x + 1$

36. $f(x) = 3x^2 - 2x - 5$

37. $y = x^3 + 6x^2 - 15x - 5$

38. $f(x) = x^3 - 6x^2 - 15x + 3$

39. $y = -3x^3 + 5x^2 + 2x - 9$

40. $y = -2x^3 + 3x^2 + 6x - 5$

41. $f(x) = x + \dfrac{5}{x}$

42. $f(x) = x + \dfrac{2}{x}$

43. $f(x) = (x - 1)^{2/3}$

44. $f(x) = (x + 2)^{2/3}$

45. $y = e^{3x}$

46. $y = e^{-3x}$

47. $g(x) = \ln(x + 1)$

48. $g(x) = \ln(x - 3)$

49. Sketch the graph of a function that satisfies the following conditions:

Domain: All real numbers
Range: All real numbers less than or equal to 5
Continuous for all real numbers
$f' > 0$ on $(-\infty, 2)$
$f' < 0$ on $(2, \infty)$
$f'' < 0$ on $(-\infty, \infty)$

50. Sketch the graph of a function that satisfies the following conditions:

Domain: All real numbers
Range: All real numbers greater than or equal to 5
Continuous for all real numbers
$f' < 0$ on $(-\infty, -3)$
$f' > 0$ on $(-3, \infty)$
$f'' > 0$ on $(-\infty, \infty)$

51. Sketch the graph of a function that satisfies the following conditions:

Domain: All real numbers
Range: All real numbers
Continuous for all real numbers
$f' > 0$ on $(-\infty, -1) \cup (3, \infty)$
$f' < 0$ on $(-1, 3)$
$f'' < 0$ on $(-\infty, 1)$
$f'' > 0$ on $(1, \infty)$

52. Sketch the graph of a function that satisfies the following conditions:

Domain: All real numbers
Range: All real numbers
Continuous for all real numbers
$f' > 0$ on $(-\infty, 0) \cup (0, \infty)$
$f'(0)$ is undefined
$f'' > 0$ on $(-\infty, 0)$
$f'' < 0$ on $(0, \infty)$

53. In Chapter 1 we saw that the graph of the quadratic function $f(x) = ax^2 + bx + c$, where a, b, and c are real numbers and $a \neq 0$, is concave up if $a > 0$ and concave down if $a < 0$. Show that this is true.

54. Show that the x-coordinate of the inflection point on the graph of any cubic polynomial of the form $f(x) = ax^3 + bx^2 + cx + d$, where a, b, c, and d are real numbers and $a \neq 0$, is given by $x = \dfrac{-b}{3a}$.

Applications

55. The Beagle Works Company has determined that its cost, in hundreds of dollars, for producing x items of its best selling product is given by

$$C(x) = x^3 - 6x^2 + 13x + 10$$

(a) Determine $C(5)$ and $C'(5)$ and interpret each.

(b) Determine intervals where *marginal cost* is increasing and where it is decreasing. Determine the relative minimum for the marginal cost function.

(c) Determine the inflection point for the graph of C.

56. The Chug-a-Mug Company has determined that its cost, in hundreds of dollars, for producing x items of its best selling product is given by

$$C(x) = x^3 - 6x^2 + 15x$$

(a) Determine $C(5)$ and $C'(5)$ and interpret each.

(b) Determine intervals where *marginal cost* is increasing and where it is decreasing. Determine the relative minimum for the marginal cost function.

(c) Determine the inflection point for the graph of C.

57. The Cool Air refrigerator company has determined that its monthly profit, in dollars, for producing and selling x items of its best-selling refrigerator, measured in hundreds, is given by

$$P(x) = -2.3x^3 + 445x^2 - 1500x - 200$$

(a) Determine $P(35)$ and $P'(35)$ and interpret each.

(b) Determine intervals where the *marginal profit* is increasing and where it is decreasing. Determine the relative extremum for marginal profit.

58. *(continuation of Exercise 57)*

(a) Determine the inflection point for the graph of P.

(b) Explain why the relative extremum for the marginal profit and the inflection point for the graph of P have the same x-value.

59. For the profit function, P, given in Exercise 57, explain why the intervals where the *marginal profit* is increasing and where it is decreasing and the intervals where the graph of P is concave up and where it is concave down are the same.

60. The Sucre Cola Company estimates that total sales of its new cola, $TS(x)$, when spending x million dollars on advertising, can be modeled by

$$TS(x) = -\frac{5}{2}x^3 + 112.5x^2 + 150x + 10{,}000 \qquad 10 \leq x \leq 30$$

Locate the point of diminishing returns for $TS(x)$ and interpret its meaning. (*Hint:* Recall the discussion following Example 6, that the *point of diminishing returns* is an inflection point.)

61. The Big Cola Company estimates that total sales of its new diet cola, $TS(x)$, when spending x million dollars on advertising, can be modeled by

$$TS(x) = -2x^3 + 90x^2 - 1200x + 10{,}000 \qquad 10 \le x \le 25$$

Locate the point of diminishing returns for $TS(x)$ and interpret its meaning. (*Hint:* Recall the discussion following Example 6, that the *point of diminishing returns* is an inflection point.)

62. Fill in the blanks.

For $B(x)$ on the interval [20, 100] in Example 6, the rate of change of sales, $B'(x)$, is increasing. This rate of change increases at a ________________ (faster/slower) rate on the interval (20, 60), whereas this rate of change increases at a ________________ (faster/slower) rate on the interval (60, 100).

63. The number of births to women 35 to 39 years old in the United States can be modeled by

$$b(x) = 208.6x^{\frac{1}{4}} \qquad 1 \le x \le 14$$

where x is the number of years since 1984 and $b(x)$ is the number of births, in thousands (*Source:* U.S. National Center for Health Statistics, www.cdc.gov/nchs).

(a) Compute $b(10)$ and $b'(10)$ and interpret each.

(b) On the interval [1, 14], show that b is increasing.

(c) Determine $b''(x)$ and show that the graph of b is concave down on the interval [1, 14].

(d) Complete the following sentence:

From 1985 to 1998, the number of births to women 35 to 39 years old was ____________ (increasing/decreasing), and it was ________________ (increasing/decreasing) at a ________________ (slower/faster) rate.

64. As arterial pressure increases, blood flow through the artery also increases. This can be modeled by

$$f(x) = 0.267e^{0.0256x} \qquad 20 \le x \le 120$$

where x represents the arterial pressure, measured in millimeters of mercury (mm Hg), and $f(x)$ is blood flow measured in milliliters per minute (mL/min).

(a) Evaluate $f(50)$ and $f'(50)$ and interpret each.

(b) Show that the graph of f is concave up on the interval [20, 120]. Explain what this means.

65. As arterial pressure increases, blood flow through the artery also increases. This can also be modeled by

$$f(x) = 0.278(1.026)^x \qquad 20 \le x \le 120$$

where x represents the arterial pressure, measured in mm Hg, and $f(x)$ is blood flow measured in mL/min.

(a) Evaluate $f(50)$ and $f'(50)$ and interpret each.

(b) Show that the graph of f is concave up on the interval [20, 120]. Explain what this means.

66. For males, the blood volume in milliliters (mL) can be approximated by

$$f(x) = -10{,}822 + 3800 \ln x \qquad 40 \le x \le 90$$

where x is weight in kilograms (kg).

(a) Compute $f(60)$ and $f'(60)$ and interpret each.

(b) Show that on the interval [40, 90] the graph of f is concave down.

67. The *cardiac index* is the cardiac output per square meter of body surface area, and its units of measure are $\dfrac{\text{liters per minute}}{\text{square meters}}$. The cardiac index can be modeled by

$$CI(x) = \frac{7.644}{\sqrt[4]{x}} \qquad 10 \le x \le 80$$

where x is an individual's age in years.

(a) Compute $CI(20)$ and $CI'(20)$ and interpret each.

(b) Determine the intervals where CI is increasing and where it is decreasing.

(c) Show that the graph of CI is concave up on $10 \le x \le 80$ and interpret what this means.

68. A certain product has a demand curve modeled by

$$q(t) = \frac{5000}{1 + 2e^{-0.2t}}$$

where $q(t)$ represents the number sold t weeks after the product was introduced to the market.

(a) Compute $q(26)$ and $q'(26)$ and interpret each.

(b) Determine $q''(t)$.

(c) On the interval [0, 52], determine where the graph of q is concave up and where it is concave down. (*Hint:* Graph $q''(t)$ and use the ZERO/ROOT command.)

(d) Assuming that advertising increases sales, based on the result from part (c), when would be a good time to begin an advertising campaign for this product? Explain.

69. The U.S. federal debt can be modeled by

$$d(x) = -1.01x^3 + 33.28x^2 - 29.14x + 999.46 \qquad 1 \le x \le 21$$

where x is the number of years since 1979 and $d(x)$ is the federal debt, in billions of dollars (*Source:* Bureau of the Public Debt, www.publicdebt.treas.gov).

(a) Use $d'(x)$ to show that on [1, 21] the federal debt is increasing.

(b) Determine the inflection point. Round coordinates to the nearest whole number and interpret.

(c) Determine the years for which the rate of change of the U.S. federal debt was increasing.

(d) Determine the years for which the rate of change of the U.S. federal debt was decreasing.

SECTION PROJECT

The number of births to women under 20 years old in the United States can be modeled by

$$b(x) = -0.088x^4 + 1.971x^3 - 15.471x^2 + 44.789x + 488.25 \quad 1 \le x \le 10$$

where x is the number of years since 1988 and $b(x)$ is the number of births, in thousands, to women under 20 years of age (*Source:* U.S. National Center for Health Statistics, www.cdc.gov/nchs).

(a) Compute $b'(9)$ and interpret.

(b) Use $b'(x)$ to determine intervals where b is increasing and where it is decreasing.

(c) Determine intervals where the rate of change of b is increasing and where it is decreasing.

(d) Determine intervals where the graph of b is concave up and where it is concave down.

(e) Use the result from part (b) to determine any relative extrema for b, and interpret each coordinate.

(f) Use the result from part (d) to determine any inflection points for the graph of b, and interpret each coordinate.

Section 15.3 Graphical Analysis and Curve Sketching

In Section 15.1 we learned how the rate function f' can be used to determine the behavior of a continuous function f. To find relative maxima and minima, determine the critical values of f and use the First Derivative Test.

- The point $(c, f(c))$ is a relative maximum if $f' > 0$ to the left of c and $f' < 0$ to the right of c.
- The point $(c, f(c))$ is a relative minimum if $f' < 0$ to the left of c and $f' > 0$ to the right of c.

In Section 15.2 we learned how the second derivative f'' gives important information about the behavior of the function f. To find intervals of concavity for a function f, determine the second derivative f'' and apply the Tests for Concavity.

- If $f'' > 0$ on an open interval, then the graph of f is concave up.
- If $f'' < 0$ on an open interval, then the graph of f is concave down.

In this section, we pull together all these ideas to analyze the behavior of functions in greater depth. We will also analyze how the **Second Derivative Test** may be used to determine relative extrema.

The Second Derivative Test

In Section 15.1 we saw how the First Derivative Test was used to locate relative extrema. At this time, we explore another method for locating relative extrema called the **Second Derivative Test**.

Example 1 Determining Relative Extrema

Determine the critical values of the function shown in Figure 15.3.1. Classify each as giving a relative maximum or a relative minimum, and determine the concavity of the graph of f at each relative extrema.

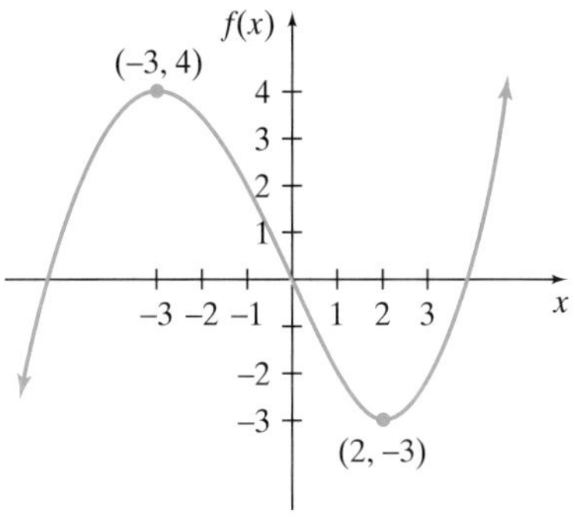

Figure 15.3.1

Solution

Since the graph of f has no breaks or sharp turns, $f'(x)$ is defined for all x. It appears that $f'(x) = 0$ at $x = -3$ and $x = 2$. So the critical values are $x = -3$ and $x = 2$. We have a relative maximum at $x = -3$, specifically at $(-3, 4)$, and a relative minimum at $x = 2$, specifically at $(2, -3)$. From the graph, it appears that at the relative maximum $(-3, 4)$ the graph is *concave down*, whereas at the relative minimum $(2, -3)$ the graph is *concave up*. ■

In Example 1, since the graph is concave down at the relative maximum $(-3, 4)$, we know that $f''(-3) < 0$. Since the graph is concave up at the relative minimum $(2, -3)$, we know that $f''(2) > 0$. This observation is formally given as the **Second Derivative Test**.

The Second Derivative Test

For a function whose second derivative exists on an open interval containing c and has a critical value $x = c$, where $f'(c) = 0$, the point $(c, f(c))$ is a

1. **Relative minimum** if $f''(c) > 0$
2. **Relative maximum** if $f''(c) < 0$

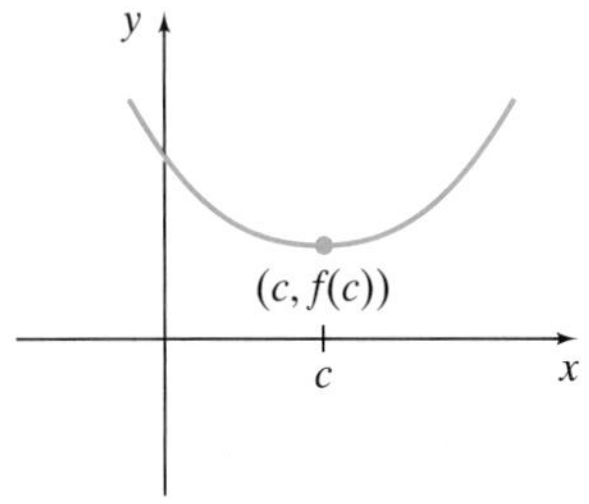

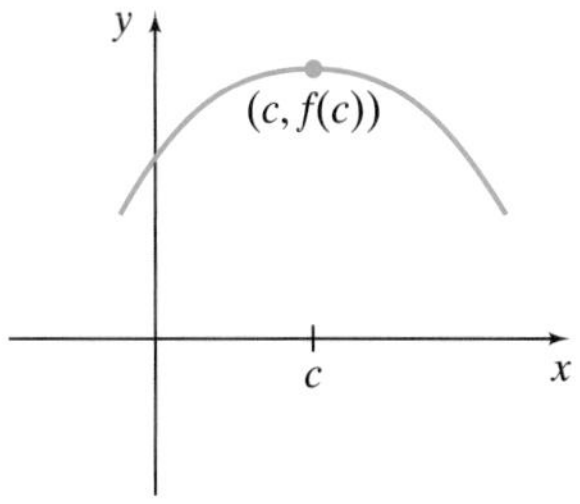

The Second Derivative Test fails if $f''(c) = 0$ or if $f''(c)$ is undefined. In this case, the First Derivative Test may be used.

Example 2 Using the Second Derivative Test to Locate Extrema

Use the Second Derivative Test to locate the relative extrema for

$$f(x) = x^3 + 3x^2 - 9x + 5$$

Solution

Interactive Activity

Use the First Derivative Test to verify the results of Example 2.

To apply the Second Derivative Test, we must first determine the critical values for f, which means that we need to determine $f'(x)$.

$$f'(x) = 3x^2 + 6x - 9$$

Since $f'(x)$ is defined for all x, the only critical values are where $f'(x) = 0$. Solving $f'(x) = 0$ yields

$$3x^2 + 6x - 9 = 0$$
$$3(x^2 + 2x - 3) = 0$$
$$3(x + 3)(x - 1) = 0$$
$$x = -3 \quad \text{or} \quad x = 1$$

So the critical values are $x = -3$ and $x = 1$. We now compute $f''(x)$ to be

$$f''(x) = 6x + 6$$

We substitute each critical value into $f''(x)$ to obtain

$$f''(-3) = 6(-3) + 6 = -12$$
$$f''(1) = 6(1) + 6 = 12$$

Since $f''(-3) < 0$, there is a relative maximum at $x = -3$. Since $f''(1) > 0$, there is a relative minimum at $x = 1$. Figure 15.3.2 shows a graph of f.

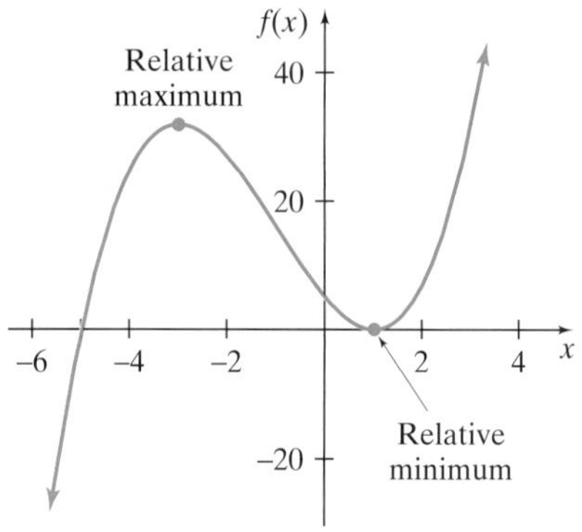

Figure 15.3.2

Checkpoint 1 Now work Exercise 19.

Synthesis

At this time, we are ready to pull together all the concepts studied so far in Chapter 15.

Example 3 Analyzing a Function and Its Graph

Consider $f(x) = 5x^4 - x^5$.

(a) Locate all relative extrema.

(b) Determine intervals where the graph of f is concave up and where it is concave down.

(c) Locate any inflection points.

(d) Using the information gathered in parts (a)–(c), sketch a graph of f and label the points found in part (a) and part (c).

Solution

(a) **Understand the Situation:** To locate relative extrema, we determine the critical values of f and apply the First Derivative Test.

The derivative is

$$f'(x) = 20x^3 - 5x^4$$

Since $f'(x)$ is defined for all x, we need to solve $f'(x) = 0$ to determine the critical values.

$$20x^3 - 5x^4 = 0$$
$$5x^3(4 - x) = 0$$
$$x = 0 \quad \text{or} \quad x = 4$$

We place the critical values on a number line and make a sign diagram using the test values $x = -1$, $x = 1$, and $x = 5$. Recall that we evaluate the derivative at these test values and place the sign of the result in our sign diagram as in Figure 15.3.3. Since $f(x) = 5x^4 - x^5$ is decreasing on $(-\infty, 0)$ and increasing on $(0, 4)$, there is a relative minimum at $x = 0$. Since $f(x) = 5x^4 - x^5$ is increasing on $(0, 4)$ and decreasing on $(4, \infty)$, there is a relative maximum at $x = 4$. Specifically, there is a relative minimum at $(0, 0)$ and a relative maximum at $(4, 256)$.

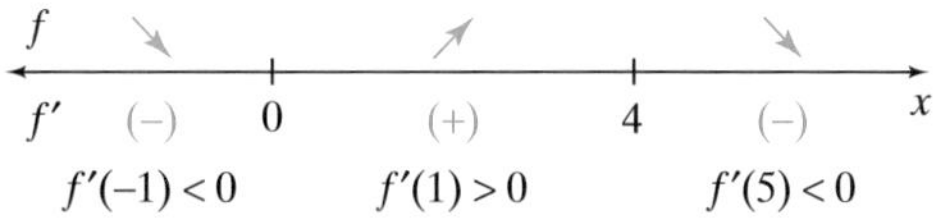

Figure 15.3.3

(b) **Understand the Situation:** To find intervals of concavity, we determine the second derivative and apply the Tests for Concavity.

The second derivative is

$$f''(x) = 60x^2 - 20x^3$$

Since $f''(x)$ is defined for all x, we need to solve $f''(x) = 0$.

$$60x^2 - 20x^3 = 0$$
$$20x^2(3 - x) = 0$$
$$x = 0 \quad \text{or} \quad x = 3$$

We place $x = 0$ and $x = 3$ on a number line and make a sign diagram (Figure 15.3.4) using the test values $x = -1$, $x = 1$, and $x = 5$. Recall that we evaluate the second derivative at these test values and place the sign of the result in our sign diagram.

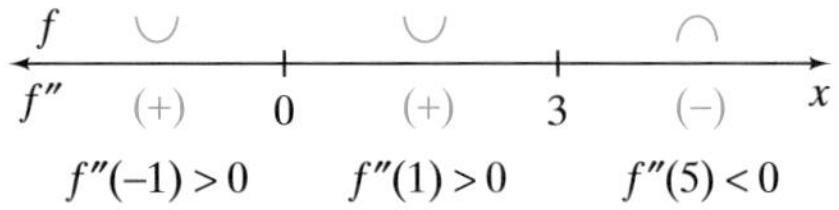

Figure 15.3.4

So the graph of $f(x) = 5x^4 - x^5$ is concave up on $(-\infty, 0) \cup (0, 3)$ and concave down on $(3, \infty)$.

(c) Recall that at an inflection point the sign of f'' changes. Since the sign of f'' changes at $x = 3$, the only inflection point occurs when $x = 3$. Specifically, it is at the point (3, 162).

(d) Figure 15.3.5 is a sketch of the function incorporating all the information gathered in parts (a)–(c).

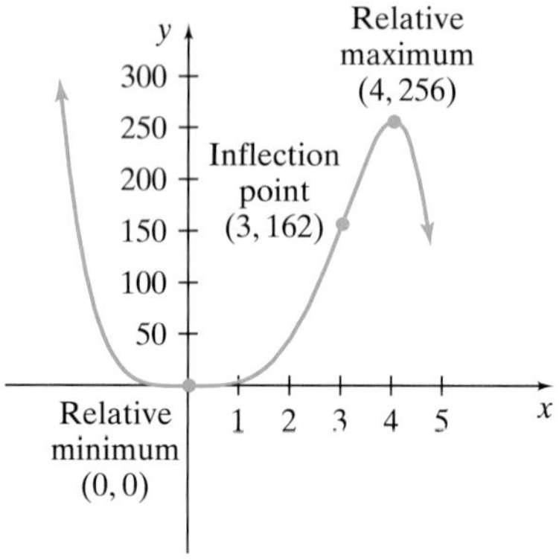

Figure 15.3.5

✓ **Checkpoint 2**

Now work Exercise 33.

Technology Option

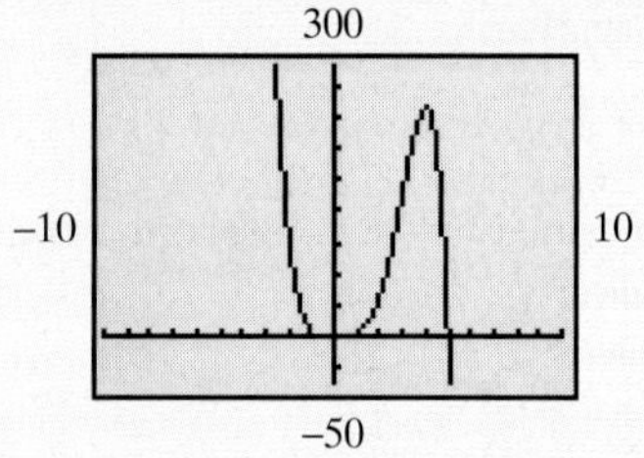

Figure 15.3.6

When doing the graphical analysis of a function, such as we did in Example 3, we can use a graphing calculator as a final step to confirm our sketch. In doing so, we notice how the calculus is helpful in determining an appropriate viewing window. For example, in Example 3 part (a), we determined that the function has a relative maximum at (4, 256) and a relative minimum at (0, 0). This tells us that a viewing window of $[-10, 10]$ by $[-50, 300]$ should show the important features of the graph. See Figure 15.3.6.

Before proceeding to Example 4, let's summarize the steps employed in Example 3.

Graphical Analysis of Function f

1. Use $f'(x)$ to:
   - Determine critical values of f.
   - Construct a sign diagram to determine intervals where f is increasing and where it is decreasing.
   - Locate relative extrema by the First Derivative Test.
2. Use $f''(x)$ to:
   - Construct a sign diagram to determine intervals where the graph of f is concave up and where it is concave down.
   - Locate inflection points.

Example 4 Analyzing a Function and Its Graph

Consider $f(x) = \dfrac{x-1}{2x-3}$.

(a) Determine any relative extrema.

(b) Determine intervals where the graph of f is concave up and where it is concave down. Also, locate any inflection points.

(c) Identify any vertical and horizontal asymptotes.

(d) Sketch a graph of f.

Solution

(a) To locate relative extrema, we determine the critical values of f and apply the First Derivative Test. Differentiating this function utilizing the **Quotient Rule** gives

$$f'(x) = \frac{1 \cdot (2x-3) - (x-1) \cdot 2}{(2x-3)^2}$$
$$= \frac{2x - 3 - 2x + 2}{(2x-3)^2} = \frac{-1}{(2x-3)^2}$$

Notice that $f'(x)$ is undefined at $x = \frac{3}{2}$ since this value makes the denominator 0. Even though $x = \frac{3}{2}$ makes the derivative undefined, it is not a critical value, because $x = \frac{3}{2}$ is not in the domain of f. Also notice that there are no values of x such that $f'(x) = 0$, that is,

$$\frac{-1}{(2x-3)^2} = 0$$

has no real solution. Thus, f has no critical values. However, we include $x = \frac{3}{2}$ on a number line when constructing our sign diagram so that we can exploit the power of Theorem 15.1. (Consult the Toolbox to the left.) See Figure 15.3.7.

From Your Toolbox

If f is continuous and $f(x) \neq 0$ for any x in the interval (a, b), then f can not change sign on (a, b).

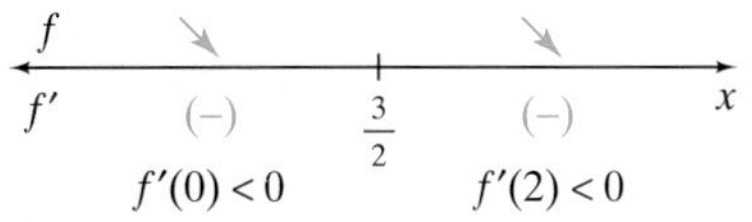

Figure 15.3.7

So $f(x) = \dfrac{x-1}{2x-3}$ is decreasing on $\left(-\infty, \frac{3}{2}\right) \cup \left(\frac{3}{2}, \infty\right)$. Utilizing the First Derivative Test using a sign diagram indicates that f has no relative extrema.

(b) To find intervals of concavity, we determine the second derivative and apply the Tests for Concavity. We have from part (a) that the first derivative is $f'(x) = \dfrac{-1}{(2x-3)^2}$. To compute $f''(x)$, we will rewrite $f'(x)$ as

$$f'(x) = -1(2x-3)^{-2}$$

and employ the Power Rule and Chain Rule. This gives

$$\begin{aligned} f''(x) &= -1[-2(2x-3)^{-3} \cdot 2] \\ &= 4(2x-3)^{-3} \\ &= \frac{4}{(2x-3)^3} \end{aligned}$$

As in part (a), $f''(x)$ is undefined at $x = \frac{3}{2}$, and $f''(x) = 0$ has no real solution. Thus, the only value that we place on the number line is $x = \frac{3}{2}$. See Figure 15.3.8.

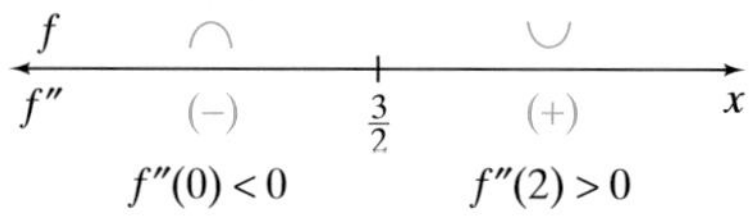

Figure 15.3.8

So the graph of $f(x) = \frac{x-1}{2x-3}$ is concave down on $\left(-\infty, \frac{3}{2}\right)$ and concave up on $\left(\frac{3}{2}, \infty\right)$. Even though the graph changes concavity at $x = \frac{3}{2}$, as stated earlier, this number is not in the domain of f. Hence, there are no inflection points for the graph of f.

(c) Utilizing information from Section 12.2, since $\lim_{x\to(3/2)^-} \frac{x-1}{2x-3} = -\infty$ and $\lim_{x\to(3/2)^+} \frac{x-1}{2x-3} = \infty$, we have a vertical asymptote at $x = \frac{3}{2}$. Also from Section 12.2, we determine the horizontal asymptote by computing

$$\begin{aligned} \lim_{x\to\infty} \frac{x-1}{2x-3} &= \lim_{x\to\infty} \frac{\frac{x}{x} - \frac{1}{x}}{\frac{2x}{x} - \frac{3}{x}} \\ &= \lim_{x\to\infty} \frac{1 - \frac{1}{x}}{2 - \frac{3}{x}} \\ &= \frac{1-0}{2-0} = \frac{1}{2} \end{aligned}$$

Similarly,

$$\lim_{x\to-\infty} \frac{x-1}{2x-3} = \frac{1}{2}$$

So the horizontal asymptote is $y = \frac{1}{2}$.

(d) Figure 15.3.9 shows a graph of f using the information gathered in parts (a)–(c).

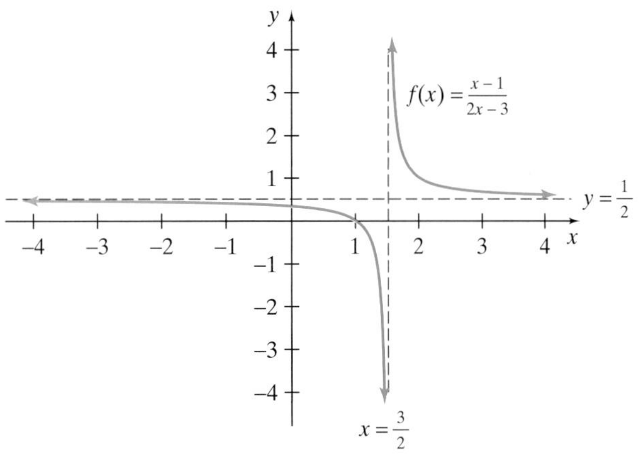

Figure 15.3.9

■

✓ **Checkpoint 3** Now work Exercise 39.

Applications

Example 5 illustrates the power of using the Second Derivative Test in locating relative extrema in some applications.

Example 5 Analyzing Timber Cut Value

The total value of timber cut in the national forest system can be modeled by

$$f(x) = 0.82x^3 - 33.09x^2 + 333.75x + 435.28 \qquad 1 \le x \le 19$$

where x represents the number of years since 1979 and $f(x)$ is the total value of timber cut in millions of dollars (*Source:* U.S. Statistical Abstract, www.census.gov/statab/www). Use the Second Derivative Test to locate any relative extrema and interpret each coordinate.

Solution

Understand the Situation: To apply the Second Derivative Test, we must first determine the critical values for f. This means that we need to determine $f'(x)$.

We compute $f'(x)$ to be

$$f'(x) = 2.46x^2 - 66.18x + 333.75$$

Since the derivative is defined for all x, the only critical value is when $f'(x) = 0$. Hence, we must solve

$$2.46x^2 - 66.18x + 333.75 = 0$$

We solve this quadratic equation by using the quadratic formula. This gives us

$$x = \frac{-b \pm \sqrt{b^2 - 4ac}}{2a} = \frac{66.18 \pm \sqrt{(-66.18)^2 - 4(2.46)(333.75)}}{2(2.46)}$$

$$= \frac{66.18 \pm \sqrt{1095.6924}}{4.92}$$

So our two solutions are, rounded to the nearest hundredth,

$$x = \frac{66.18 + \sqrt{1095.6924}}{4.92} \quad \text{or} \quad x = \frac{66.18 - \sqrt{1095.6924}}{4.92}$$

$$x = 20.18 \qquad\qquad x = 6.72$$

Notice that the solution $x = 20.18$ is not within our reasonable domain $1 \le x \le 19$. So even though it is a solution to the quadratic equation, for this problem it is not considered a critical value.

We now need to evaluate the second derivative at $x = 6.72$. The second derivative is computed to be

$$f''(x) = 4.92x - 66.18$$

Evaluating $f''(x)$ at $x = 6.72$ gives us

$$f''(6.72) = 4.92(6.72) - 66.18 \approx -33.12$$

Since $f''(6.72) \approx -33.12$, we have a relative maximum at $x = 6.72$, specifically at the point (6.72, 1432.63). (The y-coordinate has been rounded to the nearest hundredth.)

Interpret the Solution: This means that in 1985, the total value of timber cut reached a relative maximum of about \$1,432,630,000. ■

Example 6 Analyzing Timber Cut Value

In Example 5 we saw that the total value of timber cut in the national forest system can be modeled by

$$f(x) = 0.82x^3 - 33.09x^2 + 333.75x + 435.28 \qquad 1 \le x \le 19$$

where x represents the number of years since 1979 and $f(x)$ is the total value of timber cut in millions of dollars (*Source:* U.S. Statistical Abstract, www.census.gov/statab/www). Locate any inflection points and interpret each coordinate.

Solution

Understand the Situation: To find any inflection points, we need to determine intervals of concavity and apply the Tests for Concavity.

From Example 5 we determined that

$$f''(x) = 4.92x - 66.18$$

The second derivative is defined for all x, so we must solve $f''(x) = 0$. This gives

$$4.92x - 66.18 = 0$$

$$x = 13.45 \qquad \text{rounded to the nearest hundredth}$$

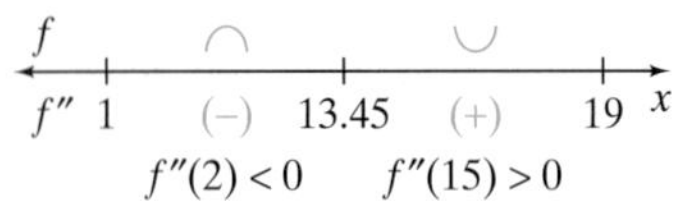

Figure 15.3.10

Placing this value on a number line, we construct the sign diagram in Figure 15.3.10. We use $x = 2$ and $x = 15$ as our test values. Evaluating the second derivative at these test values gives the completed sign diagram as shown in Figure 15.3.10.

From Figure 15.3.10 we see that the graph of f changes concavity at $x = 13.45$. So we have an inflection point at (13.45, 933.32). (The y-coordinate has been rounded to the nearest hundredth.)

Interpret the Solution: The rate at which the total value of timber cut was decreasing the greatest occurred about half way into 1992. See Figure 15.3.11.

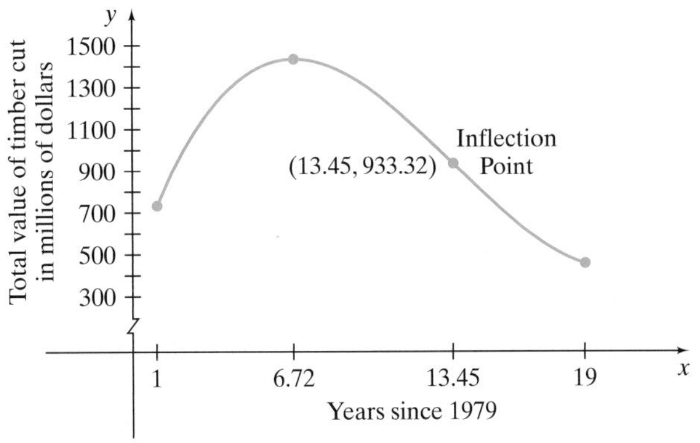

Figure 15.3.11

Checkpoint 4 Now work Exercise 67.

SUMMARY

Section 15.3 has brought together the many concepts studied in Sections 15.1 and 15.2. For a function f:

$f'(x)$

- Gives the slope of a tangent line at any point
- Gives an instantaneous rate of change
- Determines critical values
- Determines intervals of increase or decrease for f
- Determines relative extrema

$f''(x)$

- Gives the increasing or decreasing behavior of the rate of change
- Determines concavity for the graph of f
- Determines inflection points of the graph of f

In this section, we also examined the **Second Derivative Test** as a way to classify critical values as a **relative maximum** or a **relative minimum**.

SECTION 15.3 EXERCISES

Use the graph of f shown in Figure 15.3.12 to answer Exercises 1–6.

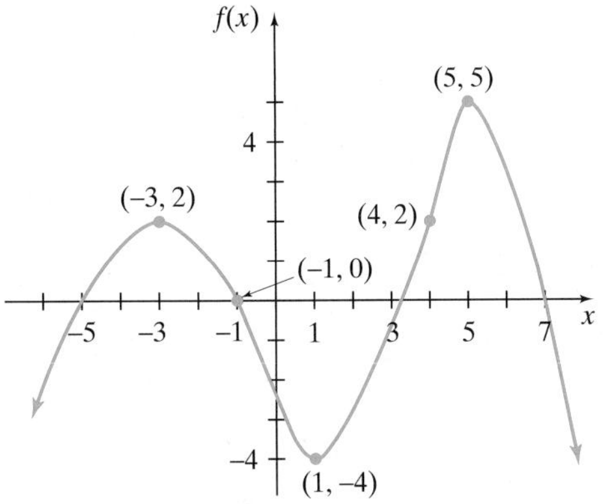

Figure 15.3.12

1. Determine intervals where f is positive and where f is negative.

2. Determine intervals where f is increasing and where f is decreasing.

3. Determine intervals where f' is positive and where f' is negative.

4. Determine intervals where f' is increasing and where f' is decreasing.

5. Determine intervals where the graph of f is concave up and where the graph of f is concave down.

6. Locate any relative extrema, and locate any inflection points.

Use the graph of f shown in Figure 15.3.13 to answer Exercises 7–12.

7. Determine intervals where f is positive and where f is negative.

8. Determine intervals where f is increasing and where f is decreasing.

9. Determine intervals where f' is positive and where f' is negative.

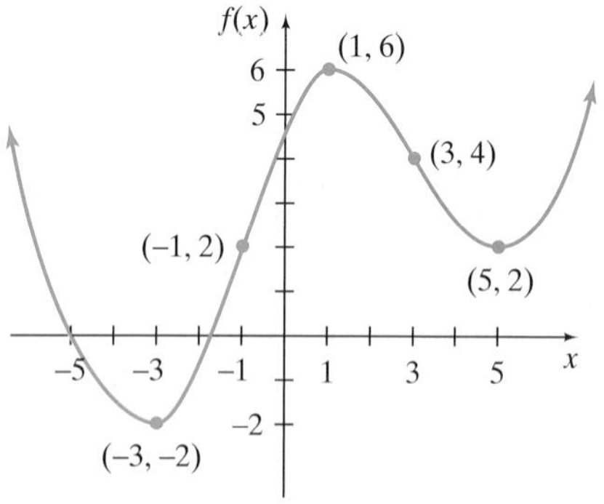

Figure 15.3.13

10. Determine intervals where f' is increasing and where f' is decreasing.

11. Determine intervals where the graph of f is concave up and where the graph of f is concave down.

12. Locate any relative extrema, and locate any inflection points.

In Exercises 13–28, use the Second Derivative Test to locate any relative extrema, if they exist. In Exercises 21 and 22, use your calculator to graph the derivative and then use the ZERO or ROOT command to approximate the solutions to $y' = 0$.

13. $f(x) = 3x^2 - 2x - 3$

14. $f(x) = -2x^2 + 3x + 2$

15. $y = x^3 - 2x^2 - 13x - 10$

16. $y = x^3 + 3x^2 - x - 3$

17. $y = \frac{1}{3}x^3 + \frac{5}{2}x^2 + 6x - 2$

18. $y = \frac{1}{3}x^3 - \frac{1}{2}x^2 - 6x + 2$

19. $f(x) = x^3 + \frac{3}{2}x^2 - 6x - 3$

20. $f(x) = -x^3 + 3x^2 - 3x + 5$

21. $y = x^4 + x^3 - 7x^2 - x + 6$

22. $y = x^4 - 3x^3 - 8x^2 + 12x + 16$

23. $g(x) = 3x^4 - 24x^2 + 16$

24. $g(x) = 2x^4 - 36x^2 - 23$

25. $f(x) = \dfrac{1}{x^2 + 1}$

26. $f(x) = \dfrac{1}{x^2 - 1}$

27. $y = 3x^6 + 9x^4 - 5$

28. $y = -\frac{1}{3}x^6 - 2x^4 + 3$

In Exercises 29–48, sketch the graph of the given function using the techniques in this section. Label all relative extrema, inflection points, and asymptotes. In Exercises 35–38, use your calculator to graph the derivative and then use the ZERO or ROOT command to approximate the solutions to $f'(x) = 0$.

29. $y = x^3 + 3x^2 - x - 3$

30. $y = x^3 - 2x^2 - 13x - 10$

31. $y = \frac{1}{3}x^3 - \frac{1}{2}x^2 - 6x + 2$

32. $f(x) = -x^3 + 3x^2 - 3x + 5$

33. $f(x) = x^3 - 6x^2 + 5$

34. $f(x) = \frac{1}{3}x^3 - 3x^2 + 5x - 2$

35. $f(x) = x^4 - 3x^3 - 8x^2 + 12x + 16$

36. $f(x) = x^4 + x^3 - 7x^2 - x + 6$

37. $f(x) = x^4 - 2x^2 + 1$

38. $f(x) = x^4 - 9x^2 + 2$

39. $y = \dfrac{x - 1}{x + 2}$

40. $y = \dfrac{3x + 1}{5x - 2}$

41. $f(x) = \dfrac{1}{x^2 + 1}$

42. $f(x) = \dfrac{x}{x^2 - 1}$

43. $y = 0.2x + 40 + \dfrac{20}{x}$

44. $y = 0.3x + 20 + \dfrac{10}{x}$

45. $f(x) = x - \ln x$

46. $y = e^x - x$

47. $f(x) = \sqrt[3]{x^2}$

48. $f(x) = \sqrt[3]{(x - 1)^2}$

In Exercises 49–52, the graph of f'' is given. Use the graph to sketch a graph of (a) f' and (b) f. There are many correct answers.

49.

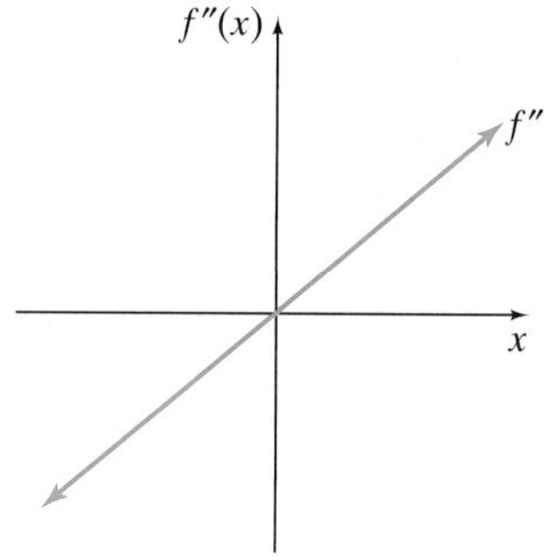

50.

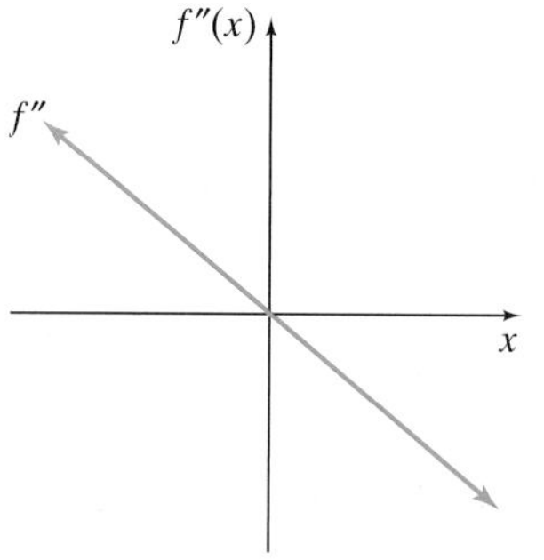

51.

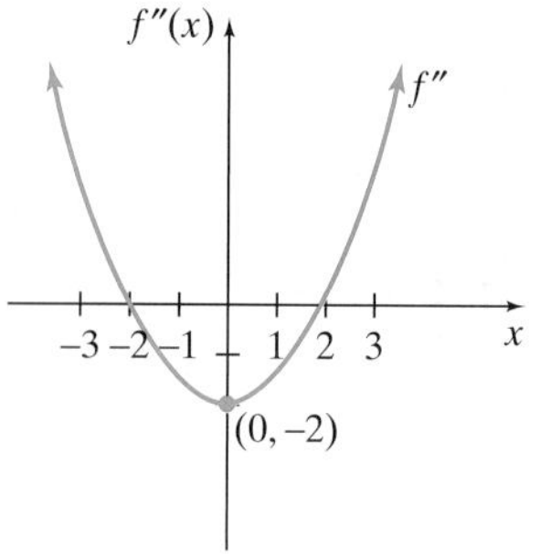

52.

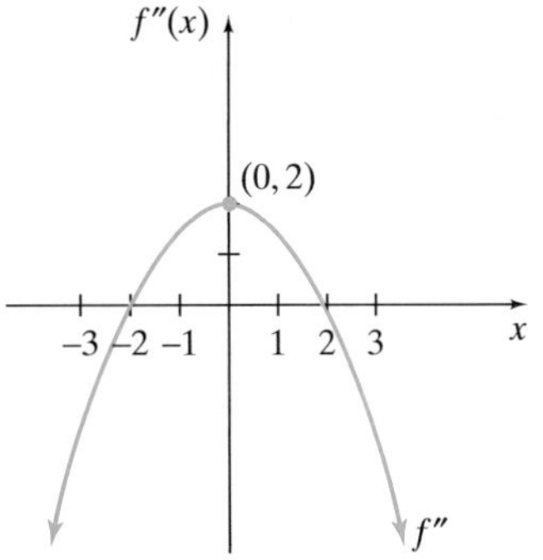

Applications

53. The marketing research department for Music Time, a manufacturer of integrated amplifiers, used a large metropolitan area to test market their new product. They determined that price, $p(x)$, in dollars per unit, and quantity demanded per week, x, was approximated by

$$p(x) = 216 - 0.08x^2 \qquad 0 \le x \le 50$$

So the weekly revenue can be approximated by

$$R(x) = p(x) \cdot x = 216x - 0.08x^3 \qquad 0 \le x \le 50$$

(a) Compute $R'(x)$ and use it to determine intervals where R is increasing and where it is decreasing.

(b) Determine intervals where the graph of R is concave up and where it is concave down.

54. Suppose that Music Time, the manufacturer in Exercise 53, has a weekly cost function, in dollars given by $C(x) = 20x + 1000$.

(a) Determine P, the weekly profit function.

(b) Using the techniques of this section, graph P. Label all extrema and inflection points, if they exist.

55. The Hanash Corporation determines that the weekly profit from producing and selling its Jog-R-Radio can be modeled by

$$P(x) = -0.01x^2 + 12x - 2000 \qquad 0 \le x \le 1000$$

where x represents the number of radios produced and sold each week and $P(x)$ is the weekly profit in dollars.

(a) The *average profit function*, denoted AP, is defined to be $AP(x) = \dfrac{P(x)}{x}$. Determine $AP(x)$.

(b) Using the techniques of this section, sketch a graph of AP on the interval $(0, 1000]$. Label all extrema and inflection points.

56. *(continuation of Exercise 55)*

(a) Determine P', the *marginal profit function*.

(b) Graph both P' and AP (from Exercise 55) in the same coordinate system on the interval $(0, 1000]$.

(c) Algebraically determine the point where the two graphs cross.

(d) Compare the x-coordinate of the result from part (c) with the x-coordinate of the relative maximum for AP. Are they the same?

57. *(continuation of Exercises 55 and 56)* Fill in the blanks illustrating an important fact from economics:

The maximum average profit occurs when ________ is equal to ________.

58. The Digital Pet Company has determined that its daily cost, in dollars, for producing x virtual pets is given by

$$C(x) = 150 + 3x + \frac{2x^2}{30} \qquad 0 \le x \le 200$$

(a) Determine the *average cost function*, $AC(x) = \dfrac{C(x)}{x}$.

(b) Using the techniques of this section, sketch a graph of AC on the interval $(0, 200]$. Label all extrema and inflection points.

59. *(continuation of Exercise 58)*

(a) Determine C', the *marginal cost function*.

(b) Graph both C' and AC (from Exercise 58) in the same coordinate system on the interval $(0, 200]$.

(c) Algebraically determine the point where the two graphs cross.

(d) Compare the x-coordinate of the result from part (c) with the x-coordinate of the relative minimum for AC. Are they the same?

60. *(continuation of Exercises 58 and 59)* Fill in the blanks illustrating an important fact from economics.

The minimum average cost occurs when ________ is equal to ________.

61. Miranda's New and Used Cars has a special program for individuals who purchase a new car. At the time of purchase, buyers can also purchase a maintenance and service contract with the dealership. This contract covers all recommended servicing of the vehicle. Tony just purchased a new car for \$11,000. The cost for the maintenance and service contract for his new car is \$500 the first year and increases \$200 per year thereafter, as long as Tony owns the car. Using the techniques of regression, he found that the total cost of the car (excluding items not covered by the service contract such as gasoline) after t years is given by

$$C(t) = 100t^2 + 400t + 11{,}000$$

(a) Determine a function for the average cost per year, $AC(t) = \dfrac{C(t)}{t}$.

(b) When is the average cost per year a minimum? Round to the nearest tenth.

(c) What is the minimum average cost per year rounded to the nearest dollar?

62. Elisa just purchased a new car at Miranda's New and Used Cars. (See Exercise 61.) The cost was \$15,000 and the cost for the maintenance and service contract for her new car is \$700 the first year and increases \$200 per year thereafter, as long as Elisa owns the car. Using the techniques of regression, she found that the total cost of the car (excluding items not covered by the service contract such as gasoline) after t years is given by

$$C(t) = 100t^2 + 600t + 15{,}000$$

(a) Determine a function for the average cost per year, $AC(t) = \frac{C(t)}{t}$.

(b) When is the average cost per year a minimum? Round to the nearest tenth.

(c) What is the minimum average cost per year rounded to the nearest dollar?

63. For males, blood volume, in milliliters (mL), can be approximated by

$$f(x) = -10{,}822 + 3800 \ln x \qquad 40 \le x \le 90$$

where x is weight in kilograms (kg).

(a) Show that f is increasing on the interval $[40, 90]$.

(b) Show that the graph of f is concave down on the interval $[40, 90]$.

64. For females, blood volume (in milliliters), can be approximated by

$$f(x) = -11{,}456 + 3888 \ln x \qquad 40 \le x \le 60$$

where x is weight in kilograms.

(a) Show that f is increasing on the interval $[40, 60]$.

(b) Show that the graph of f is concave down on the interval $[40, 60]$.

65. The movement of air is known as convection, and the removal of heat from the human body by convection air currents is called *heat loss by convection*. If we let x represent wind velocity, in miles per hour, then $f(x)$ represents the percent of total heat loss by convection and is approximated by

$$f(x) = 12 + 16.694 \ln(x + 1) \qquad 0 \le x \le 60$$

(a) Compute $f(20)$ and $f'(20)$ and interpret each.

(b) Show that f is increasing on the interval $[0, 60]$.

(c) Show that the graph of f is concave down on the interval $[0, 60]$.

(d) Sketch a graph of f on the interval $[0, 60]$.

66. In the 1990s, many municipalities stopped hauling yard waste to land fills due to lack of space. The percent of the waste generated in the United States that is yard waste is approximated by

$$f(t) = 17.27e^{-0.05t} \qquad 1 \le t \le 7$$

where t is the number of years since 1990 and $f(t)$ is the percent of the waste generated in the United States that is yard waste (*Source:* U.S. Statistical Abstract, www.census. gov/statab/www).

(a) Compute $f(6)$ and $f'(6)$ and interpret each.

(b) Compute $f'(t)$ and use it to show that on the interval $[1, 7]$ the percent of waste generated that is yard waste is decreasing.

67. The total carbon content of carbon dioxide emitted into the atmosphere by the U.S. can be modeled by

$$f(t) = -0.55t^3 + 9.03t^2 - 21.8t + 1361.01 \qquad 1 \le t \le 9$$

where t is the number of years since 1989 and $f(t)$ is the total carbon content of carbon dioxide, in millions of metric tons, emitted into the atmosphere by the U.S. (*Source:* U.S. Statistical Abstract, www.census.gov/statab/www)

(a) Evaluate $f(7)$ and $f'(7)$ and interpret each.

(b) Determine where f is increasing and where it is decreasing. Also, determine any relative extrema.

(c) Determine where the graph of f is concave up and where it is concave down. Also, determine the inflection point.

(d) Sketch a graph of f on the interval $[1, 9]$. Label any relative extrema and inflection point.

68. Evaluate $f'(t)$ in Exercise 67 at the t-coordinate of the inflection point and interpret.

SECTION PROJECT

In Table 15.3.1, participation in the labor force can be interpreted to mean "had a job."

(a) Let x represent the number of years since 1959 and let y represent the married female participation rate as a percentage in the U.S. labor force. Plot the data and determine a quadratic regression model for the data. Round coefficients to the nearest thousandth.

(b) Compute the derivative of the model. Evaluate it at $x = 35$ and interpret.

(c) Use the derivative of the model to show that the married female participation rate was increasing from 1960 to 1997.

(d) Use the second derivative to show that the graph of the model is concave down on the indicated interval. What does this result mean?

Table 15.3.1

Year	Married Female Participation Rate in the Labor Force (%)
1960	31.9
1970	40.5
1975	44.3
1980	49.8
1985	53.8
1990	58.4
1991	58.5
1992	59.3
1993	59.4
1994	60.7
1995	61
1996	61.2
1997	61.6

SOURCE: U.S. Bureau of Labor Statistics, www.bls.gov

(e) The participation rate in the U.S. labor force, as a percentage, for married males from 1960 to 1997 can be modeled by

$$f(t) = 89.18e^{-0.004t} \qquad 1 \le t \le 38$$

where t represents the number of years since 1959 and $f(t)$ represents the married male participation rate. Compute $f'(35)$ and compare with part (b).

(f) Use $f'(t)$ to show that from 1960 to 1997 the married male participation rate was decreasing.

(g) Use $f''(t)$ to show that the graph of f is concave up on $[1, 38]$. What does this result mean?

Section 15.4 Optimizing Functions on a Closed Interval

In Section 15.1 we used the first derivative and the First Derivative Test to determine the location of the *relative maximum* and *relative minimum* (plural is *relative extrema*) for a function. In this section we go one step further and determine the location of the **absolute maximum** and **absolute minimum** for a function. Intuitively, these points are on the "highest hill" and in the "lowest valley." Once we know where these **absolute extrema** are located, we will start to apply our new knowledge to a wide array of applications. Thus, we begin our voyage into what is known as **optimization**. We begin by taking a look at the tools used to find the absolute maximum and absolute minimum values on closed intervals.

Absolute Extrema on a Closed Interval

In Section 15.1 we stated that the word *relative* in *relative maximum* and *relative minimum* meant *relative to points nearby*. An **absolute maximum** is the largest value that the function attains on its domain or, stated another way, the highest point on the graph. Similarly, an **absolute minimum** is the smallest value that the function attains on its domain or, stated another way, the lowest point on the graph. For any given function, the **absolute extrema** may or may not exist. The graphs in Figure 15.4.1 show the possibilities for some functions that are defined for all values of x. These graphs suggest that absolute extrema may occur at critical points, that is, where $f'(x) = 0$ or where $f'(x)$ is undefined, which is quite similar to what we found in Section 15.1. Notice that Figure 15.4.1 (d) indicates that a critical point does not necessarily give an absolute extrema.

Now, if we consider any continuous function on a *closed interval* $[a, b]$, the function will have an absolute maximum and an absolute minimum on the interval. In fact, these absolute extrema will be located at either (1) critical points that are *within* the interval (a, b), or (2) at the endpoints of the interval $[a, b]$. Figures 15.4.2 through 15.4.4 illustrate some of the possibilities.

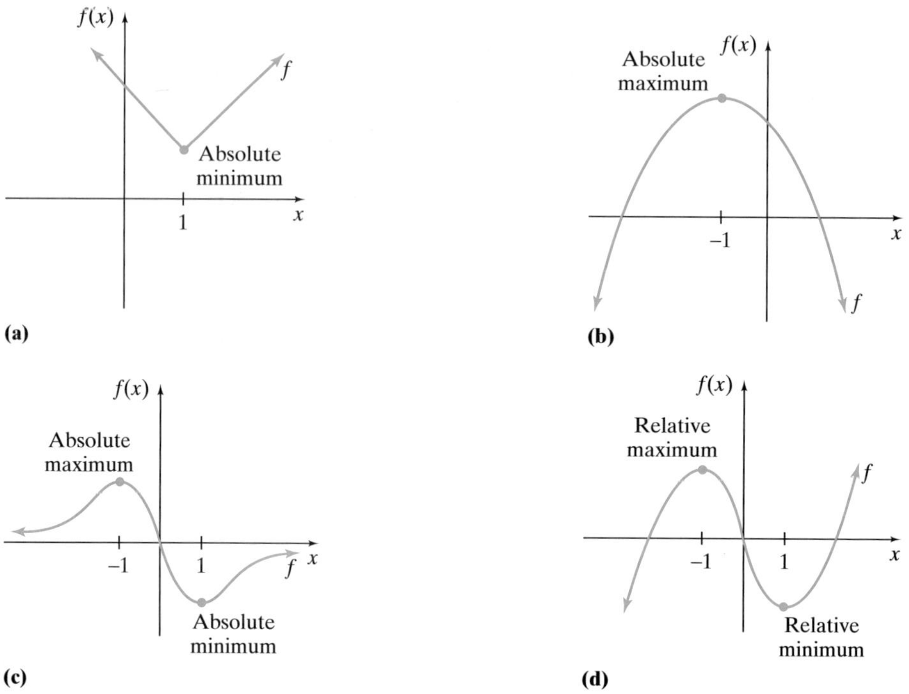

Figure 15.4.1 **(a)** Absolute minimum at $x = 1$. No absolute maximum. **(b)** Absolute maximum at $x = -1$. No absolute minimum. **(c)** Absolute maximum at $x = -1$. Absolute minimum at $x = 1$. **(d)** Relative maximum at $x = -1$. Relative minimum at $x = 1$. No absolute extrema.

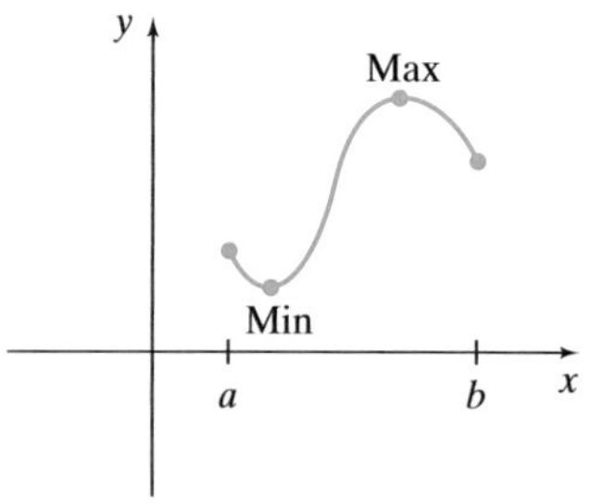

Figure 15.4.2 Absolute extrema occur at critical points.

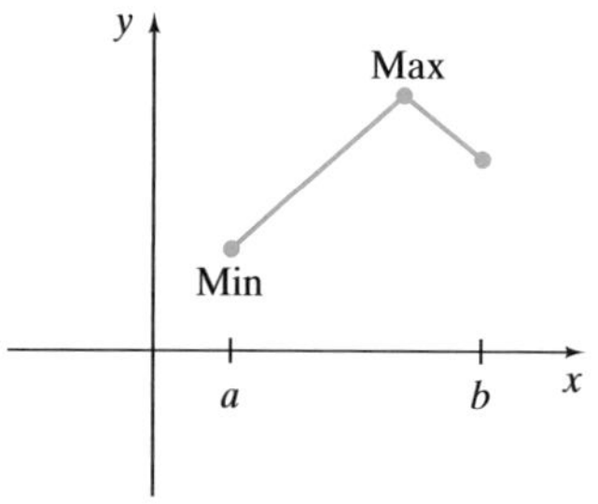

Figure 15.4.3 Absolute maximum at critical point. Absolute minimum at endpoint.

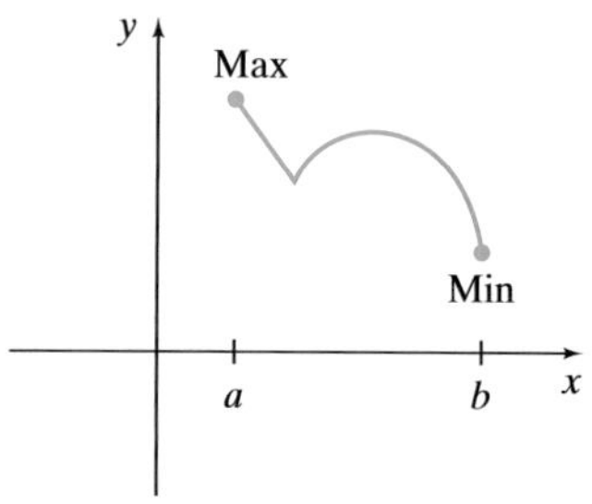

Figure 15.4.4 Absolute extrema occur at endpoints.

The prior discussion and figures lead to the following theorem.

Extreme Value Theorem

If a function is continuous on a closed interval $[a, b]$, then the function must have both an absolute maximum and an absolute minimum on $[a, b]$.

The Extreme Value Theorem tells us when an absolute maximum and absolute minimum are guaranteed to exist. Since we know that absolute extrema occur at critical points within the closed interval or at the endpoints, we supply the following procedure to locate these absolute extrema. When we determine the absolute extrema for a function, we are said to be **optimizing** the function.

Four-step Process for Locating the Absolute Extrema of f on $[a, b]$

1. Verify that f is continuous on $[a, b]$.
2. Determine the critical values of f in (a, b).
3. Evaluate $f(x)$ at the critical values and at the endpoints of the interval.
4. The largest value that you obtain from step 3 is the absolute maximum of f on $[a, b]$, and the smallest value that you obtain from step 3 is the absolute minimum of f on $[a, b]$.

Example 1 Determining Absolute Extrema

Determine the absolute extrema for $f(x) = 2x^3 + 3x^2 - 12x + 1$ on $[-3, 2]$.

Solution

Step 1: Since f is a polynomial function, we know that it is continuous on the closed interval $[-3, 2]$. By the Extreme Value Theorem, it must have an absolute maximum and an absolute minimum.

Step 2: To determine the critical values, we first need the derivative

$$f'(x) = 6x^2 + 6x - 12$$

Since $f'(x)$ is defined for all x, the only critical values are where $f'(x) = 0$. So we need to solve

$$6x^2 + 6x - 12 = 0$$
$$6(x^2 + x - 2) = 0$$
$$6(x + 2)(x - 1) = 0 \quad \text{Factor}$$
$$x = -2 \quad \text{or} \quad x = 1$$

So the critical values are $x = -2$ and $x = 1$.

Step 3: Here, we evaluate $f(x)$ at the critical values and the endpoints.

Critical values: $x = -2$ $\quad f(-2) = 2(-2)^3 + 3(-2)^2 - 12(-2) + 1 = 21$

$x = 1$ $\quad f(1) = 2(1)^3 + 3(1)^2 - 12(1) + 1 = -6$

Endpoints: $x = -3$ $\quad f(-3) = 2(-3)^3 + 3(-3)^2 - 12(-3) + 1 = 10$

$x = 2$ $\quad f(2) = 2(2)^3 + 3(2)^2 - 12(2) + 1 = 5$

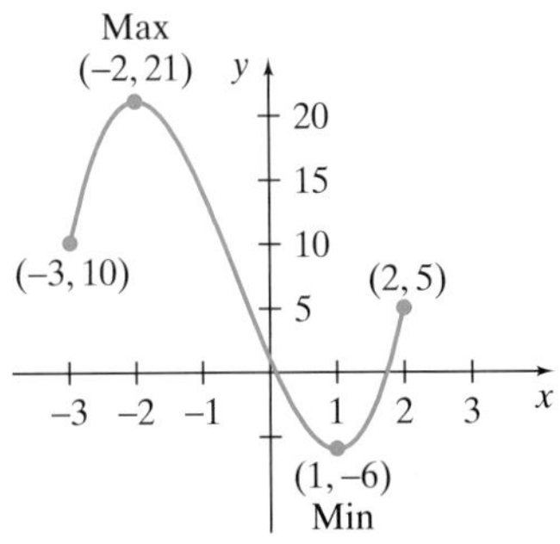

Figure 15.4.5

Step 4: Since 21 is the largest value from step 3, we conclude that the absolute maximum is 21 and it occurs at $x = -2$. The smallest value from step 3 is -6, so -6 is the absolute minimum and occurs at $x = 1$. See Figure 15.4.5. ■

Example 2 Determining Absolute Extrema

Determine the absolute extrema for $f(x) = 3x^4 + 4x^3 - 36x^2 + 1$ on $[-1, 1]$.

Solution

Step 1: Since f is a polynomial function, we know that it is continuous on the closed interval $[-1, 1]$. By the Extreme Value Theorem, it must have an absolute maximum and an absolute minimum.

Step 2: To determine the critical values, we first need the derivative

$$f'(x) = 12x^3 + 12x^2 - 72x$$

Since $f'(x)$ is defined for all x, the only critical values are where $f'(x) = 0$. So we need to solve

$$12x^3 + 12x^2 - 72x = 0$$
$$12x(x^2 + x - 6) = 0$$
$$12x(x + 3)(x - 2) = 0 \qquad \text{Factor}$$
$$x = 0, \quad x = -3, \quad \text{or} \quad x = 2$$

So the critical values are $x = 0$, $x = -3$, and $x = 2$.

Step 3: Since $x = 0$ is the only critical value in the interval $[-1, 1]$, we evaluate $f(x)$ at this critical value and at the endpoints.

Critical value: $x = 0$ $\quad f(0) = 3(0)^4 + 4(0)^3 - 36(0)^2 + 1 = 1$

Endpoints: $x = -1$ $\quad f(-1) = 3(-1)^4 + 4(-1)^3 - 36(-1)^2 + 1 = -36$

$x = 1$ $\quad f(1) = 3(1)^4 + 4(1)^3 - 36(1)^2 + 1 = -28$

Step 4: The absolute maximum is 1 and it occurs at $x = 0$. The absolute minimum is -36 and occurs at $x = -1$. See Figure 15.4.6. ■

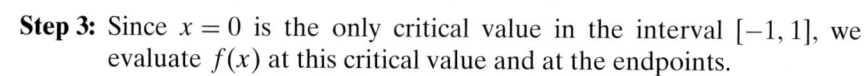

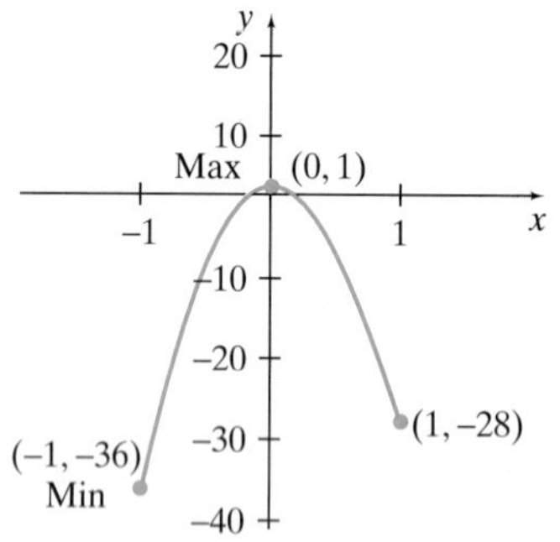

Figure 15.4.6

Checkpoint 1 Now work Exercise 9.

Applications

We now turn our attention to problems where we can maximize revenue, profit, area, or volume, as well as minimize costs, average costs, and pollution. This is a small sample of the types of problems that we may encounter. The examples that follow supply a general strategy for solving any optimization problem on a closed interval.

Example 3 Optimizing Ice Cream Consumption

The per capita consumption of ice cream in the U.S. can be modeled by

$$f(x) = 0.004x^3 - 0.111x^2 + 0.661x + 17.007 \qquad 1 \le x \le 19$$

where x is the number of years since 1979 and $f(x)$ is the per capita ice cream consumption in pounds (*Source:* U.S. Department of Agriculture, www.usda.gov). Determine the absolute extrema on $[1, 19]$ and interpret. Give the answer to the nearest year.

Solution

Understand the Situation: Even though our function models a real-world situation, we still employ the four-step process.

Step 1: Since we have a polynomial function, it is continuous on the interval [1, 19]. Thus, the Extreme Value Theorem applies, and we can expect to find an absolute maximum and an absolute minimum on the interval.

Step 2: To determine the critical values, we need the derivative, which is

$$f'(x) = 0.012x^2 - 0.222x + 0.661$$

The derivative is defined for all x, so the only critical values are where $f'(x) = 0$.

So we need to solve

$$0.012x^2 - 0.222x + 0.661 = 0$$

Using the quadratic formula gives us

$$x = \frac{-b \pm \sqrt{b^2 - 4ac}}{2a} = \frac{0.222 \pm \sqrt{(-0.222)^2 - 4(0.012)(0.661)}}{2(0.012)}$$

$$= \frac{0.222 \pm \sqrt{0.017556}}{0.024}$$

So the two solutions are

$$x = \frac{0.222 + \sqrt{0.017556}}{0.024} \approx 14.771 \quad \text{or} \quad x = \frac{0.222 - \sqrt{0.017556}}{0.024} \approx 3.729$$

Step 3: Both critical values are in the reasonable domain, so we now evaluate the function at the critical values and at the endpoints of the closed interval. (We will round to the nearest thousandth.)

Critical values: $x = 14.771$

$$f(14.771) = 0.004(14.771)^3 - 0.111(14.771)^2 + 0.661(14.771) + 17.007 \approx 15.443$$

$x = 3.729$

$$f(3.729) = 0.004(3.729)^3 - 0.111(3.729)^2 + 0.661(3.729) + 17.007 \approx 18.136$$

Endpoints: $x = 1$

$$f(1) = 0.004(1)^3 - 0.111(1)^2 + 0.661(1) + 17.007 = 17.561$$

$x = 19$

$$f(19) = 0.004(19)^3 - 0.111(19)^2 + 0.661(19) + 17.007 = 16.931$$

Step 4: We have 18.136 as the absolute maximum and it occurs at $x = 3.729$. Rounding to the nearest year gives $x = 4$. The absolute minimum is 15.443 and it occurs at $x = 14.771$. Rounding to the nearest year gives $x = 15$.

Interpret the Solution: From 1980 to 1998, the absolute maximum of ice cream consumption per capita was 18.136 pounds per person and it occurred in 1983.

The absolute minimum of ice cream consumption per capita was 15.443 pounds per person and it occurred in 1994. ■

Checkpoint 2

Now work Exercise 19.

In many optimization problems, we must first determine the function that we are trying to optimize. Example 4 illustrates how we handle these types of problems.

Example 4 Optimizing Area

One of the authors needs to build a rectangular enclosure for his beloved beagle. He has 200 feet of fence. Determine the dimensions of the rectangle that will make the area of the enclosure as large as possible.

Solution

Understand the Situation: Two rectangular regions are shown in Figure 15.4.7 that each use the 200 feet of fencing. Our goal here is to find the one rectangle that has the maximum area.

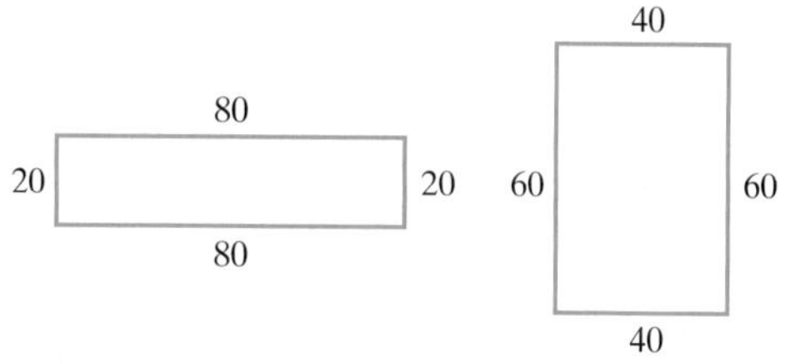

Figure 15.4.7

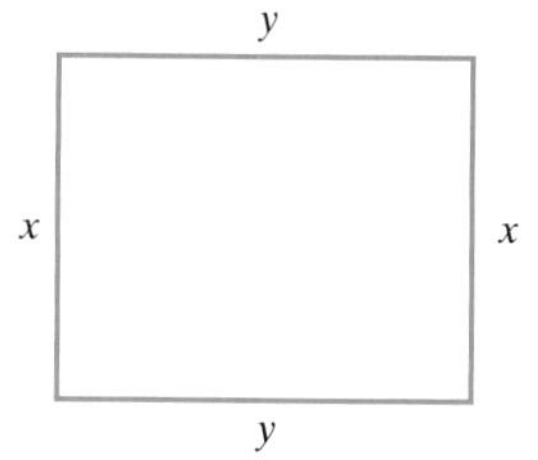

Figure 15.4.8

Instead of continuing in a "guess-and-check" fashion, let's consider a generic rectangle with a width of x and a length of y, as shown in Figure 15.4.8. The area, A, of this generic rectangle is given by

$$A = x \cdot y$$

We want to maximize the area, A, but it is written as a function of two variables. To use the tools we currently have, we need to rewrite $A = x \cdot y$ as a function of *one* variable. Since we have 200 feet of fence, we conclude that the perimeter, P, of the rectangular enclosure is 200 feet. That is,

$$\begin{aligned} P = x + y + x + y &= 200 \\ 2x + 2y &= 200 \\ 2y &= 200 - 2x \\ y &= 100 - x \end{aligned}$$

We can substitute this into our area equation and get $A = x \cdot (100 - x)$ or, more precisely,

$$A(x) = 100x - x^2$$

Since this enclosure is for an animal that must be able to move around in it and turn around in it, a reasonable domain is $10 \le x \le 90$. So we now have the situation of determining the absolute maximum of

$$A(x) = 100x - x^2, \qquad \text{on } [10, 90]$$

We now employ our four-step process just as we did in Examples 1, 2, and 3.

Step 1: Since A is a polynomial function, it is continuous on the stated interval, which means that the Extreme Value Theorem applies.

Step 2: Next, determine the critical values. The derivative of $A(x)$ is

$$A'(x) = 100 - 2x$$

Since $A'(x)$ is defined for all x, the only critical value is where $A'(x) = 0$. Solving this equation yields

$$100 - 2x = 0$$
$$x = 50$$

Step 3: Evaluating $A(x)$ at the critical value and endpoints results in

Critical value:	$x = 50$	$A(50) = 100(50) - (50)^2 = 2500$
Endpoints:	$x = 10$	$A(10) = 100(10) - (10)^2 = 900$
	$x = 90$	$A(90) = 100(90) - (90)^2 = 900$

Step 4: The largest number from step 3 is 2500 and occurs at $x = 50$. So the dimensions of the rectangular enclosure that maximizes the area for the beagle are

$$x = 50$$
$$y = 100 - x = 100 - 50 = 50$$

Interpret the Solution: This means that a rectangle with length of 50 feet and width of 50 feet maximizes the area. ■

Interactive Activity

Reread Example 4 and describe how the strategy to the right was used in the solution.

Notice that in Example 4 we needed to do some preliminary work *before* we could use our four-step process. Example 4 suggests the following procedure when solving these types of applied optimization problems.

Strategy for Solving Applied Optimization Problems on a Closed Interval

1. **Understand the Situation**. This includes:
   - Read the question carefully and, if possible, sketch a picture that represents the problem.
   - Select variables to represent the quantity to be maximized or minimized and all other unknowns.
   - Write an equation for the quantity to be maximized or minimized. If necessary, eliminate extra variables so that the quantity to be optimized is a function of one variable.
2. Apply the four-step process to determine the absolute extrema.
3. **Interpret the Solution**. This includes writing a sentence that answers the question posed in the problem.

From Your Toolbox

In Section 11.2 we presented the Problem Solving Method which we have used throughout the text. Observe that the strategy given to the right follows the general strategy given back in Section 11.2.

In Example 5 we return to a problem first encountered in the Section 15.1 Exercises.

Example 5 Determining a Maximum Medication Concentration

Researchers have determined through experimentation that the percent concentration of a certain medication t hours after it has been administered can be approximated by

$$p(t) = \frac{230t}{t^2 + 6t + 9} \qquad 1 \le t \le 20$$

where $p(t)$ is the percent concentration. How many hours after the administration of this medication is the concentration at a maximum? What is the maximum concentration? See Figure 15.4.9.

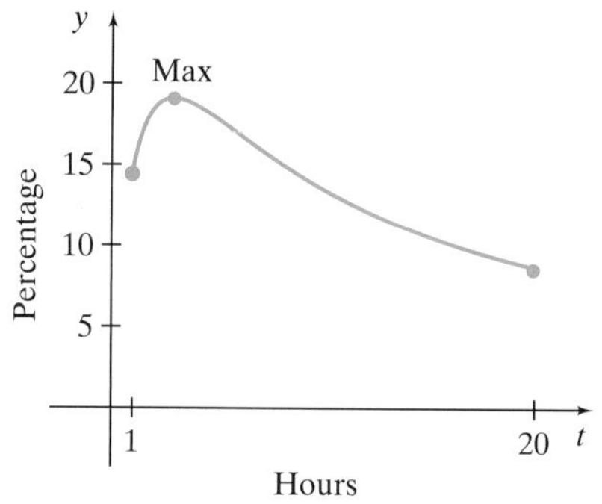

Figure 15.4.9

Solution

Understand the Situation: Since we are given the function to be maximized, we can immediately use the four-step process.

Step 1: The function p is continuous for all values of t except $t = -3$. But, since $t = -3$ is not in the stated interval, p is continuous on the stated interval. Thus, the Extreme Value Theorem guarantees an absolute maximum.

Step 2: By the Quotient Rule, we compute the derivative to be

$$p'(t) = \frac{230(t^2 + 6t + 9) - 230t(2t + 6)}{(t^2 + 6t + 9)^2} = \frac{-230t^2 + 2070}{(t^2 + 6t + 9)^2} = \frac{-230(t^2 - 9)}{(t^2 + 6t + 9)^2}$$

Notice that $p'(t)$ is undefined at $t = -3$, but, since this is not in the stated interval, we do not consider it. We now solve $p'(x) = 0$, which gives

$$\frac{-230(t^2 - 9)}{(t^2 + 6t + 9)^2} = 0$$

Since the only time that a fraction can equal zero is when the numerator equals zero, we have

$$\begin{aligned} -230(t^2 - 9) &= 0 \\ t^2 - 9 &= 0 \\ t^2 &= 9 \\ t &= \pm 3 \end{aligned}$$

Disregarding $t = -3$, we have a critical value at $t = 3$.

Step 3: Evaluating $p(t)$ at the critical value $t = 3$ and at the endpoints and rounding to the nearest hundredth gives

$$\text{Critical value:} \quad p(3) = \frac{230(3)}{(3)^2 + 6(3) + 9} \approx 19.17$$

$$\text{Endpoints:} \quad p(1) = \frac{230(1)}{(1)^2 + 6(1) + 9} \approx 14.38$$

$$p(20) = \frac{230(20)}{(20)^2 + 6(20) + 9} \approx 8.70$$

Step 4: The largest result from step 3 is 19.17 and it occurs at $t = 3$.

Interpret the Solution: This means that 3 hours after administration of this medication, the percent concentration is maximized at about 19.17%. ■

✓ Checkpoint 3

Now work Exercise 37.

In our final example, we return to the functions of business, as well as determining an equation for a line.

Example 6 Maximizing Profit

Only Beef, a local sandwich store, estimates that it can sell 500 roast beef sandwiches per week if it sets the price at \$2.50, but its weekly sales will rise by 50 sandwiches for each \$0.05 decrease in price. The company has fixed costs each week of \$525.00, and variable costs for making a roast beef sandwich are \$0.55. Let x represent the number of roast beef sandwiches made and sold each week. Determine the value of x that maximizes weekly profit, assuming that $500 \leq x \leq 1500$.

Solution

Understand the Situation: We need to employ the strategy supplied after Example 4 to attack this problem. We want to maximize profit, $P(x)$. Recall that $P(x) = R(x) - C(x)$, $R(x) = \text{price} \cdot \text{quantity}$, and $C(x) = \text{fixed costs} + \text{variable costs}$. Since fixed costs are \$525 and x is the number of sandwiches made and sold each week, the variable costs are $0.55 \cdot x$. Hence,

$$C(x) = \text{fixed costs} + \text{variable costs}$$
$$C(x) = 525 + 0.55x$$

To determine $R(x)$, we need price, $p(x)$. We know that when $x = 500$ price is \$2.50, and for each \$0.05 *decrease* in price, weekly sales *increase* by 50. To aid us in analyzing this situation, we make Table 15.4.1 with this information.

Table 15.4.1

x, Number of Sandwiches	$p(x)$, Price of Sandwich
500	2.50
550	2.45
600	2.40
650	2.35

Notice that, for each change in x of 50, $p(x)$ changes by 0.05. As x increases at a constant rate, $p(x)$ seems to be decreasing at a constant rate, which means that we should be able to model $p(x)$ with a *linear function*. We can write

$$\frac{\text{Change in } p(x)}{\text{Change in } x} = \frac{-0.05}{50} = -0.001$$

which tells us that the slope of the line is $m = -0.001$. Using the point–slope form of a line with the point (500, 2.50) gives

$$\begin{aligned} y - y_1 &= m(x - x_1) \\ y - 2.50 &= -0.001(x - 500) \\ y &= -0.001(x - 500) + 2.50 \quad \text{or} \quad y = -0.001x + 3 \end{aligned}$$

So the price function is given by

$$p(x) = -0.001x + 3$$

The revenue function is then given by

$$\begin{aligned} \text{Revenue} &= \text{price} \cdot \text{quantity} \\ R(x) &= p(x) \cdot x \\ R(x) &= (-0.001x + 3) \cdot x = -0.001x^2 + 3x \end{aligned}$$

Now that we have $R(x)$ and $C(x)$, profit, $P(x)$, is given by

$$\begin{aligned} P(x) &= R(x) - C(x) \\ P(x) &= -0.001x^2 + 3x - (525 + 0.55x) = -0.001x^2 + 2.45x - 525 \end{aligned}$$

We have determined the function that is to be optimized. In other words, our task is to determine the absolute maximum for

$$P(x) = -0.001x^2 + 2.45x - 525, \quad \text{on } [500, 1500]$$

Step 1: Since P is a polynomial function, it is continuous on the stated interval and the Extreme Value Theorem guarantees an absolute maximum.

Step 2: The derivative is

$$P'(x) = -0.002x + 2.45$$

Since $P'(x)$ is defined for all x, the only critical value is where $P'(x) = 0$. Solving this equation gives

$$\begin{aligned} -0.002x + 2.45 &= 0 \\ x &= 1225 \end{aligned}$$

Step 3: Evaluating $P(x)$ at the critical value $x = 1225$ and the endpoints results in

$$\begin{aligned} \text{Critical value:} \quad P(1225) &= -0.001(1225)^2 + 2.45(1225) - 525 = 975.625 \\ \text{Endpoints:} \quad P(500) &= -0.001(500)^2 + 2.45(500) - 525 = 450 \\ P(1500) &= -0.001(1500)^2 + 2.45(1500) - 525 = 900 \end{aligned}$$

Step 4: The largest value from step 3 is 975.625 and it occurs at $x = 1225$.

Interpret the Solution: Thus, to maximize weekly profit, Only Beef should sell 1225 roast beef sandwiches each week. ■

In Example 6, the price function, p, was determined from the information given in the problem. We recommend that you check its validity. One way of doing this is to substitute values into the price function that correspond with the table. For example, we could substitute $x = 500$ or $x = 600$ into the price function to verify that we get a price of \$2.50 and \$2.40, respectively. This type of checking is an invaluable tool in helping to secure the correct result.

Checkpoint 4 Now work Exercise 49.

SUMMARY

The main concept in this section was **optimizing**, that is, determining absolute extrema for a function on a closed interval. The **Extreme Value Theorem** stated that if f is a continuous function on a closed interval then f is guaranteed to have an absolute maximum and an absolute minimum. These absolute extrema occur at either the endpoints of an interval or at a critical point that is in the interior of the interval. We then presented a **four-step process** to locate these absolute extrema and the **Strategy for Solving Applied Optimization Problems on a Closed Interval**.

SECTION 15.4 EXERCISES

In Exercises 1–18, determine the absolute extrema of each function on the indicated interval. In Exercises 7 and 8, use a graphing calculator to graph the derivative. Then use the ZERO or ROOT command to approximate the zeros of the derivative.

1. $f(x) = x^2 - 2x - 7$ on $[-2, 1]$
2. $f(x) = 2x^2 + 3x - 1$ on $[-1, 2]$
3. $f(x) = x^3 - 2x^2 - 5x + 6$ on $[-2, 2]$
4. $f(x) = x^3 + 4x^2 + x - 6$ on $[-2, 0]$
5. $f(x) = x^3 - 3x^2$ on $[-1, 3]$
6. $f(x) = x^3 - 12x$ on $[0, 4]$
7. $f(x) = x^4 - 15x^2 - 10x + 24$ on $[-3, 3]$
8. $y = x^4 + 2x^3 - 13x^2 - 14x + 24$ on $[0, 4]$
9. $h(x) = x^4 - x^3 + 5$ on $[-2, 2]$
10. $f(x) = 2x^3 - 6x^2 + 4$ on $[-1, 4]$
11. $y = (2x^2 - 1)^4$ on $[0, 2]$
12. $y = (3x + 1)^3$ on $[-2, 1]$
13. $y = \sqrt[3]{x}$ on $[-1, 1]$
14. $f(x) = \sqrt[3]{x^2}$ on $[-1, 8]$
15. $f(x) = \dfrac{1}{x - 2}$ on $[0, 1]$
16. $f(x) = \dfrac{x}{x - 2}$ on $[3, 5]$
17. $y = \dfrac{1}{x^2 + 1}$ on $[1, 4]$
18. $y = \dfrac{1}{x^2 + 1}$ on $[-1, 1]$

Applications

19. From past records, the owner of the Sleep Cheap Motel has determined that when \$$x$ per day is charged to rent a room the daily profit, $P(x)$, is given by

$$P(x) = -x^2 + 92x - 180 \qquad 40 \le x \le 60$$

What should the owner charge to maximize profit?

20. The Wax and Wick Company sells jumbo-sized holiday scented candles with 10 in a box. The price–demand function is given by

$$p(x) = 102 - 3x$$

where x is the number of boxes and $p(x)$ is in dollars per box.

(a) Determine the revenue function, R.

(b) Determine the number of boxes sold that maximizes revenue.

21. The 40-Acre Pond is a local swimming pond for residents of Point King. Periodically, the pond is treated to control the growth of harmful bacteria. The concentration, $HB(x)$, of harmful bacteria per cubic centimeter is given by

$$HB(x) = 15x^2 - 210x + 750 \qquad 0 \le x \le 14$$

where x is the number of days after a treatment.

(a) How many days after a treatment is the concentration minimized?

(b) What is the minimum concentration?

22. The Camp-n-Swim campgrounds has a small lake for recreational swimming. Periodically, the lake is treated to control the growth of harmful bacteria. The concentration, $HB(x)$, of harmful bacteria per cubic centimeter is given by

$$HB(x) = 3x^2 - 42x + 150 \qquad 0 \le x \le 14$$

where x is the number of days after a treatment.

(a) How many days after a treatment is the concentration minimized?

(b) What is the minimum concentration?

23. From 1989 to 1999, the percent of the U.S. labor force unemployed can be modeled by

$$y = 0.016x^3 - 0.356x^2 + 2.075x + 3.32 \qquad 1 \le x \le 11$$

where x is the number of years since 1988 and y is the percent of the labor force that is unemployed (*Source:* U.S. Bureau of Labor Statistics, www.bls.gov). To the nearest year, determine when the United States had the highest unemployment percentage and when it had the lowest unemployment percentage. Determine the percentage for each.

24. The total number of public elementary and secondary school districts in the United States from 1980 to 1996 can be modeled by

$$f(x) = -0.04x^3 - 2.12x^2 - 23.08x + 16{,}063 \qquad 1 \le x \le 17$$

where x is the number of years since 1979 and $f(x)$ represents the number of districts (*Source:* U.S. Statistical Abstract, www.census.gov/statab/www). Determine the absolute maximum and the absolute minimum and interpret each.

25. The number of bowling establishments in the United States can be modeled by

$$f(x) = -3.98x^2 + 8.11x + 8617.17 \qquad 1 \le x \le 24$$

where x is the number of years since 1974 and $f(x)$ represents the number of bowling establishments in the U.S. (*Source:* U.S. Statistical Abstract, www.census.gov/statab/www). Determine the absolute maximum and the absolute minimum and interpret each. Round each to the nearest whole number.

26. The average salary of players in the National Basketball Association can be modeled by

$$f(x) = 280.8e^{0.16x} \qquad 1 \le x \le 14$$

where x represents the number of years since 1984 and $f(x)$ represents the average salary in thousands of dollars (*Source:* U.S. Statistical Abstract, www.census.gov/statab/www). Determine the absolute maximum and the absolute minimum and interpret each. Round the salary to the nearest thousand.

27. The total number of Catholic elementary schools in the United States from 1960 to 1997 can be modeled by

$$f(x) = 10{,}560e^{-0.012x} \qquad 1 \le x \le 38$$

where x represents the number of years since 1959 and $f(x)$ represents the number of Catholic elementary schools (*Source:* U.S. Statistical Abstract, www.census.gov/statab/www). Determine the absolute maximum and the absolute minimum and interpret each.

28. The total number of Catholic secondary schools in the United States from 1960 to 1997 can be modeled by

$$f(x) = 2410e^{-0.019x} \qquad 1 \le x \le 38$$

where x is the number of years since 1959 and $f(x)$ is the number of Catholic secondary schools (*Source:* U.S. Statistical Abstract, www.census.gov/statab/www). Determine the absolute maximum and the absolute minimum and interpret each.

29. Let y_1 equal the model in Exercise 27 and y_2 equal the model in Exercise 28. Define y_3 as follows:

$$y_3 = y_1 + y_2$$

(a) What does y_3 represent?

(b) Determine the absolute extrema for y_3 and interpret.

30. From 1980 to 1993, the total number of postneonatal (infants from 28 days to 11 months old) deaths in the United States can be modeled by

$$y = -1.69x^4 + 39.88x^3 - 277.12x^2 + 432.56x + 14{,}715.83 \qquad 1 \le x \le 14$$

where x is the number of years since 1979 and y represents the total number of postneonatal deaths (*Source:* U.S. National Center for Health Statistics, www.cdc.gov/nchs). Determine the absolute maximum and the absolute minimum and interpret each. (*Hint:* Graph the derivative and use the ZERO or ROOT command to approximate the zeros of the derivative.)

31. The average price of natural gas in the United States can be modeled by

$$g(x) = 0.046x^4 - 0.468x^3 + 1.429x^2 - 1.077x + 1.62 \qquad 1 \le x \le 5$$

where x is the number of years since 1994 and $g(x)$ is the average price in dollars per 1000 cubic feet (*Source:* U.S. Energy Information Administration, www.eia.doe.gov). Determine the absolute extrema and interpret each. (*Hint:* Graph the derivative and use the ZERO or ROOT command to approximate the zeros of the derivative.)

32. In Example 4 we learned that one of the authors needs to build a rectangular enclosure for his beloved beagle. His wife suggested that, if the rectangular enclosure was built so that one side was along the outside wall of the house, the 200 feet of fencing would make an even bigger enclosure. Follow the wife's advice and determine the dimensions of the rectangular enclosure that will make the enclosure as large as possible. Assume that the author has 200 feet of fencing and the side along the wall needs no fencing.

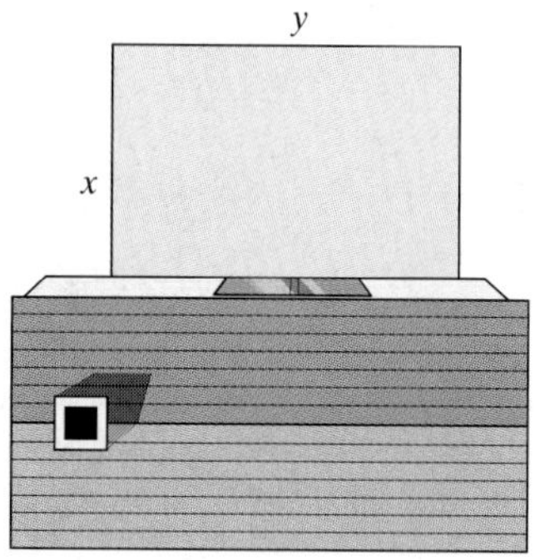

33. The recreation department in Crown Point has been authorized to construct a rectangular playground whose area is 10,000 square feet. Due to a city ordinance, the playground must be enclosed on all four sides by a fence. Let x be the length and y be the width. Given that $50 \le x \le 200$, determine the dimensions of the rectangle that minimize the number of feet of fencing required to enclose the playground.

34. One of the authors of this text recently purchased a new home. He has an extra acre of land to grow a garden. He wants the garden to be rectangular, and he wants the garden to have an area of 1440 square feet. The design of his garden incorporates an 8-foot-wide walkway on the north and south sides of the garden and a 5-foot-wide walkway on the east and west sides. See Figure 15.4.11. Determine the dimensions of the garden that will minimize the *total* area for the garden *and* the walkways. Assume that x is the length of the garden and y is the width of the garden. Also assume that $10 \le x \le 100$.

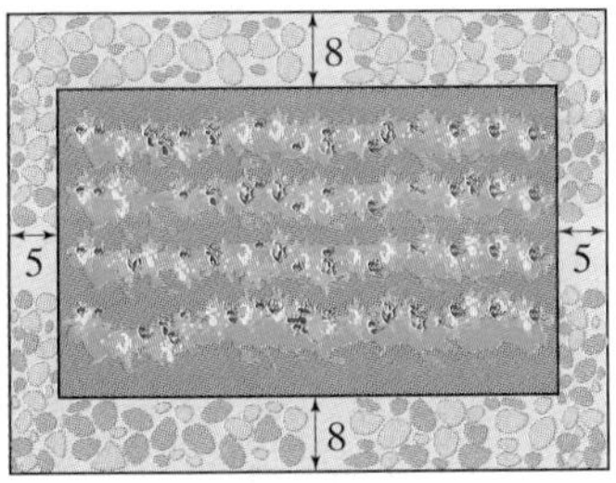

Figure 15.4.11

35. The marketing research department of *Shank*, a quarterly magazine for beginning golfers, has determined that the price–demand equation for the magazine is approximated by

$$p(x) = 2.75 - 0.01x \qquad 0 \le x \le 275$$

where x represents the number of magazines printed and sold each quarter, in hundreds, and $p(x)$ is the price, in dollars, of the magazine. The cost of printing, distributing, and advertising is given by

$$C(x) = 5 + 0.5x + 0.003x^2$$

where $C(x)$ is in hundreds of dollars.

(a) Determine the level of sales that maximizes profit.

(b) Determine the price that the magazine should sell at in order to maximize profit.

36. The concentration of a certain medication in a patient's bloodstream can be given by

$$C(t) = \frac{2.5t}{2t^2 + 7t + 4} \qquad 0 \le t \le 8$$

where $C(t)$ is in milligrams per cubic centimeter and t is the number of hours after the medication has been administered.

(a) How many hours after the medication has been administered is the concentration at a maximum?

(b) What is the maximum concentration?

37. The concentration of a certain medication in a patient's bloodstream can be given by

$$C(t) = \frac{5.3t}{t^2 + 4t + 5} \qquad 0 \le t \le 8$$

where $C(t)$ is in milligrams per cubic centimeter and t is the number of hours after the medication has been administered.

(a) How many hours after the medication has been administered is the concentration at a maximum?

(b) What is the maximum concentration?

38. The Double D Corporation analyzed the production costs for one of its products and determined that the daily cost function can be given by

$$C(x) = 0.03x^3 - 2.4x^2 + 73.16x + 102.27 \qquad 5 \le x \le 55$$

where x represents the number of units produced each day and $C(x)$ is the daily cost in dollars.

(a) Determine the average cost function, AC.

(b) Determine the production level that minimizes the average cost.

(c) Determine the average cost and the total cost at the production level found in part (b).

39. For the product in Exercise 38, the Double D Corporation gathered data for the price of a unit, in dollars, and the number of units demanded per day. The price–demand function that models these data is given by

$$p(x) = 0.14x^2 - 15.52x + 561.21 \qquad 5 \le x \le 55$$

where x represents the number of units demanded each day and $p(x)$ is the price, in dollars, that consumers pay to buy exactly x units per day.

(a) Determine R, revenue as a function of x.

(b) Determine the maximum revenue.

(c) Determine the demand that maximizes revenue.

(d) Determine the price that maximizes revenue.

40. Refer to Exercises 38 and 39. Using the model for C from Exercise 38 and the function for R from Exercise 39, determine the following:

(a) P, profit as a function of x.

(b) Maximum profit.

(c) The demand that maximizes profit.

(d) The price that maximizes the profit.

41. The total energy consumption in the United States can be modeled by

$$f(x) = 0.016x^4 - 0.355x^3 + 2.884x^2 - 8.703x + 82.433 \qquad 1 \le x \le 9$$

where x is the number of years since 1980 and $f(x)$ is the total consumption in quadrillions of Btu (*Source:* U.S.

Energy Information Administration, www.eia.doe.gov). Determine the absolute extrema and interpret each. (*Hint:* Graph $f'(x)$ and use the ZERO or ROOT command to approximate the solution to $f'(x) = 0$.)

42. The average (per household) amount spent on energy in the U.S. can be modeled by

$$f(x) = -0.07x^4 + 2.82x^3 - 38.11x^2 + 219.88x + 628.98 \qquad 1 \leq x \leq 18$$

where x is the number of years since 1979 and $f(x)$ is the average amount spent on energy, per household, in dollars (*Source:* U.S. Energy Information Administration, www.eia.doe.gov). Determine the absolute extrema and interpret each. (*Hint:* Graph $f'(x)$ and use the ZERO or ROOT command to approximate the solution to $f'(x) = 0$.)

43. The total energy consumption in the United States can be modeled by

$$f(x) = -0.018x^3 + 0.195x^2 + 1.194x + 82.573 \qquad 1 \leq x \leq 9$$

where x is the number of years since 1990 and $f(x)$ is the total consumption in quadrillions of Btu (*Source:* U.S. Energy Information Administration, www.eia.doe.gov). Determine the absolute extrema and interpret each.

44. The United States per capita consumption of beef, in pounds, can be modeled by

$$B(t) = 0.03t^3 - 0.76t^2 + 4.54t + 68.34 \qquad 1 \leq t \leq 16$$

where t is the number of years since 1979 and $B(t)$ is the per capita consumption of beef, in pounds (*Source:* U.S. Department of Agriculture, www.usda.gov). Determine the absolute maximum and the absolute minimum and interpret each.

45. The total amount of methane gas emitted into the atmosphere by the U.S. can be modeled by

$$f(x) = 0.04x^3 - 0.46x^2 + 1.42x + 28.83 \qquad 1 \leq x \leq 6$$

where x is the number of years since 1992 and $f(x)$ is the total amount of methane gas emitted into the atmosphere in millions of metric tons (*Source:* U.S. Energy Information Administration, www.eia.doe.gov). Determine the absolute extrema and interpret each.

46. As arterial pressure increases, blood flow through the artery also increases. This can be modeled by

$$f(x) = 0.267e^{0.0256x} \qquad 20 \leq x \leq 120$$

where x represents arterial pressure, measured in millimeters of mercury (mm Hg), and $f(x)$ is blood flow measured in milliliters per minute (mL/min). Determine the absolute extrema and interpret each.

47. The hematocrit of blood is the percent of blood that is cells. For example, a hematocrit of 40 means that 40% of the blood volume is cells and the rest is plasma. Viscosity is the internal friction of a fluid caused by molecular attraction, which makes it resist a tendency to flow. As hematocrit increases, the viscosity of blood increases. If we arbitrarily consider water to have a viscosity of 1, then the viscosity of whole blood at normal hematocrit is about 3 or 4. This means that three to four times as much pressure is needed to force whole blood through a given tube than to force water through the same tube. The relationship between hematocrit and viscosity is given by

$$v(h) = 0.0015h^2 - 0.019h + 1.563 \qquad 10 \leq h \leq 80$$

where h is the hematocrit and $v(h)$ represents the viscosity. (The hematocrit of a normal human is about 40.) Determine the absolute maximum and the absolute minimum and interpret each.

48. The Boogie-All-Night Company manufactures 1970s-style black lights. The monthly fixed costs are \$100 and variable costs are \$12 for each light. The company estimates that 20 lights can be sold each month if the price for a light is \$20 and that 2 more lights can be sold for each decrease of \$1 in the price. Let x represent the number of lights produced and sold each month.

(a) Determine the monthly cost function, C.

(b) Determine the monthly revenue function, R.

(c) Determine the value of x that maximizes the monthly profit, assuming that $10 \leq x \leq 40$.

✓ **49.** The Boogie-All-Night Company also manufactures authentic 1970s-style lava lamps. The monthly fixed costs are \$1000 and variable costs are \$60 for each lamp. The company estimates that 100 lamps can be sold each month if the price for a lamp is \$100 and that 7 more lamps can be sold for each decrease of \$5 in the price. Let x represent the number of lamps produced and sold each month.

(a) Determine the monthly cost function, C.

(b) Determine the monthly revenue function, R.

(c) Determine the value of x that maximizes the monthly profit, assuming that $50 \leq x \leq 200$.

50. The Tool Shack manufactures ratchet sets. The monthly fixed costs are \$15,000 and variable costs are \$180 for each ratchet set produced. The company estimates that 300 sets can be sold each month if the unit price is \$280 and that 20 more units can be sold for each decrease of \$14 in the price. Let x represent the number of ratchet sets produced and sold each month.

(a) Determine the monthly cost function, C.

(b) Determine the monthly revenue function, R.

(c) Determine the value of x that maximizes the monthly profit, assuming that $200 \leq x \leq 580$.

51. The VirtualPet Company makes handheld electronic virtual pets. The marketing department is confident that it can sell 3000 virtual pets per week at a price of \$10 each and also believes that reducing the price by \$0.17 each will increase the weekly sales by 180. Let x represent the number of virtual pets manufactured and sold each week.

(a) Determine the weekly revenue, R.

(b) Determine the value of x that maximizes weekly revenue, assuming that $3000 \leq x \leq 10{,}000$.

SECTION PROJECT

Lisa's Bakery collected the data shown in Table 15.4.2 on the price of a dozen brownies, in dollars, and the number of dozens demanded per day.

Table 15.4.2

Number of dozens demanded	Price, in dollars
14	4.00
19	3.75
21	3.50
23	3.25
26	3.00
31	2.75
35	2.50

(a) A price–demand function that models these data is

$$p(x) = -0.07x + 5.04 \qquad 14 \le x \le 40$$

where x is the number of dozens of brownies demanded per day and $p(x)$ represents the price at which people buy exactly x dozen brownies per day. Determine R, revenue as a function of x.

(b) Determine the maximum revenue and determine what price and what demand will result in the maximum revenue.

(c) If each dozen brownies costs Lisa's Bakery \$1.50 to make, determine P, profit as a function of x.

(d) Determine the maximum profit and determine what price and what demand will result in the maximum profit.

(e) A price–demand function for the data can also be modeled by

$$p(x) = 5.64e^{-0.023x} \qquad 14 \le x \le 40$$

where x is the number of dozens of brownies demanded per day and $p(x)$ represents the price at which people buy exactly x dozen brownies per day. Determine R, revenue as a function of x.

(f) Using the result from part (e), determine the maximum revenue and determine what price and what demand will result in the maximum revenue.

(g) If each dozen brownies costs Lisa's Bakery \$1.50 to make, using the result from part (f), determine P, profit as a function of x.

(h) Determine the maximum profit and determine what price and what demand will result in the maximum profit.

(i) If you owned Lisa's Bakery, which model would you use to determine the price to maximize profit, the model from part (a) or the model from part (e)? Explain.

Section 15.5 The Second Derivative Test and Optimization

In Section 15.4 we learned that if a function is continuous on a closed interval then it has an absolute maximum and an absolute minimum. We also saw how to locate the absolute extrema by a four-step process. The key to the last section was that we considered functions on *closed intervals*. In this section we consider functions on **open** and **half-open** intervals. This means that we cannot use the four-step process and must rely on something else. What we tend to rely on when optimizing a function on an open (or half-open) interval is a graph of the function, a table of values, and **the Second Derivative Test**. What we use depends on the number of critical values on the open interval.

Absolute Extrema on Open Intervals

Recall, in Section 15.4, that we used the Extreme Value Theorem to find the absolute maximum and absolute minimum on a closed interval. The key to this theorem was that it told us on a *closed interval*, a function is *guaranteed* to have an absolute maximum and an absolute minimum. If the interval is not closed, there is no guarantee that absolute extrema exist.

Figures 15.5.1 through 15.5.3 demonstrate that when the interval is open we may or may not have absolute extrema. Probably one of the best methods to determine if a function has absolute extrema on an open interval is to look at the graph of the function. However, when the function in question is continuous and has only one critical value in the indicated interval, we can employ the **Second**

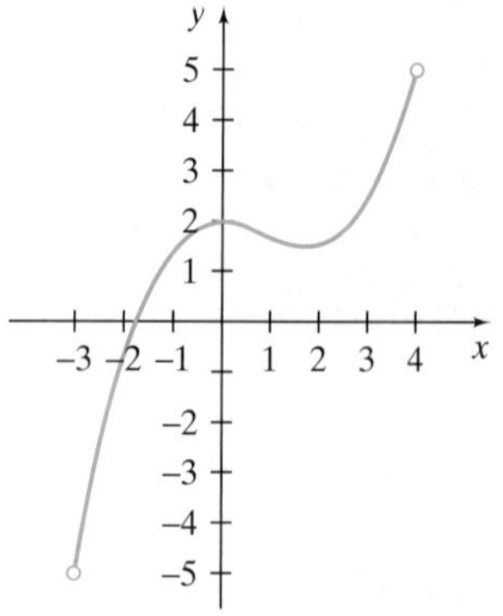

Figure 15.5.1 No absolute extrema on $(-3, 4)$.

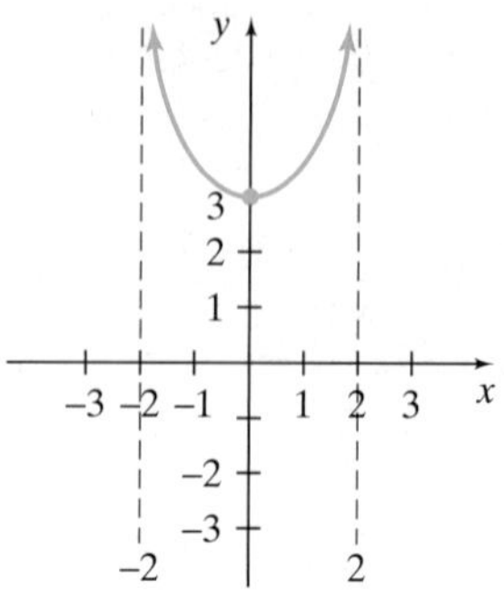

Figure 15.5.2 Absolute minimum, but no absolute maximum on $(-2, 2)$.

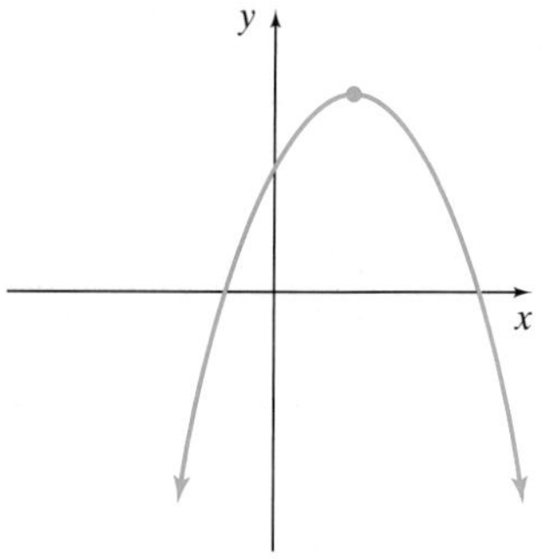

Figure 15.5.3 Absolute maximum, but no absolute minimum on $(-\infty, \infty)$.

Derivative Test to determine if an absolute maximum or an absolute minimum exists.

The Second Derivative Test may be used on *any* interval (closed, open, or half-open) that contains only one critical value. So when we have a function, f, that is continuous on an interval and $x = c$ is the only critical value in the interval, we can modify the Second Derivative Test from Section 15.3 to locate **absolute extrema** as follows.

Second Derivative Test for Absolute Extrema

For a continuous function, f, on *any* interval (open, closed, or half-open), and if $x = c$ is the *only* critical value in the interval where $f'(c) = 0$, if $f''(c)$ exists, then the point $(c, f(c))$ is

1. an **absolute minimum** if $f''(c) > 0$

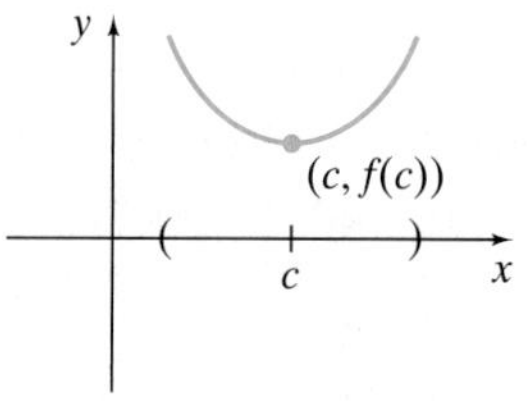

2. an **absolute maximum** if $f''(c) < 0$

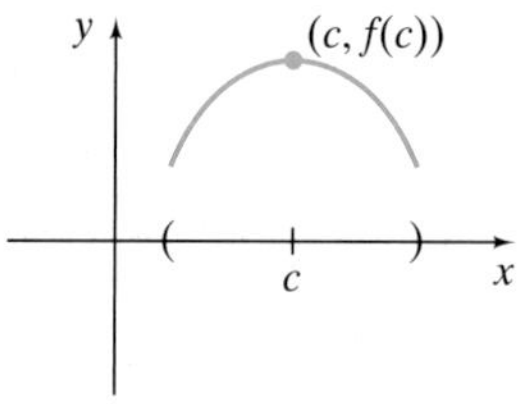

If $f''(c) = 0$, then this test fails.

Example 1 Determining an Absolute Extremum on an Interval

Determine the absolute minimum for $f(x) = 3x + \dfrac{27}{x}$ on $(0, \infty)$.

Solution

Understand the Situation: Since $(0, \infty)$ is an open interval, the techniques of Section 15.4 do not apply here. Applying the new techniques of this section requires us to first determine the derivative.

Rewriting the function algebraically and then differentiating yields

$$f(x) = 3x + \frac{27}{x} = 3x + 27x^{-1}$$

$$f'(x) = 3 - 27x^{-2} = 3 - \frac{27}{x^2}$$

Now, according to the Second Derivative Test for Absolute Extrema, we need to find the critical value $x = c$, where $f'(c) = 0$. Setting the derivative equal to zero and solving gives

$$3 - \frac{27}{x^2} = 0$$

$$\frac{3x^2 - 27}{x^2} = 0$$

$$\frac{3(x^2 - 9)}{x^2} = 0$$

$$\frac{3(x + 3)(x - 3)}{x^2} = 0$$

$$x = 3 \quad \text{or} \quad x = -3$$

So the only critical value in $(0, \infty)$, where $f'(x) = 0$, is $x = 3$. Since there is only one critical value that makes $f'(x) = 0$, we can apply the Second Derivative Test for Absolute Extrema. To apply this requires the second derivative.

$$f'(x) = 3 - 27x^{-2}$$

$$f''(x) = 54x^{-3} = \frac{54}{x^3}$$

We now evaluate the second derivative at the critical value and get

$$f''(3) = \frac{54}{(3)^3} = \frac{54}{27} = 2$$

Since $f''(3) = 2$, $f''(3) > 0$, we conclude by the Second Derivative Test for Absolute Extrema that the absolute minimum for f on $(0, \infty)$ occurs at $x = 3$, and the absolute minimum is $f(3) = 3(3) + \frac{27}{3} = 18$. See Figure 15.5.4.

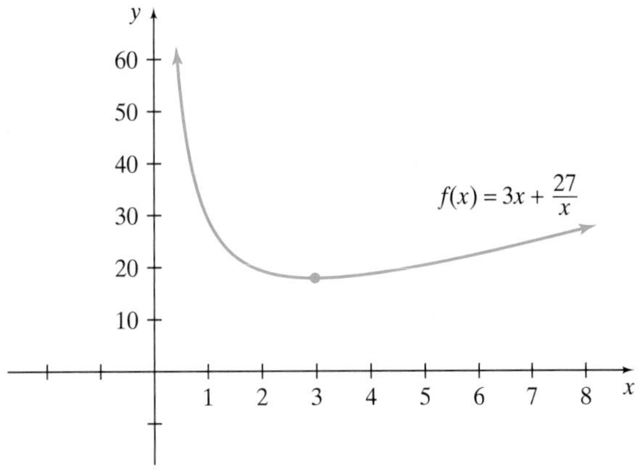

Figure 15.5.4

■

Checkpoint 1 Now work Exercise 3.

Applications

In contrast to the problems in Section 15.4, the problems in this section have intervals that are not closed. Hence, we need the techniques presented in this section.

Example 2 Maximizing Volume

Lydia would like to make an open-top box out of a square piece of corrugated cardboard that measures 12 inches on each side. To do this, she needs to cut small squares from each corner and then turn up the edges. Determine the dimensions of the resulting box if its volume is to be a maximum.

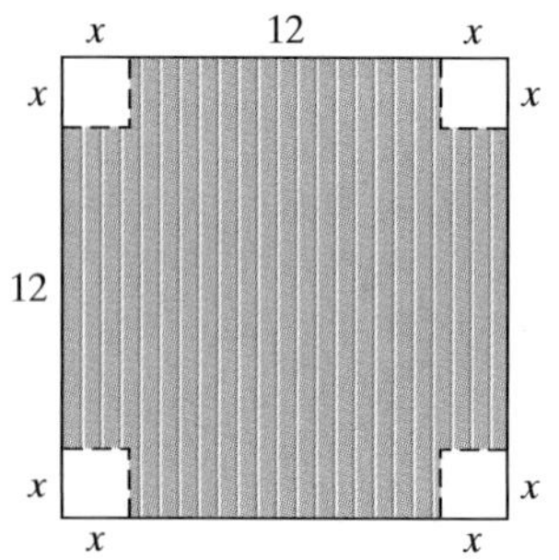

Figure 15.5.5

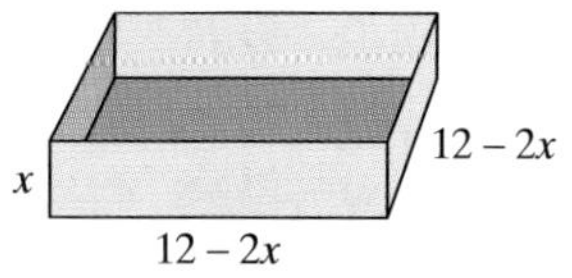

Figure 15.5.6

Solution

Understand the Situation: We begin the solution by sketching the square piece of corrugated cardboard and labeling the length and width of the cutout pieces with an x. See Figure 15.5.5. The formula for volume of a box is

$$V = l \cdot w \cdot h$$

Volume is what we want to maximize, but as it is currently written, we do not have the tools in our calculus arsenal to solve it. We need to represent the volume as a function of a single variable. To do this, we now draw a figure of what the box would look like. See Figure 15.5.6. The height, h, of the box is the same as x. The length, l, of our box is given by $(12 - 2x)$. This is due to the fact that the 12-inch square had x inches cut out from each corner. Hence, $12 - x - x$ results in the length being $12 - 2x$. By a similar argument, the width, w, of the box also measures $12 - 2x$. So we now have for the volume of the box

$$V = l \cdot w \cdot h$$
$$V(x) = (12 - 2x) \cdot (12 - 2x) \cdot x = (12 - 2x)^2 \cdot x$$

Now that we have the volume as a function of a single variable, we can apply techniques that are in our calculus arsenal to solve the problem. But first we need to consider the interval, or reasonable domain, for V. Since we want to make a box, $x \neq 0$, because $x = 0$ corresponds to making no cuts at all. Also, $x < 6$ since a square cutout of 6 inches on each side would produce four square pieces of corrugated cardboard and we could not make a box. (There would be nothing left to turn up!) So we claim that the length and width of the cutout, x, is in the interval $0 < x < 6$. Thus, we need to maximize

$$V(x) = (12 - 2x)^2 \cdot x, \quad \text{on } (0, 6)$$

The interval is an open interval, so we need the techniques of this section to solve it. We first determine the critical value as follows:

$$\begin{aligned} V'(x) &= 2(12 - 2x)(-2) \cdot x + (12 - 2x)^2 \cdot 1 && \text{Product and Chain Rules} \\ &= -4x(12 - 2x) + (12 - 2x)^2 \\ &= (12 - 2x)[-4x + (12 - 2x)] \\ &= (12 - 2x)(-6x + 12) \end{aligned}$$

Setting the derivative equal to zero and solving gives

$$(12 - 2x)(-6x + 12) = 0$$
$$x = 6 \quad \text{or} \quad x = 2$$

The only critical value that is *in* the interval $(0, 6)$ is $x = 2$. We now use the Product Rule to determine $V''(x)$.

$$V''(x) = \frac{d}{dx}[(12 - 2x)(-6x + 12)]$$
$$= -2(-6x + 12) + (12 - 2x)(-6)$$
$$= 12x - 24 - 72 + 12x = 24x - 96$$

Interactive Activity

In Example 2, expand the equation for $V(x) = (12 - 2x)^2 \cdot x$ by multiplication. Compute $V'(x)$ and solve the problem to verify that the same result is attained. Notice that by doing this, the differentiation may be easier.

Evaluating $V''(x)$ at the critical value $x = 2$ yields

$$V''(2) = 24(2) - 96 = 48 - 96 = -48$$

Since $V''(2) = -48$, $V''(2) < 0$, we have an absolute maximum at $x = 2$. Thus, the dimensions of the box that has maximum volume is

Length: $12 - 2x = 12 - 2(2) = 12 - 4 = 8$

Width: $12 - 2x = 12 - 2(2) = 12 - 4 = 8$

Height: $x = 2$

Interpret the Solution: So a length of 8 inches, width of 8 inches, and height of 2 inches maximize the volume at 128 cubic inches. ■

Example 2 suggests the following procedure when solving applied optimization problems on an open interval.

Strategy for Solving Applied Optimization Problems on an Open Interval

1. **Understand the Situation**. This includes:
   - Read the question carefully and, if possible, sketch a picture that represents the problem.
   - Select variables to represent the quantity to be maximized or minimized and other unknowns.
   - Write an equation for the quantity to be maximized or minimized and determine the interval (reasonable domain) over which the function is to be optimized. If necessary, eliminate extra variables so that the quantity to be optimized is a function of one variable.
2. If the interval is open (or half-open), you must use the techniques of this section. If the interval is closed, you can use the techniques of Section 15.4.
3. **Interpret the Solution**. This includes writing a sentence that answers the question posed in the problem.

Checkpoint 2

Now work Exercise 21.

In many settings it is very easy to become too greedy. For example, if a farmer plants too much corn on an acre of ground, the yield can actually be reduced. If a restaurant has too many tables and they are too close together, patrons may actually stay away. If an accounting firm places too many accountants in an office, their productivity may decrease. Calculus and optimization can allow us to handle the problem of overcrowding, as illustrated in Example 3.

Example 3 Maximizing a Harvest

Laurie's Lemons has determined that the annual yield per lemon tree is fairly constant at 320 pounds when the number of trees per acre is 50 or fewer. The owner of Laurie's Lemons would like to maximize the annual yield per acre. To do this, she wants to plant more lemon trees. Research has shown that for each additional tree over 50 the annual yield per tree decreases by 4 pounds due to overcrowding. How many trees should be planted on each acre to maximize the annual yield from an acre?

Solution

Understand the Situation: The total yield for an acre is represented by

$$\text{Total yield} = (\text{number of trees per acre}) \cdot (\text{yield per tree})$$

Let x represent the number of trees per acre. To aid us in determining the yield per tree, we make the chart shown in Table 15.5.1.

Table 15.5.1

Number of Trees	Yield per Tree	Total Yield
10	320	3200
20	320	6400
40	320	12,800
50	320	16,000
51	$320 - 4(51 - 50) = 316$	16,116
52	$320 - 4(52 - 50) = 312$	16,224
53	$320 - 4(53 - 50) = 308$	16,324
x	$320 - 4(x - 50)$	$x \cdot [320 - 4(x - 50)]$

The "yield per tree" column was unchanged until we hit 51 trees per acre. *Each* additional tree beyond 50 results in a decrease in yield of 4 pounds per tree. After listing a few values beyond 50, we generalized by placing an x in for the number of trees and followed the pattern that was found for 51, 52, and 53 trees. If we now let $TY(x)$ be the total yield in pounds, we can represent this scenario by

$$TY(x) = x \cdot [320 - 4(x - 50)]$$

The reasonable domain for TY is $[50, 130)$. We do not include $x = 130$ since a quick check shows that at $x = 130$ the yield would be zero! So our problem is to maximize

$$TY(x) = x \cdot [320 - 4(x - 50)], \quad \text{on } [50, 130)$$

The derivative is computed after algebraically simplifying $TY(x)$ to be

$$TY(x) = x \cdot [520 - 4x] = 520x - 4x^2$$
$$TY'(x) = 520 - 8x$$

Determining the critical value x such that $TY'(x) = 0$, gives

$$520 - 8x = 0$$
$$x = 65$$

We compute the second derivative to be

$$TY''(x) = -8$$

Evaluating the second derivative at $x = 65$ gives

$$TY''(65) = -8$$

So $TY''(65) < 0$, which means, according to the techniques of this section, that at $x = 65$ we have an absolute maximum.

Interpret the Solution: So Laurie's Lemons should plant 65 lemon trees per acre to maximize the total yield. ■

✓ Checkpoint 3

Now work Exercise 31.

The method outlined in Example 3 is used in any situation where increasing one item decreases total output by some amount. For example, adding additional work stations in a machine shop may decrease the output per station. This scenario would follow the Example 3 process.

Sustainable Harvest

Our next example shows how to **maximize a sustainable harvest**. If one is in the business of harvesting fish, possibly by aquaculture, animals, or even trees, the ability to resist overharvesting is fundamental to the long-term success of the business. Overharvesting can inhibit the ability of the population to reproduce and sustain itself. This is illustrated in Example 4.

Example 4 Maximizing a Sustainable Harvest

Sussex County allows hunters to shoot deer during a limited open hunting season. The length of the season is carefully determined by the State Department of Natural Resources to ensure a harvest that is sustainable year after year. A state biologist has determined that, in Sussex County, the deer population after one year is given by

$$f(x) = 2.1x - 0.001x^2$$

where x is the original population size of the deer measured in hundreds. Determine the optimal population size and the yearly kill that it will sustain.

Solution

Understand the Situation: We start by determining $h(x)$, the yearly harvest. The harvest is simply the difference between the deer population, $f(x)$, after one year and the original population, x. That is,

$$\begin{aligned} h(x) &= f(x) - x \\ &= 2.1x - 0.001x^2 - x \\ &= 1.1x - 0.001x^2 \end{aligned}$$

Since we have no idea of the population size, the interval on which we consider $h(x)$ is $(0, \infty)$. (We know that the population cannot be negative!) We wish to maximize $h(x)$ using the techniques in this section.

Computing the derivative gives

$$h'(x) = 1.1 - 0.002x$$

The critical value x, such that $h'(x) = 0$, is found to be

$$1.1 - 0.002x = 0$$
$$x = 550$$

The second derivative is $h''(x) = -0.002$. Evaluating the second derivative at the critical value gives

$$h''(550) = -0.002$$

Thus, we have a maximum at $x = 550$.

Interpret the Solution: So the population of deer in Sussex County should grow to 55,000. It will then sustain a yearly harvest of

$$h(550) = 1.1(550) - 0.001(550)^2 = 302.5$$

that is, a yearly sustainable harvest of 30,250 deer per year. ■

Inventory Costs

A successful retail store must pay attention to the size of its inventory. Overstocking can lead to extra interest costs, excessive warehouse rental charges, and possibly the danger of the product becoming obsolete or damaged. On the other hand, too small an inventory involves additional paperwork in reordering and extra delivery charges. There is also the danger of running out of stock.

Let's take a look at a real inventory problem. Records from previous years show that Linger's Luxury Office Furniture sells 300 executive desks a year. To plan their inventory accurately, their analysts assume that the desks sell steadily throughout the year. They could order these desks in lots of size 300, 150, or 100 or in general lots of size x. Regardless of the lot size, on average this store will have $\frac{x}{2}$ executive desks in stock on which it must pay **inventory costs**. See Figure 15.5.7. Ordering all 300 at once could result in high **storage costs**, whereas reordering several times throughout the year could result in high **reorder costs**. In Example 5 we determine the best lot size that minimizes the total of storage costs and reorder costs.

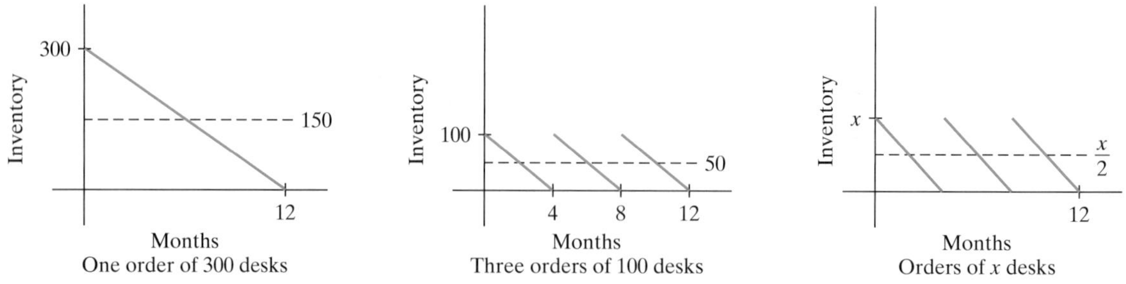

Figure 15.5.7

Example 5 Minimizing Inventory Costs

Linger's Luxury Office Furniture expects to sell 300 executive desks a year. Each desk costs the store \$400, and there is a fixed charge of \$800 per order. If it costs \$200 to store an executive desk for a year, how large should each order be and how often should orders be placed to minimize the inventory costs?

Solution

Understand the Situation: We begin by letting x represent the lot size, the number in each order. Our total inventory costs are represented by

$$\text{Total costs} = \text{storage costs} + \text{reorder costs}$$

We begin by determining the storage costs. We assume that the desks sell steadily throughout the year and that Linger's reorders x more when the stock is depleted. The inventory throughout the year would look like the graph of inventory as a function of the months in a year, as shown in Figure 15.5.8.

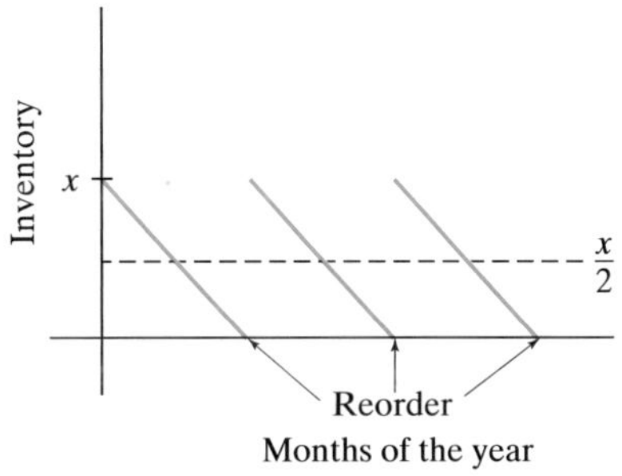

Figure 15.5.8

Since the number in stock varies from x to 0, we consider an *average inventory* of $\frac{x}{2}$. Since it costs \$200 to store a desk for a year, we determine the storage costs as follows:

$$\text{Storage costs} = (\text{storage per item}) \cdot (\text{average number of items})$$
$$= 200 \cdot \left(\frac{x}{2}\right) = 100x$$

The next piece to determine is the reorder costs. In general, reorder costs are determined by

$$\text{Reorder costs} = (\text{cost per order}) \cdot (\text{number of orders}).$$

Since each desk costs \$400, we know that ordering x desks costs $400x$. There is a fixed charge of \$800 per order, which gives the cost per order as

$$\text{Cost per order} = 400x + 800$$

A yearly supply is 300 desks, and with x desks in each order, we find the number of orders in one year to be

$$\text{Number of orders} = \frac{300}{x}$$

This means that our reorder costs are given by

$$\text{Reorder costs} = (400x + 800) \cdot \left(\frac{300}{x}\right)$$

Thus, our total inventory costs, $IC(x)$, are given by

$$IC(x) = 100x + \frac{300}{x}(400x + 800) \qquad 0 < x \leq 300$$

To minimize $IC(x)$, we employ the methods of this section. The derivative is

$$IC(x) = 100x + 120{,}000 + \frac{240{,}000}{x} = 100x + 120{,}000 + 240{,}000x^{-1}$$

$$IC'(x) = 100 - 240{,}000x^{-2} = 100 - \frac{240{,}000}{x^2}$$

We now find the critical value x, where $IC'(x) = 0$.

$$100 - \frac{240{,}000}{x^2} = 0$$

$$100 = \frac{240{,}000}{x^2}$$

$$x^2 = 2400$$

$$x \approx \pm 48.99$$

or, since x represents the lot size, we round to the nearest integer and get

$$x = \pm 49$$

We can immediately reject $x = -49$ and begin to verify that $x = 49$ minimizes inventory costs. Computing the second derivative gives

$$IC''(x) = 480{,}000x^{-3} = \frac{480{,}000}{x^3}$$

Evaluating the second derivative at the critical value $x = 49$ gives

$$IC''(49) = \frac{480{,}000}{(49)^3} \approx 4.0799$$

Since $IC''(49) > 0$, we conclude that inventory costs, $IC(x)$, are minimized at $x = 49$. If there are 49 desks per order, the yearly total of 300 would require $\frac{300}{49} \approx 6.1224$ orders per year. Since this number of orders per year is impossible, it seems reasonable to round x to 50 (the number of orders would then be 6) and answer the question as follows:

Interpret the Solution: To minimize inventory costs, each lot size should have 50 desks with orders placed 6 times per year. ■

Checkpoint 4

Now work Exercise 39.

SUMMARY

The main theme in this section was determining absolute extrema on intervals that are not closed. We illustrated how the second derivative handles this case, provided that there is only one critical value, $x = c$, on the interval where $f'(c) = 0$. Specifically, we presented the **Second Derivative Test for Absolute Extrema** and the **Strategy for Solving Applied Optimization Problems on an Open Interval.**

- **Second Derivative Test for Absolute Extrema**: For a continuous function, f, on *any* interval (open, closed, or half-open), and if $x = c$ is the *only* critical value in the interval where $f'(c) = 0$, if $f''(c)$ exists, then the point $(c, f(c))$ is an (1) **absolute minimum** if $f''(c) > 0$ or (2) **absolute maximum** if $f''(c) < 0$. If $f''(c) = 0$, then this test fails.

SECTION 15.5 EXERCISES

In Exercises 1–7, determine the absolute minimum of each function on the indicated interval.

1. $f(x) = 2x + \frac{6}{x}$ on $(0, 10)$

2. $f(x) = 4x + \frac{2}{x}$ on $(0, 10)$

✓ 3. $f(x) = 4x - 3 + \frac{2}{x}$ on $(0, 20)$

4. $f(x) = 2x - 5 + \frac{6}{x}$ on $(0, 20)$

5. $y = 3x^2 - 2 + \frac{6}{x^2}$ on $(0, 5)$

6. $y = 2x^2 - 3 + \frac{2}{x^2}$ on $(0, 5)$

7. $g(x) = 3x^2 - 1 + \frac{2}{x^2}$ on $(0, 10)$

In Exercises 8–15, determine the absolute maximum of each function on the given interval.

8. $f(x) = -2x - \frac{9}{x}$ on $(0, 20)$

9. $f(x) = -3x - \frac{2}{x}$ on $(0, 20)$

10. $y = -3x - \frac{5}{x}$ on $(0, 10)$

11. $y = -2x - \frac{3}{x}$ on $(0, 10)$

12. $g(x) = 3 - 2x^2 - \frac{3}{x^2}$ on $(0, 10)$

13. $g(x) = 2 - 3x^2 - \frac{2}{x^2}$ on $(0, 7)$

14. $f(x) = 2 - 3x - \frac{2}{x^2}$ on $(0, 10)$

15. $y = 3 - 2x - \frac{5}{x}$ on $(0, 10)$

Applications

16. Melissa has 400 feet of fencing with which to enclose two adjacent lots as shown in Figure 15.5.9. Determine the dimensions x and y that maximize the total area. What is the maximum area?

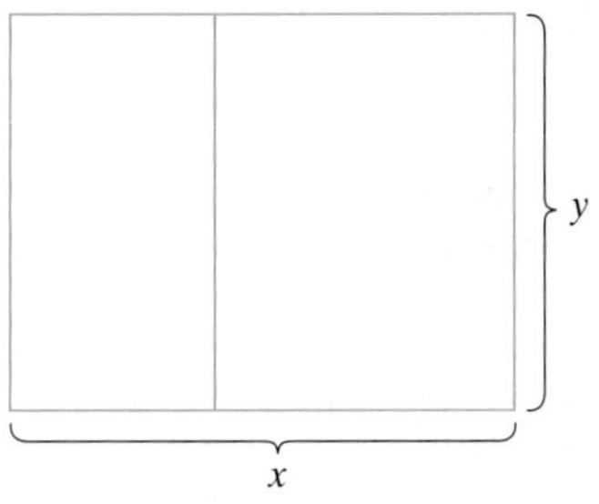

Figure 15.5.9

17. Melissa decided to put the two adjacent lots along a canal in such a way that the side labeled x in Figure 15.5.9 requires no fencing. Assuming that she has 400 feet of fencing, determine the dimensions x and y that maximize the total area. What is the maximum area?

18. A 1320-foot track encloses a rectangular region and the adjoining semicircular ends, as shown in Figure 15.5.10. Determine the dimensions x and y of the rectangle of maximum area. What is the maximum area?

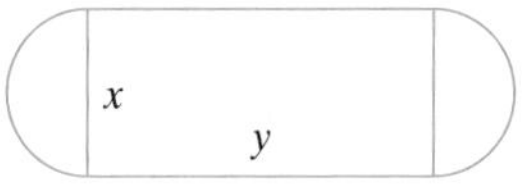

Figure 15.5.10

19. Juan is an avid gardener. He wishes to enclose two identical rectangular plots each with an area of 1500 square feet, as shown in Figure 15.5.11. To keep raccoons out of his garden, the outer boundary requires a heavy fence costing \$8.50 per foot, but the fence for the partition costs only \$3.20 per foot. Determine the dimensions x and y that minimize the total cost of the fencing.

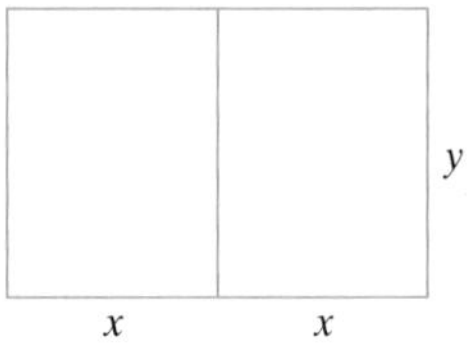

Figure 15.5.11

20. Richelle is an avid gardener. She wishes to enclose two identical rectangular plots each with an area of 1200 square feet. (Consult Figure 15.5.11 here, too.) To keep skunks out of her garden, the outer boundary requires a heavy fence costing \$6 per foot, but the fence for the partition costs only \$4 per foot. Determine the dimensions x and y that minimize the total cost of the fencing.

✓ 21. A serving tray is made from a 6-inch by 16-inch sheet of tin by cutting identical squares from the corners and then folding up the flaps. Determine the dimensions of the tray of maximum volume.

22. An open-top box used to carry small toys is to be made from a 10-inch by 12-inch piece of corrugated cardboard by cutting identical squares from the corners and then folding up the flaps. Determine the dimensions of the box of maximum volume.

23. Use Figure 15.5.12 for Exercises 23 and 24. An open-top box is made with a square base and should have a

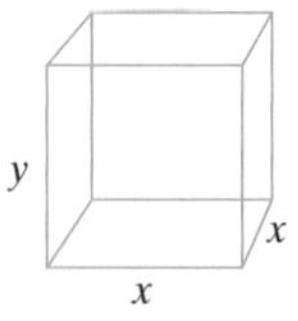

Figure 15.5.12

volume of 6000 cubic inches. If the material for the sides costs \$0.20 per square inch and the material for the base costs \$0.30 per square inch, determine the dimensions of the box that minimize the cost of the materials. (*Hint:* The surface area is equal to the area of the base plus the area of the four sides.)

24. A closed box is made with a square base and must have a volume of 343 cubic inches. The material for the sides and the top cost \$0.02 per square inch, and the material for base costs \$0.04 per square inch. Determine the dimensions of the box that minimize the cost of the materials.

25. The marketing research department of *Shank*, a quarterly magazine for beginning golfers, has determined that the cost of printing, distributing, and advertising is given by

$$C(x) = 5 + 0.5x + 0.003x^2 \qquad x > 0$$

where x represents the number of magazines printed and sold each quarter, in hundreds, and $C(x)$ is in hundreds of dollars.

(a) Determine AC, the average cost function.

(b) Determine the value of x that minimizes $AC(x)$.

26. It has been determined that the cost $C(t)$, in dollars, of owning a new car (including a service and maintenance contract) that sells for \$11,000 after t years can be approximated by

$$C(t) = 100t^2 + 400t + 11{,}000$$

Determine $AC(t)$ and the value for t that minimizes $AC(t)$.

27. It has been determined that the cost $C(t)$ in dollars, of owning a new car (including a service and maintenance contract) that sells for \$15,000 after t years can be approximated by

$$C(t) = 100t^2 + 600t + 15{,}000$$

Determine $AC(t)$ and the value for t that minimizes $AC(t)$.

28. A 200-room hotel is filled when the room rate is \$50 per day. For each \$1.50 increase in the room rate, two fewer rooms are rented. Determine the room rate that maximizes daily revenue.

29. For the hotel described in Exercise 28, it costs \$4.50 per day to clean and maintain each room occupied. Determine the room rate that maximizes daily profit.

30. Dudley's Delicious Apples determines that the annual yield per apple tree is fairly constant at 352 pounds when the number of trees per acre is 55 or fewer. For each additional tree over 55, the annual yield decreases by 5 pounds due to overcrowding. How many trees should be planted on each acre to maximize the annual yield from an acre?

✓ **31.** Pear Paradise has determined that the annual yield per pear tree is fairly constant at 160 pounds when the number of trees per acre is 40 or fewer. For each additional tree over 40, the annual yield per tree decreases by 2 pounds due to overcrowding. How many trees should be planted on each acre to maximize the annual yield from an acre?

32. The Orange Works has determined that the annual yield per orange tree is fairly constant at 270 pounds per tree when the number of trees per acre is 30 or fewer. For each additional tree over 30, the annual yield per tree decreases by 3 pounds due to overcrowding. How many trees should be planted on each acre to maximize the annual yield from an acre?

33. Walt's Walnut Grove has determined that the annual yield per walnut tree is fairly constant at 50 pounds per tree when the number of trees per acre is 30 or fewer. For each additional tree over 30, the annual yield per tree decreases by 1.25 pounds due to overcrowding. How many trees should be planted on each acre to maximize the annual yield from an acre?

34. In a certain state, hunters are allowed to shoot deer during a limited hunting season. The length of the season is carefully determined by the State Department of Natural Resources to ensure a harvest that is sustainable year after year. A state biologist has determined that the yearly growth curve for the deer population is given by

$$f(x) = 1.5x - 0.002x^2$$

where x is measured in thousands. Determine the optimal population size and the yearly kill that it will sustain.

35. In a western U.S. state, it has been estimated that the yearly growth curve for elk can be approximated by

$$f(x) = 1.75x - 0.003x^2$$

where x is measured in hundreds. Determine the optimal population size and the yearly kill that it will sustain.

36. A certain state has determined that the yearly growth curve for rabbits can be approximated by

$$f(x) = 1.3x - 0.0003x^2$$

where x is measured in hundreds. Determine the optimal population size and the yearly kill that it will sustain.

37. Essex County is renowned for its pheasant hunting. It has been determined that the growth curve for pheasant can be approximated by

$$f(x) = 2.2x - 0.01x^2$$

where x is in thousands. Determine the optimal population size and the yearly kill that it will sustain.

In Exercises 38–46, find the lot size and how often orders should be placed to minimize the inventory costs. See Example 5. Round to the nearest lot size that produces an integer value for the number of orders placed.

38. Linger's Luxury Office Furniture expects to sell 400 junior executive desks a year. Each desk costs the store

\$300, and there is a fixed charge of \$600 per order. If it costs \$200 to store a junior executive desk for a year, how large should each order be and how often should orders be placed to minimize the inventory costs?

✓ **39.** Linger's Luxury Office Furniture expects to sell 200 presidential desks in a year. Each desk costs the store \$600, and there is a fixed charge of \$600 per order. If it costs \$300 to store a presidential desk for a year, how large should each order be and how often should orders be placed to minimize inventory costs?

40. The VideoPhile Camera Store expects to sell 480 video camera carrying cases in a year. Each carrying case costs the store \$26, and there is a fixed charge of \$30 per order. If it costs \$1 to store a carrying case for a year, how large should each order be and how often should orders be placed to minimize inventory costs?

41. The VideoPhile Camera Store described in Exercise 40 has recently found a new place to store the carrying cases. Here it costs \$0.50 to store a carrying case for a year. Given this new amount, rework Exercise 40 and determine how large each order should be and how often orders should be placed to minimize inventory costs.

42. Sandkuhl's Appliances expects to sell 2500 gas stoves in a year. Each gas stove costs the store \$500, and there is a fixed charge of \$20 per order. If it costs \$10 to store a gas stove for a year, how large should each order be and how often should orders be placed to minimize inventory costs?

43. Sandkuhl's Appliances expects to sell 2400 microwave ovens in a year. Each microwave oven costs the store \$120, and there is a fixed charge of \$54 per order. If it costs \$9.00 to store a microwave oven for a year, how large should each order be and how often should orders be placed to minimize inventory costs?

44. Beagle's department store expects to sell 1200 pairs of blue jeans in a year. Each pair of blue jeans costs the store \$12, and there is a fixed charge of \$100 per order. If it costs \$4 to store a pair of blue jeans for a year, how large should each order be and how often should orders be placed to minimize the inventory costs?

45. Gene's New and Used Cars expects to sell 300 cars in a year. On average, the cars cost \$8500 each, and there is a fixed charge of \$750 per order. If it costs \$1500 to store a car for a year, how large should each order be and how often should orders be placed to minimize the inventory costs?

46. Simpson's Shoes expects to sell 1000 pairs of a certain running shoe in a year. Each pair of shoes costs the store \$15, and there is a fixed charge of \$150 per order. If it costs \$6 to store a pair of shoes for a year, how large should each order be and how often should orders be placed to minimize inventory costs?

Exercises 47–49 examine that aspect of manufacturing known as production runs. *These exercises can be handled in a manner similar to Exercises 38–46.*

47. A publisher estimates that the annual demand for a book will be 5000 copies. Each book costs \$17 to print, and setup costs are \$1800 for each printing. If storage costs are \$3 per book per year, determine how many books should be printed per run and how many printings will be needed in order to minimize costs.

48. A manufacturer estimates the demand for a new CD-ROM game to be 8000 per year. Each CD-ROM game costs \$14 to produce, plus it costs \$800 in setup costs. If a CD-ROM game can be stored for a year at a cost of \$3, how many should be produced at a time and how many production runs will be needed in order to minimize costs?

49. A toy manufacturer estimates the demand for a toy car to be 50,000 per year. Each toy car costs \$3 to make, plus setup costs of \$500 for each production run. If it costs \$2 to store a toy car for a year, how many toy cars should be manufactured at a time and how many production runs will be needed in order to minimize costs?

SECTION PROJECT

The source of the data for this project is the Federal Express web site (www.fedex.com).

(a) *Federal Express* states that any package that exceeds 165 inches in length plus girth is to be considered U.S. Domestic Freight. See Figure 15.5.13. Assuming that the front of the package is a square, determine the largest-volume package that *Federal Express* will accept and not declare it to be U.S. Domestic Freight.

(b) *Federal Express* states that the maximum length plus girth for its FedEx Overnight Freight and its FedEx 2Day Freight shipments is 300 inches. (See Figure 15.5.13.) Assuming that the front of the package is a square and the maximum length plus girth is 300 inches, determine the dimensions that will maximize the volume and determine the maximum volume.

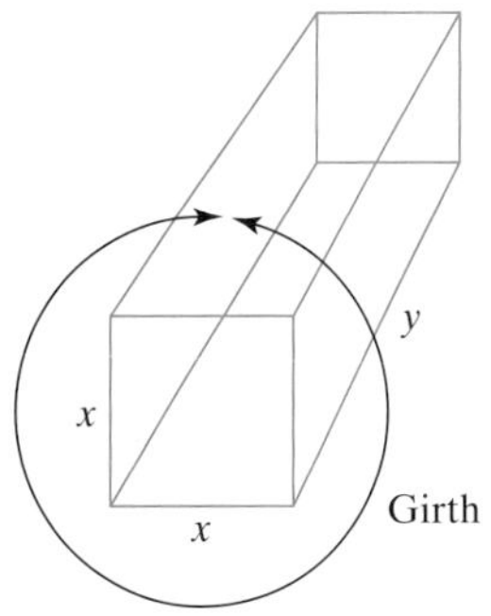

Figure 15.5.13

Why We Learned It

In this final chapter on applications of the derivative, we focused on graphical analysis of functions and optimizing functions on closed as well as open intervals. This allowed us to determine relative extrema and absolute extrema for functions. For example, if we wish to determine the production level that maximizes profit we need to determine an extremum. Likewise, if we wish to determine the production level that minimizes the average cost, we need to locate an extremum. To determine the number of days after a pond has been treated to control harmful bacteria growth when the concentration of the harmful bacteria is minimized also requires that we locate an extremum. Other applications of locating extrema include minimizing inventory costs and maximizing a sustainable harvest. All of these varied applications require using the derivative in an applied setting.

CHAPTER REVIEW EXERCISES

For each function graphed in Exercises 1–4:

(a) Determine where the function is increasing.

(b) Determine where the function is decreasing.

(c) Determine where the function is constant.

1.

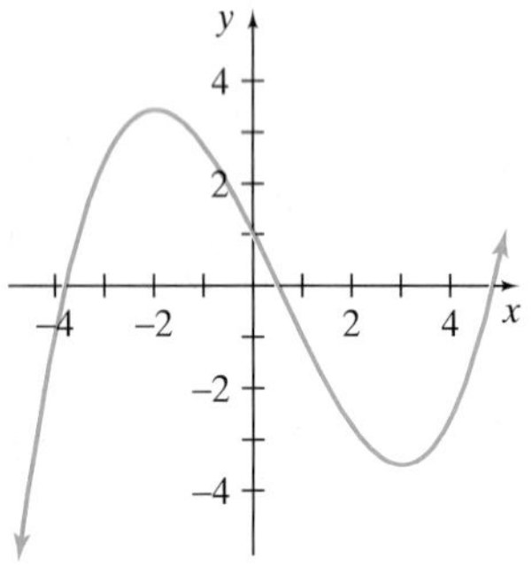

2.

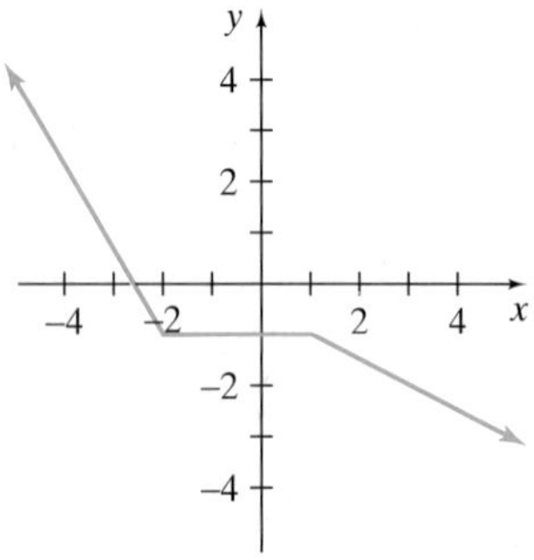

3.

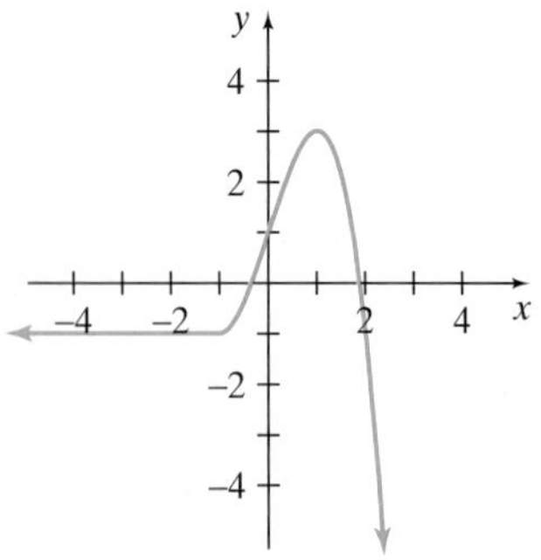

4.

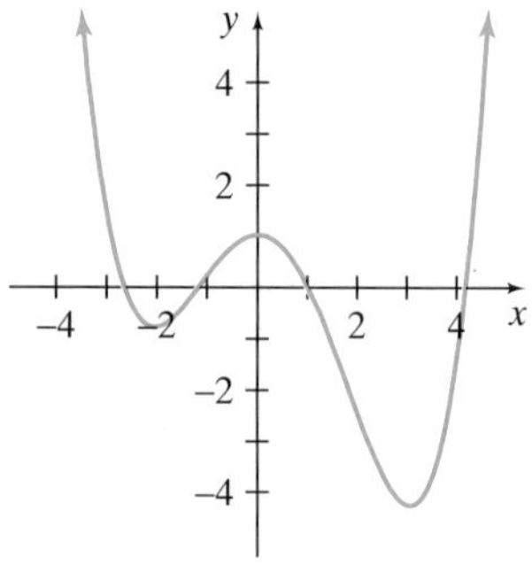

For the derivatives graphed in Exercises 5–8:

(a) List intervals where the function is increasing.

(b) List intervals where the function is decreasing.

(c) List where the function is constant.

5.

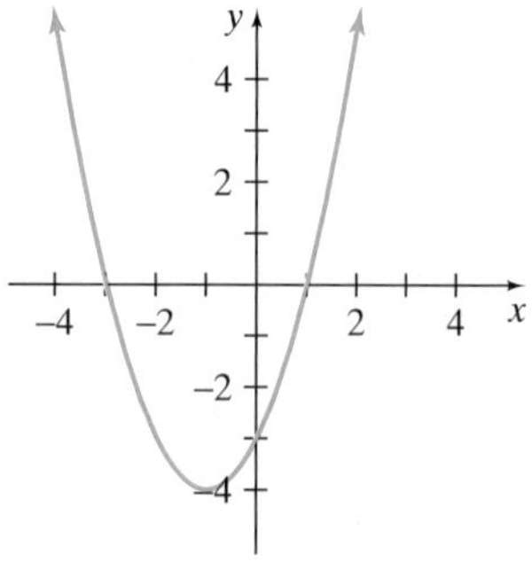

6.

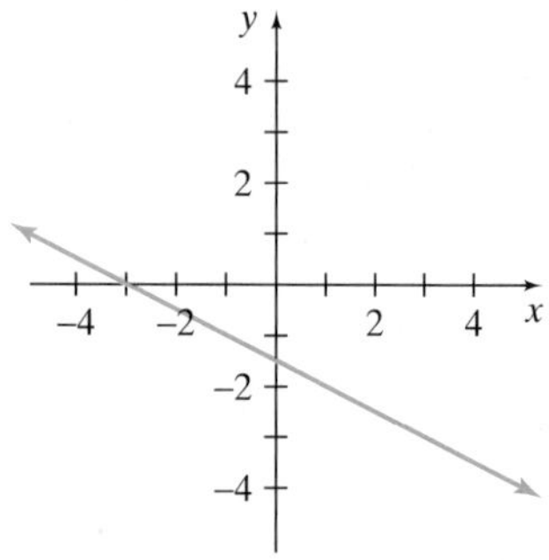

7.

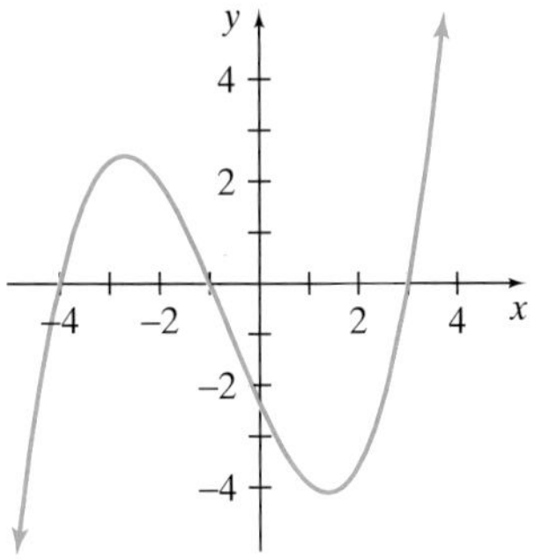

8.

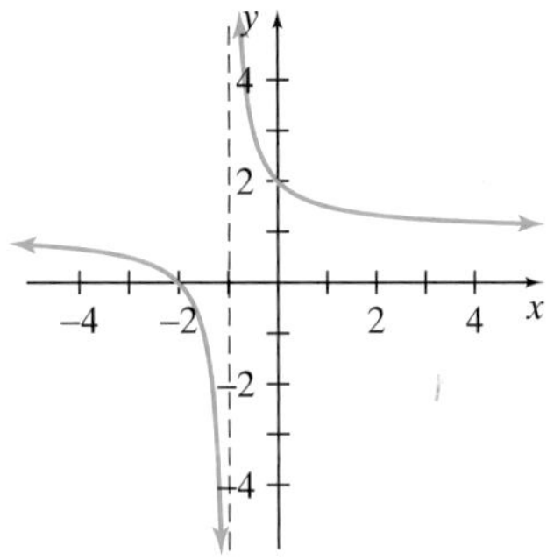

For each function in Exercises 9–14:

(a) Determine the critical values of the function.

(b) Determine intervals where the function is increasing and decreasing.

(c) Locate the relative extrema.

(d) Use a calculator to graph the function and verify your results to parts (a) and (b).

9. $f(x) = \frac{1}{2}x - 4$

10. $g(x) = -x^2 + 3x + 8$

11. $y = x^3 + 3x^2 - 24x + 20$

12. $f(x) = \sqrt{9 - x}$

13. $g(x) = (3x - 2)^4$

14. $y = \dfrac{x + 4}{3 - x}$

Use the derivatives given in Exercises 15–18 to determine the x-values where f has a relative maximum and/or a relative minimum.

15. $f'(x) = -3x + 8$

16. $f'(x) = 5x(x - 1)$

17. $f'(x) = x^2 - 9$

18. $f'(x) = (x - 4)^2(x - 1)(x + 2)^3(x + 5)$

For Exercises 19 and 20, sketch a graph of a continuous function that satisfies the given data.

19.

x	$f(x)$	$f'(x)$
0	1	−4
1	−2	−2
2	−3	0
3	−2	2
4	1	4
5	6	6

20.

x	$f(x)$	$f'(x)$
0	3	17
1	12	2
2	9	−7
3	0	−10
4	−9	−7
5	−12	2

21. The LuvYerPet Company is introducing a new brand of dog shampoo. The company has determined that the demand for the product can be modeled by

$$p(x) = 4.5e^{-0.03x}$$

where x represents the quantity demanded in thousands of containers of dog shampoo and $p(x)$ represents the price in dollars.

(a) Use the model to determine the price if there is demand for 30,000 containers of dog shampoo (that is, $x = 30$).

(b) Determine R, the revenue in thousands of dollars as a function of the quantity demanded in thousands of containers.

(c) Use the techniques of Section 15.1 to determine intervals where R is increasing and decreasing.

(d) Determine the relative maximum. Round the x- and y-coordinates to the nearest hundredth and interpret.

(e) Graph R in the viewing window [0, 100] by [0, 100] and verify your work.

22. Researchers have determined through experimentation that the percent concentration of a certain medicine during the first 20 hours after it has been administered can be approximated by

$$p(t) = \frac{150t}{t^2 + 4t + 4} \qquad 0 \le t \le 20$$

where t is the time in hours after administration of the medication and $p(t)$ is the percent concentration.

(a) Compute $p'(t)$.

(b) Determine the critical values.

(c) Determine intervals where p is increasing and decreasing.

(d) Determine the relative maximum and interpret.

23. A manufacturer of telephones has the following costs when producing x telephones in one day, where $0 < x < 500$:

Fixed costs: \$800

Unit production cost: \$15 per telephone

Equipment maintenance: $\dfrac{x^2}{18}$

Therefore, the total cost of manufacturing x telephones in one day is given by

$$C(x) = 800 + 15x + \frac{x^2}{18} \qquad 0 < x < 500$$

(a) Determine the average total cost, AC.

(b) Determine the critical value(s) for AC.

(c) Determine intervals where AC is increasing and decreasing.

(d) Determine the relative minimum and interpret.

24. The average value of farmland in the United States (excluding Alaska and Hawaii) can be modeled by

$$F(t) = 0.813t^3 - 2.63t^2 + 0.815t + 741.676 \qquad 1 \le t \le 17$$

where t is the number of years since 1979 and $F(t)$ is the average value of farmland in dollars per acre (*Source:* U.S. Department of Agriculture, www.usda.gov).

(a) Determine the critical value(s).

(b) Determine intervals where F is increasing and decreasing.

(c) Determine the relative extrema and interpret.

In Exercises 25–28, compute the first three derivatives of the given function.

25. $f(x) = x^3 - 7x^2 + 5$

26. $y = 1.3x^5 + 2.7x^4 - 6.1x^2 + 5.8x$

27. $y = \sqrt[3]{x - 5}$

28. $f(x) = x^2 + 5x - \ln 3x$

For each function in Exercises 29–35:

(a) Determine intervals where the function is concave up and where it is concave down.

(b) Determine any inflection points.

29. $f(x) = x^3 - x^2 + 5x - 7$

30. $y = x^2 - 18x + 15$

31. $y = 12 - \sqrt{x}$

32. $f(x) = \sqrt{25 - x^2}$

33. $y = 3x - \dfrac{5}{x}$

34. $g(x) = e^{-0.7x}$

35. $y = e^{5x}$

For each function in Exercises 36–39:

(a) Determine intervals where the function is increasing and where it is decreasing.

(b) Determine the relative extrema.

(c) Determine intervals where the graph of the function is concave up and where it is concave down.

(d) Determine any inflection points.

36. $y = 3x^2 + 2x - 8$

37. $f(x) = 2x^3 + 3x^2 - 36x$

38. $g(x) = 2x + \dfrac{18}{x}$

39. $y = \ln(2 - x)$

40. Sketch the graph of a function that satisfies the following conditions.

Domain: All reals

Range: All reals greater than or equal to 2

Continuous for all reals

$f' > 0$ on $(-3, 0) \cup (3, \infty)$

$f' < 0$ on $(-\infty, -3) \cup (0, 3)$

$f'(0)$ is undefined

$f'' > 0$ on $(-\infty, 0) \cup (0, \infty)$

41. The Sweet Truth Cookie Company has determined that its daily profit is given by

$$P(x) = -0.5x^3 + 3x^2 + 68x - 133$$

where x is the number of hundreds of cookies baked and sold and $P(x)$ is the daily profit in dollars.

(a) Determine $P(4)$ and $P'(4)$ and interpret each.

(b) Determine intervals where P is increasing and where it is decreasing. Determine the relative maximum and interpret each coordinate.

(c) Determine intervals where the *marginal profit* is increasing and where it is decreasing. What can you say about the graph of P on these intervals?

(d) Determine the inflection point for the graph of P.

42. Suppose that blood flow through the artery of a certain animal species can be modeled by

$$f(x) = 0.317e^{0.0103x} \qquad 20 \le x \le 120$$

where x represents arterial pressure, measured in millimeters of mercury, and $f(x)$ is blood flow, measured in milliliters per minute.

(a) Evaluate $f(40)$ and interpret.

(b) Evaluate $f'(40)$ and interpret.

(c) Show that the graph of f is concave up on the interval $[20, 120]$. Explain what this means.

43. Suppose that the number of mountain goats in a certain region can be modeled by

$$p(t) = \frac{1400}{1 + 43e^{-0.09t}}$$

where $p(t)$ represents the number of goats t years after the goats were first introduced to the region.

(a) Graph p in the viewing window $[0, 100]$ by $[0, 1500]$.

(b) Compute $p(84)$ and $p'(84)$ and interpret each.

(c) Determine $p''(t)$.

(d) On the interval $[0, 100]$, determine intervals where the graph of p is concave up and where it is concave down. (*Hint:* Graph p'' and use the ZERO/ROOT command.)

(e) When is the population of goats growing fastest?

44. The following table gives the annual number of complaints against U.S. airlines from 1988–1996.

Year	Complaints
1988	21,493
1989	10,553
1990	7,703
1991	6,106
1992	5,639
1993	4,438
1994	5,179
1995	4,629
1996	5,778

SOURCE: U.S. Census Bureau, www.census.gov

Let x represent the number of years since 1987 and let y represent the annual number of complaints against U.S. airlines.

(a) Determine a cubic regression equation for the data in the form

$$y_1 = ax^3 + bx^2 + cx + d \qquad 1 \le x \le 9$$

(b) Determine intervals where y_1 is increasing and where it is decreasing.

(c) Determine intervals where the graph of y_1 is concave up and where it is concave down.

(d) According to the model, when was the number of complaints lowest?

Use the graph of f shown to answer Exercises 45–51.

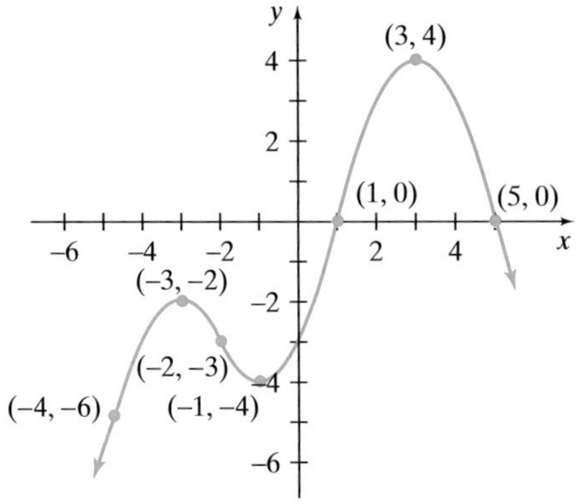

45. Determine intervals where f is positive and where f is negative.

46. Determine intervals where f is increasing and where f is decreasing.

47. Determine intervals where f' is positive and where f' is negative.

48. Determine intervals where f' is increasing and where f' is decreasing.

49. Determine intervals where the graph of f is concave up and where the graph of f is concave down.

50. Locate any relative extrema.

51. Locate any inflection points.

In Exercises 52–55, use the Second Derivative Test to locate any relative extrema, if they exist.

52. $f(x) = x^2 + 12x + 5$

53. $y = x^3 - 5x^2 - 8x$

54. $y = 3x^4 - 2x^3 - 3x^2$

55. $g(x) = \dfrac{1}{x^2 + 9}$

In Exercises 56–61, sketch the graph of the given function. Label all relative extrema, inflection points, and any asymptotes.

56. $y = x^3 - 9x^2 + 4$

57. $f(x) = 2x^3 - \dfrac{3}{2}x^2 - 9x + 5$

58. $g(x) = x^4 - 14x^2 - 24x$

59. $y = \dfrac{1}{x^2 - 4}$

60. $f(x) = 0.4x + 16 + \dfrac{8}{x}$

61. $g(x) = e^x - 4x$

In Exercises 62 and 63, the graph of f'' is given. Use the graphs to sketch graphs of f' and f. There are many correct answers.

62.

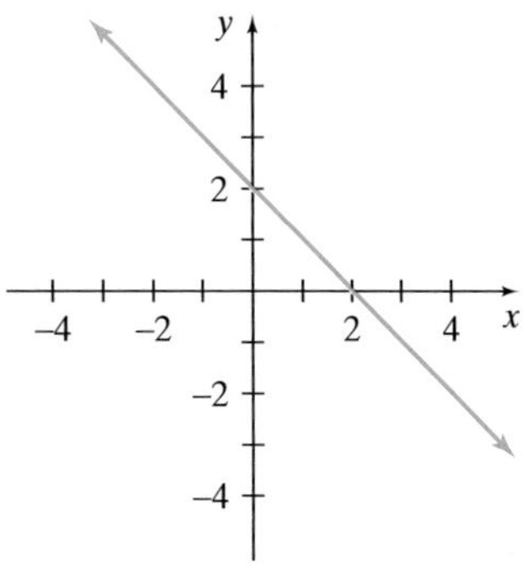

63.

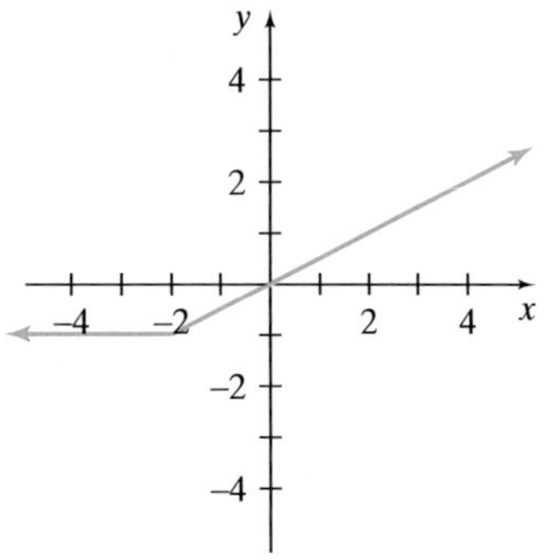

64. Frank is an artist who sells rock sculptures at local art festivals. He has determined that the relationship between price $p(x)$, in dollars per sculpture, and the quantity x demanded per day can be approximated by $p(x) = 85 - 0.12x^2$. He also knows that the cost function for the sculptures is given by $C(x) = 15x + 200$.

(a) Determine the daily revenue, R.

(b) Determine the daily profit, P.

(c) Compute $P'(x)$ and use it to determine intervals where P is increasing and where it is decreasing.

(d) Determine intervals where the graph of P is concave up and where it is concave down.

(e) Sketch the graph of P. Label all relative extrema and inflection points, if they exist.

65. *(continuation of Exercise 64)*

(a) Determine the average cost function, $AC(x)$.

(b) For what value of x is $AC(x)$ a minimum?

(c) Determine the average profit function, AP.

(d) For what value of x is $AP(x)$ a maximum?

66. The Honest Otis Auto Center offers a special maintenance and service contract for individuals who purchase a used car. This contract covers all recommended servicing of the vehicle, such as oil changes and lube jobs. Penelope just purchased a used car for \$6000. The maintenance and service contract for her car costs \$300 at the end of the first year and increases \$150 per year thereafter, as long as Penelope owns the car. Using techniques of regression, she found that the total cost of the car (excluding items not covered by the service contract, such as gasoline) after t years is given by

$$C(t) = 75t^2 + 225t + 6000$$

(a) Define a function for the average cost per year, AC.

(b) When is the average cost per year a minimum? Round to the nearest tenth.

(c) What is the minimum average cost per year? Round to the nearest dollar.

67. The number of African American members of the U.S. House of Representatives can be modeled by

$$f(t) = 15.7e^{0.058t} \qquad 1 \le t \le 15$$

where t is the number of years since 1980 and $f(t)$ is the number of African American members of the U.S. House of Representatives (*Source:* U.S. Census Bureau, www.census.gov).

(a) Compute $f(9)$ and $f'(9)$ and interpret each.

(b) Compute $f'(t)$ and use it to show that, according to the model, the number of African American members of the U.S. House of Representatives was increasing from 1981 to 1995.

(c) Compute $f''(t)$ and use it to show that the graph of f is concave up on the indicated interval. Interpret what this means.

(d) Sketch a graph of f.

In Exercises 68–74, determine the absolute extrema of each function on the indicated interval.

68. $f(x) = x^2 - 5x + 3$ on $[0, 10]$

69. $y = x^3 + 6x^2 + 9x$ on $[-5, 5]$

70. $f(x) = x^3 - 6x$ on $[0, 5]$

71. $h(x) = \frac{1}{5}x^5 - \frac{5}{3}x^3 + 4x$ on $[-5, 5]$

72. $g(x) = \sqrt{x^3}$ on $[0, 5]$

73. $f(x) = x + \frac{4}{x}$ on $[1, 10]$

74. $y = x + \frac{4}{x}$ on $[-10, -5]$

75. From 1973 to 1997, the amount of money spent by the U.S. government on national defense can be modeled by

$$y = -0.0668x^3 + 2.04x^2 - 2.41x + 71.29 \qquad 1 \le x \le 25$$

where x is the number of years since 1972 and y is the amount spent on national defense, in billions of dollars (*Source:* U.S. Statistical Abstract, www.census.gov/statab/www). Determine the absolute maximum and the absolute minimum and interpret each.

76. Mike has 60 feet of fencing that he intends to use to build a rectangular enclosure for his Scottish terrier. He plans to build the enclosure against one side of his house, so the fencing is needed on just three sides of the enclosure. Determine the dimensions that will make the enclosure have the largest possible area.

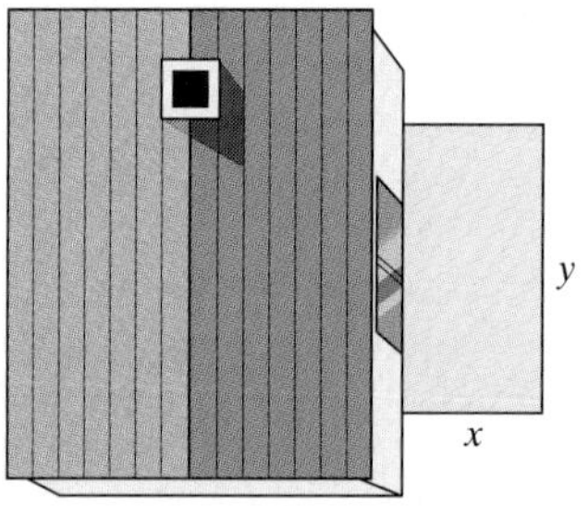

77. Your friend's family is planning to build a swimming pool. The water surface is to have an area of 1200 square feet. The pool will also have a 6-foot-wide walkway on the east and west sides and an 8-foot walkway on the north and south sides. Let x be the length of the pool (east to west), and let y be the width (north to south). Given that $10 \le x \le 100$, find the dimensions that will minimize the *total* area of the swimming pool and walkways.

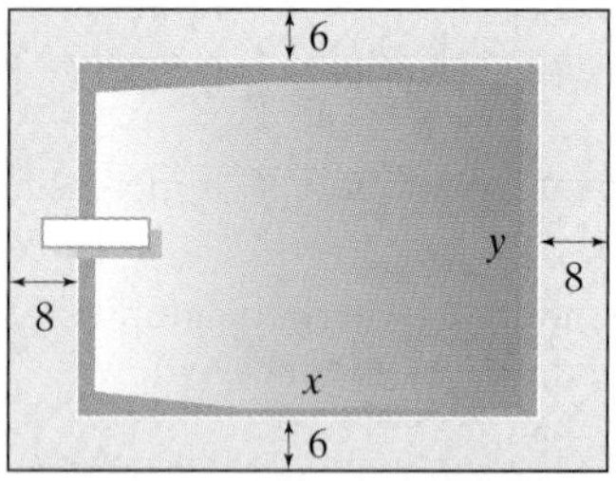

78. The GruntWorks Company analyzed the production costs for its new weight-training equipment and determined that its weekly cost function can be given by

$$C(x) = 0.001x^3 - 0.6x^2 + 217x + 7200 \qquad 50 \le x \le 550$$

where x represents the number of units produced each week and $C(x)$ is the weekly cost in dollars.

(a) Determine the average cost function, AC.

(b) Determine the production level that minimizes the average cost.

(c) Determine the average cost and the total cost at the production level found in part (b).

79. For the weight-training equipment in Exercise 78, the GruntWorks company gathered data for the price of a unit and the number of units demanded per week. The price–demand function that models these data is given by

$$p(x) = -0.002x^2 + 1.5x - 1.7 \qquad 50 \le x \le 550$$

where x represents the number of units demanded each week and $p(x)$ is the price, in dollars, that consumers pay for each unit to buy exactly x units per week.

(a) Determine R, the weekly revenue as a function of x.

(b) Determine the maximum weekly revenue.

(c) Determine the demand that maximizes weekly revenue.

(d) Determine the price that maximizes weekly revenue.

80. *(continuation of Exercise 79)* Using the functions C and R from Exercises 78 and 79, respectively:

(a) Determine P, the profit as a function of x.

(b) Determine the maximum weekly profit.

(c) Determine the demand that maximizes weekly profit.

(d) Determine the price that maximizes weekly profit.

In Exercises 81–84, determine the absolute minimum of each function on the indicated interval.

81. $y = 3x + \dfrac{6}{x}$ on $(0, 10)$

82. $f(x) = 5x - 9 + \dfrac{15}{x}$ on $(0, 6)$

83. $g(x) = 4x^2 - 2 + \dfrac{9}{x^2}$ on $(0, 5)$

84. $y = \dfrac{x^2 + 4}{x}$ on $(0, 4)$

In Exercises 85–88, determine the absolute maximum of each function on the given interval.

85. $f(x) = -4x - \dfrac{3}{x}$ on $(0, 10)$

86. $y = 3x - 5 + \dfrac{6}{x}$ on $(-5, 0)$

87. $f(x) = 5 - x^2 - \dfrac{9}{x^2}$ on $(0, 5)$

88. $g(x) = (9 - x)\sqrt{x}$ on $(2, 7)$

89. A window has the shape of a rectangle with an adjoining semicircle, as shown. The perimeter of the window is 12 feet. Determine the dimensions x and y if the area of the window is as large as possible. What is the maximum area?

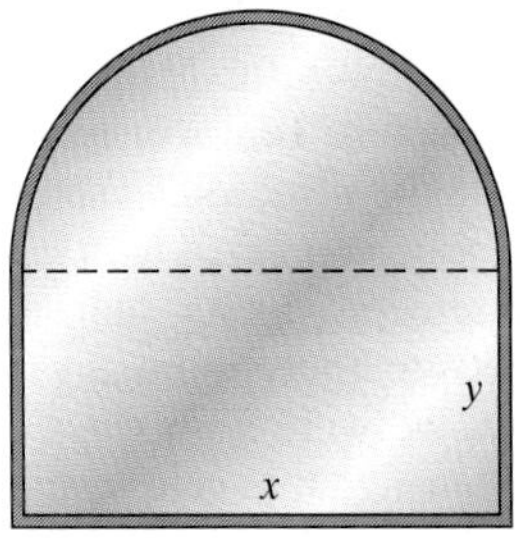

90. A newspaper publisher has determined that the daily cost of printing, distributing, and advertising is

$$C(x) = 0.0015x^2 + 0.6x + 4.5 \qquad x > 0$$

where x represents the daily circulation, in thousands, and $C(x)$ is the cost in hundreds of dollars.

(a) Determine AC, the average cost function.

(b) Determine the value of x that minimizes $AC(x)$.

91. An open-top box is made by cutting identical squares from the corners of a 15-inch by 24-inch piece of cardboard and then folding up the flaps. Determine the dimensions of the box of maximum volume.

92. The IzataFax store expects to sell 800 fax machines in a year. Each fax machine costs the store \$80, and there is a fixed cost of \$24 per order. If it costs \$62 to store a fax machine for a year, how large should each order be and how often should orders be placed to minimize inventory costs? Round to the nearest lot size that produces an integer value for the number of orders placed each year.

93. Freddy's Fruit Orchard has determined that the annual yield per fruit tree is 150 pounds per tree when the number of trees per acre is 35 or fewer. For each additional tree over 35, the annual yield per tree decreases by 4 pounds due to overcrowding. How many trees should be planted on each acre to maximize the annual yield from an acre?

94. The United Parcel Service states that a lightweight package (under 30 pounds) is *oversize* if the sum of its length and girth is 84 inches (*Source:* UPS, www.ups.com). Assuming that the front of the package is a square and the maximum length plus girth is 84 inches, determine the dimensions that will maximize the volume without requiring the customer to pay the oversize charge. What is the maximum volume?

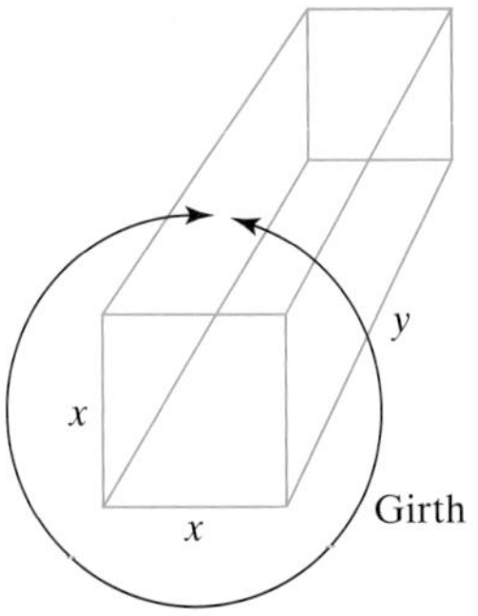

CHAPTER 15 PROJECT

Chapter 15, the last chapter devoted solely to differential calculus, offers great tools for learning about a function without having to graph it. Topics include learning what the first and the second derivative tell us about the shape of the graph. This information combined with knowing how to find the location of intercepts, limits, and asymptotes will be everything one needs to know about any function.

In this project, we will revisit the Chapter 14 project. Recall that the following table gives the population for the state of Ohio from 1800 to 2000.

Year	Population	Year	Population
1800	42,159	1910	4,767,121
1810	230,760	1920	5,759,394
1820	581,434	1930	6,646,697
1830	937,903	1940	6,907,612
1840	1,519,467	1950	7,946,627
1850	1,980,329	1960	9,706,397
1860	2,339,511	1970	10,652,017
1870	2,665,260	1980	10,797,630
1880	3,198,062	1990	10,847,115
1890	3,672,329	2000	11,353,140
1900	4,157,545		

SOURCE: U.S. Census Bureau, www.census.gov

Let x represent the number of years since 1799.

1. Find $g(x) = \dfrac{c}{1 + a \cdot e^{-b \cdot x}}$ where $g(x)$ is a logistic regression function for $1 \le x \le 201$.

2. Evaluate $g(151)$ and interpret. Compare this to the given Ohio population in 1950.

3. Evaluate $g'(151)$ and interpret.

4. If Ohio's population had continued to grow at the 1950 rate, what would the population be in 2000? Compare to $g(201)$ and to the value given in the table.

5. Find the maximum value of $g'(x)$ and interpret. Notice that this value is where the graph of g switches concavity. In general, what is such a point called?

6. In the Chapter 14 project, Ohio's population was modeled by

$$f(x) = \begin{cases} 245016.54 \cdot 1.03^x & 1 \le x < 141 \\ -54{,}195{,}506.44 + 12{,}452{,}405.31 \ln x & 141 \le x \le 201 \end{cases}$$

where x is the number of years since 1799 and $f(x)$ is Ohio's population.

Evaluate and interpret $\lim_{x \to \infty} f(x)$. Evaluate and interpret $\lim_{x \to \infty} g(x)$ where $g(x)$ is the logistic regression function from 1. Which result makes sense?

Chapter 16 Integral Calculus

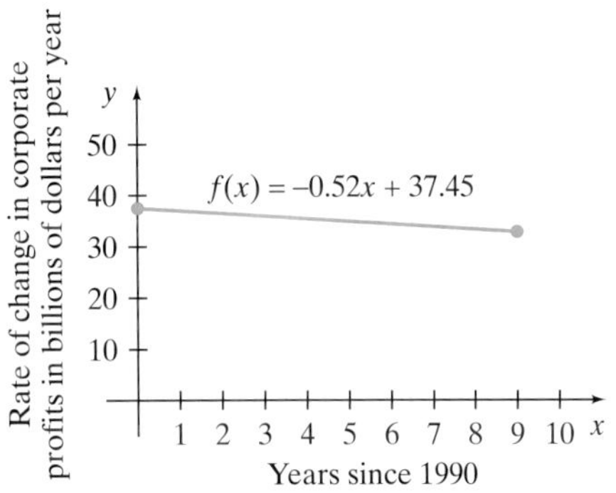

(b)

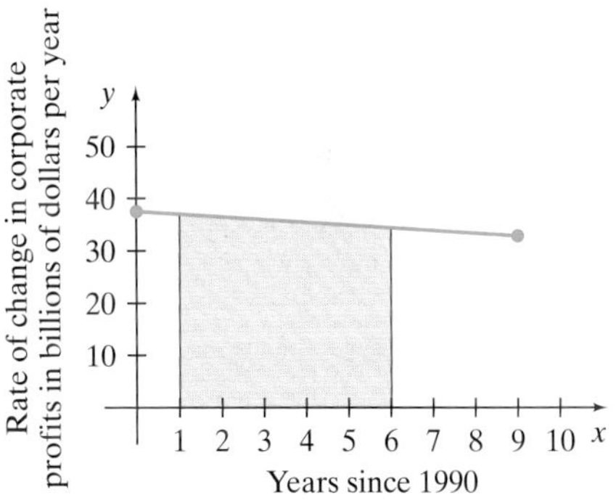

(c)

The 1990s saw the U.S. economy reach levels that had never before been reached in U.S. history. The prolonged expansion of the economy allowed many corporations in the U.S. to realize profits for many years in the 1990s. Based on data gathered at the U.S. Bureau of Economic Analysis, www.bea.doc.gov, the graph in Figure b shows the rate of change in U.S. corporate profits, after taxes, from 1990–1999. In Figure c, the shaded area represents the total increase in U.S. corporate profits, after taxes, from 1991 to 1996.

What We Know

Our focus thus far has been on the study of the derivative and instantaneous rates of change. Among other things, we have seen how we can use the derivative to determine maximum and minimum values on open and closed intervals.

Where Do We Go

We now wish to discover techniques for undoing the process of differentiation. This technique, called antidifferentiation, can be used to recover functions if the rate of change function is given. We will then study the Fundamental Theorem of Calculus, which is the glue that holds differential and integral calculus together.

Section 16.1 The Indefinite Integral

Until now, we have focused almost exclusively on how to determine the derivative and interpret the resulting rate of change function. For example, we found the derivative of a profit function and learned that this derivative is a rate of change function that measures the marginal profit. But what if we are given a **rate of change function** (that is, a rate function or derivative) and wish to find the original function? In this section, we will learn the basics of **integration**, a process that finds the original function, given its derivative, by reversing the differentiation. Then we will consider applications of integration.

The Indefinite Integral

To see how differentiation can be reversed, let's start with a common rate function from business. The WorkSharp Company determines that the marginal cost function for their designer suspenders is

$$MC(x) = C'(x) = 2x \qquad 0 \le x \le 10$$

where x is the number of designer suspenders produced in thousands and $MC(x)$ is the marginal cost in thousands of dollars.

To find the company's cost function, we need to determine a cost function with the property that $\frac{d}{dx}(C(x)) = MC(x)$. In words, we seek a cost function so that when we take its derivative we get $MC(x) = 2x$. A function such as this is referred to as an **antiderivative** of the given function MC. To gain insight into how this antidifferentiation process is done, let's review our basic differentiation rules shown in the Toolbox below.

From Your Toolbox

For differentiable functions f and g with real numbers n and k, we have:

- If $f(x) = x^n$, then $f'(x) = n \cdot x^{n-1}$. Power Rule
- If $f(x) = k$, then $f'(x) = 0$. Constant Function Rule
- If $h(x) = f(x) \pm g(x)$, then $h'(x) = f'(x) \pm g'(x)$. Sum and Difference Rule

Since the WorkSharp Company's marginal cost function is $2x$, we need an expression whose derivative is $2x$. Since the term $2x$ is linear (first degree), the Power Rule tells us that we need a quadratic function (second degree) for the antiderivative. Notice that if we choose the quadratic x^2 then $\frac{d}{dx}(x^2) = 2x$. This is just what we wanted! So is $C(x) = x^2$ the cost function we want? Not quite, because there are many functions whose derivatives are $2x$, including $x^2 + 3$, $x^2 + \frac{11}{2}$, and $x^2 - 10$. In fact, if C is any real number constant, then $x^2 + C$ is the antiderivative that we seek. We call C an **arbitrary constant**. To determine the WorkSharp Company's actual cost function, it would appear that we need a little more information.

Before continuing, let's generalize what we have done.

Antiderivative

We call F an **antiderivative** of f on an interval if $F'(x) = f(x)$ for all x in the interval. In other words,

$$\frac{d}{dx}[F(x)] = f(x)$$

We call $F(x) + C$, where C is any real number constant, the **general antiderivative** of f on an interval if

$$\frac{d}{dx}[F(x) + C] = f(x)$$

for all x in the interval.

Note: The process of determining an antiderivative is called **antidifferentiation**. Antidifferentiation is the **inverse operation** of differentiation.

Our definition provides an easy way to determine if a function is an antiderivative of another function. Example 1 illustrates how we can do this.

Example 1 Verifying General Antiderivatives

Determine if the function F is the general antiderivative of the function f.

(a) $F(x) = \frac{2}{3}x^{3/2} + 4x + C$; $f(x) = \sqrt{x} + 4$

(b) $F(x) = 2x^4 - x + C$; $f(x) = \frac{2}{3}x^3 - 1$

Solution

For each, we need to determine if $\frac{d}{dx}F(x) = f(x)$.

(a) Differentiating, we get

$$\frac{d}{dx}F(x) = \frac{d}{dx}\left(\frac{2}{3}x^{3/2} + 4x + C\right) = \frac{2}{3} \cdot \frac{3}{2}x^{1/2} + 4 + 0$$
$$= x^{1/2} + 4 = \sqrt{x} + 4 = f(x)$$

So $F(x) = \frac{2}{3}x^{3/2} + 4x + C$ is the general antiderivative of $f(x) = \sqrt{x} + 4$.

(b) Again, computing the derivative of $F(x)$ yields

$$\frac{d}{dx}F(x) = \frac{d}{dx}(2x^4 - x + C) = 8x^3 - 1 \neq f(x)$$

Since $\frac{d}{dx}F(x) \neq f(x)$, we conclude that $F(x) = 2x^4 - x + C$ is *not* the general antiderivative of f. ■

Another way to represent the general antiderivative of a function f is by

$$\int f(x)\,dx$$

which is called the **indefinite integral** of f.

Indefinite Integral

If $F'(x) = f(x)$ for all x, then

$$\int f(x)\,dx = F(x) + C$$

is called the **indefinite integral** of $f(x)$. In this notation, $\int$ is the **integral sign**, $f(x)$ is the **integrand**, and C is any real number constant. dx tells us the variable of integration.

Note: If we can determine the indefinite integral of a function, we say that the function is **integrable**. The process of determining a general antiderivative is called **integration**.

With this new notation, we can write the result of part (a) of Example 1 as

$$\int (\sqrt{x} + 4)\,dx = \frac{2}{3}x^{3/2} + 4x + C$$

Now let's look for a method to determine the indefinite integral for an expression involving x^n. First, we will consider the expression x^3. We know from the Power Rule for differentiation that the antiderivative of x^3 must be fourth degree. Since we know that $\frac{d}{dx}x^4 = 4x^3$, then $\frac{1}{4}\frac{d}{dx}(x^4) = \frac{1}{4}(4x^3) = x^3$. Thus,

$$\int x^3\,dx = \frac{1}{4}x^4 + C$$

Notice in the antiderivative $\frac{1}{4}x^4 + C$ that the power of x is one more than in x^3. Also, the antiderivative includes a constant multiple that is the reciprocal of the exponent 4.

On page 1001 we said that antidifferentiation is the inverse operation of differentiation. Here we are using the inverse operations of the Power Rule for differentiation to find an antiderivative. Instead of *subtracting* 1 from the exponent, we *add* 1, and instead of *multiplying* by the constant, we *divide*. We call this rule the **Power Rule for Integration**.

Power Rule for Integration

For any real number n, where $n \neq -1$, the indefinite integral of x^n is

$$\int x^n\,dx = \frac{1}{n+1}x^{n+1} + C$$

Note: The restriction that n cannot be -1 is because a value of -1 for n would make the denominator of the coefficient zero. In Section 16.5 we will learn the rule for the indefinite integral $\int x^{-1}\,dx = \int \frac{1}{x}\,dx$.

Example 2 Using the Power Rule for Integration

Determine the following indefinite integrals.

(a) $\int x^8\,dx$ **(b)** $\int \sqrt[4]{x}\,dx$ **(c)** $\int \frac{1}{x^5}\,dx$

Solution

(a) Applying the Power Rule for Integration, we get

$$\int x^8\,dx = \frac{1}{8+1}x^{8+1} + C = \frac{1}{9}x^9 + C = \frac{x^9}{9} + C$$

Notice that the coefficient $\frac{1}{9}$ is the reciprocal of the exponent 9.

(b) Here we use the rules of algebra to write the integrand in the form x^n. Since $\sqrt[4]{x} = x^{1/4}$, we can write

$$\int \sqrt[4]{x}\,dx = \int x^{1/4}\,dx = \left(\frac{1}{\frac{1}{4}+1}\right)\cdot x^{(1/4)+1} + C$$

$$= \frac{4}{5}x^{5/4} + C$$

Notice that the coefficient $\frac{4}{5}$ is the reciprocal of the exponent $\frac{5}{4}$.

(c) Again, the rules of algebra come into play when rewriting the integrand. Here we can write $\frac{1}{x^5}$ as x^{-5} and then compute

$$\int \frac{1}{x^5}\,dx = \int x^{-5}\,dx = \frac{1}{-5+1}x^{-5+1} + C$$

$$= \frac{1}{-4}x^{-4} + C$$

$$= -\frac{1}{4}\cdot\frac{1}{x^4} + C$$

$$= -\frac{1}{4x^4} + C$$

Notice that the coefficient $-\frac{1}{4}$ is the reciprocal of the exponent -4. ■

> **Interactive Activity**
>
> Use differentiation to show that the indefinite integral of $f(x) = e^{2x}$ is $F(x) = \frac{1}{2}e^{2x} + C$. Do the same for $f(x) = \frac{1}{x+3}$ and $F(x) = \ln(x+3) + C$. Explain why the Power Rule for Integration does not apply in these cases.
>
> WWW

✓ Checkpoint 1 Now work Exercise 17.

The next rule involves integration of a constant. To find $\int 5\,dx$, we need a function whose derivative is 5. Since we know that $\frac{d}{dx}(5x) = 5$, we can write $\int 5\,dx = 5x + C$. This suggests the following rule:

■ **Constant Rule for Integration**

If k is any real number, then the indefinite integral of k is

$$\int k\,dx = kx + C$$

Some of the properties of integration are much the same as for limits and for differentiation. Consider the integral $\int(2x-5)\,dx$. We know that $\int(2x)\,dx = x^2 + C$, and we can show from the Constant Rule for Integration that $\int 5\,dx = 5x + C$. So its seems natural that the integral of the linear function

$2x - 5$ is $\int (2x - 5)\,dx = x^2 - 5x + C$, which you should check using differentiation. We could have rewritten the expression as

$$\int (2x - 5)\,dx = \int 2x\,dx - \int 5\,dx = x^2 - 5x + C$$

Some may be concerned that since there are two integrals there must be two constants of integration C. But here we are simply adding the two constants to get a new constant, which we call C.

Sum and Difference Rule for Integration

For integrable functions f and g, we have

$$\int [f(x) \pm g(x)]\,dx = \int f(x)\,dx \pm \int g(x)\,dx$$

Note: This rule means that we are allowed to determine indefinite integrals term by term.

Example 3 Using the Sum and Difference Rule for Integration

Apply the Sum and Difference Rule for Integration to determine the indefinite integrals.

(a) $\int (x^2 + 3)\,dx$ **(b)** $\int (\sqrt[3]{x} + 5)\,dx$

Solution

Interactive Activity

Use differentiation to check the solutions to Example 3(a) and (b).

(a) The Sum and Difference Rule for Integration yields

$$\int (x^2 + 3)\,dx = \int x^2\,dx + \int 3\,dx = \frac{1}{3}x^3 + 3x + C$$

(b) We begin by writing the first term in the integrand in the form x^n to get

$$\int (\sqrt[3]{x} + 5)\,dx = \int (x^{1/3} + 5)\,dx$$

Applying the Sum and Difference Rule gives

$$= \int x^{1/3}\,dx + \int 5\,dx = \frac{1}{\frac{1}{3} + 1}x^{(1/3)+1} + 5x + C$$

$$= \frac{3}{4}x^{4/3} + 5x + C$$

■

Our next rule addresses functions with coefficients other than 1. In Chapter 12 we learned that

$$\frac{d}{dx}[k \cdot f(x)] = k \cdot \frac{d}{dx}[f(x)]$$

For example,

$$\frac{d}{dx}[3x^2] = 3 \cdot \frac{d}{dx}[x^2] = 3 \cdot 2x = 6x$$

Since integration is the inverse operation of differentiation, we know that

$$\int 6x\,dx = 3x^2 + C$$

But how would we get the general antiderivative if we did not already know it? We could try "pulling out" the constant much as we did with the constant multiple differentiation rule.

$$\int 6x\,dx = 6 \cdot \int x\,dx = 6 \cdot \frac{1}{2}x^2 + C = 3x^2 + C$$

This illustrates that when we are integrating a coefficient times a function we can put the coefficient in front of the integral sign and integrate the function.

Coefficient Rule for Integration

Given any real number coefficient c and integrable function f,

$$\int c \cdot f(x)\,dx = c \cdot \int f(x)\,dx$$

Armed with these integration rules, we can find the indefinite integral of any polynomial function. For example,

$$\int (x^2 - 3x + 4)\,dx = \int x^2\,dx - 3\int x\,dx + \int 4\,dx$$

$$= \frac{x^3}{3} - 3 \cdot \frac{x^2}{2} + 4x + C = \frac{1}{3}x^3 - \frac{3}{2}x^2 + 4x + C$$

Example 4 shows that we do not necessarily have to integrate a function with respect to x.

Example 4 **Integrating Polynomial Functions**

Determine $\int (-2t^3 + 3t + 5)\,dt$.

Solution

Understand the Situation: Here the dt indicates that we are integrating the cubic $-2t^3 + 3t + 5$ with respect to t. The same rules that we have developed in this section for the independent variable x apply here as well.

Performing the integration gives us

$$\int (-2t^3 + 3t + 5)\,dt = \int -2t^3\,dt + \int 3t\,dt + \int 5\,dt$$

$$= -2\int t^3\,dt + 3\int t\,dt + \int 5\,dt$$

$$= -2 \cdot \frac{t^4}{4} + 3 \cdot \frac{1}{2}t^2 + 5t + C$$

$$= -\frac{1}{2}t^4 + \frac{3}{2}t^2 + 5t + C$$

■

Checkpoint 2 Now work Exercise 27.

Recovering Functions from Rate Functions

Through much of our remaining study of the calculus, we will study the properties of rate functions.

Rate Function

A function that gives an instantaneous rate of change is a **rate function**. (A rate function may be considered a derivative since the derivative gives us an instantaneous rate of change.)

Note: We use rate function and rate of change function interchangeably. Also, in applications pay close attention to the units given for a rate function.

Recovering functions from rate functions is very useful in many applications in which the rate of change is given. For example, let's say that we have a derivative $f'(x) = x^2$ and we wish to know the function from which it came, that is, f. By integrating, we know that

$$f(x) = \int x^2\,dx = \frac{1}{3}x^3 + C$$

Each value of C produces a different function, as shown in Figure 16.1.1.

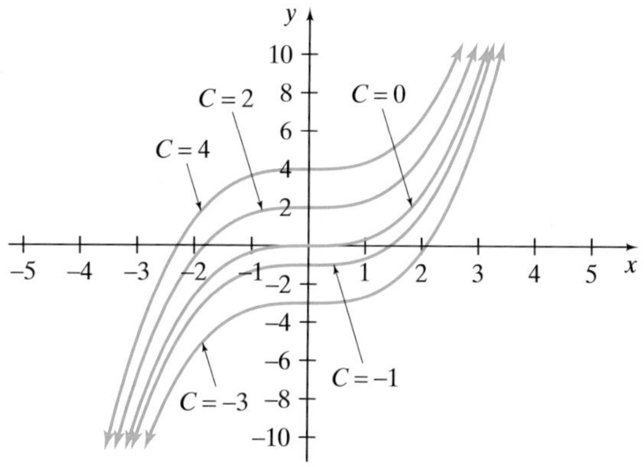

Figure 16.1.1 Graphs of $f(x) = \frac{1}{3}x^3 + C$ for $C = -3, -1, 0, 2, 4$.

Now suppose we know that an ordered pair, say (3, 11), is on the graph of f. By replacing $f(x)$ with 11 and x with 3, we can write an equation to find the value of the integration constant C.

$$f(x) = \frac{1}{3}x^3 + C$$

$$11 = \frac{1}{3}(3)^3 + C$$

$$11 = \frac{1}{3}\cdot 27 + C$$

$$11 = 9 + C$$

$$2 = C$$

So, for the derivative $f'(x) = x^2$ and point (3, 11) on the graph of f, the corresponding function f is $f(x) = \frac{1}{3}x^3 + 2$. This is illustrated in Figure 16.1.2.

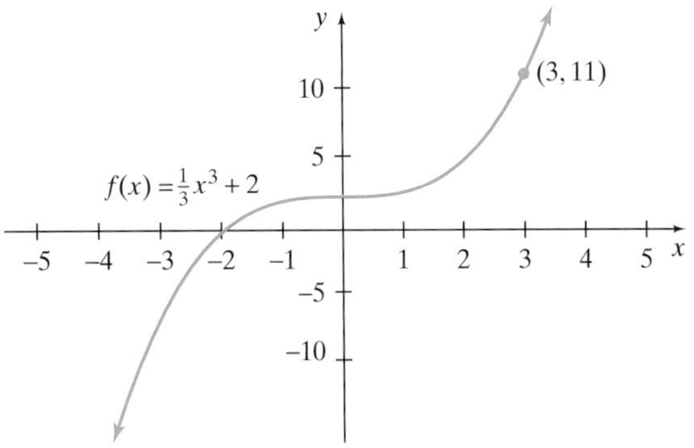

Figure 16.1.2

Example 5 Recovering a Function from a Rate Function

The HonCo marketing firm finds the rate of change of total sales of a product featured on an infomercial can be modeled by

$$R(t) = 15\sqrt[3]{t} \qquad 0 \le t \le 20$$

where t represents the number of months that the infomercial has aired and $R(t)$ represents the rate of change of total sales of the product measured in $\frac{\text{thousand units}}{\text{month}}$.

(a) Compute $R(10)$ and interpret.

(b) Knowing that there were 500 units sold in trial marketing just as the infomercial began airing, find a total sales function S for the product.

(c) Compute $S(10)$ and interpret.

Solution

(a) **Understand the Situation:** First, note that R is a rate function because it measures the rate of change in total sales with respect to time. In other words, R is the derivative of some function.

Evaluating the sales rate function at $t = 10$ yields

$$R(10) = 15\sqrt[3]{10} \approx 32.32\,\frac{\text{thousand units}}{\text{month}}$$

Interpret the Solution: This means that in the tenth month of airing the infomercial the total sales of the product were increasing at a rate of $32.32\,\frac{\text{thousand units}}{\text{month}}$, or about 32,320 units per month.

(b) **Understand the Situation:** Integrating the sales rate function R with respect to time t recovers the total sales function, S.

Proceeding with the integration gives us

$$S(t) = \int R(t)\,dt = \int 15\sqrt[3]{t}\,dt = \int 15t^{1/3}\,dt = 15\int t^{1/3}\,dt$$

$$= 15\left(\frac{1}{\frac{1}{3}+1}t^{(1/3)+1} + C\right) = 15\cdot\frac{3}{4}\cdot t^{4/3} + C = \frac{45}{4}\sqrt[3]{t^4} + C$$

Thus, the sales function has the general form $S(t) = \frac{45}{4}\sqrt[3]{t^4} + C$. Since we know that 500 units were sold when the infomercial began airing, this means that, when $t = 0$, $S(t) = 0.5$ [that is, the initial value is $(t, S(t)) = (0, 0.5)$]. Substituting these values in our general form, we get

$$S(t) = \frac{45}{4}\sqrt[3]{t^4} + C$$

$$0.5 = \frac{45}{4}\sqrt[3]{0^4} + C$$

$$0.5 = 0 + C$$

$$0.5 = C$$

Thus, the total sales function is

$$S(t) = \frac{45}{4}\sqrt[3]{t^4} + 0.5$$

where $S(t)$ represents the total sales in thousands.

(c) Evaluating the sales function at $t = 10$, we get

$$S(10) = \frac{45}{4}\sqrt[3]{10^4} + 0.5$$

$$= \frac{45}{4}\cdot\sqrt[3]{10{,}000} + 0.5 \approx 242.87 \text{ thousand units}$$

Interpret the Solution: This means that, during the tenth month of airing the infomercial the total sales were 242.87 thousand, or about 242,870 units of the product. ■

In Example 5, the rate function R is measured in $\frac{\text{thousand units}}{\text{month}}$ and the recovered sales function is measured in thousands of units. ***Generally, for a rate function, the units are given in*** $\frac{\textbf{dependent units}}{\textbf{independent unit}}$ ***and the function recovered is measured with dependent units.***

Recovering Business Functions

We have seen that given a rate function, more commonly called a derivative, we can recover the function from which it came. The rate functions, or derivatives, associated with the cost C, revenue R, and profit P functions are the **marginal functions**, denoted by MC, MR and MP, respectively. Let's review these functions in the following Toolbox.

From Your Toolbox

Let x be the number of units produced and sold of a certain product in a specific time interval.

- The **Marginal Cost Function** $MC(x) = C'(x)$ is the approximate cost of producing one additional unit at a production level x.
- The **Marginal Revenue Function** $MR(x) = R'(x)$ is the approximate loss or gain in revenue from producing and selling one additional unit at a production level x.
- The **Marginal Profit Function** $MP(x) = P'(x)$ is the approximate loss or gain in profit from producing and selling one additional unit at a production level x.

Recovering business functions involves integrating their corresponding marginal functions and evaluating at some given value. We illustrate this recovery in Example 6.

Example 6 Recovering a Profit Function

The Best Dressed Clothing Company finds that its marginal profit MP is linear and has the form $MP(q) = mq + b$, where m and b are constants. The company gets about \$171 additional profit from producing the 101st sport coat and \$169 in additional profit from producing the 151st sport coat in each production run.

(a) Determine the marginal profit function MP.

(b) Knowing that the company gets \$11,300 profit from 150 sport coats, find the profit function P.

Solution

(a) Understand the Situation: Since the marginal profit function is linear, we start by finding the slope m of the linear function. From the definition of marginal profit, we know that $MP(100) = 171$ and $MP(150) = 169$.

So the slope, or average rate of change, of the marginal profit function is

$$m = \frac{\text{change in } MP(q)}{\text{change in } q} = \frac{169 - 171}{150 - 100} = \frac{-2}{50} = -\frac{1}{25}$$

Now we can use the point–slope form with point $(q, MP(q)) = (100, 171)$.

$$y - y_1 = m(x - x_1)$$

$$y - 171 = -\frac{1}{25}(x - 100)$$

$$y = -\frac{1}{25}(x - 100) + 171$$

Replacing x with q and y with $MP(q)$ gives us

$$MP(q) = -\frac{1}{25}(q - 100) + 171$$

$$= -\frac{1}{25}q + 4 + 171 = -\frac{1}{25}q + 175$$

So the marginal profit function is $MP(q) = -\frac{1}{25}q + 175$.

(b) To recover the profit function P, we must integrate MP with respect to q. Mathematically, this means that

$$P(q) = \int MP(q)\,dq = \int \left(-\frac{1}{25}q + 175\right) dq$$
$$= -\frac{1}{25} \cdot \frac{1}{2}q^2 + 175q + C$$
$$= -\frac{1}{50}q^2 + 175q + C$$

To find the value of C, we use the point $(q, P(q)) = (150, 11{,}300)$ [that is, $q = 150$, $P(q) = 11{,}300$].

$$11{,}300 = -\frac{1}{50}(150)^2 + 175(150) + C$$
$$11{,}300 = -\frac{1}{50}(22{,}500) + 26{,}250 + C$$
$$11{,}300 = -450 + 26{,}250 + C$$
$$11{,}300 = 25{,}800 + C$$
$$-14{,}500 = C$$

Thus, the recovered profit function is $P(q) = -\frac{1}{50}q^2 + 175q - 14{,}500$. ■

SUMMARY

This section began with a discussion of the **antiderivative** and quickly led us to the **indefinite integral**. We examined how to integrate powers of functions, sums and differences of functions, and simple radical functions. Then we discussed how to recover a function when its rate function is given. We used this method to recover the business functions from the marginal business functions.

Important Integration Rules

- **Power Rule**: $\int x^n\,dx = \frac{1}{n+1}x^{n+1} + C,\ n \neq -1$
- **Constant Rule**: $\int k\,dx = kx + C$
- **Sum and Difference Rule**: $\int [f(x) \pm g(x)]\,dx = \int f(x)\,dx \pm \int g(x)\,dx$
- **Coefficient Rule**: $\int c \cdot f(x)\,dx = c \cdot \int f(x)\,dx$

SECTION 16.1 EXERCISES

In Exercises 1–10, determine if the function F is the general antiderivative of the function f.

1. $F(x) = 6x + C;\quad f(x) = 6$

2. $F(x) = \frac{1}{2}x + C;\quad f(x) = \frac{1}{2}$

3. $F(x) = x^8 + C;\quad f(x) = x^7$

4. $F(x) = \frac{1}{2}x^3 + C;\quad f(x) = x^2$

5. $F(t) = 8t + \frac{t^2}{2} + C;\quad f(t) = 8 + t$

6. $F(t) = 7t + et + C;\quad f(t) = 7 + e$

7. $F(x) = x + C;\quad f(x) = 0$

8. $F(x) = C;\quad f(x) = 0$

9. $F(x) = \frac{x^{11}}{11} + C;\quad f(x) = x^{10}$

10. $F(x) = -\dfrac{5}{6x^6} + C; \quad f(x) = \dfrac{5}{x^7}$

In Exercises 11–20, use the Power Rule for Integration to find the indefinite integrals.

11. $\int x^4\,dx$

12. $\int x^9\,dx$

13. $\int x^{2.31}\,dx$

14. $\int x^{0.27}\,dx$

15. $\int \dfrac{1}{t^3}\,dt$

16. $\int \dfrac{1}{t^{11}}\,dt$

✓ **17.** $\int \sqrt[4]{x^5}\,dx$

18. $\int \sqrt{x^5}\,dx$

19. $\int \dfrac{1}{\sqrt[3]{x}}\,dx$

20. $\int \dfrac{1}{\sqrt{x}}\,dx$

For Exercises 21–44, determine the indefinite integral.

21. $\int 0.4x^6\,dx$

22. $\int 0.5x^7\,dx$

23. $\int (2x+3)\,dx$

24. $\int (5x-2)\,dx$

25. $\int \left(\dfrac{2}{3}x+4\right)dx$

26. $\int \left(\dfrac{1}{2}t+5\right)dt$

✓ **27.** $\int (3t^2+2t+10)\,dt$

28. $\int (4x^4+5x-6)\,dx$

29. $\int (1-2x^2+3x^3)\,dx$

30. $\int (5-2x+x^3)\,dx$

31. $\int (6.21x^2+0.03x-4.01)\,dx$

32. $\int (0.03x^2-0.21x+4.02)\,dx$

33. $\int \left(\dfrac{1}{x^2}-\dfrac{3}{x^3}\right)dx$

34. $\int \left(\dfrac{3}{x^4}+6x^5\right)dx$

35. $\int (3+2\sqrt{x})\,dx$

36. $\int (\sqrt{x}-3x^{3/2})\,dx$

37. $\int \left(\dfrac{3}{x^3}+2x^{3/2}-4\right)dx$

38. $\int \left(x^{5/2}-\dfrac{4}{x^5}-\sqrt{x}\right)dx$

39. $\int \dfrac{3t^3-2t}{6t}\,dt$

40. $\int \dfrac{4x^4-5x^3}{x^2}\,dx$

41. $\int (0.1z^{-3}+2z^{-2}+z^3)\,dz$

42. $\int (5x^{-5}+3x^{-4})\,dx$

43. $\int (2t^{0.13}+5)\,dt$

44. $\int (0.1x^{0.318}+7x)\,dx$

In Exercises 45–58, solve the given rate function using the given value. Note that $f(a)=b$ corresponds to the ordered pair (a, b).

45. $f'(x)=-2; \quad f(0)=4$

46. $f'(x)=7; \quad f(0)=1$

47. $f'(x)=5x; \quad f(0)=0$

48. $f'(x)=3x; \quad f(0)=0$

49. $f'(x)=2x-3; \quad f(0)=4$

50. $g'(x)=5-6x; \quad g(0)=6$

51. $f'(t)=500-0.05t; \quad f(0)=40$

52. $\dfrac{dy}{dx}=4x^2-6x; \quad y(1)=0$

53. $s'(x)=2x^{-2}+3x^{-3}-1; \quad s(1)=2$

54. $M'(x)=\dfrac{2x^4-x}{x^3}; \quad M(1)=10$

55. $y'=\dfrac{5t+2}{\sqrt[3]{t}}; \quad (t, y)=(0, 1)$

56. $p'(t)=-\dfrac{10}{t^2}+t; \quad p=20$ when $t=1$

57. $R'(t)=\dfrac{1-t^4}{t^3}; \quad (t, R)=(1, 4)$

58. $y'(t)=\dfrac{\sqrt{t^3}-t}{(\sqrt{t})^3}; \quad y=4$ when $t=9$

Applications

59. Consider the marginal profit for producing q units of a product given by the linear function

$$MP(q)=40-0.05q$$

where $MP(q)$ is in dollars per unit.

(a) Knowing that $P=0$ when $q=0$, recover the profit function P.

(b) Use your solution from part (a) to find the total profit realized from selling 200 units of product.

60. Consider the marginal profit for producing x units of a product given by the linear function

$$MP(x)=-0.20x+40$$

where $MP(x)$ is in dollars per unit.

(a) Knowing that $P=-10$ when $x=0$, recover the profit function P.

(b) Use your solution from part (a) to find the total profit realized from selling 150 units of product.

61. The marginal revenue function for the FrontRide Bus Company is given by

$$MR(x) = R'(x) = 0.000045x^2 - 0.03x + 3.75 \qquad 0 \le x \le 500$$

(a) Knowing that $R(x) = 0$ when $x = 0$, recover the revenue function R.

(b) Find the price–demand function p for the bus company. Recall that the general formula for the revenue function is

$$\begin{aligned} R(x) &= (\text{quantity produced}) \cdot (\text{price of each unit}) \\ &= x \cdot p(x) \end{aligned}$$

(c) What should the price be when the demand is 100 passengers?

62. The daily marginal revenue function for the BlackDay Sunglasses Company is given by

$$MR(x) = R'(x) = 30 - 0.0003x^2 \qquad 0 \le x \le 540$$

where x represents the number of sunglasses produced and sold.

(a) Knowing that $R(x) = 1487.5$ when $x = 50$, recover the revenue function R.

(b) Find the price–demand function p for the sunglasses. Recall that the general formula for the revenue function is

$$\begin{aligned} R(x) &= (\text{quantity produced}) \cdot (\text{price of each unit}) \\ &= x \cdot p(x) \end{aligned}$$

(c) What should the price be when the demand is 250 sunglasses?

63. The marginal average cost function for producing x promotional banners is given by

$$MAC(x) = -\frac{100}{x^2} \qquad x > 0$$

(a) Knowing that it costs \$2.50 per banner to produce 100 banners, recover the average cost function AC.

(b) Knowing the average cost function AC from part (a), find the cost function C. Recall that, by definition, $AC(x) = \dfrac{C(x)}{x}$.

(c) Using the cost function from part (b), evaluate $C(100)$ and interpret.

64. The marginal average cost function for producing x QuickVid digital cameras is given by

$$MAC(x) = 0.03x^2 - 0.04x + 5 \qquad 0 < x \le 50$$

(a) Knowing that it costs \$422 per camera to produce 20 cameras, recover the average cost function AC.

(b) Knowing the average cost function AC from part (a), find the cost function C. Recall that, by definition, $AC(x) = \dfrac{C(x)}{x}$.

(c) Using the cost function from part (b), evaluate $C(20)$ and interpret.

65. The rate of change in the numbers of murders in the U.S. can be modeled by

$$f(x) = 0.24x^2 - 2.28x + 3.23 \qquad 1 \le x \le 7$$

where x is the number of years since 1991 and $f(x)$ is the murder rate measured in $\dfrac{\text{thousands of murders}}{\text{year}}$ (*Source:* U.S. Federal Bureau of Investigation, www.fbi.gov).

(a) Evaluate $f(5)$ and interpret.

(b) Knowing that there were 18.2 thousand murders in 1997, recover F, the function that gives the number of murders.

(c) Evaluate $F(5)$ and interpret.

66. The sales rate of Lotto tickets sold in the United States can be modeled by

$$R(x) = -8.67x^2 + 113.98x + 516.58 \qquad 1 \le x \le 20$$

where x is the number of years since 1979 and $R(x)$ is the rate of sales measured in $\dfrac{\text{millions of dollars}}{\text{year}}$ (*Source:* U.S. Statistical Abstract, www.census.gov/statab/www).

(a) Evaluate $R(17)$ and interpret.

(b) Knowing that there were \$8563 million in Lotto tickets sold in 1990, recover S, the ticket sales function.

(c) Evaluate $S(17)$ and interpret.

67. The rate of bankruptcy filings by businesses in the U.S. can be modeled by

$$b(x) = -1.89x^2 + 15.2x - 29.41 \qquad 1 \le x \le 7$$

where x is the number of years since 1992 and $b(x)$ is the rate of bankruptcy filings measured in $\dfrac{\text{thousands of businesses}}{\text{year}}$ (*Source:* U.S. Statistical Abstract, www.census.gov/statab/www).

(a) Evaluate $b(6)$ and interpret.

(b) Knowing that there were 66.428 thousand bankruptcies filed by businesses in 1993, recover B, the bankruptcy filing function.

(c) Evaluate $B(6)$ and interpret.

68. *(continuation of Exercise 66)* Graph the rate function R from Exercise 66 in the window [1, 20] by [−800, 1000]. Use the graph to analyze where the ticket sales are increasing and decreasing. Interpret.

69. *(continuation of Exercise 67)* Graph the rate function b from Exercise 67 in the window [1, 7] by [−20, 5]. Use the graph to analyze where the bankruptcy filings are increasing and decreasing. Interpret.

SECTION PROJECT

The Luv-n-Care Shop produces handcrafted art. The management of Luv-n-Care tracks its marginal costs for producing x units of art at various production levels, as shown in Table 16.1.1.

Table 16.1.1

Items Produced, x	Marginal Cost, $MC(x)$
10	50
20	56
30	62
40	67
50	73
60	77

(a) Use the regression capabilities of your calculator to compute a power regression model of the form $y = a \cdot x^b$ for the marginal cost function MC.

(b) Evaluate and interpret $MC(45)$.

(c) Evaluate $MC(40)$ and compare to the marginal cost given in the table. Explain why there is a difference in the marginal cost values.

(d) If the fixed costs are set at 1500 (that is, $C(x) = 1500$ when $x = 0$), recover the cost function C.

(e) Evaluate $C(45)$ and interpret.

Section 16.2 Area and the Definite Integral

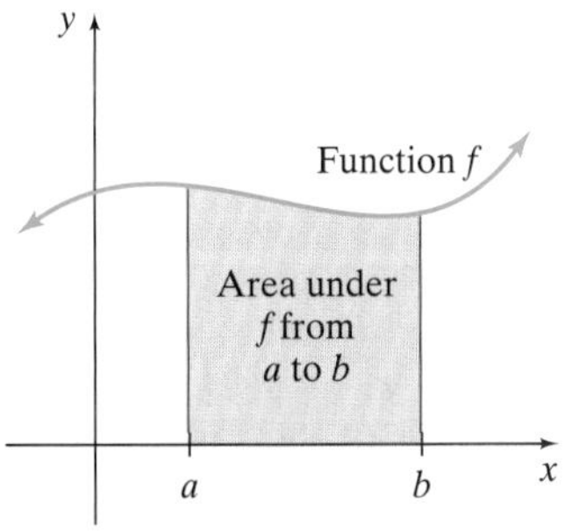

Figure 16.2.1

In Chapter 12, we learned how to complete a mathematical task that, up until then, was seemingly impossible—to find the slope of a curve at a single point. The calculus tool we used to solve this problem was the *derivative*. Now we must try to solve yet another problem that challenged mathematicians for years. To find the area A under a curve given by a function $y = f(x)$ between two endpoints a and b or, as we also say, on a *closed interval* $[a, b]$. An illustration of our challenge is shown in Figure 16.2.1.

In this section, we learn how to approximate the area as well as how to represent the area using a **definite integral**. In Section 16.3 we learn how to determine the area exactly. Then in Section 16.4 we apply the area under a curve concept to many varied applications.

Finding the Area under a Curve Numerically: The Left Sum Method

To begin to tackle the problem of finding the area under a curve, we rely on area formulas learned in geometry, such as the area of a rectangle. Recall that, for a rectangle with base b and height h, the area of the rectangle is given by

$$A = \text{base} \cdot \text{height} = b \cdot h$$

To convince ourselves that finding a definite area under a curve is possible, we will start with a continuous function $f(x) = x^2 + 1$ on a closed interval $[0, 2]$, as shown in Figure 16.2.2 on page 1014.

Since we know how to find the area of a rectangle, we can *approximate* the area under the curve by "slicing up" the whole area into thin rectangular strips, computing the area of each strip, and then adding up the areas. Let's say that we want to start with $n = 4$ of these rectangular strips. We start the process by dividing the interval $[a, b]$ into four equal parts. These parts are called **subintervals** and the width of each is called the **step size**, denoted by Δx. To get the step size, we simply divide the length of the closed interval $[a, b]$ by the number of subintervals n.

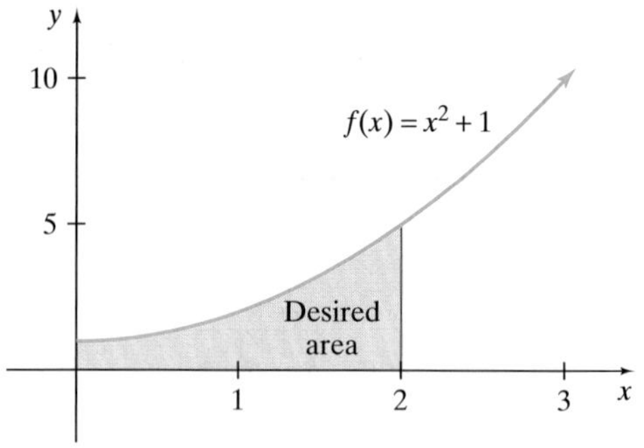

Figure 16.2.2 Area under $f(x) = x^2 + 1$ from $x = 0$ to $x = 2$.

Step Size

For a closed interval $[a, b]$ divided into n equally spaced subintervals, the **step size**, denoted by Δx, is given by

$$\Delta x = \frac{\text{length of interval } [a, b]}{\text{number of subintervals}} = \frac{b-a}{n}$$

Example 1 Computing a Step Size

For the interval $[0, 2]$ and $n = 4$:

(a) Calculate the step size Δx.

(b) List the x-coordinates, x_0, x_1, x_2, x_3, x_4 of the endpoints of the subintervals.

Solution

(a) Knowing that $a = 0$, $b = 2$, and $n = 4$, we get

$$\Delta x = \frac{b-a}{n} = \frac{2-0}{4} = \frac{1}{2}$$

(b) For the interval $[0, 2]$ and step size $\Delta x = \frac{1}{2}$, we get the coordinates shown in Table 16.2.1.

Table 16.2.1

x-Coordinate	x-Coordinate on $[0, 2]$ for $n = 4$
x_0	0
x_1	$0 + \frac{1}{2} = \frac{1}{2}$
x_2	$\frac{1}{2} + \frac{1}{2} = 1$
x_3	$1 + \frac{1}{2} = \frac{3}{2}$
x_4	$\frac{3}{2} + \frac{1}{2} = 2$

■

The coordinates in Table 16.2.1 are the x-coordinates of the endpoints of the bases of the four rectangles. We can either use the left endpoint or the right endpoint to determine the height of the rectangles. Picking the left endpoints to determine the heights is shown in Figure 16.2.3a, and choosing the right endpoints to determine the heights is shown in Figure 16.2.3b.

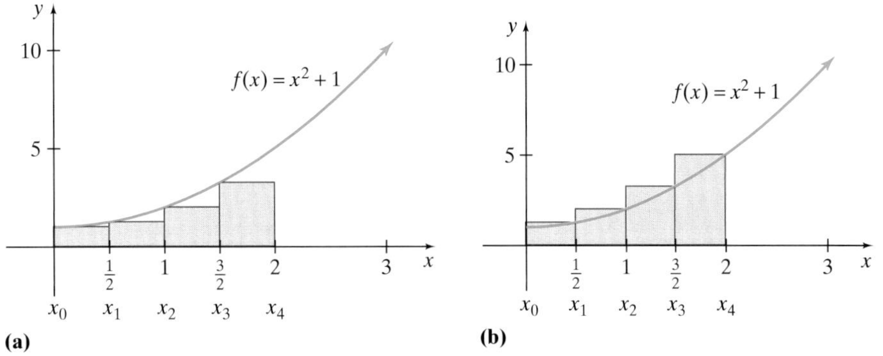

Figure 16.2.3 **(a) Left** endpoints used for the heights of rectangles. **(b) Right** endpoints used for the heights of rectangles.

For our first approximation, we will let the *left* endpoint of the rectangles determine the height. The height is found by computing the value of the function $f(x) = x^2 + 1$ at the left endpoint of each of the four rectangles. We now include these values in our table. (See Table 16.2.2.)

Table 16.2.2

x-Coordinate	x-Coordinate on $[0, 2]$ for $n = 4$	Rectangle Height $f(x_i)$
x_0	0	$f(x_0) = f(0) = 0^2 + 1 = 1$
x_1	$0 + \frac{1}{2} = \frac{1}{2}$	$f(x_1) = f\left(\frac{1}{2}\right) = \left(\frac{1}{2}\right)^2 + 1 = \frac{5}{4}$
x_2	$\frac{1}{2} + \frac{1}{2} = 1$	$f(x_2) = f(1) = 2$
x_3	$1 + \frac{1}{2} = \frac{3}{2}$	$f(x_3) = f\left(\frac{3}{2}\right) = \frac{13}{4}$

So it appears that, using these rectangles, the approximate area A under $f(x) = x^2 + 1$ on the interval $[0, 2]$ is

$$A \approx \begin{pmatrix}\text{area of first}\\ \text{rectangle}\end{pmatrix} + \begin{pmatrix}\text{area of second}\\ \text{rectangle}\end{pmatrix} + \begin{pmatrix}\text{area of third}\\ \text{rectangle}\end{pmatrix} + \begin{pmatrix}\text{area of fourth}\\ \text{rectangle}\end{pmatrix}$$

$$= \Delta x f(x_0) + \Delta x f(x_1) + \Delta x f(x_2) + \Delta x f(x_3)$$

$$= \frac{1}{2}(1) + \frac{1}{2}\left(\frac{5}{4}\right) + \frac{1}{2}(2) + \frac{1}{2}\left(\frac{13}{4}\right)$$

$$= \frac{1}{2} + \frac{5}{8} + 1 + \frac{13}{8}$$

$$= \frac{15}{4} \text{ square units} \quad \text{or} \quad 3.75 \text{ un}^2$$

This arca is shown in Figure 16.2.4. We call this way of approximating the area under the curve the **Left Sum Method**.

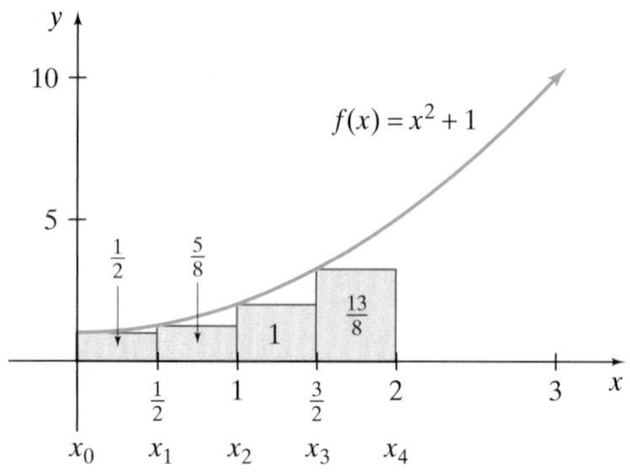

Figure 16.2.4

Left Sum Method

Given a continuous function $y = f(x)$ defined on a closed interval $[a, b]$, we compute the **Left Sum** using n equally spaced subintervals by the following steps:

1. Compute the step size Δx by calculating $\Delta x = \dfrac{b-a}{n}$.
2. Find the x-coordinates $x_0, x_1, x_2, \ldots, x_{n-1}$ of the left endpoints of each of the rectangles by starting with $x_0 = a$ and successively adding step size Δx.
3. Compute the Left Sum:

$$A = \begin{pmatrix}\text{area of first}\\ \text{rectangle}\end{pmatrix} + \begin{pmatrix}\text{area of second}\\ \text{rectangle}\end{pmatrix} + \begin{pmatrix}\text{area of third}\\ \text{rectangle}\end{pmatrix} + \cdots + \begin{pmatrix}\text{area of } n\text{th}\\ \text{rectangle}\end{pmatrix}$$
$$= \Delta x f(x_0) + \Delta x f(x_1) + \Delta x f(x_2) + \cdots + \Delta x f(x_{n-1})$$

Using **summation notation**, we could write the Left Sum Method formula in step 3 as

$$A = \sum_{i=0}^{n-1} \Delta x f(x_i) \quad \text{or} \quad A = \sum_{i=0}^{n-1} f(x_i)\Delta x$$

This method is sometimes called **rectangular approximation** and is a straightforward way to estimate the area under a curve.

The Right Sum Method

Another way to approximate area using rectangles is to use the *right* endpoint of each rectangle to determine the rectangles' heights.

Example 2 Approximating Area Using the Right Sum Method

For the function $f(x) = x^2 + 1$ on the interval $[0, 2]$, and $n = 4$:

(a) Determine the height of each rectangle using the *right* endpoint of the rectangles.

(b) Use the heights to approximate the area.

Solution

(a) Here we use the step size $\Delta x = \frac{1}{2}$ that we determined in Example 1. We get the heights shown in Table 16.2.3 for right endpoints x_1, x_2, x_3, x_4.

Table 16.2.3

x-Coordinate	x-Coordinate on $[0, 2]$ for $n = 4$	Rectangle Height $f(x_i)$
x_1	$0 + \frac{1}{2} = \frac{1}{2}$	$f(x_1) = f\left(\frac{1}{2}\right) = \left(\frac{1}{2}\right)^2 + 1 = \frac{5}{4}$
x_2	$\frac{1}{2} + \frac{1}{2} = 1$	$f(x_2) = f(1) = 2$
x_3	$1 + \frac{1}{2} = \frac{3}{2}$	$f(x_3) = f\left(\frac{3}{2}\right) = \frac{13}{4}$
x_4	$\frac{3}{2} + \frac{1}{2} = 2$	$f(x_4) = f(2) = 5$

(b) Now, approximating the area using the right endpoint gives us

$$\begin{aligned} A &\approx \begin{pmatrix}\text{area of first}\\ \text{rectangle}\end{pmatrix} + \begin{pmatrix}\text{area of second}\\ \text{rectangle}\end{pmatrix} + \begin{pmatrix}\text{area of third}\\ \text{rectangle}\end{pmatrix} + \begin{pmatrix}\text{area of fourth}\\ \text{rectangle}\end{pmatrix} \\ &= \Delta x f(x_1) + \Delta x f(x_2) + \Delta x f(x_3) + \Delta x f(x_4) \\ &= \frac{1}{2}\left(\frac{5}{4}\right) + \frac{1}{2}(2) + \frac{1}{2}\left(\frac{13}{4}\right) + \frac{1}{2}(5) \\ &= \frac{5}{8} + 1 + \frac{13}{8} + \frac{5}{2} \\ &= \frac{23}{4} \text{ square units} \quad \text{or} \quad 5.75 \text{ un}^2 \end{aligned}$$

■

Checkpoint 1 Now work Exercise 15.

Right Sum Method

Given a continuous function $y = f(x)$ defined on a closed interval $[a, b]$, we compute the **Right Sum** using n equally spaced subintervals by the following steps:

1. Compute the step size Δx by calculating $\Delta x = \dfrac{b - a}{n}$.
2. Find the x-coordinates $x_1, x_2, \ldots, x_n$ of the right endpoints of each rectangle by starting with $x_1 = a + \Delta x$ and successively adding step size Δx.
3. Compute the Right Sum:

$$\begin{aligned} A &= \begin{pmatrix}\text{area of first}\\ \text{rectangle}\end{pmatrix} + \begin{pmatrix}\text{area of second}\\ \text{rectangle}\end{pmatrix} + \begin{pmatrix}\text{area of third}\\ \text{rectangle}\end{pmatrix} + \cdots + \begin{pmatrix}\text{area of } n\text{th}\\ \text{rectangle}\end{pmatrix} \\ &= \Delta x f(x_1) + \Delta x f(x_2) + \Delta x f(x_3) + \cdots + \Delta x f(x_n) \end{aligned}$$

Using summation notation, we could write the Right Sum Method formula in step 3 as

$$A = \sum_{i=1}^{n} \Delta x f(x_i) \quad \text{or} \quad A = \sum_{i=1}^{n} f(x_i)\Delta x$$

The approximations using the two methods are compared in Figure 16.2.5 through Figure 16.2.8. Notice that in order to get better approximations for the area under the curve, we could add more and more rectangles and compute their sums.

Left Sum Method

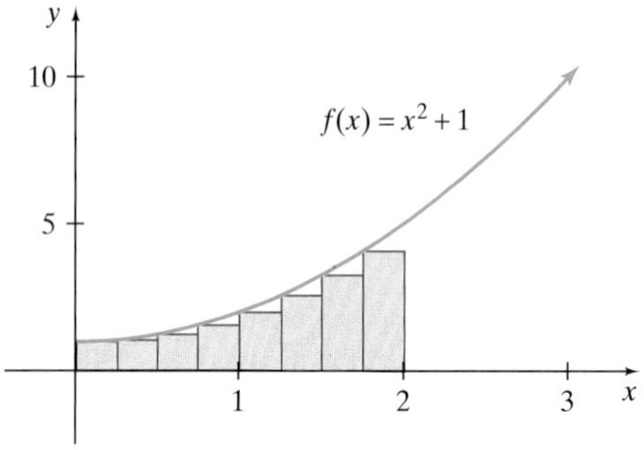

Figure 16.2.5 $n = 8$ rectangles. Area ≈ 4.1875

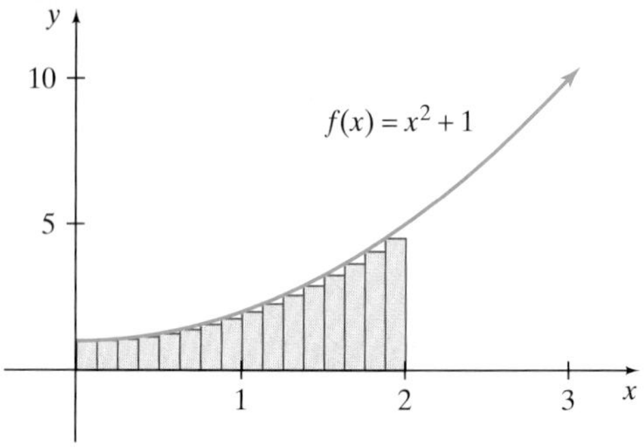

Figure 16.2.6 $n = 16$ rectangles. Area ≈ 4.421875

Right Sum Method

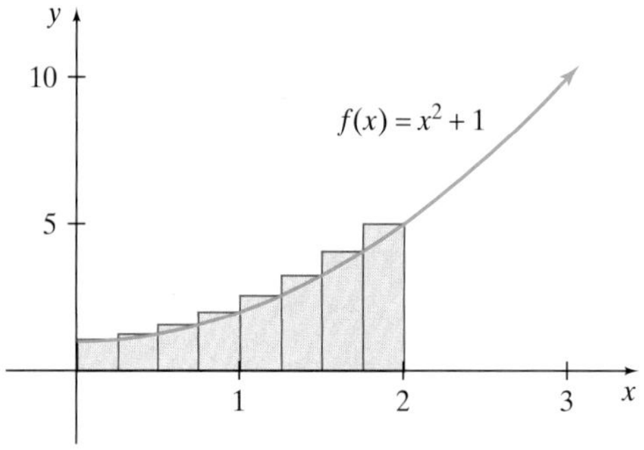

Figure 16.2.7 $n = 8$ rectangles. Area ≈ 5.1875

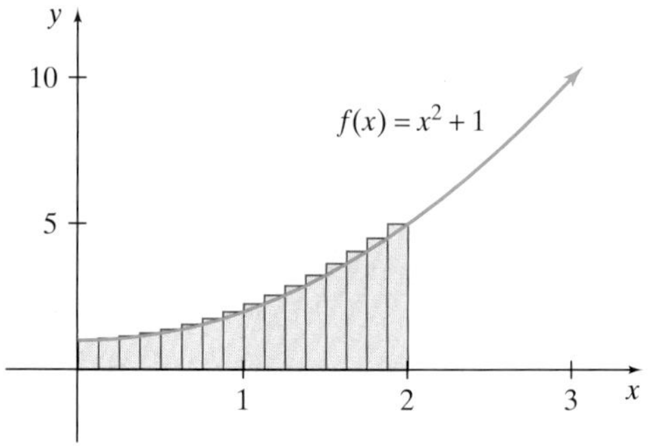

Figure 16.2.8 $n = 16$ rectangles. Area ≈ 4.921875

Technology Option

The shortcoming of adding more and more rectangles is that computing these approximations by hand can quickly become taxing. Fortunately, our calculators not only serve as graphing devices, but also are programmable handheld computers that complete these computational tasks in seconds. Following are calculator programs for the Left and Right Sum Methods written for the Texas Instruments TI-83. You may need to consult your calculator manual for programming your model of calculator.

Left Sum and Right Sum Method Calculator Programs

These programs approximate the area under the graph of a continuous function f on a closed interval $[a, b]$ for n rectangles.

The function $y = f(x)$ must be entered in Y_1 before the program is executed. When the program executes, the user is prompted to enter a left endpoint A, a right endpoint B, and number of rectangles N.

```
PROGRAM:LEFTSUM
Prompt A
Prompt B
Prompt N
(B-A)/N→D
0→T
For (I,0,N-1)
A+I* D→X
T+D*Y1→T
End
Disp "LEFT SUM APPROX",T
```

```
PROGRAM:RTSUM
Prompt A
Prompt B
Prompt N
(B-A)/N→D
0→T
For(I,1,N)
A+I*D→X
TD*Y1→T
End
Disp "RIGHT SUM APPROX",T
```

Using the Left and Right Sum Calculator Programs, we approximate the area under the graph of $f(x) = x^2 + 1$ on the interval [0, 2] in Table 16.2.4.

Table 16.2.4

N Rectangles	LEFTSUM Approximation	RTSUM Approximation
5	3.92	5.52
10	4.28	5.08
100	4.6268	4.7068
1,000	4.662668	4.670668
10,000	4.66626668	4.66706668

As the number of rectangles gets larger and larger, that is, $n \to \infty$, the approximations approach the actual area. The following formula, which results from letting $n \to \infty$, was derived by Georg Riemann, who was a disciple of the famous German mathematician Karl Friedreich Gauss. Riemann also proved that the end result is the same, even if the rectangles have different widths. This is why we name the width of the rectangle Δx_i and not just Δx.

Area Under a Curve Formula

If f is a continuous function defined on a closed interval $[a, b]$, then the area under the graph of f from a to b is

$$\left(\begin{array}{c}\text{Area under the}\\ \text{graph of } f \text{ from } a \text{ to } b\end{array}\right) = f(x_1)\Delta x_1 + f(x_2)\Delta x_2 + f(x_3)\Delta x_3 + \cdots$$

or, alternatively,

$$= \lim_{n\to\infty} \sum_{i=1}^{n} f(x_i) \cdot \Delta x_i$$

The Definite Integral

In the Area Under a Curve Formula, we saw that an alternative way to represent the area under f from a to b is

$$\lim_{n\to\infty} \sum_{i=1}^{n} f(x_i) \cdot \Delta x_i$$

We can also use the integral that has beginning and ending values, called a **definite integral** to represent this limit.

Definite Integral

For a continuous function f, the **definite integral** of f on the interval $[a, b]$ is

$$\int_a^b f(x)\,dx = \lim_{n\to\infty} \sum_{i=1}^{n} f(x_i)\cdot \Delta x_i$$

where a and b are the **limits of integration**.

Note: The definite integral is often referred to as a **limit of a sum**. Also, a definite integral can be used to represent the area under a curve.

Example 3 Applying the Definition of Definite Integral

Write a definite integral to represent the shaded area in Figure 16.2.9.

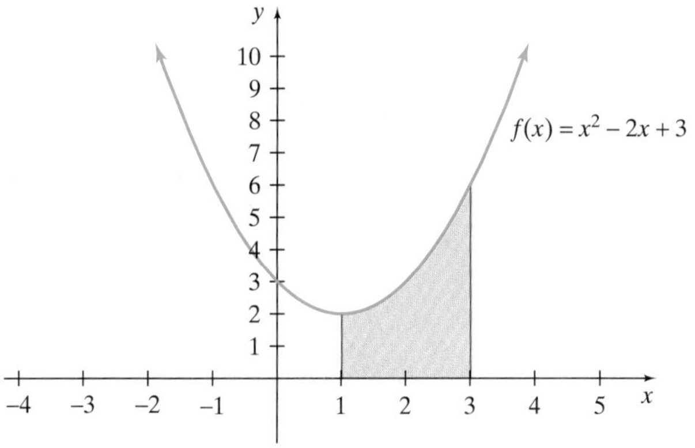

Figure 16.2.9

Solution

Understand the Situation: The shaded area is below the graph of the function $f(x) = x^2 - 2x + 3$ on the interval $[1, 3]$, which means that $a = 1$ and $b = 3$.

We represent the shaded area as

$$\int_1^3 (x^2 - 2x + 3)\,dx$$

■

In Section 16.1 we saw several properties of the indefinite integral. Many of the properties that we first saw in that section apply here, too.

Properties of the Definite Integral

1. The properties of the indefinite integral also apply to the definite integral.
   - $\int_a^b k\cdot f(x)\,dx = k\int_a^b f(x)\,dx$. This is the **Constant Rule** for definite integrals.
   - $\int_a^b [f(x) \pm g(x)]\,dx = \int_a^b f(x)\,dx \pm \int_a^b g(x)\,dx$. This is the **Sum and Difference Rule** for definite integrals.

2. If f is continuous on $[a, b]$ and there is a value c between a and b (see Figure 16.2.10), then

$$\int_a^b f(x)\,dx = \int_a^c f(x)\,dx + \int_c^b f(x)\,dx$$

This property, called the **Interval Addition Rule** for definite integrals, means that we can "break up" the limits of integration at a point $x = c$ that is between a and b.

3. $\int_a^a f(x)\,dx = 0$. This property tells us that the area under a single point on a curve is zero.

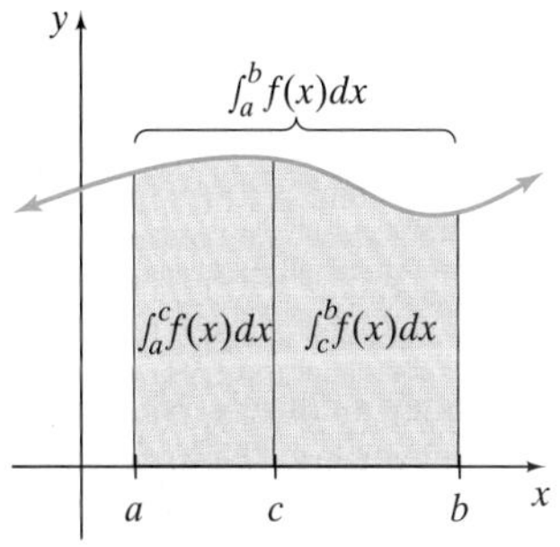

Figure 16.2.10 $\int_a^b f(x)\,dx = \int_a^c f(x)\,dx + \int_c^b f(x)\,dx.$

The Interval Addition Rule is particularly useful when integrating piecewise-defined functions. Its use is demonstrated in Example 4.

Example 4 Applying the Interval Addition Rule for Definite Integrals

Use the Interval Addition Rule to write a definite integral to represent the shaded region for the piecewise-defined function

$$f(x) = \begin{cases} x + 1, & x \le 2 \\ -2x + 7, & x > 2 \end{cases}$$

A graph of the piecewise-defined function is given in Figure 16.2.11.

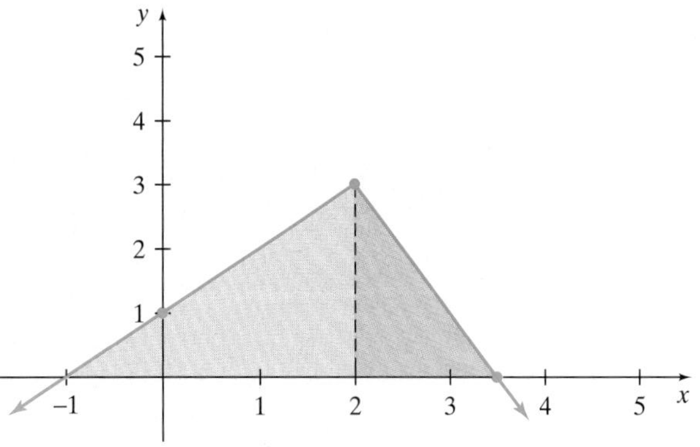

Figure 16.2.11 Graph of $f(x) = \begin{cases} x + 1, & x \le 2 \\ -2x + 7, & x > 2 \end{cases}$

Solution

We must break up the integral into two pieces. On the interval $[-1, 2]$ we integrate the expression $x+1$, and on the interval $\left[2, \frac{7}{2}\right]$ we integrate $-2x+7$. Since the "switch" in the expressions occurs at $x=2$, and 2 is between -1 and $\frac{7}{2}$, we can use the Interval Addition Rule to write

$$\int_{-1}^{7/2} f(x)\,dx = \int_{-1}^{2} (x+1)\,dx + \int_{2}^{7/2} (-2x+7)\,dx$$

■

Checkpoint 2 Now work Exercise 41.

SUMMARY

In this section we saw that the area under a curve could be approximated numerically by using the Left and Right Sum Methods. We also said that, as the number of rectangles becomes infinitely large, we get the definite integral.

- **Step Size**: $\Delta x = \dfrac{b-a}{n}$
- **Left Sum Method**: $\sum_{i=0}^{n-1} f(x_i)\Delta x$
- **Right Sum Method**: $\sum_{i=1}^{n} f(x_i)\Delta x$
- **Definite Integral**: $\int_a^b f(x)\,dx = \lim_{n\to\infty} \sum_{i=1}^{n} f(x_i)\cdot \Delta x_i$
- **Constant Rule**: $\int_a^b k\cdot f(x)\,dx = k\int_a^b f(x)\,dx$
- **Sum and Difference Rule**:

$$\int_a^b [f(x)\pm g(x)]\,dx = \int_a^b f(x)\,dx \pm \int_a^b g(x)\,dx$$

- **Interval Addition Rule**:

$$\int_a^b f(x)\,dx = \int_a^c f(x)\,dx + \int_c^b f(x)\,dx$$

SECTION 16.2 EXERCISES

In Exercises 1–10, determine the step size Δx for the given interval $[a, b]$ and number of subintervals n.

1. $[0, 2], n=4$

2. $[0, 5], n=5$

3. $[1, 4], n=6$

4. $[1, 2], n=4$

5. $[3, 5], n=4$

6. $[3, 6], n=6$

7. $[-1, 2], n=6$

8. $[-2, 2], n=4$

9. $[-3, 5], n=28$

10. $[-10, 3], n=52$

In Exercises 11–20, given f, the closed interval $[a, b]$, and n number of equally spaced subintervals:

(a) Calculate by hand the Left Sum to approximate the area under the graph of f.

(b) Calculate by hand the Right Sum to approximate the area under the graph of f.

11. $f(x)=5, [0, 2], n=4$

12. $f(x)=7, [0, 5], n=5$

13. $f(x)=3x, [1, 4], n=6$

14. $f(x)=2x, [1, 2], n=4$

✓ **15.** $f(x)=2x-1, [3, 5], n=4$

16. $f(x)=4x-2, [3, 6], n=6$

17. $f(x) = 2x^2, [0, 4], n = 4$

18. $f(x) = x^3, [0, 2], n = 4$

19. $f(x) = 4 - x^2, [-1, 2], n = 6$

20. $f(x) = x^2 + 2, [-2, 2], n = 4$

Complete the table for each of the functions f on [a, b] in Exercises 21–30 using the appropriate calculator program.

N Rectangles	LEFTSUM Approximation	RTSUM Approximation
10		
100		
1000		

21. $f(x) = 3x^2, [-2, 0]$

22. $f(x) = \dfrac{x^2}{2}, [0, 9]$

23. $f(x) = 8\sqrt{x}, [1, 4]$

24. $f(x) = 6x + \sqrt{x}, [1, 2]$

25. $f(x) = x^2 + 2x, [1, 4]$

26. $f(x) = 4x^2 - x, [0, 5]$

27. $f(x) = x^3 - 3x, [0, 4]$

28. $f(x) = x^3 + 2x, [0, 2]$

29. $f(x) = 0.2x^2 + 1.3x + 2.3, [0, 5]$

30. $f(x) = 0.3x^3 - 2.7x + 4.1, [0, 3]$

In Exercises 31–42, write a definite integral to represent the shaded area.

31.

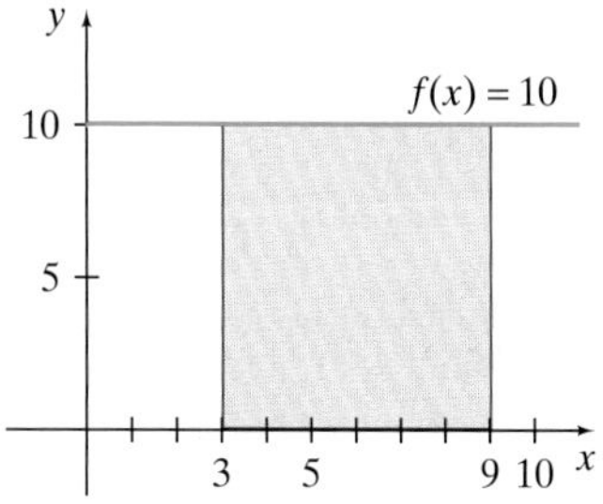

32.

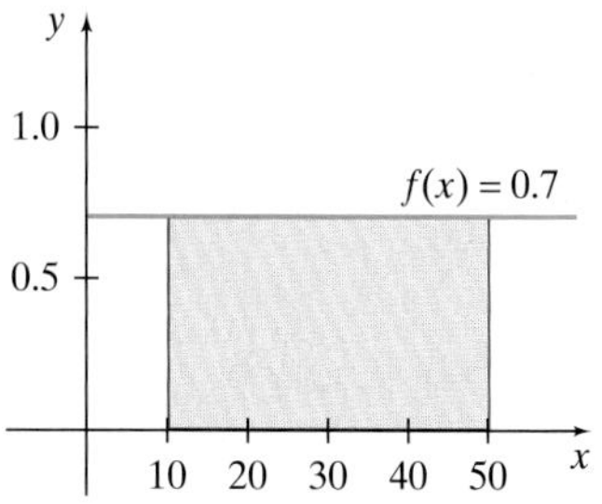

33.

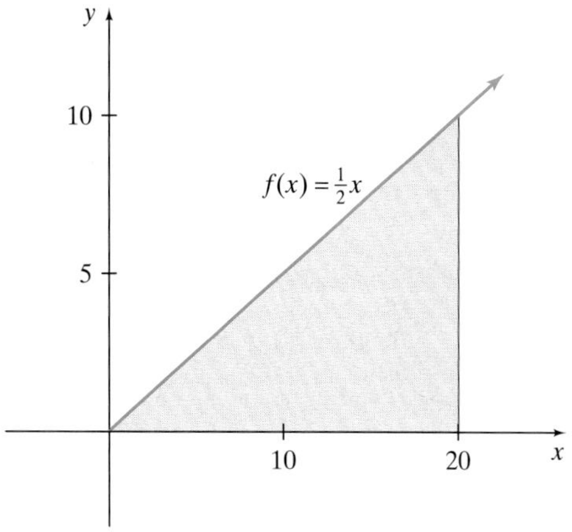

34.

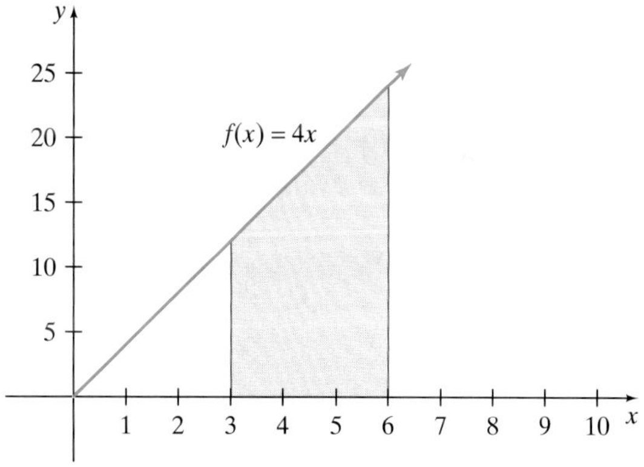

35.

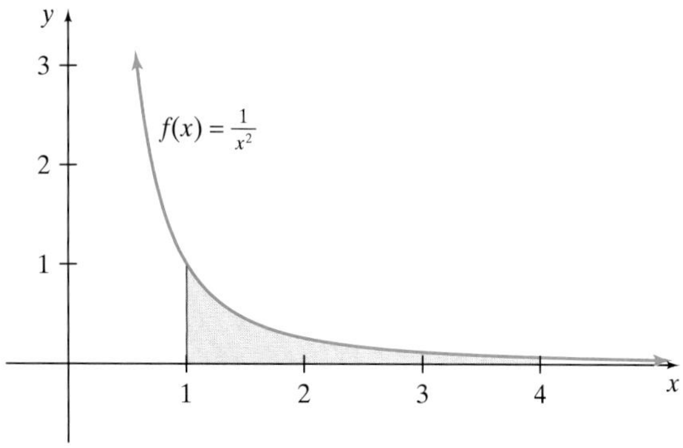

36.

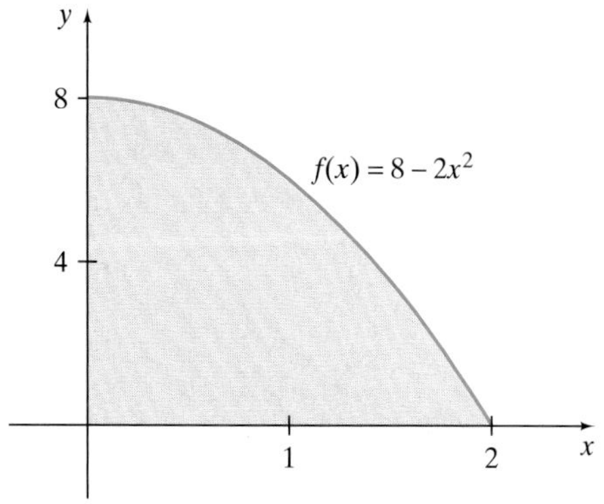

37.

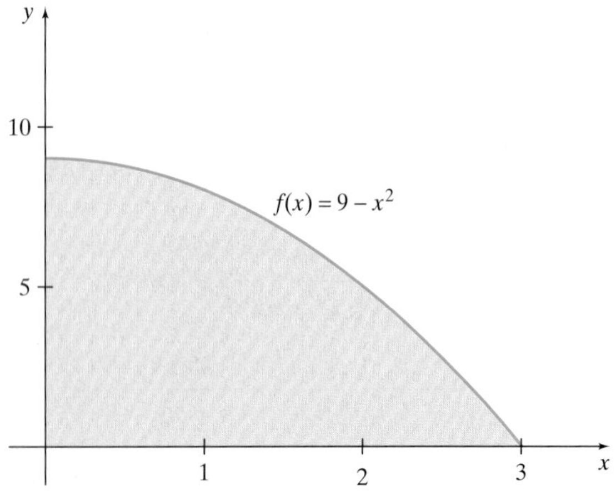

38.

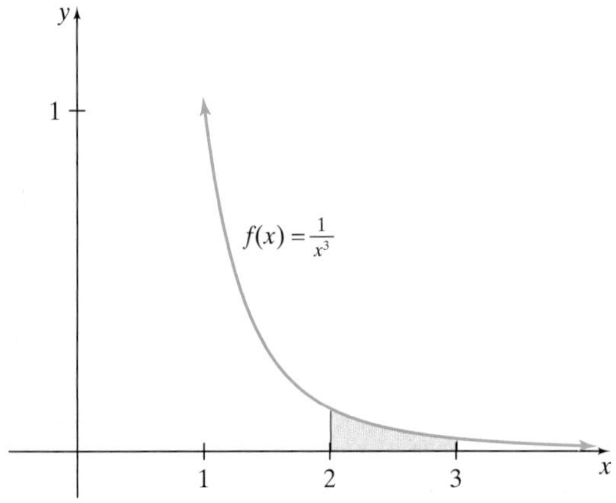

39.

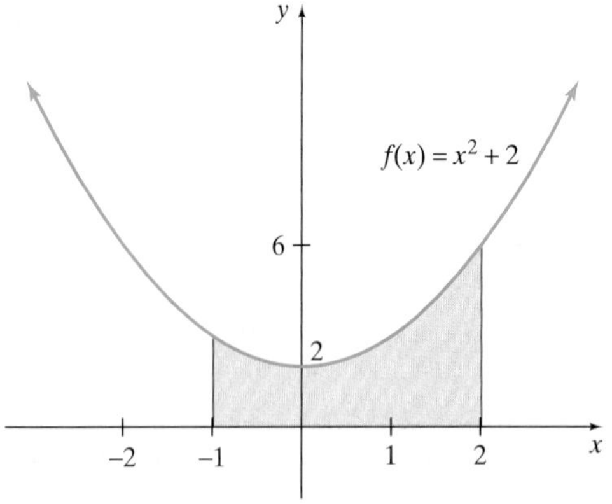

40.

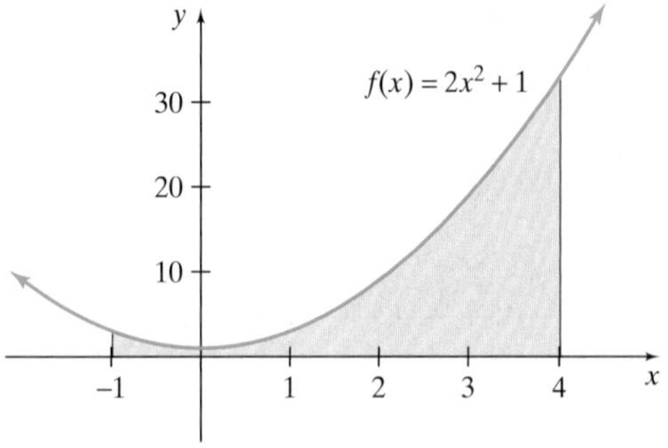

41.

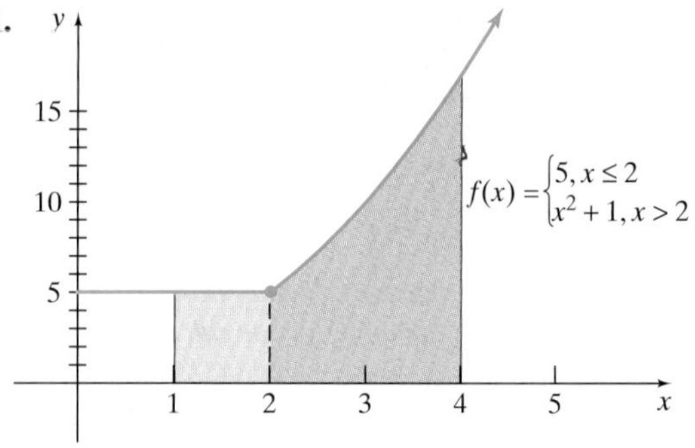

42.

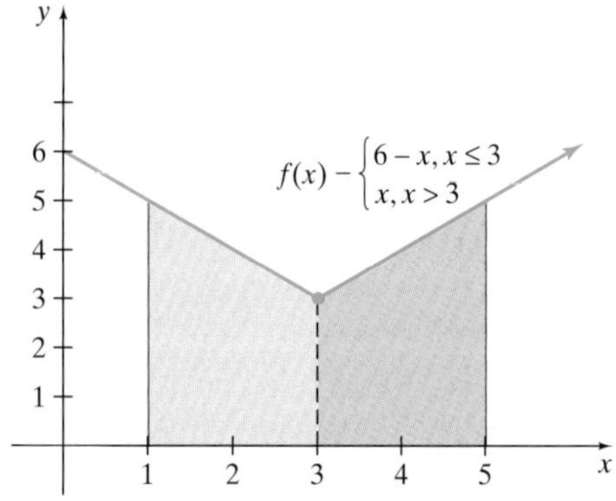

In Exercises 43–46, use a known geometric formula to determine the area under the curve for the following.

43. The curve in Exercise 31.

44. The curve in Exercise 32.

45. The curve in Exercise 33.

46. The curve in Exercise 34.

In Exercises 47–52, use the Left Sum and Right Sum Programs with $n = 100$ to approximate the area under the curve for the following.

47. The curve in Exercise 35.

48. The curve in Exercise 36.

49. The curve in Exercise 37.

50. The curve in Exercise 38.

51. The curve in Exercise 39.

52. The curve in Exercise 40.

In Exercises 53–60, sketch and shade the region given by the definite integrals.

53. $\int_0^3 x^2\,dx$ **54.** $\int_1^2 x^3\,dx$ **55.** $\int_1^4 \frac{1}{x^2}\,dx$

56. $\int_2^4 \frac{3}{x^4}\,dx$ **57.** $\int_1^5 \sqrt[3]{x}\,dx$ **58.** $\int_3^6 \sqrt{x^3}\,dx$

59. $\int_1^2 \frac{2}{\sqrt{x}}\,dx$ **60.** $\int_2^8 \frac{1}{\sqrt[3]{x}}\,dx$

SECTION PROJECT

Consider the function $f(x) = |x| + 1$.

(a) Graph the function and classify the function as symmetric to the x-axis, symmetric to the y-axis, symmetric to the origin, or not symmetric. Explain your answer.

(b) Rewrite f as a piecewise-defined function.

(c) Use a known geometric formula to determine the area under the graph of f on the interval $[-6, 6]$.

(d) Use a known geometric formula to determine the area under the graph of f on the interval $[0, 6]$. How does this answer compare to part (c)?

(e) How are the approximations in parts (c) and (d) related? How is this related to the symmetry found in part (a)? Explain.

Section 16.3 Fundamental Theorem of Calculus

In the previous section, we saw that the *area under a curve formula* and the *definite integral* are indeed the same. Mathematically, this means that

$$\lim_{n\to\infty} \sum_{i=1}^{n} f(x_i)\Delta x_i = \int_a^b f(x)\,dx$$

In this section, we will learn a powerful tool that is used to quickly evaluate a definite integral—the Fundamental Theorem of Calculus. This will allow us to quickly (and exactly) determine the area under the graph of a function on a closed interval. Then in the next section we will see why determining an area is so important.

Fundamental Theorem of Calculus

As you can imagine, it can be difficult to find the area under a curve by computing the left or right sum for thousands of rectangles. Then why is the Area under a Curve Formula so important? It turns out that the founders of calculus discovered an amazing relationship between the area under a curve and the definite integral. To understand this relationship, let's reconsider an example from Section 16.1.

Flashback

Indefinite Integrals Revisited

In part (a) of Example 3 in Section 16.1, we computed $\int (x^2 + 3)\,dx$ to be

$$F(x) = \frac{x^3}{3} + 3x + C$$

(a) For $F(x) = \dfrac{x^3}{3} + 3x + C$, evaluate $F(6) - F(3)$.

(b) Write a definite integral to represent the area under the graph of $f(x) = x^2 + 3$ on the interval $[3, 6]$. The Left Sum approximation for this definite integral is 71.9595045 for $n = 1000$ rectangles. Compare this to the result of part (a).

Flashback Solution

(a) Evaluating $F(6) - F(3)$, we get

$$F(6) - F(3) = \left[\frac{6^3}{3} + 3(6) + C\right] - \left[\frac{3^3}{3} + 3(3) + C\right]$$
$$= [72 + 18 + C] - [9 + 9 + C]$$
$$= 72 + 18 + C - 9 - 9 - C$$
$$= 72$$

(b) From the definition of definite integral, we get

$$\int_3^6 (x^2 + 3)\, dx$$

The Left Sum approximation for this definite integral, 71.9595045, for $n = 1000$ rectangles is very close to the result of part (a).

The Flashback shows that the difference $F(6) - F(3)$ is very close to the approximation for the area under the curve on the interval [3, 6]. This result is no accident. In fact, we can show that

$$\begin{pmatrix}\text{Area under}\\ f \text{ from } a \text{ to } b\end{pmatrix} = \lim_{n\to\infty} \sum_{i=1}^{n} f(x_i)\Delta x_i = F(b) - F(a)$$

Since $\lim_{n\to\infty} \sum_{i=1}^{n} f(x_i)\Delta x_i = \int_a^b f(x)\, d(x)$ (Definition of Definite Integral), it appears that we have

$$\int_a^b f(x)\, dx = F(b) - F(a)$$

where F is any antiderivative of f. If we can find an antiderivative of f, we can employ one of the most powerful theorems in all of mathematics, the **Fundamental Theorem of Calculus**. A proof of this theorem is in Appendix E.

Fundamental Theorem of Calculus

If f is a continuous function defined on a closed interval $[a, b]$ and F is an *antiderivative* of f, then

$$\int_a^b f(x)\, dx = F(b) - F(a)$$

where

- $\int_a^b f(x)\, dx$ is the **definite integral** of f on the interval $[a, b]$.
- a and b are the **limits of integration**.

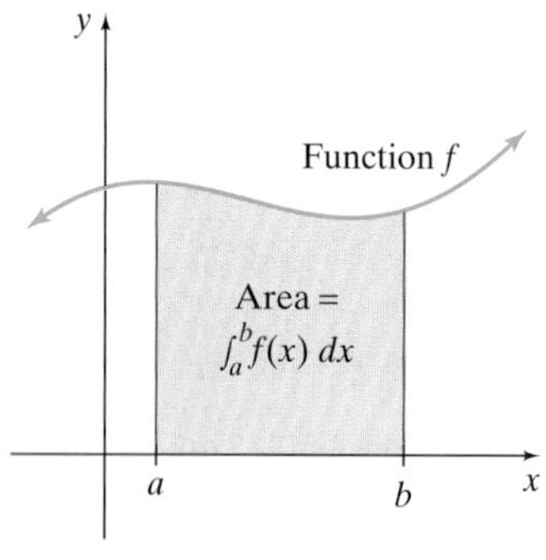

Figure 16.3.1

Before continuing, let's list a few key points of the Fundamental Theorem of Calculus.

- $\int_a^b f(x)\, dx = F(b) - F(a)$ gives the area under the graph of f from $x = a$ to $x = b$ is a special case of the Fundamental Theorem of Calculus. This is true when $f(x) \geq 0$ everywhere on $[a, b]$. See Figure 16.3.1.
- The difference $F(b) - F(a)$ is also written as $F(x)\big|_a^b$.

Example 1 Using the Fundamental Theorem of Calculus

Evaluate the definite integrals.

(a) $\int_{-2}^{0} (x^2 - 3x)\,dx$ **(b)** $\int_{1}^{5} \frac{1}{\sqrt{x}}\,dx$

Solution

(a)
$$\int_{-2}^{0} (x^2 - 3x)\,dx = \left(\frac{x^3}{3} - \frac{3x^2}{2}\right)\Bigg|_{-2}^{0}$$
$$= \left(\frac{0^3}{3} - \frac{3(0)^2}{2}\right) - \left(\frac{(-2)^3}{3} - \frac{3(-2)^2}{2}\right)$$
$$= (0 - 0) - \left(\frac{-8}{3} - 6\right)$$
$$= \frac{8}{3} + \frac{18}{3} = \frac{26}{3}$$

(b)
$$\int_{1}^{5} \frac{1}{\sqrt{x}}\,dx = \int_{1}^{5} x^{-1/2}\,dx = (2x^{1/2})\Big|_{1}^{5} = 2(\sqrt{x})\Big|_{1}^{5}$$
$$= 2(\sqrt{5} - \sqrt{1}) = 2\sqrt{5} - 2$$

✓ **Checkpoint 1**

Now work Exercise 15.

Technology Option

We can evaluate definite integrals on a graphing calculator using the FNINT command. To check our work in Example 1a, we begin by entering $f(x) = x^2 - 3x$ into Y_1. Figure 16.3.2 shows what must be entered into the calculator and Figure 16.3.3 shows the result when we execute the command.

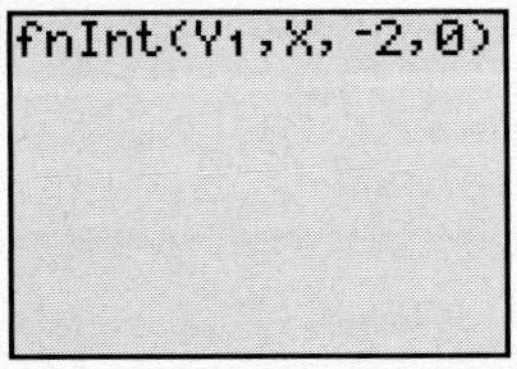

Figure 16.3.2

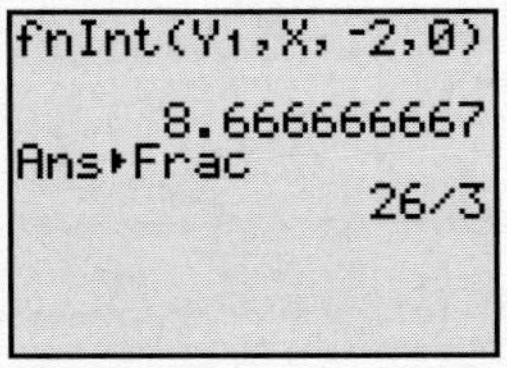

Figure 16.3.3

For more information on the FNINT command or to see how to use this command on your calculator, consult the online graphing calculator manual at www.prenhall.com/armstrong.

To see how the Fundamental Theorem of Calculus can be used to determine an area, let's take another look at an example from Section 16.2.

Example 2 Determining an Exact Area

Determine the exact area shaded in Figure 16.3.4.

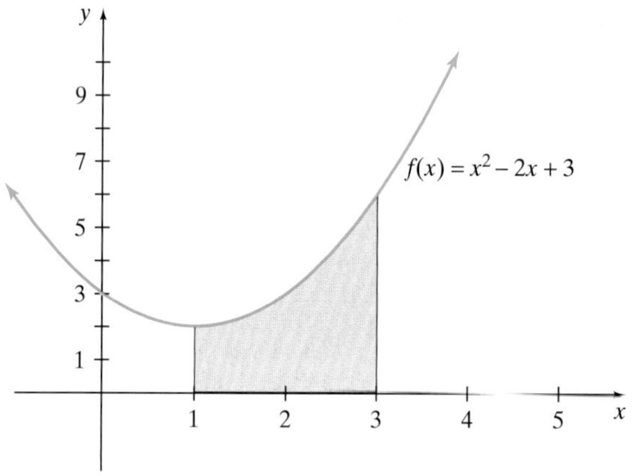

Figure 16.3.4

Solution

As we stated in Example 3 in Section 16.2, the corresponding definite integral for the shaded area is $\int_1^3 (x^2 - 2x + 3)\,dx$. Evaluating this integral yields

$$\int_1^3 (x^2 - 2x + 3)\,dx = \left(\frac{x^3}{3} - x^2 + 3x\right)\Big|_1^3$$

$$= \left[\frac{3^3}{3} - 3^2 + 3(3)\right] - \left[\frac{1^3}{3} - 1^2 + 3(1)\right]$$

$$= (9 - 9 + 9) - \left(\frac{1}{3} - 1 + 3\right)$$

$$= 9 - \frac{7}{3} = \frac{20}{3}\ \text{un}^2$$

Interpret the Solution: Thus, the area under the graph of the function $f(x) = x^2 - 2x + 3$ on the interval $[1, 3]$ is $\frac{20}{3}$ square units. ■

Checkpoint 2 Now work Exercise 33.

Technology Option

We can evaluate definite integrals on a graphing calculator using the $\int f(x)\,dx$ command. To check our work in Example 2, we begin by entering $f(x) = x^2 - 2x + 3$ into Y_1 and graph the function in the standard viewing window. See Figure 16.3.5. From the graphing window, we execute the $\int f(x)\,dx$ command. In Figure 16.3.6 we enter the lower limit of integration, in Figure 16.3.7

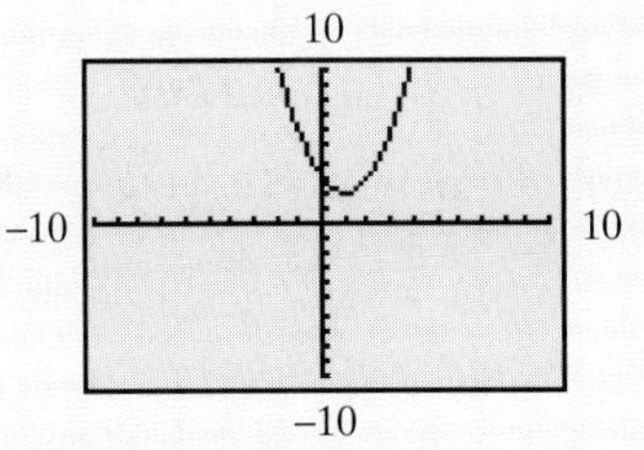

Figure 16.3.5

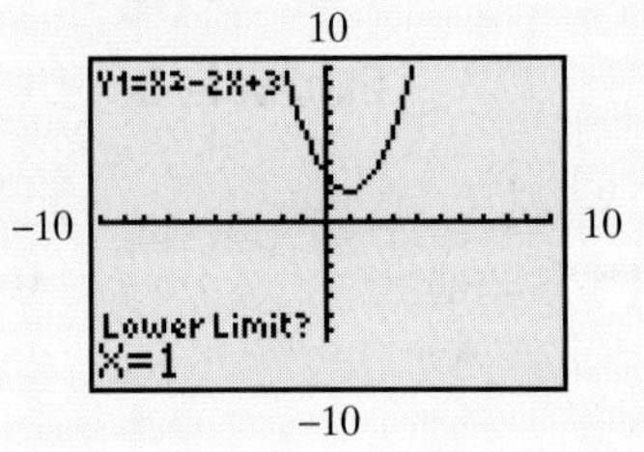

Figure 16.3.6

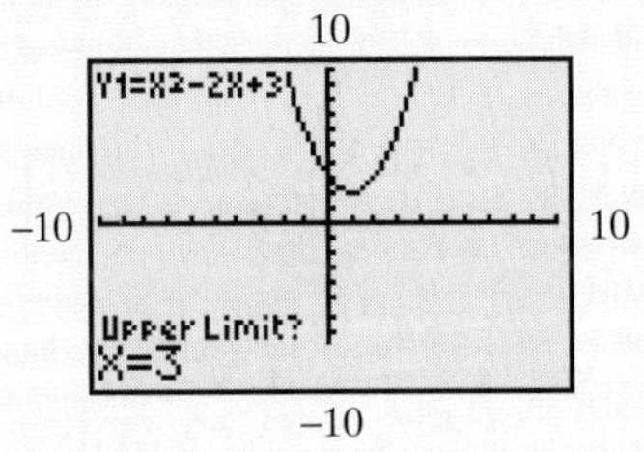

Figure 16.3.7

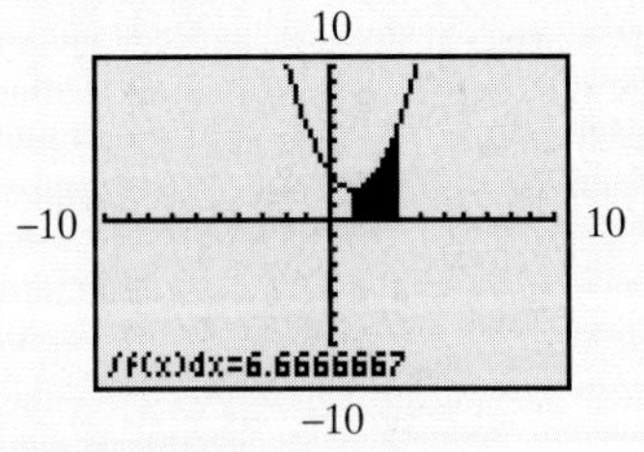

Figure 16.3.8

we enter the upper limit of integration, and in Figure 16.3.8 is the result of executing the $\int f(x)\,dx$ command. For more information on the $\int f(x)\,dx$ command or to see how to use this command on your calculator, consult the online graphing calculator manual at www.prenhall.com/armstrong.

The value of the definite integral depends not only on the limits of integration, but also on the orientation of the function relative to the x-axis.

- If $f(x) > 0$ on $[a, b]$, then $\int_a^b f(x)\,dx > 0$. Thus, if the graph of f is above the x-axis for all the values $a \le x \le b$, then the definite integral is positive.
- If $f(x) < 0$ on $[a, b]$, then $\int_a^b f(x)\,dx < 0$. Thus, if the graph of f is below the x-axis for all the values $a \le x \le b$, then the definite integral is negative.

These properties lead to the notions of net and gross area. Generally, the area between the graph of f and the x-axis on $[a, b]$ is called the **net area**. If we compute the absolute value of the region below the x-axis, we call this the **gross area**. See Figure 16.3.9.

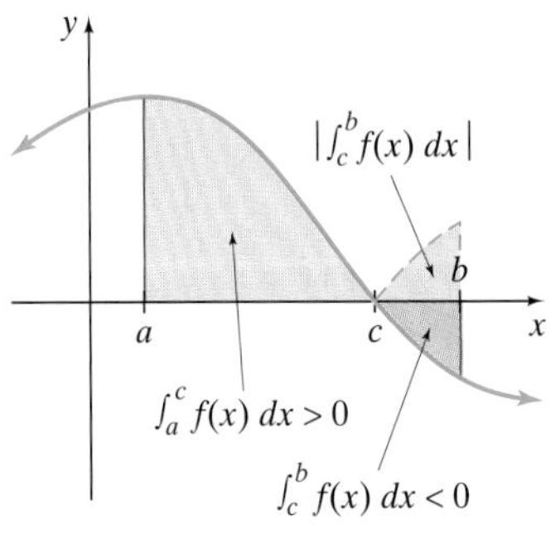

Figure 16.3.9

Example 3 Determining Net and Gross Area

Consider the cubic function $f(x) = x^3 - x^2 - 4x + 4$.

(a) Evaluate the net area by computing $\int_{-2}^{2} f(x)\,dx$. Graph the region.

(b) Calculate the gross area between the graph of f and the x-axis.

Solution

(a) Computing the net area, we get

$$\int_{-2}^{2} f(x)\,dx = \int_{-2}^{2}(x^3 - x^2 - 4x + 4)\,dx = \left(\frac{x^4}{4} - \frac{x^3}{3} - 2x^2 + 4x\right)\Bigg|_{-2}^{2}$$

$$= \left[\frac{(2)^4}{4} - \frac{(2)^3}{3} - 2(2)^2 + 4(2)\right] - \left[\frac{(-2)^4}{4} - \frac{(-2)^3}{3} - 2(-2)^2 + 4(-2)\right]$$

$$= \frac{4}{3} - \left(-\frac{28}{3}\right) = \frac{32}{3}\ \text{un}^2$$

A graph of the integrated region is shown in Figure 16.3.10

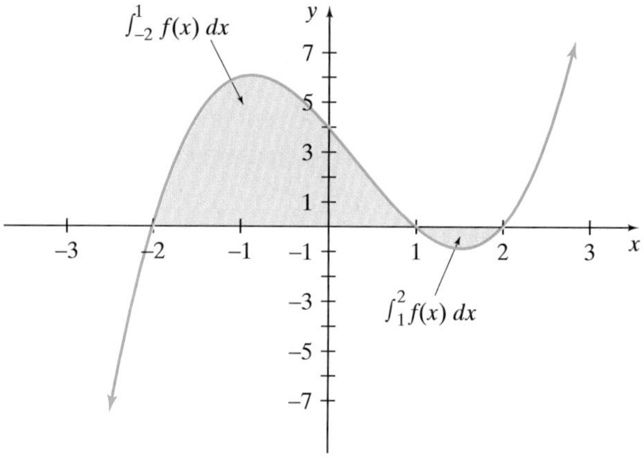

Figure 16.3.10 Net area of $\int_{-2}^{2}(x^3 - x^2 - 4x + 4)\,dx$.

(b) Understand the Situation: To get the gross area, we need to consider the interval [1, 2], where the graph of f lies below the x-axis. Since this integrated region has a negative value, we need to take the absolute value of the integral on [1, 2].

$$\text{Gross area} = \int_{-2}^{1}(x^3 - x^2 - 4x + 4)\,dx + \left|\int_{1}^{2}(x^3 - x^2 - 4x + 4)\,dx\right|$$

$$= \left(\frac{x^4}{4} - \frac{x^3}{3} - 2x^2 + 4x\right)\Bigg|_{-2}^{1} + \left|\left(\frac{x^4}{4} - \frac{x^3}{3} - 2x^2 + 4x\right)\Bigg|_{1}^{2}\right|$$

$$= \left[\frac{(1)^4}{4} - \frac{(1)^3}{3} - 2(1)^2 + 4(1)\right] - \left[\frac{(-2)^4}{4} - \frac{(-2)^3}{3} - 2(-2)^2 + 4(-2)\right]$$

$$+ \left|\left[\frac{(2)^4}{4} - \frac{(2)^3}{3} - 2(2)^2 + 4(2)\right] - \left[\frac{(1)^4}{4} - \frac{(1)^3}{3} - 2(1)^2 + 4(1)\right]\right|$$

$$= \frac{23}{12} - \left(-\frac{28}{3}\right) + \left|\frac{4}{3} - \frac{23}{12}\right| = \frac{45}{4} + \left|-\frac{7}{12}\right| = \frac{71}{6}\ \text{un}^2$$

Interpret the Solution: So the area bounded by the graph of the function f and the x-axis is $\frac{71}{6}$ un^2. This is illustrated in Figure 16.3.11.

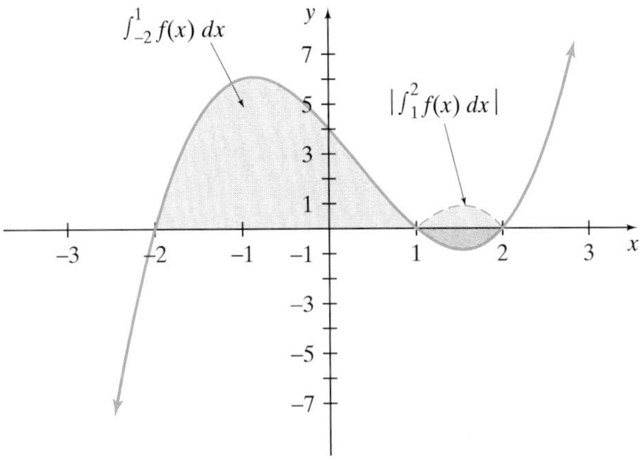

Figure 16.3.11 Gross area of $\int_{-2}^{2}(x^3 - x^2 - 4x + 4)\,dx$.

Checkpoint 3 Now work Exercise 57.

SUMMARY

In this section we were introduced to the Fundamental Theorem of Calculus and saw how we could use this theorem to evaluate definite integrals. We then applied the Fundamental Theorem of calculus to determine the exact area of a region under a curve.

- **Fundamental Theorem of Calculus**: If f is a continuous function defined on a closed interval $[a, b]$ and F is an antiderivative of f, then $\int_a^b f(x)\,dx = F(b) - F(a)$.
- If $f(x) > 0$ on $[a, b]$, then $\int_a^b f(x)\,dx > 0$.
- If $f(x) < 0$ on $[a, b]$, then $\int_a^b f(x)\,dx < 0$.

SECTION 16.3 EXERCISES

Exercises 1–24, evaluate the definite integrals.

1. $\int_1^4 3x\,dx$ **2.** $\int_1^2 2x\,dx$

3. $\int_0^2 5\,dx$ **4.** $\int_0^5 7\,dx$

5. $\int_3^5 (2x-1)\,dx$ **6.** $\int_3^6 (4x-2)\,dx$

7. $\int_0^4 2x^2\,dx$ **8.** $\int_0^2 x^3\,dx$

9. $\int_{-1}^2 (4-x^2)\,dx$ **10.** $\int_{-2}^2 (x^2+2)\,dx$

11. $\int_{-2}^0 3x^2\,dx$ **12.** $\int_0^9 \frac{x^2}{2}\,dx$

13. $\int_1^4 8\sqrt{x}\,dx$ **14.** $\int_1^2 (6x+\sqrt{x})\,dx$

✓ **15.** $\int_1^4 (x^3-3x)\,dx$ **16.** $\int_0^2 (x^3+2x)\,dx$

17. $\int_0^5 (0.2x^2+1.3x+2.3)\,dx$

18. $\int_0^3 (0.3x^3-2.7x+4.1)\,dx$

19. $\int_{-2}^{3} (5 + x - 6x^2)\,dx$ **20.** $\int_{1}^{4} (x^2 - 4x - 3)\,dx$

21. $\int_{0}^{2} (x^4 - 2x^3)\,dx$ **22.** $\int_{-2}^{3} (8x^3 + 3x - 1)\,dx$

23. $\int_{4}^{9} \frac{x-3}{\sqrt{x}}\,dx$ **24.** $\int_{-2}^{-1} \frac{2x-7}{x^3}\,dx$

In Exercises 25–30, use the FNINT *command on your calculator to evaluate the given definite integral.*

25. $\int_{1}^{4} \sqrt{x^2+3}\,dx$ **26.** $\int_{2}^{7} \sqrt{x^2+3x+1}\,dx$

27. $\int_{3}^{7} \frac{1}{x-2}\,dx$ **28.** $\int_{4}^{10} \frac{3}{x-3}\,dx$

29. $\int_{2}^{8} \frac{2x}{x^2+1}\,dx$ **30.** $\int_{2}^{6} \frac{3x^2}{x^3+1}\,dx$

In Exercises 31–42, use the Fundamental Theorem of Calculus to determine the area of the region under the graph of f on the given interval $[a, b]$.

31. $f(x) = 2x + 3$; $[2, 6]$

32. $f(x) = 3x + 2$; $[1, 5]$

✓ **33.** $f(x) = 8 - 2x^2$; $[0, 2]$

34. $f(x) = 9 - x^2$; $[0, 3]$

35. $f(x) = \frac{1}{x^2}$; $[1, 4]$

36. $f(x) = \frac{1}{x^3}$; $[1, 3]$

37. $f(x) = x^2 + 2$; $[-1, 2]$

38. $f(x) = 2x^2 + 1$; $[-1, 4]$

39. $f(x) = x^3 + 5$; $[1, 3]$

40. $f(x) = x^3 + 2$; $[0, 4]$

41. $f(x) = \sqrt{x} + 1$; $[1, 4]$

42. $f(x) = \sqrt{x} + x$; $[1, 4]$

43. Consider the absolute value function $f(x) = |x - 3|$.
(a) Graph f.
(b) Rewrite f as a piecewise-defined function.
(c) Use the Interval Addition Rule to evaluate the definite integral $\int_{-2}^{5} f(x)\,dx$.

44. Consider the absolute value function $f(x) = |x + 6|$.
(a) Graph f.
(b) Rewrite f as a piecewise-defined function.
(c) Use the Interval Addition Rule to evaluate the definite integral $\int_{-10}^{0} f(x)\,dx$.

45. Consider the absolute value function $f(x) = |x| - 4$.
(a) Graph f.
(b) Rewrite f as a piecewise-defined function.
(c) Determine the x-intercepts of the graph of f.
(d) Use the Interval Addition Rule to determine the net and gross areas of $\int_{-2}^{6} f(x)\,dx$.

46. Consider the absolute value function $f(x) = |x| - 2$.
(a) Graph f.
(b) Rewrite f as a piecewise-defined function.
(c) Determine the x-intercepts of the graph of f.
(d) Use the Interval Addition Rule to determine the net and gross areas of $\int_{0}^{9} f(x)\,dx$.

In Exercises 47–52, use the Interval Addition Rule to evaluate the indicated definite integral to find the area between the graph of g and the x-axis.

47. $\int_{-3}^{4} g(x)\,dx$, where $g(x) = \begin{cases} -x, & x < 0 \\ x, & x \geq 0 \end{cases}$

48. $\int_{-3}^{3} g(x)\,dx$, where $g(x) = \begin{cases} -x, & x \leq 1 \\ x - 2, & x > 1 \end{cases}$

49. $\int_{-1}^{2} g(x)\,dx$, where $g(x) = \begin{cases} 1 - x^2, & x < 0 \\ x + 1, & x \geq 0 \end{cases}$

50. $\int_{0}^{5} g(x)\,dx$, where $g(x) = \begin{cases} x^2, & x < 3 \\ x + 6, & x \geq 3 \end{cases}$

51. $\int_{0}^{5} g(x)\,dx$, where $g(x) = \begin{cases} 10 - x^2, & x \leq 2 \\ 6, & x > 2 \end{cases}$

52. $\int_{-4}^{0} g(x)\,dx$, where $g(x) = \begin{cases} 8, & x < -2 \\ x^2 + 4, & x \geq -2 \end{cases}$

For Exercises 53–62:

(a) Determine the x-intercepts of the graph of the function f.
(b) Integrate to determine the net area.
(c) Use the Interval Addition Rule to determine the gross areas for the indicated integral.

53. $f(x) = x + 2$; $\int_{-5}^{5} f(x)\,dx$

54. $f(x) = x - 5$; $\int_{0}^{10} f(x)\,dx$

55. $f(x) = 2\sqrt{x} - 4$; $\int_{0}^{9} f(x)\,dx$

56. $f(x) = \sqrt{x} - 1$; $\int_{0}^{4} f(x)\,dx$

✓ **57.** $f(x) = x^2 - 9$; $\int_{-1}^{4} f(x)\,dx$

58. $f(x) = x^2 - 4$; $\int_{-2}^{4} f(x)\,dx$

59. $f(x) = x^2 - x - 2$; $\int_{-1}^{4} f(x)\,dx$

60. $f(x) = x^2 + 4x + 3$; $\int_{-3}^{1} f(x)\,dx$

61. $f(x) = x^3 + x^2 - 2x$; $\int_{-2}^{2} f(x)\,dx$

62. $f(x) = x^3 - 2x^2 - 3x$; $\int_{-2}^{3} f(x)\,dx$

SECTION PROJECT

(a) Graph $f(x) = x^2 + 1$ on the interval $[-3, 3]$. Evaluate $\int_{-3}^{0}(x^2+1)\,dx$, $\int_{0}^{3}(x^2+1)\,dx$ and $\int_{-3}^{3}(x^2+1)\,dx$.

(b) Graph $f(x) = x^4 + 1$ on the interval $[-2, 2]$. Evaluate $\int_{-2}^{0}(x^4+1)\,dx$, $\int_{0}^{2}(x^4+1)\,dx$ and $\int_{-2}^{2}(x^4+1)\,dx$.

(c) Graph $f(x) = x^3$ on the interval $[-2, 2]$. Evaluate $\int_{-2}^{0} x^3\,dx$, $\int_{0}^{2} x^3\,dx$ and $\int_{-2}^{2} x^3\,dx$.

(d) Graph $f(x) = x^5$ on the interval $[-2, 2]$. Evaluate $\int_{-2}^{0} x^5\,dx$, $\int_{0}^{2} x^5\,dx$ and $\int_{-2}^{2} x^5\,dx$.

(e) The functions in parts (a) and (b) are called **even** functions. Observe that an even function is symmetric with respect to the y-axis. For f an even function, what is the relationship between $\int_{-a}^{a} f(x)\,dx$ and $\int_{0}^{a} f(x)\,dx$?

(f) The functions in parts (c) and (d) are called **odd** functions. Observe that an odd function is symmetric with respect to the origin. For f an odd function, what is the value of $\int_{-a}^{a} f(x)\,dx$?

Section 16.4 Problem Solving: Integral Calculus and Total Accumulation

In Section 16.2 we learned how to approximate area under a curve using rectangles in the Left Sum and Right Sum Method. We also introduced the **definite integral** and used it to represent the area under a curve. In Section 16.3 we presented one of the most remarkable results in mathematics, **The Fundamental Theorem of Calculus**. The Fundamental Theorem of Calculus supplied us with a way to evaluate definite integrals and determine an **exact** answer. In turn this allowed us to determine areas exactly.

In this section, we focus on why determining an area under a curve is important for us. Specifically, we focus on determining the area under the graph of a **rate of change function**, or simply a **rate function**, and interpreting what it gives us, a **total accumulation**. We will encounter many varied applications of this concept in the remainder of Chapter 16 as well as Chapter 17. Determining the area under the graph of a rate of change function is a powerful application of the calculus and specifically The Fundamental Theorem of Calculus.

Thinking About the Integral as a Continuous Sum

In Section 16.2 we stated that the definite integral can be thought of as a limit of a sum. The definite integral can also be used to represent area under the graph of a function. When a definite integral is used to represent the area under the graph of a rate of change function (which we also call a rate function), evaluating it using The Fundamental Theorem of Calculus produces a *continuous sum of a rate function over a closed interval.* To see what we mean by this statement, as well as why this observation is so important, let's consider the SafeMate Company. The SafeMate Company manufactures home emergency kits and has determined that the cost of manufacturing these kits is given by

$$C(x) = 0.02x^2 + 3x + 800 \qquad 0 \le x \le 200$$

where x represents the number of emergency kits produced daily and $C(x)$ is the cost, in dollars. In order to calculate the additional cost of increasing production from 100 to 160 emergency kits, we can compute

$$\begin{aligned} &C(160) - C(100) \\ &= [0.02(160)^2 + 3(160) + 800] - [0.02(100)^2 + 3(100) + 800] \\ &= 1792 - 1300 = 492 \end{aligned}$$

So it costs an additional \$492 to increase production from 100 to 160 units. Now let's examine this scenario using the **marginal cost function**. We know that the marginal cost function is $C'(x) = MC(x) = 0.04x + 3$. Let's determine the area under the graph of MC on the interval [100, 160]. See Figure 16.4.1.

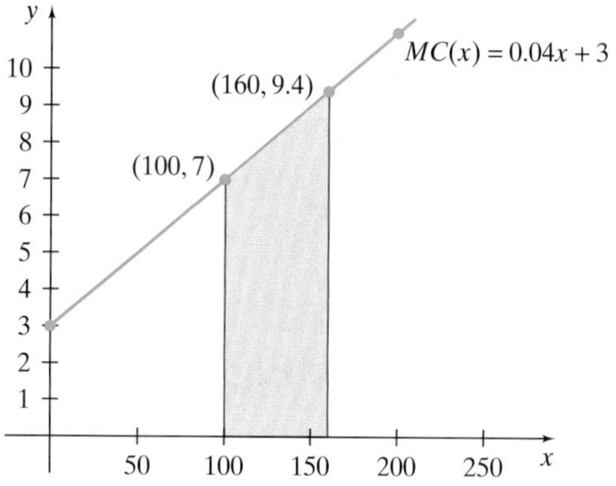

Figure 16.4.1

We can represent the area in Figure 16.4.1 with the definite integral $\int_{100}^{160} (0.04x + 3)\,dx$. We determine the exact area by utilizing the Fundamental Theorem of Calculus as follows.

$$\begin{aligned} &\int_{100}^{160} (0.04x + 3)\,dx \\ &= \left(\frac{0.04}{2}x^2 + 3x\right)\Bigg|_{100}^{160} \\ &= [0.02(160)^2 + 3(160)] - [0.02(100)^2 + 3(100)] \\ &= 992 - 500 \\ &= 492 \end{aligned}$$

Notice that this result is exactly the same solution that we had when we computed the additional cost of increasing production from 100 to 160 emergency home kits. Also notice that the marginal cost function, MC, is a type of **rate function** since it gives the instantaneous rate of change in costs. (Consult the Toolbox to the left.)

Interpreting the area under the graph of a rate function is a powerful tool in applied calculus. To ensure that we understand the power and meaning of this tool, let's try another scenario in Example 1.

From Your Toolbox

A function that gives an instantaneous rate of change is a **rate function**. (A rate function may be considered a derivative since the derivative gives us an instantaneous rate of change.)

Example 1 Analyzing Population Growth

The population of Atlanta, Georgia can be modeled by

$$F(x) = 0.098x + 2.857 \qquad 1 \le x \le 9$$

where x is the number of years since 1989 and $F(x)$ is the population in millions (*Source:* U.S. Census Bureau, www.census.gov).

(a) Evaluate and interpret $F(7) - F(2)$.

(b) Evaluate $\int_2^7 F'(x)\,dx$ and compare to part (a).

Solution

(a) Understand the Situation: Since $F(7)$ will give us the population of Atlanta in 1996 and $F(2)$ will give us the population of Atlanta in 1991, computing $F(7) - F(2)$ gives us the total increase in Atlanta's population from 1991 to 1996. Note that these populations are in millions.

We compute $F(7) - F(2)$ as

$$F(7) - F(2) = [0.098(7) + 2.857] - [0.098(2) + 2.857] = 0.49$$

Interpret the Solution: This mean that from 1991 to 1996, the total increase in Atlanta's population was about 0.49 million people or 490,000 people.

(b) Since $F(x) = 0.098x + 2.857$ we have $F'(x) = 0.098$. So we have

$$\begin{aligned}\int_2^7 F'(x)\,dx &= \int_2^7 0.098\,dx\\ &= 0.098x\Big|_2^7\\ &= 0.098(7) - 0.098(2)\\ &= 0.49\end{aligned}$$

We notice that this is exactly the same result as in part (a). Notice that the definite integral $\int_2^7 0.098\,dx$ corresponds to the region shaded in Figure 16.4.2. This illustrates that the area under the graph of a rate function can be interpreted as giving a total accumulation.

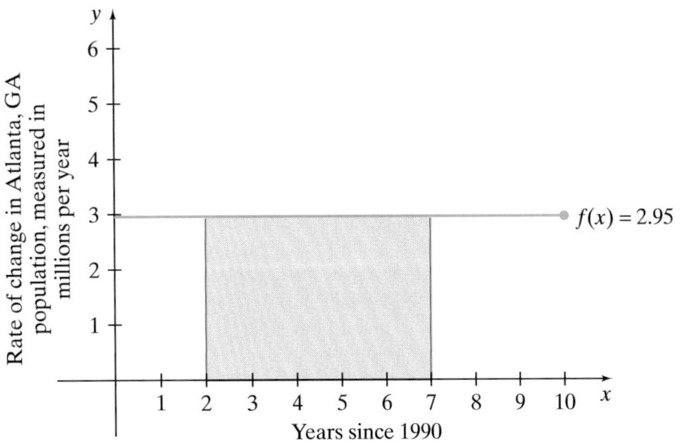

Figure 16.4.2

■

The beginning of the section, along with Example 1, illustrate an important interpretation of the area under the graph of a rate of change function. We summarize this interpretation as follows.

Interpreting the Area Under the Graph of a Rate Function

Let f be a rate function on the interval $[a, b]$ where $[a, b]$ is in the reasonable domain of f. Then the area under the graph of f on $[a, b]$ gives the total accumulation of F from a to b, where F is an antiderivative of f.

Note:
- The area under the graph of f on $[a, b]$ is given by $\int_a^b f(x)\,dx$.
- We evaluate $\int_a^b f(x)\,dx$ by using the Fundamental Theorem of Calculus. That is, $\int_a^b f(x)\,dx = F(b) - F(a)$ where F is an antiderivative.

Table 16.4.1 illustrates some of the uses of this property.

Table 16.4.1

Diagram	Accumulation
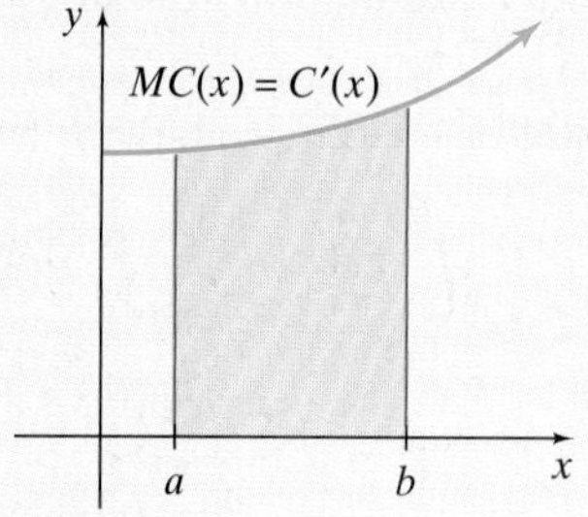 **Figure 16.4.3**	$\int_a^b MC(x)\,dx$ = Total increase in cost of producing from a to b units.
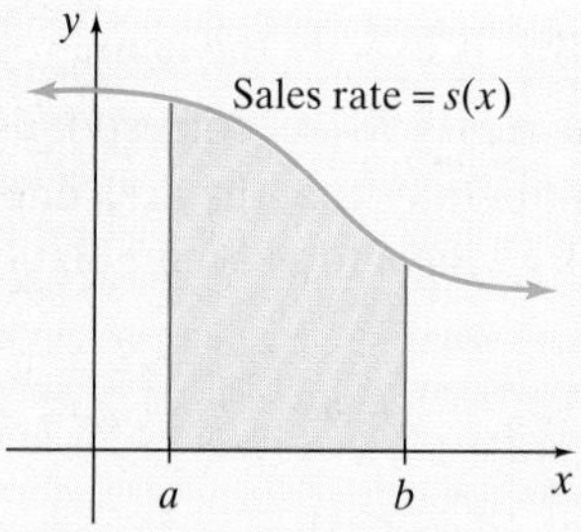 **Figure 16.4.4**	$\int_a^b s(x)\,dx$ = The total increase in sales from time a to b.
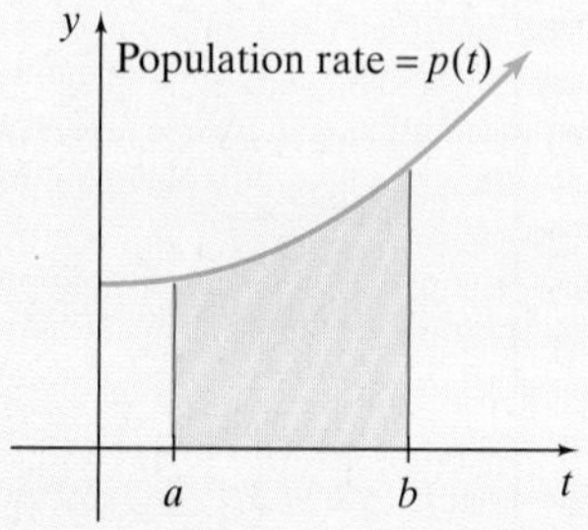 **Figure 16.4.5**	$\int_a^b p(t)\,dt$ = The total increase in population from time a to b.

Example 2 Interpreting a Total Accumulation

The rate of change in U.S. corporate profits, after taxes, can be modeled by

$$f(x) = -0.52x + 37.45 \qquad 0 \le x \le 9$$

where x represents the number of years since 1990 and $f(x)$ represent the rate of change in corporate profits measured in $\frac{\text{billions of dollars}}{\text{year}}$ (*Source:* U.S. Bureau of Economic Analysis, www.bea.doc.gov).

(a) Graph f and shade the area that represents the increase in corporate profits after taxes from 1991 to 1996.

(b) Write a definite integral to determine the shaded area in part (a). Use the Fundamental Theorem of Calculus to evaluate the definite integral and interpret.

Solution

(a) **Understand the Situation:** Since we need to find the area that represents the years 1991 ($x = 1$) to 1996 ($x = 6$), we shade below the graph of f and above the x-axis on the interval $[1, 6]$.

The desired region is shown in Figure 16.4.6.

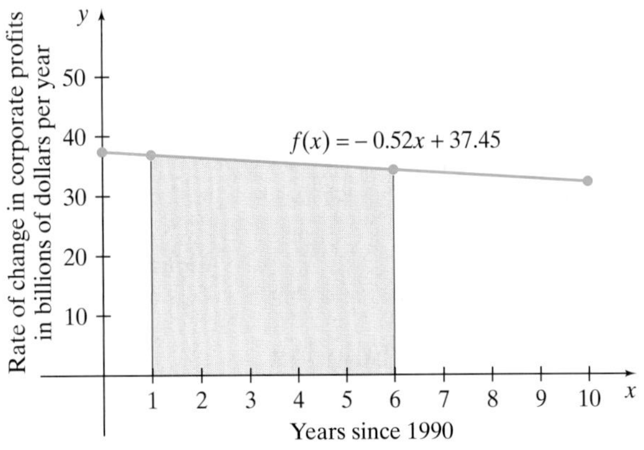

Figure 16.4.6

(b) The definite integral that represents the shaded region in Figure 16.4.6 is

$$\int_1^6 (-0.52x + 37.45)\,dx$$

Evaluating this definite integral gives us

$$\begin{aligned}\int_1^6 (-0.52x + 37.45)\,dx &= \left(\frac{-0.52}{2}x^2 + 37.45x\right)\Big|_1^6 \\ &= \left[\left(\frac{-0.52}{2}(6)^2 + 37.45(6)\right) - \left(\frac{-0.52}{2}(1)^2 + 37.45(1)\right)\right] \\ &= 215.34 - 37.19 \\ &= 178.15\end{aligned}$$

Interpret the Solution: This means that the total increase in U.S. corporate profits, after taxes, from 1991 to 1996 was about \$178.15 billion. ■

Checkpoint 1 Now work Exercise 9.

Example 3 Interpreting the Area Under the Graph of a Rate Function

The rate of change in the amount of ferrous metals that was recycled from 1990 to 2000 can be modeled by

$$f(x) = 0.06x^2 - 0.92x + 5.03 \qquad 0 \le x \le 10$$

where x represents the number of years since 1990 and $f(x)$ represents the rate of change in the amount of ferrous material recycled, measured in millions of tons per year (*Source:* U.S. Environmental Protection Agency, www.epa.gov). Determine the area of the shaded region in Figure 16.4.7. Interpret this area as a total accumulation.

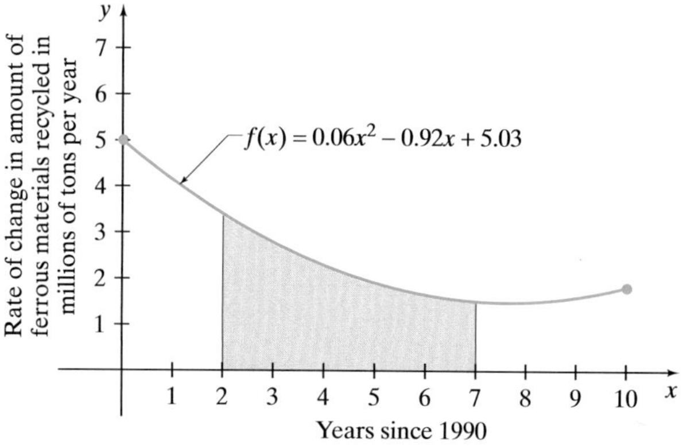

Figure 16.4.7

Solution

Understand the Situation: We first note that the interval [2, 7] corresponds to the years 1992 to 1997. The area of the shaded region is represented by the definite integral $\int_2^7 (0.06x^2 - 0.92x + 5.03)\,dx$. We integrate and use the Fundamental Theorem of Calculus to evaluate this definite integral.

Performing the integration gives us

$$\int_2^7 (0.06x^2 - 0.92x + 5.03)\,dx$$

$$= \left(\frac{0.06}{3}x^3 - \frac{0.92}{2}x^2 + 5.03x\right)\Big|_2^7$$

$$= \left[\left(\frac{0.06}{3}(7)^3 - \frac{0.92}{2}(7)^2 + 5.03(7)\right) - \left(\frac{0.06}{3}(2)^3 - \frac{0.92}{2}(2)^2 + 5.03(2)\right)\right]$$

$$= 19.53 - 8.38$$

$$= 11.15$$

Interpret the Solution: This means that the total increase in recycled ferrous materials from 1992 to 1997 was about 11.15 million tons. ■

Checkpoint 2 Now work Exercise 7.

In our next example, we determine the area bounded by a rate function where the region is below the x-axis.

Example 4 Integrating a Rate Function

The rate of change in the price per pack of cigarettes in the United States can be modeled by

$$c(x) = 0.06x^2 - 1.5x + 8.24 \qquad 0 \le x \le 35$$

where x is the number of years since 1960 and $c(x)$ is the rate of change in the price per pack measured in $\frac{\text{cents}}{\text{year}}$ (*Source:* U.S. Statistical Abstract, www.census.gov/statab/www). See Figure 16.4.8.

(a) Evaluate $\int_{23}^{28} c(x)\,dx$ and interpret.

(b) Evaluate $\int_{10}^{14} c(x)\,dx$ and interpret.

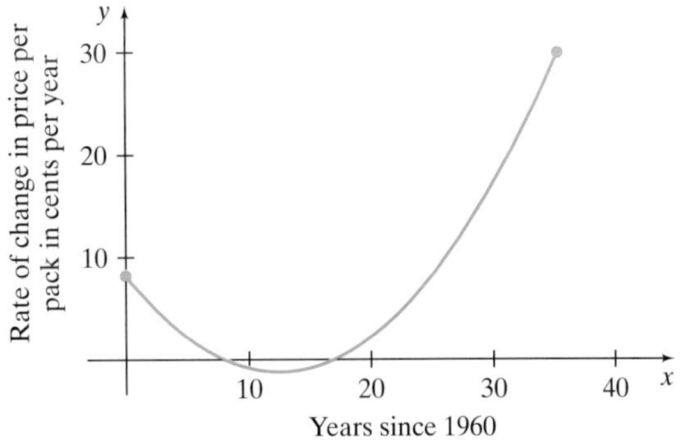

Figure 16.4.8

Solution

(a) **Understand the Situation:** Since $c(x)$ is a rate function, this integration will give us a total accumulation from 1983 ($x = 23$) to 1988 ($x = 28$). Performing the integration and using the Fundamental Theorem of Calculus gives us

$$\int_{23}^{28} c(x)\,dx = \int_{23}^{28} (0.06x^2 - 1.5x + 8.24)\,dx$$

$$= (0.02x^3 - 0.75x^2 + 8.24x)\Big|_{23}^{28}$$

$$= [(0.02(28)^3 - 0.75(28)^2 + 8.24(28)) - (0.02(23)^3 - 0.75(23)^2 + 8.24(23))]$$

$$= 45.65$$

Interpret the Solution: This means that from 1983 to 1988 the total increase in the price of a pack of cigarettes was about 45.65¢.

(b) Here, performing the integration and using the Fundamental Theorem of Calculus gives us

$$\begin{aligned}\int_{10}^{14} c(x)\,dx &= \int_{10}^{14} (0.06x^2 - 1.5x + 8.24)\,dx \\ &= (0.02x^3 - 0.75x^2 + 8.24x)\Big|_{10}^{14} \\ &= [(0.02(14)^3 - 0.75(14)^2 + 8.24(14)) - (0.02(10)^3 \\ &\quad - 0.75(10)^2 + 8.24(10))] \\ &= -4.16\end{aligned}$$

Interpret the Solution: This means that from 1970 to 1974, the total **decrease** in the price of a pack of cigarettes was about 4.16¢. See Figure 16.4.9.

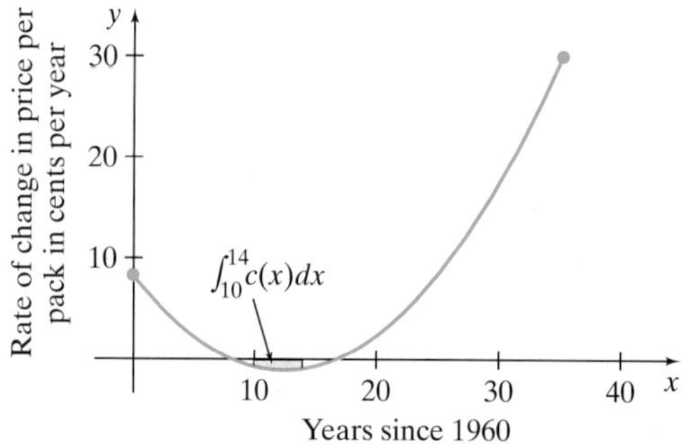

Figure 16.4.9

Notice in Example 4b that the definite integral produced a negative number. As a result, the interpretation was that it represented a total **decrease**. This reinforces that the total accumulation is not always a positive number giving us a total increase.

Checkpoint 3

Now work Exercise 25.

The integration of a rate function to get a total accumulation can also be applied to problems in the physical sciences. Before we go further, let's review a fact from physics in the Toolbox to the left.

So to determine the total movement (or **displacement**) of an object given its velocity, we can use a definite integral and the Fundamental Theorem of Calculus.

From Your Toolbox

For a position function s, the *velocity* of an object at time t is given by $v(t) = \frac{d}{dt}[s(t)]$. This means that velocity, v, can be considered a rate function.

Example 5 Finding Total Displacement of an Object

The velocity of an object is given by

$$v(t) = 8t + 3$$

where t is the time in seconds and $v(t)$ is the velocity measured in $\frac{\text{feet}}{\text{second}}$.

(a) Evaluate $v(10)$ and interpret.

(b) Determine the total movement, or displacement, of the object between 2 and 8 seconds.

Solution

(a) Evaluating, we have

$$v(10) = 8(10) + 3 = 83$$

Interpret the Solution: This means that after 10 seconds the velocity of the object is $83\ \dfrac{\text{feet}}{\text{second}}$.

(b) Understand the Situation: Since $v(t)$ is a rate function, we determine the total displacement by evaluating $\int_2^8 v(t)\,dt$.

Evaluating the definite integral gives us

$$\int_2^8 v(t)\,dt = \int_2^8 (8t+3)\,dt = (4t^2+3t)\Big|_2^8$$
$$= [(4(8)^2+3(8)) - (4(2)^2+3(2))]$$
$$= 280 - 22 = 258$$

Interpret the Solution: This means that from 2 to 8 seconds the object moved a total of 258 feet. ■

Checkpoint 4 Now work Exercise 41.

SUMMARY

This section focused on interpreting the area under the graph of a rate of change function. This concept is encountered regularly throughout the remainder of Chapter 16 as well as Chapter 17.

Interpreting the Area Under the Graph of a Rate Function

Let f be a rate function on the interval $[a, b]$ where $[a, b]$ is in the reasonable domain of f. Then the area under the graph of f on $[a, b]$ gives the total accumulation of F from a to b, where F is an antiderivative of f. The area under the graph of f on $[a, b]$ is given by $\int_a^b f(x)\,dx$. We evaluate $\int_a^b f(x)\,dx$ by using the Fundamental Theorem of Calculus. That is, $\int_a^b f(x)\,dx = F(b) - F(a)$ where F is an antiderivative.

SECTION 16.4 EXERCISES

In Exercises 1–8:

(a) Write a definite integral to determine the shaded area.

(b) Use the Fundamental Theorem of Calculus to evaluate the definite integral in part (a) and interpret.

1.

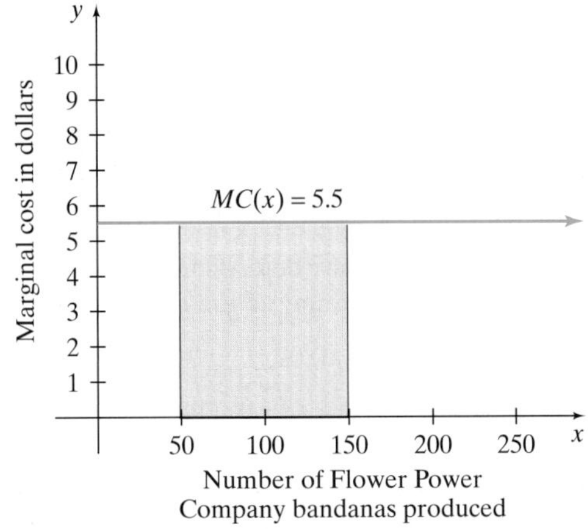

2.

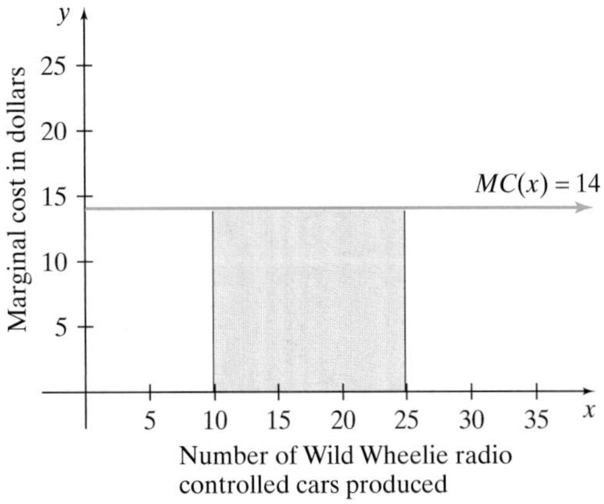

3.

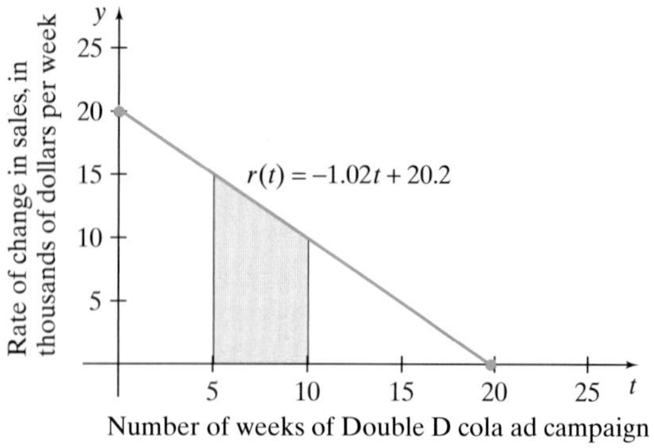

4.

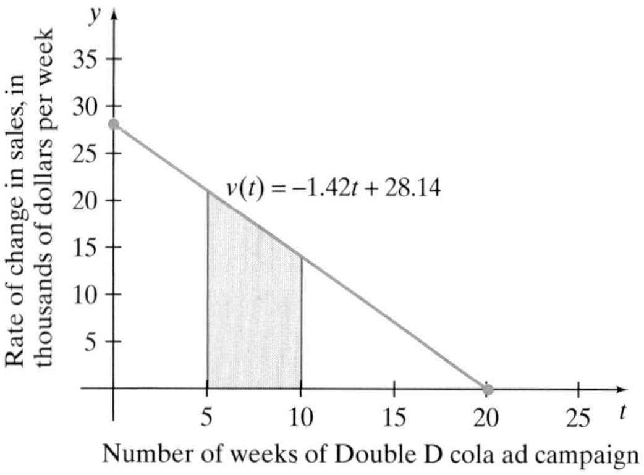

5.

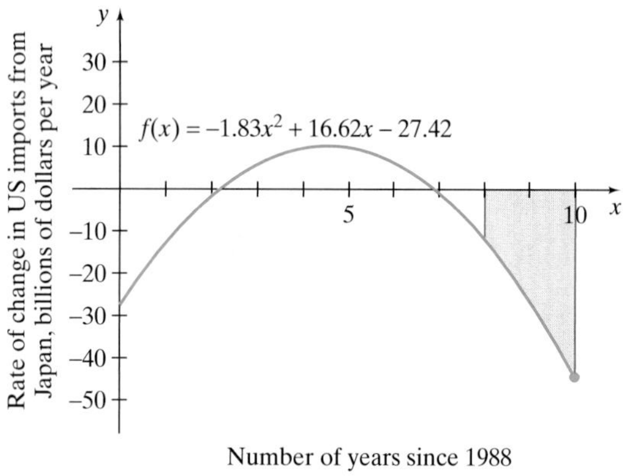

Source: U.S. Statistical Abstract, www.census.gov/statab/www

6.

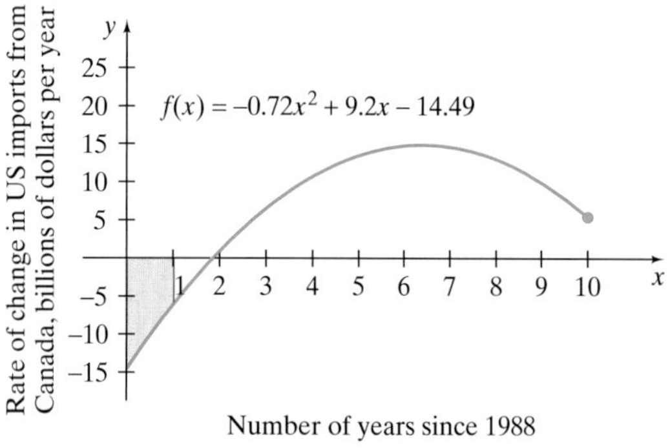

Source: U.S. Statistical Abstract, www.census.gov/statab/www

7.

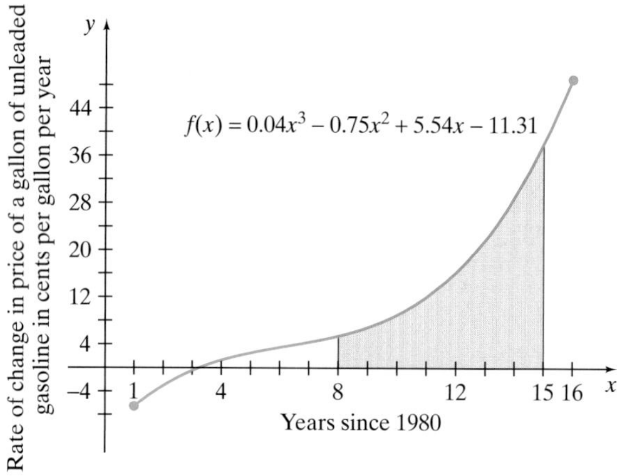

Source: U.S. Energy Information Administration, www.eia.doe.gov

8.

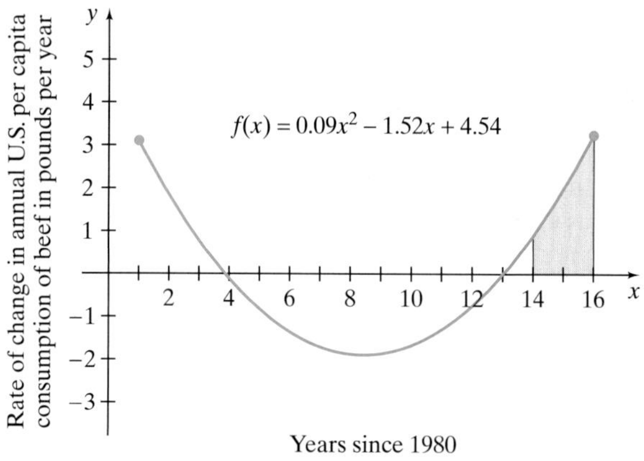

Source: U.S. Department of Agriculture, www.usda.gov

9. The TuffyToe Company determines that the marginal cost function for their new walking shoe is given by

$$MC(x) = C'(x) = 50 - 0.8x \qquad 0 \le x \le 80$$

where x represents the number of pairs of shoes produced per shift and $MC(x) = C'(x)$ represents the marginal costs measured in dollars per shoe.

(a) Graph $MC(x)$ on its reasonable domain and shade the area that represents the total cost of producing the first 50 pairs of shoes.

(b) Write a definite integral to determine the shaded area in part (a). Use The Fundamental Theorem of Calculus to evaluate the definite integral and interpret.

10. The TinyTot Toy Company determines that the marginal cost for producing their new action figure is given by

$$MC(x) = C'(x) = 4 - 0.02x \qquad 0 \le x \le 100$$

where x represents the number of toys made daily and $MC(x) = C'(x)$ represents the marginal costs measured in dollars per toy.

(a) Graph $MC(x)$ on its reasonable domain and shade the area that represents the total cost of producing the first 30 action figures.

(b) Write a definite integral to determine the shaded area in part (a). Use the Fundamental Theorem of Calculus to evaluate the definite integral and interpret.

11. The WiredWorld Audio Company determines that the marginal revenue function for producing and selling a waterproof headset radio is given by

$$MR(x) = R'(x) = 70 + x \qquad 0 \le x \le 70$$

where x is the number of headset radios produced and sold in thousands and $R'(x)$ is the marginal revenue measured in dollars per headset radio.

(a) Evaluate $\int_{10}^{40} R'(x)\,dx$ and interpret.

(b) Sketch the region determined in part (a).

12. By market analysis, the SatSet Satellite TV Company determines that the marginal revenue function for their best selling satellite dish is given by

$$MR(x) = R'(x) = 50 + 0.4x^3 \qquad 0 \le x \le 40$$

where x represents the number of units produced and sold monthly and $R'(x)$ is the marginal revenue measured in dollars per satellite dish.

(a) Evaluate $\int_{20}^{30} R'(x)\,dx$ and interpret.

(b) Sketch the region determined in part (a).

13. The marginal revenue function for the FrontRide Bus Company is given by

$$MR(x) = R'(x) = 0.000045x^2 - 0.03x + 3.75 \qquad 0 \le x \le 500$$

where x represents the number of bus tickets sold and $MR(x)$ is the marginal revenue measured in dollars per ticket. Evaluate $\int_{0}^{200} MR(x)\,dx$ and interpret.

14. The daily marginal revenue function for the BlackDay Sunglasses Company is given by

$$MR(x) = R'(x) = 30 - 0.0003x^2 \qquad 0 \le x \le 540$$

where x represents the number of sunglasses produced and sold and $MR(x)$ is the marginal revenue measured in dollars per sunglasses. Evaluate $\int_{0}^{300} MR(x)\,dx$ and interpret.

15. The ScandiTrac Company determines that their marginal profit function for producing and selling a new economy model of cross-country ski machine at a mall is given by

$$MP(x) = P'(x) = 0.3x^2 + 0.2x \qquad 0 \le x \le 30$$

where x is the number of machines produced and sold and $P'(x)$ is the marginal profit measured in dollars per ski machine. Evaluate $\int_{10}^{20} P'(x)\,dx$ and interpret.

16. The See-The-Fine-Print Company determines that their marginal profit function for producing and selling over the counter reading glasses at a regional store is given by

$$MP(x) = P'(x) = 0.0015x^2 - 0.01x \qquad 0 \le x \le 60$$

where x is the number of reading glasses sold and $P'(x)$ is the marginal profit measured in dollars per pair of glasses. Evaluate $\int_{30}^{40} P'(x)\,dx$ and interpret.

17. The marginal average cost function for producing x promotional banners is given by

$$MAC(x) = \frac{d}{dx}[AC(x)] = -\frac{100}{x^2} \qquad x > 0$$

where x represents the number of banners produced. Evaluate $\int_{10}^{30} MAC(x)\,dx$ and interpret.

18. The marginal average cost function for producing x QuickVid digital cameras is given by

$$MAC(x) = \frac{d}{dx}[AC(x)] = 0.03x^2 - 0.04x + 5 \qquad 0 < x \le 50$$

where x represents the number of digital cameras produced. Evaluate $\int_{0}^{25} MAC(x)\,dx$ and interpret.

19. The rate of arrests for drug abuse violations by U.S. adults is given by the function

$$f(x) = -0.3x^2 + 10.56x - 40.31 \qquad 0 \le x \le 25$$

where x represents the number of years since 1970 and $f(x)$ represents the arrest rate measured in $\frac{\text{thousands of arrests}}{\text{year}}$ (*Source:* U.S. Census Bureau, www.census.gov). Evaluate and interpret $\int_{10}^{20} f(x)\,dx$.

20. The sales rate of Lotto tickets sold in the U.S. is given by the rate function

$$R(x) = -14.58x^2 + 186.84x + 383.93 \qquad 0 \le x \le 16$$

where x represents the number of years since 1980 and $R(x)$ represents the rate of sales, measured in $\frac{\text{millions of dollars}}{\text{year}}$ (*Source:* U.S. Statistical Abstract, www.census.gov/statab/www). Evaluate and interpret $\int_{5}^{10} R(x)\,dx$.

21. The rate of change in the number of employees in the U.S. federal government's executive branch can be modeled by

$$f(x) = 0.36x^3 - 10.26x^2 + 78.52x - 123.18 \qquad 0 \le x \le 15$$

where x represents the number of years since 1980 and $f(x)$ represents the rate of change in the number of employees in the U.S. Federal Government's Executive

Branch measured in $\frac{\text{thousands of employees}}{\text{year}}$ (*Source:* U.S. Office of Personnel Management, www.opm.gov). Evaluate $\int_{10}^{15} f(x)\,dx$ and interpret.

22. The rate of monthly precipitation in Seattle, Washington can be modeled by

$$p(x) = 0.32x - 2.02 \qquad 1 \le x \le 12$$

where $x = 1$ corresponds to January, $x = 2$ corresponds to February, etc. and $p(x)$ is the rate of precipitation each month, measured in inches per month (*Source:* U.S. National Oceanic and Atmospheric Administration, www.noaa.gov). Evaluate $\int_7^{10} p(x)\,dx$ and interpret.

23. The rate of change in the total amount of outstanding revolving credit in the U.S. can be modeled by

$$r(x) = -0.076x^3 + 2.394x^2 - 19.28x + 53.283 \qquad 1 \le x \le 20$$

where x is the number of years since 1979 and $r(x)$ is the rate of change in the total amount of outstanding revolving credit measured in billions of dollars per year (*Source:* U.S. Statistical Abstract, www.census.gov/statab/www). Evaluate and interpret $\int_1^{11} r(x)\,dx$.

24. The rate of U.S. military sales deliveries to Israel can be modeled by

$$s(x) = 12.148x^3 - 236.115x^2 + 1402.494x - 2465.916 \qquad 1 \le x \le 12$$

where x represents the number of years since 1986 and $s(x)$ represents the rate of sales measured in $\frac{\text{millions of dollars}}{\text{year}}$ (*Source:* U.S. Statistical Abstract, www.census.gov/statab/www).

(a) Evaluate and interpret $\int_1^3 s(x)\,dx$.

(b) Evaluate and interpret $\int_4^6 s(x)\,dx$.

✓ **25.** The rate of military expenditures for NATO countries can be modeled by

$$m(x) = 0.248x^3 - 4.371x^2 + 18.616x - 7.493 \qquad 1 \le x \le 12$$

where x represents the number of years since 1985 and $m(x)$ represents the rate of sales measured in $\frac{\text{billions of dollars}}{\text{year}}$ (*Source:* U.S. Statistical Abstract, www.census.gov/statab/www).

(a) Evaluate and interpret $\int_2^6 m(x)\,dx$.

(b) Evaluate and interpret $\int_7^{11} m(x)\,dx$.

26. The rate of change in the number of Medicaid recipients in the United States can be modeled by

$$g(x) = 0.22x - 0.91 \qquad 1 \le x \le 16$$

where x represents the number of years since 1980 and $g(x)$ represents the rate of change in the number of Medicaid recipients measured in $\frac{\text{millions of recipients}}{\text{year}}$ (*Source:* U.S. Census Bureau, www.census.gov).

(a) Evaluate $\int_1^4 g(x)\,dx$ and interpret.

(b) Evaluate $\int_{10}^{15} g(x)\,dx$ and interpret.

27. The rate of change in revenue in the resource recovery industry in the United States can be modeled by

$$f(x) = 0.288x^2 - 3.488x + 8.903 \qquad 1 \le x \le 11$$

where x is the number of years since 1990 and $f(x)$ is the rate of change in revenue in the resource recovery industry measured in $\frac{\text{billions of dollars}}{\text{year}}$.

(a) Evaluate $\int_1^3 f(x)\,dx$ and interpret.

(b) Evaluate $\int_4^8 f(x)\,dx$ and interpret.

28. The rate of consumption of jet fuel in the U.S. each year can be modeled by

$$f(x) = 3.78x^2 + 25.56x - 119.48 \qquad 0 \le x \le 25$$

where x represents the number of years since 1970 and $f(x)$ represents the rate of consumption of jet fuel, measured in millions of gallons each year (*Source:* U.S. Census Bureau, www.census.gov). Determine the total increase (or decrease) in consumption of jet fuel in the U.S. from 1980 to 1990.

29. The rate of change in the total amount of outstanding automobile loans in the U.S. can be modeled by

$$c(t) = -1.04t^3 + 15.66t^2 - 61.86t + 56.15 \qquad 0 \le t \le 10$$

where t represents the number of years since 1985 and $c(t)$ represents the rate of change in the total amount of outstanding automobile loans, measured in $\frac{\text{billions of dollars}}{\text{year}}$ (*Source:* U.S. Census Bureau, www.census.gov). Determine the total increase (or decrease) in the amount of outstanding automobile loans from 1985 to 1995.

30. The import rate of crude oil into the U.S. can be modeled by

$$g(x) = -2.73x^2 + 35.08x + 22.18 \qquad 1 \le x \le 10$$

where x is the number of years since 1989 and $g(x)$ is the rate of imports annually measured in $\frac{\text{millions of barrels}}{\text{year}}$ (*Source:* U.S. Energy Information Administration, www.eia.doe.gov). Determine the total increase (or decrease) in imports of crude oil into the U.S. from 1992 to 1997.

31. The Alaskan oil field production rate can be modeled by

$$g(x) = -0.81x^2 + 5x - 27.56 \qquad 1 \le x \le 10$$

where x is the number of years since 1989 and $g(x)$ is the rate of oil production in the Alaskan oil field measured in $\frac{\text{millions of barrels}}{\text{year}}$ (*Source:* U.S. Energy Information Administration, www.eia.doe.gov). Determine the total increase (or decrease) in Alaskan oil field production from 1992 to 1997.

32. From past records, a botanist knows that a certain species of tree has a rate of growth that can be modeled by

$$f(x) = \frac{3}{2}x^{-\frac{1}{2}} \qquad 1 \le x \le 4$$

where x is the age of the tree in years and $f(x)$ is the growth rate in $\frac{\text{feet}}{\text{year}}$. Determine the total increase in the growth of the tree from the age of 1 year to 4 years.

33. From past records, a botanist knows that a certain species of tree has a rate of growth that can be modeled by

$$f(x) = 2x^{-\frac{1}{2}} \qquad 1 \le x \le 4$$

where x is the age of the tree in years and $f(x)$ is the growth rate in $\frac{\text{feet}}{\text{year}}$. Determine the total increase in the growth of the tree from the age of 1 year to 4 years.

In Exercises 34–39, the velocity function $v(t)$ for an object is given, measured in feet per second.

(a) Evaluate $v(t)$ at the point $t = t_1$.

(b) Evaluate and interpret $\int_{t_1}^{t_2} v(t)\,dt$.

34. $v(t) = 4t;\ t_1 = 2, t_2 = 8$

35. $v(t) = 0.3t;\ t_1 = 10, t_2 = 20$

36. $v(t) = 3t + 6;\ t_1 = 1, t_2 = 5$

37. $v(t) = 2t + 10;\ t_1 = 3, t_2 = 5$

38. $v(t) = -9.8t + 60;\ t_1 = 1, t_2 = 5$

39. $v(t) = -9.8t + 100;\ t_1 = 2, t_2 = 8$

40. The velocity of an object is given by

$$v(t) = -32t + 88$$

where t is the time in seconds and $v(t)$ is the velocity measured in $\frac{\text{feet}}{\text{second}}$.

(a) Evaluate $v(1)$ and interpret.

(b) Determine the total movement, or displacement, of the object between 0 and 2 seconds.

✓ **41.** The velocity of an object is given by

$$v(t) = -32t + 120$$

where t is the time in seconds and $v(t)$ is the velocity measured in $\frac{\text{feet}}{\text{second}}$.

(a) Evaluate $v(1)$ and interpret.

(b) Determine the total movement, or displacement, of the object between 0 and 3 seconds.

In Exercises 42–47, you are given a rate function that, at the present time, you are unable to antidifferentiate. (In Sections 16.5 and 16.6 we will present techniques that will allow us to determine an antiderivative.) As a result, you will need to use the FNINT *command or $\int f(x)\,dx$ on your graphing calculator.*

16.5 **42.** The CustomKey Company determines that the marginal cost of producing sterling silver key fobs is given by

$$MC(x) = \frac{10x}{\sqrt{x^2 + 10{,}000}} \qquad 0 \le x \le 100$$

where x represents the number of fobs produced daily and $MC(x)$ is the marginal cost in dollars. Use your calculator to approximate the integral $\int_0^{75} \frac{10x}{\sqrt{x^2 + 10{,}000}}\,dx$ and interpret.

16.5 **43.** The TightNut Company determine that the marginal cost of producing center punches is given by

$$MC(x) = 0.003x\sqrt{x + 1} \qquad 0 \le x \le 50$$

where x represents the number of center punches produced in a work shift and $MC(x)$ represents the marginal cost in dollars. Use your calculator to approximate the integral $\int_{35}^{48} 0.003x\sqrt{x+1}\,dx$ and interpret.

16.6 **44.** The rate of change function for foreign student enrollment in U.S. colleges can be modeled by

$$g(x) = \frac{92.43}{x} \qquad 1 \le x \le 22$$

where x represents the number of years since 1975 and $g(x)$ represents the rate of change in foreign student enrollment, measured in $\frac{\text{thousands of students}}{\text{year}}$ (*Source:* U.S. Statistical Abstract, www.census.gov/statab/www). Use your calculator to approximate $\int_1^8 \frac{92.43}{x}\,dx$ and interpret.

16.6 **45.** The rate of change in the average dollar amount awarded in the Pell Grant system can be modeled by

$$g(x) = \frac{327}{x} \qquad 1 \le x \le 20$$

where x is the number of years since 1979 and $g(x)$ is the rate of change in the average dollar amount awarded in the Pell Grant system, measured in $\frac{\text{dollars}}{\text{year}}$ (*Source:* U.S. Department of Education, www.ed.gov). Use your calculator to approximate $\int_1^{10} \frac{327}{x}\,dx$ and interpret.

16.6 **46.** The rate of change in the U.S. population can be modeled by

$$g(x) = 1.03e^{0.013x} \qquad 0 \le x \le 100$$

where x represents the number of years since 1900 and $g(x)$ represents the rate of change in U.S. population measured in $\frac{\text{millions}}{\text{year}}$ (*Source:* U.S. Census Bureau, www.census.gov). Use your calculator to approximate the total increase in the U.S. population from 1900 to 1950.

16.6 **47.** The rate of change in tuition and fees for in-state students at four year colleges can be modeled by

$$g(x) = 95.59e^{0.07x} \qquad 1 \le x \le 15$$

where x represents the number of years since 1984 and $g(x)$ represents the rate of change in tuition and fees for in-state students at four year colleges measured in $\frac{\text{dollars}}{\text{year}}$ (*Source:* U.S. National Center for Education Statistics, www.nces.ed.gov). Use your calculator to approximate the total increase in tuition and fees for in-state students at four year colleges from 1985 to 1995.

SECTION PROJECT

The rate of change in the average price of natural gas in the United States from 1995 to 1999 is shown in Table 16.4.2.

(a) Use your calculator to determine a cubic regression model in the form

$$g(x) = ax^3 + bx^2 + cx + d \qquad 1 \le x \le 5$$

where x represents the number of years since 1994 and $g(x)$ is the rate of change in the price of natural gas in the U.S. measured in $\frac{\text{dollars per 1000 cubic feet}}{\text{year}}$. Round all coefficients to the nearest thousandth.

(b) Evaluate and interpret $g(3)$ and explain why this answer is the same as the rate of change in price given in Table 16.4.2.

Table 16.4.2

Year	Rate of Change in Price
1995	0.561
1996	0.495
1997	−0.171
1998	−0.333
1999	1.113

SOURCE: U.S. Energy Information Administration, www.eia.doe.gov

(c) Compute $g'(x)$ and use it to determine when the rate of change in price was decreasing most rapidly.

(d) Evaluate and interpret $\int_1^5 g(x)\,dx$.

Section 16.5 Integration by *u*-Substitution

The rules of integration we have learned so far apply to functions that are straightforward. We have found that polynomial functions and simple root functions are easy to integrate. But what if we need to integrate a function like $\int x\sqrt[3]{4x^2+1}\,dx$? We really have no way to perform the integration with the rules that we have established. In this section we will learn how to rewrite an integral that contains a composite function so that it can be expressed in a simplified form that can easily be integrated. This is called the **method of *u*-substitution**. We will see how this method is used to determine both indefinite and definite integrals and the effect that *u*-substitution has on the limits of integration. We will also examine applications that take advantage of this technique.

u-Substitution with Indefinite Integrals

From Your Toolbox

The **differential in y**, denoted by dy, of the dependent variable is given by

$$dy = f'(x)\,dx$$

We begin by reviewing the definition of the differential in the Toolbox to the left.

Using the definition of differential, we see that if we substitute y with u we have $u = f(x)$ then $\frac{du}{dx} = f'(x)$, so the **differential in u** is $du = f'(x)\,dx$. To see how this can help us with integration, let's consider the indefinite integral $\int 4(4x+3)^5\,dx$. We could integrate by expanding $4x+3$, but we would have to raise this binomial to the fifth power! To avoid all this unnecessary algebra, let's see if a substitution in the integrand will make it simpler. If we let $u = 4x+3$ and compute the differential in u, we get

$$\begin{aligned} u &= 4x+3 \\ \frac{du}{dx} &= 4 \\ du &= 4\,dx \end{aligned}$$

Notice that we now have an expression in terms of u for every part of the integrand. Since $u = 4x+3$ and $du = 4\,dx$, we can rewrite the integral $\int \overbrace{4\underbrace{(4x+3)}_{u}^5\,dx}^{du}$ in terms of u as

$$\int u^5 du$$

Now we can use the Power Rule to find the integral with respect to u as

$$\int u^5 du = \frac{1}{6}u^6 + C$$

Since we are given the integral originally in terms of x, and knowing that $u = 4x+3$, we can resubstitute and write the solution as

$$\int 4(4x+3)^5\,dx = \frac{1}{6}(4x+3)^6 + C$$

We learned in Section 16.1 that *the derivative of the indefinite integral yields the integrand*, hence we can check our answer by differentiating the result via the Chain Rule.

$$\begin{aligned} \frac{d}{dx}\left[\frac{1}{6}(4x+3)^6 + C\right] &= \frac{1}{6}\cdot 6(4x+3)^5 \cdot \frac{d}{dx}(4x+3) + 0 \\ &= (4x+3)^5 \cdot 4 = 4(4x+3)^5 \end{aligned}$$

Now let's generalize by stating the Power Rule with u-substitution.

Power Rule with u-Substitution

If f is a differentiable function of x and $u = f(x)$, then we can write an integral of the form $\int [f(x)]^n f'(x)\,dx$ as $\int u^n\,du$ with

$$\int u^n\,du = \frac{1}{n+1}u^{n+1} + C \qquad n \neq -1$$

where $du = f'(x)\,dx$.

Example 1 Using the Power Rule with u-Substitution

Use the method of u-substitution to evaluate the given indefinite integrals.

(a) $\int \sqrt{x+2}\,dx$

(b) $\int 2(2x+1)^3\,dx$

Solution

(a) For this integration, we can let u be the radicand and write $u = x+2$. Differentiating to get an expression for du, we find that

$$\frac{du}{dx} = 1$$

$$du = 1 \cdot dx = dx$$

Since $du = dx$, the substitution is

$$\int \sqrt{x+2}\,dx = \int \sqrt{u}\,du$$

$$= \int u^{1/2}\,du = \frac{2}{3}u^{3/2} + C$$

Since $u = x + 2$, we resubstitute and get

$$\int \sqrt{x+2}\,dx = \frac{2}{3}(x+2)^{3/2} + C$$

(b) Here we will let u be the expression in the parentheses and write $u = 2x + 1$. Differentiating with respect to x, we find that

$$\frac{du}{dx} = 2$$

$$du = 2\,dx$$

Rewriting the integral in terms of u and integrating yields

$$\int 2(2x+1)^3\,dx = \int u^3 du = \frac{1}{4}u^4 + C$$

Resubstituting gives

$$\int 2(2x+1)^3\,dx = \frac{1}{4}(2x+1)^4 + C$$

■

Checkpoint 1 Now work Exercise 3.

In words, the Power Rule with u-substitution states that if an integrand can be expressed as an expression raised to a power, and the derivative of the expression is another factor in the integrand, we can apply the u-substitution. This form of the integral

$$\int (\text{expression})^{\text{power}} \cdot \frac{d}{dx}(\text{expression})\,dx$$

occurs more frequently than one might think. As we have seen in the first example, we commonly represent u as an expression raised to a power or as a radicand.

Example 2 Integrating Using a u-Substitution

Use the Power Rule with u-substitution to integrate $\int \sqrt[3]{(4x^3+x)}(12x^2+1)\,dx$.

Solution

Understand the Situation: To get $\int \sqrt[3]{(4x^3+x)}(12x^2+1)\,dx$ in the form $\int u^n\,du$, we can let u be the radicand of the cube root and get

$$u = 4x^3 + x$$

So the differential in u is

$$du = (12x^2+1)\,dx$$

Now, writing the integral with respect to u and integrating yields

$$\int \sqrt[3]{(4x^3+x)}(12x^2+1)\,dx = \int \sqrt[3]{u}\,du$$
$$= \int u^{1/3}\,du = \frac{3}{4}u^{4/3} + C$$

Knowing that $u = 4x^3 + x$, we can write the solution in terms of x as

$$\int \sqrt[3]{(4x^3+x)}(12x^2+1)\,dx = \frac{3}{4}(4x^3+x)^{4/3} + C$$
$$= \frac{3}{4}\sqrt[3]{(4x^3+x)^4} + C$$

■

Sometimes, we must alter the differential in u to get it to match the remaining part of the integrand. We can do this by multiplying or dividing both sides of the differential by a constant. Let's say that we wish to integrate $\int x^2\sqrt{x^3+9}\,dx$. If we let u represent the radicand, we have $u = x^3 + 9$. Computing du, we find that

$$\frac{du}{dx} = 3x^2$$
$$du = 3x^2\,dx$$

But we want du to just be $x^2\,dx$. We can get this desired expression by just multiplying both sides of the differential by $\frac{1}{3}$. This gives us

$$du = 3x^2\,dx$$
$$\frac{1}{3}\,du = x^2\,dx$$

Now, writing the integral in terms of u, we get

$$\int x^2\sqrt{x^3+9}\,dx = \int \left(\sqrt{u}\cdot\frac{1}{3}\right)du = \frac{1}{3}\int u^{1/2}\,du$$
$$= \frac{1}{3}\cdot\frac{2}{3}\cdot u^{3/2} + C = \frac{2}{9}u^{3/2} + C$$

In terms of x we get, by substitution,

$$= \frac{2}{9}(x^3+9)^{3/2} + C$$

Example 3 Adjusting a Constant with u-substitution

Evaluate the indefinite integral $\int \frac{1}{(4x-1)^3}\,dx$.

Solution

Understand the Situation: Here we let $u = 4x - 1$ so that $du = 4\,dx$. Since we want $du = 1 \cdot dx$, we can divide both sides of the differential by 4 to get

$$\frac{1}{4}du = dx$$

So the integral in terms of u becomes

$$\begin{aligned}\int \frac{1}{(4x-1)^3}\,dx &= \int \left(\frac{1}{4}\cdot\frac{1}{u^3}\right) du \\ &= \frac{1}{4}\int u^{-3}\,du \\ &= \frac{1}{4}\cdot\frac{1}{-2}u^{-2} + C \\ &= -\frac{1}{8}\cdot\frac{1}{u^2} + C \\ &= -\frac{1}{8(4x-1)^2} + C \qquad \text{Written in terms of } x\end{aligned}$$

Interactive Activity

Verify the solution to Example 3 by differentiating $\frac{d}{dx}\left(-\frac{1}{8(4x-1)^2} + C\right)$.

Let's summarize what we have learned so far and list some tips on integration by u-substitution.

Tips on u-Substitution

- The correct choice of u is usually a radicand, a denominator, or an expression in parentheses.
- Compute the differential in u; $du = f'(x)\,dx$.
- Multiply or divide du by a constant, if needed, to make the remaining part of the integral match the differential in u.
- After the u-substitution, no factors can contain x.
- Integrate and rewrite the integral in terms of the original independent variable (usually x).

Sometimes we must use a "trick" to complete the substitution. If the expression for u is linear, and so is the remainder of the integrand, we have to solve the expression for u in terms of x. We demonstrate this technique in Example 4.

Example 4 Solving the u-Expression for x to Complete a Substitution

Determine the indefinite integral $\int x(x-7)^2\,dx$.

Solution

If we assign u to the expression in parentheses, we see that

$$\begin{aligned}u &= x - 7 \\ du &= dx\end{aligned}$$

This doesn't help because the desired du should be $x\,dx$. However, since $u = x - 7$, we can write

$$x = u + 7$$

Now, making the substitution, we write

$$\begin{aligned}
\int x(x-7)^2\,dx &= \int (u+7)u^2\,du \\
&= \int (u^3 + 7u^2)\,du && \text{Distribute } u^2 \\
&= \frac{1}{4}u^4 + \frac{7}{3}u^3 + C && \text{Integrate with respect to } u \\
&= \frac{1}{4}(x-7)^4 + \frac{7}{3}(x-7)^3 + C && \text{Written in terms of } x
\end{aligned}$$

■

Checkpoint 2 Now work Exercise 29.

u-Substitution with Definite Integrals

We also can apply the integration technique of u-substitution to definite integrals in one of two ways.

1. Define u, perform the integration with respect to du, and rewrite the answer in terms of x. Evaluate the definite integral using the original limits of integration.
2. Define u and rewrite the integrand *and* the limits of integration in terms of the variable u. Integrate with respect to u and evaluate the integral using the new limits of integration.

We illustrate these two methods in Example 5.

Example 5 Evaluating a Definite Integral by u-Substitution

Evaluate $\displaystyle\int_0^4 \frac{x}{\sqrt{x^2+9}}\,dx$.

Solution

By letting $u = x^2 + 9$, we have $du = 2x\,dx$, so $\frac{1}{2}\,du = x\,dx$. We can now proceed in either of the ways outlined prior to Example 5.

1. $$\begin{aligned}
\int \frac{x}{\sqrt{x^2+9}}\,dx &= \int \frac{1}{\sqrt{u}} \cdot \frac{1}{2}\,du \\
&= \frac{1}{2}\int \frac{1}{\sqrt{u}}\,du = \frac{1}{2}\int u^{-(1/2)}\,du \\
&= \frac{1}{2}(2)u^{1/2} = \sqrt{u} = (x^2+9)^{1/2}
\end{aligned}$$

   Now we use the Fundamental Theorem of Calculus with the original limits to get

   $$\int_0^4 \frac{x}{\sqrt{x^2+9}}\,dx = (x^2+9)^{1/2}\Big|_0^4 = (25)^{1/2} - (9)^{1/2} = 2$$

2. Since we assign $u = x^2 + 9$, we have:

$$\text{If } x = 0, \text{ then } u = 0^2 + 9 = 9.$$
$$\text{If } x = 4, \text{ then } u = 4^2 + 9 = 25.$$

Thus, after we do the substitution, we can rewrite the limits as follows:

$$\int_0^4 \frac{x}{\sqrt{x^2+9}}\,dx = \frac{1}{2}\int_9^{25} \frac{1}{\sqrt{u}}\,du = \frac{1}{2}\int_9^{25} u^{-(1/2)}\,du$$
$$= u^{1/2}\Big|_9^{25} = (25)^{1/2} - (9)^{1/2} = 2$$ ■

Notice in Example 5 that changing the limits of integration saved a few steps in the evaluation process.

Checkpoint 3

Now work Exercises 35 and 47.

Applications

The u-substitution technique allows us to examine new families of applications. One of these is recovering business functions.

Example 6 Using u-Substitution to Recover Business Functions

A manager at the Black Box microprocessor manufacturing company finds through data gathered in research that the marginal cost function for a certain type of automobile computer chip made at the facility is given by

$$MC(x) = C'(x) = 6x\sqrt{x^2 + 11}$$

where x represents the number of auto computer chips produced each hour and $MC(x)$ represents the marginal cost. The manager also knows that it costs \$1932 to manufacture five chips.

(a) Recover the cost function, C. (To avoid confusion, we will call the arbitrary constant d.)

(b) Determine the fixed costs.

Solution

(a) **Understand the Situation:** Recall from Section 16.1 that $C(x) + d = \int MC(x)\,dx$, where d is a constant that can be found given an initial value. So we must integrate $\int 6x\sqrt{x^2+11}\,dx$.

To evaluate $\int 6x\sqrt{x^2+11}\,dx$, it is reasonable to let $u = x^2 + 11$, making

$$du = 2x\,dx$$

However, the remaining part of the integrand is $6x\,dx$. This means we must multiply the differential in u by 3 to get the parts to match.

$$3 \cdot du = 3 \cdot 2x\,dx = 6x\,dx$$

So now our substitution is

$$\int 6x\sqrt{x^2+11}\,dx = \int \sqrt{u}\cdot 3\,du = 3\int \sqrt{u}\,du$$
$$= 3\int u^{1/2}\,du = 3\cdot\left(\frac{2}{3}\right)u^{3/2} + d = 2u^{3/2} + d$$

So, in terms of x, we get the partially recovered cost function as

$$C(x) = 2(x^2+11)^{3/2} + d = 2\sqrt{(x^2+11)^3} + d$$

where d is a constant. To find d, we can use the initial value that when $x = 5$, $C(x) = 1932$ to get

$$1932 = 2\sqrt{(5^2+11)^3} + d$$
$$1932 = 2\sqrt{46{,}656} + d$$
$$1932 = 432 + d$$
$$1500 = d$$

So the recovered cost function is $C(x) = 2\sqrt{(x^2+11)^3} + 1500$.

(b) **Understand the Situation:** When the cost function was represented by a polynomial function, the fixed cost was always the same as the constant. But here the cost function is a radical function, so we need to use the definition of the cost function in its literal sense and find the cost of producing zero items. So the fixed costs are found by evaluating $C(x)$ at $x = 0$.

This gives us

$$C(0) = 2\sqrt{(0^2+11)^3} + 1500$$
$$= 2\sqrt{1331} + 1500 \approx 1572.97$$

Interpret the Solution: Thus, the fixed costs are about \$1572.97. ■

As in the previous section, we can also compute the continuous sum of a rate function using the u-substitution technique. But now we can integrate composite rate functions, as illustrated in Example 7.

Example 7 Integrating a Rate Function Using *u*-Substitution

Media consultants for the new local magazine *Rave!* have projected that the number of subscriptions will grow during the first five years at a rate given by

$$S(t) = \frac{1000}{(1+0.3t)^{3/2}} \qquad 0 \le t \le 60$$

where t is the number of months since the magazine's first issue and $S(t)$ is the rate of change in the number of subscriptions measured in $\frac{\text{subscriptions}}{\text{month}}$.

(a) Evaluate and interpret $S(12)$.

(b) Evaluate and interpret $\int_0^6 S(t)\,dt$.

Solution

(a) Understand the Situation: We first note that S is a rate function. Also, $t = 12$ corresponds to 12 months since the magazine's first issue.

Evaluating, we get

$$S(12) = \frac{1000}{(1 + 0.3(12))^{3/2}} = \frac{1000}{(4.6)^{3/2}} \approx 101.36$$

Interpret the Solution: This means that at the end of the first year the subscriptions will be growing at a rate of about 101 $\frac{\text{subscriptions}}{\text{month}}$.

(b) Understand the Situation: Since S is a rate function, integration gives us a total accumulation. Specifically, integrating from $t = 0$ to $t = 6$ gives us the total number of subscriptions after 6 months.

Here we need to integrate $\int_0^6 \frac{1000}{(1+0.3t)^{3/2}}\,dt$. If we let $u = 1 + 0.3t$, then $du = 0.3\,dt$, so $\frac{1}{0.3}\,du = dt$. Changing the limits of integration, we see that:

If $t = 0$, then $u = 1 + 0.3(0) = 1$.
If $t = 6$, then $u = 1 + 0.3(6) = 2.8$.

So, in terms of u, the integral is

$$\begin{aligned}
\int_0^6 \frac{1000}{(1+0.3t)^{3/2}}\,dt &= 1000\int_0^6 \frac{1}{(1+0.3t)^{3/2}}\,dt \\
&= 1000\int_1^{2.8} \frac{1}{u^{3/2}}\left(\frac{1}{0.3}\right)du = \frac{1000}{0.3}\int_1^{2.8}\frac{1}{u^{3/2}}\,du \\
&= \frac{1000}{0.3}\int_1^{2.8} u^{-(3/2)}\,du = \frac{1000}{0.3}(-2)\left[u^{-(1/2)}\right]\Big|_1^{2.8} \\
&= \frac{-2000}{0.3}\left[(2.8)^{-(1/2)} - (1)^{-(1/2)}\right] \approx 2682.57
\end{aligned}$$

Interpret the Solution: This means that at the end of the first six months the number of subscriptions is estimated to be about 2683. ■

Checkpoint 4 Now work Exercise 63.

SUMMARY

This section concentrated on the frequently used integration technique called ***u*-substitution**. We saw that the technique was useful in a variety of applications.

- **Power rule with *u*-substitution**:

$$\int u^n\,du = \frac{1}{n+1}u^{n+1} + C$$

- **Tips on *u*-substitution**:
  - The correct choice of u is usually a radicand, a denominator, or an expression in parentheses.
  - Compute the differential in u; $du = f'(x)\,dx$.
  - Multiply or divide du by a constant if needed to make the remaining part of the integral match the differential in u.
  - After the u-substitution, no factors can contain x.
  - Integrate and rewrite the integral in terms of the original independent variable.

SECTION 16.5 EXERCISES

In Exercises 1–12, use u-substitution to evaluate the indefinite integrals. Check by differentiating the solution.

1. $\int 3(3x+4)^2\,dx$

2. $\int 5(5x+3)^2\,dx$

3. $\int 8x(4x^2+1)^3\,dx$

4. $\int 6x(3x^2-5)^3\,dx$

5. $\int 2t\sqrt{t^2-1}\,dt$

6. $\int 10t\sqrt[3]{5t^2-11}\,dt$

7. $\int (3x^2-2)(x^3-2x)\,dx$

8. $\int (4x-1)(2x^2-x)\,dx$

9. $\int \frac{3x^2}{\sqrt[3]{x^3-5}}\,dx$

10. $\int \frac{8}{\sqrt{8x+9}}\,dx$

11. $\int \frac{4x}{(2x^2+3)^3}\,dx$

12. $\int \frac{3x^2-2}{(x^3-2x)^2}\,dx$

In Exercises 13–28, use u-substitution to evaluate the indefinite integrals.

13. $\int (4x+7)^4\,dx$

14. $\int (12x-1)^9\,dx$

15. $\int t(t^2-3)^5\,dt$

16. $\int t^2(t^3-1)^3\,dt$

17. $\int \frac{x^2}{\sqrt{x^3-5}}\,dx$

18. $\int \frac{x^3}{\sqrt{x^4-4}}\,dx$

19. $\int \frac{5x}{(10x^2-4)^2}\,dx$

20. $\int \frac{x^2}{(3x^3-8)^5}\,dx$

21. $\int \frac{1}{x^2}\sqrt{1-x^{-1}}\,dx$

22. $\int \frac{1}{\sqrt{x}}\sqrt[3]{1+\sqrt{x}}\,dx$

23. $\int (t-7)^{10}\,dt$

24. $\int (6t-10)^5\,dt$

25. $\int \frac{x+1}{(x^2+2x+5)^5}\,dx$

26. $\int \frac{2x+3}{(2x^2+6x-10)^4}\,dx$

27. $\int (6x+9)(x^2+3x-10)\,dx$

28. $\int (20x+20)(x^2+2x-9)^2\,dx$

In Exercises 29–34, determine the integrals by solving the u-expression to complete the substitution. See Example 4.

29. $\int x\sqrt{x+1}\,dx$

30. $\int x\sqrt[3]{x-2}\,dx$

31. $\int t(t-1)^5\,dt$

32. $\int 2t(t+1)^3\,dt$

33. $\int \frac{3x}{\sqrt{x-1}}\,dx$

34. $\int \frac{x}{\sqrt[3]{x+2}}\,dx$

In Exercises 35–46, evaluate the definite integrals. Do not change the limits of integration.

35. $\int_0^4 \frac{2x}{\sqrt{x^2+9}}\,dx$

36. $\int_{-2}^0 \frac{1}{(x-4)^2}\,dx$

37. $\int_0^2 x(x^2-2)^3\,dx$

38. $\int_0^1 3x(x^2-1)^2\,dx$

39. $\int_0^4 (3t-5)^2\,dt$

40. $\int_0^2 (2t+3)^3\,dt$

41. $\int_{-2}^1 \sqrt{1-x}\,dx$

42. $\int_{-5}^0 \sqrt[3]{2-x}\,dx$

43. $\int_0^1 \frac{x}{(1+3x^2)^2}\,dx$

44. $\int_{-1}^1 \frac{2x}{(1+x^2)^3}\,dx$

45. $\int_2^{10} \frac{1}{\sqrt{t-1}}\,dt$

46. $\int_0^7 \frac{1}{\sqrt[3]{t+1}}\,dt$

In Exercises 47–58, evaluate the definite integrals. Change the limits of integration to u-limits.

47. $\int_0^3 \frac{x}{(x^2+1)^2}\,dx$

48. $\int_0^1 \frac{8x}{(2x^2+1)^3}\,dx$

49. $\int_1^4 \frac{2x+1}{(x^2+x-1)^2}\,dx$

50. $\int_1^3 \frac{2x+3}{(x^2+3x)^3}\,dx$

51. $\int_0^2 \frac{3t^2}{(1+t^3)^5}\,dt$

52. $\int_{-1}^0 \frac{t^3}{(2-t^4)^4}\,dt$

53. $\int_0^2 3x^2\sqrt{x^3+1}\,dx$

54. $\int_0^1 2x^3\sqrt{x^4+4}\,dx$

55. $\int_{-1}^1 t(t^2-1)^3\,dt$

56. $\int_{-2}^0 t^2(t^3-2)\,dt$

57. $\int_2^{10} \frac{3}{\sqrt{5x-1}}\,dx$

58. $\int_0^4 \frac{x}{\sqrt{x^2+9}}\,dx$

Applications

In Exercises 59–62, determine the exact area of the shaded region.

59.

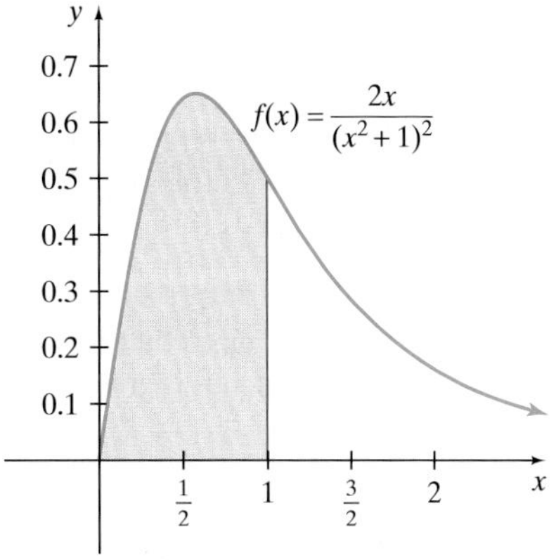

60.

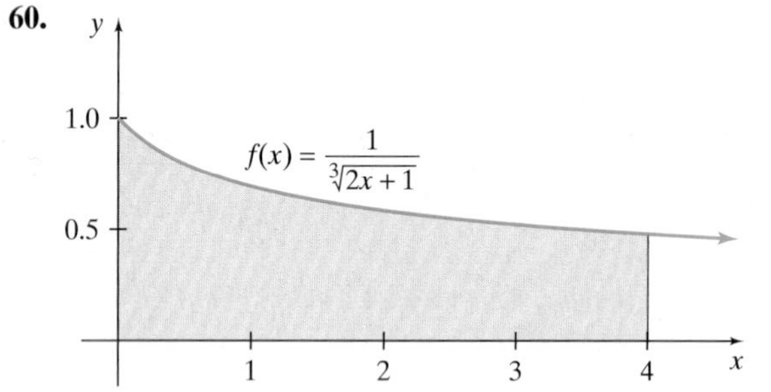

61.

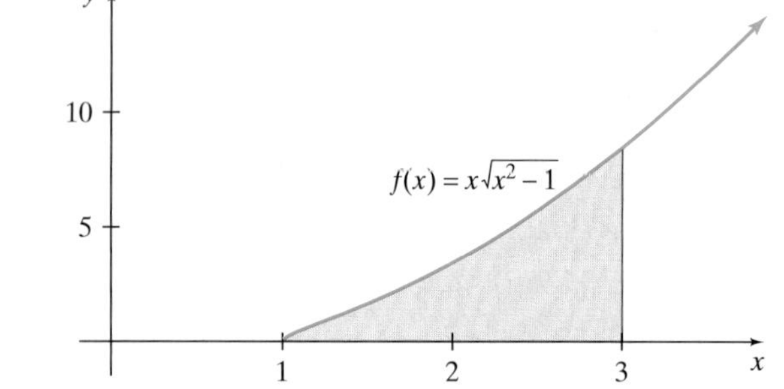

62.

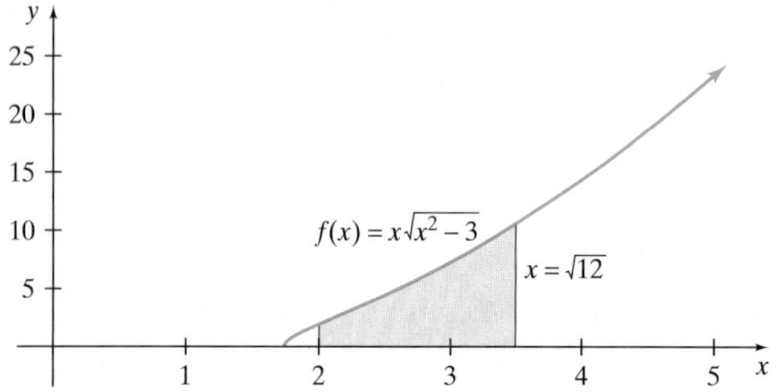

16.4 ✓ **63.** The CustomKey Company determines that the marginal cost of producing sterling silver key fobs is given by

$$MC(x) = \frac{10x}{\sqrt{x^2 + 10{,}000}} \qquad 0 \le x \le 100$$

where x represents the number of fobs produced daily and $MC(x)$ is the marginal cost in dollars.

(a) Evaluate $MC(75)$ and interpret.

(b) Evaluate $\int_0^{75} MC(x)\,dx$ and interpret.

(c) Knowing that the cost to produce 75 fobs is \$3250, recover the cost function C.

(d) Determine the fixed costs.

16.4 **64.** The TightNut Company determines that the marginal cost of producing center punches is given by

$$MC(x) = 0.003x\sqrt{x+1} \qquad 0 \le x \le 50$$

where x is the number of center punches produced in a work shift and $MC(x)$ is the marginal cost in dollars.

(a) Evaluate and interpret $MC(48)$.

(b) Evaluate $\int_{35}^{48} MC(x)\,dx$ and interpret.

(c) Knowing that it costs \$23.40 to produce 24 center punches, recover the cost function C.

(d) Determine the fixed costs.

65. The marginal profit function for seasonal flags produced by the WaveFree Company is given by

$$MP(x) = \frac{2x}{\sqrt{2x^2 - 400}} \qquad 15 \le x \le 105$$

where x is the number of flags produced and sold monthly and $MP(x)$ is the marginal profit in hundreds of dollars.

(a) Evaluate and interpret $MP(70)$.

(b) Evaluate $\int_{20}^{100} MP(x)\,dx$ and interpret.

(c) Knowing that the WaveFree Company breaks even when 20 units are sold [that is, $P(x) = 0$ when $x = 20$], recover the profit function P.

66. The FemTouch Accessory Manufacturer determines their marginal profit function for decorative scarves is

$$MP(x) = \frac{4}{\sqrt[3]{12x - 200}} \qquad 20 \le x \le 150$$

where x is the number of scarves produced and sold daily and $MP(x)$ is the marginal profit in hundreds of dollars.

(a) Evaluate $MP(100)$ and interpret.

(b) Evaluate $\int_{20}^{100} MP(x)\,dx$ and interpret.

(c) Knowing that the FemTouch accessory manufacturer breaks even when 100 units are sold [that is, $P(x) = 0$ when $x = 100$], recover the profit function P.

67. Suppose that the velocity of an object is given by the function

$$v(t) = \frac{t}{\sqrt{t^2 + 9}}$$

where t is the time in seconds and $v(t)$ is the velocity in $\frac{\text{feet}}{\text{second}}$.

(a) Determine the total distance moved between 3 and 5 seconds.

(b) Knowing that when $t = 4$ seconds $s(t) = 8$ feet, determine the position function s.

68. Suppose that the velocity of an object is given by the function

$$v(t) = \frac{2\sqrt{t+1} + 1}{\sqrt{t+1}}$$

where t is the time in seconds and $v(t)$ is the velocity in $\frac{\text{feet}}{\text{second}}$.

(a) Determine the total distance moved between 4 and 16 seconds.

(b) Knowing that when $t = 3$ seconds, $s(t) = 16$ feet, determine the position function s.

69. The number of homes being built in the computer-designed town of EastWorld is increasing at a rate of

$$h'(t) = \frac{15t^2}{\sqrt{t^3+4}} \qquad 0 \le t \le 10$$

where t is the number of years since the town was first incorporated and $h'(t)$ represents the rate of homes being built measured in $\frac{\text{homes}}{\text{year}}$.

(a) Evaluate and interpret $h'(5)$. Round your answer to the nearest whole number.

(b) Evaluate and interpret $\int_0^5 h'(t)\,dt$. Round your answer to the nearest whole number.

(c) Knowing that there were no homes completed when the town was first incorporated, determine h, the number of homes in the town after t years.

(d) Evaluate and interpret $h(5)$. Round your answer to the nearest whole number.

SECTION PROJECT

Conservationists find that the rate of growth of carp in the first year in a stocked pond can be modeled by

$$f(x) = \frac{10x}{\sqrt{x+1}} \qquad 0 \le x \le 12$$

where x is the number of months since the pond was stocked and $f(x)$ is the rate of growth in the number of carp in the pond, measured in $\frac{\text{carp}}{\text{month}}$.

(a) Evaluate and interpret $f(6)$. Round your answer to the nearest whole number.

(b) Evaluate and interpret $\int_0^{12} f(x)\,dx$. Round your answer to the nearest whole number.

(c) Knowing that the pond was stocked with 100 carp, determine $F(x)$, the number of carp in the pond after x months.

(d) Evaluate and interpret $F(5)$. Round your answer to the nearest whole number.

(e) Determine an equation of the line tangent to the graph of F when $x = 6$. Evaluate the tangent line equation at the value of $x = 12$ and interpret.

Section 16.6 Integrals That Yield Logarithmic and Exponential Functions

In Chapter 14, we discussed the derivatives of the logarithmic and exponential functions, along with their associated applications. Now we turn our attention to **integrals** that involve **exponential and logarithmic functions**. Just as we need special rules to differentiate this family of functions, we also need an additional set of integration rules. We will see how to use the u-substitution technique with these functions and check out applications that lend themselves to these functions.

Integrals That Yield Logarithmic Functions

Before examining the role that the logarithm plays in integration, let's review the applicable differentiation rules. (Consult the following Toolbox.)

From Your Toolbox

- $\frac{d}{dx}[\ln x] = \frac{1}{x}$.
- If f is a differentiable function of x, then, by the Chain Rule,

$$\frac{d}{dx}\ln[f(x)] = \frac{1}{f(x)} \cdot f'(x)$$

- The derivative of the indefinite integral yields the integrand. Mathematically, this means that $\int f(x)\,dx = F(x) + C$ if and only if

$$\frac{d}{dx}[F(x) + C] = f(x).$$

Remember that the Power Rule for Integration stated that $\int x^n\,dx = \frac{1}{n+1}x^{n+1} + C$, provided that $n \neq -1$. To find the integral of x^n when $n = -1$, we must integrate $\int x^{-1}\,dx = \int \frac{1}{x}\,dx$. Since $\frac{d}{dx}[\ln x] = \frac{1}{x}$, we can conclude that $\int \frac{1}{x}\,dx = \ln x + C$. But we need to be careful here, because the natural logarithm function is defined only for positive numbers. Let's say that x is negative, in which case $(-x)$ is positive. Then, by the Chain Rule for the logarithm function,

$$\frac{d}{dx}\ln(-x) = \frac{1}{-x}\frac{d}{dx}(-x) = \frac{1}{-x}(-1) = \frac{1}{x}$$

So we see that $\int \frac{1}{x}\,dx = \ln(-x) + C$ if x is negative. Instead of writing two integration rules for positive and negative integrands, we can use the absolute value to write one neat integration formula.

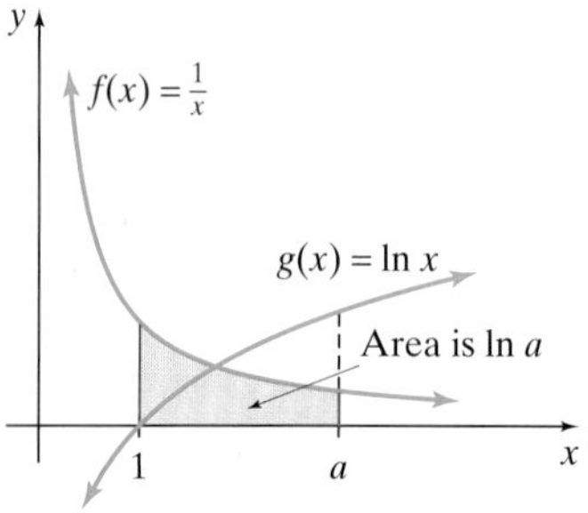

Figure 16.6.1 $\int_1^a \frac{1}{x}\,dx = \ln a$.

Integration Formula for $\frac{1}{x}$

For the function $f(x) = \frac{1}{x}$, the indefinite integral is given by

$$\int \frac{1}{x}\,dx = \ln|x| + C$$

This relationship between the functions $f(x) = \frac{1}{x}$ and $g(x) = \ln x$ is shown in Figure 16.6.1.

Note: For $a > 1$, $\int_1^a \frac{1}{x}\,dx = \ln x\Big|_1^a = \ln a - \ln 1 = \ln a - 0 = \ln a$. This means that the area under the graph of $f(x) = \frac{1}{x}$ on the interval $[1, a]$ is the same as the value of $\ln a$. Some books define the natural logarithm function exactly this way.

Example 1 Integrating Functions of the Form $f(x) = \frac{k}{x}$

Evaluate the following definite integrals.

(a) $\int_{-6}^{-2} \frac{1}{x}\,dx$

(b) $\int_1^5 \frac{6}{x}\,dx$

Solution

(a) Using the Integration Formula for $\frac{1}{x}$, we get

$$\int_{-6}^{-2} \frac{1}{x}\,dx = \ln|x|\Big|_{-6}^{-2} = \ln|-2| - \ln|-6| = \ln(2) - \ln(6) \approx -1.10$$

(b) We can begin by factoring out the constant 6.

$$\int_1^5 \frac{6}{x}\,dx = 6\int_1^5 \frac{1}{x}\,dx = 6 \cdot \ln|x|\Big|_1^5$$
$$= 6[\ln|5| - \ln|1|] = 6[\ln(5) - \ln(1)]$$
$$= 6[\ln(5) - 0] = 6\ln(5) \approx 9.66$$

■

Since we rarely must integrate $f(x) = \frac{1}{x}$, the usefulness of the integration formula for $\frac{1}{x}$ is limited. However, the u-substitution form of the formula is quite useful.

Integration Formula for $\frac{1}{u}$

If u is a differentiable function of x, then

$$\int \frac{1}{u}\,du = \ln|u| + C$$

Example 2 Integrating using u-Substitution

Determine the following indefinite integrals.

(a) $\displaystyle\int \frac{1}{t+2}\,dt$ **(b)** $\displaystyle\int \frac{4x}{x^2+5}\,dx$

Solution

(a) We can use a u-substitution here with $u = t + 2$. Differentiating both sides of the equation gives

$$du = dt$$

So, in terms of u, the integral becomes

$$\int \frac{1}{t+2}\,dt = \int \frac{1}{u}\,du = \ln|u| + C$$

In terms of t, the solution is

$$= \ln|t+2| + C$$

(b) To use the Integration Formula for $\frac{1}{u}$, we let $u = x^2 + 5$; then $du = 2x\,dx$. Since we want the expression containing dx to include $4x$, we can multiply by 2 to get $2\,du = 4x\,dx$. In terms of u, the integral becomes

$$\int \frac{4x}{x^2+5}\,dx = \int \frac{1}{u} \cdot 2\,du = 2\int \frac{1}{u}\,du = 2\ln|u| + C$$

So, in terms of x, the solution is

$$= 2\ln|x^2+5| + C = 2\ln(x^2+5) + C$$

Notice that, since the expression $x^2 + 5$ is positive for all values of x, we can remove the absolute value sign.

■

Checkpoint 1 Now work Exercise 13.

Many applications that involve integrating a rate function use our new integration rule.

Example 3 Integrating a Rate Function That Yields a Logarithmic Function

Since running a series of first-come, first-served promotions, the FineHomes Furniture Store has found that its sales rate during its three-month sales drive can be modeled by

$$s(t) = \frac{10}{t} + 2 \qquad 1 \le t \le 12$$

where t represents the number of weeks that the promotion has been running and $s(t)$ is the sales rate measured in $\frac{\text{thousands of dollars}}{\text{week}}$.

(a) Determine the total increase in sales generated from the first to the fifth week.

(b) Knowing that \$6000 was made during the first week, recover $R(t)$, the revenue generated after t weeks.

Solution

(a) **Understand the Situation:** Since we are given a rate function, we need to compute $\int_1^5 s(t)\,dt$ to obtain the total increase in sales generated from the first to the fifth week.

Evaluating the definite integral, we get

$$\int_1^5 s(t)\,dt = \int_1^5 \left(\frac{10}{t} + 2\right) dt = \int_1^5 \frac{10}{t}\,dt + \int_1^5 2\,dt$$

$$= 10\int_1^5 \frac{1}{t}\,dt + 2\int_1^5 dt = (10\ln|t| + 2t)\Big|_1^5$$

$$= (10\ln|5| + 2(5)) - (10\ln|1| + 2(1))$$

$$= 10\ln 5 + 10 - 0 - 2 = 10\ln 5 + 8$$

$$\approx 24.09 \text{ thousand dollars}$$

Interpret the Solution: So, from weeks 1 to 5 of the promotion, the total increase in sales was about \$24,090.

(b) **Understand the Situation:** To recover the revenue $R(t)$, we begin by evaluating the indefinite integral $\int s(t)\,dt$.

This gives us

$$R(t) = \int s(t)\,dt = \int \left(\frac{10}{t} + 2\right) dt$$

$$= 10\int \frac{1}{t}\,dt + \int 2\,dt$$

$$= 10\ln|t| + 2t + C$$

Since the time is represented by positive numbers only, we can omit the absolute value and write the partial solution $R(t) = 10\ln t + 2t + C$. To

Interactive Activity

Use the result of Example 3b to evaluate $R(5) - R(1)$ and compare to Example 3a. What does this illustrate?

WWW

find the constant C, we use the fact that, when $t = 1$, $R(t) = 6$, and get

$$\begin{aligned} R(t) &= 10 \ln t + 2t + C \\ 6 &= 10 \ln(1) + 2(1) + C \\ 6 &= 10(0) + 2 + C \\ 6 &= 2 + C \\ 4 &= C \end{aligned}$$

So the revenue function for the store is $R(t) = 10 \ln t + 2t + 4$. ■

Integration of Exponential Functions

From Your Toolbox

- $\frac{d}{dx}[e^x] = e^x$.
- If f is a differentiable function of x, then, by the Chain Rule,

  $\frac{d}{dx} e^{f(x)} = e^{f(x)} \cdot f'(x)$

Now we turn our attention from integrals yielding logarithmic functions to those involving exponential functions. Let's consult the Toolbox to the left to recall some facts on exponential functions.

From the first bulleted fact, we know that $\frac{d}{dx}[e^x] = e^x$. This, along with the fact that differentiation and integration are inverse operations, gives the following integration rule.

Integration Formula for e^x

For the exponential function $f(x) = e^x$,

$$\int e^x \, dx = e^x + C$$

Example 4 Integration of Functions Containing e^x

Determine the definite integral $\int_0^4 (2 - e^x) \, dx$.

Solution

Integrating, we get

$$\begin{aligned} \int_0^4 (2 - e^x) \, dx &= \int_0^4 2dx - \int_0^4 e^x \, dx \\ &= (2x - e^x)\Big|_0^4 = [2(4) - e^4] - [2(0) - e^0] \\ &= 8 - e^4 - 0 + 1 = 9 - e^4 \approx -45.60 \end{aligned}$$ ■

In practice, few integrands are simply e^x. We can easily extend the methods of u-substitution to the exponential function, as is shown in the integration formula for e^u.

Integration Formula for e^u

If u is a differentiable function of x, then

$$\int e^u \, du = e^u + C$$

Example 5 Integration of Functions Containing e^u

Determine the given indefinite integrals.

(a) $\int 10xe^{x^2}\,dx$ **(b)** $\int (2x+1)e^{2x^2+2x}\,dx$

Solution

(a) Understand the Situation: Since the integration formula for e^u suggests letting u be the expression in the exponent, we let $u = x^2$, so $du = 2x\,dx$. Multiplying by 5, we get

$$5\,du = 10x\,dx$$

So the indefinite integral is

$$\int 10xe^{x^2}\,dx = \int e^u \cdot 5\,du = 5\int e^u\,du = 5e^u + C$$

In terms of x, this gives

$$\int 10xe^{x^2}\,dx = 5e^{x^2} + C$$

(b) Understand the Situation: Again, if we let $u = 2x^2 + 2x$, then $du = (4x+2)\,dx$. Since we want the du expression to contain $2x+1$, we multiply by $\frac{1}{2}$ to get

$$\frac{1}{2}\,du = \frac{1}{2}(4x+2)\,dx = (2x+1)\,dx$$

This gives us

$$\int (2x+1)e^{2x^2+2x}\,dx = \int e^u \cdot \frac{1}{2}\,du$$

$$= \frac{1}{2}\int e^u\,du = \frac{1}{2}e^u + C$$

In terms of x, this is

$$\int (2x+1)e^{2x^2+2x}\,dx = \frac{1}{2}e^{2x^2+2x} + C$$

■

✓ Checkpoint 2 Now work Exercise 43.

In many applications, we are asked to integrate functions of the form $f(x) = e^{kx}$, where k is a real number constant. To determine this integral, we use a u-substitution with $u = kx$. Then we see that $du = k\,dx$ and that $\frac{1}{k}\,du = dx$. So, in terms of u, the integral becomes

$$\int e^{kx}\,dx = \frac{1}{k}\int e^u\,du$$

$$= \frac{1}{k}e^u + C = \frac{1}{k}e^{kx} + C$$

Integration Formula for e^{kx}

If k is a nonzero real number coefficient of x, then, for the exponential function $f(x) = e^{kx}$,

$$\int e^{kx}\,dx = \frac{1}{k}e^{kx} + C$$

Example 6 Applying the Integration Formula for e^{kx}

The rate of change in annual total assets in mutual funds in the United States can be modeled by

$$f(x) = 24.83e^{0.196x} \qquad 1 \le x \le 16$$

where x represents the number of years since 1979 and $f(x)$ represents the rate of change in total assets in mutual funds measured in $\frac{\text{billions of dollars}}{\text{year}}$. The graph of f is shown in Figure 16.6.2 (*Source:* U.S. Statistical Abstract, www.census.gov/statab/www).

(a) Evaluate $f(11)$ and interpret.

(b) Use the model to estimate the total increase in the total assets in mutual funds from 1990 to 1995.

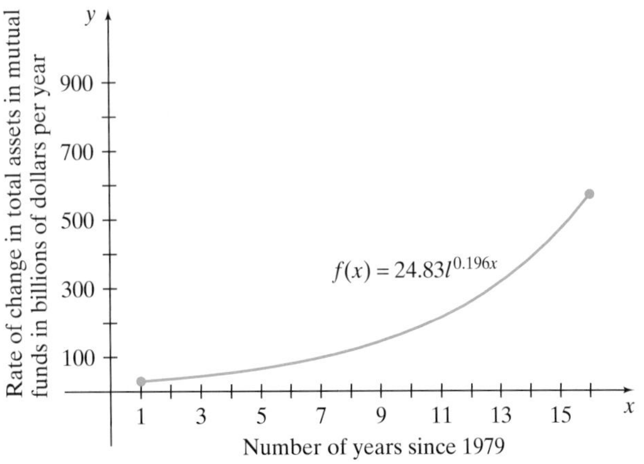

Figure 16.6.2

Solution

(a) **Understand the Situation:** We note that $x = 11$ corresponds to the year 1990. Evaluating $f(11)$ gives us

$$f(11) = 24.83e^{0.196(11)} = 24.83e^{2.156} \approx 214.44$$

Interpret the Solution: This means that in 1990, the total assets in mutual funds was increasing at a rate of about $214.44 \frac{\text{billion dollars}}{\text{year}}$.

(b) **Understand the Situation:** Since function f is a rate function and since the years 1990 and 1995 correspond to the independent variable values $x = 11$

and $x = 16$, respectively, we need to integrate $\int_{11}^{16} f(x)\,dx$ to determine the total increase.

Performing the integration gives us

$$\int_{11}^{16} 24.83e^{0.196x}\,dx = 24.83\int_{11}^{16} e^{0.196x}\,dx$$

From the integration formula for e^{kx}, we have

$$\begin{aligned} 24.83\int_{11}^{16} e^{0.196x}\,dx &= 24.83 \cdot \frac{1}{0.196}e^{0.196x}\Big|_{11}^{16} \\ &= \frac{24.83}{0.196}(e^{0.196(16)} - e^{0.196(11)}) \\ &= \frac{24.83}{0.196}(e^{3.136} - e^{2.156}) \\ &\approx 1821.092 \end{aligned}$$

Interpret the Solution: This means that from 1990 to 1995, the total increase in total assets in mutual funds was about \$1821.092 billion or \$1.821092 trillion. ■

Checkpoint 3

Now work Exercise 63.

Integrating General Exponential Functions

From Your Toolbox

The general exponential model $f(x) = a \cdot b^x$ can be *rewritten in the exponential form* as $g(x) = a \cdot e^{kx}$, where $k = \ln b$.

Thus far we have discussed how to integrate functions of the type $f(x) = e^{kx}$. But what about general exponential functions such as $f(x) = 2^x$ or $g(x) = 3^{0.6x}$? Surely they can be integrated, too. Some calculus courses derive a formula specifically for these functions, but we will rely on techniques that we have already learned. Refer to the Toolbox to the left to recall key ideas from Chapters 11 and 14 on general exponential functions.

To determine the indefinite integral whose integrand has the form b^x, such as $\int 2^x\,dx$, we rewrite the integrand in the exponential form e^{kx}, where $k = \ln b$. Thus, we can rewrite $\int 2^x\,dx$ as $\int e^{\ln 2 \cdot x}\,dx$ and then integrate using the integration formula for e^{kx}.

Example 7 Integrating a General Exponential Function

Evaluate the definite integral $\int_{-2}^{1}\left(\frac{3}{4}\right)^x dx$.

Solution

Understand the Situation: We begin by rewriting the integrand as $e^{\ln(3/4)\cdot x}$ and use the Integration Formula for e^{kx}.

$$\begin{aligned} \int_{-2}^{1}\left(\frac{3}{4}\right)^x dx &= \int_{-2}^{1} e^{\ln(3/4)\cdot x}\,dx \\ &= \frac{1}{\ln\left(\frac{3}{4}\right)}(e^{\ln(3/4)\cdot x})\Big|_{-2}^{1} \end{aligned}$$

Interactive Activity

Use the $\int f(x)\,dx$ or FNINT command on your calculator to verify the solution to Example 7.

WWW

Rewriting $e^{\ln(3/4)\cdot x}$ as $\left(\frac{3}{4}\right)^x$ yields

$$= \frac{1}{\ln\left(\frac{3}{4}\right)}\left[\left(\frac{3}{4}\right)^x\right]\Bigg|_{-2}^{1} = \frac{1}{\ln\left(\frac{3}{4}\right)}\left[\left(\frac{3}{4}\right)^1 - \left(\frac{3}{4}\right)^{-2}\right]$$

$$= \frac{1}{\ln\left(\frac{3}{4}\right)}\left(\frac{3}{4} - \frac{16}{9}\right) = \frac{1}{\ln\left(\frac{3}{4}\right)}\left(-\frac{37}{36}\right) = -\frac{37}{36\ln\left(\frac{3}{4}\right)} \approx 3.57$$

Many times we are given data that are always increasing or always decreasing on an interval. This kind of data is said to be **monotonic**. We can model these data with the general exponential mathematical model $y = a \cdot b^x$, where a is any nonzero real number and with $b > 0, b \neq 1$. To integrate functions of the type $\int a \cdot b^x\,dx$, we *pull out* the constant a and then we can determine $\int b^x\,dx$.

Example 8 Integrating a General Exponential Function Application

The rate of change in the amount of chlorofluorocarbon (CFC) gases emitted into the atmosphere by the U.S. can be modeled by

$$g(x) = -71.673(0.797)^x \qquad 1 \leq x \leq 9$$

where x is the number of years since 1989 and $g(x)$ is the rate of change in CFC gases measured in $\frac{\text{thousand metric tons}}{\text{year}}$ (*Source:* U.S. Energy Information Administration, www.eia.doe.gov).

(a) Use the model to find the total increase (or decrease) in thousands of metric tons of CFC gases emitted into the atmosphere from 1992 to 1996.

(b) Knowing that in 1993, 148 thousand metric tons of CFC gases were emitted, recover G, the emission of CFC gases function.

(c) Evaluate $G(6)$ and interpret.

Solution

(a) **Understand the Situation:** Once again, we are given a rate of change function. Integrating the rate of change function will give us the total accumulation. Here, note that 1992 and 1996 correspond to the independent variable values $x = 3$ and $x = 7$.

Proceeding with the integration gives us

$$\int_3^7 -71.673(0.797)^x\,dx = -71.673\int_3^7 (0.797)^x\,dx = -71.673\int_3^7 e^{\ln(0.797)\cdot x}\,dx$$

$$= \frac{-71.673}{\ln(0.797)}\left[e^{\ln(0.797)\cdot x}\right]\Big|_3^7$$

$$= \frac{-71.673}{\ln(0.797)}(0.797^x)\Big|_3^7$$

$$= \frac{-71.673}{\ln(0.797)}[0.797^7 - 0.797^3] \approx -95.392$$

Interpret the Solution: This means that from 1992 to 1996 the total **decrease** in CFC gases emitted into the atmosphere by the U.S. was about 95.392 thousand metric tons.

(b) **Understand the Situation:** To recover the emission function G, we begin by evaluating the indefinite integral $\int g(x)\,dx$. We then use the information that in 1993 (when $x = 4$), $G(x) = 148$ to determine the constant of integration.

Evaluating the indefinite integral $\int g(x)\,dx$ gives us

$$\begin{aligned} G(x) = \int g(x)\,dx &= \int -71.673(0.797)^x\,dx \\ &= -71.673 \int (0.797)^x\,dx \\ &= -71.673 \int_3^7 e^{\ln(0.797)\cdot x}\,dx \\ &= \frac{-71.673}{\ln(0.797)} e^{\ln(0.797)\cdot x} + C \\ &= \frac{-71.673}{\ln(0.797)}(0.797)^x + C \end{aligned}$$

We now use the information that when $x = 4$, $G(x) = 148$ to determine C.

$$\begin{aligned} G(x) &= \frac{-71.673}{\ln(0.797)}(0.797)^x + C \\ 148 &= \frac{-71.673}{\ln(0.797)}(0.797)^4 + C \\ 148 - \frac{-71.673}{\ln(0.797)}(0.797)^4 &= C \\ 20.546 &\approx C \end{aligned}$$

Interpret the Solution: So the emissions of CFC gases into the atmosphere by the U.S. can be modeled by

$$G(x) = \frac{-71.673}{\ln(0.797)}(0.797)^x + 20.546 \qquad 1 \le x \le 9$$

where x is the number of years since 1989 and $G(x)$ is the amount of CFC gases emitted by the U.S. in thousands of metric tons.

(c) Using the result from part (b), we substitute $x = 6$ and find that

$$\begin{aligned} G(x) &= \frac{-71.673}{\ln(0.797)}(0.797)^x + 20.546 \\ G(6) &= \frac{-71.673}{\ln(0.797)}(0.797)^6 + 20.546 \\ &\approx 101.506 \end{aligned}$$

Interpret the Solution: This means that in 1995, about 101.506 thousand metric tons of CFC gases were emitted into the atmosphere by the U.S. ■

SUMMARY

In this section we analyzed and applied integration formulas for the exponential and logarithmic functions.

- **Integration formula for $\frac{1}{x}$:** $\int \frac{1}{x}\,dx = \ln|x| + C$
- **Integration formula for $\frac{1}{u}$:** $\int \frac{1}{u}\,du = \ln|u| + C$
- **Integration formula for e^x:** $\int e^x\,dx = e^x + C$
- **Integration formula for e^u:** $\int e^u\,du = e^u + C$
- **Integration formula for e^{kx}:** $\int e^{kx}\,dx = \frac{1}{k}e^{kx} + C$

SECTION 16.6 EXERCISES

In Exercises 1–10, evaluate the indefinite and definite integrals.

1. $\int \frac{2}{x}\,dx$

2. $\int \frac{5}{x}\,dx$

3. $\int \frac{-2}{3t}\,dt$

4. $\int \frac{-7}{2t}\,dt$

5. $\int \frac{1+e}{x}\,dx$

6. $\int \frac{e^2-1}{x}\,dx$

7. $\int_1^5 \frac{3}{x}\,dx$

8. $\int_1^8 \frac{5}{x}\,dx$

9. $\int_2^5 \frac{1}{2}x^{-1}\,dx$

10. $\int_5^8 \frac{2}{5}x^{-1}\,dx$

In Exercises 11–24, evaluate the integrals by u-substitution.

11. $\int \frac{1}{2x+3}\,dx$

12. $\int \frac{1}{7-5x}\,dx$

✓ **13.** $\int \frac{t}{t^2+1}\,dt$

14. $\int \frac{2t}{t^2+3}\,dt$

15. $\int \frac{x-2}{x^2-4x+9}\,dx$

16. $\int \frac{3x^2-3}{x^3-3x-7}\,dx$

17. $\int \frac{x^3}{x^4+10}\,dx$

18. $\int \frac{3x^2}{x^3-4}\,dx$

19. $\int \frac{1}{x\cdot\ln 2x}\,dx$

20. $\int \frac{3}{x\cdot\ln 6x}\,dx$

21. $\int_1^2 \frac{2}{\sqrt{x}(\sqrt{x}+4)}\,dx$

22. $\int_1^2 \frac{\ln x}{x}\,dx$

23. $\int_1^3 \frac{2}{3x-2}\,dx$

24. $\int_{-1}^0 \frac{1}{4-5x}\,dx$

Applications

25. The Top-2-Bottom Dress Store has a clearance sale to remove old inventory. Past records show that the rate of sales is given by

$$S(t) = \frac{40}{t} \qquad 1 \le t \le 8$$

where t represents the number days that the sale has been running and $S(t)$ represents the sales rate, measured in $\frac{\text{dresses}}{\text{day}}$.

(a) Evaluate $S(2)$ and interpret.

(b) Determine the total increase in the number of dresses sold between days 2 and 4.

26. The Silver Spur Gun Shop runs a last chance sale based on the sales rate model

$$S(t) = \frac{82}{t} \qquad 1 \le t \le 10$$

where t represents the number days that the sale has been running and $S(t)$ represents the sales rate, measured in $\frac{\text{guns}}{\text{day}}$.

(a) Evaluate $S(4)$ and interpret.

(b) If the store owner must sell 150 guns in the first five days to break even, will the owner do so according to the model?

⇐ 16.4 **27.** The rate of change function for foreign student enrollment in U.S. colleges can be modeled by

$$g(x) = \frac{92.43}{x} \qquad 1 \le x \le 22$$

where x represents the number of years since 1975 and $g(x)$ represents the rate of change in the foreign student enrollment in U.S. colleges, measured in $\frac{\text{thousands of students}}{\text{year}}$ (*Source:* U.S. Statistical Abstract, www.census.gov/statab/www).

(a) Evaluate and interpret $g(6)$.

(b) Evaluate and interpret $\int_1^6 g(x)\,dx$.

⇐ 16.4 **28.** The rate of change in the average dollar amount awarded in the Pell Grant system can be modeled by

$$g(x) = \frac{327}{x} \qquad 1 \le x \le 20$$

where x is the number of years since 1979 and $g(x)$ is the rate of change in the average dollar amount awarded in the

Pell Grant system, measured in $\frac{\text{dollars}}{\text{year}}$ (*Source:* U.S. Department of Education, www.ed.gov).

(a) Evaluate $g(13)$ and interpret.

(b) Evaluate and interpret $\int_{13}^{19} g(x)\,dx$.

29. The rate of change function for the number of employees in the solid waste management industry is modeled by

$$g(x) = \frac{53.44}{x+1} \qquad 0 \le x \le 16$$

where x represents the number of years since 1980 and $g(x)$ represents the rate of change in employees, measured in $\frac{\text{thousands of employees}}{\text{year}}$ (*Source:* U.S. Census Bureau, www.census.gov).

(a) Evaluate and interpret $g(16)$.

(b) Knowing that there were 209.5 thousand employees in the solid waste management industry in 1990, recover the model f that represents the number of employees in the solid waste management industry. [*Hint:* f has the form $f(x) = a + b\ln x$.]

30. The rate of change in the number of public schools with interactive video disks can be modeled by

$$g(x) = \frac{15.1}{x} \qquad 1 \le x \le 7$$

where x represents the number of years since 1991 and $g(x)$ represents the rate of change in the number of public schools with interactive video disks, measured in $\frac{\text{thousands of schools}}{\text{year}}$ (*Source:* U.S. Statistical Abstract, www.census.gov/statab/www).

(a) Evaluate and interpret $g(2)$.

(b) Knowing that in 1992 there were 6.5 thousand schools with the video disks, recover the model f, where $f(x)$ represents the number of public schools with interactive video disks and f has the form $f(x) = a + b\ln x$.

In Exercises 31–40, evaluate the indefinite and definite integrals.

31. $\int 2e^x\,dx$

32. $\int 6e^x\,dx$

33. $\int (e^x - 1)\,dx$

34. $\int (e^x + 3)\,dx$

35. $\int (4e^t + t - 1)\,dt$

36. $\int (t^3 - 3e^t + t)\,dt$

37. $\int_0^1 (e^x - 1)\,dx$

38. $\int_0^1 (6 - e^x)\,dx$

39. $\int_1^2 \frac{e^x + 4}{2}\,dx$

40. $\int_1^2 \frac{3 - e^x}{6}\,dx$

In Exercises 41–58, evaluate the integrals by u-substitution.

41. $\int e^{4x+1}\,dx$

42. $\int e^{8x}\,dx$

✓ **43.** $\int 2xe^{x^2+1}\,dx$

44. $\int xe^{x^2}\,dx$

45. $\int \frac{e^x}{e^x+2}\,dx$

46. $\int \frac{e^x - e^{-x}}{e^x + e^{-x}}\,dx$

47. $\int \frac{3}{e^x(1 - e^{-x})}\,dx$

48. $\int x^2 e^{x^3}\,dx$

49. $\int \frac{e^{\sqrt{x}}}{\sqrt{x}}\,dx$

50. $\int \frac{(e^x+1)^2}{e^x}\,dx$

51. $\int_1^3 e^{-4x}\,dx$

52. $\int_0^1 e^{2x+3}\,dx$

53. $\int_0^1 \frac{1+e^x}{e^x}\,dx$

54. $\int_1^2 \left(e^{2x} - \frac{2}{x}\right)dx$

55. $\int xe^{x^2}\,dx$

56. $\int 3x^5 e^{x^6}\,dx$

57. $\int e^{4x}(e^{4x} + 4)\,dx$

58. $\int e^x(e^x + 10)\,dx$

Applications

59. The FlowStop Company determines that the marginal cost for their new line of faucet parts is given by

$$MC(x) = 1.50 + 0.04e^{0.02x}$$

(a) Evaluate $MC(20)$ and interpret.

(b) If the number of units of the part produced increases from 100 to 150, what is the total increase in cost?

60. The population of a small town is growing at a rate given by

$$R(t) = 250e^{0.04t}$$

where t represents the number of years from the present and $R(t)$ represents the town's population growth rate, measured in $\frac{\text{people}}{\text{year}}$.

(a) Evaluate and interpret $\int_0^3 R(t)\,dt$.

(b) A civil planner estimates that 3000 more people can move into the town without negatively affecting the quality of life. How many more years will it take for the population to reach that number?

61. The rate of change in the number of infant deaths per 1000 live births in Guam is given by

$$g(x) = -1.24e^{-0.042x} \qquad 1 \le x \le 34$$

where x represents the number of years since 1964 and $g(x)$ represents the rate of change in the number of infant deaths per 1000 live births in Guam, measured in $\frac{\text{infant deaths per 1000 live births}}{\text{year}}$ (*Source:* U.S. National Center for Health Statistics, www.cdc.gov/nchs).

(a) Evaluate and interpret $g(31)$.

(b) Evaluate and interpret $\int_1^{31} g(x)\,dx$.

62. The rate of change in the number of infant deaths per 1000 live births in Puerto Rico is given by

$$g(x) = -1.51e^{-0.036x} \qquad 1 \le x \le 39$$

where x represents the number of years since 1959 and $g(x)$ represents the rate of change in the number of infant deaths per 1000 live births in Puerto Rico, measured in $\frac{\text{infant deaths per 1000 live births}}{\text{year}}$ (*Source:* U.S. National Center for Health Statistics, www.cdc.gov/nchs).

(a) Evaluate and interpret $g(36)$.

(b) Evaluate and interpret $\int_6^{36} g(x)\,dx$.

16.4 **63.** The rate of change in the U.S. population can be modeled by

$$g(x) = 1.03e^{0.013x} \qquad 0 \le x \le 100$$

where x represents the number of years since 1900 and $g(x)$ represents the rate of change in the U.S. population, measured in $\frac{\text{millions}}{\text{year}}$ (*Source:* U.S. Census Bureau, www.census.gov).

(a) Evaluate and interpret $g(50)$.

(b) Determine the total increase in the U.S. population from 1900 to 1950.

16.4 **64.** The rate of change in the tuition and fees for in-state students at four-year colleges in the United States can be modeled by

$$g(x) = 95.59e^{0.07x} \qquad 1 \le x \le 15$$

where x represents the number of years since 1984 and $g(x)$ represents the rate of change in the tuition and fees for in-state students at four year colleges, measured in $\frac{\text{dollars}}{\text{year}}$ (*Source:* U.S. National Center for Education Statistics, www.nces.ed.gov).

(a) Evaluate and interpret $g(12)$.

(b) Determine the increase in the total paid in tuition and fees for in-state students at four-year colleges in the United States from 1992 to 1996.

65. *(continuation of Exercise 63)*

(a) Graph the model given in Exercise 63 in the viewing window [0, 100] by [0, 10].

(b) Use the $\int f(x)\,dx$ or the FNINT command to verify the result of Exercise 63b.

(c) Use the graph from part (a) to show that the population of the U.S. is increasing everywhere on the interval [0, 100].

66. *(continuation of Exercise 64)*

(a) Graph the model in Exercise 64 in the viewing window [1, 15] by [0, 300].

(b) Use the $\int f(x)\,dx$ or the FNINT command to verify the result of Exercise 64b.

(c) Use the graph from part (a) to show that tuition and fees for in-state students at four-year colleges is increasing everywhere on the interval [1, 15].

In Exercises 67–76, evaluate the indefinite and definite integrals.

67. $\int 5^x\,dx$

68. $\int 10^x\,dx$

69. $\int 5 \cdot \left(\frac{1}{2}\right)^x dx$

70. $\int 2 \cdot \left(\frac{5}{8}\right)^x dx$

71. $\int 7^{-x}\,dx$

72. $\int 10^{-x}\,dx$

73. $\int_1^2 10^{3x}\,dx$

74. $\int_1^5 5^{-2x}\,dx$

75. $\int_{-1}^1 2^{3x-1}\,dx$

76. $\int_1^{\sqrt{2}} x \cdot 2^{x^2}\,dx$

Applications

77. The rate of change function for the number of people confined in state or correctional facilities for more than one year can be modeled by

$$P(x) = 0.2612(1.086)^x \qquad 0 \le x \le 15$$

where x represents the number of years since 1980 and $P(x)$ represents the rate of change in the U.S. prison population, measured by $\frac{\text{100,000 inmates}}{\text{year}}$ (*Source:* U.S. Statistical Abstract, www.census.gov/statab/www).

(a) Evaluate $P(10)$ and interpret the result.

(b) Find the total change in the prison population from 1985 to 1990.

78. The rate of change in dormitory charges for in-state students at two-year colleges in the United States can be modeled by

$$f(x) = 29.28(1.036)^x \qquad 1 \le x \le 15$$

where x represents the number of years since 1984 and $f(x)$ represents the rate of change in dormitory charges in $\frac{\text{dollars}}{\text{year}}$ (*Source:* U.S. National Center for Education Statistics, www.nces.ed.gov).

(a) Evaluate $f(12)$ and interpret.

(b) Evaluate $\int_1^{12} f(x)\,dx$ and interpret.

SECTION PROJECT

The rate of change in dormitory charges for in-state students at four-year colleges in the United States is shown in Table 16.6.1. The rate is measured in $\frac{\text{dollars}}{\text{year}}$.

Table 16.6.1

Year	Rate of Change in Dorm Prices
1985	58.97
1990	74.69
1992	82.08
1994	90.22
1996	99.16
1998	108.98
1999	114.25

SOURCE: U.S. National Center for Education Statistics, www.nces.ed.gov

(a) Let x represent the number of years since 1984 and $f(x)$ represent the rate of change in dormitory prices. Use the regression capabilities of your calculator to determine a general exponential growth model in the form

$$f(x) = a \cdot b^x \qquad 1 \le x \le 15$$

(b) Write the units for $f(x)$.

(c) Evaluate and interpret $\int_1^{14} f(x)\,dx$.

(d) Let x represent the number of years since 1984 and $g(x)$ represent the rate of change in dormitory prices. Use the regression capabilities of your calculator to determine a linear model in the form

$$g(x) = ax + b \qquad 1 \le x \le 15$$

(e) Evaluate and interpret $\int_1^{14} g(x)\,dx$. Why is this result different from that found in part (c)?

Section 16.7 Differential Equations: Separation of Variables

So far in this chapter, we have seen equations of the form

$$P'(x) = \frac{x}{15} + 20 \quad \text{and} \quad \frac{dy}{dx} = 4x^2 - 3x$$

These are examples of differential equations. Quite simply, an equation that involves a function and one (or more) of its derivatives is called a **differential equation**. Entire courses and textbooks are dedicated to the study of these equations. In this section and the next, we will study **first-order differential** equations. These are differential equations that involve a first derivative but no higher derivatives. The technique we will use to solve this kind of differential equation is called **separation of variables**.

The Differential Equation $y = f'(x)$

Since the beginning of Chapter 16, we have really been solving differential equations that have the form $y = f'(x)$. When we see

$$\frac{dy}{dx} = 3x^2$$

we are saying that the derivative of some function is $3x^2$. This differential equation is solved by integrating

$$y = \int 3x^2\,dx = x^3 + C$$

Example 1 Solving a Differential Equation

Solve the differential equation $y' = x^4$.

Solution

If we replace y' with $\frac{dy}{dx}$, we have

$$\frac{dy}{dx} = x^4$$

Since $\frac{dy}{dx} = x^4$, we know that y is an antiderivative of x^4; hence,

$$y = \int x^4\,dx = \frac{1}{5}x^5 + C$$ ■

Example 1 indicates that:

> In general, a differential equation of the form $y = f'(x)$ or $\frac{dy}{dx} = f(x)$ is solved by integrating
>
> $$y = \int f(x)\,dx$$

In Example 1 we saw that the solution to $y' = x^4$ is $y = \frac{1}{5}x^5 + C$. We call $y = \frac{1}{5}x^5 + C$ the **general solution** of the differential equation $y' = x^4$ because it represents all possible solutions, one for each choice of the constant C. When the constant C is replaced with a particular value, we then have a **particular solution**. A few solutions to the differential equation $y' = x^4$ are

$$y = \frac{1}{5}x^5 + 1 \qquad C = 1$$

$$y = \frac{1}{5}x^5 \qquad C = 0$$

$$y = \frac{1}{5}x^5 - 1 \qquad C = -1$$

Notice in Figure 16.7.1 that each curve has the same shape. This is because each curve has the same derivative $y' = x^4$. For this reason, we say that the different values for C in a general solution of a differential equation produces a *family of curves*, where the general solution $y = f(x) + C$ represents the entire family.

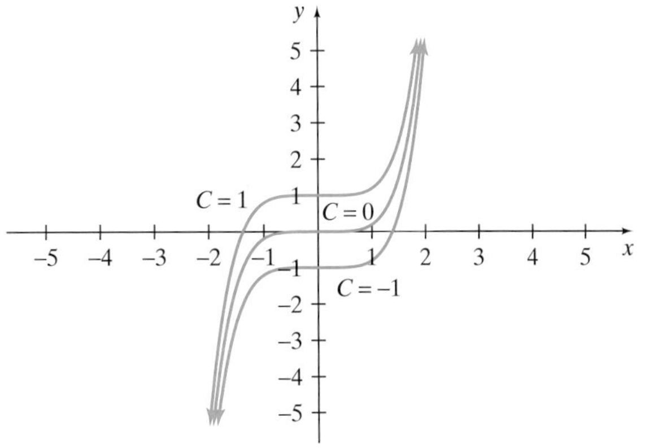

Figure 16.7.1 Graphs of $y = \frac{1}{5}x^5 + C$ for $C = -1, C = 0$, and $C = 1$.

Separation of Variables

One important type of differential equation is called a separable differential equation. We say that a differential equation is *separable* if we can move everything involving x's to one side of the equation and everything involving y's to the other side. Example 2 illustrates the process.

Example 2 **Finding a General Solution by Separation of Variables**

Determine the general solution to $\frac{dy}{dx} = \frac{x^2}{y^3}$.

Solution

The method known as separation of variables requires us to separate the x's from the y's algebraically.

$$\frac{dy}{dx} = \frac{x^2}{y^3}$$

$$dy = \frac{x^2}{y^3}\,dx \qquad \text{Multiply both sides by } dx$$

$$y^3\,dy = x^2\,dx \qquad \text{Multiply both sides by } y^3$$

We now integrate both sides of the equation and get

$$\int y^3\,dy = \int x^2\,dx \qquad \text{Integrate both sides}$$

$$\frac{1}{4}y^4 + C_0 = \frac{1}{3}x^3 + C_1 \qquad \text{Utilize the Power Rule for Integration}$$

$$\frac{1}{4}y^4 = \frac{1}{3}x^3 + C \qquad \text{Combine both constants and rewrite with one constant}$$

Solving for y yields

$$y^4 = \frac{4}{3}x^3 + 4C$$

$$y^4 = \frac{4}{3}x^3 + C \qquad \text{Four times a constant is just another constant}$$

$$y = \sqrt[4]{\frac{4}{3}x^3 + C} \qquad \text{The general solution using principal fourth root}$$

■

Notice that we can check our solution for Example 2 as follows:

$$y = \sqrt[4]{\frac{4}{3}x^3 + C} = \left(\frac{4}{3}x^3 + C\right)^{1/4}$$

$$\frac{dy}{dx} = \frac{1}{4}\left(\frac{4}{3}x^3 + C\right)^{-(3/4)} \cdot 4x^2 = \frac{x^2}{\left(\frac{4}{3}x^3 + C\right)^{3/4}} = \frac{x^2}{\left[\left(\frac{4}{3}x^3 + C\right)^{1/4}\right]^3}$$

$$= \frac{x^2}{y^3} \qquad \text{Substitute } y \text{ for } \left(\frac{4}{3}x^3 + C\right)^{1/4}$$

So we have $\frac{dy}{dx} = \frac{x^2}{y^3}$ as desired. Example 2 outlines the following process for separation of variables.

Separation of Variables

A separable first-order differential equation has the form

$$\frac{dy}{dx} = f(x) \cdot g(y) \quad \text{or} \quad \frac{dy}{dx} = \frac{f(x)}{g(y)}, \qquad g(y) \neq 0$$

It is solved by separating the variables and integrating

$$\int g(y)dy = \int f(x)\,dx$$

In Example 2 we found a general solution. Many times, we are given some additional information that allows us to determine the **particular solution**. We call the additional information an **initial condition**. Example 3 illustrates how to determine a particular solution.

Example 3 Determining a Particular Solution

Solve the differential equation $\frac{dy}{dx} = \frac{x^2}{y^3}$ with the condition that $x = 0$, $y = 1$.

Solution

From Example 2, we know that the general solution is

$$y = \sqrt[4]{\frac{4}{3}x^3 + C}$$

Given that $y = 1$, when $x = 0$, we can substitute this into the general solution and determine C.

$$y = \sqrt[4]{\frac{4}{3}x^3 + C}$$

$$1 = \sqrt[4]{\frac{4}{3}(0)^3 + C}$$

$$1 = \sqrt[4]{C}$$

$$1 = C$$

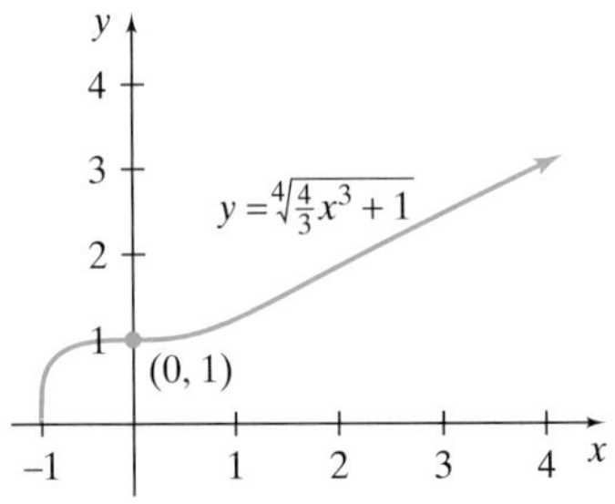

Figure 16.7.2

So the particular solution is $y = \sqrt[4]{\frac{4}{3}x^3 + 1}$. (See Figure 16.7.2.) ■

Checkpoint 1

Now work Exercise 33.

Example 3 illustrates that, when solving a separable differential equation with an initial condition, we use the following steps:

1. Separate the variables.
2. Determine the general solution.
3. Use the initial condition to determine C.

Several solutions to $\frac{dy}{dx} = \frac{x^2}{y^3}$ are shown in Figure 16.7.3, with our particular solution from Example 3 shown in blue.

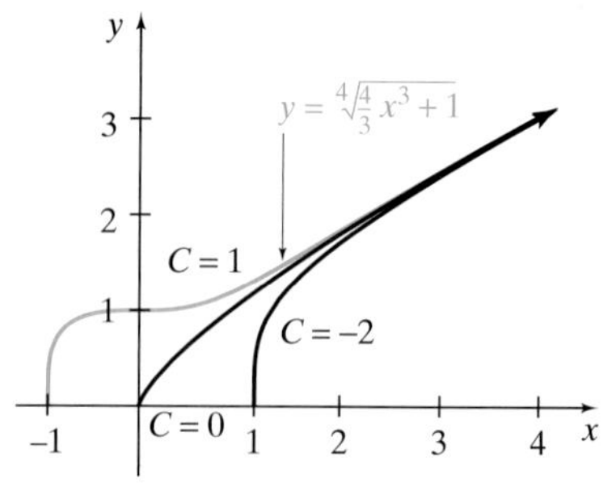

Figure 16.7.3 Graphs of $y = \sqrt[4]{\frac{4}{3}x^3 + C}$ of the solutions to $\frac{dy}{dx} = \frac{x^2}{y^3}$.

Technology Option

The computing capabilities of a graphing calculator can help us to visualize the solutions of a differential equation. For example, the differential equation $\frac{dy}{dx} = \frac{x^2}{y^3}$ in Example 3 tells us that the slope at any point (x, y) on a graph is given by the expression $\frac{x^2}{y^3}$. A graph of these slopes is called a **slope field**. Calculator programs can draw slope fields for us. A slope field program for the TI-83 is included in Appendix C.

Example 4 Finding a Particular Solution When a Logarithm Is Involved

Determine the particular solution for $\frac{dy}{dx} = 3xy$ with the condition $x = 0$, $y = 3$.

Solution

We start by separating the variables.

$$\frac{dy}{dx} = 3xy$$

$$\frac{1}{y}\frac{dy}{dx} = 3x$$

$$\frac{1}{y}\,dy = 3x\,dx$$

Integrating both sides yields

$$\int \frac{1}{y}\,dy = \int 3x\,dx$$

$$\ln y = \frac{3}{2}x^2 + C \qquad y > 0$$

T From Your Toolbox

Recall, the definition of a logarithm, $y = \ln x$ if, and only if, $e^y = x$.

Using the Toolbox to the left, we can rewrite this equation as

$$y = e^{(3/2)x^2 + C}$$
$$y = e^{(3/2)x^2} \cdot e^C$$

Since e^C represents an arbitrary constant, we can simply write C for e^C and represent the general solution as

$$y = C \cdot e^{(3/2)x^2}$$

Now, applying the initial condition $x = 0$, $y = 3$,

$$3 = C \cdot e^{(3/2)(0)^2}$$
$$3 = C \cdot e^0 = C \cdot 1 = C$$

So the particular solution is $y = 3e^{(3/2)x^2}$. ■

Checkpoint 2

Now work Exercise 63.

Sometimes we will find that there can be more than one general solution to a differential equation. This is illustrated in Example 5.

Example 5 Finding All General Solutions to a Differential Equation

Find the general solution to $y \cdot \dfrac{dy}{dx} = 2x$.

Solution

Since the x's and y's are already separated, we can simply multiply by dx and then integrate both sides.

$$y \cdot \frac{dy}{dx} = 2x$$
$$y\,dy = 2x\,dx$$
$$\int y\,dy = \int 2x\,dx$$
$$\frac{1}{2}y^2 = x^2 + C$$
$$y^2 = 2x^2 + 2C \quad \text{Multiply by 2}$$
$$y^2 = 2x^2 + C \quad \text{Since } 2C \text{ is simply another constant } C$$
$$y = \pm\sqrt{2x^2 + C} \quad \text{Take the square root of both sides}$$

So we have two square roots, one positive and the other negative. Thus, these two solutions

$$y = \sqrt{2x^2 + C}$$

and

$$y = -\sqrt{2x^2 + C}$$

together are the general solution to the differential equation. ■

Example 6 Solving a Differential Equation by *u*-Substitution

Determine the general solution for $\frac{dy}{dx} = xy - 3x$.

Solution

To separate the variables for this differential equation, we need to factor on the right side.

$$\frac{dy}{dx} = xy - 3x$$

$$\frac{dy}{dx} = x(y-3) \qquad \text{Factor } x \text{ from each term}$$

$$\frac{dy}{y-3} = x\,dx \qquad \text{Multiply by } dx \text{ and divide by } y-3$$

$$\int \frac{1}{y-3}\,dy = \int x\,dx \qquad \text{Integrate both sides}$$

$$\int \frac{1}{u}\,du = \int x\,dx \qquad \text{Substitute } u = y-3 \text{ so that } du = dy$$

$$\ln u = \frac{1}{2}x^2 + C \qquad \text{Integrate and assume that } u > 0$$

$$\ln(y-3) = \frac{1}{2}x^2 + C \qquad \text{Resubstitute } y-3 \text{ for } u$$

$$y - 3 = e^{(1/2)x^2 + C} \qquad \text{Definition of logarithm}$$

$$y - 3 = e^{(1/2)x^2} \cdot e^C = Ce^{(1/2)x^2} \qquad \text{Rule of exponents and replace } e^C \text{ by } C$$

$$y = Ce^{(1/2)x^2} + 3$$

■

Applications

Many of the applications of separable differential equations involve differential rate equations, where a rate of change function is given. Usually, we use the separation of variables technique to recover an "original" function.

Example 7 Application of Separation of Variables

The annual rate of increase in employees between 1950 and 1985 at the New-Media Company can be modeled by the differential rate equation

$$\frac{dy}{dt} = 0.3t^{0.6} \qquad 0 \le t \le 35$$

where t represents the number of years since 1950 and y represents the number of employees in hundreds.

(a) Knowing that the company had 1800 employees in 1965, find the employee growth function y for the company.

(b) Use the result from part (a) to estimate the number of employees at the company in 1983.

Solution

(a) We start by determining the general solution to the differential rate equation.

$$\frac{dy}{dt} = 0.3t^{0.6}$$

$$dy = (0.3t^{0.6})\,dt$$

$$\int dy = \int (0.3t^{0.6})\,dt$$

$$y = 0.3 \cdot \frac{1}{1.6}t^{1.6} + C$$

$$y = \frac{3}{16}t^{1.6} + C$$

Since the company had 1800 employees in 1965, we know that $y = 18$ when $t = 15$. This allows us to solve for C.

$$18 = \frac{3}{16}(15)^{1.6} + C$$

$$18 - \frac{3}{16}(15)^{1.6} = C$$

$$3.72 \approx C$$

So the employee growth of the company can be modeled by

$$y = \frac{3}{16}t^{1.6} + 3.72$$

(b) To estimate the number of employees at the company in 1983, we need to use the model found in part (a) and evaluate when $t = 33$.

$$y = \frac{3}{16}(33)^{1.6} + 3.72$$

$$y \approx 54.14$$

So, according to the model, the New Media Company had about 54.14 hundred, or about 5414, employees in 1983. ■

SUMMARY

- A differential equation of the form $y = f'(x)$ or $\frac{dy}{dx} = f(x)$ is solved by integrating $y = \int f(x)\,dx$.
- A separable first-order differential equation $\frac{dy}{dx} = f(x) \cdot g(y)$, $g(y) \neq 0$, is solved by separating the variables and integrating $\int g(y)\,dy = \int f(x)\,dx$.

SECTION 16.7 EXERCISES

In Exercises 1–14, determine if the given function f is the solution to the given differential equation by determining the derivative f′.

1. $f(x) = 5x + 3$, $y' = 5$
2. $f(x) = x^2$, $f'(x) = 2x - 7$
3. $f(x) = x^5 - 4$, $y' = x^4$
4. $f(x) = x^6 + x - 4$, $y' = 6x^5 - 1$
5. $f(x) = \frac{1}{x}$, $y' = \frac{-1}{x^2}$
6. $f(x) = \frac{1}{x}$, $xy' = -y$
7. $f(x) = 2x^3 + 10$, $y' = 3x^2$
8. $f(x) = x^2 + 1$, $3xy' = 6y - 6$
9. $f(x) = \ln x$, $xy' = 1$
10. $f(x) = \frac{1}{2}\ln x$, $2xy' = 1$
11. $f(x) = \frac{3}{e^{4x}}$, $y' + 4y = 0$
12. $f(x) = xe^x$, $xy' = y(x + 1)$
13. $f(x) = e^{3x} + 5$, $y' + 15 = 3y$
14. $f(x) = \frac{e^{2x} + x^2}{2}$, $y' - 3 = e^{2x} + x$

In Exercises 15–24, determine the general solution to the differential equation.

15. $\frac{dy}{dx} = -2$
16. $\frac{dy}{dx} = e$
17. $\frac{dy}{dx} = 1 - 3x$
18. $\frac{dy}{dx} = 2x$
19. $\frac{dy}{dx} = 5x^3 + x - 2$
20. $\frac{dy}{dx} = 10x^4 - x^2 + 2$
21. $\frac{dy}{dx} = \sqrt{x} + 2$
22. $\frac{dy}{dx} = \sqrt[4]{x} + x$
23. $\frac{dy}{dx} = \frac{5}{x^2}$
24. $\frac{dy}{dx} = x^2 - \frac{1}{x^2}$

In Exercises 25–30, graph a family of curves for the general solutions of the given differential equations using the values $C = 0$, $C = -2$, and $C = 3$.

25. $\frac{dy}{dx} = 2$
26. $\frac{dy}{dx} = -5$
27. $\frac{dy}{dx} = 3x + 1$
28. $\frac{dy}{dx} = 1 - 2x$
29. $\frac{dy}{dx} = \sqrt{x} + 3$
30. $\frac{dy}{dx} = \sqrt{x + 2}$

For Exercises 31–44, determine the particular solution to the differential equation using the given condition.

31. $\frac{dy}{dx} = 12x$, $x = 1$, $y = 8$
32. $\frac{dy}{dx} = 12x$, $x = 1$, $y = -1$
33. $\frac{dy}{dx} = 3x^2 + 4x$, $x = 1$, $y = 6$
34. $\frac{dy}{dx} = xe^{x^2}$, $x = 0$, $y = 2$
35. $\frac{dy}{dt} = 7 - 4t$, $t = 0$, $y = 3$
36. $\frac{dy}{dt} = 7 - 4t$, $t = -1$, $y = 3$
37. $\frac{dy}{dx} = x^3 - 2x$, $x = 0$, $y = -2$
38. $\frac{dy}{dx} = -x^2 - x$, $x = 1$, $y = 1$
39. $\frac{dy}{dt} = t^2 + 2t + 3$, $t = 0$, $y = 4$
40. $\frac{dy}{dx} = \sqrt[3]{x^2} - x$, $x = 0$, $y = 6$
41. $\frac{dy}{dx} = 2x^{0.3}$, $x = 0$, $y = 4$
42. $\frac{dy}{dx} = 4x^{1.7}$, $x = 0$, $y = 1$
43. $\frac{dy}{dx} = 1 - \frac{2}{x}$, $x = 1$, $y = 6$
44. $\frac{dy}{dx} = 3 - \frac{1}{x}$, $x = 1$, $y = 8$

Applications

45. The Strike Now Company determines that the marginal cost for their MaxiPak of lighters is given by the differential rate equation

$$\frac{dC}{dx} = -0.02x + 6 \qquad 0 \le x \le 300$$

where x is the number of MaxiPaks produced and C is the cost in dollars.

(a) Determine the general solution of the cost function C.

(b) Find the particular cost function C, knowing that the cost of producing 10 MaxiPaks of lighters is \$400.

46. The Sleep-4-Ever Company determines that the marginal revenue for their queen-sized mattress is given by the differential rate equation

$$\frac{dR}{dx} = 157$$

where x is the number of queen-sized mattresses sold and R is the revenue in dollars.

(a) Determine the general solution of the revenue function R.

(b) Find the particular revenue function R, knowing that there is zero revenue for zero sales.

47. Suppose that the ChocoCola Company has changed their formula for their Chocolate Cherry Cola. The public is slow in accepting this new formula, but the company is winning over Chocolate Cherry Cola drinkers over time. The company's research department determines that the rate of acceptance of the new cola can be modeled by the differential equation

$$\frac{dA}{dt} = t + 1.1 \qquad 0 \le t \le 12$$

where t is the time that the new formula for Chocolate Cherry Cola has been on the market, measured in months, and A is the percentage of the Chocolate Cherry Cola drinkers who have accepted the new formula.

(a) Evaluate $\frac{dA}{dt}$ when $t = 10$ and interpret.

(b) Determine the general solution for the differential rate equation and interpret.

(c) Find the particular solution, knowing that 5% of the drinkers had accepted the new formula as it was released. [That is, the initial condition is (0, 5).]

48. The increase in leasable area in shopping centers in the United States can be modeled by the differential rate equation

$$\frac{dA}{dt} = 0.12 \qquad 1 \le t \le 10$$

where t represents the number of years since 1989 and A represents the leasable area in shopping centers in billions of square feet (*Source:* U.S. Statistical Abstract, www.census.gov/statab/www).

(a) Determine the general solution for the differential rate equation.

(b) Knowing that there was 4.87 billion square feet of leasable area in shopping centers in 1994, determine the particular solution for the differential equation.

(c) Evaluate $A(8)$ and interpret.

49. The change in the death rate from heart disease in the United States can be modeled by the differential rate equation

$$\frac{dD}{dt} = \frac{-20.17}{t^{1.06}} \qquad 1 \le t \le 20$$

where t represents the number of years since 1979 and D is the number of deaths per 100,000 Americans caused by heart disease (*Source:* U.S. Census Bureau, www.census.gov).

(a) Evaluate $\frac{dD}{dt}$ when $t = 10$ and interpret.

(b) Determine the particular solution $D(t)$ to the differential equation, knowing that the number of deaths in 1980 was found to be 336.18 deaths per 100,000 Americans.

(c) Evaluate $D(15)$ and interpret.

50. A conservationist determines that the rate of population growth of the rare blue-spotted frog on a wildlife reserve during a year long study can be modeled by the differential rate equation

$$\frac{dp}{dt} = \frac{15}{t^{0.6}} \qquad 1 \le t \le 12$$

where t is the number of months since the start of the study and p represents the number of frogs.

(a) Evaluate $\frac{dp}{dt}$ when $t = 5$ and interpret.

(b) Determine the particular solution $p(t)$ of the differential equation knowing that the number of frogs observed after $t = 9$ months of the study was 90.

(c) Evaluate $p(5)$ and interpret.

For Exercises 51–61, determine the general solution to the separable differential equations.

51. $\frac{dy}{dx} = \frac{x}{y}$

52. $\frac{dy}{dx} = \frac{3y}{x}$

53. $\frac{dy}{dx} = 3x^2 y$

54. $\frac{dy}{dx} = 4x^3 y$

55. $\frac{dy}{dx} = e^{x+y}$ (*Hint:* Rewrite the right-hand side as two factors.)

56. $\frac{dy}{dx} = e^{x-y}$ (*Hint:* Rewrite the right-hand side as two factors.)

57. $\frac{dy}{dx} = (2x + 2)y$

58. $\frac{dy}{dx} = 4xy + 7y$

59. $\frac{dy}{dx} = \sqrt{xy}$ (*Hint:* Rewrite the right-hand side as two factors.)

60. $\frac{dy}{dx} = \frac{y^2}{x^2}$

61. $\frac{dy}{dx} = ye^x$

For Exercises 62–72, determine the particular solution to the separable differential equations using the given condition.

62. $\frac{dy}{dx} = xy, x = 0, y = -1$

✓ **63.** $\frac{dy}{dx} = 2xy, x = 0, y = 1$

64. $\frac{dy}{dx} = \frac{y}{x}, x = 1, y = 3$

65. $\frac{dy}{dx} = 2xy^4, x = 0, y = 1$

66. $\frac{dy}{dx} = \frac{x+1}{xy}, x = 1, y = 2$

67. $\frac{dy}{dx} = e^{x-y}, x = 0, y = 0$

68. $x \cdot \frac{dy}{dx} = xy, x = 0, y = 1$

69. $x \cdot \frac{dy}{dx} = x^4 y, x = 3, y = 10$

70. $\frac{dy}{dx} = xye^{x^2}, x = 0, y = e$

71. $\frac{dy}{dx} = \frac{xy}{1+x^2}, x = -1, y = 2$

72. $\frac{dy}{dx} = \frac{1+x^2}{xy}, x = 2, y = 4$

SECTION PROJECT

Consider the New Skin Company, which tracks the marginal revenue (in hundreds of dollars), at various weekly sales levels, of its designer face cream. The data are summarized in Table 16.7.1.

Table 16.7.1

Sales Level, x	Marginal Revenue, $\frac{dR}{dx}$
1	3.97
5	3.91
10	3.77
15	3.68
20	3.60
25	3.48
30	3.43

(a) Use the regression capabilities of your calculator to find a linear regression model for the marginal revenue in the form $\frac{dR}{dx} = ax + b$ for $1 \leq x \leq 30$.

(b) Knowing that there is no revenue when sales are zero, determine the revenue function R.

(c) Find the sales level x that will maximize the revenue.

(d) Determine the price–demand function p.

Section 16.8 Differential Equations: Growth and Decay

Quite often in the business, life, and social sciences, we wish to study the behavior of a quantity as it increases or decreases over time. In this section, we will study **exponential growth** and **exponential decay** phenomena that are modeled by a specific type of *differential equation*. These growth and decay models have the properties of exponential functions that we have studied so far. We will also examine how exponential growth and decay models affect both the business and life sciences. And we will examine **limited** and **logistic models** in which the growth is limited by some y-value. In Chapter 12 we called these y-values **horizontal asymptotes**.

Unlimited Growth

There are many applications in which *the rate of change increases proportionally to the amount present.* With compound interest, for example, we know that as interest in the account accumulates, the faster the account balance grows. In the social sciences, we know that as a population increases, more individuals are added to the population, and the population increases faster. These are all examples of **unlimited growth models**. If we call y the quantity at any given time t, the *exponential growth model* satisfies the equation

(rate of change in quantity) = (is proportional to)(the amount present) or, as a differential equation with constant k, we write

$$y' = ky$$

To solve this differential equation, we rely on our separation of variables technique. Since the independent variable for this kind of model is usually time t, we can write

$$\frac{dy}{dt} = ky \qquad \text{Replace } y' \text{ with } \frac{dy}{dt}$$

$$\int \frac{1}{y}\,dy = \int k\,dt \qquad \text{Divide by } y \text{ and integrate each side}$$

$$\ln y = kt + C \qquad \text{Perform the integration, assume } y > 0$$

$$y = e^{kt+C} \qquad \text{Definition of logarithm}$$

$$y = e^{kt} \cdot e^{C} = Ce^{kt} \qquad \text{Rule of exponents and replace } e^{C} \text{ with } C$$

Unlimited Growth Model

The solution to the differential equation $y' = ky$ is the **unlimited growth model**

$$y = Ce^{kt}$$

with initial value when $t = 0$, $y = C$. (See Figure 16.8.1.)

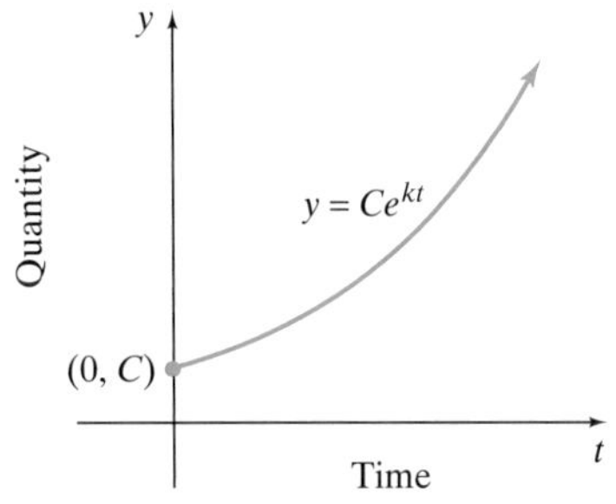

Figure 16.8.1

Notes: 1. When the rate of growth is proportional to the present size, we have a condition called **unlimited growth**. (The term unlimited means that y grows without limit.)

2. Whenever we encounter this situation, the differential equation of the form $y' = ky$, we can immediately write the solution as $y = Ce^{kt}$. Remember that C is the initial value.

3. k is called the **rate of growth**.

Example 1 Writing an Exponential Growth Model with a Given Rate

A biological researcher initially starts a culture of 200 bacteria and knows from past experience that the number of bacteria grows proportionally to the amount present. Past studies have shown that the rate of growth is 11% each hour.

(a) Write an exponential growth model for the number of bacteria present after t hours.

(b) Evaluate y, when $t = 6$ and interpret.

(c) Evaluate y' when $t = 6$ and interpret.

Solution

Interactive Activity

Using the model in Example 1, graph y and y' in the same viewing window. Is y' positive or negative for $t \geq 0$? Is y' increasing or decreasing? What does this imply about the properties of y?

WWW

(a) **Understand the Situation:** We know that when the time starts the number of bacteria is 200. Thus, when $t = 0$, $y = 200$, meaning that the initial amount $C = 200$. This gives us the exponential growth model

$$y = 200\,e^{kt}$$

Since past studies have shown the rate of growth to be 11% each hour, we have $k = 0.11$.

So the function for the number of bacteria present after t hours is given by

$$y = 200\,e^{0.11t}$$

(b) Evaluating y at $t = 6$ gives

$$y = 200\,e^{0.11(6)} = 200\,e^{0.66} \approx 386.96$$

Interpret the Solution: This means that, according to the model, there will be about 387 bacteria present after 6 hours.

(c) The derivative of $y = 200\,e^{0.11t}$ is

$$y' = \frac{d}{dt}(200\,e^{0.11t}) = 200\,e^{0.11t}\,0.11 = 22\,e^{0.11t}$$

Evaluating y' at $t = 6$ yields

$$y' = 22\,e^{0.11(6)} = 22\,e^{0.66} \approx 42.57\,\frac{\text{bacteria}}{\text{hour}}$$

Interpret the Solution: Thus, after 6 hours the number of bacteria in the culture is growing at a rate of about 42.57 bacteria per hour. ■

Checkpoint 1

Now work Exercise 3.

In practice, we are usually not given the growth rate for a differential equation as we were in Example 1. Instead, we are usually given an initial amount C and then given some other ordered pair, which we will call (t_1, y_1). We can use this ordered pair to determine the rate of growth. (This technique will be demonstrated in Example 2.)

Exponential Decay

Another family of applications relating to exponential models is called **exponential decay**. This type of model is used to exhibit behavior in which the rate of change decreases proportionally to the amount present. In this case, the differential equation for such behavior is

$$(\text{rate of change in quantity}) = (\text{is proportional to})(\text{the amount present})$$

$$f'(t) = -k \cdot f(t)$$

The exponential model associated with this differential equation is $f(t) = a \cdot e^{-kt}$, where k is the decay constant.

Rule for Exponential Decay

The general exponential model $y = a \cdot e^{-kt}$ satisfies the differential equation

$$y' = -k \cdot f(t)$$

where the rate of change decreases proportionally to the amount present. a is called the **initial amount** and k is the **rate of decay**.

Just as with growth models, exponential decay models are usually presented with an initial amount, $f(0) = a$, and some other condition, given by an ordered pair (t_1, y_1). One common application of exponential decay models is the *half-life of a radioactive substance.* Half-life is the amount of time that a radioactive substance will be reduced by half to an inert substance.

Example 2 Determining Radioactive Decay

A scientist at the Los Altos Laboratory has a 100-milligram sample of the radioactive substance cobalt 60. The half-life of cobalt 60 is known to be 5.3 years.

(a) Determine an exponential decay model for the number of milligrams of cobalt 60 present after t years.

(b) How many years will it take for the sample of cobalt 60 to decay to 10 milligrams?

Interactive Activity

Notice that for radioactive decay models the decay rate is independent of the initial amount. Solve the exponential equation $\frac{1}{2}a = a \cdot e^{-kT}$ to find a formula for the decay rate k of a radioactive substance with a half-life of T years.

WWW

Solution

(a) Understand the Situation: We know that the initial amount is 100 milligrams, so this gives us

$$f(t) = 100e^{-kt}$$

where t represents the number of years and $f(t)$ represents the amount of cobalt 60 present. Knowing that the half-life of cobalt 60 is 5.3 years tells us that, when $t_1 = 5.3$, $y_1 = \frac{1}{2}(100) = 50$ milligrams.

So we arrive at the equation

$$50 = 100e^{-k(5.3)}$$

$$\frac{1}{2} = e^{-k(5.3)}$$

$$\ln\left(\frac{1}{2}\right) = \ln e^{-k(5.3)}$$

$$\ln\left(\frac{1}{2}\right) = -k(5.3)$$

$$-\frac{\ln\left(\frac{1}{2}\right)}{5.3} = k$$

$$0.13 \approx k$$

Interpret the Solution: So the exponential decay model for the cobalt 60 sample is

$$f(t) = 100e^{-0.13t}$$

(b) **Understand the Situation:** Here we need the t-value so that $f(t) = 10$. So we need to solve the equation $10 = 100e^{-0.13t}$ for t.

Solving for t gives us

$$\begin{aligned} 10 &= 100e^{-0.13t} \\ \frac{1}{10} &= e^{-0.13t} \\ \ln\left(\frac{1}{10}\right) &= \ln e^{-0.13t} \\ \ln\left(\frac{1}{10}\right) &= -0.13t \\ -\frac{\ln\left(\frac{1}{10}\right)}{0.13} &= t \\ 17.71 &\approx t \end{aligned}$$

Interpret the Solution: This means that it will take about 17.71 years for the 100-milligram cobalt 60 sample to decay to 10 milligrams. ■

Limited Growth Models

There are some pitfalls to the unlimited growth model. For example, a town of 4 thousand that is currently growing at a rate of 18% per year has an unlimited growth model of

$$y = 4000e^{0.18t}$$

If we use the model to project the population 50 years from now, we get

$$\begin{aligned} y &= 4000e^{0.18(50)} \\ &= 4000e^{9} \approx 32{,}412{,}336 \text{ people} \end{aligned}$$

It seems rather unrealistic to believe that a town with a population of 4 thousand will swell to over 32 million in 50 years. There must be a reasonable limit, or **upper bound**, for almost every such quantity. For these kind of situations, we use the **limited growth model**. Here the rate of change is directly proportional to the distance from a limiting point, or upper bound. If we denote L as the upper bound and y as the value of the function at time t, then this distance from the upper bound is given by the difference $L - y$. So the differential equation for this situation is expressed as

$$(\text{Rate of growth}) = (\text{constant of variation}) \cdot (\text{distance from upper bound } L)$$

or, mathematically,

$$y' = k(L - y)$$

For these applications, we assume that the initial condition is $y = 0$ when $t = 0$. Let's now use the separation of variables technique to solve this differential equation. Since time t is usually our independent variable, we can write

$$\frac{dy}{dt} = k(L - y)$$

Now, separating the variables t and y, we get

$$\frac{1}{L-y}\,dy = k\,dt$$

$$\int \frac{1}{L-y}\,dy = \int k\,dt \qquad \text{Integrate both sides}$$

We need to use a u-substitution on the left-hand side. Let $u = L - y$ so that $du = -dy$ to get

$$\int \frac{1}{u}(-du) = \int k\,dt$$

$$-\int \frac{1}{u}\,du = \int k\,dt$$

$$-\ln u = kt + C$$

In terms of y this is

$$-\ln(L-y) = kt + C$$

$$\ln(L-y) = -kt - C$$

Now, using the definition of a logarithm gives us

$$L - y = e^{-kt-C}$$

$$L - y = e^{-kt}\cdot e^{-C}$$

$$-y = -L + e^{-kt}\cdot e^{-C}$$

$$y = L - e^{-kt}\cdot e^{-C}$$

Using the initial condition that $(t, y) = (0, 0)$, we can find the value of the constant e^{-C}.

$$0 = L - e^{-k(0)}\cdot e^{-C}$$

$$0 = L - e^{-C}$$

$$e^{-C} = L$$

Thus, we get the solution

$$y = L - Le^{-kt}$$

Factoring L from the right side, we get the limited growth model as a function of t to be

$$y = L(1 - e^{-kt})$$

Limited Growth Model

The solution to the differential equation $y' = k(L - y)$ is the **limited growth model** given by the function

$$y = L(1 - e^{-kt})$$

where L is the **limit of growth** and k is the **rate of growth**.

Notes: 1. $\lim_{t\to\infty} f(t) = L$. That is, the limit of growth is L.

2. y is increasing and the graph is concave down for all $t \geq 0$.

3. Applications of the limited growth model include learning curves, company growth, depreciation of equipment, and diffusion of mass media information.

Example 3 Deriving a Limited Growth Model

In a study of the memorization of unrelated shapes, a person can memorize no more than 150 of them in three hours. In a psychology class study, it is found that, on average, a subject can memorize 23 of the shapes after a 10-minute period.

(a) Write a limited growth model where t is the time used to memorize the shapes and y is the number of shapes memorized after t minutes.

(b) Estimate the number of shapes that could be memorized in 45 minutes.

Solution

(a) To determine the particular solution, we need to find the value of k. Given that $L = 150$ (the upper bound) and $y = 23$ when $t = 10$, we can get k by solving

$$
\begin{aligned}
y &= L(1 - e^{-kt}) \\
23 &= 150(1 - e^{-k \cdot 10}) \\
\frac{23}{150} &= 1 - e^{-k \cdot 10} \\
e^{-k \cdot 10} &= 1 - \frac{23}{150} \\
e^{-k \cdot 10} &= \frac{127}{150} \\
\ln(e^{-k \cdot 10}) &= \ln\left(\frac{127}{150}\right) \\
-k \cdot 10 &= \ln\left(\frac{127}{150}\right) \\
k &= -\frac{1}{10} \ln\left(\frac{127}{150}\right) \approx 0.017
\end{aligned}
$$

This gives us the learning curve model $y = 150(1 - e^{-0.017t})$. The graph is shown in Figure 16.8.2.

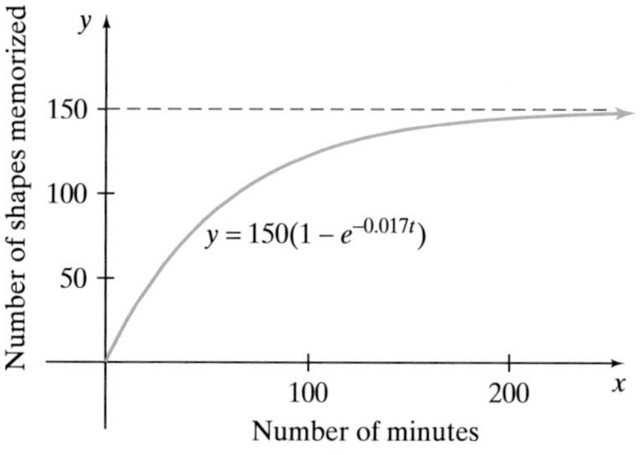

Figure 16.8.2

(b) To estimate the number of shapes memorized in 45 minutes, we need to evaluate y when $t = 45$ to get

$$y = 150(1 - e^{-0.017(45)}) = 150(1 - e^{-0.765}) \approx 80.20$$

Interpret the Solution: Thus, according to the model, about 80 shapes can be memorized in a 45-minute period. Keep in mind that the valid range for this model is the set of whole numbers from 0 to 150 since the units are measured in number of shapes. ■

Checkpoint 2 Now work Exercise 31.

Logistic Growth Model

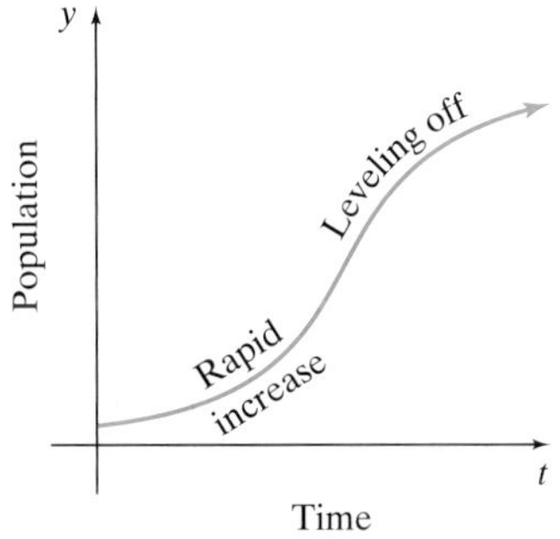

Figure 16.8.3

Many phenomena in the business, life, and social sciences follow a combination of initial rapid growth, only to flatten out for large values of t. The population of many developing countries follows this pattern of a rapid increase in population early on and then a leveling off as the country develops. (See Figure 16.8.3.) This results in a differential equation that exhibits a combination of both unlimited growth and limited growth. In this type of differential equation, the rate of change y' is proportional to the amount currently present times the distance from an upper bound L.

$$\begin{pmatrix}\text{Rate of}\\ \text{growth}\end{pmatrix} = \begin{pmatrix}\text{constant of}\\ \text{variation}\end{pmatrix} \cdot \begin{pmatrix}\text{amount currently}\\ \text{present}\end{pmatrix} \cdot \begin{pmatrix}\text{distance from}\\ \text{upper bound } L\end{pmatrix}$$

or, mathematically,

$$y' = ky(L - y)$$

By using the separation of variables technique with independent variable t, we get

$$\frac{dy}{dt} = y' = ky(L - y)$$

$$\int \frac{1}{y(L - y)}\,dy = \int k\,dt$$

Computing the general solution to this differential equation is done in the exercises. The result is the **logistic growth model**.

Logistic Growth Model

The solution to the differential equation $y' = ky(L - y)$ is the **logistic growth model** given by the function

$$y = \frac{L}{1 + ae^{-kLt}}$$

where L is the limit of growth and a and k are real number constants.

Notes: 1. To determine a logistic model, we are often given three pieces of information:

- An upper limit L
- An initial value that allows us to find the constant a
- A boundary condition to find the exponent kL in the denominator of the model

2. Applications of the logistic growth model include new product sales, dispersion of mass media information and rumors, spread of disease, and long-term population growth.
3. Notice that when $t = 0$

$$y = \frac{L}{1 + ae^{-kL(0)}} = \frac{L}{1 + a}$$

so $\left(0, \frac{L}{1 + a}\right)$ is the initial value. Many of the properties of the logistic growth model are verified in the exercises.

One common application of logistic growth models occurs in the *harvest models* of trees. The model applies in this situation because, as the population of trees in a forest or tree farm increases, the room for roots to spread becomes limited, the amount of sunlight hitting the trees can become restricted, and the number of seedlings that have room and light to grow in becomes less and less. This results in a tapering of the number of trees that can thrive in a fixed area. Consequently, conservationists use a logistic model for the population size for trees.

Example 4 Deriving a Logistic Growth Model

Suppose that commercial tree growers estimate that the maximum number of pine trees that can be supported at a large nursery is 15,000. The nursery was started with 1200 harvestable trees on the land. After 14 years, they find 6000 of the trees are ready to be harvested. Write a differential equation for the rate of growth in population of the nursery trees; then determine a logistic model for the number of trees that are harvestable after t years.

Solution

Understand the Situation: In the differential equation $y' = ky(L - y)$, we know that the upper bound is $L = 15{,}000$, so we get $y' = ky(15{,}000 - y)$. To get the harvest function, we know that its general solution has the form

$$y = \frac{15{,}000}{1 + ae^{-k \cdot 15{,}000t}}$$

To simplify our computations, we replace $k \cdot 15{,}000$ with another constant b.

$$y = \frac{15{,}000}{1 + ae^{-bt}}$$

Since the nursery started with 1200 trees, we know that $y = 1200$ when $t = 0$.

$$1200 = \frac{15{,}000}{1 + ae^{0}}$$

$$1200 = \frac{15{,}000}{1 + a}$$

$$1 + a = \frac{15{,}000}{1200}$$

$$a = \frac{15{,}000}{1200} - 1 = 11.5$$

This gives us the general solution

$$y = \frac{15{,}000}{1 + 11.5\,e^{-bt}}$$

To get the proper value of b, we use the condition that $y = 6000$ when $t = 14$. This gives us

$$\begin{aligned}
6000 &= \frac{15{,}000}{1 + 11.5\,e^{-b(14)}} \\
1 + 11.5\,e^{-b(14)} &= \frac{15{,}000}{6000} \\
11.5\,e^{-b(14)} &= \frac{15{,}000}{6000} - 1 \\
e^{-b(14)} &= \frac{1}{11.5}\left(\frac{15{,}000}{6000} - 1\right) \\
-b(14) &= \ln\left[\frac{1}{11.5}\left(\frac{15{,}000}{6000} - 1\right)\right] \\
b &= -\frac{1}{14}\left\{\ln\left[\frac{1}{11.5}\left(\frac{15{,}000}{6000} - 1\right)\right]\right\} \approx 0.15
\end{aligned}$$

Interactive Activity

Conservationists define the *maximum sustained yield* (MSY) as the greatest harvest that can be taken indefinitely. This MSY is found at the *inflection point* of the logistic model. Determine the MSY for the nursery in Example 4 and plot the MSY on the graph. The inflection point can be found by graphing the second derivative on your calculator and using the ZERO command to determine the x-intercept of the graph of f''.

WWW

Interpret the Solution: Thus, the logistic model for the number of trees that are harvestable after t years is

$$y = \frac{15{,}000}{1 + 11.5\,e^{-0.15t}}$$

The graph of the model for the first five decades is displayed in Figure 16.8.4.

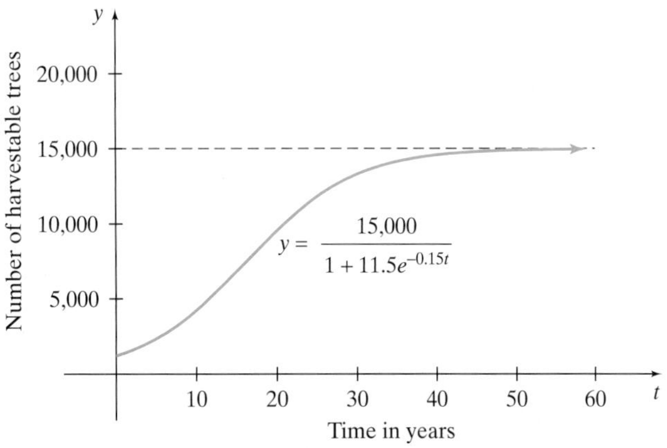

Figure 16.8.4

■

Many times, we are not given information such as the upper bound or growth rate for an application. As a matter of fact, these are the pieces of information that are expected to be discovered. In practice, logistic model applications come from collected data and are generated by regression techniques. Example 5 is one such application.

Example 5 Modeling the Logistic Curve

The number of deaths occurring annually from acquired immune deficiency syndrome (AIDS) from 1981 to 1994 can be modeled by the logistic function

$$y = \frac{48{,}171}{1 + 68.53\,e^{-0.49t}} \qquad 1 \le t \le 14$$

where t represents the number of years since 1980 and y represents the number of deaths occurring annually from AIDS (*Source:* Centers for Disease Control and Prevention, www.cdc.gov).

(a) Graph the function and determine $\lim_{t\to\infty} \frac{48{,}171}{1 + 68.53\,e^{-0.49t}}$ and interpret.

(b) Compute y when $t = 10$ and interpret.

Solution

(a) The graph of the model is shown in Figure 16.8.5. Computing the limit, we get

$$\lim_{t\to\infty} \frac{48{,}171}{1 + 68.53\,e^{-0.49t}} = \frac{48{,}171}{1 + 68.53(0)}$$
$$= 48{,}171$$

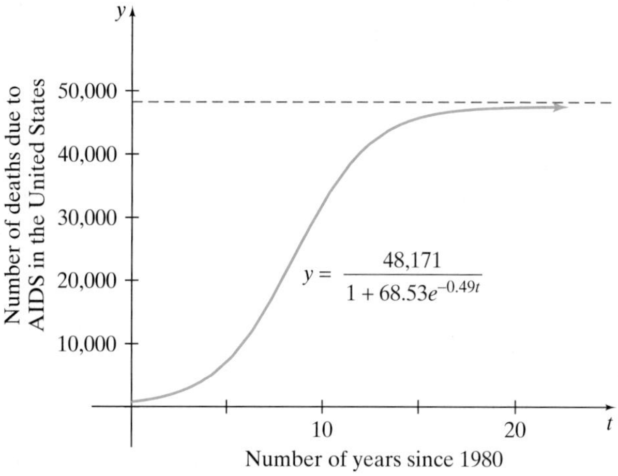

Figure 16.8.5 Logistic model of U.S. AIDS deaths.

Interpret the Solution: This means that according to the logistic model, the upper bound for the number of deaths occurring from AIDS is about 48,171 each year.

(b) Evaluating the model at $t = 10$ yields

$$y = \frac{48{,}171}{1 + 68.53\,e^{-0.49(10)}} \approx 31{,}894.68$$

Interpret the Solution: This means that according to the model, in 1990 approximately 31,895 people died due to AIDS. ■

SUMMARY

In this section, we focused on four different types of growth and decay models. We found that each differential equation from which the model came could be found by separation of variables.

- The **exponential growth model** $y = Ce^{kt}$ is the solution to the differential equation $y' = ky$.
- The **exponential decay model** $y = Ce^{-kt}$ is the solution to the differential equation $y' = -ky$.
- The **limited growth model** $y = L(1 - e^{-kt})$ is the solution to the differential equation $y' = k(L - y)$.
- The **logistic growth model** $y = \dfrac{L}{1 + a \cdot e^{-kLt}}$ is the solution to the differential equation $y' = ky(L - y)$.

SECTION 16.8 EXERCISES

Applications

1. The population of Sensenville is increasing at a rate of 5% per year. The current population of the town is 50,000. Assume that the rate of change in the population is increasing proportionally to the population size.

(a) Write a differential equation and an initial condition for the population of Sensenville.

(b) Write an exponential growth model for the population of Sensenville in t years.

(c) Use the model to estimate the population in 10 years.

2. The population of Burgentown is increasing at a rate of 2% per year. The current population of the city is 250,000. Assume that the rate of change in the population is increasing proportionally to the population size.

(a) Write a differential equation and an initial condition for the population of Burgentown.

(b) Write an exponential growth model for the population of Burgentown in t years.

(c) Use the model to estimate the population in 10 years.

✓ **3.** The number of bacteria in a culture starts at 5000 and increases at a rate of 12% per hour. Assume that the rate of change in the number of bacteria is increasing proportionally to the number of bacteria present.

(a) Write a differential equation for the number of bacteria present after t hours.

(b) Write an exponential growth model of the bacteria present after t hours.

(c) Find the number of bacteria after $t = 3$, 12, and 24 hours.

4. The number of bacteria in a culture starts at 500 and increases at a rate of 7% per hour. Assume that the rate of change in the number of bacteria is increasing proportionally to the number of bacteria present.

(a) Write a differential equation for the number of bacteria present after t hours.

(b) Write an exponential growth model of the bacteria present after t hours.

(c) Find the number of bacteria after $t = 0$, 3, 6, and 12 hours.

5. In 1970 the consumption of timber in the United States was 39.9 million board feet, and in 1994 the consumption rose to 60.4 million board feet (*Source:* U.S. Department of Agriculture, www.usda.gov).

(a) Write a natural resource function $N(t) = a \cdot e^{kt}$, where t represents the number of years since 1970 and $N(t)$ represents the amount of timber consumed in millions of board feet.

(b) Compute and graph $\int_0^{20} N(t)\,dt$ to determine the total amount of timber consumed between 1970 and 1990.

6. In 1970 the human consumption of fish in the United States was 11.3 billion pounds, and in 1995 the consumption increased to 16.6 billion pounds (*Source:* U.S. Statistical Abstract, www.census.gov/statab/www).

(a) Write a natural resource function $N(t) = a \cdot e^{kt}$, where t represents the number of years since 1970 and $N(t)$ represents the amount of fish consumed in billions of pounds.

(b) Compute and graph $\int_0^{25} N(t)\,dt$ to determine the total amount of fish consumed between 1970 and 1995.

7. The percentage of wood waste generated in the United States can be modeled by the exponential growth model

$$y = 6.0\,e^{0.034t} \qquad 0 \le t \le 9$$

where t represents the number of years since 1990 and y represents the percentage of wood waste generated (*Source:* U.S. Statistical Abstract, www.census.gov/statab/www).

(a) Write a differential equation and an initial condition for the percentage of wood waste generated in the United States.

(b) Evaluate y and y' when $t = 3$ and interpret.

8. The percentage of paper waste generated in the United States can be modeled by the exponential growth model

$$f(t) = 35.29\,e^{0.022t} \qquad 0 \le t \le 9$$

where t represents the number of years since 1990 and $f(t)$ represents the percentage of paper waste generated (*Source:* U.S. Statistical Abstract, www.census.gov/statab/www).

(a) Write a differential equation and an initial condition for the percentage of paper waste generated in the United States.

(b) Evaluate y and y' when $t = 6$ and interpret.

9. Suppose a scientist has a 3000-milligram sample of the radioactive substance phosphorus 32. The half-life of phosphorus 32 is known to be 14.2 days.

(a) Determine an exponential decay model for the number of milligrams of phosphorus 32 present after t days.

(b) How many days will it take for the sample of phosphorus 32 to decay to 2000 milligrams?

10. Suppose that a scientist has a 300-gram sample of the radioactive substance radon. The half-life of radon is known to be 3.82 days.

(a) Determine an exponential decay model for the number of grams of radon present after t days.

(b) How many days will it take for the sample of radon to decay to 50 grams?

11. The population of Devonburg was 920,000 in 1980. Because of changing economic conditions in the community, the population had decreased to 700,000 in 1995. Assume that the rate of change in the population is decreasing proportionally to the population size.

(a) Write a differential equation and an initial condition for the population of Devonburg.

(b) Write an exponential decay model for the population of Devonburg in t years.

(c) Use the model to estimate the population in 1998.

12. The population of Grandland was 604,000 in 1985. Due to urban flight, the population decreased to 500,000 in 1990. Assume that the rate of change in the population is decreasing proportionally to the population size.

(a) Write a differential equation and an initial condition for the population of Grandland.

(b) Write an exponential decay model for the population of Grandland in t years.

(c) Use the model to estimate the population in 2000.

13. The percentage of Caucasians in the U.S. labor force who are unemployed can be modeled by

$$y = 6.54\,e^{-0.09t} \qquad 0 \le t \le 8$$

where t represents the number of years since 1990 and $f(t)$ represents the percentage of Caucasians in the U.S. labor force who are unemployed (*Source:* U.S. Bureau of Labor Statistics, www.bls.gov).

(a) Write a differential equation and an initial condition for the percentage of Caucasians in the U.S. labor force who are unemployed.

(b) Evaluate y and y' when $t = 3$ and interpret.

14. The number of new car sales that were imports can be modeled by

$$y = 2.54\,e^{-0.09t} \qquad 0 \le t \le 9$$

where t represents the number of years since 1990 and y represents the number of new car sales that were imports, measured in millions (*Source:* U.S. Census Bureau, www.census.gov).

(a) Write a differential equation and an initial condition for the number of new car sales which were imports.

(b) Evaluate y and y' when $t = 1$ and interpret.

For Exercises 15–26:

(a) Write the corresponding differential equation $y' = k(L - y)$ using the definition of the limited growth model.

(b) Evaluate y when $t = 0$ to determine the initial value and identify the limit of growth L. Use the definition of the limited growth model.

15. $y = 10(1 - e^{-t})$

16. $y = 5(1 - e^{-t})$

17. $y = 8(1 - e^{-2t})$

18. $y = 8(1 - e^{-3t})$

19. $y = 20(1 - e^{-0.5t})$

20. $y = 16(1 - e^{-0.3t})$

21. $y = 12 - 12e^{-0.1t}$

22. $y = 24 - 24e^{-0.2t}$

23. $y = 1.5(1 - e^{-0.2t})$

24. $y = 0.01(1 - e^{-0.3t})$

25. $y = 15.5 - 15.5e^{-0.1t}$

26. $y = 62 - 62e^{-0.02t}$

Applications

27. Suppose that the Wesson and Selverstone Investment Group has introduced a new computer system for its 3000 employees. The developer of the software has determined that the employees will learn the new system according to the model

$$y = 3000(1 - e^{-0.3t}) \qquad t \ge 0$$

where t represents the number of weeks since the system has been installed and y represents the number of employees that have learned the new system.

(a) Evaluate y and y' when $t = 2$ and interpret.

(b) According to the model, how many weeks will it take for at least 2500 employees to learn the new system?

28. Consider a college dormitory of 600 students where a flu epidemic is infecting students according to the model

$$y = 600(1 - e^{-0.22t}) \qquad t \geq 0$$

where t represents the number of days since the flu was first diagnosed and y represents the number of dorm residents infected.

(a) Evaluate y and y' when $t = 4$ and interpret.

(b) According to the model, how many days will it take so that at least 400 students have been infected by the flu?

29. The MidWest Fabricating Company installs a new metal press for its 300 employees. The company that makes the press has determined through experience that employees learn to use the new press in a way that follows the limited growth model

$$y = L(1 - e^{-kt}) \qquad t \geq 0$$

where t is the number of weeks since the new press was installed and y represents the number of employees who have learned to use the new equipment. They find that after the first week 40 employees have learned to use the new metal press.

(a) Identify the limit of growth L.

(b) Solve the equation $40 = 300(1 - e^{-k(1)})$ to determine the rate of growth k.

(c) Graph the model for the first half-year (that is, 26 weeks) after installation.

(d) If the company has set a goal of having half of the employees skilled at using the machine in the first month (that is, in the first 4 weeks), will the company reach its goal?

30. A municipal park is being overrun by 200 deer, so park officials decide to open the park to hunting for a short time in an attempt to curb the deer population. The officials believe that the number of deer killed by the hunters will follow the limited growth model

$$y = L(1 - e^{-kt}) \qquad t \geq 0$$

where t is the number of days since the hunting season began and y represents the number of deer killed. After the first two days of hunting, 24 deer were killed.

(a) Identify the limit of growth L.

(b) Solve the equation $24 = 200(1 - e^{-k(2)})$ to determine the rate of growth k.

(c) Graph y to display the number of deer killed if the hunting season lasted 30 days.

(d) If the officials plan to call an end to the hunting season when the deer population is cut in half, how many days will this take?

✓ **31.** The Newportage Company conducts an aptitude test in which a maximum of 75 manufacturing procedures are learned. Studies show that, on the average, 15 of the procedures can be memorized in a 1.5-day period. Assume that the rate of learning is proportional to the distance from the maximum number of learning procedures. (That is, the number of procedures learned can be modeled by a limited growth model.)

(a) Identify the limit of growth L and determine the rate of growth k.

(b) Write a limited growth model in the form

$$y = L(1 - e^{-kt}) \qquad t \geq 0$$

where t is the number of days that an employee has been memorizing the procedures and y represents the number of procedures memorized. Graph the model for $0 \leq t \leq 30$.

(c) Evaluate y' when $x = 5$ and compare to y' when $x = 25$.

For Exercises 32–41:

(a) Write the corresponding differential equation $y' = ky(L - y)$ for the model. Use the definition of the logistic growth model.

(b) Evaluate y when $t = 0$ to determine the initial value and identify the limit of growth L.

32. $y = \dfrac{30}{1 + e^{-t}}$

33. $y = \dfrac{15}{1 + e^{-t}}$

34. $y = \dfrac{50}{1 + 2e^{-t}}$

35. $y = \dfrac{10}{1 + 5e^{-t}}$

36. $y = \dfrac{150}{1 + e^{-2t}}$

37. $y = \dfrac{90}{1 + e^{-3t}}$

38. $y = \dfrac{400}{1 + 2e^{-3t}}$

39. $y = \dfrac{2500}{1 + 5e^{-4t}}$

40. $y = \dfrac{1000}{1 + 4.5e^{-0.9t}}$

41. $y = \dfrac{4000}{1 + 6e^{-1.2t}}$

Applications

42. According to estimates from the Whitehead Research Group, the percentage of households that own video disk recorders in a certain county is given by the logistic model

$$y = \frac{90}{1 + 22e^{-0.7t}} \qquad 0 \le t \le 10$$

where t represents the number of years since 1990 and $f(t)$ represents the percentage of households that own video disk recorders.

(a) According to the model, what percentage of the households in the county owned video disk recorders in 1999?

(b) Evaluate y' when $t = 6$ and interpret.

(c) According to the model, what is the maximum number of households in the county that are expected to own a video disk recorder?

43. On a certain university campus, the 12,000 students who attend home basketball games are quickly swept by the new fad of tossing toilet paper onto the basketball court after the first basket is made for the home team. Research students in the sociology department think that the fad started with about 100 fans and that the number of students at the home games who throw toilet paper follow the logistic model

$$y = \frac{12{,}000}{1 + 119e^{-0.8x}} \qquad 0 \le x \le 12$$

where x is the number of home games completed and y represents the number of students expected to toss toilet paper onto the basketball court after the first basket is made for the home team.

(a) Evaluate y when $x = 8$ and interpret.

(b) If the university's athletic director threatens to search all students coming to the game for toilet paper when at least half of the student fans are engaging in the activity, how many home games will this take?

44. The Top Cove Fishery, established a harvesting area that is designed to hold a maximum of 7500 catfish. The fishery was initially stocked with nearly 1000 catfish. After two years, the number of catfish in the fishery had grown to nearly 2500.

(a) What is the limit of growth L?

(b) Solve the equation $2500 = \dfrac{7500}{1 + 6.5e^{-b(2)}}$ to determine the rate of growth where $b = kL$.

(c) Write the logistic function in the form $y = \dfrac{L}{1 + ae^{-bt}}$ for the first 5 years, since the harvesting area was established where $b = kL$.

(d) Conservationists define the maximum sustained yield (MSY) as the largest number of the population that can be removed while sustaining the population. For logistic functions, the MSY is found at the inflection point of the curve that is the point $\left(\dfrac{\ln a}{b}, \dfrac{L}{2}\right)$, where $b = kL$. Determine the MSY for the harvesting area.

45. Consider a town with a population of 30,000 whose city council president has just resigned. He announces the resignation at a closed session of the 10-member city council. It takes only 6 hours for 27,300 citizens to learn about the resignation.

(a) What is the limit of growth L?

(b) Solve the equation $27{,}300 = \dfrac{30{,}000}{1 + 3000e^{-b(6)}}$ to determine the rate of growth, where $b = kL$.

(c) Write the logistic function in the form $y = \dfrac{L}{1 + ae^{-bt}}$ for the first 7 hours after the mayor's announcement.

46. The percentage of African Americans 25 years old and over who completed 4 years or more of high school can be modeled by

$$y = \frac{94.3}{1 + 4e^{-0.07t}} \qquad 1 \le t \le 40$$

where t represents the number of years since 1959 and y represents the percentage of African Americans 25 years old and over who completed 4 years or more of high school (*Source:* U.S. Census Bureau, www.census.gov).

(a) Evaluate and interpret y when $t = 16$.

(b) Write the corresponding differential equation $y' = ky(L - y)$ for the model. (*Hint:* $b = kL = 0.07$.)

(c) Evaluate y' when $t = 16$ and interpret.

(d) Determine $\lim_{t \to \infty} \left(\dfrac{94.3}{1 + 4e^{-0.07t}}\right)$ and interpret.

47. Consider the differential equation corresponding to the logistic growth model $y' = ky(L - y)$.

(a) Use the separation of variables technique to write an equation with expressions involving y on the left side and the constant k on the right.

(b) Verify that the derivative of the expression $\dfrac{1}{L}\ln\left(\dfrac{y}{L - y}\right)$ is $\dfrac{1}{y(L - y)}$ using the differentiation rules. Keep in mind that L is a constant.

(c) Show that when we solve the equation $\dfrac{1}{L}\ln\left(\dfrac{y}{L - y}\right) = kt + C$ for y we get $y = \dfrac{L}{1 + ae^{-kLt}}$, where $a = e^{-CL}$. (*Hint:* Isolate y and divide the numerator and denominator of the resulting fraction by ae^{kLt}.)

(d) Verify that $y = \dfrac{L}{1 + ae^{-kLt}}$ satisfies the differential equation $y' = ky(L - y)$, and explain why $y' > 0$ for any $t > 0$. This shows that the logistic function is increasing over its domain.

(e) Use the properties of limits and exponential functions to explain why $\lim_{t \to \infty} \left(\dfrac{L}{1 + ae^{-kLt}}\right) = L$.

SECTION PROJECT

Consider the U.S. population figures from 1790 to 1980 given in Table 16.8.1.

Table 16.8.1

Year	Population (in millions)	Year	Population (in millions)
1790	3.929	1890	62.980
1800	5.308	1900	76.212
1810	7.240	1910	92.228
1820	9.638	1920	106.022
1830	12.861	1930	123.203
1840	17.063	1940	132.165
1850	23.192	1950	151.326
1860	31.443	1960	179.324
1870	50.189	1970	203.302
1880	50.189	1980	226.549

SOURCE: U.S. Census Bureau, www.census.gov

(a) Make a scatterplot of the data, where t is the number of years past 1800 and $-10 \leq t \leq 180$.

(b) Use your calculator to get a logistic regression model y that represents the U.S. population and t is the number of years past 1800. Now graph the scatterplot along with the logistic model y for $-10 \leq t \leq 300$ and then for $0 \leq t \leq 370$.

(c) Determine the year in which the rate of growth in the U.S. population started to decrease. (*Hint:* Find the inflection point.) What do you notice about this answer and the answer to part (b)?

(d) According to the model, what is the largest that the U.S. population can reach?

Why We Learned It

In Chapter 16, we started by learning the mechanics of antidifferentiation. This skill was immediately applied to recovering a function if we knew its derivative. For example, if we know the marginal cost function for a product and the cost to produce 1000 units of the product, then antidifferentiation allows us to recover the cost function. We then learned that integration was another name for antidifferentiation. After we were introduced to the indefinite integral we then saw how the definite integral could be used to represent the area under a curve. The Fundamental Theorem of Calculus then supplied us with the necessary tool to evaluate definite integrals. A main area of application of definite integrals and the Fundamental Theorem of Calculus was in the interpretation of the area under the graph of a rate function. We saw that this interpretation was basically a total accumulation. For example, if we know the rate function for the population of a country from 1950 to 2000 then integrating the rate function over the interval corresponding to the years 1950 to 2000 gives us the total increase (or decrease) in the country's population. Another application of the total accumulation concept is with the marginal business functions. If we know the marginal profit for a certain product, then integrating the marginal profit function over an interval, say from 50 to 100, gives the total increase in profit when selling from 50 to 100 units of the product. This idea of a total accumulation is quite powerful and we will see even more applications of it in Chapter 17. In this chapter we also learned an integration technique called u-substitution as well as integrals that produce natural logarithms. Integrals involving exponentials were also addressed. Again we saw that many rate functions utilize these functions. Hence knowing their antiderivatives was imperative when we needed to determine a total accumulation.

CHAPTER REVIEW EXERCISES

In Exercises 1–4, determine if the function F is an antiderivative of the function f.

1. $F(x) = x^2 + C;\ f(x) = x$

2. $F(x) = x^3 - 5x + C;\ f(x) = 3x^2 - 5$

3. $F(t) = \frac{1}{5}t^5 + \frac{1}{4}t^4 + \frac{1}{3}t^3 + C;\ f(t) = t^4 + t^3 + t^2$

4. $F(x) = \frac{2}{x^3} + 4x + C;\ f(x) = \frac{6}{x^4} + 4$

In Exercises 5–8, use the Power Rule for Integration to find the indefinite integrals.

5. $\int x^7\,dx$

6. $\int x^{5.13}\,dx$

7. $\int \frac{1}{t^5}\,dt$

8. $\int \frac{1}{\sqrt[4]{x^3}}\,dx$

For Exercises 9–12, determine the indefinite integral.

9. $\int (0.5x^7 - 6x)\,dx$

10. $\int (6t^5 + 4t^4 - 3)\,dt$

11. $\int (x^2 + 0.4x - \sqrt{x^5})\,dx$

12. $\int \frac{5x^2 + 3x}{x}\,dx$

In Exercises 13 and 14, the rate function f′ is shown. Use the graph to determine the intervals where its corresponding function f is increasing or decreasing and where the graph of f is concave up or down and where the relative maximum and minimum values occur.

13.

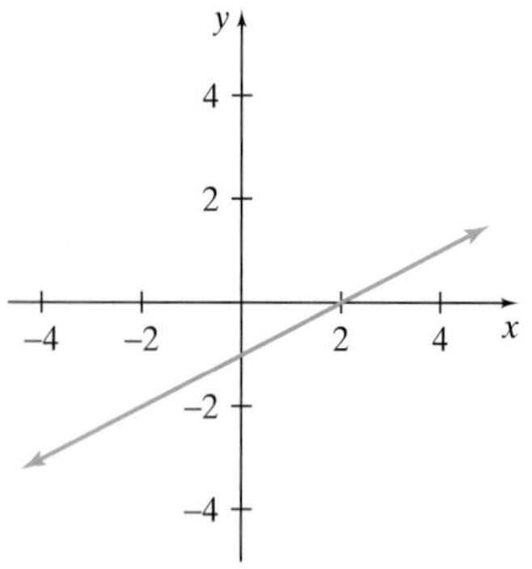

14.

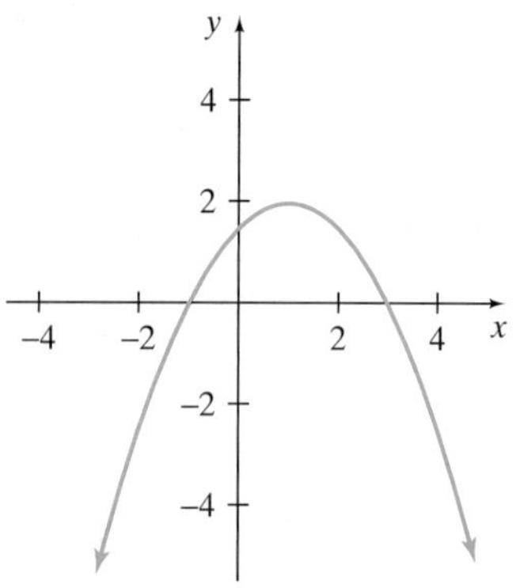

In Exercises 15–18, solve the given problem given the rate function and the given value. Note that $f(a) = b$ corresponds to the ordered pair (a, b).

15. $f'(x) = 3x - 4;\ f(0) = 11$

16. $\frac{dy}{dx} = x^2 - 5x;\ y = 10$ when $x = 6$

17. $y' = \sqrt{x};\ (x, y) = (9, -5)$

18. $P'(t) = 0.5 + 0.15\sqrt[3]{t^4};\ P(27) = 3.1$

19. Consider the marginal profit (in dollars per unit) for producing q units of a product given by the linear function

$$MP(q) = 48 - 0.03q$$

(a) Knowing that $P = 1600$ when $q = 100$, recover the profit function P.

(b) Use your solution from part (a) to find the total profit realized from selling 500 units of product.

20. The monthly marginal revenue function for the Byrne Rubber Tire Company is given by

$$MR(x) = R'(x) = -0.3x^2 - 1.4x + 35.2$$

where x is the number of thousands of tires produced and sold and $MR(x)$ is the marginal revenue in thousands of dollars per thousand tires.

(a) Knowing that $R(x) = 0$ when $x = 0$, recover the revenue function R.

(b) Find the price–demand function p for the tires. Recall that the general formula for the revenue function is

$$R(x) = (\text{quantity produced}) \cdot (\text{price of each unit}) = x \cdot p(x)$$

(c) What should the price be when the demand is 6000 tires (that is, $x = 6$)?

In Exercises 21–24, determine the step size Δx for the given interval $[a, b]$ and number of subintervals n.

21. $[0, 9],\ n = 27$

22. $[-5, 5],\ n = 10$

23. $[7, 11],\ n = 32$

24. $[-4, 9],\ n = 39$

In Exercises 25–28, calculate (a) the Left Sum and (b) the Right Sum to approximate the area under the graph of f on the closed interval $[a, b]$ using n equally spaced subintervals.

25. $f(x) = 5x$, $[0, 5]$, $n = 5$

26. $f(x) = 3x - 8$, $[6, 8]$, $n = 4$

27. $f(x) = x^2$, $[0, 3]$, $n = 6$

28. $f(x) = 3x^2 + 4$, $[-1, 1]$, $n = 4$

Complete the table for each of the functions f on $[a, b]$ in Exercises 29–32.

N Rectangles	LEFTSUM Approximation	RTSUM Approximation
10		
100		
1000		

29. $f(x) = 5x^2 - 3$, $[0, 8]$

30. $f(x) = \sqrt{x - 2}$, $[3, 5]$

31. $f(x) = 3x^4 - 5x^3$, $[0, 4]$

32. $f(x) = 0.6x^2 - 1.3x + 7.3$, $[0, 10]$

In Exercises 33–38, evaluate the definite integrals.

33. $\int_3^7 6x\,dx$

34. $\int_0^3 (3x - 4)\,dx$

35. $\int_{-2}^2 (10 - x^2)\,dx$

36. $\int_{-2}^8 (x^5 + 9x^2 - 2)\,dx$

37. $\int_0^5 (2.7x^2 - 1.4x + 0.8)\,dx$

38. $\int_1^{64} x(\sqrt{x} - \sqrt[3]{x})\,dx$

In Exercises 39–41, determine the area of the shaded region.

39.

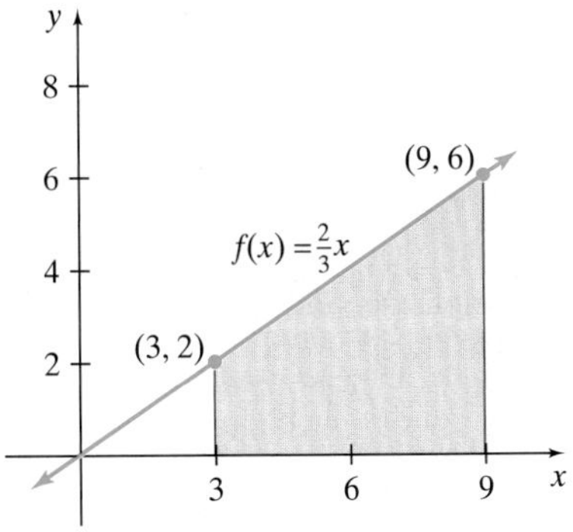

40.

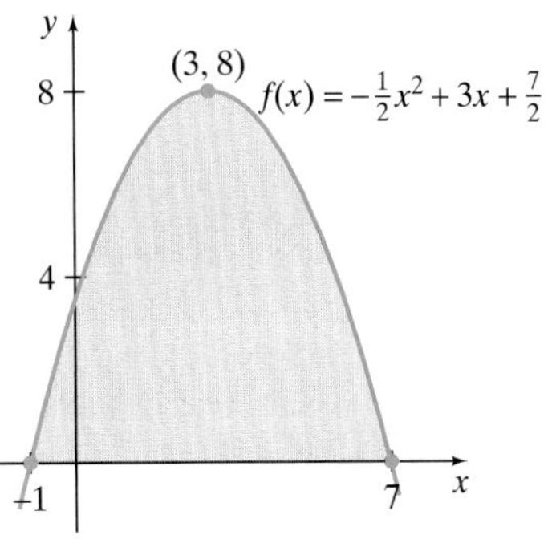

41.

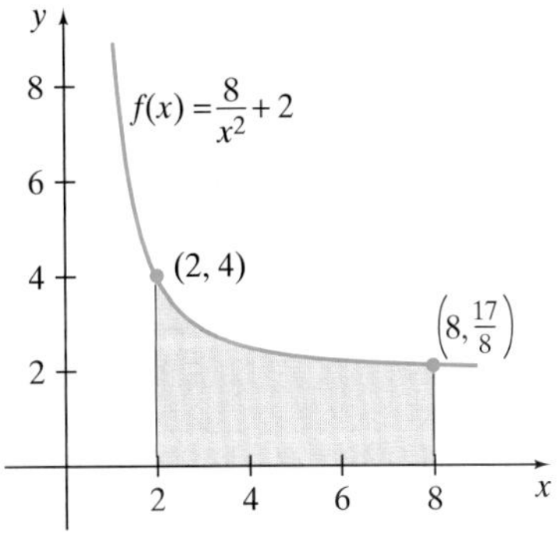

42. Consider the areas under the graphs of f and g.

(a) Determine the area of the shaded region in the figure.

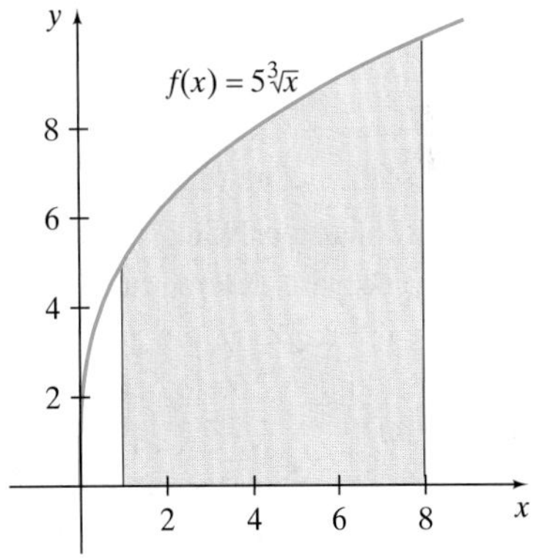

(b) Determine the area of the shaded region in the figure.

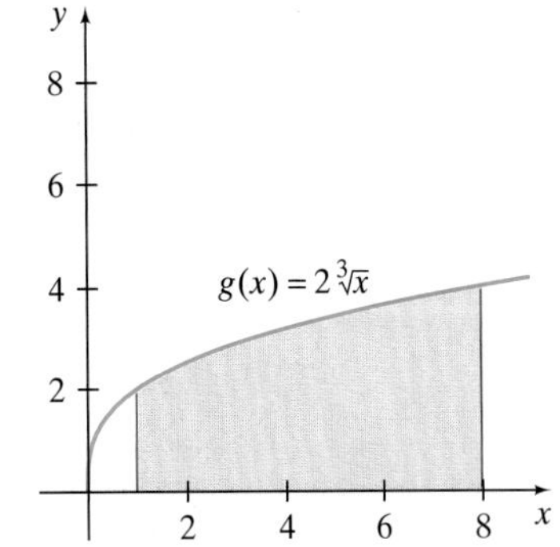

43. Consider the areas under the graphs of f and g.

(a) Determine the area of the shaded region in the figure.

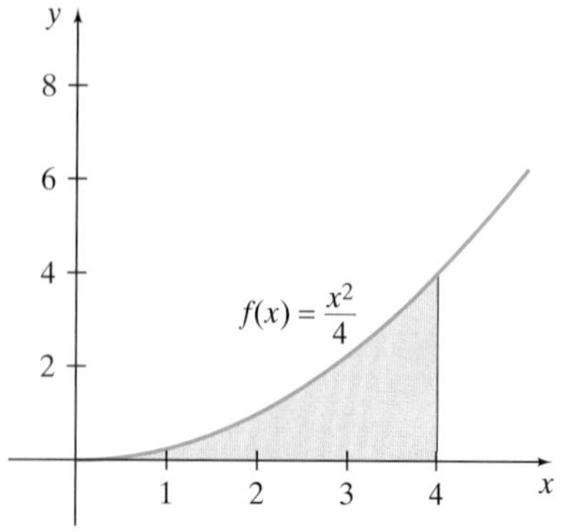

(b) Determine the area of the shaded region in the figure.

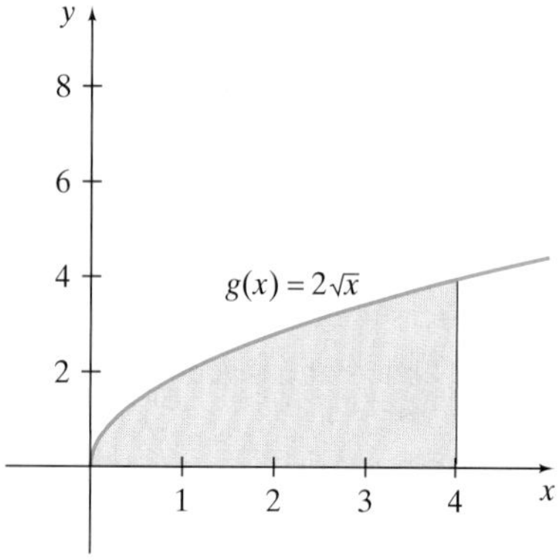

44. Consider the absolute value function $f(x) = |x| - 3$.

(a) Graph f in the standard viewing window.

(b) Rewrite f as a piecewise-defined function.

(c) Determine the x-intercepts of the graph of f.

(d) Use the interval addition rule to determine the net and gross areas of $\int_{-2}^{5} f(x)\,dx$.

For Exercises 45–48:

(a) Graph the piecewise-defined function g.

(b) Use the interval addition rule to evaluate the indicated definite integral to find the area between the graph of g and the x-axis.

45. $g(x) = \begin{cases} -2x, & x < 0 \\ 3x, & x \geq 0 \end{cases}; \quad \int_{-5}^{2} g(x)\,dx$

46. $g(x) = \begin{cases} 6 - 3x, & x < 1 \\ 3x, & x \geq 1 \end{cases}; \quad \int_{0}^{6} g(x)\,dx$

47. $g(x) = \begin{cases} x^2 - 5, & x \leq -3 \\ x + 7, & x > -3 \end{cases}; \quad \int_{-5}^{2} g(x)\,dx$

48. $g(x) = \begin{cases} 5, & x < 4 \\ x^2 - 11, & x \geq 4 \end{cases}; \quad \int_{0}^{10} g(x)\,dx$

For Exercises 49–52:

(a) Determine the x-intercepts of the graph of the function f.

(b) Use the interval addition rule to determine the net and gross areas of the indicated integral.

49. $f(x) = x - 3; \quad \int_{0}^{10} f(x)\,dx$

50. $f(x) = 2 - \sqrt{x}; \quad \int_{0}^{10} f(x)\,dx$

51. $f(x) = x^2 - x - 2; \quad \int_{-1}^{5} f(x)\,dx$

52. $f(x) = x^3 + 2x^2 - 3x; \quad \int_{-5}^{5} f(x)\,dx$

53. The AddEmUp Calculator Company determines that the marginal cost function for a new calculator is given by

$$MC(x) = C'(x) = \frac{1}{20}x + 12 \qquad 0 \leq x \leq 300$$

where x is the number of calculators produced per day and $MC(x) = C'(x)$ is the marginal cost in dollars per calculator.

(a) Evaluate $C'(160)$ and interpret.

(b) Evaluate $\int_{0}^{200} C'(x)\,dx$ and interpret.

54. The Voll-E Ball Company determines that the marginal profit function for selling a new kind of football is

$$MP(x) = P'(x) = 0.0002x^2 - 0.01x + 2.85 \qquad 0 \leq x \leq 100$$

where x is the number of footballs made and sold and $P'(x)$ is the marginal profit function in dollars per football.

(a) Knowing that the company breaks even if 20 footballs are sold (that is, the profit for making and selling 20 footballs is \$0), recover the profit function P.

(b) Evaluate $\int_{50}^{100} P'(x)\,dx$ and interpret.

55. The rate of U.S. government expenditures on defense- and space-related research and development is given by

$$r(t) = -0.0167t^3 + 0.448t^2 - 0.85t + 12.5 \qquad 1 \leq t \leq 21$$

where t represents the number of years since 1974 and $r(t)$ represents the rate of expenditures, measured in $\frac{\text{billions of dollars}}{\text{year}}$ (*Source:* U.S. Statistical Abstract, www.census.gov/statab/www).

(a) Evaluate and interpret $r(7)$.

(b) Determine the total defense- and space-related research and development from 1979 to 1985.

56. The velocity function for an object, in feet per second, is $v(t) = 6t - 9$.

(a) Evaluate $v(t)$ at the point $t = 2$.

(b) Evaluate and interpret $\int_{50}^{100} v(t)\,dt$.

In Exercises 57–60, use the u-substitution method to evaluate the indefinite integrals. Check by differentiating the solution.

57. $\int 5(5x-7)^3\,dx$

58. $\int 3t^2(t^3+5)^4\,dt$

59. $\int (2x-5)(x^2-5x+3)\,dx$

60. $\int \frac{2x}{\sqrt{x^2-5}}\,dx$

In Exercises 61–64, use the u-substitution method to evaluate the indefinite integrals.

61. $\int (3x+6)^8\,dx$

62. $\int t^4(2t^5-7)^3\,dt$

63. $\int \frac{x+2}{x^2+4x+6}\,dx$

64. $\int (8x-28)(x^2-7x+9)^3\,dx$

In Exercises 65–68, determine the integrals by solving the u-expression to complete the substitution.

65. $\int 3t(t+4)^2\,dt$

66. $\int x(x-1)^6\,dx$

67. $\int \frac{x}{\sqrt{x+3}}\,dx$

68. $\int x\sqrt{x-5}\,dx$

In Exercises 69–72, evaluate the definite integrals, and evaluate the limits of integration in terms of x.

69. $\int_{10}^{17} 2x\sqrt{x^2-64}\,dx$

70. $\int_0^{12} (x-4)^3\,dx$

71. $\int_2^{11} \frac{3x^2}{\sqrt{x^3-4}}\,dx$

72. $\int_5^{20} \sqrt[4]{21-t}\,dt$

In Exercises 73–76, evaluate the definite integrals. Change the limits of integration to u-limits.

73. $\int_1^4 \frac{2x-5}{x^2-5x+1}\,dx$

74. $\int_0^2 (3t^2-4)(t^3-4t+7)^4\,dt$

75. $\int_4^8 3x^2\sqrt{x^3+17}\,dx$

76. $\int_5^9 \frac{2x}{\sqrt{x^2+144}}\,dx$

77. Determine the exact area of the shaded region.

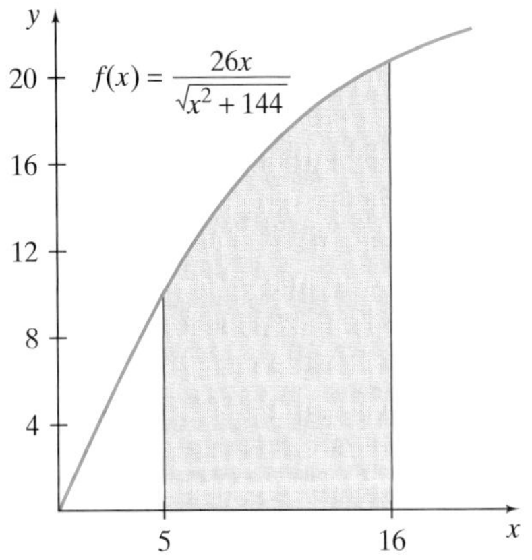

78. The marginal profit function for hard disks produced by the Elephant Media Company is given by

$$MP(x) = \frac{80x}{\sqrt{x^2+576}} \qquad 0 \le x \le 100$$

where x is the number of hard disks produced in a work shift and $MP(x)$ is the marginal profit.

(a) Evaluate and interpret $MP(7)$.

(b) Evaluate $\int_7^{45} MP(x)$ and interpret.

(c) Knowing that the Elephant Media Company breaks even when 32 hard disks are sold [that is, $P(x)=0$ when $x=32$], recover the profit function P.

79. Suppose that the velocity of an object is given by

$$v(t) = (t^3-5t)^3(3t^2-5)$$

where t is the time in seconds and $v(t)$ is the velocity in $\frac{\text{meters}}{\text{second}}$.

(a) Determine the total distance moved between $t=3$ and 6 seconds.

(b) Knowing that when $t=3$ seconds $s(t)=736$ meters, meters, determine the position function s.

In Exercises 80–87, evaluate the indefinite and definite integrals.

80. $\int \frac{3}{x}\,dx$

81. $\int \frac{3}{4}t^{-1}\,dt$

82. $\int 3e^x\,dx$

83. $\int (5e^x - 3x)\,dx$

84. $\int \left(\frac{1}{2}\right)^t dt$

85. $\int_3^5 \frac{6}{x}\,dx$

86. $\int_0^5 \frac{e^t - 5}{3}\,dt$

87. $\int_0^2 5^x\,dx$

In Exercises 88–95, evaluate the integrals using u-substitution.

88. $\int \frac{2x+5}{x^2+5x-3}\,dx$

89. $\int \frac{5}{\sqrt{x}(\sqrt{x}+4)}\,dx$

90. $\int e^{3t-5}\,dt$

91. $\int e^x(5-e^x)^3\,dx$

92. $\int 3x^2 \cdot 5^{x^3}\,dx$

93. $\int_0^3 \frac{2x}{x^2+16}\,dx$

94. $\int_3^7 e^{2t-6}\,dt$

95. $\int_{-2}^2 2^{3x}\,dx$

96. The PaperCut Discount Stationery Store is having a sale on pen and pencil sets. The rate of sales is given by

$$S(t) = \frac{60}{t} \qquad 1 \le t \le 10$$

where t represents the number of days that the sale has been running and $S(t)$ represents the sales rate, measured in $\frac{\text{pen and pencil sets}}{\text{day}}$.

(a) Evaluate $S(4)$ and interpret.

(b) Use an integral to determine the total increase in the number of pen and pencil sets sold between days 4 and 8.

97. The rate of change in the number of U.S. billed international telephone calls can be modeled by

$$r(t) = 63.8e^{0.798t} \qquad 1 \le t \le 11$$

where t represents the number of years since 1984 and $r(t)$ represents the rate of change in the number of U.S. billed international telephone calls, measured in $\frac{\text{millions of calls}}{\text{year}}$ (*Source:* U.S. Statistical Abstract, www.census.gov/statab/www).

(a) Evaluate and interpret $r(6)$.

(b) Evaluate and interpret $\int_1^{11} r(t)\,dt$.

In Exercises 98–101, write an exponential growth model in the form $f(t) = a \cdot e^{kt}$ that satisfies the given differential equation and initial condition.

98. $f'(t) = 0.7 \cdot f(t);\ f(0) = 300$

99. $f'(t) = 1.37 \cdot f(t);\ f(0) = 8.2$

100. $f'(t) = -0.07 \cdot f(t);\ f(t) = 5$ when $t = 0$

101. $f'(t) = -0.87 \cdot f(t);\ f(0) = 384$

102. The population of Popuville is increasing at a rate of 11% per year. The current population of the town is 42,000. Assume that the rate of change in the population is increasing proportionally to the population size.

(a) Write a differential equation and an initial condition for the population of Popuville.

(b) Write an exponential growth model for the population of Popuville after t years.

(c) Use the model to estimate the population after 15 years.

103. The number of bacteria in a culture starts at 300 and increases at a rate of 8% per hour. Assume that the rate of change in the number of bacteria is increasing proportionally to the number of bacteria present.

(a) Write a differential equation for the number of bacteria present after t hours.

(b) Write an exponential growth model for the number of bacteria present after t hours.

(c) Complete the table.

Number of Hours, t	Number of Bacteria, $f(t)$
0	
2	
4	
10	
20	

(d) Graph the exponential growth model for the values $0 \le t \le 20$.

104. In 1970, 8.6 million tons of solid waste were recovered for recycling in the United States. By 1995, this number had increased to 56.2 million tons (*Source:* U.S. Statistical Abstract, www.census.gov/statab/www).

(a) Considering recycled waste as a "natural resource," write a natural resource function

$$N(t) = a \cdot e^{kt}$$

where t represents the number of years since 1970 and $N(t)$ represents the number of millions of tons of solid waste recovered each year for recycling.

(b) Evaluate and interpret $\int_0^{25} N(t)\,dt$.

CHAPTER 16 PROJECT

One of the most important results in mathematics is presented in Chapter 16: The Fundamental Theorem of Calculus. This theorem ties together everything that has been learned over the past six chapters: from limits to differentiation to integration. Not only is integration an important operation of a function, its interpretation is even more so.

Consider the following example:

Good Times is a maker of relaxing heated seat massagers for the home, car, or office. The rate of change of revenue in dollars, $MR(x)$, of the seat massagers is given by the following chart:

Number of Seat Massagers	Rate of Change of Revenue	Number of Seat Massagers	Rate of Change of Revenue
0	0.00	12	627.02
1	21.00	13	593.10
2	63.93	14	520.83
3	124.81	15	407.53
4	197.10	16	249.45
5	275.80	17	39.90
6	356.92	18	−222.82
7	436.84	19	−547.27
8	508.65	20	−937.11
9	567.99	21	−1,395.10
10	612.52	22	−1,926.65
11	634.12	23	−2,540.19

where x represents the number of seat massagers sold in one week with $0 \leq x \leq 23$. It is also known that the rate of change of cost is increasing at a constant rate of $139.50 per unit sold.

1. Determine a cubic regression model of the form $MR(x) = ax^3 + bx^2 + cx + d$ for the given data. Round coefficients to the nearest hundredth.

2. For what value(s) of x does $MR(x) = 0$? What does this imply about the graph of the revenue function R? For what value(s) of x does $MR(x)$ have a maxima? What does this imply about the graph of the revenue function R?

3. Determine a model for the marginal cost of the form $MC(x) = a$. What type of function does this imply the cost function is?

4. Find and interpret $MR(x) - MC(x)$.

5. Find and interpret $\int MR(x)\,dx$ given no revenue is made when no seat massagers are sold.

6. Find and interpret $\int MC(x)\,dx$ given there is a fixed cost of $255.

7. Use 5 and 6 to find and interpret $\int [MR(x) - MC(x)]\,dx$ given $255 is lost when no seat massagers are sold.

8. Use 7 to evaluate and interpret $\int_0^{12} [MR(x) - MC(x)]\,dx$.

9. Find the profit made by selling the twelfth seat massager.

Chapter

17 Applications of Integral Calculus

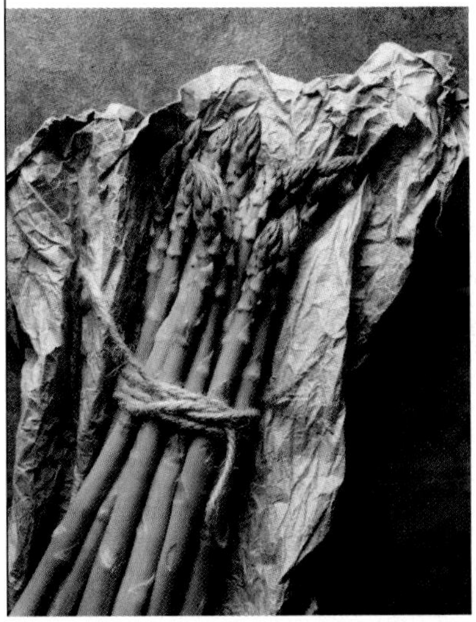

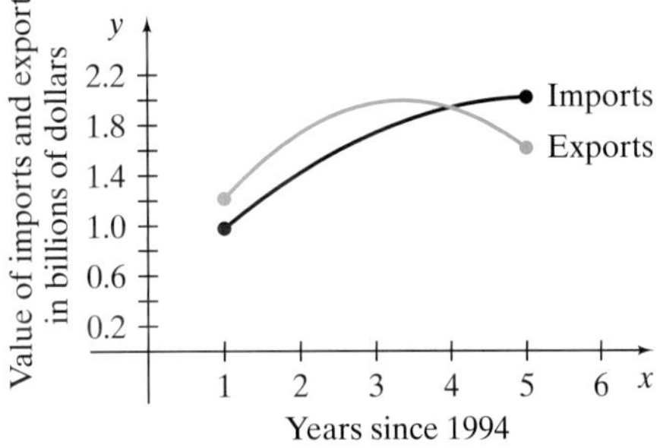

(b)

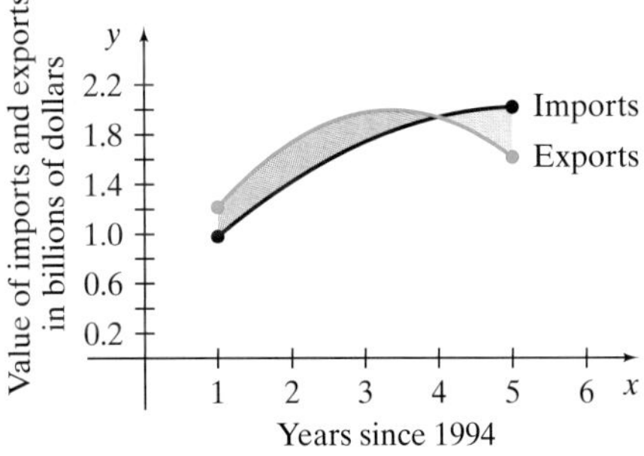

(c)

When fresh asparagus appears in local grocery stores during the winter months, it may have been grown in Peru. The issue of the balance of trade with Latin American nations, including Peru, is a hot economic issue. Figure b shows the U.S. exports and imports with Peru since 1994. An analysis of the data indicates that the U.S. had a total accumulated trade surplus with Peru of about 0.72 billion dollars from 1995 to 1998. From 1998 to 1999 the U.S. had a total accumulated trade deficit with Peru of about 0.187 billion dollars. The total accumulated trade surplus and the total accumulated trade deficit is found by determining the orange shaded area and the yellow shaded area, respectively, as shown in Figure c.

What We Know

In Chapter 16 we saw how the definite integral can be used to give a total accumulation. We also saw how integration and differentiation were related through the Fundamental Theorem of Calculus.

Where Do We Go

In this chapter we will see even more applications of integration. Included applications will be average value, perpetuities, the Gini coefficient, consumers' and producers' surplus, and many others. We continue to strengthen the concepts set forth in Chapter 16, especially an accumulation.

Section 17.1 Average Value of a Function and the Definite Integral in Finance

In this section we examine the **average value** of a function. Compound interest will be revisited as we examine what is known as an **income stream**. The process of computing an average value for a function over an interval will be exploited throughout Chapter 17, and the concept of an income stream will also reappear in later sections of this chapter. We encourage you to work to understand these concepts as they arise, to insure your continued success in these later sections.

Average Value of a Function

Before we analyze how to determine an average value of a function, let's first recall that the average (or mean) of n numbers, $y_1, y_2, y_3, \ldots, y_n$, is simply the sum of the numbers divided by n. That is,

$$\bar{y} = \frac{1}{n}\sum_{i=1}^{n} y = \frac{y_1 + y_2 + y_3 + \cdots + y_n}{n}$$

In the following Flashback, we review how to determine the average of a set of numbers. The Flashback also provides the necessary foundation for the definition of the **average value of a function**.

Flashback

Carbon Dioxide (CO_2) Emissions Revisited

In Section 15.3 we analyzed the carbon content of CO_2 emissions, measured in millions of metric tons, into the atmosphere by the United States from 1990 to 1998. The data are given in Table 17.1.1. We can represent the data graphically by plotting the data points and then making a bar graph as shown in Figure 17.1.1.

Table 17.1.1

Year	Carbon Content
1990	1347.0
1991	1358.8
1992	1379.5
1993	1388.6
1994	1409.9
1995	1423.8
1996	1471.5
1997	1490.4
1998	1495.5

SOURCE: U.S. Energy Information Administration, www.eia.doe.gov

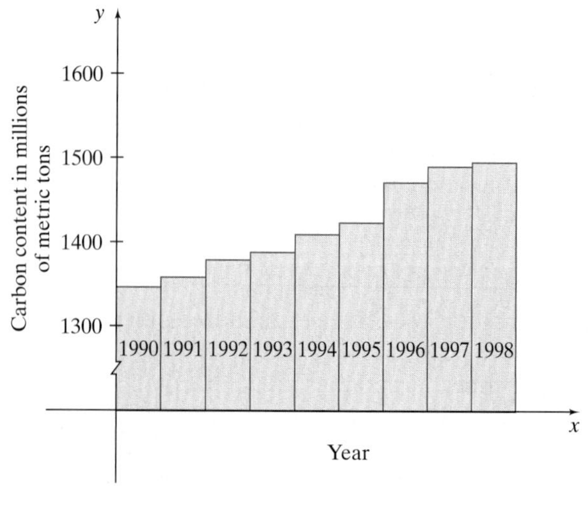

Figure 17.1.1

(a) Using the data values, determine the average value of the carbon content of CO_2 emissions over the 7-year period. Round to the nearest tenth.

(b) Determine the total area of the nine rectangles in Figure 17.1.1.

Flashback Solution

(a) The average value of the carbon content of the CO_2 emissions into the atmosphere by the United States over the 9-year period is given by

$$\bar{y} = \frac{1}{n} = \sum_{i=1}^{n} y$$

$$= \frac{1347.0 + 1358.8 + 1379.5 + 1388.6 + 1409.9 + 1423.8 + 1471.5 + 1490.4 + 1495.5}{9}$$

$$= \frac{12{,}765}{9} \approx 1418.3 \text{ million metric tons per year}$$

(b) The base of the first rectangle is 1, and the height is the carbon content, 1347.0. So the area of this rectangle is $1 \cdot 1347.0 = 1347.0$, which is just the carbon content for that year. Repeating this process for the other rectangles, we see that the total area of the nine rectangles is simply the sum of the values of the carbon content, 12,765, which we determined in part (a).

If we divide the area of the nine rectangles by the number of rectangles, 9, and round to the nearest tenth, we get

$$\frac{12{,}765}{9} \approx 1418.3$$

which is the average value we found in the Flashback, part (a). Thus, we can interpret this result to mean

$$\frac{\text{area under the bar graph}}{\text{width of the bar graph}} = \text{average value}$$

Now, if a continuous function f models the data, we can find the average value of the function over the interval $[a, b]$. Using the preceding result for bar graphs, it can be shown that

$$\frac{\text{area under the curve}}{\text{width of the interval}} = \text{average value}$$

Since we know that the area under a curve is given by $\int_a^b f(x)\,dx$, these results suggest the following definition for the average value of a continuous function f on the interval $[a, b]$.

Average Value of a Continuous Function on $[a, b]$

The **average value** of a continuous function, f, on the interval $[a, b]$ is given by

$$\frac{1}{b-a} \int_a^b f(x)\,dx$$

Example 1 Determining an Average Value

The carbon content, in million of metric tons, of CO_2 emissions into the atmosphere by the United States can be modeled by

$$y = f(x) = -0.3x^3 + 5.26x^2 - 6.52x + 1351.63 \qquad 1 \le x \le 9$$

where $x = 1$ corresponds to January 1, 1990 and y is the carbon content, in millions of metric tons, of CO_2 emissions into the atmosphere by the United States. Use the model to calculate the average value of CO_2 emissions for the period January 1, 1990 to January 1, 1998 (*Source:* U.S. Energy Information Administration, www.eia.doe.gov).

Solution

Understand the Situation: We first observe that $x = 1$ corresponds to January 1, 1990 and $x = 9$ corresponds to January 1, 1998. Thus, we need to determine the average value of $y = f(x) = -0.3x^3 + 5.26x^2 - 6.52x + 1351.63$ on the interval $[1, 9]$.

So we compute the average value to be

$$\frac{1}{9-1}\int_1^9 (-0.3x^3 + 5.26x^2 - 6.52x + 1351.63)\,dx$$

$$= \frac{1}{8}\left[\frac{-0.3}{4}x^4 + \frac{5.26}{3}x^3 - \frac{6.52}{2}x^2 + 1351.63x\right]\Bigg|_1^9$$

$$\approx 1417.1 \text{ (rounded to the nearest tenth)}$$

Interpret the Solution: This means that from January 1, 1990 to January 1, 1998, the average annual carbon content of CO_2 emissions into the atmosphere by the United States was approximately 1417.1 million metric tons. ■

Notice that the average value determined in Example 1 was fairly close to the actual average value that we determined in the Flashback. Some discrepancy is expected, since the function in Example 1 is a model that *approximates* the data.

Example 2 Determining an Average Value

Determine the average value of $f(x) = x^3 - 2x^2 - 5x + 11$ on the interval $[-2, 4]$.

Solution

Here we compute the average value to be

$$\frac{1}{b-a}\int_a^b f(x)\,dx = \frac{1}{4-(-2)}\int_{-2}^4 (x^3 - 2x^2 - 5x + 11)\,dx$$

$$= \frac{1}{6}\left[\frac{x^4}{4} - \frac{2x^3}{3} - \frac{5x^2}{2} + 11x\right]\Bigg|_{-2}^4$$

$$= \frac{1}{6}\left[\left(64 - \frac{128}{3} - 40 + 44\right) - \left(4 + \frac{16}{3} - 10 - 22\right)\right]$$

$$= \frac{1}{6}\left[\frac{76}{3} - \left(\frac{-68}{3}\right)\right] = \frac{1}{6}(48) = 8$$

So the average value of $f(x) = x^3 - 2x^2 - 5x + 11$ on $[-2, 4]$ is 8. ■

Checkpoint 1 Now work Exercise 5.

The average value can be thought of as giving the typical function value for a function on the closed interval $[a, b]$. At this time, we offer the following analogy and comparison between a discrete average value and a continuous average value.

Types of Values	Discrete	Continuous
Sum of values	$\sum_{i=1}^{n} y$	$\int_a^b f(x)\,dx$
Number of "points"	n	$b-a$
Average	$\frac{1}{n}\sum_{i=1}^{n} y$	$\frac{1}{b-a}\int_a^b f(x)\,dx$

Example 3 Determining an Average Value

The annual U.S. per capita consumption (adult population) of alcoholic beverages can be modeled by

$$f(x) = 43.75e^{-0.012x} \qquad 1 \le x \le 16$$

where x is measured in years ($x = 1$ corresponds to January 1, 1980), and $f(x)$ is the annual per capita consumption of alcoholic beverages in gallons. Use the model to determine the average per capita consumption of alcoholic beverages, rounded to the nearest tenth, during the 1980s (*Source:* U.S. Department of Agriculture, www.usda.gov).

Solution

Understand the Situation: When we say "during the 1980s," we mean from January 1, 1980, to January 1, 1990 (or January 1, 1980, through December 31, 1989). This means that we are to determine the average value of f on the interval $[1, 11]$.

Computing this integral yields

$$\frac{1}{b-a}\int_a^b f(x)\,dx = \frac{1}{11-1}\int_1^{11}(43.75e^{-0.012x})\,dx = \frac{1}{10}\left[\frac{43.75}{-0.012}e^{-0.012x}\right]\Bigg|_1^{11}$$

$$= \frac{1}{10}\left[\frac{43.75}{-0.012}e^{-0.012(11)}\right] - \frac{1}{10}\left[\frac{43.75}{-0.012}e^{-0.012(1)}\right]$$

$$\approx \frac{1}{10}[-3194.993 - (-3602.345)]$$

$$= \frac{1}{10}[407.352] = 40.7 \text{ rounded to the nearest tenth.}$$

Interpret the Solution: This means that during the 1980s the annual average alcoholic beverage consumption per capita was about 40.7 gallons. ■

Checkpoint 2 Now work Exercise 15.

Compound Interest

So far we have examined how the average value of a function is used in the life and social sciences. Now we turn our attention to economics, finance, and the managerial sciences. To do this, we need to recall a formula first encountered in Chapter 11 that is reviewed in the Toolbox to the left.

Example 4 Determining an Average Balance and Computing a Bonus

(a) Wilma deposits \$5000 into a money market account earning 6.5% interest compounded continuously. Determine Wilma's average balance, to the nearest cent, over one year.

(b) Suppose that the institution where Wilma has deposited her money also pays a bonus at the end of the year of 1.25% of the average balance in the account during the year. Compute the size of Wilma's bonus.

> **From Your Toolbox**
>
> **Continuous Compound Interest Formula**
>
> If a principal of P dollars is invested into an account earning annual interest rate r (in decimal form) compounded continuously, then the amount A in the account at the end of t years is given by
>
> $$A(t) = Pe^{rt}$$

Solution

(a) **Understand the Situation:** From the formula for continuous compound interest, we have

$$A(t) = 5000e^{0.065t}$$

We are asked to compute the average value of $A(t) = 5000e^{0.065t}$ on $[0, 1]$. So

$$\begin{aligned}\frac{1}{b-a}\int_a^b A(t)\,dt &= \frac{1}{1-0}\int_0^1 5000e^{0.065t}\,dt\\ &= \frac{5000}{0.065}e^{0.065t}\Big|_0^1\\ &= \frac{5000}{0.065}e^{0.065(1)} - \frac{5000}{0.065}e^{0.065(0)}\\ &\approx 5166.08\end{aligned}$$

Interpret the Solution: So Wilma's average balance over one year was \$5166.08.

(b) Computing 1.25% of the \$5166.08 average balance from part (a) yields a bonus of

$$0.0125(5166.08) \approx \$64.58$$

■

In Example 5, we review how to determine a total amount when given a rate of change function, as well as computing an average value.

Example 5 Determining an Account Change and an Average Balance

Stella deposits \$7000 into an account where the rate of change of the amount in the account t years after the initial deposit is given by $A'(t) = 490e^{0.07t}$.

(a) Determine by how much the account changes from the end of the third year to the end of the fifth year. Round to the nearest cent.

(b) Determine the average balance, to the nearest cent, in the account over the first 5 years.

Solution

(a) Understand the Situation: Notice that we are given a **rate function**, $A'(t) = 490e^{0.07t}$. Recall from Chapter 16 that the total change in the account from the end of the third year to the end of the fifth year can be found by the definite integral.

That is,

$$A(5) - A(3) = \int_3^5 490e^{0.07t}\,dt = \frac{490}{0.07}e^{0.07t}\Big|_3^5$$
$$= 7000e^{0.07(5)} - 7000e^{0.07(3)} \approx 1297.73$$

Interpret the Solution: Stella's account changed by about \$1297.73 from the end of the third year to the end of the fifth year.

(b) Understand the Situation: Here we need to determine the average value of A on $[0, 5]$. The first order of business is to determine $A(t)$.

To do this, we integrate

$$A(t) = \int A'(t)\,dt = \int 490e^{0.07t}\,dt$$
$$= \frac{490}{0.07}e^{0.07t} + C = 7000e^{0.07t} + C$$

So $A(t) = 7000e^{0.07t} + C$. To determine the constant C, we note that, at $t = 0$, $A(t) = 7000$. (An initial deposit corresponds to $t = 0$.) Substituting gives

$$A(0) = 7000e^{0.07(0)} + C = 7000$$
$$7000 + C = 7000$$
$$C = 0$$

Since $A(t) = 7000e^{0.07t}$, we compute the average value of $A(t)$ on $[0, 5]$ as follows:

$$\frac{1}{b-a}\int_0^5 7000e^{0.07t}\,dt = \frac{1}{5-0}\left[\frac{7000}{0.07}e^{0.07t}\right]\Big|_0^5$$
$$= \frac{1}{5}\left[100{,}000e^{(0.07\cdot 5)} - 100{,}000e^{(0.07\cdot 0)}\right] \approx 8381.35$$

Interpret the Solution: The average amount in Stella's account during the first 5 years was \$8381.35. ■

Continuous Stream of Income

The prior two examples reviewed the concept of continuous compounding. Continuous compounding does not mean that the bank or other institution where the money is invested is continuously placing money (the interest earned) into our account. However, we compute what is in our account as if it were. We can think of our account as being a **continuous income stream**. In fact, this is very similar to the electric company, the natural gas company, or the water company, which has a meter on our house that continuously monitors and totals the amount of electricity, natural gas, or water that we are consuming. Example 6 describes a continuous stream of income.

Example 6 Determining Total Income from an Income Stream

Suppose that the rate of change of income, in thousands of dollars per year, for an oil field is projected to be $f(t) = 350e^{-0.09t}$, where t is the number of years from when oil extraction began. See Figure 17.1.2. Determine the total amount of money generated from this oil field during the first two years of operation.

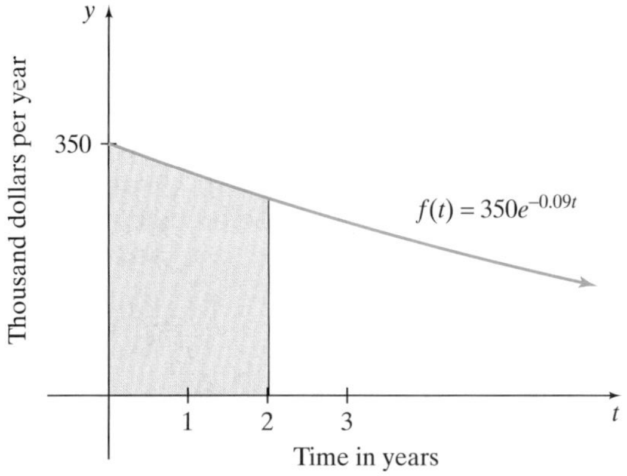

Figure 17.1.2 The area of the shaded region can be interpreted as giving the total money generated in the first 2 years.

Solution

Understand the Situation: Once again we recognize that we have a rate function. To determine the total amount of money for the first 2 years, we need to determine the area under the curve on the interval [0, 2].

Integrating yields

$$\int_0^2 350e^{-0.09t}\,dt = \left(\frac{350}{-0.09}e^{-0.09t}\right)\Bigg|_0^2$$
$$= \frac{350}{-0.09}e^{-0.09(2)} - \frac{350}{-0.09}e^{-0.09(0)}$$
$$\approx 640.61584$$

Interpret the Solution: So the first 2 years of operation will generate about \$640,615.84 in income. ■

In reality, income from the oil field in Example 6 is not received in one lump sum payment at the end of 2 years. The income is probably collected on a regular basis, possibly every month or quarter. In situations like Example 6, however, it is convenient to assume that the income is actually received in a *continuous stream*; that is, we assume that the income is a continuous function of time. The rate of change of income, $f(t) = 350e^{-0.09t}$ in Example 6, is called a **rate of flow function**.

Total Income for a Continuous Income Stream

If f is the rate of flow function of a continuous income stream, then the **total income** produced from time $t = a$ to $t = b$ is found by

$$\int_a^b f(t)\,dt$$

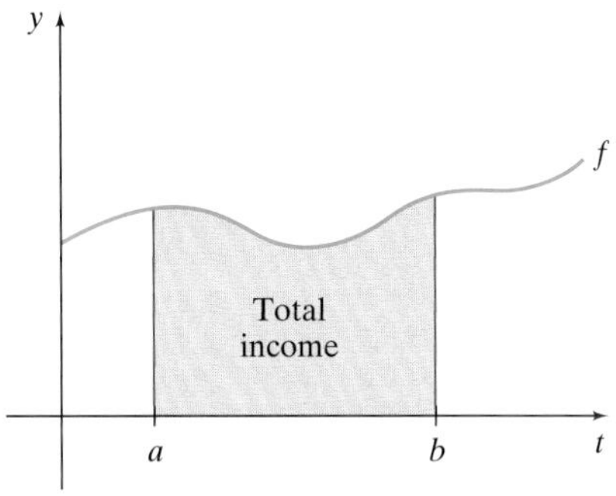

Figure 17.1.3

Checkpoint 3

Now work Exercise 39.

Annuities

We conclude this section with a look at what is known as an annuity. An **annuity** is defined as a sequence of equal payments made at equal time intervals. For example, if money is borrowed to finance the purchase of a home, a car, or even an education, the loan usually is repaid in equal monthly payments for a specified length of time. This sequence of equal monthly payments forms an annuity. Also, payments that one receives at retirement from a pension plan, a 401-k, a 403-b, or an IRA are other examples of annuities.

Assume that Maria deposits \$2000 at the beginning of each year into an IRA and that her investment earns an annual interest rate of 6% compounded continuously. To compute the amount in her account after 4 years requires that we compute what happens to the initial deposit, the second deposit, the third deposit, and so on, for the 4 years. For example, using $A(t) = Pe^{rt}$:

- The initial deposit ($t = 0$) of \$2000 is in the account for 4 years and grows to $2000e^{(0.06)(4)}$.
- The second deposit ($t = 1$) is in the account for 3 years ($4 - 1$) and grows to $2000e^{(0.06)(4-1)} = 2000e^{(0.06)(3)}$.

Continuing in this manner, the total amount S in the account at the end of 4 years can be represented by

$$S = \begin{pmatrix}\text{value of initial payment}\\ t = 0\end{pmatrix} + \begin{pmatrix}\text{value of second payment}\\ t = 1\end{pmatrix} + \begin{pmatrix}\text{value of third payment}\\ t = 2\end{pmatrix} + \begin{pmatrix}\text{value of final payment}\\ t = 3\end{pmatrix}$$

$$S = 2000e^{0.06(4)} + 2000e^{0.06(4-1)} + 2000e^{0.06(4-2)} + 2000e^{0.06(4-3)}$$

$$S = 2000e^{0.06(4)} + 2000e^{0.06(3)} + 2000e^{0.06(2)} + 2000e^{0.06(1)}$$

$$S \approx 2542.50 + 2394.43 + 2254.99 + 2123.67 = \$9315.59$$

To see how we can approximate the sum, S, with a definite integral, we plot each term of the sum as shown in Figure 17.1.4.

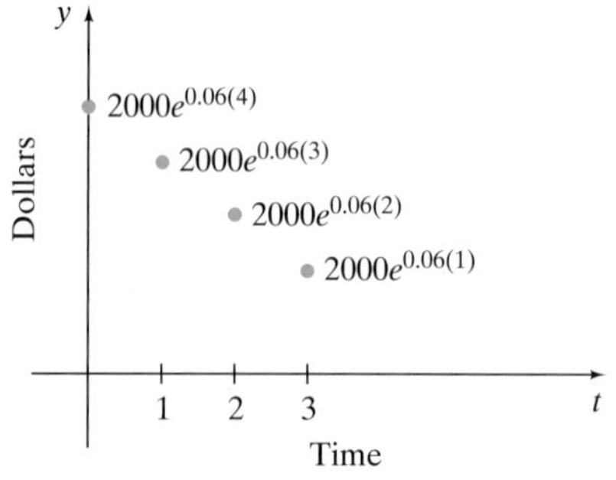

Figure 17.1.4

The sum of the amount in the IRA can be thought of as being equal to the sum of the four rectangles shown in Figure 17.1.5, since the base (width) of each rectangle is 1. Figure 17.1.6 shows the graph of $A(t) = 2000e^{0.06(4-t)}$ with the four rectangles. What we are about to do should look quite familiar by now. The area under A on the interval [0, 4] is approximately equal to the area of the four rectangles. Hence, the area under the curve can be used to approximate the amount or sum in the IRA at the end of 4 years. Example 7 illustrates the process.

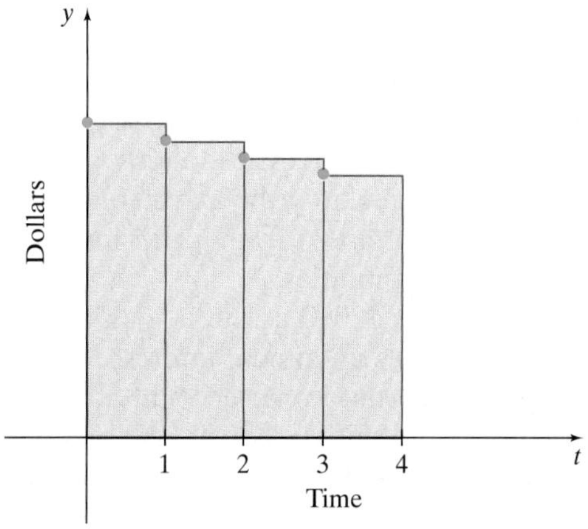

Figure 17.1.5

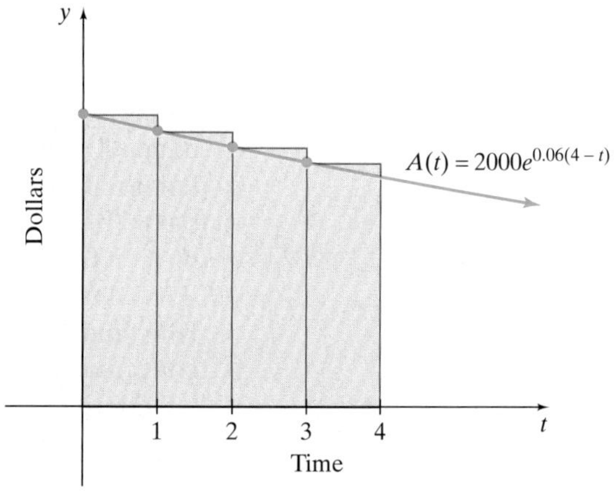

Figure 17.1.6

Example 7 Applying the Definite Integral

Write and evaluate a definite integral to find the area under A in Figure 17.1.6 that approximates the sum in the IRA at the end of 4 years.

Solution

The definite integral that determines the area under A is

$$\int_0^4 2000e^{0.06(4-t)}\,dt$$

We evaluate the definite integral using u-substitution as follows:

$$\text{Let } u = 4 - t; \text{ then } du = -dt \text{ or } -du = dt.$$

By u-substitution, the integral may be written as

$$-\int 2000e^{0.06u}\,du = -\left(\frac{2000}{0.06}e^{0.06u}\right) = -\left(\frac{2000}{0.06}e^{0.06(4-t)}\right)\Bigg|_0^4$$

$$= -\left(\frac{2000}{0.06}e^{0.06(4-4)} - \frac{2000}{0.06}e^{0.06(4-0)}\right)$$

$$\approx -(33{,}333.33 - 42{,}374.97) = \$9041.64$$

Example 7 indicates that the definite integral gives an *approximation* to the actual sum. With this in mind, the explanation preceding Example 7, as well as Example 7, outlines a procedure that we can now generalize to give an approximation for S, the **future value of an annuity**.

Future Value of an Annuity

If n equal payments of P dollars is deposited into an annuity at an annual interest rate r (in decimal form), then the **future value**, S, for a duration of the annuity T is given by

$$S \approx \int_0^T Pe^{r(T-t)}\,dt$$

Note: It is very important to keep the units of time for T, r, and P consistent when using the formula to approximate the future value of an annuity, S. Example 8 illustrates.

Example 8 Determining a Future Value

Interactive Activity

Try integrating the definite integral that approximates the future value of an annuity in the most general case, that is, $\int_0^T Pe^{r(T-t)}\,dt$, to determine a formula for the antiderivative that you may use in the exercises. Keep in mind that P, r, and T are constants when integrating with respect to t.

WWW

Irene has just started a new job teaching. At the beginning of each month she deposits \$25 into her 403-b plan, which earns 6.6% annual interest compounded continuously. Approximate how much Irene will have in her account after 20 years.

Solution

We approximate the amount in Irene's 403-b plan after 20 years by using

$$S \approx \int_0^T Pe^{r(T-t)}\,dt$$

If we have $P = \$25$ per month, then r and T must also be in terms of months. Thus,

$$T = 12(20) = 240 \text{ months} \quad \text{and} \quad r = \frac{0.066}{12} = 0.0055 \text{ per month}$$

So the approximation for S is

$$S \approx \int_0^T Pe^{r(T-t)}\,dt = \int_0^{240} 25e^{0.0055(240-t)}\,dt$$

Using a u-substitution yields $u = 240 - t$ and $du = -dt$ or $-du = dt$, and thus

$$\begin{aligned}\int_0^{240} 25e^{0.0055(240-t)}\,dt &= -\int 25e^{0.0055u}\,du \\ &= -\left(\frac{25}{0.0055}e^{0.0055u}\right) = -\left(\frac{25}{0.0055}e^{0.0055(240-t)}\right)\Big|_0^{240} \\ &= -\left(\frac{25}{0.0055}e^{0.0055(240-240)} - \frac{25}{0.0055}e^{0.0055(240-0)}\right) \\ &\approx -(4545.45 - 17{,}015.55) \\ &= \$12{,}470.10\end{aligned}$$

So the total amount in the account after 20 years is approximately \$12,470.10. ■

Checkpoint 4 Now work Exercise 45.

SUMMARY

In this introductory section on applications of the integral, we examined how recognizing a definite integral as a limit of a sum and recognizing the area under a curve as a definite integral were very important. The former was utilized to determine a formula for the average value, and the latter was used in determining a formula to approximate the future value of an annuity. We also saw how integrating a rate function gives a total accumulation. In this section, it was seen again in the context of **income streams**.

- The **average value** of a continuous function, f, on the interval $[a, b]$ is given by

$$\frac{1}{b-a}\int_a^b f(x)\,dx$$

- If n equal payments of P dollars are deposited into an annuity at an annual interest rate r (in decimal form), then the **future value**, S, for a duration of the annuity T is given by

$$S \approx \int_0^T Pe^{r(T-t)}\,dt$$

- If f is the **rate of flow function** of a continuous income stream, then the **total income** produced from time $t = a$ to $t = b$ is found by

$$\int_a^b f(t)\,dt$$

SECTION 17.1 EXERCISES

In Exercises 1–10, determine the average value of the given function on the stated interval.

1. $f(x) = 2x + 5$	$[0, 7]$
2. $f(x) = 3x + 1$	$[1, 4]$
3. $f(x) = 3x^2 - 2x$	$[-1, 2]$
4. $g(x) = 4x - 2x^2$	$[0, 3]$
✓ **5.** $y = x^3 + 2x^2 - x + 1$	$[0, 2]$
6. $y = -2x^3 + x + 2$	$[-2, 1]$
7. $g(x) = x^{2/3}$	$[1, 8]$
8. $g(x) = \sqrt{x+1}$	$[3, 7]$
9. $y = 4e^{0.2x}$	$[0, 10]$
10. $y = 2e^{-0.15x}$	$[0, 5]$

For Exercises 11–14,

(a) Use your **RTSUM** program, with $n = 100$, to approximate $\int_a^b f(x)\,dx$.

(b) Use your result from part (a) to approximate an average value of f on $[a, b]$.

11. $f(x) = \dfrac{2x+1}{x^2+1}$	$[-1, 1]$
12. $f(x) = \dfrac{x}{x+1}$	$[0, 4]$
13. $f(x) = \ln(1 + x^2)$	$[0, 2]$
14. $f(x) = \dfrac{1}{x^2+4}$	$[0, 5]$

Applications

15. The amount of money spent on home health care in the United States can be approximated by

$$f(x) = 0.13x^2 - 0.39x + 2.79 \qquad 1 \le x \le 16$$

where x is measured in years ($x = 1$ corresponds to January 1, 1980) and $f(x)$ is the amount of money, in billions of dollars. Use the model to compute the average amount spent on home health care in the United States during the 1980s (*Source:* U.S. Health Care Financing Administration, www.hcfa.gov).

16. The number of Medicare enrollees in the United States can be modeled by

$$f(x) = 0.61x + 27.68 \qquad 1 \le x \le 16$$

where x is measured in years ($x = 1$ corresponds to January 1, 1980) and $f(x)$ is the number of Medicare enrollees, in millions, in the United States. Use the model to compute the average number of Medicare enrollees during the 1980s (*Source:* U.S. Health Care Financing Administration, www.hcfa.gov).

17. The number of Medicaid recipients in the United States can be modeled by

$$f(x) = 0.11x^2 - 0.91x + 22.59 \qquad 1 \le x \le 16$$

where x is measured in years ($x = 1$ corresponds to January 1, 1980) and $f(x)$ is the number of Medicaid recipients, in millions, in the United States. Use the model to compute the average number of Medicaid recipients during the 1980s (*Source:* U.S. Health Care Financing Administration, www.hcfa.gov).

18. The annual U.S. per capita consumption, in pounds, of beef can be modeled by

$$B(t) = 0.03t^3 - 0.76t^2 + 4.54t + 68.34 \qquad 1 \le t \le 16$$

where t is measured in years ($t = 1$ corresponds to January 1, 1980) and $B(t)$ is the per capita consumption, in pounds, of beef. Determine the average per capita consumption of beef during the 1980s (*Source:* U.S. Department of Agriculture, www.usda.gov).

19. The revenue, in billions of dollars, in the resource recovery industry in the United States can be modeled by

$$f(x) = 0.096x^3 - 1.744x^2 + 8.903x + 5.834 \qquad 1 \le x \le 11$$

where x is measured in years ($x = 1$ corresponds to January 1, 1990). Determine the average revenue in the resource recovery industry during the 1990s (*Source:* U.S. Statistical Abstract, www.census.gov/statab/www).

20. The cost, in cents per gallon, of unleaded regular gasoline in the United States, can be modeled by

$$f(x) = 0.01x^4 - 0.25x^3 + 2.77x^2 - 11.31x + 133.29 \qquad 1 \le x \le 16$$

where x is measured in years ($x = 1$ corresponds to January 1, 1980) (*Source:* U.S. Energy Information Administration, www.eia.doe.gov).

(a) Determine the average price, in cents per gallon, for unleaded regular gasoline in the United States during the 1980s.

(b) Determine the average price, in cents per gallon, for unleaded regular gasoline in the United States during the first half of the 1990s.

21. The 1998 prospectus of a certain mutual fund states that during 1997 the amount of money that was kept in cash reserves was approximately linear and can be represented by $CR(x) = 0.8x + 3$, where x is the number of months after the first of the year and $CR(x)$ is the cash reserves in millions of dollars.

(a) Compute the average cash reserves on the interval [0, 3], the first quarter of 1997.

(b) Compute the average cash reserves on the interval [9, 12], the last quarter of 1997.

22. The 1998 prospectus of a certain mutual fund states that during 1997 the amount of money that was kept in cash reserves was approximately linear and can be represented by

$$CR(x) = -0.9x + 12.5$$

where x is the number of months after the first of the year and $CR(x)$ is the cash reserves in millions of dollars.

(a) Compute the average cash reserves on the interval [0, 3], the first quarter of 1997.

(b) Compute the average cash reserves on the interval [9, 12], the last quarter of 1997.

23. The price–demand function for Linguini's Pizza Palace all-you-can-eat pizza buffet is given by

$$p(x) = -0.02x + 8.3$$

where $p(x)$ is the price in dollars and x is the quantity demanded. Determine the average price on the demand interval [120, 200].

24. The price–demand function for Linguini's Pizza Palace specialty pizzas is given by

$$p(x) = 15.22e^{-0.015x}$$

where $p(x)$ is the price in dollars and x is the quantity demanded. Determine the average price on the demand interval [40, 80].

25. Researchers have determined through experimentation that the percent concentration of a certain medication during the first 20 hours after it has been administered can be approximated by

$$p(t) = \frac{200t}{2t^2 + 5} \qquad 0 \le t \le 20$$

where t is the time in hours after the medication is taken and $p(t)$ is the percent concentration. Determine the average percent concentration on the interval $[0, 20]$.

26. Researchers have determined through experimentation that the percent concentration of a certain medication during the first 20 hours after it has been administered can be approximated by

$$p(t) = \frac{300t}{6t^2 + 5} \qquad 0 \le t \le 20$$

where t is the time in hours after the medication is taken and $p(t)$ is the percent concentration.

(a) Determine the average percent concentration on the interval $[0, 10]$.

(b) Determine the average percent concentration on the interval $[10, 20]$.

(c) Is the average percent concentration greater during the first 10 hours after it has been taken or during the second 10 hours after it has been taken? Explain.

27. Dixco Engines has determined that the cost for producing x diesel engines is given by

$$C(x) = 60{,}000 + 300x$$

where $C(x)$ is the cost in dollars.

(a) Determine the average value of the cost function over the interval $[0, 500]$.

(b) Determine the average cost function, AC.

$\left[\text{Recall that } AC(x) = \frac{C(x)}{x}.\right]$ Evaluate $AC(500)$.

(c) Explain the differences between parts (a) and (b).

28. Digital Pet has determined that the cost for producing x virtual pets in one day is given by

$$C(x) = 150 + 3x + \frac{2x^2}{30}$$

where $C(x)$ is the cost in dollars.

(a) Determine the average value of the cost function over the interval $[0, 200]$.

(b) Determine the average cost function, AC.

$\left[\text{Recall that } AC(x) = \frac{C(x)}{x}.\right]$ Evaluate $AC(200)$.

(c) Explain the differences between parts (a) and (b).

29. Jan deposits \$3000 into a money market account earning 5.85% interest compounded continuously. Determine Jan's average balance, to the nearest cent, over 5 years.

30. Eugene deposits \$12,000 into a money market account earning 6.5% interest compounded continuously. Determine Eugene's average balance, to the nearest cent, over 3 years.

31. Lisa deposits \$6000 into a bank account that earns 4.5% interest compounded continuously. Determine Lisa's average balance, to the nearest cent, over 1 year.

32. Wilhelm deposits \$9000 into a money market account earning 7% interest compounded continuously. The institution where Wilhelm has deposited his money also pays a bonus at the end of the year of 1% of the average balance in the account during the year.

(a) Compute Wilhelm's average balance, to the nearest cent, over 1 year.

(b) Compute the size of Wilhelm's bonus.

33. Elmer deposits \$3500 into a bank account that earns 3.8% interest compounded continuously. The bank also pays a bonus at the end of the year of 0.75% of the average balance in the account during the year.

(a) Compute Elmer's average balance, to the nearest cent, over 1 year.

(b) Compute the size of Elmer's bonus.

34. Rusty has \$4000 to invest in a money market account. At Cold Cash Company, Rusty is offered an account that earns 5% interest compounded continuously with a 1% bonus at the end of the year on his average balance in the account during the year. At Money Time Company, Rusty is offered an account that earns 4.9% interest compounded continuously with a 1.2% bonus at the end of the year on his average balance in the account during the year. If Rusty's goal is to achieve the largest bonus at the end of the year, into which account should he place his money? Explain.

35. Shannon deposits \$8000 into an account where the rate of change of the amount in the account is given by $A'(t) = 480e^{0.06t}$, t years after the initial deposit.

(a) Determine by how much the account changes from the end of the second year to the end of the fifth year. Round to the nearest cent.

(b) Determine the average balance, to the nearest cent, in the account over the first 5 years.

36. Kevan deposits \$2000 into an account where the rate of change of the amount in the account is given by $A'(t) = 110e^{0.055t}$, t years after the initial deposit.

(a) Determine by how much the account changed from the end of the first year to the end of the third year. Round to the nearest cent.

(b) Determine the average balance, to the nearest cent, in the account over the first 3 years.

37. Muggy is a company that produces coffee mugs. The research department at Muggy determined the following marginal cost function:

$$MC(x) = 1 + \frac{x}{20}$$

Here $MC(x)$ is in dollars per mug and x is the number of coffee mugs produced per day.

(a) Compute the increase in cost going from a production level of 25 mugs per day to 75 mugs per day.

(b) If daily fixed costs are \$50, compute the average value of $C(x)$ over the interval [25, 75] and interpret.

38. Binky Inc. is a company that produces pacifiers. The research department at Binky Inc. has determined the following marginal cost function:

$$MC(x) = 0.06 + \frac{x}{300}$$

Here $MC(x)$ is in dollars per pacifier and x is the number of pacifiers produced per day.

(a) Compute the increase in cost going from a production level of 200 pacifiers per day to 300 pacifiers per day.

(b) If daily fixed costs are \$150, compute the average value of $C(x)$ over the interval [200, 300] and interpret.

✓ **39.** Geologists estimate that an oil field will produce oil at a rate given by $f(t) = 600e^{-0.1t}$ thousand barrels per month, t months into production.

(a) Write a definite integral to estimate the total production for the first year of production.

(b) Evaluate the integral from part (a) to estimate the total production for the first year of operation. Round to the nearest whole number.

40. The rate of change of income, in thousands of dollars per year, for the oil field in Exercise 39, is projected to be $f(t) = 8400e^{-0.1t}$, where t is the number of months into production.

(a) Write a definite integral to estimate the total amount of money generated from this oil field for the first year of operation.

(b) Evaluate the integral from part (a) to estimate the total amount of money generated from this oil field for the first year of operation. Round to the nearest cent.

41. Smart Ones Car Wash generates income in the first 3 years of operation at a rate given by

$$f(t) = 2 \qquad 0 \le t \le 10$$

where t is the number of years in operation and $f(t)$ is in millions of dollars per year. Determine the total income produced in the first 3 years of operation.

42. The rate of change of income produced by a vending machine located in a college dorm is given by

$$f(t) = 2500e^{0.05t} \qquad 0 \le t \le 10$$

where t is time in years since the installation of the vending machine and $f(t)$ is in dollars per year. Determine the total income generated by the vending machine during the first 3 years of operation.

43. The rate of change of income produced by a vending machine located in a busy airport is given by

$$f(t) = 6000e^{0.08t} \qquad 0 \le t \le 10$$

where t is time in years since the installation of the vending machine and $f(t)$ is in dollars per year. Determine the total income generated by the vending machine during the first 3 years of operation.

44. Rich's Lawn Service has determined that his new landscaping company has a projected rate of change of income given by

$$f(t) = 30e^{0.15t} \qquad 0 \le t \le 10$$

where t is time in years since starting the business and $f(t)$ is in thousands of dollars per year. Determine the total income generated by Rich's Lawn Service during the first 2 years of operation.

✓ **45.** At the beginning of each month, Antonio deposits \$250 into his 403-b plan, which earns 6.5% annual interest compounded continuously. Determine how much Antonio will have in his account after 25 years. Round to the nearest cent.

46. At the beginning of each month, Susan deposits \$300 into her 403-b plan, which earns 7.25% annual interest compounded continuously. Determine how much Susan will have in her account after 30 years. Round to the nearest cent.

47. At the age of 22, Vicki opened an IRA. At the beginning of each year, she deposits \$1800 into her IRA, which earns 8% annual interest compounded continuously. Determine how much Vicki has in her account in 43 years when she retires at age 65.

48. *(continuation of Exercise 47)* Determine how much of Vicki's final amount is interest.

49. At the age of 25, Jeff opened an IRA. At the beginning of each year, he deposits \$2000 into his IRA, which earns 8% annual interest compounded continuously. Determine how much Jeff has in his account in 40 years when he retires at age 65.

50. *(continuation of Exercise 49)* Determine how much of Jeff's final amount is interest.

51. Comparing Exercises 47 and 49:

(a) How much did Vicki contribute to her IRA?

(b) How much did Jeff contribute to his IRA?

(c) By how much did Vicki's final amount surpass Jeff's final amount? Explain how Vicki's final amount could surpass Jeff's final amount if Jeff contributed more to his IRA than did Vicki.

52. On the birth of their new son Dylan, Bill and Lisa opened a college savings account. They deposit \$50 at the beginning of each month into an account earning 5.5% annual interest compounded continuously. Determine how much is in the account in 18 years when Dylan starts college.

53. On the birth of their new son Randon, Donald and Melissa opened a college savings account. They deposit $75 at the beginning of each month into an account earning 5.25% annual interest compounded continuously. Determine how much is in the account in 18 years when Randon starts college.

54. On the birth of their new daughter Hannah, Steve and Felicia opened a college savings account. They deposit $100 at the beginning of each month into an account earning 5% annual interest compounded continuously. Determine how much is in the account in 18 years when Hannah starts college.

SECTION PROJECT

Maria has decided to open an IRA for retirement purposes. She has found an account that earns 7% annual interest compounded continuously. She has decided to invest $1800 each year into the IRA. She can invest this $1800 in one of two ways: either $1800 at the beginning of each year or $150 at the beginning of each month. Maria plans on investing for 30 years, at which time she will retire. To help Maria decide between the two investment options:

(a) Determine how much is in Maria's account in 30 years if she invests $1800 at the beginning of each year.

(b) Determine how much is in Maria's account in 30 years if she invests $150 at the beginning of each month.

(c) Based on parts (a) and (b), which plan would you advise Maria to take? Explain.

Section 17.2 Area between Curves and Applications

In Chapter 16, as well as in Section 17.1, we saw that the area under a curve on a closed bounded interval can be determined by the definite integral. In this section we extend this concept to compute the area between two curves on a closed interval. We will also see how to determine the closed interval over which the area trapped between two curves can be calculated. The latter will require us to find points of intersection of two curves.

Area between Curves on a Closed Interval $[a, b]$

We begin this section utilizing what was learned in Chapter 16, specifically, that the definite integral gives the area under a curve. Our first example reviews this important concept.

Example 1 Representing Area by a Definite Integral

(a) Write a definite integral to represent the shaded area in Figure 17.2.1.

(b) Write a definite integral to represent the shaded area in Figure 17.2.2.

(c) Use the results of parts (a) and (b) to write a definite integral to represent the shaded area in Figure 17.2.3.

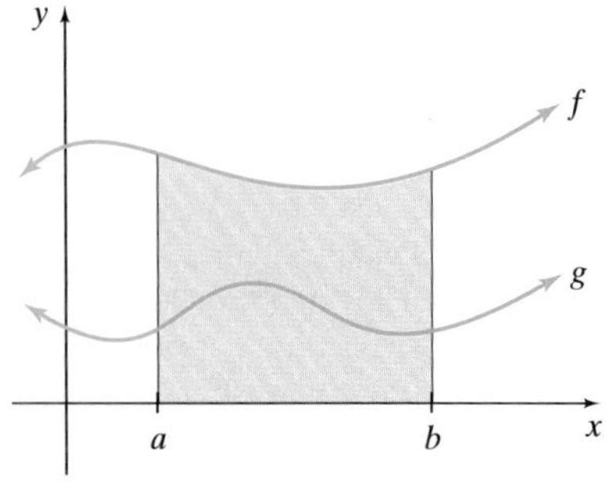

Figure 17.2.1

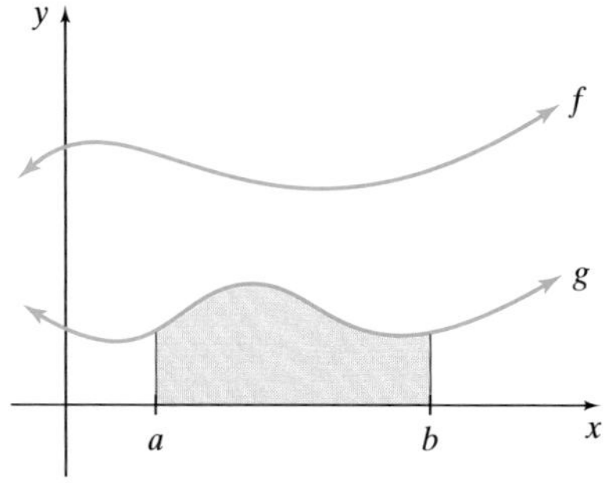

Figure 17.2.2

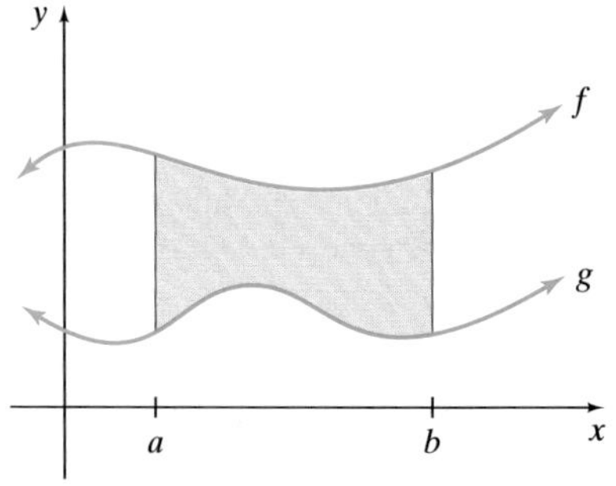

Figure 17.3.3

Solution

(a) Since the shaded area is below the graph of f on the interval $[a, b]$, the definite integral that represents the shaded area is $\int_a^b f(x)\,dx$.

(b) Since the shaded area is below the graph of g on the interval $[a, b]$, the definite integral that represents the shaded area is $\int_a^b g(x)\,dx$.

(c) Here it appears that the shaded area is simply

Area under the graph of f – area under the graph of g

$$\int_a^b f(x)\,dx - \int_a^b g(x)\,dx$$

We can now use a property from Section 16.1 and write these integrals as

$$\int_a^b [f(x) - g(x)]\,dx$$

In Example 1c, what we really found was the area between the graphs of f and g on the interval $[a, b]$. In fact, the process outlined in Example 1 can be summarized as follows.

Area between Two Curves

On a closed interval $[a, b]$, the area between two continuous functions f and g is given by:

1. If $f(x) \geq g(x)$, then the area between the graphs of f and g is $\int_a^b [f(x) - g(x)]\,dx$. See Figure 17.2.4a.
2. If $g(x) \geq f(x)$, then the area between the graphs of f and g is $\int_a^b [g(x) - f(x)]\,dx$. See Figure 17.2.4b.

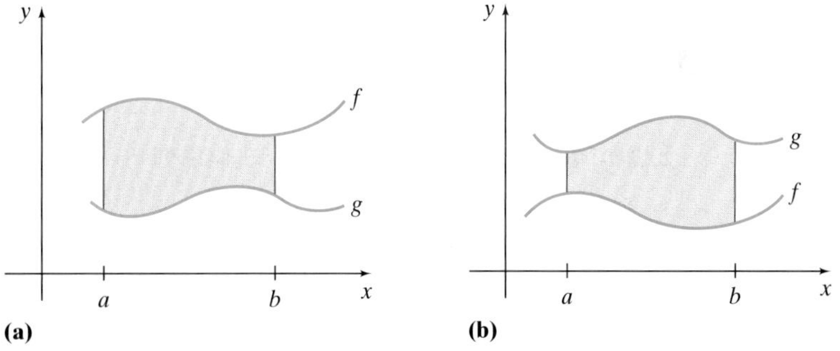

Figure 17.2.4 **(a)** Area of shaded region is given by $\int_a^b [f(x) - g(x)]\,dx$. **(b)** Area of shaded region is given by $\int_a^b [g(x) - f(x)]\,dx$.

Note: An easy way to handle *both* cases listed is to remember the following: To determine the area between two curves on $[a, b]$, integrate

$$\int_a^b [(\text{top curve} - \text{bottom curve})]\,dx$$

Example 2 Determining the Area between Two Curves

Determine the area between the x-axis and $f(x) = x^2 - x - 6$ on the interval $[1, 3]$.

Solution

Understand the Situation: We first graph the function and shade the area as an aid in setting up the correct integral. See Figure 17.2.5.

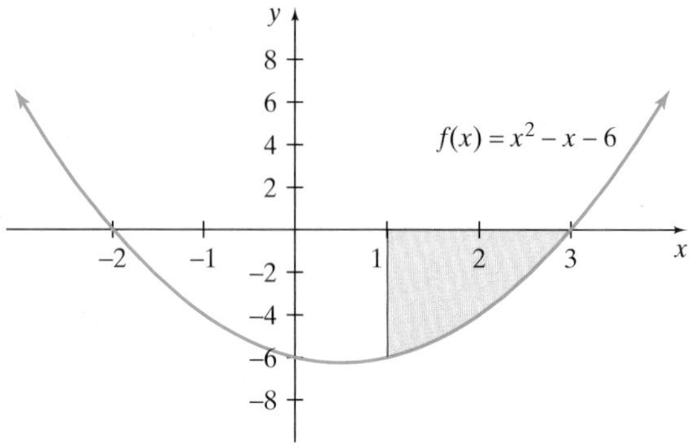

Figure 17.2.5

If we recall that the x-axis can be represented by the function $y = 0$, we can apply the techniques of this section. Thus,

$$\int_1^3 [\text{top curve} - \text{bottom curve}]\,dx = \int_1^3 [0 - (x^2 - x - 6)]\,dx$$

$$= \int_1^3 (-x^2 + x + 6)\,dx = \left(-\frac{1}{3}x^3 + \frac{1}{2}x^2 + 6x\right)\Big|_1^3$$

$$= \left[\left(-\frac{1}{3}(3)^3 + \frac{1}{2}(3)^2 + 6(3)\right) - \left(-\frac{1}{3}(1)^3 + \frac{1}{2}(1)^2 + 6(1)\right)\right]$$

$$= 7\frac{1}{3} \text{ square units}$$

■

Example 3 Determining Area between Two Curves

Determine the area between $f(x) = -x^2 + 4$ and $g(x) = 2x + 7$ on the interval $[-3, 2]$.

Solution

Understand the Situation: As in Example 2, we begin with a graph of f and g and shade the area. See Figure 17.2.6. Since $g(x) \geq f(x)$ for all x on the interval $[-3, 2]$, the graph of g is the top curve and the graph of f is the bottom curve.

$$\int_{-3}^2 [\text{top curve} - \text{bottom curve}]\,dx = \int_{-3}^2 [(2x + 7) - (-x^2 + 4)]\,dx$$

$$= \int_{-3}^2 (x^2 + 2x + 3)\,dx$$

$$= \left(\frac{1}{3}x^3 + x^2 + 3x\right)\Big|_{-3}^2$$

$$= \left(\frac{8}{3} + 4 + 6\right) - (-9 + 9 - 9)$$

$$= \frac{65}{3} \text{ square units}$$

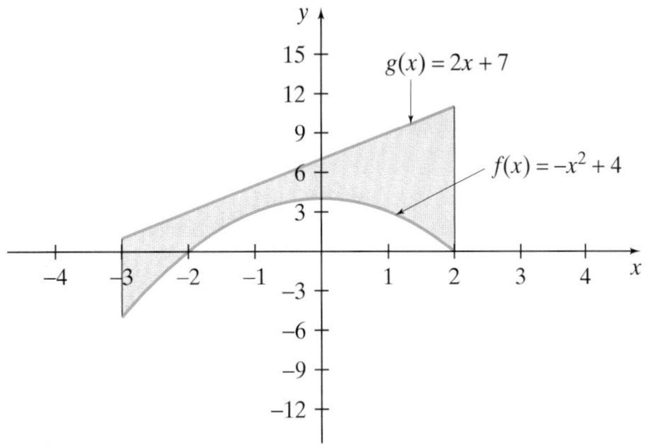

Figure 17.2.6

Checkpoint 1 Now work Exercise 15.

In Example 4, we extend the concept set forth in Chapter 16 that integrating a rate of change function gives an accumulation.

Example 4 Determining Plant Growth

From past records, a botanist knows that a certain species of tree grows at a rate of $\frac{3}{2}x^{-1/2}$ feet per year, where x is the age of the tree in years. However, if a special nitrogen-rich nutrient is given to the tree, it grows at a rate of $2x^{-1/2}$ feet per year. On the interval $[1, 4]$, how many more feet in growth would result from the special nitrogen-rich nutrient?

Solution

Understand the Situation: In Figure 17.2.7 we have graphed the two curves representing the two different growth rates. To determine how many more feet of growth results from the special nitrogen-rich nutrient, we need to integrate the difference of the rates from $x = 1$ to $x = 4$. In other words, we need to determine the area between the curves.

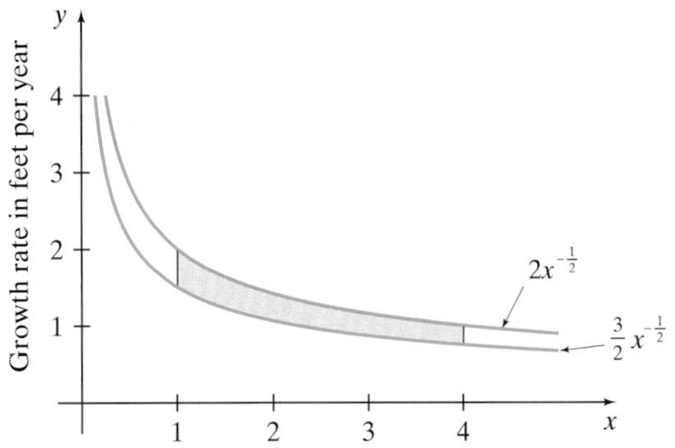

Figure 17.2.7 Additional growth is given by the shaded region.

The area between the curves is found by computing

$$\int_1^4 \left(2x^{-1/2} - \frac{3}{2}x^{-1/2}\right) dx = \int_1^4 \frac{1}{2}x^{-1/2}\, dx$$
$$= x^{1/2}\Big|_1^4 = (4)^{1/2} - (1)^{1/2} = 1$$

Interpret the Solution: So on the interval [1, 4], that is, from year 1 to year 4, the nitrogen-rich nutrient would result in an additional 1 foot of growth. ■

Checkpoint 2

Now work Exercise 47.

Area Bounded by Two Curves

So far in this section we have presented examples in which the interval of integration has been given. We now consider how to determine the area of a region that is trapped or enclosed by two curves and when no interval is given. In these situations, where the curves completely enclose an area, we need to determine where the curves intersect.

Example 5 Determining the Area Enclosed by Two Curves

Determine the area bounded by the curves $y = x^2 + x - 5$ and $y = 2x + 1$.

Solution

Interactive Activity

Graph $y_1 = 2x + 1$ and $y_2 = x^2 + x - 5$, and use the INTERSECT command to verify the points of intersection found algebraically in Example 5.

WWW

Understand the Situation: As always, our first step is to graph the curves so that we can see the region whose area we want. See Figure 17.2.8. For this region, the top curve is $y = 2x + 1$ and the bottom curve is $y = x^2 + x - 5$.

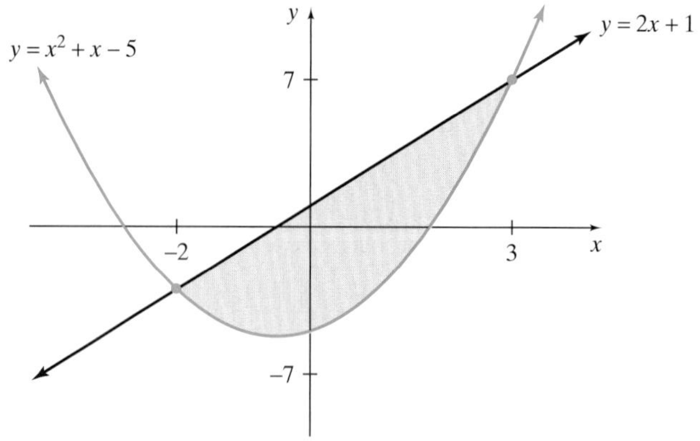

Figure 17.2.8

To determine the limits of integration, we must find the points where the curves intersect. Algebraically, this is handled by setting the two equations equal to each other and solving for x.

$$2x+1=x^2+x-5$$
$$x^2-x-6=0$$
$$(x-3)(x+2)=0$$
$$x=3 \quad \text{or} \quad x=-2$$

So the curves intersect at x-values of $x=3$ and $x=-2$. The area enclosed by the two curves is then found by integrating

$$\int_{-2}^{3}[(2x+1)-(x^2+x-5)]\,dx=\int_{-2}^{3}(-x^2+x+6)\,dx$$
$$=\left(-\frac{1}{3}x^3+\frac{1}{2}x^2+6x\right)\Big|_{-2}^{3}$$
$$=13.5-\left(-7\frac{1}{3}\right)$$
$$=20\frac{5}{6}, \quad \text{or approximately 20.83 square units.}$$ ■

In our next example, we revisit marginal revenue and marginal cost.

Example 6 Computing a Total Profit

The FrigAir Company knows that its marginal cost to produce x refrigerators is given by $C'(x)=4x+23$ and the marginal revenue is given by $R'(x)=-0.6x+460$, where both marginals are in dollars per unit. Compute the total profit from $x=0$ to $x=95$, the production level where profit is maximized.

Solution

Understand the Situation: Figure 17.2.9 has a graph of R' and C'. In Chapter 16 we saw that to determine total profit we needed to integrate P', the marginal profit function. But, since $P(x)=R(x)-C(x)$, we have $P'(x)=R'(x)-C'(x)$. Hence, to find total profit, we need to integrate

$$\int_{0}^{95}P'(x)\,dx=\int_{0}^{95}[R'(x)-C'(x)]\,dx$$

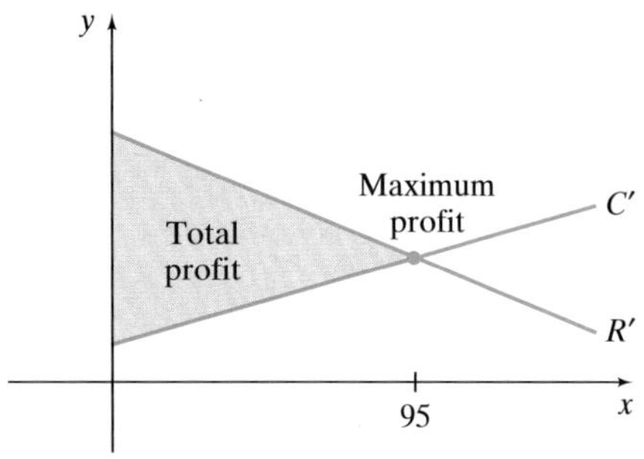

Figure 17.2.9

From Figure 17.2.9 we notice that the total profit is simply the area between the graphs of R' and C'. Continuing, we have

$$\begin{aligned}\int_0^{95}[R'(x)-C'(x)]\,dx &= \int_0^{95}[(-0.6x+460)-(4x+23)]\,dx\\ &= \int_0^{95}(-4.6x+437)\,dx = \left(\frac{-4.6}{2}x^2+437x\right)\Bigg|_0^{95}\\ &= \left[\frac{-4.6}{2}(95)^2+437(95)\right]-\left[\frac{-4.6}{2}(0)^2+437(0)\right]\\ &= 20{,}757.5\end{aligned}$$

Interpret the Solution: So the total profit from a production level of $x = 0$ to $x = 95$ is \$20,757.50. ■

Note: In Example 6 we stated that profit is maximized at $x = 95$. As shown in Figure 17.2.9, that is where $R'(x) = C'(x)$. In general, **profit is maximized when $R'(x) = C'(x)$.**

Checkpoint 3 Now work Exercise 53.

Area between Curves That Cross on $[a, b]$

The final scenario we consider in this section is what happens if two curves cross each other on a given interval. When this occurs, what was a top curve may become a bottom curve, as we illustrate in Example 7.

Example 7 Determining the Area between Curves That Cross on an Interval

Determine the area between the curves $y = x + 1$ and $y = \frac{2}{x}$ on the interval $\left[\frac{1}{2}, 3\right]$.

Solution

The first thing we do is graph the two curves on the stated interval to see which is the top curve and which is the bottom curve. We will also be able to see if the two curves cross each other in the interval. Figure 17.2.10 shows the graph, and we have shaded the area that we want to determine.

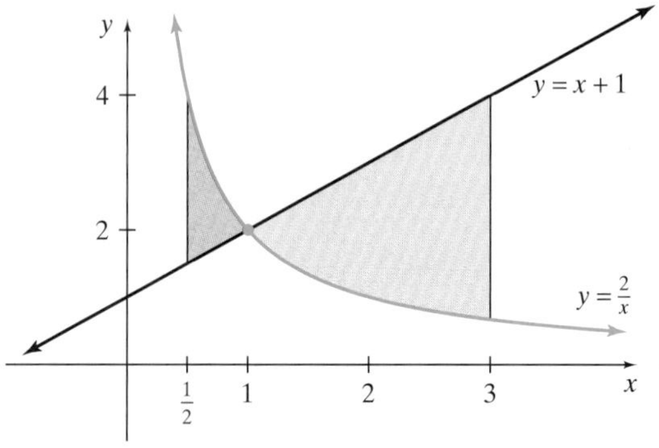

Figure 17.2.10

To the left of the point of intersection, $y = \frac{2}{x}$ is the top curve, while to the right it is the bottom curve. At this time we need to determine where the curves cross, or intersect if you prefer. Setting the functions equal to each other and solving for x gives us

$$\begin{aligned} x + 1 &= \frac{2}{x} \\ x^2 + x &= 2 && \text{Multiply each side by } x, \text{ provided } x \neq 0 \\ x^2 + x - 2 &= 0 && \text{Subtract 2 from each side} \\ (x+2)(x-1) &= 0 && \text{Factor} \\ x = -2 \quad \text{or} \quad x &= 1 \end{aligned}$$

Since $x = -2$ is not in the interval $\left[\frac{1}{2}, 3\right]$, we can ignore it.

Interactive Activity

Use the FNINT command to verify the result of Example 7.

Since the curves cross each other at $x = 1$, we integrate over the intervals $\left[\frac{1}{2}, 1\right]$ and $[1, 3]$ separately and then add the results to determine the area between the curves. This gives

$$\int_{1/2}^{1} \left[\frac{2}{x} - (x+1)\right] dx + \int_{1}^{3} \left(x + 1 - \frac{2}{x}\right) dx$$

$$= \int_{1/2}^{1} \left(\frac{2}{x} - x - 1\right) dx + \int_{1}^{3} \left(x + 1 - \frac{2}{x}\right) dx$$

$$= \left(2\ln|x| - \frac{1}{2}x^2 - x\right)\Bigg|_{1/2}^{1} + \left(\frac{1}{2}x^2 + x - 2\ln|x|\right)\Bigg|_{1}^{3} \qquad \text{Recall, } \int \frac{1}{x}\,dx = \ln|x| + c$$

$$= \left[\left(2\ln 1 - \frac{1}{2}(1)^2 - 1\right) - \left(2\ln\frac{1}{2} - \frac{1}{2}\left(\frac{1}{2}\right)^2 - \frac{1}{2}\right)\right]$$

$$+ \left[\left(\frac{1}{2}(3)^2 + 3 - 2\ln 3\right) - \left(\frac{1}{2}(1)^2 + 1 - 2\ln 1\right)\right] \approx 4.314$$

So the area between the curves $y = x + 1$ and $y = \frac{2}{x}$ on the interval $\left[\frac{1}{2}, 3\right]$ is about 4.31 square units. ■

Checkpoint 4

Now work Exercise 25.

Technology Option

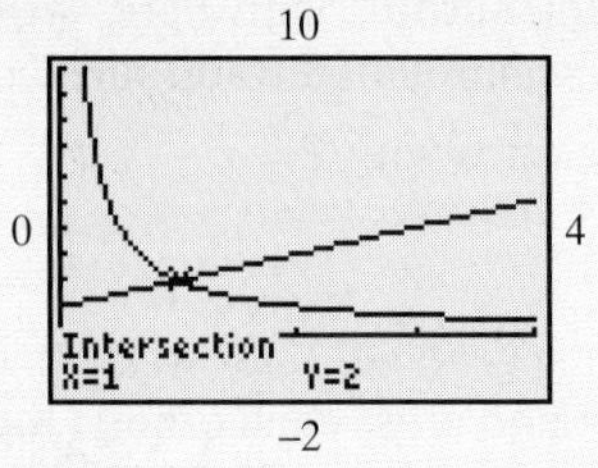

Figure 17.2.11

We could also use the INTERSECT command on the calculator to determine the point of intersection for the two curves in Example 7. Figure 17.2.11 shows the result of using the INTERSECT command.

Our final example in this section analyzes U.S. trade with Peru. In this example, we will determine when the United States had a **trade deficit**. A trade deficit occurs when imports from Peru exceed exports to Peru. We will also determine when the United States had a **trade surplus**. A trade surplus occurs when exports to Peru exceed imports from Peru. Finally, we will determine the U.S. **trade balance** with Peru.

Example 8 Determining a Trade Balance

The following represent the U.S. imports and exports with Peru.

$$\text{Imports: } f(x) = -0.06x^2 + 0.62x + 0.42 \qquad 1 \le x \le 5$$

$$\text{Exports: } g(x) = -0.14x^2 + 0.94x + 0.42 \qquad 1 \le x \le 5$$

Here x represents the number of years since 1994 and $f(x)$ and $g(x)$ represents the value of imports and exports, respectively, in billions of dollars.

See Figure 17.2.12 (*Source for both models:* U.S. Statistical Abstract, www.census.gov/statab/www).

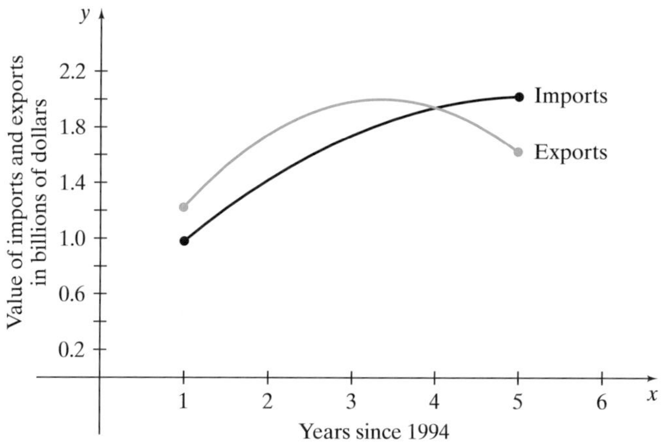

Figure 17.2.12

(a) Determine for what years the United States had a trade deficit with Peru and determine for what years the United States had a trade surplus with Peru.

(b) For the years that the United States had a trade deficit with Peru, write and evaluate an integral to determine the total accumulated trade deficit.

(c) For the years that the United States had a trade surplus with Peru, write and evaluate an integral to determine the total accumulated trade surplus.

(d) From 1995 to 1999, determine the U.S. trade balance with Peru.

Solution

(a) **Understand the Situation:** Figure 17.2.12 shows a graph of the two models. There is a value for x where the curves intersect and cross each other. Once we determine this value, we can then give intervals where the U.S. had a trade surplus as well as a trade deficit with Peru.

To determine the point of intersection, we set the models equal to each other and solve for x. This gives us

$$\begin{aligned} -0.06x^2 + 0.62x + 0.42 &= -0.14x^2 + 0.94x + 0.42 \\ 0.08x^2 - 0.32x &= 0 \\ x(0.08x - 0.32) &= 0 \\ x = 0 \quad \text{or} \quad x &= 4 \end{aligned}$$

Note that $x = 0$ is not in the interval $1 \leq x \leq 5$. So the x-coordinate of the point of intersection is $x = 4$. Note that on the interval [1, 4], the exports curve is above the imports curve whereas on the interval [4, 5] the imports curve is above the exports curve.

Interpret the Solution: This means that the United States had a trade surplus with Peru from 1995 to 1998 and a trade deficit with Peru from 1998 to 1999.

(b) **Understand the Situation:** To determine the total accumulated trade deficit, we need to compute the area of the yellow shaded region between the curves shown in Figure 17.2.13.

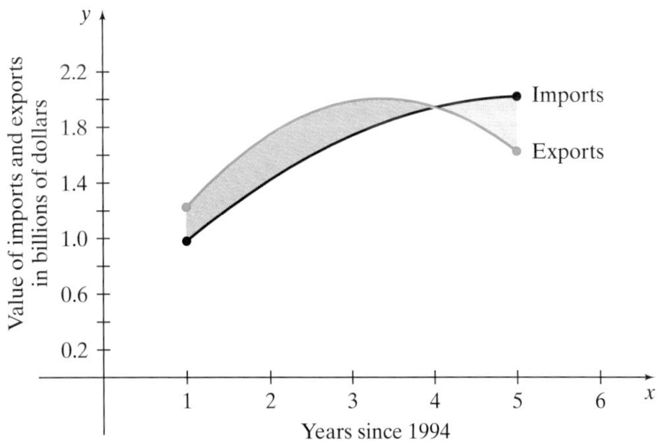

Figure 17.2.13

The total accumulated trade deficit from 1998 to 1999 is found by integrating

$$\begin{aligned} &\int_4^5 (\text{imports} - \text{exports})\, dx \\ &= \int_4^5 [(-0.06x^2 + 0.62x + 0.42) - (-0.14x^2 + 0.94x + 0.42)]\, dx \\ &= \int_4^5 (0.08x^2 - 0.32x)\, dx = \left(\frac{0.08}{3}x^3 - \frac{0.32}{2}x^2\right)\Big|_4^5 \\ &= \frac{14}{75} \approx 0.187 \end{aligned}$$

Interpret the Solution: This means that the U.S. total accumulated trade deficit with Peru from 1998 to 1999 was approximately 0.187 billion dollars.

(c) **Understand the Situation:** To determine the total accumulated trade surplus from 1995 to 1998, we need to compute the area of the orange shaded region between the two curves shown in Figure 17.2.13.

The total accumulated trade surplus from 1995 to 1998 is found by integrating

$$\int_1^4 (\text{exports} - \text{imports})\,dx$$
$$= \int_1^4 [(-0.14x^2 + 0.94x + 0.42) - (-0.06x^2 + 0.62x + 0.42)]\,dx$$
$$= \int_1^4 (-0.08x^2 + 0.32x)\,dx = \left(-\frac{0.08}{3}x^3 + \frac{0.32}{2}x^2\right)\Bigg|_1^4$$
$$= 0.72$$

Interpret the Solution: This means that the U.S. total accumulated trade surplus with Peru from 1995 to 1998 was about 0.72 billion dollars.

Interactive Activity

Determine the Peruvian trade balance with the United States from 1995 to 1999.

(d) **Understand the Situation:** The result from part (b) gives the accumulated trade deficit, while the result from part (c) gives the accumulated trade surplus. Notice that the result from part (c) is larger than the result from part (b). This implies that from 1995 to 1999 the U.S. trade balance with Peru is a positive number indicating an overall surplus.

The overall surplus can be found by

$$\int_1^4 (\text{exports} - \text{imports})\,dx - \int_4^5 (\text{imports} - \text{exports})\,dx \approx 0.72 - 0.187 = 0.533$$

Interpret the Solution: So from 1995 to 1999 the United States had an accumulated trade surplus with Peru of about 0.533 billion dollars. ■

SUMMARY

In this section we saw how to compute the area between two curves and how to apply this concept. To determine the area between two curves, we offer the following guidelines:

1. Graph the two curves to see the region whose area we want.
2. If an interval is not given, or if the curves cross each other on a given interval, determine the appropriate limits of integration by determining where the curves intersect each other.
3. Integrate top curve minus bottom curve on each interval.

SECTION 17.2 EXERCISES

In Exercises 1–8, write a definite integral to represent the area of the shaded region.

1.

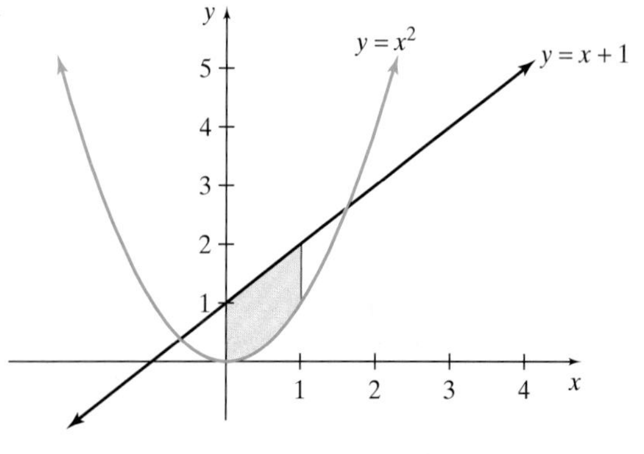

2.

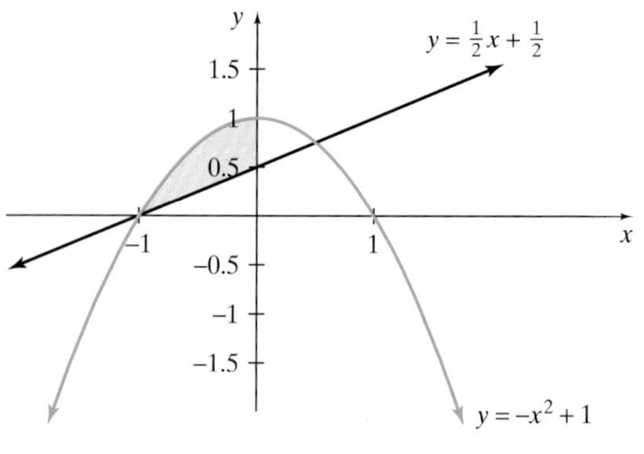

3.

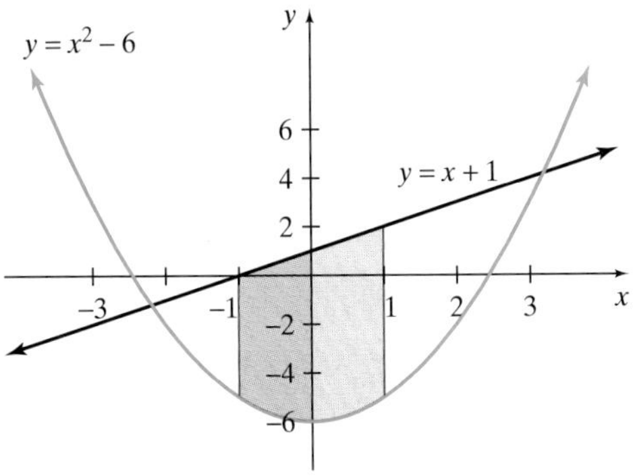

4.

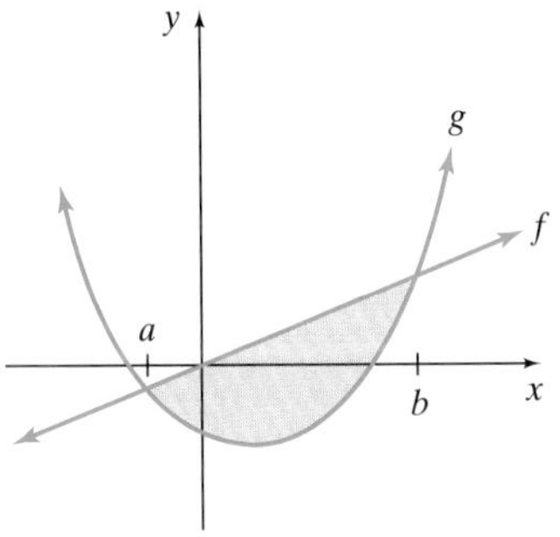

5.

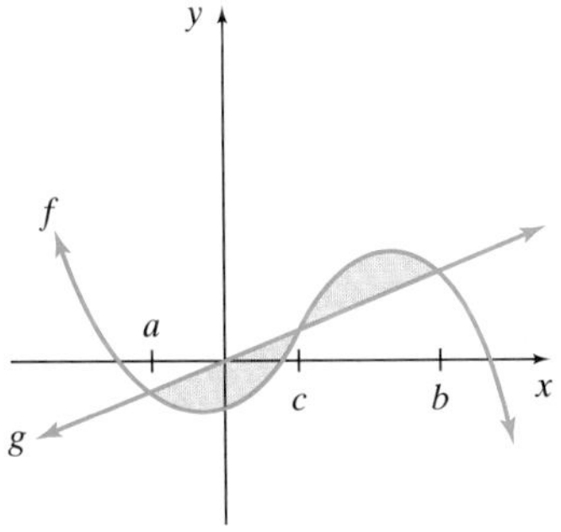

6.

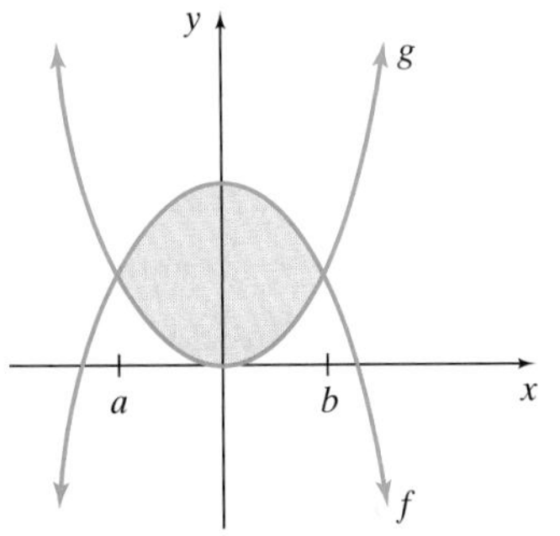

7.

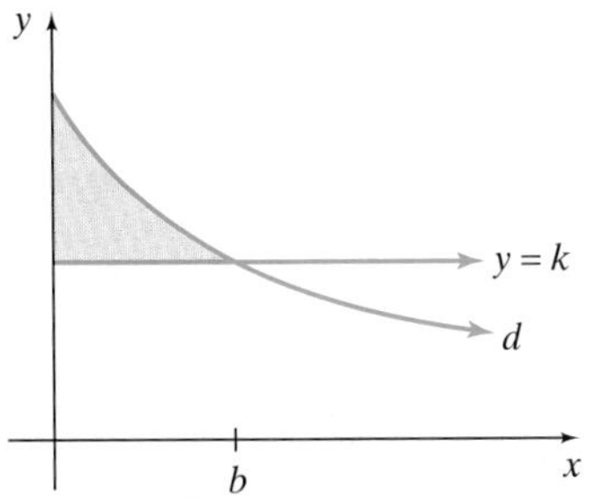

8.

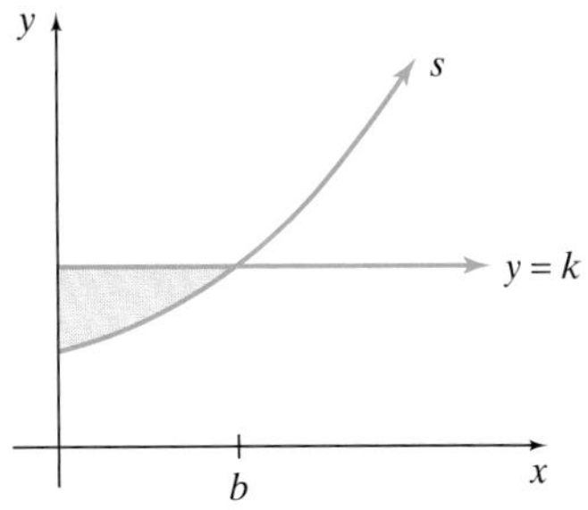

In Exercises 9–26, determine the area of the region bounded by the given conditions.

9. $f(x) = 3x + 1$ and the x-axis on $[0, 4]$

10. $f(x) = 2x + 3$ and the x-axis on $[1, 3]$

11. $f(x) = 2x^2 + x + 1$ and the x-axis on $[-2, 1]$

12. $f(x) = x^2 + 2x + 3$ and the x-axis on $[-3, 2]$

13. $f(x) = -x^2 + x + 2$ and the x-axis on $[1, 4]$

14. $f(x) = -2x^2 - x + 1$ and the x-axis on $[-2, 3]$

✓ 15. $f(x) = 8 - x^2$ and $y = 4$ on $[-2, 2]$

16. $f(x) = x^2 - 2$ and $y = 2$ on $[-2, 2]$

17. $f(x) = -x + 4$ and $g(x) = -x^2 + 1$ on $[-2, 2]$

18. $f(x) = 2x + 3$ and $g(x) = -x^2 + 1$ on $[-1, 2]$

19. $f(x) = x - 3$ and $g(x) = x^2 + x + 1$ on $[-2, 1]$

20. $f(x) = \frac{1}{2}x - 1$ and $g(x) = x^2 + 2x - 1$ on $[1, 5]$

21. $f(x) = 16 - \sqrt{x}$ and $g(x) = 8 - \sqrt[3]{x}$ on $[1, 5]$

22. $f(x) = x^2$ and $g(x) = \sqrt{x}$ on $[0, 1]$

23. $f(x) = 5e^{-0.15x}$ and $g(x) = 10e^{-0.08x}$ on $[0, 5]$

24. $f(x) = 7.2e^{-0.07x}$ and $g(x) = 2.34e^{-0.11x}$ on $[1, 4]$

✓ 25. $f(x) = \frac{3}{x}$ and $g(x) = \frac{4}{x^2}$ on $[2, 8]$

26. $f(x) = \frac{3}{x}$ and $g(x) = \frac{4}{x^2}$ on $[1, 8]$

In Exercises 27–34, determine the area bounded by the given curves.

27. $f(x) = 4 - x^2$ and $g(x) = -x + 2$

28. $f(x) = -x - 3$ and $g(x) = 9 - x^2$

29. $f(x) = -2x^2 - 5x + 3$ and $g(x) = 2x^2 + 3x - 2$

30. $f(x) = x^2 - 6$ and $g(x) = 12 - x^2$

31. $f(x) = 3x^2 - 2x$ and $g(x) = x^3$

32. $f(x) = x^2$ and $g(x) = x^3$

33. $f(x) = x^3$ and $g(x) = x$

34. $f(x) = 2x^3$ and $g(x) = -x^3 + x^2 + 2x$

Applications

35. The average wage, $W_c(t)$, in dollars per hour, for individuals in construction in the United States can be modeled by

$$W_c(t) = 0.38t + 13.01 \qquad 1 \le t \le 10$$

where t is the number of years since 1989. The average wage, $W_m(t)$, in dollars per hour, for individuals in manufacturing in the United States can be modeled by

$$W_m(t) = 0.34t + 10.41 \qquad 1 \le t \le 10$$

where t is the number of years since 1989 (*Source for both models:* U.S. Bureau of Labor Statistics, www.bls.gov). During the period 1990 through 1999, how much more money has a worker in construction earned than a worker in manufacturing? Assume that each worked 40 hours per week, made the average wage, and earned a paycheck 52 weeks each year.

36. The average wage, $W_{mi}(t)$, in dollars per hour, for individuals in mining in the United States can be modeled by

$$W_{mi}(t) = 0.38t + 13.18 \qquad 1 \le t \le 10$$

where t is the number of years since 1989. The average wage, $W_{ma}(t)$, in dollars per hour, for an individual in manufacturing in the United States can be modeled by

$$W_{ma}(t) = 0.34t + 10.41 \qquad 1 \le t \le 10$$

where t is the number of years since 1989 (*Source for both models:* U.S. Bureau of Labor Statistics, www.bls.gov). During the period 1990 through 1999, how much more money has a worker in mining earned than a worker in manufacturing? Assume that each worked 40 hours per week, made the average wage, and earned a paycheck 52 weeks each year.

37. The receipts, in billions of dollars, in the National Income and Products Accounts for State and Local Governments can be modeled by

$$f(x) = 1.13x^2 + 23.51x + 337.88 \qquad 1 \le x \le 16$$

where x is the number of years since 1979. The expenditures, in billions of dollars, in the National Income and Products Accounts for State and Local Governments can be modeled by

$$g(x) = 1.28x^2 + 18.77x + 283.09 \qquad 1 \le x \le 16$$

where x is the number of years since 1979 (*Source:* U.S. Statistical Abstract, www.census.gov/statab/www).

(a) Graph f and g in the viewing window [0, 17] by [200, 1100].

(b) Write an integral to represent the area between the two curves on the interval [1, 16].

(c) Evaluate the integral from part (b) and interpret. (When receipts exceed expenditures, a surplus exists, whereas when expenditures exceed receipts, a deficit exists.)

38. The size of a certain bacteria culture grows at a rate of $4t^{5/2}$ milligrams per minute. If a special nutrient is introduced into the culture, it grows at a rate of $4t^{18/5}$ milligrams per minute.

(a) Set up an integral to determine the area between the two curves on [0, 1].

(b) Evaluate the integral from part (a) and interpret.

39. *(continuation of Exercise 38)* The size of a certain bacteria culture grows at a rate of $4t^{5/2}$ milligrams per minute. If a special nutrient is introduced into the culture, it grows at a rate of $4t^{18/5}$ milligrams per minute.

(a) Set up an integral to determine the area between the two curves on [1, 2].

(b) Evaluate the integral from part (a) and interpret.

(c) On the interval [0, 2], how many more bacteria result from introducing the special nutrient?

40. The imports and exports for the United States with Japan can be modeled by

$$\text{Imports: } I(x) = -0.61x^3 + 8.31x^2 - 27.42x + 114.37$$
$$\text{Exports: } E(x) = -0.12x^4 + 2.28x^3 - 13.93x^2 + 32.78x + 23.22$$

where x is the number of years since 1988 and imports and exports are in billions of dollars. Determine the area between the two curves on the interval [1, 8] and interpret (*Source for both models:* U.S. Statistical Abstract, www.census.gov/statab/www).

41. The imports and exports for the United States with South Korea can be modeled by:

$$\text{Imports: } I(x) = 1.28x^2 - 6.15x + 29.38 \qquad 1 \le x \le 5$$
$$\text{Exports: } E(x) = -0.33x^2 + 1.27x + 24.69 \qquad 1 \le x \le 5$$

where x is the number of years since 1994 and the imports and exports are in billions of dollars (*Source for both models:* U.S. Statistical Abstract, www.census.gov/statab/www).

(a) In what year did the U.S. imports from South Korea first exceed the U.S. exports to South Korea?

(b) Round the x-coordinate of the point of intersection to $x = 3.85$. Evaluate and interpret $\int_1^{3.85} (E(x) - I(x))\,dx$.

(c) Round the x-coordinate of the point of intersection to $x = 3.85$. Evaluate and interpret $\int_{3.85}^{5} (I(x) - E(x))\,dx$.

(d) Determine the U.S. trade balance with South Korea from 1995 to 1999.

42. The imports and exports for the United States with Canada can be modeled by

$$\text{Imports: } I(x) = -0.24x^3 + 4.6x^2 - 14.49x + 99.96$$
$$\text{Exports: } E(x) = -0.17x^3 + 3.14x^2 - 8.14x + 85.58$$

where x is the number of years since 1988 and imports and exports are in billions of dollars. Determine the area between the two curves on the interval [1, 8] and interpret.

43. The rate at which the number of public schools in the United States with interactive videodisk units has grown according to $\frac{15.1}{x}$ thousand schools per year, where x is the number of years since 1991. From 1992 through 1996, how many schools received these interactive videodisk units? (*Source:* U.S. Statistical Abstract, www.census.gov/statab/www.)

44. The rate at which the number of public schools in the United States with computer networks has grown according to $\frac{17.58}{x}$ thousand schools per year, where x is the number of years since 1991. From 1992 through 1996, how many schools received computer networks? (*Source:* U.S. Statistical Abstract, www.census.gov/statab/www.)

45. Double D cola is planning to release a new Double Dose Caffeine Cola on the market. In the past, Double D cola ran a 20-week ad campaign for any new product that it released on the market. For example, when Double D ran a 20-week ad campaign for its new caffeine-free cola, weekly sales $s(t)$ in thousands of dollars per week t weeks after the campaign began were modeled by

$$s(t) = -0.51t^2 + 20.20t + 49.49$$

An agent for a TV personality has approached Double D cola and claims that, if Double D uses his client in the 20-week ad campaign, the weekly sales $w(t)$ in thousands of dollars per week t weeks after the campaign begins will be approximated by

$$w(t) = -0.71t^2 + 28.14t + 57.22$$

(a) On the interval [0, 20], determine the area between s and w and interpret.

(b) If the endorsement of the TV personality costs Double D cola \$1,000,000, will this endorsement be paid off in the first 20 weeks of the campaign? Explain.

46. A market analyst for Chocolate Time estimates that with no promotion the annual sales of its candy bar, Chocoloco Dream, can be modeled by

$$s_1(t) = 0.76t + 8.42$$

million dollars per year, t years from now. This same analyst estimates that, with a modest promotional campaign, annual sales can be modeled by

$$s_2(t) = 11.4e^{0.05t}$$

During the first 5 years, what will the total increase in sales be in response to the promotion?

✓ **47.** A market analyst for Chocolate Time estimates that with no promotion the annual sales of its candy bar, Chocoholic Dream, can be modeled by

$$s_1(t) = 0.76t + 8.42$$

million dollars per year, t years from now. This same analyst estimates that, with a modest promotional campaign, annual sales can be modeled by

$$s_2(t) = 12.15e^{0.04t}$$

During the first 5 years, what will the total increase in sales be in response to the promotion?

48. A psychologist has determined that people can memorize digits at the rate of $4.4e^{-0.2x}$ digits per minute, where x is the time in minutes that the individuals have been memorizing. She discovered that if a scent of peppermint oil is present while people are attempting to memorize and when asked to recall the digits memorized, the rate of learning appeared to be $5.3e^{-0.15x}$ digits per minute.

(a) Graph both rates in the viewing window [0, 5] by [0, 6].

(b) On the interval [1, 5], determine the area between the two curves and interpret.

49. A psychologist has determined that the rate at which a child learns to recognize new words is given by $25t + 50\sqrt{t}$ words per year. If the child listens to special vocabulary-building tapes, the rate at which the child recognizes new words is given by $35t + 70\sqrt{t}$. In both cases, t represents the child's age in years.

(a) Graph both rates in the viewing window [0, 10] by [0, 600].

(b) On the interval [4, 8], determine the area between the two curves and interpret.

50. Dynatronics has determined that its costs, in millions of dollars, have been increasing, mainly due to inflation, at a rate given by $1.25e^{0.12x}$, where x is time measured in years since its new telephone equipment began being produced. At the beginning of the third year of production, a breakthrough in the production process resulted in costs increasing at a rate given by $0.85\sqrt{x}$. Determine the total savings in costs on the interval [3, 6] due to the breakthrough.

51. Quality Chips has determined that its costs, in millions of dollars, have been increasing, mainly due to inflation, at a rate given by $2.67e^{0.09x}$, where x is time measured in years since its snack food started in production. At the beginning of the third year, a breakthrough in the production process resulted in costs increasing at a rate given by $0.68x^{2/3}$. Determine the total savings in costs on the interval [3, 8] due to the breakthrough.

52. See It Inc. determines that the marginal revenue to produce x graphing calculators is given by $R'(x) = -0.3x + 100$, while the marginal cost is given by $C'(x) = 0.5x + \frac{750}{x+1}$, where both marginals are in dollars per unit. Determine the total profit from $x = 10$ to $x = 100$.

✓ **53.** Monkey Works determines that the marginal revenue to produce x squeegees is given by

$R'(x) = -0.2x + 40$, while the marginal cost is given by $C'(x) = \frac{300}{x+1}$, where both marginals are in dollars per unit. Determine the total profit from $x = 10$ to $x = 100$.

54. Music Time determines that the marginal revenue to produce x integrated stereo amplifiers is given by $R'(x) = 216 - 0.24x^2$, while marginal cost is given by $C'(x) = 20$, where both marginals are in dollars per unit. Determine the total profit from $x = 10$ to $x = 30$.

55. Java Buddy determines that the marginal revenue to produce x coffee mugs is given by $R'(x) = 5 - 0.05x$, while marginal cost is given by $C'(x) = 1 + \frac{x}{40}$, where both marginals are in dollars per unit. Determine the total profit from $x = 5$ to $x = 40$.

56. A factory worker can assemble units at the rate of $-3x^2 + 13x + 11$ units per hour during the first 4 hours on the job, where x is the number of hours on the job. Determine how many units can be assembled by this worker during the first 4 hours.

57. *(continuation of Exercise 56)* If the factory worker in Exercise 56 has breakfast along with two cups of coffee, then she can assemble units at the rate of $-2x^2 + 11x + 13$ units per hour during the first 4 hours on the job, where x is the number of hours on the job.

(a) In this case, how many units can she assemble during the first 4 hours?

(b) How many more units can she assemble in the first 4 hours if she has breakfast and 2 cups of coffee?

58. A botanist knows from past records that a certain species of oak tree grows at a rate of $\frac{4x^2 + 16x + 9}{2x + 4}$ feet per year, where x is the age of the tree in years. Determine how much growth takes place on the interval [3, 8].

59. *(continuation of Exercise 58)* The botanist in Exercise 58 knows that restricting the amount of light that the oak tree receives inhibits its growth. When restricting the light, the oak tree grows at a rate of $\frac{2x^2 + 8x + 9}{2x + 4}$ feet per year, where x is the age of the tree in years. On the interval [3, 8], how many fewer feet in growth will result from restricting the amount of light that the tree receives?

SECTION PROJECT

(*Source for all models:* U.S. Census Bureau, www.census.gov)

(a) The average net income for a General and Family practice physician in the United States is modeled by

$$f(x) = -0.09x^3 + 1.52x^2 - 1.54x + 77.95 \qquad 1 \le x \le 10$$

thousands of dollars per year, where x is the number of years since 1984. Determine the area between the graph of f and the x-axis on the interval [1, 9] and interpret.

(b) The average net income for a surgeon in the United States is modeled by

$$s(x) = -0.04x^3 - 0.24x^2 + 18.97x + 136.5 \qquad 1 \le x \le 10$$

thousands of dollars per year, where x is the number of years since 1984. Determine the area between the graph of s and the x-axis on the interval [1, 9] and interpret.

(c) The average net income for a pediatrician in the United States is modeled by

$$k(x) = -0.19x^3 + 3x^2 - 5.9x + 79.52 \qquad 1 \le x \le 10$$

thousands of dollars per year, where x is the number of years since 1984. Determine the area between the graph of k and the x-axis on the interval [1, 9] and interpret.

Assume from 1985 to 1993 that Dylan is a family practice physician earning the average net income each year, Lisa is a surgeon earning the average net income each year, and Austin is a pediatrician earning the average net income each year.

(d) From 1985 to 1993, how much more money does Lisa earn than Dylan?

(e) From 1985 to 1993, how much more money does Lisa earn than Austin?

(f) From 1985 to 1993, how much more money does Austin earn than Dylan?

Section 17.3 Economic Applications of Area between Two Curves

In Section 17.2 we introduced how to determine the area between two curves and considered many applications of this concept. In this section we focus mainly on two specific applications of area between two curves: **consumers' and producers' surplus** and **income distributions**. Our discussion of consumers' and producers' surplus will also include a look at **market price**, **total consumer expenditure**, **revenue**, **market demand**, and **equilibrium point**. Our discussion of income distribution will include a look at **Lorenz curves** and the **Gini Index**.

Price–Demand Function Revisited

In Chapter 11, we were first introduced to what is known as the price–demand function. We review this function in the Toolbox to the left.

A common way to represent the price–demand function in economics is with the letter d. Quite simply, $d(x)$ yields the price at which exactly x units of a product will be sold.

From Your Toolbox

Price–Demand Function

$p(x)$ gives the price at which people buy exactly x units of a product.

Note: This section contains applications that are also taught in many economics courses. Since many of you will see the letter d for the price–demand function in your economics course, we will use this notation in this section. In this section, we will use d to represent the price–demand function, which we will simply call the **demand function**.

As shown in Figure 17.3.1, the demand function is a decreasing function. Intuitively, this seems reasonable. As the price decreases, more people will buy the product. Economists refer to the curve in Figure 17.3.1 simply as a **demand curve**. This is because it expresses the price per unit as a function of the quantity, x, in demand.

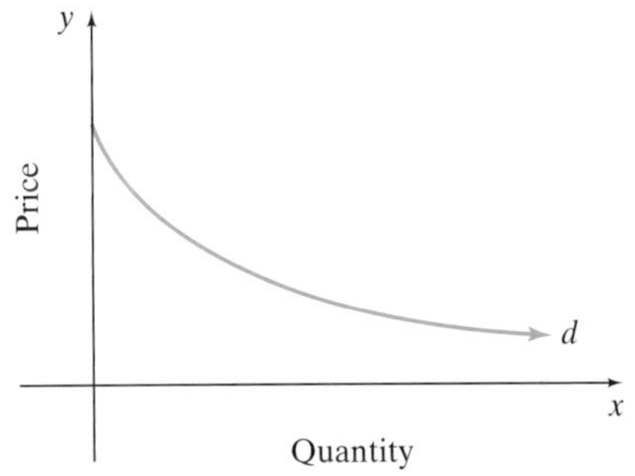

Figure 17.3.1

Example 1 Evaluating a Demand Function

Skinner Bikes, a bicycle retailer, has determined that a demand function for a certain brand of mountain bike is given by $d(x) = -0.3x + 330$, where x is the quantity demanded each month and $d(x)$ is the price per bicycle in dollars. Compute $d(200)$, $d(250)$, and $d(300)$ and interpret each.

Solution

Since $d(x) = -0.3x + 330$, we have

$$d(200) = -0.3(200) + 330 = \$270 \text{ per bicycle}$$
$$d(250) = -0.3(250) + 330 = \$255 \text{ per bicycle}$$
$$d(300) = -0.3(300) + 330 = \$240 \text{ per bicycle}$$

The interpretation of these results are, respectively,

200 bicycles will be sold each month at a price of \$270 per bicycle.
250 bicycles will be sold each month at a price of \$255 per bicycle.
300 bicycles will be sold each month at a price of \$240 per bicycle.

■

Consumers' Surplus

For the sake of argument, let's assume that you really like the bicycle discussed in Example 1 and were willing to pay \$275 for it. If it currently costs, say, \$200, then in a way you "saved" \$75; the \$275 you were willing to pay minus the \$200 price. Now consider *any* price above \$200 that some consumer is willing to pay minus the actual price of \$200. The sum of these savings over all possible prices above \$200 is known as the **consumers' surplus** for this bicycle. In general, consumers' surplus measures the benefit that consumers get from an economy where competition keeps prices down.

The actual price at which any item is sold, called the **market price**, is influenced by many factors. However, when a market price is determined, we can compute the quantity that consumers demand using the demand function. Returning to our bicycle retailer in Example 1, we note that at a price of \$255 per bicycle 250 bicycles are demanded each month.

In Figure 17.3.2 is a shaded rectangle whose area represents what economists call the *actual consumer expenditure*. In general, the actual consumer expenditure is the amount of money that consumers spend for x bicycles at price $d(x)$ dollars per bicycle. In this case, the actual consumer expenditure is computed to be

$$(\$255)(250) = \$63{,}750$$

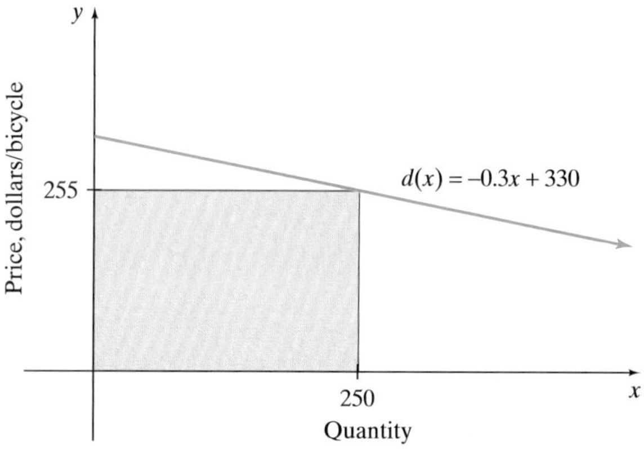

Figure 17.3.2 Actual consumer expenditure.

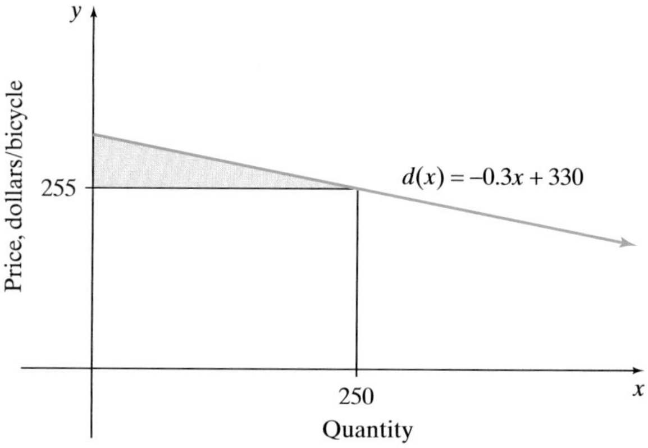

Figure 17.3.3 Shaded area represents consumers' surplus.

The area of the rectangle in Figure 17.3.2 is the **actual consumer expenditure**, while the area below the demand curve $d(x) = -0.3x + 330$ and above the rectangle gives a *total savings* that consumers realize by buying at \$255 per bicycle. See Figure 17.3.3. This total savings is what is known as **consumers' surplus**.

Consumers' Surplus

The **consumers' surplus** is the total amount saved by consumers who are willing to pay more than market price, d_{mp}, for a product, yet are able to purchase the product for d_{mp}.

Example 2 Using a Definite Integral to Determine Consumers' Surplus

(a) Set up a definite integral to determine the area of the shaded region in Figure 17.3.3, which represents the consumers' surplus for the bicycle from Example 1.

(b) Evaluate the integral from part (a) to determine the area of the shaded region, as well as determining the consumers' surplus.

Solution

(a) Here we need to determine the area between two curves. We have top curve is $d(x) = -0.3x + 330$ and bottom curve is $y = 255$

From our work in Section 17.2, we represent the area of the shaded region as

$$\int_0^{250} [(-0.3x + 330) - 255]\, dx = \int_0^{250} (-0.3x + 75)\, dx$$

(b) We evaluate the integral from part (a) to be

$$\int_0^{250} (-0.3x + 75)\, dx = \left(\frac{-0.3}{2}x^2 + 75x\right)\Bigg|_0^{250} = \$9375$$

So the consumers' surplus is \$9375 each month. ■

Interactive Activity

There is another way to determine the area representing consumers' surplus in Example 2. We can take the area under the graph of d from $x = 0$ to $x = 250$ and subtract the area of the rectangle that represents actual consumer expenditure. This corresponds to integrating $\int_0^{250} (-0.3x + 330)\, dx$ and subtracting \$63,750 (the actual consumer expenditure). Perform this integration and subtraction to verify the consumers' surplus determined in Example 2.

WWW

Example 2 illustrates how we can determine consumers' surplus for any situation.

Computing Consumers' Surplus

For a demand function d and a point (x_m, d_{mp}) on the graph of d, where x_m is called the **market demand** and d_{mp} is called the **market price**, we determine the consumers' surplus by integrating

$$\int_0^{x_m} [d(x) - d_{mp}]\, dx$$

Example 3 Computing Consumers' Surplus

Assume that the market price of the bicycle in Example 1 is \$210 per bicycle. Determine the consumers' surplus.

Solution

Understand the Situation: Recall that $d(x) = -0.3x + 330$; its graph is shown in Figure 17.3.4. We know that the market price is \$210, but we do not have the market demand. Since we need the market demand in our definite integral (it is the upper limit of integration), we must solve

$$\begin{aligned} 210 &= -0.3x + 330 \\ -120 &= -0.3x \\ 400 &= x \end{aligned}$$

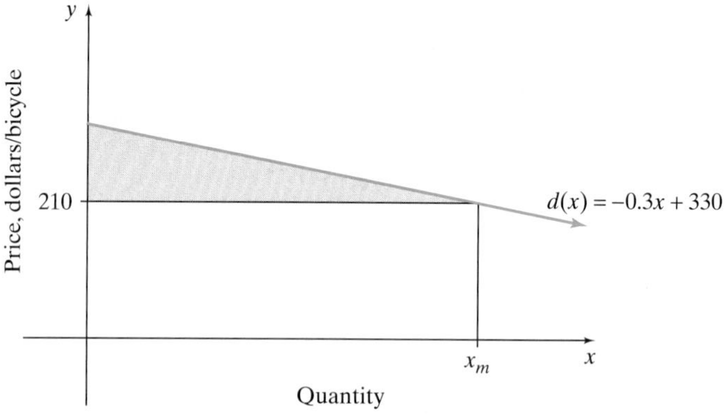

Figure 17.3.4

To find the consumers' surplus, we need to integrate

$$\begin{aligned}\int_0^{400} [(-0.3x + 330) - 210]\,dx &= \int_0^{400} (-0.3x + 120)\,dx \\ &= \left(\frac{-0.3}{2}x^2 + 120x\right)\Bigg|_0^{400} \\ &= \$24{,}000\end{aligned}$$

Interpret the Solution: So the consumers' surplus is \$24,000 each month. ■

Checkpoint 1 Now work Exercise 15.

Supply Function

Now let's take a look at things from the producers' perspective. We know that when prices go up consumers usually demand less. However, manufacturers tend to respond to higher prices by *supplying more*. So a typical supply curve, denoted s, that relates the price per unit $s(x)$ as a function of quantity, x, supplied should be an increasing function, as shown in Figure 17.3.5. Notice that as the quantity x increases so does the price.

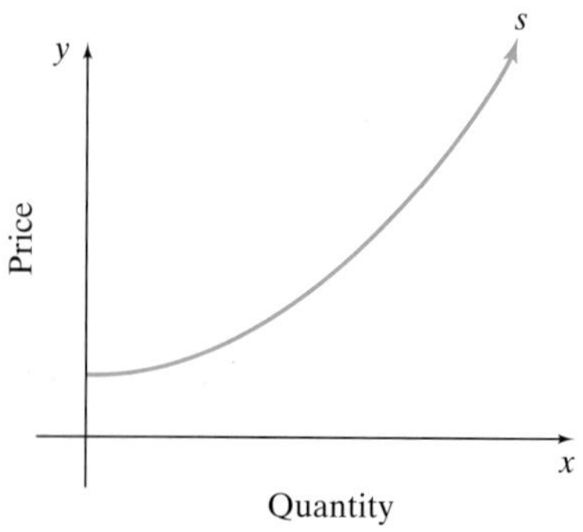

Figure 17.3.5

The supply function is defined as follows:

Supply Function

The **supply function**, s, for any product gives the price $s(x)$ at which exactly x units of the product will be supplied.

Example 4 Evaluating a Supply Function

Suppose that the supplier of the bicycle that we have discussed in this section has a supply function given by $s(x) = 0.2x + 75$, where x is the quantity supplied each month and $s(x)$ is the price per bicycle in dollars. Compute $s(100)$, $s(200)$, and $s(300)$ and interpret each.

Solution

Since $s(x) = 0.2x + 75$, we have

$$s(100) = 0.2(100) + 75 = \$95 \text{ per bicycle}$$
$$s(200) = 0.2(200) + 75 = \$115 \text{ per bicycle}$$
$$s(300) = 0.2(300) + 75 = \$135 \text{ per bicycle}$$

The interpretations of these results are, respectively,

The supplier will supply 100 bicycles each month at a price of \$95 per bicycle.
The supplier will supply 200 bicycles each month at a price of \$115 per bicycle.
The supplier will supply 300 bicycles each month at a price of \$135 per bicycle.

Producers' Surplus

Let's return to the scenario where the market price for a bicycle is \$200 per bicycle. Suppose that the bicycle supplier is willing to stay in business if the price per bicycle dropped to \$150. The fact that bicycles sell for \$200 gives a "gain" of \$50 per bicycle to the supplier. Now consider any price below \$200 at which some supplier is willing to supply these bicycles, subtracted from the actual price of \$200. The sum of these gains over all possible prices below \$200 is known as the **producers' surplus**.

However, once a market price, d_{mp}, is determined for an item, we can compute the quantity that producers will supply by the supply function. Our bicycle supplier in Example 4 supplies 100 bicycles each month when the price is \$95 per bicycle. The shaded region in Figure 17.3.6 gives the actual amount that the supplier received or, if you prefer, the *revenue*. (From the consumers' perspective, this amount is called *consumer expenditure*.) We can quickly compute this amount to be $\$95(100) = \$9500$.

Figure 17.3.6 shows the actual amount (or revenue) received, while Figure 17.3.7 shows the area above the supply curve, $s(x) = 0.2x + 75$, and below the line $y = 95$. This area in Figure 17.3.7 is the extra amount that suppliers receive over what they are willing to receive. This extra amount is called the **producers' surplus**.

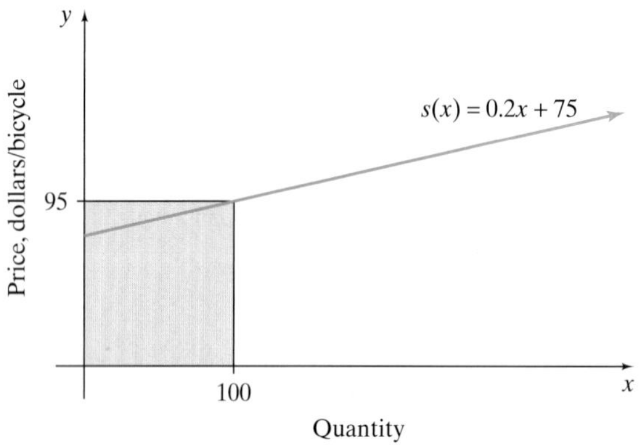

Figure 17.3.6 Shaded region represents revenue.

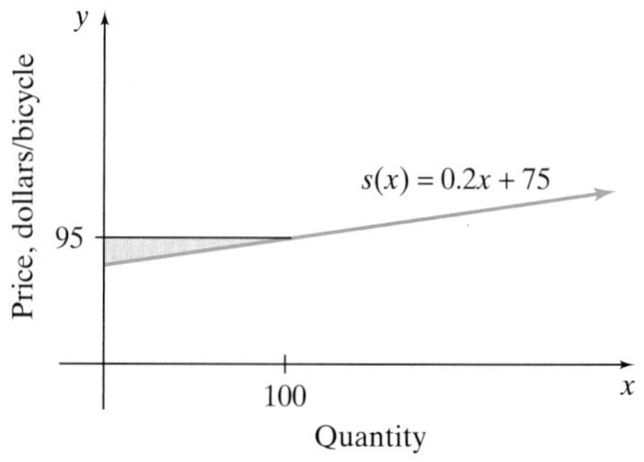

Figure 17.3.7 Shaded region represents producer's surplus.

Producer's Surplus

Producers' surplus is the total amount gained by producers who are willing to receive less than market price, d_{mp}, for a product, yet are able to receive d_{mp} for the product.

To determine the area of the shaded region in Figure 17.3.7, and therefore the producers' surplus, we need to determine the area between the two curves $y = 95$ and $s(x) = 0.2x + 75$. This is found by integrating

$$\int_0^{100} [95 - (0.2x + 75)]\,dx = \int_0^{100} (-0.2x + 20)\,dx$$

Notice that, just like computing consumers' surplus, determining producers' surplus is analogous to determining the area between two curves. The following shows how to determine producers' surplus in any situation.

Computing Producers' Surplus

For a supply function s and a point (x_m, d_{mp}) on the graph of s, where x_m is called the **market demand** and d_{mp} is called the **market price**, we determine the producers' surplus by integrating

$$\int_0^{x_m} [d_{mp} - s(x)]\,dx$$

Example 5 Determining Producers' Surplus

Assume the supply function from Example 4 for the bicycle to be $s(x) = 0.2x + 75$, where $s(x)$ is the price per bicycle in dollars and x is the number of bicycles. Determine the producers' surplus if the market price is \$210 per bicycle.

Solution

Understand the Situation: Since $s(x) = 0.2x + 75$ and the market price is \$210 per bicycle, to determine producers' surplus we must first determine the quantity supplied, or the market demand. To determine the market demand, we need to solve

$$210 = 0.2x + 75$$
$$135 = 0.2x$$
$$x = 675$$

To find producers' surplus, we integrate

$$\int_0^{675} [210 - (0.2x + 75)]\,dx = \int_0^{675} (135 - 0.2x)\,dx$$
$$= \left(135x - \frac{0.2x^2}{2}\right)\Big|_0^{675}$$
$$= \$45{,}562.50 \text{ each month.}$$

Interpret the Solution: So the producers' surplus is \$45,562.50 each month. ■

Checkpoint 2

Now work Exercise 29.

Interactive Activity

There is another way to determine the area representing producers' surplus in Example 5. We can take the area under the graph of s from $x = 0$ to $x = 675$, and subtract it from the area of the rectangle that represents the actual amount that producers' receive to supply 675 bicycles. This corresponds to integrating $\int_0^{675} (0.2x + 75)\,dx$ and subtracting the result from $675(210) = \$141{,}750$. Perform the integration and subtraction to verify the producers' surplus determined in Example 5.

Equilibrium Point

If we graph a demand function, d, and a supply function, s, together on the same set of axes, we have the situation shown in Figure 17.3.8. The demand, x, at which the demand and supply curves intersect is called the *equilibrium demand*, while the price is called the *equilibrium price*. The point of intersection for d and s is called the **equilibrium point**. Also notice in Figure 17.3.8 that both the consumers' surplus and producers' surplus can be shown together.

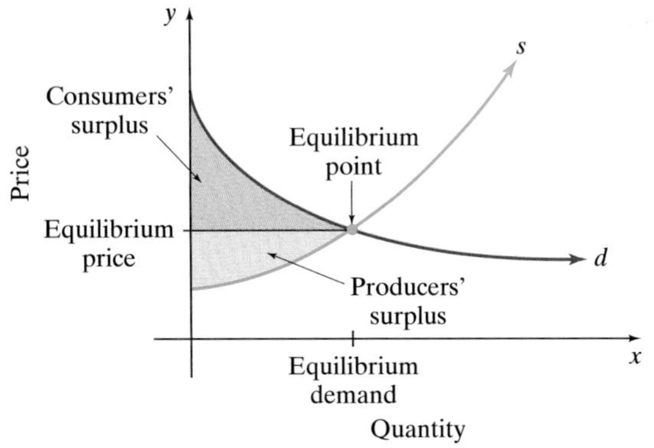

Figure 17.3.8

Example 6 Locating an Equilibrium Point

Continuing with our bicycle scenario, if the demand function is given by $d(x) = -0.3x + 330$ and the supply function is given by $s(x) = 0.2x + 75$:

(a) Determine the equilibrium point.

(b) Determine the consumers' surplus at the equilibrium demand.

(c) Determine the producers' surplus at the equilibrium demand.

Solution

(a) **Understand the Situation:** Figure 17.3.9 has a graph of d and s. We determine the equilibrium point algebraically by setting $d(x) = s(x)$ and solving for x. Once we have a value for x, the equilibrium demand, we can substitute it into either $d(x)$ or $s(x)$ to determine the equilibrium price.

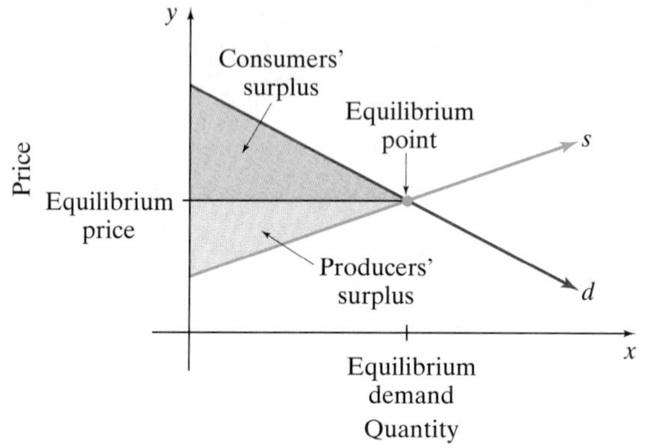

Figure 17.3.9

Setting $d(x) = s(x)$ and solving for x gives us

$$\begin{aligned} -0.3x + 330 &= 0.2x + 75 \\ 255 &= 0.5x \\ 510 &= x \end{aligned}$$

We have the equilibrium demand to be 510 bicycles each month. The equilibrium price is found by using either $d(x)$ or $s(x)$. Using $s(x)$, we get

$$s(510) = 0.2(510) + 75 = 177$$

The equilibrium price is \$177, which means that the equilibrium point is (510, \$177).

(b) As shown in Figure 17.3.9, the consumers' surplus is found by integrating

$$\begin{aligned} \int_0^{510} [(-0.3x + 330) - 177]\,dx &= \int_0^{510} (-0.3x + 153)\,dx \\ &= \left(\frac{-0.3}{2}x^2 + 153x\right)\Bigg|_0^{510} \\ &= \left[\frac{-0.3}{2}(510)^2 + 153(510)\right] - \left[\frac{-0.3}{2}(0)^2 + 153(0)\right] \\ &= \$39{,}015 \text{ each month.} \end{aligned}$$

So the consumers' surplus is \$39,015 each month.

(c) As shown in Figure 17.3.9, the producers' surplus is determined by integrating

$$\begin{aligned} \int_0^{510} [177 - (0.2x + 75)]\,dx &= \int_0^{510} (102 - 0.2x)\,dx = \left(102x - \frac{0.2}{2}x^2\right)\Bigg|_0^{510} \\ &= \left[102(510) - \frac{0.2}{2}(510)^2\right] - \left[102(0) - \frac{0.2}{2}(0)^2\right] \\ &= \$26{,}010 \text{ each month} \end{aligned}$$

So the producers' surplus is \$26,010 each month. ■

Checkpoint 3 Now work Exercise 41.

Lorenz Curves and Gini Index

It is a fact that in our society some individuals make more money than others. Economists measure the gap between the rich and the poor by examining the proportion of the total income that is earned by the lowest 20% of the population, and then the proportion earned by the lowest 40% of the population, and so on. For example, data for income distribution in the United States in 1990 are shown in Table 17.3.1.

The table tells us that in 1990 the lowest 40% of the population earns only $0.135 = 13.5\%$ of the total income. The graph of $y = L(x)$ through the data points in Figure 17.3.10 is called a **Lorenz curve**. A Lorenz curve gives us the proportion of total income earned by the lowest proportion, x, of the population. The domain for L is $[0, 1]$ and the range is $[0, 1]$.

Table 17.3.1

Proportion of Population	Proportion of Income
0.20	0.039
0.40	0.135
0.60	0.294
0.80	0.534
1.00	1.00

SOURCE: U.S. Census Bureau, www.census.gov

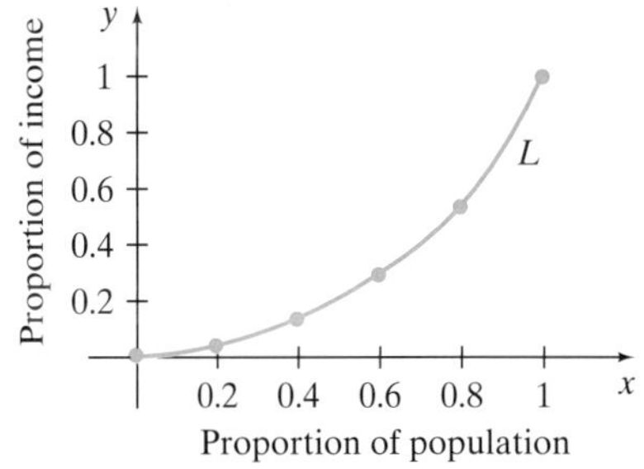

Figure 17.3.10

There are two extreme cases of income distribution, which means that there are two extreme cases for Lorenz curves. They are:

1. Absolute equality of income distribution
2. Absolute inequality of income distribution

Absolute equality of income distribution means that everyone earns the same income. The lowest 20% earns 20% of the income, the lowest 40% earns 40% of the income, and so on. The Lorenz curve for absolute equality is $L(x) = x$, as shown in Figure 17.3.11. Absolute inequality of income distribution means that no one earns any income except one person who earns it all. This would yield the Lorenz curve shown in Figure 17.3.12.

All Lorenz curves that we will analyze lie between these two extremes. See Figure 17.3.13. To measure how the actual income distribution differs from

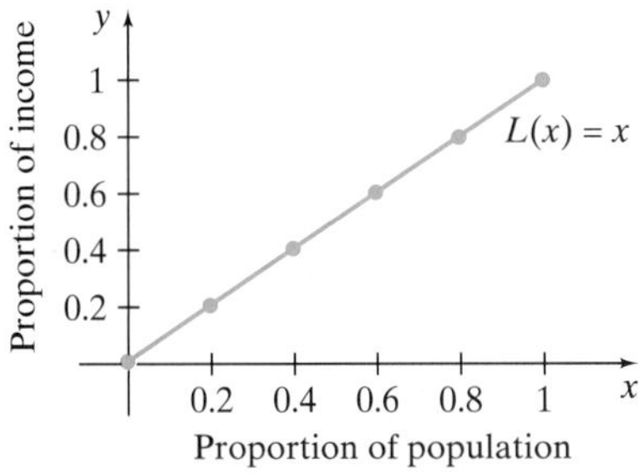

Figure 17.3.11 Lorenz curve for absolute equality.

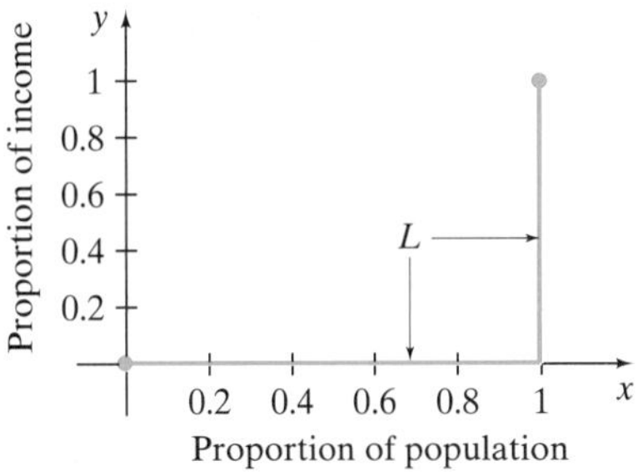

Figure 17.3.12 Lorenz curve for absolute inequality.

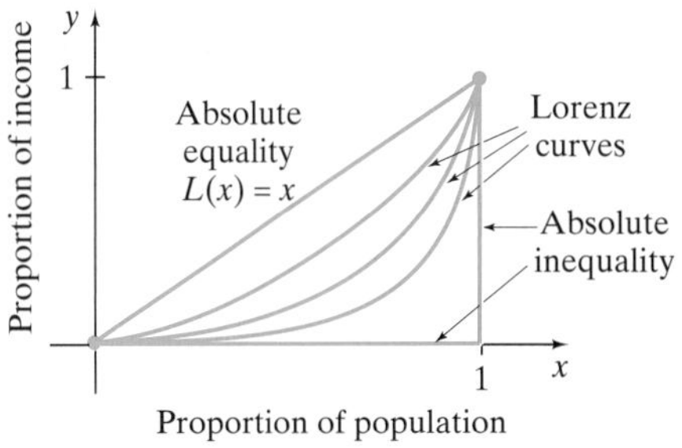

Figure 17.3.13

absolute equality, we will calculate the area between $L(x) = x$ (absolute equality) and the Lorenz curve modeling our income distribution.

Since the domain and range of any Lorenz curve are [0, 1], the area between absolute equality and absolute inequality is $\frac{1}{2}$. (The region between these two extremes is $\frac{1}{2}$ of a square whose area is 1 square unit.) It follows that the area between any Lorenz curve and absolute equality is at most $\frac{1}{2}$. However, economists multiply this area by 2 to get a number between 0 and 1, with 0 being absolute equality and 1 being absolute inequality.

The area between a Lorenz curve and absolute equality gives a measure that economists call the **Gini Index** or the **Gini coefficient**. A larger Gini Index means more area between the Lorenz curve and absolute equality, which then means a greater inequality in income distribution.

Example 7 Computing a Gini Index

In Utopia Land, the Lorenz curve for income distribution is modeled by $L(x) = x^2$.

(a) Graph $L(x) = x$, absolute equality, and $L(x) = x^2$, the Lorenz curve for Utopia Land on the same set of axes.

(b) Compute the Gini Index for Utopia Land.

Solution

(a) Figure 17.3.14 gives the required graphs.

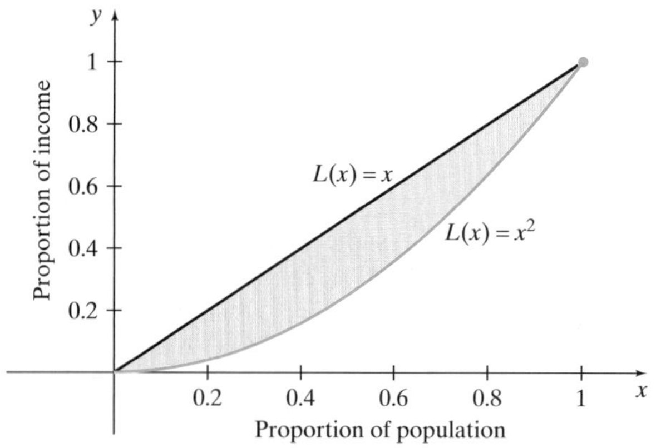

Figure 17.3.14

(b) The shaded region shown in Figure 17.3.14 is the area between two curves. We compute it to be

$$\int_0^1 (x - x^2)\,dx = \left(\frac{1}{2}x^2 - \frac{1}{3}x^3\right)\Bigg|_0^1 = \left[\frac{1}{2}(1)^2 - \frac{1}{3}(1)^3\right] - \left[\frac{1}{2}(0)^2 - \frac{1}{3}(0)^3\right] = \frac{1}{6}$$

We now multiply by 2 to get the Gini Index of

$$2\left(\frac{1}{6}\right) = \frac{1}{3}$$

So the Gini Index for Utopia Land is $\frac{1}{3}$ or approximately 0.33. ■

We can generalize the work done in Example 7 as follows:

Computing a Gini Index

For a Lorenz curve, L, the **Gini Index** is found by

$$2 \cdot \int_0^1 [x - L(x)]\,dx$$

Example 8 Computing a Gini Index

The Lorenz curve for the distribution of income in the United States in 1990 can be modeled by

$$L(x) = 2.03x^4 - 3.15x^3 + 2.22x^2 - 0.1x \qquad 0 \le x \le 1$$

Compute the Gini Index for the United States in 1990 (*Source:* U.S. Census Bureau, www.census.gov).

Solution

The Gini index is found by

$$\begin{aligned} 2 \cdot \int_0^1 [x - L(x)]\,dx &= 2 \cdot \int_0^1 [x - (2.03x^4 - 3.15x^3 + 2.22x^2 - 0.1x)]\,dx \\ &= 2 \cdot \int_0^1 (-2.03x^4 + 3.15x^3 - 2.22x^2 + 1.1x)\,dx \\ &= 2 \cdot \left[\left(\frac{-2.03}{5}x^5 + \frac{3.15}{4}x^4 - \frac{2.22}{3}x^3 + \frac{1.1}{2}x^2 \right) \Big|_0^1 \right] \\ &= 2 \cdot [0.1915] \\ &= 0.383 \end{aligned}$$

So the Gini index in 1990 was about 0.383.

Checkpoint 4 Now work Exercise 65.

SUMMARY

In this section we studied some applications for area between two curves. **Consumers' surplus** and **producers' surplus** are simply areas between two curves, and we stress that they should be viewed that way, as opposed to simply memorizing a formula. We also discussed where to locate an **equilibrium point** and the relationship between **consumer expenditure** and revenue. The final topic in this section dealt with income distributions and a measure of income distribution called the **Gini Index**. Again, this is an application of area between two curves.

- **Computing consumers' surplus**: For demand function d and point (x_m, d_{mp}) on the graph of d, where x_m is the **market demand** and d_{mp} is the **market price**, we determine the consumers' surplus by integrating $\int_0^{x_m} [d(x) - d_{mp}]\,dx$.
- **Computing producers' surplus**: For supply function s and a point (x_m, d_{mp}) on the graph of s, where x_m is the **market demand** and d_{mp} is the **market price**, we determine the producers' surplus by integrating $\int_0^{x_m} [d_{mp} - s(x)]\,dx$.
- **Computing a Gini Index**: For a Lorenz curve, $y = L(x)$, the Gini Index is found by $2 \cdot \int_0^1 [x - L(x)]\,dx$.

SECTION 17.3 EXERCISES

Exercises 1–4 refer to Figure 17.3.15.

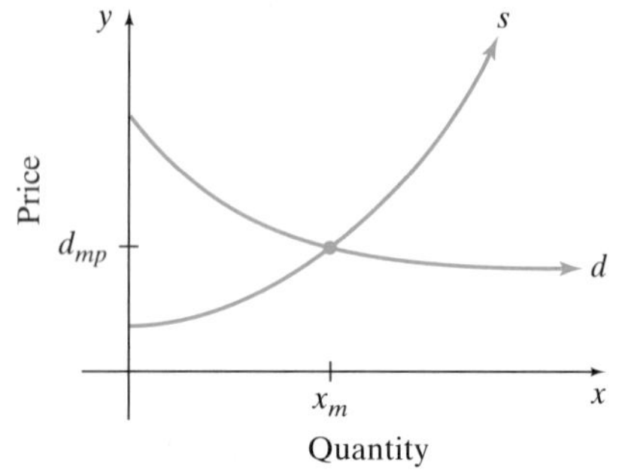

Figure 17.3.15

1. Shade the region that corresponds to the consumers' surplus. Write an integral to determine consumers' surplus.

2. Shade the region that corresponds to the producers' surplus. Write an integral to determine producers' surplus.

3. Shade the region that corresponds to the actual consumer expenditure. Determine the actual consumer expenditure.

4. Shade the region that corresponds to the revenue. Determine the actual revenue.

Exercises 5 and 6 refer to Figure 17.3.16.

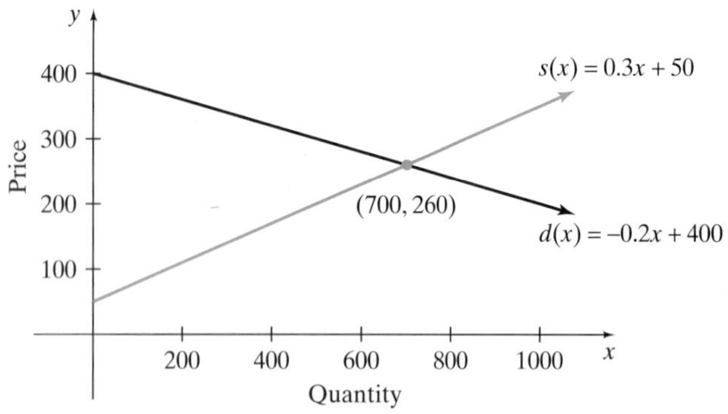

Figure 17.3.16

5. Shade the region that corresponds to the producers' surplus. Write an integral to determine producers' surplus.

6. Shade the region that corresponds to the consumers' surplus. Write an integral to determine consumers' surplus.

Exercises 7 and 8 refer to Figure 17.3.17.

7. Shade the region that corresponds to the producers' surplus. Write an integral to determine producers' surplus.

8. Shade the region that corresponds to the consumers' surplus. Write an integral to determine consumers' surplus.

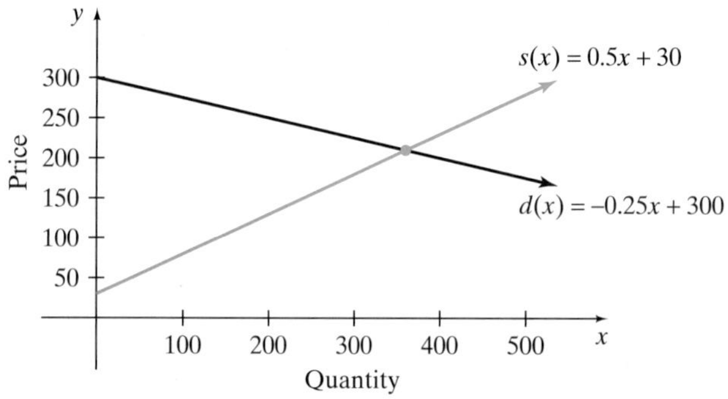

Figure 17.3.17

Exercises 9 and 10 refer to Figure 17.3.18.

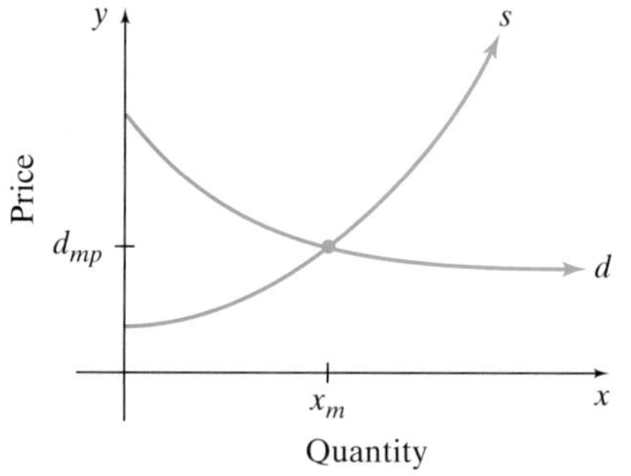

Figure 17.3.18

9. Shade the region that corresponds to the consumers' surplus. Write an integral to determine consumers' surplus.

10. Shade the region that corresponds to the producers' surplus. Write an integral to determine producers' surplus.

In Exercises 11–24, for each demand function given and demand level x, determine the consumers' surplus.

11. $d(x) = 3500 - 3x,\ x = 200$

12. $d(x) = 3500 - 3x,\ x = 500$

13. $d(x) = 220 - \frac{1}{3}x,\ x = 100$

14. $d(x) = 330 - \frac{1}{5}x,\ x = 200$

✓ **15.** $d(x) = 600 - 0.6x,\ x = 150$

16. $d(x) = 540 - 0.4x,\ x = 300$

17. $d(x) = 1610 - 0.08x^2,\ x = 100$

18. $d(x) = 2720 - 0.05x^2,\ x = 200$

19. $d(x) = 400 - \frac{1}{3}x^2,\ x = 30$

20. $d(x) = 300 - \frac{1}{4}x^2, x = 30$

21. $d(x) = 600e^{-0.01x}, x = 300$

22. $d(x) = 550e^{-0.02x}, x = 150$

23. $d(x) = 230e^{-0.02x}, x = 175$

24. $d(x) = 320e^{-0.01x}, x = 100$

In Exercises 25–38, for each supply function given and demand level x, determine the producers' surplus.

25. $s(x) = 0.2x + 50, x = 75$

26. $s(x) = 0.3x + 66, x = 90$

27. $s(x) = 0.15x + 80, x = 100$

28. $s(x) = 0.27x + 60, x = 80$

✓ **29.** $s(x) = \frac{1}{3}x + 55, x = 60$

30. $s(x) = \frac{1}{2}x + 40, x = 70$

31. $s(x) = 0.05x^2 + 20, x = 100$

32. $s(x) = 0.01x^2 + 25, x = 200$

33. $s(x) = 0.1x^2 + 10, x = 150$

34. $s(x) = 0.2x^2 + 15, x = 110$

35. $s(x) = 10e^{0.02x}, x = 90$

36. $s(x) = 20e^{0.01x}, x = 120$

37. $s(x) = 13e^{0.01x}, x = 110$

38. $s(x) = 25e^{0.03x}, x = 55$

For Exercises 39–48:

(a) Determine the equilibrium point. In Exercises 47 and 48, use the INTERSECT command on your calculator and round equilibrium demand and equilibrium price to the nearest whole number.

(b) Determine the consumers' surplus at equilibrium demand.

(c) Determine the producers' surplus at equilibrium demand.

39. $d(x) = 3553 - 13x; s(x) = 5.70x$

40. $d(x) = 329 - \frac{1}{5}x; s(x) = 0.27x$

✓ **41.** $d(x) = 398 - 0.4x; s(x) = 0.5x + 11$

42. $d(x) = 89 - 0.25x; s(x) = 0.85x + 19.7$

43. $d(x) = 2743 - 0.04x^2; s(x) = 0.06x^2 + 20.5$

44. $d(x) = 2528 - 0.01x^2; s(x) = 0.02x^2 + 5$

45. $d(x) = 500 - 0.2x^2; s(x) = 0.2x^2 + 10$

46. $d(x) = 1426 - 0.02x^2; s(x) = 0.01x^2 + 13.33$

47. $d(x) = 600e^{-0.01x}; s(x) = 1.125e^{0.02x}$

48. $d(x) = 550e^{-0.02x}; s(x) = 20e^{0.01x}$

Applications

49. Lindsay's Department Store has determined that the demand function for a new type of nonsticking frying pan is given by

$$d(x) = -1.4x + 25$$

where x is the number of pans demanded each day and $d(x)$ is the price per pan in dollars.

(a) Assuming that the equilibrium price is \$11 per pan, determine the equilibrium demand.

(b) Determine the consumers' surplus at equilibrium demand.

50. The demand function for a new crock pot is given by

$$d(x) = 98 - 7\sqrt{x}$$

where x is the number of crock pots demanded each day and $d(x)$ is the price per crock pot in dollars.

(a) Assuming that the equilibrium price is \$28 per crock pot, determine the equilibrium demand.

(b) Determine the consumers' surplus at equilibrium demand.

51. Frickel's Department Store has determined that the demand function for an extra-wide toaster is given by

$$d(x) = 27.3e^{-0.09x}$$

where x is the number of toasters demanded each day and $d(x)$ is the price per toaster in dollars. Assuming that the equilibrium price is \$12.14, determine the consumers' surplus at equilibrium demand. Round equilibrium demand to the nearest whole number.

52. The Hacker's Delight has determined that the demand function for the new Doppler Don Driving Iron is given by

$$d(x) = -x^2 - x + 650$$

where x is the number of driving irons demanded each month and $d(x)$ is the price per driving iron in dollars. Assuming that the equilibrium price is \$230, determine the consumers' surplus.

53. Balata Inc., a producer of golf balls, has determined that the supply function for the new Xtrah golf ball is given by

$$s(x) = 0.24x + 3.70$$

where x is the number of dozens supplied each month and $s(x)$ is the price per dozen.

(a) If the equilibrium price is \$19.30 per dozen, determine the equilibrium demand.

(b) Determine the producers' surplus at equilibrium demand.

54. Linguini's Pizza Palace has determined that the supply function for a large pizza is given by

$$s(x) = 0.02x^2 + 0.78$$

where x is the number of pizzas supplied each day and $s(x)$ is the price per pizza.

(a) If the equilibrium price of a pizza is \$8 per pizza, determine the equilibrium demand.

(b) Determine producers' surplus at equilibrium demand.

55. Balata Inc., a producer of golf balls, has determined that the supply function for the new Equalizer golf ball is given by

$$s(x) = 6.7e^{0.02x}$$

where x is the number of dozens supplied each month and $d(x)$ is the price per dozen. Assuming that the equilibrium price is \$19.75 per dozen, determine the producers' surplus. Round equilibrium demand to the nearest whole number.

56. Baker's Bake Shoppe, a supplier of specialty baking pans, has determined that the supply function for a certain birthday cake baking pan is given by

$$s(x) = 4e^{0.05x}$$

where x is the number of these baking pans supplied each month and $s(x)$ is the price per pan in dollars. Assuming that the equilibrium price is \$8 per pan, determine the producers' surplus. Round equilibrium demand to the nearest whole number.

57. BuildIt, a local home improvement store, has determined that the demand function for a 30-gallon trash can is given by

$$d(x) = 13 - 0.01x^2$$

while the related producer supply function is given by

$$s(x) = 0.1x + 1$$

where x is the daily quantity and $d(x)$ and $s(x)$ are in dollars per can.

(a) Determine the equilibrium point.

(b) Determine the consumers' surplus at equilibrium.

(c) Determine the producers' surplus at equilibrium.

58. Fresh Paint, a local paint supply and paint accessories store, has determined that the demand function for a step ladder is given by

$$d(x) = -0.2x^2 + 100$$

while the related producer supply function is given by

$$s(x) = x + 1$$

where x is the weekly quantity and $d(x)$ and $s(x)$ are in dollars per ladder.

(a) Determine the equilibrium point. Round the equilibrium demand to the nearest whole number and round the equilibrium price to the nearest dollar.

(b) Determine the consumers' surplus at equilibrium.

(c) Determine the producers' surplus at equilibrium.

59. Office House, an office supply store, has determined that the demand function for a floppy disk storage case is given by

$$d(x) = 30 - x$$

while the related producer supply function is given by

$$s(x) = \sqrt{x}$$

where x is the weekly quantity and $d(x)$ and $s(x)$ are in dollars per storage case.

(a) Determine the equilibrium point.

(b) Determine the consumers' surplus at equilibrium.

(c) Determine the producers' surplus at equilibrium.

60. Just Smell It, a scented candle shop, has determined that the demand function for a large vanilla-scented candle is given by

$$d(x) = 20 - \frac{1}{2}x$$

while the related producer supply function is given by

$$s(x) = \frac{1}{2}\sqrt{x}$$

where x is the monthly quantity and $d(x)$ and $s(x)$ are in dollars per candle.

(a) Determine the equilibrium point. Round the equilibrium demand to the nearest whole number and round the equilibrium price to the nearest dollar.

(b) Determine the consumers' surplus at equilibrium.

(c) Determine the producers' surplus at equilibrium.

In Exercises 61–66, the techniques of regression were used on data from the Census Bureau (www.census.gov) *to obtain Lorenz curves for selected years. Recall that x represents the proportion of the population and L(x) represents the proportion of income. Determine the Gini Index for each year. Where appropriate, round to the nearest ten-thousandth.*

61. In 1988, $L(x) = 1.93x^4 - 2.96x^3 + 2.12x^2 - 0.09x$

62. In 1990, $L(x) = 2.03x^4 - 3.15x^3 + 2.22x^2 - 0.1x$

63. In 1992, $L(x) = 2.01x^4 - 3.07x^3 + 2.18x^2 - 0.11x$

64. In 1994, $L(x) = 2.396x^4 - 3.65x^3 + 2.39x^2 - 0.14x$

✓ **65.** In 1996, $L(x) = 2.5x^4 - 3.84x^3 + 2.5x^2 - 0.16x$

66. In 1998, $L(x) = 2.55x^4 - 3.92x^3 + 2.52x^2 - 0.16x$

67. The Gini Index is similar to a snapshot in time. The index does not depend on time. However, a collection of indices over some period of time produces a "moving picture" of income distribution. Exercises 61–66 are a collection over a 10-year period of time, specifically, 1988 to 1998. Analyze the Gini Indices over this 10-year period of time. Does the Gini Index appear to be increasing or decreasing over this time period? What does this mean about income distribution in the United States over this time period?

68. Use the data in Table 17.3.2 for U.S. income distribution in 1995.

(a) Let x represent the proportion of the population and let y represent the proportion of income. Enter the data into your calculator and determine a quartic regression equation to model the data. Round all coefficients to the nearest hundredth.

Table 17.3.2

Proportion of Population	Proportion of Income
0	0
0.2	0.044
0.6	0.303
0.8	0.535
1	1

SOURCE: U.S. Census Bureau, www.census.gov

(b) Compute the Gini Index for 1995. Round to the nearest ten-thousandth.

(c) Compare the Gini Index for 1995 with the Gini Index for 1990. See Exercise 62. Interpret the results.

SECTION PROJECT

An office supply store has collected the following data for a certain brand of briefcase.

Number of Briefcases Demanded per Week	Price ($ per Briefcase)
2	79.99
5	61.99
7	51.99
9	43.99
12	33.99
15	25.99

Meanwhile, the supplier of the briefcases has collected the following data:

Number Supplied Each Week	Price ($ per Briefcase)
1	23.99
3	29.99
6	39.99
8	47.49
11	61.99
13	70.99
16	89.99

(a) Let x represent the quantity demanded per week. Enter the data into your calculator, plot the data, and determine a quadratic regression model for d. Round coefficients to the nearest thousandth.

(b) Now let x represent the quantity supplied each week. Enter the data into your calculator, plot the data, and determine a quadratic regression model for s. Round coefficients to the nearest thousandth.

(c) Graph d and s in the viewing window [0, 20] by [0, 70].

(d) Determine the equilibrium point by using the INTERSECT command. Round the equilibrium demand to the nearest whole number and round the equilibrium price to the nearest dollar.

(e) Determine the consumers' surplus at equilibrium.

(f) Determine the producers' surplus at equilibrium.

Section 17.4 Integration by Parts

In this section we present another technique of integration called **integration by parts**. Integration by parts is somewhat analogous to the Product Rule for differentiating. Applications involving this new technique will include total production from oil fields and a return to continuous income streams, first presented in Section 17.1. Here we will analyze the **present value** of a continuous income stream.

Integration by Parts

In many real-life situations, managers encounter integrals that cannot be evaluated with the integration techniques that we have learned so far. Other integration techniques are needed to solve these problems. For example, the manager of a taxicab fleet knows that the variable cost per mile for maintaining a cab that has been in service for x years since 2000 is modeled by

$$C(x) = 8.3 + 0.88 \ln x$$

To find the average variable cost per mile for a cab that has been in service for 2 years since 2000, the manager would need to evaluate the integral

$$\frac{1}{2}\int_0^2 (8.3 + 0.88 \ln x)\,dx$$

We cannot evaluate this integral because we have not yet determined $\int \ln x\,dx$. (We know how to *differentiate* $\ln x$.) A new method of integration, called **integration by parts**, can help to solve this problem.

From Your Toolbox

The differential of y, dy, is defined to be $dy = f'(x)\,dx$.

To see where the formula for integration by parts comes from, let's assume that we have two differentiable functions, u and v. We need to recall the definition of the differential from Section 13.1. (Consult the Toolbox to the left.) So in differential notation, for our two functions u and v, we have

$$du = u'\,dx \quad \text{and} \quad dv = v'\,dx$$

The derivative of $u \cdot v$ is determined by the Product Rule to be

$$\frac{d}{dx}(u \cdot v) = u'v + uv'$$

Now, if we integrate both sides of this equation with respect to x, we get

$$\int \frac{d}{dx}(u \cdot v)\,dx = \int (u'v + uv')\,dx$$

$$u \cdot v = \int u'v\,dx + \int uv'\,dx$$

Since $du = u'\,dx$ and $dv = v'\,dx$, we rewrite this as

$$u \cdot v = \int v\,du + \int u\,dv$$

We now solve this equation for $\int u\,dv$ and get

$$\int u\,dv = u \cdot v - \int v\,du$$

This last equation is the formula for the integration technique known as **integration by parts**.

Integration by Parts

For two differentiable functions u and v,

$$\int u\,dv = u \cdot v - \int v\,du$$

When applying the integration by parts formula, we are really performing a *double substitution*. Example 1 shows how this works. After Example 1, we will supply some general guidelines for using integration by parts.

Example 1 Using the Integration by Parts Formula

Determine $\int xe^x\,dx$.

Solution

Understand the Situation: Our integration by parts formula has four parts that must be determined: u, du, v, and dv. Our goal is to turn our original problem, $\int xe^x\,dx$, into an integral of the form $\int u\,dv$. This is where the double substitution comes in. We will let u equal some part of $xe^x\,dx$, and then dv will equal the rest of $xe^x\,dx$.

We decide to select $u = x$, since $du = 1 \cdot dx$, and $dv = e^x\,dx$, since $\int e^x\,dx = e^x$. So we have

$$u = x \qquad dv = e^x\,dx$$

$$du = 1 \cdot dx = dx \qquad v = \int e^x\,dx = e^x$$

Applying the integration by parts formula yields

$$\int u\,dv = u \cdot v - \int v\,du$$

$$\int xe^x\,dx = xe^x - \int e^x\,dx$$

$$= xe^x - e^x + C$$

■

Note: 1. Notice that the differentials du and dv include the dx.

2. When we integrate dv to get v, we can omit the constant of integration C. One constant C at the very end is enough.

The key step in using integration by parts is selecting the *two* substitutions u and dv. We rewrite the original integral in the form $\int u\,dv$ by substituting u for part of the original integrand and dv for the rest. We choose u and dv so that the resulting expression is simpler than the original. In Example 1 we chose u and dv so that the resulting expression has $\int e^x\,dx$, which we can easily integrate. We offer the following guidelines for choosing u and dv.

Guidelines for the Selection of u and dv

- Either select dv to be the most complicated part of the integrand that is easily integrated. Then u is the rest.
- Or select u so that its derivative, du, is a simpler function than u. Then dv is the rest.

Example 2 Using the Integration by Parts Formula

Determine $\int x \ln x\,dx$.

Solution

Understand the Situation: In the integrand $x \ln x$, we can easily integrate x, but not $\ln x$. With this in mind, we select $u = \ln x$ and $dv = x\,dx$, which means that

Interactive Activity

Check the result of Example 2 by differentiating the answer, that is,

$$\frac{d}{dx}\left[\frac{1}{2}x^2 \ln x - \frac{1}{4}x^2 + C\right]$$

WWW

the four required pieces are

$$u = \ln x \qquad dv = x\,dx$$
$$du = \frac{1}{x}\,dx \qquad v = \int x\,dx = \frac{1}{2}x^2$$

Applying the integration by parts formula yields

$$\int u\,dv = u \cdot v - \int v\,du$$
$$\int x \ln x\,dx = (\ln x) \cdot \frac{1}{2}x^2 - \int \frac{1}{2}x^2 \cdot \frac{1}{x}\,dx$$
$$= \frac{1}{2}x^2 \ln x - \frac{1}{2}\int x\,dx$$
$$= \frac{1}{2}x^2 \ln x - \frac{1}{2} \cdot \frac{1}{2}x^2 + C = \frac{1}{2}x^2 \ln x - \frac{1}{4}x^2 + C$$

■

Checkpoint 1

Now work Exercise 3.

Example 3 Using the Integration by Parts Formula

Determine $\int \ln x\,dx$.

Solution

Interactive Activity

Verify the result of Example 3 by differentiating the answer.

Understand the Situation: This one looks a bit unusual in that there does not appear to be anything multiplied by $\ln x$. If we imagine $\ln x$ to be $1 \cdot \ln x$, then we can employ our guidelines. We urge caution here. It may be tempting to let $u = 1$, but this would force $dv = \ln x\,dx$. We do not know the integral of $\ln x$—that is what we are trying to determine! Basically, we have no choice but to select $u = \ln x$ and $dv = 1 \cdot dx$. This gives

$$u = \ln x \qquad dv = 1 \cdot dx$$
$$du = \frac{1}{x}\,dx \qquad v = \int 1\,dx = x$$

The integration by parts formula gives

$$\int u\,dv = u \cdot v - \int v\,du$$
$$\int \ln x\,dx = x \ln x - \int x \cdot \frac{1}{x}\,dx$$
$$= x \ln x - \int 1\,dx$$
$$= x \ln x - x + C$$

■

Example 4 Using the Integration by Parts Formula

Determine $\int xe^{0.2x}\,dx$.

Solution

From the guidelines we select $u = x$ and $dv = e^{0.2x}\,dx$. This gives

$$u = x \qquad dv = e^{0.2x}\,dx$$

$$du = dx \qquad v = \int e^{0.2x}\,dx = \frac{1}{0.2}e^{0.2x} = 5e^{0.2x}$$

So, via the integration by parts formula, we have

$$\int u\,dv = u \cdot v - \int v\,du$$

$$\int xe^{0.2x}\,dx = 5xe^{0.2x} - \int 5e^{0.2x}\,dx = 5xe^{0.2x} - 5\int e^{0.2x}\,dx$$

$$= 5xe^{0.2x} - 5\left(\frac{1}{0.2}e^{0.2x}\right) + C = 5xe^{0.2x} - 25e^{0.2x} + C$$ ■

Note: Many times, when evaluating an integral using integration by parts and part of the integrand involves e^{ax}, choose $dv = e^{ax}\,dx$. We then immediately have $v = \frac{1}{a}e^{ax}$.

Checkpoint 2 Now work Exercise 21.

Example 5 **Evaluating an Indefinite Integral**

Determine $\int 2xe^{x^2}\,dx$.

Solution

Do not be fooled! This is nothing more than a u-substitution first encountered in Section 16.4. To integrate $\int 2xe^{x^2}\,dx$, we let $u = x^2$; then $du = 2x\,dx$ and by u-substitution we have

$$\int 2xe^{x^2}\,dx = \int e^u\,du = e^u + C = e^{x^2} + C$$ ■

Example 5 serves as a reminder that, even though we are currently learning a new integration technique in integration by parts, we should not forget our previously learned techniques. Our final example, before we consider some applications, requires two uses of the integration by parts formula.

Example 6 **Utilizing the Integration by Parts Formula Twice**

Determine $\int x^2e^{2x}\,dx$.

Solution

From the guidelines and the Note following Example 4, we select $u = x^2$ and $dv = e^{2x}\,dx$. This gives

$$u = x^2 \qquad dv = e^{2x}\,dx$$

$$du = 2x\,dx \qquad v = \frac{1}{2}e^{2x}$$

The integration by parts formula gives

$$\int x^2 e^{2x}\,dx = \frac{1}{2}x^2e^{2x} - \int \frac{1}{2}e^{2x}\cdot 2x\,dx = \frac{1}{2}x^2e^{2x} - \int xe^{2x}\,dx$$

This time, the integration by parts formula did not produce an integral that we could quickly evaluate. However, the new integral is simpler than the original, and it also looks similar to the one that we did in Example 4. So let's use integration by parts again. For the integral $\int xe^{2x}\,dx$, we select $u = x$ and $dv = e^{2x}\,dx$. This gives

$$u = x \qquad dv = e^{2x}\,dx$$
$$du = dx \qquad v = \frac{1}{2}e^{2x}$$

Integration by parts yields

$$\begin{aligned}\int xe^{2x}\,dx &= \frac{1}{2}xe^{2x} - \int \frac{1}{2}e^{2x}\,dx = \frac{1}{2}xe^{2x} - \frac{1}{2}\int e^{2x}\,dx\\ &= \frac{1}{2}xe^{2x} - \frac{1}{2}\left(\frac{1}{2}e^{2x}\right) + C\\ &= \frac{1}{2}xe^{2x} - \frac{1}{4}e^{2x} + C\end{aligned}$$

Putting it all together results in

$$\begin{aligned}\int x^2e^{2x}\,dx &= \frac{1}{2}x^2e^{2x} - \int xe^{2x}\,dx\\ &= \frac{1}{2}x^2e^{2x} - \left(\frac{1}{2}xe^{2x} - \frac{1}{4}e^{2x} + C\right)\\ &= \frac{1}{2}x^2e^{2x} - \frac{1}{2}xe^{2x} + \frac{1}{4}e^{2x} + C\end{aligned}$$

■

Checkpoint 3 Now work Exercise 33.

Applications

Most of the applications that we consider in this section require us to evaluate a definite integral. We will disregard whatever limits of integration we have until after we have found an indefinite integral, using whatever technique is required.

Example 7 Estimating Oil Field Production

Geologists have estimated that an oil field will produce oil at a rate given by

$$B(t) = 5te^{-0.2t}$$

where $B(t)$ is thousands of barrels per month t months from the start of operation. Estimate the total production in the first year of operation.

Solution

Understand the Situation: Since B is a rate function we know that the total production in the first year of operation is given by a definite integral. So we need to evaluate

$$\int_0^{12} 5te^{-0.2t}\,dt$$

As mentioned prior to Example 7, we will ignore the limits of integration until after we have found an antiderivative. Using the guidelines for integration by parts, we select $u = 5t$ and $dv = e^{-0.2t}\,dt$. This gives

$$u = 5t \qquad dv = e^{-0.2t}\,dt$$

$$du = 5\,dt \qquad v = \frac{1}{-0.2}e^{-0.2t} = -5e^{-0.2t}$$

So, by integration by parts, we have

$$\begin{aligned}\int 5te^{-0.2t}\,dt &= 5t(-5e^{-0.2t}) - \int(-5e^{-0.2t})\cdot 5\,dt\\ &= -25te^{-0.2t} + 25\int e^{-0.2t}\,dt\\ &= -25te^{-0.2t} + 25\left(\frac{1}{-0.2}e^{-0.2t}\right) + C\\ &= -25te^{-0.2t} - 125e^{-0.2t} + C\end{aligned}$$

Now to evaluate the *definite* integral and answer the question, we evaluate from $t = 0$ to $t = 12$ and get

$$\begin{aligned}(-25te^{-0.2t} - 125e^{-0.2t})\Big|_0^{12} &= [(-25(12)e^{-0.2(12)} - 125e^{-0.2(12)})\\ &\quad - (-25(0)e^{-0.2(0)} - 125e^{-0.2(0)})]\\ &\approx 86.44\end{aligned}$$

Interpret the Solution: So, in the first year of operation, the total production is estimated to be 86.44 thousand barrels. ■

✓ Checkpoint 4

Now work Exercise 49.

Flashback

Continuous Income Stream Revisited

In Section 17.1 we introduced the concept of a continuous income stream. Recall that a continuous income stream is very similar to the electric company, natural gas company, or water company that has a meter on our house that continuously monitors and totals the amount of electricity, natural gas, or water that we are consuming. Suppose that the rate of change of income in thousands of dollars per year for the oil field in Example 7 is projected to be

$$f(t) = 500e^{-0.09t}$$

where t is the number of years from when oil extraction began. See Figure 17.4.1. Find the total income from this field during the first year of operation.

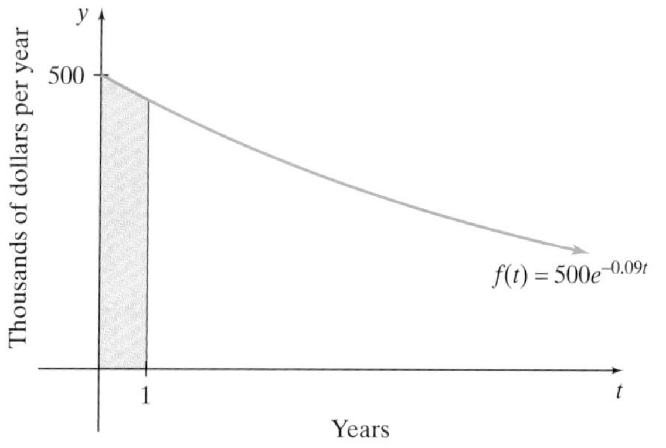

Figure 17.4.1 Shaded area gives total income.

Flashback Solution

Again, since we have a rate function, determining the total income generated during the first year of operation is equivalent to determining the area under the curve from $t = 0$ to $t = 1$. Integrating yields

$$\int_0^1 500e^{-0.09t}\,dt = \left(\frac{500}{-0.09}e^{-0.09t}\right)\Big|_0^1$$

$$= \frac{500}{-0.09}e^{-0.09(1)} - \frac{500}{-0.09}e^{-0.09(0)}$$

$$\approx 478.1600818$$

So in the first year of operation this oil field will generate approximately \$478,160.08 in income.

Recall from Section 17.1 that the rate of change of income in the Flashback is called the *rate of flow function*. Also in Section 17.1 we stated that in general the *total income* produced from time $t = a$ to $t = b$ is found by $\int_a^b f(t)\,dt$.

Now that we have properly reviewed continuous income streams, we turn our attention to the **present value of a continuous income stream**.

Present Value of a Continuous Income Stream

Suppose that we have a continuous income stream from an oil field and we know the rate of flow function. The ability to determine its *present value* is very important if we wish to sell (or buy) this oil field. We need to recall that, if P dollars are invested at an annual interest rate r compounded continuously, then the amount A after t years is given by $A = Pe^{rt}$. Solving this equation for P yields

$$P = Ae^{-rt}$$

which we say is the **present value** of A. This means that P is the amount that needs to be invested *now* at interest rate r compounded continuously so that we have A dollars t years from now.

We now generalize the present value concept to continuous income streams by revisiting rectangles and the definite integral. First, we assume that the present money can be invested at the given rate, r, compounded continuously over the given time interval. Even though this assumption may sound unrealistic, managers make these assumptions regularly based on historical data.

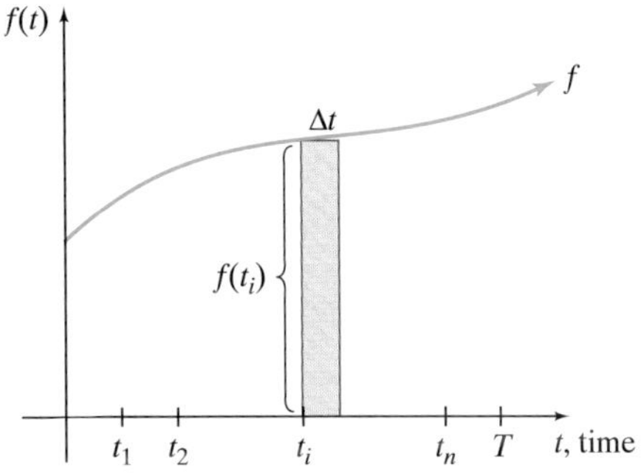

Figure 17.4.2

Suppose that f is the rate of flow function for a continuous income stream. Divide the interval $[0, T]$ into n equal subintervals of length Δt, as shown in Figure 17.4.2. In a typical subinterval, we can pick any point t_i. Let's say that it is the left-hand endpoint of the subinterval. *The total income produced over this subinterval is approximately equal to the area of the rectangle shaded in Figure 17.4.2.* This area is given by

$$f(t_i)\,\Delta t$$

Using the present value formula, $P = Ae^{-rt}$, with $A = f(t_i)\,\Delta t$ and $t = t_i$, the present value of the income received over this subinterval, P_i, is approximately equal to

$$P_i \approx f(t_i)\,\Delta t \cdot e^{-rt_i} = f(t_i)e^{-rt_i}\,\Delta t$$

To get the total present value on the interval $[0, T]$ requires us to sum *all* these present values. We know from our work in Chapter 16 that this is given by the definite integral $\int_0^T f(t)e^{-rt}\,dt$.

Present Value of a Continuous Income Stream

If f is the rate of flow function for a continuous income stream, then the **present value**, P, at annual interest rate, r, compounded continuously for T years is given by

$$P = \int_0^T f(t)e^{-rt}\,dt$$

Example 8 Determining the Present Value of a Continuous Income Stream

A window washing business generates income at the rate of $3t$ thousand dollars per year, where t is the number of years from now.

(a) Determine the present value of this continuous income stream for the next 7 years at 8% compounded continuously.

(b) Determine the total amount (income plus interest) produced by the window washing business over this 7-year period.

Solution

(a) **Understand the Situation:** We know that $f(t) = 3t$, so the present value for the next 7 years at 8% compounded continuously is

$$P = \int_0^T f(t)e^{-rt}\,dt = \int_0^7 3te^{-0.08t}\,dt$$

Since this definite integral requires integration by parts, we will ignore the limits of integration until we have determined an antiderivative. Here we select $u = 3t$ and $dv = e^{-0.08t}\,dt$. This yields

$$u = 3t \qquad dv = e^{-0.08t}\,dt$$

$$du = 3\,dt \qquad v = \frac{1}{-0.08}e^{-0.08t} = -12.5e^{-0.08t}$$

Using integration by parts, we have

$$\begin{aligned}\int 3te^{-0.08t}\,dt &= 3t(-12.5e^{-0.08t}) - \int(-12.5e^{-0.08t})\cdot 3\,dt\\ &= -37.5te^{-0.08t} + 37.5\int e^{-0.08t}\,dt\\ &= -37.5te^{-0.08t} + 37.5\left(\frac{1}{-0.08}e^{-0.08t}\right) + C\\ &= -37.5te^{-0.08t} - 468.75e^{-0.08t} + C\end{aligned}$$

We now evaluate this from $t = 0$ to $t = 7$ and get

$$(-37.5te^{-0.08t} - 468.75e^{-0.08t})\Big|_0^7 \approx 51.053372$$

Interpret the Solution: So the present value of this continuous income stream over the next 7 years is about $51,053.37.

(b) **Understand the Situation:** The total amount (income plus interest) earned over the next 7 years is equivalent to the amount earned by investing the present value $51,053.37 in an account earning 8% annual interest compounded continuously for 7 years.

So we use the compound interest formula and get

$$A = Pe^{rt} = 51{,}053.37e^{0.08(7)} \approx \$89{,}377.73$$

Interpret the Solution: So the total amount (income plus interest) produced by this window washing business over this 7-year period is $89,377.73. ■

✓ Checkpoint 5 Now work Exercise 59.

Example 8 illustrates how to determine the present value of a continuous income stream and also how to determine the total amount (income plus interest) generated by the continuous income stream. If we wish to compute just the total income generated by the continuous income stream, we would evaluate $\int_0^7 3t\,dt$.

SUMMARY

In this section we presented an integration technique known as **integration by parts**. We suggest becoming familiar with the guidelines for selecting u and dv, as well as *when* we need to use integration by parts. A mathematical application of integration by parts was the determination of an antiderivative for $\ln x$ to be $\int \ln x\,dx = x\ln x - x + C$. The section concluded with some applications involving integration by parts. We also determined a formula for the present value of a continuous income stream.

- **Integration by Parts Formula**: $\int u\,dv = uv - \int v\,du$.
- **Guidelines for the Selection of u and dv**: (1) Select dv to be the most complicated part of the integral that is easily integrated. Then u is the rest. (2) Or select u so that its derivative, du, is a simpler function than u. Then dv is the rest.
- **Present Value of a Continuous Income Stream Formula**: $P = \int_0^T f(t)e^{-rt}\,dt$.

SECTION 17.4 EXERCISES

Evaluate the given integrals in Exercises 1–26. If integration by parts is required, consult the guidelines for selecting u and dv given in this section.

1. $\int 2xe^{3x}\,dx$

2. $\int 8xe^{5x}\,dx$

3. $\int xe^{4x}\,dx$

4. $\int xe^{6x}\,dx$

5. $\int xe^{-x}\,dx$

6. $\int xe^{-3x}\,dx$

7. $\int xe^{-0.03x}\,dx$

8. $\int xe^{-0.07x}\,dx$

9. $\int_0^2 xe^{-x}\,dx$

10. $\int_0^1 5xe^{-2x}\,dx$

11. $\int_0^5 xe^{-0.04x}\,dx$

12. $\int_0^6 xe^{-0.03x}\,dx$

13. $\int_1^e x\ln x\,dx$

14. $\int_1^3 x\ln x\,dx$

15. $\int x^2\ln x\,dx$

16. $\int x^3\ln x\,dx$

17. $\int \ln 2x\,dx$

18. $\int \ln 3x\,dx$

19. $\int (x+3)e^x\,dx$ (*Hint:* Let $u = x+3$)

20. $\int (x-5)e^x\,dx$ (*Hint:* Let $u = x-5$)

21. $\int 6te^{-0.1t}\,dt$

22. $\int_0^3 6te^{-0.1t}\,dt$

23. $\int_0^5 5te^{-0.06t}\,dt$

24. $\int 5te^{-0.06t}\,dt$

25. $\int 1.2te^{-0.08t}\,dt$

26. $\int 2.3te^{-0.05t}\,dt$

In Exercises 27–30, evaluate the integral of the form $\int (x+c)(x+d)^n\,dx$ by letting $u = x+c$ and $dv = (x+d)^n$ and applying the integration by parts formula.

27. $\int (x+2)(x+1)^5\,dx$

28. $\int (x+2)(x-3)^6\,dx$

29. $\int (x-2)(x+3)^6\,dx$

30. $\int (x-1)(x-2)^4\,dx$

Exercises 31–40 require two (or more) applications of integration by parts.

31. $\int x^2e^{3x}\,dx$

32. $\int x^2e^x\,dx$

33. $\int x^2e^{-x}\,dx$

34. $\int x^2e^{4x}\,dx$

35. $\int x^2e^{5x}\,dx$

36. $\int (x+3)^2e^x\,dx$

37. $\int (\ln x)^2\,dx$

38. $\int x(\ln x)^2\,dx$

39. $\int x(\ln x)^3\,dx$

40. $\int (\ln x)^3\,dx$

In Exercises 41–48, determine the method needed to evaluate the given integral. If the method selected is integration by parts, state the choice for u and dv. If the method selected is u-substitution, simply state the choice for u.

41. $\int xe^{-x}\,dx$

42. $\int xe^{-x^2}\,dx$

43. $\int \frac{1}{x\ln x}\,dx$

44. $\int \frac{\ln x}{x}\,dx$

45. $\int (x+6)e^x\,dx$

46. $\int (x-5)e^x\,dx$

47. $\int 3xe^{2x}\,dx$

48. $\int 3x^2e^{x^3}\,dx$

Applications

49. It is estimated that an oil field will produce oil at a rate given by $B(t) = 6te^{-0.15t}$ thousand barrels per month, t months into production.

(a) Write a definite integral to estimate the total production for the first year of operation.

(b) Evaluate the integral from part (a) to estimate the total production for the first year of operation.

50. It is estimated that an oil field will produce oil at a rate given by $B(t) = 7.5te^{-0.13t}$ thousand barrels per month, t months into production.

(a) Write a definite integral to estimate the total production for the first year of operation.

(b) Evaluate the integral from part (a) to estimate the total production for the first year of operation.

51. The BrenKev Corporation has determined that its marginal profit function is given by

$$P'(t) = 2te^{0.2t}$$

where t is time in years and $P'(t)$ is in millions of dollars per year. Assuming that $P(0) = 0$, determine $P(t)$.

52. The VivaMix Corporation has determined that its marginal profit function is given by

$$P'(t) = 3te^{0.15t}$$

where t is time in years and $P'(t)$ is in millions of dollars per year. Assuming that $P(0) = 0$, determine $P(t)$.

53. The BrenKev Corporation has determined that its marginal cost function is given by

$$C'(t) = te^{-0.15t}$$

where t is time in years and $C'(t)$ is in millions of dollars per year. Assuming that $C(0) = 2$, determine $C(t)$.

54. The VivaMix Corporation has determined that its marginal cost function is given by

$$C'(t) = 0.3te^{-0.1t}$$

where t is time in years and $C'(t)$ is in millions of dollars per year. Assuming that $C(0) = 1.5$, determine $C(t)$.

55. Medical researchers have determined that the rate of absorption of a certain medication is given by $4te^{-0.51t}$ milligrams per hour, where t is the number of hours since the medication was taken. Determine the total amount of the medication absorbed during the first 6 hours.

56. Medical researchers have determined that the rate of absorption of a certain medication is given by $6te^{-0.42t}$ milligrams per hour, where t is the number of hours since the medication was taken. Determine the total amount of the medication absorbed during the first 8 hours.

57. Dena's Car Wash generates income at the rate of $2t$ million dollars per year, where t is the number of years that the car wash has been in business. Determine the present value of this continuous income stream over the first 5 years at a continuous compound interest rate of 6%.

58. SterilizeIt is a medical instrument company that generates income at the rate of $500{,}000 + 30{,}000t$ dollars per year, where t is the number of years that the company has been in operation. Determine the present value of this continuous income stream over the first 7 years at a continuous compound interest rate of 6.5%.

59. Suppose that an oil well produces income at the rate of $200e^{-0.1t}$ thousand dollars per year for t years. Determine the present value of this continuous income stream over the first 4 years of operation assuming a continuous compound interest rate of 5.75%.

60. Suppose that an oil well produces income at the rate of $150e^{-0.15t}$ thousand dollars per year for t years. Determine the present value of this continuous income stream over the first 3 years of operation assuming a continuous compound interest rate of 5.5%.

61. Reggie has a choice of two investments. Each choice requires the same initial investment and each produces a continuous income stream at an interest rate of 8% compounded continuously. The first investment is an oil well that generates income at a rate of $50t$ thousand dollars per year, where t is the number of years from now. The second investment is a natural gas well that generates income at a rate of $30 + 40t$ thousand dollars per year, where t is the number of years from now. Compare the present value of each investment to determine which is the better choice over the next 5 years.

62. Repeat Exercise 61 except here determine which is the better investment over the next 10 years.

63. The college enrollment for males, in millions, in the United States can be approximated by

$$f(x) = 2.18 + 1.1\ln x \qquad 1 \le x \le 38$$

where x is in years ($x = 1$ corresponds to January 1, 1960) and $f(x)$ is the male enrollment in millions (*Source:* U.S. Census Bureau, www.census.gov).

(a) Determine the average value of f on the interval $[1, 11]$ and interpret.

(b) Determine the average value of f on the interval $[11, 21]$ and interpret.

(c) Determine the average value of f on the interval $[21, 31]$ and interpret.

64. The college enrollment for females, in millions, in the United States can be approximated by

$$g(x) = 0.22x + 0.94 \qquad 1 \leq x \leq 38$$

where x is in years ($x = 1$ corresponds to January 1, 1960) and $g(x)$ is the female enrollment in millions (*Source:* U.S. Census Bureau, www.census.gov).

(a) Determine the average value of g on the interval $[1, 11]$ and interpret.

(b) Determine the average value of g on the interval $[11, 21]$ and interpret.

(c) Determine the average value of g on the interval $[21, 31]$ and interpret.

65. The percent of solid waste generated in the United States that is paper and paperboard can be modeled by

$$f(x) = 35.8 + 1.96 \ln x \qquad 1 \leq x \leq 6$$

where x is in years ($x = 1$ corresponds to January 1, 1991) and $f(x)$ is the percent of solid waste that is paper and paperboard. Determine the average value of f on $[1, 6]$ and interpret (*Source:* U.S. Statistical Abstract, www.census.gov/statab/www).

66. The revenue generated in the solid waste management industry in the United States is approximated by

$$R(x) = 8.13 + 8.5 \ln x \qquad 1 \leq x \leq 16$$

where x is in years ($x = 1$ corresponds to January 1, 1980) and $R(x)$ is the revenue in billions of dollars (*Source:* U.S. Statistical Abstract, www.census.gov/statab/www).

(a) Determine the average value of R on the interval $[1, 11]$ and interpret.

(b) Determine the average value of R on the interval $[11, 16]$ and interpret.

67. The total output of motor gasoline of U.S. refineries can be modeled by

$$g(x) = 2506.74 + 148.92 \ln x \qquad 1 \leq x \leq 11$$

where x is in years ($x = 1$ corresponds to January 1, 1990) and $g(x)$ is the total output of motor gasoline in millions of barrels (*Source:* U.S. Energy Information Administration, www.eia.doe.gov). Determine the average value of g on the interval $[1, 11]$ and interpret.

68. The annual per capita consumption of light and skim milk can be modeled by

$$m(x) = 10.12 + 2 \ln x \qquad 1 \leq x \leq 16$$

where x is in years ($x = 1$ corresponds to January 1, 1980) and $m(x)$ is the annual per capita consumption in gallons (*Source:* U.S. Department of Agriculture, www.usda.gov).

(a) Determine the average value of m on the interval $[1, 11]$ and interpret.

(b) Determine the average value of m on the interval $[11, 16]$ and interpret.

69. The annual per capita consumption of cheese (excluding cottage cheese) is approximated by

$$h(x) = 17.13 + 3.4 \ln x \qquad 1 \leq x \leq 16$$

where x is in years ($x = 1$ corresponds to January 1, 1980) and $h(x)$ is the annual per capita consumption in pounds (*Source:* U.S. Department of Agriculture, www.usda.gov).

(a) Determine the average value of h on the interval $[1, 11]$ and interpret.

(b) Determine the average value of h on the interval $[11, 16]$ and interpret.

70. The amount of money spent annually for admission to spectator sports in the United States is modeled by

$$s(x) = 2.1 + 1.3 \ln x \qquad 1 \leq x \leq 12$$

where x is in years ($x = 1$ corresponds to January 1, 1985) and $s(x)$ is the amount spent in billions of dollars (*Source:* U.S. Bureau of Economic Analysis, www.bea.doc.gov).

(a) Determine the average value of s on the interval $[1, 12]$ and interpret.

(b) Evaluate $\int_1^{12} s(x)\,dx$ and interpret.

71. The variable cost (gas, oil, maintenance, and the like) per mile of owning and operating an automobile in the United States is approximated by

$$v(x) = 8.3 + 0.88 \ln x \qquad 1 \leq x \leq 6$$

where x is in years ($x = 1$ corresponds to January 1, 1990) and $v(x)$ is the variable cost in cents per mile. Determine the average value on the interval $[1, 6]$ and interpret (*Source:* U.S. Statistical Abstract, www.census.gov/statab/www).

72. The average annual salary for a high school principal in the United States can be modeled by

$$f(x) = 40.9 + 12.1 \ln x \qquad 1 \leq x \leq 15$$

where x is in years ($x = 1$ corresponds to January 1, 1985) and $f(x)$ is the average salary in thousands of dollars per year. Suppose that Fred is a high school principal who earned the average salary from January 1, 1985, to January 1, 2000 (*Source:* U.S. Statistical Abstract, www.census.gov/statab/www).

(a) Write an integral to determine Fred's total salary from January 1, 1985, to January 1, 2000.

(b) Evaluate the integral from part (a) and interpret.

SECTION PROJECT

The data in Table 17.4.1 give the average annual salaries, in thousands, for certain employees in the U.S. public school systems. The year column corresponds to January 1 of the given year.

Table 17.4.1

Year	Superintendent	Business Administrator	Teacher (K-12)
1985	56.954	40.344	23.587
1990	75.425	52.354	31.278
1994	87.717	59.997	36.531
1995	90.198	61.323	37.264
1996	94.229	63.840	38.706
1997	98.106	65.797	39.580
1998	101.519	67.724	40.133
1999	106.122	71.387	41.351

SOURCE: U.S. Statistical Abstract, www.census.gov/statab/www

(a) Let x represent years ($x = 1$ correspond to January 1, 1985) and y_1 represent the average salary in thousands of a superintendent. Determine a linear regression model for the data. Round coefficients to the nearest hundredth.

(b) Let x represent years ($x = 1$ correspond to January 1, 1985) and y_2 represent the average salary in thousands of a business administrator. Determine a linear regression model for the data. Round coefficients to the nearest hundredth.

(c) Let x represent years ($x = 1$ correspond to January 1, 1985) and y_3 represent the average salary in thousands of a teacher. Determine a natural logarithmic regression model for the data. Round coefficients to the nearest hundredth.

(d) Suppose that Christine is a superintendent who earned the average salary from January 1, 1985, to January 1, 2000. Write an integral to determine Christine's total salary from January 1, 1985, to January 1, 2000.

(e) Evaluate the integral from part (d) and interpret.

(f) Suppose that Joe is a business administrator who earned the average salary from January 1, 1985, to January 1, 2000. Write an integral to determine Joe's total salary from January 1, 1985, to January 1, 2000.

(g) Evaluate the integral from part (f) and interpret.

(h) Suppose that Rich is a teacher who earned the average salary from January 1, 1985, to January 1, 2000. Write an integral to determine Rich's total salary from January 1, 1985, to January 1, 2000.

(i) Evaluate the integral from part (h) and interpret.

Section 17.5 Numerical Integration

In Chapters 16 and 17 we presented several techniques of integration. In spite of all the varied techniques of integration, we still encounter integrals that cannot be antidifferentiated by any of our methods. For example, $\int e^{-x^2}\,dx$ has no elementary antiderivative and cannot be evaluated using the Fundamental Theorem of Calculus. (Incidentally, $\int e^{-x^2}\,dx$ is related to the bell-shaped curve seen in probability and statistics courses.)

So how do we evaluate $\int_0^1 e^{-x^2}\,dx$? We approximate it numerically by interpreting it as an area under a curve. We have already seen how to approximate a definite integral using rectangles. Hence, we are familiar with some numerical methods. In this section, we explore the Trapezoidal Rule as another numerical method. The main use, and beauty, of the Trapezoidal Rule is that it is ideally suited for discrete data.

Trapezoidal Rule

To gain an understanding of the Trapezoidal Rule, we need to review the process behind rectangular approximations.

Example 1 Reviewing Rectangular Approximations

Consider $f(x) = \frac{1}{3}x^3 - 2.5x^2 + 4x + 4$. Use four left-endpoint rectangles to approximate $\int_2^6 \left(\frac{1}{3}x^3 - 2.5x^2 + 4x + 4\right) dx$, that is, the area under the curve from $x = 2$ to $x = 6$.

Solution

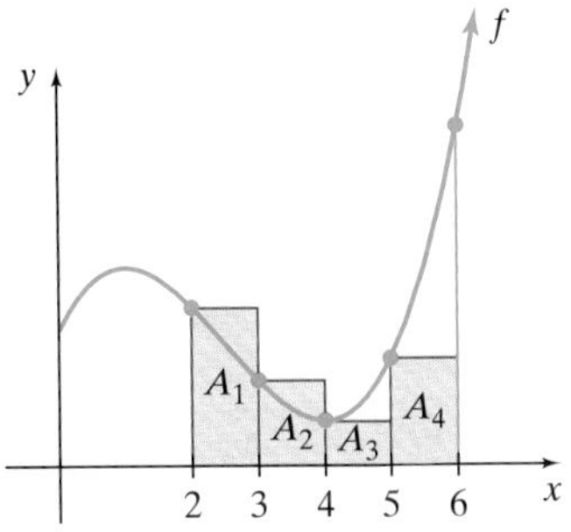

Figure 17.5.1

Figure 17.5.1 shows the graph of f along with the four rectangles. First, recall that:

- a is the left endpoint of the interval.
- b is the right endpoint of the interval.
- n is the number of rectangles.
- Δx is the width, or base, of each rectangle.

So here we have $a = 2$, $b = 6$, and $n = 4$. Then Δx is

$$\Delta x = \frac{b-a}{n} = \frac{6-2}{4} = 1$$

Each rectangle has a base $= \Delta x = 1$, and the height is given by f (left endpoint). With this in mind, we find the area of rectangle A_1 to be

$$\begin{aligned} A_1 &= \text{base} \cdot \text{height} \\ &= \Delta x \cdot f(\text{left endpoint}) \\ &= 1 \cdot f(2) = 1 \cdot \frac{14}{3} = \frac{14}{3} \end{aligned}$$

For rectangle A_2 we get

$$A_2 = \Delta x \cdot f(\text{left endpoint}) = 1 \cdot f(3) = 1 \cdot \frac{5}{2} = \frac{5}{2}$$

In a similar fashion we compute the area of rectangles A_3 and A_4 to be

$$A_3 = \frac{4}{3} \quad \text{and} \quad A_4 = \frac{19}{6}$$

The sum of the four rectangles gives the rectangular approximation. This yields

$$\begin{aligned} \int_2^6 \left(\frac{1}{3}x^3 - 2.5x^2 + 4x + 4\right) dx &\approx A_1 + A_2 + A_3 + A_4 \\ &= \frac{14}{3} + \frac{5}{2} + \frac{4}{3} + \frac{19}{6} = \frac{35}{3} \end{aligned}$$

■

Interactive Activity

Evaluate $\int_2^6 \left(\frac{1}{3}x^3 - 2.5x^2 + 4x + 4\right) dx$ by determining an antiderivative and using the Fundamental Theorem of Calculus.

WWW

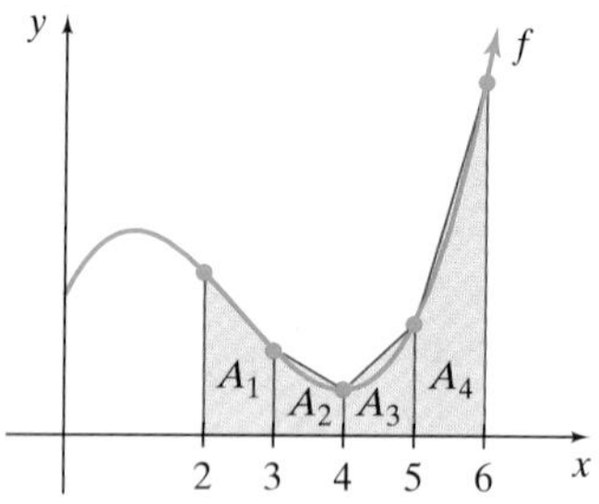

Figure 17.5.2

Recall that in Chapter 16 we improved on this approximation by placing smaller and smaller rectangles under the curve. Now we will investigate another numerical technique that may yield a better approximation with the same value for n. To do this we will use *trapezoids* rather than rectangles.

Let's look at Figure 17.5.2. Here we have redrawn $f(x) = \frac{1}{3}x^3 - 2.5x^2 + 4x + 4$, but instead of four rectangles we have drawn four trapezoids. It appears that the four trapezoids will yield a better approximation than the four rectangles did. All we need is to recall the formula for the area of a trapezoid:

$$A_{\text{trap}} = \frac{1}{2}(h_1 + h_2)b$$

where h_1 and h_2 represents the two heights of a trapezoid and b represents the base. *Remember that h_1 is parallel to h_2.* Let's rework the function in Example 1 with four trapezoids.

Example 2 Approximating an Integral with Trapezoids

Consider $f(x) = \frac{1}{3}x^3 - 2.5x^2 + 4x + 4$ from Example 1. Use four trapezoids to approximate $\int_2^6 f(x)\,dx$, that is, the area under the graph of f from $x = 2$ to $x = 6$.

Solution

Just as with rectangles we have $a = 2$, $b = 6$, $n = 4$, and $\Delta x = \dfrac{b-a}{n} = \dfrac{6-2}{2} = 1$. So each trapezoid has a base $= \Delta x = 1$. Notice that h_1 and h_2 are found to be f (left endpoint) and f (right endpoint), respectively. The areas of the four trapezoids are

$$\begin{aligned}
A_1 &= \frac{1}{2}(h_1 + h_2)b \\
&= \frac{1}{2}[f(\text{left endpoint}) + f(\text{right endpoint})] \cdot b \\
&= \frac{1}{2}[f(2) + f(3)] \cdot \Delta x \\
&= \frac{1}{2}\left[\frac{14}{3} + \frac{5}{2}\right] \cdot 1 = \frac{43}{12}
\end{aligned}$$

$$\begin{aligned}
A_2 &= \frac{1}{2}(h_1 + h_2)b \\
&= \frac{1}{2}[f(\text{left endpoint}) + f(\text{right endpoint})] \cdot b \\
&= \frac{1}{2}[f(3) + f(4)] \cdot \Delta x \\
&= \frac{1}{2}\left[\frac{5}{2} + \frac{4}{3}\right] \cdot 1 = \frac{23}{12}
\end{aligned}$$

$$\begin{aligned} A_3 &= \frac{1}{2}(h_1 + h_2)b \\ &= \frac{1}{2}[f(\text{left endpoint}) + f(\text{right endpoint})] \cdot b \\ &= \frac{1}{2}[f(4) + f(5)] \cdot \Delta x \\ &= \frac{1}{2}\left[\frac{4}{3} + \frac{19}{6}\right] \cdot 1 \\ &= \frac{9}{4} \end{aligned}$$

$$\begin{aligned} A_4 &= \frac{1}{2}(h_1 + h_2)b \\ &= \frac{1}{2}[f(\text{left endpoint}) + f(\text{right endpoint})] \cdot b \\ &= \frac{1}{2}[f(5) + f(6)] \cdot \Delta x \\ &= \frac{1}{2}\left[\frac{19}{6} + 10\right] \cdot 1 \\ &= \frac{79}{12} \end{aligned}$$

The sum of the four trapezoids gives the approximation

$$\begin{aligned} \int_2^6 \left(\frac{1}{3}x^3 - 2.5x^2 + 4x + 4\right) dx &\approx A_1 + A_2 + A_3 + A_4 \\ &= \frac{43}{12} + \frac{23}{12} + \frac{9}{4} + \frac{79}{12} = \frac{43}{3} \end{aligned}$$ ■

As you discovered in the Interactive Activity following Example 1, the exact answer to $\int_2^6 \left(\frac{1}{3}x^3 - 2.5x^2 + 4x + 4\right) dx$ is $\frac{40}{3}$. We can see that four trapezoids is more accurate than four rectangles. Also notice that, just as we did with rectangular approximations, our subintervals are of equal length when using trapezoids.

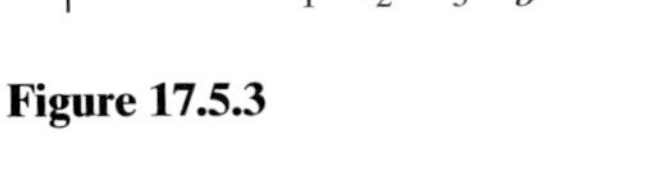

Figure 17.5.3

We want to look back at our work in Example 2 and refer to Figure 17.5.3 in order to generalize the trapezoid process. To find the area of the trapezoids, we calculated

$$A_1 = \frac{1}{2}[f(a) + f(x_1)] \cdot \Delta x$$

$$A_2 = \frac{1}{2}[f(x_1) + f(x_2)] \cdot \Delta x$$

$$A_3 = \frac{1}{2}[f(x_2) + f(x_3)] \cdot \Delta x$$

$$A_4 = \frac{1}{2}[f(x_3) + f(b)] \cdot \Delta x$$

The total area of the four trapezoids is the sum

$$A_1 + A_2 + A_3 + A_4 = \frac{1}{2}[f(a) + f(x_1)] \cdot \Delta x + \frac{1}{2}[f(x_1) + f(x_2)] \cdot \Delta x + \frac{1}{2}[f(x_2) + f(x_3)] \cdot \Delta x + \frac{1}{2}[f(x_3) + f(b)] \cdot \Delta x$$

Factoring out a $\frac{1}{2}\Delta x$ yields

$$= \frac{\Delta x}{2}[f(a) + f(x_1) + f(x_1) + f(x_2) + f(x_2) + f(x_3) + f(x_3) + f(b)]$$

$$= \frac{\Delta x}{2}[f(a) + 2f(x_1) + 2f(x_2) + 2f(x_3) + f(b)]$$

The last equation is the Trapezoidal Rule. In general, if n trapezoids are used, then the approximate value of the definite integral is given by the trapezoidal rule, as follows:

Trapezoidal Rule

If f is continuous on $[a, b]$, then

$$\int_a^b f(x)\,dx \approx \frac{\Delta x}{2}[f(a) + 2f(x_1) + 2f(x_2) + \cdots + 2f(x_{n-1}) + f(b)]$$

where a is the left endpoint, b is the right endpoint, n is the number of trapezoids, and $\Delta x = \dfrac{b-a}{n}$. Notice that $x_1 = a + \Delta x$, $x_2 = x_1 + \Delta x$, and so on.

As with rectangular approximations, the larger the value of n the more trapezoids we will use and the smaller Δx will be. Also, the result of using larger values for n is a better approximation to the definite integral.

Note: There is some error involved in the trapezoidal approximation. The formula to compute the error is beyond the scope of this text. However, it is worth noting that doubling the number of trapezoids reduces the maximum error by a factor of 4.

Example 3 Using the Trapezoidal Rule

Use the trapezoidal rule with four trapezoids to approximate $\displaystyle\int_0^2 \frac{1}{x^2+16}\,dx$.

Solution

We have $a = 0, b = 2, n = 4$, and $\Delta x = \dfrac{2-0}{4} = \dfrac{1}{2}$. So by the trapezoid rule we have

$$\int_0^2 \frac{1}{x^2+16}\,dx \approx \frac{\frac{1}{2}}{2}\left[f(0) + 2f\left(\frac{1}{2}\right) + 2f(1) + 2f\left(\frac{3}{2}\right) + f(2)\right]$$

$$\approx \frac{1}{4}[0.0625 + 0.1230 + 0.1176 + 0.1096 + 0.05] \approx 0.1157$$

See Figure 17.5.4.

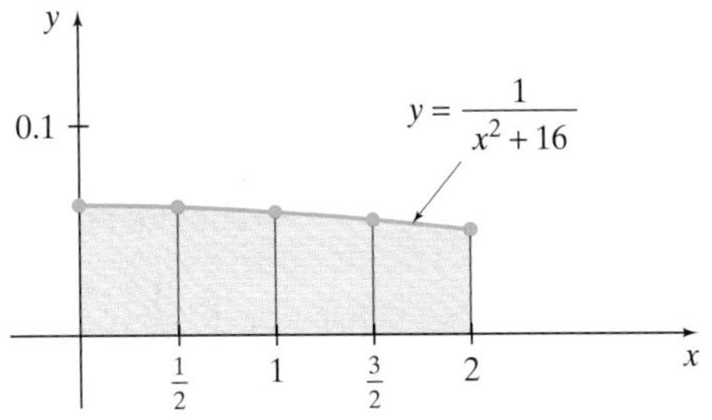

Figure 17.5.4

■

Checkpoint 1 Now work Exercise 3.

Technology Option

Because of the structure of the Trapezoid Rule, it can be easily programmed into our calculator (or computer). We supply the following program for the TI-83. As always, if you do not have a TI-83, your code may be different. For this reason, we also supply a line-by-line explanation of what the code is performing. Also, consult www.prenhall.com/armstrong for the program for your calculator.

■ TI-83 Program for Trapezoid Rule When Given $f(x)$

Disp"ENTER FUNCTION IN Y_1"	Stores function in Y_1.
Prompt A, B, N	Asks for A, B, and N, the left endpoint, right endpoint, and number of trapezoids.
$(B - A)/N \to D$	Computes the base of each trapezoid.
$Y_1(A) + Y_1(B) \to S$	Evaluates function in Y_1 at A and B; adds the results and stores sum in S.
$A + D \to X$	Left endpoint plus base stored in X.
Lbl 1	Beginning of loop.
$2*Y_1(X) + S \to S$	Two times function evaluated at X; result added to sum in S; result stored in S.
$X + D \to X$	Current X increased by base; result in X.
If $X < B$	If current $X <$ right endpoint, execute next line.
Then	Prior line true; execute next line.
Goto 1	Return to beginning of loop.
Else	If $X \geq B$, execute next line.
$S*(D/2) \to S$	Current sum in S multiplied by D/2; result in S.
Disp"TRAP APPROX IS", S	Display trapezoid approximation.

To use the trapezoidal rule program to approximate $\int_{-1}^{1} e^{-x^2}\,dx$ using $n = 10$, 50, and 100 trapezoids, we first enter $y_1 = e^{-x^2}$ into our calculator. Next we execute the program. Results are listed in Table 17.5.1. Note that $a = -1$ and $b = 1$. We conclude that $\int_{-1}^{1} e^{-x^2}\,dx \approx 1.49$.

Table 17.5.1

n	Trapezoidal Approximation to $\int_{-1}^{1} e^{-x^2}\,dx$
10	1.48873668
50	1.493452053
100	1.493599214

Trapezoidal Rule for Discrete Data

We have now seen how to use the Trapezoidal Rule to approximate $\int_a^b f(x)\,dx$. Many times, however, we have a set of data and we cannot determine a model for the data. In other words, we have no $f(x)$. In these circumstances, the trapezoidal rule is extremely powerful, as shown in Example 4.

Example 4 Using the Trapezoidal Rule for Discrete Data

Table 17.5.2

x	Rate of Change of Revenue
1	3.3
2	2
3	2.5
4	2.8
5	2.9
6	2.2
7	2

Table 17.5.2 shows the rate of change of revenue in millions of dollars per year for a bowling alley, where x represents the number of years since 1989.

(a) Let $x = 1$ correspond to January 1, 1990, and y represent the rate of change of revenue in millions of dollars per year. Plot the data.

(b) Approximate the bowling alley's revenue from January 1, 1990, to January 1, 1996, by using the trapezoid rule with the given data.

Solution

(a) Figure 17.5.5 gives a plot of the data.

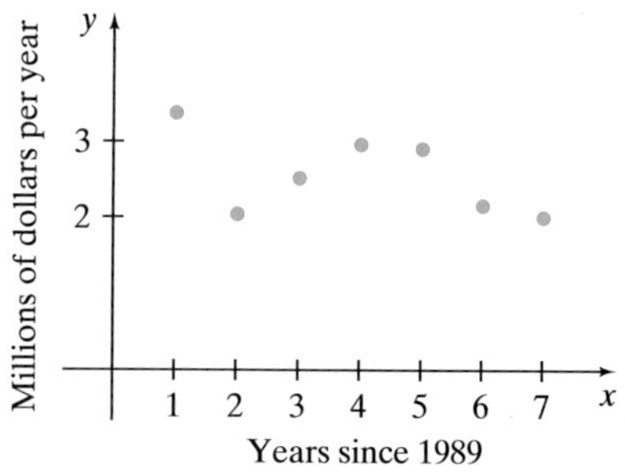

Figure 17.5.5

(b) Understand the Situation: Figure 17.5.6 is a plot of the data with the trapezoids sketched in as well. Here is where the power of the trapezoid rule comes through. Without a model for the data, we can use the data as is and

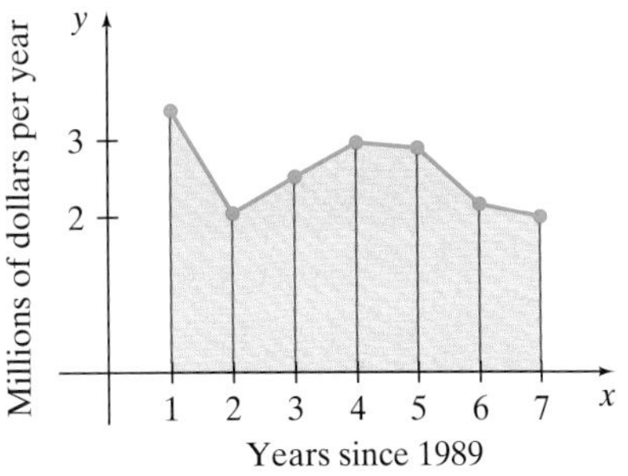

Figure 17.5.6

approximate the revenue. To use the trapezoid rule, note that $a = 1$, $b = 7$, $n = 6$, and $\Delta x = 1$. So by the trapezoid rule we have

$$\frac{1}{2}[f(1) + 2f(2) + 2f(3) + 2f(4) + 2f(5) + 2f(6) + f(7)]$$

Now, instead of a formula for our function, we use the functional values from the data in our table! For example, from the table we know that $f(1) = 3.3$, $f(2) = 2$, and so on.

Proceeding, we get the approximation to be

$$\frac{1}{2}[3.3 + 2(2) + 2(2.5) + 2(2.8) + 2(2.9) + 2(2.2) + 2] = 15.05$$

So the bowling alley's revenue from January 1, 1990, to January 1, 1996, was approximately 15.05 million dollars. ■

Checkpoint 2 Now work Exercise 23.

Technology Option

Applying the trapezoidal rule to a set of data can be a bit tedious, especially if there are many data points. We supply the following program that performs the trapezoidal rule on a set of data. Again, the code is for a TI-83 and the program makes use of Lists. In L_1 (List 1) we shall place the values for the independent variable, while in L_2 (List 2) we place the values of the dependent variable. Consult the online graphing calculator manual at www.prenhall.com/armstrong to learn more about LISTS.

TI-83 Program for the Trapezoidal Rule When Given Data Points

Disp "NUMBER DATA POINTS"	Asks for the number of data points.
Prompt P	Enter number of data points.
$(L_1(P) - L_1(1))/(P - 1) \rightarrow D$	Computes the base of each trapezoid; stores in D.
$L_2(1) + L_2(P) \rightarrow S$	Sums first and last dependent variable values; stores sum in S.

$2 \to I$	2 stored in I.
Lbl 1	Beginning of loop.
$2^*L_2(I) + S \to S$	Two times dependent variable value in current I added to value in S; result stored in S.
$I + 1 \to I$	Increment size of I.
If $I < P$	If updated $I < P$, execute next line.
Then	If previous line true, execute next line.
Goto 1	Return to beginning of loop.
Else	If $I \geq P$, execute next line.
$S^*(D/2) \to S$	Current value in S times (D/2) stored in S.
Disp"TRAP APPROX IS",S	Display trapezoidal approximation.

Table 17.5.3

x	Rate of Change of Revenue
1	2.5
2	2.8
3	1.2
4	0.9
5	2.7
6	2.1
7	3.1

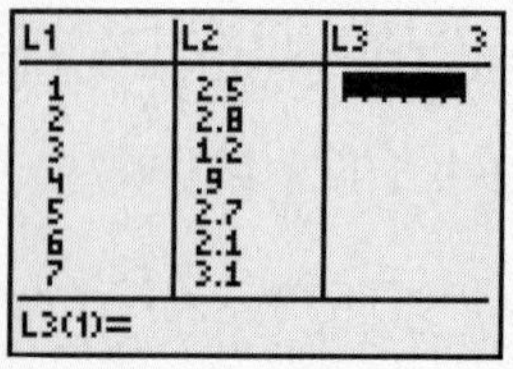

Figure 17.5.7

Table 17.5.3 shows the rate of change of revenue in millions of dollars per year for a restaurant, where x represents the number of years since 1989. To approximate the restaurant's revenue from 1990 to 1996 using the trapezoidal rule program for data points, we first enter the data in the x column into L_1 and the rate of change of revenue column into L_2. Figure 17.5.7 shows our screen after doing this. We now run our trapezoidal rule program for data points. We are first asked how many data points and we enter a 7. The program returns 12.5. So we conclude that from 1990 to 1996 the restaurant's revenue was approximately 12.5 million dollars.

Note: Both trapezoidal rule programs are designed for subintervals of equal length. When using the trapezoidal rule program for data points, the independent variable values should be equally spaced.

Example 5 Using the Trapezoidal Rule to Analyze Sulfur Dioxide Emissions

Table 17.5.4 shows the U.S. sulfur dioxide emissions in thousands of tons from 1940 to 1990. Figure 17.5.8 gives a plot of the data, where $x = 1$ corresponds to

Table 17.5.4

x	Sulfur Dioxide Emissions
1	19,953
11	22,358
21	22,227
31	31,161
41	25,905
51	22,433

SOURCE: U.S. Environmental Protection Agency, www.epa.gov

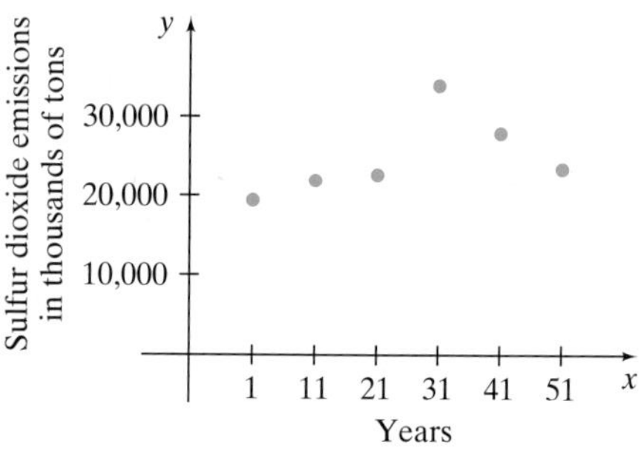

Figure 17.5.8

the beginning of 1940 and y represents the sulfur dioxide emissions in thousands of tons.

(a) Approximate the total amount of sulfur dioxide emissions by the United States from 1940 to 1990.

(b) Approximate the average amount of sulfur dioxide emissions on a yearly basis from 1940 to 1990.

Solution

(a) **Understand the Situation:** By looking at Figure 17.5.8 we see that in order to use the trapezoidal rule, we have $a = 1, b = 51, n = 5$ and $\Delta x = 10$. So by the trapezoidal rule we have

$$\frac{1}{2}[f(1) + 2f(11) + 2f(21) + 2f(31) + 2f(41) + f(51)] \cdot \Delta x$$

As in Example 4, we use the functional values from Table 17.5.4 to determine that $f(1) = 19{,}953$, $f(11) = 22{,}358$ and so on. Proceeding we get the approximation to be

$$\frac{1}{2}[19{,}953 + 2(22{,}358) + 2(22{,}227) + 2(31{,}161) + 2(25{,}905) + 22{,}433] \cdot 10$$

$$= 1{,}228{,}440$$

Interpret the Solution: This means that from 1940 to 1990 the amount of U.S. sulfur dioxide emissions was approximately 1,228,440 thousand tons or 1,228,440,000 tons.

(b) To compute this average value, we take the result from part (a) and divide by $b - a$, just as we learned in Section 17.1. This yields

$$\frac{1{,}228{,}440{,}000}{51 - 1} = \frac{1{,}228{,}440{,}000}{50}$$

$$= 24{,}568{,}800$$

We conclude that from 1940 to 1990 the annual average value of the U.S. sulfur dioxide emissions was approximately 24,568,800 tons. ■

✓ Checkpoint 3

Now work Exercise 35.

Simpson's Rule

Another popular numerical integration technique is called **Simpson's Rule**. Simpson's Rule utilizes adjacent parabolic segments over a closed interval $[a, b]$. The sum of the areas below these parabolic segments approximates $\int_a^b f(x)\,dx$. Approximations using Simpson's Rule are very accurate, but it requires an *even* number n of equal subintervals. This could be limiting if our data do not have an even number of data points. For the curious, Simpson's Rule can be found in almost any engineering calculus textbook. We will not pursue this numerical technique in this text.

SUMMARY

In this section we introduced our final numerical technique for integration, the **trapezoidal rule**. At this time the numerical techniques that we have discussed include:

- Rectangular approximations
- Built-in numerical integration on the calculator
- Trapezoidal approximations

We supplied two programs for the trapezoidal rule: one if we are given a function and another if we do not have a function, or we cannot determine one, for a set of data. Applications utilizing the trapezoidal rule for data points were discussed. We want to stress that if some data are *not* easily modeled, the trapezoidal rule is a powerful weapon to analyze the data. The Trapezoidal Rule assumes that the independent variable values of the data points are equally spaced.

SECTION 17.5 EXERCISES

In Exercises 1–10, use the Trapezoidal Rule to approximate the given definite integral using $n = 4$ trapezoids.

1. $\int_{-1}^{3} x^3\,dx$ **2.** $\int_{0}^{2} x^4\,dx$

3. $\int_{1}^{3} \frac{5x}{5+2x^2}\,dx$ **4.** $\int_{0}^{2} \frac{1}{(1+2x)^2}\,dx$

5. $\int_{0}^{1} x^2e^x\,dx$ **6.** $\int_{0}^{1} x^5e^x\,dx$

7. $\int_{3}^{5} (\ln x)^2\,dx$ **8.** $\int_{2}^{5} (\ln x)^3\,dx$

9. $\int_{0}^{2} x^4\,dx$ **10.** $\int_{-1}^{3} x^3\,dx$

In Exercises 11–22, use the calculator program for the Trapezoidal Rule when given $f(x)$ to approximate the given definite integral using $n = 10$, $n = 50$, and $n = 100$ trapezoids.

11. $\int_{0}^{2} \frac{1}{(1+2x)^2}\,dx$ **12.** $\int_{1}^{3} \frac{5x}{5+2x^2}\,dx$

13. $\int_{1}^{4} \frac{e^{2x}}{1+e^{4x}}\,dx$ **14.** $\int_{1}^{2} \frac{e^x-e^{-x}}{2}\,dx$

15. $\int_{1}^{2} \frac{e^x+e^{-x}}{2}\,dx$ **16.** $\int_{0}^{1} x^2e^x\,dx$

17. $\int_{0}^{1} x^5e^x\,dx$ **18.** $\int_{3}^{5} (\ln x)^2\,dx$

19. $\int_{2}^{5} (\ln x)^3\,dx$ **20.** $\int_{0}^{2} \frac{2x+3}{\sqrt{x^2+2x+5}}\,dx$

21. $\int_{1}^{3} \frac{x+2}{\sqrt[3]{(6x-x^2)}}\,dx$ **22.** $\int_{0}^{3} e^{x^2}\,dx$

Applications

In Exercises 23–32, use the Trapezoidal Rule.

23. The following data give the rate of change of revenue in thousands of dollars per week for Sparkle Time Car Wash. Here x represents the number of weeks after April 1. Approximate Sparkle Time's revenue for this period of time.

x	Rate of Change of Revenue
1	2.35
2	3.75
3	1.25
4	2.5
5	1.75

24. The following data give the rate of change of sales, in millions of dollars per month, for Spring Brook Mall. Here x corresponds to the month of the year. Approximate Spring Brook Mall's total sales for this period of time.

x	Rate of Change of Sales
1	3.35
2	2.69
3	1.72
4	1.11
5	2.35
6	2.51

25. The following data give the marginal cost for different levels of production at Muggy Inc. Here x represents the number of coffee mugs produced and $C'(x)$ is in dollars per mug. Approximate the total cost in going from a production level of 10 mugs to 50 mugs.

x	10	20	30	40	50
$C'(x)$	1.5	2	2.4	3.1	3.5

26. The following data give the marginal cost for different levels of production at Binky Inc. Here x represents the number of pacifiers produced and $C'(x)$ is in dollars per pacifier. Approximate the total cost in going from a production level of 220 pacifiers to 300 pacifiers.

x	220	240	260	280	300
$C'(x)$	1.18	1.25	1.36	1.48	1.55

27. The Beef Place keeps a monthly record of the rate of increase in profits, $P'(x)$, during the first 5 months of the year. The following are the data where x is the month of the year and $P'(x)$ is measured in thousands of dollars per month. Approximate the total profit during this period of time.

x	1	2	3	4	5
$P'(x)$	0.62	0.77	0.78	0.85	1.03

28. The amount of energy produced, in quadrillion Btu, by the United States from the beginning of 1985 to the beginning of 1993 is listed in the following table. Here $x = 1$ corresponds to the beginning of 1985.

x	Btu Produced
1	84.55
3	84.73
5	86.85
7	89.31
9	88.52

SOURCE: U.S. Census Bureau, www.census.gov

Approximate the total amount of energy produced by the United States from the beginning of 1985 to the beginning of 1993.

29. The following data give the rate of change of revenue, in thousands of dollars per week, for Timber!, a local tree trimming company. Here x represents the number of weeks after May 1. Approximate the company's revenue for this time period.

x	Rate of Change of Revenue	x	Rate of Change of Revenue
1	2.25	6	2.12
2	3.75	7	3.11
3	1.35	8	2.11
4	2.55	9	1.50
5	1.85	10	3.12

30. Anthony has been keeping a monthly record of a company's rate of increase in profits, $P'(x)$, over a 1-year period. The following are the data that Anthony collected, where x is the month of the year and $P'(x)$ is in thousands of dollars per month. Approximate the total profit for this time period.

x	$P'(x)$	x	$P'(x)$
1	6.2	7	10.7
2	7.7	8	11.1
3	7.8	9	12.1
4	8.5	10	12.3
5	10.3	11	12.1
6	10.2	12	12.5

31. Tina has been asked to determine the marginal cost for different levels of production. She collected the following data, where x represents the number of items produced and $C'(x)$ is in hundreds of dollars per item. Use the data that Tina collected to determine the total cost of producing 20 items.

x	0	4	8	12	16	20
$C'(x)$	7	13	18	24	31	39

32. Vanessa is the leader of a sales management team for a company, and her team has been asked to prepare a year-end report. Vanessa's team gathered data showing the rate of change of sales, $s'(t)$, in millions of dollars per month, for the prior year, where t is measured in months. The following are the data that were collected.

t	$s'(t)$
0	0.8
1	0.7
2	1.13
3	1.44
4	0.68
5	0.22
6	1.01
7	1.51
8	2.13
9	1.78
10	1.42
11	0.77
12	0.96

Use the data that Vanessa's team gathered to approximate the total sales for the year.

For the remainder of the exercises, use the calculator program for the Trapezoidal Rule when given data points, where appropriate.

33. The following data give the number of females, single and married, in the U.S. labor force, in thousands, from 1960 to 1995. Here x is in years where $x = 1$ corresponds to the beginning of 1960.

x	Single	Married
1	5,410	12,893
6	5,976	14,829
11	7,265	18,475
16	9,125	21,484
21	11,865	24,980
26	13,163	27,894
31	14,612	30,901
36	15,467	33,359

SOURCE: U.S. Bureau of Labor Statistics, www.bls.gov

(a) Approximate the average number of single females, on a yearly basis, in the U.S. labor force from the beginning of 1960 to the beginning of 1995.

(b) Approximate the average number of married females, on a yearly basis, in the U.S. labor force from the beginning of 1960 to the beginning of 1995.

In Exercises 34, 35, and 36, use the following data for U.S. air pollution emissions from 1940 to 1990 measured in thousands of tons per year. Here x is in years where $x = 1$ corresponds to the beginning of 1940.

x	Nitrogen Dioxides	Volatile Organic Compounds	Carbon Monoxide
1	7,374	17,161	93,615
11	10,093	20,936	102,609
21	14,140	24,459	109,745
31	20,625	30,646	128,079
41	23,281	25,893	115,625
51	23,038	23,599	100,650

SOURCE: U.S. Environmental Protection Agency, www.epa.gov

34. (a) Approximate the total amount of nitrogen dioxide emissions in the United States from the beginning of 1940 to the beginning of 1990.

(b) Approximate the average amount of nitrogen oxide emissions on a yearly basis from the beginning of 1940 to the beginning of 1990.

35. (a) Approximate the total amount of volatile organic compounds emissions in the United States from the beginning of 1940 to the beginning of 1990.

(b) Approximate the average amount of volatile organic compounds on a yearly basis from the beginning of 1940 to the beginning of 1990.

36. (a) Approximate the total amount of carbon monoxide emissions in the United States from the beginning of 1940 to the beginning of 1990.

(b) Approximate the average amount of carbon monoxide emissions on a yearly basis from the beginning of 1940 to the beginning of 1990.

In Exercises 37 and 38, use the following data on work stoppages in the United States from 1960 to 1995. Here x is in years where $x = 1$ corresponds to the beginning of 1960, and the number of workers involved are in thousands.

x	Number of Work Stoppages	Number of Workers Involved
1	222	896
6	268	999
11	381	2468
16	235	965
21	187	795
26	54	324
31	44	185
36	31	192

SOURCE: U.S. Statistical Abstract, www.census.gov/statab/www

37. (a) Approximate the total number of work stoppages in the United States from the beginning of 1960 to the beginning of 1995.

(b) Approximate the average number of work stoppages, on a yearly basis, in the United States from the beginning of 1960 to the beginning of 1995.

38. (a) Approximate the total number of workers involved in work stoppages in the United States from the beginning of 1960 to the beginning of 1995.

(b) Approximate the average number of workers involved in work stoppages, on a yearly basis, in the United States from the beginning of 1960 to the beginning of 1995.

In Exercises 39 and 40, use the following data for the number of high school graduates, male and female, in thousands from 1960 to 1992. Here x is in years where $x = 1$ corresponds to the beginning of 1960.

x	Males	Females	x	Males	Females
1	756	923	21	1500	1589
5	997	1148	25	1429	1583
9	1184	1422	29	1334	1339
13	1420	1541	33	1216	1182
17	1450	1537			

SOURCE: U.S. Census Bureau, www.census.gov

39. (a) Approximate the total number of males who graduated from high school from the beginning of 1960 to the beginning of 1992.

(b) Approximate the average number of males who graduated from high school, on a yearly basis, from the beginning of 1960 to the beginning of 1992.

40. (a) Approximate the total number of females who graduated from high school from the beginning of 1960 to the beginning of 1992.

(b) Approximate the average number of females who graduated from high school, on a yearly basis, from the beginning of 1960 to the beginning of 1992.

41. The following data give the cost in constant 1987 dollars for the regulation and monitoring of pollution abatement in the United States. Here x is in years where $x = 1$ corresponds to the beginning of 1985, and the regulation and monitoring costs are in billions of constant 1987 dollars.

x	1	3	5	7	9
Costs	1.361	1.519	1.657	1.654	1.656

SOURCE: U.S. Bureau of Economic Analysis, www.bea.doc.gov

Approximate the total cost in constant 1987 dollars spent on regulation and monitoring from the beginning of 1985 to the beginning of 1993.

42. Crude oil imports, in millions of barrels per year, into the United States from 1970 to 1995 are given in the following table. Here x is in years where $x = 1$ corresponds to the beginning of 1970.

x	Number of Barrels (in millions)
1	483
6	1498
11	1926
16	1168
21	2151
26	2608

SOURCE: U. S. Energy Information Administration, www.eia.doe.gov

(a) Approximate the total number of barrels of oil imported into the United States from the beginning of 1970 to the beginning of 1995.

(b) Approximate the average number of barrels, on a yearly basis, of oil imported into the United States from the beginning of 1970 to the beginning of 1995.

(c) A barrel contains 42 gallons. How many total gallons of oil were imported into the United States from the beginning of 1970 to the beginning of 1995?

43. The following data give the number of pregnancies in the United States from 1976 to 1996. Here x is in years where $x = 1$ corresponds to the beginning of 1976, and the number of pregnancies is in millions.

x	Number	x	Number
1	5.002	13	6.341
3	5.433	15	6.668
5	5.912	17	6.484
7	6.024	19	6.373
9	6.019	21	6.240
11	6.129		

SOURCE: U.S. National Center for Health Statistics, www.cdc.gov/nchs

(a) Approximate the total number of pregnancies in the United States from the beginning of 1976 to the beginning of 1996.

(b) Approximate the average number of pregnancies, on a yearly basis, in the United States from the beginning of 1976 to the beginning of 1996.

In Exercises 44 and 45, use the following data for the first-year enrollment of students pursuing a registered nursing degree, in thousands, where the degree is either a 4-year baccalaureate or a 2-year associate degree. Here x is in years where $x = 1$ corresponds to the beginning of 1985.

x	4-year Degree	2-year Degree
1	39.573	63.776
3	28.026	54.330
5	29.042	63.973
7	33.437	69.869
9	41.290	75.382
11	43.451	76.016

SOURCE: U.S. Statistical Abstract, www.census.gov/statab/www

44. Approximate the average number of first-year registered nursing students in a 4-year baccalaureate degree program from the beginning of 1985 to the beginning of 1995.

45. Approximate the average number of first-year registered nursing students in a 2-year associate degree program from the beginning of 1985 to the beginning of 1995.

SECTION PROJECT

The following data give the average tuition and fees as well as room and board charges, in dollars, from 1985 to 1999 for public and private 2- and 4-year colleges. Here x is in years where $x = 1$ corresponds to the beginning of 1985.

	Public			
	Tuition and Fees		Room and Board	
x	2-Year Colleges	4-Year Colleges	2-Year Colleges	4-Year Colleges
1	584	1386	2223	2513
3	660	1651	2328	2819
5	730	1846	2453	3059
7	824	2159	2644	3425
9	1025	2604	2774	3838
11	1194	2982	2955	4100
13	1276	3323	3128	4469
15	1328	3644	3293	4985

SOURCE: U.S. National Center for Education Statistics, www.nces.ed.gov

	Private			
	Tuition and Fees		Room and Board	
x	2-Year Colleges	4-Year Colleges	2-Year Colleges	4-Year Colleges
1	3485	6843	2718	3400
3	3684	8118	2700	4160
5	4817	9451	3149	4622
7	5570	11,379	3733	5124
9	6059	13,055	3845	5843
11	6865	14,510	4194	6500
13	7236	16,552	4718	6968
15	7815	18,237	5437	7106

SOURCE: U.S. National Center for Education Statistics, www.nces.ed.gov

(a) Approximate the average tuition and fees on a yearly basis from 1985 to 1999 at public 2-year colleges.

(b) Approximate the average tuition and fees on a yearly basis from 1985 to 1999 at private 2-year colleges.

(c) Approximate the average room and board on a yearly basis from 1985 to 1999 at public 2-year colleges.

(d) Approximate the average room and board on a yearly basis from 1985 to 1999 at private 2-year colleges.

(e) Approximate the average tuition and fees on a yearly basis from 1985 to 1999 at public 4-year colleges.

(f) Approximate the average tuition and fees on a yearly basis from 1985 to 1999 at private 4-year colleges.

(g) Approximate the average room and board on a yearly basis from 1985 to 1999 at public 4-year colleges.

(h) Approximate the average room and board on a yearly basis from 1985 to 1999 at private 4-year colleges.

(i) Assume the average yearly amount determined in parts (e) and (f) for a 4-year degree. How much more does it cost in tuition and fees for 4 years at a private 4-year college than at a public 4-year college?

(j) Assume the average yearly amount determined in parts (a) and (b) for a 2-year degree. How much more does it cost in tuition and fees for 2 years at a private 2-year college than at a public 2-year college?

Section 17.6 Improper Integrals

So far all the definite integrals we have considered have been on a closed bounded interval $[a, b]$. In this section, we relax this condition and define integrals over an unbounded interval such as $(-\infty, b]$, $[a, \infty)$, or even $(-\infty, \infty)$. Integrals defined on unbounded intervals are called **improper integrals** and have many applications. To understand what is happening with improper integrals, we need a brief review of limits at infinity.

Limits at Infinity

In Chapter 12 we discussed the limit concept. We saw that the definitions of the derivative and the definite integral demonstrate how central the limit concept is to calculus. Let's revisit limits at infinity and recall some concepts from our work in Chapter 12. Consult the Toolbox to the left.

Let's quickly review some limits. Remembering that the limit of a constant is that constant yields

$$\lim_{b\to\infty}\left(3-\frac{1}{b^2}\right) = \lim_{b\to\infty} 3 - \lim_{b\to\infty}\frac{1}{b^2} = 3 - 0 = 3$$

Employing other results in the Toolbox gives

$$\lim_{b\to\infty}(e^{-2b} - 3) = \lim_{b\to\infty} e^{-2b} - \lim_{b\to\infty} 3 = 0 - 3 = -3$$

From Your Toolbox

- For $f(x) = b^x$, b a real number where $b > 1$,
  (a) $\lim_{x\to\infty} b^x = \infty$ and $\lim_{x\to-\infty} b^x = 0$.
  (b) $\lim_{x\to\infty} b^{-x} = 0$ and $\lim_{x\to-\infty} b^{-x} = \infty$.
- $\lim_{x\to\infty} \ln x = \infty$.
- $\lim_{x\to\infty} \frac{k}{x^n} = 0$ and $\lim_{x\to-\infty} \frac{k}{x^n} = 0$, where k is any real constant and n is a positive real number.

Improper Integrals

At this time one may be wondering, "Why are we discussing limits at this point in the book?" The answer becomes apparent as we consider Examples 1 and 2.

Example 1 **Evaluating Definite Integrals Using the Fundamental Theorem of Calculus**

Consider $f(x) = \dfrac{1}{x^3}$. Determine the following.

(a) $\displaystyle\int_1^5 \frac{1}{x^3}\,dx$ **(b)** $\displaystyle\int_1^{10} \frac{1}{x^3}\,dx$ **(c)** $\displaystyle\int_1^b \frac{1}{x^3}\,dx$

Solution

(a) Evaluation of $\int_1^5 \frac{1}{x^3}\,dx$ will give us the area of the shaded region shown in Figure 17.6.1.

$$\int_1^5 \frac{1}{x^3}\,dx = \int_1^5 x^{-3}\,dx = -\frac{1}{2}x^{-2}\Big|_1^5 = -\frac{1}{2x^2}\Big|_1^5$$

$$= -\frac{1}{2(5)^2} - -\frac{1}{2(1)^2} = \frac{24}{50}$$

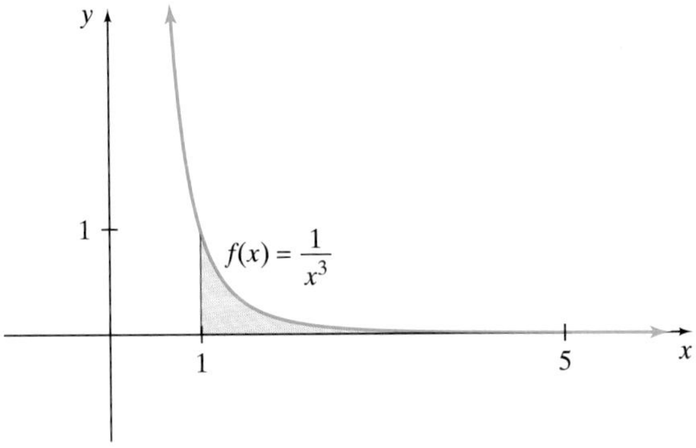

Figure 17.6.1

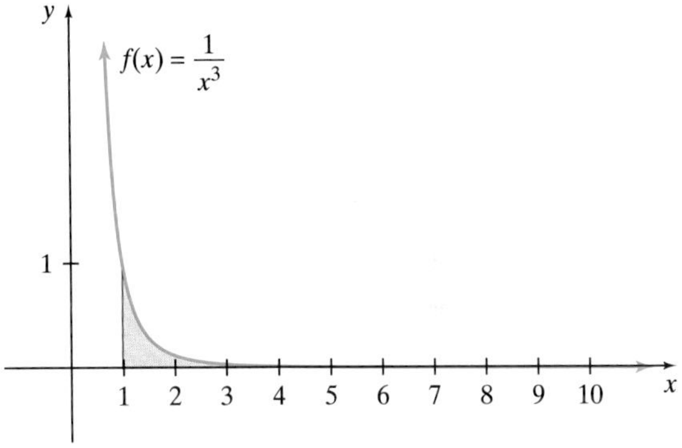

Figure 17.6.2

(b) Evaluation of $\int_1^{10} \frac{1}{x^3}\,dx$ will give us the area shown shaded in Figure 17.6.2.

$$\int_1^{10} \frac{1}{x^3}\,dx = \int_1^{10} x^{-3}\,dx = -\frac{1}{2}x^{-2}\Big|_1^{10} = -\frac{1}{2x^2}\Big|_1^{10}$$

$$= -\frac{1}{2(10)^2} - -\frac{1}{2(1)^2} = \frac{99}{200}$$

(c) Here we are asked to determine the area under $f(x) = \frac{1}{x^3}$ from $x = 1$ to some arbitrary point $x = b$, as shown in Figure 17.6.3. The process employed

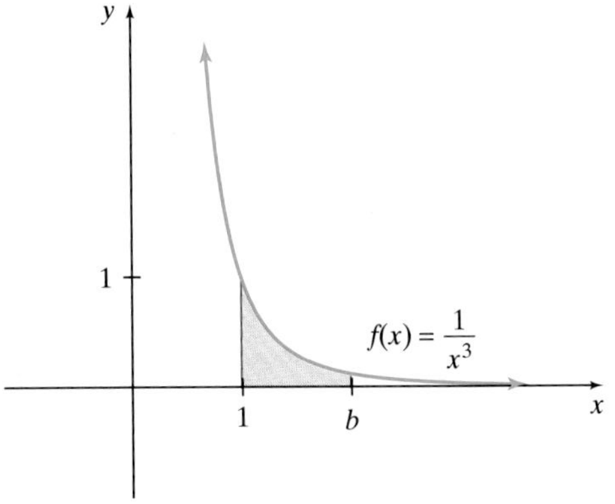

Figure 17.6.3

in parts (a) and (b) does not change.

$$\int_1^b \frac{1}{x^3}\,dx = \int_1^b x^{-3}\,dx = -\frac{1}{2}x^{-2}\Big|_1^b = -\frac{1}{2x^2}\Big|_1^b$$

$$= -\frac{1}{2(b)^2} - -\frac{1}{2(1)^2} = -\frac{1}{2b^2} + \frac{1}{2}$$ ■

We are now ready to make sense out of integrating over an unbounded region.

Example 2 Evaluating an Improper Integral

Consider the graph of $f(x) = \frac{1}{x^3}$ and the shaded region shown in Figure 17.6.4. The shaded region is unbounded; that is, it continues indefinitely to the right. We represent the area of this unbounded region by

$$\int_1^\infty \frac{1}{x^3}\,dx$$

which is an example of an *improper integral*. Evaluate this integral.

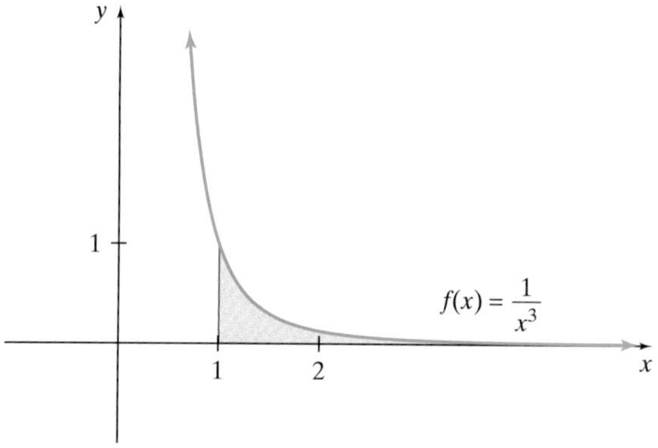

Figure 17.6.4

Solution

To do this, we employ a strategy almost identical to Example 1c. That is, we pick an arbitrary point to the right of $x = 1$, call it b, and integrate $\int_1^b \frac{1}{x^3}\,dx$. We know from Example 1c that

$$\int_1^b \frac{1}{x^3}\,dx = -\frac{1}{2b^2} + \frac{1}{2}$$

Now we take the limit as $b \to \infty$ of this result and get

$$\lim_{b\to\infty}\left(-\frac{1}{2b^2} + \frac{1}{2}\right) = \lim_{b\to\infty}\left(-\frac{1}{2b^2}\right) + \lim_{b\to\infty}\frac{1}{2}$$

$$= 0 + \frac{1}{2} = \frac{1}{2}$$

What we really did in Example 2 is compute the following:

$$\lim_{b\to\infty}\left[\int_1^b \frac{1}{x^3}\,dx\right]$$

Notice that this evaluation took place in two stages: first we integrated using the Fundamental Theorem of Calculus; then we did the limit. ■

It may contradict our intuition that an unbounded region, as in Figure 17.6.4, has a finite area of $\frac{1}{2}$ square unit, but our intuition is somewhat clouded when we encounter the concept of infinity. Do not despair. Some of the greatest minds in history have struggled with the concept of infinity.

At this time we generalize what we did in Example 2.

Improper Integral $\int_a^\infty f(x)\,dx$

Let f be continuous and nonnegative for any $x \geq a$. Then

$$\int_a^\infty f(x)\,dx = \lim_{b\to\infty}\int_a^b f(x)\,dx$$

provided the limit exists. If the limit exists, we say the improper integral $\int_a^\infty f(x)\,dx$ **converges**. If the limit does not exist, we say the improper integral $\int_a^\infty f(x)\,dx$ **diverges**.

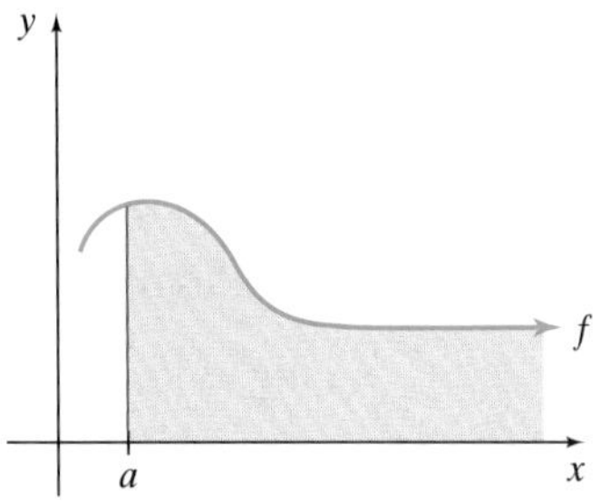

Figure 17.6.5

Example 3 Evaluating an Improper Integral

Evaluate.

(a) $\int_1^\infty \frac{1}{x^2}\,dx$ **(b)** $\int_1^\infty \frac{1}{x}\,dx$

Solution

(a) Evaluating $\int_1^\infty \frac{1}{x^2}\,dx$ will give us the area of the shaded region in Figure 17.6.6. We begin by rewriting the integral as

$$\int_1^\infty \frac{1}{x^2} = \lim_{b\to\infty}\left[\int_1^b \frac{1}{x^2}\,dx\right]$$

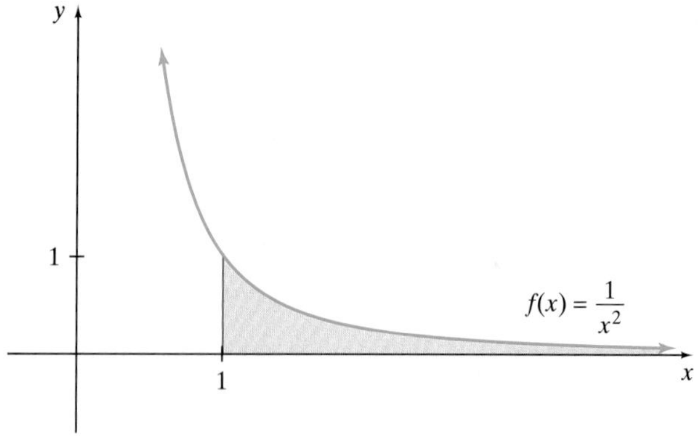

Figure 17.6.6

We perform the integration as follows:

$$\int_1^b \frac{1}{x^2}\,dx = \int_1^b x^{-2}\,dx = -x^{-1}\Big|_1^b$$

$$= -\frac{1}{x}\Big|_1^b = -\frac{1}{b} + 1$$

Next we evaluate the limit to get

$$\lim_{b\to\infty}\left(-\frac{1}{b}+1\right) = \lim_{b\to\infty}\left(-\frac{1}{b}\right) + \lim_{b\to\infty} 1 = 0 + 1 = 1$$

So the improper integral, $\int_1^\infty \frac{1}{x^2}\,dx$, converges and has a value of 1.

(b) Evaluating $\int_1^\infty \frac{1}{x}\,dx$ will give us the area shaded in Figure 17.6.7. Proceeding as in part (a) gives

$$\int_1^\infty \frac{1}{x}\,dx = \lim_{b\to\infty}\left[\int_1^b \frac{1}{x}\,dx\right]$$

Integrating gives

$$\int_1^b \frac{1}{x}\,dx = \ln x\Big|_1^b = \ln b - \ln 1 = \ln b$$

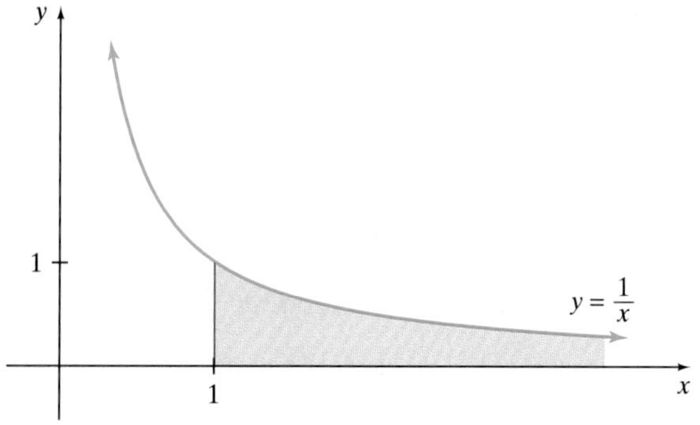

Figure 17.6.7

Evaluating the limit of this result gives

$$\lim_{b\to\infty}(\ln b) = \infty$$

This means the improper integral, $\int_1^\infty \frac{1}{x}\,dx$, diverges. ■

Checkpoint 1 Now work Exercise 5.

We will discuss two other types of improper integrals in this section. Consider the shaded areas in Figures 17.6.8 and 17.6.9. To determine the area under $f(x) = \frac{1}{x^2}$ on the interval $(-\infty, -1]$, as shown in Figure 17.6.8, requires that we evaluate the **improper integral** $\int_{-\infty}^{-1} \frac{1}{x^2}\,dx$. To determine the area under $f(x) = \frac{1}{\sqrt{2\pi}}e^{-(x^2/2)}$ on the interval $(-\infty, \infty)$, as shown in Figure 17.6.9, requires that we evaluate the **improper integral** $\int_{-\infty}^{\infty} \frac{1}{\sqrt{2\pi}}e^{-(x^2/2)}\,dx$. (The graph in Figure 17.6.9 is called the *normal curve*, which is seen frequently in statistics and probability.) We can evaluate these improper integrals in a manner that is similar to evaluating $\int_a^\infty f(x)\,dx$.

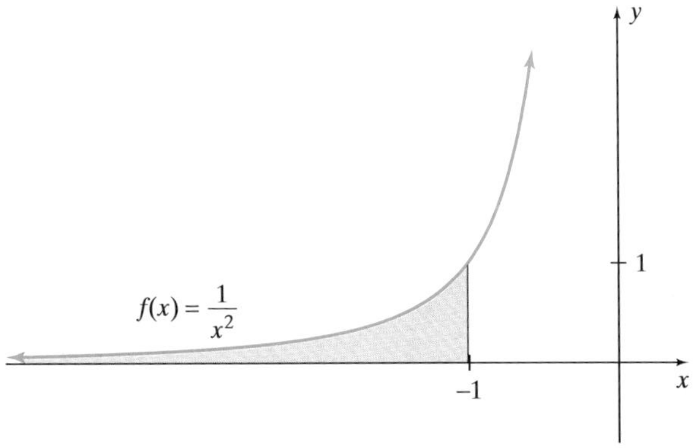

Figure 17.6.8

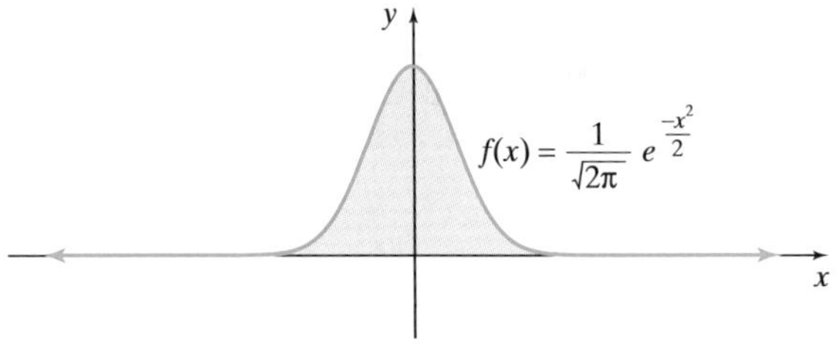

Figure 17.6.9

Improper Integrals $\int_{-\infty}^{b} f(x)\,dx$ and $\int_{-\infty}^{\infty} f(x)\,dx$

Let f be continuous and nonnegative on the indicated interval. Then we have:

1. $\int_{-\infty}^{b} f(x)\,dx = \lim_{a\to-\infty}\int_a^b f(x)\,dx$
2. $\int_{-\infty}^{\infty} f(x)\,dx = \int_{-\infty}^{c} f(x)\,dx + \int_c^\infty f(x)\,dx$

In 2, c is any point in $(-\infty, \infty)$, provided that both improper integrals on the right exist. Again, if the limits exist we say that the improper integral **converges**. If the limit does not exist, we say the improper integral **diverges**.

Note: If either integral on the right in the second formula in the box diverges, then the improper integral diverges.

Example 4 Evaluating an Improper Integral

Evaluate $\int_{-\infty}^{0} e^x\,dx$.

Solution

First we rewrite the improper integral as

$$\int_{-\infty}^{0} e^x\,dx = \lim_{a\to-\infty}\left[\int_a^0 e^x\,dx\right]$$

Just as we did in the previous example, we will integrate first.

$$\int_a^0 e^x\,dx = e^x\Big|_a^0 = e^0 - e^a = 1 - e^a$$

Next we take the limit of this result and get

$$\lim_{a\to-\infty}(1-e^a) = \lim_{a\to-\infty} 1 - \lim_{a\to-\infty} e^a = 1 - 0 = 1$$

So the improper integral $\int_{-\infty}^{0} e^x\,dx$ converges and has a value of 1. ■

Checkpoint 2 Now work Exercise 17.

Example 5 Evaluating an Improper Integral

Evaluate $\displaystyle\int_{-\infty}^{\infty} \frac{4x^3}{(1+x^4)^2}\,dx$.

Solution

We begin by rewriting the improper integral as

$$\begin{aligned}\int_{-\infty}^{\infty} \frac{4x^3}{(1+x^4)^2}\,dx &= \int_{-\infty}^{0} \frac{4x^3}{(1+x^4)^2}\,dx + \int_{0}^{\infty} \frac{4x^3}{(1+x^4)^2}\,dx\\ &= \lim_{a\to-\infty}\int_a^0 \frac{4x^3}{(1+x^4)^2}\,dx + \lim_{b\to\infty}\int_0^b \frac{4x^3}{(1+x^4)^2}\,dx\end{aligned}$$

To integrate $\displaystyle\int_a^0 \frac{4x^3}{(1+x^4)^2}\,dx$, we use u-substitution. Let $u = 1 + x^4$; then $du = 4x^3\,dx$. (We opt to leave off limits until we have an antiderivative.) Proceeding, we have

$$\begin{aligned}\int \frac{4x^3}{(1+x^4)^2}\,dx &= \int \frac{1}{u^2}\,du = \int u^{-2}\,du = -u^{-1} = -\frac{1}{u}\\ &= -\frac{1}{(1+x^4)}\Big|_a^0 = -1 + \frac{1}{1+a^4}\end{aligned}$$

We now take the limit and get

$$\lim_{a\to-\infty}\left(-1+\frac{1}{1+a^4}\right)=\lim_{a\to-\infty}(-1)+\lim_{a\to-\infty}\left(\frac{1}{1+a^4}\right)$$
$$=-1+0=-1$$

Integrating $\int_0^b \frac{4x^3}{(1+x^4)^2}\,dx$ will result in the same antiderivative as just computed, since it has the same integrand, just different limits. Hence, we can immediately write

$$\int_0^b \frac{4x^3}{(1+x^4)^2}\,dx=-\frac{1}{(1+x^4)}\bigg|_0^b=-\frac{1}{1+b^4}+1$$

Taking the limit of this result yields

$$\lim_{b\to\infty}\left(-\frac{1}{1+b^4}+1\right)=\lim_{b\to\infty}\left(-\frac{1}{1+b^4}\right)+\lim_{b\to\infty}1$$
$$=0+1=1$$

Putting this all together results in

$$\int_{-\infty}^{\infty}\frac{4x^3}{(1+x^4)^2}\,dx=\int_{-\infty}^{0}\frac{4x^3}{(1+x^4)^2}\,dx+\int_0^{\infty}\frac{4x^3}{(1+x^4)^2}\,dx$$
$$=\lim_{a\to-\infty}\int_a^0\frac{4x^3}{(1+x^4)^2}\,dx+\lim_{b\to\infty}\int_0^b\frac{4x^3}{(1+x^4)^2}\,dx$$
$$=-1+1=0$$

So the improper integral converges. ■

Checkpoint 3 Now work Exercise 31.

Applications

One family of applications of improper integrals is in estimating the total amount of oil or natural gas that will be produced by a well, given its production rate. Example 6 illustrates this.

Example 6 **Analyzing Total Oil Production**

Petroleum engineers estimate that an oil field will produce oil at a rate given by $f(t)=600e^{-0.1t}$ thousand barrels per month, t months into production. Estimate the total amount of oil produced by this well.

Solution

Understand the Situation: Since f is a rate function, integration gives us the total amount of oil produced. So the total amount of oil produced by the oil well in T months of production is given by

$$\int_0^T 600e^{-0.1t}\,dt$$

Since we want to know the potential output of the well, we assume that the well will operate indefinitely. Thus, the total amount of oil produced is given by

$$\int_0^{\infty} 600e^{-0.1t}\,dt = \lim_{T\to\infty}\left[\int_0^T 600e^{-0.1t}\,dt\right]$$

As we have seen in this section, we integrate first and then apply the limit concept. Integration yields

$$\int_0^T 600e^{-0.1t}\,dt = \frac{600}{-0.1}e^{-0.1t}\Big|_0^T = -6000e^{-0.1t}\Big|_0^T$$

$$= -6000e^{-0.1T} + 6000e^{-0.1(0)}$$

$$= -6000e^{-0.1T} + 6000$$

Applying the limit results in

$$\lim_{T\to\infty}(-6000e^{-0.1T} + 6000) = 0 + 6000 = 6000$$

Interpret the Solution: So the total production of the oil well is estimated to be 6000 thousand, or 6,000,000 barrels. ■

Capital Value

We begin our study of capital value by reconsidering the idea of the present value of a continuous income stream.

Flashback

Continuous Income Stream Revisited

In Section 17.4 we analyzed the present value of a continuous income stream. Recall that if $f(t)$ represents the rate of flow of a continuous income stream then the present value, P, at annual interest rate r compounded continuously for T years is given by

$$P = \int_0^T f(t)e^{-rt}\,dt$$

The **capital value** of a continuous income stream is simply the present value on the interval $[0, \infty)$; that is,

$$\text{capital value} = \int_0^{\infty} f(t)e^{-rt}\,dt$$

In other words, capital value gives the worth of an investment that generates income forever. B. K. O'Neill has just discovered that some land he inherited has a huge oil deposit on it. He decides to lease the oil rights to Shannon Oil for an indefinite annual payment of \$50,000. Determine the capital value of this lease at an annual interest rate of 8% compounded continuously.

Flashback Solution

We first note that the annual payments create a continuous income stream given by

$$f(t) = 50{,}000$$

that lasts indefinitely, or forever. So the capital value is given by

$$\int_0^{\infty} f(t)e^{-rt}\,dt = \int_0^{\infty} 50{,}000e^{-0.08t}\,dt$$

$$= \lim_{T\to\infty}\left[\int_0^{T} 50{,}000e^{-0.08t}\,dt\right]$$

First, the integration gives

$$\int_0^{T} 50{,}000e^{-0.08t}\,dt = \frac{50{,}000}{-0.08}e^{-0.08t}\Big|_0^T = -625{,}000e^{-0.08t}\Big|_0^T$$

$$= -625{,}000e^{-0.08T} + 625{,}000$$

Evaluating the limit of this result gives

$$\lim_{T\to\infty}(-625{,}000e^{-0.08T} + 625{,}000) = 625{,}000$$

So the capital value is \$625,000.

Endowments and Perpetuities

A fund that creates a steady income forever (think of a continuous income stream) may also be called a *permanent endowment* or a **perpetuity**. Some scholarship funds, such as the one in Example 7, and endowed chairs at universities are examples of perpetuities.

Example 7 Funding a Scholarship

Maria Lopez, a wealthy alumna of Old State University, wants to establish a nursing student scholarship in her name at her alma mater. If the annual proceeds of the scholarship are to be \$10,000 and the annual rate of interest is 6% compounded continuously, determine the amount, to the nearest cent, that Maria needs to fund this scholarship.

Solution

Understand the Situation: We first note that the annual proceeds determine a continuous income stream given by

$$f(t) = 10{,}000$$

that is to last forever. So the amount necessary to fund this scholarship is given by

$$\int_0^{\infty} 10{,}000e^{-0.06t}\,dt = \lim_{T\to\infty}\left[\int_0^{T} 10{,}000e^{-0.06t}\,dt\right]$$

Integrating first produces

$$\int_0^{T} 10{,}000e^{-0.06t}\,dt = \frac{10{,}000}{-0.06}e^{-0.06t}\Big|_0^T = -166{,}666.67e^{-0.06T} + 166{,}666.67$$

Evaluating the limit yields

$$\lim_{T\to\infty}(-166{,}666.67e^{-0.06T} + 166{,}666.67) = 166{,}666.67$$

Interpret the Solution: We conclude that the amount Maria needs to fund this scholarship indefinitely is \$166,666.67. ■

Checkpoint 4

Now work Exercise 53.

Probability Density Functions

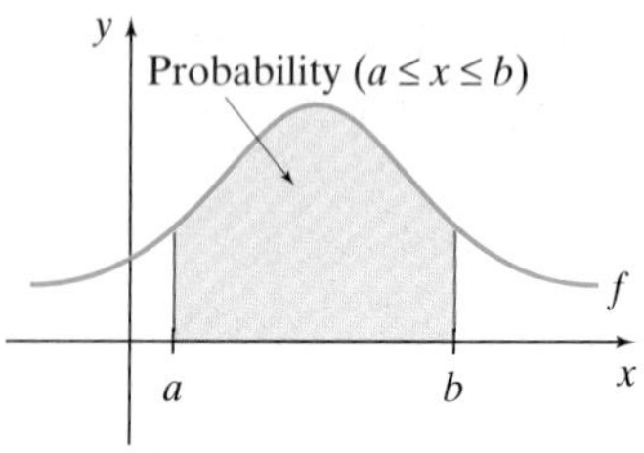

Figure 17.6.10

We conclude this section with a brief, intuitive, and informal look at how to use improper integrals with **probability density functions**. Consider an experiment that is designed so that any real number x in the interval $[a, b]$ is a possible outcome. An example of such an experiment might be to list the height or weight of a person selected at random.

Under some circumstances it is possible to determine a function f that will determine the probability that x assumes a value on a stated subinterval of $(-\infty, \infty)$. Functions that do this are called **probability density functions**. A probability density function must satisfy three conditions (see Figure 17.6.10):

1. For all x in $(-\infty, \infty)$, $f(x) \ge 0$.
2. The area under the graph of f is 1. That is, $\int_{-\infty}^{\infty} f(x)\,dx = 1$.
3. If $[a, b]$ is a subinterval of $(-\infty, \infty)$ then

$$\text{Probability}\,(a \le x \le b) = \int_a^b f(x)\,dx$$

Example 8 Calculating a Probability

The length of telephone calls in minutes from a pay phone in a college dormitory has a probability density function of

$$f(x) = \begin{cases} 0.5e^{-0.5x}, & x \ge 0 \\ 0, & x < 0 \end{cases}$$

Calculate the probability that a call selected at random will last between 1 and 2 minutes.

Interactive Activity

Graph f from Example 8 to verify that it satisfies condition 1 of probability density functions. Also, verify that

$$\int_{-\infty}^{\infty} 0.5e^{-0.5x}\,dx = 1,$$

showing that condition 2 is satisfied. [*Hint:* Since $f(x) = 0$ for $x < 0$, $\int_{-\infty}^{0} f(x) = 0$.]

Solution

The first two conditions of a probability density function are left to you to verify in the Interactive Activity to the left of this example. The probability that a call selected at random will last between 1 and 2 minutes is

$$\text{Probability}\,(1 \le x \le 2) = \int_1^2 0.5e^{-0.5x}\,dx = \frac{0.5}{-0.5}e^{-0.5x}\Big|_1^2$$

$$= -e^{-0.5x}\Big|_1^2 = -e^{-0.5(2)} + e^{-0.5(1)} \approx 0.2387$$ ■

Example 9 Calculating a Probability

For the probability density function given in Example 8, determine the probability that a telephone call selected at random is 4 or more minutes in length.

Solution

Understand the Situation: The probability that a call selected at random is 4 or more minutes in length is determined by integrating

$$\text{Probability } (x \geq 4) = \int_4^\infty 0.5e^{-0.5x}\,dx = \lim_{b\to\infty}\left[\int_4^b 0.5e^{-0.5x}\,dx\right]$$

As always, we do the integration first. This yields

$$\int_4^b 0.5e^{-0.5x}\,dx = -e^{-0.5x}\Big|_4^b = -e^{-0.5b} + e^{-2}$$

Now we determine the limit as follows:

$$\lim_{b\to\infty}(-e^{-0.5b} + e^{-2}) = e^{-2} \approx 0.1353$$

Interpret the Solution: So we have Probability $(x \geq 4) \approx 0.1353$. ■

Checkpoint 5 Now work Exercise 63.

SUMMARY

In this section we saw that an integral in which a limit of integration was ∞ (or $-\infty$) is called an **improper integral**. We saw how to use the limit concept to evaluate improper integrals. In particular, we determined that if f is continuous and nonnegative we have:

- $\int_a^\infty f(x)\,dx = \lim_{b\to\infty}\int_a^b f(x)\,dx$
- $\int_{-\infty}^b f(x)\,dx = \lim_{a\to-\infty}\int_a^b f(x)\,dx$
- $\int_{-\infty}^\infty f(x)\,dx = \int_{-\infty}^c f(x)\,dx + \int_c^\infty f(x)\,dx$

If the limit exists, we said that the improper integral **converges**, and if the limit does not exist, we said that the improper integral **diverges**. We applied improper integrals to determine the total output of an oil well, to assess a capital value, to determine the size of a **perpetuity**, and to briefly explore **probability density functions**. Recall that the **capital value** is determined by

$$\text{capital value} = \int_0^\infty f(t)e^{-rt}\,dt$$

where $f(t)$ is a continuous income stream.

SECTION 17.6 EXERCISES

In Exercises 1–16, evaluate each improper integral or state that it diverges.

1. $\int_2^\infty \frac{1}{x^4}\,dx$ **2.** $\int_4^\infty \frac{1}{x^3}\,dx$

3. $\int_3^\infty \frac{1}{\sqrt{x}}\,dx$ **4.** $\int_5^\infty \frac{3}{\sqrt{x}}\,dx$

✓ **5.** $\int_2^\infty \frac{1}{x^{1.5}}\,dx$ **6.** $\int_3^\infty \frac{1}{x^{2.5}}\,dx$

7. $\int_1^\infty \frac{1}{x^{2/3}}\,dx$ **8.** $\int_1^\infty \frac{1}{x^{3/4}}\,dx$

9. $\int_1^\infty e^{-x}\,dx$ **10.** $\int_0^\infty e^{-x}\,dx$

11. $\int_1^\infty e^{-0.03x}\,dx$ **12.** $\int_1^\infty e^{0.03x}\,dx$

13. $\int_1^\infty \frac{\ln x}{x}\,dx$ **14.** $\int_{10}^\infty \ln x\,dx$

15. $\int_1^\infty \frac{2x}{x^2+3}\,dx$ **16.** $\int_0^\infty \frac{x^4}{(x^5+2)^2}\,dx$

In Exercises 17–28, evaluate each improper integral or state that it diverges.

✓ **17.** $\int_{-\infty}^0 e^{2x}\,dx$ **18.** $\int_{-\infty}^0 e^{-2x}\,dx$

19. $\int_{-\infty}^0 e^{-x}\,dx$ **20.** $\int_{-\infty}^0 e^{2.3x}\,dx$

21. $\int_{-\infty}^0 e^{0.2x}\,dx$ **22.** $\int_{-\infty}^0 e^{x+1}\,dx$

23. $\int_{-\infty}^{-4} x^{-2}\,dx$ **24.** $\int_{-\infty}^{-2} x^{-4}\,dx$

25. $\int_{-\infty}^{-2} \frac{x^3}{(x^4-1)^2}\,dx$

26. $\int_{-\infty}^{0} \frac{3x}{(x^2+5)^4}\,dx$

27. $\int_{0}^{\infty} 2xe^{-x^2}\,dx$

28. $\int_{-\infty}^{0} 2xe^{-x^2}\,dx$

In Exercises 29–36, evaluate the given improper integral or state that it diverges.

29. $\int_{-\infty}^{\infty} \frac{4x^3}{(x^4+3)^2}\,dx$

30. $\int_{-\infty}^{\infty} \frac{4x^3}{(x^4+1)^2}\,dx$

✓ 31. $\int_{-\infty}^{\infty} 2xe^{-x^2}\,dx$

32. $\int_{-\infty}^{\infty} 2xe^{x^2}\,dx$

33. $\int_{-\infty}^{\infty} e^{-x}\,dx$

34. $\int_{-\infty}^{\infty} e^{x}\,dx$

35. $\int_{-\infty}^{\infty} \frac{2x}{x^2+1}\,dx$

36. $\int_{-\infty}^{\infty} \frac{1}{\sqrt{(x^2+1)}}\,dx$

Applications

37. Petrohas Engineering estimates that an oil well produces oil at a rate of $B(t) = 60e^{-0.05t} - 60e^{-0.1t}$ thousand barrels per month, where t is the number of months from now. Estimate the total amount of oil produced by this well.

38. A geologist estimates that an oil well produces oil at a rate of $B(t) = 85e^{-0.02t} - 85e^{-0.1t}$ thousand barrels per month, where t is the number of months from now. Estimate the total amount of oil produced by this well.

39. GeoSurveys Inc. estimates that a natural gas well has a rate of production given by $CF(t) = te^{-0.25t}$, where $CF(t)$ is millions of cubic feet per month, t months from now. Estimate the total amount of natural gas produced.

40. EarthWorks Inc. estimates that a natural gas well has a rate of production given by $CF(t) = 2te^{-0.37t}$, where $CF(t)$ is millions of cubic feet per month, t months from now. Estimate the total amount of natural gas produced.

41. The rate at which a certain hazardous chemical is being released into a lake from an abandoned dump is given by $350e^{-0.3t}$ tons per year, t years from now. Determine the total amount of the hazardous chemical that will be released into the lake if the leak continues indefinitely.

42. The rate at which a certain toxic chemical is being released into a river from an abandoned dump is given by $300e^{-0.25t}$ tons per year, t years from now. Determine the total amount of the toxic chemical that will be released into the river if the leak continues indefinitely.

43. It is a fact that when people take medication the human body does not absorb all the medication. Researchers at a pharmaceutical company know that one way to determine the amount of medication absorbed by the body is to determine the rate at which the body eliminates the medication. For a certain arthritis medication, researchers have determined that the rate at which the body eliminates the medication is given by $5e^{-0.02t} - 5e^{-0.035t}$ milliliters per minute, t minutes after the medication was administered. Determine the amount of the medication that was eliminated from the body.

44. For a certain asthma medication, the rate at which the body eliminates the medication is given by $6.5e^{-0.025t} - 6.5e^{-0.05t}$ milliliters per minute, t minutes after the medication was administered. Determine the amount of the medication that was eliminated from the body.

45. Elle owns a rental property that generates an indefinite annual rent of \$10,000. Determine the capital value of this property at an annual interest rate of 6% compounded continuously.

46. Ginger owns a rental property that generates an indefinite annual rent of \$6000. Determine the capital value of this property at an annual interest rate of 8.5% compounded continuously.

47. Austin has created a new computer game. He decides to lease the rights to his computer game to Macrohard for an indefinite annual payment of \$30,000. Determine the capital value of this lease at an annual interest rate of 7.25% compounded continuously.

48. Nikita has created a new computer game. She decides to lease the rights to her computer game to Star Systems for an indefinite annual payment of \$15,000. Determine the capital value of this lease at an annual interest rate of 9.75% compounded continuously.

49. Dylan has inherited some land that has a huge gold deposit on it. He decides to lease the rights to the gold to Hoofer Mining Company for an indefinite annual payment of \$70,000. Determine the capital value of this lease at an annual interest rate of 6.5% compounded continuously.

50. Emily owns a property that has a huge silver deposit on it. She decides to lease the rights to the silver to Hoofer Mining Company for an indefinite annual payment of \$25,000. Determine the capital value of this lease at an annual interest rate of 8.8% compounded continuously.

51. Lisa owns a rental property that generates an indefinite annual rent of \$12,000. Determine the capital value of this property at an annual interest rate of 5.5% compounded continuously.

52. Gene owns a rental property that generates an indefinite annual rent of \$2000. Determine the capital value of this property at an annual interest rate of 7% compounded continuously.

✓ **53.** If the annual proceeds from the Emma Lou Smith scholarship fund will be \$8000 indefinitely and the annual interest rate is 7.5% compounded continuously, how much should be invested to fund this scholarship?

54. If the annual proceeds from the Count Moncheech de Squeeg cinematography scholarship fund will be \$5000 indefinitely and the annual interest rate is 7% compounded continuously, how much should be invested to fund this scholarship?

55. The Shannon Oil Company has decided to provide an annual scholarship to individuals studying environmental science in the amount of \$3000. If the annual rate of interest is 6.6% compounded continuously, how much should be invested to fund this scholarship perpetually?

56. The Beagle Works Company has decided to provide an annual scholarship to individuals studying veterinary medicine in the amount of \$10,000. If the annual rate of interest is 7.75% compounded continuously, how much should be invested to fund this scholarship perpetually?

57. You wish to leave a legacy at your alma mater in the form of a scholarship that bears your name. You want the scholarship to be an annual award of \$5000. Assume that the annual interest rate is 6.25% compounded continuously.

(a) How much should you invest to fund this annual \$5000 scholarship for the next 75 years?

(b) How much should you invest to fund this annual \$5000 scholarship indefinitely?

58. You wish to leave a legacy at your alma mater in the form of a scholarship that bears your name. You want the scholarship to be an annual award of \$8000. Assume that the annual interest rate is 6.25% compounded continuously.

(a) How much should you invest to fund this annual \$8000 scholarship for the next 75 years?

(b) How much should you invest to fund this annual \$8000 scholarship indefinitely?

59. The length of telephone calls in minutes from a pay phone in a college dormitory has a probability density function of

$$f(x) = \begin{cases} 0.3e^{-0.3x}, & x \ge 0 \\ 0, & x < 0 \end{cases}$$

(a) Verify that $\int_{-\infty}^{\infty} f(x)\,dx = 1$.

(b) Calculate the probability that a call selected at random lasts between 3 and 4 minutes.

(c) Calculate the probability that a call selected at random lasts 3 minutes or longer.

60. The length of telephone calls in minutes from a pay phone in a college cafeteria has a probability density function of

$$f(x) = \begin{cases} 0.6e^{-0.6x}, & x \ge 0 \\ 0, & x < 0 \end{cases}$$

(a) Verify that $\int_{-\infty}^{\infty} f(x)\,dx = 1$.

(b) Calculate the probability that a call selected at random lasts between 1 and 2 minutes.

(c) Calculate the probability that a call selected at random lasts 3 minutes or longer.

61. The length of time for which a calculus book on a 4-hour reserve at a college library is checked out has a probability density function of

$$f(x) = \begin{cases} \frac{1}{8}x, & 0 \le x \le 4 \\ 0, & \text{otherwise} \end{cases}$$

(a) Verify that $\int_{-\infty}^{\infty} f(x)\,dx = 1$.

(b) Compute the probability that a student selected at random has the book checked out from 1 to 2 hours.

(c) Compute the probability that a student selected at random has the book checked out from 30 minutes to 1 hour.

62. The length of time for which a solutions manual for a certain calculus book on a 6-hour reserve at a college library is checked out has a probability density function of

$$f(x) = \begin{cases} \frac{1}{18}x, & 0 \le x \le 6 \\ 0, & \text{otherwise} \end{cases}$$

(a) Verify that $\int_{-\infty}^{\infty} f(x)\,dx = 1$.

(b) Compute the probability that a student selected at random has the solutions manual checked out from 2 to 4 hours.

(c) Compute the probability that a student selected at random has the solutions manual checked out from 1 to 2 hours.

✓ **63.** The length of a prison term for convicted felons has a probability density function of

$$f(x) = \begin{cases} 0.1e^{-0.1x}, & x \ge 0 \\ 0, & x < 0 \end{cases}$$

(a) Verify that $\int_{-\infty}^{\infty} f(x)\,dx = 1$.

(b) Determine the probability that a randomly selected felon has a prison term of 5 to 10 years.

(c) Determine the probability that a randomly selected felon has a prison term of 1 to 5 years.

(d) Determine the probability that a randomly selected felon has a prison term of 8 years or more.

64. The number of months until failure for a certain part on a copying machine has a probability density function of

$$f(x) = \begin{cases} 0.02e^{-0.02x}, & x \ge 0 \\ 0, & x < 0 \end{cases}$$

(a) Verify that $\int_{-\infty}^{\infty} f(x)\,dx = 1$.

(b) Compute the probability that the part will fail in the first year.

(c) Compute the probability that the part will fail in the second year.

65. The number of months until failure for a certain computer chip has a probability density function of

$$f(x) = \begin{cases} 0.03e^{-0.03x}, & x \geq 0 \\ 0, & x < 0 \end{cases}$$

(a) Verify that $\int_{-\infty}^{\infty} f(x)\,dx = 1$.

(b) Compute the probability that the part will fail in the first year.

(c) Compute the probability that the part will fail in the second year.

66. At Robotics Inc. the number of minutes that a randomly selected telephone customer is on hold has a probability density function of

$$f(x) = \begin{cases} 0.6e^{-0.6x}, & x \geq 0 \\ 0, & x < 0 \end{cases}$$

(a) Verify that $\int_{-\infty}^{\infty} f(x)\,dx = 1$.

(b) Compute the probability that a customer is on hold from 1 to 4 minutes.

(c) Compute the probability that a customer is on hold for 2 minutes or longer.

67. For a certain variation of the flu, the amount of time in days for a complete recovery has a probability density function of

$$f(x) = \begin{cases} 0.025e^{-0.025x}, & x \geq 0 \\ 0, & x < 0 \end{cases}$$

(a) Verify that $\int_{-\infty}^{\infty} f(x)\,dx = 1$.

(b) Determine the probability that an infected individual recovers between 3 and 6 days.

(c) Determine the probability that an infected individual recovers between 7 and 10 days.

(d) Determine the probability that an infected individual takes 10 days or more to recover.

SECTION PROJECT

Peggy never finishes her calculus lecture before the end of the period, but always finishes her lectures within 5 minutes of the end of the period. The length of time that elapses between the end of the period and the end of her lecture has a probability density function of

$$f(x) = \begin{cases} \dfrac{3}{125}x^2, & 0 \leq x \leq 5 \\ 0, & \text{otherwise} \end{cases}$$

(a) Verify that $\int_{-\infty}^{\infty} f(x)\,dx = 1$.

(b) Determine the probability that the lecture continues for 1 to 2 minutes beyond the end of the period.

(c) Compute the probability that the lecture ends within 1 minute of the end of the period.

Donald never starts his calculus lecture at the beginning of a period, but always starts his lecture within 2 to 9 minutes from the beginning of the period. The length of time that elapses from 2 minutes after the beginning of the period and the beginning of his lecture is given by the probability density function of

$$f(x) = \begin{cases} \dfrac{3}{721}x^2, & 2 \leq x \leq 9 \\ 0, & \text{otherwise} \end{cases}$$

(d) Verify that $\int_{-\infty}^{\infty} f(x)\,dx = 1$.

(e) Compute the probability that his lecture begins within 3 to 5 minutes from the beginning of the period.

(f) Compute the probability that his lecture begins within 7 to 9 minutes from the beginning of the period.

Why We Learned It

This chapter focused on the many varied applications of the integral. The chapter began with a discussion of determining the average value of a function on a closed interval. We used this concept to compute the average revenue, on a yearly basis, in the resource recovery industry during the 1980s. We then turned our attention to determining the area between two curves. If we are given two growth rates for a tree, one without a special nutrient and the other with a special nutrient, the area between the graph of the two growth rates on a closed interval gives us the total additional growth that the special nutrient provides. We also analyzed trade balances, consumers' surplus, producers' surplus and the Gini index. All of these were applications of the area between two curves. A new integration technique, called integration by parts, was introduced and used to integrate functions involving the natural logarithm. The trapezoidal rule provided us with another numerical technique for approximating definite integrals. This rule is especially powerful when we have data, yet no model for the data. The chapter concluded with a brief discussion about improper integrals. These types of integrals were instrumental in analyzing the capital value of a continuous income stream. We also used improper integrals in our brief introduction to probability.

CHAPTER REVIEW EXERCISES

In Exercises 1–4, determine the average value of the given function on the stated interval.

1. $f(x) = 3x + 4$ $[0, 6]$

2. $f(x) = x^2 - 2x$ $[1, 5]$

3. $g(x) = 4x^3 - 3x^2 + 2$ $[0, 5]$

4. $y = 3e^{-0.2x}$ $[0, 2]$

5. For $f(x) = \ln(1 + x)$:

(a) Use your RTSUM program to approximate $\int_a^b f(x)\,dx$ on $[0, 8]$. Use $n = 100$

(b) Use your approximation from part (a) to determine an (approximate) average value for $f(x)$ on $[0, 8]$.

6. The price–demand function for the 5-minute special lunch at the Reggie's Veggies Restaurant is given by

$$p(x) = 9.10 - 0.07x$$

where $p(x)$ is the price in dollars and x is the quantity demanded. Determine the average price in the demand interval $[24, 60]$.

7. The Tinnitus Phone Factory has determined that the cost for producing x telephones is given by

$$C(x) = 1300 + 18x$$

where $C(x)$ is the cost in dollars.

(a) Determine the average value of the cost function over the interval $[0, 400]$.

(b) Determine the average cost function, $AC(x) = \dfrac{C(x)}{x}$. Evaluate $AC(400)$.

(c) Explain the differences between parts (a) and (b).

8. Frederick deposits \$1400 into a money market account earning 5.25% interest compounded continuously. Determine Frederick's average balance, to the nearest cent, over 1 year.

9. Penultimate Savings and Loan offers a money market account that earns 4.7% interest compounded continuously. At the end of the year, the account also pays a bonus of 0.85% of the average balance in the account during the year. Carlene deposits \$2400 into a Penultimate money market account.

(a) Compute Carlene's average balance, to the nearest cent, over 1 year.

(b) Compute the size of Carlene's bonus.

10. Sylvia deposits \$800 into an account where the rate of change of the amount in the account is given by $A'(t) = 56e^{0.07t}$, t years after the initial deposit.

(a) Determine by how much the account will change from the end of the first year to the end of the fourth year. Round to the nearest cent.

(b) Determine the average balance, to the nearest cent, in the account over the first 4 years.

11. The rate of change of the total income produced by a pinball machine located in a college dorm is given by

$$f(t) = 4000e^{0.09t}$$

where t is the time in years since the installation of the pinball machine and $f(t)$ is in dollars per year. Determine the total income generated by the pinball machine during its first 5 years of operation.

12. Calvin contributes \$2000 to his IRA account at the beginning of each year for 30 years. The account earns 6.2% annual interest compounded continuously. Determine how much money Calvin has in his account after 30 years.

13. Upon adopting their daughter Felicia, Chris and Pat opened a college savings account. They deposit \$65 at the beginning of each month into an account earning 4.8% annual interest, compounded continuously. Determine how much is in the account in 16 years, when Felicia starts college.

14. The percentage of wage and salary workers in the United States who are members of unions can be modeled by

$$f(x) = -0.0069x^3 + 0.18x^2 - 1.68x + 21.61 \qquad 1 \le x \le 14$$

where $x = 1$ corresponds to January 1, 1983, and $f(x)$ is the percentage of wage and salary workers in the United States who were members of unions at time x. Use the model to compute the average percentage of salary workers in the United States who were members of unions between

January 1, 1983, and January 1, 1996 (*Source:* U.S. Bureau of Labor Statistics, www.bls.gov).

For Exercises 15 and 16, write a definite integral to represent the area of the shaded region.

15.

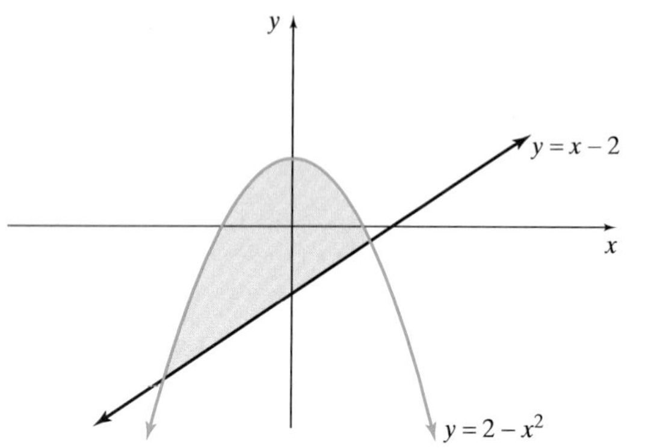

16.

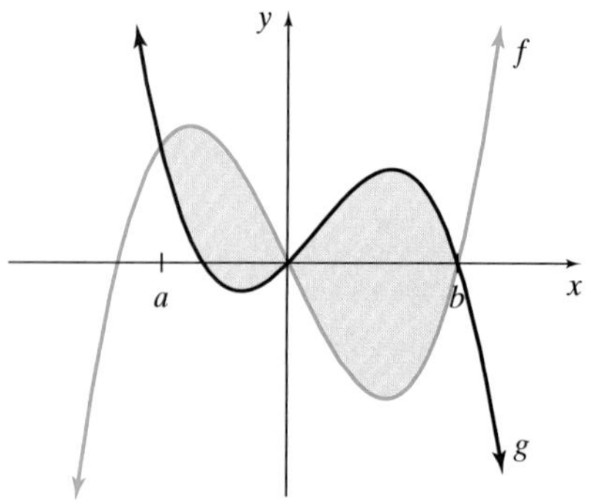

In Exercises 17–24, determine the area of the region bounded by the given conditions.

17. $f(x) = 9 - x^2$ and the x-axis on $[-3, 3]$

18. $f(x) = x^3 - 5x^2 - 2x + 3$ and the x-axis on $[1, 5]$

19. $f(x) = x^2 - 4$ and $g(x) = x + 2$ on $[-2, 3]$

20. $f(x) = x^3 + 2x^2 - 20$ and $g(x) = -3x^2 + 4x$ on $[-5, 2]$

21. $f(x) = 4x + 6$ and $g(x) = x^2 - 6$ on $[-2, 6]$

22. $f(x) = 4 + \sqrt{x}$ and $g(x) = \sqrt[3]{x}$ on $[0, 6]$

23. $f(x) = 5e^x$ and $g(x) = 3.7^x$ on $[-2, 5]$

24. $f(x) = \dfrac{8}{x^3}$ and $g(x) = \dfrac{2}{x}$ on $[1, 4]$

In Exercises 25–28, determine the area of the region bounded by the given curves.

25. $f(x) = x^2$ and $g(x) = x + 6$

26. $f(x) = x^3 + 2x^2 - 5x - 6$ and the x-axis

27. $f(x) = x^3$ and $g(x) = 2x^2 + 8x$

28. $f(x) = \sqrt[3]{x}$ and $g(x) = \dfrac{1}{4}x$

29. The Krazy Dog Company is planning to sell a new tricycle for dogs. It estimates that its monthly sales, $s(t)$, in units per month, t months after the product is released, will be given by

$$s(t) = 1.7t^2 - 16.1t + 160$$

An agent for Lucky, the star of the famous movie *Rover's Revenge*, claims that if Lucky is featured in an ad campaign the monthly sales, $m(t)$, in units per month, t months after the product is released, will be

$$m(t) = 164e^{0.09t}$$

On the interval [0, 15], determine the area between the graphs of s and m and interpret.

30. The size of bacteria culture A grows at a rate of $20x^{1.5}$ bacteria per minute and the size of bacteria culture B grows at a rate of $10x^{2.5}$ bacteria per minute.

(a) Set up an integral to determine the area between the two curves on [0, 2].

(b) Evaluate the integral from part (a) and interpret.

(c) Set up an integral to determine the area between the two curves on [2, 4].

(d) Evaluate the integral from part (c) and interpret.

(e) On the interval [0, 4], determine which culture has more bacterial growth and by how many bacteria.

31. The Loose Nail Furniture Company has determined that its costs, in thousands of dollars, have been increasing mainly due to inflation at a rate given by $215e^{0.14x}$, where x is time measured in years since a certain item started being produced. At the beginning of the fourth year, a breakthrough in the production process resulted in costs increasing at a rate given by $80x^{3/4}$. Determine the total savings in the interval [4, 8] due to the breakthrough.

32. The Lost Sole Shoe Company determines that the marginal revenue to produce x pairs of shoes is given by $R'(x) = 85 - 0.005x^2$, while the marginal cost is given by $C'(x) = 32$, where both marginals are in dollars per pair of shoes. Determine the total profit from $x = 45$ to $x = 84$.

33. Fred and Ginger are assembly workers at the Clobini Factory. Fred assembles units at the rate of $f(x) = -5x^2 + 23x + 31$ units per hour and Ginger assembles units at the rate of $g(x) = -4x^2 + 25x + 32$, where x is the number of hours on the job, during the first 4 hours on the job.

(a) How many units does Fred assemble during his first 4 hours?

(b) How many more units does Ginger assemble than Fred during the same time period?

34. The imports and exports for the United States with Belgium can be modeled by

Imports: $I(x) = 0.55x + 4.11 \quad 1 \le x \le 5$

Exports: $E(x) = -0.37x^3 + 3.42x^2 - 8.38x + 15.08 \quad 1 \le x \le 5$

where x is the number of years since 1991 and imports and exports are in billions of dollars. Determine the area between the two curves on $[1, 5]$ and interpret (*Source:* U.S. Statistical Abstract, www.census.gov/statab/www).

Exercises 35–38 refer to the accompanying figure.

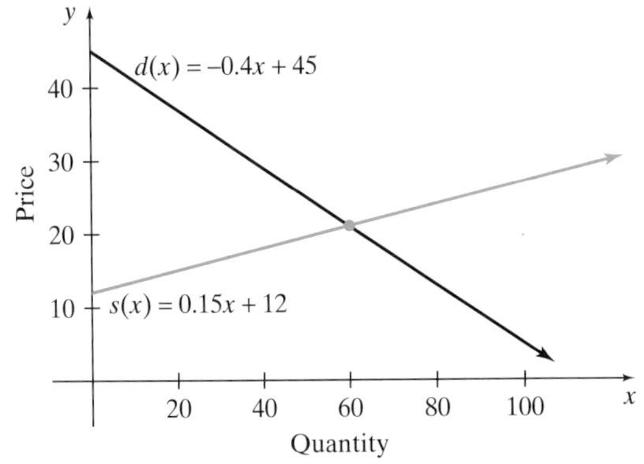

35. Shade the region that corresponds to the consumers' surplus. Write an integral to determine the consumers' surplus.

36. Shade the region that corresponds to the producers' surplus. Write an integral to determine the producers' surplus.

37. Shade the region that corresponds to the actual consumer expenditure. Determine the actual consumer expenditure.

38. Shade the region that corresponds to the actual revenue. Determine the actual revenue.

In Exercises 39–42, for each demand function given and demand level x, determine the consumers' surplus.

39. $d(x) = 450 - 0.8x,\ x = 100$

40. $d(x) = 3600 - 0.02x^2,\ x = 300$

41. $d(x) = 100 - \sqrt{x},\ x = 36$

42. $d(x) = 225e^{-0.01x},\ x = 250$

In Exercises 43–46, for each supply function given and demand level x, determine the producers' surplus.

43. $s(x) = 0.5x + 1200,\ x = 400$

44. $s(x) = \frac{1}{15}x + 10,\ x = 150$

45. $s(x) = 0.02x^2 + 25,\ x = 50$

46. $s(x) = 84e^{0.02x},\ x = 200$

For Exercises 47–50:

(a) Determine the equilibrium point. Round equilibrium demand and equilibrium price to the nearest whole number.

(b) Determine the consumers' surplus at equilibrium demand.

(c) Determine the producers' surplus at equilibrium demand.

47. $d(x) = 250 - 0.3x;\ s(x) = 60 + 0.2x$

48. $d(x) = 1280 - 0.8x;\ s(x) = 272 + 1.6x$

49. $d(x) = 4500 - 0.04x^2;\ s(x) = 0.03x^2 + 1700$

50. $d(x) = 700e^{-0.05x};\ s(x) = 50e^{0.06x}$

51. Knutsen Bolts Hardware has determined that the demand function for a new type of electric drill is given by

$$d(x) = 22 - 2\sqrt{x}$$

where x is the number of drills demanded each day and $d(x)$ is the price per drill in dollars.

(a) Assuming that the equilibrium price is \$14 per drill, determine the equilibrium demand. Round to the nearest whole number.

(b) Determine the consumers' surplus at equilibrium demand.

52. Carpets Galore has determined that the demand function for a certain kind of carpet is given by

$$d(x) = -0.5x + 400$$

where x is the number of square yards sold per week and $d(x)$ is the price per square yard in dollars. Assuming that the equilibrium price is \$18 per square yard, determine the consumers' surplus.

53. The Heavy Stuff Sporting Goods company has determined that the supply function for its bowling ball is given by

$$s(x) = 18e^{0.007x}$$

where x is the number of bowling balls supplied each month and $s(x)$ is the price per bowling ball in dollars. Assuming that the equilibrium price is \$47 per bowling ball, determine the producers' surplus.

54. A cushion manufacturer has determined that the supply function for a certain kind of cushion is given by

$$s(x) = 0.04x^2 + 11$$

where x is the number of these cushions supplied each week and $s(x)$ is the price per cushion in dollars. Assuming that the equilibrium price is \$23.96, determine the producers' surplus.

55. Springs 'n' Things has determined that the demand function for its deluxe clipboard is given by

$$d(x) = 12.48 - 0.005x^2$$

while the related producer supply function is given by

$$s(x) = \sqrt{x}$$

where x is the weekly quantity and $d(x)$ and $s(x)$ are in dollars per clipboard.

(a) Determine the equilibrium point.

(b) Determine the consumers' surplus at equilibrium.

(c) Determine the producers' surplus at equilibrium.

56. The following Lorenz curve models income distribution in the United States in 1992. Recall that x represents the proportion of the population and $L(x)$

represents the proportion of income (*Source:* U.S. Census Bureau, www.census.gov).

$$L(x) = 2.65x^4 - 4.32x^3 + 2.92x^2 - 0.25x$$

Determine the Gini Index for 1992. Round to the nearest ten-thousandth.

Evaluate the given integrals in Exercises 57–62. If integration by parts is required, consult the guidelines for selecting u and dv given in Section 17.4.

57. $\int 4xe^{2x}\,dx$ **58.** $\int xe^{-0.05x}\,dx$

59. $\int x^4 \ln x\,dx$ **60.** $\int (x-2)e^x\,dx$

61. $\int 8te^{-0.2t}\,dt$ **62.** $\int (x+5)(x-2)^7\,dx$

Exercises 63–66 require two (or more) applications of integration by parts.

63. $\int x^2e^{-x}\,dx$ **64.** $\int (x+1)^2e^x\,dx$

65. $\int (\ln x)^4\,dx$ **66.** $\int x(\ln x)^4\,dx$

In Exercises 67–70, determine the method needed to evaluate the given integral. If the method selected is integration by parts, state the choice for u and dv. If the method selected is u-substitution, simply state the choice for u.

67. $\int x^2e^{-x^3}\,dx$ **68.** $\int 2x\ln x\,dx$

69. $\int xe^{5x}\,dx$ **70.** $\int \frac{(\ln x)^5}{x}\,dx$

71. It is estimated that an oil field will produce oil at a rate given by $B(t) = 8te^{-0.17t}$ thousand barrels per month, t months into production.

(a) Write a definite integral to estimate the total production for the first year of operation.

(b) Evaluate the integral from part (a) to estimate the total production for the first year of operation.

72. Hoppin Pepper Inc. has determined that its marginal profit function is given by $P'(t) = 42te^{0.12t}$, where t is time in years and $P'(t)$ is in thousands of dollars per year. Assuming that $P(0) = 0$, determine $P(t)$.

73. Kevin's business generates income at the rate of $90t$ thousand dollars per year, where t is the number of years since the business started. Determine the present value of this continuous income stream over the first 4 years at a continuous compound interest rate of 7%.

74. Suppose that a gold mine produces income at the rate of $140e^{-0.08t}$ thousand dollars per year, where t is the number of years since mining began. Determine the present value of this continuous income stream over the first 6 years of operation, assuming a continuous compound interest rate of 5%.

75. The average sales price for a home in the United States can be approximated by

$$h(x) = 5.63x + 39.25 \qquad 1 \le x \le 21$$

where $x = 1$ corresponds to October 1, 1976, and $h(x)$ is the average selling price, in thousands of dollars (*Source:* U.S. Census Bureau, www.census.gov).

(a) Determine the average value of h on the interval [1, 6] and interpret.

(b) Determine the average value of h on the interval [8, 12] and interpret.

(c) Determine the average value of h on the interval [16, 21] and interpret.

76. The average annual expenditures on food by U.S. households can be modeled by

$$f(x) = 4.14 + 0.15\ln x \qquad 1 \le x \le 7$$

where $x = 1$ corresponds to January 1, 1989, and $f(x)$ is the amount spent on food by the average household in a year, in thousands of dollars (*Source:* U.S. Census Bureau, www.census.gov).

(a) Determine the average value of f on the interval [1, 4] and interpret.

(b) Determine the average value of f on the interval [4, 7] and interpret.

In Exercises 77–80, use the Trapezoidal Rule to approximate the given definite integral using n = 4 trapezoids.

77. $\int_{-3}^{1} x^5\,dx$ **78.** $\int_{4}^{8} (\ln x)^4\,dx$

79. $\int_{0}^{1} (x^2 - 2x)\,dx$ **80.** $\int_{0}^{8} \frac{1}{x+1}\,dx$

In Exercises 81–84, use the Trapezoidal Rule to approximate the given definite integral using n = 10, n = 50, and n = 100 trapezoids.

81. $\int_{1}^{11} \sqrt{x^2+3x+2}\,dx$ **82.** $\int_{3}^{8} x^2e^{0.5x}\,dx$

83. $\int_{1}^{5} \frac{(\ln x)^3}{x}\,dx$ **84.** $\int_{0}^{10} \frac{3x+4}{x^2+2}\,dx$

85. The following data give the rate of change of the year-to-date revenues in dollars per month for Jack's Handcrafted Birdhouses. Here x represents the number of months after May 1.

x	Rate of Change of Year-to-Date Revenue
1	2047
2	3124
3	2387
4	1937
5	2183
6	2706

Use the Trapezoidal Rule to approximate the revenues for Jack's Handcrafted Birdhouses for this period of time.

86. The following data give the marginal revenue for different levels of production at David's DVD Division. Here x represents the number of DVDs produced and $R'(x)$ is in dollars per DVD.

x	$R'(x)$
100	30
200	20
300	15
400	10
500	8

Use the Trapezoidal Rule to approximate the total revenue increase that would be obtained by going from a production level of 100 to 500 DVDs.

87. The GHI company has determined the marginal cost for different levels of production. The following data have been collected, where x represents the number of items produced and $C'(x)$ is in thousands of dollars per item.

x	$C'(x)$	x	$C'(x)$
0	3.8	14	5.8
2	3.6	16	6.9
4	3.2	18	7.8
6	3.4	20	9.4
8	3.9	22	12.7
10	4.2	24	18.5
12	4.7		

Assume that $C(0) = 0$ and use the Trapezoidal Rule to approximate the total cost of producing 24 items.

88. The following data give the annual amount spent by the U.S. federal government on human resources, in billions of dollars, from 1973 to 1997. Here, $x = 1$ corresponds to the beginning of 1973.

x	Amount spent (billions of dollars)
1	119.5
5	221.9
9	362.0
13	471.8
17	568.7
21	827.5
25	1019.4 (est.)

SOURCE: U.S. Census Bureau, www.census.gov

Use the Trapezoidal Rule to approximate the total amount spent from the beginning of 1973 to the beginning of 1997.

In Exercises 89 and 90, use the following information and the Trapezoidal Rule to answer the questions.

The following data give the number of new incorporations (in thousands) and the number of business failures (in thousands) in the United States from 1985 to 1995. Here $x = 1$ represents the beginning of 1985.

x	Number of New Incorporations (thousands)	Business Failures (thousands)
1	664	57
3	686	61
5	677	50
7	629	88
9	707	86
11	768	71

SOURCE: U.S. Census Bureau, www.census.gov

89. (a) Approximate the total number of businesses that incorporated from the beginning of 1985 to the beginning of 1995.

(b) Approximate the average number of businesses that incorporated, on a yearly basis, from the beginning of 1985 to the beginning of 1995.

90. (a) Approximate the total number of business failures that occurred from the beginning of 1985 to the beginning of 1995.

(b) Approximate the average number of business failures that occurred, on a yearly basis, from the beginning of 1985 to the beginning of 1995.

In Exercises 91–100, evaluate each improper integral or state that it is divergent.

91. $\int_5^\infty \frac{1}{x^2}\,dx$

92. $\int_1^\infty \frac{4}{\sqrt[3]{x}}\,dx$

93. $\int_{10}^\infty \frac{3}{x}\,dx$

94. $\int_0^\infty (8x - 2)\,dx$

95. $\int_0^\infty e^{0.1x}\,dx$

96. $\int_5^\infty \frac{4x}{2x^2 + 3}\,dx$

97. $\int_{-\infty}^0 e^{3x}\,dx$

98. $\int_{-\infty}^0 xe^{-0.2x^2}\,dx$

99. $\int_{-\infty}^\infty e^{0.05x}\,dx$

100. $\int_{-\infty}^\infty \frac{2x}{\sqrt{x^2 + 5}}\,dx$

101. An oil well has been estimated to produce oil at a rate of $B(t) = 43e^{-0.03t} - 43e^{-0.08t}$ thousand barrels per month, where t is the number of months from now. Estimate the total amount of oil that will be produced by this well.

102. Kevin owns a rental property that generates an indefinite annual rent of $14,000. Determine the capital value of this property at an annual interest rate of 5.25% compounded continuously.

103. A scholarship fund is to provide an annual scholarship in the amount of $8000. If the annual rate of interest is 9%, compounded continuously, how much should be invested to fund this scholarship perpetually?

104. The rate at which a certain toxic chemical is being released into an ocean from an abandoned dump is given by $260e^{-0.2t}$ tons per year t years from now. Determine the total amount of the toxic chemical that will be released into the ocean if the leak continues indefinitely.

105. The number of days for which a book is checked out of the Springfield Public Library has a probability density function of

$$f(x) = \begin{cases} \frac{2}{441}x, & 0 \le x \le 21 \\ 0, & \text{otherwise} \end{cases}$$

(a) Verify that $\int_{-\infty}^{\infty} f(x)\,dx = 1$.

(b) Compute the probability that the length of time for which a book is checked out is from 7 to 14 days.

(c) Compute the probability that the length of time for which a book is checked out is from 14 to 21 days.

106. In a certain country, the number of days for a mailed letter to reach its destination has a probability function of

$$f(x) = \begin{cases} 0.08e^{-0.08x}, & x \ge 0 \\ 0, & x < 0 \end{cases}$$

(a) Verify that $\int_{-\infty}^{\infty} f(x)\,dx = 1$.

(b) Compute the probability that a letter takes 2 to 5 days to reach its destination.

(c) Compute the probability that a letter takes 10 or more days to reach its destination.

CHAPTER 17 PROJECT

SOURCE: U.S. Census Bureau, www.census.gov

Many topics of integration were discussed in Chapter 17; everything from integration by parts to improper integrals. Two of these topics will be the focus of the Chapter 17 project: average value and area between curves.

Let x represent the number of years since 2000. The projected rate of change in the number of deaths in the United States from 2000 to 2050 can be modeled by

$$f(x) = 23.24(1.01)^x$$

where $0 \le x \le 50$ and $f(x)$ is measured in thousands of deaths per year. During the same time period, the projected rate of change in the number of births in the United States can be modeled by

$$g(x) = -0.92x + 41.75$$

where $0 \le x \le 50$ and $g(x)$ is measured in thousands of births per year.

1. Graph f and g in the viewing window [0, 50] by [−10, 45].

2. Evaluate $f(5) - g(5)$ and interpret.

3. Evaluate $f(35) - g(35)$ and interpret.

4. How fast is f changing at $x = 5$? How fast is g changing at $x = 5$?

5. Integrate f on the interval $5 \le x \le 10$ and interpret. Integrate g on the interval $5 \le x \le 10$ and interpret.

6. Evaluate $\frac{1}{10-5}\int_5^{10} f(x)\,dx$ and $\frac{1}{10-5}\int_5^{10} g(x)\,dx$ and interpret.

7. Determine the point of intersection of f and g and interpret each coordinate.

8. Evaluate $\int_5^{10}[f(x) - g(x)]\,dx$ and interpret.

9. Evaluate $\int_{30}^{40}[f(x) - g(x)]\,dx$ and interpret. What does the sign of the solution tell us?

Chapter

18 Calculus of Several Variables

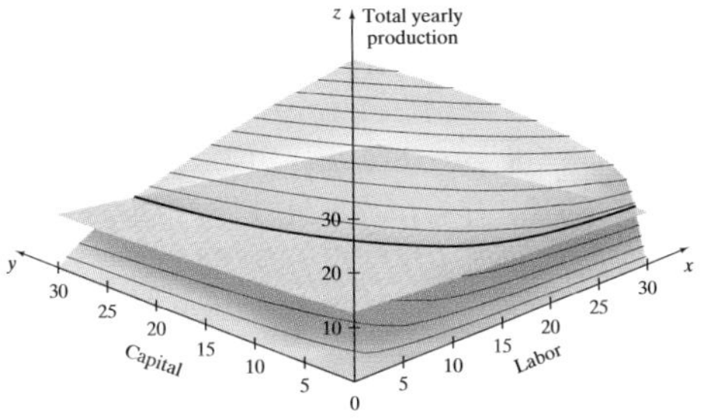

(b)

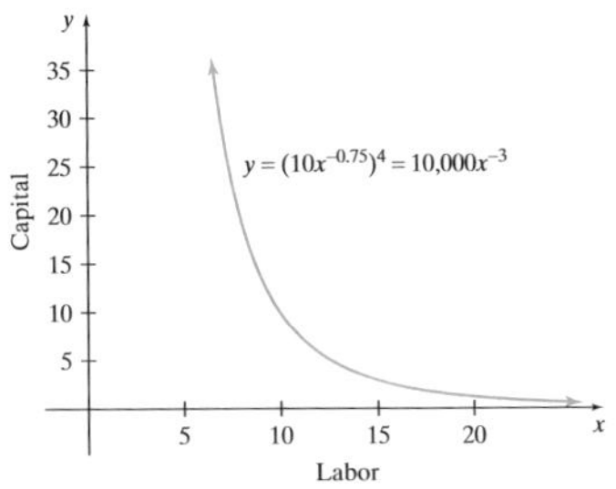

(c)

When manufacturing a product such as a golf cart, there are many inputs used in the production process. Economists tend to reduce all of the different inputs to two, labor and capital. This allows economists to determine a production function that relates the quantity produced as a function of labor and capital. The graph of a production function is a surface in three dimensions. In Figure b is the graph of a production function for the manufacturing of golf carts. In Figure b the surface of the graph of the production function has been sliced with a horizontal plane. The curve that results from where the plane and the graph of the production function intersect is shown in Figure c. Economists call such a curve an isoquant. An isoquant gives all combinations of labor and capital that yield a fixed quantity to be produced. Knowing all combinations of labor and capital that yield a fixed quantity to be produced is important for any business manager.

What We Know

In the previous seven chapters of this text, we saw how the derivative and the integral can be used in a variety of settings. In all these settings the functions that we applied calculus to were functions of a single independent variable.

Where Do We Go

In this chapter we extend our calculus knowledge to functions that involve two independent variables. We will still see that rates of change are central to the derivative concept in the many varied applied settings.

Section 18.1 Functions of Several Independent Variables

In this section we introduce the concept of a function of two or more independent variables and the coordinate system used to analyze the graphs of such functions. Some calculators can graph a function of two independent variables. At this time we suggest that you consult the owner's manual of your calculator to determine if your calculator does this.

Functions of Two or More Independent Variables

Many quantities in the world depend on several independent variables. For example, the wind chill factor and the heat index, two numbers that we may hear on the local weather report, each depend on two variables. The wind chill depends on the air temperature and the wind speed, while the heat index depends on the air temperature and the relative humidity. In mathematical terms, if WCI represents the wind chill index, then we may write

$$\text{WCI} = f(v, t)$$

where v is the wind speed in miles per hour and t is the air temperature in degrees Fahrenheit. We can evaluate a function that has more than one independent variable. Before we show how to do this, we first present a definition for a function of two independent variables.

Function of Two Independent Variables

An equation of the form $z = f(x, y)$ represents a **function of two independent variables** if, for each ordered pair of real numbers (x, y), the equation determines a unique real number z.

Note: The variables x and y are the **independent variables**, and z is the **dependent variable**.

The **domain** of a function of two variables is the set of all ordered pairs (x, y) for which z is defined. Example 1 illustrates how to evaluate a function of two independent variables.

Example 1 Evaluating a Function of Two Independent Variables

Consider $f(x, y) = 2x^2 - 3y^3$. Determine the following:

(a) $f(2, 1)$ **(b)** $f(-1, 2)$ **(c)** $f(1, -2)$

Solution

(a) Evaluating a function of two independent variables is just like evaluating a function with only one independent variable. That is, we substitute 2 in for x and 1 in for y. This produces

$$f(x, y) = 2x^2 - 3y^3$$
$$f(2, 1) = 2(2)^2 - 3(1)^3 = 8 - 3 = 5$$

So we have $f(2, 1) = 5$.

(b) Evaluating as we did in part (a) yields

$$f(x, y) = 2x^2 - 3y^3$$
$$f(-1, 2) = 2(-1)^2 - 3(2)^3 = 2 - 24 = -22$$

So we have $f(-1, 2) = -22$.

(c) Evaluating as we did in parts (a) and (b) gives

$$f(x, y) = 2x^2 - 3y^3$$
$$f(1, -2) = 2(1)^2 - 3(-2)^3 = 2 + 24 = 26$$

So we have $f(1, -2) = 26$. ■

✓ Checkpoint 1 Now work Exercise 7.

In Example 2, we illustrate how a function of two variables can be constructed to determine a cost function.

Example 2 Determining a Cost Function

DynaBall Corporation manufactures two types of golf balls. One has a balata cover favored by professionals and other low-handicap players. The other has a durable cover favored by high-handicap players. The cost to make a balata ball is \$0.97, whereas the cost to make a durable-cover ball is \$0.89.

(a) Determine a cost function, C as a function of x, for producing x balata balls.

(b) Determine a cost function, C as a function of y, for producing y durable-cover balls.

(c) Determine a cost function, C as a function of x and y, for the balata and durable-cover balls.

Solution

(a) The cost function for the balata balls is simply

$$C(x) = (\text{unit cost}) \cdot (\text{quantity}) = 0.97x$$

(b) For the durable-cover balls, the cost function is

$$C(y) = (\text{unit cost}) \cdot (\text{quantity}) = 0.89y$$

(c) **Understand the Situation:** Notice here, in the language of functions, that we are asked to determine cost as a function of x *and* y. Intuitively, this should be the sum of the cost functions from parts (a) and (b).

This gives us

$$C(x, y) = 0.97x + 0.89y$$ ■

The cost function in Example 2c, $C(x, y) = 0.97x + 0.89y$, gives us the cost of producing x balata balls and y durable-cover balls. This function is an example of a **function of two independent variables x and y**. If this company decides to expand its golf ball line by producing and selling a special 80 compression ball for senior citizens, we would introduce a third independent variable and rewrite our cost function as a *function of three independent variables*. For the most part, we will be discussing functions of two independent variables in this chapter. For many of the problems that we encounter, we may have to place restrictions on the domain to determine the realistic domain. Example 3 illustrates this idea.

Example 3 Determining a Domain and Evaluating a Function of Two Variables

In Example 2 we determined that the cost function to produce x balata balls and y durable-cover balls is given by $C(x, y) = 0.97x + 0.89y$.

(a) Determine the reasonable domain for $C(x, y)$.

(b) Evaluate $C(250, 700)$ and interpret.

Solution

(a) At first glance, we notice that $C(x, y)$ is defined for all values of x and y. In other words, any ordered pair (x, y) is in the domain. But since $C(x, y)$ represents a *cost function*, we claim that it is unrealistic for x or y to be negative. (How does one produce -3 golf balls?) It is possible for x or y to be 0, so we could write that the domain is all ordered pairs (x, y) with real numbers x and y such that $x \geq 0$ and $y \geq 0$.

(b) We evaluate $C(250, 700)$ by simply substituting as follows:

$$C(x, y) = 0.97x + 0.89y$$
$$C(250, 700) = 0.97(250) + 0.89(700) = 865.50$$

Interpret the Solution: This means that it costs \$865.50 to produce 250 balata balls and 700 durable-cover balls. ■

Cartesian Coordinates in 3-Space

So far in this text we have considered graphs of functions where the function has one independent variable and one dependent variable. The graph of this type of function is determined by points in the xy-plane that we called **ordered pairs**. To graph a function of two independent variables and one dependent variable, we simply extend this concept. Since we now have a total of three variables, two independent and one dependent, we need three coordinate axes and we plot points determined by **ordered triples** (x, y, z). In Figure 18.1.1 we have three coordinate axes that meet, at right angles to each other, at a point called the **origin** $(0, 0, 0)$. Think of our familiar xy-plane as being horizontal, like the floor of a room, while the z-axis extends vertically above and below the plane. See Figure 18.1.2.

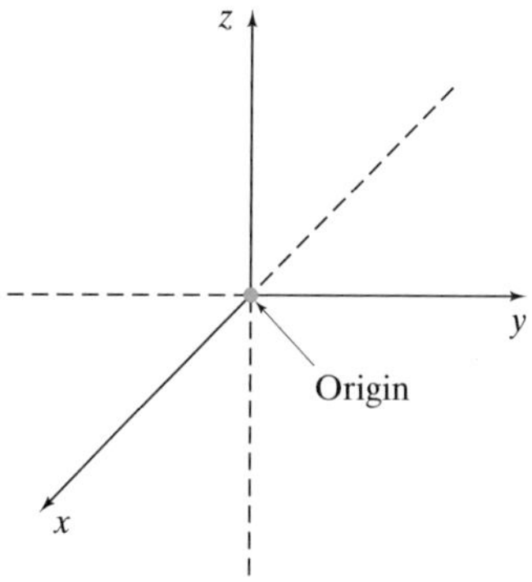

Figure 18.1.1

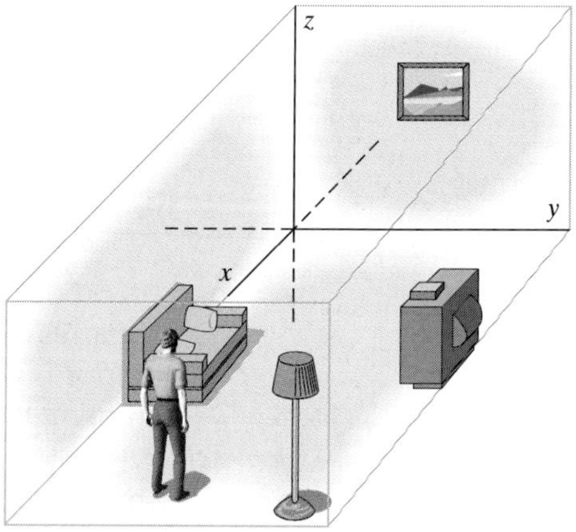

Figure 18.1.2

In Figure 18.1.1 we have labeled the x-, y-, and z-axes. Consider the solid portion of each axis to be the positive portion of the axis, while the dashed part corresponds to the negative portion of each axis. These three axes may be used to determine the location of any point in three-dimensional space.

Any point in three-dimensional space, or **3-space**, can be represented with an ordered triple (x, y, z). When considering points in space, imagine the first two coordinates of an ordered triple as indicating where to go in the xy-plane. The third coordinate, the z-coordinate, tells us whether to go up or down or to stay put. Example 4 shows how to plot points in space.

Example 4 Plotting Points in Space

Plot the following points:

(a) $(2, 3, 1)$ **(b)** $(-2, -4, -2)$

Solution

(a) As shown in Figure 18.1.3a, we first locate the point $(2, 3)$ in the xy-plane. We start at the origin and move 2 units along the positive x-axis and then 3 units to the right (in the positive direction), taking care to remain parallel to the y-axis. We use a "✗" to locate $(2, 3)$ in the xy-plane. Since our z-coordinate is a *positive* 1, we move *up* 1 unit directly above this location, taking care to remain parallel to the z-axis, and plot the point $(2, 3, 1)$ as shown in Figure 18.1.3b.

(b) In a manner similar to part (a), we first locate $(-2, -4)$ in the xy-plane. See Figure 18.1.4a. Here our z-coordinate is a *negative* 2, so we move *down* 2 units from this location and plot the point $(-2, -4, -2)$, as shown in Figure 18.1.4b.

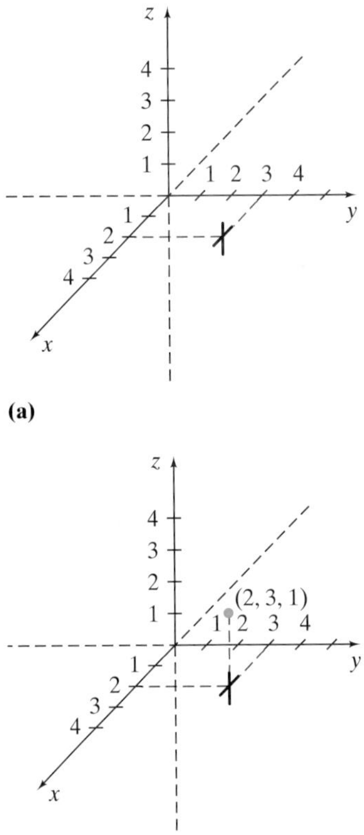

(a)

(b)

Figure 18.1.3

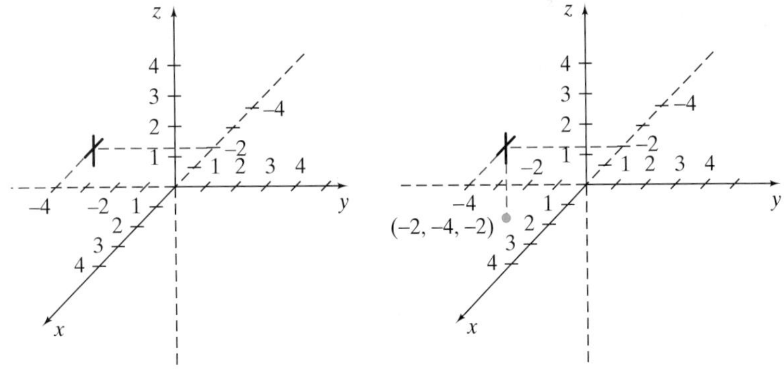

(a) (b)

Figure 18.1.4

■

Checkpoint 2 Now work Exercise 31.

Graphing in 3-Space

Recall that for a function f of one independent variable the graph of $y = f(x)$ is a curve that is above or below the x-axis. This curve is a collection of all points (x, y) such that $y = f(x)$. Also note that the curve lies in the xy-plane.

For a function of two independent variables, say $z = f(x, y)$, the graph of $f(x, y)$ is a collection of all points (x, y, z) such that $z = f(x, y)$. The graph of $z = f(x, y)$ is called a **surface** and is either above or below the xy-plane. Recall that an equation of the form $z = f(x, y)$ represents a function of two independent variables if for each ordered pair of real numbers (x, y) the equation determines a unique real number z. We can think of the domain of $z = f(x, y)$ as being any ordered pair (x, y) for which z is defined.

As we can imagine, graphing functions of two variables is difficult since it involves drawing three-dimensional graphs. Computers and some calculators can generate a 3D graph fairly quickly. Figure 18.1.5 shows several graphs of surfaces.

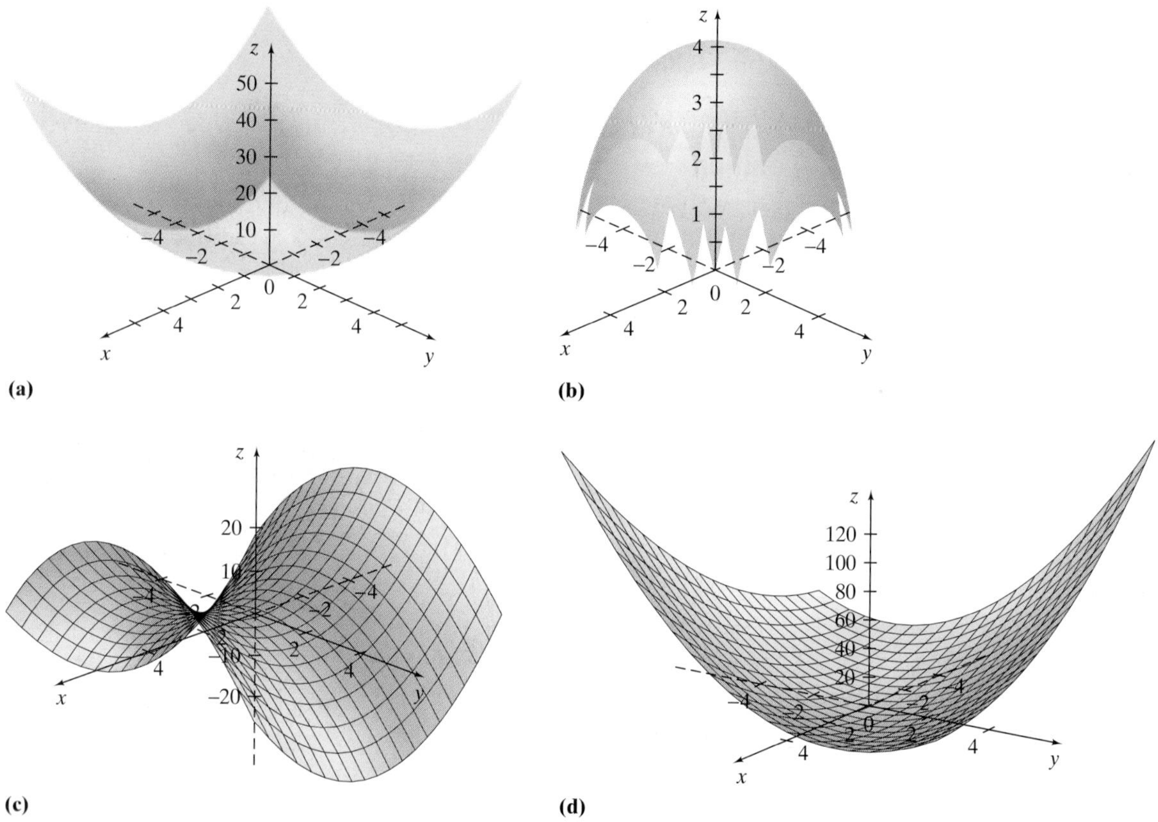

Figure 18.1.5 **(a)** Graph of $z = x^2 + y^2$. **(b)** Graph of $z = \sqrt{16 - x^2 - y^2}$. **(c)** Graph of $z = x^2 - y^2$. **(d)** Graph of $z = x^2 - 2xy + 2y^2$.

In the next section we will use two-dimensional techniques to aid us in visualizing the three-dimensional (or 3D) graphs. In this section we direct our focus on the domains of functions of two variables and some very basic 3D graphs. The most elementary 3D graphs are **planes**. Some planes are illustrated in Example 5.

Example 5 Graphing Planes

Sketch a graph of the following planes.

(a) $z = 2$ **(b)** $x = 2$ **(c)** $y = 3$

Solution

(a) The equation $z = 2$ is satisfied by all ordered triples having the form $(x, y, 2)$. To find a point on this graph, we locate any point (x, y) in the xy-plane and then move up 2 units. The result is a plane 2 units above the xy-plane and parallel to the xy-plane. The surface is shown in Figure 18.1.6.

(b) The equation $x = 2$ is satisfied by all ordered triples of the form $(2, y, z)$. To find any point on this graph, we locate any point of the form $(2, y)$ in the xy-plane or, equivalently, any point on the line $x = 2$ in the xy-plane and plot all points above, on, or below this line. The graph of $x = 2$ is a plane parallel to the yz-plane, as shown in Figure 18.1.7.

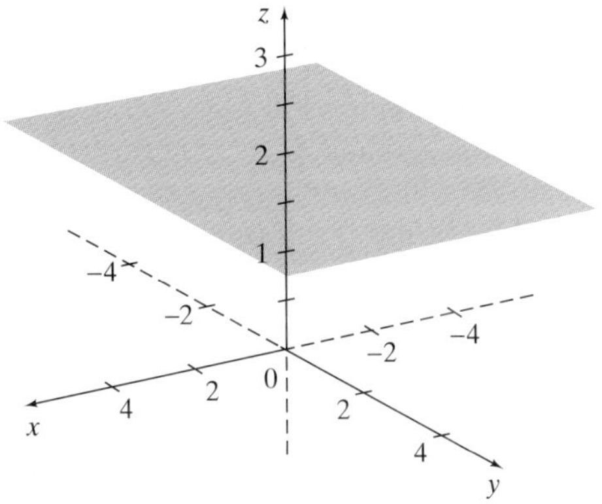

Figure 18.1.6 Graph of $z = 2$.

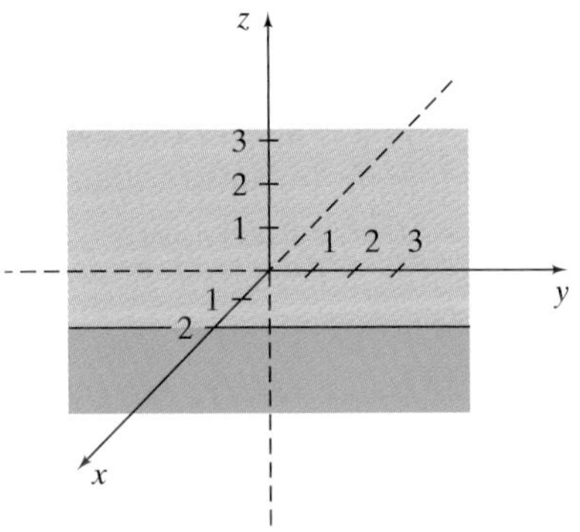

Figure 18.1.7 Graph of $x = 2$.

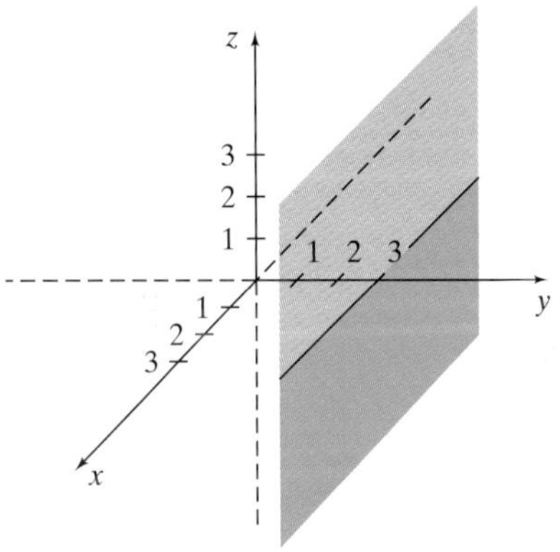

Figure 18.1.8 Graph of $y = 3$.

(c) The equation $y = 3$ is satisfied by all ordered triples of the form $(x, 3, z)$. To find any point on this graph, we locate any point of the form $(x, 3)$ in the xy-plane or, equivalently, any point on the line $y = 3$ in the xy-plane and plot all points above, on, or below this line. The graph of $y = 3$ is a plane parallel to the xz-plane, as shown in Figure 18.1.8. ■

In Example 5, all the planes that we graphed had only one variable in the equation. We notice that when this occurs the resulting graph is a plane that is parallel to one of the **coordinate planes**. The three coordinate axes determine three coordinate planes, which are the xy-, the xz-, and the yz-plane. See Figure 18.1.9.

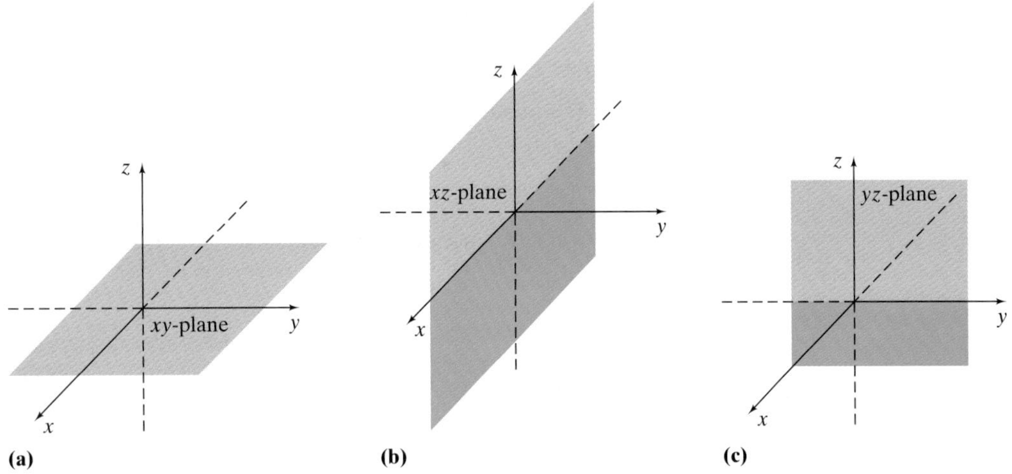

Figure 18.1.9 **(a)** Graph of the xy-plane. **(b)** Graph of the xz-plane. **(c)** Graph of the yz-plane.

Planes parallel to the coordinate planes are used in the next section when we look at *level curves* and *cross sections* as part of our two-dimensional analysis of a 3D surface. Since this will be used in future work, we offer the following:

For any real constant c, we have the following:

- The graph of $z = c$ is parallel to the xy-plane.
- The graph of $x = c$ is parallel to the yz-plane.
- The graph of $y = c$ is parallel to the xz-plane.

Checkpoint 3

Now work Exercise 47.

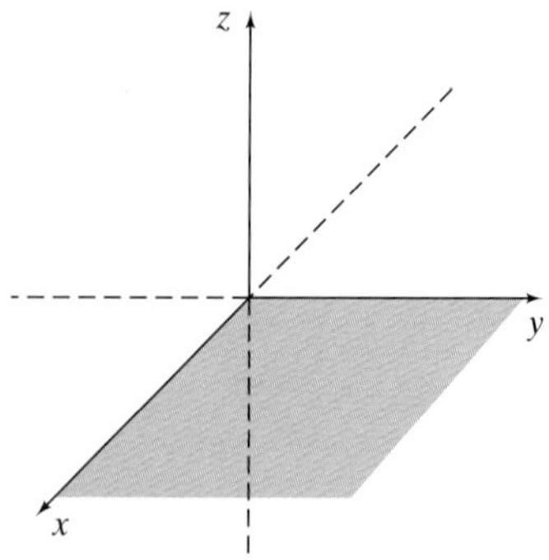

Figure 18.1.10

Most applications that we encounter in this chapter will have the domain restricted to quadrant 1 in the xy-plane, that is, $x \geq 0$ and $y \geq 0$. Figure 18.1.10 shows the domain for most applications that we will examine. The surface of a 3D graph is either entirely above, entirely below, or has portions above and portions below this quadrant of the xy-plane. The last part of this section focuses on one such application.

Cobb–Douglas Production Function

When producing any item, many inputs are used in the production process. Mathematically, if we let Q represent the output, then we have

$$Q = f(x_1, x_2, x_3, \ldots, x_k)$$

where each $x_1, x_2, \ldots$ is a different input. Clearly, Q is a function of several variables. However, economists tend to reduce all of these different inputs in this **production function** to two, L and K, and use

$$Q = f(L, K)$$

where Q is the number of units of output, L is the number of units of **labor**, and K is the number of units of **capital**. Capital includes many items, such as buildings, equipment, and insurance. This production function was developed and made popular by the mathematician Charles Cobb and the economist Paul Douglas and is called the Cobb–Douglas production function. The most general form of a Cobb–Douglas production function is given in the following box.

Cobb–Douglas Production Function

If L represents the units of labor and K represents the units of capital, then the total production Q is given by the **Cobb–Douglas production function**

$$Q = aL^bK^c$$

where a, b, and c are constants.

Example 6 Analyzing a Cobb–Douglas Production Function

The Cobb–Douglas production function was introduced in 1928. It was originally constructed for all the manufacturing output in the United States for the years 1899 to 1922. The production function for all U.S. manufacturing output from

1899 to 1922 is given by

$$Q = 1.01L^{0.75}K^{0.25}$$

where Q is the total yearly production, K is the capital investment, and L is the total labor force. Using x for labor and y for capital, we can rewrite this production function as

$$Q = f(x, y) = 1.01x^{0.75}y^{0.25}$$

(a) Determine the reasonable domain of $Q = f(x, y)$.

(b) Determine if the graph of Q is above or below the xy-plane.

(c) Evaluate Q when $x = 50$ and $y = 60$ and interpret.

Solution

(a) Since x represents the total labor force and y represents the total capital investment, we conclude that realistically neither can be negative. So the reasonable domain is all ordered pairs (x, y) such that $x \geq 0$ and $y \geq 0$. Graphically, it is the region shaded in Figure 18.1.11.

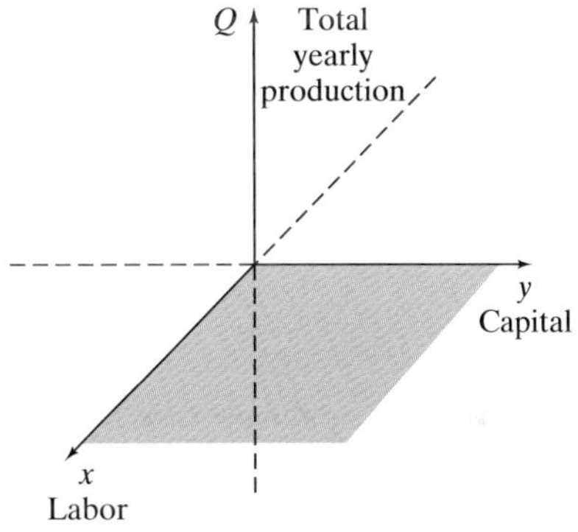

Figure 18.1.11

(b) From part (a) we know that $x \geq 0$ and $y \geq 0$. If $x = 0$ or if $y = 0$, we have $Q = 0$. If $x \neq 0$ and $y \neq 0$, then $Q > 0$. So we conclude that the graph of Q, the surface, lies above the xy-plane. (It intersects the x-axis and the y-axis.) Figure 18.1.12 gives two different perspectives of the graph of Q. In Figure 18.1.12a the axes are in a standard position. In Figure 18.1.12b the axes are rotated 180°.

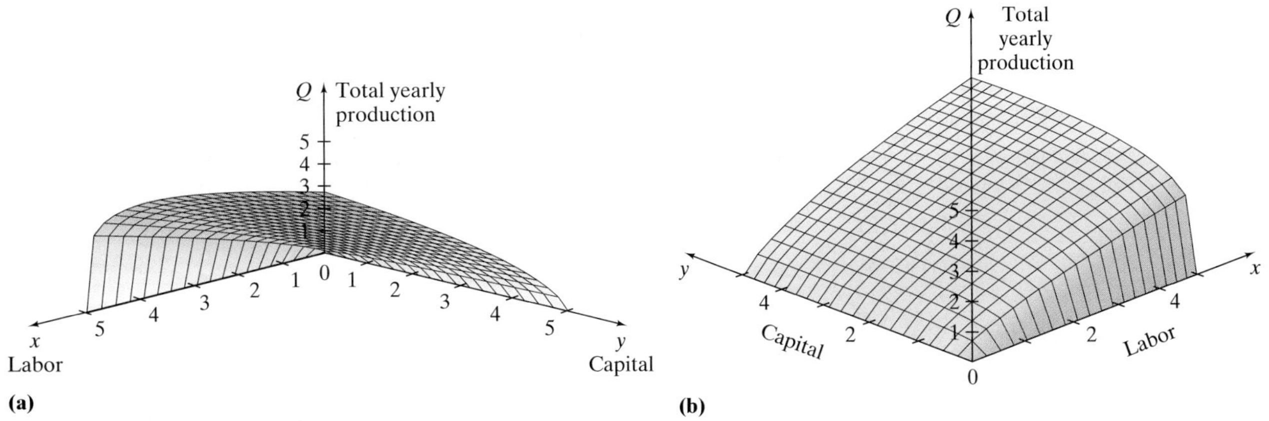

Figure 18.1.12 Graph of $Q = 1.01x^{0.75}y^{0.25}$.

Interactive Activity

How many units above the xy-plane is the point on the surface of $Q = 1.01x^{0.75}y^{0.25}$ when $x = 50$ and $y = 60$? Explain.

WWW

(c) To evaluate Q when $x = 50$ and $y = 60$, we simply substitute these values for x and y into the production function, as follows:

$$Q = 1.01x^{0.75}y^{0.25} = 1.01(50)^{0.75}(60)^{0.25} \approx 52.86$$

Interpret the Solution: Thus, when 50 units of labor and 60 units of capital are used, the number of units produced is approximately 52.86. ■

Note: For future graphs of production functions we will use the perspective in Figure 18.1.12b unless stated otherwise. This perspective shows the important features of the surface of a production function quite nicely.

When Cobb and Douglas first introduced their production function, they gave the more specific form

$$Q = aL^bK^{1-b}$$

They used this form because they restricted the results to what economists call **constant return to scales**. Return to scales is discussed in economics theory, but for our purposes we just want to notice that, to have constant return to scales, the sum of the exponents is equal to 1. In fact, this is the only difference between Cobb and Douglas's specific form and the general form presented prior to Example 6.

SUMMARY

In this section we introduced the concept of a function of two independent variables.

- **Function of two independent variables**: An equation of the form $z = f(x, y)$ represents a function of two independent variables if for each ordered pair of real numbers (x, y) the equation determines a unique real number z.

We saw that the domain of a function of two variables is in the xy-plane and the graph is in **3-space**. We plotted points in 3-space, looked at graphs of some surfaces in space, and graphed some very basic surfaces in space called **planes**.

- For any real constant c, we have the following:

  The graph of $z = c$ is parallel to the xy-plane. For example, see Figure 18.1.6.

  The graph of $x = c$ is parallel to the yz-plane. For example, see Figure 18.1.7.

  The graph of $y = c$ is parallel to the xz-plane. For example, see Figure 18.1.8.

SECTION 18.1 EXERCISES

In Exercises 1–4, evaluate $f(x, y) = x^2 - 2xy + 3y^2$ for the indicated values.

1. $f(2, 3)$ **2.** $f(-1, 4)$

3. $f(-2, -1)$ **4.** $f(3, -2)$

In Exercises 5–8, evaluate $f(x, y) = \dfrac{x^2 + y^2}{x + y}$ for the indicated values.

5. $f(3, 1)$ **6.** $f(-1, 2)$

7. $f(-2, -3)$ **8.** $f(1, -2)$

In Exercises 9–12, evaluate $g(x, y) = \dfrac{3x - y + 1}{x^2 - y^2}$ for the indicated values.

9. $g(1, 4)$ **10.** $g(-2, -5)$

11. $g(-2, 2)$ **12.** $g(2, 0)$

In Exercises 13–16, evaluate $f(x, y) = y + x\ln x + xe^y$ for the indicated values.

13. $f(1, 2)$ **14.** $f(1, 0)$

15. $f(e, 0)$ **16.** $f(e^2, 1)$

For Exercises 17–22, determine the domain of the given function. Recall that here the domain is the set of all ordered pairs (x, y) in the xy-plane.

17. $f(x, y) = x + 2y$ **18.** $g(x, y) = 2x - y$

19. $g(x, y) = \dfrac{1}{x + y}$ **20.** $f(x, y) = \dfrac{3x}{x - y}$

21. $f(x, y) = \dfrac{3}{x - 4y}$ **22.** $g(x, y) = e^y + x\ln x$

For Exercises 23–36, plot the given points.

23. $(4, 4, 2)$ **24.** $(-3, 2, -1)$

25. $(-2, 1, 2)$
26. $(-3, 4, 1)$
27. $(-3, -1, 1)$
28. $(2, -2, 3)$
29. $(1, -3, -2)$
30. $(2, -3, -2)$
✓ **31.** $(4, 2, -1)$
32. $(2, 5, -3)$
33. $(-1, -3, -2)$
34. $(-2, -4, -1)$
35. $(2, -4, 0)$
36. $(-3, 1, 0)$

For Exercises 37–44, without plotting the point, determine if the given point is above or below the xy-plane. Explain.

37. $(2, 2, 3)$
38. $(3, 1, 2)$
39. $(-3, -2, 5)$
40. $(-1, -2, 4)$
41. $(4, -2, -1)$
42. $(2, 0, -5)$
43. $(0, 3, -0.5)$
44. $(1, -1, 0.1)$

In Exercises 45–50, sketch the graph of the given plane in 3-space.

45. $x = 3$
46. $x = -2$
✓ **47.** $z = 3$
48. $z = -2$
49. $y = -1$
50. $y = 2$

Applications

51. Seamount Boats spends x thousand dollars each week on newspaper advertising and y thousand dollars each week on radio advertising. The company has weekly sales, in tens of thousands of dollars, given by

$$S(x, y) = 2x^2 + y$$

(a) Determine $S(5, 3)$ and interpret.
(b) Determine $S(3, 5)$ and interpret.

52. Lakeway Boats spends x thousand dollars each week on radio advertising and y thousand dollars each week on television advertising. The company has weekly sales, in tens of thousands of dollars, given by

$$S(x, y) = 3x + 2y^3$$

(a) Determine $S(2, 6)$ and interpret.
(b) Determine $S(6, 2)$ and interpret.

53. Tube Town, a recently opened water park, spends x thousand dollars on radio advertising and y thousand dollars on television advertising. The park has weekly ticket sales, in tens of thousands of dollars, given by

$$TS(x, y) = 1.5x^2 + 3.2y^3$$

(a) Determine $TS(1, 0.5)$ and interpret.
(b) Determine $TS(0.5, 1)$ and interpret.

54. In their study of human groupings, anthropologists often use an index that indicates the shape of the head, the *cephalic index*. The cephalic index is given by

$$C(W, L) = 100 \cdot \frac{W}{L}$$

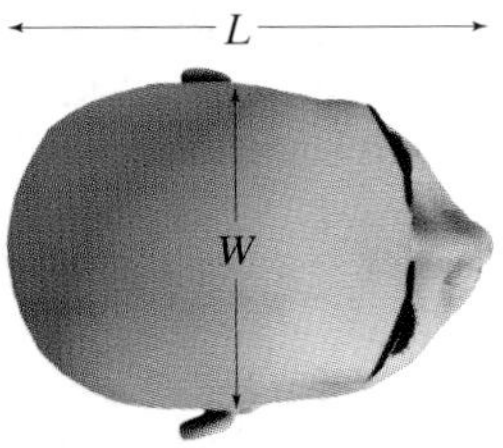

where W is the width and L is the length of an individual's head. Both measurements are made across the top of the head and are in inches. Determine $C(6, 8.2)$ and $C(8.8, 9.7)$.

55. Poiseuille's law states that the resistance, R, for blood flowing in a blood vessel is given by

$$R(L, r) = K \cdot \frac{L}{r^4}$$

where K is a constant, L is the length of the blood vessel, and r is the radius of the blood vessel. Determine $R(36, 1)$ and $R(36, 2)$.

56. An individual's body surface area is approximated by

$$BSA(w, h) = 0.007184w^{0.425}h^{0.725}$$

where BSA is in square meters, w is weight in kilograms, and h is height in centimeters. Determine $BSA(70, 160)$ and interpret. This formula is known as the DuBois and DuBois formula (*Source:* www-users.med.cornell.edu/~spon/picu/bsacalc.htm).

57. The wind chill index is modeled by

$$f(v, T) = 35.74 + 0.6215T - 35.75v^{0.16} + 0.4275Tv^{0.16}$$

where v represents the velocity of the wind in miles per hour, T is the actual air temperature in degrees Fahrenheit, and $f(v, T)$ is the wind chill index in degrees Fahrenheit. (Wind chill is the temperature that it actually feels like.) Evaluate $f(25, 10)$ and interpret. Round the answer to the nearest integer (*Source:* National Oceanic and Atmospheric Administration, www.noaa.gov).

58. The *Doyle log rule* is one method used to determine the yield of a log, measured in board-feet. In English, the rule states

> Deduct 4 inches from the diameter of the log as an allowance for slab; square one-quarter of the remainder and multiply the result by the length of the log in feet. (*Source:* http://www.forestry.uga.edu.docs/950-measurementscaling.html.)

Mathematically, this is translated as

$$f(d, L) = \left(\frac{d - 4}{4}\right)^2 \cdot L$$

where d is the diameter in inches, L is the length in feet, and $f(d, L)$ is the number of board-feet. Determine $f(30, 12)$ and interpret.

59. The Leaf Eater Company manufactures and sells leaf blowers and a special 10-foot blower attachment to clean gutters. The monthly cost function, in dollars, for the company is given by

$$C(x, y) = 1000 + 35x + 1.5y$$

where x is the number of leaf blowers produced each month and y is the number of 10-foot blower attachments produced each month. Determine $C(50, 30)$ and interpret.

60. The Leaf Eater Company from Exercise 59 has determined the following:

$p = 120 - 0.8x - 0.1y,$ the price in dollars for a leaf blower

$q = 52.3 - 0.15x + 0.015y,$ the price in dollars for a 10-foot attachment

where x is the number of leaf blowers sold each month and y is the number of 10-foot blower attachments sold each month.

(a) Determine the revenue function $R(x, y)$.

(b) Determine $R(50, 30)$ and interpret.

61. Refer to Exercises 59 and 60.

(a) Using the cost function from Exercise 59 and the revenue function from Exercise 60, determine the profit function $P(x, y)$.

(b) Determine $P(50, 30)$ and interpret.

62. The Chalet Bicycle Company manufactures 21-speed racing bicycles and 21-speed mountain bicycles. Let x represent the weekly demand for a 21-speed racing bicycle and y represent the weekly demand for a 21-speed mountain bicycle. The weekly price–demand equations are given by

$p = 350 - 4x + y,$ the price in dollars for a 21-speed racing bicycle

$q = 450 + 2x - 3y,$ the price in dollars for a 21-speed mountain bicycle

The cost function is given by

$$C(x, y) = 390 + 95x + 100y$$

(a) Determine the weekly revenue function $R(x, y)$.

(b) Evaluate $R(15, 20)$ and interpret.

(c) Determine the weekly profit function $P(x, y)$.

(d) Evaluate $P(15, 20)$ and interpret.

63. A T-shirt maker produces two types of tie-dyed T-shirts. The full-rainbow T-shirt costs \$5 each to produce and the partial-rainbow T-shirt costs \$4 each to produce. She sells the full-rainbow T-shirts for \$12 each and the partial-rainbow T-shirts for \$9.50 each.

(a) Determine the cost function $C(x, y)$ for making x full-rainbow T-shirts and y partial-rainbow T-shirts. Assume that the fixed costs are \$50.

(b) Determine the revenue function $R(x, y)$.

(c) Determine the profit function $P(x, y)$.

64. If \$5000 is invested at an annual interest rate of r (in decimal form) compounded quarterly for t years, the total amount accumulated is given by

$$f(r, t) = 5000\left(1 + \frac{r}{4}\right)^{4t}$$

Determine $f(0.075, 15)$ and interpret.

65. If an amount P dollars is invested at an annual interest rate of 6.125% compounded monthly for t years, the total amount accumulated is given by

$$f(P, t) = P\left(1 + \frac{0.06125}{12}\right)^{12t}$$

Determine $f(2000, 20)$ and interpret.

66. If an amount P dollars is invested at an annual interest rate of 6% compounded continuously for t years, the total amount accumulated is given by

$$f(P, t) = Pe^{0.06t}$$

Determine $f(3000, 10)$ and interpret.

67. If \$2000 is invested at an annual interest rate of r (in decimal form) compounded continuously for t years, the total amount accumulated is given by

$$f(r, t) = 2000e^{r \cdot t}$$

Determine $f(0.0725, 25)$ and interpret.

68. A golf club manufacturer has a Cobb–Douglas production function given by

$$Q = f(x, y) = 21x^{0.3}y^{0.75}$$

where x is utilization of labor and y is utilization of capital. Determine the number of units of golf clubs produced when 200 units of labor and 75 units of capital are used.

69. A sports shoe manufacturer has a Cobb–Douglas production function given by

$$Q = f(x, y) = 42x^{0.37}y^{0.66}$$

where x is utilization of labor and y is utilization of capital. Determine the number of units of sports shoe produced when 300 units of labor and 100 units of capital are used.

70. The volume of a cylinder, such as a soup can, is given by $\pi r^2 h$, where r is the radius and h is the height. Since the volume is a function of the radius and the height, we can write

$$V(r, h) = \pi r^2 h$$

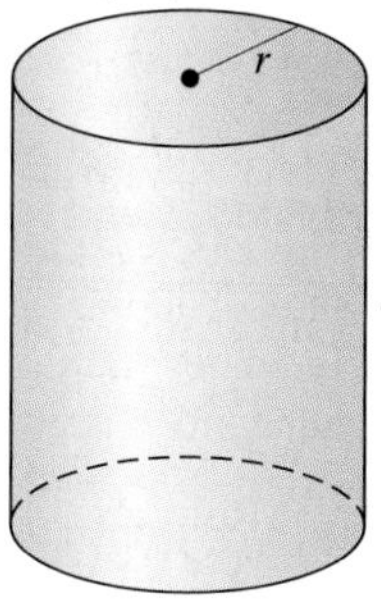

(a) Determine $V(2, 6)$ and interpret. (Each is measured in inches.)

(b) If the radius and height are equal, we can express the volume of the cylinder as a function of one variable, either r or h. Determine $V(r)$ and $V(h)$.

71. An individual's IQ is defined to be

$$\text{IQ} = f(a, m) = \frac{100m}{a}$$

where m is the individual's mental age (as determined by a test) in years and a is the individual's actual age in years.

(a) Evaluate $f(9, 12)$ and interpret.

(b) Evaluate $f(12, 9)$ and interpret.

(c) Evaluate $f(12, 12)$ and interpret.

(d) Determine IQ if $a = m$. In your own words, what does this tell us about an IQ of 100?

72. A bicycle seat manufacturer has a Cobb–Douglas production function given by

$$Q = f(x, y) = 22x^{0.75}y^{0.25}$$

where x is utilization of labor and y is utilization of capital. Determine the number of units of bicycle seats produced when 100 units of labor and 50 units of capital are used.

73. *(continuation of Exercise 72)* Suppose that we wish to know what combinations of labor and capital would result in 2000 units of bicycle seats being produced. In other words, we want to know what values of x and y satisfy

$$2000 = 22x^{0.75}y^{0.25}$$

If we solve this equation for y, we get

$$\frac{2000}{22x^{0.75}} = y^{0.25}$$

$$(90.91x^{-0.75})^4 = y$$

(a) In the viewing window [0, 200] by [0, 1000], graph $y = (90.91x^{-0.75})^4$. The curve that you see is called an *isoquant*. We will discuss isoquants in Section 18.2.

(b) Use the TRACE command to approximate the number of units of capital needed if 70 units of labor are used to produce 2000 units of bicycle seats.

(c) Use the TRACE command to approximate the number of units of capital needed if 100 units of labor are used to produce 2000 units of bicycle seats.

(d) Determine three combinations of labor and capital that will produce 2000 units of bicycle seats. (Answers may vary.)

74. The O'Neill Corporation has 10 soft drink bottling plants located in the United States. In a recent year the data for each plant gave the number of labor hours (in thousands), capital (total net assets, in millions), and the total quantity produced (in thousands of gallons). The data are shown in Table 18.1.1. The plants all use the same technology, so a production function can be determined. A Cobb–Douglas production function modeling these data is given by

$$Q = f(L, K) = 1.64L^{0.623}K^{0.357}$$

Table 18.1.1

Labor	Capital	Quantity
100	11	68
100	13	72
110	14	79
125	16	89
133	17	95
140	20	105
151	23	114
152	23	115
160	24	120
166	26	127

(a) Determine $f(125, 18)$ and interpret.

(b) Determine $f(70, 30)$ and interpret.

SECTION PROJECT

The program in Appendix C that performs multiple regression is needed to do this Section Project.

The Basich Company has eight plants in the United States producing handheld calculators. In a recent year the data for each plant gave the number of labor hours (in thousands), capital (in millions), and total quantity produced. The data are shown in Table 18.1.2. The plants all use the same technology, so a production function can be determined.

(a) Enter the data into your calculator and execute the program to determine a Cobb–Douglas production function to model the data. Round all values to the nearest hundredth.

Table 18.1.2

Labor	Capital	Quantity
96	15	12,234
103	20	13,907
104	21	14,187
109	22	14,903
111	27	15,915
121	30	17,507
122	31	17,768
127	36	19,047

(b) Suppose that Richard, CEO of the Basich Company, wants to know all combinations of labor and capital that produce 15,000 calculators. To do this, use the production function from part (a) and rename L with an x and K with a y. Substitute 15,000 for Q and solve the resulting equation for y.

(c) Graph the equation found in part (b) in an appropriate viewing window. Use the TRACE command to determine three realistic combinations of labor and capital that will produce 15,000 calculators.

Section 18.2 Level Curves, Contour Maps, and Cross-sectional Analysis

In Section 18.1 we were introduced to functions of two independent variables and saw that the graphs of these functions are surfaces in 3-space. Figure 18.2.1 shows a graph of a production function.

As mentioned in Section 18.1, graphing surfaces in 3-space is very difficult without the aid of a computer. For this reason **cross-sectional analysis** can aid in understanding the behavior of a surface by analyzing the surface using graphs in two dimensions. For example, consider Figure 18.2.2, which shows the surface in Figure 18.2.1 sliced with a horizontal plane $z = 4$.

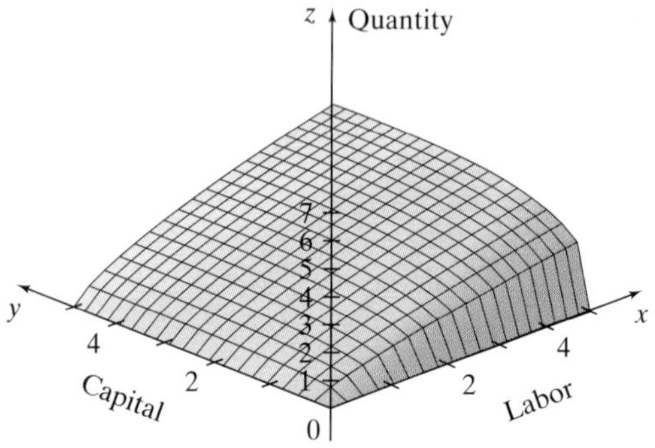

Figure 18.2.1 Graph of $z = 1.3x^{0.75}y^{0.3}$.

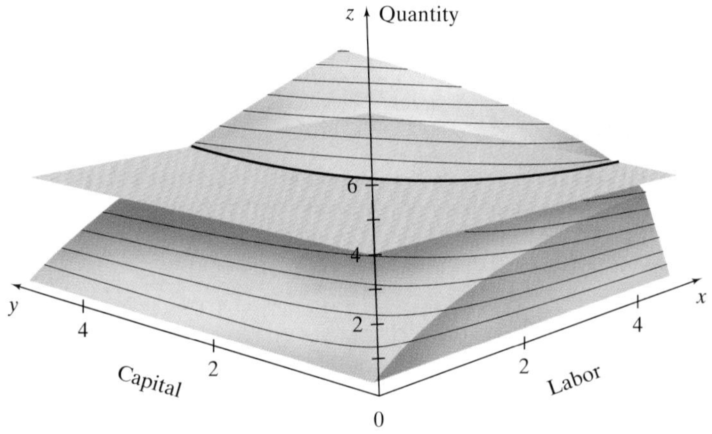

Figure 18.2.2 Graph of $z = 1.3x^{0.75}y^{0.3}$ and the horizontal plane $z = 4$.

The horizontal plane $z = 4$ intersects the surface to form a curve that is parallel to the xy-plane, which means that we can graph the curve in the xy-plane. Slicing of surfaces with horizontal plane(s) and graphing the resulting curve(s) in a single xy-plane is where we begin our analysis of surfaces in 3-space.

Horizontal Cross Sections, Level Curves, and Isoquants

To have an idea of what we will be doing, let's consider the **topographical map** shown in Figure 18.2.3. A *topographical map* is simply a two-dimensional graph of a three-dimensional surface, such as a mountain. The lines we see on the topographical map connect points with the same elevation. In mathematics, we call lines that connect points of equal elevation **contour lines** or **level curves**. Also, we call a collection of contour lines a **contour map**. Assuming that the elevation between the contour lines on the topographical map changes by a constant amount, we see that the more closely packed the lines are, the steeper the terrain. The more spread apart the contour lines are, the flatter the terrain. A topographical map gives a good overall picture of the terrain, indicating where hills

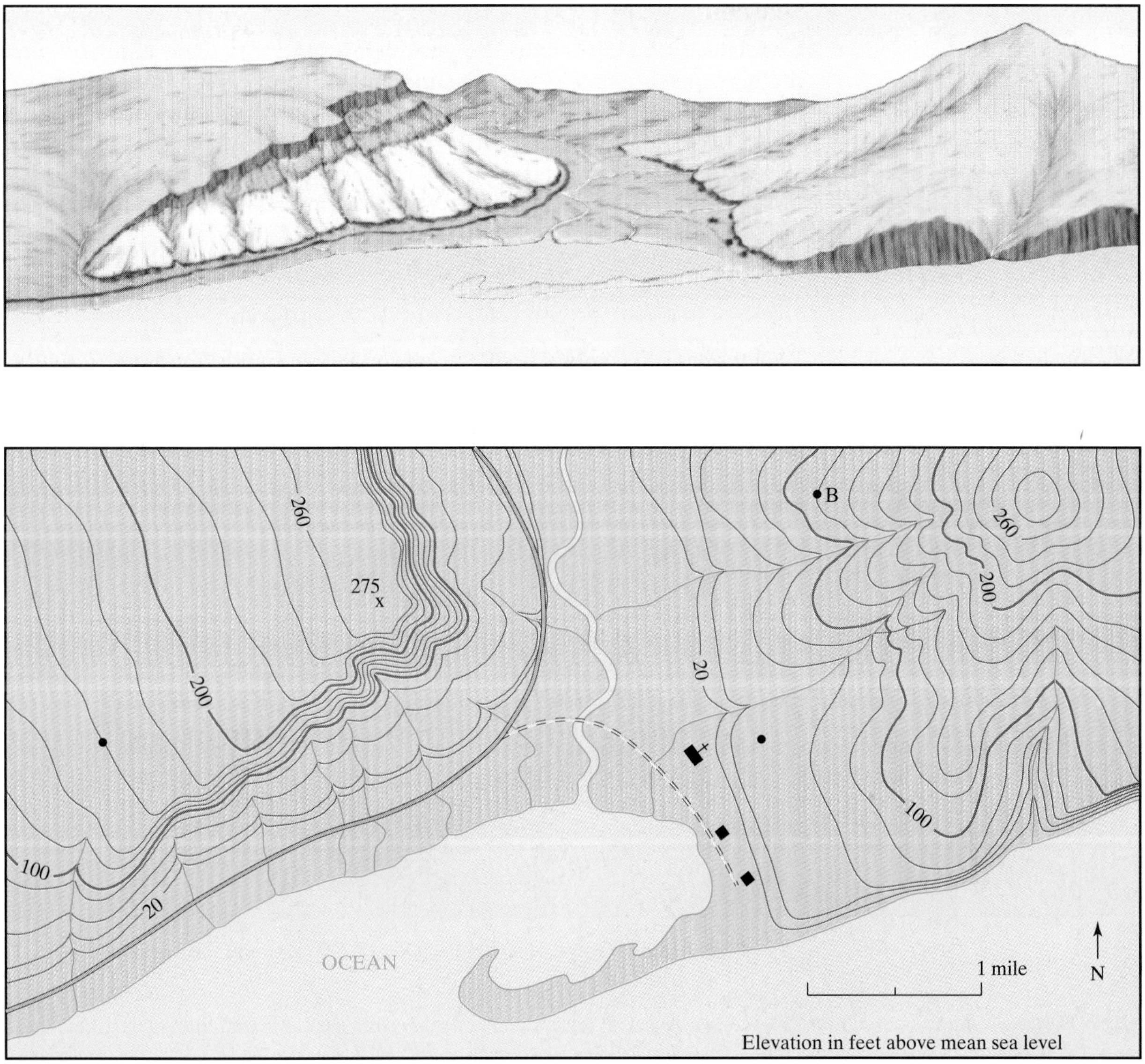

Figure 18.2.3

and flat areas are located. In Example 1 we employ the concept illustrated in the topographical map to a surface in 3-space.

Example 1 Determining a Level Curve

An economist for the Linger Golf Cart Corporation has computed a production function for the manufacture of their golf carts to be

$$f(x, y) = Q = 1.3x^{0.75}y^{0.25}$$

where Q is the number of golf carts produced each week, x is the number of labor hours each day, and y is the daily usage of capital investment. What combinations of labor hours each day and daily usage of capital investment will result in 13 golf carts being produced each week?

Solution

Understand the Situation: We are being asked to determine *all* possible combinations of labor hours (x) and daily usage of capital investment (y) such that $Q = 13$. In other words, we want to know for what values of x and y does

$$13 = 1.3x^{0.75}y^{0.25}$$

Solving this equation for y gives

$$10 = x^{0.75}y^{0.25}$$

$$10x^{-0.75} = y^{0.25}$$

$$y = (10x^{-0.75})^4 = 10{,}000x^{-3}$$

To determine all combinations of x and y, we can simply graph $y = (10x^{-0.75})^4 = 10{,}000x^{-3}$ in the xy-plane. Figure 18.2.4 shows a graph of $f(x, y) = Q = 1.3x^{0.75}y^{0.25}$, and Figure 18.2.5 shows a graph of $y = 10{,}000x^{-3}$. The graph in Figure 18.2.5 answers the question.

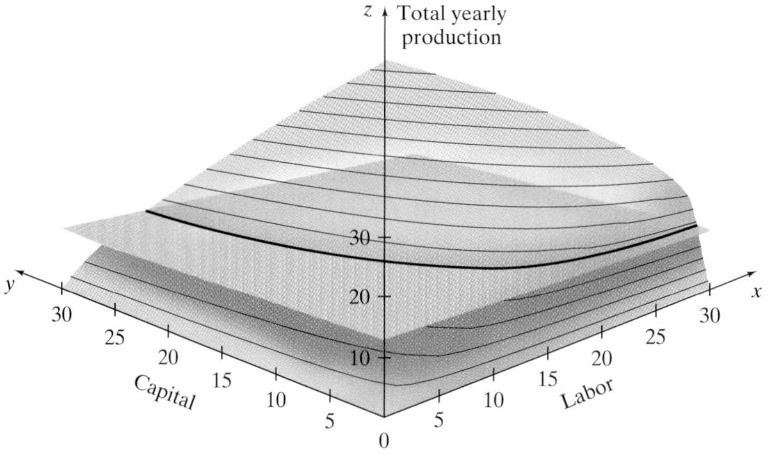

Figure 18.2.4 Graph of $f(x, y) = Q = 1.3x^{0.75}y^{0.25}$ and the horizontal plane $Q = 13$.

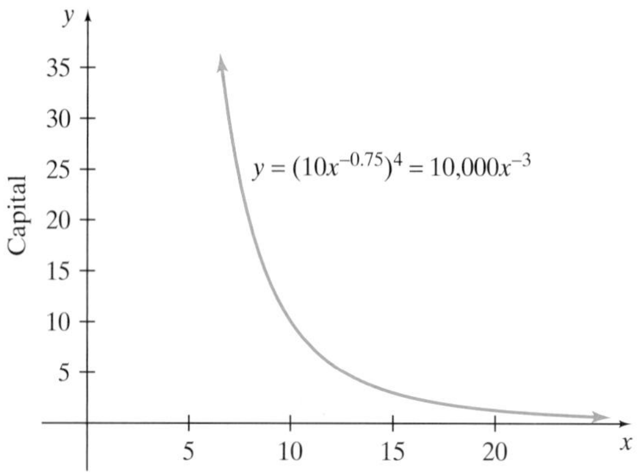

Figure 18.2.5

Economists call the graph in Figure 18.2.5 an **isoquant** (*iso* means same and *quant* is short for quantity). It gives all combinations of x and y that yield a production of 13 golf carts each week. Graphically, it is the result of slicing the surface given by $f(x, y) = Q = 1.3x^{0.75}y^{0.25}$ with the horizontal plane $Q = 13$. (We can think of the variable Q as behaving like the variable z in our xyz-coordinate system.)

Technology Option

To see the isoquant shown in Figure 18.2.5 on your calculator, simply graph $y = 10{,}000x^{-3}$ in the viewing window [0, 20] by [0, 30]. See Figure 18.2.6

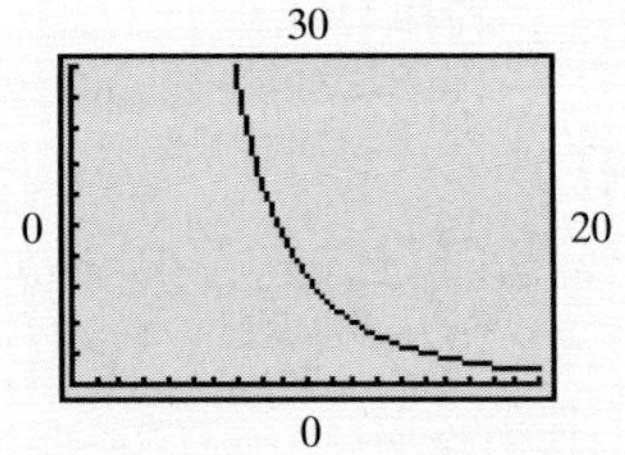

Figure 18.2.6

If we take the curve in Figure 18.2.5 and lift it 13 units above the xy-plane, we get a picture of the behavior of the *surface* 13 units above the xy-plane. Figure 18.2.7 illustrates this.

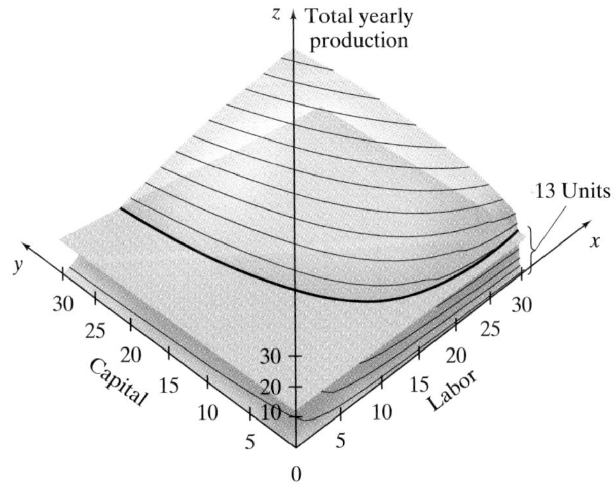

Figure 18.2.7

As we saw with the topographical map, the general mathematical word for an isoquant is *level curve*. It has this name since every point on the curve has the same dependent variable value. Also, recall that another name for level curve is a *contour line*. In general, we have the following:

Level Curve or Contour Line

A **level curve**, or **contour line**, is obtained from a surface $z = f(x, y)$ by slicing it with a horizontal plane $z = c$. The equation for the level curve at height c is given by

$$c = f(x, y)$$

A collection of level curves is called a **contour map** or **contour diagram**. Example 2 illustrates how to construct a contour map.

Example 2 Constructing a Contour Map

Let $z = f(x, y) = x^2 + y^2$. Construct a contour map using $c = 2, 4$, and 6. Compare the contour map to the graph of the surface.

Solution

The level curve at any height c is given by

$$c = x^2 + y^2$$

This is simply the equation of a circle, centered at the origin, with a radius of $\sqrt{c}$. So the level curve at $c = 2$ is found by graphing

$$2 = x^2 + y^2$$

For $c = 4$, we graph $4 = x^2 + y^2$, and for $c = 6$, we graph $6 = x^2 + y^2$. The contour map is shown in Figure 18.2.8, and the graph of $z = x^2 + y^2$ is shown in Figure 18.2.9. Notice that the graph of $z = x^2 + y^2$ gets steeper as we move farther away from the origin. This can be observed on the contour map in Figure 18.2.8, since the level curves become more tightly packed together as we move away from the origin.

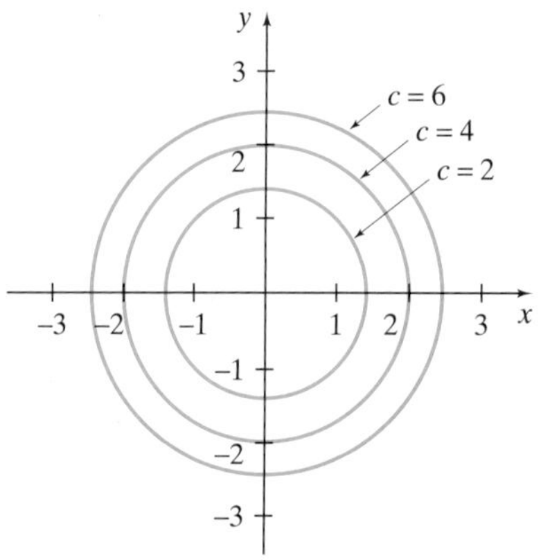

Figure 18.2.8

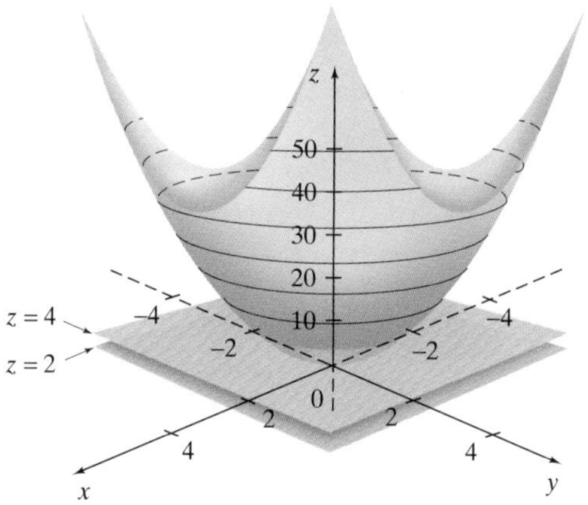

Figure 18.2.9

Checkpoint 1 Now work Exercise 5.

Example 3 Constructing a Contour Map in an Application

Recall that the golf cart manufacturer in Example 1 has a production function of $Q = f(x, y) = 1.3x^{0.75}y^{0.25}$, where Q is the number of golf carts produced each week, x is the number of labor hours each day, and y is the daily usage of capital investment. Construct a contour map (here, a collection of isoquants) using $c = 10, 20, 30$, and 40.

Solution

Recall that we can think of Q as being the same as z. The isoquant (level curve) at any production level (height) is given by

$$c = 1.3x^{0.75}y^{0.25}$$

Solving this equation for y gives

$$\frac{c}{1.3} = x^{0.75}y^{0.25}$$

$$\frac{c}{1.3}x^{-0.75} = y^{0.25}$$

$$y = \left(\frac{c}{1.3}x^{-0.75}\right)^4 = \frac{c^4}{2.8561}x^{-3}$$

We now substitute the given values of c into this equation and graph the resulting equations, as shown in Figure 18.2.10. The graph of the surface is in Figure 18.2.11.

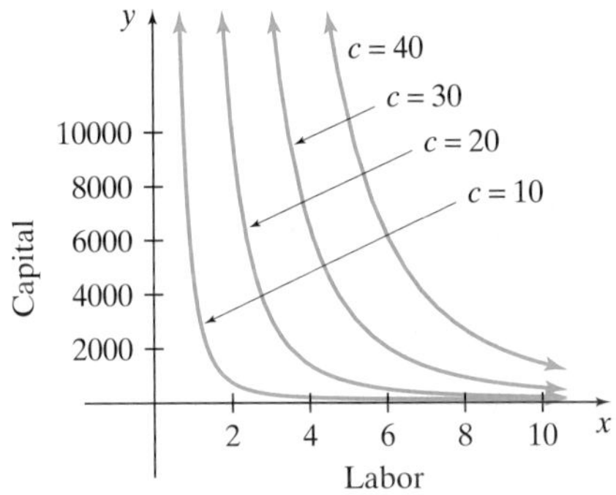

Figure 18.2.10

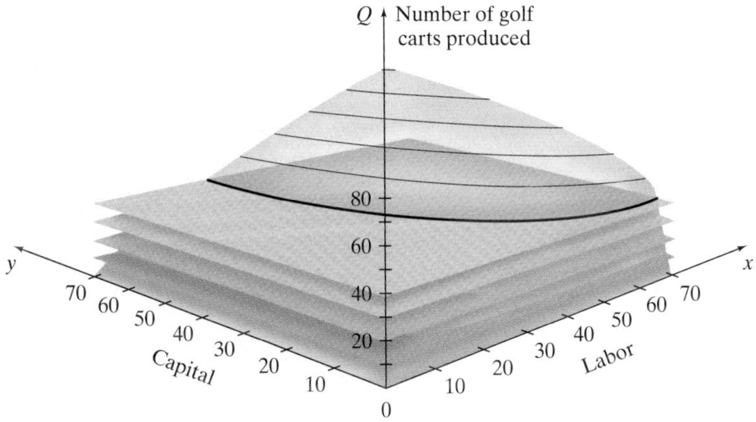

Figure 18.2.11

Notice in Examples 2 and 3 that the contour map is constructed on one xy-plane. Also notice that the values of c in both examples are equally spaced. Having equally spaced c-values and constructing a contour map on a single xy-plane are necessary if we want the contour map to give us an idea of the behavior of the surface.

Indifference Curves

So far we have seen that the level curves of a production function are called isoquants. We now look at another level curve from economics that is called an **indifference curve**. First we need a brief discussion on what economists call a **utility function**. Utility functions and indifference curves are the basis of the modern theory of consumer behavior.

We say that an individual derives *satisfaction* or *utility* from commodities consumed during a given time period. In the time period, the individual will consume a large variety of different commodities. Economists refer to this collection of different commodities as a **commodity bundle**. For different commodity bundles, economists assume that each individual compares alternative commodity bundles and states a preference. Example 4 illustrates this idea.

Example 4 Analyzing Commodity Bundles

Tom enjoys eating cheeseburgers and drinking colas. He was asked to rank the following commodity bundles (of cheeseburgers and colas) for a typical week, with a preferred bundle assigned a higher number. His rankings are given in Table 18.2.1.

Table 18.2.1

Bundle	Colas (x)	Cheeseburgers (y)	Rank
A	3	5	3
B	4	3	3
C	5	2	3
D	1	4	1
E	2	2	1
F	3	1	1

(a) Which bundle(s) is(are) most preferred by Tom? Least preferred?

(b) Make a plot of the data, placing quantity of colas on the x-axis and quantity of cheeseburgers on the y-axis. Connect with a smooth curve those points that have the same rank.

Solution

(a) It appears that Tom prefers bundles A, B, and C the most since he assigned each a rank of 3. He is said to be *indifferent* among these three bundles. Tom prefers bundles D, E, and F the least since he assigned each a rank of 1. Tom is said to be *indifferent* among these three bundles.

(b) Figure 18.2.12 gives a plot of the data.

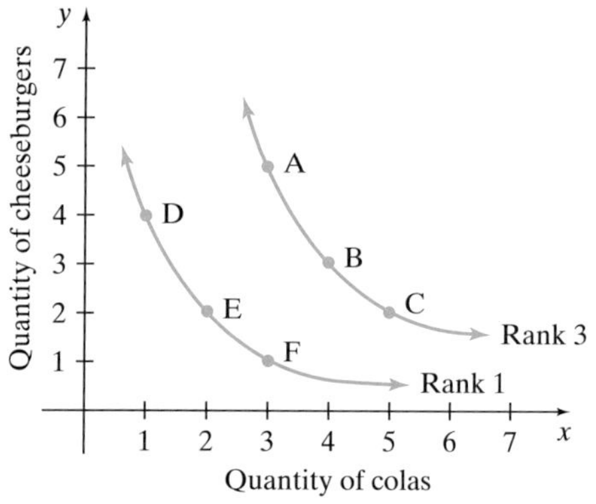

Figure 18.2.12

■

The two curves seen in Figure 18.2.12 are called **indifference curves**, since the consumer in question, Tom, is indifferent to any bundle on each curve. In other words, on the rank 3 indifference curve, Tom would receive the same utility from bundle A as from bundle B or C.

Since the indifference curves in Figure 18.2.12 are labeled with rank 1 and rank 3, we can easily imagine that the *indifference curves are level curves* of some function. The function in question is called a **utility function** and is denoted

$$U(x, y) = f(x, y)$$

The utility function assigns a numerical value (or utility level) to commodity bundles for goods x and y. Without going too far into economics theory, all that we require of the utility function is that it reflect the same rankings that a consumer assigns to alternative commodity bundles. So if a consumer prefers bundle A to bundle D, the utility function has to assign a *larger* numerical value to bundle A than to bundle D. Example 5 shows how this is done.

Example 5 Sketching an Indifference Map

Suppose that the utility from consuming x colas and y cheeseburgers is given by $U(x, y) = \sqrt{xy}$. Draw a contour map for $c = 1, 2, 3,$ and 4. Here we will have four indifference curves. A collection of indifference curves is also known as an **indifference map**.

Solution

As we have seen so far in this section, we can algebraically set this up as

$$\begin{aligned} U(x, y) &= \sqrt{xy} \\ c &= \sqrt{xy} \\ c^2 &= xy \\ y &= \frac{c^2}{x} \end{aligned}$$

Now we substitute 1, 2, 3, and 4 in for c and graph the following in the xy-plane.

$$y = \frac{1}{x}, \quad y = \frac{4}{x}, \quad y = \frac{9}{x} \quad \text{and} \quad y = \frac{16}{x}$$

Figure 18.2.13 is the contour map and Figure 18.2.14 is a graph of the utility function. Again notice that the curves in Figure 18.2.13 are the result of slicing the surface in Figure 18.2.14 with horizontal planes 1, 2, 3, and 4 units, respectively, above the xy-plane.

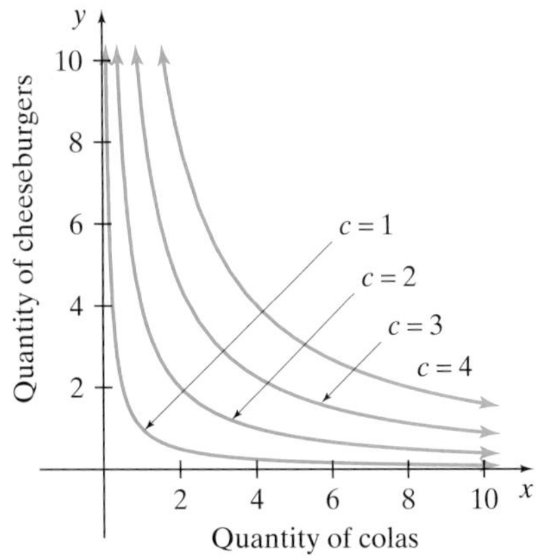

Figure 18.2.13

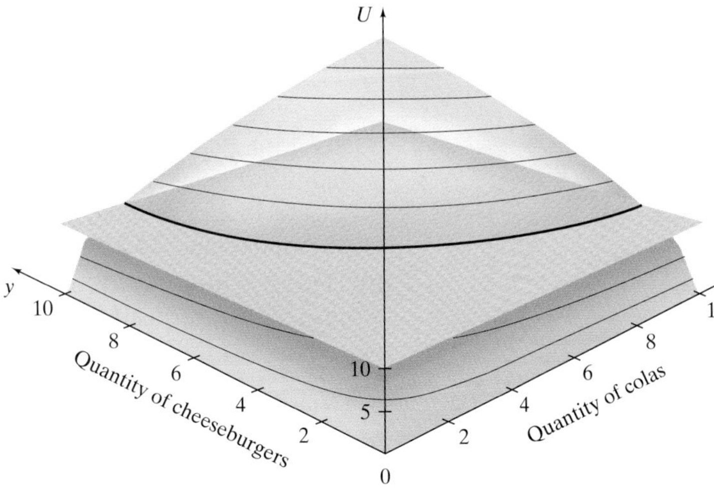

Figure 18.2.14

■

✓ Checkpoint 2

Now work Exercise 17.

Vertical Cross Sections

So far we have seen how slicing a surface with a horizontal plane produces a level curve that can be drawn in the xy-plane. A collection of several slices at different heights produces a collection of level curves that gives us an idea of the behavior of the surface in 3-space. However, if we look at **vertical cross sections** of the form $x = c$ and $y = c$, we can sometimes improve on our 3-space visualization.

In Example 2, we sliced the surface with planes of the form $z = c$. See Figure 18.2.9. Notice that these planes are perpendicular to the z-axis and that the collection of level curves, that is, the contour map, was graphed on the xy-plane. The procedure for vertical cross-sections is basically the same. If we slice a surface with planes of the form $x = c$ (these will be vertical planes), the planes are perpendicular to the x-axis, and the resulting cross sections are graphed on the yz-plane. Similarly, if we slice a surface with planes of the form $y = c$ (again, vertical planes), the planes are perpendicular to the y-axis, and the resulting cross sections are graphed on the xz-plane. We demonstrate this process in Example 6.

Example 6 Constructing Vertical Cross Sections

Consider $z = y^2 - x^2$.

(a) Sketch the cross sections on the xz-plane with y fixed at 0, ± 1, and ± 2.

(b) Sketch the cross sections on the yz-plane with x fixed at 0, ± 1, and ± 2.

Solution

(a) Here y is fixed, which means that we are slicing the surface with vertical planes. The vertical planes are perpendicular to the y-axis and are of the form $y = 0$, $y = \pm 1$, and $y = \pm 2$. So we need to graph the following curves in the xz-plane.

$$\begin{aligned} z &= -x^2 && (y = 0) \\ z &= 1 - x^2 && (y = \pm 1) \\ z &= 4 - x^2 && (y = \pm 2) \end{aligned}$$

To graph these curves, we simply need to treat z as the dependent variable and x as the independent variable. The graphs of these curves are given in Figure 18.2.15, and the graph of the surface with a vertical cross section is shown in Figure 18.2.16.

(b) Here x is fixed, which means that we are slicing the surface with vertical planes. The vertical planes are perpendicular to the x-axis and are of the

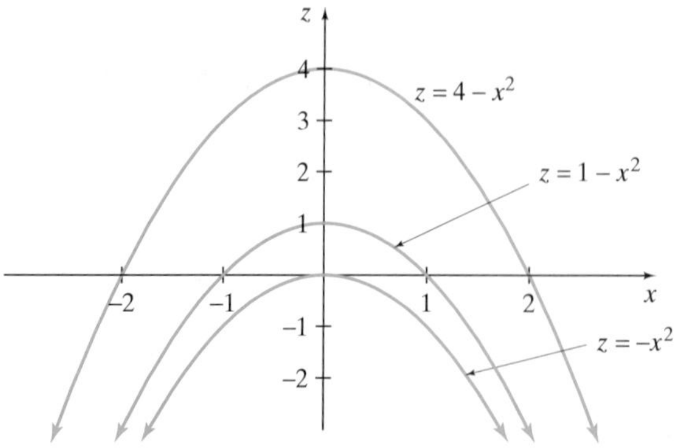

Figure 18.2.15

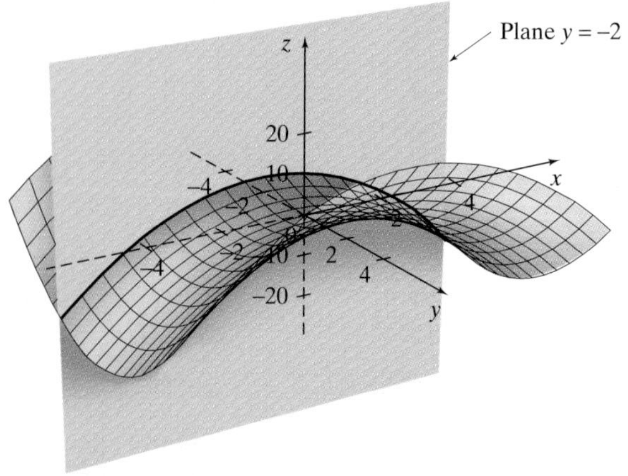

Figure 18.2.16 Vertical plane $y = -2$ intersects the surface in a curve.

form $x = 0$, $x = \pm 1$, and $x = \pm 2$. Hence, we need to graph the following curves in the yz-plane.

$$z = y^2 \qquad (x = 0)$$
$$z = y^2 - 1 \qquad (x = \pm 1)$$
$$z = y^2 - 4 \qquad (x = \pm 2)$$

Notice that here z is the dependent variable and y is the independent variable. The graphs of these cross sections are given in Figure 18.2.17, and the graph of the surface with a vertical cross section is shown in Figure 18.2.18.

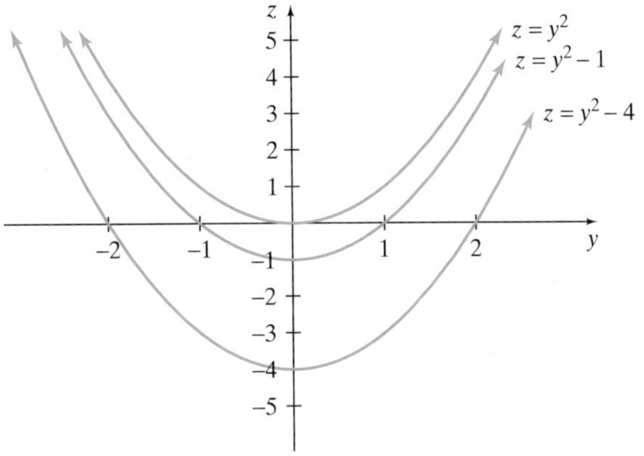

Figure 18.2.17

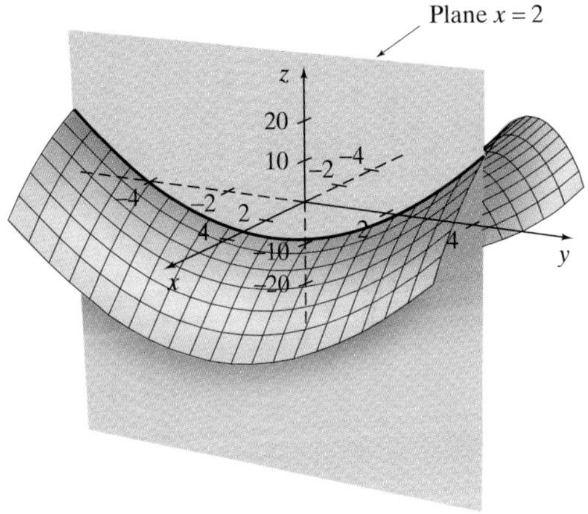

Figure 18.2.18 Vertical plane $x = 2$ intersects the surface in a curve.

Notice the upward opening parabolas in the y-direction, which matches ourwork in part (b). Also, notice the downward opening parabolas in the x-direction, which matches our work in part (a). ■

Technology Option

We can use our graphing calculator to reproduce the vertical cross sections in Example 6. To get the cross section shown in Figure 18.2.15, since z is the dependent variable it acts like y for the calculator purposes. Here, x is still the independent variable. So we enter and graph $y_1 = 4 - x^2$, $y_2 = 1 - x^2$ and $y_3 = -x^2$ to reproduce Figure 18.2.15. See Figure 18.2.19.

To get the cross section in Figure 18.2.17, since z is the dependent variable it acts like y for the calculator purposes. Here, since y is the independent variable, it acts like x for the calculator. So we enter and graph $y_1 = x^2$, $y_2 = x^2 - 1$ and $y_3 = x^2 - 4$ to reproduce Figure 18.2.17. See Figure 18.2.20.

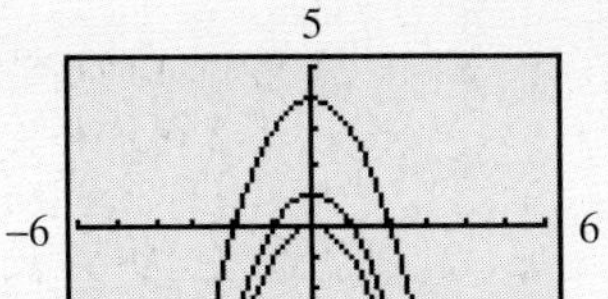

Figure 18.2.19

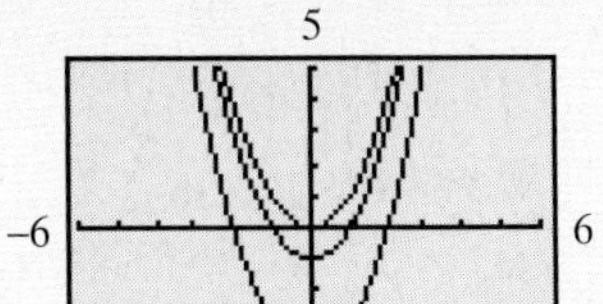

Figure 18.2.20

Interactive Activity

Construct a contour map for the function in Example 6 for c-values of $c = 0, \pm 1$, and ± 2. Using these horizontal cross sections along with the vertical cross section from parts (a) and (b) in Example 6, do you "see" the surface as shown in Figure 18.2.18?

Before we look at our concluding example, we want to mention that our work with vertical cross sections will be very important in Section 18.3 when we tackle derivatives of functions of two variables. These derivatives, called *partial derivatives*, are related to vertical cross sections.

Example 7 Applying Cross-sectional Analysis to a Production Function

Consider the production function for the Linger Golf Cart Corporation in Example 1. It is

$$f(x, y) = Q = 1.3x^{0.75}y^{0.25}$$

where Q is the number of golf carts produced per week, x is the number of labor hours each day, and y is the daily usage of capital investment. Construct a cross section on an xz-plane with y fixed at $y = 20$. Approximate how many labor hours are required each day to produce 10 golf carts per week.

Solution

Interactive Activity

Graphically verify the result of Example 7 by using the INTERSECT command on your calculator.

WWW

We begin by replacing the dependent variable Q with z so that we have

$$z = 1.3x^{0.75}y^{0.25}$$

Again, since y is fixed, we are slicing the surface with a vertical plane perpendicular to the y-axis of the form $y = 20$. So we need to graph in an xz-plane the following:

$$z = 1.3x^{0.75}(20)^{0.25} \approx 2.75x^{0.75}$$

The graph of the cross section is given in Figure 18.2.21, and the graph of the surface is given in Figure 18.2.22.

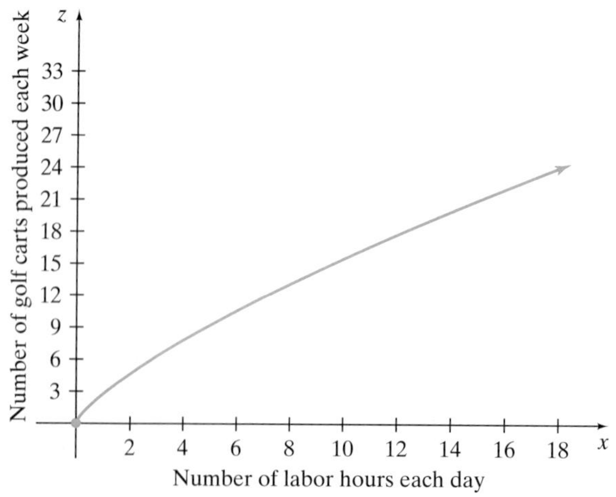

Figure 18.2.21 Graph of $z = 2.75x^{0.75}$ representing the vertical cross section when y is fixed at $y = 20$.

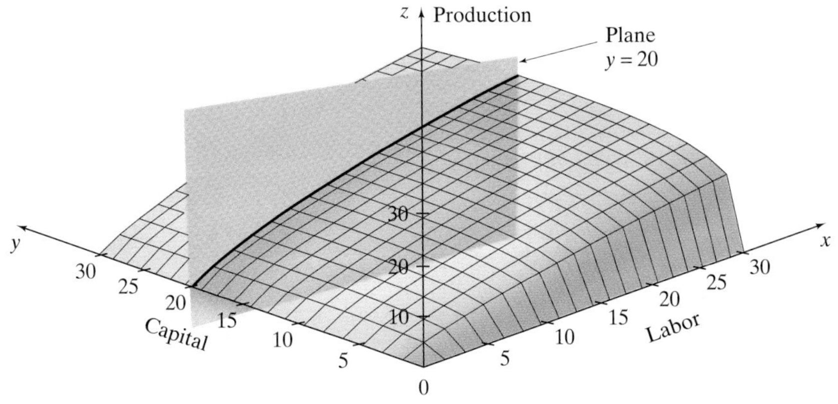

Figure 18.2.22

Now to approximate how many labor hours are required each day to produce 10 golf carts per week requires us to substitute $z = 10$ into $z = 2.75x^{0.75}$ and solve for x. This gives us

$$z = 2.75x^{0.75}$$

$$10 = 2.75x^{0.75} \quad \text{Substitute } z = 10$$

$$\frac{10}{2.75} = x^{0.75} \quad \text{Divide both sides by 2.75}$$

$$\left(\frac{10}{2.75}\right)^{4/3} = (x^{0.75})^{4/3} \quad \text{Raise both sides to the } \frac{4}{3} \text{ power}$$

$$5.59 \approx x \quad \text{On the right, 0.75 times } \frac{4}{3} \text{ equals 1.}$$

Interpret the Solution: This means that approximately 5.6 labor hours each day will produce a quantity of 10 golf carts per week when the capital investment is fixed at $y = 20$ units each day. ■

✓ Checkpoint 3 Now work Exercise 37.

SUMMARY

We began this section by analyzing a surface using horizontal cross sections, called **level curves**. We then analyzed surfaces in space using vertical cross sections. We sliced a surface with planes of the form $x = c$ and $y = c$ and graphed the resulting cross sections on the yz-plane and the xz-plane, respectively. Notice that each type of cross section tells us something about the shape of the surface.

- If the surface has an equation of the form $z = f(x, y)$, a level curve is obtained by slicing the surface with a horizontal plane $z = c$. The equation for the level curve at height c is given by $c = f(x, y)$.

SECTION 18.2 EXERCISES

In Exercises 1–8, make a contour map of the given function for the specified c-values.

1. $f(x, y) = x + y + 1$ $\quad c = 0, 1, 2, 3$

2. $f(x, y) = 1 - x - y$ $\quad c = 0, 3, 6, 9$

3. $f(x, y) = 2x + 3y + 6$ $\quad c = 0, 2, 4, 6$

4. $f(x, y) = 2y - 3x + 1$ $\quad c = 0, 2, 4, 6$

✓ **5.** $f(x, y) = \sqrt{16 - x^2 - y^2}$ $\quad c = 0, 2, 4$

6. $f(x, y) = \sqrt{25 - x^2 - y^2}$ $c = 1, 3, 5$

7. $f(x, y) = e^x + y$ $c = 0, 2, 4, 6$

8. $f(x, y) = \ln x + y - 1$ $c = 0, 2, 4, 6$

In Exercises 9–14, make a contour map of the given production functions for the specified c-values.

9. $Q = 24x^{0.5}y^{0.5}$ $c = 10, 20, 30, 40$

10. $Q = 36.5x^{0.5}y^{0.5}$ $c = 100, 200, 300, 400$

11. $Q = 5.6x^{0.65}y^{0.41}$ $c = 10, 20, 30, 40$

12. $Q = 6.3x^{0.7}y^{0.35}$ $c = 20, 40, 60, 80$

13. $Q = 180x^{0.7}y^{0.3}$ $c = 100, 200, 300, 400$

14. $Q = 120x^{0.75}y^{0.25}$ $c = 100, 200, 300, 400$

In Exercises 15–20, make a contour map of the given utility function for the specified c-values.

15. $U(x, y) = \sqrt{xy}$ $c = 2, 4, 6, 8$

16. $U(x, y) = x^{0.25}y^{0.75}$ $c = 1, 2, 3, 4$

17. $U(x, y) = x^{0.75}y^{0.25}$ $c = 1, 2, 3, 4$

18. $U(x, y) = x^{0.65}y^{0.35}$ $c = 1, 2, 3, 4$

19. $U(x, y) = x^{0.55}y^{0.45}$ $c = 1, 2, 3, 4$

20. $U(x, y) = x^{0.55}y^{0.45}$ $c = 2, 4, 6, 8$

In Exercises 21–28, sketch on the appropriate plane, either the xz-plane or yz-plane, cross sections for the given function for the specified vertical plane values. See Example 6.

21. $z = 15x^{0.5}y^{0.5}$ $x = 2, 4, 6$

22. $z = 15x^{0.5}y^{0.5}$ $x = 3, 6, 9$

23. $z = 15x^{0.5}y^{0.5}$ $y = 2, 4, 6$

24. $z = 15x^{0.5}y^{0.5}$ $y = 3, 6, 9$

25. $z = 120x^{0.75}y^{0.25}$ $x = 10, 20, 30$

26. $z = 120x^{0.75}y^{0.25}$ $x = 20, 40, 60$

27. $z = 120x^{0.75}y^{0.25}$ $y = 20, 40, 60$

28. $z = 120x^{0.75}y^{0.25}$ $y = 10, 20, 30$

Applications

29. Cranky Corporation has three crankshaft producing plants in the United States and a Cobb–Douglas production function of $Q = f(x, y) = 1.2x^{0.7}y^{0.3}$, where x represents the number of labor hours (in thousands), y represents the capital (total net assets in dollars), and Q represents the quantity produced (in thousands). Sketch isoquants for $Q = 20$, 40, and 60.

30. Ling Incorporated has four pretzel making plants in the United States and a Cobb–Douglas production function of $Q = f(x, y) = 2.3x^{0.2}y^{0.8}$, where x represents the number of labor hours (in hundreds), y represents the capital (total net assets in dollars), and Q represents the quantity produced (in hundreds of pounds). Sketch isoquants for $Q = 30$, 60, and 90.

31. Suppose that the utility from consuming x colas and y slices of pizza in a typical week is given by $U(x, y) = x^{2/3}y^{1/3}$. Make an indifference map for $c = 1$, 2, and 3.

32. Suppose that the utility from consuming x chocolate milkshakes and y cheeseburgers in a typical week is given by $U(x, y) = x^{1/3}y^{2/3}$. Make an indifference map for $c = 1$, 2, and 3.

33. In their study of human groupings, anthropologists often use an index called the **cephalic index**. The cephalic index is given by $z = C(x, y) = 100 \cdot \frac{x}{y}$, where x is the width and y is the length of an individual's head. Both measurements are made across the top of the head and both are in inches. Construct a contour map for $c = 75$, 80, 85, and 90.

34. The **Doyle log rule** is one method used to determine the yield of a log, measured in board feet. It is given by

$$f(x, y) = \left(\frac{x - 4}{4}\right)^2 \cdot y$$

where x is the diameter of the log in inches, y is the length of the log in feet, and $f(x, y)$ is the number of board-feet (*Source:* www.forestry.uga.edu/docs/950-measurementscaling.html).

(a) Construct a contour map for $c = 400$, 500, 600, and 700.

(b) In your own words, describe what the level curve for $c = 500$ represents.

35. The O'Neill Corporation has 10 soft drink bottling plants in the United States and a Cobb–Douglas production function of $Q = f(x, y) = 1.64x^{0.6}y^{0.4}$, where x is the number of labor hours (in thousands), y is the capital (total net assets in millions), and Q is the quantity produced (in thousands of gallons). Sketch isoquants for $Q = 80$, 100, 120, and 140.

36. *(continuation of Exercise 35)* James, the CEO of the O'Neill Corporation, wants to keep the number of labor hours at each plant fixed at $x = 110$.

(a) By fixing labor at $x = 110$, is James looking at a vertical or horizontal cross section?

(b) Graph the cross section when $x = 110$.

(c) Approximate how much capital is required to produce 100.

(d) In your own words, describe what the graph in part (b) represents.

37. *(continuation of Exercise 35)* James, the CEO of the O'Neill Corporation, wants to keep capital at each plant fixed at $y = 18$.

(a) By fixing capital at $y = 18$, is James looking at a vertical or horizontal cross section?

(b) Graph the cross section when $y = 18$.

(c) Approximate how many labor hours would be required to produce 100.

(d) In your own words, describe what the graph in part (b) represents.

38. A golf club manufacturer has a Cobb–Douglas production function given by

$$Q = f(x, y) = 21x^{0.3}y^{0.75}$$

where x is the utilization of labor, y is the utilization of capital, and Q is the number of units of golf clubs produced. Sketch the isoquants for $Q = 500, 1000, 1500$, and 2000.

39. *(continuation of Exercise 38)* The CEO of the golf club manufacturer wants to keep labor fixed at $x = 100$ units.

(a) By fixing labor at $x = 100$ units, is the CEO looking at a vertical or horizontal cross section?

(b) Graph the cross section when $x = 100$.

(c) Approximate how many units of capital would be required to produce 1500 units of golf clubs.

(d) In your own words, describe what the graph in part (b) represents.

40. *(continuation of Exercise 38)* The CEO of the golf club manufacturer wants to keep capital fixed at $y = 100$ units.

(a) By fixing capital at $y = 100$ units, is the CEO looking at a vertical or horizontal cross section?

(b) Graph the cross section when $y = 100$.

(c) Approximate how many units of labor would be required to produce 1500 units of golf clubs.

(d) In your own words, describe what the graph in part (b) represents.

41. The Brody Corporation has 11 plants worldwide and a Cobb–Douglas production function given by

$$Q = f(x, y) = 1.7x^{0.8}y^{0.2}$$

where x is the number of units of labor (in thousands), y is the number of units of capital (in millions), and Q is the number of units of quantity produced. Construct isoquants for $c = 300, 400, 500$, and 600.

42. *(continuation of Exercise 41)* Cheryl, the CEO of the Brody Corporation, wants to keep labor fixed at $x = 350$ units.

(a) Graph the vertical cross section when $x = 350$.

(b) Approximate how much capital is required to produce 560 units.

43. *(continuation of Exercise 41)* Cheryl, the CEO of the Brody Corporation, wants to keep capital fixed at $y = 50$ units.

(a) Graph the vertical cross section when $y = 50$.

(b) Approximate how many units of labor are required to produce 560 units.

44. The volume of a cylinder with radius r and height h is given by $V(r, h) = \pi r^2 h$. Substituting x for r and y for h gives $V(x, y) = \pi x^2 y$. Make a contour map for $c = 10, 20, 30$, and 40.

45. Suppose that the height of the cylinder in Exercise 44 is fixed at 4 inches, that is, $y = 4$.

(a) Graph the vertical cross section when $y = 4$.

(b) Approximate the radius necessary to produce a cylinder with a volume of approximately 113 cubic inches.

(c) Use the graph to approximate the volume if the radius is 6 inches.

(d) In your own words, what does the graph in part (a) represent?

46. IQ is defined to be $\text{IQ} = f(x, y) = \dfrac{100y}{x}$, where y is the individual's mental age in years and x is the individual's actual age in years.

(a) Make a contour map for $c = 90, 100, 110$, and 120. Keep in mind that $x > 0$ and $y > 0$.

(b) Give three combinations of actual age and mental age that give an IQ of 120.

47. For the IQ function in Exercise 46, consider a vertical cross section by keeping mental age fixed at 20, that is, $y = 20$. Our function then becomes

$$\text{IQ} = \frac{100 \cdot 20}{x} = \frac{2000}{x}$$

(a) To accommodate our calculator, substitute y for IQ. This yields $y = \dfrac{2000}{x}$. Graph this in the window [0, 60] by [0, 300].

(b) The point $\left(15, 133\frac{1}{3}\right)$ is on the graph. Interpret what this means.

(c) The point (16, 125) is on the graph. Interpret what this means.

48. For the IQ function in Exercise 46, consider a vertical cross section by keeping actual age fixed at 20, that is, $x = 20$. Our function then becomes

$$\text{IQ} = \frac{100y}{20} = 5y$$

(a) To accommodate our calculator, substitute y for IQ and x for y. This gives $y = 5x$. Graph this in the viewing window [0, 60] by [0, 300].

(b) The point (30, 150) is on the graph. Interpret what this means.

(c) The point (31, 155) is on the graph. Interpret what this means.

(d) The point (29, 145) is on the graph. Interpret what this means.

(e) Fill in the blanks.

When actual age is fixed at 20, each 1-year increase in mental age ______________ (increases/decreases) IQ by ______________ units.

SECTION PROJECT

The program in Appendix C that performs multiple regression is needed to do this section project.

The Riblet Engine Company has six plants around the United States to produce small engines. In a recent year the data for each plant gave the number of labor hours (in thousands), capital (in millions), and total quantity produced, as follows:

Labor	Capital	Quantity
42	5	27,825
44	6	31,058
47	7	34,633
49	9	39,762
50	10	42,201
52	13	48,621

(a) The plants all operate with the same technology, so a production function can be determined. Enter the data into your calculator and execute the program to determine a Cobb–Douglas production function to model the data. Round all values to the nearest hundredth.

(b) Graph isoquants for $q = 20{,}000, 30{,}000, 40{,}000$, and $50{,}000$.

(c) The CEO of the company, Gene, wants to keep labor fixed at 48. Is Gene asking for a vertical or horizontal cross section?

(d) Graph the cross section when labor is fixed at 48.

(e) Use the graph to approximate how much capital is required to produce a quantity of 35,000.

Section 18.3 Partial Derivatives and Second-order Partial Derivatives

In this section we discuss how to measure the rate of change of a function of two independent variables. We will observe how the vertical cross sections discussed in Section 18.2 are central to the study of this rate of change, which is known as **partial derivatives**. Partial derivatives simply measure the rate of change of a function of two independent variables, and they will be compared to the derivative that was discussed in Section 12.4. We also will discuss *second-order* partial derivatives. Second-order partial derivatives are analogous to second derivatives, which we discussed in Section 15.2.

From Your Toolbox

- Recall that for a given function f the derivative of f at x is

$$f'(x) = \lim_{h \to 0} \frac{f(x+h) - f(x)}{h}$$

- f' gives the rate of change of f.

Partial Derivatives

We have encountered many interpretations and applications of the derivative of a function of one independent variable, f.

Naturally, we would like to extend this rate of change concept to a function of two variables, $f(x, y)$. To do this, we consider how the function values change as x changes, that is, when y is kept constant, and we consider how the function values change as y changes, that is, when x is kept constant. Geometrically, we are talking about the **vertical cross sections** introduced in Section 18.2.

Flashback

O'Neill Corporation Revisited

In Section 18.2 we were introduced to the O'Neill Corporation. Recall that the O'Neill Corporation has 10 soft drink bottling plants and a production function of

$$Q = f(x, y) = z = 1.64x^{0.6}y^{0.4}$$

where x is the number of labor hours (in thousands), y is the capital (total net assets in millions), and Q is the quantity produced (in thousands of gallons). James, the CEO of the O'Neill Corporation, wants to keep the number of labor hours at each plant fixed at $x = 110$. Graph the cross section when $x = 110$ and describe what the graph represents.

Flashback Solution

Figure 18.3.1 gives the graph of the surface and the plane $x = 110$, and Figure 18.3.2 is a graph of the cross section. Notice that the cross section is graphed on the yz-plane.

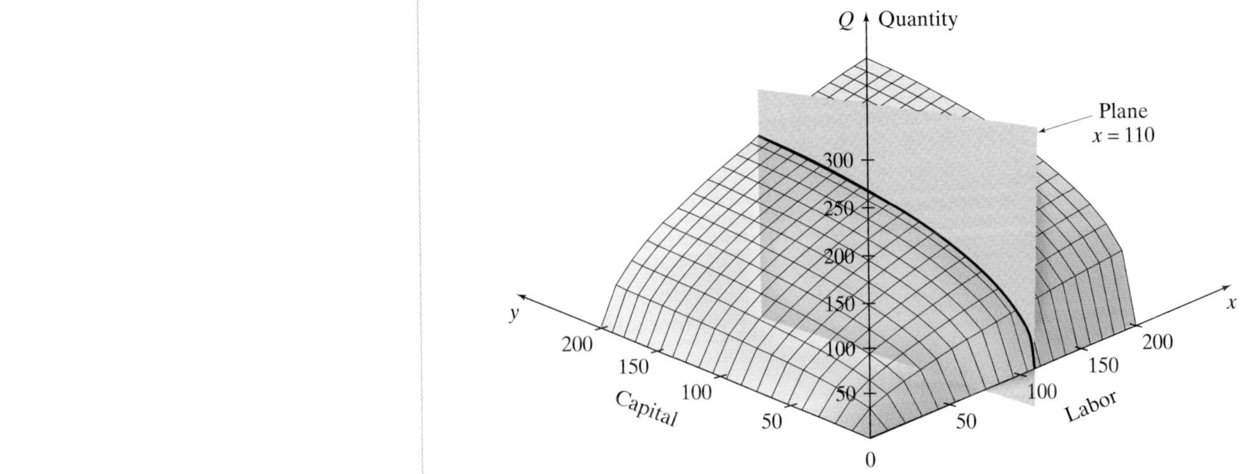

Figure 18.3.1

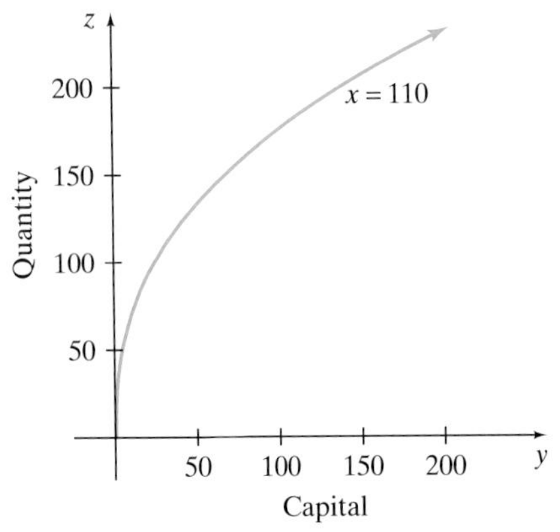

Figure 18.3.2

The graph of the cross section shows us that quantity is a function of capital when labor is held constant at $x = 110$. Specifically, we have

$$\begin{aligned} Q &= f(x, y) = 1.64x^{0.6}y^{0.4} \\ &= f(110, y) = 1.64(110)^{0.6}y^{0.4} \end{aligned}$$

So, by holding labor constant at $x = 110$, we now have quantity as a function of one variable, capital or y.

Let's take another look at Figure 18.3.2. Notice that when labor is fixed at $x = 110$ the vertical cross section is a function of *one* variable. (All along the curve in Figure 18.3.2, x is held constant at $x = 110$.) As we saw in the Flashback, this function of one variable has the form

$$f(110, y) = 1.64(110)^{0.6}y^{0.4}$$

From our work with derivatives, we know that we can find the rate at which $f(110, y) = 1.64(110)^{0.6}y^{0.4}$ changes, as y changes, by taking the derivative. What exactly does all this tell us? Example 1 provides the answer.

Example 1 Determining a Rate of Change

For the O'Neill Corporation production function, we determined in the Flashback that when labor is held constant at $x = 110$ we get a function of one variable

$$f(110, y) = 1.64(110)^{0.6}y^{0.4}$$

(a) Determine $\frac{d}{dy}[1.64(110)^{0.6}y^{0.4}]$.

(b) If the number of labor hours is fixed at $x = 110$, how fast is the number of gallons produced increasing as the capital increases from a level of 20?

Solution

(a) The derivative is

$$\frac{d}{dy}[1.64(110)^{0.6}y^{0.4}] = 1.64(110)^{0.6}(0.4)y^{-0.6}$$

(b) Evaluating the derivative in part (a) at $y = 20$ gives

$$1.64(110)^{0.6}(0.4)(20)^{-0.6} \approx 1.8244$$

Evaluating the derivative at $y = 20$ gives the *slope of the line tangent to the curve at* $y = 20$. See Figure 18.3.3. The curve in Figure 18.3.3 is the graph of the cross section when $x = 110$, so it is the graph of quantity as a function of capital when labor is fixed at 110. This tells us that, when the number of labor hours is held constant at $x = 110$ and capital is at 20 units, as capital increases 1 unit, production is increasing at a rate of about $1.8 \frac{\text{thousand gallons}}{\text{unit of capital}}$!

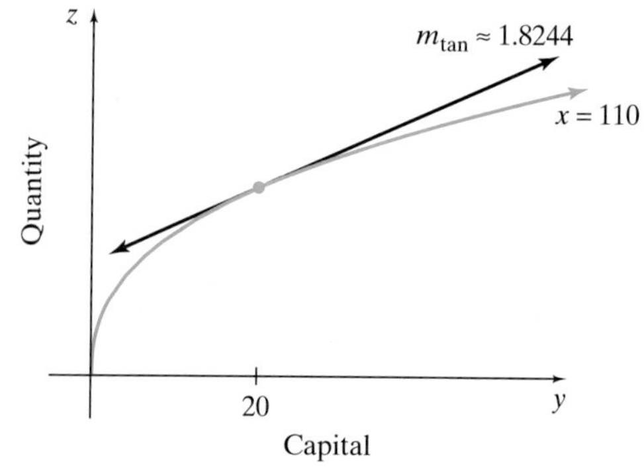

Figure 18.3.3

■

Let's summarize what was done in the Flashback and Example 1.

- When x is held constant at $x = 110$, we produce a **vertical cross section** of the surface.
- The graph of this cross section in the yz-plane shows production as a function of *one* variable.
- We can differentiate the function of one variable to give us a rate of change.

Definition of Partial Derivatives

The two-dimensional analysis that we did in the Flashback and in Example 1 tells us something remarkable about the three-dimensional surface. Since x is constant at $x = 110$ and we looked specifically at the scenario when $y = 20$, this means that we were at the point (110, 20, 91.22) on the graph of the surface (z-coordinate rounded to nearest hundredth) of our function. If we are at the point (110, 20, 91.22) on the surface and move *in the y-direction*, that is, parallel to the y-axis, the surface is changing at a rate of approximately $1.8 \frac{\text{units of } z}{\text{unit of } y}$. This is exactly the same as the slope of the tangent line to the curve that resulted from slicing the surface with the plane $x = 110$! See Figure 18.3.4.

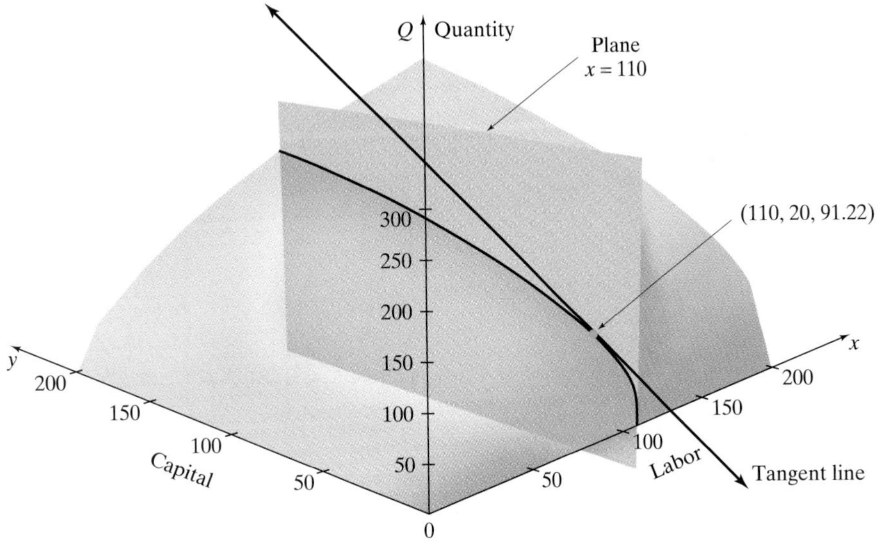

Figure 18.3.4

The process of holding x constant at any value $x = a$, moving in the y-direction (that is, parallel to the y-axis), and determining the rate of change of the surface is one interpretation of the **partial derivative of $f(x, y)$ with respect to y**. In a completely analogous manner, the process of holding y constant at any value $y = b$, moving in the x-direction (that is, parallel to the x-axis), and determining the rate of change of the surface is one interpretation of the **partial derivative of $f(x, y)$ with respect to x**. See Figures 18.3.5a–c.

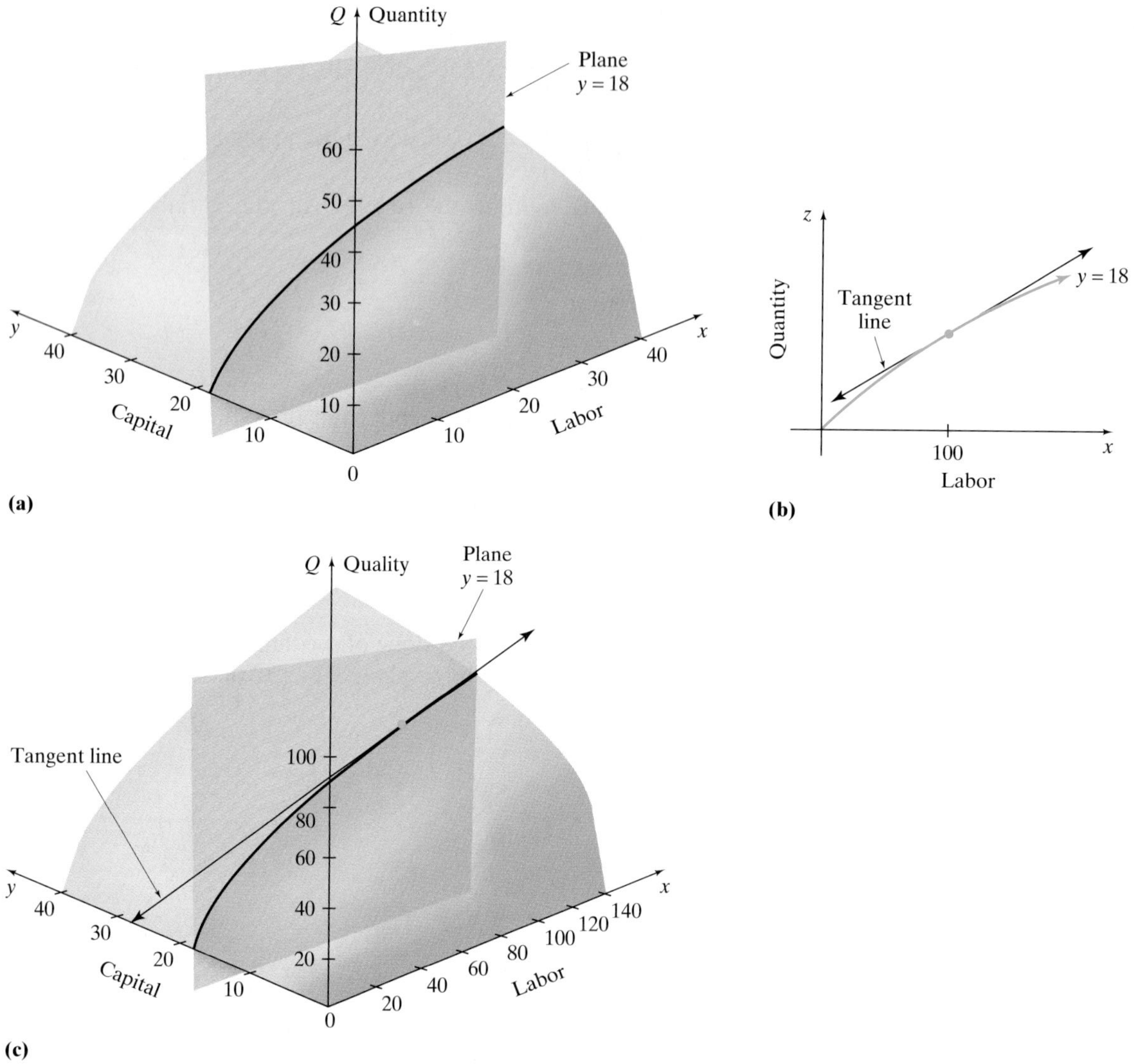

Figure 18.3.5

The previous explanation serves as a basis for the following formal definition:

Partial Derivative

Let $f(x, y)$ be a function of two variables. The **partial derivative of $f(x, y)$ with respect to x** is

$$f_x(x, y) = \lim_{h \to 0} \frac{f(x+h, y) - f(x, y)}{h}$$

The **partial derivative of $f(x, y)$ with respect to y** is

$$f_y(x, y) = \lim_{h \to 0} \frac{f(x, y+h) - f(x, y)}{h}$$

Of course, each of these partial derivatives exists only if the appropriate limit exists.

The important practical point on how to compute these partial derivatives is given in the following note.

Note:

1. The **partial derivative of $f(x, y)$ with respect to x** is found by treating y as a constant and performing our ordinary differentiation techniques. The notation for this partial derivative is $f_x(x, y)$ or $\frac{\partial f}{\partial x}$. The units of this partial derivative are units of f per unit of x.
2. The **partial derivative of $f(x, y)$ with respect to y** is found by treating x as a constant and performing our ordinary differentiation techniques. The notation for this partial derivative is $f_y(x, y)$ or $\frac{\partial f}{\partial y}$. The units of this partial derivative are units of f per unit of y.

Example 2 Computing a Partial Derivative

Consider $f(x, y) = x^2y - y^2$.

(a) Determine $f_x(x, y)$. **(b)** Determine $f_y(x, y)$.

Solution

(a) **Understand the Situation:** To determine $f_x(x, y)$, we treat y as a constant and take the derivative with respect to x.

This gives

$$f(x, y) = x^2y - y^2$$
$$f_x(x, y) = \frac{\partial}{\partial x}(x^2y) - \frac{\partial}{\partial x}(y^2)$$
$$f_x(x, y) = 2x \cdot y - 0 = 2xy$$

Observe that $\frac{\partial}{\partial x}(y^2) = 0$ because y is treated as a constant.

(b) **Understand the Situation:** To compute $\frac{\partial}{\partial y}$, we treat x as a constant and take the derivative with respect to y.

This yields

$$f(x, y) = x^2y - y^2$$
$$f_y(x, y) = \frac{\partial}{\partial y}(x^2y) - \frac{\partial}{\partial y}(y^2)$$
$$f_y(x, y) = x^2 \cdot 1 - 2y = x^2 - 2y$$

■

Checkpoint 1 Now work Exercise 3.

Example 3 Computing Partial Derivatives

Consider $f(x, y) = x^2y + y^3x - 2xy + y$.

(a) Determine $f_x(x, y)$. **(b)** Determine $f_y(x, y)$.

Interactive Activity

In your own words, explain why there was no usage of the Product Rule for derivatives when computing the partial derivatives in Example 3.

WWW

Solution

(a) **Understand the Situation:** To determine $f_x(x, y)$, we treat y as a constant and differentiate with respect to x.

$$f(x, y) = x^2y + y^3x - 2xy + y$$
$$f_x(x, y) = (2 \cdot x)y + y^3 \cdot 1 - 2 \cdot 1 \cdot y + 0$$
$$= 2xy + y^3 - 2y$$

(b) **Understand the Situation:** To find $f_y(x, y)$, we treat x as a constant and differentiate with respect to y.

$$f(x, y) = x^2y + y^3x - 2xy + y$$
$$f_y(x, y) = x^2 \cdot 1 + (3y^2) \cdot x - 2x \cdot 1 + 1$$
$$= x^2 + 3y^2x - 2x + 1$$

■

Checkpoint 2 Now work Exercise 7.

Example 4 Evaluating and Interpreting a Partial Derivative

Let $f(x, y) = 2x^3y^4 + \frac{1}{3}y^3 + e^{3x} - e^{2y}$.

(a) Determine $f_x(x, y)$.

(b) Evaluate $f_x(x, y)$ at $\left(1, \frac{3}{2}\right)$ and interpret.

Solution

(a) To determine $f_x(x, y)$, we treat y as a constant and differentiate with respect to x.

$$f(x, y) = 2x^3y^4 + \frac{1}{3}y^3 + e^{3x} - e^{2y}$$
$$f_x(x, y) = (6x^2)y^4 + 0 + 3e^{3x} - 0 = 6x^2y^4 + 3e^{3x}$$

(b) Evaluating $f_x(x, y)$ at $\left(1, \frac{3}{2}\right)$ yields

$$f_x\left(1, \frac{3}{2}\right) = 6(1)^2\left(\frac{3}{2}\right)^4 + 3e^{3(1)}$$
$$= 30.375 + 3e^3$$
$$\approx 90.63$$

Interpret the Solution: This means that if we are at the point $\left(1, \frac{3}{2}, 11.25\right)$ on the surface and move along the surface in the x-direction, parallel to the x-axis, the function is changing at a rate of approximately $90.63 \frac{\text{units of } f}{\text{unit of } x}$. Equivalently, we say that at the point $\left(1, \frac{3}{2}, 11.25\right)$ the slope of the surface in the x-direction is approximately 90.63. See Figure 18.3.6.

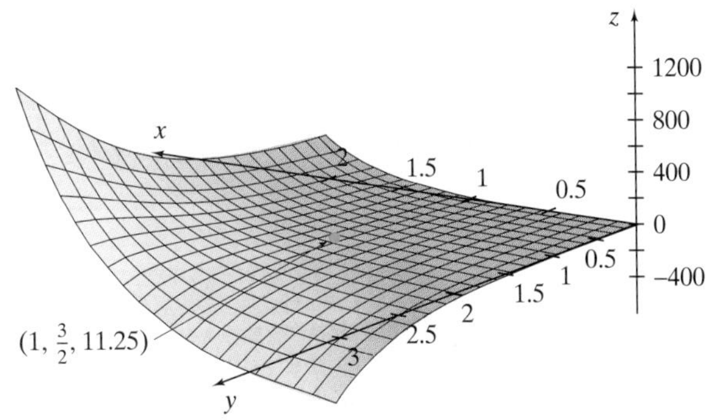

Figure 18.3.6

■

Checkpoint 3 Now work Exercise 33.

Applications

In the next example, we analyze the relationship between a partial derivative and **marginal analysis**.

Example 5 **Analyzing Marginal Revenue**

A company sells leaf blowers and a special 10-foot attachment to the leaf blower to clean gutters. Let x and y be the number of leaf blowers sold per month and the number of 10-foot attachments sold per month, respectively. Suppose that

$p = 1200 - 8x - y,$ the price in dollars for a leaf blower

$q = 523 - 1.5x - 0.15y,$ the price in dollars for an attachment

(a) Determine the revenue function $R(x, y)$.

(b) Determine $R_x(x, y)$ and $R_y(x, y)$. Interpret each.

(c) Evaluate and interpret $R_x(50, 25)$.

Interactive Activity

Evaluate and interpret $R_y(50, 25)$ for the revenue function in Example 5.

WWW

Solution

(a) Since revenue is price times quantity, we have

$$\begin{aligned} R(x, y) &= x(1200 - 8x - y) + y(523 - 1.5x - 0.15y) \\ &= 1200x - 8x^2 - xy + 523y - 1.5xy - 0.15y^2 \\ &= 1200x - 8x^2 - 2.5xy + 523y - 0.15y^2 \end{aligned}$$

(b) To determine $R_x(x, y)$, we treat y as a constant and differentiate with respect to x.

$$\begin{aligned} R_x(x, y) &= 1200 \cdot 1 - 8 \cdot 2x - 2.5(1)(y) + 0 - 0 \\ &= 1200 - 16x - 2.5y \end{aligned}$$

Interpret the Solution: $R_x(x, y)$ gives the amount that each additional leaf blower adds to the total revenue. In other words, it gives the *marginal revenue* for leaf blowers.

To determine $R_y(x, y)$, we treat x as a constant and differentiate with respect to y.

$$\begin{aligned} R_y(x, y) &= 0 - 0 - 2.5(x)(1) + 523(1) - 0.15(2y) \\ &= -2.5x + 523 - 0.3y \end{aligned}$$

Interpret the Solution: $R_y(x, y)$ gives the amount that each additional 10-foot attachment adds to the total revenue. In other words, it is the *marginal revenue* for the 10-foot attachments.

(c) Since $x = 50$ and $y = 25$, we have

$$\begin{aligned} R_x(50, 25) &= 1200 - 16(50) - 2.5(25) \\ &= 337.5 \end{aligned}$$

Interpret the Solution: This means that when 50 leaf blowers and 25 of the 10-foot attachments have been sold, the company receives \$337.50 for each *additional* leaf blower sold. ■

Example 6 Computing Partial Derivatives of a Volume Formula

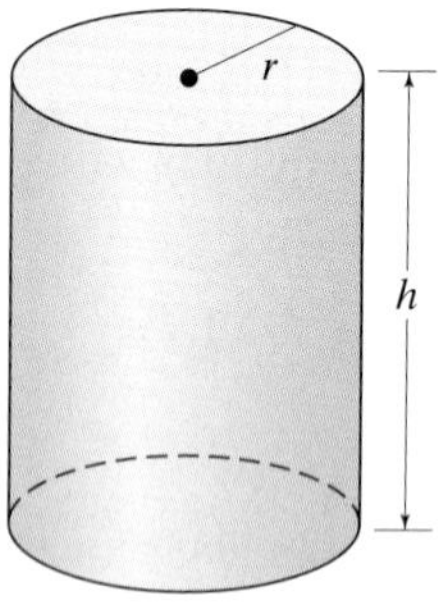

Recall that the formula for the volume of a cylinder is given by $V(r, h) = \pi r^2 h$, where r is the radius and h is the height.

(a) Determine $V_r(r, h)$ and $V_h(r, h)$.

(b) Evaluate each partial derivative when $r = 2$ centimeters (cm) and $h = 3$ cm and interpret.

Solution

(a) **Understand the Situation:** Keep in mind that r and h are the independent variables. Thus, $V_r(r, h)$ is found by keeping h constant.

This gives

$$V_r(r, h) = 2\pi r h$$

To find $V_h(r, h)$, we keep r constant, which gives

$$V_h(r, h) = \pi r^2$$

(b) Evaluating $V_r(r, h)$ when $r = 2$ cm and $h = 3$ cm yields

$$\begin{aligned} V_r(2, 3) &= 2\pi(2)(3) \\ &= 12\pi \end{aligned}$$

Interpret the Solution: This means that when $r = 2$ cm and $h = 3$ cm, if the radius increases by 1 cm while the height remains constant at $h = 3$ cm, the volume of the cylinder increases by approximately 12π cm^3.

Evaluating $V_h(r, h)$ when $r = 2$ cm and $h = 3$ cm yields

$$\begin{aligned} V_h(2, 3) &= \pi(2)^2 \\ &= 4\pi \end{aligned}$$

Interpret the Solution: Thus, when $r = 2$ cm and $h = 3$ cm, if the height increases by 1 cm while the radius remains constant at $r = 2$ cm, the volume of the cylinder increases by approximately 4π cm^3. ■

Marginal Productivity of Labor and Capital

Example 1 foreshadowed the concept known as **marginal productivity of labor** and **marginal productivity of capital**. These two partial derivatives are very important in economics.

Marginal Productivity of Labor and Capital

For any production function of the form $Q = f(x, y) = ax^m y^n$, where a, m, and n are positive constants and x represents units of labor and y represents units of capital, we have the following:

- $f_x(x, y)$ gives the approximate change in productivity per unit change in labor and is called **marginal productivity of labor.**
- $f_y(x, y)$ gives the approximate change in productivity per unit change in capital and is called **marginal productivity of capital.**

Example 7 Computing Marginal Productivity of Labor and Capital

Recall that the production function for the O'Neill Corporation is given by

$$\begin{aligned} Q &= f(x, y) \\ &= 1.64x^{0.6}y^{0.4} \end{aligned}$$

where x is the number of labor hours in thousands, y is capital in millions, and Q is the quantity in thousands of gallons.

(a) Compute the marginal productivity of labor and the marginal productivity of capital.

(b) Evaluate the marginal productivity of labor at $x = 100$ and $y = 18$ and interpret.

(c) Evaluate the marginal productivity of capital at $x = 110$ and $y = 20$ and interpret.

Interactive Activity

Compare the result of Example 7c with the result from Example 1.

Solution

(a) Since marginal productivity of labor is $f_x(x, y)$, we have

$$\begin{aligned} f_x(x, y) &= 1.64(0.6)x^{-0.4}y^{0.4} \\ &= 0.984x^{-0.4}y^{0.4} \end{aligned}$$

Marginal productivity of capital is $f_y(x, y)$, which yields

$$\begin{aligned} f_y(x, y) &= 1.64x^{0.6}(0.4)y^{-0.6} \\ &= 0.656x^{0.6}y^{-0.6} \end{aligned}$$

(b) Since $x = 100$ and $y = 18$, we evaluate $f_x(100, 18)$, which gives

$$f_x(100, 18) = 0.984(100)^{-0.4}(18)^{0.4} \approx 0.5$$

Interpret the Solution: This means that when the O'Neill Corporation uses 100 units of labor and 18 units of capital and keeps capital fixed at 18 units, production is increasing at a rate of approximately $0.5 \frac{\text{thousand gallons}}{\text{unit of labor}}$.

(c) Since $x = 110$ and $y = 20$, we need to evaluate $f_y(110, 20)$, which yields

$$f_y(110, 20) = 0.656(110)^{0.6}(20)^{-0.6} \approx 1.8$$

Interpret the Solution: Thus, when the O'Neill Corporation uses 110 units of labor and 20 units of capital and keeps labor fixed at 110 units, production is increasing at the rate of approximately $1.8 \frac{\text{thousand gallons}}{\text{unit of capital}}$. ■

For some amounts currently invested in labor and in capital, a natural question to ask is whether production would increase more rapidly if additional resources were invested in labor or in capital. We can approximate the change in productivity by simply doing the following:

- The change in productivity with respect to labor is the product of

$$f_x(x, y) \cdot (\text{change in } x)$$

- The change in productivity with respect to capital is the product of

$$f_y(x, y) \cdot (\text{change in } y)$$

Example 8 Allocating Additional Resources

Suppose that a production function is given by

$$f(x, y) = 4.23x^{0.37}y^{0.66}$$

where x represents dollars in millions spent on labor and y represents dollars in millions spent on capital equipment. Currently, $x = 5$ and $y = 1$. Would production increase more by spending an additional \$1 million on labor or \$500,000 on capital equipment?

Solution

Understand the Situation: We need to compute and compare $f_x(5, 1) \cdot (1)$ and $f_y(5, 1) \cdot (0.5)$. Notice that, since capital, y, is in millions, \$500,000 is the same as \$0.5 million.

Proceeding with the partial derivatives, we have

$$f_x(x, y) = 4.23(0.37)x^{-0.63}y^{0.66} = 1.5651x^{-0.63}y^{0.66}$$
$$f_x(5, 1) = 1.5651(5)^{-0.63}(1)^{0.66} \approx 0.568$$

The change in productivity with respect to labor is $f_x(5, 1) \cdot (1) \approx 0.568$. Continuing, we get for capital

$$f_y(x, y) = 4.23(0.66)x^{0.37}y^{-0.34} = 2.7918x^{0.37}y^{-0.34}$$
$$f_y(5, 1) = 2.7918(5)^{0.37}(1)^{-0.34} \approx 5.06$$

The change in productivity with respect to capital is $f_y(5, 1) \cdot (0.5) \approx 2.53$.

Interpret the Solution: Since the change in productivity with respect to capital is greater than the change in productivity with respect to labor, production increases more by spending \$500,000 on capital equipment than by spending \$1 million on labor. ■

Second-order Partial Derivatives

Just like the second derivatives for a function of one variable, there are **second-order partial derivatives**. These second-order partial derivatives are very important for Section 18.4, when we discuss locating relative extrema on the surface of $z = f(x, y)$.

Second-order Partial Derivatives

If $z = f(x, y)$, then the four possible second-order partial derivatives are as follows:

$$f_{xx}(x, y) = \frac{\partial}{\partial x}\left(\frac{\partial f}{\partial x}\right) \qquad f_{yy}(x, y) = \frac{\partial}{\partial y}\left(\frac{\partial f}{\partial y}\right)$$

$$f_{xy}(x, y) = \frac{\partial}{\partial y}\left(\frac{\partial f}{\partial x}\right) \qquad f_{yx}(x, y) = \frac{\partial}{\partial x}\left(\frac{\partial f}{\partial y}\right)$$

Note: In subscript notation, we differentiate from left to right. That is, to compute f_{yx}, we first compute f_y and then differentiate this result with respect to x.

Example 9 Computing a Second-order Partial Derivative

Determine all four second-order partial derivatives of

$$f(x, y) = x^2y + y^3x - 2xy + y.$$

Solution

First we calculate $f_x(x, y)$. It is given by

$$f_x(x, y) = 2xy + y^3 - 2y$$

We now can compute $f_{xx}(x, y)$ from this result by finding the derivative of $f_x(x, y)$ with respect to x.

$$\begin{aligned} f_{xx}(x, y) &= \frac{\partial}{\partial x}[f_x(x, y)] \\ &= 2y + 0 - 0 = 2y \end{aligned}$$

Also, we can compute $f_{xy}(x, y)$ from $f_x(x, y)$ by finding the derivative of $f_x(x, y)$ with respect to y.

$$\begin{aligned} f_{xy}(x, y) &= \frac{\partial}{\partial y}[f_x(x, y)] \\ &= 2x + 3y^2 - 2 \end{aligned}$$

To determine the other two second-order partial derivatives, we return to our original function, $f(x, y) = x^2y + y^3x - 2xy + y$, and compute $f_y(x, y)$. It is

$$f_y(x, y) = x^2 + 3y^2x - 2x + 1$$

From this result, we have

$$f_{yy}(x, y) = \frac{\partial}{\partial y}[f_y(x, y)] = 6yx$$

$$f_{yx}(x, y) = \frac{\partial}{\partial x}[f_y(x, y)] = 2x + 3y^2 - 2$$

■

Notice in Example 9 that $f_{xy}(x, y) = f_{yx}(x, y)$. This is true for many functions that we encounter, but not all functions.

Checkpoint 4 Now work Exercise 45.

SUMMARY

In this section we saw how the vertical cross sections introduced in Section 18.2 are key to our understanding of a partial derivative. We supplied the following note on computing partial derivatives:

- The **partial derivative of $f(x, y)$ with respect to x** is found by treating y as a constant and performing our ordinary differentiation techniques. The notation for this partial derivative is $f_x(x, y)$ or $\frac{\partial f}{\partial x}$.
- The **partial derivative of $f(x, y)$ with respect to y** is found by treating x as a constant and performing our ordinary differentiation techniques. The notation for this partial derivative is $f_y(x, y)$ or $\frac{\partial f}{\partial y}$.

We encountered several applications of partial derivatives, including marginal productivity of labor and marginal productivity of capital. These were defined as follows:

- For any production function of the form $Q = f(x, y) = ax^m y^n$:

$f_x(x, y)$ gives the approximate change in productivity per unit change in labor and is called **marginal productivity of labor**.

$f_y(x, y)$ gives the approximate change in productivity per unit change in capital and is called **marginal productivity of capital**.

The section concluded with a look at second-order partial derivatives.

- If $z = f(x, y)$, then the four possible **second-order partial derivatives** are as follows:

$$f_{xx}(x, y) = \frac{\partial}{\partial x}\left(\frac{\partial f}{\partial x}\right)$$

$$f_{yy}(x, y) = \frac{\partial}{\partial y}\left(\frac{\partial f}{\partial y}\right)$$

$$f_{xy}(x, y) = \frac{\partial}{\partial y}\left(\frac{\partial f}{\partial x}\right)$$

$$f_{yx}(x, y) = \frac{\partial}{\partial x}\left(\frac{\partial f}{\partial y}\right)$$

SECTION 18.3 EXERCISES

In Exercises 1–30, determine $f_x(x, y)$ and $f_y(x, y)$.

1. $f(x, y) = 5x^2 - 6y^3$

2. $f(x, y) = 3x + 2y + 10$

3. $f(x, y) = 2xy - y^2 + 1$

4. $f(x, y) = 3x^4 + 3x + 5y^2$

5. $f(x, y) = x^3 y^2$

6. $f(x, y) = x^2 y^4$

7. $f(x, y) = x^2 + 3x^2 y^3 - 2y^3 - xy$

8. $f(x, y) = 2x^4 + x^2 y^2 - 3y^2 - y$

9. $f(x, y) = 2x^3 - y^2 + 2x - 3$

10. $f(x, y) = 3x^4 - 2y^3 + 3x^2 - 5xy$

11. $f(x, y) = (2x + 3y)^3$

12. $f(x, y) = (3x - 5y)^4$

13. $f(x, y) = 37.21x^{0.15} y^{0.87}$

14. $f(x, y) = 2.41x^{0.27} y^{0.71}$

15. $f(x, y) = e^x \ln y$

16. $f(x, y) = e^y \ln x$

17. $f(x, y) = \frac{y}{x}$

18. $f(x, y) = \frac{x}{y}$

19. $f(x, y) = 3x^2 - 2x^3 y^4 + 7$

20. $f(x, y) = 4y^3 - 5x^2 y + 2$

21. $f(x, y) = (x^2 + y^3)^2$

22. $f(x, y) = (x^3 - y^2)^4$

23. $f(x, y) = \ln(x^2 + y^3)$

24. $f(x, y) = \ln(y^2 - x)$

25. $f(x, y) = \frac{x^2 y}{y + x}$

26. $f(x, y) = \frac{xy^2}{y - x}$

27. $f(x, y) = 3x^2 y^3 + e^{x+y}$

28. $f(x, y) = -2x^3 y - e^{x-y}$

29. $f(x, y) = e^{xy} + \ln y$

30. $f(x, y) = y^2 e^{xy} + \ln x$

In Exercises 31–40, for the given function, evaluate the stated partial derivative and interpret. See Example 4.

31. $f(x, y) = 4y^3 - 5x^2 y + 2$ Determine $f_x(1, 2)$.

32. $f(x, y) = 3x^4 - 2y^3 + 3x^2 - 5xy$ Determine $f_x(2, 1)$.

33. $f(x, y) = 2x^4 + x^2 y^2 - 3y^2 - y$ Determine $f_y(2, 3)$.

34. $f(x, y) = \ln(y^2 - x)$ Determine $f_y(2, 5)$.

35. $f(x, y) = \ln(y^2 - x)$ Determine $f_x(2, 5)$.

36. $f(x, y) = 37.21x^{0.15} y^{0.87}$ Determine $f_y(5, 2)$.

37. $f(x, y) = 37.21x^{0.15} y^{0.87}$ Determine $f_x(5, 2)$.

38. $f(x, y) = 3x^2 y^3 + e^{x+y}$ Determine $f_y(1, 2)$.

39. $f(x, y) = -2x^3 y - e^{x-y}$ Determine $f_x(2, 4)$.

40. $f(x, y) = \frac{\ln x}{x^2 - y^2}$ Determine $f_x(4, 3)$.

In Exercises 41–56, determine $f_{xx}(x, y)$, $f_{xy}(x, y)$, and $f_{yy}(x, y)$.

41. $f(x, y) = 4y^3 - 5x^2 y + 2$

42. $f(x, y) = 2x^3 - x^2 y^3 + 2y^4$

43. $f(x, y) = 3x^2 - 2x^3 y^2 + 2x^3$

44. $f(x, y) = 3x^4 + 2x^3 y^2 - y$

45. $f(x, y) = 5xy - 6x^3 + 7y$

46. $f(x, y) = 5xy^3 - 7x^3 y$

47. $f(x, y) = y^5 - 2x^3y^2 + 7x^2$

48. $f(x, y) = y^3 + 3x^2y - 8x^3$

49. $f(x, y) = y^2e^{xy} + \ln x$ **50.** $f(x, y) = x^3e^{xy} + \ln y$

51. $f(x, y) = 5.2x^{0.65}y^{0.4}$ **52.** $f(x, y) = 2.41x^{0.27}y^{0.71}$

53. $f(x, y) = \dfrac{2y}{x}$ **54.** $f(x, y) = \dfrac{y^2}{x}$

55. $f(x, y) = ye^{xy}$ **56.** $f(x, y) = y^2e^{xy}$

57. For $f(x, y) = x^2 + y^2 - xy + 3y$, determine values for x and y such that $f_x(x, y) = 0$ and $f_y(x, y) = 0$ simultaneously.

58. For $f(x, y) = x^2 + y^2 - 8x + 2y + 7$, determine values for x and y such that $f_x(x, y) = 0$ and $f_y(x, y) = 0$ simultaneously.

Applications

59. Seamego Boats spends x thousand dollars each week on newspaper advertising and y thousand dollars each week on radio advertising. The company has weekly sales, in tens of thousands of dollars, given by

$$S(x, y) = 2x^2 + y$$

(a) Determine $S_x(x, y)$ and $S_y(x, y)$.

(b) Determine $S_x(3, 5)$ and $S_y(3, 5)$ and interpret each.

60. Hilltop Boats spends x thousand dollars each week on radio advertising and y thousand dollars each week on television advertising. The company has weekly sales, in tens of thousands of dollars, given by

$$S(x, y) = 3x + 2y^3$$

(a) Determine $S_x(x, y)$ and $S_y(x, y)$.

(b) Determine $S_x(2, 6)$ and $S_y(2, 6)$ and interpret each.

61. Tube Town, a recently opened water park, spends x thousand dollars on radio advertising and y thousand dollars on television advertising. The park has weekly ticket sales, in tens of thousands of dollars, of

$$TS(x, y) = 1.5x^2 + 3.2y^2$$

(a) Determine $TS_x(x, y)$ and $TS_y(x, y)$.

(b) Determine $TS_x(1, 0.5)$ and $TS_y(1, 0.5)$ and interpret each.

62. In Section 18.1 we saw that the wind chill index is given by

$$f(v, T) = 35.74 + 0.6215T - 35.75v^{0.16} + 0.4275Tv^{0.16}$$

where v is the wind speed in miles per hour and T is the actual air temperature in degrees Fahrenheit (*Source:* National Oceanic and Atmospheric Administration, www.noaa.gov).

(a) Determine $f_T(v, T)$ and $f_v(v, T)$.

(b) Evaluate $f(25, 5)$, $f_T(25, 5)$, and $f_v(25, 5)$ and interpret each.

63. If an amount P dollars is invested at an annual interest rate of 6% compounded continuously for t years, the total amount accumulated is given by $f(P, t) = Pe^{0.06t}$. Determine $f_P(P, t)$ and $f_t(P, t)$ and interpret each.

64. In human cells, sodium ions are transported from inside the cell to outside the cell; at the same time, potassium ions are transported from outside the cell to inside. When this takes place across a nerve fiber in a nerve cell, a slight negative electric charge is left behind in the nerve fiber. The amount of voltage that will develop across a membrane is given by the Nernst equation, $f(x, y) = -61 \log \frac{x}{y}$, where x is the concentration of potassium ions inside the nerve fiber, y is the concentration of potassium ions outside the nerve fiber, and $f(x, y)$ is measured in millivolts. Determine $f_x(x, y)$ and $f_y(x, y)$ and interpret each.

65. The formula for a cone is given by

$$V(r, h) = \frac{1}{3}\pi r^2 h$$

where r is the radius and h is the height.

(a) Determine $V_r(r, h)$ and $V_h(r, h)$.

(b) Compute $V(3, 5)$, $V_r(3, 5)$, and $V_h(3, 5)$ and interpret each.

66. Poiseuille's law states that the resistance, R, for blood flowing in a blood vessel is given by

$$R(L, r) = k \cdot \frac{L}{r^4}$$

where k is a constant, L is the length of the blood vessel, and r is the radius of the blood vessel.

(a) Determine $R_L(L, r)$ and $R_r(L, r)$.

(b) Evaluate $R_L(6, 0.3)$ and $R_r(6, 0.3)$ and interpret each.

67. In their study of human groupings, anthropologists often use an index called the cephalic index. The cephalic index is given by

$$C(W, L) = 100 \cdot \frac{W}{L}$$

where W is the width and L is the length of an individual's head. Both measurements are made across the top of the head and are in inches. Determine $C_W(6, 8.2)$ and $C_L(6, 8.2)$ and interpret each.

68. An individual's body surface area is approximated by

$$BSA(w, h) = 0.007184w^{0.425}h^{0.725}$$

where BSA is in square meters, w is weight in kilograms, and h is height in centimeters. This formula is known as the DuBois and DuBois formula (*Source:* www-users.med.cornell.edu/~spon/picu/bsacalc.htm).

(a) Determine $BSA_w(w, h)$ and $BSA_h(w, h)$.

(b) Evaluate $BSA(70, 160)$, $BSA_w(70, 160)$, and $BSA_h(70, 160)$ and interpret each.

69. The Chalet Bicycle Company manufactures 21-speed racing bicycles and 21-speed mountain bicycles. Let x represent the weekly demand for a 21-speed racing bicycle and y represent the weekly demand for a 21-speed mountain bicycle. The weekly price–demand equations are given by

$p = 350 - 4x + y$, the price in dollars for a 21-speed racing bicycle

$q = 450 + 2x - 3y$, the price in dollars for a 21-speed mountain bicycle.

(a) Determine the revenue function $R(x, y)$.

(b) Determine $R_x(x, y)$ and $R_y(x, y)$. Interpret each.

(c) Evaluate $R(15, 20)$ and $R_x(15, 20)$ and interpret each.

70. *(continuation of Exercise 69)* The cost function for Chalet Bicycle Company is given by

$$C(x, y) = 390 + 95x + 100y$$

(a) Determine the weekly profit function $P(x, y)$.

(b) Determine $P_x(x, y)$ and $P_y(x, y)$ and interpret each.

(c) Evaluate $P(15, 20)$ and $P_x(15, 20)$ and interpret each.

71. The Kerr Company produces two types of graphing calculators, x units of a 2D graphing calculator and y units of a 3D graphing calculator each month. The monthly revenue, in dollars, is given by

$$R(x, y) = 70x + 95y + 0.5xy - 0.04x^2 - 0.04y^2$$

(a) Determine the marginal revenue function for the 2D graphing calculator.

(b) Determine the marginal revenue function for the 3D graphing calculator.

72. *(continuation of Exercise 71)* The Kerr Company knows that the cost function for producing the two types of calculators is

$$C(x, y) = 4x + 5y + 3200$$

(a) Determine the marginal cost function for the 2D graphing calculator.

(b) Determine the marginal cost function for the 3D graphing calculator.

73. The Smolki Engine Company produces two types of small engines, x thousand two-stroke engines and y thousand four-stroke engines. The vice president of the company, Jamie, has determined that the revenue and cost functions for a year are, in thousands of dollars,

$$R(x, y) = 42x + 39.5y$$
$$C(x, y) = 0.5x^2 + 0.8xy + y^2 + 5x + 4y$$

(a) Determine the profit function $P(x, y)$.

(b) Determine $P_x(x, y)$ and $P_y(x, y)$ and interpret each.

(c) Compute $P(18, 10)$ and $P_x(18, 10)$ and interpret each.

(d) Compute $P_y(18, 10)$ and interpret.

74. A golf club manufacturer has a Cobb–Douglas production function given by

$$Q = f(x, y) = 21x^{0.3}y^{0.75}$$

where x is the utilization of labor (in millions), y is the utilization of capital (in millions), and Q is the number of units of golf clubs produced.

(a) Compute $f_x(x, y)$ and $f_y(x, y)$.

(b) If the golf club manufacturer is currently using 150 units of labor and 100 units of capital, determine the marginal productivity of labor and the marginal productivity of capital.

(c) Would production increase more by spending an additional \$1 million on labor or \$500,000 on capital? Explain.

75. The Kevshan Company has a production function of

$$Q = f(x, y) = 161x^{0.8}y^{0.25}$$

where x is the number of labor hours in thousands, y is the capital equipment in millions, and Q is the quantity of calculators produced.

(a) Compute $f_x(x, y)$ and $f_y(x, y)$.

(b) If the Kevshan Company is now using 111 units of labor and 25 units of capital, determine the marginal productivity of labor and the marginal productivity of capital and interpret each.

76. The Baker Compact Disc Company has eight plants around the United States that manufacture compact disc players. The Cobb–Douglas production function for the eight plants is given by

$$Q = f(x, y) = 3.14x^{0.6}y^{0.45}$$

where x is the dollars (in millions) spent on labor, y is the dollars (in millions) spent on capital, and Q is the quantity produced (in thousands).

(a) Compute $f_x(x, y)$ and $f_y(x, y)$.

(b) If the Baker Compact Disc Company is currently using 1.7 units of labor and 5 units of capital, determine the marginal productivity of labor and the marginal productivity of capital and interpret each.

(c) Would production increase more by spending an additional \$1 million on labor or \$500,000 on capital? Explain.

77. The Toth Engine Company has a production function given by

$$Q = f(x, y) = 1434.33x^{0.6}y^{0.45}$$

where x is the number of labor hours in thousands, y is the capital in millions, and Q is the total quantity produced.

(a) Compute $f_x(x, y)$ and $f_y(x, y)$.

(b) If the Toth Engine Company is currently using 47 units of labor and 8 units of capital, determine the marginal productivity of labor and the marginal productivity of capital and interpret each.

78. Anastasia's, a manufacturer of skis, has a monthly production function given by

$$Q = f(x, y) = 210x^{0.6}y^{0.4}$$

where Q is the number of pairs of skis produced, x is the amount of labor used, and y is the amount of capital used.

(a) Compute $f_x(x, y)$ and $f_y(x, y)$.

(b) If Anastasia's is currently using 15 units of labor and 23 units of capital, determine the marginal productivity of labor and the marginal productivity of capital and interpret each.

79. The Jendrag Motorcycle Company has a production function given by

$$Q = f(x, y) = 60x^{0.55}y^{0.5}$$

where x is the dollars (in millions) spent on labor, y is the dollars (in millions) spent on capital, and Q is the quantity produced (in thousands).

(a) Compute $f_x(x, y)$ and $f_y(x, y)$.

(b) If the Jendrag Company is currently using 220 units of labor and 140 units of capital, determine the marginal productivity of labor and the marginal productivity of capital and interpret each.

(c) Would production increase more by spending an additional $1 million on labor or $1.5 million on capital? Explain.

SECTION PROJECT

The program in Appendix C that performs multiple regression is needed to do the section project.

The IronWorks Company has six plants in the U.S. Midwest that make fireplace accessories. Data for each plant give the dollars (in millions) spent on labor, the dollars (in millions) spent on capital, and the total quantity produced (in thousands) for 1999. The plants all operate at the same level of technology, so a production function can be determined.

Labor	Capital	Quantity
2.2	1.7	10.7
2.3	1.6	10.8
2.5	1.75	11.8
2.9	1.8	13.3
3	1.85	13.7
3.2	2	14.7

(a) Enter the data into your calculator and execute the program to determine a Cobb–Douglas production function. Round all values to the nearest hundredth.

(b) On average, IronWorks is currently investing 2.7 units of labor and 1.8 units of capital. Find the marginal productivity of labor and marginal productivity of capital at these levels of labor and capital investment and interpret each.

(c) At its current levels of capital and labor investment, would production increase more by spending, on average, an additional $800,000 on labor or $400,000 on capital?

Section 18.4 Maxima and Minima

We are now ready for a brief analysis of extreme values for a function of two variables. If we can locate the high points (**relative maxima**) and the low points (**relative minima**) on a surface, then we can optimize the quantity represented by the surface. For example, we could determine a maximum for profit or a minimum for cost. This should sound familiar, because we optimized functions of one variable in Sections 15.4 and 15.5. There we learned how the **Second Derivative Test** aided us in locating the relative extrema. We shall see in this section that there is a Second Derivative Test for relative extrema of a function of two variables. Just as we did with a function of one variable, we must first discuss the concept of **critical points**.

Relative Extrema, Critical Points, and the Second Derivative Test

We begin our analysis by making an assumption. Given $z = f(x, y)$, we assume that all second-order partial derivatives exist for $f(x, y)$ in some circular region in the xy-plane. This assumption guarantees that we are dealing with what is known as a **smooth surface**. A **smooth surface** is a one that has no edges (like a shoe box) or sharp points (like the point of a nail) or breaks (like the San Andreas fault line). These surfaces are illustrated in Figure 18.4.1.

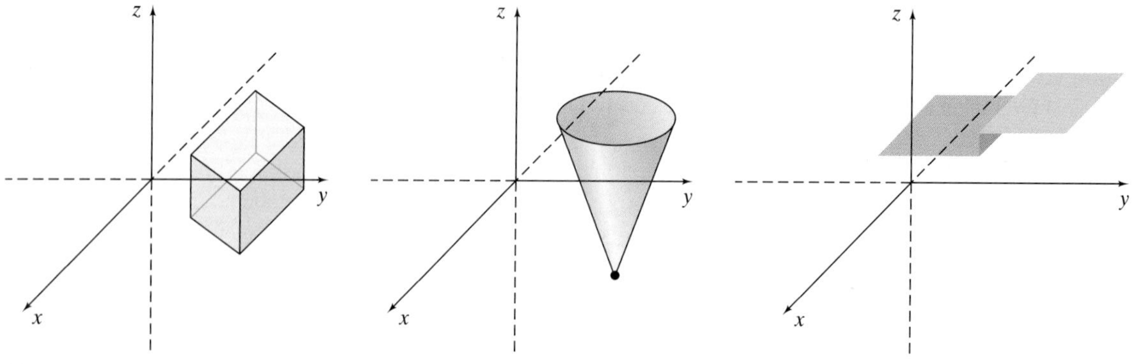

Figure 18.4.1 **(a)** Surface has an edge. **(b)** Surface has a sharp point. **(c)** Surface has a break.

Finally, we are not going to concern ourselves with boundary points. Thus, we are not going to concern ourselves with absolute extrema.

In Section 15.1 we defined relative extrema for a function of one variable as follows:

T From Your Toolbox

Definition of Relative Extrema

For some open interval containing c:

1. f has a **relative maximum** at c if $f(c) \geq f(x)$ for all x in the interval.
2. f has a **relative minimum** at c if $f(c) \leq f(x)$ for all x in the interval.

We extend this definition to functions of two variables as follows:

Definition of Relative Extrema in 3-Space

Let $f(x, y)$ be a function of two variables. The value $f(a, b)$ is:

1. A **relative maximum** if $f(a, b) \geq f(x, y)$ for all points (x, y) in some circular region in the xy-plane around (a, b), where (a, b) is the center of the circular region.
2. A **relative minimum** if $f(a, b) \leq f(x, y)$ for all points (x, y) in some circular region in the xy-plane around (a, b), where (a, b) is the center of the circular region.

Note: The circular region described in these definitions should be small. If it is too large, it may contain some values of $f(x, y)$ that are larger or smaller than our relative extremum.

Figure 18.4.2 illustrates a relative maximum and Figure 18.4.3 illustrates a relative minimum.

Figure 18.4.4 illustrates a **saddle point**, which is neither a relative maximum nor a relative minimum. No matter how small a circular region we draw with (a, b) as the center, there are always values of $f(x, y)$ such that $f(x, y) > f(a, b)$ and $f(x, y) < f(a, b)$.

Look carefully at Figures 18.4.2 and 18.4.3. Recall that a vertical cross section of the form $x = c$ is parallel to the y-axis and provides the geometric

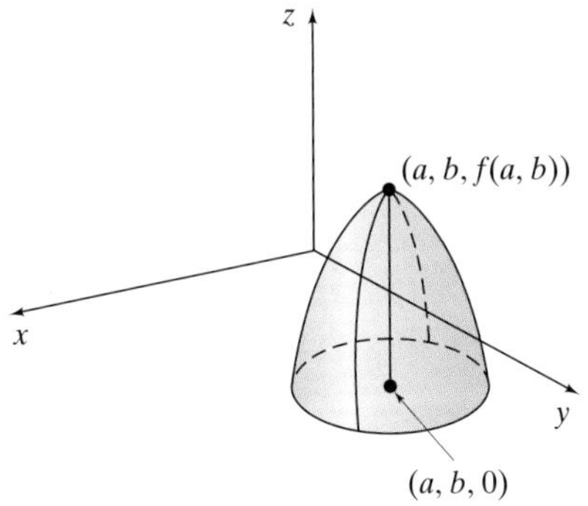

Figure 18.4.2 Relative maximum.

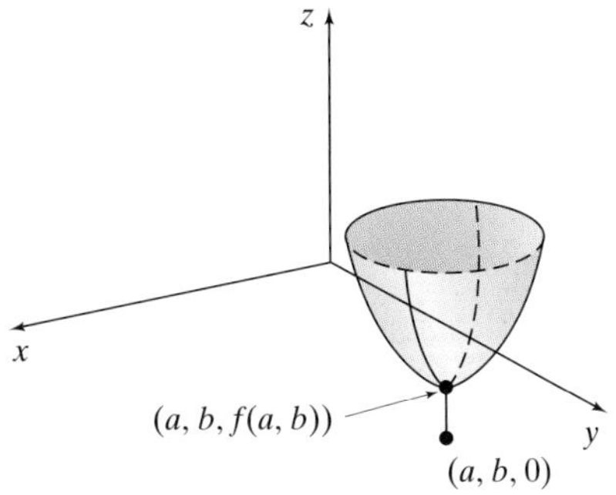

Figure 18.4.3 Relative minimum.

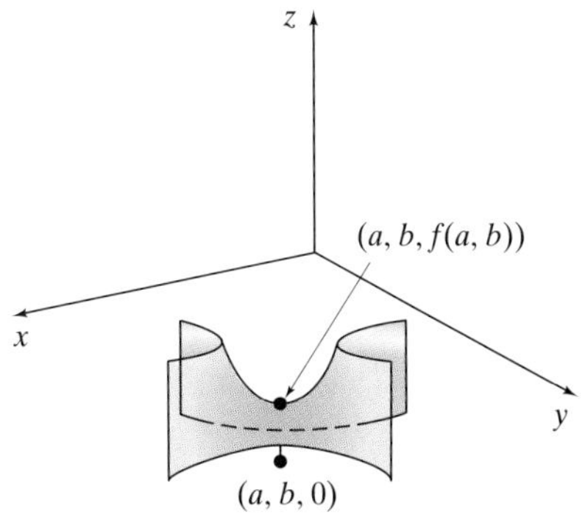

Figure 18.4.4 Saddle point.

meaning of $f_y(x, y)$. Likewise, a vertical cross section of the form $y = c$ is parallel to the x-axis and provides the geometric meaning of $f_x(x, y)$. Now in Figures 18.4.2 and 18.4.3 it appears that *any* vertical cross section will have a horizontal tangent at the relative extreme, in particular, both $f_x(x, y) = 0$ and $f_y(x, y) = 0$.

Recall our definition of the **critical value** of a function of one variable:

T From Your Toolbox

Definition of Critical Values

A **critical value** for f is an x-value in the domain of f for which (1) $f'(x) = 0$ or (2) $f'(x)$ is undefined.

The possibilities for a function of two variables are beyond the scope of this text. We shall restrict our discussion to **critical points** for which $f_x(x, y) = 0$ and $f_y(x, y) = 0$.

Critical Point

The point (a, b) is a **critical point** for $f(x, y)$ if $f_x(a, b) = 0$ **and** $f_y(a, b) = 0$.

Note: We call it a critical *point* because the domain of $f(x, y)$ is a region in the xy-plane.

Example 1 Determining Critical Points

Determine the critical points for $f(x, y) = x^2 + y^2 - 8x + 2y + 7$.

Solution

Understand the Situation: Critical points are determined by setting the partial derivatives $f_x(x, y) = 0$ and $f_y(x, y) = 0$ and solving this system of equations for x and y.

The partial derivatives are

$$f_x(x, y) = 2x - 8 \quad \text{and} \quad f_y(x, y) = 2y + 2$$

Setting each equal to zero and solving yields

$$\begin{aligned} 0 &= 2x - 8 \quad &\text{and} \quad 0 &= 2y + 2 \\ 4 &= x \quad &\text{and} \quad -1 &= y \end{aligned}$$

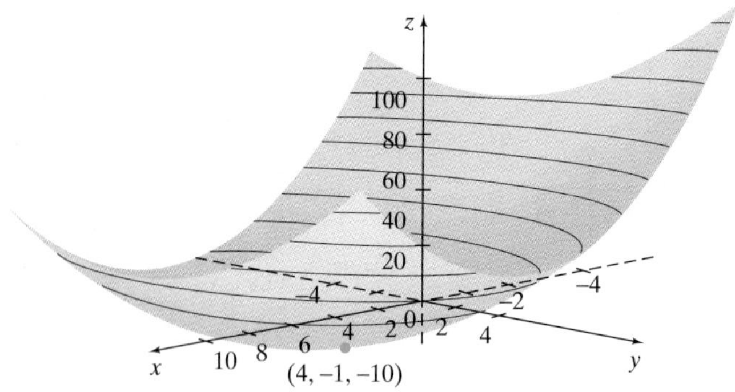

Figure 18.4.5

So the only critical point is the point $(4, -1)$. Notice that this point is in the xy-plane. If we wanted to know the point on the *surface*, we would find the z-coordinate by evaluating $f(4, -1)$. This yields a z-coordinate of -10. So the point on the surface would be $(4, -1, -10)$. See Figure 18.4.5. ■

Checkpoint 1 Now work Exercise 3.

Example 2 Determining Critical Points

Determine the critical points for $f(x, y) = x^2 + y^2 - xy + 3y$.

Solution

The partial derivatives are

$$f_x(x, y) = 2x - y \quad \text{and} \quad f_y(x, y) = 2y - x + 3$$

Setting each to zero gives

$$0 = 2x - y \quad \text{and} \quad 0 = 2y - x + 3$$

Here we have a system of two equations. We need to determine values of x and y that satisfy *both* equations. There are several methods that we could use to solve this system. We use an algebraic technique known as *substitution*. For a review of the substitution method for solving a system of equations, look back to Section 1.5. This method yields

$$0 = 2x - y \quad \text{and} \quad 0 = 2y - x + 3$$
$$y = 2x$$

We substitute $y = 2x$ into the second equation for y and get

$$0 = 2(2x) - x + 3$$
$$0 = 4x - x + 3$$
$$0 = 3x + 3$$
$$-1 = x$$

We simply substitute $x = -1$ into either of the original equations and solve for y. We choose the first equation and this gives

$$0 = 2x - y$$
$$0 = 2(-1) - y$$
$$-2 = y$$

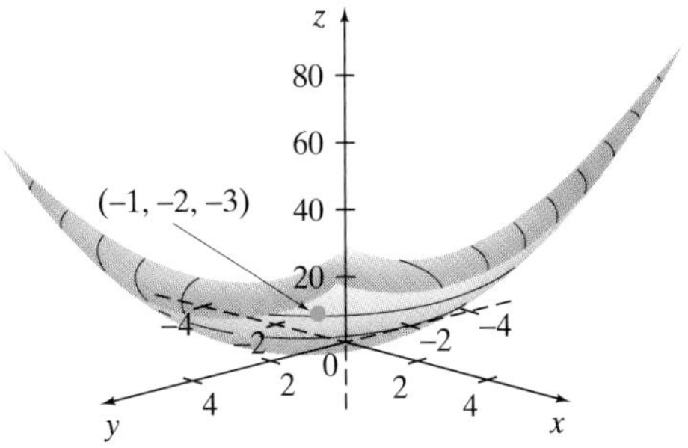

Figure 18.4.6

So $x = -1$ and $y = -2$ are the solutions to both equations. Thus, our critical point is $(-1, -2)$. See Figure 18.4.6. ■

Checkpoint 2 Now work Exercise 7.

Now that we know how to find critical points, our next task is to determine whether we have a relative extrema at a given critical point. Just as not all critical values of a function of one variable give relative extrema, there is no guarantee that we will find a relative extrema at critical points of a function of two variables. For example, the saddle point in Figure 18.4.4 is a critical point that does not produce a relative extremum. The following **Second Derivative Test** enables us to determine the shape of the surface in the vicinity of critical points, which in turn tells us if we have a relative extremum or not.

Second Derivative Test

Let $z = f(x, y)$ be a function of two variables such that $f_{xx}(x, y)$, $f_{yy}(x, y)$, and $f_{xy}(x, y)$ all exist. If $f_x(a, b) = 0$ and $f_y(a, b) = 0$, that is, (a, b) is a critical point, then we define a number D to be

$$D = f_{xx}(a, b) \cdot f_{yy}(a, b) - [f_{xy}(a, b)]^2$$

The Second Derivative Test has the following form:

1. If $D > 0$ and $f_{xx}(a, b) < 0$, then $f(a, b)$ is a **relative maximum.**
2. If $D > 0$ and $f_{xx}(a, b) > 0$, then $f(a, b)$ is a **relative minimum.**
3. If $D < 0$, then $f(a, b)$ is a **saddle point.**
4. If $D = 0$, the test gives no information about $f(a, b)$.

Although this test appears complicated, it is fairly easy to apply, as Example 3 illustrates.

Example 3 Determining Relative Extrema

Determine any relative extrema for $f(x, y) = x^2 + y^2 - xy + 3y$.

Solution

This is the function from Example 2. There we determined the partial derivatives to be

$$f_x(x, y) = 2x - y \quad \text{and} \quad f_y(x, y) = 2y - x + 3$$

and the only critical point occurred at $x = -1$ and $y = -2$, that is, at $(-1, -2)$. To use the Second Derivative Test, we need the following second-order partial derivatives:

$$f_{xx}(x, y) = 2, \quad f_{yy}(x, y) = 2, \quad \text{and} \quad f_{xy}(x, y) = -1$$

We now evaluate our critical point $(-1, 2)$ at each second-order partial derivative.

$$f_{xx}(-1, -2) = 2, \quad f_{yy}(-1, -2) = 2, \quad \text{and} \quad f_{xy}(-1, -2) = -1$$

We then define D to be

$$\begin{aligned} D &= f_{xx}(-1, -2) \cdot f_{yy}(-1, -2) - [f_{xy}(-1, -2)]^2 \\ &= 2 \cdot 2 - [-1]^2 = 4 - 1 = 3 \end{aligned}$$

Since $D > 0$ and $f_{xx}(-1, -2) > 0$, the Second Derivative Test asserts that there is a *relative minimum* at $(-1, -2)$. Specifically, on the *surface* there is a relative minimum at $(-1, -2, -3)$. See Figure 18.4.6. ■

We offer the following guidelines on applying the Second Derivative Test:

1. Determine all critical points (a, b).
2. Determine $f_{xx}(x, y)$, $f_{yy}(x, y)$, and $f_{xy}(x, y)$.
3. Evaluate the second-order partial derivatives at each critical point.
4. Determine D.
5. Apply the Second Derivative Test.

Before we address some applications, let's do Example 4 to ensure that the process is understood.

Example 4 Determining Relative Extrema

Determine the relative extrema for $f(x, y) = x^3 + 3x^2 - y^2 + 2y + 4$.

Solution

We begin by locating the critical points. First, the partial derivatives are

$$f_x(x, y) = 3x^2 + 6x \quad \text{and} \quad f_y(x, y) = -2y + 2$$

Setting each equal to zero and solving yields

$$\begin{aligned} 0 &= 3x^2 + 6x \quad &\text{and} \quad 0 &= -2y + 2 \\ 0 &= 3x(x + 2) & 1 &= y \\ x &= 0, \quad x = -2 & & \end{aligned}$$

So here we have two critical points, when $x = 0$ and $y = 1$ as well as when $x = -2$ and $y = 1$. This gives the points $(0, 1)$ and $(-2, 1)$ as the critical points.

Next we determine the second-order partial derivatives to be

$$f_{xx}(x, y) = 6x + 6, \quad f_{yy}(x, y) = -2, \quad \text{and} \quad f_{xy}(x, y) = 0$$

Evaluating the second-order partial derivatives at the critical point $(0, 1)$ gives

$$f_{xx}(0, 1) = 6(0) + 6 = 6, \quad f_{yy}(0, 1) = -2, \quad \text{and} \quad f_{xy}(0, 1) = 0$$

For the critical point $(0, 1)$, we determine D to be

$$\begin{aligned} D &= f_{xx}(0, 1) \cdot f_{yy}(0, 1) - [f_{xy}(0, 1)]^2 \\ &= 6(-2) - [0]^2 = -12 \end{aligned}$$

Since $D < 0$ at the critical point $(0, 1)$, the Second Derivative Test asserts that we have a *saddle point* at $(0, 1)$. Specifically, we have a saddle point at $(0, 1, 5)$.

Evaluating the second-order partial derivatives at the critical point $(-2, -1)$ gives

$$f_{xx}(-2,-1) = 6(-2) + 6 = -6, \quad f_{yy}(-2,-1) = -2, \quad \text{and} \quad f_{xy}(-2,-1) = 0$$

For the critical point $(-2, -1)$, we now determine D to be

$$\begin{aligned} D &= f_{xx}(-2,-1) \cdot f_{yy}(-2,-1) - [f_{xy}(-2,-1)]^2 \\ &= (-6)(-2) - [0]^2 = 12 \end{aligned}$$

Since $D > 0$ at the critical point $(-2, -1)$ and $f_{xx}(-2, -1) < 0$, the Second Derivative Test tells us that we have a *relative maximum* at $x = -2$, $y = -1$. Specifically, we have a *relative maximum* at $(-2, -1, 9)$. Figure 18.4.7 shows the surface with the saddle point and the relative maximum.

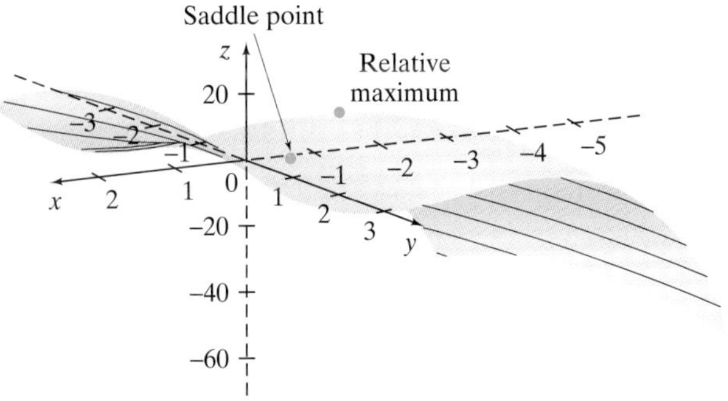

Figure 18.4.7

■

Checkpoint 3

Now work Exercise 21.

Applications

The final two examples of this section illustrate how, through applications, the Second Derivative Test is a very powerful method in locating relative extrema.

Example 5 Maximizing Revenue

The All Clear Company sells two types of car windshield wiper fluid: *regular*, which can be used at temperatures above 10°F, and a *no-freeze*, which can be used at temperatures above −30°F. Let x represent the number of gallons of no-freeze sold each year, in millions, and let y represent the number of gallons of regular sold each year, also in millions. Suppose that

$p = 3.9 - 0.5x - 0.1y$, the price in dollars of a gallon of no-freeze

$q = 3.9 - 0.1x - 0.8y$, the price in dollars of a gallon of regular

(a) Determine the revenue function $R(x, y)$.

(b) Determine x and y such that revenue is maximized and find the maximum revenue.

Solution

(a) As always, revenue equals price times quantity, which gives

$$\begin{aligned} R(x, y) &= x(3.9 - 0.5x - 0.1y) + y(3.9 - 0.1x - 0.8y) \\ &= 3.9x - 0.5x^2 - 0.1xy + 3.9y - 0.1xy - 0.8y^2 \\ &= -0.5x^2 - 0.8y^2 + 3.9x + 3.9y - 0.2xy \end{aligned}$$

(b) Since we want to maximize revenue, first locate any critical points of $R(x, y)$. Computing $R_x(x, y)$ and $R_y(x, y)$ yields

$$R_x(x, y) = -x + 3.9 - 0.2y \quad \text{and} \quad R_y(x, y) = -1.6y + 3.9 - 0.2x$$

Setting each equal to zero and solving gives

$$0 = -x + 3.9 - 0.2y \quad \text{and} \quad 0 = -1.6y + 3.9 - 0.2x$$

We opt to solve by using the substitution method, as we did in Example 2. From the first equation we have

$$x = 3.9 - 0.2y$$

Substituting this into the second equation yields

$$\begin{aligned} 0 &= -1.6y + 3.9 - 0.2(3.9 - 0.2y) \\ 0 &= -1.6y + 3.9 - 0.78 + 0.04y \\ 0 &= -1.56y + 3.12 \\ y &= 2 \end{aligned}$$

Substituting $y = 2$ into the first equation gives us

$$\begin{aligned} 0 &= -x + 3.9 - 0.2(2) \\ x &= 3.5 \end{aligned}$$

So we have a critical point at $x = 3.5$ and $y = 2$, or more precisely at $(3.5, 2)$. We now proceed to determine if the critical point yields a relative extremum. First we compute the second-order partial derivatives to be

$$R_{xx}(x, y) = -1, \quad R_{yy}(x, y) = -1.6, \quad \text{and} \quad R_{xy}(x, y) = -0.2$$

Evaluating our critical point at these second-order partial derivatives gives

$$R_{xx}(3.5, 2) = -1, \quad R_{yy}(3.5, 2) = -1.6, \quad \text{and} \quad R_{xy}(3.5, 2) = -0.2$$

Now, for the critical point $(3.5, 2)$ we define D as

$$\begin{aligned} D &= R_{xx}(3.5, 2) \cdot R_{yy}(3.5, 2) - [R_{xy}(3.5, 2)]^2 \\ &= (-1)(-1.6) - [-0.2]^2 = 1.56 \end{aligned}$$

Since $D > 0$ for the critical point $(3.5, 2)$ and $R_{xx}(3.5, 2) < 0$, we have a *relative maximum* at $x = 3.5$, $y = 2$. So All Clear needs to sell 3.5 million gallons of the no-freeze and 2 million gallons of the regular to maximize revenue for the year. The maximum revenue is

$$\begin{aligned} R(3.5, 2) &= -0.5(3.5)^2 - 0.8(2)^2 + 3.9(3.5) + 3.9(2) - 0.2(3.5)(2) \\ &= 10.725 \end{aligned}$$

So the maximum revenue is $10.725 million dollars. ■

Interactive Activity

Determine the critical point in Example 5 by setting each partial derivative equal to zero, solving for y, graphing, and using the INTERSECT command.

WWW

Example 6 Maximizing Volume

Figure 18.4.8

North-South Airlines allows each passenger to carry up to three suitcases as long as the sum of the width, length, and height of each suitcase is less than or equal to 60 inches. Determine the dimensions of a suitcase of maximum volume that a passenger may carry with this restriction.

Solution

Understand the Situation: First we sketch a picture of a generic suitcase. See Figure 18.4.8. We represent length with an x, width by y, and height by z. The airline's restriction is that

$$x + y + z \leq 60$$

The volume of our suitcase is given by

$$V = x \cdot y \cdot z$$

Notice that the volume is a function of three independent variables. We need to somehow rewrite this as a function of two variables so that we may use the techniques learned in this section. Since we believe that the largest volume for the suitcase results when $x + y + z = 60$, we solve this equation for z and get

$$z = 60 - x - y$$

We substitute this into our equation for volume and get a function of two variables,

$$V(x, y) = xy(60 - x - y) = 60xy - x^2y - xy^2$$

To maximize the volume, we proceed by determining the critical points and applying the Second Derivative Test. First, we determine $V_x(x, y)$ and $V_y(x, y)$.

$$V_x(x, y) = 60y - 2xy - y^2 \quad \text{and} \quad V_y(x, y) = 60x - x^2 - 2xy$$

Setting each equal to zero yields

$$0 = 60y - 2xy - y^2 \quad \text{and} \quad 0 = 60x - x^2 - 2xy$$
$$0 = y(60 - 2x - y) \quad \text{and} \quad 0 = x(60 - x - 2y)$$

Thus,

$$0 = y \quad \text{or} \quad 0 = 60 - 2x - y \quad \text{and} \quad 0 = x \quad \text{or} \quad 0 = 60 - x - 2y$$

We can immediately reject the situation when $x = 0$ and $y = 0$ since this would produce a suitcase with a volume of 0. We are then left with solving

$$0 = 60 - 2x - y \quad \text{and} \quad 0 = 60 - x - 2y$$

Using the substitution method, we solve the first equation for y and get

$$y = 60 - 2x$$

Substituting this into our second equation gives

$$0 = 60 - x - 2(60 - 2x)$$
$$0 = 60 - x - 120 + 4x$$
$$0 = -60 + 3x$$
$$x = 20$$

Substituting $x = 20$ into the first equation to determine y gives

$$y = 60 - 2(20) = 20$$

So our critical point is $x = 20$ and $y = 20$. Next we need the second-order partial derivatives.

$$V_{xx}(x, y) = -2y, \quad V_{yy}(x, y) = -2x, \quad \text{and} \quad V_{xy}(x, y) = 60 - 2x - 2y$$

Evaluating each second-order partial derivative at the critical point (20, 20) yields

$$V_{xx}(20, 20) = -40$$
$$V_{yy}(20, 20) = -40$$
$$V_{xy}(20, 20) = 60 - 2(20) - 2(20) = 60 - 40 - 40 = -20$$

Interactive Activity

How was the height of the suitcase in Example 6 computed to be 20 inches?

For the critical point $x = 20$, $y = 20$, we determine D to be

$$D = V_{xx}(20, 20) \cdot V_{yy}(20, 20) - [V_{xy}(20, 20)]^2$$
$$= (-40)(-40) - [-20]^2 = 1200$$

Since $D > 0$ for the critical point (20, 20) and $V_{xx}(20, 20) < 0$, we have a relative maximum at $x = 20$, $y = 20$.

Interpret the Solution: So the dimensions that would maximize volume would be 20 by 20 by 20 inches. ■

SUMMARY

In this section we defined the **relative extrema** of a function of two variables.

- Let $f(x, y)$ be a function of two variables. The value $f(a, b)$ is:

  A **relative maximum** if $f(a, b) \geq f(x, y)$ for all points (x, y) in some circular region in the xy-plane around (a, b), where (a, b) is the center of the circular region.

  A **relative minimum** if $f(a, b) \leq f(x, y)$ for all points (x, y) in some circular region in the xy-plane around (a, b), where (a, b) is the center of the circular region.

We also defined critical points for a function of two variables as follows:

- The point (a, b) is a **critical point** for $f(x, y)$ if $f_x(a, b) = 0$ and $f_y(a, b) = 0$.

We presented the **Second Derivative Test**, which allows us to determine if the critical point produces a relative maximum, a relative minimum, a saddle point, or none of these.

- If (a, b) is a critical point, then we define a number D to be

$$D = f_{xx}(a, b) \cdot f_{yy}(a, b) - [f_{xy}(a, b)]^2$$

The Second Derivative Test has the following form:

1. If $D > 0$ and $f_{xx}(a, b) < 0$, the $f(a, b)$ is a **relative maximum**.
2. If $D > 0$ and $f_{xx}(a, b) > 0$, then $f(a, b)$ is a **relative minimum**.
3. If $D < 0$, then $f(a, b)$ is a **saddle point**.
4. If $D = 0$, the test gives no information about $f(a, b)$.

SECTION 18.4 EXERCISES

In Exercises 1–10, determine the critical points for the given function.

1. $f(x, y) = x^2 + y^2 - 4x + 6y - 4$

2. $f(x, y) = x^2 + y^2 - 6x - 4y + 3$

✓ **3.** $f(x, y) = x^2 + y^2 + 2x - 4y - 3$

4. $f(x, y) = x^2 + y^2 - 8x + 2y - 2$

5. $f(x, y) = 2x^2 + 3y^2 - 8x + 6y + 5$

6. $f(x, y) = x^2 + y^2 + xy - 6y + 1$

✓ **7.** $f(x, y) = x^2 + y^2 - xy - 3y + 5$

8. $f(x, y) = x^2 + y^2 - xy - 3x - 3y + 2$

9. $f(x, y) = x^2 - y^2 + xy + 2x - 9y - 5$

10. $f(x, y) = 2x^2 + 3y^2 + 2xy + 4x - 8y + 3$

In Exercises 11–30, use the Second Derivative Test to locate any relative extrema and saddle points.

11. $f(x, y) = x^2 + y^2 - 6x - 4y + 3$

12. $f(x, y) = x^2 + y^2 - 4x + 6y - 4$

13. $f(x, y) = x^2 + y^2 - 6x + 4y + 2$

14. $f(x, y) = 3x^2 - y^2 - 4x + 6y + 1$

15. $f(x, y) = -x^2 - 3y^2 - 4x - 4y - 1$

16. $f(x, y) = x^2 + y^2 - xy - 3y + 5$

17. $f(x, y) = x^2 + y^2 + xy - 6y + 1$

18. $f(x, y) = x^2 - y^2 + xy + 2x - 9y - 5$

19. $f(x, y) = x^2 + y^2 - xy - 3x - 3y + 2$

20. $f(x, y) = 3x^2 + y^2 + xy + 3y + 4$

✓ 21. $f(x, y) = \frac{4}{3}x^3 - y^2 - 4x^2 + 2y - 1$

22. $f(x, y) = 4x^3 - 6x^2 - 24x + 2y^2 - 4y + 6$

23. $f(x, y) = 2y^3 - 6y^2 - 18y + 2x^2 - 4x + 12$

24. $f(x, y) = x^3 + 3xy - y^3$

25. $f(x, y) = x^3 - 3xy + y^3$

26. $f(x, y) = 2x^3y - 2x + 16y - 5$

27. $f(x, y) = e^{-x^2-y^2}$

28. $f(x, y) = e^{xy}$

29. $f(x, y) = 6xy + \frac{12}{x} - \frac{3}{y}$

30. $f(x, y) = 4xy + \frac{8}{x} - \frac{2}{y}$

Applications

31. The annual cost for labor and specialized robotics equipment for an automobile manufacturer, in millions of dollars, is given by

$$C(x, y) = 5x^2 + 5xy + 7.5y^2 - 40x - 45y + 135$$

where x is the amount, in millions of dollars, spent each year on labor and y is the amount, in millions of dollars, spent each year on robotics equipment.

(a) Determine how much should be spent on each, per year, to minimize cost.

(b) Determine the minimum cost.

32. New Jeans, a specialty blue jeans manufacturer, produces two types of blue jeans each day, x pairs of straight leg and y pairs of wide leg. The daily profit function, in dollars, is given by

$$P(x, y) = 78x + 3xy - 3x^2 - y^2 - 2y$$

(a) How many straight-leg jeans and how many wide-leg jeans should be produced and sold each day to maximize profit?

(b) What is the maximum profit?

33. The Wokon Company produces two types of woks each day, x number of regular sized and y number of jumbo-sized. The daily profit from the sale of x number of regular-sized woks and y number of jumbo-sized woks is given by

$$P(x, y) = -0.3y^3 - 0.2x^2 + 6xy - 2$$

where $P(x, y)$ is measured in hundreds of dollars.

(a) How many regular-sized woks and how many jumbo-sized woks should be sold each day to maximize profit?

(b) What is the maximum profit?

34. The BigBang Engine Company produces two types of engines, x thousand two-stroke engines and y thousand four-stroke engines. Vice-president Kathy determined the revenue and cost functions for a year to be (in millions of dollars)

$$R(x, y) = 42x + 39.5y$$
$$C(x, y) = 0.5x^2 + 0.8xy + y^2 + 5x + 4y$$

(a) Determine how many two-stroke engines and how many four-stroke engines should be produced each year to maximize profit.

(b) What is the maximum profit?

35. Crispy Chips produces two types of potato chips each year, x 18-ounce bags (in millions) of sour cream and chives and y 18-ounce bags (in millions) of barbecue. The revenue and cost functions for the year, in millions of dollars, are

$$R(x, y) = 1.5x + 2y$$
$$C(x, y) = x^2 - xy + 2y^2 + 2.5x - 9y + 2$$

(a) Determine the profit function $P(x, y)$.

(b) How many 18-ounce bags of each type of chip should be produced each year to maximize profit?

(c) What is the maximum profit?

36. The Palton Company produces two types of graphing calculators each month, x units of a 2D graphing calculator and y units of a 3D graphing calculator. The monthly revenue and cost functions, in dollars, are

$$R(x, y) = 75x + 101y + 0.05xy - 0.04x^2 - 0.04y^2$$
$$C(x, y) = 13.98x + 15.02y + 32{,}000$$

(a) Determine how many 2D graphing calculators and how many 3D graphing calculators should be sold each month to maximize profit.

(b) What is the maximum profit?

37. Sammy's Cycles manufactures 21-speed touring bicycles and 21-speed mountain bicycles. Let x represent the weekly demand for a 21-speed touring bicycle, and let y represent the weekly demand for a 21-speed mountain bicycle. The

weekly price–demand equations are given by

$p = 349 - 4x + y$, the price in dollars for a 21-speed touring bicycle

$q = 446 + 2x - 3y$, the price in dollars for a 21-speed mountain bicycle

(a) Determine the revenue function $R(x, y)$.

(b) How many of each type of bicycle should be produced each week to maximize revenue?

(c) What is the maximum revenue?

38. *(continuation of Exercise 37)* Sammy's Cycles has a cost function given by

$$C(x, y) = 390 + 95x + 96y$$

(a) Determine the profit function $P(x, y)$.

(b) How many of each type of bicycle should be produced to maximize profit?

(c) What is the maximum profit?

39. The Leaf Chewer Company manufactures and sells leaf blowers and a special 8-foot blower attachment to clean gutters. Let x represent the number of leaf blowers produced and sold each month, and let y represent the number of special 8-foot attachments produced and sold each month. The monthly price–demand equations are given by

$p = 228 - 8x - y$, the price in dollars for a leaf blower

$q = 31 - x - 0.15y$, the price in dollars for a 8-foot attachment

(a) Determine the revenue function $R(x, y)$.

(b) Determine how many leaf blowers and how many 8-foot attachments should be produced and sold each month to maximize revenue.

40. *(continuation of Exercise 39)* The monthly cost function for the Leaf Chewer Company is given by

$$C(x, y) = 1000 + 28x + 5y$$

(a) Determine the monthly profit function $P(x, y)$.

(b) Determine how many leaf blowers and how many 8-foot attachments should be sold each month to maximize profit.

(c) What is the maximum profit?

41. You have been challenged to design a rectangular box with no top and two parallel partitions that holds 125 cubic inches. Determine the dimensions that will require the least amount of material to do this.

SECTION PROJECT

(a) *Federal Express* states that the maximum length plus girth for its FedEx Overnight Freight and its FedEx 2Day Freight shipments is 300 inches (*Source:* Federal Express, www.fedex.com). See Figure 18.4.9. Determine the dimensions of the largest-volume package that can be sent by FedEx Overnight Freight and FedEx 2Day Freight. Assume that the length is y.

(b) To avoid U.S. Domestic Freight Service charges, any package sent by *Federal Express* must have a length plus girth not exceeding 165 inches (*Source:* Federal Express, www.fedex.com). See Figure 18.4.9. Determine the dimensions of the largest-volume package that can be sent by *Federal Express* without incurring Domestic Freight Service charges. Assume that the length is y.

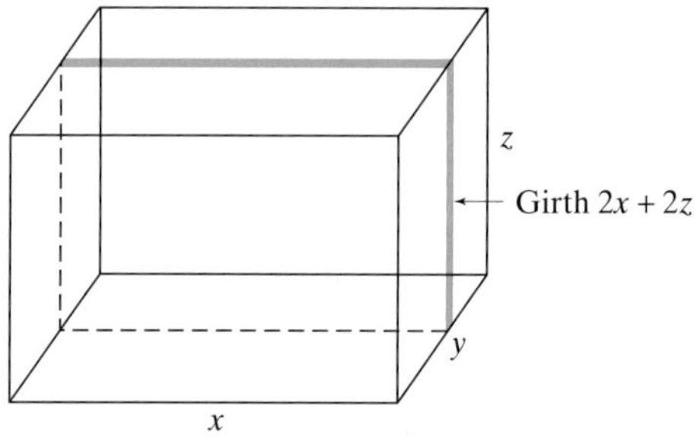

Figure 18.4.9

Section 18.5 Lagrange Multipliers

Many optimization problems that we encounter are actually *constrained* by some external circumstances. For example, North-South Airlines in Example 6 from Section 18.4 had a *constraint* that length plus width plus height of a suitcase cannot exceed 60 inches. We maximized the volume of a suitcase, $V = x \cdot y \cdot z$, subject to the constraint $x + y + z \leq 60$. Other examples include maximizing an area of an enclosure given a finite amount of fence or maximizing production given a budget constraint. Each of these problems may be solved by using the **Method of Lagrange Multipliers**.

Method of Lagrange Multipliers

The method that we are about to study is for solving maximum or minimum problems when dealing with restrictions on the independent variables. The method is named after the French mathematician Joseph Louis Lagrange (1736–1813), who discovered the method at the age of 19. We introduce the method through a Flashback.

Flashback

The Beloved Beagle Revisited

In Section 15.4, we saw that one of the authors needs to build an enclosure for his beloved beagle. He has 200 feet of fence and wants to make a rectangular enclosure along a straight wall of his house (see Figure 18.5.1).

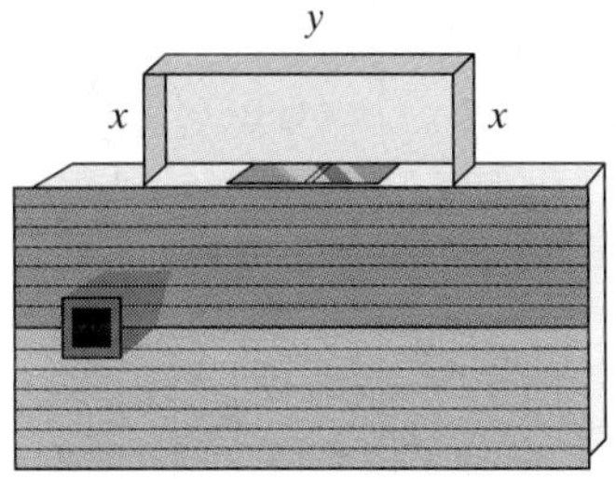

Figure 18.5.1

(a) Determine a function of two independent variables for the area of the enclosure.

(b) Determine any restrictions on x and y, assuming that we want to use all 200 feet of fence to maximize the area of the enclosure.

Flashback Solution

(a) We know that area is equal to length times width. For the enclosure shown in Figure 18.5.1, this gives

$$A(x, y) = xy$$

(b) We only have 200 feet of fence and we wish to use all of it to make this enclosure. From Figure 18.5.1 we determine that this means that x and y are such that

$$2x + y = 200$$

This equation represents the restrictions placed on the independent variables x and y. These restrictions placed on x and y are called **constraints**.

In short, what we are asked to do in situations like the Flashback is to

Maximize $\quad A(x, y) = xy$

Subject to $\quad 2x + y = 200 \quad$ or $\quad 2x + y - 200 = 0$

This is a specific example of the more general class of maximum/minimum problems of the form

Maximize/Minimize	$z = f(x, y)$
Subject to	$g(x, y) = 0$

In Section 15.4 we determined the dimensions that maximized the area of the enclosure in the Flashback by solving the *constraint equation* for y and substituted the result into the area equation. This gave us a function of one variable, which we solved using methods developed in Section 15.4. So why do we need a new method to solve this problem? There are several reasons.

- Sometimes our constraint equation is complicated and cannot be solved for one variable.
- The *Method of Lagrange Multipliers* generalizes to functions of several variables subject to one or more constraints.

Theorem of Lagrange

The relative extrema of the function $z = f(x, y)$ subject to the constraint equation $g(x, y) = 0$ occur among those points (x, y) for which there exists a value of λ (*lambda*),

$$F(x, y, \lambda) = f(x, y) + \lambda \cdot g(x, y)$$

where we have the following:

$$F_x(x, y, \lambda) = 0, \quad F_y(x, y, \lambda) = 0, \quad \text{and} \quad F_\lambda(x, y, \lambda) = 0$$

provided all partial derivatives exist.

Note: The λ (Greek letter *lambda*) in the definition of $F(x, y, \lambda)$ is called the *Lagrange multiplier*. To organize our work when using the method of Lagrange, we use the following steps.

How to Use the Method of Lagrange

1. Write the situation in the form

Maximize (or minimize)	$z = f(x, y)$
Subject to	$g(x, y) = 0$

2. Define $F(x, y, \lambda) = f(x, y) + \lambda \cdot g(x, y)$.
3. Determine the partial derivatives

$$F_x(x, y, \lambda), \quad F_y(x, y, \lambda), \quad \text{and} \quad F_\lambda(x, y, \lambda)$$

4. Solve the system of equations

$$F_x(x, y, \lambda) = 0, \quad F_y(x, y, \lambda) = 0, \quad F_\lambda(x, y, \lambda) = 0$$

5. The maximum (or minimum) of $f(x, y)$ is among the values in step 4. Simply evaluate $z = f(x, y)$ at each.

Example 1 Utilizing the Method of Lagrange

Use the Method of Lagrange to determine the maximum area of an enclosure for the beloved beagle in the Flashback.

Solution

We will break down the method of Lagrange into the five steps just given.

Step 1: Write the situation in the appropriate form.

$$\begin{aligned} &\text{Maximize} && A(x, y) = xy \\ &\text{Subject to} && 2x - y - 200 = 0 \end{aligned}$$

Step 2: Define function F using the Lagrange multiplier λ.

$$F(x, y, \lambda) = A(x, y) + \lambda \cdot g(x, y) = xy + \lambda \cdot (2x + y - 200)$$

Step 3: Determine the partial derivatives.

$$\begin{aligned} F_x(x, y, \lambda) &= y + \lambda(2 + 0 - 0) = y + 2\lambda \\ F_y(x, y, \lambda) &= x + \lambda(0 + 1 - 0) = x + \lambda \\ F_\lambda(x, y, \lambda) &= 0 + 1 \cdot (2x + y - 200) = 2x + y - 200 \end{aligned}$$

Step 4: Solve the system. (Solutions to the system are critical points for F.)

$$\begin{aligned} y + 2\lambda &= 0 \\ x + \lambda &= 0 \\ 2x + y - 200 &= 0 \end{aligned}$$

To solve this system, notice from the first two equations that we have

$$y = -2\lambda \quad \text{and} \quad x = -\lambda$$

If we substitute these values for x and y into the third equation and solve for λ, we have

$$\begin{aligned} 2(-\lambda) + (-2\lambda) - 200 &= 0 \\ -4\lambda - 200 &= 0 \\ -4\lambda &= 200 \\ \lambda &= -50 \end{aligned}$$

Substituting $\lambda = -50$ into the first two equations yields

$$\begin{aligned} &y = -2\lambda && \text{and} && x = -\lambda \\ &y = -2(-50) && \text{and} && x = -(-\lambda) \\ &y = 100 && \text{and} && x = 50 \end{aligned}$$

Step 5: Here it is important to realize exactly what the Theorem of Lagrange states. Quite simply, the method of Lagrange finds only critical points. It does not tell whether the function is maximized, minimized, or neither at the critical point. In this problem, as in each problem that we do, we must know that the maximum (or minimum) does exist. It then follows that the maximum (or minimum) must occur at a critical point as determined by Lagrange. For this situation, there is certainly an optimal area enclosure for the beagle. Since there is only one possibility from step 4, it follows that the dimensions are $x = 50$ feet by $y = 100$ feet, and the maximum enclosure is $A(50, 100) = 5000$ square feet. ■

Interactive Activity

Refer to Section 15.4, Exercise 32, to verify that we achieved the same result as well as to compare the two different methods.

Example 2 Applying the Method of Lagrange

$$\begin{aligned} &\text{Minimize} && f(x, y) = x^2 + 3y^2 \\ &\text{Subject to} && x + y = 2 \end{aligned}$$

Solution

Step 1: Write in an appropriate form by writing the constraint equation as $g(x, y) = 0$.

$$\begin{aligned} &\text{Minimize} && f(x, y) = x^2 + 3y^2 \\ &\text{Subject to} && g(x, y) = x + y - 2 = 0 \end{aligned}$$

Step 2: Define function F using the Lagrange multiplier λ.

$$F(x, y, \lambda) = f(x, y) + \lambda \cdot g(x, y) = x^2 + 3y^2 + \lambda(x + y - 2)$$

Step 3: Determine the partial derivatives.

$$\begin{aligned} F_x(x, y, \lambda) &= 2x + \lambda \\ F_y(x, y, \lambda) &= 6y + \lambda \\ F_\lambda(x, y, \lambda) &= x + y - 2 \end{aligned}$$

Step 4: Solve the system

$$\begin{aligned} 2x + \lambda &= 0 \\ 6y + \lambda &= 0 \\ x + y - 2 &= 0 \end{aligned}$$

From the first two equations, we have

$$2x = -\lambda \quad \text{and} \quad 6y = -\lambda$$

Setting these equal to each other yields

$$\begin{aligned} 2x &= 6y \\ y &= \frac{1}{3}x \end{aligned}$$

Since we have y in terms of x, we can return to our constraint equation, $x + y - 2 = 0$ and substitute for y, which yields

$$\begin{aligned} x + \frac{1}{3}x - 2 &= 0 \\ \frac{4}{3}x &= 2 \\ x &= \frac{3}{2} \end{aligned}$$

From $y = \frac{1}{3}x$ we also have

$$y = \frac{1}{3}\left(\frac{3}{2}\right) = \frac{1}{2}$$

Step 5: There is only one solution from step 4, $x = \frac{3}{2}$ and $y = \frac{1}{2}$. So the function is minimized at $f\left(\frac{3}{2}, \frac{1}{2}\right)$. This gives

$$f\left(\frac{3}{2}, \frac{1}{2}\right) = \left(\frac{3}{2}\right)^2 + 3\left(\frac{1}{2}\right)^2 = 3$$

The minimum value of $f(x, y) = x^2 + 3y^2$ subject to $x + y = 2$ is 3. ■

Checkpoint 1 Now work Exercise 7.

Geometric Analysis

Now let's look at what is happening geometrically. If we add a *constraint*, $g(x, y) = 0$, in the xy-plane, it gives the sketch in Figure 18.5.2. Now we look at only that part of the surface that is directly above our constraint equation. This yields our *constrained maximum*, (x_0, y_0, z_0), which we notice in Figure 18.5.2 is different from our *unconstrained (or free) maximum*, (x, y, z).

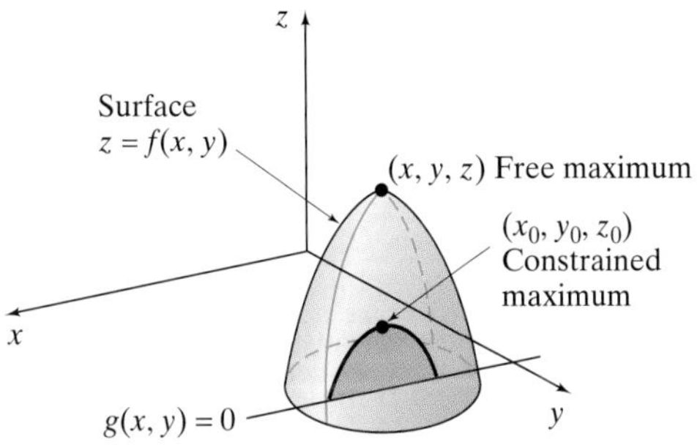

Figure 18.5.2

To gain another perspective, let's return to our two-dimensional setting and look at the level curves that we discussed in Section 18.2. Recall that a level curve is the result of slicing the surface with a horizontal plane.

Example 3 Analyzing Level Curves and a Constraint Equation

In Example 1 we

$$\text{Maximized} \quad A(x, y) = xy$$
$$\text{Subject to} \quad g(x, y) = 2x + y - 200 = 0$$

Sketch level curves for $A(x, y)$ for c-values of 3000, 5000, and 7000, and graph the constraint equation in the xy-plane.

Solution

Recall that to sketch level curves for $A(x, y)$, we graph equations of the form

$$c = xy \quad \text{or} \quad y = \frac{c}{x}$$

So for $c = 3000$, $c = 5000$, and $c = 7000$, we need to graph

$$y = \frac{3000}{x}, \quad y = \frac{5000}{x}, \quad \text{and} \quad y = \frac{7000}{x}$$

We also need to graph the constraint equation

$$g(x, y) = 2x + y - 200$$

When $g(x, y) = 0$, the graph is a line in the xy-plane. Thus, we need to graph

$$2x + y - 200 = 0 \quad \text{or} \quad y = 200 - 2x$$

In Figure 18.5.3 we show the graphs of $y = \dfrac{3000}{x}$, $y = \dfrac{5000}{x}$, $y = \dfrac{7000}{x}$, and $y = 200 - 2x$.

Interactive Activity

In the viewing window [0, 200] by [0, 250], graph $y = \frac{3000}{x}$, $y = \frac{5000}{x}$, $y = \frac{7000}{x}$, and $y = 200 - 2x$. Use the INTERSECT command to confirm the results of Figure 18.5.3.

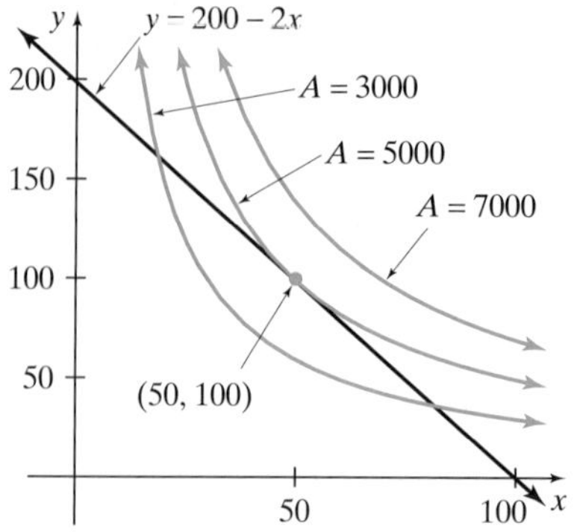

Figure 18.5.3

From Figure 18.5.3, we can make the following observations.

1. The constraint line shows all combinations of x and y that we could use to exhaust our 200 feet of fence.
2. The level curve corresponding to $A = 7000$ is not possible since it does not intersect our constraint equation.
3. The maximum value for A on the constraint equation occurs where the constraint is tangent to the level curve $A = 5000$.

Example 3 demonstrates that the critical points of our function $F(x, y, \lambda)$ in the Lagrange multipliers method occur at points where a level curve of $f(x, y)$ is tangent to the constraint curve $g(x, y)$.

Example 4 Maximizing a Production Function

Recall that the O'Neill Corporation has a production function of

$$Q = f(x, y) = 1.64x^{0.6}y^{0.4}$$

where x is labor hours, y is capital, and $f(x, y)$ is thousands of gallons produced. Each unit of labor is \$15,000 and each unit of capital is \$7000, and the total expense for both (a budget constraint) is limited to \$1.8 million. (These figures are for each of the O'Neill Corporation's 10 plants.) Determine the number of units of labor and capital needed to maximize production and give the maximum production.

Solution

This appears to be a maximization problem requiring the Method of Lagrange multipliers.

Step 1: We want to maximize $f(x, y) = 1.64x^{0.6}y^{0.4}$. Since each unit of labor is \$15,000 and each unit of capital is \$7000, we have the constraint equation $15{,}000x + 7000y = 1{,}800{,}000$. So our problem has the form

$$\begin{aligned} &\text{Maximize} && f(x, y) = 1.64x^{0.6}y^{0.4} \\ &\text{Subject to} && g(x, y) = 15{,}000x + 7000y - 1{,}800{,}000 = 0 \end{aligned}$$

Step 2: Define $F(x, y, \lambda) = f(x, y) + \lambda \cdot g(x, y)$ using the Lagrange multiplier λ.

$$F(x, y, \lambda) = 1.64x^{0.6}y^{0.4} + \lambda(15{,}000x + 7000y - 1{,}800{,}000)$$

Step 3: Determine the partial derivatives.

$$F_x(x, y, \lambda) = 0.984x^{-0.4}y^{0.4} + 15{,}000\lambda$$
$$F_y(x, y, \lambda) = 0.656x^{0.6}y^{-0.6} + 7000\lambda$$
$$F_\lambda(x, y, \lambda) = 15{,}000x + 7000y - 1{,}800{,}000$$

Step 4: Solve the system

$$0.984x^{-0.4}y^{0.4} + 15{,}000\lambda = 0$$
$$0.656x^{0.6}y^{-0.6} + 7000\lambda = 0$$
$$15{,}000x + 7000y - 1{,}800{,}000 = 0$$

To solve this nasty looking system, let's solve the first two equations for $-\lambda$. This gives

$$0.984x^{-0.4}y^{0.4} = -15{,}000\lambda \quad \text{and} \quad 0.656x^{0.6}y^{-0.6} = -7000\lambda$$
$$\frac{0.984x^{-0.4}y^{0.4}}{15{,}000} = -\lambda \quad \text{and} \quad \frac{0.656x^{0.6}y^{-0.6}}{7000} = -\lambda$$

Now we can set these equal to each other to give

$$\frac{0.984x^{-0.4}y^{0.4}}{15{,}000} = \frac{0.656x^{0.6}y^{-0.6}}{7000}$$

Now for the "trick." Multiply both sides of the equation by $x^{0.4}y^{0.6}$.

$$x^{0.4}y^{0.6}\left(\frac{0.984x^{-0.4}y^{0.4}}{15{,}000}\right) = x^{0.4}y^{0.6}\left(\frac{0.656x^{0.6}y^{-0.6}}{7000}\right)$$
$$\frac{0.984x^{-0.4+0.4}y^{0.4+0.6}}{15{,}000} = \frac{0.656x^{0.6+0.4}y^{-0.6+0.6}}{7000}$$
$$\frac{0.984y}{15{,}000} = \frac{0.656x}{7000}$$
$$y = \frac{15{,}000}{0.984}\left(\frac{0.656x}{7000}\right) \approx 1.43x$$

Substituting this back into the budget constraint equation for y gives

$$15{,}000x + 7000(1.43x) - 1{,}800{,}000 = 0$$
$$25{,}010x = 1{,}800{,}000$$
$$x \approx 71.97$$

From $y \approx 1.43x$, we also determine that

$$y \approx 1.43(71.97) \approx 102.92$$

Step 5: Since there is only one solution from step 4, $x \approx 71.97$ and $y \approx 102.92$, we conclude that $x \approx 71.97$ units of labor and $y \approx 102.92$ units of capital maximize production. The maximum production is approximately

$$f(71.97, 102.92) = 1.64(71.97)^{0.6}(102.92)^{0.4}$$
$$\approx 136.19 \text{ thousand gallons}$$

■

Checkpoint 2 Now work Exercise 25.

Marginal Productivity of Money

In Example 4 we did not determine the value of $-\lambda$. In economics, $-\lambda$ has a practical meaning. To determine $-\lambda$ in Example 4, we could use either

$$\frac{0.984x^{-0.4}y^{0.4}}{15{,}000} = -\lambda \quad \text{or} \quad \frac{0.656x^{0.6}y^{-0.6}}{7000} = -\lambda$$

from step 4 and substitute in our values for x and y to give

$$-\lambda \approx 7.57 \times 10^{-5} = 0.0000757$$

The first thing we observe is that $-\lambda$ is positive. Now let's investigate the meaning of $-\lambda$ by looking at what happens when the budget for the O'Neill Corporation increases from \$1.8 million to \$1.9 million. How does this change x, y, and Q? Example 5 shows us.

Example 5 Maximizing a Production Function

If the budget for the O'Neill Corporation has been increased from \$1.8 million to \$1.9 million, what combinations of x and y maximize Q?

Interactive Activity

Work out the details of steps 2 through 4 in Example 5 to confirm our results.

WWW

Solution

$$\text{Maximize} \qquad Q = f(x, y) = 1.64x^{0.6}y^{0.4}$$
$$\text{Subject to} \qquad g(x, y) = 15{,}000x + 7000y - 1{,}900{,}000 = 0$$

The process outlined in steps 2 through 4 of Example 4 is the same here, except 1,800,000 is replaced with 1,900,000. We save the details for you to work out in the Interactive Activity to the left. The solution is $x \approx 75.97$ and $y \approx 108.64$. The maximum production with $x \approx 75.97$ units of labor and $y \approx 108.64$ units of capital is about $Q \approx 143.76$ thousand gallons. ■

Notice that our value for Q in Example 5 is approximately equal to our value for Q from Example 4 plus $10^5 \cdot (-\lambda)$. In other words,

$$Q \text{ from Example 5} \approx Q \text{ from Example 4} + 10^5 \cdot (-\lambda)$$
$$\approx 136.19 + 10^5(7.57 \times 10^{-5}) \approx 143.76$$

Thus, in a production function setting:

- $-\lambda$ gives the extra production achieved by increasing the budget by *one* unit. Since the budget increased by \$100,000 = 10^5 in Example 5, we multiplied $-\lambda$ by 10^5.
- The value of $-\lambda$ approximates the increase in the optimal value of Q when the budget is increased by *one* unit.

In the language of calculus this means that $-\lambda$ **gives the rate of change of the optimal value of Q as the budget increases.**

Marginal Productivity of Money

Let $Q = f(x, y)$ be a production function and $g(x, y) = 0$ be a budget constraint. For

$$F(x, y, \lambda) = f(x, y) + \lambda \cdot g(x, y)$$

$-\lambda$ is the **marginal productivity of money** at (x, y), which is the extra production achieved by increasing the budget one unit.

Note: $-\lambda$ in this setting is always a positive number.

Now, from our work in Example 5 and the discussion immediately following Example 5, we can observe where the following formula from economics arises.

Formula for Optimal Increase in Production

Assume that $-\lambda$ is the marginal productivity of money and P additional dollars is available to the budget. The **optimal increase in production** is given by $-\lambda P$.

Example 6 ties together elements from Examples 4 and 5 and the marginal productivity of money concept.

Example 6 Maximizing Production and Determining Optimal Increase

RollyOn Tires has a Cobb–Douglas production function of

$$Q = f(x, y) = 250x^{0.25}y^{0.75}$$

where x represents units of labor and y represents units of capital. Suppose that each unit of labor is \$100 and each unit of capital is \$250. Assume that the budget is limited to \$100,000.

(a) Determine x and y that maximize production and determine the maximum production.

(b) Determine the marginal productivity of money for the division of labor and capital determined in part (a) and interpret. Find the optimal increase in production if an additional \$15,000 is budgeted.

Solution

(a) Step 1: We want to

$$\text{Maximize} \quad f(x, y) = 250x^{0.25}y^{0.75}$$

$$\text{Subject to} \quad g(x, y) = 100x + 250y - 100{,}000 = 0$$

Step 2: Define $F(x, y, \lambda) = f(x, y) + \lambda g(x, y)$ using the Lagrange multiplier λ.

$$F(x, y, \lambda) = 250x^{0.25}y^{0.75} + \lambda(100x + 250y - 100{,}000)$$

Step 3: Determine the partial derivatives.

$$F_x(x, y, \lambda) = 62.5x^{-0.75}y^{0.75} + 100\lambda$$

$$F_y(x, y, \lambda) = 187.5x^{0.25}y^{-0.25} + 250\lambda$$

$$F_\lambda(x, y, \lambda) = 100x + 250y - 100{,}000$$

Step 4: As we did in Example 4, we solve the first two equations for $-\lambda$.

$$62.5x^{-0.75}y^{0.75} = -100\lambda \quad \text{and} \quad 187.5x^{0.25}y^{-0.25} = -250\lambda$$

$$\frac{62.5x^{-0.75}y^{0.75}}{100} = -\lambda \quad \text{and} \quad \frac{187.5x^{0.25}y^{-0.25}}{250} = -\lambda$$

Setting these equal to each other and multiplying both sides of the equation by $x^{0.75}y^{0.25}$ gives

$$x^{0.75}y^{0.25}\left(\frac{62.5x^{-0.75}y^{0.75}}{100}\right) = x^{0.75}y^{0.25}\left(\frac{187.5x^{0.25}y^{-0.25}}{250}\right)$$

$$\frac{62.5y}{100} = \frac{187.5x}{250}$$

$$y = \frac{100}{62.5}\left(\frac{187.5x}{250}\right) = 1.2x$$

Substituting this into the budget constraint equation for y gives

$$100x + 250(1.2x) - 100{,}000 = 0$$

$$400x = 100{,}000$$

$$x = 250$$

From $y = 1.2x$ we also determine that $y = 300$.

Step 5: So $x = 250$ units of labor and $y = 300$ units of capital maximize production at $f(250, 300) = 71{,}658.2$ units. We round this to 71,658 units.

(b) The marginal productivity of money at $x = 250$ and $y = 300$ is

$$-\lambda = \frac{62.5x^{-0.75}y^{0.75}}{100} = \frac{62.5(250)^{-0.75}(300)^{0.75}}{100} \approx 0.717$$

Thus, for each additional dollar available to the budget, production increases by about 0.717 unit. The optimal increase in production is $-\lambda P$. Since $P = \$15{,}000$, $-\lambda P \approx 0.717(15{,}000) = 10{,}755$. Hence, we now have

71,658	(maximum production with a budget of \$100,000)
+10,755	(additional production from extra \$15,000 added to the budget)
82,413	

This 82,413 is the optimal production possible with \$115,000. ■

Checkpoint 3 Now work Exercise 27.

SUMMARY

In this section we have seen how to maximize or minimize problems that arise when we have **constraints** on our independent variables. The **Method of Lagrange Multipliers** was used to solve these problems.

- **How to Use the Method of Lagrange**
  1. Write the situation in the form

     Maximize (or minimize) $z = f(x, y)$

     Subject to $g(x, y) = 0$
  2. Define $F(x, y, \lambda) = f(x, y) + \lambda \cdot g(x, y)$.
  3. Determine the partial derivatives $F_x(x, y, \lambda)$, $F_y(x, y, \lambda)$, and $F_\lambda(x, y, \lambda)$.
  4. Solve the system of equations

     $F_x(x, y, \lambda) = 0, \quad F_y(x, y, \lambda) = 0, \quad F_\lambda(x, y, \lambda) = 0$
  5. The maximum (or minimum) of $f(x, y)$ is among the values in step 4. Simply evaluate $z = f(x, y)$ at each.

The section concluded with problems from the world of economics, specifically marginal productivity of money.

- $-\lambda$ is the **marginal productivity of money** at (x, y), which is the extra production achieved by increasing the budget one unit.
- Assume that $-\lambda$ is the marginal productivity of money and P additional dollars is available to the budget. The **optimal increase in production** is given by $-\lambda P$.

SECTION 18.5 EXERCISES

In Exercises 1–18, use the method of Lagrange multipliers to solve each problem.

1. Maximize $f(x, y) = 3xy$ Subject to $x + y = 1$

2. Maximize $f(x, y) = 2xy$ Subject to $2x + y = 20$

3. Maximize $f(x, y) = xy$ Subject to $2x + y = 12$

4. Maximize $f(x, y) = 2xy - 5$ Subject to $x + y = 12$

5. Minimize $f(x, y) = xy$ Subject to $x - y = -8$

6. Maximize $f(x, y) = 3xy + x$ Subject to $x + y = 1$

✓ **7.** Minimize $f(x, y) = x^2 + 2y^2$ Subject to $2x + y = 8$

8. Minimize $f(x, y) = x^2 + y^2 - 7$ Subject to $x + 2y = 10$

9. Maximize $f(x, y) = 2xy - 4x$ Subject to $x + y = 12$

10. Maximize $f(x, y) = 25 - x^2 - y^2$
Subject to $2x + y = 10$

11. Maximize $f(x, y) = 1.64x^{0.6}y^{0.4}$
Subject to $12{,}000x + 5000y = 1{,}100{,}000$

12. Maximize $f(x, y) = 125x^{0.75}y^{0.25}$
Subject to $100x + 250y = 100{,}000$

13. Minimize $f(x, y) = x^2 + y^2 - xy - 4$
Subject to $x + y - 6 = 0$

14. Maximize $f(x, y) = 16 - x^2 - y^2$
Subject to $x + 2y - 6 = 0$

15. Maximize $f(x, y) = xy$ Subject to $x^2 + y^2 = 8$

16. Minimize $f(x, y) = xy$ Subject to $x^2 + y^2 = 8$

17. Maximize $f(x, y) = e^{xy}$ Subject to $x^2 + y^2 = 8$

18. Minimize $f(x, y) = e^{x^2+y^2}$ Subject to $x + y = 5$

Applications

19. On-the-Go Music produces two types of walkman's each day, x units of its regular model and y units of its joggers' model. The cost, in dollars, is given by

$$C(x, y) = 0.1x^2 + 0.2y^2$$

Due to limitations on the supply of parts, it is necessary that

$$x + y = 180$$

(a) Determine how many of each type of walkman should be produced to minimize cost.

(b) Determine the minimum cost.

20. DeeDee's produces two types of diapers each week, x packages of its regular absorbent and y packages of its extra absorbent. The cost, in dollars, is given by

$$C(x, y) = 0.1x^2 + 0.5y^2$$

Due to limited materials, it is necessary that

$$x + y = 1200$$

(a) Determine how many of each type of diaper should be produced to minimize cost.

(b) Determine the minimum cost.

21. SalPal produces two types of personal security devices each day, x units of its flashlight and pepper spray device and y units of its flashlight and siren device. The cost, in dollars, is given by

$$C(x, y) = 0.1x^2 + 0.2y^2$$

Due to shipping restrictions, it is necessary that

$$x + y = 120$$

(a) Determine how many of each type of security device should be produced to minimize cost.

(b) Determine the minimum cost.

22. Rikki's Bicycle Company makes two models of bicycles, 10-speeds and 21-speeds. Each week, the cost to make x 10-speeds and y 21-speeds is given by

$$C(x, y) = 20{,}000 + 100x + 140y - xy$$

where $C(x, y)$ is in dollars. Determine the number of 10-speeds and the number of 21-speeds that should be produced each week to minimize costs if the total number produced each week is 200 bicycles.

23. The Riblet Engine Company has installed a newer, more efficient production line, which is alongside an older production line that is still in use. Let x represent the number of units produced on the older line, and let y represent the number of units produced on the new line. The cost of a production run is given by

$$C(x, y) = 2x^2 - xy + y^2 + 250$$

where $C(x, y)$ is in dollars. A production run has been scheduled for which both lines will be used to complete an order of 104 units.

(a) Determine the number of units produced by each line in order to minimize the cost.

(b) Determine the minimum production cost.

24. Sumption Sonic Incorporated has a production function for the manufacture of solar cells of

$$Q = f(x, y) = 2.5x^{0.7}y^{0.3}$$

where Q is the number of solar cells produced per week, x is the number of units of labor each week, and y is the number of units of capital. A unit of labor costs \$400, a unit of capital costs \$500, and the total budget for both is \$180,000.

(a) Determine the number of units of labor and the number of units of capital required to maximize production.

(b) Determine the maximum production.

✓ **25.** The Linger Golf Cart Corporation has a production function for the manufacture of golf carts given by

$$Q = f(x, y) = 13x^{0.75}y^{0.3}$$

where Q is the number of golf carts produced per week, x is the number of labor hours per week, and y is the weekly capital investment. A unit of labor is \$15 and a unit of capital is \$50, and the total budget for both is limited to \$18,000.

(a) Determine the number of units of labor and the number of units of capital required to maximize production.

(b) Determine the maximum production.

26. The $(RB)^2$ Company has a production function of

$$Q = f(x, y) = 1610x^{0.8}y^{0.25}$$

where Q is the quantity of handheld calculators produced per year, x is the number of labor hours in thousands, and y is the capital equipment. A unit of labor is \$15,000 and a unit of capital is \$4000, and the total budget for both is limited to \$1.5 million.

(a) Determine the number of units of labor and the number of units of capital required to maximize production.

(b) Determine the maximum production.

✓ **27.** The Arnold Engine Company has a production function of

$$Q = f(x, y) = x^{0.6}y^{0.45}$$

where x is the number of units of labor, y is the number of units of capital, and Q is the total quantity produced. Each unit of labor is \$16,000 and each unit of capital is \$4000, and the total budget for both is limited to \$1.7 million.

(a) Determine the number of units of labor and capital needed to maximize production.

(b) Determine the marginal productivity of money for the labor and capital found in part (a) and interpret.

(c) Determine the optimal increase in production if an additional \$200,000 is budgeted.

28. Hogrefe's Incorporated has a production function given by

$$Q = f(x, y) = x^{0.7}y^{0.3}$$

where x is the number of units of labor, y is the number of units of capital, and Q is the total quantity produced. Each unit of labor is \$1200, each unit of capital is \$800, and the total budget for both is limited to \$480,000.

(a) Determine the number of units of labor and capital needed to maximize production.

(b) Determine the marginal productivity of money for the labor and capital found in part (a) and interpret.

(c) Determine the optimal increase in production if an additional \$20,000 is budgeted.

29. Hoffman Laboratories has a production function given by

$$Q = f(x, y) = 5.6x^{0.65}y^{0.41}$$

where Q is the quantity of pain-killing tablets produced (in thousands) per week, x is the weekly units of labor, and y is the weekly units of capital. Each unit of labor is \$2000, each unit of capital is \$1500, and the budget for both is limited to \$800,000.

(a) Determine the number of units of labor and capital needed to maximize production.

(b) Determine the marginal productivity of money for the labor and capital found in part (a) and interpret.

(c) Determine the optimal increase in production if an additional \$50,000 is budgeted.

30. Blair's Inc. has a production function given by

$$Q = f(x, y) = 25x^{0.8}y^{0.2}$$

where Q is the total quantity produced, x is the number of units of labor used, and y is the number of units of capital used. Each unit of labor costs \$30, each unit of capital costs \$60, and the budget for both is \$300,000.

(a) Determine the number of units of labor and capital needed to maximize production.

(b) Determine the marginal productivity of money for the labor and capital found in part (a) and interpret.

(c) Determine the optimal increase in production if an additional \$40,000 is budgeted.

31. New River Outfitters has a production function given by

$$Q = f(x, y) = 100x^{0.6}y^{0.4}$$

where x is the number of units of labor, y is the number of units of capital, and Q is the quantity produced. Each unit of labor costs \$100, each unit of capital is \$150, and the budget for both is \$60,000.

(a) Determine the number of units of labor and capital needed to maximize production.

(b) Determine the marginal productivity of money for the labor and capital found in part (a) and interpret.

(c) Determine the optimal increase in production if an additional \$10,000 is budgeted.

32. The Bowen Company has a production function given by

$$Q = f(x, y) = 25x^{0.65}y^{0.35}$$

where x is the number of units of labor, y is the number of units of capital, and Q is the quantity produced. Each unit of labor costs \$200, each unit of capital costs \$250, and the budget for both is \$130,000.

(a) Determine the number of units of labor and capital needed to maximize production.

(b) Determine the marginal productivity of money for the labor and capital found in part (a) and interpret.

(c) Determine the optimal increase in production if an additional \$20,000 is budgeted.

33. A company wishes to enclose a rectangular parking lot using an existing building as one side of the boundary and adding fencing for the other boundaries. If 625 feet of fencing is available, determine the dimensions of the largest parking lot that can be enclosed.

34. Determine the dimensions of a rectangular garden with an area of 5000 square feet that minimizes the cost of fencing if one side costs three times as much as the other three sides.

35. *Federal Express* states that the maximum length plus girth for its FedEx Overnight Freight and its FedEx 2Day Freight shipments is 300 inches (*Source:* Federal Express, www.fedex.com). See Figure 18.5.4. Use the method of Lagrange multipliers to determine the dimensions of the largest-volume package that can be sent by FedEx Overnight Freight and FedEx 2Day Freight. Assume that the length is y.

36. To avoid U.S. Domestic Freight Services charges, any packages sent by *Federal Express* must have the length plus girth not exceed 165 inches (*Source:* Federal Express, www.fedex.com). See Figure 18.5.4. Use the method of Lagrange multipliers to determine the dimensions of the largest-volume package that can be sent by *Federal Express* without incurring Domestic Freight Service charges. Assume that the length is y.

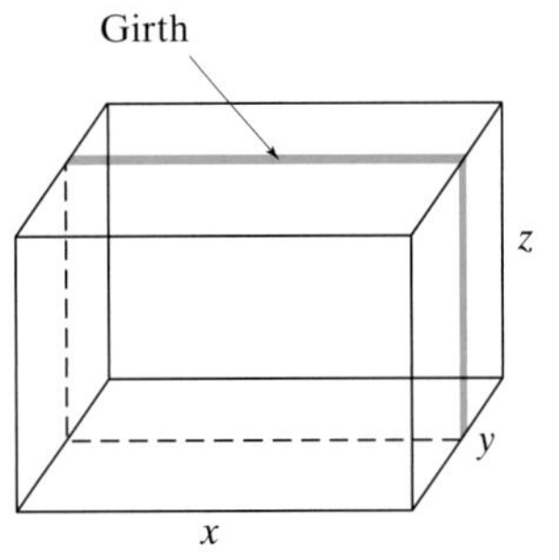

Figure 18.5.4

SECTION PROJECT

This section project synthesizes information studied in Chapter 18 and uses the following data. Also, the program in Appendix C that performs multiple regression is needed to do the section project.

The Gantzler Corporation has six plants around the United States that produce plastic tubing. In a recent year the following data for each plant gave the number of labor hours (in thousands), capital (in millions), and quantity produced.

Labor	Capital	Quantity
38	3.1	3440
39	3.2	3533
41	3.7	3800
42	3.8	3894
42	3.81	3896
43	3.9	3987

(a) The plants have the same technology, so a production function can be determined. Enter the data into your calculator and execute the program to determine a Cobb–Douglas production function to model the data. Use x for labor and y for capital and round all values to the nearest hundredth.

(b) Compute $f_x(x, y)$ and $f_y(x, y)$.

(c) If, on average, the Gantzler Corporation is now using 41 units of labor and 3.6 units of capital, determine the marginal productivity of labor and the marginal productivity of capital and interpret each.

(d) Suppose that each unit of labor costs \$12,000 and each unit of capital costs \$2000. The budget for both is \$1.1 million. Determine the number of units of labor and capital needed to maximize production.

(e) Determine the marginal productivity of money for the division of labor and capital found in part (d).

(f) Determine the optimal increase in production if an additional \$100,000 is budgeted.

Section 18.6 Double Integrals

In this chapter we have seen how to generalize the derivative concept to functions of two variables. A natural question to ask is whether we can do the same with integration. The answer is yes, and we give a brief description of the process in this section. To fully understand the process, we begin the section by looking at **partial antidifferentiation**.

Partial Antidifferentiation and Iterated Integrals

In Chapter 16 we saw that antidifferentiation is the reverse operation of differentiation. In other words, antidifferentiation can "undo" differentiation. Remember that at that time we were dealing with functions of a single variable. To antidifferentiate a function of two variables with respect to one variable, we simply treat all other variables as if they are constants. This means that when we see

$$\int f(x, y)\,dx \qquad \text{Notice the } dx \text{ in this integral}$$

we antidifferentiate $f(x, y)$ with respect to x and treat y as a constant. Likewise, when we see

$$\int f(x, y)\,dy \qquad \text{Notice the } dy \text{ in this integral}$$

we antidifferentiate $f(x, y)$ with respect to y and treat x as a constant. Example 1 illustrates the process.

Example 1 Antidifferentiating Functions of Two Variables

Determine the following indefinite integrals:

(a) $\int (x^2y + 3y^2x)\,dx$ **(b)** $\int (x^2y + 3y^2x)\,dy$

Solution

(a) Understand the Situation: Here we are antidifferentiating with respect to x, so we treat y as a constant.

We simply apply some properties of antidifferentiation from Chapter 16. Consult the Toolbox below.

From Your Toolbox

- $\int [f(x) + g(x)]\,dx = \int f(x)\,dx + \int g(x)\,dx$
- $\int k \cdot f(x)\,dx = k \cdot \int f(x)\,dx$, where k is a constant

$$\begin{aligned}\int (x^2y + 3y^2x)\,dx &= \int x^2y\,dx + \int 3y^2x\,dx \\ &= y\int x^2\,dx + 3y^2\int x\,dx \qquad y \text{ is constant} \\ &= y\left(\frac{1}{3}x^3\right) + 3y^2\left(\frac{1}{2}x^2\right) + C(y) \\ &= \frac{1}{3}x^3y + \frac{3}{2}x^2y^2 + C(y)\end{aligned}$$

Notice that the "constant" is not just a C. It can be *any function of* y, since for any function of y we know

$$\frac{\partial}{\partial x}C(y) = 0$$

Interactive Activity

Check the result in Example 1b by computing the partial derivative

$$\frac{\partial}{\partial y}\left[\frac{1}{2}x^2y^2 + xy^3 + C(x)\right]$$

WWW

(b) **Understand the Situation:** Here we antidifferentiate with respect to y, so we treat x as a constant.

This gives

$$\begin{aligned}\int (x^2y + 3y^2x)\,dy &= \int x^2y\,dy + \int 3y^2x\,dy \\ &= x^2\int y\,dy + x\int 3y^2\,dy \qquad x \text{ is constant} \\ &= x^2\left(\frac{1}{2}y^2\right) + x(y^3) + C(x) \\ &= \frac{1}{2}x^2y^2 + xy^3 + C(x)\end{aligned}$$

Notice here that our antiderivative has as its arbitrary "constant" $C(x)$. ■

In Chapter 16 we encouraged checking our antidifferentiation by differentiating the answer. We still encourage this type of checking. In Example 1 our result can be checked by partial differentiation. To check the result in Example 1a, we compute the partial derivative

$$\begin{aligned}\frac{\partial}{\partial x}\left[\frac{1}{3}x^3y + \frac{3}{2}x^2y^2 + C(y)\right] &= \left(\frac{1}{3}\cdot 3x^2\right)y + \left(\frac{3}{2}\cdot 2x\right)y^2 + 0 \\ &= x^2y + 3xy^2\end{aligned}$$

Example 1 shows us that we can easily antidifferentiate functions of two variables as long as we remember which variable is being treated as a constant. In Example 2 we extend the process to evaluating definite integrals.

Example 2 Evaluating Partial Antiderivatives

Evaluate the following definite integrals.

(a) $\int_0^1 (x^2y + 3y^2x)\,dx$

(b) $\int_0^1 (x^2y + 3y^2x)\,dy$

Solution

Interactive Activity

In Example 2a, what did we choose $C(y)$ to equal?

In Example 2b, what did we choose $C(x)$ to equal?

(a) Using the result from Example 1a, we have

$$\begin{aligned}\int_0^1 (x^2y + 3y^2x)\,dx &= \left(\frac{1}{3}x^3y + \frac{3}{2}x^2y^2\right)\Bigg|_{x=0}^{x=1} \\ &= \left(\frac{1}{3}(1)^3y + \frac{3}{2}(1)^2y^2\right) - \left(\frac{1}{3}(0)^3y + \frac{3}{2}(0)^2y^2\right) \\ &= \frac{1}{3}y + \frac{3}{2}y^2\end{aligned}$$

Notice that this definite integral produces a function of y.

(b) Using the result from Example 1b, we have

$$\int_0^1 (x^2y + 3y^2x)\,dy = \left(\frac{1}{2}x^2y^2 + xy^3\right)\Bigg|_{y=0}^{y=1}$$

$$= \left(\frac{1}{2}x^2(1)^2 + x(1)^3\right) - \left(\frac{1}{2}x^2(0)^2 + x(0)^3\right)$$

$$= \frac{1}{2}x^2 + x$$

Notice that this definite integral produces a function of x. ■

Checkpoint 1

Now work Exercise 9.

In Example 2, we noticed the following:

1. $\displaystyle\int_0^1 (x^2y + 3y^2x)\,dx = \frac{1}{3}y + \frac{3}{2}y^2 = f(y)$
2. $\displaystyle\int_0^1 (x^2y + 3y^2x)\,dy = \frac{1}{2}x^2 + x = f(x)$

So it appears that when we integrate and evaluate a definite integral of the form

$$\int_a^b f(x, y)\,dx$$

the result is a function of a single variable, y. (It could also produce a constant.) Likewise, when we integrate and evaluate a definite integral of the form

$$\int_c^d f(x, y)\,dy$$

the result is a function of a single variable, x. (It could also produce a constant.) Since each of these produces a function of a single variable, they in turn could be an integrand of a second integral. In Example 3, we evaluate what are called **iterated integrals**. The word iterated simply means repeated.

Example 3 Evaluating Iterated Integrals

Evaluate the following:

(a) $\displaystyle\int_0^1 \left[\int_0^1 (x^2y + 3y^2x)\,dx\right] dy$

(b) $\displaystyle\int_0^1 \left[\int_0^1 (x^2y + 3y^2x)\,dy\right] dx$

Solution

(a) **Understand the Situation:** The brackets indicate the order in which the definite integrals are to be evaluated. Hence, we first evaluate the inside integral.

From Example 2a we have

$$\int_0^1 (x^2y + 3y^2x)\,dx = \frac{1}{3}y + \frac{3}{2}y^2$$

This result is now the integrand for the outer integral. This gives

$$\int_0^1\left[\int_0^1 (x^2y+3y^2x)\,dx\right]dy = \int_0^1\left(\frac{1}{3}y+\frac{3}{2}y^2\right)dy$$
$$= \left(\frac{1}{6}y^2+\frac{1}{2}y^3\right)\Big|_0^1$$
$$= \left(\frac{1}{6}+\frac{1}{2}\right)-(0+0) = \frac{2}{3}$$

(b) From Example 2b the inside integral is

$$\int_0^1 (x^2y+3y^2x)\,dy = \frac{1}{2}x^2+x$$

This result is now the integrand for the outer integral. This gives

$$\int_0^1\left[\int_0^1 (x^2y+3y^2x)\,dy\right]dx = \int_0^1\left(\frac{1}{2}x^2+x\right)dx$$
$$= \left(\frac{1}{6}x^3+\frac{1}{2}x^2\right)\Big|_0^1$$
$$= \left(\frac{1}{6}+\frac{1}{2}\right)-(0+0) = \frac{2}{3}$$

■

Double Integrals

It is no accident that the results in Examples 3a and b are identical. Examples 1 through 3 suggest the following definition of a **double integral**.

Double Integral

The **double integral** of $f(x, y)$ over a rectangular region, denoted R, is

$$\iint f(x,y)\,dA,$$
$$R = \{(x,y) | a \le x \le b, c \le y \le d\}$$

This may be rewritten as

$$\int_a^b \int_c^d f(x,y)\,dy\,dx$$

or

$$\int_c^d \int_a^b f(x,y)\,dx\,dy.$$

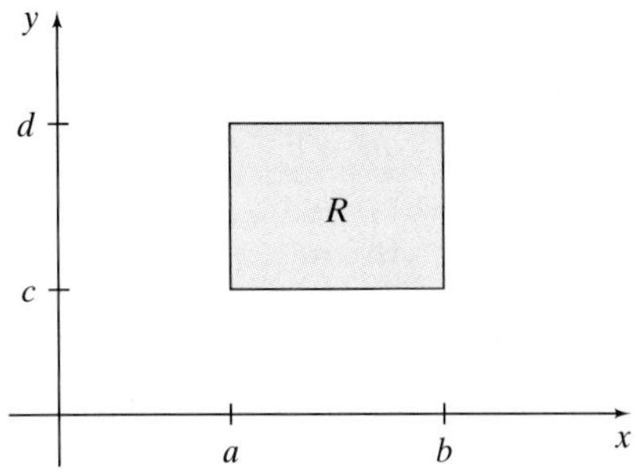

Note: In the double integral, $\iint f(x, y)\,dA$, R is called the **region of integration**. Notice that this region is in the xy-plane. Also, dA indicates that either order of integration, $dy\,dx$ or $dx\,dy$, may be used.

We must mention that a more general definition of double integrals exists. The one that we supply is more applicable to the functions that we will study.

Example 4 Writing and Evaluating a Double Integral

For $\iint (2x + y)\,dA$, where $R = \{(x, y)|1 \le x \le 3, 0 \le y \le 2\}$:

(a) Sketch the region of integration R in the xy-plane.

(b) Write and evaluate a double integral with $dy\,dx$ order of integration.

(c) Write and evaluate a double integral with $dx\,dy$ order of integration.

Solution

(a) The region of integration is shown in Figure 18.6.1.

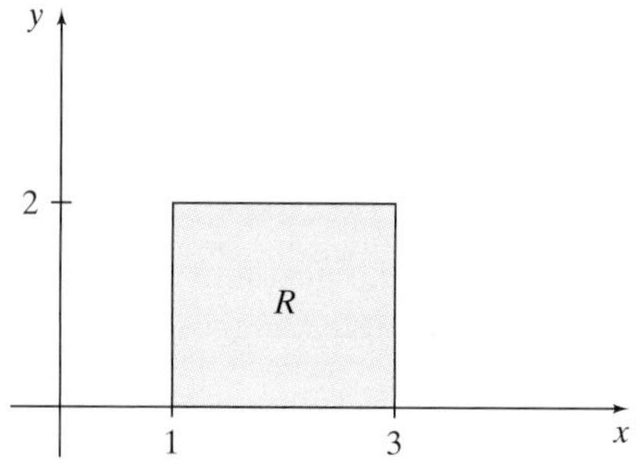

Figure 18.6.1

(b) **Understand the Situation:** A $dy\,dx$ order of integration means that the innermost integral has limits determined by the bounds for y and the outermost integral has limits determined by the bounds for x.

This gives

$$\int_1^3 \int_0^2 (2x + y)\,dy\,dx$$

We evaluate by performing the innermost integration first. This gives

$$\int_1^3 \left[\int_0^2 (2x + y)\,dy\right] dx = \int_1^3 \left[\left(2xy + \frac{1}{2}y^2\right)\Big|_{y=0}^{y=2}\right] dx$$

$$= \int_1^3 \left[\left(2x(2) + \frac{1}{2}(2)^2\right) - \left(2x(0) + \frac{1}{2}(0)^2\right)\right] dx$$

$$= \int_1^3 (4x + 2)\,dx = (2x^2 + 2x)\Big|_1^3$$

$$= (18 + 6) - (2 + 2) = 20$$

(c) **Understand the Situation:** A $dx\,dy$ order of integration means that the innermost integral has limits determined by the bounds for x and the outermost integral has limits determined by the bounds for y.

This gives

$$\int_0^2 \int_1^3 (2x + y)\, dx\, dy$$

Evaluating, starting with the innermost integral, yields

$$\int_0^2 \left[\int_1^3 (2x + y)\, dx\right] dy = \int_0^2 \left[(x^2 + xy)\Big|_{x=1}^{x=3}\right] dy$$
$$= \int_0^2 [(9 + 3y) - (1 + y)]\, dy$$
$$= \int_0^2 (8 + 2y)\, dy = (8y + y^2)\Big|_0^2$$
$$= (16 + 4) - (0 + 0) = 20$$

Again Example 4 illustrates that, as long as our function is continuous, either order of integration ($dy\,dx$ or $dx\,dy$) yields the same result.

Checkpoint 2

Now work Exercise 15.

Example 5 Evaluating a Double Integral

Given $\int_{-1}^{1} \int_0^2 x^2 e^{-y} dx\, dy$, sketch the region of integration and evaluate the double integral.

Solution

Interactive Activity

Evaluate $\int_0^2 \int_{-1}^{1} x^2 e^{-y}\, dy\, dx$ to verify the result of Example 5.

WWW

From the limits of integration and the $dx\, dy$ order of integration, we have

$$-1 \le y \le 1 \quad \text{and} \quad 0 \le x \le 2$$

The region is shown in Figure 18.6.2.

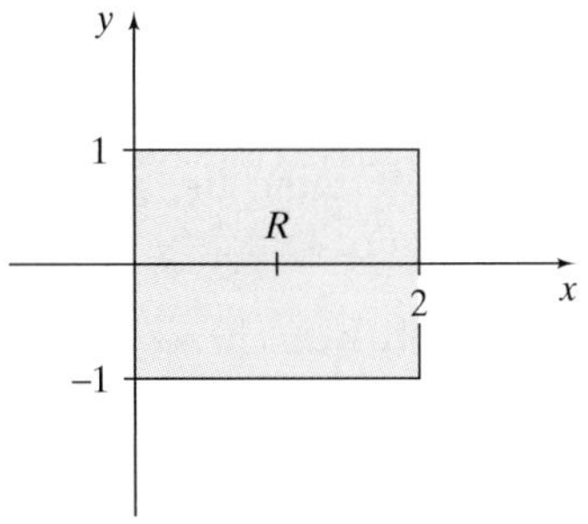

Figure 18.6.2

Evaluating, starting with the innermost integral, gives

$$\int_{-1}^{1} \left[\int_0^2 x^2 e^{-y} dx\right] dy = \int_{-1}^{1} \left[\left(\frac{1}{3}x^3 e^{-y}\right)\Bigg|_{x=0}^{x=2}\right] dy$$
$$= \int_{-1}^{1} \left[\frac{1}{3}(2)^3 e^{-y} - \frac{1}{3}(0)^3 e^{-y}\right] dy$$
$$= \int_{-1}^{1} \frac{8}{3} e^{-y}\, dy = \left(-\frac{8}{3}e^{-y}\right)\Bigg|_{-1}^{1} = -\frac{8}{3}e^{-1} + \frac{8}{3}e$$

Checkpoint 3 Now work Exercise 25.

Average Value over Rectangular Regions

In Section 17.1 we defined the average value of a function, f, over an interval $[a, b]$. We can extend the average value concept to functions of two variables.

From Your Toolbox

The *average value* of a continuous function, f, on the interval $[a, b]$ is given by

$$\frac{1}{b-a}\int_a^b f(x)\,dx$$

Average Value over Rectangular Regions

The **average value** of $f(x, y)$ over rectangular region R is defined to be

$$\frac{1}{(b-a)(c-d)}\iint f(x, y)\,dA$$

or

$$\frac{1}{\text{area of } R}\iint f(x, y)\,dA$$

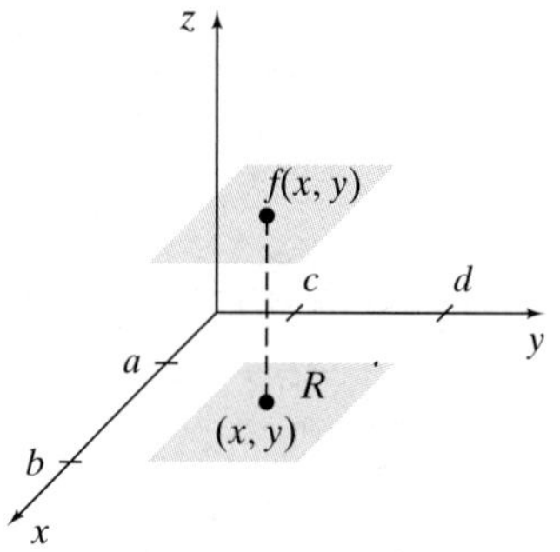

Note: Observe that $(b-a)(c-d)$ is simply the area of rectangular region R.

Example 6 Computing an Average Value

Determine the average value of $f(x, y) = x + y$ over the region R shown in Figure 18.6.3.

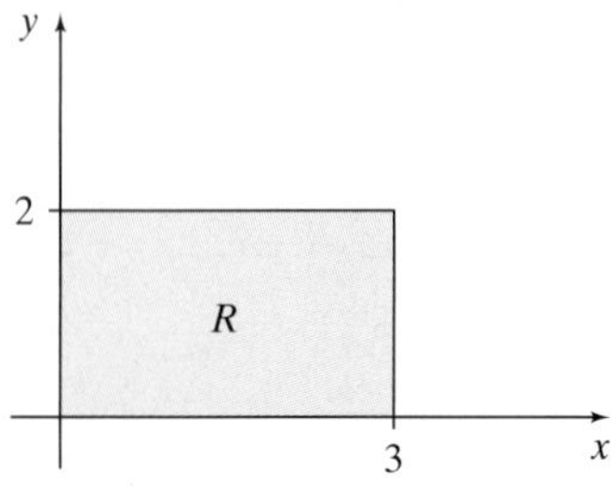

Figure 18.6.3

Solution

From Figure 18.6.3 we see that R is simply

$$0 \le x \le 3 \quad \text{and} \quad 0 \le y \le 2$$

Hence, the area of R is 6. Using a $dy\,dx$ order of integration, we compute the average value to be

$$\frac{1}{6}\int_0^3\int_0^2 (x+y)\,dy\,dx = \frac{1}{6}\int_0^3\left[\int_0^2 (x+y)\,dy\right]dx = \frac{1}{6}\int_0^3\left[\left(xy+\frac{1}{2}y^2\right)\Big|_{y=0}^{y=2}\right]dx$$

$$= \frac{1}{6}\int_0^3 [(2x+2)-(0+0)]\,dx$$

$$= \frac{1}{6}\int_0^3 (2x+2)\,dx = \frac{1}{6}\left[(x^2+2x)\Big|_0^3\right]$$

$$= \frac{1}{6}[(9+6)-(0+0)] = \frac{1}{6}(15) = \frac{15}{6} = \frac{5}{2} = 2.5$$

Interpret the Solution: Thus, 2.5 is simply the average of the z-values for $z = f(x, y) = x + y$ over region R. See Figure 18.6.4.

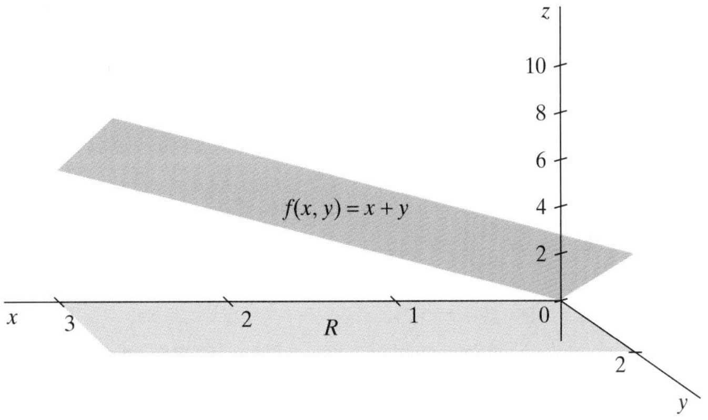

Figure 18.6.4

Example 7 Computing an Average Value of a Production Function

The Boroff Engine Company has a production function given by

$$Q = f(x, y) = x^{0.6}y^{0.45}$$

where Q is the number of engines produced per week, x is the number of employees at the company, and y is the weekly operating budget in thousands. Because the company uses a temporary labor agency, it uses anywhere from 40 to 50 employees each week, and its operating budget is anywhere from \$10,000 to \$15,000 each week. Determine the average number of engines that the company can produce each week.

Solution

Understand the Situation: Since the Boroff Engine Company uses anywhere from 40 to 50 employees each week, we know that

$$40 \le x \le 50$$

Also, since its operating budget is anywhere from \$10,000 to \$15,000 each week,

$$10 \le y \le 15 \quad \text{Recall that } y \text{ is in thousands}$$

Thus, we need to determine the average value of $Q = f(x, y) = x^{0.6}y^{0.45}$ over the rectangular region shown in Figure 18.6.5.

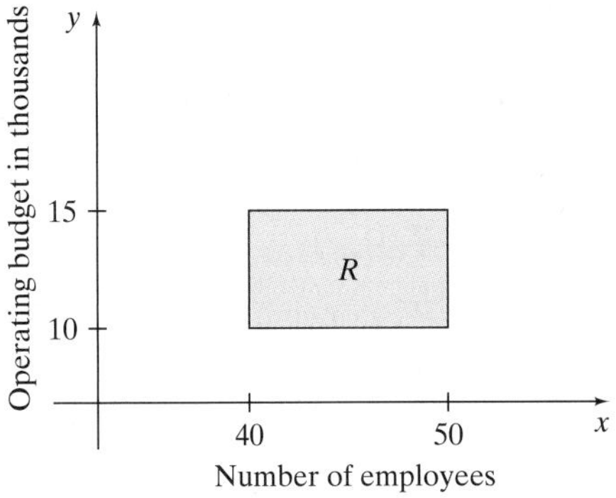

Figure 18.6.5

The area of the region is found to be

$$(50 - 40)(15 - 10) = 50$$

The average value is

$$\frac{1}{50}\int_{40}^{50}\int_{10}^{15} x^{0.6}y^{0.45}\,dy\,dx = \frac{1}{50}\int_{40}^{50}\left[\int_{10}^{15} x^{0.6}y^{0.45}\,dy\right]dx$$

$$= \frac{1}{50}\int_{40}^{50}\left[\left(\frac{1}{1.45}x^{0.6}y^{1.45}\right)\Bigg|_{10}^{15}\right]dx$$

$$= \frac{1}{50}\int_{40}^{50}\left(\frac{1}{1.45}x^{0.6}(15)^{1.45} - \frac{1}{1.45}x^{0.6}(10)^{1.45}\right)dx$$

$$\approx \frac{1}{50}\int_{40}^{50}(34.99x^{0.6} - 19.44x^{0.6})\,dx$$

$$= \frac{1}{50}\int_{40}^{50} 15.55x^{0.6}\,dx$$

$$= \frac{15.55}{50}\int_{40}^{50} x^{0.6}\,dx = \frac{15.55}{50}\left[\frac{1}{1.6}x^{1.6}\Bigg|_{40}^{50}\right]$$

$$= \frac{15.55}{50}\left[\frac{1}{1.6}(50)^{1.6} - \frac{1}{1.6}(40)^{1.6}\right] \approx 30.51$$

Interpret the Solution: To the nearest engine, the Boroff Engine Company can produce an average of 31 engines each week. Figure 18.6.6 gives a graph of $Q = f(x, y) = x^{0.6}y^{0.45}$ over the region R. Again, the result of 31 is simply the average of the Q-values over R.

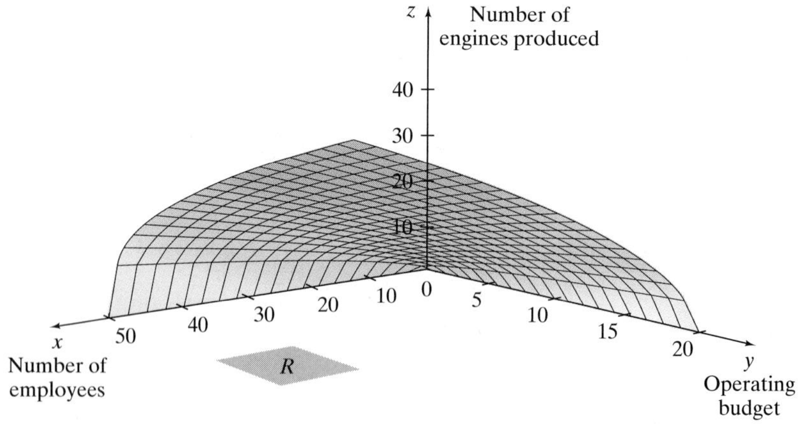

Figure 18.6.6

■

Volume

We conclude this section with a discussion on volume. In Chapter 16 we saw that the area of the region under a curve given by $y = f(x)$ on the interval $[a, b]$ is determined by $\int_a^b f(x)\,dx$, as long as $f(x)$ is nonnegative on $[a, b]$. See Figure 18.6.7. Here, we can determine the *volume* of the *solid* under a surface given by $f(x, y)$ over rectangular region R using double integrals. See Figure 18.6.8.

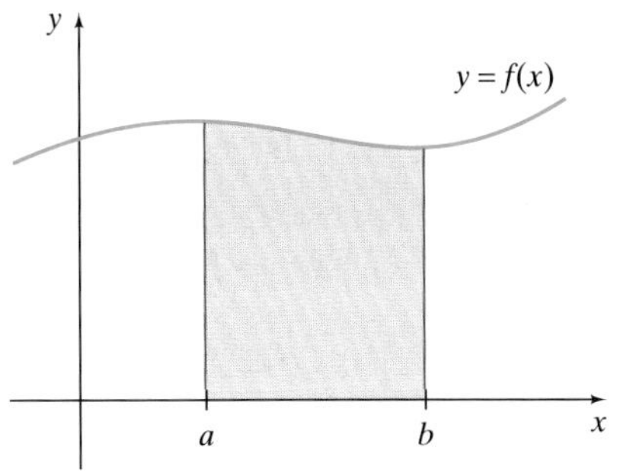

Figure 18.6.7

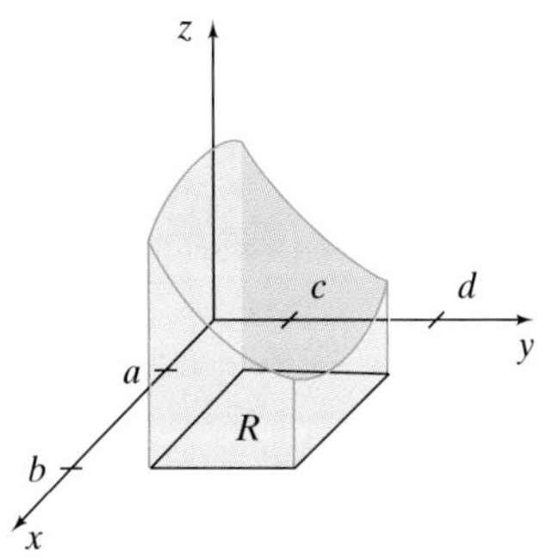

Figure 18.6.8 Volume $= \int\int f(x, y)\, dA$.

Volume under a Surface

If $z = f(x, y)$ is nonnegative and continuous ($f(x, y) \geq 0$) over rectangular region R, $R = \{(x, y)|a \leq x \leq b, c \leq y \leq d\}$, then the volume of the solid under $f(x, y)$ over R is given by

$$V = \int\int f(x, y)\, dA$$

Example 8 Computing a Volume

Determine the volume under $f(x, y) = 4 - x^2 - y^2$ above the region

$$R = \{(x, y)|0 \leq x \leq 1, 0 \leq y \leq 1\}.$$

Solution

Figure 18.6.9 shows the region, and Figure 18.6.10 shows the volume that we are trying to determine.

The volume is

$$V = \int\int (4 - x^2 - y^2)\, dA$$

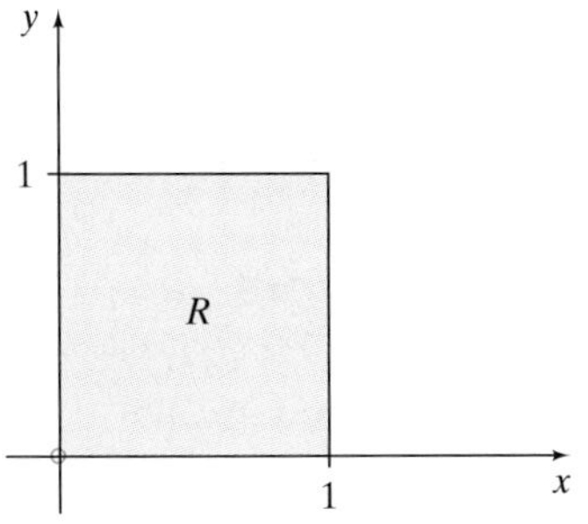

Figure 18.6.9

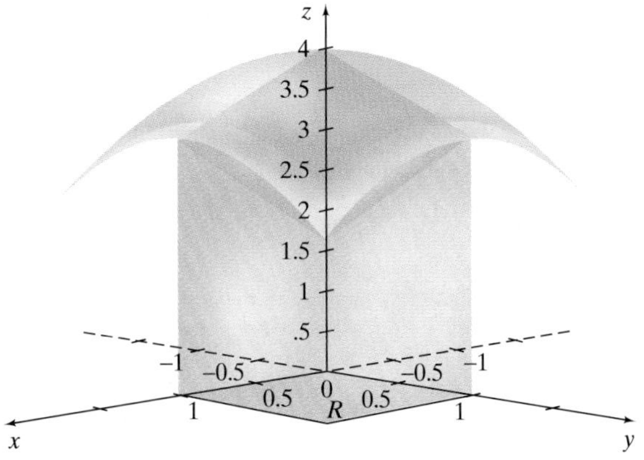

Figure 18.6.10

We select a $dy\,dx$ order of integration, which gives

$$V = \int_0^1 \int_0^1 (4 - x^2 - y^2)\,dy\,dx = \int_0^1 \left[\left(4y - x^2y - \frac{1}{3}y^3\right)\Big|_{y=0}^{y=1} \right] dx$$

$$= \int_0^1 \left(4 - x^2 - \frac{1}{3}\right) dx = \int_0^1 \left(\frac{11}{3} - x^2\right) dx$$

$$= \left(\frac{11}{3}x - \frac{1}{3}x^3\right)\Big|_0^1 = \left(\frac{11}{3} - \frac{1}{3}\right) - (0 - 0)$$

$$= \frac{10}{3} \text{ cubic units}$$

■

Checkpoint 4 Now work Exercise 39.

SUMMARY

In this section we have seen how to extend the concept of antidifferentiation to functions of two variables. Partial antidifferentiation led to iterated integrals, which in turn led us to double integrals. We then extended the concept of average value to functions of two variables. We concluded the section with a brief look at how a double integral can be used to determine the volume of a solid under a surface over a rectangular region.

- The **double integral** of $f(x, y)$ over a rectangular region, denoted R, is

$$\iint f(x, y)\,dA, \quad R = \{(x, y) | a \le x \le b, c \le y \le d\}$$

This may be rewritten as $\int_a^b \int_c^d f(x, y)\,dy\,dx$ or $\int_c^d \int_a^b f(x, y)\,dx\,dy$.

- The **average value** of $f(x, y)$ over rectangular region R is defined to be

$$\frac{1}{(b-a)(c-d)} \iint f(x, y)\,dA$$

or

$$\frac{1}{\text{area of } R} \iint f(x, y)\,dA$$

- If $z = f(x, y)$ is nonnegative and continuous ($f(x, y) \ge 0$) over rectangular region R, $R = \{(x, y) | a \le x \le b, c \le y \le d\}$, then the volume of the solid under $f(x, y)$ over R is given by $V = \int\int f(x, y)\,dA$.

SECTION 18.6 EXERCISES

In Exercises 1–12, evaluate the given integral.

1. $\int_1^3 3x^2y^2\,dy$ **2.** $\int_1^4 2x^3y^3\,dy$

3. $\int_1^3 3x^2y^2\,dx$ **4.** $\int_1^4 2x^3y^3\,dx$

5. $\int_0^2 (2x + 3y)\,dy$ **6.** $\int_0^3 (3x - 2y)\,dy$

7. $\int_0^2 (2x + 3y)\,dx$ **8.** $\int_0^3 (3x - 2y)\,dx$

✓ **9.** $\int_1^2 (x^3y^2 - 2xy)\,dy$ **10.** $\int_2^4 (x^2y^3 - 3xy)\,dy$

11. $\int_1^2 (x^3y^2 - 2xy)\,dx$ **12.** $\int_2^4 (x^2y^3 - 3xy)\,dx$

For Exercises 13–22:

(a) Sketch the region of integration.

(b) Write and evaluate a double integral with a $dy\,dx$ order of integration.

(c) Write and evaluate a double integral with a $dx\,dy$ order of integration.

13. $\iint 2xy\,dA$ $R = \{(x, y) | 0 \le x \le 2, 0 \le y \le 3\}$

14. $\iint 3xy\,dA$ $R = \{(x, y) | 0 \le x \le 2, 0 \le y \le 3\}$

✓ **15.** $\iint (3x + y)\,dA$ $R = \{(x, y) | 0 \le x \le 1, 0 \le y \le 2\}$

16. $\iint (4x - y)\,dA$ $R = \{(x, y) | 2 \le x \le 4, 1 \le y \le 2\}$

17. $\iint_R \sqrt{xy}\,dA$ $\quad R = \{(x, y)|1 \le x \le 16, 1 \le y \le 4\}$

18. $\iint_R \sqrt{xy}\,dA$ $\quad R = \{(x, y)|1 \le x \le 4, 1 \le y \le 16\}$

19. $\iint_R x^2y^2\,dA$ $\quad R = \{(x, y)|1 \le x \le 2, 0 \le y \le 1\}$

20. $\iint_R x^2y^2\,dA$ $\quad R = \{(x, y)|0 \le x \le 1, 1 \le y \le 2\}$

21. $\iint_R \left(2 - \frac{1}{2}x^2 + y^2\right) dA$

$R = \{(x, y)|0 \le x \le 1, 0 \le y \le 1\}$

22. $\iint_R \left(3 + x^2 - \frac{1}{2}y^2\right) dA$

$R = \{(x, y)|0 \le x \le 1, 0 \le y \le 1\}$

In Exercises 23–28, sketch the region of integration and evaluate the given double integral.

23. $\int_0^1 \int_0^2 xy\,dy\,dx$

24. $\int_1^3 \int_1^2 5xy\,dy\,dx$

25. $\int_1^2 \int_1^3 (x^2y + y)\,dx\,dy$

26. $\int_1^3 \int_1^2 (x^2y - y)\,dx\,dy$

27. $\int_{-1}^1 \int_0^2 x^2e^y\,dx\,dy$

28. $\int_{-1}^1 \int_0^2 x^2e^y\,dy\,dx$

In Exercises 29–36, determine the average value of the function over the region R.

29. $f(x, y) = x^2 + y^2$ $\quad R = \{(x, y)|0 \le x \le 1, 0 \le y \le 1\}$

30. $f(x, y) = x^2 + y^2$ $\quad R = \{(x, y)|0 \le x \le 2, 0 \le y \le 2\}$

31. $f(x, y) = 9 - x^2 - y^2$

$R = \{(x, y)|0 \le x \le 1, 0 \le y \le 1\}$

32. $f(x, y) = 9 - x^2 - y^2$

$R = \{(x, y)|0 \le x \le 2, 0 \le y \le 2\}$

33. $f(x, y) = 2xy$ $\quad R = \{(x, y)|0 \le x \le 2, 0 \le y \le 2\}$

34. $f(x, y) = 2xy$ $\quad R = \{(x, y)|0 \le x \le 1, 0 \le y \le 1\}$

35. $f(x, y) = x^2e^y$ $\quad R = \{(x, y)|0 \le x \le 2, 0 \le y \le 1\}$

36. $f(x, y) = x^2e^y$ $\quad R = \{(x, y)|0 \le x \le 1, 0 \le y \le 2\}$

In Exercises 37–44, determine the volume under $f(x, y)$ above the given region R.

37. $f(x, y) = 9 - x^2 - y^2$ $\quad R = \{(x, y)|0 \le x \le 1, 0 \le y \le 2\}$

38. $f(x, y) = 12 - x^2 - y^2$

$R = \{(x, y)|0 \le x \le 1, 0 \le y \le 2\}$

39. $f(x, y) = x^2 + 2y^2$

$R = \{(x, y)|0 \le x \le 1, 0 \le y \le 2\}$

40. $f(x, y) = 2x^2 + y^2$ $\quad R = \{(x, y)|0 \le x \le 2, 0 \le y \le 2\}$

41. $f(x, y) = 3xy$ $\quad R = \{(x, y)|0 \le x \le 2, 0 \le y \le 1\}$

42. $f(x, y) = 2xy$ $\quad R = \{(x, y)|0 \le x \le 1, 0 \le y \le 1\}$

43. $f(x, y) = x^2e^y$ $\quad R = \{(x, y)|0 \le x \le 2, 0 \le y \le 1\}$

44. $f(x, y) = x^2e^y$ $\quad R = \{(x, y)|0 \le x \le 1, 0 \le y \le 2\}$

Applications

45. Sumption Sonic Inc. has a production function given by

$$Q = f(x, y) = 250x^{0.7}y^{0.3}$$

where Q is the number of solar cells produced each week, x is the number of employees at the company, and y is the weekly operating budget in thousands. Because the company uses a temporary labor agency, it uses anywhere from 30 to 40 employees each week, and its operating budget is anywhere from \$150,000 to \$180,000 each week. Determine the average number of solar cells that the company can produce each week.

46. The Linger Golf Cart Corporation has a production function given by

$$Q = f(x, y) = 1.3x^{0.75}y^{0.3}$$

where Q is the number of golf carts produced each week, x is the number of employees at the company, and y is the weekly operating budget in thousands. Because the company uses a temporary labor agency, it uses anywhere from 10 to 15 employees each week, and its operating budget is anywhere from \$12,000 to \$16,000 each week. Determine the average number of golf carts that the company can produce each week.

47. The Basich Company has a production function given by

$$Q = f(x, y) = 161x^{0.8}y^{0.25}$$

where Q is the number of hand-held calculators produced each week, x is the number of employees at the company, and y is the weekly operating budget in thousands. Because the company use a temporary labor agency, it uses anywhere from 70 to 80 employees each week, and its operating budget is anywhere from \$20,000 to \$30,000 each week. Determine the average number of hand-held calculators that the company can produce each week.

48. Hoffman Laboratories has a production function given by

$$Q = f(x, y) = 5.6x^{0.65}y^{0.41}$$

where Q is the number of pain-killing tablets produced (in thousands) each week, x is the number of employees at the company, and y is the weekly operating budget in thousands. Because the company use a temporary labor agency, it uses anywhere from 25 to 35 employees each week, and its operating budget is anywhere from \$40,000 to \$50,000 each week. Determine the average number of pain-killing tablets that the company can produce each week.

49. The temperature, in degrees Fahrenheit, x miles east and y miles north of a reporting weather station is given by

$$f(x, y) = 36 + 2x - 3y$$

Determine the average temperature over the region shown in Figure 18.6.11.

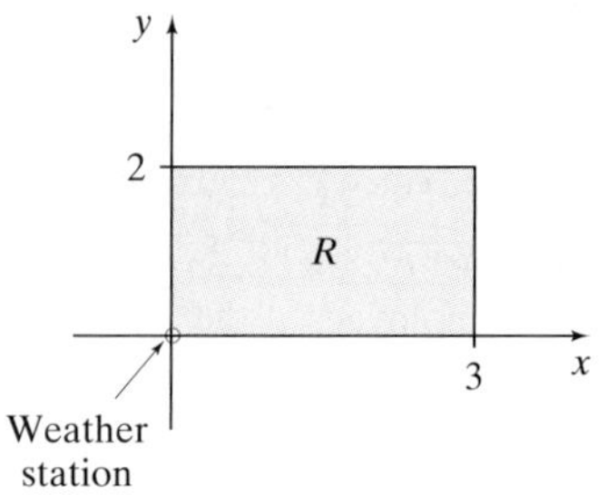

Figure 18.6.11

50. The temperature, in degrees Fahrenheit, x miles east and y miles north of a reporting weather station is given by

$$f(x, y) = 45 + x - 2y$$

Determine the average temperature over the region shown in Figure 18.6.12.

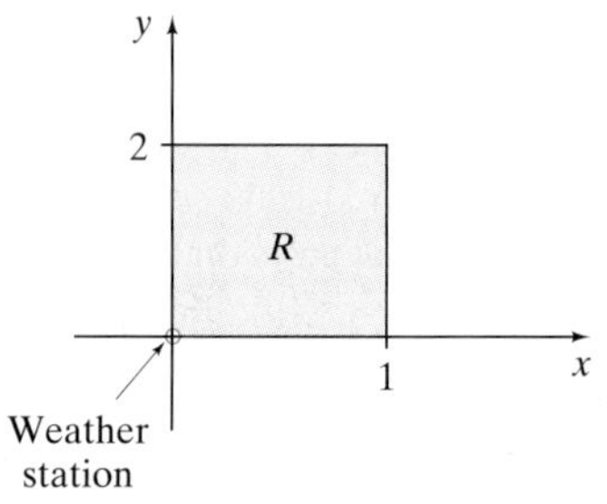

Figure 18.6.12

51. Cedar Falls is a town roughly shaped like a square, as shown in Figure 18.6.13. The **population density**, in people per square mile, x miles east and y miles north of the southwest corner of the town is given by

$$h(x, y) = 30{,}000e^{-y}$$

Determine the average population density for Cedar Falls.

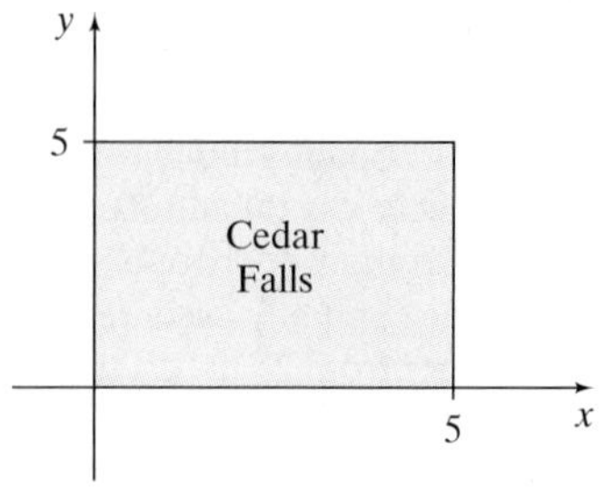

Figure 18.6.13

52. Bucksville is a town situated along a fairly straight river and is fairly rectangular, as shown in Figure 18.6.14. The population density, in people per square mile, x miles east and y miles north of the southwest corner of the town is given by

$$h(x, y) = 15{,}000(5 - y^2)$$

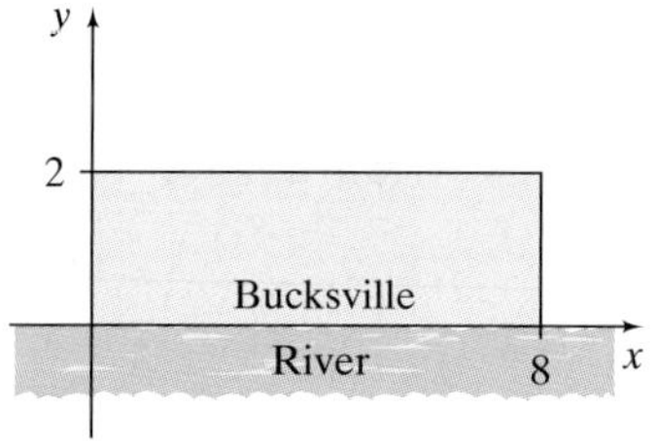

Figure 18.6.14

Determine the average population density for Bucksville.

53. Cedar Falls is a town roughly shaped like a square, as shown in Figure 18.6.13. The population density, in people per square mile, x miles east and y miles north of the southwest corner of the town is given by

$$h(x, y) = 30{,}000e^{-y}$$

Determine the *total population* of Cedar Falls. (To do this, integrate the population density over region R. This illustrates that a double integral can determine a *continuous sum*.)

54. Bucksville is a town situated along a fairly straight river and is fairly rectangular, as shown in Figure 18.6.14. The population density, in people per square mile, x miles east and y miles north of the southwest corner of the town is given by

$$h(x, y) = 15{,}000(5 - y^2)$$

Determine the total population of Bucksville.

55. A heavy industrial complex in Bakerstown emits pollution into the atmosphere. Due to the prevailing winds, the air pollution (in parts per million) x miles east and y miles north of the complex is given by

$$f(x, y) = 125 - 10x^2 - 10y^2$$

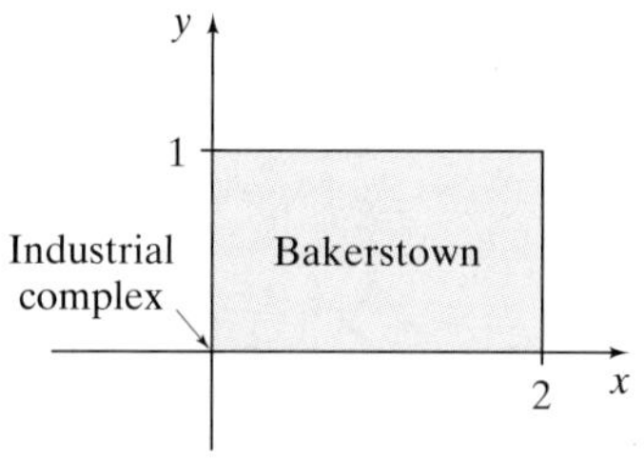

Figure 18.6.15

Determine the average concentration of air pollution in the northeast part of Bakerstown. See Figure 18.6.15.

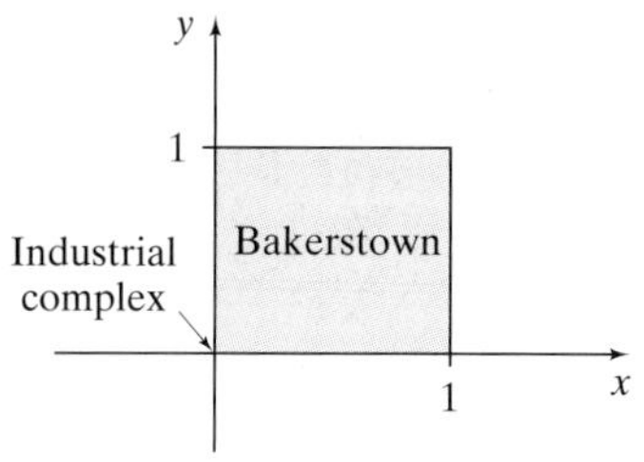

Figure 18.6.16

56. Repeat Exercise 55 for the portion of Bakerstown shown in Figure 18.6.16.

57. An individual's IQ is defined to be

$$\text{IQ} = f(a, m) = \frac{100m}{a}$$

where m is the individual's mental age in years and a is the individual's actual age in years. In a calculus class, the mental age varies from 23 to 32 years, and the actual age varies from 19 to 25 years. Determine the average IQ for this calculus class.

58. Repeat Exercise 57, except here the mental age varies from 23 to 45 years and the actual age varies from 22 to 32 years.

SECTION PROJECT

In their study of human groupings, anthropologists often use an index called the *cephalic index*. The cephalic index is given by

$$C(W, L) = 100 \cdot \frac{W}{L}$$

where W is the width and L is the length of an individual's head. Both measurements are made across the top of the head and are in inches. In a calculus class the width of heads varies from 6 to 8.5 inches and the length of heads varies from 8 to 10 inches.

(a) Determine the average value of the cephalic index for this class.

(b) Repeat part (a), except here the width of heads varies from 6.5 to 8.2 inches and the length of heads varies from 7.6 to 9.8 inches.

Why We Learned It

In this chapter we extended the function concept to functions of more than one independent variable. If we have a business that produces two items, we can use a function of two variables to represent the total costs, revenue and profit for these two items. After introducing functions of more than one variable, we looked at level curves and vertical cross sections. Two economic applications of level curves were isoquants and indifference curves. Vertical cross sections introduced us to partial derivatives. Partial derivatives allowed us to analyze rates of change of a function of more than one variable. We saw some familiar applications of partial derivatives including marginal revenue and marginal profit. Another frequently seen application in this chapter was the Cobb–Douglas production function. Partial derivatives allowed us to analyze the marginal productivity of labor as well as the marginal productivity of capital. Once these marginal derivatives were introduced and analyzed, we then used them to determine how to allocate additional resources most effectively. The chapter concluded with a look at determining relative extrema for functions of more than one variable and Lagrange multipliers. This then allowed us to maximize a profit function as well as minimize a cost function. Double integrals was our final topic and with it we returned to some old applications such as the average value concept.

CHAPTER REVIEW EXERCISES

In Exercises 1–4, evaluate $f(x, y) = \dfrac{x^2 + 3}{x - y^2}$ *for the indicated values.*

1. $f(3, 5)$

2. $f(-2, 4)$

3. $f(10, 3)$

4. $f(20, -4)$

In Exercises 5 and 6, determine the domain of the given function. Recall that here the domain is the set of all ordered pairs (x, y) *in the xy-plane.*

5. $f(x, y) = \sqrt{x + y}$

6. $f(x, y) = \dfrac{5x}{x - 2y}$

In Exercises 7–10, plot the given points.

7. $(5, 0, 0)$

8. $(3, 8, -5)$

9. $(0, -4, 3)$ **10.** $(-2, -6, 5)$

In Exercises 11–14, without plotting the point, determine if the given point is above or below the xy-plane. Explain.

11. $(-3, -4, 2)$ **12.** $(4, 3, -8)$

13. $(6, -2, -1)$ **14.** $(-5, 9, 4)$

In Exercises 15 and 16, sketch the graph of the given plane in 3-space.

15. $x = -5$ **16.** $z = 2$

17. The Moondoe Coffee Company has determined that its monthly profit, in dollars, is given by

$$P(x, y) = 0.8x + 1.2y - 6000$$

where x is the number of regular cups of coffee sold in a month and y is the number of cups of cappuccino sold in a month. Determine $P(4000, 3000)$ and interpret.

18. The Stellar Stereo Store spends x thousand dollars each month on television advertising and y thousand dollars each month on direct mail advertising. The company has monthly sales given by

$$S(x, y) = x^2 + 3y$$

(a) Determine $S(4, 7)$ and interpret.

(b) Determine $S(7, 4)$ and interpret.

19. The WriteNow Company sells a regular calligraphy set for \$12 and a deluxe calligraphy set for \$21. The cost to produce these items is \$7 for the regular set and \$13 for the deluxe set. The company also has fixed monthly expenses totaling \$5000.

(a) Determine the cost $C(x, y)$ of producing x regular sets and y deluxe sets in a month.

(b) Determine the revenue function $R(x, y)$.

(c) Determine the profit function $P(x, y)$.

20. The Digital Calculator Company manufactures scientific calculators and graphing calculators. Let x represent the weekly demand for scientific calculators, and let y represent the demand for graphing calculators. The weekly price–demand equations are given by

$p = 60 - 0.3x + 0.1y$, the price in dollars for a scientific calculator

$q = 100 + 0.2x - 0.5y$, the price in dollars for a graphing calculator

The cost function is given by

$$C(x, y) = 3000 + 25x + 35y$$

(a) Determine the weekly revenue function $R(x, y)$.

(b) Evaluate $R(40, 60)$ and interpret.

(c) Determine the weekly profit function $P(x, y)$.

(d) Evaluate $P(40, 60)$ and interpret.

21. If an amount P dollars is invested at an annual interest rate of r (in decimal form) compounded continuously for 10 years, the total amount accumulated is given by

$$f(P, r) = Pe^{10r}$$

determine $f(4500, 0.045)$ and interpret.

22. The volume of a pyramid with a square base is given by

$$V(x, h) = \frac{1}{3}x^2h$$

where x is the length of one side of the square base and h is the height of the pyramid. Assume that x and h are measured in inches.

(a) Determine $V(8, 12)$ and interpret.

(b) If the height is equal to the side length of the square base, the volume can be expressed as a function of one variable, either x or h. Determine $V(x)$ and $V(h)$.

In Exercises 23–26, make a contour map of the given function for the specified c-values.

23. $f(x, y) = 2x - y + 3$ $c = 0, 2, 4, 6$

24. $f(x, y) = x^2 - y$ $c = -2, 0, 2, 4$

25. $f(x, y) = 16 - x^2 - y^2$ $c = 0, 4, 8, 12$

26. $f(x, y) = x - e^y$ $c = 0, 1, 2, 3$

In Exercises 27–30, make a contour map of the given production functions for the specified c-values.

27. $Q = 10x^{0.5}y^{0.5}$ $c = 10, 20, 30, 40$

28. $Q = 2.7x^{0.3}y^{0.75}$ $c = 20, 40, 60, 80$

29. $Q = 8.9x^{0.62}y^{0.42}$ $c = 100, 200, 300, 400$

30. $Q = 80x^{0.21}y^{0.79}$ $c = 50, 100, 150, 200$

In Exercises 31–34, make a contour map of the given utility function for the specified c-values.

31. $U = \sqrt[3]{xy^2}$ $c = 2, 4, 6, 8$

32. $U = x^{0.4}y^{0.6}$ $c = 1, 2, 3, 4$

33. $U = x^{0.15}y^{0.85}$ $c = 2, 4, 6, 8$

34. $U = x^{0.3}y^{0.7}$ $c = 1, 2, 3, 4$

In Exercises 35–38, sketch on the appropriate plane, either the xz-plane or yz-plane, cross sections for the given function for the specified vertical plane values. See Example 6 in Section 18.2.

35. $z = 20x^{0.3}y^{0.7}$ $x = 2, 4, 6$

36. $z = 20x^{0.3}y^{0.7}$ $y = 2, 4, 6$

37. $z = x^2 - y$ $y = 0, 1, 2, 3$

38. $z = x^2 - y$ $x = 0, 1, 2, 3$

39. The McGillicutty Corporation has 12 manufacturing plants and has a Cobb–Douglas production function of $Q = f(x, y) = 1.87x^{0.7}y^{0.3}$, where x is the number of labor-hours (in thousands), y is the capital (total net assets in millions), and Q is the quantity produced (in thousands of units). Sketch isoquants for $Q = 60, 80, 100,$ and 120.

40. Refer to Exercise 39. Lucille, the CEO of the McGillicutty Corporation, wants to keep the capital at each plant fixed at $y = 4.1$.

(a) By fixing capital at $y = 4.1$, is Lucille looking at a vertical or horizontal cross section?

(b) Graph the cross section when $y = 4.1$.

(c) Approximate how many labor-hours would be required to produce a quantity of 150,000 units (that is, $Q = 150$).

41. Refer to Exercises 39 and 40. Instead of keeping the capital fixed, suppose that Lucille, the CEO of the McGillicutty Corporation, wants to keep the number of labor-hours at each plant fixed at $x = 240$.

(a) By fixing labor at $x = 240$, is Lucille looking at a vertical or horizontal cross section?

(b) Graph the cross section when $x = 240$.

(c) Approximate how much capital is required to produce 150,000 units (that is, $Q = 150$).

42. The MaxiSound Company has seven plants around the United States that produce speakers for computers. In a recent year, the data for each plant gave the number of labor-hours (in thousands), capital (in millions), and total quantity of speakers produced, as follows:

Labor	Capital	Quantity
24	4.3	13,016
26	4.9	14,565
29	5.6	16,500
31	6.1	17,852
38	5.8	18,600
45	6.4	21,052
46	7.3	23,084

(a) The plants all use the same technology, so a production function can be determined. Enter the data into your calculator and execute the multiple regression program in Appendix C to determine a Cobb–Douglas production function to model the data. Round all values to the nearest hundredth.

(b) Construct isoquants for $Q = 12{,}000$, 16,000, 20,000, and 24,000.

In Exercises 43–50, determine $f_x(x, y)$ and $f_y(x, y)$.

43. $f(x, y) = 5x - 3y^2$

44. $f(x, y) = 2x^3 - 5xy + y^8$

45. $f(x, y) = (8y^2 - 3x)^5$

46. $f(x, y) = 5.83x^{0.38}y^{0.61}$

47. $f(x, y) = \dfrac{x + y}{3x}$

48. $f(x, y) = \ln(5x - y^3)$

49. $f(x, y) = e^{x-y} + 5x^3y$

50. $f(x, y) = \ln xy - e^{x/y}$

In Exercises 51–54, for the given function, evaluate the stated partial derivative and interpret.

51. $f(x, y) = 5x^2 + 3xy^4 - 2y^5$; determine $f_y(3, 5)$.

52. $f(x, y) = \ln(x^2 + 5y)$; determine $f_x(2, 7)$.

53. $f(x, y) = 12.85x^{0.38}y^{0.62}$; determine $f_y(3, 2)$.

54. $f(x, y) = \dfrac{\ln 2y}{x^2 + y}$; determine $f_x(-2, 5)$.

In Exercises 55–58, determine $f_{xx}(x, y)$, $f_{xy}(x, y)$, and $f_{yy}(x, y)$.

55. $f(x, y) = -4x^3 + 10xy + xy^5$

56. $f(x, y) = 8x^2y^7 - x^3y^5 + 12x^2$

57. $f(x, y) = (x + 2y)^2$

58. $f(x, y) = \ln(3x - y)$

59. For $f(x, y) = 6x^2 - 4xy + 5y^2 + x - 3y + 17$, determine values for x and y such that $f_x(x, y) = 0$ and $f_y(x, y) = 0$ simultaneously.

60. The Heavy Stuff Athletic Club spends x thousand dollars each year on television advertising and y thousand dollars each year on newspaper advertising. The club has weekly revenues (in thousands of dollars) given by

$$R(x, y) = 1.5x + 0.6x^2$$

(a) Determine $R_x(x, y)$ and $R_y(x, y)$ and interpret.

(b) Determine $R_x(12, 8)$ and interpret.

(c) Determine $R_y(12, 8)$ and interpret.

61. Pixel Power, Inc., manufactures 15-inch computer monitors and 17-inch computer monitors. Let x represent the weekly demand for a 15-inch monitor, and let y represent the weekly demand for a 17-inch monitor. The weekly price–demand equations are given by

$p = 480 - 0.6x + 0.2y$, the price in dollars for a 15-inch monitor

$q = 720 + 0.3x - 0.7y$, the price in dollars for a 17-inch monitor

(a) Determine the weekly revenue function $R(x, y)$.

(b) Determine $R_x(x, y)$ and $R_y(x, y)$ and interpret each.

(c) Determine $R_x(640, 850)$ and $R_y(640, 850)$ and interpret each.

62. *(continuation of Exercise 61)* The weekly cost function for Pixel Power, Inc., is given by

$$C(x, y) = 12{,}000 + 128x + 163y$$

(a) Determine the weekly profit function $P(x, y)$.

(b) Determine $P_x(x, y)$ and $P_y(x, y)$ and interpret each.

(c) Determine $P_x(640, 850)$ and $P_y(640, 850)$ and interpret each.

63. The WatchIt Company produces two types of videocassette recorders, x hundred regular VCRs and y hundred stereo VCRs each month. The monthly revenue and cost functions are, in hundreds of dollars,

$$R(x, y) = -3.5x^2 - 3.5xy + 1.2y^2 + 175x + 215y$$
$$C(x, y) = 15xy + 65x + 85y + 600$$

(a) Determine the monthly profit function $P(x, y)$.

(b) Determine $P_x(x, y)$ and $P_y(x, y)$ and interpret each.

(c) Determine $P_x(8, 11)$ and $P_y(8, 11)$ and interpret each.

(d) Determine $P(8, 11)$ and interpret.

64. The Superior Gizmo Company has a production function given by

$$Q = f(x, y) = 5.64x^{0.25}y^{0.8}$$

where x is the number of labor-hours in thousands, y is the capital in millions, and Q is the total quantity produced in thousands.

(a) Compute $f_x(x, y)$ and $f_y(x, y)$.

(b) If the Superior Gizmo Company is currently using 38 units of labor and 1.6 units of capital, determine the marginal productivity of labor and the marginal productivity of capital and interpret each.

(c) Would production increase more by increasing the labor by 1200 labor-hours or by increasing the capital by \$40,000?

In Exercises 65–68, determine the critical points for the given function.

65. $f(x, y) = x^2 + y^2 + 8x - 6y + 8$

66. $f(x, y) = x^2 - y^2 + 4x + 10y - 3$

67. $f(x, y) = x^2 + y^2 - 5xy + 6x - 8y + 7$

68. $f(x, y) = 3x^2 + 5y^2 - 8xy + 2x + 6y + 3$

In Exercises 69–76, use the Second Derivative Test to locate any relative extrema and saddle points.

69. $f(x, y) = x^2 + y^2 - 2x + 8y - 7$

70. $f(x, y) = 2x^2 + 5y^2 + 6x - 2y + 12$

71. $f(x, y) = x^2 - 2y^2 + xy + 3x - 3y$

72. $f(x, y) = \frac{4}{3}y^3 + x^2 - 4x - 8y - 17$

73. $f(x, y) = 2x^3 - 2x^2 + 4xy + y^2$

74. $f(x, y) = \frac{1}{5}x^5 - y^2 - 16x + 6y - 8$

75. $f(x, y) = e^{x^2 - y^2 + 4y}$

76. $f(x, y) = 4xy + \frac{1}{x} + \frac{2}{y}$

77. Roberta's Hair Salon offers a basic haircut and a deluxe haircut, which includes shampoo and styling. Let x represent the daily demand for basic haircuts, and let y represent the daily demand for deluxe haircuts. The daily price–demand equations are given by

$p = 12 - 0.3x + 0.1y$, the price in dollars for a basic haircut

$q = 20 + 0.1x - 0.2y$, the price in dollars for a deluxe haircut

(a) Determine the revenue function $R(x, y)$.

(b) How many basic haircuts and how many deluxe haircuts should be given per day in order to maximize revenue?

(c) What is the maximum daily revenue?

78. *(continuation of Exercise 77)* Roberta's Hair Salon has a daily cost function given by

$$C(x, y) = 200 + 4x + 5y$$

(a) Determine the profit function $P(x, y)$.

(b) How many basic haircuts and how many deluxe haircuts should be given per day in order to maximize profit?

(c) What is the maximum daily profit?

79. The WriteNow Company makes marking pens with either a fine tip or a large tip. Let x represent the number of thousands of fine-tip pens, and let y represent the number of thousands of large-tip pens produced each day. The daily profit function is given by

$$P(x, y) = -3x^2 - 2y^2 + 2xy + 800x + 1200y - 400$$

where $P(x, y)$ is the profit in dollars for producing x thousand fine-tip pens and y thousand large-tip pens.

(a) How many fine-tip pens and how many large-tip pens should be sold each day to maximize profit?

(b) What is the maximum profit?

80. You have been challenged to design a rectangular box with no top and three parallel partitions that holds 80 cubic inches. Determine the dimensions that will require the least amount of material to do this.

In Exercises 81–86, use the method of Lagrange multipliers to solve each problem.

81. Maximize $f(x, y) = 5xy$ Subject to $x + y = 10$

82. Minimize $f(x, y) = 3xy$ Subject to $x - 2y = 6$

83. Maximize $f(x, y) = 16 - 2x^2 + 5xy - 18y^2$ Subject to $x + 3y = 6$

84. Minimize $f(x, y) = 3 + x^2 + y^2$ Subject to $2x - y = 10$

85. Maximize $f(x, y) = 3.8x^{0.35}y^{0.6}$ Subject to $120x + 260y = 2500$

86. Minimize $f(x, y) = xy$ Subject to $x^2 + 4y^2 = 8$

87. The Liberty Bell Telephone Company has a production function for the manufacture of portable telephones of

$$Q = f(x, y) = 28x^{0.6}y^{0.4}$$

where Q is the number of thousands of telephones produced per week, x is the number of units of labor each week, and y is the number of units of capital. A unit of labor is \$500 and a unit of capital is \$800. The total expense for both is limited to \$4000.

(a) Determine the number of units of labor and the number of units of capital required to maximize production.

(b) Determine the maximum production.

88. The Dizzy Dog Company has a production function for the manufacture of dog toys given by

$$Q = f(x, y) = 4.3x^{0.55}y^{0.5}$$

where Q is the quantity of dog toys produced (in thousands) per week, x is the number of units of labor, and y is the number of units of capital. Each unit of labor is \$600 and each unit of capital is \$400. The budget for both is limited to \$24,000.

(a) Determine the number of units of labor and the number of units of capital required to maximize production.

(b) Determine the marginal productivity of money for the division of labor and capital found in part (a) and interpret.

(c) Determine the optimal increase in production if an additional \$2000 is allotted toward the budget.

89. A family plans to plant a garden with an area of 300 square feet. Three sides of the garden will have fencing, and the fourth side will have a wall that costs five times as much (per linear foot) as the fencing. Use the method of Lagrange multipliers to determine the dimensions of the garden that will minimize the cost.

90. Packages weighing less than 15 pounds sent by Priority Mail are charged at the standard rate (based on the weight of the package), provided that the length plus girth is no more than 84 inches. Use the method of Lagrange multipliers to determine the dimensions of the largest volume package that can be sent via Priority Mail at the standard rate. Assume that the length is y (*Source:* United States Postal Service, www.usps.gov).

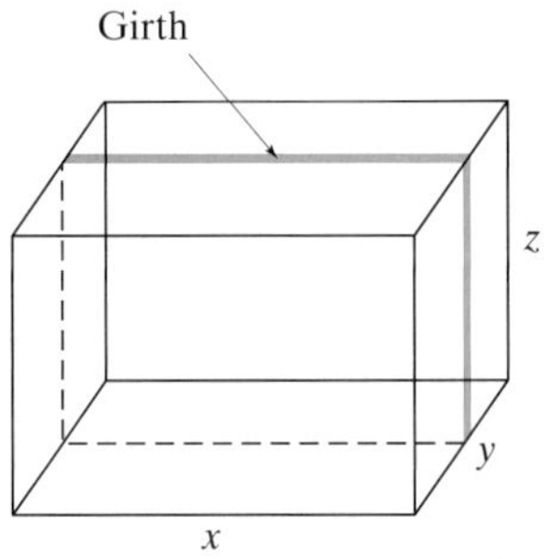

In Exercises 91–94, evaluate the given integral.

91. $\int_5^8 2x^2y^3\,dx$

92. $\int_0^5 (3x - 4y)\,dy$

93. $\int_2^6 (x^2y - 4y^3)\,dy$

94. $\int_3^9 (x^2y^3 + 4xy)\,dx$

For Exercises 95 and 96:

(a) Sketch the region of integration.

(b) Write and evaluate a double integral with a $dy\,dx$ order of integration.

(c) Write and evaluate a double integral with a $dx\,dy$ order of integration.

95. $\int (3x - 4xy)\,dA \quad R = \{(x, y)|0 \le x \le 5, 0 \le y \le 2\}$

96. $\int \sqrt[3]{xy}\,dA \quad R = \{(x, y)|0 \le x \le 8, 1 \le y \le 27\}$

For Exercises 97 and 98:

(a) Sketch the region of integration.

(b) Evaluate the given double integral.

97. $\int_0^4 \int_1^3 3xy^2\,dy\,dx$

98. $\int_{-2}^6 \int_0^3 (2x + ye^x)\,dx\,dy$

In Exercises 99 and 100, determine the volume under $f(x, y)$ above the given region R.

99. $f(x, y) = 25 - x^2 - y^2 \quad R = \{(x, y)|0 \le x \le 3, 0 \le y \le 4\}$

100. $f(x, y) = \dfrac{e^x}{y} \quad R = \{(x, y)|0 \le x \le 4, 1 \le y \le e\}$

In Exercises 101–104, determine the average value of the function over the region R.

101. $f(x, y) = x^2 + 3xy - y^2 \quad R = \{(x, y)|0 \le x \le 3, 0 \le y \le 4\}$

102. $f(x, y) = 3x + 4y^3 \quad R = \{(x, y)|0 \le x \le 5, 0 \le y \le 2\}$

103. $f(x, y) = x^3y \quad R = \{(x, y)|1 \le x \le 5, 3 \le y \le 7\}$

104. $f(x, y) = e^{x+y/2} \quad R = \{(x, y)|0 \le x \le 1, 0 \le y \le 2\}$

105. The Cripton Company has a production function given by

$$Q = f(x, y) = 11.8x^{0.68}y^{0.45}$$

where Q is the number of radios produced (in thousands) each week, x is the number of employees at the company, and y is the weekly operating budget in thousands. Because the company uses a temporary labor agency, it has anywhere from 18 to 30 employees each week, and its operating budget is anywhere from \$16,000 to \$24,000 each week. Determine the average number of radios that the company produces each week.

106. The temperature, in degrees Fahrenheit, x miles east and y miles north of a reporting weather station is given by

$$f(x, y) = 68 + 2x - 3y$$

Determine the average temperature over the region shown in the figure.

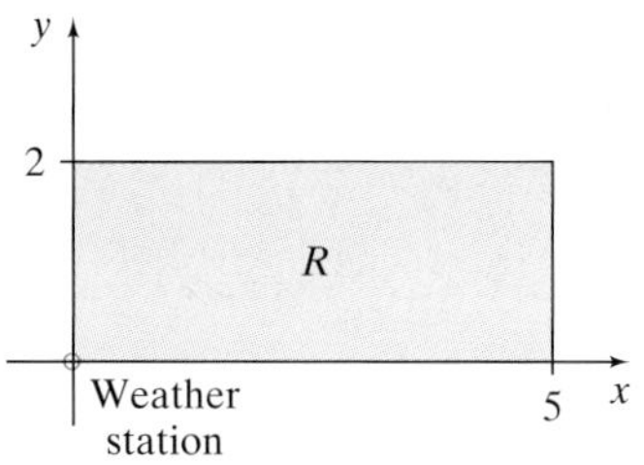

107. Brandonville is a town whose shape is fairly rectangular, as shown. A heavy industrial complex located at the southwest corner of the town emits pollution into the atmosphere. The air pollution (in parts per million) x miles east and y miles north of the complex is given by

$$f(x, y) = 150 - x^2 - 2y^2$$

Determine the average concentration of air pollution in Brandonville.

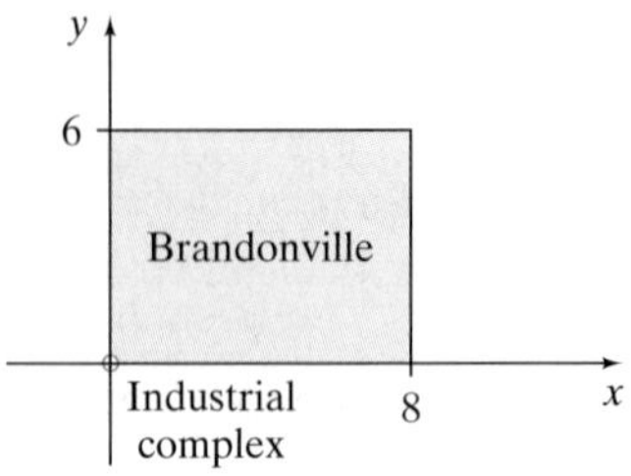

108. Refer to Exercise 107. The population density of Brandonville, in people per square mile, x miles east and y miles north of the southwest corner of the town is given by

$$h(x, y) = 8000(3 + x + y^2)$$

Determine the average population of Brandonville.

CHAPTER 18 PROJECT

SOURCE: US Census Bureau, www.census.gov

Wind chill can be described as the temperature it really feels like outside. On November 1, 2001, the National Weather Service updated the Wind Chill Temperature Index. As you will see below, the index factors in wind speed. Wind speed, which used to be measured from a height of 33 feet off the ground, is now measured at a height of 5 feet off the ground—the height of an average human's face. One of the new features that the index includes is that it takes into acount skin tissue resistance. Currently, the index does not factor in the effects of the sun on various sky conditions like sunny or cloudy. The National Weather Service plans to revise this in 2002.

In this project, we will compare the former and new Wind Chill Temperature Index. Both functions are of two variables, which makes them fit nicely into Chapter 18.

Let f represent the former Wind Chill Temperature Index and let g represent the new Wind Chill Temperature Index, both are measured in dgreees Fahrenheit. That is, let

$$f(v, T) = 91.4 - (0.4474677 - 0.020425v + 0.303107\sqrt{v}) \times (91.4 - T)$$

and

$$g(v, T) = 35.74 + 0.6215T - 35.75v^{0.16} + 0.4275Tv^{0.16}$$

where v represents the wind speed in miles per hour and T represents the air temperature.

1. Evaluate, interpret and compare $f(45, 5)$ and $g(45, 5)$.

2. Find $f_v(v, T)$, $f_T(v, T)$, $g_v(v, T)$, and $g_T(v, T)$.

3. Evaluate, interpret and compare, $f_v(45, 5)$ and $g_v(45, 5)$.

4. Using $g(v, T)$, find all combinations of wind speed and air temperature that will result in wind chill of $-12°$F.

5. For both functions, find how fast the wind chill changes with respect to air temperature at a wind speed of 10 miles per hour when the air temperature is $20°$F.

Appendices

Appendix A Prerequisites

Section A.1 Percents, Decimals, and Fractions

In many applications, we are given information in the form of a **percentage**. When we hear facts such as "15.5% of the U.S. military in 1998 was made up of females" or "In 1997, 25.6% of all the lumber exported from the United States went to Japan" we are using percentages. Whenever we perform mathematical calculations involving percentages, we convert them to **decimal** form. Since the word **percent** means "per 100," we convert a percent into a decimal form by dividing by 100. For example, 25.6% in decimal form is determined by

$$25.6\% = \frac{25.6}{100} = 0.256$$

This illustrates the following shortcut used in converting percents to decimals.

Converting Percents to Decimals

To convert a percent into decimal form, we move the decimal point two places to the left and remove the percent sign.

Example 1 Converting Percents to Decimals

Rewrite each of the following in decimal form.

(a) 73.1% **(b)** 3.1%

Solution

(a) When we move the decimal two places to the left and remove the percent sign, we see that

$$73.1\% = 0.731$$

(b) Since there is only one digit to the left of the decimal place, we insert a zero to get

$$3.1\% = 0.031$$

Checkpoint 1

Now work Exercise 3 in the A.1 Exercises.

Let's see how percent to decimal conversion is used in an application.

Example 2 Applying Percent to Decimal Conversion

In 1997, about 24% of the doctorates conferred in earth science were awarded to females (*Source:* National Science Foundation, www.nsf.gov). If 862 earth science doctorates were awarded in 1997, how many went to females?

Solution

Understand the Situation: To determine the number of females receiving earth science doctorates, we must take 24% of those receiving doctorates.

This is determined by performing

$$24\% \text{ of } \begin{pmatrix} \text{total earth} \\ \text{science doctorates} \end{pmatrix} = 0.24 \cdot 862 = 206.88 \approx 207$$

Interpret the Solution: This means that of the 862 students receiving earth science doctorates in 1997 about 207 of them were female. ■

Converting Fractions to Percents

Another often used tool in many of our applications is the conversion of fractions into percents. To do this, we first convert the fraction to a decimal and then convert the decimal into a percent.

Converting Fractions to Percents

To convert a fraction into a percent, first convert the fraction into a decimal by dividing the numerator of the fraction by its denominator. Then multiply the decimal number by 100 and add on a percent sign.

Note: Converting decimals to percents is equivalent to moving the decimal point to the right two places and then adding on a percent sign.

Example 3 Converting Fractions to Percents

Convert each of the following fractions into a percent.

(a) $\frac{4}{5}$ **(b)** $\frac{47}{200}$

Solution

(a) We begin by writing the fraction in decimal form as follows.

$$\frac{4}{5} = 4 \div 5 = 0.80$$

Now we multiply the quotient by 100 and add on a percent sign.

$$0.80 \cdot 100 = 80\%$$

(b) Dividing the numerator by the denominator gives us

$$\frac{47}{200} = 47 \div 200 = 0.235$$

Multiplying this decimal number by 100 gives us

$$0.235 \cdot 100 = 23.5\%$$

■

Checkpoint 2

Now work Exercise 15 in the A.1 Exercises.

Many of the applications that we study involve finding what percentage of a whole is made up by a small group. In these cases, converting the quantity to a fraction can make the solution easier to understand. Now let's look at an application of fraction to percentage conversion.

Example 4 Converting Fractions to Percents in Applications

Interactive Activity

The information in Example 4 shows that the number of farms in 1992 was about 1,925,000, with 283,000 being 50- to 99-acre farms. Use these numbers to convert to a percent. Is the answer the same as in Example 4? Explain.

WWW

In 1992 there were about 1925 thousand farms in the United States and 283 thousand of these had sizes between 50 and 99 acres (*Source:* U.S. Department of Agriculture, www.usda.gov). Convert $\frac{283}{1925}$ to a percentage and interpret.

Solution

Dividing this fraction gives us

$$\frac{283}{1925} \approx 0.147$$

Now converting to a percent gives us

$$0.147 \cdot 100 = 14.7\%$$

Interpret the Solution: This means that in 1992 about 14.7% of all U.S. farms had a size between 50 and 99 acres. ■

SECTION A.1 EXERCISES

In Exercises 1–8, write each of the following in decimal form.

1. 73%
2. 56%
3. 4%
4. 8%
5. 59.1%
6. 46.4%
7. 0.7%
8. 0.9%

9. In 1990 the average daily cost of hospital care was \$1104, of which 47.5% of the charges were reimbursed by private health insurance. How much per day was covered by private health insurance? (*Source:* U.S. Census Bureau, www.census.gov).

10. In 1997 the average daily cost of hospital care was \$2131, of which 49.9% of the charges were reimbursed by private health insurance. How much per day was covered by private health insurance? (*Source:* U.S. Census Bureau, www.census.gov).

11. In 1980 about 9866 million board feet of lumber was imported to the United States from Canada. If this represents 97.5% of all the lumber imported that year, how many board feet does this represent? (*Source:* U.S. Census Bureau, www.census.gov).

12. In 1998 there were about 272 trillion cubic feet of natural gas produced worldwide. If the United States produced 23.1% of all the natural gas in 1998, how many trillion cubic feet of natural gas does this represent? (*Source:* U.S. Census Bureau, www.census.gov).

In Exercises 13–20, convert each of the following fractions to a percent.

13. $\frac{1}{5}$
14. $\frac{3}{4}$
15. $\frac{1}{8}$
16. $\frac{7}{8}$
17. $\frac{267}{500}$
18. $\frac{117}{250}$
19. $\frac{3}{7}$
20. $\frac{4}{11}$

21. The 2008 projected school enrollment of students in kindergarten through eighth grade is about 38 million, of which 4.6 million are projected to be in private schools. Convert $\frac{4.6}{38}$ to a percent and interpret. (*Source:* U.S. National Center for Education Statistics, www.nces.ed.gov).

22. In 1988 the school enrollment of students in kindergarten through eighth grade was about 38.4 million, of which about 4 million were enrolled in private schools. Convert $\frac{4}{38.4}$ to a percent and interpret. (*Source:* U.S. National Center for Education Statistics, www.nces.ed.gov).

23. In 1985 the United States spent about \$49,887 million on research and development, of which \$2725 million was spent on technology research and development. Convert $\frac{2725}{49{,}887}$ to a percent and interpret. (*Source:* U.S. Census Bureau, www.census.gov).

24. In 1999 the United States spent about \$77,225 million on research and development, of which \$8037 million was spent on technology research and development. Convert $\frac{8037}{77{,}225}$ to a percent and interpret. (*Source:* U.S. Census Bureau, www.census.gov).

Section A.2 Evaluating Expressions and Order of Operations

In mathematics an **expression** can be thought of as a combination of arithmetic operations (addition, subtraction, multiplication, and division), along with exponents and parentheses. An expression can contain numbers and variables. By **evaluating** an expression, we mean to perform as many of these operations as possible. To evaluate expressions in a consistent manner, we use the **order of operations** to perform the simplifications.

Order of Operations

Given a mathematical expression, we perform all operation in the following order.

1. **Parentheses**: Simplify all expressions in parentheses or other grouping symbols such as fraction bars. Calculate from the innermost parentheses out. In the case of a fraction bar, calculate the numerator and denominator separately and then perform the division.
2. **Exponents**: Simplify all expressions containing exponents by raising all numbers to their indicated powers.
3. **Multiplication and Division**: Perform all multiplications and divisions, reading from left to right, in the order that they appear.
4. **Addition and Subtraction**: Perform all additions and subtractions, reading from left to right, in the order that they appear.

Note: Some recall the order of operations by remembering the pneumonic **P**lease **E**xcuse **M**y **D**ear **A**unt **S**ally.

Example 1 Performing the Order of Operations

Evaluate the following expressions:

(a) $4 - 36 \div 4 + 3$

(b) $(2 + 13) \cdot 2 - 3^3 + 6$

(c) $\dfrac{6 - 3}{4}$

(d) $-2^2 + 3(5)$

Solution

(a) Since no parentheses or exponents are involved, we start by performing the division.

$$4 - \mathbf{36 \div 4} + 3 = 4 - 9 + 3$$

Now we can perform the subtraction and the addition, going from left to right.

$$\mathbf{4 - 9} + 3 = -5 + 3 = -2$$

(b) Completing the operation within the parentheses gives

$$\mathbf{(2 + 13)} \cdot 2 - 3^3 + 6 = 15 \cdot 2 - 3^3 + 6$$

Then we simplify the exponential part of the expression to get

$$15 \cdot 2 - \mathbf{3^3} + 6 = 15 \cdot 2 - 27 + 6$$

Multiplying yields

$$\mathbf{15 \cdot 2} - 27 + 6 = 30 - 27 + 6$$

Now we go from left to right, performing the addition and subtraction.

$$\mathbf{30 - 27} + 6 = 3 + 6 = 9$$

(c) For this fraction, we first simplify the numerator and then divide.

$$\frac{6 - 3}{4} = \frac{3}{4} = 0.75$$

(d) For this expression, we begin by squaring the 2. Since we can write -2 as $-1 \cdot 2$, -2^2 really means $-1 \cdot 2^2$. Notice that the result is still negative.

$$\mathbf{-2^2} + 3(5) = -4 + 3(5)$$

Now we perform the multiplication implied by the parentheses and finally the addition to get

$$-4 + \mathbf{3(5)} = \mathbf{-4 + 15} = 11$$

■

Technology Option

We can evaluate expressions on a graphing calculator fairly easily. Figures A.2.1 through A.2.4 show the results of Example 1, parts (a)–(d), respectively.

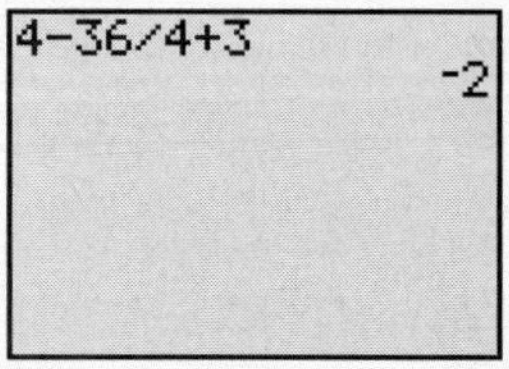

Figure A.2.1

Figure A.2.2

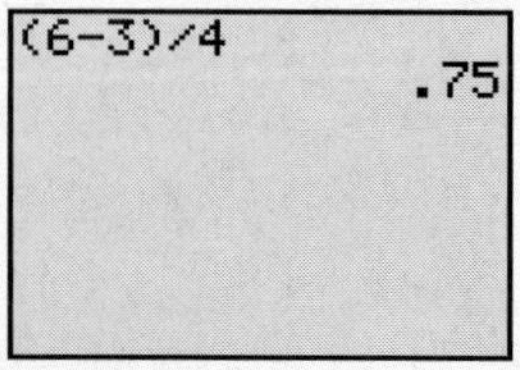

Figure A.2.3

Figure A.2.4

To see how expressions are evaluated on your calculator, consult the on-line calculator manual at www.prenhall.com/armstrong.

Checkpoint 1 Now work Exercise 5 in the A.2 Exercises.

Many applications involve evaluating an expression that comes from a given formula. Example 2 illustrates how the order of operations can be used in an application.

Example 2 **Using the Order of Operations in an Application**

In 1998 about 77% of all Asian Americans living in the United States were anywhere from $31.9 - 1.2(20.1)$ to $31.9 + 1.2(20.1)$ years old. (*Source:* U.S. Census Bureau, www.census.gov). Simplify these expressions and interpret.

Solution

When simplifying these expressions, we must first perform the multiplication implied by the parentheses and then complete the addition and subtraction.

$$31.9 - 1.2(20.1) = 31.9 - 24.12 = 7.78 \qquad 31.9 + 1.2(20.1) = 31.9 + 24.12 = 56.02$$

Interpret the Solution: This means that in 1997 about 77% of all Asian Americans living in the United States were anywhere from 7.78 to 56.02 years old. ■

SECTION A.2 EXERCISES

In Exercises 1–8, evaluate the following expressions without the use of a calculator.

1. $3 + 4 \cdot 7$

2. $2 + 5 \cdot 8$

3. $9 - 8 \div 4 + 1$

4. $5 + 8 \div 2 + 7$

✓ **5.** $(5 - 2)^2 - 5$

6. $(1 + 3)^2 - 1$

7. $2(3) + 4(1)$

8. $6(8) + 2(4)$

In Exercises 9–18, evaluate the following expressions using a calculator.

9. $\dfrac{7-3}{0.5}$

10. $\dfrac{10-6}{0.2}$

11. $3(-2)^2 - 4(-3)^2$

12. $2(-1)^2 - 6(-3)^2$

13. $12(0.6)^2(0.4)^3$

14. $8(0.7)^4(0.3)^2$

15. $5^2 + 2 \cdot 9 - (4 - 6 \div 3)$

16. $2^3 + 3 \cdot 4 - (6 - 15 \div 5)$

17. $(0.7)(0.3) + (0.4)(0.6)$

18. $(0.1)(0.9) + (0.5)(0.5)$

19. In 1996 the middle 50% of the unmarried mothers in the United States had ages from $23.3 - 0.67(5.9)$ to $23.3 + 0.67(5.9)$ years old (*Source:* U.S. National Center for Health Statistics, www.cdc.gov/nchs). Simplify these expressions and interpret.

20. The middle 60% of the families who had a child born during 1998 had family incomes anywhere from $29.44 - 0.84(20.33)$ to $29.44 + 0.84(20.33)$ (*Source:* U.S. Census Bureau, www.census.gov). Simplify these expressions and interpret.

Section A.3 Properties of Integer Exponents

Integer exponents are used as an efficient way to write repeated multiplication, just as multiplication is used as an efficient way to write repeated addition. Many of the properties of integer exponents can be derived directly from the way that we define the integer exponent.

From Your Toolbox

- By an **integer**, we mean the set of numbers comprised of the whole numbers and their opposites. The set of integers looks like $\{\ldots, -3, -2, -1, 0, 1, 2, 3, \ldots\}$.
- A **real number** is any number on the real number line.

Exponent

If b is a real number and n is a positive integer, then b^n is the product of n factors of b. That is

$$b^n = \underbrace{b \cdot b \cdot \cdots \cdot b}_{n \text{ factors of } b}$$

where n is the **exponent** and b is the **base**.

To see how we can use this definition to develop some shortcuts for simplifying exponents, let's consider a product such as $3^2 \cdot 3^4$. Using the definition of exponent, we see that

$$3^2 \cdot 3^4 = (3 \cdot 3) \cdot (3 \cdot 3 \cdot 3 \cdot 3) = 3 \cdot 3 \cdot 3 \cdot 3 \cdot 3 \cdot 3 = 3^6$$

Notice that if we simply add the exponents we get

$$3^2 \cdot 3^4 = 3^{2+4} = 3^6$$

This illustrates the **multiplication rule** for exponents. The other basic properties of integer exponents can be verified in a similar manner.

Basic Properties of Exponents

Let m and n be integers and let b represent any real number.

Rule	Meaning	Example
1. Multiplication Rule: $b^m \cdot b^n = b^{m+n}$	When multiplying numbers with the same base, add the exponents.	$5^4 \cdot 5^3 = 5^{4+3} = 5^7$
2. Division Rule: $\frac{b^m}{b^n} = b^{m-n}, \quad b \neq 0$	When dividing numbers with the same base, subtract the exponents.	$\frac{9^{11}}{9^5} = 9^{11-5} = 9^6$
3. Power-to-Power Rule $(b^m)^n = b^{m \cdot n}$	When an exponential expression is raised to a power, multiply the exponents.	$(6^2)^4 = 6^{2 \cdot 4} = 6^8$

Note: We can show that these properties, along with the other properties in this section, also apply to real number exponents. However, we shall restrict our discussion to only integer exponents.

Example 1 Simplifying Expressions with Integer Exponents

Use the basic properties of exponents to simplify the following and write the solution in the form b^n. Then evaluate the simplified expression if possible.

(a) $\frac{4^{10}}{4^8}$ **(b)** $(1 + a)^6(1 + a)^3$ **(c)** $(0.5^2)^3$

Solution

(a) Using the division rule, we get

$$\frac{4^{10}}{4^8} = 4^{10-8} = 4^2 = 16$$

(b) In this case the base is the expression $1 + a$. The multiplication rule gives us

$$(1+a)^6(1+a)^3 = (1+a)^{6+3} = (1+a)^9$$

(c) To simplify this expression, we use the power rule to get

$$(0.5^2)^3 = 0.5^{2\cdot 3} = 0.5^6 = 0.015625$$

■

Checkpoint 1 Now work Exercises 9 and 13 in the A.3 Exercises.

Now we examine a few special cases of exponents. Let's say that we have the fraction $\frac{3^4}{3^4}$. If we use the definition of exponent to rewrite the numerator and denominator and then cancel the common factors, we get

$$\frac{3^4}{3^4} = \frac{3\cdot 3\cdot 3\cdot 3}{3\cdot 3\cdot 3\cdot 3} = 1$$

However, using the Division Rule, this fraction can be simplified as $\frac{3^4}{3^4} = 3^{4-4} = 3^0$. So it appears that $3^0 = 1$. Since we can use any nonzero base and get a similar result, this illustrates that any nonzero real number raised to the zero power is equal to 1.

To see what other properties can be shown, let's examine the fraction $\frac{6^2}{6^5}$. If we factor and cancel the common factors, the reduced form of the fraction is

$$\frac{6^2}{6^5} = \frac{6\cdot 6}{6\cdot 6\cdot 6\cdot 6\cdot 6} = \frac{1}{6\cdot 6\cdot 6} = \frac{1}{6^3}$$

From the division rule, we know that $\frac{6^2}{6^5} = 6^{2-5} = 6^{-3}$. So it appears that $6^{-3} = \frac{1}{6^3}$.

To conclude these properties, let's now consider the expression $(4x)^3$. In this case, the exponent is 3 and the base is $4x$. If we use the definition of exponent to write the factored form, we get

$$(4x)^3 = 4x\cdot 4x\cdot 4x = (4\cdot 4\cdot 4)\cdot(x\cdot x\cdot x) = 4^3\cdot x^3 = 64x^3$$

This illustrates the property that if a product, like $4x$, is raised to a power we raise each factor to that power. A similar argument can be made to show that $\left(\frac{5}{x}\right)^3 = \frac{5^3}{x^3} = \frac{125}{x^3}$. Now let's gather this list of integer exponent properties.

■ Properties of Integer Exponents

Let m and n be integers and let a and b represent any real numbers.

1. $b^0 = 1, \quad b \neq 0$
2. $b^{-n} = \frac{1}{b^n}, \quad b \neq 0$
3. $(ab)^n = a^n b^n$
4. $\left(\frac{a}{b}\right)^n = \frac{a^n}{b^n}, \quad b \neq 0$

Example 2 Using the Properties of Integer Exponents

Simplify the following:

(a) 4^{-2} **(b)** $(5x^3)^2$ **(c)** $\left(\frac{2}{5}\right)^3$

Solution

(a) Here we have a base of 4 and the exponent is negative. Property 2 gives us

$$4^{-2} = \frac{1}{4^2} = \frac{1}{16}$$

(b) Here the base is $5x^3$, so we have to square 5 and square x^3 using Property 3.

$$(5x^3)^2 = 5^2(x^3)^2 = 25x^6$$

Observe how we used the Power-to-Power Rule to simplify $(x^3)^2$.

(c) To simplify this expression, we must take the numerator and the denominator to the third power by using Property 4 to get

$$\left(\frac{2}{5}\right)^3 = \frac{2^3}{5^3} = \frac{8}{125}$$

■

Many of the applications that we encounter, particularly when studying the mathematics of finance in Chapter 5, make use of integer exponents. We also see these exponents used in the study of probability.

Example 3 Applying Integer Exponents

Shawna buys a big-screen digital television and wants to pay back the \$4600 price in 18 months. The electronics store is running a financing promotion and determines that the cost of each monthly payment is given by the expression $4600\frac{0.015}{1-(1.015)^{-18}}$. Evaluate this expression and interpret.

Solution

Using the properties of integer exponents, we can rewrite the denominator and simplify.

$$4600\frac{0.015}{1-(1.015)^{-18}} = 4600\frac{0.015}{1-\frac{1}{1.015^{18}}} \approx 4600 \cdot \frac{0.015}{0.235} \approx 293.62$$

This means that Shawna will pay \$293.62 each month for 18 months to pay off the television. ■

SECTION A.3 EXERCISES

In Exercises 1–14, simplify using the basic properties of exponents and write the solution in the form b^n. Then evaluate the simplified expression, if possible.

1. $3 \cdot 3^5$

2. $5 \cdot 5^4$

3. $2^{10} \cdot 2^5$

4. $9^8 \cdot 9^2$

5. $\frac{10^8}{10^2}$

6. $\frac{7^7}{7^5}$

7. $\frac{3^8}{3^7}$

8. $\frac{8^{12}}{8^{11}}$

✓ **9.** $(6^2)^3$

10. $(8^3)^4$

11. $(-3^3)^5$

12. $(-2)^7$

✓ **13.** $(x+3)^4(x+3)^2$

14. $(5-x)^5(5-x)$

In Exercises 15–30, simplify using the properties of integer exponents.

15. 5^0

16. 3^0

17. $(-5)^0$

18. $(-3)^0$

19. 6^{-2}

20. 3^{-3}

21. $\left(\frac{1}{6}\right)^{-2}$

22. $\left(\frac{1}{3}\right)^{-3}$

23. $(2x^3)^4$

24. $(4x^5)^3$

25. $(3x^5)^{-1}$

26. $(8x^2)^{-1}$

27. $\left(\frac{3}{4}\right)^2$

28. $\left(\frac{2}{3}\right)^3$

29. $\left(\frac{1}{8}\right)^{-2}$

30. $\left(\frac{1}{10}\right)^{-3}$

31. During 1995, statistics show that if 20 women who just had babies were selected at random the chance that 2 of them were African American is given by the expression $190(0.16)^2(0.84)^{20-2}$ (*Source:* U.S. Census Bureau, www.census.gov). Evaluate this expression, convert to a percent, and interpret.

32. U.S. Coast Guard statistics show that if the records of 30 oil spills from 1994 are chosen at random the chance that 27 of them were spills of less than 100 gallons is given by the expression $4060(0.94)^{27}(0.06)^{30-27}$ (*Source:* U.S. Coast Guard, www.uscg.mil). Evaluate this expression, convert to a percent, and interpret.

33. Candy is considering the purchase of a new car from West Side Auto Sales. She chooses a car costing \$22,000 and is told by the salesperson that, if monthly payments are made for 5 years, the cost of each payment is given by the expression $22{,}000\frac{0.065}{1-(1.065)^{-60}}$. Evaluate this expression and interpret.

34. The Punderson family is shopping for a new dining room table and hutch. They find that the table and hutch that they want costs \$9600. They decide to finance the purchase and learn from their banker that, in order to pay off the purchase in 2 years, each monthly payment will be $9600\frac{0.08}{1-(1.08)^{-24}}$. Evaluate this expression and interpret.

Section A.4 Rational Exponents and Radicals

Fractions that have integers in the numerator and denominator are called **rational numbers**. Let's see what happens if an expression contains an exponent that is a rational number. Using the multiplication rule for exponents, we know that, for example,

$$3^{\frac{1}{2}} \cdot 3^{\frac{1}{2}} = 3^{\frac{1}{2}+\frac{1}{2}} = 3^1 = 3$$

But we also know that $\sqrt{3} \cdot \sqrt{3} = \sqrt{9} = 3$. So it appears that we can say that $3^{\frac{1}{2}} = \sqrt{3}$. This illustrates that we can take an expression written in radical form and write it as a rational exponent, and vice versa.

Rational Exponent

Let m and n be integers, $n \neq 0$, and let b represent any real number. Then $\frac{m}{n}$ is a **rational exponent** and we can write

$$b^{\frac{m}{n}} = \sqrt[n]{b^m} = \left(\sqrt[n]{b}\right)^m$$

Example 1 Writing Radical Expression in Exponential Form

Write each of the following in the form $b^{\frac{m}{n}}$.

(a) $\sqrt[3]{x^2}$

(b) $\frac{1}{\sqrt{x}}$

Solution

(a) Using the definition of rational exponent with $n = 3$ and $m = 2$, we get

$$\sqrt[3]{x^2} = x^{\frac{2}{3}}$$

(b) Here we see that, since the radical expression is in the denominator, the exponent will be negative.

$$\frac{1}{\sqrt{x}} = \frac{1}{x^{\frac{1}{2}}} = x^{-\frac{1}{2}}$$ ■

Checkpoint 1 Now work Exercise 9 in the A.4 Exercises.

Example 2 Writing Expressions in Radical Form

Write each of the following expressions in radical form.

(a) $3x^{\frac{3}{2}}$ **(b)** $x^{-\frac{1}{4}}$

Interactive Activity

Write the expression $(3x)^{\frac{3}{2}}$ in radical form. How does this answer differ from that in part (a) of Example 2?

Solution

(a) Here the exponent is $\frac{3}{2}$ and the base is x, so we get

$$3x^{\frac{3}{2}} = 3\sqrt{x^3}$$

(b) We first make the exponent positive to get

$$x^{-\frac{1}{4}} = \frac{1}{x^{\frac{1}{4}}} = \frac{1}{\sqrt[4]{x}}$$ ■

SECTION A.4 EXERCISES

In Exercises 1–10, write each of the following in the form $b^{\frac{m}{n}}$.

1. $\sqrt{x}$
2. $\sqrt[3]{x}$
3. $\sqrt[7]{x^4}$
4. $\sqrt[5]{x^3}$
5. $\sqrt[4]{5x^2}$
6. $\sqrt[6]{10x^5}$
7. $\frac{1}{\sqrt[5]{x}}$
8. $\frac{1}{\sqrt[8]{x}}$
9. ✓ $\frac{8}{\sqrt[7]{x^2}}$
10. $\frac{11}{\sqrt[5]{x^4}}$

In Exercises 11–20, write each of the following expressions in radical form.

11. $x^{\frac{1}{4}}$
12. $x^{\frac{1}{6}}$
13. $x^{\frac{5}{8}}$
14. $x^{\frac{3}{4}}$
15. $(6x^2)^{\frac{1}{4}}$
16. $(4x^4)^{\frac{1}{2}}$
17. $x^{-\frac{1}{3}}$
18. $x^{-\frac{1}{10}}$
19. $5x^{-\frac{2}{3}}$
20. $9x^{-\frac{5}{7}}$

Appendix B
Algebra Review

Section B.1 Multiplying Polynomial Expressions

A **polynomial** in one variable is an algebraic expression that contains terms of the form $a_n x^n$ that are combined using addition and subtraction. The first part of the term, a_n, is called the **coefficient** and can be any real number. The **variable part** of the term is x^n, and the number n is the **degree** of the term and can be any whole number. We name some polynomials by the number of terms that they have. See Table B.1.1.

Table B.1.1

Number of terms	Name	Example
1	Monomial	$4x^2$
2	Binomial	$x^2 - 7x$
3	Trinomial	$3x^3 + 2x - 7$
4	Four-term polynomial	$7x^3 - 0.2x^2 + 6x - 4$
$\vdots$	$\vdots$	$\vdots$

From Your Toolbox

The **Multiplication Rule** for integer exponents tells us that when we are multiplying and the bases are the same we add the exponents. That is, $b^m \cdot b^n = b^{m+n}$.

To see how polynomials are multiplied together, let's consider several cases.

Case 1: Multiplying a Monomial by a Monomial

To Multiply: Use the Multiplication Rule for integer exponents to multiply the two variable parts (consult the Toolbox to the left) and then multiply the coefficients.

Example 1 Multiplying a Monomial by a Monomial

Multiply the following monomials.

(a) $4x^3 \cdot 12x$ **(b)** $(-3x^2)(x^4)$ **(c)** $-6(5x^4)$

Solution

(a) Here the respective coefficients are 4 and 12, and the variable parts are x^3 and x. So we get

$$4x^3 \cdot 12x = (4 \cdot 12)(x^3 \cdot x) = 48x^{3+1} = 48x^4$$

(b) We can write the term x^4 as $1x^4$ to get the product

$$(-3x^2)(x^4) = (-3 \cdot 1)(x^2 \cdot x^4) = -3x^6$$

(c) Since we know that $x^0 = 1$, we can write the first terms as $-6x^0$ to give us

$$-6(5x^4) = (-6 \cdot 5)(x^0 \cdot x^4) = -30x^4$$

■

✓ Checkpoint 1

Now work Exercise 3 in the B.1 Exercises.

Case 2: Multiplying a Monomial by a Polynomial

To Multiply: Use the Distributive Property.

■ **Distributive Property**

1. $a(b + c) = ab + ac$
2. $a(b - c) = ab - ac$
3. $a(b + c + d + \cdots + n) = ab + ac + ad + \cdots + an$

The distributive property means that we can multiply the monomial by each term of the polynomial in the parentheses. After multiplying, we can sometimes combine the **like terms**. Two terms are like terms if their variable parts are the same and they have the same degree. To combine like terms, we combine their coefficients and write the common variable part.

Example 2 Multiplying a Monomial by a Polynomial

Multiply the following and simplify if possible.

(a) $6x(3x^2 - 7x + 4)$ **(b)** $2x(x + 3) + (x^2 + 7x)$

Solution

(a) Using the distributive property, we get

$$6x(3x^2 - 7x + 4) = 6x(3x^2) - 6x(7x) + 6x(4)$$
$$= 18x^3 - 42x^2 + 24x$$

(b) We begin by distributing $2x$ over the binomial $x + 3$.

$$2x(x + 3) + (x^2 + 7) = 2x(x) + 2x(3) + (x^2 + 7x)$$
$$= 2x^2 + 6x + (x^2 + 7x)$$

Now let's remove the parentheses and combine the like terms.

$$= 2x^2 + 6x + x^2 + 7x$$
$$= 3x^2 + 13x$$

■

Case 3: Multiplying a Binomial by a Binomial

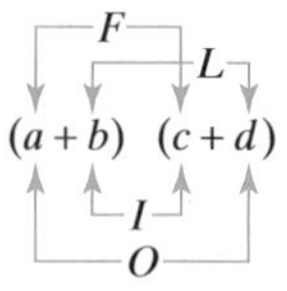

Figure B.1.1

To Multiply: Use the FOIL Method.

The FOIL method reminds us that when we multiply two binomials together we multiply the **F**irst two terms, the **O**utside terms, the **I**nside terms, and the two **L**ast terms. See Figure B.1.1.

Example 3 Multiplying Two Binomials

Multiply and combine like terms.

(a) $(x - 3)(2x + 1)$ **(b)** $(0.3x^2 + 5)(2x^2 + 6)$

Solution

(a) Let's begin by identifying the terms used for the FOIL method.

$$\begin{aligned}(x-3)(2x+1) &= \overset{\textbf{F}}{x(2x)} + \overset{\textbf{O}}{x(1)} - \overset{\textbf{I}}{3(2x)} - \overset{\textbf{L}}{3(1)} \\ &= 2x^2 + x - 6x - 3 \\ &= 2x^2 - 5x - 3\end{aligned}$$

(b) Using the FOIL method, we get

$$\begin{aligned}(0.3x^2+5)(2x^2+6) &= (0.3x^2)(2x^2) + (0.3x^2)(6) + (5)(2x^2) + (5)(6) \\ &= 0.6x^4 + 1.8x^2 + 10x^2 + 30 \\ &= 0.6x^4 + 11.8x^2 + 30\end{aligned}$$

■

Checkpoint 2 Now work Exercise 15 in the B.1 Exercises.

Case 4: Multiplying a Polynomial by Another Polynomial

To Multiply: Use the general distributive property.

General Distributive Property

If p is a polynomial expression, then

1. $(b + c)p = bp + cp$
2. $(b + c + d + \cdots + n)p = bp + cp + dp + \cdots + np$

The general distributive property means that we can distribute a polynomial over each term of another polynomial. Let's see how this property is used to multiply two polynomials together.

Example 4 Multiplying Two Polynomials

Multiply and combine like terms.

(a) $(3x - 2)(x^2 - 6x + 4)$ **(b)** $(2x^2 - x + 6)(x^2 + 3x - 1)$

Solution

(a) Using the general distributive property, we start by distributing the terms $3x$ and -2 over the trinomial $x^2 - 6x + 4$. Then we combine the like terms.

$$\begin{aligned}(3x-2)(x^2-6x+4) &= 3x(x^2-6x+4) - 2(x^2-6x+4) \\ &= 3x^3 - 18x^2 + 12x - 2x^2 + 12x - 8 \\ &= 3x^3 - 20x^2 + 24x - 8\end{aligned}$$

(b) When we multiply these two trinomials, we distribute the terms $2x$, $-x$, and 6 over the trinomial $x^2 + 3x - 1$ to get

$$\begin{aligned}(2x^2 - x + 6)(x^2 + 3x - 1) &= 2x^2(x^2 + 3x - 1) - x(x^2 + 3x - 1) + 6(x^2 + 3x - 1)\\ &= 2x^4 + 6x^3 - 2x^2 - x^3 - 3x^2 + x + 6x^2 + 18x - 6\\ &= 2x^4 + 5x^3 + x^2 + 19x - 6\end{aligned}$$

■

Interactive Activity

Polynomials can also be multiplied using the vertical method. The solution to part (a) of Example 4 using the vertical method is

$$\begin{array}{rrrr} & x^2 & -\ 6x & +\ 4\\ & & 3x & -\ 2\\ \hline & -\ 2x^2 & +\ 12x & -\ 8\\ 3x^3 & -\ 18x^2 & +\ 12x & \\ \hline 3x^3 & -\ 20x^2 & +\ 24x & -\ 8\end{array}$$

Verify the solution to part (b) of Example 4 using this method.

SECTION B.1 EXERCISES

In Exercises 1–6, multiply the following monomials.

1. $5x^3 \cdot 2x^2$
2. $7x^3 \cdot 9x^7$
3. $(-8x^9)(x^3)$
4. $(-2x^2)(x^2)$
5. $-10(15x^4)$
6. $-9(6x^3)$

In Exercises 7–14, multiply the following:

7. $2x(x - 5)$
8. $3x(x + 1)$
9. $-4x^2(x^2 - 3x)$
10. $-9x^3(x^2 - 2x)$
11. $(4x^5 - 6x)(-3x)$
12. $(2x^3 - 7x)(-6x)$
13. $-2x^2(x^2 - 7x + 5)$
14. $-5x^4(2x^2 + 9x - 1)$

In Exercises 15–22, multiply and combine like terms.

15. $(x + 3)(x + 5)$
16. $(x + 4)(x + 7)$
17. $(1.2x - 5)(x - 4)$
18. $(1.4x - 1)(x - 3)$
19. $(3x + 4)(3x - 8)$
20. $(8x - 1)(2x + 6)$
21. $(4x^2 + 5)(2x^2 + 3)$
22. $(6x^2 + 3)(3x^2 + 2)$

In Exercises 23–32, multiply and combine like terms.

23. $(x + 1)(x^2 + 2x + 3)$
24. $(x + 2)(x^2 + 2x + 5)$
25. $(0.2x - 3)(3x^2 - 7x + 1)$
26. $(0.3x - 2)(2x^2 + x - 3)$
27. $(x^2 + x - 2)(4x^2 + 3)$
28. $(x^2 + 2x - 1)(5x^2 + 2)$
29. $(x^2 + 3x - 3)(x^2 + 9x + 2)$
30. $(x^2 - 5x + 2)(x^2 + 2x + 8)$
31. $(2x^3 - 2x^2 + 3)(x^2 + 5x + 2)$
32. $(x^3 + 5x^2 - 6)(x^2 + 9x + 8)$

Section B.2 Factoring Trinomials

Factoring polynomials can be thought of as the opposite of multiplying them. Factoring can be a useful tool for solving equations. When we say factoring, we are referring to factoring polynomials so that their factors contain integer coefficients and constants. Several techniques can be used in factoring, depending on the characteristics of a given polynomial. Let's consider each case by case.

Case 1: Factoring Out the Greatest Common Monomial

To Factor: Use the reverse of the distributive property. That is, we use

$$\overset{\text{Factor out } a}{ab + ac = a(b + c)}$$

or, in general,

$$ab + ac + ad + \cdots + an = a(b + c + d + \cdots + n)$$

The **greatest common monomial** is made up of a coefficient part and a variable part. The coefficient part is the greatest common factor of all the coefficients. The variable part contains the smallest power of the variable that appears in every term of the polynomial.

Example 1 Factoring Out the Greatest Common Monomial

Factor the following:

(a) $36x^3 + 24x^2$ **(b)** $26x^5 + 13x^3 - 39x^2$

Solution

(a) The coefficients are 36 and 24, so the greatest common factor of these numbers is 12. The smallest power of x in each term is 2, so the greatest common monomial is $12x^2$.

$$36x^3 + 24x^2 = 12x^2(3x + 2)$$

(b) For this trinomial, the greatest common monomial is $13x^2$. Factoring this monomial from each term yields

$$26x^5 + 13x^3 - 39x^2 = 13x^2(2x^3 + x - 3)$$

■

Case 2: Difference of Squares

To Factor: Use the formula $a^2 - b^2 = (a + b)(a - b)$

We call the binomial $a^2 - b^2$ a difference of squares because each term is a **perfect square** (such as 4, 9, 16, 25 or $x^2, x^4, x^6, \ldots$), and, by subtracting, we are taking their **difference**. To verify the difference of squares formula, let's use the FOIL method to multiply $(a + b)(a - b)$.

$$(a + b)(a - b) = a \cdot a - ab + ab - b \cdot b = a^2 - b^2$$

Notice that the two middle terms cancel one another.

Example 2 Factoring a Difference of Squares

Factor the following completely.

(a) $x^2 - 16$ **(b)** $x^4 - 81$

Solution

(a) We see that both terms are perfect squares since they can be written in the form $(x)^2$ and $(4)^2$, respectively. So we get

$$x^2 - 16 = (x)^2 - (4)^2 = (x + 4)(x - 4)$$

(b) Knowing that $x^4 = (x^2)^2$ and $81 = (9)^2$, we get

$$x^4 - 81 = (x^2)^2 - (9)^2 = (x^2 + 9)(x^2 - 9)$$

But now notice that the binomial $x^2 - 9$ is also a difference of squares, so we can factor this binomial to give us the completely factored form

$$x^4 - 81 = (x^2 + 9)(x^2 - 9) = (x^2 + 9)(x + 3)(x - 3)$$ ■

Case 3: Factoring by Grouping

To Factor: Using the general distributive property, factor $ap + bp = (a + b)p$, where p is a polynomial.

Factoring by grouping is used when a polynomial expression has four terms. We first factor the greatest common monomial from the two first terms and then repeat the process for the next two terms. Finally, we factor using the general distributive property.

Example 3 Factoring by Grouping

Factor the following completely.

(a) $8x^3 - 12x^2 + 6x - 9$ **(b)** $x^3 + 2x^2 - x - 2$

Solution

> **Interactive Activity**
>
> Verify the answer to part (a) of Example 3 by multiplying $(4x^2 + 3)(2x - 3)$ using the FOIL method.

(a) We can factor $4x^2$ from the two first terms and factor 3 from the last two terms. This gives us

$$8x^3 - 12x^2 + 6x - 9 = 4x^2(2x - 3) + 3(2x - 3)$$

Using the general distributive property, we get

$$\underset{a}{4x^2}\underset{p}{(2x - 3)} + \underset{b}{3}\underset{p}{(2x - 3)} = \underset{(a+b)}{(4x^2 + 3)}\underset{p}{(2x - 3)}$$

(b) For this polynomial, we begin by factoring x^2 from the two first terms.

$$x^3 + 2x^2 - x - 2 = x^2(x + 2) - x - 2$$

To complete the factoring by grouping process, we factor -1 from the last two terms.

$$x^2(x + 2) - x - 2 = x^2(x + 2) - 1(x + 2) = (x^2 - 1)(x + 2)$$

Now we see that the first binomial $x^2 - 1$ is a difference of squares. So the complete factored form of the polynomial is

$$x^3 + 2x^2 - x - 2 = (x^2 - 1)(x + 2) = (x + 1)(x - 1)(x + 2)$$ ■

The final two cases involve the techniques that are used to factor **quadratic trinomials**. There are two cases that depend on whether the leading coefficient is 1 or the leading coefficient is some integer other than 1.

Case 4: Factoring Trinomials of the Form $x^2 + bx + c$

To Factor: Factor the expression $x^2 + bx + c$ into $(x + n_1)(x + n_2)$ under the conditions that $n_1 \cdot n_2 = c$ and $n_1 + n_2 = b$.

To see how this factoring method works, let's multiply $(x+n_1)(x+n_2)$ using the FOIL method.

$$(x+n_1)(x+n_2) = x^2 + n_2x + n_1x + n_1 \cdot n_2$$

If we factor x from the two middle terms, we get

$$x^2 + \underbrace{(n_1+n_2)}_{b}x + \underbrace{n_1 \cdot n_2}_{c} = x^2 + bx + c$$

Example 4 Factoring a Trinomial of the Form $x^2 + bx + c$

Factor the following completely.

(a) $x^2 + 3x - 10$ **(b)** $x^4 - 13x^2 - 48$

Solution

(a) For this trinomial, we need two numbers, n_1 and n_2, so that $n_1 \cdot n_2 = -10$, while $n_1 + n_2 = 3$. That is, we need two numbers such that when they are multiplied together we get -10, and when we add the two numbers, we get 3. Notice that if we let $n_1 = 5$ and $n_2 = -2$ we get $n_1 \cdot n_2 = (5)(-2) = -10$ and $n_1 + n_2 = 5 + (-2) = 3$. This means that we can write the factored form of the polynomial as

$$\mathbf{x^2 + bx + c = (x + n_1)(x + n_2)}$$
$$x^2 + 3x - 10 = (x+5)\ (x-2)$$

This solution can be checked using the FOIL method.

(b) Even though this is a fourth-degree polynomial, we can rewrite it in quadratic form by letting $x^2 = y$. This gives us

$$x^4 - 13x^2 - 48 = y^2 - 13y - 48$$

Now we need to two numbers n_1 and n_2 such that their product is $n_1 \cdot n_2 = -48$ and their sum is $n_1 + n_2 = -13$. We can find these values by listing the factors of -48. See Table B.2.1.

Table B.2.1

Factors of −48	Sum of Factors	
−48(1)	$-48+1=-47$	
−24(2)	$-24+2=-22$	
−16(3)	$-16+3=-13$	Desired values
−12(4)	$-12+4=-8$	
−8(6)	$-8+6=-2$	
−6(8)	$-6+8=2$	
−4(12)	$-4+12=8$	
−3(16)	$-3+16=13$	
−2(24)	$-2+24=22$	
−1(48)	$-1+48=47$	

Now we see that $n_1 = -16$ and $n_2 = 3$. This gives us the factored form

$$y^2 - 13y - 48 = (y-16)(y+3)$$

Rewriting the factored form in terms of x (recall that we let $x^2 = y$) gives us

$$x^4 - 13x^2 - 48 = (x^2 - 16)(x^2 + 3)$$

But notice that the binomial $x^2 - 16$ is a difference of squares. Factoring this binomial, we get

$$x^4 - 13x^2 - 48 = (x^2 - 16)(x^2 + 3) = (x + 4)(x - 4)(x^2 + 3)$$

The expression is now factored completely. ■

Checkpoint 1

Now work Exercise 31 in the B.2 Exercises.

Case 5: Factoring a Trinomial of the Form $ax^2 + bx + c$, $a \neq 1$

To Factor: One can use the following process:

1. Determine numbers n_1 and n_2 so that their product is $n_1 \cdot n_2 = a \cdot c$ and their sum is $n_1 + n_2 = b$.
2. Rewrite the trinomial as a four-term polynomial of the form

$$ax^2 + n_1x + n_2x + c$$

where $n_1x + n_2x = bx$.

3. Factor by grouping.

Let's see how this process works in the next example.

Example 5 Factoring a Trinomial of the Form $ax^2 + bx + c$, $a \neq 1$

Factor the following completely.

(a) $2x^2 - 5x - 3$ **(b)** $6x^2 + 11x - 10$

Solution

Interactive Activity

In part (b) of Example 5, suppose that we let the two numbers be $n_1 = -4$ and $n_2 = 15$. Will we get the same answer? Try it and see.

WWW

(a) For this part of the example, we will outline the three steps.

1. We need two numbers such that $n_1 \cdot n_2 = a \cdot c = (2)(-3) = -6$ and $n_1 + n_2 = b = -5$. We can see that both of these conditions are satisfied when $n_1 = -6$ and $n_2 = 1$.
2. Rewriting the trinomial as a four-term polynomial using the number that we found in step 1 gives us

$$2x^2 - 5x - 3 = 2x^2 - 6x + x - 3$$

3. Using the factoring by grouping technique, we begin by factoring $2x$ from the two first terms and 1 from the last two terms.

$$\begin{aligned} 2x^2 - 6x + x - 3 &= 2x(x - 3) + 1(x - 3) \\ &= (2x + 1)(x - 3) \end{aligned}$$

(b) For this trinomial we need $n_1 \cdot n_2 = (6)(-10) = -60$ and $n_1 + n_2 = 11$. We can check to see that the two numbers are $n_1 = 15$ and $n_2 = -4$. Using these values, we rewrite the trinomial and factor by grouping to give us

$$\begin{aligned} 6x^2 + 11x - 10 &= 6x^2 + 15x - 4x - 10 \\ &= 3x(2x + 5) - 2(2x + 5) \\ &= (3x - 2)(2x + 5) \end{aligned}$$

■

We have this final comment about factoring. Many students opt to use a trial and error method to factor trinomials, once any common factors have been factored. This method is perfectly fine and will yield the same completely factored form.

SECTION B.2 EXERCISES

In Exercises 1–8, factor out the greatest common monomial.

1. $2x - 2$ **2.** $5x + 5$

3. $35x - 10$ **4.** $28x - 7$

5. $12x^2 - 26x^3$ **6.** $2x^3 + 3x^2$

7. $21x^3 + 4x^2 - 14$ **8.** $15x^3 + 3x - 5$

In Exercises 9–18, factor the difference of squares completely.

9. $x^2 - 81$ **10.** $x^2 - 49$

11. $9x^2 - 1$ **12.** $4x^2 - 25$

13. $(2x - 1)^2 - 4$ **14.** $(x + 1)^2 - 100$

15. $x^4 - 1$ **16.** $x^4 - 16$

17. $16x^4 - 81$ **18.** $256x^4 - 1$

In Exercises 19–26, factor by grouping.

19. $x^2 + 4x + x + 4$

20. $x^2 + 5 + x + 5$

21. $2x^3 + x^2 + 6x + 1$

22. $3x^3 + 5x^2 + 3x + 5$

23. $2x^3 - 2x^2 + 4x - 4$

24. $9x^3 + 18x^2 - 5x - 10$

25. $6x^3 - 24x - 3x^2 + 12$

26. $5x^3 - 45x + x^2 - 9$

In Exercises 27–36, factor completely. In Exercises 35 and 36, first factor out the greatest common monomial.

27. $x^2 + 8x + 15$ **28.** $x^2 + x + 9$

29. $x^2 + 2x + 1$ **30.** $x^2 + 12x + 36$

✓ **31.** $x^2 - 4x - 5$ **32.** $x^2 - 5x - 6$

33. $x^2 - 9x + 81$ **34.** $x^2 - 10x - 200$

35. $2x^2 + 10x - 48$ **36.** $4x^2 - 4x - 8$

In Exercises 37–46, factor completely using the three-step process or trial and error.

37. $2x^2 + 3x + 1$ **38.** $3x^2 + 7x + 2$

39. $3x^2 - 8x + 4$ **40.** $2x^2 - 9x + 4$

41. $7x^2 - 12x + 5$ **42.** $5x^2 + 36x + 7$

43. $7x^2 + 4x - 11$ **44.** $5x^2 - 54x - 11$

45. $15x^2 + 10x - 40$ **46.** $3x^2 - 21x + 36$

Section B.3 Solving Linear Equations

A **linear equation** is an equation that can be written in the form $ax + b = c$, where a is the real number coefficient of the variable x ($a \neq 0$), and b and c represent any real numbers. Once a linear equation is in this form, we can solve for x using the following steps:

$$ax + b = c \qquad \text{Given equation}$$

$$ax = b - c \qquad \text{Subtract } c \text{ from both sides}$$

$$x = \frac{b - c}{a} \qquad \text{Divide both sides by } a$$

Example 1 Solving a Linear Equation

Solve the following equations for x.

(a) $3x - 4 = 6$ **(b)** $0.25x + 1 = 3$

Solution

(a) Using the steps for solving a linear equation, we get

$$3x - 4 = 6 \qquad \text{Given equation}$$
$$3x = 6 + 4 \qquad \text{Add 4 to both sides.}$$
$$3x = 10 \qquad \text{Simplify.}$$
$$x = \frac{10}{3} \qquad \text{Divide both sides by 3.}$$

(b) Even though the coefficient of x is a decimal, we solve the equation in the same manner.

$$0.25x + 1 = 3$$
$$0.25x = 3 - 1$$
$$0.25x = 2$$
$$x = \frac{2}{0.25}$$
$$x = 8$$

■

There are times when a linear equation must be rewritten or simplified so that it is in the form $ax + b = c$. This could involve simplifying each side of the equation, collecting all the variable terms on one side of the equation, or the use of the distributive property. Let's see how we can solve these kinds of linear equations.

Example 2 Simplifying and Solving a Linear Equation

Simplify and solve the following equations for x.

(a) $5x - 2(3x + 12) = 10$ **(b)** $6x - (5x + 1) = 2(x + 3)$

Solution

(a) We begin by distributing -2 over the binomial $3x + 12$.

$$5x - 2(3x + 12) = 10 \qquad \text{Given equation}$$
$$5x - 6x - 24 = 10 \qquad \text{Distribute } -2.$$
$$-x - 24 = 10 \qquad \text{Combine like terms.}$$
$$-x = 34 \qquad \text{Add 24 to both sides.}$$
$$x = -34 \qquad \text{Divide both sides by } -1.$$

(b) In this case the variable x appears on both sides of the equation.

$$6x - (5x + 1) = 2(x + 3) \qquad \text{Given equation}$$
$$6x - 5x - 1 = 2x + 6 \qquad \text{Distribute } -1 \text{ (left side) and 2 (right side).}$$
$$x - 1 = 2x + 6 \qquad \text{Combine like terms.}$$
$$-x - 1 = 6 \qquad \text{Subtract } 2x \text{ from both sides.}$$
$$-x = 7 \qquad \text{Add 1 to both sides.}$$
$$x = -7 \qquad \text{Divide both sides by } -1.$$

■

Checkpoint 1 Now work Exercise 11 in the B.3 Exercises.

When studying applications of linear equations, we often solve equations that are based on **linear models**. A linear model is a mathematical expression derived from real-world data. When solving equations that involve linear models, we write a sentence that **interprets** the mathematical solution. Let's examine one of these applications.

Example 3 Solving an Equation Based on a Linear Model

The year that the number of violent crimes in the United States was about 7.7 million can be determined by solving the equation

$$-0.78x + 13.94 = 7.7$$

where x represents the number of years since 1990 and the number of violent crimes is expressed in millions (*Source:* U.S. Bureau of Justice Statistics, www.ojp.usdoj.gov/bjs). Solve for x and interpret.

Solution

Solving this equation for x gives us

$$\begin{aligned} -0.78x + 13.94 &= 7.7 \\ -0.78x &= 7.7 - 13.94 \\ -0.78x &= -6.24 \\ x &= \frac{-6.24}{-0.78} \\ x &= 8 \end{aligned}$$

Since x represents the number of years since 1990, we see that $x = 8$ corresponds to the year 1998.

Interpret the Solution: This means that in 1998 about 7.7 million violent crimes were reported in the United States. ■

SECTION B.3 EXERCISES

In Exercises 1–10, solve the equation for x.

1. $x + 6 = 8$ **2.** $x - 7 = 3$

3. $2x = 6$ **4.** $3x = 15$

5. $\frac{1}{4}x = 7$ **6.** $\frac{1}{6}x = 8$

7. $3x + 1 = 3$ **8.** $8x - 2 = 13$

9. $0.2x + 1.3 = 0.8$ **10.** $0.7x - 1.2 = 0.9$

In Exercises 11–20, simplify and solve the following:

✓ **11.** $3x - 11 = 7 - 2x$

12. $x + 18 = 6x - 3$

13. $2(2x + 1) = 5 + 3x$

14. $3(x - 4) = y + 6$

15. $4(x + 1) = -2(4 - x)$

16. $5(x + 4) = 2(3 - x)$

17. $3(x - 5) + 10 = 2(x + 4)$

18. $2(5x + 2) + 1 = 3(3x - 2)$

19. $0.22(4x - 3) = 1.37$

20. $0.85(5x + 10) = 2.55$

21. The year that the annual expenditures on pollution abatement in the United States totaled 60.17 billion dollars can be determined by solving the equation

$$2x + 48.17 = 60.17$$

where x represents the number of years since 1972 (*Source:* U.S. Census Bureau, www.census.gov). Solve the equation and interpret.

22. The year that United States industries spent 3 billion dollars on research and development can be determined by solving the equation

$$0.19x + 1.67 = 3$$

where x represents the number of years since 1972 (*Source:* U.S. Census Bureau, www.census.gov). Solve the equation and interpret.

23. The year that worldwide military expenditures were 887.92 billion dollars can be determined by solving the equation

$$-50.77x + 1091 = 887.92$$

where x represents the number of year since 1990 (*Source:* U.S. Census Bureau, www.census.gov). Solve the equation and interpret.

24. The year that the total expenditures for colleges and universities in the United States was 221.7 billion dollars can be determined by solving the equation

$$5.05x + 191.4 = 221.7$$

where x represents the number of years since 1990 (*Source:* U.S. National Center for Education Statistics, www.nces.ed.gov). Solve the equation and interpret.

Section B.4 Solving Quadratic Equations

So far we have examined solutions to linear equations. Now we turn our attention to **quadratic equations**. These are equations that include a second-degree polynomial and have the form

$$ax^2 + bx + c = 0, \qquad \text{where } a \neq 0$$

We call this form the **standard form** of a quadratic equation. Quadratic equations are useful in such applications as physical motion problems, geometry problems, and mathematical modeling. We will study two methods for solving quadratic equations. These are solving by factoring and solving by the quadratic formula.

Solving Quadratic Equations by Factoring

If a trinomial in the form $ax^2 + bx + c$ is factorable, then its corresponding standard form equation $ax^2 + bx + c = 0$ can be solved by using the **zero product property**.

Zero Product Property

Let a and b represent any two real numbers. If the product

$$a \cdot b = 0$$

then either $a = 0$, $b = 0$, or both equal zero.

Note: The zero product property tells us that if a product is equal to zero then at least one of its factors is equal to zero.

Let's say that we have an equation such as $(3x - 1)(x - 7) = 0$. We can use the zero product property to set each linear factor equal to zero and solve.

$$
\begin{aligned}
(3x-1)(x-7) &= 0 && \text{Given equation}\\
3x-1=0 \quad \text{or} \quad x-7 &= 0 && \text{By the zero product property}\\
3x=1 \quad \text{or} \quad x &= 7 && \text{Solving each equation for } x\\
x=\frac{1}{3} \quad \text{or} \quad x &= 7 && \text{The solution to the equation}
\end{aligned}
$$

So the solutions to the equation $(3x-1)(x-7)=0$ are $x=\frac{1}{3}$ and $x=7$. If a quadratic equation is factorable over the real numbers, then the solution consists of either two real number solutions or one repeated number solution.

Solving a Quadratic Equation by Factoring

1. If needed, write the quadratic equation in the standard form $ax^2+bx+c=0$ by setting one side equal to zero.
2. Factor the polynomial completely.
3. Use the zero product property to set each linear factor equal to zero.
4. Solve the linear equations found in step 3.

Example 1 Solving a Quadratic Equation by Factoring

Solve the following equations by factoring.

(a) $x^2+x=6$ **(b)** $2x^2-7x-4=0$

Solution

(a) We begin by writing the equation in standard form.

$$
\begin{aligned}
x^2+x &= 6\\
x^2+x-6 &= 0
\end{aligned}
$$

Next we factor the trinomial on the left side of the equation.

$$
\begin{aligned}
x^2+x-6 &= 0 && \text{Equation in standard form}\\
(x+3)(x-2) &= 0 && \text{Factor the polynomial completely.}\\
x+3=0 \quad \text{or} \quad x-2 &= 0 && \text{Use the zero product property.}\\
x=-3 \quad \text{or} \quad x &= 2 && \text{Solve the linear factors for } x.
\end{aligned}
$$

So the solutions are $x=-3$ and $x=2$.

(b) We can see that there is no greatest common monomial to factor out. So we factor and solve as in part (a).

$$
\begin{aligned}
2x^2-7x-4 &= 0 && \text{Given equation}\\
(2x+1)(x-4) &= 0 && \text{Factor by trial and error.}\\
2x+1=0 \quad \text{or} \quad x-4 &= 0 && \text{Use the zero product property.}\\
x=-\frac{1}{2} \quad \text{or} \quad x &= 4 && \text{Solve the linear factors for } x.
\end{aligned}
$$

So the solutions are $x=-\frac{1}{2}$ and $x=4$. ■

Technology Option

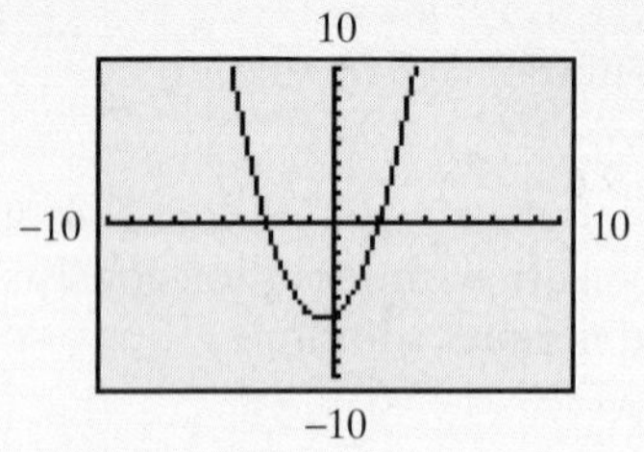

Figure B.4.1

We can verify the solution to part (a) of Example 1 by graphing $y = x^2 + x - 6$ on the calculator. See Figure B.4.1. Notice that the graph crosses the x-axis at the points $(-3, 0)$ and $(2, 0)$. Explain why this shows that $x = -3$ and $x = 2$ are the solutions to $x^2 + x - 6 = 0$.

Solving Quadratic Equations Using the Quadratic Formula

If a quadratic equation is not factorable, then we must use another method to determine the solution. The general solution to the quadratic equation $ax^2 + bx + c = 0$ for $a \neq 0$ is given by the **quadratic formula**

$$x = \frac{-b \pm \sqrt{b^2 - 4ac}}{2a}$$

This formula has two important features. First is the symbol $\pm$ (read "plus or minus"), which means that the quadratic formula is really two formulas that represent the two solutions of a quadratic equation.

$$x = \frac{-b - \sqrt{b^2 - 4ac}}{2a} \quad \text{or} \quad x = \frac{-b + \sqrt{b^2 - 4ac}}{2a}$$

The other important feature is the expression under the radical sign. This expression, $b^2 - 4ac$, is called the **discriminant** and can be used to determine what type of solution a quadratic equation will have. See Table B.4.1.

Table B.4.1

Value of Discriminant $b^2 - 4ac$	Type of Solution
Positive, meaning $b^2 - 4ac > 0$	Two distinct real number solutions
Zero, meaning $b^2 - 4ac = 0$	One repeated real number solution
Negative, meaning $b^2 - 4ac < 0$	No real number solutions

Solving a Quadratic Equation Using the Quadratic Formula

1. If needed, write the quadratic equation in the standard form $ax^2 + bx + c = 0$ by setting one side equal to zero.
2. Identify the values of a, b, and c.
3. Substitute the values from step 2 into the quadratic formula

$$x = \frac{-b \pm \sqrt{b^2 - 4ac}}{2a}$$

4. Evaluate and simplify to obtain the solutions.

Note: If after step 4 we see that the discriminant is negative, then we write *no real solutions*. The solutions are not real-number solutions because we cannot take the square root of a negative number. Such solutions are complex-number solutions and involve the imaginary unit i.

Example 2 Using the Quadratic Formula

Find the real-number solutions of the following equations, if possible, using the quadratic formula.

(a) $x^2 + 4x + 5 = 0$ **(b)** $4x^2 = 11 - 4x$

Solution

Interactive Activity

For the two solutions in Example 2(b), convert to decimal approximations. Check the results in the original equation.

(a) The equation is given in standard form, so we can see that $a = 1$, $b = 4$, and $c = 5$. Substituting these values into the quadratic equation yields

$$x = \frac{-b \pm \sqrt{b^2 - 4ac}}{2a} = \frac{-4 \pm \sqrt{4^2 - 4(1)(5)}}{2(1)} = \frac{-4 \pm \sqrt{16 - 20}}{2} = \frac{-4 \pm \sqrt{-4}}{2}$$

Since the discriminant $b^2 - 4ac = -4$ is negative, this equation has *no real solutions*.

(b) We begin by writing the equation in standard form.

$$4x^2 = 11 - 4x$$
$$4x^2 + 4x - 11 = 0$$

This gives $a = 4$, $b = 4$, and $c = -11$. Substituting these values in the quadratic formula gives us

$$x = \frac{-4 \pm \sqrt{4^2 - 4(4)(-11)}}{2(4)} = \frac{-4 \pm \sqrt{16 + 176}}{8} = \frac{-4 \pm \sqrt{192}}{8}$$

Since $192 = 64 \cdot 3 = 8^2 \cdot 3$, we can simplify the solutions by writing

$$x = \frac{-4 \pm \sqrt{192}}{8} = \frac{-4 \pm \sqrt{8^2 \cdot 3}}{8} = \frac{-4 \pm 8\sqrt{3}}{8} = \frac{4(-1 \pm 2\sqrt{3})}{8} = \frac{-1 \pm 2\sqrt{3}}{2}$$

So the solutions to the equation $4x^2 = 11 - 4x$ are $x = \dfrac{-1 - 2\sqrt{3}}{2}$ and $x = \dfrac{-1 + 2\sqrt{3}}{2}$. ■

Quadratic equations are often used in applications that involve **mathematical models**. When solving a quadratic equation based on a mathematical model, we must be sure that the solution is part of the **reasonable values of *x***.

These values, denoted in the form $l \leq x \leq u$, are the values on which the model is based.

Example 3 Solving Quadratic Models

The year that the number of calories consumed each day per person in the United States was 3361 can be represented by the equation

$$0.89x^2 - 1.93x + 3306.27 = 3361, \qquad 0 \leq x \leq 15$$

where x represents the number of years since 1974 (*Source:* U.S. Census Bureau, www.census.gov).

(a) Interpret the meaning of $0 \leq x \leq 15$.

(b) Solve the equation for x and interpret.

Solution

(a) The statement $0 \leq x \leq 15$ gives the reasonable values of x on which the model is based. Since x represents the number of years since 1974, this model is valid for the years 1974 to 1989.

(b) Writing the equation in standard form, we get

$$0.89x^2 - 1.93x - 54.73 = 0$$

Knowing that $a = 0.89$, $b = -1.93$, and $c = -54.73$, we have

$$x = \frac{1.93 \pm \sqrt{(-1.93)^2 - 4(0.89)(-54.73)}}{2(0.89)} = \frac{1.93 \pm \sqrt{198.5637}}{1.78}$$

Using a calculator to approximate these solutions, we get

$$x = \frac{1.93 - \sqrt{198.5637}}{1.78} \approx -6.83 \quad \text{or} \quad x = \frac{1.93 + \sqrt{198.5637}}{1.78} \approx 9.00$$

Since -6.83 is not between $0 \leq x \leq 15$, the only approximate solution is $x \approx 9$.

Interpret the Solution: This means that in 1983 the number of calories consumed each day per person in the United States was 3361. ■

Checkpoint 1 Now work Exercise 23 in the B.4 Exercises.

SECTION B.4 EXERCISES

In Exercises 1–12, solve the equation by factoring.

1. $4x^2 + 8x = 0$

2. $5x^2 - 10x = 0$

3. $x^2 - 9 = 0$

4. $x^2 - 36 = 0$

5. $x^2 + 8x + 15 = 0$

6. $x^2 + 7x + 12 = 0$

7. $x^2 = 5x - 6$

8. $x^2 = 3x - 2$

9. $2x^2 + 3x + 1 = 0$

10. $3x^2 + 10x + 3 = 0$

11. $3x^2 = x + 4$

12. $2x^2 = 7x + 4$

In Exercises 13–20, solve the equation by using the quadratic formula. If the discriminant is negative, write **no real solutions**.

13. $4x^2 + x - 2 = 0$

14. $3x^2 + 7x + 3 = 0$

15. $x^2 - 3x + 3 = 0$

16. $2x^2 + 3x + 3 = 0$

17. $4x^2 + 7 = 12x$

18. $3x^2 = 2 - 6x$

19. $2x^2 = 6x - 1$

20. $5x^2 = 1 - 2x$

21. The year that the membership in the Boy Scouts of America was 5433 thousand can be found by solving the equation

$$21.82x^2 - 143.75x + 5555.57 = 5433, \qquad 0 \le x \le 7$$

where x represents the number of years since 1989. (*Source:* U.S. Census Bureau, www.census.gov). Solve the equation for x and interpret.

22. The year that the annual total U.S. exports was $442.5 billion can be found by solving the equation

$$5.18x^2 + 5.86x + 410.6 = 442.5, \qquad 1 \le x \le 6$$

where x represents the number of years since 1990 (*Source:* U.S. Census Bureau, www.census.gov). Solve the equation for x and interpret.

23. The year that the number of lawyers in private practice was 195 thousand can be found by solving the equation

$$-0.37x^2 - 1.43x + 190.34 = 195, \qquad 0 \le x \le 30$$

where x represents the number of years since 1960. (*Source:* U.S. Census Bureau, www.census.gov). Solve the equation for x and interpret. (*Hint:* Begin by multiplying each side of the equation by -1.)

24. The year that the number of aggravated assaults in the United States was 297 per 100,00 people can be found by solving the equation

$$-3.08x^2 + 40.35x + 305.89 = 297, \qquad 0 \le x \le 10$$

where x represents the number of years since 1986 (*Source:* U.S. Census Bureau, www.census.gov). Solve the equation for x and interpret. (*Hint:* Begin by multiplying each side of the equation by -1.)

Section B.5 Properties of Logarithms

We can think of subtraction as a way to undo addition and division as a way to undo multiplication. A logarithm can be thought of as a way to undo exponentiation. To see how, let's take a look at the definition of logarithm.

Logarithm

We can define a **logarithm base *b* of *x*** by saying that

$$y = \log_b x \quad \text{if and only if} \quad b^y = x, \qquad b \ne 1,\ x > 0$$

Note: From the definition we see that we cannot take the logarithm of a negative number.

This definition shows that an expression written in logarithmic form can be written in exponential form, and vice versa. Table B.5.1 lists some examples.

Table B.5.1

Logarithmic Form	Exponential Form
$\log_2 8 = 3$	$2^3 = 8$
$\log_5 625 = 4$	$5^4 = 625$
$\log_{10} 10{,}000 = 4$	$10^4 = 10{,}000$
$\log_{\frac{1}{2}} \frac{1}{32} = 5$	$(\frac{1}{2})^5 = \frac{1}{32}$
$\log_{\frac{1}{4}} 64 = -3$	$(\frac{1}{4})^{-3} = 64$

Some logarithms have special bases. If $b = 10$, then $\log_{10} x = \log x$ is a **common logarithm**. Many applications in the area of finance use the number e as the base. This is the **natural logarithm** and is denoted by $\log_e x = \ln x$.

The definition of logarithm can be used to prove many of the properties of logarithms. Let's take a look at some of these properties.

Properties of Logarithms

Let $b > 0$ with $b \neq 1$ and let m and n represent positive real numbers with r being any real number.

1. $\log_b 1 = 0$
2. $\log_b b = 1$
3. $\log_b(m \cdot n) = \log_b m + \log_b n$
4. $\log_b(\frac{m}{n}) = \log_b m - \log_b n$
5. $\log_b m^r = r \log_b m$
6. $\log_b b^r = r$

Example 1 Using the Properties of Logarithms

Use the properties of logarithms to write the following as a sum, difference, or product of logarithms.

(a) $\log \dfrac{(x-3)y}{x}$ **(b)** $\ln((x+7)y)^4$

Solution

(a) $\log \dfrac{(x-3)y}{x}$	Given expression
$= \log(x-3)y - \log x$	Logarithm property 4
$= \log(x-3) + \log y - \log x$	Logarithm property 3
(b) $\ln((x+7)y)^4$	Given expression
$= 4\ln((x+7)y)$	Logarithm property 5
$= 4(\ln(x+7) + \ln y)$	Logarithm property 3

■

Checkpoint 1 Now work Exercise 5 in the B.5 Exercises.

At times we need to solve **logarithmic equations**. Sometimes we can use the definition of a logarithm to solve equations. Other times we need to use the method based on the following property.

$$\text{If } \boldsymbol{\log_b m = \log_b n}, \quad \text{then } \boldsymbol{m = n}.$$

Here m and n are called the **arguments** of the logarithmic expressions. Another useful property is

$$\text{If } \boldsymbol{m = n}, \quad \text{then } \boldsymbol{\log_b m = \log_b n}.$$

In mathematics, when we have this "two-way street," we can express both properties by writing

$$\log_b m = \log_b n \quad \text{if and only if} \quad m = n.$$

Note: This assumes, from the definition, that $m > 0$ and $n > 0$.

Part (a) of the next example illustrates using the definition of a logarithm for solving, while part (b) shows how the aforementioned property is used.

Example 2 Solving Logarithmic Equations

Solve each of the following for x.

(a) $\log(2x + 1) = 4$ **(b)** $\ln(x + 6) - \ln(x + 2) = \ln x$

Solution

(a) Here we use the definition of logarithm to solve. This gives us

$\log(2x + 1) = 4$	Given equation
$10^4 = 2x + 1$	Definition of logarithm
$10{,}000 = 2x + 1$	Simplification
$9999 = 2x$	Subtracting 1 from both sides
$\frac{9999}{2} = x$	Dividing both sides by 2

(b) For this equation we need to rewrite the left side as a single logarithmic term using property 4 in order to apply the property $\log_b m = \log_b n$ if and only if $m = n$.

$\ln(x + 6) - \ln(x + 2) = \ln x$	Given equation
$\ln\left(\frac{x+6}{x+2}\right) = \ln x$	Logarithm property 4
$\frac{x+6}{x+2} = x$	$\log_b m = \log_b n$ if and only if $m = n$
$x + 6 = x(x + 2)$	Multiply both sides by $x + 2$
$x + 6 = x^2 + 2x$	Distribute x.
$0 = x^2 + x - 6$	The quadratic equation in standard form
$0 = (x - 2)(x + 3)$	Factor the right side of the equation.
$x = 2$ or $x = -3$	Solve for x.

Since we cannot take the logarithm of a negative number, we see that $x = -3$ is not a solution. Thus, $x = 2$ is the only solution to the equation. ■

Logarithms are a useful tool for solving **exponential equations**. These are equations that have a form such as $b^y = x$, where the variable that we are solving for is the exponent. Since the property $\log_b m = \log_b n$ if and only if $m = n$ is a two-way street, we can solve exponential equations by taking the logarithm of each side of the equation and using logarithm property 5. Since it does not matter which base of logarithm we use when solving these equations, we usually use the natural logarithm.

Example 3 Solving Exponential Equations

Solve the equation $3(5^x) = 21$ for x.

Solution

$$3(5^x) = 21 \quad \text{Given equation}$$
$$5^x = 7 \quad \text{Divide both sides by 3.}$$
$$\ln(5^x) = \ln 7 \quad \mathbf{log}_b\boldsymbol{m} = \mathbf{log}_b\boldsymbol{n} \text{ if and only if } \boldsymbol{m} = \boldsymbol{n}$$
$$x \ln 5 = \ln 7 \quad \text{Logarithm property 5}$$
$$x = \frac{\ln 7}{\ln 5} \approx 1.21 \quad \text{Divide both sides by ln 5.}$$

■

SECTION B.5 EXERCISES

In Exercises 1–8, use the properties of logarithms to write the following as a sum, difference, or product of logs.

1. $\log(2x \cdot y)$

2. $\log\left(\frac{3x}{5y}\right)$

3. $\log\left(\frac{2y}{5x}\right)^3$

4. $\log\left(\frac{10x}{y}\right)^7$

✓ **5.** $\ln\left(\frac{xy^2}{x+2}\right)$

6. $\ln\left(\frac{x^3 y}{x-10}\right)$

7. $\ln\sqrt[3]{\frac{5x^3}{9y}}$

8. $\ln\sqrt{\frac{y^4}{x^3 y}}$

In Exercises 9–16, solve the logarithmic equation for x.

9. $\log(x-2) = 2$

10. $\log(x+5) = 3$

11. $\ln(4x-2) = 3$

12. $\ln(5x+1) = 5$

13. $\log(x-2) = \log(x-7) + \log 4$

14. $\log_2 x - \log_2(x+1) = \log_2 5$

15. $\ln(4x+5) - \ln(x+3) = \ln 3$

16. $\ln(4x-2) + \ln 4 = \ln(x-2)$

In Exercises 17–22, solve the exponential equation for x.

17. $3^x = 6$

18. $4^x = 12$

19. $2^{5x+2} = 8$

20. $10^{3x-7} = 5$

21. $5(2^{3x}) = 8$

22. $0.4(3^x) = 0.8$

23. The year that 18.89% of all waste generated in the United States was yard waste can be determined by solving the exponential equation

$$19.92(0.994^x) = 18.89, \qquad 1 \le x \le 6$$

where x represents the number of years since 1990 (*Source:* U.S. Census Bureau, www.census.gov). Solve the equation for x and interpret.

24. The year that the pupil–teacher ratio in the United States was 19 students for each teacher can be determined by solving the exponential equation

$$25.34(0.987^x) = 19, \qquad 1 \le x \le 36$$

where x represents the number of years since 1959 (*Source:* U.S. National Center for Education Statistics, www.nces.ed.gov). Solve the equation for x and interpret.

Section B.6 Geometric Sequence and Series

In our everyday use of the word *sequence*, we usually mean that there is a collection of some objects with an order to the collection. For example, if we were to list in order the U.S. presidents since the end of World War II, our list would have an identified first member (Harry S. Truman), an identified second member

(Dwight D. Eisenhower), an identified third member (John F. Kennedy), and so on.

In mathematics, we use the word *sequence* in a manner similar to our everyday usage of the word. The main difference is that in mathematics, our collection of objects is usually numbers. At this time, we limit our study to **geometric sequences**.

Geometric Sequence

A finite **geometric sequence** has a finite number of terms where each successive term is found by multiplying the preceding term by a number, denoted r. (Think of r as being short for *ratio*.) In general, a finite geometric sequence looks like the following:

Finite Geometric Sequence

A **finite geometric sequence** has the form

$$a, ar, ar^2, ar^3, \ldots, ar^{n-1}$$

where a is the first term and r is the **common ratio** ($r \neq 1$), and there are n terms.

Example 1 Identifying Elements of a Finite Geometric Sequence

For the following geometric sequences, identify a, r, and n.

(a) 3, 9, 27, 81

(b) $3, \frac{3}{10}, \frac{3}{10^2}, \frac{3}{10^3}, \frac{3}{10^4}$

Solution

(a) The first term is $a = 3$. The common ratio can be found by dividing any two consecutive terms. This gives us $r = \frac{9}{3} = 3$. The number of terms is $n = 4$.

(b) Here it may be beneficial to rewrite the sequence as follows:

$$3, \frac{3}{10}, \frac{3}{10^2}, \frac{3}{10^3}, \frac{3}{10^4}$$

$$3, 3\left(\frac{1}{10}\right), 3\left(\frac{1}{10^2}\right), 3\left(\frac{1}{10^3}\right), 3\left(\frac{1}{10^4}\right)$$

$$3, 3\left(\frac{1}{10}\right), 3\left(\frac{1}{10}\right)^2, 3\left(\frac{1}{10}\right)^3, 3\left(\frac{1}{10}\right)^4$$

The first term is $a = 3$. The common ratio is $r = \frac{1}{10}$. The number of terms is $n = 5$. Notice how rewriting the sequence made determining r relatively easy. Also notice that the number of terms, $n = 5$, is one more than the last exponent. ■

Geometric Series

When we sum the terms of a sequence, we form what is called a **series**. A finite **geometric series** has the following definition:

Finite Geometric Series

A **finite geometric series** has the form

$$a + ar + ar^2 + ar^3 + \cdots + ar^{n-1}$$

where a is the first term and r is the common ratio ($r \neq 1$), and there are n terms.

Note: 1. Notice that we are simply adding the terms of a finite geometric sequence.
2. The $\cdots$ means that there may be other terms that follow the indicated pattern.

Example 2 Identifying Elements of a Finite Geometric Series

For the following geometric series, identify a, r, and n.

(a) $3 + 9 + 27 + 81$

(b) $3 + \frac{3}{10} + \frac{3}{10^2} + \cdots + \frac{3}{10^{50}}$

Solution

(a) As in Example 1a, we have $a = 3$, $r = 3$ and $n = 4$.

(b) As in Example 1b, we can rewrite the series as follows.

$$3 + \frac{3}{10} + \frac{3}{10^2} + \cdots + \frac{3}{10^{50}}$$

$$3 + 3\left(\frac{1}{10}\right) + 3\left(\frac{1}{10}\right)^2 + \cdots + 3\left(\frac{1}{10}\right)^{50}$$

We have $a = 3$ and $r = \frac{1}{10}$. The number of terms is 51, one more than the last exponent. ■

Each finite geometric series in Example 2 has a sum. We can quickly compute that the sum in Example 2(a) is 120, whereas the sum in Example 2(b) would require more effort. At this time we would like to develop a formula to find the sum of **any** finite geometric series. (This formula is used in Chapter 5.) To do this, we will call the sum S_n. This gives us

$$S_n = a + ar + ar^2 + ar^3 + \cdots + ar^{n-1} \qquad \text{(I)}$$

Next we multiply both sides of the equation by r to give us

$$rS_n = ar + ar^2 + ar^3 + ar^4 + \cdots + ar^n \qquad \text{(II)}$$

Now we subtract Equation (II) from Equation (I).

$$S_n - rS_n = (a + ar + ar^2 + ar^3 + \cdots + ar^{n-1}) - (ar + ar^2 + ar^3 + ar^4 + \cdots + ar^n)$$
$$= a - ar^n$$

Finally, we factor and solve for S_n to get

$$S_n(1 - r) = a(1 - r^n)$$
$$S_n = \frac{a(1 - r^n)}{1 - r}, \qquad r \neq 1$$

Sum of Finite Geometric Series

A finite geometric series has a sum of

$$a + ar + ar^2 + ar^3 + \cdots + ar^{n-1} = \frac{a(1 - r^n)}{1 - r}, \qquad r \neq 1$$

Example 3 Using the Sum of a Finite Geometric Series Formula

Which would produce more money? $10,000 each day in the month of April or 1¢ the first day, 2¢ the second day, 4¢ the third day, and so on, with the money received doubling each day.

Solution

Since April has 30 days, the first choice would generate 10,000(30) = 300,000 or $300,000. The second choice would generate, in cents,

$$\text{Total} = 1 + 2 \cdot 1 + 2^2 \cdot 1 + 2^3 \cdot 1 + 2^4 \cdot 1 + \cdots + 2^{29} \cdot 1$$

This is a finite geometric series with $a = 1$, $r = 2$, and $n = 30$. Using our sum formula yields

$$\text{total} = \frac{a(1 - r^n)}{1 - r} = \frac{1(1 - 2^{30})}{1 - 2} = \frac{-1{,}073{,}741{,}823}{-1} = 1{,}073{,}741{,}823¢$$

In dollars, this is equal to $10,737,418.23. Clearly, the second choice is much better! ■

SECTION B.6 EXERCISES

In Exercises 1–8, find the sum of the given finite geometric series.

1. $3 + 3\left(\frac{1}{10}\right) + 3\left(\frac{1}{10}\right)^2 + \cdots + 3\left(\frac{1}{10}\right)^{50}$

2. $4 + 4\left(\frac{1}{10}\right) + 4\left(\frac{1}{10}\right)^2 + \cdots + 4\left(\frac{1}{10}\right)^{50}$

3. $1 + 2 + 2^2 + 2^3 + \cdots + 2^{20}$

4. $1 + 3 + 3^2 + 3^3 + \cdots + 3^{20}$

5. $2 + 2(5) + 2(5)^2 + 2(5)^3 + \cdots + 2(5)^9$

6. $5 + 5(2) + 5(2)^2 + 5(2)^3 + \cdots + 5(2)^9$

7. $2 - 2(3) + 2(3)^2 - 2(3)^3 + \cdots + 2(3)^8$

8. $3 - 3(2) + 3(2)^2 - 3(2)^3 + \cdots + 3(2)^8$

Applications

9. An NBA player has recently been fined for fighting with a fan during a game. The league office has decided to fine him according to the following: $1000 for the first day after its decision; $2000 for the second day after its decision; $4000 for the third day, and so on, with the amount doubling each day until the player pays the fines. Find the total fine for 12 days, when the player decides to pay the fine.

10. The NBA player in Exercise 9 decides to appeal his fine. The appeal board gave the player the option of paying the total fine as determined in Exercise 9 or to pay $100 to every fan in attendance at the team's next home game (maximum capacity of the basketball arena is 19,000). As his agent, what would you advise him to do?

11. Your very wealthy and very eccentric Aunt Liz has offered you the following choices as a graduation present: She picks the month and gives you $100,000 each day of the month *or* you pick the month and she gives you 1¢ the first day, 2¢ the second day, 4¢ the third day, and so on, doubling the amount received each day. Which would you choose and why?

12. Suppose that Aunt Liz in Exercise 11 slightly changes your choices as follows: You pick the month and she gives you $100,000 each day of the month *or* she picks the month and gives you 1¢ the first day, 2¢ the second day, 4¢ the third day, and so on, doubling the amount received each day. Which would you choose and why?

Appendix C
Calculator Programs

Simplex Method Program

The following program, SIMPLEX, utilizes the simplex method to solve linear programming problems as examined in Chapter 4. As with all programs in Appendix C, this program is written for the Texas Instruments TI-83 family of calculators. If you have a calculator different from this, consult the text's companion Web site, www.prenhall.com/armstrong, for the program for your model of calculator.

```
PROGRAM:SIMPLEX
ClrHome
dim([A])→L1
L1(1)-1→dim(L2)
rowSwap([A],1,L1(1))→[A]
For(J,1,L1(2)-1)
0→C:0→M
For(I,2,L1(1))
C+abs([A](I,J))→C
If [A] (I,J)=1
Then
I→R:1→M
End:End
If C=1 and [A](1,J)>0 and M=1
Then
*row+(-[A] (1,J), [A], R, 1)→[A]
End:End
Lbl LP
-1→C
Matr▶list([A]ᵀ,1,L3)
L1(2)-1→dim(L3)
For(I,1,L1(2)-1)
If L3(I)=min(L3) and L3(I)<0
Then
I→C
End:End
If C=-1
Then
Goto QT
End
For(I,2,L1(1))
If [A] (I,C)>0
```

```
Then
[A] (I,L1(2))/[A] (I,C)→L2(I-1)
Else:1E6→L2(I-1)
End:End
For(I,1,dim(L2))
If L2(I)=min(L2)
Then
I+1→R
End:End
*row(1/[A] (R,C),[A],R)→[A]
For(I,1,L1(1))
If I≠R
Then
*row+(-[A] (I,C),[A],R,I)→[A]
End:End
Goto LP
Lbl QT
round([A],4)→[A]
[A] (1,L1(2))→Z
Disp "Z="
Output(1,5,Z)
Disp ""
rowSwap([A],1,L1(1))→[A]
L1(2)-1→dim(L4)
For(J,1,L1(2)-1)
Matr▶list([A],J,L2)
If sum(abs(L2))>1
Then
0→L4(J)
Else
For(I,1,L1(1))
If [A] (I,J)=1
Then
[A] (I,L1(2))→L4(J)
End:End:End:End
L4
```

Example **Executing the Simplex Program**

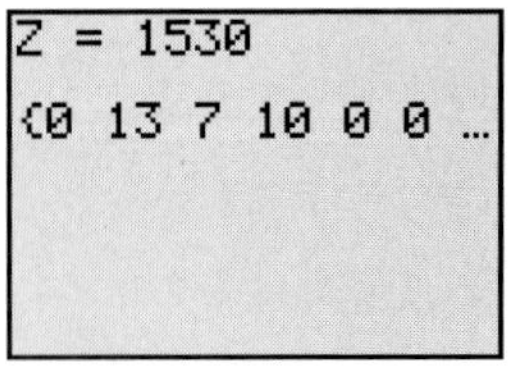

Figure C.1

For the situation in Section 4.2, Example 3, we enter the initial simplex tableau into matrix A and execute the SIMPLEX program. The output of the program is shown in Figure C.1. The optimal solution of 1530 is shown, as well as the values of all variables. You can scroll to the right to see the value of all decision and slack variables. ■

Future Value of an Annuity Program

The following program calculates the future value of an annuity.

```
PROGRAM:ANNFV
Prompt P,I,N
P*((1+I)^N-1)/I→F
Disp "FUTURE VALUE IS"
Disp F
```

Example Executing the ANNFV Program

Figure C.2

For the situation in Section 5.3, Example 3, the input and output for executing the future value of an annuity program are shown in Figure C.2. Notice that the I must be entered as the periodic rate in decimal form. ■

Payment to Fund an Annuity Program

The following program determines the periodic payment amount needed to fund an annuity. Classic examples are sinking funds.

```
PROGRAM:ANNPMT
Prompt F,I,N
(F*I)/((1+I)^N-1)→P
Disp "PAYMENT IS"
Disp P
```

Example Executing the ANNPMT Program

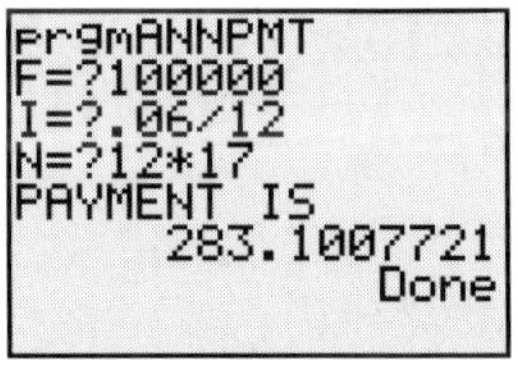

Figure C.3

For the situation in Section 5.3, Example 4, the input and output for executing the payment to fund an annuity program are shown in Figure C.3. Pay particular attention to how I and N were entered. ■

Present Value of an Annuity Program

The following program calculates the present value of an annuity.

```
PROGRAM:ANNPV
Prompt P,I,N
P*((1-(1+I)^-N)/I)→A
Disp "PRESENT VALUE IS"
Disp A
```

Example Executing the ANNPV Program

Figure C.4

For the situation in Section 5.4, Example 1, the input and output for executing the present value of an annuity program are shown in Figure C.4. Pay particular attention to how I and N are entered. ■

Program for Payment to Amortize a Loan

The following program calculates the periodic payment necessary to amortize a loan.

```
PROGRAM:AMORTPAY
Prompt P,I,N
P*(I/(1-(1+I)^-N))→T
Disp "PAYMENT IS"
Disp T
```

Example Executing the AMORTPAY Program

Figure C.5

For the situation in Section 5.4, Example 4, the input and output for executing the payment to amortize a loan program are shown in Figure C.5. Pay particular attention to how I and N are entered. ■

Bayes Theorem Program

This program is used to compute

$$P(A) = P(B_1) \cdot P(A \mid B_1) + P(B_2) \cdot P(A \mid B_2) + \cdots + P(B_k) \cdot P(A \mid B_k)$$

and

$$P(B_i \mid A) = \frac{P(B_i) \cdot P(A \mid B_i)}{P(B_1) \cdot P(A \mid B_1) + P(B_2) \cdot P(A \mid B_2) + \cdots + P(B_k) \cdot P(A \mid B_k)}$$

for any value i, where $1 \le i \le k$.

The values of $P(B_1), P(B_2), \ldots, P(B_k)$ are entered in L_1 and the values of $P(A \mid B_1), P(A \mid B_2), \ldots, P(A \mid B_k)$ are entered in L_2 before executing the program.

```
PROGRAM: BAYES
Input "NO. BRANCHES ",K
0→T
For (J,1,K)
L1(J)*L2(J)→S
T+S→T
End
Disp "P(A)= ",T
Input "ITH BRANCH ",I
(L1(I)*L2(I))→N
N/T→P
Disp "PROBABILITY",P
```

Example Executing the BAYES Program

For the situation in Section 6.7, Example 4, the input and output for executing the Bayes program are shown in Figures C.6 to C.8.

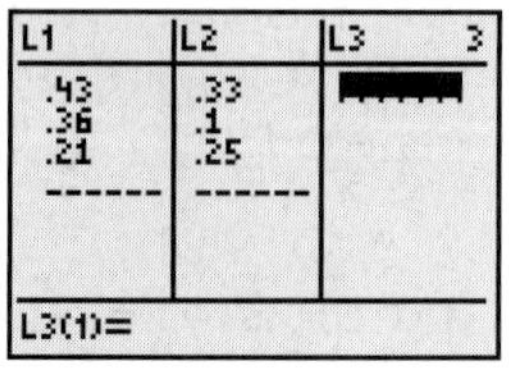

Figure C.6

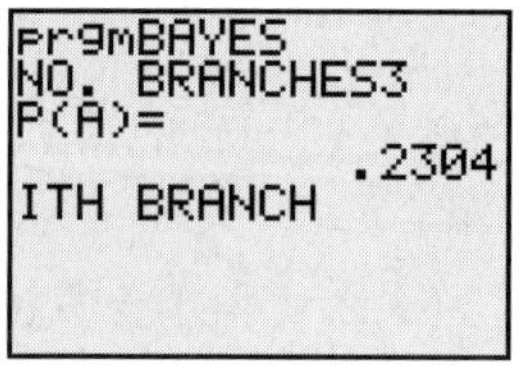

Figure C.7

Figure C.8 ■

Mean and Standard Deviation Program

For a given frequency distribution, this program computes the mean and standard deviation where the midpoints of the class are in L_1 and the frequencies are in L_2.

```
PROGRAM:MEANSD
sum(L1*L2)/sum(L2)→M
ClrList L3
(L1-M)²→L3
√((sum(L3*L2))/(sum(L2)-1))→S
Disp "MEAN", M
Disp "STD DEV", S
```

To see the input and output for this program, consult the Technology Option box after Example 3 in Section 7.4.

Binomial Probability Distribution Program

This program displays a table of $X = x$ and $P(X = x)$ values of the binomial distribution for given values of n and p.

```
PROGRAM: BINOMPD
Input "N ",N
Input "P ",P
ClrList L1,L2
For (I, 0, N)
I→L1(I+1)
binompdf(N,P,I)→L2(I+1)
End
```

Example **Executing the BINOMPD Program**

For the situation in Section 8.3, Example 3, the input for the program is in Figure C.9 and the probability distribution is in Figure C.10. Recall that for this example $n = 7$ and $p = 0.71$.

Figure C.9

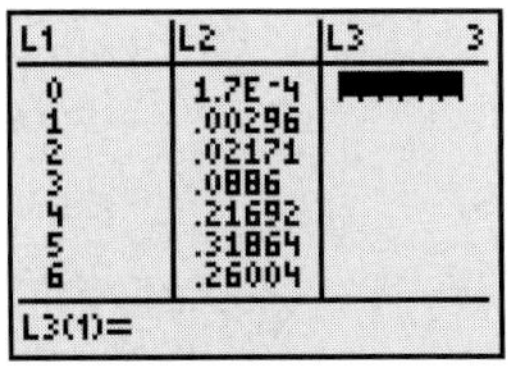

Figure C.10

■

Secant Line Slope Program

This program allows us to quickly compute the slope of secant lines over smaller and smaller intervals. The function that you are computing the secant line slope over smaller and smaller intervals for, must first be entered into Y_1 in your calculator. This program then allows us to approximate the tangent line slope at a point.

```
PROGRAM:MSEC
Input"X-VALUE",X
For(I,0,3)
1/(10^I)→H
(Y1(X+H)-Y1(X))/H→M
Disp M
End
```

```
Disp"PRESS A KEY"
Disp"TO CONTINUE"
Pause
For(J,0,3)
-1/(10^J)→H
(Y1(X+H)-Y1(X))/H→M
Disp M
End
Stop
```

Example Use the Quadratic Formula Program to approximate the roots of the quadratic equation $2x^2 - 4x = 3$.

Solution

Figure C.11

First we need to write the equation in the standard form $2x^2 - 4x - 3 = 0$. When we enter $a = 2$, $b = -4$, and $c = -3$ after the program is executed, we get the output shown in Figure C.11. ■

Newton's Method Program

We can extend the ideas of tangent lines and linear approximations discussed in Section 13.1 to derive what is known as Newton's Method, named after one of the founding fathers of the calculus, Isaac Newton. This method can be used to numerically determine zeros (or roots) of differentiable functions. For the method, we let f be a differentiable function and suppose that r is a real number zero. If x_n is an approximation to r, then the next approximation is given by $x_{n+1} = x_n - \dfrac{f(x_n)}{f'(x_n)}$, provided that $f'(x_n) \neq 0$, where $n = 0, 1, 2, \ldots$. Before running the program, $f(x)$ must be stored in Y_1 and its derivative $f'(x)$ stored in Y_2. Upon execution of the program, an initial guess $X = x_0$ is prompted.

```
PROGRAM:NEWTON
Prompt X
Lbl A
X-Y1(X)/Y2(X)→X
Disp X
Pause
Goto A
```

Example For the function $f(x) = x^2 - 4x - 2$, compute the first three iterations of an approximation to a real root of f using the Newton's Method program. Use the initial guess $x = -2$.

Solution

Figure C.12

First we enter $f(x)$ in Y_1 and $\dfrac{d}{dx}(x^2 - 4x - 2) = 2x - 4$ in Y_2 and then execute the program. The first three iterations are displayed in Figure C.12. The ENTER key must be pressed to get the next approximation. ■

Left/Right Sum Program

This program combines the left sum and right sum methods, which were discussed in Section 16.2. To run the program, the integrand of $\int_a^b f(x)\,dx$ must be entered in Y_1. When executed, the program prompts the user to enter values for the limits of integration a and b and the desired number of rectangles n.

```
PROGRAM:LFTRTSUM
Prompt A,B,N
(B-A)/N→D
0→T
0→Z
For (I,0,N-1)
A+I*D→X
T+D*Y1→T
End
Disp"LEFT SUM APPROX",T
For (J,1,N)
A+J*D→X
Z+D*Y1→Z
End
Disp"RIGHT SUM APPROX",Z
Disp"DIFFERENCE",abs(T-Z)
```

Example Use the Left/Right Sum Program to approximate $\int_0^3 \frac{1}{1+x}\,dx$ using $n = 20$ rectangles.

Solution

Here we enter $\frac{1}{1+x}$ in Y_1. After the program is executed, we enter $a = 0$, $b = 3$, and $n = 20$. The output is shown in Figure C.13. ■

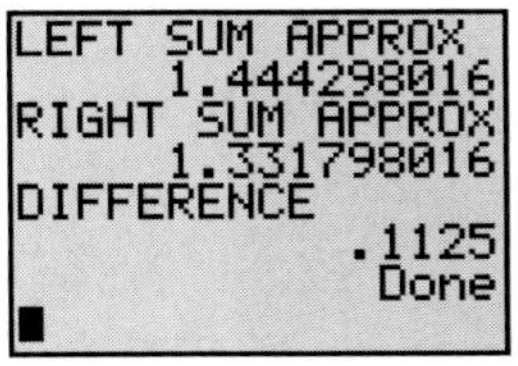

Figure C.13

Midsum Program

This program is another way to approximate the area under a curve based on the rectangular approximation methods discussed in Section 16.2. It uses the midpoint rule, which states that the area under the graph of f on the closed interval $[a, b]$ can be approximated by

$$\text{Area} \approx \sum_{i=1}^{n} f\left(\frac{x_i + x_{i-1}}{2}\right) \Delta x$$

where the height of each rectangle is determined by the middle of its base. The integrand of $\int_a^b f(x)\,dx$ must be entered in Y_1. When executed, the program prompts the user to enter values for the limits of integration a and b and the number of desired rectangles n.

```
PROGRAM:MIDSUM
Prompt A,B,N
0→T
(B-A)/N→H
```

```
For(I,1,N,1)
ACI*H→F
A+(I-1)*H→G
(F+G)/2→X
T+Y1*H→T
End
Disp"APPROX AREA",T
```

Example Use the Midsum Program to approximate $\int_0^3 \frac{1}{1+x}\,dx$ using $n = 20$ rectangles.

Solution

Here we enter $\frac{1}{1+x}$ in Y_1. When we enter $a = 0$, $b = 3$, and $n = 20$ after the program is executed, we get the output shown in Figure C.14. ■

Figure C.14

Slope Field Program

This program sketches the slope field, or family of solution curves, for the differential equation $\frac{dy}{dx} = f(x, y)$. Here $\frac{dy}{dx}$ is entered in Y_1.

```
PROGRAM:SLPFILD
Clr Draw Fnoff
7*(Xmax-Xmin)/83→H
7*(Ymax-Ymin)/55→K
6.25/H^2→A
6.25/K^2→B
Xmin+0.5*H→X
Ymin+0.5*K→V
1→L
Lbl 1
V→Y
1→J
Lbl 2
Y1→T
1/√(A+B*T^2)→M
T*M→N
Line (X-M,Y-N,X+M,Y+N)
Y+K→Y
IS>(J,8)
Goto 2
X+H→X
IS>(L,12)
Goto 1
```

Example Graph the slope field for the differential equation $\frac{dy}{dx} = \frac{x}{2y}$ in the viewing window $[-3, 3]$ by $[-3, 3]$.

Solution

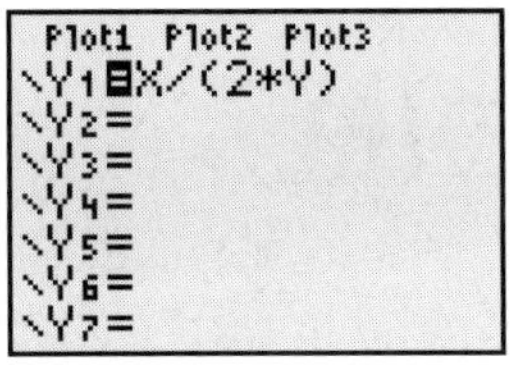

Figure C.15

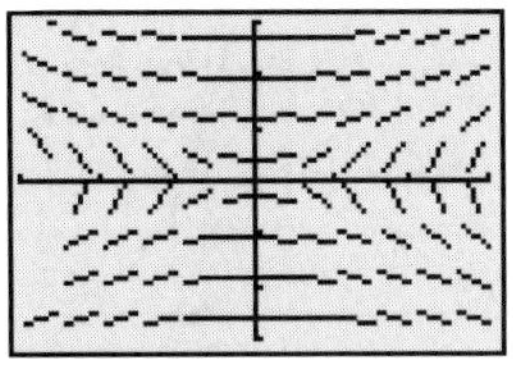

Figure C.16

We enter the differential equation as shown in Figure C.15. When the program is executed, we get the output shown in Figure C.16. ■

Multiple Regression Program

This program models the Cobb–Douglas production function $Q(L, K) = kL^a K^b$, where the labor data (L) is entered in list L_1, the capital data in L_2, and the quantity data in L_3. Upon execution, the program prompts the user to enter the number of data points used.

```
PROGRAM:MULTIREG
Disp"NUMBER OF DATA POINTS"
Prompt N
log(L1)→L1
log(L2)→L2
log(L3)→L3
(sum(L1))/N→P
(sum(L2))/N→Q
(sum(L3))/N→R
sum(L1*L3)-N*P*R→A
sum(L2*L3)-N*Q*R→B
sum((L1)^2)-N*P^2→C
sum(L1*L2)-N*P*Q→D
sum((L2)^2)-N*Q^2→E
(A*E-B*D)/(C*E-D^2)→F
(B*C-A*D)/(C*E-D^2)→G
R-F*P-G*Q→H
Disp"EXPONENT FOR L IS",F
Disp"EXPONENT FOR K IS",G
Disp"CONSTANT IS",10^H
```

Example The Brandy Corporation has 11 plants worldwide. In a recent year, the data for each plant gave the number of labor hours (in thousands), capital (total net assets, in millions), and total quantity produced, as follows:

Labor	250	270	300	320	350	400	440	440	450	460	460
Capital	30	34	44	50	70	76	84	86	104	110	116
Quantity	245	240	300	320	390	440	520	520	580	600	600

The plants all use the same technology, so a production function can be determined. Enter the data and execute the program to determine a Cobb–Douglas production function to model the data.

Solution

Figure C.17

We enter the labor, capital, and quantity data in lists L_1, L_2, and L_3, respectively. The output is displayed in Figure C.17.

So, based on the data, the Cobb–Douglas production function can be modeled by $Q(L, K) = 0.73L^{0.83}K^{0.35}$. ■

Appendix D
Standard Normal Table

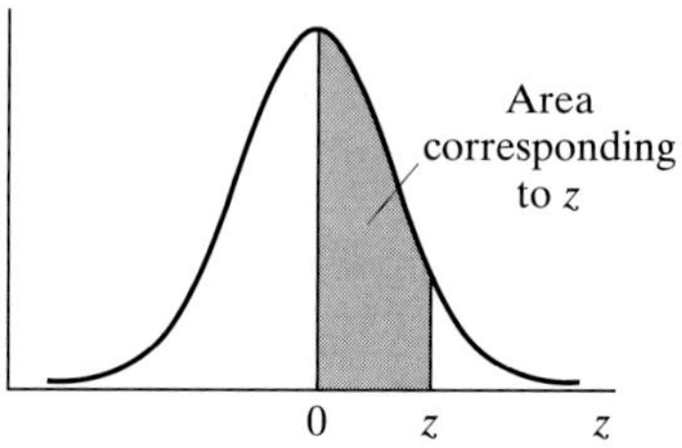

Area Under the Standard Normal Curve

z	.00	.01	.02	.03	.04	.05	.06	.07	.08	.09
0.0	0.0000	0.0040	0.0080	0.0120	0.0160	0.0199	0.0239	0.0279	0.0319	0.0359
0.1	0.0398	0.0438	0.0478	0.0517	0.0557	0.0596	0.0636	0.0675	0.0714	0.0753
0.2	0.0793	0.0832	0.0871	0.0910	0.0948	0.0987	0.1026	0.1064	0.1103	0.1141
0.3	0.1179	0.1217	0.1255	0.1293	0.1331	0.1368	0.1406	0.1443	0.1480	0.1517
0.4	0.1554	0.1591	0.1628	0.1664	0.1700	0.1736	0.1772	0.1808	0.1844	0.1879
0.5	0.1915	0.1950	0.1985	0.2019	0.2054	0.2088	0.2123	0.2157	0.2190	0.2224
0.6	0.2257	0.2291	0.2324	0.2357	0.2389	0.2422	0.2454	0.2486	0.2517	0.2549
0.7	0.2580	0.2611	0.2642	0.2673	0.2704	0.2734	0.2764	0.2794	0.2823	0.2852
0.8	0.2881	0.2910	0.2939	0.2967	0.2995	0.3023	0.3051	0.3078	0.3106	0.3133
0.9	0.3159	0.3186	0.3212	0.3238	0.3264	0.3289	0.3315	0.3340	0.3365	0.3389
1.0	0.3413	0.3438	0.3461	0.3485	0.3508	0.3531	0.3554	0.3577	0.3599	0.3621
1.1	0.3643	0.3665	0.3686	0.3708	0.3729	0.3749	0.3770	0.3790	0.3810	0.3830
1.2	0.3849	0.3869	0.3888	0.3907	0.3925	0.3944	0.3962	0.3980	0.3997	0.4015
1.3	0.4032	0.4049	0.4066	0.4082	0.4099	0.4115	0.4131	0.4147	0.4162	0.4177
1.4	0.4192	0.4207	0.4222	0.4236	0.4251	0.4265	0.4279	0.4292	0.4306	0.4319
1.5	0.4332	0.4345	0.4357	0.4370	0.4382	0.4394	0.4406	0.4418	0.4429	0.4441
1.6	0.4452	0.4463	0.4474	0.4484	0.4495	0.4505	0.4515	0.4525	0.4535	0.4545
1.7	0.4554	0.4564	0.4573	0.4582	0.4591	0.4599	0.4608	0.4616	0.4625	0.4633
1.8	0.4641	0.4649	0.4656	0.4664	0.4671	0.4678	0.4686	0.4693	0.4699	0.4706
1.9	0.4713	0.4719	0.4726	0.4732	0.4738	0.4744	0.4750	0.4756	0.4761	0.4767
2.0	0.4772	0.4778	0.4783	0.4788	0.4793	0.4798	0.4803	0.4808	0.4812	0.4817
2.1	0.4821	0.4826	0.4830	0.4834	0.4838	0.4842	0.4846	0.4850	0.4854	0.4857
2.2	0.4861	0.4864	0.4868	0.4871	0.4875	0.4878	0.4881	0.4884	0.4887	0.4890
2.3	0.4893	0.4896	0.4898	0.4901	0.4904	0.4906	0.4909	0.4911	0.4913	0.4916
2.4	0.4918	0.4920	0.4922	0.4925	0.4927	0.4929	0.4931	0.4932	0.4934	0.4936
2.5	0.4938	0.4940	0.4941	0.4943	0.4945	0.4946	0.4948	0.4949	0.4951	0.4952
2.6	0.4953	0.4955	0.4956	0.4957	0.4959	0.4960	0.4961	0.4962	0.4963	0.4964
2.7	0.4965	0.4966	0.4967	0.4968	0.4969	0.4970	0.4971	0.4972	0.4973	0.4974
2.8	0.4974	0.4975	0.4976	0.4977	0.4977	0.4978	0.4979	0.4979	0.4980	0.4981
2.9	0.4981	0.4982	0.4982	0.4983	0.4984	0.4984	0.4985	0.4985	0.4986	0.4986
3.0	0.4987	0.4987	0.4987	0.4988	0.4988	0.4989	0.4989	0.4989	0.4990	0.4990
3.1	0.4990	0.4991	0.4991	0.4991	0.4992	0.4992	0.4992	0.4992	0.4993	0.4993
3.2	0.4993	0.4993	0.4994	0.4994	0.4994	0.4994	0.4994	0.4995	0.4995	0.4995
3.3	0.4995	0.4995	0.4995	0.4996	0.4996	0.4996	0.4996	0.4996	0.4996	0.4997
3.4	0.4997	0.4997	0.4997	0.4997	0.4997	0.4997	0.4997	0.4997	0.4997	0.4998
3.5	0.4998	0.4998	0.4998	0.4998	0.4998	0.4998	0.4998	0.4998	0.4998	0.4998
3.6	0.4998	0.4998	0.4999	0.4999	0.4999	0.4999	0.4999	0.4999	0.4999	0.4999
3.7	0.4999	0.4999	0.4999	0.4999	0.4999	0.4999	0.4999	0.4999	0.4999	0.4999
3.8	0.4999	0.4999	0.4999	0.4999	0.4999	0.4999	0.4999	0.4999	0.4999	0.4999
3.9	0.5000	0.5000	0.5000	0.5000	0.5000	0.5000	0.5000	0.5000	0.5000	0.5000

Table entries represent the area under the standard normal curve from 0 to z, $z \geq 0$.

Taken from Barnett/Ziegler/Byleen Finite Math 8/e (Prentice Hall)

Appendix E
Selected Proofs

Quadratic Formula

The solution of the quadratic equation $ax^2 + bx + c = 0$ is given by
$x = \dfrac{-b \pm \sqrt{b^2 - 4ac}}{2a}$.

Proof We will solve the quadratic equation $ax^2 + bx + c = 0$ by completing the square.

$ax^2 + bx = -c$	Subtract c
$x^2 + \dfrac{b}{a}x = -\dfrac{c}{a}$	Divide by a
$x^2 + \dfrac{b}{a}x + \left(\dfrac{b}{2a}\right)^2 = \left(\dfrac{b}{2a}\right)^2 - \dfrac{c}{a}$	Complete the square
$x^2 + \dfrac{b}{a}x + \left(\dfrac{b}{2a}\right)^2 = \dfrac{b^2}{4a^2} - \dfrac{c}{a}$	Simplify the right side
$\left(x + \dfrac{b}{2a}\right)^2 = \dfrac{b^2}{4a^2} - \dfrac{c}{a}$	Write the left side as a squared binomial
$x + \dfrac{b}{2a} = \pm\sqrt{\dfrac{b^2}{4a^2} - \dfrac{c}{a}}$	Take the square root of both sides
$x + \dfrac{b}{2a} = \pm\sqrt{\dfrac{b^2 - 4ac}{4a^2}}$	Write the radicand as a single fraction
$x + \dfrac{b}{2a} = \pm\dfrac{\sqrt{b^2 - 4ac}}{2a}$	Take the square root of the denominator on the right side
$x = -\dfrac{b}{2a} \pm \dfrac{\sqrt{b^2 - 4ac}}{2a}$	Subtract $-\dfrac{b}{2a}$
$x = \dfrac{-b \pm \sqrt{b^2 - 4ac}}{2a}$	Write the right side as a single fraction ■

Properties of Logarithms

Here we will prove three of the properties of logarithms listed in Appendix A.

Property: $\log_b(m \cdot n) = \log_b m + \log_b n$

Proof Let $x = \log_b m$ and $y = \log_b n$. Then, by the definition of logarithm, we may write $b^x = m$ and $b^y = n$. Multiplying these equations gives

$$b^x \cdot b^y = m \cdot n = b^{x+y}$$

Using the definition of logarithm, we may write the equation $m \cdot n = b^{x+y}$ as

$$\log_b(m \cdot n) = x + y$$

Since $x = \log_b m$ and $y = \log_b n$, we get

$$\log_b(m \cdot n) = \log_b m + \log_b n$$

■

Property: $\log_b\left(\dfrac{m}{n}\right) = \log_b m - \log_b n$

Proof Let $x = \log_b m$ and $y = \log_b n$. Then, by the definition of logarithm, we may write $b^x = m$ and $b^y = n$. Dividing these equations gives

$$\frac{b^x}{b^y} = \frac{m}{n} = b^{x-y}$$

Using the definition of logarithm, we may write the equation $\dfrac{m}{n} = b^{x-y}$ as

$$\log_b\left(\frac{m}{n}\right) = x - y$$

Since $x = \log_b m$ and $y = \log_b n$, we get

$$\log_b\left(\frac{m}{n}\right) = \log_b m - \log_b n$$

■

Property: $\log_b m^r = r \log_b m$

Proof Let $x = \log_b m$. Then, by the definition of logarithm, we may write the equation in exponential form as $b^x = m$. If we raise both sides of this equation to the rth power, we get

$$(b^x)^r = m^r \qquad b^{xr} = m^r$$

By the definition of logarithm, we can write the equation $b^{xr} = m^r$ as

$$xr = \log_b m^r$$

Since $x = \log_b m$, we have $r \log_b m = \log_b m^r$.

■

Power Rule

If $f(x) = x^n$, where n is any real number, then $f'(x) = nx^{n-1}$. Equivalently, $\dfrac{d}{dx}[x^n] = nx^{n-1}$, and if $y = x^n$, then $\dfrac{dy}{dx} = nx^{n-1}$.

Proof Here we will prove the Power Rule for n being a positive integer. By the definition of derivative, we have $\dfrac{d}{dx}[x^n] = \lim\limits_{h \to 0} \dfrac{(x+h)^n - x^n}{h}$. From the Binomial

Theorem in algebra, we know that $(x+h)^n$ can be written as

$$(x+h)^n = x^n + nx^{n-1}h + \frac{n(n-1)}{2}x^{n-2}h^2 + \cdots + nxh^{n-1} + h^n$$

Substituting this expression into the definition of derivative, we get

$$\begin{aligned}
\frac{d}{dx}[x^n] &= \lim_{h\to 0} \frac{[x^n + nx^{n-1}h + \frac{n(n-1)}{2}x^{n-2}h^2 + \cdots + nxh^{n-1} + h^n] - x^n}{h} \\
&= \lim_{h\to 0} \frac{nx^{n-1}h + \frac{n(n-1)}{2}x^{n-2}h^2 + \cdots + nxh^{n-1} + h^n}{h} \\
&= \lim_{h\to 0} \left[nx^{n-1} + \frac{n(n-1)}{2}x^{n-2}h + \cdots \begin{pmatrix}\text{other terms} \\ \text{with } x \text{ and} \\ h \text{ factors}\end{pmatrix} \cdots + nxh^{n-2} + h^{n-1} \right]
\end{aligned}$$

Since every term in the brackets except the first contains h as a factor, every term except the first approaches zero as $h \to 0$, and we get

$$\frac{d}{dx}[x^n] = nx^{n-1}$$

■

Constant Multiple Rule

If $f(x) = k \cdot g(x)$, where k is any real number, then $f'(x) = k \cdot g'(x)$, assuming that g is differentiable. Equivalently, $\frac{d}{dx}[k \cdot g(x)] = k \cdot g'(x)$.

Proof Recall that the definition of the derivative of a function f is $\frac{d}{dx}[f(x)] = \lim_{h\to 0} \frac{f(x+h) - f(x)}{h}$. Using this definition for $f(x) = k \cdot g(x)$, we get

$$\frac{d}{dx}[k \cdot g(x)] = \lim_{h\to 0} \frac{k \cdot g(x+h) - k \cdot g(x)}{h}$$ Using the definition of derivative

$$= \lim_{h\to 0} \frac{k[g(x+h) - g(x)]}{h}$$ Factor k from the numerator

$$= k \cdot \lim_{h\to 0} \frac{g(x+h) - g(x)}{h}$$ From Limit Theorem 3 in Section 12.1

$$= k \cdot g'(x)$$ By definition, $\lim_{h\to 0} \frac{g(x+h) - g(x)}{h} = g'(x)$

■

Sum Rule

If $h(x) = f(x) + g(x)$, where f and g are differentiable functions, then $h'(x) = f'(x) + g'(x)$. Equivalently, $\frac{d}{dx}[f(x) + g(x)] = \frac{d}{dx}[f(x)] + \frac{d}{dx}[g(x)] = f'(x) + g'(x)$.

Proof We want to show that $\frac{d}{dx}[f(x)+g(x)] = f'(x) + g'(x)$, so, by the definition of derivative,

$\frac{d}{dx}[f(x)+g(x)] = \lim_{h\to 0} \frac{[f(x+h)+g(x+h)]-[f(x)+g(x)]}{h}$	Using the definition of derivative
$= \lim_{h\to 0} \frac{f(x+h)+g(x+h)-f(x)-g(x)}{h}$	Removing the parentheses
$= \lim_{h\to 0} \frac{f(x+h)-f(x)+g(x+h)-g(x)}{h}$	Interchanging the two middle terms in the numerator
$= \lim_{h\to 0} \left[\frac{f(x+h)-f(x)}{h} + \frac{g(x+h)-g(x)}{h}\right]$	Rewriting the difference quotient as two fractions
$= \lim_{h\to 0} \frac{f(x+h)-f(x)}{h} + \lim_{h\to 0} \frac{g(x+h)-g(x)}{h}$	Limit Theorem 4 in Section 12.1
$= f'(x) + g'(x)$	By the definition of derivative ■

Difference Rule

If $h(x) = f(x) - g(x)$, where f and g are differentiable functions, then

$$h'(x) = f'(x) - g'(x).$$

Equivalently,

$$\frac{d}{dx}[f(x)-g(x)] = \frac{d}{dx}[f(x)] - \frac{d}{dx}[g(x)] = f'(x) - g'(x).$$

Proof We want to show that $\frac{d}{dx}[f(x)-g(x)] = f'(x) - g'(x)$, so, by the definition of derivative,

$\frac{d}{dx}[f(x)-g(x)] = \lim_{h\to 0} \frac{[f(x+h)-g(x+h)]-[f(x)-g(x)]}{h}$	Using the definition of derivative
$= \lim_{h\to 0} \frac{f(x+h)-g(x+h)-f(x)+g(x)}{h}$	Removing the parentheses
$= \lim_{h\to 0} \frac{f(x+h)-f(x)-g(x+h)+g(x)}{h}$	Interchanging the two middle terms in the numerator
$= \lim_{h\to 0} \frac{[f(x+h)-f(x)]-[g(x+h)-g(x)]}{h}$	Factor −1 from the last two terms in the numerator
$= \lim_{h\to 0} \left[\frac{f(x+h)-f(x)}{h} - \frac{g(x+h)-g(x)}{h}\right]$	Rewriting the difference quotient as two fractions
$= \lim_{h\to 0} \frac{f(x+h)-f(x)}{h} - \lim_{h\to 0} \frac{g(x+h)-g(x)}{h}$	Limit Theorem 4 in Section 12.1
$= f'(x) - g'(x)$	By the definition of derivative ■

Product Rule

If $h(x) = f(x) \cdot g(x)$, where f and g are differentiable functions, then

$$h'(x) = f'(x) \cdot g(x) + f(x) \cdot g'(x).$$

Equivalently,

$$\frac{d}{dx}[f(x) \cdot g(x)] = \frac{d}{dx}[f(x)] \cdot g(x) + f(x) \cdot \frac{d}{dx}[g(x)] = f'(x) \cdot g(x) + f(x) \cdot g'(x).$$

Proof We want to show that $\frac{d}{dx}[f(x) \cdot g(x)] = f'(x) \cdot g(x) + f(x) \cdot g'(x)$, so, by the definition of derivative,

$$\frac{d}{dx}[f(x) \cdot g(x)] = \lim_{h \to 0} \frac{f(x+h) \cdot g(x+h) - f(x) \cdot g(x)}{h}$$

Now we subtract and add the term $f(x) \cdot g(x+h)$ in the numerator to get

$$= \lim_{h \to 0} \frac{f(x+h) \cdot g(x+h) - f(x) \cdot g(x+h) + f(x) \cdot g(x+h) - f(x) \cdot g(x)}{h}$$

Factoring $g(x+h)$ from the first two terms and $f(x)$ from the last two terms gives

$$= \lim_{h \to 0} \frac{[f(x+h) - f(x)] \cdot g(x+h) + f(x) \cdot [g(x+h) - g(x)]}{h}$$

Writing the difference quotient as two fractions yields

$$= \lim_{h \to 0} \left[\frac{f(x+h) - f(x)}{h} \cdot g(x+h) \right] + \lim_{h \to 0} \left[f(x) \cdot \frac{g(x+h) - g(x)}{h} \right]$$
$$= f'(x) \cdot g(x) + f(x) \cdot g'(x)$$

■

Quotient Rule

If $h(x) = \dfrac{f(x)}{g(x)}$, where f and g are differentiable functions, then

$$h'(x) = \frac{f'(x) \cdot g(x) - f(x) \cdot g'(x)}{[g(x)]^2} \quad \text{where } g(x) \neq 0.$$

Equivalently,

$$\frac{d}{dx}\left[\frac{f(x)}{g(x)}\right] = \frac{\frac{d}{dx}[f(x)] \cdot g(x) - f(x) \cdot \frac{d}{dx}[g(x)]}{[g(x)]^2}.$$

Proof We want to show that $\frac{d}{dx}\left[\frac{f(x)}{g(x)}\right] = \frac{f'(x) \cdot g(x) - f(x) \cdot g'(x)}{[g(x)]^2}$, so, by the definition of derivative,

$$\frac{d}{dx}\left[\frac{f(x)}{g(x)}\right] = \lim_{h \to 0} \frac{\frac{f(x+h)}{g(x+h)} - \frac{f(x)}{g(x)}}{h}$$

Multiplying the numerator and denominator by $g(x+h) \cdot g(x)$ gives

$$= \lim_{h \to 0} \frac{f(x+h) \cdot g(x) - f(x) \cdot g(x+h)}{h \cdot g(x+h) \cdot g(x)}$$

Subtracting and adding the term $f(x) \cdot g(x)$ in the numerator yields

$$= \lim_{h\to 0} \frac{f(x+h)\cdot g(x) - f(x)\cdot g(x) + f(x)\cdot g(x) - f(x)\cdot g(x+h)}{h\cdot g(x+h)\cdot g(x)}$$

$$= \lim_{h\to 0} \frac{[f(x+h) - f(x)]g(x) - f(x)[-g(x) + g(x+h)]}{h\cdot g(x+h)\cdot g(x)}$$

If we write the numerator in the form of two difference quotients and take the limit, we get

$$= \lim_{h\to 0}\left[\frac{\dfrac{f(x+h)-f(x)}{h}\cdot g(x) - f(x)\cdot\dfrac{g(x+h)-g(x)}{h}}{g(x+h)\cdot g(x)}\right]$$

$$= \frac{f'(x)\cdot g(x) - f(x)\cdot g'(x)}{[g(x)]^2}$$ ■

Differentiability Implies Continuity

If f is differentiable at $x = c$, then f is continuous at $x = c$.

Proof Let f be differentiable at $x = c$. So we know that $f'(c)$ exists, and we need to show that $\lim_{x\to c} f(x) = f(c)$ or, equivalently, $\lim_{x\to c}[f(x) - f(c)] = 0$. Assuming $x \neq c$, we begin by multiplying and dividing the expression in the limit argument by $(x - c)$ to get

$$\lim_{x\to c}[f(x) - f(c)] = \lim_{x\to c}\left[(x-c)\cdot\frac{f(x)-f(c)}{x-c}\right]$$

By Limit Theorem 5 in Section 12.1, we have

$$\lim_{x\to c}[f(x) - f(c)] = \lim_{x\to c}(x-c)\cdot\lim_{x\to c}\frac{f(x)-f(c)}{x-c}$$

$$= 0\cdot f'(c) = 0$$

Hence, $\lim_{x\to c} f(x) = f(c)$ and f is continuous at $x = c$. ■

Fundamental Theorem of Calculus

If f is a continuous function defined on a closed interval $[a, b]$ and F is an antiderivative of f, then $\int_a^b f(x)dx = F(b) - F(a)$.

Proof Let f be a continuous function with $f(x) > 0$ for all x in $[a, b]$. Now we define an area function A, which represents the area under the graph of f from a to x. See Figure E.1.

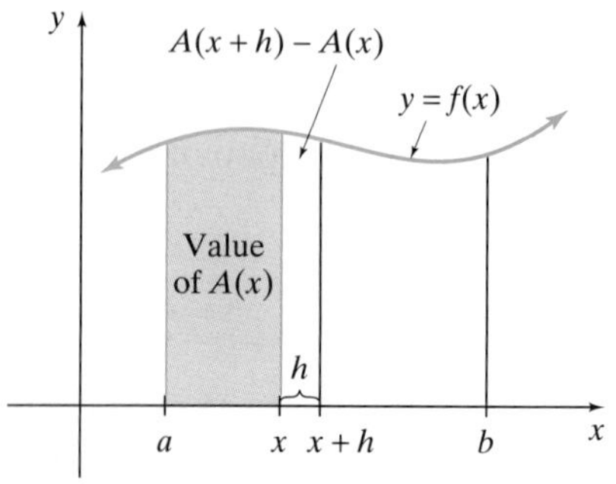

Figure E.1

We need to first show that $\begin{pmatrix}\text{Area under the graph of } f\\ \text{on } [a, b]\end{pmatrix} = A(b) - A(a)$. Since f is continuous on $[a, b]$, we know that $\int_a^x f(t)\,dt$ exists, and so we define $A(x) = \int_a^x f(t)\,dt$. We see that $A(a) = \int_a^a f(t)\,dt = 0$, and so $A(b) = A(b) - 0 = A(b) - A(a)$. So we have $\begin{pmatrix}\text{Area under the graph of}\\ f \text{ on } [a, b]\end{pmatrix} = A(b) - A(a)$. Now we need to show that A is an antiderivative of f. By the definition of derivative, we have $A'(x) = \lim_{h\to 0} \dfrac{A(x+h) - A(x)}{h}$. Analyzing this limit, we see the following:

- $A(x+h)$ is the area under the graph of f between a and $x+h$.
- $A(x)$ is the area under the graph of f between a and x.
- $A(x+h) - A(x)$ is the area between x and $x+h$.

For a small value of h, x and $x+h$ are values close to one another, and since f is continuous, $f(x+h)$ is close in value to $f(x)$. In other words,

$$A(x+h) - A(x) \approx \begin{pmatrix}\text{Area of rectangle with height}\\ f(x) \text{ and width } h\end{pmatrix} = f(x)\cdot h$$

So, for a small value of h, $\dfrac{A(x+h) - A(x)}{h} \approx \dfrac{f(x)\cdot h}{h} = f(x)$. Thus, as $h \to 0$ we see that $A'(x) = \lim_{h\to 0} \dfrac{A(x+h) - A(x)}{h} = f(x)$. Hence, A is an antiderivative of f. ■

Derivation of the Least-squares Formulas

Given a collection of data points $(x_1, y_1), (x_2, y_2), (x_3, y_3), \ldots, (x_n, y_n)$, the linear regression model for the data has the form $y = mx + b$, where

$$m = \frac{n\sum xy - \sum x \cdot \sum y}{n\sum x^2 - \left(\sum x\right)^2} \quad \text{and} \quad b = \frac{\sum y \cdot \sum x^2 - \sum x \cdot \sum xy}{n\sum x - \left(\sum x\right)^2}$$

Proof Let $(x_1, y_1), (x_2, y_2), (x_3, y_3), \ldots, (x_n, y_n)$ be a set of data points. To determine a regression line, we need to find values m and b so that the sum of the squares of the residuals $d = y - (mx + b)$ is a minimum. If we write m and b in the form of a function of two independent variables, we see that we need to minimize

$$\begin{aligned} f(m, b) &= \sum d^2 = \sum[y - (mx + b)]^2 \\ &= (y_1 - mx_1 - b)^2 + (y_2 - mx_2 - b)^2(y_3 - mx_3 - b)^2 + \cdots + (y_n - mx_n - b)^2 \end{aligned}$$

The partial derivative of f with respect to m is

$$\begin{aligned} \frac{\partial f}{\partial m} &= 2(y_1 - mx_1 - b)(-x_1) + 2(y_2 - mx_2 - b)(-x_2) \\ &\quad + 2(y_3 - mx_3 - b)(-x_1) + \cdots + 2(y_n - mx_n - b)(-x_n) \\ &= 2[(-x_1y_1 - x_2y_2 - x_3y_3 - \cdots - x_ny_n) + m(x_1^2 + x_2^2 + x_3^2 + \cdots + x_n^2) \\ &\quad + b(x_1 + x_2 + x_3 + \cdots + x_n] \\ &= 2\left[-\sum xy + m\sum x^2 + b\sum x\right] \end{aligned}$$

The partial derivative of f with respect to b is

$$\begin{aligned}\frac{\partial f}{\partial b} &= 2(y_1 - mx_1 - b)(-1) + 2(y_2 - mx_2 - b)(-1) \\ &\quad + 2(y_3 - mx_3 - b)(-1) + \cdots + 2(y_n - mx_n - b)(-1) \\ &= 2[(-y_1 - y_2 - y_3 - \cdots - y_n) + m(x_1 + x_2 + x_3 + \cdots + x_n) \\ &\quad + b(1 + 1 + 1 + \cdots + 1) \\ &= 2\left[-\sum y + m\sum x + bn\right]\end{aligned}$$

Now, if we set $\dfrac{\partial f}{\partial m} = 0$ and $\dfrac{\partial f}{\partial b} = 0$ and solve the system of equations, we get

$$\begin{cases} -\sum xy + m\sum x^2 + b\sum x = 0 \\ -\sum y + m\sum x + bn = 0 \end{cases} \quad \text{or} \quad \begin{aligned} m\sum x^2 + b\sum x &= \sum xy \quad \text{(I)} \\ m\sum x + bn &= \sum y \quad \text{(II)} \end{aligned}$$

To solve for m, we multiply equation (I) by n and equation (II) by $-\sum x$ and then add the equations to get

$$\begin{aligned} nm\sum x^2 - m\left(\sum x\right)^2 &= n\sum xy - \sum x \sum y \\ m\left[n\sum x^2 - \left(\sum x\right)^2\right] &= n\sum xy - \sum x \sum y \\ m &= \frac{n\sum xy - \sum x \sum y}{n\sum x^2 - \left(\sum x\right)^2} \end{aligned}$$

To solve for b, we multiply the equation (I) by $-\sum x$ and equation (II) by $\sum x^2$ and then add the equations to get

$$\begin{aligned} bn\sum x - b\left(\sum x\right)^2 &= \sum x^2 \sum y - \sum x \sum xy \\ b\left[n\sum x - \left(\sum x\right)^2\right] &= \sum x^2 \sum y - \sum x \sum xy \\ b &= \frac{\sum y \cdot \sum x^2 - \sum x \cdot \sum xy}{n\sum x - \left(\sum x\right)^2} \end{aligned}$$

■

Example 1

Use the least-squares formulas to determine a linear regression function of the form $y = mx + b$ for the following sample of data. Use the regression capabilities of your calculator to verify your answer.

x	19	23	21	15	16	18
y	55	7	20	123	88	76

Solution

When we compute the necessary sums, we get

$$\begin{aligned} n &= 6 \\ \sum x &= 19 + 23 + \cdots + 18 = 112 \\ \sum y &= 55 + 7 + \cdots + 76 = 369 \\ \sum x^2 &= 19^2 + 23^2 + \cdots + 18^2 = 2136 \\ \sum xy &= 19 \cdot 55 + 23 \cdot 7 + \cdots + 18 \cdot 76 = 6247 \end{aligned}$$

Evaluating the derived formulas for m and b gives

$$m = \frac{6(6247) - (112)(369)}{6(2136) - (112)^2} \approx -14.14,$$

$$b = \frac{(369)(2136) - (112)(6247)}{6(112) - (112)^2} \approx 325.44$$

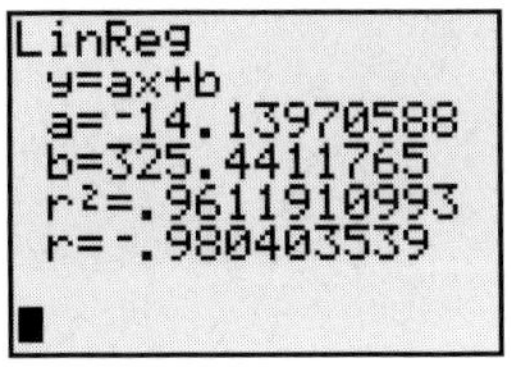

Figure E.2

So the linear regression model for the data is $y = -14.14x + 325.44$. Entering the data in the calculator and using the LINREG command gives the output shown in Figure E.2. ■

Appendix F
Photo and Illustration Credits

PAGE xxix: Author photos, courtesy of authors. **PAGE 1:** Statue of Liberty, New York / Getty Images, Inc. **PAGE 71:** Reticulated buses cross a busy intersection in downtown Seattle, Washington / Jonathan Nourok / PhotoEdit; agricultural landscape of lettuce growing in a field / Pete Turner, Inc. / Getty Images, Inc. **PAGE 139:** Exterior of a McDonald's restaurant with golden arches / Michael Newman / PhotoEdit. **PAGE 191:** Empty chair and computer in an office / Robert Daly / Getty Images, Inc. **PAGE 255:** An open package of cigarettes / C. Sherbume / Getty Images, Inc.; exterior of US Bank building / Eric R. Berndt / Unicorn Stock Photos. **PAGE 307:** Tiger Woods, USPGA Championship 1997, Winged Foot Golf Course, NY, August 14–17 / Jamie Squire / Getty Images, Inc. **PAGE 397:** Rural barn and field / Gary Vestal / Index Stock Imagery, Inc. **PAGE 461:** Man squeezing a grip strengthener at his computer / Ryan McVay / Getty Images, Inc. **PAGE 525:** Inner city elementary school classroom, Long Beach, California / A. Ramey / Woodfin Camp & Associates. **PAGE 567:** Retail establishment / Laima Druskis / Pearson Education / PH College. **PAGE 601:** Carolyn Schaefer / Liaison Agency, Inc. **PAGE 711:** K. Knudson / PhotoDisc, Inc. **PAGE 789:** Michael Newman / PhotoEdit. **PAGE 823:** Robert Brenner / PhotoEdit. **PAGE 875:** Michael Newman / PhotoEdit. **PAGE 951:** N. Hashimoto / Sygma Photo News. **PAGE 1037:** Stewart Cohen / Tony Stone Images. **PAGE 1135:** Illustration by Dennis Tasa. From *Earth, An Introduction to Physical Geology*, sixth edition, Edward J. Tarbuck, Frederick K. Lutgens, copyright 1999 by Prentice-Hall, Inc.

Answers

Chapter 1

Section 1.1

1. $x + 10 = 14$ **3.** $x - 7 = 28$ **5.** $5x = 45$

7. $6x - 1 = 29$ **9.** $3x + 11 = 44$

11. **(a)** It will cost \$267 to produce ten picture frames.
(b) $C = 1.50x + 252$

13. **(a)** It costs \$11,250 to produce 110 portable cribs.
(b) $C = 100x + 250$

15. **(a)** $220x + 80y = 1400$ **(b)** $6x + 4y = 40$

17. **(a)** $x + y = 640$ **(b)** $0.75x + 2y = 1080$

19. **(a)** Let x = number of student tickets sold and let y = number of adult tickets sold.

$$\begin{aligned} x + y &= 1600 \\ 3x + 5y &= 5760 \end{aligned}$$

(b) $$\begin{aligned} x + y &= 1600 \\ 3x + 5y &= 5980 \end{aligned}$$ **(c)** $$\begin{aligned} x + y &= 1600 \\ 3x + 5y &= 5500 \end{aligned}$$

21. **(a)** Let C = monthly checking costs of City account and let x = fee per check.
$C = 5 + 0.15x$
(b) Let C = monthly checking costs of Urban account and let x = fee per check.
$C = 4 + 0.2x$

23. **(a)** Let C = attorney costs at Frank & Earnst Law Firm and let x = amount of settlement.
$C = 1500 + 0.25x$
(b) Let C = attorney costs at Mackenzie Law Firm and let x = amount of settlement.
$C = 2000 + 0.20x$

25. **(a)** $0.5x + 0.8y + 1.1z = 450$
(b) $0.6x + 0.7y + 0.9z = 390$
(c) $0.2x + 0.3y + 0.6z = 190$

27. **(a)** $x + y + z = 50{,}000$ **(b)** $z = \frac{1}{2}x + 1000$
(c) $0.06x + 0.07y + 0.08z = 3410$

29. **(a)** $240x + 215y + 336z = 65{,}790$
(b) $148x + 98y + 171z = 31{,}860$
(c) $42x + 69y + 147z = 21{,}630$

31. For parts (a), (b), and (c), let x = amount of Basic!, let y = amount of Vaca, and let z = amount of Secra.
(a) $20x + 10y + 30z = 340$ **(b)** $10x + 5y + 20z = 205$
(c) $20x + 15y + 10z = 225$

33. For parts (a), (b), and (c), let x = quantity produced at Plant I, let y = quantity produced at Plant II, and let z = quantity produced at Plant III.
(a) $18x + 20y + 16z = 5900$
(b) $12x + 9y + 11z = 3580$
(c) $22x + 18y + 17z = 6350$

Section 1.2

1. Ordered pairs are (0, 1), (3, 5), and (6, 7).

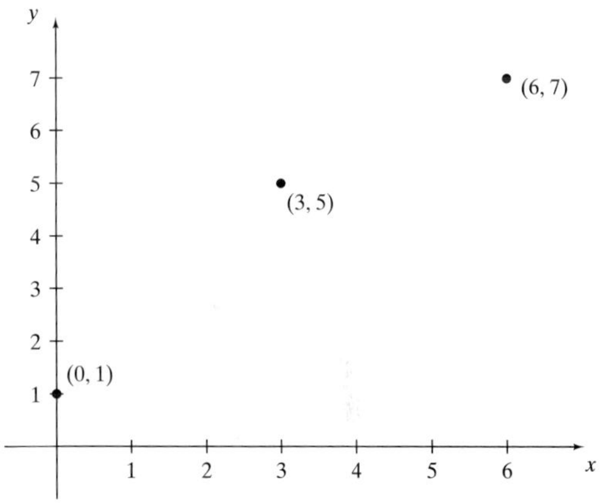

3. Ordered pairs are (−2, 3), (−1, 5), and (4, −4).

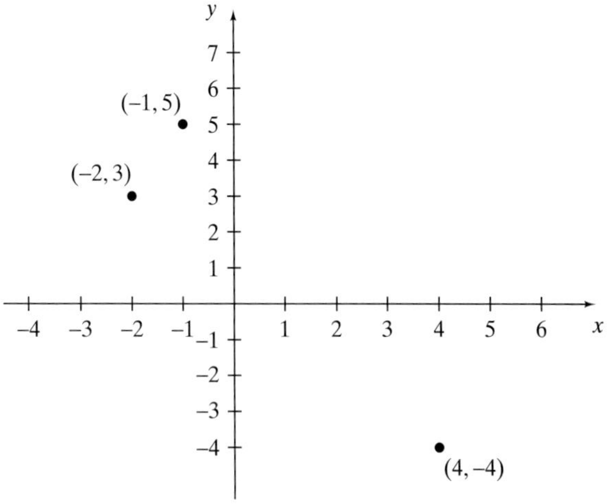

5. Ordered pairs are (−4, −5), (−3.5, −6), (−2, −8), and (2, −11).

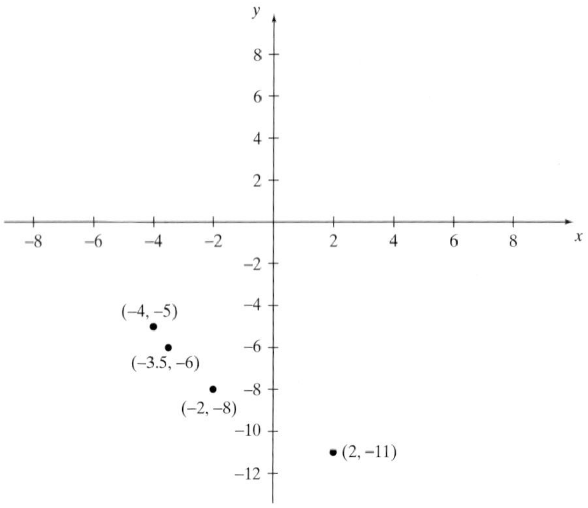

7. Ordered pairs are (0, 1.1), (1, 1.5), (2, 1.8), and (3, 2).

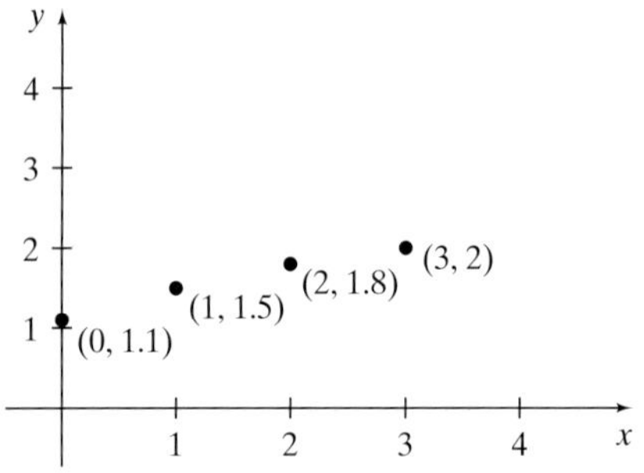

9. Domain: {0, 3, 6}
Range: {1, 5, 7}

11. Domain: {−2, −1, 4}
Range: {−4, 3, 5}

13. Domain: {−4, −3.5, −2, 2}
Range: {−11, −8, −6, −5}

15. Domain: {0, 1, 2, 3}
Range: {1.1, 1.5, 1.8, 2}

17. **(a)** Independent variable: x
Dependent variable: $f(x)$

(b)

x	$f(x)$
2	−3
−3	−8

19. **(a)** Independent variable: x
Dependent variable: $g(x)$

(b)

x	$g(x)$
−1	−2.5
0	0.5
5	15.5

21. **(a)** Independent variable: x
Dependent variable: y

(b)

x	y
0	6
5	7.5
10	9

23. **(a)** Independent variable: t
Dependent variable: $f(t)$

(b)

t	$f(t)$
0	3.5
1	3.35
3	3.05
6	2.6

25. **(a)** Independent variable: x
Dependent variable: $f(x)$

(b)

x	$f(x)$
1	1
4	2.2
6	3

27. The table represents a function; there is a unique y-value for every x-value.

29. The table does not represent a function; there are two y-values for the x-value of −2.

31. The graph represents a function.

33. The graph does not represent a function.

35. **(a)**

x	Expression	y
0	$3(0) + 1$	1
2	$3(2) + 1$	7
1	$3(1) + 1$	4

(b) $f(x) = 3x + 1$

37. **(a)**

x	Expression	y
0	$6 - 3(0)$	6
1	$6 - 3(1)$	3
2	$6 - 3(2)$	0
3	$6 - 3(3)$	−3

(b) $f(x) = 6 - 3x$

39. $m = \frac{3}{10}$; increasing

41. $m = -3$; decreasing

43. $m = 0$; horizontal

45. m is undefined; vertical

47. $m = -\frac{3}{2}$; decreasing

49. y-intercept: (0, −10)
x-intercept: (5, 0)

51. y-intercept: (0, −2)
x-intercept: (3, 0)

53. y-intercept: (0, −4)
x-intercept: (8, 0)

55. y-intercept: $\left(0, -\frac{5}{2}\right)$
x-intercept: $\left(\frac{1}{2}, 0\right)$

57. y-intercept: (0, 10.3)
x-intercept: (−51.5, 0)

59. $C(x) = 70x + 10{,}000$

61. $C(x) = 134x + 14{,}500$

63. **(a)** Fixed costs are \$1250 and the cost per unit is \$3.
(b) $C(400) = 2450$. The cost to produce 400 quick-replace spokes is \$2450.
(c) 90 quick-replace spokes can be produced for \$830.

65. **(a)** The fixed costs are \$10 and the cost per unit is \$32.
(b) $C(15) = 490$
To arrange and deliver 15 get-well bouquets, the cost is \$490.
(c) 250 get-well bouquets can be arranged and delivered for \$8010.

67. $m = 2.42$
From 1989 to 1999, public expenditures at private universities and colleges in the United States increased at an average rate of 2.42 billion dollars per year.

Section 1.3

1. $\frac{2}{3}x - y = \frac{19}{3}$ **3.** $3x + y = 12$ **5.** $4x - y = 7$

7. $1.375x - y = -3.725$ **9.** $x = 0$ **11.** $x = 9$

13. $\frac{2}{3}x + y = 2$ **15.** $f(x) = 3x - 20$ **17.** $f(x) = -\frac{5}{8}x + \frac{29}{8}$

19. $f(x) = 5x - 13$ **21.** $f(x) = \frac{2}{11}x + \frac{31}{11}$

23. $f(x) = 0.1825x + 0.0875$ **25.** $f(x) = -\frac{1}{4}x + \frac{7}{4}$

27. $f(x) = \frac{1}{8}x + \frac{3}{2}$ **29.** $f(x) = -7.25x + 32.55$

31. $y = \frac{1}{6}x + \frac{8}{3}$ **33.** $y = -\frac{2}{7}x + \frac{22}{7}$

35. **(a)** Let x = age of photocopier, in years. Let y = value of photocopier, in dollars.

$y = 25{,}000 - 1250x$

(b) The photocopier will be worth \$10,000 in 12 years.

37. **(a)** Let x = age of machine press, in years. Let y = value of machine press, in dollars.

$y = 90{,}000 - 6000x$

(b) The machine press will be worth half of its original value in 7.5 years.

39. **(a)** The amount of money spent on pollution abatement is generally increasing at an average rate of \$2 billion per year.

(b) $f(12) = 72.17$

In 1984, \$72.17 billion were spent on pollution abatement.

(c) In 1992.

41. **(a)** $f(50) = 163.15$

In 1950, the population of the United States was about 163.15 million people.

(b) $x \approx 43.15$

In 1943, the population of the United States was about 150 million people.

43. **(a)** Spending in college bookstores increases \$190,000,000 each year, on average.

(b) In 1985, \$2.24 billion was spent in college bookstores in the United States.

(c) In 1990, \$3.19 billion was spent in college bookstores in the United States.

(d) \$0.95 billion more was spent in 1990 than in 1985.

45. **(a)** $g(x) = y = 0.12x + 4.24$

(b) In 1994, there were 4.72 million children in the United States with physical disabilities.

47. **(a)** $y = 1776.71(x - 5) + 25{,}483.95$

(b) During the year 1997, the average price for an import car was \$30,000.

49. **(a)** $m = 4.36$

The average rate of change in the median price from 1989 to 1999 was \$4360 per year.

(b) $y = 4.36x + 96.4$

(c) During 1994 the median price of a single family house in the southern United States was \$118.2 thousand.

51. **(a)** $m = 1.52$

The average rate of change in the adult civilian labor force in the United States from 1990 to 2000 was increasing at a rate of 1.52 million people per year.

(b) $f(x) = 1.52x + 125.50$

(c) $f(4) = 131.58$

In 1994, the size of the adult civilian labor force in the United States was 131.58 million people.

53. **(a)** $m = 2.94$

The number of motor vehicle registrations in the United States from 1990 to 2000 was increasing at an average rate of 2.94 million vehicles per year.

(b) $f(x) = 2.94x + 188.34$

(c) The y-intercept is (0,188.34). In 1990, the number of motor vehicle registrations was 188.34 million.

55. **(a)** $11x - 100y = -1375$

(b) $x = 5$

There were 14.3 million students enrolled in higher education institutions in the United States in 1995.

Section 1.4

1. Yes **3.** No **5.** No **7.** No **9.** No

11. The solution to the system is the ordered pair $(3, -1)$.

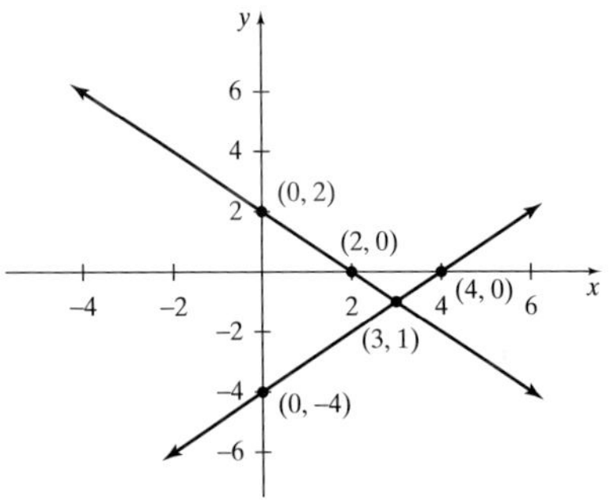

13. The solution to the system is the ordered pair $(0, 4)$.

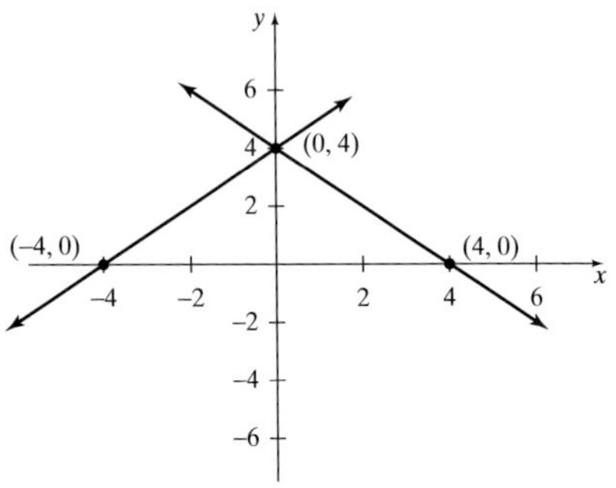

15. The solution to the system is the ordered pair $(0, -3)$.

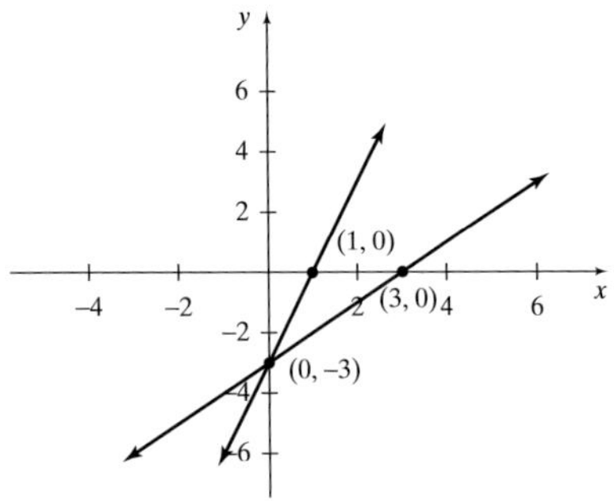

17. Parallel lines; no solution.

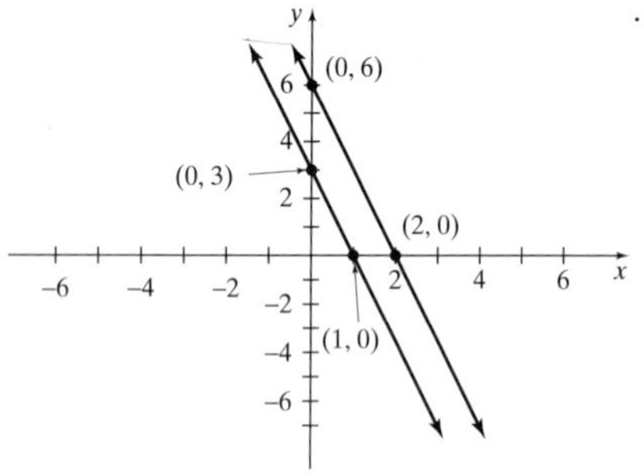

19. Graphs coincide; infinitely many solutions.

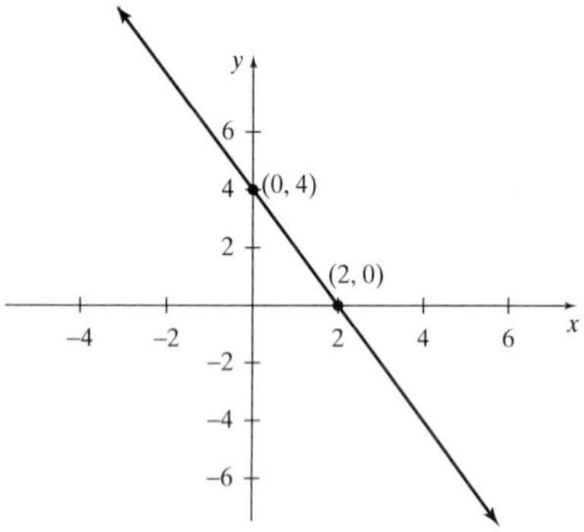

21. The intersection is at (2, −1).

23. The intersection is at (1, 0).

25. (6, −2)

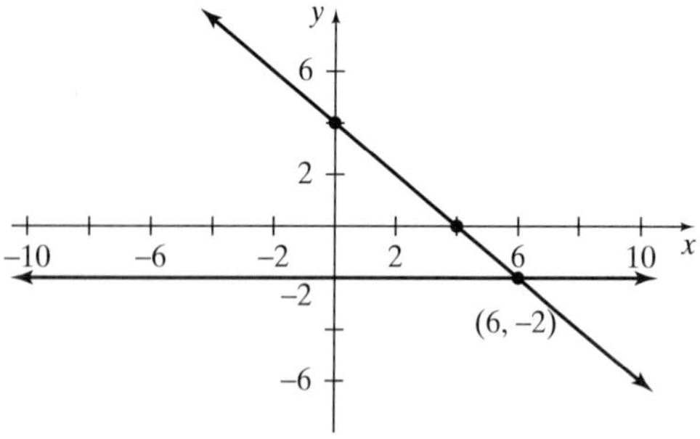

27. (4, −4)

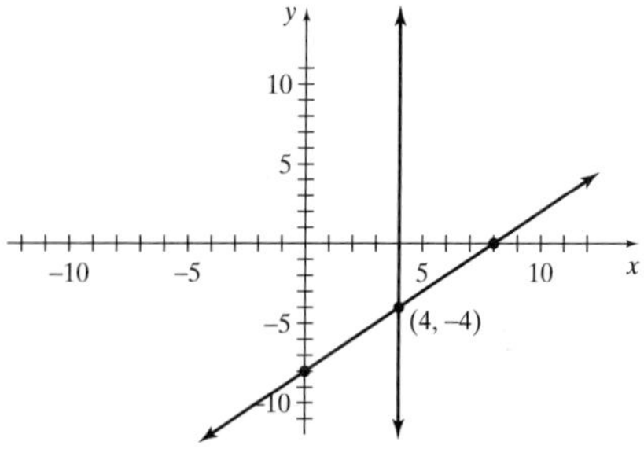

29. (3, −2)

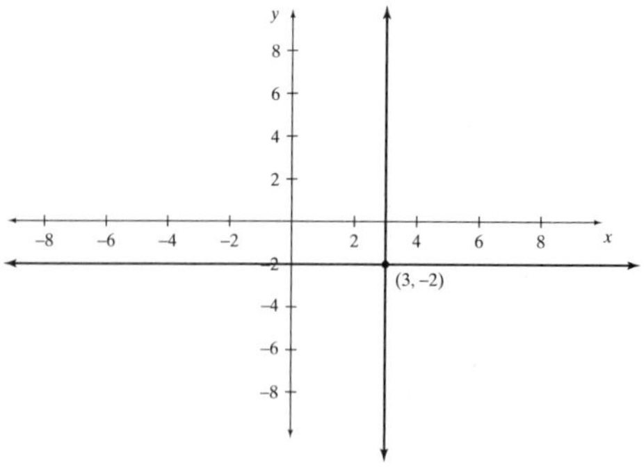

31. (2, −1)

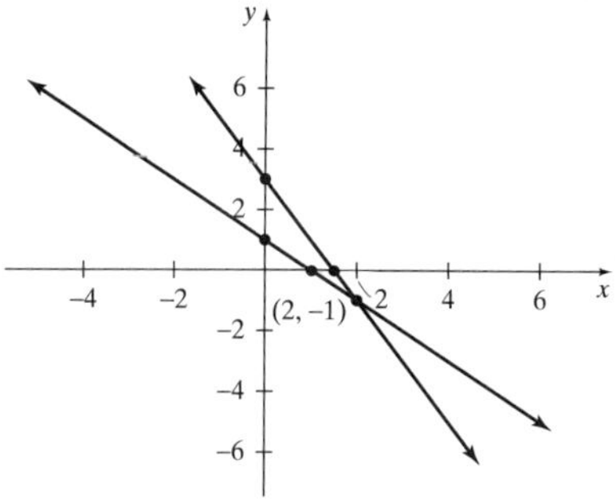

33. The solution to the system is the ordered pair (−3, −2).

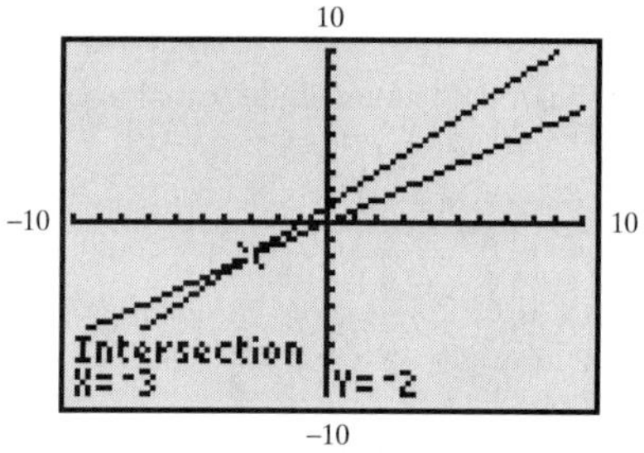

35. The system has no solution.

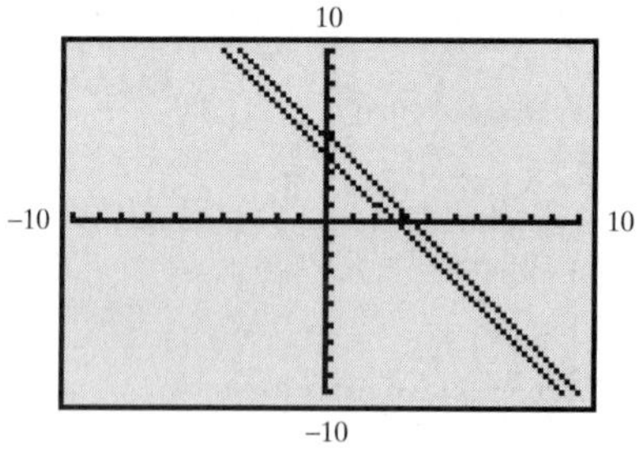

37. The solution to the system of equations is the ordered pair (0, 3).

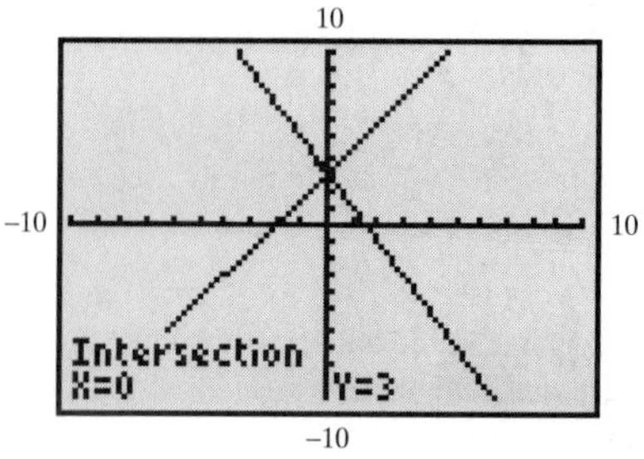

39. $k = 3$. **41.** $k = 6$.

43. Answers vary. They are in the form $k = z - \frac{1}{2}y$, where z is any number, $z \neq 2$.

45. (a) x-intercept: (133.33, 0)
y-intercept: (0, 80)

(b) $s(0) = 0$
$s(130) = 52$

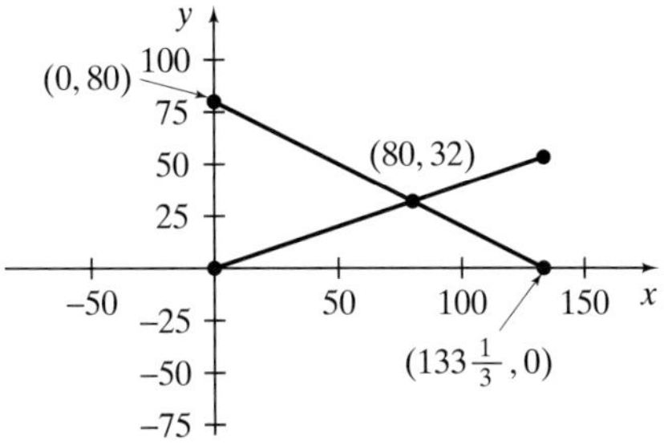

(c) The intersection is at the point (80, 32). The supply will meet the demand when 80 teleconference mini-suites are supplied at a price of \$32 per unit.

47. (a) x-intercept: (750, 0)
y-intercept: (0, 150)

(b) $s(0) = 0$
$s(400) = 160$

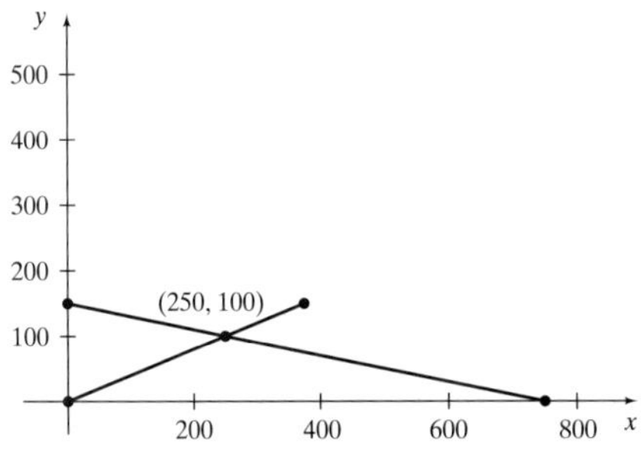

(c) The intersection is at the point (250, 100). The supply will meet the demand when 250 TK-2 model color televisions are supplied at a cost of \$100 per unit.

49. (a) $d(x) = y = 75 - \frac{1}{4}x$ $\quad s(x) = y = \frac{1}{2}x$

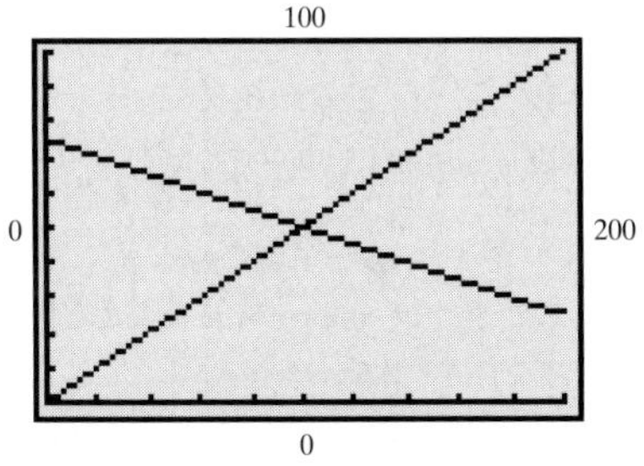

(b) The intersection point is (100, 50).

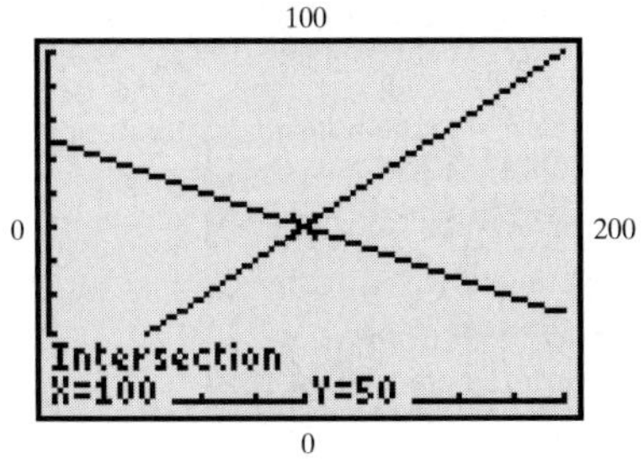

49. (b) *(continued)* The daily supply will meet demand when 100 pup tents are supplied at a cost of \$50 per tent.

51. (a) $C(x) = y = 252 + 1.5x$
$R(x) = y = 5.5x$

(b) The lines intersect at about (60, 350). This is the break-even point. The x-coordinate is the quantity of frames produced and the y-coordinate is the amount, in dollars, of the cost and revenue.

53. Let x = the quantity of cribs produced, $C(x)$ = the cost, in dollars, to produce the cribs, and $R(x)$ = the revenue, in dollars, from sales of the cribs.
$C(x) = 250 + 100x$
$R(x) = 125x$

55. Let x = quantity of bumper stickers sold, $C(x)$ = the cost, in dollars, to produce the bumper stickers, and $R(x)$ = the revenue, in dollars, from sales.
$C(x) = 200 + 3x$
$R(x) = 5x$

57. Let x = the quantity of radios, $C(x)$ = the cost to produce the radios, and $R(x)$ = the revenue, in dollars, from sales.
$C(x) = 20{,}000 + 30x$
$R(x) = 50x$

59. Let x = the quantity of pressure washers, $C(x)$ = the cost, in dollars, to produce the pressure washers, and $R(x)$ = the revenue, in dollars, from sales.
$C(x) = 100{,}000 + 150x$
$R(x) = 250x$

61. (a) $f(0) = 2.51$
$f(8) = 3.47$
$g(0) = 2.84$
$g(8) = 2.68$

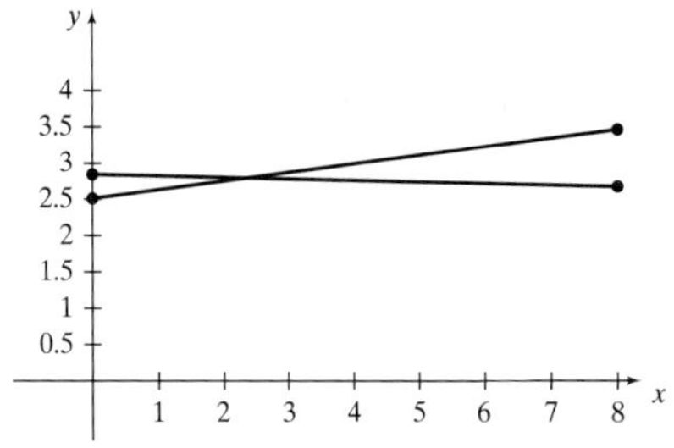

(b) The point of intersection is approximately (2.5, 3). In 1992, there were approximately 3 million children under the age of 5 who were of Hispanic origin and 3 million children under the age of 5 who were of African-American origin in the United States.

63. (a) $f(0) = 2.335$
$f(8) = 2.527$
$g(0) = 2.629$
$g(8) = 2.445$

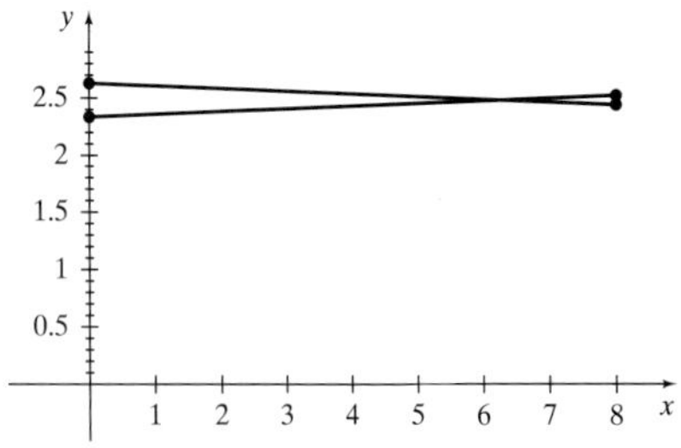

(b) The intersection point is approximately (6.5, 2.5). In 1996, there were approximately 2.5 million people 25 to 29 years of age who were of Hispanic origin and 2.5 million people 25 to 29 years of age who were of African-American origin in the United States.

Section 1.5

1. $(x, y) = (2, 2)$ **3.** $(x, y) = (1, 3)$ **5.** $(x, y) = (5, 1)$

7. $(x, y) = (4, 1)$ **9.** $(x, y) = (-1, -2)$

11. There is no solution to the system. **13.** $(x, y) = (3, 7)$

15. $(x, y) = (x, 3x - 5)$ **17.** $(x, y) = \left(\frac{8}{3}, \frac{7}{3}\right)$

19. $(x, y) = (0.2, -0.3)$ **21.** $(x, y) = (2, 4)$

23. $(x, y) = (5, 2)$ **25.** $(x, y) = (-4, -2)$

27. $(x, y) = (5, -6)$ **29.** $(x, y) = (1, 3)$

31. There is no solution to the system of equations.

33. $(x, y) = \left(-1, \frac{5}{2}\right)$ **35.** $(x, y) = \left(\frac{12}{5}, \frac{3}{5}\right)$

37. $(x, y) = (2, 3)$ **39.** $(x, y) = \left(\frac{7}{25}, -\frac{1}{25}\right)$

41. $(x, y, z) = \left(-\frac{11}{27}, \frac{79}{27}, \frac{22}{27}\right)$ **43.** $(x, y, z) = (1, -3, 3)$

45. $(x, y, z) = (-3, 0, 4)$

47. There is no solution to the system of equations.

49. $(x, y, z) = (4, -5, 8)$

51. **(a)** Let x = price of a coat and y = price of the slacks.

$2x + y = 120$

$x + 2y = 120$

(b) $(x, y) = (40, 40)$

The price for the coats was \$40 and the price of the slacks were also \$40.

53. $(x, y) \approx (4.4, 250.29)$

In 1984, both the District of Columbia and South Dakota had about 250 thousand households.

55. $(x, y) \approx (4.78, 555.29)$

In 1994, population of the District of Columbia and the population of Alaska were equal at about 555.29 thousand people.

57. **(a)** $x + y + z = 100$

(b) $4x + 6y + 7z = 520$

$6x + 9y + 12z = 810$

(c) $(x, y, z) = (50, 30, 20)$

The company should manufacture 50 boxes of standard clips, 30 boxes of large clips, and 20 boxes of jumbo clips.

59. **(a)** $2x + 3y + z = 14$

$x + 2y + z = 9$

$2x + y + 2z = 9$

(b) $(x, y, z) = (2, 3, 1)$

Two servings of Food A, three servings of Food B, and one serving of Food C are needed to provide the necessary number of grams of fat, carbohydrates and protein.

Chapter 1 Review Exercises

1. $x + 7 = 12$ **3.** $3x = 36$ **5.** $9x + 5 = 41$

7. **(a)** $7x + 5y = 70$ **(b)** $4x + 4y = 48$

9. **(a)** $x + y + z = 20{,}000$ **(b)** $y = \frac{1}{4}x + 8000$

(c) $0.06x + 0.07y + 0.05z = 1280$

11.

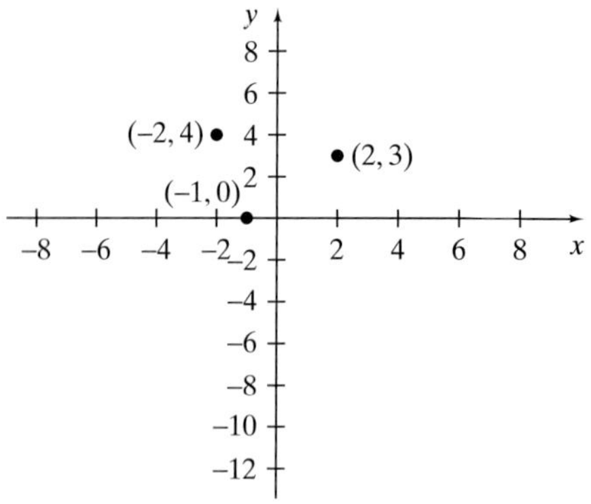

13. Domain: $\{-2, -1, 2\}$
Range: $\{0, 3, 4\}$

15. **(a)** x is the independent variable.

$g(x)$ is the dependent variable.

(b)

x	$g(x)$
-2	$-\frac{13}{4}$
4	$\frac{35}{4}$
10	$\frac{83}{4}$

17. This table does not represent a function.

19. This graph represents a function.

21. **(a)**

x	Expression	y
0	$2(0) - 1$	-1
4	$2(4) - 1$	7
$\frac{1}{2}$	$2(\frac{1}{2}) - 1$	0
2	$2(2) - 1$	3

(b) $y = 2x - 1$

23. $m = \frac{1}{4}$
Increasing

25. The x-intercept is the point $(7, 0)$.
The y-intercept is the point $(0, -14)$.

27. $C(x) = 17.33x + 210$

29. **(a)** The fixed cost is \$700, and the per unit cost is \$1.50.

(b) $C(50) = 775$

It costs \$775 to produce 50 baseballs.

(c) 75 baseballs can be produced for \$812.50.

31. $3x - 4y = -19$ **33.** $x = -5$ **35.** $y = 2x - 6$

37. $g(x) = -3x - 5$ **39.** $y = -3x + 2$

41. **(a)** Let y represent the value of the car in dollars after x years.

$y = -800x + 15{,}525$

(b) In four years the car will have a value of \$12,325.

43. **(a)** The total pounds of garbage produced per year is increasing at a rate of 50 pounds per year.

(b) 5,200 pounds

(c) $f(5) = 5450$

In the year 2000 the housing community was producing 5450 pounds of garbage per year.

45. **(a)** $y = 2.07(x - 2) + 20.15$

(b) The average price of a fishing license exceeded \$24.50 during the year 1994.

47. The given ordered pair is a solution to the system.

49. There are infinite solutions because the graphs are the same line.

51. The solution is the point of intersection $(2, -3)$.

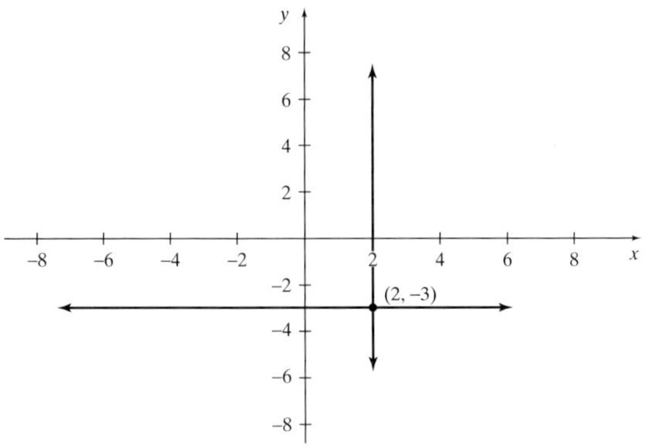

53. The lines are parallel; there is no solution to the system.

55. $k = 12$

57. **(a)** $d(x) = 100 - \frac{1}{5}x$

$s(x) = \frac{3}{5}x$

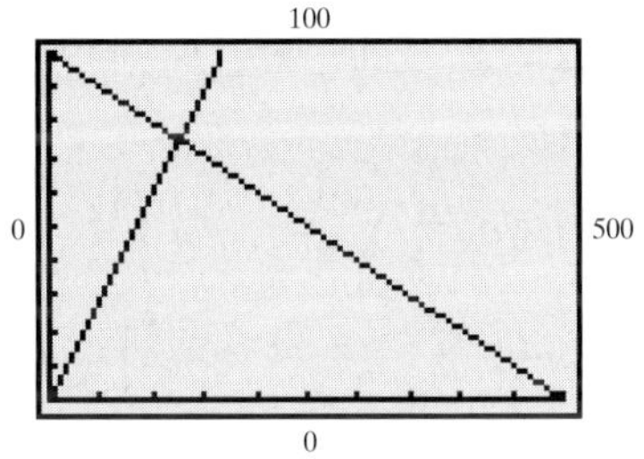

(b) The point of intersection is $(125, 75)$.
The supply will meet the demand when 125 shoes are supplied at a price of \$75 per shoe.

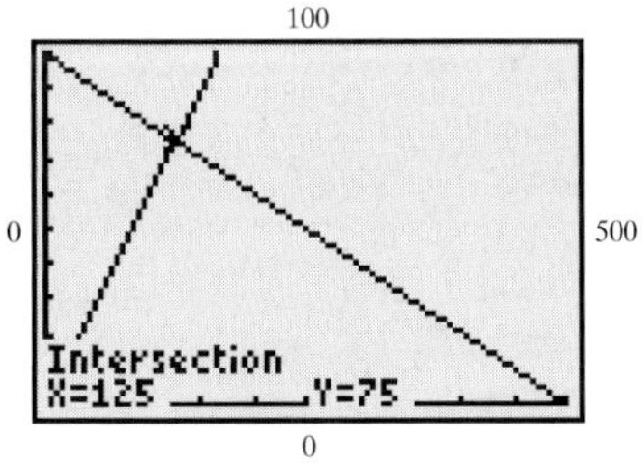

59. **(a)** Let x = the quantity of sweaters produced, $C(x)$ = the cost to produce x sweaters, and $R(x)$ = the revenue from producing x sweaters.

$C(x) = 14x + 65$
$R(x) = 30x$

(b) The intersection point is approximately (4,122). Molly needs to produce and sell approximately 4 sweaters for \$122 in order to break even.

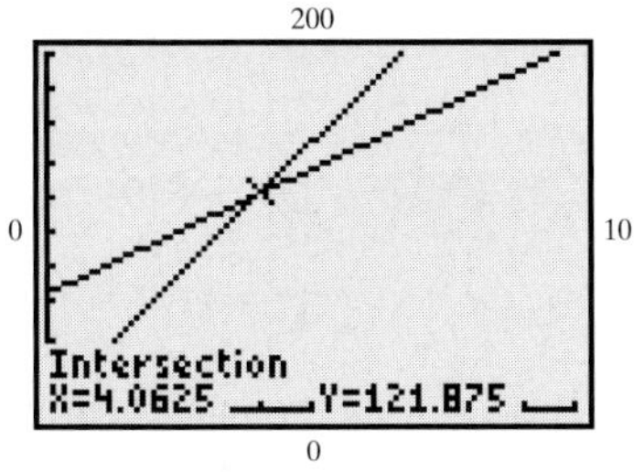

61. The solution to the system is $(-2, 1)$.

63. The solution to the system is $(0, -3)$.

65. The solution to the system is $(10, -2)$.

67. There are infinitely many solutions. The solution in parameter form is $(3y + y, y)$.

69. The solution to the system of equations is $(1, 1, -2)$.

71. **(a)** Let x = the cost of the T-shirts and y = the cost of the jeans.

$$3x + 2y = 90$$
$$x + 4y = 130$$

(b) $x, y = (10, 30)$

The T-shirts were \$10 each and the jeans were \$30 each.

73. The point of intersection is approximately (7.22, 3.91). Approximately 7.22 years after July 1990 (about September 1997) the populations of Colorado and Kentucky were about equal at 3.91 million people in each state.

75. **(a)**
$$4x + 2y + z = 14$$
$$3x + y + 2z = 15$$
$$2x + 2y + 3z = 18$$

(b) $x, y, z = (2, 1, 4)$

Two servings of food A, one serving of food B, and four servings of food C are required to satisfy requirements.

Chapter 2

Section 2.1

1. 2×3 **3.** 3×2 **5.** 3×4 **7.** 4×1 **9.** $b_{3,1} = 5$

11. $d_{2,3} = -2$ **13.** $f_{2,3} = 7$ **15.** $h_{1,3} = 2$

17. $\left[\begin{array}{cc|c} 3 & -7 & 4 \\ 2 & 5 & 8 \end{array}\right]$ **19.** $\left[\begin{array}{cc|c} 1 & 3 & -11 \\ 2 & 5 & -17 \end{array}\right]$

21. $\left[\begin{array}{ccc|c} 1 & 1 & 1 & 1 \\ 3 & 1 & -1 & -2 \\ 2 & -1 & 3 & 4 \end{array}\right]$ **23.** $\left[\begin{array}{ccc|c} 3 & 1 & 10 & 38 \\ 5 & 10 & 1 & 22 \\ 7 & -1 & -12 & -15 \end{array}\right]$

25. $2x - 3y = 7$, $-2x + 4y = 3$ **27.** $x = 8$, $y = -3$ **29.** $3x - y + 4z = 7$, $2x + 8y - 3z = 2$, $-x + 4y + 5z = -3$

31. $x = 8$, $y = -2$, $z = 7$ **33.** $\left[\begin{array}{cc|c} 1 & -1 & -7 \\ 3 & -1 & -4 \end{array}\right]$ **35.** $\left[\begin{array}{cc|c} 1 & 3 & -2 \\ 0 & 10 & -1 \end{array}\right]$

37. $\left[\begin{array}{ccc|c} 1 & -1 & -\frac{3}{4} & -\frac{1}{2} \\ -5 & 18 & 4 & 22 \\ 7 & 31 & -2 & 37 \end{array}\right]$ **39.** $\left[\begin{array}{ccc|c} 1 & -1 & -1 & -2 \\ 0 & 3 & 2 & 5 \\ 0 & -6 & 3 & 9 \end{array}\right]$

41. The solution is $x = 13$, $y = 11$.

43. The solution is $x = 11$, $y = 5$.

45. The solution is $x = 4$, $y = 3$.

47. The solution is $x = -\frac{41}{27}$, $y = \frac{22}{27}$.

49. The solution is $x = 1$, $y = 0$, $z = -1$.

51. The solution is $x = 6$, $y = -2$, $z = 3$.

53. The solution is $x = \frac{21}{4},\ y = 0,\ z = -\frac{1}{4}$.

55. The solution is $x = \frac{39}{35},\ y = -\frac{44}{35},\ z = -\frac{8}{5}$.

57. The solution is $x = -\frac{1}{2},\ y = \frac{11}{10},\ z = -\frac{13}{10},\ w = -\frac{9}{5}$

59. $x = 1.5738$, $y = -1.5229$ **61.** $x = 1.9668$, $y = -0.0922$

63. $x = 1.2752$, $y = -1.0035$, $z = 0.1508$ **65.** $x = -0.1178$, $y = 0.2013$, $z = -1.0281$, $w = -0.1270$

67. (a) $5x + 4y = 40$
$200x + 80y = 1200$

(b) Max can make 4 end tables and 5 night stands.

69. The company should make 544 two-person models and 96 four-person models to operate at full capacity.

71. Six boxes should be ordered from Tidy Trader wholesaler and 2 boxes from the Trade-it-Now wholesaler.

73. (a) $0.5x + 0.8y + 1.1z = 450$
$0.6x + 0.7y + 0.9z = 390$
$0.2x + 0.3y + 0.6z = 190$

(b) Twenty-two-person tents, 396 four-person tents, and 112 six-person tents can be made in one week.

75. (a) $x + y + z = 50{,}000$
$0.06x + 0.07y + 0.08z = 3410$
$-0.5x + z = 1000$

(b) The corporation borrowed \$20,000 at 6%, \$19,000 at 7%, and \$11,000 at 8%.

77. $\frac{8}{7}$ quarts of the 21% acid solution and $\frac{6}{7}$ quarts of the 14% acid solution should be used to make 2 quarts of an 18% acid solution.

79. The patient should eat 4 ounces of Basic!, 5 ounces of Vaca, and 7 ounces of Secra.

81. (a) It is not possible for him to use all his resources completely.

(b) With the additional amount of nitrogen, Juan can use all resources completely. He should plant 30 acres of asparagus, 210 acres of cauliflower, and 40 acres of celery.

83. $\left[\begin{array}{ccc|c} 0.8 & 1.1 & 1.7 & 32.1 \\ 0.6 & 1.3 & 1.5 & 29.7 \\ 1.2 & 0.8 & 1.3 & 33.0 \end{array}\right]$

In reduced row echelon form:

$$\left[\begin{array}{ccc|c} 1 & 0 & 0 & 15 \\ 0 & 1 & 0 & 9 \\ 0 & 0 & 1 & 6 \end{array}\right]$$

Tammy K's should admit 15 students to the medical training, 9 students to the small business training, and 6 students to the executive secretary training.

Section 2.2

1. $\begin{bmatrix} 2 & -1 & 0 \\ 1 & 4 & 5 \end{bmatrix}$ **3.** $A + B$ cannot be added. **5.** $\begin{bmatrix} 5 & -4 \\ 1 & -7 \end{bmatrix}$

7. $\begin{bmatrix} 4 & 0 & -6 \\ -2 & 6 & -8 \end{bmatrix}$ **9.** $[4 \quad -4 \quad 16 \quad -12]$

11. $\begin{bmatrix} 2 & -2 & 3 \\ 3 & 5 & 14 \end{bmatrix}$ **13.** $\begin{bmatrix} -13 & -4 \\ -21 & 25 \\ 23 & -14 \end{bmatrix}$ **15.** $\begin{bmatrix} 14 & 0 \\ 18 & -18 \\ -18 & 16 \end{bmatrix}$

17. *B* and *C*. **19.** *E*.

21. *D* and *G* cannot be added because they do not have the same dimension. *D* is a 2×3 matrix and *G* is a 3×2 matrix.

23. $D^T = \begin{bmatrix} 4 & 0 \\ -1 & 7 \\ -3 & 1 \end{bmatrix}; 3 \times 2$ **25.** $C^T = \begin{bmatrix} -4 & 3 \\ 1 & 5 \end{bmatrix}; 2 \times 2$

27. $\begin{bmatrix} 0 & 0 & 0 \\ 0 & 0 & 0 \end{bmatrix}$ **29.** $\begin{bmatrix} 0 & 0 \\ 0 & 0 \\ 0 & 0 \end{bmatrix}$ **31.** $x = -1;\ y = 4;\ z = 3$

33. $x = 5,\ y = 3,\ z = 1$ **35.** $x = 5,\ y = 13,\ z = 3$

37. $x = 27,\ y = 4,\ z = 2,\ w = 3,\ p = \frac{1}{2}$

39. $\begin{bmatrix} 9.43 & 13.9 & 2.6 \\ 7.29 & 5.92 & 10.2 \end{bmatrix}$ **41.** $\begin{bmatrix} 26.97 & 37.03 & 6.61 \\ 18.73 & 14.98 & 24.99 \end{bmatrix}$

43. $\begin{bmatrix} 1.716 & 6.071 & 1.547 \\ 4.082 & 3.614 & 7.293 \end{bmatrix}$ **45.** $\begin{bmatrix} 1.7497 & 9.4988 & 2.5806 \\ 6.7363 & 6.0548 & 12.5154 \end{bmatrix}$

47. No answer; verification.

49. (a) $J = \begin{bmatrix} 13 & 7 \\ 5 & 12 \end{bmatrix},\ Y = \begin{bmatrix} 12 & 9 \\ 7 & 14 \end{bmatrix},\ A = \begin{bmatrix} 14 & 11 \\ 3 & 19 \end{bmatrix}$

(b) $J + Y + A = \begin{bmatrix} 39 & 27 \\ 15 & 45 \end{bmatrix}$

51. $\begin{bmatrix} 13 & 9 \\ 5 & 15 \end{bmatrix}$

This represents the average that was sold at Krizan's Northside dealership for the months of June, July, and August.

53. (a) $\begin{bmatrix} 26.5 & 37 \\ 12.5 & 26 \end{bmatrix}$

This represents the average of material and labor costs for the East and West Coast plants.

(b) $1.1E = \begin{bmatrix} 27.5 & 38.5 \\ 11 & 27.5 \end{bmatrix},\ 0.95W = \begin{bmatrix} 26.6 & 37.05 \\ 14.25 & 25.65 \end{bmatrix}$

55. (a) $I = 0.08A = \begin{bmatrix} 1600 & 2000 \\ 1200 & 2800 \end{bmatrix}$

(b) $A + I = \begin{bmatrix} 21{,}600 & 27{,}000 \\ 16{,}200 & 37{,}800 \end{bmatrix}$

57. (a) $MCR = \begin{bmatrix} 32.3 \\ 33.2 \\ 33.1 \\ 33.9 \\ 34.7 \\ 35.2 \\ 35.6 \end{bmatrix}$ **(b)** $MCD = \begin{bmatrix} 24.3 \\ 29.4 \\ 31.7 \\ 31.6 \\ 31.9 \\ 31.5 \\ 29.0 \end{bmatrix}$

(c) $TG = MCR + MCD = \begin{bmatrix} 56.6 \\ 62.6 \\ 64.8 \\ 65.5 \\ 66.6 \\ 66.7 \\ 64.6 \end{bmatrix}$

This represents the number of people, in millions, covered by Medicare and Medicaid insurance from 1990 through 1997.

59. **(a)** $m_{3,2} = 31.4$; 31.4% of males in the United States between the ages of 25 and 34 are smokers. $f_{3,2} = 28.8$; 28.8% of females in the United States between the ages of 25 and 34 are smokers.

(b) $M - F = \begin{bmatrix} -2.4 & 6.2 & 6.1 & 3.5 & 6.1 \\ 4.1 & 3.4 & 9.7 & 4.5 & 3.1 \\ 4.6 & 2.6 & 6.4 & 5.5 & 2.1 \end{bmatrix}$

61. **(a)** $M = \begin{bmatrix} 57.2 & 51.3 \\ 58.0 & 47.3 \\ 54.0 & 41.6 \\ 53.4 & 42.6 \end{bmatrix}$ **(b)** $F = \begin{bmatrix} 48.5 & 36.8 \\ 45.0 & 34.7 \\ 40.7 & 26.4 \\ 35.7 & 21.9 \end{bmatrix}$

(c) $M - F = \begin{bmatrix} 8.7 & 14.5 \\ 13.0 & 12.6 \\ 13.3 & 15.2 \\ 17.7 & 20.7 \end{bmatrix}$

Section 2.3

1. **(a)** 3×3 **(b)** 2×2 **3.** **(a)** 2×2 **(b)** 4×4

5. **(a)** Does not exist. **(b)** 5×4

7. **(a)** Does not exist. **(b)** Does not exist.

9. $\begin{bmatrix} -3 \\ -2 \end{bmatrix}$ **11.** $\begin{bmatrix} 2 \\ 10 \end{bmatrix}$ **13.** $\begin{bmatrix} 10 & -10 \\ 7 & -9 \end{bmatrix}$ **15.** $\begin{bmatrix} 28 & 1 \\ -8 & 17 \end{bmatrix}$

17. Matrix product does not exist.

19. $\begin{bmatrix} 5 & -6 \\ -2 & 1 \end{bmatrix}$ **21.** $\begin{bmatrix} 6 & -1 & 4 \\ 0 & 3 & 2 \\ 1 & 4 & 5 \end{bmatrix}$ **23.** $\begin{bmatrix} 440 \\ 212 \\ 620 \\ 694 \end{bmatrix}$

25. $\begin{bmatrix} 19.48 & 12.83 \\ 43.6 & 28.28 \end{bmatrix}$ **27.** $\begin{bmatrix} 1 & 0 \\ 0 & 1 \end{bmatrix}$ **29.** $\begin{bmatrix} 93{,}300 \\ 43{,}600 \\ 78{,}700 \end{bmatrix}$

31. $\begin{bmatrix} 1 & 0 & 0 \\ 0 & 1 & 0 \\ 0 & 0 & 1 \end{bmatrix}$

33. **(a)** $A^2 = \begin{bmatrix} 0.671 & 0.329 \\ 0.6674 & 0.3326 \end{bmatrix}$

$A^5 = \begin{bmatrix} 0.669811064 & 0.330188936 \\ 0.6698118416 & 0.3301881584 \end{bmatrix}$

$A^{10} = \begin{bmatrix} 0.6698113208 & 0.3301886792 \\ 0.6698113208 & 0.3301886792 \end{bmatrix}$

$A^{20} = \begin{bmatrix} 0.6698113208 & 0.3301886792 \\ 0.6698113208 & 0.3301886792 \end{bmatrix}$

(b) The values of $a_{1,1}$ and $a_{2,1}$ are becoming equal as the matrix is raised to increasingly higher powers. The same thing is happening to the entries $a_{2,1}$ and $a_{2,2}$.

35. $\begin{bmatrix} 3 & -2 \\ 2 & -5 \end{bmatrix}\begin{bmatrix} x \\ y \end{bmatrix} = \begin{bmatrix} 14 \\ 8 \end{bmatrix}$ **37.** $\begin{bmatrix} 1 & 1 \\ -3 & -2 \end{bmatrix}\begin{bmatrix} x \\ y \end{bmatrix} = \begin{bmatrix} 4 \\ -5 \end{bmatrix}$

39. $\begin{bmatrix} 1 & 2 & 2 \\ 1 & -1 & -1 \\ 2 & 5 & 9 \end{bmatrix}\begin{bmatrix} x \\ y \\ z \end{bmatrix} = \begin{bmatrix} 11 \\ -4 \\ 39 \end{bmatrix}$

41. $\begin{bmatrix} -2 & 3 & 4 \\ 3 & -5 & 2 \\ 4 & 2 & -3 \end{bmatrix}\begin{bmatrix} x \\ y \\ z \end{bmatrix} = \begin{bmatrix} 10 \\ 7 \\ -2 \end{bmatrix}$

43. **(a)** $MP = \begin{bmatrix} 965{,}300 & 1{,}436{,}200 \\ 1{,}000{,}100 & 1{,}487{,}800 \end{bmatrix}$

(b) $mp_{1,1}$ is the wholesale value of the three most popular selling SUVs at the East dealership. $mp_{1,2}$ is the retail value of the three most popular selling SUVs at the East dealership. $mp_{2,1}$ is the wholesale value of the three most popular selling SUVs at the West dealership. $mp_{2,2}$ is the retail value of the three most popular selling SUVs at the West dealership.

45. **(a)** $LW = \begin{bmatrix} 17.5 & 18.2 \\ 23.9 & 25.0 \\ 33.8 & 35.5 \end{bmatrix}$ **(b)** $18.20 per hour

(c) $33.80 per hour

(d) It costs $17.50 per hour to manufacture a two-person tent on the East coast, and $18.20 per hour to manufacture a two-person tent on the West coast. For the four-person tent, it costs $23.90 per hour to manufacture it on the East coast and $25.00 per hour to manufacture it on the West coast. For the six-person tent, the manufacturing costs are $33.80 per hour on the East coast and $35.50 per hour on the West coast.

47. **(a)** $A = \begin{bmatrix} 8 \\ 5 \\ 9 \end{bmatrix}$ **(b)** $B = \begin{bmatrix} 4 \\ 5 \\ 7 \end{bmatrix}$

(c) $FA = \begin{bmatrix} 480 \\ 285 \\ 325 \end{bmatrix}$

At lunchtime, the patient consumed 480 units of calcium, 285 units of vitamin A, and 325 units of vitamin C

(d) $FB = \begin{bmatrix} 340 \\ 205 \\ 225 \end{bmatrix}$

At dinner, the patient consumed 340 units of calcium, 205 units of vitamin A, and 225 units of vitamin C.

49. **(a)** $FP = \begin{bmatrix} 46{,}400 \\ 36{,}770 \\ 15{,}520 \end{bmatrix}$

(b) He will need 46,400 pounds of nitrogen, 36,770 pounds of phosphates, and 15,520 pounds of potash for the entire growing season.

51. $BA = [21{,}429.92 \quad 21{,}545.72 \quad 18{,}728.24]$

In 1985, there were 21,429.92 thousands of tons of these pollutants released into the atmosphere. In 1990, there were 21,545.72 thousands of tons of these pollutants released into the atmosphere, and in 1995, there were 18,728.24 thousands of tons of these pollutants released into the atmosphere. In all 3 years, pollutants caused by electric companies.

53. **(a)** $AP = \begin{bmatrix} 41{,}100 \\ 49{,}850 \end{bmatrix}$

(b) On Dec. 31, 2000, Denyse's stock holdings were valued at $41,100; Jamie's were valued at $49,850.

55. **(a)** $\frac{1}{3}(SC) = \begin{bmatrix} 82.67 \\ 87.0 \\ 91.3 \\ 86.67 \\ 81.67 \end{bmatrix}$

This matrix represents the average grade that each student received for the three tests given. Josh had an average score of 82.67, Juan's average score was 87.0, Maria's average score was 91.3, Diane's average was 86.67, and Zach's average score for the three tests was 81.67.

(b) $\begin{bmatrix} 87.4 & 82 & 88 \end{bmatrix}$

This matrix represents the class average for each test given. For the first test, the class average was 87.4. On the second test, the class earned an average of 82, and for the third test the average grade for the class was 88.

57. **(a)** $NF = \begin{bmatrix} 220 & 150 & 140 \\ 570 & 405 & 390 \\ 750 & 525 & 500 \end{bmatrix}$

(b) There are 150 grams per ounce of protein in Mix II.

(c) There are 570 grams per ounce of carbohydrates in Mix I.

(d) In Mix I there are 220 grams per ounce of protein, 570 grams per ounce of carbohydrates, and 750 grams per ounce of fat. In Mix II, there are 150 grams per ounce of protein, 405 grams per ounce of carbohydrates, and 525 grams per ounce of fat. In Mix III, there are 140 grams per ounce of protein, 390 grams per ounce of carbohydrates, and 500 grams per ounce of fat.

59. $BA = \begin{bmatrix} 89.778 & 18.531 & 79.3835 \end{bmatrix}$

Of the eligible voters, 89,778,000 will vote Democratic, 18,531,000 will vote Independent, and 79,383,500 will vote Republican.

Section 2.4

1. They are inverses of each other.

3. They are not inverses of each other.

5. They are inverses of each other.

7. They are not inverses of each other.

9. $\begin{bmatrix} -\frac{1}{5} & \frac{3}{10} \\ \frac{2}{5} & -\frac{1}{10} \end{bmatrix}$ **11.** $\begin{bmatrix} -\frac{1}{30} & -\frac{7}{30} \\ \frac{2}{15} & -\frac{1}{15} \end{bmatrix}$

13. $\begin{bmatrix} -\frac{11}{10} & -\frac{1}{5} & \frac{7}{10} \\ -\frac{9}{10} & \frac{1}{5} & \frac{3}{10} \\ \frac{3}{5} & \frac{1}{5} & -\frac{1}{5} \end{bmatrix}$ **15.** $\begin{bmatrix} -\frac{1}{9} & -\frac{1}{3} & \frac{2}{9} \\ \frac{11}{9} & \frac{2}{3} & -\frac{4}{9} \\ -\frac{2}{3} & 0 & \frac{1}{3} \end{bmatrix}$

17. $\begin{bmatrix} -1 & 1 & 1 & 0 \\ -11 & \frac{54}{5} & \frac{38}{5} & \frac{7}{5} \\ -6 & \frac{28}{5} & \frac{21}{5} & \frac{4}{5} \\ 4 & -\frac{19}{5} & -\frac{13}{5} & -\frac{2}{5} \end{bmatrix}$ **19.** $\begin{bmatrix} \frac{2}{7} & \frac{3}{28} \\ -\frac{1}{7} & \frac{1}{14} \end{bmatrix}$

21. The matrix does not have an inverse; it is singular.

23. $\begin{bmatrix} \frac{2}{11} & \frac{3}{11} \\ -\frac{1}{11} & -\frac{7}{11} \end{bmatrix}$ **25.** $\begin{bmatrix} \frac{7}{10} & \frac{1}{10} \\ \frac{2}{5} & \frac{1}{5} \end{bmatrix}$

27. No answer, verification.

29. $\begin{bmatrix} -\frac{21}{52} & 1 & -\frac{21}{52} \\ \frac{41}{52} & 0 & -\frac{11}{52} \\ -\frac{5}{26} & -1 & \frac{21}{26} \end{bmatrix}$ **31.** $\begin{bmatrix} -\frac{42}{113} & \frac{19}{113} & -\frac{31}{226} & \frac{79}{226} \\ \frac{22}{113} & \frac{7}{226} & \frac{27}{226} & -\frac{18}{113} \\ -\frac{13}{113} & \frac{1}{226} & \frac{10}{113} & \frac{11}{226} \\ \frac{38}{113} & -\frac{29}{226} & -\frac{15}{226} & \frac{10}{113} \end{bmatrix}$

33. $\begin{bmatrix} \frac{5}{22} & -\frac{5}{44} & \frac{3}{11} & -\frac{1}{44} \\ \frac{1}{11} & -\frac{13}{44} & \frac{31}{22} & \frac{37}{44} \\ \frac{3}{22} & -\frac{7}{22} & \frac{19}{22} & \frac{7}{11} \\ -\frac{5}{22} & \frac{4}{11} & -\frac{17}{22} & -\frac{5}{22} \end{bmatrix}$

35. $\begin{bmatrix} \frac{159}{1213} & -\frac{757}{1213} & -\frac{325}{1213} & -\frac{385}{1213} & \frac{428}{1213} \\ \frac{293}{1213} & -\frac{716}{1213} & -\frac{309}{1213} & -\frac{534}{1213} & \frac{392}{1213} \\ \frac{86}{1213} & -\frac{28}{1213} & \frac{137}{1213} & \frac{13}{1213} & -\frac{5}{1213} \\ -\frac{229}{1213} & \frac{2021}{1213} & \frac{552}{1213} & \frac{1531}{1213} & -\frac{1242}{1213} \\ -\frac{34}{1213} & -\frac{807}{1213} & -\frac{167}{1213} & -\frac{795}{1213} & \frac{679}{1213} \end{bmatrix}$

37. The solution is $x = \frac{13}{10}$, $y = \frac{19}{10}$.

39. The solution is $x = \frac{16}{15}$, $y = \frac{11}{15}$.

41. The solution is $x = -\frac{17}{80}$, $y = -\frac{133}{80}$, $z = \frac{31}{40}$.

43. The solution is $x = \frac{11}{9}$, $y = \frac{5}{9}$, $z = -\frac{2}{3}$.

45. The solution is $x = -3$, $y = -\frac{153}{5}$, $z = -\frac{81}{5}$, $w = \frac{58}{5}$.

47. The solution is $x = -\frac{45}{13}$, $y = \frac{112}{13}$, $z = -\frac{14}{13}$.

49. The solution is $x = \frac{53}{22}$, $y = \frac{70}{11}$, $z = \frac{111}{22}$, $w = -\frac{97}{22}$.

51. The solution is $x = \frac{3890}{1213}$, $y = \frac{4277}{1213}$, $z = \frac{357}{1213}$, $v = -\frac{12,728}{1213}$, $w = \frac{6347}{1213}$.

53. **(a)** $\begin{bmatrix} 1 & 2 & 2 \\ 3 & 2 & 3 \\ 2 & 2 & 1 \end{bmatrix} \cdot \begin{bmatrix} x \\ y \\ z \end{bmatrix} = \begin{bmatrix} 24 \\ 45 \\ 27 \end{bmatrix}$

(b) $\begin{bmatrix} x \\ y \\ z \end{bmatrix} = \begin{bmatrix} 8 \\ 3 \\ 5 \end{bmatrix}$

Daily orders are 8 loaves of Lemon Poppyseed glaze, 3 loaves of Pumpkin glaze, and 5 loaves of Banana Walnut glaze.

55. **(a)** Thursday: $x + y = 1600$, $3x + 5y = 5760$

Friday: $x + y = 1600$, $3x + 5y = 5980$

Saturday: $x + y = 1600$, $3x + 5y = 5500$

(b) Thursday: $\begin{bmatrix} 1 & 1 \\ 3 & 5 \end{bmatrix}\begin{bmatrix} x \\ y \end{bmatrix} = \begin{bmatrix} 1{,}600 \\ 5{,}760 \end{bmatrix}$

Friday: $\begin{bmatrix} 1 & 1 \\ 3 & 5 \end{bmatrix}\begin{bmatrix} x \\ y \end{bmatrix} = \begin{bmatrix} 1{,}600 \\ 5{,}980 \end{bmatrix}$

Saturday: $\begin{bmatrix} 1 & 1 \\ 3 & 5 \end{bmatrix}\begin{bmatrix} x \\ y \end{bmatrix} = \begin{bmatrix} 1600 \\ 5500 \end{bmatrix}$

(c) On Thursday night, 1120 three dollar tickets were sold and 480 five dollar tickets were sold. On Friday night, 1010 three dollar tickets were sold and 590 five dollar tickets were sold. On Saturday night, 1250 three dollar tickets were sold and 350 five dollar tickets were sold.

57. **(a)** $18x + 20y + 16z = 5900$

$12x + 9y + 11z = 3580$

$22x + 18y + 17z = 6350$

(b) $\begin{bmatrix} 18 & 20 & 16 \\ 12 & 9 & 11 \\ 22 & 18 & 17 \end{bmatrix} \cdot \begin{bmatrix} x \\ y \\ z \end{bmatrix} = \begin{bmatrix} 5900 \\ 3580 \\ 6350 \end{bmatrix}$

(c) $\begin{bmatrix} x \\ y \\ z \end{bmatrix} = \begin{bmatrix} 130 \\ 90 \\ 110 \end{bmatrix}$

To fill the order, Plant I should schedule 130 hours, Plant II should schedule 90 hours, and Plant III should schedule 110 hours.

59. $\begin{bmatrix} x \\ y \\ z \end{bmatrix} = \begin{bmatrix} 20 \\ 396 \\ 112 \end{bmatrix}$

The company can produce 20 two-person tents, 396 four-person tents, and 112 six-person tents operating at full capacity.

61. (a) $\begin{bmatrix} 1 & 1 & 1 \\ 0.06 & 0.075 & 0.085 \\ -\frac{1}{3} & 0 & 1 \end{bmatrix} \cdot \begin{bmatrix} x \\ y \\ z \end{bmatrix} = \begin{bmatrix} 120{,}000 \\ 8{,}350 \\ 5{,}000 \end{bmatrix}$

(b) $\begin{bmatrix} x \\ y \\ z \end{bmatrix} = \begin{bmatrix} 60{,}000 \\ 35{,}000 \\ 25{,}000 \end{bmatrix}$

The corporation borrowed $60,000 at 6.0%, $35,000 at 7.5%, and 25,000 at 8.5%.

63. $\begin{bmatrix} x \\ y \\ z \end{bmatrix} = \begin{bmatrix} 5 \\ 4 \\ 7 \end{bmatrix}$

To exactly meet the dietary requirements, Mona should use 5 ounces of Lacto, 4 ounces of Fracto, and 7 ounces of Hima.

65. To use all of his resources, Zach should plant 120 acres of head lettuce, 100 acres of cantaloupe, and 80 acres of bell peppers.

Section 2.5

1. 1 unit of agriculture requires 0.27 units of agriculture, 0.05 units of manufacturing, 0.06 units of trade, and 0.12 units of service.

3. 1 unit of trade requires 0 units of agriculture, 0 units of manufacturing, 0.36 units of trade, and 0.2 units of service.

5. $TX = \begin{bmatrix} 18{,}250 \\ 10{,}500 \end{bmatrix}$

The internal demands for utilities and transportation are $18,250 and $10,500, respectively.

7. $TX = \begin{bmatrix} 4{,}800 \\ 15{,}850 \\ 20{,}650 \end{bmatrix}$

The internal demands for agriculture, construction, and utilities are $4,800, $15,850 and $20,650, respectively.

9. To produce $1.00 worth of construction, 0.2 units of construction and 0.4 units of services are required.

11. $I - T = \begin{bmatrix} 0.8 & -0.1 \\ -0.4 & 0.4 \end{bmatrix}$ **13.** $X = \begin{bmatrix} \frac{145}{14} \\ \frac{160}{7} \end{bmatrix}$

$(I - T)^{-1} = \begin{bmatrix} \frac{10}{7} & \frac{5}{14} \\ \frac{10}{7} & \frac{20}{7} \end{bmatrix}$

15. To produce $1.00 worth of M, 0.2 units of A, 0.5 units of M, and 0 units of T are required.

17. $(I - T) = \begin{bmatrix} 0.8 & -0.2 & -0.1 \\ -0.4 & 0.5 & -0.2 \\ -0.1 & 0 & 0.8 \end{bmatrix}$ **19.** $\begin{bmatrix} \frac{6150}{247} \\ \frac{12{,}020}{247} \\ \frac{5400}{247} \end{bmatrix}$

21. The total output for agriculture and energy is $1.1 billion and $4.75 billion respectively.

23. The total output is $1955.88 million in agriculture and $293,227.26 million in manufacturing.

25. (a) $T = \begin{array}{c} \\ A \\ M \\ T \end{array} \begin{array}{c} \begin{array}{ccc} A & M & T \end{array} \\ \begin{bmatrix} 0.27 & 0.39 & 0 \\ 0.05 & 0.15 & 0 \\ 0.06 & 0.07 & 0.36 \end{bmatrix} \end{array}$ (b) $D = \begin{bmatrix} 8 \\ 23 \\ 10 \end{bmatrix}$

(c) $26.23 billion, $28.47 billion, and $21.09 billion for agriculture, manufacturing, and trade, respectively.

27. $73,894.20 million, $755,004.27 million, and $117,761.22 million for agriculture, manufacturing, and trade, respectively.

Chapter 2 Review Exercises

1. 3×2 **3.** -1 **5.** $\left[\begin{array}{ccc|c} -2 & 3 & -1 & 5 \\ 1 & 1 & 1 & 13 \\ 4 & -5 & 2 & -1 \end{array}\right]$

7. $x = -1,\ y = 2$ **9.** $(-5z + 4,\ 2z - 1,\ z)$

11. $x = -3.8248,\ y = -0.4023$

13. (a) $6x + 8y = 40$
$60x + 80y = 400$
(b) Four 4-wt rods, two 5-wt rods

15. (a) $0.3x + 0.4y + 0.6z = 310$
$0.4x + 0.5y + 0.8z = 405$
$0.1x + 0.2y + 0.3z = 150$

(b) 100 small, 250 medium, 300 large

17. 10 oz Alive, 8 oz Gday, 5 oz Cardio

19. $\begin{bmatrix} 3 & 4 & 0 \\ -1 & -8 & 0 \\ 5 & 0 & -1 \end{bmatrix}$ **21.** $\begin{bmatrix} 3 & 9 & -15 \\ 0 & -24 & -12 \\ 6 & -3 & 9 \end{bmatrix}$

23. $\begin{bmatrix} -1 & 7 & -25 \\ 2 & -24 & -20 \\ 0 & -5 & -13 \end{bmatrix}$ **25.** $\begin{bmatrix} 1 & -1 & 1 \\ -2 & 8 & 2 \\ 10 & 8 & 5 \end{bmatrix}$

27. $x = 5,\ y = 3,\ z = -7,\ w = 7,\ p = 4$

29. $\begin{bmatrix} -0.76 & -3.19 \\ -36.95 & -3.67 \\ 3.23 & -12.14 \end{bmatrix}$

31. (a) $I = \begin{bmatrix} 2250 & 1800 \\ 1350 & 2700 \end{bmatrix}$ (b) $\begin{bmatrix} 27{,}250 & 21{,}800 \\ 16{,}350 & 32{,}700 \end{bmatrix}$

(c) CD: 6%; TN: 8%

33. $A_{2,2} = 30{,}000$ and represents the amount of money in treasury notes at Hilltop Bank; $I_{1,2} = 1800$ and represents the amount of interest earned after 1 year on $20,000 worth of treasury notes at First Century.

35. (a) 5×5 (b) 2×2 **37.** $\begin{bmatrix} 9 \\ 2 \end{bmatrix}$

39. $\begin{bmatrix} -13 & -8 & 7 \\ 32 & 40 & 4 \\ 19 & 10 & -12 \end{bmatrix}$ **41.** $\begin{bmatrix} 49 \\ -21 \\ 3 \\ -27 \end{bmatrix}$

43. (a) $\begin{bmatrix} 0.3895 & 0.6105 \\ 0.3774 & 0.626 \end{bmatrix}, \begin{bmatrix} 0.3820324245 & 0.6179675755 \\ 0.3820163194 & 0.61798336806 \end{bmatrix},$

$\begin{bmatrix} 0.3820224721 & 0.6179775279 \\ 0.3820224718 & 0.6179775282 \end{bmatrix},$

$\begin{bmatrix} 0.3820224719 & 0.6179775281 \\ 0.3820224719 & 0.6179775281 \end{bmatrix}$

(b) As n increases, A^n approaches $\begin{bmatrix} 0.3820224719 & 0.6179975281 \\ 0.3820224719 & 0.6179975281 \end{bmatrix}$.

45. $\begin{bmatrix} 1 & 3 & -2 \\ -2 & 4 & -5 \\ -1 & -1 & 1 \end{bmatrix}\begin{bmatrix} x \\ y \\ z \end{bmatrix} = \begin{bmatrix} 4 \\ -1 \\ 7 \end{bmatrix}$

47. (a) $A = \begin{bmatrix} 5 \\ 8 \\ 12 \end{bmatrix}$ **(b)** $B = \begin{bmatrix} 6 \\ 10 \\ 16 \end{bmatrix}$

(c) $\begin{bmatrix} 251 \\ 285 \\ 129 \end{bmatrix}$; $fa_{1,1}$ represents the total units of iron consumed at breakfast from all three food types. $fa_{2,1}$ represents the total units of vitamin C consumed at breakfast from all three food types. $fa_{3,1}$ represents the total units of vitamin B consumed at breakfast from all three food types.

(d) $\begin{bmatrix} 318 \\ 360 \\ 164 \end{bmatrix}$; $fb_{1,1}$ represents the total units of iron consumed at lunch from all three food types. $fb_{2,1}$ represents the total units of vitamin C consumed at lunch from all three food types. $fb_{3,1}$ represents the total units of vitamin B consumed at lunch from all three food types.

49. (a) $\begin{bmatrix} 500 \\ 400 \\ 600 \end{bmatrix}$ **(b)** $\begin{bmatrix} 860 \\ 640 \end{bmatrix}$ **(c)** 860 people **(d)** 640 people

51. Not inverses

53. $\begin{bmatrix} -\frac{2}{5} & \frac{3}{10} & \frac{3}{20} \\ \frac{2}{5} & \frac{1}{5} & -\frac{2}{5} \\ \frac{1}{5} & \frac{1}{10} & \frac{1}{20} \end{bmatrix}$ **55.** $\begin{bmatrix} \frac{3}{13} & -\frac{2}{13} \\ \frac{1}{13} & \frac{3}{26} \end{bmatrix}$

57. $\begin{bmatrix} \frac{93}{239} & -\frac{40}{239} & -\frac{13}{239} \\ \frac{40}{239} & \frac{119}{239} & -\frac{59}{239} \\ -\frac{123}{239} & -\frac{109}{239} & \frac{102}{239} \end{bmatrix}$ **59.** $x = -1,\ y = 4$

61. $x = \frac{80}{239},\ y = -\frac{479}{239},\ z = -\frac{260}{239}$

63. (a) $X = \begin{bmatrix} x \\ y \\ z \end{bmatrix}$; $\begin{bmatrix} 1 & 3 & 3 \\ 2 & 3 & 4 \\ 1 & 2 & 3 \end{bmatrix}\begin{bmatrix} x \\ y \\ z \end{bmatrix} = \begin{bmatrix} 135 \\ 180 \\ 125 \end{bmatrix}$

(b) $\begin{bmatrix} x \\ y \\ z \end{bmatrix} = \begin{bmatrix} 15 \\ 10 \\ 30 \end{bmatrix}$ **(c)** 30 large pizzas

65. (a) $10x + 12y + 8z = 2800$
$15x + 7y + 12z = 2880$
$6x + 12y + 15z = 2970$

(b) $\begin{bmatrix} 10 & 12 & 8 \\ 15 & 7 & 12 \\ 6 & 12 & 15 \end{bmatrix}\begin{bmatrix} x \\ y \\ z \end{bmatrix} = \begin{bmatrix} 2800 \\ 2880 \\ 2970 \end{bmatrix}$

(c) $\begin{bmatrix} x \\ y \\ z \end{bmatrix} = \begin{bmatrix} 80 \\ 120 \\ 70 \end{bmatrix}$; plant I should operate 80 hr, plant II 120 hr, and plant III 70 hr.

67. (a) $\begin{bmatrix} 0.511 & 0.48 & 0.474 \\ 0.382 & 0.423 & 0.41 \\ 0.07 & 0.091 & 0.107 \end{bmatrix}\begin{bmatrix} x \\ y \\ z \end{bmatrix} = \begin{bmatrix} 9815 \\ 7930 \\ 1669 \end{bmatrix}$

(b) $\begin{bmatrix} x \\ y \\ z \end{bmatrix} \approx \begin{bmatrix} 10{,}020 \\ 5312 \\ 4525 \end{bmatrix}$, approximately 10,020,000, 5,312,000, and 4,525,000 people voted in California, Florida, and Ohio, respectively.

69. 0.18 unit of agriculture, 0.15 unit of energy, 0 units of chemicals

71. Util.: \$13,500; Trans.: \$45,000

73. 0.1 unit of C, 0.2 unit of M, 0.1 unit of T

75. $\begin{bmatrix} 0.9 & -0.3 & -0.1 \\ -0.2 & 0.9 & -0.3 \\ -0.4 & -0.1 & 0.5 \end{bmatrix}$

77. C: 19.6 million; M: 22.2 million; T: 40.1 million

79. A: 12.3 million; E: 21.7 million

81. (a) $T = \begin{array}{c} \quad A \quad\ \ C \quad\ \ E \\ \begin{bmatrix} 0.15 & 0.02 & 0.04 \\ 0.04 & 0.12 & 0.18 \\ 0.05 & 0.25 & 0.10 \end{bmatrix} \end{array}\begin{matrix} A \\ C \\ E \end{matrix}$ **(b)** $D = \begin{bmatrix} 12 \\ 8 \\ 15 \end{bmatrix}$

(c) A: 15.5 billion; C: 14.2 billion; E: 21.5 billion

Chapter 3

Section 3.1

1.

x	$x > 6$	Statement
−12	$-12 > 6$	False
2	$2 > 6$	False
6	$6 > 6$	False
10	$10 > 6$	True

3.

x	$x \le -4$	Statement
−8	$-8 \le -4$	True
−6	$-6 \le -4$	True
−4	$-4 \le -4$	True
0	$0 \le -4$	False

5.

x	$x \ge -1$	Statement
−2	$-2 \ge -1$	False
−1	$-1 \ge -1$	True
0	$0 \ge -1$	True
1	$1 \ge -1$	True

7. $x + 5 > 4$ **9.** $x - 8 < -18$ **11.** $9x \ge 54$

13. $3x - 15 \le -20$ **15.** $8x + 1 \ne 14$

17. $x > 7$

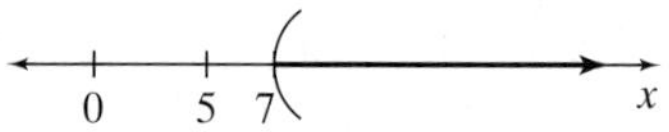

19. $x < 5$

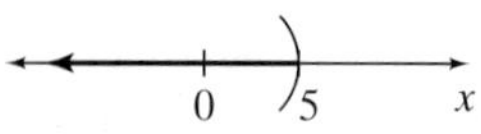

21. $x \leq 8$

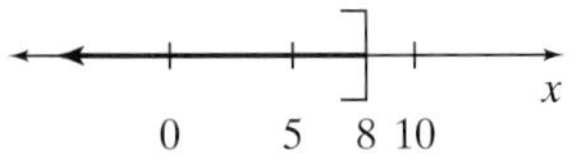

23. $x \geq 5$

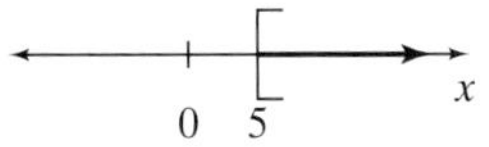

25. $x < -5$

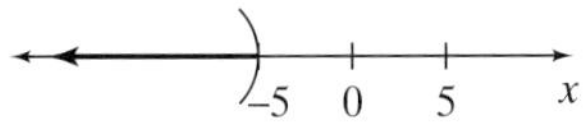

27. $x \geq 3$

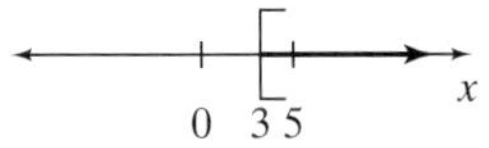

29. $2x + 3y \leq 32$
$2.5x + 2.5y \leq 30$
$x \geq 0,\ y \geq 0$

31. $4x + 3y \leq 240$
$x + 2y \leq 140$
$x \geq 0,\ y \geq 0$

33. $x + y \leq 200$
$110x \geq 11{,}000$
$35y \geq 2{,}205$

35. **(a)** Let x = quantity of two-person tents produced and y = quantity of four-person tents produced.
$x \geq 0,\ y \geq 0$

(b) $6x + 8y \leq 580$
$2x + 4y \leq 260$

37. **(a)** Let x = quantity of WOW 1000s produced and y = quantity of WOW 1500s produced.
$x \geq 0,\ y \geq 0$

(b) $3x + 6y \leq 150$

$3x + 305y \leq 100$

39. **(a)** Let x = acres of cauliflower to plant and y = acres of head lettuce to plant.
$x \geq 0,\ y \geq 0$

(b) $x + y \leq 210$
$215x + 173y \leq 38{,}380$
$98x + 122y \leq 22{,}240$

41. **(a)** Let x = number of students in the medical field training program, y = number of students in the small business program, and z = number of students in the executive secretary program.
$x \geq 0,\ y \geq 0,\ z \geq 0$

(b) $0.8x + 1.1y + 1.7z \leq 32.1$
$0.6x + 1.3y + 1.5z \leq 29.7$
$1.2x + 0.8y + 1.3z \leq 33$

Section 3.2

1. The correct graph is (b).

3. The correct graph is (d).

5.

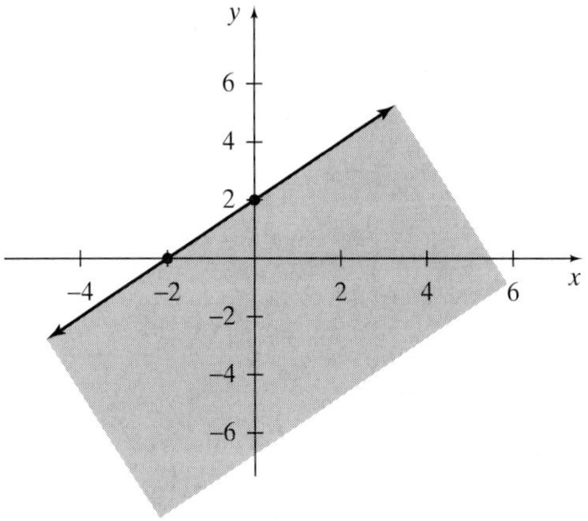

7.

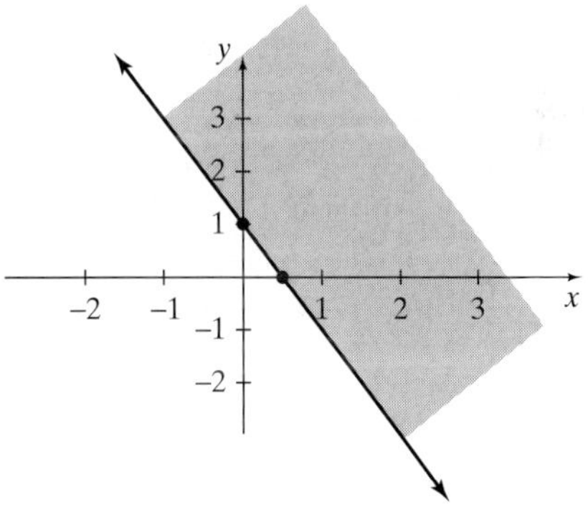

9.

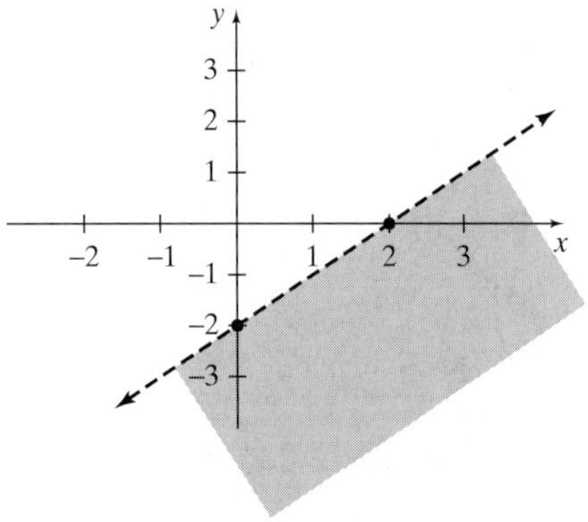

11.

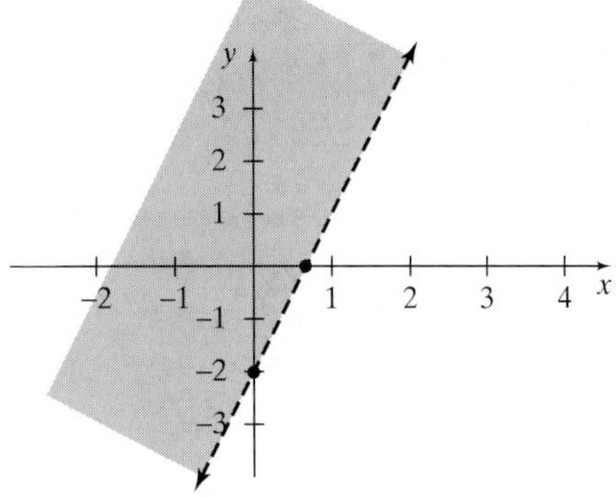

13.

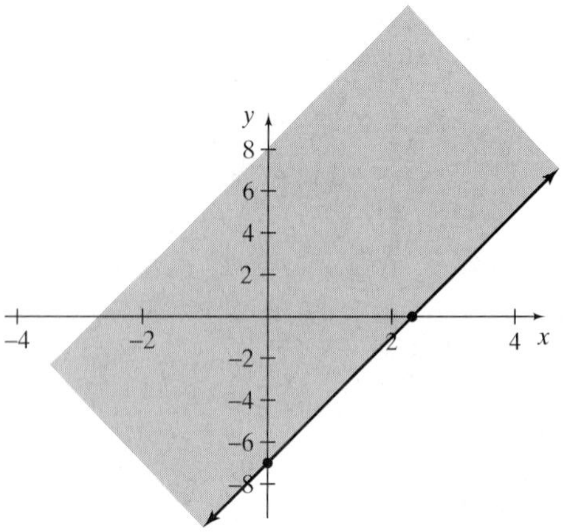

15.

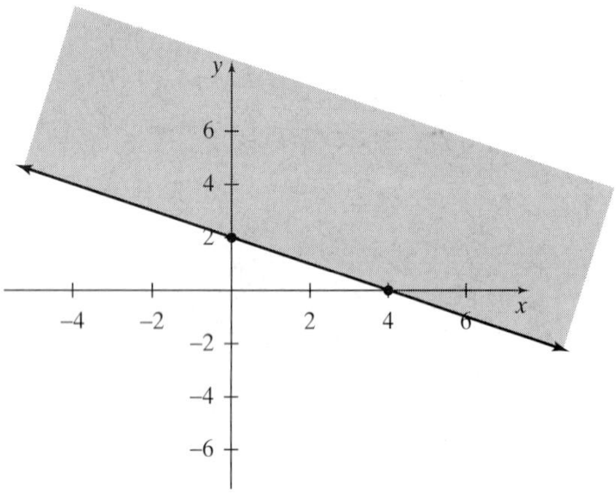

17.

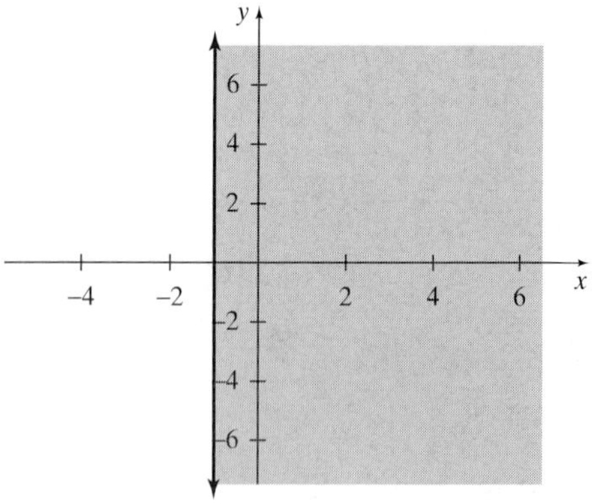

19.

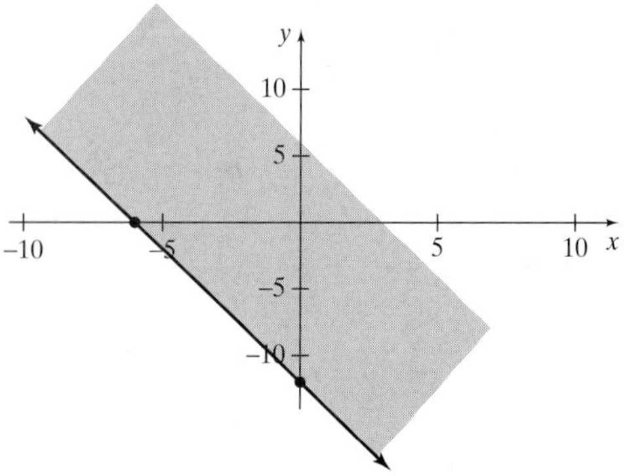

21.

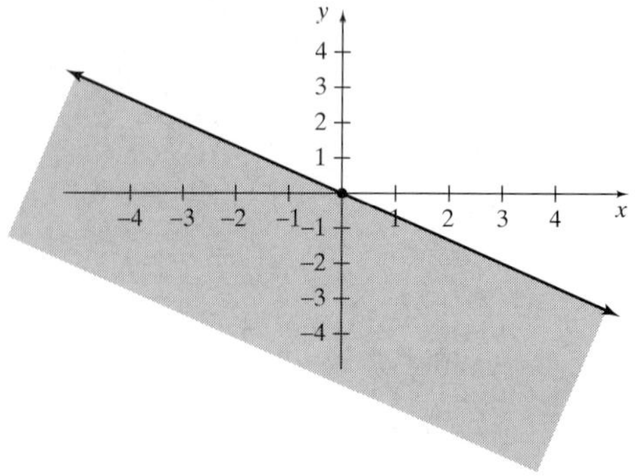

23.

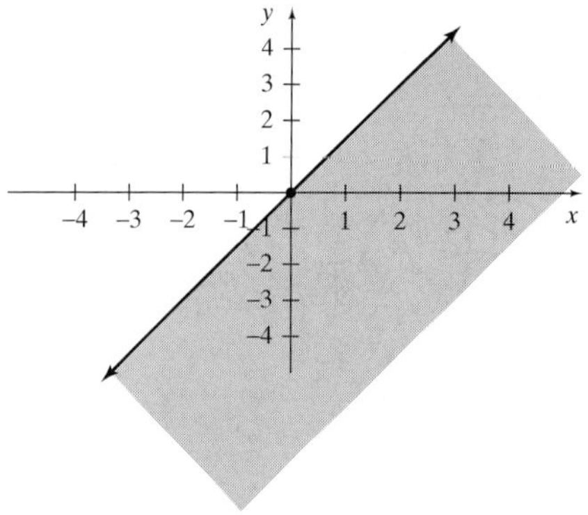

25. The system of inequalities corresponds to region II.

27. The system of inequalities corresponds to region III.

29. The region is unbounded.

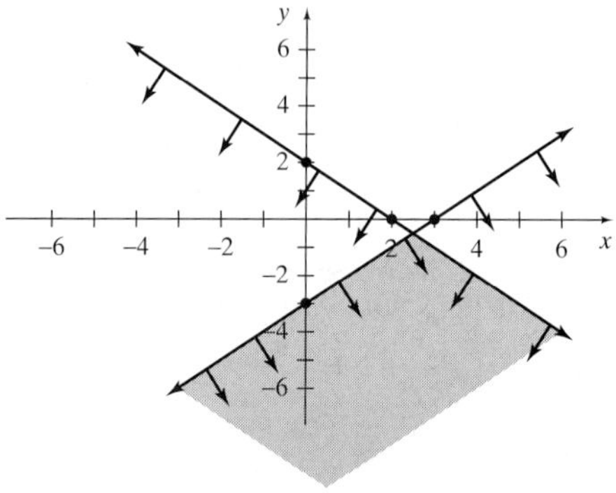

31. The region is unbounded.

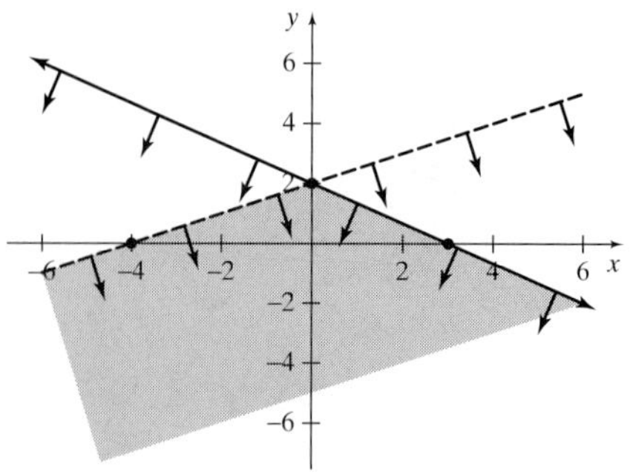

33. The lines are parallel, and there is no feasible solution to the system of inequalities.

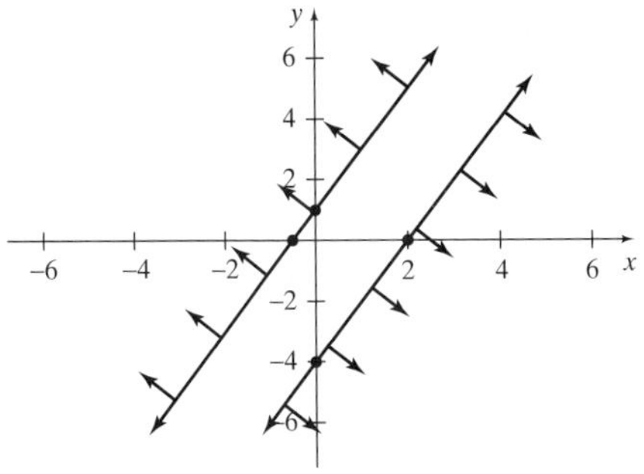

35. The region is bounded.

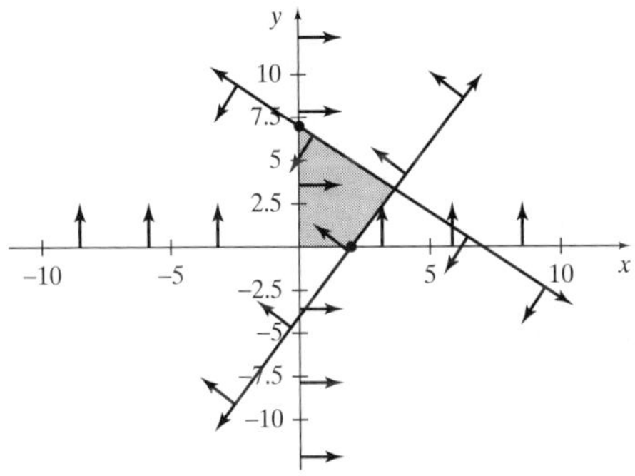

37. The region is bounded.

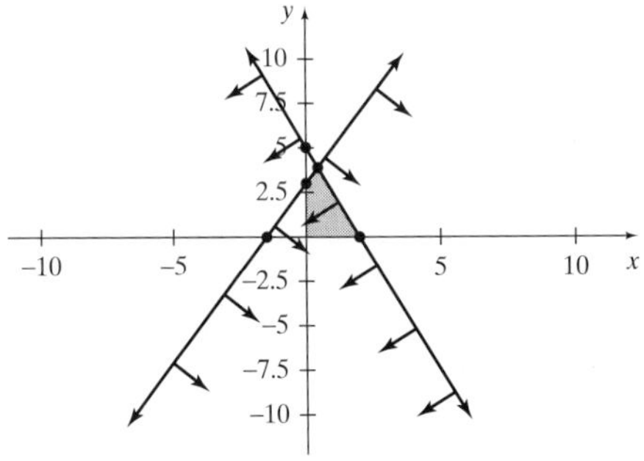

39.

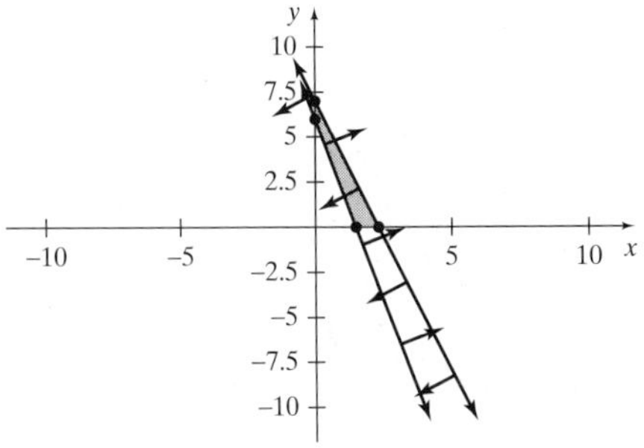

The region is bounded.

41. The region is bounded.

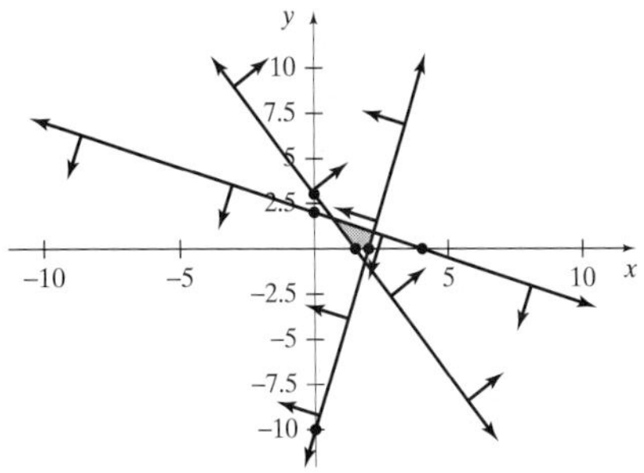

43.

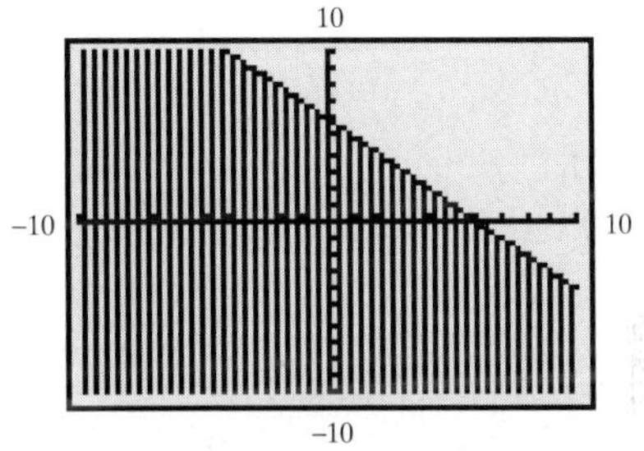

45.

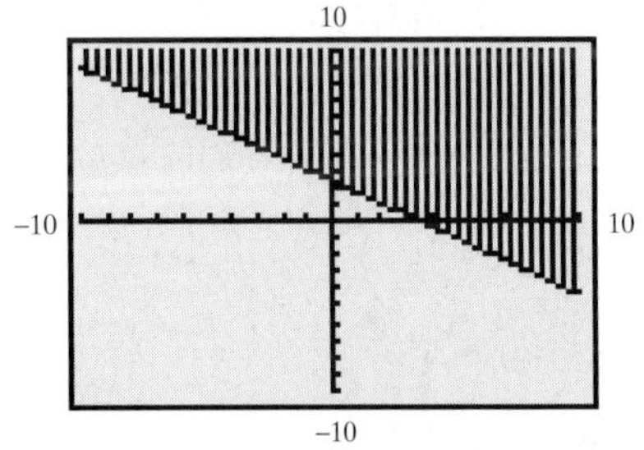

47.

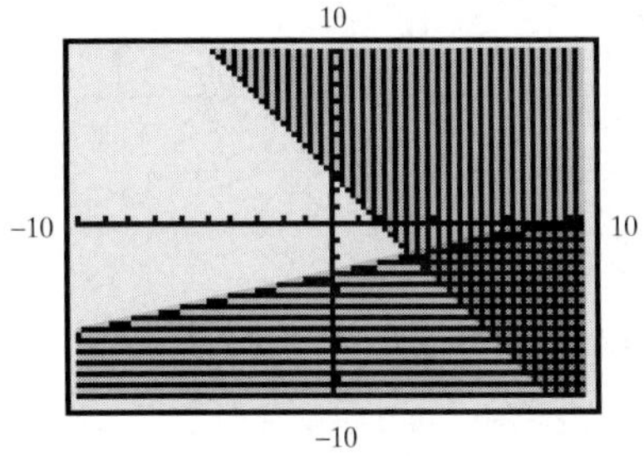

49. **(a)** $2x + 3y \leq 32$
$2.5x + 2.5y \leq 30$
$x \geq 0,\ y \geq 0$

(b) Every point in the feasible region satisfies the system of inequalities.

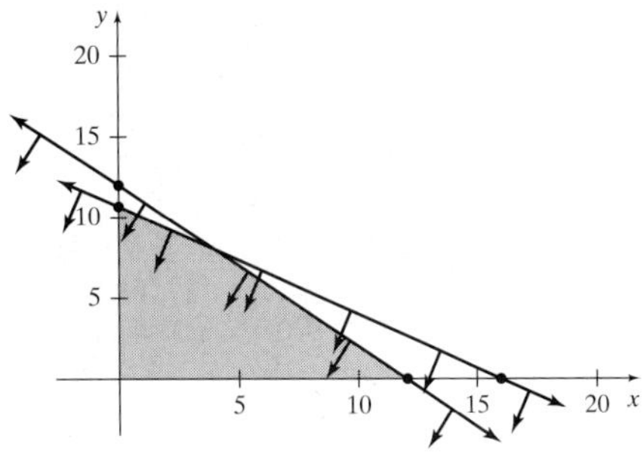

51. **(a)** Let x = the quantity of switch A produced and y = the quantity of switch B produced.

$4x + 3y \le 240$
$x + 2y \le 140$
$x \ge 0,\ y \ge 0$

(b) Every point in the feasible region satisfies the system of inequalities.

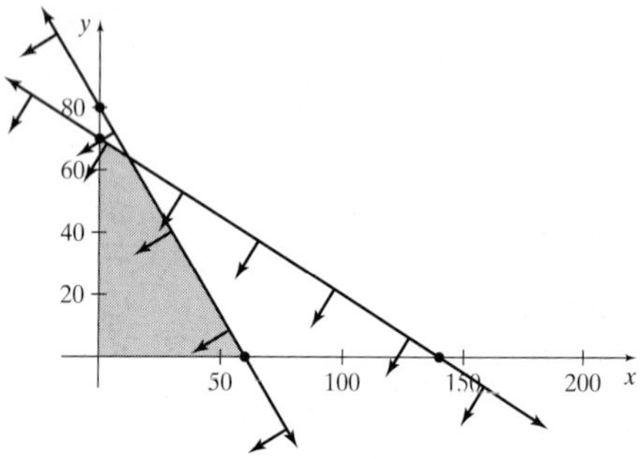

53. **(a)** Let x = acres of corn to plant and y = acres of wheat to plant.

$x + y \le 200$
$110x \ge 11{,}000$
$35y \ge 2205$

(b) Every point in the feasible region satisfies the system of inequalities.

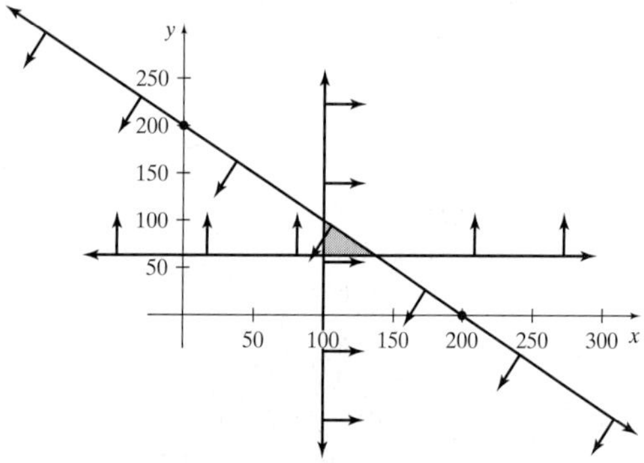

55. **(a)** Let x = small orders of French fries and y = garden salads.

$3x + 7y \ge 12$
$x \ge 0,\ y \ge 0$

(b) Every point in the feasible region satisfies the system of inequalities.

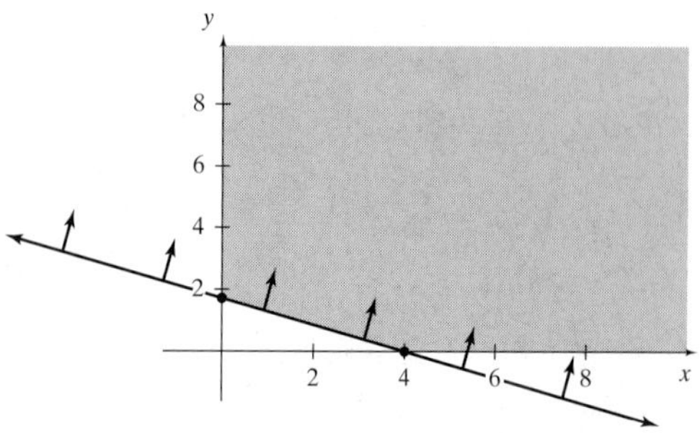

57. **(a)** $3x + 7y \ge 12$
$10x + 6y \le 25$
$x \ge 0,\ y \ge 0$

(b) Every point in the feasible region satisfies the system of inequalities.

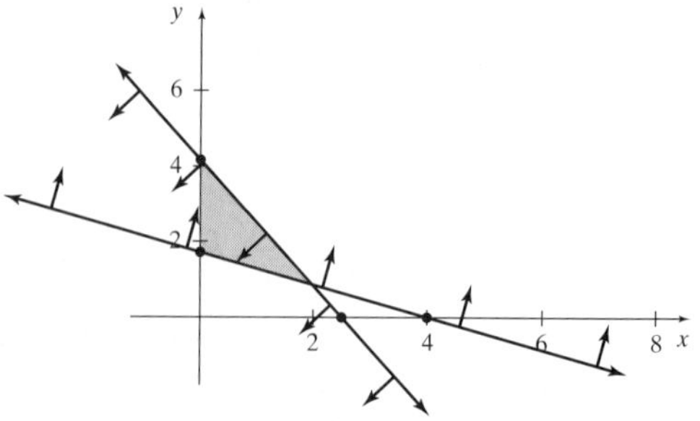

59. **(a)** No answer given; verification.

(b)

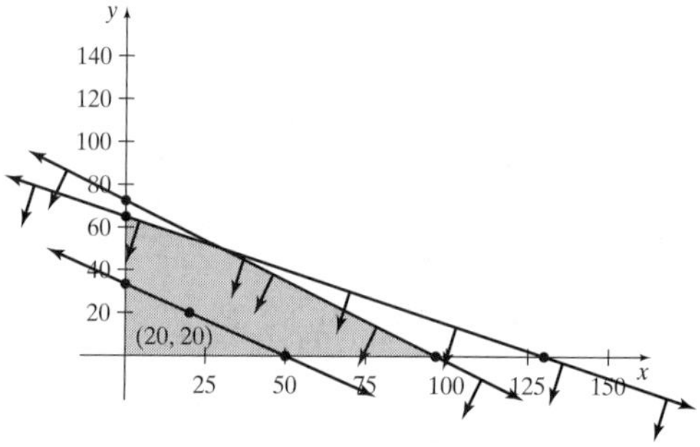

(c) $40x + 60y = 2400$

(d) The slope of the line $40x + 60y = 2000$ is $-\frac{2}{3}$, which is equal to the slope of the line $40x + 60y = 2400$. The lines are parallel.

Section 3.3

1. The minimum value of $z = 2x + 3y$ is 8 at the corner point (1, 2); the maximum value is 39 at the corner point (6, 9).
3. The minimum value of $z = 2x + 5y$ is 0 at the corner point (0, 0); the maximum value is 30 at the corner point (0, 6).
5. The minimum value of $z = x + 4y$ is 5 at the corner point (5, 0); the region is unbounded and no maximum exists.
7. The minimum value of $z = 0.4x + 0.3y$ is 0 at the corner point (0, 0); the maximum value is 4.5 at the corner point (3, 11).
9. The maximum value of $z = 4x + 3y$ is 11 at the corner point (2, 1).
11. The minimum value of $z = x + 2y$ is 0 at the corner point (0, 0).
13. The maximum value of $z = 2x + 2y$ is 14 at the corner points (7, 0) and (0, 7). This is a multiple optimal solution, and all points on the line segment connecting (7, 0) and (0, 7) will produce a maximum value of 14.
15. The minimum value of $z = 3x + 2y$ is 0 at the corner point (0, 0), and the maximum value is 11 at the corner point (1, 4).
17. The minimum value of $z = x + 4y$ is 0 at the corner point (0, 0), and the maximum value is 48 at the corner point (0, 12).

19. The minimum value of $z = 2x + 5y$ is 0 at the corner point $(0, 0)$, and the maximum value is 20 at the corner point $(0, 4)$.

21. The minimum value of $z = 3x + 5y$ is 7 at the corner point $\left(\frac{3}{2}, \frac{1}{2}\right)$, and the maximum value is 47 at the corner point $(4, 7)$.

Section 3.4

1. (a) Let P = profit, x = quantity of two-person tents produced and y = quantity of four-person tents produced.

Maximize: $P = 40x + 60y$

Subject to: $6x + 8y \le 580$
$2x + 4y \le 260$
$x \ge 0,\ y \ge 0$

(b) To maximize weekly profit and stay within department hour restriction, the Yuen Company should produce and sell 30 two-person tents and 50 six-person tents. This generates a maximum weekly profit of \$4200.

3. (a) Let P = profit, x = quantity Zap 1000s produced, and y = quantity of Zap 1200s produced.

Maximize: $P = 300x + 450y$

Subject to: $3x + 6y \le 150$
$3x + 3.5y \le 100$
$x \ge 0,\ y \ge 0$

(b) To maximize weekly profit and stay within department hour restrictions, Denyse and Mike should produce and sell 10 Zap 1000s and 20 Zap 1200s. This will generate a maximum weekly profit of \$12,000.

5. (a) Let C = cost, x = quantity of slick tires produced, and y = quantity of rain tires produced.

Minimize: $C = 50x + 60y$

Subject to: $4x + 12y \ge 18{,}000$
$x + y \ge 2000$
$y \ge 800$

(b) To minimize daily costs and stay within labor union restrictions, HotRubber, Inc., should produce and sell 750 slick tires and 1250 rain tires per day. The minimum daily costs will be \$112,500.

7. To maximize return on investment, Kevan should invest approximately \$16,666.67 in the growth mutual fund and \$33,333.33 in the income fund. Kevan will realize a \$5000 return on his investment.

9. (a) To fulfill the order while minimizing daily operating costs, the Findlay refinery should operate for 60 days and the Ft. Wayne refinery should operate for 150 days.

(b) The minimum cost will be \$6,630,000.

11. (a) To maximize profit while following EPA guidelines, the Roskos Chemical Co. should produce 500 gallons using the old process and 5500 gallons using the new process.

(b) The maximum daily profit is \$1080.

13. (a) To maximize revenue, 39 hardback books and 95 softback books should be shipped.

(b) The maximum revenue is \$1521.66.

15. (a) To maximize revenue, one 1200 series laptop and seven 1700 series laptops should be shipped.

(b) The maximum potential revenue is \$11,592.

17. (a) To maximize revenue and have enough ingredients, Dominic should produce and sell 8 pancakes and 14 crepes.

(b) The maximum revenue is \$101.

19. (a) To minimize daily costs and meet minimum nutrition requirements, Anita should use 50 ounces of HiGlo and 300 ounces of Sheen.

(b) The minimum daily cost is \$11.50.

21. (a) To minimize daily food costs and meet minimum nutritional requirements, Mike should eat 3 cheeseburgers and 10 small orders of onion rings daily.

(b) Mike's minimum daily food costs are \$13.47.

23. (a) To minimize daily fat consumption while meeting minimum daily nutritional requirements, Austin should eat 6 Filet-o-Fish™ sandwiches and 4 large orders of French fries daily.

(b) Minimum daily total fat intake is 260 grams of fat.

25. (a) To maximize profit, Frances should plant 40 acres of artichokes and 100 acres of asparagus.

(b) Maximum profit is \$21,600.

Chapter 3 Review Exercises

1.

x	$x > 5$	Statement
−14	$-14 > 5$	False
5	$5 > 5$	False
7	$7 > 5$	True
10	$10 > 5$	True

3. $x + 8 = 12$ **5.** $5x - 7 \le 9$

7. $x > 8$

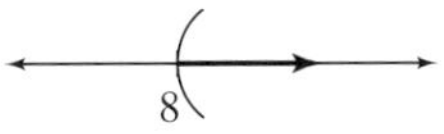

9. $x \ge 4$

11. $x \ge 0,\ y \ge 0$
$2x + 3y \le 37$
$3x + 4y \le 51$

13. (a) Let x = number of day packs that can be made each week and y = number of multiday packs that can be made each week; $x \ge 0,\ y \ge 0$.

(b) $4x + 8y \le 340$
$2x + 6y \le 230$

15. (a) Let x = number of students admitted each month for the drum workshop, y = number of students admitted each month for the guitar workshop, and z = number of students admitted each month for the piano workshop; $x \ge 0,\ y \ge 0,\ z \ge 0$.

(b) $2x + 3y + 5z \le 57$
$4x + 5y + 2z \le 54$
$x + 3y + z \le 24$

17. **(a)**

19.

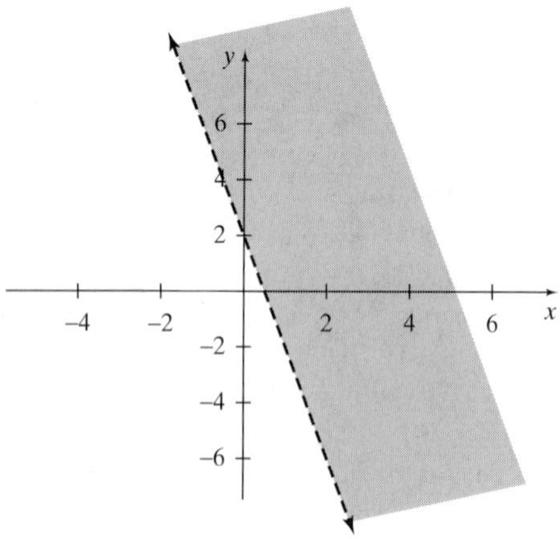

21.

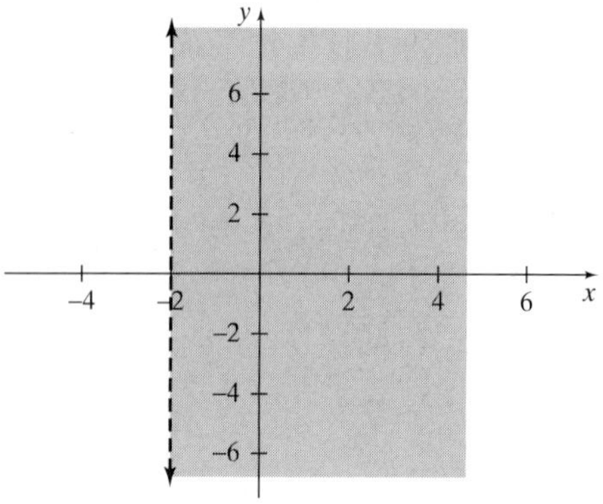

23. Region I

25. Unbounded

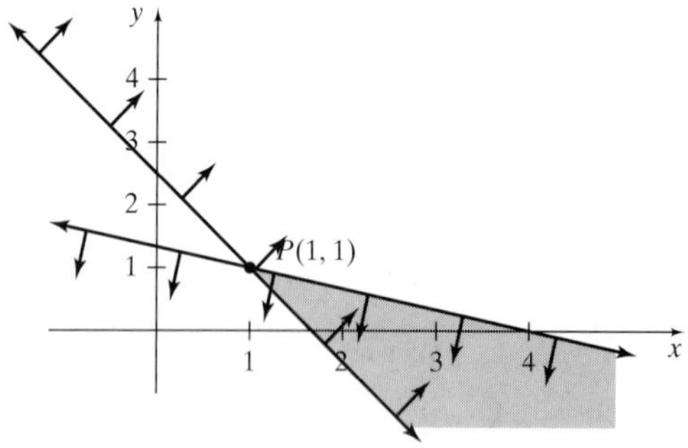

27. Unbounded

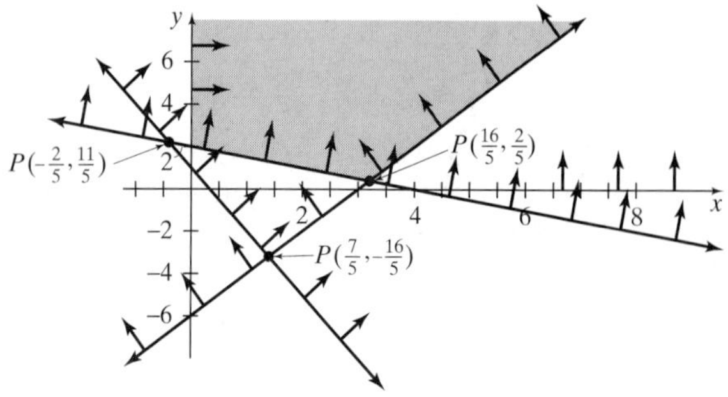

29.

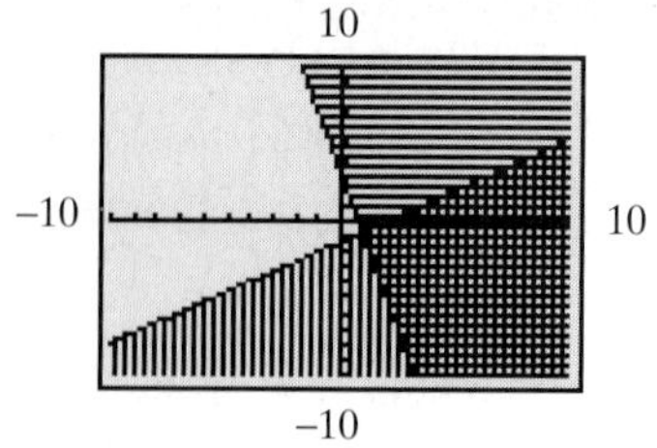

31. **(a)** $x \geq 0,\ y \geq 0$
$2x + 3y \leq 37$
$3x + 4y \leq 51$
$x + y \leq 14$

(b) Every point in the feasible region is a feasible solution to the problem.

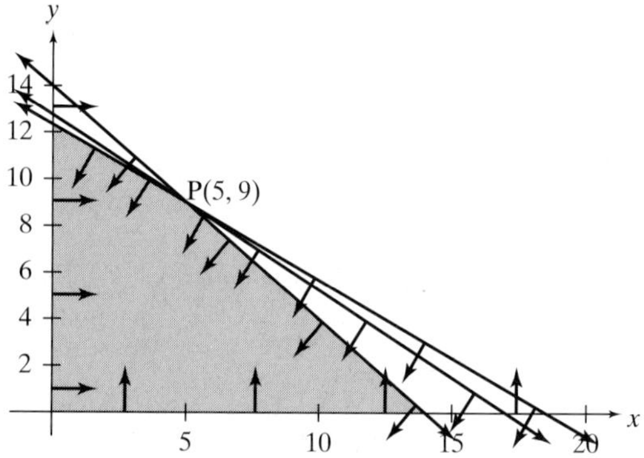

33. **(a)** Let x = number of 6-inch turkey subs, and y = number of 6-inch veggie delite subs.

$x \geq 0,\ y \geq 0$
$16x + 7y \geq 25$

(b) Every point in the feasible region is a feasible solution to the problem.

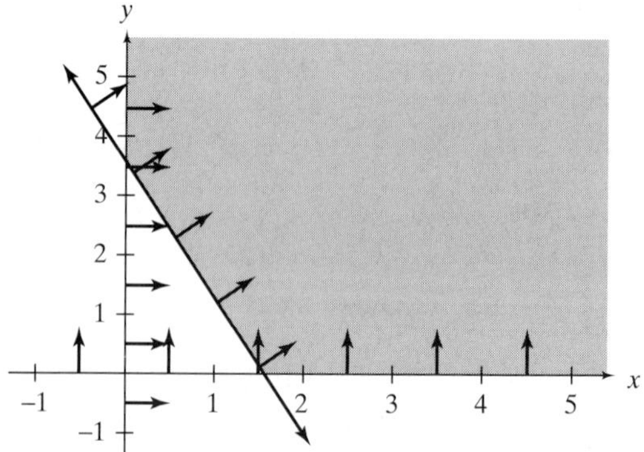

35. **(a)** $x \geq 0, y \geq 0$
$16x + 7y \geq 25$
$3.5x + 2.5y \leq 8$

(b) Every point in the feasible region is a feasible solution to the problem.

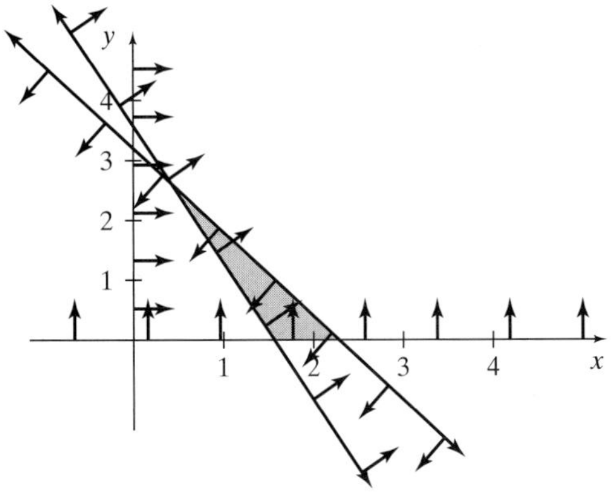

37. Maximum: 23; minimum: 0

39. Maximum: none; minimum: 10

41. Maximum: 6 at (2, 0) **43.** Minimum: $\frac{86}{5}$ at $\left(\frac{8}{5}, \frac{14}{5}\right)$

45. Minimum: 6 at $\left(\frac{3}{2}, 0\right)$; maximum: 32 at $\left(\frac{14}{3}, \frac{8}{3}\right)$

47. (a) 17 drum students and 0 guitar students **(b)** \$425

49. (a) Aztec: 10 days; Buena Vista: 282 days **(b)** \$8,780,000

Chapter 4

Section 4.1

1. $x_1 + x_2 + s_1 = 8$
$2x_1 - x_2 + s_2 = 5$

3. $2x_1 + x_2 - x_3 + s_1 = 5$
$-3x_1 + 2x_2 + x_3 + s_2 = 7$
$4x_1 - x_2 + 2x_3 + s_3 = 9$

5. $2x_1 - x_2 + s_1 = 4$
$-3x_1 + 2x_2 + s_2 = 7$
$x_1 - 4x_2 + s_3 = 8$

7. $2x_1 + 3x_2 - x_3 + s_1 = 8$
$-4x_1 + x_2 + x_3 + s_2 = 5$

9.
$$\begin{array}{ccccc|c} x_1 & x_2 & s_1 & s_2 & z & \\ 1 & 1 & 1 & 0 & 0 & 6 \\ 1 & -1 & 0 & 1 & 0 & 2 \\ \hline -2 & -1 & 0 & 0 & 1 & 0 \end{array}$$

11.
$$\begin{array}{ccccc|c} x_1 & x_2 & s_1 & s_2 & z & \\ 1 & 1 & 1 & 0 & 0 & 7 \\ 2 & 3 & 0 & 1 & 0 & 16 \\ \hline -1 & -2 & 0 & 0 & 1 & 0 \end{array}$$

13.
$$\begin{array}{ccccc|c} x_1 & x_2 & s_1 & s_2 & z & \\ 1 & 1 & 1 & 0 & 0 & 3 \\ 1 & -2 & 0 & 1 & 0 & 0 \\ \hline -2 & -2 & 0 & 0 & 1 & 0 \end{array}$$

15.
$$\begin{array}{cccccc|c} x_1 & x_2 & s_1 & s_2 & s_3 & z & \\ 3 & -2 & 1 & 0 & 0 & 0 & 0 \\ 1 & 1 & 0 & 1 & 0 & 0 & 5 \\ 2 & 1 & 0 & 0 & 1 & 0 & 6 \\ \hline -2 & -2 & 0 & 0 & 0 & 1 & 0 \end{array}$$

17.
$$\begin{array}{ccccccc|c} x_1 & x_2 & s_1 & s_2 & s_3 & s_4 & z & \\ -1 & 2 & 1 & 0 & 0 & 0 & 0 & 10 \\ 2 & 1 & 0 & 1 & 0 & 0 & 0 & 15 \\ 5 & 1 & 0 & 0 & 1 & 0 & 0 & 8 \\ -1 & 1 & 0 & 0 & 0 & 1 & 0 & 1 \\ \hline -5 & -1 & 0 & 0 & 0 & 0 & 1 & 0 \end{array}$$

19.
$$\begin{array}{cccccc|c} x_1 & x_2 & x_3 & s_1 & s_2 & z & \\ 1 & -1 & 2 & 1 & 0 & 0 & 4 \\ 2 & 1 & -3 & 0 & 1 & 0 & 7 \\ \hline -2 & -1 & -3 & 0 & 0 & 1 & 0 \end{array}$$

21.
$$\begin{array}{ccccccc|c} x_1 & x_2 & x_3 & s_1 & s_2 & s_3 & z & \\ 2 & 1 & 4 & 1 & 0 & 0 & 0 & 8 \\ 4 & -1 & 3 & 0 & 1 & 0 & 0 & 15 \\ 12 & 2 & 7 & 0 & 0 & 1 & 0 & 12 \\ \hline -3 & -1 & -5 & 0 & 0 & 0 & 1 & 0 \end{array}$$

23.
$$\begin{array}{cccccc|c} x_1 & x_2 & x_3 & s_1 & s_2 & z & \\ 2 & 0 & 3 & 1 & 0 & 0 & 8 \\ 3 & 2 & 0 & 0 & 1 & 0 & 5 \\ \hline -4 & -2 & -3 & 0 & 0 & 1 & 0 \end{array}$$

25.
$$\begin{array}{cccccc|c} x_1 & x_2 & x_3 & s_1 & s_2 & z & \\ 15 & 3 & 0 & 1 & 0 & 0 & 9 \\ 0 & 6 & 1 & 0 & 1 & 0 & 7 \\ \hline -5 & -1 & -3 & 0 & 0 & 1 & 0 \end{array}$$

27. (a) Basic variables: x_1, s_1, z
Nonbasic variables: x_2, s_2

(b) $x_1 = 5, x_2 = 0, s_1 = 2, s_2 = 0, z = 10$

29. (a) Basic variables: s_1, s_2, z
Nonbasic variables: x_1, x_2

(b) $x_1 = 0, x_2 = 0, s_1 = 3, s_2 = 7, z = 9$

31. (a) Basic variables: s_1, s_2, s_3, z
Nonbasic variables: x_1, x_2

(b) $x_1 = 0, x_2 = 0, s_1 = 3, s_2 = 8, s_3 = 15, z = 0$

33. (a) Basic variables: x_1, x_2, z
Nonbasic variables: x_3, s_1, s_2

(b) $x_1 = 7, x_2 = 12, x_3 = 0, s_1 = 0, s_2 = 0, z = 27$

35. (a) Basic variables: x_2, x_3, s_2, z
Nonbasic variables: x_1, x_4, s_1, s_3

(b) $x_1 = 0, x_2 = 15, x_3 = 20, x_4 = 0, s_1 = 0, s_2 = 10, s_3 = 0, z = 100$

37. (a)
$$\begin{array}{ccccc|c} x_1 & x_2 & s_1 & s_2 & z & \\ 1 & 0 & -\frac{1}{2} & \frac{5}{2} & 0 & 4 \\ 0 & 1 & \frac{1}{2} & \frac{1}{2} & 0 & 1 \\ \hline 0 & 0 & \frac{5}{2} & \frac{13}{2} & 1 & 15 \end{array}$$

(b) $x_1 = 4, x_2 = 1, s_1 = 0, s_2 = 0, z = 15$

39. (a)
$$\begin{array}{ccccc|c} x_1 & x_2 & s_1 & s_2 & z & \\ 1 & -\frac{3}{2} & \frac{1}{2} & 0 & 0 & \frac{3}{2} \\ 0 & 12 & -2 & 1 & 0 & 1 \\ \hline 0 & \frac{7}{2} & \frac{1}{2} & 0 & 1 & \frac{21}{2} \end{array}$$

(b) $x_1 = \frac{3}{2}, x_2 = 0, s_1 = 0, s_2 = 1, z = \frac{21}{2}$

41. (a)

$$\begin{array}{cccccc|c} x_1 & x_2 & x_3 & s_1 & s_2 & z & \\ \end{array}$$

$$\left[\begin{array}{cccccc|c} -\frac{7}{4} & 0 & -\frac{1}{4} & 1 & 0 & 0 & \frac{3}{2} \\ \frac{3}{8} & 1 & \frac{1}{8} & 0 & 0 & 0 & \frac{3}{4} \\ \frac{41}{8} & 0 & -\frac{5}{8} & 0 & 1 & 0 & \frac{17}{4} \\ \hline \frac{5}{8} & 0 & \frac{7}{8} & 0 & 0 & 1 & \frac{21}{4} \end{array}\right]$$

(b) $x_1 = 0,\ x_2 = \frac{3}{4},\ x_3 = 0,\ s_1 = \frac{3}{2},\ s_2 = \frac{17}{4}, z = \frac{21}{4}$

43. (a)

$$\begin{array}{cccccc|c} x_1 & x_2 & x_3 & s_1 & s_2 & z & \\ \end{array}$$

$$\left[\begin{array}{cccccc|c} 7 & \frac{5}{2} & 0 & 1 & -\frac{1}{2} & 0 & 70 \\ \frac{3}{2} & \frac{3}{4} & 1 & 0 & \frac{1}{4} & 0 & 10 \\ \hline 10 & 3 & 0 & 0 & 2 & 1 & 80 \end{array}\right]$$

(b) $x_1 = 0, x_2 = 0, x_3 = 10, s_1 = 70, s_2 = 0, z = 80$

45. (a)

$$\begin{array}{cccccccc|c} x_1 & x_2 & x_3 & x_4 & s_1 & s_2 & s_3 & z & \\ \end{array}$$

$$\left[\begin{array}{cccccccc|c} 1 & 0 & -\frac{5}{2} & 0 & -\frac{11}{2} & -\frac{25}{2} & -\frac{39}{2} & 0 & -\frac{117}{2} \\ 0 & 0 & -\frac{3}{2} & 1 & -\frac{11}{2} & -\frac{13}{2} & -\frac{25}{2} & 0 & -\frac{55}{2} \\ 0 & 1 & \frac{1}{2} & 0 & \frac{3}{2} & \frac{7}{2} & \frac{9}{2} & 0 & \frac{27}{2} \\ \hline 0 & 0 & 1 & 0 & 4 & 7 & 13 & 1 & 107 \end{array}\right]$$

(b) $x_1 = -\frac{117}{2},\ x_2 = \frac{27}{2},\ x_3 = 0,\ x_4 = -\frac{55}{2},$
$s_1 = 0,\ s_2 = 0,\ s_3 = 0,\ z = 107$

47. (b) $x_1 = 0,\ x_2 = 2.2,\ s_1 = 0.6,\ s_2 = 0,\ z = 21.6$

49. (b) $x_1 = 0,\ x_2 = 0,\ x_3 = 13,\ s_1 = 0,\ s_2 = -11,\ z = 130$

51. (a) Let x_1 = gallons of chemical produced using older process and x_2 = gallons of chemical produced using newer process.

Maximize: $z = 0.40x_1 + 0.16x_2$
Subject to: $4x_1 + 2x_2 \le 10{,}000$
$3x_1 + x_2 \le 7000$
$x_1 \ge 0, x_2 \ge 0$

(b)
$$4x_1 + 2x_2 + s_1 = 10{,}000$$
$$3x_1 + x_2 + s_2 = 7000$$
$$-0.4x_1 - 0.16x_2 + z = 0$$

(c)

$$\begin{array}{ccccc|c} x_1 & x_2 & s_1 & s_2 & z & \\ \end{array}$$

$$\left[\begin{array}{ccccc|c} 4 & 2 & 1 & 0 & 0 & 10{,}000 \\ 3 & 1 & 0 & 1 & 0 & 7000 \\ \hline -0.4 & -0.16 & 0 & 0 & 1 & 0 \end{array}\right]$$

53. (a) Let x_1 = quantity of WOW 1000s produced and x_2 = quantity of WOW 1500s produced.

Maximize: $P = 100x_1 + 150x_2$
Subject to: $3x_1 + 6x_2 \le 150$
$3x_1 + 3.5x_2 \le 100$
$x_1 \ge 0, x_2 \ge 0$

(b)
$$3x_1 + 6x_2 + s_1 = 150$$
$$3x_1 + 3.5x_2 + s_2 = 100$$
$$-100x_1 - 150x_2 + z = 0$$

(c)

$$\begin{array}{ccccc|c} x_1 & x_2 & s_1 & s_2 & z & \\ \end{array}$$

$$\left[\begin{array}{ccccc|c} 3 & 6 & 1 & 0 & 0 & 150 \\ 3 & 3.5 & 0 & 1 & 0 & 100 \\ \hline -100 & -150 & 0 & 0 & 1 & 0 \end{array}\right]$$

55. (a) Let x_1 = acres of artichokes and x_2 = acres of asparagus.

Maximize: $z = 140x_1 + 160x_2$
Subject to: $197x_1 + 240x_2 \le 31{,}880$
$123x_1 + 42x_2 \le 9120$
$x_1 \ge 0, x_2 \ge 0$

(b)
$$197x_1 + 240x_2 + s_1 = 31{,}880$$
$$123x_1 + 42x_2 + s_2 = 9120$$
$$-140x_1 - 160x_2 + z = 0$$

(c)

$$\begin{array}{ccccc|c} x_1 & x_2 & s_1 & s_2 & z & \\ \end{array}$$

$$\left[\begin{array}{ccccc|c} 197 & 240 & 1 & 0 & 0 & 31{,}880 \\ 123 & 42 & 0 & 1 & 0 & 9120 \\ \hline -140 & -160 & 0 & 0 & 1 & 0 \end{array}\right]$$

57. (a) Let x_1 = quantity of two-person tents, x_2 = quantity of four-person tents, and x_3 = quantity of six-person tents.

Maximize: $z = 30x_1 + 50x_2 + 80x_3$
Subject to: $0.5x_1 + 0.8x_2 + 1.1x_3 \le 420$
$0.6x_1 + 0.7x_2 + 0.9x_3 \le 310$
$0.2x_1 + 0.3x_2 + 0.6x_3 \le 180$
$x_1 \ge 0, x_2 \ge 0, x_3 \ge 0$

(b)
$$0.5x_1 + 0.8x_2 + 1.1x_3 + s_1 = 420$$
$$0.6x_1 + 0.7x_2 + 0.9x_3 + s_2 = 310$$
$$0.2x_1 + 0.3x_2 + 0.6x_3 + s_3 = 180$$
$$-30x_1 - 50x_2 - 80x_3 + z = 0$$

(c)

$$\begin{array}{ccccccc|c} x_1 & x_2 & x_3 & s_1 & s_2 & s_3 & z & \\ \end{array}$$

$$\left[\begin{array}{ccccccc|c} 0.5 & 0.8 & 1.1 & 1 & 0 & 0 & 0 & 420 \\ 0.6 & 0.7 & 0.9 & 0 & 1 & 0 & 0 & 310 \\ 0.2 & 0.3 & 0.6 & 0 & 0 & 1 & 0 & 180 \\ \hline -30 & -50 & -80 & 0 & 0 & 0 & 1 & 0 \end{array}\right]$$

59. (a) Let x_1 = acres of asparagus, x_2 = acres of cauliflower, and x_3 = acres of celery.

Maximize: $z = 160x_1 + 90x_2 + 80x_3$
Subject to: $240x_1 + 215x_2 + 336x_3 \le 65{,}970$
$148x_1 + 98x_2 + 171x_3 \le 31{,}860$
$42x_1 + 69x_2 + 147x_3 \le 21{,}630$
$x_1 \ge 0, x_2 \ge 0, x_3 \ge 0$

(b)
$$240x_1 + 215x_2 + 336x_3 + s_1 = 65{,}970$$
$$148x_1 + 98x_2 + 171x_3 + s_2 = 31{,}860$$
$$42x_1 + 69x_2 + 147x_3 + s_3 = 21{,}630$$
$$-160x_1 - 90x_2 - 80x_3 + z = 0$$

(c)

$$\begin{array}{ccccccc|c} x_1 & x_2 & x_3 & s_1 & s_2 & s_3 & z & \\ \end{array}$$

$$\left[\begin{array}{ccccccc|c} 240 & 215 & 336 & 1 & 0 & 0 & 0 & 65{,}970 \\ 148 & 98 & 171 & 0 & 1 & 0 & 0 & 31{,}860 \\ 42 & 69 & 147 & 0 & 0 & 1 & 0 & 21{,}630 \\ \hline -160 & -90 & -80 & 0 & 0 & 0 & 1 & 0 \end{array}\right]$$

Section 4.2

1. $x_1 = 2,\ x_2 = 0,\ s_1 = 3,\ s_2 = 0,\ z = 18$

3. $x_1 = 5,\ x_2 = 0,\ s_1 = 1,\ s_2 = 0,\ z = 29$

5. $x_1 = 0, x_2 = \frac{3}{4}$,
$s_1 = \frac{3}{2}, s_2 = 0, s_3 = \frac{17}{4}$,
$z = 5.25$

7. $x_1 = 0,\ x_2 = 0,\ x_3 = 10,\ s_1 = 70,\ s_2 = 0,\ z = 80$

9. $x_1 = 0,\ x_2 = \frac{9}{5},\ x_3 = \frac{117}{5},\ x_4 = \frac{38}{5}$,
$s_1 = 0,\ s_2 = \frac{38}{5},\ s_3 = 0,\ z = 107$

11. (a)
$$\begin{array}{ccccc|c} x_1 & x_2 & s_1 & s_2 & z & \\ \hline 1 & 1 & 1 & 0 & 0 & 6 \\ 1 & -1 & 0 & 1 & 0 & 2 \\ \hline -2 & -1 & 0 & 0 & 1 & 0 \end{array}$$

(b) $x_1 = 4,\ x_2 = 2,\ s_1 = 0,\ s_2 = 0,\ z = 10$

13. (a)
$$\begin{array}{ccccc|c} x_1 & x_2 & s_1 & s_2 & z & \\ \hline 3 & 2 & 1 & 0 & 0 & 8 \\ 1 & -1 & 0 & 1 & 0 & 7 \\ \hline -2 & -3 & 0 & 0 & 1 & 0 \end{array}$$

(b) $x_1 = 0,\ x_2 = 4,\ s_1 = 0,\ s_2 = 11,\ z = 12$

15. (a)
$$\begin{array}{cccccc|c} x_1 & x_2 & s_1 & s_2 & s_3 & z & \\ \hline 2 & 1 & 1 & 0 & 0 & 0 & 32 \\ 1 & 1 & 0 & 1 & 0 & 0 & 18 \\ 1 & 3 & 0 & 0 & 1 & 0 & 36 \\ \hline -1 & -4 & 0 & 0 & 0 & 1 & 0 \end{array}$$

(b) $x_1 = 0,\ x_2 = 12,\ s_1 = 20,\ s_2 = 6,\ s_3 = 0,\ z = 48$

17. (a)
$$\begin{array}{cccccc|c} x_1 & x_2 & s_1 & s_2 & s_3 & z & \\ \hline 3 & -2 & 1 & 0 & 0 & 0 & 0 \\ 1 & 1 & 0 & 1 & 0 & 0 & 5 \\ 2 & 1 & 0 & 0 & 1 & 0 & 6 \\ \hline -2 & -2 & 0 & 0 & 0 & 1 & 0 \end{array}$$

(b) $x_1 = 1,\ x_2 = 4,\ s_1 = 5,\ s_2 = 0,\ s_3 = 0,\ z = 10$
(solution not unique; multiple optimal solution)

19. (a)
$$\begin{array}{ccccccc|c} x_1 & x_2 & s_1 & s_2 & s_3 & s_4 & z & \\ \hline -1 & 2 & 1 & 0 & 0 & 0 & 0 & 10 \\ 2 & 1 & 0 & 1 & 0 & 0 & 0 & 15 \\ 5 & 1 & 0 & 0 & 1 & 0 & 0 & 8 \\ -1 & 1 & 0 & 0 & 0 & 1 & 0 & 1 \\ \hline -5 & -1 & 0 & 0 & 0 & 0 & 1 & 0 \end{array}$$

(b) $x_1 = \frac{8}{5},\ x_2 = 0,\ s_1 = \frac{58}{5}$,
$s_2 = \frac{59}{5},\ s_3 = 0,\ s_4 = \frac{13}{5}$,
$z = 8$

21. (a)
$$\begin{array}{cccccc|c} x_1 & x_2 & x_3 & s_1 & s_2 & z & \\ \hline 2 & 1 & 1 & 1 & 0 & 0 & 10 \\ 5 & -1 & 2 & 0 & 1 & 0 & 8 \\ \hline -3 & -2 & -4 & 0 & 0 & 1 & 0 \end{array}$$

(b) $x_1 = 0,\ x_2 = 4,\ x_3 = 6$,
$s_1 = 0,\ s_2 = 0,\ z = 32$

23. (a)
$$\begin{array}{ccccccc|c} x_1 & x_2 & x_3 & s_1 & s_2 & s_3 & z & \\ \hline 5 & 2 & 1 & 1 & 0 & 0 & 0 & 20 \\ 1 & 3 & 5 & 0 & 1 & 0 & 0 & 13 \\ 4 & 1 & 3 & 0 & 0 & 1 & 0 & 15 \\ \hline -2 & -5 & -1 & 0 & 0 & 0 & 1 & 0 \end{array}$$

(b) $x_1 = \frac{34}{13},\ x_2 = \frac{45}{13},\ x_3 = 0$,
$s_1 = 0,\ s_2 = 0,\ s_3 = \frac{14}{13},\ z = \frac{293}{13}$

25. (a)
$$\begin{array}{cccccc|c} x_1 & x_2 & x_3 & s_1 & s_2 & z & \\ \hline 2 & 0 & 3 & 1 & 0 & 0 & 8 \\ 3 & 2 & 0 & 0 & 1 & 0 & 5 \\ \hline -4 & -2 & -3 & 0 & 0 & 1 & 0 \end{array}$$

(b) $x_1 = 0,\ x_2 = \frac{5}{2},\ x_3 = \frac{8}{3}$,
$s_1 = 0,\ s_2 = 0,\ z = 13$

27. To maximize profit, the Young Furniture Company should produce 30 tables and 40 chairs. This yields a maximum profit of $4100.

29. To maximize weekly profit and stay within EPA regulations, the Henderson Chemical Company should produce 2000 gallons using the older process and 1000 gallons using the newer process. This will yield a maximum weekly profit of approximately $960.

31. To maximize weekly profit, 10 units of the WOW 1000 and 20 units of the WOW 1500 should be produced. This will yield a maximum profit of $4000.

33. (a) To maximize profit, Franco should plant 40 acres of artichokes and 100 acres of asparagus.

(b) Maximum profit is $21,600.

35. To maximize profit, the Smolko Tent Company should produce and sell 0 two-person tents, 160 four-person tents, and 220 six-person tents.

37. To maximize profit, Eldrick should plant about 215 acres of asparagus, 0 acres of cauliflower, and 0 acres of celery. This will yield a maximum profit of approximately $34,443. The nonzero values for s_1 and s_2 indicate that there will be approximately 14,305 pounds of nitrogen and 12,589 pounds of potash, respectively, that will not be used.

39. To maximize total grams of protein, Meagan should eat 7.2 servings of Cookie's 'n' Creme candies, and no servings of Reese's Cups or York Peppermint. She will consume 14.4 grams of protein. The nonzero value for s_1 indicates that she will consume 52 fewer calories than the maximum amount (700). The nonzero value for s_2 indicates that she will consume 8 fewer grams of carbohydrates than the maximum amount (80 grams).

41. To maximize profit, Julia should make 8 loaves of Lemon Poppy Seed Glaze bread, 3 loaves of Pumpkin Glaze bread and 5 loaves of Banana Walnut Glaze bread. This will yield a maximum profit of $20.

Section 4.3

1. $A^T = \begin{bmatrix} 2 & 4 \\ 1 & 1 \\ 3 & 2 \end{bmatrix}$ **3.** $A^T = \begin{bmatrix} 5 & 3 & 0 \\ 7 & 2 & 1 \\ 6 & 1 & 1 \\ 1 & 0 & 2 \end{bmatrix}$ **5.** $A^T = \begin{bmatrix} 4 & 1 & 8 \\ 1 & 2 & 1 \\ 3 & 7 & 9 \end{bmatrix}$

7. Maximize: $z = 4x_1 + 7x_2$
Subject to: $5x_1 + 3x_2 \le 2$
$2x_1 + x_2 \le 3$
$x_1 \ge 0, x_2 \ge 0$

9. Maximize: $z = 7x_2 + 11x_3$
Subject to: $2x_1 + 8x_2 + 5x_3 \le 1$
$x_1 + x_2 + 2x_3 \le 3$
$4x_1 + x_2 + 7x_3 \le 4$
$x_1 \ge 0, x_2 \ge 0, x_3 \ge 0$

11. Maximize: $z = 8x_1 + 7x_2 + 11x_3$
Subject to: $5x_1 + 3x_2 \le 2$
$x_1 + x_3 \le 1$
$8x_2 + x_3 \le 3$
$x_1 \ge 0, x_2 \ge 0, x_3 \ge 0$

13. (a) Minimum is $w = 6$ at $y_1 - 6,\ y_2 = 0$.
(b) Maximum is $z = 6$ at $x_1 = 1,\ x_2 = 0$.

15. (a) Minimum is $w = 24$ at $y_1 = 12,\ y_2 = 0$.
(b) Maximum is $z = 24$ at $x_1 = 2,\ x_2 = 0,\ x_3 = 0$.

17. (a) Maximize: $z = 15x_1 + 4x_2$
Subject to: $8x_1 + x_2 \le 7$
$10x_1 + x_2 \le 3$
$x_1 \ge 0, x_2 \ge 0$
(b) Minimum is $w = 12$ at $y_1 = 0,\ y_2 = 4$.

19. (a) Maximize: $z = 5x_1 + 6x_2$
Subject to: $x_1 + x_2 \le 1$
$2x_1 + 3x_2 \le 4$
$x_1 \ge 0, x_2 \ge 0$
(b) Minimum is $w = 6$ at $y_1 = 6,\ y_2 = 0$.

21. (a) Maximize: $z = 8x_1 + 4x_2$
Subject to: $x_1 + x_2 \le 2$
$x_1 + 2x_2 \le 1$
$x_1 \ge 0, x_2 \ge 0$
(b) Minimum is $w = 8$ at $y_1 = 0,\ y_2 = 8$.

23. (a) Maximize: $z = 29x_1 + 55x_2 + 18x_3$
Subject to: $2x_1 + 5x_2 + 2x_3 \le 4$
$3x_1 + 4x_2 + x_3 \le 5$
$x_1 \ge 0, x_2 \ge 0, x_3 \ge 0$
(b) Minimum is $w = 53$ at $y_1 = 7,\ y_2 = 5$.

25. (a) Maximize: $z = 29x_1 + 20x_2 + 78x_3$
Subject to: $15x_1 + x_2 + 8x_3 \le 2$
$x_1 + 6x_2 + 7x_3 \le 3$
$x_1 \ge 0, x_2 \ge 0, x_3 \ge 0$
(b) Minimum is $w = 22$ at $y_1 = 8,\ y_2 = 2$.

27. (a) Maximize: $z = 6x_1 + 9x_2$
Subject to: $x_1 + 2x_2 \le 14$
$x_1 + x_2 \le 8$
$3x_1 + x_2 \le 20$
$x_1 \ge 0, x_2 \ge 0$
(b) Minimum is $w = 66$ at $y_1 = 3,\ y_2 = 3,\ y_3 = 0$.

29. (a) Maximize: $z = 6x_1 + 9x_2 + 3x_3$
Subject to: $x_1 + x_2 + 3x_3 \le 4$
$3x_1 + x_2 + x_3 \le 10$
$x_1 + 3x_2 + x_3 \le 9$
$x_1 \ge 0, x_2 \ge 0, x_3 \ge 0$
(b) Minimum is $w = \frac{63}{2}$ at $y_1 = \frac{9}{2},\ y_2 = 0,\ y_3 = \frac{3}{2}$.

31. (a) Minimize: $C = 0.08y_1 + 0.09y_2$
Subject to: $5y_1 + 10y_2 \ge 50$
$y_1 + 2y_2 \ge 5$
$y_1 \ge 0, y_2 \ge 0$
(b) Maximize: $z = 50x_1 + 5x_2$
Subject to: $5x_1 + x_2 \le 0.08$
$10x_1 + 2x_2 \le 0.09$
$x_1 \ge 0, x_2 \ge 0$
(c) The solution to the primal problem is $C = 0.45$ at $y_1 = 0,\ y_2 = 5$.
To minimize cost, Richie should take 0 pills of the AllDay supplement and 5 pills of the Live! supplement. Minimum daily cost is $0.45.

33. (a) Minimize: $C = 0.30y_1 + 0.40y_2$
Subject to: $210y_1 + 240y_2 \ge 2115$
$150y_1 + 380y_2 \ge 3460$
$y_1 \ge 0, y_2 \ge 0$
(b) Maximize: $z = 2115x_1 + 3460x_2$
Subject to: $210x_1 + 150x_2 \le 0.30$
$240x_1 + 380x_2 \le 0.40$
$x_1 \ge 0, x_2 \ge 0$
(c) The solution to the primal problem is $C = \frac{346}{95} \approx 3.64$ at $y_1 = 0,\ y_2 = \frac{173}{19} \approx 9.11$.
To minimize cost, Elle should feed her beagles 0 pounds of Shine dog food and approximately 9.11 pounds of Lustre dog food. Minimum daily cost will be about $3.64.

35. (a) Let y_1 = number of days Mine I should operate, y_2 = number of days Mine II should operate.
Minimize: $C = 12y_1 + 10y_2$
Subject to: $3y_1 + 2y_2 \ge 13$
$3y_1 + y_2 \ge 9$
$4y_1 + 6y_2 \ge 24$
$y_1 \ge 0, y_2 \ge 0$
(b) Maximize: $z = 13x_1 + 9x_2 + 24x_3$
Subject to: $3x_1 + 3x_2 + 4x_3 \le 12$
$2x_1 + x_2 + 6x_3 \le 10$
$x_1 \ge 0, x_2 \ge 0, x_3 \ge 0$
(c) The solution to the primal problem is $C = 56$ at $y_1 = 3,\ y_2 = 2$.
To minimize daily cost, AJA Mining Co. should operate Mine I for 3 days and Mine II for 2 days. Minimum daily cost will be $56,000.

37. (a) To minimize cost, and fulfill the minimum order, the refinery in Findlay should operate for 60 days and the refinery in Ft. Wayne should operate for 150 days.
(b) The minimum cost is $6,630,000.

39. (a) To minimize cost, and meet dietary requirements, Anita should use 50 ounces of Flakes and 300 ounces of Woday.

(b) The minimum daily cost is \$11.50.

41. (a) To minimize fat intake, and meet dietary requirements, Warren should eat 2 servings of Burrito Supreme^R-Beef and 3 servings of Double Burrito Supreme^R-Steak.

(b) Minimum daily fat intake is 90 grams of fat.

43. (a) The cost of a milligram of vitamin C is \$0.009 and the cost of a milligram of niacin is 0.

(b) New total cost is about \$0.46.

45. (a) A unit of protein is worth zero and a unit of fat is worth $\frac{1}{950}$ of a dollar.

(b) New total cost is about \$3.64.

Section 4.4

1. $3x_1 + 2x_2 + s_1 = 12$
$x_1 + 4x_2 - s_2 = 7$

3. $x_1 + 4x_2 + 3x_3 + s_1 = 9$
$2x_1 + x_2 + 5x_3 - s_2 = 10$
$3x_1 + 5x_2 + 2x_3 + s_3 = 8$

5. $x_1 + 2x_2 + x_3 + s_1 = 60$
$2x_1 + 3x_3 - s_2 = 18$
$3x_2 + 2x_3 - s_3 = 30$

7. (a) Maximize: $z = -w = -8x_1 - 3x_2$
Subject to: $4x_1 + 2x_2 \le 8$
$x_1 + 4x_2 \ge 7$
$x_1 \ge 0, x_2 \ge 0$

(b) $4x_1 + 2x_2 + s_1 = 8$
$x_1 + 4x_2 - s_2 = 7$
$8x_1 + 3x_2 + z = 0$

(c)
$$\begin{array}{ccccc|c} x_1 & x_2 & s_1 & s_2 & z & \\ 4 & 2 & 1 & 0 & 0 & 8 \\ 1 & 4 & 0 & -1 & 0 & 7 \\ \hline 8 & 3 & 0 & 0 & 1 & 0 \end{array}$$

9. (a) Maximize: $z = -w = -2x_1 - x_2 - 3x_3$
Subject to: $5x_1 + 6x_2 + x_3 \le 20$
$2x_1 + x_2 + 5x_3 \ge 10$
$7x_2 + 4x_3 \ge 20$
$x_1 \ge 0, x_2 \ge 0, x_3 \ge 0$

(b) $5x_1 + 6x_2 + x_3 + s_1 = 20$
$2x_1 + x_2 + 5x_3 - s_2 = 10$
$0x_1 + 7x_2 + 4x_3 - s_3 = 20$
$2x_1 + x_2 + 3x_3 + z = 0$

(c)
$$\begin{array}{ccccccc|c} x_1 & x_2 & x_3 & s_1 & s_2 & s_3 & z & \\ 5 & 6 & 1 & 1 & 0 & 0 & 0 & 20 \\ 2 & 1 & 5 & 0 & -1 & 0 & 0 & 10 \\ 0 & 7 & 4 & 0 & 0 & -1 & 0 & 20 \\ \hline 2 & 1 & 3 & 0 & 0 & 0 & 1 & 0 \end{array}$$

11. The solution is $x_1 = 18,\ x_2 = 0,\ s_1 = 7,\ s_2 = 12,\ s_3 = 0,\ z = 90$.

13. The solution is $x_1 = 0, x_2 = 10, s_1 = 5, s_2 = 0, s_3 = 18, z = 30$.

15. The solution is $x_1 = 0,\ x_2 = 15,\ x_3 = 45,\ s_1 = 45,\ s_2 = 0,\ s_3 = 0,\ z = 150$.

17. The solution is $x_1 = 2,\ x_2 = 0,\ s_1 = 8,\ s_2 = 0,\ w = 2$.

19. The solution is $x_1 = 0,\ x_2 = 1,\ s_1 = 16,\ s_2 = 9,\ s_3 = 0,\ w = 4$.

21. The solution is $x_1 = 0, x_2 = \frac{8}{5}, x_3 = \frac{13}{5}, s_1 = 7, s_2 = 0, s_3 = 0, w = \frac{34}{5}$.

23. The solution is $x_1 = 0, x_2 = 12, x_3 = 0, s_1 = 10, s_2 = 30, s_3 = 0, w = -12$.

25. (a) To maximize profit, the Tentsmith Co. should make and sell 30 two-person tents and 50 six-person tents. This will yield a profit of \$4200.

(b) $s_1 = 0, s_2 = 0$ indicate that all hours in the fabric cutting and assembly and packaging departments have been used. $s_3 = 60$ indicates that the company has exceeded the minimum number of tents required by 60.

27. (a) To maximize profit, Mike and Denyse should produce and sell 10 Deluxe 1000 models and 20 Deluxe 1200 models. Maximum profit will be \$12,000.

(b) $s_1 = s_2 = 0$ indicate that all hours in the assembly and testing department are being used. $s_3 = 22$ indicates that they will exceed the minimum 8 microcomputers required by 22.

29. (a) To minimize daily food cost and meet minimum nutritional requirements, Mike should eat 3 cheeseburgers and 10 small orders of onion rings.

(b) Minimum daily cost will be \$13.47.

31. (a) To minimize operating cost and fulfill the minimum order, the Findlay refinery should operate for 60 days and the Ft. Wayne refinery should operate for 150 days. The minimum cost is \$6,630,000.

(b) $s_1 = 138{,}000$ indicates that they will exceed the minimum order of 60,000 barrels of low-octane gasoline by 138,000 barrels. $s_4 = 90$ indicates that the refineries will operate 90 days less than the 300-day maximum restriction, and the zero values for s_2 and s_3 indicate that they will exactly meet the minimum requirements of high-octane and diesel, respectively.

33. (a) To minimize cost and meet nutritional requirements, Anita should use 50 ounces of Goli and 300 ounces of Zarobila. Minimum daily costs are \$11.50.

(b) $s_2 = 150$ indicates that the minimum requirement of 500 units of minerals will be exceeded by 150 units. $s_4 = 100$ indicates that Anita will use 100 ounces less than the maximum 450 ounces allowed. The zero values for the remaining slack and surplus variables indicate that these restrictions are being met exactly.

35. To minimize transportation costs, warehouse A should ship 110 DVD players to retailer I and zero DVD players to retailer II. Warehouse B should ship 150 DVD players to retailer II and zero DVD players to retailer I. The minimum transportation costs will be \$3860.

Chapter 4 Review Exercises

1. $$\begin{aligned} x_1 + x_2 + s_1 &= 14 \\ -2x_1 + 5x_2 + s_2 &= 6 \end{aligned}$$

3. $$\begin{array}{c} \begin{array}{cccccc} x_1 & x_2 & s_1 & s_2 & z & \end{array} \\ \left[\begin{array}{ccccc|c} 1 & 2 & 1 & 0 & 0 & 8 \\ 1 & -1 & 0 & 1 & 0 & 4 \\ \hline -1 & -3 & 0 & 0 & 1 & 0 \end{array}\right] \end{array}$$

5. $$\begin{array}{c} \begin{array}{ccccccc} x_1 & x_2 & x_3 & s_1 & s_2 & z & \end{array} \\ \left[\begin{array}{cccccc|c} 1 & 2 & 3 & 1 & 0 & 0 & 14 \\ -1 & 4 & -2 & 0 & 1 & 0 & 3 \\ \hline -2 & -3 & -1 & 0 & 0 & 1 & 0 \end{array}\right] \end{array}$$

7. (a) Basic: x_1, s_1, z; nonbasic: x_2, s_2

(b) $x_1 = 14, x_2 = 0, s_1 = 6, s_2 = 0, z = 5$

9. (a) $$\begin{array}{c} \begin{array}{ccccccc} x_1 & x_2 & s_1 & s_2 & s_3 & z & \end{array} \\ \left[\begin{array}{cccccc|c} \frac{11}{4} & 0 & 1 & -\frac{3}{4} & 0 & 0 & 0 \\ -\frac{1}{4} & 1 & 0 & \frac{1}{4} & 0 & 0 & 1 \\ \frac{1}{2} & 0 & 0 & \frac{1}{2} & 1 & 0 & 14 \\ \hline \frac{17}{4} & 0 & 0 & -\frac{5}{4} & 0 & 1 & -3 \end{array}\right] \end{array}$$

(b) $x_1 = 0, x_2 = 1, s_1 = 0, s_2 = 0, s_3 = 14, z = -3$

11. (a) Let x_1 = number of students admitted each month for the drum workshop, and let x_2 = number of students admitted each month for the guitar workshop.

Maximize: $P = 25x_1 + 30x_2$

Subject to:
$$\begin{aligned} 2x_1 + 4x_2 &\le 84 \\ 4x_1 + 2x_2 &\le 78 \\ x_1 \ge 0, x_2 &\ge 0 \end{aligned}$$

(b) $$\begin{aligned} 2x_1 + 4x_2 + s_1 &= 84 \\ 4x_1 + 2x_2 + s_2 &= 78 \\ -25x_1 - 30x_2 + P &= 0 \end{aligned}$$

(c) $$\begin{array}{c} \begin{array}{cccccc} x_1 & x_2 & s_1 & s_2 & P & \end{array} \\ \left[\begin{array}{ccccc|c} 2 & 4 & 1 & 0 & 0 & 84 \\ 4 & 2 & 0 & 1 & 0 & 78 \\ \hline -25 & -30 & 0 & 0 & 1 & 0 \end{array}\right] \end{array}$$

13. (a) Let x_1 = number of + 15 bags made each week, and let x_2 = number of – 15 bags made each week.

Maximize: $P = 60x_1 + 70x_2$

Subject to:
$$\begin{aligned} 3x_1 + 6x_2 &\le 240 \\ 3x_1 + 3x_2 &\le 180 \\ x_1 \ge 0, x_2 &\ge 0 \end{aligned}$$

(b) $$\begin{aligned} 3x_1 + 6x_2 + s_1 &= 240 \\ 3x_1 + 3x_2 + s_2 &= 180 \\ -60x_1 - 70x_2 + P &= 0 \end{aligned}$$

(c) $$\begin{array}{c} \begin{array}{cccccc} x_1 & x_2 & s_1 & s_2 & P & \end{array} \\ \left[\begin{array}{ccccc|c} 3 & 6 & 1 & 0 & 0 & 240 \\ 3 & 3 & 0 & 1 & 0 & 180 \\ \hline -60 & -70 & 0 & 0 & 1 & 0 \end{array}\right] \end{array}$$

15. (a) Let x_1 = number of acres allotted for head lettuce, x_2 = number of acres allotted for cantaloupe, and x_3 = number of acres allotted for bell peppers.

Maximize: $P = 120x_1 + 100x_2 + 140x_3$

Subject to:
$$\begin{aligned} x_1 + x_2 + x_3 &\le 220 \\ 173x_1 + 175x_2 + 212x_3 &\le 34{,}700 \\ 122x_1 + 161x_2 + 108x_3 &\le 27{,}845 \\ 76x_1 + 39x_2 + 72x_3 &\le 13{,}350 \\ x_1 \ge 0, x_2 \ge 0, x_3 &\ge 0 \end{aligned}$$

(b) $$\begin{aligned} x_1 + x_2 + x_3 + s_1 &= 220 \\ 173x_1 + 175x_2 + 212x_3 + s_2 &= 34{,}700 \\ 122x_1 + 161x_2 + 108x_3 + s_3 &= 27{,}845 \\ 76x_1 + 39x_2 + 72x_3 + s_4 &= 13{,}350 \\ -120x_1 - 100x_2 - 140x_3 + P &= 0 \end{aligned}$$

(c) $$\begin{array}{c} \begin{array}{cccccccccc} x_1 & x_2 & x_3 & s_1 & s_2 & s_3 & s_4 & P & \end{array} \\ \left[\begin{array}{cccccccc|c} 1 & 1 & 1 & 1 & 0 & 0 & 0 & 0 & 220 \\ 173 & 175 & 212 & 0 & 1 & 0 & 0 & 0 & 34{,}700 \\ 122 & 161 & 108 & 0 & 0 & 1 & 0 & 0 & 27{,}845 \\ 76 & 39 & 72 & 0 & 0 & 0 & 1 & 0 & 13{,}350 \\ \hline -120 & -100 & -140 & 0 & 0 & 0 & 0 & 1 & 0 \end{array}\right] \end{array}$$

17. $x_1 = 5, x_2 = 0, s_1 = 23, s_2 = 0, s_3 = 17, z = 20$

19. $x_1 = 0, x_2 = 1, x_3 = \frac{12}{5}, x_4 = \frac{41}{5}, s_1 = 0, s_2 = 0, s_3 = 0, z = \frac{149}{5}$

21. (a) $$\begin{array}{c} \begin{array}{ccccccc} x_1 & x_2 & s_1 & s_2 & s_3 & z & \end{array} \\ \left[\begin{array}{cccccc|c} 4 & -2 & 1 & 0 & 0 & 0 & 0 \\ 1 & 1 & 0 & 1 & 0 & 0 & 7 \\ 2 & 1 & 0 & 0 & 1 & 0 & 8 \\ \hline -2 & -2 & 0 & 0 & 0 & 1 & 0 \end{array}\right] \end{array}$$

(b) Maximum is 14 at $x_1 = 1, x_2 = 6, s_1 = 8$, and $s_2 = s_3 = 0$ (may be multiple solutions).

23. (a) $$\begin{array}{c} \begin{array}{ccccccc} x_1 & x_2 & x_3 & s_1 & s_2 & z & \end{array} \\ \left[\begin{array}{cccccc|c} 1 & -1 & 1 & 1 & 0 & 0 & 5 \\ 2 & 1 & 4 & 0 & 1 & 0 & 4 \\ \hline -3 & -1 & -1 & 0 & 0 & 1 & 0 \end{array}\right] \end{array}$$

(b) Maximum is $z = 6$ at $x_1 = 2, x_2 = x_3 = s_2 = 0$, and $s_1 = 3$.

25. (a) $$\begin{array}{c} \begin{array}{cccccc} x_1 & x_2 & s_1 & s_2 & z & \end{array} \\ \left[\begin{array}{ccccc|c} 1 & 1 & 1 & 0 & 0 & 9 \\ 2 & 2 & 0 & 1 & 0 & 4 \\ \hline -8 & -2 & 0 & 0 & 1 & 0 \end{array}\right] \end{array}$$

(b) Maximum is $z = 16$ at $x_1 = 2, x_2 = s_2 = 0$, and $s_1 = 7$.

27. (a) A maximum monthly profit of \$660 is obtained by admitting 8 drum students and 12 guitar students each month. The value of 2 for s_3 indicates that 2 hours of available music recording instruction are unused.

(b) Answers will vary.

29. A maximum profit of \$23,000 is obtained by allotting 150 acres for head lettuce and 50 acres for cantaloupe. The value of 20 for s_1 indicates that there are 20 available acres unused. The value of 1495 for s_3 indicates that there are 1495 available pounds of phosphate unused.

31. $$\begin{bmatrix} 3 & 2 & 4 \\ 1 & 0 & 1 \\ 0 & 3 & 8 \\ 2 & 7 & 9 \end{bmatrix}$$

33. Maximize: $z = 10x_1 + 5x_2$
Subject to: $4x_1 + x_2 \le 6$
$2x_1 + 3x_2 \le 4$
$x_1 \ge 0, x_2 \ge 0$

35. (a) Minimum is 4 at $y_1 = 4, y_2 = 0$.

(b) Maximum is 4 at $x_1 = 1, x_2 = 0$.

37. (a) Maximize: $z = 6x_1 + 9x_2$
Subject to: $4x_1 + 3x_2 \le 3$
$2x_1 + 4x_2 \le 4$
$x_1 \ge 0, x_2 \ge 0$

(b) Minimum is 9 at $y_1 = 3, y_2 = 0$.

39. (a) Maximize: $z = 5x_1 + 12x_2$
Subject to: $x_1 + 2x_2 \le 16$
$2x_1 + x_2 \le 7$
$3x_1 + x_2 \le 22$
$x_1 \ge 0, x_2 \ge 0$

(b) Minimum is 84 at $y_1 = 0, y_2 = 12, y_3 = 0$.

41. (a) Minimize: $C = 0.08y_1 + 0.07y_2$
Subject to: $6y_1 + 6y_2 \ge 60$
$2y_1 + y_2 \ge 12$
$y_1 \ge 0, y_2 \ge 0$

(b) Maximize: $z = 60x_1 + 12x_2$
Subject to: $6x_1 + 2x_2 \le 0.08$
$6x_1 + x_2 \le 0.07$
$x_1 \ge 0, x_2 \ge 0$

(c) A minimum cost of \$0.72 is obtained by taking 2 Everyday pills and 8 Allday pills daily.

43. (a) Aztec: 50 days; Buena Vista: 125 days

(b) \$5,800,000

45. (a) $x_1 =$ the value of each mg of vitamin C, $x_2 =$ the value of each mg of Niacin; $x_1 = 0.01, x_2 = 0.01$.

(b) \$0.74

47. $4x_1 + x_2 + 3x_3 + s_1 = 12$
$x_1 + 2x_2 + 5x_3 - s_2 = 9$
$3x_1 + x_2 + 7x_3 + s_3 = 4$

49. (a) Maximize: $z = -w = -x_1 - 3x_2 - 8x_3$
Subject to: $x_1 + 2x_2 + 5x_3 \le 7$
$x_1 + x_2 + 8x_3 \ge 10$
$4x_2 + 2x_3 \ge 24$
$x_1 \ge 0, x_2 \ge 0, x_3 \ge 0$

(b) $x_1 + 2x_2 + 5x_3 + s_1 = 7$
$x_1 + x_2 + 8x_3 - s_2 = 10$
$4x_2 + 2x_3 - s_3 = 24$
$x_1 + 3x_2 + 8x_3 + z = 0$

(c)
$$\begin{array}{c} \begin{matrix} x_1 & x_2 & x_3 & s_1 & s_2 & s_3 & z & \end{matrix} \\ \left[\begin{array}{ccccccc|c} 1 & 2 & 5 & 1 & 0 & 0 & 0 & 7 \\ 1 & 1 & 8 & 0 & -1 & 0 & 0 & 10 \\ 0 & 4 & 2 & 0 & 0 & -1 & 0 & 24 \\ \hline 1 & 3 & 8 & 0 & 0 & 0 & 1 & 0 \end{array}\right] \end{array}$$

51. Maximum is $z = 64$ at $x_1 = 16, x_2 = s_3 = 0, s_1 = 6, s_2 = 22$.

53. Minimum is $w = 4$ at $x_1 = s_2 = 0, x_2 = 2, s_1 = 20$.

55. Minimum is $w = 10$ at $x_1 = 5, x_2 = x_3 = s_2 = 0, s_1 = 25, s_3 = 15$.

57. (a) Day packs: 90; Multi-day packs: 0; \$3150

(b) $s_1 = 0$ indicates that no available hours in fabric cutting are unused, $s_2 = 80$ indicates that 80 available hours in assembly and packaging are unused, $s_3 = 75$ indicates that 75 more backpacks are made weekly than the minimum requirement of 15.

59. (a) 6-inch roasted chicken sandwiches: 3; 6-inch Subway Club® sandwiches: 2

(b) \$19.40

(c) Answers will vary.

Chapter 5

Section 5.1

1. \$187.50 **3.** \$2750 **5.** \$318.75 **7.** \$11,000

9. \$2280 **11.** \$13,250 **13.** \$7350 **15.** \$32,675

17. 7.5% **19.** 9% **21.** 9.51% **23.** 10.52%

25. \$4310.34 **27.** \$7920.79 **29.** \$3814.06 **31.** \$13,300.08

33. (a) $A = 8000 + 520t$

(b)

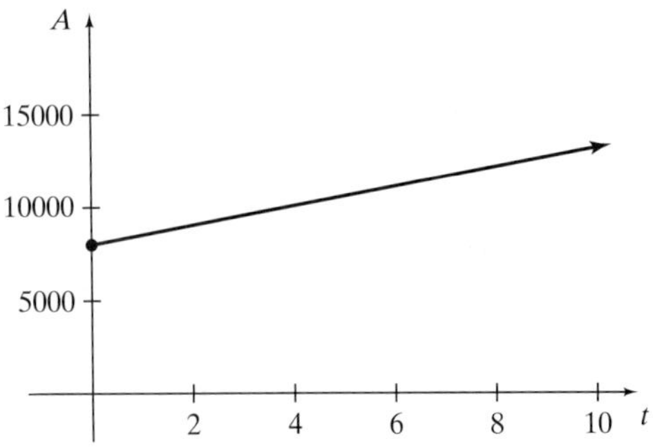

$m = 520$

Each year, Sue earns \$520 in interest on her \$8000 deposit.

35. (a) $A = 4000 + 210t$

(b)

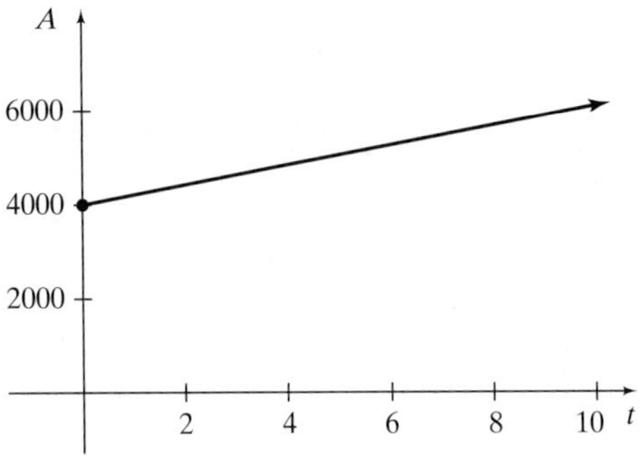

$m = 210$

Each year, Rich earns \$210 in interest on his \$4000 deposit.

37. It will take 10 years for a deposit to double if invested at 10% simple interest.

39. It will take approximately 11.76 years for a deposit to double if invested at 8.5% simple interest.

41. It will take approximately 17.39 years for a deposit to double if invested at 5.75% simple interest.

43. The discount is $46.67 and the proceeds are $953.33.

45. The discount is $650 and the proceeds are $4350.

47. The discount is $2887.50 and the proceeds are $7112.50.

49. $1048.95 51. $5747.13 53. $14,059.75

55. The amount of interest due on the loan is $18,000.

57. The amount of interest due on the loan is $160.

59. The amount of interest due on the loan is $18 million.

61. Ardra repaid $25,630 to her aunt.

63. Denyse repaid $23,625 to her grandfather.

65. The discount is $1190 and the proceeds are $5810.

67. Brian should borrow $8433.73.

69. Melissa should invest approximately $2643.17.

71. Bill should invest approximately $16,293.28.

73. Francine will pay 8.25% interest.

75. Lenny will pay 7.3% interest.

77. Dylan should invest $3948.06.

79. Franco should invest $1864.80.

81. The annual simple interest rate is 3.7%.

83. Alicia should pay $4773.27.

85. (a) The loan to value ratio is $32,800.
(b) Bill will be able to take a $2788 tax deduction in a given year due to interest on the home equity loan.

Section 5.2

1. Future value of approximately $3828.84 after 5 years and a future value of approximately $4432.37 after 8 years.

3. Future value of approximately $2623.30 after 5 years and a future value of approximately $3087.02 after 8 years.

5. Future value of approximately $11,458.08 after 5 years and a future value of approximately $14,214.29 after 8 years.

7. Future value of approximately $21,001.72 after 5 years and a future value of approximately $25,701.19 after 8 years.

9. Future value of approximately $14,560.77 after 5 years and a future value of approximately $18,579.40 after 8 years.

11. Future value of $9424.55 after 3 years and $11,102.76 after 6 years.

13. Future value of $11,877.82 after 3 years and $14,108.27 after 6 years.

15. The present value is approximately $4113.51.

17. The present value is approximately $2287.19.

19. The present value is approximately $5523.22.

21. The present value is approximately $8115.48.

23. The present value is approximately $3639.06.

25. The present value is approximately $6995.44.

27. The present value is approximately $5817.44.

29. Graph (c) 31. Graph (a)

33. It will take approximately 4 years.

35. It will take approximately 18 compounding periods (about 9 years).

37. It will take approximately 37 compounding periods (about $9\frac{1}{4}$ years).

39. It will take approximately 92 compounding periods (about $7\frac{1}{2}$ years).

41. The effective rate is 8.775%.

43. The effective rate is 6.168%.

45. The effective rate is 5.319%.

47. The effective rate is 6.963%.

49. The effective rate is 7.714%.

51. The effective rate is 7.229%.

53. The effective rate is 6.348%.

55. The effective rate is 5.904%.

57. (a) The nominal rate of 6.25% compounded quarterly is 6.109%.
(b) The nominal rate of 6.25% compounded monthly is 6.078%.

59. (a) The nominal rate of 5.5% compounded quarterly is 5.39%.
(b) The nominal rate of 5.5% compounded monthly is 5.366%.

61. (a) The nominal rate of 5.75% compounded quarterly is 5.63%.
(b) The nominal rate of 5.75% compounded monthly is 5.604%.

63. (a) The nominal rate of 7.5% compounded quarterly is 7.298%.
(b) The nominal rate of 7.5% compounded monthly is 7.254%.

65. 8% compounded monthly is the better investment.

67. 8.25% compounded quarterly is the better investment.

69. If the rate holds true, in 5 years a gallon of milk will cost about $2.62.

71. The account will be worth about $2962.37 in 4 years.

73. If the rate holds true, in 7 years the house will be worth about $166,033.

75. Alec and Lexi should invest about $16,074.45 now at 5.5% interest compounded quarterly to have $20,000 in 4 years.

77. Sam and Diane should invest $16,368.75 now at 8.5% interest compounded monthly to have $25,000 in 5 years.

79. It will take 37 compounding periods, or 37 quarters, for the \$10,000 to grow to at least \$16,000.

81. **(a)** The effective rate of 4.5% compounded monthly is 4.594%.
(b) The effective rate of 4.4% compounded daily is 4.498%.
(c) The account at First Internet bank paying 4.50% interest compounded monthly is a better investment; the effective rate is higher.

83. Paige's grandparents should pay \$11,546.49 for the bond.

85. The annual compound rate of return is 9.746%.

87. The interest rate, compounded annually, is 25.137%.

Section 5.3

1. The future value of the annuity is \$24,581.90.

3. The future value of the annuity is \$206,798.81.

5. The future value of the annuity is \$191,086.80.

7. The future value of the annuity is \$79,240.57.

9. The future value of the annuity is \$105,997.82.

11. The future value of the annuity is \$110,943.44.

13. The amount of interest is \$6581.90.

15. The amount of interest is \$146,798.81.

17. The amount of interest is \$131,086.80.

19. The periodic payment amount is \$1517.36 annually.

21. The periodic payment amount is \$263.55 monthly.

23. The periodic payment amount is \$700 quarterly.

25. The annual payment needed is \$1193.84.

27. The monthly payment needed is \$258.14.

29. The quarterly payment needed is \$1129.35.

31. **(a)** \$15,173.60 **(b)** \$4,826.40

33. **(a)** \$47,439 **(b)** \$32,561

35. **(a)** \$22,400 **(b)** \$7600

37. The annual interest rate is 6.66%.

39. The annual interest rate is 6.92%.

41. The annual interest rate is 8.12%.

43. The account will be worth \$15,322.29 after 10 years.

45. Teri deposited \$12,000 in the account and earned \$3322.29 in interest.

47. The account will be worth \$71,924.44 after 30 years.

49. The account will be worth \$41,693.54 after 10 years.

51. The account will be worth \$18,838.56 after 4 years.

53. They should deposit \$356.76 monthly.

55. The quarterly deposit needed is \$1014.92.

57. Irene should invest \$208.81 each month.

59. **(a)** \$5355
(b) 4 years of college tuition at \$5355 per year will be \$21,420.
(c) \$80.14 should be invested monthly at 5% interest compounded monthly to have \$21,420 after 15 years.

61. **(a)** \$188,921.57 **(b)** \$272,615.08 **(c)** \$398,041.76

63. **(a)** On her 65th birthday, Debbie's account will be worth \$367,264.71.
(b) Bubba's account will be worth \$291,901.24 on his 65th birthday.
(c) Debbie deposited \$20,000 into her account, and earned \$347,264.71 in interest. Bubba deposited \$66,000 into his account and earned \$225,901.24 in interest. The amount that Debbie deposited was considerably less than the amount that Bubba deposited, but she had more in her account on her 65th birthday. The moral is to begin making regular deposits into a savings account as early as possible!

65. **(a)** \$13,286.89 **(b)** Monthly investment is \$3286.89.
(c) The value of the account after 30 years is \$3,301,640.04.

Section 5.4

1. \$60,876.17 **3.** \$111,718.92 **5.** \$21,994.72

7. \$66,753.57 **9.** \$103,671.89 **11.** \$15,586.87

13. \$296.75 **15.** \$716.41 **17.** \$761.87 **19.** \$410.33

21. \$820.95 **23.** \$707.29

25. **(a)** The total amount paid over the life of the loan is \$35,610.
(b) The total amount of interest paid over the life of the loan is \$10,610.

27. **(a)** The total amount paid over the life of the loan is \$257,907.60.
(b) The total amount of interest paid over the life of the loan is \$157,907.60.

29. **(a)** The total amount paid over the life of the loan is \$15,237.40.
(b) The total amount of interest paid over the life of the loan is \$2237.40.

31. Teri needs to deposit a lump sum of \$9537.64 now at 4.75% interest compounded monthly for 10 years to yield the same amount as the annuity.

33. **(a)** The monthly payment will be \$359.45.
(b) The total interest paid over the life of the loan is \$15,934.

35. Assuming that the account continues to earn 6.5% interest compounded monthly, Irene can withdraw \$1493.80 each month for 25 years.

37. **(a)** Irene deposited \$72,000 into her account.
(b) Nick deposited \$72,000 into his account.
(c) Irene receives \$17,925.60 annually while Nick receives \$16,994.73 annually. Irene's account earned more interest than Nick's due to the more frequent deposits and the monthly compounding of interest.

39. Malika needs to deposit \$310.26 into the annuity each month for 35 years to be able to withdraw \$3000 each month for 25 years.

41. **(a)** The lump sum payment will be \$1,835,310.77.
(b) The lump sum payment will be \$1,654,538.78.

43. The cost of the house is $138,566.64.

45. (a) The down payment is $36,000. The financed amount is $144,000.

(b) The monthly payment is $995.06.

47. (a) They will pay $214,221.60 in interest if they take out the 30-year mortgage at 7.38%.

(b) They will pay $88,975.80 in interest if they take out the 15-year mortgage at 7.0%.

49. (a) After 10 years of payments, the unpaid balance is $124,651.82.

(b) After 15 years of payments, the unpaid balance is $108,134.44.

(c) After 15 years of payments, the unpaid balance is $84,272.55.

51. (a) $y = 995.06\left[\frac{1-(1.00615)^{-(360-x)}}{0.00615}\right]$

(b)

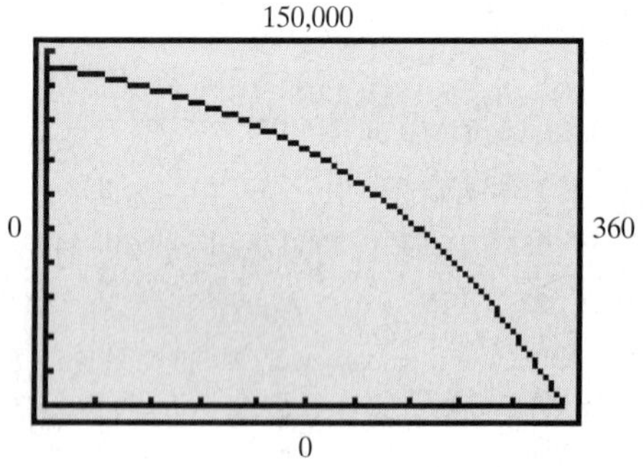

(c) The calculator approximation of the unpaid balance is $124,651.82, which is the same result as in Exercise 49a.

(d) The calculator approximation of the unpaid balance is $108,134.44, which is the same result as in Exercise 49b.

(e) The calculator approximation of the unpaid balance is $84,272.55, which is the same result as in Exercise 49c.

53.

Payment Number	Amount of Payment	Interest for Period	Reduction of Unpaid Balance	Unpaid Balance
0				10,000
1	874.51	75.00	799.51	9,200.49
2	874.51	69.00	805.51	8,394.98
3	874.51	62.96	811.55	7,583.43
4	874.51	56.88	817.63	6,765.80
5	874.51	50.74	823.77	5,942.03
6	874.51	44.57	829.94	5,112.09
7	874.51	38.34	836.17	4,275.92
8	874.51	32.07	842.44	3,433.48
9	874.51	25.75	848.76	2,548.72
10	874.51	19.39	855.12	1,693.60
11	874.51	12.70	861.81	831.79
12	838.03	6.24	831.79	0

Notice that the final month's payment will usually need to be adjusted to bring the unpaid balance to zero.

Chapter 5 Review Exercises

1. (a) $2970 **(b)** $14,970 **3.** 7.5% **5.** $6439.39

7. (a) $A = 6000 + 375t$

(b)

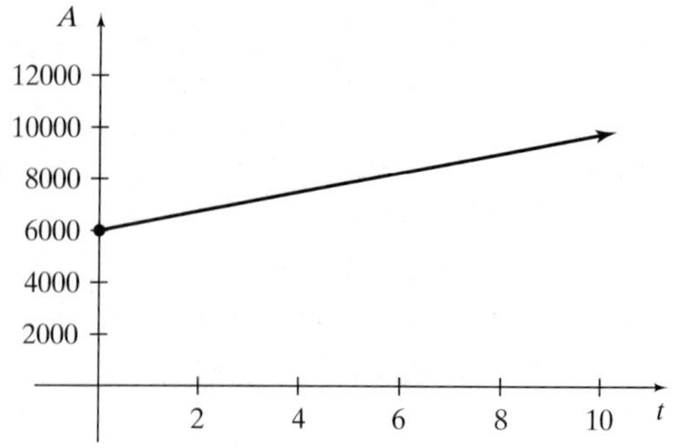

(c) $m = 375$; Amara earns $375 in interest each year on her $6000 deposit.

9. 16 years **11.** $D = \$4200$, $p = \$7800$ **13.** $29,937.63

15. $20,860 **17.** 10% **19.** $9128.98, $12,722.65

21. $10,064.76, $11,902.67 **23.** $5008.01 **25.** $8929.13

27. 10 years **29.** 7.978% **31. (a)** 5.630% **(b)** 5.604%

33. $1.87 **35.** $6997.61 **37.** $14,280.11 **39.** $270,949.97

41. $180,949.97 **43.** $861.97

45. (a) $11,606.40 **(b)** $8393.60 **47.** 4.47%

49. $6918.28 **51.** $312.38 **53.** $94,891.55

55. $46,269.22 **57.** $265.47 **59.** $511.63

61. (a) $47,784.60 **(b)** $17,784.60

63. (a) $291.04 **(b)** $22,387.20

65. (a) $1849.27

(b) If Lindsay deposits $240 a month for 25 years, she deposits a total of $72,000 and receives equal monthly payments of $1421.11 for the last 20 years. Whereas, if Lindsay deposits $400 each month for 15 years, she deposits a total of $73,000 and receives monthly payments of $1849.27 for the last 20 years.

67.

Payment Number	Amount of Payment	Interest for Period	Reduction of Unpaid Balance	Unpaid Balance (principal)
0				$6000
1	$521.93	$40.00	$481.93	$5518.07
2	521.93	36.79	485.14	5032.93
3	521.93	33.55	488.38	4544.55
4	521.93	30.30	491.63	4052.92
5	521.93	27.02	494.91	3558.01
6	521.93	23.72	498.21	3059.80
7	521.93	20.40	501.53	2558.27
8	521.93	17.06	504.87	2053.40
9	521.93	13.69	508.24	1545.16
10	521.93	10.30	511.63	1033.53
11	521.93	6.89	515.04	518.49
12	521.95	3.46	518.49	0

69. (a) Simple interest **(b)** 4.44%

71. (a) Compound interest **(b)** $9971.65

Chapter 6

Section 6.1

1. This is an example of probability.
3. This is an example of statistical inference.
5. This is an example of probability.
7. This is an example of probability.
9. This process represents a probability experiment.
11. This process does not represent a probability experiment.
13. All the questions that have an oval to fill in as a response represent a probability experiment.
15. Order is important. 17. Order is important.
19. Order is not important. 21. Order is not important.
23. Order is not important.
25. The events can happen at the same time.
27. The events cannot happen at the same time.
29. The events can happen at the same time.
31. The events cannot happen at the same time.
33. The events cannot happen at the same time.

In Exercises 35–45, the answers will vary.

35. The outcome is tails. 37. It will be light outside.
39. The person has perfect hearing.
41. The person has blonde hair.
43. The student receives a passing grade in geography.
45. The card is a queen.
47. The outcome of the flood influences the postponement of the picnic.
49. The outcome of the first event does not influence the second event.
51. The outcome of the first event influences the second event.
53. The outcome of the first event does not influence the second event.
55. The outcome of the first event influences the second event.

In Exercises 57–67, the answers will vary.

57. There will be a full moon on Wednesday night.
59. The child watches Sesame Street in the afternoon.
61. The outcome of the next roll is an odd number.
63. The event that the person will have a sunburn.
65. The event that Tammy gives a speech at the ceremony.
67. The event that Ron fails the political science exam.

Section 6.2

1. $A = \{1, 2, 3, 4, 5, 6, 7\}$
3. $C = \{2, 4, 6, 8, 10, 12, 14, 16, 18, 20, 22, 24\}$
5. $E = \{\text{September, October, November, December}\}$
7. $A = \{x \mid x \text{ is a whole number and } 2 \le x \le 6\}$
9. $C = \{x \mid x \text{ is a multiple of 5 and } 1 \le x \le 30\}$
11. $E = \{x \mid x \text{ is a day of the week that starts with the letter T}\}$
13. $\emptyset, \{2\}$ 15. $\emptyset$, {Honda}, {Ford}, {Honda, Ford}
17. $\emptyset, \{8\}, \{9\}, \{10\}, \{8, 9\}, \{8, 10\}, \{9, 10\}, \{8, 9, 10\}$
19. $\supset$ 21. $\subseteq$ 23. $\subseteq$ 25. $A \cap B = \{2, 4\}$
27. $B \cap C = \emptyset$ 29. $A \cup B = \{1, 2, 3, 4, 5, 6, 8, 10\}$
31. $B \cup C = \{2, 4, 6, 8, 10, 11, 12, 13, 14, 15\}$
33. $A^C = \{6, 7, 8, 9, 10, 11, 12, 13, 14, 15\}$
35. $B \cap C^C = \{2, 4, 6, 8, 10\}$ 37. $U^C = \emptyset$
39. $(A \cap C)^C = \{1, 2, 3, 4, 5, 6, 7, 8, 9, 10, 11, 12, 13, 14, 15\}$
41. $A^C \cup C = \{6, 7, 8, 9, 10, 11, 12, 13, 14, 15\}$
43. $A \cup \emptyset = \{1, 2, 3, 4, 5\}$ 45. $\emptyset \cap B^C = \emptyset$
47. $A = \{a, b, c, d, e\}$
49. $U = \{a, b, c, d, e, f, g, h, i, j, k, l, m, n\}$
51. $A \cap B = \{b, e\}$ 53. $A^C = \{f, g, h, i, j, k, l, m, n\}$
55. $(A \cup B)^C = \{h, j, k, l, m, n\}$
57. $A \cup (B \cap C) = \{2, 4, 5, 6, 8\}$

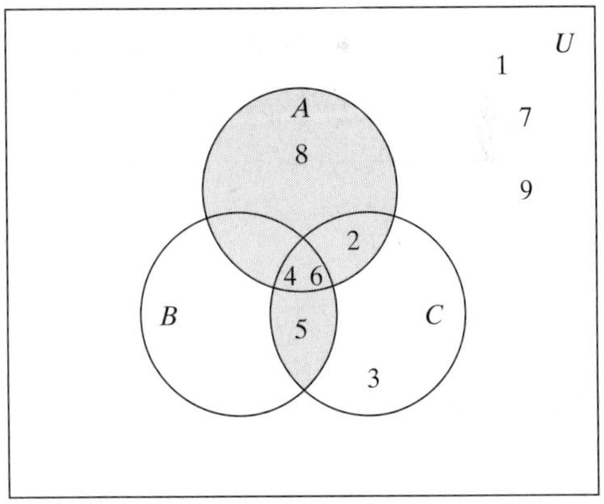

59. $A \cap (B \cup C)^C = \{8\}$

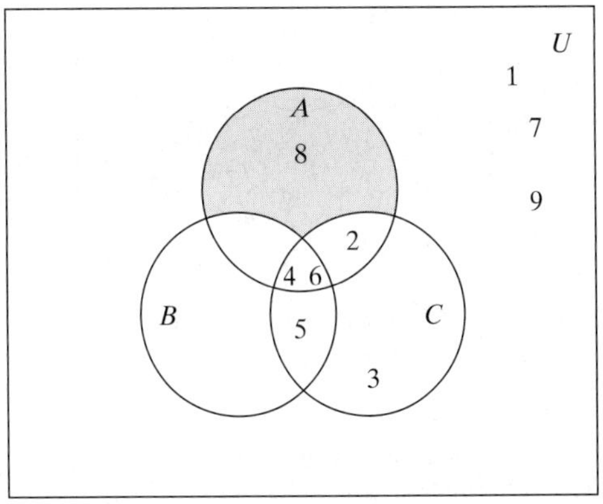

61. $(A \cup B) \cap (A \cup C) = \{2, 4, 5, 6, 8\}$

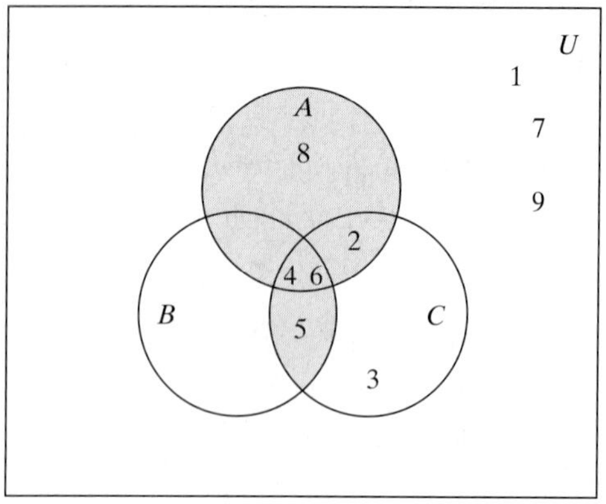

63. $A^C \cup B^C = \{1, 2, 3, 5, 7, 8, 9\}$

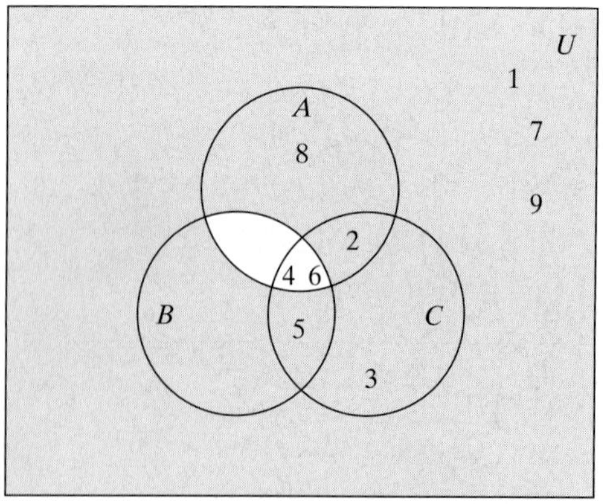

65. $(A \cap B)^C = \{1, 2, 3, 5, 7, 8, 9\}$

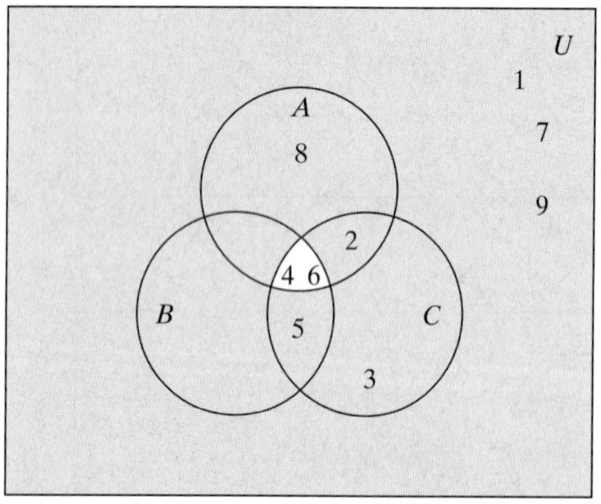

67. $n(A) = 250$

During 1996, the value of arm transfers from the United States to India and Pakistan was $250 million.

69. $n(B \cap D) = 70$

During 1996, the value of arms transfers from the United Kingdom to Pakistan was $70 million.

71. $n(B \cup D) = n(B) + n(D) - n(B \cap D)$
$= 70 + 210 - 70 = 210$

During 1996, the values of arms transfers either sent by the United Kingdom or received by Pakistan was $210 million.

73. $n(A^C) = n(U) - n(A)$
$= 84.9 - 53.8 = 31.1$

There are about 31.1 million households in the Midwest and southern United States whose household income is not under $35,000.

75. $n(A \cup D) = 71.8$

For the southern and midwestern United States, there are about 71.8 million households who are in the southern U.S. or whose household income is under $35,000.

77. $n(A^C \cup D) = 49.6$

For the southern and midwestern United States, there are 49.6 million households located in the southern U.S or whose household income is not under $35,000.

Section 6.3

1. (a)

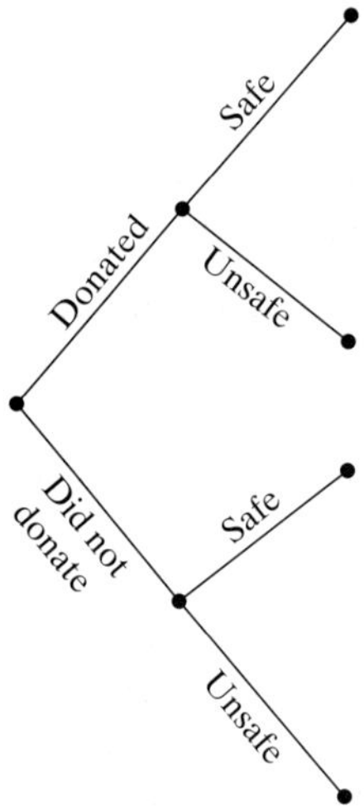

(b) There are four possible ways a respondent could answer the questions.

3. (a)

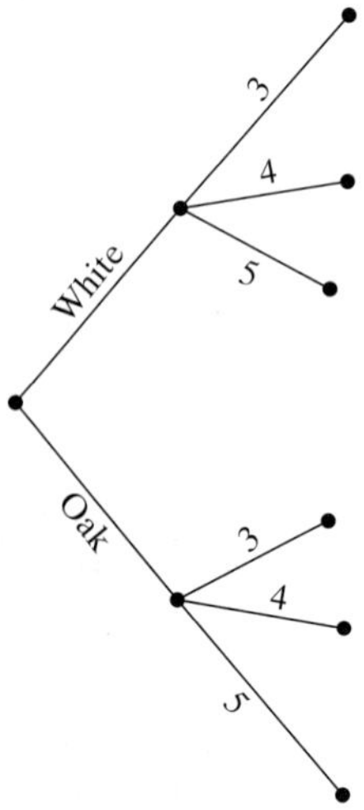

(b) There are six possible ways to build the bookshelf.

5. **(a)**

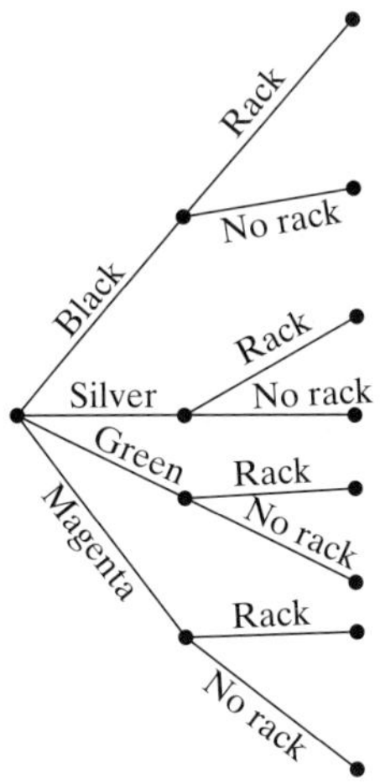

(b) There are eight possible ways to create a motorcycle package.

7. There are 192 ways that a person can construct a package deal.

9. There are 1512 ways that the awards can be given.

11. There are 720 possible ways to order the meal.

13. The textbooks can be chosen in 4550 ways.

15. The car can be ordered in 60 ways.

17. There are 30,240 possible combinations for the lock.

19. There are 1,125, 899, 906, 842, 624 different ways to answer the 25 questions.

21. 5040 **23.** 60 **25.** 5040 **27.** 1

29. ${}_{50}P_{20} \approx 1.15 \times 10^{32}$ **31.** ${}_{40}P_{15} \approx 5.26 \times 10^{22}$

33. ${}_{90}P_{30} \approx 1.79 \times 10^{56}$

35. There are 1320 different ways that the company officers can be selected.

37. There are 5814 ways that the winner, first runner-up, and second runner-up can be selected from the 19 entrants.

39. There are 303,600 different ways that the restaurants could be ranked.

41. 95,040 different messages are possible.

43. There are 362,880 ways that the batting order could be arranged.

45. 10 **47.** 210 **49.** 792 **51.** 7 **53.** 1 **55.** 1

57. ${}_{50}C_{20} \approx 4.71 \times 10^{13}$ **59.** ${}_{40}C_{15} \approx 4.02 \times 10^{10}$

61. ${}_{90}C_{30} \approx 6.73 \times 10^{23}$

63. There are 20 ways that the judges can be selected.

65. There are 15,504 different ways that the research team can be formed.

67. There are 118,755 different ways that five of the companies could be chosen.

69. In the Maine state lottery *Megabucks*, numbers can be chosen in 5,245,786 different ways.

71. 76,904,685 different sampler cups are possible.

73. There are 350 different ways that the subcommittee could be formed with 2 Democrats and 4 Republicans.

75. There are 8,953,560 ways that a sample of 10 could be chosen so that 2 are defective and 8 are not.

77. There are 8400 different ways that the committee of 7 could be formed with 2 Republicans, 2 Democrats, and 3 Green Party members.

Section 6.4

1. $S = \{1, 2, 3, 4, 5, 6, 7, 8, 9\}$
$n(S) = 9$

3. $S = \{2, 3, 4, 5, 6, 7, 8, 9, 10\}$
$n(S) = 9$

5. $S = \{\text{conservative, moderate, liberal}\}$
$n(S) = 3$

7. $S = \{(\text{good}, 1), (\text{good}, 2), (\text{good}, 3)$
$(\text{good}, 4), (\text{excellent}, 1),$
$(\text{excellent}, 2), (\text{excellent}, 3),$
$(\text{excellent}, 4), (\text{superior}, 1),$
$(\text{superior}, 2), (\text{superior}, 3),$
$(\text{superior}, 4)\}$
$n(S) = 12$

9. $S = \{(\text{better, AM}), (\text{better, FM}),$
$(\text{worse, AM}), (\text{worse, FM})\}$
$n(S) = 4$

11. $P(A) = \frac{5}{9},\ P(B) = \frac{4}{9}$ **13.** $P(A) = \frac{5}{9},\ P(B) = \frac{4}{9}$

15. $P(A) = 0,\ P(B) = 0.2$

17. $P(A) = \frac{4}{12} \approx 0.33,\ P(B) = \frac{3}{12} = 0.25$

19. $P(A) = \frac{4}{16} = 0.25,\ P(B) = \frac{7}{16} = 0.4375$

21. **(a)** ${}_{30}C_2 = 435$ **(b)** 105

(c) $P(A) = \dfrac{105}{435} \approx 0.24$

There is about a 24% chance that both cookies will be oatmeal raisin.

23. **(a)** 4060 **(b)** 455

(c) $P(A) = \dfrac{455}{4,060} \approx 0.112$

There is about an 11.2% chance that the 3 marbles will be blue.

25. **(a)** 924 **(b)** 350

(c) $P(A) = \dfrac{350}{924} \approx 0.3788$

There is about a 37.9% probability that the subcommittee will be formed with 2 Democrats and 4 Republicans.

27. **(a)** 30,045,015 **(b)** 8,953,560

(c) $P(A) = \dfrac{8,953,560}{30,045,015} \approx 0.298$

There is about a 29.8% probability that in a sample of 10 car stereos 2 will be defective.

29. $P(A) = \dfrac{8400}{77,520} \approx 0.108$

There is about a 10.8% chance that a committee of 7 will be formed with 2 Republicans, 2 Democrats, and 3 Green Party members.

31. (a) Let A represent the event that the periodical was a quarterly.

$P(A) = \frac{1444}{8904} \approx 0.162$

The probability that the periodical was a quarterly is approximately 16.2%

(b) Let B represent the event that the periodical was a weekly.

$P(B) = \frac{1716}{8904} \approx 0.193$

The probability that the periodical was a weekly is approximately 19.3%

33. (a) Let A represent the event that the crime involved a gas station.

$P(A) = \frac{13}{459} \approx 0.0283$

The probability that the crime involved a gas station is about 2.83%.

(b) Let B represent the event that the crime involved a bank.

$P(B) = \frac{11}{459} \approx 0.024$

The probability that the crime involved a bank is about 2.4%.

35. (a) Let A represent the event that the person was in the Army.

$P(A) = \frac{162.3}{380.1} \approx 0.427$

The probability that a randomly selected military personnel was in the Army is about 42.7%.

(b) Let B represent the event that the person was in the Navy.

$P(B) = \frac{93.3}{380.1} \approx 0.2455$

The probability that a randomly selected military personnel was in the Navy is about 24.55%.

37. (a) Let A represent the event that the scientist is a life scientist.

$P(A) = \frac{180.0}{665.8} \approx 0.270$

The probability that a randomly selected scientist, employed in the United States in 1996, is a life scientist is about 27.0%.

(b) The probability that a randomly selected scientist, employed in the United States in 1996, is not a life scientist is about 73.0%.

39. (a) Let A represent the event that the house sold for anywhere between $120,000 and $149,000.

$P(A) = \frac{183{,}000}{891{,}000} = 0.205$

The probability that a house sold in 1998, selected at random, sold for anywhere between $120,000 and $149,000 is about 20.5%.

(b) Let B represent the event that the house sold for $120,000 or more.

$P(B) = \frac{639{,}000}{891{,}000} \approx 0.7172$

The probability that a house sold in 1998, selected at random, sold for $120,000 or more is about 20.5%.

41. Subjective **43.** Empirical **45.** Classical

Section 6.5

1. The events are not mutually exclusive.

3. The events are mutually exclusive.

5. The events are not mutually exclusive.

7. The events are mutually exclusive.

9. The events are mutually exclusive.

11. $\frac{1}{5}$ **13.** $\frac{1}{10}$ **15.** $\frac{11}{15}$ **17.** $\frac{1}{3}$

19. $P(A) = \frac{175}{500} = 0.35$

The probability that a Canal Junction water treatment plant worker, chosen at random, took a 3-week vacation last year is 35.0%.

21. $P(A \cap B) = \frac{50}{500} = 0.10$

The probability that a Canal Junction water treatment plant worker, chosen at random, took a 3-week vacation and worked more than 5 weeks of overtime last year is 10.0%.

23. $P(A \cup B) = 0.50$

The probability that a worker, selected at random, took a 3-week vacation or worked more than 5 weeks of overtime is 50%.

25. $P(B) = 0.35$

The probability that a senior, selected at random, had been placed on the Dean's List for at least one semester is 35.0%.

27. $P(A \cap B)^C = 0.95$

The probability that a senior, selected at random, was not placed on the dean's list and was not placed on academic probation for at least one semester last year is 95.0%.

29. $P(A \cup B) = 0.55$

The probability that a senior, selected at random, was either placed on academic probation or was placed on the dean's list is 55.0%.

31. $P(A) = \frac{3113}{6639} \approx 0.469$

The probability that the doctorate is in social sciences is about 46.9%.

33. $P(A)^C = 0.531$

The probability that the doctorate is not in social sciences is about 53.1%.

35. $P(A \cap C) = \frac{2353}{6639} \approx 0.354$

The probability that the doctorate is in social sciences and is conferred on a U.S. citizen is about 35.4%.

37. $P(A \cap B) = 0$

The events are mutually exclusive.

39. $P(A \cup D) \approx 0.636$

The probability that the doctorate is in social sciences or is conferred on foreign-born citizen is about 63.6%.

41. $P(B) = 52.4\%$

In 1996, the probability that a U.S. adult, selected at random, exercised regularly is 52.4%.

43. $P(B^C) = 47.6\%$

In 1996, the probability that a U.S. adult, selected at random, did not exercise regularly is 47.6%.

45. $P(B \cap D) = 26.8$

The probability that, in 1996, a U.S. adult, selected at random, exercised regularly and is a female is 26.8%.

47. $P(C \cap D) = 0$

The events are mutually exclusive.

49. $P(B \cup C) = 74.4\%$.

The probability that in 1996 a U.S. adult, selected at random, either exercised regularly or is male is 74.4%.

51. **(a)** The odds in favor of event A occurring are 4 to 1.

(b) The odds against event A occurring are 1 to 4.

53. **(a)** The odds in favor of event A occurring are 3 to 7.

(b) The odds against event A occurring are 7 to 3.

55. $P(A) = \frac{5}{8}$

57. The probability that event A will not occur is $\frac{2}{7}$.

59. The odds that a tail will come up in a single toss of a coin is 1 to 1.

61. The odds that at least one tail will come up in two coin tosses is 3 to 1.

63. The odds against a head coming up in a single toss of a coin are 1 to 1.

65. The odds against 2 tails coming up in two tosses of a coin are 1 to 3.

67. **(a)** The odds that it will rain tomorrow are 3 to 7.

(b) The odds that it will not rain tomorrow are 7 to 3.

69. The probability that the Dallas Cowboys will win the 2002 Super Bowl is $\frac{1}{76}$ or about 1.32%.

71. The probability that the Los Angeles Dodgers will win the 2001 World Series is $\frac{1}{26}$ or about 3.84%.

73. The probability that the Los Angeles Lakers will win the 2002 NBA Championship is $\frac{5}{12}$ or about 41.7%.

75. No answer given; verification

Section 6.6

1. The events are dependent.

3. The events are independent.

5. The events are dependent.

7. The events are independent.

9. The events are dependent.

11. $\frac{1}{9}$ **13.** $\frac{25}{81}$

15. **(a)** The events are independent. The opinion of one randomly selected person does not influence the opinion of another randomly selected person.

(b) $P(A \cap B) = 0.0196$

There is a 1.96% probability that, in the year 2000, two Americans selected at random believe that the most important problem facing the country is the decline of ethics, morals, and family values.

(c) $P(A^C) = P(B^C) = 0.86$

The probability that, in 2000, the first person does not believe that the most important problem facing the county is the decline of ethic, morals, and family values is 86%. (Same probability for the second person also.)

(d) $P(A^C \cap B^C) = 0.7396$

The probability that, in 2000, two randomly selected Americans did not believe that the biggest problem facing the country is the decline of ethic, morals, and family values is 73.96%.

17. $(0.25)^4 \approx 0.004$

The probability that, in March 2000, four randomly selected Americans all reported their doctor had made an error is about 0.4%.

19. **(a)**

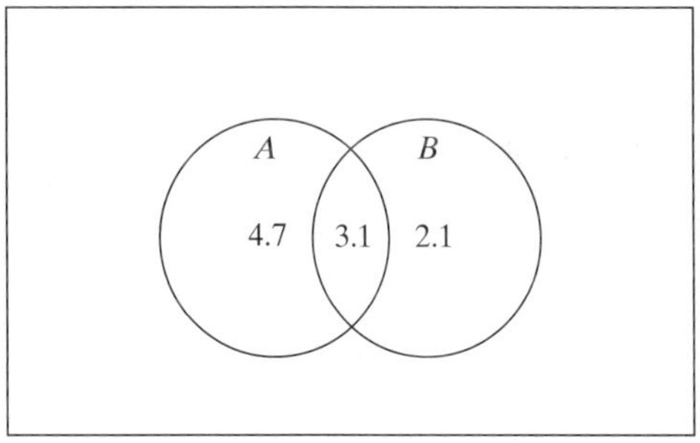

(b) $P(A \cap B) = \dfrac{3.1}{9.9} \approx 0.31$

The probability that a randomly selected American, who uses her computer for more than 16 hours per week, uses the computer for personal and business use is about 31%.

(c) $P(B|A) = \dfrac{\frac{3.1}{9.9}}{\frac{7.8}{9.9}} = \dfrac{3.1}{7.8} \approx 0.397$

The probability that a randomly selected computer owner, who uses her computer for more than 16 hours a week, uses the computer for business use, given that the computer is for personal use, is about 39.7%.

21. $P(A \cap B) = \dfrac{4}{27}$

$P(B \mid A) = \dfrac{4}{9} \approx 0.44$

23. $P(A \cap B) = \dfrac{12}{40} = 0.3$

$P(B \mid A) = \dfrac{12}{20} = 0.6$

25. $P(A \cap B) = \dfrac{1}{18}$

$P(B \mid A) = \dfrac{1}{3}$

27. $P(A \cap B) = 0.38$

$P(B \mid A) = \dfrac{0.38}{0.61} \approx 0.623$

29. $P(A \cap C) = \dfrac{40}{200} = 0.2$

$P(C \mid A) = \dfrac{40}{95} \approx 0.421$

31. $P(A \cap C) = \dfrac{87}{250} = 0.348$

$P(C \mid A) = \dfrac{87}{148} \approx 0.588$

33. $P(B \cap D) = \dfrac{2}{5}$

$P(D \mid B) = \dfrac{4}{7}$

35. $P(B \cap D) = 0.39$

$P(D \mid B) = \dfrac{0.39}{0.63} \approx 0.619$

37. (a) $\frac{4}{15}$ **(b)** $\frac{2}{15}$ **39. (a)** $\frac{1}{120}$ **(b)** $\frac{7}{40}$

41. $P(C \cap D) = 0$

The events are mutually exclusive.

43. $P(B \cap D) = \dfrac{1111}{6639} \approx 0.167$

45. $P(D \mid B) = \dfrac{\frac{1111}{6639}}{\frac{3526}{6639}} = \dfrac{1111}{3526} \approx 0.315$

There is 31.5% probability that the doctorate was conferred on a foreign-born citizen, given that the doctorate is in physical sciences.

47. $P(A \mid D) = \dfrac{\frac{760}{6639}}{\frac{1871}{6639}} = \dfrac{760}{1871} \approx 0.406$

49. $P(A \cap B) = 0$

The events are mutually exclusive.

51. $P(A \cap C) = 22.0$

The probability that the person is male and did not exercise regularly is 22.0%.

53. $P(C \mid A) = \dfrac{22.0}{47.6} \approx 0.462$

The probability that the person is male, given that the person did not exercise regularly, is about 46.2%.

55. $P(B \mid C) = \dfrac{25.6}{47.6} \approx 0.538$

The probability that the person exercised regularly, given that the person is male, is about 53.8%.

Section 6.7

1. $P(D \mid B) = \dfrac{1}{5}$ $P(B \mid D) = \dfrac{1}{4}$

3. $P(D \mid B) = \dfrac{16}{27}$ $P(B \mid D) = \dfrac{8}{11}$

5. $P(C \mid A) = \dfrac{2}{3}$ $P(A \mid C) = \dfrac{1}{2}$

7. $P(C \mid A) = \dfrac{0.22}{0.64} = 0.34375$ $P(A \mid C) = \dfrac{0.22}{0.37} \approx 0.595$

9. (a) $P(D|B) = \dfrac{1111}{3526} \approx 0.315$

(b) $P(B|D) = \dfrac{1111}{1871} \approx 0.594$

11. $P(A) = 0.17$

$P(B \mid A) = \dfrac{(0.3)(0.1)}{0.17} \approx 0.176$

$P(B^C \mid A) = \dfrac{(0.7)(0.2)}{0.17} \approx 0.824$

13. $P(A) = \dfrac{1}{4}$

$P(B \mid A) = \dfrac{(\frac{1}{4})(\frac{2}{5})}{\frac{1}{4}} = \dfrac{2}{5}$

$P(B^C \mid A) = \dfrac{(\frac{3}{4})(\frac{1}{5})}{\frac{1}{4}} = \dfrac{3}{5}$

15. $P(A) = 0.1454$

$P(B \mid A) = \dfrac{(0.23)(0.13)}{0.1454} \approx 0.206$

$P(B^C \mid A) = \dfrac{(0.77)(0.15)}{0.1454} \approx 0.794$

17. (a) $P(A) = \left(\dfrac{300}{500}\right)\left(\dfrac{150}{300}\right) + \left(\dfrac{200}{500}\right)\left(\dfrac{150}{200}\right) = \dfrac{3}{5}$

The probability that a randomly selected factory worker favored St Patrick's day for a paid holiday is 60%.

(b) $P(B \mid A) = \frac{1}{2}$

The probability that a randomly selected factory worker works day shift, given that the worker favors St. Patrick's day as a paid holiday, is 50%.

19. (a) $P(A) = 0.2586$

There is a 25.86% probability that a randomly selected Ketterson College student favors switching from quarters to semesters.

(b) $P(B \mid A) \approx 0.842$

The probability that a Ketterson College student is an underclassperson, given that he favors switching from quarters to semesters is about 84.2%.

21. (a) $P(A) = 0.6842$

The probability that a randomly selected respondent thought that Tiger Woods was the greatest active golfer is about 68.4%.

(b) $P(B \mid A) \approx 0.308$

There is about a 30.8% probability that the respondent is a golf fan, given that he/she felt that Tiger Woods is the greatest active golfer.

(c) $P(B^C \mid A) \approx 0.692$

There is about a 69.2% probability that the respondent is not a golf fan, given that he/she felt that Tiger Woods is the greatest active golfer.

23. (a) $P(A) = 0.3795$

There is a 37.95% probability that a randomly selected respondent felt that Shoeless Joe Jackson should be eligible for the Hall of Fame.

(b) $P(B \mid A) \approx 0.522$

The probability that the respondent is a baseball fan, given that he/she felt that Shoeless Joe Jackson should be eligible for the Hall of Fame, is about 52.2%.

(c) $P(B^C \mid A) \approx 0.478$

There is about a 47.8% probability that the respondent is not a baseball fan, given that he/she feels that Shoeless Joe Jackson should be eligible for the Hall of Fame.

25. (a) $P(A) = 0.4826$

The probability that the respondent is willing to pay higher prices for organic foods is 48.26%.

(b) $P(B \mid A) \approx 0.62$

The probability that the respondent eats organic foods regularly, given that he/she is willing to pay higher prices for organic foods, is about 62%.

(c) $P(B^C \mid A) \approx 0.38$

The probability that the respondent does not eat organic foods regularly, given that he/she is willing to pay higher prices for organic foods, is about 38%.

27. **(a)** $P(A) = 0.612$

There is a 61.2% probability that the respondent felt that kids whose parents are not involved in school sometimes get shortchanged.

(b) $P(B \mid A) \approx 0.624$

There is about a 62.4% probability that the respondent is a parent, given that he/she felt that kids whose parents are not involved in school sometimes get shortchanged.

(c) $P(B^C \mid A) \approx 0.353$

There is about a 35.3% probability that the respondent is a teacher, given that the respondent felt that kids whose parents are not involved is school sometimes get shortchanged.

29. $P(B_1) + P(B_2) = 1$

The events are collectively exhaustive.

31. $P(B_1) + P(B_2) \neq 1$

The events are not collectively exhaustive.

33. $P(B_1) + P(B_2) + P(B_3) = 1$

The events are collectively exhaustive.

35. $P(A) = 0.24$
$P(B_1|A) \approx 0.333$
$P(B_3|A) \approx 0.167$

37. $P(A) = \dfrac{37}{240}$
$P(B_1|A) = \dfrac{20}{37}$
$P(B_3|A) = \dfrac{12}{37}$

39. $P(A) = 0.22$
$P(B_1|A) = \dfrac{9}{22}$
$P(B_3|A) = \dfrac{1}{11}$

41. **(b)** $P(A) = 0.023$

The probability that the light fixture is defective is about 2.3%

(c) $P(B_2|A) \approx 0.217$

The probability that the light fixture came from factory II, given that the fixture is defective, is about 21.7%.

(d) $P(B_3|A) \approx 0.609$

The probability that the light fixture came from factory III, given that the light fixture is defective, is about 60.9%.

43. **(b)** $P(A) = 0.277$

There is a 27.7% probability that the randomly selected woman has never married or cohabitated.

(c) $P(B_2|A) \approx 0.196$

There is about a 19.6% probability that the woman is between the ages of 25 and 34, given that she has never married or cohabitated.

(d) $P(B_3|A) \approx 0.078$

There is about a 7.8% probability that the woman is between the ages of 35 and 44, given that she has never married or cohabited.

Chapter 6 Review Exercises

1. Statistical inference **3.** Yes **5.** Yes **7.** Yes

9. No **11.** No **13.** No

15. **(a)** Answers may vary. **(b)** Answers may vary.

17. $B = \{$New Hampshire, New Jersey, New Mexico, New York$\}$

19. $B = \{x | x$ is a state that starts with the letter A$\}$

21. Ø, {McKinley}, {Pikes Peak}, {Washington}, {McKinley, Pikes Peak}, {McKinley, Washington}, {Pikes Peak, Washington}, {McKinley, Pikes Peak, Washington}

23. $A \cap B = \{17, 19\}$

25. $(A \cup C)^C = \{21, 22, 23, 24, 25\}$ **27.** $A^C \cap B = \{6, 7, 8\}$

29. $A^C \cap (B \cup C) = \{2, 4\}$

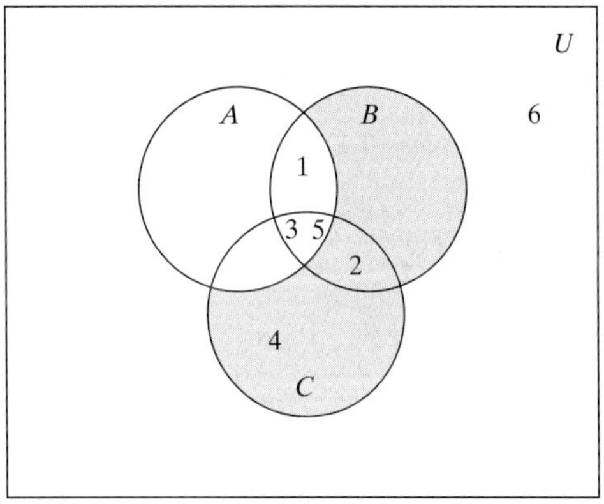

31. $n(A \cup C) = 25{,}508{,}000$; in 1997, 25,508,000 students were either in grades 9 to 12 at public and private schools or were at a public school in grades 9 to 12 and college.

33. 528 ways **35.** 1024 ways **37.** 30 **39.** 2.74×10^{17}

41. 70 **43.** 4.19×10^{15} **45.** 28 ways **47.** 5,565,120 ways

49. $S = \{$A, B, C, D, E, F, G, H, I, J, K, L$\}$; $n(S) = 12$

51. $S = \{$(Yes, public), (Yes, private), (No, public), (No, private)$\}$; $n(S) = 4$

53. $P(A) = \dfrac{1}{3}$; $P(B) = \dfrac{1}{6}$ **55.** $P(A) = \dfrac{1}{6}$; $P(B) = \dfrac{5}{36}$

57. **(a)** 2002 ways **(b)** 560

(c) $\frac{560}{2002} \approx 0.2797$; there is about a 28% probability that a subcommittee is formed with 3 Democrats and 2 Republicans.

59. **(a)** About 0.763 **(b)** About 0.237 **61.** Classical

63. Not mutually exclusive **65.** $\dfrac{25}{47} \approx 0.532$

67. $\frac{210}{400} = 0.525$; if we select a student from the survey group at random, the probability that the student is an upperclassman is 0.525 or 52.5%.

69. $\frac{320}{400} = 0.80$; if we select a student from the survey group at random, the probability that the student is not an upperclassman with season football tickets is 80%.

71. $\frac{150}{400} = 0.375$ **73.** 0.557 **75.** **(a)** $\frac{2}{3}$ **(b)** $\frac{3}{2}$

77. $\frac{2}{3}$ **79.** $\frac{1}{6}$

81. Dependent; if Rebecca's car breaks down, the chances that she will need to ride her bike to work are increased, so the events are dependent.

83. (a)

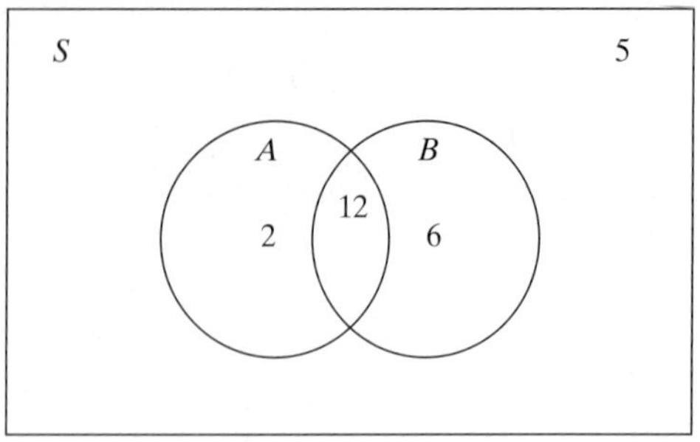

(b) $\frac{12}{25} = 0.48$; if one of the sampled students is selected at random, the probability that the student studied for at least 3 hours and received a grade of 80% or higher is 0.48 or 48%.

(c) $\frac{12}{14} \approx 0.857$; if one of the sampled students is selected at random, the probability that the student received a grade of 80% or higher, given that the student studied at least 3 hours, is about 85.7%.

85. $P(A \cap C) = \frac{15}{105} \approx 0.143$; $P(C|A) = \frac{15}{40} = 0.375$

87. (a) $\frac{24}{91}$ **(b)** $\frac{15}{91}$

89. 0; if a person living in California or Texas is selected at random, the probability that the person lives in California and Texas is zero.

91. 0.5 **93.** $P(D|B) = \frac{10}{40} = 0.25$; $P(B|D) = \frac{10}{25} = 0.40$

95. (a) About 0.519; if a person who participated in the survey is selected at random, the probability that the person is a man, given that the person is afraid of heights, is about 0.519 or 51.9%.

(b) 0.624; if a person who participated in the survey is selected at random, the probability that the person is afraid of heights, given that the person is a man, is 0.624 or 62.4%.

97. $P(A) = \frac{2}{5}$; $P(B|A) = \frac{1}{4}$; $P(B^C|A) = \frac{3}{4}$

99. (a) About 0.570; if a respondent is selected at random, the probability that the respondent supports the attacks is about 0.570 or 57%.

(b) About 0.739

(c) About 0.261; if a respondent is selected at random, the probability that the respondent was not aware of the attacks prior to the survey, given that the respondent supports the attacks after being made aware of them, is about 0.261 or 26.1%.

101. Collectively exhaustive

103. $P(A) = 0.26$; $P(B_1|A) \approx 0.077$; $P(B_3|A) \approx 0.615$

105. (a)

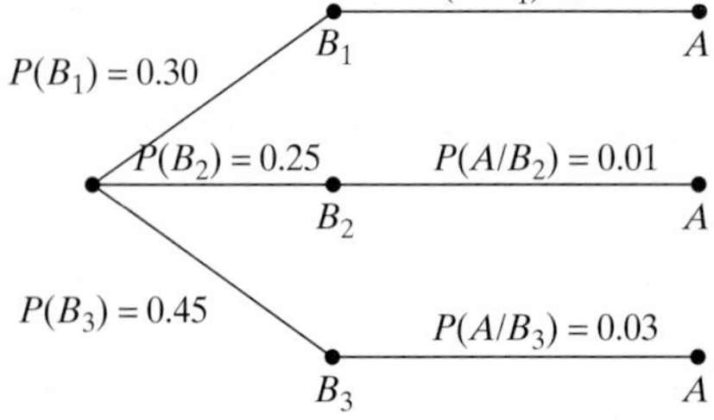

(b) 0.022; if a ski binding is selected at random, the probability that the binding has a defect is 0.022 or 2.2%.

(c) About 0.114

(d) About 0.614; if a ski binding is selected at random, the probability that the binding came from factory III, given that the binding is defective, is about 0.614 or 61.4%.

Chapter 7

Section 7.1

1. The data set is a population. **3.** The data set is a sample.

5. The data set is a population. **7.** The data set is a sample.

9. The population is all adult Americans; the sample is the 568 adult Americans who were surveyed.

11. The population is all baseball fans; the sample is the 467 baseball fans that were surveyed.

13. The population is all Internet users; the sample is the 19,026 respondents.

15. Qualitative **17.** Quantitative **19.** Qualitative

21.

Cash Receipts From Top Farm Livestock Commodities in the United States, 1998.

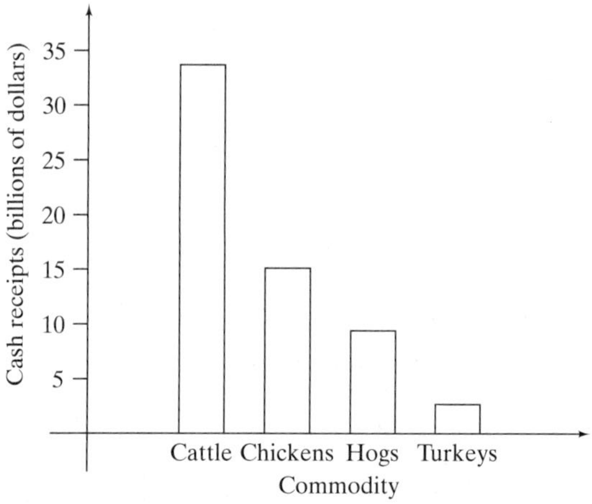

23.

Selected Service Industries in the United States, 1992.

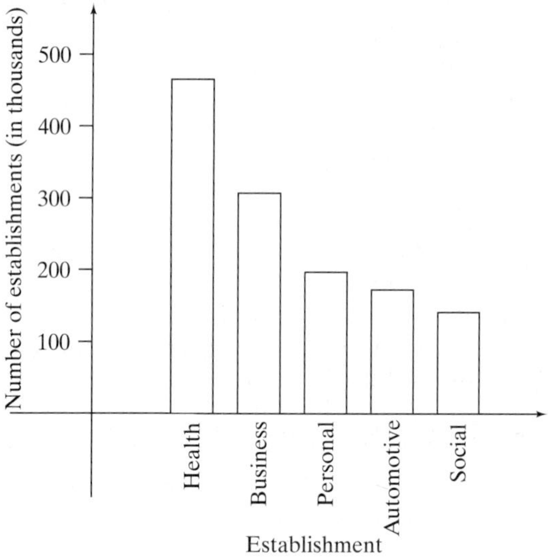

25. Recorded Audio Media Shipped in the United States, 1990.

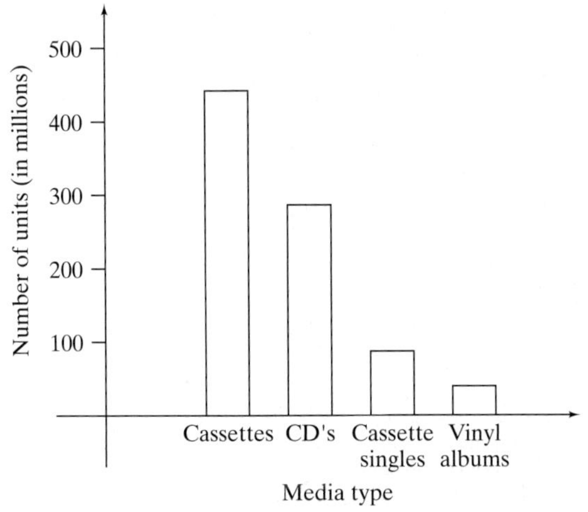

27. Books in various Subjects Imported to the United States, 1990.

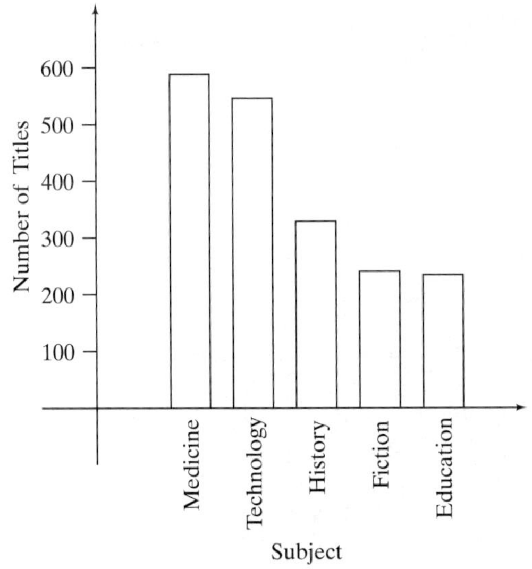

29. Foreign Students who came to Study in the United States, 1995.

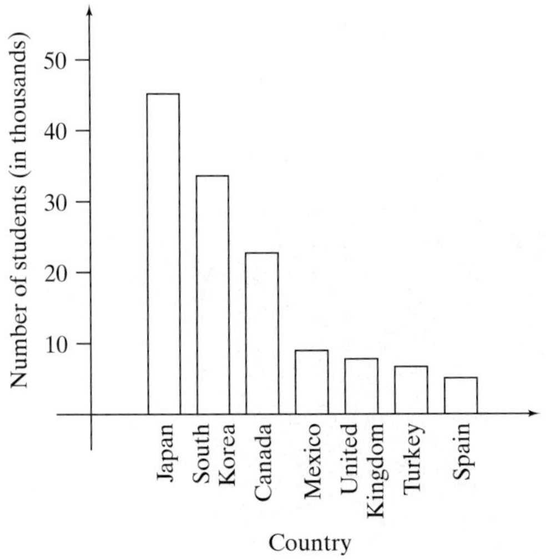

31. **(a)** United States, United Kingdom
(b) Singapore, Germany

33. **(a)** Sulfur dioxide, carbon monoxide, nitrogen oxide and lead
(b) There are about 740 more monitoring stations for particulates than there are for lead.

35. **(a)** About 65,000 households took trips for rest and relaxation.
(b) About 40,000 more households took trips for rest and relaxation than households that took trips for sightseeing.

37. **(a)** In 1995, 58.6% of family debt was from home secured debt and installment.
(b) In 1995, 63.6% of family debt was from credit card balances and installment.

39. **(a)** In 1996, buses were the least used mode of motorized transportation in France.
(b) About 8.9% were motorcycles.

41. **(a)** Life scientist
(b) About 39.6% of the scientists were social scientists.

43. **(a)** Premium cable
(b) About 23.8% of the time was spent watching basic cable.

45. **(a)** About 45,000 gas station and convenience store crimes were reported in 1996.
(b) About 12.4% of crimes reported in 1996 were residence robbery crimes.

Section 7.2

1. **(a)** The class width is 500.
(b) In 1997, there were 10,680,000 single family homes with square footage from 2000 to 2499 square feet.

3. **(a)** The class width is 10.
(b) The 75 to 84 age group had the smallest amount of financial debt in 1995 and the 45 to 54 age group had the largest amount of financial debt in 1995.

5. **(a)** In 1992, 908,000 farms were owned by people who were 55 or older.
(b) In 1992, 990,000 farms were owned by people who were between 25 and 54 years of age.

7. **(a)** 204,700,000 people, aged 15 to 74 years were covered by private or government health insurance in 1997.
(b) The age group from 35 to 44 years had the highest frequency of Americans covered by private or government health insurance in 1997.

9. **(a)** 4,755,000 northeastern homeowners had monthly housing costs of less than $500.
(b) 4,648,000 northeastern homeowners had monthly housing costs of $300 or more but less than $700.

11. **(a)**

Class Boundaries
500.5 to 999.5
999.5 to 1499.5
1499.5 to 1999.5
1999.5 to 2499.5
2499.5 to 2999.5
2999.5 to 3499.5
3499.5 to 3999.5

(b)

Square Footage for Single-Family Homes in the United States, 1997.

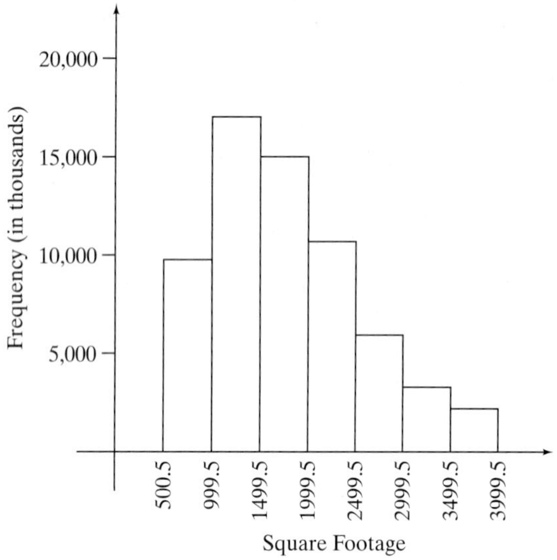

13. (a)

Class Boundaries
24.5 to 34.5
34.5 to 44.5
44.5 to 54.5
54.5 to 64.5
64.5 to 74.5
74.5 to 84.5

(b)

Financial Debt Held by Families in the United States, 1995.

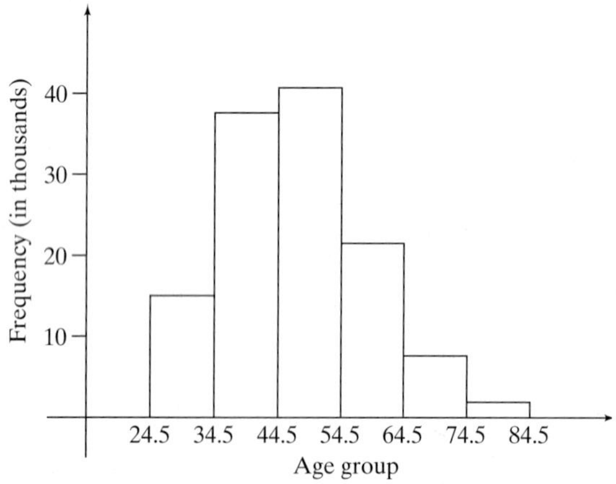

15. (a)

Class Boundaries
14.5 to 24.5
24.5 to 34.5
34.5 to 44.5
44.5 to 54.5
54.5 to 64.5
64.5 to 74.5

(b)

Number of Farms Owned by Americans, 1992.

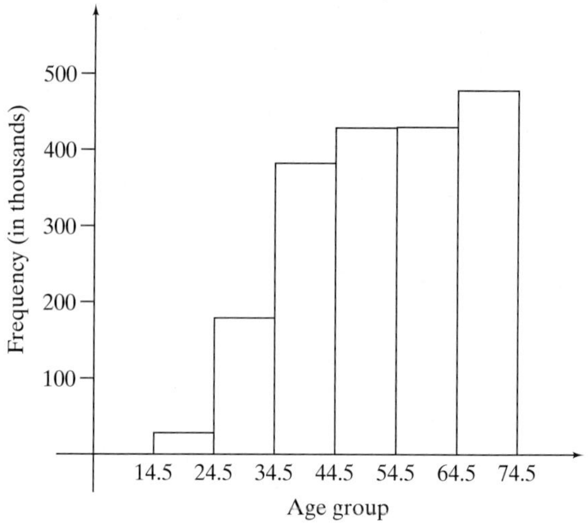

17. (a)

Class Boundaries
14.5 to 24.5
24.5 to 34.5
34.5 to 44.5
44.5 to 54.5
54.5 to 64.5
64.5 to 74.5

(b)

Class	Frequency	Relative Frequency
15 to 24	32.3	0.16
25 to 34	39.3	0.19
35 to 44	44.5	0.22
45 to 54	34.1	0.17
55 to 64	22.3	0.11
65 to 74	32.1	0.16

(c)

Relative Frequency of Americans, Aged 15 to 74, Covered by either Private or Government Health Insurance, 1997.

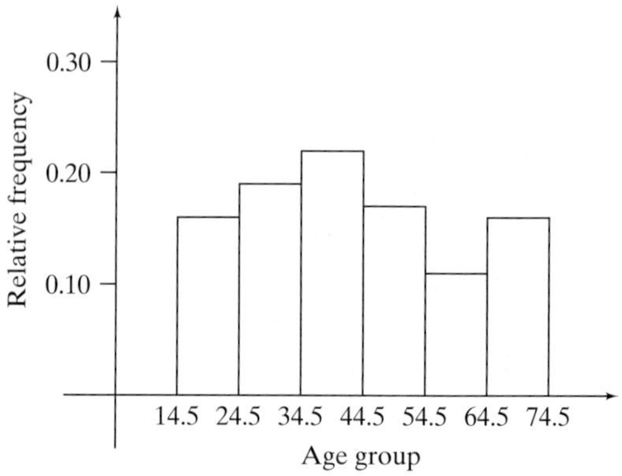

19. (a)

Class Boundaries
199.5 to 299.5
299.5 to 399.5
399.5 to 499.5
499.5 to 599.5
599.5 to 699.5
699.5 to 799.5
799.5 to 899.5
899.5 to 999.5

(b)

Class	**Frequency**	**Relative Frequency**
200 to 299	1923	0.23
300 to 399	1493	0.17
400 to 499	1339	0.16
500 to 599	967	0.11
600 to 699	849	0.10
700 to 799	765	0.09
800 to 899	482	0.06
900 to 999	722	0.08

(c)

Relative Frequency of Monthly Housing Costs under 1000 for Homeowners in the Northeastern United States

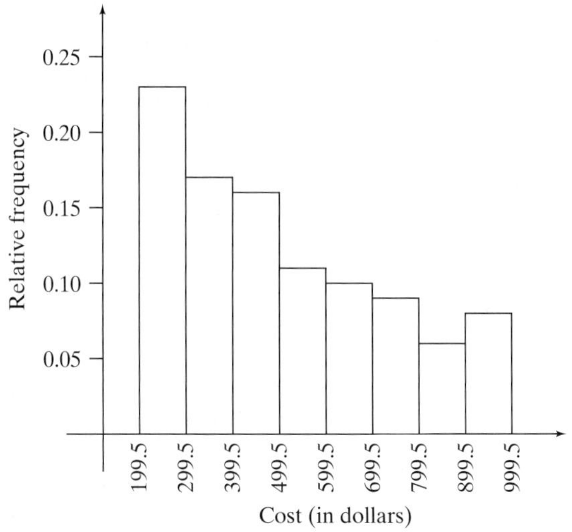

21. (a)

Class Limits	**Frequency**
0 to 79	4
80 to 159	5
160 to 239	5
240 to 319	10
320 to 400	6

(b)

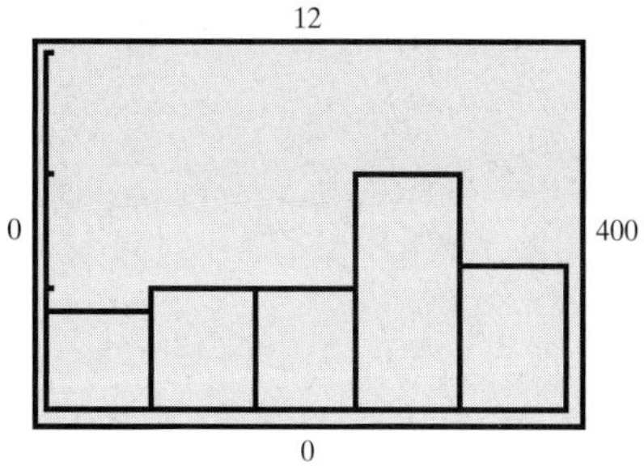

23. (a)

Class Limits	**Frequency**
75 to 197	18
198 to 320	6
321 to 443	0
444 to 566	1
567 to 689	2
690 to 813	2

(b)

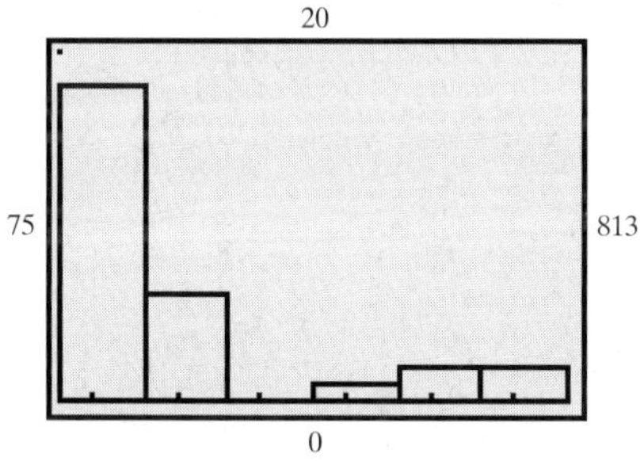

25. (a)

Class Limits	**Frequency**
649 to 1027	12
1028 to 1406	5
1407 to 1785	3
1786 to 2164	3
2165 to 2543	2

(b)

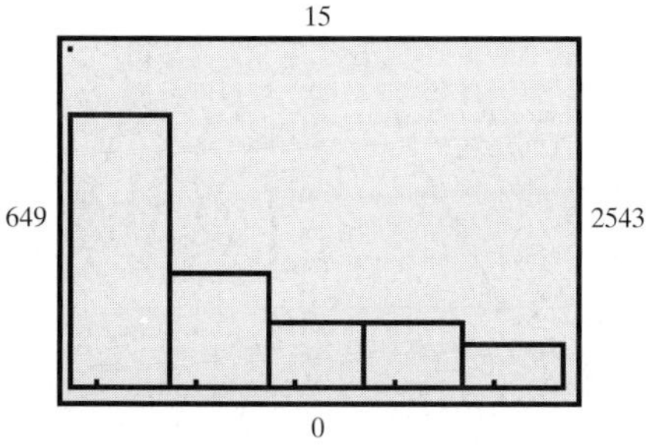

27. (a)

Class	**Midpoint**	**Frequency**
20 to 29	24.5	15
30 to 39	34.5	50
40 to 49	44.5	56
50 to 59	54.5	30
60 to 69	64.5	18

(b) Employee Ages at New Finish Corporation

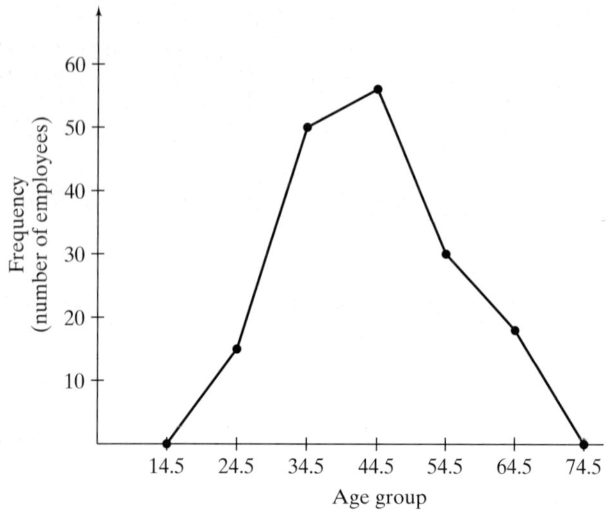

29. (a)

Class	Midpoint	Frequency
59 to 61	60	4
62 to 64	63	8
65 to 67	66	12
68 to 70	69	13
71 to 73	72	21
74 to 76	75	15
77 to 79	78	12
80 to 82	81	9
83 to 85	84	4
86 to 88	87	2

(b) Weights of 100 Sixth-graders at P. H. Johnson Elementary School

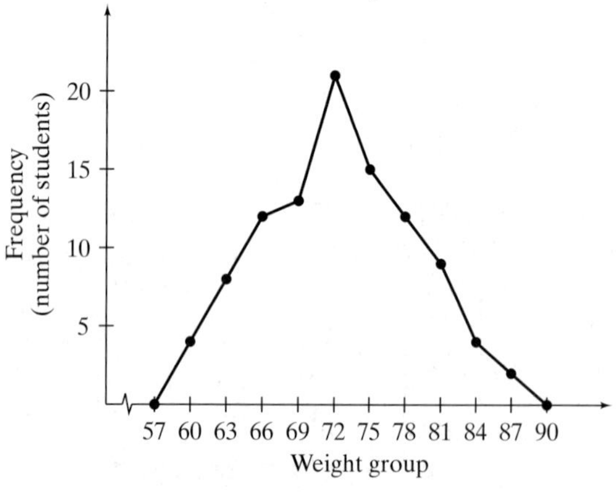

31. (a)

Class	Midpoint
501 to 999	750
1000 to 1499	1249.5
1500 to 1999	1749.5
2000 to 2499	2249.5
2500 to 2999	2749.5
3000 to 3499	3249.5
3500 to 3999	3749.5

(b) Square Footage of Single-family Homes in the United States, 1997.

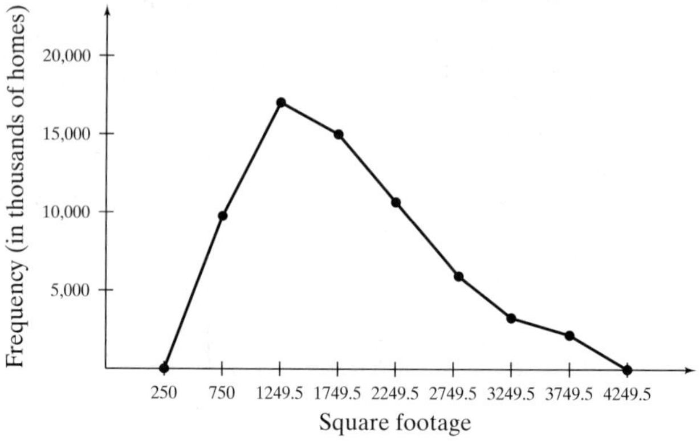

33. (a)

Class	Midpoint
25 to 34	29.5
35 to 44	39.5
45 to 54	49.5
55 to 64	59.5
65 to 74	69.5
75 to 84	79.5

(b) Average Amount of United States Family Debt, 1995.

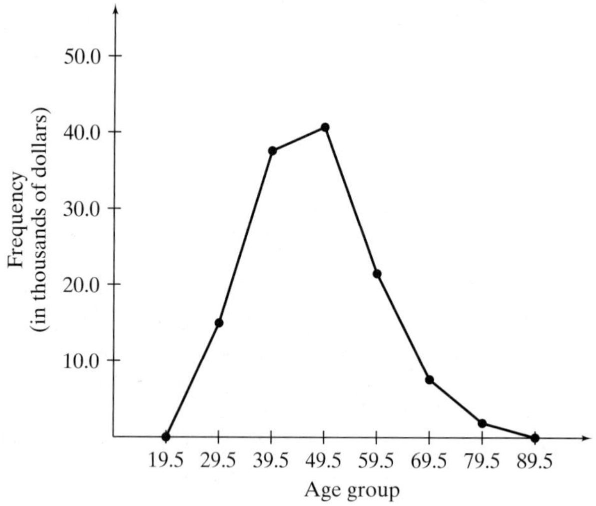

35. (a)

Class	Cumulative Frequency
20 to 29	15
30 to 39	65
40 to 49	121
50 to 59	151
60 to 69	169

(b)

Employees Ages at New Finish Corporation

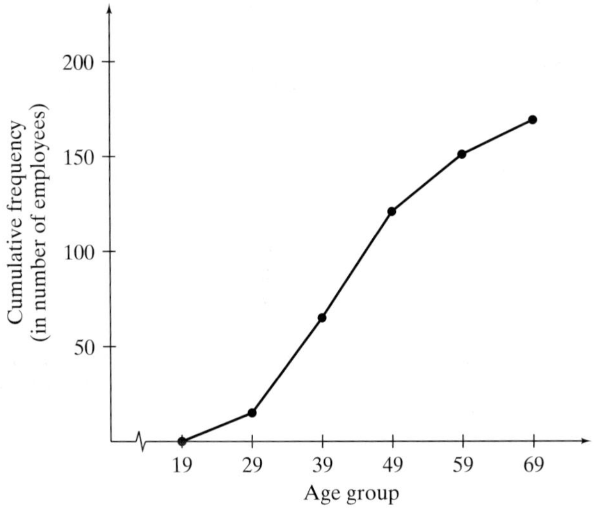

37. (a)

Class	Cumulative Frequency
59 to 61	4
62 to 64	12
65 to 67	24
68 to 70	37
71 to 73	58
74 to 76	73
77 to 79	85
80 to 82	94
83 to 85	98
86 to 88	100

(b)

Weights of 100 Sixth-graders at P. H. Johnson Elementary School

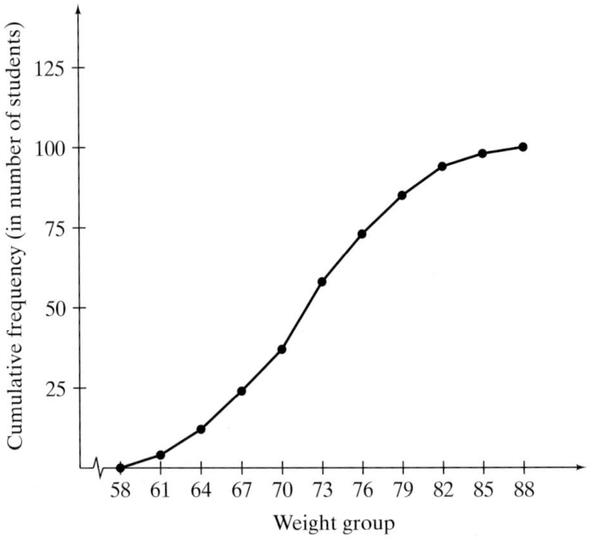

39. (a)

Class (age group)	Cumulative Frequency
15 to 24	21
25 to 34	149
35 to 44	520
45 to 54	987
55 to 64	1414
65 to 74	1911

(b)

Farm Ownership by Americans, 1997.

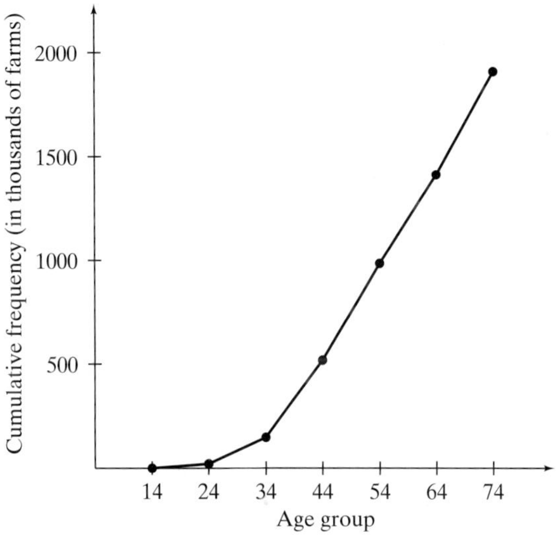

41. The 24.5 to 29.5 age group had the most newborns.

43. The class width is 5.

45. The 24.5 to 29.5 age group had the most newborns.

47. The class width is 5. **49.** The class width is 3.

51. The second class (parents of children ages 4 to 6) spent over $1300 annually on child care and education.

53. The class width is 10.

55. In the seventh and eighth classes, the number of deaths caused by cancer is close to 160,000; this corresponds to the age groups of 65 to 74 and 75 to 84, respectively.

57. The lower limit of the first class is 15 and the upper limit is 24.

59. About 25,300 drivers aged 34.5 years and below were involved in alcohol-related fatal crashes in 1998.

61. The lower limit of the first class is 5 and the upper limit is 14.

63. About 54,300 individuals aged 54.5 years and below died from accidental and adverse effects in 1997.

Section 7.3

1. $\bar{x} \approx 6.17$, $M_0 = 6$ **3.** $\bar{x} = 3.6$, $M_0 = 1$ and 4 **5.** $\bar{x} \approx 11.3$, There is no mode.

7. $\bar{x} = 15.16$, $M_0 = 11.2$ and 15.8 **9.** $\bar{x} = 0.5$, There is no mode. **11.** $\mu = 5.7$

13. $\mu \approx 154.7$ 15. $\mu = 54.0375$ 17. $M = 5$, $MR = 7$

19. $M = 5$, $MR = 6$ 21. $M = 19.5$, $MR = 20$ 23. $M = 0.44$, $MR = 0.51$

25. $M = 205$, $MR = 353$ 27. The mode is medium.

29. Blue and brown are the modes. 31. $\bar{x} \approx 6.64$

33. $\bar{x} = 30$ 35. $\bar{x} = \frac{94.5}{25} = 3.78$ 37. $\bar{x} \approx 93.3$

39. Symmetric 41. Skewed left

43. Skewed left 45. Symmetric

47. (a) $\bar{x} = 40.3$
In May 2000, the average number of visits to the top six (most visited) Web sites was 40.3 million.
(b) $M = 40.7$
In May 2000, the median number of visits to the top six (most visited) web sites was 40.7 million.

49. (a) $\bar{x} \approx 93.2$
There was an average of 93.2 billion barrels of oil reserves in the top nine countries (with the largest oil reserves) in 1999.
(b) $M_0 = 93$
The most frequently occurring amount of oil reserves is 93 million barrels.

51. (a) $M = 8.3$
In 1998, in the top 15 countries in terms of personal computer usage, the median number of PCs in use was 8.3 million.
(b) $MR = \frac{129 + 4.6}{2} = 66.8$
In 1998, in the top 15 countries in terms of personal computer usage, the midrange number of computers was 66.8 million.

53. (a) $n = 95.5$
This means that in 1997 the number of emergency room visits in the United States by people aged 5 to 84 totaled about 95.5 million.
(b) The mean is about 38.73. This means that the average age of people going to emergency rooms in the United States in 1997 was about 38.73 years.

55. (a) $\bar{x} \approx 37.76$
In 1997, the average age of a U.S. prisoner sentenced to death was about 37.76 years.
(b) In general, the data are skewed right.

Section 7.4

1. $R = 6$ 3. $R = 8$ 5. $R = 17$ 7. $R = 10.5$

9. $R = 0.82$ 11. $\sigma \approx 3.07$ 13. $\sigma \approx 22.63$

15. $\sigma \approx 14.42$ 17. $s \approx 3.11$ 19. $s \approx 2.34$

21. $s \approx 6.48$ 23. $s \approx 0.247$ 25. $s \approx 92.3$

27. $s \approx 3.65$ 29. $s \approx 10.3$ 31. $s \approx 0.55$

33. $s \approx 36.3$

35. (a) $R = 691$
The range of selected bodies of water in the Pacific Coast region is 691 square miles.
(b) $s \approx 326.71$
This means that the areas of these selected bodies of water varied on average by about 326.71 square miles from the mean area of 328.5 square miles.

37. $\sigma \approx 9$
The ages of the U.S. prisoners sentenced to death in 1980 varied on average by about 9 years from the mean age of about 30.5 year.

39. $\sigma \approx 21.94$
This means that the ages of people who visited emergency rooms in the United States in 1997 varied on average by about 21.94 years from the mean age of 38.73 years.

41. $SK \approx -6.43$
The distribution is skewed left.

43. $SK = 0$
The distribution is symmetric.

45. $SK = 0.25$
The distribution is skewed right.

Chapter 7 Review Exercises

1. The population consists of the responses, or possible responses, of all American adults. The sample is made up of the 1060 American adults who participated in the survey.

3. Quantitative 5. Qualitative

7. Number of pounds of domestic fish and shell fish caught, 1998.

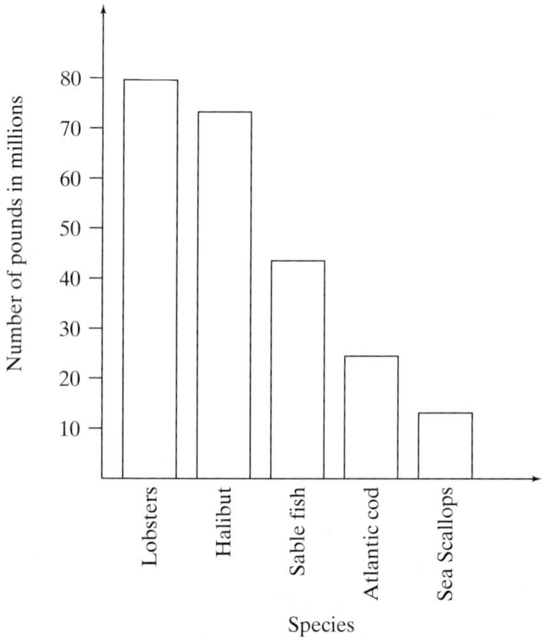

9. **(a)** Bachelor's, master's, doctorate
(b) About $40,000

11. **(a)** About 200,000
(b) About 650,000

13. **(a)** Dangerous nonnarcotic drugs
(b) 77.0%

15. **(a)** 10 **(b)** 21.9 million adults

17. **(a)** 24.5 to 34.5, 34.5 to 44.5, 44.5 to 54.5, 54.5 to 64.5, 64.5 to 74.5
(b)

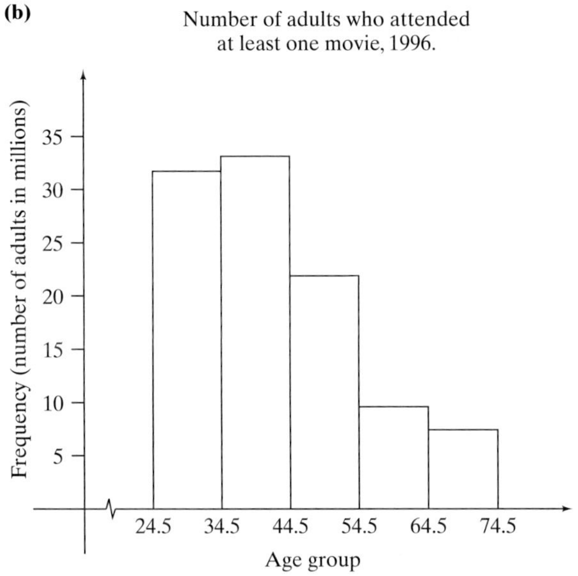

19. **(a)**

Class (number of minutes exercising)	Frequency, f (number of adults)
0 to 17	10
18 to 35	7
36 to 53	6
54 to 71	3
72 to 89	1
90 to 107	1
108 to 125	2

(b)

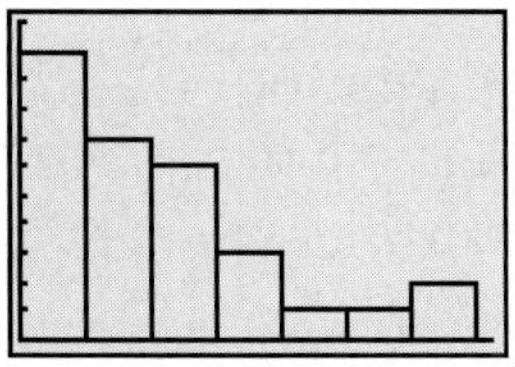

21. **(a)**

Class (age group)	Class midpoint, m
25 to 34	29.5
35 to 44	39.5
45 to 54	49.5
55 to 64	59.5
65 to 74	69.5

(b)

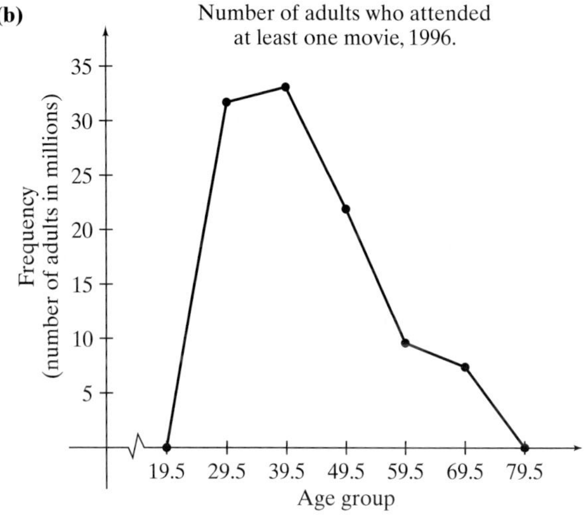

23. **(a)**

Class (age group)	Cumulative Frequency
25 to 34	31.7
35 to 44	64.8
45 to 54	86.7
55 to 64	96.3
65 to 74	103.7

(b)

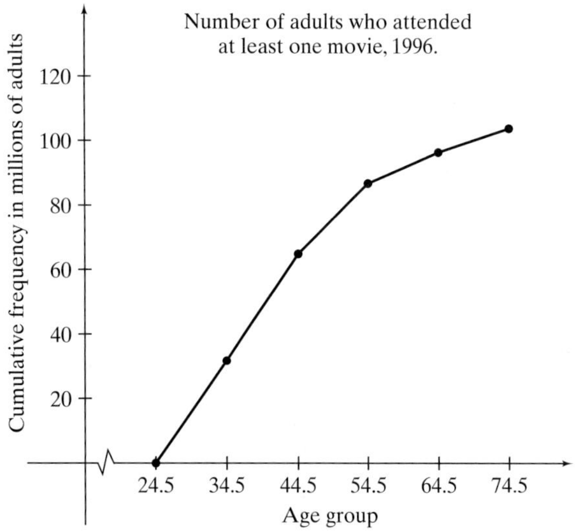

25. 10 **27.** 55 to 64 **29.** 10 **31.** 5.1; 2, 8 **33.** 133

35. 18.5; 20.5 **37.** First **39.** Approximately 8.39

41. Skewed right

43. **(a)** Approximately 381.9; the average domestic gross, in millions, per film of the top six summer films of all time was about $381.9 million.

(b) 378.45; the median domestic gross, in millions, of the top six summer films of all time was $378.45 million. Half of the top six summer films grossed more than $378.45 million and half grossed less than $378.45 million.

45. **(a)** Approximately 15.67; an estimate of the average age of a 10 to 18 year-old arrested in 1999 was 15.67 years old.

(b) Skewed left

47. 38 **49.** Approximately 4.20 **51.** Approximately 2.87

53. Approximately 4.87 **55.** Approximately 4.48

57. **(a)** 334; of the seven counties listed in the table, the difference between the most and least populated counties is 334,000.

(b) Approximately 116.26; the population of the seven counties varied on average by about 116,260 from the mean.

59. Approximately 19.46; skewed right

Chapter 8

Section 8.1

1. $P(X = x)$ meets both criteria; therefore, it is a probability mass function.

3. Since $P(X = x)$ meets both criteria, it is a probability mass function.

5. Both criteria are met; therefore, it is a probability mass function.

7. Since $P(X = 10) = -0.1$, $P(X = x)$ is not a probability mass function.

9. Since $P(X = x)$ meets both criteria, it is a probability mass function.

11. **(a)**

$X = x$	$P(X = x) = \frac{1}{6}x$
$X = 0$	$P(X = 0) = 0$
$X = 1$	$P(X = 1) = \frac{1}{6}$
$X = 2$	$P(X = 2) = \frac{1}{3}$
$X = 3$	$P(X = 3) = \frac{1}{2}$

(b) Both of the criteria have been met, so $P(X = x)$ is a probability mass function.

13. **(a)**

$X = x$	$P(X = x) = \frac{1}{4}$
$X = 1$	$P(X = 1) = \frac{1}{4}$
$X = 2$	$P(X = 2) = \frac{1}{4}$
$X = 3$	$P(X = 3) = \frac{1}{4}$
$X = 4$	$P(X = 4) = \frac{1}{4}$

(b) Both criteria have been met, so $P(X = x)$ is a probability mass function.

15. **(a)**

$X = x$	$P(X = x) = \frac{1}{15}x^2$
$X = 0$	$P(X = 0) = 0$
$X = 1$	$P(X = 1) = \frac{1}{15}$
$X = 2$	$P(X = 2) = \frac{4}{15}$
$X = 3$	$P(X = 3) = \frac{9}{15}$
$X = 4$	$P(X = 4) = \frac{16}{15}$

(b) Since $P(X = 4) > 1$, $P(X = x)$ is not a probability mass function.

17. **(a)**

$X = x$	$P(X = x) = 0.05x$
$X = 2$	$P(X = 2) = 0.10$
$X = 3$	$P(X = 3) = 0.15$
$X = 4$	$P(X = 4) = 0.20$
$X = 6$	$P(X = 6) = 0.30$

(b) Both criteria have not been met, so $P(X = x)$ is not a probability mass function.

19. **(a)**

$X = x$	$P(X = x) = \frac{x+2}{10}$
$X = 0$	$P(X = 0) = \frac{1}{5}$
$X = 1$	$P(X = 1) = \frac{3}{10}$
$X = 2$	$P(X = 2) = \frac{2}{5}$
$X = 3$	$P(X = 3) = \frac{1}{2}$

(b) Both criteria have not been met, so $P(X = x)$ is not a probability mass function.

21. $k = 6$ **23.** $k = \frac{1}{50}$ **25.** $k = 21$

27.

$X = x$	$P(X = x)$
$X = 1$	$P(X = 1) = \frac{5}{20} = 0.25$
$X = 2$	$P(X = 2) = \frac{5}{20} = 0.25$
$X = 3$	$P(X = 3) = \frac{10}{20} = 0.50$

29.

$X = x$	$P(X = x)$
$X = 1$	$P(X = 1) = \frac{10}{75} \approx 0.13$
$X = 2$	$P(X = 2) = \frac{20}{75} \approx 0.27$
$X = 3$	$P(X = 3) = \frac{15}{75} = 0.20$
$X = 4$	$P(X = 4) = \frac{5}{75} \approx 0.07$
$X = 5$	$P(X = 5) = \frac{25}{75} \approx 0.33$

31.

$X = x$	$P(X = x)$
$X = 0$	$P(X = 0) = \frac{150}{1350} \approx 0.11$
$X = 1$	$P(X = 1) = \frac{600}{1350} \approx 0.44$
$X = 2$	$P(X = 2) = \frac{400}{1350} \approx 0.30$
$X = 3$	$P(X = 3) = \frac{200}{1350} \approx 0.15$

33. (a)

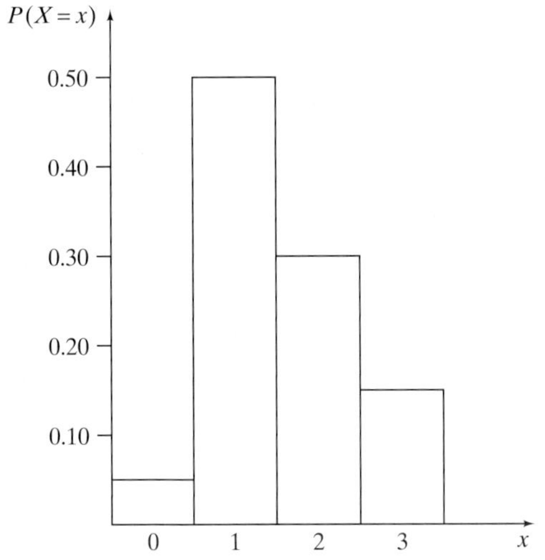

(b) $P(1 \le X \le 2) = 0.80$

The probability that a customer will rent one or two videos is 80%.

35. (a)

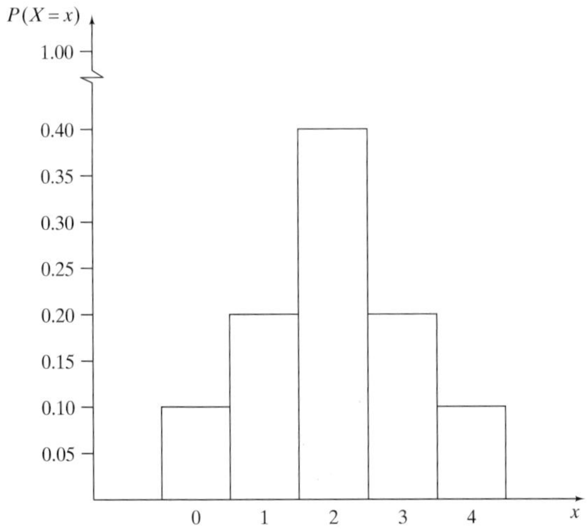

(b) $P(2 \le X \le 4) = 0.7$

The probability that 2, 3, or 4 women are hired is 70%.

37. (a)

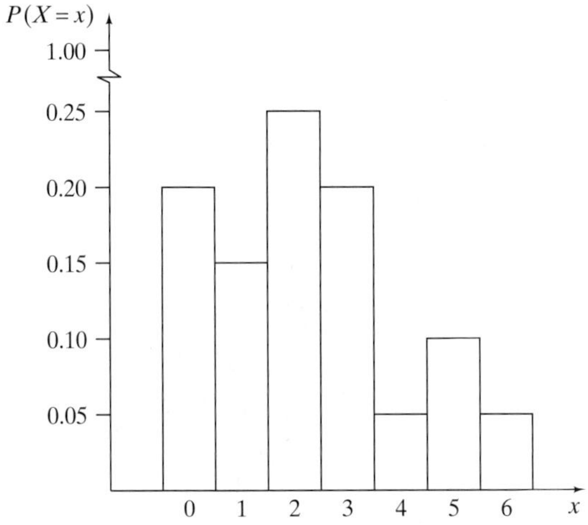

(b) $P(X \le 3) = 0.80$

There is an 80% probability that the number of commercials aired per show during after school television will be 3 or fewer.

39. (a) $P(X \le 14) = 0.65$

The probability that 12, 13, or 14 students will drop out of introductory psychology during any given quarter is 65%.

(b) 12 is the least likely number of students to drop out of introductory psychology during any given quarter.

41. (a) $P(X \ge 4) = 0.62$

The probability that the size of the family will be at least 4 is 62%.

(b) 7 is the least likely size that the chosen family will be.

43. (a)

$X = x$	$P(X = x)$
0	$P(X=0) = \frac{7800}{19{,}000} \approx 0.41$
1	$P(X=1) = \frac{5500}{19{,}000} \approx 0.29$
2	$P(X=2) = \frac{3200}{19{,}000} \approx 0.17$
3	$P(X=3) = \frac{2500}{19{,}000} \approx 0.13$

(b) $P(X = 3) \approx 0.13$

There is about a 13% probability that a randomly selected student has completed 3 years of college.

45. (a)

$X = x$	$P(X = x)$
$X = 0$	$P(X=0) = \frac{53}{665} \approx 0.08$
$X = 1$	$P(X=1) = \frac{127}{665} \approx 0.19$
$X = 2$	$P(X=2) = \frac{280}{665} \approx 0.42$
$X = 3$	$P(X=3) = \frac{128}{665} \approx 0.19$
$X = 4$	$P(X=4) = \frac{77}{665} \approx 0.12$

(b) $P(X = 3) \approx 0.19$

There is about a 19% probability that 3 passengers will miss the overnight flight from Tokyo to Honolulu.

47. (a)

$X = x$	$P(X = x)$
$X = 16$	$P(X=16) = \frac{1.6}{24.2} \approx 0.07$
$X = 17$	$P(X=17) = \frac{2.2}{24.2} \approx 0.09$
$X = 18$	$P(X=18) = \frac{2.4}{24.2} \approx 0.10$
$X = 19$	$P(X=19) = \frac{2.7}{24.2} \approx 0.11$
$X = 20$	$P(X=20) = \frac{2.8}{24.2} \approx 0.12$
$X = 21$	$P(X=21) = \frac{2.9}{24.2} \approx 0.12$
$X = 22$	$P(X=22) = \frac{2.9}{24.2} \approx 0.12$
$X = 23$	$P(X=23) = \frac{3.1}{24.2} \approx 0.13$
$X = 24$	$P(X=24) = \frac{3.6}{24.2} \approx 0.15$

(b) $P(X = 20) \approx 0.12$

There is about a 12% probability that the driver involved in the accident was 20 years old.

(c) $P(X \leq 20) = 0.49$

There is a 49% probability that the driver involved in the accident was 20 years old or younger.

49. (a) $P(X = 2) = 0.197$

The probability that at least one parent participated in 2 school activities is 19.7%.

(b) $P(X \geq 2) = 0.891$

There is an 89.1% probability that at least one parent participated in 2 or more school activities.

Section 8.2

1. $E(X) = 0.75$ **3.** $E(X) = 2.65$ **5.** $E(X) = 0.34$

7. $E(X) = 29$ **9.** $E(X) = 10.25$ **11.** $E(X) = 1$

13. $E(X) = 4.2$ **15.** $E(X) = \frac{7}{3}$ **17.** $E(X) = 3.75$

19. $E(X) = 2.5$ **21.** $E(X) = \frac{18}{7}$ **23.** $E(X) = 4.5$

25. $E(X) = 1$ **27.** $\sigma_X = \sqrt{0.6875} \approx 0.829$

29. $\sigma_X = \sqrt{1.4275} \approx 1.195$ **31.** $\sigma_X = \sqrt{69} \approx 8.31$

33. $\sigma_X = \sqrt{0.556} \approx 0.745$ **35.** $\sigma_X = \sqrt{1.25} \approx 1.12$

37. $\sigma_X = \sqrt{0.388} \approx 0.623$ **39.** $\sigma_X = \sqrt{\frac{73}{70}} \approx 1.02$

41. $E(X) = 1.55$

The average number of video rentals per customer is 1.55 videos.

43. (a) $E(X) = 2.25$

The average number of commercials per show aired during after school television is 2.25.

(b) The most frequently occurring number of commercials on any given show is 2.

45. The number of students expected to drop out of introductory psychology is 14.14 students.

47. $E(X) = -8$

The expected value of one ticket is −$8.

49. (a) $E(X) \approx 1.53$

The number of inspections required to discover a defective spring is expected to be 1.53.

(b) $\sigma_X \approx 0.72$

The number of inspections required to discover a defective spring varied an average by about 0.72 from the mean of 1.53.

51. (a) $E(X) = 178.81$

The expected average weight of an American male aged 70 to 79, from 1988 to 1994, is 178.81 pounds.

(b) $\sigma_X \approx 31$

The expected average weight of an American male aged 70 to 79 varied on average by approximately 31 pounds from the mean of 178.81 pounds.

53. The insurance company's expected gain is −$1624.

55. (a) The expected value of Site I is $7 million. The expected value of Site II is $4 million.

(b) Based on part (a), advise the company to select Site I.

(c) Site I has less risk.

57. (a) $P(X = x) = \frac{1}{10}, x = 1, 2, 3, 4, 5, 6, 7, 8, 9, 10$

(b) $\mu_X = 5.5$

The expected number of times that the printing machine is expected to break down during a shift is 5.5 times.

(c) $\sigma_X \approx 2.87$

The expected number of times that the machine breaks down varied on average by about 2.87 from the mean of 5.5.

(d) $P(X \leq 3) = 0.3$

There is a 30% probability that during any given shift the machine will break down 3 or fewer times.

59. (a) $P(X = x) = \frac{1}{25}, \quad 1 \leq x \leq 25$

(b) $\mu_X = 13$

The expected hardness of the plastic pieces manufactured is 13 on the Rockford hardness scale.

(c) $\sigma_X \approx 7.21$

The expected hardness of the plastic pieces varied on average by about 7.21 from the mean of 13.

(d) $P(X \geq 20) = 0.24$

The probability that a plastic piece, selected at random, will have a rating on the Rockford hardness scale of more than 20 is 24%.

Section 8.3

1. $P(X = 2) = 0.375$ **3.** $P(X = 0) = 0.32768$

5. $P(X = 6) \approx 0.2731$ **7.** $P(X = 5) \approx 0.0264$

9. $P(X = 10) \approx 0.1762$

11. This is a binomial experiment.

$p = 0.50 \quad n = 30$

13. This is not a binomial experiment. There are not exactly two outcomes.

15. This is a binomial experiment.

$p = 0.90 \quad n = 12$

17. This is not a binomial experiment. There is not a definite number of trials.

19. This is a binomial experiment.

$p = 0.20 \quad n = 150$

21. (a)

$X = x$	$P(X = x)$
$X = 0$	${}_3C_0\left(\frac{1}{2}\right)^0\left(1 - \frac{1}{2}\right)^{3-0} = 0.125$
$X = 1$	${}_3C_1\left(\frac{1}{2}\right)^1\left(1 - \frac{1}{2}\right)^{3-1} = 0.375$
$X = 2$	${}_3C_2\left(\frac{1}{2}\right)^2\left(1 - \frac{1}{2}\right)^{3-2} = 0.375$
$X = 3$	${}_3C_3\left(\frac{1}{2}\right)^3\left(1 - \frac{1}{2}\right)^{3-3} = 0.125$

(b)

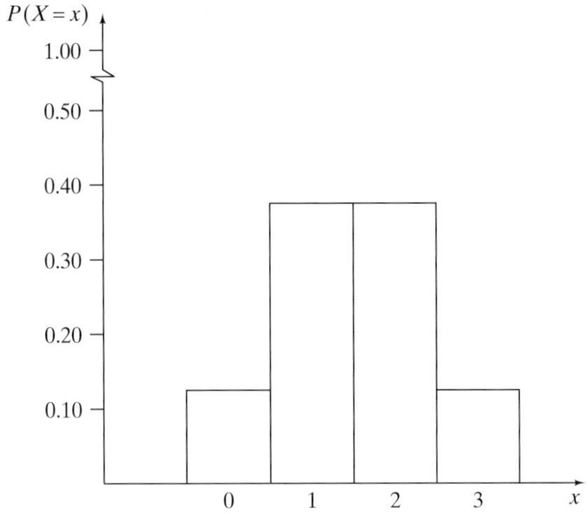

23. (a)

$X = x$	$P(X = x)$
$X = 0$	${}_5C_0(0.3)^0(1 - 0.3)^{5-0} = 0.16807$
$X = 1$	${}_5C_1(0.3)^1(1 - 0.3)^{5-1} = 0.36015$
$X = 2$	${}_5C_2(0.3)^2(1 - 0.3)^{5-2} = 0.3087$
$X = 3$	${}_5C_3(0.3)^3(1 - 0.3)^{5-3} = 0.1323$
$X = 4$	${}_5C_4(0.3)^4(1 - 0.3)^{5-4} = 0.02835$
$X = 5$	${}_5C_5(0.3)^5(1 - 0.3)^{5-5} = 0.00243$

(b)

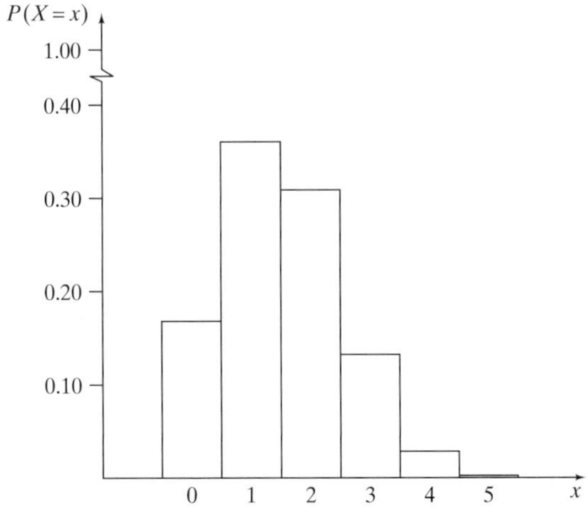

25. (a)

$X = x$	$P(X = x)$
$X = 0$	${}_5C_0(0.7)^0(1 - 0.7)^{5-0} = 0.00243$
$X = 1$	${}_5C_1(0.7)^1(1 - 0.7)^{5-1} = 0.02835$
$X = 2$	${}_5C_2(0.7)^2(1 - 0.7)^{5-2} = 0.1323$
$X = 3$	${}_5C_3(0.7)^3(1 - 0.7)^{5-3} = 0.3087$
$X = 4$	${}_5C_4(0.7)^4(1 - 0.7)^{5-4} = 0.36015$
$X = 5$	${}_5C_5(0.7)^5(1 - 0.7)^{5-5} = 0.16807$

(b)

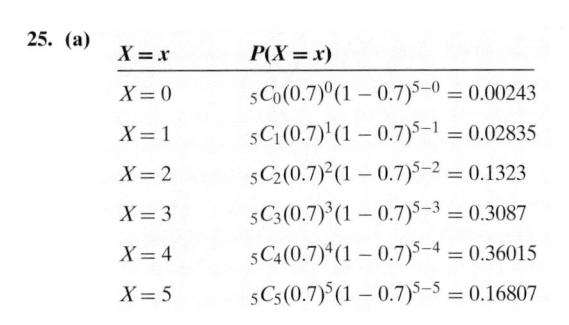

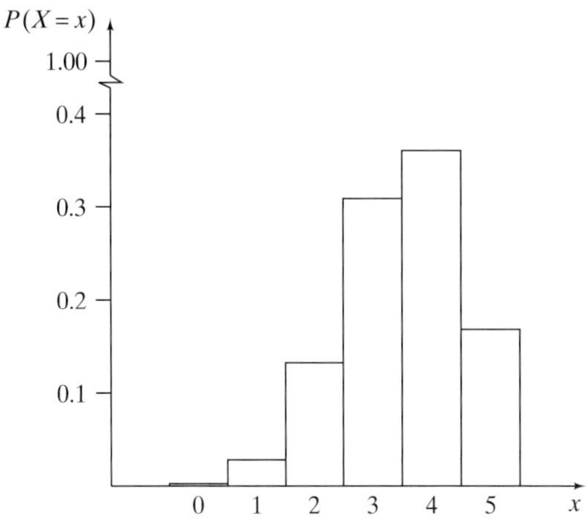

27. b **29.** c **31.** f

33. $E(X) = \mu_X = 6$
$\sigma_X = \sqrt{3} \approx 1.732$

35. $E(X) = \mu_X = 5$
$\sigma_X = \sqrt{3.75} \approx 1.94$

37. $E(X) = \mu_X = 1.5$
$\sigma_X = \sqrt{1.125} \approx 1.06$

39. $E(X) = \mu_X = 2.25$
$\sigma_X = \sqrt{\frac{63}{32}} \approx 1.40$

41. (a) There are two outcomes: a prospective employee either passes or does not pass the exam. The probability is the same (0.40). There are a definite number of trials (7), which are independent of each other.

(b) $n = 7$, $p = 0.40$

$P(X = 5) \approx 0.077$

There is a 7.7% probability that 5 prospective employees will pass the exam.

43. (a) $P(X = 4) \approx 0.213$

The probability that 4 of the 12 sweepstakes entrants also ordered a magazine is about 21.3%.

(b) $P(4 \le X \le 5) \approx 0.44$

The probability that 4 or 5 of the 12 sweepstakes entrants also ordered a magazine is about 44%.

45. (a) $P(X = 3) \approx 0.24$

There is about a 24% probability that 3 of the 15 Republicans responded Abraham Lincoln.

(b) $P(X \le 3) \approx 0.535$

The probability that 3 or fewer of the 15 Republicans responded Abraham Lincoln is about 53.5%.

47. (a) $P(X = 7) \approx 0.267$

The probability that 7 of the 8 households had cable is abut 26.7%.

(b) $P(X \le 5) \approx 0.320$

The probability that at most 5 of the 8 households had cable is about 32%.

49. (a) There is about a 34.8% probability that one of the 50 people was unemployed.

(b) There is about a 59% probability that at most one of the 50 people was unemployed.

51. **(a)** The probability that 3 of the 13 were cigar smokers is about 20.6%.

(b) The probability that at most 3 of the 13 were cigar smokers is about 38%.

53. **(a)**

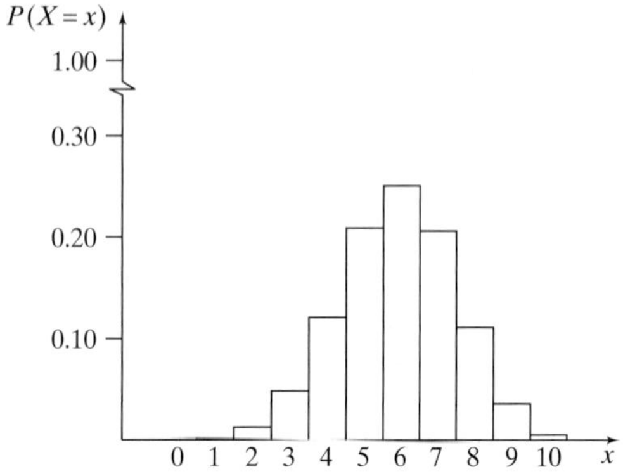

(b) $\mu_X = 5.9$

Out of 10 African Americans in New York City it is expected that 5.9 will not have an automobile.

$\sigma_X = \sqrt{2.419} \approx 1.56$

As the experiment is repeated, in samples of 10, the number of African Americans in New York without a car varies on average by about 1.56 from the expected value of 5.9.

55. **(a)** $\mu_X = 9.8$

It is expected, out of 20 prisoners, that 9.8 of them will not have high school diplomas.

(b)

$X = x$	$P(X = x)$
$X = 0$	$P(X = 0) \approx 0.0000014$
$X = 1$	$P(X = 1) \approx 0.000027$
$X = 2$	$P(X = 2) \approx 0.00025$
$X = 3$	$P(X = 3) \approx 0.00143$
$X = 4$	$P(X = 4) \approx 0.00585$
$X = 5$	$P(X = 5) \approx 0.01799$
$X = 6$	$P(X = 6) \approx 0.04321$
$X = 7$	$P(X = 7) \approx 0.08302$
$X = 8$	$P(X = 8) \approx 0.12962$
$X = 9$	$P(X = 9) \approx 0.16605$
$X = 10$	$P(X = 10) \approx 0.17549$
$X = 11$	$P(X = 11) \approx 0.15328$
$X = 12$	$P(X = 12) \approx 0.11045$
$X = 13$	$P(X = 13) \approx 0.06531$
$X = 14$	$P(X = 14) \approx 0.03137$
$X = 15$	$P(X = 15) \approx 0.01206$
$X = 16$	$P(X = 16) \approx 0.00363$
$X = 17$	$P(X = 17) \approx 0.00082$
$X = 18$	$P(X = 18) \approx 0.00013$
$X = 19$	$P(X = 19) \approx 0.000013$
$X = 20$	$P(X = 20) \approx 0.00000064$

The most likely number of prisoners to not have a high school diploma is 10.

57. **(a)** $E(X) = \mu_X = 9.78$

Out of 30 women, selected at random, who gave birth in 1994, 9.78 of them are expected to be unmarried.

(b) $P(X \leq 10) \approx 0.618$

There is about a 61.8% probability that at most 10 of the 30 women were unmarried.

Section 8.4

1. **(a)** For the Jupiter, $z = 1$. For the Atlantis, $z \approx 0.608$

(b) The Jupiter got better mileage relative to the rest of the cars of the same make.

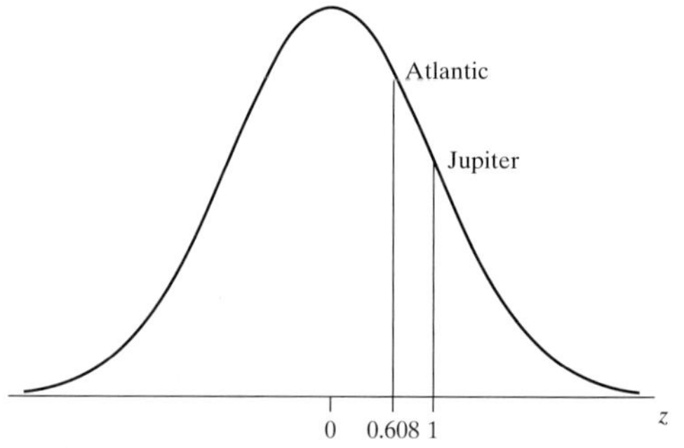

3. **(a)** 1999–2000: $z \approx -1.51$

2000–2001: $z \approx -1.26$

(b) The student who took the test in 2000–2001 scored better on the test relative to the rest of the students who took the test that year.

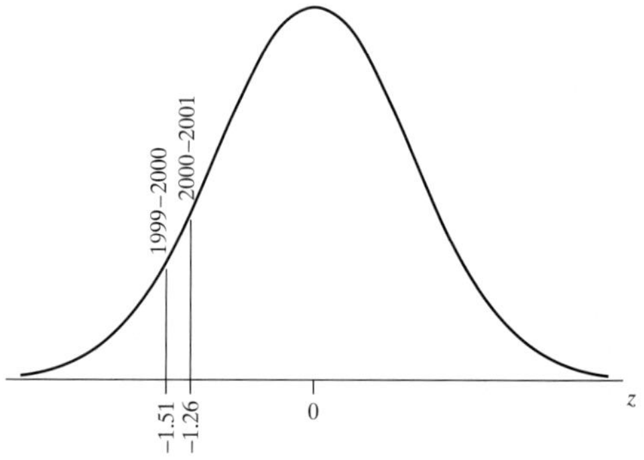

5. **(a)** MegaStar: $z = 2.5$. HemelGraph: $z = 2.3$

(b) HemelGraph's sale division traveled less, relative to the rest of the months.

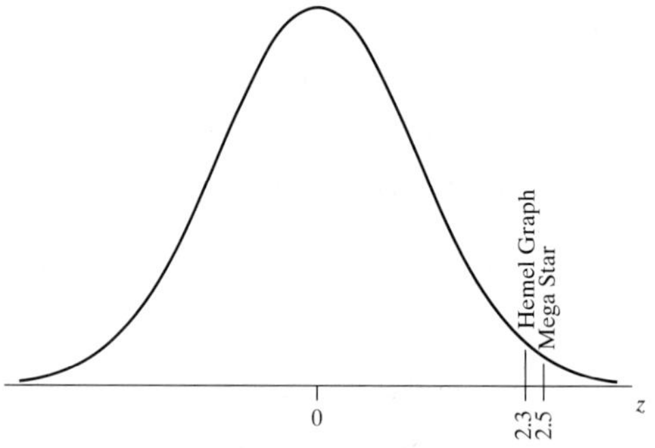

7. 0.4842 **9.** 0.2371 **11.** 0.0721 **13.** 0.5759

15. 0.3894 **17.** 0.4332 **19.** 0.2422 **21.** 0.6198

23. 0.2807 **25.** 0.3191 **27.** $P(Z > 0) = 0.5000$

29. $P(Z > 1.30) = 0.0968$ **31.** $P(Z \leq -1.76) = 0.0392$

33. $P(Z \geq -0.73) = 0.7673$ **35.** $P(Z < 1.96) = 0.9750$

37. $P(-0.50 < Z < 0.50) = 0.3830$

39. $P(-0.91 \leq Z \leq -0.12) = 0.2708$

41. (a) $z = \frac{15-15}{0.15} = 0$

$P(X > 15) = P(Z > 0) = 0.50$

There is a 50% probability that the shuttle bus will have made more than 15 runs from the airport to the local hotels.

(b) $P(X < 15) = P(Z < 0) = 0.50$

There is a 50% probability that the shuttle bus will have made less than 15 runs from the airport to the local hotels.

43. (a) $P(0.12 \leq X \leq 0.16) = P(0 \leq Z \leq 2) = 0.4772$

The probability that the wire is between 0.12 and 0.16 inch is about 47.72%.

(b) $P(0.08 \leq X \leq 0.12) = P(-2 \leq Z \leq 0) = 0.4772$

The probability that the wire is between 0.08 and 0.12 inch is about 47.72%.

45. (a) $P(X \leq 260) \approx P(Z \leq 0.52) = 0.50 + 0.1985 = 0.6985$

The probability that the player will weigh less than 260 pounds is about 69.85%.

(b) $P(200 \leq X \leq 260) \approx P(-0.71 \leq Z \leq 0.52)$
$= 0.2611 + 0.1985 = 0.4596$

The probability that the player will weigh between 200 and 260 pounds is about 45.96%.

47. (a) $P(X \leq 30) \approx P(Z \leq -0.77) = 0.2206$

The probability that the average monthly temperature was below 30°F is about 22%.

(b) $P(30 \leq X \leq 40) \approx P(-0.77 \leq Z \leq 1.31) = 0.6843$

The probability that the average monthly temperature was between 30° and 40°F is about 68.43%.

49. (a) $P(1.10 \leq X \leq 1.50) \approx P(0.22 \leq Z \leq 1.09) = 0.275$

There is about a 27.5% probability that the selected minimum wage earner made between \$1.10 and \$1.50 per hour.

(b) $P(0.90 \leq X \leq 1.50) \approx P(-0.22 \leq Z \leq 1.09) = 0.4492$

There is about a 45% probability that the selected minimum wage earner made between \$0.90 and \$1.50 per hour.

51. (a) $P(X \geq 40) \approx P(Z \geq -0.31) = 0.6217$

The probability that the patient is at least 40 years old is about 62.17%.

(b) $X = 50: \quad z = \frac{50 - 45.73}{18.68} \approx 0.23$

$P(0 \leq X \leq 50) \approx P(-2.45 \leq Z \leq 0.23) = 0.5839$

The probability that the patient is at most 50 years old is about 58.39%.

53. (a) $P(16 \leq X \leq 19) \approx P(-0.54 \leq Z \leq -0.40) = 0.05$

The probability that the patient was between 16 and 19 years old is about 5%.

(b) $P(20 \leq X \leq 40) \approx P(-0.35 \leq Z \leq 0.60) = 0.3625$

The probability that the patient was between 20 and 40 years old is about 36.25%.

55. (a) $P(X \geq 85) \approx P(Z \geq 0.45) = 0.3264$

The probability that the resident was older than 85 is about 32.64%.

(b) $X = 72: \quad z = \frac{72-81.77}{7.25} \approx -1.35$
$P(X \geq 72) \approx P(Z \geq -1.35) = 0.9155$

The probability that the resident was older than 72 is about 91.55%.

Chapter 8 Review Exercises

1. Does not meet criteria

3. Does not meet criteria

5. (a)

$X = x$	$P(X = x)$
$X = 1$	$P(X = 1) = \frac{1}{30}$
$X = 2$	$P(X = 2) = \frac{2}{15}$
$X = 3$	$P(X = 3) = \frac{3}{10}$
$X = 4$	$P(X = 4) = \frac{8}{15}$

(b) Does meet criteria

7. $k = 56$

9.

$X = x$	$P(X = x)$
$X = 1$	$P(X = 1) = \frac{10}{54} \approx 0.185$
$X = 2$	$P(X = 2) = \frac{12}{54} \approx 0.222$
$X = 3$	$P(X = 3) = \frac{18}{54} \approx 0.333$
$X = 4$	$P(X = 4) = \frac{14}{54} \approx 0.259$

11. (a)

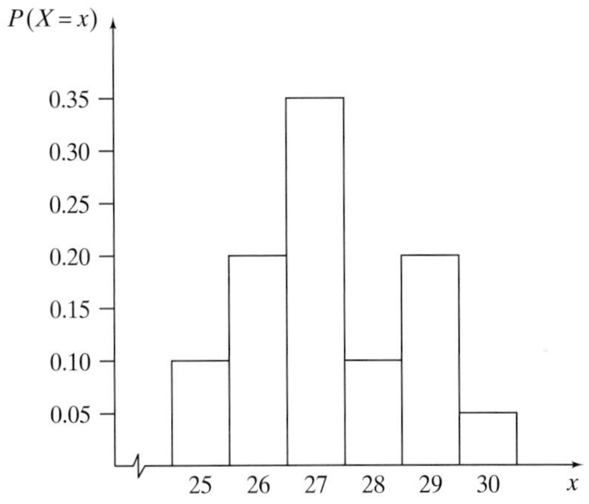

(b) 0.65; the probability that a randomly selected class at Salem High School will have between 26 and 28 students inclusive is 0.65 or 65%.

13. (a) 0.63 **(b)** 5 memberships

15. (a)

$X = x$	$P(X = x)$
$X = 1997$	$P(X = 1997) = \frac{56}{244} \approx 0.230$
$X = 1998$	$P(X = 1998) = \frac{51}{244} \approx 0.209$
$X = 1999$	$P(X = 1999) = \frac{58}{244} \approx 0.238$
$X = 2000$	$P(X = 2000) = \frac{79}{244} \approx 0.324$

(b) 0.324; the probability that a person selected at random from the group of people who were involved in unprovoked shark attacks between 1997 and 2000 was involved in an attack in 2000 is 0.324 or 32.4%.

(c) 0.561; the probability that a person selected at random from the group of people who were involved in unprovoked shark attacks between 1997 and 2000 was involved in an attack in 1999 or 2000 is 0.561 or 56.1%.

17. $E(X) = 0.505$ **19.** $E(X) = 3.5$

21. $E(X) \approx 4.33$ **23.** $\sigma_X \approx 1.12$

25. $E(X) = 27.25$; if classes at Salem High School are selected at random, in the long run, the average number of students per class is 27.25.

27. (a) 5.87 **(b)** Approximately 1.29

29. (a) $P(X = x) = \frac{1}{4}, x = 1, 2, 3, 4$

(b) $\mu_X = 2.5$; the average number of times KTTV radio gives the weather each hour is 2.5 times.

(c) Approximately 1.19 **(d)** 0.5

31. $\frac{64}{125}$ **33.** Not satisfied

35. (a)

$X = x$	$P(X = x)$
$X = 0$	$P(X = 0) \approx 0.670$
$X = 1$	$P(X = 1) \approx 0.287$
$X = 2$	$P(X = 2) \approx 0.041$
$X = 3$	$P(X = 3) \approx 0.002$

(b)

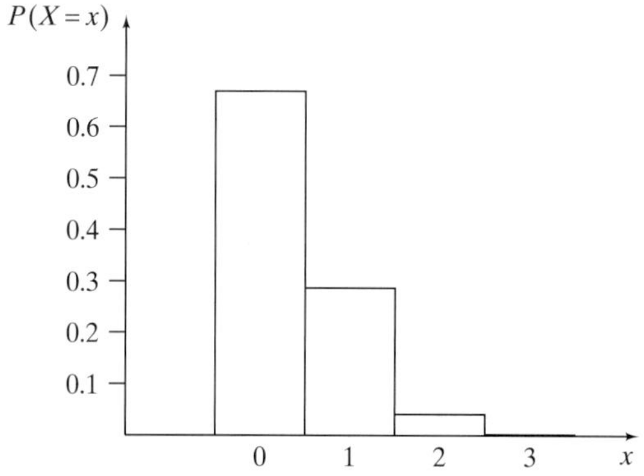

37. b **39.** $E(X) = 3.5$; $\sigma_X \approx 1.62$

41. (a) 1. Success: member of sample passes exam; failure: member of sample fails exam

2. $P = 0.75$ **3.** $n = 8$ **4.** Trials are independent.

(b) $P(X = 5) \approx 0.21$; if 8 students at Hawthorne Middle School are selected at random, the probability that 5 of them pass their physical fitness test is about 0.21 or 21%.

43. (a) Approximately 0.19 **(b)** Approximately 0.53

45. (a) 2.32 **(b)** Approximately 0.82 **47.** 0.3810

49. 0.5123 **51.** 0.3790 **53.** 0.1003 **55.** 0.3310

57. (a) 0.9876; if a business day is selected at random, there is about a 98.8% chance that between 8000 and 12,000 copies are made.

(b) 0.4938; if a business day is selected at random, there is about a 49.4% chance that between 10,000 and 12,000 copies are made.

59. (a) 0.0446 **(b)** 0.9195

Chapter 9

Section 9.1

1. It is a transition matrix. It is a 2×2 matrix, with the properties that the sum of each row is equal to 1 and all the elements of the matrix are between 0 and 1.

3. It is not a transition matrix. The sum of the second row is greater than 1.

5. It is not a transition matrix. All the elements are not between 0 and 1.

7. It is a transition matrix. It is a 3×3 matrix, with the properties that the sum of each row is equal to 1 and all the elements of the matrix are between 0 and 1.

9. It is a transition matrix. It is a 3×3 matrix, with the properties that the sum of each row is equal to 1 and all the elements of the matrix are between 0 and 1.

11. $P = \begin{bmatrix} 0.6 & 0.4 \\ 0.2 & 0.8 \end{bmatrix}$ **13.** $P = \begin{bmatrix} 0.1 & 0.9 \\ 1.0 & 0.0 \end{bmatrix}$

15. $P = \begin{bmatrix} 0.3 & 0.3 & 0.4 \\ 0.2 & 0.6 & 0.2 \\ 0.3 & 0.1 & 0.6 \end{bmatrix}$ **17.** $P = \begin{bmatrix} 1.0 & 0 & 0 \\ 0 & 0.5 & 0.5 \\ 0.6 & 0.4 & 0 \end{bmatrix}$

19. $P = \begin{bmatrix} 0.6 & 0.4 & 0 & 0 \\ 0.3 & 0.5 & 0 & 0.2 \\ 0.1 & 0 & 0.3 & 0.6 \\ 0 & 0 & 0.2 & 0.8 \end{bmatrix}$

21.

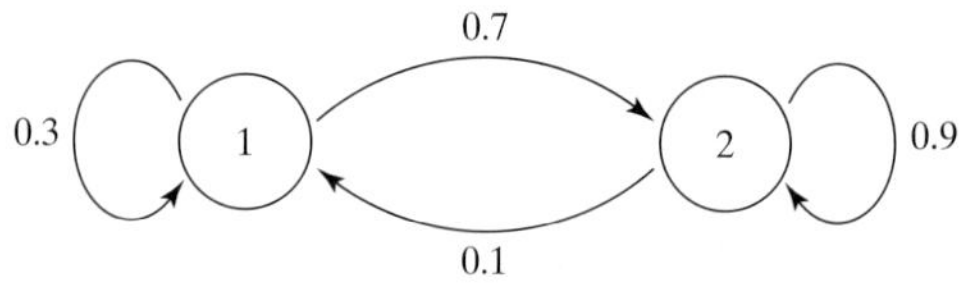

23.

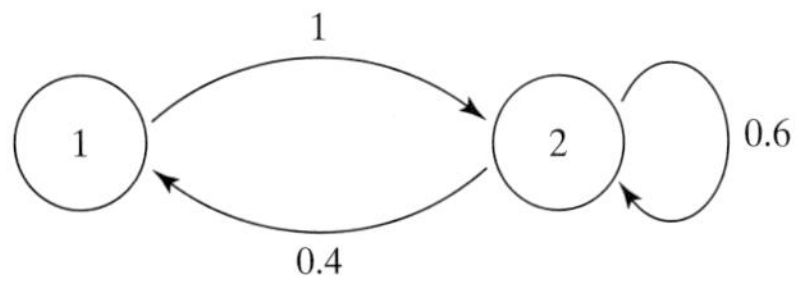

25.

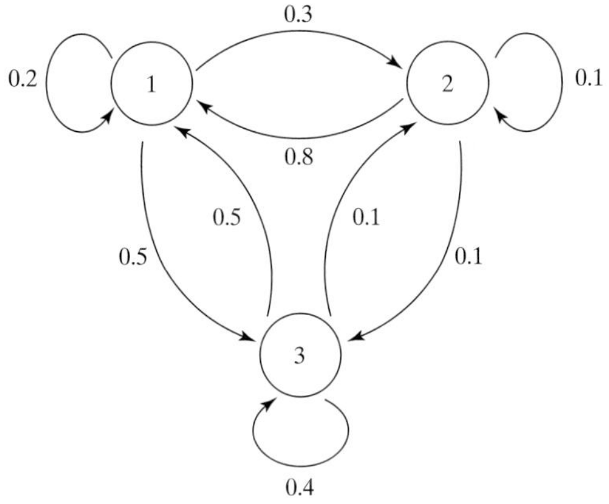

27.

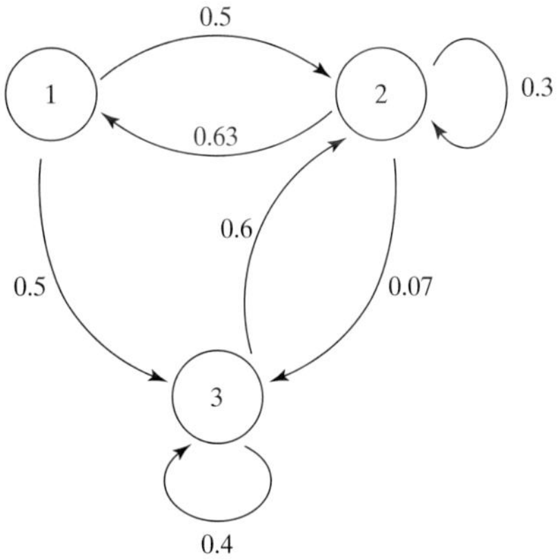

29.

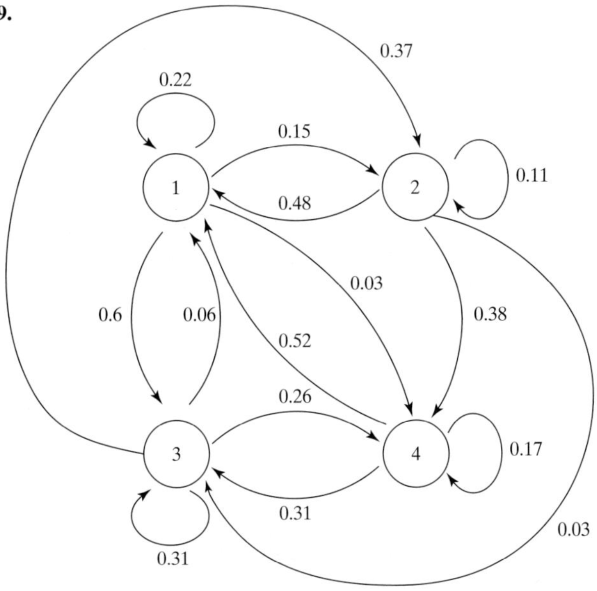

31. **(a)** State 1: owns Blade Shades
State 2: owns Optifilters

(b)

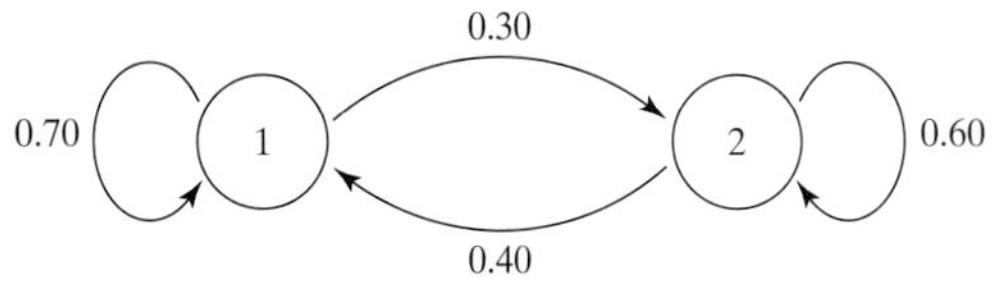

(c) $P = \begin{bmatrix} 0.70 & 0.30 \\ 0.40 & 0.60 \end{bmatrix}$

33. **(a)** State 1: hot summer
State 2: moderate summer

(b)

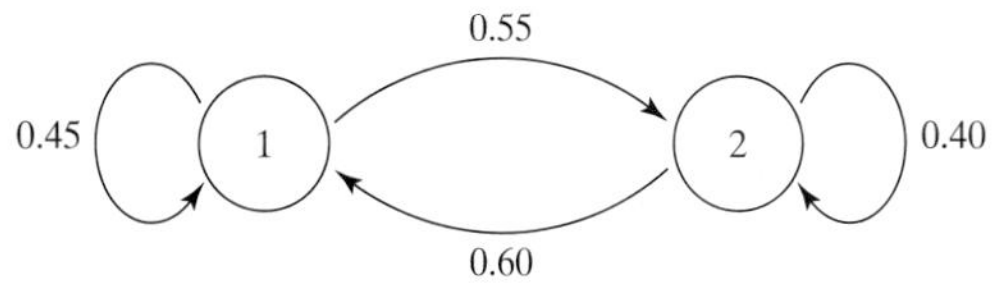

(c) $P = \begin{bmatrix} 0.45 & 0.55 \\ 0.60 & 0.40 \end{bmatrix}$

35. **(a)** State 1: Democrat in office
State 2: Republican in office

(b)

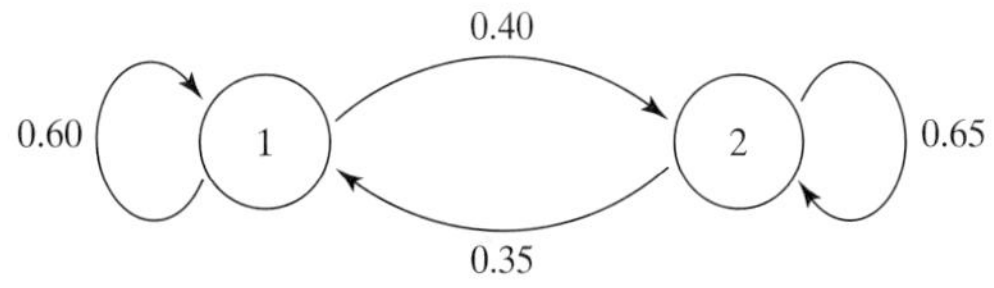

(c) $P = \begin{bmatrix} 0.60 & 0.40 \\ 0.35 & 0.65 \end{bmatrix}$

37. **(a)** State 1: rented in Adamstown
State 2: rented in Breston
State 3: rented in Carpersville

(b)

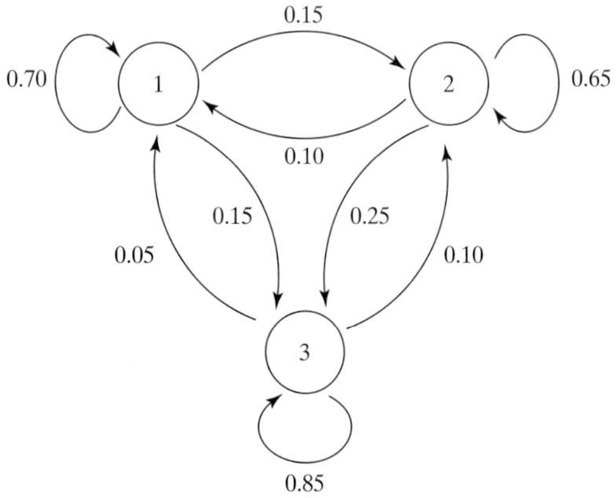

(c) $P = \begin{bmatrix} 0.70 & 0.15 & 0.15 \\ 0.10 & 0.65 & 0.25 \\ 0.05 & 0.10 & 0.85 \end{bmatrix}$

39. (a) State 1: Republican Party
State 2: Democratic Party
State 3: Green Party

(b)

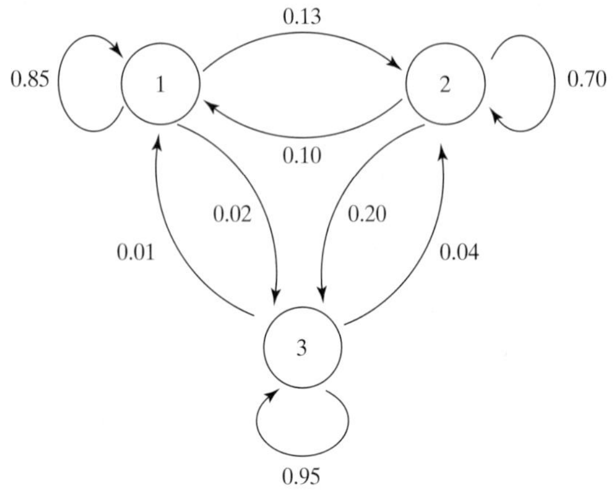

(c) $P = \begin{bmatrix} 0.85 & 0.13 & 0.02 \\ 0.10 & 0.70 & 0.20 \\ 0.01 & 0.04 & 0.95 \end{bmatrix}$

41. (a) $S_0 = [\,0.60 \quad 0.40\,]$

(b) $S_3 = [\,0.5722 \quad 0.4278\,]$

After 3 years, Blade Shades will have about 57% of the market share and Optifilter will have about 43% of the market share.

43. (a) Currently, a 40% chance of a severely hot summer and a 60% chance of a moderate summer.

(b) $S_2 = [\,0.519 \quad 0.481\,]$

After two years there will be about 52% chance of a severely hot summer and about a 48% chance of a moderate summer.

45. (a) A 40% chance that the next mayor is a Democrat and a 60% chance that the next mayor is a Republican.

(b) $S_5 \approx [\,0.467 \quad 0.533\,]$

After 10 years about a 47% chance that the mayor is a Democrat and about a 53% chance that the mayor is a Republican.

47. (a) $S_0 = [\,800 \quad 1200 \quad 400\,]$

(b) $S_2 = [\,622 \quad 792 \quad 986\,]$

After 2 years the Adamstown store will have 622 trucks to rent, the Breston store will have 792 trucks to rent, and the Carpersville store will have 986 trucks to rent.

49. (a) Initially, 62% of the voters were members of the Republican Party, 37% were members of the Democratic Party, and 1% were members of the Green Party.

(b) $S_4 \approx [\,0.4317 \quad 0.2766 \quad 0.2917\,]$

After 4 elections about 43% of the voters will declare themselves as members of the Republican Party, about 28% as members of the Democratic Party, and about 29% as members of the Green Party.

Section 9.2

1. $P^4 = \begin{bmatrix} 0.6424 & 0.3576 \\ 0.4023 & 0.5977 \end{bmatrix}$

$P^{16} \approx \begin{bmatrix} 0.5310 & 0.4690 \\ 0.5277 & 0.4723 \end{bmatrix}$

$P^{32} \approx \begin{bmatrix} 0.5294 & 0.4706 \\ 0.5294 & 0.4706 \end{bmatrix}$

$P^{64} \approx \begin{bmatrix} 0.5294 & 0.4706 \\ 0.5294 & 0.4706 \end{bmatrix}$

This is a regular transition matrix

$P^n = \begin{bmatrix} 0.5294 & 0.4706 \\ 0.5294 & 0.4706 \end{bmatrix}$ for large values of n.

3. $P^4 \approx \begin{bmatrix} 0.5670 & 0.4330 \\ 0.5196 & 0.4804 \end{bmatrix}$

$P^{16} \approx \begin{bmatrix} 0.5455 & 0.4545 \\ 0.5455 & 0.4545 \end{bmatrix}$

$P^{32} \approx \begin{bmatrix} 0.5455 & 0.4545 \\ 0.5455 & 0.4545 \end{bmatrix}$

$P^{64} \approx \begin{bmatrix} 0.5455 & 0.4545 \\ 0.5455 & 0.4545 \end{bmatrix}$

This is a regular transition matrix.

$P^n = \begin{bmatrix} 0.5455 & 0.4545 \\ 0.5455 & 0.4545 \end{bmatrix}$ for large values of n.

5. $P^4 \approx \begin{bmatrix} 1 & 0 \\ 0.985 & 0.015 \end{bmatrix}$

$P^{16} \approx \begin{bmatrix} 1 & 0 \\ 1 & 0 \end{bmatrix}$

$P^{32} \approx \begin{bmatrix} 1 & 0 \\ 1 & 0 \end{bmatrix}$

$P^{64} \approx \begin{bmatrix} 1 & 0 \\ 1 & 0 \end{bmatrix}$

This is a regular transition matrix.

$P^n = \begin{bmatrix} 1 & 0 \\ 1 & 0 \end{bmatrix}$ for large values of n.

7. The matrix is a regular transition matrix.

$P^n = \begin{bmatrix} 0.1188 & 0.1881 & 0.6931 \\ 0.1188 & 0.1881 & 0.6931 \\ 0.1188 & 0.1881 & 0.6931 \end{bmatrix}$ for large values of n.

9. The matrix is a regular transition matrix.

$P^n = \begin{bmatrix} 0.3636 & 0.4545 & 0.1818 \\ 0.3636 & 0.4545 & 0.1818 \\ 0.3636 & 0.4545 & 0.1818 \end{bmatrix}$ for large values of n.

11. The steady-state vector is

$[\,x \quad y\,] = [\,\frac{1}{9} \quad \frac{8}{9}\,]$

$P^n = \begin{bmatrix} \frac{1}{9} & \frac{8}{9} \\ \frac{1}{9} & \frac{8}{9} \end{bmatrix}$ for large values of n.

13. The steady-state vector is

$[\,x \quad y\,] = [\,\frac{7}{16} \quad \frac{9}{16}\,]$

$P^n = \begin{bmatrix} \frac{7}{16} & \frac{9}{16} \\ \frac{7}{16} & \frac{9}{16} \end{bmatrix}$ for large values of n.

15. The steady-state vector is

$S = \begin{bmatrix} \frac{5}{6} & \frac{1}{6} \end{bmatrix}$

$P^n = \begin{bmatrix} \frac{5}{6} & \frac{1}{6} \\ \frac{5}{6} & \frac{1}{6} \end{bmatrix}$ for large values of n.

17. The steady-state vector is

$S = \begin{bmatrix} \frac{17}{32} & \frac{15}{32} \end{bmatrix}$

$P^n = \begin{bmatrix} \frac{17}{32} & \frac{15}{32} \\ \frac{17}{32} & \frac{15}{32} \end{bmatrix}$ for large values of n.

19. The steady-state vector is

$\begin{bmatrix} x & y & z \end{bmatrix} = \begin{bmatrix} \frac{3}{11} & \frac{5}{11} & \frac{3}{11} \end{bmatrix}$

$P^n = \begin{bmatrix} \frac{3}{11} & \frac{5}{11} & \frac{3}{11} \\ \frac{3}{11} & \frac{5}{11} & \frac{3}{11} \\ \frac{3}{11} & \frac{5}{11} & \frac{3}{11} \end{bmatrix}$ for large values of n.

21. The steady-state vector is

$\begin{bmatrix} x & y & z \end{bmatrix} = \begin{bmatrix} \frac{1}{5} & \frac{2}{5} & \frac{2}{5} \end{bmatrix}$

$P^n = \begin{bmatrix} \frac{1}{5} & \frac{2}{5} & \frac{2}{5} \\ \frac{1}{5} & \frac{2}{5} & \frac{2}{5} \\ \frac{1}{5} & \frac{2}{5} & \frac{2}{5} \end{bmatrix}$ for large values of n.

23. The steady-state vector is

$\begin{bmatrix} x & y & z \end{bmatrix} = \begin{bmatrix} \frac{3}{7} & \frac{8}{21} & \frac{4}{21} \end{bmatrix}$

$P^n = \begin{bmatrix} \frac{3}{7} & \frac{8}{21} & \frac{4}{21} \\ \frac{3}{7} & \frac{8}{21} & \frac{4}{21} \\ \frac{3}{7} & \frac{8}{21} & \frac{4}{21} \end{bmatrix}$ for large values of n.

25. The steady-state vector is

$\begin{bmatrix} x & y & z \end{bmatrix} = \begin{bmatrix} \frac{1}{10} & \frac{19}{55} & \frac{61}{110} \end{bmatrix}$

$P^n = \begin{bmatrix} \frac{1}{10} & \frac{19}{55} & \frac{61}{110} \\ \frac{1}{10} & \frac{19}{55} & \frac{61}{110} \\ \frac{1}{10} & \frac{19}{55} & \frac{61}{110} \end{bmatrix}$ for large values of n.

27. (a) State 1: the sample passes the quality test
State 2: the sample fails the quality test

$P = \begin{bmatrix} 0.80 & 0.20 \\ 0.70 & 0.30 \end{bmatrix}$

(b) $P^n = \begin{bmatrix} \frac{7}{9} & \frac{2}{9} \\ \frac{7}{9} & \frac{2}{9} \end{bmatrix}$ for large values of n.

(c) In the long run, the probability that the sample will pass the quality test is 7/9 (approx, 78%).

29. (a)

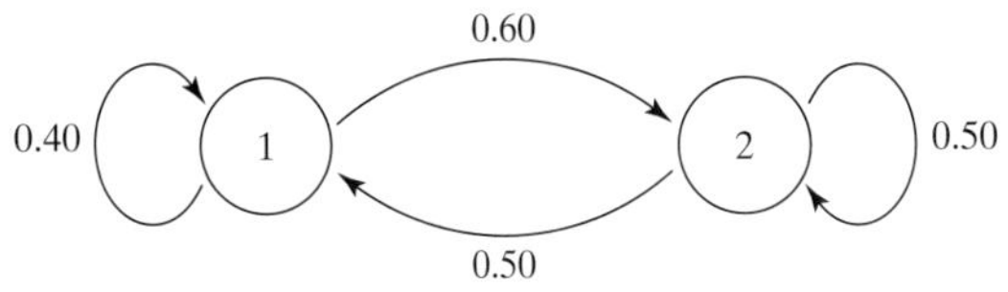

$P = \begin{bmatrix} 0.40 & 0.60 \\ 0.50 & 0.50 \end{bmatrix}$

(b) $P^n = \begin{bmatrix} \frac{5}{11} & \frac{6}{11} \\ \frac{5}{11} & \frac{6}{11} \end{bmatrix}$

(c) In the long run about 9091 motorists will be classified as low risk and about 10,909 will be classified as high risk.

31. (a) $P^n = \begin{bmatrix} \frac{4}{7} & \frac{3}{7} \\ \frac{4}{7} & \frac{3}{7} \end{bmatrix} \approx \begin{bmatrix} 0.5714 & 0.4286 \\ 0.5714 & 0.4286 \end{bmatrix}$ for large values of n.

(b) In the long run Blade Shades will have about 57% of the market share and Optifilters will have about 43% of the market share.

33. (a)

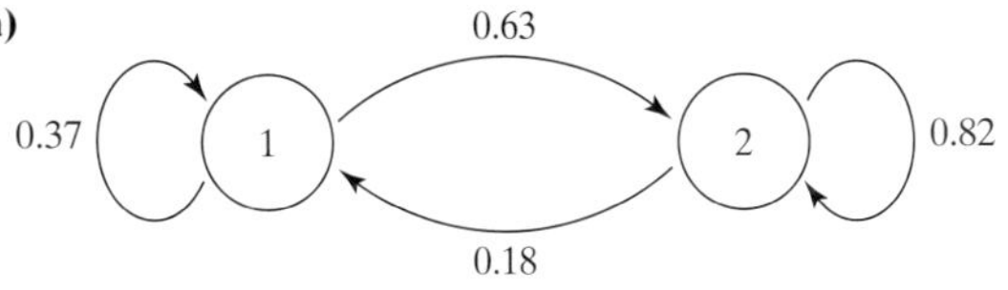

(b) $P^n = \begin{bmatrix} \frac{2}{9} & \frac{7}{9} \\ \frac{2}{9} & \frac{7}{9} \end{bmatrix} \approx \begin{bmatrix} 0.2222 & 0.7778 \\ 0.2222 & 0.7778 \end{bmatrix}$ for large values of n.

In the long run the college can expect to have about 22% of the students declared business majors and about 78% declared general studies majors.

35. (a) $P^n = \begin{bmatrix} \frac{7}{15} & \frac{8}{15} \\ \frac{7}{15} & \frac{8}{15} \end{bmatrix} \approx \begin{bmatrix} 0.4667 & 0.5333 \\ 0.4667 & 0.5333 \end{bmatrix}$ for large values of n.

(b) About 47% of mayoral election Democrats are expected to win in the long run.

37. (a) $P^n \approx \begin{bmatrix} 0.1774 & 0.2419 & 0.5806 \\ 0.1774 & 0.2419 & 0.5806 \\ 0.1774 & 0.2419 & 0.5806 \end{bmatrix}$ for large values of n.

(b) In the long run about 18% of the trucks will be returned to Adamstown, about 24% of the trucks to Breston, and about 58% to Carpersville.

39. (a) $P^n \approx \begin{bmatrix} 0.1511 & 0.1577 & 0.6911 \\ 0.1511 & 0.1577 & 0.6911 \\ 0.1511 & 0.1577 & 0.6911 \end{bmatrix}$ for large values of n.

(b) In the long run about 69% of the voters are expected to be members of the Green Party.

Section 9.3

1. There is not an absorbing state in this matrix.

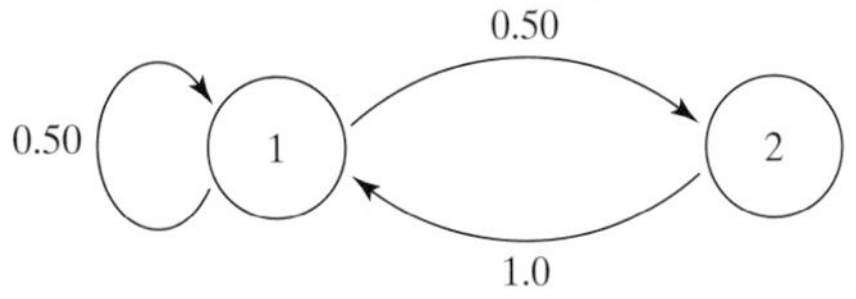

3. States 2 and 3 are absorbing states.

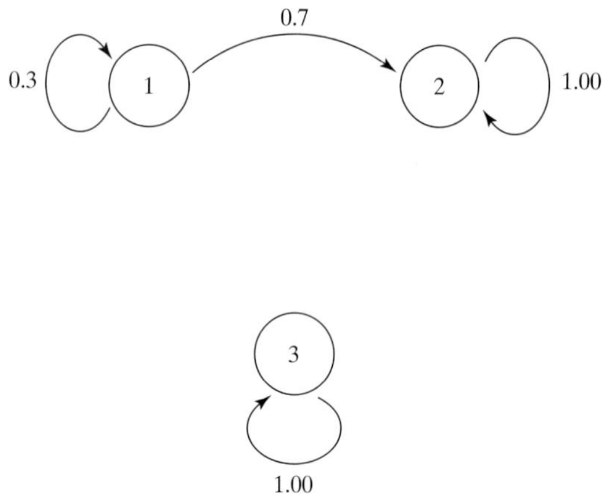

5. State 3 is an absorbing state.

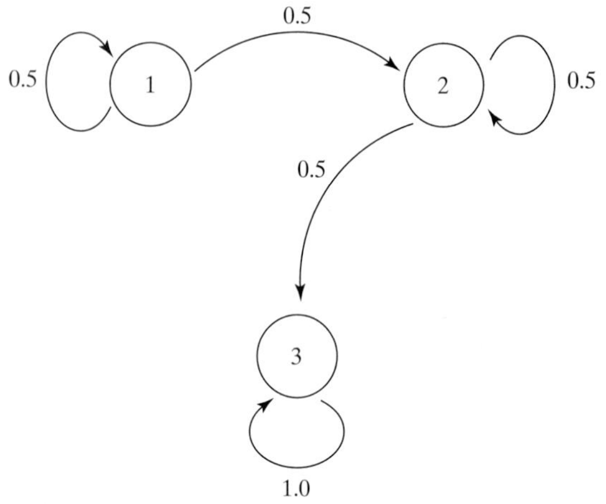

7. States 1 and 3 are absorbing states.

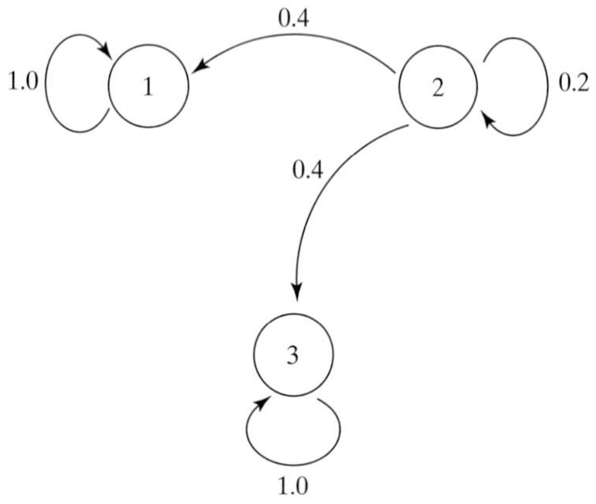

9. State 4 is an absorbing state.

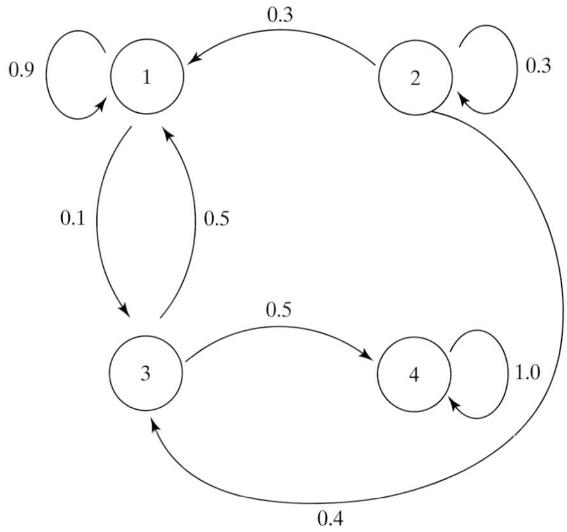

11. State 1 is an absorbing state. The diagram represents an absorbing Markov chain.

13. States 1 and 3 are absorbing states. The diagram represents an absorbing Markov chain.

15. State 2 is an absorbing state. The diagram represents an absorbing Markov chain.

17. States 2 and 4 are absorbing states. The diagram represents an absorbing Markov chain.

19. State 5 is an absorbing state. The diagram represents an absorbing Markov chain.

21. **(a)** $F = \begin{bmatrix} 2 & 0 \\ 2 & \frac{4}{3} \end{bmatrix}$ **(b)** $FR = \begin{bmatrix} 1 \\ 1 \end{bmatrix}$

23. **(a)** $F = \begin{bmatrix} 2 & 0 \\ 2 & \frac{4}{3} \end{bmatrix}$ **(b)** $FR = \begin{bmatrix} 1 \\ 1 \end{bmatrix}$

25. **(a)** $F = \begin{bmatrix} \frac{20}{9} & \frac{14}{9} \\ \frac{10}{9} & \frac{16}{9} \end{bmatrix}$ **(b)** $FR = \begin{bmatrix} 1 \\ 1 \end{bmatrix}$

27. Singular matrix, therefore no inverse exists.

29. **(a)** $F = \begin{bmatrix} \frac{10}{7} & 0 \\ \frac{10}{7} & 1 \end{bmatrix}$ **(b)** $FR = \begin{bmatrix} 1 \\ 1 \end{bmatrix}$

31. Singular matrix, therefore no inverse exists.

33. **(a)** $F = \begin{bmatrix} \frac{40}{27} & \frac{10}{27} \\ \frac{25}{27} & \frac{40}{27} \end{bmatrix}$ **(b)** $FR = \begin{bmatrix} \frac{8}{27} & \frac{19}{27} \\ \frac{5}{27} & \frac{22}{27} \end{bmatrix}$

35. **(a)** $F = \begin{bmatrix} \frac{5620}{4343} & \frac{1900}{4343} & \frac{1220}{4343} \\ \frac{1080}{4343} & \frac{5620}{4343} & \frac{1780}{4343} \\ \frac{1640}{4343} & \frac{2100}{4343} & \frac{5920}{4343} \end{bmatrix}$ **(b)** $FR = \begin{bmatrix} \frac{2405}{4343} & \frac{1938}{4343} \\ \frac{2085}{4343} & \frac{2258}{4343} \\ \frac{2201}{4343} & \frac{2142}{4343} \end{bmatrix}$

37. **(a)** State 1: zoned residential
State 2: zoned commercial
State 3: zoned public

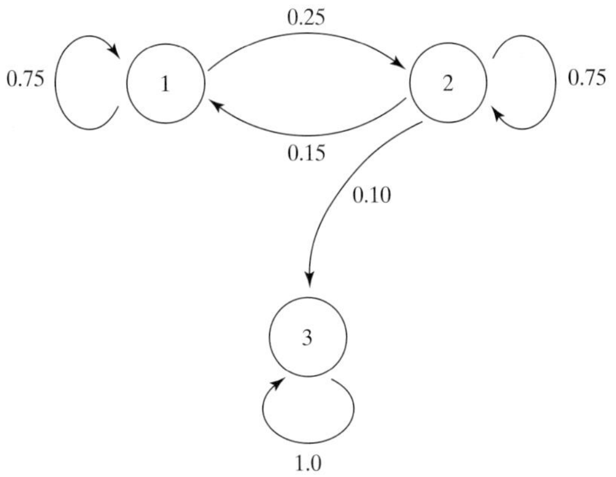

(b) $P = \begin{bmatrix} 0.75 & 0.25 & 0 \\ 0.15 & 0.75 & 0.10 \\ 0 & 0 & 1 \end{bmatrix}$

$F = \begin{bmatrix} 10 & 10 \\ 6 & 10 \end{bmatrix}$

(c) $f_{1,2} = 10$; land that was initially zoned as residential will spend about 10 years zoned as commercial until it is eventually zoned for public use.

39. (a) State 1: freshman
State 2: sophomore
State 3: leave the school

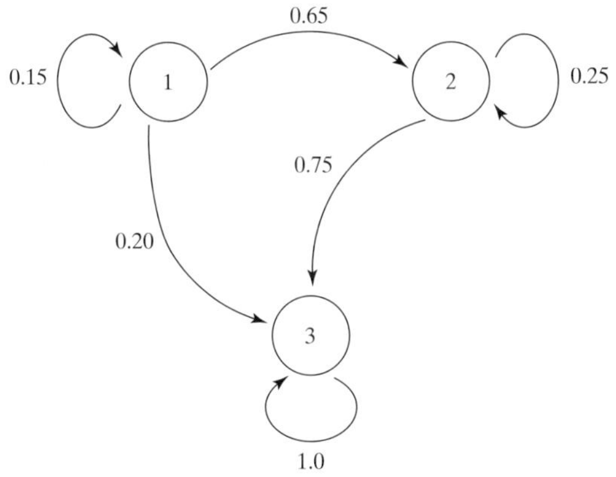

(b) $P = \begin{bmatrix} 0.15 & 0.65 & 0.20 \\ 0 & 0.25 & 0.75 \\ 0 & 0 & 1 \end{bmatrix}$

$F = \begin{bmatrix} \frac{20}{17} & \frac{52}{51} \\ 0 & \frac{4}{3} \end{bmatrix}$

(c) $f_{1,2} = \frac{52}{51}$; this means that a student who is initially a freshman will spend a little more than 1 year as a sophomore before eventually leaving school.

41. (a) State 1: account in good standing
State 2: probationary
State 3: pay off loan
State 4: repossession

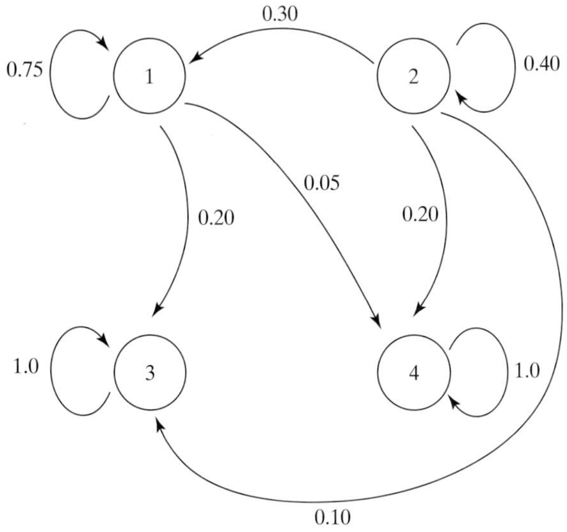

(b) $P = \begin{bmatrix} 0.75 & 0 & 0.20 & 0.05 \\ 0.30 & 0.40 & 0.10 & 0.20 \\ 0 & 0 & 1 & 0 \\ 0 & 0 & 0 & 1 \end{bmatrix}$ $\quad F = \begin{bmatrix} 4 & 0 \\ 2 & \frac{5}{3} \end{bmatrix}$

(c) The sum of the entries in row 1 is 4. This indicates that we can expect an account in good standing to be either paid off or enter into repossession in four months.

The sum of the entries in row 2 is about 3.7. This indicates that we can expect an account that is in a probationary state to be either paid off or enter into repossession in about 3.7 months.

(d) $FR = \begin{bmatrix} \frac{4}{5} & \frac{1}{5} \\ \frac{17}{30} & \frac{13}{30} \end{bmatrix}$

An account in good standing has an 80% chance of remaining in good standing and a 20% chance of becoming probationary.

43. (a) State 1: trainee
State 2: technician
State 3: supervisor
State 4: leave the company

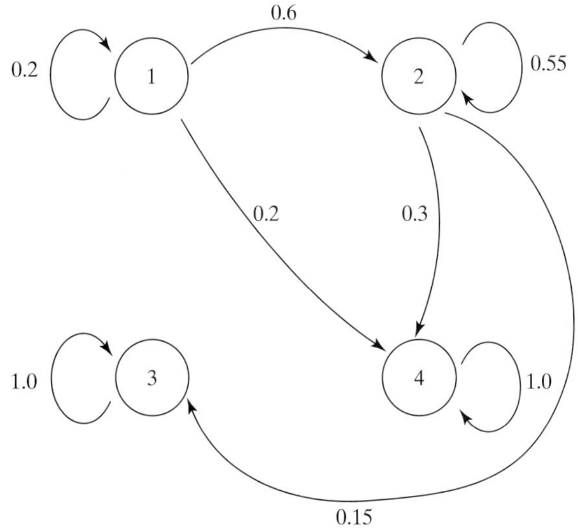

(b) $P = \begin{bmatrix} 0.20 & 0.60 & 0 & 0.20 \\ 0 & 0.55 & 0.15 & 0.30 \\ 0 & 0 & 1 & 0 \\ 0 & 0 & 0 & 1 \end{bmatrix}$ $\quad F = \begin{bmatrix} \frac{5}{4} & \frac{5}{3} \\ 0 & \frac{20}{9} \end{bmatrix}$

(c) The sum of the entries in row 1 is approximately 2.92. The sum of the entries in row 2 is approximately 2.22. This indicates that a trainee can expect to be with the company about 3 years and a technician can expect to be with the company about 2.2 years before becoming a supervisor.

45. (a) State 1: freshman
State 2: sophomore
State 3: junior
State 4: senior
State 5: withdraw
State 6: graduate

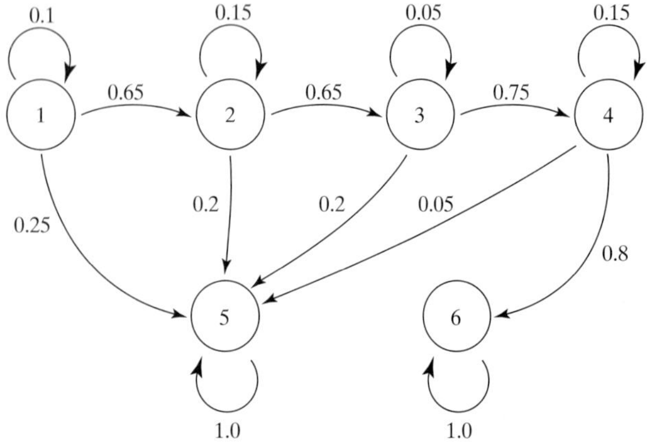

(b) $P = \begin{bmatrix} 0.10 & 0.65 & 0 & 0 & 0.25 & 0 \\ 0 & 0.15 & 0.65 & 0 & 0.20 & 0 \\ 0 & 0 & 0.05 & 0.75 & 0.20 & 0 \\ 0 & 0 & 0 & 0.15 & 0.05 & 0.80 \\ 0 & 0 & 0 & 0 & 1 & 0 \\ 0 & 0 & 0 & 0 & 0 & 1 \end{bmatrix}$

$F \approx \begin{bmatrix} \frac{10}{9} & \frac{130}{153} & \frac{1690}{2907} & 0.513 \\ 0 & \frac{20}{17} & \frac{260}{323} & \frac{3900}{5491} \\ 0 & 0 & \frac{20}{19} & \frac{300}{323} \\ 0 & 0 & 0 & \frac{20}{17} \end{bmatrix}$

The sum of row 1 is approximately 3.055.
The sum of row 2 is approximately 2.692.
The sum of row 3 is approximately 1.981.
The sum of row 4 is approximately 1.176.

This means that a freshman, sophomore, junior, and senior can expect to be at the university 3.055, 2.692, 1.981, and 1.176 years, respectively.

Chapter 9 Review Exercises

1. Transition matrix **3.** Transition matrix

5. $\begin{bmatrix} 0.8 & 0 & 0.2 \\ 0.1 & 0.1 & 0.8 \\ 0.6 & 0.3 & 0.1 \end{bmatrix}$

7.

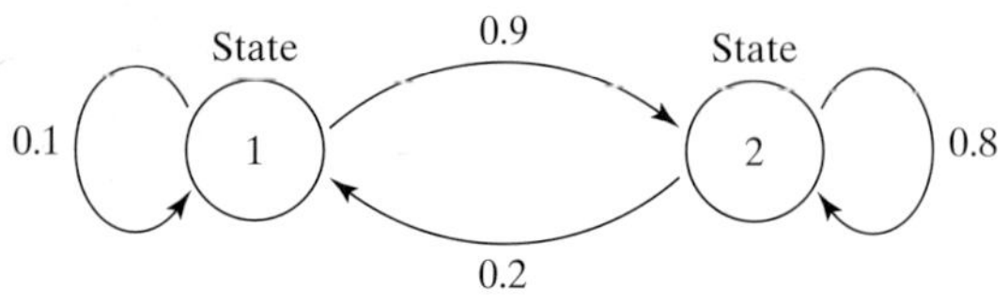

9.

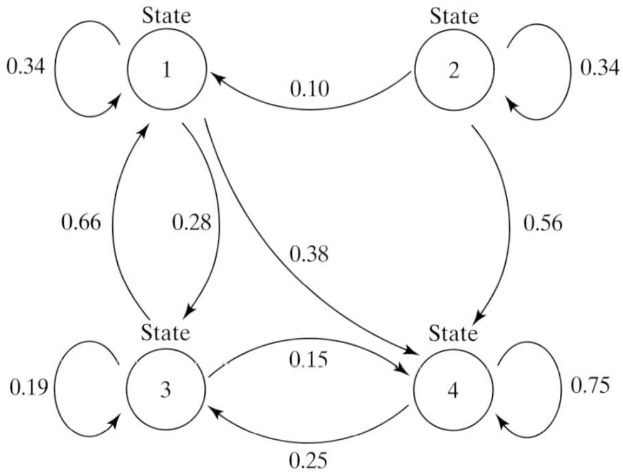

11. (a) State 1: bike is returned to the library; state 2: bike is returned to the YMCA, state 3: bike is returned to the grocery store.

(b)

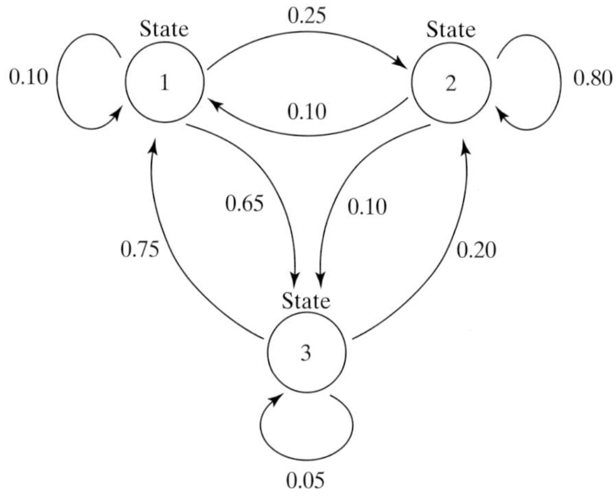

(c) $\begin{bmatrix} 0.10 & 0.25 & 0.65 \\ 0.10 & 0.80 & 0.10 \\ 0.75 & 0.20 & 0.05 \end{bmatrix}$

13. (a) $S_0 = [20 \quad 24 \quad 18]$

(b) $[16.795 \quad 29.975 \quad 15.230]$; on Wednesday morning there will be about 17 bikes at the library, 30 bikes at the YMCA, and 15 bikes at the grocery store.

15. Regular; $P^n = \begin{bmatrix} 0.1250 & 0.8750 \\ 0.1250 & 0.8750 \end{bmatrix}$ for large values of n.

17. Regular; $P^n = \begin{bmatrix} 0.3162 & 0.2222 & 0.4615 \\ 0.3162 & 0.2222 & 0.4615 \\ 0.3162 & 0.2222 & 0.4615 \end{bmatrix}$ for large values of n.

19. $\begin{bmatrix} \frac{2}{7} & \frac{5}{7} \end{bmatrix}$; $\begin{bmatrix} \frac{2}{7} & \frac{5}{7} \\ \frac{2}{7} & \frac{5}{7} \end{bmatrix}$

21. $\begin{bmatrix} \frac{4}{13} & \frac{4}{13} & \frac{5}{13} \end{bmatrix}$; $\begin{bmatrix} \frac{4}{13} & \frac{4}{13} & \frac{5}{13} \\ \frac{4}{13} & \frac{4}{13} & \frac{5}{13} \\ \frac{4}{13} & \frac{4}{13} & \frac{5}{13} \end{bmatrix}$

23. $\begin{bmatrix} \frac{52}{151} & \frac{45}{302} & \frac{153}{302} \end{bmatrix}$; $\begin{bmatrix} \frac{52}{151} & \frac{45}{302} & \frac{153}{302} \\ \frac{52}{151} & \frac{45}{302} & \frac{153}{302} \\ \frac{52}{151} & \frac{45}{302} & \frac{153}{302} \end{bmatrix}$

25. **(a)** State 1: off probation; state 2: on probation;

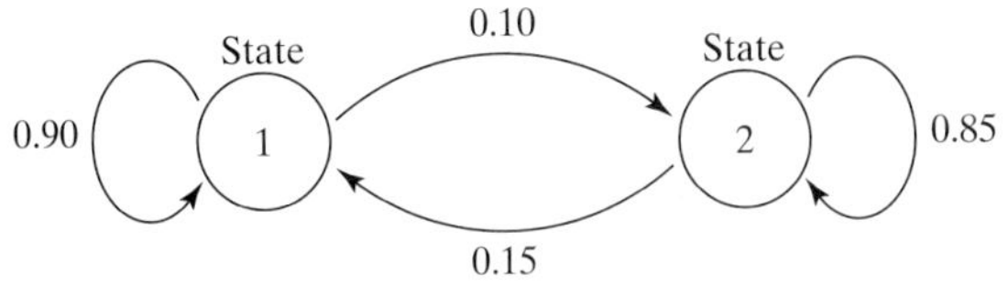

$\begin{bmatrix} 0.90 & 0.10 \\ 0.15 & 0.85 \end{bmatrix}$

(b) $\begin{bmatrix} \frac{3}{5} & \frac{2}{5} \\ \frac{3}{5} & \frac{2}{5} \end{bmatrix}$

(c) Off probation: 720 students; on probation: 480 students

27. **(a)** $\begin{bmatrix} \frac{68}{277} & \frac{147}{277} & \frac{62}{277} \\ \frac{68}{277} & \frac{147}{277} & \frac{62}{277} \\ \frac{68}{277} & \frac{147}{277} & \frac{62}{277} \end{bmatrix}$ **(b)** About 0.53 or 53%

29. Absorbing state: state 2

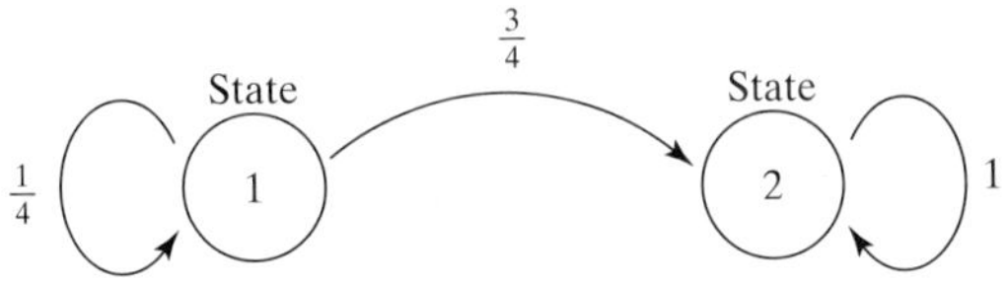

31. Absorbing states: states 3 and 4

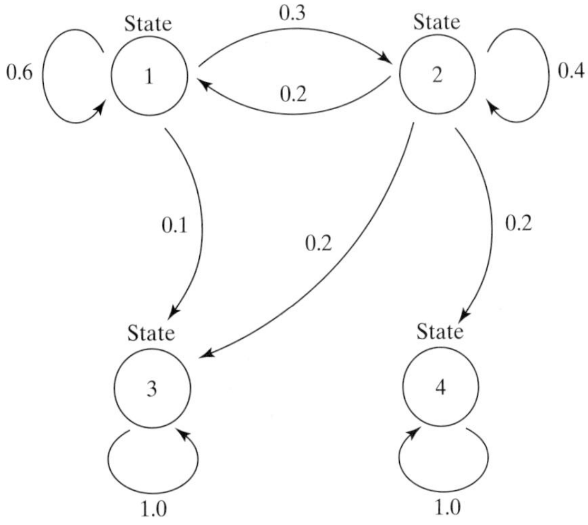

33. Absorbing states: states 2 and 3; does not represent an absorbing Markov chain

35. Absorbing states: states 4, 5, and 6; does not represent an absorbing Markov chain

37. $\begin{bmatrix} 1.25 & 0 \\ 0.3125 & 2.5 \end{bmatrix}$

39. $\begin{bmatrix} \frac{5}{3} & \frac{5}{3} \\ \frac{5}{6} & \frac{5}{2} \end{bmatrix}$

41. **(a)** State 1: first string, State 2: second string, State 3: leave the team;

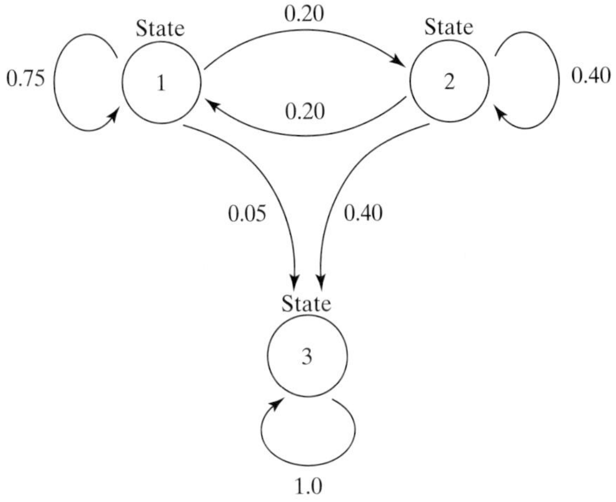

(b) $\begin{bmatrix} 0.75 & 0.20 & 0.05 \\ 0.20 & 0.40 & 0.40 \\ 0 & 0 & 1 \end{bmatrix}$; entries rounded to the nearest hundredth, $\begin{bmatrix} 5.45 & 1.82 \\ 1.82 & 2.27 \end{bmatrix}$

(c) A first-string player (state 1) will spend about 1.82 seasons as a second-string player (state 2) before leaving the team (state 3).

43. **(a)** State 1: stage I; state 2: stage II; state 3: patient transfers; state 4: patient completes program;

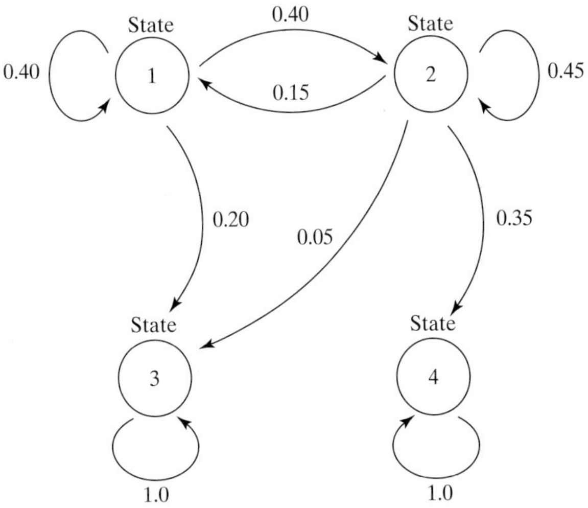

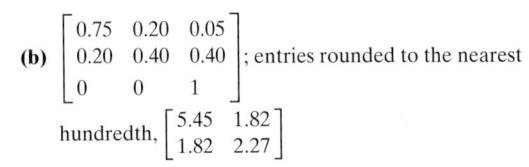

(b)
$$\begin{bmatrix} 0.40 & 0.40 & 0.20 & 0 \\ 0.15 & 0.45 & 0.05 & 0.35 \\ 0 & 0 & 1 & 0 \\ 0 & 0 & 0 & 1 \end{bmatrix}$$; entries rounded to the nearest

hundredth: $\begin{bmatrix} 2.04 & 1.48 \\ 0.56 & 2.22 \end{bmatrix}$

(c) Row 1: about 3.52, a patient in stage I can be expected to be in the program for about 3.52 weeks before transferring or completing the program; row 2: about 2.78, a patient in stage II can be expected to be in the program for about 2.78 weeks before transferring or completing the program.

45. (a) State 1: freshman; state 2: sophomore; state 3: junior; state 4: senior; state 5: withdraw; state 6: graduate;

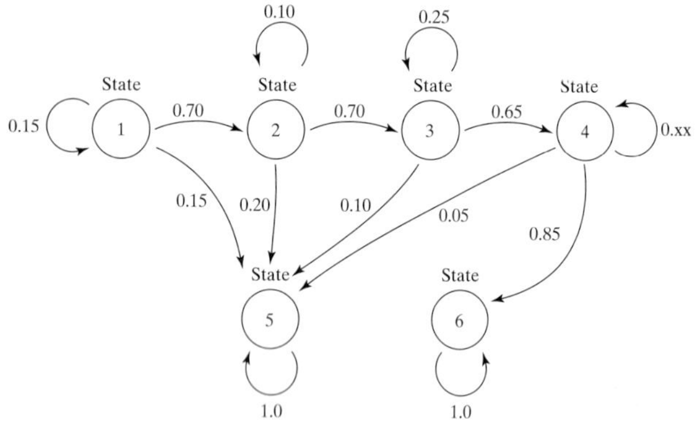

(b)
$$\begin{bmatrix} 0.15 & 0.70 & 0 & 0 & 0.15 & 0 \\ 0 & 0.10 & 0.70 & 0 & 0.20 & 0 \\ 0 & 0 & 0.25 & 0.65 & 0.10 & 0 \\ 0 & 0 & 0 & 0.10 & 0.05 & 0.85 \\ 0 & 0 & 0 & 0 & 1 & 0 \\ 0 & 0 & 0 & 0 & 0 & 1 \end{bmatrix}$$; entries rounded

to the nearest hundredth:
$$\begin{bmatrix} 1.18 & 0.92 & 0.85 & 0.62 \\ 0 & 1.11 & 1.04 & 0.75 \\ 0 & 0 & 1.33 & 0.96 \\ 0 & 0 & 0 & 1.11 \end{bmatrix}$$

(c) Freshman: about 3.57 years; sophomore: about 2.9 years; junior: about 2.29 years; senior: about 1.11 years

(d) About 0.53 or 53%

(e) About 0.36 or 36%

Chapter 10

Section 10.1

1. Jenna wins $2.   **3.** Elle wins $2.

5. Jenna wins $6.   **7.** Elle wins $3.

9. Don $\begin{matrix} & \text{Bill} \\ & \begin{bmatrix} -3 & 2 \\ 3 & 4 \end{bmatrix} \end{matrix}$

11. Patrice $\begin{matrix} & \text{Quincy} \\ & \begin{bmatrix} 3 & -4 & -4 \\ -1 & 6 & -6 \\ -2 & -3 & 8 \end{bmatrix} \end{matrix}$

13. (a) The saddle point is the entry $a_{1,1} = 2$.

(b) The optimal strategy for Bill is to play row 1 and the optimal strategy for Al is to play column 1.

(c) The value of the game is 2 and the game favors Bill.

15. There is no saddle point in this game. This is not a strictly determined game.

17. (a) The saddle points are the entries $a_{1,1}$ and $a_{1,3}$, which each have a value of 2.

(b) The optimal strategy for Khalid is to play row 1. The optimal strategy for Scoonie is to choose column 1 or 3.

(c) The value of the game is 2 and the game favors Khalid.

19. (a) The saddle point is the entry $a_{2,1} = 5$.

(b) The optimal strategies are for Austin to play row 2 and for Dylan to play column 1.

(c) The value of the game is 5 and the game favors Austin.

21. (a) The saddle point is the entry $a_{3,2} = 3$.

(b) The optimal strategies are for Brock to play row 3 and for Ash to play column 2.

(c) The value of the game is 3 and the game favors Brock.

23. (a) The saddle points are the entries $a_{2,2}$ and $a_{4,2}$, which both are equal to zero.

(b) The optimal strategies are for Brian to play row 2 or 4 and for Mike to play column 2.

(c) The value of the game is 0 and it is a fair game.

25. (a) The saddle point is the entry $a_{1,2} = -100,000$. The best strategy for The Chasm is to remodel (row 1).

27. (a) This is a strictly determined game.

(b) The saddle point is the entry $a_{2,2} = 25$. The optimal strategy for each bank is to open a branch in Kerrton.

(c) The value of the game is 25. The game favors Bank Two.

29. (a)
$$\begin{matrix} & & \text{Erog} \\ & & \text{City} \quad \text{Suburb} \\ \text{Redan} & \begin{matrix} \text{City} \\ \text{Suburb} \end{matrix} & \begin{bmatrix} 2 & -6 \\ 8 & -4 \end{bmatrix} \end{matrix}$$

(c) Optimal strategy for each store is to build in the suburbs.

31. (a)
$$\begin{matrix} & & \text{Noflab} \\ & & \text{MA} \quad \text{SR} \\ \text{Gutbuster} & \begin{matrix} \text{MA} \\ \text{SR} \\ \text{TW} \end{matrix} & \begin{bmatrix} 5 & -5 \\ 5 & 0 \\ 10 & 15 \end{bmatrix} \end{matrix}$$

(c) The optimal strategy for Gutbuster is to start a twenty-something fit-for-life program. The optimal strategy for Noflab is to start a middle-age spread program.

33. (a)
$$\begin{matrix} & & \text{Media} \\ & & \text{Critical} \quad \text{Noncritical} \\ \text{Candidate} & \begin{matrix} \text{Reveals} \\ \text{Doesn't reveal} \end{matrix} & \begin{bmatrix} -35 & 7 \\ -150 & 0 \end{bmatrix} \end{matrix}$$

(c) The optimal strategy is for the candidate to reveal the skeleton.

Section 10.2

1. The expected value is $\frac{1}{6}$.

3. The expected value is $\frac{9}{8}$.

5. The expected value is 0.41.

7. The expected value is 2.85.

7. The expected value is 2.85.

9. The expected value is 0.84.

11. (a) $E = [0.25]$ **(b)** $E = \left[\frac{13}{12}\right]$ **(c)** $E = \left[\frac{26}{35}\right]$

(d) $E = \left[\frac{23}{24}\right]$

Strategy (b) is the most favorable to the row player.

13. $p_{\text{opt}} = \begin{bmatrix} \frac{7}{16} & \frac{9}{16} \end{bmatrix}$ $q_{\text{opt}} = \begin{bmatrix} \frac{9}{16} \\ \frac{7}{16} \end{bmatrix}$

15. This is a strictly determined game. Optimal strategy for the row player is row 1; for the column player it is column 2.

17. $p_{\text{opt}} = \begin{bmatrix} \frac{4}{7} & \frac{3}{7} \end{bmatrix}$ $q_{\text{opt}} = \begin{bmatrix} \frac{3}{7} \\ \frac{4}{7} \end{bmatrix}$

19. $p_{\text{opt}} = \begin{bmatrix} \frac{1}{2} & \frac{1}{2} \end{bmatrix}$ $q_{\text{opt}} = \begin{bmatrix} \frac{5}{9} \\ \frac{4}{9} \end{bmatrix}$

21. $p_{\text{opt}} = \begin{bmatrix} \frac{1}{2} & \frac{1}{2} \end{bmatrix}$ $q_{\text{opt}} = \begin{bmatrix} \frac{5}{12} \\ \frac{7}{12} \end{bmatrix}$

23. (a) $v = -\frac{1}{16}$ **(b)** The game favors the column player.

25. (a) $v = \frac{12}{7}$ **(b)** The game favors the row player.

27. (a) $v = \frac{1}{12}$ **(b)** The game favors the row player.

31. (a) $p_{\text{opt}} = \begin{bmatrix} \frac{19}{42} & \frac{23}{42} \end{bmatrix}$

This is the optimal strategy for the Stitz Co.

$q_{\text{opt}} = \begin{bmatrix} \frac{10}{21} \\ \frac{11}{21} \end{bmatrix}$

This is the optimal strategy for the Stumpf Co.

(b) $v = \frac{31}{105}$

The game favors the Stitz Co.

33. (a) Bridge

	Economy: Strong	Economy: Recession
Remodel	400	−150
Don't remodel	−125	−100

(b) $p_{\text{opt}} = \begin{bmatrix} \frac{1}{23} & \frac{22}{23} \end{bmatrix}$

35. (a) Stock choice

	Economy: Growing	Economy: Recession
Tech.	0.35	−0.15
Pharm.	0.15	0.16

(b) $p_{\text{opt}} = \begin{bmatrix} \frac{1}{51} & \frac{50}{51} \end{bmatrix}$

The optimal strategy for the Riblets is to invest about $980 in technology stock and about $49,020 in pharmaceutical stock.

(c) The Riblets can expect about a 15.39% annual gain. So after one year the Riblets can expect a profit of about $7695.

37. (a) Toth

	Williams: Rural	Williams: Urban
Rural	0.60	0.40
Urban	0.35	0.65

(b) $p_{\text{opt}} = [0.6 \quad 0.4]$

The optimal strategy for Toth is to campaign 60% of the time in rural areas and 40% of the time in urban areas.

$q_{\text{opt}} = [0.5 \quad 0.5]$

The optimal strategy for Williams is to campaign 50% of the time in rural areas and 50% of the time in urban areas.

Section 10.3

1. (a) Optimal strategy for the column player is $Q_{\text{opt}} = \begin{bmatrix} \frac{1}{7} \\ \frac{6}{7} \end{bmatrix}$.

Optimal strategy for the row player is $P_{\text{opt}} = \begin{bmatrix} \frac{5}{7} & \frac{2}{7} \end{bmatrix}$.

(b) The value of the game is $\frac{19}{7}$.

3. (a) Column player's optimal strategy is $Q_{\text{opt}} = \begin{bmatrix} \frac{1}{5} \\ \frac{4}{5} \end{bmatrix}$.

Row player's optimal strategy is $p_{\text{opt}} = \begin{bmatrix} \frac{2}{5} & \frac{3}{5} \end{bmatrix}$.

(b) The value of the game is $\frac{43}{5}$.

5. (a) Row player's optimal strategy is row 1; column player's is column 1. (This is a strictly determined game).

(b) Value of game is 2.

7. (a) Column player's optimal strategy is $Q_{\text{opt}} = \begin{bmatrix} \frac{3}{5} \\ \frac{2}{5} \end{bmatrix}$.

The row player's optimal strategy is $P_{\text{opt}} = \begin{bmatrix} 0 & \frac{1}{5} & \frac{4}{5} \end{bmatrix}$.

(b) The value of the game is $\frac{17}{5}$.

9. (a) The column player's optimal strategy is $Q_{\text{opt}} = \begin{bmatrix} \frac{2}{3} \\ 0 \\ \frac{1}{3} \end{bmatrix}$.

The row player's optimal strategy is $P_{\text{opt}} = \begin{bmatrix} \frac{2}{3} & \frac{1}{3} \end{bmatrix}$.

(b) The value of the game is $\frac{8}{3}$.

11. (a) The column player's optimal strategy is $Q_{\text{opt}} = \begin{bmatrix} \frac{3}{8} \\ \frac{5}{8} \end{bmatrix}$.

The row player's optimal strategy is $P_{\text{opt}} = \begin{bmatrix} \frac{1}{2} & \frac{1}{2} \end{bmatrix}$.

13. The column player's optimal strategy is $Q_{\text{opt}} = \begin{bmatrix} \frac{3}{7} \\ \frac{4}{7} \end{bmatrix}$.

The row player's optimal strategy is $P_{\text{opt}} = \begin{bmatrix} \frac{13}{21} & \frac{8}{21} \end{bmatrix}$.

15. The column player's optimal strategy is $Q_{\text{opt}} = \begin{bmatrix} 0 \\ \frac{1}{2} \\ \frac{1}{2} \end{bmatrix}$.

The row player's optimal strategy is $P_{\text{opt}} = \begin{bmatrix} \frac{1}{2} & \frac{1}{2} \end{bmatrix}$.

17. The column player's optimal strategy is $Q_{opt} = \begin{bmatrix} \frac{43}{116} \\ \frac{9}{29} \\ \frac{37}{116} \end{bmatrix}$.

The row player's optimal strategy is $P_{opt} = \begin{bmatrix} \frac{27}{58} & \frac{10}{29} & \frac{11}{58} \end{bmatrix}$.

19. The column player's optimal strategy is $Q_{opt} = \begin{bmatrix} \frac{4}{7} \\ \frac{3}{7} \end{bmatrix}$.

The row player's optimal strategy is $P_{opt} = \begin{bmatrix} \frac{4}{7} & 0 & \frac{3}{7} \end{bmatrix}$.

21. (a) The column player's optimal strategy is $Q_{opt} = \begin{bmatrix} \frac{2}{5} \\ \frac{3}{5} \end{bmatrix}$.

The optimal strategy for the Hoffman company is to have $\frac{2}{5}$ of their advertising on television and $\frac{3}{5}$ on radio.

The row player's optimal strategy is $P_{opt} = \begin{bmatrix} \frac{2}{5} & \frac{3}{5} \end{bmatrix}$.

The optimal strategy for the Baker Company is to have $\frac{2}{5}$ of their advertising on television and $\frac{3}{5}$ on radio.

(b) The value of the game is $\frac{1}{5}$. The game favors the Baker Company.

23. (a) Jessie (rows) vs. Jamie (columns)

Jessie \ Jamie	L	R
L	−4	3
R	6	−4

(b) Jamie's optimal strategy will be to guess the left hand 7 out of 17 times and the right hand 10 out of 17 times. Jessie's optimal strategy is to hide the coin in the left hand 10 out of 17 times and in the right hand 7 out of 17 times.

(c) The value of the game is $\frac{2}{17}$. The game favors Jessie.

25. The doctor's optimum strategy would be to prescribe treatment A.

Chapter 10 Review Exercises

1. −3, Amy loses \$3, Grace wins \$3.

3. 6, Amy wins \$6, Grace loses \$6.

5. Sharon (rows) vs. Ron (columns): $\begin{bmatrix} -2 & 1 \\ 5 & 3 \end{bmatrix}$

7. (a) $a_{2,2} = 2$ **(b)** Mike: row 2; Jen: column 2

(c) 2; favors Mike

9. (a) $a_{4,1} = 3$ **(b)** Lindsay: row 4; Mark: column 1

(c) 3; favors Lindsay

11. −8000; don't move

13. (a) Willard Mountain (rows) vs. Gorham Mountain (columns)

Willard Mountain \ Gorham Mountain	2-hour pass	3-hour pass
2-hour pass	−2	4
3-hour pass	0	1
4-hour pass	−1	−2

(b) $a_{2,1} = 0$ is a saddle point.

(c) Willard Mountain: offer a 3-hour pass; Gorham Mountain: offer a 2-hour pass

15. (a) Bookstore (rows) vs. Economy (columns)

Bookstore \ Economy	Strong	Recession
Expand	50,000	−100,000
Don't expand	−75,000	−80,000

(b) $a_{2,2} = -80{,}000$ is a saddle point. **(c)** Don't expand

17. 1.78

19. (a) 0 **(b)** $\frac{5}{4} = 1.25$ **(c)** $-\frac{7}{9} \approx -0.78$ **(d)** $-\frac{3}{8} = -0.375$

Strategy (c) is the most favorable to the column player.

21. $[1 \quad 0]$; $\begin{bmatrix} 0 \\ 1 \end{bmatrix}$

23. (a) $-\frac{1}{3} \approx -0.33$ **(b)** Column player

25. (a) $\frac{1}{7} \approx 0.14$ **(b)** Row player

27. (a) Bookstore (rows) vs. Economy (columns)

Bookstore \ Economy	Strong	Recession
Expand	50	−100
Don't expand	−125	−80

(b) $\begin{bmatrix} \frac{3}{13} & \frac{10}{13} \end{bmatrix}$ **(c)** Don't expand

29. (a) Kris (rows) vs. Rachel (columns)

Kris \ Rachel	All school meetings	Football games
All school meetings	0.55	0.40
Football games	0.45	0.60

(b) $\begin{bmatrix} \frac{1}{2} & \frac{1}{2} \end{bmatrix}$; $\begin{bmatrix} \frac{2}{3} \\ \frac{1}{3} \end{bmatrix}$

31. (a) $\begin{bmatrix} \frac{1}{2} & \frac{1}{2} \end{bmatrix}$; $\begin{bmatrix} \frac{3}{8} \\ \frac{5}{8} \end{bmatrix}$ **(b)** $\frac{11}{2}$

33. (a) $\begin{bmatrix} 0 & \frac{2}{3} & \frac{1}{3} \end{bmatrix}$; $\begin{bmatrix} \frac{1}{3} \\ \frac{2}{3} \end{bmatrix}$ **(b)** $\frac{10}{3}$

35. (a) $[1 \quad 0]$; $\begin{bmatrix} 1 \\ 0 \\ 0 \end{bmatrix}$ **(b)** 2

37. $\begin{bmatrix} \frac{12}{19} & \frac{7}{19} \end{bmatrix}$; $\begin{bmatrix} \frac{14}{19} \\ \frac{5}{19} \end{bmatrix}$

39. $\begin{bmatrix} \frac{2}{21} & \frac{10}{21} & \frac{3}{7} \end{bmatrix}$; $\begin{bmatrix} \frac{10}{21} \\ \frac{19}{42} \\ \frac{1}{14} \end{bmatrix}$

41. (a) $\begin{bmatrix} \frac{2}{7} & \frac{5}{7} \end{bmatrix}$; $\begin{bmatrix} \frac{3}{7} \\ \frac{4}{7} \end{bmatrix}$ **(b)** Yes; the Babcock Company

43. (a) $S = (a + d) - (b + c) = (3 + 1) - [-2 + (-1)] = 4 + 3 = 7$;

$$p_1 = \frac{d - c}{S} = \frac{1 - (-1)}{7} = \frac{2}{7};\ p_2 = 1 - p_1 = 1 - \frac{2}{7} = \frac{5}{7};$$

$$q_1 = \frac{d - b}{S} = \frac{1 - (-2)}{7} = \frac{3}{7};\ q_2 = 1 - q_1 = 1 - \frac{3}{7} = \frac{4}{7}$$

(b) $\frac{1}{7}$

45. $\begin{bmatrix} 0 & \frac{5}{11} & \frac{6}{11} \end{bmatrix}$; treatment C

Chapter 11

Section 11.1

1. The ordered pairs are (−2, 3), (1, 5), and (4, 4). The scatterplot of the points is:

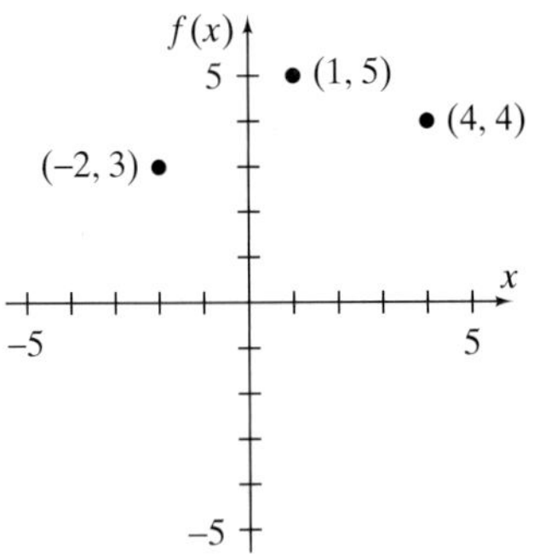

3. The ordered pairs are (−4, −5), (−3.5, −6), (−2, −8), and (2, −11). The scatterplot is:

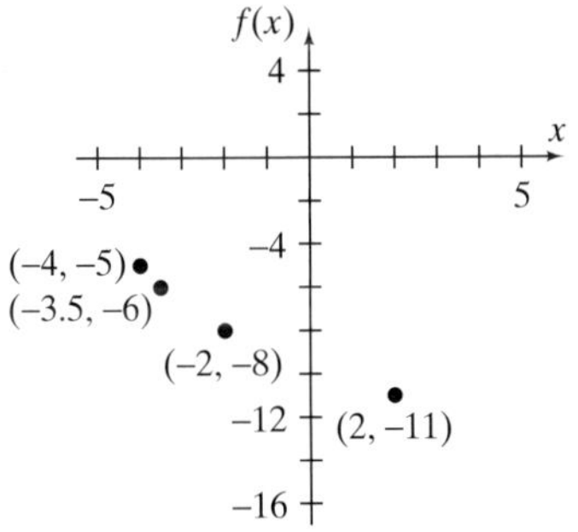

5.

x	y
2	6
4	5
6	4
8	2
10	0

The ordered pairs are (2, 6), (4, 5), (6, 4), (8, 2), and (10, 0).

7.

x	y
10	60
20	50
30	40
40	20
50	10

The ordered pairs are (10, 60), (20, 50), (30, 40), (40, 20), and (50, 10).

9.

x	y
−3	−30
−2	−10
−1	0
0	10
2	10

The ordered pairs are (−3, −30), (−2, −10), (−1, 0), (0, 10), and (2, 10).

11.

Years Since 1980	Expenditures at Public Colleges and Universities (in billions of dollars)
0	88
5	99
10	125
13	133
16	141
19	154

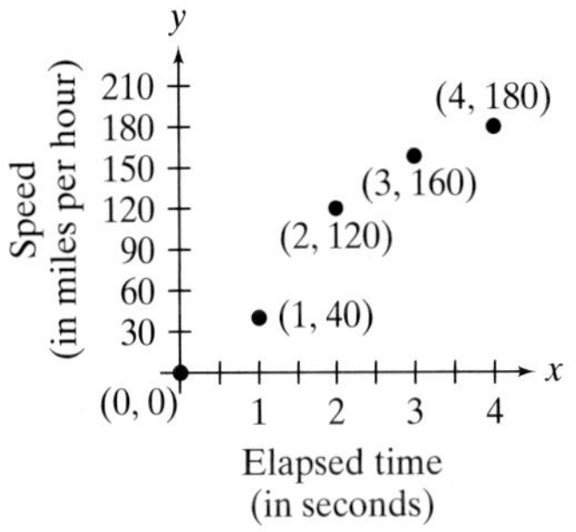

13.

Time	Speed (in miles per hour)
0	0
1	40
2	120
3	160
4	180

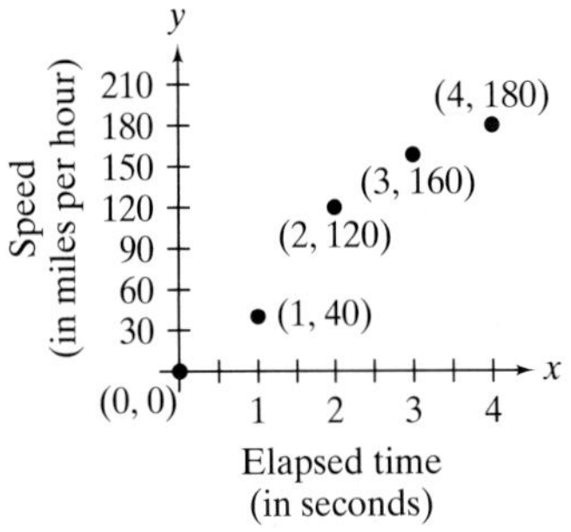

15. Function

17. Not a function

19. Function

21. Not a function

23. Function **25.** (2, 3)

27. (1, 1) **29.** (3, 9)

31. (0, 0)

33. $\left(\frac{1}{2}, \frac{1}{4}\right)$

35. (−0.25, 0.0625)

37. **(a)** t **(b)** $y = f(t)$ **(c)** 10 **(d)** 20 **(e)** $[0, \infty)$ **(f)** $[0, \infty)$

39.

Interval Notation	Inequality Notation	Number Line
$[-\frac{1}{2}, 5)$	$-\frac{1}{2} \le x < 5$	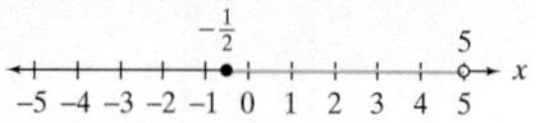
$[-2, \infty)$	$x \ge -2$	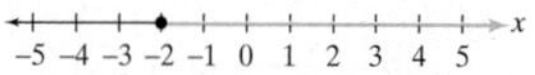
$(-\infty, -3)$	$x < -3$	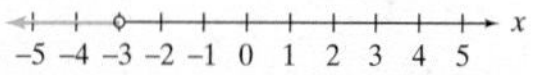
$(-1, 10)$	$-1 < x < 10$	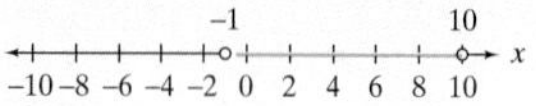
$(-\infty, 3) \cup (3, \infty)$	$x < 3$ or $x > 3$	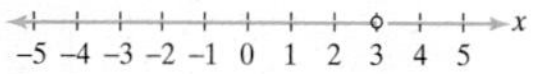

41. $(-\infty, \infty)$ **43.** $(-\infty, \infty)$ **45.** $(-\infty, 5) \cup (5, \infty)$

47. $(-\infty, 6]$ **49.** $(-\infty, \infty)$ **51.** $(-\infty, 0) \cup (0, 2) \cup (2, \infty)$

53. The domain is $(-\infty, \infty)$; the range is $[\frac{1}{2}, \infty)$.

55. The domain is $(-\infty, \infty)$; the range is $(-0.43, 0.1)$.

57. **(a)**

x	0	5	10	15	20	25	30	35
$p(x)$	41.25	40	38.75	37.5	36.25	35	33.75	32.50

(b)

(c) $[32.5, 41.25]$

(d) If 22 units of Teddy Bear Designer Lingerie are sold per day, the price will be $35.75 per unit.

59. $36.75 **61.** The domain is $(-\infty, \infty)$; the range is $(-\infty, \infty)$.

63. The domain is $(-5, \infty)$; the range is $(-\infty, 7]$.

65. The domain is $(-50, 100]$; the range is $[0, 10)$.

Section 11.2

1. $C(300) = 2150$. The cost of producing 300 quick-replace spokes is $2150.

3. **(a)** $C(30) = 22{,}720.5$. The cost of producing 30 units is $22,720.50.

(b) 231 units can be produced at a cost of $24,197.85.

5. **(a)** $C(110) = 14{,}453.5$. The cost of producing 110 units is $14,453.50.

(b) 440 units can be made at a cost of $16,714.

7. $0 \le x \le 18$ **9.** $1 \le x \le 6$ **11.** $1 \le x \le 100$

13. Since x represents the number of years since 1979, the model is defined for the years 1980 to 1999.

15. Since x represents the number of years since 1960, the model is defined for the years 1960 to 1995.

17. Since x represents the number of years since 1979, the model is defined for the years 1980 to 1995.

19. **(a)** $f(12) = 72.17$. This means that in 1984 the annual total expenditures on pollution abatement in the U.S. (in constant 1987 dollars) was about $72.17 billion.

(b) The year 1997 corresponds to an x-value of 25, which is not in the reasonable domain of the model.

21. **(a)** $f(20) = 408.22$; this means that in 1994 the number of milligrams of cholesterol consumed each day per person was about 408.22.

(b) The year 1974 corresponds to an x-value of zero which is not in the reasonable domain of the model.

23. **(a)** $f(20) = 6.3$; this means that in 1995 the percentage of people taking the SAT whose intended area of study was engineering was about 6.3%.

(b) The year 1999 corresponds to an x-value of 24 which is not in the reasonable domain of the model.

25. $x = 5$; in 1977 the annual total expenditures on pollution abatement was about $58.17 billion.

27. $x = 7$; in 1993 the total payment by U.S. travelers to foreign countries was about $53.27 billion.

29. $t = 15$; since 15 is not in the reasonable domain, we conclude that there is no year in which the average weekly earnings was about $599.36.

Section 11.3

1. -5 **3.** $-\frac{5}{6}$ **5.** 1.1 **7.** 1 **9.** $-\frac{5}{3}$ **11.** -4 **13.** $\frac{17}{4}$

15. $y - 6 = -3(x - 2)$

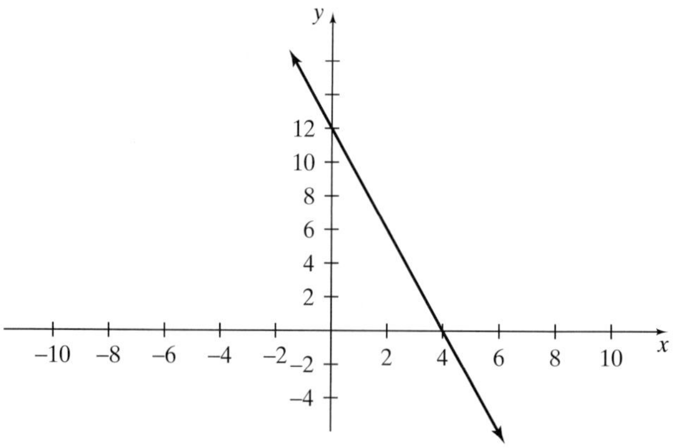

17. $y - 5 = 4(x - 3)$

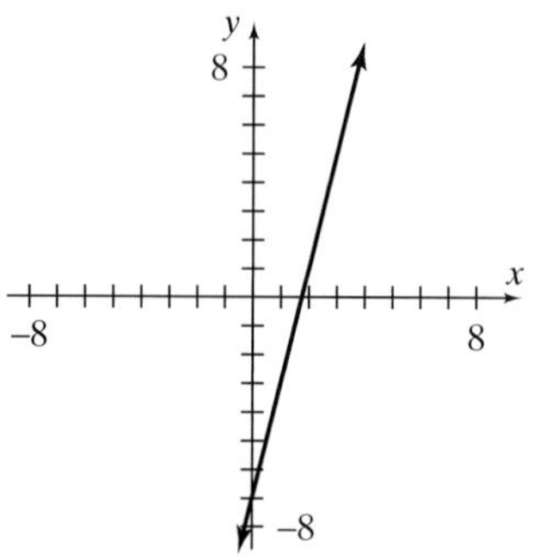

19. $x = 0$

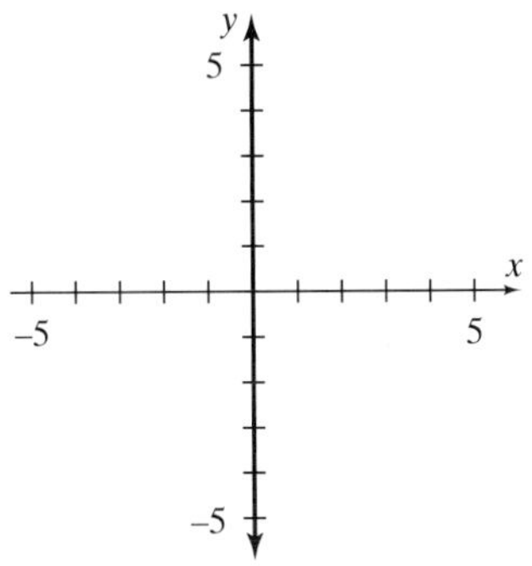

21. $x = -2$

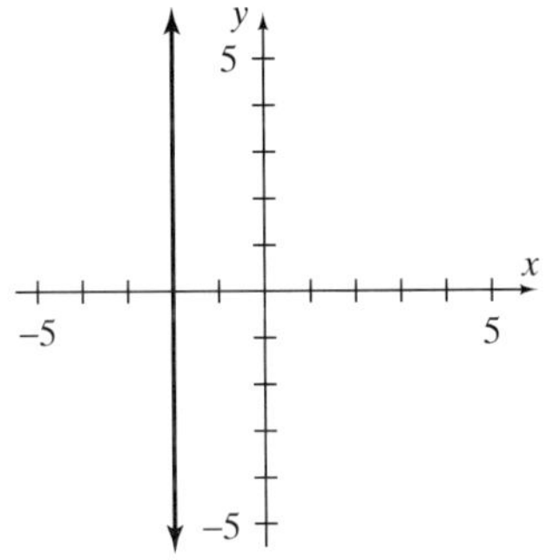

23. $y = 3(x - 8) + 4$ **25.** $y = 5(x - 0) - 13$

27. $y = 0.3(x + 4)$ **29.** $y = -\frac{1}{4}(x + 1) + 2$

31. $y = \frac{1}{8}(x + 12)$ **33.** $y = -1.175(x - 0) + 6$

35. x-intercept: $(1.25, 0)$; y-intercept: $(0, 5)$

37. x-intercept: $(400, 0)$; y-intercept: $(0, 120)$

39. x-intercept: $(-44.9, 0)$; y-intercept: $(0, 8.98)$

41. **(a)** $f(x) = -1250x + 25{,}000$ **(b)** Twelve years

43. **(a)** $f(x) = -6000x + 90{,}000$ **(b)** 7.5 years

45. The slope of the line is -4.5; decreasing.

47. The slope of the line is $+4$; increasing.

49. The slope of the line is $+4$; increasing.

51. Slope $= \frac{1}{3}$; increasing **53.** Slope $= 14$; increasing

55. $C(x) = 147.75x + 2700$ **57.** $C(x) = 575x + 1250$

59. $C(x) = 80x + 650$

61. **(a)** The fixed cost is \$550 and the variable costs are \$1.25 per bottle.
(b) $C(70) = 637.5$; it costs \$637.50 to produce 70 bottles.
(c) 115 bottles **(d)** \$1.25 per bottle **(e)** \$1.25

63.

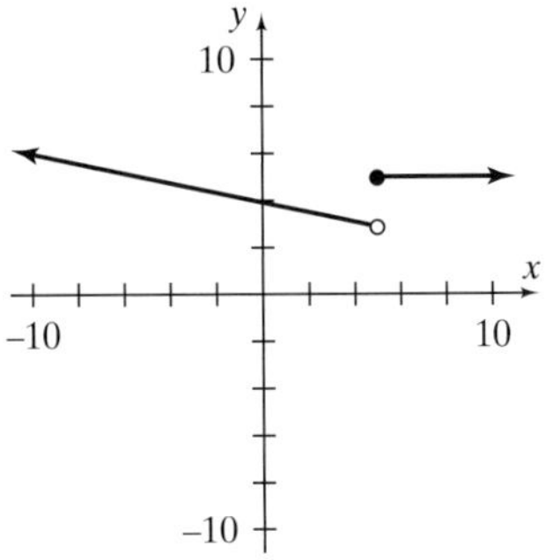

65.

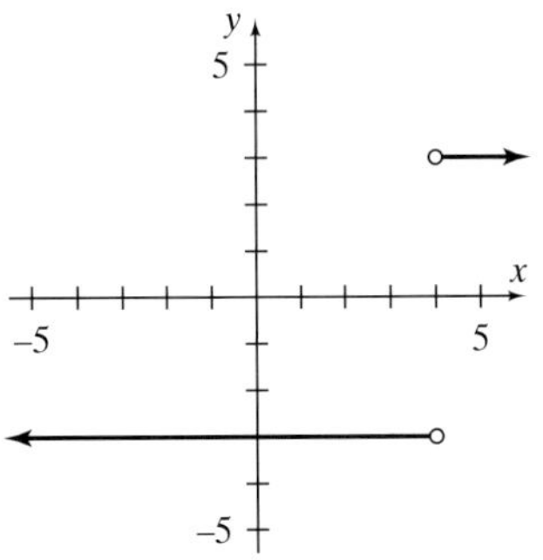

67.

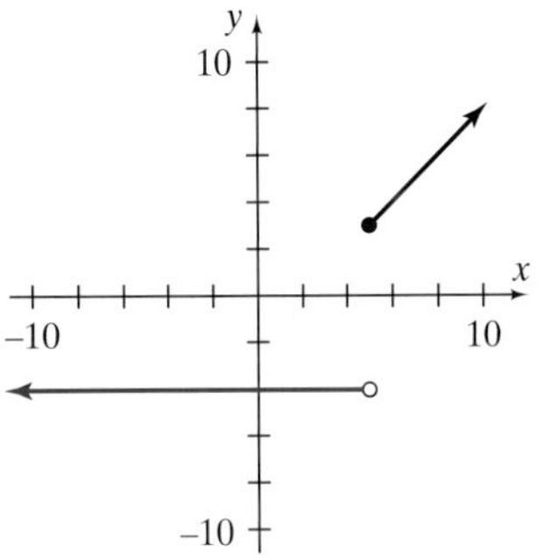

69.

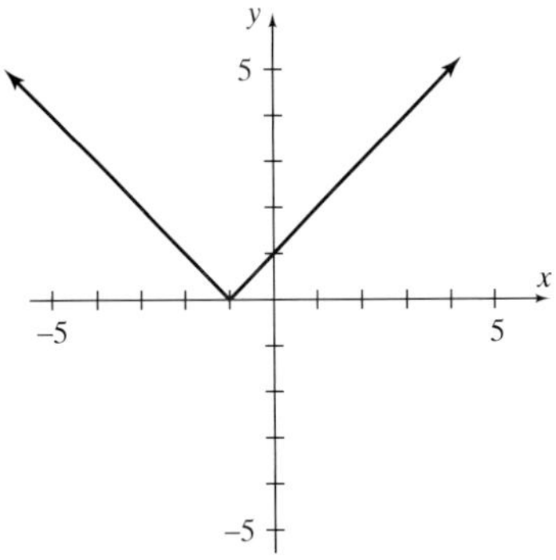

$$f(x) = |x + 1| = \begin{cases} -x - 1 & x < -1 \\ x + 1 & x \geq -1 \end{cases}$$

71.

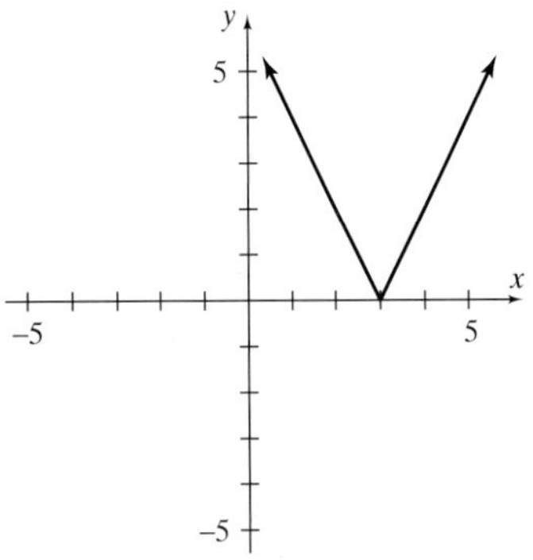

$$g(x) = |6 - 2x| = \begin{cases} 6 - 2x & x < 3 \\ -6 + 2x & x \geq 3 \end{cases}$$

73.

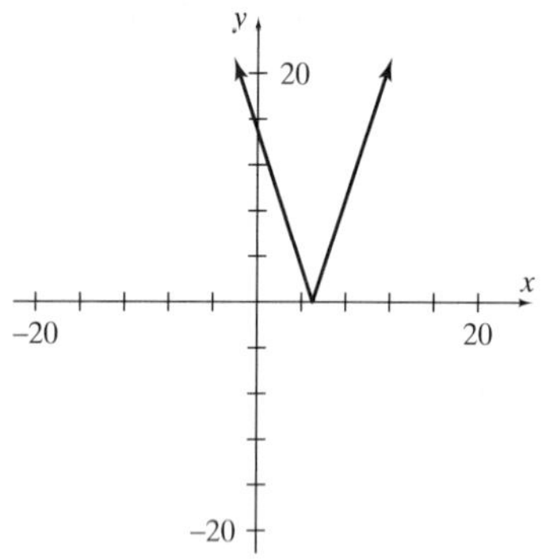

$g(x) = |3x - 15|$

$= \begin{cases} -3x + 15 & x < 5 \\ 3x - 15 & x \geq 5 \end{cases}$

75. **(a)** The fixed costs are \$20. The variable costs are \$0.15 per mile.
(b) $C(x) = 0.15x + 20$ **(c)** $(0, \infty)$
(d) $C(120) = 38$; it cost \$38 to rent a truck and drive 120 miles.
(e) \$0.15; this is called the marginal cost.

77. **(a)** Since the fixed cost is equal to $C(0)$, then the fixed cost is \$50.
(b) It costs \$15 per hour for each additional hour of labor. This is called the marginal cost.
(c) $C(x) = 15x + 50$ **(d)** \$117.50

79. **(a)** Spending is increasing at a rate of \$0.19 billion per year.
(b) \$2.24 billion
(c) \$4.14 billion
(d) \$1.9 billion more was spent in 1995 than in 1985.

Section 11.4

1. **(a)** $(-3, -4)$
(b) Increasing on $(-3, \infty)$ and decreasing on $(-\infty, -3)$

3. **(a)** $(3, 15)$
(b) Increasing on $(-\infty, 3)$ and decreasing on $(3, \infty)$

5. **(a)** $(-0.6, -4.8)$
(b) Increasing on $(-0.6, \infty)$ and decreasing on $(-\infty, -0.6)$

7. **(a)** $\left(-\frac{25}{14}, -\frac{209}{280}\right) \approx (-1.79, -0.75)$
(b) Increasing on $\left(-\frac{25}{14}, \infty\right)$ and decreasing on $\left(-\infty, -\frac{25}{14}\right)$

9. **(a)** $(-1, 1.4)$
(b) Increasing on $(-\infty, -1)$ and decreasing on $(-1, \infty)$

11. $x = 4$ or $x = -4$ **13.** $x = 0$ or $x = -2$ **15.** $x = 2$ or $x = 3$

17. $x = \frac{10}{3}$ or $x = -2.5$ **19.** $x = \dfrac{1 + \sqrt{5}}{2}$ and $x = \dfrac{1 - \sqrt{5}}{2}$

21. $x = \dfrac{7 + \sqrt{5}}{22}$ and $x = \dfrac{7 - \sqrt{5}}{22}$ **23.** No real roots

25. $x = -4$ and $x = 6$ **27.** $x = 1.5$ **29.** No real zeros

31. **(a)** -5 **(b)** -5 **(c)** -5

33. **(a)** 5 **(b)** 3 **(c)** 7 **35.** **(a)** 3 **(b)** 2 **(c)** 1.5

37. 101; this means that during the five-year period, the number of parking tickets issued increased at an average rate of $101 \frac{\text{tickets}}{\text{year}}$.

39. **(a)** $s(3) = 96$ feet; this is the height of the rock 3 seconds after it was thrown.
(b) 16 feet per second; the rock traveled an average speed of 16 feet per second over the time interval $[1, 3]$.

41. **(a)** $g(3) = 2033$; this means that after 3 hours, there were 2033 bacteria in the colony.
(b) 17 bacteria per hour; the bacteria were growing at an average rate of 17 bacteria per hour over the time interval $[3, 6]$.

43. **(a)** $f(2) = 374.27$; this means that in 1988, there were 374.27 aggravated assaults per 100,000 people.
(b) The greatest assault rate occurred some time during the middle of 1992.
(c) 0.31 aggravated assaults per 100,000 people per year

45. **(a)** $f(8) = 419.74$; this means that in 1982, the number of milligrams of cholesterol consumed each day per person in the United States was about 419.74 milligrams.
(b) Vertex at about (18, 408); this means that the lowest amount of cholesterol consumed each day per person was about 408 milligrams, and this occurred in 1992.
(c) $m_{\text{sec}} = -2.28$; this means that from 1975 to 1989, the number of milligrams of cholesterol consumed each day per person in the United States decreased at an average rate of $2.28 \frac{\text{milligrams}}{\text{year}}$.
(d) $m_{\text{sec}} = 0.58$; this means that from 1993 to 1997, the number of milligrams of cholesterol consumed each day per person in the United States increased at an average rate of $0.58 \frac{\text{milligrams}}{\text{year}}$.

47. **(a)** $f(22) = 337.96$; this means that in 1982, there were about 337.96 thousand lawyers in private practice in the U.S.
(b)

x	0	5	10	15	25	30
$f(x)$	190.34	192.44	213.04	252.14	385.84	480.44

(c) $m_{\text{sec}} = 9.67$; this means that from 1970 to 1980, the number of lawyers in private practice in the U.S. increased at an average rate of $9.67 \frac{\text{thousand lawyers}}{\text{year}}$.
(d) $m_{\text{sec}} = 17.07$; this means that from 1980 to 1990, the number of lawyers in private practice in the U.S. increased at an average rate of $17.07 \frac{\text{thousand lawyers}}{\text{year}}$. This rate is about 7.4 thousand lawyers per year more than from 1970 to 1980.

Section 11.5

1. **(a)** $(f + g)(x) = 10x + 2$; the domain is $(-\infty, \infty)$.
(b) $(f - g)(x) = 2x + 4$; the domain is $(-\infty, \infty)$.
(c) $(f \cdot g)(x) = 24x^2 + 6x - 3$; the domain is $(-\infty, \infty)$.
(d) $\left(\dfrac{f}{g}\right)(x) = \dfrac{6x + 3}{4x - 1}$; the domain is $(-\infty, 0.25) \cup (0.25, \infty)$.

3. **(a)** $(f + g)(x) = x^2 + \sqrt{x + 5} + 5$; the domain is $[-5, \infty)$.
(b) $(f - g)(x) = -x^2 + \sqrt{x + 5} - 5$; the domain is $[-5, \infty)$.
(c) $(f \cdot g)(x) = x^2\sqrt{x + 5} + 5\sqrt{x + 5}$; the domain is $[-5, \infty)$.
(d) $\left(\dfrac{f}{g}\right)(x) = \dfrac{\sqrt{x + 5}}{x^2 + 5}$; the domain is $[-5, \infty)$.

5. 1 **7.** 336 **9.** -11 **11.** $\frac{15}{13}$

13. **(a)** −245; the Handi-Neighbor Hardware Store makes \$245 less than the U-Do-It Store when they both sell 20 hammers.

(b) 635; both stores will make \$635 combined if each sells 20 hammers.

15. **(a)** 57.53% **(b)** $\left(\dfrac{x+35}{4x+45}\right)\cdot 100\%$

17. **(a)**

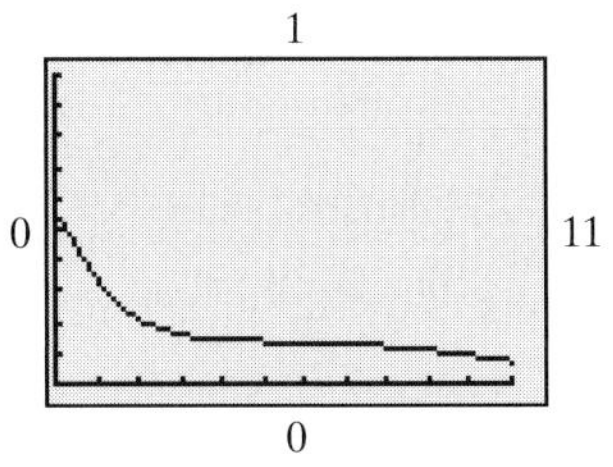

(b) $[1, 3]$

19. **(a)**

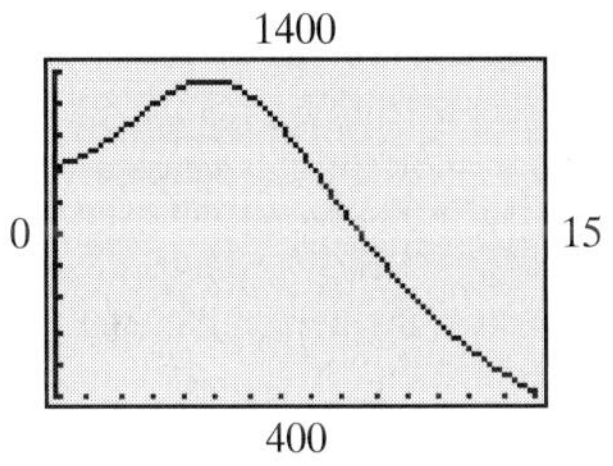

(b) 1989

21. $R(x) = 87.1x$ **23.** $R(x) = -0.19x^2$ **25.** $R(x) = \dfrac{-x^2}{2000} + 3x$

27. $R(x) = -0.12x^2 + 30x$ **29.** **(a)** (625, 5000) **(b)** \$5000

31. **(a)** (0.05, 5.00) and (79.95, 1603) **(b)** \$5 or \$1603

33. **(a)**

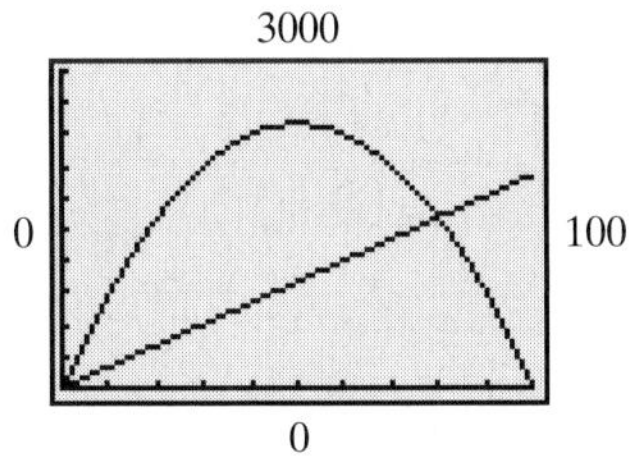

(b) (0.05, 5.00) and (79.95, 1603)
(c) \$5 or \$1603

35. **(a)**

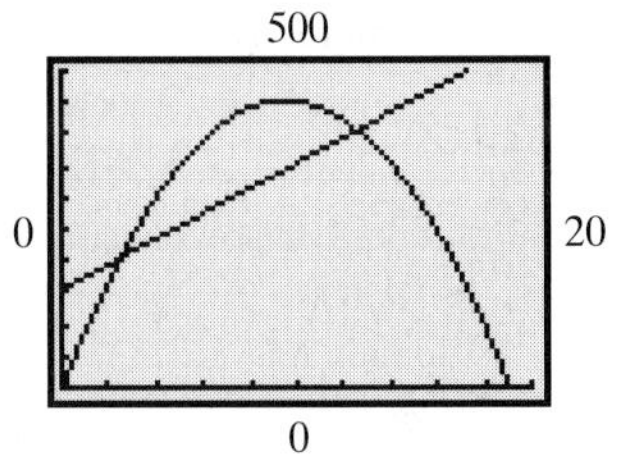

(b) (2.47, 204.31) and (12.57, 406.49)
(c) \$204.31 or \$406.49

37. **(a)** $P(x) = -37x^2 + 1052x - 4300$

(b) $x \approx 4.9$ or $x \approx 23.5$; this means that when either 4.9 or 23.5 units are produced, the break-even point is reached.

(c) Vertex at about (14.2, 3177.7); this means that the maximum profit will be about \$3177.70 when the production level is at about 14.2 units.

39. **(a)** $P(x) = -2.1x^2 + 420x - 9500$

(b) $x \approx 26$ or $x \approx 174$; this means that when either 26 or 174 items are produced, the break-even point is reached.

(c) Vertex at about (100, 11,500); this means that the maximum profit will be about \$11,500 when the production level is at about 100 units.

41. **(a)** $P(x) = -x^2 + 8.8x - 10.15$

(b) (4.4, 9.21); maximum profit of \$9.21 is attained when $x = 4.4$.

43. **(a)** $P(x) = -10x^2 + 540x - 6650$

(b) (27, 640); maximum profit of \$640 is attained when $x = 27$.

45. **(a)** $P(x) = -x^2 + 25x - 60$

(b) (12.5, 96.25); maximum profit of \$96.25 is attained when $x = 12.5$.

47. $x \approx 1.37$ and $x \approx 7.43$; this means that the break-even points occur at production levels of about 1.37 and 7.43 units.

49. $x \approx 2.69$ and $x \approx 22.31$; this means that the break-even points occur at production levels of about 2.69 and 22.31 units.

51. Polynomial function; linear

53. Not a polynomial function

55. Polynomial function; cubic

57. As $x \to -\infty$, $f(x) \to \infty$; as $x \to \infty$, $f(x) \to -\infty$.

59. As $x \to -\infty$, $g(x) \to -\infty$; as $x \to \infty$, $g(x) \to \infty$.

61. As $x \to -\infty$, $g(x) \to \infty$; as $x \to \infty$, $g(x) \to \infty$.

63. As $x \to -\infty$, $f(x) \to -\infty$; as $x \to \infty$, $f(x) \to \infty$.

65. As $x \to -\infty$, $g(x) \to \infty$; as $x \to \infty$, $g(x) \to \infty$.

67. **(a)** 44; the ball is 44 feet above the ground after 2 seconds.

(b) −10; the average rate of change of height in the first 2.5 seconds is −10 feet per second.

69. **(a)** 2664.38; the sale of 250 Boomer headphones generates \$2664.38 in revenue.

(b) 10.4521; during sales in the range of 100 to 300 Boomer headphones, the revenue increased by an average of \$10.45 per Boomer headphone.

71. **(a)** $f(7) = 20.06$; this means that in 1982, about 20.06% of the people taking the SAT intended to study business.

(b) $m_{\text{sec}} = 0.65$; this means that from 1977 to 1987, the percent of people taking the SAT who intended to study business increased at an average rate of about 0.65 percent per year.

(c) $m_{\text{sec}} = -0.95$; this means that from 1987 to 1997, the percent of people taking the SAT who intended to study business decreased at an average rate of about 0.95 percent per year.

(d) In the period from 1977 to 1987, the average rate of change in the percent of people taking the SAT who intended to study business increased, while from 1987 to 1997, the average rate of change decreased.

73. **(a)** As $x \to \infty$, $f(x) \to -\infty$ and as $x \to -\infty$, $f(x) \to -\infty$.

(b) f has a peak at (−5, 46) and has no valleys.

(c) f is increasing on $(-\infty, -5)$ and decreasing on $(-5, \infty)$.

75. **(a)** As $x \to \infty$, $f(x) \to -\infty$ and as $x \to -\infty$, $f(x) \to -\infty$.

(b) f has a peak at about (1.26, 19.56) and has no valleys.

(c) f is increasing on $(-\infty, 1.26)$ and decreasing on $(1.26, \infty)$.

77. **(a)** As $x \to \infty$, $f(x) \to \infty$ and as $x \to -\infty$, $f(x) \to \infty$.

(b) f has a valley at (3, −2) and has no peaks.

(c) f is decreasing on $(-\infty, 3)$ and increasing on $(3, \infty)$.

79. **(a)** As $x \to \infty$, $f(x) \to \infty$ and as $x \to -\infty$, $f(x) \to \infty$.
(b) f has valleys at $(-1, -\frac{1}{2})$ and at $(1, -\frac{1}{2})$ and a peak at $(0, 0)$.
(c) f is decreasing on $(-\infty, -1)$ and $(0, 1)$ and is increasing on $(-1, 0)$ and $(1, \infty)$.

Section 11.6

1. **(a)** $(-\infty, 0) \cup (0, \infty)$ **(b)** Hole at $x = 0$

3. **(a)** $(-\infty, 2) \cup (2, \infty)$ **(b)** Vertical asymptote at $x = 2$

5. **(a)** $(-\infty, -2) \cup (-2, \infty)$ **(b)** Hole at $x = -2$

7. **(a)** $(-\infty, -3) \cup (-3, -1) \cup (-1, \infty)$
(b) Hole at $x = -1$ and a vertical asymptote at $x = -3$

9. **(a)** $(-\infty, -1.5) \cup (-1.5, 2) \cup (2, \infty)$
(b) Hole at $x = -1.5$ and a vertical asymptote at $x = 2$

11. No horizontal asymptote

13. $y = 3$ **15.** $y = 3$

17. $y = 1$ **19.** $y = 2$

21. y-intercept is $(0, 0)$; x-intercept is $(0, 0)$

23. y-intercept is $(0, -\frac{3}{4})$; x-intercept is $(0.5, 0)$

25. y-intercept is $(0, -0.5)$; x-intercept is $(-2, 0)$

27. y-intercept is $(0, -\frac{2}{9})$; x-intercept is $(2, 0)$

29. y-intercept is $(0, \frac{4}{3})$; x-intercepts are $(-2, 0)$ or $(2, 0)$

31. y-intercept is $(0, 5)$; no x-intercepts

33. y-intercept is $(0, -\frac{2}{3})$; x-intercept is $\approx (0.632, 0)$

35. y-intercept is $(0, 0.5)$; x-intercept is $\approx (-0.657, 0)$

37.

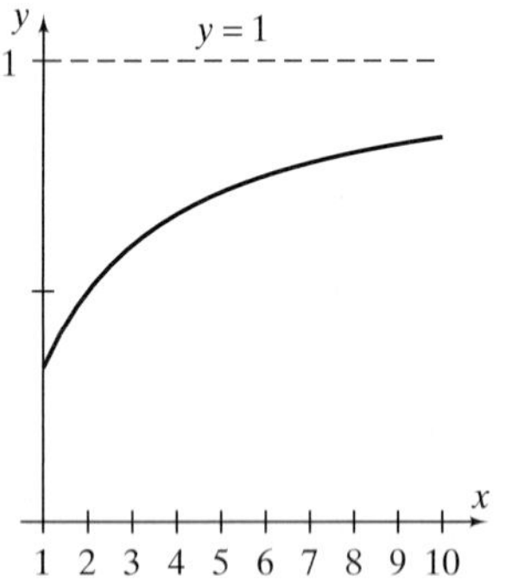

39.

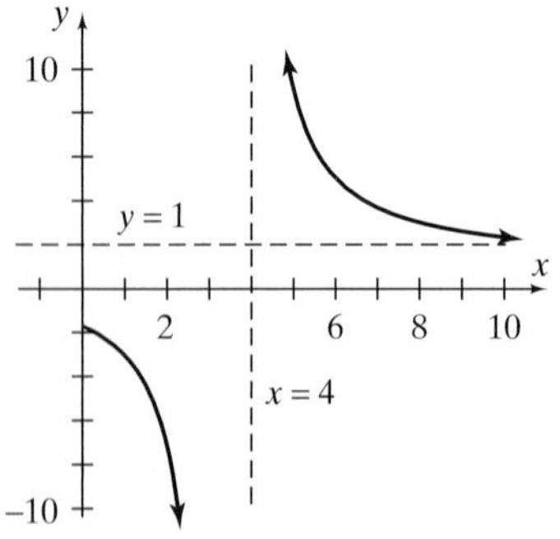

41.

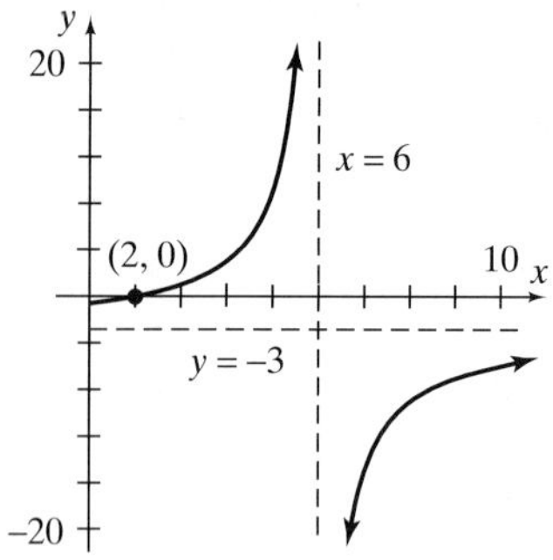

43. **(a)** $f(85) = 170$; it will cost \$170 million to remove 85% of the pollutants.

(b)

x	5	50	70	90	95
$f(x)$	≈ 1.6	30	70	270	570

(c) \$29,400,000,000

(d) $f(100)$ is undefined.

45. **(a)** $f(10) = 18.2$; after 10 days, the person had about 18 out of 20 items remembered.
(b) $f(100) = 19.82$; after 100 days, the person had about 20 out of 20 items remembered.

47. **(a)** $S(10) \approx 59.44$; if the company spends \$10,000 on advertising, the income from the sales will be about \$5,944,000.

(b)

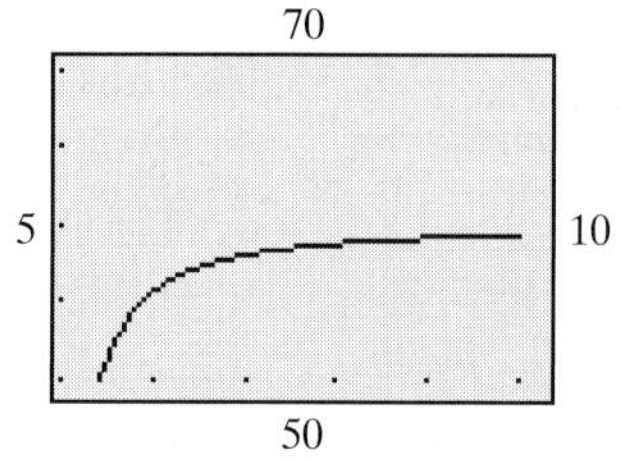

(c) The function has a horizontal asymptote of $y = \frac{120}{2}$ or $y = 60$. This means that the most income the company can generate by advertising is \$6,000,000.

49. $f(x) = (2x + 3)^{1/2}$; $[-1.5, \infty)$

51. $f(x) = (5x - 8)^{1/3}$; $(-\infty, \infty)$

53. $g(x) = (6x + 1)^{5/4}$; $[-\frac{1}{6}, \infty)$

55. $f(x) = 3(7x - 2)^{-1/2}$; $(\frac{2}{7}, \infty)$

57. $f(x) = \sqrt{8x - 9}$; $[1.125, \infty)$

59. $f(x) = \sqrt[3]{7x + 9}$; $(-\infty, \infty)$

61. $f(x) = \sqrt{(4x - 5)^3}$; $[1.25, \infty)$

63. $g(x) = \frac{1}{\sqrt[3]{7x-2}}$; $(-\infty, \frac{2}{7}) \cup (\frac{2}{7}, \infty)$

65. **(a)** $g(70) \approx 11.71$; if the tower is 70 feet tall, an observer can see about 11.71 miles into the forest.
(b) ≈ 429 feet high

67. (a) $g(300) \approx 188$; there are about 188 plant species in 300 square miles of rain forest.

(b) 0.0599; from 1728 square miles to 2197 square miles the average increase in plant species is about 0.0599 species per square mile.

69. (a) $f(3) \approx 16{,}787$; in 1992, there were about 16,787 grocery stores in the United States.

(b) Between 1992 and 1995, the number of grocery stores decreased at an average rate of 107.55 stores per year.

71. (a) $f(19) \approx 26.62$; this means that in 1998, there were about 26.62 million retired workers in the U.S. that received Social Security benefits.

(b) $m_{\text{sec}} \approx 0.19$; this means that from 1991 to 1998, the number of retired workers in the U.S. that received Social Security benefits increased at an average rate of about $0.19 \frac{\text{million people}}{\text{year}}$.

73. (a)

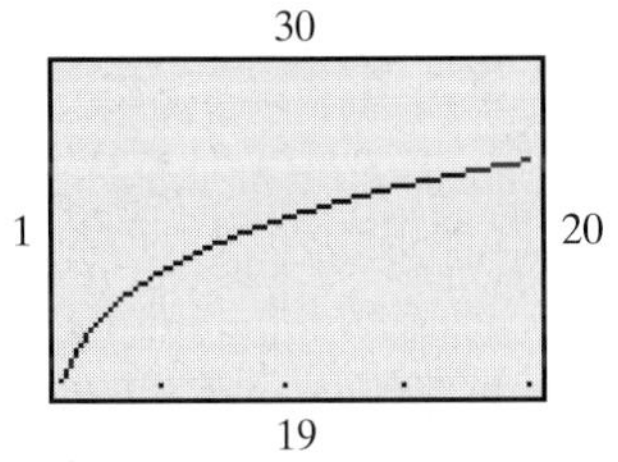

From the graph, we see that as x increases, so does $f(x)$, so the number of retired workers in the U.S. that received Social Security benefits is generally increasing.

(b) From the graph, f is 25 when $x \approx 10.85$. So the year that the number of retired workers in the U.S. that received Social Security benefits exceeded 25 million was 1989.

Section 11.7

1. (a)

x	−5	−4	−3	−2	−1	0
$(\frac{1}{3})^x$	243	81	27	9	3	1

x	1	2	3	4	5
$(\frac{1}{3})^x$	$\frac{1}{3}$	$\frac{1}{9}$	$\frac{1}{27}$	$\frac{1}{81}$	$\frac{1}{243}$

(b)

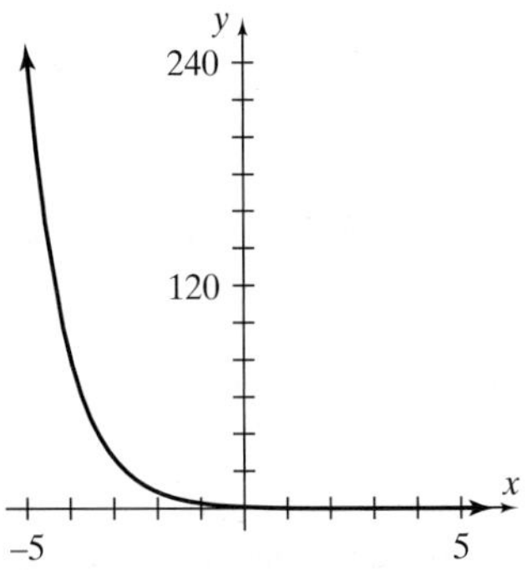

The function is an exponential decay function.

3. (a)

x	−5	−4	−3	−2	−1	0
4^x	$\frac{1}{1024}$	$\frac{1}{256}$	$\frac{1}{64}$	$\frac{1}{16}$	$\frac{1}{4}$	1

x	1	2	3	4	5
4^x	4	16	64	256	1024

(b)

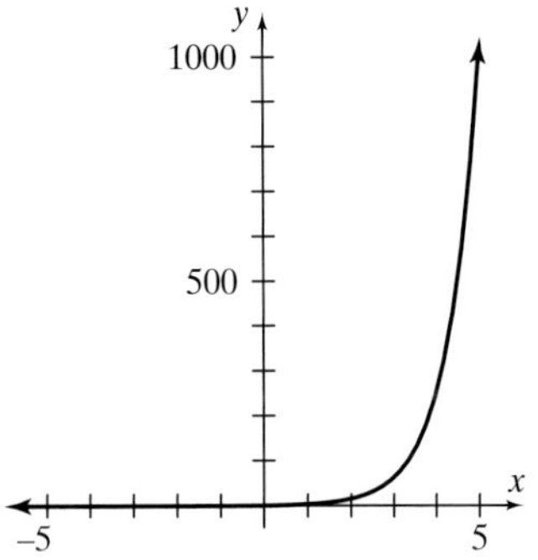

The function is an exponential growth function.

5. (a) All values are rounded to the nearest hundredth or to 2 significant digits.

x	−5	−4	−3	−2	−1	0
$(2.3)^x$	0.016	0.036	0.082	0.19	0.43	1

x	1	2	3	4	5
$(2.3)^x$	2.3	5.29	12.17	27.98	64.36

(b)

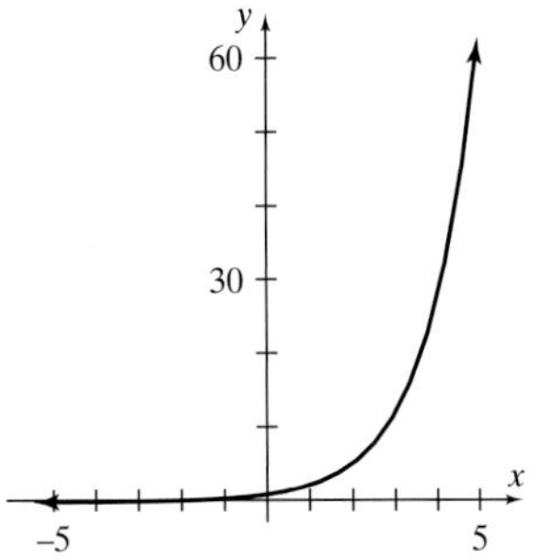

The function is an exponential growth function.

7. (a) All values are rounded to the nearest hundredth.

x	−5	−4	−3	−2	−1	0
$(0.7)^x$	5.95	4.16	2.92	2.04	1.43	1

x	1	2	3	4	5
$(0.7)^x$	0.7	0.49	0.34	0.24	0.17

(b)

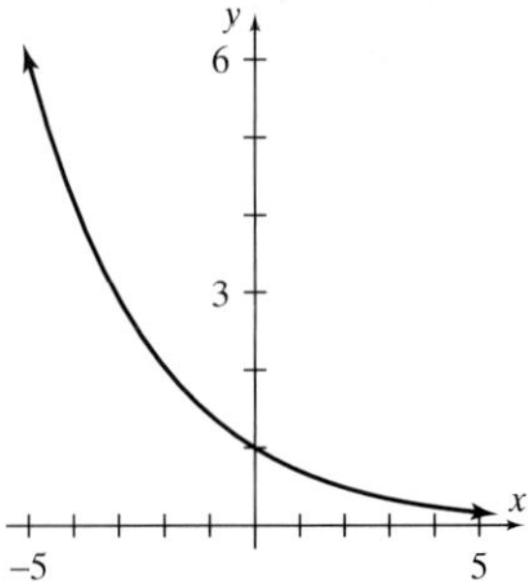

The function is an exponential decay function.

9. (a) All values are rounded to the nearest hundredth or to 2 significant digits.

x	-5	-4	-3	-2
e^{2x}	0.000045	0.00034	0.0025	0.018

x	-1	0	1	2
e^{2x}	0.14	1	7.39	54.60

x	3	4	5
e^{2x}	403.43	2980.96	22,026.47

(b)

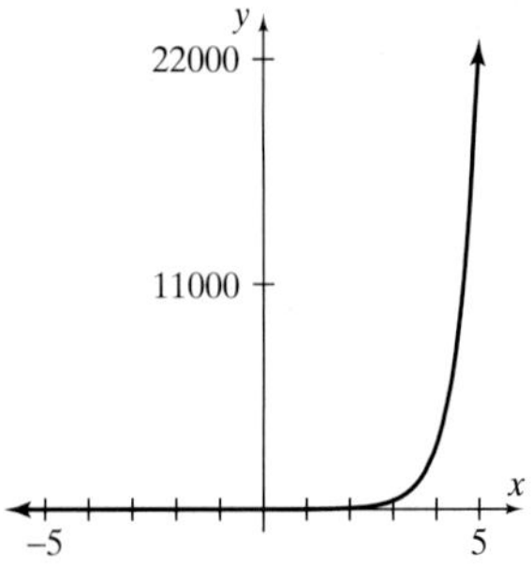

The function is an exponential growth function.

11. (a) All values are rounded to the nearest hundredth.

x	-5	-4	-3	-2	-1	0
$e^{0.3x}$	0.22	0.30	0.41	0.55	0.74	1

x	1	2	3	4	5
$e^{0.3x}$	1.35	1.82	2.46	3.32	4.48

(b)

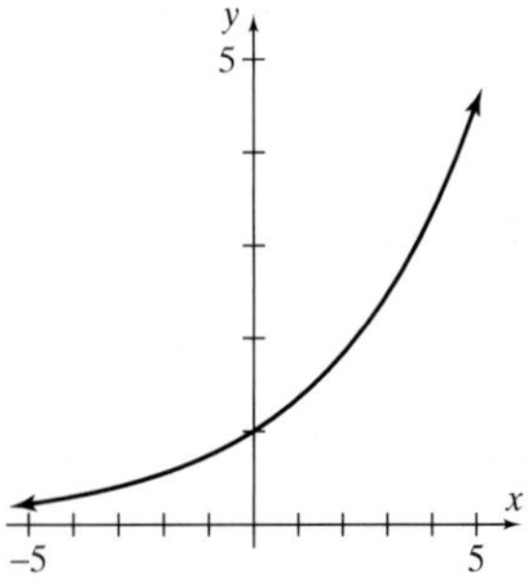

The function is an exponential growth function.

13. (a) All values are rounded to the nearest hundredth or 2 significant digits.

x	-5	-4	-3	-2
$e^{-1.6x}$	2980.96	601.85	121.51	24.53

x	-1	0	1	2
$e^{-1.6x}$	4.95	1	0.20	0.041

x	3	4	5
$e^{-1.6x}$	0.0082	0.0017	0.00034

(b)

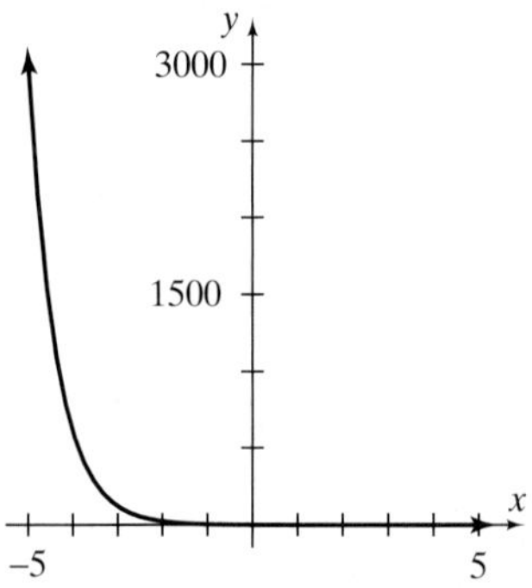

The function is an exponential decay function.

15. (a) $f(7) \approx 41.94$; on the seventh day, the wound is about 42 cm^2.

(b) On the third day, the wound is approximately half its original size.

(c) $m_{\text{sec}} \approx -18.06$; between the first and eighth days, the wound was shrinking at an average rate of 18.06 square centimeters per day.

17. (a) $g(5) = 4237$; by the fifth day, 4237 students had heard the rumor.

(b) $m_{\text{sec}} = 779.15$; between the first and fifth days, the rumor was spreading at an average rate of 779.15 students per day.

19. (a)

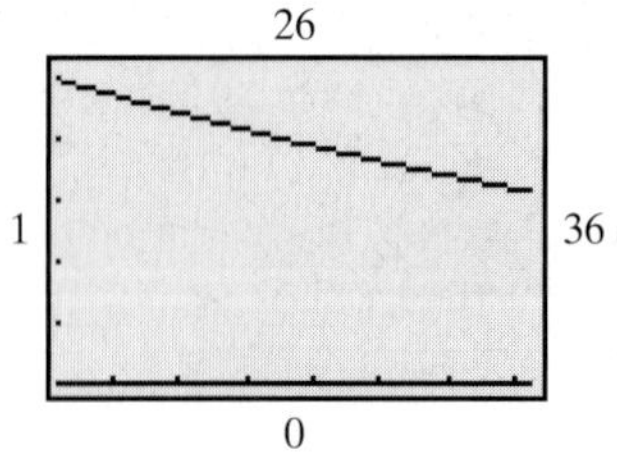

(b) When $f(x) = 21$, $x \approx 14.36$; this means that in 1973 the pupil-student ratio became less than 21 $\frac{\text{pupils}}{\text{teacher}}$.

21. (a) $f(2) \approx 17.04$; this means that in 1992, the percentage of waste generated that is yard waste was about 17.04%.

(b) $m_{\text{sec}} \approx -0.90$; this means that from 1991 to 1996, the percentage of waste generated that is yard waste decreased at an average rate of 0.90 percent per year.

23. (a)

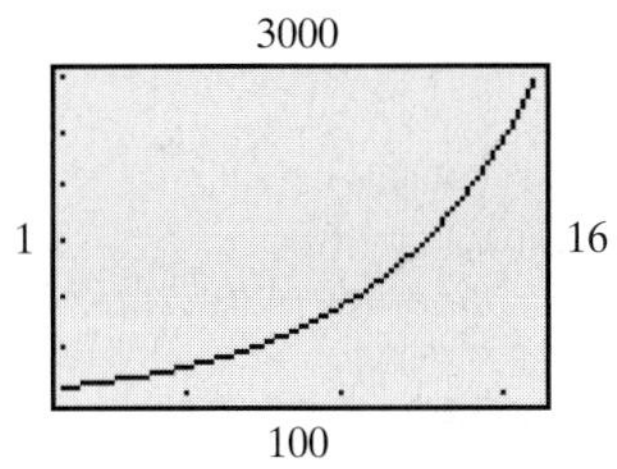

(b) When $f(x) = 1000$, $x \approx 10.52$; this means that the first year that the annual total assets in mutual funds exceeded \$1000 billion was 1989.

25. \$520 **27.** \$12,915.78 **29.** \$1374.55

31. (a) \$9385.69 **(b)** \$9560.92 **(c)** \$9573.82

33. (a) \$14,462.52 **(b)** \$14,589.82 **(c)** \$14,655.81

35. $A(5) = \$8834.69$; \$2434.69 in interest made

37. $A(5) = \$10{,}242.97$; \$2742.97 in interest made

39. (a) $A(t) = 4000\left(1 + \frac{0.0575}{12}\right)^{12t}$

(b)

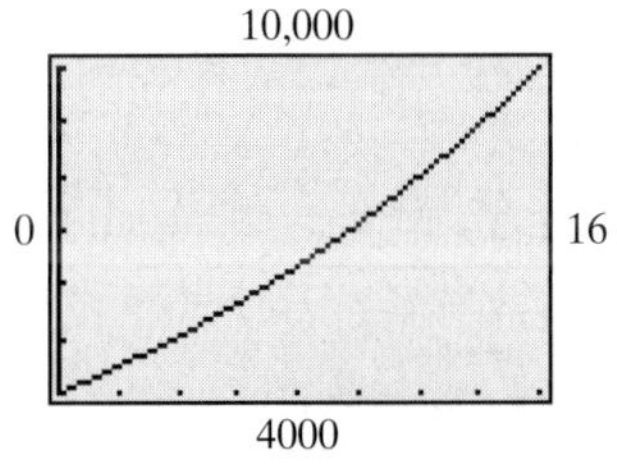

(c)

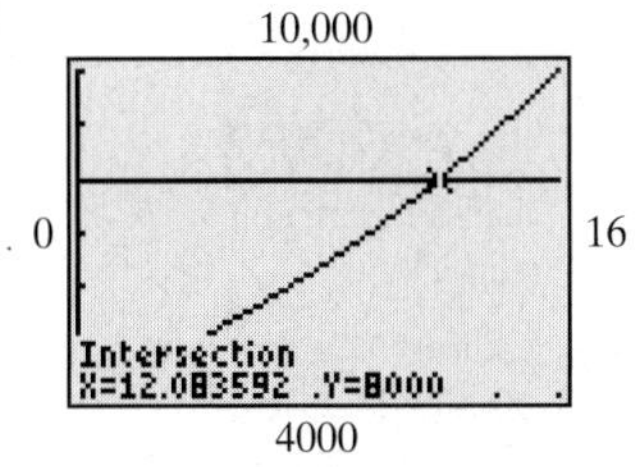

41. (a) $f(5) \approx 3789831$; after 5 minutes, there are approximately 3,789,831 bacteria in the colony.

(b) $m_{\text{sec}} \approx 1{,}076{,}901.90$; in the first 10 minutes, the colony was growing at an average rate of about 1,076,901.90 bacteria per minute.

43. (a) $S(5) \approx 16{,}321$; after 5 years on the market, approximately 16,321 Flashfast Cigarette Lighters had been sold.

(b) $m_{\text{sec}} \approx 1264.24$; during the first 5 years, sales increased at an average rate of about 1264.24 lighters per year.

45. (a) $f(12) \approx 824.50$; in 1991, there were approximately 824.5 thousand WWII veterans receiving compensation for service-connected disabilities.

(b) Exponential decay

(c) $m_{\text{sec}} \approx -38.31$; between 1980 and 1990, the number of WWII veterans receiving compensation decreased at an average rate of about 38,310 veterans per year.

47. (a) The function is an exponential growth model.

(b) $g(21) \approx 218.96$; in 1970, the federal government outlays were about \$218.96 billion.

(c) $m_{\text{sec}} \approx 66.76$; between 1980 and 1990, the federal government outlays increased at an average rate of about \$66.76 billion per year.

49. $A(7) \approx 1469.61$; after 7 years, there will be \$1469.61 in the account.

51. $A(7) \approx 30{,}226.90$; after 7 years, there will be \$30,226.90 in the account.

53. $A(7) \approx 14{,}103.76$; after 7 years, there will be \$14,103.76 in the account.

55. (a) $f(3) \approx 0.2326$; in 2000, about 23.26% of the households in the United States will own a DVD player.

(b) $m_{\text{sec}} \approx 0.0547$; between the years 2000 and 2002, the number of households that own a DVD player increased at an average rate of 5.47% per year.

57. (a) $f(0) \approx 5$; when the influenza initially first broke out, 5 students were infected.

(b) $m_{\text{sec}} \approx 13$; between the third and seventh day after the outbreak, the number of infected students increased at an average rate of about 13 students per day.

59. (a) The initial size of the seed colony was 8 lizards.

(b) After 8 years, there are 723 lizards in the colony.

(c) $m_{\text{sec}} \approx 89$; in the first 8 years, the population grew at an average rate of 89 lizards per year.

Section 11.8

1. $f(g(3)) = 54$; $f(g(-2)) = -36$

3. $f(g(0)) = 2$; $g(f(0)) = 1$

5. $g(f(x)) = 24x - 6$; $f(g(x)) = 24x + 5$

7. $g(f(x)) = 25x^2 - 15x + 4$; $f(g(x)) = 5x^2 + 15x + 17$

9. $g(f(x)) = \dfrac{1}{x^3}$; $f(g(x)) = \dfrac{1}{x^3}$

11. $g(f(x)) = x + 3$; $f(g(x)) = \sqrt{x^2 + 3}$

13. Let $g(x) = x + 3$ and $f(x) = x^3$.

15. Let $g(x) = \dfrac{1}{x+3}$ and $f(x) = x^2$.

17. Let $g(x) = x - 2$ and $f(x) = \sqrt[4]{x}$.

19. Let $g(x) = \sqrt{x}$ and $f(x) = 2 - 3x$.

21. (a) $r(t) = 1.3t$

(b) $V(r(t)) = \frac{8.788}{3}\pi t^3$; $V(t)$ is the volume of the sphere as a function of time.

(c) $V(6) \approx 1987.80$; after 6 seconds, the volume of the sphere is 1987.80 in^3.

23. $f(g(x)) = f(\frac{1}{8}x) = 8(\frac{1}{8}x) = x$

25. $f(g(x)) = f\left(\dfrac{x+10}{7}\right) = 7\left(\dfrac{x+10}{7}\right) - 10 = x + 10 - 10 = x$

27. $f(g(x)) = f(\sqrt[3]{x-6}) = (\sqrt[3]{x-6})^3 + 6 = (x-6) + 6 = x$

29. $f(g(x)) = f(x^2 + 1)$
$= \sqrt{(x^2+1) - 1} = \sqrt{x^2} = |x| = x$ since $x \geq 0$

31. $5 = \log_2(32)$ **33.** $-3 = \log_2\left(\frac{1}{8}\right)$ **35.** $2 = \log_{1/2}\left(\frac{1}{4}\right)$

37. $\frac{3}{4} = \log_{81}(27)$ **39.** $5 = \log(100{,}000)$ **41.** $1 = \ln(e)$

43. $\log_2(3) - \log_2(5)$ **45.** $\log(8) + \log(20)$ **47.** $\frac{1}{2}\ln(26)$

49. $\log_3(4) - 1.5$ **51.** 1.63 **53.** 2.67 **55.** 3.43 **57.** −2.95

59. **(a)** $f(4) \approx 8.89$; in 1996, 8.89% of the unemployed workers in the United States labor force were Hispanic.

(b) $x \approx 2$; $x \approx 2$ corresponds to 1994, hence after 1994, the number of unemployed workers in the United States labor force that were Hispanic dropped below 10%.

61. **(a)**

x	1	3	5	7	9
$f(x)$	4.21	5.63	6.29	6.72	7.04

(b) $f(6) \approx 6.52$; this means that in 1995, the total annual personal expenditures for admission to spectator sports was about \$6.52 billion.

(c) $m_{sec} \approx 0.35$; this means that from 1990 to 1998, total annual personal expenditures for admission to spectator sports increased at an average rate of about \$0.35 billion per year.

63. **(a)** $f(10) \approx 26.1$; this means that in 1994, the total school expenditures in higher education by the federal government was about \$26.1 billion.

(b) $m_{sec} \approx 0.29$; this means that from 1994 to 1997, the total school expenditures in higher education by the federal government increased at an average rate of about \$0.29 billion per year.

Section 11.9

1. Quadratic Model **3.** Cubic Model

5. A Logarithmic Model first and then a Power Model as a second choice.

7. Exponential Model **9.** Logistic Model

11. $y = 0.57x + 18.26$ **13.** $y = 15{,}557.13 + 2259.10 \ln x$

15. **(a)** $f(10) = 30.06$; this means that in 1994, about 30.06% of births were to unmarried mothers.

(b) $f(22) = -7.14$; this means that in 2006, about −7.14% of births will be to unmarried mothers.

(c) The solution to part (b) does not make sense since a percentage cannot be negative. Note that $x = 22$ is not in the reasonable domain.

Chapter 11 Review Exercises

1.

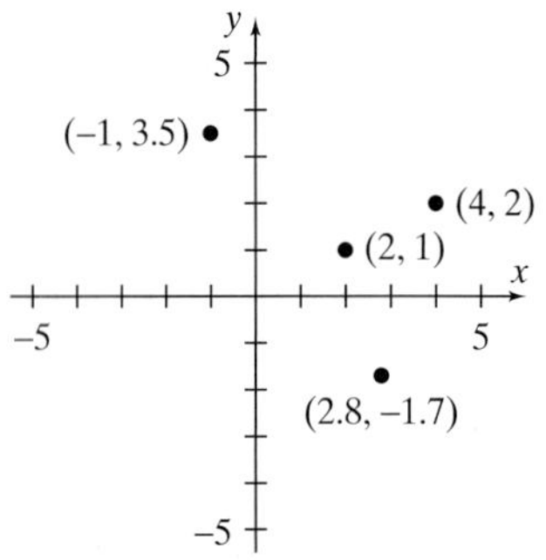

3. r is the independent variable: A is the dependent variable.

5. Function **7.** (3, 5)

9. **(a)** x is the independent variable.

(b) y is the dependent variable.

(c) 2 **(d)** 2 **(e)** [1, 4] **(f)** [1, 5] **11.** $[2, \infty)$

13.

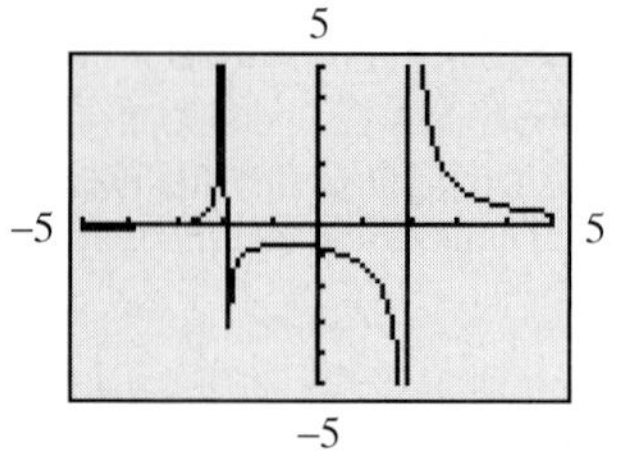

Domain: $(-\infty, -2) \cup (-2, 2) \cup (2, \infty)$
Range: $(-\infty, -0.655) \cup (-0.095, \infty)$

15. The domain is $(-2, \infty)$; the range is $[-3, \infty)$. **17.** $\frac{2}{3}$

19. $y - 5 = \frac{5}{4}(x - 0)$

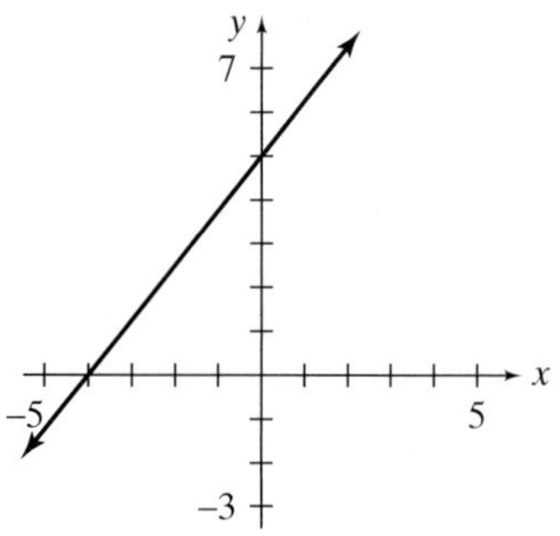

21. $y = \frac{3}{4}x - 7$

23. **(a)** $C(x) = -40x + 800$

(b) In 5 years, the bike will be worth \$600.

25. $C(x) = 75x + 2600$

27.

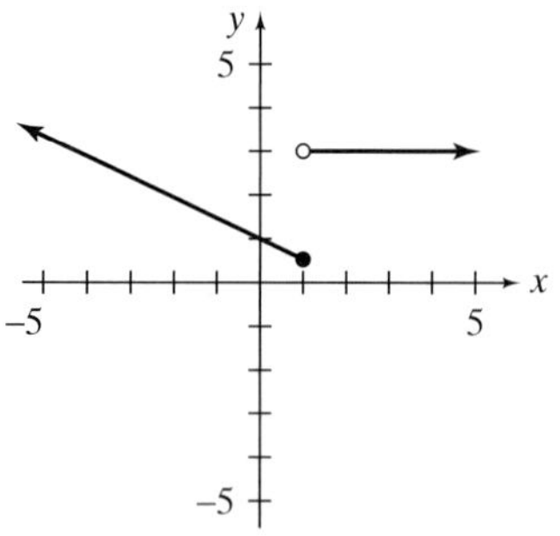

29. **(a)** Assuming that the cost function is linear, the y-coordinate of the y-intercept will correspond to the fixed costs. Since (0, 2500) is the y-intercept, then the fixed costs are \$2500.

(b) The marginal cost, the cost to produce one additional ceiling fan, is \$60 per fan.

(c) $C(x) = 60x + 2500$

(d) It will cost \$7540 to produce 84 ceiling fans.

31. **(a)** $\left(-\frac{3}{2},\ 16.5\right)$ **(b)** $(-\infty,\ 16.5]$

33. **(a)** (1.6, −5.24)

(b) The function is increasing on $(-\infty,\ 1.6)$ and decreasing on $(1.6, \infty)$.

35. $x = 2.5$ or $x = -4$

37.

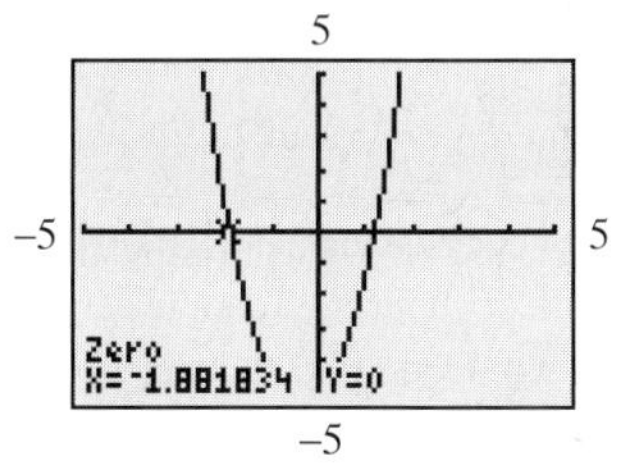

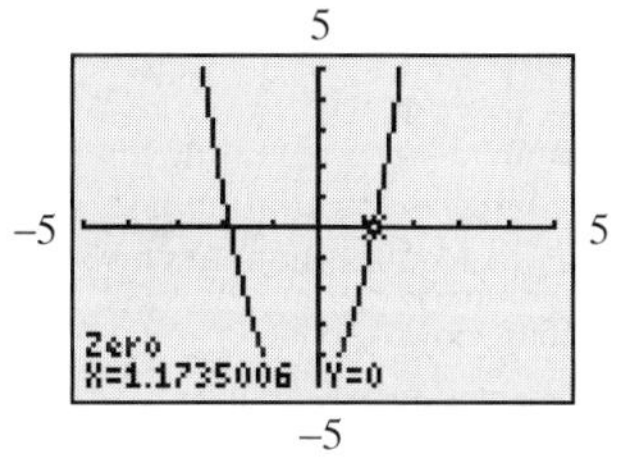

Roots are $x \approx -1.88$ and $x \approx 1.17$

39. **(a)** $m_{\text{sec}} = -1$ **(b)** $m_{\text{sec}} = 1$ **(c)** $m_{\text{sec}} = 5$

41. **(a)** $f(4) = 82.7$; in 1994, the average American spent 82.7 hours per year watching premium cable television programming.

(b)

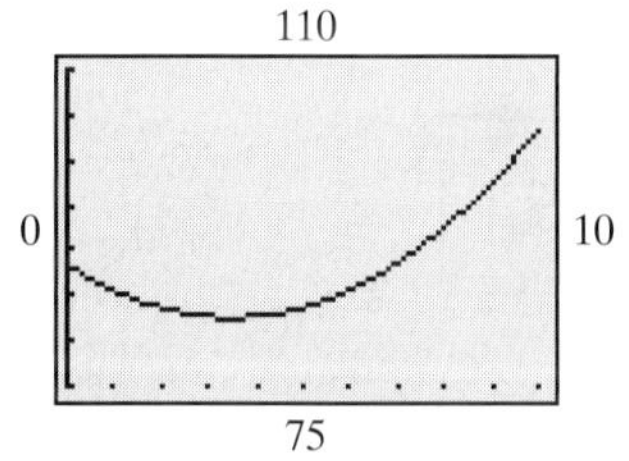

(c) [82.57, 103.4]

(d)

x	$f(x)$
0	88.5
1	85.58
2	83.64
3	82.68
4	82.7
5	83.7
6	85.68
7	88.64
8	92.58
9	97.5
10	103.4

(e) In 1992 through 1995 ($x = 2, 3, 4$, and 5), the average American spent less than 85 hours watching premium cable television programming.

43. **(a)** $(f+g)(x) = \sqrt{x+2} + x - 2$; the domain is $[-2, \infty)$.

(b) $(f-g)(x) = \sqrt{x+2} - x + 2$; the domain is $[-2, \infty)$.

(c) $(f \cdot g)(x) = (\sqrt{x+2})(x-2)$; the domain is $[-2, \infty)$.

(d) $\left(\dfrac{f}{g}\right)(x) = \dfrac{\sqrt{x+2}}{x-2}$; the domain is $[-2, 2) \cup (2, \infty)$.

45. $(f-g)(4) = -29$

47. $\left(\dfrac{f}{g}\right)(3) = 0.16$

49. $R(x) = 145x - 0.15x^2$

51. $P(x) = -0.08x^2 + 16x - 400$; vertex: (100, 400); since a is negative, P has a maximum profit of \$400 when $x = 100$ units are produced.

53. **(a)** Polynomial; quadratic

(b) Polynomial; constant

(c) Polynomial; linear

(d) Polynomial; quartic

55. As $x \to -\infty$, $g(x) \to \infty$ and as $x \to \infty$, $g(x) \to \infty$.

57. **(a)** As $x \to -\infty$, $f(x) \to -\infty$ and as $x \to \infty$, $f(x) \to \infty$.

(b)

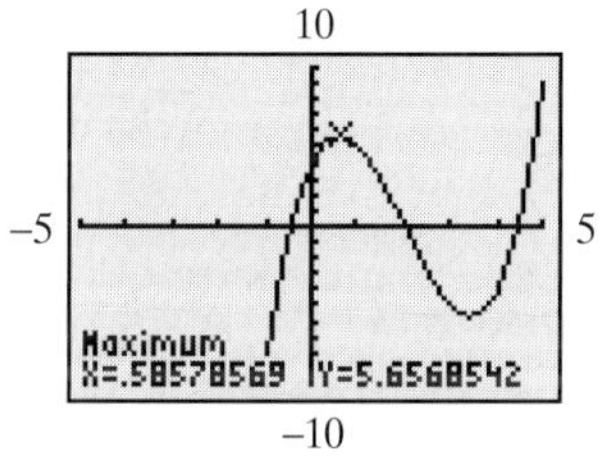

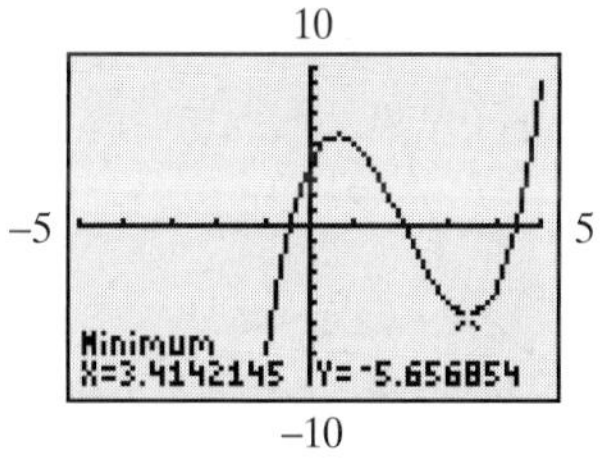

Thus, we have a "peak" at (0.59, 5.66) and a "valley" at (3.41, −5.66).

(c) $f(x)$ is decreasing on (0.59, 3.41) and increasing on $(-\infty, 0.59) \cup (3.41, \infty)$.

59. **(a)** $(-\infty, -1.5) \cup (-1.5, \infty)$

(b) Vertical asymptote at $x = -1.5$

(c) No horizontal asymptote

61. **(a)** $(-\infty, -3) \cup (-3, 3) \cup (3, \infty)$

(b) Vertical asymptote at $x = -3$; "hole" at $x = 3$

(c) Horizontal asymptote of $y = 2$

63. y-intercept is $(0, -\frac{16}{9})$; x-intercept is (4, 0)

65. y-intercept is (0, 2); no x-intercepts

67.

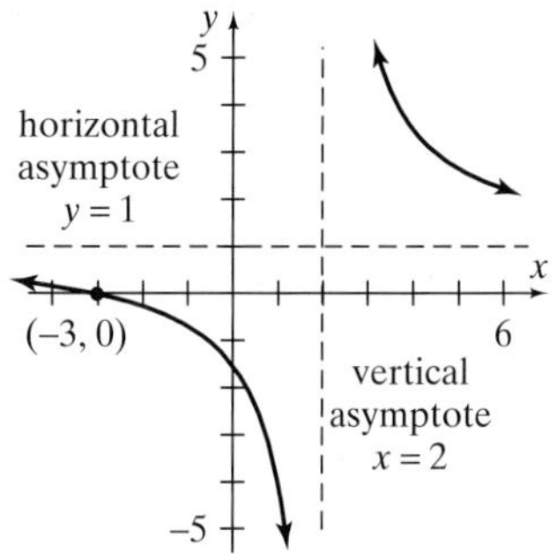

69. $f(x) = (x - 2)^{3/4}$; the domain is $[2, \infty)$.

71. (a) $f(64) = 2$; it will take 2 seconds for an object to fall 64 feet.

(b) It will take 3 seconds for an object to fall 144 feet.

73. (a) $f(12) \approx 123.65$; in 1987, about $123.65 billion was spent on research and development in the United States.

(b) $m_{\text{sec}} \approx 8.01$; between 1983 and 1987, the amount spent on research and development was increasing at an average rate of about $8.01 billion per year.

75. (a)

x	$f(x)$
−5	≈ 0.0155
−4	≈ 0.0357
−3	≈ 0.0822
−2	≈ 0.189
−1	≈ 0.435
0	1

x	$f(x)$
1	2.3
2	5.29
3	≈ 12.17
4	≈ 27.98
5	≈ 64.36

(b)

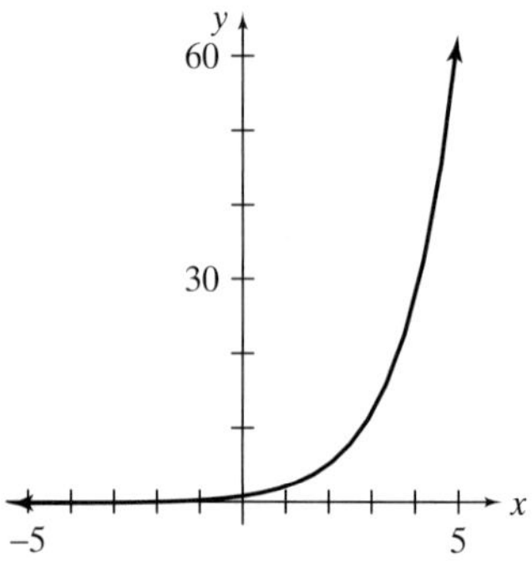

(c) This is an exponential growth function.

77. (a)

x	$f(x)$
−5	≈ 729,416
−4	≈ 49,021
−3	≈ 3294.5
−2	≈ 221.41
−1	≈ 14.88
0	≈ 1

x	$f(x)$
1	≈ 0.0672
2	≈ 0.0045
3	$\approx 3 \times 10^{-4}$
4	$\approx 2 \times 10^{-5}$
5	$\approx 1 \times 10^{-6}$

(b)

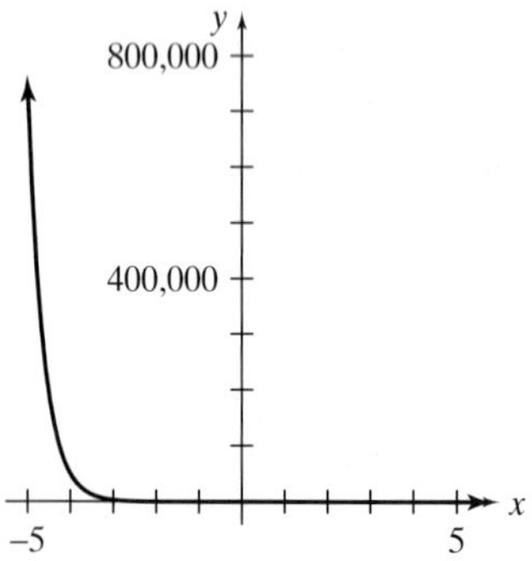

(c) This is an exponential decay function.

79. After 7 years, $2461.60 will be in the account.

81. (a) $A(t) = 8000(1.016875)^{4t}$; this is the amount in the account as a function of time.

(b) $A(6) \approx 11{,}953.96$; after 6 years, $11,953.96 will be in the account.

(c) $3953.96

83. (a) $f(4) \approx 75{,}105$; in 1984, there were about 75,105 people in Granaco City.

(b) $m_{\text{sec}} \approx 3322.13$; between 1985 and 1988, the population was growing at an average rate of about 3322 people per year.

(c) In 1994, the population exceeded 110,000 people.

85. (a) Effective rate ≈ 0.0852 or 8.52%; so, investing in an account that pays 8.2% compounded monthly is equivalent to investing in an account that pays 8.52% compounded annually.

(b) Effective rate ≈ 0.0693 or 6.93%; so, investing in an account that pays 6.7% compounded continuously is equivalent to investing in an account that pays 6.93% compounded annually.

87. $f(g(4)) = 0$

89. (a) Their domains are all real numbers or $(-\infty, \infty)$.
(b) $f(g(x)) = 5x^2 + 3$ **(c)** $g(f(x)) = 25x^2 + 30x + 9$

91. Let $f(x) = x^5$ and $g(x) = x^2 + 2$.

93. Let $f(x) = \sqrt{x}$ and $g(x) = 6 - x$.

95.

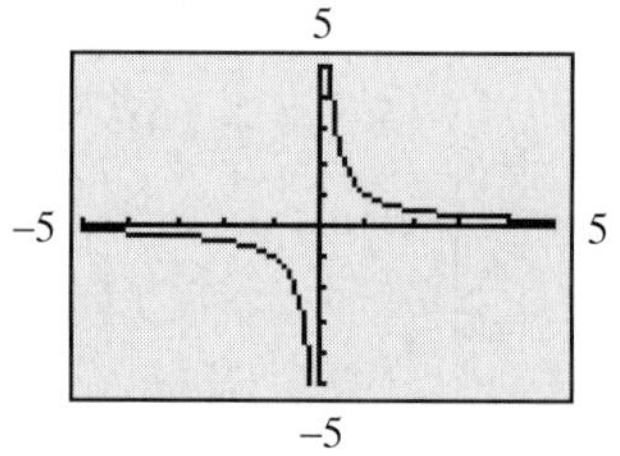

The function is one-to-one.

97. (a) $f(g(x)) = f(x^3 - 5) = \sqrt[3]{(x^3 - 5) + 5} = \sqrt[3]{x^3} = x$

(b)

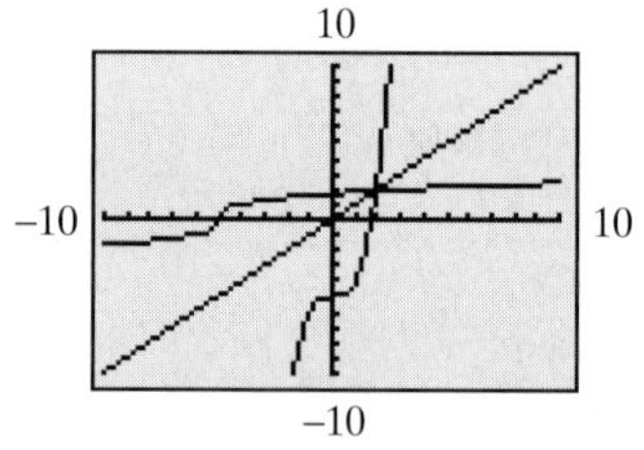

99. $-4 = \log(0.0001)$ **101.** $2\log_2(5) - \log_2(7)$ **103.** 2.52

105. A Logarithmic Model first and then a Power Model as a second choice.

107. $y = 2.9x + 234.867$

109. $y = (2.862 \times 10^{-9})x^{5.452}$

Chapter 12

Section 12.1

1. (a) −1 **(b)** −1 **(c)** −1 **3. (a)** 48 **(b)** 48 **(c)** 48

5. (a) 2 **(b)** 2 **(c)** 2 **7. (a)** 12 **(b)** 12 **(c)** 12

9. **(a)** 1 **(b)** 1 **(c)** 1 **11.** −6 **13.** 3 **15.** 4

17. $\frac{1}{4}$ **19.** −5 **21.** 13 **23.** 6 **25.** 0 **27.** 0

29. −2 **31.** $\sqrt{3}$ **33.** 0 **35.** 0 **37.** 1125

39. 5 **41.** 1 **43.** $\frac{1}{4}$

45. **(a)** 1 **(b)** 1 **(c)** 1 **(d)** 2

(e) Does not exist **(f)** 2 **(g)** 3

47. $2x$ **49.** $3x$ **51.** $4x^2$ **53.** $-3x^2$

55. **(a)** $3h+2$ **(b)** 3 **57.** **(a)** h^2+2h+3 **(b)** 2

59. **(a)** h^2+5h+3 **(b)** 5 **61.** **(a)** $\dfrac{1}{h+1}$ **(b)** −1

63. **(a)** $N(10) = 1948$; when $10,000 is spent on advertising, 194,800 items are sold.

(b) $\lim_{x\to 10} N(x) = 1948$; as the level of advertising expenditures approaches $10,000, the number of items sold approaches 194,800.

65. **(a)** $AC(10) = 2257.35$; the average cost of producing 10 units of a product is $2257.35 per unit.

(b) $\lim_{x\to 10} AC(x) = 2257.35$; as the level of production approaches 10 units, the average cost approaches $2257.35 per unit.

67. **(a)** $c(100) = 30$; rental charges for driving 100 miles will be $30.

(b) $\lim_{m\to 100} c(m) = 30$

(c) $c(200) = 60$; rental charges for driving 200 miles will be $60.

(d) $\lim_{m\to 200} c(m)$ does not exist.

(e) The charges for $m \geq 200$ were incurred during a second day of driving.

Section 12.2

1. **(a)** $-\infty$ **(b)** ∞ **(c)** Does not exist

3. **(a)** ∞ **(b)** ∞ **(c)** ∞ **5.** **(a)** $-\frac{1}{5}$ **(b)** $-\frac{1}{5}$ **(c)** $-\frac{1}{5}$

7. **(a)** $-\infty$ **(b)** ∞ **(c)** Does not exist

9. Does not exist **11.** $-\frac{1}{5}$ **13.** $\frac{1}{6}$ **15.** Does not exist

17. **(a)** Exponential growth

(b) $\lim_{x\to\infty} f(x) = \infty$; $\lim_{x\to-\infty} f(x) = 0$

19. **(a)** Exponential growth

(b) $\lim_{x\to\infty} f(x) = \infty$; $\lim_{x\to-\infty} f(x) = 0$

21. **(a)** Exponential decay

(b) $\lim_{x\to\infty} f(x) = 0$; $\lim_{x\to-\infty} f(x) = \infty$

23. **(a)** Exponential decay

(b) $\lim_{x\to\infty} f(x) = 0$; $\lim_{x\to-\infty} f(x) = \infty$

25. 2 **27.** $\frac{3}{2}$ **29.** 0 **31.** 0 **33.** $y = 2$ **35.** $y = 0$

37. $-\infty$ **39.** Does not exist **41.** 1 **43.** 0

45. **(a)** $0 \leq x < 100$

(b) $C(40) = 62$; the cost of removing 40% of the pollutants is $62,000.

(c) $\lim_{x\to 100^-} C(x) = \infty$; as the amount of pollutants removed approaches 100%, the cost grows without bound.

47. 3200; as the amount of money spent on advertising grows without bound, the number of items sold approaches 320,000.

49. If $1.5 \leq x \leq 3.75$, then $2700 \leq N(x) \leq 3000$.

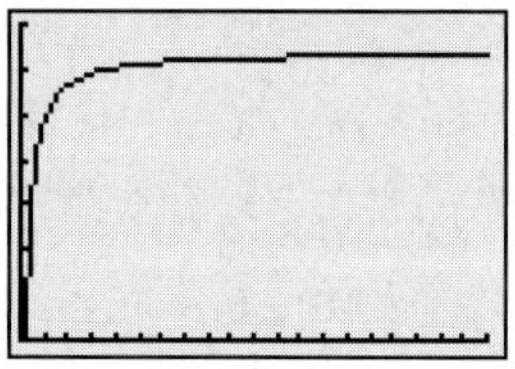

51. 4.85; as the number of units produced increases without bound, the average cost of production approaches $4.85 per item.

53. **(a)** 2.5; this means that initially, 2.5 milliliters of medication is administered.

(b) 0; as the amount of time since the medication was administered increases without bound, the amount present in the body approaches 0.

55. $t \approx 4.58$

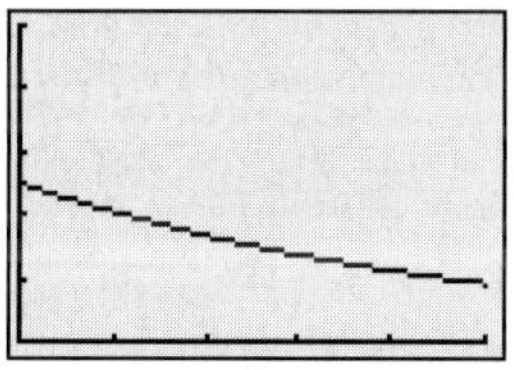

57. **(a)** 758 bass

(b) 3500; as time increases without bound, the bass population approaches 3500.

61. **(a)** 2 people

(b) 1000; as time increases without bound, the number of people affected by influenza approaches 1000, the entire population.

Section 12.3

1. Instantaneous rate of change **3.** Average rate of change

5. Instantaneous rate of change **7.** Average rate of change

9. Average rate of change **11.** Instantaneous rate of change

13. $\dfrac{f(4)-f(1)}{4-1} = 6$ **15.** $\dfrac{f(4)-f(2)}{4-2} = 59$

17. $\dfrac{f(1)-f(-2)}{1-(-2)} = -1$ **19.** $\dfrac{g(9)-g(1)}{9-1} = \dfrac{3}{4}$

21. $\dfrac{f(20)-f(0)}{20-0} \approx 5.63$ **23.** $\dfrac{f(15)-f(0)}{15-0} \approx -0.13$

25. $\dfrac{f(10)-f(1)}{10-1} \approx 0.26$

27. **(a)** Positive: A, B and D, E; negative: B, C; zero: C, D

(b) Positive: B, C; negative: C, D; zero: A, B

29. 3 **31.** 7 **33.** 99 **35.** $\frac{3}{4}$

37. ≈ 0.63 **39.** ≈ -0.23

41. **(a)** $m_{sec} = 44.924$; from 1980 to 1990, membership in the Girl Scouts of America increased at an average rate of 44,924 Girl Scouts per year.

(b) $m_{tan} \approx 55.68$; in 1990, membership in the Girl Scouts of America was increasing at a rate of about 55,684 Girl Scouts per year.

43. **(a)** $m_{sec} = 32.5$; when production changes from 100 to 150 rain coats produced, the cost increased at an average rate of \$32.5 per rain coat.

(b) $m_{tan} \approx 31$; when production is at 200 rain coats, the cost is increasing at a rate of about \$31 per rain coat.

45. $m_{tan} \approx 0.3$; in 1990, the poverty threshold in the U.S. was increasing at a rate of about \$300 per year.

47. $m_{tan} \approx 5$; when 50 magazine subscriptions are sold each month, the profit is increasing at a rate of about \$5 per month.

Section 12.4

1. 2 **3.** −6 **5.** −6 **7.** $\frac{1}{2}$ **9.** −13.6

11. $y = -6(x+2)+8 = -6x-4$ **13.** $y = -6(x-4)-8 = -6x+16$

15. $y = -13.6(x-7)-38.5 = -13.6x+56.7$ **17.** $f'(x) = 2$

19. $g'(x) = -2$ **21.** $f'(x) = 2x$ **23.** $h'(x) = 2x-2$

25. $g'(x) = 4.2x+3.2$ **27.** $f'(x) = -4x+3$

29. $f'(x) = -4.6x+3.1$ **31.** $g'(x) = 3x^2+2x$

33. $x = -1$: $m_{tan} = 2$; $x = 2$: $m_{tan} = 2$

35. $x = -3$: $m_{tan} = -8$; $x = 2$: $m_{tan} = 2$

37. $x = 0$: $m_{tan} = 3.2$; $x = 2$: $m_{tan} = 11.6$

39. $x = 3$: $m_{tan} = -9$; $x = 6$: $m_{tan} = -21$

41. $x = -1$: $m_{tan} = 1$; $x = 1$: $m_{tan} = 5$

43. **(a)** $P'(q) = 1000-4q$

(b) $P'(200) = 200$; $P'(300) = -200$; when 200 modems are produced and sold, the profit is increasing at a rate of \$200 per modem; when 300 modems are produced and sold, the profit is decreasing at a rate of \$200 per modem.

45. **(a)** $R(50) = 7500$; $R(150) = 7500$; when 50 sets of Junior Golf Clubs are produced and sold, the revenue is \$7500; when 150 sets of Junior Golf Clubs are produced and sold, the revenue is \$7500.

(b) $R'(q) = 200-2q$

(c) $R'(50) = 100$; $R'(150) = -100$; when 50 sets of Junior Golf Clubs are sold, revenue is increasing at a rate of \$100 per set; when 150 sets of Junior Golf Clubs are sold, revenue is decreasing at a rate of \$100 per set.

47. **(a)** $C'(x) = 2x+15$

(b) $C'(40) = 95$; $C'(100) = 215$; when 40 refrigerators are produced, the weekly cost is increasing at a rate of \$95 per refrigerator; when 100 refrigerators are produced, the weekly cost is increasing at a rate of \$215 per refrigerator.

49. **(a)** $f'(x) = 4.78x+47.05$

(b) $f'(3) = 61.39$; $f'(4) = 66.17$; $f'(5) = 70.95$; imports are increasing at a rate of \$61.39 billion per year in 1993, \$66.17 billion per year in 1994, and \$70.95 billion per year in 1995.

(c) Increasing

51. **(a)** $f'(x) = -0.04x+0.8$

(b) $f'(5) = 0.6$; $f'(10) = 0.4$; $f'(15) = 0.2$. Diet carbonated soft drink consumption was increasing at a rate of 0.6 gallons per capita per year in 1984, 0.4 gallons per capita per year in 1989 and 0.2 gallons per capita per year in 1994.

(c) Increasing

53. **(a)** $A(7) \approx 3107.97$; after 7 years, there will be \$3107.97 in the account.

(b) $A'(7) \approx 195.72$; after 7 years, the account was increasing at a rate of about \$195.72 per year.

55. **(a)** $N(15) = 1458$; $N'(15) = 2.8$; when \$15,000 is spent on advertising, sales will be 145,800 items, and sales are increasing at a rate of 280 items per thousand advertising dollars.

(b) $N(25) = 1474.8$; $N'(25) = 1.008$; when \$25,000 is spent on advertising, sales will be 147,480 items, and sales are increasing at a rate of about 101 items per thousand advertising dollars.

(c) $\lim_{x\to\infty} N'(x) = 0$; decreasing

(d)

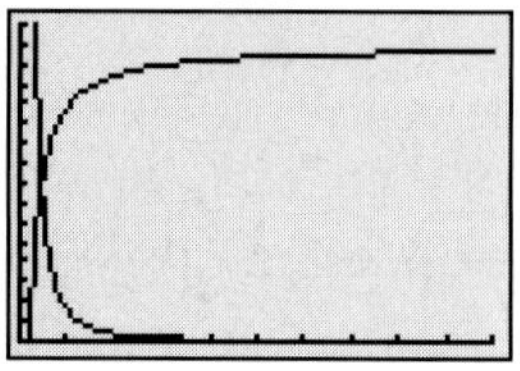

the result is reasonable.

57. Tangent line t_1 is associated with the faster growth rate since its slope is larger than the slope of tangent line t_2.

Section 12.5

1. $f'(x) = 0$ **3.** $f'(t) = 0$ **5.** $f'(x) = 6x^5$

7. $y' = -12x^3$ **9.** $g'(x) = \frac{4}{3\sqrt[3]{x}}$ **11.** $f'(x) = \frac{1}{\sqrt[3]{x^4}}$

13. $h'(x) = \frac{8x^3}{3}$ **15.** $y' = -\frac{2\sqrt[3]{x^2}}{3}$

17. $f'(x) = 6x^2+8x-7$ **19.** $g'(x) = 6x-2$

21. $y' = -10x-6$ **23.** $y' = -15x^2+7$

25. $f'(x) = \frac{3}{2}x^2+\frac{6}{5}x-\frac{2}{3}$ **27.** $g'(x) = 2.62x+2.05$

29. $d'(x) = -0.4x+10.5x^2-1.6x^3$

31. $y' = 3.45x^2-4.6x+2.53$ **33.** $f'(x) = \frac{3}{2\sqrt{x}}+\frac{1}{2}-10x$

35. $y' = \frac{1}{3\sqrt[3]{x^2}}+2x-9x^2$ **37.** $g'(x) = \frac{2}{3\sqrt[3]{x}}+\frac{2}{\sqrt{x^3}}$

39. $f'(x) = 3.1725x^{0.35}$ **41.** $AC'(x) = -\frac{10}{x^3}$

43. $y'(x) = 4x-\frac{1}{x^2}$ **45.** $f'(x) = -\frac{9}{2\sqrt{x^5}}-\frac{2}{\sqrt{x}}$

47. $y' = -\dfrac{1.175}{\sqrt{x^3}} + \dfrac{23}{15\sqrt[3]{x^5}}$

49. **(a)** $(-\infty, \infty)$ **(b)** $f(x) = \dfrac{x^3}{4} - \dfrac{3x^2}{4} + \dfrac{1}{3}$

(c) $f'(x) = \dfrac{3x^2}{4} - \dfrac{3x}{2}$ **(d)** $(-\infty, \infty)$

51. **(a)** $(-\infty, 0) \cup (0, \infty)$ **(b)** $f(x) = 2x^2 + 3x - 1 + 3x^{-1}$

(c) $f'(x) = 4x + 3 - \dfrac{3}{x^2}$ **(d)** $(-\infty, 0) \cup (0, \infty)$

53. **(a)** $(-\infty, 0) \cup (0, \infty)$ **(b)** $y = 7x^2 - 50 + x^{-1}$

(c) $y' = 14x - \dfrac{1}{x^2}$ **(d)** $(-\infty, 0) \cup (0, \infty)$

55. **(a)** $(0, \infty)$ **(b)** $h(x) = 2x^{5/2} - 7x^{3/2} + \frac{3}{2}x^{-1/2}$

(c) $h'(x) = 5x^{3/2} - \dfrac{21}{2}x^{1/2} - \dfrac{3}{4x^{3/2}}$ **(d)** $(0, \infty)$

57. **(a)** $f'(x) = 3x^2$ **(b)** $f'(-1) = 3$ **(c)** $y = 3x + 2$

(d)

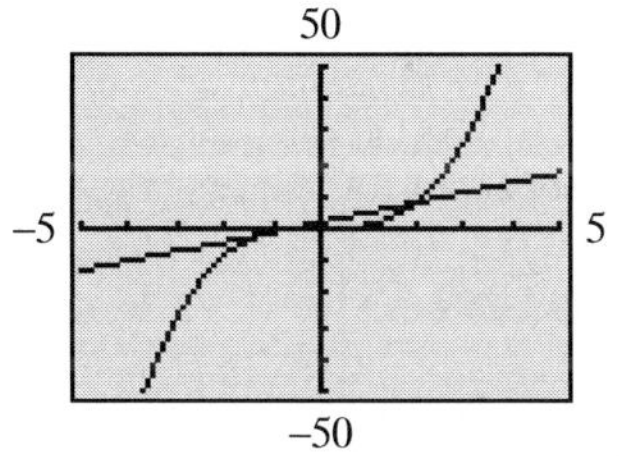

59. **(a)** $y' = -\dfrac{1}{x^2}$ **(b)** $y'(3) = -\frac{1}{9}$ **(c)** $y = -\frac{1}{9}x + \frac{2}{3}$

(d)

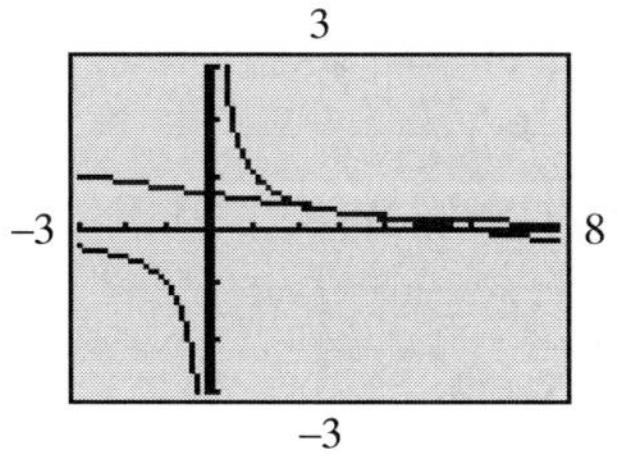

61. **(a)** $f'(x) = \dfrac{2}{3}x^{-1/3} = \dfrac{2}{3\sqrt[3]{x}}$ **(b)** $f'(8) = \frac{1}{3}$ **(c)** $y = \frac{1}{3}x + \frac{4}{3}$

(d)

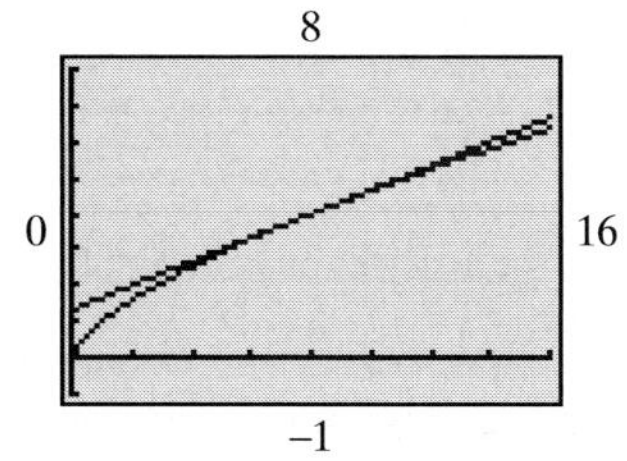

63. **(a)** $N'(x) = \dfrac{520}{x^2}$

(b) $N(10) = 2448$; $N'(10) = 5.2$; when \$10,000 is spent on advertising, 244,800 CDs are sold and sales are increasing at a rate of 520 CDs per thousand dollars spent.

65. $C(300) \approx 6334.64$; $C'(300) \approx 11.58$; when 300 coats are made per week, the total cost is about \$6334.64 and the costs are increasing at a rate of about \$11.58 per coat.

67. $SD(1) = 93.21$; $SD'(1) = -186.42$; one mile downwind, the amount of sulphur dioxide is 93.21 ppm and decreasing at a rate of 186.42 ppm per mile.

69. **(a)** $f(3) = 1362.03$; $f(7) = 1462.23$; in 1992 the total carbon content of carbon dioxide emitted into the atmosphere by the U.S. was 1362.03 million metric tons; in 1996 it was 1462.23 million metric tons.

(b) $f'(3) = 17.53$; $f'(7) = 23.77$; in 1992, the total carbon content of carbon dioxide emitted into the atmosphere by the U.S. was increasing at a rate of 17.53 million metric tons per year; in 1996, it was increasing at a rate of 23.77 million metric tons per year.

71. **(a)** $\dfrac{N(4) - N(1)}{4 - 1} = 62$; between 1 and 4 minutes, the average rate of change in the size of bacteria is 62 milligrams per minute.

(b) $N'(4) = 120$; after 4 minutes, the size of bacteria was increasing at a rate of 120 milligrams per minute.

73. **(a)** $v(t) = s'(t) = -32t$

(b) $s(1) = 134$; $s(3) = 6$; after 1 second, the egg is 134 feet above ground; after 3 seconds, it is 6 feet above ground.

(c) $\dfrac{s(3) - s(1)}{3 - 1} = -64$; -64 feet per second

(d) $v(1) = -32$; $v(3) = -96$; after 1 second, the instantaneous velocity is -32 feet per second; after 3 seconds, it is -96 feet per second.

(e) $t \approx 3.1$

75. **(a)** $f'(x) = -0.042x^2 + 0.78x - 3.11$

(b) $f'(12) = 0.202$; $f'(17) = -1.988$; the number of tuberculosis cases was increasing at a rate of 202 per year in 1991 and was decreasing at a rate of 1988 per year in 1996.

77. **(a)** $C(11) = 413.89$; $C(21) = 408.69$; in 1985, the number of milligrams of cholesterol consumed each day per person was 413.89; in 1995, it was 408.69.

(b) $C'(11) = -1.62$; $C'(21) = 0.58$; the number of milligrams of cholesterol consumed each day per person was decreasing at a rate of 1.62 milligrams per day per person per year in 1985, and was increasing at a rate of 0.58 milligrams per day per person per year in 1995.

79. **(a)** $S(23) = 244.30$; after 23 weeks, weekly sales were \$244,300.

(b) $S'(23) = -3.26$; after 23 weeks, weekly sales were decreasing at a rate of \$3260 per week.

(c) \$244,300; decreasing; \$3260 per week

81. **(a)** $g(14) = 3332.086$; in 1993, there were 3,332,086 members of the Girl Scouts of America.

(b) $g'(t) = 0.344t^3 - 11.49t^2 + 111.94t - 243.23$

(c) $g'(14) = 15.826$; in 1993, membership in the Girl Scouts of America was increasing at a rate of 15,826 members per year.

83. Boy Scouts of America; the instantaneous rate of change for the Boy Scouts was greater than the instantaneous rate of change for the Girl Scouts.

Section 12.6

1. $f'(x) = 2x(3x + 1)$ **3.** $f'(x) = x^2(15x^2 + 8x - 15)$

5. $y' = 3x^3(12x^2 - 45x + 4)$ **7.** $y' = -5x(15x^3 + 15x - 14)$

9. $f'(x) = 12x + 5$

11. $y' = 60x^3 + 57x^2 + 12x + 5$

13. $g'(x) = 24x^3 + 33x^2 - 58x + 19$

15. $y' = 24x + 9\sqrt{x} - \dfrac{4}{\sqrt{x}} - 25$

17. $f'(x) = \frac{172}{5}x^{28/15} - \frac{160}{3}x^{5/3} + \frac{66}{5}x^{6/5} - 20x - 18x^{1/5} + 25$

19. $f'(x) = -\dfrac{5}{\sqrt{x}} + \dfrac{15}{2x^{7/2}} - \dfrac{15}{x^4} + 6$

21. $h'(x) = -\dfrac{2}{x^3} - \dfrac{4}{3x^{7/3}} - \dfrac{2}{x^2} - \dfrac{1}{3x^{4/3}}$

23. $f'(x) = -\dfrac{1}{(x+1)^2}$ **25.** $y' = \dfrac{10}{(2x+1)^2}$

27. $f'(x) = \dfrac{34x^2 + 2x - 13}{(5x^2 + 3x + 2)^2}$

29. $f'(x) = \dfrac{-9x^{1/2} - \frac{3}{2}x^{-1/2} + 30}{(6x - 1)^2}$

31. $y' = \dfrac{2(x^3 - 3x^2 + 3x + 2)}{(x-1)^2}$

33. $f'(x) = \dfrac{25x^6 - 120x^5 + 45x^4 - 240x^3 + 2x^2 - 16x - 6}{(x-4)^2}$

35. $g'(x) = 6x^2 + 2x - \frac{3}{2}$

37. **(a)** $f'(x) = 2x(2x^2 - 5)$

(b) $y = -6x + 2$

39. **(a)** $f'(x) = x(5x^3 + 3x + 2)$

(b) $y = 10x - 6$

41. **(a)** $y' = -\dfrac{3}{(x-1)^2}$

(b) $y = -3x + 10$

43. **(a)** $g'(x) = \dfrac{-6(x^2 - 3x + 1)}{(-2x + 3)^2}$

(b) $y = -\frac{6}{5}x - \frac{1}{5}$

45. **(a)** $y = 20x - 36$

(b)

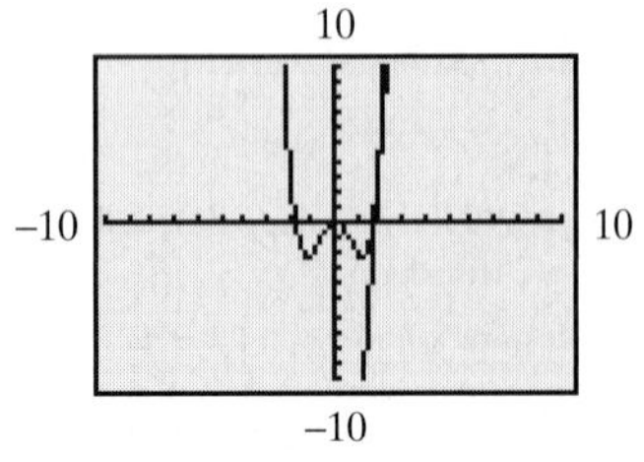

(c)

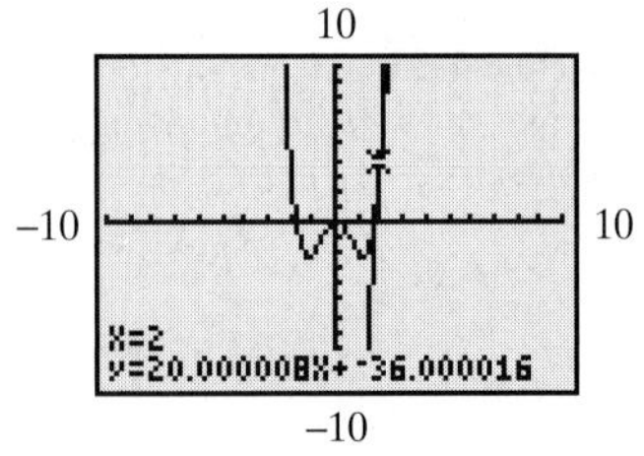

47. **(a)** $y = 7x + 10$

(b)

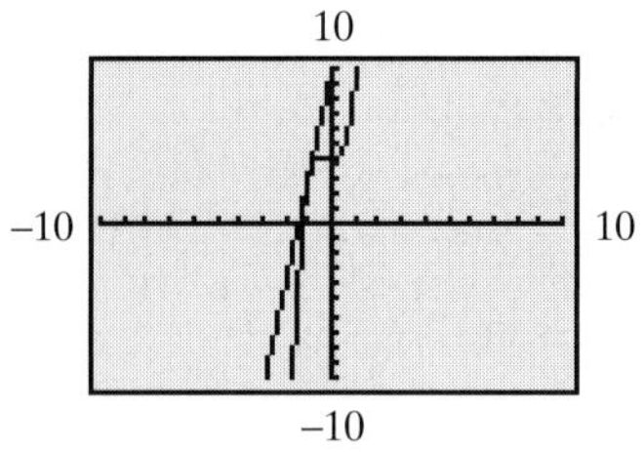

(c)

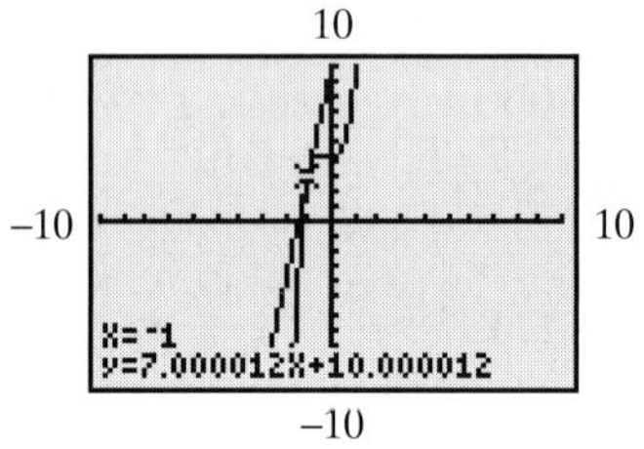

49. **(a)** $y = 6.92x - 8.64$

(b)

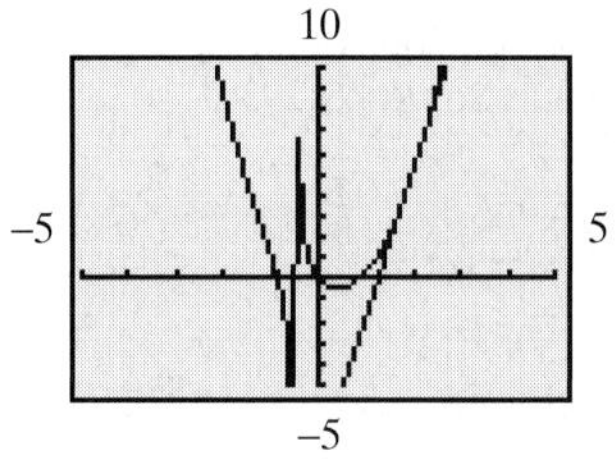

(c)

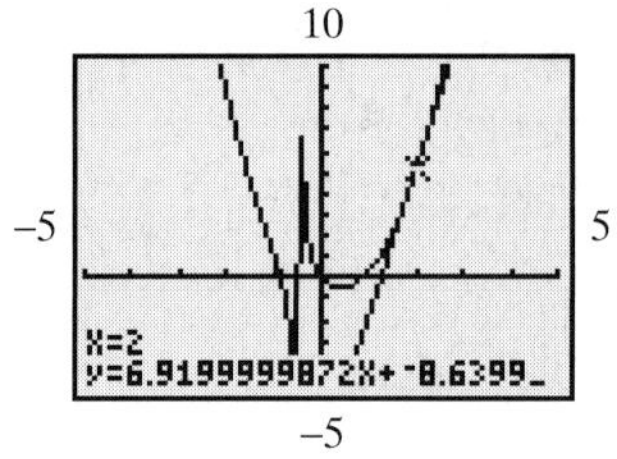

51. $f'(x) = 3x^2 + 8x - 7$ **53.** $f'(x) = 24x^3 + 42x^2 - 2x - 21$

55. $y' = 21x^{5/2} - \dfrac{15}{2}x^{3/2} + 6\sqrt{x} - \dfrac{1}{\sqrt{x}}$

57. $q(3) = 85.5$; $q'(3) = 27$; three months after it hit the market, monthly sales of the new computer were 8550 units and sales were increasing at the rate of 2700 units per month.

59. **(a)** $R(t) = q(t) \cdot p(t) = (30t - 0.5t^2)(2200 - 34t^2)$ in hundreds of dollars.

(b) $R'(t) = (30 - t)(2200 - 34t^2) + (30t - 0.5t^2)(-68t)$

(c) $R(3) = 161{,}937$; $R'(3) = 33{,}696$; three months after it hit the market, monthly revenues for the new computer were \$16,193,700 and increasing at the rate of \$3,369,600 per month.

61. $p(6) = 976$; $p'(6) = -408$; six months after it hit the market, the retail price of the new computer was \$976 and it was decreasing at the rate of \$408 per month.

63. **(a)** $R'(t) = (30 - t)(2200 - 34t^2) + (30t - 0.5t^2)(-68t)$
$= (66{,}000 - 1020t^2 - 2200t + 34t^3) + (-2040t^2 + 34t^3)$
$= 68t^3 - 3060t^2 - 2200t + 66{,}000$

(b)

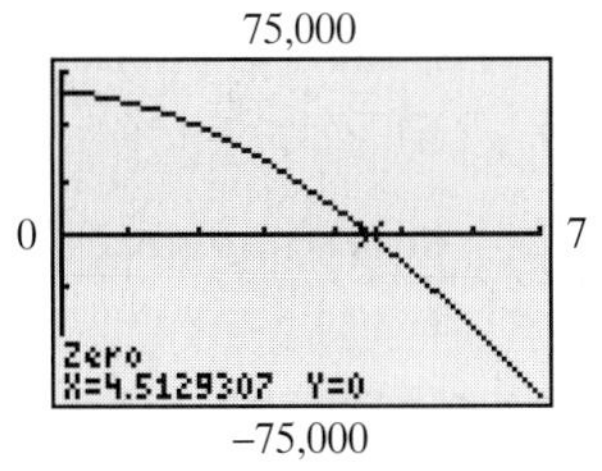

$x = 4.5129307$

(c)

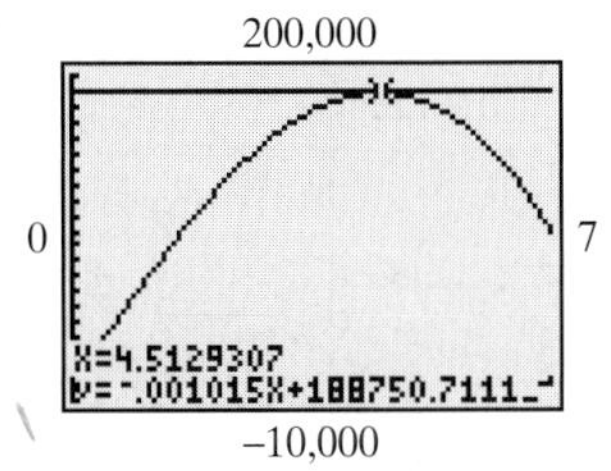

$\frac{dy}{dx} = -0.001015$

(d) Maximized; to the left of $t = 4.5129307$ the values of $R'(t)$ are positive so the slope of $R(t)$ is positive and the function is increasing. To the right of $t = 4.5129307$ the values of $R'(t)$ are negative so the slope of $R(t)$ is negative and the function is decreasing. So, $R(4.5129307) \approx 188{,}751$ must be a maximum.

65. $p(3) = 211$; $p'(3) = -6$; three months after being introduced, the retail price of the DVD drive is \$211 and the price is decreasing at the rate of \$6 per month.

67. (a) $\frac{f(3)}{g(3)} \approx 35.79$; the average weekly hours in 1992 was about 35.79.

(b) $h(t)$ gives the average weekly hours.

(c) $h(3) \approx 35.79$; $h'(3) \approx 0.03$; in 1992, the average weekly hours was 35.79 and was increasing at the rate of about $0.03 \frac{\text{hours per week}}{\text{year}}$.

69. $C(50) = 50$; $C'(50) = 2$; the cost of removing 50% of the city's pollutants is \$50,000 and is increasing at the rate of \$2000 per %.

71. $P(5) \approx 818.18$; $P'(5) \approx 112.40$; five months after stocking the lake, the bass population is about 818 and is increasing at the rate of about 112 bass per month.

73. (a) $SD(5) = 22.27$; $CP(5) = 245.76$; in 1989, the amount of sulphur dioxide emitted into the atmosphere by the U.S. was 22,270,000 tons and the U.S. civilian population was 245,760,000 people.

(b) $\frac{SD(5)}{CP(5)} \approx 0.09$; in 1989, the amount of sulphur dioxide emitted into the atmosphere by the U.S. was about 0.09 tons per person.

(c) $H'(t) = \frac{-0.1028t^3 - 13.2807t^2 + 125.771t - 346.139}{(2.57t + 232.91)^2}$

(d) $H'(5) \approx -0.001$; in 1989, the per capita emission of sulphur dioxide into the atmosphere was decreasing at the rate of about 0.001 tons per year.

Section 12.7

1. Yes **3.** No; $\lim_{x\to 3} f(x)$ does not exist.

5. Yes; $f(1)$ is defined, $\lim_{x\to 1} f(x)$ exists, and $\lim_{x\to 1} f(x) = f(1)$.

7. Yes; $f(3)$ is defined, $\lim_{x\to 3} f(x)$ exists, and $\lim_{x\to 3} f(x) = f(3)$.

9. g continuous at $x = -2$ **11.** h continuous at $x = 3$

13. g not continuous at $x = 3$ **15.** f not continuous at $x = 5$

17. f not continuous at $x = 1$ **19.** $(-\infty, \infty)$

21. $(-\infty, -5) \cup (-5, \infty)$ **23.** $(-\infty, -1) \cup (-1, 3) \cup (3, \infty)$

25. $(-\infty, \infty)$ **27.** $(-3, \infty)$ **29.** $(-\infty, \infty)$ **31.** $[-\frac{3}{2}, \infty)$

33. $x = -2$ function not continuous
$x = 2$ function not continuous

35. $x = -2$ function has corner
$x = 1$ function has corner

37. $x = 2$ function has sharp turn

39. No values where derivative does not exist

41. (a) $[0, \infty)$ **(b)** $f'(x) = \frac{1}{2\sqrt{x}}$

(c) Yes; $x = 0$ **(d)** Vertical tangent at $x = 0$.

43. (a) $(-\infty, \infty)$ **(b)** $f'(x) = \frac{2}{3x^{1/3}}$

(c) Yes; $x = 0$ **(d)** Sharp turn

45. (a) $(-\infty, -1) \cup (-1, \infty)$ **(b)** $f'(x) = \frac{x^2 + 2x + 1}{(x + 1)^2} = 1$

(c) No; however $f'(-1)$ is undefined. **(d)** Discontinuous

47. (a) $(-\infty, 2) \cup (2, \infty)$ **(b)** $f'(x) = \frac{x^2 - 4x + 4}{(x - 2)^2} = 1$

(c) No; however $f'(2)$ is undefined. **(d)** Discontinuous

49. The charges for $m \ge 200$ were incurred during a second day of driving.

51. (a) $C(x) = \begin{cases} 1.50x, & 0 < x < 2 \\ 1.00x, & 2 \le x \le 6 \end{cases}$

(b)

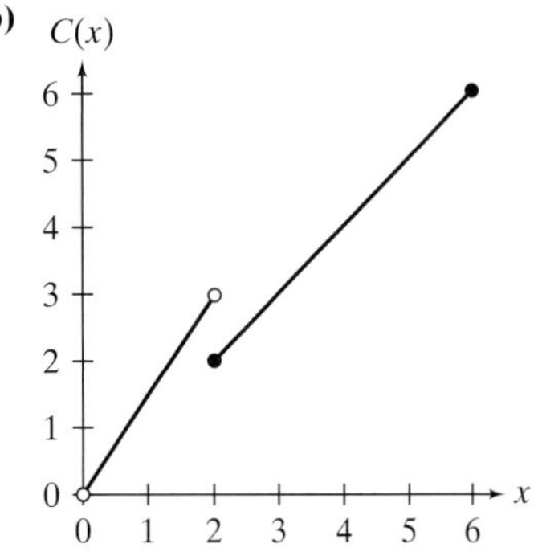

(c) $C(1.5) = 2.25$; 1.5 pounds of hamburger costs \$2.25.

(d) $C(2) = 2$; 2 pounds of hamburger costs \$2.00.

(e) No; $\lim_{x\to 2^-} C(x) \ne \lim_{x\to 2^+} C(x)$

(f) No

Chapter 12 Review Exercises

1. (a) 1 **(b)** 1 **(c)** 1

3. 23.591 **5.** 36 **7.** 8 **9.** Does not exist

11. **(a)** 2 **(b)** 0 **(c)** $-\frac{3}{2}$ **(d)** Does not exist **(e)** 1 **(f)** 1

13. $2x^2 - 9$ **15.** **(a)** $3(2+h)^2$ **(b)** 12

17. **(a)** $C(400) = 38{,}000$; when 400 teddy bears are made, the costs are \$38,000.

(b) $\lim_{x\to 100} C(x) = 37{,}000$; as the number of teddy bears produced approaches 100, the costs approach \$37,000.

(c) $AC(25) = 1460$; when 25 teddy bears are made, the average cost per bear is \$1460.

19. **(a)**

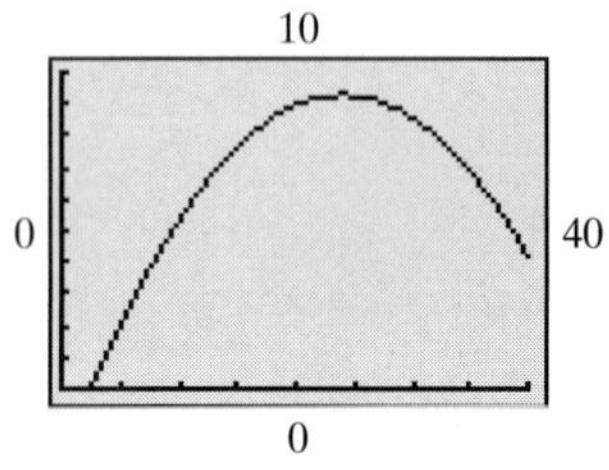

(b) $R(20) = 8.96$; $R(25) = 9.26$; $R(30) = 8.56$

(c) 0 **(d)** 24 tablespoons

21. **(a)** ∞ **(b)** ∞ **(c)** ∞

23. **(a)** $(-\infty, 0) \cup (0, \infty)$ **(b)** Does not exist

25. **(a)** Exponential growth **(b)** ∞ **(c)** 0

27. **(a)** Exponential growth **(b)** ∞ **(c)** 0

29. -2 **31.** $-\infty$ **33.** $y = 0$

35. No; $\lim_{x\to 0} f(x)$ does not exist

37. $g(x)$ is continuous at $x = 9$ since $g(9)$ defined, $\lim_{x\to 9} g(x)$ exists, and $\lim_{x\to 9} g(x) = g(9)$; $g(x)$ is not continuous at $x = -3$ since $g(-3)$ is not defined.

39. $(-\infty, -3) \cup (-3, 5) \cup (5, \infty)$ **41.** **(a)** 0 **(b)** ∞

43. **(a)** 2.4; on April 1, the mosquito population is 2,400,000.

(b) ∞; as time increases without bound, so does the population.

(c) About 92 days

45. **(a)** \$21,875

(b) 25; as time increases without bound, the annual profit approaches \$25,000.

47. $\dfrac{g(9) - g(4)}{9 - 4} = \dfrac{4}{5}$ **49.** $\dfrac{f(7.9) - f(6.2)}{7.9 - 6.2} = -\dfrac{50}{413} \approx -0.1211$

51. 2.5 **53.** ≈ 0.4610 **55.** $f'(7) = 12$ **57.** $f'(3) = -\frac{2}{27}$

59. $f'(x) = 3$ **61.** $h'(x) = 6.8x + 1.9$ **63.** $m_{\tan} = 6$

65. $m_{\tan} = -\frac{1}{49}$

67. **(a)** $A(6) = 4894.40$; after 6 years, there is \$4894.40 in the account.

(b) $A'(6) = 399.29$; after 6 years, the account is increasing at the rate of about \$399.29 per year.

69. **(a)** $P'(4) = 5$ **(b)** $y = 5t + 40$

(c) $m_{\tan} = 5$ means the fish population is growing at the rate of 5 fish per year after 4 years.

71. $g'(x) = \dfrac{3}{5x^{4/5}}$ **73.** $f'(x) = -\frac{14}{15}x^{1/6}$

75. $h'(x) = -27x^2 + 1$ **77.** $g'(x) = 12.46x + 1.98$

79. $f'(x) = \dfrac{1}{\sqrt{x}} + 14x - \dfrac{3}{2}x^2$ **81.** $h'(x) = 7.8832x^{2.79}$

83. **(a)** $(-\infty, 0) \cup (0, \infty)$

(b) $f(x) = 6x^2 + 25x - \dfrac{9}{x} + \dfrac{11}{x^2}$

(c) $f'(x) = 12x + 25 + \dfrac{9}{x^2} - \dfrac{22}{x^3}$

(d) $(-\infty, 0) \cup (0, \infty)$

85. **(a)** $f'(x) = \dfrac{3}{4}x^{-1/4} = \dfrac{3}{4x^{1/4}}$

(b) $m_{\tan} = \frac{3}{8}$

(c) $y = \frac{3}{8}x + 2$

(d)

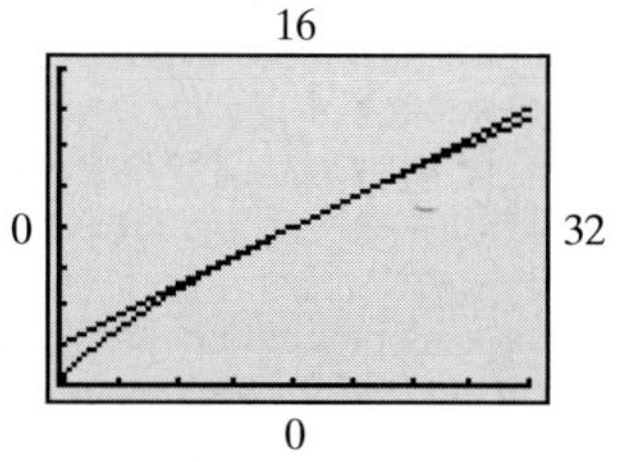

87. $h'(4) = 1600$; after 4 seconds, the rate of change of altitude (i.e., the velocity) is 1600 feet per second.

89. **(a)** $S(t) = 1.172(2)^{t/1.5}$, where t is the number of years after 1985.

(b) $S'(13) \approx 220.084$; in 1998, clock speed is increasing at a rate of 220.084 MHz per year.

91. $g'(x) = -3x(30x^3 - 8x^2 + 27x + 20)$

93. $f'(x) = 10x + \dfrac{9}{2}\sqrt{x} - 14x^{2/5} - \dfrac{27}{5x^{1/10}} - \dfrac{27}{2\sqrt{x}} - 45$

95. $h'(x) = -\dfrac{5}{(x-1)^2}$ **97.** $g'(x) = \dfrac{-9x + 12\sqrt{x} + 3}{2\sqrt{x}(3x+1)^2}$

99. **(a)** $f'(x) = -3x(9x + 10)$ **(b)** $y = -168x + 204$

(c)

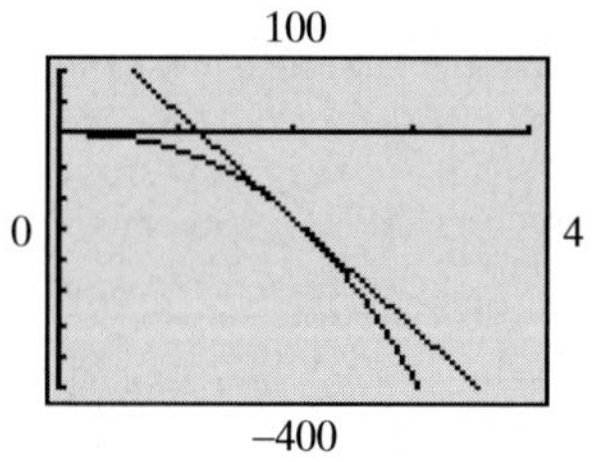

(d)

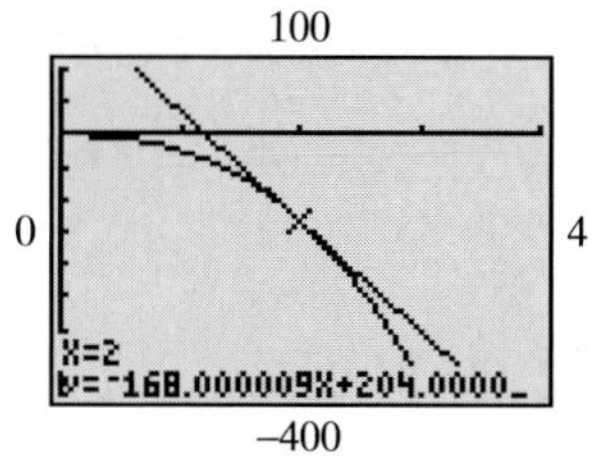

101. $x = 1$

103. **(a)** $[0, \infty)$ **(b)** $f'(x) = \frac{5}{8}x^{-3/8} = \frac{5}{8x^{3/8}}$

(c) Yes; $x = 0$ **(d)** Vertical tangent at $x = 0$.

105. $f'(x) = \frac{33}{2}x^{7/4} + \frac{175}{4}x^{3/4} - \frac{27}{4x^{1/4}}$

Chapter 13

Section 13.1

1. $dy = 6\,dx$ **3.** $dy = (-6x + 2)\,dx$ **5.** $dy = \frac{-5}{(x-1)^2}\,dx$

7. $dy = \frac{1}{(x+1)^2}\,dx$ **9.** $dy = \left(\frac{1}{2\sqrt{x}} - \frac{2}{x^2}\right)dx$

11. $dy = \left(-\frac{1}{2\sqrt{x^3}} + \frac{2}{3\sqrt[3]{x}}\right)dx$ **13.** $dy = \frac{-4x}{(x^2-1)^2}\,dx$

15. $dy = \left(5.1x^{0.7} + \frac{5.6}{x^{0.2}}\right)dx$

17. $\Delta y = 0.21$; $dy = (2x - 2)\,dx = 0.2$

19. $\Delta y = -398$; $dy = (5 - 4x)\,dx = -390$

21. $\Delta y \approx 1.58$; $dy = \frac{270}{x^2}\,dx \approx 1.67$

23. $\Delta y \approx 0.0349$; $dy = \frac{1}{2\sqrt{x}}\,dx \approx .0354$

25. $\Delta y \approx -0.0802$; $dy = \frac{-4x}{(x^2-1)^2}\,dx \approx -0.0889$

27. $\Delta y = 1.4606$; $dy = (24x^3 - 12x^2)\,dx = 1.2$

29. Linear approximation: $\frac{51}{10} = 5.1000$; calculator value: 5.0990

31. Linear approximation: $\frac{80}{27} \approx 2.9630$; calculator value: 2.9625

33. Linear approximation: 1.9938; calculator value: 1.9937

35. Linear approximation: $\frac{97}{343} \approx 0.2828$; calculator value: 0.2828

37. $dA = 2\pi r\,dr = 2\pi \approx 6.28$

39. $dy = \frac{55}{\sqrt{x}}\,dx \approx \$13.91$

41. $dy = (0.66x^2 - 4.7x + 14.32)\,dx = \$1{,}401{,}960$

43. $dy = (120 - 4.8x)\,dx = 7200$ volleyballs

45. $\Delta y = 6960$ volleyballs, $dy = 7200$ volleyballs

47. **(a)** $y = 0.358x + 5.496$

(b) When $x = 17$, the $y -$ value on the tangent line is about $y = 11.582$. This means that if the percentage of males between the ages of 45 to 64 who lived alone in the U.S. continued to increase at the 1995 rate, that about 11.582% would be living alone in 1996.

Section 13.2

1. **(a)** $MC(x) = 23$

(b) $MC(10) = 23$. The approximate cost of producing the next, or 11th, unit is \$23.

(c) $C(11) - C(10) = \$23$

3. **(a)** $MC(x) = x + 12.7$

(b) $MC(11) = 23.7$. The approximate cost of producing the next, or 12th, unit is \$23.70.

(c) $C(12) - C(11) = \$24.20$

5. **(a)** $MC(x) = 0.6x^2 - 6x + 50$

(b) $MC(30) = 410$. The approximate cost of producing the next, or 31st, unit is \$410.

(c) $C(31) - C(30) = \$425.20$

7. **(a)** $R(x) = 6x$; $P(x) = x - 500$

(b) $MP(x) = 1$; $MC(x) = 5$

9. **(a)** $R(x) = -\frac{x^2}{20} + 15x$; $P(x) = -\frac{3x^2}{50} + 8x - 1000$

(b) $MP(x) = -\frac{3x}{25} + 8$; $MC(x) = \frac{x}{50} + 7$

11. **(a)** $R(x) = -0.005x^2 + 7x$;
$P(x) = 0.001x^3 - 0.005x^2 + 3x - 100$

(b) $MP(x) = 0.003x^2 - 0.01x + 3$; $MC(x) = -0.003x^2 + 4$

13. $MC(x) = 40 - 0.002x$; $MC(200) = 39.6$. This means that the next, or 201st, tire to be produced is about \$39.60.

15. $MC(x) = 35 - 0.02x$; $MC(200) = 31$. This means that the 201st rain coat produced costs about \$31.

17. **(a)** $AC(x) = \frac{100 + 40x - 0.001x^2}{x}$; $AC(200) = 40.3$. At a production level of 200 tires per day, it costs the Country Day Company, on average, \$40.30 to produce each lawn tractor tire.

(b) $MAC(x) = -0.001 - \frac{100}{x^2}$; $MAC(200) = -0.0035$. When 200 lawn tractor tires have been produced, the average cost per tire decreases by 0.35¢ for each additional tire produced.

19. $MC(x) = -0.02x + 12$; $MC(500) = 2$; this means that the profit from producing and selling the 501st radio is about \$2.

21. $MP(x) = 5 + \frac{1}{2\sqrt{x}}$; $MP(55) = 5 + \frac{\sqrt{55}}{110} \approx 5.067$. At a sales level of 55 magazine subscriptions, there is about a \$5.07 profit in selling the 56th magazine subscription.

23. **(a)** $AP(x) = 5 + \frac{1}{\sqrt{x}}$; $AP(55) = 5 + \frac{\sqrt{55}}{55} \approx 5.13$. At 55 magazine subscriptions, the profit from selling each magazine is approximately \$5.13.

(b) $MAP(x) = -\frac{1}{2x^{3/2}}$; $MAP(55) = -\frac{\sqrt{55}}{6050} \approx -.00123$. Once 55 magazine subscriptions have been sold, the average profit per subscription decreases by 0.12¢.

25. **(a)** $C(q) = 12q + 1200$; $MC(q) = 12$

(b) $MC(100) = 12$; $MC(150) = 12$. This means that producing the 101st or 151st pocket pager has an approximate cost of \$12.

(c) Since $MC(q) = 12$, producing the 101st pocket pager or 151st pocket pager has exactly the same additional cost—\$12.

27. **(a)** $P(x) = -\frac{x^2}{95} + \frac{39x}{2} - 5500$

(b) $P(500) = \frac{30{,}750}{19} \approx 1618.42$. The New Joy Company makes a total profit of \$1618.42 when it sells 500 items.

(c) $MP(500) = \frac{341}{38} \approx 8.97$. At a production level of 500 units, the New Joy Company would realize about a \$8.97 profit for selling the 501st unit.

29. (a) $R(x) = -\frac{x^2}{30} + 300x$; $C(x) = 30x + 150{,}000$

(b) $P(x) = -\frac{x^2}{30} + 270x - 150{,}000$; smallest production level with profit: 601; largest production level with profit: 7499

(c) $P'(1000) = MP(1000) = \frac{610}{3} \approx 203.33$. This means that at a production level of 1000 hand-held computer devices, there is about \$203.33 profit for selling the 1001st hand held computer device.

31. $C(2001) \approx 57{,}070 + 30(1) = 57{,}100$

Section 13.3

1. (a) 2 (b) $\varepsilon_x = \pm\frac{2}{12} = \pm\frac{1}{6}$ (c) 16.67%

3. (a) 150 (b) $\varepsilon_x = \pm\frac{150}{4150} = \pm\frac{3}{83}$ (c) 3.61%

5. (a) 1500 (b) $\varepsilon_x = \pm\frac{1500}{43{,}500} = \pm\frac{1}{29}$ (c) 3.45%

7. $\varepsilon_y = \pm 0.3$

9. $\varepsilon_y \approx \pm 0.1125$

11. $\varepsilon_y \approx \pm 0.0125$

13. $\varepsilon_y \approx \pm 0.035$

15. $\varepsilon_y = \pm 0.57$

17. The predicted percentage of votes could be 3% more or 3% less for either candidate. Thus, the margin of error does not rule out the possibility that the first candidate garners only 49.5% of the votes while the second candidate wins with 50.5% of the votes.

19. 25% **21.** 8.3%

23. (a) 12.5%

(b) $R(5) \approx 2236$; $MR(5) \approx 671$. This means that the additional revenue generated by selling the next 1000 bolts after 5000 bolts are sold is approximately \$671.

(c) $R(6) \approx 2236 + 671 = \$2907$ (by linear approximation)

(d) $R(6) \approx 2939$. Relative error of approximation is ± 0.011 or $\pm$ 1.1%

25. $Rel(x) = \frac{2}{x}$, $Rel(5) = \frac{2}{5}$

27. $Rel(x) = \frac{1}{x - 100}$, $Rel(125) = \frac{1}{25}$

29. $Rel(x) = \frac{1}{2x}$, $Rel(8) = \frac{1}{16}$

31. $Rel(x) = \frac{6x + 5}{3x^2 + 5x}$, $Rel(2) = \frac{17}{22}$

33. (a) $f(30) = 6.59$. This means that in 1990, the poverty threshold in the U.S. was about \$6590.

(b) $f'(x) = 0.008x + 0.057$; $f'(30) = 0.297$. This means that in 1990, the poverty threshold in the U.S. was increasing at a rate of about $0.297 \frac{\text{thousand dollars}}{\text{year}}$.

(c) $Rel(30) \approx 0.045$. This means that in 1990, the poverty threshold in the U.S. was increasing at about 4.5% per year.

Chapter 13 Review Exercises

1. $dy = 4\,dx$ **3.** $dy = -\frac{8}{(x-5)^2}\,dx$

5. $dy = \left(\frac{1}{2\sqrt{x}} + \frac{12}{x^5}\right)dx$ **7.** $dy = \left(4x^3 - 2 + \frac{2}{3\sqrt[3]{x}}\right)dx$

9. $dy = (20x^4 - 21)\,dx$ **11.** $dy = \left(3.4x^{0.7} - \frac{4}{x^{0.2}}\right)dx$

13. $\Delta y = 2.34$; $dy = 2.3$ **15.** $\Delta y = 0.1$; $dy \approx 0.1051$

17. $\Delta y \approx -0.340$; $dy = -0.4$

19. Linear approximation: 8.0625; calculator: 8.0623

21. Linear approximation: 2.0094; calculator: 2.0093

23. (a) $dy = (90 - 5.4x)\,dx$

(b) An advertising increase of \$400 to \$500 would increase sales by approximately 6840 T-shirts.

25. For exercise 23, $\Delta y = 6570$ T-shirts, 270 fewer than the approximation. For exercise 24, $\Delta y = 6593$ T-shirts, 187 fewer than the approximation.

27. (a) $MC(x) = 18$

(b) $MC(12) = 18$. It costs approximately \$18 to produce the 13th unit.

(c) $C(13) - C(12) = 18$. The answers are exactly the same.

29. (a) $MC(x) = 26.7$

(b) $MC(8) = 26.7$. It costs \$26.70 to produce the 9th unit.

(c) $C(9) - C(8) = 26.7$. The answers are exactly the same.

31. (a) $MC(x) = \frac{1}{2}x + 12$

(b) $MC(31) = \frac{55}{2} = 27.5$. It costs \$27.50 to produce the 32nd unit.

(c) $C(32) - C(31) = 27.75$. The answers vary by \$0.25.

33. (a) $R(x) = 11x$ (b) $P(x) = 4x - 250$

(c) $MP(x) = 4$ (d) $MC(x) = 7$; $MR(x) = 11$

(e) $MR(x) - MC(x) = 4$; $MP(x) = MR(x) - MC(x)$ exactly.

35. (a) $R(x) = -\frac{1}{15}x^2 + 50x$ (b) $P(x) = -\frac{1}{6}x^2 + 47x - 850$

(c) $MP(x) = -\frac{1}{3}x + 47$

(d) $MC(x) = \frac{1}{5}x + 3$; $MR(x) = -\frac{2}{15}x + 50$

(e) $MR(x) - MC(x) = -\frac{1}{3}x + 47$; $MP(x) = MR(x) - MC(x)$ exactly.

37. (a) $R(x) = -0.005x^2 + 10x$

(b) $P(x) = 0.001x^3 - 0.005x^2 + 2x - 100$

(c) $MP(x) = 0.003x^2 - 0.01x + 2$

(d) $MC(x) = -0.003x^2 + 8$; $MR(x) = -0.01x + 10$

(e) $MR(x) - MC(x) = 0.003x^2 - 0.01x + 2$; $MP(x) = MR(x) - MC(x)$ exactly.

39. (a) $C(q) = 85q + 5600$ (b) $MC(q) = 85$

41. **(a)** $MAC(q) = -\dfrac{5600}{q^2}$

(b) $MAC(250) = -0.0896$. This means that when 250 bicycles have been produced, the average cost per bicycle decreases by about \$0.09 for each additional bicycle produced.

43. **(a)** $P(q) = -0.02q^2 + 65q - 5600$

(b) At least 1500 bicycles should be manufactured to maximize profit.

45. **(a)** $MC(250) = 125$; $MC(500) = 125$; at production levels of 250 and 500 bicycles, it costs about \$125 to produce one additional bicycle.

(b) They are equal because the marginal cost function is a constant function.

47. **(a)** $R(q) = -0.022q^2 + 180q$

(b) $MR(q) = -0.044q + 180$

(c) $MR(500) = 158$. This means that the revenue gained from producing and selling the 501st bicycle is about \$158.

49. **(a)** $P(x) = R(x) - C(x) = (6x) - (0.0002x^2 + 2x + 1250)$
$= -0.0002x^2 + 4x - 1250$

(b) $P(3000) = 8950$. If the Coloraura Company produces and sells 3000 water color sets, it will make a total profit of \$8950.

(c) $MP(x) = -0.0004x + 4$

(d) $MP(3000) = 2.8$. The approximate profit by producing the 3001st water color set is \$2.80.

(e) $P(3001) \approx P(3000) + MP(3000) = \$8952.80$

(f) The error is 0.0002 or 0.000002%.

51. $$MAC(x) = \frac{d}{dx}\left[\frac{C(x)}{x}\right] = \frac{C'(x)\cdot x - C(x)\cdot 1}{x^2} = \frac{C'(x) - \frac{C(x)}{x}}{x} = \frac{MC(x) - AC(x)}{x}$$

53. **(a)** $C(x) = 150x + 10{,}000$; $R(x) = -\dfrac{x^2}{75} + 250x$

(b) $P(x) = -\dfrac{x^2}{75} + 100x - 10{,}000$

(c) Smallest: 102 bookshelves. Largest: 7398 bookshelves.

(d) $MP(x) = -\frac{2}{75}x + 100$

(e) $MP(2500) = \frac{100}{3} \approx 33.33$. At a production level of 2500, there is a \$33.33 gain in profit in selling the 2501st bookshelf.

55. $C(201) \approx C(200) + MC(200) = 30{,}830 + 80 = \$30{,}910$

57. $R(101) \approx R(100) + MR(100) = 38{,}000 + 360 = \$38{,}360$

59. $R(301) \approx R(300) + MR(300) = 102{,}000 + 280 = \$102{,}280$

61. **(a)** $MAP(x) = -6000x - 50 + \dfrac{123{,}450}{x^2}$

(b) $MAP(1.5) \approx 45{,}816.67$. At a production level of 1500 printers, the average profit for the next 1000 printers is about \$45,816.67.

63. $\varepsilon_y = \pm 0.05$ **65.** $\varepsilon_y = \pm 0.152$ **67.** $\varepsilon_y = \pm 0.1125$

69. **(a)** 215 **(b)** $\varepsilon_x \approx \pm 0.163$ **(c)** $\pm 16.3\%$

71. **(a)** 250 **(b)** $\varepsilon_x = \pm 0.016$ **(c)** $\pm 1.6\%$

73. **(a)** 94,250 **(b)** $\varepsilon_x \approx \pm 0.00106$ **(c)** $\pm 0.1\%$

75. 54% to 46%

77. **(a)** 5.1%

(b) $C(88) = 5{,}374{,}400$. $MC(88) = 91{,}600$. It will cost the company \$91,600 to produce an additional 1000 modems after 88,000 have already been made.

(c) $C(89) \approx C(88) + MC(88) = 5{,}466{,}000$. The total cost of producing 89,000 modems is about \$5,466,000.

(d) $C(89) = 5{,}466{,}350$. Relative error is about -0.0064%.

79. Highest: 80%. Lowest: 76%

81. $Rel(x) = \dfrac{21x^2 - 2}{7x^3 - 2x}$, $Rel(5) \approx 0.605$

83. $Rel(x) = -\dfrac{x - 8}{x^2 - 4x}$, $Rel(8) = 0$

85. $Rel(x) = \dfrac{1}{2x}$, $Rel(2) = \dfrac{1}{4}$

87. $Rel(x) = \dfrac{7}{-7x + x^2}$, $Rel(2) = \dfrac{-7}{10}$

89. **(a)** $f(18) = 1288.20$. This means that in 1988, there were about 1,288,200 United States households with a male householder and related children under 18.

(b) $f'(x) = 3.04x + 19.9$, $f'(18) = 74.62$. This means that if the number of United States households with a male householder and related children under 18 continued at 1988 rates, such households would increase at a rate of 74,620 per year.

(c) $Rel(18) \approx 0.0579$. This means that the number of such households was growing at a rate of about 5.79% in 1988.

Chapter 14

Section 14.1

1. $2(x + 1)$ **3.** $3(x - 5)^2$ **5.** $-2(2 - x)$ **7.** $6(2x + 4)^2$

9. $-10(5 - 2x)^4$ **11.** $30x(3x^2 + 7)^4$

13. $2(x^3 - 2x^2 + x)(3x^2 - 4x + 1)$ **15.** $27x^2(x^3 - 4)^2$

17. $50(5x^2 - 3x - 1)^9(10x - 3)$

19. $55(4x^2 - x - 4)^{54}(8x - 1)$ **21.** $(2x - 4)^{-1/2}$

23. $\frac{1}{3}(x^2 + 2x)^{-2/3}(2x + 2)$ **25.** $-10(5x - 2)^{-3}$

27. $-(x^2 + 2x + 4)^{-3/2}(x + 1)$

29. $-\frac{1}{4}(3x^3 - x)^{-5/4}(9x^2 - 1)$

31. $-\dfrac{3}{(3x + 4)^2}$ **33.** $\dfrac{-10}{(x - 2)^3}$ **35.** $\dfrac{-2(2x + 2)}{(x^2 + 2x + 3)^2}$

37. $y = 6x - 5$ **39.** $y = -4x + 5$ **41.** $y = 256x - 496$

43. $x = 2$ **45.** $\dfrac{x}{\sqrt{x^2 + 5}}$ **47.** $\dfrac{2}{3}(2x - 1)^{-2/3}$ or $\dfrac{2}{3(\sqrt[3]{2x - 1})^2}$

49. $-5(2x - 8)^{-3/2}$ or $-\dfrac{5}{(\sqrt{2x - 8})^3}$

51. $-\frac{64}{3}(5x^2 - 6x + 3)^{-4/3}(10x - 6)$

53. $4(x-4)^2(x-1)$ **55.** $(x^2+3x)^{1/2} + \dfrac{x(2x+3)}{2(x^2+3x)^{1/2}}$

57. $\dfrac{3x^3-24x^2}{(3x-8)^3}$ **59.** $(x+3)^2(2x-1)(10x+9)$

61. $\dfrac{1}{2\sqrt{\frac{x+3}{x-3}}} \cdot \dfrac{-6}{(x-3)^2}$ **63.** $0.23(4x^2+5x+6)^{-0.77}(8x+5)$

65. $1.03(x+3)^{0.03}$ **67.** $1.7568(x+1)^{0.22}$

69. The number of persons expected to survive to a given age (near age 70) is decreasing at a rate of about $36.5\,\frac{\text{persons}}{\text{year of age}}$.

71. Ten years after the beginning of the study, enrollment is increasing at a rate of $192.09\,\frac{\text{students}}{\text{year}}$.

73. (a) $4.5(3x+6)^{0.5}$

(b) With production at a level of 500, the cost of producing the next 100 is about $20,622.

75. (a) $\dfrac{3}{2\sqrt{3x+5}}$

(b) With production at a level of $15\,\frac{\text{cameras}}{\text{shift}}$, the cost of producing the 16th camera is about $21.20.

77. (a) $-\dfrac{25}{\sqrt{22{,}500-50x}}$

(b) At a production level of 2500 dolls, the price is decreasing at a rate of about $0.171\,\frac{\text{dollars}}{\text{hundred dolls}}$.

79. (a) $\dfrac{-125}{(\sqrt{2x+5})^3}$

(b) At a production level of 2000 units, the price is decreasing at a rate of about $0.414\,\frac{\text{dollars}}{\text{hundred units}}$.

(c) $\dfrac{125x}{\sqrt{2x+5}}$ **(d)** $\dfrac{125x+625}{(\sqrt{2x+5})^3}$

(e) When the production level is 2000, an increase of 100 units produced and sold increases revenue by about $10.35.

81. (a) It will take the subject 72 minutes to learn a list of 12 items.

(b) $2\sqrt{x-3} + \dfrac{x}{\sqrt{x-3}}$

(c) When the list has 12 items, the time to learn additional items is increasing at a rate of $10\,\frac{\text{min}}{\text{item}}$.

83. (a) There will be about 2040 persons surviving 75 years.

(b) $-\dfrac{200}{\sqrt{101-t}}$

(c) The number of persons expected to survive to a given age (near age 75), is decreasing at a rate of about $39.2\,\frac{\text{persons}}{\text{year of age}}$.

85. (a) There are about 17,864 bacteria present 8 hours after the first observation.

(b) $73.5(10+0.5t)^{1.1}$

(c) Eight hours after the first observation, the number of bacteria present is increasing at a rate of about $1340\,\frac{\text{bacteria}}{\text{hour}}$.

Section 14.2

1. $\dfrac{5}{x}$ **3.** $\dfrac{6}{x}$ **5.** $12x^2\ln x + 4x^2$ **7.** $\dfrac{15x^4\ln x - 3x^4}{(\ln x)^2}$

9. $-\dfrac{12}{x}$ **11.** $\dfrac{1}{x+7}$ **13.** $\dfrac{2}{2x-5}$ **15.** $\dfrac{2x}{x^2+3}$

17. $\dfrac{1}{2x+5}$ **19.** $\dfrac{6(\ln x)^5}{x}$ **21.** $\dfrac{1}{2\sqrt{x}}\ln\sqrt{x} + \dfrac{1}{2\sqrt{x}}$

23. $\dfrac{2x+2}{\ln(x+5)} - \dfrac{x^2+2x+3}{[\ln(x+5)]^2(x+5)}$ **25.** $\dfrac{1}{x\ln 10}$ **27.** $\dfrac{6}{x\ln 3}$

29. $2x\log_9 x + \dfrac{x}{\ln 9}$ **31.** $\dfrac{5}{(5x+3)\ln 2}$

33. $\dfrac{1}{\ln 10(x+3)} - \dfrac{2x}{\ln 10(x^2+1)}$ **35.** $y = \dfrac{x}{2} + \ln 2 - 1$

37. $y = x-1$ **39.** $y = 4x-4$ **41.** $y = \dfrac{6}{e}x - 5$

43. (a) $\dfrac{12}{t}$

(b) Twelve hours after the start of the experiment there are about 780 bacteria and the number is growing at a rate of about $1\,\frac{\text{bacterium}}{\text{hour}}$.

45. (a) $\dfrac{5}{x\ln 2}$

(b) The rate of growth after 2 years is about $3.61\,\frac{\text{thousands of prescriptions}}{\text{year}}$ while the growth rate after 10 years is $0.72\,\frac{\text{thousands of prescriptions}}{\text{year}}$.

47. (a) $\dfrac{1.75}{x}$

(b) In 1972 the life expectancy of African American females is about 70.33 years and is increasing at a rate of about $0.58\,\frac{\text{years}}{\text{birth year}}$.

(c) $y = 0.5833x + 68.58$; in 1984 the life expectancy of African American females is 73.15 years according to the model and 77.33 years according to the tangent line approximation.

49. (a) $\dfrac{2}{x}$

(b) In 1984, the consumption growth rate was $0.4\,\frac{\text{gallons/capita}}{\text{year}}$, while in 1989 it was $0.2\,\frac{\text{gallons/capita}}{\text{year}}$.

51. (a)

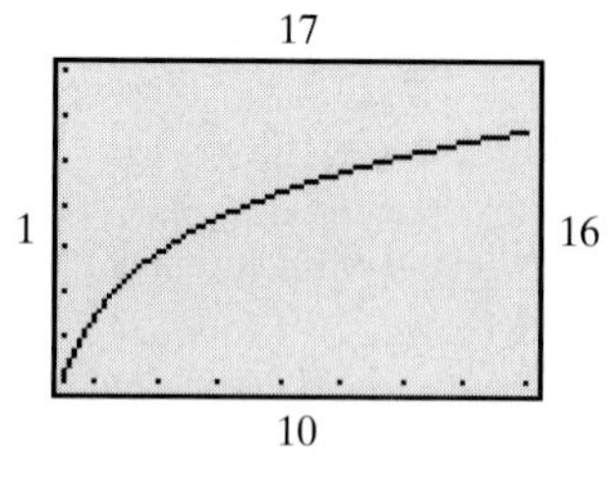

(b)

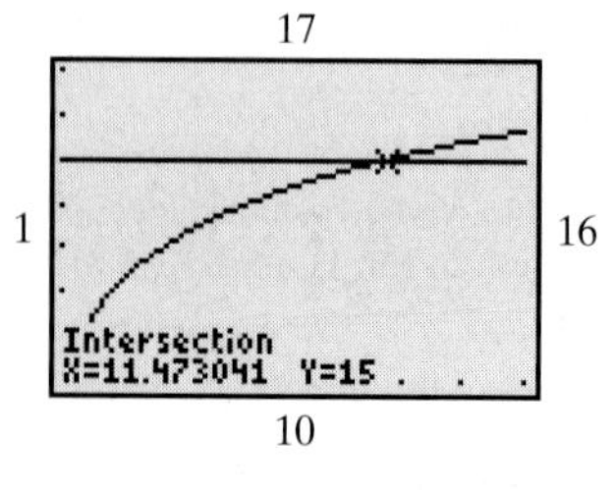

$x \approx 11.473041$. This means that in 1990, the annual per capita consumption of light and skim milk exceeded 15 gallons.

(c) At $x = 5$, $\frac{dy}{dx} \approx 0.40000001$; at $x = 10$, $\frac{dy}{dx} = 0.2$. The calculator's approximation is very close to the exact value found via calculus for $x = 5$ and gives the same value when $x = 10$.

Section 14.3

1. $7e^x$ **3.** $8 + 2e^x + 2xe^x$ **5.** $\dfrac{10e^x}{(5 - e^x)^2}$

7. $8xe^x + 4x^2e^x$ **9.** $-\dfrac{e^x}{2\sqrt{12 - e^x}}$

11. $\dfrac{e^x(x^3 - 1) - 3x^2(e^x - 10)}{(x^3 - 1)^2}$ **13.** $\dfrac{-2e^x}{(e^x - 1)^2}$

15. $2e^x + 2xe^x - 1$ **17.** $2e^{2x-1}$ **19.** $\dfrac{e^{\sqrt{x}}}{2\sqrt{x}}$ **21.** 1

23. $5e^{2x} + 10xe^{2x}$ **25.** 0 **27.** $\dfrac{2x - e^{-x}}{x^2 + e^{-x}}$

29. $10^x \ln 10$ **31.** $-3^{-x} \ln 3$

33. $3x^2 \cdot 0.3^x + x^3 \cdot 0.3^x \ln(0.3)$ **35.** $10^{x+3} \ln 10$

37. $\dfrac{-9^{1/x}}{x^2} \ln 9$ **39.** $e^x + xe^x - 2 \cdot 5^{2x} \ln 5$

41. $\dfrac{1}{x} 5^{x^2} + 2(\ln 5x)5^{x^2} x \ln 5$

43. (a) $0.42e^{.2t}$

(b) Three days after exposure the tumor diameter is growing at a rate of about $0.765 \frac{\text{millimeters}}{\text{day}}$.

45. (a) $12(0.8)^t \ln 0.8$

(b) One year after the factory opened the population was decreasing at a rate of about 214 per year compared to a rate of about 45 per year 8 years after opening.

47. (a) Growth model because $0.16 > 0$ and $e > 1$

(b) $44.928e^{0.16x}$

(c) In 1987, the average salary was increasing at a rate of about $72{,}607 \frac{\text{dollars}}{\text{year}}$, while in 1997 it was increasing at a rate of about $359{,}625 \frac{\text{dollars}}{\text{year}}$.

49. (a) $130e^{0.065t}$

(b) The amount accumulated after 5 years is \$2768.06 and is growing at a rate of \$179.92/year.

51. (a)

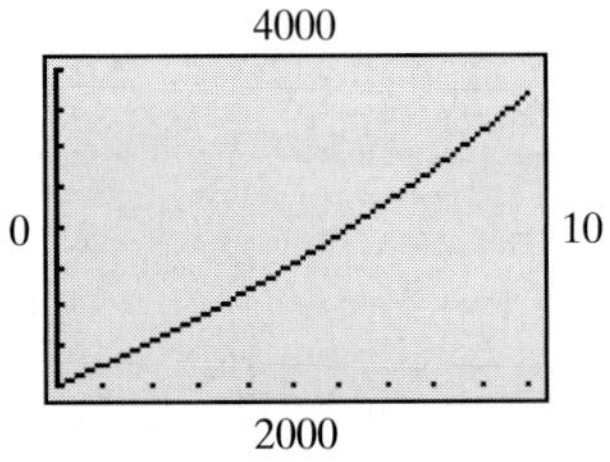

(b) The amount accumulated reaches \$3000 after about 6.24 years.

(c) $A'(2) \approx 148.05 \frac{\text{dollars}}{\text{year}}$

53. (a) $f'(x) = 1.4123e^{0.145x}$

(b) $f'(3) \approx 2.18$. This means that in 1982, the trading volume on the NYSE was increasing at a rate of about 2.18 billion shares per year.

(c) $f'(17) \approx 16.61$. This means that in 1996, the trading volume on the NYSE was increasing at a rate of about 16.61 billion shares per year.

55. (a) Growth model because $1 > 0$ and $1.27 > 1$

(b) $f'(x) = 115.9 \cdot 1.27^x \ln 1.27$. The amount invested was increasing at a rate of about $35.18 \frac{\text{billion dollars}}{\text{year}}$ in 1990 and $238.09 \frac{\text{billion dollars}}{\text{year}}$ in 1998.

(c) $g(x) = 115.9e^{0.24x}$

Section 14.4

1. -2 **3.** $-\dfrac{1}{6y}$ **5.** $\dfrac{-x}{y}$ **7.** $\dfrac{4x^3 - 15x^2}{3y^2}$ **9.** $\left(\dfrac{y}{x}\right)^{3/4}$

11. $-\dfrac{2x + y}{x}$ **13.** $\dfrac{x^2 + 4y}{y^2 - 4x}$ **15.** $\dfrac{6x - y}{x + y}$ **17.** $\dfrac{-y}{x(\ln x + 1)}$

19. $\dfrac{e^y + 2x}{-xe^y + 2y}$ **21.** $\dfrac{3x^2}{5^y \ln 5}$ **23.** $\dfrac{1}{10^{y-2} \ln 10}$ **25.** $6x(\sqrt[3]{y})^2$

27. $-\frac{1}{2}$ **29.** $\dfrac{4x^3 - y}{x}$ **31.** $\dfrac{y - 2xy^2}{x}$ **33.** $y = -\frac{3}{2}x + \frac{13}{2}$

35. $y = 2$ **37.** $y = 2x - 1$ **39.** $y = 2x - 2$

41. (a) $-2x$

(b) At the demand level of 1100 frames, the price is decreasing at a rate of \$22 per hundred frames.

43. (a) $-\dfrac{p + 2}{x}$

(b) At the demand level of 2000 tents, the price is decreasing at a rate of \$2.50 per hundred tents.

45. $2\dfrac{dx}{dt} + 3\dfrac{dy}{dt} = 0$ **47.** $2x\dfrac{dx}{dt} - 3\dfrac{dy}{dt} = 0$

49. $2x\dfrac{dx}{dt} + 2y\dfrac{dy}{dt} = 5\dfrac{dx}{dt}$ **51.** $5\dfrac{dx}{dt}y + 5x\dfrac{dy}{dt} + 4y^3\dfrac{dy}{dt} = \dfrac{dx}{dt}$

53. $-\frac{2}{7}$ **55.** -4 **57.** $\frac{3}{2}$ **59.** $\dfrac{3}{\pi} \dfrac{\text{inches}}{\text{minute}}$

61. $\dfrac{5}{3} \dfrac{\text{inches}}{\text{minute}}$ **63.** $0.32\pi \dfrac{\text{inches}^2}{\text{minute}}$ **65.** $\dfrac{32}{3} \dfrac{\text{feet}}{\text{second}}$

67. (a) $\dfrac{dR}{dt} = 250\dfrac{dx}{dt} - \dfrac{4}{5}x\dfrac{dx}{dt}$ **(b)** $34{,}000 \dfrac{\text{dollars}}{\text{month}}$

69. (a) $2x\dfrac{dx}{dt} + 2y\dfrac{dy}{dt} = 0$ **(b)** $-6.87 \dfrac{\text{feet}}{\text{second}}$

71. (a) $\dfrac{dy}{dt} = -\dfrac{2500}{(1 + x)^2}\dfrac{dx}{dt}$ **(b)** $-160 \dfrac{\text{bass}}{\text{year}}$ **73.** $620 \dfrac{\text{dollars}}{\text{week}}$

Section 14.5

1. $\frac{11}{10}$ **3.** $\frac{16}{15}$ **5.** $\frac{13}{19}$

7. (a) One thousand units will be sold at a price of 5, while 800 units will be sold at a price of 6.

(b) $E_a = 1$; the relative changes in price and quantity sold are equal.

9. Since $E_a = 1.9125 > 1$, the demand is elastic.

11. Since $E_a \approx 2.01 > 1$, the demand is elastic.

13. $x = \dfrac{-p}{100} + 6$ **15.** $x = \sqrt{\dfrac{300}{p - 10}}$ **17.** $x = -25p + 300$

19. $x = \sqrt{300 - p^2}$ **21.** $x = -10 \ln\left(\frac{p}{100}\right)$

23. 0.294, inelastic **25.** 0.941, inelastic **27.** 1.0, unitary

29. $\frac{3}{4}$, inelastic **31.** 2.0, elastic **33.** 4.0, elastic

35. 0.0350, inelastic **37.** 2.0, elastic

39. **(a)** $E(4) = \frac{1}{3}$, inelastic **(b)** Raised

41. **(a)** $E(15) = 3$, elastic **(b)** Lowered

43. **(a)** 0.190, raise **(b)** \$12.50

45. **(a)** 0.58, raise **(b)** \$1.83

47. **(a)** 0.511, raise **(b)** \$116.71

Chapter 14 Review Exercises

1. $3(x+2)^2$ **3.** $-3(8-x)^2$ **5.** $8(2x+5)^3$

7. $6(x^2 - 5x + 3)(2x - 5)$ **9.** $\frac{1}{3}(2x^2 - 5x + 7)^{-2/3}(4x - 5)$

11. $-\dfrac{6}{(2x+9)^2}$ **13.** $-\dfrac{18}{(3x+5)^4}$ **15.** $y = 400x + 368$

17. $y = 32x - 80$ **19.** $\dfrac{7}{2\sqrt{7x - 12}}$ **21.** $\dfrac{-6}{(\sqrt{4x} + 5)^3}$

23. $(x^2+5)^3 + 6x^2(x^2+5)^2$ **25.** $\dfrac{3}{\sqrt{5x-6}} - \dfrac{5}{2}\dfrac{3x-7}{(\sqrt{5x-6})^3}$

27. $0.67(3x^2 - x + 1)^{-0.33}(6x - 1)$

29. $-0.7(x^3 - x^2 + 5x + 1)^{-1.7}(3x^2 - 2x + 5)$

31. **(a)** $15(10x - 8)^{0.5}$

(b) With production at a level of 700 windows, the cost is increasing at a rate of about $11{,}811 \frac{\text{dollars}}{\text{hundred windows}}$.

(c) $\dfrac{(10x - 8)^{1.5} + 480}{x}$

(d) $15\dfrac{(10x-8)^{0.5}}{x} - \dfrac{(10x-8)^{1.5} + 480}{x^2}$

(e) With production at a level of 700 windows, the average cost is decreasing at a rate of about $289 \frac{\text{dollars}}{\text{hundred windows}}$.

33. $\dfrac{5}{x}$ **35.** $\dfrac{6x}{3x^2 - 5}$ **37.** $\dfrac{2x+5}{\ln(x+4)} - \dfrac{x^2 + 5x - 2}{[\ln(x+4)]^2(x+4)}$

39. $12x^2 \log_2 x + \dfrac{4x^2}{\ln 2}$ **41.** $\dfrac{1}{\ln 4}\left(\dfrac{2x}{x^2+5} - \dfrac{2}{2x-3}\right)$

43. $\dfrac{1}{2\sqrt{(e^x - 5)}}e^x$ **45.** $2e^{2x-13}$

47. $2\dfrac{e^{2x}}{\ln(x-4)} - \dfrac{e^{2x}}{(\ln(x-4))^2(x-4)}$

49. **(a)** $y = 4e^{-4}x + 12$

(b)

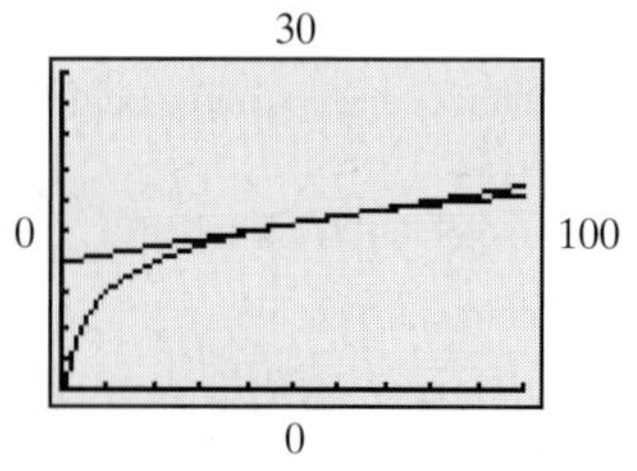

51. **(a)** $y = 2x$

(b)

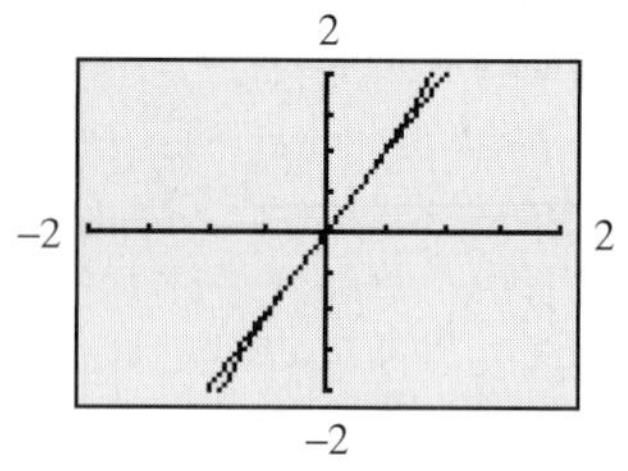

53. $-5^{-x} \ln 5$ **55.** $\frac{1}{3}\sqrt[3]{10^x} \ln 10$ **57.** $\dfrac{0.4^{\sqrt{x}}}{2\sqrt{x}} \ln(0.4)$

59. $\dfrac{1}{x \ln 7} - \dfrac{1}{x \ln 3}$

61. **(a)** Growth **(b)** $f'(x) = 2149.6 \cdot (1.036)^x \ln 1.036$

(c) In 1976 the number of stations was increasing at a rate of about $97 \frac{\text{stations}}{\text{year}}$ while in 1981 it was increasing at a rate of about $116 \frac{\text{stations}}{\text{year}}$.

63. $\frac{3}{5}$ **65.** $\dfrac{2x - 30x^5}{3y^2}$ **67.** $\dfrac{2xy}{3y^2 - x^2}$ **69.** $\dfrac{y \ln y}{y - x}$

71. $-\frac{2}{3}$ **73.** $\dfrac{2 - 3x^2y}{x^3 - 1}$ **75.** $y = \frac{5}{4}x - \frac{9}{4}$

77. $y = \frac{4}{3}x - \frac{1}{3}$

79. **(a)** $-\dfrac{p + 10}{x}$

(b) At a demand level of 15,000 lamps and a price of \$30, the price is decreasing at a rate of $2.67 \frac{\text{dollars}}{\text{thousand lamps}}$.

81. $3x^2\dfrac{dx}{dt} + 3y^2\dfrac{dy}{dt} = 6\dfrac{dx}{dt}$

83. $\dfrac{dx}{dt}y + x\dfrac{dy}{dt} - 3y^2\dfrac{dy}{dt} = 2\dfrac{dx}{dt}$ **85.** 20 **87.** −11

89. **(a)** $\dfrac{dS}{dt} = 8\pi r\dfrac{dr}{dt}$ **(b)** $140.7 \dfrac{\text{centimeters}^2}{\text{second}}$

91. 1.25 **93.** 1.60

95. Since $E_a = 1.38 > 1$, the demand is elastic.

97. $\dfrac{12}{p - 5}$ **99.** $-50.0 \ln \dfrac{p}{800}$

101. **(a)** $E(p) = \dfrac{2p^2}{400 - p^2}$

(b) $E(12) = 1.125$, elastic

103. **(a)** $E(p) = \dfrac{3p}{(120 - 3p)\ln(120 - 3p)}$

(b) $E(5) = 0.0307$, inelastic

105. **(a)** $E(p) = \dfrac{5p}{(150 - 5p)\ln(150 - 5p)}$; $E(25) = 1.55$, lower

(b) $p = 23.34$

Chapter 15

Section 15.1

1. Increasing: $(-1, 2)$; decreasing: $(-\infty, -1) \cup (2, \infty)$; constant: $x = -1$ and $x = 2$

3. Increasing: $(-\infty, -1)$; constant: $[-1, 1]$; decreasing: $(1, \infty)$

5. Increasing: $(-2, 0) \cup (1, \infty)$; decreasing: $(-\infty, -2) \cup (0, 1)$; constant: $x = -2$, $x = 0$ and $x = 1$

7. **(a)** $(-\infty, -2) \cup (0, 2)$

(b) $(-2, 0) \cup (2, \infty)$

(c) Zero: $x = -2$, $x = 0$ and $x = 2$

9. **(a)** $(-\infty, -2) \cup (0, 1) \cup (2, \infty)$

(b) $(-2, 0)$

(c) Zero: $x = 0$; undefined: $x = -2$ and $x = 2$

11. **(a)** $(-\infty, 2) \cup (2, \infty)$

(b) None

(c) undefined: $x = 2$

13. None **15.** $x = \frac{5}{2}$ **17.** $x = -5$ and $x = 3$

19. $x = -\frac{1}{2}$ **21.** None **23.** None **25.** $x = -1$

27. None

29. **(a)** Increasing: $(\frac{5}{2}, \infty)$
Decreasing: $(-\infty, \frac{5}{2})$

(b) Relative minimum at $x = \frac{5}{2}$

(c)

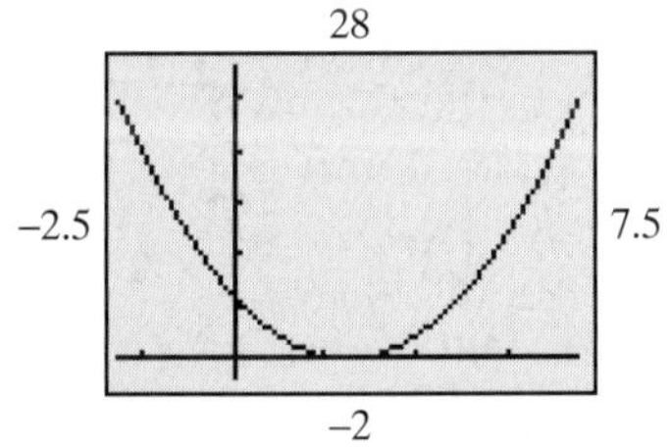

31. **(a)** Increasing: $(-\infty, -5) \cup (3, \infty)$
Decreasing: $(-5, 3)$

(b) Relative maximum at $x = -5$
Relative minimum at $x = 3$

(c)

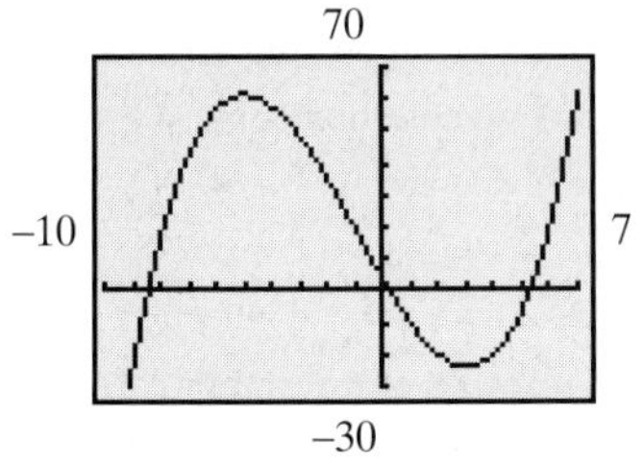

33. **(a)** Increasing: $(-\frac{1}{2}, \infty)$

(b) None

(c)

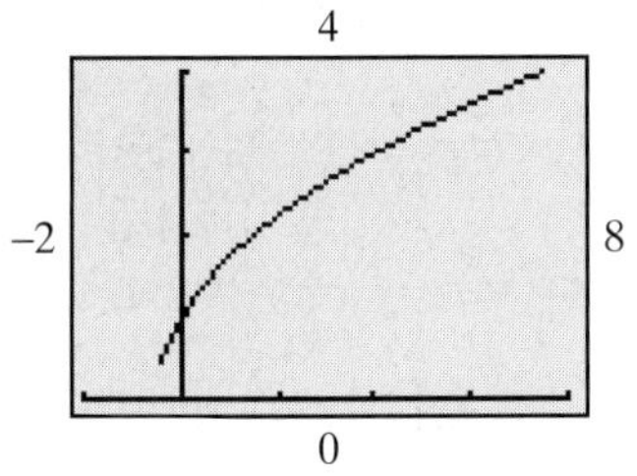

35. **(a)** Increasing: $(0, \infty)$

(b) None

(c)

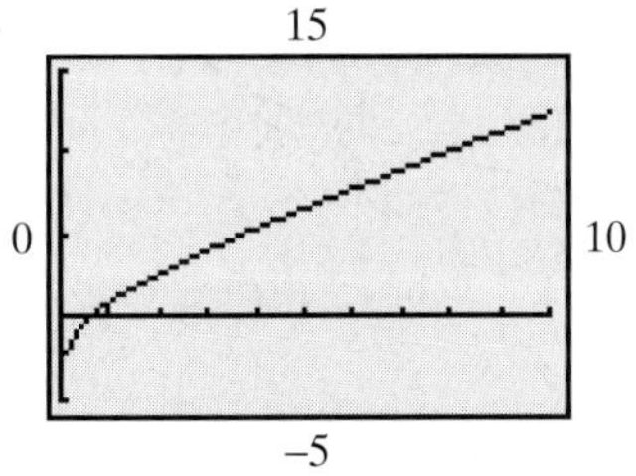

37. **(a)** Increasing: $(-\infty, \infty)$

(b) None

(c)

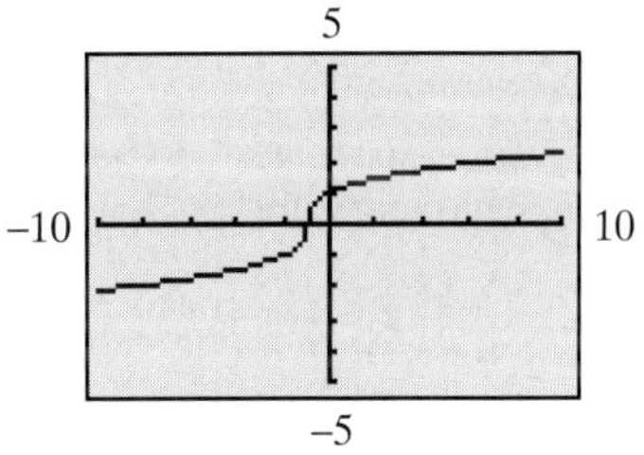

39. Relative maximum at $x = \frac{5}{2}$

41. Relative maximum at $x = -1$,
Relative minimum at $x = 0$

43. Relative maximum at $x = -3$,
Relative minimum at $x = 1$

45.

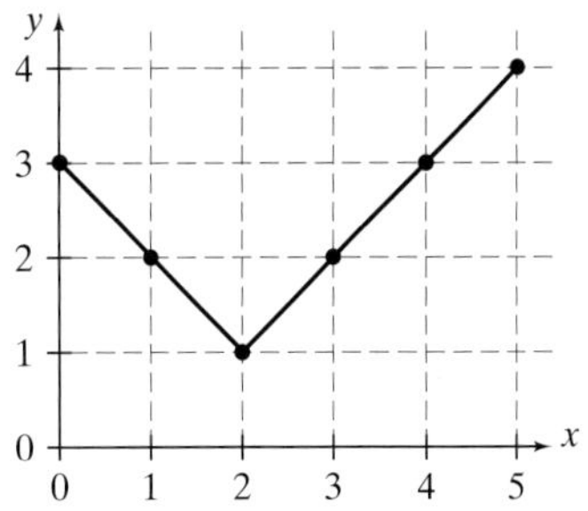

47.

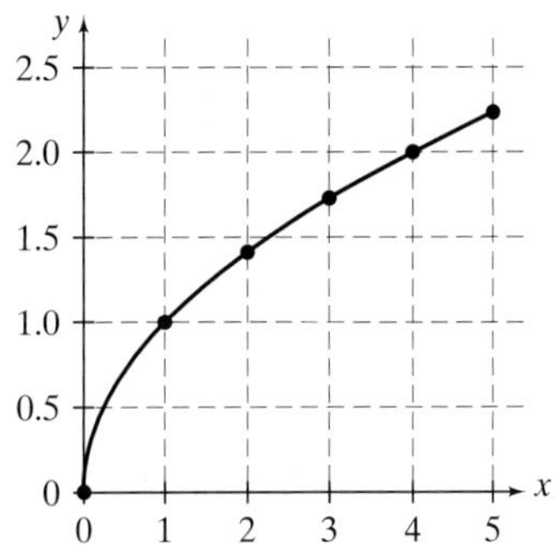

49. (a)

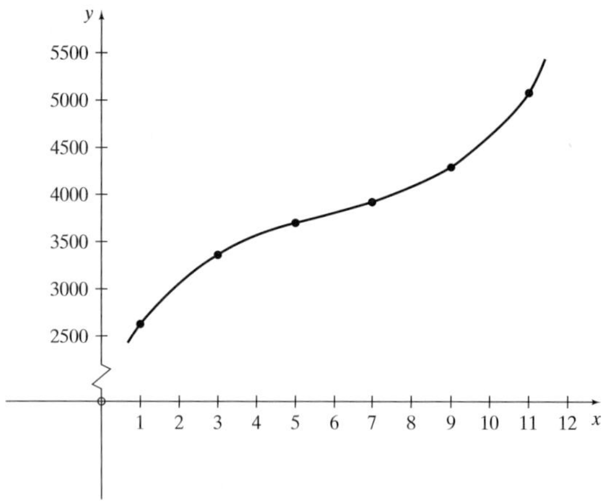

(b) $\frac{\text{number of liver transplants}}{\text{year}}$

51. (a) Increasing: (0, 500); decreasing: (500, 1000)

(b) Relative maximum = (500, 250,000). When 500 golf clubs are produced and sold, a maximum profit of \$250,000 is realized.

53. (a) Increasing: (1, 13.76); decreasing: (13.76, 19)

(b) Relative minimum ≈ (13.76, 235.73). In 1992 the U.S. egg consumption was at a minimum of about 236 eggs consumed per person.

55. (a) Increasing: $(1, 8.67) \cup (19, 20)$; decreasing: (8.67, 19)

(b) Relative maximum ≈ (8.67, 15.14); relative minimum ≈ (19, 14.04). In 1987 U.S. fish and shellfish consumption was at a maximum of about 15.14 pounds consumed per person and in 1998 it was at a minimum of about 14.04 pounds consumed per person.

57. $f'(x) = -0.81x^2 + 5x - 27.56$; since $f'(x)$ is negative for all values of x in the interval $1 \le x \le 10$ this means the graph of f is decreasing in the interval $1 \le x \le 10$. Therefore, Alaskan oil field production was decreasing from 1990 to 1999.

59. (a) \$6.67

(b) $R(x) = x \cdot p(x) = 15.22xe^{-0.015x}$

(c) Increasing: (0, 67); decreasing: $(67, \infty)$

(d) Relative maximum ≈ (67, 373.27). This means that the maximum revenue of \$373.29 is obtained when 67 pizzas are sold.

61. (a) $p'(t) = \dfrac{-230(t-3)}{(t+3)^3}$; critical value: $t = 3$

(b) Increasing: [0, 3); decreasing: (3, 20]

(c) Relative maximum ≈ (3, 19.17). The concentration of medicine is highest at about 19.17% 3 hours after administration.

63. (a) $AC(x) = \dfrac{50}{x} + 1 + \dfrac{x}{40}$; critical value: $x \approx 44.72$

(b) Increasing: (44.72, 100]; decreasing: (0, 44.72)

(c) Relative minimum ≈ (44.72, 3.24). This means that average production costs are lowest when about 45 coffee mugs are made, averaging about \$3.24 per mug.

65. (a) $P'(x) = R'(x) - C'(x)$ so $P'(x) = 0$ when $R'(x) - C'(x) = 0$; that is, when $R'(x) = C'(x)$.

(b) When $R'(x) > C'(x)$, then $R'(x) - C'(x) = P'(x) > 0$ so $P(x)$ is increasing.

(c) When $R'(x) < C'(x)$, then $R'(x) - C'(x) = P'(x) < 0$ so $P(x)$ is decreasing.

Section 15.2

1. $f'(x) = -20x^4 - 18x^2 + 7$; $f''(x) = -80x^3 - 36x$; $f'''(x) = -240x^2 - 36$

3. $y' = 21x^2 - 6x + 4$; $y'' = 42x - 6$; $y''' = 42$

5. $f'(x) = e^x$; $f''(x) = e^x$; $f'''(x) = e^x$

7. $f'(x) = \dfrac{1}{2\sqrt{x}}$; $f''(x) = -\dfrac{1}{4\sqrt{x^3}}$; $f'''(x) = \dfrac{3}{8\sqrt{x^5}}$

9. $f'(x) = \dfrac{1}{x}$; $f''(x) = -\dfrac{1}{x^2}$; $f'''(x) = \dfrac{2}{x^3}$

11. (a) Concave up: $(1, \infty)$; concave down: $(-\infty, 1)$

(b) f' increasing: $(1, \infty)$; f' decreasing: $(-\infty, 1)$

13. (a) Concave up: $(-\infty, \infty)$ **(b)** f' increasing: $(-\infty, \infty)$

15. (a) Concave up: $(-\infty, -1) \cup (1, \infty)$; concave down: $(-1, 1)$

(b) f' increasing: $(-\infty, -1) \cup (1, \infty)$; f' decreasing: $(-1, 1)$

17. Concave up: $(-2, \infty)$; concave down: $(-\infty, -2)$; inflection point: $(-2, -25)$

19. Concave up: $(-\infty, \frac{1}{6})$; concave down: $(\frac{1}{6}, \infty)$; inflection point: $(\frac{1}{6}, \frac{-134}{9})$

21. Concave up: $(-\infty, \infty)$; no inflection point(s)

23. Concave down: $(-\infty, \infty)$; no inflection point(s)

25. Concave down: $(-\infty, 0)$; concave up: $(0, \infty)$; no inflection point(s)

27. Concave down: $(-\infty, 0) \cup (0, \infty)$; no inflection point(s)

29. Concave up: $(-\infty, \infty)$; no inflection point(s)

31. Concave up: $(-\infty, \infty)$; no inflection point(s)

33. Concave down: $(-1, \infty)$; no inflection point(s)

35. (a) Increasing: $(-\frac{3}{4}, \infty)$; decreasing: $(-\infty, -\frac{3}{4})$

(b) Relative minimum at $x = -\frac{3}{4}$

(c) Concave up: $(-\infty, \infty)$ **(d)** No inflection points

37. (a) Increasing: $(-\infty, -5) \cup (1, \infty)$; decreasing: $(-5, 1)$

(b) Relative maximum at $x = -5$, relative minimum at $x = 1$

(c) Concave down: $(-\infty, -2)$; concave up: $(-2, \infty)$

(d) $(-2, 41)$

39. (a) Increasing: $\left(\dfrac{5-\sqrt{43}}{9}, \dfrac{5+\sqrt{43}}{9}\right)$;

decreasing: $\left(-\infty, \dfrac{5-\sqrt{43}}{9}\right) \cup \left(\dfrac{5+\sqrt{43}}{9}, \infty\right)$

(b) Relative minimum at $x = \dfrac{5-\sqrt{43}}{9}$; relative maximum at $x = \dfrac{5+\sqrt{43}}{9}$

(c) Concave down: $(\frac{5}{9}, \infty)$; concave up: $(-\infty, \frac{5}{9})$

(d) $x = \left(\frac{5}{9}, \frac{-1667}{243}\right)$

41. **(a)** Increasing: $(-\infty, -\sqrt{5}) \cup (\sqrt{5}, \infty)$; decreasing: $(-\sqrt{5}, 0) \cup (0, \sqrt{5})$

(b) Relative maximum at $x = -\sqrt{5}$; relative minimum at $x = \sqrt{5}$

(c) Concave up: $(0, \infty)$; concave down: $(-\infty, 0)$

(d) None

43. **(a)** Increasing: $(1, \infty)$; decreasing: $(-\infty, 1)$

(b) Relative minimum at $x = 1$

(c) Concave down: $(-\infty, 1) \cup (1, \infty)$

(d) None

45. **(a)** Increasing: $(-\infty, \infty)$ **(b)** None

(c) Concave up: $(-\infty, \infty)$ **(d)** None

47. **(a)** Increasing: $(-1, \infty)$ **(b)** None

(c) Concave down: $(-1, \infty)$ **(d)** None

49.

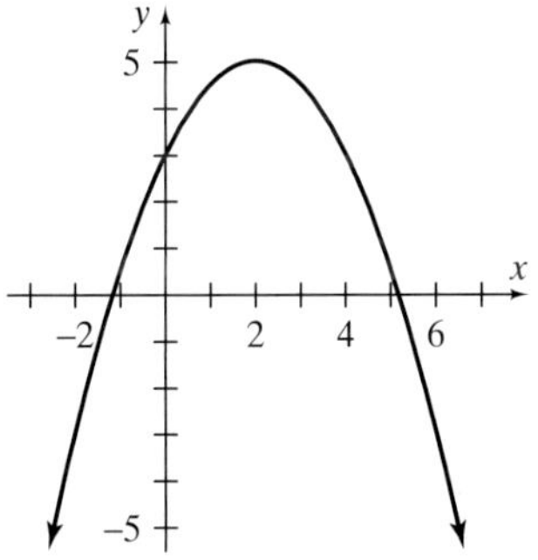

Graphs may vary.

51.

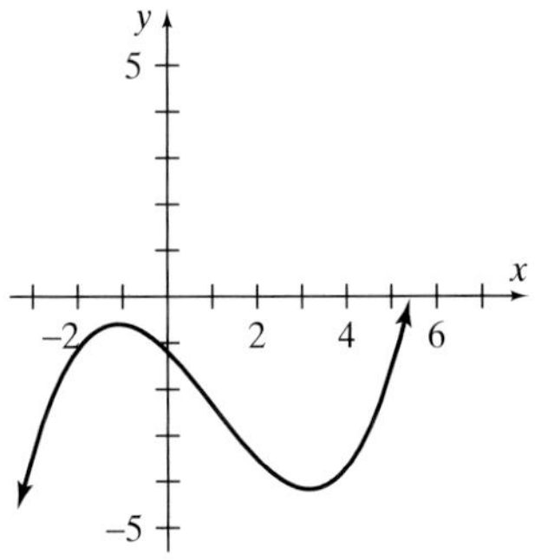

Graphs may vary.

53. If $a > 0$, then $f''(x) = 2a > 0$ for all x, so f is concave up on $(-\infty, \infty)$. A similar argument shows that $a < 0$ implies f is concave down on $(-\infty, \infty)$.

55. **(a)** $C(5) = 50$. It costs \$5000 to produce 5 units; $C'(5) = 28$. It costs approximately \$2800 to produce the sixth unit.

(b) Decreasing: $(0, 2)$; increasing: $(2, \infty)$; relative minimum: $(2, 1)$

(c) Inflection point for graph of C is $(2, 20)$.

57. **(a)** $P(35) = 393{,}812.50$. The profit from selling 3500 refrigerators is \$393,812.50; $P'(35) = 21{,}197.50$. When the production level is 3500 refrigerators, the profit gained from selling the next 100 refrigerators is about \$21,197.50.

(b) Increasing: $(0, 64.49)$; decreasing: $(64.49, \infty)$; marginal profit is maximized at about 6449 refrigerators.

59. The marginal profit is increasing where $\frac{d}{dx}MP(x) > 0$. Since $MP(x) = P'(x)$, $\frac{d}{dx}MP(x) = \frac{d}{dx}P'(x) = P''(x) > 0$. Hence P is concave up. A similar argument shows that P is concave down where the marginal profit is decreasing.

61. (15,5500). The point at which the rate of change of total sales will no longer increase as a result of advertising is \$15,000,000; This is the same point at which the graph of TS changes from concave up to concave down.

63. **(a)** $b(10) \approx 370.949$. The total number of births to women 35 to 39 years old in the United States in 1994 was about 370,949; $b'(10) \approx 9.274$. In 1994 the total number of births to women 35 to 39 years old in the United States was increasing at a rate of about 9274 $\frac{\text{births}}{\text{year}}$.

(b) On $[1, 14]$, $b'(x) = 52.15x^{-3/4} > 0$ so b is increasing on $[1, 14]$.

(c) $b''(x) = -\frac{3129}{80x^{(7/4)}}$. Over the interval $[1, 14], b''(x) < 0$. Therefore, b is concave down on the entire interval.

(d) Increasing, increasing, slower

65. **(a)** $f(50) \approx 1.00$. At 50 mm-Hg, blood flow is about 1.00 mL/min; $f'(50) \approx 0.026$. When arterial pressure is 50 mm-Hg, blood flow is increasing at a rate of about $0.026 \frac{\text{mL/min}}{\text{mm-Hg}}$.

(b) $f''(x) = 0.278(1.026)^x[\ln(1.026)]^2$ which is always greater than zero on the interval $[20, 120]$. This means that f is concave up on the entire interval, which implies that the rate of blood flow is increasing on the entire interval.

67. **(a)** $CI(20) \approx 3.61$. This means that a 20-year-old has a cardiac index of about $\frac{3.61 \text{ L/min}}{m^2}$; $CI'(20) \approx -0.045$. At the age of 20, one's cardiac index is decreasing at a rate of approximately $0.045 \frac{\frac{\text{L/min}}{m^2}}{\text{year}}$.

(b) Decreasing: $[10, 80]$

(c) $CI''(x) = \frac{2.38875}{x^{(9/4)}}$, which is positive over the interval $[10, 80]$ and therefore CI is concave up. This means that CI is decreasing progressively more slowly over the entire interval.

69. **(a)** $d'(x) = -3.03x^2 + 66.56x - 29.14$. Since $d'(x) > 0$ for all x in the interval $[1, 21]$, this means that d, the federal debt, is increasing on the interval $[1, 21]$.

(b) Inflection point $\approx (11, 3355)$. In 1990, the rate of change of the federal debt was at its greatest.

(c) $d''(x) = -6.06x + 66.56$. $d''(x) > 0$ on the interval $[1, 11]$. Since $d''(x)$ is the first derivative of $d'(x)$, the rate of change of the federal debt, we know that $d''(x) > 0$ implies $d'(x)$ was increasing. The interval $[1, 11]$ tells us that the rate of change of the federal debt was increasing from 1980 to 1990.

(d) $d''(x) = -6.06x + 66.56$. $d''(x) < 0$ on the interval [11, 21]. Since $d''(x)$ is the first derivative of $d'(x)$, the rate of change of the federal debt, we know that $d''(x) < 0$ implies $d'(x)$ was decreasing. The interval [11, 21] tells us that the rate of change of the federal debt was decreasing from 1990 to 2000.

Section 15.3

1. Positive: $(-5, -1) \cup (13/4, 7)$; negative: $(-\infty, -5) \cup (-1, 13/4) \cup (7, \infty)$

3. Positive: $(-\infty, -3) \cup (1, 5)$; negative: $(-3, 1) \cup (5, \infty)$

5. Concave up: $(-1, 4)$; concave down: $(-\infty, -1) \cup (4, \infty)$

7. Positive: $(-\infty, -5) \cup (-\frac{5}{3}, \infty)$; negative: $(-5, -\frac{5}{3})$

9. Positive: $(-3, 1) \cup (5, \infty)$; negative: $(-\infty, -3) \cup (1, 5)$

11. Concave up: $(-\infty, -1) \cup (3, \infty)$; concave down: $(-1, 3)$

13. Relative minimum at $x = \frac{1}{3}$

15. Relative maximum at $x = \dfrac{2 - \sqrt{43}}{3}$; relative minimum at $x = \dfrac{2 + \sqrt{43}}{3}$

17. Relative maximum at $x = -3$; relative minimum at $x = -2$

19. Relative maximum at $x = -2$; relative minimum at $x = 1$

21. Relative minimum at $x \approx -2.25$ and $x \approx 1.57$; relative maximum at $x \approx -0.07$

23. Relative minimum at $x = -2$ and $x = 2$; relative maximum at $x = 0$

25. Relative maximum at $x = 0$

27. Relative minimum at $x = 0$

29.

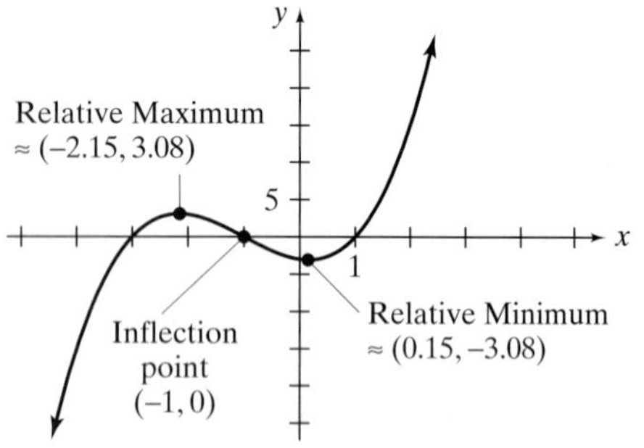

31.

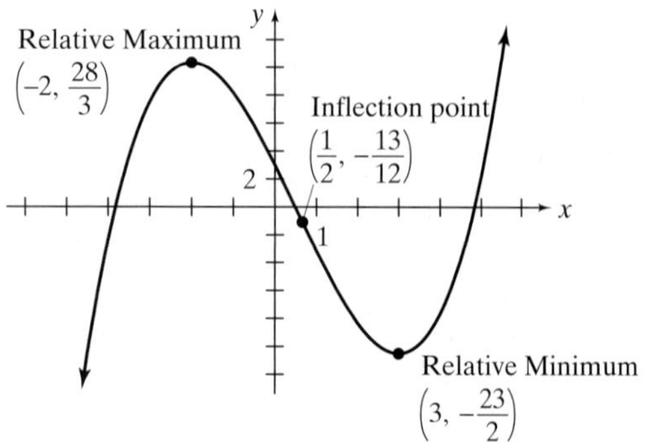

33.

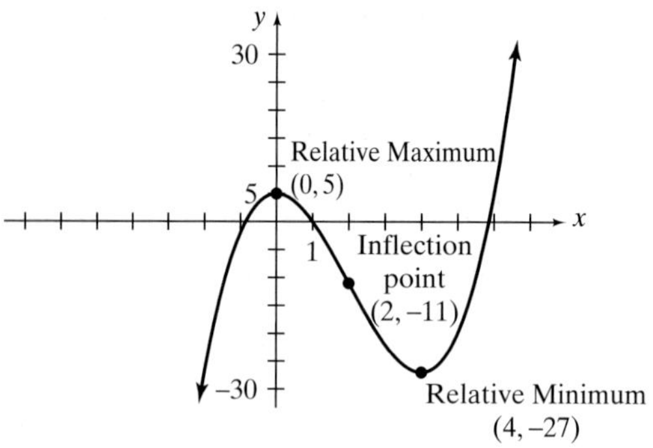

35.

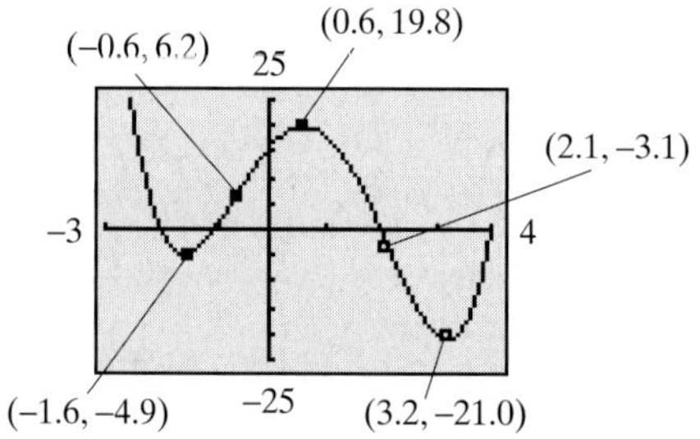

37.

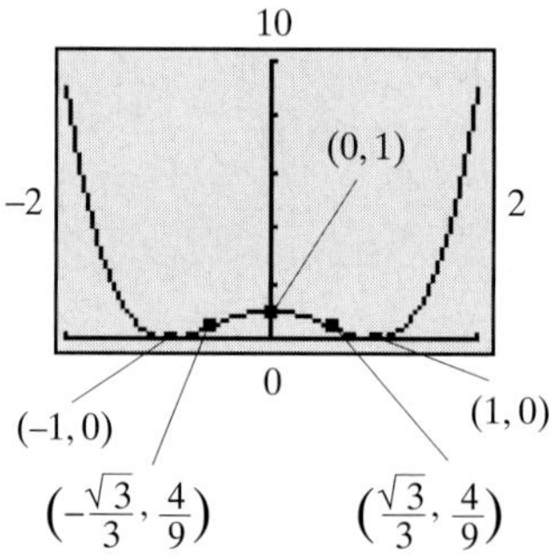

39.

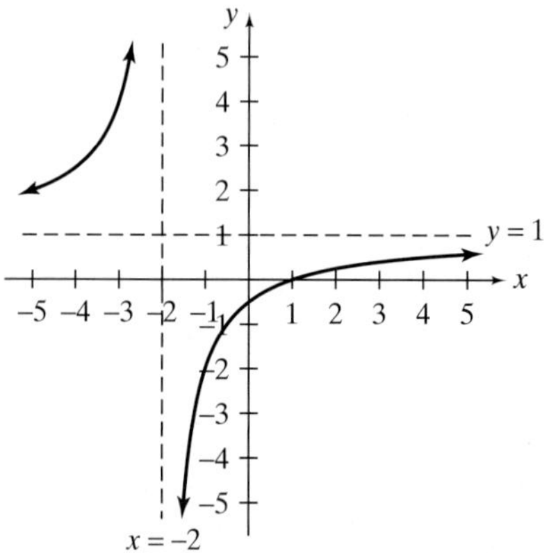

41.

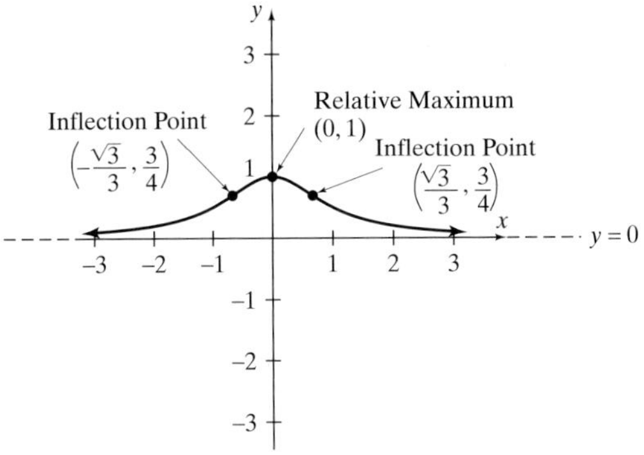

43.

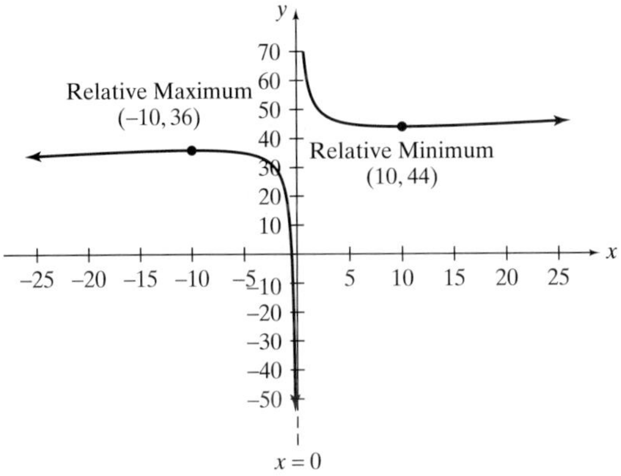

45.

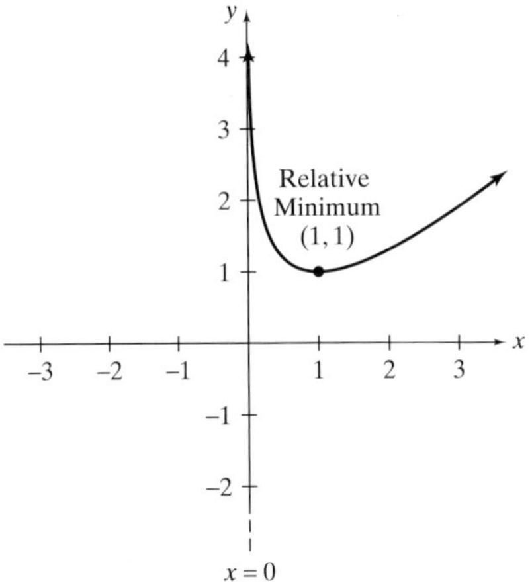

47.

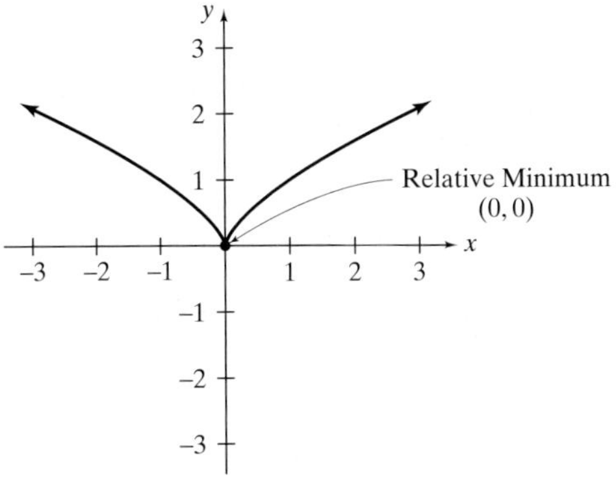

49. Answers will vary. Samples:

(a)

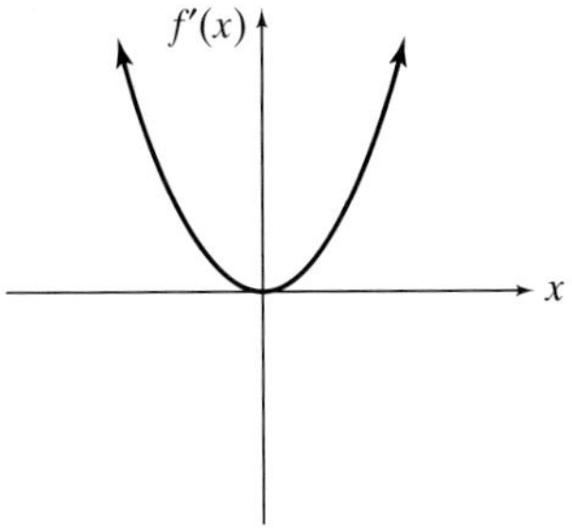

(b)

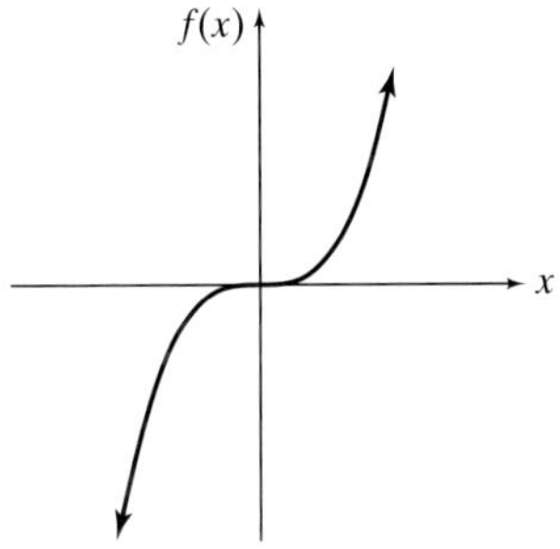

51. Answers will vary. Samples:

(a)

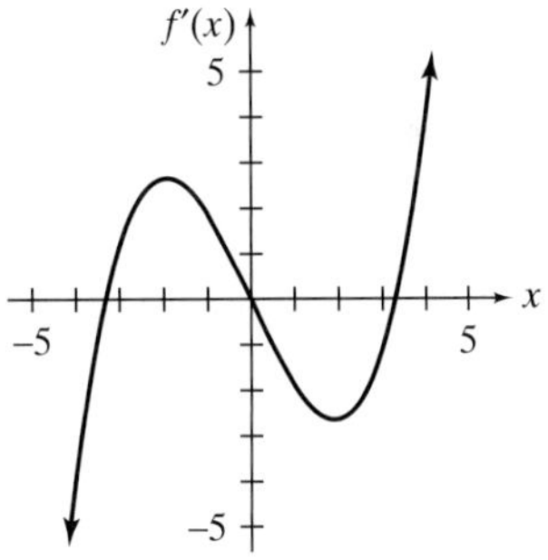

(b)

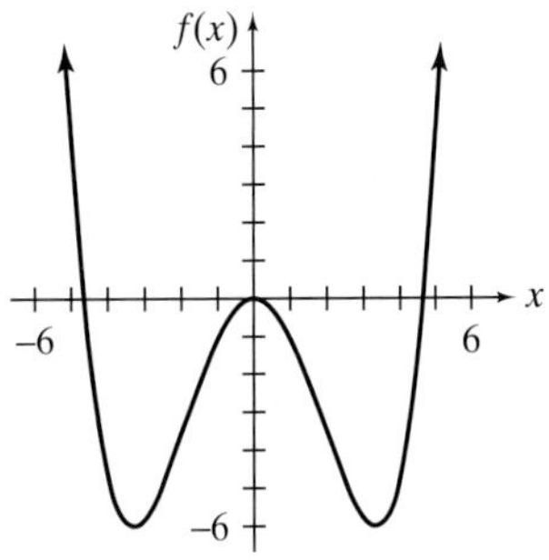

53. (a) $R'(x) = 216 - 0.24x^2$; increasing: [0, 30); decreasing: (30, 50]

(b) Concave down: (0, 50)

55. (a) $AP(x) = -0.01x + 12 - \dfrac{2000}{x}$

(b)

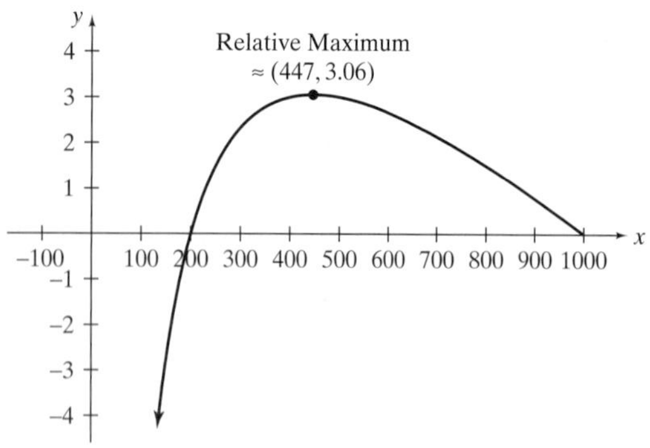

57. $AP(x) = P'(x)$

59. (a) $C'(x) = 3 + \frac{2x}{15}$

(b)

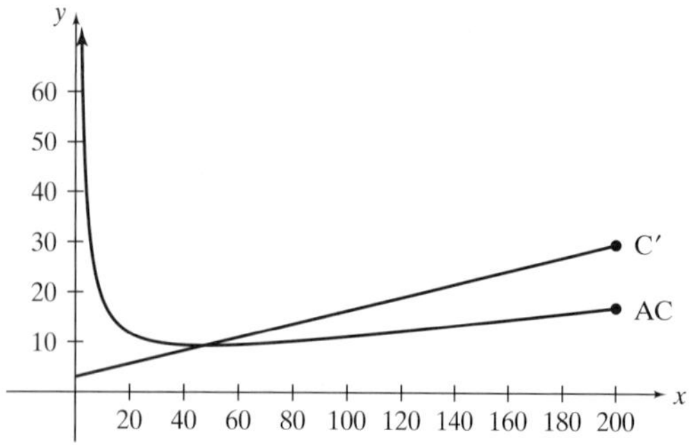

(c) (47.4, 9.32) **(d)** Yes

61. (a) $AC(t) = 100t + 400 + \frac{11,000}{t}$ **(b)** 10.5 **(c)** \$2498

63. (a) $f'(x) = \frac{3800}{x}$. Since $f'(x) > 0$ on the interval [40, 90], f is increasing.

(b) $f''(x) = -\frac{3800}{x^2}$. Since $f''(x) < 0$ on the interval [40, 90], the graph of f is concave down.

65. (a) $f(20) \approx 62.8$. At winds of 20 mph, 62.8% of total heat loss occurs by convection; $f'(20) \approx 0.79$. At winds of 20 mph, the percent of heat loss by convection is increasing by about 0.79% per mph.

(b) $f'(x) = \frac{16.694}{x+1}$. Since $f'(x) > 0$ on the interval [0, 60], f is increasing.

(c) $f''(x) = -\frac{16.694}{(x+1)^2}$. Since $f''(x) < 0$ on the interval [0, 60], the graph of f is concave down.

(d)

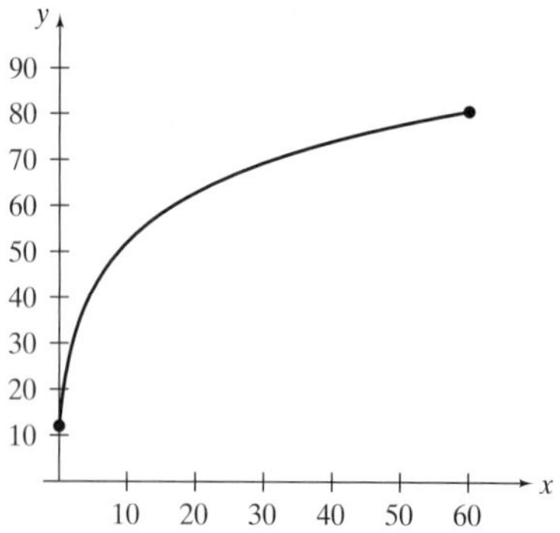

67. (a) $f(7) = 1462.23$. In 1996, the total carbon content of carbon dioxide emitted into the atmosphere by the U.S. was 1462.23 million metric tons. $f'(7) = 23.77$. In 1996 the total carbon content of carbon dioxide emitted into the atmosphere by the U.S. was increasing at a rate of 23.77 million metric tons per year.

(b) Increasing: (1.38, 9]; decreasing: [1, 1.38). Relative minimum at about $t \approx 1.38$.

(c) Concave up: (5.47, 9]; concave down: [1, 5.47). Inflection point at about (5.47, 1421.93).

(d)

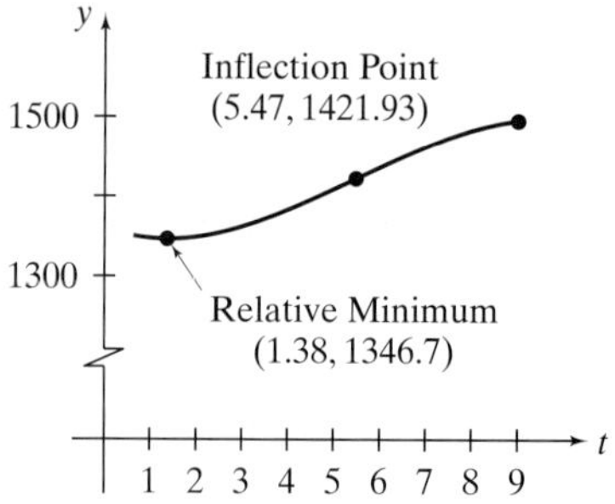

Section 15.4

1. Absolute minimum: (1, −8); absolute maximum: (−2, 1)

3. Absolute minimum: (2, −4); absolute maximum: $\left(\frac{2-\sqrt{19}}{3}, 8.21\right)$

5. Absolute minimum: (−1, −4) and (2, −4); absolute maximum: (0, 0) and (3, 0)

7. Absolute minimum: (2.89, −60.42); absolute maximum: (−0.34, 25.68)

9. Absolute minimum: (0.75, 4.89); absolute maximum: (−2, 29)

11. Absolute minimum: $\left(\frac{\sqrt{2}}{2}, 0\right)$; absolute maximum: (2, 2401)

13. Absolute minimum: (−1, −1); absolute maximum: (1, 1)

15. Absolute minimum: (1, −1); absolute maximum: $(0, -\frac{1}{2})$

17. Absolute minimum: $(4, \frac{1}{17})$; absolute maximum: $(1, \frac{1}{2})$

19. \$46 per day

21. (a) 7 days **(b)** $\frac{15 \text{ harmful bacteria}}{\text{cm}^3}$

23. Highest: 1992—6.95%; lowest: 1999—4.36%

25. Absolute maximum: 8621, in year 1975; absolute minimum: 6519, in year 1998

27. Absolute maximum: 10,434, in year 1960; absolute minimum: 6693, in year 1997

29. (a) The number of Catholic elementary *and* secondary schools in the United States from 1960 to 1997

(b) Absolute maximum: 12,799, in year 1960; absolute minimum: 7864, in year 1997

31. Absolute minimum: (1, 1.55). Natural gas was cheapest in 1995 at an average price of \$1.55 per 1000 cubic feet; absolute maximum: (2.73, 2.36). Natural gas was most expensive in 1996 at an average price of \$2.36 per 1000 cubic feet.

33. $100' \times 100'$

35. **(a)** 8654 magazines **(b)** \$1.88/issue

37. **(a)** 2.24 hours **(b)** $\frac{0.63 \text{ mg}}{\text{cm}^3}$

39. **(a)** $R(x) = 0.14x^3 - 15.52x^2 + 561.21x$
(b) Maximum revenue is \$7211.05.
(c) Demand is 55 units/day.
(d) Price is \$131.11/unit.

41. Absolute minimum: (2.46, 73.78); in 1982, the U.S. total energy consumption was at a minimum of about 73.78 quadrillion Btu. Absolute maximum: (9, 83.891); in 1989, the U.S. total energy consumption was at a maximum of about 83.891 quadrillion Btu.

43. Absolute minimum: (1, 83.944); in 1991, the U.S. total energy consumption was at a minimum of about 83.944 quadrillion Btu. Absolute maximum: (9, 95.992); in 1999, the U.S. total energy consumption was at a maximum of about 95.992 quadrillion Btu.

45. Absolute minimum: (5.52, 29.37); in 1997, the amount of methane gas emitted by the U.S. was at a minimum of about 29.37 million metric tons. Absolute maximum: (2.14, 30.15); in 1994, the amount of methane gas emitted by the U.S. was at a maximum of about 30.15 million metric tons.

47. Absolute minimum: (10, 1.52). When blood volume is 10% cells, it takes 1.52 times as much pressure to force blood through a tube as to force water through the same tube; absolute maximum: (80, 9.64). When blood volume is 80% cells, it takes 9.64 times as much pressure to force blood through a tube as to force water through the same tube.

49. **(a)** $C(x) = 1000 + 60x$
(b) $R(x) = -\frac{5}{7}x^2 + \frac{1200}{7}x$
(c) $x = 78$ maximizes profit

51. **(a)** $R(x) = -\frac{17}{18,000}x^2 + \frac{77}{6}x$
(b) Revenue is maximized at $x = 6794$.

Section 15.5

1. $(\sqrt{3}, 4\sqrt{3})$ **3.** $\left(\frac{\sqrt{2}}{2}, 4\sqrt{2} - 3\right) \approx (0.71, 2.66)$

5. $(\sqrt[4]{2}, 6\sqrt{2} - 2) \approx (1.19, 6.49)$

7. $\left(\sqrt[4]{\frac{2}{3}}, 2\sqrt{6} - 1\right) \approx (0.90, 3.90)$

9. $\left(\frac{\sqrt{6}}{3}, -2\sqrt{6}\right) \approx (0.82, -4.90)$

11. $\left(\frac{\sqrt{6}}{2}, -2\sqrt{6}\right) \approx (1.22, -4.90)$

13. $\left(\sqrt[4]{\frac{2}{3}}, 2 - 2\sqrt{6}\right) \approx (0.90, -2.90)$

15. $\left(\frac{\sqrt{10}}{2}, 3 - 2\sqrt{10}\right) \approx (1.58, -3.32)$

17. $x = 200$ ft, $y = 66.67$ ft, area $= 13,333.3$ ft^2

19. $x \approx 29.9$ ft, $y \approx 50.2$ ft **21.** $\frac{4}{3} \times \frac{40}{3} \times \frac{10}{3}$

23. $20 \times 20 \times 15$

25. **(a)** $AC(x) = \frac{5}{x} + 0.5 + 0.003x$
(b) $x \approx 41$

27. $AC(t) = 100t + 600 + \frac{15,000}{t}$, $t \approx 12.25$

29. \$102.50 per room

31. 60 trees per acre

33. 35 trees per acre

35. Optimal population: 12,500 deer; sustainable harvest: 4688 deer

37. Optimal population: 60,000 pheasants; sustainable harvest: 36,000 pheasants

39. To minimize inventory costs, each order should have 25 desks, placed 8 times per year.

41. To minimize inventory costs, each order should have 240 carrying cases, placed semi-annually.

43. To minimize inventory costs, each order should have 160 microwave ovens, placed 15 times per year.

45. To minimize inventory costs, each order should have 15 (or 20) cars, placed 20 (or 15) times per year.

47. To minimize costs, the publisher should print 2500 books per run, twice a year.

49. To minimize costs, the toy manufacturer should make 5000 toys per run, 10 times per year.

Chapter 15 Review Exercises

1. **(a)** $(-\infty, -2) \cup (3, \infty)$
(b) $(-2, 3)$
(c) $x = -2$ and $x = 3$

3. **(a)** $(-1, 1)$
(b) $(1, \infty)$
(c) $(-\infty, -1]$ and $x = 1$

5. **(a)** $(-\infty, -3) \cup (1, \infty)$
(b) $(-3, 1)$
(c) $x = -3$ and $x = 1$

7. **(a)** $(-4, -1) \cup (3, \infty)$
(b) $(-\infty, -4) \cup (-1, 3)$
(c) $x = -4, x = -1$ and $x = 3$

9. **(a)** No critical values
(b) Increasing: $(-\infty, \infty)$
(c) No relative extrema
(d)

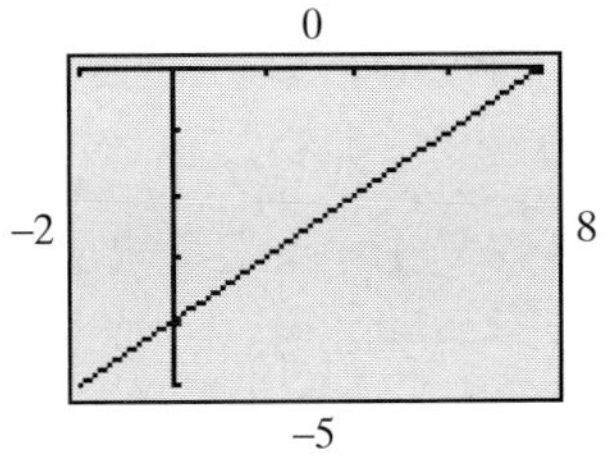

11. **(a)** $x = -4$ and $x = 2$
(b) Increasing: $(-\infty, -4) \cup (2, \infty)$
Decreasing: $(-4, 2)$
(c) Relative maximum: $(-4, 100)$
Relative minimum: $(2, -8)$

(d)

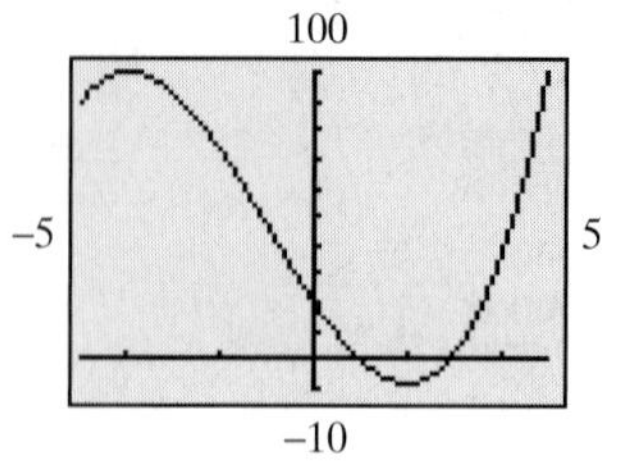

13. **(a)** $x = \frac{2}{3}$

(b) Increasing: $(\frac{2}{3}, \infty)$
Decreasing: $(-\infty, \frac{2}{3})$

(c) Relative minimum: $(\frac{2}{3}, 0)$

(d)

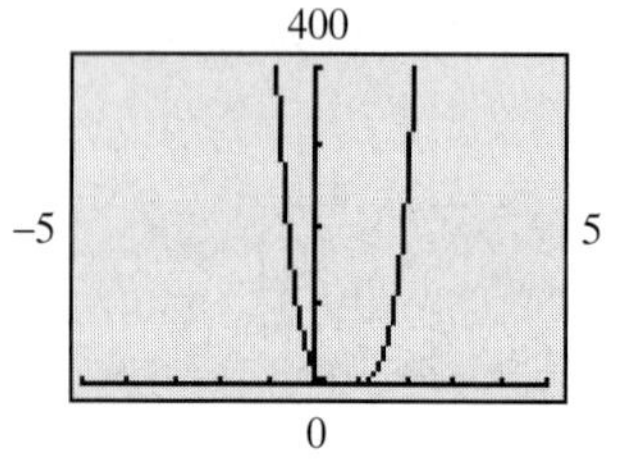

15. Relative maximum at $x = \frac{8}{3}$

17. Relative maximum at $x = -3$; relative minimum at $x = 3$

19.

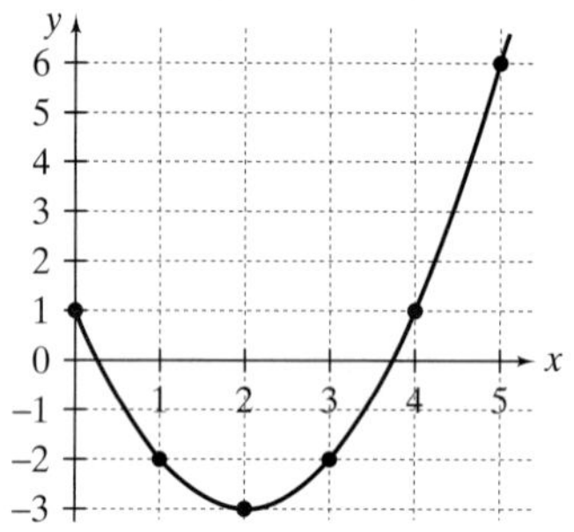

21. **(a)** $p(30) \approx 1.83$

(b) $R(x) = 4.5xe^{-0.03x}$

(c) Increasing: $(0, 33.33)$; decreasing: $(33.33, \infty)$

(d) Relative maximum $\approx (33.33, 55.18)$; if 33,330 bottles of dog shampoo are sold, \$55,180 will be generated in revenue.

(e)

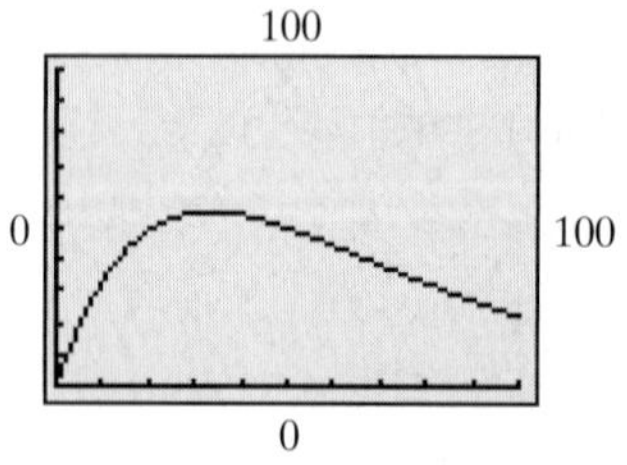

23. **(a)** $AC(x) = \dfrac{800}{x} + 15 + \dfrac{x}{18}$

(b) $x = 120$

(c) Increasing: $(120, 500)$; decreasing: $(0, 120)$

(d) Relative minimum $\approx (120, 28.33)$. At 120 telephones per day, the average cost is minimized—\$28.33 per phone.

25. $f'(x) = 3x^2 - 14x$; $f''(x) = 6x - 14$; $f'''(x) = 6$

27. $y' = \dfrac{1}{3(x-5)^{2/3}}$; $y'' = -\dfrac{2}{9(x-5)^{5/3}}$; $y''' = \dfrac{10}{27(x-5)^{8/3}}$

29. **(a)** Concave down: $(-\infty, \frac{1}{3})$; concave up: $(\frac{1}{3}, \infty)$

(b) Inflection point: $(\frac{1}{3}, -\frac{146}{27})$

31. **(a)** Concave up: $(0, \infty)$

(b) No inflection point(s)

33. **(a)** Concave up: $(-\infty, 0)$; concave down: $(0, \infty)$

(b) No inflection point(s)

35. **(a)** Concave up: $(-\infty, \infty)$

(b) No inflection points

37. **(a)** Increasing: $(-\infty, -3) \cup (2, \infty)$; decreasing: $(-3, 2)$

(b) Relative maximum: $(-3, 81)$; relative minimum: $(2, -44)$

(c) Concave up: $(-\frac{1}{2}, \infty)$; concave down: $(-\infty, -\frac{1}{2})$

(d) Inflection point: $(-\frac{1}{2}, \frac{37}{2})$

39. **(a)** Decreasing: $(-\infty, 2)$

(b) No relative extrema

(c) Concave down: $(-\infty, 2)$

(d) No inflection point(s)

41. **(a)** $P(4) = 155$. When 400 cookies are baked and sold, a \$155 profit is made; $P'(4) = 68$. When 400 cookies are sold, the profit for selling the next 100 cookies is about \$68.

(b) Increasing $\approx (0, 9.02)$; decreasing $\approx (9.02, \infty)$; relative maximum $\approx (9.02, 357.51)$. If 902 cookies are baked and sold, profits will be maximized at \$357.51.

(c) Increasing: $(0, 2)$; decreasing: $(2, \infty)$. The concavity of P changes a $x = 2$.

(d) $(2, 11)$

43. **(a)**

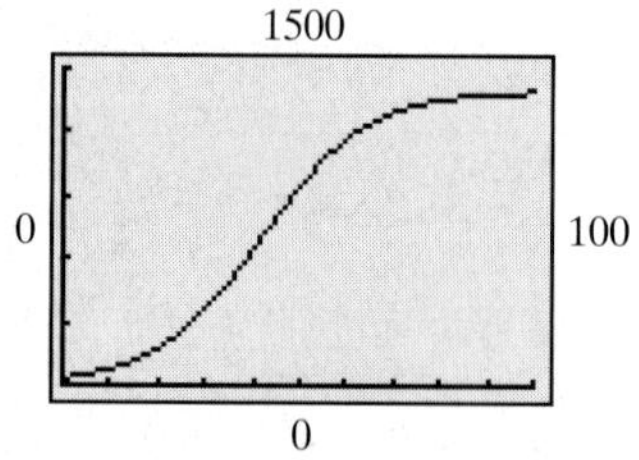

(b) $p(84) \approx 1369.33$. Eighty-four years after first being introduced, there are about 1369 goats in the region. $p'(84) \approx 2.7$. Eighty-four years after first being introduced, the number of goats is increasing at a rate of about $2.7 \frac{\text{goats}}{\text{year}}$.

(c) $p''(t) = \dfrac{487.62(43e^{-.18x} - e^{-.09x})}{(1 + 43e^{-.09x})^3}$

(d) Concave up: $(0, 41.79)$; concave down: $(41.79, 100)$

(e) About 42 years after they were first introduced

45. $f > 0$: $(1, 5)$; $f < 0$: $(-5, 1)$

47. $f' > 0$: $(-5, -3) \cup (-1, 3)$; $f' < 0$: $(-3, -1) \cup (3, 5)$

49. Concave up: $(-2, 1)$; concave down: $(-5, -2) \cup (1, 5)$

51. Inflection points: $(-2, -3)$ and $(1, 0)$

53. Relative maximum: $(-\frac{2}{3}, \frac{76}{27})$; relative minimum: $(4, -48)$

55. Relative maximum: $(0, \frac{1}{9})$

57.

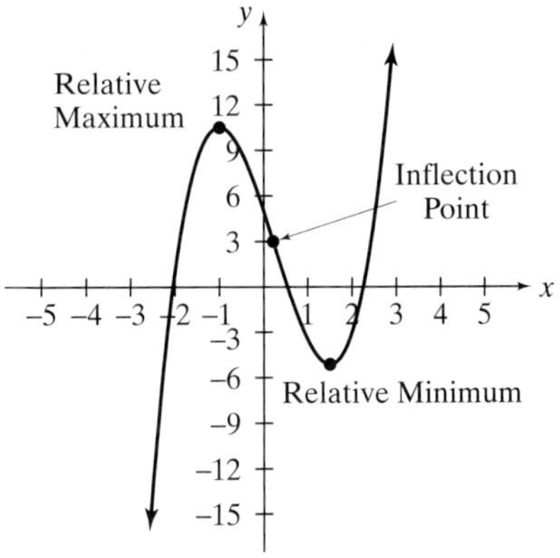

59.

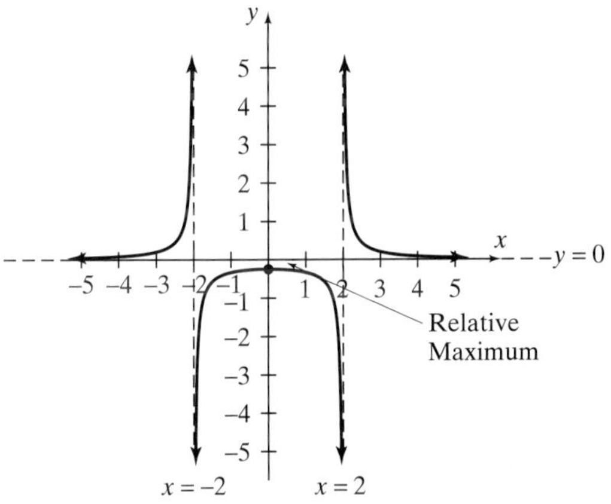

61.

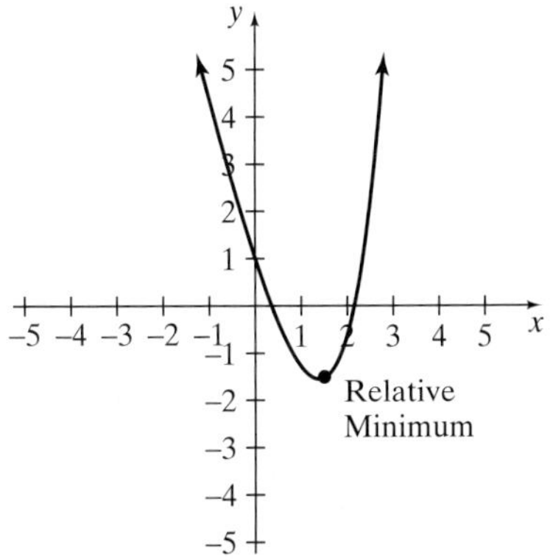

63. Answers will vary. Samples:

(a)

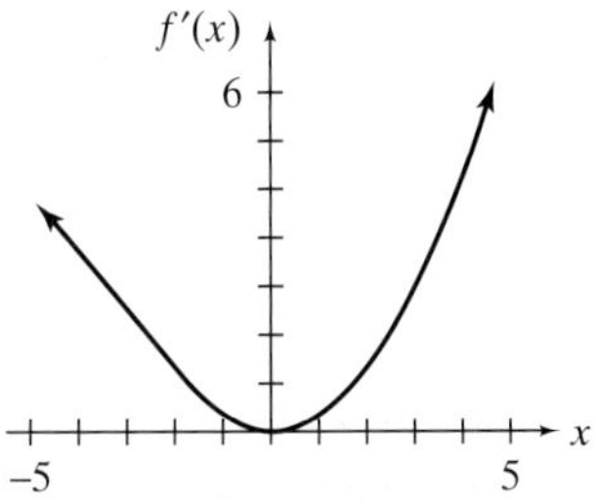

(b)

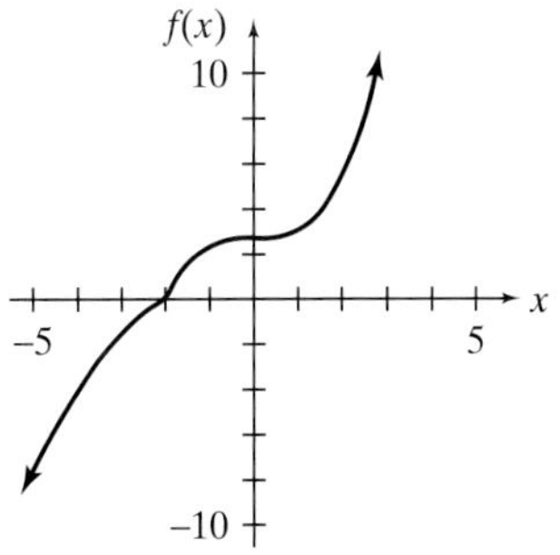

65. **(a)** $AC(x) = 15 + \dfrac{200}{x}$

(b) $AC(x)$ is always decreasing, so it has no minimum.

(c) $AP(x) = 70 - 0.12x^2 - \dfrac{200}{x}$

(d) $x \approx 9$

67. **(a)** $f(9) \approx 26$. In 1989, there were 26 African-American representatives. $f'(9) \approx 1.5$. In 1989 the number of African-American representatives was increasing at a rate of about $1.5 \frac{\text{members}}{\text{year}}$.

(b) $f'(t) = 0.9106e^{0.058t}$. Over the interval [1, 15], $f'(t)$ is always > 0. Therefore f, which represents the number of African-American representatives, is always increasing.

(c) $f''(t) \approx 0.053e^{0.058t}$; $f''(t) > 0$ over the interval [1, 15]. Therefore, f is concave up and is increasing at a faster rate.

(d)

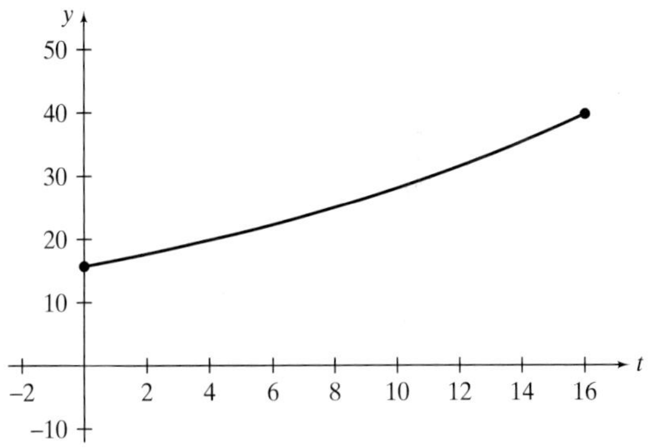

69. Absolute maximum: (5, 320); absolute minimum: (−5, −20)

71. Absolute maximum: $\left(5, \frac{1310}{3}\right)$; absolute minimum: $\left(-5, -\frac{1310}{3}\right)$

73. Absolute maximum: $\left(10, \frac{52}{5}\right)$; absolute minimum: (2, 4)

75. Absolute maximum: (19.75, 304.81). In 1991, \$304.81 billion were spent on defense; absolute minimum: (1, 70.85). In 1973, \$70.85 billion were spent on defense.

77. 40 ft × 30 ft

79. **(a)** $R(x) = -0.002x^3 + 1.5x^2 - 1.7x$

(b) \$124,150 per week

(c) 500 units per week

(d) \$248.30 per unit

81. $(\sqrt{2}, 6\sqrt{2}) \approx (1.41, 8.49)$ **83.** $\left(\dfrac{\sqrt{6}}{2}, 10\right) \approx (1.22, 10)$

85. $\left(\frac{\sqrt{3}}{2}, -4\sqrt{3}\right) \approx (0.87, -6.93)$ **87.** $(\sqrt{3}, -1) \approx (1.73, -1)$

89. ≈ 10 ft^2 **91.** $18 \times 9 \times 3$ in **93.** 36 trees

Chapter 16

Section 16.1

1. Yes **3.** No **5.** Yes **7.** No **9.** Yes

11. $\frac{x^5}{5} + C$ **13.** $\frac{x^{3.31}}{3.31} + C$ **15.** $-\frac{1}{2t^2} + C$

17. $\frac{4}{9}x^{9/4} + C$ **19.** $\frac{3}{2}x^{2/3} + C$ **21.** $\frac{2}{35}x^7 + C$

23. $x^2 + 3x + C$ **25.** $\frac{1}{3}x^2 + 4x + C$ **27.** $t^3 + t^2 + 10t + C$

29. $x - \frac{2}{3}x^3 + \frac{3}{4}x^4 + C$ **31.** $2.07x^3 + 0.015x^2 - 4.01x + C$

33. $-\frac{1}{x} + \frac{3}{2x^2} + C$ **35.** $3x + \frac{4}{3}(\sqrt{x})^3 + C$

37. $-\frac{3}{2x^2} + \frac{4}{5}x^{5/2} - 4x + C$ **39.** $\frac{1}{6}t^3 - \frac{1}{3}t + C$

41. $-\frac{1}{20z^2} - \frac{2}{z} + \frac{z^4}{4} + C$ **43.** $\frac{200}{113}t^{1.13} + 5t + C$

45. $f(x) = -2x + 4$ **47.** $f(x) = \frac{5}{2}x^2$

49. $f(x) = x^2 - 3x + 4$ **51.** $f(t) = 500t - \frac{t^2}{40} + 40$

53. $S(x) = -2x^{-1} - \frac{3}{2}x^{-2} - x + \frac{13}{2}$

55. $y = 3(\sqrt[3]{t})^5 + 3(\sqrt[3]{t})^2 + 1$

57. $R(t) = -\frac{t^{-2}}{2} - \frac{t^2}{2} + 5$

59. **(a)** $P(q) = 40q - 0.025q^2$
(b) \$7000

61. **(a)** $R(x) = 0.000015x^3 - 0.015x^2 + 3.75x$
(b) $p(x) = 0.000015x^2 - 0.015x + 3.75$
(c) \$2.40

63. **(a)** $AC(x) = \frac{100}{x} + 1.50$
(b) $C(x) = 100 + 1.50x$
(c) The cost of producing 100 banners is \$250.

65. **(a)** $f(5) = -2.17$; this means that in 1996 the number of murders in the U.S. was decreasing at a rate of about 2170 murders per year.
(b) $F(x) = 0.08x^3 - 1.14x^2 + 3.23x + 22.58$
(c) $F(5) = 20.23$; this means that in 1996, there were about 20,230 murders in the U.S.

Section 16.2

1. 0.5 **3.** 0.5 **5.** 0.5 **7.** 0.5 **9.** $\frac{2}{7}$

11. **(a)** 10.0 **(b)** 10.0

13. **(a)** 20.25 **(b)** 24.75

15. **(a)** 13.0 **(b)** 15.0

17. **(a)** 28.0 **(b)** 60.0

19. **(a)** 9.625 **(b)** 8.125

21.

N	LEFTSUM	RTSUM
10	9.24	6.84
100	8.1204	7.8804
1000	8.012004	7.988004

23.

N	LEFTSUM	RTSUM
10	36.11837	38.51837
100	37.21318	37.45318
1000	37.32133	37.34533

25.

N	LEFTSUM	RTSUM
10	32.895	39.195
100	35.68545	36.31545
1000	35.9685	36.0315

27.

N	LEFTSUM	RTSUM
10	30.24	51.04
100	38.9664	41.0464
1000	39.89606	40.10406

29.

N	LEFTSUM	RTSUM
10	33.25	39
100	35.79625	36.37125
1000	36.05459	36.11209

31. $\int_3^9 10\,dx$ **33.** $\int_0^{10} \frac{1}{2}x\,dx$ **35.** $\int_1^4 \frac{1}{x^2}\,dx$

37. $\int_0^3 (9 - x^2)\,dx$ **39.** $\int_{-1}^2 (x^2 + 2)\,dx$

41. $\int_1^2 5\,dx + \int_2^4 (x^2 + 1)\,dx$

43. Rectangle formula; 60 un^2

45. Triangle formula; 100 un^2

47. Left Sum ≈ 0.764 un^2; right Sum ≈ 0.736 un^2

49. Left Sum ≈ 18.135 un^2; right Sum ≈ 17.865 un^2

51. Left Sum ≈ 8.955 un^2; right Sum ≈ 9.045 un^2

53. $y = x^2$

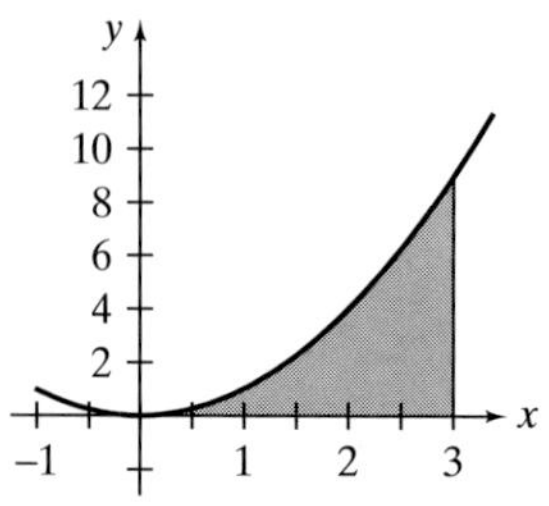

55. $y = \frac{1}{x^2}$

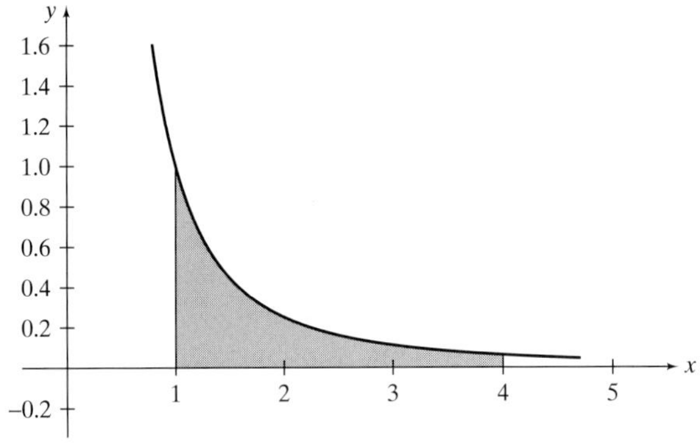

57. $y = \sqrt[3]{x}$

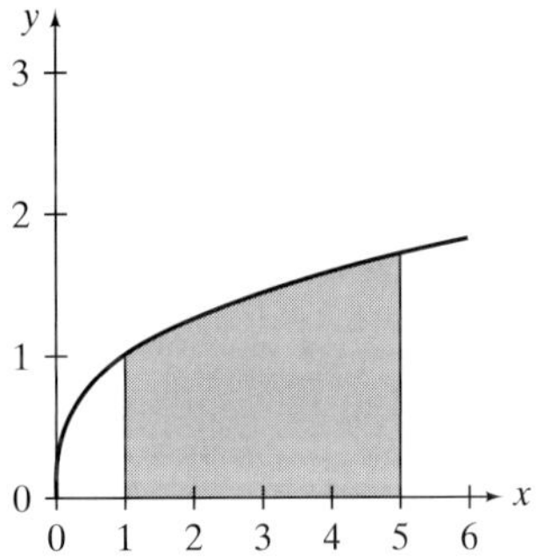

59. $y = \frac{2}{\sqrt{x}}$

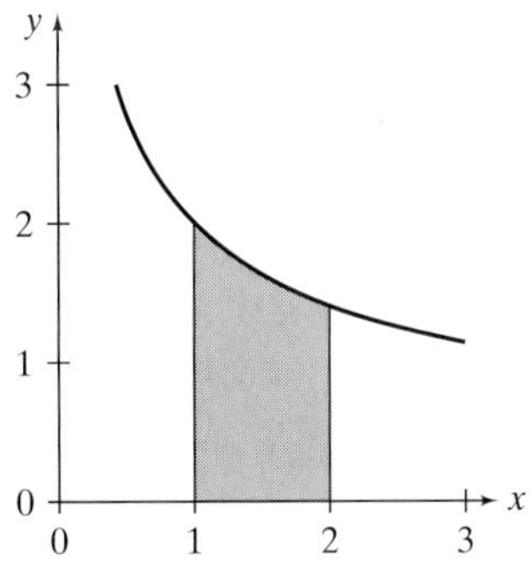

Section 16.3

1. $\frac{45}{2}$ **3.** 10 **5.** 14 **7.** $\frac{128}{3}$ **9.** 9 **11.** 8

13. $\frac{112}{3}$ **15.** $\frac{165}{4}$ **17.** $\frac{433}{12}$ **19.** $-\frac{85}{2}$ **21.** $-\frac{8}{5}$

23. $\frac{20}{3}$

25. Area ≈ 9.25 un^2

27. Area ≈ 1.61 un^2

29. Area ≈ 2.56 un^2

31. Area $= 44$ un^2

33. Area $= \frac{32}{3}$ un^2

35. Area $= \frac{3}{4}$ un^2

37. Area $= 9$ un^2

39. Area $= 30$ un^2

41. Area $= \frac{23}{3}$ un^2

43. **(a)**

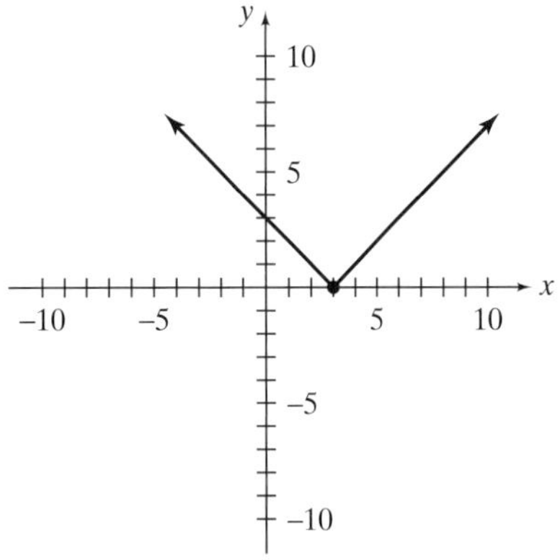

(b) $f(x) = \begin{cases} -x + 3, & x < 3 \\ x - 3, & x \geq 3 \end{cases}$

(c) $\frac{29}{2}$

45. **(a)**

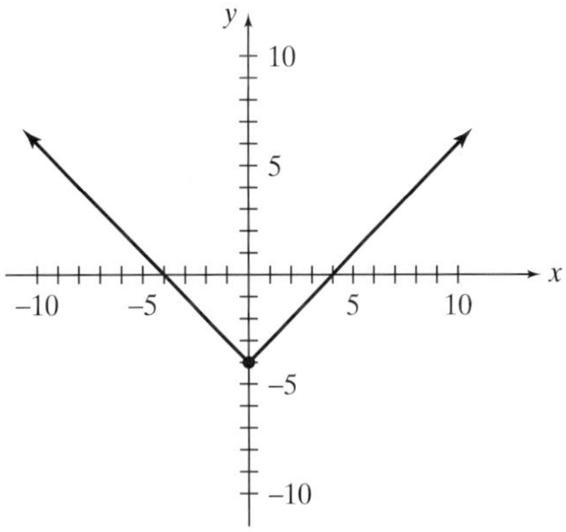

(b) $f(x) = \begin{cases} -x - 4, & x < 0 \\ x - 4, & x \geq 0 \end{cases}$

(c) $-4, 4$

(d) Net area $= -12$, gross area $= 16$

47. $\frac{25}{2}$ **49.** $\frac{14}{3}$ **51.** $\frac{106}{3}$ **53.** **(a)** -2 **(b)** 20 **(c)** 29

55. **(a)** 4 **(b)** 0 **(c)** $\frac{32}{3}$ **57.** **(a)** $-3, 3$ **(b)** $-\frac{70}{3}$ **(c)** 30

59. **(a)** $-1, 2$ **(b)** $\frac{25}{6}$ **(c)** $\frac{79}{6}$

61. **(a)** $-2, 0, 1$ **(b)** $\frac{16}{3}$ **(c)** $\frac{37}{6}$

Section 16.4

1. **(a)** $\int_{50}^{150} 5.5\, dx$

(b) 550; this means that the total increase in cost of producing 50 to 150 bandanas is $550.

3. **(a)** $\int_5^{10} (-1.02t + 20.2)\,dt$

(b) 62.75; this means that as the number of weeks of the Double D cola ad campaign increases from 5 to 10 weeks, the total increase in sales is \$62.75 thousand.

5. **(a)** $\int_8^{10} (-1.83x^2 + 16.62x - 27.42)\,dx$

(b) −53.36; this means that from 1996 to 1998, the total decrease in U.S. imports from Japan was \$53.36 billion.

7. **(a)** $\int_8^{15} (0.04x^3 - 0.75x^2 + 5.54x - 11.31)\,dx$

(b) 116.34; this means that from 1988 to 1995 the total increase in the price of a gallon of unleaded gasoline was 116.34 cents per gallon.

9. **(a)**

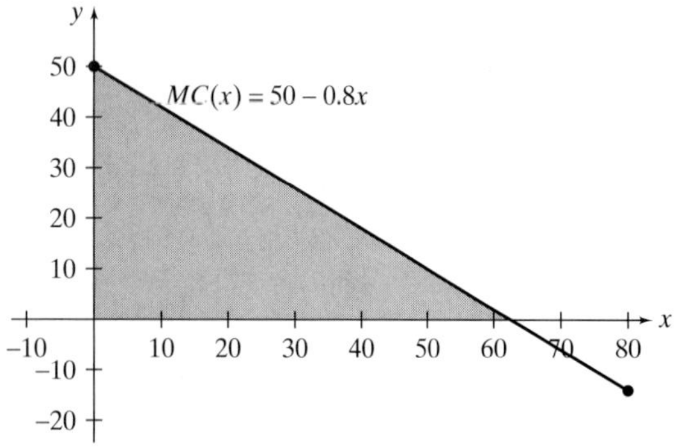

(b) $\int_0^{50} (50 - 0.8x)\,dx = 1500$; this means that the cost of producing the first 50 pairs of walking shoes is \$1500.

11. **(a)** 2850; this means that as the production and sales of the headset radios increase from 10 to 40 thousand, the total increase in revenue is \$2850.

(b)

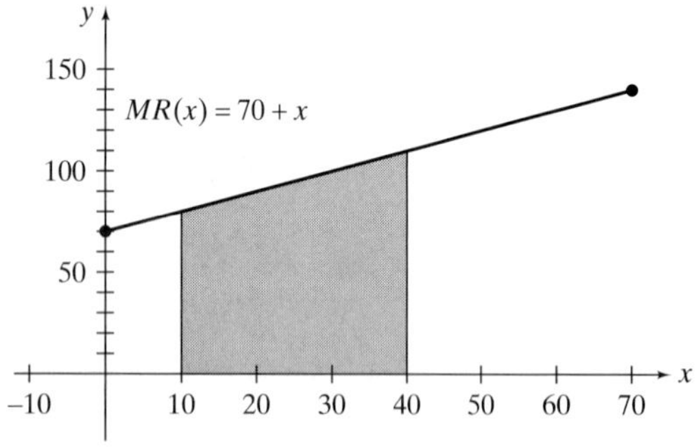

13. 270; this means that the revenue generated from selling the first 200 tickets is \$270.

15. 730; this means that as production and sales increase from 10 to 20 ski machines, the total increase in profit is \$730.

17. −6.67; this means that as the number of promotional banners increases from 10 to 30, the total decrease in average cost is \$6.67.

19. 480.9; this means that from 1980 to 1990, the total increase in the number of arrests for drug abuse violations was 480.9 thousand.

21. −174.65; this means that from 1990 to 1995, the total decrease in the number of employees in the executive branch of the U.S. federal government was 174.65 thousand.

23. 159.21; this means that from 1980 to 1990, the total increase in the amount of outstanding revolving credit was \$159.21 billion.

25. **(a)** 44.188; this means that from 1987 to 1991, the total increase in the amount of military expenditures by NATO countries was \$44.188 billion.

(b) −40.432; this means that from 1992 to 1996, the total decrease in the amount of military expenditures by NATO countries was \$40.432 billion.

27. **(a)** 6.35; this means that from 1991 to 1993, the total increase in revenue in the resource recovery industry in the U.S. was \$6.35 billion.

(b) −5.092; this means that from 1994 to 1998, the total decrease in revenue in the resource recovery industry in the U.S. was \$5.092 billion.

29. Increase of \$88.5 billion

31. Decrease of 131.25 million barrels

33. 4 feet

35. **(a)** $v(10) = 3$; this means that after 10 seconds, the velocity is 3 feet per second.

(b) 45; this means that from 10 to 20 seconds, the total displacement of the object was 45 feet.

37. **(a)** $v(3) = 16$; this means that after 3 seconds, the velocity is 16 feet per second.

(b) 36; this means that from 3 to 5 seconds, the total displacement of the object was 36 feet.

39. **(a)** $v(2) = 80.4$; this means that after 2 seconds, the velocity is 80.4 feet per second.

(b) 306; this means that from 2 to 8 seconds, the total displacement of the object was 306 feet.

41. **(a)** $v(1) = 88$; this means that after 1 second, the velocity of the object is 88 feet per second.

(b) 216; this means that in the first 3 seconds, the total displacement of the object was 216 feet.

43. 10.58; this means that the total increase in cost from 35 to 48 center punches, is \$10.58.

45. 752.95; this means that from 1980 to 1989, the total increase in the average dollar amount awarded in the Pell Grant system was \$752.95.

47. 1484.73; this means that from 1985 to 1995, the total increase in tuition and fees for in-state students at 4-year colleges was \$1484.73.

Section 16.5

1. $\frac{1}{3}(3x + 4)^3 + C$ **3.** $\frac{(4x^2 + 1)^4}{4} + C$

5. $\frac{2}{3}(\sqrt{(t^2 - 1)})^3 + C$ **7.** $\frac{(x^3 - 2x)^2}{2} + C$

9. $\frac{3}{2}(\sqrt[3]{x^3-5})^2+C$ 11. $-\frac{1}{2(2x^2+3)^2}+C$

13. $\frac{1}{20}(4x+7)^5+C$ 15. $\frac{1}{12}(t^2-3)^6+C$

17. $\frac{2}{3}\sqrt{(x^3-5)}+C$ 19. $-\frac{1}{8(5x^2-2)}+C$

21. $\frac{2}{3}(\sqrt{1-x^{-1}})^3+C$ 23. $\frac{1}{11}(t-7)^{11}+C$

25. $-\frac{1}{8(x^2+2x+5)^4}+C$ 27. $\frac{3}{2}(x^2+3x-10)^2+C$

29. $-\frac{2}{3}(\sqrt{x+1})^3+\frac{2}{3}(\sqrt{x+1})^5+C$

31. $\frac{(t-1)^6}{6}+\frac{(t-1)^7}{7}+C$ 33. $6\sqrt{x-1}+2(\sqrt{x-1})^3+C$

35. 4 37. 0 39. 52 41. $2\sqrt{3}$ 43. $\frac{1}{8}$ 45. 4

47. $\frac{9}{20}$ 49. $\frac{18}{19}$ 51. $\frac{1640}{6561}$ 53. $\frac{52}{3}$ 55. 0 57. $\frac{24}{5}$

59. $\frac{1}{2}$ 61. $\frac{16}{3}\sqrt{2}$

63. (a) When daily production is 75 fobs, it costs about \$6 to produce the 76th fob.
(b) The total increase in daily cost of producing from 0 to 75 fobs is \$250.
(c) $C(x)=10\sqrt{(x^2+10{,}000)}+2000$
(d) \$3000

65. (a) When monthly production and sales is 70 flags, the profit from the 71st flag is about \$144.40.
(b) The total increase in profit of producing and selling from 20 to 100 flags each month is \$12,000.
(c) $P(x)=\sqrt{(2x^2-400)}-20$

67. (a) ≈ 1.59 feet (b) $s(t)=\sqrt{(t^2+9)}+3$

69. (a) Five years after the town was incorporated, homes were being built at a rate of about $33\frac{\text{homes}}{\text{year}}$.
(b) Ninety-four homes were built during the first 5 years after the town was incorporated.
(c) $h(t)=10\sqrt{(t^3+4)}-20$
(d) There were 94 homes in the town after 5 years.

Section 16.6

1. $2\ln|x|+C$ 3. $-\frac{2}{3}\ln|t|+C$ 5. $(1+e)\ln|x|+C$

7. $3\ln 5$ 9. $\frac{1}{2}(\ln 5-\ln 2)$ 11. $\frac{1}{2}\ln|2x+3|+C$

13. $\frac{1}{2}\ln(t^2+1)+C$ 15. $\frac{1}{2}\ln|x^2-4x+9|+C$

17. $\frac{1}{4}\ln(x^4+10)+C$ 19. $\ln|\ln 2x|+C$

21. $4\ln(\sqrt{2}+4)-4\ln 5$ 23. $\frac{2}{3}\ln 7-\frac{2}{3}\ln 2$

25. (a) The sales rate after 2 days is $20\frac{\text{dresses}}{\text{day}}$.
(b) $40\ln 2\approx 28$

27. (a) In 1981, enrollment was increasing by $15.405\frac{\text{thousands of students}}{\text{year}}$.
(b) From 1976 to 1981, foreign enrollment increased by about 165.61 thousand students.

29. (a) In 1996, the number of employees was increasing at a rate of $3.1435\frac{\text{thousands of employees}}{\text{year}}$.
(b) $f(x)=81.356+53.44\ln(x+1)$

31. $2e^x+C$ 33. e^x-x+C 35. $4e^t+\frac{1}{2}t^2-t+C$

37. $e-2$ 39. $\frac{1}{2}(e^2-e+4)$ 41. $\frac{1}{4}e^{4x+1}+C$

43. $e^{x^2+1}+C$ 45. $\ln(e^x+2)+C$ 47. $3\ln|1-e^{-x}|+C$

49. $2e^{\sqrt{x}}+C$ 51. $-\frac{1}{4}[e^{-12}-e^{-4}]$ 53. $2-e^{-1}$

55. $\frac{1}{2}e^{x^2}+C$ 57. $\frac{1}{8}(e^{4x}+4)^2+C$

59. (a) At a production level of 20 faucets, the cost to produce the 21st faucet is about \$1.56.
(b) \$100.39

61. (a) In 1995, infant deaths were decreasing at a rate of about $0.34\frac{\text{infant deaths per 1000 live births}}{\text{year}}$.
(b) From 1965 to 1995 the total decrease in infant mortality was about $20.28\frac{\text{infant deaths}}{\text{1000 live births}}$.

63. (a) In 1950 the United States population was increasing at a rate of about $1.973\frac{\text{millions}}{\text{year}}$.
(b) 72.539 million

65. (a)

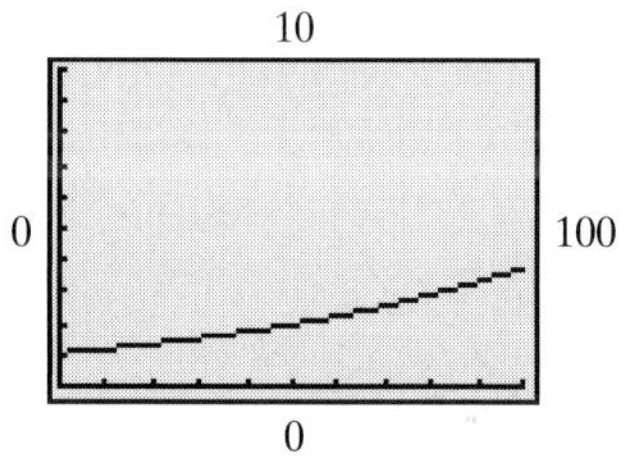

(c) The graph of g is above the x-axis on the reasonable domain, which means $g>0$ for all x in the reasonable domain. Since g is rate of change of population and $g>0$, this means that population is increasing on the reasonable domain.

67. $\frac{1}{\ln 5}5^x+C$ 69. $-\frac{5}{\ln 2}\left(\frac{1}{2}\right)^x+C$ 71. $-\frac{1}{\ln 7}7^{-x}+C$

73. $\frac{333{,}000}{\ln 10}$ 75. $\frac{21}{16\ln 2}$

77. (a) In 1990 the prison population was increasing at a rate of $59{,}603\frac{\text{inmates}}{\text{year}}$.
(b) About 244,198 inmates

Section 16.7

1. Yes 3. No 5. Yes

7. No 9. Yes 11. Yes

13. Yes 15. $y=-2x+C$

17. $y=x-\frac{3}{2}x^2+C$

19. $y=\frac{5}{4}x^4+\frac{1}{2}x^2-2x+C$

21. $y=\frac{2}{3}(\sqrt{x})^3+2x+C$

23. $y=-\frac{5}{x}+C$

25.

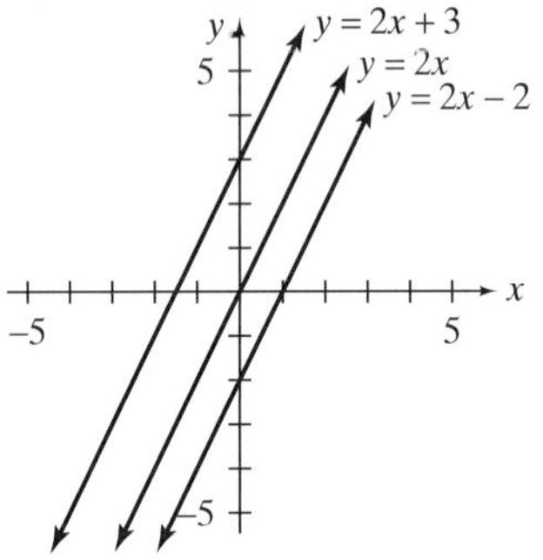

27.

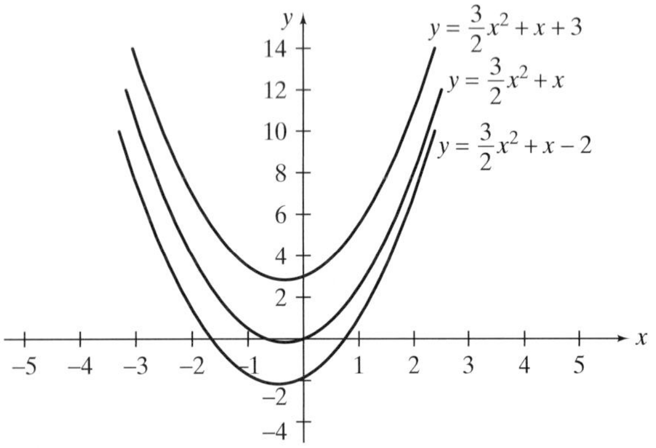

29.

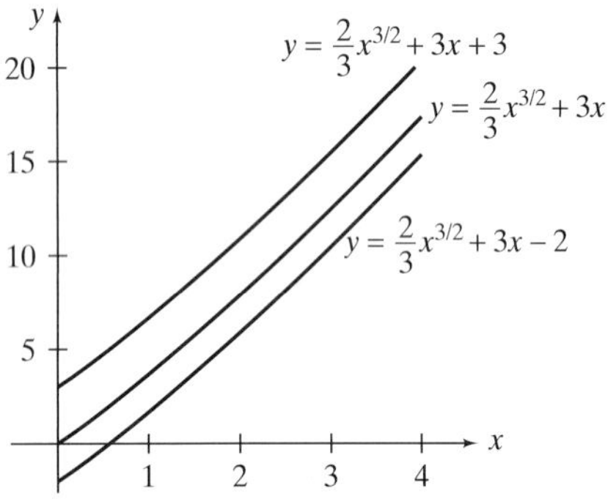

31. $y = 6x^2 + 2$ **33.** $y = x^3 + 2x^2 + 3$

35. $y = 7t - 2t^2 + 3$ **37.** $y = \frac{1}{4}x^4 - x^2 - 2$

39. $y = \frac{1}{3}t^3 + t^2 + 3t + 4$ **41.** $y = \frac{2}{1.3}x^{1.3} + 4$

43. $y = x - 2\ln|x| + 5$

45. **(a)** $C(x) = -0.01x^2 + 6x + \text{constant}$

(b) $C(x) = -0.01x^2 + 6x + 341.0$

47. **(a)** Ten months after its introduction, the percentage of drinkers accepting the new formula is increasing at a rate of $11.1\,\frac{\text{percent}}{\text{month}}$.

(b) The percentage of drinkers accepting the new formula after t months is given by $A(t) = 0.5t^2 + 1.1t + C$ where C is determined by initial conditions.

(c) $A(t) = 0.5t^2 + 1.1t + 5$

49. **(a)** In 1989, deaths from heart disease were decreasing at a rate of $1.7567\,\frac{\text{deaths}}{100{,}000\text{ Americans}}$.

(b) $D(t) = 336.17t^{-0.06} + .01$

(c) $D(15) \approx 285.76$; in 1994, the number of deaths from heart disease was about $285.76\,\frac{\text{deaths}}{100{,}000\text{ Americans}}$.

51. $y^2 = x^2 + C$ **53.** $y = Ce^{x^3},\ y > 0$

55. $e^{-y} = C - e^x$ **57.** $y = Ce^{x(x+2)},\ y > 0$

59. $y^{1/2} = \frac{1}{3}x^{3/2} + C$ **61.** $y = Ce^{e^x}$ **63.** $y = e^{x^2}$

65. $\frac{1}{y^3} = -3x^2 + 1$ **67.** $e^y = e^x$ or $y = x$

69. $y = 10e^{(x^4-81)/4}$ **71.** $y = \sqrt{2(1+x^2)}$

Section 16.8

1. **(a)** $P' = 0.05P,\ P(0) = 50{,}000$

(b) $P = 50{,}000e^{0.05t}$

(c) 82,436

3. **(a)** $P' = 0.12P$

(b) $P = 5000e^{0.12t}$

(c) $P(3) = 7167,\ P(12) = 21{,}103,\ P(24) = 89{,}071$

5. **(a)** $N(t) = 39.9e^{0.01728t}$

(b) 953.25 millions of board feet

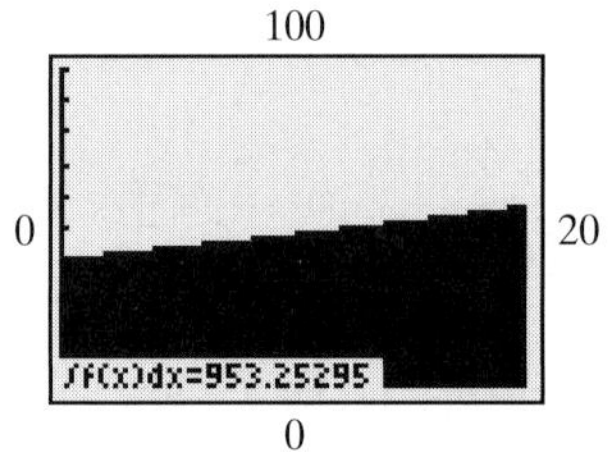

7. **(a)** $y' = 0.034y,\ y(0) = 6$

(b) In 1993 the percentage of wood waste generated in the United States was 6.64 and was growing at a rate of $0.2259\,\frac{\text{percent}}{\text{year}}$.

9. **(a)** $f(t) = 3000e^{(-\ln 2/14.2)t}$ **(b)** 8.31

11. **(a)** $P' = -0.01822P,\ P(0) = 920{,}000$

(b) $P = 920{,}000e^{-0.01822t}$

(c) 662,760

13. **(a)** $y' = -0.09y,\ y(0) = 6.54$

(b) In 1993 the percentage of Caucasians unemployed was 4.99 and was decreasing at a rate of about $0.449\,\frac{\text{percent}}{\text{year}}$.

15. **(a)** $y' = (10 - y)$

(b) $f(0) = 0,\ L = 10$

17. **(a)** $y' = 2(8 - y)$

(b) $f(0) = 0,\ L = 8$

19. **(a)** $y' = 0.5(20 - y)$

(b) $f(0) = 0,\ L = 20$

21. **(a)** $y' = 0.1(12 - y)$

(b) $f(0) = 0,\ L = 12$

23. **(a)** $y' = 0.2(1.5 - y)$

(b) $f(0) = 0,\ L = 1.5$

25. **(a)** $y' = 0.1(15.5 - y)$

(b) $f(0) = 0,\ L = 15.5$

27. **(a)** Two weeks after installation about 1354 employees have learned the new system and about $494\,\frac{\text{employees}}{\text{week}}$ are learning it.

(b) About 5.97, or 6, weeks

29. (a) 300

(b) $-\ln \frac{15}{13}$

(c)

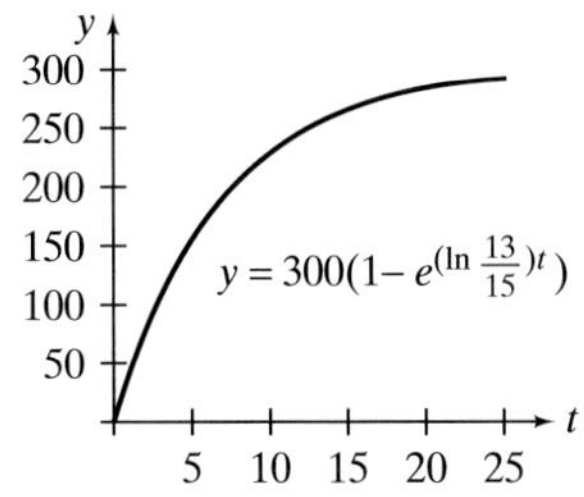

(d) No

31. (a) $L = 75, k \approx 0.14876$

(b) $y = 75(1 - e^{-0.14876t})$

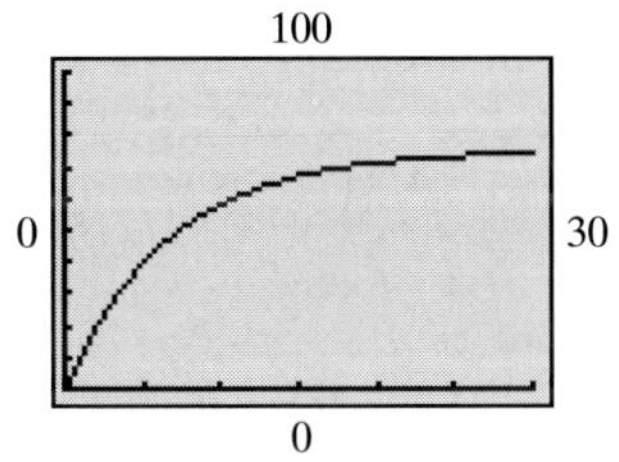

(c) $y'(5) = 5.303, y'(25) = 0.27065$

33. (a) $y' = \frac{1}{15}y(15 - y)$

(b) $y(0) = \frac{15}{2}, L = 15$

35. (a) $y' = \frac{1}{10}y(10 - y)$

(b) $y(0) = \frac{5}{3}, L = 10$

37. (a) $y' = \frac{1}{30}y(90 - y)$

(b) $y(0) = 45, L = 90$

39. (a) $y' = \frac{1}{625}y(2500 - y)$

(b) $y(0) = \frac{1250}{3}, L = 2500$

41. (a) $y' = \frac{3}{10{,}000}y(4000 - y)$

(b) $y(0) = \frac{4000}{7}, L = 4000$

43. (a) After 8 completed games, 10,019 students can be expected to toss toilet paper.

(b) 6

45. (a) 30,000 **(b)** 5.7333×10^{-5}

(c) $y = \dfrac{30{,}000}{1 + 3000^{-1.72t}}$

47. (a) $\dfrac{dy}{y(L-y)} = k\,dt$

(b) $$\frac{d}{dt}\left[\frac{1}{L}\ln\left(\frac{y}{L-y}\right)\right] = \frac{1}{L}\cdot\frac{1}{\left(\frac{y}{L-y}\right)}\cdot\frac{1(L-y) - y(-1)}{(L-y)^2} = \frac{1}{y(L-y)}$$

(c) $$\frac{1}{L}\ln\left(\frac{y}{L-y}\right) = kt + C \Rightarrow \frac{y}{L-y} = e^{kLt+CL}$$
$$\Rightarrow y = \frac{Le^{kLt+CL}}{1 + e^{kLt+CL}} = \frac{L}{e^{-kLt-CL} + 1} = \frac{L}{1 + ae^{-kLt}}$$

where $a = e^{-CL}$

(d) $$y' = \frac{L}{(1 + ae^{-kLt})^2}\cdot ae^{-kLt}\cdot(-kL) = k\left(\frac{L}{1 + ae^{-kLt}}\right)\left(\frac{Lae^{-kLt}}{1 + ae^{-kLt}}\right) = k\left(\frac{L}{1 + ae^{-kLt}}\right)\left(L - \frac{L}{1 + ae^{-kLt}}\right) = ky(L - y);$$

Since $k > 0$ and $0 < y < L$, we have $ky(L - y) = y' > 0$.

(e) As $t \to \infty$, $ae^{-kLt} \to 0$. Hence $\lim_{t\to\infty}\left(\frac{L}{1 + ae^{-kLt}}\right) = \frac{L}{1} = L.$

Chapter 16 Review Exercises

1. No **3.** Yes **5.** $\frac{x^8}{8} + C$ **7.** $-\frac{1}{4t^4} + C$

9. $0.0625x^8 - 3x^2 + C$ **11.** $\frac{x^3}{3} + 0.2x^2 - \frac{2}{7}x^{7/2} + C$

13. $(-\infty, 2)$: decreasing; $(2, \infty)$: increasing; concave up with relative minimum at $x = 2$.

15. $f(x) = \frac{3}{2}x^2 - 4x + 11$ **17.** $y = \frac{2}{3}(\sqrt{x})^3 - 23$

19. (a) $P(q) = 48q - 0.015q^2 - 3050.0$ **(b)** \$17,200.00

21. $\frac{1}{3}$ **23.** $\frac{1}{8}$ **25. (a)** 50 **(b)** 75

27. (a) 6.875 **(b)** 11.375

29.

N	LEFTSUM	RTSUM
10	705.6	961.6
100	816.576	842.176
1000	828.0538	830.6138

31.

N	LEFTSUM	RTSUM
10	211.8298	391.0298
100	285.5104	303.4304
1000	293.5047	295.2967

33. 120 **35.** $\frac{104}{3}$ **37.** 99 **39.** 24 **41.** 15

43. (a) $\frac{16}{3}$ **(b)** $\frac{32}{3}$

45. (a)

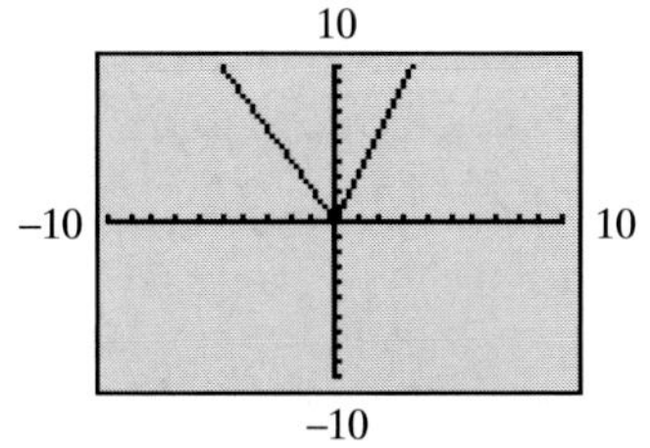

(b) 31

47. (a)

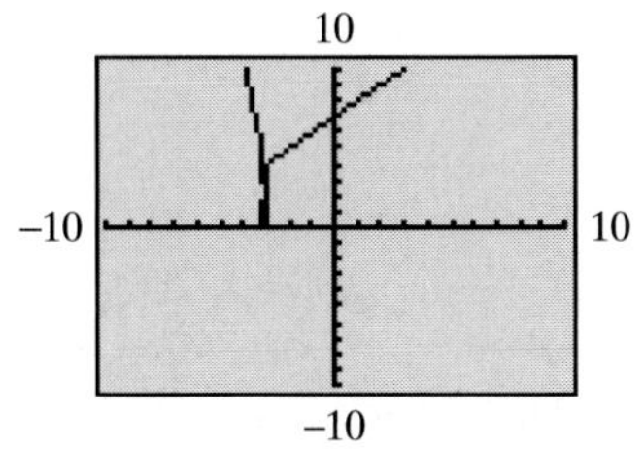

(b) $\frac{331}{6}$

49. **(a)** x-int: 3 **(b)** Net area $= 20$, gross area $= 29$

51. **(a)** x-int: $-1, 2$ **(b)** Net area $= 18$, gross area $= 27$

53. **(a)** At a production level of 160 calculators, the cost of producing the 161st is \$20.

(b) The total cost of producing the first 200 calculators is \$3400.

55. **(a)** In 1981, spending was increasing at a rate of $22.774 \frac{\text{billions of dollars}}{\text{year}}$.

(b) \$155.78 billion

57. $\frac{1}{4}(5x-7)^4 + C$ **59.** $\frac{1}{2}(x^2-5x+3)^2 + C$

61. $\frac{1}{27}(3x+6)^9 + C$ **63.** $\frac{1}{2}\ln(x^2+4x+6) + C$

65. $3\left(\frac{(t+4)^4}{4} - \frac{4(t+4)^3}{3}\right) + C$

67. $\frac{2}{3}(\sqrt{x+3})^3 - 6\sqrt{x+3} + C$ **69.** 2106 **71.** $2\sqrt{1327} - 4$

73. 0 **75.** $\frac{22.876}{3}$ **77.** 182

79. **(a)** 299,215,620 meters **(b)** $s = \frac{(t^3-5t)^4}{4} - 4448$

81. $\frac{3}{4}\ln|t| + C$ **83.** $5e^x - \frac{3}{2}x^2 + C$ **85.** $6\ln\frac{5}{3}$ **87.** $\frac{24}{\ln 5}$

89. $10\ln(\sqrt{x}+4) + C$ **91.** $-\frac{1}{4}(5-e^x)^4 + C$

93. $2\ln 5 - 4\ln 2$ **95.** $\frac{1365}{64\ln 2}$

97. **(a)** In 1990 United States billed international calls were increasing at the rate of $7659.9 \frac{\text{millions of calls}}{\text{year}}$.

(b) From 1985 to 1995 the total increase in United States billed international calls was 518,688 million.

99. $f(t) = 8.2e^{1.37t}$ **101.** $f(t) = 384e^{-0.87t}$

103. **(a)** $P' = 0.08P$ **(b)** $P(t) = 300e^{0.08t}$

(c)

t	$f(t)$
0	300
2	352
4	413
10	668
20	1486

(d)

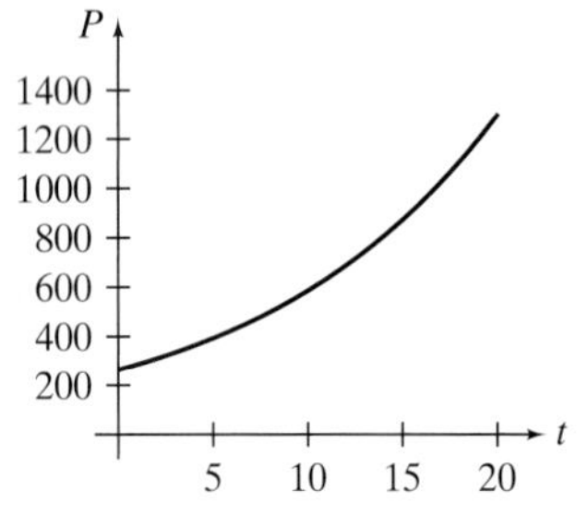

Chapter 17

Section 17.1

1. 12 **3.** 2 **5.** $\frac{14}{3}$ **7.** $\frac{93}{35}$ **9.** $2(e^2-1) \approx 12.78$

11. Answers will vary.
(a) 1.59076 **(b)** 0.79538

13. Answers will vary.
(a) 1.44929 **(b)** 0.72464

15. During the 1980s; the annual average amount spent on home health care in the U.S. was about \$6.2 billion.

17. During the 1980s; the annual average number of Medicaid recipients was about 22 million.

19. During the 1990s; the annual average revenue in the resource recovery industry was about \$17.07 billion.

21. **(a)** \$4.2 million **(b)** \$11.4 million

23. \$5.10 **25.** About 12.7%

27. **(a)** \$135,000

(b) $AC = \frac{60{,}000}{x} + 300$; $AC(500) = 420$

(c) Part (b) is the average cost per diesel engine if 500 engines are built. Part (a) is the average total cost for building 500 engines.

29. \$3484.85 **31.** \$6137.05

33. **(a)** \$3567.35
(b) \$26.76

35. **(a)** \$1778.90
(b) \$9329.57

37. **(a)** \$175

(b) \$167.71. Between 25 and 75 mugs, total average cost is \$167.71.

39. **(a)** $\int_0^{12} 600e^{-0.1t}\,dt$ **(b)** 4,193,000 barrels

41. \$6 million **43.** About \$20,344 **45.** About \$188,235

47. About \$679,207 **49.** About \$588,313

51. **(a)** \$77,400

(b) \$80,000

(c) \$90,894. Since Vicki started contributing at a younger age, Vicki's contributions had more time to earn interest.

53. About \$26,963

Section 17.2

1. $\int_0^1 [x+1-x^2]\,dx$ **3.** $\int_{-1}^1 [x+7-x^2]\,dx$

5. $\int_a^c [g(x)-f(x)]\,dx + \int_c^b [f(x)-g(x)]\,dx$

7. $\int_0^b [d(x)-k]\,dx$ **9.** 28 **11.** $\frac{15}{2}$ **13.** $\frac{59}{6}$

15. $\frac{32}{3}$ **17.** $\frac{52}{3}$ **19.** 15 **21.** About 30.88

23. About 23.62 **25.** About 2.66 **27.** $\frac{9}{2}$ **29.** 18

31. $\frac{1}{2}$ **33.** $\frac{1}{2}$ **35.** About \$52,790

37. **(a)**

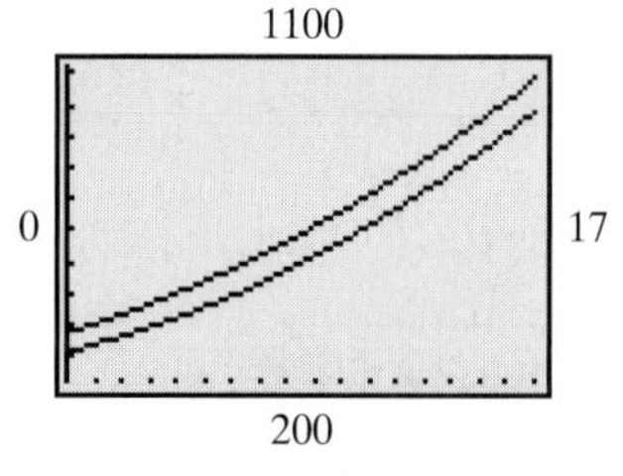

(b) $\int_1^{16} [-0.15x^2 + 4.74x + 54.79]\,dx$

(c) 1221.45. This means that receipts exceeded expenditures by over \$1.2 trillion between 1980 and 1996.

39. (a) $\int_1^2 4[t^{18/5} - t^{5/2}]\,dt$

(b) About 8.43. Between the first and second minute, the use of the nutrient increases growth by about 8.43 mg.

(c) About 8.16 mg

41. (a) In 1997

(b) About 7.83. Exports exceeded imports by about \$7.83 billion from 1995 to about October 1997.

(c) About 4.09. Imports exceeded exports by about \$4.09 billion from about October 1997 to 1999.

(d) The U.S. had a trade surplus with South Korea of about \$3.74 billion from 1995 to 1999.

43. About 24,303 schools

45. (a) About 1209. If the TV personality is used for the ad campaign, sales during the first 20 weeks will be increased by about \$1,209,000.

(b) Yes. Sales will increase by \$1,209,000, but the personality's salary is only \$1 million.

47. About \$15.7 million

49. (a)

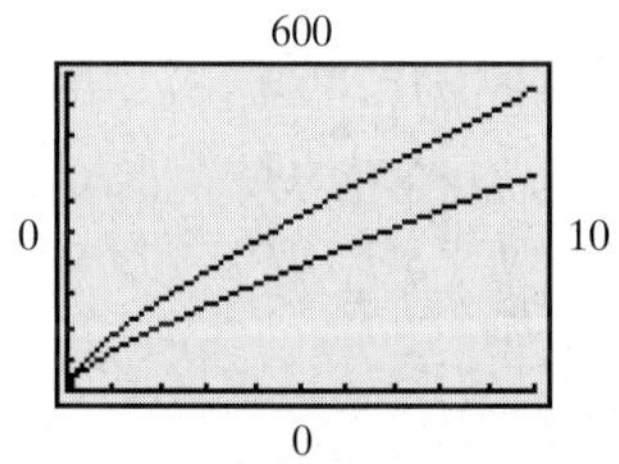

(b) About 435. Between ages 4 and 8, the child will learn an additional 435 words by listening to the tapes.

51. Between years 3 and 8, the breakthrough saved Quality Chips about \$11.6 million.

53. About \$1944.83 **55.** About \$80.94

57. (a) About 97 units **(b)** About 13 units

59. About 34.73 fewer feet of growth

Section 17.3

1. $\int_0^{xm} [d(x) - d_{mp}]dx$

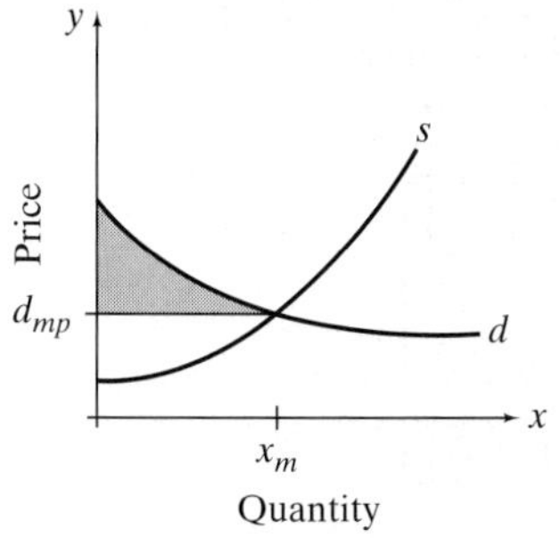

3. $d_{mp} \cdot x_m$

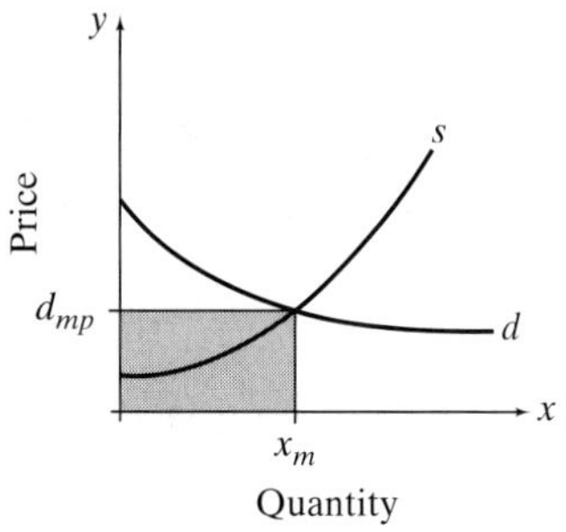

5. $\int_0^{700} [210 - 0.3x]\,dx$

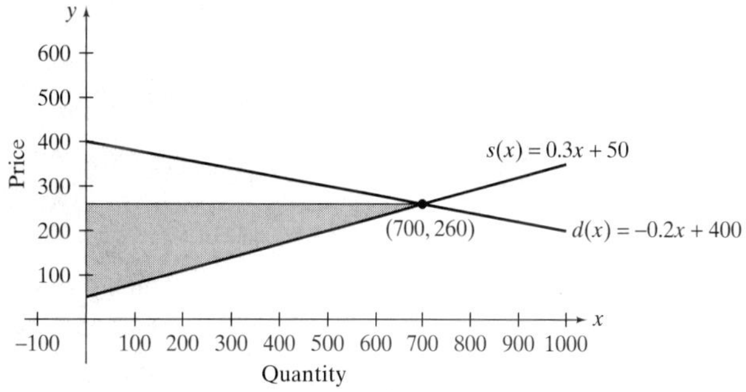

7. $\int_0^{360} [180 - 0.5x]\,dx$

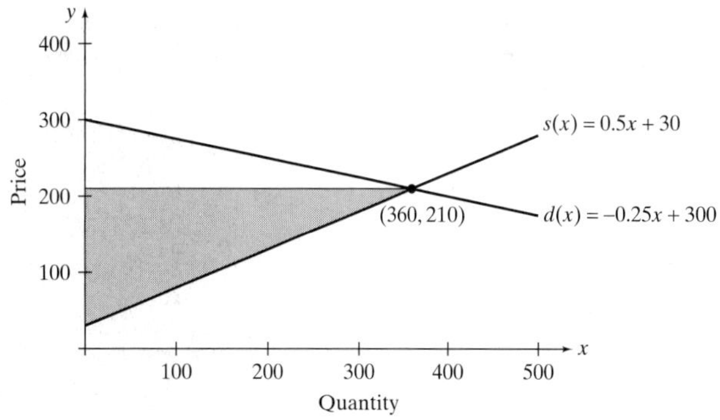

9. $\int_0^{x_m} [d(x) - d_{mp}]\,dx$

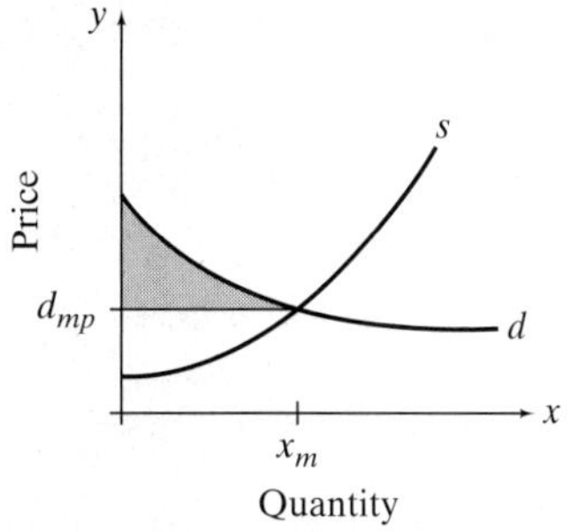

11. \$60,000 **13.** About \$1666.67 **15.** \$6750

17. About \$53,333.33 **19.** \$6000 **21.** About \$48,051.10

23. About \$9937.29 **25.** \$562.50 **27.** \$750

29. \$600 **31.** About \$33,333.33 **33.** \$225,000

35. About \$2919.86 **37.** About \$1690.54

39. **(a)** (190, 1083) **(b)** \$234,650 **(c)** \$102,885

41. **(a)** (430, 226) **(b)** \$36,980 **(c)** \$46,225

43. **(a)** (165, 1654) **(b)** \$119,790 **(c)** \$179,685

45. **(a)** (35, 255) **(b)** About \$5716.67 **(c)** About \$5716.67

47. **(a)** About (209, 74) **(b)** About \$37,112.77 **(c)** About \$11,845.42

49. **(a)** 10 **(b)** \$70

51. About \$59.13 **53.** **(a)** 65 dozen golf balls **(b)** \$507

55. About \$415.03

57. **(a)** (30, 4) **(b)** \$180 **(c)** \$45

59. **(a)** (25, 5) **(b)** \$312.50 **(c)** About \$41.67

61. 0.3847 **63.** 0.3877 **65.** 0.4133

67. Increasing; inequality in income distribution became greater.

Section 17.4

1. $e^{3x}(\frac{2}{3}x - \frac{2}{9}) + C$ **3.** $e^{4x}(\frac{1}{4}x - \frac{1}{16}) + C$

5. $e^{-x}(-x-1) + C$ **7.** $e^{-.03x}(-\frac{100}{3}x - \frac{10{,}000}{9}) + C$

9. $-\frac{3}{e^2} + 1 \approx 0.594$ **11.** $-\frac{750}{e^{1/5}} + 625 \approx 10.95$

13. $\frac{1}{4}(e^2+1) \approx 2.097$ **15.** $x^3(\frac{1}{3}\ln x - \frac{1}{9}) + C$

17. $x(\ln|2x| - 1) + C$ **19.** $e^x(x+2) + C$

21. $e^{-0.1t}(-60t - 600) + C$ **23.** $\frac{1250}{9}\left(10 - \frac{13}{e^{3/10}}\right) \approx 51.300$

25. $e^{-0.8t}(-15t - \frac{375}{2}) + C$ **27.** $(x+1)^6\left(\frac{6x+13}{42}\right) + C$

29. $(x+3)^7\left(\frac{7x-19}{56}\right) + C$ **31.** $e^{3x}\left(\frac{9x^2-6x+2}{27}\right) + C$

33. $e^{-x}(-x^2-2x-2) + C$ **35.** $e^{5x}\left(\frac{25x^2-10x+2}{125}\right) + C$

37. $x(\ln x)^2 - 2x\ln x + 2x + C$

39. $\frac{4x^2(\ln x)^3 - 6x^2(\ln x)^2 + 6x^2\ln x - 3x^2}{8} + C$

41. $u = x, dv = e^{-x}\,dx$ (Parts) **43.** $u = \ln x$ (Substitution)

45. $u = x+6, dv = e^x\,dx$ (Parts)

47. $u = 3x, dv = e^{2x}\,dx$ (Parts)

49. **(a)** $\int_0^{12} 6te^{-0.15t}\,dt$ **(b)** About 143,243 barrels

51. $P(t) = e^{0.2t}(10t - 50) + 50$

53. $C(t) = e^{-0.15t}\left(\frac{-60t-400}{9}\right) + \frac{418}{9}$

55. About 12.45 mg **57.** About \$20.5 million

59. About \$593,534

61. Oil well: about \$480,874; natural gas: about \$508,330; the natural gas well is a better investment.

63. **(a)** About 3.98 million. On average, there were 3.98 million men annually enrolled in college in the United States during the 1960s.

(b) About 5.21 million. On average, there were 5.21 million men annually enrolled in college in the United States during the 1970s.

(c) About 5.76 million. On average, there were 5.76 million men annually enrolled in college in the United States during the 1980s.

65. About 38.05. Between 1991 and 1996, paper and paperboard annually averaged 38.05% of all solid waste in the United States.

67. About 2751. The total output of motor gasoline of U.S. refineries annually averaged about 2751 million barrels during the 1990s.

69. **(a)** About 22.7. During the 1980s, per capita consumption of cheese averaged about 22.7 pounds annually.

(b) About 26.0. During the first half of the 1990s, per capita consumption of cheese averaged about 26.0 pounds annually.

71. About 9.3. During the first half of the 1990s, variable costs of operating an automobile annually averaged 9.3¢ per mile.

Section 17.5

1. 22 **3.** About 1.48 **5.** About 0.76

7. About 3.83 **9.** 7.0625

11. $n = 10$: about 0.4128; $n = 50$: about 0.4005; $n = 100$: about 0.4001

13. $n = 10$: about 0.0690; $n = 50$: about 0.0672; $n = 100$: about 0.0671

15. $n = 10$: about 2.4537; $n = 50$: about 2.4517; $n = 100$: about 2.4517

17. $n = 10$: about 0.4091; $n = 50$: about 0.3961; $n = 100$: about 0.3957

19. $n = 10$: about 6.1785; $n = 50$: about 6.1725; $n = 100$: about 6.1723

21. $n = 10$: about 4.0511; $n = 50$: about 4.0500; $n = 100$: about 4.0499

23. About \$9.55 thousand **25.** About \$100

27. About \$3225 **29.** About \$21,025 **31.** About \$43,600

33. **(a)** About 10,349,214 **(b)** About 23,098,429

35. **(a)** About 1,223,140 thousands of tons

(b) About 24,463 thousands of tons

37. **(a)** About 6478 work stoppages

(b) About 185 work stoppages annually

39. (a) About 41.2 million

(b) About 1,287,500 million annually

41. ≈ $12.677 billion (in 1987 dollars)

43. (a) About 122,008,000 pregnancies

(b) About 6,100,400 pregnancies annually

45. About 66,690 nursing students annually were pursuing a 2-year degree

Section 17.6

1. $\frac{1}{24}$ **3.** Diverges **5.** $\sqrt{2} \approx 1.41$ **7.** Diverges

9. $\frac{1}{e} \approx 0.368$ **11.** $\frac{100}{3e^{3/100}} \approx 32.35$ **13.** Diverges

15. Diverges **17.** $\frac{1}{2}$ **19.** Diverges **21.** 5

23. $\frac{1}{4}$ **25.** $-\frac{1}{60} \approx -0.0167$ **27.** 1 **29.** 0 **31.** 0

33. Diverges **35.** Diverges **37.** 600 thousand barrels

39. 16 million ft^3 **41.** About 1166.67 tons

43. About 107 ml **45.** About $166,667

47. About $413,793 **49.** About $1,076,923

51. About $218,182 **53.** About $106,667

55. About $45,455

57. (a) About $79,263
(b) $80,000

59. (b) About 0.1054
(c) About 0.4066

61. (b) 0.1875
(c) 0.046875

63. (b) About 0.2387
(c) About 0.2983
(d) About 0.4493

65. (b) About 0.3023
(c) About 0.2109

67. (b) About 0.0670
(c) About 0.0607
(d) About 0.7788

Chapter 17 Review Exercises

1. 13 **3.** 102

5. Answers will vary.
(a) About 11.8624 **(b)** About 1.4828

7. (a) $4900

(b) $AC(x) = \frac{1300}{x} + 18$; $AC(400) = \$21.25$

(c) Part (a) is the average total cost; part (b) is the average cost of one phone when 400 are produced.

9. (a) $2457.29 **(b)** $20.89

11. About $25,258 **13.** About $18,776

15. $\int_{-\frac{1+\sqrt{17}}{2}}^{\frac{-1+\sqrt{17}}{2}} [4 - x^2 - x]\,dx$ **17.** 36 **19.** $\frac{125}{6} \approx 20.83$

21. $\frac{256}{3} \approx 85.33$ **23.** ≈ 211.43 **25.** $\frac{125}{6} \approx 20.83$

27. $\frac{149}{3} \approx 49.33$

29. About 2705. If Lucky is used for the ad campaign, total sales in the first 15 months will increase by 2705 tricycles.

31. About $795,766

33. (a) 201 units **(b)** 41 units

35. $\int_0^{60} [-0.4x + 24]\,dx$

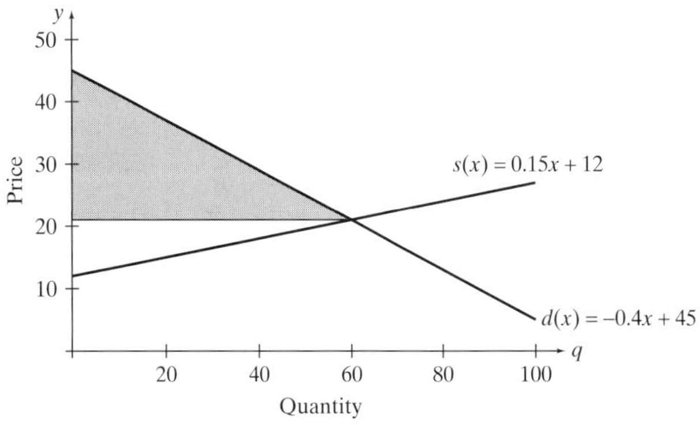

37. $1260

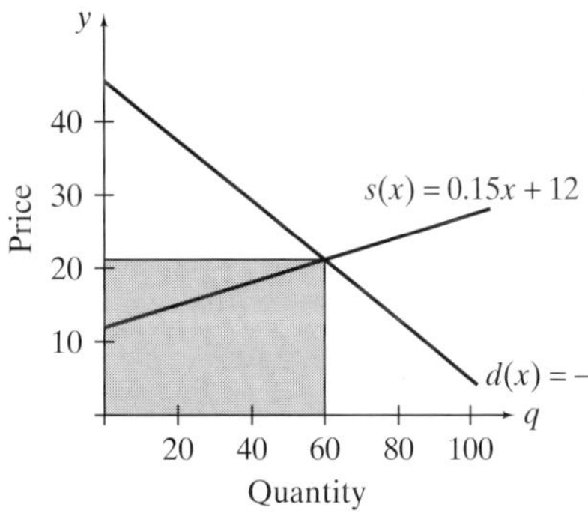

39. $4000 **41.** $72 **43.** $40,000 **45.** About $1667

47. (a) (380, 136)
(b) $21,660
(c) $14,440

49. (a) (200, 2900)
(b) About $213,333
(c) $160,000

51. (a) 16 drills **(b)** About $42.67

53. About $2301

55. (a) (36, 6) **(b)** $155.52 **(c)** $72

57. $e^{2x}(2x - 1) + C$ **59.** $x^5\left(\frac{5 \ln x - 1}{25}\right) + C$

61. $e^{-0.2t}(-200 - 40t) + C$

63. $e^{-x}(-x^2 - 2x - 2) + C$

65. $x((\ln x)^4 - 4(\ln x)^3 + 12(\ln x)^2 - 24(\ln x) + 24) + C$

67. $u = -x^3$ (Substitution) **69.** $u = x,\ dv = e^{5x}$ (Parts)

71. (a) $\int_0^{12} 8te^{-0.17t}\,dt$ **(b)** About 167,395 barrels

73. About $598,717

75. (a) 58,955. The average sales price for a home in the United States between 1976 and 1981 was $58,955.

(b) 95,550. The average sales price for a home in the United States between 1983 and 1987 was $95,550.

(c) 143,405. The average sales price for a home in the United States between 1991 and 1996 was $143,405.

77. −154 **79.** About −0.6563

81. $n = 10$: about 74.797; $n = 50$: about 74.798; $n = 100$: about 74.798

83. $n = 10$: about 1.679; $n = 50$: about 1.677; $n = 100$: about 1.677

85. About $12,008 **87.** About $153,500

89. **(a)** About 6,830,000 **(b)** About 683,000

91. $\frac{1}{5}$ **93.** Diverges **95.** Diverges **97.** $\frac{1}{3}$

99. Diverges **101.** About 895,833 barrels

103. About $88,889 **105.** **(b)** About 0.333 **(c)** About 0.556

Chapter 18

Section 18.1

1. 19 **3.** 3 **5.** $\frac{5}{2}$ **7.** $-\frac{13}{5}$ **9.** 0 **11.** Undefined

13. $2 + e^2$ **15.** $2e$ **17.** All ordered pairs (x, y)

19. All orderd pairs (x, y) such that $x + y \neq 0$

21. All ordered pairs (x, y) such that $x - 4y \neq 0$

23.

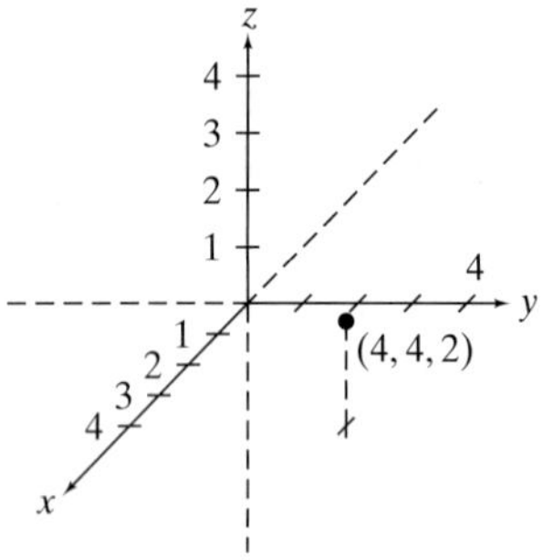

25.

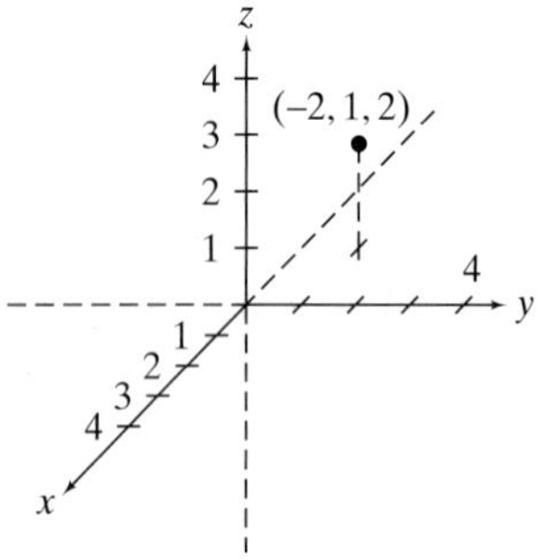

27.

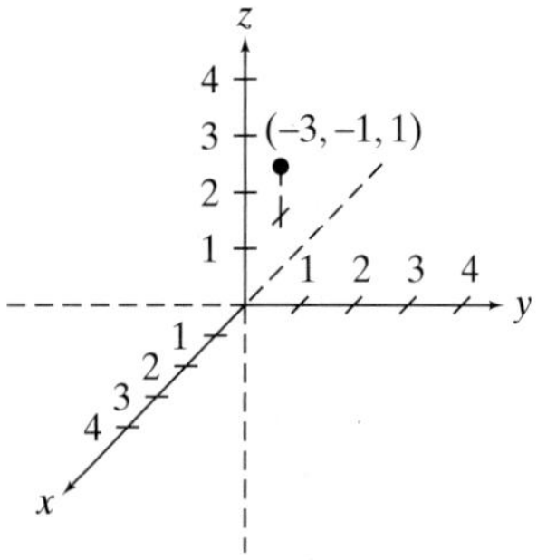

29.

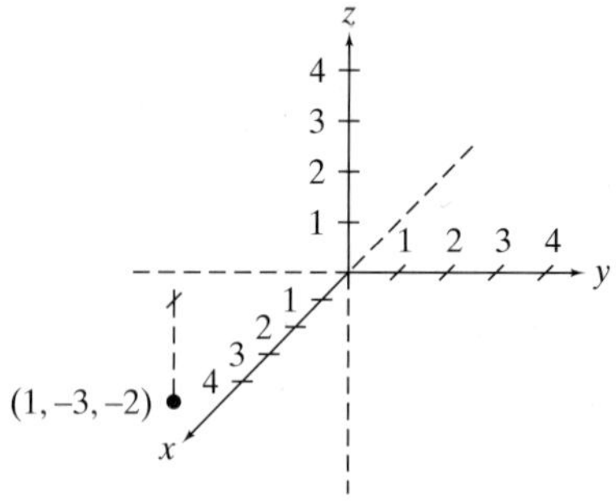

31.

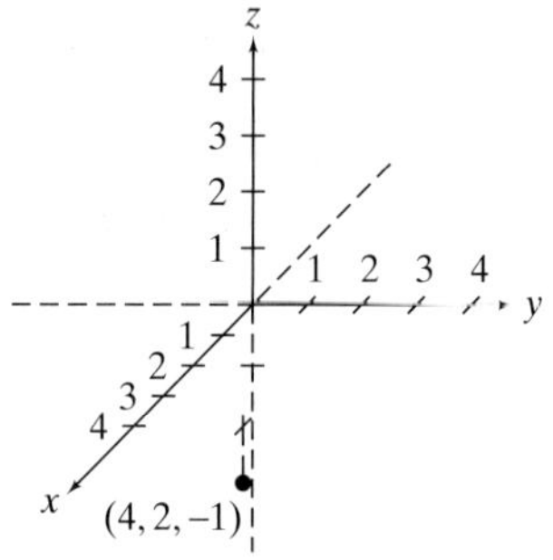

33.

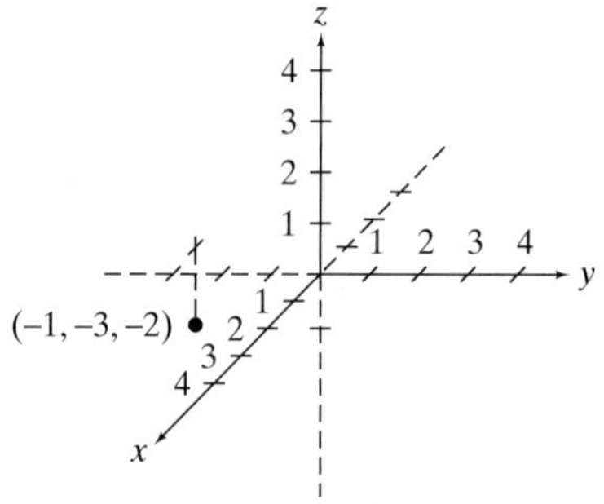

35.

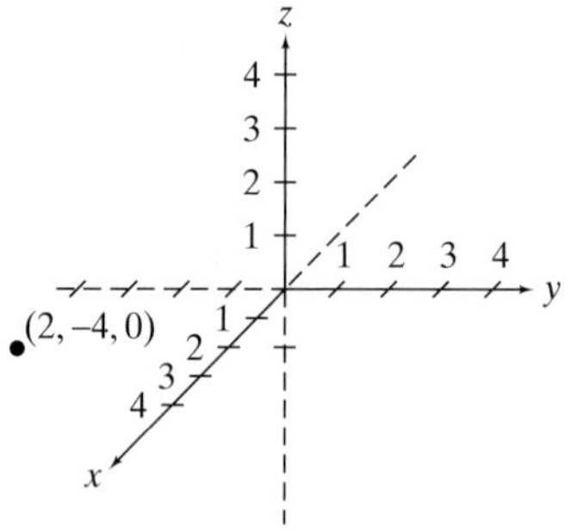

37. Above **39.** Above **41.** Below **43.** Below

45.

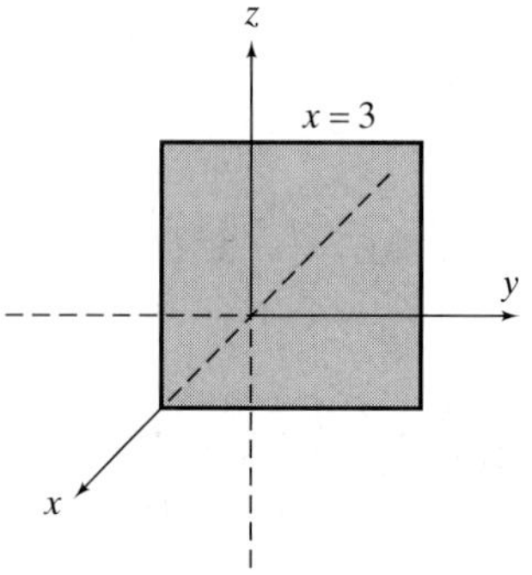

47.

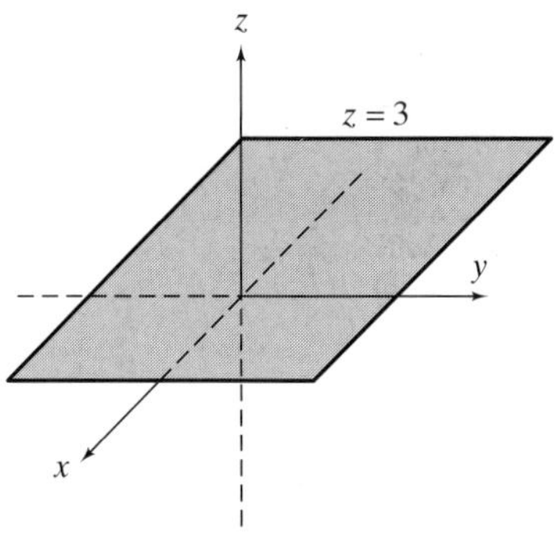

49.

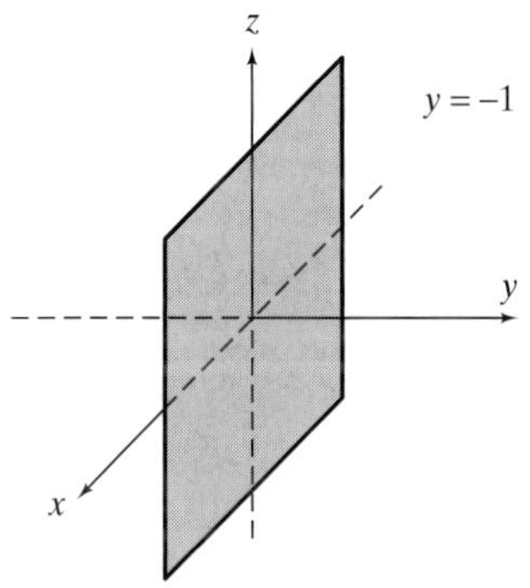

51. **(a)** $S(5, 3) = 53$; when Seamount spends \$5000 each week on newspaper advertising and \$3000 each week on radio advertising, weekly sales are \$530,000.

(b) $S(3, 5) = 23$; when Seamount spends \$3000 each week on newspaper advertising and \$5000 each week on radio advertising, weekly sales are \$230,000.

53. **(a)** $TS(1, 0.5) = 1.9$; when Tube Town spends \$1000 on radio advertising and \$500 on television advertising, weekly ticket sales are \$19,000.

(b) $TS(0.5, 1) = 3.575$; when Tube Town spends \$500 on radio advertising and \$1000 on television advertising, weekly ticket sales are \$35,750.

55. $R(36, 1) = 36K,\ R(36, 2) = \frac{9}{4}K$

57. $f(25, 10) \approx -11$; when the air temperature is 10°F and the wind velocity is 25 miles per hour, it feels like it is 11° below zero.

59. The monthly cost of producing 50 leaf blowers per month and 30 attachments per month is \$2795.00.

61. **(a)** $P(x, y) = 85x - 0.8x^2 - 0.25xy + 50.8y + 0.015y^2 - 1000.0$

(b) The monthly profit from sales of 50 leaf blowers per month and 30 attachments per month is \$2412.5.

63. **(a)** $C(x, y) = 5x + 4y + 50$

(b) $R(x, y) = 12x + 9.5y$

(c) $P(x, y) = 7x + 5.5y - 50$

65. $f(2000, 20) = 6787.15$; \$2000 invested at an annual interest rate of 6.125% compounded monthly for 20 years yields \$6787.15.

67. $f(0.0725, 25) = 12{,}251.49$; \$2000 invested at an annual interest rate of 7.25% compounded continuously for 25 years yields \$12,251.49.

69. About 7241

71. **(a)** A 9 year-old person with a 12 year-old mental age has an IQ of 133.33.

(b) A 12 year-old person with a 9 year-old mental age has an IQ of 75.00.

(c) A 12 year-old person with a 12 year-old mental age has an IQ of 100.00.

(d) If $a = m$, then $IQ = 100$. An IQ of 100 represents a person with mental age equal to actual age.

73. **(a)**

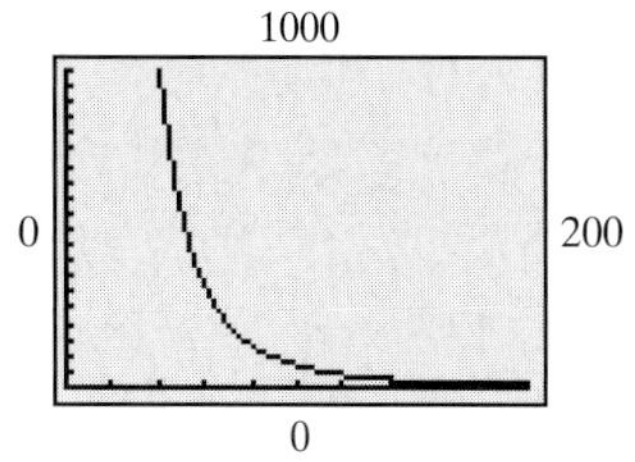

(b) About 199 **(c)** About 68

(d) Answers may vary. Sample answers: (119, 40), (81, 129), (40, 1067)

Section 18.2

1.

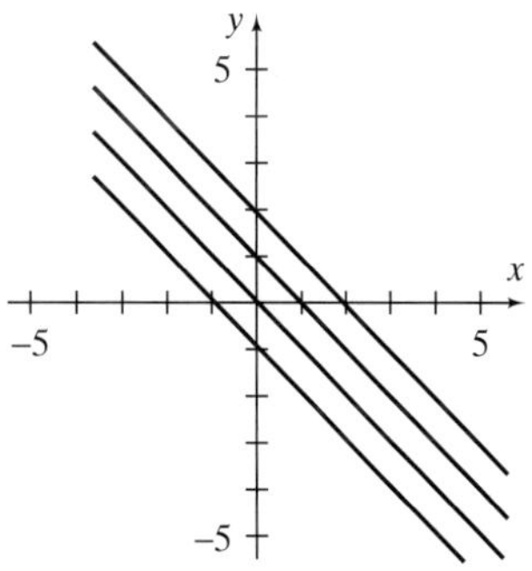

3.

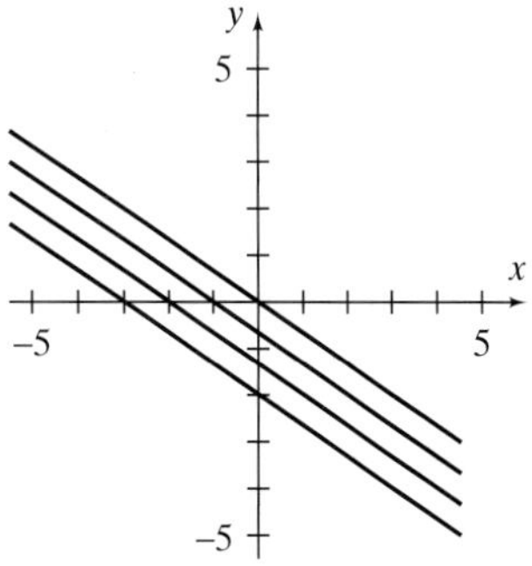

5.

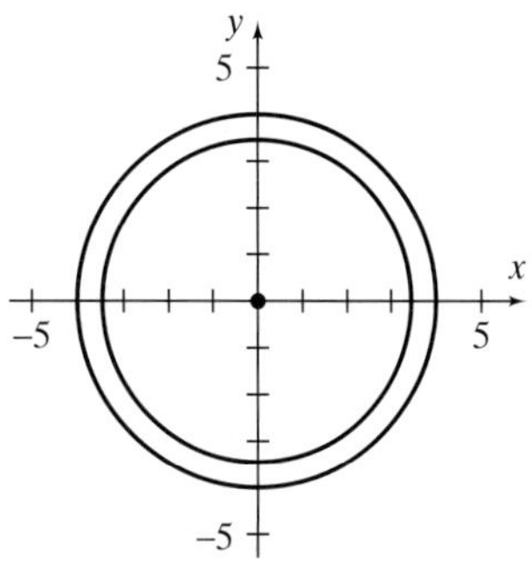

7.

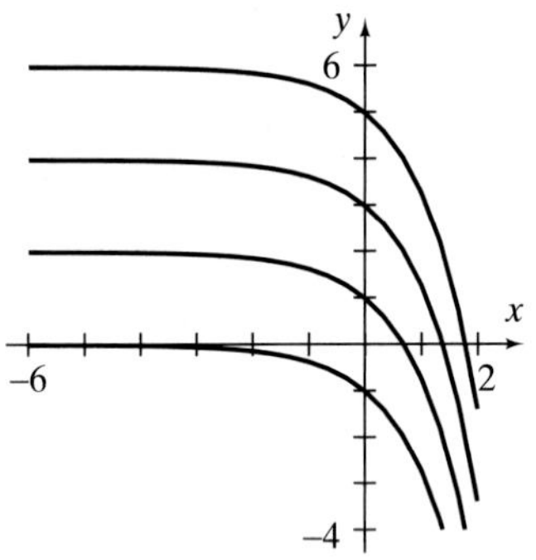

9.

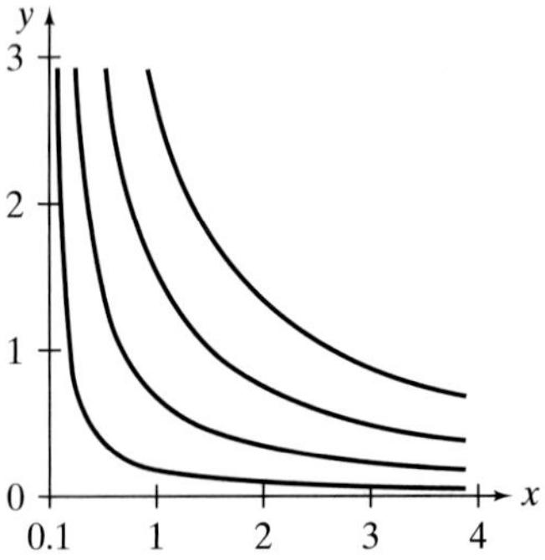

11.

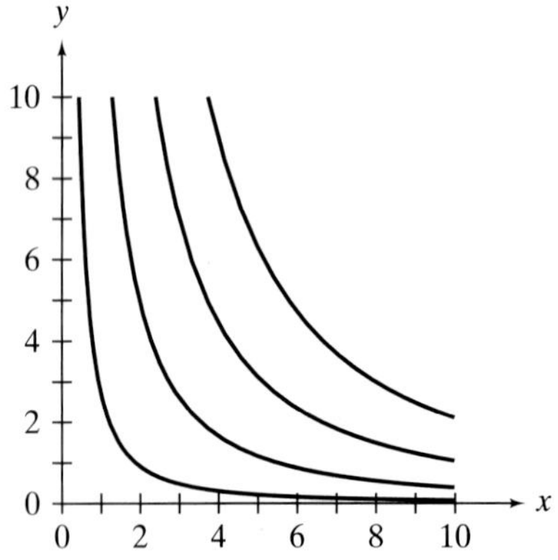

13.

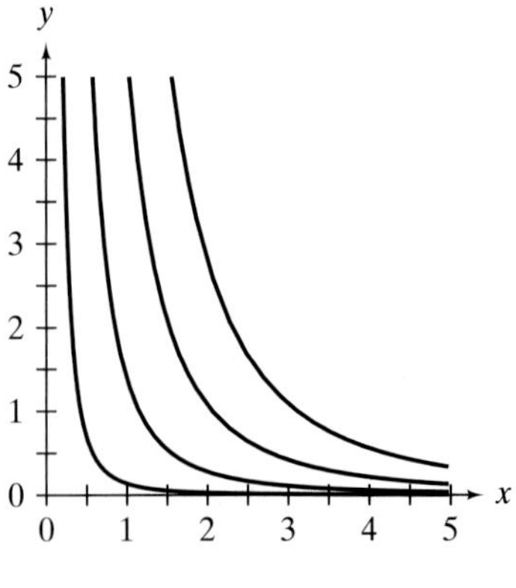

15.

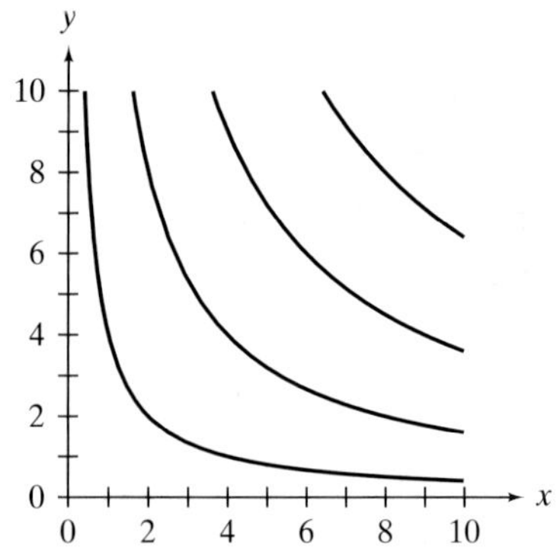

17.

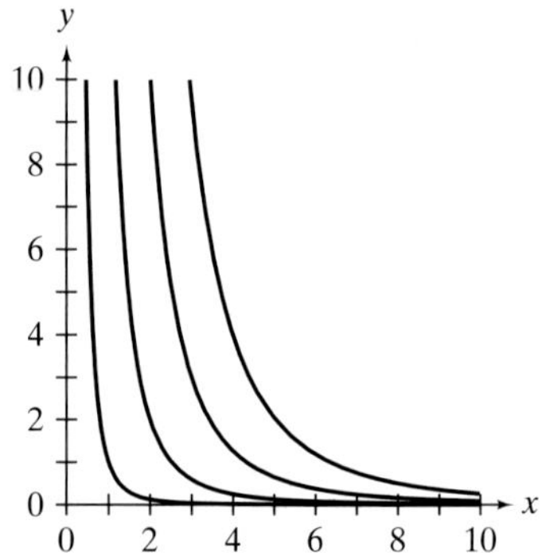

19.

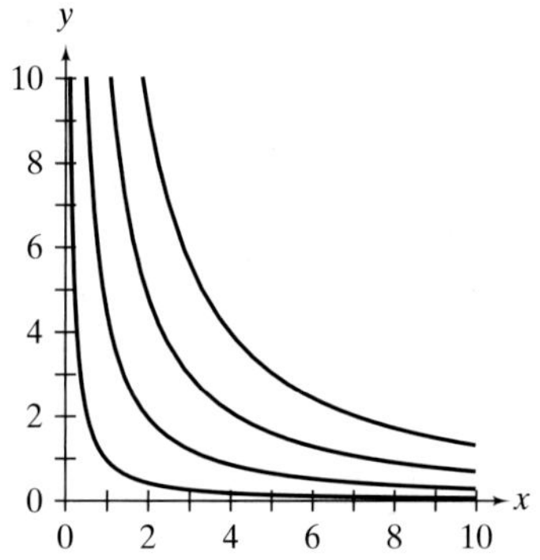

21.

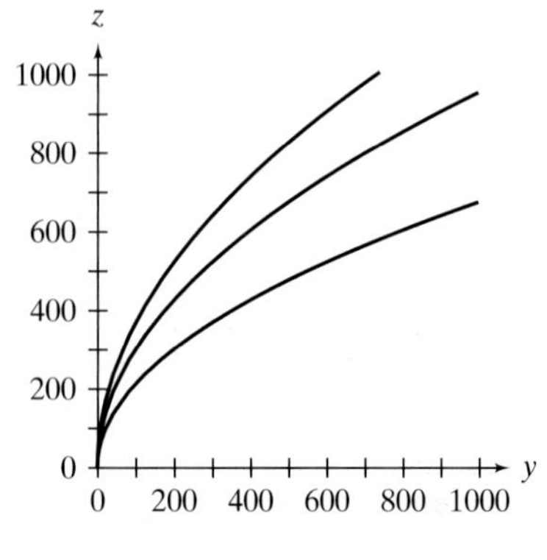

23.

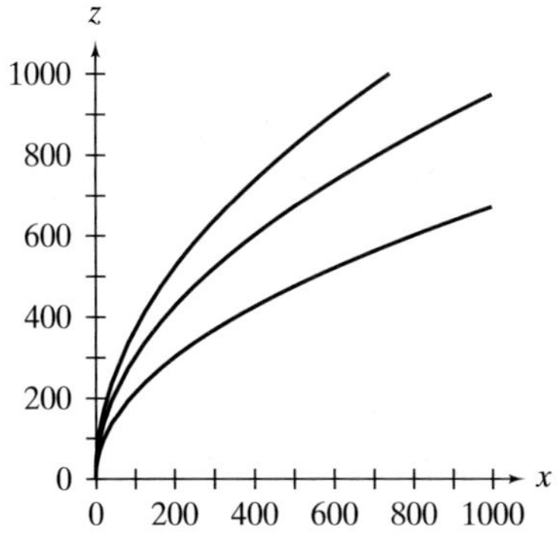

25.

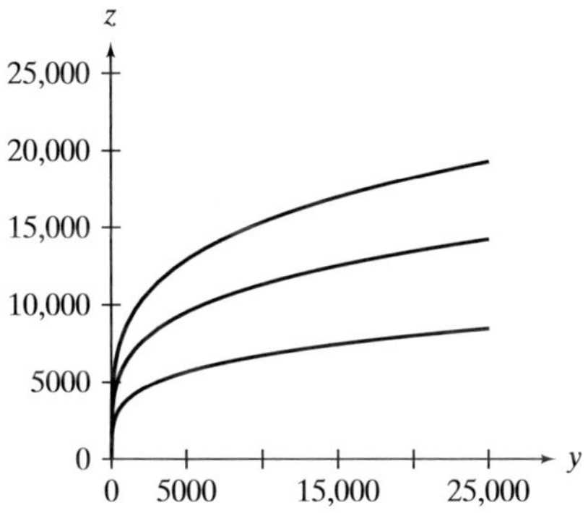

27.

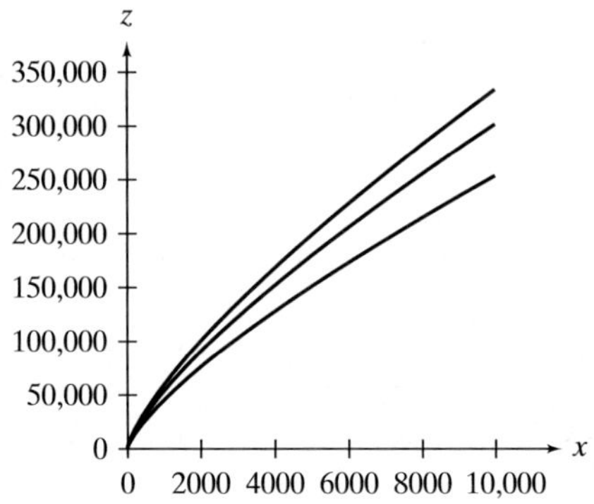

29.

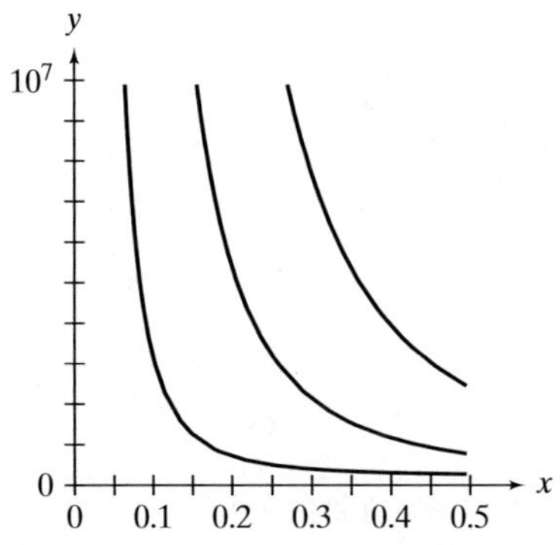

31.

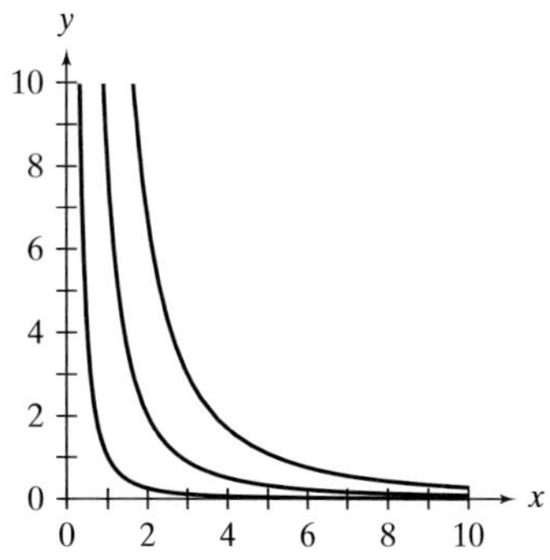

33.

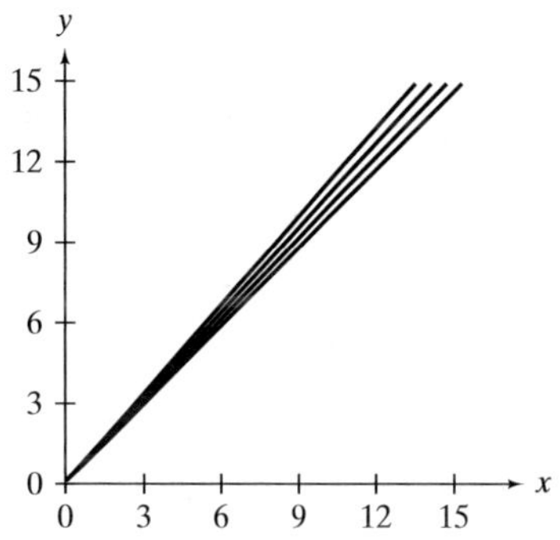

35.

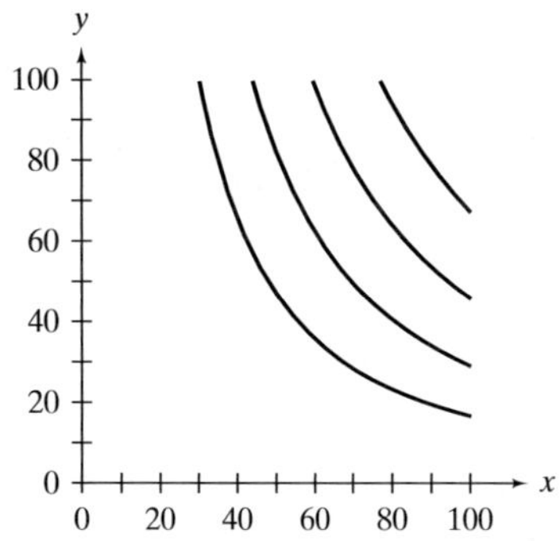

37. **(a)** Vertical

(b)

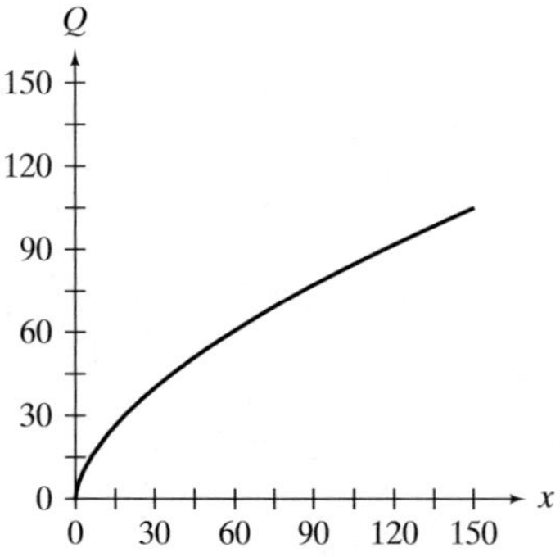

(c) About 137 thousand labor hours

(d) The graph shows the relationship between production and the number of labor hours when capital is held fixed at \$18 million.

39. (a) Vertical

(b)

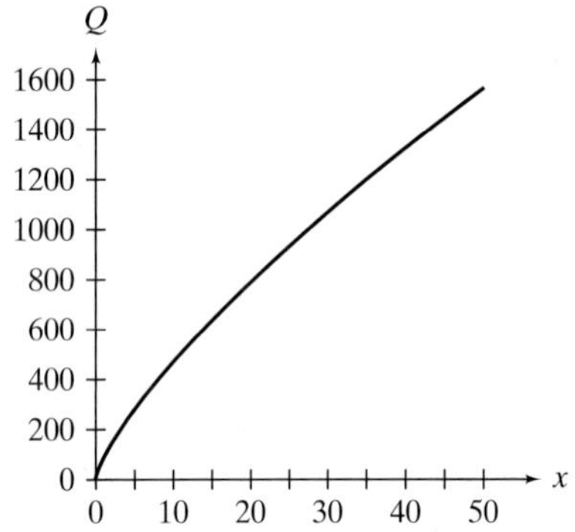

(c) 47

(d) The graph shows the relationship between production and the utilization of capital when labor utilization is held fixed at 100.

41.

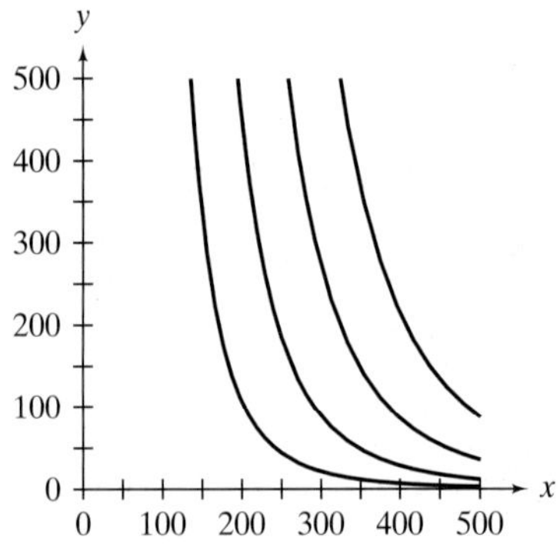

43. (a)

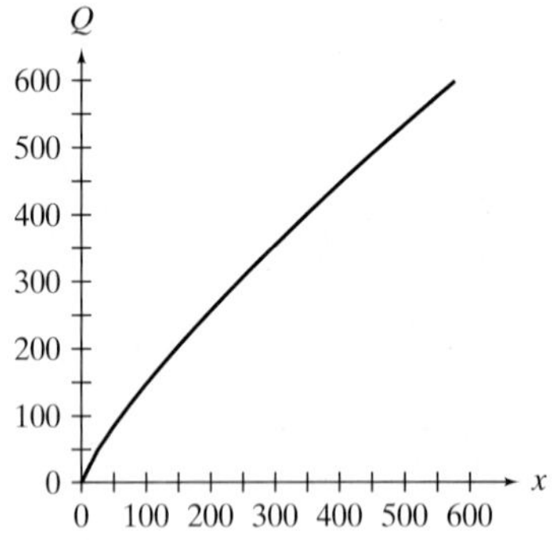

(b) About 528 thousand units of labor

45. (a)

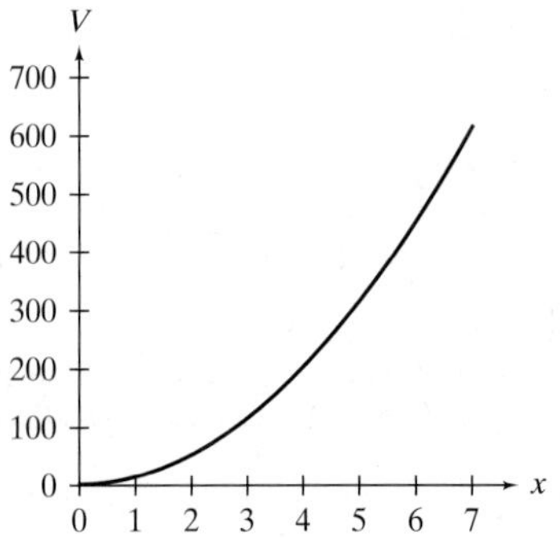

(b) 3

(c) About 452

(d) The graph shows the relationship between the volume and the radius of a cylinder with the height held fixed at 4 inches.

47. (a)

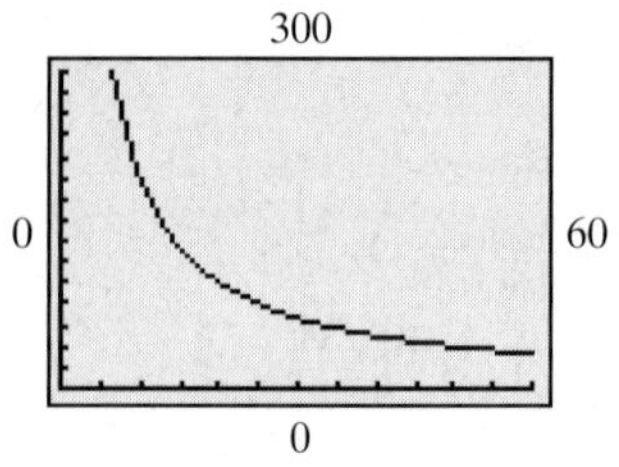

(b) A 15 year-old with a mental age of 20 has an IQ of $133\frac{1}{3}$.

(c) A 16 year-old with a mental age of 20 has an IQ of 125.

(d) $V(3, 5) = 15\pi$; the volume of a cone with radius of 3 and height of 5 is 15π; $V_r(3, 5) = 10\pi$; for a cone with a radius of 3 units and a height of 5 units, if the radius increases by 1 unit while the height remains constant at 5 units, the volume will increase by approximately 10π units; $V_h(3, 5) = 3\pi$; for a cone with a radius of 3 units and a height of 5 units, if the height increases by 1 unit while the radius remains constant at 3 units, the volume will increase by approximately 3π units.

Section 18.3

1. $f_x(x, y) = 10x$, $f_y(x, y) = -18y^2$

3. $f_x(x, y) = 2y$, $f_y(x, y) = 2x - 2y$

5. $f_x(x, y) = 3x^2y^2$, $f_y(x, y) = 2x^3y$

7. $f_x(x, y) = 2x + 6xy^3 - y$, $f_y(x, y) = 9x^2y^2 - 6y^2 - x$

9. $f_x(x, y) = 6x^2 + 2$, $f_y(x, y) = -2y$

11. $f_x(x, y) = 6(2x + 3y)^2$, $f_y(x, y) = 9(2x + 3y)^2$

13. $f_x(x, y) = 5.5815x^{-0.85}y^{0.87}$, $f_y(x, y) = 32.3727x^{0.15}y^{-0.13}$

15. $f_x(x, y) = e^x \ln y$, $f_y(x, y) = \dfrac{e^x}{y}$

17. $f_x(x, y) = -\dfrac{y}{x^2}$, $f_y(x, y) = \dfrac{1}{x}$

19. $f_x(x, y) = 6x - 6x^2y^4$, $f_y(x, y) = -8x^3y^3$

21. $f_x(x, y) = 4x(x^2 + y^3)$, $f_y(x, y) = 6y^2(x^2 + y^3)$

23. $f_x(x, y) = \dfrac{2x}{x^2 + y^3}$, $f_y(x, y) = \dfrac{3y^2}{x^2 + y^3}$

25. $f_x(x, y) = \dfrac{2xy}{y + x} - \dfrac{x^2y}{(y + x)^2}$, $f_y(x, y) = \dfrac{x^2}{y + x} - \dfrac{x^2y}{(y + x)^2}$

27. $f_x(x, y) = 6xy^3 + e^{x+y}$, $f_y(x, y) = 9x^2y^2 + e^{x+y}$

29. $f_x(x, y) = ye^{xy}$, $f_y(x, y) = xe^{xy} + \dfrac{1}{y}$

31. If we are at the point (1, 2, 24) on the surface and move along the surface in the x-direction parallel to the x-axis, the function is decreasing at a rate of $20 \frac{\text{units of } f}{\text{units of } x}$.

33. If we are at the point (2, 3, 38) on the surface and move along the surface in the y-direction parallel to the y-axis, the function is increasing at a rate of $5 \frac{\text{units of } f}{\text{units of } y}$.

35. If we are at the point $(2, 5, \ln 23)$ on the surface and move along the surface in the x-direction parallel to the x-axis, the function is decreasing at a rate of $\frac{1}{23} \frac{\text{units of } f}{\text{units of } x}$.

37. If we are at the point $(5, 2, 86.58)$ on the surface and move along the surface in the x-direction parallel to the x-axis, the function is increasing at a rate of $2.5973 \frac{\text{units of } f}{\text{units of } x}$.

39. If we are at the point $(2, 4, -64 - e^2)$ on the surface and move along the surface in the x-direction parallel to the x-axis, the function is decreasing at a rate of $-96 - e^{-2} \frac{\text{units of } f}{\text{units of } x}$.

41. $f_{xx}(x, y) = -10y$, $f_{yy}(x, y) = 24y$, $f_{xy}(x, y) = -10x$

43. $f_{xx}(x, y) = 6 - 12xy^2 + 12x$, $f_{yy}(x, y) = -4x^3$, $f_{xy}(x, y) = -12x^2y$

45. $f_{xx}(x, y) = -36x$, $f_{yy}(x, y) = 0$, $f_{xy}(x, y) = 5$

47. $f_{xx}(x, y) = -12xy^2 + 14$, $f_{yy}(x, y) = 20y^3 - 4x^3$, $f_{xy}(x, y) = -12x^2y$

49. $f_{xx}(x, y) = y^4e^{xy} - \dfrac{1}{x^2}$, $f_{yy}(x, y) = 2e^{xy} + 4yxe^{xy} + y^2x^2e^{xy}$, $f_{xy}(x, y) = 3y^2e^{xy} + y^3xe^{xy}$

51. $f_{xx}(x, y) = -1.183x^{-1.35}y^{0.4}$, $f_{yy}(x, y) = -1.248x^{0.65}y^{-1.6}$, $f_{xy}(x, y) = 1.352x^{-0.35}y^{-0.6}$

53. $f_{xx}(x, y) = \dfrac{4y}{x^3}$, $f_{yy}(x, y) = 0$, $f_{xy}(x, y) = -\dfrac{2}{x^2}$

55. $f_{xx}(x, y) = y^3e^{xy}$, $f_{yy}(x, y) = 2xe^{xy} + yx^2e^{xy}$, $f_{xy}(x, y) = 2ye^{xy} + y^2xe^{xy}$

57. $x = -1$, $y = -2$

59. **(a)** $S_x(x, y) = 4x$, $S_y(x, y) = 1$

(b) $S_x(3, 5) = 12$; if \$3000 are spent each week on newspaper advertising and \$5000 are spent each week on radio advertising and the amount spent on radio advertising is kept fixed at \$5000 per week then sales will be increasing at a rate of $\$120{,}000 \frac{\text{in sales per week}}{\text{thousands of dollars spent on newspaper advertising per week}}$; $S_y(3, 5) = 1$; if \$3000 are spent each week on newspaper advertising and \$5000 are spent each week on radio advertising and the amount spent on newspaper advertising is kept fixed at \$3000 then sales will be increasing at a rate of $\$10{,}000 \frac{\text{in sales per week}}{\text{thousands of dollars spent on radio advertising per week}}$.

61. **(a)** $TS_x(x, y) = 3x$, $TS_y(x, y) = 6.4y$

(b) $TS_x(1, 0.5) = 3$; if \$1000 are spent each week on radio advertising and \$500 are spent each week on television advertising and the amount spent on television advertising is kept fixed at \$500 per week then sales will be increasing at a rate of $\$30{,}000 \frac{\text{in sales per week}}{\text{thousands of dollars spent on radio advertising per week}}$; $TS_y(1, 0.5) = 3.2$; if \$1000 are spent each week on radio advertising and \$500 are spent each week on television advertising and the amount spent on radio advertising is kept fixed at \$1000 per week then sales will be increasing at a rate of $\$32{,}000 \frac{\text{in sales per week}}{\text{thousands of dollars spent on television advertising per week}}$.

63. The rate of change of the total amount accumulated when the point in time is held fixed and the amount of the initial investment is increased is given by $f_P(P, t) = e^{0.06t}$; the rate of change of the total amount accumulated when the amount of the initial investment is held fixed and time is increasing is given by $f_t(P, t) = 0.06Pe^{0.06t} \frac{\text{dollars}}{\text{year}}$.

65. **(a)** $V_r(r, h) = \frac{2}{3}\pi rh$, $V_h(r, h) = \frac{1}{3}\pi r^2$

67. $C_w(6, 8.2) \approx 12.2$; for a head with a width of 6 inches and a length of 8.2 inches, if the width of the head increases by 1 inch while the length remains constant at 8.2 inches, the cephalic index will increase by approximately 12.2 units; $C_L(6, 8.2) \approx -8.92$; for a head with a width of 6 inches and a length of 8.2 inches, if the length of the head increases by 1 inch while the width remains constant at 6 inches, the cephalic index will decrease by approximately 8.92 units.

69. **(a)** $R(x, y) = 350x - 4x^2 + 3xy + 450y - 3y^2$

(b) The marginal revenue from an increase in sales of racing bikes is given by $350 - 8x + 3y \frac{\text{dollars}}{\text{bikes}}$; the marginal revenue from an increase in sales of mountain bikes is given by $3x + 450 - 6y \frac{\text{dollars}}{\text{bikes}}$.

(c) $R(15, 20) = 13{,}050$; with a weekly demand of 15 racing bikes and 20 mountain bikes, the revenue is \$13,050; $R_x(15, 20) = 290$; with a weekly demand of 15 racing bikes and 20 mountain bikes, the marginal revenue from an increase in sales of racing bikes when the demand for mountain bikes is held fixed at 20 is $290 \frac{\text{dollars}}{\text{bike}}$.

71. **(a)** Marginal revenue for a 2D calculator is $R_x(x, y) = 70 + 0.5y - 0.08x$.

(b) Marginal revenue for a 3D calculator is $R_y(x, y) = 95 + 0.5x - 0.08y$.

73. **(a)** $P(x, y) = 37x + 35.5y - 0.5x^2 - 0.8xy - y^2$

(b) $P_x(x, y) = 37 - x - 0.8y$; the marginal profit from an increase in sales of two-stroke engines when the sales of four-stroke engines is held fixed is given by $37 - x - 0.8y \frac{\text{thousands of dollars}}{\text{thousand two-stroke engine}}$; $P_y(x, y) = 35.5 - 0.8x - 2y$; the marginal profit from an increase in sales of four-stroke engines when the sales of two-stroke engines is held fixed is given by $35.5 - 0.8x - 2y \frac{\text{thousands of dollars}}{\text{thousand two-stroke engine}}$.

(c) $P(18, 10) = 615$; the profit from sales of 18 thousand two-stroke engines and 10 thousand four-stroke engines is 615 thousand dollars; $P_x(18, 10) = 11$; with sales of 18 thousand two-stroke engines and 10 thousand four-stroke engines, the marginal profit from an increase in sales of two-stroke engines when sales of four-stroke engines is fixed at 10 thousand is $11 \frac{\text{thousand dollars}}{\text{thousand engines}}$.

(d) $P_y(18, 10) = 1.1$; with sales of 18 thousand two-stroke engines and 10 thousand four-stroke engines, the marginal profit from an increase in sales of four-stroke engines when sales of two-stroke engines is fixed at 18 thousand is $1.1 \frac{\text{thousand dollars}}{\text{thousand engines}}$.

75. **(a)** $f_x(x, y) = 128.8x^{-0.2}y^{0.25}$, $f_y(x, y) = 40.25x^{0.8}y^{-0.75}$

(b) If the company is now using 111 units of labor and 25 units of capital and keeps capital fixed at 25 units, production is increasing at the rate of $112.29 \frac{\text{calculators}}{\text{thousands of hours of labor}}$; if the company is now using 111 units of labor and 25 units of capital and keeps labor fixed at 111 units, production is increasing at the rate of $155.8 \frac{\text{calculators}}{\text{capital equipment in millions}}$.

77. **(a)** $f_x(x, y) = 860.598x^{-0.4}y^{0.45}$, $f_y(x, y) = 645.4485x^{0.6}y^{-0.55}$

(b) If the company is now using 47 units of labor and 8 units of capital and keeps capital fixed at 8 units, production is increasing at the rate of $470.27 \frac{\text{engines}}{\text{thousands of hours of labor}}$; if the company is now using 47 units of labor and 8 units of capital and keeps labor fixed at 47 units, production is increasing at the rate of $2072.1 \frac{\text{engines}}{\text{capital equipment in millions}}$.

79. **(a)** $f_x(x, y) = 33x^{-0.45}y^{0.5}$, $f_y(x, y) = 30x^{0.55}y^{-0.5}$

(b) If the company is now using 220 units of labor and 140 units of capital and keeps capital fixed at 140 units, production is increasing at the rate of $34.474 \frac{\text{motorcycles}}{\text{labor in millions of dollars}}$; if the company is now using 220 units of labor and 140 units of capital and keeps labor fixed at 220 units, production is increasing at the rate of $49.248 \frac{\text{motorcycles}}{\text{capital in millions of dollars}}$.

(c) Production would increase more with increased spending on capital since the change in productivity with respect to capital is larger than the change in productivity with respect to labor.

Section 18.4

1. $(2, -3)$ **3.** $(-1, 2)$ **5.** $(2, -1)$

7. $(1, 2)$ **9.** $(1, -4)$

11. The point $(3, 2, -10)$ is a relative minimum.

13. The point $(3, -2, -11)$ is a relative minimum.

15. The point $(-2, -\frac{2}{3}, \frac{13}{3})$ is a relative maximum.

17. The point $(-2, 4, -11)$ is a relative minimum.

19. The point $(3, 3, -7)$ is a relative minimum.

21. The point $(0, 1, 0)$ is a relative maximum and the point $(2, 1, -\frac{16}{3})$ is a saddle point.

23. The point $(1, -1, 20)$ is a saddle point and the point $(1, 3, -44)$ is a relative minimum.

25. The point $(0, 0, 0)$ is a saddle point and the point $(1, 1, -1)$ is a relative minimum.

27. The point $(0, 0, 1)$ is a relative maximum.

29. The point $(-2, \frac{1}{2}, -18)$ is a relative maximum.

31. **(a)** 3 million on labor, 2 million on robotics equipment

(b) \$30 million

33. **(a)** 1500 regular woks, 100 jumbo woks

(b) \$14,999,800

35. **(a)** $P(x, y) = -x + 11y - x^2 + xy - 2y^2 - 2$

(b) 1.0 million bags of sour cream and chives and 3.0 million bags of barbecue

(c) 14.0 million dollars

37. **(a)** $R(x, y) = 349x - 4x^2 + 3xy + 446y - 3y^2$

(b) 88 touring bikes, 118 mountain bikes

(c) \$41,744

39. **(a)** $R(x, y) = 228x - 8x^2 - 2xy + 31y - 0.15y^2$

(b) 8 leaf blowers, 50 8-foot attachments

41. 10 in × 5 in × 2.5 in

Section 18.5

1. $\frac{3}{4}$ **3.** 18 **5.** −16 **7.** $\frac{128}{9}$ **9.** 50

11. 108.86 **13.** 5 **15.** 4 **17.** e^4

19. **(a)** $120 \frac{\text{regular models}}{\text{day}}$, $60 \frac{\text{joggers' models}}{\text{day}}$

(b) \$2160.00

21. **(a)** $80 \frac{\text{pepper spray devices}}{\text{day}}$, $40 \frac{\text{siren devices}}{\text{day}}$

(b) \$960.00

23. **(a)** 39 units by the old line, 65 units by the new line

(b) \$4982.00

25. **(a)** 857.1 labor hours per week, 102.9 units of capital investment per week

(b) 8268 golf carts per week

27. **(a)** 60.7 units of labor, 182.1 units of capital

(b) For each additional dollar value available to the budget, production increases by about 7.5492×10^{-5} units.

(c) 15.1 units

29. **(a)** 245.3 weekly units of labor, 206.3 weekly units of capital

(b) For each additional dollar value available to the budget, production increases by about 2.3584×10^{-3} units.

(c) 117.9 units

31. **(a)** 360 units of labor, 160 units of capital

(b) For each additional dollar value available to the budget, production increases by about 0.43379 units.

(c) 4338 units

33. $\frac{625}{2}$ feet by $\frac{625}{4}$ feet

35. 50 in × 50 in × 100 in

Section 18.6

1. $26x^2$ **3.** $26y^2$ **5.** $4x + 6$ **7.** $4 + 6y$

9. $\frac{7}{3}x^3 - 3x$ **11.** $\frac{15}{4}y^2 - 3y$

13. **(a)**

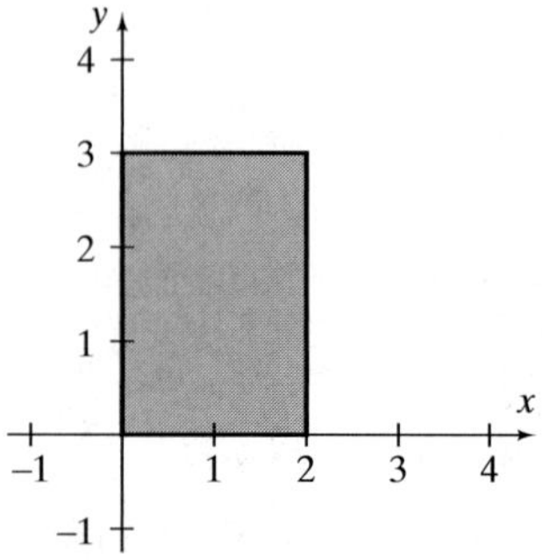

(b) $\int_0^2 \int_0^3 2xy\,dy\,dx = 18$ **(c)** $\int_0^3 \int_0^2 2xy\,dx\,dy = 18$

15. (a)

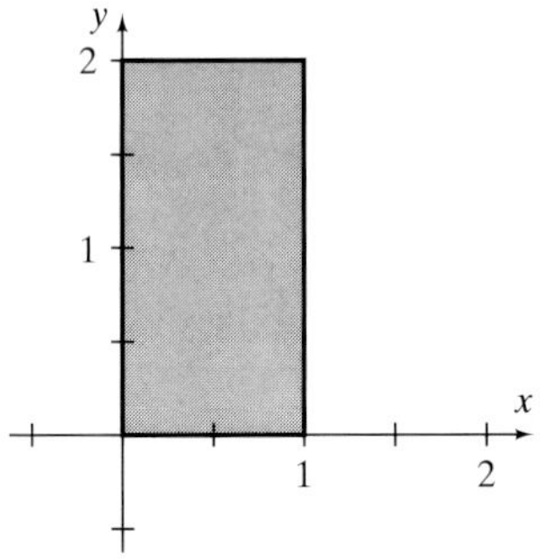

(b) $\int_0^1 \int_0^2 (3x + y)\, dy\, dx = 5$ **(c)** $\int_0^2 \int_0^1 (3x + y)\, dx\, dy = 5$

17. (a)

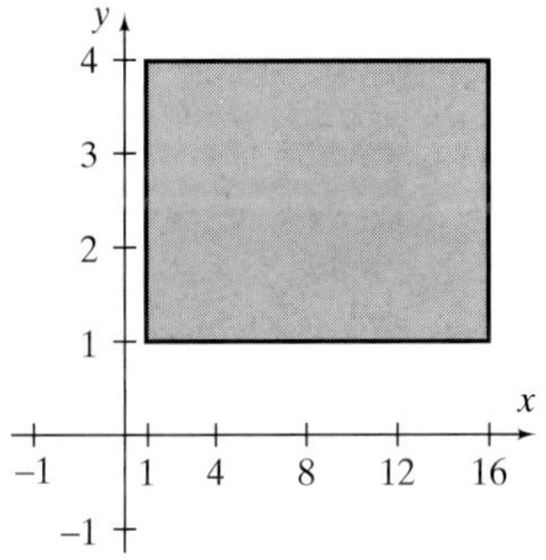

(b) $\int_1^{16} \int_1^4 \sqrt{xy}\, dy\, dx = 196$ **(c)** $\int_1^4 \int_1^{16} \sqrt{xy}\, dx\, dy = 196$

19. (a)

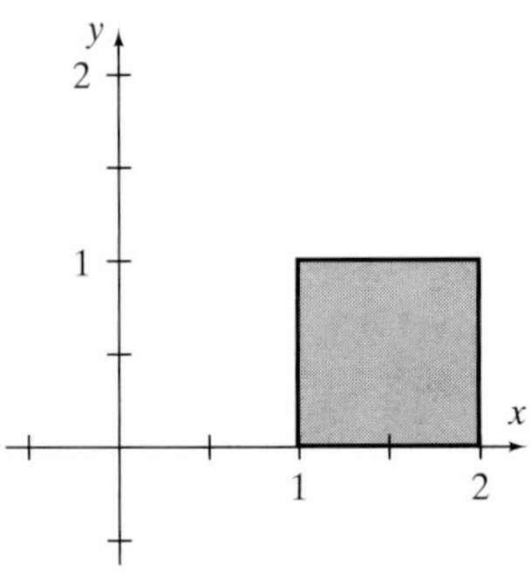

(b) $\int_1^2 \int_0^1 x^2y^2\, dy\, dx = \frac{7}{9}$ **(c)** $\int_0^1 \int_1^2 x^2y^2\, dx\, dy = \frac{7}{9}$

21. (a)

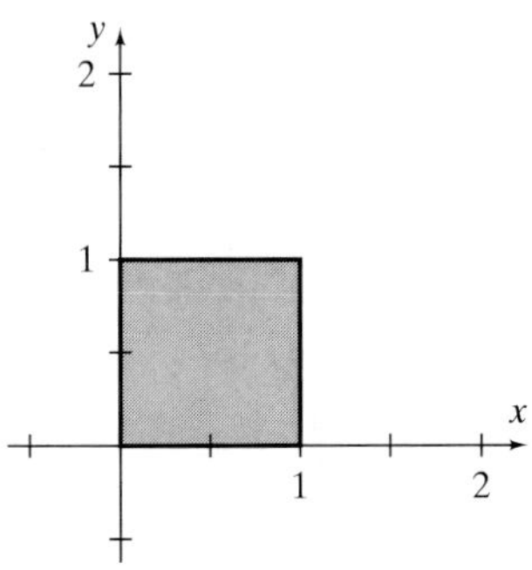

(b) $\int_0^1 \int_0^1 (2 - \frac{1}{2}x^2 + y^2)\, dy\, dx = \frac{13}{6}$

(c) $\int_0^1 \int_0^1 (2 - \frac{1}{2}x^2 + y^2)\, dx\, dy = \frac{13}{6}$

23. $\int_0^1 \int_0^2 xy\, dy\, dx = 1$

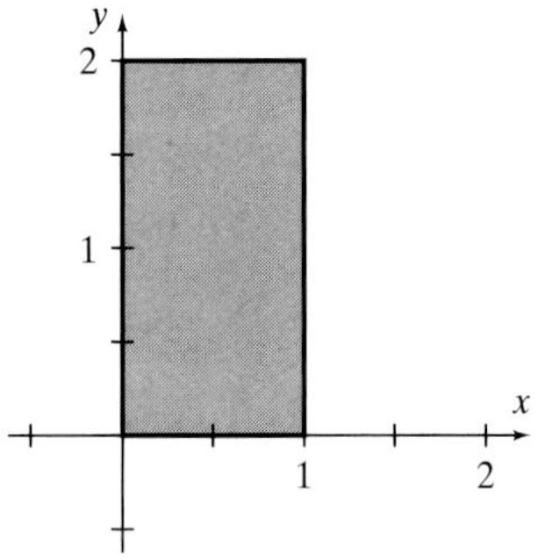

25. $\int_1^2 \int_1^3 (x^2y + y)\, dy\, dx = 16$

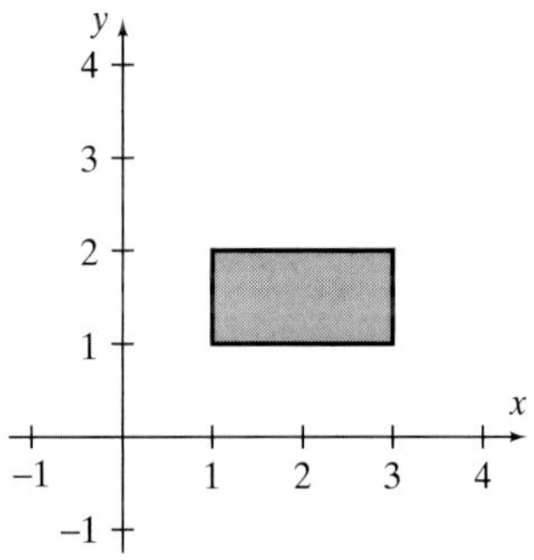

27. $\int_{-1}^1 \int_0^2 x^2e^y\, dx\, dy = \frac{8}{3}e - \frac{8}{3}e^{-1}$

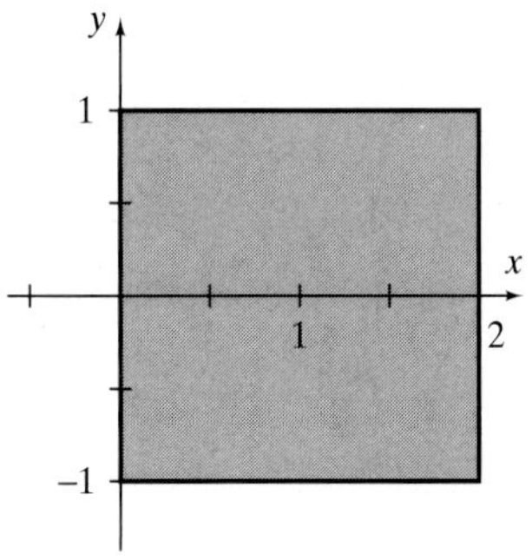

29. $\frac{2}{3}$ **31.** $\frac{25}{3}$ **33.** 2 **35.** $\frac{4}{3}e - \frac{4}{3}$

37. $\frac{44}{3}$ **39.** 6 **41.** 3 **43.** $\frac{8}{3}e - \frac{8}{3}$

45. 13918.6 **47.** 11,370 **49.** 36

51. $5959.6 \frac{\text{people}}{\text{mile}^2}$ **53.** 148,989

55. $\frac{325}{3}$ parts per million **57.** 125.78

Chapter 18 Review Exercises

1. $-\frac{6}{11}$ **3.** 103

5. All ordered pairs (x, y) such that $x + y \geq 0$

7.

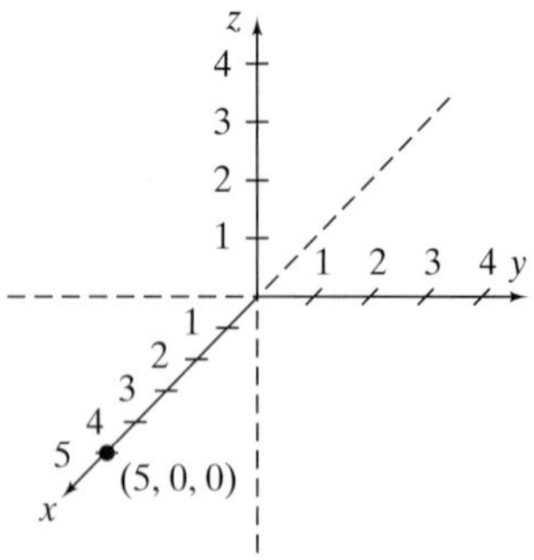

9.

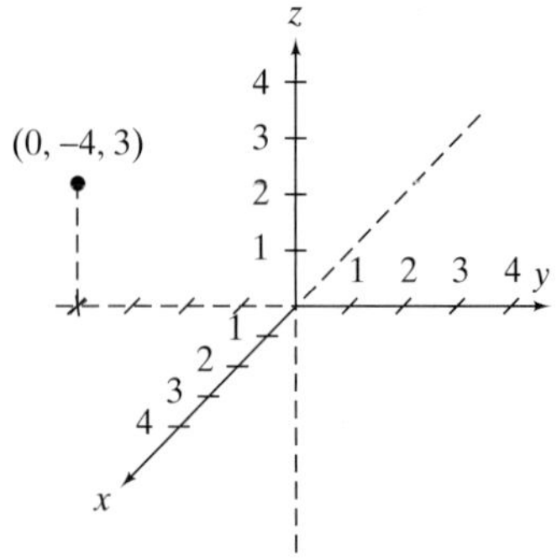

11. Above **13.** Below

15.

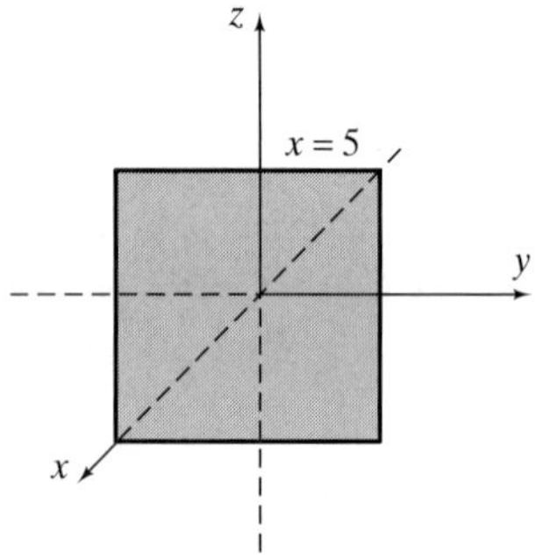

17. With sales $4000\,\frac{\text{cups of coffee}}{\text{month}}$ and $3000\,\frac{\text{cups of cappuccino}}{\text{month}}$ there are monthly profits of \$800.

19. **(a)** $C = 7x + 13y + 5000$ **(b)** $R = 12x + 21y$

(c) $P = 5x + 8y - 5000$

21. If \$4500 is invested at an annual rate of 4.5% compounded continuously for 10 years, the total amount accumulated is \$7057.40.

23.

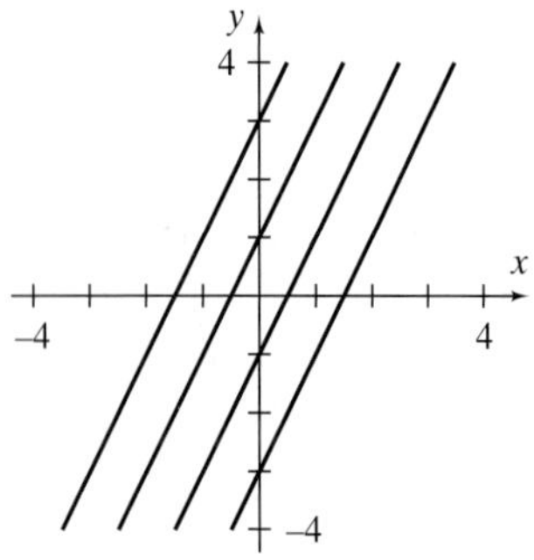

25.

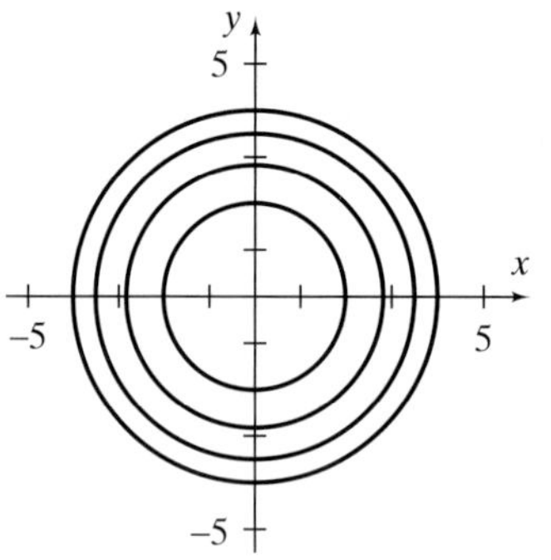

27.

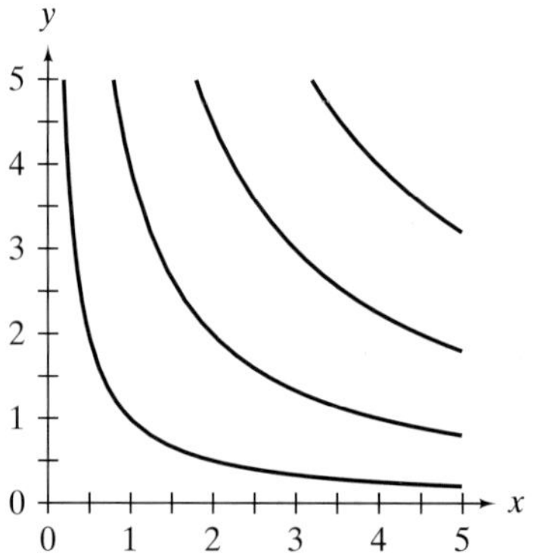

29.

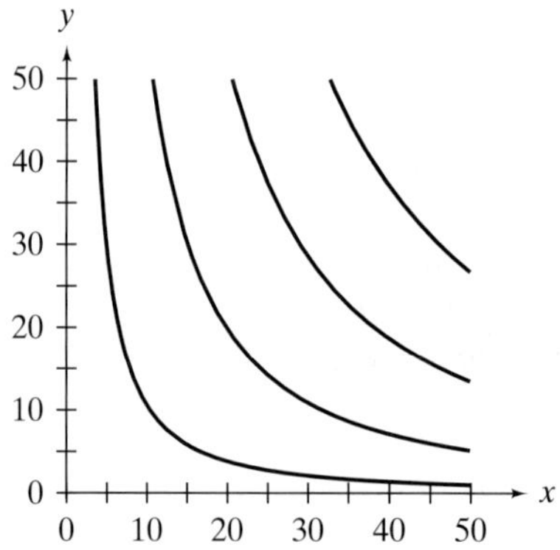

31.

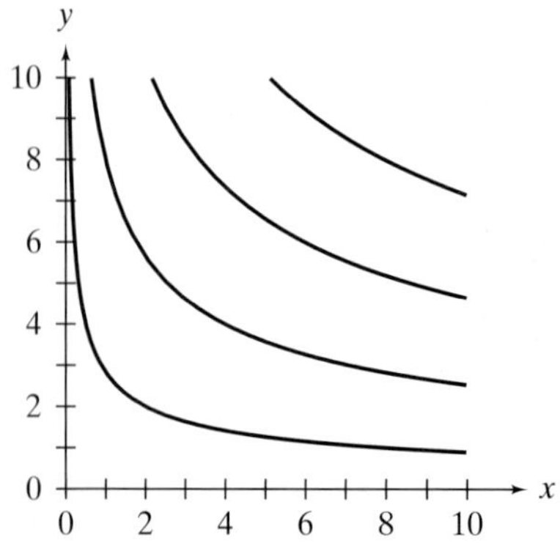

33.

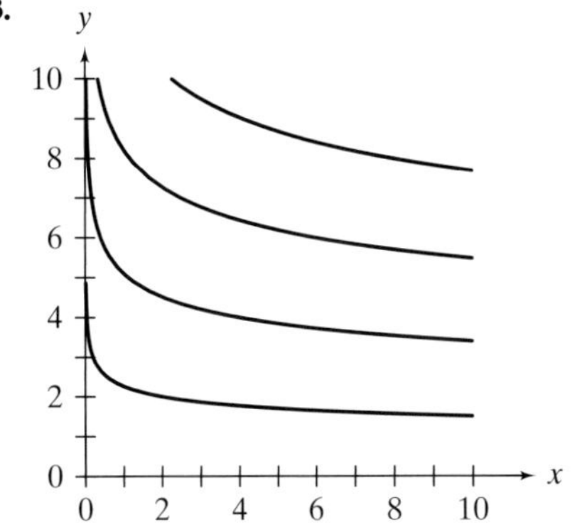

35.

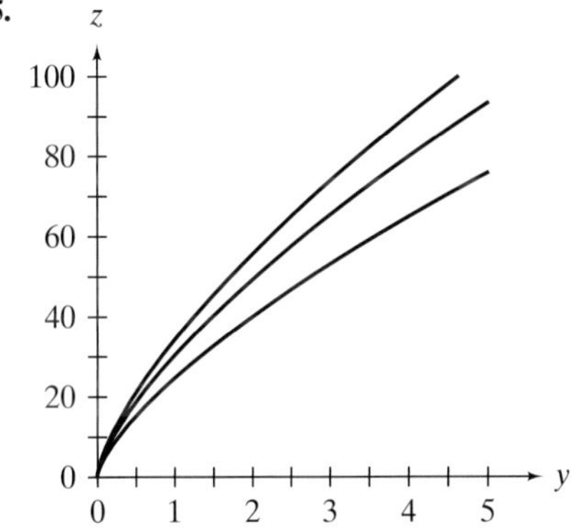

37.

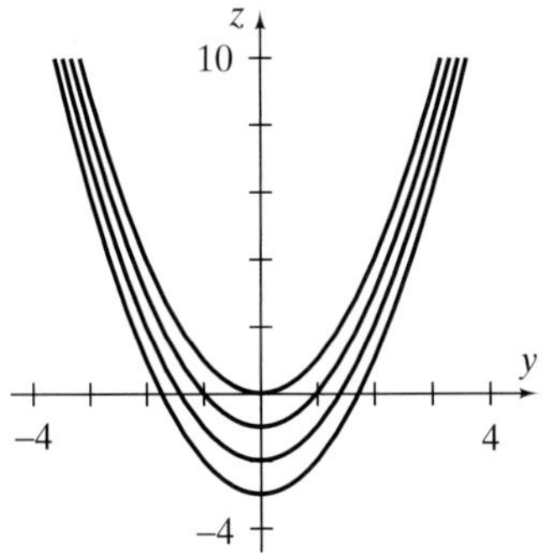

39.

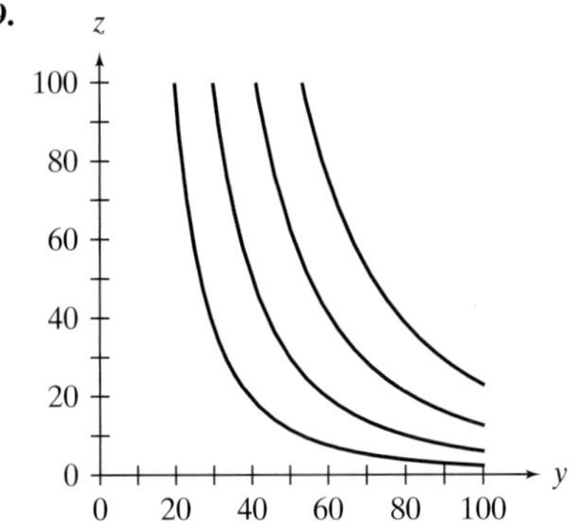

41. (a) Vertical

(b)

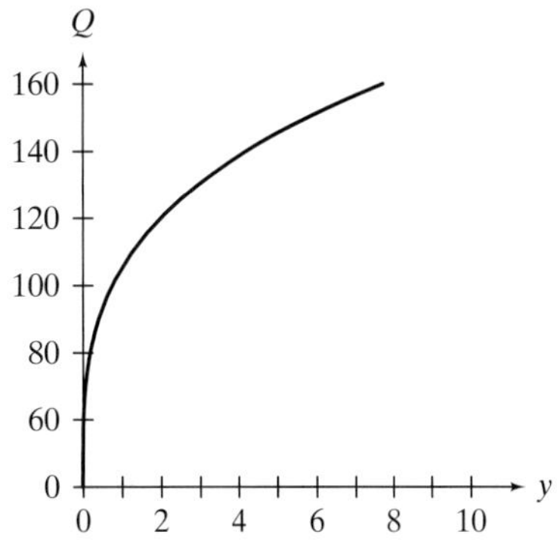

(c) \$6.2 million

43. $f_x(x, y) = 5,\ f_y(x, y) = -6y$

45. $f_x(x, y) = -15(8y^2 - 3x)^4,\ f_y(x, y) = 80(8y^2 - 3x)^4 y$

47. $f_x(x, y) = -\dfrac{y}{3x^2},\ f_y(x, y) = \dfrac{1}{3x}$

49. $f_x(x, y) = e^{x-y} + 15x^2 y,\ f_y(x, y) = -e^{x-y} + 5x^3$

51. If we are at the point (3, 5) on the surface and move along the surface in the y-direction parallel to the y-axis, the function is changing at a rate of $-1750 \frac{\text{units of } f}{\text{units of } y}$.

53. If we are at the point (3, 2) on the surface and move along the surface in the y-direction parallel to the y-axis, the function is changing at a rate of about $9.29 \frac{\text{units of } f}{\text{units of } y}$.

55. $f_{xx}(x, y) = -24x,\ f_{xy}(x, y) = 10 + 5y^4,\ f_{yy}(x, y) = 20xy^3$

57. $f_{xx}(x, y) = 2,\ f_{xy}(x, y) = 4,\ f_{yy}(x, y) = 8$

59. $x = \frac{1}{52},\ y = \frac{4}{13}$

61. (a) $R = 480.0x - 0.6x^2 + 0.5xy + 720.0y - 0.7y^2$

(b) The marginal revenue from an increase in sales of 15-inch monitors is given by $480.0 - 1.2x + 0.5y \frac{\text{dollars}}{\text{monitor}}$; the marginal revenue from an increase in sales of 17-inch monitors is given by $0.5x + 720.0 - 1.4y \frac{\text{dollars}}{\text{monitor}}$.

(c) When 640 15-inch monitors and 850 17-inch monitors have been sold the company receives \$137.00 for the 641st 15-inch monitor sold; when 640 15-inch monitors and 850 17-inch monitors have been sold the company loses \$150.00 for the 851st 17-inch monitor sold.

63. (a) $P = -3.5x^2 - 18.5xy + 1.2y^2 + 110.0x + 130y - 600$

(b) The marginal profit from an increase in sales of regular VCRs is given by $-7.0x - 18.5y + 110.0 \frac{\text{hundreds of dollars}}{\text{hundreds of VCRs}}$; the marginal profit from an increase is sales of stereo VCRs is given by $-18.5x + 2.4y + 130.0 \frac{\text{hundreds of dollars}}{\text{hundreds of VCRs}}$.

(c) With sales of 8 hundred regular VCRs and 11 hundred stereo VCRs, the marginal profit from an increase in sales of regular VCRs is $-149.5 \frac{\text{hundreds of dollars}}{\text{hundreds of VCRs}}$; with sales of 8 hundred regular VCRs and 11 hundred stereo VCRs, the marginal profit from an increase in sales of stereo VCRs is $8.40 \frac{\text{hundreds of dollars}}{\text{hundreds of VCRs}}$.

(d) The profit from sales of 8 hundred regular VCRs and 11 hundred stereo VCRs is 320 dollars.

65. $(-4, 3)$ **67.** $\left(-\frac{4}{3}, \frac{2}{3}\right)$

69. The point $(1, -4, -24)$ is a relative minimum.

71. The point $(-1, -1, -1)$ is a saddle point.

73. The point $(2, -4, -8)$ is a relative minimum and the point $(0, 0, 0)$ is a saddle point.

75. The point $(0, 2, e^4)$ is a saddle point.

77. **(a)** $R(x, y) = 12.0x - 0.3x^2 + 0.2xy + 20.0y - 0.2y^2$

(b) 44 basic haircuts, 72 deluxe haircuts

(c) \$984.00

79. **(a)** 280,000 fine-tip pens, 440,000 large-tip pens

(b) \$375,600

81. 125 **83.** -5 **85.** 22.888

87. **(a)** About 4.8 units of labor, about 2.0 units of capital

(b) About 94,692 telephones

89. 10 feet by 30 feet **91.** $258y^3$ **93.** $16x^2 - 1280$

95. **(a)**

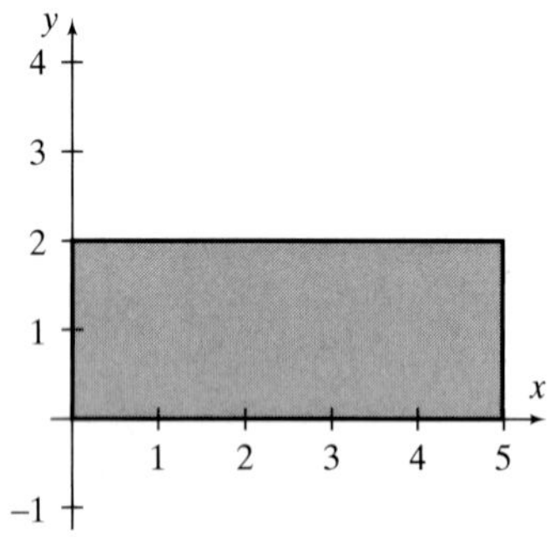

(b) $\int_0^5 \int_0^2 (3x - 4xy)\,dy\,dx = -25$

(c) $\int_0^2 \int_0^5 (3x - 4xy)\,dx\,dy = -25$

97. **(a)**

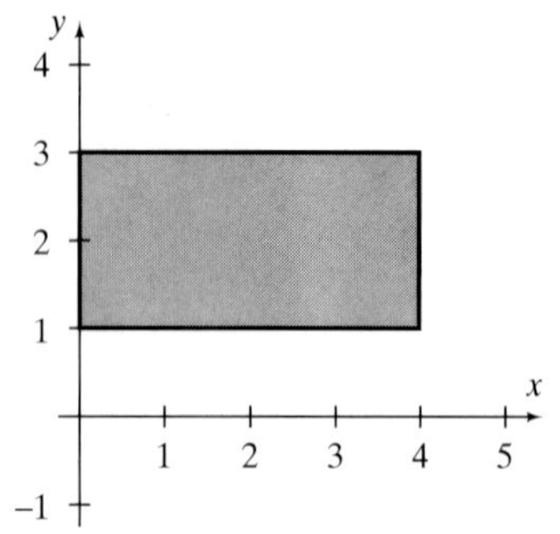

(b) 208

99. 200 **101.** $\frac{20}{3}$ **103.** 195 **105.** 392.8

107. 104.667 parts per million

Appendices

Appendix A.1

1. 0.73 **3.** 0.04 **5.** 0.591 **7.** 0.007

9. \$524.40 was covered by private health insurance in 1990.

11. About 10,119 million board feet. **13.** 20% **15.** 12.5%

17. 53.4% **19.** About 42.9%

21. It is projected that in 2008 approximately 12.1% of the students will be enrolled in private schools.

23. In 1985 approximately 5.46% of the budget was spent on technology research and development.

Appendix A.2

1. 31 **3.** 8 **5.** 4 **7.** 10 **9.** 8 **11.** -24

13. 0.27648 **15.** 41 **17.** 0.45

19. $23.3 - 0.67(5.9) = 19.347$
$23.3 + 0.67(5.9) = 27.253$

The ages of the middle 50% of the unmarried mothers in the United States in 1996 ranged from about 19 to 27.

Appendix A.3

1. $3^6 = 729$ **3.** $2^{15} = 32{,}768$ **5.** $10^6 = 1{,}000{,}000$

7. $3^1 = 3$ **9.** $6^6 = 46{,}656$ **11.** $-3^{15} = -14{,}348{,}907$

13. $(x + 3)^6$ **15.** 1 **17.** 1 **19.** $\frac{1}{36}$ **21.** 36

23. $16x^{12}$ **25.** $\frac{1}{3x^5}$ **27.** $\frac{9}{16}$ **29.** 64

31. ≈ 0.2109
There is about a 21.1% chance that 2 out of 20 were African American.

33. The payment will be \$1,463.45 each month for 5 years.

Appendix A.4

1. $x^{\frac{1}{2}}$ **3.** $x^{\frac{4}{7}}$ **5.** $5^{\frac{1}{4}}x^{\frac{2}{4}} = 5^{\frac{1}{4}}x^{\frac{1}{2}}$ **7.** $x^{-\frac{1}{5}}$ **9.** $8x^{-\frac{2}{7}}$

11. $\sqrt[4]{x}$ **13.** $\sqrt[8]{x^5}$ **15.** $\sqrt[4]{6x^2}$ **17.** $\frac{1}{\sqrt[3]{x}}$ **19.** $\frac{5}{\sqrt[3]{x^2}}$

Appendix B.1

1. $10x^5$ **3.** $-8x^{12}$ **5.** $-150x^4$ **7.** $2x^2 - 10x$

9. $-4x^4 + 12x^3$ **11.** $-12x^6 + 18x^2$ **13.** $-2x^4 + 14x^3 - 10x^2$

15. $x^2 + 8x + 15$ **17.** $1.2x^2 - 9.8x + 20$ **19.** $9x^2 - 12x - 32$

21. $8x^4 + 22x^2 + 15$ **23.** $x^3 + 3x^2 + 5x + 3$

25. $0.6x^3 - 10.4x^2 + 21.2x - 3$ **27.** $4x^4 + 4x^3 - 5x^2 + 3x - 6$

29. $x^4 + 12x^3 + 26x^2 - 21x - 6$

31. $2x^5 + 8x^4 - 6x^3 - x^2 + 15x + 6$

Appendix B.2

1. $2(x - 1)$ **3.** $5(7x - 2)$ **5.** $2x^2(6 - 13x)$

7. This expression cannot be factored. **9.** $(x + 9)(x - 9)$

11. $(3x + 1)(3x - 1)$ **13.** $(2x + 1)(2x - 3)$

15. $(x^2 + 1)(x + 1)(x - 1)$ **17.** $(4x^2 + 9)(2x + 3)(2x - 3)$

19. $(x + 1)(x + 4)$ **21.** Expression cannot be factored.

23. $(2x^2 + 4)(x - 1)$ **25.** $(6x - 3)(x + 2)(x - 2)$

27. $(x + 3)(x + 5)$ **29.** $(x + 1)^2$ **31.** $(x - 5)(x + 1)$

33. Expression is not factorable. **35.** $2(x + 8)(x - 3)$

37. $(2x+1)(x+1)$ **39.** $(3x-2)(x-2)$ **41.** $(7x-5)(x-1)$

43. $(7x+11)(x-1)$ **45.** $(5)(3x-4)(x+2)$

Appendix B.3

1. $x=2$ **3.** $x=3$ **5.** $x=28$ **7.** $x=\frac{2}{3}$

9. $x=-\frac{0.5}{0.2}=-\frac{5}{2}$ **11.** $x=\frac{18}{5}$ **13.** $x=3$

15. $x=-6$ **17.** $x=13$ **19.** $x=\frac{203}{88}$

21. $x=6$. In 1978 the annual expenditure on pollution abatement in the United States totaled \$60.17 billion.

23. $x=4$. In 1994, worldwide military expenditures were \$887.92 billion.

Appendix B.4

1. $x=0$ or $x=-2$ **3.** $x=-3$ or $x=3$

5. $x=-5$ or $x=-3$ **7.** $x=6$ or $x=-1$

9. $x=-\frac{1}{2}$ or $x=-1$ **11.** $x=\frac{4}{3}$ or $x=-1$

13. $x=\dfrac{-1\pm\sqrt{33}}{8}$ **15.** No real solution

17. $x=\dfrac{3\pm\sqrt{2}}{2}$ **19.** $x=\dfrac{3\pm\sqrt{7}}{2}$

21. In 1990 and 1994, membership in the Boy Scouts of America was 5,433,000.

23. No real solution.

Appendix B.5

1. $\log 2+\log x+\log y$ **3.** $3(\log 2+\log y-\log 5-\log x)$

5. $\ln x+2\ln y-\ln(x+2)$ **7.** $\frac{1}{3}(\ln 5+3\ln x-\ln 9-\ln y)$

9. $x=102$ **11.** $x=\frac{e^3+2}{4}\approx 5.52$ **13.** $x=\frac{26}{3}$

15. $x=4$ **17.** $x\approx 1.63$ **19.** $x=\frac{1}{5}$ **21.** $x\approx 0.23$

23. $x\approx 8.82$; in 1998, 18.89% of all waste generated in the United States was yard waste.

Appendix B.6

1. $\frac{10}{3}$ **3.** 2,097,151 **5.** 4,882,812 **7.** 9842

9. \$4,095,000

11. You pick the month choice. Pick a 31-day month and the doubling effect would amount to \$21,474,836.47. A 31-day month during which you receive \$100,000 each day results in a total of \$3,100,000.

Index

Travel

Weather